*Edited by*
*Gero Decher and*
*Joseph B. Schlenoff*

**Multilayer Thin Films**

## Related Titles

Pacchioni, G., Valeri, S. (eds.)

**Oxide Ultrathin Films**

Science and Technology

2012

ISBN: 978-3-527-33016-4

Friedbacher, G., Bubert, H. (eds.)

**Surface and Thin Film Analysis**

A Compendium of Principles, Instrumentation, and Applications

2011

ISBN: 978-3-527-32047-9

Knoll, W., Advincula, R. C. (eds.)

**Functional Polymer Films**

2 Volume Set

2011

ISBN: 978-3-527-32190-2

Kumar, C. S. S. R. (ed.)

**Polymeric Nanomaterials**

2011

ISBN: 978-3-527-32170-4

Urban, M. W. (ed.)

**Handbook of Stimuli-Responsive Materials**

2011

ISBN: 978-3-527-32700-3

Fernandez-Nieves, A., Wyss, H., Mattsson, J., Weitz, D. A. (eds.)

**Microgel Suspensions**

Fundamentals and Applications

2011

ISBN: 978-3-527-32158-2

Samori, P., Cacialli, F. (eds.)

**Functional Supramolecular Architectures**

for Organic Electronics and Nanotechnology

2011

ISBN: 978-3-527-32611-2

Chujo, Y. (ed.)

**Conjugated Polymer Synthesis**

Methods and Reactions

2011

ISBN: 978-3-527-32267-1

Leclerc, M., Morin, J.-F. (eds.)

**Design and Synthesis of Conjugated Polymers**

2010

ISBN: 978-3-527-32474-3

*Edited by*
*Gero Decher and Joseph B. Schlenoff*

# Multilayer Thin Films

Sequential Assembly of Nanocomposite Materials

*Second, Completely Revised and Enlarged Edition*

WILEY-VCH Verlag GmbH & Co. KGaA

**The Editors**

*Prof. Gero Decher*
Université de Strasbourg
Insitut Charles Sadron, UPR 22
23, rue du Loess
67034 Strasbourg Cedex 2
France

*Prof. Joseph B. Schlenoff*
Florida State University
Chemistry Department
Tallahassee, FL 32306-4390
USA

■ All books published by **Wiley-VCH** are carefully produced. Nevertheless, authors, editors, and publisher do not warrant the information contained in these books, including this book, to be free of errors. Readers are advised to keep in mind that statements, data, illustrations, procedural details or other items may inadvertently be inaccurate.

**Library of Congress Card No.:** applied for

**British Library Cataloguing-in-Publication Data**
A catalogue record for this book is available from the British Library.

**Bibliographic information published by the Deutsche Nationalbibliothek**
The Deutsche Nationalbibliothek lists this publication in the Deutsche Nationalbibliografie; detailed bibliographic data are available on the Internet at http://dnb.d-nb.de.

© 2012 Wiley-VCH Verlag & Co. KGaA, Boschstr. 12, 69469 Weinheim, Germany

All rights reserved (including those of translation into other languages). No part of this book may be reproduced in any form – by photoprinting, microfilm, or any other means – nor transmitted or translated into a machine language without written permission from the publishers. Registered names, trademarks, etc. used in this book, even when not specifically marked as such, are not to be considered unprotected by law.

**Composition**   Thomson Digital, Noida, India
**Printing and Binding**   betz-druck GmbH, Darmstadt
**Cover Design**   Schulz Grafik-Design, Fußgönheim

Printed in the Federal Republic of Germany
Printed on acid-free paper

**Print ISBN:**   978-3-527-31648-9
**ePDF ISBN:**   978-3-527-64677-7
**oBook ISBN:**   978-3-527-64674-6
**ePub ISBN:**   978-3-527-64676-0

# Contents

**List of Contributors** XXV

**Volume 1**

**1 Layer-by-Layer Assembly (Putting Molecules to Work)** 1
*Gero Decher*
1.1 The Whole is More than the Sum of its Parts 1
1.2 From Self-Assembly to Directed Assembly 1
1.3 History and Development of the Layer-by-Layer Assembly Method 4
1.4 LbL-Assembly is the Synthesis of Fuzzy Supramolecular Objects 6
1.5 Reproducibility and Choice of Deposition Conditions 7
1.6 Monitoring Multilayer Build-up 10
1.7 Spray- and Spin-Assisted Multilayer Assembly 13
1.8 Recent Developments 14
1.8.1 Self-patterning LbL-Films 14
1.8.2 Deposition of LbL-Films on Very Small Particles 15
1.8.3 Purely Inorganic LbL-"Films" 17
1.9 Final Remarks 18
References 19

**Part I Preparation and Characterization** 23

**2 Layer-by-Layer Processed Multilayers: Challenges and Opportunities** 25
*Michael F. Rubner and Robert E. Cohen*
2.1 Introduction 25
2.2 Fundamental Challenges and Opportunities 25
2.2.1 LbL Assembly on Nanoscale Elements and in Confined Geometries 25
2.2.2 Living Cells as Functional Elements of Polyelectrolyte Multilayers 28
2.2.3 Multilayer Cellular Backpacks 28
2.2.4 Direct LbL Processing of Living Cells 29
2.3 Technological Challenges and Opportunities 31
2.3.1 Improving Processing Time and Versatility 31
2.3.2 Towards Mechanically Robust Multilayer Coatings 32
2.4 The Path Forward 36
References 36

| | | |
|---|---|---|
| **3** | **Layer-by-Layer Assembly: from Conventional to Unconventional Methods** 43 | |
| | *Guanglu Wu and Xi Zhang* | |
| 3.1 | Introduction 43 | |
| 3.2 | Conventional LbL Methods 44 | |
| 3.2.1 | Electrostatic LbL Assembly 44 | |
| 3.2.2 | Hydrogen-Bonded LbL Assembly 47 | |
| 3.2.3 | LbL Assembly Driven by Coordination Interaction 50 | |
| 3.2.4 | To Combine LbL Assembly and Post-Chemical Reaction for the Fabrication of Robust Thin Films 50 | |
| 3.3 | Unconventional LbL Methods 52 | |
| 3.3.1 | Electrostatic Complex for Unconventional LbL Assembly 52 | |
| 3.3.1.1 | Nanoreactors with Enhanced Quantum Yield 53 | |
| 3.3.1.2 | "Ion Traps" for Enhancing the Permselectivity and Permeability 53 | |
| 3.3.1.3 | Surface Imprinted LbL Films 54 | |
| 3.3.1.4 | Cation-Selective µCP Based on SMILbL Film 58 | |
| 3.3.2 | Hydrogen-Bonded Complex for Unconventional LbL Assembly 58 | |
| 3.3.3 | Block Copolymer Micelles for Unconventional LbL Assembly 61 | |
| 3.3.4 | π–π Interaction Complex for Electrostatic LbL Assembly 62 | |
| 3.4 | Summary and Outlook 64 | |
| | References 64 | |
| | | |
| **4** | **Novel Multilayer Thin Films: Hierarchic Layer-by-Layer (Hi-LbL) Assemblies** 69 | |
| | *Katsuhiko Ariga, Qingmin Ji, and Jonathan P. Hill* | |
| 4.1 | Introduction 69 | |
| 4.2 | Hi-LbL for Multi-Cellular Models 70 | |
| 4.3 | Hi-LbL for Unusual Drug Delivery Modes 72 | |
| 4.4 | Hi-LbL for Sensors 75 | |
| 4.4.1 | Mesoporous Carbon Hi-LbL 75 | |
| 4.4.2 | Mesoporous Carbon Capsule Hi-LbL 76 | |
| 4.4.3 | Graphene/Ionic-Liquid Hi-LbL 78 | |
| 4.5 | Future Perspectives 79 | |
| | References 80 | |
| | | |
| **5** | **Layer-by-Layer Assembly Using Host-Guest Interactions** 83 | |
| | *Janneke Veerbeek, David N. Reinhoudt, and Jurriaan Huskens* | |
| 5.1 | Introduction 83 | |
| 5.2 | Supramolecular Layer-by-Layer Assembly 84 | |
| 5.3 | 3D Patterned Multilayer Assemblies on Surfaces 85 | |
| 5.4 | 3D Supramolecular Nanoparticle Crystal Structures 88 | |
| 5.5 | Porous 3D Supramolecular Assemblies in Solution 90 | |
| 5.6 | Conclusions 95 | |
| | References 95 | |

| 6 | **LbL Assemblies Using van der Waals or Affinity Interactions and Their Applications** *99* |
|---|---|
| | *Takeshi Serizawa, Mitsuru Akashi, Michiya Matsusaki, Hioharu Ajiro, and Toshiyuki Kida* |
| 6.1 | Introduction *99* |
| 6.2 | Stereospecific Template Polymerization of Methacrylates by Stereocomplex Formation in Nanoporous LbL Films *100* |
| 6.2.1 | Introduction *100* |
| 6.2.2 | LbLit-PMMA/st-PMMA Stereocomplex Uultrathin Film *102* |
| 6.2.3 | Fabrication of Template Nanospaces in Films *104* |
| 6.2.4 | Polymerization within Template Nanospaces Using a Double-Stranded Assembly *106* |
| 6.2.5 | Studies on the Porous Structure Obtained by LbL Assembly *111* |
| 6.3 | Preparation and Properties of Hollow Capsules Composed of Layer-by-Layer Polymer Films Constructed through van der Waals Interactions *113* |
| 6.3.1 | Introduction *113* |
| 6.3.2 | Preparation of Hollow Nanocapsules Composed of Poly(methyl methacrylate) Stereocomplex Films *114* |
| 6.3.3 | Preparation and Fusion Properties of Novel Hollow Nanocapsules Composed of Poly(lactic acid)s Stereocomplex Films *116* |
| 6.4 | Fabrication of Three-Dimensional Cellular Multilayers Using Layer-by-Layer Protein Nanofilms Constructed through Affinity Interaction *120* |
| 6.4.1 | Introduction *120* |
| 6.4.2 | Hierarchical Cell Manipulation *121* |
| 6.4.3 | High Cellular Activities Induced by 3D-Layered Constructs *124* |
| 6.4.4 | Quantitative 3D-Analyses of Nitric Oxide Diffusion in a 3D-Artery Model *124* |
| 6.5 | Conclusion *129* |
| | References *129* |
| | |
| 7 | **Layer-by-Layer Assembly of Polymeric Complexes** *135* |
| | *Junqi Sun, Xiaokong Liu, and Jiacong Shen* |
| 7.1 | Introduction *135* |
| 7.2 | Concept of LbL Assembly of Polymeric Complexes *136* |
| 7.2.1 | LbL Assembly of Polyelectrolyte Complexes for the Rapid and Direct Fabrication of Foam Coatings *136* |
| 7.2.2 | LbL Assembly of Hydrogen-Bonded Polymeric Complexes *138* |
| 7.2.3 | LbL Assembly of Polyelectrolyte-Surfactant Complexes *139* |
| 7.3 | Structural Tailoring of LbL-Assembled Films of Polymeric Complexes *140* |
| 7.3.1 | Mixing Ratio of PECs *140* |
| 7.3.2 | LbL Codeposition of PECs and Free Polyelectrolytes *142* |
| 7.3.3 | Salt Effect *143* |
| 7.4 | LbL-Assembled Functional Films of Polymeric Complexes *144* |

| | | |
|---|---|---|
| 7.4.1 | Self-Healing Superhydrophobic Coatings | 144 |
| 7.4.2 | Mechanically Stable Antireflection- and Antifogging-Integrated Coatings | 146 |
| 7.4.3 | Transparent and Scratch-Resistant Coatings with Well-Dispersed Nanofillers | 147 |
| 7.5 | Summary | 149 |
| | References | 149 |

**8 Making Aqueous Nanocolloids from Low Solubility Materials: LbL Shells on Nanocores** 151

*Yuri Lvov, Pravin Pattekari, and Tatsiana Shutava*

| | | |
|---|---|---|
| 8.1 | Introduction | 151 |
| 8.2 | Formation of Nanocores | 153 |
| 8.2.1 | Stabilizers versus Layer-by-Layer Polyelectrolyte Shell | 153 |
| 8.3 | Ultrasonication-Assisted LbL Assembly | 154 |
| 8.3.1 | Top-Down Approach | 154 |
| 8.3.2 | Bottom-Up Approach | 158 |
| 8.4 | Solvent-Assisted Precipitation Into Preformed LbL-Coated Soft Organic Nanoparticles | 159 |
| 8.5 | Washless (Titration) LbL Technique | 161 |
| 8.6 | Formation of LbL Shells on Nanocores | 163 |
| 8.7 | Drug Release Study | 165 |
| 8.8 | Conclusions | 168 |
| | References | 168 |

**9 Cellulose Fibers and Fibrils as Templates for the Layer-by-Layer (LbL) Technology** 171

*Lars Wågberg*

| | | |
|---|---|---|
| 9.1 | Background | 171 |
| 9.2 | Formation of LbLs on Cellulose Fibers | 172 |
| 9.3 | The use of LbL to Improve Adhesion between Wood Fibers | 176 |
| 9.4 | The Use of LbL to Prepare Antibacterial Fibers | 179 |
| 9.5 | The use of NFC/CNC to Prepare Interactive Layers Using the LbL Approach | 182 |
| 9.6 | Conclusions | 185 |
| | References | 186 |

**10 Freely Standing LbL Films** 189

*Chaoyang Jiang and Vladimir V. Tsukruk*

| | | |
|---|---|---|
| 10.1 | Introduction | 189 |
| 10.2 | Fabrication of Freely Standing Ultrathin LbL Films | 189 |
| 10.2.1 | LbL Films with Gold Nanoparticles and Silver Nanowires | 191 |
| 10.2.2 | LbL Films with Quantum Dots | 193 |
| 10.2.3 | LbL Films with Nanocrystals | 194 |
| 10.2.4 | LbL Films with Graphene Oxides and Carbon Nanotubes | 195 |

| | | |
|---|---|---|
| 10.2.5 | LbL with Biomaterials | 198 |
| 10.2.6 | Mechanical and Optical Properties | 199 |
| 10.2.7 | Surface Morphology and Internal Microstructure | 201 |
| 10.3 | Porous and Patterned Freely Standing LbL Films | 204 |
| 10.3.1 | Nanoporous LbL Films | 204 |
| 10.3.2 | Patterned LbL Films | 205 |
| 10.3.3 | Sculptured LbL Films | 207 |
| 10.4 | Freely Standing LbL Films with Weak Interactions | 208 |
| 10.4.1 | Responsive LbL Films | 208 |
| 10.4.2 | Anisotropic LbL Microcapsules | 209 |
| 10.4.3 | LbL-Shells for Cell Encapsulation | 210 |
| | References | 214 |
| **11** | **Neutron Reflectometry at Polyelectrolyte Multilayers** | **219** |
| | *Regine von Klitzing, Ralf Köhler, and Chloe Chenigny* | |
| 11.1 | Introduction | 219 |
| 11.2 | Neutron Reflectometry | 219 |
| 11.2.1 | Specular Reflectometry | 220 |
| 11.2.2 | Contrast Variation: Interesting Features for PEMs | 222 |
| 11.2.2.1 | Superlattice Structure | 222 |
| 11.2.2.2 | Block Structure | 223 |
| 11.2.3 | Fitting | 223 |
| 11.3 | Preparation Techniques for Polyelectrolyte Multilayers | 224 |
| 11.3.1 | Dipping | 224 |
| 11.3.2 | Spraying | 228 |
| 11.3.3 | Spin-Coating | 231 |
| 11.3.4 | Comparison of the Techniques and Model | 231 |
| 11.4 | Types of Polyelectrolytes | 233 |
| 11.4.1 | A Strong Polyanion and a Weak Polycation | 233 |
| 11.4.2 | Two Strong Polyelectrolytes | 235 |
| 11.4.3 | Two Weak Polyelectrolytes | 237 |
| 11.4.4 | Multilayers with Non-Electrostatic Interactions | 238 |
| 11.5 | Preparation Parameters | 238 |
| 11.5.1 | Ion Effects | 238 |
| 11.5.1.1 | Ionic Strength | 238 |
| 11.5.1.2 | Counterions | 239 |
| 11.5.2 | pH | 241 |
| 11.5.3 | Temperature | 241 |
| 11.6 | Influence of External Fields After PEM Assembly | 242 |
| 11.6.1 | Scalar Fields: Ionic Strength | 242 |
| 11.6.2 | Scalar Fields: Temperature | 244 |
| 11.6.3 | Scalar Fields: Water | 246 |
| 11.6.3.1 | Swelling in Water | 247 |
| 11.6.3.2 | Models for Determination of the Water Content | 252 |

| | | |
|---|---|---|
| 11.6.3.3 | Effect of Polymer Charge Density and Ions During Preparation on Swelling in Water | *254* |
| 11.6.3.4 | Odd-Even Effect | *256* |
| 11.6.4 | Directed External Fields: Mechanical and Electrical Fields | *257* |
| 11.6.4.1 | Mechanical Stress | *257* |
| 11.6.4.2 | Electric Field | *258* |
| 11.7 | PEM as a Structural Unit | *259* |
| 11.7.1 | Cushion | *259* |
| 11.7.1.1 | Cell-Membrane Mimic Systems | *260* |
| 11.7.1.2 | Multilayers Beyond Polyelectrolytes | *261* |
| 11.7.2 | Matrix | *261* |
| 11.8 | Conclusion and Outlook | *261* |
| | References | *262* |
| | | |
| **12** | **Polyelectrolyte Conformation in and Structure of Polyelectrolyte Multilayers** | ***269*** |
| | *Stephan Block, Olaf Soltwedel, Peter Nestler, and Christiane A. Helm* | |
| 12.1 | Introduction | *269* |
| 12.2 | Results | *270* |
| 12.2.1 | The First Polyelectrolyte Layer: Brush and/or Flatly Adsorbed Chains | *270* |
| 12.2.2 | Adsorption of Additional Layers and Interdiffusion | *275* |
| 12.2.3 | The Outermost Layer: From Odd–Even to Even–Odd Effect | *277* |
| 12.3 | Conclusion and Outlook | *279* |
| | References | *280* |
| | | |
| **13** | **Charge Balance and Transport in Ion-Paired Polyelectrolyte Multilayers** | ***281*** |
| | *Joseph B. Schlenoff* | |
| 13.1 | Introduction | *281* |
| 13.2 | Association Mechanism: Competitive Ion Pairing | *283* |
| 13.2.1 | Intrinsic vs. Extrinsic Charge Compensation | *285* |
| 13.2.1.1 | Key Equilibria | *285* |
| 13.2.1.2 | Interaction Energies | *287* |
| 13.2.1.3 | Multilayer Decomposition | *290* |
| 13.2.1.4 | Doping-moderated Mechanical Properties | *292* |
| 13.3 | Surface versus Bulk Polymer Charge | *292* |
| 13.3.1 | Distribution of Surface Charge in Layer-by-Layer Build-up: Mechanism | *297* |
| 13.4 | Polyelectrolyte Interdiffusion | *301* |
| 13.4.1 | Equilibrium versus Non-Equilibrium Conditions for Salt and Polymer Sorption | *304* |
| 13.5 | Ion Transport Through Multilayers: the "Reluctant" Exchange Mechanism | *305* |

| | | |
|---|---|---|
| 13.5.1 | Practical Consequences: Trapping and Self-Trapping | *313* |
| 13.6 | Concluding Remarks | *315* |
| | References | *315* |

## 14 Conductivity Spectra of Polyelectrolyte Multilayers Revealing Ion Transport Processes  *321*
*Monika Schönhoff and Cornelia Cramer*

| | | |
|---|---|---|
| 14.1 | Introduction to Conductivity Studies of LbL Films | *321* |
| 14.2 | PEM Spectra: Overview | *323* |
| 14.3 | DC Conductivities of PEMs | *324* |
| 14.4 | Modeling of PEM Spectra | *328* |
| 14.5 | Ion Conduction in Polyelectrolyte Complexes | *329* |
| 14.6 | Scaling Principles in Conductivity Spectra: From Time–Temperature to Time–Humidity Superposition | *332* |
| 14.6.1 | General Aspects of Scaling | *332* |
| 14.6.2 | Time–Temperature Superposition Principle in PEC Spectra | *333* |
| 14.6.3 | Establishment of a Time–Humidity Superposition Principle | *334* |
| | References | *335* |

## 15 Responsive Layer-by-Layer Assemblies: Dynamics, Structure and Function  *337*
*Svetlana Sukhishvili*

| | | |
|---|---|---|
| 15.1 | Introduction | *337* |
| 15.2 | Chain Dynamics and Film Layering | *338* |
| 15.2.1 | Lessons from PECs | *338* |
| 15.2.1.1 | Phase Diagrams of PECs and LbL Film Deposition | *338* |
| 15.2.1.2 | Polyelectrolyte Type and Equilibrium and Dynamics in PECs | *340* |
| 15.2.2 | pH-Induced Chain Dynamics and Film Structure | *342* |
| 15.3 | Responsive Swellable LbL Films | *348* |
| 15.3.1 | LbL-Derived Hydrogels | *348* |
| 15.3.1.1 | Preparation and Swelling | *348* |
| 15.3.1.2 | Mechanical Properties | *350* |
| 15.3.1.3 | LbL Hydrogels as Matrices for Controlled Release of Bioactive Molecules | *351* |
| 15.3.2 | LbL Films of Micelles with Responsive Cores | *353* |
| 15.4 | Conclusion and Outlook | *358* |
| | References | *359* |

## 16 Tailoring the Mechanics of Freestanding Multilayers  *363*
*Andreas Fery and Vladimir V. Tsukruk*

| | | |
|---|---|---|
| 16.1 | Introduction | *363* |
| 16.2 | Measurements of Mechanical Properties of Flat LbL Films | *364* |
| 16.2.1 | Micromechanical Properties from Bulging Experiments | *364* |

| 16.2.2 | Buckling Measurements of LbL Films   367 |
| --- | --- |
| 16.2.3 | Local Mechanical Properties Probed with Force Spectroscopies   369 |
| 16.2.4 | Summary of Mechanical Properties of Flat LbL Films   370 |
| 16.3 | Mechanical Properties of LbL Microcapsules   372 |
| 16.3.1 | Theory and Measurement of Mechanical Properties of Microcapsules   373 |
| 16.3.2 | Basic Concepts of Shell Theory   373 |
| 16.3.3 | Experimental Measurements   374 |
| 16.3.4 | Small Deformation Measurements   375 |
| 16.4 | Prospective Applications Utilizing Mechanical Properties   378 |
| 16.4.1 | Flat Freestanding LbL Films for Sensing Applications   378 |
| 16.4.2 | Microcapsules for Controlled Delivery Processes   382 |
| 16.4.2.1 | Exploding Capsules   382 |
| 16.4.2.2 | Asymmetric Capsules   382 |
| 16.4.2.3 | Mechanical Stability and Intracellular Delivery   385 |
|  | References   386 |

**17  Design and Translation of Nanolayer Assembly Processes: Electrochemical Energy to Programmable Pharmacies**   393
*Md. Nasim Hyder, Nisarg J. Shah, and Paula T. Hammond*

| 17.1 | Introduction   393 |
| --- | --- |
| 17.2 | Controlling Transport and Storing Charge in Multilayer Thin Films: Ions, Electrons and Molecules   395 |
| 17.2.1 | LbL Systems for Proton Exchange Fuel Cells   395 |
| 17.2.2 | Charge Storage Using LbL Assemblies   398 |
| 17.2.2.1 | Carbon Nanotube LbL Electrodes   398 |
| 17.2.2.2 | Carbon Nanotube/Conjugated Polymer System   400 |
| 17.2.3 | Composite Systems   402 |
| 17.2.3.1 | Virus Battery   402 |
| 17.2.3.2 | DSSC LbL Electrodes   402 |
| 17.2.3.3 | Multicolor Electrochromic Devices   404 |
| 17.3 | LbL Films for Multi-Agent Drug Delivery – Opportunities for Programmable Release   406 |
| 17.3.1 | Passive Controlled Release   407 |
| 17.3.1.1 | Release of Proteins   408 |
| 17.3.1.2 | Small Molecule Delivery   412 |
| 17.3.2 | Multi-Agent Delivery and Control of Sequence   418 |
| 17.4 | Automated Spray-LbL – Enabling Function and Translation   422 |
| 17.4.1 | Means of Achieving New Architectures   425 |
| 17.4.2 | Coating of Complex Surfaces   427 |
| 17.4.3 | Using Kinetics to Manipulate Composition   429 |
| 17.5 | Concluding Remarks   431 |
|  | References   431 |

## Contents | XIII

| | | |
|---|---|---|
| **18** | **Surface-Initiated Polymerization and Layer-by-Layer Films** *437* | |
| | Nicel Estillore and Rigoberto C. Advincula | |
| 18.1 | Introduction *437* | |
| 18.2 | Overview of Surface-Grafted Polymer Brushes *438* | |
| 18.2.1 | Synthesis of Polymer Brushes *439* | |
| 18.2.2 | Surface-Initiated Atom Transfer Radical Polymerization (SI-ATRP) *439* | |
| 18.3 | Layer-by-Layer (LbL) Self-Assembly *440* | |
| 18.3.1.1 | Water-Soluble ATRP Macroinitiators *440* | |
| 18.4 | Combined LbL-SIP Approach *441* | |
| 18.4.1 | Single Adsorption of a Polyelectrolyte ATRP Macroinitiator onto a Charged Surface *442* | |
| 18.4.2 | Adsorption of Oppositely Charged ATRP Macroinitiator(s) onto an LbL Deposited Multilayer Assembly *443* | |
| 18.4.3 | Multiple Adsorption of Oppositely Charged ATRP Macroinitiators for Enhanced Initiator Density *447* | |
| 18.5 | Applications of the Combined LbL-SIP Approach *449* | |
| 18.5.1 | Dual Stimuli-Response of the LbL-SIP Brush Modified Surfaces *449* | |
| 18.5.2 | Functional Free-Standing Brush Films *450* | |
| 18.5.3 | Cell Adhesion *451* | |
| 18.5.4 | Stimuli Responsive Free-Standing Films *451* | |
| 18.6 | Concluding Remarks *453* | |
| | References *453* | |
| | | |
| **19** | **Quartz Crystal Resonator as a Tool for Following the Build-up of Polyelectrolyte Multilayers** *455* | |
| | Mikko Salomäki and Jouko Kankare | |
| 19.1 | Introduction *455* | |
| 19.2 | Basic Concepts *456* | |
| 19.3 | Growth Processes *461* | |
| 19.3.1 | Effect of Temperature *462* | |
| 19.4 | Experimental Techniques *463* | |
| 19.5 | Analysis of QCR Data *465* | |
| | References *469* | |

**Volume 2**

**Part II  Applications** *471*

| | | |
|---|---|---|
| **20** | **Electrostatic and Coordinative Supramolecular Assembly of Functional Films for Electronic Application and Materials Separation** *473* | |
| | Bernd Tieke, Ashraf El-Hashani, Kristina Hoffmann, and Anna Maier | |
| 20.1 | Introduction *473* | |
| 20.2 | Polyelectrolyte Multilayer Membranes *474* | |
| 20.2.1 | Water Desalination *474* | |
| 20.2.2 | Size- and Charge-Selective Transport of Aromatic Compounds *476* | |

| | | |
|---|---|---|
| 20.2.3 | Ion Separation from Polyelectrolyte Blend Multilayer Membranes *479* | |
| 20.2.4 | Membranes Containing Macrocyclic Compounds *480* | |
| 20.2.4.1 | p-Sulfonato-Calix[n]arene-Containing Membranes *480* | |
| 20.2.4.2 | Azacrown Ether-Containing Membranes *483* | |
| 20.2.4.3 | Membranes Containing Hexacyclen-Hexaacetic Acid *486* | |
| 20.2.5 | LbL-Assembled Films of Prussian Blue and Analogues *487* | |
| 20.2.6 | Coordinative Assembly of Functional Thin Films *493* | |
| 20.2.6.1 | Films of Coordination Polymers Based on Schiff-Base-Metal Ion Complexes *493* | |
| 20.2.6.2 | Films of Coordination Polymers Based on Terpyridine-Metal Ion Complexes *495* | |
| 20.2.6.3 | Films of Coordination Polymers Based on Bisimidazolylpyridine-Metal Ion Complexes *502* | |
| 20.3 | Summary and Conclusions *504* | |
| | References *506* | |

**21 Optoelectronic Materials and Devices Incorporating Polyelectrolyte Multilayers** *511*

*H.D. Robinson, Reza Montazami, Chalongrat Daengngam, Ziwei Zuo, Wang Dong, Jonathan Metzman, and Randy Heflin*

| | | |
|---|---|---|
| 21.1 | Introduction *511* | |
| 21.2 | Second Order Nonlinear Optics *512* | |
| 21.3 | Plasmonic Enhancement of Second Order Nonlinear Optical Response *515* | |
| 21.4 | Nonlinear Optical Fibers *519* | |
| 21.5 | Optical Fiber Biosensors *521* | |
| 21.6 | Antireflection Coatings *525* | |
| 21.7 | Electrochromic Devices *527* | |
| 21.8 | Electromechanical Actuators *530* | |
| | References *533* | |

**22 Nanostructured Electrodes Assembled from Metal Nanoparticles and Quantum Dots in Polyelectrolytes** *539*

*Lara Halaoui*

| | | |
|---|---|---|
| 22.1 | Introduction *539* | |
| 22.2 | Nanostructured Pt Electrodes from Assemblies of Pt Nanoparticles in Polyelectrolytes *540* | |
| 22.2.1 | Assembly of Polyacrylate-capped Pt NPs in Polyelectrolytes *540* | |
| 22.2.2 | Surface Characterization by Hydrogen Underpotential Deposition *541* | |
| 22.2.3 | $H_2O_2$ Sensing at Arrays of Pt NPs in Polyelectrolyte at Low Surface Coverage *545* | |
| 22.2.4 | Biosensing at Assemblies of Glucose Oxidase Modified Pt NPs in Polyelectrolytes *548* | |

| | | |
|---|---|---|
| 22.2.5 | Surface Oxidation and Stability of Pt NP Assemblies | 549 |
| 22.2.6 | Oxygen Reduction at Pt NP Assemblies | 550 |
| 22.3 | Nanostructured Photoelectrodes from Assemblies of Q-CdS in Polyelectrolytes | 552 |
| 22.3.1 | Dip versus Dip–Spin Assembly of Q-CdS in Polyelectrolytes | 553 |
| 22.3.2 | Photoelectrochemistry at PDDA/Q-CdS Assembly in the Presence of Hole Scavengers | 555 |
| 22.3.3 | Photocurrent Polarity-Switching at Q-CdS Photoelectrodes | 556 |
| 22.4 | Conclusions | 558 |
| | References | 559 |

**23 Record Properties of Layer-by-Layer Assembled Composites 573**
*Ming Yang, Paul Podsiadlo, Bong Sup Shim, and Nicholas A. Kotov*

| | | |
|---|---|---|
| 23.1 | Introduction | 573 |
| 23.2 | LbL Assemblies of Clays | 574 |
| 23.2.1 | Structure and Properties of Clay Particles | 574 |
| 23.2.2 | Structural Organization in Clay Multilayers | 575 |
| 23.2.3 | Clay Multilayers as High-Performance Nanocomposites | 576 |
| 23.2.4 | Applications of Clay Multilayers in Biotechnology | 578 |
| 23.2.5 | Anisotropic Transport in Clay Multilayers | 580 |
| 23.2.6 | Clay Multilayers for Optical and Electronic Applications | 581 |
| 23.3 | LBL Assemblies of Carbon Nanotubes | 582 |
| 23.3.1 | Structure and Properties of CNTs | 582 |
| 23.3.2 | Structural Organization and Mechanical Properties in Multilayers of Carbon Nanotubes | 583 |
| 23.3.3 | Electrical Conductor Applications | 584 |
| 23.3.4 | Sensor Applications | 585 |
| 23.3.5 | Fuel Cell Applications | 587 |
| 23.3.6 | Nano-/Micro-Shell LbL Coatings | 587 |
| 23.3.7 | Biomedical Applications | 588 |
| 23.4 | Conclusions and Perspectives | 589 |
| | References | 590 |

**24 Carbon Nanotube-Based Multilayers 595**
*Yong Tae Park and Jaime C. Grunlan*

| | | |
|---|---|---|
| 24.1 | Introduction | 595 |
| 24.2 | Characteristics of Carbon Nanotube Layer-by-Layer Assemblies | 596 |
| 24.2.1 | Growth of Carbon Nanotube-Based Multilayers | 596 |
| 24.2.2 | Electrical Properties | 600 |
| 24.2.3 | Mechanical Behavior | 601 |
| 24.3 | Applications of Carbon Nanotube Layer-by-Layer Assemblies | 602 |
| 24.3.1 | Transparent Electrodes | 602 |
| 24.3.2 | Sensor Applications | 604 |
| 24.3.3 | Energy-Related Applications | 607 |

| | | |
|---|---|---|
| 24.3.4 | Biomedical Applications | 608 |
| 24.4 | Conclusions | 609 |
| | References | 609 |

| | | |
|---|---|---|
| **25** | **Nanoconfined Polyelectrolyte Multilayers: From Nanostripes to Multisegmented Functional Nanotubes** | **613** |

*Cécile J. Roy, Cédric C. Buron, Sophie Demoustier-Champagne, and Alain M. Jonas*

| | | |
|---|---|---|
| 25.1 | Introduction | 613 |
| 25.2 | Estimation of the Size of Polyelectrolyte Chains in Dilute Solutions | 614 |
| 25.3 | Confining LbL Assembly on Flat Surfaces | 618 |
| 25.3.1 | LbL Assembly Templated by Chemically Nanopatterned Surfaces | 619 |
| 25.3.2 | LbL Lift-Off | 621 |
| 25.4 | Confining LbL Assembly in Nanopores | 624 |
| 25.4.1 | Peculiarities of LbL Assembly in Nanopores | 627 |
| 25.4.2 | Multisegmented LbL Nanotubes | 630 |
| 25.5 | Conclusions | 633 |
| | References | 634 |

| | | |
|---|---|---|
| **26** | **The Design of Polysaccharide Multilayers for Medical Applications** | **637** |

*Benjamin Thierry, Dewang Ma, and Françoise M. Winnik*

| | | |
|---|---|---|
| 26.1 | Introduction | 637 |
| 26.2 | Polysaccharides as Multilayered film Components: An Overview of Their Structure and Properties | 638 |
| 26.2.1 | Polycations | 639 |
| 26.2.2 | Polyanions | 641 |
| 26.3 | Multilayers Formed by Assembly of Weak Polyanions and Chitosan or Chitosan Derivatives | 642 |
| 26.3.1 | Hyaluronan | 642 |
| 26.3.2 | Polygalacturonic Acid | 646 |
| 26.3.3 | Alginate | 647 |
| 26.4 | Multilayers Formed by Assembly of Strong Polyanions and Chitosan or Chitosan Derivatives | 647 |
| 26.4.1 | Heparin | 647 |
| 26.4.2 | Chondroitin Sulfate | 649 |
| 26.5 | Cardiovascular Applications of Polysaccharide Multilayers | 650 |
| 26.5.1 | Anti-Thrombogenic Properties of Polysaccharide Multilayers | 650 |
| 26.5.2 | Preparation of Polysaccharide Multilayers on Blood Vessels | 651 |
| 26.5.3 | Drug Delivery to the Vascular Wall from Polysaccharide Multilayers | 652 |
| 26.6 | Conclusions | 654 |
| | References | 655 |

| 27 | **Polyelectrolyte Multilayer Films Based on Polysaccharides: From Physical Chemistry to the Control of Cell Differentiation** 659 |
|---|---|
| | *Thomas Boudou, Kefeng Ren, Thomas Crouzier, and Catherine Picart* |
| 27.1 | Introduction 659 |
| 27.2 | Film Internal Composition and Hydration 660 |
| 27.2.1 | Film Growth 660 |
| 27.2.2 | Hydration and Swellability 662 |
| 27.2.3 | Internal Composition 665 |
| 27.3 | Film Cross-Linking: Relation Between Composition and Mechanical Properties 666 |
| 27.3.1 | Ionic Pairing 666 |
| 27.3.2 | Covalent Amide Bonds 668 |
| 27.3.3 | Mechanical Properties of Films and Correlation with Cross-Link Density 669 |
| 27.4 | Cell Adhesion onto Cross-Linked Films: Cell Adhesion, Cytoskeletal Organization and Comparison with Other Model Materials 671 |
| 27.4.1 | Film Cross-Linking Modulates Myoblast Attachment, Proliferation and Cytoskeletal Organization 671 |
| 27.4.2 | Correlation Between the Cell Spreading Area and Young's Modulus $E_0$ 675 |
| 27.4.3 | Comparison with Other Material Substrates: Polyacrylamide (PA) and Polydimethysiloxane (PDMS) 677 |
| 27.5 | Cell Differentiation: ESC and Myoblasts 679 |
| 27.5.1 | Film Cross-Linking Influences the Myogenic Differentiation Process Morphologically but not the Expression of Muscle-Specific Proteins 679 |
| 27.5.2 | Film Cross-Linking Drives the Fate of Mouse Embryonic Stem 681 |
| 27.6 | Conclusions 684 |
| | References 685 |
| | |
| 28 | **Diffusion of Nanoparticles and Biomolecules into Polyelectrolyte Multilayer Films: Towards New Functional Materials** 691 |
| | *Marc Michel and Vincent Ball* |
| 28.1 | Introduction 691 |
| 28.2 | LBL Films in Which Nanoparticles are Incorporated Step-By-Step 693 |
| 28.3 | LBL Films Made Uniquely From Nanoparticles 693 |
| 28.4 | Nanoparticles Produced by Post-treatment of Deposited Films 694 |
| 28.5 | Diffusion of Colloids in Already Deposited Films 698 |
| 28.5.1 | Permeability of LBL Films Towards Ions, Small Drugs and Dyes 698 |
| 28.5.2 | Diffusion of Nanoparticles in and out of LBL Films 700 |
| 28.6 | Emerging Properties of Films Filled with Nanoparticles by the Post-incubation Method 705 |
| 28.6.1 | Production of Anisotropically-Coated Colloids 705 |

| | | |
|---|---|---|
| 28.6.2 | Sensing and Controlled Release Applications | 705 |
| 28.6.3 | Electrical Conductivity | 705 |
| 28.7 | Conclusions and Perspectives | 706 |
| | References | 707 |

**29 Coupling Chemistry and Hybridization of DNA Molecules on Layer-by-Layer Modified Colloids** 711
*Jing Kang and Lars Dähne*

- 29.1 Introduction 711
- 29.2 Materials and Methods 712
- 29.2.1 Materials 712
- 29.2.2 Methods 713
- 29.2.2.1 Coupling of Oligonucleotides onto LbL-Coated Colloidal Particles 713
- 29.2.2.2 Hybridization of Complementary Oligos Onto the LbL-Oligo Particles 716
- 29.3 Results 716
- 29.3.1 One-Step versus Two-Step Coupling Process by EDC 716
- 29.3.2 Selection of Polyelectrolytes 716
- 29.3.2.1 Selection of Polycations 717
- 29.3.2.2 Selection of Polyanions 718
- 29.3.3 Non-Specific Binding and Its Minimization 718
- 29.3.4 Coupling by 1.1′-Carbonyldiimidazol in Organic Solvent 721
- 29.3.5 Number of Coating Layers 722
- 29.3.6 Coupling and Hybridization Efficiency 722
- 29.3.6.1 Coupling Efficiency 722
- 29.3.6.2 Hybridization Efficiency 723
- 29.3.7 Comparison of LbL Particles with Conventional Carboxylated Particles 724
- 29.3.8 Arrangement of the Coupled Oligo Molecules 724
- 29.3.8.1 Analysis by FRET Investigations 724
- 29.3.8.2 FRET of dsOligos Free in Solution 725
- 29.3.8.3 FRET of dsOligos on LbL Particles 726
- 29.3.9 Stability of the LbL-Oligo Particles 727
- 29.4 Summary 727
- References 729

**30 A "Multilayered" Approach to the Delivery of DNA: Exploiting the Structure of Polyelectrolyte Multilayers to Promote Surface-Mediated Cell Transfection and Multi-Agent Delivery** 731
*David M. Lynn*

- 30.1 Introduction 731
- 30.2 Surface-Mediated Delivery of DNA: Motivation and Context, Opportunities and Challenges 732
- 30.3 Films Fabricated Using Hydrolytically Degradable Cationic Polymers 734

| | | |
|---|---|---|
| 30.4 | Toward Spatial Control: Release of DNA from the Surfaces of Implants and Devices *736* | |
| 30.5 | Toward Temporal Control: Tunable Release and Sequential Release *739* | |
| 30.5.1 | Approaches Based on Incorporation of Different Hydrolytically Degradable Polyamines *740* | |
| 30.5.2 | Approaches Based on Incorporation of Cationic "Charge-Shifting" Polymers *742* | |
| 30.5.3 | New Approaches to Rapid Release *744* | |
| 30.6 | Concluding Remarks *745* | |
| | References *746* | |
| | | |
| **31** | **Designing LbL Capsules for Drug Loading and Release** *749* | |
| | *Bruno G. De Geest and Stefaan C. De Smedt* | |
| 31.1 | Introduction *749* | |
| 31.2 | Engineering Microparticulate Templates to Design LbL Capsules for Controlled Drug Release *750* | |
| 31.3 | Engineering the Shell to Design LbL Capsules for Controlled Drug Release *753* | |
| 31.4 | Interaction of LbL Capsules with Living Cells *In Vitro* and *In Vivo* *759* | |
| 31.5 | Conclusions *761* | |
| | References *761* | |
| | | |
| **32** | **Stimuli-Sensitive LbL Films for Controlled Delivery of Proteins and Drugs** *765* | |
| | *Katsuhiko Sato, Shigehiro Takahashi, and Jun-ichi Anzai* | |
| 32.1 | Introduction *765* | |
| 32.2 | Avidin-Containing LbL Films *765* | |
| 32.3 | Concanavalin A-containing LbL Films *768* | |
| 32.4 | Dendrimer-Containing LbL Films *771* | |
| 32.5 | Insulin-Containing LbL Films *772* | |
| 32.6 | Conclusions *774* | |
| | References *776* | |
| | | |
| **33** | **Assembly of Multilayer Capsules for Drug Encapsulation and Controlled Release** *777* | |
| | *Jinbo Fei, Yue Cui, Qiang He, and Junbai Li* | |
| 33.1 | Introduction *777* | |
| 33.2 | Magnetically Sensitive Release *779* | |
| 33.3 | Ultrasound-Stimulated Release *780* | |
| 33.4 | Photo-Stimulated Release *781* | |
| 33.5 | Thermo-Stimulated Release *783* | |
| 33.6 | pH-Sensitive Release *785* | |
| 33.7 | Redox-Controlled Release *787* | |

| 33.8 | Bio-Responsive Release  788 |
| 33.9 | Extension  792 |
| 33.10 | Concluding Remarks  794 |
| | References  794 |

**34  Engineered Layer-by-Layer Assembled Capsules for Biomedical Applications  801**
*Angus P.R. Johnston, Georgina K. Such, Sarah J. Dodds, and Frank Caruso*

| 34.1 | Introduction  801 |
| 34.2 | Template Selection  801 |
| 34.3 | Material Assembly  804 |
| 34.4 | Loading  809 |
| 34.4.1 | Preloading on/in Template  810 |
| 34.4.2 | Loading Within Layers  811 |
| 34.4.3 | Post-Loading  813 |
| 34.5 | Degradation and Release  813 |
| 34.6 | Applications  816 |
| 34.6.1 | Microreactors  816 |
| 34.6.2 | Targeting  818 |
| 34.6.3 | Therapeutic Delivery  819 |
| 34.6.3.1 | Small Molecules  819 |
| 34.6.3.2 | Vaccines  821 |
| 34.6.3.3 | DNA  822 |
| 34.7 | Conclusions  823 |
| | References  824 |

**35  Assembly of Polymer Multilayers from Organic Solvents for Biomolecule Encapsulation  831**
*Sebastian Beyer, Jianhao Bai, and Dieter Trau*

| 35.1 | Introduction  831 |
| 35.1.1 | Bio-Template-Based LbL Encapsulation in the Aqueous Phase  831 |
| 35.1.2 | Loading-Based LbL Biomolecule Encapsulation in the Aqueous Phase  832 |
| 35.1.3 | Diffusion-Based LbL Biomolecule Encapsulation in the Aqueous Phase  833 |
| 35.2 | Limitations of LbL-Based Biomolecule Encapsulation in Aqueous Phase  834 |
| 35.3 | LbL Biomolecule Encapsulation in the Organic Phase  835 |
| 35.3.1 | Reverse-Phase LbL  836 |
| 35.3.1.1 | Mechanism  836 |
| 35.3.1.2 | Technique  839 |
| 35.3.1.3 | Encapsulation of Biomolecules  840 |
| 35.3.2 | "Inwards Build-Up Self-Assembly" of Polymers for Biomolecule Encapsulation in the Organic Phase  845 |
| 35.3.2.1 | Mechanism and Technique  845 |

| 35.3.2.2 | Encapsulation of Biomolecules  846 |
|---|---|
| 35.4 | Conclusion and Outlook  847 |
| | References  849 |

**36 Stimuli-Responsive Polymer Composite Multilayer Microcapsules and Microchamber Arrays**  851
*Maria N. Antipina, Maxim V. Kiryukhin, and Gleb B. Sukhorukov*

| 36.1 | Introduction  852 |
|---|---|
| 36.2 | Fabrication of Stimuli-Responsive LbL Microcapsules  853 |
| 36.2.1 | pH-Responsive Capsules  855 |
| 36.2.2 | Salt-Responsive Capsules and Capsule Fusion  858 |
| 36.2.3 | Redox-Responsive Capsules  862 |
| 36.2.4 | Chemical-Responsive Capsules  862 |
| 36.2.4.1 | Solvent  862 |
| 36.2.4.2 | Glucose  863 |
| 36.2.4.3 | $CO_2$  863 |
| 36.2.4.4 | Enzymes  864 |
| 36.2.5 | Temperature-Responsive Capsules  865 |
| 36.2.6 | Remote Responsive Capsules  866 |
| 36.2.6.1 | Magnetic LbL Capsules  866 |
| 36.2.6.2 | Ultrasound-Triggered Release  867 |
| 36.2.6.3 | Optically Addressable Capsules  868 |
| 36.2.7 | Mechanical Addressing of Individual Capsules  869 |
| 36.2.8 | Patterning Polyelectrolyte Capsules  872 |
| 36.3 | Microchamber Arrays  873 |
| 36.3.1 | Fabrication of Microchambers by LbL Assembly on Imprinted Surfaces  873 |
| 36.3.1.1 | LbL assembly of the PEMs in confined geometries  874 |
| 36.3.1.2 | Dissolving the template and making a free-standing PEM film with an array of standing hollow microchambers  874 |
| 36.3.2 | Microchamber Loading with Substances of Interest  877 |
| 36.3.3 | Responsiveness of Chambers to Light and Mechanical Load  879 |
| 36.4 | Conclusion  881 |
| | References  882 |

**37 Domain-Containing Functional Polyelectrolyte Films: Applications to Antimicrobial Coatings and Energy Transfer**  891
*Aurélie Guyomard, Bernard Nysten, Alain M. Jonas, and Karine Glinel*

| 37.1 | Introduction  891 |
|---|---|
| 37.2 | Polyelectrolyte Films Incorporating Randomly Distributed Hydrophobic Nanodomains for Antimicrobial Applications  893 |
| 37.2.1 | Hydrophobic Nanodomains in Hydrophilic PEMs with Amphiphilic Macromolecules  893 |
| 37.2.2 | Entrapment of an Antibacterial Peptide in Hydrophobic Nanodomains  895 |

| | | |
|---|---|---|
| 37.3 | Multicompartmentalized Stratified Polyelectrolyte Films for Control of Energy Transfer  898 | |
| 37.3.1 | How Precisely Can the Stratification of a LbL Film Be Controlled?  898 | |
| 37.3.2 | Fabrication of Dye-Impermeable Polyelectrolyte Barriers  901 | |
| 37.3.3 | Control of a Cascade of Events in a Multicompartmentalized Stratified Polyelectrolyte Film  902 | |
| 37.4 | Conclusions and Perspectives  903 | |
| | References  904 | |
| | | |
| **38** | **Creating Functional Membranes Through Polyelectrolyte Adsorption  907** | |
| | *Merlin L. Bruening* | |
| 38.1 | Introduction  907 | |
| 38.2 | Functionalization of the Interior of Membranes  908 | |
| 38.2.1 | Deposition of PEMs in Porous Media  908 | |
| 38.2.2 | Functionalization of Membranes with Proteins  910 | |
| 38.2.2.1 | Protein Adsorption in Membranes  910 | |
| 38.2.2.2 | Trypsin-Containing Polyelectrolyte Films for Protein Digestion  911 | |
| 38.2.3 | Catalytic Films and Membranes  914 | |
| 38.2.3.1 | Catalytic, Nanoparticle-Containing Films  914 | |
| 38.2.3.2 | Catalytic Membranes  915 | |
| 38.3 | LBL Films as Membrane Skins  918 | |
| 38.3.1 | Early Studies  918 | |
| 38.3.2 | Removal of Dyes and Small Organic Molecules from Water  918 | |
| 38.3.3 | Selective Rejection of $F^-$ and Phosphate  920 | |
| 38.3.4 | Variation of PSS/PDADMAC film Properties with the Number of Adsorbed Layers  921 | |
| 38.4 | Challenges  922 | |
| | References  922 | |
| | | |
| **39** | **Remote and Self-Induced Release from Polyelectrolyte Multilayer Capsules and Films  925** | |
| | *Andre G. Skirtach, Dmitry V. Volodkin, and Helmuth Möhwald* | |
| | References  940 | |
| | | |
| **40** | **Controlled Architectures in LbL Films for Sensing and Biosensing  951** | |
| | *Osvaldo N. Oliveira Jr., Pedro H.B. Aoki, Felippe J. Pavinatto, and Carlos J.L. Constantino* | |
| 40.1 | Introduction  951 | |
| 40.2 | LbL-Based Sensors and Biosensors  952 | |
| 40.2.1 | Optical Detection Methods  952 | |
| 40.2.2 | Mass Change Methods  954 | |
| 40.2.3 | Electrochemical Methods  955 | |

| | | |
|---|---|---|
| 40.2.4 | Methods Involving Electrical Measurements | 956 |
| 40.2.5 | E-Tongues and E-Noses | 958 |
| 40.2.6 | Extending the Concept of E-Tongue to Biosensing | 963 |
| 40.3 | Special Architectures for Sensing and Biosensing | 964 |
| 40.4 | Statistical and Computational Methods to Treat the Data | 969 |
| 40.4.1 | Artificial Neural Networks and Regression Methods | 970 |
| 40.4.2 | Optimization of Biosensing Performance Using Multidimensional Projections | 971 |
| 40.5 | Conclusions and Perspectives | 977 |
| | References | 978 |

**41 Patterned Multilayer Systems and Directed Self-Assembly of Functional Nano-Bio Materials** *985*

*Ilsoon Lee*

| | | |
|---|---|---|
| 41.1 | New Approaches and Materials for Multilayer Film Patterning Techniques | 985 |
| 41.2 | Cell Adhesion and Patterning Using PEMs | 988 |
| 41.3 | PEMs Incorporating Proteins and Their Patterning | 990 |
| 41.4 | Metal/Graphene Conductive Patterning via PEM Films | 992 |
| 41.5 | Ordered and Disordered Particles on PEMs | 995 |
| 41.6 | Mechanical Aspects of PEM Films and Degradable Films | 997 |
| | References | 999 |

**42 Electrochemically Active LbL Multilayer Films: From Biosensors to Nanocatalysts** *1003*

*Ernesto. J. Calvo*

| | | |
|---|---|---|
| 42.1 | Introduction | 1003 |
| 42.2 | Electrochemical Response | 1004 |
| 42.3 | Dynamics of Charge Exchange | 1012 |
| 42.3.1 | Propagation of Redox Charge (Electron Hopping) | 1012 |
| 42.3.2 | Ion Exchange | 1018 |
| 42.3.3 | Applications | 1024 |
| 42.3.4 | Biosensors | 1026 |
| 42.3.5 | Core–Shell Nanoparticles | 1030 |
| 42.3.6 | Nanoreactors | 1030 |
| 42.3.7 | Biofuel Cell Cathodes | 1031 |
| 42.4 | Conclusions | 1033 |
| | References | 1034 |

**43 Multilayer Polyelectrolyte Assembly in Feedback Active Coatings and Films** *1039*

*Dmitry G. Shchukin and Helmuth Möhwald*

| | | |
|---|---|---|
| 43.1 | Introduction. The Concept of Feedback Active Coatings | 1039 |
| 43.2 | Polyelectrolyte-Based Self-Healing Anticorrosion Coatings | 1040 |
| 43.2.1 | Passive Protection Activity of Polyelectrolyte Multilayers | 1042 |

| | | |
|---|---|---|
| 43.2.2 | Controlled Release of Inhibiting Agents | 1044 |
| 43.3 | Coatings with Antibacterial Activity | 1045 |
| 43.4 | Conclusions and Outlook | 1050 |
| | References | 1050 |

**Index** 1053

# List of Contributors

**Rigoberto C. Advincula**
University of Houston
Departments of Chemistry and
Chemical and Biomolecular
Engineering
136 Fleming Building
Houston, TX 77204-5003
USA

**Hioharu Ajiro**
Osaka University
Department of Applied Chemistry
2-1 Yamada-oka
Suita, Osaka 565-0871
Japan

**Mitsuro Akashi**
Osaka University
Department of Applied Chemistry
2-1 Yamada-oka
Suita, Osaka 565-0871
Japan

**Maria N. Antipina**
Agency for Science, Technology and
Research (A*STAR)
Institute of Materials Research and
Engineering
3 Research Link
Singapore 117602
Singapore

**Jun-ichi Anzai**
Tohoku University
Graduate School of
Pharmaceutical Sciences
Aramaki, Aoba-ku
Sendai 980-8578
Japan

**Pedro H.B. Aoki**
Universidade Estadual Paulista
(UNESP)
Faculdade de Ciências e Tecnologia
19060-900 Presidente Prudente, SP
Brazil

**Katsuhiko Ariga**
National Institute for Materials Science
(NIMS)
World Premier International (WPI)
Research Center for Materials
Nanoarchitectonics (MANA)
1-1 Namiki
Tsukuba 305-0044
Japan

and

JST
CREST
1-1 Namiki
Tsukuba 305-0044
Japan

**Jianhao Bai**
National University of Singapore
Division of Bioengineering
Singapore 117576
Singapore

**Vincent Ball**
Centre de Recherche Public
Henri Tudor
Department of Advanced Materials and
Structures
66 rue de Luxembourg
4002 Esch-sur-Alzette
Luxembourg

**Sebastian Beyer**
National University of Singapore
Division of Bioengineering
Singapore 117576
Singapore

and

National University of Singapore
Graduate School for Integrative
Sciences and Engineering
Singapore 11756
Singapore

**Stephan Block**
Ernst-Moritz-Arndt Universität
Institut für Physik
Felix-Hausdorff-Str. 6
17487 Greifswald
Germany

**Thomas Boudou**
Grenoble Institute of Technology and
Centre National de la Recherche
Scientifique
CNRS UMR 5628, LMGP, MINATEC
3 Parvis Louis Néel
38016 Grenoble
France

**Merlin L. Bruening**
Michigan State University
Department of Chemistry
East Lansing, MI 48824
USA

**Cédric C. Buron**
Université Catholique de Louvain
Institute of Condensed Matter and
Nanosciences – Bio & Soft Matter
Croix du Sud 1
1348 Louvain-la-Neuve
Belgium

**Ernesto J. Calvo**
Universidad de Buenos Aires
Departamento de Química Inorgánica
Electrochemistry Group, INQUIMAE
Pabellón 2, Ciudad Universitária
Buenos Aires 1428
Argentina

**Frank Caruso**
The University of Melbourne
Department of Chemical and
Biomolecular Engineering
Building 173
Melbourne, Victoria 3010
Australia

**Chloe Chevigny**
TU Berlin
Department of Chemistry
Straße des 17. Juni 124
10623 Berlin
Germany

**Robert E. Cohen**
Massachusetts Institute of Technology
Departments of Materials Science and
Engineering and Chemical Engineering
77 Massachusetts Avenue
Cambridge, MA 02139
USA

## List of Contributors

**Carlos J.L. Constantino**
Universidade Estadual Paulista
(UNESP)
Faculdade de Ciências e Tecnologia
19060-900 Presidente Prudente, SP
Brazil

**Cornelia Cramer**
University of Münster
Institute of Physical Chemistry
Corrensstr. 28/30
48149 Münster
Germany

**Thomas Crouzier**
Grenoble Institute of Technology and
Centre National de la Recherche
Scientifique
CNRS UMR 5628, LMGP, MINATEC
3 Parvis Louis Néel
38016 Grenoble
France

**Yue Cui**
Chinese Academy of Sciences
Institute of Chemistry
Beijing National Laboratory for
Molecular Sciences (BNLMS)
Zhongguancun North First Street 2
Beijing 100190
China

**Chalongrat Daengngam**
Virginia Polytechnic Institute and
State University
Department of Physics
Robeson Hall (0435)
Blacksburg, VA 24061-0435
USA

**Bruno G. De Geest**
Ghent University
Department of Pharmaceutics
Harelbekestraat 72
9000 Ghent
Belgium

**Sophie Demoustier-Champagne**
Université Catholique de Louvain
Institute of Condensed Matter and
Nanosciences – Bio & Soft Matter
Croix du Sud 1
1348 Louvain-la-Neuve
Belgium

**Stefaan C. De Smedt**
Ghent University
Department of Pharmaceutics
Harelbekestraat 72
9000 Ghent
Belgium

**Wang Dong**
Virginia Polytechnic Institute and
State University
Department of Physics
Robeson Hall (0435)
Blacksburg, VA 24061-0435
USA

**Ashraf El-Hashani**
University of Cologne
Department of Chemistry
Luxemburger Str. 116
50939 Köln
Germany

**Nicel Estillore**
University of Houston
Departments of Chemistry and
Chemical and Biomolecular
Engineering
136 Fleming Building
Houston, TX 77204-5003
USA

**Jinbo Fei**
Chinese Academy of Sciences
Institute of Chemistry
Beijing National Laboratory for
Molecular Sciences (BNLMS)
Zhongguancun North First Street 2
Beijing 100190
China

**Andreas Fery**
University of Bayreuth
Department of Physical Chemistry II
95440 Bayreuth
Germany

**Karine Glinel**
Université de Rouen, CNRS
Laboratoire Polymères, Biopolymères,
Surfaces
Bd M. de Broglie
7681 Mont Saint Aignan
France

and

Université Catholique de Louvain
Institute of Condensed Matter and
Nanosciences – Bio & Soft Matter
Croix du Sud 1
1348 Louvain-la-Neuve
Belgium

**Jaime C. Grunlan**
Texas A&M University
Department of Mechanical Engineering
College Station, TX 77843-3123
USA

**Aurélie Guyomard**
Université de Rouen, CNRS
Laboratoire Polymères, Biopolymères,
Surfaces
Bd M. de Broglie
7681 Mont Saint Aignan
France

**Lara Halaoui**
American University of Beirut
Department of Chemistry
Beirut
Lebanon

**Paula T. Hammond**
Massachusetts Institute of Technology
Department of Chemical Engineering
77 Massachusetts Avenue
Cambridge, MA 02139
USA

**Qiang He**
Harbin Institute of Technology
Micro/Nano Technology Research
Centre
Harbin 150080
China

**Randy Heflin**
Virginia Polytechnic Institute and
State University
Department of Physics
Robeson Hall (0435)
Blacksburg, VA 24061-0435
USA

**Christiane A. Helm**
Ernst-Moritz-Arndt Universität
Institut für Physik
Felix-Hausdorff-Str. 6
17487 Greifswald
Germany

**Jonathan P. Hill**
National Institute for Materials Science
(NIMS)
World Premier International (WPI)
Research Center for Materials
Nanoarchitectonics (MANA)
1-1 Namiki
Tsukuba 305-0044
Japan

and

JST
CREST
1-1 Namiki
Tsukuba 305-0044
Japan

*Kristina Hoffmann*
University of Cologne
Department of Chemistry
Luxemburger Str. 116
50939 Köln
Germany

*Jurriaan Huskens*
University of Twente
MESA+ Institute for Nanotechnology
Molecular Nanofabrication Group
7500 AE Enschede
The Netherlands

*Md Nasim Hyder*
Massachusetts Institute of Technology
Department of Chemical Engineering
77 Massachusetts Avenue
Cambridge, MA 02139
USA

*Qingmin Ji*
National Institute for Materials Science (NIMS)
World Premier International (WPI) Research Center for Materials Nanoarchitectonics (MANA)
1-1 Namiki
Tsukuba 305-0044
Japan

*Chaoyang Jiang*
University of South Dakota
Chemistry Department
414 East Clark Street
Vermillion, SD 57069
USA

*Alain M. Jonas*
Université Catholique de Louvain
Institute of Condensed Matter and Nanosciences – Bio & Soft Matter
Croix du Sud 1
1348 Louvain-la-Neuve
Belgium

*Jouko Kankare*
University of Turku
Department of Chemistry
Laboratory of Materials Chemistry and Chemical Analysis
20014 Turku
Finland

*Toshiyuki Kida*
Osaka University
Department of Applied Chemistry
2-1 Yamada-oka
Suita, Osaka 565-0871
Japan

*Maxim V. Kiryukhin*
Agency for Science, Technology and Research (A*STAR)
Institute of Materials Research and Engineering
3 Research Link
Singapore 117602
Singapore

*Ralf Köhler*
TU Berlin
Department of Chemistry
Straße des 17. Juni 124
10623 Berlin
Germany

*Nicholas A. Kotov*
University of Michigan
Departments of Chemical Engineering, Materials Science and Engineering, and Biomedical Engineering
2300 Hayward Street
Ann Arbor, MI 48109
USA

**Ilsoon Lee**
Michigan State University
Department of Chemical Engineering
and Materials Science
2527 Engineering Building
East Lansing, MI 48824
USA

**Junbai Li**
Chinese Academy of Sciences
Institute of Chemistry
Beijing National Laboratory for
Molecular Sciences (BNLMS)
Zhongguancun North First Street 2
Beijing 100190
China

**Xiaokong Liu**
Jilin University
College of Chemistry
State Key Laboratory of Supramolecular
Structure and Materials
Changchun 130012
China

**Yuri Lvov**
Louisiana Tech University
Institute for Micromanufacturing
911 Hergot Avenue
Ruston, LA 71272
USA

**David M. Lynn**
University of Wisconsin – Madison
Department of Chemical and Biological
Engineering
1415 Engineering Drive
Madison, WI 53706
USA

**Dewang Ma**
Université de Montréal
Faculté de Pharmacie
Succursale Centre-Ville
Montréal, Quebec H3C 3J7
Canada

**Anna Maier**
University of Cologne
Department of Chemistry
Luxemburger Str. 116
50939 Köln
Germany

**Michiya Matsusaki**
Osaka University
Department of Applied Chemistry
2-1 Yamada-oka
Suita, Osaka 565-0871
Japan

**Jonathan Metzman**
Virginia Polytechnic Institute and
State University
Department of Materials Science and
Engineering
Robeson Hall (0435)
Blacksburg, VA 24061-0435
USA

**Marc Michel**
Centre de Recherche Public
Henri Tudor
Department of Advanced Materials and
Structures
66 rue de Luxembourg
4002 Esch-sur-Alzette
Luxembourg

**Helmuth Möhwald**
Max-Planck Institute of
Colloids and Interfaces
Fraunhofer Institute of
Biomedical Technology
Research Campus Golm
Am Mühlenberg 1
14424 Potsdam-Golm
Germany

**Reza Montazami**
Virginia Polytechnic Institute and
State University
Department of Materials Science and
Engineering
Robeson Hall (0435)
Blacksburg, VA 24061-0435
USA

**Peter Nestler**
Ernst-Moritz-Arndt Universität
Institut für Physik
Felix-Hausdorff-Str. 6
17487 Greifswald
Germany

**Bernard Nysten**
Université Catholique de Louvain
Institute of Condensed Matter and
Nanosciences – Bio & Soft Matter
Croix du Sud 1
1348 Louvain-la-Neuve
Belgium

**Osvaldo N. Oliveira Jr.**
Universidade de São Paulo
Instituto de Física de São Carlos
13560-970 São Carlos, SP
Brazil

**Yong Tae Park**
Texas A&M University
Department of Mechanical Engineering
College Station, TX 77843-3123
USA

**Pravin Pattekari**
Louisiana Tech University
Institute for Micromanufacturing
911 Hergot Avenue
Ruston, LA 71272
USA

**Felippe J. Pavinatto**
Universidade de São Paulo
Instituto de Física de São Carlos
13560-970 São Carlos, SP
Brazil

**Catherine Picart**
Grenoble Institute of Technology and
Centre National de la Recherche
Scientifique
CNRS UMR 5628, LMGP, MINATEC
3 Parvis Louis Néel
38016 Grenoble
France

**Paul Podsiadlo**
University of Michigan
Department of Chemical Engineering
2300 Hayward Street
Ann Arbor, MI 48109
USA

**David N. Reinhoudt**
University of Twente
MESA+ Institute for Nanotechnology
Molecular Nanofabrication Group
7500 AE Enschede
The Netherlands

**Kefeng Ren**
Grenoble Institute of Technology and
Centre National de la Recherche
Scientifique
CNRS UMR 5628, LMGP, MINATEC
3 Parvis Louis Néel
38016 Grenoble
France

**H.D. Robinson**
Virginia Polytechnic Institute and
State University
Department of Physics
Robeson Hall (0435)
Blacksburg, VA 24061-0435
USA

**Cécile J. Roy**
Université Catholique de Louvain
Institute of Condensed Matter and
Nanosciences – Bio & Soft Matter
Croix du Sud 1
1348 Louvain-la-Neuve
Belgium

**Michael F. Rubner**
Massachusetts Institute of Technology
Departments of Materials Science and
Engineering and Chemical Engineering
77 Massachusetts Avenue
Cambridge, MA 02139
USA

**Mikko Salomäki**
University of Turku
Department of Chemistry
Laboratory of Materials Chemistry and
Chemical Analysis
20014 Turku
Finland

**Katsuhiko Sato**
Tohoku University
Graduate School of
Pharmaceutical Sciences
Aramaki, Aoba-ku
Sendai 980-8578
Japan

**Joseph B. Schlenoff**
Florida State University
Department of Chemistry and
Biochemistry
95 Chieftan Way
Tallahassee, FL 32306-4390
USA

**Monika Schönhoff**
University of Münster
Institute of Physical Chemistry
Corrensstr. 28/30
48149 Münster
Germany

**Takeshi Serizawa**
University of Tokyo
Research Center for Advanced Science
and Technology
4-6-1 Komaba, Meguro-ku
Tokyo 153-8904
Japan

**Nisarg Shah**
Massachusetts Institute of Technology
Department of Chemical Engineering
77 Massachusetts Avenue
Cambridge, MA 02139
USA

**Dmitry G. Shchukin**
Max-Planck Institute of
Colloids and Interfaces
Research Campus Golm
Am Mühlenberg 1
14424 Potsdam-Golm
Germany

**Jiacong Shen**
Jilin University
College of Chemistry
State Key Laboratory of Supramolecular
Structure and Materials
Changchun 130012
China

**Bong Sup Shim**
University of Michigan
Department of Chemical Engineering
2300 Hayward Street
Ann Arbor, MI 48109
USA

**Tatsiana Shutava**
Louisiana Tech University
Institute for Micromanufacturing
911 Hergot Avenue
Ruston, LA 71272
USA

**Andre G. Skirtach**
Max-Planck-Institute of
Colloids and Interfaces
Fraunhofer Institute of
Biomedical Technology
Research Campus Golm
Am Mühlenberg 1
14476 Potsdam-Golm
Germany

**Olaf Soltwedel**
Ernst-Moritz-Arndt Universität
Institut für Physik
Felix-Hausdorff-Str. 6
17487 Greifswald
Germany

**Svetlana Sukhishvili**
Stevens Institute of Technology
Department of Chemistry, Chemical
Biology and Biomedical Engineering
1 Castle Point on Hudson
Hoboken, NJ 07030
USA

**Gleb B. Sukhorukov**
Queen Mary University of London
School of Engineering and
Materials Science
Mile End Road
London E1 4NS
UK

**Junqi Sun**
Jilin University
College of Chemistry
State Key Laboratory of Supramolecular
Structure and Materials
Changchun 130012
China

**Shigehiro Takahashi**
Tohoku University
Graduate School of
Pharmaceutical Sciences
Aramaki, Aoba-ku
Sendai 980-8578
Japan

**Benjamin Thierry**
University of South Australia
Ian Wark Research Institute
Mawson Lakes Campus
Mawson Lakes, South Australia 5095
Australia

**Bernd Tieke**
University of Cologne
Department of Chemistry
Luxemburger Str. 116
50939 Köln
Germany

**Dieter Trau**
National University of Singapore
Division of Bioengineering
Singapore 117576
Singapore

and

National University of Singapore
Department of Chemical &
Biomolecular Engineering
Singapore 11756
Singapore

**Vladimir V. Tsukruk**
Georgia Institute of Technology
Materials Science and Engineering
771 Ferst Drive N.W.
Atlanta, GA 30332-0245
USA

**Janneke Veerbeek**
University of Twente
MESA+ Institute for Nanotechnology
Molecular Nanofabrication Group
7500 AE Enschede
The Netherlands

**Dmitry V. Volodkin**
Max-Planck-Institute of
Colloids and Interfaces
Fraunhofer Institute of
Biomedical Technology
Research Campus Golm
Am Mühlenberg 1
14476 Potsdam-Golm
Germany

**Regine von Klitzing**
TU Berlin
Department of Chemistry
Straße des 17. Juni 124
10623 Berlin
Germany

**Lars Wågberg**
KTH – Royal Institute of Technology
School of Chemical Science and
Engineering
Fibre and Polymer Technology and
The Wallenberg Wood Science Centre
Teknikringen 56–58
10044 Stockholm
Sweden

**Françoise M. Winnik**
Université de Montréal
Faculté de Pharmacie and
Département de Chimie
Succursale Centre-Ville
Montréal, Quebec H3C 3J7
Canada

and

National Institute for Materials Science
(NIMS)
World Premier International (WPI)
Research Center for Materials
Nanoarchitectonics (MANA)
1-1 Namiki
Tsukuba 305-0044
Japan

**Guanglu Wu**
Tsinghua University
Department of Chemistry
116 Hetian Building
Beijing 100084
China

**Ming Yang**
University of Michigan
Department of Chemical Engineering
2300 Hayward Street
Ann Arbor, MI 48109
USA

**Xi Zhang**
Tsinghua University
Department of Chemistry
308 Hetian Building
Beijing 100084
China

**Ziwei Zuo**
Virginia Polytechnic Institute and
State University
Department of Physics
Robeson Hall (0435)
Blacksburg, VA 24061-0435
USA

# 1
## Layer-by-Layer Assembly (Putting Molecules to Work)
*Gero Decher*

### 1.1
#### The Whole is More than the Sum of its Parts

The properties of a material arise from the structural arrangement of its constituents and their dynamics.

In nature, a hierarchical organization assures that a required property of a material is realized with a minimum of constituents and a minimum of complexity through processes of evolutionary selection. In the living world functional entities exist on any length scale ranging from atoms to whole organisms, complex properties of "biological materials" such as cellular life emerging at the upper end of the nanoscale (Figure 1.1).

Chemistry, Physics and Materials Science are increasingly approaching maturity on the molecular and macroscopic length scales, shifting importance toward new nanoscale materials and nanoorganized systems. New materials properties, a prerequisite for responding to the emerging problems of the world population, will require materials design at the nanoscale, forcing the development of new multi-material nanocomposites.

In complex systems, new properties appear that are not observed for each individual component. While it is trivial that electrons and nuclei form atoms (sub-Ångstrom scale), that atoms form molecules (Ångstrom scale) or that monomers can be transformed into polymers (early nanometer scale), we are just beginning to explore the potential of supramolecular assemblies of multifunctional objects. Figure 1.1 summarizes how complex materials properties evolve hierarchically over a length scale of several orders of magnitude.

### 1.2
#### From Self-Assembly to Directed Assembly

Imagine a self-assembly experiment involving several different chemical species and obtaining equilibrium. Chances are that the result of this experiment will have produced a material with less-than-optimal properties. The generally adopted strategy for improvements is then to re-engineer the chemical structure of the

*Multilayer Thin Films: Sequential Assembly of Nanocomposite Materials*, Second Edition.
Edited by Gero Decher and Joseph B. Schlenoff.
© 2012 Wiley-VCH Verlag GmbH & Co. KGaA. Published 2012 by Wiley-VCH Verlag GmbH & Co. KGaA.

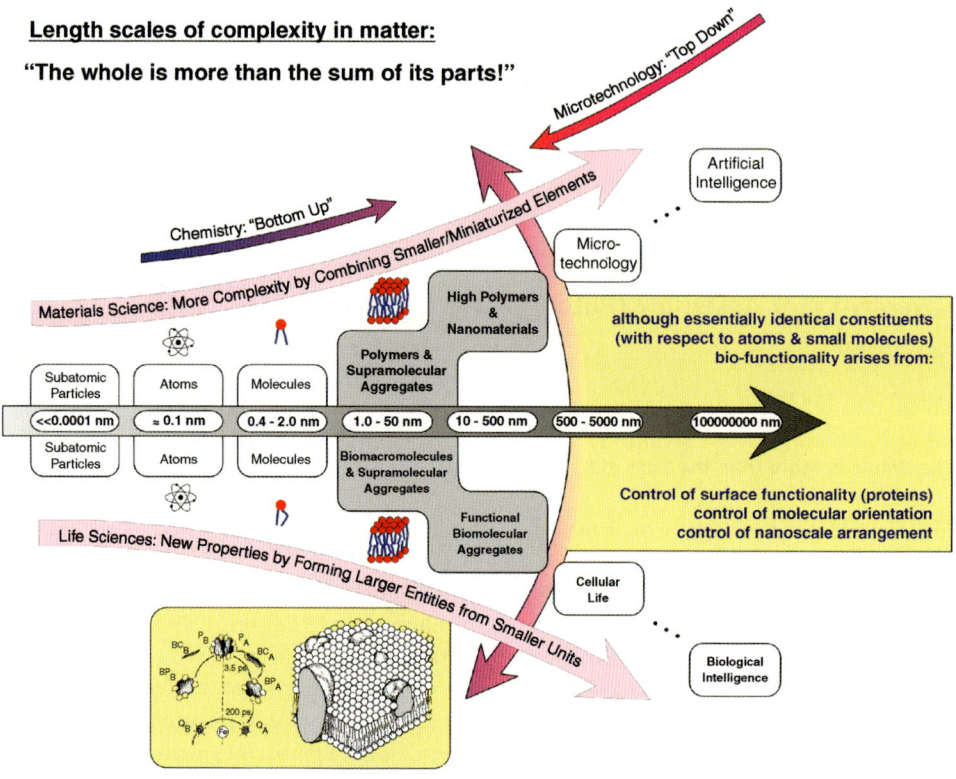

**Figure 1.1** Complexity as a function of length scale. Materials Science is not yet as far advanced as the evolutionary developments present everywhere in nature. The photosystem is just one example of the precise spatial assembly of a functional molecular machine. The drawing is adapted from [1].

molecules (objects) involved, to test different assembly conditions and to hope for a better outcome in the next experiment. This is a tedious approach as several optimization cycles are frequently required. This approach will necessarily become utterly hopeless when larger numbers of components are being involved and when hierarchically organized materials are targeted.

Another strategy to prepare a desired hierarchically organized composite material is to use an assembly procedure that bypasses equilibrium by trapping every compound kinetically in a predetermined spatial arrangement. At present there are only very few approaches in this direction, likely because as in nature, multiple assembly steps are required to arrive at the desired target structure. Most of the work has been carried out for rather simple cases, for example by assembling different materials with one-dimensional order in a multilayer film. For about 65 years the molecularly controlled fabrication of nanostructured multilayers has been dominated by the conceptually elegant Langmuir–Blodgett (LB) technique, in which monolayers are formed on a water surface [2] and subsequently transferred onto a solid support [3, 4]. The pioneering work on synthetic nanoscale *multi*composites of organic molecules was carried out by H. Kuhn and colleagues in the late 1960s using the LB technique [5]. His experiments with donor and acceptor dyes in different layers of LB-films provided direct proof of distance-dependent Förster energy transfer on the nanoscale. These were also the first true nanomanipulations as they allowed mechanical handling of individual molecular layers such as separation and contact formation with Ångstrom precision [6]. Despite being conceptually very elegant, the LB-technique is rather limited with respect to the set of molecular components suitable for LB-deposition, and because molecules are often not firmly trapped and frequently rearrange after or even during deposition.

The so-called layer-by-layer (LbL) deposition technique [1, 7–16] (Figure 1.2) is a more recent approach toward nanoscale multimaterial films that allows one to choose

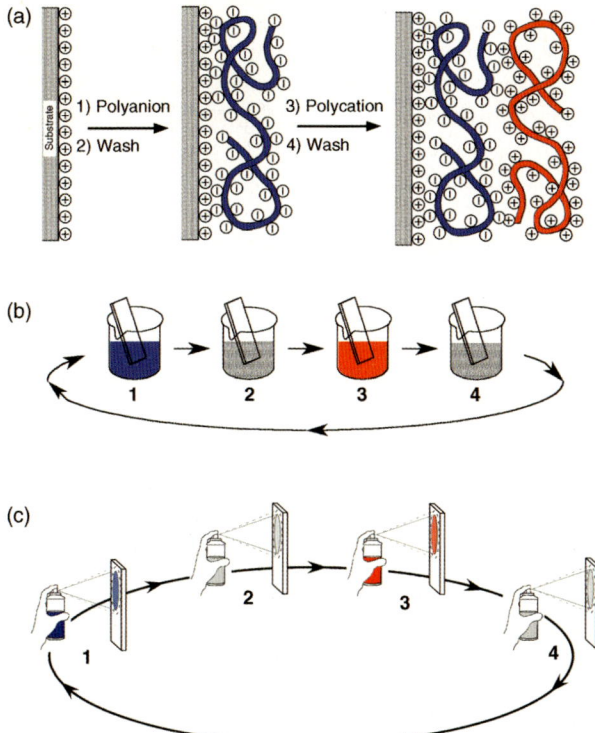

**Figure 1.2** (a) Simplified molecular conception of the first two adsorption steps depicting film deposition starting with a positively charged substrate. The polyion conformation and layer interpenetration are an idealization of the surface charge reversal with each adsorption step which is the base of the electrostatically driven multilayer build-up depicted here. Counterions are omitted for clarity. (b) Schematic of the film deposition process using glass slides and beakers. Steps 1 and 3 represent the adsorption of a polyanion and polycation respectively, and steps 2 and 4 are washing steps. The four steps are the basic build-up sequence for the simplest film architecture $(A/B)_n$ where $n$ is the number of deposition cycles. The construction of more complex film architectures is trivial and just requires additional beakers and an extended deposition sequence (see Figure 1.3). (c) Instead of bringing the surface into contact with the liquid that contains the adsorbing species by immersion, the liquid is sprayed against the receiving surface onto which the polyelectrolyte multilayer is deposited.

from an unprecedented variety of different components as constituents of the multilayer films and falls into the category of template assisted assembly.

The fabrication of multicomposite films by the LbL procedure means literally the nanoscopic assembly of hundreds of different materials in a single device using environmentally friendly, ultra-low-cost techniques. The materials can be small organic molecules or inorganic compounds, macromolecules including biomacromolecules, such as proteins or DNA or even colloids (metallic or oxidic colloids or latex particles). The choice of different components for LbL-assembly has become so large that the hundreds of different published articles describing that work cannot be meaningfully cited. The positive side of this aspect points to a key strength of LbL-assembly, the preparation of multicomponent nanomaterials. At present most publications in the field describe films composed of only two to five different components, however, the door is wide open for assembling more complex systems. Looking at the complexity of "materials" found in biological systems it seems clear that future LbL-devices will certainly contain more components in better controlled spatial arrangement (barriers, compartments, ...) with precise response to stimuli (degradation, opening/closing of pores, renewable surfaces, ...).

The LbL-technique can be applied to solvent accessible surfaces of almost any kind and any shape, the more exotic ones being colloids, fruit, textiles, paper or, even biological cells.

One of the key advantages of LbL-assembly is that LbL-films often display close to identical properties after deposition of the first few layers, even if films are deposited on very different surfaces. This means that it is frequently possible to prepare films for analytical purposes on (e.g., silicon) wafers while the films are put to work on surfaces that prevent thorough physico-chemical characterization (e.g., paper, textiles). Another interesting consequence is that different materials can be equipped with close to identical surface functionalities (e.g., anti-coagulation coatings on different surfaces in contact with blood).

## 1.3
### History and Development of the Layer-by-Layer Assembly Method

It is now a little over 20 years since layer-by-layer assembly was introduced as yet another method to functionalize surfaces and to fabricate thin films. Starting with simple bola-shaped amphiphiles [7] it was quickly extended to simple polyelectrolytes [8] and functional macromolecules including proteins [17, 18] or DNA [19, 20]. Eventually, in 1994, biological nanoparticles [21] and inorganic nanoparticles [22, 23], including magnetic [24] and gold [25] nanoparticles, were added to the list of possible multilayer film constituents. The field took off in the late 1990s, after it became clear that stratified multilayer architectures could be prepared [12, 26, 27] and that multilayer films can also be assembled on non-planar surfaces, namely on microparticles, as first demonstrated by T. Mallouk in 1995 [28] and then brought to a sheer explosion by H. Möhwald and his team a few years later [29, 30]. Polyelectrolyte multilayers were even deposited on nanoparticles [31, 32] a process that was further enhanced [33, 34] and which led to very interesting and very small multifunctional objects [35, 36]. Today, multilayer films are even prepared on surfaces such as textiles, paper, or on the skin of fruits. Due to its simplicity and the unprecedented choice of different components for polyelectrolyte multilayers and related systems, the method is today well established in materials science and making its way into the life sciences as well. Both developments have led to a solid and continuous growth of this field with no sign of reaching a peak or a plateau yet. In 2010, there were about 1000 articles published in the field.

Looking back at the historic development of this method over the last 20 years is quite interesting. When we started to work on this topic in 1989 (the first graduate student Jong-Dal Hong–now professor in Korea–finished his thesis in 1991) we were not aware of any other previous work on multilayer fabrication using electrostatic interactions. It was only much later that we learned about the work of R. K. Iler of Du Pont de Nemours & Co., on "Multilayers of colloidal particles" that was published in *J. Colloid and Interface Science* already back in 1966 [37]. In the early 1990s, the inversion of the zeta potential of charged colloids after adsorption of an oppositely charged polyelectrolyte had already been experimentally observed, for example, from work in the field of flocculation [38]. However, we were initially not aware of work by C. Gölander *et al.* [39] on "heparin layer formation" through sequential adsorption of a heparin complex and heparin which also demonstrated charge reversal, nor of a proposal of P. Fromherz [40] outlining similar ideas, or work of R. Aksberg *et al.* on the adsorption of a polyanion on cellulosic fibers with pre-adsorbed polycations [41]. We also did not know about a somewhat similar SILAR method (*S*uccessive *I*onic *L*ayer *A*dsorption and *R*eaction) for preparing polycrystalline inorganic films first reported by Nicolau in 1985 [42]. Neither seemed any of these research teams to know about each other.

Nowadays there is even "*unintentional*" use of LbL-assembly principles. Since about 1991 one increasingly finds protocols in biology in which "*two-component coatings*" are used or recommended for certain experiments (e.g., poly-L-lysine or poly-D-lysine and laminin for controlling cell attachment), nobody being aware that they are using layer-by-layer principles for their experiments. Imagine how much such experiments could benefit from knowing what is happening at the molecular scale and how the field of for example cell biology could gain advantage from engineering more suitable coatings for certain experiments. Today, the impetus for improvements in this field is coming from colleagues in the materials sciences (see many chapters in this book) who are addressing cell biology issues by introducing LbL-assembly to the life science community.

By 1988, when I started to work on layer-by-layer assembled films (unfortunately the first attempts during my time as a postdoctoral fellow using 4,4′-terphenyl disulfonic acid failed), the field of thin organic multilayer films was dominated by the so-called Langmuir-Blodgett technique (see above). Already in the early 80s, the field of LB-films had grown so large that the community started a meeting series "International Conference on Langmuir-Blodgett Films" the first of which being held in Durham, England in 1982, followed by meetings in Scenectady (1985) and Göttingen (1987). The meeting in Durham was of particular interest also with respect to layer-by-layer assembled films. At this meeting G.L. Gaines contributed two important papers "On the history of Langmuir-Blodgett films" [43] and "Deposition of colloidal particles in monolayers and multilayers" [44]. The abstract of his latter contribution reads as follows:

> "Preliminary observations are reported on the deposition of alumina, silica, zinc sulfide and gold colloidal particles on solid surfaces. The alumina, which is positively charged, can induce subsequent deposition of the other colloids (which are negative) onto glass, as pointed out by Iler in 1966. **However, more uniform, reproducible and rapid deposition occurs on two monolayers of docosylamine sulfate applied to the glass by the Langmuir-Blodgett technique.** The zinc sulfide sol, which is not stable to flocculation, deposits as three-dimensional aggregates. The deposition of colloidal gold was followed by optical absorption measurements, and layers containing a substantial fraction of the close-packed limit were obtained."

So Iler's early experiments were in fact known to the thin film community in the early 1980s, they were even taken up and presented by one of the leading researchers of the field. With this in mind, one can only conclude that Langmuir-Blodgett films seemed so overwhelmingly promising at that time that the electrostatically driven assembly of inorganic particles must have looked like a scientific dead-end street. In the meantime this has changed, in the recent meetings of this conference series layer-by-layer assembled films have taken an ever growing fraction of the scientific program, due to the growing number of developments and applications.

## 1.4
### LbL-Assembly is the Synthesis of Fuzzy Supramolecular Objects

For most cases a LbL film has a unique layer sequence that is strictly defined by the deposition sequence (Figure 1.3). This indicates that LbL deposition should be considered as an analog to a chemical reaction sequence. While a chemical reaction takes part between different synthons and typically yields a unique molecule after each step of a given synthesis, layer-by-layer deposition involves the adsorption of a single species or mixture in each adsorption step and yields a multilayer film with a defined layer sequence at the end of the film assembly. While molecules are synthesized in several consecutive reaction steps, a multi-composite film is fabricated in several subsequent adsorption steps.

The reagents in classic synthesis are typically molecules, in layer-by-layer deposition they can be chosen from a wide range of materials. While today most of the multilayer films have been fabricated using mainly electrostatic attraction as the driving force for multilayer build-up, this is by no means a prerequisite. There are many other interactions that have successfully been used for multilayer deposition including: donor/acceptor interactions, hydrogen bridging, adsorption/drying cycles, covalent bonds, stereocomplex formation or specific recognition.

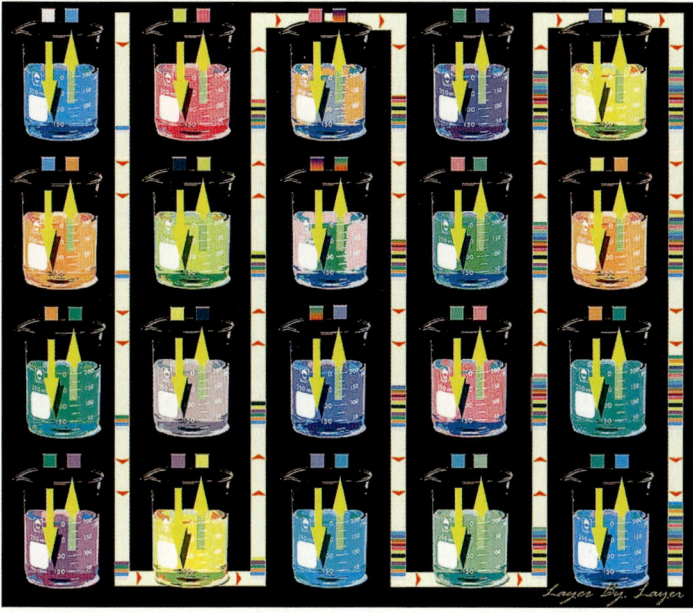

**Figure 1.3** An artists rendition of layer-by-layer assembly has recently appeared on the cover of a review published by K. Ariga *et al.* [45]. It nicely depicts that the architecture of a multicomponent film with a defined layer sequence conveniently depends on the sequence of immersions in different beakers containing the different species.

In general one needs just any interaction (this may be one or several different interactions) between two species "reagents" in order to incorporate them into a multilayer film. The interaction can easily be tested in solution prior to carrying out the deposition if both film constituents are soluble in the same solvent. When both solutions are mixed and flocculation occurs it is a good sign that multilayer fabrication will be possible. However this is only a very crude test for the "complementarity" of different compounds, multilayer formation may also be possible in the absence of flocculation.

It is possible to coat almost any solvent-accessible surface starting with nanoparticles or nanochannels up to the inside of tubings, or even objects with a surface of several square meters. Like a chemical reaction, the precise structure of each layer and the properties of the whole film depend on a set of control parameters, such as concentration, adsorption times, ionic strength, pH or temperature, but in general the processing window is rather broad.

## 1.5
### Reproducibility and Choice of Deposition Conditions

The question of reproducibility arises immediately when we draw the analogy between a chemical reaction and layer-by-layer adsorption. On first sight one may say that molecules are unique species and multilayer films are "only" fuzzy supramolecular objects. This is essentially the same argument that has downgraded for years macromolecular chemistry in comparison to organic chemistry. Today it is generally accepted that "ill-defined" macromolecules are also unique species that can indeed be well described by distributions and average properties like polydispersity or degree of polymerization. The situation is similar for multilayer films as they are characterized by a sequence of layers in which each layer has its individual structure and properties (Figure 1.3). While the sequence of layers is as strict as the arrangement of atoms in a molecule, the properties of each layer can only be described as an average over a certain area or over a certain distance along the layer normal. The most obvious property of an individual layer is its thickness, which is dependent on the nature of the underlying surface and on the deposition conditions. Parameters presumed to be important with respect to the underlying surface are, for example, nature and density of charged groups, their local mobility (in the case of a polymeric surface), and the surface roughness. Other important parameters are: solvent, concentration of adsorbing species, adsorption time, temperature, nature and concentration of added salt, rinsing time, humidity of the surrounding air, drying, agitation during adsorption or rinsing, dipping speed, and so forth. Typically, polymer and salt concentrations and deposition times are well described in the literature, however, some teams have already seen reproducibility issues when trying in colder weather to repeat results obtained during a hot summer.

How do we choose the right deposition conditions? This is not an easy question, because experimental constraints are very different for different investigations or applications. Film parameters, such as target thickness, target roughness, target functionalities, must be decided and are the result of an optimized deposition procedure. However, there are some guiding principles for polyelectrolyte multilayer deposition:

1) Longer adsorption times lead to more reproducible results. Everybody is aware that the plateau of adsorption is reached as a function of the concentration of the adsorbing component and of the adsorption time. While concentrations are easy to reproduce with little error, it is much more difficult to reproduce adsorption times in the range of seconds, especially if far from the plateau of adsorption or if adsorbing onto large or irregular shaped objects. Even a small difference in the adsorption times may lead to a large difference in the adsorbed mass in the initial phase of adsorption, whereas even a much larger difference in the adsorption time will lead only to a small difference in the adsorbed mass close to the plateau of adsorption. Adsorption kinetics are conveniently followed *in situ* for example, by quartz crystal microbalance (Figure 1.6).

2) The rinsing volume is important for avoiding cross-contamination of deposition solutions. While many multilayer films grow well even without rinsing, one should carefully calculate the required rinsing volumes to avoid cross-contaminations in the case of LbL-assembly by "dipping". Of course this is only relevant when the rinsing solution is in a beaker into which the substrate is immersed (e.g., "dipping robot"). When the substrate is withdrawn from a deposition solution a thin film of the deposition solution will adhere to its surface, the

volume of the adhering liquid can be estimated from the surface area of the immersed object, assuming a thickness of the adhering liquid film of a few microns. The dilution factor of the first rinsing bath is calculated by dividing the volume of the first rinsing bath by the estimated volume of the adhering liquid. Each further rinsing bath will increase the dilution factor correspondingly. The number and volume of the rinsing baths should be chosen such that the overall dilution factor is at least $1:10^6$, otherwise the liquid adhering on the surface of the substrate will contaminate the following deposition solution. *Cross-contamination and the depletion of the concentration of the adsorbing molecules are frequently underestimated, especially with large surface areas and with a large number of deposited layers.*

3) The surface coverage of functional groups is a key parameter for reproducibility. While most LbL-films show linear growth, which is likely associated with densities of functional groups that are independent of the layer number, this is not always the case. Superlinear growth may result from increasing surface coverage (or densities) of functional groups (or from the reservoir effect brought about by molecules diffusing into the whole film) and sub-linear growth results from decreasing densities of functional groups which will finally lead to stagnation of layer growth. In general, stagnation of layer growth is more likely with molecules or objects that possess only few functional groups, especially when the adsorbed geometries permit orientation of all functional groups toward the surface. For molecules with a large number of functional groups (i.e., high degree of polymerization) such unfavorable orientations for layer growth are much less likely. However, reproducible layer-by-layer assembly can be performed even with molecules containing only two functional groups (e.g., [7, 46]).

While the LbL-technique works generally very well due to the fact that the processing window is rather large, it is highly recommended to keep the deposition conditions as constant as possible in order to get highly reproducible results. If this is done rigorously, one obtains films composed of tens of layers whose thickness for example, differs by about 1%.

When comparing data, one should not overlook that one must not only maintain exactly the deposition conditions, but also the conditions under which the measurements were taken. Figure 1.4 shows an example of how the film thickness of a (PSS/PAH)$_8$ multilayer film, for which both polyions were deposited from solutions containing 2 M sodium chloride, depends on the temperature and on the relative humidity at the time of the measurement [47].

Often it is said that polyelectrolyte multilayer films are independent of the underlying substrate. This is an oversimplified statement, of course there is a dependence on the underlying surface, as stated above. However, since polyanion and polycation adsorption is often repeated consecutively, each polyanion adsorbs onto a polycation covered surface and vice versa. This means that, after a few layers, the structure and properties of each layer are often governed by the choice of the respective polyanion/polycation pair and by the deposition conditions, and that the influence of the substrate is typically lost after a few deposition cycles.

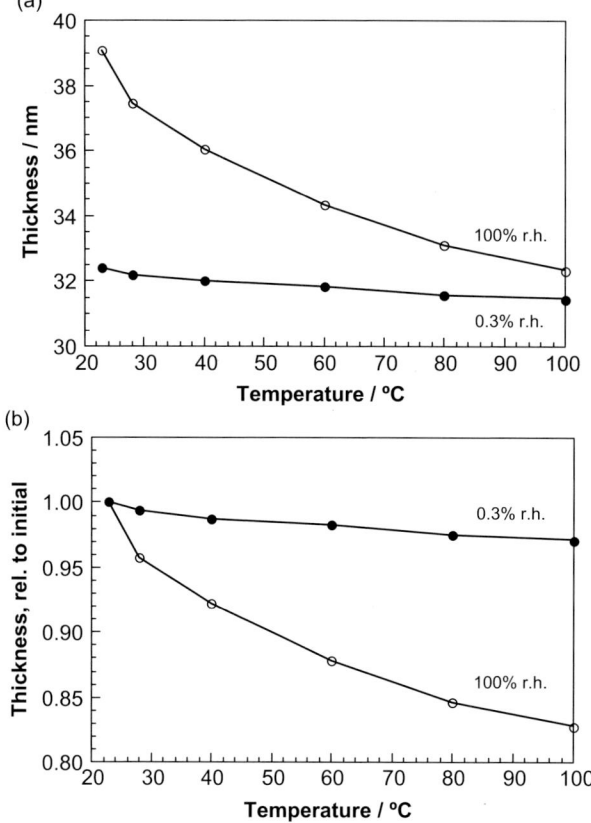

**Figure 1.4** (a) Film thickness of the same multilayer specimen as a function of temperature and of relative humidity (r.h.). The differences in thickness at identical temperatures are entirely due to a difference in water content within the film and not to a negative thermal expansion coefficient. This difference becomes less pronounced at elevated temperature, when the water is driven out of the film. (b) The same data as in (a), but represented normalized with respect to the initial film thickness. It becomes obvious that even small differences in temperature or humidity can easily account for changes in film thickness of the order of 5–10% depending on the swellability of the film.

## 1.6
### Monitoring Multilayer Build-up

The easiest way to follow multilayer build-up is probably by UV/Vis spectroscopy which works for all colored materials. Figure 1.5 is an example for poly(styrene sulfonate)/poly(allyl amine) $(PSS/PAH)_n$ films [8] which constitute probably the best characterized system as of today.

Equivalent to measuring the optical absorbance, one can also determine the film thickness by ellipsometry or X-ray reflectometry. These characterization methods are straightforward and widely available, but they require interruption of the deposition

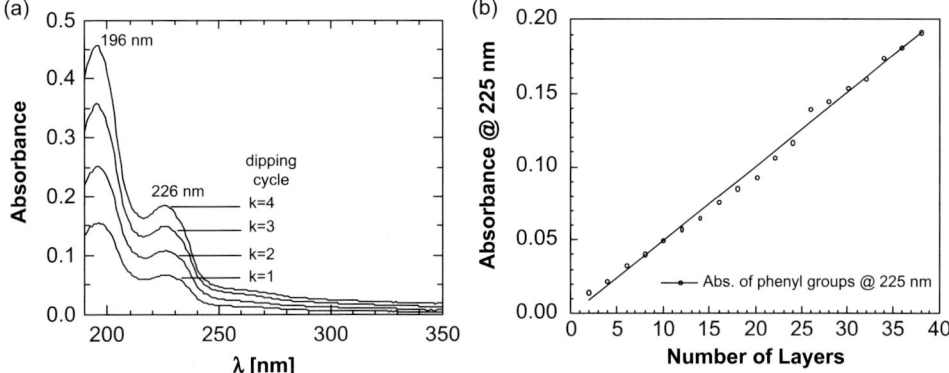

**Figure 1.5** (a) UV/Vis spectra taken after different numbers of adsorption cycles (k) during the preparation of a PSS/PAH multilayer. The bands at 195 nm and 226 nm originate from the aromatic chromophore of the styrene monomer unit of PSS. The absorbance increases regularly with the number of PSS layers. (b) Plot of the absorbance of the PSS band at 225 nm versus the number of layers deposited. The numerical fit to the data (solid line) shows that the increase of absorbance per layer is constant. The absorbance per layer is less than in (a) because the salt concentration in the deposition solutions was different. The slight deviation from a straight increase after 26 layers is due to the interruption of the deposition overnight.

process for taking the measurement. Not only are the measurements an interruption, they also have to be taken in the dry which may not be desirable in some cases.

*In-situ* methods are available for samples that cannot be dried. Depending on their time resolution, such methods also allow one to follow the kinetics of adsorption (Figure 1.6) and/or multilayer reorganization. Besides measurements of the zeta potential (Figure 1.7) and results obtained by quartz crystal microbalance (Figure 1.6), typical *in-situ* methods include surface plasmon spectroscopy, OWLS (optical waveguide lightmode spectroscopy), optical reflectometry in stagnation point flow cells, scanning angle reflectometry (SAR), *in-situ* ellipsometry, *in-situ* AFM, attenuated total reflection Fourier transform infrared spectroscopy (ATR-FTIR), surface forces measurements, X-ray and neutron reflectometry or second harmonic generation (SHG). Quartz crystal microbalance is ideally suited for screening the adsorption kinetics for new components and for optimizing the adsorption conditions.

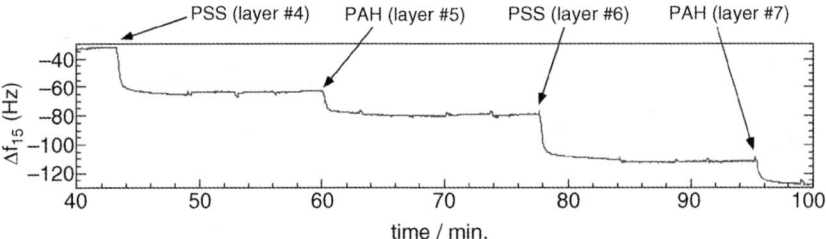

**Figure 1.6** Continuous QCM-trace of the third harmonic at 15 MHz (raw data) during the deposition of PSS (large displacements) and PAH (small displacements) during 4 arbitrary adsorption steps (layer numbers 4, 5, 6 and 7) of a longer deposition sequence.

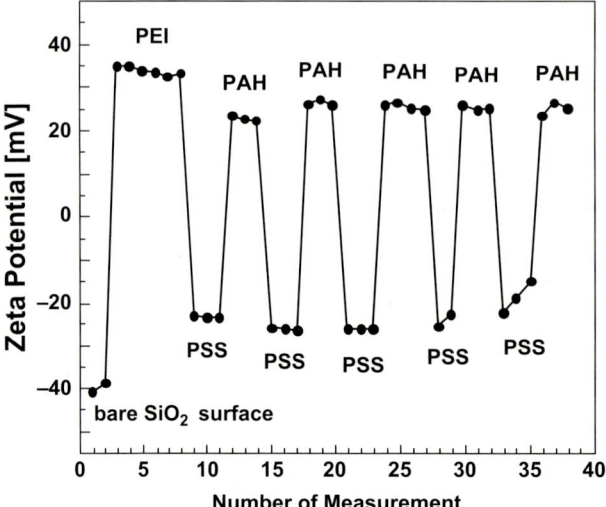

**Figure 1.7** Streaming potential measurement showing the surface charge reversal during multilayer buildup *in situ*. The first layer was poly(ethylene imine) (PEI) followed by 5 deposition cycles PSS and PAH [48].

A prerequisite, however, is that the multilayer being deposited should be rather rigid in order to evaluate frequency displacements as adsorbed mass using the Sauerbrey equation. If the multilayer has to be treated as a viscous film, a more sophisticated QCM instrument and data evaluation is needed. The data in Figure 1.6 on the consecutive deposition of poly(styrene sulfonate) (PSS) and poly(allyl amine) (PAH) were taken with a so-called QCM-D instrument, that also registers viscoelastic components of adsorbed films (data not shown). It is seen nicely that adsorption kinetics are rather fast (about 1–2 minutes per layer), that the thickness of each monolayer is autolimited (plateau) and that there is no visible desorption after rinsing (plateau). In this plateau region the QCM cell was rinsed three times for each adsorption step, once with the solution containing the respective polyion and twice with the buffer in which the polyions were dissolved. Please also note that there is a fast and a slow component in data like that shown in Figure 1.6 and that the structure and properties of a multilayer assembly may depend on processes occurring after the rapid adsorption step. Figure 1.6 arbitrarily shows the adsorption of layer numbers 4, 5, 6 and 7 out of a longer deposition sequence. The diagram shows that the film build-up is very regular.

In the early days of layer-by-layer assembly, there was just the idea of layer build-up driven by electrostatics, the surface potential of polyelectrolyte covered surfaces was only discussed to some extent, mostly in colloid science. In the meantime several Zeta potential measurements have been published, the diagram shown in Figure 1.7 results from a measurement using a quartz capillary that was carried out in collaboration with the groups of P. Schaaf and J. C. Voegel [48]. It demonstrates nicely that the adsorption of each polyelectrolyte layer leads to an overcompensation of the previous surface charge, just as we had assumed earlier on and schematically drawn in illustrations such as Figure 1.2. However, newer data suggest that the alternation of the zeta potential may not be a strict prerequisite for layer-by-layer assembly (see further down).

The theoretical description of polyelectrolyte complex formation including polyelectrolyte multilayers has considerably progressed recently [49]. However, measurements such as shown in Figure 1.7 are not a proof that multilayer build-up is entirely driven by electrostatic attraction (incoming layer) and electrostatic repulsion (auto-limitation to a single layer). Such measurements only show that there is a contribution of electrostatics in the case of multilayer build-up using positively and negatively charged components. Note that the release of the counterions also plays an important role as a driving force of layer-by-layer assembly. Depending on the chemical nature of the polyions and/or colloids employed for deposition, the importance of the electrostatic contribution should vary and other interactions, such as van der Waals, hydrogen bonding or charge transfer may more or less be involved as well.

A rather curious case turned up recently, when we studied the deposition of poly (sodium phosphate) (PSP) with a much longer polycation, poly(allylamine hydrochloride) (PAH) [50]. Despite the fact that both components form a polyelectrolyte complex in bulk, it is *very* difficult to prepare polyelectrolyte multilayers from these constituents. We finally succeeded by using spray assembly, but it turned out that multilayer growth depends strongly on the polymer concentrations, on the pH and on the ionic strength, even the growth regime can change as a function of these parameters. Most interestingly, in the case of linear growth at $10^{-4}$ M concentrations of PAH and PSP at pH 6.7 and in 0.15 M NaCl the zeta potential does not alternate between positive and negative when changing from polycation to polyanion. While the zeta potential starts out at $+60$ mV, it decreases to 0 mV during the deposition of the first 75 "layers" then becoming negative, finally stabilizing at $-20$ mV for layer numbers of 150 and above. The decline is not smooth, consecutively adsorbed layers lead to a difference in the surface potential of about 10 mV, but a classic alternation is not observed [50]. At present this behavior is not understood.

## 1.7
## Spray- and Spin-Assisted Multilayer Assembly

In a variation to the deposition by adsorption from solution, the application of layers by spraying was introduced by L. Winterton [51] and J. Schlenoff [52] and the use of spin-coaters was demonstrated by J.-D. Hong [53, 54] and also by H.-L. Wang [55, 56]. Both spraying and spin coating have the advantage that only small amounts of liquids are needed to coat large surface areas. More importantly, both techniques can lead to an enormous gain in deposition speed. We have recently shown that this reduction in deposition speed by spray-assisted assembly [15] does lead to well ordered LbL-films [16] and does not lead to a degradation of the nanostratification of such spray-assembled films (Figure 1.8).

Again, this underlines that the deposition conditions play an important role with respect to the final film characteristics. Spraying and spin coating extend the

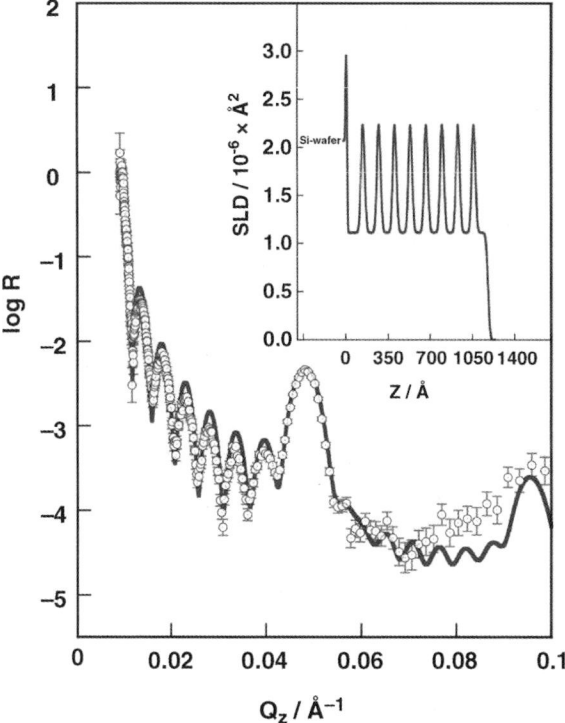

**Figure 1.8** Neutron reflectometry data of spray-assembled LbL-films of Si/SiO$_2$/PEI/[(PSSh$_7$/PAH)$_4$/(PSSd$_7$/PAH)$_1$]$_8$/(PSSh$_7$/PAH)$_4$. The data points are raw data, the solid line is the expected trace for the scattering length densities displayed in the inset [16]. As previously reported, spray-assembled films were found to be thinner than films assembled by dipping.

parameter space of LbL-deposition even further. It is to be expected, however, that both methods will contribute to the general acceptance of the LbL-technology.

## 1.8
## Recent Developments

### 1.8.1
### Self-patterning LbL-Films

While the patterning of LbL-films has already been described using classic micro- and nanostructuration techniques, polyelectrolyte multilayers are typically fairly smooth structures. In some cases, especially after changing the pH or the ionic strength, some surface corrugations were observed (sometimes in the context of changing the porosity of the film), but the controlled preparation of nanoscale surface patterns of polyelectrolyte multilayer films is hard to achieve.

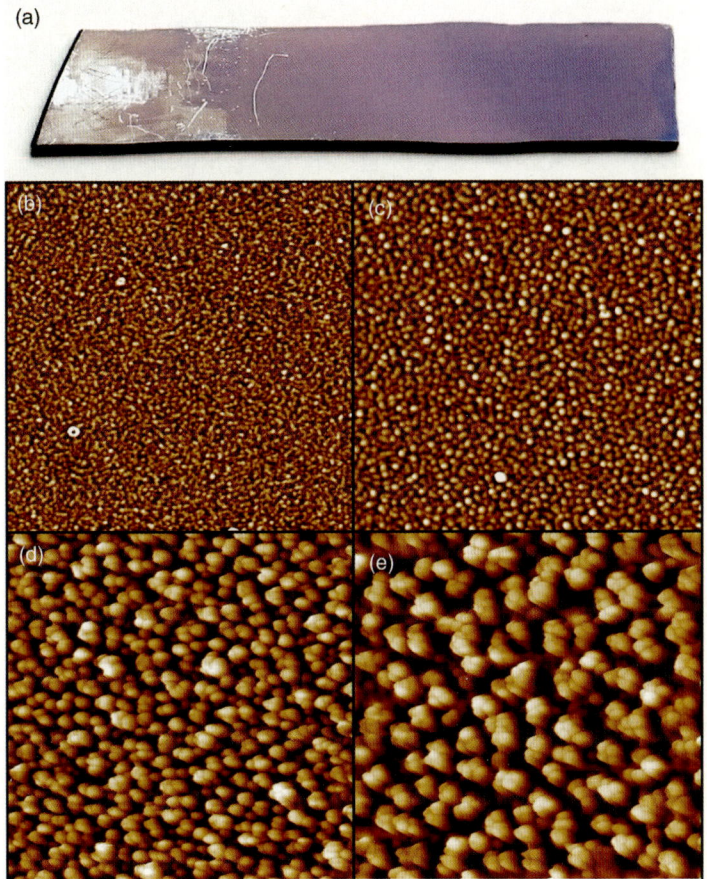

**Figure 1.9** Photographic image of a (PSP/PAH)$_n$ "film" with $n = 75$ on a silicon wafer (a) and AFM topographic images with $n = 3$ (b), $n = 10$ (c), $n = 30$ (d) and $n = 55$ (e), the dimensions of each image being $2 \times 2$ microns. Quantitative evaluation shows that the surface roughness of the "films" scales linearly with the "film" thickness [50].

It turns out that the poly(sodium phosphate)/poly(allylamine hydrochloride) system mentioned above allows one to precisely control the feature size of nanoscale surface patterns in LbL-assembled deposits (Figure 1.9) [50].

### 1.8.2
### Deposition of LbL-Films on Very Small Particles

While LbL-assembly is easy to envision on macroscopic objects, it is much more difficult to imagine how very small objects that cannot be handled individually can be coated. In the case of very small objects it is required to bring the objects in question in contact with an oppositely charged macromolecule in the same solution, a situation that is well know as a classic condition for bridging flocculation. Such

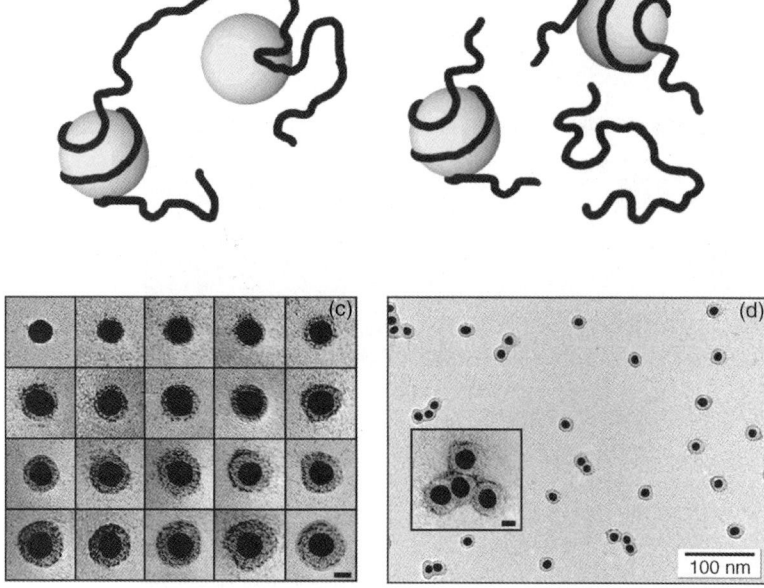

**Figure 1.10** (a,b) Schematic showing bridging flocculation of two nanoparticles connected by a single chain of a long polymer (a) and two nanoparticles wrapped individually with short polymers that are in stoichiometric excess (b) [34]. (c,d) Electron microscopic images of a (PAH/PSS)$_n$ film stepwise assembled around gold nanoparticles with a diameter of 13 nm (c). The scale bar is 10 nm. (d) is an electron micrograph of a suspension of particles coated with 13 layers that demonstrates that the majority of particles are individually dispersed [33]. This is very important for therapeutic applications (e.g., [36]) where precise particle diameters are required for passive tissue targeting which is proposed for tumor therapy by extravasation of nanoparticles through increased permeability of the tumor vasculature and ineffective lymphatic drainage (EPR effect) [57].

investigations were started in 1995 by T. Mallouk, who LbL-assembled redox-active components in an onion-like fashion around Cab-O-Sil SiO$_2$ particles [28]. In 1998 the team around H. Möhwald prepared the first micron-sized hollow spheres, a procedure that was termed "colloidal templating" [29, 30] and that rapidly then developed a dynamic of its own (several chapters in this book). A little bit later the team of F. Caruso coated gold nanoparticles with LbL-films [31, 32]. Avoiding bridging flocculation in coating nanoparticles with oppositely charged polyelectrolytes turned out to be a very difficult task when assembling larger numbers of layers on nanoscale particles [33, 34]. Figure 1.10 shows the deposition of a total of 20 polyelectrolyte layers on gold nanoparticles with a size of 13 nm and that these particles remain predominantly individually dispersed. These experiments led later to the proof of a well ordered layer structure even around small particles through distance dependent quenching of fluorescence [35], and to the potential application of such particles as multifunctional therapeutic agents [36]. We were in fact able to prepare for the first time nanoparticles equipped with dual functionality: capable of releasing a cytotoxic drug and stealthy toward THP-1 cells.

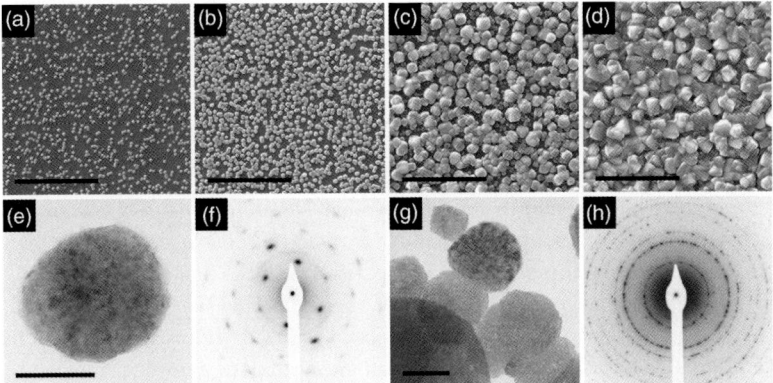

**Figure 1.11** Top view scanning electron micrographs of films composed of CaF$_2$ (a–d) at various stages of film growth. The numbers of spray cycles for each sample are as follows: 3 (a), 10 (b), 50 (c) and 200 (d). The scale bar represents 10 μm. Electron micrographs and diffraction patterns were obtained by transmission electron microscopy from CaF$_2$ crystals after one spraying cycle (e, f) and three spraying cycles (g, h). The scale bars represent 100 nm for image e and 200 nm for image g) [58].

### 1.8.3
### Purely Inorganic LbL-"Films"

Very recently we opened the field toward another important class of LbL-assembled films. Based on the alternate spraying of complementary inorganic salt solutions against a receiving surface we described the formation of purely inorganic films ("precipitation coating") [58]. The method applies whenever the solubility of the deposited material is smaller than that of the salts in the solutions of the reactants. The film thickness is controlled from nanometers to hundreds of micrometers simply by varying the number of spraying steps; 200 spray cycles, corresponding to less than 15 min deposition time, yield films with thicknesses exceeding one micrometer and reaching tens of micrometers in some cases.

CaF$_2$ films constitute a case in which the individual crystals are monocrystalline and form a film in which they grow together with an increasing number of spraying cycles, finally forming a quite dense film with few pores (Figure 1.11). Other inorganic materials yield different structures and morphologies. The approach is also compatible with conventional layer-by-layer assembly and permits the fabrication of multimaterial sandwich-like coatings. This solution-based spray-assembly process is similar to the so-called SILAR method for preparing polycrystalline inorganic films which was first reported by Nicolau in 1985 [42]. The obvious advantage of a spray process over immersion is that it can easily be adapted to different surfaces without needing large baths for large objects. Evidently then, the spray-assembly of complementary species also reduces the risk for cross-contamination as baths are not being used.

## 1.9
## Final Remarks

It has been a pleasure to see Layer-by-Layer assembly grow to the state it has reached today. The chapters that follow in the new edition of this book are a very clear indication that LbL-assembly has reached a certain maturity (despite quite a few remaining open questions) and that a new area has begun in which polyelectrolyte multilayers are more and more being used as materials or in devices that can frequently not be prepared otherwise. In comparison with the first edition, the book has grown in volume since many new fields of application have been "popping up". Instead of giving a detailed overview of the field, as was my intention in my chapter in the first edition of this book, I wanted to focus on presenting the guiding principles of layer-by-layer assembly and also on its historical context (as far as I am aware of it) and on the development of the technique.

Last year we had the occasion to organize at our institute the international LbL-Symposium 2011: "20 Years Layer-by-Layer Assembly: New Frontiers for Fundamental Science and for Applications". While over 160 participants attended the meeting (see Figure 1.12), the tight agendas of today caused some invitations for presentations to be declined as was unfortunately also the case with some invitations for chapters in the second edition of this book. Nevertheless, during this meeting a phrase was coined by Helmut Ringsdorf who expressed his amazement about the exceptional development of the field. In the context

**Figure 1.12** Conference photograph taken on occasion of the international LbL-Symposium 2011: "20 Years Layer-by-Layer Assembly: New Frontiers for Fundamental Science and for Applications", that was held March 10–12, 2011 in our institute in Strasbourg, France.

of delivering his presentation with the title: "Art is I – Science is We" he remarked that Layer-by-Layer assembly benefitted strongly in its development from a unique community that distinguishes itself from others through *"Competitive Collaboration"*.

## References

1 Decher, G. (1996) Layered nanoarchitectures via directed assembly of anionic and cationic molecules, in *Templating, Self-Assembly and Self-Organization* (eds J.-P. Sauvage and M.W. Hosseini), Pergamon Press, Oxford, pp. 507–528.

2 Langmuir, I. and Blodgett, K.B. (1935) Über einige neue methoden zur untersuchung von monomolekularen Filmen. *Kolloid-Z.*, **73**, 257–263.

3 Blodgett, K.B. (1934) Monomolecular films of fatty acids on glass. *J. Am. Chem. Soc.*, **56**, 495.

4 Blodgett, K.B. and Langmuir, I. (1937) Build-up films of barium stearate and their optical properties. *Phys. Rev.*, **51**, 964–982.

5 Kuhn, H. and Möbius, D. (1971) Systems of Monomolecular Layers—Assembling and Physico-Chemical Behavior. *Angew. Chem. Int. Ed.*, **10**, 920–637.

6 Inacker, O., Kuhn, H., Möbius, D., and Debuch, G. (1976) Manipulation in molecular dimensions. *Z. Phys. Chem. Neue Folge*, **101**, 337–360.

7 Decher, G. and Hong, J.-D. (1991) Buildup of ultrathin multilayer films by a self-assembly process: I. Consecutive adsorption of anionic and cationic bipolar amphiphiles. *Makromol. Chem., Macromol. Symp.*, **46**, 321–327.

8 Decher, G., Hong, J.-D., and Schmitt, J. (1992) Buildup of ultrathin multilayer films by a self-assembly process: III. Consecutively alternating adsorption of anionic and cationic polyelectrolytes on charged surfaces. *Thin Solid Films*, **210/211**, 831–835.

9 Decher, G., Eßler, F., Hong, J.-D., Lowack, K., Schmitt, J., and Lvov, Y. (1993) Layer-by-layer adsorbed films of polyelectrolytes, proteins or DNA. *Polymer Preprints*, **34**, 745.

10 Schmitt, J., Grunewald, T., Decher, G., Pershan, P.S., Kjaer, K., and Losche, M. (1993) Internal structure of Layer-by-layer adsorbed polyelectrolyte films - a neutron and X-ray reflectivity study. *Macromolecules*, **26**, 7058–7063.

11 Decher, G. (1996) Multilayer films (polyelectrolytes), in *The Polymeric Materials Encyclopedia: Synthesis, Properties, and Applications* (ed. J.C. Salamone), CRC Press Inc., Boca Raton, pp. 4540–4546.

12 Decher, G. (1997) Fuzzy nanoassemblies: Toward layered polymeric multicomposites. *Science*, **277**, 1232–1237.

13 Decher, G., Eckle, M., Schmitt, J., and Struth, B. (1998) Layer-by-layer assembled multicomposite films. *Curr. Opin. Colloid Interface Sci.*, **3**, 32–39.

14 Losche, M., Schmitt, J., Decher, G., Bouwman, W.G., and Kjaer, K. (1998) Detailed structure of molecularly thin polyelectrolyte multilayer films on solid substrates as revealed by neutron reflectometry. *Macromolecules*, **31**, 8893–8906.

15 Izquierdo, A., Ono, S.S., Voegel, J.C., Schaaf, P., and Decher, G. (2005) Dipping versus spraying: Exploring the deposition conditions for speeding up layer-by-layer assembly. *Langmuir*, **21**, 7558–7567.

16 Felix, O., Zheng, Z.Q., Cousin, F., and Decher, G. (2009) Are sprayed LbL-films stratified? A first assessment of the nanostructure of spray-assembled multilayers by neutron reflectometry. *C. R. Chim.*, **12**, 225–234.

17 Decher, G., Lehr, B., Lowack, K., Lvov, Y., and Schmitt, J. (1994) New nanocomposite films for biosensors - layer-by-layer adsorbed films of polyelectrolytes, proteins or DNA. *Biosens. Bioelectron.*, **9**, 677–684.

18 Lvov, Y., Ariga, K., and Kunitake, T. (1994) Layer-by-layer assembly of alternate protein polyion ultrathin films. *Chem. Lett.*, 2323–2326.

19 Lvov, Y., Decher, G., and Sukhorukov, G. (1993) Assembly of thin films by means of successive deposition of alternate layers of DNA and poly(allylamine). *Macromolecules*, **26**, 5396–5399.

20 Sukhorukov, G.B., Möhwald, H., Decher, G., and Lvov, Y.M. (1996) Assembly of polyelectrolyte multilayer films by consecutively alternating adsorption of polynucleotides and polycations. *Thin Solid Films*, **284/285**, 220–223.

21 Lvov, Y., Haas, H., Decher, G., Möhwald, H., Michailov, A., Mtchedlishvily, B., Morgunova, E., and Vainshtain, B. (1994) Successive deposition of alternate layers of polyelectrolytes and a charged virus. *Langmuir*, **10**, 4232–4236.

22 Kleinfeld, E.R. and Ferguson, G.S. (1994) Stepwise formation of multilayered nanostructural films from macromolecular precursors. *Science*, **265**, 370–373.

23 Keller, S.W., Kim, H.-N., and Mallouk, T.E. (1994) Layer-by-layer assembly of intercalation compounds and heterostructures on surfaces: Towards molecular "beaker" epitaxy. *J. Am. Chem. Soc.*, **116**, 8817–8818.

24 Kotov, N.A., Dékány, I., and Fendler, J.H. (1995) Layer-by-layer self-assembly of polyelectrolyte-semiconductor nanoparticle composite films. *J. Phys. Chem.*, **99**, 13065–13069.

25 Schmitt, J., Decher, G., Dressik, W.J., Brandow, S.L., Geer, R.E., Shashidhar, R., and Calvert, J.M. (1997) Metal nanoparticle/polymer superlattice films: Fabrication and control of layer structure. *Adv. Mater.*, **9**, 61–65.

26 Schmitt, J., Grünewald, T., Kjær, K., Pershan, P., Decher, G., and Lösche, M. (1993) The internal structure of Layer-by-layer adsorbed polyelectrolyte films: A neutron and X-ray reflectivity study. *Macromolecules*, **26**, 7058–7063.

27 Lösche, M., Schmitt, J., Decher, G., Bouwman, W.G., and Kjaer, K. (1998) Detailed structure of molecularly thin polyelectrolyte multilayer films on solid substrates as revealed by neutron reflectometry. *Macromolecules*, **31**, 8893–8906.

28 Keller, S.W., Johnson, S.A., Brigham, E.S., Yonemoto, E.H., and Mallouk, T.E. (1995) Photoinduced charge separation in multilayer thin films grown by sequential adsorption of polyelectrolytes. *J. Am. Chem. Soc.*, **117**, 12879–12880.

29 Caruso, F., Caruso, R.A., and Möhwald, H. (1998) Nanoengineering of inorganic and hybrid hollow spheres by colloidal templating. *Science*, **282**, 1111–1114.

30 Donath, E., Sukhorukov, G.B., Caruso, F., Davis, S.A., and Möhwald, H. (1998) Novel hollow polymer shells by colloid-templated assembly of polyelectrolytes. *Angew. Chem. Int. Ed.*, **37**, 2202–2205.

31 Gittins, D.I. and Caruso, F. (2000) Multilayered polymer nanocapsules derived from gold nanoparticle templates. *Adv. Mater.*, **12**, 1947.

32 Gittins, D.I. and Caruso, F. (2001) Tailoring the polyelectrolyte coating of metal nanoparticles. *J. Phys. Chem. B*, **105**, 6846–6852.

33 Schneider, G. and Decher, G. (2004) From functional core/shell nanoparticles prepared via layer-by-layer deposition to empty nanospheres. *Nano Lett.*, **4**, 1833–1839.

34 Schneider, G. and Decher, G. (2008) Functional core/shell nanoparticles via layer-by-layer assembly. investigation of the experimental parameters for controlling particle aggregation and for enhancing dispersion stability. *Langmuir*, **24**, 1778–1789.

35 Schneider, G., Decher, G., Nerambourg, N., Praho, R., Werts, M.H.V., and Blanchard-Desce, M. (2006) Distance-dependent fluorescence quenching on gold nanoparticles ensheathed with layer-by-layer assembled polyelectrolytes. *Nano Lett.*, **6**, 530–536.

36 Schneider, G.F., Subr, V., Ulbrich, K., and Decher, G. (2009) Multifunctional cytotoxic stealth nanoparticles. A model approach with potential for cancer therapy. *Nano Lett.*, **9**, 636–642.

37 Iler, R.K. (1966) Multilayers of colloidal particles. *J. Colloid Interface Sci.*, **21**, 569–594.

38 Ries, H.E. and Meyers, B.L. (1971) Microelectrophoresis and electron-microscope studies with polymeric flocculants. *J. Appl. Polym. Sci.*, **15**, 2023–2034.

39 Gölander, C.-G., Arwin, H., Eriksson, J.C., Lundstrom, I., and Larsson, R. (1982) Heparin surface film formation through adsorption of colloidal particles studied by ellipsometry and scanning electron microscopy. *Colloids Surf.*, **5**, 1–16.

40 Fromherz, P. (1980) Assembling proteins at lipid monolayers, in *Electron Microscopy at Molecular Dimensions* (eds W. Baumeister and W. Vogell), Springer Verlag, Berlin, Heidelberg, pp. 338–349.

41 Aksberg, R. and Ödberg, L. (1990) Adsorption of an anionic polyacrylamide on cellulosic fibers with pre-adsorbed cationic polyelectrolytes. *Nord. Pulp Pap. Res. J.*, **5**, 168–171.

42 Nicolau, Y.F. (1985) Solution deposition of thin solid compound films by a successive ionic-layer adsorption and reaction process. *Appl. Surf. Sci.*, **22-3**, 1061–1074.

43 Gaines, G.L. (1983) On the history of Langmuir-Blodgett films. *Thin Solid Films*, **99**, R9–R13.

44 Gaines, G.L. (1983) Deposition of colloidal particles in monolayers and multilayers. *Thin Solid Films*, **99**, 243–248.

45 Ariga, K., Hill, J.P., and Ji, Q.M. (2007) Layer-by-layer assembly as a versatile bottom-up nanofabrication technique for exploratory research and realistic application. *PCCP Phys. Chem. Chem. Phys.*, **9**, 2319–2340.

46 Gill, R., Mazhar, M., Félix, O., and Decher, G. (2010) Covalent Layer-by-layer assembly and solvent memory of multilayer films from homobifunctional poly(dimethoxysilane). *Angew. Chem. Int. Ed*, **49**, 6116–6119.

47 Decher, G. and Schlenoff, J.B. (eds) (2003) *Multilayer Thin Films: Sequential Assembly of Nanocomposite Materials*, Wiley-VCH, Weinheim.

48 Ladam, G., Schaad, P., Voegel, J.C., Schaaf, P., Decher, G., and Cuisinier, F. (2000) In-situ determination of the structural properties of initially deposited polyelectrolyte multilayers. *Langmuir*, **16**, 1249–1255.

49 Gucht, J.v.d., Spruijt, E., Lemmers, M., and Stuart, M.A.C. (2011) Polyelectrolyte complexes: Bulk phases and colloidal systems. *J. Colloid Interface Sci.*, **361**, 407–422.

50 Cini, N., Tulun, T., Decher, G., and Ball, V. (2010) Step-by-step assembly of self-patterning polyelectrolyte films violating (almost) all rules of layer-by-layer deposition. *J. Am. Chem. Soc.*, **132**, 8264–8265.

51 Winterton, L., Vogt, J., Lally, J., and Stockinger, F. (1999) Coating of polymers, Novartis AG, World Patent WO9935520.

52 Schlenoff, J.B., Dubas, S.T., and Farhat, T. (2000) Sprayed polyelectrolyte multilayers. *Langmuir*, **16**, 9968–9969.

53 Cho, J., Char, K., Hong, J.D., and Lee, K.B. (2001) Fabrication of highly ordered multilayer films using a spin self-assembly method. *Adv. Mater.*, **13**, 1076.

54 Lee, S.S., Hong, J.D., Kim, C.H., Kim, K., Koo, J.P., and Lee, K.B. (2001) Layer-by-layer deposited multilayer assemblies of ionene-type polyelectrolytes based on the spin-coating method. *Macromolecules*, **34**, 5358–5360.

55 Chiarelli, P.A., Johal, M.S., Casson, J.L., Roberts, J.B., Robinson, J.M., and Wang, H.L. (2001) Controlled fabrication of polyelectrolyte multilayer thin films using spin-assembly. *Adv. Mater.*, **13**, 1167.

56 Chiarelli, P.A., Johal, M.S., Holmes, D.J., Casson, J.L., Robinson, J.M., and Wang, H.L. (2002) Polyelectrolyte spin-assembly. *Langmuir*, **18**, 168–173.

57 Peer, D., Karp, J.M., Hong, S., Farokhzad, O.C., Margalit, R., and Langer, R. (2007) Nanocarriers as an emerging platform for cancer therapy. *Nature Nanotechnol.*, **2**, 751–760.

58 Popa, G., Boulmedais, F., Zhao, P., Hemmerle, J., Vidal, L., Mathieu, E., Felix, O., Schaaf, P., Decher, G., and Voegel, J.C. (2010) Nanoscale precipitation coating: The deposition of inorganic films through step-by-step spray-assembly. *ACS Nano*, **4**, 4792–4798.

# Part I
# Preparation and Characterization

*Multilayer Thin Films: Sequential Assembly of Nanocomposite Materials,* Second Edition.
Edited by Gero Decher and Joseph B. Schlenoff.
© 2012 Wiley-VCH Verlag GmbH & Co. KGaA. Published 2012 by Wiley-VCH Verlag GmbH & Co. KGaA.

# 2
# Layer-by-Layer Processed Multilayers: Challenges and Opportunities

*Michael F. Rubner and Robert E. Cohen*

## 2.1
## Introduction

It has now been well established that essentially any material with suitable secondary bonding abilities (the ability to form ionic or hydrogen bonds, for example) that can be dissolved or dispersed in aqueous solutions, can be assembled into multilayer constructs by using the layer-by-layer (LbL) processing approach. For example, the LbL process has been successfully accomplished with polymers [1–3], proteins [4], small molecules [5], nanoparticles [6], block copolymer micelles [7–9], living cells [10], viruses [11], clay nanosheets [12] and carbon nanotubes [13], to name just a few. This enormous versatility has allowed the creation and exploration of a wide range of functional thin film multilayer heterostructures with nanoscale controllable thicknesses, layered architectures and physical/chemical properties.

An interesting question to ask at this moment in time is whether or not the field of layer-by-layer processed multilayers has matured to the point that only incremental advances will be made in the future. After about 20 years of research worldwide, it is clear that the versatility of this process, in terms of both the types of materials that can be manipulated and the manner by which they are manipulated, is enormous, perhaps more so than any other thin film processing approach yet described. However, there still remain many unresolved, and largely unexplored, fundamental and technological issues that, if resolved, could result in an explosion of new research vectors and, ultimately, unique technological advances. This chapter will focus on a small subset of these issues, keeping in mind that issues are oftentimes opportunities waiting to be discovered.

## 2.2
## Fundamental Challenges and Opportunities

### 2.2.1
### LbL Assembly on Nanoscale Elements and in Confined Geometries

In the fundamental arena, it is to be recognized that the deposition of LbL processed materials inevitably involves the sequential adsorption of relatively weakly interacting

*Multilayer Thin Films: Sequential Assembly of Nanocomposite Materials*, Second Edition.
Edited by Gero Decher and Joseph B. Schlenoff.
© 2012 Wiley-VCH Verlag GmbH & Co. KGaA. Published 2012 by Wiley-VCH Verlag GmbH & Co. KGaA.

molecules or nanoelements (nanoparticles, micelles, etc.) from dilute aqueous solutions. A key advantage of a process based on molecular adsorption is that conformal multilayer constructs can be assembled onto any surface that is accessible to the repeated application of assembly and rinsing solutions. This means that, in principle, it is possible to create functional, structurally well defined (i.e., complex z-direction heterostructures) LbL coatings on the surfaces of typically hard to access substrates, such as nanochannels [14], nanotextured elements [15] and nanoparticles [16]. In all of these cases, one begins to enter an adsorption regime that is quite interesting and not well understood. Specifically, a regime is entered where the dimensions of the adsorbing species start to approach and exceed at least one of the dimensions of the substrate of interest. In the case of adsorbing polymers, for example, this means that the surface features/elements are in the size range of the radius of gyration of the polymer. The deformability of the adsorbing material (hydrated polymer versus rigid nanoparticle), and geometric constraints, such as the effective surface area it uses to bind to the surface, become particularly important in this situation.

While both charged macromolecules and charged nanoparticles may contain identical pH-sensitive substrate-binding groups, only the macromolecule, owing to its configurational flexibility, can place multiple copies of these binding groups simultaneously on the surface. The placement of these binding sites can occur over an area that is very large compared to the size of the binding site itself, and that area may be copiously interpenetrated by binding groups from other macromolecules. Desorption requires the unlikely simultaneous release of these multiple binding interactions. On the other hand, the rigid nanoparticle is essentially restricted to point contact with the substrate and the single weak binding event leads to easy desorption, possibly followed by aggregation of neighboring nanoparticles, with subsequent adsorption of these aggregates on the substrate.

A particularly interesting case is the situation where the adsorbing polymer or nanoparticle must enter a geometry confined by multiple surfaces, such as in the attempted LbL coating of deep nanochannels or trenches. In this case, surface charge and size exclusion effects can lead to a concentration depletion of the adsorbing species that strongly influences the resultant multilayer growth behavior [14,17]. Also, depending on the deposition conditions, materials and substrate feature sizes, multilayers can assemble conformally onto the surfaces contained within the confined geometries and/or bridge the nanoscale features, thereby preventing further deposition within the channels. A survey of the literature published on this subject [18–22] reveals that anything is possible and limited theoretical [23] or experimental guidance is available to predict behavior, particularly when polymer assembly is involved. Clearly, unresolved fundamentally important questions in this area include an understanding of the factors that control the assembly of multilayers in confined geometries and how they can be utilized to create predictable coating outcomes.

To address these questions, we have been investigating LbL assembly within the confined geometries of model $Si/SiO_2$ nanochannels [14, 24, 25]. These model nanochannels are 10–15 μm deep and have channel widths in the range of 800 to

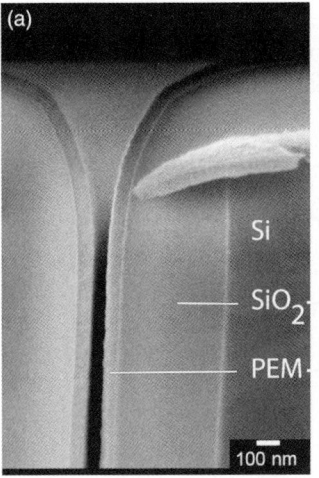 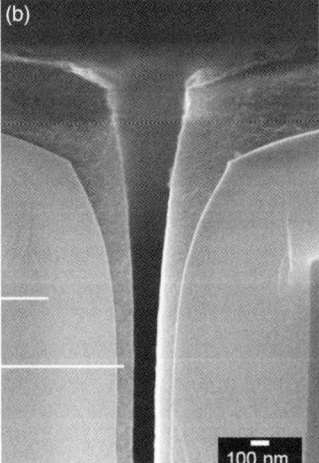 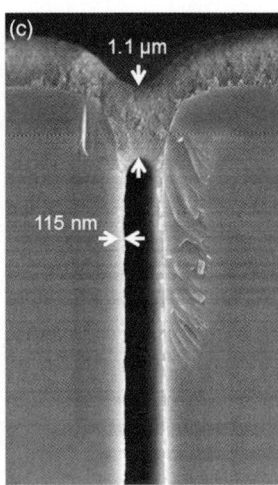

**Figure 2.1** Scanning electron micrographs showing conformal nanochannel coatings of (a) a 40 layer pair PEM of poly(allylamine hydrochloride)/poly(styrene sulfonic acid), (b) a 120 layer pair PEM of 6 nm diameter $TiO_2$ nanoparticles/poly(vinyl sulfonic acid), (c) a 60 layer pair PEM of 6 nm diameter $TiO_2$ nanoparticles/15 nm diameter $SiO_2$ nanoparticles (note bridging of the channel opening by the all-nanoparticle multilayer). In all cases, the PEM coating uniformly covers the walls and bottom of the 10–15 μm deep channel.

200 nm. As revealed in Figure 2.1, it is possible to conformally coat the walls of these nanochannels from top to bottom with uniform thickness polymer/polymer [14], polymer/nanoparticle [24] and nanoparticle/nanoparticle mutlilayers [25]. In the case of the nanoparticle/nanoparticle multilayer assembly (Figure 2.1c), the conditions utilized produced very thin conformal coatings within the channels and ultimately resulted in the assembly of nanoparticle multilayers that uniformly overcoated (bridged) the entrance of the nanochannel. The smallest gap created within a nanochannel was realized with polymer/polymer multilayer assembly, in which case the original gap size could be reduced down to about 11 nm (wet-state gap). The radius of gyration of the polymers used in this assembly process was of the order of 30–50 nm. Amazingly, all of these nanochannel coatings were accomplished by a simple static dipping process in which the adsorbing species diffused readily into the channels during each deposition step.

The SEM images presented in Figure 2.1 suggest a few of the numerous opportunities presented by the conformal LbL coating of nanochannel constructs. First, this approach can be used to systematically reduce nanochannel gaps to sizes not easily obtained by usual processing methodologies. Second, all of the rich surface functionalization demonstrated over the years by LbL assembly can be applied to the nanochannel walls, including controlling surface charge and chemistry, as has been done with microfluidic channels [26, 27].

For example, single-stranded DNA molecules or antibodies can be covalently attached to the wall surfaces to provide a selective molecular capture and release capability. The assembly of temperature and pH-driven stimuli-response multilayers onto the channel walls further provides an ability to create nanoscale devices with dynamically tunable surfaces and channel gap sizes [28]. Third, the coatings themselves can be nanoporous, as in the case of the polymer/nanoparticle and nanoparticle/nanoparticle multilayers of Figure 2.1. In the filtration area, this could provide opportunities to create nanoscale size-exclusion effects and high surface areas for enhanced separations. Fourth, LbL-coated nanochannel membranes can be used as sacrificial templates to create polyelectrolyte multilayer nanotubes and related nanoscale objects [29]. Finally, multilayer bridges can be formed that could provide additional control over the filtration and selection of molecules passing through or into the channels.

### 2.2.2
### Living Cells as Functional Elements of Polyelectrolyte Multilayers

Suitably engineered polyelectrolyte multilayers (PEMs) have long been recognized as versatile biomaterial platforms due to their similarity to naturally occurring, highly hydrated systems, ease of biochemical functionalization, high level of biocompatibility and tunable mechanical compliance [30–34]. These attributes can be realized with synthetic polymers, and more importantly, with naturally occurring polymers, including DNA, polypeptides and polysaccharides. Tremendous progress has been made in creating biofunctional multilayer systems that can, for example, promote or inhibit the attachment and proliferation of living cells and bacteria [35–48]. Although, in this regard, the interface between living cells and PEMs has been extensively explored, issues associated with polycation toxicity and the narrow processing window of living cells have limited perhaps the more interesting possibility of using living cells as functional building blocks in conjunction with multilayers. This would include the LbL incorporation of living cells into multilayers, as well as the attachment of functional multilayers to the surfaces of cells.

### 2.2.3
### Multilayer Cellular Backpacks

From the fundamental standpoint, the newly emerging concept of functionalizing living cells with synthetic elements [49, 50] is a promising new direction for polyelectrolyte multilayer research. For example, by combining the native functions of living immune system cells, such as their ability to target disease sites, with on-demand drug-releasing multilayers, it may be possible to enhance significantly the body's ability to overcome specific diseases. We have recently found that it is possible to attach multifunctional PEM "backpacks" to a small fraction of the membrane surface area of living immune system cells, including lymphocytes, phagocytes and dendritic cells (DCs) [51–53]. Since the PEM backpack does not completely occlude the cellular surface from its environment, this technique leaves the cell free to perform its native functions, including the ability to migrate and spread on surfaces. This process has been used to apply backpacks to murine B and human T lympho-

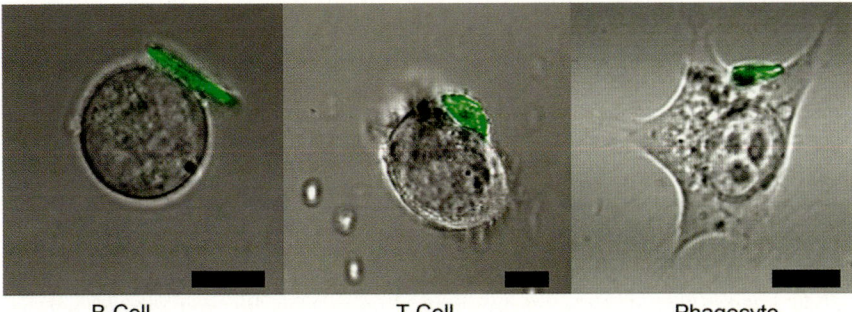

B-Cell    T-Cell    Phagocyte

**Figure 2.2** Confocal laser scanning microscopy images of fluorescent multilayer backpacks attached to three different cell types; a CH27 B-lymphocyte, a HuT 78 T cell and a murine DC2.4 dendritic cell. The green fluorescence is from a FTIC-labeled polycation (PAH) in the multilayer. Scale bars are 10 μm. The multilayer backpacks contain magnetic iron oxide nanoparticles for cell manipulation, and chitosan/hyaluronic acid surface bilayers to promote binding to CD44 cell-surface receptor sites (Adapted with permission from Swiston, A. J., Cheng, C., Um, S. H., Irvine, D. J., Cohen, R. E., & Rubner, M. F. Surface Functionalization of Living Cells with Multilayer Patches. Nano Lett. 8(12), 4446–4453 (2008). Copyright (2008) American Chemical Society.).

cytes, as well as murine dendritic cells, and mouse macrophages (see Figure 2.2). When backpacks are attached to T cells, they do not interfere with the cell's ability to migrate on a 2D, ICAM-coated surface. For example, T cells were observed to migrate for several hours, moving the backpack around its surface to always be in the trailing end of the polarized cell. When a backpack is attached to the surface of a phagocyte, such as a murine DC or mouse macrophage, the cell amazingly does not endocytose the backpack. This observation leads to interesting possibilities in the design of phagocytosis-resistant materials that may serve in advanced vaccine therapies, such as by serving as adjuvants or local antigen depots.

Possible payloads within the PEM cellular backpacks include drugs, vaccine antigens, thermally responsive polymers, and nanoparticles, all of which may be delivered by the cell to a site for beneficial drug release and/or diagnostic bioimaging. Furthermore, backpack-modified cells can reveal fundamental insights into cell behavior changes due to surface binding events, such as the clustering of surface receptors, how membrane mechanics influence the cell's biochemical state, or how cells respond to very localized mechanical and chemical cues. This approach has broad potential for applications in bioimaging, single-cell functionalization, immune system and tissue engineering, and cell-based therapeutics.

### 2.2.4
### Direct LbL Processing of Living Cells

In the arena of tissue engineering and cell-based biosenors and bioreactors, the possibility of incorporating living cells into the LbL process to create cell-containing multilayers (cellular multilayers) with well controlled matrix components/molecular environments and complex 3D cellular architectures is quite intriguing [54].

Early work is this area demonstrated [55] the ability to fabricate multilayer thin films containing hepatocyte–(chitosan/DNA)$_x$–hepatocyte cell assemblies, hepatocyte–(chitosan/DNA)$_x$–endothelial cell assemblies, and hepatocyte–(chitosan/DNA)$_x$–fibroblast cell assemblies, with an objective of constructing multilayer constructs that mimic the structure of liver tissues *in vivo*. A follow-on to this study [56] utilized chitosan/hyaluronic acid multilayers to construct liver mimic architectures. 3D cellular constructs assembled with fibronectin/gelatin multilayers have also been reported [57, 58], along with results demonstrating the importance of such architectures in determining cellular stability and function. The complications of assembling potentially cytotoxic polycations onto living cells has, however, limited the more general applicability of this approach [59]. The use of hydrogen-bonded multilayers [60] and spray-assembled multilayers [61] look to be promising directions to pursue when one needs to assemble polyelectrolyte layers with living cells.

In our work, we have taken advantage of the binding interaction of polysaccharides, such as hyaluronic acid, with specific cell surface receptor sites, to attach multilayers to immune system cells (the cellular backbacks described above) or to attach B cells to patterned surfaces [62]. The specific binding of interest in this case is a result of a hyaluronic acid–CD44 receptor interaction. This approach has also led to a robust system for fabricating 3D cellular architectures comprised of B cells [63]. Figure 2.3

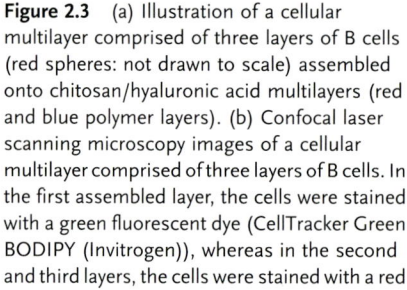
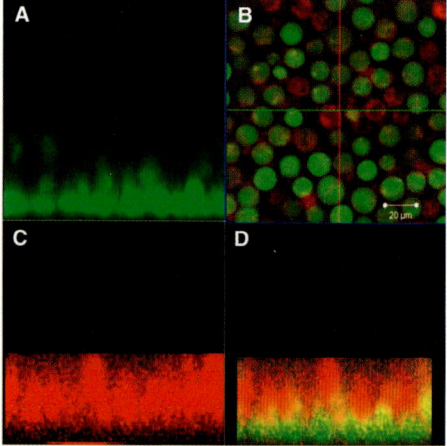

**Figure 2.3** (a) Illustration of a cellular multilayer comprised of three layers of B cells (red spheres: not drawn to scale) assembled onto chitosan/hyaluronic acid multilayers (red and blue polymer layers). (b) Confocal laser scanning microscopy images of a cellular multilayer comprised of three layers of B cells. In the first assembled layer, the cells were stained with a green fluorescent dye (CellTracker Green BODIPY (Invitrogen)), whereas in the second and third layers, the cells were stained with a red fluorescent dye (CellTracker Red CMTPX (Invitrogen)). (A) Reconstructed edge view from a sample that was excited with a 488 nm Argon laser to excite the green fluorescent dye. (B) A single z-slice of the cellular multilayer showing a majority of green labeled cells from the bottom layer and some red labeled cells from the top layer. (C) Reconstructed edge view from a sample that was excited with a 543 nm HeNe laser to excite the red fluorescent dye. (D) Overlay of (A) and (C).

shows a multilayer thin film comprised of three B cell layers assembled in the usual LbL process by using hyaluronic acid/chitosan multilayers both to anchor the cells to the surface and to enable subsequent multilayer growth. The LbL assembly process produces robust, B cell-containing multilayers only when hyaluronic acid/chitosan layer pairs are used as the molecular glue to bind additional B cell layers to the surface. This early work points to the need to identify and utilize naturally occurring biopolymers and cell surface receptors as the key elements used to assemble typically non-anchorage-dependent cells onto surfaces and into functional multilayer films.

## 2.3
## Technological Challenges and Opportunities

### 2.3.1
### Improving Processing Time and Versatility

The LbL process as it is typically practiced involves the assembly of molecules, polymers and nanoparticles onto surfaces from dilute aqueous solutions, with rinsing steps between each adsorption step. This is most often accomplished by sequentially immersing substrates (dip assembly) for a few minutes into assembly component and rinsing solutions. In many cases, such as the coating of complex nanochannels, mircofluidic devices and colloidal particles, this may be the only viable processing scheme available. From an industrial standpoint, however, this process as applied to more conventional substrates may be considered slow and not well suited to the large-scale production of functional coatings. This is particularly true for large area optical coatings for architectural elements, glass panes, windshields and solar panels, where, to be economically competitive, high optical quality coatings must be produced quickly and in high yield. New developments in the spray-assembly of LbL coatings look very promising in this regard, and much effort is currently devoted to the development of this process. Since the initial publication [64] showing that multilayers comparable to those obtained by dip assembly can be realized much faster by simply spraying the various solutions onto surfaces, this area has expanded rapidly [61, 65–80] and is quickly moving toward technologically viable commercial products [81].

Bragg stacks (dielectric mirrors), comprised of many nanoscale, alternating regions of high and low refractive index, represent a particularly challenging multilayer heterostructure to fabricate over large areas in reasonably fast processing times. These so-called structural color coatings require the reproducible deposition of a number of materials into well-defined nanoscale regions, and they contain many internal interfaces that need to be of low enough roughness to prevent optical scattering. Current techniques used to fabricate dielectric mirrors utilize vacuum deposition techniques that can limit the size of the substrate to be coated and require expensive and complicated processing equipment. We have previously demonstrated that the LbL process can be used to create high quality Bragg stacks assembled with all-polymer or nanoparticle/polymer layer pairs [82–86]. To create these structures,

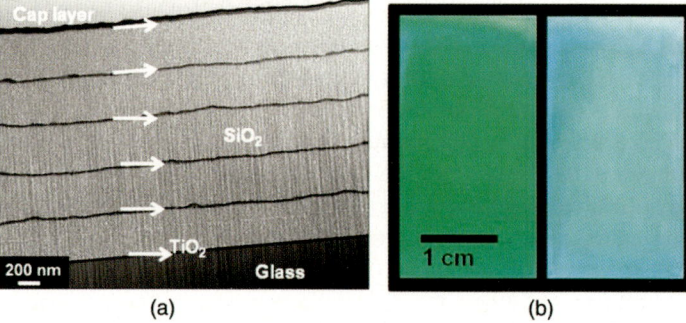

**Figure 2.4** (a) Cross-sectional TEM image of an 11-stack Bragg reflector made of alternating regions of $TiO_2$ (dark gray) and $SiO_2$ (light gray) nanoparticles assembled via spray LbL. (b) Photographic image of an 11-stack [$TiO_2$/$SiO_2$] multilayer spray assembled on a microscope glass slide. Both photographs were taken of the same sample but at different angles to show the iridescence of this multiple-stack film. (Adapted with permission from Nogueira, G. M., Banerjee, D., Cohen, R. E., & Rubner, M. F. Spray-Layer-by-Layer Assembly Can More Rapidly Produce Optical-Quality Multistack Heterostructures. Langmuir 27(12), 7860–7867 (2011). Copyright (2011) American Chemical Society.)

upwards of 1400 individual adsorbed polymer or nanoparticle layers were deposited using conventional dipping procedures over the course of about 3–5 days. Thus, although high quality coatings with interesting functionality were possible (for example dynamically tunable structural color and self-cleaning effects), the processing time required to achieve these coatings was impractical from a technological standpoint.

Figure 2.4 shows a structural color coating created by an automated LbL spray assembly technique. This particular multilayer heterostructure was assembled with 840 individual layers by the alternate spraying of polymer, nanoparticle and rinsing solutions onto a substrate, typically with 6 s spraying times [65]. When compared to an identical multilayer heterostructure fabricated by an LbL immersion process, the optical quality and uniformity is as good or better, and the processing time is reduced from days to hours. This approach can also be used to fabricate high quality multistack anti-reflection coatings with broadband capability (greater than 98% transmission over the wavelength range of 400–800 nm). In both cases, low interfacial roughness and nanoscale thickness control over regions of low and high refractive index result in high quality coatings with optical profiles that can be designed on paper and realized through assembly from aqueous solutions.

### 2.3.2
### Towards Mechanically Robust Multilayer Coatings

The limited mechanical robustness of some ultrathin films of polyelectrolyte multilayers, and their inability oftentimes to handle a range of aggressive mechanical/chemical environments is perhaps one of the most important unmet challenges preventing the wide-scale adoption of PEMs by the traditional coatings industry. For many important applications of PEMs, such as their use as functional biomaterials, the mechanical properties typically realized through LbL assembly are more than adequate to ensure the levels of mechanical and chemical robustness needed for the intended application. If needed, further enhancement of these attributes can be readily realized by using solution crosslinking chemistries well known in the community, such as aqueous-based NHS/carbodiimide and glutaraldehyde crosslinking chemistry. The demanding requirements of some coatings, however, are not easily realized by PEMs, particularly in their ultrathin form (typically about 50–100 nm). Such coatings would include anti-reflection, anti-fogging, anti-static, and scratch resistant coatings, to name a few. Thus, although it is possible to enable remarkable functionality with exceptionally thin coatings based on PEMs, the limited mechanical robustness of any ultrathin film coating (metal, polymer, etc.) including PEM coatings must be recognized and addressed.

PEMs can be assembled from a wide variety of synthetic polymers, biopolymers, small molecules and nanoparticles. From the standpoint of realizing robust thin film coatings, nanoparticle-containing multilayers (polymer/nanoparticle or all-nanoparticle [87]) are particularly attractive. When nanoparticles are utilized to create PEMs, it is possible to significantly enhance mechanical robustness and durability by activating post-assembly processes that partially fuse the nanoparticles together, or that form covalent bonds between the nanoparticle and polymer components (see Figure 2.5).

If high temperatures can be tolerated, the particles can be partially fused together by thermal calcination reactions, typically at temperatures in the range of 500–550 °C. This process, however, removes any polymer component present and clearly cannot be used with plastic substrates. To preserve the benefits and function of a polymer matrix and/or utilize plastic substrates, hydrothermal treatments can be carried out at

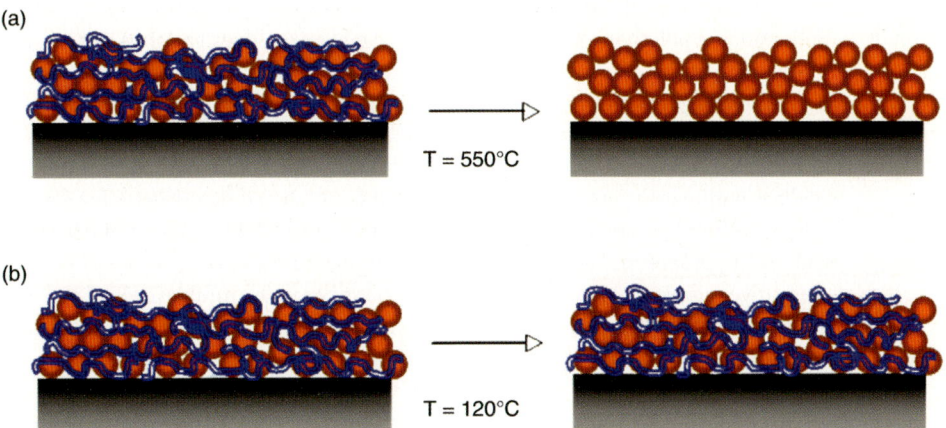

**Figure 2.5** (a) thermal calcination improves mechanical robustness of polymer/nanoparticle coatings but removes any polymer component. (b) Hydrothermal treatment or thermal crosslinking improves mechanical robustness without eliminating the polymer components.

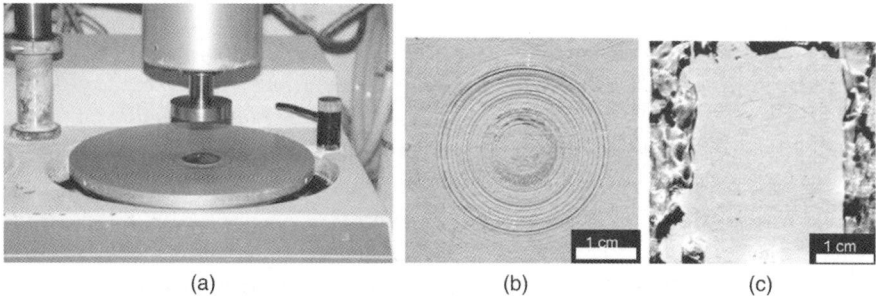

**Figure 2.6** Photographic images of (a) the metal polishing instrument modified to perform a quantitative abrasion test. In this test, the bottom wheel was held stationary while the top wheel was rotated at 150 rpm. (b) A sample of bare polycarbonate after abrasion testing under a 25 kPa normal stress. (c) Bare polycarbonate coated with ~100 nm thick, autoclaved (124 °C, 1 h) poly(diallyldimethylammonium chloride)/ silica nanoparticle PEM after abrasion testing under a 25 kPa normal stress; the edges are covered with glue (Adapted with permission from Gemici, Z., Shimomura, H., Cohen, R. E., & Rubner, M. F. Hydrothermal Treatment of Nanoparticle Thin Films for Enhanced Mechanical Durability. Langmuir 24(5), 2168–2177 (2008). Copyright (2008) American Chemical Society).

temperatures as low as 120 °C. We have found that this process produces coatings as good as or better than (in terms of mechanical robustness) those created with high temperature thermal calcination reactions [88]. In essence, mechanical robustness is achieved by simply autoclaving the coating after LbL assembly. Figure 2.6 shows how the mechanical robustness (abrasion resistance) of a nanoporous anti-reflection PEM coating deposited onto a polycarbonate substrate can be improved by a post-assembly hydrothermal treatment. As revealed in Figure 2.6, the surface of bare polycarbonate is severely damaged by the abrasion test, whereas a polycarbonate surface coated with only a 100 nm thick PEM remains essentially undamaged. At the low temperatures employed, the polymer component is not degraded or removed and thus can be used to provide additional functionality to the coating.

Another possible advantage of using nanoparticles to reinforce the mechanical properties of LbL thin films is the ability to carry out post-assembly processes based on capillary condensation principles [89]. If assembly conditions are adjusted to produce nanoporous structures, it is possible to condense liquids into the porous regions of the film and further modify properties, including mechanical, chemical and wetting properties. The unique thermodynamic habitat established in the zones of negative radius of curvature between nanoparticles in the multilayer film makes it possible to condense liquids under conditions (temperature/pressure) that would not normally support condensation on flat surfaces.

As illustrated in Figure 2.7, the level of pore filling can be controlled by using nanoparticles of different sizes, with smaller nanoparticles favoring complete filling of the pores, and larger nanoparticles only allowing the condensation of a thin layer of condensate on the particle surfaces. Since this is largely a thermodynamic process, many different types of molecules can be condensed into the pores, including

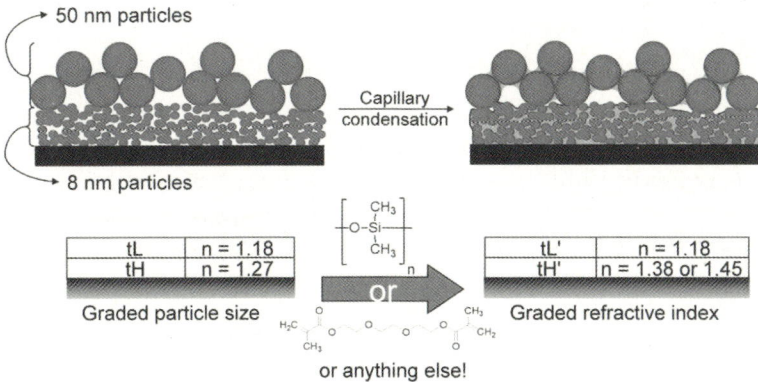

**Figure 2.7** Illustrations and refractive index profiles of two-stack, graded-index anti-reflection coatings before and after PDMS- and TEGDMA- functionalization. The two stacks were fabricated from PAH/SiO$_2$ multilayers (reprinted with permission from Gemici, Z., Schwachulla, P. I., Williamson, E. H., Rubner, M. F., & Cohen, R. E. Targeted Functionalization of Nanoparticle Thin Films Via Capillary Condensation. Nano Letters 9(3), 1064–1070 (2009). Copyright (2009) American Chemical Society.).

hydrophilic and hydrophobic molecules. For example, we have successfully condensed oligomers of poly(dimethyl siloxane) (PDMS) and triethylene glycol dimethacrylate monomers (TEGDMA) into the porous regions of nanoparticle-containing multilayer heterostructures comprised of separate regions of 8 and 50 nm particles. The pores of the smaller particle regions were essentially filled whereas the larger particles were only coated with the condensate. This made it possible to modify the refractive index profile of the heterostructure to produce a broadband anti-reflection capability. When PDMS was condensed, the multilayer film was rendered hydrophobic and resistant to water condensation. When TEGDMA was condensed, the mechanical properties of the resultant hydrophilic multilayer could be further modified by a crosslinking reaction of the methacrylate groups.

Clearly, more work is needed in the development of mechanically robust multilayers. New approaches to reinforcing PEMs during assembly and post-assembly that significantly enhance mechanical robustness without diminishing targeted functionality should be explored [90]. When used in conjunction with the mechanical reinforcement provided by nanoelements, such as nanoclay sheets [12] or nanoparticles, it should be possible to produce ultrathin coatings that better meet the needs of many different types of coating applications. It is unlikely that scratch resistant coatings with properties comparable to the multi-micron thick nanoparticle reinforced scratch resistant coatings currently used in industry will be realized from ultrathin PEM coatings only 100 nm in thickness. However, with suitable effort, it will no doubt be possible to better exploit the unique functionality provided by ultrathin PEM coatings by designing and fabricating multilayer systems with sufficient mechanical integrity to be viable in many different and important coating applications.

## 2.4
## The Path Forward

The field of LbL-processed materials is indeed alive and well. The focus of future research and development will no doubt be directed at multilayer systems that provide unique functionality but also cannot be easily or economically fabricated by other means. There are certainly other approaches that can be used to create ultrathin film coatings with high functionality (grafted polymer layers, self-assembled monolayers, chemical vapor deposited layers, etc.). Thus, when one considers utilizing the LbL approach to construct new types of coatings, it is a worthwhile exercise to ask the question: does this coating derive benefit from the unique capabilities provided by LbL processing ? Such capabilities include water-based processing, the ability to manipulate a wide-range of quite different materials into well defined thin film structures, the ability to compartmentalize materials in thin film heterostructures, the ability to control at the nanoscale the spacing of complementary or interacting materials in the z-direction and the ability to conformally coat micro- [91] and nanoscale objects and topographies with complex multilayer heterostructures. There are also exciting new possibilities for creating biofunctional coatings that take full advantage of these unique capabilities. It is in this latter area that the next 20 years of LbL research and development will most likely enable the most exciting and innovative new ideas and material systems.

## References

1 Kozlovskaya, V., Kharlampieva, E., Erel, I., and Sukhishvili, S.A. (2009) Multilayer-derived, ultrathin, stimuli-responsive hydrogels. *Soft Matter*, **5** (21), 4077–4087.
2 Ariga, K., Hill, J.P., and Ji, Q.M. (2007) Layer-by-layer assembly as a versatile bottom-up nanofabrication technique for exploratory research and realistic application. *Phys. Chem. Chem. Phys.*, **9** (19), 2319–2340.
3 Hammond, P.T. (2004) Form and function in multilayer assembly: new applications at the nanoscale. *Adv. Mater.*, **16** (15), 1271–1293; Such, G.K., Johnston, A.P.R., and Caruso, F. (2011) Engineered hydrogen-bonded polymer multilayers: from assembly to biomedical applications. *Chem. Soc. Rev.*, **40** (1), 19–29.
4 Lvov, Y., Ariga, K., Ichinose, I., and Kunitake, T. (1995) Assembly of multicomponent protein films by means of electrostatic layer-by-layer adsorption. *J. Am. Chem. Soc.*, **117** (22), 6117–6123.
5 Ariga, K., Lvov, Y., and Kunitake, T. (1997) Assembling alternate dye-polyion molecular films by electrostatic layer-by-layer adsorption. *J. Am. Chem. Soc.*, **119** (9), 2224–2231.
6 Ariga, K., Ariga, K., Lvov, Y., Onda, M., Ichinose, I., and Kunitake, T. (1997) Alternately assembled ultrathin film of silica nanoparticles and linear polycations. *Chem. Lett. (Jpn.)* (2), 125–126;Lvov, Y., Ariga, K., Onda, M., Ichinose, I., and Kunitake, T. (1997) Alternate assembly of ordered multilayers of $SiO_2$ and other nanoparticles and polyions. *Langmuir*, **13** (23), 6195–6203.
7 Tan, W.S., Cohen, R.E., Rubner, M.F., and Sukhishvili, S.A. (2010) Temperature-induced, reversible swelling transitions in multilayers of a cationic triblock copolymer and a polyacid. *Macromolecules*, **43** (4), 1950–1957.
8 Kim, B.S., Park, S.W., and Hammond, P.T. (2008) Hydrogen-

bonding layer-by-layer assembled biodegradable polymeric micelles as drug delivery vehicles from surfaces. *ACS Nano.*, **2** (2), 386–392.

9 Ma, N., Zhang, H.Y., Song, B., Wang, Z.Q., and Zhang, X. (2005) Polymer micelles as building blocks for layer-by-layer assembly: An approach for incorporation and controlled release of water-insoluble dyes. *Chem. Mater.*, **17** (20), 5065–5069.

10 Rajagopalan, P., Shen, C.J., Berthiaume, F., Tilles, A.W., Toner, M., and Yarmush, M.L. (2006) Polyelectrolyte nano-scaffolds for the design of layered cellular architectures. *Tissue Eng.*, **12** (6), 1553–1563.

11 Nam, K.T., Kim, D.W., Yoo, P.J., Chiang, C.Y., Meethong, N., Hammond, P.T., Chiang, Y.M., and Belcher, A.M. (2006) Virus-enabled synthesis and assembly of nanowires for lithium ion battery electrodes. *Science*, **312** (5775), 885–888.

12 Tang, Z.Y., Kotov, N.A., Magonov, S., and Ozturk, B. (2003) Nanostructured artificial nacre. *Nature Mater.*, **2** (6), 413–418.

13 Olek, M., Ostrander, J., Jurga, S., Mohwald, H., Kotov, N., Kempa, K., and Giersig, M. (2004) Layer-by-layer assembled composites from multiwall carbon nanotubes with different morphologies. *Nano. Lett.*, **4** (10), 1889–1895.

14 DeRocher, J.P., Mao, P., Han, J.Y., Rubner, M.F., and Cohen, R.E. (2010) Layer-by-layer assembly of polyelectrolytes in nanofluidic devices. *Macromolecules*, **43** (5), 2430–2437.

15 Lin, Y.H., Jiang, C., Xu, J., Lin, Z.Q., and Tsukruk, V.V. (2007) Sculptured layer-by-layer films. *Adv. Mater.*, **19** (22), 3827.

16 Schneider, G. and Decher, G. (2004) From functional core/shell nanoparticles prepared via layer-by-layer deposition to empty nanospheres. *Nano. Lett.*, **4** (10), 1833–1839.

17 Bohmer, M.R., Evers, O.A., and Scheutjens, J. (1990) Weak polyelectrolytes between two surfaces – adsorption and stabilization. *Macromolecules*, **23** (8), 2288–2301.

18 Alem, H., Blondeau, F., Glinel, K., Demoustier-Champagne, S., and Jonas, A.M. (2007) Layer-by-layer assembly of polyelectrolytes in nanopores. *Macromolecules*, **40** (9), 3366–3372.

19 Wang, Y.J., Yu, A.M., and Caruso, F. (2005) Nanoporous polyelectrolyte spheres prepared by sequentially coating sacrificial mesoporous silica spheres. *Angew Chem. Int. Edit.*, **44** (19), 2888–2892.

20 Ali, M., Yameen, B., Cervera, J., Ramirez, P., Neumann, R., Ensinger, W., Knoll, W., and Azzaroni, O. (2010) Layer-by-layer assembly of polyelectrolytes into ionic current rectifying solid-state nanopores: insights from theory and experiment. *J. Am. Chem. Soc.*, **132** (24), 8338–8348.

21 Lazzara, T.D., Lau, K.H.A., Abou-Kandil, A.I., Caminade, A.M., Majoral, J.P., and Knoll, W. (2010) Polyelectrolyte layer-by-layer deposition in cylindrical nanopores. *ACS Nano.*, **4** (7), 3909–3920.

22 Roy, C.J., Dupont-Gillain, C., Demoustier-Champagne, S., Jonas, A.M., and Landoulsi, J. (2010) Growth mechanism of confined polyelectrolyte multilayers in nanoporous templates. *Langmuir*, **26** (5), 3350–3355.

23 Carrillo, J.M.Y. and Dobrynin, A.V. (2011) Layer-by-layer assembly of charged nanoparticles on porous substrates: molecular dynamics simulations. *ACS Nano.*, **5** (4), 3010–3019.

24 Kim, J.Y., DeRocher, J.P., Mao, P., Han, J., Cohen, R.E., and Rubner, M.F. (2010) Formation of nanoparticle-containing multilayers in nanochannels via layer-by-layer assembly. *Chem. Mater.*, **22** (23), 6409–6415.

25 DeRocher, J.P., Kim, J.Y., Mao, P., Cohen, R.E., and Rubner, M.F. (2012) Formation of all nanoparticle-containing multilayers in nanochannels. *ACS Appl. Mater. Interfaces*, **4** (1), 391–396.

26 Sui, Z.J. and Schlenoff, J.B. (2003) Controlling electroosmotic flow in microchannels with pH-responsive polyelectrolyte multilayers. *Langmuir*, **19** (19), 7829–7831.

27 Barker, S.L.R., Ross, D., Tarlov, M.J., Gaitan, M., and Locascio, L.E. (2000) Control of flow direction in microfluidic devices with polyelectrolyte

multilayers. *Anal. Chem.*, **72** (24), 5925–5929.

28 Lee, D., Nolte, A.J., Kunz, A.L., Rubner, M.F., and Cohen, R.E. (2006) pH-induced hysteretic gating of track-etched polycarbonate membranes: swelling/deswelling behavior of polyelectrolyte multilayers in confined geometry. *J. Am. Chem. Soc.*, **128** (26), 8521–8529.

29 Chia, K.K., Rubner, M.F., and Cohen, R.E. (2009) pH-responsive reversibly swellable nanotube arrays. *Langmuir*, **25** (24), 14044–14052.

30 Jang, Y., Park, S., and Char, K. (2011) Functionalization of polymer multilayer thin films for novel biomedical applications. *Korean J. Chem. Eng.*, **28** (5), 1149–1160.

31 Detzel, C.J., Larkin, A.L., and Rajagopalan, P. (2011) Polyelectrolyte multilayers in tissue engineering. *Tissue Eng. Part B*, **17** (2), 101–113.

32 Boudou, T., Crouzier, T., Ren, K., Blin, G., and Picart, C. (2010) Multiple functionalities of polyelectrolyte multilayer films: New biomedical applications. *Adv. Mater.*, **22** (4), 441–467.

33 Jewell, C.M. and Lynn, D.M. (2008) Multilayered polyelectrolyte assemblies as platforms for the delivery of DNA and other nucleic acid-based therapeutics. *Adv. Drug Deliv. Rev.*, **60** (9), 979–999.

34 Tang, Z., Wang, Y., Podsiadlo, P., and Kotov, N.A. (2006) Biomedical applications of layer-by-layer assembly: from biomimetics to tissue engineering. *Adv. Mater.*, **18** (24), 3203–3224.

35 Lichter, J.A., Van Vliet, K.J., and Rubner, M.F. (2009) Design of antibacterial surfaces and interfaces: polyelectrolyte multilayers as a multifunctional platform. *Macromolecules*, **42** (22), 8573–8586.

36 Picart, C. (2008) Polyelectrolyte multilayer films: from physico-chemical properties to the control of cellular processes. *Curr. Med. Chem.*, **15** (7), 685–697.

37 Berg, M.C., Yang, S.Y., Hammond, P.T., and Rubner, M.F. (2004) Controlling mammalian cell interactions on patterned polyelectrolyte multilayer surfaces. *Langmuir*, **20** (4), 1362–1368.

38 Boulmedais, F., Frisch, B., Etienne, O., Lavalle, P., Picart, C., Ogier, J., Voegel, J.C., Schaaf, P., and Egles, C. (2004) Polyelectrolyte multilayer films with pegylated polypeptides as a new type of anti-microbial protection for biomaterials. *Biomaterials*, **25** (11), 2003–2011.

39 Chua, P.H., Neoh, K.G., Kang, E.T., and Wang, W. (2008) Surface functionalization of titanium with hyaluronic acid/chitosan polyelectrolyte multilayers and RGD for promoting osteoblast functions and inhibiting bacterial adhesion. *Biomaterials*, **29** (10), 1412–1421.

40 Elbert, D.L., Herbert, C.B., and Hubbell, J.A. (1999) Thin polymer layers formed by polyelectrolyte multilayer techniques on biological surfaces. *Langmuir*, **15** (16), 5355–5362.

41 Fu, J.H., Ji, J., Yuan, W.Y., and Shen, J.C. (2005) Construction of anti-adhesive and antibacterial multilayer films via layer-by-layer assembly of heparin and chitosan. *Biomaterials*, **26** (33), 6684–6692.

42 Lichter, J.A., Thompson, M.T., Delgadillo, M., Nishikawa, T., Rubner, M.F., and Van Vliet, K.J. (2008) Substrata mechanical stiffness can regulate adhesion of viable bacteria. *Biomacromolecules*, **9** (6), 1571–1578.

43 Lichter, J.A. and Rubner, M.F. (2009) Polyelectrolyte multilayers with intrinsic antimicrobial functionality: the importance of mobile polycations. *Langmuir*, **25** (13), 7686–7694.

44 Richert, L., Lavalle, P., Payan, E., Shu, X.Z., Prestwich, G.D., Stoltz, J.F., Schaaf, P., Voegel, J.C., and Picart, C. (2004) Layer by layer buildup of polysaccharide films: Physical chemistry and cellular adhesion aspects. *Langmuir*, **20** (2), 448–458.

45 Mendelsohn, J.D., Yang, S.Y., Hiller, J., Hochbaum, A.I., and Rubner, M.F. (2003) Rational design of cytophilic and cytophobic polyelectrolyte multilayer thin films. *Biomacromolecules*, **4** (1), 96–106.

46 Schneider, A., Francius, G., Obeid, R., Schwinte, P., Hemmerle, J., Frisch, B., Schaaf, P., Voegel, J.C., Senger, B., and Picart, C. (2006) Polyelectrolyte multilayers with a tunable young's

modulus: Influence of film stiffness on cell adhesion. *Langmuir*, **22** (3), 1193–1200.

47 Thompson, M.T., Berg, M.C., Tobias, I.S., Rubner, M.F., and Van Vliet, K.J. (2005) Tuning compliance of nanoscale polyelectrolyte multilayers to modulate cell adhesion. *Biomaterials*, **26** (34), 6836–6845.

48 Yang, S.Y., Mendelsohn, J.D., and Rubner, M.F. (2003) New class of ultrathin, highly cell-adhesion-resistant polyelectrolyte multilayers with micropatterning capabilities. *Biomacromolecules*, **4** (4), 987–994.

49 Stephan, M.T. and Irvine, D.J. (2011) Enhancing cell therapies from the outside in: cell surface engineering using synthetic nanomaterials. *Nano. Today*, **6** (3), 309–325.

50 Stephan, M.T., Moon, J.J., Um, S.H., Bershteyn, A., and Irvine, D.J. (2010) Therapeutic cell engineering with surface-conjugated synthetic nanoparticles. *Nat. Med.*, **16** (9), 1035–1135.

51 Swiston, A.J., Cheng, C., Um, S.H., Irvine, D.J., Cohen, R.E., and Rubner, M.F. (2008) Surface functionalization of living cells with multilayer patches. *Nano. Lett.*, **8** (12), 4446–4453.

52 Swiston, A.J., Gilbert, J.B., Irvine, D.J., Cohen, R.E., and Rubner, M.F. (2010) Freely suspended cellular "backpacks" lead to cell aggregate self-assembly. *Biomacromolecules*, **11** (7), 1826–1832.

53 Doshi, N., Swiston, A.J., Gilbert, J.B., Alcaraz, M.L., Cohen, R.E., Rubner, M.F., and Mitragotri, S. (2011) Cell-based drug delivery devices using phagocytosis-resistant backpacks. *Adv. Mater.*, **23** (12), H105–H109.

54 Guillame-Gentil, O., Semenov, O., Roca, A.S., Groth, T., Zahn, R., Voeroes, J., and Zenobi-Wong, M. (2010) Engineering the extracellular environment: strategies for building 2D and 3D cellular structures. *Adv. Mater.*, **22** (48), 5443–5462.

55 Rajagopalan, P., Shen, C.J., Berthiaume, F., Tilles, A.W., Toner, M., and Yarmush, M.L. (2006) Polyelectrolyte nano-scaffolds for the design of layered cellular architectures. *Tissue Eng.*, **12** (6), 1553–1563.

56 Kim, Y., Larkin, A.L., Davis, R.M., and Rajagopalan, P. (2010) The design of in vitro liver sinusoid mimics using chitosan-hyaluronic acid polyelectrolyte multilayers. *Tissue Eng. Part A*, **16** (9), 2731–2741.

57 Kadowaki, K., Matsusaki, M., and Akashi, M. (2010) Three-dimensional constructs induce high cellular activity: Structural stability and the specific production of proteins and cytokines. *Biochem. Biophys. Res. Commun.*, **402** (1), 153–157.

58 Matsusaki, M., Kadowaki, K., Nakahara, Y., and Akashi, M. (2007) Fabrication of cellular multilayers with nanometer-sized extracellular matrix films. *Angew Chem. Int. Edit.*, **46** (25), 4689–4692.

59 Kadowaki, K., Matsusaki, M., and Akashi, M. (2010) Control of cell surface and functions by layer-by-layer nanofilms. *Langmuir*, **26** (8), 5670–5678.

60 Kozlovskaya, V., Harbaugh, S., Drachuk, I., Shchepelina, O., Kelley-Loughnane, N., Stone, M., and Tsukruk, V.V. (2011) Hydrogen-bonded LbL shells for living cell surface engineering. *Soft Matter*, **7** (6), 2364–2372.

61 Grossin, L., Cortial, D., Saulnier, B., Felix, O., Chassepot, A., Decher, G., Netter, P., Schaaf, P., Gillet, P., Mainard, D., Voegel, J.C., and Benkirane-Jessel, N. (2009). Step-by-step build-up of biologically active cell-containing stratified films aimed at tissue engineering. *Adv. Mater.*, **21** (6), 650.

62 Vasconcellos, F.C., Swiston, A.J., Beppu, M.M., Cohen, R.E., and Rubner, M.F. (2010) Bioactive polyelectrolyte multilayers: hyaluronic acid mediated b lymphocyte adhesion. *Biomacromolecules*, **11** (9), 2407–2414.

63 Endale, G., Gilbert, J., Cohen, R., and Rubner, M. (2011) results to be published.

64 Schlenoff, J.B., Dubas, S.T., and Farhat, T. (2000) Sprayed polyelectrolyte multilayers. *Langmuir*, **16** (26), 9968–9969.

65 Nogueira, G.M., Banerjee, D., Cohen, R.E., and Rubner, M.F. (2011) Spray-layer-by-layer assembly can more rapidly produce

optical-quality multistack heterostructures. *Langmuir*, **27** (12), 7860–7867.
66 Kyung, K.-H., Fujimoto, K., and Shiratori, S. (2011) Control of structure and film thickness using spray layer-by-layer method: application to double-layer anti-reflection film. *Jpn. J. Appl. Phys.*, **50** (3), 035803-1–035803-5.
67 Krogman, K.C., Lowery, J.L., Zacharia, N.S., Rutledge, G.C., and Hammond, P.T. (2009) Spraying asymmetry into functional membranes layer-by-layer. *Nature Mater.*, **8** (6), 512–518.
68 Chunder, A., Etcheverry, K., Wadsworth, S., Boreman, G.D., and Zhai, L. (2009) Fabrication of anti-reflection coatings on plastics using the spraying layer-by-layer self-assembly technique. *J. Soc. Inf. Display*, **17** (4), 389–395.
69 Kolasinska, M., Krastev, R., Gutberlet, T., and Warszynski, P. (2009) Layer-by-layer deposition of polyelectrolytes. Dipping versus spraying. *Langmuir*, **25** (2), 1224–1232.
70 Felix, O., Zheng, Z., Cousin, F., and Decher, G. (2009) Are sprayed lbl-films stratified? A first assessment of the nanostructure of spray-assembled multilayers by neutron reflectometry. *C. R. Chim.*, **12** (1–2), 225–234.
71 Krogman, K.C., Zacharia, N.S., Schroeder, S., and Hammond, P.T. (2007) Automated process for improved uniformity and versatility of layer-by-layer deposition. *Langmuir*, **23** (6), 3137–3141.
72 Hong, J. and Park, H. (2011) Fabrication and characterization of block copolymer micelle multilayer films prepared using Dip-, Spin- and spray-assisted layer-by-layer assembly deposition. *Colloid Surf. A*, **381** (1–3), 7–12.
73 Fukao, N., Kyung, K.-H., Fujimoto, K., and Shiratori, S. (2011) Automatic spray-LbL machine based on in-situ QCM monitoring. *Macromolecules*, **44** (8), 2964–2969.
74 Stewart-Clark, S.S., Lvov, Y.M., and Mills, D.K. (2011) Ultrasonic nebulization-assisted layer-by-layer assembly for spray coating of multilayered, multicomponent, bioactive nanostructures. *J. Coating Tech. Res.*, **8** (2), 275–281.
75 Ashcraft, J.N., Argun, A.A., and Hammond, P.T. (2010) Structure-property studies of highly conductive layer-by-layer assembled membranes for fuel cell PEM applications. *J. Mater. Chem.*, **20** (30), 6250–6257.
76 Suzuki, Y., Pichon, B.P., D'Elia, D., Beauger, C., and Yoshikawa, S. (2009) Preparation and microstructure of titanate nanowire thin films by spray layer-by-layer assembly method. *J. Ceram. Soc. Jpn.*, **117** (1363), 381–384.
77 Lefort, M.L.M., Boulmedais, F., Jierry, L., Gonthier, E., Voegel, J.C., Hemmerle, J., Lavalle, P., Ponche, A., and Schaaf, P. (2011) Simultaneous spray coating of interacting species: general rules governing the poly(styrene sulfonate)/poly(allylamine) system. *Langmuir*, **27** (8), 4653–4660.
78 Ladhari, N., Hemmerle, J., Ringwald, C., Haikel, Y., Voegel, J.C., Schaaf, P., and Ball, V. (2008) Stratified PEI-(PSS-PDADMAC)-PSS-(PDADMAC-TiO$_2$) multilayer films produced by spray deposition. *Colloid. Surf. A*, **322** (1–3), 142–147.
79 Izquierdo, A., Ono, S.S., Voegel, J.C., Schaaf, P., and Decher, G. (2005) Dipping versus spraying: exploring the deposition conditions for speeding up layer-by-layer assembly. *Langmuir*, **21** (16), 7558–7567.
80 Porcel, C.H., Izquierdo, A., Ball, V., Decher, G., Voegel, J.C., and Schaaf, P. (2005) Ultrathin coatings and (poly (glutamic acid)/polyallylamine) films deposited by continuous and simultaneous spraying. *Langmuir*, **21** (2), 800–802.
81 Svaya Nanotechnology, http://www.svaya-nano.com/index-2.html
82 Zhai, L., Nolte, A.J., Cohen, R.E., and Rubner, M.F. (2004) pH-gated porosity transitions of polyelectrolyte multilayers in confined geometries and their application as tunable Bragg reflectors. *Macromolecules*, **37** (16), 6113–6123.
83 Kurt, P., Banerjee, D., Cohen, R.E., and Rubner, M.F. (2009) Structural color via layer-by-layer deposition: layered nanoparticle arrays with near-UV and

visible reflectivity bands. *J. Mater. Chem.*, **19** (47), 8920–8927.
84 Olugebefola, S.C., Kuhlman, W.A., Rubner, M.F., and Mayes, A.M. (2008) Photopatterned nanoporosity in polyelectrolyte multilayer films. *Langmuir*, **24** (9), 5172–5178.
85 Wu, Z., Lee, D., Rubner, M.F., and Cohen, R.E. (2007) Structural color in porous, superhydrophilic, and self-cleaning $SiO_2/TiO_2$ Bragg stacks. *Small*, **3** (8), 1445–1451.
86 Nolte, A.J., Rubner, M.F., and Cohen, R.E. (2004) Creating effective refractive index gradients within polyelectrolyte multilayer films: molecularly assembled rugate filters. *Langmuir*, **20** (8), 3304–3310.
87 Lee, D., Rubner, M.F., and Cohen, R.E. (2006) All-nanoparticle thin-film coatings. *Nano. Lett.*, **6** (10), 2305–2312.
88 Gemici, Z., Shimomura, H., Cohen, R.E., and Rubner, M.F. (2008) Hydrothermal treatment of nanoparticle thin films for enhanced mechanical durability. *Langmuir*, **24** (5), 2168–2177.
89 Gemici, Z., Schwachulla, P.I., Williamson, E.H., Rubner, M.F., and Cohen, R.E. (2009) Targeted functionalization of nanoparticle thin films via capillary condensation. *Nano. Lett.*, **9** (3), 1064–1070.
90 Dafinone, M.I., Feng, G., Brugarolas, T., Tettey, K.E., and Lee, D. (2011) Mechanical reinforcement of nanoparticle thin films using atomic layer deposition. *ACS Nano.*, **5** (6), 5078–5087.
91 Caruso, F. (2001) Nanoengineering of particle surfaces. *Adv. Mater.*, **13**, 11–22.

# 3
# Layer-by-Layer Assembly: from Conventional to Unconventional Methods

*Guanglu Wu and Xi Zhang*

## 3.1
## Introduction

Layer-by-layer (LbL) assembly is a technique for depositing multilayers. The first work on LbL can be traced back to 1966, when Iler reported the fabrication of multilayer films by alternating deposition of positively and negatively charged colloidal particles [1]. However, the really extraordinary growth of this field was initiated in the beginning of the 1990s by Decher *et al.*, who rediscovered that anionic and cationic bipolar amphiphiles or polyelectrolytes could be consecutively adsorbed on a charged surface to form multilayer thin films [2–5]. Since then, the unique features of LbL assembly, such as simplicity, controllability, universality and diversity, have driven extensive research on LbL thin films from fundamental aspects to practical applications.

In principle, various building blocks can be incorporated into multilayer thin films with controlled architectures and functions, including bolaform amphiphiles [6], colloids and nanoparticles [7, 8], dyes [9, 10], dendrimers [11, 12], proteins and enzymes [13–18], DNA [19, 20], viruses [21], and so on. All of these building blocks can be fabricated into multilayer thin films simply by alternating deposition at the liquid/solid interface, the so-called conventional LbL method. However, there are other building blocks, such as singly charged or water-insoluble species, that cannot be incorporated into multilayer thin films by the conventional LbL method. To solve this problem, we have developed a series of unconventional LbL methods. The main idea of these methods includes more than one step in the assembly process, as shown in Figure 3.1. For example, the building blocks can self-assemble in solution to form a supramolecular complex, and the supramolecular complex can be further used as one of the building blocks for LbL assembly at a liquid/solid interface. In this way, those building blocks which cannot be fabricated by the conventional LbL method can be assembled by this unconventional method, endowing the multilayer thin films with new structures and functions [22].

*Multilayer Thin Films: Sequential Assembly of Nanocomposite Materials*, Second Edition.
Edited by Gero Decher and Joseph B. Schlenoff.
© 2012 Wiley-VCH Verlag GmbH & Co. KGaA. Published 2012 by Wiley-VCH Verlag GmbH & Co. KGaA.

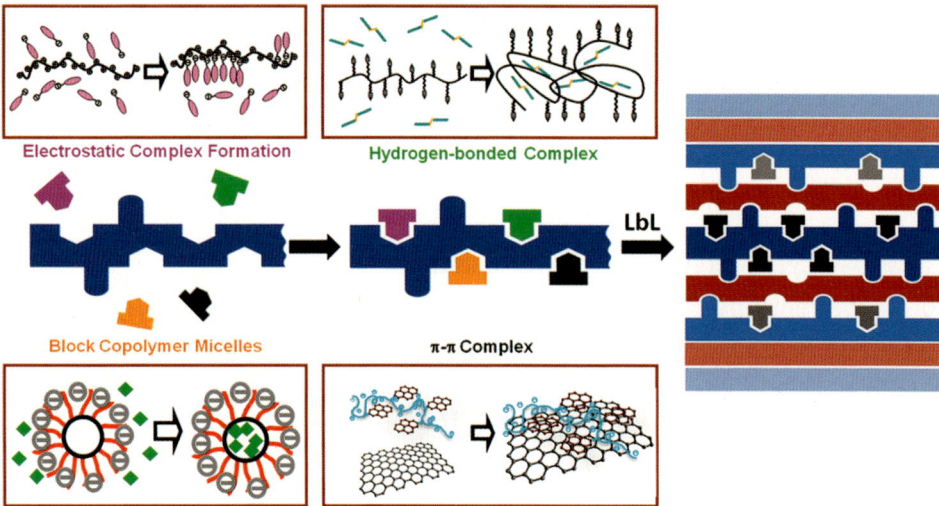

**Figure 3.1** Unconventional LbL method includes two steps: (1) Building blocks self-assemble in solution to form supramolecular complex; (2) the supramolecular complex is used as one of the building blocks for LbL assembly at the liquid/solid interface.

## 3.2
## Conventional LbL Methods

The conventional LbL method refers to the direct alternating deposition of building blocks at a liquid/solid interface, which can be driven by different intermolecular interactions, such as electrostatic interaction, hydrogen bonding, halogen bonding [23], and coordination interactions. In this section we will introduce a few conventional LbL methods and discuss how to employ these methods for the construction of functional thin films.

### 3.2.1
### Electrostatic LbL Assembly

The electrostatic LbL assembly requires the building blocks to be water soluble and multi-charged species. One of the most popular candidates for electrostatic LbL assembly is a polyelectrolyte, including polyanions and polycations with one charged group per monomer unit. The alternating deposition of polyanions and polycations does not require an exact positional matching of the charged points. It should be noted, however, that there is no clear interface structure for the polyelectrolyte multilayer. In many cases, X-ray reflectograms have exhibited only so-called Kiessig fringes that arise from the interference of X-ray beams reflected at the substrate/film and the film/air interfaces. In other words, the adjacent layers are interdigitated to some extent in polyelectrolyte multilayers. It is found, however, that Bragg diffraction can be clearly observed on multilayer thin films fabricated from a polyelectrolyte and a bolaform amphiphile bearing mesogenic groups, suggesting that the order of multilayer thin films can be improved to some extent in this way [24].

Taking advantage of the penetrability of polyelectrolyte multilayer thin films, we have first demonstrated the formation of a rough surface with micro- and nanostructures on an LbL film through electrochemical deposition of gold or silver clusters [25, 26]. As shown in Figure 3.2a, a flat ITO electrode was modified with six layer pairs of poly(diallyldimethylammonium chloride) (PDDA) and poly(4-stylene sulfonate) (PSS), and was then immersed in a mixture of $H_2SO_4$ (0.5 M) and $KAuCl_4$ (1 mg mL$^{-1}$). Electrochemical deposition was conducted at $-200$ mV in a single potential time base mode, using a platinum electrode as the counter electrode and Ag/AgCl as the reference electrode. From SEM images (Figure 3.2b), gold clusters were found under these conditions, exhibiting interesting dendritic structures with nanoscale protuberances. It was shown that the surface density and the size of the gold clusters increased with increasing deposition time (Figure 3.2c). A LbL polyelectrolyte multilayer film is crucial for the formation of the dendritic

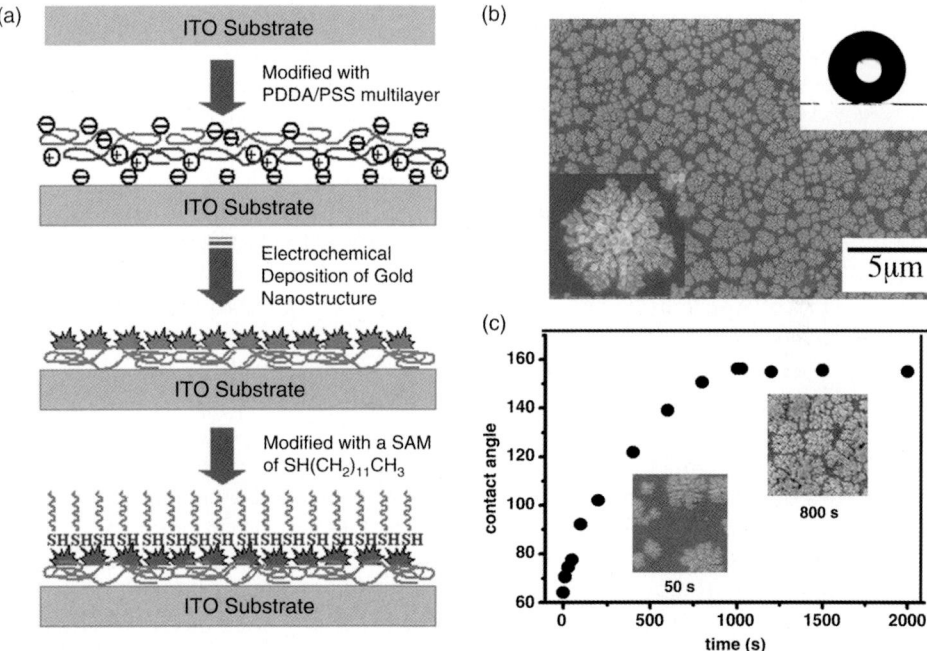

**Figure 3.2** Multilayer thin films as the matrix for electrochemical deposition of gold clusters achieving superhydrophobic surfaces: (a) Schematic illustration of the fabrication of superhydrophobic surfaces; (b) SEM images of dendritic gold clusters formed on an ITO electrode modified with a PDDA/PSS multilayer film by electrochemical deposition at $-200$ mV (vs Ag/AgCl) (single potential time base mode) for 200 s; (c) The contact angles of the surface as a function of the electrochemical deposition time. (Adapted with permission from [25]. Copyright 2011 American Chemical Society).

structure of gold clusters, and it is a guarantee for the strong adhesion between the ITO surface and the formed gold aggregates. By further chemisoption of a surface-active molecule such as *n*-dodecanethiol onto the above rough surface, a superhydrophobic surface with a contact angle greater than 170° can be obtained. The surface showed a tilted angle as low as 1.5°, indicating that this study provides a new avenue for the fabrication of a self-clean surface.

This method for realizing a superhydrophobic surface is independent of the size and shape of the substrate, which means that a superhydrophobic surface can be achieved on a planar substrate as well as on a nonplanar substrate. For example, we have fabricated a superhydrophobic surface on gold threads, and confirmed that this superhydrophobic surface can be used not only for enhancing supporting force but also for decreasing the fluidic drag. It is well known that the legs of water striders are equipped with suprerhydrophobic coatings, allowing them to walk easily on the surface of water. Taking the deformation force into consideration, it is found that the total supporting force provided by the normally hydrophobic coating (contact angle ≈ 110°) is strong enough to support the water strider when it merely floats on the water surface. Since nature selects a superhydrophobic coating (contact angle > 150°) for water striders, it must have its reasons. In an attempt to understand why a surperhydrophobic surface is needed by water striders, gold threads were chosen to mimic water striders' legs [27]. Our study suggests that one reason is to provide an additional buoyancy force by wrapping its body with a thin layer of air. The additional buoyancy force is caused by the displacement of the air layer. When a water strider penetrates the water surface and is immersed beneath the water surface, the additional buoyancy force will provide enough upward force to help the water strider to pull itself back out through the water surface and refloat on the surface.

The other reason is to reduce the fluidic drag during movement on the water surface. A superhydrophobic coating guarantees a much faster movement than that of a hydrophobic coating at the water surface. To confirm this hypothesis, a smart experiment has been designed. Pt aggregates were deposited electrochemically on one end of the gold threads. Since the Pt aggregates can catalyze the decomposition of $H_2O_2$ and release oxygen, this process can provide the prepared gold threads with a method of propulsion in a solution of $H_2O_2$ (Figure 3.3a). Both the superhydrophobic and normally hydrophobic gold threads were placed carefully on the surface of the solution of $H_2O_2$. An interesting finding is that the average rate of movement of the two gold threads was measured as $26.0\,\text{cm}\,\text{min}^{-1}$ for the superhydrophobic gold threads, and $15.6\,\text{cm}\,\text{min}^{-1}$ for the normal hydrophobic ones (Figure 3.3b). This indicates that the superhydrophobic coating is indispensable and can decrease the fluidic drag by introducing a thin layer of air surrounding the surface, thereby making the water strider move faster.

To construct a smart surface, other stimuli-responsive surface reactive molecules should be used instead of *n*-dodecanethiol. For example, pH-responsive 2-(11-mercaptoundecanamido) benzoic acid was used to modify the rough gold surface. Interestingly, the surface wettabilities can be manipulated from near superhydrophobicity to superhydrophilicity by adjusting the pH. Thus, the flotation and motion of gold threads equipped with such smart coatings can be feasibly controlled by pH [28].

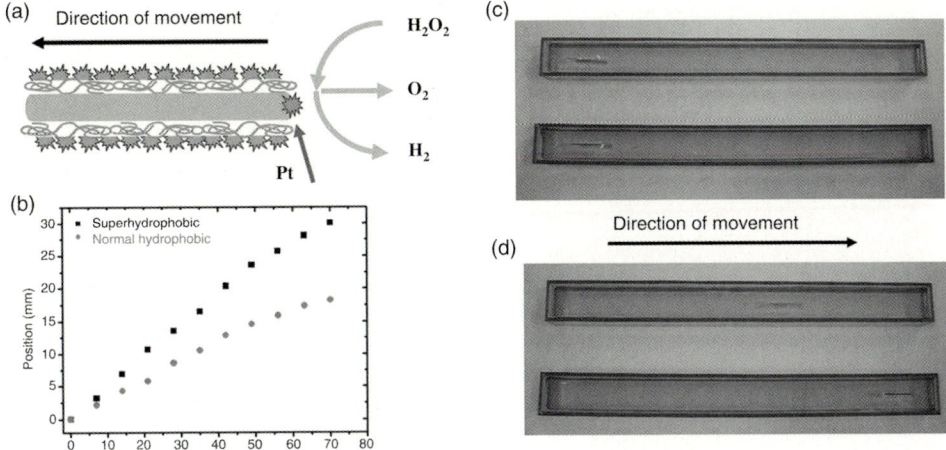

**Figure 3.3** (a) Schematic illustration of the movement of a superhydrophobic gold thread with Pt aggregates deposited on one end. (b) The position of the superhydrophobic and normal hydrophobic gold threads versus time of movement. (c) The initial picture when both of the superhydrophobic and normal hydrophobic gold threads with Pt aggregates on one of their ends were simultaneously put onto the surface of 30% $H_2O_2$ solution in two troughs. (d) 70 s later than (c) when the superhydrophobic gold thread has reached the other end of the trough. (Adapted with permission from [27]. Copyright 2011 Wiley-VCH Verlag GmbH & Co. KGaA).

### 3.2.2
### Hydrogen-Bonded LbL Assembly

The hydrogen bond as a driving force for LbL assembly was first demonstrated by our group and Rubner *et al.* simultaneously in 1997 [29, 30]. Polymers bearing hydrogen bond donors and acceptors are employed as building blocks for LbL assembly. For example, poly(4-vinylpyridine) (PVP) as a hydrogen bond acceptor can alternately deposit with a hydrogen bond donor such as poly(acrylic acid) (PAA) [30] or a copolymer of *p*-(hexaflouro-2-hydroxylisopropyl)-α-methylstyrene and styrene [31] to fabricate multilayer thin films in organic solvent on the basis of hydrogen bonding, which has been identified by FT-IR spectroscopy. In addition, carboxyl-terminated poly-ether dendrimer (DEN-COOH) can also interact with PVP for hydrogen-bonded LbL assembly [32]. Since the carboxyl group can act as both a hydrogen bond donor and acceptor, a uniform film by self-deposition of DEN-COOH has been successfully realized [12].

A hydrogen bond can be formed under one set of conditions and broken under other conditions. To take advantage of this unique feature, one can employ a hydrogen-bonded LbL assembly for construction of erasable thin films [33, 34] and microporous films [35]. It has been demonstrated that a microporous film can be obtained by immersion of the above PVP/PAA LbL film into a basic aqueous solution. The AFM images in Figure 3.4 of the 25-layer LbL films, after immersion in pH 13 NaOH aqueous solutions at 25 °C for different periods of time, show that the microporous films can be produced by prolonged immersion of the LbL film in a

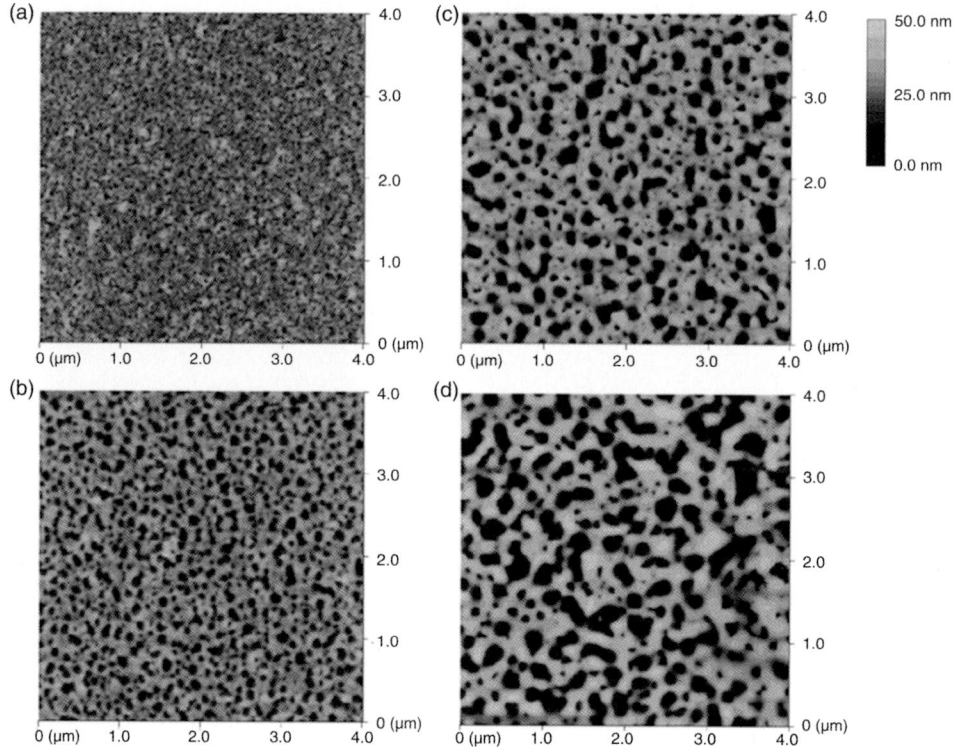

**Figure 3.4** The microporous films can be produced by prolonged immersion of the LbL film in a basic solution: AFM height images of 25-layer PAA/PVP LbL films after immersion in pH 13 NaOH aqueous solution at 25 °C for 10 (a), 40 (b), 100 (c), and 180 min (d). (Adapted with permission from [35]. Copyright 2011 American Chemical Society).

basic solution. During the basic treatment, the characteristic FT-IR band of the carboxyl group gradually disappeared, suggesting the dissolution of PAA. However, it is worth noting that the release of PAA should not be directly responsible for the formation of the microporous film. As a control, a spin-coated film from a mixture of PAA and PVP solution was prepared. Although the FT-IR clearly indicated similar removal of PAA by a basic solution, only a nonporous film was formed.

It is proposed that the morphology variation is a result of the reconformation or folding of PVP induced by the basic solution. The above hypothesis was further supported by single molecule force spectroscopy (SMFS). Desorption of PVP chains assembled on a $NH_2$-modified substrate in methanol was measured by SMFS. Sawtooth-like force profiles were obtained (Figure 3.5) which was attributed to the detachment of polymer loops from the substrate. If PVP chains are folded in a basic solution as proposed, the number of PVP attached points per unit length in a basic solution would be less than that in methanol; in other words, the average length of loops in the basic solution should be greater. Estimated by dividing the curve length

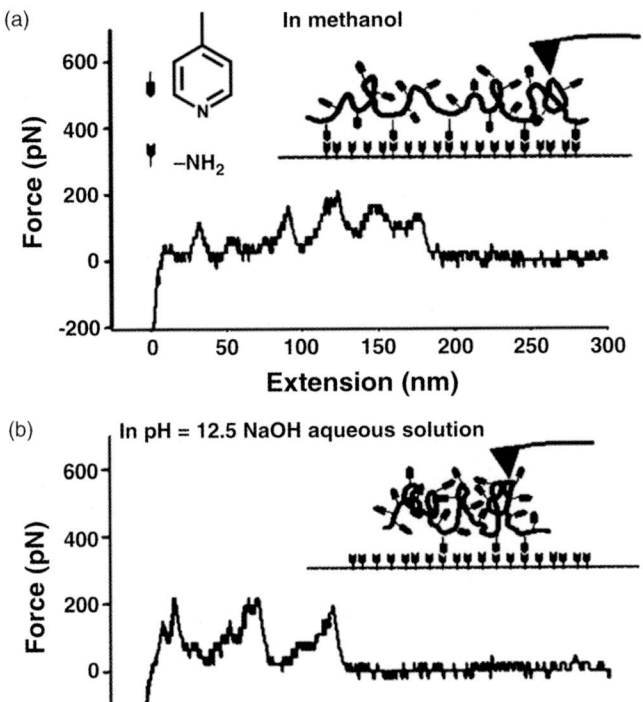

**Figure 3.5** Typical force extension curves of PVP in methanol (a) and in pH 12.5 NaOH aqueous solution (b). The insets are the corresponding schematic drawings of SMFS. (Adapted with permission from [35]. Copyright 2011 American Chemical Society).

by the number of peaks, the average lengths of loops in methanol and in the basic solution were 27.9 and 35.8 nm, respectively, as shown in Figure 3.6. As a result, the average length of loops in the basic solution is greater than that in methanol, in agreement with our hypothesis that the reconformation of PVP polymer chains is responsible for the formation of microporous films [35].

One of the applications of the microporous film is as an ink reservoir for multiple printing in soft lithography. For example, Xu et al. fabricated the above microporous film from PVP/PAA hydrogen-bonded LbL film on the surface of polydimethlysiloxane (PDMS) stamps for microcontact printing (μCP). The porous structures on the top of the stamp can act as a so-called "ink reservoir" for the adsorption of inks, such as proteins and particles. Multiple printing of these large ensembles is not possible with conventional μCP. However, the microporous film-modified PDMS stamps allow multiple printing of proteins and particles without the need to re-ink, therefore improving a nanotechnology for fabrication of functional surfaces [36].

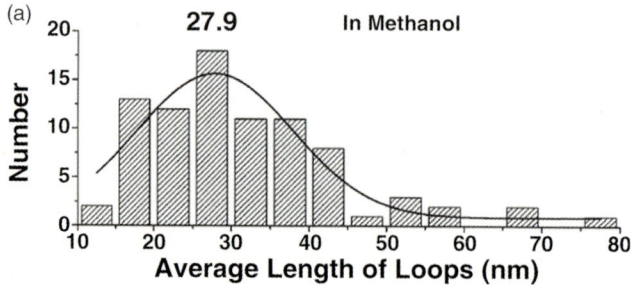
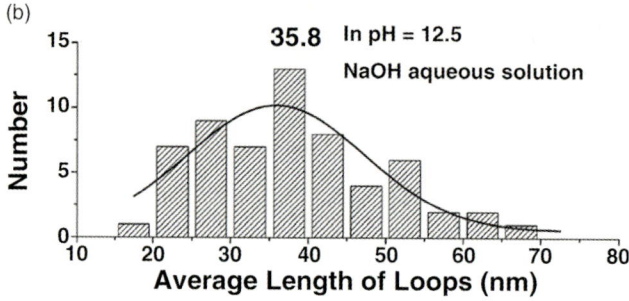

**Figure 3.6** Histogram of the average length of PVP loops in methanol (a) and in pH 12.5 NaOH aqueous solution (b) (Adapted with permission from [35]. Copyright 2011 American Chemical Society).

### 3.2.3
### LbL Assembly Driven by Coordination Interaction

Besides weak intermolecular interaction, much stronger coordination interaction is also available for a conventional LbL method. The $Cu^{2+}$ neutralized polyelectrolyes, poly(copper styrene 4-sulfonate) ($PSS(Cu)_{1/2}$), and ligand containing polymer PVP have been utilized to construct LbL films, driven by the coordination interaction between $Cu^{2+}$ and pyridine [37]. Similar methods have been extended to many other coordinating systems. For example, nanoparticles like CdS can be directly incorporated into multilayer films by coordination interaction [38].

To incubate the LbL films of $PSS(Cu)_{1/2}$/PVP in $H_2S$ gas, an *in situ* chemical reaction was carried out yielding an organic–inorganic hybrid multilayer film, PSS/$Cu_2S$/PVP. Since the reaction is confined between the layers, $Cu_2S$ nanoparticles with a size of only a few nanometers are obtained. Therefore, the combination of LbL assembly driven by coordination interaction and post-chemical reaction opens a new avenue for the formation of organic–inorganic hybrid films.

### 3.2.4
### To Combine LbL Assembly and Post-Chemical Reaction for the Fabrication of Robust Thin Films

Although LbL assembly on the basis of different interactions has been developed, electrostatic interaction is one of the major driving forces for the formation of

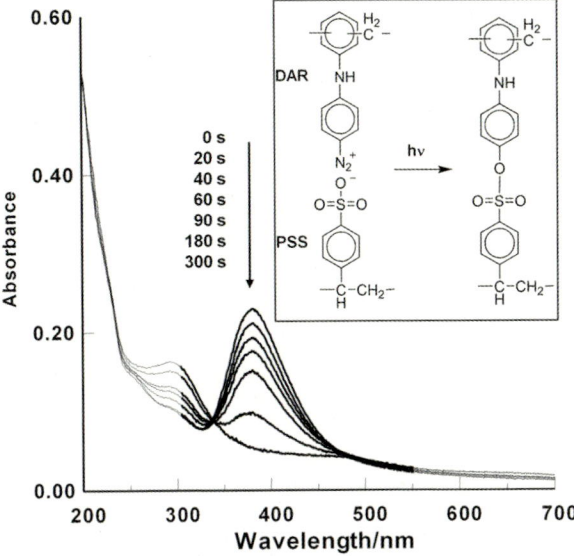

**Figure 3.7** UV–VIS absorption spectra of eight layer pairs of DAR/PSS upon irradiation with UV light for (a) 0, (b) 20, (c) 40, (d) 60, (e) 90, (f) 180 and (g) 300 s. The inset is the schematic representation for structure changes of multilayers from ionically to covalently attach upon UV irradiation. (Adapted with permission from [40]. Copyright 2011 Wiley-VCH Verlag GmbH & Co. KGaA).

multilayer thin films. However, the electrostatic interaction can be easily influenced by many environmental factors, such as the ionic strength of the solution, the pH of the solvent, temperature, and so on, and so is the stability of the resulting multilayer films. In other words, these films are not robust enough for use in some rigorous environments, especially in solutions with a high ionic strength, high pH, or low pH.

To meet the requirement of stability, we have proposed a method that combines LbL deposition and post-photochemical reaction. Diazoresin (DAR), a reactive and water-soluble polycation, was employed as one of the building blocks for alternating deposition with negatively charged building blocks, for example, PSS, yielding a multilayer thin film of DAR/PSS. Upon UV irradiation, DAR was converted to its phenyl cationic form, followed by an $S_N1$ attack by sulfonate to form a cross-linked network [39, 40], as shown in Figure 3.7. In this way, the ionic interaction between the layers is converted into much stronger covalent bonds, which significantly improves the stability of the multilayer thin films.

It should be pointed out that this is a general method for fabrication of robust multilayer thin films. In addition to electrostatic interaction, DAR derivatives can be used for constructing thin films on the basis of hydrogen bonding. Moreover, the photoreaction is not only limited to DAR with sulfonate-bearing building blocks like PSS, but also works with building blocks bearing carboxyl, phosphate, or hydroxy groups. In addition to the strong polyelectrolyte, DAR has been employed to stabilize weak polyelectrolytes, PAA [41], oligo-charged dyes [42, 43], and inorganic nanoparticles [8].

Taking advantage of the solubility difference of the photoreactive multilayer films before and after UV irradiation, a robust multilayer patterning of surfaces can be produced by photolithographic techniques [44]. If we further modify the patterned surface by interfacial chemistry, the patterned polyelectrolyte multilayer thin films can be used as a good matrix for selective adsorption of polystyrene nanoparticles with appropriate modification [45].

In order to meet the requirement for stabilizing positively charged building blocks, we have employed a photoreactive polyanion, PAA-$N_3$, a PAA derivative grafted with azido groups, and provided an effective way to improve the stability of multilayer thin films also by post-photoreaction. This polyanion can construct multilayer films with various positively charged building blocks, including weak or strong polycations, and small molecules with multiple positively charged sites. Moreover, besides electrostatic LbL assembly, this polyanion could also be employed for multilayer fabrication on the basis of hydrogen bonding in organic solvents [46].

## 3.3
### Unconventional LbL Methods

As mentioned previously, unconventional LbL methods are proposed for incorporating building blocks which cannot be incorporated into multilayer thin films through the above conventional LbL methods, such as singly charged or water-insoluble species. A typical unconventional LbL method usually includes two assembly processes. First, in bulk solution, the building blocks self-assemble to form a supramolecular complex. Diverse complexes have been achieved based on different interactions, including electrostatic complexes, hydrogen-bonded complexes, block copolymer micelles and π–π complexes. Second, at a liquid/solid interface, the supramolecular complex is subsequently used as a building block for LbL assembly, which can also be driven by different forces. As a result, the singly charged or water-insoluble building blocks become self-assembling for the construction of multilayer thin films, leading to the development of new functions of surface imprinting, nanocontainers, nanoreactors and biosensors.

### 3.3.1
### Electrostatic Complex for Unconventional LbL Assembly

One typical example for this type of unconventional LbL method is the incorporation of singly charged molecules, for example, sodium 9-anthracenepropionate (SANP) into multilayer thin films [47], as shown in Figure 3.8. Driven by electrostatic interaction in solution, this negatively charged SANP combines with positively charged PDDA to form a supramolecular complex (PDDA-SANP), and multilayer films are fabricated by alternating deposition of the PDDA–SANP complex with PSS at a liquid/solid interface, also driven by electrostatic interaction. It is well known that small molecules, such as SANP, can diffuse into conventional LbL films of PDDA/PSS. However, the amount of SANP assembled by the unconventional method is

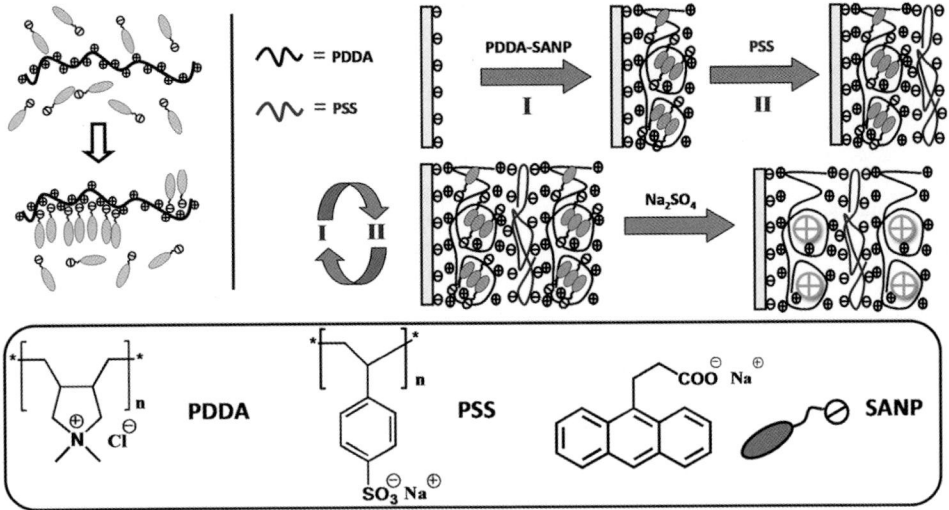

**Figure 3.8** Schematic illustration of the unconventional LbL method by electrostatic complex formation: Step I, SANP tethered to PDDA in solution; Step II, LbL assembly of PDDA-SANP/PSS. Multilayer film with charge selectivity can be obtained by removing SANP from the film.

much larger than that would be included by diffusion, and moreover, a controllable amount of SANP can be incorporated by adjusting the initial concentration of SANP in the PDDA–SANP complex solution.

### 3.3.1.1 Nanoreactors with Enhanced Quantum Yield

Can LbL films act as a nanoreactor? To answer this question, the above LbL film of PDDA-SANP/PSS is a nice model system, since the anthracene moiety in SANP can undergo photocycloaddition under UV irradiation. As shown in Figure 3.9, the characteristic absorbance of anthracene between 250 and 425 nm decreases with UV irradiation, while at the same time the absorbance of benzene around 205 nm increases, which indicates that SANP moieties incorporated in the LbL film undergo photocycloaddition to produce a photocyclomer. Interestingly, the quantum yield of photocycloaddition is about four times higher than that in the solution. The reason why such photocycloaddition occurs with an enhanced quantum yield should be correlated with the aggregation of SANP in the LbL films which facilitates the reaction.

### 3.3.1.2 "Ion Traps" for Enhancing the Permselectivity and Permeability

The combination of supramolecular complexes and LbL deposition allows not only incorporation of single charged moieties into LbL films, but also controlled release of them from LbL films, leading to so-called "ion traps". For example, when immersing a multilayer film of PDDA–SANP/PSS into an aqueous solution of $Na_2SO_4$, the SANP can be released from the film quickly, depending on the ionic strength of the solution. An interesting finding is that, after the release of SANP, the LbL film is endowed with

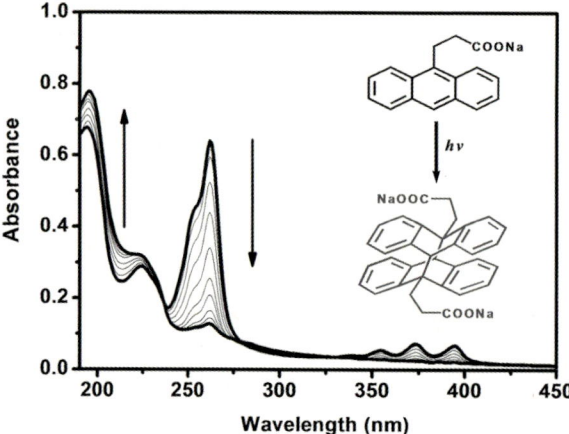

**Figure 3.9** The characteristic absorbance of SANP as a function of UV irradiation time: SANP undergoes photocycloaddition to produce a photocyclomer. The quantum yield of photocycloaddition is about four times higher than that in solution. (Adapted with permission from [47]. Copyright 2011 American Chemical Society).

charge selectivity. That is to say, the as-prepared LbL film can re-adsorb only negatively charged moieties, whereas it repels positively charged moieties. As a control experiment, small molecules can diffuse into normal LbL films of PDDA/PSS, however, either positively charged or negatively charged species can be equally incorporated, indicative of no charge selectivity. In addition, the loading capacity of SANP in a PDDA–SANP/PSS film is seven times higher than that in a PDDA/PSS film. The same method can be extended to incorporate positively charged building blocks and to fabricate films with abilities to re-adsorb only positively charged moieties and repel negatively charged ones [48]. Therefore, the LbL films fabricated by this unconventional LbL method can be used as materials of permselectivity.

Not all small molecules are suitable templates for fabrication of LbL films that can trap ions of one kind of charge while repelling those of opposite charge. We have tried different cations and anions and realized that singly charged molecules bearing condensed aromatic structures are good candidates. The reasons are: (i) Singly charged molecules can form complexes with polyelectrolytes and also unbind easily, which is an important factor for successful incorporation into LbL films, as mentioned above. Molecules with two or more charges can hardly unbind from the polyelectrolytes. (ii) The small molecules used in our experiment have a hydrophilic group and a hydrophobic group with a condensed aromatic moiety. When forming a complex in aqueous solution, the aromatic hydrophobic groups might get together due to hydrophobic interaction as well as the π–π stacking interaction.

### 3.3.1.3 Surface Imprinted LbL Films

Combining the above properties of an ion trap with previously discussed approaches for robust films, gives a new concept of surface molecular imprinted LbL (SMILbL)

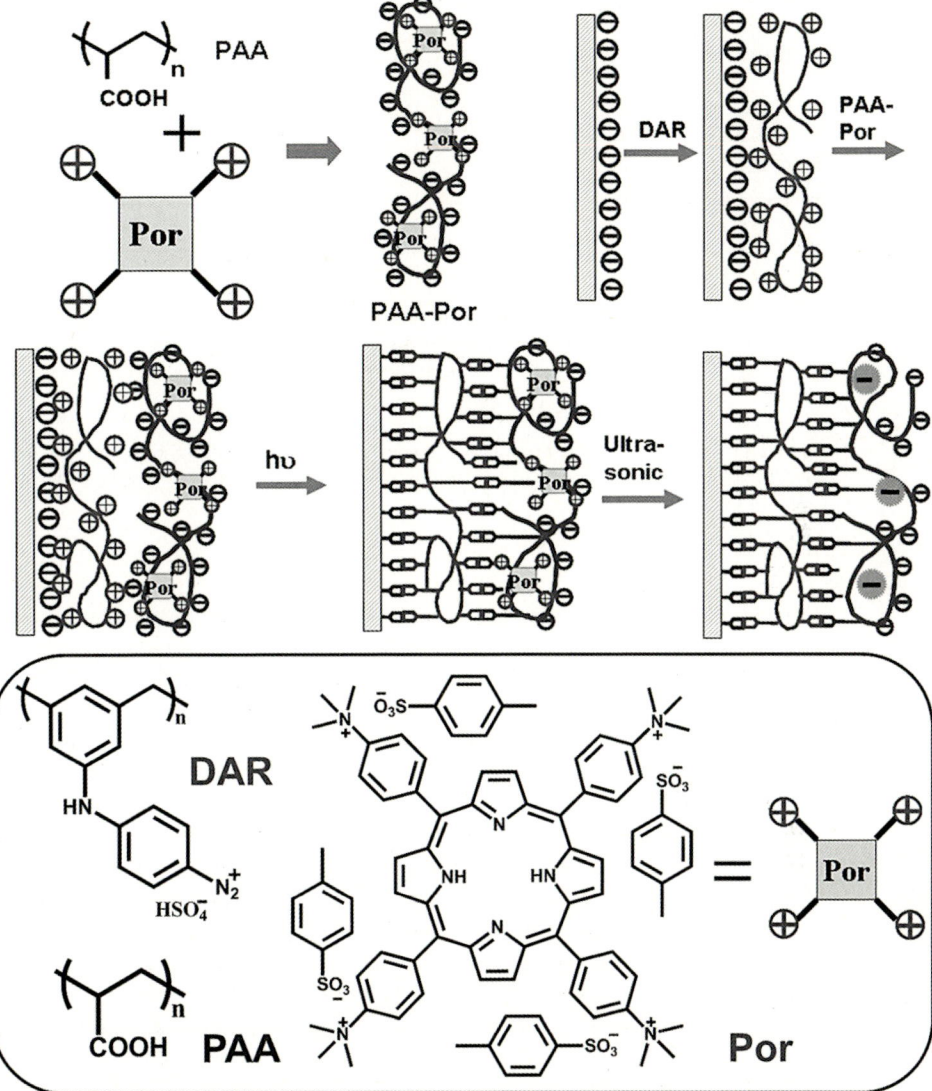

**Figure 3.10** Schematic illustration of the formation of the imprinting complexes (PAA-Por) and the experimental procedure for the formation of SMILbL films.

films. A general procedure for the preparation of SMILbL films includes four steps. Take the generation of imprinted sites for the porphyrin derivative (Por) as an example, as shown in Figure 3.10. First, an electrostatically stabilized complex between the positively charged porphyrin and PAA was formed in aqueous solution. Second, a multilayer thin film was fabricated by alternating deposition of the complex PAA-Por and photoreactive DAR. Third, the layered structure was photo cross-linked

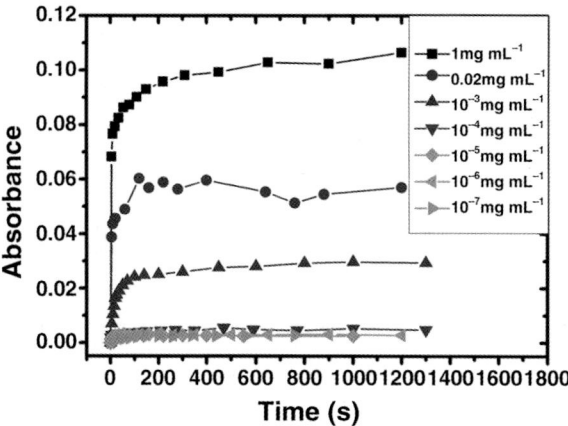

**Figure 3.11** Time-dependent absorbance of the Por association to the imprinted film at different bulk concentrations of Por. (Adapted with permission from [49]. Copyright 2011 Wiley-VCH Verlag GmbH & Co. KGaA).

to yield the covalent bridging of the layers by UV irradiation. In the final step, the template porphyrin molecules were washed off from the film, yielding the surface imprinted matrix [49].

The SMILbL films have advantages over other methods in terms of thermodynamics and kinetics. As shown in Figure 3.11, the binding rate for loading Por back to the imprinted film is very fast, and the loading process reaches a saturated value within 2 min. Moreover, the saturated absorbance of Por increases upon elevating the bulk concentration of Por, indicating a concentration-dependent loading process. That is to say, the association of Por is an equilibrium process, with a binding constant of about $2 \times 10^5 \, M^{-1}$.

To give convincing proof that specific imprinted sites, rather than electrostatic interactions, play a more important role in binding template molecules, the positively charged $Ru(NH_3)_6^{3+}$ is utilized as an electrochemical label to probe its association with LbL films before and after exclusion of the positively charged template, Por. When the multilayer film was loaded with Por, no redox response was observed for the $Ru(NH_3)_6^{3+}$ label, Figure 3.12 (curve 1), implying that the film insulates the interfacial electron transfer to the redox label. Exclusion of the template resulted in an electrical response of $Ru(NH_3)_6^{3+}$, Figure 3.12 (curve 2). That is, unoccupied negatively charged sites, which come from the exclusion of positively charged templates, were occupied by $Ru(NH_3)_6^{3+}$, making the film permeable to the redox label. Significantly, treating the $Ru(NH_3)_6^{3+}$-loaded film with the imprinted template, Por, resulted in a decrease in the electrical response of $Ru(NH_3)_6^{3+}$, Figure 3.12 (curves 3 and 4), implying that the redox label was competitively displaced by Por. That is, Por exhibited a substantially higher affinity for the polymer as compared to $Ru(NH_3)_6^{3+}$. Therefore, the binding sites of SMILbL films here are specific imprinted sites rather than simple ion traps.

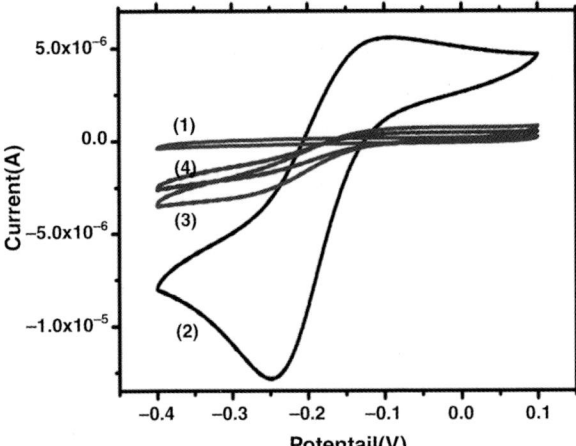

**Figure 3.12** Cyclic voltammograms of the gold electrodes modified by (DAR/PAA-Por)$_5$DAR multilayer films in the presence of an electrolyte solution consisting of 3 mM Ru(NH$_3$)$_6^{3+}$ and 0.1 M KCl; (1) after UV irradiation and in the presence of bound Por; (2) removal of Por from the film by ultrasonic agitation in the ternary solution; (3, 4) after immersing the unloaded film in a mixed solution of 0.2 mg mL$^{-1}$ Por, 2.4 mM Ru(NH$_3$)$_6^{3+}$, and 0.08 M KCl for 2 and 20 min, respectively. (Adapted with permission from [49]. Copyright 2011 Wiley-VCH Verlag GmbH & Co. KGaA).

In addition to photoreactive DAR, the stabilized matrix can also be obtained by forming cross-linked disulfide bonds between the layers. Again, the unconventional LbL method was adopted to form SMILbL films based on the disulfide cross-linking. The polycation, poly(allylamine hydrochloride) (PAH), preassembles in aqueous solution to a PAH-TPPS complex with TPPS, another porphyrin derivative with negative charges. Subsequently, the preassembled PAH-TPPS complex undergoes an LbL assembly at the liquid/solid interface with PAASH60, a PAA with 60% of its carboxylic acid grafted by thiol groups. After oxidative cross-linking to form disulfide bonds between the layers, the multilayer with preassembly of the PAH-TPPS complex allowed the release and loading of TPPS in a reproducible way. The release of TPPS from the loaded film is a pH-controlled process. Moreover, the release of TPPS could also be achieved by cleaving the cross-linking through reduction of disulfide bonds, and the release rates could be controlled by the reductive efficiency of the reductants in the media. In this way, the release of TPPS is pH/reductant dually controllable, thereby facilitating a new route to multi-stimuli controllable materials [50].

To improve the selectivity of SMILbL films, the cooperativity of various specific interactions has been introduced within the binding sites. Theophylline derivatives are chosen as the model template molecules to investigate the feasibility of our method in the fabrication of SMILbL films. Theophylline-7-acetic acid (THAA) is covalently conjugated to polyelectrolyte PAA with a cystamine bridge by amide linkage to form precursor assemblies PAAtheo15, which is a PAA with 15% of its carboxylic acid grafted by THAA. The disulfide bond moiety of the cystamine bridge can be reduced to thiol groups, and the thiol group is able to form a hydrogen bond with the hydroxy group of the guest molecules, offering an extra recognition site. Cooperatively combined with other hydrogen-bonded interactions established through template incorporation and film construction, this additional mercapto recognition site renders selectivity for the nanostructured binding cavities in the LbL film [51].

#### 3.3.1.4 Cation-Selective μCP Based on SMILbL Film

The SMILbL film is very attractive for chemical- and biological-sensing applications because it can reduce the response time by significantly shortening the diffusion path for the reagents. Similarly to the ink reservoir from microporous LbL films, we have combined the selectivity of SMILbL films with the μCP technique to selectively load positively charged chemical species and transfer them to various substrates [52].

A multilayer thin film of (PAA-Por/DAR) was constructed on the surface of a patterned PDMS through an unconventional LbL method. After cross-linking the multilayer film between DAR and PAA through UV irradiation and subsequent washing, most of the porphyrins were removed to generate binding sites that could selectively reload cations. A principal factor of μCP is that the interaction between the ink and the substrate should be stronger than between the ink and the stamp. Alcian Blue 8GX, a phthalocyanate derivative with four positive quaternary ammonium groups and four sulfide groups ($Pc^{4+}$), was chosen as the cation candidate, of which the sulfide groups rendered strong interactions with Au substrates to facilitate the transfer of cations from the stamps to the substrates.

In order to demonstrate the successful transfer of the very thin monolayer of cations by μCP, the unpatterned area was backfilled with octadecanethiol, which is a wet etch-resist. After acidic etching of the $Pc^{4+}$ monolayers, line patterns were clearly seen in the AFM image of the film (Figure 3.13a), indicating that the SMILbL-film-modified PDMS stamp reloaded the $Pc^{4+}$ during the absorption step, and successfully transferred the $Pc^{4+}$ to the gold substrate. When the same procedure was applied with the structurally analogous but negatively charged $Por^{4-}$, only a flat surface without any pattern was observed in the AFM image (Figure 3.13c), implying that during the step of inking, no $Por^{4-}$ ink molecules were loaded. As a control, almost no ink molecules were loaded by LbL-film-modified PDMS, which was modified by a conventional LbL film without specific binding sites (Figure 3.13b and d).

### 3.3.2
### Hydrogen-Bonded Complex for Unconventional LbL Assembly

Along the same line of research, the unconventional method of LbL assembly on the basis of hydrogen bonding has been developed. It involves hydrogen-bonded complexation in the solution and hydrogen-bonded LbL assembly at the liquid/solid interface. The solvent used could be organic, which favors the formation of a

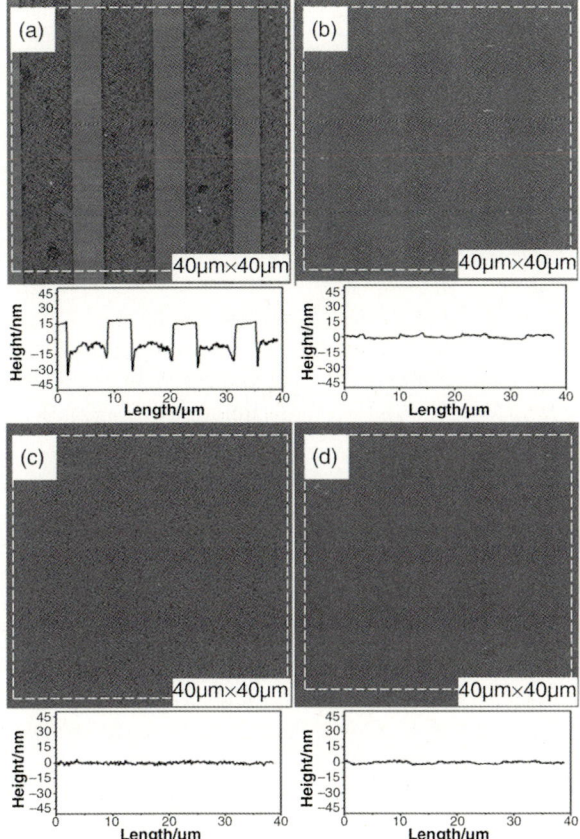

**Figure 3.13** AFM height images and section analyses of etched gold patterns produced by μCP (2 min): (a) μCP of $Pc^{4+}$ by the SMILbL-PDMS stamp; (b) μCP of $Pc^{4+}$ by the LbL-PDMS stamp; (c) μCP of $Por^{4-}$ by the SMILbL-PDMS stamp; (d) μCP of $Por^{4-}$ by the LbL-PDMS stamp. After being backfilled with ODT (15 s), the samples were subsequently etched in an Fe(III)/thiourea etching bath for 3 min at 50 °C. The ink concentration was 0.25 mM in all cases. (Adapted with permission from [52]. Copyright 2011 Wiley-VCH Verlag GmbH & Co. KGaA).

hydrogen bond. In this way, some water-insoluble small organic molecules can be loaded into multilayer thin films.

An example of hydrogen-bonded unconventional LbL assembly [53] is shown in Figure 3.14. First, a small organic molecule, bis-triazine (DTA), was mixed with PAA in methanol to form a hydrogen-bonded complex (PAA-DTA); second, LbL assembly was performed between the methanol solutions of PAA-DTA and DAR, driven by hydrogen-bonded interaction. In this way, DTA was loaded into the LbL film in a convenient and well-controlled manner. Since DAR is a photoreactive polycation, one can irradiate the film with UV light to convert the hydrogen bond into a covalent bond, thus forming a robust multilayer film.

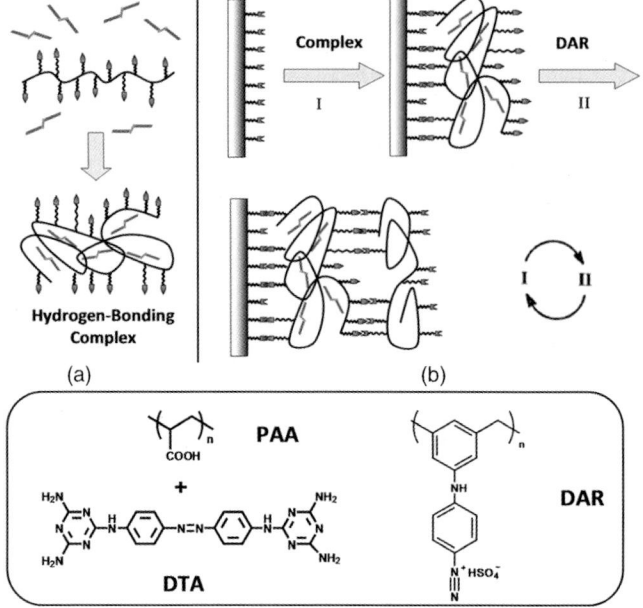

**Figure 3.14** Schematic illustration of the unconventional LbL method on the basis of hydrogen bonding: Step I, formation of hydrogen-bonded PAA-DTA complexes (a); Step II, LbL assembly of PAA-DTA and DAR (b).

The loading amount of DTA can be adjusted by several parameters, such as the concentration of DTA, the solvent composition, temperature, and so on. For example, no DTA could be loaded into the film if a methanol solvent containing more than 20% DMSO was used for PAA-DTA complexation. From another aspect, DMSO can be utilized for the release of DTA from the cross-linked film. After release, DTA could be reloaded upon further immersion of the film into a methanol solution of DTA. Compared with normal PAA/DAR multilayer films, the loading capacity of PAA-DTA/DAR films could be much larger than the normal ones, even five times larger at certain concentrations. Thus it is demonstrated that DTA can serve as a template to fabricate multilayer films with enhanced loading capacity of DTA. This is conceptually similar to molecular imprinting.

This method has been applied to a series of structurally related molecules with an increasing number of hydrogen bond donors and acceptors to find out the structural demand of the method. Our conclusion is that only the molecules that can form multiple and strong hydrogen bonds with PAA are suitable for our method. One simple technique to test whether molecules can interact with PAA strongly is as follows: when mixing the molecules with PAA in solution, if floccules appear, it means that there exists a strong interaction between the molecule and PAA. Therefore, those molecules are usually suitable for this unconventional LbL assembly.

### 3.3.3
### Block Copolymer Micelles for Unconventional LbL Assembly

Amphiphilic block copolymers are able to self-assemble into core–shell micellar structures in a selective solvent. To take advantage of hydrophobic cores of the block copolymer micelles, water-insoluble molecules, for example, pyrene, were incorporated into the hydrophobic micellar cores of poly(styrene-b-acrylic acid), and then the loaded block copolymer micelles were employed as building blocks for LbL assembly [54]. As shown in Figure 3.15, the block copolymer micelles of poly(styrene-b-acrylic acid) with acrylic acid on the shell functioned as polyanions, allowing LbL assembly by alternating deposition with polycations. This is certainly another unconventional LbL assembly that involves micellar formation in solution and the use of loaded micelles for LbL deposition at the liquid/solid interface. In this way, small water-insoluble molecules can be incorporated.

The same concept can be extended to incorporate different water-insoluble molecules, such as azobenzene, for LbL assembly [55, 56]. It is well known that azobenzene can undergo a reversible photoisomerization under UV irradiation, but the rate of photoisomerization is faster in solution than in solid films. For a multilayer film of azobenzene-loaded poly(styrene-b-acrylic acid) micelles and PDDA, it was found, interestingly, that the photoisomerization of the azobenzene in the multilayer film completed within several minutes, which was much faster than in normal solid films, but similar to that in dilute solutions, suggesting a method for enhancing the photophysical properties in the LbL films.

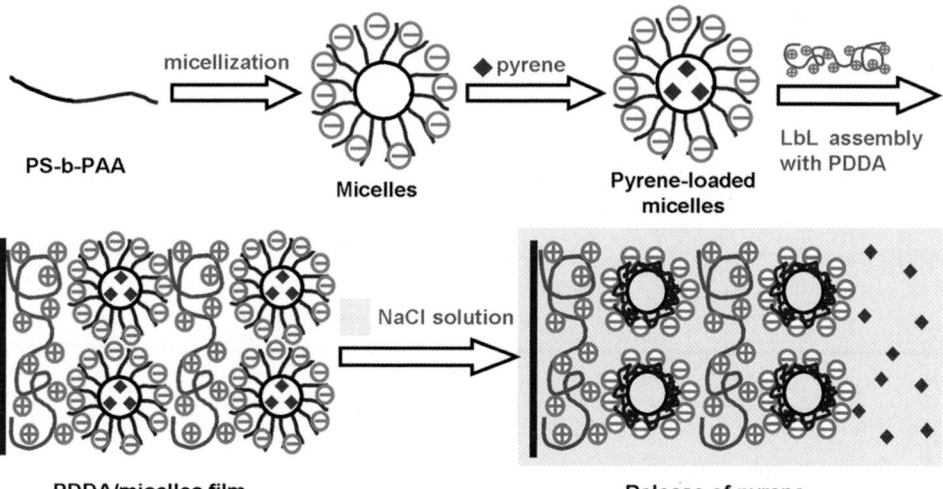

**Figure 3.15** Schematic illustration of the unconventional LbL method using block copolymer micelles: Step 1, incorporation of pyrene into block copolymer micelles; Step 2, LbL deposition of loaded micelles with PDDA; Step 3, the release of pyrene from the multilayer thin film.

In principle, amphiphilic block copolymer micelles can be used to incorporate different water-insoluble molecules of suitable size and shape. Cationic block copolymer micelles can be alternately deposited with anionic block copolymer micelles directly to form a multilayer thin film of micelles [57]. Micelles formed by low molecular weight surfactant are less stable than block copolymer micelles, which usually cannot be used for LbL deposition. To improve the stability of micelles, Sun *et al.* have proposed a strategy to solve the problem, and it involves the use of polyelectrolyte to stabilize the micelles, and the stabilized micelles can then be used for LbL assembly [58].

### 3.3.4
### π–π Interaction Complex for Electrostatic LbL Assembly

Extensive attention has been paid to graphene because of its superior electron conductivity and single atom thickness. Graphene-modified electrodes have been demonstrated to show not only excellent electrocatalytic activity for $H_2O_2$, $O_2$, NADH, and other important electroactive species, but also greatly improved performance in

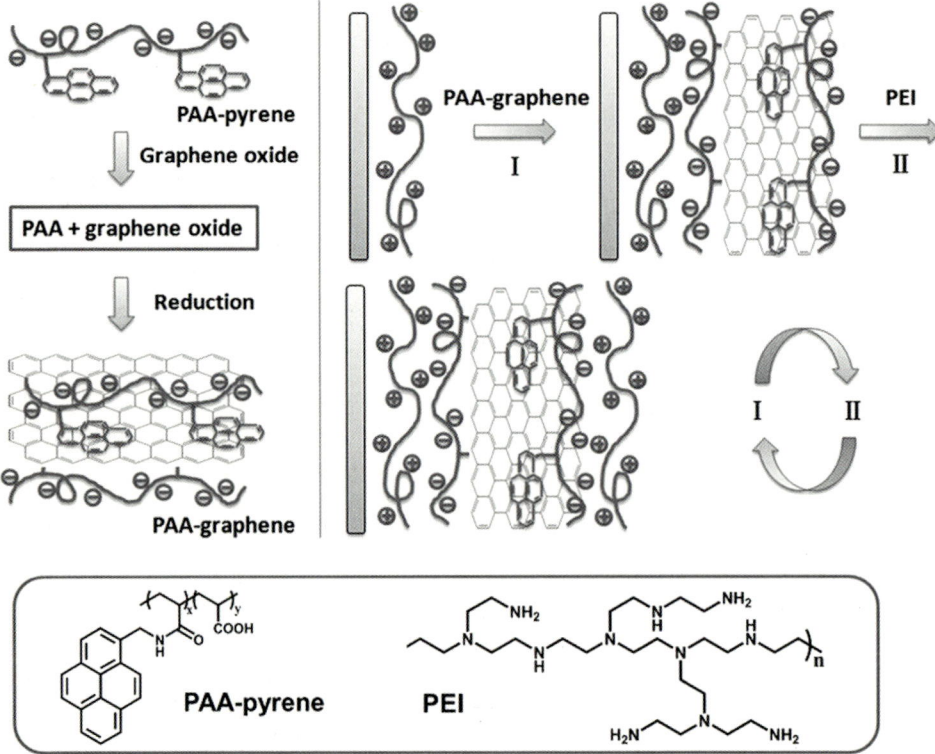

**Figure 3.16** Schematic illustration of the unconventional LbL method on the basis of π–π interaction for the immobilization of graphene on the surface.

enzyme-based biosensing. An unconventional method of LbL assembly has been applied for immobilization of graphene onto the surface [59]. As shown in Figure 3.16, chemically reduced graphene oxide was modified by pyrene-grafted PAA (PAA-pyrene) in aqueous solution on the basis of π–π stacking as well as van der Waals interactions. Then PAA-modified graphene (PAA-graphene) was LbL assembled with poly(ethyleneimine) (PEI).

It is found that graphene can promote electron transfer and facilitates the redox reaction of the probe. Based on the (PEI/PAAgraphene)$_n$ electrode with high electrocatalytic activity toward $H_2O_2$, more complicated and integrated structures

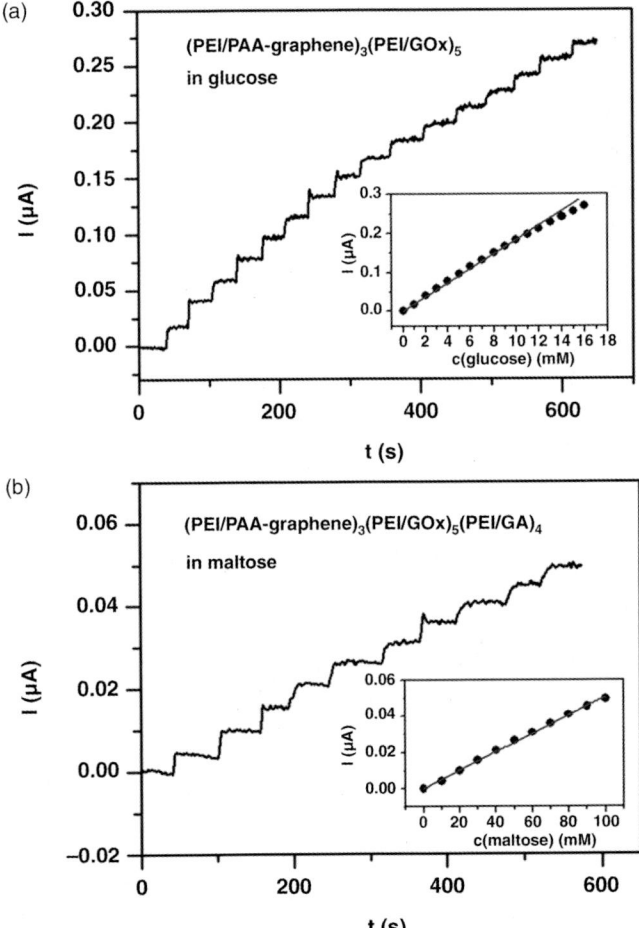

**Figure 3.17** Current–time curves measured at +0.9 V for a (PEI/PAA-graphene)$_3$(PEI/GOx)$_5$ electrode (a) and a (PEI/PAA-graphene)$_3$(PEI/GOx)$_5$(PEI/GA)$_4$ electrode (b) with successive addition of glucose and maltose, respectively. Inset: calibration curves. Electrolyte: $O_2$-saturated and magnetically stirred 0.05 M pH 7.0 PBS. (Adapted with permission from [59]. Copyright 2011 American Chemical Society).

can be designed on the electrode for enzyme-based biosensors. We have employed the bienzyme system of glucose oxidase (GOx) and glucoamylase (GA) as a proof of principle. In a multilayer film successively assembled with graphene, GOx, and GA, GA catalyzes the hydrolysis of maltose to produce glucose, which is subsequently oxidized by GOx to produce $H_2O_2$. Then, $H_2O_2$ is detected by the graphene-multilayer-film-modified electrode.

We start from a monoenzyme system for glucose sensing. As shown in Figure 3.17a, the amperometric response of a GOx multilayer film upon titration of 1 mM glucose was recorded. As shown by the calibration curve, the current increased linearly with glucose concentration within 10 mM. The detection limit and sensitivity are determined to be 0.168 mM (S/N = 3) and $0.261\,\mu A\,mM^{-1}\,cm^{-2}$, respectively. The above results demonstrate that a glucose biosensor based on glucose oxidase has been fabricated successfully.

To construct a bienzyme system for maltose sensing, a GA multilayer film was assembled on top of the GOx multilayer film and its amperometric response to the successive addition of 10 mM maltose was examined (Figure 3.17b). A perfect calibration curve was obtained. As the concentration of maltose increased, the response to maltose addition became slower. The time before reaching the steady state increased from 5 to 20 s as the maltose concentration increased from 10 to 100 mM. The detection limit and sensitivity were determined to be 1.37 mM (S/N = 3) and $0.00715\,\mu A\,mM^{-1}\,cm^{-2}$, respectively.

## 3.4
**Summary and Outlook**

In this chapter, we have summarized our work on the conventional and unconventional LbL methods for the construction of a great variety of building blocks with controlled architectures and compositions. Although each method has its own scope of applications as well as limitations, it is highly anticipated that a combination of different methods can facilitate the fabrication of multilayer thin films with complex and elaborate structures for the integration of functionalities.

## References

1 Iler, R.K. (1966) Multilayers of colloidal particles. *J. Colloid Interface Sci.*, **21** (6), 569–594.

2 Decher, G. and Hong, J.D. (1991) Buildup of ultrathin multilayer films by a self-assembly process: I. Consecutive adsorption of anionic and cationic bipolar amphiphiles. *Makromol. Chem., Macromol. Symp.*, **46** (1), 321–327.

3 Decher, G. and Hong, J.D. (1991) Buildup of ultrathin multilayer films by a self-assembly process: II. Consecutive adsorption of anionic and cationic bipolar amphiphiles and polyelectrolytes on charged surfaces. *Ber. Bunsenges. Phys. Chem.*, **95** (11), 1430–1434.

4 Decher, G., Hong, J.D., and Schmitt, J. (1992) Buildup of ultrathin multilayer films by a self-assembly process: III. Consecutively alternating adsorption of anionic and cationic polyelectrolytes on

charged surfaces. *Thin Solid Films*, **210/211** (2), 831–835.

5 Wu, T. and Zhang, X. (2001) Self-assembled ultrathin films: from layered nanoarchitechtures to functional assemblies. *Chem. J. Chin. Univ.*, **22** (6), 1057–1065.

6 Gao, M.L., Kong, X.X., Zhang, X., and Shen, J.C. (1993) Preparation and characterization of functional ultrathin ordered molecular deposition film. *Chem. J. Chin. Univ.*, **14** (8), 1182–1183.

7 Gao, M.Y., Yang, Y., Yang, B., Bian, F.L., and Shen, J.C. (1994) Synthesis of PbS nanoparticles in polymer matrices. *J. Chem. Soc., Chem. Commun.* (24), 2779–2780.

8 Fu, Y., Xu, H., Bai, S.L., Qiu, D.L., Sun, J.Q., Wang, Z.Q., and Zhang, X. (2002) Fabrication of a stable polyelectrolyte/Au nanoparticles multilayer film. *Macromol. Rapid Commun.*, **23** (4), 256–259.

9 Zhang, X., Gao, M.L., Kong, X.X., Sun, Y.P., and Shen, J.C. (1994) Build-up of a new type of ultrathin film of porphyrin and phthalocyanine based on cationic and anionic electrostatic attraction. *J. Chem. Soc., Chem. Commun.* (9), 1055–1056.

10 Sun, Y.P., Zhang, X.;., Sun, C.Q., Wang, Z.Q., Shen, J.C., Wang, D.J., and Li, T.L. (1996) Supramolecular assembly of alternating porphyrin and phthalocyanine layers based on electrostatic interactions. *Chem. Commun.* (20), 2379–2380.

11 Crespo-Biel, O., Dordi, B., Reinhoudt, D.N., and Huskens, J. (2005) Supramolecular layer-by-layer assembly: alternating adsorptions of guest- and host-functionalized molecules and particles using multivalent supramolecular interactions. *J. Am. Chem. Soc.*, **127** (20), 7594–7600.

12 Huo, F.W., Xu, H.P., Zhang, L., Fu, Y., Wang, Z.Q., and Zhang, X. (2003) Hydrogen-bonding based multilayer assemblies by self-deposition of dendrimer. *Chem. Commun.* (7), 874–875.

13 Kong, W., Wang, L.P., Gao, M.L., Zhou, H., Zhang, X., Li, W., and Shen, J.C. (1994) Immobilized bilayer glucose isomerase in porous trimethylamine polystyrene based on molecular deposition. *J. Chem. Soc., Chem. Commun*, (11), 1297–1298.

14 Kong, W., Zhang, X., Gao, M.L., Zhou, H., Li, W., and Shen, J.C. (1994) A new kind of immobilized enzyme multilayer based on cationic and anionic interaction. *Makromol. Rapid Commun.*, **15** (5), 405–409.

15 Lvov, Y., Ariga, K., and Kunitake, T. (1994) Layer-by-layer assembly of alternate protein/polyion ultrathin films. *Chem. Lett.*, **23** (12), 2323–2326.

16 Sun, Y.P., Zhang, X., Sun, C.Q., Wang, B., and Shen, J.C. (1996) Fabrication of ultrathin film containing bienzyme of glucose oxidase and glucoamylase based on electrostatic interaction and its potential application as a maltose sensor. *Macromol. Chem. Phys.*, **197** (1), 147–153.

17 Lvov, Y. and Möhwald, H. (2000) *Protein Architectures: Interfacing Molecular Assemblies and Immobilization Biotechnology*, Marcel Dekker, New York.

18 Sun, J.Q., Sun, Y.P., Wang, Z.Q., Sun, C.Q., Wang, Y., Zhang, X., and Shen, J.C. (2001) Ionic self-assembly of glucose oxidase with polycation bearing Os complex. *Macromol. Chem. Phys.*, **202** (1), 111–116.

19 Lvov, Y., Decher, G., and Sukhorukov, G. (1993) Assembly of thin films by means of successive deposition of alternate layers of DNA and poly(allylamine). *Macromolecules*, **26** (20), 5396–5399.

20 Shchukin, D.G., Patel, A.A., Sukhorukov, G.B., and Lvov, Y.M. (2004) Nanoassembly of biodegradable microcapsules for DNA encasing. *J. Am. Chem. Soc.*, **126** (11), 3374–3375.

21 Lvov, Y., Haas, H., Decher, G., Möhwald, H., Mikhailov, A., Mtchedlishvily, B., Morgunova, E., and Vainshtein, B. (1994) Successive deposition of alternate layers of polyelectrolytes and a charged virus. *Langmuir*, **10** (11), 4232–4236.

22 Zhang, X., Chen, H., and Zhang, H.Y. (2007) Layer-by-layer assembly: from conventional to unconventional methods. *Chem. Commun.* (14), 1395–1405.

23 Wang, F., Ma, N., Chen, Q.X., Wang, W.B., and Wang, L.Y. (2007) Halogen bonding as a new driving force for layer-by-layer assembly. *Langmuir*, **23** (19), 9540–9542.

24 Zhang, X., Sun, Y.P., Gao, M.L., Kong, X.X., and Shen, J.C. (1996) Effects of pH on the supramolecular structure of polymeric molecular deposition films. *Macromol. Chem. Phys.*, **197** (2), 509–515.

25 Zhang, X., Shi, F., Yu, X., Liu, H., Fu, Y., Wang, Z.Q., Jiang, L., and Li, X.Y. (2004) Polyelectrolyte multilayer as matrix for electrochemical deposition of gold clusters: toward super-hydrophobic surface. *J. Am. Chem. Soc.*, **126** (10), 3064–3065.

26 Zhao, N., Shi, F., Wang, Z.Q., and Zhang, X. (2005) Combining layer-by-layer assembly with electrodeposition of silver aggregates for fabricating superhydrophobic surfaces. *Langmuir*, **21** (10), 4713–4716.

27 Shi, F., Niu, J., Liu, J.L., Liu, F., Wang, Z.Q., Feng, X.Q., and Zhang, X. (2007) Towards understanding why a superhydrophobic coating is needed by water striders. *Adv. Mater.*, **19** (17), 2257–2261.

28 Chen, X.X., Gao, J., Song, B., Smet, M., and Zhang, X. (2010) Stimuli-responsive wettability of nonplanar substrates: pH-controlled floatation and supporting force. *Langmuir*, **26** (1), 104–108.

29 Stockton, W.B. and Rubner, M.F. (1997) Molecular-level processing of conjugated polymers. 4. layer-by-layer manipulation of polyaniline via hydrogen-bonding interactions. *Macromolecules*, **30** (9), 2717–2725.

30 Wang, L.Y., Wang, Z.Q., Zhang, X., and Shen, J.C. (1997) A new approach for the fabrication of an alternating multilayer film of poly(4-vinylpyridine) and poly(acrylic acid) based on hydrogen bonding. *Macromol. Rapid Commun.*, **18** (6), 509–514.

31 Wang, L.Y., Cui, S.X., Wang, Z.Q., Zhang, X., Jiang, M., Chi, L.F., and Fuchs, H. (2000) Multilayer assemblies of copolymer PSOH and PVP on the basis of hydrogen bonding. *Langmuir*, **16** (26), 10490–10494.

32 Zhang, H.Y., Fu, Y., Wang, D., Wang, L.Y., Wang, Z.Q., and Zhang, X. (2003) Hydrogen-bonding-directed layer-by-layer assembly of dendrimer and poly(4-vinylpyridine) and micropore formation by post-base treatment. *Langmuir*, **19** (20), 8497–8502.

33 Sukhishvili, S.A. and Granick, S. (2000) Layered, erasable, ultrathin polymer films. *J. Am. Chem. Soc.*, **122** (39), 9550–9551.

34 Sukhishvili, S.A. and Granick, S. (2002) Layered, erasable polymer multilayers formed by hydrogen-bonded sequential self-assembly. *Macromolecules*, **35** (1), 301–310.

35 Fu, Y., Bai, S.L., Cui, S.X., Qiu, D.L., Wang, Z.Q., and Zhang, X. (2002) Hydrogen-bonding-directed layer-by-layer multilayer assembly: reconformation yielding microporous films. *Macromolecules*, **35** (25), 9451–9458.

36 Xu, H.P., Gomez-Casado, A., Liu, Z.H., Reinhoudt, D.N., Lammertink, R.G.H., and Huskens, J. (2009) Porous multilayer-coated PDMS stamps for protein printing. *Langmuir*, **25** (24), 13972–13977.

37 Xiong, H.M., Cheng, M.H., Zhou, Z., Zhang, X., and Shen, J.C. (1998) A new approach to the fabrication of a self-organizing film of heterostructured polymer/$Cu_2S$ nanoparticles. *Adv. Mater.*, **10** (7), 529–532.

38 Hao, E.C., Wang, L.Y., Zhang, J.H., Yang, B., Zhang, X., and Shen, J.C. (1999) Fabrication of polymer/inorganic nanoparticles composite films based on coordinative bonds. *Chem. Lett.*, **28** (1), 5–6.

39 Sun, J.Q., Wu, T., Sun, Y.P., Wang, Z.Q., Zhang, X., Shen, J.C., and Cao, W.X. (1998) Fabrication of a covalently attached multilayer via photolysis of layer-by-layer self-assembled films containing diazo-resins. *Chem. Commun.* (17), 1853–1854.

40 Sun, J.Q., Wang, Z.Q., Wu, L.X., Zhang, X., Shen, J.C., Gao, S., Chi, L.F., and Fuchs, H. (2001) Investigation of the covalently attached multilayer architecture based on diazo-resins and poly(4-styrene sulfonate). *Macromol. Chem. Phys.*, **202** (7), 967–973.

41 Sun, J.Q., Wu, T., Liu, F., Wang, Z.Q., Zhang, X., and Shen, J.C. (2000) Covalently attached multilayer assemblies by sequential adsorption of polycationic diazo-resins and polyanionic poly(acrylic acid). *Langmuir*, **16** (10), 4620–4624.

42 Sun, J.Q., Wang, Z.Q., Sun, Y.P., Zhang, X., and Shen, J.C. (1999) Covalently attached multilayer assemblies of diazo-resins and porphyrins. *Chem. Commun.* (8), 693–694.

43 Sun, J.Q., Wu, T., Zou, B., Zhang, X., and Shen, J.C. (2001) Stable entrapment of small molecules bearing sulfonate groups in multilayer assemblies. *Langmuir*, **17** (13), 4035–4041.

44 Shi, F., Dong, B., Qiu, D.L., Sun, J.Q., Wu, T., and Zhang, X. (2002) Layer-by-layer self-assembly of reactive polyelectrolytes for robust multilayer patterning. *Adv. Mater.*, **14** (11), 805–809.

45 Shi, F., Wang, Z.Q., Zhao, N., and Zhang, X. (2005) Patterned polyelectrolyte multilayer: surface modification for enhancing selective adsorption. *Langmuir*, **21** (4), 1599–1602.

46 Wu, G.L., Shi, F., Wang, Z.Q., Liu, Z., and Zhang, X. (2009) Poly(acrylic acid)-bearing photoreactive azido groups for stabilizing multilayer films. *Langmuir*, **25** (5), 2949–2955.

47 Chen, H., Zeng, G.H., Wang, Z.Q., Zhang, X., Peng, M.L., Wu, L.Z., and Tung, C.H. (2005) To combine precursor assembly and layer-by-layer deposition for incorporation of single-charged species: Nanocontainers with charge-selectivity and nanoreactors. *Chem. Mater.*, **17** (26), 6679–6685.

48 Chen, H., Zeng, G.H., Wang, Z.Q., and Zhang, X. (2007) To construct "ion traps" for enhancing the permselectivity and permeability of polyelectrolyte multilayer films. *Macromolecules*, **40** (3), 653–660.

49 Shi, F., Liu, Z., Wu, G.L., Zhang, M., Chen, H., Wang, Z.Q., Zhang, X., and Willner, I. (2007) Surface imprinting in layer-by-layer nanostructured films. *Adv. Funct. Mater.*, **17** (11), 1821–1827.

50 Niu, J., Shi, F., Liu, Z., Wang, Z.Q., and Zhang, X. (2007) Reversible disulfide cross-linking in layer-by-layer films: preassembly enhanced loading and pH/reductant dually controllable release. *Langmuir*, **23** (11), 6377–6384.

51 Niu, J., Liu, Z.H., Fu, L., Shi, F., Ma, H.W., Ozaki, Y., and Zhang, X. (2008) Surface imprinted nanostructured layer-by-layer film for molecular recognition of theophylline-derivatives. *Langmuir*, **24** (20), 11988–11994.

52 Liu, Z.H., Yi, Y., Xu, H.P., Zhang, X., Ngo, T.H., and Smet, M. (2010) Cation-selective microcontact printing based on surface-molecular-imprinted layer-by-layer films. *Adv. Mater.*, **22** (24), 2689–2693.

53 Zeng, G.H., Gao, J., Chen, S.L., Chen, H., Wang, Z.Q., and Zhang, X. (2007) Combining hydrogen-bonding complexation in solution and hydrogen-bonding-directed layer-by-layer assembly for the controlled loading of a small organic molecule into multilayer films. *Langmuir*, **23** (23), 11631–11636.

54 Ma, N., Zhang, H.Y., Song, B., Wang, Z.Q., and Zhang, X. (2005) Polymer micelles as building blocks for layer-by-layer assembly: an approach for incorporation and controlled release of water-insoluble dyes. *Chem. Mater.*, **17** (20), 5065–5069.

55 Ma, N., Wang, Y.P., Wang, Z.Q., and Zhang, X. (2006) Polymer micelles as building blocks for the incorporation of azobenzene: enhancing the photochromic properties in layer-by-layer films. *Langmuir*, **22** (8), 3906–3909.

56 Ma, N., Wang, Y.P., Wang, B.Y., Wang, Z.Q., Zhang, X., Wang, G., and Zhao, Y. (2007) Interaction between block copolymer micelles and azobenzene-containing surfactants: from coassembly in water to layer-by-layer assembly at the interface. *Langmuir*, **23** (5), 2874–2878.

57 Cho, J., Hong, J., Char, K., and Caruso, F. (2006) Nanoporous block copolymer micelle/micelle multilayer films with dual optical properties. *J. Am. Chem. Soc.*, **128** (30), 9935–9942.

58 Liu, X.K., Zhou, L., Geng, W., and Sun, J.Q. (2008) Layer-by-layer-assembly multilayer films of polyelectrolyte-stabilized surfactant micelles for the incorporation of noncharged organic dyes. *Langmuir*, **24** (22), 12986–12989.

59 Zeng, G.H., Xing, Y.B., Gao, J., Wang, Z.Q., and Zhang, X. (2010) Unconventional layer-by-layer assembly of graphene multilayer films for enzyme-based glucose and maltose biosensing. *Langmuir*, **26** (18), 15022–15026.

# 4
# Novel Multilayer Thin Films: Hierarchic Layer-by-Layer (Hi-LbL) Assemblies

*Katsuhiko Ariga, Qingmin Ji, and Jonathan P. Hill*

## 4.1
## Introduction

Many functional complexes in biology have highly hierarchic structures, as can be seen in various systems such as organelles, cells, tissues, and organs. In these systems, hierarchy is important in relaying events at different structural dimensions resulting in harmonized functions. Therefore, the development of biological mimics with hierarchic structures is believed to be useful for construction of highly functional materials. We can create functional units with artificial motifs through approaches based on biomimetic chemistry that successfully create various important biomimic systems, such as artificial enzymes and cell membrane mimics. However, their integratation into well-defined structures to attain bio-like higher functionality with good efficiencies and specificities is not always easy.

In biological systems and processes, hierarchic structures are constructed through spontaneous self-assembly that is achieved through molecular design and the appropriate selection of components. Nature accomplishes these processes only under optimized conditions that have evolved over billions of years. However, it would be very difficult to mimic all available natural processes by using non-biochemical approaches. One possible approach for constructing functional hierarchic structures would be a multi-step process where nanostructured materials are first synthesized and then further assembled into organized structures of higher order. In order to satisfy this strategy for various kinds of target structures the assembly process should be universal, that is, applicable to a wide range of materials, and it should even be possible to assemble unusual nanomaterials by conventional procedures.

As one of the most versatile techniques for the construction of organized structures, alternate layer-by-layer (LbL) assembly has been paid much attention [1–6]. LbL assembly is applicable to a huge variety of target materials, including simple and functional polymers [7–9], inorganic nanostructures [10, 11], molecular assemblies [12–14], biomaterials themselves [15–20], and even viruses [21]. The driving force for LbL assembly was originally limited to simple electrostatics but has been expanded to include hydrogen bonding [22, 23], metal coordination [24, 25],

---

*Multilayer Thin Films: Sequential Assembly of Nanocomposite Materials*, Second Edition.
Edited by Gero Decher and Joseph B. Schlenoff.
© 2012 Wiley-VCH Verlag GmbH & Co. KGaA. Published 2012 by Wiley-VCH Verlag GmbH & Co. KGaA.

stereo-complex formation [26, 27], charge transfer [28], bio-specific interactions [29, 30], covalent bonding [31, 32], and electrochemical coupling [33]. Its implementation is simple and inexpensive so that the films can be assembled only using solution dipping with beakers and tweezers, however, various modifications, including spin-coating [34–36], spraying [37], nanofabrication [38, 39], and automated machine processes [40, 41] have been made. Assembly morphologies are not limited to thin films. Multilayered three-dimensional objects, including spheres, capsules, and tubes can be fabricated by combining LbL assembly with appropriate template syntheses [42, 43]. The most important feature of the LbL technology is the wide freedom of choice in the resulting layered structures, with layer thickness (the number of layers) and layer sequences being easily tuned.

The LbL technique can be used in the fabrication of functional hierarchic structures using pre-synthesized nanostructured objects. In this chapter, several examples for the construction of hierarchic structures by LbL assembly are described. The related technique is here referred to as the hierarchic layer-by-layer (Hi-LbL) method. Several demonstrations of the Hi-LbL method are described in three categories, multi-cellular models, unusual drug delivery, and sensor applications.

## 4.2
### Hi-LbL for Multi-Cellular Models

Formation of cell membrane mimics, that is, artificial cell membranes, has been investigated as one of the most popular targets in biomimetic chemistry. Cell mimicry requires spherical structures that can form an inner space as a container. Liposomes and lipid layer pair vesicles are known as models of a spherical cell membrane possessing a direct mimic of a unicellular membrane. Unfortunately, their limited mechanical stability is often disadvantageous for some kinds of practical application. Structurally stable cell-like vesicle structures have been investigated using silica-related materials. For example, Caruso, Möhwald and coworkers reported the formation of hollow silica vesicles through LbL assembly on colloidal nanoparticle templates [44]. Polyelectrolytes and smaller silica particles were initially assembled on a colloidal core by the LbL technique. Subsequent destruction of template colloidal particles resulted in a hollow silica/polymer hybrid vesicle, and further calcination of the hybrid vesicles left a hollow vesicle composed of silica.

Katagiri et al. pursued the covalent linkage of a siloxane framework to a lipid layer pair vesicle, and sophisticated organization of multi-cellular systems. The resulting vesicles have a siloxane network covalently attached to the layer pair membrane surface and are called cerasomes (ceramics + soma) (Figure 4.1) [45, 46]. Alkoxysilane-bearing lipids were first dispersed in an acidic aqueous solution using a vortex mixer. The dispersion behavior depended significantly on the solution pH. Upon mixing 10 mmol of the alkoxysilane-bearing lipid with aqueous HCl (pH 1, 5.0 ml), precipitation occurred immediately, probably because of an excessive rate of hydrolysis and condensation. In contrast, basic conditions resulted in an uneven dispersion of aqueous lipid solution. Medium acidic conditions (pH 3) were found to be

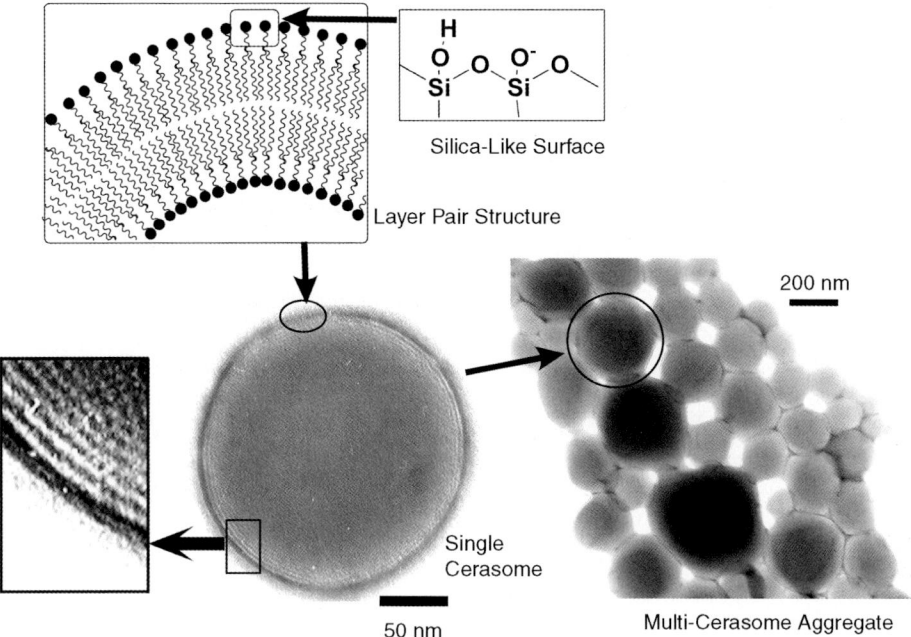

**Figure 4.1** Cell-mimicry, cerasome, and its aggregates.

most appropriate for the stable formation of cerasomes. Under the optimized conditions, a stable dispersion was obtained after vortex mixing for 15 min at room temperature. Formation of vesicular structures was confirmed by transmission electron microscopy (TEM) with the aid of hexaammonium heptamolybdate tetrahydrate as a contrast agent. The image of the multi-lamellar cerasomes with a layer pair thickness of ca. 4 nm and vesicular diameter of 150 nm was clearly visible. Interestingly, the TEM image of the vesicular aggregates was observed in the same specimen, where the collapse and fusion of the cerasome were suppressed, probably through the formation of the intra- and intermembrane siloxane network.

Subjecting the cerasome structure to LbL techniques could result in pre-designed multi-cellular mimics. Hi-LbL assemblies between cationic polyelectrolyte (poly (diallyldimethylammonium chloride), PDDA) and anionic vesicles (Figure 4.2a) have been investigated using a quartz crystal microbalance (QCM) that provides sensitive mass detection of materials deposited on the surface from its frequency shifts [47]. LbL assembly between the anionic cerasome and cationic PDDA exhibited reasonable frequency shifts of the QCM response for the formation of multi-cellular films. The steps in the frequency shifts upon cerasome adsorption were in good agreement with the cerasome deposition in accordance with their spherical structure. The morphology of the assembled structure was also confirmed by atomic force microscopy (AFM).

Cerasomes with cationic surface charge can also be prepared using another alkoxysilane-bearing lipid. Using both the anionic and cationic cerasomes, direct

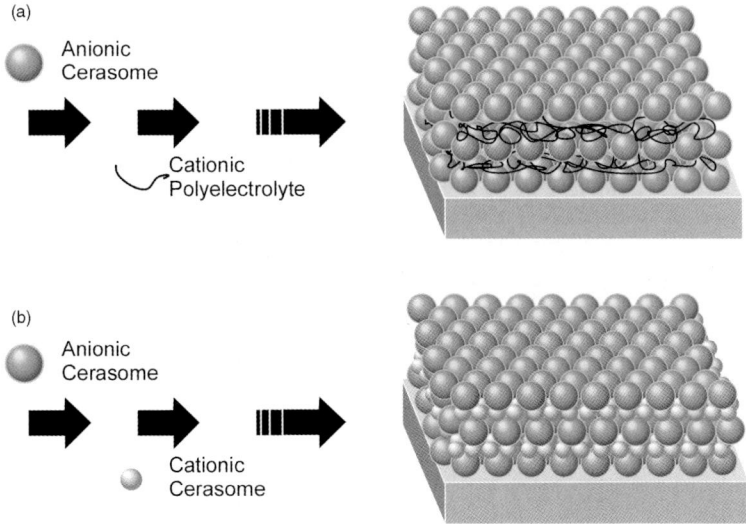

**Figure 4.2** Hi-LbL of cerasomes: (a) anionic cerasome and cationic polyelectrolyte; (b) anionic cerasome and cationic cerasome.

LbL assembly of cerasome structures in the absence of polyelectrolyte counterions became possible (Figure 4.2b) [48]. According to TEM observation, cationic cerasomes are smaller in diameter (20–100 nm) than the anionic cerasomes (70–300 nm). When Hi-LbL assembly was conducted between cationic and anionic cerasomes, large and small steps in the frequency changes were observed with the odd- and even-number steps, corresponding to adsorption of the anionic and cationic cerasome, respectively. The presence of closely packed cerasome particles in both layers was clearly confirmed by AFM observation of the surface of the assembled structures. The hierarchically structured assembly obtained can be regarded as a multi-cellular mimic and could be used as a bioreactor or biosensor. Further functionalization of the cerasome surface using various biomolecules, such as enzymes and antibodies, through covalent linkage indicates a great potential for creating various kinds of biomimetic silica nanohybrids.

## 4.3
### Hi-LbL for Unusual Drug Delivery Modes

Further increase in the mechanical stability of assembled components using the Hi-LbL process is useful for a wide range of applications. We used mesoporous capsules for Hi-LbL in place of cerasomes. The resulting hierarchic structures fabricated by Hi-LbL assembly of mesoporous capsules are called mesoporous nanocompartment films [49, 50]. The major components of the mesoporous nanocompartment films are hollow silica capsules containing hierarchical micro- and nanospaces: an internal void microspace ($1000 \times 700 \times 300 \, nm^3$) in the interior, and mesopores (average diameter 2.2 nm) at the silica walls. Core–shell silica particles were first prepared on pseudohexagonal prismatic zeolite cores using octadecyltrimethoxysilane as the strucure-directing reagent. Aluminum was next incorporated into the silica (ZCMS) framework by an impregnation method to yield ZCMS aluminosilicate with strong acid sites for the polymerization of phenol and paraformaldehyde. With the ZCMS aluminosilicate, phenol incorporation and reaction with paraformaldehyde, followed by dissolution of the ZCMS silicate using dilute HF solution, generated mesoporous carbon capsules. In the final process, the mesoporous carbon capsules were filled with tetraethyl orthosilicate and exposed to HCl vapor to induce hydrolysis and condensation of the silica. Finally, the silica–carbon nanocomposite was heated in air to remove the carbon framework, resulting in mesoporous silica capsules.

Anionic silica capsules were deposited on a QCM resonator using LbL assembly (Figure 4.3). The Hi-LbL assembly between the hollow capsules and PDDA was performed with the aid of anionic silica nanoparticles as co-adsorbers of the capsule component. QCM frequency changes during a LbL process involving alternate immersion of the QCM resonator in a solution of a mixture of silica particles and silica capsules (w/w, 10: 1) and a solution of PDDA, resulted in the required structure. Scanning electron microscopic (SEM) images of the silica particle/capsule compartment films reveal that the silica capsules are dispersed among the silica particles.

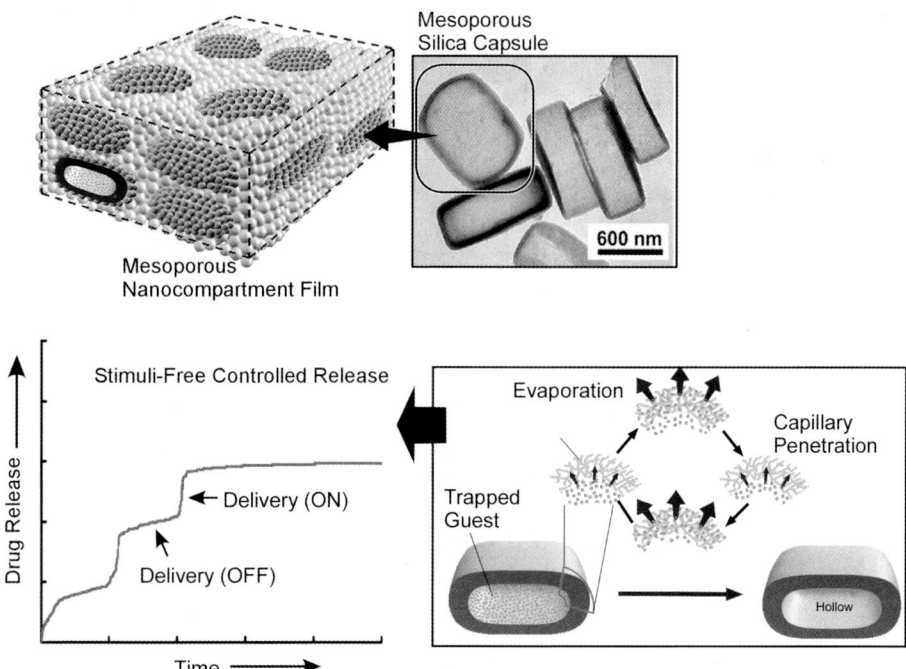

**Figure 4.3** Mesoporous nanocompartment film and its stimuli-free controlled release behavior.

The top and cross-sectional views of the nanocompartment film show that the silica capsules are embedded within the smooth film with retention of their capsular morphology.

We performed quantitative analyses of water evaporation from the compartment films under a variety of conditions. After immersing the compartment film on the QCM resonator in water and drying under nitrogen flow, the net change in weight of the film after each cycle was measured in air using the QCM. An increase in QCM frequency corresponds to a decrease in mass and, in this case, is symptomatic of water release from the interior of the silica capsule within the compartment film. Surprisingly, the frequency shifts upon water evaporation from the mesoporous nanocompartment films show a stepwise profile, even though no external stimulus was applied.

A plausible mechanism for automodulated stepwise water release from the mesoporous nanocompartment films is discussed below. In the proposed mechanism, stepwise release is assumed to originate from a combination of two processes: water evaporation from the pores and capillary penetration into the pores. Judging from the release profiles observed under different encapsulation conditions, the number of release steps seems to be related to the ratio of the water volume to the mesopore volume in the capsule wall. The interior volume of a silica capsule is four times as large as its mesoporous wall. If one step of water release originates in water release from the mesopore region then the number of steps becomes one, two, three, or, four, depending on the intial water content. When the volume of water contained in the capsule is less than that contained in the pore wall ($<26\%$), release in a single step is observed. Two-step release observed in the QCM frequency profile corresponds to an encapsulated water content of 26–52%, three-step release to an encapsulated water content of 52–78%, while greater degrees of encapsulation ($>78\%$) result in multiply-stepped release. These expectations are in good agreement with the experimental observation.

Initially, water entrapped in mesopore channels evaporates to the exterior, which is observed as the first step of water release. After most of the water has evaporated from the mesopore channels, water enters that region from the capsule interior, probably by rapid capillary penetration. Subsequently, water again evaporates from the mesopore channels to the exterior and is apparent as the second evaporation step. This mechanism explains why the quantity of each step is almost identical, and the number of steps is determined by the ratio between the amount of entrapped water and the mesopore volume. In addition, the water evaporation rate at each step can be controlled by several factors, such as temperature and the co-adduct materials (silica particles and polymer).

This release profile was used in a demonstration of the controlled release of various fluid drugs, such as fragrance molecules. The results obtained in these experiments suggest that the stepped release mode appears to be due to the non-equilibrated concurrent evaporation of material from the pore channels to the exterior, and capillary penetration from the interior into the mesoporous channels. The mesoporous nanocompartment films containing hollow mesoporous capsules will introduce potential applications based on their excellent chemical storage/release mechanism.

## 4.4 Hi-LbL for Sensors

Most of the currently available control release systems perform modulation in release using some external stimulus. However, we have presented here a rare example of *a stimulus-free controlled release medium*, which operates in a stepwise manner with prolonged release efficiency, a feature useful for controlled-release drug delivery. This new system has been shown to possess features of controlled loading/release, which are of great utility for the development of energy-less and clean *stimulus-free* controlled drug release applications.

## 4.4
## Hi-LbL for Sensors

### 4.4.1
### Mesoporous Carbon Hi-LbL

We have also demonstrated sensor applications of the Hi-LbL structures prepared from mesoporous carbon materials and polyelectrolytes. Although conventional mesoporous carbon materials, such as CMK-3, do not possess surface charges sufficient for successful Hi-LbL assembly, surface oxidation of carbon using ammonium persulfate enabled us to introduce negative carboxylate groups to mesoporous carbon (CMK-3). Hi-LbL assembly of oxidized CMK-3 was performed using polycation PDDA on a QCM plate (Figure 4.4) [51]. Continuous film growth was confirmed for the oxidized CMK-3, and SEM images also revealed an increasing

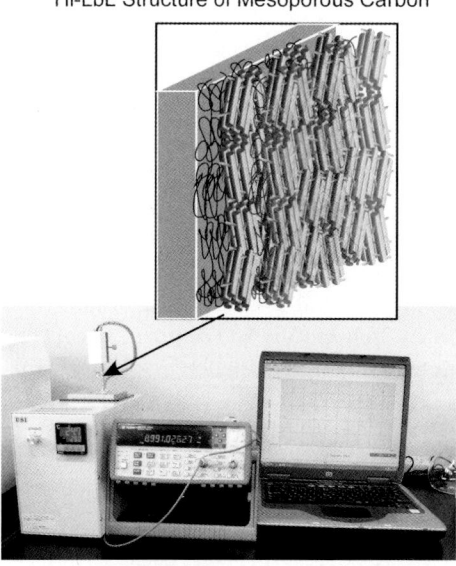

**Figure 4.4** Hi-LbL assembly of mesoporous carbon for sensor use.

surface coverage of the QCM plate as LbL assembly proceeds. Because imperfect surface coverage is expected to be advantageous for diffusion of guest molecules, two-layered LbL films were used for subsequent sensing applications.

Sensing performances were investigated in aqueous solution where a QCM plate covered with the Hi-LbL film of CMK-3 was immersed and a sensing target was injected. A frequency shift upon adsorption of tannic acid onto the CMK-3 LbL films was observed immediately after injection of tannic acid. For reference, adsorption of tannic acid onto the SAM surface of octadecanethiol was similarly measured, but resulted in much less significant frequency shifts. LbL films composed only of polyelectrolyte (PDDA and poly(sodium stryrenesulfonate) (PSS)) also displayed a poor adsorption capacity for tannic acid, even at the highest concentration examined. The frequency shifts upon adsorption of tannic acid greatly exceed those for catechin and caffeine. The resulting sensitivity ratios of tannic acid to catechin or caffeine are circa 3.9 and 13.6, respectively. The superior adsorption capacity for tannic acid likely originates in its molecular structure, that is, multiple phenyl rings of the tannic acid molecule can interact with the carbon surface through $\pi$–$\pi$ interactions and hydrophobic effects. In addition, size fitting of tannic acid (a roughly circular molecule with approximate diameter 3 nm) to the CMK-3 nanochannel may result in enhanced interactions between the guests themselves and/or the guest and carbon surface. A comparable frequency shift for adsorbed tannic acid and that for immobilized CMK-3 suggests filling of most of the pores by tannic acid molecules, as judged from the pore volume of $1 \text{ cm}^3 \text{ g}^{-1}$.

The amounts of tannic acid adsorbed onto the CMK Hi-LbL film at equilibrium exhibited a sigmoidal profile at low concentrations. This cooperative binding profile was absent for the adsorption of tannic acid onto the SAM surface of octadecanethiol. Highly cooperative behavior might result from confinement effects during adsorption. Although cooperative adsorption isotherms have been explained by citing structural changes of the host materials or phase transitions of guest–host complexes, neither of these phenomena can explain the sigmoidal adsorption isotherm observed in the present work because of the rigid carbon framework of CMK-3. The observed behavior may be similarly explained by enhanced guest–guest interaction, since the adsorbed tannic acid can have effective $\pi$–$\pi$ and/or hydrophobic interactions when confined. The entropically favored release of the clustered water from mesopores upon guest inclusion might induce this peculiar behavior for the guest adsorption. These observations will also promote our understanding of molecular interactions within nanospaces, especially non-specific interactions in aqueous media, a full exploration of which might clarify important phenomena, including those of biological systems.

### 4.4.2
#### Mesoporous Carbon Capsule Hi-LbL

As described previously, carbon capsules were synthesized using zeolite crystals as templates. The synthesized capsules have homogeneous dimensions ($1000 \times 700 \times 300 \text{ nm}^3$) with 35-nm-thick mesoporous walls with a uniform pore size distribu-

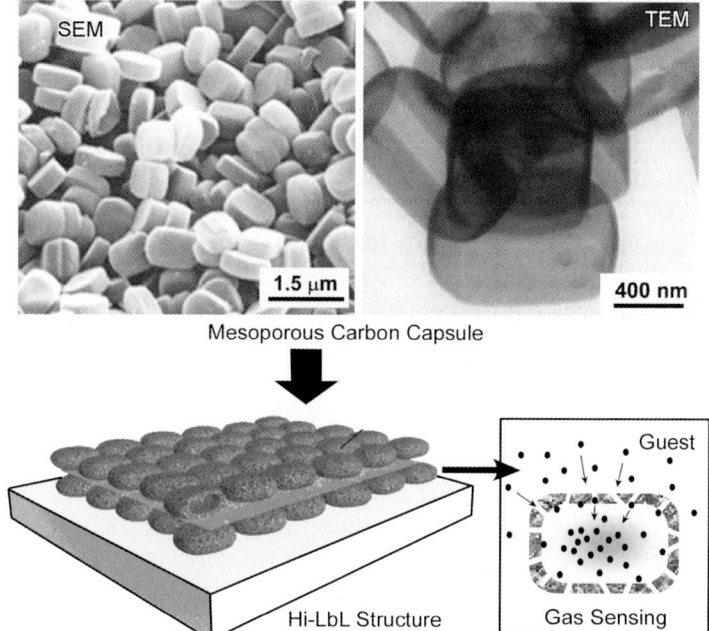

**Figure 4.5** Mesoporous carbon capsule and its Hi-LbL assembly.

tion centered at 4.3 nm in diameter, and a specific surface area of $918\,m^2\,g^{-1}$. Surfactant-covering enables us to assemble non-charged substances in the Hi-LbL process with the aid of counterionic polyelectrolyte. Hi-LbL films to detect specific molecules in the gas phase were prepared from mesoporous carbon capsules and polyelectrolyte (Figure 4.5) [52].

Adsorption of various volatile substances onto the carbon capsule Hi-LbL films in vapor-saturated atmospheres was investigated by *in situ* frequency decrease of the QCM resonator used as the film support. Generally, aromatic hydrocarbons, such as benzene and toluene, are better detected in this sensing system than aliphatic hydrocarbons, such as cyclohexane. In particular, the amount of benzene adsorbed at equilibrium is circa five times larger than that of cyclohexane, despite their very similar vapor pressures, molecular weights, and structures, and indicating the crucial role of $\pi$–$\pi$ interactions on the adsorption of volatiles in the carbon capsule film. Selectivity could be easily tuned by impregnation with additional recognition components that can be introduced after film preparation. The carbon capsule film impregnated with lauric acid showed the greatest affinity for non-aromatic amines and the second highest affinity for acetic acid. FT-IR spectra of the lauric acid-impregnated capsule film indicate that ammonia adsorption causes a shift in $\nu(C=O)$ from $1702\,cm^{-1}$ (original) to $1736\,cm^{-1}$ (after adsorption) with the appearance of new peaks at $1670\,cm^{-1}$ ($\delta(NH)$, H-bonded)) and $1559\,cm^{-1}$ ($COO^-$), suggesting strong entrapment of amines through acid–base interactions. In contrast, impregnation of dodecylamine into the carbon capsule films resulted in a strong preference for acetic acid.

The prepared hierarchic layer-by-layer films with dual pore carbon capsules exhibit excellent adsorption capabilities for volatile guests such as aromatic hydrocarbons. In addition, the selectivity of the gas adsorption can be controlled flexibly by impregnation with second recognition sites. Such designed materials will find widespread applications as sensors or filters because of their designable guest selectivity. As the carbon materials used are stable in water, this system could also be used for removal of toxic materials from water.

### 4.4.3
### Graphene/Ionic-Liquid Hi-LbL

Graphene is one of the most attractive materials prepared in recent times. Pieces of graphene can be disassembled from graphite then re-assembled into hierarchic structures through the Hi-LbL technique. In our research, graphene oxide sheet (GOS) was first prepared by oxidation of graphite under acidic conditions, followed by its reduction to graphene sheet (GS) in the presence of ionic liquids in water. Composites of grapheme sheet/ionic liquid (GS–IL) behave as charge-decorated nanosheets and were assembled alternately with poly(sodium styrenesulfonate) (PSS) by Hi-LbL adsorption on appropriate solid supports to provide layered assemblies of GS-IL composite with PSS on the surface of a QCM resonator (Figure 4.6) [53].

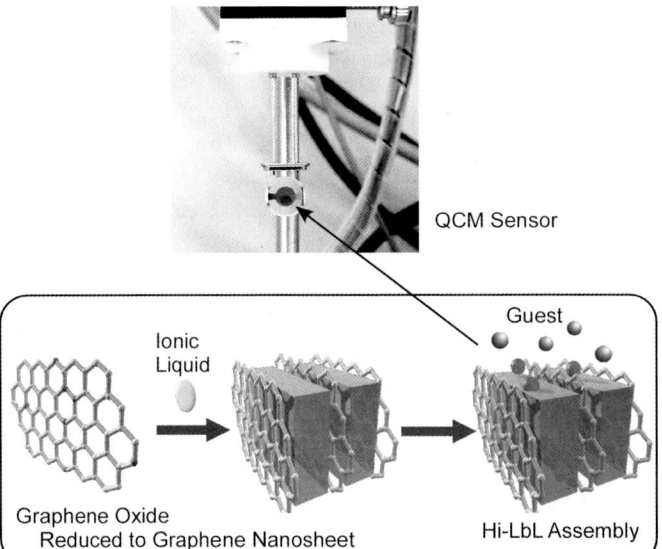

**Figure 4.6** Hi-LbL assembly of graphene nanosheet and ionic liquid for sensor application.

Exposure of the composite films to various saturated vapors (25 °C), after equilibration under an ambient atmosphere caused an *in situ* decrease in frequency of the QCM due to gas adsorption. The amount of adsorbed benzene clearly observed for the GS–IL films depends significantly on the type of ionic liquid used, while GS–water (prepared without ionic liquid component) showed almost no adsorption capability. This behavior is a striking indication of the highly selective detection of aromatic guests within the well-defined π-electron-rich nanospace in the GS–IL films. For example, some GS–IL films showed significantly higher selectivity (more than 10 times) for benzene vapor over cyclohexane, despite their similar molecular sizes, molecular weights, and vapor pressures. Detection of vapors can be repeated through alternate exposure and removal of the subject solvents. Interestingly, gradual degradation of the ON/OFF response was noted for benzene detection, probably caused by the strong interactions between aromatic compounds and the graphene layer, while the response to cyclohexane was fully reversible. Responses to mixtures of benzene and cyclohexane at different mole ratios showed an approximately linear relation with small cooperative deviations, facilitating estimation of gas fractions in mixtures. Control of the electronic properties of the GS–IL films upon gas adsorption was also demonstrated. The electrical resistance of GS–IL film is about $178\,\Omega\,\text{sq}^{-1}$ and is sensitively influenced by gas adsorption leading to, for example, $163\,\Omega\,\text{sq}^{-1}$ upon benzene adsorption. This enables us to convert gas detection to electrical signals. In addition, the GS–IL films have a variety of potential practical applications, including environment remediation through the capture of atmospheric $CO_2$. Adsorption of $CO_2$ vapors from a saturated sodium hydrocarbonate solution into the GS–IL films showed enhanced adsorption volume compared to the GS films without intercalated ionic liquids.

As modern hirerarchic structures, GS–IL films are formed by reduction of graphene oxide sheets in the presence of room temperature ionic liquids and subsequent LbL assembly. Clear enhancement of aromatic gas adsorption based on increases in graphene layer spacing results in highly selective sensing. The experimental results also suggest great potential for practical usage of GS-IL films.

## 4.5
### Future Perspectives

In this chapter, we briefly introduced several examples of the fabrication of hierarchic structures by the Hi-LbL method. The LbL technique is highly versatile when compared with other preparation methods. Therefore, unusual objects such as cerasomes, mesoporous capsules, and composites of graphene nanosheets and ionic liquids can be assembled into organized structures. The prepared hierarchic structures have novel properties and potential including, as shown, in stimuli-free automodulated drug delivery and selectivity-tuned molecular sensing, as well as multi-cellular models, the latter being expected to open up a new area in biomimetic chemistry. However, the examples presented are still at a relatively primitive stage since they use very simple sensing devices such as QCM. In the future, integration of the Hi-LbL structures into advanced micro-fabricated systems will allow expression of more sophisticated functions. The high versatility of LbL assembly will allow us to prepare hierarchic structures even within integrated circuits and micro-fluidic structures. In addition, application of the Hi-LbL structures in human biomedical applications is anticipated. Using biological elements in the Hi-LbL assembly would be very useful for the construction and/or regeneration of bio-like hierarchic structures with real biological components.

## Acknowledgements

We thank the World Premier International Research Center Initiative (WPI Initiative), MEXT, Japan, and the Core Research for Evolutional Science and Technology (CREST) program of the Japan Science and Technology Agency (JST), Japan.

## References

1 Decher, G. (1997) *Science*, **277**, 1232.
2 Jaber, J.A. and Schlenoff, J.B. (2006) *Curr. Opin. Colloid Interface Sci.*, **11**, 324.
3 Ariga, K., Hill, J.P., and Ji, Q. (2007) *Phys. Chem. Chem. Phys.*, **9**, 2319.
4 Ariga, K., Hill, J.P., and Ji, Q. (2008) *Macromol. Biosci.*, **8**, 981.
5 Ariga, K., Hill, J.P., Lee, M.V., Vinu, A., Charvet, R., and Acharya, S. (2008) *Sci. Technol. Adv. Mater.*, **9**, 014109.
6 Ariga, K., McShane, M., Lvov, Y.M., Ji, Q., and Hill, J.P. (2011) *Expert Opin. Drug Deliv.*, **8**, 633.
7 Lvov, Y., Ariga, K., Onda, M., Ichinose, I., and Kunitake, T. (1999) *Colloid Surf. A: Physicochem. Eng. Asp.*, **146**, 337.
8 Fujii, N., Fujimoto, K., Michinobu, T., Akada, M., Hill, J.P., Shiratori, S., Ariga, K., and Shigehara, K. (2010) *Macromolecules*, **43**, 3947.
9 Lvov, Y., Onda, M., Ariga, K., and Kunitake, T. (1998) *J. Biomater. Sci., Polym. Ed.*, **9**, 345.
10 Lvov, Y., Ariga, K., Ichinose, I., and Kunitake, T. (1996) *Langmuir*, **12**, 3038.
11 Lvov, Y., Ariga, K., Onda, M., Ichinose, I., and Kunitake, T. (1997) *Langmuir*, **13**, 6195.
12 Ramsden, J.J., Lvov, Y.M., and Decher, G. (1995) *Thin Solid Films*, **254**, 246.
13 Lvov, Y., Essler, F., and Decher, G. (1993) *J. Phys. Chem.*, **97**, 13773.
14 Ariga, K., Lvov, Y., and Kunitake, T. (1997) *J. Am. Chem. Soc.*, **119**, 2224.
15 Lvov, Y., Ariga, K., and Kunitake, T. (1994) *Chem. Lett.*, 2323.
16 Lvov, Y., Ariga, K., Ichinose, I., and Kunitake, T. (1995) *J. Am. Chem. Soc.*, **117**, 6117.
17 Onda, M., Lvov, Y., Ariga, K., and Kunitake, T. (1996) *J. Ferment. Bioeng.*, **82**, 502.
18 Onda, M., Lvov, Y., Ariga, K., and Kunitake, T. (1996) *Biotechnol. Bioeng.*, **51**, 163.
19 Caruso, F., Furlong, D.N., Ariga, K., Ichinose, I., and Kunitake, T. (1998) *Langmuir*, **14**, 4559.
20 Onda, M., Ariga, K., and Kunitake, T. (1999) *J. Biosci. Bioeng.*, **87**, 69.
21 Lvov, Y., Haas, H., Decher, G., Möhwald, H., Mikhailov, A., Mtchedlishvily, B., Morgunova, E., and Vainstein, B. (1994) *Langmuir*, **10**, 4232.
22 Stockton, W.B. and Rubner, M.F. (1997) *Macromolecules*, **30**, 2717.
23 Sukhishvili, S.A. and Granick, S. (2000) *J. Am. Chem. Soc.*, **122**, 9550.
24 Lee, H., Kepley, L.J., Hong, H.G., and Mallouk, T.E. (1988) *J. Am. Chem. Soc.*, **110**, 618.
25 Wanunu, M., Vaskevich, A., Cohen, S.R., Cohen, H., Arad-Yellin, R., Shanzer, A., and Rubinstein, I. (2005) *J. Am. Chem. Soc.*, **127**, 17877.

26 Serizawa, T., Hamada, K., Kitayama, T., Fujimoto, N., Hatada, K., and Akashi, M. (2000) *J. Am. Chem. Soc.*, **122**, 1891.

27 Serizawa, T., Hamada, K., and Akashi, M. (2004) *Nature*, **429**, 52.

28 Shimazaki, Y., Mitsuishi, M., Ito, S., and Yamamoto, M. (1998) *Langmuir*, **14**, 2768.

29 Lvov, Y., Ariga, K., Ichinose, I., and Kunitake, T. (1995) *J. Chem. Soc., Chem. Commun.*, 2313.

30 Hoshi, T., Akase, S., and Anzai, J. (2002) *Langmuir*, **18**, 7024.

31 Chen, J., Huang, L., Ying, L., Luo, G., Zhao, X., and Cao, W. (1999) *Langmuir*, **15**, 7208.

32 Such, G.K., Quinn, J.F., Quinn, A., Tjipto, E., and Caruso, F. (2006) *J. Am. Chem. Soc.*, **128**, 9318.

33 Li, M., Ishihara, S., Akada, M., Liao, M., Sang, L., Hill, J.P., Krishnan, V., Ma, Y., and Ariga, K. (2011) *J. Am. Chem. Soc.*, **133**, 7348.

34 Lee, S.-S., Hong, J.-D., Kim, C.H., Kim, K., Koo, J.P., and Lee, K.-B. (2001) *Macromolecules*, **34**, 5358.

35 Lefaux, C.J., Zimberlin, J.A., Dobrynin, A.V., and Mather, P.T. (2004) *J. Polym. Sci. B*, **42**, 3654.

36 Lee, M.-H., Chung, W.J., Park, S.K., Kim, M., Seo, H.S., and Ju, J.J. (2005) *Nanotechnology*, **16**, 1148.

37 Izquierdo, A., Ono, S.S., Voegel, J.-C., Schaaf, P., and Decher, G. (2005) *Langmuir*, **21**, 7558.

38 Hua, F., Cui, T., and Lvov, Y.M. (2004) *Nano Lett.*, **4**, 823.

39 Hammond, P.T. (2004) *Adv. Mater.*, **16**, 1271.

40 Shiratori, S.S., Ito, T., and Yamada, T. (2002) *Colloids Surf. A*, **198**, 415.

41 Clark, S.L. and Hammond, P.T. (1998) *Adv. Mater.*, **10**, 1515.

42 Sukhorukov, G.B., Donath, E., Davis, S., Lichtenfeld, H., Caruso, F., Popov, V.I., and Möhwald, H. (1998) *Polym. Adv. Technol.*, **9**, 759.

43 Donath, E., Sukhorukov, G.B., Caruso, F., Davis, S.A., and Möhwald, H. (1998) *Angew. Chem. Int. Ed.*, **37**, 2202.

44 Caruso, F., Caruso, R.A., and Möhwald, H. (1998) *Science*, **282**, 1111.

45 Katagiri, K., Ariga, K., and Kikuchi, J. (1999) *Chem. Lett.*, 661.

46 Katagiri, K., Hashizume, M., Ariga, K., Terashima, T., and Kikuchi, J. (2007) *Chem. Eur. J.*, **13**, 5272.

47 Katagiri, K., Hamasaki, R., Ariga, K., and Kikuchi, J. (2002) *Langmuir*, **18**, 6709.

48 Katagiri, K., Hamasaki, R., Ariga, K., and Kikuchi, J. (2002) *J. Am. Chem. Soc.*, **124**, 7892.

49 Ji, Q., Miyahara, M., Hill, J.P., Acharya, S., Vinu, A., Yoon, S.B., Yu, J.-S., Sakamoto, K., and Ariga, K. (2008) *J. Am. Chem. Soc.*, **130**, 2376.

50 Ji, Q., Acharya, S., Hill, J.P., Vinu, A., Yoon, S.B., Yu, J.-S., Sakamoto, K., and Ariga, K. (2009) *Adv. Funct. Mater.*, **19**, 1792.

51 Ariga, K., Vinu, A., Ji, Q., Ohmori, O., Hill, J.P., Acharya, S., Koike, J., and Shiratori, S. (2008) *Angew. Chem. Int. Ed.*, **47**, 7254.

52 Ji, Q., Yoon, S.B., Hill, J.P., Vinu, A., Yu, J.-S., and Ariga, K. (2009) *J. Am. Chem. Soc.*, **131**, 4220.

53 Ji, Q., Honma, I., Paek, S.-M., Akada, M., Hill, J.P., Vinu, A., and Ariga, K. (2010) *Angew. Chem. Int. Ed.*, **49**, 9737.

# 5
# Layer-by-Layer Assembly Using Host-Guest Interactions

*Janneke Veerbeek, David N. Reinhoudt, and Jurriaan Huskens*

## 5.1
## Introduction

The huge interest in nanomaterials has become an important line of research in nanotechnology for the generation of functional molecular assemblies. For the construction of molecule-based functional devices, it is necessary to develop methods for integrating these molecular components into well-ordered assemblies with a well-defined supramolecular architecture. Therefore, it is essential to study and develop methods for the controlled assembly of multicomponent nanostructures. While systems with monomolecular films on surfaces have been studied extensively, it is attractive to extend this approach to multilayer films, to enhance their properties and create materials with functional groups at controlled sites in three-dimensional arrangements [1].

In recent years, layer-by-layer (LbL) assembly has emerged as a promising method for fabricating structured and functional thin films on solid substrates, since it can be used to construct a film on top of a substrate of almost any composition or topology by alternating its exposure to solutions containing species of complementary affinities. While most multilayer assemblies are prepared by electrostatic interactions, we have used supramolecular host–guest interactions to grow LbL films [2]. Host–guest interactions offer unique properties, such as tunable binding affinity, controlled orientation, reversibility, and so on, and can be compatible with and orthogonal to other interaction motifs. Compared to the well-known electrostatic LbL films, supramolecular multilayer films may offer additional benefits, such as incorporation of neutral molecules and/or biomolecules, implementation of host–guest motifs at the interface of nanomaterials to be assembled in a LbL scheme, straightforward control over the growth process and layer thickness by structural and steric design of the building blocks, and so on.

In this chapter, the research on supramolecular layer-by-layer assembly will be reviewed. The first section explains the principle of supramolecular LbL assembly in general. The second section evaluates the combination of LbL assembly with nanoimprint lithography (NIL). This combination of bottom-up and top-down

---

*Multilayer Thin Films: Sequential Assembly of Nanocomposite Materials*, Second Edition.
Edited by Gero Decher and Joseph B. Schlenoff.
© 2012 Wiley-VCH Verlag GmbH & Co. KGaA. Published 2012 by Wiley-VCH Verlag GmbH & Co. KGaA.

techniques is essential for the fabrication of 3D structures to accommodate the need for device applications. Furthermore, highly ordered 3D supramolecular nanoparticle structures were produced by a combination of transfer printing and LbL assembly, which could be used to create freestanding particle bridges. This will be discussed in the third section. Finally, the fourth section focuses on the production of porous 3D supramolecular assemblies that can be used to encapsulate guest molecules.

## 5.2
### Supramolecular Layer-by-Layer Assembly

Within our group, cyclodextrin (CD) self-assembled monolayers (SAMs) have been prepared on gold [3] and silicon oxide [4] surfaces on which stable positioning and patterning of molecules has been achieved. These CD SAMs can be used as molecular printboards for the positioning of thermodynamically and kinetically stable assemblies of multivalent systems, for example, dendrimers [5–7]. In solution, CD-modified gold particles and adamantyl (Ad)-functionalized dendrimers appeared to aggregate, in part controlled by the ligand valency [8]. Crespo-Biel et al. combined these results and developed a method for the stepwise construction of a novel kind of self-assembled organic/inorganic multilayers based on multivalent supramolecular interactions [2]. This supramolecular LbL assembly was used for the alternating adsorption of guest-functionalized dendrimers and host-modified gold nanoparticles. These components were chosen because the CD–adamantyl interaction is one of the strongest, the number of adamantyl units (ranging from 4 to 64) can be easily varied, and the spherical shape of the dendrimers and particles allows the multivalent display of these host/guest functionalities in a 3D manner.

The LbL assembly of the Ad-functionalized dendrimers and CD-Au nanoparticles is shown in Figure 5.1. The dendrimers were stable in an aqueous solution (pH 2) by complexation of the adamantyl end groups with a slight excess of cyclodextrin and by protonation of the dendrimer core amine functionalities. While the dendrimers precipitated at pH>7, the Au nanoparticles are not stable in acidic solutions. Therefore, the LbL assembly was typically accomplished by alternately dipping the substrate into a solution of dendrimers in CD (pH 2), followed by rinsing the substrate with the same CD solution (pH 2), and then into an aqueous CD-Au nanoparticle solution, followed by rinsing with water [2].

UV/Vis absorption spectroscopy showed a linear increase in absorption with the number of layer pairs. The growth of the gold nanoparticle plasmon absorption band corresponded to approximately a dense monolayer of gold nanoparticles per layer pair. Ellipsometry and atomic force microscopy (AFM) scratching experiments were used to measure the development of the film thickness with the number of layer pairs, confirming linear growth and a thickness increase of approximately 2 nm per layer pair. The deposition process was monitored by surface plasmon resonance spectroscopy. When the concentration of both dendrimers and CD-Au nanoparticles was 10 times higher, the adsorption was only 1.5 times higher, which clearly confirms the supramolecular specificity of binding [2].

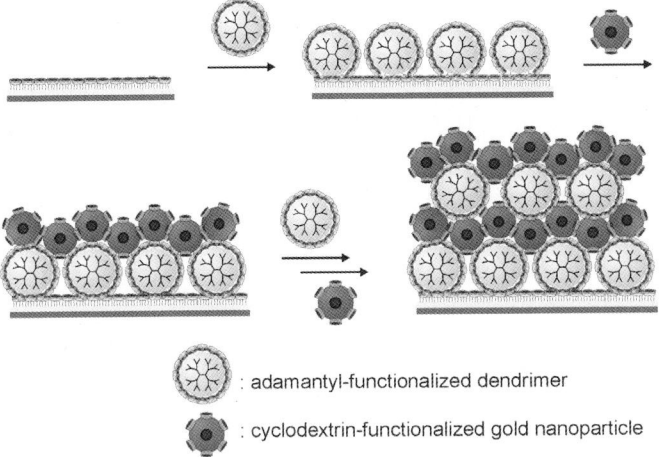

**Figure 5.1** Layer-by-layer assembly scheme using supramolecular host–guest interactions [2], here represented by the alternating adsorption of adamantyl-terminated dendrimers and CD-Au nanoparticles onto CD SAMs.

In conclusion, a supramolecular procedure was developed for the stepwise construction of multilayer thin films. This procedure is based on LbL assembly of guest-functionalized dendrimers and CD-Au nanoparticles, resulting in multilayer thin films with thickness control at the nm level. It was shown that the multilayer formation was well-defined, had an accurate thickness control, and needed specific host–guest interactions [2].

## 5.3
## 3D Patterned Multilayer Assemblies on Surfaces

The fabrication of 3D nanostructures with tunable, sub-100 nm dimensions in all three directions is a key issue of nanotechnology. When producing thin films by LbL assembly for practical applications, it is necessary to achieve high spatial resolution while retaining the interfacial properties and specificity of the components. Therefore, different patterning methods have been developed. Crespo-Biel et al. compared various strategies to create patterns of supramolecular LbL assemblies made by using multivalent supramolecular interactions between dendritic guest molecules and CD-modified gold nanoparticles [9]. Because the supramolecular specificity was not sufficient for multistep directed assembly onto patterned CD SAMs, due to non-specific interactions induced by the dendrimers, nanotransfer printing (nTP) was employed to transfer complete LbL assemblies and NIL to provide topographical masks for LbL assembly. This section focuses on the production of patterned multilayer thin films by the combination of NIL and LbL assembly.

The integration of NIL and LbL assembly is advantageous due to the high resolution of NIL, and the fact that the polymer protects the substrate from

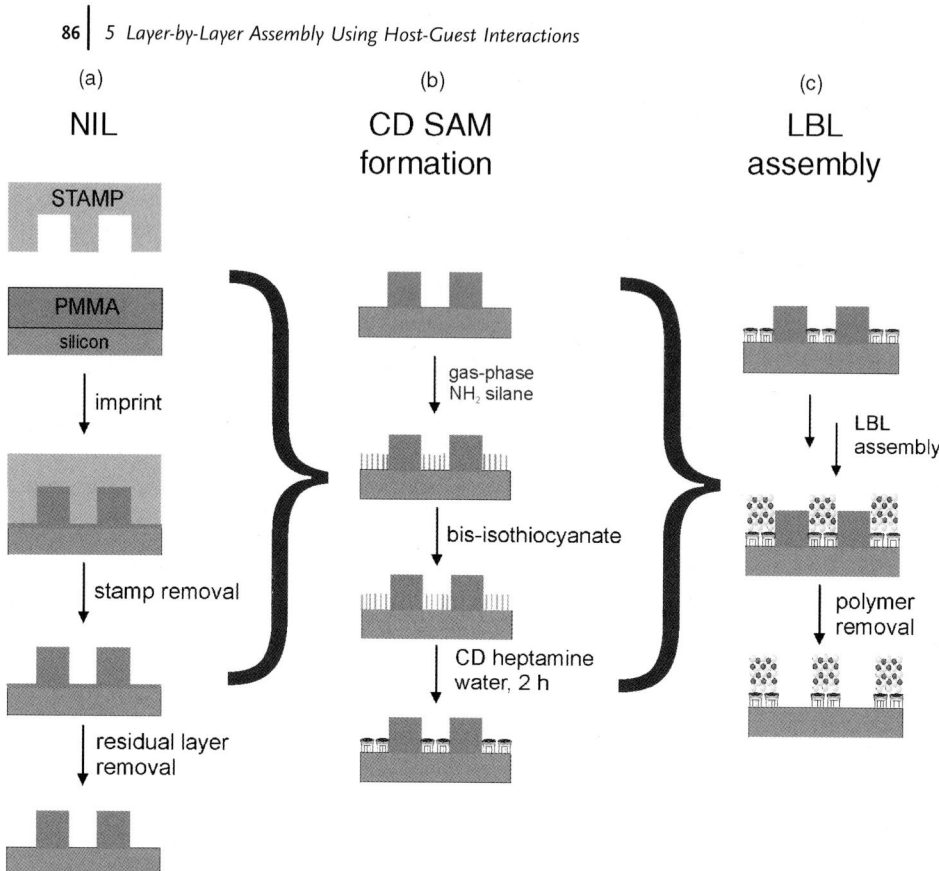

**Figure 5.2** Preparation of 3D, NIL-patterned, LbL assemblies using NIL (a), CD monolayer formation (b), and LbL assembly and lift-off (c) [9].

nonspecific adsorption [10–12]. Furthermore, the $x,y$ control by NIL can be combined with the $z$ control by LbL assembly, resulting in a versatile 3D nanofabrication methodology of 10–40 process steps. The integrated scheme of the three multistep processes, encompassing NIL, multistep monolayer formation, and LbL assembly, is shown in Figure 5.2. Figure 5.2a shows the NIL process, resulting in patterned poly(methyl methacrylate) (PMMA) structures with bare silicon oxide substrate areas in between. Figure 5.2b shows the three-step process for the fabrication of CD monolayers on the bare substrate areas, resulting in patterned CD areas. The PMMA structures are used as a physical barrier for the CD pattern. The LbL assembly on the NIL-patterned substrates and the polymer removal are shown in Figure 5.2c, resulting in patterned LbL structures with a desired thickness. The LbL assembly makes use of supramolecular host–guest interactions between adamantyl guest-functionalized dendrimers and cyclodextrin-modified gold nanoparticles, as described in the previous section.

The LbL assembly method is fully compatible with the NIL-prepared PMMA structures on the substrate, since aqueous solutions are used for the LbL process, which do not dissolve or degrade the PMMA structures. Furthermore, the possible nonspecific adsorption of the LbL components onto the PMMA structures does not pose any problems, since the polymer is dissolved in the final lift-off step, together with the material adsorbed on it [13].

AFM images of various NIL-patterned LbL structures showed that the thickness increase was about 1.1 nm per layer pair, which is slightly lower than LbL assembly on nonpatterned substrates [9]. This is probably due to different wetting and mass transport limitations of the LbL components on the PMMA-structured substrates. Nevertheless, the observed linear growth shows the potential of NIL for structuring supramolecular LbL assemblies.

Maury *et al.* demonstrated the versatility of the process by using nanoparticles of different size and nature [12]. They developed high-resolution patterning of supramolecular LbL assemblies by nanostructuring LbL assemblies at the sub-200 nm scale. LbL assembly of CD-functionalized Au nanoparticles (CD-Au NPs) was performed on a high-resolution patterned CD layer, resulting in a large number of attached layer pairs. Furthermore, supramolecular LbL assembly was carried out with silica ($SiO_2$) nanoparticles by using capillary assisted vertical deposition to increase the particle density and order the attached NPs. Both nanoparticles were assembled in combination with adamantyl-terminated dendrimers.

Furthermore, periodic patterns of single nanoparticles were obtained [12]. It was possible to attach CD-functionalized $SiO_2$ nanoparticles in the holes after polymer removal. Particles were physically confined in the shape of the features, as heptagons in the case of 60 nm particles confined in 200 nm holes, and as single particles in 60 nm holes. Single particle attachment is of high importance, for instance, for application in the catalyst industry, sensing, and drug screening.

Ling *et al.* further investigated the combination of NIL and LbL assembly by using ferrocenyl-functionalized silica (Fc-$SiO_2$) as guest-functionalized nanoparticles [14] and CD-Au and CD-$SiO_2$ as host-functionalized nanoparticles [15]. The effects that are induced by the order of the nanoparticle assembly steps, going from large to small and from small to large nanoparticles, were compared. Different sizes and core materials were investigated in order to extend the supramolecular LbL assembly to the build-up of multicomponent hybrid (organic–metallic–inorganic) nanoobjects.

To investigate the influence of the order of the assembly steps, Ling *et al.* assembled dendrimers, CD-Au, Fc-$SiO_2$ and CD-$SiO_2$, from small to large, respectively, resulting in a total height of about 450 nm [15]. All nanostructures showed good packing density, but order was lacking. The resulting cross-sectional height profiles corresponded well to the accumulated diameters of the deposited nanoparticles. Moreover, the specific supramolecular assembly of nanoparticles was self-limited, that is, one nanoparticle layer was adsorbed per assembly step. This indicates that control over the thickness of the supramolecular hybrid nanostructure can be achieved by choosing the nanoparticle size, irrespective of their core material. When the multilayer stack was reversed, that is, LbL assembly with decreasing nanoparticle size, the roughness of the surface was increased, which was attributed to the reverse order and, in particular, to the corrugations on the surface introduced by the first layer of (large) particles.

In conclusion, it has been shown that the integration of top-down and bottom-up nanofabrication processes is possible. By combining the $x,y$ control of NIL and the $z$ control of LbL assembly, 3D structures of supramolecular LbL assemblies on CD molecular printboards have been created. The methodologies can be used in other nanofabrication schemes and may lead to well-defined, high-resolution 3D nanostructures of a large variety of materials. In addition, the high-resolution patterning of CD layers with single and multilayer particle nanostructuring was possible, and single particle attachment was achieved.

## 5.4
### 3D Supramolecular Nanoparticle Crystal Structures

When particles are integrated into devices, accurate placement of the particles is often required. This can be achieved by combining top-down nanofabrication techniques with the self-assembly of particles, as explained in the previous section. Microcontact printing can be extended to transfer printing in order to transfer the assembled nanostructure onto another substrate. Wolf *et al.* have reported the transfer printing of nanoparticles with a sub-100 nm particle resolution [16]. In many cases, only a single layer of particles was transferred in each step, since transfer printing was carried out via conformal contact. Therefore, the success of printing was strongly dependent on the adhesion between the transferred objects and the target surface.

While most chemically directed particle assembly is focused on the formation of 2D particle arrays, Ling *et al.* used a combination of transfer printing and self-assembly of particles to develop a new strategy of forming 3D supramolecular particle structures with macroscopic robustness and improved order [17–19]. The conventional LbL assembly of oppositely charged polyelectrolytes generally produces nanometer-thick coatings via attractive electrostatic forces between the polyelectrolytes. Ling *et al.* used supramolecular LbL assembly of guest-functionalized organic molecules and host-functionalized metallic NPs, providing a thin, stable coating on and within a preassembled crystal of CD-functionalized polystyrene (PS) particles. The constituents act as complementary supramolecular glues that strengthen the cohesion between the individual PS-CD particles by thermodynamically stable multivalent host–guest interactions.

Ling *et al.* have thus constructed mechanically robust and crystalline supramolecular particle structures by decoupling nanoparticle assembly and supramolecular glue infiltration into a sequential process. In this way, the advantages of convective assembly (in ordering particles) and supramolecular chemistry (in chemically bonding adjoining particles by infiltration with supramolecular glues) are combined into one process. In the first step of the process, CD-functionalized polystyrene particles ($d \sim 500$ nm) were assembled on a CD-functionalized surface by using convective assembly. This resulted in highly ordered, but mechanically unstable, particle crystals. In the second step, the crystals were infiltrated by a solution of adamantyl-functionalized dendrimers, functioning as a supramolecular glue to bind neighboring particles together and to couple the entire particle crystal to the CD

## 5.4 3D Supramolecular Nanoparticle Crystal Structures

surface, both in a noncovalent way. The supramolecular particle crystals could withstand agitation by ultrasonication, so they are highly robust [18, 19].

When the dendrimer-infiltrated particle crystals were assembled on a polydimethylsiloxane (PDMS) stamp, they could be transfer-printed onto a CD-functionalized target surface via host–guest interactions. The patterned PDMS stamp functions as a template to provide spatial and geometrical confinement to the CD-functionalized polystyrene particle crystal. The geometry and size of the PDMS stamps were varied, resulting in single particle lines, interconnected particle rings, and V-shaped particle assemblies. The layer thickness and lateral dimensions of the transferred crystal could be conveniently controlled by the design of the template. Furthermore, the number of layers of the particle crystal could be controlled by tuning the individual particle diameter [18].

The printed particle structures can be addressed as functional building blocks for further assembly of molecules or nanostructures. Divalent adamantyl guest molecules were printed orthogonally onto the particle lines. Fluorescence microscopy of the particle structure printed with these guests showed only squares that were fluorescent, equivalent to the size of the contact areas between the particle structures and the PDMS stamp. This indicates that the supramolecular host property of the individual particles remains intact after the transfer printing process. The supramolecular particle structures can thus potentially act as a "pathway for the transportation of molecular information" [19].

Ling *et al.* investigated the fabrication of freestanding particle bridges by the combination of the bottom-up self-assembly of nanoparticles, supramolecular LbL assembly and transfer printing (Figure 5.3a) [17, 19]. To achieve this, the bonding

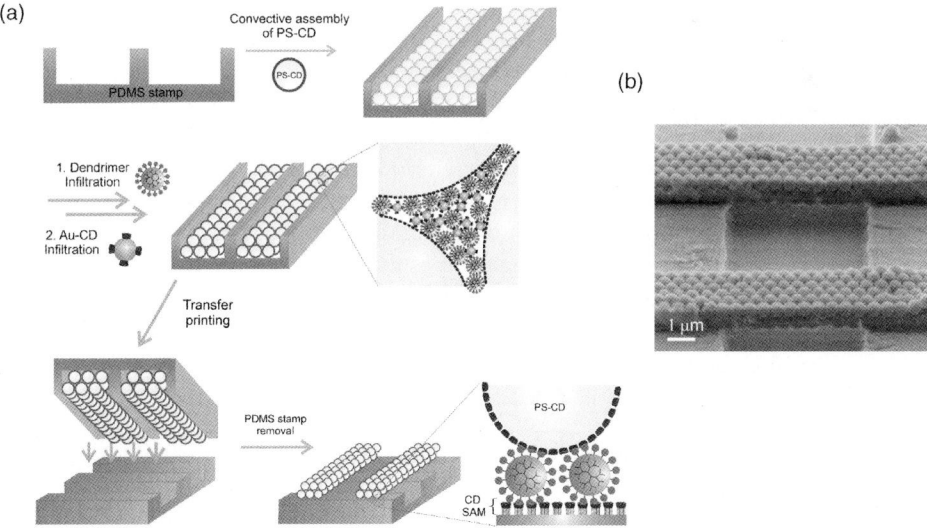

**Figure 5.3** The formation of free-standing 3D supramolecular particle crystal bridges by a combination of convective assembly, supramolecular LbL assembly, and transfer printing onto a topographically patterned target substrate [19] (a), and SEM image of a freestanding particle bridge made by such a procedure (b).

between the nanoparticles had to be further strengthened by using Ad dendrimers and CD-functionalized Au NPs as the respective guest- and host-functionalized glues. CD-functionalized particle crystals were first convectively assembled onto a patterned PDMS stamp. The glues were assembled within the particle crystal in a supramolecular LbL way, yielding a stable and ordered 3D particle composite on the PDMS stamp. The dendrimers and Au-CD were alternately assembled with up to 30 LbL cycles in total. The particle composite was perpendicularly printed onto topographically patterned CD SAM surfaces via host–guest interactions, yielding 3D freestanding and ordered hybrid polystyrene nanoparticle bridges, as shown in Figure 5.3b. Ten layer pairs appeared to be sufficient to get reliable and high-yielding free-standing structures.

The freestanding particle structures bridge micrometer distances between the vertical posts of the substrate, resulting in single-span particle microbridges. The yield of these particle bridges was typically in the range of 50–90% across the entire substrate. The mechanical robustness and rigidity of the particle bridges can be controlled by varying the number of LbL cycles of supramolecular glues of Au NPs and dendrimers. AFM-based microbending measurements showed that the particle bridges fulfilled the classical supported-beam characteristics, with bending moduli varying between 0.8 and 1.1 GPa, depending on the degree of filling by the supramolecular glues [17, 19]. This approaches the bending modulus of pure PS, which shows that the glue has a strength that is comparable to the PS core material.

In conclusion, host–guest-based LbL assembly has been shown to be useful as a supramolecular glue, allowing the formation of mechanically robust 3D structures. In particular, the high stability and chemical versatility allow decoupling of the particle assembly step, performed under conditions in which strong attractive interactions are absent to allow ordered particle lattice formation, from the glueing infiltration step, in which the host–guest interactions are installed, providing stabilization of the structure. It is thus also possible to produce freestanding structures of desired shapes and sizes, as determined mainly by the design of the stamp used as a mold in the first assembly step. The tunable supramolecular LbL assembly method could be used to fine-tune the stiffness of the particle bridge from semiflexible to rigid. This ability to control the mechanical properties could initiate new perspectives for the development of the assembly and patterning of particle crystals for applications in optical devices, electronic devices, and microelectromechanical systems (MEMS) [17, 19].

## 5.5
### Porous 3D Supramolecular Assemblies in Solution

There is an emerging interest in fabricating well-defined 3D structured materials, ranging from nanometer to micrometer length scale, since they can be used as basic components for applications in information storage, electronics and sensor devices. It is important to control the morphology, dimensions, composition and functionality of such 3D structures, since this allows direct manipulation of the physical and chemical properties of the resulting device. As stated before, LbL assembly can be used to form 3D mesoscopic structures with tunable morphology, size and composition. In this section, two new strategies for the production of 3D supramolecular assemblies with controlled size and geometry will be discussed [20, 21]. These assembled structures can be transferred to solution and can be used to encapsulate external guest molecules.

The first technique [20] consists of a new strategy to produce discrete supramolecular 3D hybrid structures with controllable morphology and dimensions. These discrete, monodisperse, and robust supramolecular assemblies are dispersible in solution. The structures are formed by combining supramolecular chemistry and nanofabrication techniques on surfaces, and thus constitute an alternative to solution-based supramolecular aggregation, which generally lacks size control.

The process to fabricate 3D hybrid structures consists of three key steps (Figure 5.4) [20]. In the first step, nanoimprint lithography was used for the formation of patterned PMMA, which served as a morphological and dimensional template for

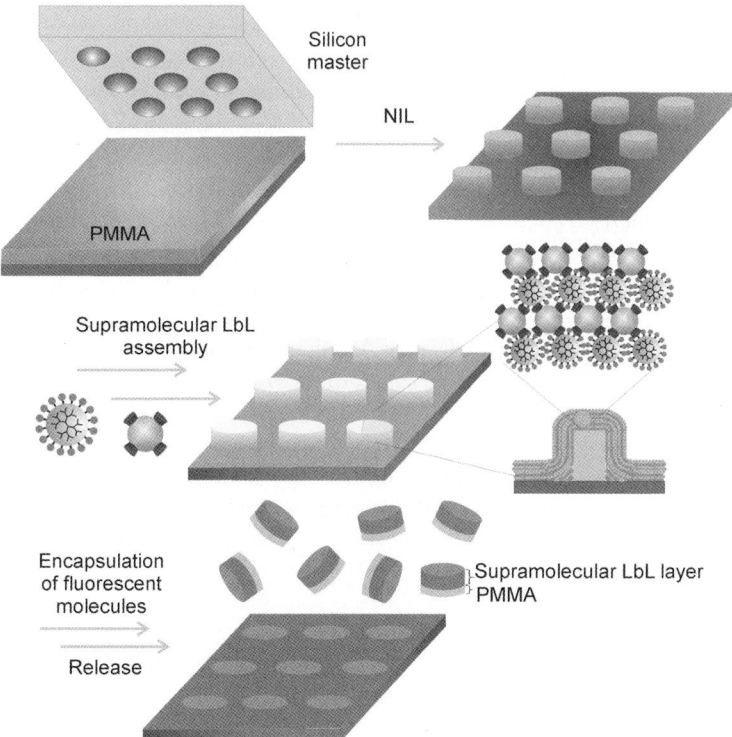

**Figure 5.4** 3D supramolecular hybrid material structures prepared by three key steps: (i) NIL fabrication of a polymer template, (ii) supramolecular LbL assembly of multilayered Ad dendrimers and Au-CD nanoparticles onto the polymer pattern, and (iii) a lift-off process to release the structures into solution. Fluorescent molecules can be encapsulated within the structures to form discrete fluorescent hybrid objects [20].

forming hexagonal and square 3D hybrid structures. Thereafter, alternating LbL assembly of Ad dendrimers and Au-CD nanoparticles led to a supramolecular hybrid multilayer thin film on the PMMA template. Subsequently, the hybrid supramolecular structure-PMMA objects were released from the silicon substrate by a lift-off process, resulting in freely suspended hybrid structures in water. The supramolecular recognition properties of the 3D structures can be used to encapsulate external guest molecules and nanoparticles.

Scanning electron microscope (SEM) imaging showed that the entire substrate was covered with the supramolecular multilayers before lift-off. After releasing the hexagonally shaped hybrid structures, only a honeycomb-shaped ring network of the supramolecular layers remained on the substrate. After deposition of the released structures onto another substrate, imaging showed that almost all hybrid structures had retained their original hexagonal shape and size. The AFM height profile showed a thickness of about 800 nm, comparable to the original 800-nm thick PMMA pattern. Notably, this is much higher than the supramolecular multilayer of Au nanoparticles and dendrimers, which is only 100 nm thick. Accordingly, approximately 100 nm PMMA seemed to dissolve during the lift-off process. This shows that the PMMA layer and the hybrid structures remained integrated, even after the lift-off process. This composite structure was evident from the images as the two different sides of the hybrid objects, that is, exposing either the supramolecular multilayers or the PMMA, could be clearly distinguished from the different roughnesses on the respective surfaces.

When different NIL molds were used, discrete 20 μm hybrid hexagons and 2 μm squares cpuld be produced [20]. These structures were slightly smaller than the NIL silicon mold, which is probably due to the reduction of the polymer pattern size during the polymer residual layer removal process. This demonstrates the key aspect of the process, namely the possibility of varying the dimensions of the discrete supramolecular hybrid structures at will by the design of the NIL mold and by the LbL process, leading to structures with a very narrow (<5%) size distribution.

The strong multivalent host–guest interactions between the Au NPs and dendrimers bind the entire structure together. As a consequence, the free-standing structures are robust in preserving their shape and dimensions during the lift-off process and when suspended in solution. The mesoscopic structures possess anisotropic (Janus) properties, since one face consists of organic–metallic layers with supramolecular recognition functionalities, while the other face exposes the PMMA layer which functions as a rigid backbone for the supramolecular layers. The supramolecular material still allows binding of other entities, biomolecules for example, in order to provide targeting properties to the thus-assembled nanomaterials.

Since the dendrimers within the hybrid structures contain voids and protonated tertiary amines, they can encapsulate small negatively charged guest molecules. When fluorescent molecules are encapsulated, discrete fluorescent objects can be produced. For example, the negatively charged dye Rhodamine was embedded within the positively charged core of the adamantyl dendrimers by using electrostatic interactions [20]. This Rhodamine loading was achieved by immersing the substrate in the Rhodamine solution before lift-off and a subsequent release of the discrete hybrid structures. The Rhodamine molecules were firmly and selectively encapsulated within the patterned LbL structures. Other fluorescent molecules could also be bound within the hybrid structures while retaining their fluorescent properties. It was observed that the PMMA layer did not hinder the fluorescence intensity of the encapsulated molecules. When hybrid structures with different encapsulated dyes were mixed in solution, the difference in molecular guest information could still be clearly addressed. This indicates that, by varying the size, shape and loading of the hybrid structures, libraries of discrete molecularly encoded hybrid structures can be created.

The other technique to produce porous 3D supramolecular assemblies in solution includes a double-templating strategy [21]. The sizes and geometries of these stable and ordered free-standing assemblies can be controlled at different length scales. This approach extends the use of supramolecular chemistry in forming particle composite bridges with macroscopic robustness (as described in Section 5.4) to stand-alone and freely-suspended porous nanoparticle structures with specific functionalities.

The fabrication procedure of free-standing hybrid particle crystals and of the porous assemblies derived thereof is shown in Figure 5.5 [21]. A 3D crystal structure within a PMMA template with Au-CD nanoparticles and Ad dendrimers as glues was produced in the same way as described in Section 5.3, thus using a combination of NIL and supramolecular LbL assembly. The PMMA template was used as a carrier, in order to release the entire particle composites from the silicon substrate into water. The resulting floating PMMA template at the water/air interface, including the hybrid nanoparticle composites, was picked up by a target substrate for visualization. The PS cores of the hybrid particle structure and the PMMA template were simultaneously removed by rinsing with dichloromethane, resulting in interconnected porous capsules. The size and shape of these capsules are determined by the PS core size and the design of the polymer template. The integrity, order and functionality were fully preserved.

Upon release of the PMMA template with embedded supramolecular structures from the starting substrate, the PS-CD particles in 3 µm line structures remained adhered to the PMMA template, probably due to Van der Waals forces and electrostatic interactions between the positively charged Ad dendrimers and the negatively charged oxidized PMMA surface. The PMMA-particle structure is flexible owing to its thickness of <1 µm. After dissolving the PS core and PMMA template, ribbons of interconnected hollow hybrid capsules were observed, with a length of up to 500 µm. The internal cohesion by the supramolecular glues remained intact and the majority of the hollow capsules retained their original spherical shape.

Ling et al. also investigated the ability of the sub-millimeter hollow capsule ribbons to encapsulate fluorescent molecules, including 8-anilino-1-naphthalene sulfonic acid (ANS) [21]. ANS is a fluorescent probe that is highly sensitive toward its microenvironment, as its intensity depends on the polarity of the medium. Thus severe fluorescence quenching is observed in water. The as-prepared particle composites were dipped into an ANS solution. The negatively charged ANS penetrates into the

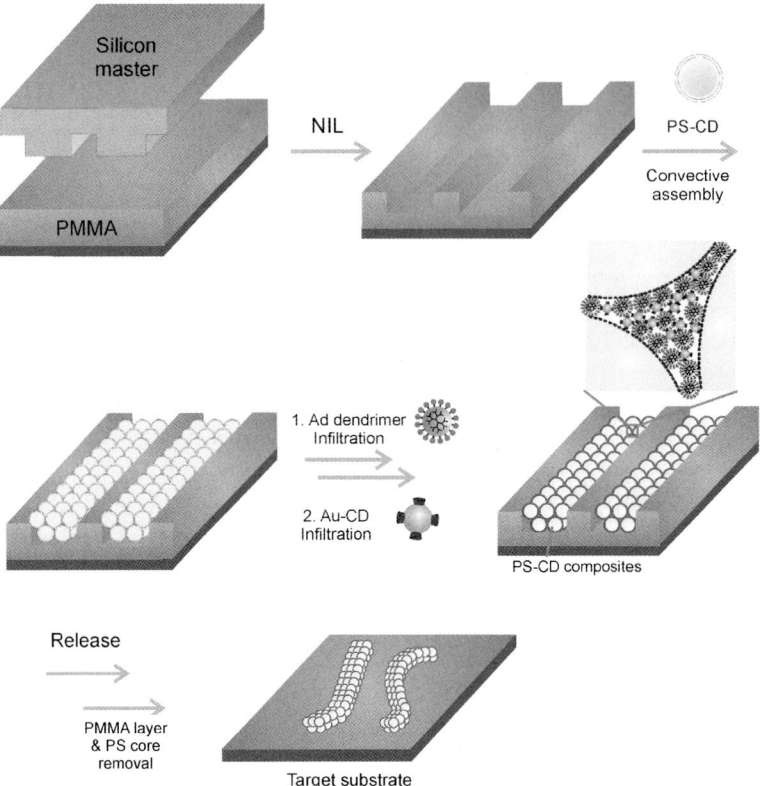

**Figure 5.5** The preparation of a PMMA template and the release-and-transfer of free-standing hydrid particle composites, followed by template and PS core removal [21].

core of the Ad dendrimers via electrostatic interactions to restore its fluorescence intensity. After removal of the PS cores and the PMMA layer, the Ad dendrimers, and thus the complete capsule ribbons, could be observed by fluorescence microscopy. This indicates that the ANS molecules are stored within the shells of the hollow capsules and remain in the structure upon release, transfer and rinsing.

It was also possible to produce other structures, for example, a network structure, by changing the design of the PMMA template [21]. This means that the individual size of the hollow capsules can be manipulated by the choice of the size of the PS particles, while the overall shape and geometry of the entire capsule structure can be designed by the geometry and size of the PMMA template.

Summarizing, the two strategies described above allow the fabrication of 3D supramolecular assemblies with controllable size and geometry. These assemblies can be transferred to solution and can be used to encapsulate external guest molecules. Therefore, the structures can potentially be used as molecular information carriers to store and release molecular information.

## 5.6
## Conclusions

Nanotechnology requires new methodologies for the assembly of molecular- to micrometer-scale objects onto substrates in predetermined arrangements for the production of sensors, electrical and optical devices, MEMS and photonic systems. It is important for these methodologies to be able to control the lateral dimensions, deposit films of different compositions onto surfaces, and to construct 3D devices. This chapter has shown the versatility of layer-by-layer assembly using host–guest interactions, which can be used to achieve these goals.

By using guest-functionalized dendrimers and cyclodextrin-modified gold particles, multilayer thin films with thickness control at the nm level can be produced. This LbL assembly can be patterned by combining the $x,y$ control of, for example, nanoimprint lithography with the $z$ control of LbL assembly. As a consequence, 3D nanostructures with tunable, sub-100 nm dimensions in all three directions can be produced. This method may lead to well-defined, high-resolution 3D nanostructures of a large variety of materials.

When layer-by-layer assembly is combined with transfer printing, 3D supramolecular nanoparticle crystal structures can be produced. These printed particle structures can be addressed as functional building blocks for further assembly of molecules or nanostructures, or may be used as a pathway for the transportation of molecular information. By strengthening the supramolecular interactions within a particle crystal, freestanding particle bridges can be obtained. The tunable supramolecular LbL assembly method can be used to control the mechanical properties of the particle bridges.

Finally, two novel strategies to produce 3D supramolecular assemblies with controlled size and geometry have been reviewed. The thus-produced assemblies can be transferred to solution and, at the same time, allow the encapsulation of external guest molecules. Therefore, the structures can potentially be used as molecular information carriers to store and release molecular information. In general, the strategies described here could initiate new perspectives for the development of the assembly and patterning of particle structures and crystals for applications in electronic devices, sensing and MEMS, while the high degree of order achieved in the crystal structures could prove useful for application in photonic bandgap materials and optical circuits.

## References

1 Crespo-Biel, O., Ravoo, B.J., Reinhoudt, D.N., and Huskens, J. (2006) Noncovalent nanoarchitectures on surfaces: from 2D to 3D nanostructures. J. Mater. Chem., **16** (41), 3997–4021.
2 Crespo-Biel, O., Dordi, B., Reinhoudt, D.N., and Huskens, J. (2005) Supramolecular layer-by-layer assembly: alternating adsorptions of guest- and host-functionalized molecules and particles using multivalent supramolecular interactions. J. Am. Chem. Soc., **127** (20), 7594–7600.

3 de Jong, M.R., Huskens, J., and Reinhoudt, D.N. (2001) Influencing the binding selectivity of self-assembled cyclodextrin monolayers on gold through their architecture. *Chem.-Eur. J.*, **7** (19), 4164–4170.

4 Onclin, S., Mulder, A., Huskens, J., Ravoo, B.J., and Reinhoudt, D.N. (2004) Molecular printboards: monolayers of β-cyclodextrins on silicon oxide surfaces. *Langmuir*, **20** (13), 5460–5466.

5 Ludden, M.J.W., Reinhoudt, D.N., and Huskens, J. (2006) Molecular printboards: versatile platforms for the creation and positioning of supramolecular assemblies and materials. *Chem. Soc. Rev.*, **35** (11), 1122–1134.

6 Huskens, J., Mulder, A., Auletta, T., Nijhuis, C.A., Ludden, M.J.W., and Reinhoudt, D.N. (2004) A model for describing the thermodynamics of multivalent host-guest interactions at interfaces. *J. Am. Chem. Soc.*, **126** (21), 6784–6797.

7 Huskens, J., Deij, M.A., and Reinhoudt, D.N. (2002) Attachment of molecules at a molecular printboard by multiple host-guest interactions. *Angew. Chem. Int. Edit.*, **41** (23), 4467–4471.

8 Crespo-Biel, O., Jukovic, A., Karlsson, M., Reinhoudt, D.N., and Huskens, J. (2005) Multivalent aggregation of cyclodextrin gold nanoparticles and adamantyl-terminated guest molecules. *Isr. J. Chem.*, **45** (3), 353–362.

9 Crespo-Biel, O., Dordi, B., Maury, P., Peter, M., Reinhoudt, D.N., and Huskens, J. (2006) Patterned, hybrid, multilayer nanostructures based on multivalent supramolecular interactions. *Chem. Mater.*, **18** (10), 2545–2551.

10 Crespo-Biel, O., Dordi, B., Maury, P., Péter, M., Reinhoudt, D.N., and Huskens, J. (2006) Patterned, hybrid, multilayer nanostructures based on multivalent supramolecular interactions. *Chem. Mater.*, **18** (10), 2545–2551.

11 Huskens, J., Maury, P., Crespo-Biel, O., Péter, M., and Reinhoudt, D.N. (2005) Fabrication of three-dimensional hybrid nanostructures by an integrated process comprising nanoimprint lithography and layer-by-layer assembly. Proceedings of the institution of mechanical engineers, Part N. *J. Nanoeng. Nanosyst.*, **219** (4), 157–163.

12 Maury, P., Peter, M., Crespo-Biel, O., Ling, X.Y., Reinhoudt, D.N., and Huskens, J. (2007) Patterning the molecular printboard: patterning cyclodextrin monolayers on silicon oxide using nanoimprint lithography and its application in 3D multilayer nanostructuring. *Nanotechnology*, **18** (4), 044007

13 Maury, P., Crespo-Biel, O., Péter, M., Reinhoudt, D.N., and Huskens, J. (2005) Integration of top-down and bottom-up nanofabrication schemes. *Mater. Res. Soc. Symp. Proc.*, **901**, 441–449.

14 Ling, X.Y., Reinhoudt, D.N., and Huskens, J. (2006) Ferrocenyl-functionalized silica nanoparticles: preparation, characterization, and molecular recognition at interfaces. *Langmuir*, **22** (21), 8777–8783.

15 Ling, X.Y., Phang, I.Y., Reinhoudt, D.N., Vancso, G.J., and Huskens, J. (2008) Supramolecular layer-by-layer assembly of 3D multicomponent nanostructures via multivalent molecular recognition. *Int. J. Mol. Sci.*, **9** (4), 486–497.

16 Kraus, T., Malaquin, L., Schmid, H., Riess, W., Spencer, N.D., and Wolf, H. (2007) Nanoparticle printing with single-particle resolution. *Nat. Nanotech.*, **2** (9), 570–576.

17 Ling, X.Y., Phang, I.Y., Schönherr, H., Reinhoudt, D.N., Vancso, G.J., and Huskens, J. (2009) Freestanding 3D supramolecular particle bridges: Fabrication and mechanical behavior. *Small*, **5** (12), 1428–1435.

18 Ling, X.Y., Phang, I.Y., Reinhoudt, D.N., Vancso, G.J., and Huskens, J. (2009) Transfer-printing and host-guest properties of 3D supramolecular particle structures. *ACS Appl. Mater. Interface*, **1** (4), 960–968.

19 Ling, X.Y., Phang, I.Y., Maijenburg, W., Schönherr, H., Reinhoudt, D.N., Vancso, G.J., and Huskens, J. (2009) Free-standing 3D supramolecular hybrid particle structures. *Angew. Chem. Int. Edit.*, **48** (5), 983–987.

**20** Ling, X.Y. and Huskens, J. (2009) Fabrication of 3D supramolecular hybrid particle microstructures with controllable morphology and dimensions. *Chem. Commun.*, (37), 5521–5523.

**21** Ling, X.Y., Phang, I.Y., Reinhoudt, D.N., Vancso, G.J., and Huskens, J. (2009) Free-standing porous supramolecular assemblies of nanoparticles made using a double-templating strategy. *Faraday Discuss.*, **143**, 117–127.

# 6
# LbL Assemblies Using van der Waals or Affinity Interactions and Their Applications

*Takeshi Serizawa, Mitsuru Akashi, Michiya Matsusaki, Hioharu Ajiro, and Toshiyuki Kida*

## 6.1
## Introduction

In 1991, Decher *et al.* found that stepwise immersion of substrates such as mica and glass into aqueous solutions of positively and negatively charged polymers produced multilayered ultrathin polymer films with a controllable nanometer thickness [1–6]. This method is called layer-by-layer (LbL) assembly, and both basic research and applications have been widely developed in the field of polymer science. Because this technique involves the simple immersion of a substrate into an oppositely charged polymer solution, researchers extended LbL assembly to include deposition not only of water-soluble linear charged-polymers but also of viruses [7], proteins [8–11], silica colloids [12, 13], metal nanoparticles [14–17], dyes [18–20], metal oxides [21–23], amphiphiles [24, 25], clays [26–28], and polystyrene nanospheres [29–31]. In the beginning, electrostatic interactions were essentially utilized for LbL assembly. A while later, other interactions such as hydrogen-bonding [32–34] and charge transfer [35, 36] were used to facilitate polymer association to give ultrathin film deposition. The most important concept for LbL assembly is how to use interactions between polymeric materials. It follows that other weak interactions between macromolecules should be crucial for further development of LbL research. Van der Waals interaction in physical chemistry is the sum of the attractive or repulsive forces between molecules, or between parts of the same molecule, other than those due to covalent bonds or to the electrostatic interactions of ions with one another or with neutral molecules. This weak interaction between molecules plays a fundamental role in the field of supramolecular chemistry, structural biology, and polymer science. Affinity interaction, such as interactions between proteins in a biosystem is also an attractive target for LbL research.

In this chapter, we introduce LbL assemblies using van der Waals and affinity interactions and their applications. Poly(methyl methacrylate) stereocomplex and poly(lactic acid) stereocomplex are typical polymer complexes using van der Waals interaction to facilitate LbL assembly of films on a substrate. First, we achieved stereoreospecific template polymerization of methacrylates by stereocomplex

*Multilayer Thin Films: Sequential Assembly of Nanocomposite Materials*, Second Edition.
Edited by Gero Decher and Joseph B. Schlenoff.
© 2012 Wiley-VCH Verlag GmbH & Co. KGaA. Published 2012 by Wiley-VCH Verlag GmbH & Co. KGaA.

formation in nanoporous LbL stereocomplex films. Secondly, we induced hollow nanocapsules composed of LbL stereocomplex films on silica particles. In the field of tissue engineering, three-dimensional cellular chips have been strongly desired. Finally, we developed a simple and unique bottom-up approach, "hierarchical cell manipulation using nanofilms", which used the LbL technique for fibronectin-gelatin films as a nano-ECM (extracellular matrix).

## 6.2
## Stereospecific Template Polymerization of Methacrylates by Stereocomplex Formation in Nanoporous LbL Films

### 6.2.1
### Introduction

Natural polymers, such as polynucleotides and proteins, are always produced with a highly regulated polymer structure through "template polymerization", which is based on the polymer–polymer interaction with multivalent and non-covalent interactions composed of hydrogen bonds and van der Waals interactions. The template polymerization system makes it possible to control the arrangement of the varied monomers along the polymer chain backbone, and the three-dimensional higher order structure as well as the first order structure. Since it is a very attractive polymerization method for controlled structure, it has been applied to synthetic polymers, such as methacrylate polymers [37–46], polyesters [47, 48], and poly(amino acid)s [49–51].

It is known that isotactic (it) and syndiotactic (st) poly(methyl methacrylate)s (PMMAs) (see Figure 6.1) form a stereocomplex in certain solvents, or in films, due to the structurally well-defined synthetic polymers with structural fitting between the polymer chains or between lateral functional groups with van der Waals contacts. The it-PMMA/st-PMAA stereocomplex forms a double [52–55] or triple [56] stranded helix in polar organic solvents, such as acetonitrile and dimethylformamide (DMF), on the basis of structural fitting with van der Waals interactions [37–41], in which complex it-PMMAs are surrounded by twice the length of st-PMMAs [52–55].

Researchers are highly interested in the double-stranded moiety of PMMA stereocomplexes as an adequate assembly system to perform *in situ* free radical template syntheses of stereoregular PMMAs in the presence of a single PMMA [57–66] Although the stoichiometry of the complex is different from that of nucleic acids, template polymerization is inspired by template syntheses of polynucleotides and proteins under mild physiological conditions. However, the structural transcription of templates to polymers synthesized by this polymerization method has thus far been insufficient. Polymerization has been demonstrated under restrictive conditions such as low conversion, low temperatures, and high molecular weights of templates [57].

Since biological systems synchronously utilize sophisticated nanospaces produced by appropriate enzymes as well as template effects, solvated polymers with random

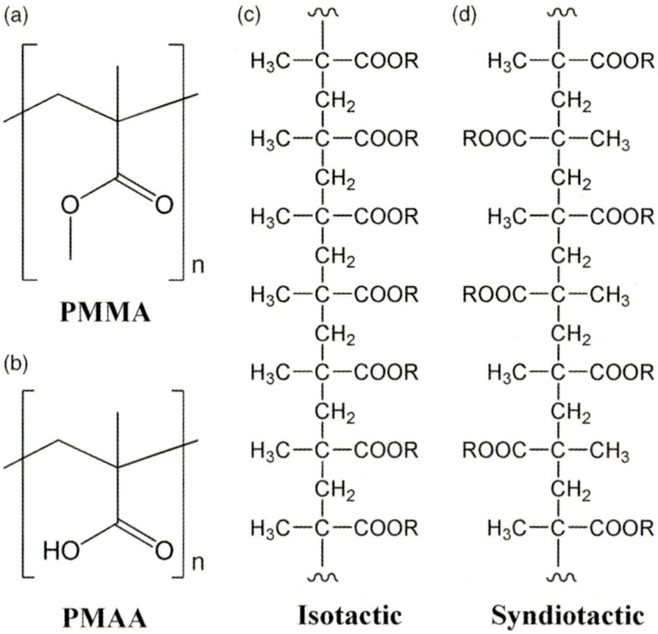

**Figure 6.1** Chemical structure of PMMA (a) and PMAA (b). Schematic illustration of isotactic structure (c) and syndiotactic structure (d).

and dynamic conformations are not suitable polymerization templates. In order to overcome this problem, it was proposed to use a free radical polymerization process within nanospaces fabricated by polymer templates such as porous ultrathin films, based on the double stranded or van der Waals contacted stereocomplex formed between it-PMMA and st-poly(methacrylic acid) (st-PMAA) [67, 68].

The aforementioned approach for precision template polymerization is comprised of the following three steps: (i) The most important step is the fabrication of ultrathin films composed of stereocomplexes formed between it-PMMA and st-PMAA. (ii) The selective extraction of a single component from the films, to fabricate porous films with template nanospaces. (iii) The free radical polymerization of monomers within the nanospaces and the characterization of the selectively extracted resulting polymers. To fabricate ultrathin stereocomplex films on surfaces, we applied stereocomplex formation to layer-by-layer (LbL) assembly, which is normally used for the fabrication of polyelectrolyte multilayers through electrostatic polyion complex formation. The concept of LbL assembly means that we can assemble certain polymers by stabilizing them on a substrate after adsorption due to polymeric interactions. Thus, this process should be potentially applicable to the formation of stereoregular PMMA stereocomplexes.

Although detailed studies of stereocomplex formation between structurally regular polymers have been performed [37–41], the stereocomplex characteristics obtained by the stepwise fabrication of ultrathin polymer assemblies on a substrate should be different from the bulk stereocomplex samples. The stepwise stereocomplex assembly on a substrate would require the structural rearrangement of a pre-adsorbed polymer, possibly from a random conformation during the subsequent adsorption process of the second stereoregular PMMA, in order to form regular nanostructures. In this case, the polymers would have huge steric requirements during formation. The *in situ* assembly of ultrathin stereocomplex films has more general implications regarding dynamic events involving polymeric interactions at ultrathin polymer film surfaces. In other words, the stereocomplex formation suggests that a dynamic conformational change in polymers adsorbed on a film's surface is available for the ultrathin film assembly process. It is interesting to investigate whether polymers at the interface can interact with other polymers exhibiting conformational changes using the LbL system.

In this section, a highly efficient stereoregular polymerization process is demonstrated, using free radical initiators within template nanospaces, which can be prepared by the selective extraction of a single component from structural regular ultrathin stereocomplex films composed of it-PMMA and st-PMAA.

## 6.2.2
### LbLit-PMMA/st-PMMA Stereocomplex Uultrathin Film

For the first fundamental combination, the assembly of it-PMMA and st-PMMA was selected [69]. Stereocomplex formation between it-PMMA and st-PMMA in organic solutions is known to be dependent on solvent species. As strong complexing solvents, acetonitrile, acetone, and DMF were selected [39]. Figure 6.2 shows the dependence of frequency shifts on assembly steps, when the quartz crystal microbalance (QCM) was alternately immersed in it-PMMA ($M_n$ 20 800, PDI 1.26, $mm:mr:rr = 97:2:1$) and st-PMMA ($M_n$ 22 700, PDI 1.26, $mm:mr:rr = 0:11:89$) solutions in three solvents for 15 min at ambient temperature. It is clear that acetonitrile was the best solvent for PMMA assembly in large amounts, as shown in Figure 6.2a. When the QCM was immersed in it-PMMA or st-PMMA solution alone for a much longer time, the frequency shift saturated to a level obtained during the first step of the assembly process. This observation suggests that the stereocomplex of stereoregular PMMAs was produced on the QCM. In the initial two steps, larger frequency shifts were observed, possibly due to the direct influence of the QCM gold substrate on the assembly amount. The frequency shift after a 20-step assembly was $-416$ Hz with an experimental error of $\pm 5\%$. The shift corresponded to an adsorbtion of 362 ng, and the film thickness was estimated to be 9.7 nm, assuming a density for solid PMMA of $1.188 \, \text{g cm}^{-3}$. The estimated thickness was consistent with the thickness that was obtained by scratching the assembly on the QCM substrate using an atomic force microscope (AFM) tip ($9.7 \pm 0.3$ nm). When the assembly was demonstrated without drying, the total frequency shift was not affected, indicating that the drying process is not essential for the stepwise assembly. In the case of the acetonitrile solvent, the mean ratio of st-PMMA and it-PMMA was determined as $2.0 \pm 0.4$. The present data suggest that solvent selection is a key factor in the stepwise assembly of

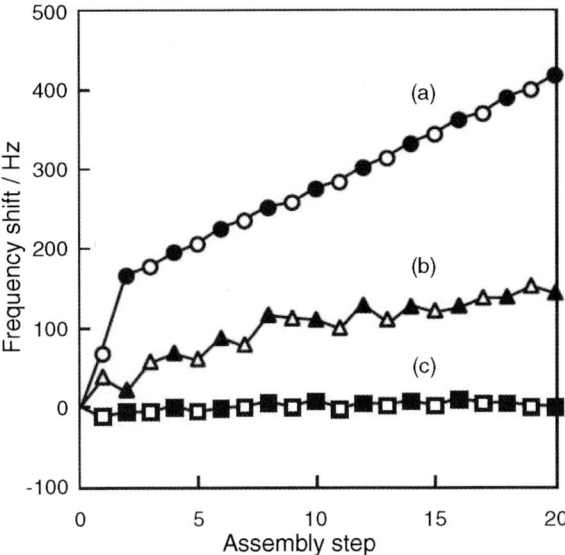

**Figure 6.2** Frequency shift of QCM by the stepwise assembly from organic solutions of it-PMMA and st-PMMA at a concentration of 1.7 mg mL$^{-1}$ in acetonitrile (a), acetone (b), DMF (c). Open and closed symbols show it-PMMA and st-PMMA steps, respectively. Reprinted with permission from [69], T. Serizawa et al., J. Am. Chem. Soc. **2000**, 122, 1891. © 2000, American Chemical Society.

stereoregular PMMAs. A system using acetonitrile solutions was used in the following characterization of the stereocomplex films.

Static contact angles on the air side of film surfaces of it- and st-PMMA are significantly different from each other due to the selective accumulation of functional groups at the surface. The mean contact angle at odd it-PMMA assembly steps was 63.6 ± 0.3°. This value was comparable to the adsorbed bulk it-PMMA film on the bare QCM substrate [70], while that of a bare gold surface of a QCM is 44 ± 1° in air. On the other hand, the mean value at even st-PMMA assembly steps was 71.2 ± 0.4°, which was slightly smaller than a previously reported value of the st-PMMA film surface [70], and smaller than that (73.2 ± 0.8°) of a physically adsorbed st-PMMA film on a QCM, within experimental error. This observation implies that the more hydrophilic ester group in st-PMMA should point outwards in the molecular structure of the stereocomplex [55]. Therefore, it is concluded that the surface composition of the ultrathin PMMA film was altered by the stepwise assembly of stereoregular PMMAs, possibly by the physical adsorption of it-PMMA without stereocomplex formation at it-PMMA steps and by the stereocomplex formation at st-PMMA steps on the surfaces (Figure 6.3).

The stereocomplex formation was evidenced with further analyses. The reflection adsorption spectrum (RAS) was measured for the 20-step assembled PMMA films. The main peak in CH$_2$-rocking adsorption for the assembly was at approximately 860 cm$^{-1}$, together with a shoulder at approximately 840 cm$^{-1}$. The peak positions

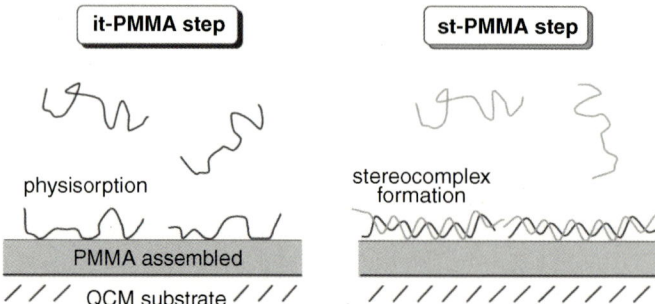

**Figure 6.3** Schematic representation of the PMMA stereocomplex assembly on surfaces. Reprinted with permission from [69], T. Serizawa et al., J. Am. Chem. Soc. **2000**, *122*, 1891. © 2000, American Chemical Society.

were significantly different from those of cast films of it-PMMA and st-PMMA, thereby suggesting stereocomplex formation between it-PMMA and st-PMMA on the substrate. When the AFM was employed for the analysis of assembled PMMA films, the image showed a domain-like structure that was different from those on a bare QCM electrode. The mean diameter and height of the domain were $60 \pm 10$ and $11 \pm 2$ nm, respectively, and the $R_a$ was 4.3 nm, which was more than three times greater than that of the bare QCM, but was nonetheless relatively smooth. Furthermore, the same analytical results of stereocomplex formation were obtained with the other syndiotactic methacrylate polymers, such as st-poly(ethyl methacrylate) (st-PEMA) [71], st-poly(propyl methacrylate) (st-PPMA) [71], and st-poly(methacrylic acid) (st-PMAA) [72], with it-PMMA, because the methyl ester groups of st-PMMA are oriented toward the outside of the stereocomplexes. Especially, st-PMAA was important due to the differential solubility of it-PMMA, which can be utilized as a host film against stereoregular polymers of methacrylates, and nanospaces for stereoregular polymerization.

### 6.2.3
### Fabrication of Template Nanospaces in Films

After the stereocomplex assembly of it-PMMA ($M_n$ 20 800, PDI 1.26, *mm:mr: rr* = 97 : 1 : v2) and st-PMAA ($M_n$ 37 900, PDI 1.26, *mm:mr:rr* = 1 : 2 : 97) on the QCM surface, the immersion of the assembly in a 10 mM sodium hydroxide aqueous solution for 5 min resulted in a drastic frequency increase, thus indicating desorption of some of the polymers from the assembly (Figure 6.4) [73]. There was no additional frequency change after immersion for several hours, thereby suggesting the selective extraction of st-PMAA, which can be readily dissolved in an aqueous alkaline solution. The extracted amount was consistent with that of st-PMAA assembled on the substrate, possibly indicating a 100% extraction of st-PMAA. St-PMAA was successfully extracted from the stereocomplex film assembled with it-PMMA and st-PMAA. Subsequently, st-PMAA was incorporated into the resulting it-PMAA film from the solution (0.017 M) with a complexing efficiency of approximately 80%.

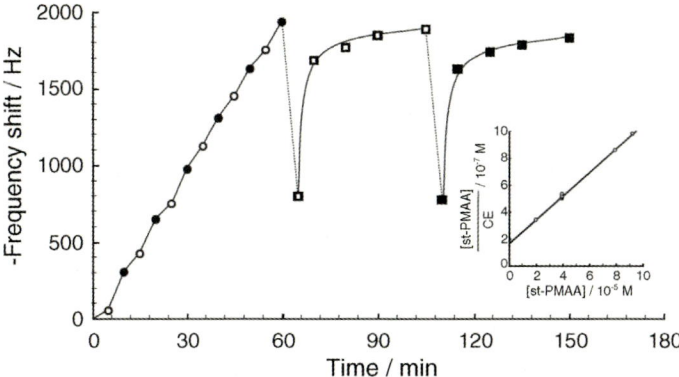

**Figure 6.4** QCM analysis of the LbL assembly, the selective extraction, and the subsequent st-PMAA incorporation. The it-PMMA(○) in acetonitrile and st-PMAA (●) in an acetonitrile/water (4/6, v/v) were alternately assembled on a QCM substrate for a 12-step assembly at a concentration of 0.017 M at 25 °C, following our previous study [65]. A dotted line shows the selective extraction process of st-PMAA in a 10 mM NaOH aqueous solution for 5 min, in order to obtain the designed host it-PMMA film. St-PMAA was subsequently incorporated in the it-PMMA film from the acetonitrile/water (4/6, v/v) solution at a concentration of 0.017 M at 25 °C (□). The extraction and incorporation was repeated again (■). The inset shows the apparent Langmuir plot, in order to obtain the maximum complexing efficiency and the apparent incorporation constant. Reprinted with permission from [73], T. Serizawa et al., *Angew. Chem. Int. Ed.*. **2003**, *42*, 1118. © 2003, Wiley-VCH.

The incorporation saturated over time. The extraction from the stereocomplex assembly was repeated at least three times, and the subsequent incorporation of st-PMAA was also observed with the same complexing efficiency at the same concentration. Atactic PMAA was not incorporated into the porous it-PMMA film, indicating that the above incorporation occurred by complementary stereocomplex formation.

RAS in the carbonyl vibration band regions was used for the detection of complex formation and to obtain conformational information (Figure 6.5). There were two peaks for the assembled film at 1737 and 1725 cm$^{-1}$, which were assigned to it-PMMA and st-PMAA, respectively (Figure 6.5a) Only a single peak was observed at 1739 cm$^{-1}$ after extraction using an alkaline solution (Figure 6.5b), meaning that a single polymer component was completely extracted from the assembled film. The two peaks were also observed after immersion in an st-PMAA solution (Figure 6.5c), thereby resulting in recovery of the complex assembly. It is significant that the peak intensity corresponding to st-PMAA was smaller than that for the assembled film, which is reasonable because the complexing efficiency was smaller and around 80% of the incorporating st-PMAA concentration, as shown by the QCM analysis. Regarding the AFM analysis, it is surprising that the thickness was 44.1 ± 3.8 nm, and this was not changed by the extraction, strongly suggesting that an it-PMMA film with a macromolecular porous structure was obtained. Note that st-PMMA was

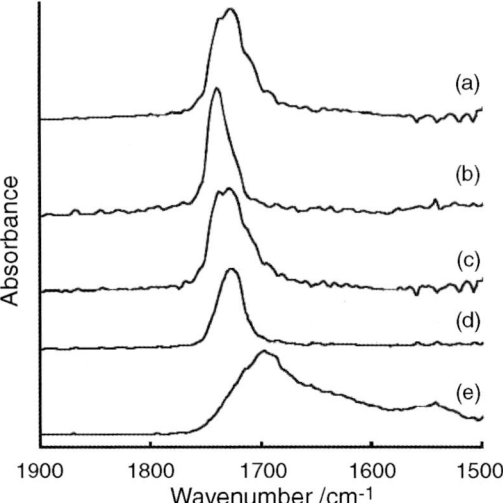

**Figure 6.5** RAS of the LbL assembly of it-PMMA and st-PMAA for a 12-step assembly (a), the extracted film (b), the st-PMAA incorporated film (c), a cast film of it-PMMA (d), and a cast film of st-PMAA (e). Films were similarly prepared on a gold-sputtered poly(ethylene terephthalate) film, as shown above. Reprinted with permission from [73], T. Serizawa et al., Angew. Chem. Int. Ed.. **2003**, 42, 1118. © 2003, Wiley-VCH.

incorporated into the host with moderate efficiency, and that the complexing efficiency was small for st-PEMA and st-PPMA.

### 6.2.4
### Polymerization within Template Nanospaces Using a Double-Stranded Assembly

LbL assembly of these polymers led successfully to the fabrication of ultrathin double-stranded stereocomplex films with expected stoichiometries, as described above. st-PMAA was selectively extracted from assembled films in aqueous solutions of sodium hydroxide, followed by the fabrication of porous it-OMMA films. We applied the porous it-PMMA films for template syntheses of st-PMAAs, as shown schematically in Figure 6.6. Polymerizations from MAAs (1.7 mg mL$^{-1}$) in porous it-PMMA ($M_n$ 20 800, PDI 1.2, $mm:mr:rr = 97:2:1$) films (mean thickness, 44 nm) at 40 °C in the presence of the free radical initiator, 2,2′-azobis-($N,N$-dimethyleneisobutyramidine)dihydrochloride (VA-044) (5.0 mg mL$^{-1}$) ([monomer]/[initiator] = 20/1, mol/mol) in 10 mL acetonitrile/water (4/6, v/v) were preliminarily analyzed by immersing the QCM substrate coated with porous films. After 2 h polymerization periods in both solution and film, the PMAAs synthesized were incorporated into porous films at an 80% fill rate (where, 100% indicates that st-PMAA extracted is equal to incorporated PMAAs). Attenuated total reflection (ATR) spectra in carbonyl vibration bands demonstrated the emergence of a newly observed peak at 1725 cm$^{-1}$ corresponding to PMAA, in addition to the original peak of template it-PMMA at 1737 cm$^{-1}$ [67, 68]. In order to characterize the

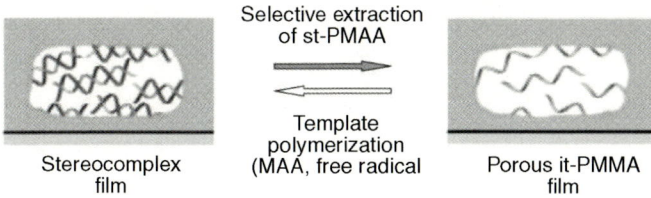

**Figure 6.6** Schematic representation of template polymerization of st-PMAA using ultrathin porous it-PMMA films based on double-stranded stereocomplexes. Reprinted with permission from [68], T. Serizawa et al., Macromolecules **2005**, 38, 6759. © 2005, American Chemical Society.

synthesized PMMA, porous it-PMMA films were prepared on silica colloids with greater surface areas (10 g, mean diameter: 1.6 μm). 1.5 g MAA was similarly polymerized in 500 mL solvent in the presence of coated silica. The PMAAs were extracted in an aqueous 10 mM sodium hydroxide solution. Approximately 0.2 g of PMAA was recovered, and yields were analyzed to be 13%. PMAAs were methylated using diazomethane to obtain PMMAs, and characterized by $^1$H NMR. Figure 6.7 shows an NMR trace for the PMMA prepared using template it-PMMAs. Chemical shifts for α-methyl groups at 1–1.5 ppm, which can be used as signals for analyzing tacticity, were almost single peaks at 1.2 ppm, thereby indicating that >96% syndiotacitc PMAA had been prepared.

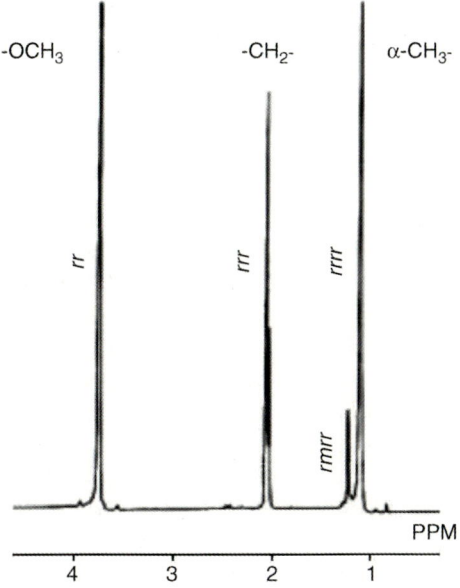

**Figure 6.7** Typical $^1$H NMR spectrum of st-PMAA (mm:mr:rr. 0:2:98) methylated from st-PMAAs polymerized in porous it-PMMA template films. Reprinted with permission from [68], T. Serizawa et al., Macromolecules **2005**, 38, 6759. © 2005, American Chemical Society.

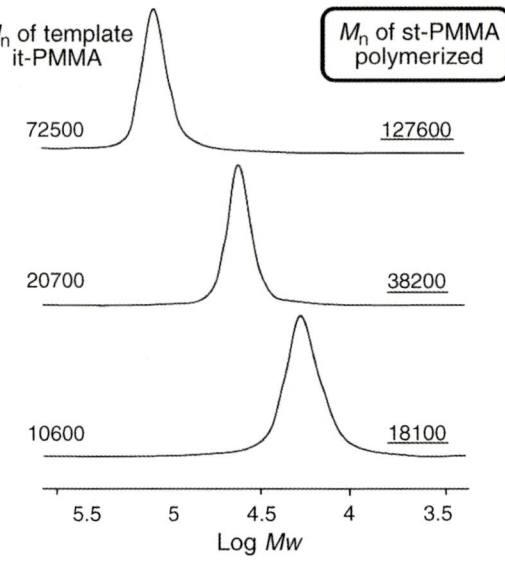

**Figure 6.8** SEC charts of st-PMMAs polymerized in porous template films from it-PMMAs with various molecular weights. Reprinted with permission from [68], T. Serizawa et al., Macromolecules **2005**, *38*, 6759. © 2005, American Chemical Society.

The molecular weights of synthesized st-PMMAs were measured by size exclusion chromatography (SEC), shown in Figure 6.8. The $M_n$s of the st-PMMAs were approximately twice those of the template it-PMMAs in all cases. It is surprising that the mass ratio was potentially consistent with the stoichiometry of double-stranded 1:2 (it-PMMA:st-PMAA) stereocomplexes. The molecular weights of PMMAs simultaneously prepared in solution ($M_n$ 2300, PDI 4.0) were obviously independent of template it-PMMAs. A pot polymerization in the presence of mixed silica coated with template it-PMMAs with different $M_n$ (PDI) (11 000 (1.5), 22 800 (1.2), and 68 900 (1.5)) resulted in the synthesis of st-PMMA mixtures with $M_n$ of 19 000, 37 900, and 121 500, respectively, as shown in Figure 6.9 [68]. The observations indicate that pot polymerizations prepare mixtures of stereoregular methacrylate polymers with various molecular weights. The aforementioned 1:2 stereocomplex film composed of it-PMMA and st-PMAA gave macromolecularly porous it-PMMA film, by the selective extraction of st-PMAA from the stereocomplex, however, it-PMMA was not extracted from the film. Next, the formation of a stereocomplex between it-PMMA and st-PMAA with a 1:1 length stoichiometry was examined (Figure 6.10). To obtain an assembly with a 1:1 stoichiometry, DMF and DMF/water (2/3, v/v) were selected as solvents for it-PMMA and st-PMAA, respectively. LbL assembly in both solutions produced a film with the objective stoichiometry of $1.0 \pm 0.2$ (st-PMAA/it-PMMA length ratio). The immersion of the assembled film in both chloroform and 10 mM sodium hydroxide aqueous solution resulted in the desorption of approximately

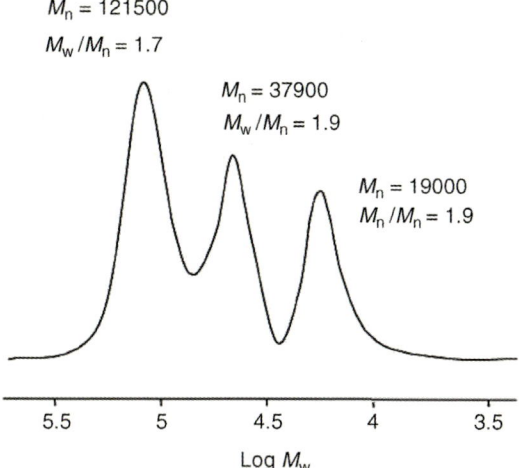

**Figure 6.9** SEC chart of a pot polymerized st-PMMA in porous template films from it-PMMAs with $M_n$ of 11 000, 22 800, and 68 900. Reprinted with permission from [68], T. Serizawa et al., *Macromolecules* **2005**, *38*, 6759. © 2005, American Chemical Society.

half of the polymers. The amount desorbed was consistent with the amount of each polymer totally assembled, thereby suggesting that it-PMMA and st-PMAA were selectively extracted from the film [67].

For template polymerization the QCM substrate coated with the porous st-PMAA film was immersed in a 10 mL DMF solution of MMA (1.7 mg mL$^{-1}$) for 3 h at 70 °C in the presence of a free radical initiator, 2,2′-azobisisobutyronitrile (5 mg mL$^{-1}$).

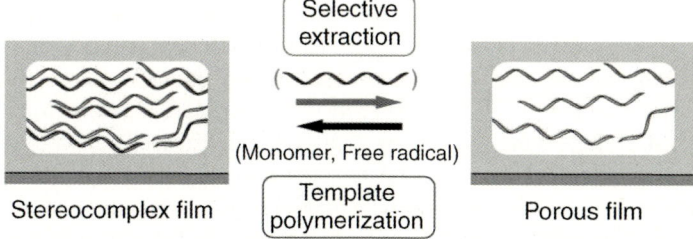

**Figure 6.10** Schematic representation of template polymerization using ultrathin porous films. LbL assembly prepares the ultrathin film of a stereocomplex comprised of it-PMMA and st-PMAA. Two polymer chains are drawn by rigid helical rods, and should be partially distorted and entangled in the film. There must be a disordered region of the stereocomplex. A single component is selectively extracted from the film, resulting in the preparation of the porous film with regular nanospaces. Then, the porous film is used as a reaction mold for free radical template polymerization of MMA or MAA, followed by the regeneration of the stereocomplex film. Reprinted with permission from [67], T. Serizawa et al., *Nature* **2004**, *429*, 52. © 2004, Nature Publishing Group.

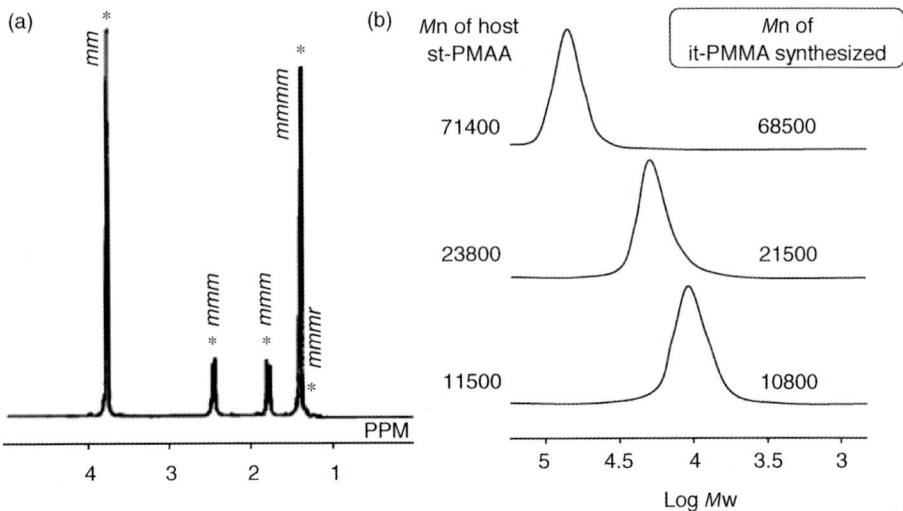

**Figure 6.11** Characterization of it-PMMAs polymerized in porous st-PMAA films for 3 h at 70 °C. (a) A typical 1H NMR chart of it-PMMA polymerized in a porous st-PMAA (Mn 23 800) film. (b) SEC curves of it-PMMAs polymerized in porous films of st-PMAAs with various molecular weights. Reprinted with permission from [67], T. Serizawa et al., Nature **2004**, 429, 52. © 2004, Nature Publishing Group.

After rinsing with DMF, approximately 90% of the it-PMMA previously extracted was recovered. The ATR spectrum of the obtained film showed the same band peaks as those observed for the stereocomplex film, possibly suggesting that it-PMMA was prepared, and that it formed a stereocomplex with st-PMAA in the film. In order to analyze the stereoregularity and the molecular weight of the PMMA, the porous film was prepared on silica particles (10 g, mean diameter: 1.6 μm). Then, MMA was polymerized in DMF, and the PMMA was extracted with chloroform. Approximately 0.4 g PMMA was obtained from feeding 1.5 g MMA/500 mL DMF, which yielded approximately 25% under these polymerization conditions. A $^1$H NMR chart of the PMMA clearly demonstrated the isotactic specific polymerization of MMA in the porous st-PMAA film (Figure 6.11a), indicating that an isotactic specific template polymerization was achieved on the basis of stereocomplex formation with the st-PMAA. SEC curves potentially demonstrate the control of molecular weights within a narrow distribution (Figure 6.11b). Isotacticities were greater than 92% in all cases. The $M_n$ of the host st-PMAA controlled the $M_n$ of the it-PMMA, and was almost the same. As a consequence, the control of the stereoregularity and of the molecular weight of it-PMMA was realized by the template polymerization method using porous template films.

The porous film of it-PMMA was similarly prepared from the 1 : 1 complex film to polymerize st-PMAA. Silica particles coated with the porous it-PMMA film were used for free radical polymerization of MMA in DMF/water (2/3, v/v) in the presence of a water-soluble free radical initiator, VA-044 for 2 h at 40 °C. St-PMAA was successfully prepared by using the template it-PMMA film, and the molecular weight of the host it-PMMA similarly regulated that of the st-PMAA. These observations indicate an alternative utilization of porous films for template polymerization. Note that conventionally prepared films did not realize the above template polymerization. Template polymerization of the combination of it-PMMA and st-PMAA in the solution only demonstrated the acceleration of polymerization rates, since the association is relatively weaker than that of stereoregular PMMAs. Accordingly, the potential of the present porous films was confirmed [74].

### 6.2.5
### Studies on the Porous Structure Obtained by LbL Assembly

The nanoporous films were prepared for the template polymerization field. The stability [75–79] of the films and the polymerization mechanism [80–82] were clarified, using the porous it-PMMA films. Figure 6.12 shows the results of X-ray diffraction (XRD) of LbL films [75]. A 100-step alternative immersion process into an it-PMMA acetonitrile solution and a st-PMAA acetonitrile/water (4/6, v/v)

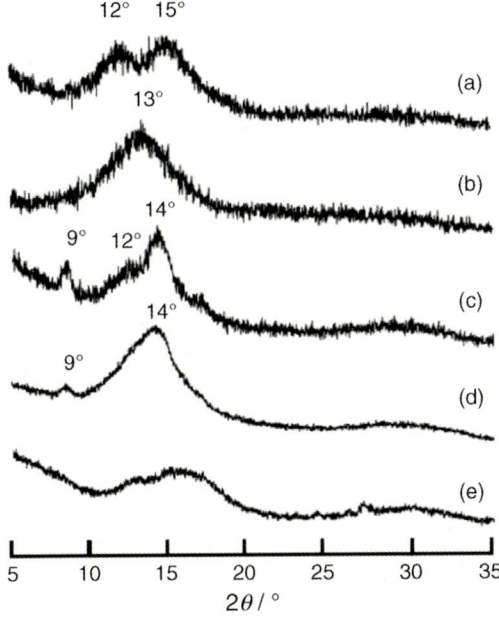

**Figure 6.12** XRD patterns of (a) it-PMMA/st-PMAA stereocomplex films, (b) porous it-PMMA films, (c) st-PMAA incorporated films (after 600 min of immersion), (d) it-PMMA powder, and (e) st-PMAA powder. Reprinted with permission from [75], D. Kamei et al., Chem. Lett. **2008**, 37, 332. © 2008, The Chemical Society of Japan.

mixed solution of 0.017 M was performed to generate stereocomplexes of it-PMMA/st-PMAA on a glass substrate at 25 °C. Next, 10 mM NaOH aq. was used to extract the st-PMAA from the stereocomplex by simple immersion for 30 min, which resulted in porous it-PMMA thin films. Subsequently, st-PMAA was incorporated into the porous films by immersing the films in the same st-PMAA solution at 25 °C. The XRD patterns were taken by Rigaku RINT2000. Ni-filtered CuK$\alpha$ ($\lambda = 0.154$ nm) was used as the X-ray source and operated with a Rigaku ultraX18 (40 kV, 200 mA). Films were examined in the scanning angle range from 5° to 35° at a scan rate of 0.5° min$^{-1}$.

The observed reflections of the multilayered polymer films are ascribed to the packing of the polymer chains. The XRD patterns from the LbL assembly of it-PMMA and st-PMAA for a 100-step assembly (Figure 6.12a) showed two characteristic peaks of it-PMMA/st-PMAA stereocomplex ($2\theta = 12°$ and $15°$, $d = 0.74$ and $0.59$ nm, respectively) [45], none of which are present in it-PMMA or st-PMAA. This result clearly indicates that it-PMMA/st-PMAA stereocomplex films were formed on the glass surface. st-PMAA was then selectively extracted from the stereocomplex thin films to form porous it-PMMA thin films, supported by IR spectra [73]. Surprisingly, the peak from the porous films (Figure 6.12b) shifted from that of the stereocomplex films. The broad reflection at $2\theta = 13°$ ($d = 0.6$ nm) suggested that the crystallinity in the films had decreased. The distance between the polymer chains in the films also expanded as compared to it-PMMA powder (Figure 6.12d) [83], suggesting that a porous structure was obtained, and that stereoregular (st-PMAA) nanospaces could be fabricated in the thin films. In a previous report using AFM [73], the surface roughness of the porous films was found to increase after the extraction of st-PMAA, supporting the molecular-level extraction in this result. Next, st-PMAA was incorporated into porous it-PMMA films. In Figure 6.12c, the XRD pattern of st-PMAA incorporated films (after 600 min immersion) showed a shoulder peak characteristic of the stereocomplex ($2\theta = 12°$) with two peaks of crystalline it-PMMA ($2\theta = 9°$ and $14°$, $d = 0.96$ and $0.62$ nm, respectively) [83]. It is known that st-PMAA was incorporated into about 80% of the it-PMMA in the porous films analyzed by a quartz crystal microbalance (12-step assembly) [73]. The present result indicates that the stereocomplexes were partially formed, and the residual semicrystalline it-PMMA crystallized. Thus, the molecular motion to assemble, caused by polymer–polymer interactions, could be a driving force to crystallize, even at the solid/liquid interface, regardless of whether it is it-PMMA itself or the it-PMMA/st-PMAA stereocomplex.

For example, the solvent effect is summarized in Figure 6.13. The stereocomplex formation would be generated without any structural change of it-PMMA in water (Figure 6.13A). In contrast, after immersion in acetonitrile/water (4/6, v/v) for 10 and 40 h, the incorporation was reduced to around 50% and 35%, respectively [76]. The decreased values suggest that crystallized it-PMMA in the thin films lost its stereocomplex formation capability as it may have shortened the packing distance (Figure 6.13b). The aforementioned assemblies were also observed with AFM [76] and SEM [77], and these kind of changes were clearly influenced by the LbL formation conditions [78, 79].

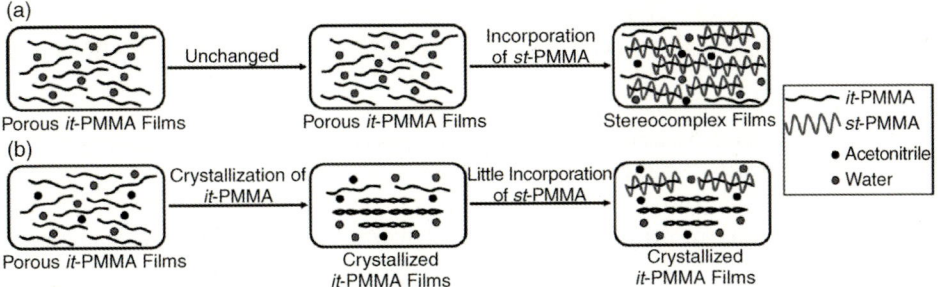

**Figure 6.13** Schematic illustration of the molecular motion of it-PMMA in multilayered thin films. Reprinted with permission from [76], D. Kamei et al., Langmuir **2009**, 25, 280. © 2009, American Chemical Society.

## 6.3
## Preparation and Properties of Hollow Capsules Composed of Layer-by-Layer Polymer Films Constructed through van der Waals Interactions

### 6.3.1
### Introduction

Hollow capsules of nano- to micrometer size have attracted much attention in the biomedical and pharmaceutical fields [84], since they can be utilized as drug carriers and containers in which sensitive biomolecules can be preserved. The construction of hollow capsules of uniform size and controlled film thickness has been efficiently carried out by the deposition of polymer films onto the surface of colloidal core templates via the LbL technique and the subsequent removal of the template cores [85, 86]. In most cases, this method has been applied to the preparation of hollow capsules composed of polyelectrolyte multilayers based on electrostatic interactions [87, 88], except for several reports where hollow capsules were prepared using the hydrogen bonding interactions between the uncharged polymers [89–92], such as poly (methacrylic acid) and poly(N-vinylpyrrolidone) acting as the hydrogen bonding donor and acceptor, respectively. We also prepared biodegradable multilayered hollow nanocapsules via the LbL assembly of the biodegradable polyelectrolytes, chitosan and dextran sulfate, on silica nanoparticles as a template, with the subsequent removal of the template [93] to study the usefulness of these multilayered hollow capsules as drug delivery and controlled release carriers [94, 95]. On the other hand, there has been no report on hollow capsules composed of nonionic multilayers constructed based on their van der Waals interactions, possibly because these hollow capsules have been believed to be easily ruptured after the removal of the template core, due to the much weaker van der Waals interactions. It can be expected that hollow capsules composed of nonionic multilayers possess a unique shell permeability and morphology, unlike conventional hollow capsules composed of polyelectrolyte multilayers. In this section, we describe the preparation and properties of hollow nanocapsules composed of LbL stereocomplex films, such as poly(methyl methacrylate) stereocomplex films and poly(lactic acid)s stereocomplex films.

### 6.3.2
### Preparation of Hollow Nanocapsules Composed of Poly(methyl methacrylate) Stereocomplex Films [96]

Poly(methyl methacrylate) (PMMA) has found many applications in the biomedical fields, due to its excellent biocompatibility. The stereoregular PMMA stereocomplex, which is a double-stranded helical assembly formed with van der Waals interactions between isotactic (*it*) and syndiotactic (*st*) PMMAs [37, 55], has been utilized as a component of hollow fiber membranes for artificial dialysis [97]. Recently, our research group prepared stable ultrathin films composed of double-stranded PMMA stereocomplex by the alternate LbL assembly of it- and st-PMMAs on a solid substrate [69]. The application of these it-/st-PMMA stereocomplex films to the hollow capsule shell will allow us to construct novel hollow nanocapsules composed of nonionic multilayers, based on their van der Waals interactions. Here, we describe the preparation of hollow nanocapsules composed of PMMA stereocomplex multilayer shells by a combination of the alternate LbL assembly of it- and st-PMMAs and the silica template method.

The fabrication of it-/st-PMMA stereocomplex hollow capsules was carried out according to the process shown in Figure 6.14. We chose silica nanoparticles, which can be easily removed from those coated with polymer films by treatment with aqueous HF [89, 93, 98], as a template core. Acetonitrile/water (9 : 1) was chosen as an immersion solvent, since this solvent allows deposition of a relatively large amount of the LbL assembly composed of PMMA stereocomplex ont the silica particles, according to the previous result on the alternate LbL assembly of it- and st-PMMAs on an Au plate [69]. Silica nanoparticles with a diameter of 330 nm were alternately immersed in a 25 mL acetonitrile/water (9 : 1) solution of it-PMMA ($M_n = 20\,400$, $M_w/M_n = 1.21$, $mm{:}mr{:}rr = 99 : 1 : 0$) and st-PMMA ($M_n = 73\,200$, $M_w/M_n = 1.20$, $mm{:}mr{:}rr = 1 : 9 : 90$). The immersion process was continued for 10 cycles to afford 10 layer pairs of it- and st-PMMAs. The resulting particles were then treated with 2.3% aqueous HF to remove the silica cores. Inductively coupled plasma (ICP) emission

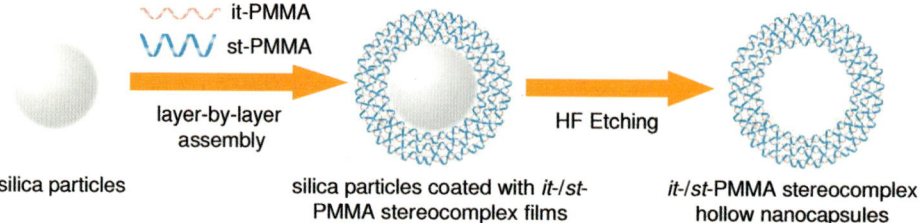

**Figure 6.14** Schematic illustration of the fabrication process of PMMA stereocomplex hollow nanocapsules. Reprinted with permission from [96], T. Kida *et al.*, *Angew. Chem. Int. Ed.* **2006**, *45*, 7534. © 2006, Wiley-VCH.

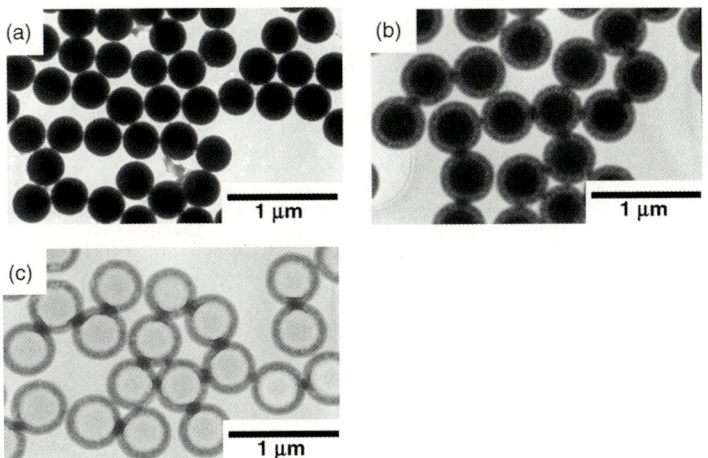

**Figure 6.15** TEM images of (a) silica particles, (b) silica particles coated with PMMA stereocomplex films, and (c) PMMA stereocomplex hollow nanocapsules. Reprinted with permission from [96], T. Kida et al., Angew. Chem. Int. Ed. **2006**, 45, 7534. © 2006, Wiley-VCH.

analysis of the obtained particles indicated that more than 99.8% of the Si atoms were removed from the PMMA film-coated silica particles by HF etching.

Figure 6.15b and c show the transmission electron microscopic (TEM) images of PMMA film-coated silica particles and PMMA hollow capsules, respectively. These images clearly indicate that the PMMA films were constructed on the surface of the silica particles, and PMMA hollow capsules were successfully fabricated by the removal of the silica core from the core–shell particles without any damage to the shell. The PMMA hollow capsules are spherical with an average diameter of 510 nm, and their shell thickness is approximately 90 nm. On the other hand, the self-assembled multilayer films of it- or st-PMMA alone were not formed on the silica particles, in accordance with the previous result on the LbL assembly of it- or st-PMMA alone on an Au plate [96]. This finding shows that the stereocomplex formation between it- and st-PMMAs is a crucial driving force for the construction of the PMMA multilayered film on a silica particle.

XRD analysis and IR spectroscopy are effective methods for evaluating the stereocomplex formation between stereoregular PMMAs. The XRD patterns of the particles before and after the HF etching of PMMA film-coated silica particles showed two peaks characteristic of the PMMA stereocomplex ($2\theta = 11°$ and $15°$) [99], which were significantly different from those of it- and st-PMMAs. This result clearly indicates that PMMA stereocomplex films were formed on the silica surface by the LbL assembly of it- and st-PMMAs. This stereocomplex structure was maintained even after the removal of the silica template. In the FT-IR/ATR spectra of the PMMA hollow capsules obtained after the removal of the silica core, absorption peaks corresponding to the C=O stretching vibrations and the main-chain $CH_2$ rocking vibrations of the PMMA stereocomplex were observed at around 1750 and 860 $cm^{-1}$,

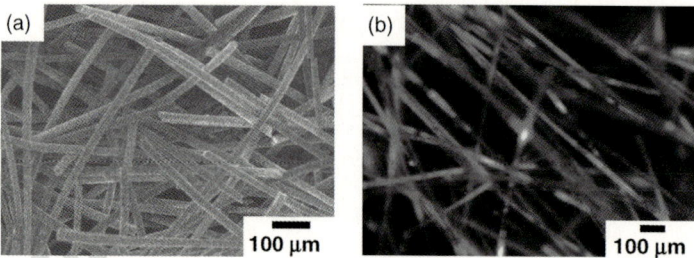

**Figure 6.16** (a) SEM images of it-/st-PMMA stereocomplex hollow tubes. (b) Fluorescence microscopic images of Rhodamine 6G-encapsulated PMMA hollow tubes. Fluorescence microscopic image was taken with the fluorescence filter for Rhodamine 6G. Wiley-VCH.

respectively [37, 55, 69, 100–103], strongly supporting the formation of the PMMA stereocomplex hollow capsules.

The combination of the LbL assembly of it- and st-PMMAs and the silica template method can offer a significant advantage over conventional fabrication methods for PMMA hollow materials, such as water-in-oil-in-water (W/O/W) emulsion polymerization [104–106]. The former method makes it possible to fabricate hollow materials with a variety of shapes in addition to the spherical shape, depending on the shape of the silica template employed. Figure 6.16a shows the SEM images of the PMMA hollow fibers that were constructed by the LbL assembly of it- and st-PMMAs on silica fibers (diameter about 10 μm, length about 10 mm) and the subsequent removal of silica fiber. Fluorescent microscopic images of the Rhodamine 6G-encapsulated hollow fibers are also shown in Figure 6.16b. These results demonstrate that PMMA hollow materials with a variety of shapes can be prepared simply by changing the shape of the silica template.

We believe that these PMMA stereocomplex hollow capsules are potentially applicable as drug carriers and containers. The replacement of the st-PMMA by st-poly(methacrylic acid) (st-PMAA) in the stereocomplex shell will allow us to construct pH-responsive hollow capsules, since the st-PMAA is easily and selectively extracted from the it-PMMA/st-PMAA stereocomplex shell by an alkaline aqueous solution [73].

### 6.3.3
### Preparation and Fusion Properties of Novel Hollow Nanocapsules Composed of Poly(lactic acid)s Stereocomplex Films [107]

Poly(lactic acid)s (PLAs) have been widely used as biomaterials due to their excellent biocompatibility, biodegradability, and mechanical strength [108, 109]. The mixture of poly(L-lactic acid) (PLLA) and poly(D-lactic acid) (PDLA) in polar organic solvents forms triclinic racemic crystals (β-form), called a stereocomplex, in which the left- and right-handed $3_1$ helices pack side by side via van der Waals interactions [110, 111]. The stereocomplex formation has been utilized as the driving force for the preparation of LbL polymer thin films [69]. In this section, we mention the preparation of

## 6.3 Preparation and Properties of Hollow Capsules Composed of Layer-by-Layer Polymer Films

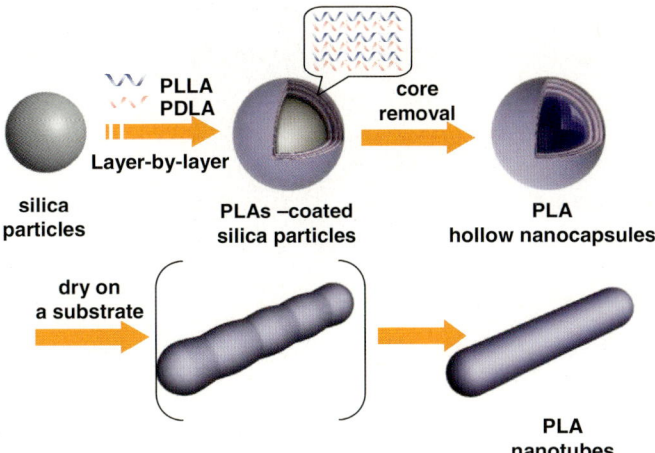

**Figure 6.17** Schematic illustration of formation of poly(lactic acid)s hollow nanocapsules and nanotubes. Reprinted with permission from [107], K. Kondo et al., J. Am. Chem. Soc. **2010**, *132*, 8236. © 2010, American Chemical Society.

hollow capsules composed of PLA stereocomplex multilayer shells by a combination of the alternate LbL assembly of PLLA and PDLA and the silica template method. In addition, the unique one-dimensional fusion properties of the resulting PLA hollow capsules to give nanotubes are described (Figure 6.17).

The preparation of hollow nanocapsules composed of PLLA/PDLA stereocomplex multilayer shells was carried out in a similar manner to the preparation of PMMA hollow nanocapsules [96]. Here, PLLAs with two different molecular weights ($M_w = 30\,000$, $M_w/M_n = 2.4$ and $M_w = 5500$, $M_w/M_n = 2.0$) and PDLAs ($M_w = 26\,000$, $M_w/M_n = 2.0$ and $M_w = 5800$, $M_w/M_n = 2.3$) were employed. Silica nanoparticles with a diameter of 300 nm were alternately immersed in acetonitrile solutions (5 mL) of PLLA and PDLA at 50 °C. The immersion process was continued for 10 cycles to afford 10 layer pairs of PLLA and PDLA. The resulting particles were then treated with 2.3% aqueous HF to remove the silica core. Figure 6.18 shows TEM images of

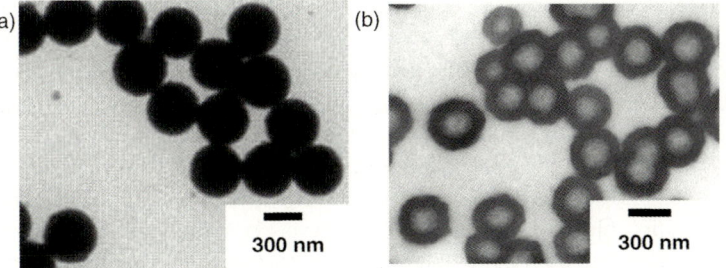

**Figure 6.18** TEM images of (a) silica particles coated with $(PLLA/PDLA)_{10}$ and (b) $(PLLA/PDLA)_{10}$ hollow capsules obtained after HF etching. Reprinted with permission from [107], K. Kondo et al., J. Am. Chem. Soc. **2010**, *132*, 8236. © 2010, American Chemical Society.

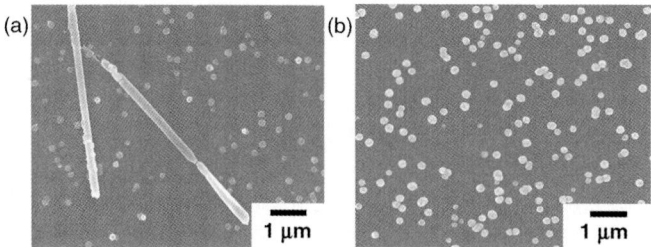

**Figure 6.19** SEM images of nanostructures obtained after evaporation of water from the water dispersion of PLA hollow nanocapsules on a PET substrate: (a) $M_{w\ PLLA} = 5500$, $M_{w\ PDLA} = 5800$; (b) $M_{w\ PLLA} = 30\ 000$, $M_{w\ PDLA} = 26\ 000$. Reprinted with permission from [107], K. Kondo et al., J. Am. Chem. Soc. **2010**, *132*, 8236. © 2010, American Chemical Society.

silica particles coated with a (PLLA/PDLA)$_{10}$ film, where the combination of the higher molecular weights of PLLA ($M_w = 30\ 000$) and PDLA ($M_w = 26\ 000$) was used, and the (PLLA/PDLA)$_{10}$ hollow nanocapsules obtained after the HF treatment. These images clearly indicate that (PLLA/PDLA)$_{10}$ hollow capsules were successfully fabricated by the removal of the silica core from the core–shell particles. The TEM energy dispersive X-ray spectroscopy (TEM-EDX) and FT-IR/ATR spectra also confirmed the complete removal of the silica core. The (PLLA/PDLA)$_{10}$ hollow capsules are spherical with a diameter of $320 \pm 20$ nm, and their shell thickness is approximately 60 nm. In the electron diffraction (ED) pattern of these hollow particles, the crystalline patterns corresponding to (010) and (200) of the PLLA/PDLA stereocomplex were observed [112]. The Bragg spacing values of the diffraction spots $d_{010}$ and $d_{100}$, calculated by Au calibration, were 0.339 nm and 0.424 nm, respectively. This result demonstrates that the hollow nanocapsules were composed of PLLA/PDLA stereocomplex films. In the course of the fabrication of hollow nanocapsules composed of LbL stereocomplex films of PLLA and PDLA, we found that tubular nanostructures were formed by drying a water dispersion of the hollow nanocapsules on a substrate. When evaporating water at ambient temperature from a water dispersion of the hollow capsules composed of the lower molecular weight PLAs ($M_{w\ PLLA} = 5500$, $M_{w\ PDLA} = 5800$) on a polyethylene terephthalate (PET) substrate, the formation of tubular assemblies was observed (Figure 6.19a). These tubular assemblies have an average diameter of 300 nm and lengths of 2–5 μm. On the other hand, tubular assemblies were not formed with hollow capsules composed of the higher molecular weight PLLA ($M_w = 30\ 000$) and PDLA ($M_w = 26\ 000$) (Figure 6.19b). It is well known that the stability of stereocomplex polymer films increases with increase in the molecular weight of the polymer components [113]. These results suggest that the formation of nanotubes is related to the stability of the PLAs stereocomplex films comprising the hollow capsule shell. Figure 6.20 shows TEM images of PLA nanotubes and hollow capsules composed of lower molecular weights PLAs ($M_{w\ PLLA} = 5500$, $M_{w\ PDLA} = 5800$). The hollow feature of these tubular structures was confirmed by the TEM images. These images clearly indicate that two or more hollow nanocapsules are one-dimensionally fused with each other.

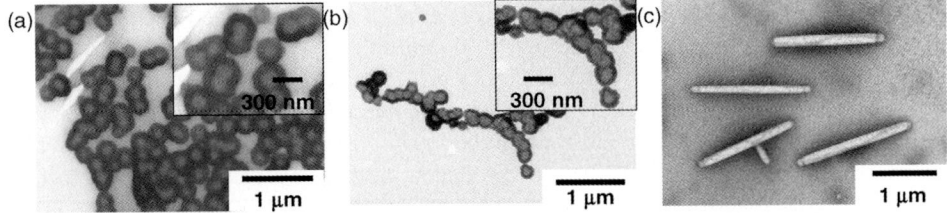

**Figure 6.20** TEM images of (a, b) intermediates of morphology transition and (c) the final formed nanotubes composed of lower molecular weight PLAs ($M_{w\ PLLA} = 5500$, $M_{w\ PDLA} = 5800$). Reprinted with permission from [107], K. Kondo et al., *J. Am. Chem. Soc.* **2010**, *132*, 8236. © 2010, American Chemical Society.

These observations reveal that the PLA nanotubes were formed through the continuous one-dimensional fusion of PLA hollow nanocapsules.

In order to fabricate nanotubes more efficiently from hollow nanocapsules, silica nanoparticles coated with (PLLA/PDLA)$_{10}$ films were densely deposited onto the PET substrate by a vertical deposition technique [114] in advance, and then the nanoparticles deposited on the PET substrate were immersed into 2.3% aqueous HF to remove the silica core. SEM images of silica nanoparticles coated with (PLLA/PDLA)$_{10}$ ($M_{w\ PLLA} = 5500$, $M_{w\ PDLA} = 5800$) deposited on the PET substrate and the nanostructures obtained after the HF treatment are shown in Figure 6.21a and b, respectively. These images clearly show that efficient formation of nanotubes was achieved by the dense deposition of hollow capsules on the substrate. The diameters and lengths of the formed nanotubes were $400 \pm 70$ nm and $3.0 \pm 2.0$ μm,

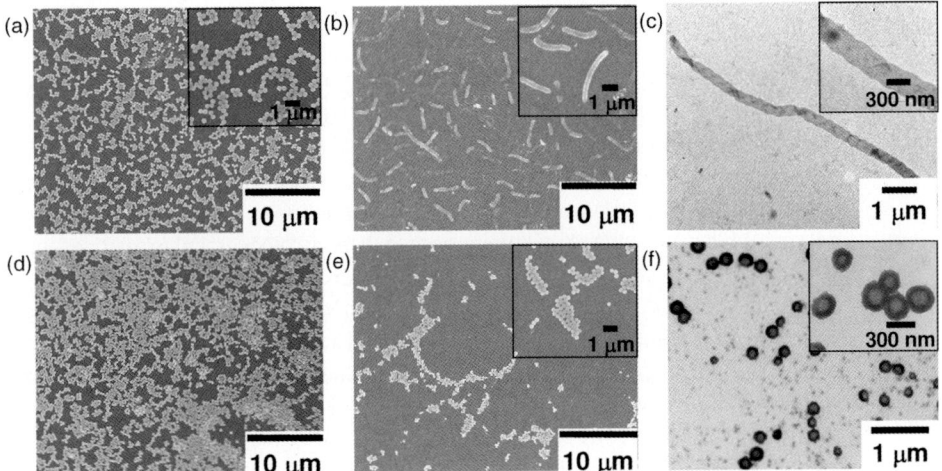

**Figure 6.21** SEM images of (a, d) silica particles coated with (PLLA/PDLA)$_{10}$ and (b, e) nanotubes and hollow capsules obtained after HF etching. TEM images of (c) nanotube and (f) hollow capsules. Molecular weights of PLAs are (a, b, c) $M_{w\ PLLA} = 5500$, $M_{w\ PDLA} = 5800$, (d, e, f) $M_{w\ PLLA} = 30\,000$, $M_{w\ PDLA} = 26\,000$. Reprinted with permission from [107], K. Kondo et al., *J. Am. Chem. Soc.* **2010**, *132*, 8236. © 2010, American Chemical Society.

respectively. On the other hand, when silica particles coated with the higher molecular weight PLAs ($M_{w\ PLLA} = 30\ 000$, $M_{w\ PDLA} = 26\ 000$) were treated with the same procedure, nanotube formation was not observed (Figure 6.21d and e). Taken together these findings with the above-mentioned tendency for tube formation suggest that the formation of nanotubes is closely related to the stability of the PLAs stereocomplex film; less stable PLA hollow capsules would lead to more efficient transformation into PLA nanotubes. We believe that nanotube formation through the one-dimensional fusion of LbL hollow capsules may provide a new strategy for creating nanotubes with a defined diameter.

## 6.4
## Fabrication of Three-Dimensional Cellular Multilayers Using Layer-by-Layer Protein Nanofilms Constructed through Affinity Interaction

### 6.4.1
### Introduction

In the body, nearly all tissue cells reside in the fibrous nano-meshwork of the extracellular matrix (ECM; 10 to 300 nm diameter and several micrometers in length). The ECM is typically composed of fibronectin (FN) and collagen, and provides complex biochemical and physical signals [115–117]. The *in vitro* development of highly-organized, three-dimensional engineered tissue constructs composed of not only multiple types of cells, but also ECM fiber scaffolds which possess a similar structure and function to natural tissues, is a key challenge for implantable tissues in tissue engineering, and for model tissues in pharmaceutical assays [118]. Recently, the topological control of biodegradable porous scaffolds [119, 120], especially nanofiber scaffolds by electrospinning [121] or self-assembling amphiphilic peptides [122], has attracted much attention due to their high porosity and the controlled alignment of the fibers to control cellular function and the development of 3D-engineered tissues [123]. However, 3D-engineered tissues which precisely control the cell type, cell alignment, and cell–cell interactions in three dimensions have not yet been developed. A conventional approach using biodegradable matrices, such as hydrogels or fiber scaffolds, seems to have several limitations in developing 3D-tissue constructs which satisfy the above requirements. A bottom-up approach using multiple cell types as pieces of tissue is expected to solve these problems. Currently, various technologies, such as a cell sheet [124] or magnetic liposomes [125], have been reported in constructing a multilayered cell sheet. These methods are intriguing examples of a bottom-up approach, but have limitations due to the complicated manipulation of the fragile cell sheet or the remains of magnetite particles in the cells.

Recently, we have developed a simple and unique bottom-up approach, "hierarchical cell manipulation using nanofilms," which uses nanometer-sized LbL FN-gelatin (FN-G) films as a nano-ECM [126]. The FN-G nanofilms were prepared directly on the cell surface, and these nanofilms acted as a stable adhesive surface for adhesion of the second cell layer. In this section, we describe the fabrication of 3D-cell multilayers by our technique and their applications as an *in vitro* human 3D-tissue model for analyzing a drug response or tissue function.

## 6.4.2
### Hierarchical Cell Manipulation

In order to obtain a 3D-cell multilayer, preparation of a cell-adhesive surface like an ECM onto the surface of a cell membrane is necessary, because there is not enough ECM on the surface of a cell membrane. We focused on the LbL technique, which is an appropriate method to prepare nanometer-sized films on a substrate by their alternate immersion into interactive polymer solutions [1, 4]. The preparation of nanometer-sized multilayer films composed of ECM components on the surface of the first layer of cells will provide a cell-adhesive surface for the second layer of cells. Rajagopalan *et al.* demonstrated a layer pair structure composed of hepatocytes and other cells by preparing a polyelectrolyte multilayer consisting of chitosan and DNA on the hepatocyte surface [127]. However, chitosan cannot dissolve in neutral buffer, and fabrication of polyelectrolyte multilayers on a cell surface has a limitation due to the cytotoxicity of polycations [128, 129]. The appropriate choice of natural ECM components for the nanofilms will be significant, because the typical ECM presents with cell-adhesive moieties such as RGD (arginine-glycine-aspartic acid) and other amino acid sequences for cellular functions [130]. We selected FN and G to prepare nano-ECM films on the cell surface. FN is a flexible multifunctional glycoprotein, and plays an important role in cell attachment, migration, differentiation, and so on [116, 131]. It is well known to interact not only with a variety of ECM proteins, such as collagens (gelatins) and glycosaminoglycans, but also with the $\alpha_5\beta_1$ integrin receptor on the cell surface [132]. We reported FN-based protein multilayers composed of FN and ECM components, such as gelatin, heparin, and elastin, constructed by LbL assembly [133]. Although FN and G have a negative charge under physiological conditions, they interacted with each other because FN has a collagen-binding domain [116]. Thus, the FN-G nanofilms are expected to provide a suitable cell-adhesive surface, similar to the natural ECM, for the second layer of cells.

The fabrication of 3D-cellular multilayers composed of cells and FN-G nano-ECM films was performed according to the process shown in Figure 6.22. The LbL assembly of FN and G on the cell surface was analyzed quantitatively using a quartz crystal microbalance (QCM) as the assembly substrate, and with a phospholipid layer pair membrane as a model cell membrane (Figure 6.23). A phospholipid layer pair composed of 1,2-dipalmitoyl-sn-glycero-3-phosphatidylcholine (DPPC) and 1,2-dipalmitoyl-sn-glycero-3-phosphate (DPPA) was prepared on the base layer, a four-step assembly of poly(diallyldimethylammonium chloride) (PDDA) and poly(styrenesulfate sodium) (PSS), according to Krishna's report [134]. The mean thickness of the LbL assembly after 1, 7, and 23 steps was calculated to be 2.3, 6.2, and 21.1 nm, respectively. Furthermore, the top and cross-sections of the confocal laser scanning microscopy (CLSM) 3D-merged images suggested a homogeneous assembly of fluorescently labeled FN-G nanofilms on the mouse L929 fibroblast cell surface. We fabricated a layer pair of mouse L929 fibroblast cells with or without

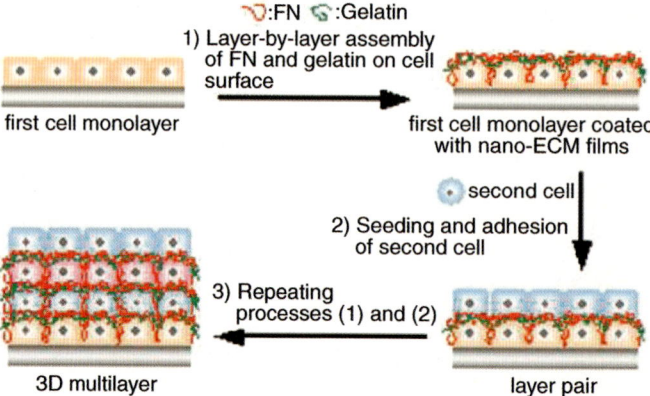

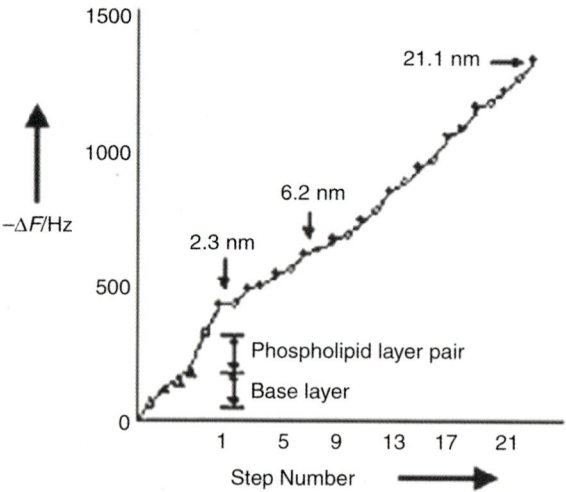

**Figure 6.22** Fabrication process of 3D-cellular multilayers composed of cells and nano-ECM films. The outermost surfaces of all films were FN, which allowed cell adhesion. Reprinted with permission from [126], M. Matsusaki et al., Angew. Chem. Int. Ed. **2007**, 46, 4689. © 2007, Wiley-VCH.

FN-G nanofilms by using a cover glass as a substrate. When seven-step assembled FN-G nanofilms were prepared on the surface of the first L929 cell layer, the second layer cells were then observed on the first cell layer (Figure 6.24a and b). However, when the nanofilm was not prepared or the one-step-assembled nanofilm (only FN) was assembled on the first cell layer, then the layer pair architecture was not

**Figure 6.23** QCM analysis of the LbL assembly of FN and G onto a phospholipid layer pair (DPPC/DPPA 4:1). Open and closed circles show the assembly steps of G and FN, respectively. The base layer was a four-step assembly of PDDA and PSS. Reprinted with permission from [126], M. Matsusaki et al., Angew. Chem. Int. Ed. **2007**, 46, 4689. © 2007, Wiley-VCH.

## 6.4 Fabrication of Three-Dimensional Cellular Multilayers Using Layer-by-Layer Protein Nanofilms

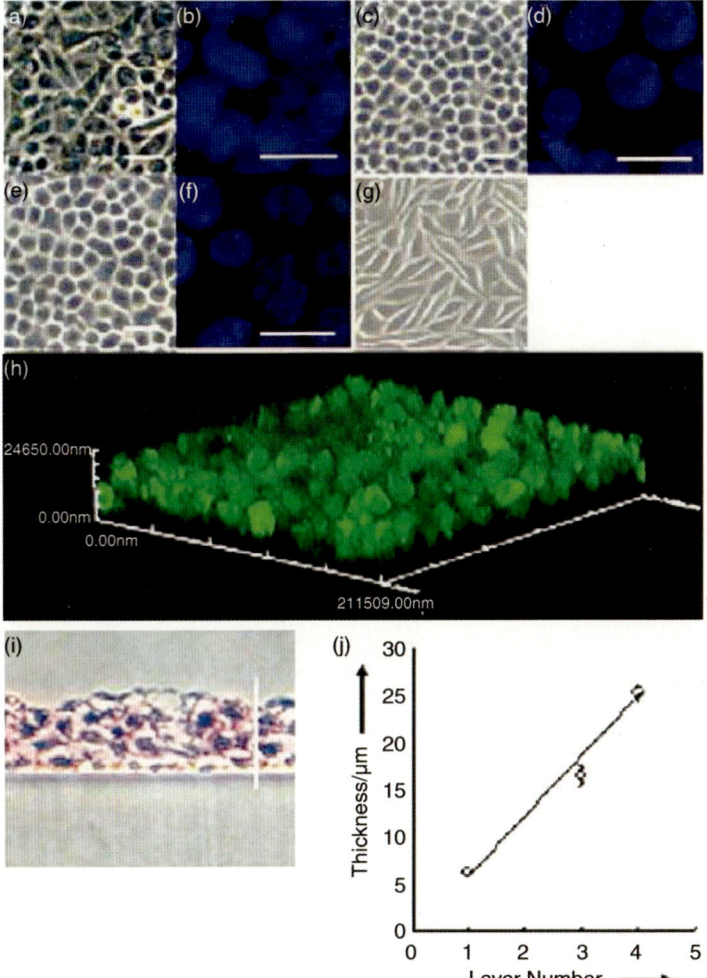

**Figure 6.24** (a,c,e) Phase-contrast (Ph) and (b, d,f) fluorescence (Fluo) microscope images of L929 fibroblast cell layer pairs with a seven-step-assembled (a,b), without a seven-step-assembled (c,d), or with a one-step-assembled (e,f) FN-G nanofilm on the surface of the first cell layer. (g) Ph image of a L929 cell monolayer as a control. The Fluo images were obtained by nuclei labeling with DAPI. The scale bars in (a–g) are 20 μm (Ph) and 10 μm (Fluo). (h) 3D-reconstructed CLSM cross-section image of four-layered L929 cells. The cells were labeled with cell tracker green. (i) Hematoxylin and eosin (HE) staining image of four-layered L929 cells. The scale bar is 30 μm. (j) Relationship between the L929 cell layer number and the mean thickness estimated from 3D-CLSM images ($n = 3$). Reprinted with permission from [126], M. Matsusaki et al., Angew. Chem. Int. Ed. **2007**, 46, 4689. © 2007, Wiley-VCH.

observed (Figure 6.24c–f). These results suggested 2.3 nm of FN film was inadequate, and at least approximately 6 nm of FN-G nanofilm was required as a stable adhesive surface for the second cell layer. The four layered cellular structures were clearly observed after four repetitions of these steps, as shown in Figure 6.24h,i.

Furthermore, blood vessel multilayers composed of human umbilical artery smooth muscle cells (UASMC) and human umbilical vascular endothelial cells (HUVEC) were successfully fabricated. This present methodology can be applied as one of the biomedical applications of LbL assembly to fabricate various cellular multilayers composed of target cells and ECMs.

### 6.4.3
**High Cellular Activities Induced by 3D-Layered Constructs**

Studies on the functions of layered cellular architectures as compared with cell monolayer are valuable, not only for understanding how a 3D environment composed of cells, ECM, and signaling molecules regulates functions similar to natural tissues, but also for creating 3D-artificial tissues resembling natural tissues. Recently, some researchers have reported the functions of layered cellular architectures *in vitro* [127, 135–138]. However, the basic properties induced by 3D-cellular structures, such as the layer number or the cell types, have not yet been clarified. We evaluated the structural stability of layered constructs consisting of human fibroblast cells (FCs) and HUVECs in relation to their layer number [139].

Interestingly, the ECs adhered homogeneously onto four-layered (4L) FCs, and tight-junction formation was widely observed at the centimeter scale, while heterogeneous EC domain structures were observed on the monolayered (1L) FCs (Figure 6.25). The production of heat shock protein70 (Hsp70) and interleukin-6 (IL-6) from the cellular structures was investigated to elucidate any 3D-structural effect on cellular function (Figure 6.26). The Hsp70 expression of the ECs decreased after adhesion onto the 4L-FC structure as compared with the EC monolayer. Surprisingly, the Hsp70 production response to heat shock increased drastically, approximately 10-fold, as compared with a non-heat shock by 3D structure formation, whereas the monolayer structures showed no change. Moreover, the production of the inflammatory cytokine IL-6 decreased significantly, depending on the layer number of FCs. These results suggested that 4L-FC provided a more favorable environment for ECs than a cell culture plastic disk to induce high thermosensitivity, and to suppress the inflammatory response from the substrate. Since the FN-G films prepared on the cell surface did not affect cellular functions [140], we concluded that the layered construct would be an analogous environment to natural tissues. These findings could be important not only for tissue engineering, but also for basic cell biology.

### 6.4.4
**Quantitative 3D-Analyses of Nitric Oxide Diffusion in a 3D-Artery Model**

A blood vessel is crucial, not only for circulatory diseases and treatments, but also for the biological evaluation of drug diffusion to target tissues, the penetration of cancer cells or pathogens, and the control of blood pressure. A blood vessel is generally composed of three distinct layers: the intima, an inner single layer of endothelial cells (ECs); the media, medium layers of smooth muscle cells (SMCs); and the adventitia,

## 6.4 Fabrication of Three-Dimensional Cellular Multilayers Using Layer-by-Layer Protein Nanofilms

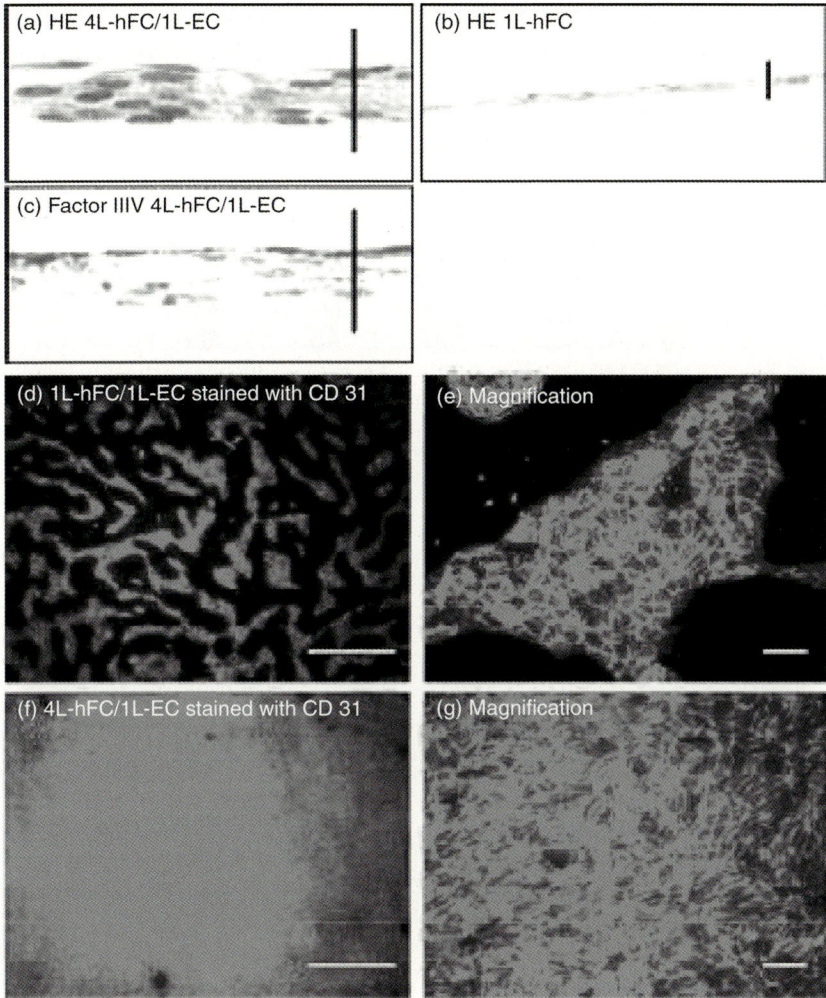

**Figure 6.25** Histological images: (a) stained with HE and (c) immunostained for factor VIII of the 4L-FC/1L-EC construct as compared to (b) HE images of the 1L-FC construct. Fluorescence immunostaining images with an anti-CD31 antibody of (d and e) 1L-FC/1L-EC and (f and g) 4L-FC/1L-EC constructs. The scale bars are (a and c) 40 μm, (b) 20 μm, (d and f) 3000 μm, and (e and g) 100 μm, respectively. Reprinted with permission from [139], K. Kadowaki *et al.*, *Biochem. Biophys. Res. Commun.* **2010**, *402*, 153. © 2010, ELSEVIER.

an outer layer of fibroblast cells [141]. The nitric oxide produced from ECs diffuses into the SMCs through their cell membranes, and activates guanylate cyclase to produce intracellular cyclic guanosine monophosphate (cGMP), which induces a signaling pathway, mediated by kinase proteins, leading to SMC relaxation [142]. Accordingly, quantitative, kinetic, and spatial analyses of the extracellular delivery of NO molecules from the EC layer to the SMC layers upon drug stimulation are crucial for pharmaceutical and biomedical evaluations of hypertension and diabetes. So far,

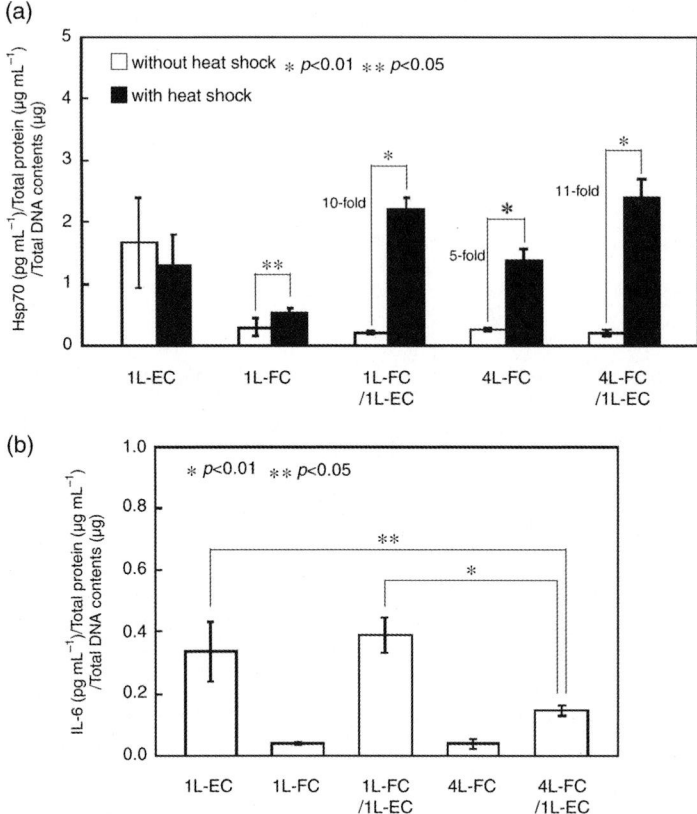

**Figure 6.26** (a) HSP70 production versus the total protein from non-heat and heat-shocked layered structures composed of FCs and ECs ($n = 4$). The heat shock condition is defined as 20 min of incubation at 45 °C, followed by a 2 h recovery period. (b) IL-6 production versus the total protein from layered structures composed of FCs and ECs ($n = 3$). Reprinted with permission from [139], K. Kadowaki et al., *Biochem. Biophys. Res. Commun.* **2010**, *402*, 153. © 2010, ELSEVIER.

pharmaceutical assays of NO production have been performed by *in vivo* animal experiments, but low reproducibility and differences in NO production depending on animal types are unsolved issues. Thus, development of a convenient and versatile method for the *in vitro* quantitative and spatial analyses of NO diffusion inside the artery wall instead of animal experiments is important for biological and pharmaceutical applications.

We reported biocompatible and highly sensitive NO sensor particles prepared by LbL assembly. The mesoporous, micrometer-sized silica particles encapsulating the 4,5-diaminofluorescein (DAF-2), NO fluorescent indicator dye, were covered with biocompatible chitosan (CT)-dextran sulfate (DS) LbL films to provide cytocompatibility, and to inhibit leakage of the DAF-2. The NO sensor particles (SPs) showed high NO-sensitivity at 5–500 nM, which is sufficient to detect NO at concentrations of

### 6.4 Fabrication of Three-Dimensional Cellular Multilayers Using Layer-by-Layer Protein Nanofilms

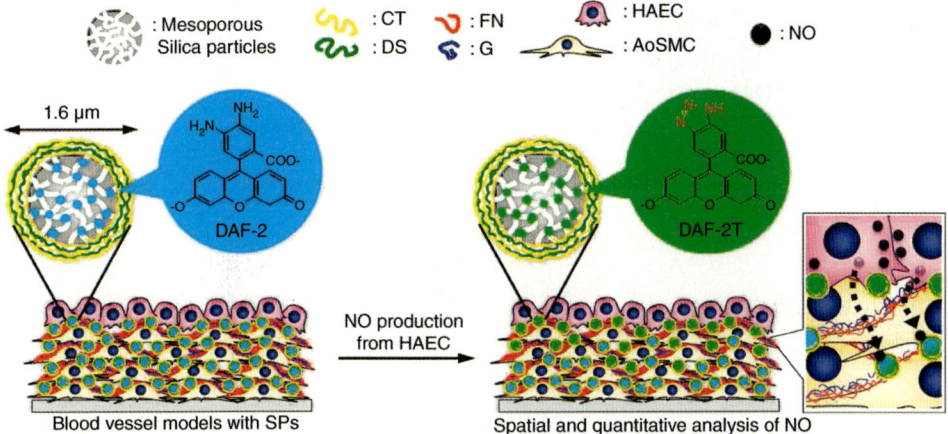

**Figure 6.27** Schematic illustration of the *in vitro* spatial and quantitative analyses of NO diffusion from the uppermost EC layer to the SMC layers in a 3D-artery model by SPs, which were allocated into each layer.

hundreds of nM (EC production level [143]) [144]. If an artificial 3D-artery model allocating these SPs can be developed, the extracellular diffusion of NO from the EC layer to the SMC layers with chemical and physical stimuli is expected to be observed *in vitro* fluorescently by using CLSM. We successfully developed the 3D-blood vessel models consisting of human ECs and SMCs, and their morphology and histology were evaluated in detail [145]. We combined both the SP technique and hierarchical cell manipulation to develop five layered (5L)-artery models including SPs using human aortic smooth muscle cells (AoSMCs) and human aortic endothelial cells (HAECs), as shown in Figure 6.27. To clarify the effect of the layered structure consisting of AoSMCs and HAECs on NO production, the NO concentrations from various 2D- or 3D-structures consisting of AoSMCs and HAECs after 48 h of culture were evaluated by a horseradish peroxidase (HRP) assay [146] Interestingly, the layer pair structure of HAECs with AoSMCs promoted a twofold higher NO production as compared to 1L-HAECs, whereas the heterogeneous monolayered co-culture with AoSMCs did not show any effect. A vertical interaction between upper HAECs and lower AoSMCs like a native artery may effectively stimulate NO production, because the ECs are known to have high polarity. We demonstrated a quantitative 3D-analysis of NO diffusion from the uppermost 1L-HAECs to the underlying 4L-AoSMCs in 5L-artery models using the SPs. The SPs were allocated into each layer as shown in Figure 6.27.

Figure 6.28a shows 3D-reconstructed CLSM images of the 5L-constructs after 48 h of culture, where the green and blue colors represent the SPs and nuclei, respectively. The averaged local NO concentrations in each layer were quantified from the cross-sectional image (Figure 6.28b) and summarized in Figure 6.28c. The NO concentrations in first (HAEC) and second (AoSMC) layers were about 530~550 nM, and then gradually decreased with increasing layer number. The NO concentration reached about 50% (290 nM) in the fifth layer (32 μm in depth). Furthermore, the distance of NO diffusion from the top HAEC layer to the underlying AoSMC layers was

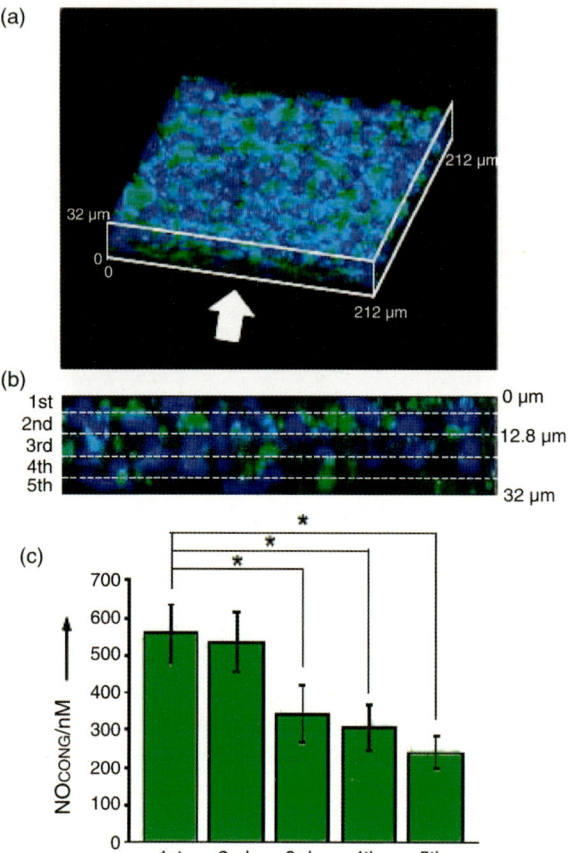

**Figure 6.28** (a) 3D-reconstructed CLSM image of the 5L-artery model containing SPs (green) in each layer after 48 h of incubation. The first layer is HAECs, and the second to fifth layers are AoSMCs. The nuclei (blue) were labeled with 4′,6-diamidino-2-phenylindole dihydrochloride (DAPI). The image area is 212.1 × 212.1 μm$^2$, and the height is 32.0 μm. (b) Cross-sectional CLSM image of (a) in the white arrow direction. The dashed lines indicate brief interfaces of each layer. (c) The localized NO concentrations in each cellular layer were analyzed by SPs. The fluorescence intensities of the SPs were measured from each layer (6.4 × 212.1 μm$^2$) in (b) ($n = 3$, over 20 SPs per image). Asterisk denotes statistically significant difference using a two sample $t$-test ($^*p < 0.01$) for each comparison.

estimated at approximately 60 μm from the equation of NO diffusion distance obtained from Figure 6.28c, which shows the correlation between the NO concentration and the distance from the HAECs. Although this value is slightly lower than the reported *in vivo* NO diffusion distance of approximately 100 μm [147, 148], the obtained value is reasonable because it is known that about 37% of the NO produced is consumed in chemical reactions in the artery walls [148]. These results suggest that this technique will be a useful method for *in vitro* artery assays instead of *in vivo* animal experiments.

## 6.5
## Conclusion

In this chapter, we have shown that weak interactions between polymers or macromolecules are useful for constructing LbL assemblies in an analogous way to electrostatic interaction. Making good use of van der Waals interactions between stereoregular poly(methyl methacrylate)s or poly(lactic acid)s not only stimulated new development in LbL studies, but also resolved an outstanding problem in polymer chemistry. Actually, template polymerization has been one of the goals in synthetic polymer chemistry for a long time. Then peculiar hollow nanotube formation from hollow nanocapsules prepared by the LbL technique was discovered, although the mechanism is not clarified at present. Because these interactions are weak enough in polymer materials, recombination of polymer–polymer interaction should be induced in mild ambient conditions. Moreover, we believe that application of LbL assemblies for 3D cell architectures is an epoch-making study. As affinity interaction in biosystems is a key factor for life maintenance, LbL tissue engineering using the interaction between biomolecules is expected to develop into a new field.

In conclusion, the LbL assembly technique using weak interactions tends to allow polymers to be easily arranged into the most stable conformation structure of a polymer chain, as well as the construction of a thin film. Such effects could be caused by a kind of self-sorting based on the small energetic barrier among polymer conformations in a LbL film. Therefore, the technique enabled LbL film to produce various functional materials, such as the precise polymerization field, the fusion of hollow nanoparticles, and the biomimetic field for cell culture. It is expected that the LbL system with weak polymer interaction will create further applications, because it exerts molecularly controlled structure in the films.

## References

1 Decher, G. and Hong, J.D. (1991) *Makromol. Chem. Macromol. Symp.*, **46**, 321.
2 Decher, G. and Hong, J.D. (1991) *Ber. Bunsen-Ges. Phys. Chem.*, **95**, 1430.
3 Decher, G. (1996) *Compr. Supramol. Chem.*, **9**, 507.
4 Decher, G. (1997) *Science*, **277**, 1232.
5 Decher, G. and Schlenoff, J.B. (eds) (2003) *Multilayer Thin Films*, Wiley-VCH, Weinheim.
6 Tauk, L., Schröder, A.P., Decher, G., and Giuseppone, N. (2009) *Nat. Chem.*, **1**, 649.
7 Lvov, Y., Haas, H., Decher, G., Möhwald, H., Mikhailov, A., Mtchedlishvily, B., Morgunova, E., and Vainshtein, B. (1994) *Langmuir*, **10**, 4232.
8 Lvov, Y., Ariga, K., and Kunitake, T. (1994) *Chem. Lett.*, 2323.
9 Lvov, Y., Ichinose, I., Ariga, K., and Kunitake, T. (1995) *J. Am. Chem. Soc.*, **117**, 6117.
10 Ariga, K., McShane, M., Lvov, Y., Ji, Q., and Hill, J.P. (2011) *Expert Opin. Drug Deliv.*, **8**, 633.
11 Komatsu, T., Qu, X., Ihara, H., Fujihara, M., Azuma, H., and Ikeda, H. (2011) *J. Am. Chem. Soc.*, **133**, 3246.
12 Ariga, K., Lvov, Y., Onda, M., Ichinose, I., and Kunitake, T. (1997) *Chem. Lett.*, 125.

13 Lvov, Y., Ariga, K., Onda, M., Ichinose, I., and Kunitake, T. (1997) *Langmuir*, **13**, 6195.

14 Yang, H.C., Aoki, K., Hong, H.-G., Sackett, D.D., Arendt, M.F., Yau, S.-L., Bell, C.M., and Mallouk, T.E. (1993) *J. Am. Chem. Soc.*, **115**, 11855.

15 Yonezawa, T., Onoue, S., and Kunitake, T. (1998) *Adv. Mater.*, **10**, 414.

16 Shchukin, D.S., Ustinovich, E.A., Sukhorukov, G.B., Möhwald, H., and Sviridov, D.V. (2005) *Adv. Mater.*, **17**, 468.

17 Anders, C.M. and Kotov, N.A. (2010) *J. Am. Chem. Soc.*, **132**, 14496.

18 Sun, Y., Zhang, X., Sun, C., Wang, Z., Shen, J., Wang, D., and Li, T. (1996) *J. Chem. Soc., Chem. Commun.*, 2379.

19 Ariga, K., Lvov, Y., and Kunitake, T. (1997) *J. Am. Chem. Soc.*, **119**, 2224.

20 Linford, M.R., Auch, M., and Möhwald, H. (1998) *J. Am. Chem. Soc.*, **120**, 178.

21 Caruso, F., Caruso, R.A., and Möhwald, H. (1998) *Science*, **282**, 1111.

22 Zebi, B., Susha, A.S., Sukhorukov, G.B., Rogach, A.L., and Parak, W.J. (2005) *Langmuir*, **21**, 4262.

23 Hu, S.H., Tsai, C.H., Liao, C.F., Liu, D.M., and Chen, S.Y. (2008) *Langmuir*, **24**, 11811.

24 Ichinose, I., Fujiyoshi, K., Mizuki, S., Lvov, Y., and Kunitake, T. (1996) *Chem. Lett.*, 257.

25 Sohling, U. and Schouten, A.J. (1996) *Langmuir*, **12**, 3912.

26 Keller, S.W., Kim, H.-N., and Mallouk, T.E. (1994) *J. Am. Chem. Soc.*, **116**, 8817.

27 Kotov, N.A., Haraszti, T., Turi, L., Zavala, G., Geer, R.E., Dékány, I., and Fendler, J.H. (1997) *J. Am. Chem. Soc.*, **119**, 12184.

28 Hua, F., Cui, T., and Lvov, Y. (2004) *Nano Lett.*, **4**, 823.

29 Serizawa, T. and Akashi, M. (1997) *Chem. Lett.*, 809.

30 Serizawa, T., Kamimura, S., and Akashi, M. (2000) *Colloids Surf.*, **164**, 237.

31 Correa-Duarte, M.A., Kosiorek, A., Kandulski, W., Giersig, M., and Liz-Marzán, L.M. (2005) *Chem. Mater.*, **17**, 3268.

32 Stockton, W.B. and Rubner, M.F. (1997) *Macromolecules*, **30**, 2717.

33 Wang, L., Wang, Z.Q., Zhang, X., Shen, J.C., Chi, L.F., and Fuchs, H. (1997) *Macromol. Rapid Commun.*, **18**, 509.

34 Kharlampieva, E., Kozlovskaya, V., and Sukhishvili, S.A. (2009) *Adv. Mater.*, **21**, 3053.

35 Shimazaki, Y., Mitsuishi, M., Ito, S., and Yamamoto, M. (1997) *Langmuir*, **13**, 1385.

36 Wang, F., Ma, N., Chen, Q., Wang, W., and Wang, L. (2007) *Langmuir*, **23**, 9540.

37 Spěváček, J. and Schneider, B. (1987) *Adv. Colloid Interface Sci.*, **27**, 81.

38 Watanabe, W.H., Ryan, C.F., Fleischer, P.C.Jr., and Garrett, B.S. (1961) *J. Phys. Chem.*, **65**, 896.

39 Liquori, A.M., Anzuino, G., Coiro, V.M., D'Alagni, M., Santis, P.D., and Savino, M. (1965) *Nature*, **206**, 358.

40 Feitsma, E.L., De Boer, A., and Challa, G. (1975) *Polymer*, **16**, 515.

41 De Boer, A. and Challa, G. (1976) *Polymer*, **17**, 633.

42 Bosscher, F., Keekstra, D., and Challa, D. (1981) *Polymer*, **22**, 124.

43 Kitayama, T., Fujimoto, N., Terawaki, Y., and Hatada, K. (1990) *Polym. Bull.*, **23**, 279.

44 Lohmeyer, J.H.G.M., Kransen, G., and Tan, Y.Y. (1975) *J. Polym. Sci., Polym. Lett. Ed.*, **13**, 725.

45 Lohmeyer, J.H.G.M., Tan, Y.Y., Lako, P., and Challa, G. (1978) *Polymer*, **19**, 1171.

46 Hatada, K., Shimizu, S., Terawaki, Y., Ohta, K., and Yuki, H. (1981) *Polym. J.*, **13**, 811.

47 Ikada, Y., Jamshidi, K., Tsuji, H., and Hyon, S.-H. (1987) *Macromolecules*, **20**, 904.

48 Grenier, D. and Prud'homme, R.E. (1984) *J. Polym. Sci., Polym. Phys. Ed.*, **22**, 577.

49 Fukuzawa, T. and Uematsu, I. (1974) *Polym. J.*, **6**, 537.

50 Takahashi, T., Tsutsumi, A., Hikichi, K., and Kaneko, M. (1974) *Macromolecules*, **7**, 806.

51 Nomori, H., Tsuchihashi, N., Takagi, S., and Hatano, M. (1975) *Bull. Chem. Soc. Jpn.*, **48**, 2522.

52 Bosscher, F., Brinke, G.T., and Challa, G. (1982) *Macromolecules*, **15**, 1442.

## References

53 Brinke, G.T., Schomaker, E., and Challa, G. (1985) *Macromolecules*, **18**, 1925.
54 Spěváček, J., Schneider, B. and Straka, J. (1990) *Macromolecules*, **23**, 304.
55 Schomaker, E. and Challa, G. (1989) *Macromolecules*, **22**, 3337.
56 Kumaki, J., Kawauchi, T., Okoshi, K., Kusanagi, H., and Yashima, E. (2007) *Angew. Chem. Int. Ed.*, **46**, 5348.
57 Buter, R., Tan, Y.Y., and Challa, G. (1972) *J. Polym. Sci. A-1*, **10**, 1031.
58 Buter, R., Tan, Y.Y., and Challa, G. (1973) *J. Polym. Sci., Polym. Chem. Ed.*, **11**, 1003.
59 Buter, R., Tan, Y.Y., and Challa, G. (1973) *J. Polym. Sci., Polym. Chem. Ed.*, **11**, 1013.
60 Buter, R., Tan, Y.Y., and Challa, G. (1973) *J. Polym. Sci., Polym. Chem. Ed.*, **11**, 2975.
61 Gons, J., Vorenkamp, E.J., and Challa, G. (1975) *J. Polym. Sci., Polym. Chem. Ed.*, **13**, 1699.
62 Gons, J., Slagter, W.O., and Challa, G. (1977) *J. Polym. Sci., Polym. Chem. Ed.*, **15**, 771.
63 Gons, J., Vorenkamp, E.J., and Challa, G. (1977) *J. Polym. Sci., Polym. Chem. Ed.*, **15**, 3031.
64 Gons, J., Straatman, L.J.P., and Challa, G. (1978) *J. Polym. Sci., Polym. Chem. Ed.*, **16**, 427.
65 Lohmeyer, J.H.G.M., Tan, Y.Y., and Challa, G. (1980) *J. Macromol. Sci., Chem.*, **A14**, 945.
66 Szumilewicz, J. (2000) *Macromol. Symp.*, **161**, 183.
67 Serizawa, T., Hamada, K., and Akashi, M. (2004) *Nature*, **429**, 52.
68 Serizawa, T., Hamada, K., and Akashi, M. (2005) *Macromolecules*, **38**, 6759.
69 Serizawa, T., Hamada, K., Kitayama, T., Fujimoto, N., Hatada, K., and Akashi, M. (2000) *J. Am. Chem. Soc.*, **122**, 1891.
70 Tretinnikov, O.N. (1997) *Langmuir*, **13**, 2988.
71 Hamada, K., Serizawa, T., Kitayama, T., Fujimoto, N., Hatada, K., and Akashi, M. (2001) *Langmuir*, **17**, 5513.
72 Serizawa, T., Hamada, K., Kitayama, T., Katsukawa, K.-I., Hatada, K., and Akashi, M. (2000) *Langmuir*, **16**, 7112.
73 Serizawa, T., Hamada, K., Kitayama, T., and Akashi, M. (2003) *Angew. Chem. Int. Ed.*, **42**, 1118.
74 Serizawa, T. and Akashi, M. (2006) *Polym. J.*, **38**, 311.
75 Kamei, D., Ajiro, H., Hongo, C., and Akashi, M. (2008) *Chem. Lett.*, **37**, 332.
76 Kamei, D., Ajiro, H., Hongo, C., and Akashi, M. (2009) *Langmuir*, **25**, 280.
77 Kamei, D., Ajiro, H., and Akashi, M. (2010) *Polym. J.*, **42**, 131.
78 Kamei, D., Ajiro, H., and Akashi, M. (2010) *J. Polym. Sci. Part A: Polym. Chem.*, **48**, 3651.
79 Ajiro, H., Maegawa, M., and Akashi, M. (2010) *J. Polym. Sci. Part A: Polym. Chem.*, **48**, 3265.
80 Ajiro, H., Kamei, D., and Akashi, M. (2008) *J. Polym. Sci. Part A: Polym. Chem.*, **46**, 5879.
81 Ajiro, H., Kamei, D., and Akashi, M. (2009) *Polym. J.*, **41**, 90.
82 Ajiro, H., Kamei, D., and Akashi, M. (2009) *Macromolecules*, **42**, 3019.
83 Kusy, R.P. (1976) *J. Polym. Sci., Polym. Chem. Ed.*, **14**, 1527.
84 Meier, W. (2000) *Chem. Soc. Rev.*, **29**, 295. and references therein.
85 Caruso, F. (2000) *Chem. Eur. J.*, **6**, 413.
86 Hammond, P.T. (2004) *Adv. Mater.*, **16**, 1271.
87 Antipov, A.A. and Sukhorukov, G.B. (2004) *Adv. Colloid Interface Sci.*, **111**, 49.
88 Dähne, L. and Peyratout, C.S. (2004) *Angew. Chem. Int. Ed.*, **43**, 3762.
89 Zhang, Y., Guan, Y., Yang, S., Xu, J., and Han, C.C. (2003) *Adv. Mater.*, **15**, 832.
90 Kozlovskaya, V., Ok, S., Sousa, A., Libera, M., and Sukhishvili, S.A. (2003) *Macromolecules*, **36**, 8590.
91 Kharlampieva, E., Kozlovskaya, V., Tyutina, J., and Sukhishvili, S.A. (2005) *Macromolecules*, **38**, 10523.
92 Kharlampieva, E., Kozlovskaya, V., and Sukhishvili, S.A. (2009) *Adv. Mater.*, **21**, 3053.
93 Itoh, Y., Matsusaki, M., Kida, T., and Akashi, M. (2004) *Chem. Lett.*, **33**, 1552.
94 Itoh, Y., Matsusaki, M., Kida, T., and Akashi, M. (2006) *Biomacromolecules*, **7**, 2715.
95 Itoh, Y., Matsusaki, M., Kida, T., and Akashi, M. (2008) *Biomacromolecules*, **9**, 2202.
96 Kida, T., Mouri, M., and Akashi, M. (2006) *Angew. Chem. Int. Ed.*, **45**, 7534.

97 Bikson, B., Nelson, J.K., and Muruganandam, N. (1994) *J. Membr. Sci.*, **94**, 313.
98 Schuetz, P. and Caruso, F. (2003) *Adv. Funct. Mater.*, **13**, 929.
99 Vorenkamp, E.J., Bosscher, F., and Challa, G. (1979) *Polymer*, **20**, 59.
100 Spěváček, J. and Schneider, B. (1974) *J. Polym. Sci. Part C, Polym. Lett. Ed.*, **12**, 349.
101 Dybal, J., Stokr, J., and Schneider, B. (1983) *Polymer*, **24**, 971.
102 Spěváček, J., Schneider, B., Dybal, J., Stokr, J., Baldrian, J., and Pelzbauer, Z. (1984) *J. Polym. Sci. Part B, Polym. Phys. Ed.*, **22**, 617.
103 Tretinnikov, O.N., Nakano, K., Ohta, K., and Iwamoto, R. (1996) *Macromol. Chem. Phys.*, **197**, 753.
104 Florence, A.T. and Whitehill, D. (1982) *Int. J. Pharm.*, **11**, 277.
105 Itou, N., Masukawa, T., Ozaki, I., Hattori, M., and Kasai, K. (1999) *Colloid Surf. A*, **153**, 311.
106 Kim, J.-W., Joe, Y.-G., and Suh, K.-D. (1999) *Colloid Polym. Sci.*, **277**, 252.
107 Kondo, K., Kida, T., Arikawa, Y., Ogawa, Y., and Akashi, M. (2010) *J. Am. Chem. Soc.*, **132**, 8236.
108 Kalkarni, R.K., Pani, K.G., Neuman, G., and Leonard, F. (1966) *Arch. Surg.*, **93**, 839.
109 Heino, A., Naukkarinen, A., Kulju, T., Törmälä, P., Pohjonen, T., and Mäkeäl, E.A. (1996) *J. Biomed. Mater. Res.*, **30**, 187.
110 Ikada, Y., Jamshidi, K., Tsuji, H., and Hyon, S.H. (1987) *Macromolecules*, **20**, 906.
111 Tsuji, H., Horii, F., Nakagawa, M., Ikada, Y., Odani, H., and Kitamaru, R. (1992) *Macromolecules*, **25**, 4114.
112 Hu, J., Tang, Z., Qiu, X., Pang, X., Yang, Y., Chen, X., and Jing, X. (2005) *Biomacromolecules*, **6**, 2843.
113 de Jong, S.J., van Dijk-Wolthuis, W.N.E., Kettenes-van Bosch, J.J., Schuyl, P.J.W., and Hennink, W.E. (1998) *Macromolecules*, **31**, 6397.
114 Jiang, P., Bertone, F., Hwang, K.S., and Colvin, V.L. (1999) *Chem. Mater.*, **11**, 2132.
115 Boudreau, N.J. and Jones, P.L. (1999) *Biochem. J.*, **339**, 481.
116 Hynes, R.O. (1990) *Fibronectins*, Springer-Verlag Inc., New York.
117 Raines, E.W. (2000) *Int. J. Exp. Pathol.*, **81**, 173.
118 Langer, R. and Vacanti, J.P. (1993) *Science*, **260**, 920.
119 Lee, J., Cuddihy, M.J., and Kotov, N.A. (2008) *Tissue Eng. Part B*, **14**, 61.
120 Lee, K.Y. and Mooney, D.J. (2001) *Chem. Rev.*, **101**, 1869.
121 Dzenis, Y. (2004) *Science*, **304**, 1917.
122 Zhang, S. (2003) *Nature Biotech.*, **21**, 1171.
123 Stevens, M.M. and George, J.H. (2005) *Science*, **310**, 1135.
124 Sasagawa, T., Shimizu, T., Sekiya, S., Haraguchi, Y., Yamato, M., Sawa, Y., and Okano, T. (2010) *Biomaterials*, **31**, 1646.
125 Akiyama, H., Ito, A., Kawabe, Y., and Kamihira, M. (2010) *Biomaterials*, **31**, 1251.
126 Matsusaki, M., Kadowaki, K., Nakahara, Y., and Akashi, M. (2007) *Angew. Chem. Int. Ed.*, **46**, 4689.
127 Rajagopalan, P., Shen, C.J., Berthiaume, F., Tilles, A.W., Toner, M., and Yarmush, M.L. (2006) *Tissue Eng.*, **12**, 1553.
128 Fischer, D., Li, Y., Ahlemeyer, B., Krieglstein, J., and Kissel, T. (2003) *Biomaterials*, **24**, 1121.
129 Chanana, M., Gliozzi, A., Diaspro, A., Chodnevskaja, I., Huewel, S., Moskalenko, V., Ulrichs, K., Galla, H.-J., and Krol, S. (2005) *Nano Lett.*, **5**, 2605.
130 Kleinman, H.K., Phlip, D., and Hoffman, M.P. (2003) *Curr. Opin. Biotechnol.*, **14**, 526.
131 Yamada, K.M. (1983) *Annu. Rev. Biochem.*, **52**, 761.
132 Ruoslahti, E. and Pierschbacher, M.D. (1987) *Science*, **238**, 491.
133 Nakahara, Y., Matsusaki, M., and Akashi, M. (2007) *J. Biomater. Sci. Polymer Edn.*, **18**, 1565.
134 Krishna, G., Shutava, T., and Lvov, Y. (2005) *Chem. Commun.*, 2796.
135 Ito, A., Jitsunobu, H., Kawabe, Y., and Kamihira, M. (2007) *J. Biosci. Bioeng.*, **104**, 371.
136 Harimoto, M., Yamato, M., Hirose, M., Takahashi, C., Isoi, Y., Kikuchi, A., and

Okano, T. (2002) *J. Biomed. Mater. Res.*, **62**, 464.

**137** Ohno, M., Motojima, K., Okano, T., and Taniguchi, A. (2009) *J. Biochem.*, **145**, 591.

**138** L'Heureux, N., Stoclet, J.-C., Auger, F.A., Lagaud, G.J., Gemain, L., and Andriantsitohaina, R. (2001) *FASEB J.*, **15**, 515.

**139** Kadowaki, K., Matsusaki, M., and Akashi, M. (2010) *Biochem. Biophys. Res. Commun.*, **402**, 153.

**140** Kadowaki, K., Matsusaki, M., and Akashi, M. (2010) *Langmuir*, **26**, 5670.

**141** Isenberg, B.C. and Wong, J.Y. (2006) *Mater. Today*, **9**, 54.

**142** Alderton, W.K., Cooper, C.E., and Knowles, R.G. (2001) *Biochem. J.*, **357**, 593.

**143** Nakatsubo, N., Kojima, H., Kikuchi, K., Nagoshi, H., Hirata, Y., Maeda, D., Imai, Y., Irimura, T., and Nagano, T. (1998) *FEBS Lett.*, **427**, 263.

**144** Amemori, S., Matsusaki, M., and Akashi, M. (2010) *Chem. Lett.*, **39**, 42.

**145** Matsusaki, M., Kadowaki, K., Adachi, E., Sakura, T., Yokoyama, U., Ishikawa, Y., and Akashi, M. (2012) *J. Biomater. Sci.* **23**, 63–79.

**146** Kikuchi, K., Nagano, T., and Hirobe, M. (1996) *Biol. Pharm. Bull.*, **19**, 649.

**147** Malinski, T. and Taha, Z. (1992) *Nature*, **358**, 676.

**148** Malinsk, T., Taha, Z., and Grunfeld, S. (1993) *Biochem. Biophys. Res. Commun.*, **193**, 1076.

# 7
# Layer-by-Layer Assembly of Polymeric Complexes

*Junqi Sun, Xiaokong Liu, and Jiacong Shen*

## 7.1
## Introduction

Interest in the layer-by-layer (LbL) assembly technique, which involves the alternate deposition of species with complementary chemical interactions for the fabrication of composite films, has been continuously growing since Decher and coworkers put forward this method in the early 1990s [1]. Compared with other film preparation methods, the LbL assembly is unique in precisely controlling the chemical composition and structure of the films because all species are deposited in a predesigned LbL fashion. Meanwhile, LbL assembly is particularly suitable for film fabrication on nonflat surfaces with large areas [2]. The LbL assembly is generally regarded as a group of methods for ultrathin film fabrication (i.e., films with thickness less than 100 nm) because the deposition of one layer of film usually takes several tens of minutes and several hundreds of layers are often required to reach a thickness of one micrometer [3]. Compared with ultrathin films, the LbL assembled films with micrometer thickness have the irreplaceable advantages of high loading capacity, enhanced mechanical robustness, convenience in tailoring micro- and nanoscaled hierarchical structures, and integrating multiple functions into one film. Although several methods such as spin LbL assembly [4], spray LbL assembly [5] and exponential LbL (e-LbL) assembly [6] have been developed to fabricate LbL-assembled thick films or speed up the LbL assembly process, novel facile methods for rapid construction of LbL-assembled thick films are still highly desirable.

Polymeric complexes represent a large variety of supramolecular assemblies held together by multiple weak interactions [7], which include polyelectrolyte–polyelectrolyte complexes (PECs), polyelectrolyte–surfactant complexes, non-charged polymer–polymer complexes, polymeric–inorganic hybrid complexes, and so forth (Figure 7.1). Compared with uncomplexed polymers, polymeric complexes have an abundance of compositions, relatively large dimensions, and diverse structures, which can all be easily tailored by changing parameters such as the mixing ratio, solution pH, ionic strength, temperature, and so on during or after the formation of polymeric complexes. All of these characteristics make polymeric complexes ideal

*Multilayer Thin Films: Sequential Assembly of Nanocomposite Materials*, Second Edition.
Edited by Gero Decher and Joseph B. Schlenoff.
© 2012 Wiley-VCH Verlag GmbH & Co. KGaA. Published 2012 by Wiley-VCH Verlag GmbH & Co. KGaA.

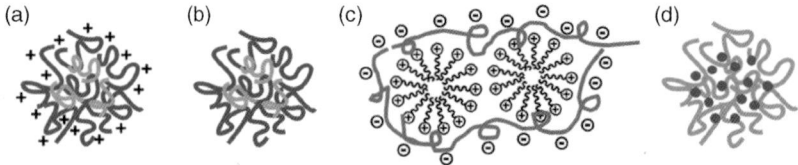

**Figure 7.1** Illustration of different types of polymeric complexes. (a) Polyelectrolyte–polyelectrolyte complexes. (b) Non-charged polymer–polymer complexes. (c) Polyelectrolyte–sufactant complexes. (d) Polymeric–inorganic complexes.

building blocks for the LbL assembly of composite films [8]. Although previous studies proved that water-dispersible PECs can be LbL assembled with oppositely charged polyelectrolyte partners for the fabrication of composite polymeric films [9], the importance of using polymeric complexes for LbL film fabrication was not fully explored. Our recent work has established that the LbL assembly of polymeric complexes can integrate the merits of both polymeric complexes and the LbL assembly for the fabrication of functional polymeric composite films [8]. In this chapter, the LbL assembly of various kinds of polymeric complexes for the fabrication of functional films, as well as their structural tailoring is presented.

## 7.2
### Concept of LbL Assembly of Polymeric Complexes

Nonstoichiometric polymeric complexes have an abundance of uncomplexed groups on their outer surface. These uncomplexed groups make polymeric complexes suitable for LbL assembly with complementary partner species to produce composite films. Taking PECs, polyelectrolyte–surfactant complexes and hydrogen-bonded polymer–polymer complexes, for examples, we demonstrate that the LbL assembly of polymeric complexes can enable the rapid fabrication of LbL-assembled films as well as convenient tailoring of film composition and structure.

### 7.2.1
#### LbL Assembly of Polyelectrolyte Complexes for the Rapid and Direct Fabrication of Foam Coatings

In general, the LbL-assembled porous films, especially the thick ones, can hardly be fabricated without using phase separation or templating methods [10, 11]. The inability of LbL assembly for straightforward porous film fabrication originates from the self-adjusting capability of polymer building blocks during the LbL assembly process, which tends to close up any pre-formed pores and make the film smooth and compact. The self-adjusting capability originates from the soft nature of polymers employed for LbL film fabrication. Water-dispersible PECs are electrostatically cross-linked polyelectrolyte colloidal particles, having larger dimension and higher rigidity than their corresponding uncomplexed polyelectrolytes. The self-adjusting capability

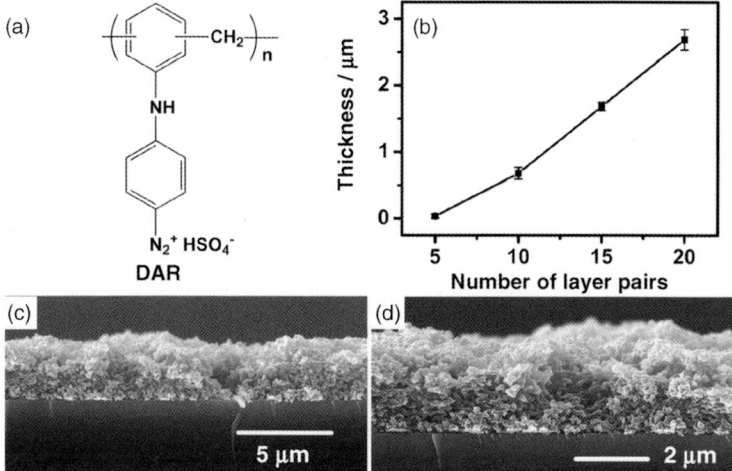

**Figure 7.2** (a) Chemical structure of DAR. (b) Dependence of the thickness of UV-irradiated PAA-DAR/DAR-PSS coatings on the number of deposition cycles. (c) Cross-sectional SEM image of a UV-irradiated (PAA-DAR/DAR-PSS)*20 coating. (d) Magnified SEM image of (c). Reproduced with permission [12]. Copyright 2009, Royal Society of Chemistry.

of rigid PECs during the LbL assembly process can be considerably suppressed. Therefore, oppositely charged PECs are employed for the direct LbL fabrication of thick porous coatings [12].

Negatively charged PECs of poly(acrylic acid) (PAA) and diazoresin (DAR, the structure is shown in Figure 7.2a) (denoted as PAA-DAR) with an average hydrodynamic diameter of ∼178 nm, and positively charged PECs of DAR and PSS (noted as DAR-PSS) with an average hydrodynamic diameter of ∼125 nm were LbL assembled for the rapid fabrication of porous coatings. No drying step was used in the coating deposition procedure unless it was in the last layer. After UV irradiation, covalently cross-linked PAA-DAR/DAR-PSS coatings were finally obtained, because the diazonium groups of DAR decomposed to produce cationic phenyl groups, which reacted with the nucleophile groups of carboxylate and sulfonate in PAA and PSS to form carboxylate and sulfonate esters [13]. The thickness of the UV-irradiated PAA-DAR/DAR-PSS coatings increases almost linearly with increasing number of coating deposition cycles (Figure 7.2b). The UV-irradiated (PAA-DAR/DAR-PSS)*20 coating has an average thickness of $2.69 \pm 0.15$ μm. The cross-sectional scanning electron microscopy (SEM) images of the UV-irradiated (PAA-DAR/DAR-PSS)*20 coating in Figure 7.2c and d clearly disclose the foam structure, where macropores distribute within the whole coating. In sharp contrast, the LbL-assembled (DAR/PAA)*20 and (DAR/PSS)*20 films are compact with a thickness of only ∼40 nm. The loose stacking of rigid sphere-like PECs accounts for the formation of foam coatings. The rapid foam coating fabrication originates from the large dimensions of PAA-DAR and DAR-PSS complexes. It should be noted that the non-drying LbL assembly is critically important for rapid and direct fabrication of foam coatings, because $N_2$ drying after each layer deposition can flatten the sphere-like PECs and produce thin and compact films. The LbL assembly of oppositely charged PECs for the fabrication of macroporous foam coatings promises that the LbL assembly of PECs is unique in realizing rapid film fabrication and controlling film structures [12].

### 7.2.2
### LbL Assembly of Hydrogen-Bonded Polymeric Complexes

Polymeric complexes can also be formed by the complexation of polymers driven by hydrogen bonds [14]. Hydrogen-bonded polymeric complexes of poly(vinylpyrrolidone) (PVPON) and PAA with a feed monomer molar ratio of PVPON to PAA of 1: 1.8 and 1: 1.2 (denoted as PVPON-PAA_1.8 and PVPON-PAA_1.2) were LbL assembled with poly(methacrylic acid) (PMAA) to demonstrate that hydrogen-bonded polymeric complexes are highly useful building blocks for LbL film fabrication [8a]. No drying step was used in the film deposition procedure to ensure the rapid fabrication of PVPON-PAA/PMAA films. PVPON-PAA_1.8 and PVPON-PAA_1.2 complexes have hydrodynamic diameters of ∼670 and ∼238 nm, respectively. The PVPON-PAA/PMAA films are constructed by the hydrogen-bonding interaction between them. As indicated in Figure 7.3a, the thickness of the PVPON-PAA/PMAA films shows an exponential increase with the number of film deposition cycles. With the same number of deposition cycles, the PVPON-PAA complexes with a larger dimension produce a thicker PVPON-PAA/PMAA film. As a comparison, the LbL-assembled PVPON/PMAA films show a linear deposition behavior, with much smaller thickness when the number of depositions is equal to those of PVPON-PAA/PMAA films. The thicknesses of (PVPON-PAA_1.8/PMAA)*30, (PVPON-PAA_1.2/PMAA)*30 and

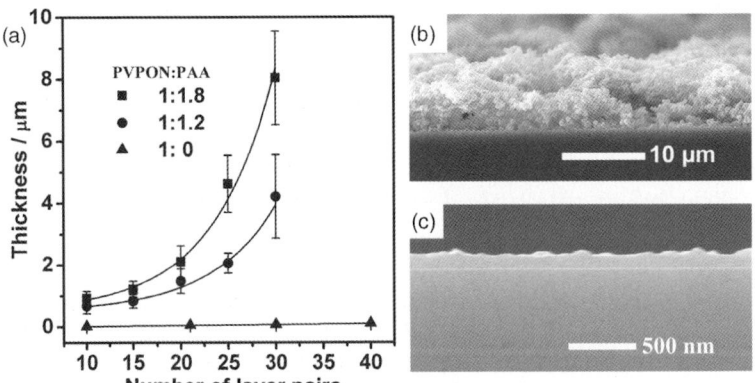

**Figure 7.3** (a) Dependence of the thickness of the PVPON-PAA/PMAA and PVPON/PMAA films on the number of deposition cycles. (b and c) Cross-sectional SEM images of a (PVPON-PAA/PMAA-1.8)*30 film (b) and a (PVPON/PMAA)*30 film (c). Reproduced with permission [8a]. Copyright 2009, Royal Society of Chemistry.

(PVPON/PMAA)*30 films are 8.0 ± 1.5 μm (Figure 7.3b), 4.2 ± 1.3 μm and 71 ± 6 nm (Figure 7.3c), respectively. These results affirmatively confirm that hydrogen-bonded PVPON-PAA complexes can be LbL assembled for rapid fabrication of micrometer-thick polymeric films.

The as-prepared PVPON-PAA/PMAA films are composed of aggregated particles of PVPON-PAA complexes glued together by PMAA (Figure 7.3b). The surface roughness of the PVPON-PAA/PMAA films increases with increasing number of film deposition cycles, which permits more PVPON-PAA complexes and PMAA to deposit than in the previous deposition cycle. Therefore, the LbL-assembled PVPON-PAA/PMAA films are fabricated in an exponential way. The thickness and surface roughness of the PVPON-PAA/PMAA films can be well-tailored by varying the complexing ratios of PVPON-PAA complexes and the number of deposition cycles. The rapidly fabricated PVPON-PAA/PMAA coatings with well tailored roughness are highly useful to fabricate superhydrophobic coatings after chemical vapor deposition of a layer of fluoroalkylsilane [8a].

### 7.2.3
### LbL Assembly of Polyelectrolyte-Surfactant Complexes

Surfactant micelles have a high capability to incorporate water-insoluble species. The complexation between polyelectrolytes and surfactant micelles stabilizes the resultant surfactant micelles in aqueous solution. Polyelectrolyte-stabilized surfactant micelles, as carriers for noncharged species, can be LbL assembled with oppositely charged partner polyelectrolytes to incorporate noncharged species into multilayer films. To demonstrate the success of this method, PAA-stabilized cetyltrimethylammonium bromide (CTAB) micelles loaded with pyrene (Py) were LbL assembled with poly(diallyldimethylammonium chloride) (PDDA) to fabricate Py-incorporated polymeric multilayer films [15].

As shown schematically in Figure 7.4a, hydrophobic molecules such as Py can be encapsulated into the hydrophobic cores of CTAB micelles. Py-loaded CTAB micelles (noted as Py@CTAB) can be complexed with negatively charged PAA polyelectrolyte to produce PAA-stabilized Py@CTAB complexes (noted as PAA-(Py@CTAB)) because of the electrostatic interaction between them. PAA-(Py@CTAB) complexes are negatively charged with an average hydrodynamic diameter of 237 nm. As depicted in Figure 7.4b, the negatively charged PAA-(Py@CTAB) complexes can be LbL assembled with polycation PDDA to produce PAA-(Py@CTAB)/PDDA multilayer films. UV–vis absorption spectra show that the amount of the incorporated Py molecules increases with increasing number of film deposition cycles. The fluorescence spectrum confirms that Py molecules exist in the hydrophobic cores of CTAB micelles in LbL-assembled PAA-(Py@CTAB)/PDDA films. Controlled experiments reveal that CTAB micelles cannot be alternately deposited with polyanions such as PSS to produce a multilayer film because CTAB micelles disassemble during the LbL deposition process. Therefore, the PAA-stabilized CTAB micelles are essential for realizing the incorporation of Py into LbL-assembled films. It is anticipated that the LbL assembly of polyelectroyte-stabilized surfactant micelles

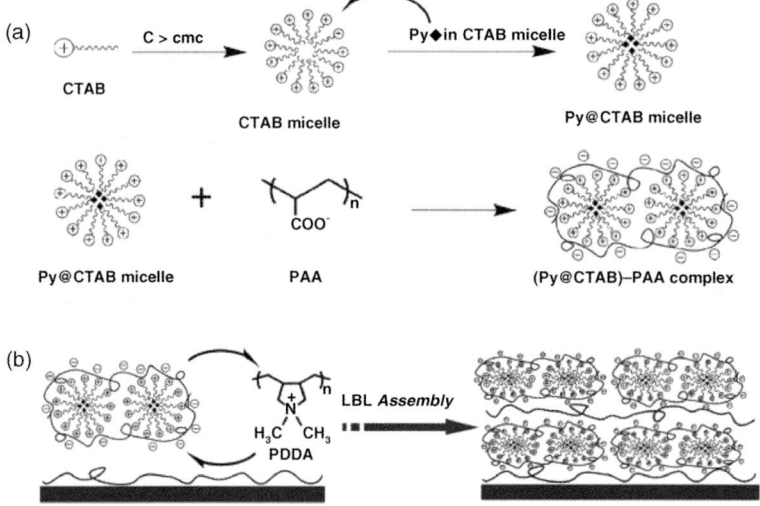

**Figure 7.4** (a) Preparative process of PAA-stabilized pyrene@CTAB micelles. (b) LbL deposition process for fabrication of PAA-(pyrene@CTAB)/PDDA multilayer films. Adapted with permission [15]. Copyright 2008, American Chemical Society.

will open a general and cost-effective avenue for the fabrication of advanced film materials containing noncharged species, such as organic molecules, nanoparticles and so forth.

## 7.3
## Structural Tailoring of LbL-Assembled Films of Polymeric Complexes

The structural tailoring of LbL-assembled films is essential for their functionalization. The diverse and highly controllable structures of polymeric complexes facilitate the precise structural tailoring of LbL-assembled films of polymeric complexes, which can be accomplished by changing the structure of the polymeric complexes in dipping solution or post-processing the LbL-assembled films. In general, a $N_2$ drying step after each layer deposition produces a lateral shearing force which can spread the deposited polymeric complexes and produce thin, flat and compact films [8a,12]. The structural tailoring of LbL-assembled films of PECs was demonstrated to show the convenience in controlling the film structures of LbL-assembled polymeric complexes [16–18].

### 7.3.1
### Mixing Ratio of PECs

Mixing ratio is one of the primary parameters which determine the composition and structure of PECs. By varying the mixing ratios, the electrostatic cross-linking density,

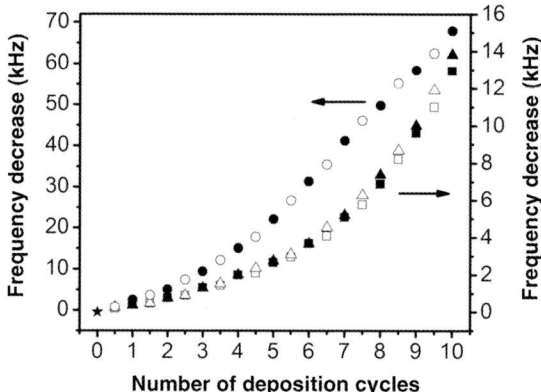

**Figure 7.5** QCM frequency decrease (-$\Delta F$) of alternate deposition of PAA-PAA complexes (hollow symbols) and PSS (solid symbols): PAH-PAA$_{0.25}$/PSS (squares), PAH-PAA$_{0.5}$/PSS (triangles), and PAH-PAA$_{0.75}$/PSS (circles) films. Reproduced with permission [16]. Copyright 2009, American Chemical Society.

size, and rigidity of the PECs can be systematically tailored. Water-dispersible PECs of poly(allylamine hydrochloride) (PAH) and PAA with monomer ratios of PAH to PAA of 1: 0.25, 1: 0.5 and 1: 0.75 (denoted as PAH-PAA$_{0.25}$, PAH-PAA$_{0.5}$, PAH-PAA$_{0.75}$, respectively) have average hydrodynamic diameters of 145, 177, and 735 nm, respectively. They are positively charged with decreasing $\zeta$-potentials with increasing ratio of PAA components in PAH-PAA complexes. PAH-PAA complexes were LbL assembled with polyanion PSS to fabricate PAH-PAA/PSS films [16]. Quartz crystal microbalance (QCM) measurements were employed to monitor the film deposition process. As shown in Figure 7.5, the PAH-PAA$_{0.25}$/PSS and PAH-PAA$_{0.5}$/PSS films show a quite similar deposition behavior because the PAH-PAA$_{0.25}$ and PAH-PAA$_{0.5}$ complexes have quite similar sizes and surface charge density. In contrast, the increased size of PAH-PAA$_{0.75}$ complexes leads to more PAH-PAA complexes and PSS deposition when the number of film deposition layers is equal to those of PAH-PAA$_{0.25}$/PSS and PAH-PAA$_{0.5}$/PSS films. Surface morphology investigation reveals that PAH-PAA$_{0.75}$/PSS films have a rougher surface than that of the PAH-PAA$_{0.25}$/PSS and PAH-PAA$_{0.5}$/PSS films. The cross-sectional SEM images of the (PSS/PAH-PAA$_{0.25}$)*20, (PSS/PAH-PAA$_{0.5}$)*20 and (PSS/PAH-PAA$_{0.75}$)*20 films reveal an average film thickness of 853.5 ± 29.2, 966.1 ± 15.5 and 2048 ± 676 nm, respectively. PAH-PAA complexes with low mixing ratios of PAA to PAH contain a large amount of free amine groups and are flexible with a slightly cross-linked structure. In comparison, the PAH-PAA$_{0.75}$ complexes have larger sizes and more rigid spherical structure than that of PAH-PAA$_{0.25}$ and PAH-PAA$_{0.5}$ complexes because of the heavy cross-linking between the PAH and PAA chains. The larger sizes and higher rigidity of the PAH-PAA$_{0.75}$ complexes explain the larger thickness and higher roughness of the PSS/PAH-PAA$_{0.75}$ films. Therefore, by varying the mixing ratios, the size, rigidity and surface charge density of the PECs can be controlled to well tailor the structures of the LbL-assembled films of PECs.

## 7.3.2
### LbL Codeposition of PECs and Free Polyelectrolytes

Free polyelectrolytes usually exist in the dispersions of PECs because there is a balance between the PECs and free polyelectrolytes, especially for those with one kind of polyelectrolyte being in excess [7b]. It is important to systematically investigate the deposition behavior and structural tailoring of the LbL-assembled films when PECs and free polyelectrolytes coexist in the dipping solutions. To do so, DAR-PAA complex dispersion with a feed monomer molar ratio of DAR to PAA of 1: 0.4 was prepared which contains ~46% uncomplexed DAR. Aqueous DAR-PAA complex dispersion and PAA solution were used as dipping solutions for the LbL assembly of PAA/(DAR-PAA + DAR) films [17]. No drying step was used during the film deposition procedure until it was in the last layer. The as-prepared polymeric films were then cross-linked under UV irradiation [12]. As shown in Figure 7.6a and b, the [PAA/(DAR-PAA + DAR)]*50 film is recognized by its bilayer structures, which comprise the underlying continuous layer and the top discontinuous layer. The continuous films originate from the LbL deposition of PAA with free DAR in the aqueous DAR-PAA dispersion, whereas the discontinuous films with hierarchical structures correspond to the alternate deposition of PAA with the positively charged DAR-PAA complexes. The deposition of PAA/DAR films takes place simultaneously with the deposition of PAA/DAR-PAA discontinuous films, but proceeds more slowly than that of PAA/PAA-DAR because of the smaller dimensions of free DAR than those of DAR-PAA complexes. Therefore, PAA/DAR films occupy the inter-

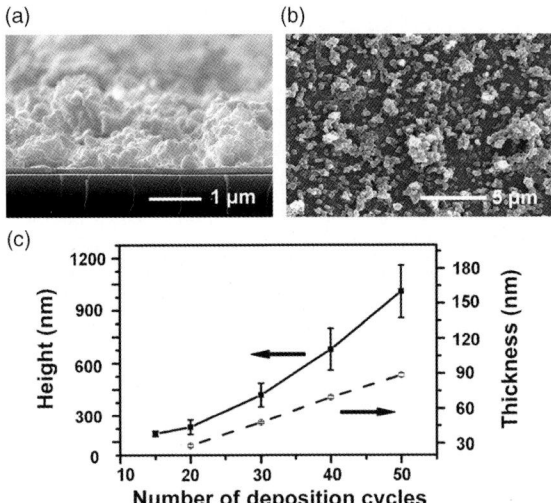

**Figure 7.6** (a) Cross-sectional and (b) top-view SEM images of UV irradiated [PAA/(DAR-PAA + DAR)] *50 coatings. (c) Dependence of the height of the hierarchical structures (■) and the thickness of the continuous films (○) of the PAA/(DAR-PAA + DAR) films as a function of the number of coating deposition cycles. Reproduced with permission [17]. Copyright 2010, American Chemical Society.

stices of PAA/DAR-PAA coatings. As a result, polymeric coatings with the hierarchical structures of PAA/DAR-PAA rooting in the underlying continuous PAA/DAR films are produced. The thickness of the underlying PAA/DAR and height of the PAA/DAR-PAA hierarchical structures in PAA/(DAR-PAA + DAR) films increase with increasing number of film deposition cycles, as indicated in Figure 7.6c. By varying the ratio of free DAR to DAR-PAA complexes in the dipping solution, the layer pair structures of the films can be well tailored.

### 7.3.3
### Salt Effect

The structure and properties of PECs depend strongly on the total ionic strength of the solution. Therefore, the structural tailoring of the LbL-assembled films of PECs can be accomplished by varying the ionic strength of the aqueous dispersions of PECs. Salt-containing PAH-PAA complexes with mixing molar ratio of PAH to PAA of 1: 0.75 were LbL assembled with PSS to investigate the ionic-strength-dependent fabrication of polymeric films [18]. As indicated in Figure 7.7a, the hydrodynamic diameters of PAH-PAA complexes increase with increasing concentration of NaCl in their aqueous dispersions. In the first stage, when the concentration of NaCl is lower than 80 mM, the hydrodynamic diameters of the PECs slightly increase from ~120 to

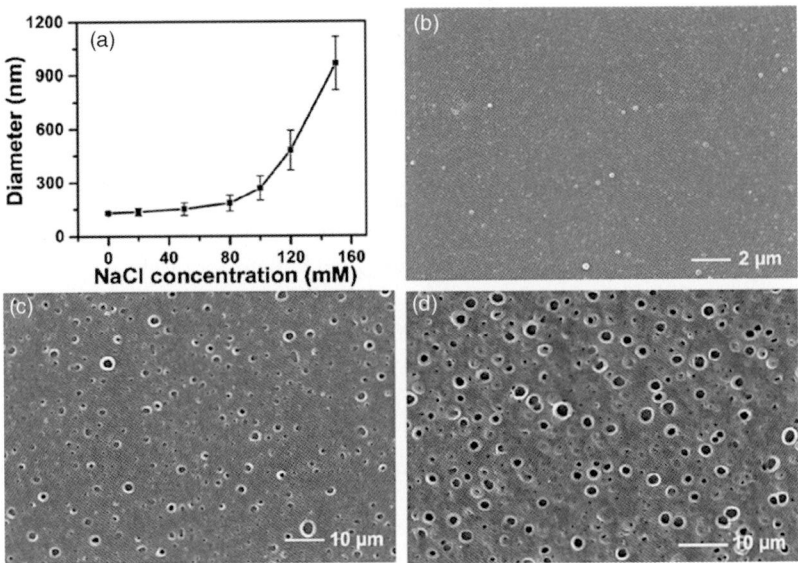

**Figure 7.7** (a) Dependence of the hydrodynamic diameter of PAH-PAA complexes on NaCl concentration in aqueous PAH-PAA dispersions. (b-d) SEM images of (PSS/PAH-PAAm)*30 films with the concentration of NaCl in the dipping solutions of PAH-PAA complexes being (b) 0, (c) 100, and (d) 120 mM. Adapted with permission [18]. Copyright 2011, American Chemical Society.

~180 nm, illustrating slight swelling of the PECs particles. In the second stage, a pronounced salt-induced size increase of PAH-PAA complexes is observed when the concentration of NaCl is higher than 80 mM, suggesting the aggregation of PAH-PAA complexes. Meanwhile, the $\zeta$-potentials of the salt-containing PAH-PAA complexes decrease with the addition of NaCl.

No drying step was used in the LbL deposition of PSS/PAHPAA$_m$ films (herein $m$ refers to the concentration of NaCl added to the complex dispersions, with $m$ being 0, 100 or 120 mM) unless it was in the last layer. With the same number of deposition cycles, the thickness of the PSS/PAH-PAA films increases with increasing concentration of NaCl in the corresponding aqueous PAH-PAA dispersions. The surface of the (PSS/PAH-PAA$_0$)*30 film is full of tiny particles and is compact (Figure 7.7b). Pores are observed on the surface of the (PSS/PAH-PAA$_{100}$)*30 and (PSS/PAH-PAA$_{120}$)*30 films (Figure 7.7c and d). Some of the pores can run through the whole film. A higher concentration of NaCl leads to greater porosity in (PSS/PAH-PAA$_m$)*30 films. Taking a (PSS/PAH-PAA$_{120}$)*30 film for instance, *in situ* atomic force microscopy (AFM) measurements disclose that microphase separation takes place for the film in the wet state, but no pores as in the dried film can be observed. Upon drying the wet (PSS/PAH-PAA$_{120}$)*30 film with N$_2$ flow, porous (PSS/PAH-PAA$_{120}$)*30 film can be obtained. Therefore, the dewetting process is considered to lead to the formation of pores in LbL-assembled films of salt-containing PAH-PAA complexes. The microphase separation in the wet (PSS/PAH-PAA$_{120}$)*30 film can be considered as the nucleation of the dewetting pores, while the N$_2$ drying process is the growth stage of the dewetting pores [19]. The LbL assembly of salt-containing PECs cannot only enable the fine structural tailoring of the LbL-assembled films of PECs, but also provide a way for the fabrication of porous polymeric films.

## 7.4
### LbL-Assembled Functional Films of Polymeric Complexes

Along with the rapid fabrication of micrometer-thick films with well-tailored film structures, the LbL assembly of polymeric complexes is expected to build up novel functional films which are difficult or impossible to fabricate by LbL assembly of uncomplexed polymers. Self-healing superhydrophobic coatings, antireflection- and antifogging-integrated coatings, and mechanically reinforced coatings enabled by the homogeneous dispersion of nanofillers are fabricated to show the uniqueness of the LbL assembly of polymeric complexes for functional film preparation.

### 7.4.1
#### Self-Healing Superhydrophobic Coatings

Superhydrophobic coatings promise a wide range of applications, from self-cleaning surfaces to corrosion-resistant, anti-adhesive, and drag-reducing coatings [20]. However, the artificial superhydrophobic coatings lose their unique properties due to sun bleaching or mechanical actions. Endowing artificial superhydrophobic coatings

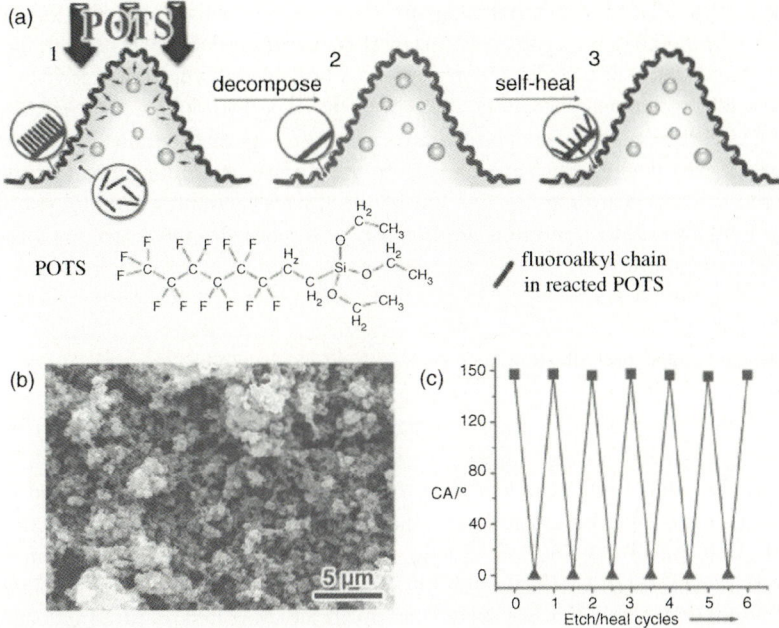

**Figure 7.8** (a) Working principle of self-healing superhydrophobic coatings. (b) Top-view SEM images of a (PAH–SPEEK/PAA)*60.5 coating on a silicon wafer. (c) Contact angle (CA) of $O_2$ plasma-damaged superhydrophobic coating (▲) and the coating after self-healing (■). Adapted with permission [21]. Copyright 2010, Wiley-VCH.

with a self-healing ability can solve the durability problem. As shown in Figure 7.8a and b, porous polymer coatings with micro- and nanoscaled hierarchical structures fabricated by LbL assembly of PECs of PAH and sulfonated poly(ether ether ketone) (SPEEK, sulfonation degree ≈82%) with PAA were employed to fabricate self-healing superhydrophobic coatings [21]. The as-prepared (PAH-SPEEK/PAA)*60.5 coating has an average thickness of $2.7 \pm 0.4$ μm. Thermal cross-linking produces amide bonds between the carboxylate and amine groups and enhances the mechanical robustness of the coatings. Upon chemical vapor deposition of 1H, 1H, 2H, 2Hperfluorooctyltriethoxysilane (POTS), superhydrophobic coatings with a water contact angle of 157° and a sliding angle as low as 1° were finally obtained. Water droplets roll off easily from such coatings. During the chemical vapor deposition process, POTS molecules not only deposit on the coating surface, but also penetrate into the whole coating, because of the porosity of the coating. The porous PAH–SPEEK/PAA coatings act as a reservoir to accommodate an abundance of POTS healing agents. Once the primary top POTS layer is decomposed or scratched away, the preserved POTS can migrate to the coating surface under a slightly humid environment to lower surface energy through rearrangement of polyelectrolyte chains. In this way, the damaged superhdrophobicity of the coating is healed. As indicated in Figure 7.8c, the $O_2$ plasma-treated coating became superhydrophilic.

After being transferred to an ambient environment with a relative humidity (RH) of 40% for 4 h, the coating regained its original superhydrophobicity. The etching–healing process can be repeated many times without decreasing the superhydrophobicity of the self-healed coatings. Moreover, the superhydrophobic (PAA/PAH-SPEEK)*60.5 coatings are scratch-resistant and their superhydrophobicity can be self-healed, even when scratches are made on the surface. It is anticipated that the introduction of a self-healing function into artificial superhydrophobic coatings will open a new avenue to extending the lifespan of superhydrophobic coatings for practical applications.

### 7.4.2
### Mechanically Stable Antireflection- and Antifogging-Integrated Coatings

The antireflection- and antifogging-integrated coatings are widely useful in daily life because they can effectively enhance the transmission of light and, meanwhile, considerably reduce water condensation. Highly nanoporous films comprised of hydrophilic materials with low refractive index meet the requirement of integrated antireflection and antifogging coatings [8b]. To fabricate antireflection- and antifogging-integrated coatings with extremely high transmittance and mechanical stability, complexes of PDDA and sodium silicate (PDDA-silicate) were first prepared and then LbL assembled with PAA to fabricate PAA/PDDA-silicate hybrid films (Figure 7.9). The positively charged PDDA-silicate complexes have an average size of about 13.2 nm, as revealed by transmission electron microscopy (TEM) measurements. The PAA/PDDA-silicate films show a linear deposition behavior, with a thickness

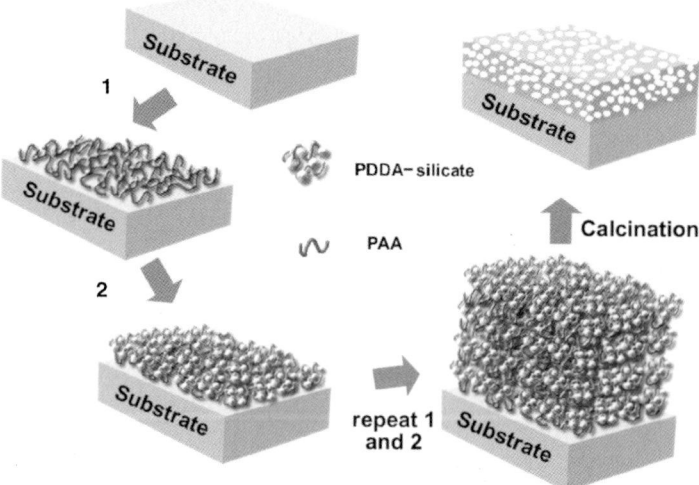

**Figure 7.9** Schematic illustration of the LbL deposition of PDDA-silicate complexes and PAA for the fabrication of antireflective and antifogging coatings. Reproduced with permission [8b]. Copyright 2008, American Chemical Society.

increment of 21.4 nm per deposition cycle. The as-prepared (PAA/PDDA-silicate)*12 film has a compact film structure, with a constant thickness of ~285.7 nm. Calcination produces highly porous silica coatings because the PDDA and PAA porogens are burned out. Meanwhile, calcination cross-links the porous silica coatings via the formation of stable siloxane bridges, which endows the coatings with high mechanical stability and excellent adhesion to the substrates. The thickness of the calcinated (PAA/PDDA-silicate)*12 film is ~121.8 nm, with a reduced thickness of ~55% because of the removal of organic components. The bare quartz substrate has a transmittance below ~93% in the spectral range between 300 and 800 nm. The quartz with both sides covered with calcinated (PAA/PDDA-silicate)*12 coatings has a maximum transmittance of 99.86% at a wavelength of 570 nm. Controlled experiments show that the LbL-assembled binary silicate/polycation films after calcination have a maximum transmittance of about 97.5% in the visible region. Compared with the uncomplexed silicate, the PDDA-silicate complexes allow the introduction of a higher ratio of organic components in the LbL-assembled hybrid films, and high porosity after calcinations. Therefore, PDDA-silicate complexes are indispensable for the fabrication of porous antireflection coatings with extremely high transmittance.

The highly porous silica coatings derived from the calcinated (PAA/PDDA-silicate)*12 films are superhydrophilic, which can significantly suppress the fogging behavior because condensed water droplets spread flat, almost instantaneously, to form a thin sheet-like water membrane. In this way, light scattering by the condensed water droplets is eliminated. The pencil hardness of the porous silica coatings was higher than 5H. The long-term application of the antireflection and antifogging coatings under conditions of daily maintenance is guaranteed. The LbL assembly of PDDA-silicate complexes not only provides a facile and cost-effective method for the fabrication of mechanically stable antireflection- and antifogging-integrated coatings, but also extends the concept of LbL assembly of polymeric complexes to polymeric–inorganic complexes [8b].

### 7.4.3
**Transparent and Scratch-Resistant Coatings with Well-Dispersed Nanofillers**

Uniform dispersion of nanofillers into polymer materials provides a very effective strategy to enhance their mechanical properties. However, the uniform dispersion of the nanofillers in polymer composites is always challenging because the nanosized fillers possess extremely large surface area and have a strong tendency to agglomerate [22]. The recently developed e-LbL assembly enables the fabrication of micrometer-thick polyelectrolyte coatings in a rapid way because of the "in-and-out" diffusion of polyelectrolytes [6]. An innovative and straightforward strategy to homogeneously disperse nanofillers into micrometer-thick polymeric coatings was developed by making use of the "in-and-out" diffusion of polyelectrolytes during e-LbL assembly of polyelectrolyte–nanofiller complexes [23].

The complexes of PAA and *in situ* synthesized $CaCO_3$ nanoparticles (denoted as PAA-$CaCO_3$) were alternately deposited with PAH to fabricate hybrid PAA-$CaCO_3$/

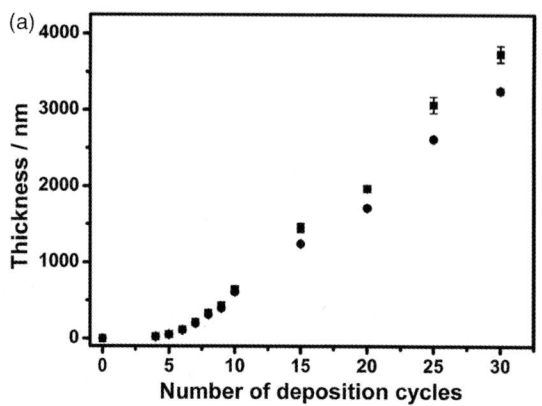 

**Figure 7.10** (a) The dependence of the thickness of the as-prepared and thermally cross-linked PAA-CaCO$_3$/PAH coatings on the number of deposition cycles. ■ and ● represent the as-prepared and the cross-linked coatings, respectively. (b) Photographs of azobenzene-containing films with (right) and without (left) the (PAA-CaCO$_3$/PAH)*20 protecting coating after repeatedly rubbing. Adapted with permission [23]. Copyright 2010, Royal Society of Chemistry.

PAH coatings. As indicated in Figure 7.10a, the PAA-CaCO$_3$/PAH coatings exhibit a typical exponential deposition behavior because of the "in-and-out" diffusion of PAH during the coating fabrication process. The thermally cross-linked (PAA-CaCO$_3$/PAH)*20 coating has a constant thickness of $1.72 \pm 0.19$ μm. The loading content of CaCO$_3$ nanofillers in the hybrid coatings was determined by thermogravimetric analysis to be 4.2 wt%. TEM measurements reveal that ultrafine CaCO$_3$ nanoparticles of ~2 nm diameter are homogeneously distributed in the PAA-CaCO$_3$/PAH coatings. The "in-and-out" diffusion of polyelectrolytes and the strong interaction of CaCO$_3$ nanoparticles with PAA enable the homogeneous dispersion of the CaCO$_3$ nanofillers in the exponentially growing PAA-CaCO$_3$/PAH coatings. The extremely smooth surface and uniformly dispersed CaCO$_3$ nanofillers make the thermally cross-linked (PAA-CaCO$_3$/PAH)*20 coatings highly transparent. The thermally cross-linked PAA-CaCO$_3$/PAH coatings, which have greatly enhanced hardness and Young's elastic modulus because of the well-dispersed CaCO$_3$ nanofillers, are highly scratch-resistant. Such PAA-CaCO$_3$/PAH coatings are very useful as scratch-protection layers of other functional films. For instance, an azobenzene-containing film was covered with a thermally cross-linked (PAA-CaCO$_3$/PAH)*20 coating and repeatedly scratched with a piece of ramie cloth under an applied pressure of $1.12 \times 10^4$ Pa. As depicted in Figure 7.10b, the azobenzene film was seriously worn away after 40 cycles of rubbing. However, there are no scratches observed in the azobenzene film protected by thermally cross-linked (PAA-CaCO$_3$/PAH)*20 coating after 1100 cycles of rubbing. These results clearly demonstrate that cross-linked PAA-CaCO$_3$/PAH coatings of high transparency can be used as excellent scratch-resistant layers of other functional films. The e-LbL assembly of polyelectrolyte–nanofiller complxes is believed to provide a general way to well disperse nanofillers in rapidly growing polymeric composite films for fabricating mechanically robust functional coatings.

## 7.5
## Summary

We have demonstrated that the LbL assembly of polymeric complexes can integrate the merits of both polymeric complexes and the LbL assembly technique for the fabrication of functional polymeric composite films. The large dimensions of polymeric complexes enable the rapid fabrication of LbL-assembled thick films. The structural diversity of polymeric complexes combined with the stepwise film deposition pattern facilitates the structural tailoring of LbL-assembled films of polymeric complexes. Therefore, the LbL assembly of polymeric complexes can build up novel functional films that are difficult or impossible to fabricate by the LbL assembly of uncomplexed polymers, as exemplified by the fabrication of foam coatings, self-healing superhydrophobic coatings, mechanically reinforced coatings with uniformly dispersed nanofillers, and so forth. We believe that the LbL assembly of polymeric complexes will certainly enrich the structures and, therefore, functionalities of the LbL-assembled films.

## References

1 Decher, G. and Hong, J.-D. (1991) *Makromol. Chem., Macromol. Symp.*, **46**, 321–327.
2 (a) Decher, G. (1997) *Science*, **277**, 1232–1237; (b) Hammond, P.T. (2004) *Adv. Mater.*, **16**, 1271–1293; (c) Zhang, X., Chen, H., and Zhang, H.Y. (2007) *Chem. Commun.*, 1395–1405.
3 Tang, Z.Y., Kotov, N.A., Magonov, S., and Ozturk, B. (2003) *Nat. Mater.*, **2**, 413–418.
4 Cho, J., Char, K., Hong, J.-D., and Lee, K.-B. (2001) *Adv. Mater.*, **13**, 1076–1078.
5 (a) Schlenoff, J.B., Dubas, S.T., and Farht, T. (2000) *Langmuir*, **16**, 9968–9969; (b) Izquierdo, A., Ono, S.S., Voegel, J.-C., Schaaf, P., and Decher, G. (2005) *Langmuir*, **21**, 7558–7567.
6 Picart, C., Mutterer, J., Richert, L., Luo, Y., Prestwich, G.D., Schaaf, P., Voegel, J.-C., and Lavalle, P. (2002) *Proc. Natl. Acad. Sci. USA*, **99**, 12531–12535.
7 (a) Tsuchida, E. and Abe, K. (1982) *Adv. Polym. Sci.*, **45**, 1–119; (b) Thünemann, A.F., Müller, M., Dautzenberg, H., Joanny, J.-F., and Löwen, H. (2004) *Adv. Polym. Sci.*, **166**, 113–171; (c) Kabanov, V. (2003) *Multilayer Thin Films: Sequential Assembly of Nanocomposite Material* (eds G. Decher and J.B. Schlenoff), Wiley-VCH Verlag GmbH, Weinheim, Germany, pp. 47–86; (d)Philipp, B., Dautzenberg, H., Linow, K.J., Kötz, J., and Dawydoff, W. (1989) *Prog. Polym. Sci.*, **14**, 91–172.
8 (a) Liu, X.K., Dai, B.Y., Zhou, L., and Sun, J.Q. (2009) *J. Mater. Chem.*, **19**, 497–504; (b) Zhang, L.B., Li, Y., Sun, J.Q., and Shen, J.C. (2008) *Langmuir*, **24**, 10851–10857.
9 (a) Schuetz, P. and Caruso, F. (2002) *Colloids Surf. A*, **207**, 33–40; (b) Reihs, T., Müller, M., and Lunkwitz, K. (2003) *Colloids Surf. A*, **212**, 79–95.
10 (a) Mendelsohn, J.D., Barrett, C.J., Chan, V.V., Pal, A.J., Mayes, A.M., and Rubner, M.F. (2000) *Langmuir*, **16**, 5017–5023; (b) Li, Q., Quinn, J.F., Wang, Y., and Caruso, F. (2006) *Chem. Mater.*, **18**, 5480–5485.
11 Zimnitsky, D., Shevchenko, V.V., and Tsukruk, V.V. (2008) *Langmuir*, **24**, 5996–6006.
12 Zhang, L. and Sun, J.Q. (2009) *Chem. Commun.*, 3901–3903.
13 (a) Sun, J.Q., Tao, W., Sun, Y.P., Wang, Z.Q., Zhang, X., Shen, J.C. and Cao, W.X. (1998) *Chem. Commun.*,

1853–1854; (b) Sun, J.Q., Wu, T., Liu, F., Wang, Z.Q., Zhang, X., and Shen, J.C. (2000) *Langmuir*, **16**, 4620–4624.

14 Nurkeeva, Z.S., Mun, G.A., and Khutoryanskiy, V.V. (2003) *Macromol. Biosci.*, **3**, 283–295.

15 Liu, X.K., Zhou, L., Geng, W., and Sun, J.Q. (2008) *Langmuir*, **24**, 12986–12989.

16 Guo, Y.M., Geng, W., and Sun, J.Q. (2009) *Langmuir*, **25**, 1004–1010.

17 Zhang, L. and Sun, J.Q. (2010) *Macromolecules*, **43**, 2413–2420.

18 Zhang, L., Zheng, M., Liu, X.K., and Sun, J.Q. (2011) *Langmuir*, **27**, 1346–1352.

19 Petrov, J.G., Ralston, J., and Hayes, R.A. (1999) *Langmuir*, **15**, 3365–3373.

20 Verho, T., Bower, C., Andrew, P., Franssila, S., Ikkala, O., and Ras, R.H.A. (2011) *Adv. Mater.*, **23**, 673–678.

21 Li, Y., Li, L., and Sun, J.Q. (2010) *Angew. Chem. Int. Ed.*, **49**, 6129–6133.

22 Rong, M.Z., Zhang, M.Q., and Ruan, W.H. (2006) *Mater. Sci. Technol.*, **22**, 787–796.

23 Liu, X.K., Zhou, L., Liu, F., Ji, M.Y., Tang, W.G., Pang, M.J., and Sun, J.Q. (2010) *J. Mater. Chem.*, **20**, 7721–7727.

# 8
# Making Aqueous Nanocolloids from Low Solubility Materials: LbL Shells on Nanocores

*Yuri Lvov, Pravin Pattekari, and Tatsiana Shutava*

## 8.1
## Introduction

Many organic and inorganic compounds with good application potential are limited in admixing/dispersing in liquids because of their poor (lower than <0.005 mg mL$^{-1}$) or moderate solubility in aqueous and nonaqueous media, thus excluding needed composite formulations [1]. For liquid hydrophobic materials, such as oils, there is a well-established procedure of emulsification, which allows the stable dispersion of microdroplets in water [2, 3]. We have successfully converted solid materials into stable aqueous nanocolloids with a particle size of less than 200 nm through simultaneous powerful ultrasonication and layer-by-layer (LbL) polyelectrolyte coating (sonication-assisted layer-by-layer (SLbL) encapsulation). This was a development of LbL microencapsulation introduced by Sukhorukov, Donath, Caruso, Möhwald et al. [4–10]. Most of their works exploit the formation of nano/engineered polyelectrolyte shells on pre-formed microtemplates with much larger diameters of 2 to 5 µm [11–18].

There are a number of publications on micronizing drug or dye particles and building LbL shells, typically containing 4 to 10 polyelectrolyte bilayers, and allowing a long particle dissolution time (from minutes up to 10 h) through adjustable capsule wall thickness (20–50 nm) [8–10, 19]. LbL-shell-coated dye microparticles have been used as paint additives, and LbL cellulose microfibers in paper production. Soluble drugs, such as furosemide, nifedipine, naproxen, biotin, vitamin K3 and insulin, were mechanically crushed into a dry powder and used as cores for LbL shell assembly at a pH where they have low solubility, in order to preserve the drug microcores from dissolution during the preparation [12–15]. The typical particle sizes of such a formulation were 2–10 µm. In another approach, LbL microcapsules were assembled on sacrificed microcores (a few micrometers diameter $CaCO_3$, $MnCO_3$, or silica). Then these cores were dissolved and the empty shells were loaded with proteins or drugs through a pH-controlled capsule wall opening [13–15]. Induced drug release is also possible with light responsive capsule opening [14]. Contrary to the case of solid drug cores, these microshells contained a relatively low amount of loaded materials (1–5 vol%).

*Multilayer Thin Films: Sequential Assembly of Nanocomposite Materials*, Second Edition.
Edited by Gero Decher and Joseph B. Schlenoff.
© 2012 Wiley-VCH Verlag GmbH & Co. KGaA. Published 2012 by Wiley-VCH Verlag GmbH & Co. KGaA.

Here, we restrict ourselves to nanoparticles with diameters of 100–200 nm coated with very thin LbL-shells (two or three polyelectrolyte bilayers of 10–15 nm thickness). If such cores are of poorly-soluble drugs (often anticancer drugs), they are suitable for intravenous blood injection, while LbL coating provides a strong surface charge (zeta-potential of about ±40 mV), producing stable aqueous nanocolloids from these initially poorly-soluble materials. If the cores are of dye pigments, one may expect better colloidal stability of the polar solvent based paint loaded with these core–shell nanoparticles.

As a consequence of working with nanometer-sized cores, one faces many technological limitations in analytical and physico-chemical methods of analysis. For example, laser confocal microscopy is very efficient for the detailed study of the structure of LbL microcapsules, proving the location of the loaded drugs and demonstrating their penetration into cells [16]. However, such a technique cannot be used successfully for the nanometer-diameter size. An application of SEM and TEM to nanocore–shell capsules is necessary.

Centrifugation and filtration, which are commonly applied to 1–5 μm particles in the step-wise polyelectrolyte deposition in LbL encapsulation, are difficult for 100 nm capsules. Centrifuge precipitation leads to large losses of the sample and aggregation. Therefore, use of a non-washing LbL assembly technique to form a consistent number of layers in the shell is needed. To produce the needed good colloidal stability surface potential, two to three polyelectrolyte bilayers are sufficient. For poorly-soluble nanocores, one does not need to build thick shells, contrary to the common LbL microcapsule concept where walls serve as an adjustable diffusion barrier. With low solubility drugs (tamoxifen, paclitaxel, and curcumin), the main task is to produce nanocolloids with a concentration of 4–5 mg mL$^{-1}$ that are stable in biological buffers. For such drugs, we do not need thicker capsule walls because the core materials already have low solubility; and even with two bilayer polycation/polyanion shells, the core dissolution in a volume sufficient for its complete solubilization (sink conditions) takes 4–20 h. Additional coating with sequential polyelectrolyte layers allows the building of a sophisticated capsule wall architecture for advanced properties such as, targeting with specific proteins (cell-penetrating peptide (Tat), monoclonal nucleosome-specific 2C5 antibody (mAb 2C5)), antifouling properties, long-time salt and buffer stability, lower toxicity and immunogenicity, altered distribution in the body (covalently attached PEG and dextran macromolecules), and anticoagulation properties.

Our approach for 100–200 nm diameter capsules is based on the powerful sonication of powders of poorly-soluble materials in the presence of polyelectrolytes and amphiphiles, which are adsorbed on newly developed surfaces, charging particles and preventing smaller and smaller pieces from re-aggregation. Nanocolloids of organic and inorganic poorly-soluble materials with content up to 80 wt% and concentration up to 5 mg mL$^{-1}$ were prepared with SLbL via alternate adsorption of oppositely-charged natural polyelectrolytes [20–24]. This method allowed the nanoparticulation of low solubility anticancer drugs, anticorrosion agents, MRI contrast improving agent, insoluble dyes, and inorganic salts.

## 8.2
## Formation of Nanocores

Experimental manufacturing techniques that allow a high degree of material dispersion can be divided into two classes: top-down and bottom-up approaches [25–28]. Top-down disintegration involves high-pressure homogenization, ultrasonication or pearl/ball milling of a coarse material in either water or nonaqueous media. Pressure and very high speed rotation of the milling pearls create powerful disruptive forces (shearing, collision, and cavitation) which disintegrate the bulk powder into nanoparticles [25–27]. The particle size of the resulting nanocrystals depends on the amount of initial substance, stabilizer, and technological parameters of the material processing, which are variable and highly dependent on the instrumentation used. High-pressure homogenization and ball milling are widely used in industry for preparing drug nanocrystals [27, 28], dye microparticles [17, 25] and inorganic crystals [25]. Several drug nanocrystal formulations such as Triglige®, Rapamune®, Emend®, Tricore®, Megas ES®, and Invega® have been approved by the FDA for medical usage, and several others are in clinical trials [27, 28].

The application of ultrasound technology has increased over the last decades. The ultrasound cavitation bubbles in a liquid expand and collapse. A temperature of several thousand degrees and pressure exceeding one thousand atmospheres are developed in the confined volume of a collapsing cavitation bubble, crushing particles to smaller pieces [29–34].

Bottom-up techniques, or nano-precipitation methods, start from real solutions of substances of interest in an appropriate organic solvent, such as acetone, tetrahydrofuran, DMSO, ethanol, N-methyl-2-pyrrolidone, and so on. Seed development is initiated by adding this highly concentrated solution to a miscible anti-solvent (often water) in the presence of a stabilizer, and then slowly increasing the amount of anti-solvent in the mixture. The latter can be achieved by means of a secondary affect of powerful sonication (elevated temperature of the reaction media), which enhances evaporation of an organic solvent, increases the percentage of water in the media that leads to slow oversaturation, multiple seeds, and, finally, nanocrystal formation [20–24]. Growth of such seeded nanocrystals is limited by the adsorption of amphiphilic molecules and polyelectrolytes, resulting in 100–200 nm cores. In another approach, a sonication method of preparation of aqueous 50-nm beta-lactoglobulin particles based on thermal denaturation of the protein was proposed [35, 36].

## 8.2.1
## Stabilizers versus Layer-by-Layer Polyelectrolyte Shell

When particle dimensions are at the nanoscale range, the surface area of the nanoparticles increases, thus increasing the Gibbs free energy of the system [26, 37]. The system prefers to reduce it by agglomeration, especially in concentrated suspensions [38]. This results in particle size increase and poorer colloidal stability. To minimize the surface energy and prevent re-aggregation, all the aforementioned techniques require the presence of surfactants, both ionic (e.g., sodium lauryl sulfate, sodium docusate) and non-ionic (polysorbates, polyethylene glycols, polyvinyl pyrrolidone) [26, 39, 40]. Among ionic polymeric stabilizers, several belong to the class of polyelectrolytes, such as: carboxymethyl cellulose, chitosan and its quaternized derivatives, carrageenans, amine derivatives of dextran, and so on. For better performance, surfactants are applied in a combination that provides both electrostatic and steric repulsions between nanoparticles [39, 40]. Typical formulations contain up to 5–10 wt% of stabilizers in supernatant that cannot be removed or exchanged without losing colloidal stability [28, 39, 40].

LbL polyelectrolyte-stabilized nanoparticles are different. After adsorption of the first, usually polycationic, layer, one has a colloidal dispersion of materials coated with a layer, which provides a high surface $\xi$-potential, usually of $+35$ mV. Deposition of the second anionic polyelectrolyte ties chains of the first polyelectrolyte together, preventing their detachment from the surface while bringing the nanoparticle $\xi$-potential to a negative value (ca. $-40$ mV). Such multilayer polyelectrolyte shells provide colloidal stability for months.

## 8.3
### Ultrasonication-Assisted LbL Assembly

We have applied top-down and bottom-up approaches in ultrasonication-assisted LbL assembly [20–24]. We prepared stable 100–200 nm colloids of inorganic and organic low solubility drugs, dye pigments and anticorrosion agents (Figure 8.1). These core–shell nanoparticles have up to 80 wt% of the loaded compound; there are no free stabilizers in the solution, and they can be concentrated up to 5 mg mL$^{-1}$ without agglomeration.

### 8.3.1
### Top-Down Approach

In a typical experimental set-up for ultrasonication-assisted LbL assembly (Figure 8.2), insoluble crystals (curcumin, paclitaxel) or amorphous micropowder (resveratrol, tamoxifen) are dispersed in water to attain an initial concentration of 0.5–2 mg mL$^{-1}$. A buffer (e.g., PBS) or water-miscible organic solvent (such as glycerol, PEG400) can also be used. The crude dispersion is sonicated using a powerful ultrasonic processor (20 kHz, 50–100 W) with the temperature controlled at 25 °C. A polyelectrolyte bearing charge opposite to that of the core material is added to the dispersion before or during the sonication at a typical ratio of $\sim$0.2 g per 1 g of nanoparticles. This amount of polyelectrolyte (usually, polycation) has been found experimentally to be enough to coat completely the nanoparticles' surface and reverse the zeta-potential to an opposite value.

After 1 h of intense sonication, the size of drug particles reaches 200–250 nm (Table 8.1). At that stage one can proceed with the addition of oppositely charged polyelectrolyte (e.g., polyanion) and sonicate for 20 min to obtain nanoparticles coated with a polycation/polyanion bilayer. More layers may be deposited in a similar

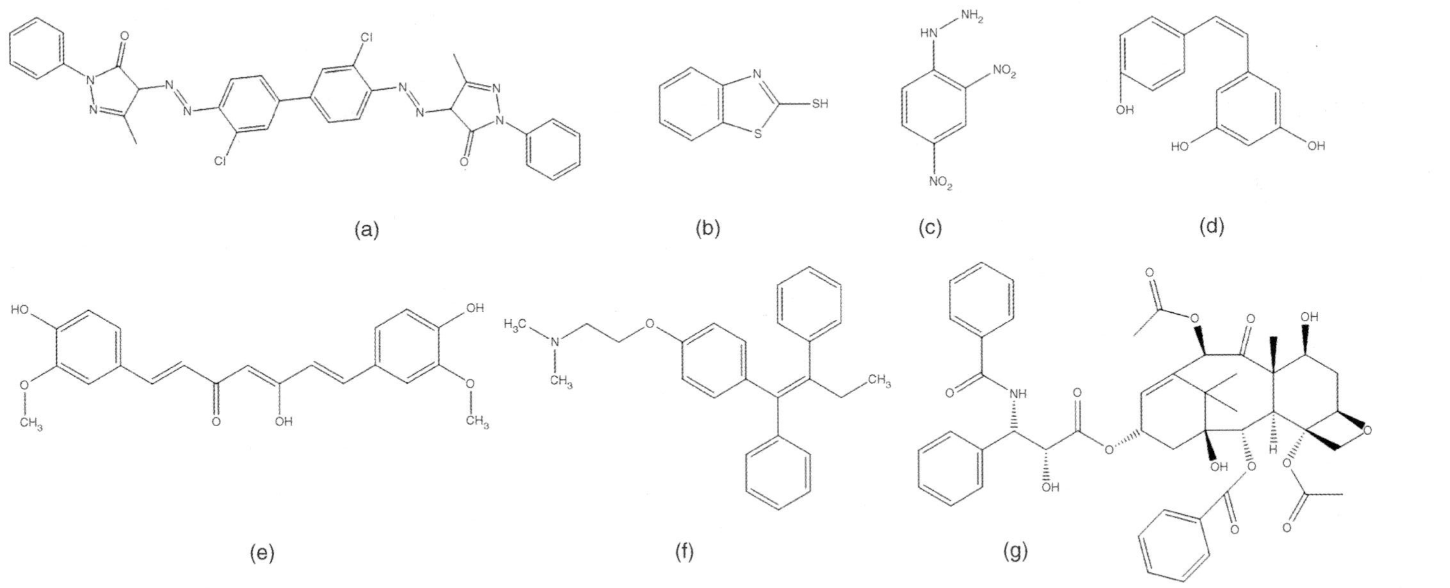

**Figure 8.1** Chemical structures of some poorly-soluble organic compounds: (a) pigment Orange-13, (b) 2-mercaptobenzothiazole, (c) 2,4-dinitrophenylhydrazine, (d) resveratrol, (e) curcumin, (f) tamoxifen, and (g) paclitaxel.

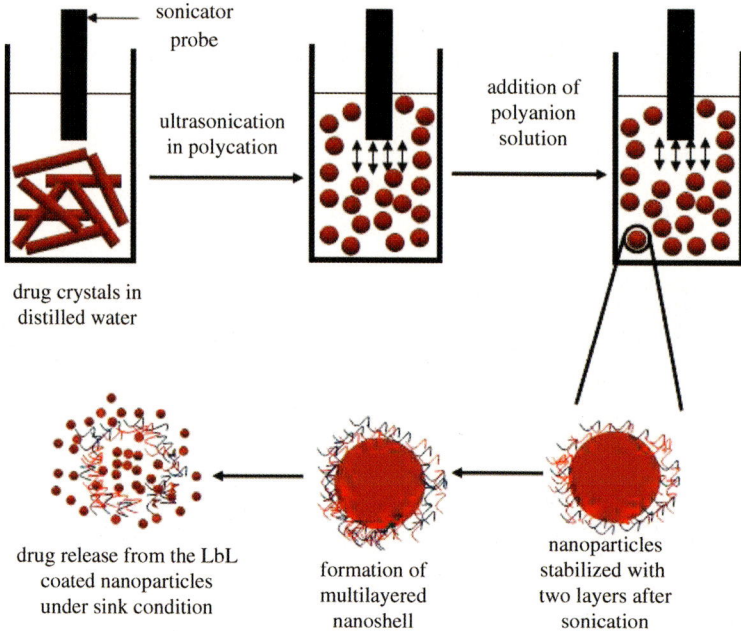

**Figure 8.2** Schematic representation of the ultrasonication-assisted top-down LbL assembly technique.

way. Being predominantly negatively charged, during SLbL processing, many materials were coated with positively charged poly(allylamine hydrochloride) (PAH) or polyethyleneimine (PEI) followed by negatively charged poly(sodium styrene sulfonate) (PSS). For *in vitro* release studies [22–24], bio/polyelectrolytes (alginic acid, chitosan, heparin, poly-L-lysine (PLL)) and proteins (protamine sulfate (PS), bovine serum albumin (BSA)) were used for the LbL process. To eliminate the yield of

**Table 8.1** Experimental details for ultrasonication-assisted top-down LbL assembly of poorly-soluble organic compounds.

| Organic compounds | Initial particle size (μm) | First layer | Second layer | Diameter after SLbL coating (nm) |
|---|---|---|---|---|
| paclitaxel | $15 \pm 5$ | PAH, chitosan, PLL | PSS, alginic acid, heparin | $211 \pm 40$ |
| tamoxifen | $12 \pm 7$ | PAH, PEI | PSS, PAA | $220 \pm 32$ |
| curcumin | $18 \pm 5$ | PAH, PS | PSS, BSA | $107 \pm 17$ |
| resveratrol | $180 \pm 40$ | chitosan | alginic acid | $200 \pm 30$ |
| pigment Orange 13 | $19 \pm 6$ | PAH, PS | PSS, BSA | $264 \pm 23$ |
| 2,4-dinitrophenylhydrazine | $82 \pm 20$ | PAH, PEI | PSS | $216 \pm 29$ |
| 2-mercaptobenzothiazole | $27 \pm 8$ | PAH, PS | PSS, BSA | $290 \pm 33$ |

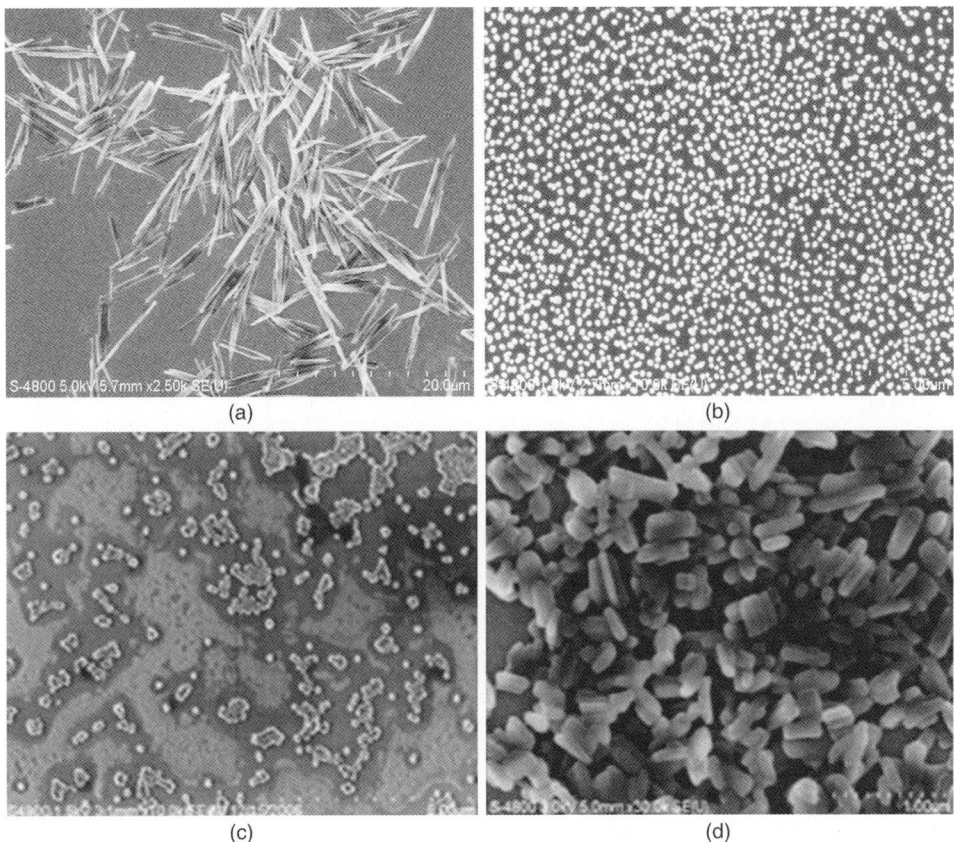

**Figure 8.3** SEM images of (a) original paclitaxel microcrystals and (b–d) nanocolloids after sonication-assisted LbL assembly using a top-down approach: (b) paclitaxel coated with (PLL/heparin)$_4$, (c) tamoxifen coated with (PDDA/PSS)$_2$, (d) paclitaxel coated with (PAH/BSA)$_2$.

bigger particles and traces of titanium coming from the sonication probe during the process, all samples were centrifuged at a low centrifugation speed. Figures 8.3 and 8.4 show images of such organic nanoparticles prepared using the SLbL approach.

Similarly to organic compounds, we elaborated a preparation scheme for nanocolloids of poorly-soluble inorganic compounds. Figure 8.5 and Table 8.2 show a reduction of the original particle size from 10 to 50 μm to an average diameter of 200 nm in nanocolloids.

Sulfides and oxides are the most insoluble inorganic compounds, but their crystalline structure and hardness vary widely as, for example, for iron (II) sulfide minerals. We assume that the efficiency of SLbL for inorganic compounds depends on their appearance in an amorphous or crystalline state. Van-der-Waals and hydrogen bonds are less strong and easier to break than covalent bonds. Amorphous compounds, or those having an asymmetric crystalline structure (black carbon) are

**Figure 8.4** SEM of (a) microcrystals of pigment Orange 13 and (b) PAH/PSS – coated nanoparticles of the pigment after SLbL.

easy to break; but our attempts to efficiently prepare nanocolloids from monocrystal silicon failed. Forming stable polyelectrolyte-coated nanocolloids from different solid materials using the aforementioned SLbL technique may find industrial applications in composite materials [25]. However, a bottom-up approach allows one to achieve a higher degree of disintegration, thus getting smaller particle diameters.

### 8.3.2
**Bottom-Up Approach**

This ultrasonication-assisted LbL assembly allows a variety of conditions for nanoparticulation. All of them start from highly concentrated solutions of the substance of interest in a "good" solvent (ca. 10–50 mg mL$^{-1}$). For example [23], the anticancer drug paclitaxel was completely dissolved in 60% ethanol, and an aqueous polycation (PAH) was slowly added to the solution under sonication. With increasing water

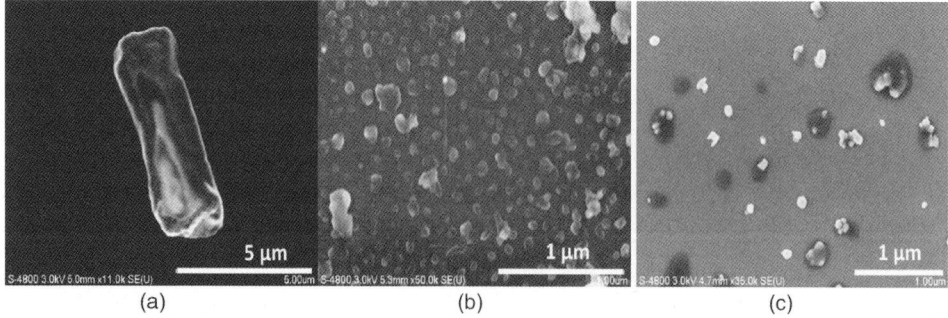

**Figure 8.5** SEM images of poorly-soluble inorganic materials: cupric oxide before (a) and after (b) SLbL coating with PAH/PSS layer pairs, and (c) nanoparticles of barium sulfate coated with (PAH/PSS)$_2$.

**Table 8.2** Experimental details for ultrasonication-assisted layer-by-layer assembly of poorly-soluble inorganic compounds.

| Inorganic compounds | Initial particle size (μm) | First layer | Second layer | Particle size after SLbL coating (nm) |
|---|---|---|---|---|
| CuO | $15 \pm 23$ | PAH, PS | PSS, BSA | $175 \pm 20$ |
| CuCO$_3$ | $50 \pm 17$ | PAH, PEI | PSS | $199 \pm 25$ |
| BaSO$_4$ | $6 \pm 3$ | PAH | PSS | $120 \pm 50$ |
| FeS | $40 \pm 8$ | PAH | PSS | $160 \pm 30$ |
| MnS | $50 \pm 20$ | PAH | PSS | $230 \pm 20$ |
| Al(OH)$_3$ | $30 \pm 15$ | PAH | PSS | $225 \pm 25$ |

concentration, the solubility of paclitaxel decreased, reached saturation, and then nucleation began, resulting in the formation of nano-sized particles. A powerful sonication prevents the formation of larger particles and polycation adsorption provides a surface charge sufficient for colloidal stability. After 45 min of the ultrasound treatment needed to evaporate the organic solvent, an anionic polyelectrolyte layer (PSS) was added. Paclitaxel nanoparticles with an average particle size of about 100 nm were obtained. One has to achieve a balance between the low polyelectrolyte solubility in organic solvent/water mixtures (with greater than 60% organic solvent) and the diminishing solubility of many drugs with increasing water content. Therefore, the conditions of the assembly should be chosen carefully, and such an approach is difficult for most of the natural polyelectrolytes.

In another bottom-up experimental set-up, paclitaxel was dissolved in acetone or ethanol at a 50 mg mL$^{-1}$ concentration and less than 150 μL of the solution was added under constant sonication to a 40 times larger volume of an aqueous solution containing surfactant (sodium docusate, polysorbate 80) and polycation (chitosan, poly-L-lysine). In the presence of both polycation and surfactants, the diameter of paclitaxel particles was two times smaller than in the presence of surfactants alone and reached 180 nm (Figure 8.6). Parikh [39] and Muller et al. [40] showed that mixtures of ionic and non-ionic surfactants and polymers in homogenization techniques are more efficient and produce smaller particles.

The top-down SLbL method is more efficient in the preparation of large quantities of about 200 nm diameter colloids, and the bottom-up approach produces smaller nanocolloids of about 100 nm diameter.

## 8.4 Solvent-Assisted Precipitation Into Preformed LbL-Coated Soft Organic Nanoparticles

An encapsulation of poorly water-soluble drugs and dyes into preformed microcapsules with the interior filled with polyelectrolytes was proposed by Sukhorukov [41, 42] and developed by others [43, 44]. That approach was based on

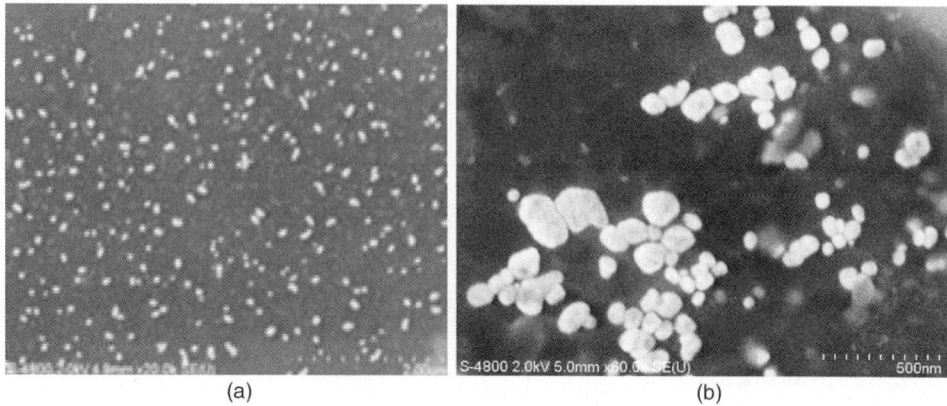

**Figure 8.6** SEM images of paclitaxel nanocolloids after sonication-assisted LbL using a bottom-up approach and coated with (a) (PLL/heparin)$_4$ and (b) (docusate + Polysorbate 80)/(PLL-block-PEG/heparin)$_{3.5}$.

the gradually changing polarity between microcapsule interior and external solution which contains a miscible organic solvent. Despite better control over the size distribution of the particulated material, increased colloidal stability of the particles, and easy surface modification, there are only a few examples of poorly-soluble compound precipitation/adsorption into preformed soft nanocores (protein or polysaccharide possessing high affinity for the drugs [45]).

Among techniques useful for bulk batch oil-free nanoparticle synthesis, there are desolvation of gelatin, albumin, or lactoglobulin with acetone or ethanol [46–48], pH [49] or sulfate salt [50] driven precipitation of proteins and polysaccharides, thermally assisted denaturation of proteins [35, 36], interpolyelectrolyte and polyelectrolyte/protein complexation. Biopolymer-based nanoparticles have been proven to be safe and effective nonviral gene delivery vehicles with a prolonged *in vivo* circulation time and high accumulation at the tumor tissues [35, 47–50].

Gelatin-based 200 nm nanoparticles consisting of a soft gel-like interior, strongly binding the poorly-soluble compounds of the polyphenol family, were developed. These nanocores were coated with LbL shells (e.g., poly-L-glutamic acid/poly-L-lysine, dextran sulfate/protamine sulfate, carboxymethyl cellulose/gelatin) (Figure 8.7). These LbL nanocapsules encased a wide range of polyphenolic compounds with demonstrated anticancer potential [51]. To overcome the low solubility of theaflavin and curcumin in water, they were loaded from 25% acetone or 25% ethanol: 12.5% acetone in water, while EGCG and tannic acid were adsorbed from aqueous solutions. Adsorption of the polyphenols into gelatin nanoparticles equalizes their extremely different solubility, and makes the features of polyphenol loading and release a function of the general rules of polyphenol/protein interaction [45]. Adsorption of polyphenols with higher molecular weights and a larger number of phenolic −OH groups was found to be higher. For theaflavin having the highest molecular weight, it reached 70% of the mass of nanoparticulated material on a dry weight basis. The degree of tannic acid and (−)-epigallacathechin gallate (EGCG) encapsulation was

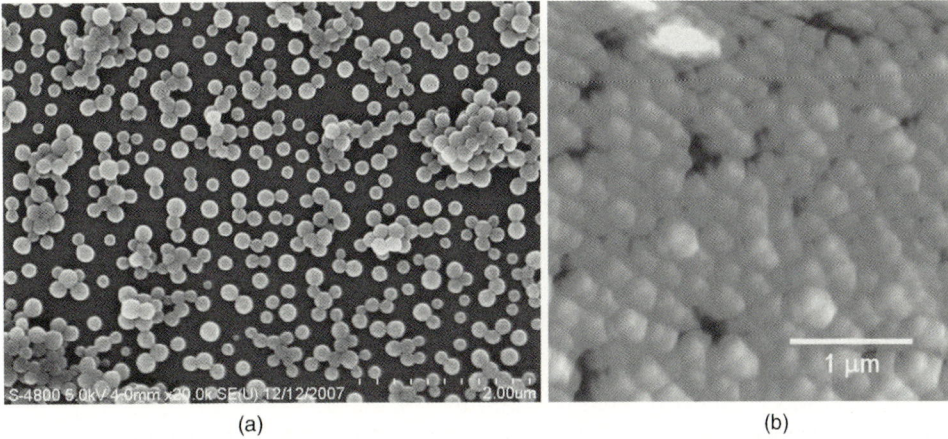

**Figure 8.7** SEM (a) and AFM (b) images of 200 nm (PGA/PLL)$_2$-coated gelatin nanoparticles loaded with polyphenol; PGA – poly-L-glutamic acid, PLL – poly-L-lysine.

lower (25–30%). All encapsulated polyphenols retained their antioxidant properties, as was proved with the ABTS cation-radical assay, and biological activity [51].

Beta-lactoglobulin-based nanoparticles of 50 nm diameter have also been proposed for encapsulation of natural polyphenols, such as EGCG and resveratrol [36, 52]. Complexation with beta-lactoglobulin nanoparticles enhances the hydrosolubility of the compounds and increases the stability of the drug loaded into the protein-based nanoparticles.

## 8.5
## Washless (Titration) LbL Technique

The fundamentals of the LbL deposition technique on microcores are well described [53–55]. An application of this process for colloid coating was first mentioned by Mallouk *et al.* for polyelectrolyte "onion-like" structures [56], and was elaborated for polyelectrolyte microshells with intermediate washing and centrifugation in the Max Planck Institute, Golm [4–15, 57–60]. LbL-encapsulation using a separation of colloids from the excess of polyelectrolyte by filtration allowed a larger production scale [61]. A method for LbL coating of colloids without washing was mentioned by Sukhorukov, Caruso *et al.* [57–59, 62]. It is based on adding polyelectrolytes in the amounts that will be adsorbed completely on the colloid/water interface. In this way, the need to remove the excess polycation before adding the next oppositely charged polyanion is avoided. It has also been exploited recently for dye emulsions, but without details of the coating technique [63].

We have elaborated a non-washing LbL procedure for different nano-dispersions: TiO$_2$ cores, nanocrystals of low solubility drugs (paclitaxel, tamoxifen, and curcumin), and cellulose microfibers [64]. The procedure of washless LbL assembly is

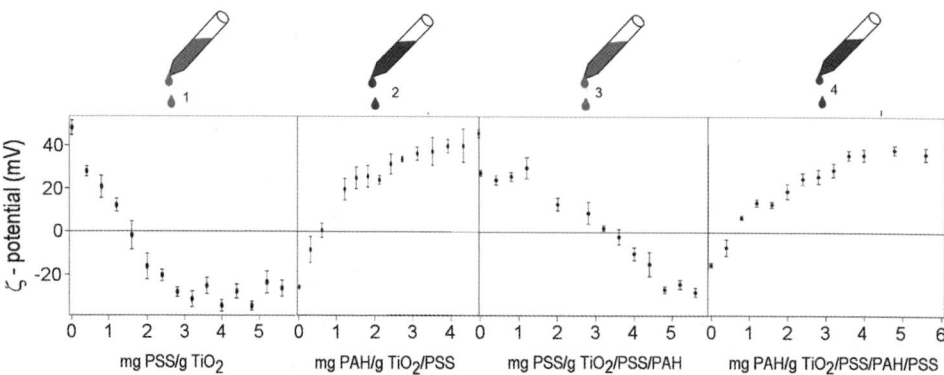

**Figure 8.8** ζ-potential versus polyelectrolyte concentration for a 20 mg mL$^{-1}$ TiO$_2$ nanocolloid, pH 6.5. Four steps of polyanion/polycation deposition are shown: First step: the onset of the ζ-potential plateau is approximately at 3.2 mg PSS added to 1 g TiO$_2$. Second step: the onset of the ζ-potential plateau is at 2 mg PAH added to 1 g TiO$_2$ coated with PSS. Third step: the onset of the ζ-potential plateau is at 4.9 mg PSS added to 1 g TiO$_2$ coated with PSS/PAH layer pair. Fourth step: The onset of the ζ-potential plateau is at 3.7 mg PAH added to 1 g TiO$_2$ coated with PSS/PAH/PSS.

based on the permanent monitoring of the particle surface potential as a polyanion is gradually added so as to find the moment when the particle recharging is completed. Immediately after this, we switch to addition of a polycation until positive surface charge saturation is reached, and so on. Figure 8.8 shows the ζ-potential changes for 300 nm TiO$_2$ particles in the process of stepwise addition of sodium poly(styrene sulfonate) (PSS) and poly(allylamine hydrochloride) (PAH). An addition of polyelectrolytes under constant sonication allowed one to avoid aggregation of drug nanoparticles and to decrease the sonication time to 1–2 min.

The washless approach scales up the LbL-process using a highly concentrated nanoparticles suspension (about 25 mg mL$^{-1}$ for TiO$_2$ particles), making the process easier. The procedure is more economical because less material is used and the separation colloidal particles from the supernatant is avoided. This also allows usage of smaller amounts of polyelectrolytes for coating, precluding the need to recycle excessive materials. This is important for the modified polyelectrolytes (e.g., PEGylated) used in pharmaceutical applications that are often expensive or only available in small amounts. While one needs only 2.5 mg PSS per 1 g TiO$_2$ to completely reverse the charge of the suspension, the amount used to change its charge after centrifugation and re-dispersion is much larger, about 20 mg. Both centrifugation and washless methods give the same thickness of polyelectrolyte shell formed on the nanocores [62, 64].

In a washless LbL coating of 200 nm diameter paclitaxel particles with poly-L-lysine and heparin layers, an addition of polyelectrolytes under constant sonication avoided aggregation, but we had a significant size increase if a traditional centrifugation-based LbL procedure with a threefold excess of polyelectrolytes was used. Application of the washless LbL method for porous substrates, such as lignocelluloses microfibers, has an advantage in coating only the external fiber surface and not using polyelectrolytes for internal pores and lumen treatment if it is superfluous [65, 66].

## 8.6
## Formation of LbL Shells on Nanocores

LbL assembly of polyelectrolytes and enzymes on nanocores generally follows the same rules as that for microcores [4–15]. However, several size-related problems should be mentioned. The smallest cores successfully utilized for LbL assembly are the 13.5 nm gold nanoparticles used by Decher [67] for coating with 20 alternated layers of 15 kDa PAH and 13 kDa PSS. The 27 nm final diameter of the polyelectrolyte shell on the gold cores was confirmed by TEM. These gold core–shell particles were easily centrifuged and re-dispersed without aggregation. Using lower molecular weight polyelectrolytes was essential. Specificity of LbL assembly on small nanoparticles relates to limited polyelectrolyte segment adsorption on a nanoparticle and to the profound influence of the solution ionic strength. As a result, extended polyelectrolyte tails protrude into solution [68] favoring the formation of aggregates [69]. For 300 nm $TiO_2$ nanoparticles [64], the use of shorter polymers decreased their aggregation. The use of higher molecular weight PSS increased aggregation, but it was reduced in the next step by using lower molecular weight PAH (still this was worse than using both polyelectrolytes of low molecular weight).

Among silica nanoparticles of different diameters, stronger adsorption of lysozyme was observed for larger particles, ultimately resulting in a qualitative change of adsorption behavior from stoichiometric protein–silica conjugates for 4 nm particles, through monolayer adsorption on 20 nm particles to, finally, multilayer adsorption on 100 nm particles [70]. Large particles allow polyelectrolyte to spread to the same extent as on a flat surface [68], however, such stretching leads to protein unfolding, resulting in partial lost of enzymatic activity [70].

We have carried out LbL assembly of different bio/polyelectrolytes on the surface of 100–200 nm diameter gelatin-based gel cores [51] and on 100–250 nm nanocolloids of poorly-soluble drugs [22–24] to produce stabilized nanoparticles. A better stability of the latter in salt and buffer solutions and cell binding resistance was achieved using pre-PEGylated polymers. Figure 8.9 shows alternation of the $\xi$-potential of paclitaxel nanoparticles in the processes of washless LbL coating with PLL/heparin and PLL-block-PEG/heparin layers. The 4 mg mL$^{-1}$ paclitaxel nanocolloids coated with 3–4 layers of PEGylated polycations were stable in phosphate buffered saline for up to 150 h.

Modification of the nanocores' surfaces with polyelectrolyte layer-by-layer shells allows modulation of the nanoparticle cell uptake rate, increasing nanoparticle colloidal stability, and controlling loading/release characteristics, and provides a template for their modification with tumor-targeting agents [71–73].

A very intriguing aspect of LbL assembly on nanocores is the possibility and, in many cases, the necessity to adsorb charged polyelectrolytes from solutions with an excess of uncharged, sometimes polymeric, pharmaceutical excipients used upon

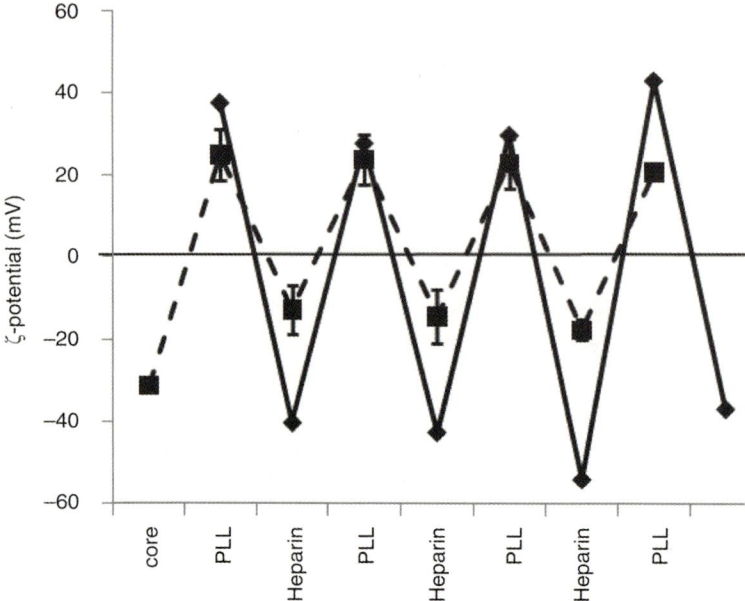

**Figure 8.9** ξ-potential of 170 nm paclitaxel nanoparticles coated with poly-L-lysine and heparin using the washless LbL method. Solid line: top-down approach; sonication in the presence of PLL followed by LbL assembly. Dashed line: bottom-up method; cores produced by adding paclitaxel solution in acetone to a docusate and Polysorbate 80 solution, followed by LbL assembly using PLL-block-PEG.

preparation or storage stabilization of micro/nanoparticles (e.g., polyvinylpyrrolidone, polyethylene glycol, sucrose, glycerol, etc.) [74, 75]. We successfully deposited 3–10 layers of bio/polyelectrolytes (chondroitin sulfate, poly-L-glutamic acid, poly-L-aspartic acid, dextran sulfate, carboxymethyl cellulose, poly-L-lysine hydrochloride, poly-L-arginine hydrochloride, and protamine sulfate) on quartz flat supports and on PROMAXX® insulin microcores of 1 μm diameter in aqueous solutions composed of 16% polyethylene glycol MW 3350 kD (PEG 3350) and NaCl (Figure 8.10). A typical layer pair thickness was 4–5 nm. The dissolution properties of insulin were controlled by alternating polyelectrolyte layers on the PROMAXX® particles' surface [74, 75].

In another experiment, an alternate adsorption of PEI and PSS on the surface of quartz crystal took place, even at a PEG concentration of 30 wt%, without sufficient influence on the thickness of the formed PEI/PSS multilayer films. LbL polyelectrolyte assembly from up to 70% aqueous glycerol was also demonstrated. The possibility to form LbL multilayers in the presence of high concentrations of water-soluble pharmaceutical polymers allows stabilization and modification of colloids produced by different homogenization techniques.

Using PEGylated polyelectrolytes in LbL assembly to increase the stability of the nanocapsules of paclitaxel was also efficient. For example, an alternate deposition of linear block-copolymers of poly-L-lysine and PEG (e.g., PLL 16kDa-PEG 5kDa) and

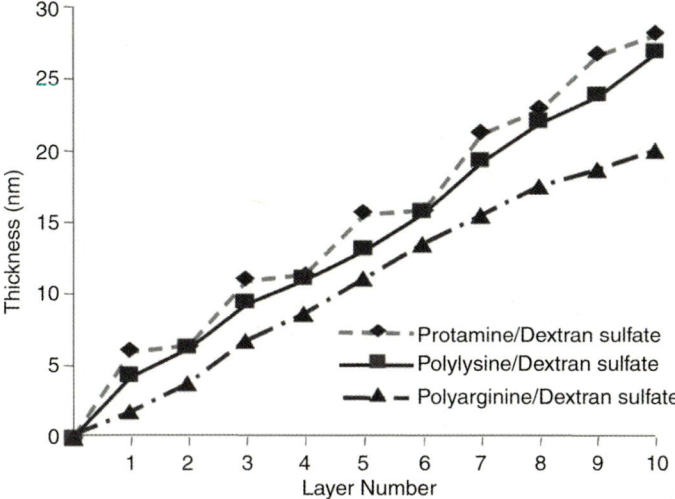

**Figure 8.10** Quartz crystal microbalance data for LbL assembly of different polyelectrolytes from 1 mg mL$^{-1}$ solutions in a 16% PEG-0.7% NaCl buffer, pH 5.8 at room temperature. Experimental errors are ±0.5 nm.

heparin on 170 nm paclitaxel cores allowed the formation of stable in PBS buffer of 5 mg mL$^{-1}$ drug colloids. Interestingly, better results on colloidal stability were obtained while using PEG-PLL block-copolymer at every layer pair with the shell architecture of (PLL-PEG/heparin)$_3$ rather than using PEGylated polyelectrolyte only in the outermost layer of the capsule.

## 8.7
## Drug Release Study

LbL modification of the surface of organic crystals and microparticles to extend release time has been explored by many researchers [4–16, 19]. Beginning with Antipov *et al.*'s paper [76] on fluorescein crystals coated with a PAH/PSS shell of 5–15 layer pairs, there have been a number of reports showing a possibility to vary the release time within 1–10 h with increasing wall thickness of the polyelectrolyte microcapsules [11, 12, 16]. Biocompatible microshells based on 10–12 chitosan/ dextran sulfate and chitosan/alginic acid layer pairs have been reported to decrease the release rate for ibuprofen microparticles from seconds to a few hours [8, 77].

For poorly-soluble compounds, which many anticancer drugs are, low dissolution rate is not a problem. Typically, the complete drug release takes many hours and may be reached only if the amount of the drug in a volume is lower than its solubility limit (sink conditions). The *in vitro* release studies for poorly-soluble nanoparticles often have to be done in the presence of surfactants, proteins, or directly in serum to increase drug solubility to a level where detection is possible. The release profiles of paclitaxel from its LbL nanocolloids coated with a different number of poly-L-lysine/

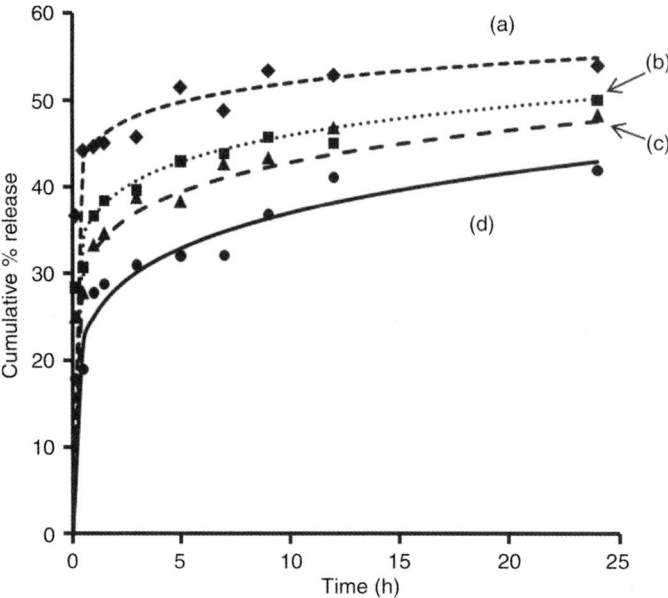

**Figure 8.11** Drug release of paclitaxel from 300 nm nanocolloids prepared using top-down SLbL method and coated with 0.5 (a), 4 (b), 8 (c), 12 (d) poly-L-lysine/heparin layer pairs (in water at 37 °C).

heparin layers are shown in Figure 8.11. 4, 8 and 12 polyelectrolyte layer pair coatings decreased the release time as compared with nanocolloids coated with only one poly-L-lysine layer. In another experiment, for 170 nm paclitaxel nanoparticles prepared via a bottom-up approach and PLL-block-PEG copolymer and heparin for coating, no distinguishable influence on release was found for shells consisting of less than 3.5 layer pairs.

For LbL shells on soft insulin PROMAXX® cores, we combined two concepts: interpolyelectrolyte complexation and the formation of an organized multilayer shell for prolonged release. Formation of 10–50 nm LbL multilayers on the soft permeable insulin cores was not sufficient to provide a tight physical barrier to slow down insulin diffusion into aqueous solution. We established conditions for complexation of the insulin core with the first polyelectrolyte layer, which played a critical role for sustained release. After a complex of insulin in the core conjugated with the first layer of polycation is formed, the particle is stabilized and the addition of a polyanion layer is performed. Balancing the strong interaction of the first polyelectrolyte layer with the protein core, and easing this interaction by depositing the second oppositely-charged polyelectrolyte allowed us to reach a sustained 3-month insulin release (Figure 8.12) [74, 75]. With this, we converted the concept where a LbL multilayer was considered as a diffusion barrier with adjustable thickness of 5–12 layer pairs [4–8, 12–16], to a modified approach where the main role in the controlled release from protein micro/nano cores is given to adjustable interpolyelectrolyte complex formation controlled by alternate polycation/polyanion coating with 2–3 monolayers.

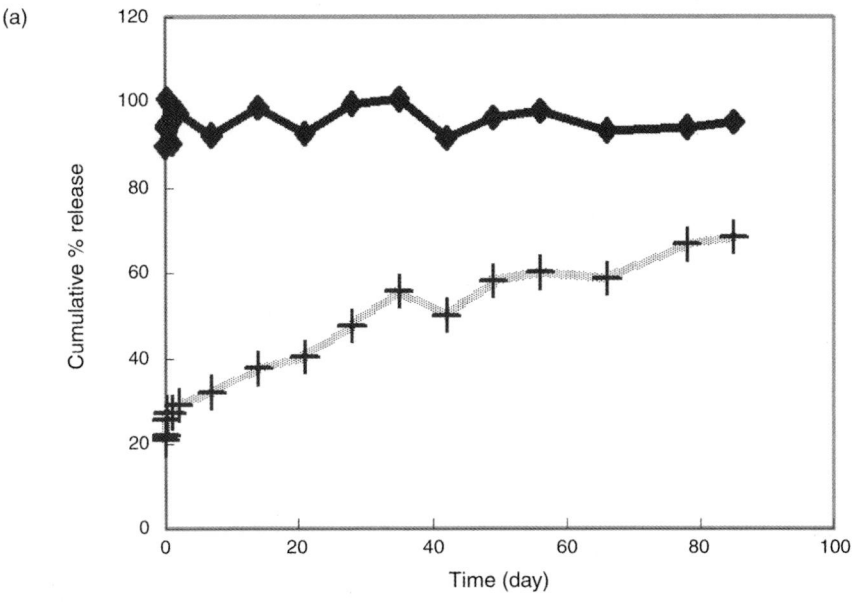

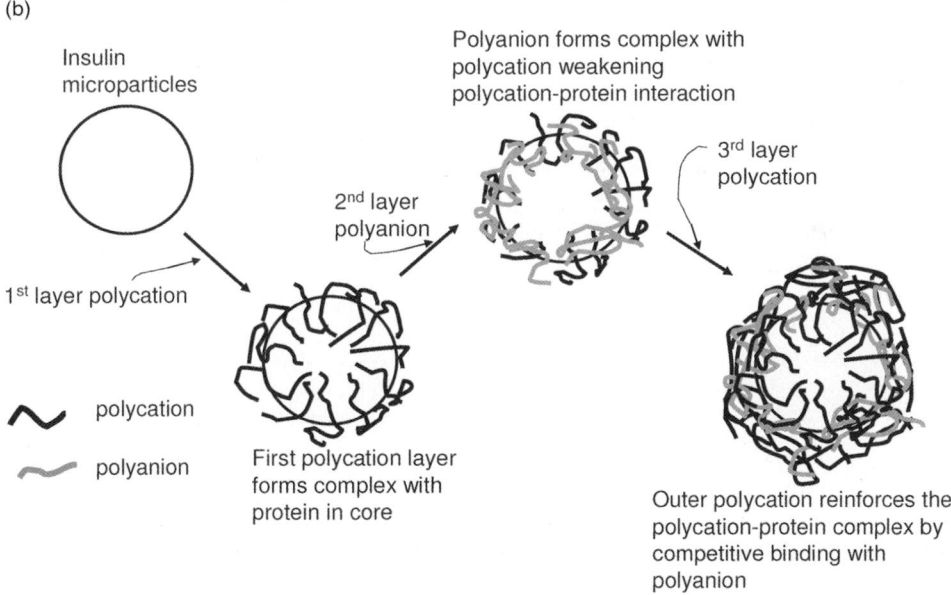

**Figure 8.12** (a) *In vitro* release profiles of insulin from PROMAXX® particles; black – uncoated, and gray – coated with protamine sulfate/alginic acid shell; (b) scheme of the assembly of polyelectrolyte layers on PROMAXX® insulin particles.

## 8.8
## Conclusions

Nanoencapsulation of poorly-soluble or insoluble materials was developed with sonication-assisted layer-by-layer polyelectrolyte coating. We concentrated on the formation of core–shell structures with diameters of 100–200 nm, which are much smaller than traditional LbL microcapsules. We changed the strategy of LbL-encapsulation from making microcapsules with many layers in the shell walls for encasing highly-soluble materials to using very thin polycation/polyanion coating on low-soluble nanoparticles to provide good colloidal stability. This technique works for low-solubility drugs, dyes, and even for insoluble inorganic salts.

In the top-down approach, nanocolloids were prepared by rupturing powdered material using ultrasonication and simultaneous sequential adsorption of oppositely-charged polyelectrolytes (often with a non-washing procedure). In the bottom-up approach, the drugs were dissolved in an organic solvent (ethanol or acetone), and drug nucleation was initiated by gradually worsening the solution with the addition of an aqueous polyelectrolyte, assisted by ultrasonication. For example, SLbL encapsulation of the cancer drug paclitaxel resulted in stable 170 nm diameter colloids with a high surface $\xi$-potential (−45 mV) and drug content of 80 wt%. The release rates from such nanocapsules were controlled by assembling multilayer shells with variable thicknesses, and are in the range of 10–20 h. The small real nano-size of the particles, stability in bio-buffers at high concentration (5 mg mL$^{-1}$), and the use of PEGylated biodegradable natural polyelectrolytes in the shell allowed the development of new drug nanoformulations which have a potential use in intravenous administration. A similar SLbL approach for inorganic materials allowed the production of stable nanocolloids of pigments, corrosion inhibitors and inorganic salts for their better dispersion in aqueous solutions (for example, anticorrosion benzotriazole and medical contrast BaSO$_4$).

## References

1 Hampsey, J., De Castro, C., McCaughey, B., Wang, D., Mitchell, B., and Lu, Y. (2004) *J. Am. Ceram. Soc.*, **87**, 1280.
2 Nilsson, L. and Bergenstahl, B. (2007) *J. Colloid Interface Sci.*, **308**, 508.
3 Guzey, D. and McClements, D. (2006) *Adv. Colloid Interface Sci.*, **128**, 227.
4 Sukhorukov, G., Donath, E., Davis, S., Lichtenfeld, H., Caruso, F., Popov, V., and Möhwald, H. (1998) *Polym. Adv. Technol.*, **9**, 759.
5 Donath, E., Sukhorukov, G., Caruso, F., Davis, S., and Möhwald, H. (1998) *Angew. Chem. Int. Ed.*, **37**, 2202.
6 Antipov, A. and Sukhorukov, G. (2004) *Adv. Colloid Interface Sci.*, **111**, 49.
7 Caruso, F. (2001) *Adv. Mater.*, **13**, 11.
8 Qiu, X., Leporatti, S., Donath, E., and Möhwald, H. (2001) *Langmuir*, **17**, 5375.
9 Lvov, Y., Antipov, A., Mamedov, A., Möhwald, H., and Sukhorukov, G. (2001) *Nano Lett.*, **1**, 125.
10 Sukhorukov, G. (2002) *Dentrimers*, MML Series, vol. **5** (eds R. Arshady and A. Guyot), Citrus Books, NY, pp. 111–147.
11 Ai, H., Jones, S., de Villiers, M., and Lvov, Y. (2003) *J. Controlled Release*, **86**, 59.
12 Sheney, D. and Sukhorukov, G. (2004) *Eur. J. Phar. Biopharm.*, **58**, 521.

13 De Geest, B., Sukhorukov, G., and Möhwald, H. (2009) *Expert Opin. Drug Deliv.*, **6**, 613.
14 Bédard, M., De Geest, B., Skirtach, A., Möhwald, H., and Sukhorukov, G. (2010) *Adv. Colloid Interface Sci.*, **158**, 2.
15 De Cock, L., De Koke, S., De Geest, B., Grooten, J., Vervaet, C., Remon, Je., Sukhorukov, G., and Antipina, M. (2010) *Angew. Chem.*, **122**, 9820.
16 Palankar, R., Skirtach, A., Kreft, O., Bedard, M., Garstka, M., Gould, K., Möhwald, H., Sukhorukov, G., Winterhalter, M., and Springer, S. (2009) *Small*, **5**, 2168.
17 Yuan, J., Zhou, S., You, B., and Wu, L. (2005) *Chem. Mater.*, **17**, 3587.
18 Katsuhiko, A., Qingmin, J., and Jonathan, P.H. (2010) *Adv. Polym. Sci.*, **229**, 51.
19 Pargaonkar, N., Lvov, Y., Li, N., Steenekamp, J., and de Villiers, M. (2005) *Pharm. Res.*, **22**, 826.
20 Agarwal, A., Lvov, Y., Sawant, R., and Torchilin, V. (2008) *J. Control Release*, **128**, 255.
21 Lvov, Y., Agarwal, A., Sawant, R., and Torchilin, V. (2008) *Pharma Focus Asia*, **7**, 36; Lvov, Y., Shutava, T., Arapov, K., Torchilin, V., and DeVilliers, M. (2011) *Pharma Focus Asia*, **15**, 28.
22 Zheng, Z., Zhang, X., Carbo, D., Clark, C., Nathan, C.-A., and Lvov, Y. (2010) *Langmuir*, **26**, 7679.
23 Pattekari, P., Zheng, Z., Zhang, X., Levchenko, T., Torchilin, V., and Lvov, Y. (2011) *Phys. Chem. Chem. Phys.*, **13**, 9014.
24 Lvov, Y., Pattekari, P., Zhang, X., and Torchilin, V. (2011) *Langmuir*, **27**, 1212.
25 Tsuzuki, T. (2009) *Int. J. Nanotech.*, **6**, 567.
26 Sivasankar, M. and Kumar, B.P. (2010) *Int. J. Res. Pharm. Biomed. Sci.*, **1**, 41.
27 Chen, H., Khemtong, C., Yang, X., Chang, X., and Gao, J. (2011) *Drug Discov. Today*, **16**, 354.
28 Keck, C.M., Kobierski, S., Mauludin, R., and Muller, R.H. (2008) *Dosis*, **2**, 124.
29 Suslick, K. (1990) *Science*, **247**, 1439.
30 Suslick, K. Jr., Cline, R., and Hammerton, D. (1986) *J. Am. Chem. Soc.*, **108**, 5641.
31 Suslick, K., Flint, E., Grinstaff, M., and Kemper, K. (1993) *J. Phys. Chem.*, **97**, 3098.
32 Borkent, B., Gekle, S., Prosperetti, A., and Lohse, D. (2009) *Phys. Fluids*, **21**, 102003.
33 Dibbern, E., Toublan, F., and Suslick, K. (2006) *J. Am. Chem. Soc.*, **128**, 6540.
34 Santos, H.M., Lodeiro, C., and Capello-Martinez, J.L. (2009) *Ultrasound in Chemistry: Analytical Applications* (ed. J.L. Capelo-Martinez), Wiley-VCH Verlag GmbH, Weinheim, pp. 1–16.
35 Livney, Y.D. (2010) *Curr. Opin. Colloid. Interface*, **15**, 73.
36 Shpigelman, A., Israeli, G., and Livney, Y.D. (2010) *Food Hydrocolloid*, **24**, 735.
37 Lee, J., Choi, J.Y., and Park, C.H. (2008) *Int. J. Pharm.*, **355**, 328.
38 Kahlweit, M. (1975) *Adv. Colloid Interface Sci.*, **5**, 1.
39 Parikh, I., (1999) Composition and method of preparing microparticles of water-insoluble substances US Patent 5922355. *filed* Sep. 29, 1997 and issued Jul. 13, 1999.
40 Muller, R.H., Jacobs, C., and Kayser, O. (2000) *Pharmaceutical Emulsions and Suspensions. Drugs and the Pharmaceutical Sciences*, vol. 105 (eds F. Nielloud and G. Marti-Mestres), Marcel Dekker, pp. 383–408.
41 Radtchenko, I., Sukhorukov, G., and Möhwald, H. (2002) *Int. J. Pharm.*, **242**, 219.
42 Sukhorukov, G., Döhne, L., Hartmann, J., Donath, E., and Möhwald, H. (2000) *Adv. Mater.*, **12**, 112.
43 Wang, Y., Yan, Y., Cui, J., Hosta-Rigau, L., Heath, J., Nice, E., and Caruso, F. (2010) *Adv. Mater.*, **22**, 4293.
44 Shi, X., Wang, S., Chen, X., Meshinchi, S., and Baker, J. (2006) *Mol. Pharm.*, **3**, 144.
45 Haslam, E. (1996) *J. Nat. Prod.*, **59**, 205.
46 Zillies, J.C., Zwiorek, K., Winter, G., and Coester, C. (2007) *Anal. Chem.*, **79**, 4574.
47 Kommareddy, S. and Amiji, M. (2005) *Bioconjugate Chem.*, **16**, 1423.
48 Das, S., Banerjee, R., and Bellare, J. (2005) *Trends Biomater. Artif. Organs*, **18**, 203.
49 Zhang, H., Oh, M., Allen, C., and Kumacheva, E. (2004) *Biomacromolecules*, **5**, 2461.
50 Lu, Z., Yeh, T.K., Tsai, M., Au, J.L.S., and Wientjes, M.G. (2004) *Clin. Cancer Res.*, **10**, 7677.

51 Shutava, T.G., Balkundi, S.S., Vangala, P., Steffan, J.J., Bigelow, R.L., Cardelli, J.A., O'Neal, D.P., and Lvov, Y.M. (2009) *ACS Nano*, **37**, 1877.

52 Liang, L., Tajmir-Riahi, H.A., and Subirade, M. (2008) *Biomacromolecules*, **9**, 50.

53 Decher, G. (2003) *Multilayer Thin Films: Sequential Assembly of Nanocomposite Materials* (eds G. Decher and J. Schlenoff), Wiley-VCH Verlag GmbH, Weinheim, pp. 1–46.

54 Lvov, Y., Ariga, K., Ichinose, I., and Kunitake, T. (1995) *J. Am. Chem. Soc.*, **117**, 6117.

55 Lvov, Y., Decher, G., and Möhwald, H. (1993) *Langmuir*, **9**, 481.

56 Keller, S., Johnson, S., Brigham, E., Yonemoto, E., and Mallouk, T. (1995) *J. Am. Chem. Soc.*, **117**, 12879.

57 Sukhorukov, G., Donath, E., Davis, S., Lichtenfeld, H., Caruso, F., Popov, V., and Möhwald, H. (1998) *Polym. Adv. Technol.*, **9**, 759.

58 Sukhorukov, G., Donath, E., Lichtenfeld, H., Knippel, E., Knippel, M., Budde, A., and Möhwald, H. (1998) *Colloids Surf. A*, **137**, 253.

59 Caruso, F., Caruso, R., and Möhwald, H. (1998) *Science*, **282**, 1111.

60 Schlenoff, J., Dubas, S., and Farhat, T. (2000) *Langmuir*, **16**, 9968.

61 Voigt, A., Lichtenfeld, H., Sukhorukov, G., Zastrow, H., Donath, E., Baumler, H., and Möhwald, H. (1999) *Ind. Eng. Chem. Res.*, **38**, 4037.

62 Caruso, F. and Sukhorukov, G. (2003) *Multilayer Thin Films: Sequential Assembly of Nanocomposite Materials* (eds G. Decher and J. Schlenoff), Wiley-VCH Verlag GmbH, Weinheim, pp. 331–362.

63 Ogawa, S., Decker, E., and McClements, D. (2004) *J. Agric. Food Chem.*, **52**, 3595.

64 Bantchev, G., Lu, Z., and Lvov, Y. (2009) *J. Nanosci. Nanotech.*, **9**, 396.

65 Zheng, Z., McDonald, J., Khillan, R., Shutava, T., Grozdits, G., and Lvov, Y. (2006) *J. Nanosci. Nanotech.*, **6**, 624.

66 Lvov, Y., Zheng, Z., and Grozdits, G. (2006) *Nord. Pulp Pap. Res. J.*, **21**, 87.

67 Schneider, G. and Decher, G. (2004) *Nano Lett.*, **4**, 1833.

68 Chodanowski, P. and Stoll, S. (2001) *J. Chem. Phys.*, **115**, 4951.

69 Li, Y., Xia, J., and Dubin, P.L. (1994) *Macromolecules*, **27**, 7049.

70 Vertegel, A.V., Siegel, R.W., and Dordick, J.S. (2004) *Langmuir*, **20**, 6800.

71 Ai, H., Jones, S.A., and Lvov, Y.M. (2003) *Cell Biochem. Biophys.*, **39**, 23.

72 Zahr, A.S. and Pishko, M.V. (2007) *Biomacromolecules*, **8**, 2004.

73 Ai, H., Pink, J.J., Shuai, X., Boothman, D.A., and Gao, J. (2005) *J. Biomed. Mater. Res.*, **73**, 303.

74 Rashba–Step, J., Darvari, R., Lin, Q., Kelly, J., Shutava, T., Lvov, Y., and Scott, T. (2006) 33rd Annual Meeting and Exposition of the Controlled Release Society, 75/477.

75 Rashba-Step, J., Scott, T., Darvari, R., Lvov, Y., and Shutava, T. (2006) Surface-modified microparticles and method of forming the same. US Patent 20060260777. filed Apr. 27, 2006 and issued Nov. 23, 2006.

76 Antipov, A.A., Sukhorukov, G.B., Donath, E., and Möhwald, H. (2001) *J. Phys. Chem. B*, **105**, 2281.

77 Ye, S.Q., Wang, C.Y., Liu, X.X., and Tong, Z. (2005) *J. Control Release*, **106**, 319.

# 9
# Cellulose Fibers and Fibrils as Templates for the Layer-by-Layer (LbL) Technology
*Lars Wågberg*

## 9.1
## Background

Cellulose fibers from wood have been used in a variety of different products for more than 100 years with a huge benefit for both consumers and commercial companies. They constitute a fantastic raw material and, depending on the source of the fibers, they naturally have different properties. Bleached chemical softwood fibers from Scandinavian spruce have a cylindrical shape with a length of about 2 mm, a width of 20 µm, and a wall thickness of 4 µm, and there are about 5 million fibers per gram They have a cellulose content of about 83%, a hemicelluloses content of about 16% and the remaining 1% consists of residual lignin, extractives and impurities. The cellulose molecules are arranged in fibrils with a cross-section of $4 \times 4 \, \text{nm}^2$ and a length of more than 1 µm. In delignified fibers these fibrils are aggregated in clusters with a cross-section of $20 \times 20 \, \text{nm}^2$ and the crystallinity of the cellulose in these fibrils is about 60%. In the wet state, the bleached fibers have a specific surface area of $100-200 \, \text{m}^2 \, \text{g}^{-1}$, but in the dry state the fibrils are totally joined together, the fiber wall is compact and the specific surface area is only about $1 \, \text{m}^2 \, \text{g}^{-1}$ [1]. This means that the wet fiber wall is a fantastic template for various types of modifications, provided the application for which the fibers are intended has been identified.

The most common use of fibers today is in the manufacture of various paper products, but large quantities of fibers are also used in hygiene products, and in fiber-reinforced composites. In the preparation of different types of papers, it is very common to use cationic and anionic polyelectrolytes to increase the wet and dry strength of the paper, and to improve the efficiency of the papermaking process [2]. The polyelectrolytes are often added in sequence [3], primarily to increase the adsorption of the polyelectrolytes to the fibers. This means that layer-by-later deposition of polyelectrolytes on wood fibers already existed before the LbL-concept was discovered, since the fibers are recycled in the process and treated several times with polyelectrolytes. However, to the knowledge of the author, nobody detected this LbL formation on the fibers in the papermaking process, and nobody considered the huge benefits of using this technology to tailor fibers for specific end-uses.

*Multilayer Thin Films: Sequential Assembly of Nanocomposite Materials*, Second Edition.
Edited by Gero Decher and Joseph B. Schlenoff.
© 2012 Wiley-VCH Verlag GmbH & Co. KGaA. Published 2012 by Wiley-VCH Verlag GmbH & Co. KGaA.

After the initial work of Decher [4, 5] and Hoogeven [6], it was realized that the LbL technique could be used to tailor wood fibers [7–12] for use in more advanced products, such as fiber-reinforced biocomposites and/or totally new fiber-based products. The technique is extremely interesting since it enables the tailoring of fibers without the need for the covalent modification that had earlier been the chosen method for fiber modification [12]. When using covalent modification, it is often necessary to use harsh, oxidative reaction conditions, including organic solvents [13] that may also break down the cellulose chains, leading to a decrease in molecular mass, which is negative for the material properties of, for example, fiber-reinforced biocomposites. With the LbL technique it is, on the contrary, possible to use water, a neutral pH and room temperature. Since the renewable wood fibers can be used to form fibrous networks, they are ideal for the manufacture of new, interactive materials with a 3D structure.

During the last decade, there has been a large focus on using nanofibrillated cellulose (NFC) [14] and cellulose nanocrystals (CNC) [15] as building blocks of cellulose in new materials. NFC and CNC emanate from the cellulose fibrils inside the wood fiber wall, as described earlier, and they are liberated in different ways using both chemical treatment and high pressure homogenization [14, 15] to separate the fibrils from each other, and to stabilize them in aqueous dispersion. The methods used to prepare CNC lead to the removal of the amorphous cellulose from the fibrils, leaving a material with a higher crystallinity than NFC but with a much lower aspect ratio. Depending on the preparation conditions, it is also possible to create CNCs and NFCs with different charge densities and charge signs [14] on the nanomaterial surfaces. This naturally affects the stability of the dispersions, their self-associative properties and their interactions with other nanomaterials. It has, for example, been shown that it is possible to form well-ordered LbLs of anionic CNC and cationic polyelectrolytes [16, 17], and NFC and different types of cationic polyelectrolytes [18, 19].

The present chapter will focus on different aspects of the LbL and cellulose fibers. First, the way in which it has been established that LbLs are actually formed on wood fibers will be described. This will be followed by different examples of how fibers and fibrils can be modified to generate cellulose surfaces with totally new properties without the need for the traditional covalent modification that usually demands organic solvents and harsh reactions conditions. Finally, the LbL technique using NFC/CNC and cationic polyelectrolytes will be given some attention, considering the large promise these combinations have for the development of future interactive films.

## 9.2
### Formation of LbLs on Cellulose Fibers

As was mentioned earlier, the delignified wood fibers constitute a highly porous fibrillar network, as shown schematically in Figure 9.1. The interior of the fibrils consists of crystalline cellulose whereas their external parts are less ordered, as

**Figure 9.1** A schematic representation of the open fibrillar structure of a wet cellulose fiber wall where the lignin and most of the hemicelluloses have been removed. The diameter of these fibril aggregates is about 20 nm and the pores between the fibrils are between 10 and 50 nm. Owing to the carboxyl groups on the hemicelluloses, the fibrils have an anionic charge when dispersed in water above a pH of about 2.5.

determined with, for example, solid state NMR [20]. Since the hemicellulose content of most delignified fibers is between 5 and 15% it should be realized and stressed that the fibrils in this image have a layer of hemicelluloses on their external surface. It is also these hemicelluloses that give the cellulose fibers/fibrils an anionic charge due to the presence of carboxyl groups in some of the hemicelluloses [1]. The cellulose as such contains no charge.

When this open network is exposed to oppositely charged polyelectrolytes, it is important to choose a polyelectrolyte that can penetrate the porous fiber wall if the intention is to completely cover all the fibrils with a single layer or multilayers of polyelectrolyte [21]. If the polyelectrolyte is too highly charged [21], or has too large a radius of gyration it will only be adsorbed on the external surface of the fibers [22]. This means that LbL films can be formed at different structural levels in the fiber wall, and this might be of great importance for the end-use applications of the fibers.

However, since fibers are geometrically inhomogeneous, depending on their growth position in the tree and due to the processing of the fibers, it is not a simple task to establish that LbL structures are actually formed on or within the fibers. Lvov [10] used fluorescent labeling to study the formation of LbLs on single wood fibers, and Lingström et al. [23] used Cahn-balance measurements for single fibers to show the change in wettability of the fibers depending on the polyelectrolyte adsorbed in the outermost layer. A schematic description of this experimental arrangement is shown in Figure 9.2, together with the contact angles calculated from the wetting

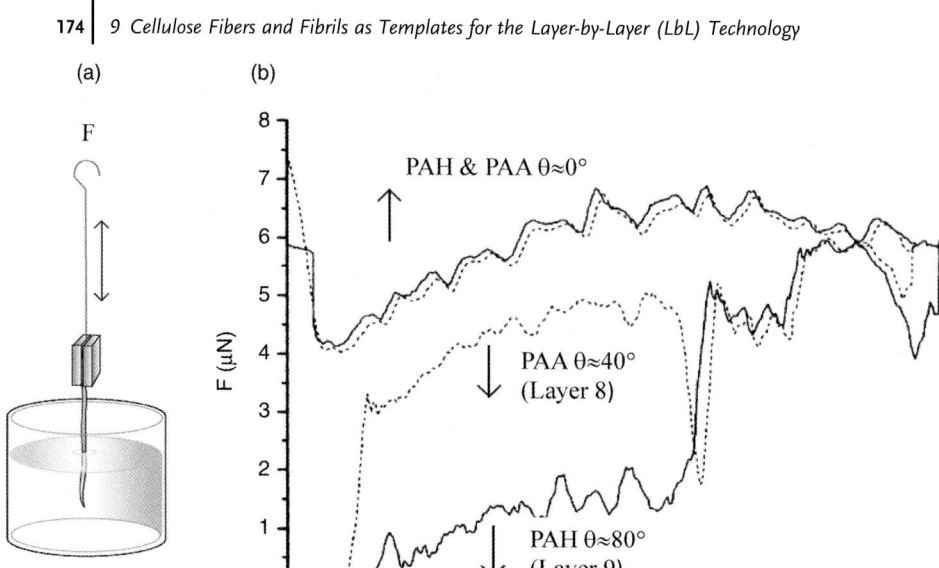

**Figure 9.2** (a) Schematic description of the Cahn balance technique where a single fiber is subjected to LbL treatment by moving it between beakers with cationic and anionic polyelectrolytes with intermediate rinsing with pure water. Only a fraction of the fiber is treated, leaving the rest of the fiber untreated. (b) Two force traces from the rinsing step with pure water are shown with PAH (8th layer) and PAA (9th layer), respectively, in the outer layer of the multilayer [23]. The LbL treatment extended only to a height of 0.9 mm, and above this height the fibers were untreated. The pH was 5 for the deposition of both PAH and PAA in these experiments.

force determined with the Cahn-balance for a LbL system of polyallylaminehydrochloride (PAH) and polyacrylicacid (PAA). In these experiments, the weight average molecular mass was 15 000 and 8000 for the PAH and the PAA, respectively. A very useful feature of this latter technique is that a treated and a non-treated part of the same fiber can be compared, thus eliminating the influence of the fiber geometry on the wetting behavior since the multilayers are deposited only on the lower parts of the fiber.

As shown in Figure 9.2, there is a large difference in the wetting force, and hence in the calculated contact angle, for the advancing measurements (arrow pointing downwards), when PAH or PAA is in the outer layer. With PAH in the outer layer the contact angle was 80° and with PAA it was 40°. The exact molecular reason for this difference has not been fully clarified, but it is suggested that it is due to a more hydrophobic nature of the PAH chain than that of the PAA. In the case of the receding measurements, the result was independent of which polymer was present in the external layer of the multilayer, and this was attributed to the fact that the fibers were completely wet and the receding measurements were thus made against a water-wet surface. It is, nevertheless, striking to see the great difference in wetting made by a single molecular layer on the surface of the cellulose fibers.

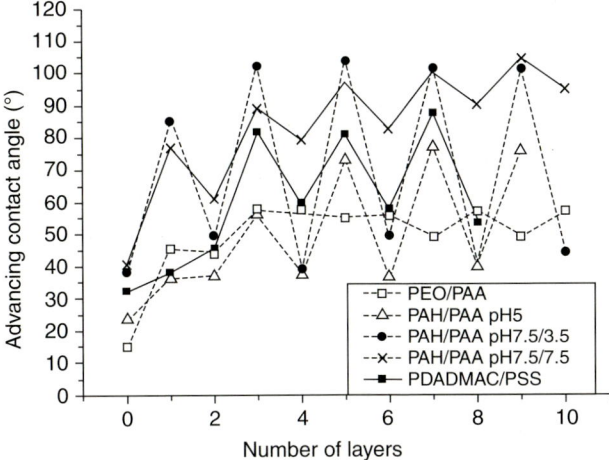

**Figure 9.3** Influence of the LbL treatment on the advancing contact angle of water on cellulose fibers [23] as measured by the Cahn balance technique. Apart from combinations of PAH/PAA ($M_w$ 15 000 and 5000, respectively) at different pH conditions, the figure also shows data for polydiallyldimethylammoniumchloride (PDADMAC) and polystyrenesulfonate (PSS) and PEO/PAA. Layer 0 corresponds to a treatment with a cationic polymer followed by an anionic polymer. In the case of PEO/PAA the first layer was PDADMAC followed by a layer of PAA at pH 7 and then a layer of PEO at pH 2, followed by consecutive treatments with PAA and PEO at pH 2.

The measurements were repeated for different combinations of polyelectrolytes and for a combination of PEO (polyethylene oxide) and PAA (Figure 9.3). A similar pattern was found for all the LBLs based on cationic and anionic polyelectrolytes, the contact angle being greater when the cationic polyelectrolyte was in the external layer.

This figure also shows that the pH conditions at which the layers were prepared had a large influence on the advancing contact angles. When, for example, the PAH was adsorbed at pH 7.5 and the PAA at pH 3.5, the difference in contact angle was even larger (105°/40°) than when both polymers were adsorbed at pH 5 (80°/40°). It has not been possible to establish an exact molecular mechanism for this behavior, but similar investigations using silicon wafers [24] showed that the pH strategy used in the formation of this type of multilayer of PAH and PAA had a considerable influence on the wetting properties. and on the thickness of the LbL films. With the PEO/PAA system, no difference in contact angle could be detected depending on the pH.

With this wetting force technique and with high resolution environmental scanning electron microscopy [25] it was, however, established that an LbL structure is actually formed on cellulose fibers and, depending on the molecular mass [26], the multilayers can be formed either throughout the entire fiber wall (low molecular mass polyelectrolytes), see Figure 9.1, or solely on the external surface of the fibers (high molecular mass polyelectrolytes).

## 9.3
### The use of LbL to Improve Adhesion between Wood Fibers

It was discovered at a fairly early stage that LbL modification of cellulose fibers with PAH and PAA led to a large increase in the tensile strength of fibrous networks, that is, papers, made from these treated fibers [27–29]. It was also shown that both the dry and wet tensile strengths of the networks were improved by this physical treatment method [28]. Since the mechanical properties of fibrous networks are built up from the fiber strength, the joint strength between the fibers and the number of joints per unit volume, it was necessary to establish that the LbL treatment also had an effect on the joint strength between the fibers, measured as the force necessary to break a single fiber/fiber cross [30]. As shown in Figure 9.4 the LbL treatment has a similar effect on both the stress at the break of the fibrous networks [29] and the joint strength between the fibers.

This certainly indicates that the improvement in mechanical properties of fibrous networks due to the LbL treatment is caused by an improved adhesion between the fibers. It is also interesting to note that the stress at the break of the networks, as well as the joint strength, is higher when the PAH is in the external layer of the multilayers when the PAH and PAA were adsorbed at pH 7.5 and 3.5, respectively. This is similar to the difference that was observed in the wetting experiments.

On a molecular level the improved strength of the fiber/fiber joint can be caused either by increased molecular adhesion or by an increased molecular contact area in

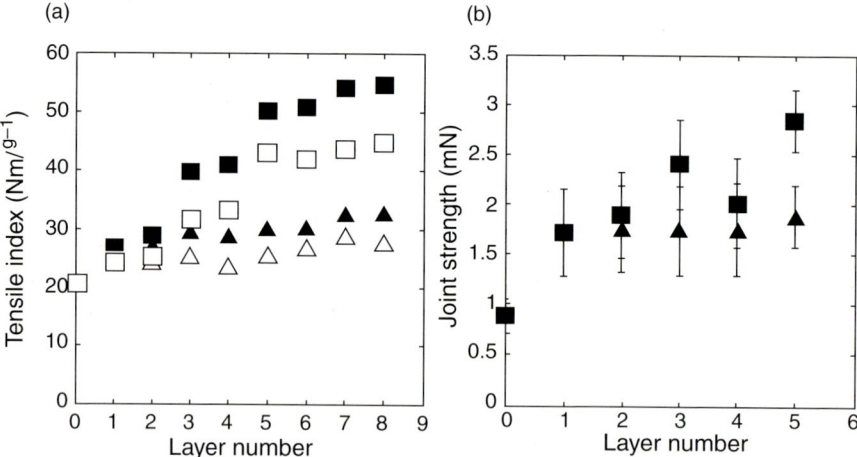

**Figure 9.4** (a) Stress at break of a fibrous network [29] and (b) the joint strength of a fiber/fiber cross [30] as a function of LbL treatment, that is, layer number, with PAH and PAA ($M_w$ of 15 000 and 8000, respectively). Odd numbers refer to PAH and even numbers to PAA treatment. Squares refer to adsorption of PAH at pH 7.5 and PAA at pH 3.5 whereas triangles refer to adsorption of both polyelectrolytes at pH 7.5. Open symbols refer to conventionally dried samples (i.e., 85 °C maximum temperature) whereas filled symbols refer to heat-treated samples where prepared sheets and fiber/fiber crosses were heated to 160 °C for 30 min. The confidence intervals in (b) indicate 95% confidence.

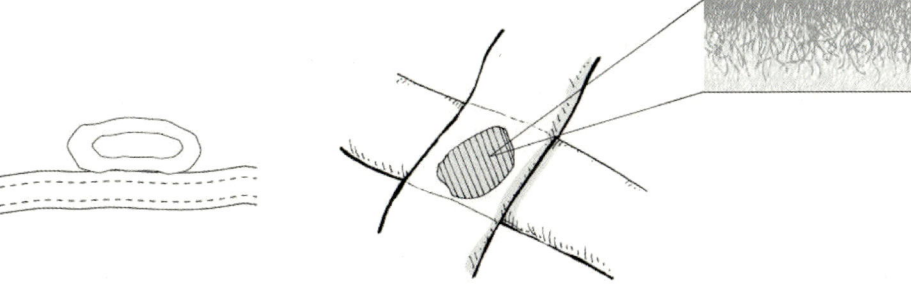

**Figure 9.5** Schematic illustration of a fiber/fiber cross where the molecular contact zone is illustrated by the hatched area of the crossed fibers and where, in the case of the LbL-treated fibers, it can be anticipated that the adsorbed multilayers on each fiber (insert to the far right in the figure) can mix to create an improved molecular adhesion between the treated fibers.

the fiber/fiber cross. This is illustrated schematically in Figure 9.5 where the molecular contact zone is represented by the hatched area and the increased molecular adhesion is illustrated by the intermixed polymer layers indicated in the insert in the figure.

In order to study the possible interdiffusion of the multilayers formed on the fibers, model experiments with colloidal probe atomic force microscopy (AFM) were conducted. In these experiments, a borosilicate colloidal probe and a carefully oxidized silicon wafer were both covered with multilayers of PAH and PAA, and the pull-off force was measured as a function of the number of layers and of the time during which the surfaces were kept in full contact, that is, under maximum load, before they were separated [31]. The same types of PAH and PAA as those used in the experiments shown in Figures 9.2 to 9.4 were used in these experiments, together with a PAH/PAA combination with a higher molecular mass (70 000/240 000). As shown in Figure 9.6a and b, the normalized pull-off force was higher when the PAH was in the outer layer of the multilayer and, furthermore, the pull-off force was much higher when the surfaces were left in contact under maximum load for a longer time.

A comparison of the data in Figure 9.6a and b shows that the time under maximum load had a greater influence on the LbLs made from a PAH/PAA of lower molecular mass, and that the pull-off force was also higher for this polyelectrolyte combination. It is suggested that this is due to an entanglement of the multilayers on opposite surfaces and that, in order to achieve the same effect with the higher molecular mass combination, it would have been necessary to apply pressure for longer times. However, after longer times, the pull-off forces became so high that a change of cantilevers would have been necessary, and this would have made it impossible to compare different measuring series. Nevertheless, these results strongly support the hypothesis that it is an intermixing of the polymers in the adsorbed layers that increases the molecular adhesion in the wet state, and that this is maintained in the dry state since both the fiber/fiber joint strength and the fibrous network strength follow the same trends as the AFM colloidal probe measurements. It should also be mentioned that similar experiments with the colloidal probe technique and poly-

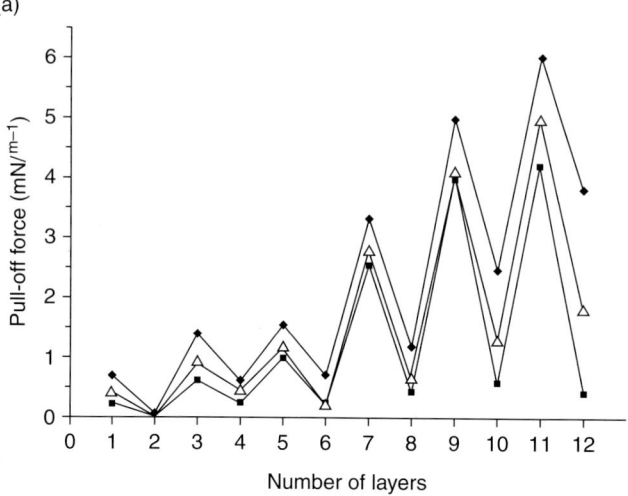

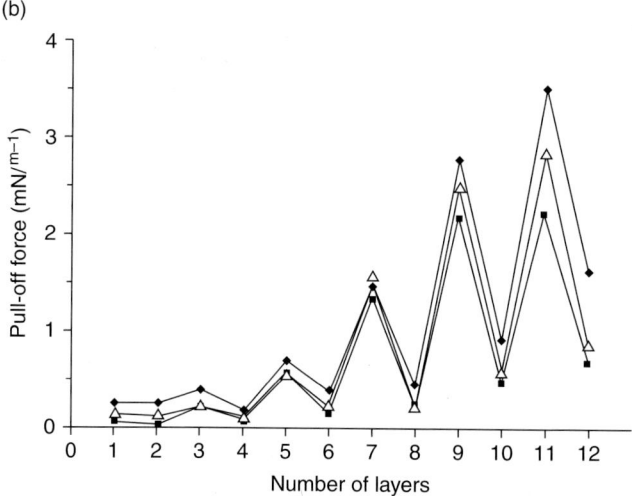

**Figure 9.6** Normalized pull-off force as a function of layer number and time at maximum load for a borosilicate colloidal probe and a silicon oxide surface both covered with LbLs of PAH and PAA of different molecular mass. (a) Molecular mass 15 000/8000 and (b) 70 000/240 000 for PAH and PAA, respectively. The pH during the assembly of both layers was 7.5. Filled squares = 0 s, open triangles = 1 s and filled diamonds = 5 s time under maximum load [31].

electrolyte multilayers have been performed by other groups, although in these investigations a bare colloidal probe was used and in this experimental set-up the electrostatic interactions predominate [32].

It is, naturally, difficult to find an exact molecular reason for the increased pull-off force when the PAH is in the external layer of the multilayer, but experiments with the QCM-D (quartz crystal microbalance with dissipation) [33] have given some insight

## 9.4 The Use of LbL to Prepare Antibacterial Fibers

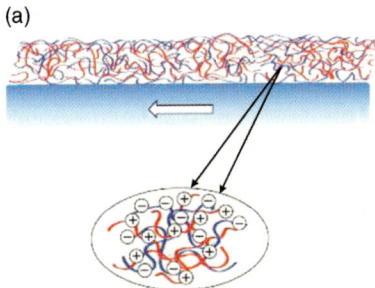
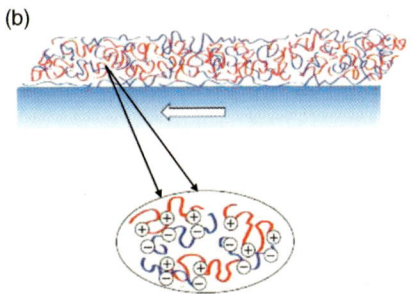
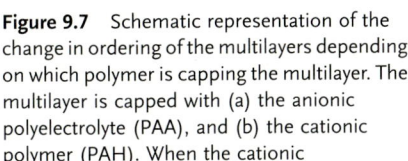

**Figure 9.7** Schematic representation of the change in ordering of the multilayers depending on which polymer is capping the multilayer. The multilayer is capped with (a) the anionic polyelectrolyte (PAA), and (b) the cationic polymer (PAH). When the cationic polyelectrolyte is in the external layer the viscosity and the modulus of the entire LbL structure are reduced. The arrow in the figures are thought to indicate the movement of the crystal in the QCM-D equipment. Redrawn from [33].

into this. In these experiments, the low molecular mass combination of PAH/PAA was used, and with both pH 7.5 for both layers and pH 3.5 for the PAA and pH 7.5 for the PAH layer, the viscoelastic properties of the adsorbed multilayer were significantly dependent on which polyelectrolyte was in the external layer. With PAH in the external layer, the modulus and viscosity of the entire multilayer were much lower than when PAA was in the external layer [33]. This indicates that the ordering of the entire layer is less when the multilayer is capped with PAH and that this promotes the intermixing of the layers of two surfaces that are pressed against each other, as shown schematically in Figure 9.7.

When fiber/fiber crosses are formed in water the fibers are pushed together by the capillary forces created between the fibers as the water is gradually removed. This can explain why the same trends are found with the AFM colloidal probe technique as with the fiber/fiber crosses. Since fibrous networks are formed from numerous fiber/fiber crosses, it is logical that the networks show the same trends in strength development as the other measurements.

It has been suggested [34] that the surface structure of the LbL, depending on the pH strategy for the formation of the LbL assembly and the type of polyelectrolyte in the external layer, is the dominating factor for the properties of the multilayer. However, the results from the AFM colloidal probe measurements [31], as well as from the fiber/fiber joint strength measurements [30] and the fiber network testing [29], show that this is probably not the explanation of the improved adhesion when the PAH is capping the polyelectrolyte multilayer. More work is needed to establish the exact origin of the detected trends.

## 9.4
### The Use of LbL to Prepare Antibacterial Fibers

The use of the LbL-approach to engineer surfaces is naturally very appealing, due both to its simplicity and to the use of aqueous solutions/dispersions and mild reaction conditions [5]. It has thus attracted considerable attention for both nanocomposites [35, 36] and for biointeractive surfaces [37]. Recently, there has been a great interest in preparing antibacterial surfaces with the aid of active polymers [38–41], in order to find remedies for the development of multiresistant bacteria. In these cases [38–41] the polymers are usually covalently linked to the surfaces in order to secure a high surface density (g m$^{-2}$) [41] of the polymers. This usually requires the use of organic solvents and relatively harsh reaction conditions. However, since most of the antibacterial polymers carry a significant cationic charge, as one of the most important ingredients, they would be ideal materials to incorporate in polyelectrolyte multilayers, to increase the amount of cationic polymer per m$^2$, and several authors have shown recently that the LbL technique can actually be used to prepare surfaces that bind bacteria and prevent bacterial growth [42–45]. In this respect, the treatment of wood fibers [42, 43] has a more general interest since they are very similar to cotton fibers that are used in textiles of different types, where this anti-bacterial treatment could be of considerable interest. The polyelectrolytes used in [42, 43] were hydrophobically modified polyvinylamine (PVAm) which consisted of almost completely hydrolyzed polyvinylformamide with C6, C8 and C12 alkyl substituents and degrees of substitution of 30% (C6) and 10% (C8 and C12), respectively. The general structure of the PVAm polyelectrolytes is shown in Figure 9.8. The degree of hydrolysis was 91% for the PVAm used in the investigations [42, 43].

In [42, 43], the PVAm-C6 polyelectrolytes were also adsorbed onto glass surfaces in a single layer [42] or combined with PAA to form LbL structures on oxidized cellulose model surfaces [43]. The C6- and C8-modified PVAm polymers were found to be most efficient in killing bacteria when they were mixed with the bacteria in solution [42]. When the PVAm-C6 polyelectrolytes were used in LbL to modify silicon oxide (single layer) [42] or cellulose surfaces (multilayers) it was shown that bacterial growth was basically prevented especially when 5.5 layer pairs (PVAm/PAA) were adsorbed onto the cellulose surfaces [43]. However, it was difficult to detect any dead bacteria on these polyelectrolyte-treated surfaces. There may be several reasons for the difference in the action of the polyelectrolytes in solution and when they are

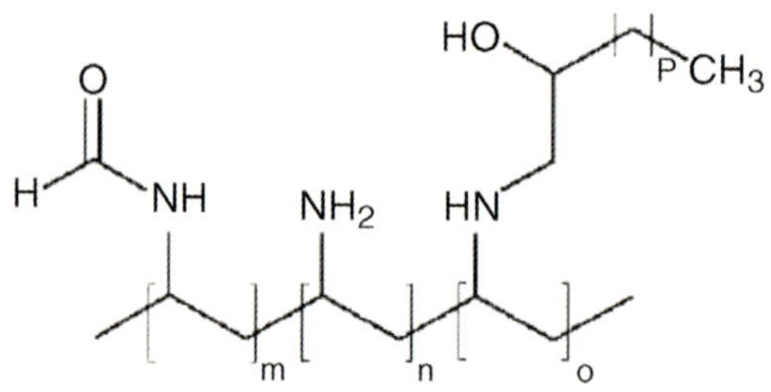

**Figure 9.8** Structure of the PVAm used as an antibacterial polyelectrolyte in LbL treatment of cellulose and silicon oxide surfaces [43]. The alkyl chain length ($p$) was varied between C6, C8 and C12 and the degrees of substitution (by mole) were 30% (C6) and 10% (C8 and C12).

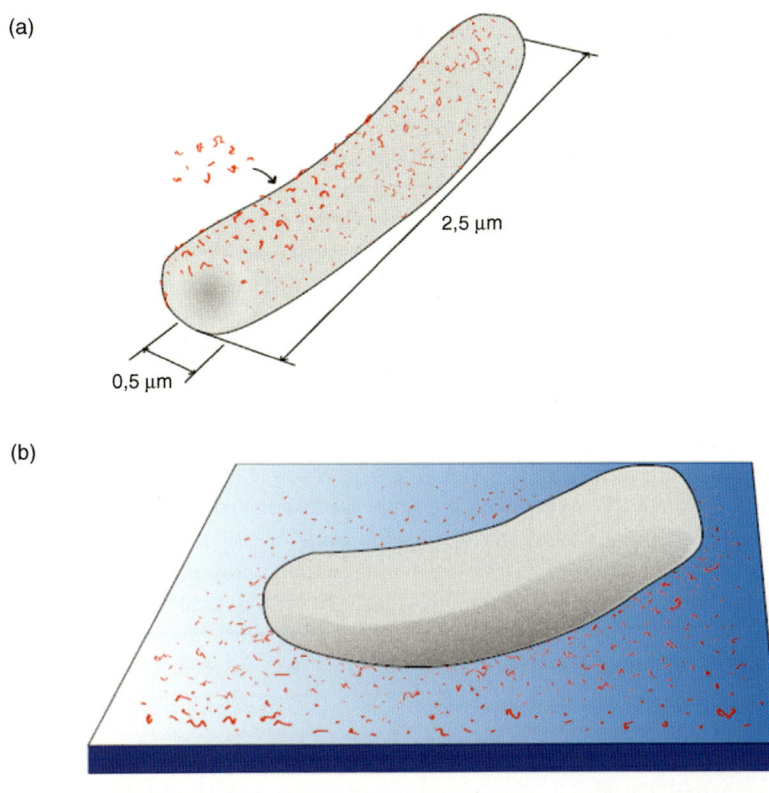

**Figure 9.9** Schematic representation of the difference between the manner in which the bacteria will be affected (a) by polyelectrolytes in solution and (b) by polyelectrolytes adsorbed on a solid surface. The polymers and the bacteria are drawn to scale, and, as can be seen, the bacteria will be surrounded by polyelectrolyte when they are mixed in solution, whereas there will only be a thin layer (5–10 nm for a single layer) affecting the bacteria when they are adsorbed on a solid substrate.

adsorbed on a solid surface, but one very obvious difference is the geometrical constraints of the different situations. When the polyelectrolytes are adsorbed onto the bacteria in solution they will saturate the surface of the bacteria and the entire cell wall of the bacteria will be affected, as is illustrated schematically in Figure 9.9a where the polyelectrolytes and the bacteria are drawn to scale. The size of the PVAm polyelectrolytes was determined to be between 50 and 70 nm, using dynamic light scattering (DLS), at salt concentrations relevant to the testing of the antibacterial action.

On the other hand, when the polyelectrolytes are adsorbed onto a solid surface, only a very small part of the cell wall of the bacteria will be affected by the thin adsorbed layer (between 5–10 nm for a single layer of polyelectrolyte). This naturally means that the efficiency in killing the bacteria will be much lower when the polymers are fixed at an interface, and this can also explain why LbL structures will be more efficient since they are softer and can also be structured depending on the conditions under which they are formed [33, 34, 44, 45]. Another important factor is the cationic surface charge that is created when the polyelectrolyte is adsorbed onto the surfaces, that is, the degree of overcharging. In an elegant work [41], Murata et al. showed that the concentration of cationic charges on the surface should exceed $5 \times 10^{15}$ charges $cm^{-2}$ in order for the surface to kill a single layer of *E-Coli*. For a C6-modified PVAm, the adsorption onto a $SiO_2$ surface is about $0.5 \, mg \, m^{-2}$ in a single layer, and assuming that a large fraction (50%) of the PVAm is neutralized by the surface charges, it can be estimated that the cationic overcharging of the surface would correspond to $1.5 \times 10^{14}$ charges $cm^{-2}$, which is low if the treated surface is to be able to kill bacteria. This means that the adsorption on the surface must be increased, and this can either be achieved by oxidation of the surface or by LbL treatment of the surface. In [43], cellulose surfaces were oxidized and this increased the adsorption of PVAm C6. With the same assumption of charge neutralization of the surface charges, as mentioned earlier, the available cationic charge on the surface was $4.5 \times 10^{15}$ charges $cm^{-2}$, which is about the limit indicated by Murata et al. [41]. With the LbL treatment, the amount of cationic charges is naturally increased, but all these charges will not be available for interaction with the bacteria. Since the efficiency to prevent bacterial growth increases with increasing layer number, it seems reasonable to suggest that at least some of these cationic charges are available for interaction with the bacteria.

It has also been shown recently [46] that when bacteria were incorporated into fibrous assemblies, that is, fiber pads where the fibers had been treated with PVAm and modified PVAm, the efficiency in the prevention of bacterial growth was greater than with flat model surfaces saturated with PVAm. This is believed to be because the bacteria are then completely surrounded by fibers with PVAm layers, and that this is more efficient in preventing bacterial growth. Naturally, more work is necessary to establish the exact mechanism behind the delayed bacterial growth, and it will also be necessary to quantify the balance between bacterial killing and prevention of bacterial growth.

## 9.5
### The use of NFC/CNC to Prepare Interactive Layers Using the LbL Approach

As was mentioned in Section 9.1, there has been considerable interest over the last five to ten years in using NFC or CNC as building blocks in new materials and new devices [14, 15, 47]. This has been brought about by the interesting properties and renewability of the NFC/CNC, new efficient processes for their preparation, developments in nanotechnology, and the need for new markets for the wood natural resource due to the decreased consumption of graphical papers.

As was shown in Figure 9.1, the wet, delignified fiber wall is composed of a rather open network of cellulose fibrils with a diameter of about 20 nm and a length

## 9.5 The use of NFC/CNC to Prepare Interactive Layers Using the LbL Approach

exceeding 1 μm. These fibril aggregates are created during the removal of lignin and hemicelluloses from the fiber wall. In the production of NFC, the fiber wall is treated both enzymatically (endoglucanase) and mechanically before the fibers are treated in a high-pressure homogenizer to liberate the fibrils from the fiber wall [48]. The fibrils thus liberated have a diameter of between 15 and 18 nm and a length of over 1 μm. Instead of using the enzymatic/mechanical pretreatment, the cellulose can be oxidized to increase the charge on the fibers and this will in turn increase the fiber swelling to avoid clogging during high-pressure homogenization of the fibers [18]. The fibrils thus prepared have a diameter of about 5 nm and a length of more than 1 μm, corresponding to the original fibrils of the fiber wall before aggregation has occurred during fiber processing.

In the preparation of CNC, the fibers are usually boiled in rather concentrated HCl or $H_2SO_4$ to remove the amorphous part of the cellulose [49]. After this treatment, the remaining fibers are thoroughly washed and then subjected to some mechanical treatment to liberate the CNC. Revol *et al.* [49] used ultrasonication and, since $H_2SO_4$ was used, the CNC also contained sulfate groups which helped the liberation of the CNC in water. The crystallinity of this material was found to be about 80%, Cellulose I, which can be compared with a crystallinity of about 60% for the NFC [50]. The dimensions of the CNC vary depending on the liberation process but with the sulfuric acid method the diameter of CNC from spruce is about 10 nm and the length is about 200 nm [49]. A schematic image of the two different materials is given in Figure 9.10.

(a)
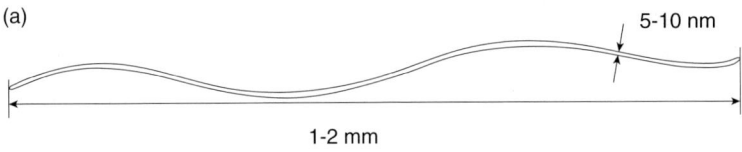

(b)
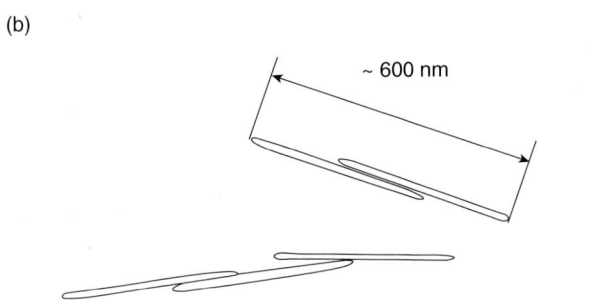

**Figure 9.10** Schematic representation of the dimensions of NFC and CNC. The crystallinity, Cellulose I, of the NFC is about 60% and for the CNC about 80%. The width of the NFC is about the same as for the CNC.

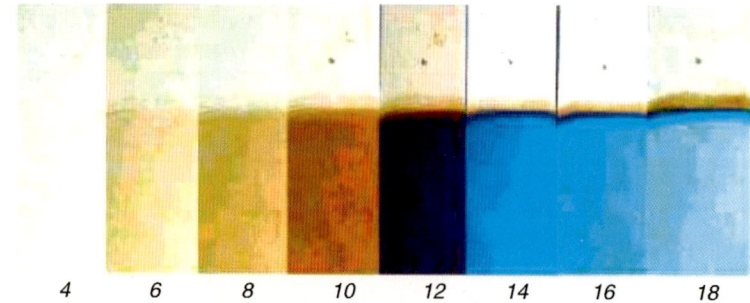

**Figure 9.11** The development of interference colors as the number of layers of PEI and NFC is increased [18]. The numbers in the diagram correspond to the total number of layers and 18 thus means 9 PEI layers and 9 NFC layers.

Both the NFC and the CNC carry anionic charges and, due to their dimensions, they have properties somewhere between those of soluble polyelectrolytes and colloidal particles [51]. The NFC contains carboxyl groups, either from residual hemicelluloses or from oxidation of the cellulose hydroxyl groups, whereas the CNC contains sulfate groups from the boiling of the cellulose fibers in concentrated sulfuric acid. The charges of the CNC induce a self-organization of the crystals as the concentration of the crystals increases, and the remaining cellulose film attains a chiral nematic order. This means that the charge of the CNC, the salt concentration, and the type of ions can be used to control the optical properties of the prepared films.

The presence of an anionic charge also means that both NFC and CNC can be used as a component in LbL structures [16–18, 52] and the layers formed are so well defined that they show bright interference colors as the number of layers is increased [16, 18]. This is exemplified in Figure 9.11 where NFC, with a charge $515\,\mu eq.\,g^{-1}$, is combined with a cationic polyethyleneimine (PEI), with a molecular mass of 25 000, to form LbL structures on oxidized silicon wafers.

Since the NFC is hygroscopic, the films change color when they are subjected to moisture, and this has been explored to develop non-contact moisture sensors based on NFC/PEI multilayers [53].

It has also been shown that the properties of the NFC, that is, charge and size distribution, and the types of polyelectrolytes used have a significant effect on the build-up of the LbLs. It was shown [18] that the use of a 3D-polyelectrolyte (PEI) leads to much thicker layer pairs than when a linear cationic polymer (PAH) is used. However, as the salt concentration is increased during the adsorption of the cationic polyelectrolyte, the linear polyelectrolyte starts to approach the properties created by the 3D polyelectrolyte, indicating that the geometrical structure of the polyelectrolyte has a profound effect on the properties of the LbL. It has also been shown that the charge of the NFC has a large effect on the properties of the layer pairs formed with NFC. As expected, the films formed with NFC of a lower charge have NFC/PEI layer pairs with a more open structure, whereas a higher charge leads to much denser films [50].

## 9.6 Conclusions

This chapter has shown that it is possible to form multilayers of polyelectrolytes, nanoparticles and non-charged polymers on anionically charged cellulose fibers. By carefully selecting the dimensions of the polyelectrolytes/nanoparticles it is possible to treat either the surface of the fibers or the entire fiber wall due to the nanoporous structure of the fibers. Single fiber wetting experiments were used to show how the wetting of the fibers was altered when the number of polyelectrolyte layers were changed. Together with ESEM-images this showed that LbL structures formed on the fiber surface.

It has also been shown that the adhesion between fibers can be greatly improved by treating the fibers with LbL. From model experiments with AFM colloidal probe measurements it was shown that this could be due to entanglements of polyelectrolytes adsorbed onto the fibers as the fibers are brought together during drying of single fiber/fiber joints and fibrous networks.

In an application example it was shown that the LbL treatment of fibers could be used to create antibacterial fibers. It was found that the selection of polyelectrolyte and the pre-treatment of the fibers are essential in order to achieve an efficient bacterial binding to the fibers and killing of the immobilized bacteria. Since wood fibers are very similar to cotton fibers this shows a great potential for antibacterial treatment of fiber surfaces using the LbL technique. In another example it was shown that it is possible to form LbL structures of cationic polyelectrolytes and CNC and/or NFC, that is, cellulose crystals and cellulose fibrils liberated from the fiber wall. It has been found that the layer pairs of these materials are so well defined that the layers formed show clear interference colors depending on the thickness of the LbL film. These types of films can be used as sensors or, for example, in security papers.

## Acknowledgements

The author wishes to extend his warmest gratitude to present and former PhD students and Post Docs that have helped in all the experiments summarized in this chapter. Your hard work, your intelligence, and great company have made this work possible and the time spent together such fun. I also want to thank BiMaC Innovation at KTH, and especially Professor Tom Lindström, for continuous financial and scientific support. The Wallenberg Wood Science Center is thanked for financing parts of the work and for financing the writing of this chapter. I also would like to extend a special thank you to Professor Gero Decher for our cooperation on LbL and NFC. Vinnova, SCA AB and BASF are thanked for financing the work on antibacterial fibers, Dr Hjalmar Granberg for fantastic cooperation in the preparation of LbL films with tailored optical properties, Dr Sven Forsberg for continuous support in LbL work with fibers and Dr Mats Rundlöf, owner of Capisco AB, for the fantastic illustrations and for great scientific collaboration. Finally, Dr Anthony Bristow is thanked for the linguistic adjustment of the text.

## References

1 Wågberg, L. and Annergren, G. (1997) *Fundamentals of Papermaking Raw Materials* (ed. C.F. Baker), Transactions of the 11th Fundamental Research Symposium, Cambridge, UK, pp. 1–82.
2 Lindström, T., Wågberg, L., and Larsson, T. (2005) (Sept.) *Advances in Paper Science and Technology*, Transactions of the 13th Fundamental Research Symposium, Cambridge, UK, **1**, pp. 457–562.
3 Aksberg, R. and Ödberg, L. (1990) *Nord. Pulp Paper Res. J.*, **4**, 168–171.
4 Decher, G. and Hong, J.D. (1991) *Ber. Bunsen-Ges. Phys. Chem.*, **95** (11), 1430–1434.
5 Decher, G. (1997) *Science*, **277**, 1232–1237.
6 Hoogeveen, N.G., Cohen Stuart, M.A., Fleer, G.J., and Böhmer, M.R. (1996) *Langmuir*, **12**, 3675–3681.
7 Wågberg, L., Forsberg, S., and Juntti, P. (2002) *J. Pulp Paper Sci.*, **28** (2), 222.
8 Eriksson, M., Notley, S.M., and Wågberg, L. (2005) *J. Colloid Interface Sci.*, **292**, 38–45.
9 Lingström, R., Wågberg, L., and Larsson, P.T. (2006) *J. Colloid Interface Sci.*, **296**, 396–408.
10 Agarwal, M., Lvov, Y., and Varahramyan, K. (2006) *Nanotechnology*, **17**, 5319.
11 Agarwal, M., Xing, Q., Shim, B.S., Kotov, N., and Varahramyan, K., and Lvov, Y. (2009) *Nanotechnology*, **20**, 215602.
12 Renneckar, S. and Zhou, Y. (2009) *Appl. Mater. Interf.*, **1** (3), 559–566.
13 Lindström, T. and Wågberg, L. (2002) *Sustainable Natural and Polymeric Composites- Science and Technology* (eds H. Lilholt, B. Madsen, H.L. Toftegaard, E. Cendre, M. Megnis, L.P. Mikkelsen, and B.F. Sørensen), Risø National Laboratory, Roskilde, Denmark, pp. 35–59.
14 Klemm, D., Kramer, F., Moritz, S., Lindström, T., Ankerfors, M., Gray, D., and Dorris, A. (2011) *Angew. Chem. Int. Ed.*, **50**, 2–31.
15 Eichhorn, S.J., Dufresne, A., Aranguren, M., Marcovich, N.E., Capadona, J.R., Rowan, S.J., Weder, C., Thielemans, W., Roman, M., Renneckar, S., Gindl, W., Veigel, S., Keckes, J., Yano, H., Abe, K., Nogi, M., Nakagaito, A.N., Mangalam, A., Simonsen, J., Benight, A.S., Bismarck, A., Berglund, L.A., and Peijs, T. (2010) *Rev. J. Mater. Sci.*, **45** (1), 1–33.
16 Gray, D.G. and Cranston, E.D. (2006) *Biomacromolecules*, **7**, 2522–2530.
17 Cranston, E.D. and Gray, D.G. (2010) *Model Cellulose Surfaces*, ACS Symposium Series (ed. M. Roman), American Chemical Society, Washington DC, USA, pp. 95–114.
18 Wågberg, L., Decher, G., Norgren, M., Lindström, T., Ankerfors, M., and Axnäs, K. (2008) *Langmuir*, **24** (3), 784–795.
19 Aulin, C., Varga, I., Claesson, P.M., Wågberg, L., and Lindström, T. (2008) *Langmuir*, **24**, 2509–2518.
20 Larsson, P.T., Hult, E.-L., Wickholm, K., Pettresson, E., and Iversen, T. (1999) *Solid State Nucl. Mag.*, **15**, 31–40.
21 Horvath, A.T., Horvath, A.E., Lindström, T., and Wågberg, L. (2008) *Langmuir*, **24**, 10797–10806.
22 Wågberg, L. and Hägglund, R. (2001) *Langmuir*, **17**, 1096–1103.
23 Lingström, R. and Wågberg, L. (2007) *J. Colloid Interface Sci.*, **314**, 1–9.
24 Yoo, D., Shiratori, S.S., and Rubner, M.F. (1998) *Macromolecules*, **31**, 4309–4318.
25 Lingström, R., Wågberg, L., and Larsson, P.T. (2006) *J. Colloid Interface Sci.*, **296**, 396–408.
26 Lingström, R. and Wågberg, L. (2008) *J. Colloid Interface Sci.*, **328**, 233–242.
27 Forsberg, S. and Wågberg, L.(30th Nov. 1998) Intern. Pat. Appl. PCT WO 00/32702, Priority Date.
28 Wågberg, L., Forsberg, S., Johansson, A., and Juntti, P. (2002) *J. Pulp Paper Sci.*, **28** (7), 222–228.
29 Eriksson, M., Notley, S.M., and Wågberg, L. (2005) *J. Colloid Interface Sci.*, **292**, 38–45.
30 Eriksson, M., Torgnysdotter, A., and Wågberg, L. (2006) *Ind. Eng. Chem. Res.*, **45**, 5279–5286.

31 Johansson, E., Blomberg, E., Lingström, R., and Wågberg, L. (2009) *Langmuir*, **25**, 2887–2894.
32 Gong, H., Garcia-Turiel, J., Vasilev, K., and Vinogradova, O.I. (2005) *Langmuir*, **21**, 7545–7550.
33 Notley, S.M., Eriksson, M., and Wågberg, L. (2005) *J. Colloid Interface Sci.*, **292**, 29–37.
34 Rubner, M.F. (2003) *Multilayer Thin Films-Sequential Assembly of Nanocomposites* (eds G. Decher and J. Schlenoff), Wiley-VCH Verlag GmbH, ISBN 3-527-30440-1 pp. 133–154.
35 Decher, G. (2003) *Multilayer Thin Films-Sequential Assembly of Nanocomposites* (eds G. Decher and J. Schlenoff), Wiley-VCH Verlag GmbH, ISBN 3-527-30440-1 pp. 1–46.
36 Hammond, P.T. (2004) *Adv. Mater.*, **16**, 1271–1293.
37 Boudou, T., Crouzier, T., Ren, K., Blin, G., and Picart, C. (2010) *Adv. Mater.*, **22**, 441–467.
38 Kenawy, E.-R., Worley, S.D., and Broughton, R. (2007) *Biomacromolecules*, **8**, 1359–1384.
39 Tiller, J.C., Lee, S.B., Lewis, K., and Klibanov, A.M. (2002) *Biotechnol. Bioeng.*, **79** (4), 465–471.
40 Klibanov, A.M. (2003) *Biotechnol. Bioeng.*, **83** (2), 168–172.
41 Murata, H., Koepsel, R.R., Matyjaszewski, K., and Russel, A.J. (2007) *Biomaterials*, **28** (32), 4870–4879.
42 Wågberg, L., Westman, E.-H., Ek, M., and Enarsson, L.-E. (2009) *Biomacromolecules*, **10**, 1478–1483.
43 Ek, M., Westman, E.-H., and Wågberg, L. (2009) *Holzforschung*, **63**, 33–39.
44 Lichter, J.A. and Rubner, M.F. (2009) *Langmuir*, **25**, 7686–7693.
45 Lichter, J., Van Vliet, K.J., and Rubner, M.F. (2009) *Macromolecules*, **42**, 8573–8586.
46 Illergård, J., PhD Thesis, KTH, March 2012.
47 Ikkala., O., Ras, R.H.A., Houbenov, H., Ruokolainen, J., Pääkkö, M., Laine, J., Leskelä, M., Berglund, L.A., Lindström, T., ten Brinke, G., Iatrou, H., Hadjichristidis, N., and Faul, C.F.J. (2009) *Faraday Discuss.*, **143**, 95–107.
48 Henriksson, M., Henriksson, G., Berglund, L.A., and Lindström, T. (2007) *Eur. Polym. J.*, **43**, 3434–3441.
49 Revol, J.-F., Bradford, H., Giasson, J., Marchessault, R.H., and Gray, D.G. (1992) *Int. J. Biol. Macromol.*, **14**, 170–172.
50 Aulin, A., Ahola, S., Josefsson, P., Nishino, T., Hirose, Y., Österberg, M., and Wågberg, L. (2009) *Langmuir*, **25**, 7675–7685.
51 Fall, A., Lindström, S., Ödberg, L., and Wågberg, L. (2011) *Langmuir*, **27**, 11332–11338.
52 Podisalo, P., Choi, S.-Y., Shim, B., Lee, J., Cuddihy, M., and Kotov, N. (2005) *Biomacromolecules*, **6**, 2914–2918.
53 Granberg, H., Coppel, L., Eita, M., Antunez de Mayolo, Arwin, H. and Wågberg, L. Accepted for publication in Nordic Pulp and paper Research Journal 2012.

# 10
# Freely Standing LbL Films
*Chaoyang Jiang and Vladimir V. Tsukruk*

## 10.1
## Introduction

Freely standing flexible and robust ultrathin films fabricated by layer-by-layer (LbL) assembly are important nanomaterials with promising applications in a variety of fields, such as thermomechanical sensing, controlled release, optical detection, chemical sensors, microactuators, and controlled drug delivery. Freely standing ultrathin LbL films have several advantages over traditional ultrathin LbL films assembled on solid substrates including two-side accessibility, adjustable shape and dimensions, controlled permeability, and, in particular, high sensitivity to external stimuli such as mechanical pressure [1].

In this chapter, the authors' work on freely standing LbL films over the past eight years will be briefly summarized with some relevant results from recent literature also discussed. We will first introduce freely standing LbL films which contain a variety of nanostructures, such as gold nanoparticles, silver nanowires, semiconductor quantum dots, lanthanide-doped upconversion inorganic nanocrystals, carbon nanotubes, graphene oxides, silk fibroins, and other biomaterials and nanoparticles. Furthermore, more complex and non-uniform freely standing LbL films will also be presented, including nanoporous, patterned, and sculptured LbL films. Finally, we will overview some recent results from the authors' labs on freely standing LbL films with weak interactions, such as responsive hydrogel LbL films, anisotropic microcapsules with ultrathin LbL shells, and LbL-designed non-toxic ultrathin shells for cell encapsulation.

## 10.2
## Fabrication of Freely Standing Ultrathin LbL Films

The freely standing LbL structures and films can be assembled from purely polymeric compounds as well as with various additional reinforcing components according to the modified LbL assembly procedure on sacrificial substrates which will be mostly

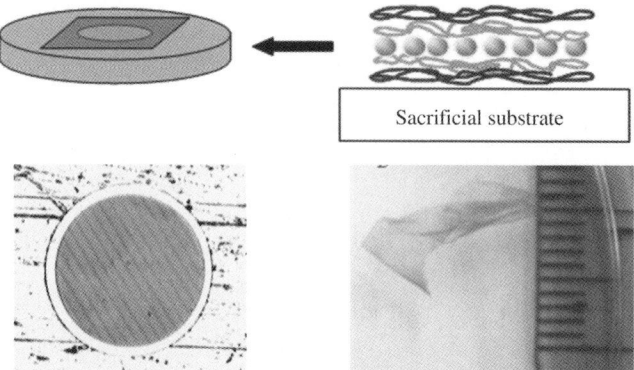

**Figure 10.1** Freely standing ultrathin LbL films can be fabricated from polyelectrolytes and functional nanoscale fillers. By dissolving the sacrificial substrate, the LbL films can be suspended in liquid or picked up onto a holey substrate. Adapted from [25, 51, 60].

discussed in this chapter. Robust and compliant ultrathin LbL films can be fabricated from polyelectrolytes, dendrimers, inorganic nanoparticles, and biomolecules usually reinforced with nanoparticles by commonly utilizing a combination of spin coating and LbL techniques (Figure 10.1). For instance, Jiang *et al.* applied spin-assisted LbL (SA-LbL) assembly for the fabrication of purely polymeric and nanocomposite LbL films from the polyelectrolytes poly(allylamine hydrochloride) (PAH) and poly(sodium 4-styrenesulfonate) (PSS) [1]. In this study, the authors also exploited gold nanoparticles with a diameter of about 13 nm to assemble the robust ultrathin LbL films, which have an overall thickness ranging from 20 to over 100 nm. The central layer of gold nanoparticles used in the ultrathin LbL film had a large optical extinction and unique plasmonic properties. For fabricating the freely standing nanocomposite ultrathin LbL film, a cellulose acetate layer was used as a sacrificial substrate for the LbL assembly. LbL films containing gold nanoparticles and polyelectrolytes deposited on this substrate were then submerged in acetone to dissolve the cellulose acetate and release the films to the liquid (Figure 10.1). With that, freely suspended multilayer LbL films were achieved and ready for additional exploration (Figure 10.1).

To facilitate further characterizations and studies of physical properties, the freely standing ultrathin LbL films could be picked up with a variety of solid substrates and holey supports. These holey substrates include, but are not limited to, microfabricated silicon holders, copper grids, and a copper holder with a center hole of several hundred nanometers in diameter. Figure 10.1 shows the microphotograph of a freely standing LbL film released onto a water surface before transfer to a solid substrate. A top-view micrograph a $(PAH-PSS)_n PAH/Au/(PAH-PSS)_n PAH$ LbL film collected on a copper grid is shown in Figure 10.2 [2]. The ultrathin LbL films are both mechanically and chemically stable to sustain transfer and drying procedures.

A long shelf life of many months is reported for these ultrathin LbL films on the holey substrates. It is believed that such outstanding mechanical properties and long-

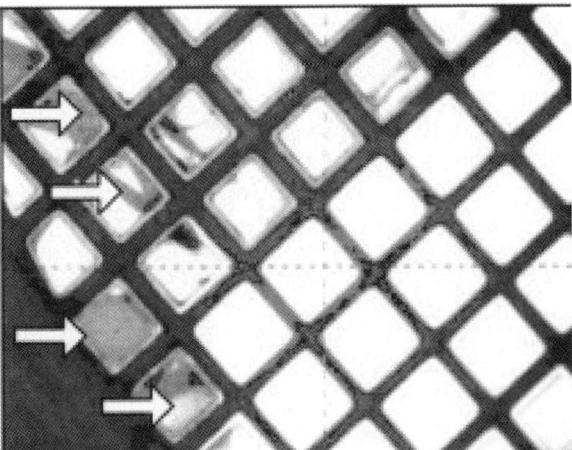

**Figure 10.2** Top-view photograph of a SA-LbL film picked up by a 100 μm copper grid. The white arrows indicate the broken pieces of the film. Reprinted from [1].

term stability originate from the high level of in-plane, bi-axial orientation of the polyelectrolyte chains during the spin-coating process, combined with a reinforcement contribution from the nanoparticle fillers.

### 10.2.1
### LbL Films with Gold Nanoparticles and Silver Nanowires

Ultrathin nanocomposite LbL films containing gold nanoparticles are one of the most intensively studied subjects in the authors' groups. Jiang and coworkers have conducted systematic research on such freely standing LbL ultrathin films, including their fabrication, characterization, and application as highly sensitive sensors [2]. The LbL films with thicknesses of 25–70 nm are fabricated at a molecular precision by the time-efficient SA-LbL assembly. A central layer containing gold nanoparticles was placed between multilayered films composed of alternating polyelectrolytes, PAH and PSS (Figure 10.1). The color of the LbL films in solution after release was light blue, which can be explained by a broad absorption around 600 nm due to the plasmon resonance of gold nanoparticles and their aggregates.

AFM imaging revealed a uniform and smooth surface of the ultrathin LbL films and there are no signs of polymer aggregations or significant nanoparticle clustering during assembly. The freely standing LbL thin films containing gold nanoparticles demonstrated extreme sensitivity with large deflections upon external mechanical/pressure stimulation. Both the results of bulging tests for large deflection and AFM experiments for nanoscale deflection indicated that the nanocomposite LbL films possess outstanding robustness and flexibility. Additional experiments showed that such membranes also have outstanding recovery abilities [3]. Such an autorecovery phenomenon can be considered as a novel mechanism for the self-healing of

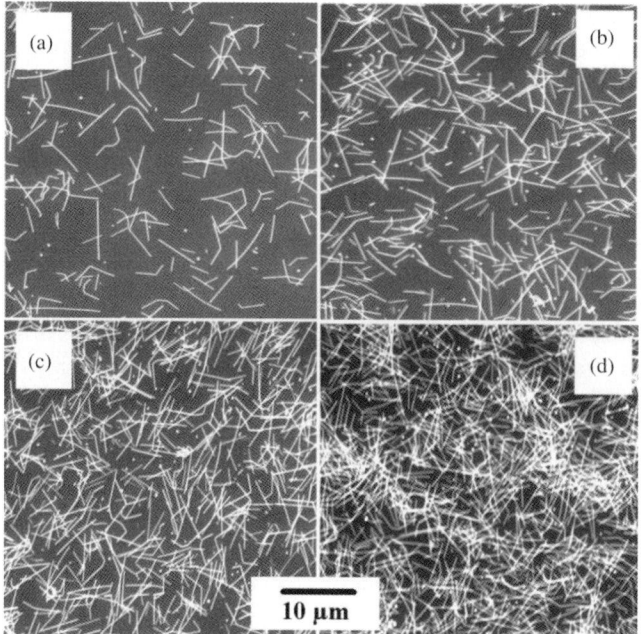

**Figure 10.3** Scanning electron microscopy images of silver nanowires on LbL multilayers. The samples have different volume fractions of silver nanowires of (a) 2.5, (b) 7.5, (c) 15, and (d) 22.5%, respectively. Reprinted from [5].

composite nanomaterials, and thus can serve as a safeguard against the event of overstretching, providing a fast recovery path and facilitating a long lifetime for sensing devices.

Freely standing nanocomposite LbL films with silver nanorods and nanowires can be fabricated by sandwiching the layer of silver nanostructures into a multilayer structure [4]. The SEM images of silver nanowires on an LbL film surface show uniform surface coverage (Figure 10.3). The majority of the nanowires form random monolayers and there is only modest aggregation, even at the highest surface coverage. Fascinating properties of one-dimensional silver nanostructures bring several interesting properties into the ultrathin LbL films, such as excellent conductivity when the density of silver nanowires exceeds the percolation limit, significant mechanical toughening, and the ability for surface enhanced Raman scattering (SERS). In the LbL films, the diameter of the silver nanowires is about 70 nm, which is thicker than that of LbL films (total of about 50 nm). AFM investigation revealed that a peculiar conformal morphology was formed with the nanowires protruding from the planar LbL films. It is not surprising that such asymmetric structures also influence the electrical conductivity of the ultrathin nanocomposite LbL films.

Indeed, LbL films with low silver nanowire content were insulating, while a 22.5% volume fraction of silver nanowires induced overall conductivity. The final calculated experimental percolation threshold, $8\% < \emptyset_c < 12.0\%$, is consistent with the theoretical prediction for a 2D array of the silver nanowires. Furthermore, a sixfold difference in in-plane conductivity for bottom-down and bottom-up orientation of the ultrathin nanocomposite LbL films was observed. Since the conductivity is inversely proportional to the thickness of an insulating layer, the conductivity difference might be explained by a thicker polyelectrolyte matrix on silver nanowires at the bottom of the LbL, an assumption that was confirmed with AFM investigation.

A detailed study of the mechanical properties was further conducted on freely standing nanocomposite LbL films containing silver nanowires [5]. The ultrathin LbL films with silver nanowires possess elastic moduli ranging from 2 to 4.6 GPa. The modulus increased with the increasing fraction of the silver nanowires. The ultimate strain was found to decrease slightly with increasing volume fraction of the silver nanowires, which indicates the higher stiffness. The toughness of the ultrathin LbL films with silver nanowires ranges from 400 to 1000 kJ m$^{-3}$, with a trend towards lower values for stiffer films and purely polymeric films. The maximum toughness of about 1000 kJ m$^{-3}$ is achieved for the LbL films with an intermediate composition of 7.5% silver nanowires, which can be related to a combination of higher strength induced by nanowires and high ultimate strain remaining.

## 10.2.2
### LbL Films with Quantum Dots

Semiconductor nanocrystals such as colloidal quantum dots (QDs) exhibit unique size-dependent strong luminescence and are currently of great interest for various applications in photovoltaic devices, opto-electronic, and biolabeling and bioimaging. Zimnitsky *et al.* fabricated freely standing ultrathin nanocomposite LbL films encapsulating a monolayer of quantum dots [6]. Core–shell CdSe/ZnS QDs were prepared according to the known method and a ligand exchange with thioacetic acid was conducted for LbL assembly. In this work, freely standing LbL films were studied with different QD densities by using solutions with different concentrations. Although the LbL films were stable and robust for process, transfer, and sample preparation, the overall elastic modulus only increases slightly due to the low volume content of QD nanoparticles. All the ultrathin LbL films showed luminescence emission with a single peak centered about 630 nm, which is close to that of the QD solutions. For LbL films with low QD content, the luminescence intensity was directly proportional to the number of QDs.

Substrate- and time-dependent photoluminescence of semiconductor QDs inside ultrathin LbL thin films deposited onto several different types of substrates were studied in detail. Zimnitsky and coworkers found that the luminescence intensity of QD-containing LbL films depends strongly upon the distance between the QDs and the silicon substrate [7]. The increase in the luminescence intensity with the increasing distance between the QD monolayer and the silicon surface can be related to the reduction in the energy transfer between different components. The fact that an effective quench phenomenon can be observed at a distance as high as 30 nm can be explained by the fact that the polyelectrolytes in LbL films can act as a charge conductor. A complete suppression of the luminescence for the QD-contain-

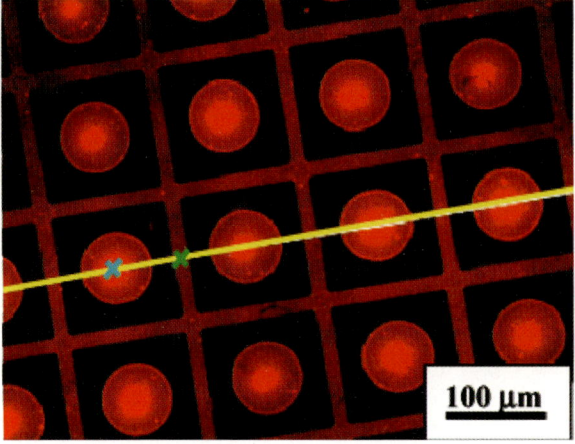

**Figure 10.4** Fluorescent image of a LbL film containing QDs. The LbL film is deposited on a microfabricated wafer with cylindrical microscopic captivities. Reprinted from [6].

ing LbL films was observed on conductive metal surfaces (copper and gold). QD-containing LbL films on glass substrates were observed to have 40% higher luminescence intensity than that on silicon and PDMS substrates.

The QD-containing ultrathin free-standing films showed a very interesting phenomenon when assembled onto a silicon substrate with microfabricated cavities. A regular distribution of the photoluminescence was observed which followed the patterned array beneath (Figure 10.4). The photoluminescent intensity of freely suspended multilayer films is about four times higher than for those films attached to the substrate. It was also found that a less dramatic contrast in luminescence intensity occurred for QD-containing LbL films transferred on PDMS substrates with similar microscopic cavities. Besides the quenching effect mentioned above, it is also believed that the high intensity of freely standing films on silicon cavities can be due to the reflection of incident light from the highly reflective bottom of the cavities. The optimization of the ultrathin LbL films containing QDs can greatly enhance their photoluminescence appearance.

### 10.2.3
### LbL Films with Nanocrystals

Lanthanide-doped nanocrystals can be integrated into freely standing ultrathin nanocomposite LbL films that exhibit the fascinating optical properties of upconversion luminescence. Jiang and coworkers synthesized citrate-coated $NaYF_4$: 17% Yb, 3% Er nanocrystals using a high-boiling-point solvent method and integrated them into ultrathin LbL films [8]. These nanocrystals, with diameter around 50 nm, can absorb near-infrared (NIR) photons and emit higher energy photons in the visible range, so-called upconversion (UC). Prior to the LbL assembly, a ligand-exchange process was conducted on the UC nanocrystals (UCNCs) which resulted in hydrophilic UCNCs with a large negative zeta-potential (below $-50.8\,\text{mV}$).

Both AFM and TEM revealed that the UCNCs are homogeneously distributed in the SA-LbL thin films, which is a significant improvement as compared to UCNCs in PMMA thin films. Robust ultrathin LbL films with total thicknesses below 70 nm can be released from the cellulose acetate layer and freely suspended over holey substrates for months. Only a slight decrease in the upconversion efficiency was found with hydrophilic UCNCs after the surface modification. Such slight luminescence quenching could attribute to the increase in the number of OH oscillators at the surface of the UCNCs. Time evolution studies on the upconversion emission confirmed that the upconversion efficiency of UCNCs in LbL films is similar to that of hydrophilic UCNCs in solutions. The freely standing upconversion nanocomposite LbL films possess reasonable long-term luminescence stability and have potential for NIR imaging devices.

## 10.2.4
### LbL Films with Graphene Oxides and Carbon Nanotubes

Ultrathin LbL films containing carbon nanotubes (CNTs), graphenes and their derivatives attract increasing interest due to the unique optical, electrical, mechanical, and thermal properties of the carbon-based nanostructures [9, 10]. Carbon nanotubes can act as reinforcement fillers and thus strengthen the mechanical properties of the freely standing ultrathin nanocomposite LbL films with sandwiched CNTs. Jiang *et al.* fabricated LbL films with CNTs via the SA-LbL assembly and systematically studied the internal stress and structural reorganization in the course of their responsive behavior under mechanical stimulation [11]. In their work, multilayer films with a general formula of $(\text{PAH-PSS})_n\text{PAH/CNT/(PAH-PSS)}_n\text{PAH}$ were freely suspended over copper plates. The overall volume fraction of CNTs in LbL films was estimated to be about 1%. An enhanced elastic modulus of such nanocomposite thin films was measured to be 8.8 GPa from the bulging tests. In the case when the fraction of the CNTs is reduced by using a micropatterning process, the elastic modulus of the ultrathin nanocomposite thin films with CNT patterns is 5.9 GPa, a slight decrease compared to the non-patterned ones [11].

Strain-sensitive Raman modes of CNTs in freely suspended LbL nanomembranes were monitored for a better understanding of the reinforcement role of the CNTs inside the LbL films. Since CNTs have a relatively large Raman cross-section, the feature of the CNT Raman spectra in the ultrathin LbL films can be obtained without any metal nanoparticle enhancement. The CNT-containing LbL films have Raman peaks such as the D band at $1352\,\text{cm}^{-1}$ and the G band at 1571 ($G^-$) and 1599 ($G^+$) $\text{cm}^{-1}$ with a width about $20\,\text{cm}^{-1}$. For the maximum deflection at the pressure of 98 mbar, the G bands were shifted to 1589 and $1564\,\text{cm}^{-1}$, respectively, with the position of the D band unchanged (Figure 10.5). The changes in the peak position under mechanical deformation are reversible and repeatable. Analysis of the experimental data revealed that the Raman position shifts were quite different for small and large deflections, where a very insignificant shift occurs below $2\,\mu\text{m}$ deflection

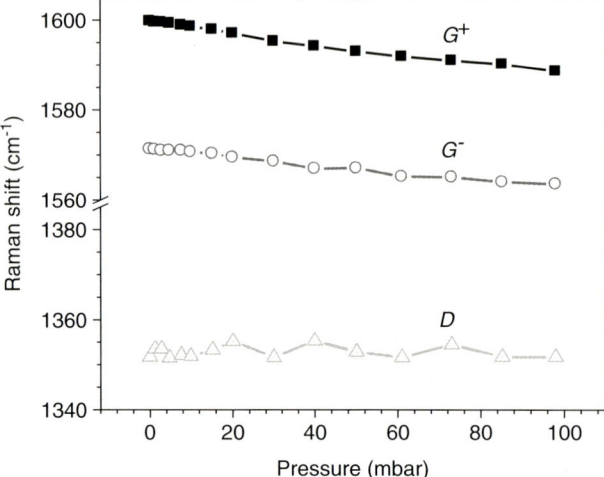

**Figure 10.5** Shifts of Raman bands of carbon nanotubes in 9CNT9 nanomembranes with the pressure applied to the films. Reprinted from [11].

and dramatic shifts happen for larger deflection of the LbL nanomembranes. This behavior can be explained by a model of two distinctive deformational regimes which include the bending and buckling deformation of the CNTs in the initial stages and a local collapse of the CNT walls at higher deformation.

Assembling graphene-based materials into LbL films with minimum folding and wrinkling can be achieved by the combination of LbL assembly with the Langmuir–Blodgett technique (Figure 10.6). Kulkarni et al. recently successfully fabricated

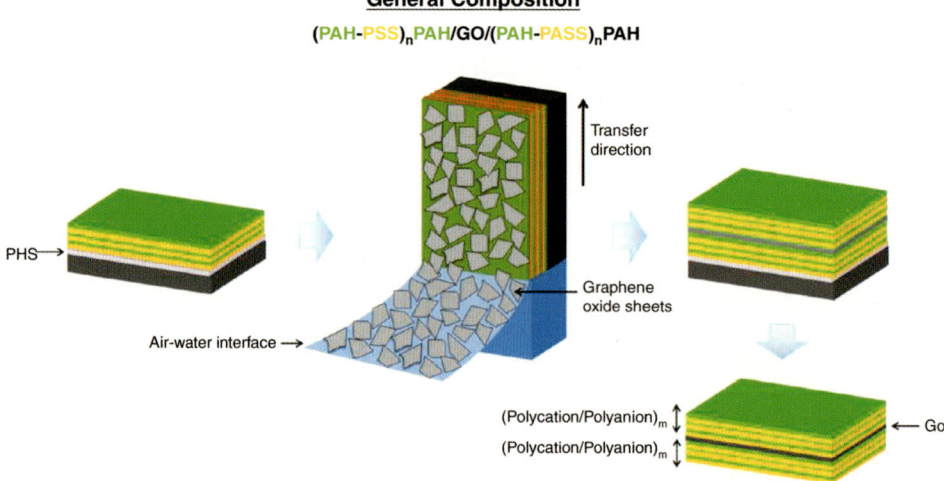

**Figure 10.6** Assembly of freely standing LbL film containing graphene oxide sheets. Reprinted from [12].

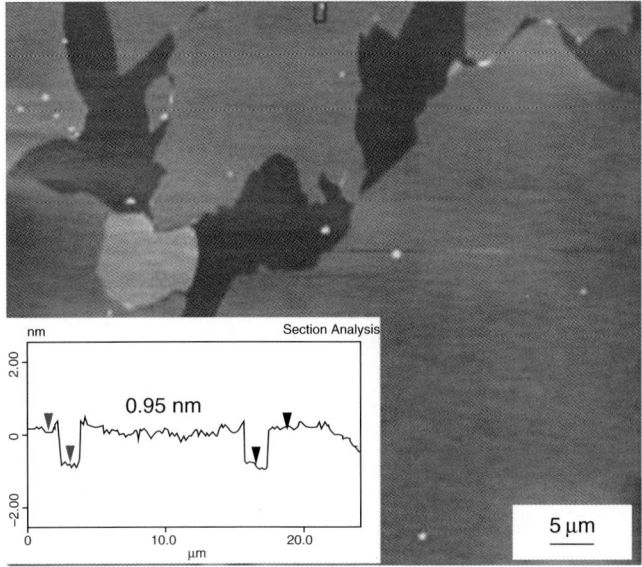

**Figure 10.7** AFM image showing graphene oxide LB sheets used for LbL assembly (inset shows the sectional image), Z-scale is 5 nm. Reprinted from [12].

graphene oxide-polyelectrolyte LbL nanomembranes with an overall thickness of about 50 nm, where the graphene oxide (GO) sheets were uniformly incorporated inside the LbL polyelectrolyte matrix [12]. GO sheets prepared by the oxidative exfoliation of graphite flakes provided a homogeneous dispersion with most of the exfoliated flakes being single layers and layer pairs. By controlling the surface pressure at the air/water interface, a GO monolayer with a high density of 90% can be obtained with only occasional wrinkles. Such monolayers are ready to be integrated into freely standing ultrathin LbL films. AFM images indicated the GO sheets followed the morphology of the polyelectrolyte layers (Figure 10.7). The contrast between the GO and polymer can be observed due to the difference in their functionalities and stiffness.

The elastic modulus of ultrathin LbL films containing GO sheets was studied with both buckling instability tests and bulging tests. The buckling instability tests result in a very uniform buckling pattern. The spacing of the wrinkles increased from 2.5 to 5.0 μm with increasing concentration of the GO sheets in the LbL films. Accordingly, the Young's modulus of the nanocomposite LbL films increased from 1.5 to about 4.0 GPa. On the other hand, bulging tests show higher elastic values, from 4 to about 20 GPa with increasing GO loading from 1.7 to 8% in the LbL films. The disagreement between the results of bulging and buckling measurements is quite unusual for bulk composite materials and might be related to the unique structures of the ultrathin films containing monolayer GO sheets. Due to the slippage and buckling, the stress applied is not completely transferred to the filler, and thus the reinforcing contribution of stiff, but easily pliable GO sheets is significantly undermined.

During the bulging tests, the nanocomposite LbL films are subjected to tensile stress, which is evenly transferred across the thickness of the films. With that, all of the GO sheets contribute to the reinforcement fully with minimum slippage, thus providing a significant toughness, about 2 MJ m$^{-3}$, which is almost five times that of the films without embedded GO sheets. The maximum toughness was reached at a very small content of the GO sheets due to the combination of higher mechanical strength and ultimate strain.

### 10.2.5
### LbL with Biomaterials

The assembly of ultrathin LbL films from biomaterials (proteins, enzymes, cells) can be of great interest for a wide range of applications for those requiring robust mechanical properties, such as excellent elasticity combined with biocompatible and biodegradable properties.

Jiang *et al.* exploited SA-LbL assembly to fabricate thin (500 nm) and ultrathin (20–100 nm) silk fibroin films, and these films were tested for tensile and compressive mechanical properties [13]. The silk films were characterized for elastic modulus, elongation to break, and toughness. The mechanical parameters in both compressive and tensile modes have been revealed for such ultrathin silk films. Structural control of the silk protein was gained through physical crosslinks ($\beta$-sheets), resulting in robust and stable ultrathin material coatings with excellent mechanical characteristics that do not require specific chemical or photoinitiated crosslinking reactions (Figure 10.8). This study demonstrates that the approach, which is well-known for bulk silk material, can be successfully applied to ultrathin silk films with the thickness comparable to chain dimensions.

The FTIR-ATR spectra of silk films with different treatments confirmed that silk II with $\beta$-sheet structure can be formed in both the thin and ultrathin silk films in the process of the fabrication, even without additional methanol treatment. The presence of the $\beta$-sheets in the films was additionally confirmed by X-ray diffraction, which shows the presence of two preferential intermolecular packings of the backbones.

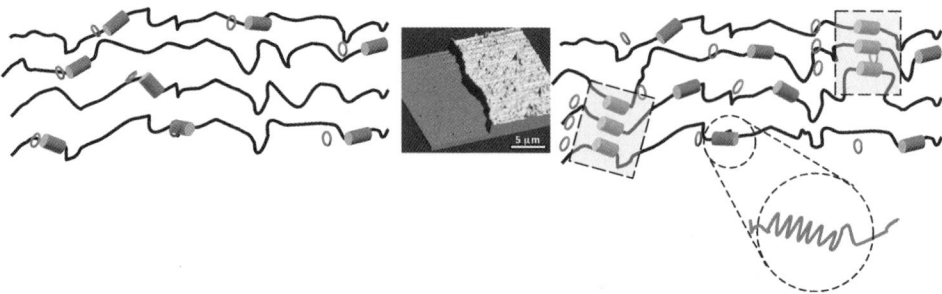

**Figure 10.8** Schematics of $\beta$-sheet formation in the silk fibroin LbL nanomembrane, in which the LbL process can cause an increase in $\beta$-sheet content. Reprinted from [13].

These results indicate the coexistence of an amorphous matrix and poorly ordered domains with crystalline structure formed by $\beta$–sheets (silk II form) [14, 15]. Silk II is an antiparallel $\beta$-sheet in which the polypeptide chains are aligned and adjacent chains are locked with hydrogen bonds. The amino acid repeat unit in B. mori silkworm cocoon silk fibroin that is responsible for the formation of β-sheets consists of the sequence GAGAGSGAAG[SGAGAG]$_8$Y (where G = glycine, A = alanine, S = serine, Y = tyrosine) separated by less regular domains with a random conformation [16].

The $\beta$–sheet secondary structure was more pronounced in the ultrathin LbL films and could be further induced by the drying process [17–19]. Methanol treatment of the ultrathin LbL films resulted in increased $\beta$–sheet content (Figure 10.8). The formation of these crystalline domains results in reinforcement with a network of physical crosslinks, thus increasing elastic modulus, ultimate strength, and film toughness. The high values of elastic modulus (3–4 GPa for untreated, as-prepared silk films and 6–9 GPa for methanol-treated silk films) and the ultimate strength were obtained in this and other studies [20, 21].

Even more robust ultra-thin silk-inorganic nanocomposite membranes with enhanced mechanical and tunable optical properties have been reported by Kharlampieva et al. [22]. LbL assembly of silk protein with clay nanoplatelets resulted in the highly transparent films with significantly enhanced mechanical properties, many-fold higher than that for pristine silk nanomaterials. The ultrathin reinforced silk films demonstrated outstanding elastic modulus reaching 20 GPa and toughness reaching 1000 kJ m$^{-3}$. On the other hand, incorporation of highly reflective and densely packed silver nanoplates caused similar enhancement of mechanical properties but, in contrast, created highly-reflective flexible silk films. The LbL films composed of a pre-crossinked silk fibroin matrix with incorporated silica nanoparticles with silsesquioxane cores have also been studied by Kharlampieva et al. [23]. These nanoparticle-reinforced silk membranes have enhanced mechanical properties as compared to traditional silk-based nanocomposites. Many-fold increase in elastic modulus and toughness was also found for these nanocomposites.

### 10.2.6
**Mechanical and Optical Properties**

The ultrathin nanocomposite LbL films exhibit unique physical characteristics, including their outstanding mechanical and optical properties, which enable their great potential as active membranes in sensor, detector, and eletromechanical systems. As we know, one of the advantages of the LbL assembly technique is that functional nanoscale building blocks can be introduced into the LbL films without significant alteration of their multilayer structure but with enhanced ultimate properties [24]. The versatility of such LbL nanomaterials, together with the ability of the fine structural tuning, ensures a broad range of prospective applications.

Traditional methods, such as tensile experiments, three-point load tests, and indentation measurements, are used to handle macroscale materials and, therefore, are not suitable for studying the mechanical properties of ultrathin films

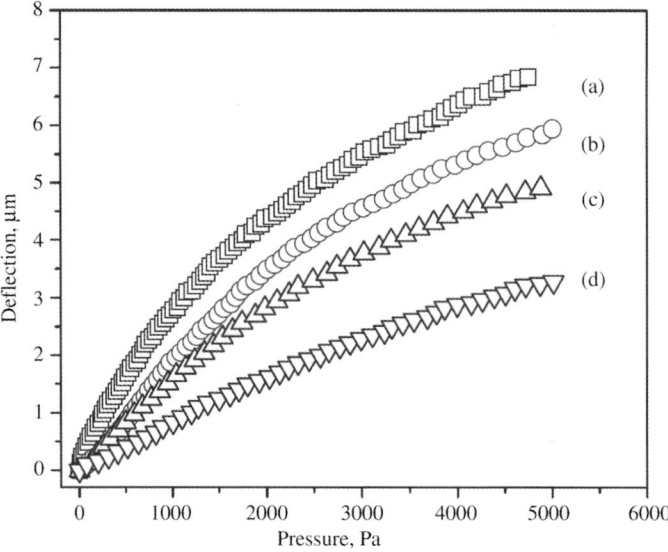

**Figure 10.9** Deflection versus pressure measurements for freely standing LbL films containing silver nanowires with various volume fractions, (a) 2.5, (b) 7.5, (c) 15.0, and (d) 22.5%. Reprinted from [5].

(see Chapter 16). There are only a few examples on using micro-tensile testing to study the mechanical strength of LbL films with thickness of several hundred nanometers [25]. Markutsya and coworkers performed a bulging test to study the mechanical properties of free-standing ultrathin LbL films on a home-designed interferometer [26]. In the bulging test, the pressure difference between the two sides of the free-standing LbL films can be regulated by an automatic syringe pump. The deflections of the ultrathin LbL films are calculated from the changes in the interference patterns. Figure 10.9 shows typical deflection–pressure curves for ultrathin LbL films containing silver nanowires. The experimental results of bulging tests clearly demonstrated the increase in the Young's modulus of the ultrathin LbL film with increasing content of the nanoscale fillers [5].

Besides the bulging tests, mechanical properties can also be studied by using buckling instability tests and nanoscale indentation experiments. Buckling patterns due to the mismatch of the elastic modulus of the ultrathin films and the soft substrates can be easily recorded with optical microscopy and analyzed to obtain the Young's modulus of the thin film. The wrinkles of the non-patterned ultrathin LbL films are quite uniform while more complex patterns can be observed for patterned LbL films [27].

In addition to the excellent mechanical properties, freely standing ultrathin nanocomposite LbL films also have interesting optical properties, such as absorption, fluorescence, and surface-enhanced Raman scattering (SERS). The optical behaviors of the ultrathin nanocomposite LbL films are closely associated with the optical properties of the embedded nanostructures. The sandwiched gold nanoparticles reserve their strong surface plasmon resonance (SPR) in the visible spectrum. The positions and shapes of SPR peaks can be significantly affected by the nanostructures in the LbL films, such as the density of the gold nanoparticles in the monolayer, the number of gold nanoparticle layers and their interlayer distances. Jiang et al. reported that in the absorption spectra extinction peaks are due to the SPR of individual gold nanoparticles and the collective SPR from both intra- and interlayer interparticle plasmon coupling [28].

Lin et al. reported the fabrication of purely polymeric freely standing ultrathin LbL films from conjugated polyelectrolytes with strong fluorescence properties without any nanofillers [29]. In that work, a water soluble conjugated polymer, poly(2,5-methoxypropyloxy sulfonate phenylene vinylene) (MPS-PPV) was synthesized and assembled into LbL multilayers with a positively charge polyelectrolyte, PAH. Smooth LbL films with microroughness below 5 nm were obtained. The conjugate nature of the MPS-PPV chains brings an intense fluorescence with an emission maximum in the visible range. The authors also revealed that the intense fluorescence is extremely stable, even under large mechanical deformation. The robust fluorescent freely standing conjugated LbL films can be considered for applications such as flexible lightweight displays and large luminescent panels.

Interesting optical properties of the ultrathin nanocomposite LbL films can also be related to the interaction of the sandwiched nanoscale building blocks and the polymeric matrix, for example Raman spectra of polyelectrolytes that were enhanced by the embedded gold nanoparticle arrays in the LbL films. Gold nanoparticles encapsulated in the freely standing ultrathin LbL films allow the use of SERS to monitor the conformational changes under deformation (Figure 10.10). After optimizing the design of gold nanoparticles forming chainlike aggregates, Jiang and coworkers conducted an *in situ* SERS study of the Raman spectra of the ultrathin nanocomposite LbL films in the course of the membrane deformation [30]. The authors observed that the position of the main Raman peak was gradually shifted to a higher frequency and the trend changed to the opposite for strain above 0.06%. The authors suggested that the results are due to the deformation of the ultrathin nanocomposite LbL films initially reorienting the side groups and then stretching the polymer backbone segments (Figure 10.10). The authors believe that chain bridging spread over a number of adjacent gold nanoparticles creates an effective toughening mechanism.

## 10.2.7
### Surface Morphology and Internal Microstructure

Numerous studies of LbL films with AFM since the introduction of LbL assembly have focused mostly on routine verification of the surface morphology to assure its uniformity and integrity, monitoring polyelectrolyte or nanoparticle adsorption and aggregation, imaging dewetting behavior, estimating the surface microroughness for LbL films, and measuring the thickness of the films (usually in combination with ellipsometry and X-ray reflectivity) [31–34].

(a)

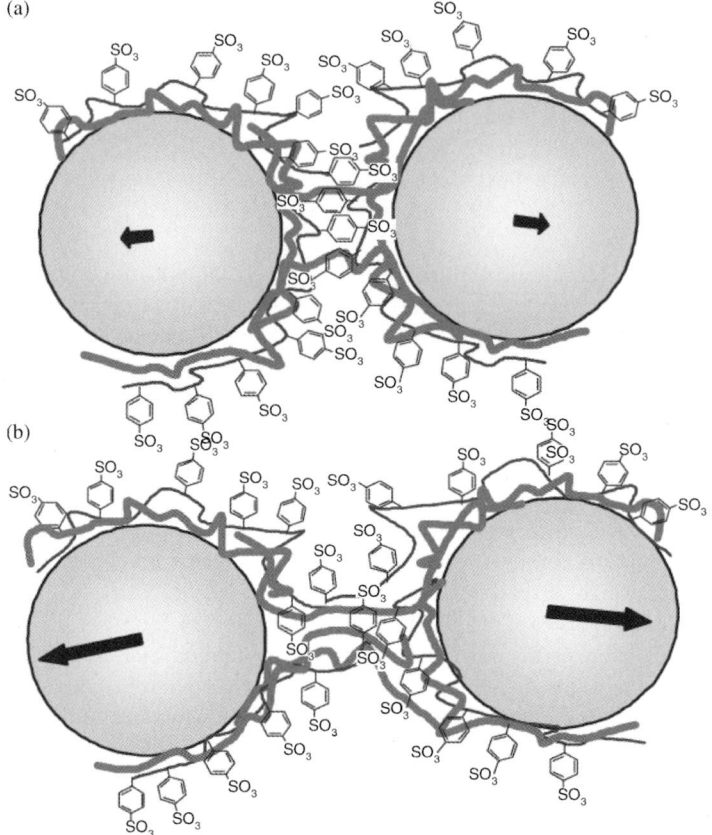

(b)

**Figure 10.10** Side groups and backbone segments for low deformation of LbL nanomembrane (a) followed by the formation of bridging chain segments between gold nanoparticles for large deformation (b). Reprinted from [30].

Generally, despite the relatively high roughness of sacrificial substrates, freely standing LbL films show very uniform surface morphology with microroughness (usually measured within $1\,\mu m^2$ surface area) below 0.5 nm, which is the typical diameter of polymeric backbones [35]. In contrast, various weakly-assembled and heterogeneous LbL films, such as hydrogen-bonded films, nanocomposite films, porous films, exponentially-grown films, and LbL films with crosslinked components, usually show much higher surface microroughness and larger scale irregularities, exceeding 10–20 nm in some cases. This is due to the localized nonhomogeneous distribution of components, significant intermixing or demixing, local dewetting and surface-induced roughening.

AFM was also employed as an imaging tool for *ex situ* and *in situ* monitoring of the LbL assembly processes under variable environments. For instance, in earlier studies, Tsukruk *et al.* exploited AFM for the investigation of monolayers and layer pairs of polyionic materials deposited on charged self-assembled monolayers

(SAMs) [36]. *Ex situ* AFM was used to monitor PSS adsorption on the charged surfaces of amine-terminated SAMs and PAH adsorption on a PSS monolayer. Observations of the PSS layer at various stages of electrostatic deposition revealed highly inhomogeneous adsorption and chain assembly at the earliest stages of deposition. For longer deposition times, local coverage increases significantly in the areas of submicron defects, such as the edges of atomic planes. Even longer deposition times result in an equilibration of polymer chains and the formation of a homogeneous monolayer.

Neutron reflectivity was utilized to study the internal structure of dip-assisted LbL films made by the traditional route of adsorption from solution of strong polyelectrolytes combined with selectively deuterated polyelectrolyte layers [37–40]. In a recent study to reveal the internal organization of freely standing spin-assisted LbL films, Kharlampieva *et al.* applied neutron reflectivity for contrasted LbL films (Figure 10.11) [41]. They found that the level of the stratification and the degree of layer intermixing can be controlled by varying the type and concentration of salt during SA-LbL assembly. They observed well-defined layered structure in SA-LbL films when deposited from salt-free solutions. These films feature 2-nm-thick layer pairs, which are three times thicker than those in conventional LbL films. SA-LbL films obtained from buffer solutions were more stratified than the highly intermixed layers seen in conventional LbL films assembled from buffer solutions.

Neutron reflectivity for the contrasted LbL films is shown in Figure 10.12. The distinct and periodic Bragg peaks observed indicate a high degree of stratification. The scattering-length-density (SLD) profile reveals the well-defined natural silicon oxide layer (Figure 10.12). The SLD profile also reveals a sequence of three equally spaced labeled *d*PSS layers with thickness of 1.6 nm. The marker layers are distinct and do not vary with distance from the substrate. In contrast, a dip-assisted LbL film drawn from salt-free solutions featured smaller layer pair thickness and fast deterioration of the segregated structures (Figure 10.12). Addition of phosphate

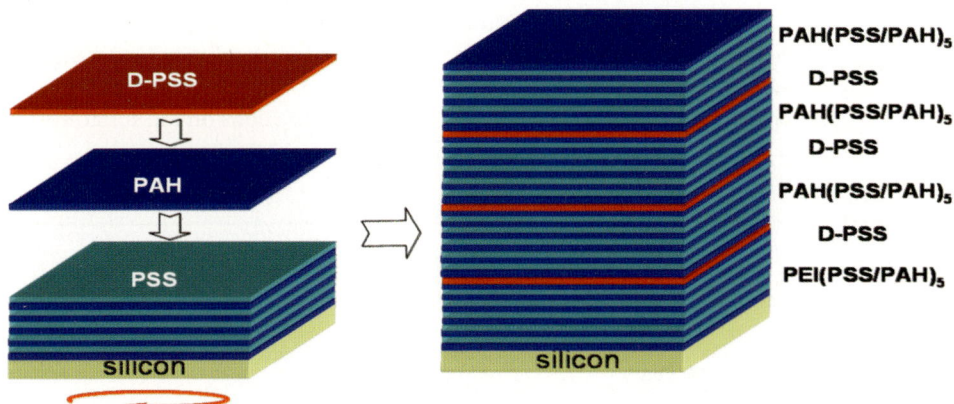

**Figure 10.11** Spin-assisted LbL assembly and resulting multilayer architecture with selectively labeled *d*PSS layers for neutron reflectivity. Reprinted from [41].

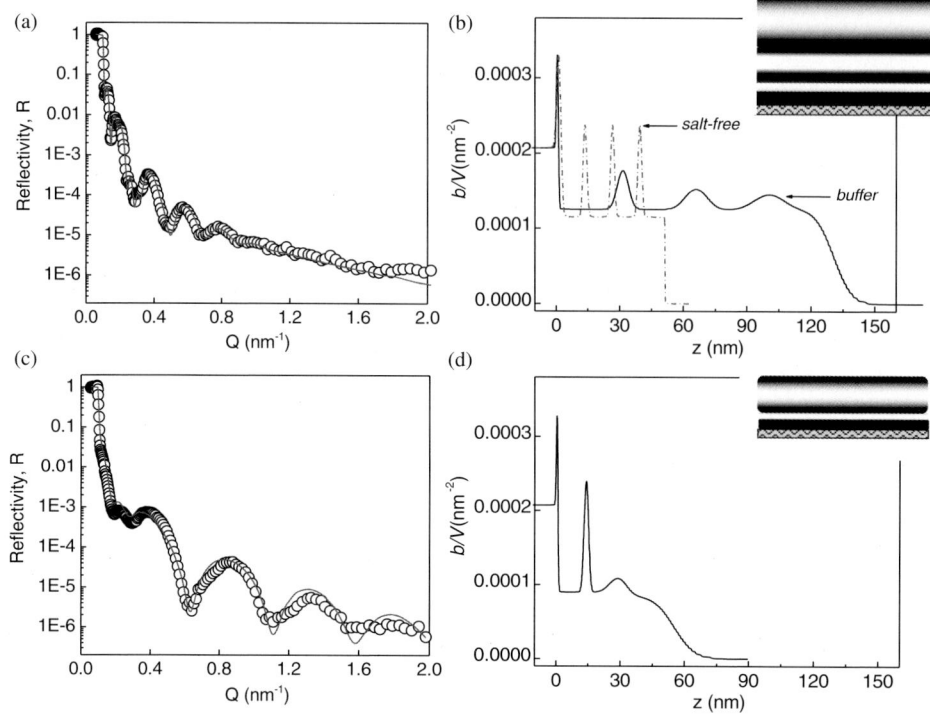

**Figure 10.12** Neutron reflectivity (a,c), the corresponding scattering-length-density profile and schematic (b,d) for spin-assisted (a,b) and dip-assisted (c,d) LbL films. Adapted from [41].

buffer results in an increase in film thickness for both SA and conventional LbL films, but induces a significant layer intermixing. The stratification in SA-LbL films can then be restored in the presence of NaCl in phosphate buffer solutions. Finally, the SA-LbL films made from buffer solutions display a better resolved layered structure than the conventional LbL films assembled from the buffer solution. Apparently, in conventional LbL dipping deposition, the absence of salt promotes thinner polyelectrolyte layers due to mostly uncoiled chain conformations on the surfaces with local repulsion caused by a lack of the charge screening.

## 10.3
## Porous and Patterned Freely Standing LbL Films

### 10.3.1
### Nanoporous LbL Films

Synthetic membranes with pore sizes close to molecular dimensions are considered to be critical for prospective applications in membrane science and technology

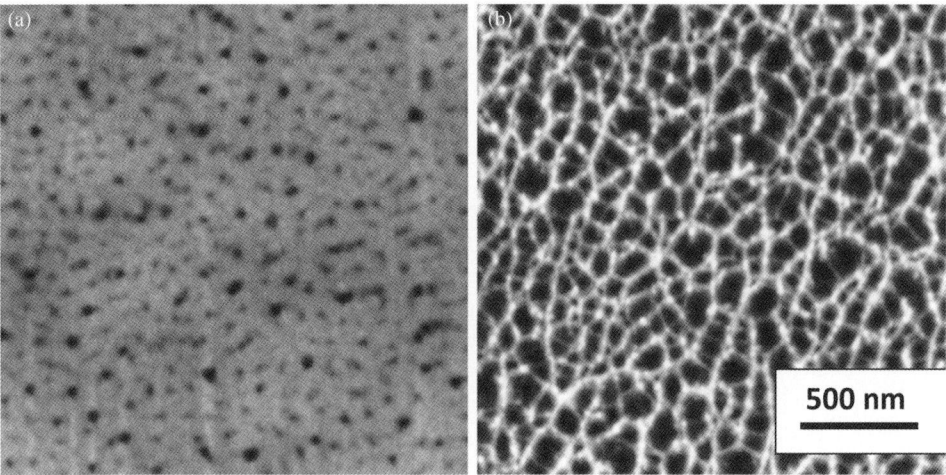

**Figure 10.13** AFM images of the "top" (a) and "bottom" (b) side of the freely suspended (PAH-PSS)$_{15}$ film. Reprinted from [48].

[42, 43]. In recent studies, Rubner and coworkers discovered that porous multilayered films can be formed by immersing poly(acrylic acid)/PAH (PAA-PAH) LBL films in acidic solution [44–47]. The resulting films may undergo a secondary reorganization in neutral water, leading to morphology with discrete, through pores. An alternative approach to the formation of porous LbL membranes was reported by Zimnitsky *et al.* [48]. In this approach, the dewetting of polyelectrolyte solutions on hydrophobic PVA surfaces, coupled with the fast drying during spin-casting, promotes partial phase separation leading to porous morphology of freely standing LbL films.

It was found that during the spin-assisted deposition of the LbL film on the PVA layer, the random porous and network-like morphology can be formed with a high concentration of nanopores, reaching 30–40% (Figure 10.13). This surface morphology is characterized by high microroughness reaching 4 nm within a 1 μm$^2$ surface area, and pores which propagate through the entire thickness of the film. The surface density of pores with 25–50 nm in diameter reaches $10^{10}$ cm$^{-2}$, which is about an order of magnitude higher than those for regular track membranes. The overall film morphology is controlled by the number of polyelectrolyte layer pairs with decreasing pore concentration observed for thicker LbL films. The elastic modulus of porous films with 30% porosity was estimated from buckling tests to be around 2 GPa, which is on a par with solid LbL films [26, 35].

### 10.3.2
### Patterned LbL Films

Freely standing patterned LbL films have unique properties that are not achievable for usual planar LbL films. There are several methods to precisely control the in-plane composition of the nanocomposite LbL films, including lithographic patterning, microcontact printing, and polymer-on-polymer stamping [24, 49, 50]. Here, we will discuss the authors' recent work on the fabrication, characterization, and properties of patterned LbL films by microcontact printing of a sacrificial template. Tsukruk and coworkers fabricated patterned gold nanoparticles, carbon nanotubes, or silver nanowires into the freely standing nanocomposite LbL films, as will be discussed briefly below.

The fabricating procedure of the patterned freely standing LbL films is described in the literature [51, 52]. In the case of patterned LbL films with gold nanoparticles, the PDMS stamp was applied to transfer the PAH ink onto the PSS layer. This PAH surface was used to selectively attach gold nanoparticles. After that, a desired number of polyelectrolyte layers were deposited. The sacrificial layer can be dissolved in a suitable solvent and the patterned freely standing nanocomposite LbL films can be transferred onto holey substrates for further characterization. The patterned structures in the ultrathin nanomembrane were stable and could be kept on the shelf for months under ambient conditions. For the patterned CNT-containing LbL films, a mask of polystyrene was printed for the CNT patterning. Accordingly, robust patterned LbL films can be obtained. Mechanical tests indicated that the elastic modulus of the patterned LbL films was slightly lower than that of non-patterned LbL films.

Surface-enhanced Raman scattering spectra of the polyelectrolyte matrix with the patterned gold nanoparticle arrays were recorded and the Raman intensity distribution confirmed the patterned composition of LbL films (Figure 10.14). The lateral

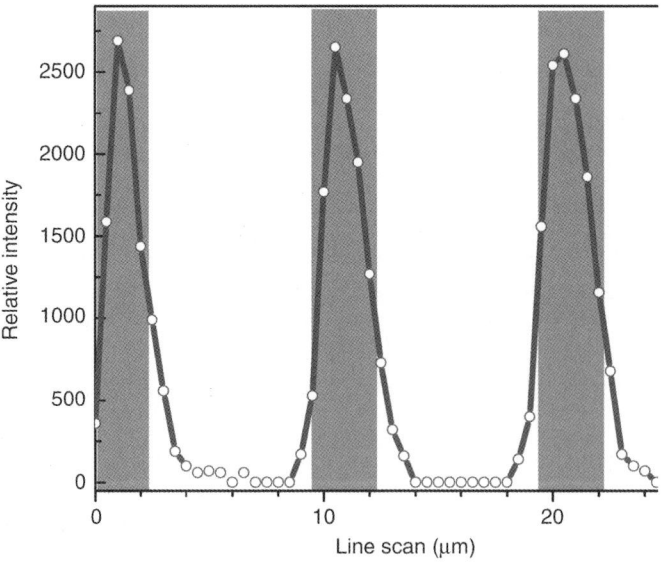

**Figure 10.14** The intensity variation of the 1590 cm$^{-1}$ peak across several stripes of the gold nanoparticle regions in the patterned LbL films. The locations of the stripes are marked by a gray background. Reprinted from [52].

resolution of the confocal system allows a clear differentiation of both the 2.5 μm wide gold-rich strips and the 7.5 μm wide gold-free regions. An intensity ratio of over 100 can be easily obtained. The patterned gold nanoparticle LbL films provide a unique example of a microscopic SERS grating with controlled spacing, high uniformity, and excellent contrast. Similar Raman mapping can also be observed from the samples of patterned CNT-containing LbL films [51].

### 10.3.3
### Sculptured LbL Films

Lin *et al.* fabricated sculptured LbL films with 3D microscopic structures by using conjugated polyelectrolytes on a sacrificial modulated substrate [53]. The conjugated polyelectrolyte MPS-PPV was used as a component in this study. It is critical to use robust conjugated polyelectrolytes since the mechanical stability is a key issue for the 3D LbL structures.

Both SEM and optical microscopy show uniform patterned structures over a size of several hundred micrometers (Figure 10.15). An effective thickness of 60 nm and a

**Figure 10.15** Sculptured LbL film on a TEM grid: (a) optical image, (b) fluorescence image with fluorescence spectrum (inset), (c) SEM image of the damaged piece of the 3D LbL film with characteristic fracturing (top view), (d) high-resolution side-view SEM image of freely standing sculptured LbL film. Reprinted from [53].

160 nm peak-to-peak value were found for the sculptured LbL films. The micromechanical properties of the sculptured LbL films were studied with both compressive and tensile methods. In bulging tests, the elastic modulus of the LbL films is around 0.4 GPa, which is an order smaller than that of usual PPV-containing LbL films. This value of the elastic modulus implies a very different mechanism of the elastic deformation for sculptured LbL films during the bulging tests, in which an expansion (or unfolding) of the 3D structures might occur. Similar values (0.3 GPa) were obtained from the buckling experiments. Another interesting phenomenon is that the collapsed films are under the (11) direction of the patterns, which is different to the preferential fracturing direction, (10) or (01), for the bulging tests.

The sculptured LbL films are capable of diffracting light due to the modulation of local shape with microscopic periodicity. White light can be reflected by the sculptured LbL films with the intense color changing from blue to red while adjusting the view angle. It is remarkable that such intense color of diffracted light is observed, considering that the physical thickness is only 60 nm. Similarly, a 2D diffraction pattern can be observed under an incident illumination from a laser beam. The diffraction pattern in a reciprocal space is defined by the symmetry and spacing of the spatial modulations of LbL films.

## 10.4
### Freely Standing LbL Films with Weak Interactions

### 10.4.1
#### Responsive LbL Films

Responsive LbL films have been exploited for stimuli-responsive polymer coatings [54], adaptive roughness, wettability, biocompatibility, and optical appearance [55–59]. In a recent study, Kozlovskaya *et al.* reported pH-responsive ultrathin plasmonic LbL membranes of (PMAA-Au NRs)$_{20}$ via post-inclusion of gold nanorods into highly swollen (PMAA)$_{20}$ layered hydrogel films (Figure 10.16) [60, 61]. These films showed a dramatic pH response with a significant shift in an easily detectable plasmon band position, in contrast to most of the pH-responsive large-scale structures known to date. In this study, the PMAA films were synthesized via chemical crosslinking of hydrogen-bonded (poly(*N*-vinylpyrrolidone) (PVPON)/PMAA)$_{20}$ films with ethylenediamine (EDA) (Figure 10.16). The number of nanorods included in the LbL film at pH 8 is less than that at pH 3. This difference is caused by free CTAB molecules from the nanorod solution at pH 8, which can be electrostatically attracted to the swollen, negatively charged hydrogel film along with the CTAB-coated nanorods. Indeed, the estimation of the spatial distribution of surface-tethered and embedded gold nanorods confirmed the inclusion of the nanorods within the hydrogel, with at least half of them buried deep inside hydrogel films, well beneath the topmost hydrogel surface layer.

It has been demonstrated that the longitudinal plasmon peak from (PMAA-Au NRs)$_{20}$ hydrogel LbL films blue shifts in response to a pH change from 8 to 5 (and

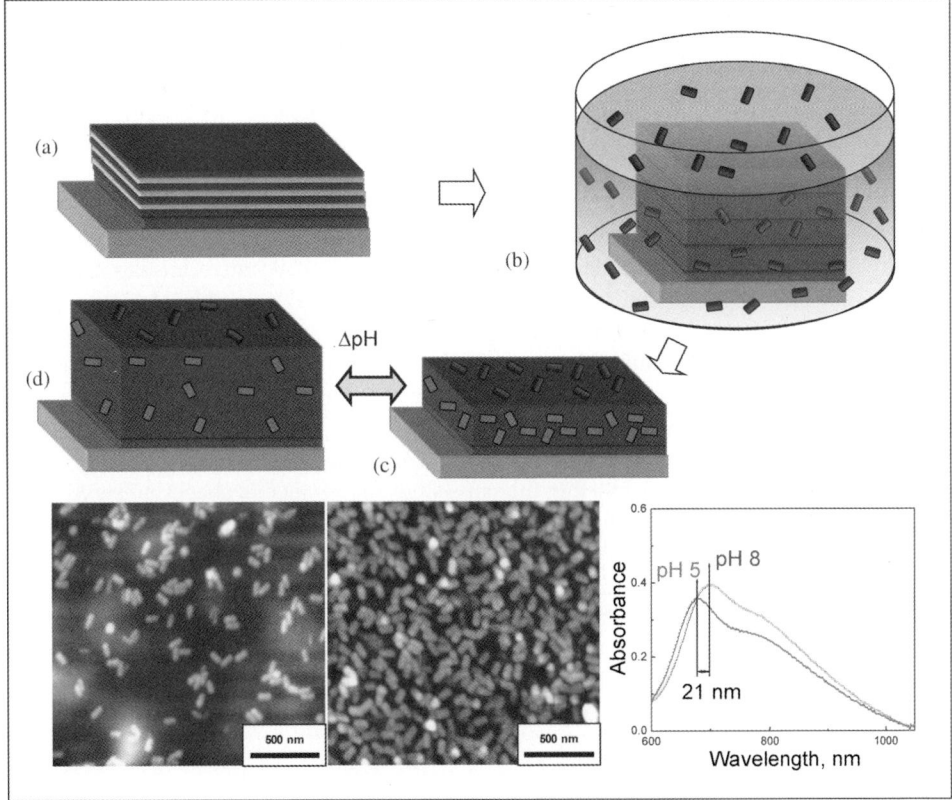

**Figure 10.16** Preparation of pH-sensitive LbL hydrogels: (a) LbL assembly of PMAA/PVPON films; (b) swelling cross-linked hydrogel in gold nanoparticles solution; (c, d) swelling of nanorod/LbL film can be controlled by pH, resulting in a shift in the longitudinal SPR peak. AFM images show gold nanorods embedded into LbL film and UV plot shows the shift of the SPR band. Adapted from [60].

back) due to de-swelling of the film, resulting in stronger interactions of the gold nanorods (Figure 10.16). Thus, the use of closely-packed gold nanorods as opposed to spherical particles in a combination with pH-responsive LbL films facilitates significant shifts in longitudinal plasmon bands.

### 10.4.2
### Anisotropic LbL Microcapsules

Hollow synthetic microcapsules are of great interest because of their potential for controlled storage and release, formation of ordered "opal" materials, and synthetic cell-like structures [62, 63]. Organic microcapsules find applications in enzymatic catalysis and as a platform for the construction of artificial cells and organelles [64, 65].

Successful demonstration of LbL microcapsules ignited recent intense efforts focused on the construction of a variety of spherical hollow structures [66–68]. Recently, significant efforts have also been made toward the fabrication of more complex microcapsules, such as multicompartmental capsules [69]. For anisotropic hollow microcapsules with sharp edges only a few examples have been reported to date [70–73].

Cubic microcapsules have been fabricated with hydrogen-bonded LbL shells on cubic $CdCO_3$ crystals, as suggested by Sukhishvili *et al.* [74]. In a very recent study, Schelepina *et al.* have demonstrated anisotropic hollow microcapsules from LbL shells with cubic and tetrahedral shapes that are stable at physiological pH and capable of retaining their anisotropic shape under different pHs [75]. Tetrahedral shapes, including sub-micron capsules introduced in this study in comparison with cubic shapes, present a challenge in the preparation of shaped microcapsules due to additional stresses at the edges and corners.

Moreover, dramatic scaling down of microcapsule size to submicron scale (below 200 nm) was demonstrated in the same study for tetrahedral shapes preserved after removal of SnS cores (Figure 10.17). SEM images directly confirmed the presence of hollow tetrahedral LbL structures preserved in colloidal solution, with characteristic dimensions and shapes replicating those for initial crystals. Corresponding computer modeling suggested that introducing sharp edges and vertices acting as a reinforcing frame can potentially prevent random buckling and collapse, thus enhancing microcapsule stability under osmotic pressure variation.

### 10.4.3
### LbL-Shells for Cell Encapsulation

The advantages of this approach for cell surface modification include conformal coating of geometrically challenged templates, a precise control of the membrane nanoscale thickness, and tunability of membrane functionalities and properties [24, 76–78]. The ability to tailor permeability properties by the LbL approach is of particular importance for encapsulation of living cells, as cell viability depends critically on the diffusion of nutrients through the polymer shell [79, 80]. Despite successful examples, cytotoxicity of the polycations posed severe limitations on this approach for cell surface engineering [81–83]. As suggested, the overall toxicity of the PAH/PSS LbL shells originates from the positive charge of cationic polyelectrolytes, which causes cell membrane pore formation followed by cell damage and death [84, 85]. For instance, poly(L-lysine)/PSS, poly(ethyleneimine)/PSS, and PAH/PSS resulted in significant decreases in viability of mammalian cells [86].

Kozlovskaya *et al.* reported a different design of cytocompatible synthetic shells, from highly permeable, hydrogen-bonded LbL multilayers, for cell surface engineering with preservation of long-term cell function [87, 88]. In contrast to traditional polyelectrolyte systems, the shells suggested in this study were based on hydrogen bonding, allowing gentle cell encapsulation using non-ionic components such as poly(*N*-vinylpyrrolidone) (PVPON) and tannic acid (TA). The yeast cells were

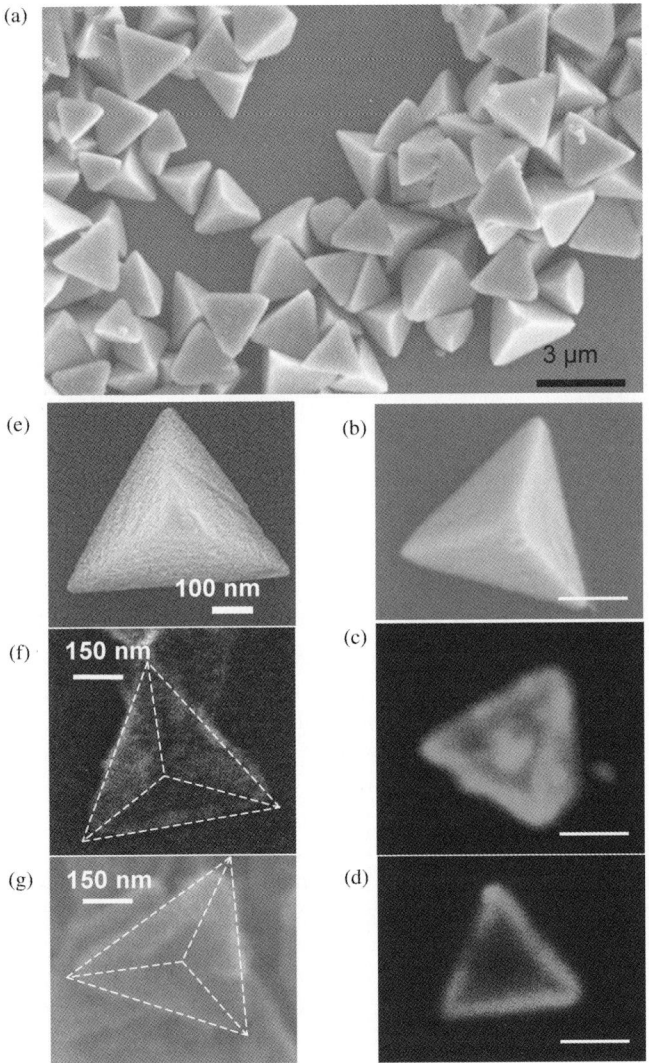

**Figure 10.17** SEM images of tetrahedral SnS crystals (a) with micrometer (b) and submicrometer (e) dimensions; confocal microscopy images of hollow (TA/PVPON)$_3$ tetrahedral microcapsules produced by selective core dissolution: side view (c) of 3D reconstructed capsules and cross-section (d); SEM images of individual collapsed hollow capsules produced from submicron tetrahedral cores (f, g). Dashed lines are added to emphasis edges and vertices. Scale bars for (b, c, d) are 1 μm. Reprinted from [75].

sequentially coated with (TA/PVPON)$_n$ LbL shells through hydrogen bonding under deposition conditions which preserved cell functioning (Figure 10.18) To ensure the adhesion of the following multilayers, the cell surfaces can be primed with a poly (ethyleneimine) (PEI) monolayer [89].

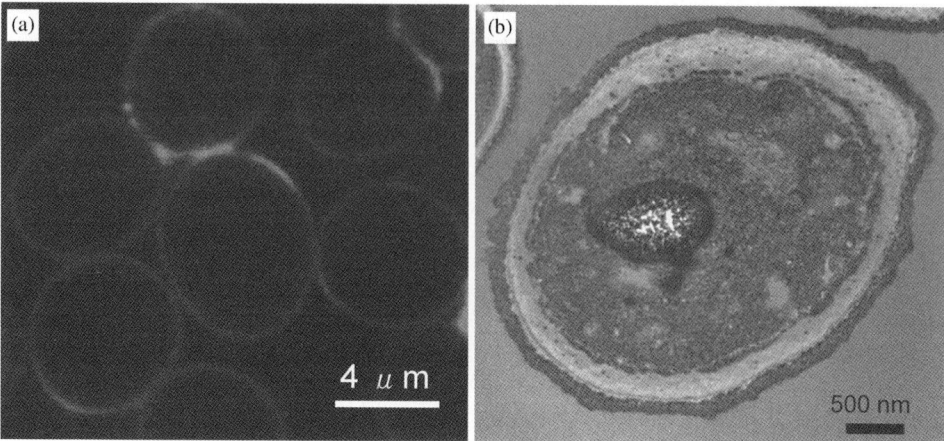

**Figure 10.18** Confocal images of yeast cells encapsulated with LbL shells (a); TEM images of freeze-dried LbL-coated yeast cells (b). Reprinted from [87].

Figure 10.18 demonstrates homogeneous fluorescence from AlexaFluor532-tagged PVPON, confirming formation of the uniform shells. TEM demonstrates the integrity of the cell membrane upon adsorption of the PEI(TA/PVPON)$_6$ hydrogen-bonded shell (Figure 10.18). AFM images show grainy surface morphology of the hollow shells with domain dimensions of about 50 nm. The thickness of the shells increased linearly to 22 nm, indicating a regular growth mode with the average TA/PVPON layer pair thickness of 4.0 nm. Concurrently, the surface microroughness of the LbL films increased significantly as a function of layer pair number due to weak aggregation of hydrogen-bonded components.

Preserved cell function and viability is indicated by their ability to bud after the cell surfaces were modified with the LbL shells. Figure 10.19 demonstrates the characteristic S-shaped cell growth of the control cells and LbL-coated cells. During the initial lag phase the rate of growth, or cell division, is slow in all cases. Red fluorescence in confocal images of the LbL-coated cells taken in the lag phase witnesses a homogenous polymer coating around the cells (Figure 10.19). Moreover, the green fluorescence from the yEGFP-reporter produced by the yeast cells confirms that the functional capacity of the cells was not adversely influenced by the shell presence. This stage is followed by the exponential growth mode, when cell division accelerates and unicellular organisms duplicate, that is, one cell produces two (see divided cells as indicated by arrows in Figure 10.19). During this phase very rapid multiplication of yeast cells is observed. The exponential phase then proceeds to a stationary phase where there is no discernible change in cell concentration.

While original yeast cells go into the exponential phase after 8 h, duplicating every 4 h, there is a delay of the exponential phase for the (TA/PVPON)-coated cells which is dependent on the thickness (Figure 10.19). Confocal imaging revealed that during exponential growth coated cells are able to break the polymer shell (Figure 10.19). This is similar to polycation/polyanion coatings and is due to the

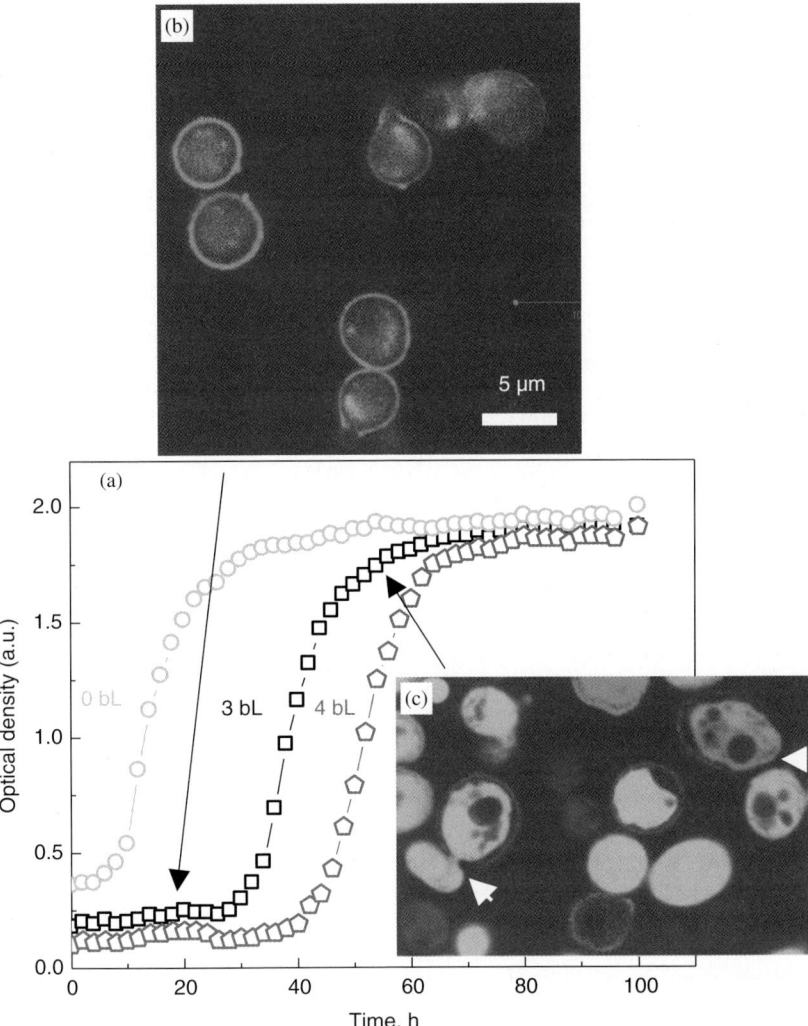

**Figure 10.19** Growth of PEI(TA/PVPON)-coated YPH501cells after yEGFP expression was induced (a). Confocal microscopy images of LbL-coated cells after 10 h (b) and 46 h (c) of the yEGFP expression. Reprinted from [87].

comparable rigidity of the (TA/PVPON) coatings which exhibit high Young's modulus in the dry state. Overall, this study demonstrated that hydrogen-bonded LbL shells with higher diffusion facilitate outstanding cell survivability, reaching 90% (for LbL shells without PEI prelayer [90]) in contrast to the only 20% viability level achieved with traditional LbL coatings. The authors suggested that the drastic increase in cell viability and preservation of cell functioning after coating with synthetic shells stems from the minimal exposure of the cells to toxic polycations, and high shell permeability.

## Acknowledgements

The activity in this field is continuously supported with funding provided by the National Science Foundation, Air Force Office of Scientific Research, Air Force Research Laboratory, Department of Energy, National Aeronautics and Space Administration, and Semiconductor Research Corporation.

## References

1 Jiang, C.Y., Markutsya, S., and Tsukruk, V.V. (2004) Compliant, robust, and truly nanoscale free-standing multilayer films fabricated using spin-assisted layer-by-layer assembly. *Adv. Mater.*, **16** (2), 157–161.

2 Jiang, C.Y., Markutsya, S., Pikus, Y., and Tsukruk, V.V. (2004) Freely suspended nanocomposite membranes as highly sensitive sensors. *Nat. Mater.*, **3** (10), 721–728.

3 Jiang, C.Y., Rybak, B.M., Markutsya, S., Kladitis, P.E., and Tsukruk, V.V. (2004) Self-recovery of stressed nanomembranes. *Appl. Phys. Lett.*, **86** (12), 121912.

4 Gunawidjaja, R., Jiang, C., Ko, H., and Tsukruk, V.V. (2006) Freestanding 2D arrays of silver nanorods. *Adv. Mater.*, **18** (21), 2895–2899.

5 Gunawidjaja, R., Jiang, C., Peleshanko, S., Ornatska, M., Singamaneni, S., and Tsukruk, V.V. (2006) Flexible and robust 2D arrays of silver nanowires encapsulated within freestanding layer-by-layer films. *Adv. Funct. Mater.*, **16** (15), 2024–2034.

6 Zimnitsky, D., Jiang, C., Xu, J., Lin, Z., Zhang, L., and Tsukruk, V.V. (2007) Photoluminescence of a freely suspended monolayer of quantum dots encapsulated into layer-by-layer films. *Langmuir*, **23** (20), 10176–10183.

7 Zimnitsky, D., Jiang, C., Xu, J., Lin, Z., and Tsukruk, V.V. (2007) Substrate- and time-dependent photoluminescence of quantum dots inside the ultrathin polymer LbL film. *Langmuir*, **23** (8), 4509–4515.

8 Bao, Y., Luu, Q.A.N., Lin, C., Schloss, J.M., May, P.S., and Jiang, C. (2010) Layer-by-layer assembly of freestanding thin films with homogeneously distributed upconversion nanocrystals. *J. Mater. Chem.*, **20** (38), 8356–8361.

9 Mamedov, A.A., Kotov, N.A., Prato, M., Guldi, D.M., Wicksted, J.P., and Hirsch, A. (2002) Molecular design of strong single-wall carbon nanotube/polyelectrolyte multilayer composites. *Nat. Mater.*, **1** (3), 190–194.

10 Shen, J., Hu, Y., Li, C., Qin, C., Shi, M., and Ye, M. (2009) Layer-by-layer self-assembly of graphene nanoplatelets. *Langmuir*, **25** (11), 6122–6128.

11 Jiang, C., Ko, H., and Tsukruk, V.V. (2005) Strain-sensitive Raman modes of carbon nanotubes in deflecting freely suspended nanomembranes. *Adv. Mater.*, **17** (17), 2127–2131.

12 Kulkarni, D.D., Choi, I., Singamaneni, S.S., and Tsukruk, V.V. (2010) Graphene oxide–polyelectrolyte nanomembranes. *ACS Nano*, **4** (8), 4667–4676.

13 Jiang, C., Wang, X., Gunawidjaja, R., Lin, Y.H., Gupta, M.K., Kaplan, D.L., Naik, R.R., and Tsukruk, V.V. (2007) Mechanical properties of robust ultrathin silk fibroin films. *Adv. Funct. Mater.*, **17** (13), 2229–2237.

14 Valluzzi, R. and Jin, H.-J. (2004) X-ray Evidence for a "super"-secondary structure in silk fibers. *Biomacromolecules*, **5** (3), 696–703.

15 Um, I.C., Ki, C.S., Kweon, H., Lee, K.G., Ihm, D.W., and Park, Y.H. (2004) Wet spinning of silk polymer: II. Effect of drawing on the structural characteristics and properties of filament. *Int. J. Biological Macromolecules*, **34** (1–2), 107–119.

16 Mita, K., Ichimura, S., and James, T.C. (1994) Highly repetitive structure and its organization of the silk fibroin gene. *J. Mol. Evolution*, **38** (6), 583–592.

17 Valluzzi, R., Gido, S.P., Zhang, W., Muller, W.S., and Kaplan, D.L. (1996) Trigonal crystal structure of bombyx mori silk incorporating a threefold helical chain conformation found at the air–water interface. *Macromolecules*, **29** (27), 8606–8614.

18 Magoshi, J., Magoshi, Y., Becker, M.A., and Nakamura, S. (1996) in *Materials Encyclopedia* (ed. J.C. Salamone), CRC Press, NewYork, p. 667.

19 Nam, J. and Park, Y.H. (2001) Morphology of regenerated silk fibroin: Effects of freezing temperature, alcohol addition, and molecular weight. *J. Appl. Polym. Sci.*, **81** (12), 3008–3021.

20 Ebenstein, D.M. and Wahl, K. (2006) . J., Anisotropic nanomechanical properties of Nephila clavipes dragline silk. *J. Mater. Res.*, **21** (8), 2035–2044.

21 Ebenstein, D.M., Park., J., Kaplan, D.L., and Wahl, K.J. (eds) (2005) *MRS Symp. Series*, **841**, 57.

22 Kharlampieva, E., Kozlovskaya, V., Gunawidjaja, R., Shevchenko, V.V., Vaia, R., Naik, R.R., Kaplan, D.L., and Tsukruk, V.V. (2010) Flexible silk–inorganic nanocomposites: from transparent to highly reflective. *Adv. Funct. Mater.*, **20** (5), 840–846.

23 Kharlampieva, E., Kozlovskaya, V., Wallet, B., Shevchenko, V.V., Naik, R.R., Vaia, R., Kaplan, D.L., and Tsukruk, V.V. (2010) Co-cross-linking silk matrices with silica nanostructures for robust ultrathin nanocomposites. *ACS Nano*, **4** (12), 7053–7063.

24 Hammond, P.T. (2004) Form and function in multilayer assembly: New applications at the nanoscale. *Adv. Mater.*, **16** (15), 1271–1293.

25 Tang, Z., Kotov, N.A., Magonov, S., and Ozturk, B. (2003) Nanostructured artificial nacre. *Nat. Mater.*, **2** (6), 413–418.

26 Markutsya, S., Jiang, C., Pikus, Y., and Tsukruk, V.V. (2005) Freely suspended layer-by-layer nanomembranes: Testing micromechanical properties. *Adv. Funct. Mater.*, **15** (5), 771–780.

27 Jiang, C., Singamaneni, S., Merrick, E., and Tsukruk, V.V. (2006) Complex buckling instability patterns of nanomembranes with encapsulated gold nanoparticle arrays. *Nano Lett.*, **6** (10), 2254–2259.

28 Jiang, C., Markutsya, S., and Tsukruk, V.V. (2004) Collective and individual plasmon resonances in nanoparticle films obtained by spin-assisted layer-by-layer assembly. *Langmuir*, **20** (3), 882–890.

29 Lin, Y.-H., Jiang, C., Xu, J., Lin, Z., and Tsukruk, V.V. (2007) Robust, fluorescent, and nanoscale freestanding conjugated films. *Soft Matter*, **3** (4), 432–436.

30 Jiang, C., Lio, W.Y., and Tsukruk, V.V. (2005) Surface enhanced Raman scattering monitoring of chain alignment in freely suspended nanomembranes. *Phys. Rev. Lett.*, **95** (11), 115503.

31 Tsukruk, V.V. (1997) Assembly of supramolecular polymers in ultrathin films. *Progr. Polym. Sci.*, **22** (2), 247–311.

32 Lvov, Y., Ariga, K., Onda, M., Ichinose, I., and Kunitake, T. (1997) Alternate assembly of ordered multilayers of SiO2 and other nanoparticles and polyions. *Langmuir*, **13** (23), 6195–6203.

33 Lvov, Y., Ariga, K., Onda, M., Ichinose, I., and Kunitake, T. (1999) A careful examination of the adsorption step in the alternate layer-by-layer assembly of linear polyanion and polycation. *Colloid Surf. Physicochem. Eng. Aspect*, **146** (1–3), 337–346.

34 Lobo, R.F.M., Pereira-da-Silva, M.A., Raposo, M., Faria, R.M., and Oliveira, O.N. (2003) The morphology of layer-by-layer films of polymer/polyelectrolyte studied by atomic force microscopy. *Nanotechnology*, **14** (1), 101–108.

35 Jiang, C. and Tsukruk, V.V. (2005) Organized arrays of nanostructures in freely suspended nanomembranes. *Soft Matter*, **1** (5), 334–337.

36 Tsukruk, V.V., Bliznyuk, V.N., Visser, D., Campbell, A.L., Bunning, T.J., and Adams, W.W. (1997) Electrostatic deposition of polyionic monolayers on charged surfaces. *Macromolecules*, **30** (21), 6615–6625.

37 Schmitt, J., Gruenewald, T., Decher, G., Pershan, P.S., Kjaer, K., and Loesche, M. (1993) Internal structure of layer-by-layer adsorbed polyelectrolyte films: a neutron and X-ray reflectivity study. *Macromolecules*, **26** (25), 7058–7063.

38 Kellogg, G.J., Mayes, A.M., Stockton, W.B., Ferreira, M., Rubner, M.F., and Satija, S.K. (1996) Neutron reflectivity investigations of self-assembled conjugated polyion multilayers. *Langmuir*, **12** (21), 5109–5113.

39 Lösche, M., Schmitt, J., Decher, G., Bouwman, W.G., and Kjaer, K. (1998) Detailed structure of molecularly thin polyelectrolyte multilayer films on solid substrates as revealed by neutron reflectometry. *Macromolecules*, **31** (25), 8893–8906.

40 Korneev, D., Lvov, Y., Decher, G., Schmitt, J., and Yaradaikin, S. (1995) Neutron reflectivity analysis of self-assembled film superlattices with alternate layers of deuterated and hydrogenated polysterenesulfonate and polyallylamine. *Physica B: Condensed Matter*, **213–214**, 954–956.

41 Kharlampieva, E., Kozlovskaya, V., Chan, J., Ankner, J.F., and Tsukruk, V.V. (2009) Spin-assisted layer-by-layer assembly: Variation of stratification as studied with neutron reflectivity. *Langmuir*, **25** (24), 14017–14024.

42 Nishizawa, M., Menon, V.P., and Martin, C.R. (1995) Metal nanotubule membranes with electrochemically switchable ion-transport selectivity. *Science*, **268** (5211), 700–702.

43 Hofnung, M. (1995) An intelligent channel (and more). *Science*, **267** (5197), 473–474.

44 Mendelsohn, J.D., Barrett, C.J., Chan, V.V., Pal, A.J., Mayes, A.M., and Rubner, M.F. (2000) Fabrication of microporous thin films from polyelectrolyte multilayers. *Langmuir*, **16** (11), 5017–5023.

45 Hiller, J.A., Mendelsohn, J.D., and Rubner, M.F. (2002) Reversibly erasable nanoporous anti-reflection coatings from polyelectrolyte multilayers. *Nat. Mater.*, **1** (1), 59–63.

46 Zhai, L., Nolte, A.J., Cohen, R.E., and Rubner, M.F. (2004) pH-Gated porosity transitions of polyelectrolyte multilayers in confined geometries and their application as tunable Bragg reflectors. *Macromolecules*, **37** (16), 6113–6123.

47 Berg, M.C., Zhai, L., Cohen, R.E., and Rubner, M.F. (2005) Controlled drug release from porous polyelectrolyte multilayers. *Biomacromolecules*, **7** (1), 357–364.

48 Zimnitsky, D., Shevchenko, V.V., and Tsukruk, V.V. (2008) Perforated, freely suspended layer-by-layer nanoscale membranes. *Langmuir*, **24** (12), 5996–6006.

49 Hua, F., Cui, T., and Lvov, Y.M. (2004) Ultrathin cantilevers based on polymer–ceramic nanocomposite assembled through layer-by-layer adsorption. *Nano Lett.*, **4** (5), 823–825.

50 Jiang, X., Zheng, H., Gourdin, S., and Hammond, P.T. (2002) Polymer-on-polymer stamping: Universal approaches to chemically patterned surfaces. *Langmuir*, **18** (7), 2607–2615.

51 Ko, H., Jiang, C., Shulha, H., and Tsukruk, V.V. (2005) Carbon nanotube arrays encapsulated into freely suspended flexible films. *Chem. Mater.*, **17** (10), 2490–2493.

52 Jiang, C., Markutsya, S., Shulha, H., and Tsukruk, V.V. (2005) Freely suspended gold nanoparticle arrays. *Adv. Mater.*, **17** (13), 1669–1673.

53 Lin, Y.H., Jiang, C., Xu, J., Lin, Z., and Tsukruk, V.V. (2007) Sculptured layer-by-layer films. *Adv. Mater.*, **19** (22), 3827–3832.

54 Luzinov, I., Minko, S., and Tsukruk, V.V. (2004) Adaptive and responsive surfaces through controlled reorganization of interfacial polymer layers. *Prog. Polym. Sci.*, **29** (7), 635–698.

55 Lupitskyy, R., Roiter, Y., Tsitsilianis, C., and Minko, S. (2005) From smart polymer molecules to responsive nanostructured surfaces. *Langmuir*, **21** (19), 8591–8593.

56 Sidorenko, A., Krupenkin, T., Taylor, A., Fratzl, P., and Aizenberg, J. (2007) Reversible switching of hydrogel-actuated nanostructures into complex

micropatterns. *Science*, **315** (5811), 487–490.

57 Sidorenko, A., Krupenkin, T., and Aizenberg, J. (2008) Controlled switching of the wetting behavior of biomimetic surfaces with hydrogel-supported nanostructures. *J. Mater. Chem.*, **18** (32), 3841–3846.

58 Krsko, P. and Libera, M. (2005) Biointeractive hydrogels. *Mater. Today*, **8** (12), 36–44.

59 Houbenov, N., Minko, S., and Stamm, M. (2003) Mixed polyelectrolyte brush from oppositely charged polymers for switching of surface charge and composition in aqueous environment. *Macromolecules*, **36** (16), 5897–5901.

60 Kozlovskaya, V., Kharlampieva, E., Khanal, B.P., Manna, P., Zubarev, E.R., and Tsukruk, V.V. (2008) Ultrathin layer-by-layer hydrogels with incorporated gold nanorods as pH-sensitive optical materials. *Chem. Mater.*, **20** (24), 7474–7485.

61 Kozlovskaya, V., Kharlampieva, E., Chang, S., Muhlbauer, R., and Tsukruk, V.V. (2009) pH-responsive layered hydrogel microcapsules as gold nanoreactors. *Chem. Mater.*, **21** (10), 2158–2167.

62 Caruso, F., Caruso, R.A., and Möhwald, H. (1998) Nanoengineering of inorganic and hybrid hollow spheres by colloidal templating. *Science*, **282** (5391), 1111–1114.

63 Donath, E., Sukhorukov, G.B., Caruso, F., Davis, S.A., and Möhwald, H. (1998) Novel hollow polymer shells by colloid-templated assembly of polyelectrolytes. *Angew. Chem. Int. Ed.*, **37** (16), 2201–2205.

64 Sukhorukov, G.B., Rogach, A.L., Garstka, M., Springer, S., Parak, W.J., Muñoz-Javier, A., Kreft, O., Skirtach, A.G., Susha, A.S., Ramaye, Y., Palankar, R., and Winterhalter, M. (2007) Multifunctionalized polymer microcapsules: Novel tools for biological and pharmacological applications. *Small*, **3** (6), 944–955.

65 Stadler, B., Price, A.D., Chandrawati, R., Hosta-Rigau, L., Zelikin, A.N., and Caruso, F. (2009) Polymer hydrogel capsules: en route toward synthetic cellular systems. *Nanoscale*, **1** (1), 68–73.

66 Shchepelina, O., Kozlovskaya, V., Singamaneni, S., Kharlampieva, E., and Tsukruk, V.V. (2010) Replication of anisotropic dispersed particulates and complex continuous templates. *J. Mater. Chem.*, **20** (32), 6587–6603.

67 Johnston, A.P.R., Cortez, C., Angelatos, A.S., and Caruso, F. (2006) Layer-by-layer engineered capsules and their applications. *Curr. Opin. Colloid Interface Sci.*, **11** (4), 203–209.

68 Jiang, C.Y. and Tsukruk, V.V. (2006) Freestanding nanostructures via layer-by-layer assembly. *Adv. Mater.*, **18** (7), 829–840.

69 Delcea, M., Yashchenok, A., Videnova, K., Kreft, O., Möhwald, H., and Skirtach, A.G. (2010) Multicompartmental micro- and nanocapsules: Hierarchy and applications in biosciences. *Macromol. Biosci.*, **10** (5), 465–474.

70 Caruso, F. (2000) Hollow capsule processing through colloidal templating and self-assembly. *Chem.-Eur. J.*, **6** (3), 413–419.

71 Ai, H., Jones, S., and Lvov, Y. (2003) Biomedical applications of electrostatic layer-by-layer nano-assembly of polymers, enzymes, and nanoparticles. *Cell Biochem. Biophys.*, **39** (1), 23–43.

72 Decher, G. and Schlenoff, J.B. (eds) (2003) *Multilayer Thin Films*, Wiley-VCH Verlag GmbH, Weinheim.

73 Holt F B, Lam, R., Meldrum, F.C., Stoyanov, S.D., and Paunov, V.N. (2007) Anisotropic nano-papier mache microcapsules. *Soft Matter*, **3** (2), 188–190.

74 Kozlovskaya, V., Yakovlev, S., Libera, M., and Sukhishvili, S.A. (2005) Hydrogen-bonded multilayers of thermoresponsive polymers. *Macromolecules*, **38**, 4828–4836.

75 Shchepelina, O., Kozlovskaya, V., Kharlampieva, E., Mao, W., Alexeev, A., and Tsukruk, V.V. (2010) Anisotropic micro- and nano-capsules. *Macromol. Rapid Commun.*, **31** (23), 2041–2046.

76 Quinn, J.F., Johnston, A.P.R., Such, G.K., Zelikin, A.N., and Caruso, F. (2007) Next generation, sequentially assembled

ultrathin films: beyond electrostatics. *Chem. Soc. Rev.*, **36** (5), 707–718.

77 del Mercato, L.L., Rivera-Gil, P., Abbasi, A.Z., Ochs, M., Ganas, C., Zins, I., Sonnichsen, C., and Parak, W.J. (2010) LbL multilayer capsules: recent progress and future outlook for their use in life sciences. *Nanoscale*, **2** (4), 458–467.

78 Veerabadran, N.G., Goli, P.L., Stewart-Clark, S.S., Lvov, Y.M., and Mills, D.K. (2007) Nanoencapsulation of stem cells within polyelectrolyte multilayer shells. *Macromol. Biosci.*, **7** (7), 877–882.

79 Diaspro, A., Silvano, D., Krol, S., Cavalleri, O., and Gliozzi, A. (2002) Single living cell encapsulation in nano-organized polyelectrolyte Shells. *Langmuir*, **18** (13), 5047–5050.

80 Krol, S., Nolte, M., Diaspro, A., Mazza, D., Magrassi, R., Gliozzi, A., and Fery, A. (2004) Encapsulated living cells on microstructured surfaces. *Langmuir*, **21** (2), 705–709.

81 De Koker, S., De Geest, B.G., Cuvelier, C., Ferdinande, L., Deckers, W., Hennink, W.E., De Smedt, S.C., and Mertens, N. (2007) In vivo Cellular uptake, degradation, and biocompatibility of polyelectrolyte microcapsules. *Adv. Funct. Mater.*, **17** (18), 3754–3763.

82 Städler, B., Chandrawati, R., Price, A.D., Chong, S.-F., Breheney, K., Postma, A., Connal, L.A., Zelikin, A.N., and Caruso, F. (2009) A microreactor with thousands of subcompartments: Enzyme-loaded liposomes within polymer capsules. *Angew. Chem. Int. Ed.*, **48** (24), 4359–4362.

83 Veerabdran, N.G., Goli, P.L., Stewart-Clark, S.S., Lvov, Y.M., and Mills, D.K. (2007) Nanoencapsulation of stem cells within polyelectrolyte multilayer shells. *Macromol. Biosci.*, **7** (7), 877–882.

84 Bieber, T., Meissner, W., Kostin, S., Niemann, A., and Elsasser, H.-P. (2002) Intracellular route and transcriptional competence of polyethylenimine-DNA complexes. *J. Control. Release*, **82** (2–3), 441–454.

85 Godbey, W.T., Wu, K.K., and Mikos, A.G. (1999) Size matters: Molecular weight affects the efficiency of poly(ethylenimine) as a gene delivery vehicle. *J. Biomed. Mater. Res.*, **45** (3), 268–275.

86 Germain, M., Balaguer, P., Nicolas, J.-C., Lopez, F., Esteve, J.-P., Sukhorukov, G.B., Winterhalter, M., Richard-Foy, H., and Fournier, D. (2006) Protection of mammalian cell used in biosensors by coating with a polyelectrolyte shell. *Biosens. Bioelectron.*, **21** (8), 1566–1573.

87 Kozlovskaya, V., Harbaugh, S., Drachuk, I., Shchepelina, O., Kelley-Loughnane, N., Stone, M., and Tsukruk, V.V. (2011) Hydrogen-bonded LbL shells for living cell surface engineering. *Soft Matter*, **7** (6), 2364–2372.

88 Kozlovskaya, V., Kharlampieva, E., Drachuk, I., Cheng, D., and Tsukruk, V.V. (2010) Responsive microcapsule reactors based on hydrogen-bonded tannic acid layer-by-layer assemblies. *Soft Matter*, **6** (15), 3596–3608.

89 Brunot, C., Ponsonnet, L., Lagneau, C., Farge, P., Picart, C., and Grosgogeat, B. (2007) Cytotoxicity of polyethyleneimine (PEI), precursor base layer of polyelectrolyte multilayer films. *Biomaterials*, **28** (4), 632–640.

90 Carter, J.L., Drachuk, I., Harbaugh, S., Kelley-Loughnane, N., Stone, M., and Tsukruk, V.V. (2011) Truly non-ionic polymer shells for encapsulation of living cells. *Macromol. Biosci.*, **11** (9), 1244–1253.

# 11
# Neutron Reflectometry at Polyelectrolyte Multilayers

*Regine von Klitzing, Ralf Köhler, and Chloe Chenigny*

## 11.1
## Introduction

Functionalized materials have a high impact with respect to both applications (e.g., in catalysis, electronic or optical devices, or sensors) and basic research in, for example, material and life sciences. To be accessible, functional units are usually located at interfaces.

Due to their self-assembling and self-organizing properties, thin polymer films are advantageous for the design of interface structures hosting functional groups. Different compounds can be combined with ease by self-organization in order to get materials with new structural and dynamic features.

Properties, like sensitivity to external parameters or stability, are highly correlated with the internal structure of polyelectrolyte multilayers (PEMs). Therefore, understanding and control of the internal structure are essential for tuning their physico-chemical properties. Macroscopic properties of PEMs, like mechanics or permeability, are often easier to study at capsules, while the internal structure and dynamics of the multilayers are better accessible at planar interfaces, for example, by neutron reflectometry (NR).

This chapter focuses on topics on PEMs which are studied by NR. A unique advantage of NR is that the internal structure can be studied in a non-invasive way. The relationship between internal structure, preparation conditions and post-treatment will be shown. A certain focus is on the swelling behavior in water, which is also strongly related to the internal structure.

## 11.2
## Neutron Reflectometry

Reflection of light at a surface is a well-known phenomenon since the seventeenth century, caused by the change in refractive index during the passage through an interface. Later, it was found that the same optics laws apply for X-ray reflectivity, only with different refractive indices, depending on the number of electrons per unit

*Multilayer Thin Films: Sequential Assembly of Nanocomposite Materials*, Second Edition.
Edited by Gero Decher and Joseph B. Schlenoff.
© 2012 Wiley-VCH Verlag GmbH & Co. KGaA. Published 2012 by Wiley-VCH Verlag GmbH & Co. KGaA.

volume. In 1944, Fermi and Zimm demonstrated that neutron reflectivity works in the same way, but again with new specific refractive indices, depending on the nuclei number and on the scattering power of each atom. A scattering length is defined for every atom and quantifies its ability to scatter neutrons. This scattering length is sensitive to different isotopes. Each medium is then characterized by a scattering length density $N_b$:

$$N_b = \sum_i b_i n_i$$

with $n_i$ as the number of a certain nuclei type per volume unit and $b_i$ as its scattering length. As with visible light, total reflection occurs when neutrons pass from a medium of higher refractive index to one of lower refractive index, this index $n$ is linked to $N_b$ by:

$$n = 1 - \delta - ik \quad \text{with} \quad \delta = \frac{\lambda^2}{2\pi} N_b$$

including the neutron wavelength $\lambda$.

The imaginary part $ik$ represents absorption and will be neglected in this chapter. $\delta$ is of the order of $10^{-6}$; the scattering length $b_i$ can be either positive or negative, as can $N_b$ and $\delta$. The refractive index for neutrons can then be greater than or less than 1. If less than 1, total external reflection is observed. For instance, H (hydrogen) has a negative $b$, and D (deuterium) a positive one, leading to total reflection at the $Si/D_2O$ interface but not at the $Si/H_2O$ interface. As for visible light, interferences can be produced between reflected waves from the two interfaces of a thin layer and create interference fringes, called "Kiessig fringes", in the reflectivity profile.

During the past 20 years neutron reflectivity has developed to a very powerful and non-destructive technique to access buried interfaces. It has the unique possibility of contrast variation via selective use of different isotopes. The most common being the use of H and D. The use of deuterated compounds, the simplest being $D_2O$ allows contrast optimization. This gives either the largest possible contrast, that is, better statistics, or a contrast matching in order to mask selected parts of the sample. Reflectivity is widely used to obtain refined concentration profiles on surfaces, such as surfactants, lipids, proteins or adsorbed polymers at interfaces, which can be liquid or solid. Specular reflectivity gives information along the $z$-direction, that is, perpendicular to the surface. For studies of the lateral shape of interfaces, off-specular measurements have to be carried out.

This chapter addresses primarily specular neutron scattering adapted to studies of PEMs. For more details of the method itself, the work of Daillant and Gibaud [1] on X-rays and neutron reflectivity, or of Névot and Croce [2], or Fragneto and Cubitt [3] should be consulted.

### 11.2.1
**Specular Reflectometry**

Figure 11.1 shows a simple reflectivity geometry: in air or vacuum, a neutron beam is incident with a wave vector $\vec{k}_i$ on a surface characterized by a scattering length density

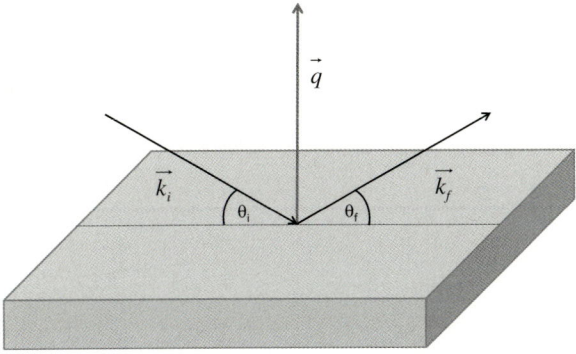

**Figure 11.1** Specular reflectivity geometry: $\theta_f = \theta_i$.

$N_b$. The beam is reflected with a wave vector $\vec{k}_f$. The reflection is characterized by the momentum transfer $\vec{q} = \vec{k}_f - \vec{k}_i$. For elastic scattering, that is, no energy loss, $|\vec{k}_i| = |\vec{k}_f|$ and under specular conditions, $\theta_i = \theta_f = \theta$ is valid. Then, the momentum transfer is given by

$$|\vec{q}| = q_z = \frac{4\pi}{\lambda}\sin(\theta).$$

This formula shows already the two options for performing reflectivity studies. Either the angle of incidence is varied in a monochromatic experiment or the surface is radiated by an energy spectrum at one or two fixed angles in a time of flight (TOF) experiment. The measured reflectivity $R = I/I_0$ is the reflected intensity $I$ normalized by the intensity $I_0$ of the incident beam, and can be described as a function of $q$ and $q_c$, the critical wave vector, below which reflection is total:

$$R = \left[\frac{q - (q^2 - q_c^2)^{1/2}}{q + (q^2 - q_c^2)^{1/2}}\right] \quad \text{with} \quad q_c = \sqrt{16\pi N_b}$$

If the sample consists of several layers, like a polymer film on a substrate, or even multilayers, the matrix formalism from optics is used, to get the reflectivity from one layer to the other [4].

Another important feature is the roughness of the interfaces, that is, any irregularity or interdiffusion between adjacent layers as is the case for instance in PEMs. Indeed, if the interface is not perfectly smooth, the reflectivity is reduced by a Debye–Waller-like coefficient:

$$R \approx \left(\frac{16\pi^2}{q^4} N_b^2\right) e^{-q_z^2 \sigma^2}$$

where $\sigma$ is the characteristic size of the imperfections, but does not contain any information about the type of roughness (irregularities or diffuse interface). A diffuse interface and a rough one, with the same $\sigma$ will then give the same specular reflectivity profile; but the rough one might create off-specular reflectivity whereas the diffuse one will not. For PEMs $\sigma$ characterizes (i) the interdigitation between two polyelectrolyte layers, and (ii) the surface irregularities, coming from polymer coiling, aggregates on the surface and so on, when defined between the film and the external medium (air or water). In the first case it is referred to as "internal roughness" and in the latter case "external roughness". The external roughness is also accessible by X-rays reflectivity or atomic force microscopy (AFM).

### 11.2.2
### Contrast Variation: Interesting Features for PEMs

Neutron reflectivity (NR) has two main principal characteristics which make it particularly suitable for the study of PEMs: the ability to perform measurements in water, and the option for contrast variation via selective deuteration. These two features are impossible to realize with X-rays, which are mainly used for "quick" characterizations of thickness and external roughness. Neutrons are then a unique technique which allows direct insight into the internal structure of PEMs, like the individual layer thicknesses and interdigitation, in both the dry and the water-swollen state.

#### 11.2.2.1 Superlattice Structure
The multilayering preparation process of PEMs allows specific labeling of only some of the layers: the so-called "superlattice structure". Every $n$ layers (protonated), one of the layers is going to be deuterated, creating a superlattice and ordering on a larger scale, as described in Figure 11.2. In most experiments the deuterated layer is poly(styrene sulfonate) because of its commercial availability. There will then be two characteristic sizes of the sample: the total thickness $D$, which gives regular Kiessig fringes, and the repeat unit thickness $d$, which gives so-called "Bragg peaks". Without any fitting of the curve, these distances can be found by simply translating the characteristic width in reciprocal space $\Delta q$ between two adjacent Kiessig fringes or two adjacent Bragg peaks into the real space distance with the usual relation $\Delta q(\text{Bragg}) = 2\pi/d$ or $\Delta q(\text{Kiessig}) = 2\pi/D$. Due to the interdigitation/diffusion between adjacent layers, the Bragg peaks are not as small as for crystalline structures, but broadened. The width gives information on the interdigitation length. Often only 1–3 Bragg peaks can be observed in the available $q$-regime. The decrease in intensity of adjacent Bragg peaks is again described by the Debye–Waller factor and gives information about the position correlation of the deuterated layers. It is worth pointing out that with X-ray reflectometry (XRR) only Kiessig fringes, but no Bragg peaks can be observed for polyanion/polycation multilayers, since the electron density of both polyelectrolyte types exhibits not sufficient contrast.

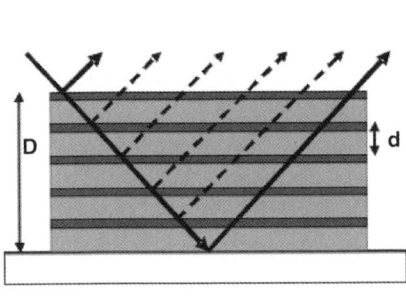
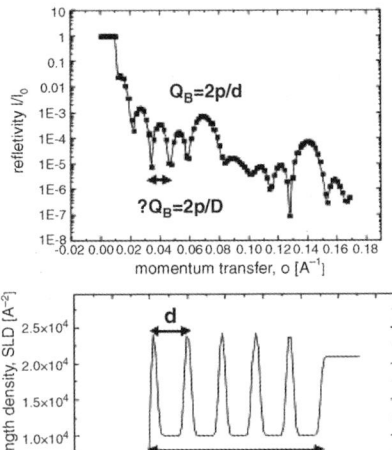

**Figure 11.2** Scheme of the superlattice set-up commonly used to observe the innner structure of PEMs (a). Every n protonated layers (light shade), a deuterated layer (dark shade) is incorporated to create long-distance ordering and distinguishable features in reflectivity curves: Total thickness, $D$, and thickness of repeat unit, $d$. (b) Characteristics reflectivity curve with Kiesig fringes and Bragg peaks. (c) According scattering length density profile.

#### 11.2.2.2 Block Structure

Recently, a different multilayer set-up was studied: a block structure. The multilayer is then made of two blocks, a block of deuterated double layers (either d-PSS and h-PAH, widely interpenetrated [5], or fully deuterated bilayers of d-PSS and deuterated poly (4vinyl-benzylsulfanyl)-pyridinium chloride, d-PS-Pyr [6]), plus a block of protonated ones. By keeping the total number of layers constant but changing the position of the interface, precise information is gained about the evolution of the interdigitation length and the average layer thickness along the direction normal to the substrate.

### 11.2.3
### Fitting

Because detectors cannot follow the frequency of neutrons, the intensity is recorded instead of the amplitude. This leads to a loss in phase information. As a consequence, several profiles correspond to the same reflectivity. To fit the data, one has first to assume a model of the interface, consisting of a set of parallel layers of different homogeneous materials. Each layer is characterized by a scattering length density (SLD) $N_b$ and a thickness $t$, and if needed each interface can be characterized by a roughness $\sigma$. The substrate and the external medium (air or water) are considered to have infinite thickness. From this model a theoretical reflectivity is calculated by the optical matrix method (Parratt or Abeles) [4], and compared with the measured reflectivity. The model is then adjusted, by changing every $N_b$, $t$ or $\sigma$ which seems relevant, to optimize the fit of the data. Fits are mostly done using programs dedicated specifically to reflectivity, developed at facilities (neutron sources or synchrotons).

Without any deuterated material, most PEMs can be modeled by one layer, because of the small contrast, that is, the difference in scattering length density between different polyelectrolytes. For fitting the data of experiments against aqueous half-space, several layers may be needed to describe properly the different amounts of water swelling the film, since often the films swell inhomogeneously perpendicular to the substrate surface. If some of the layers are deuterated, the number of layers, which has to be included into the model can rise dramatically. Here, a simplification of the model has to be found. Because of the isotope sensitivity of neutrons, some uncertainties about the exact profile may be resolved by using several contrasts when immersed in water, usually by measuring in both $D_2O$ and $H_2O$. Contrast matching, that is, measuring against a mixture of solvents (here $D_2O/H_2O$) which masks one of the layers' reflectivity (by equalizing the scattering lengths densities), is theoretically possible but experimentally difficult with PEMs because of their "sponge-like" structure and the inhomogeneous swelling behavior.

The problem with some NR studies is that they are carried out under undefined humidity. In order to calculate the amount of water it is essential to use a controlled environment. Dry film can be either studied in a vacuum chamber or against a drying agent like$P_2O_5$. The relative humidity can be controlled by $N_2/H_2O$ vapor mixtures or against aqueous solutions of different salt content. PEMs against liquid are studied in a *flow cell*. The latter gives the opportunity to perform *in situ* adsorption and to investigate the effect of salt, pH or other additives on the structure of the PEMs.

## 11.3
### Preparation Techniques for Polyelectrolyte Multilayers

PEMs are prepared via alternating deposition of oppositely charged macromolecules. Different methods are used for preparation, like dipping, spraying and spin-coating. There is controversy in the literature as to whether or not the type of preparation technique has an effect on the structure and response to external stimuli. This section compares the results of the different techniques and aims to find a unifying concept to explain the partially contradicting observations. Since most of the studies are done at PSS/PAH multilayers, the methods will be mainly compared on the basis of the PSS/PAH system.

### 11.3.1
### Dipping

When the fabrication of polyelectrolyte multilayer films was invented in 1992 [7, 8], it was also the first time that a new method for preparation of molecularly thin

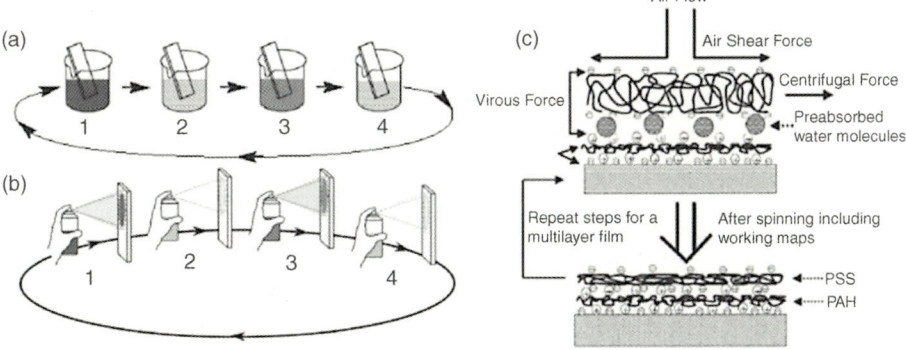

**Figure 11.3** Preparation techniques. Three layer by layer deposition techniques are mainly used for preparation of PEMs: Dipping (a), spraying (b), spin-coating (c). Taken from ref. [8–10].

films was used. The primary layer-by-layer (LbL) technique was the *dipping* method. It seems to still be the most prominent and frequently used technique. Thereby, the substrate is dipped alternately into aqueous polyanion and polycation solutions. Most of the substrates used for preparation of PEMs carry an excess charge (e.g., silicone or glass) which is the precondition for the adsorption of oppositely charged polymer chains. By using special pretreatment, for example, plasma cleaning, other substrates, for instance silicone rubber (PDMS), can also be modified to allow adsorption of PEMs. For a better attachment very often the branched PE polyethylenimine (PEI) is used as an intermediate layer between the substrate and the linear chains of the first PE layer. After each step of PE adsorption the sample is usually immersed into a water bath to remove excess PE (Figure 11.3). In principle, this procedure can be repeated as often as wished. A limiting factor is the exposure time needed for an adsorption step, which in most of the studies is 15–20 min. This makes dipping a time-consuming preparation process.

The first NR study was performed by Schmitt *et al.* in 1993 [11]. This study proved the remarkably high potential of NR for soft matter research. Besides the determination of the thickness of the PEM film, the study used the option of contrast variation by fabrication of a superlattice, as shown schematically in Figure 11.2. As basic polyelectrolytes, poly[allylamine hydrochloride] (PAH) and sodium poly[styrene-4-sulfonate] (PSS) were used. Building up the PEM using deuterated PSS (dPSS) for every third PSS layer yielded a superstructure of the film which causes Bragg peaks in the reflectometry curve (curve A2 in Figure 11.4).

In this way Schmitt could prove that the PEM film consists of interdigitating but stratified layers of one PE species. For multilayers prepared at 2 M NaCl, the internal roughness was determined to be 19 Å, while the thickness of a PSS/PAH layer pair was about 51 Å (PSS: about 30 Å and PAH: about 20 Å) (see Table 11.1).

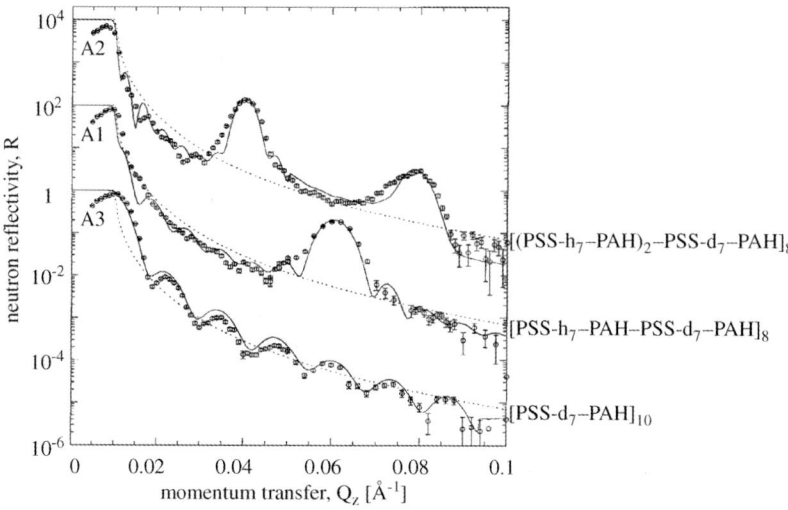

**Figure 11.4** Changing the number of intercalating layers in PSS/PAH: Direct effect on the Bragg peaks. From top to bottom: 5, 3 and 1 intercalated protonated layer(s). Taken from ref. [13], data for sample A2 published first in [11].

The limit of interdigitation of polyionic layers was addressed by Korneev et al. [12] and Lösche et al. [13] via investigating the Bragg peak of differently deuterated samples. The authors deuterated either every, every second, or every third PSS layer. In both latter cases Bragg peaks could be observed, indicating a separation of the deuterated layers from each other, as shown in Figure 11.4. For the first sample the strong interdigitation of dPSS and PAH-layers smears out the SLD contrast completely, related to an internal roughness which is about 1/3 to 1/2 of the layer pair thickness.

From the fit of the Bragg peaks one finds evidence for a variation of the layer pair thicknesses with respect to the layer number. The inner layers close to the solid surface are thinner compared to the subsequently adsorbed layers. After 4–5 molecular layers the equilibrium layer pair thickness is reached, leading to a constant growth [11]. This analysis became an essential element of the model of the *linear growth* of PSS/PAH multilayers (see [8] and references therein). In agreement with Schmitt et al. [11], Lösche et al. showed an increasing roughening of successively adsorbed layers close to the substrate, which provides an increasing number of adsorption sites resulting in thicker layers [13]. This region, where the substrate has a pronounced influence on the PEM structure, occurs in many systems and has to be excluded, if one wants to categorize the system with respect to the type of growth, such as linear or exponential growth (see Section 11.4.2).

Recently, a study by Soltwedel presented the opposite behavior of the internal roughness [5]. Approaching the opposite surface, that is, the film/air interface instead of the substrate/film interface, the inner roughness of PSS/PAH multilayers decreases linearly with smaller distance from the film/air interface. A similar

**Table 11.1** Several structure parameters of PEM obtained with NR. Comparison of the different preparation techniques: Dipping (D), spraying (S), and spin coating (SC); conditions during preparation: salt concentration and temperature; and the experiment: relative humidity (RH). Data obtained with other techniques: atomic force microscopy (AFM), ellipsometry (Elli), and X-ray reflectometry (XRR).

| Author, ref. | Polycation | Preparation | Exp. Conditions | $\sigma_{s/f}$ [Å] | $\sigma_{int}$ [Å] | $\sigma_{f/a}$ [Å] | $d_{bl}$ [Å] |
|---|---|---|---|---|---|---|---|
| Schmitt [11] | PAH | D 2 M +buffer | amb.c. | 3.9 ± 0.5 XRR | 19 ± 1 | 13…15 XRR | 51.0 ± 0.6 |
| Korneev [12] | PAH | D 2 M | amb.c. | — | — | — | 53.1 |
| Lösche [13] | PAH | D 3 M | amb.c. | 5 | 28 | — | 76 |
| Lösche [13] | PAH | D 2 M | amb.c. | 5 | 24 | — | 59 |
| Lösche [13] | PAH | D 1 M | amb.c. | 5 | 19 | — | 45 |
| Lösche [13] | PAH | D 0.5 M | amb.c. | 5 | 15 | — | 35 |
| Steitz [14] | PAH | D 1 M | XRR, amb.c. | 5 ± 0.3 | — | 12 ± 2.5 | 51.7 ± 0.6 |
| v. Klitzing [15] | PAH | D 0.1 M | 33–45% RH | — | — | — | 27 |
| v. Klitzing [15] | PDADMAC | D 0.05 M | 33–45% RH | — | — | — | 40.5 |
| Jomaa [16] | PDADMAC | D 0.1 M | amb.c. | — | 12 | 21 AFM | 44.3 |
| Izquierdo [9] | PAH | D 0.5 M | XRR, amb.c. | — | — | 10 | 26.9 |
| Izquierdo [9] | PAH | S 0.5 M | XRR, amb.c. | — | — | 9.9 | 20.7 |
| Krogman [17] | PDADMAC | D 0.1 M | Elli, amb.c. | — | — | — | 30.9 |
| Krogman [17] | PDADMAC | S 0.1 M | Elli, amb.c. | — | — | — | 26.7 |
| Felix [18] | PAH | D 0.5 M | amb.c. | 9 [PEI] | 18 | 18 | 31.2 ± 0.2 |
| Felix [18] | PAH | S 0.5 M | amb.c. | 9 [PEI] | 13 | 13 | 23.6 ± 0.1 |
| Kolasinska [19] | PAH | D 0.15 M | XRR, 0% RH | — | — | — | 21.7 ± 1.4 |
| Kolasinska [19] | PAH | S 0.15 M | XRR, 0% RH | — | — | — | 16.0 ± 1.4 |
| Köhler [20] | PAH | S 0.5 M | 0% RH | 5 | — | 5 | 30.0 ± 0.3 |
| Kharlampieva [21] | PAH | SC no salt | amb.c. | — | 16 | 3 | 20.9 |
| Kharlampieva [21] | PAH | D no salt | Elli, amb.c. | — | — | — | 6 |
| Kharlampieva [21] | PAH | SC buffer | amb.c. | — | 50…130 | 160 | 54.9 |
| Kharlampieva [21] | PAH | D buffer | amb.c. | — | 19…74 | 180 | 23.0 |
| Soltwedel [5] | PAH | D 1 M, 40 °C | 0% RH | — | 12–26 | 11 | 46 |
| Soltwedel [5] | PDADMAC | D 0.1M, 20 °C | 0% RH | — | 31–52 | 14 | 46 |
| Dodoo [22] | PDADMAC | D 0.5 M | 0% RH, vac. | 0 | — | 22 ± 5 | 80.2 ± 0.5 |
| Dodoo [22] | PDADMAC | D 0.25 M | 0% RH, vac. | 0 | — | 11 ± 4 | 44.5 ± 1 |
| Dodoo [22] | PDADMAC | D 0.1 M | 0% RH, vac. | — | — | 10 ± 4 | 24.0 ± 0.2 |
| Früh [23] | PDADMAC | S 0.25 M | 30% RH | 2 | 10 | 10 | 24.9 |

observation was made by Chevigny et al. [6] for another polymer system (see Section 11.4.2).

The paradox between the two contradicting observations on roughness can be resolved by the following explanation: The increase in roughness for the first layers close to the substrate is due to a change in surface topology, such as the development of PE aggregates, and so on with increasing deposition steps. It is a *topological roughness*, this type of roughness increase leads to an increase in thickness increments, since more adsorption sites are offered, and levels off after a few deposition steps.

The second type of roughness, which decreases with increasing distance from the substrate, is an *interfusion roughness*. Due to the multiple immersion, the first adsorbed layers have a longer integrated contact time with the aqueous salt–polymer solution than the layers which are adsorbed later. This leads to stronger chain diffusion of the layer adsorbed at the beginning close to the substrate in comparison to those which are deposited later. This type of internal roughness does not affect the film thickness.

Drying during the preparation process causes irreversible changes of PEM with respect to a continuation of the adsorption steps. Subsequent adsorption of new layers after drying results in a thinner new layer [13]. In order to get high reproducibility most of the studies in the last decade have been done without any intermediate drying.

Up to now, most of the studies on and with PEM have been performed on multilayer thin films produced in the *classical* dipping technique, even though it is quite time consuming. The advantage of he dipping technique is that the material is not lost, as is the case with spraying or spin-coating.

## 11.3.2
### Spraying

A second technique for LbL-film preparation of PEM was established by Schlenoff et al. They used spraying devices to moisten solid surfaces/substrates with solutions of oppositely charged polyions, subsequently [24]. This LbL spraying technique offers the possibility to obtain a PEM in a fraction of the time required for preparation with the dipping technique. Typically, the samples are exposed twice to the same solution for several seconds, with a short time in between. The sample is then sprayed with pure water to remove excess material. In doing so, the preparation time for one layer is reduced to a few tens of seconds instead of the tens of minutes needed for the dipping technique.

Recently, the spraying method was varied by Lefort et al. to reduce the preparation time by simultaneous application of the polymeric counterions [25]. However, strictly speaking this is no longer an LbL technique. Therefore, it will not be considered further in the following.

When the spraying technique was first introduced, the authors claimed a fairly similar structure and properties for the films obtained when the exposure time was comparable [24]. Using ellipsometry, voltametry and FTIR, Schlenoff and coworkers found an identical chemical composition and conductive behavior for sprayed and dipped samples of PSS/PDADMAC, even when the contact time was only 10 s.

This was confirmed in a study by Izquierdo *et al.* who compared, for the first time, the obtained structure of dipped and sprayed PEM (PSS/PAH) when varying a large number of parameters of the spraying technique [9]. They tested several different cycles of spraying for their output in film quality and thickness. Spraying with intermediate rinsing steps yields thinner films than dipping, which in turn produces thinner films than produced by spraying without rinsing. The external roughness of dipped and sprayed samples is similar, about 10 Å in ambient conditions. If the exposure to PSS was repeated after a rinsing step the film thickness always increased. The thickness of the sprayed multilayer increases with increasing polyelectrolyte concentration, while the spraying time above 1 s has no pronounced effect. In contrast to dipped multilayers, intermediate drying has no effect, while intermediate rinsing leads to a higher increment than without rinsing. Dipping a film for a similar total time to that needed for the sprayed film, that is, a factor 40–400 faster than usual, the multilayer becomes very thin and much rougher ($\sigma > 300$ Å).

In 2007, Krogman *et al.* reported an ellipsometric study to compare (PSS/PDADMAC) multilayers sprayed or dipped from solutions containing 0.1 M NaCl [17]. In both cases the multilayers grew linearly with the same increment, but the inital growth (below five layer pairs) is smaller for dipped multilayers. This is due to an inhomogeneous coverage for the first dipped layers, while in the case of spraying a homogeneous coverage was obtained from the beginning, as proven by AFM measurements. This difference is explained by kinetic trapping of polyelectrolyte chains for spraying and diffusive processes during dipping. In contrast to this, other polyelectrolyte pairs, such as dendrimer PAMAM/PAA[1]) show the same small initial growth rate for dipping and spraying. This supports the important role of the competition between sticking at the surface and the lateral mobility of polyelectrolytes.

To get deeper insight into the inner structure of sprayed PEMs Felix *et al.* performed NR measurements on several PSS/PAH films, with and without superstructure, prepared with 0.5 M NaCl [18]. The study confirms that sprayed films are also stratified. The dipped multilayer is about 30% thicker than the sprayed one. The data were fitted with fixed values for the internal and external roughness, which gave the best fits: 13 Å for all sprayed multilayers and 18 Å for the dipped multilayer. The values for the SLD were also kept fixed: $1.11 \times 10^{-6}$ Å$^{-2}$ for hydrogenated layers and $2.75 \times 10^{-6}$ Å$^{-2}$ for deuterated layers, independently of whether the samples were dipped or sprayed. Interestingly, the resulting fit quality was better for a sprayed sample than for a dipped one. In this case a model for a super lattice structure with equidistant layers and constant internal roughness was used. The poorer quality of the fits with this model for dipped samples supports findings of varying internal roughnesses for dipped samples [5, 6, 26].

---

1) Poly-(amido amine), poly(acrylic acid).

Because the SLD and the roughness were kept constant and the same value for internal and external roughness was used for a certain sample, it is difficult to analyze these values. Lösche *et al.* obtained 16 Å as internal roughness of dipped PSS/PAH multilayers prepared with 0.5 M NaCl [13].

Kolasinska *et al.* performed NR measurements at PSS/PAH multilayers prepared with 0.15 M NaCl [19]. The dipped multilayers are again thicker (by 30–40%) than the sprayed ones after rinsing, which is in good agreement with the results mentioned above [9, 18]. The scattering length densities obtained by both preparation techniques are similar. Since no superlattice was used, only the external roughness could be obtained, which is, in contrast to [18], higher for the sprayed films, but against liquid water and water vapor (5–10 Å for dipped vs. 15–24 Å for sprayed). In this study the roughness was a fitting parameter and not fixed like in [18]. On the other hand, the values for the roughness are not very reliable in [19], since the roughnesses of the dry films are higher than against water, which is amazing due to the higher surface tension in the case of dry films. The roughness against $N_2$ is only given for the dipped film and is between 11 and 15 Å, which is similar to [18]. To summarize, in both studies [18, 19], the fit results seem to be not very sensitive to changes in external roughness.

Sprayed PSS/PAH multilayers seem to be less stable to external pH variation than dipped samples [19].

Spraying also allows polyelectrolyte adsorption at a low degree of polymer charge, which is not possible with dip coating. Dip coating of PSS and 68% charged P(DADAMC–*stat*-NMVA) (DADMAC: diallyldimethyl ammonium chloride, NMVA: N-methyl-N-vinylacetamide) gives very thin layers [28], while alternating spraying of PSS and 60% charged P(DADAMC–*stat*-NMVA) gives much higher but irregularly adsorbed amount of polymer [29] Figure 11.5. A similar observation was made at reduced charge density (75% charged PDADMAC) and high salt concentration (0.5 M NaCl). While no sucessive adsorption takes place in a dipping process, a film can be sprayed [20]. The effect of charge density will be discussed in more detail in

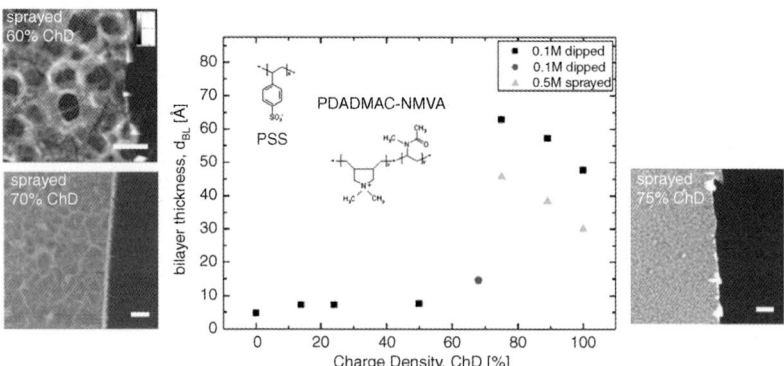

**Figure 11.5** Comparision between dipped and sprayed PSS/PDADMAC multilayers with different charge density of P(DADMAC–stat–NMVA). Multilayers dipped with 0.1 M NaCl [27, 28] and sprayed with 0.5 M NaCl [20]. The AFM micrographs show sprayed PEMs of 60, 70 and 75%-ly charged P(DADMAC–stat–NMVA), respectively. Scale bar, 5μm [20, 29].

Section 11.4.2. Here, the difference between dipped and sprayed multilayers will be analyzed. It is known that a decrease in polymer charge and an increase in concentration lead to more mobile polyelectrolyte chains due to the reduced opportunity to form complexation sites. During spraying the structure sticks very fast to the surface, has no time to rearrange with the environment, and is "frozen". During dipping the chains have more time to desorb from the surface, if insufficient complexation sites are formed.

### 11.3.3
### Spin-Coating

Spinning was already used for drying of dipped samples [30]. It improves the surface quality of the PEM by reducing the roughness, possibly due to the directed drainage and the shear forces at the interface. There are hints that the lower roughness, that is, lower porosity leads to less swelling in subsequent feeding solution. In 2001 a new approach for preparation of multilayer films was introduced by Chiarelli et al. [31]. The authors used a spin-coating technique to produce PEM from PEI and PAZO.[2] The film thickness increases linearly with increasing number of layers. The first three layer pairs close to the substrate are denser. The more commonly used PAH/PSS system was characterized by Char and coworkers [32, 33]. The PEM showed higher layer pair thickness with low roughness (24 Å vs. 3 Å) than for dipping (8 Å vs. 4 Å).

Recently, Kharlampieva et al. published a neutron reflectometry study of spin-assembled PEM [21]. They investigated the PSS/PAH model system and compared the effect of dipping and spin-coating preparation on the internal structure of the PEM. The samples were rotated at 3000 rpm for 20 s and rinsed twice with water. The contact time during the dipping procedure was 10 min. By forming a PEM superlattice, where every sixth PSS layer was deuterated, Bragg peaks were observed, which are an indication of a stratified multilayer. All multilayers grow linearly, while the thickness is about three times larger than for the dipping procedure. Without any additional salt the external roughness is very low (3 Å), which can be explained by pronounced elongation of the polymer chains due to high shear rate. The internal roughness is about 16 Å, a value between those for dipped and sprayed samples at 0.5 M NaCl [18]. The external roughness is smaller by trend for spin-coated samples, but scattering of the roughness values is quite high. PEMs spin-coated from 0.01 M buffer show an external roughness of 120–160 Å (dipped: 180–250 Å) and an internal roughness 50–130 Å (dipped: 19–7 Å). These values are about one order of magnitude higher than in all other studies of spraying and dip coating (see above and Table 11.1).

### 11.3.4
### Comparison of the Techniques and Model

Spraying and spin-coating are very efficient techniques for preparing PEMs. The preparation time is speeded up by at least a factor of 20–400 compared to the dipping

---

[2] PEI, poly[1-[4-(3-carboxy-4-hydroxy phenyl-azo)benzensulfonamido]-1,2-ethanediyl sodium].

technique. A disadvantage is that the material is lost after spraying and spin-coating. Both techniques are restricted more or less to planar interfaces, although recently a spray coating procedure for tissues (fiber mat) was reported [34].

The adsorption process of polymers consists of several steps, such as delivery (diffusion/flow) of polyelectrolyte to the surface, the adsorption kinetics and the rearrangement, including possible removal/desorption of excess material. These steps have different features in the three considered preparation techniques due to differences in contact time with the preparation solution, the drainage and flow and, related to that, different surface topologies and porosity/permeability.

There are just a few studies, where spraying and dipping are systematically compared at for example, the same salt concentrations [9, 18, 19, 21]. Another serious problem is that many neutron reflectometry measurements are carried out under ambient laboratory conditions without controlling the relative humidity. The conditions between different studies vary, which makes it difficult to compare the effect of the different methods. Spraying itself contains a lot of parameters, like spraying, rinsing and waiting times, spraying distance, and spraying pressure. So far, there is no unified procedure and only trends can be recognized. The comparison between dipping on the one hand and spraying and spin-coating on the other cannot be done properly, since the preconditions with respect to the kinetics are different. Nevertheless, in the following some essentials will be worked out.

First, the polyelectrolyte concentration and the contact time with the polyelectrolyte solution, that is, the amount of polyelectrolyte which is offered within a certain time frame, are crucial for the multilayer formation. In the case of dipping, one can estimate the number of polyelectrolyte chains which will reach the substrate within the dipping time, via the diffusion coefficient and the known concentration. In the case of spraying the amount of polyelectrolyte within the contact time is not evident, since the experimental parameters, such as spraying pressure and the angle of aperture, determine the number of polyelectrolyte chains/time reaching the surface. To find the same conditions for spraying and dipping with respect to the integrated amount of polyelectrolyte will be almost impossible. Analogous problems occur for the comparison between spin-coating and dipping. Here, the offered polymer density is determined by the concentration of the solution and the rotation speed and time. Spraying and spin coating exhibit a linear growth of PSS/PAH systems from the first layer on, but dipping gives a smaller increment for the first layers. This could be explained again by the amount of polymer/time sticking at the surface. Spraying the first layers results in a full coverage, while dipping leads often to an incomplete coverage during the first deposition steps.

Secondly, the adsorption kinetics is different. The driving force for complexation is the entropy due to release of small ions. Due to the long contact time in the dipping process, the polyelectrolyte chains have more time to diffuse into the formerly adsorbed layers in order to form complexes, to rearrange, and perhaps also to desorb than in the case of spraying or spin coating. This would lead to higher internal roughnesses than for spraying, as was observed in the literature [18]. The multiple relatively long contact with polyelectrolyte solutions during dipping allows the polyelectrolyte chains closer to the substrate stronger diffusion into the adjacent layers and increases the internal roughness. This leads to a decrease in internal roughness towards the outer film surface (film/air or film/water) [5]. The internal roughness of sprayed multilayers seems to be more uniform perpendicular to the substrate surface than for dipped PEMs. Another consequence of the short contact time in the case of spraying and spin-coating is the more "frozen" structure.

Due to the reduced possibility for rearrangement, the sprayed layers seem to be less stable in comparison to the dipped ones, which is proved by loss in material due to intermediate rinsing leading to the order in thickness: "spraying without rinsing" > "dip coating with rinsing" > "spraying with rinsing" [9] and by the change in the external pH [19]. It is worth mentioning also that the dipped multilayers are not in thermodynamic equilibrium, which was proved, for instance, by heating after preparation. A clear annealing process could be identified [35], which is discussed in Section 11.6.2.

The external roughness is a critical parameter, since often fits of neutron reflectivity data are not very sensitive to the external roughness at the film/air or film/water interface. In addition, the outcome depends strongly on the roughness which was assumed for the substrate/film surface. The values should be compared with the results of independent AFM studies. In this context, the outcome of the work of Izquierdo *et al.* that the external roughnesses of sprayed and dipped samples are similar seems to be reliable [9].

Besides the short contact time, spraying and spin-coating have the directed flow in common. In the case of spin-coating it is driven by the centrifugal force, and during spraying a vertical flow occurs. This leads to different orientations of the polymers than in dipped multilayers, and leads to less pronounced swelling of sprayed and spin-coated films. This point is discussed in Section 11.6.3. The very difference between spraying and spinning is that spinning means transportation of polymer/water mixture and evaporation. The film becomes steadily thinner until the (excess) macromolecules can no longer be transported, which could lead to larger thicknesses.

Unless stated otherwise, the multilayers addressed in the following sections are prepared by the dipping technique.

## 11.4
### Types of Polyelectrolytes

The type of polyelectrolytes affects the properties of the PEM, such as thickness, roughness, and type of growth. The differences are caused by interactions between polyanion and polycation, the stiffness of the polyelectrolytes, or the balance between hydrophobicity and hydrophilicity. PSS for example has a rather hydrophobic backbone [36].

### 11.4.1
#### A Strong Polyanion and a Weak Polycation

The first polyanion/polycation combination ever used to form polyelectrolytes multilayers was made of a weak polycation, poly(allylamine hydrochloride) PAH

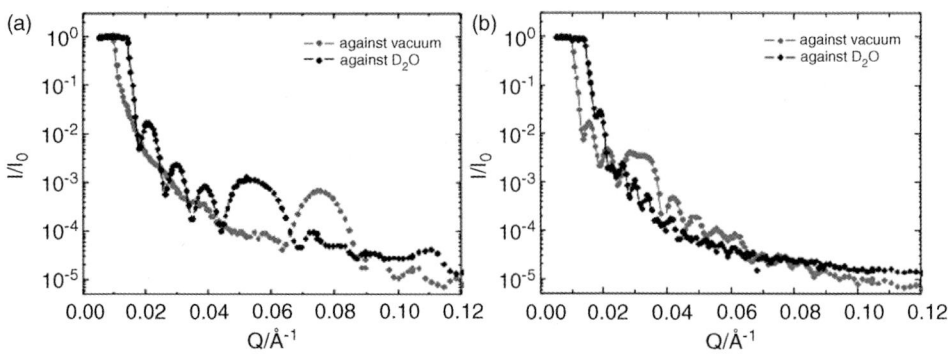

**Figure 11.6** (a) PSS/PAH multilayer ([(PSS/PAH)$_2$/d-PSS/PAH]$_6$ prepared with 0.1 M NaCl, (b) PSS/PDADMAC multilayer [(PSS/PDADMAC)$_3$/d-PSS/PDADMAC]$_6$ prepared with 0.05 M. Both systems are studied in vacuum and D$_2$O. The fact that the Bragg peaks vanish in the case of PSS/PDADMAC against D$_2$O is due to contrast matching. Taken from ref. [15].

and a strong polyanion, poly(styrene sulfonate) PSS [7]. Because of its versatility and accessibility this system remains the most investigated. Some features with respect to different preparation methods have already been described in Section 11.3. Here, a short summary is given in order to compare PSS/PAH PEMs with other types of PEMs. The interpenetration of polyelectrolyte chains between adjacent layers is extended over 1/3 to 1/2 of the layer pair thickness, which leads to the absence of Bragg peaks if every PSS layer is deuterated. If the distance between the deuterated layers becomes larger, Bragg-reflexes occur, which are resistant to salt during [13] or after preparation (see Section 11.6.1) and to drying and reswelling in water vapor, as shown in Figure 11.6.

The positions of the peaks also allowed an estimation of the layer thicknesses much more precisely than with only the overall thickness (obtained by XRR): layers closer to the substrate were found to be less thick [11] because they are in the "substrate-influence" zone [37]. Out of this zone, the thickness increases linearly with increasing number of dipping cycles.

The "reverse" samples, using a deuterated polycation as spacer instead of deuterated polyanion, were also investigated [38, 39]. As PAH is unavailable in its deuterated form, poly(p-phenylene vinylene) (d-PPV) is used instead. The same type of reflectivity curve, with clear Bragg peaks, is obtained, and the internal roughness estimated to be the same as for the d-PSS samples: around 15–20 Å.

The experiments of Soltwedel et al. on PSS/PAH prepared in 1 M NaCl and with a deuterated and a non-deuterated block (Section 11.2.2), confirmed the inhomogeneity in layer thicknesses distribution [5]. They are smaller near the substrate (70% of the equilibrium value), and have a constant layer pair equilibrium value estimated to be 46 Å per layer after the first 4 to 5 layer pairs. As already mentioned in Section 11.3 the interdigitation decreases continuously from the substrate to the outer interface, which is attributed to interdiffusion taking place during each dipping step. Since the layers closer to the substrate are more often exposed to the feeding solution than those near the air/multilayer interface, the polymers inside have more time to rearrange and to interdigitate.

It is important for NR to be aware of the replacement of H versus D. Steitz *et al.* [40] and Ahrens *et al.* [41] found that 1 H is exchanged by 1 D. Later, Ivanova *et al.* found that, on the timescale of a day, at every PAH monomer unit 2–3 mobile protons (H) are replaced by deuterium [42]. This time dependence could not be confirmed by Steitz *et al.* and the reason for the mismatch is still unclear. The deuterium cannot be exchanged in $H_2O$ and remains 100% bound (D) [40, 42].

To conclude, PSS/PAH is a system which forms highly ordered multilayers, with moderate internal and external roughnesses. The growth of the system is linear, and the film thicknesses are then moderate. But these general features can be tuned, for example, by changing the ionic strength. As PAH is a weak polyelectrolyte, pH can also be used as a tuning agent. This will be detailed in Section 11.5, which focuses on preparation parameters.

Kellogg *et al.* addressed the inner multilayer structure of weak polyelectrolytes using PAH and partially deuterated SPAn[3] [26]. The report presents, in contrast to former work of Schmitt *et al.* (PAH/PSS), a linear interfacial roughening away from the solid substrate.

## 11.4.2
### Two Strong Polyelectrolytes

The most common system using two strong polyelectrolytes (and the only one observed with neutrons) is composed of PSS and, PDADMAC.[4] This association is characterized by a higher thickness and larger external and internal roughness than the PEM from PAH/PSS.

Multilayers of PSS and 100% charged PDADMAC show a linear growth during multilayer build-up. PDADMAC is highly hydrophilic, and has a stiffer backbone than PAH [43]: the persistence length is about 5 Å, whereas it is only 1 Å for PSS and PAH. The intra-charges distance is also longer for PDADMAC (3.6 Å) than for PSS or PAH (2.5 Å). These differences may be caused by the ring in the PDADMAC structure, which makes the whole polymer stiffer. Because of this rigidity, both due to the electrostatic repulsion and the backbone stiffness, there are less "complexation points" between PSS and PDADMAC than between PSS and PAH, which makes PSS/PDADMAC a more mobile sytem [44]. This higher mobility causes a stronger diffusion of the polyelectrolyte into the adjacent layers during exposure to the polyelectrolyte solution, leading to an increase in internal roughness. Another reason for the high internal roughness could be the quite high external roughness after each adsorption step, which increases with increasing number of layers.

In order to study the internal structure the usual superlattice set-up can be used. Due to stronger chain coiling and interdigitation, a larger number of protonated spacer layers has to be used to observe Bragg peaks: with four layer pairs, peaks

---
[3] Sulfonated polyaniline.
[4] Poly(diallyl dimethyl ammonium chloride).

are observed in 0.1 M NaCl [16], while with three, even the Kiessig fringes are blurred [45]. If the ionic strength is decreased to 0.05 M NaCl, with 2 protonated layer pair spacers a fuzzy Bragg peak can be observed, becoming more obvious with 3 layer pair spacers [15, 46] (see Figure 11.6). Interpretation of these data remains particularly tricky because the repeat unit thickness (from Bragg peaks) does not correspond to the total thickness (from Kiessig fringes), indicating an irregular inner structure. More adapted is the block set-up, which allows a more precise determination of the internal roughness. For PSS/PDADMAC in 0.1 M NaCl, the same global features as for PAH/PSS are found: smaller thickness near the substrate (50%), and this time also larger near the outer surface (145%), with constant equilibrium layer pair thickness in the core zone (46 Å) and internal roughness which increases when closer to the substrate [5]. The internal roughness for PSS/PDADMAC is, in similar ionic conditions, about a factor of 2 higher than for PSS/PAH and interdigitation takes place over more than one layer.

In PSS/PDADMAC multilayers no H can be exchanged by D, and the NR curves in vacuum after exposure to $D_2O$ and $H_2O$, respectively, coincide perfectly [22].

As mentioned in Section 11.2.2, recently, the block set-up was used to investigate another strong system, PSS/PS-Pyr[5], this combination has the advantage of being fully deuterable [6]. In 0.1 M NaCl, an average layer pair thickness of 20 Å is found, but layers close to the substrate (the first 3 layer pairs) are 17% thinner. The variation of internal roughness is quite small (this roughness is itself small, between 18 and 23 Å), but again $\sigma_{int}$ is found to increase when closer to the substrate. The values here are more comparable with PSS/PAH, because the polycation PS-Pyr has a very similar structure to PSS, and the same intra-charges distance and persistence length as PSS or PAH: It is a stiffer system, less mobile and with more intrinsic charge compensation. This is in good agreement with the type of growth which is linear, as reported for PSS/PAH. A scheme of this specific block set-up with the corresponding reflectivity curves obtained for PSS/PS-Pyr is reproduced in Figure 11.7.

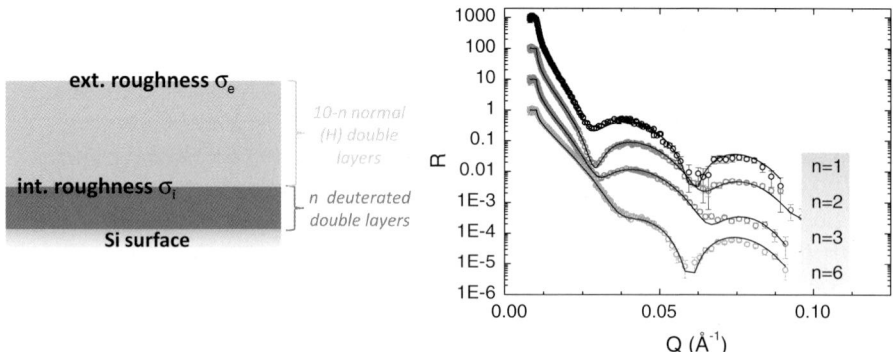

**Figure 11.7** The "block" set-up to observe the inner structure of PEMs via NR. Changing the position of the interface between d-layers and h-layers allows one to scan the whole internal structure of PEMs. From [6].

5) PS-Pyr: poly(4vinyl-benzylsulfanyl)-pyridinium chloride.

The charge density of PDADMAC polymer can be tuned via copolymerization with neutral NVMA, *N*-methyl-*N*-vinylacetamide. In this case the intra-charge distance increases and the number of "complexation points" with PSS is reduced as the charge density decreases, until there are too few to form stable films: for charge density below 68%, no multilayers can be formed [28, 47] in the case of dip coating (see Figure 11.5). Reducing the charge density will promote more coiled structures caused by weaker electrostatic and intramolecular repulsion. The mobility of the system is increased due to the reduced number of complexation points [44]. Recently, the structure of sprayed PSS/PDADMAC multilayers with charge densities varying between 75 and 100% for PDADMAC was studied using NR, giving access to the thickness, and external and internal roughness of the PEM [20]. The measurements confirmed the increasing thickness, external and internal roughness when decreasing the PDADMAC charge density, coming from the more coiled conformations of the polymers and increased mobility (Figure 11.5). The internal roughness is higher than the external one, and the effect is stronger when the charge density decreases (from the supporting information data in [20]). First hints for this behavior were found in the experiments by Steitz *et al.* [46], who measured two superlattices of PSS/PDADMAC, with the same number of spacer layers and the same ionic strength, the only difference being the PDADMAC charge density: for 100%, Bragg peaks are observable, while for 75%-charged PDADMAC, no Bragg peaks occur, clearly indicating an increased interdigitation and mobility in the latter case.

### 11.4.3
### Two Weak Polyelectrolytes

Only a few papers deal with the association weak/weak PEMs and neutron reflectivity, and there are two simple reasons for that: first, no weak polyanion or polycation is available in its deuterated form, which greatly lowers the interest in studying their structure with NR, and second, these systems are pH-sensitive and pH can be tricky to keep stable. Indeed, weak polyelectrolytes are highly sensitive to pH, which changes their charge density, and react differently to a change in the pH value: the polyanion is charged in the basic regime and uncharged in the acidic regime, but for the polycation the opposite is true [48, 49]. Therefore, such a system is not widely used. We can mention the work of Tanchak *et al.* [50, 51], and of Elzbieciak *et al.* [52], both on the PAH/PAA system. Here, nonlinear growth could be observed. Due to the unavailability of a deuterated form of any of the polyions, NR is not so well-adapted for structural studies, and simply provides together thickness, density and external roughness. NR is, however, very useful for investigations in water, to study pH influence, water or ions localization, which will be detailed later in this chapter [cross ref].

In general the high mobility of weak PE in the multilayer structure and during the adsorption cannot yield defined and layered structures [37, 53]. In contrast to that, Kharlampieva *et al.* showed that it is possible to obtain stratified layers with deuterated weak PEs [54]. The study was performed with $[(PMAA/QPVP)_4(dPMAA/QPVP)]_4$.[6]

---

6) Poly-methacrylic acid/quaternized poly-4-vinyl pyridine with 20% of charged units.

### 11.4.4
### Multilayers with Non-Electrostatic Interactions

It is also possible to form multilayers with uncharged compounds. In this case, rather than electrostatic interactions, van der Waals forces or hydrogen bonds are responsible for polymer attraction and multilayer build-up. A few systems have been investigated by NR. Kharlampieva *et al.* [55] systematically studied the inner structure of multilayers formed by PMAA[7] and several water-soluble polymers (among them PEO[8] and PNIPAM[9]), attracted by H-bonds, by combining NR and superlattice structure (a layer of deuterated PMAA is intercalated every five layer pairs). It was found that weak intermolecular interactions lead to high internal roughness and nonlinear growth. In the case of PEO/PMAA films the interpenetration length was too high for the occurrence of Bragg peaks. More recently, systems based on cellulose/xyloglucan have been investigated [56, 57]. NR allowed probing of the internal structure, revealing a dramatic change in internal structure with increasing xyloglucan concentration, caused by a change in conformation already in solution, from the dilute to the semi-dilute and entangled regime.

## 11.5
## Preparation Parameters

As already mentioned in previous sections, the chain mobility is one important feature of the specific PE, which is responsible for the structure of polyelectrolyte multilayers. It mainly depends on the density of complexation sites between the polyanions and polycations. Besides the charge density of the polyelectrolytes, other preparation conditions, like ionic strength, type of ion and pH, tune the strength of complexation.

### 11.5.1
### Ion Effects

During the preparation of polyelectrolyte solutions, salt is usually added for a better control of the multilayer thickness. Changing this salt, or even simply changing salt concentration, is the most common way to modify a PEM's properties: this will alter the polyelectrolyte's conformation in solution, and the interactions between polyelectrolytes, therefore altering the whole construction of the PEM. Ionic effects can be divided into electrostatic effects, related to the ionic strength, and ion-specific effects. Both have a strong influence on chain conformation in solution and, therefore, also on conformation after adsorption in PEMs.

#### 11.5.1.1 Ionic Strength
Lösche *et al.* [13], using the superlattice structure and NR for PAH/PSS multilayers at different concentrations in NaCl, highlighted first a direct proportionality between

---

7) Poly(methacrylic acid)
8) Poly(ethylene oxide)
9) Poly(*N*-isopropyl acrylamide)

the salt concentration and the inner structure of the PEM. Because of the Bragg peaks, which give a precise value of the double-layer thickness, they were able to extract a formula linearly linking the ionic strength $I$ (from 0.5 to 3 M NaCl) to the double-layer thickness $d$: $d = (27.6 + 15.9 \times I)$ (Figure 11.8). A similar proportionality was found between internal roughness and ionic strength: $\sigma_{\text{int}} = 0.4 \times d$.

Lower ionic strengths (below 0.5 M NaCl) were investigated later by Steitz et al. [14] at PSS/PAH multilayers formed in two steps: a common six-layer-pairs precursor film (1 M NaCl), on top of which six additional layer pairs are deposited, at variable ionic strength (0, 0.2, 0.5 and 1 M NaCl). The increase in thickness upon addition of salt is still obvious in the reflectivity curves (Kiessig fringes get smaller) but the dependence of the thickness is on $\sqrt{I}$ and not $I$. These observations were confirmed, for a PSS/PAH system, with a superlattice structure studied by NR [58]. Figure 11.8 shows the average thickness of a PSS/PAH layer pair in dependence on the NaCl concentration of the preparation solution. Above a salt concentration of 0.5 M a linear relation is found. Below this concentration the thickness is proportional to $\sqrt{I}$. The values of different studies also fit quantitatively after correction with respect to the swelling in ambient conditions.

Charge compensation via polycation–polyanion complexation is called intrinsic charge compensation [60]. Increasing the salt concentration will increase the extrinsic charge compensation, which reduces the number of complexation points between two polyelectrolytes, allowing them more possibilities to coil or move through the film [44], which gives rougher and thicker structures.

### 11.5.1.2 Counterions

Besides the obvious influence of ionic strength, the ion type is also able to change the structure of a PEM, although the electrostatic interactions according to Debye–

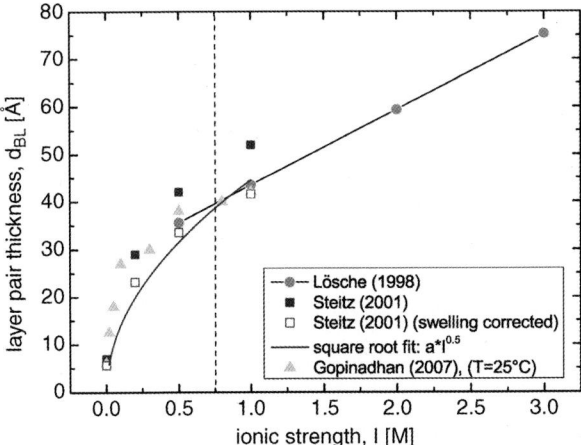

**Figure 11.8** Thickness of a PSS/PAH layer pair in dependence on the ionic strength of the polyelectrolyte solution. The correction with respect to swelling in ambient conditions is done under the assumption that the PEM swells by about 20% [59]. The data are taken from [13, 14, 58].

Hückel theory should be the same. This can be understood by classifying the ions into the Hofmeister series [43, 61]: smaller ions have a relatively small polarizability, high electric fields at short distances, and keep their water of hydration. They have a well-ordered large hydration shell and are called cosmotropic ions, that is, water structure makers. In contrast, chaotropic ions are larger with a significant polarizability, a weak electric field and their water of hydration can be easily removed. These characteristics can affect the structure of the PEM.

Gopinadhan et al. [58] studied the influence of the cation (NaCl, KCl) on the structure of PSS/PAH films, with NR at 0% relative humidity. No obvious influence of the cation type was found on the thickness or internal roughness of the PEM, except a higher amount of bound water for NaCl, which is coherent with the Hofmeister's classification: $Na^+$ has a larger hydration shell than $K^+$, it is then logical that more water is bound. In contrast PSS/PAMPS multilayers show an obvious effect of the cation type on thickness and roughness at a salt concentration above 0.1 M.

The anion effect on PSS/PDADMAC films was recently studied by NR [22]. It is known that the effect should be much higher than the cation one [62]. The reflectivity curves confirmed the trend in thickness and roughness as a function of counteranion type: the larger the anion, the thicker and rougher the film. Larger ions have higher polarizability, and therefore interact more strongly with polyelectrolytes, promoting more coiled structures, which gives thicker and rougher films. The stronger interaction ions/polyions will also increase the polymer mobility inside the PEM [44], by decreasing the complexation points between polyanions and polycations. Actually, the effect of increasing anion size ($F^-$, $Cl^-$ and $Br^-$ used as counteranions) is qualitatively the same as increasing ionic strength (0.1, 0.25 and 0.5 M NaCl), as shown in Figure 11.9 which compares the effect of these two parameters on PSS/PDADMAC films, observed with reflectivity.

Typically, ion specific effects are observed at an ionic strength above 0.1 M. Below this concentration the interactions are dominated by electrostatics [62]. The phenomenon has been observed in many other systems.

Kharlampieva et al. studied the competition between ions of different multivalence [21]. The addition of 0.01 M phosphate buffer to the polyelectrolyte solutions

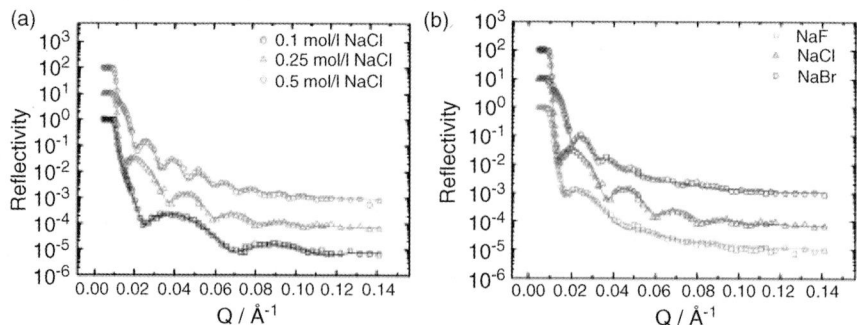

**Figure 11.9** Effect of ionic strength (a) and counteranion (b) on a PSS/PDADMAC film, observed under vacuum with NR. Increasing anion size (from F to Br at 0.25 M) or ionic strength (from 0.1 to 0.5 M NaCl) gives a similar change in the Kiessig fringes pattern. Taken from [22].

during preparation leads to a tremendous increase in thickness (by a factor 2–3), but the addition of 0.1 M NaCl to the buffer has no further effect. This can be explained by the fact that multivalent ions are more efficient in interaction with polyelectrolytes than monovalent. Here electrostatics also comes into play.

Depending on the preparation parameters, the linear growth, which was observed for 100% charged PDADMAC, can be transferred into a nonlinear, a so-called exponential one. With decreasing charge density (minimum 75%), increasing salt concentration and addition of ions with a small hydration shell the multilayer formation develops from linear via nonlinear growth to unstable (zig-zag) growth [59]. This reflects the increase in chain mobility [44].

### 11.5.2
### pH

The influence of pH in the build-up of PEM only has a meaning if at least one of the polyelectrolytes is weak. In this case, depending on the pH range, the weak polyion will dissociate partially or totally (weakly or strongly charged). As the charge density changes the polyelectrolyte conformation, which then changes the multilayers build-up and structure, pH is an important parameter to be taken into account. In a recent paper [52] on PAH/PAA films Kiessig fringes could only observed at pH 11, where PAH is weakly charged and PAA strongly charged. Data obtained from PAH/PAA multilayers prepared at other pH values or pH combinations could not be analyzed. This was explained by a too high external roughness, coming from the chains coiling due to low charge densities.

### 11.5.3
### Temperature

As chain conformation and the solubility of the polyelectrolytes depend strongly on the temperature, this parameter is important in controlling the PEM properties during preparation. When heating a solution of polyelectrolytes, the polymer–solvent interactions are changing: for most polyelectrolytes the quality of water as solvent decreases, meaning that the chains become more coiled as the temperature increases. Eventually, above a transition temperature, the polyelectrolyte precipitates, for PSS this precipitation temperature is around 60 °C.

The effect of changing solution temperature (from 5 to 60 °C) on the inner structure of PSS/PAH multilayers, observed by NR, is essentially studied in references [41, 58, 63]. The usual superlattice structure is used, allowing precise determination of the layer thickness and interdigitation length. In a first experiment the salt concentration is fixed at 1 M KCl, in order to have a strong screening of electrostatic interactions. As reproduced in Figure 11.10, there is an obvious effect of the temperature on internal structure: upon heating, the Bragg peak is shifted towards the left, indicating an increase in layer thickness, and its amplitude is reduced, indicating an increase in internal roughness. This structure change of the PEM is linked to the structure change of the polyelectrolytes in solution: upon heating, they adopt a pearl-necklace (or even a globular) structure in solution,

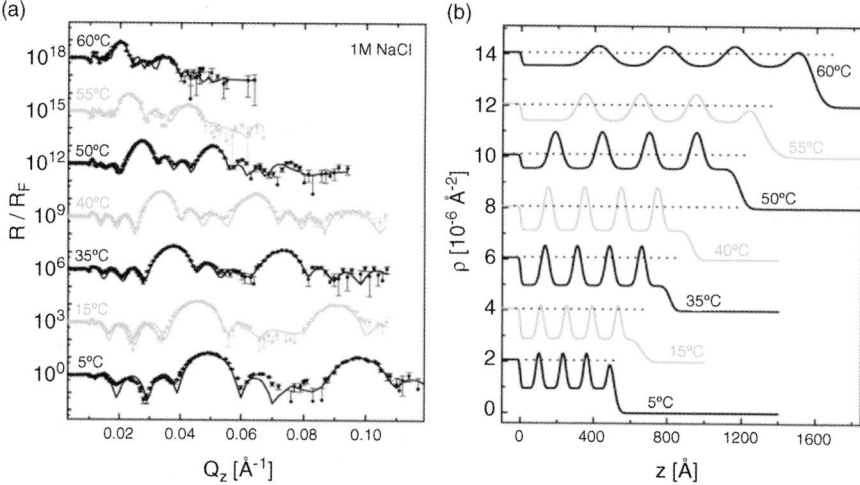

**Figure 11.10** Changing the dipping solution temperature: effect on PEM structure for a PSS/PAH film at 0.1 M NaCl. (a) Reflectivity pattern ($R/R_F$) with Bragg peaks getting closer to each other and smaller with increasing temperature. (b) The corresponding density profile of the film, showing a clear roughening and thickening upon temperature increase. Taken from [58].

and these more coiled conformations in solution are maintained inside the PEM. These conclusions were confirmed by other experiments, where both the type of salt (NaCl and KCl) and the ionic strength (from 0 to 1 M) were varied. Whereas no clear counterion effect is found, the importance of ionic strength is confirmed: below 0.05 M, temperature has no effect because electrostatic interactions are too strong and overwhelm any other effects. Above this threshold, the temperature effect can be seen, getting stronger while screening (via increasing ionic strength) increases.

## 11.6
### Influence of External Fields After PEM Assembly

This section addresses the different parameters which can affect the PEM structure after multilayer preparation. The response to outer stimuli has a high impact with respect to applications, for example, in sensors. More or less the same parameters which are important *during* preparation can be used to manipulate the structure and dynamics *after* preparation. Another wide field is the swelling in water. Here, especially, NR gives unique insights. This section is divided into scalar fields, like ionic strength, temperature and water and vector fields, like mechanical stress and electric fields.

### 11.6.1
### Scalar Fields: Ionic Strength

Swelling in salt and buffer solutions was observed very early [64, 65]. After increase of the thickness and roughness the PEM finally decompose if the salt concentration has risen to a critical value. What happens below this critical salt concentration?

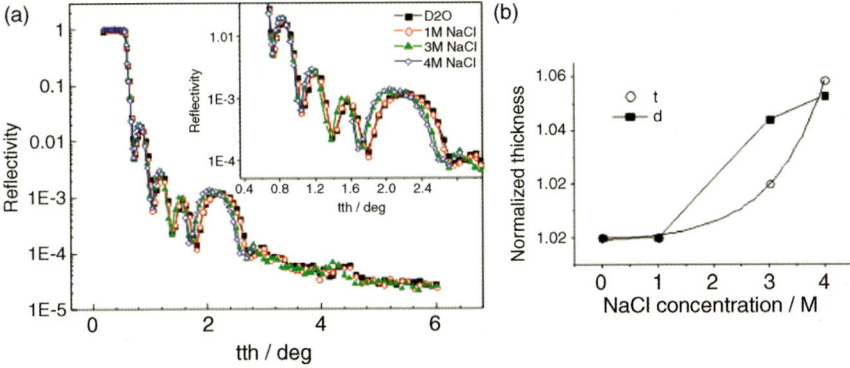

**Figure 11.11** (a) Reflectivity data of a (PSS/PAH)$_2$/d-PSS/PAH multilayer after exposure to aqueous solutions of different NaCl concentrations. (b) Total film thickness (t) and distance between the repetition units (d) in dependence on different external NaCl concentrations. Data provided by [66].

Steitz *et al.* [66] studied a [(PSS/PAH)$_2$/d-PSS/PAH]$_6$ multilayer in aqueous solutions of different salt concentrations. Figure 11.11 shows that the exposure to 1 M NaCl has no effect on the structure. Both, Kiessig fringes and Bragg peaks do not shift with respect to the curves obtained with pure water. With further increase in external salt concentration to 3 and 4 M the Kiessig fringes are shifted to lower q-values, indicating a swelling of the film. The Bragg peaks are still present even at 4 M NaCl during at least 12 h incubation, and no decomposition can be observed. The average distance between internal repetition units and the total film thickness swell differently. This is explained by a non-homogeneous swelling of the PEM perpendicular to the substrate.

In contrast, PSS/PDADAMAC multilayers show an internal "melting" after exposure to 0.8 M NaCl [16]. The intensity of the Bragg peaks decreases with increasing incubation time, as shown in Figure 11.12. Opposite to the former described PSS/PAH study, both Bragg peaks and Kiessig fringes remain at the same positions. That means that the thickness does not change significantly. It is worth noting that the study shown in Figure 11.12 was carried out at a salt concentration where no change in thickness is observed for the PSS/PAH system.

The difference between the PSS/PAH and the PSS/PDADMAC multilayer can be explained by a different chain mobility, as mentioned in Section 11.4.2. The diffusion coefficient within the PSS/PDADMAC matrix is higher than in the PSS/PAH one, due to the lower density of complex sites. The complexes can be partially resolved by small ions, which increase the mobility and the interdigitation between adjacent layers. Obviously, the complex sites in the rather glassy PSS/PAH multilayer are less sensitive to salt.

If the sample was rinsed in salt solution instead of pure water during sample formation a superstructure was not formed [45].

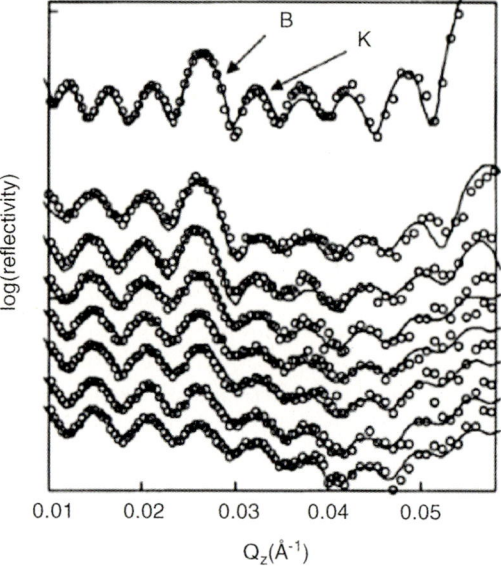

**Figure 11.12** Reflectivity data (open circles) and fit (solid line) of a [(PDADMAC/PSS)$_4$ (PDADMAC/d-PSS)]$_5$(PDADMAC/PSS) multilayer. Uppermost curve: as deposited sample measured in ambient. Lower curves, measured under argon (from top to bottom): after annealing (for 10, 25, 55, 110, 170, 260 min) in 0.8 M NaCl. A final anneal in 1 M NaCl for 120 min (lowest spectrum) almost completely removes the Bragg peak, B, to leave a Kiessig fringe, K. Taken from [16].

### 11.6.2
### Scalar Fields: Temperature

Varying the temperature of the PEM after preparation may cause several effects. One of the most prominent would be the passing of the glass transition point of the PEM in one or the other direction, which would drastically alter the chain mobility. Another effect of temperature change is the variation of the hydrogen-bonding interaction, which is reduced when the temperature rises. To the best of our knowledge, the first example of deliberate temperature variation monitored with NR was the transformation of a PPV[10] layer, sandwiched between PEM blocks, into a conjugated polymer [38], the spacer block (3.5 PSS/PAH layer pairs) shrank by 2% upon heating for 12 h at 200 °C in vacuum. Simultaneously, the PPV layer lost about 13% of its thickness. Another experiment using temperature to modify an adsorbed polymer was the melting of wax particles on top of a PSS/PAH multilayer at 60 °C [67]. The heat treatment showed no noticeable variations of PSS/PAH before and after annealing, but the capping with a wax layer effectively reduced water penetration and water distribution in the PEM.

10) Poly(phenylene vinylene) an organic conductor and light emitting polymer.

In 2002, Steitz et al. studied the effect of temperature on PEMs consisting of a [PSS/PAH]$_4$ precursor block and an in situ adsorbed (ad) block of (I) [PSS/PDAD-MAC]$_3$ and (II) [PSS/PAH]$_3$, respectively [35]. The samples were immersed in water (D$_2$O) and tempered subsequently at (I) 20–40–10 °C or (II) 20–40–20 °C. The precursor block, which was dried after preparation, shrank by $6 \pm 3\%$ on average, but the ad-blocks did not vary within experimental error. Although no, or only minor, thickness change was observed, the SLD of the first PE system showed a significant decay of the SLD which was attributed to a loss of heavy water during heating: System (I) by 12% on average and system (II) by 4% on average. In other words, especially the PDADMAC-containing system (I) was densified in a pronounced way. This means that conformational changes are the origin of the thickness and density changes in the PEM, rather than just compression of the film. In parallel the Kiessig fringes at high $q$-values became more pronounced, indicating an annealing process. An experimental report from 2003 supports the interpretation of a molecular rearrangement upon heating [68]. The Bragg peak of the superstructure of a deuterated PSS/PAH multilayer vanished upon heating to 80 °C. This indicates the destruction of the given superstructure by partial mixing of the mobilized polymeric chains upon heating.

The former observed shrinkage in a tempered water bath was also confirmed for heating PSS/PAH films in saturated D$_2$O vapor by Ahrens et al. [41].

Heating of hollow PSS/PAH capsules of several micrometers diameter leads to an extreme shrinkage of about 70% [69]. This is possibly because a hollow capsule does not experience any geometrical restriction, for example, pinning or confinement effects, so the shrinkage could be observed here in its purest form. This was confirmed by comparing experiments with hollow capsules and capsules which still contain the template (polystyrene latex, blood cell or porous silica particle) [70–72]. It was shown by small angle neutron scattering (SANS) that the freestanding multilayer capsules are thicker and contain more water than the precursor PEM still adsorbed on the particle template [72].

In general, heating provides the energy to overcome the kinetic barrier and allow rearrangement in an energetically more favored state. The driving force is the hydrophobic effect, which minimizes the water/polymer contact of the hydrophobic polymeric units [73].

In order to increase the temperature sensitivity, poly-N-isopropyl acrylamide (PNIPAM), which has a lower critical solution temperature (LCST) of 32 °C was embedded in a polyelectrolyte multilayer. It was shown that it is possible to form multilayers from PDADMAC and a diblock–copolymer which consist of a PSS block (11%) and a PNIAPM block (89%). Although the average charge of 11% is well below the charge minimum which is needed to build up a PEM (see Section 11.4.2), a strongly charged short block is sufficient for adsorption [47]. Figure 11.13a shows NR curves of a (PSS-$b$-PNIPAM/PDADMAC)$_{10}$ multilayer in D$_2$O at different temperatures.

With increasing temperature up to 60 °C the thickness decreases by about 5% [74]. Simultaneously, the oscillations at high $q$-values become more pronounced, which indicates a decrease in roughness and a kind of annealing, as already mentioned

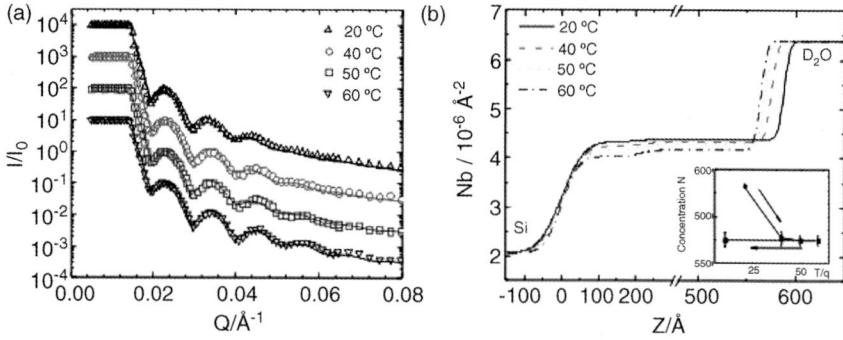

**Figure 11.13** Neutron reflectometry of a (PSS–b–PNIPAM/PDADMAC)$_{10}$ multilayer on Si/PEI against D$_2$O at different temperatures: (a) reflectivity data, solid lines are the fits by the Parratt algorithm; (b) corresponding density profiles from the data fits; Inset: thickness obtained from data fits in dependence on temperature. Taken from [74].

above. Both the shrinking and the annealing are not reversible by cooling to 10 °C. The observed effect is much less pronounced than in 3D microgels. This is thought to be due to the de-coupling of the PNIPAM layers by the PDADMAC layers in the interface bound multilayer. The structural changes found in the functionalized multilayers are irreversible on the timescale of the experiments. This is not in line with the fully reversible phase transition observed for the 3D microgels. An explanation could be the strong interdigitation between adjacent layers, which might become even stronger during the annealing process.

Of course there are other strategies to design thermosensitive coatings, as described in [74].

### 11.6.3
### Scalar Fields: Water

Water in polyelectrolyte multilayers is a very important, interesting, and complex topic. On the one hand the preparation of a PEM is water based, that is, water as the solvent at least partially governs the formation process of the multilayers. On the other hand, water is an active component inside the PEM, which could mediate transport through the multilayer and lead to shrinkage/swelling, which is an important feature for drug uptake and release. The reason for the sensitivity to water is the need of the "waterborne species" to establish an equilibrium between the solid-like film and the adjacent liquid or humidity, by regulating differences in the chemical potential or the osmotic pressure. What can NR do, to clarify the influence of water on the PEM? Of course neutron scattering is not the preferred method for short time experiments. Data on water uptake/release kinetics on a timescale of seconds or minutes is difficult to obtain. The strength of using neutron scattering is the opportunity to determine structural properties, like thickness, internal structure and roughness, and to quantify the amount of incorporated material via the detection of material-specific SLD.

Already, in the first NR experiment Schmitt et al. reported a quite high water content of about 27% inside the PEM structure (PSS/PAH) [11]. They found that the PSS unit is surrounded by about four molecules of water at ambient conditions. The ambient humidity was not controlled. This problem applies to several of the early studies, which makes it difficult to compare results of different studies. In 1998 Lösche found that six water molecules are attached to a PSS monomer unit and only one to a PAH monomer unit in fully hydrated films. Experiments with water vapor and $P_2O_5$ were performed. Under the assumption that the volume of a PSS monomer is three times larger than PAH, two times more water (by volume) was associated with PSS than with PAH [13].

The phenomenon of high amounts of adsorbed water was also found in other PEM. Plech et al. investigated very thin double or quadlayers of PEI/PSS on a glass substrate [75]. They state that every PE monomer is associated with 1.5–2 water molecules. This is 20 wt% of water (or roughly 30% of the volume). This value might be a little underestimated in this case, since the authors use a sample in ambient $H_2O$-containing atmosphere as the reference for a *dry* PEM.

### 11.6.3.1 Swelling in Water

To our knowledge the first NR study which mentioned a volume change caused by a variation in the amount of incorporated solvent molecules (water) was the work of Bijsma et al. in 1997 [76]. A reduction in thickness was traced back to a drying process, additionally the authors state a non-uniform distribution of water in the PVP/PSS multilayer,[11] two layers with different "water densities" were observed.

In 1998 Hong et al. claimed that "after a year of exposure to Jerusalem air" PSS/PAH/D-PPV[12] films swell by about 2% due to air moisture. The authors report that swelling during preparation can be significantly reduced by spinning the sample, which leads to smoother surfaces (see Section 11.3). Less rough surfaces minimize water penetration, at least on the time scale of the next dipping step, that is, 10 min [30].

Also in 1998, the swelling and water content of PEM were investigated, exploiting the unique option for the variation of the contrast of the scattering length density (SLD). Performing swelling experiments with $D_2O$ and $H_2O$ vapor Lösche et al. obtained two independent sets of data which allowed quantification of the water content in the PEM [13]. The volume share of water in a PAH/PSS multilayer system in saturated water vapor (100% RH) is $\geq 40\%$. The swelling process caused a thickness change of about 29% (for a 32 layer pair thick film). Other authors observed a water content of 20 wt% in an ultrathin (2 layer pair) PEI/PSS film. The swelling was 10% minimum [75].

In 2000, the first NR measurements of PEMs in liquid water were performed [14]. Steitz et al. showed that water content and volume change is not a material constant but depends on the preparation conditions, such as salt content, and sample history (Figure 11.14). A precursor of a PSS/PAH multilayer which was dried previously,

---

11) Poly(1-methyl-2-vinylpyridine)/poly(p-styrene sulfonate).
12) D-PPV = deuterated poly(p-phenylene vinylene).

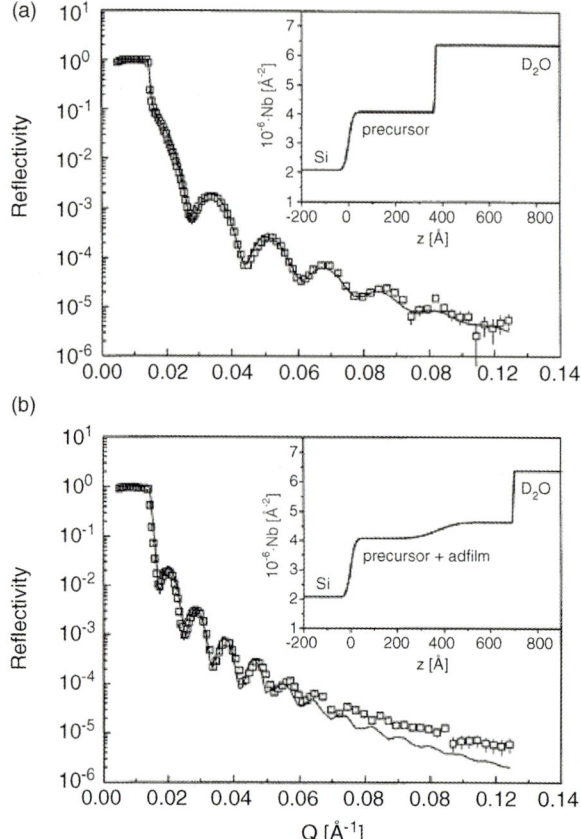

**Figure 11.14** Structure and water content: Two structural units of one film adsorbed under the same salt conditions but after a drying times show different swelling and water uptake when immersed in water. (a) Reflectivity curve and Scattering length density profile of precursor film. (b) Same of the freshly added outer zone, which has a higer water content than the zone close to the substrate [14].

contains about 42 vol% of water with exposure to liquid water and shrinks by about 10% upon drying. An (ad)-PSS/PAH multilayer which was adsorbed *in situ* and never dried after preparation contains 56% water, and shrinks by 32% upon drying.

The *dry* film, that is, in ambient conditions, has a water content of 35%, which is a value between the values found by other authors mentioned above [11, 13].

The first NR measurements with different relative humidities were carried out by Kügler *et al.* [77]. They showed a nonlinear swelling behavior of PSS/PAH multilayers with increase in vapor pressure (Figure 11.15). The maximum swelling was about 30% in saturated $H_2O$ or $D_2O$. The PSS/PAH multilayer proved very hydrophobic with a Flory–Huggins parameter, $\chi$, of 0.91. This means that despite all the charged sites of the single PE molecules, complexed PE in compounds are rather hydrophobic and tend to phase separate when exposed to high partial pressures of water vapor.

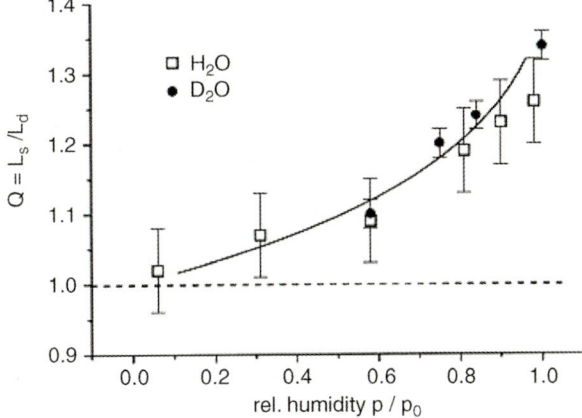

**Figure 11.15** Swelling in $H_2O$ and $D_2O$ vapors. Homogeneous swelling and water distribution in PEM, according to a one-box model. Nonlinear swelling and Flory–Huggins fit. Taken from ref. [77].

A nonlinear swelling with increasing ambient humidity was confirmed by Wong et al. [59]. In addition the authors report a distinct reduction in the water content of the PEM with increasing film thickness from 50% over 30 to 23% for 6, 12, or 26 layer pairs, respectively. The water content was determined via the volume change (*volume model* see below, Section 11.6.3). This would mean that a stack of layer pairs close to the film/air interface, which the authors assume to be of minor density, would attract and store more water than the layers close to the substrate. A thin film is in total loosely packed and becomes more dense by adsorption of polyelectrolyte layers, where the outer layers remain loosely packed. Here, the idea of a structure-based, multi-zone model of swelling comes into play.

A higher water content in the outer film zone was confirmed by Tanchak et al., who came to the conclusion of a continuous but inhomogeneous water distribution in the PEM in contact with liquid water ($D_2O$ or *zero-scattering water*)[13] and saturated vapor. The amount of water, incorporated in a PEM-structure made of the weak polyelectrolytes PAA/PAH decays almost linearly from the outer PEM/water interface, ending in a drastic reduction and a depletion zone close to the solid/PEM interface (Figure 11.16): The PEM exhibits 35% swelling in water, the outer layers contain 54% water, the water content in the core of the film lies between 30 and 45%, but the inner, interfacial zone stores approximately 18% water only. Tanchak et al. present an explanation which is based mainly on kinetic arguments: Clusters of water are formed which effectively depress transport processes towards the film close to the substrate by blocking *microchannels*. Thus the region close to the PEM/water interface stores distinctly more water than the buried regions of the film. The concept

---

13) A $H_2O/D_2O$ mixture of 92:8 has a scattering length density of zero. Thus the water does not contribute to the SLD-profile. A change in SLD of the PEM film can be directly attributed to a network expansion.

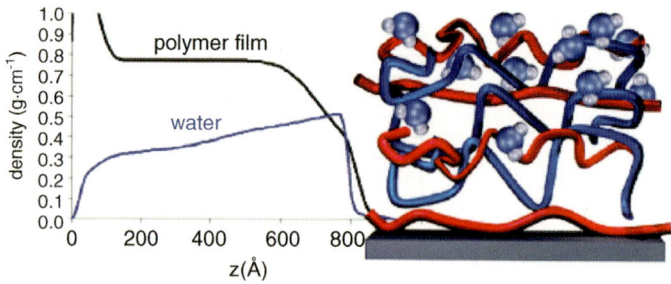

**Figure 11.16** Water distribution in PEM. Widely continuous and linear decay of the amount of water inside PEM with increasing distance from the PEM/water interface. Taken from ref. [50].

of a water distribution varying with distance to the interfaces is opposite to the ideas of almost constant water distribution [11, 13, 76]. Later, Tanchak used the same experimental method to determine the distribution of ions in a PEM of weak PEs [51].

Köhler *et al.* confirmed the nonlinear swelling behavior reported from Kugler *et. al.* [77] also for strong polyelectrolyte systems (PSS/dPDADMAC) prepared by the spraying technique [20]. They describe the nonlinear swelling as a two-stage mechanism: hydration of the PEM, characterized by a small increase in water content and volume, followed by a significantly higher volume increase with respect to the amount of the incorporated water (Figure 11.17). Besides this they identified that the volume increase and water content are not identical or proportionally related. They explain the two-stage swelling by a phase separation, which is also indicated by

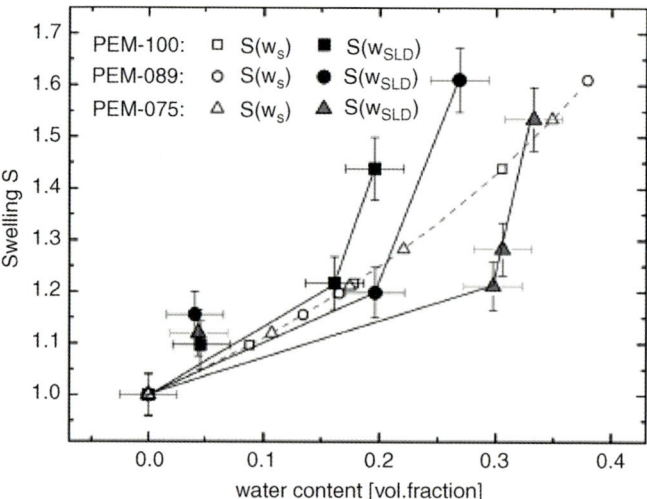

**Figure 11.17** Swelling and water content. The graph shows the mismatch between the water content either calculated via the *displacement model* (dashed line, empty symbols) or with the obtained SLDs *density model* (full symbols). Taken from ref. [20].

the high Flory–Huggins parameter, $\chi = 0.85$. A similar value to that found for dipped PSS/PAH [77]. The maximum swelling of the sprayed PSS/dPDADMAC[14] is 44% with a water content of 20%. Studying the conductivity in PEM Cramer *et al.* found an exponential increase in ion mobility upon increasing the surrounding relative humidity [78], which might explain the nonlinear swelling behavior.

Except for Figure 11.17 [20] all results referred to in this section obtained using up to now are dipping technique. However, as asserted in Section 11.3, other preparation techniques yield different inner structures which, in turn, could result in different or deviating mechanisms of water incorporation and volume change.

Kolasinska *et al.* observed that dipped samples (6.5, 7 layer pairs from 0.15 M NaCl solutions) swell about 9% more than sprayed when immersed in liquid $D_2O$ [19]. This finding is supported by the data of Dodoo and Köhler who found, for sprayed and dipped PSS/PDADMAC multilayers, both prepared in 0.5 M sodium chloride solution, 44 and 55% swelling, respectively [20, 22]. A summary of the data is shown in Figure 11.18a. An explanation could be that in the case of spraying the chains are "frozen" in a rather flat conformation due to the shear force and drainage parallel to the surface. In a dipped PEM the orientation of polymer segments is a bit more isotropically oriented due to diffusion into the adjacent layers. This allows a stronger stretching of the chains, that is, swelling, perpendicular to the surface than after spraying.

Some studies report a difference in swelling between exposure to vapor and to liquid water [19, 41, 79]. Other studies observed similar swelling under both

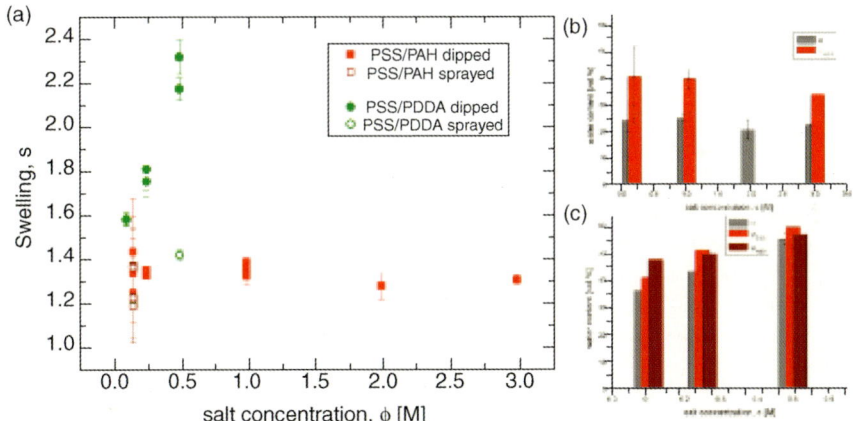

**Figure 11.18** Swelling and water content in the PEM according to three different model calculations. (a) Swelling of PSS/PAH and PSS/PDADMAC systems in water or saturated water vapor ($H_2O$ or $D_2O$) prepared by the dipping or spraying techniques with different salt (NaCl) concentrations. (b) Water content in dipped PSS/PAH-PEM according to the *volume model*, and the *percolation model*. (c) Water content in dipped PSS/PDADMAC-PEM according to the same model calculations as in (b). Error bars of (a) and (b) are the standard deviations. All data are included in Table 11.2.

14) PEM-100 with 100% charge density.

conditions [66], which is expected, since the chemical potential of liquid water and saturated vapor is the same. It is important to point out that it is quite difficult to reach 100% r.h., and especially at high r.h. (90–100%) the sensitivity of swelling to slight changes in r.h. is very high. In addition, it is very likely that water will condense at the sample surface and form a thin liquid film at r.h. > 95%.

#### 11.6.3.2 Models for Determination of the Water Content

There have been many swelling studies, carried out with different methods. Beside NR, often ellipsometry or scanning force microscopy (SFM) are used. This subsection gives an overview of the different models used to calculate the water content. The different results for NR are summarized in Table 11.2. The swelling, $S$, is defined as the ratio between the thickness of the dry, $d_0$, and the swollen PEM, $d_s$:

$$S = \frac{d_s}{d_0}$$

Three models are used in the literature for the estimation of the water content.

1) The simplest model (*volume model*) refers to the change in volume. The amount of incorporated water, $\phi$, is calculated by:

$$\phi = \frac{d_s - d_0}{d_s} = 1 - S^{-1} \tag{11.1}$$

This model is the most popular model, since it can be applied to data obtained by many different methods. In contrast, both the following models can only be applied to NR.

2) The second model (*density model*) takes into account that the change in the scattering length density of the polymer originates from the incorporation of an additional scatterer, water in this case, with its own scattering length density and the dilatation of the polymeric network due to swelling, that is inversely proportional to the swelling, $S$. The model calculates the fraction of water, $w$, inside a swollen PEM, from the thickness of the dry and the swollen film, $d_0$ and $d_s$, and the SLD of the solvent, $\varrho_{water}$ ($H_2O$ or $D_2O$), and the dry and the swollen films, $\varrho_0$ and $\varrho_s$ [20]:

$$w = \frac{\varrho_s - \varrho_0/S}{\varrho_{water} - \varrho_0/S} \tag{11.2}$$

3) The third model (*percolation model*) assumes that there are vacuum *voids* in the dry PEM. During exposure to water, the volume fraction of the voids is filled with water, which contributes to the change in scattering length but not to the change in swelling. This water fraction is called *void water*, $\phi_{void}$. The larger amount of water contributes to both swelling and change in scattering length density and is called *swelling water*, $\phi_{swell}$, which corresponds to the relative change in volume $\phi$ in Equation 11.1. The total amount of water, $\psi$, is then the sum of both water fractions [22, 79]:

**Table 11.2** Calculation of water content of a fully swollen PEM according to the *volume* and the *percolation model*. Conditions: PEM spraying in prepared (S), swollen in water (l) or saturated vapour (v), salt concentration during preparation. Swelling: thickness and scattering length density of dry film $d_0$, $\varrho_0$, thickness and SLD of swollen film $d_{100}$, $\varrho_{100}$. (*value corrected according personal communication).

| Author, Ref. | System | Conditions | $S = d_{100}/d_0$ | $\varrho_0$ [$10^{-6}$ Å$^{-2}$] | $\varrho_{100}$ [$10^{-6}$ Å$^{-2}$] | $\phi = \frac{d_{100}-d_0}{d_{100}}$ [%/100] | $\psi D_2O/\psi H_2O$ [%/100] |
|---|---|---|---|---|---|---|---|
| Lösche [13] | ABS(PSS/PAH)$_{32}$ | D$_2$Ol | 1.29 ± 0.01 | — | — | 0.22 ± 0.01 | — |
| Kugler [77] | PEI(PSS/PAH)$_4$ | H$_2$Ov | 1.26 ± 0.06 | 1.36 ± 0.05 | 0.95 ± 0.05 | 0.21 ± 0.04 | —/0.23 |
| Kugler [77] | PEI(PSS/PAH)$_{10}$ | D$_2$Ov | 1.29 ± 0.02 | 1.55 ± 0.05 | 3.37 ± 0.05 | 0.23 ± 0.01 | 0.34/— |
| Wong [59] | PEI(PSS/PAH)$_8$PSS | H$_2$Ov | 1.31 | — | — | 0.237 | — |
| Wong [59] | PEI(PSS/PAH)$_9$ | H$_2$Ov | 1.335 | — | — | 0.251 | — |
| Ivanova [42] | PEI(PSS/PAH)$_8$PSS | D$_2$Ov | 1.35 | 1.04 | 3.1 | 0.26 | 0.37/— |
| Ivanova [42] | PEI(dPSS/PAH)$_8$dPSS | D$_2$Ov | 1.31 | 3.5 | 5.23 | 0.24 | 0.40/— |
| Kolasinska [19] | PEI(PSS/PAH)$_6$ | D$_2$Ov | 1.33 ± 0.25 | 2.5 ± 0.1 | 3.8 ± 0.2 | 0.25 ± 0.14 | 0.30/— |
| Kolasinska [19] | PEI(PSS/PAH)$_6$ | D$_2$Ol | 1.42 ± 0.25 | 2.5 ± 0.1 | 5.1 ± 0.2 | 0.29 ± 0.12 | 0.52/— |
| Kolasinska [19] | PEI(PSS/PAH)$_6$ | S D$_2$Ov | 1.17 ± 0.17 | — | 3.6 ± 0.2 | 0.15 ± 0.12 | — |
| Kolasinska [19] | PEI(PSS/PAH)$_6$ | S D$_2$Ol | 1.36 ± 0.18 | — | 5.1 ± 0.1 | 0.26 ± 0.10 | — |
| Kolasinska [19] | PEI(PSS/PAH)$_6$PSS | D$_2$Ov | 1.23 ± 0.14 | 2.6 ± 0.1 | 4.2 ± 0.1 | 0.19 ± 0.09 | — |
| Kolasinska [19] | PEI(PSS/PAH)$_6$PSS | D$_2$Ol | 1.32 ± 0.16 | 2.6 ± 0.1 | 5.0 ± 0.1 | 0.24 ± 0.09 | 0.32/— |
| Kolasinska [19] | PEI(PSS/PAH)$_6$PSS | S D$_2$Ov | 1.21 ± 0.19 | — | 4.3 ± 0.3 | 0.17 ± 0.13 | 0.48— |
| Kolasinska [19] | PEI(PSS/PAH)$_6$PSS | S D$_2$Ol | 1.35 ± 0.17 | — | 5.1 ± 0.3 | 0.26 ± 0.10 | — |
| Köhler [20] | PEI(PSS/PDDA)$_6$ | S H$_2$Ov | 1.41 ± 0.03 | 2.50 ± 0.02 | 1.29 ± 0.02 | 0.29 ± 0.01 | —/0.88 |
| Delajon [79] | PEI(PSS/PAH)$_6$ | D$_2$Ov | 1.31 ± 0.04 | 3.33 ± 0.17 | 4.95 ± 0.18 | 0.24 ± 0.02 | 0.38/— |
| Delajon [79] | PEI(PSS/PAH)$_6$ | H$_2$Ov | 1.33 ± 0.05 | 3.33 ± 0.17 | 2.30 ± 0.17 | 0.25 ± 0.03 | —/0.37 |
| Delajon [79] | PEI(PSS/PAH)$_6$PSS | D$_2$Ov | 1.37 ± 0.02 | 3.4 | 5.33 ± 0.16 | 0.27 ± 0.01 | 0.45/— |
| Delajon [79] | PEI(PSS/PAH)$_6$PSS | H$_2$Ov | 1.33 ± 0.03 | 3.4 | 2.25 ± 0.07 | 0.25 ± 0.02 | —/0.56 |
| Dodoo [22] | PEI(PSS/PDDA)$_6$ | H$_2$Ov 0.1 M | 1.57 ± 0.03 | 1.33 ± 0.02 | 0.58 ± 0.01 | 0.36 ± 0.01 | —/0.48 |
| Dodoo [22] | PEI(PSS/PDDA)$_6$ | H$_2$Ov 0.25 M | 1.74 ± 0.07 | 1.34 ± 0.03 | 0.49 ± 0.02 | 0.43 ± 0.02 | —/0.5 |
| Dodoo [22] | PEI(PSS/PDDA)$_6$ | H$_2$Ov 0.5 M | 2.16 ± 0.05 | 0.74 ± 0.04 | 0.02 ± 0.01* | 0.54 ± 0.01 | —/0.57 |
| Dodoo [22] | PEI(PSS/PDDA)$_6$ | D$_2$Ov 0.1 M | 1.57 ± 0.03 | 1.33 ± 0.02 | 3.47 ± 0.01 | 0.36 ± 0.01 | 0.41/— |
| Dodoo [22] | PEI(PSS/PDDA)$_6$ | H$_2$Ov 0.25 M | 1.80 ± 0.10 | 1.34 ± 0.03 | 4.00 ± 0.02 | 0.44 ± 0.03 | 0.51/— |
| Dodoo [22] | PEI(PSS/PDDA)$_6$ | H$_2$Ov 0.5 M | 2.31 ± 0.08 | 0.74 ± 0.04 | 4.11 ± 0.01 | 0.57 ± 0.01 | 0.60— |

$$\psi = \phi_{\text{void}} + \phi$$
$$= (1-x)(1-\phi) + \phi \quad \text{with} \quad x = \frac{\varrho_0}{\varrho_{\text{water}}} - \frac{\varrho_s - \phi\,\varrho_{\text{water}}}{(1-\phi)\varrho_{\text{water}}} + 1 \quad (11.3)$$

In the following, for the sake of clarity, only analysis by the first and the third models is compared. Figure 11.18a shows a very different swelling behavior for the PSS/PAH and the PSS/PDADMAC system: The PSS/PAH is almost insensitive to the change in salt concentration. In contrast, PSS/PDADMAC multilayers show a pronounced swelling with increasing salt concentration used during preparation condition (dipping/spraying) shows only an effect for the PSS/PDADMAC PEM. Within experimental error the sprayed PSS/PAH sample swells to the same extent as the dipped ones.

The similarity of the swelling of PSS/PAH PEMs applies also to the water content Figure 11.18b. The percentage of incorporated water does not differ with the salt concentration of the preparation, but the amount of water in the PSS/PDADMAC system increases with increasing salt concentration. The effect of ionic strength on the water content and the different swelling behavior of PSS/PAH and PSS/PDADAMAC multilayers will be discussed in Section 11.6.3. Under the same conditions (0.1 M, dipped) PSS/PAH PEM contains about 40% water including 15% of void water and PSS/PDADMAC has about 45% water including about 10% of void water.

The percolation model (3) gives similar amounts of incorporated water for heavy and light water, that is, there is no isotopic effect observed for PSS/PDADMAC [22] irrespective of the ionic strength and type of salt used for preparation. Differences in swelling in $D_2O$ and $H_2O$, reported in other studies, are minor and nonsystematic. Kügler found a lower swelling in $H_2O$ vapor than in $D_2O$, 26% versus 29%, in liquid $D_2O$, it was 27% [77]. For 6 layer pair (PAH/dPSS) 31% swelling in $D_2O$ vapor and 33% swelling in $H_2O$ was observed [79]; for 6.5 layer pairs it was 37% in $D_2O$ and 33% in $H_2O$ vapor.

From time to time an isotopic effect is pointed out in the literature. As shown above, the deviations are either within experimental error or dominated by the odd/even effect (see Section 11.6.3.4) which also gives a difference of several % in the amount of swelling.

### 11.6.3.3 Effect of Polymer Charge Density and Ions During Preparation on Swelling in Water

*Polymer Charge Density* The water content of P(DADMAC–stat–NMVA) multilayers increases with decreasing charge density (ChD). For sprayed PEMs in saturated water vapor the water content is 0.2 (100% ChD), 0.27 (89% ChD), and 0.33 (75% ChD) (Figure 11.17). In addition, it was found that the water content and outer osmotic pressure do not depend linearly on each other [20]. This is explained by a reduction in density of the complexation sites within the multilayer which leads to increasing mobility of the chains [44]. The multilayer can be considered as a sponge, by decreasing the density of complexation sites larger pores are created, and the multilayer can swell in a more pronounced way.

*Ionic Strength* Figure 11.18a shows that the water uptake of PSS/PDADMAC multilayers increases with increasing salt concentration during preparation [22]. This can be explained by a reduction in complexation sites related to a transition form intrinsic to more extrinsic charge compensation (see Section 11.5.1), which leads to higher chain mobility [44]. The same study shows that the diffusion coefficient in PSS/PAH multilayers is one to two orders of magnitude lower than for PSS/PADMAC multilayers, which might explain the lower water content and its insensitivity to the salt concentration during preparation.

The amount of *hydration* water, that is, water which does not leave the PEM even at 0% r.h., depends on the preparation conditions, among others the concentration and the kind of the salt during preparation, for 1 M salt solution more bound water is found for NaCl than for KCl, 3.6 or 2.3 water molecules per PAH/PSS complex [58]. Other studies showed that there is no hydration water left in the multilayers [40]. The contradiction might be explained by the sample treatment. In [58] 0% r.h. is obtained by exposure to $N_2$ and in [40] by creating a vacuum ($10^{-6}$–$10^{-5}$ bar).

In a system of weak PE (PAA/PAH) increase in the pD significantly increases the amount of water in the film, the same is true for the decrease in pD from pD = 6.3 [51].

*Type of Ion* The effect of ions during preparation on the water content of PSS/PADAMC multilayers was studied by Dodoo et al. [22]. The measurements were carried out in vacuum ($10^{-6}$–$10^{-5}$ bar) after exposure to liquid $D_2O$ and $H_2O$ and in liquid $D_2O$ and $H_2O$. The authors describe two types of incorporated water, water that does not contribute to the swelling of the PEM, "void water", and water which contributes or causes volume change, "swelling water" (see Section 11.6.3). Part of the Hofmeister-series was tested ($F^-$, $Cl^-$, $Br^-$). The results are summarized in Figure 11.19. The void water decreases with increasing salt concentration in the order $F^- > Cl^- > Br^-$. Thus the density of the dry multilayer increases in the

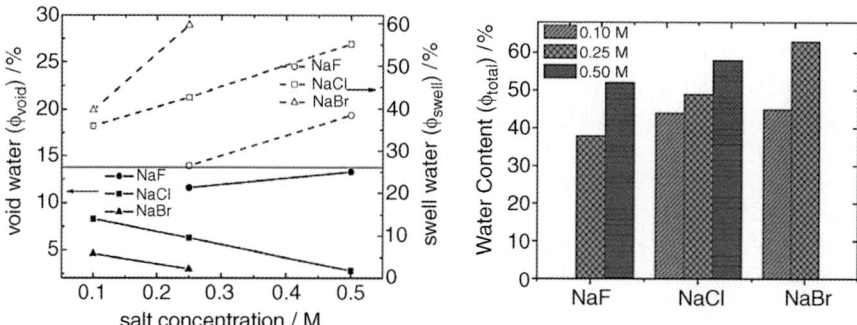

**Figure 11.19** Water content. The model of "void and swell water" (percolation model) delivers the total amount of incorporated water and differentiates between water which does not contribute to swelling and water which does. PEM of 6 BL PSS/PDADMAC prepared in different salt solutions of varying ionic strength. Taken from ref. [22].

order $F^- < Cl^- < Br^-$ and with increasing salt concentration. This means that a stronger coiling of the chains due to either stronger anion/PDADMAC interaction or increasing ionic strength leads to a more compact structure. The decreasing density of complex sites makes it easier for the system to rearrange and to adapt, leading to higher density. On the other hand the reduction of complex sites allows the system to swell more in water, that is, there is a higher amount of swelling water (upper part of Figure 11.19a). The total amount of water, that is, the sum of void water and swelling water increases with increasing ionic strength and anion size (Figure 11.19b).

Guzman et al. also investigated PSS/PDADMAC multilayers. The water content calculated via the volume change (without a correction for the void water) was found to be dependent on the salt concentration upon preparation: 0.3 (0.1 M NaCl, 24 BL), 0.6 (0.5 M NaCl, 24 BL) [45]. These values are close to the data for the swelling water presented by Doodo et al. [22].

#### 11.6.3.4 Odd-Even Effect

At high r.h. (98%) PSS/PAH multilayers with PSS as the terminated layer are thicker than those with PAH in the outermost layer [59], which leads to a kind of zig–zag increase in thickness with increasing number of deposited layers. In contrast, the respective dry PEMs (3% r.h.) show a monotonic increase in thickness, which means that no material is lost during the adsorption of PAH. The results indicate that water has to be pressed out of the multilayer when PAH is adsorbed onto the PSS surface of the multilayer and that again more water diffuses into the multilayer on adsorption of the next PSS layer. This behavior, often referred to as *odd-even effect* was, to our knowledge, first reported in 1997 by Hsieh et al. [80]. Schwarz and Schönhoff found an increased mobility and/or water content in the outer shell of PEM-coated silica particles by using NMR [81, 82]. Thus, in a PEM of two polyelectrolyte species the surface properties change for the even or odd layer. Later this issue was addressed by neutron reflectometry [83, 84]. Solvent surface density ($D_2O$) was shown to be about 40% higher for PSS than for PAH as the terminated layer [83] and studies of the amount of swelling support this trend [84]. Carriere et al. studied the odd-even effect in dependence on the concentration of the preparative salt [85] and the number of deposited layers. For 1 M NaCl (13–16 BL) the effect is significant: thickness, surface roughness and scattering length density of the film oscillate with layer number. The increase/decrease of solvent fraction upon the subsequent adsorption step was roughly 30%. In contrast, at 0.25 M NaCl (27–30 BL) almost no odd-even effect was observed. An overview of the results of different studies is given in Figure 11.20.

An explanation of the effect could be that the outer layer and the inner/buried interface form a capacitor with water as the dielectric fluid. It is worth mentioning that the contact angle also shows an odd-even effect, where PSS-terminated layers show a lower contact angle, that is, more hydrophilic behavior than PAH-capped PEMs. This would lead to a stronger attraction of water and stronger swelling, as observed in the experiments.

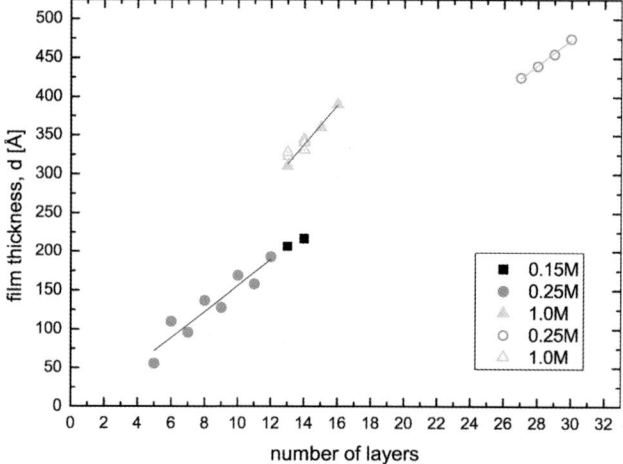

**Figure 11.20** Odd-even effect. PSS/PAH films of varying numbers of layers swell differently in water (or 100% water vapor) when the terminating layer is either PSS or PAH. The effect seems to cancel out with increasing number of layers. Data from [19, 59, 79, 85].

### 11.6.4
### Directed External Fields: Mechanical and Electrical Fields

Beside the scalar fields, vector fields, such as mechanical load or electric fields, may also alter the structure and the properties of the PEM. This topic was only recently addressed by neutron reflectometry. Thus, the following section is rather a kind of outlook.

#### 11.6.4.1 Mechanical Stress

In view of future applications, the mechanical properties of PEM, for example, stiffness, resistance, or fatigue, are of high importance. But for fundamental science the interplay of internal interactions and external macroscopic behavior is also an interesting topic. The first studies of mechanics were done on microcapsules [86, 87]. For planar films the topic was first addressed by a group from Strasbourg [88]. Mertz and coworkers used stretched PEM films on silicone rubber and observed cracks and film rupture, but also softening of the films under the influence of water as plasticizer. The impact of water on the elasticity, for example, the Young's modulus, of different PEM was studied in detail by Nolte *et al.* [89, 90].

With a newly adapted reflectometry technique Früh *et al.* were able to detect the response of the PEM on the molecular scale [91]. Sprayed samples of PSS/PDAD-MAC (0.25 M NaCl) were deformed, the change in thickness was observed with NR [23]. They found a transition between elastic and plastic behavior for an elongation, $\varepsilon$, of about 0.2%. This quite low value equals the limit of elasticity of metals such as copper or aluminum. After elongation, PEM showed a sudden increase in thickness (within seconds or minutes), then later the thickness decayed significantly. This was observed with XRR measurement within 1–2 h of the stress being applied (Figure 11.21). This might originate in reorientation and relaxation

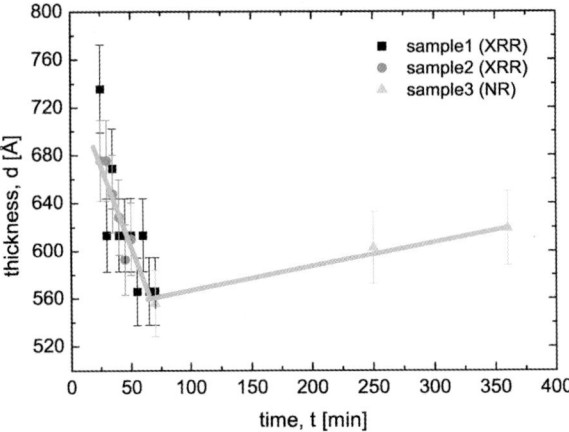

**Figure 11.21** Evolution of film thickness with time after applying mechanical load. The measurements were performed at 30% relative humidity. Taken from ref. [91].

processes in the PEM after the initial mechanical disturbance. NR measurements which probe the sample on a longer timescale show a constant film thickness within experimental error, or only a very moderate increase in thickness. Ellipsometric measurements gave a hint that a slight increase in film thickness could be caused by water uptake and swelling in the ambient atmosphere (30% r.h.). Samples prepared with a superstructure of partially deuterated layers showed irreversible degradation of the Bragg peak during elongation. This occured even when the deformation was below the transition from elastic to plastic behavior.

#### 11.6.4.2 Electric Field

The first attempts to exploit the electrochemical potential manipulation of PEM were made in the late 1990s with multilayers produced from PSS, PBV[15] and PAH-FC, GOx,[16] respectively [92, 93].

Thin PEM films containing reversibly oxidizable/reducible blocks can swell and deswell via exchange of ions and/or solvent between the polymer film and the adjacent liquid phase. If they are electrochemically active a continuous stimulus can be applied by changing the electric field [94, 95]. Ferrocyanides can be incorporated after build-up of PEM. Varying the electric field leads to swelling of the multilayer, a variation in the Young's modulus, and the permeability for ions, small molecules, and the solvent. The electrochemical swelling originates from a change in the osmotic pressure in the film upon change in the redox state of the active sites. Transport of counterions is necessary for compensation of the electrostatic and osmotic imbalance [96].

Zahn *et al.* investigated PGA/PAH[17] multilayers with incorporated ferrocyanides and monitored the electrochemical swelling with neutron reflectometry [97].

---

15) Poly(butanyl viologen).
16) PAH-ferrocene, glucose oxidase.
17) Poly(L-glutamic acid)/PAH.

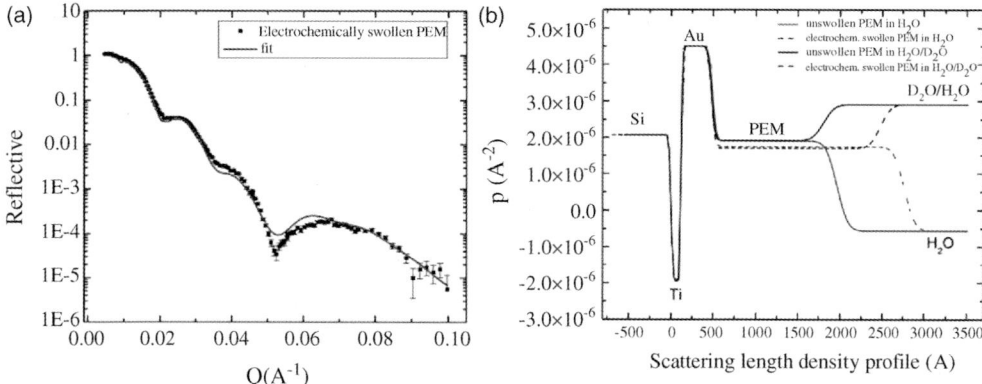

**Figure 11.22** Electrochemical swelling. Application of an electrical potential causes oxidation of incorporated ferro cyanide (FC). The osmotic pressure is equilibrated by water uptake and swelling. (a) Reflectivity curve of electrochemical swollen PEM. (b) Comparision of Scattering length density profiles of electrochemical swollen and water swollen PEMs. Taken from [96].

The PEM is un-swollen as long as no potential is applied. A low positive potential oxidizes the FC and yields to a subsequent swelling of the PEM structure. A volume increase of 4% only was observed at low ionic strength (0.1 M KCl), but at high ionic strength (0.3 M KCl) the film swelled about 40%.

Experiments in water mixtures of different scattering length density, that is, by varying the $H_2O/D_2O$ ratio, lead to the conclusion that the increase in volume upon change in the electrical potential is mainly caused by water uptake and not by incorporation of counterions (Figure 11.22). One can expect that neutron scattering techniques will contribute to a deeper understanding of the complex electrochemical swelling behavior in future.

## 11.7
## PEM as a Structural Unit

PEMs are widely used as supports, templates or matrices for more complex structures. These structures can also be characterized with neutrons, using the same two main features as for simple PEMs: selective deuteration to refine the structure determination, and water or hydration measurements.

### 11.7.1
### Cushion

A widespread use of PEM is as a substrate, that is, as a soft cushion on which to deposit something. The deposited objects are often of biological interest, like lipid layers, peptides or bacteria. The PEMs allow a better investigation of these thin layers and soften the surface, which is important for biological systems.

### 11.7.1.1 Cell-Membrane Mimic Systems

Lipid-bilayer membranes supported on solid substrates are often used as cell-surface models that connect biological and artificial materials. They can be placed either directly on solids, or on ultrathin polymer supports that mimic the generic role of the extracellular matrix [97]: we will focus on this last set-up. Neutron reflectivity is useful to determine the specific thickness and shape of the deposited layer pair: moreover, a single layer pair being too thin to be investigated by NR, thus, the PEM has also an essential role as a spacer. For example, it was found that, depending on the charge of the terminating layer, phospholipids (dimyristoylphosphatidylcholin DMPC) could form a smooth layer pair (PSS-terminated) or only roughly adsorb on a few spots (PAH) [98, 99]. This can be explained by the less flat surface of PSS-terminated PEMs, more favorable to adsorption, by the higher hydrophilicity of PSS, favoring the attraction of the hydrated polar headgroups of the phospholipids, or simply by electrostatic interactions, the DMPC being zwitterionic but with a slightly positive headgroup which could be attracted to the negatively-charged PSS. The adsorption of phospholipids (1,2-dipalmitoyl-sn-glycero-3-phosphocholine, DPPC) on the same PEM cushion (PAH/PSS) can lead to a decrease in water content of this supporting cushion, because of reorganization of the system and water diffusing into the new phospholipid layer [100]. More complex set-ups can also be formed by adding another multilayer on top of the lipidic structure, which then acts as a barrier inside the PEM [98, 101]. Reflectivity is again essential to characterize the internal lipid layer pair.

Besides lipid layer pairs, other biological macromolecules were used to mimic the cell wall. Toca-Herrera et al. [102, 103] used subunits (proteins) from bacterial cell surface layers, called S-layers, and studied the conditions for recrystallization on PEM. The recrystallization was studied by AFM, but NR in $D_2O$ was necessary to understand the process. Because of a higher roughness, PAH-terminated PEMs could not support recrystallization, the S-layers just adsorbed densely at the surface, excluding $D_2O$. On the contrary, on the flatter PSS-terminated surface, the proteins formed a smooth film at low-packing density (corresponding to crystallization) filled with water [103], as is reproduced in Figure 11.23. The influence of temperature on

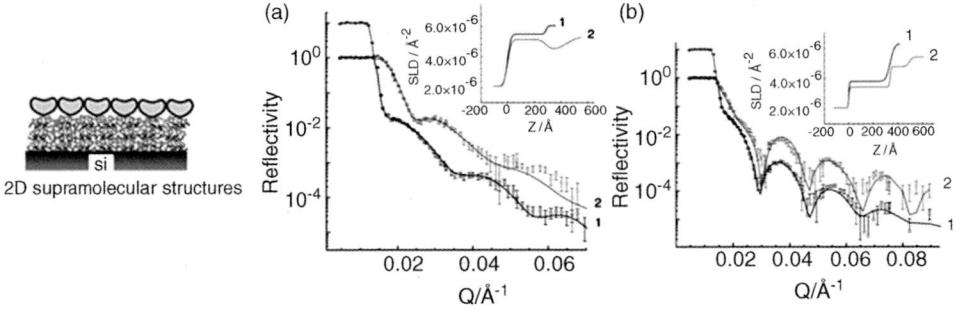

**Figure 11.23** Adsorption of S-layers on PAH-terminated (a) or PSS-terminated (b) multilayers. Taken from ref. [103].

these structures showed an irreversible loss of order, with the S-layers penetrating the PEMs above 50 °C [102], as evidenced with neutrons.

#### 11.7.1.2 Multilayers Beyond Polyelectrolytes

Lots of other complexes of interest can be formed on PEMs. Peptide and protein adsorption on multilayers was studied by NR. This allows the determination of the thickness of the adsorbed peptide layer [104], and the structure of deposited protein layers. Patches (high roughness observed) or smooth layers of proteins can be formed [105]. The deposition of micelles [106] and of a specific polymer (poly(o-methoxyaniline)) [107] has also been investigated.

### 11.7.2 Matrix

PEMs can be used as a matrix to design composite materials by, for example, dispersion of small fillers, like clay platelets or nanoparticles, in them. NR was, for example, used to quantify the salt diffusion in polyelectrolytes/montmorillonite multilayers [108]: after the construction of 6 layer pairs of PSS/PAH, a thin layer of clay is adsorbed, and six new layer pairs (this time with d-PSS) are deposited on top. The samples are then immersed in a 2 M NaCl aqueous solution. The experiments proved the efficiency of the clay layer as a barrier for ions.

Krasteva *et al.* [109, 110] worked with specific multilayers made from gold nanoparticles and dendrimers. Vapor sorption of different solvents (water, toluene, methanol) was studied by NR. Refined profiles of the solvent distribution inside the films were obtained, which are linked to their hydrophilicity: water penetrates in the whole film, whereas hydrophobic toluene only forms a thin wetting layer at the interface.

## 11.8 Conclusion and Outlook

Neutron reflectometry has the unique advantage of being able to determine the internal structure of films, besides the external parameters like thickness and roughness. By controlled deuteration the internal interfaces can be analyzed. Another advantage is that the water content can be precisely determined, even the water fraction which does not contribute to changes in thickness. This gives a powerful tool to investigate and to better understand the complex behavior of PEMs.

The internal structure shows clear hints for a stratification of the polyelectrolyte multilayers due to the layerwise preparation method, but the layers are more or less smeared out, which leads to a lower ordering of the system. The internal roughness can be considered as a diffusion length, which increases with *diffusion time* and *diffusion coefficient*. This chapter shows how these two parameters can be tuned by preparation and experimental conditions. Starting with the preparation method, dipping with longer contact time of the PEM with the polyelectrolyte solution gives a higher internal roughness than spraying, due to the longer diffusion time. Longer diffusion times lead also to larger internal roughnesses close to the substrate, which is the reason for inhomogeneous internal roughnesses across a PEM. A pronounced rearrangement of the polymer chains within the PEM gives more stable films.

The driving force for the formation of polyanion–polycation complexes is the gain in entropy due to release of counterions. With decreasing density of complex sites the diffusion coefficient of the multilayer system increases. First, the choice of the polymer type plays an important role. The stiffness, charge density and ratio between the charge densities of both polyelectrolytes are important parameters. For instance, a decrease in charge density leads to a higher mobility. In addition, the complexation sites can be tuned with respect to their number density and strength via the preparation conditions. Increasing ionic strength and decreasing ion radius of the added salt reduces the density of complexation sites due to screening of the polymer charges (extrinsic charge compensation) leading to a higher diffusion coefficient. Both parameters can affect the system in the same direction, but after preparation.

Multilayers with a high mobility swell more strongly in water than stiffer systems. The lower density of complexes in mobile systems leads to a kind of sponge-like behavior. In contrast to that, the more mobile systems are denser under dry conditions, since the more flexible chains can adjust better than stiff chains.

Recently, besides the scalar external fields mentioned above, vector fields, like mechanical stress or electric fields, have also been applied to polyelectrolytes in order to study their response. In future, these types of study should be extended. They give interesting insights with respect to applications for such as actuators or optoelectronic devices. With respect to fundamental research they might give new insights into the properties of soft matter materials. A lack of knowledge is the dynamics of the PEM. A deeper insight would probably help to better understand, for example, the uptake and release of water.

## References

1 Daillant, J. and Gibaud, A. (1999) *X-Ray and Neutron Reflectivity: Principles and Applications*, Springer.
2 Nevot, L. and Croce, P. (1980) Characterization of surfaces by grazing x-ray reflection–application to the study of polishing of some silicate glasses. *Rev. Phys. Appl.*, **15** (3), 761–780.
3 Cubitt, R. and Fragneto, G. (2002) Neutron Reflection: Principles and Examples of Applications. In *Scattering*, Pike, R. and Sabatier, P., eds., Academic Press: London, Chapter 2.8.3, pp 1198–1208.
4 Parratt, L.G. (1954) Surface studies of solids by total reflection of X-rays. *Phys. Rev.*, **95** (2), 359.
5 Soltwedel, O., Ivanova, O., Nestler, P., Müller, M., Köhler, R., and Helm, C.A. (2010) Interdiffusion in polyelectrolyte multilayers. *Macromolecules*, **43** (17), 7288–7293.
6 Chevigny, C. and Klitzing, R.v. (November 2010) Substrate influence on the swelling behaviour and internal roughness of polyelectrolytes multilayers. Experimental Report LLB, EROS-10328.

7 Decher, G., Hong, J.-D., and Schmitt, J. (1992) Build-up of ultrathin multilayer films composed of alternating layers of anionic and cationic polyelectrolytes on charged surfaces. *Thin Solid Films*, **210/211**, 831–835.

8 Decher, G. (1997) Fuzzy nanoassemblies: Toward layered polymeric multicomposites. *Science*, **277** (5330), 1232–1237.

9 Izquierdo, A., Ono, S.S., Voegel, J.-C., Schaaf, P., and Decher, G. (2005) Dipping versus spraying: Exploring the deposition conditions for speeding up layer-by-layer assembly. *Langmuir*, **21** (16), 7558–7567.

10 Cho, J., Char, K., Hong, J.D., and Lee, K.B. (2001) Fabrication of highly ordered multilayer films using a spin self-assembly method. *Adv. Mater.*, **13** (14), 1076.

11 Schmitt, J., Grunewald, T., Decher, G., Pershan, P.S., Kjaer, K., and Losche, M. (1993) Internal structure of layer-by-layer adsorbed polyelectrolyte films - a neutron and x-ray reflectivity study. *Macromolecules*, **26** (25), 7058–7063.

12 Korneev, D., Lvov, Y., Decher, G., Schmitt, J., and Yaradaikin, S. (1995) Neutron reflectivity analysis of self-assembled film superlattices with alternate layers of deuterated and hydrogenated polysterenesulfonate and polyallylamine. *Physica B*, **213**, 954–956; Yamada Conference XXXXI/International Conference on Neutron Scattering (ICNS 94), Sendai, Japan, OCT 11-14, 1994.

13 Lösche, M., Schmitt, J., Decher, G., Bouwman, W.G., and Kjaer, K. (1998) Detailed structure of molecularly thin polyelectrolyte multilayer films on solid substrates as revealed by neutron reflectometry. *Macromolecules*, **31** (25), 8893–8906.

14 Steitz, R., Leiner, V., Siebrecht, R., and Klitzing, R.v. (2000) Influence of the ionic strength on the structure of polyelectrolyte films at the solid/liquid interface. *Colloid. Surf. A*, **163** (1), 63–70.

15 von Klitzing, R. and Steitz, R. (2003) Interdigitation between polyelectrolyte layers in multilayer systems at the solid/liquid and at the solid/air interface. Technical report, ILL Institute Laue Langevine, Grenoble, France.

16 Jomaa, H.W. and Schlenoff, J.B. (2005) Salt-induced polyelectrolyte interdiffusion in multilayered films: A neutron reflectivity study. *Macromolecules*, **38** (20), 8473–8480.

17 Krogman, K.C., Zacharia, N.S., Schroeder, S., and Hammond, P.T. (2007) Automated process for improved uniformity and versatility of layer-by-layer deposition. *Langmuir*, **23** (6), 3137–3141.

18 Felix, O., Zheng, Z., Cousin, F., and Decher, G. (2009) Are sprayed lbl-films stratified? a first assessment of the nanostructure of spray-assembled multilayers by neutron reflectometry. *C. R. Chim.*, **12** (1–2), 225–234.

19 Kolasinska, M., Krastev, R., Gutberlet, T., and Warszynski, P. (2009) Layer-by-layer deposition of polyelectrolytes. dipping versus spraying. *Langmuir*, **25** (2), 1224–1232.

20 Köhler, R., Dönch, I., Ott, P., Laschewsky, A., Fery, A., and Krastev, R. (2009) Neutron reflectometry study of swelling of polyelectrolyte multilayers in water vapors: Influence of charge density of the polycation. *Langmuir*, **25** (19), 11576–11585. PMID: 19788217.

21 Kharlampieva, E., Kozlovskaya, V., Chan, J., Ankner, J.F., and Tsukruk, V.V. (2009) Spin-assisted layer-by-layer assembly: Variation of stratification as studied with neutron reflectivity. *Langmuir*, **25** (24), 14017–14024.

22 Dodoo, S., Steitz, R., Laschewsky, A., and von Klitzing, R. (2011) Effect of ionic strength and type of ions on the structure of water swollen polyelectrolyte multilayers. *Phys. Chem. Chem. Phys.* doi: 10.1039/C0CP01357A

23 Früh, J., Rühm, A., Möhwald, H., Krastev, R., and Köhler, R. (2011) Neutron and X-ray reflectometry on curved substrates (in prep.).

24 Schlenoff, J.B., Dubas, S.T., and Farhat, T. (2000) Sprayed polyelectrolyte multilayers. *Langmuir*, **16** (26), 9968–9969.

25 Lefort, M., Boulmedais, F., Jierry, L., Gonthier, E., Voegel, J.C., Hemmerle, J., Lavalle, P., Ponche, A., and Schaaf, P. (2011) Simultaneous spray coating of interacting species: General rules governing the poly (styrene sulfonate)/poly(allylamine) system. *Langmuir*, **27**, 4653–4660.

26 Kellogg, G.J., Mayes, A.M., Stockton, W.B., Ferreira, M., Rubner, M.F., and Satija, S.K. (1996) Neutron reflectivity investigations of self-assembled conjugated polyion multilayers. *Langmuir*, **12**, 5109–5113.

27 Steitz, R., Jaeger, W., and Klitzing, R.v. (2001) Influence of charge density and ionic strength on the multilayer formation of strong polyelectrolytes. *Langmuir*, **17** (15), 4471–4474.

28 Voigt, U., Jaeger, W., Findenegg, G.H., and Klitzing, R.v. (2003) Charge effects on the formation of multilayers containing strong polyelectrolytes. *J. Phys. Chem. B*, **107**, 5273–5280.

29 Dönch, I. (2008) Mechanische Eigenschaften von Polyelektrolyt-Multilagen bei verschiedenen Ladungsdichten und Hydratationszuständen. PhD thesis, Universität Potsdam, Deutschland.

30 Hong, H.P., Steitz, R., Kirstein, S., and Davidov, D. (1998) Superlattice structures in poly(phenylenevinylene)-based self-assembled films. *Adv. Mater.*, **10** (14), 1104.

31 Chiarelli, P.A., Johal, M.S., Casson, J.L., Roberts, J.B., Robinson, J.M., and Wang, H.-L. (2001) Controlled fabrication of polyelectrolyte multilayer thin films using spin-assembly. *Adv. Mater.*, **13** (15), 1167–1171.

32 Char, K., Kim, S., Cho, J., Sohn, H., and Jang, H. (2002) Characteristics and micropatterning of spin self-assembled ultrathin multilayers. *Int. J. Nanosci.*, **1** (5–6), 375–381.

33 Cho, J. and Char, K. (2004) Effect of layer integrity of spin self-assembled multilayer films on surface wettability. *Langmuir*, **20** (10), 4011–4016. PMID: 15969392.

34 Bruening, M. and Dotzauer, D. (2009) Polymer films: Just spray it. *Nature Mater.*, **8**, 449–450.

35 Steitz, R., Leiner, V., Tauer, K., Khrenov, V., and Klitzing, R.v. (2002) Temperature-induced changes in polyelectrolyte films at the solid-liquid interface. *Appl. Phys. A-Mater.*, **74** (Suppl. 1), s519–s521.

36 Klitzing, R.v. (2006) Internal structure of polyelectrolyte multilayer assemblies. *Phys. Chem. Chem. Phys.*, **8**, 5012–5033.

37 Ladam, G., Schaad, P., Voegel, J.C., Schaaf, P., Decher, G., and Cuisinier, F. (2000) In situ determination of the structural properties of initially deposited polyelectrolyte multilayers. *Langmuir*, **16** (3), 1249–1255.

38 Tarabia, M., Hong, H., Davidov, D., Kirstein, S., Steitz, R., Neumann, R., and Avny, Y. (1998) Neutron and X-ray reflectivity studies of self-assembled heterostructures based on conjugated polymers. *J. Appl. Phys.*, **83**, 725–732.

39 Kirstein, S., Hong, H.P., Steitz, P., and Davidov, D. (1999) Super-lattice structure in ppv-based self-assembled films. *Synth. Met.*, **102**, 519–521.

40 Steitz, R. and von Klitzing, R. (2011) unpublished results.

41 Ahrens, H., Büscher, K., Eck, D., Förster, S., Luap, C., Papastavrou, G., Schmitt, J., Steitz, R., and Helm, C.A. (2004) Poly(styrene sulfonate) self-organization: electrostatic and secondary interactions. *Macromol. Symp.*, **211** (1), 93–106.

42 Ivanova, O., Soltwedel, O., Gopinadhan, M., Köhler, R., Steitz, R., and Helm, C.A. (2008) Immobile light water and proton-deuterium exchange in polyelectrolyte multilayers. *Macromolecules*, **41** (19), 7179–7185.

43 Klitzing, R.v., Wong, J.E., Jaeger, W., and Steitz, R. (2004) Short range interactions in polyelectrolyte multilayers. *Curr. Opin. Colloid Interface*, **9**, 158–162.

44 Nazaran, P., Bosio, V., Jaeger, W., Anghel, D.F., and von Klitzing, R. (2007) Lateral mobility of polyelectrolyte chains in multilayers. *J. Phys. Chem. B*, **111** (29), 8572–8581; 6th International Symposium on Polyelectrolytes, Dresden, Germany, SEP, 2006.

45 Guzman, E., Ritacco, H., Rubio, J.E.F., Rubio, R.G., and Ortega, F. (2009)

Salt-induced changes in the growth of polyelectrolyte layers of poly(diallyldimethylammonium chloride) and poly(4-styrene sulfonate of sodium). *Soft Matter*, **5**, 2130–2142.

46 Klitzing, R.v. and Steitz, R.(November 2002) Interdigitation between polyelectrolytes layers in a multilayer system at the solid/liquid interface. Experimental Report ILL, ADAM-9-11-894.

47 Voigt, U., Khrenov, V., Tauer, K., Hahn, M., Jaeger, W., and von Klitzing, R. (2003) The effect of polymer charge density and charge distribution on the formation of multilayers. *J. Phys. Condens. Matter*, **15** (1), 213–218.

48 Shiratori, S.S. and Rubner, M.F. (2000) ph-dependent thickness behavior of sequentially adsorbed layers of weak polyelectrolytes. *Macromolecules*, **33**, 4213–4219.

49 Yoo, D., Shiratori, S.S., and Rubner, M.F. (1998) Controlling bilayer composition and surface wettability of sequentially adsorbed multilayers of weak polyelectrolytes. *Macromolecules*, **31**, 4309–4318.

50 Tanchak, O.M., Yager, K.G., Fritzsche, H., Harroun, T., Katsaras, J., and Barrett, C.J. (2006) Water distribution in multilayers of weak polyelectrolytes. *Langmuir*, **22** (11), 5137–5143.

51 Tanchak, O.M., Yager, K.G., Fritzsche, H., Harroun, T., Katsaras, J., and Barrett, C.J. (2008) Ion distribution in multilayers of weak polyelectrolytes: A neutron reflectometry study. *J. Chem. Phys.*, **129** (8), 084901.

52 Elzbieciak, M., Kolasinska, M., Zapotoczny, S., Krastev, R., Nowakowska, M., and Warszynski, P. (2009) Nonlinear growth of multilayer films formed from weak polyelectrolytes. *Colloid. Surf. A*, **343**, 89–95.

53 Porcel, C., Lavalle, P., Ball, V., Decher, G., Senger, B., Voegel, J.-C., and Schaaf, P. (2006) From exponential to linear growth in polyelectrolyte multilayers. *Langmuir*, **22** (9), 4376–4383.

54 Kharlampieva, E., Ankner, J.F., Rubinstein, M., and Sukhishvili, S.A. (2008) ph-induced release of polyanions from multilayer films. *Phys. Rev. Lett.*, **100** (12), 128303.

55 Kharlampieva, E., Kozlovskaya, V., Ankner, J.F., and Sukhishvili, S.A. (2008) Hydrogen-bonded polymer multilayers probed by neutron reflectivity. *Langmuir*, **24**, 11346–11349.

56 Jean, B., Heux, F., Dubreuil, L., Chambat, G., and Cousin, F. (2009) Non-electrostatic building of biomimetic cellulose-xyloglucan multilayers. *Langmuir*, **25**, 3920–3923.

57 Cerclier, C., Cousin, F., Bizot, H., Moreau, C., and Cathala, B. (2010) Elaboration of spin-coated cellulose-xyloglucan multilayered thin films. *Langmuir*, **26**, 17248–17255.

58 Gopinadhan, M., Ivanova, O., Ahrens, H., Gunther, J.-U., Steitz, R., and Helm, C.A. (2007) The influence of secondary interactions during the formation of polyelectrolyte multilayers: Layer thickness, bound water and layer interpenetration. *J. Phys. Chem. B*, **111** (29), 8426–8434.

59 Wong, J.E., Rehfeldt, F., Hanni, P., Tanaka, M., and Klitzing, R.v. (2004) Swelling behavior of polyelectrolyte multilayers in saturated water vapor. *Macromolecules*, **37** (19), 7285–7289.

60 Schlenoff, J. and Dubas, S. (2001) Mechanism of polyelectrolyte multilayer growth: Charge overcompensation and distribution. *Macromolecules*, **34** (3), 592–598.

61 Salomaki, M., Tervasmaki, P., Areva, S., and Kankare, J. (2004) The hofmeister anion effect and the growth of polyelectrolyte multilayers. *Langmuir*, **20** (9), 3679–3683.

62 Wong, J.E., Zastrow, H., Jaeger, W., and von Klitzing, R. (2009) Specific ion versus electrostatic effects on the construction of polyelectrolyte multilayers. *Langmuir*, **24**, 14061–14070.

63 Gopinadhan, M., Ahrens, H., Gunther, J.U., Steitz, R., and Helm, C.A. (2005) Approaching the precipitation temperature of the deposition solution and the effects on the internal order of polyelectrolyte multilayers. *Macromolecules*, **38**, 5228–5235.

64 Dubas, S.T. and Schlenoff, J.B. (2001) Swelling and smoothing of polyelectrolyte multilayers by salt. *Langmuir*, **17** (25), 7725–7727.

65 Dubas, S.T. and Schlenoff, J.B. (2001) Polyelectrolyte multilayers containing a weak polyacid: Construction and deconstruction. *Macromolecules*, **34** (11), 3736–3740.

66 Steitz, R. and von Klitzing, R. (2003) unpublished results.

67 Glinel, K., Prevot, M., Krustev, R., Sukhorukov, G.B., Jonas, A.M., and Mohwald, H. (2004) Control of the water permeability of polyelectrolyte multilayers by deposition of charged paraffin particles. *Langmuir*, **20** (12), 4898–4902.

68 Estrela-Lopis, I., Donath, E., Ibarz, G., and Krastev, R. (2003) Thermodynamic stability of polyelectrolyte multilayers. Technical report, BENSC, Hahn-Meitner-Institut Berlin, Germany.

69 Kohler, K., Shchukin, D.G., Sukhorukov, G.B., and Mohwald, H. (2004) Drastic morphological modification of polyelectrolyte microcapsules induced by high temperature. *Macromolecules*, **37** (25), 9546–9550.

70 Estrela-Lopis, I., Leporatti, S., Moya, S., Brandt, A., Donath, E., and Moehwald, H. (2002) SANS studies of polyelectrolyte multilayers on colloidal templates. *Langmuir*, **18**, 7861–7866.

71 Estrela-Lopis, I., Leporatti, S., Typlt, E., Clemens, D., and Donath, E. (2007) Small angle neutron scattering investigations (SANS) of polyelectrolyte multilayer capsules templated on human red blood cells. *Langmuir*, **23**, 7209–7215.

72 Estrela-Lopis, I., Leporatti, S., Clemens, D., and Donath, E. (2009) Polyelectrolyte multilayer hollow capsules studied by small-angle neutron scattering (SANS). *Soft Matter*, **5**, 214–219.

73 Kohler, K., Shchukin, D.G., Mohwald, H., and Sukhorukov, G.B. (2005) Thermal behavior of polyelectrolyte multilayer microcapsules. 1. the effect of odd and even layer number. *J. Phys. Chem. B*, **109** (39), 18250–18259.

74 Burmistrova, A., Steitz, R., and von Klitzing, R. (2010) Temperature response of PNIPAM derivatives at planar surfaces: Comparison between polyelectrolyte multilayers and adsorbed microgels. *Chemphyschem*, **11** (17), 3571–3579.

75 Plech, A., Salditt, T., Munster, C., and Peisl, J. (2000) Investigation of structure and growth of self-assembled polyelectrolyte layers by x-ray and neutron scattering under grazing angles. *J. Colloid. Interface Sci.*, **223**, 74–82.

76 Bijlsma, R., vanWell, A.A., and Stuart, M.A.C. (1997) Characterization of self-assembled multilayers of polyelectrolytes. *Physica B*, **234**, 254–255.

77 Kugler, R., Schmitt, J., and Knoll, W. (2002) The swelling behavior of polyelectrolyte multilayers in air of different relative humidity and in water. *Macromol. Chem. Phys.*, **203** (2), 413–419.

78 Cramer, C., De, S., and Schönhoff, M. (2011) Time-humidity-superposition principle in electrical conductivity spectra of ion-conducting polymers. *Phys. Rev. Lett.*, **107** (2), 028301.

79 Delajon, C., Gutberlet, T., Moehwald, H., and Krastev, R. (2009) Absorption of light and heavy water vapours in polyelectrolyte multilayer films. *Colloid. Surf. B-Biointerfaces*, **74**, 462–467.

80 Hsieh, M., Farris, R., and McCarthy, T. (1997) Surface "priming" for layer-by-layer deposition: Polyelectrolyte multilayer formation on allylamine plasma-modified poly (tetrafluoroethylene). *Macromolecules*, **30** (26), 8453–8458.

81 Schwarz, B. and Schönhoff, M. (1998) Surface potential driven swelling of polyelectrolyte multilayers. *Macromolecules*, **31**, 8893.

82 Schwarz, B. and Schönhoff, M. (2002) A 1h nmr relaxation study of hydration water in polyelectrolyte mono and multilayers adsorbed to colloidal particles. *Colloid. Surf. A*, **198–200**, 293–304.

83 Schönhoff, M., Carriere, D., Delajon, C., and Krastev, R. (2003) Polyelectrolyte multilayers: hydration and odd-and-even

effects. Technical report, BENSC, Hahn-Meitner-Institut Berlin, Germany.
84 Klitzing, R.v., Wong, J., Schemmel, S., and Steitz, R. (2003) Is there an odd/even effect of the water content of polyelectrolyte multilayers? Technical report, BENSC, Hahn-Meitner-Institut Berlin, Germany.
85 Carriere, D., Krastev, R., and Schonhoff, M. (2004) Oscillations in solvent fraction of polyelectrolyte multilayers driven by the charge of the terminating layer. *Langmuir*, **20** (26), 11465–11472.
86 Dubreuil, F., Elsner, N., and Fery, A. (2003) Elastic properties of polyelectrolyte capsules studied by atomic-force microscopy and ricm. *Eur. Phys. J. E: Soft Matter Biol. Phys.*, **12** (2), 215–221.
87 Lulevich, V.V., Radtchenko, I.L., Sukhorukov, G.B., and Vinogradova, O.I. (2003) Deformation properties of nonadhesive polyelectrolyte microcapsules studied with the atomic force microscope. *J. Phys. Chem. B*, **107** (12), 2735–2740.
88 Mertz, D., Hemmerle, J., Boulmedais, F., Voegel, J.-C., Lavalle, P., and Schaaf, P. (2007) Polyelectrolyte multilayer films under mechanical stretch. *Soft Matter*, **3** (11), 1413–1420.
89 Nolte, A.J., Rubner, M.F., and Cohen, R.E. (2005) Determining the young's modulus of polyelectrolyte multilayer films via stress-induced mechanical buckling instabilities. *Macromolecules*, **38** (13), 5367–5370.
90 Nolte, A.J., Treat, N.D., Cohen, R.E., and Rubner, M.F. (2008) Effect of relative humidity on the young's modulus of polyelectrolyte multilayer films and related nonionic polymers. *Macromolecules*, **41** (15), 5793–5798.
91 Früh, J. (2011) Structural Change of Polyelectrolyte Multilayers under Mechanical Stress. PhD thesis, Universität Potsdam, Deutschland.
92 Laurent, D. and Schlenoff, J.B. (1997) Multilayer assemblies of redox polyelectrolytes. *Langmuir*, **13** (6), 1552–1557.
93 Hodak, J., Etchenique, R., Calvo, E.J., Singhal, K., and Bartlett, P.N. (1997) Layer-by-layer self-assembly of glucose oxidase with a poly(allylamine)ferrocene redox mediator. *Langmuir*, **13** (10), 2708–2716.
94 Zahn, R., Voros, J., and Zambelli, T. (2010) Swelling of electrochemically active polyelectrolyte multilayers. *Curr. Opin. Colloid Interface.*, **15** (6), 427–434.
95 Calvo, EJ, Wolosiuk, A. (2002) Donnan permselectivity in layer-by-layer self-assembled redox polyelectrolye thin films. *J. Am. Chem. Soc*, **124**, 8490–8497.
96 Grieshaber, D., Voros, J., Zambelli, T., Ball, V., Schaaf, P., Voegel, J.C., and Boulmedais, F. (2008) Swelling and contraction of ferrocyanide-containing polyelectrolyte multilayers upon application of an electric potential. *Langmuir*, **24**, 13668–13676.
97 Zahn, R., Köhler, R., Bickel, K., Vörös, J., and Zambelli, T. (2010) Water distribution in electrochemically swellable polyelectrolyte multilayers. Technical report, Helmholtz Center Berlin for Materials and Energy GmbH, Berlin, Germany.
98 Tanaka, M. and Sackmann, E. (2005) Polymer-supported membranes as models of the cell surface. *Nature*, **437**, 656–663.
99 Delajon, C., Gutberlet, T., Steitz, R., Mohwald, H., and Krastev, R. (2005) Formation of polyelectrolyte multilayer architectures with embedded dmpc studied in situ by neutron reflectometry. *Langmuir*, **21**, 8509–8514.
100 Gromelski, S., Saraiva, A.M., Krastev, R., and Brezesinski, G. (2009) The formation of lipid bilayers on surfaces. *Colloid. Surf. B*, **74**, 477–483.
101 Perez-Salas, U.A., Faucher, K.M., Majkrzak, C.F., Berk, N.F., Krueger, S., and Chaikof, E.L. (2003) Characterization of a biomimetic polymeric lipid bilayer by phase sensitive neutron reflectometry. *Langmuir*, **19**, 7688–7694.
102 Chen, J.S., Kohler, R., Gutberlet, T., Moehwald, H., and Krastev, R. (2009) Asymmetric lipid bilayer sandwiched in polyelectrolyte multilayer films through layer-by-layer assembly. *Soft Matter*, **5**, 228–233.

103 Delcea, M., Krastev, R., Gutberlet, T., Pum, D., Sleytr, U.B., and Toca-Herrera, J.L. (2008) Thermal stability, mechanical properties and water content of bacterial protein layers recrystallized on polyelectrolyte multilayers. *Soft Matter*, **4**, 1414–1421.

104 Toca-Herrera, J.L., Krastev, R., Bosio, V., Kupcu, S., Pum, D., Fery, A., Sara, M., and Sleytr, U.B. (2005) Recrystallization of bacterial s-layers on flat polyelectrolyte surfaces and hollow polyelectrolyte capsutes. *Small*, **1**, 339–348.

105 Rocha, S., Krastev, R., Thunemann, A.F., Pereira, M.C., Moehwald, H., and Brezesinski, G. (2005) Adsorption of amyloid beta-peptide at polymer surfaces: A neutron reflectivity study. *Chemphyschem*, **6**, 2527–2534.

106 Jackler, G., Czeslik, C., Steitz, R., and Royer, C.A. (2005) Spatial distribution of protein molecules adsorbed at a polyelectrolyte multilayer. *Phys. Rev. E*, **71**, 041912.

107 Voets, I.K., de Vos, W.A., Hofs, B., de Keizer, A., Stuart, M.A.C., Steitz, R., and Lott, D. (2008) Internal structure of a thin film of mixed polymeric micelles on a solid/liquid interface. *J. Phys. Chem. B*, **112**, 6937–6945.

108 Ribeiro, P.A., Steitz, R., Lopis, I.E., Haas, H., Souza, N.C., Oliveira, O.N., and Raposo, M. (2006) Thermal stability of poly(o-methoxyaniline) layer-by-layer films investigated by neutron reflectivity and uv-vis spectroscopy. *J. Nanosci. Nanotech.*, **6**, 1396–1404.

109 Struth, B., Eckle, M., Decher, G., Oeser, R., Simon, P., Schubert, D.W., and Schmitt, J. (2001) Hindered ion diffusion in polyelectrolyte/montmorillonite multilayers: Toward compartmentalized films. *Eur. Phys. J. E*, **6**, 351–358.

110 Krasteva, N., Krustev, R., Yasuda, A., and Vossmeyer, T. (2003) Vapor sorption in self-assembled gold nanoparticle/dendrimer films studied by specular neutron reflectometry. *Langmuir*, **19** (19), 7754–7760.

111 Krasteva, N., Möhwald, H., and Krastev, R. (2009) Structural changes in stimuli-responsive nanoparticle/dendrimer composite films upon vapor sorption. *C. R. Chim.*, **12** (1–2), 129–137.

# 12
# Polyelectrolyte Conformation in and Structure of Polyelectrolyte Multilayers

*Stephan Block, Olaf Soltwedel, Peter Nestler, and Christiane A. Helm*

## 12.1
## Introduction

"The ability to engineer surfaces that present multiple functionalities when and where they are needed could lead to important advances in electrooptical devices, separations, and biomaterials [1]. The layer-by-layer (LbL) [2] electrostatic assembly technique via alternating adsorption of positively and negatively charged species from aqueous solutions is ideally suited for such applications because it allows absolute control over the sequence in which multiple functional elements are incorporated into a growing film [3–6]. Polymer organic and organic/inorganic thin films formed using the LbL technique may contain a number of different functional groups, including electro-optic, electroluminescent, conducting, and dielectric layers. These achievements were obtained by empirically optimizing the building conditions (deposition process, post-preparation treatment).

The linear LbL growth mode of polyelectrolyte multilayers (PEMs) is an inherently nonequilibrium process, yielding a nonequilibrium structure. In the adsorption process itself, individual molecules attach to the surface of a growing PEM via many ion pairing contacts. Understanding the polyelectrolyte (PE) molecular conformation, both that of the freshly adsorbed monolayer, and the ensuing changes during PEM formation, allows one to optimize the PEMs. To pursue this goal, we follow different approaches.

It is well known that PEs adsorbed from pure water form a flat compact monolayer and cause surface charge reversal [7–9]. The next adsorption step is again accompanied by charge reversal. These results were obtained with the surface forces apparatus [10]. However, PEMs are very often formed in salt solutions. Increasing the ion concentration in the deposition solution beyond 0.1 M yields an increase in the thickness per deposited polycation/polyanion layer pair, suggesting a strong involvement of electrostatic intermolecular interactions [11–13]. The conformation of single polyelectrolyte layers adsorbed from 1 M salt was only determined recently [14–16]. To understand the respective contribution of electrostatic and steric forces it was

*Multilayer Thin Films: Sequential Assembly of Nanocomposite Materials*, Second Edition.
Edited by Gero Decher and Joseph B. Schlenoff.
© 2012 Wiley-VCH Verlag GmbH & Co. KGaA. Published 2012 by Wiley-VCH Verlag GmbH & Co. KGaA.

necessary to image the surface topology, and to identify domains of brush phase and of flatly adsorbed PEs. To achieve that, the colloidal probe technique was optimized.

Early publications on PEMs from $(PAH/PSS)_n$ with selectively deuterated layers demonstrate a stratified structure [2]. However, not all PEMs with the linear LbL growth mode show this stratified structure, an example is $(PDADMAC/PSS)_n$ built from solutions with high salt content [5, 6]. To monitor the internal structure of PEMs and its changes due to interdiffusion, we used neutron reflectivity with a dedicated film architecture. A PEM consists of two blocks, one with deuterated layers, the second with only protonated layers [17–19]. The number of layer pairs is kept constant, while the thickness of the respective blocks is varied. Thus, the position and the width of a sequence of internal interfaces can be probed systematically.

Usually, the LbL technique is used at high salt conditions. Then, the top PE layer extends into solution, in a coiled or brush conformation. With ellipsometry and neutron reflectivity, it was shown that the kind of PE in the outermost layer determines the water content and thickness of the PEM (the so-called odd–even effect [5, 6]). Using multiangle null-ellipsometry, we explored how polycation and polyanion thickness are affected by the thickness of the supporting PEM, as well as by a change of deposition conditions.

Summarizing, to understand the internal structure of PEMs, we determine the PE conformation in the first layer, the position-dependent layer interdiffusion and the thickness/water-content of the top layer as a function of the degree of polymerization $N$, the temperature $T$ and the salt concentration $I_{Ads}$ in the deposition solution. Different techniques are adapted for PEMs which yield complementary information. Systematic variation of parameters is necessary to understand the subtle interplay of the molecular interactions which determine the PE coverage and conformation.

## 12.2
## Results

### 12.2.1
### The First Polyelectrolyte Layer: Brush and/or Flatly Adsorbed Chains

The surface forces between two PSS layers adsorbed from pure water onto an oppositely charged surface (silicon is functionalized with 3-aminopropyldimethylethoxysilane) are shown in Figure 12.1a. The surface forces are consistent with DLVO theory [8]. The electrostatic repulsion decays exponentially with increasing surface separation. The decay length $\gamma$ is the Debye length, its signature is its decrease on increase of the bulk NaCl concentration $I$ ($\gamma = 1/\kappa = 0.304 \times I^{-0.5}$ nm, with $I$ in M). Actually, the measured decay lengths agree quantitatively with the theoretical prediction, indicating a flat, compact PSS adsorption layer [14].

However, the surface forces differ qualitatively and quantitatively (cf. Figure 12.1b), if PSS is adsorbed from high salt conditions ($I_{Ads} = 1$ M): (i) The magnitude of the forces is larger; at a selected surface separation and bulk concentration $I$, the difference is one to two orders of magnitude. (ii) At large separations, the

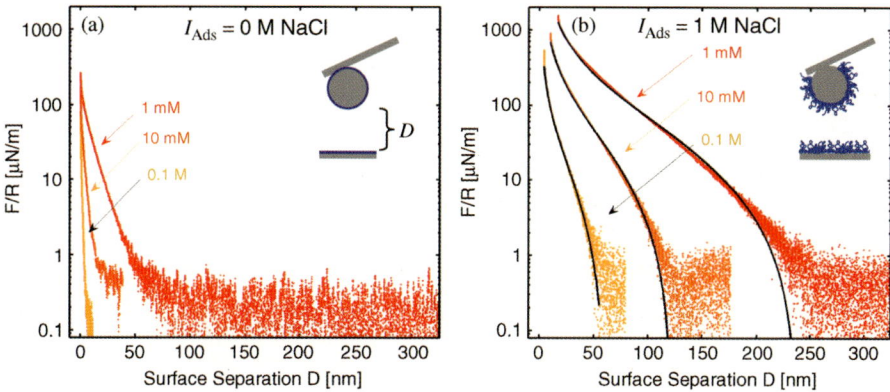

**Figure 12.1** Surface forces between PSS ($N = 1800$) layers adsorbed from pure water (a) and from 1 M NaCl solution (b), measured at salt concentration $I$ as indicated. (a) The forces increase exponentially on approach of the surfaces. As expected for two flat surfaces of equal charge, the decay length $\gamma$ is the Debye-length ($\gamma = 1/\kappa = 0.304 \times I^{-0.5}$ nm). (b) The theory of Alexander–de Gennes for neutral brushes (black lines) describes the surface forces: typical is the sharp increase when the brushes touch. On further approach, the repulsive force increases exponentially. The decay length is $\gamma = 2L/\pi$. $L$, the brush thickness, increases on dilution.

forces increase steeply. On further approach, an almost exponential increase occurs. (iii) The decay length $\gamma$ increases when the bulk salt concentration $I$ is lowered. The surface forces can be described by the theory of Alexander and de Gennes for neutral polymer brushes [20, 21]. Basically, this theory contains two independent parameters: the brush length $L$ and the area per brush $s^2$. With both surfaces covered with PEs and at intermediate surface separations, when the surface forces can be almost described as an exponential function, the amplitude is proportional to $L \times s^{-3}$, and the decay length $\gamma$ equals $2L/\pi$. It seems counterintuitive to describe the surface forces caused by polyelectrolyte brushes with the theories derived for neutral brushes. However, this approach is valid, since all counterions are incorporated into the salted polyelectrolyte brush [22, 23]. Indeed, the brush length $L$ scales with the product of $I$ and $s^2$ with the exponent $-1/3$ (cf. Figure 12.2a,b), a behavior predicted theoretically and found experimentally for the salted brush phase [24–26]. This scaling seems to be general, Figure 12.2a presents the results for PLL and PAH [14]. To summarize: when polyelectrolytes are adsorbed from high salt concentrations ($I_{Ads} = 1$ M), they generate surface forces, as known from neutral brushes; however, the brush length scales as a salted polyelectrolyte brush.

To visualize the swelling/shrinking in different salt bulk solutions, the surfaces are imaged. However, a sharp AFM tip (radius 5–50 nm) penetrates the brush [15, 27]. Therefore, a large radius is necessary ($\approx\mu$m), and we use colloid probe tapping mode (CPTM). When PSS is adsorbed only onto a gold step (cf. Figure 12.3), the change in step height can be unambiguously attributed to the swelling/shrinking of the adsorbed PSS layer. The brush thickness changes by a factor of four (from 20 to 80 nm, the contour length of a single chain is 425 nm). This behavior is very different

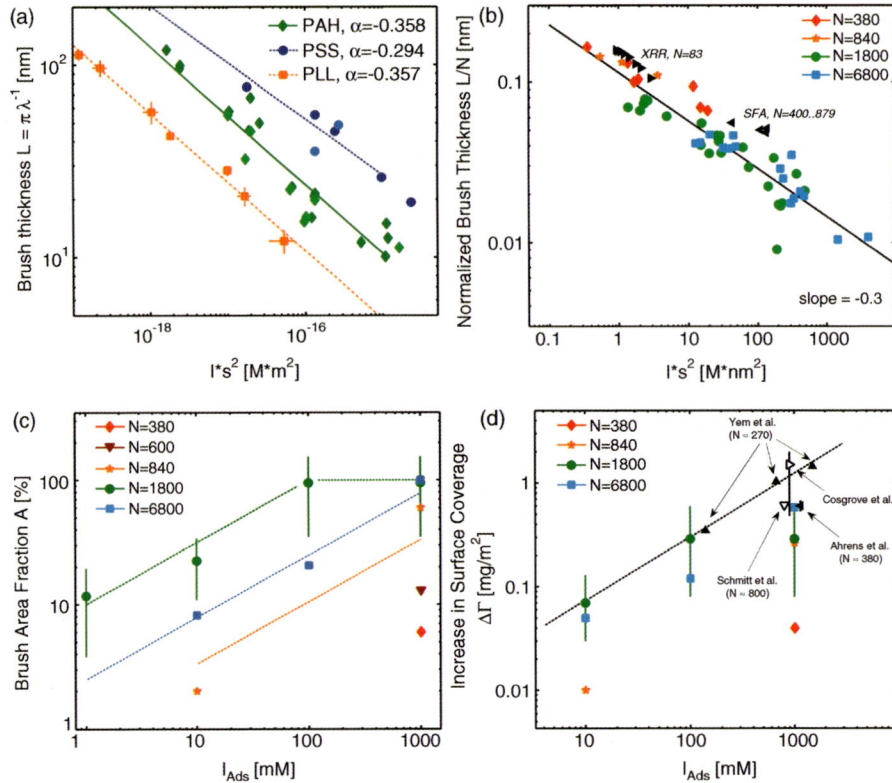

**Figure 12.2** Parameters influencing the brush of adsorbed PSS chains. (a) Brush thickness as derived from the decay length $\gamma = \lambda^{-1}$ of the force law (cf. Figure 12.1b) for different physisorbed polyelectrolytes versus the product of the salt concentration $I$ and the area per chain $s^2$ ($I_{Ads} = 1$ M NaCl. Contour length of PAH, PSS and PLL is 220, 425, 350 nm, respectively.). (b) Brush thickness of PSS normalized with respect to the degree of polymerization $N$ versus $I\,s^2$. Also shown are results characterizing end-grafted PSS chains from X-ray reflectivity [24] and from SFA measurements [25]. (c) Brush area fraction $A$ versus the ion concentration $I_{Ads}$ in the deposition solution. (d) Molecular weight per unit area of the fraction of PSS-chains protruding into solution (determined from brush thickness $L$, area per chain $s^2$ and the brush area fraction $A$) as function of $I_{Ads}$. For comparison, additional surface coverage determined by neutron [28, 29] and X-ray reflectivity [30], as well as UV–vis [31] are shown, using different substrates. For all measurements with PSS, $N$ is varied as indicated.

from polyelectrolyte layers within a PEM, where a swelling of only 30–50% is observed [5, 17].

Interestingly, if shorter PSS is used, and/or the salt concentration in the adsorption solution is lowered (to $I_{Ads} = 0.01$ M), the adsorbed PSS layer is no longer homogeneous (cf. Figure 12.3). Islands can be observed, which obviously consist of a phase similar to polyelectrolyte brushes, since they swell and shrink with changing adsorption conditions. Between the islands, a flat compact polyelectrolyte layer is found [16].

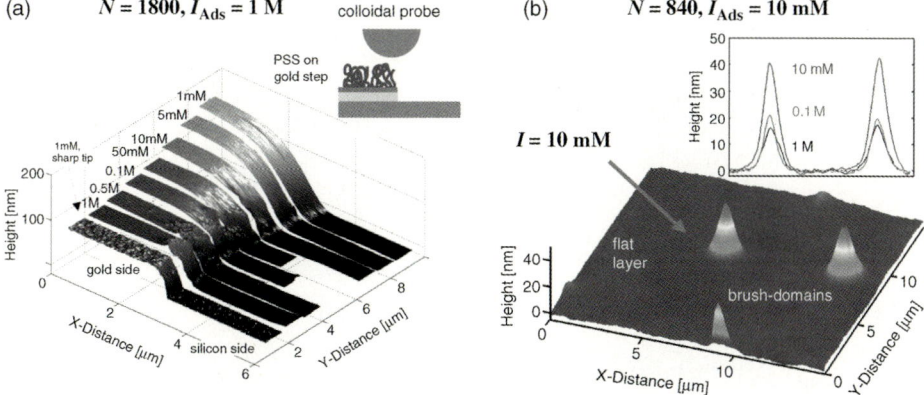

**Figure 12.3** Colloid probe tapping mode images of the swelling/shrinking of non-flat adsorbed PSS layers in salt solutions with concentration $I$ (as indicated). (a) A homogeneous layer ($N = 1800$, $I_{Ads} = 1$ M) is adsorbed onto a silanized gold step (height 50 nm). With sharp tips, the brush cannot be resolved, since the tip penetrates the brush. Therefore, the step height as obtained with the sharp tip is independent of the brush, and only one typical image is shown. (b) The inhomogeneous layer ($N = 840$, $I_{Ads} = 0.01$ M, $I$ indicated) is imaged on a silanized silica slide. The observed islands can be attributed to PSS chains in a brush-like conformation (cf. swelling/shrinking in the inset). Apparently, if PSS with low molecular weight is adsorbed at small salt concentrations, the adsorbed PSS layer consists of two distinct phases.

We started to investigate the factors which influence the phase coexistence of physisorbed polyelectrolyte layers. In a first step, the degree of polymerization $N$ is systematically varied, while the ion concentration in the adsorption solution is kept high and constant (1 M) [16]. For degree of polymerization $N = 380$, the brush area fraction is 6% (cf. Figure 12.4). With increasing degree of polymerization, the brush area fraction and the domain radii increase (from 50 nm to 1.5 μm). Lateral homogeneous brush layers are found for a degree of polymerization exceeding 1100. The surface forces are a superposition of steric and electrostatic forces; their respective contribution is determined by the brush area fraction. Therefore, once the area fraction of the brush-like phase is known, its parameters (brush length $L$ and area per chain $s^2$) can be determined from the surface forces.

Independent of the area fraction, for each degree of polymerization $N$ the brush-like phase scales like a polyelectrolyte brush. Actually, if the brush length is normalized with respect to the degree of polymerization, the data superimpose and collapse onto master curves known for salted PSS brushes (cf. Figure 12.2b). The normalized brush length is somewhat shorter than observed for real end-grafted PSS chains [24, 25], demonstrating that a small fraction of the chains is adsorbed onto the substrate. Note that both the brush length and the Debye length scale with the surrounding ion concentration $I$. However, the brush length changes less, it scales with $-1/3$, while the former scales with $-1/2$ (cf. Figure 12.2b).

Afterward, by increasing the salt content in the deposition solution, $I_{Ads}$, the changes in PSS surface coverage are explored while the degree of polymerization,

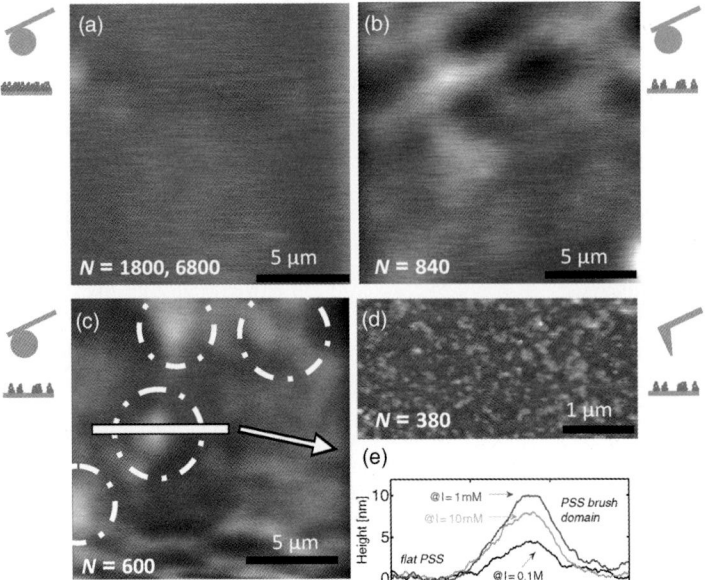

**Figure 12.4** Colloidal probe tapping mode images of PSS layers physisorbed from high salt solution ($I_{Ads} = 1$ M) onto silanized silica, the degree of polymerization $N$ is varied. The images are obtained in aqueous solution at low salt conditions ($I = 1$ mM). (a) For large $N$ (>1100) the layer is homogeneous. (b) For intermediate $N$ (= 840), holes are observed, (c) and for small $N$ (= 600), isolated islands. (d) For very small $N$ (= 380), also the domains are small. (e) Height scan of the indicated island in (c) at different ionic strengths, as indicated. The swelling/shrinking is reversible. The radius of the colloid probe is 3 µm, except for (d) (≈100 nm).

$N$, is kept constant [32]. Even the addition of very little salt ($I_{Ads} > 1$ mM) leads to two coexisting PSS phases (cf. Figure 12.3b). On raising the salt content in the adsorption solution the area fraction increases until the surface is covered homogeneously with the brush-like phase (cf. Figure 12.2c). Again, the surface forces can be described as a superposition of electrostatic and steric forces, each force contributes according to its respective area fraction. Surprisingly, the molecular properties of the brush phase (i.e., the area per chain $s^2$ and the brush thickness $L$) depend only on the degree of polymerization, and not on the salt content of the adsorption solution (1 mM < $I_{Ads}$ < 1 M). Once the PSS layer is formed, the brush thickness $L$ can be controlled by the salt content $I$ of the solution. The observed increase in brush-like physisorbed PSS chains provides an explanation for the increase in PSS surface coverage on addition of salt in the deposition solution, which is widely reported in the literature [28–31].

However, quantitative comparison between the laterally averaged literature data (cf. Figure 12.2d) and our experiments shows deviations. Since in each experiment, chemically different substrates were used, different surface coverage may be attributed to different monomer/substrate interactions. Extending these studies to the top layers of polyelectrolyte multilayers will help our understanding of PEMs.

## 12.2.2
### Adsorption of Additional Layers and Interdiffusion

How much a polyelectrolyte layer within a PEM diffuses determines spatially organized structures, as well as the interaction with neighboring layers. It is thought that linearly growing PEMs interact mostly with two or three adjacent layers [2]. However, diffusing polyelectrolytes are assumed to be able to rapidly diffuse throughout PEMs' architectures during assembly, resulting in exponential film growth and poorly organized, blended structures [4]. To verify and eventually quantify this concept, the internal structure and interdiffusion need to be measured. We start with PEMs showing linear growth, using neutron reflectometry.

Figure 12.5 presents results obtained with PAH/PSS and PDADMA/PSS films. Each film consists of a protonated and a deuterated block, built from $x$ protonated and $y$ deuterated polycation/polyanion layer pairs, respectively. The number of layer pairs $N = x + y$ is kept constant; the position of the interface between the blocks is varied

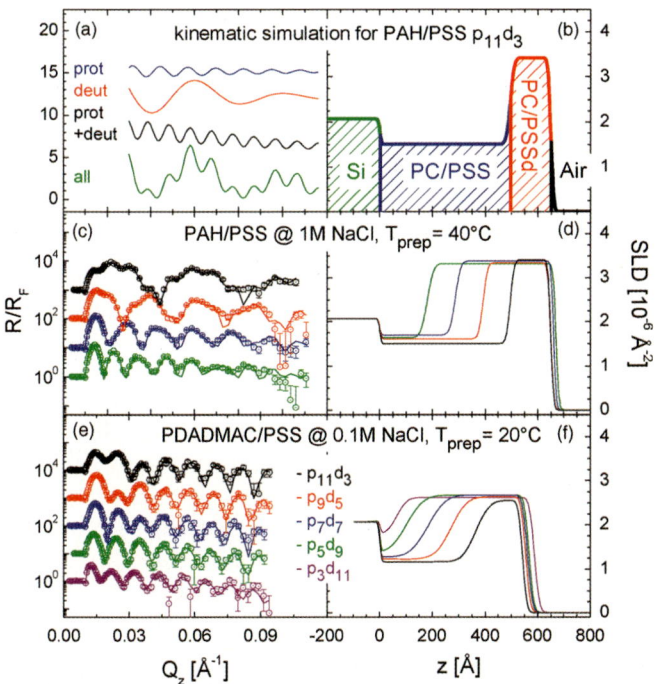

**Figure 12.5** Normalized neutron reflectivity curves (a, c, e) and corresponding scattering length density profiles (b, d, f). (a, b) Concept of film architecture: the calculated reflectivity (green) is a superposition of the interference fringes of the protonated block $p_{14-n}$ ($14 - n$ polycation/PSS layer pairs, blue), the deuterated block $d_n$ ($n$ polycation/PSS-d layer pairs, red) and the total film $p_{14-n}d_n$ (black). (c–f) Experiments on PAH/PSS and PDADMA/PSS at 0% r.h. For clarity, the reflectivity curves are shifted relative to each other. The deposition conditions are standard for PDADMA/PSS (0.1 M NaCl, 20 °C); they were adjusted to get a similar film thickness for PAH/PSS (1 M NaCl, 40 °C).

systematically. This film architecture is advantageous because the reflectivity curve can be described with only seven independent parameters, that is, the thickness and scattering length density of each block and the roughness of the three adjacent interfaces. The focus of interest is the width and position of the interface between the blocks (its roughness is $\sigma_{int}$) [18].

Counting from the top, the internal roughness increases with the number of deposited layer pairs (cf. Figure 12.5). For PAH/PSS, it grows from 10 to 25 Å, the latter number is measured with 15 layer pairs beneath the film/air interface. Possibly, diffusion stops far away from the film/air interface [18]. Obviously, each deposition step promotes interlayer diffusion. Mathematically, it is a one-dimensional diffusion, an error function describes the segment density profile [33]. One obtains $\sigma_{int}^2 = 2Dt$ (with $D$ the diffusion coefficient and $t$ time). We assume that interdiffusion takes place whenever the internal interface under investigation is immersed into the deposition solution. Thus, each deposition step adds another 30 min interdiffusion time. With these assumptions, the diffusion coefficient for PAH/PSS is $1.1 \times 10^{-18}$ cm$^2$ s$^{-1}$ and for PDADMA/PSS $1.1 \times 10^{-17}$ cm$^2$ s$^{-1}$.

For PDADMA/PSS we observe that, after at most seven layer pairs beneath the film/air interface, an equilibrium internal roughness is reached. For this polycation/polyanion pair, the influence of the salt concentration in the deposition solution and the molecular weight of PDADMAC are explored [19].

The neutron reflectivity curves of films with the block architecture have distinct features. In the kinematic approximation the reflected intensity caused by the single blocks decays proportionally to $\exp(-Q_z^2 \sigma_{int}^2/2)$. Therefore, for a large internal roughness $\sigma_{int}$, the blocks do not contribute to the reflected intensity at large $Q_z$. The measurements in Figure 12.6a, b illustrate this concept with $p_x d_5$ PEMs, built from PDADMAC with different molecular weights. The largest internal roughness is found for the $p_9 d_5$ film made from the heavy PDADMAC ($M_{w,PDADMAC} = 510$ kDa; $\sigma_{int} = 48$ Å; $\sigma_{air} = 14$ Å). With the exception of a strongly damped second maximum (at $Q_z \approx 0.025$ Å$^{-1}$), the neutron reflectivity curve could be described as Kiessig fringes from a homogeneous thin film. In contrast, the film made from the lowest molecular weight PDADMAC ($M_{w,PDADMAC} = 35$ kDa, $\sigma_{int} = 20$ Å) is, even at large $Q_z$, a superposition of three different interference patterns, since the three interfaces involved all have comparable small roughnesses. The film made from intermediate weight PDADMAC ($M_{w,PDADMAC} = 322$ kDa, $\sigma_{int} = 29$ Å) has an intermediate internal roughness, therefore the single blocks only contribute to the reflected intensity for $Q_z < 0.065$ Å$^{-1}$.

Figure 12.6c, d presents the position-dependent internal roughness $\sigma_{int}$. It is smallest next to the film/air interface and increases with the number of layer pairs away from the film/air interface until an equilibrium value is reached. Both the equilibrium internal roughness and the interdiffusion constant grow with the salt concentration in the deposition solution, the latter by about a factor of ten. This is attributed to a reduction of ion pairing contacts between monomers. On increase of the PDADMAC molecular weight, the internal roughness and the PSS diffusion constant also increase. This is very counterintuitive, since the diffusion constant of a polymer decreases with its weight. Therefore, we suggest intralayer diffusion and PDADMA segment relaxation. Furthermore, since we measure PSS, not PDADMA

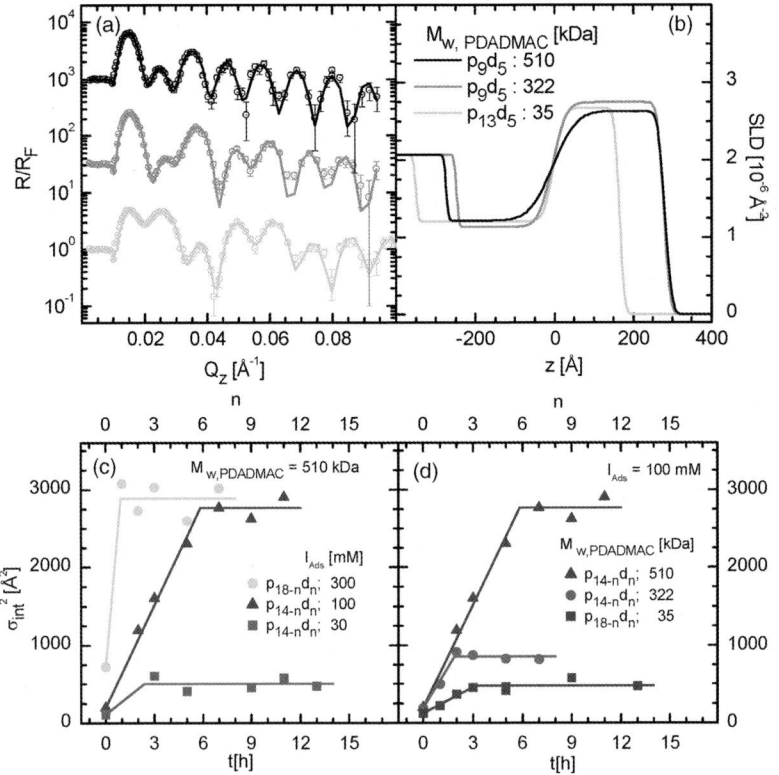

**Figure 12.6** (a, b) Normalized neutron reflectivity curves (a) and corresponding SLD profiles (b) of PEI/PSS/(PDADMA/PSS)$_{x-1}$/(PDADMA/PSS-d)$_5$ films, measured at 0% r.h. $x$ is either 9 or 13, as indicated. Varied is $M_{W,PDADMAC}$, all films are prepared at 0.1 M NaCl and 20 °C. For the scattering length density profiles, $z = 0$ is set to the interface between the deuterated and the protonated block. (c, d) Square of internal roughness, $\sigma_{int}^2$, of PEI/PSS/(PDADMA/PSS)$_{N-n}$/(PDADMA/PSS-d)$_n$ films versus the position of the investigated interface, that is, $n$ layer pairs below the film/air interface (or versus the immersion time in the deposition solution). The film/air interface is set to $n = 0$. Films were prepared from different NaCl concentration $I_{Ads}$ ((c) with $M_{W,PDADMAC} = 510$ kDa) or from different PDADMA polymer weight ((d) with 100 mM NaCl). The lines are fits to $\sigma_{int}^2 = 2Dt$, while $\sigma_{int}$ is changing.

profiles, PSS does not diffuse independently, but cooperative effects occur. One can imagine that segments or strongly coupled fragments of PDADMA/PSS chains diffuse together.

### 12.2.3
### The Outermost Layer: From Odd–Even to Even–Odd Effect

Complementary to the first layer on the substrate, we investigate the final layer with null-ellipsometry [34, 35]. To minimize the interdiffusion with the PEM, PAH/PSS

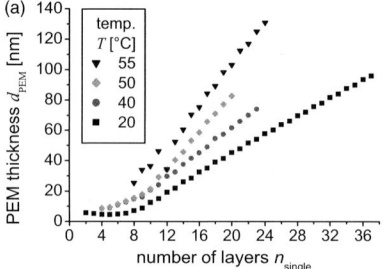 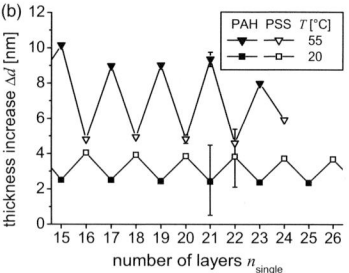

**Figure 12.7** (a) PSS/PAH film thickness $d_{PEM}$ determined *in situ* with null-ellipsometry as function of number of adsorbed layers $n_{single}$ ($I_{Ads} = 1$ M NaCl at temperatures indicated). After the adsorption of about twelve polyelectrolyte layers, the film grows linearly with $n_{single}$. (b) PEM thickness increase $\Delta d = d_{PEM}(n_{single}) - d_{PEM}(n_{single} - 1)$ as function of the number of layers $n_{single}$ in the linear regime. The thickness increase $\Delta d$ shows an oscillating pattern, depending on the charge of the terminating layer. Open symbols are used, when the polyanion PSS (even $n_{single}$) is the terminating layer; full symbols indicate the polycation PAH (odd $n_{single}$). At room temperature, terminating PSS layers are thicker than PAH layers ("odd–even effect"). At 55 °C, the effect is reversed; the polycation layers are thicker ("even–odd effect").

layers are used, since the internal roughness between the top layer and the supporting PEM is very low [18]. We adsorb from high concentration salt solutions since we intend to understand the nonflat adsorption of PEs better [14–16, 32]. Furthermore, to explore non-electrostatic interactions, the temperature of the deposition solution is varied. As a method multiple-angle null-ellipsometry is chosen, because for films thicker than 10 nm the thickness and the index of refraction can be measured independently [36, 37]. The PEMs are always immersed in 1 M NaCl solution.

Figure 12.7a demonstrates very moderate film growth for the first eight layers, as often described in the literature [5]. After the deposition of the first twelve layers, the film growth becomes linear. The film thickness $d_{PEM}$ increases with the temperature, as described earlier [13, 38].

On close inspection of Figure 12.7, one can observe evidence of the odd–even effect. Considering the film prepared at 20 °C in the linear growth regime, two consecutive layers always have a similar thickness, then a more pronounced increase in $d_{PEM}$ occurs. To quantify this observation, the thickness increase $\Delta d = d_{PEM}(n_{single}) - d_{PEM}(n_{single} - 1)$ is shown in Figure 12.7b. The large thickness increase occurs for the layers with an even number, that is, the polyanion layers. Obviously, each PSS layer is thicker than the previous or consecutive PAH layer. With each deposition step, charge compensation occurs. Since for both the top PSS and PAH layer the same index of refraction is found, the larger thickness of the PSS layer is explained with its larger monomer volume (about a factor of two [13, 18]). Within the resolution of Figure 12.7a for preparation temperatures of 40 and 50 °C, the odd–even effect cannot be discerned. However, at 55 °C, again every second adsorbed layer contributes more strongly to the film thickness. The quantitative data analysis as shown in Figure 12.7b demonstrates that now the odd layers are thicker than the even ones. The odd–even effect is reversed, an even–odd effect occurs. The index of refraction of the outermost PSS layers is larger than that of the corresponding PAH layers, suggesting a very compact PSS structure and a rather fluffy PAH layer with a large water content. The reason may be the hydrophobic nature of PSS, which causes almost a collapse of PSS layers. The PAH partially protrudes in the PSS layers after adsorption, until the surface charge is neutralized. On drying these PEMs destabilize, their roughness increases and, with AFM, a random pattern of holes is observed, if the thickness of a layer pair in solution exceeds 8.6 nm [13, 35].

## 12.3
## Conclusion and Outlook

With the aim to understand the internal structure of PEMs, we investigated the first and last layer of a PEM in detail, as well as the internal structure and its changes on deposition of additional layers.

For the first PE layers adsorbed from high salt conditions, strong steric forces are found. Domains of PE brushes coexist with flatly adsorbed layers, as colloid probe imaging shows. For PSS, the surface forces can be quantified as a superposition of steric and electrostatic forces. The area fraction of the brush domains depends on the degree of polymerization and the ion content in the adsorption solution.

The outermost layer of PEMs from PAH and PSS varies in thickness and water content. Either the polycation or the polyanion is thicker, as was found with null-ellipsometry. At low temperature the PSS is thicker, because of the larger monomer volume. On temperature increase, the odd–even effect is inversed to an even–odd effect, which is attributed to a shift in the balance of electrostatic and secondary forces.

The internal roughness of PEMs changes with the distance from the film surface, indicating progressive film interdiffusion during dipping cycles. For PDADMA/PSS we find that the internal roughness and the interdiffusion constant increase with the salt content in the deposition solution. Eventually, an equilibrium internal roughness is reached. On increase of the PDADMAC molecular weight, the internal roughness and the interdiffusion constant also grow. It is suggested that PDADMA segment relaxation occurs after deposition and augments the PSS mobility.

We are very much at the beginning of our studies. Obviously, one needs to adapt and optimize existing physical methods to characterize conformation and interactions of the PEs within a PEM. To understand the first layer, we needed to combine surface forces with colloidal probe imaging. For the last layer, multiple-angle null-ellipsometry provided new insights. To measure the internal roughness within a PEM with neutron reflectivity, a suitable film architecture was found. Now, relevant parameters as well as different kinds of PEs need to be explored.

## References

1. Hammond, P.T. (2004) *Adv. Mater.*, **16**, 1271.
2. Decher, G. (1997) *Science*, **277**, 1232.
3. Schlenoff, J.B. (2009) *Langmuir*, **25**, 14007.
4. Picart, C., Lavalle, P., Hubert, P., Cuisinier, F.J.G., Decher, G., Schaaf, P., and Voegel, J.C. (2001) *Langmuir*, **17**, 7414.
5. Schönhoff, M. (2003) *Curr. Opin. Colloid Interface Sci.*, **8**, 86.
6. Klitzing, R.v. (2006) *Phys. Chem. Chem. Phys.*, **8**, 5012.
7. Berndt, P., Kurihara, K., and Kunitake, T. (1992) *Langmuir*, **8**, 2486.
8. Lowack, K. and Helm, C.A. (1998) *Macromolecules*, **41**, 823.
9. Claesson, P.M., Poptoshev, E., Blomberg, E., and Dedinaite, A. (2005) *Adv. Colloid Interface.*, **114**, 173.
10. Israelachvili, J.N. (1973) *J. Chem. Soc., Faraday Trans. 2*, **11**, 1729.
11. Decher, G., Hong, J.D., and Schmitt, J. (1992) *Thin Solid Films*, **210/211**, 831.
12. Gopinadhan, M., Ivanova, O., Ahrens, H., Günther, J.U., Steitz, R., and Helm, C.A. (2007) *J. Phys. Chem. B*, **111**, 8426.
13. Cornelsen, M., Helm, C.A., and Block, S. (2010) *Macromolecules*, **43**, 4300.
14. Block, S. and Helm, C.A. (2007) *Phys. Rev. E*, **76**, 030801(R).
15. Block, S. and Helm, C.A. (2008) *J. Phys. Chem. B*, **112**, 9318.
16. Block, S. and Helm, C.A. (2009) *Macromolecules*, **42**, 6733.
17. Ivanova, O., Soltwedel, O., Gopinadhan, M., Köhler, R., Steitz, R., and Helm, C.A. (2008) *Macromolecules*, **41**, 7179.
18. Soltwedel, O., Ivanova, O., Nestler, P., Müller, M., Köhler, R., and Helm, C.A. (2010) *Macromolecules*, **43**, 7288.
19. Soltwedel, O., Nestler, P., Neumann, H.-G., Köhler, R., and Helm, C.A., *Macromolecules*, subm.
20. Alexander, S. (1977) *J. Phys.*, **38**, 983.
21. de Gennes, P.-G. (1987) *Adv. Colloid Interface Sci*, **27**, 189.
22. Biesheuvel, P.M. (2004) *J. Colloid Interface Sci.*, **275**, 97.
23. Zhulina, E.B., Birshtein, T.M., and Borisov, O.V. (1995) *Macromolecules*, **28**, 1491.
24. Ahrens, H., Förster, S., and Helm, C.A. (1998) *Phys. Rev. Lett.*, **81**, 4172.
25. Balastre, M., Li, F., Schorr, P., Yang, J.C., Mays, J.W., and Tirrell, M.V. (2002) *Macromolecules*, **35**, 9480.
26. Rühe, J., Ballauff, M., Biesalski, M., Dziezok, P., Grohn, F., Johannsmann, D., Houbenov, N., Hugenberg, N., Konradi, R., Minko, S., Motornov, M., Netz, R.R., Schmidt, M., Seidel, C., Stamm, M., Stephan, T., Usov, D., and Zhang, H.N. (2004) Polyelectrolyte brushes, in: *Polyelectrolytes with Defined Molecular Architecture I*, Springer Verlag, Berlin, **165**, p. 79.
27. Halperin, A. and Zhulina, E.B. (2010) *Langmuir*, **26**, 8933.
28. Cosgrove, T., Obey, T.M., and Vincent, B. (1986) *J. Colloid Interface Sci.*, **111**, 409.
29. Yim, H., Kent, M., Matheson, A., Ivkov, R., Satija, S., Majewski, J., and Smitth, G.S. (2000) *Macromolecules*, **33**, 6126.
30. Ahrens, H., Baltes, H., Schmitt, J., Möhwald, H., and Helm, C.A. (2001) *Macromolecules*, **34**, 4504.
31. Schmitt, J. (1996) Aufbau und Strukturelle Charakterisierung von Multilagen aus Polyelektrolyten, Reaktivpolymeren und Kolloiden. Dissertation, Johannes Gutenberg-Universität.
32. Block, S. and Helm, C.A. (2011) *J. Phys. Chem. B*, **115**, 7301.
33. Karim, A., Mansour, A., Felcher, G.P., and Russell, T.P. (1989) *Physica B*, **156 & 157**, 430.
34. Nestler, P. (2010) Diploma Thesis, Greifswald University.
35. Nestler, P., Block, S., and Helm, C.A. (2012) *J. Phys. Chem. B*, **116**, 1234.
36. Ibrahim, M.M. and Bashara, N.M. (1971) *J. Opt. Soc. Am.*, **61**, 1622.
37. McCrackin, F.L., Passaglia, E., Stromberg, R.R., and Steinberg, H.L. (1963) *J. Res. NBS A Phys. Chem.*, **A a67**, 363.
38. Büscher, K., Ahrens, H., Graf, K., and Helm, C.A. (2002) *Langmuir*, **18**, 3585.

# 13
# Charge Balance and Transport in Ion-Paired Polyelectrolyte Multilayers

*Joseph B. Schlenoff*

## 13.1
### Introduction

Polyelectrolyte multilayers, PEMUs, are one of several possible morphologies of polyelectrolyte complexes. A thorough characterization of solution-precipitated polyelectrolyte complexes, PECs, made from highly-charged polyelectrolytes was undertaken in the 1960s by Alan Michaels and coworkers [1–3], who described the following properties: insolubility in common solvents [2], infusibility [2], amorphous structure [2], blending of constituents on a molecular level [4], virtually no salt counterions within the complex [2], high dielectric constant when wet [5], permeable to water and electrolytes [2], impermeable to macrosolutes [2], and swelling and plasticizing by aqueous electrolytes [2]. Principal projected uses [2] were as membranes for dialysis, ultrafiltration, battery separators, fuel cell membranes, electrically conductive coatings, and medical devices, including contact lenses and chemical sensors. Processing of PECs is possible by dissolving them in special ternary solvents, then casting them [2], or extruding them if they are sufficiently plasticized by added salt ("saloplastics") [6, 7].

Quasisoluble polyelectrolyte complexes, qPECs, colloidal dispersions of a complex stabilized by excess surface charge, were researched in depth by the Moscow State group [8]. Conditions favoring the formation of qPECs are dilute solutions, a strong excess of one polyelectrolyte, and large differences in molecular weights between positive and negative macromolecules [9]. Kabanov *et al.* [8] elegantly demonstrated place-exchange reactions between certain types of polyelectrolyte. This labile behavior was contrasted with non-labile complexation between other pairings of polymers, illustrating the intricate kinetics of polyelectrolyte association and dissociation.

Coacervates are the most diffuse, in character and composition, of the complexes [10, 11]. Coacervates are often formed between polysaccharides or proteins, making them important in the food industry [12]. They are strongly hydrated liquid gels which show clear phase separation from solution. Usually, coacervates are not dense, scattering precipitates, but there is no sharp boundary, at least in current

*Multilayer Thin Films: Sequential Assembly of Nanocomposite Materials*, Second Edition.
Edited by Gero Decher and Joseph B. Schlenoff.
© 2012 Wiley-VCH Verlag GmbH & Co. KGaA. Published 2012 by Wiley-VCH Verlag GmbH & Co. KGaA.

definitions, between "coacervate" and "complex", that is, a "dense coacervate" may be the same as a "diffuse complex."

In each of these morphologies pairing between charged repeat units, with loss of counterions, drives association. The differences are in the density of pairing, the stoichiometry and the amount of water. Many other interactions between polymers exist which are not attributable to ion-pairing, such as hydrogen bonding, charge transfer interactions, host–guest complexation, metal–organic complexation, and hydrophobic interactions not coupled to ion-pairing. These interactions also drive multilayer build-up and are presented in more detail in other chapters. The current chapter is limited to ion pairing as a mechanism.

The majority of multilayer studies have involved only two components alternately adsorbed. Nanoparticle/polyelectrolyte combinations, where polymer "glues" the particles together, often require some subtle optimization of parameters. The assembly of an ultrathin film using highly charged synthetic polyelectrolytes alone is more straightforward and yields uniform blends. Thus, multilayers made from polyelectrolyte components only (PEMUs) are typically amorphous, with neighboring layers extensively interpenetrated [13–17]. Because of this intermingling, actual stratification, or composition modulation in PEMUs is observed only if components are separated by several "layers" of material which include a third component [13, 14, 17]. The definition of a "layer" is rather loose, and, for the majority of films, really means a thickness increment following an adsorption step. Although the lack of layering may make the thin film less intriguing from a structural point of view, it simplifies the theoretical treatment, since the film is represented by an average composition with a mean field response to external stimuli.

One may ask what PEMUs have to offer that PECs do not. The main differences between PECs and PEMUs are in the area of processing. In the latter, each adsorption step ("layer") leads to surface charge reversal, and the adsorbed amount is self-limiting. These two properties induce layer-by-layer propagation, as detailed in Decher's pioneering work two decades ago [15, 18]. Because of the sequential adsorption approach, PEMUs form ultrathin, uniform, smooth, contour-following films. The exceptional thinness, not accessible with solvent cast PECs, of defect-free PEMU membranes provides high flux for selective separations. PEMUs are better suited to surface modification, where uniformity is at a premium and a minimal thickness of a mechanically weak film is advantageous. A few nanometers of PEMU controls wetting [19], biocompatibility [20], and surface charge in microchannels for microfluidics [21, 22].

Further distinguishing features of PEMUs are evidenced by the demonstration of actual layering in PEMUs and related systems. For example, ultrathin strata of polymers may be isolated by layers of clay nanosheets for light-emitting devices [23]. Layers of magnetic nanoparticles have been separated by clay minerals [24]. Layered systems composed of polymers only may be obtained when one of the components is a liquid crystalline polyelectrolyte [25]. We have prepared multilayered "onion skins" that decompose to yield free membranes [26].

Some degree of molecular orientation is possible in PEMUs – an additional advantage. Second order nonlinear optical responses [27–29] are a clear indication of molecular ordering in PEMUs comprising azobenzene polyelectrolytes. From a scale-up perspective, organic solvents and high salt conditions used to process PEC into films are environmentally less acceptable than are pure aqueous salt solutions. Finally, a PEC processing solvent may also be incompatible with the substrate or materials, such as enzymes [30, 31], that must be incorporated.

The system treated here, composed of two synthetic polyelectrolytes, salt ions and water, is the simplest of multicomponent thin films. In this chapter we track all PEMU ingredients and show that focusing on the way charged components contribute to overall electrical neutrality at the surface and in the bulk helps to understand the formation and function of PEMUs. The importance of salt, which controls build-up, permeability and stability, is immediately apparent, but will be no surprise to scientists acquainted with the peculiarities of charged polymers: charged microions are expected to modify electrostatic interactions between polymers [32].

The balance of this chapter is organized in three parts. First, we will consider the forces holding constituent polymers together and the way salt counterions govern these forces and participate in charge balance (13.2). Then, a view of how excess surface charge, key to propagating layer-by-layer build-up, is spread out through the multilayer will provide a semiquantitative model for their construction (13.3, 13.4). Finally, we will show how the internal counterion population, regulated by external salt, controls the transport of charged species through PEMUs (13.5). This chapter is mainly limited to discussion of "linear" PEMUs, where the build-up is strongly kinetically controlled. "Exponential" growth, which has been explored over the past several years, is discussed in other chapters.

## 13.2
### Association Mechanism: Competitive Ion Pairing

Consideration of the response of polyelectrolyte systems to salt usually starts with electrostatic theory [32, 33]. The local electrostatic potential caused by the local charge density is determined by the Poisson equation. Using a mean-field approximation, local ionic concentrations are then expressed as a Boltzmann distribution. Typically, the Poisson–Boltzmann equation is linearized by assuming low charge density (Debye–Hückel approximation) or more elaborate nonlinear solutions are sought. For polyelectrolytes, concepts of local stiffness, or persistence length, and excluded volume have been introduced, as well as sophisticated scaling concepts. Interactions are not specific and there is no dependence of interaction energy on the chemical identity of the charges. The interaction energy between like and opposite charges is modified by screening, the characteristic decay length (Debye length) of electric potential being shorter in solutions of higher salt concentration [32, 33].

In this chapter, a more "chemical" perspective is taken in dealing with salt ions, where they are to be found, and how they modify polymer–polymer interactions. In spite of the imposing precision and complexity of electrostatic theory, it does not take into account the polarizibility of functional groups, their degree of hydration and,

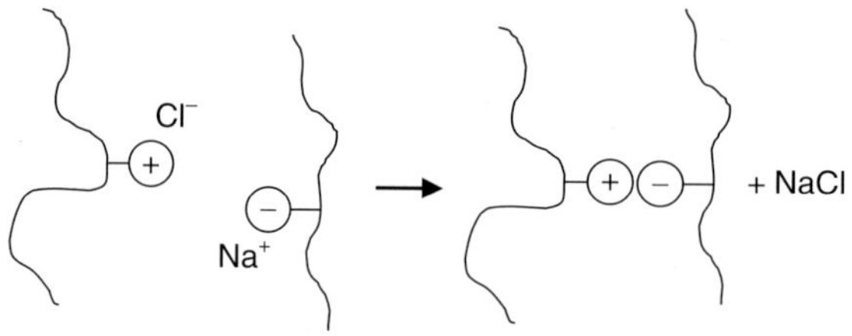

**Figure 13.1** Complexation between positive polyelectrolyte segment and negative polyelectrolyte segment releasing counterions.

more specifically, differential hydration. Here, precise values of interactions, obtained from electrostatic first principles, are forgone in favor of a combination of equilibrium constants that must be experimentally obtained, but can be rationalized and categorized *post hoc* by intuitive arguments based on entropy, hydrophobicities, polarizabilities, polymer excluded volume and other fundamental physical-chemical constructs.

The complexation of polyelectrolytes is known to be virtually athermal [2, 3]. The overall free energy change of complexation is derived mainly from the release of counterions [34] ("escaping tendency", that is, increase in entropy [3]). Small free energy changes per segment are additive, to a point [35], due to the cooperative nature of polyelectrolyte complexation, yielding strong association on a per-molecule basis. Since ions are well hydrated, the release of waters of hydration, specifically the differential levels of solvation between free, solvated polymer and associated complex, should also be considered in the net energy equation [36]. The term "hydrophobicity" is used here to denote the relative numbers of waters of hydration released [36].

The equilibria considered here are variations of place-exchange or ion-exchange reactions. In its most generic form, the interaction between charged polymer segments is represented by Figure 13.1.

Upon complexation, polyelectrolyte segments form ion pairs and relinquish their counterions. Electrostatic neutrality is maintained and no chemical bonds are formed in the reshuffling of charged species. The term "ion pairing" is used to describe an energetically favorable pairing of polymer segments, driven by the loss of water and counterions [36].

From the perspective of a single polyelectrolyte segment, an oppositely-charged segment and a counterion compete to balance the charge. There are two arenas where this competition for charge takes place. The first is in the initial formation of the PEMU (or PEC). Individual solution phase macromolecules come together, expelling their counterions as they complex. There may be some conceptual fuzziness, with a long history [32], as to exactly where the counterions are and how closely they are associated with the polymer chain, but for simplicity we assume they are predominantly "condensed" [37] on the polymer chain. For a polyelectrolyte adsorbing on an oppositely-charged surface, the surface counterions are displaced by adsorbing polymer segments, a process that can be directly tracked by radiochemical labeling techniques [38].

In the second arena, the chemical potential of salt in solution can cause swelling or infiltration of ions into a bulk polyelectrolyte complex *after* it has been formed. For assembled multilayers there is a clear phase change on going from complexed polymer to solution. Waters of hydration can be added into the equation, as is done below.

The situation cannot be described simply by vacuum electrostatics between naked, isolated charges, which would yield enormous free energy changes. The driving force is ion pairing between polyelectrolyte segments, driven by release of counterions and water. Additional salt ions modify the overall free energy of interaction by competing for polymer charge. Given this, the mechanism is better defined as competitive ion pairing [39].

Systems without added salt are at the theoretical maximum strength of interaction, but are also poorly defined, since the ionic strength is that of the polyelectrolyte and its counterions. Multilayers have been created using small components of low charge, but, in the absence of specific interactions, must be assembled at low ionic strength [40–42].

## 13.2.1
### Intrinsic vs. Extrinsic Charge Compensation

#### 13.2.1.1 Key Equilibria

In constructing a multilayer, there are a host of experimental options, each with potential impact on formation [39]. Variables include exposure time, temperature, polymer type, polymer molecular weight, polymer concentration, salt concentration, salt identity, pH, and solvent composition. Fortunately, since almost all work is performed with high molecular weight material at room temperature in aqueous NaCl solutions the only significant variables are polymer type, and salt concentration, and also pH if one or more of the polymers is a weak acid/base [19, 43]. Even then, the process is extraordinarily robust (forgiving) and some form of multilayering is easily obtained (as long as the substrate is properly cleaned!).

We begin with a multilayer that has been formed with a pair of oppositely charged polymers. Following sequential adsorption of polyelectrolytes to the desired film thickness, the PEMU is typically rinsed in pure water to remove excess polymer and salt.

For a complex with 1:1 charge stoichiometry, one can envisage two scenarios, shown in Figure 13.2, of how net charge neutrality is maintained, by a combination of polymer repeat units and salt ions, within the as-made film. In one case, termed *intrinsic* compensation, polymer positive charge is balanced by a negative charge, also on a polymer [44]. Alternatively, polymer charge is balanced by salt counterions derived from the bathing solution used to construct the multilayers (*extrinsic* charge compensation) [44]. A continuum of intrinsic → extrinsic composition is available, with the proviso that fully extrinsic multilayers are not realistic since, in the absence of other interaction mechanisms, the PEMU would decompose back into isolated molecules.

The presence of salt ions would have a substantial impact on the performance of PEMUs in certain applications. For electronic devices, mobile counterions would

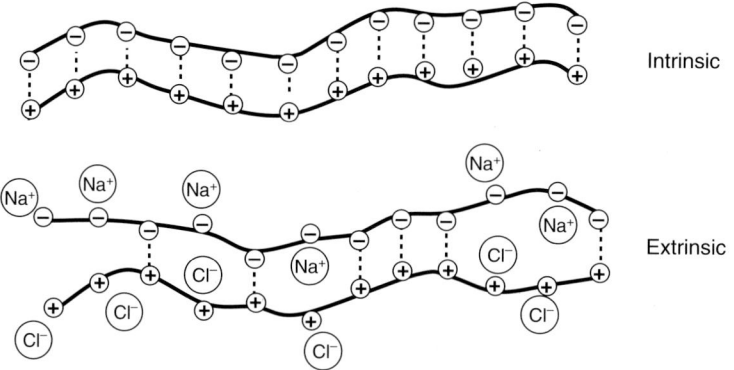

**Figure 13.2** Comparison of intrinsic and extrinsic charge compensation in polyelectrolyte multilayers. Polyelectrolyte–polyelectrolyte ion pairs are shown as dotted lines.

control the dielectric constant and the conductivity of thin film PEMU insulators, and would degrade electrical contacts. Intrinsic compensation controls transport of charged species through PEMUs when they are employed as membranes, as discussed below.

An interesting physiological question arises when considering intrinsic compensation: how can the polymers undergo the contortions required to match each positive segment with each negative segment and obviate the need for salt ions, especially if the spacings of charged units along the polyelectrolyte backbone are very different? In considering this question for PECs, Michaels [2, 3] concluded that polyelectrolytes do not complex as "ladders," where a polymer is associated with only one other, complementary molecule (DNA-like), rather, they form amorphous "scrambled salts." Our experiments, described below, confirm the intrinsic compensation mode for the bulk of PEMUs.

Although as-made PEMUs contain no salt ions it is possible to introduce them by increasing their external (solution) concentration (chemical potential). In the following equilibrium, polyelectrolyte/polyelectrolyte ion pairs are replaced by polyelectrolyte/salt ion pairs. The process is a type of ion exchange [45], the difference being that the polyelectrolyte segments are not free to leave the PEMU matrix. The system, believed to be under rapid, reversible equilibrium control for NaCl [46], as it involves the transport of small ions with minimal displacement of polyelectrolyte segments, is represented by

$$\text{Pol}^+\text{Pol}^-_m + \text{Na}^+_{aq} + \text{Cl}^-_{aq} = \text{Pol}^+\text{Cl}^-_m + \text{Pol}^-\text{Na}^+_m \tag{13.1}$$

where $\text{Pol}^+$ and $\text{Pol}^-$ are, respectively, positive and negative polyelectrolyte repeat units. The subscript "m" refers to components in the multilayer phase. The equilibrium constant may be written as

$$K' = \frac{y^2}{(1-y)[\text{NaCl}]^2_{aq}} = \left(\frac{y^2}{[\text{NaCl}]^2_{aq}}\right)_{y \to 0} \tag{13.2}$$

where $y$ is the fraction of the multilayer in the extrinsic form and $1-y$ is the intrinsic fraction [46]. Of course, salt activities are more properly used in lieu of concentration. It is clear, from this equilibrium, how salt directly controls the degree of association or interaction between polymer segments. Note that the number of positive and negative counterions is equal. These ions are, themselves, available for exchange with other ions. We have termed these systems, having matched positive and negative immobile ion pairs in their as-made state, "reluctant exchangers," since exchangeable ions can be forced into them with external chemical potential, as in the equilibrium above [46].

"Interaction energy" means, in the present context, the free energy of ion pairing between polyelectrolyte segments, which is the reverse of Equilibrium 13.1. Different pairs of polyelectrolytes have different strengths of interaction [36].

### 13.2.1.2 Interaction Energies

Multilayers expand as ions (and additional waters of hydration) enter the film [47], although this is not always the case [48]. In early experiments, Sukhorukov et al. measured the expansion of PEMUs by salt *ex situ* using X-ray reflectivity on dried films [49]. We have employed atomic force microscopy (AFM) to determine swelling *in situ* of representative pairs of polyelectrolytes [50]. The first measurement for each PEMU was made on the "dry" film (exposed to ambient atmosphere of approximately 30% relative humidity). Pure water was then added to the cell, followed by salt solutions. All films expanded on immersion in water. Very different swelling behavior, Figure 13.3, is observed for the three combinations of "typical" polyelectrolytes: poly(styrene sulfonate), PSS/poly(acrylic acid), PAA; PSS/poly(diallyldimethylammonium), PDADMA; PSS/poly(allylamine hydrochloride), PAH. These differences reflect the relative strengths of association of oppositely charged polyelectrolyte segments. Polymer pairs which form more hydrophobic complexes associate more strongly in contact with aqueous solutions and are thus less prone to swelling by salt ($K'$ smaller). Swelling curves were reproducible for the same sample provided $[NaCl]_{aq}$ did not exceed 2 M (for PSS/PDADMA) or 0.3 M (for PAA/PDADMA), after which concentrations the films started to decompose.

The ratio of the "wet" to "dry" thicknesses, given in the caption for Figure 13.3, may be used as a relative index of hydrophobicity of polyelectrolyte pairs. The water content for PSS/PDADMA in contact with liquid water is in approximate agreement with prior estimates [51]. A quantitative measure of salt infiltration can be provided by defining a volume swelling coefficient, $Q_{swell}$ (the slope from Figure 13.3),

$$Q_{swell} = \frac{\% \text{ swelling}}{[\text{salt}]} \quad (13.3)$$

$Q_{swell}$ values for PAA/PDADMA, PSS/PDADMA, and PSS/PAH are about 400, 20, and $<1 \, M^{-1}$, respectively. Swelling involves sorption of both salt and additional waters of hydration, since polyelectrolyte segments paired with salt counterions will be more hydrophilic. Interaction energies under nonstandard conditions (i.e., for salt concentrations other than 1 M) are given by $\Delta G = \Delta G^o + RT\ln[NaCl]_{aq}^2$.

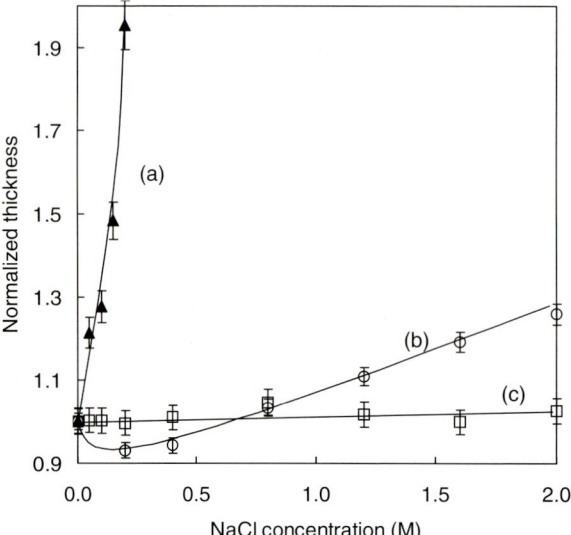

**Figure 13.3** Thickness (normalized to [NaCl]$_{aq}$ = 0) as a function of NaCl concentration for three multilayers: (a) PAA/PDADMA; (b) PSS/PDADMA; (c) PSS/PAH. For these PEMUs, "wet" (immersed in pure water)/"dry" (ca. 30% r.h.) thickness (Å): (a) 2330/640 = 3.64; (b) 3550/1750 = 2.03; (c) 600/490 = 1.22. Solid lines are a guide to the eye.

The detailed behavior of each PEMU is of interest. PSS/PDADMA exhibits a (reproducible) decrease in thickness at low salt concentrations. This is due to the removal of water in pores or voids ("liquid water") under increasing external osmotic pressure, as illustrated by the addition of a neutral osmotic stressing agent such as PEG [36]. PAA/PDADMA exhibits a "hyperswelling" behavior, consistent with the observation that the films actually decompose at intermediate salt concentrations [52]. Different salts yield different $Q_{swell}$, depending on their hydrophobicity and charge.

Swelling is reversible and reproducible. Salt concentration jump experiments on the permeability of redox active probe ions through PSS/PDADMA multilayers have demonstrated that swelling is rapid and the equilibrium swelling level is also attained quickly [46]. During the normal course of multilayer build-up, where the substrate is alternately immersed in salt solutions and salt-free rinse water, the multilayer is undergoing cyclic swelling and shrinking as ions rush in and out of the film under changing conditions of solution salt concentration – a phenomenon we term "ion breathing."

Calorimetry experiments on polyelectrolyte complexes have confirmed the small heats of complexation [34, 53]. In order to compare the relative importance of enthalpic and entropic driving forces in polyelectrolyte association, $\Delta G$ is needed ($\Delta G = \Delta H - T\Delta S$). Doping constants, the inverse of association equilibrium constants, were determined using *in situ* FTIR spectroscopy of multilayers doped with infrared-active ions [36]. The slopes of the doping curves in Figure 13.4 were used to

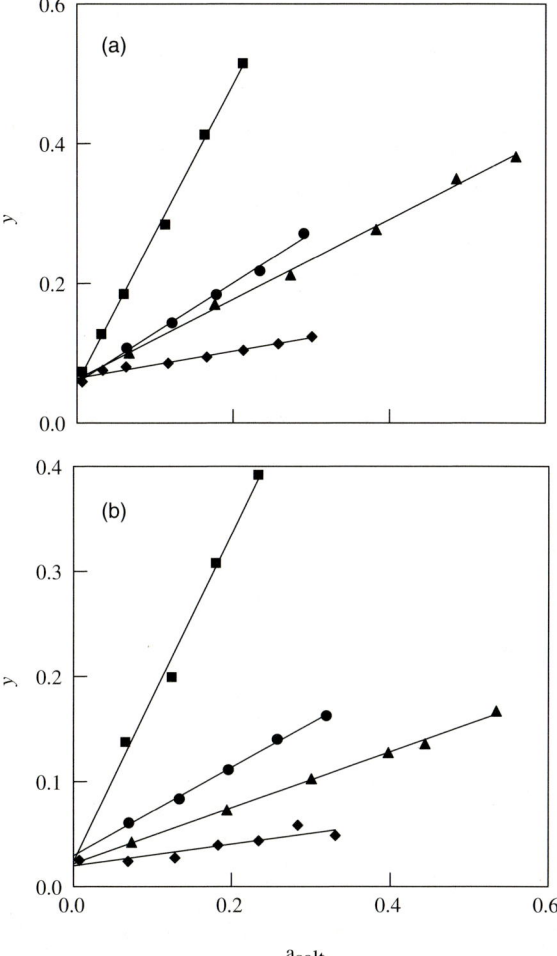

**Figure 13.4** Mole ratio of salt/sulfonate, representing the fraction of extrinsic sites or doping level, y, for each of the IR active ions versus solution activity of the salts: ■, perchlorate; ● thiocyanate; ▲, nitrate; ◆, azide. (a) (PDADMA/PSS)$_{12.5}$, and (b) (P4VMP/PSS)$_{167.5}$. The intercept on the y-axis suggests residual anion content of 6 and 2% in the respective PEMUs.

determine $K'$, then $\Delta G$ was calculated from $\Delta G = -RT\ln K'$. The thermodynamics, see Table 13.1, were shown to be largely driven by entropy [36].

Doping equilibrium constants follow the Hofmeister series [36] – the least hydrated ions are the strongest dopants. The Hofmeister series was also demonstrated for controlling the rate of multilayer growth in 1999 using *ex situ* ellipsometric measurements [39] and in 2004 using *in situ* quartz crystal microbalance measurements [54]. As seen from Table 13.2, thicker multilayers are obtained with less

**Table 13.1** $K_{dop}$ and $\gamma_{residual}$ determined from Figure 13.4 and $\Delta G_{association}$, calculated from $\Delta G_a = RT \ln K_{dop}$.

| Salt | $K_{dop,PDADMA/PSS}$ | $\gamma_{residual}$ | $\Delta G_a$ (kJ mol$^{-1}$) |
|---|---|---|---|
| NaClO$_4$ | 3.8 | 0.06 | 3.3 |
| NaSCN | 0.42 | 0.05 | −2.1 |
| NaNO$_3$ | 0.27 | 0.06 | −3.2 |
| NaN$_3$ | 0.031 | 0.06 | −8.6 |
| | $K_{dop,\,P4VMP/PSS}$ | | |
| NaClO$_4$ | 2.9 | 0.02 | 2.6 |
| NaSCN | 0.18 | 0.03 | −4.2 |
| NaNO$_3$ | 0.071 | 0.02 | −6.6 |
| NaN$_3$ | 0.011 | 0.02 | −11 |

a) An entropy driven association is also demonstrated by the temperature independence of K (Figure 13.5).

hydrated ions. Interestingly, the cation series, H$^+$, Li$^+$, Na$^+$, K$^+$, Rb$^+$, Cs$^+$, shows a stronger trend than the anion series, F$^-$, Cl$^-$, Br$^-$.

### 13.2.1.3 Multilayer Decomposition

Figure 13.4 suggests it is possible to reverse polyelectrolyte association at sufficiently high salt concentrations. At some critical salt concentration, [NaCl]$_{crit}$, the extrinsic compensation is pushed to the point, $y_{crit}$, where the few remaining polymer/polymer ion pairs are no longer capable of holding the PEMU together, and the

**Table 13.2** Multilayer deposition of 10 layer pairs of PDADMAC and PSS, 1 mM each, on Si disk at 600 rpm. Deposition time, 5 min. 3 × 1-minute rinses in fresh water between each layer.

| Salt | $a/\gamma/M$[a] | $n_{cation}$[b] | $n_{anion}$[c] | Multilayer Thickness,[d] Å |
|---|---|---|---|---|
| NaF | 0.33/0.632/0.52 | 3.5 | 2.7 | 430 |
| NaCl | 0.33/0.681/0.48 | 3.5 | 2.0 | 903 |
| NaBr | 0.33/0.697/0.46 | 3.5 | 1.9 | 900 |
| NaI | 0.33/0.723/0.45 | 3.5 | 1.6 | n/a[e] |
| HCl | 0.33/0.757/0.43 | 2.7[f] | 2.0 | 705 |
| LiCl | 0.33/0.739/0.44 | 5.2 | 2.0 | 820 |
| NaCl | 0.33/0.681/0.48 | 3.5 | 2.0 | 903 |
| KCl | 0.33/0.649/0.50 | 2.6 | 2.0 | 1200 |
| RbCl | 0.33/0.634/0.52 | 2.4 | 2.0 | 1466 |
| CsCl | 0.33/0.606/0.54 | 2.1 | 2.0 | 1420 |

a) $a/\gamma/M$ = activity/activity coefficient/concentration.
b) Hydration number of cation in aqueous solution.
c) Hydration number of anion in aqueous solution.
d) Ellipsometric thickness, +/− 3%.
e) PDADMAC solution precipitates in NaI.
f) Hydration number for H$_3$O$^+$ ion.

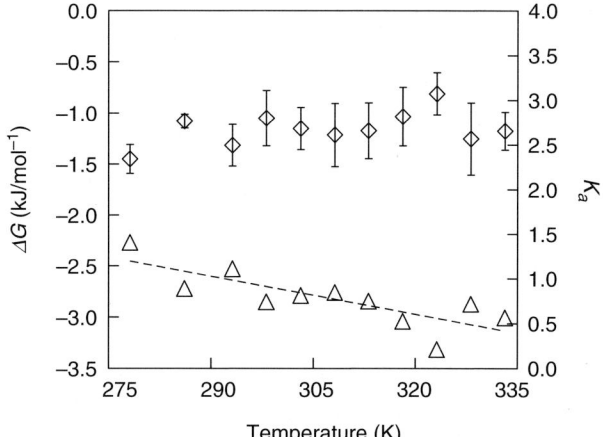

**Figure 13.5** Association constant for ion pairing, $K_a$, in a 30-layer PSS/PDADMA PEMU doped with sodium nitrate versus temperature (5–60 °C) (◇); corresponding $\Delta G$s in kJ mol$^{-1}$ (△).

individual polyelectrolytes surrender to the entropic forces driving dissociation. Figure 13.6 demonstrates this effect for a PAA/PDADMA system [52]. The multilayer is stable to salt concentrations up to about 0.4 M, after which point the system decomposes. The deconstruction of PEMUs may actually yield very loosely associated complexes, that is, not true solution of individual polyelectrolyte molecules.

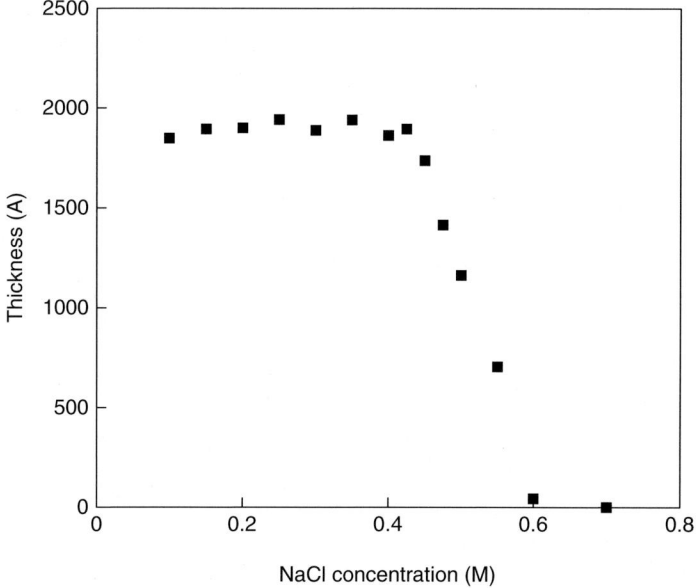

**Figure 13.6** Thickness of 20-layer PAA/PDADMA film, prepared with 84 500 $M_w$ PAA, as a function of salt concentration. The deposition, soaking and rinsing solutions were maintained at pH 11.

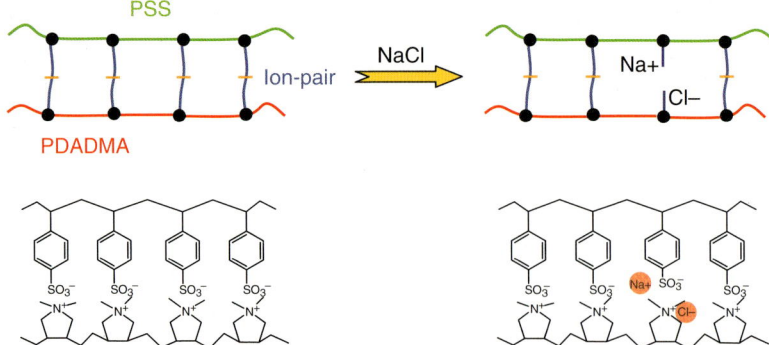

**Figure 13.7** Complexed chains of PDADMA and PSS. Ion pair crosslinks between PDADMA and PSS are shown as blue lines. On the addition of salt, crosslinks are broken.

The concept of minimum charge density along the backbone needed to form a stable multilayer has been addressed previously [55, 56]. We add here the proviso that this minimum charge density depends on the salt concentration and salt type, and also very strongly on the chemical identity of the charged units being relied on to form the thin film. We have used the preferential decomposition of "weak" multilayers, inserted between more salt-resilient ones, to prepare onionskin PEMUs that, on exposure to salt or pH changes, delaminate to yield free ultrathin membranes [26].

#### 13.2.1.4 Doping-moderated Mechanical Properties

Direct evidence that doping actually disrupts polymer/polymer ion pairing (as opposed to simply swelling the complex) is available from measurements of the mechanical properties of multilayers [57, 58]. As shown in Figure 13.7, each ion pair functions as a physical crosslink between polyelectrolytes. Crosslinks are progressively broken as the multilayer is doped. Micromechanical measurements on micron-thick samples of PEMU illustrate the classical behavior expected when crosslinks are broken: the modulus decreases [57].

Stress–strain plots for PSS/PDADMA PEMUs are depicted in Figure 13.8. It is clear that doping by salt leads to a strong decrease in bulk modulus. In fact, classical theories of rubber elasticity were followed quite well when the modulus was correlated to the doping level and thus crosslink density [57].

### 13.3
### Surface versus Bulk Polymer Charge

Figure 13.9 depicts a typical growth versus layer number for a strongly dissociated pair of polyelectrolytes (in this case, PSS and poly(*N*-methyl-4-vinylpyridine) PNM4VP[44] These particular polymers have been radiolabeled,[38, 59, 60] making it a simple task to accurately track their surface concentration. A strong dependence

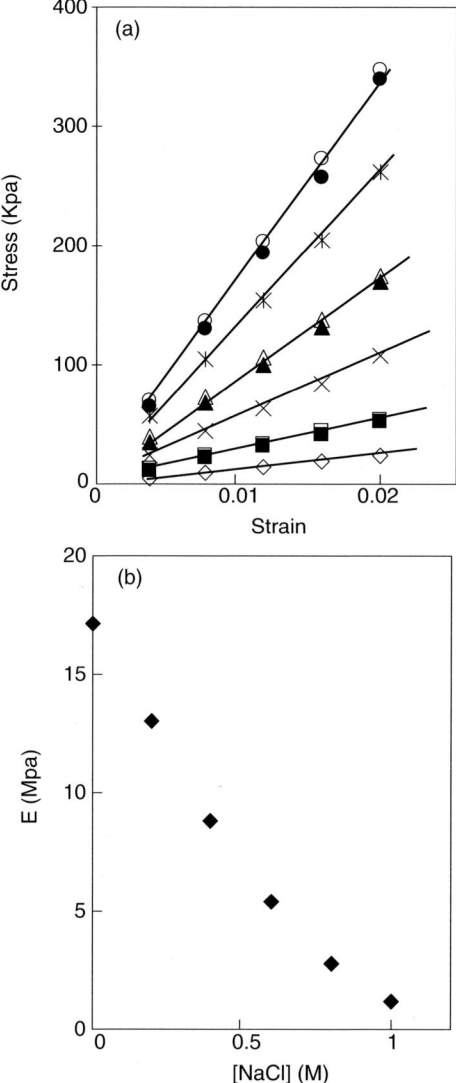

**Figure 13.8** (a) Stress–strain curves of a (PDADMA/PSS)$_{250}$ multilayer at different salt concentrations. Open circles, asterisk, triangles, crosses, squares and diamonds correspond to the stretching cycle (in increasing order of elongation) with salt concentrations of 0.0, 0.2, 0.4, 0.6, 0.8 and 1.0 M NaCl respectively. Solid circles, triangles and squares indicate a decreasing elongation cycle at 0.0, 0.4, 0.8 M. A similar trend was observed at the other salt concentrations (b) Elastic modulus obtained from the slope of the curves in (a). At 0.0 M NaCl, $E = 17$ MPa. Dimensions were corrected to account for swelling.

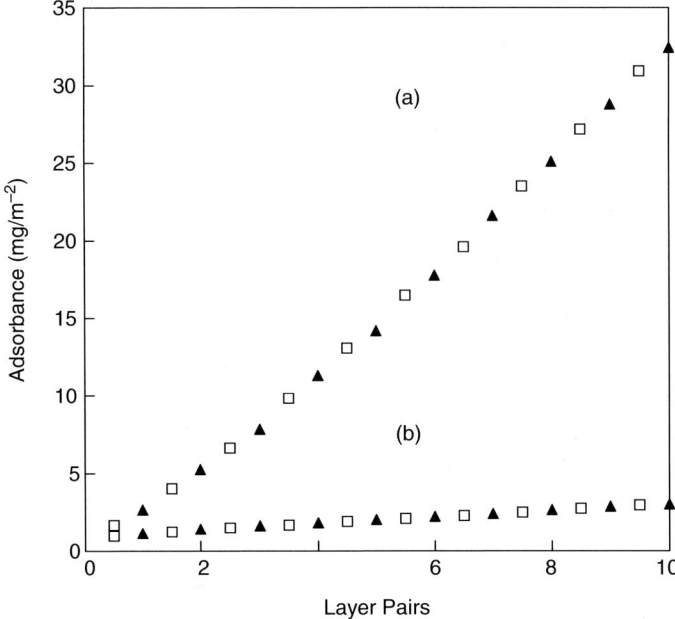

**Figure 13.9** Build-up of a multilayer of PM2VP (squares) and PSS (triangles) from 0.1 M NaCl$_{aq}$ (a) and pure water (b).

of the amount deposited on the solution salt concentration is observed. Other typical multilayer growth features include a slight upwards curvature of the amount per layer towards the beginning of the plot (for 0.1 M salt) followed by linear behavior. The linear portion of the curve is evidence that the eventual thickness increment per cycle does not depend on the amount already deposited, and thus the controlling features of multilayer growth must be related to properties at or near the surface. The total molar charge of positive and negative polyelectrolytes balance for Figure 13.9 [44], which points towards intrinsic bulk compensation, but does not prove it conclusively.

The non-intuitive reversal of surface charge that occurs at each adsorption step, critical for multilayer build-up, was recognized at the outset [18, 61]. The sign of surface charge and its reversal are shown by electroosmotic flow experiments on capillaries coated with multilayers [21, 22]. The direction of electroosmotic flow in these microfluidic systems depends on the sign of the wall charge. As shown in Figure 13.10, the flow is reversed as each new polymer layer is added. The apparent charge density, which is taken to be the net charge density presented to the solution/multilayer interface, is much less than the total integrated areal excess (net) polymer charge density and is also remarkably constant. Surface charge reversal is also seen in the conceptually inverted arrangement of PEMU-coated particles moving under an electric field [62].

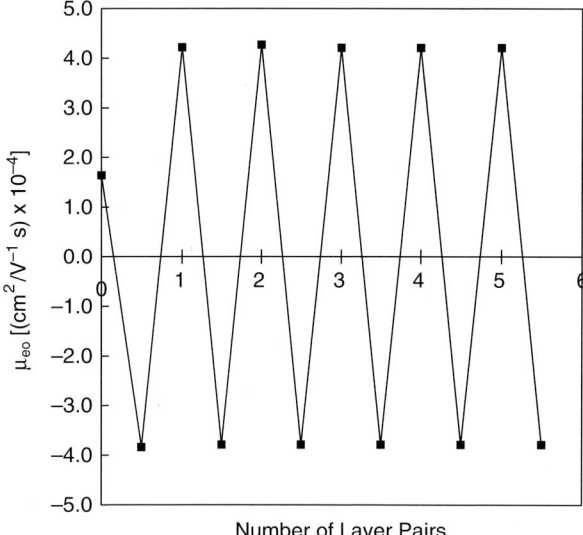

**Figure 13.10** Demonstration of surface charge reversal using capillary electrophoresis. PDADMA and PSS were alternately deposited from 0.5 M NaCl$_{aq}$. Acetone was used as a neutral marker to measure electroosmotic flow, with detection at 254 nm. Capillary length 37 cm, 50 μm I.D. column, 15 kV applied voltage. 20 mM phosphate buffer.

Direct evaluation of the total areal excess charge density of a multilayer was accomplished using radiolabeled counterions and multilayers terminated with either a positive or a negative polyelectrolyte [44]. For both types of multilayer, ion self-exchange experiments indicate the presence of positive or negative salt ions within the film. For the example in Figure 13.11 a PSS/PDADMA multilayer of thickness 245 Å, having either PSS or PDADMA as the "capping" layer, was exposed to a solution of radiolabeled calcium ions ($^{45}$Ca$^{2+}$, details on this type of radiochemical experiment may be found in references [38, 60, 63–65]) At the point indicated, an excess of *unlabeled* calcium ions was added, which would self-exchange with any labeled calcium in the film. For the film capped with PSS (negative surface) an excess positive polymer charge was detected from the self-exchange, but the positive-capped film has no detectable salt cations. Symmetrical results were obtained for a negative radiolabeled probe ion ($^{14}$CH$_3$COO$^-$) [44]. These results were interpreted to show that there are no salt ions within the bulk of an as-made multilayer (within the limit of detection of about 100 ppm. Salt concentrations at the low ppm level may be inferred from AC dielectric measurements [66]) and only one type of salt ion exists at the surface. The absence of salt ions contradicts the results from FTIR seen in Figure 13.4 above. However, the PEMUs for FTIR measurements had to be thick and so were constructed with 1.0 M NaCl. It appears nonstoichiometry or residual extrinsic sites may be induced with higher salt concentrations.

The absence of salt ions within the bulk of as-made multilayers is consistent with prior, but less sensitive, studies using XPS [16, 67], and with thermal gravimetric

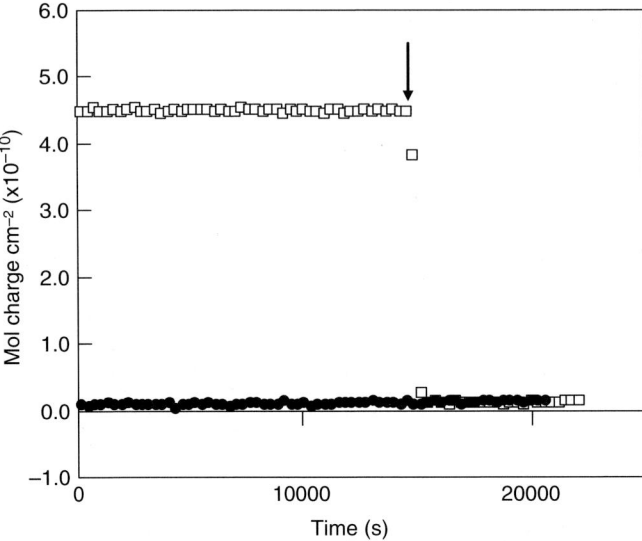

**Figure 13.11** Salt cation content, in terms of areal charge density, of a PDADMA/PSS multilayer deposited from 0.1 M NaCl, with PDADMA as the top layer (circles, 10.5 layer pairs) or PSS as the top layer (squares, 10 layer pairs). $^{45}Ca^{2+}$ ($10^{-4}$ M) was used as a probe. At the point indicated an excess of unlabeled $Ca^{2+}$ was added to self-exchange salt cations in the PEMU. The multilayer contains exchangeable positive charge, at the surface only, when PSS is the top layer.

analysis [51], which shows no inorganic residue on thermolysis of multilayers. It is also consistent with the strong exclusion of ions when multilayers are used as membranes. Of course, salt added to the external solution contacting the multilayer can introduce extrinsic charge, as in Equation 13.1 and Figure 13.4.

The absence of extrinsic charge within as-made multilayers is an important and interesting property. It confirms the driving forces leading to complete ion pairing, and supports strong interpenetration [14] of "layers" and the scrambled or amorphous structure [2, 3]. It seems to be a property of a bulk polyelectrolyte complex that excess polymer charge is *extruded* toward the surface, much like a bulk conductor localizes excess electronic charge on the surface. Charge extrusion is observed for PEMUs containing weak acid polyelectrolytes when the bulk charge balance is perturbed by changes in pH [68].

Despite the fact that ion pairing defeats many attempts to induce actual layering of material within PEMUs [55], several strategies may be invoked to create extrinsic charge and, thereby, layering in two-component systems. (Layering in three-component polyelectrolyte systems is possible provided the spacing of one of the components, such as a deuterated layer, is greater than a few layer thicknesses [14, 17]). In two-component polyelectrolyte systems, a fundamental requirement for layering with periodicity greater than the width of a polymer chain is the presence of extrinsic compensation – layering is not possible for completely mixed intrinsic materials. Layering with particle/polymer combinations is observed due to the inability of the components to mix.

Post-assembly processing steps can induce excess polymer charge of one type within a multilayer. Such steps include thermolytic or photolytic elimination of charged groups, electrochemical injection of charge into the multilayer [16, 44, 69, 70], or the introduction of charge via pH changes when the multilayer includes weakly acidic polyelectrolytes [19, 43]. For example [44], the polycationic precursor to poly(phenylene vinylene), PPV, when incorporated into a multilayer, will, on heating, eliminate thiophene and a positive charge.

Extrinsic charges produced by mechanisms other than "reluctant" swelling may be thermodynamically unstable and may relax towards intrinsic compensation with *extrusion* to the surface and possible *expulsion* of excess polyelectrolyte charge from the multilayer. Immobilization of polymers, for example by crosslinking, may be required to prevent this type of relaxation.

## 13.3.1
### Distribution of Surface Charge in Layer-by-Layer Build-up: Mechanism

Surface excess polyelectrolyte charge is known to be compensated, then overcompensated, on the addition of oppositely-charged polymer [55, 61, 71, 72]. It is understood that the increment per deposition cycle depends on the magnitude of this excess surface charge. For the PEMU pair of PSS and PDADMA, commonly employed in multilayer studies, we have shown [39] that the film thickness is an approximately linear function of the salt concentration used in the deposition solution, $[NaCl]_{build}$. This implies that the surface excess charge must also be proportional to the salt concentration. We have verified this relationship using radiolabeled probes ($^{35}SO_4^{2-}$) of the net areal surface charge density [73]. As shown in Figure 13.12, the areal surface charge density of a multilayer capped with PDADMA shows a roughly linear dependence on $[NaCl]_{build}$, whereas the same multilayer capped with PSS contains neither surface nor bulk negative counterions.

If deposition is performed from concentrated salt solutions, several equivalent monolayers of polymer accrue on a single deposition step [39]. This large polymer excess charge is believed to be accommodated in the substrate-perpendicular direction over a length scale equal to several "layer thicknesses [39, 44]." Such a spreading of polymer charge is depicted in Scheme 13.1, which shows contributions to volume charge density from both polyelectrolytes, with an exponential decay of positive excess polymer charge from the surface to the bulk having a characteristic length $l_{cp}$. The excess polymer charge density is mirrored by a counteranion profile, required to maintain net neutrality.

Scheme 13.1 incorporates several elements discussed above and found in the literature. First, the bulk of the PEMU is intrinsically compensated, in contrast to the region near the surface, where salt counterions balance excess polymer charge [44]. Second, both polymers are present at all points, since they are interpenetrating. [14, 15] This coexistence extends to the surface, allowing excess polymer to be distributed smoothly into the multilayer, rather than remain a thick surface layer of loops and tails.

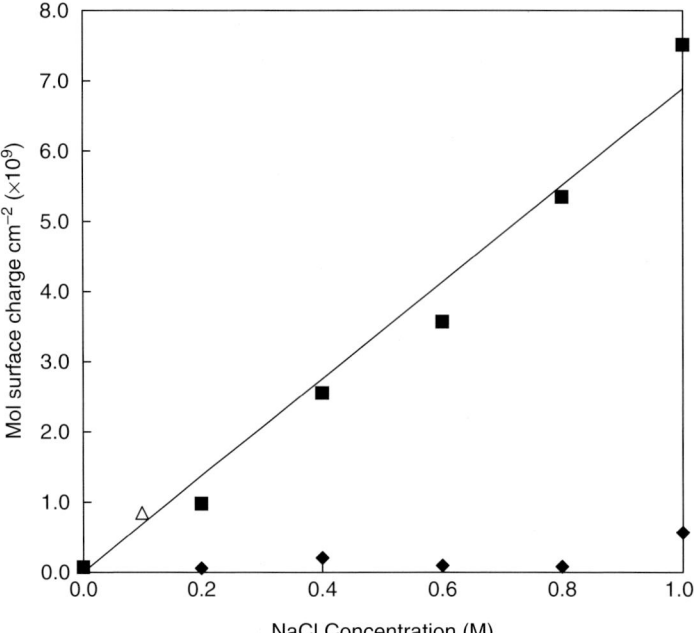

**Figure 13.12** Positive surface areal charge density for 15-layer PSS/PDADMA multilayers (PDADMA-capped, squares), and 14-layer multilayers (PSS-capped, diamonds) made from solutions containing 1 mM polymer and different [NaCl]$_{build}$. The probe for the surface charge was $^{35}$S-labeled sulfate ion. No surface or bulk positive polymer excess charge is observed for the PSS-capped multilayer (i.e., the multilayer contains no negative salt ions). The solid line is a guide to the eye. The open triangle is from a similar strongly-dissociated polyelectrolyte system. Only for multilayer assembled from 1.0 M salt does the PEMU contain bulk anions, as also seen in the FTIR results of Figure 13.4

At the multilayer/solution interface the charge of the last-added polymer compensates the previous one by a factor $\phi$. If $\phi = 1$ the opposite polymer charges are in exact stoichiometric ratio. $\phi - 1$ is the surface overcompensation level. In our model, overcompensation, essential for multilayer growth, is driven by entropy gains due to the adsorbing polymer preserving some configurational degrees of freedom despite being immobilized on the multilayer surface.

The exact reason for the magnitude of $l_{cp}$ for different conditions is not clear. We initially believed $l_{cp}$ to be controlled by the increasing density, and repulsion, of polyelectrolyte segments going into the PEMU. On the other hand, the mechanism for formation of a layer, illustrated below, emphasizes the kinetic nature of a polyelectrolyte diffusing into the surface. In any case, in the model, at increasing lengths $l$ from the interface, overcompensation decreases in an exponential fashion (decay rate $l_{cp}$).

In order to estimate the amount of polymer added in each deposition step, the area under the exponential polymer excess charge profile is integrated and doubled,

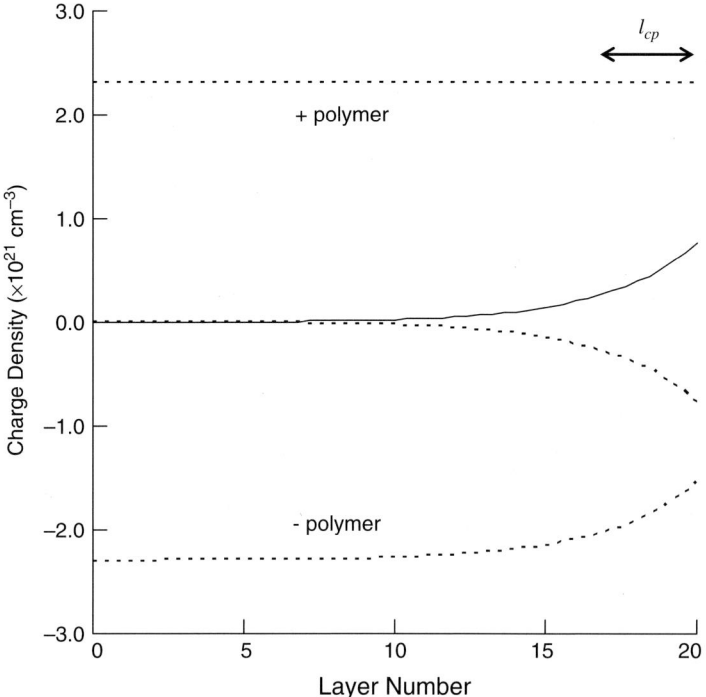

**Scheme 13.1** Sketch of charge density versus distance from the substrate (in terms of nominal layer number) for an as-made 20 layer pair PSS/PDADMA multilayer measured under dry conditions. The curves were modeled according to Equations 13.4 and 13.5 with $l_{cp} = 200$ Å and $\phi = 1.5$. The system assumes intrinsic compensation within the bulk of the multilayer (layers 0 to 10), that is, 1:1 polymer charge stoichiometry and no salt ions, as is generally found for "linearly" growing systems. The final deposition is from PDADMAC. Contributions from positive and negative polymer are shown as dotted lines and the net polymer charge density is shown as the solid line. The area under the latter curve would represent the polymer excess "surface" charge. Chloride counterions are represented by the dashed line. Charge density is estimated assuming a bulk density of 1.2 g cm$^{-3}$ and a segment ion pair molecular weight of 310 g mole$^{-1}$.

assuming the adding polymer compensated the multilayer then overcompensates by the same amount.

Assuming we are modeling the build-up of a PSS/PDADMA multilayer, thickness $t$, from 1.0 M NaCl, shown in Figure 13.13, the thickness increments, $t_n^+$ and $t_n^-$, for odd (PDADMA) and even (PSS) layers, respectively, are

$$t_n^+ = 2l_{cp}\left(\frac{\phi-1}{\phi}\right)f^+\left(1-e^{-t/l_{cp}}\right) \tag{13.4}$$

$$t_n^- = 2l_{cp}\left(\frac{\phi-1}{\phi}\right)f^-\left(1-e^{-t/l_{cp}}\right) \tag{13.5}$$

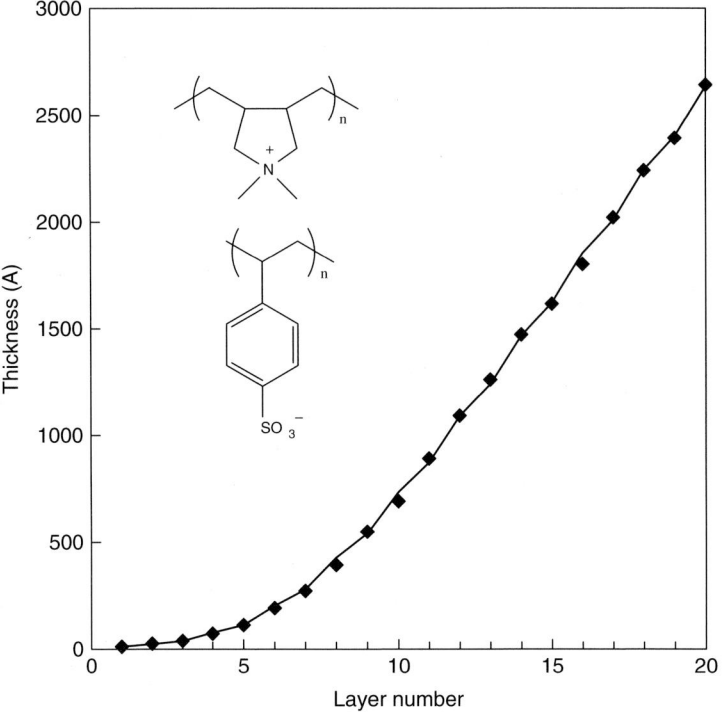

**Figure 13.13** Thickness as a function of the number of layers for a PSS/PDADMA multilayer deposited on silicon wafer from polymer solutions containing 1.0 M [NaCl]$_{build}$. Odd layers are from PDADMA; even from PSS. Polymer concentration 10 mM; 5 min deposition time with three water rinses in between layers. Solid line is a fit to Equations 13.4–13.6 with $\phi = 1.86$ and $l_{cp} = 425$ Å. The thickness of the first layer, $t_1$, was measured to be 12 Å. Exponential growth is followed by linear growth after the thickness $> l_{cp}$.

The thickness increment from each layer is limited, at the beginning of build-up, by truncation of the excess charge profile by the substrate, represented by the rightmost term in Equations 13.4 and 13.5. At sufficient thicknesses, or smaller $l_{cp}$s, the growth is a linear function of layer number. The overall thickness of a film is the sum of thickness increments

$$t = \sum t_n = t_1 + \sum t_n^+ + \sum t_n^- \tag{13.6}$$

Where $t_1$ is the anomalous, but small, thickness of the first layer.

An iterative computer program [73] yields thicknesses for specific values of $\phi$ and $l_{cp}$. The model using the optimum pair of $\phi$ and $l_{cp}$ is also presented in Figure 13.13. The experimental data are reproduced well with these two parameters. Note that the linear portion of the curve yields the product $\phi l_{cp}$. Clear separation between $\phi$ and $l_{cp}$ can only be obtained if there is initial curvature to the data, since $l_{cp}$ effectively delays the onset of steady-state growth.

The model shows how excess charge is spread over several "layers" ($l_{cp}$ is about 2.5 nominal "layers") so that $\phi$ does not need to be high, even for very "thick" films. A significant preliminary finding was that values of $\phi$ for the PSS/PDADMA multilayer system were almost independent of salt concentration, and film growth is almost entirely controlled by $l_{cp}$. For salt concentration M, the approximate relationship is $l_{cp}(\text{Å}) = 400M$ in the present case. The independence of $\phi$ on salt concentration is consistent with results from electroosmotic flow measurements of capillaries coated with multilayers [21], where it was found that the apparent surface charge remained strictly constant as a function of the salt concentration (layer thickness).

In "exponentially growing" films[74] $l_{cp}$ is $> t$ (i.e., the adding polymer overcompensates the whole film). From Equations 13.4 and 13.5, if $l_{cp}$ for both positive and negative layers is $> t$

$$t_n^+ = \left(\frac{\phi^+ - 1}{\phi^+}\right) f^+ t \tag{13.7a}$$

and

$$t_n^- = \left(\frac{\phi^- - 1}{\phi^-}\right) f^- t \tag{13.7b}$$

In other words, each "layer" overcompensates all the PEMU of thickness $t$ by $\varphi^+ - 1$ or $\varphi^- - 1$, so the film grows exponentially. In current theories of exponential growth [74–78], one polymer is more mobile and overcompensates the whole film whereas the other polymer simply compensates excess charge from the mobile one. In this case

$$t_n = 2\left(\frac{\phi - 1}{\phi}\right) t \tag{13.8}$$

Where $\varphi - 1$ is the overcompensation level for the mobile polymer.

## 13.4
## Polyelectrolyte Interdiffusion

During exposure of our freshly-prepared PEMUs to salt solutions a significant decrease in the surface roughness was observed [50]. Smoothing of the surface was especially rapid in solutions of high salt concentration. Higher salt concentrations are known to enhance the mobility of charge-paired polyelectrolyte chains [63, 79–81], presumably by freeing up "frozen" segments via Equation 13.1. As a PEMU surface smooths, to reduce the PEMU/solution interfacial contact area, the material in the "peaks" diffuses into the "valleys." For a PSS/PDADMA multilayer prepared from 1.0 M NaCl, a characteristic peak-valley dimension, $\Delta$, of 150 nm was found. This value, inserted into the well known [82] relationship $\Delta = (2Dt)^{1/2}$, provides values for the polyelectrolyte interdiffusion coefficient, $D$, of order $6 \times 10^{-15}$ and $5 \times 10^{-14}$ cm$^2$ s$^{-1}$ for 0.5 and 1.0 M salt, respectively. The strongly nonlinear effect of salt in

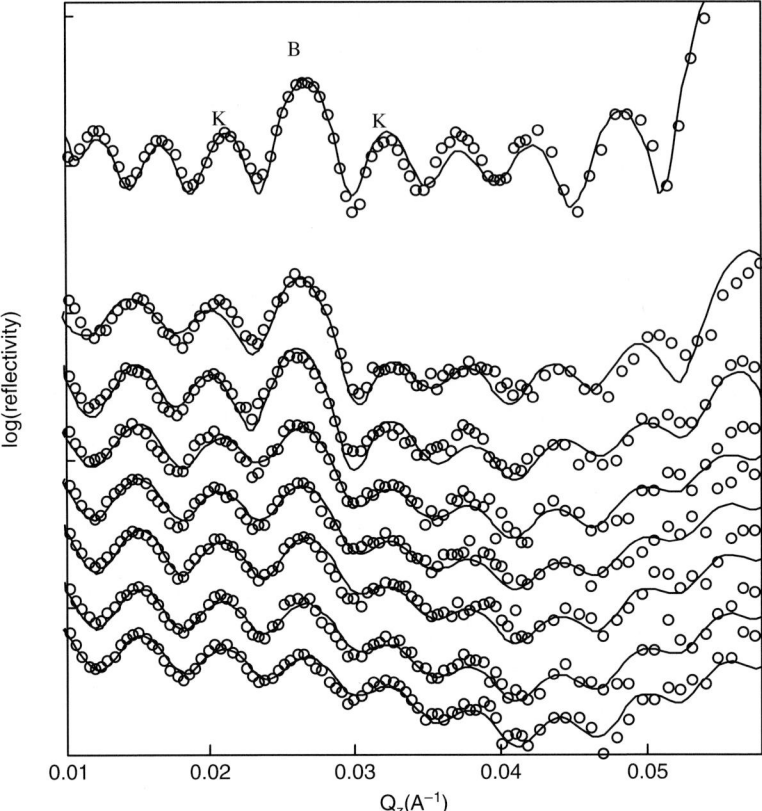

**Figure 13.14** Plot of log (reflectivity) versus $Q$. Experimental data (open circles) and fit (solid line) of [(PDADMA/PSS)$_4$(PDADMA/d-PSS)]$_5$(PDADMA/PSS) multilayer. Uppermost curve: as-deposited sample measured in ambient. Lower curves, measured under argon, (from top to bottom): after annealing (for 10, 25, 55, 110, 170, 260 min) in 0.8 M NaCl. A final anneal in 1 M NaCl for 120 min (lowest spectrum) almost completely removes the Bragg peak, B, to leave a Kiessig fringe, K [85].

"lubricating" the motions of one charged polymer against another, as discussed for the diffusion of a polyelectrolyte into an oppositely-charged gel [83, 84], indicates the cooperative properties of polyelectrolyte segments in retarding diffusion [50].

Samples not used immediately after preparation displayed an induction period before the smoothing effect started, possibly a result of the aggregation of hydrophobic domains within the multilayer. For practical purposes, a brief annealing period in high salt concentration is generally recommended for applications where PEMU roughness is undesired. The optimum annealing conditions are probably a function of $Q_{swel}$ Evidence for salt-induced mobility of bulk polyelectrolytes *inside* PEMUs was obtained from neutron reflectivity measurements [85]. A three-component PEMU with deuterated PSS (d-PSS) layers every five layer pairs was constructed: [(PDADMA/PSS)$_4$(PDADMA/d-PSS)]$_5$(PDADMA/PSS). Neutron reflectivity curves (Figure 13.14)

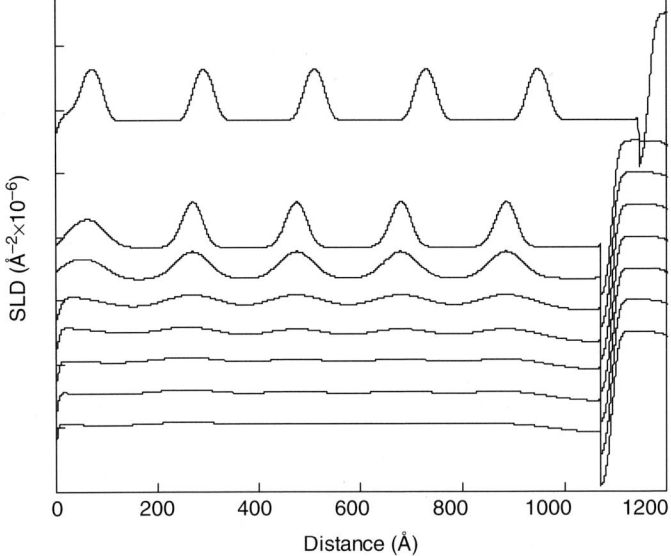

**Figure 13.15** Plot of the scattering length density versus distance from the air/PEMU interface. Samples as in Figure 13.14. Increases in SLD are due to deuterated layers. The deuterated stratum closest to the surface dissipates faster than the rest, consistent with the finding of enhanced surface diffusion.

showed a Bragg peak corresponding to the spacing of the deuterated layers. On the addition of 0.8 M salt, the polyelectrolytes interdiffused and the Bragg peak was slowly lost. Scattering length density fits to the reflectivity curves as a function of "annealing" time in 0.8 M NaCl are shown in Figure 13.15, where it is clear that deuterated polymer diffuses into neighboring undeuterated regions.

The role of salt in promoting polyelectrolyte interdiffusion is in separating polymer segments. Movement of part or all a polyelectrolyte molecule is difficult. For two rod-like oppositely charged polyelectrolytes, all ion pairs would have to be disengaged simultaneously in order to move a polymer chain. For flexible polymers, such as the present case, short runs of adjacent repeat units are able to act in a localized fashion, and if a sufficient number of ion pairs are simultaneously unpaired, a correlated group of units might be able to move in concert, as illustrated by Scheme 13.2. In Scheme 13.2, a group of adjacent intrinsic ion pairs has sequentially dissociated, moved, and reassociated. As the probability for each dissociation step is low, the probability of achieving the simultaneous dissociated condition shown in the transition state in the scheme is much lower. In Scheme 13.2b is a representation of interdiffusion if extrinsic charge enters the picture: one of the positive polyelectrolyte charges is extrinsically compensated by a salt anion (such as chloride). The chloride ion facilitates the local rearrangement of the polyelectrolyte by hopping with the polyelectrolyte segment as it moves. The transition in Scheme 13.2b merely suggests a way that the energy barrier for hopping can be minimized by concerted participation of four charges species (connected by dotted lines).

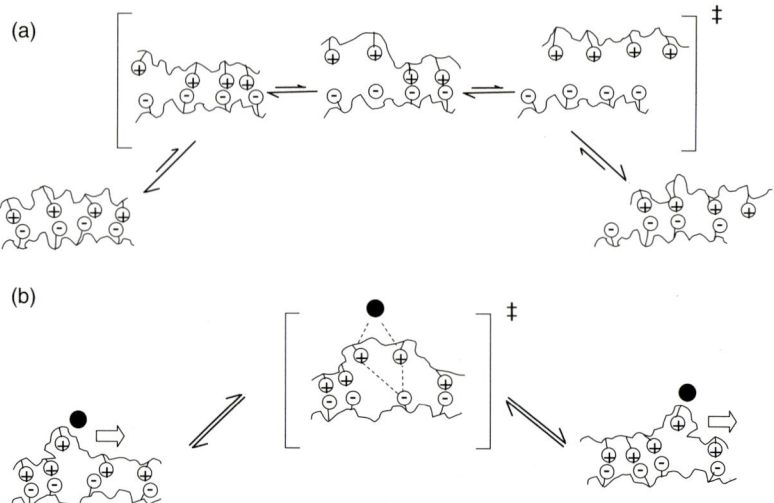

**Scheme 13.2** Movement of polyelectrolyte segments within a polyelectrolyte complex. In (a) several ion pairing contacts must break before the segment moves. In (b) participation of an extrinsic charge (solid dot, a negative counterion in this example) allows localized place-exchange of individual repeat units, or a small group of them. Suggested transition-states are depicted.

### 13.4.1
**Equilibrium versus Non-Equilibrium Conditions for Salt and Polymer Sorption**

The rapid, irreversible adsorption of polyelectrolytes, generally acknowledged from the outset of research into multilayered systems [55, 61, 71, 72], was demonstrated using radiolabeled polyelectrolyte [39], which, when incorporated into a PEMU as the top layer, did not exchange with unlabeled solution polyelectrolyte on the timescale of the assembly. Irreversibility is an essential and fortuitous property. If desorption were not kinetically limited, multilayer polymer could be stripped from the surface on exposure to a solution of its oppositely charged partner [52]. Indeed, polymers with sufficiently low molecular weight can be prised off the surface by this process, suppressing PEMU growth [86]. Polymer stripping using a large excess of oppositely charged polyelectrolyte may yield soluble complexes, which, having more configurational and translational entropy, are more stable than precipitated complexes. Salt, again, mediates the removal of surface polyelectrolyte by loosening the interactions between polymers via competitive ion exchange. Adsorbed polyelectrolyte conformational changes are known to be accelerated by the addition of salt [63, 79]. For successful PEMU growth the competition between polymer addition to the surface and the loss of complex to solution favors the former on the timescale of the experiment. In the presence of salt, the kinetic balance is tipped towards polymer addition by sufficient molecular weight [56].

The above explanation holds for "linear" systems, where overcompensation is limited to near the surface. In exponential systems, where overcompensation occurs throughout the film, there is evidence that bulk polymer exchanges with solution polymer [87]. For labile systems, there may be loss of complex to the solution even if molecular weights match [56, 88]. Chapter 15 draws further comparisons between traditional PECs and PEMUs, emphasizing the importance of kinetic limitations in multilayer build-up versus qPEC formation.

## 13.5
## Ion Transport Through Multilayers: the "Reluctant" Exchange Mechanism

PEMUs have been employed as selective membranes for the separation of gases [89, 90] and dissolved species [91–94]. Multilayers offer several advantages for use as membranes: they are uniform, easily prepared on a variety of substrates, continuous, resistant to protein adsorption [21], fairly rugged, the thickness is reproducible, very thin layers allow high permeation rates, and there is a wide range of possible compositions of polymers and particles. Separation by pervaporation of volatiles [95–97], for example water and ethanol, has achieved high separation factors. Membrane selectivity for ions of low charge has been demonstrated using conductivity [92] or redox electrochemical [93, 94] methods. Post-deposition thermal crosslinking of polyamine/polyacrylic acid PEMs significantly decreases ion permeability [93]. The ion flux through a dendrimer multilayer composite may be controlled by the external pH [98]. The reader is referred to Chapters 20 and 38 on the design and use of PEMUs as separation membranes.

In the remainder of this chapter, ionic transport through PEMUs will be rationalized in terms of the kinetics and distribution of charged species within multilayers. The latter is elucidated by considering various equilibria governing reluctant exchange.

We employ a rotating disk electrode, RDE, coated with a multilayer, to evaluate the permeation of electrochemically active ionic probes [46]. For reversible redox species, such as ferrocyanide, $Fe(CN)_6^{4-}$, this apparatus yields an S-shaped redox wave with a steady state limiting current, $i_l$, at the plateau [99]. Figure 13.16 depicts, for example, the response of a RDE coated with 700 Å of PSS/PDADMA to a solution of 1 mM ferrocyanide with various $[NaCl]_{aq}$. A striking and unconventional feature of the electrochemistry at the PEMU-coated RDE is the increase in current as the $[NaCl]_{aq}$ increases.

The flux, or current, to the membrane-coated RDE electrode is limited by convective diffusion through a stagnant layer of electrolyte next to the membrane and by diffusion through the membrane [100, 101]. The effect of these two series resistances to mass transport is expressed as follows:

$$\frac{1}{i_l} = \frac{1}{i_m} + \frac{1}{i_{bare}} \tag{13.9}$$

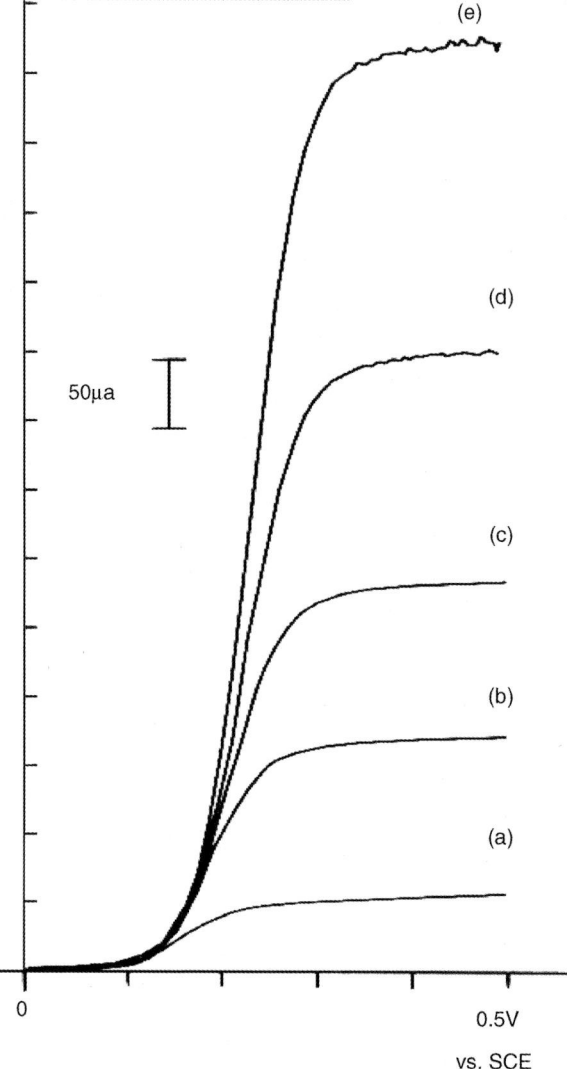

**Figure 13.16** Linear scan voltammograms of a rotating disc electrode; uncoated (e), and coated with a PSS/PDADMA multilayer of thickness 700 Å (a–d). Polymer solutions used to make the 20-layer membrane were 1 mM in polyelectrolyte and 0.25 M salt. The last layer (PSS) was deposited from 10 mM salt to provide a pseudo-neutral surface. The electrolyte for the voltammograms contained 1 mM potassium ferrocyanide and 1.2 M (a); 1.6 M (b); 1.8 M (c); 2.0 M (d) salt. Electrode was 8 mm diameter platinum, temperature (22 ± 0.1) °C, rotation rate 1000 rpm.

where $i_m$ is the membrane-limited current and $i_{bare}$ is the current at the bare electrode, area $A$. The latter is given, with precision for this system with well-defined hydrodynamics, by the Levich equation [99]

$$i_{bare} = 0.620\, nFAD_{aq}^{2/3}\omega^{1/2}v^{-1/6}C_{aq} \tag{13.10}$$

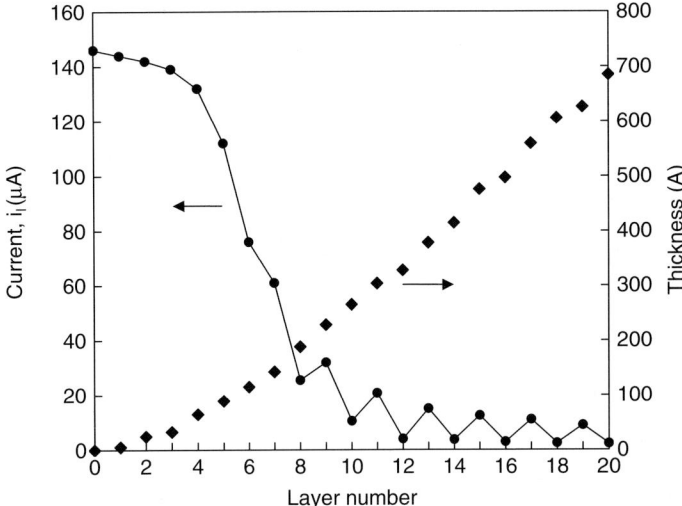

**Figure 13.17** Current and thickness versus layer number for the build-up of a 20-layer PEMU from 0.25 M salt, as in Figure 13.16 The flux of ferrocyanide through the membrane after the addition of each layer was measured in 1 mM Fe(CN)$_6^{4-}$ and 0.6 M NaCl. Rotation rate 1000 rpm.

where $n$ is the number of electrons transferred (1 in this case), $F$ is Faraday's constant, $D_{aq}$ is the solution diffusion coefficient of electroactive species, $\omega$ is the angular velocity, $\nu$ is the kinematic viscosity of the electrolyte (approximately 0.01 cm$^2$ s$^{-1}$ for dilute aqueous solutions) and $C_{aq}$ is the solution concentration (mol cm$^{-3}$). With plots of $i_l$ versus $\omega^{1/2}$ on coated and bare electrodes, $i_m$ is easily extracted.

Steady-state diffusion-limited flux, $J_m$, through a membrane of thickness $d$ is given by [45]

$$J_m = -\frac{\bar{D}\bar{C}}{d} \tag{13.11}$$

where $\bar{D}$ and $\bar{C}$ are, respectively, the diffusion coefficient and concentration of the probe species within the membrane. Note that the flux, which is converted to current by $i_m = J_m nFA$, is purely diffusive, that is, it does not contain an electric field driven, or migration, component. Evidence that this is the case for the electrochemical system in Figure 13.16 comes from the fact that the limiting plateau is flat instead of ramped. The electric field is suppressed by the presence of salt ions within the multilayer. Given Equation 13.11 one would expect current to decrease for thicker films. This is indeed observed (see Figure 13.17).

The distribution of charged species within the multilayer is governed by the chemical potential of external salt. The competition between intrinsic and extrinsic compensation in the presence of pure salt solutions is characterized, for the ferrocyanide system used in our studies, by the following equilibria

$$\text{Pol}^+\text{Pol}^-_{(m)} + \text{Cl}^-_{(aq)} + \text{Na}^+_{(aq)} = \text{Pol}^+\text{Cl}^-_{(m)} + \text{Pol}^-\text{Na}^+_{(m)} \tag{13.12}$$

$$4\text{Pol}^+\text{Pol}^-_{(m)} + \text{Fe(CN)}_6^{4-}{}_{(aq)} + 4\text{Na}^+_{(aq)} = \text{Pol}_4^{4+}\text{Fe(CN)}_6^{4-}{}_{(aq)} + 4\text{Pol}^-\text{Na}^+_{(m)}$$
(13.13)

A third equilibrium represents competition between solution anion species for positive polymer segments.

$$4\text{Pol}^+\text{Cl}^-_{(m)} + \text{Fe(CN)}_6^{4-}{}_{(aq)} = \text{Pol}_4^{4+}\text{Fe(CN)}_6^{4-}{}_{(m)} + 4\text{Cl}^-_{(aq)}$$
(13.14)

Assuming activity coefficients are unity

$$K_1 = \frac{[\text{Pol}^+\text{Cl}^-]_m [\text{Pol}^-\text{Na}^+]_m}{[\text{Pol}^+\text{Pol}^-]_m [\text{Cl}^-]_{aq} [\text{Na}^+]_{aq}}$$
(13.15)

$$K_2 = \frac{[\text{Pol}_4^{4+}\text{Fe(CN)}_6^{4-}]_m [\text{Pol}^-\text{Na}^+]_m^4}{[\text{Pol}^+\text{Pol}^-]_m^4 [\text{Fe(CN)}_6^{4-}]_{aq} [\text{Na}^+]_{aq}^4}$$
(13.16)

$$K_3 = \frac{[\text{Pol}_4^{4+}\text{Fe(CN)}_6^{4-}]_m [\text{Cl}^-]_{aq}^4}{[\text{Pol}^+\text{Cl}^-]_m^4 [\text{Fe(CN)}_6^{4-}]_{aq}}$$
(13.17)

where $K_3 = K_2/K_1^4$.

At this point, it is worth comparing the differences in ion exchange behavior between salt-swollen PEMUs and classical ion exchanger resins. Resins are typically prepared with one type of polyelectrolyte, charge balanced completely via exchangeable salt ions, and crosslinked to prevent the polymer from dissolving [45]. These classical exchangers exhibit permselective Donnan exclusion of like-charged mobile species and inclusion of oppositely charged ones [44]. This selectivity sets up a concentration gradient and, therefore, chemical potential across the membrane/solution interface. Competition between solution species for a fixed number of ion exchanging sites in classical exchangers is represented by an equilibrium of the type shown in Equation 13.14. The difference with reluctant exchangers is that additional exchanging sites are also *created* by external ions (Equations 13.12 and 13.13). The number of sites is controlled by the external ion concentration.

There are some examples of reluctant exchangers in the literature. It is possible to synthesize resins containing equal numbers of *fixed* positive and negative groups (so-called "amphoteric" exchangers) [102] which, if intrinsically compensated, would be reluctant. More relevant examples are the "snake cage" exchangers developed by Dow in the 1950s [103]. These interesting systems are prepared by polymerizing ionic counterions (making "snakes") inside an ion exchanger matrix (the "cage"). If polymerization were complete, the system would be fully intrinsic as made. Snake cage resins also exhibit the characteristic reluctant properties of swelling in electrolyte (unlike Donnan exchangers, which shrink) and non-permselectivity [103]. There are examples where solution-precipitated PECs have been formed into membranes.[32, 104, 105] These materials, having similar morphologies to PEMUs also exhibit selective transport for ions of low charge.

Some unusual oscillating features (Figure 13.17), also seen in other work [72], are observed in the current as the PEMU is built layer by layer. For our negatively charged probe ion, multilayers terminated with a positive layer yield larger membrane currents than those ending with a negative layer. These features may be explained by Donnan inclusion of ferrocyanide into the positively charged surface layer. This inclusion increases the membrane concentration $\bar{C}$, leading to greater $J_m$, as predicted by Equation 13.11.

Manipulation of the equilibrium expressions above in the limit of high $[NaCl]_{aq}$ leads to some important conclusions. If the solution concentration of NaCl is much greater than that of the ferrocyanide, we make the following approximations: $[Pol^+Cl^-]_m \gg [Pol_4^{4+}(Fe(CN)_6^{4-}]_m$ and $[Pol^+Cl^-]_m \approx [Pol^-Na^+]_m$, and $[Cl^-]_{aq} = [Na^+]_{aq}$. In the limit of low swelling, $(1-y) \to 1$ and $[Pol^+Pol^-]_m \approx C_t$, where $C_t$ is the total concentration of intrinsically plus extrinsically compensated polymer, which remains constant if the volume is constant, which is the case for small $y$. Thus,

$$\bar{C} = [Pol_4^{4+}Fe(CN)_6^{4-}]_m = K_3 K_1^2 C_t^2 [Fe(CN)_6^{4-}]_{aq} \tag{13.18}$$

The linear dependence of $J_m$ on $[Fe(CN)_6^{4-}]_{aq}$ through $\bar{C}$ was verified experimentally. From Equation 13.2

$$y = [Cl^-]_{aq} \sqrt{K'} \tag{13.19}$$

Two important facts emerge from the foregoing treatment of the high salt concentration limit: (i) ferrocyanide concentration in the PEM is independent of solution salt – the ferrocyanides in the multilayer displaced by chloride ions are compensated by the fact that the chlorides *also* open up an equal number of new, reluctant, ferrocyanide-occupied anion exchanging sites; (ii) the total exchanger capacity ($y$) is directly proportional to the external salt ion concentration, at least for low $y$. Given finding (i), that is, $\bar{C}$ is constant as $[NaCl]_{aq}$ increases, *all enhancement of flux by $[NaCl]_{aq}$ is due to an increase in $\bar{D}$*.

The constancy of redox probe ion concentration, $\bar{C}$, in the PEMU as a function of $[NaCl]_{aq}$ is a critical feature of the "reluctant" ion exchange mechanism and the ultimate control of the flux by $\bar{D}$ (Equation 13.11). (This constancy may be generalized for all combinations of probe ion and salt ion charge.) We have obtained direct experimental evidence for the constancy of $\bar{C}$ [46].

Membrane currents extracted from $i_l$ using Equation 13.9 are plotted as a function of the activity of NaCl (determined by the product of $[NaCl]_{aq}$ and a tabulated activity coefficient) in Figure 13.18 for ferro- and ferricyanide. The dependence of flux on salt activity (or $[NaCl]_{aq}$) is clearly nonlinear. On inspection of the data, we established the following empirical fit for transport of ions of charge $n$ through PEMUs

$$J_m = k a_{NaCl}^n = cy^n \text{ (since $y$ scales with $a_{NaCl}$)} \tag{13.20}$$

where $c$ is another constant. The fits with $n = 3$ or 4 for ferri- or ferrocyanide, respectively, are shown in Figure 13.18.

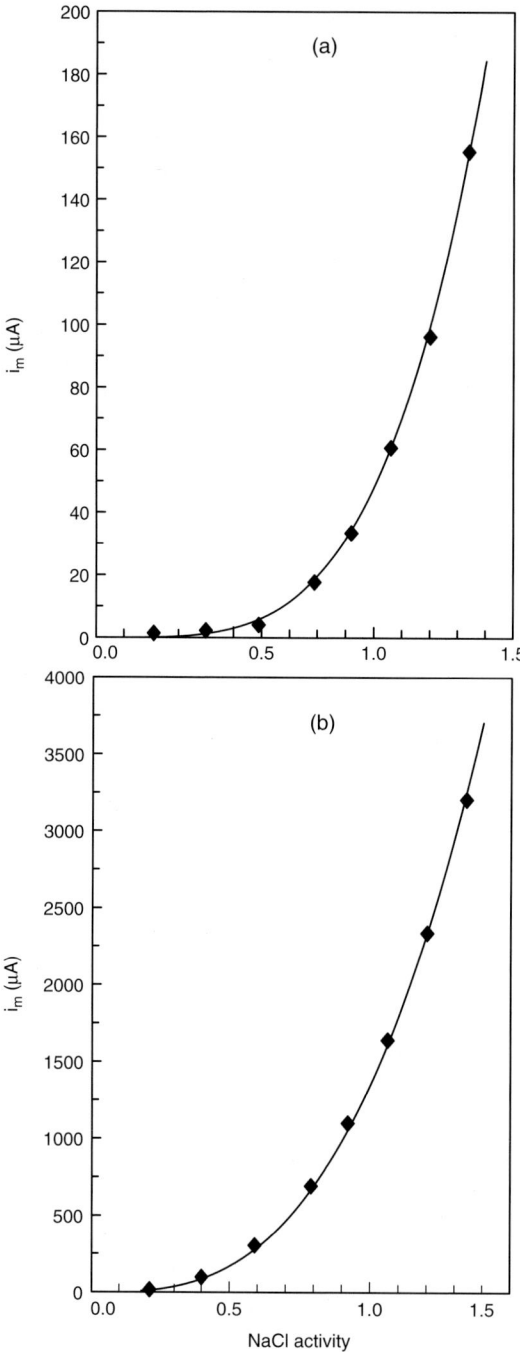

**Figure 13.18** Membrane flux, in terms of $i_m$ versus activity of salt $a_{NaCl}$. The solid line represents an empirical fit of $J_m = ka_{NaCl}^4$ for 1 mM ferrocyanide ($k = 48$ (a)); and $J_m = ka_{NaCl}^3$ for 1 mM ferricyanide ($k = 1350$ (b)).

## 13.5 Ion Transport Through Multilayers: the "Reluctant" Exchange Mechanism

The nonlinear control of diffusion coefficient by external salt represents a fascinating new paradigm for the transport of charge through "soft" condensed matter. The phenomenological trends in Figure 13.18 suggest transport of ions within the PEMU occurs via hopping from site to site [106]. Each site that is occupied by a small ion (mostly $Cl^-$ and $Na^+$) acts as a charge carrier. Positive charge carriers, $[Pol^+Cl^-]_m$ sites, are introduced, or "doped," into the PEM by the external chemical potential of the salt solution. The $n^{th}$ power relationship to the site exchanger site concentration, $[Pol^+Cl^-]_m$ (or normalized concentration, $y$) is reasonable, since each ion requires the simultaneous displacement of $n$ adjacent chloride ions from $Pol^+$ sites, as shown in Scheme 13.3. The probability that this occurs is a function of $y^n$. A hopping-type mechanism would be thermally activated, as observed [46].

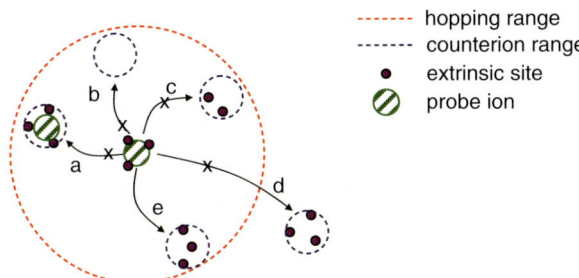

**Scheme 13.3** Hopping model. Triply charged species (green) transported through the membrane can hop within a certain range (dashed red line) only if a small cluster (blue dashed line) of three ion exchange sites (violet) is available (attempt e). All other attempts fail because: sites are already occupied (a); there are insufficient sites (b and c); sites are too far away (d). The resulting membrane flux is thus a strongly nonlinear function of salt concentration.

A further implication of the charge carrier argument for ion transport through reluctant exchangers is that polyelectrolyte multilayers will be much more permeable to species of lower charge. This trend has been observed previously [91, 93, 94], and is confirmed in a quantitative fashion for the ferro/ferricyanide data (Figure 13.18). Membrane selectivity for the species of lower charge is evident from the data in Figure 13.19, and the transport mechanism in Scheme 13.3.

Monte Carlo simulations of reluctant exchanger films with extrinsic ions support the selectivity according to charge and the $y^n$ dependence of flux [106]. In these simulations, probe ions of charge $n$ are assumed to hop only if $n$ oppositely charged sites are generated, by random movement of extrinsic salt counterions, within a specified range.

An interesting comparison may be made of the mechanism of ion transport in the system described with electron transport. Some electron transport materials, such as conducting polymers, exhibit strongly nonlinear increases in conductivity with doping level [107, 108]. However, since electrons, or holes, are of unit charge, a

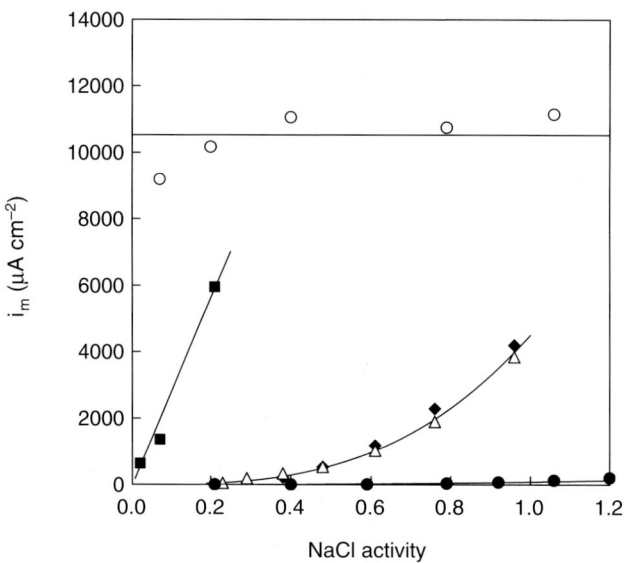

**Figure 13.19** Membrane current density, as a function of salt activity, through a 700 Å thick multilayer of PSS/PDADMA, using species of different charge, all 1 x 10$^{-3}$ M concentration. Hydroquinone (open circles) uncharged; iodide (charge −1, squares); ferricyanide (−3, diamonds); ferric ion (+3, triangles); and ferrocyanide (−4, solid circles). Solid lines are fits using the scaling behavior: flux ~ $a_{NaCl}^{charge}$.

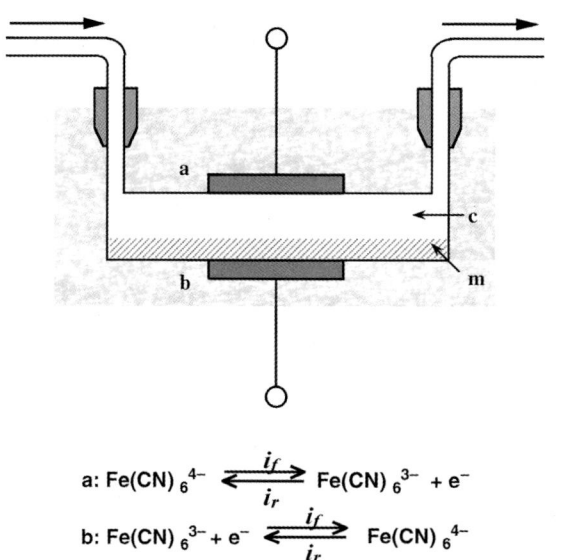

a: $Fe(CN)_6^{4-}$ $\underset{i_r}{\overset{i_f}{\rightleftarrows}}$ $Fe(CN)_6^{3-}$ + $e^-$

b: $Fe(CN)_6^{3-}$ + $e^-$ $\underset{i_r}{\overset{i_f}{\rightleftarrows}}$ $Fe(CN)_6^{4-}$

**Figure 13.20** The salt gated "transistor." Two electrodes, one coated with a polyelectrolyte multilayer (m), and the other bare, are separated by a channel (c) containing a constant concentration of redox-active probe ions and a variable concentration of salt. An electrochemical current flowing in the forward, $i_f$, direction, is carried through the multilayer by ferricyanide ions, whereas in the reverse direction it is carried by ferrocyanide ions. The multilayer is comprised of a complex of poly (diallyldimethylammonium) and poly(styrene sulfonate).

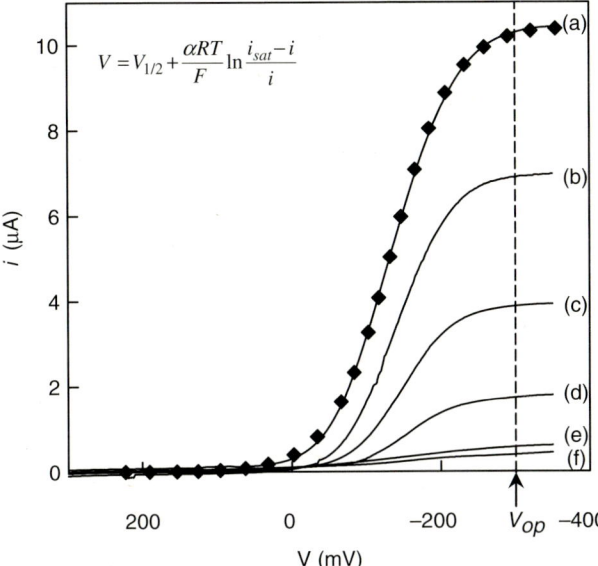

**Figure 13.21** A salt-gated family of $i$–$V$ curves. Current flows through the multilayer according to the equation shown, where $i_{sat}$ is the saturation current, $V_{1/2}$ the half-wave potential, $R$ is the gas constant, $T$ the temperature, $F$ Faraday's constant and $\alpha$ a shape factor between 1 and 2. Increasing concentrations of salt (f → a, 0.1, 0.2, 0.4, 0.6, 0.8, 1.0 M) lead to higher doping of the multilayer, which produces more sites for ion transport. The concentration of ferri- (ferro)cyanide probe ions was constant ($10^{-3}$ M).

dependence of transport on the charge of the carrier is not an issue. The diffusion constant of ferricyanide transport through PSS/PDADMA PEMUs was found to be strongly thermally activated, with an activation energy of 98 kJ mol$^{-1}$ [109].

An example of exploiting the differential flux of ions to make a device is shown in Figure 13.20. In the salt gated "transistor" one side of a channel is coated with PEMU [110]. The channel is filled with a mixture of ferri- and ferrocyanide. As seen in Figure 13.21, current passes in one direction through the PEMU much more easily than the other, producing a rectifying junction.

### 13.5.1
### Practical Consequences: Trapping and Self-Trapping

Many practical considerations follow from the control of permeability and selectivity by salt. For membrane separations with PEMUs, the selectivity for species of differing charge, in addition to being high, may be tuned by salt. As is commonly encountered in membrane separations, enhanced selectivity comes at the cost of flux. For the trapping of molecules, such as drugs and biomolecules, by PEMU microcapsules [62, 111–113], detailed by other contributors to this volume, salt will play a

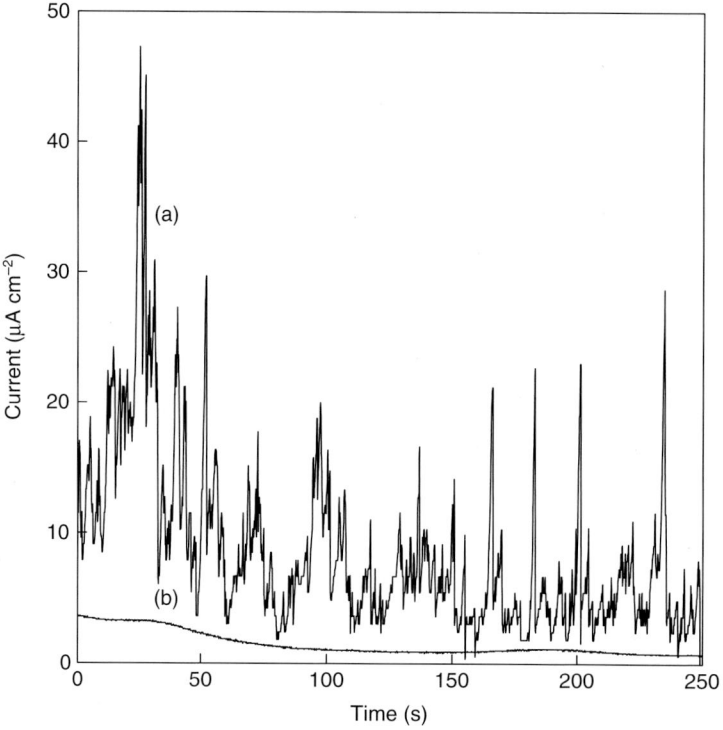

**Figure 13.22** Current versus time in the metastable pitting region (0.6 V vs. SCE) for (a) a bare steel wire; (b) wire coated with 70 nm PDADMA/PSS in 0.1 M $NaCl_{aq}$.

strong role in loading and in the rate of release. Selective transportation of species with lower charge provides a convenient mechanism for loading PEMU membranes or capsules: under conditions of high salt, PEMU capsules, rendered significantly more permeable, may be loaded with polyvalent molecules. When the system is exposed to fresh water, ions of low charge rush out, leaving the lumbering highly-charged species trapped within the capsule. "Self-trapping" is possible for single ionic species only (i.e., no other salts are added) when the overall ionic content of a film falls to a value that does not support diffusion. Self-trapping for ions of higher charge is a nonlinear (power law) positive feedback mechanism, as fewer ions within the film mean significantly decreased $\bar{D}$. Thus, notably more polyvalent ions will be self-trapped, as is probably the case for multiply charged dye probes [42]. Self-trapping implies that it is impossible to rid a multilayer of all salt ions on immersion in pure water, although their concentration may be below the detection limit.

As an example of another potential application, we have shown that PEMUs are useful in preventing corrosion [114]. The effectiveness of PEMUs in suppressing the corrosion of steel in salt water, surprising in light of the fact that the films contain appreciable amounts of water and salt, was attributed partially to the exclusion of "liquid" water, or water that is not occupied in hydrating ion pairs. In addition, the partial mobility, in the presence of salt, of polymeric material in PEMUs, evidenced in Figures 13.13 and 13.14, was thought to allow small defects in the ultrathin film to heal over. Figure 13.22 depicts corrosion current versus time in the metastable pitting region of stainless steel. A 70 nm layer of PEMU is extremely effective at suppressing pitting events.

## 13.6
## Concluding Remarks

Many properties of polyelectrolyte complexes depend on the content, balance and transport of ions, including mechanical properties, permeability, and selectivity. This chapter has presented the thread and major findings of our work in this area. It is not intended to be a comprehensive review of all the work in the area, as the reader will discover after perusing the rest of the fine contributions to this volume set.

## References

1 Michaels, A.S. and Miekka, R.G. (1961) Polycation-polyanion complexes: preparation and properties of poly(vinylbenzyltrimethylammonium styrenesulfonate). *J. Phys. Chem.*, **65**, 1765–1773.
2 Michaels, A.S. (1965) Polyelectrolyte complexes. *J. Ind. Eng. Chem.*, **57** (10), 32–40.
3 Bixler, H.J. and Michaels, A.S. (1969) Polyelectrolyte complexes, in *Encyclopedia of Polymer Science and Technology*, (eds H.F. Mark, N.G. Gaylord, and N.M. Bikales), Wiley (Interscience), New York, vol. 10, 765–780.
4 Oyama, H.T. and Frank, C.W. (1986) Structure of the polyion complex between poly(sodium p-styrene sulfonate) and poly(diallyl dimethyl ammonium chloride). *J. Polym. Sci., Part B Polym. Phys.*, **24** (8), 1813–1821.
5 Michaels, A.S., Falkenstein, G.L., and Schneider, N.S. (1965) Dielectric properties of polyanion–polycation complexes. *J. Phys. Chem.*, **69** (5), 1456–1465.
6 Porcel, C.H. and Schlenoff, J.B. (2009) Compact polyelectrolyte complexes: "Saloplastic" candidates for biomaterials. *Biomacromolecules*, **10** (11), 2968–2975.
7 Hariri, H.H. and Schlenoff, J.B. (2010) Saloplastic macroporous polyelectrolyte complexes: Cartilage mimics. *Macromolecules*, **43** (20), 8656–8663.
8 Kabanov, V.M. and Zezin, A.B. (1984) Soluble interpolymeric complexes as a new class of synthetic polyelectrolyte. *Pure Appl. Chem.*, **56**, 343–354.
9 Dautzenberg, H. *et al.* (1996) Stoichiometry and structure of polyelectrolyte complex particles in diluted solutions. *Ber. Bunsen-Ges. Phys. Chem.*, **100** (6), 1024–1032.
10 Ahmed, L.S. *et al.* (1994) Stoichiometry and the mechanism of complex-formation in protein-polyelectrolyte coacervation. *J. Macromol. Sci., Part A Pure Appl. Chem.*, **31** (1), 17–29.
11 Magnin, D. *et al.* (2004) Physicochemical and structural characterization of a polyionic matrix of interest in biotechnology, in the pharmaceutical and biomedical fields. *Carbohyd. Polym.*, **55** (4), 437–453.
12 Weinbreck, F., Tromp, R.H., and de Kruif, C.G. (2004) Composition and structure of whey protein/gum arabic coacervates. *Biomacromolecules*, **5** (4), 1437–1445.
13 Schmitt, J. *et al.* (1993) Internal structure of layer-by-layer adsorbed polyelectrolyte films - a neutron and X-ray reflectivity study. *Macromolecules*, **26** (25), 7058–7063.

14 Loesche, M. et al. (1998) Detailed structure of molecularly thin polyelectrolyte multilayer films on solid substrates as revealed by neutron reflectometry. *Macromolecules*, **31** (25), 8893–8906.

15 Decher, G. (1997) Fuzzy nanoassemblies: Toward layered polymeric multicomposites. *Science*, **277** (5330), 1232–1237.

16 Laurent, D. and Schlenoff, J.B. (1997) Multilayer assemblies of redox polyelectrolytes. *Langmuir*, **13** (6), 1552–1557.

17 Decher, G., Lvov, Y., and Schmitt, J. (1994) Proof of multilayer structural organization in self-assembled polycation-polyanion molecular films. *Thin Solid Films*, **244** (1–2), 772–777.

18 Decher, G., Hong, J.D., and Schmitt, J. (1992) Buildup of ultrathin multilayer films by a self-assembly process .3. Consecutively alternating adsorption of anionic and cationic polyelectrolytes on charged surfaces. *Thin Solid Films*, **210** (1–2), 831–835.

19 Yoo, D., Shiratori, S.S., and Rubner, M.F. (1998) Controlling bilayer composition and surface wettability of sequentially adsorbed multilayers of weak polyelectrolytes. *Macromolecules*, **31** (13), 4309–4318.

20 Mendelsohn, J.D. et al. (2003) Rational design of cytophilic and cytophobic polyelectrolyte multilayer thin films. *Biomacromolecules*, **4** (1), 96–106.

21 Graul, T.W. and Schlenoff, J.B. (1999) Capillaries modified by polyelectrolyte multilayers for electrophoretic separations. *Anal. Chem.*, **71** (18), 4007–4013.

22 Barker, S.L.R. et al. (2000) Plastic Microfluidic Devices Modified with Polyelectrolyte Multilayers. *Anal. Chem.*, **72** (20), 4899–4903.

23 Eckle, M. and Decher, G. (2001) Tuning the performance of layer-by-layer assembled organic light emitting diodes by controlling the position of isolating clay barrier sheets. *Nano Lett.*, **1** (1), 45–49.

24 Mamedov, A. et al. (2000) Stratified assemblies of magnetite nanoparticles and montmorillonite prepared by the layer-by-layer assembly. *Langmuir*, **16** (8), 3941–3949.

25 Arys, X., Laschewsky, A., and Jonas, A.M. (2001) Ordered polyelectrolyte "multilayers". 1. Mechanisms of growth and structure formation: A comparison with classical fuzzy "multilayers". *Macromolecules*, **34** (10), 3318–3330.

26 Dubas, S.T., Farhat, T.R., and Schlenoff, J.B. (2001) Multiple membranes from "true" polyelectrolyte multilayers. *J. Am. Chem. Soc.*, **123** (22), 5368–5369.

27 Laschewsky, A. et al. (1996) A new route to thin polymeric, non-centrosymmetric coatings. *Thin Solid Films*, **285**, 334–337.

28 Roberts, M.J. et al. (1998) Thermally stable nonlinear optical films by alternating polyelectrolyte deposition on hydrophobic substrates. *J. Am. Chem. Soc.*, **120** (43), 11202–11203.

29 Heflin, J.R. et al. (1999) Thickness dependence of second-harmonic generation in thin films fabricated from ionically self-assembled monolayers. *Appl. Phys. Lett.*, **74** (4), 495–497.

30 Onda, M. et al. (1996) Sequential actions of glucose oxidase and peroxidase in molecular films assembled by layer-by-layer alternate adsorption. *Biotechnol. Bioeng.*, **51** (2), 163–167.

31 Lvov, Y.M. et al. (1998) Direct electrochemistry of myoglobin and cytochrome p450(cam) in alternate layer-by-layer films with DNA and other polyions. *J. Am. Chem. Soc.*, **120** (17), 4073–4080.

32 Dautzenberg, H. et al. (1994) *Polyelectrolytes: Formation, Characterization and Applications*, Hanser., Munich.

33 Radeva, T. (ed.) (2001) *Physical Chemistry of Polyelectrolytes. Surfactant Science*, vol. 99, M. Dekker, New York.

34 Bucur, C.B., Sui, Z., and Schlenoff, J.B. (2006) Ideal mixing in polyelectrolyte complexes and multilayers: Entropy driven assembly. *J. Am. Chem. Soc.*, **128** (42), 13690–13691.

35 Tsuchida, E. and Osada, H. (1974) *Makromol. Chem.*, **175**, 593–601.

36 Schlenoff, J.B., Rmaile, A.H., and Bucur, C.B. (2008) Hydration contributions to association in polyelectrolyte multilayers and complexes: Visualizing hydrophobicity. *J. Am. Chem. Soc.*, **130** (41), 13589–13597.

37 Manning, G.S. (1969) Limiting laws and counterion condensation in polyelectrolyte solutions I. Colligative properties. *J. Chem. Phys.*, **51**, 924–933.

38 Schlenoff, J.B. and Li, M. (1996) Kinetics and multilayering in the adsorption of polyelectrolytes to a charged surface. *Ber. Bunsen-Ges. Phys. Chem. Chem. Phys.*, **100** (6), 943–947.

39 Dubas, S.T. and Schlenoff, J.B. (1999) Factors controlling the growth of polyelectrolyte multilayers. *Macromolecules*, **32** (24), 8153–8160.

40 Ariga, K., Lvov, Y., and Kunitake, T. (1997) Assembling alternate dye-polyion molecular films by electrostatic layer-by-layer adsorption. *J. Am. Chem. Soc.*, **119** (9), 2224–2231.

41 Kim, J. et al. (1999) Novel layer-by-layer complexation technique and properties of the fabricated films. *Chem. Mater.*, **11** (8), 2250–2256.

42 Tedeschi, C. et al. (2000) Adsorption and desorption behavior of an anionic pyrene chromophore in sequentially deposited polyelectrolyte-dye thin films. *J. Am. Chem. Soc.*, **122** (24), 5841–5848.

43 Shiratori, S.S. and Rubner, M.F. (2000) pH-dependent thickness behavior of sequentially adsorbed layers of weak polyelectrolytes. *Macromolecules*, **33** (11), 4213–4219.

44 Schlenoff, J.B., Ly, H., and Li, M. (1998) Charge and mass balance in polyelectrolyte multilayers. *J. Am. Chem. Soc.*, **120** (30), 7626–7634.

45 Helfferich, F. (1962) *Ion Exchange*, McGraw-Hill., New York.

46 Farhat, T.R. and Schlenoff, J.B. (2001) Ion transport and equilibria in polyelectrolyte multilayers. *Langmuir*, **17** (4), 1184–1192.

47 Ruths, J. et al. (2000.) Polyelectrolytes I: Polyanion/polycation multilayers at the air/monolayer/water interface as elements for quantitative polymer adsorption studies and preparation of hetero-superlattices on solid surfaces. *Langmuir*, **16** (23), 8871–8878.

48 Jaber, J.A. and Schlenoff, J.B. (2007) Counterions and water in polyelectrolyte multilayers: A tale of two polycations. *Langmuir*, **23** (2), 896–901.

49 Sukhorukov, G.B., Schmitt, J., and Decher, G. (1996) Reversible swelling of polyanion/polycation multilayer films in solutions of different ionic strength. *Ber. Bunsen-Ges. Phys. Chem. Chem.Phys.*, **100** (6), 948–953.

50 Dubas, S.T. and Schlenoff, J.B. (2001) Swelling and smoothing of polyelectrolyte multilayers by salt. *Langmuir*, **17** (25), 7725–7727.

51 Farhat, T. et al. (1999) Water and ion pairing in polyelectrolyte multilayers. *Langmuir*, **15** (20), 6621–6623.

52 Dubas, S.T. and Schlenoff, J.B. (2001) Polyelectrolyte multilayers containing a weak polyacid: Construction and deconstruction. *Macromolecules*, **34** (11), 3736–3740.

53 Laugel, N. et al. (2006) Relationship between the growth regime of polyelectrolyte multilayers and the polyanion/polycation complexation enthalpy. *J. Phys. Chem. B*, **110** (39), 19443–19449.

54 Salomaeki, M. et al. (2004) The Hofmeister anion effect and the growth of polyelectrolyte multilayers. *Langmuir*, **20** (9), 3679–3683.

55 Bertrand, P. et al. (2000) Ultrathin polymer coatings by complexation of polyelectrolytes at interfaces: suitable materials, structure and properties. *Macromol. Rapid Commun.*, **21** (7), 319–348.

56 Hoogeveen, N.G. et al. (1996) Formation and stability of multilayers of polyelectrolytes. *Langmuir*, **12** (15), 3675–3681.

57 Jaber, J.A. and Schlenoff, J.B. (2006) Mechanical properties of reversibly cross-linked ultrathin polyelectrolyte complexes. *J. Am. Chem. Soc.*, **128** (9), 2940–2947.

58 Jaber, J.A. and Schlenoff, J.B. (2006) Dynamic viscoelasticity in polyelectrolyte multilayers: Nanodamping. *Chem. Mater.*, **18** (24), 5768–5773.

59 Wang, R.M. and Schlenoff, J.B. (1998) Adsorption of a radiolabeled random hydrophilic/hydrophobic copolymer at the liquid/liquid interface: Kinetics, isotherms, and self-exchange. *Macromolecules*, **31** (2), 494–500.

60 Graul, T.W., Li, M., and Schlenoff, J.B. (1999) Ion exchange in ultrathin films. *J. Phys. Chem. B*, **103** (14), 2718–2723.

61 Sauvage, J.P. and Hosseini, M.W. (eds) (1996) Chapter 14, in *Comprehensive Supramolecular Chemistry*, vol. **9**, Pergamon Press, Oxford.

62 Caruso, F., Donath, E., and Möhwald, H. (1998) Influence of polyelectrolyte multilayer coatings on Forster resonance energy transfer between 6-carboxyfluorescein and rhodamine B-labeled particles in aqueous solution. *J. Phys. Chem. B*, **102** (11), 2011–2016.

63 Dubin, P.L. and Farinato, R. (eds) (1999) *Radiochemical Methods for Polymer Adsorption. Colloid-Polymer Interactions*, Wiley, New York.

64 Wang, R. and Schlenoff, J.B. (1998) Adsorption of a radiolabeled random hydrophilic/hydrophobic copolymer at the liquid/liquid interface: Kinetics, isotherms, and self-exchange. *Macromolecules*, **31** (2), 494–500.

65 Li, M. and Schlenoff, J.B. (1994) Ion exchange using a scintillating polymer with a charged surface. *Anal. Chem.*, **66** (6% R doi: 10.1021/ac00078a011): 824–829.

66 Durstock, M.F. and Rubner, M.F. (2001) Dielectric properties of polyelectrolyte multilayers. *Langmuir*, **17** (25), 7865–7872.

67 Korneev, D. et al. (1995) Neutron reflectivity analysis of self-assembled film superlattices with alternate layers of deuterated and hydrogenated polysterenesulfonate and polyallyamine. *Physica B*, **213**, 954–956.

68 Sui, Z.J. and Schlenoff, J.B. (2004) Phase separations in pH-responsive polyelectrolyte multilayers: Charge extrusion versus charge expulsion. *Langmuir*, **20** (14), 6026–6031.

69 Stepp, J. and Schlenoff, J.B. (1997) Electrochromism and electrocatalysis in viologen polyelectrolyte multilayers. *J. Electrochem. Soc.*, **144** (6), L155–L157.

70 Schlenoff, J.B. et al. (1998) Redox-active polyelectrolyte multilayers. *Adv. Mater.*, **10** (4), 347–349.

71 Decher, G. and Schmitt, J. (1992) *Prog. Coll. Polym. Sci.*, **89**, 160–164.

72 Lvov, Y. and Möhwald, H. (eds) (2000) *Protein Architecture: Interfacing Molecular Assemblies and Immobilization Biotechnology*, Dekker., New York, N.Y., p. 394.

73 Schlenoff, J.B. and Dubas, S.T. (2001) Mechanism of polyelectrolyte multilayer growth: Charge overcompensation and distribution. *Macromolecules*, **34** (3), 592–598.

74 Lavalle, P. et al. (2004) Modeling the buildup of polyelectrolyte multilayer films having exponential growth. *J. Phys. Chem. B*, **108** (2), 635–648.

75 Hubsch, E. et al. (2004) Controlling the growth regime of polyelectrolyte multilayer films: changing from exponential to linear growth by adjusting the composition of polyelectrolyte mixtures. *Langmuir*, **20** (5), 1980–1985.

76 Schaaf, P. et al. (2007) Influence of the polyelectrolyte molecular weight on exponentially growing multilayer films in the linear regime. *Langmuir*, **23** (4), 1898–1904.

77 Porcel, C. et al. (2006) From exponential to linear growth in polyelectrolyte multilayers. *Langmuir*, **22** (9), 4376–4383.

78 Picart, C. et al. (2002) Molecular basis for the explanation of the exponential growth of polyelectrolyte multilayers. *Proc. Natl. Acad. Sci. USA*, **99** (20), 12531–12535.

79 Pefferkorn, E., Jeanchronberg, A.C., and Varoqui, R. (1990) Conformational relaxation of polyelectrolytes at a solid liquid interface. *Macromolecules*, **23** (6), 1735–1741.

80 Karibyants, N. and Dautzenberg, H. (1998) Preferential binding with regard to chain length and chemical structure in the reactions of formation of quasi-soluble polyelectrolyte complexes. *Langmuir*, **14** (16), 4427–4434.

81 Dautzenberg, H. (1997) Polyelectrolyte complex formation in highly aggregating

systems. 1. Effect of salt: Polyelectrolyte complex formation in the presence of NaCl. *Macromolecules*, **30** (25), 7810–7815.

82 Crank, J. (1975) *The Mathematics of Diffusion*, Clarendon Press., Oxford.

83 Zezin, A., Rogacheva, V., and Kabanov, V.A. (1997) *Macromol. Symp.*, **126**, 123–141.

84 Schlenoff, J.B. and Li, M. (1995) Diffusion of a charged polymer into an oppositely-charged matrix - the polyelectrolyte clock. *PMSE Prepr. ACS Proc.*, **72**, 269–270.

85 Jomaa, H.W. and Schlenoff, J.B. (2005) Salt-induced polyelectrolyte interdiffusion in multilayered films: A neutron reflectivity study. *Macromolecules*, **38** (20), 8473–8480.

86 Sui, Z.J., Salloum, D., and Schlenoff, J.B. (2003) Effect of molecular weight on the construction of polyelectrolyte multilayers: Stripping versus sticking. *Langmuir*, **19** (6), 2491–2495.

87 Lavalle, P. et al. (2004) Direct evidence for vertical diffusion and exchange processes of polyanions and polycations in polyelectrolyte multilayer films. *Macromolecules*, **37** (3), 1159–1162.

88 Sukhishvili, S.A., Kharlampieva, E., and Izumrudov, V. (2006) Where polyelectrolyte multilayers and polyelectrolyte complexes meet. *Macromolecules*, **39** (26), 8873–8881.

89 Stroeve, P. et al. (1996) Gas transfer in supported films made by molecular self-assembly of ionic polymers. *Thin Solid Films*, **285**, 708–712.

90 Levasalmi, J.M. and McCarthy, T.J. (1997) Poly(4-methyl-1-pentene)-supported polyelectrolyte multilayer films: Preparation and gas permeability. *Macromolecules*, **30** (6), 1752–1757.

91 Harris, J.J. and Bruening, M.L. (2000) Electrochemical and in situ ellipsometric investigation of the permeability and stability of layered polyelectrolyte films. *Langmuir*, **16** (4), 2006–2013.

92 Krasemann, L. and Tieke, B. (2000) Selective ion transport across self-assembled alternating multilayers of cationic and anionic polyelectrolytes. *Langmuir*, **16** (2), 287–290.

93 Harris, J.J., DeRose, P.M., and Bruening, M.L. (1999) Synthesis of passivating, nylon-like coatings through cross-linking of ultrathin polyelectrolyte films. *J. Am. Chem. Soc.*, **121** (9), 1978–1979.

94 Harris, J.J., Stair, J.L., and Bruening, M.L. (2000) Layered polyelectrolyte films as selective, ultrathin barriers for anion transport. *Chem. Mater.*, **12** (7), 1941–1946.

95 Krasemann, L. and Tieke, B. (1998) Ultrathin self-assembled polyelectrolyte membranes for pervaporation. *J. Membr. Sci.*, **150** (1), 23–30.

96 van Ackern, F., Krasemann, L., and Tieke, B. (1998) Ultrathin membranes for gas separation and pervaporation prepared upon electrostatic self-assembly of polyelectrolytes. *Thin Solid Films*, **329**, 762–766.

97 Krasemann, L. and Tieke, B. (2000) Highly efficient composite membranes for ethanol-water pervaporation. *Chem. Eng. Technol.*, **23** (3), 211–213.

98 Liu, Y.L. et al. (1997) pH-switchable, ultrathin permselective membranes prepared from multilayer polymer composites. *J. Am. Chem. Soc.*, **119** (37), 8720–8721.

99 Bard, A.J. and Faulkner, L.R. (1980) *Electrochemical Methods*, Wiley., New York.

100 Gough, D.A. and Leypoldt, J.K. (1979) Membrane-covered, rotated disk electrode. *Anal. Chem.*, **51** (3), 439–444.

101 Ikeda, T. et al. (1982) Permeation of electroactive solutes through ultrathin polymeric films on electrode surfaces. *J. Am. Chem. Soc.*, **104** (10), 2683–2691.

102 Friedlander, H.Z. (1968) Membranes, in *Encyclopedia of Polymer Science and Technology*, (eds H.F. Mark, N.G. Gaylord, and N.M. Bikales), Wiley (Interscience), New York, vol. 8, 620–638.

103 Hatch, M.J., Dillon, J.A., and Smith, H.B. (1957) *Ind. Eng. Chem.*, **49**, 1812.

104 Bromberg, L.E. (1991) *J. Membr. Sci.*, **62**, 131–143.

105 Urairi, M. et al. (1992) Bipolar reverse-osmosis membrane for separating monovalent and divalent ions. *J. Membr. Sci.*, **70** (2–3), 153–162.

106 Farhat, T.R. and Schlenoff, J.B. (2003) Doping-controlled ion diffusion in polyelectrolyte multilayers: Mass transport in reluctant exchangers. *J. Am. Chem. Soc.*, **125** (15), 4627–4636.

107 Chien, J.C.W. (1984) *Polyacetylene. Chemistry, Physics, and Materials Science*, Academic Press., Orlando.

108 Skotheim, T.A. (ed.) (1986) *Handbook of Conducting Polymers*, Academic Press, New York.

109 Ghostine, R.A. and Schlenoff, J.B. (2011) Ion diffusion coefficients through polyelectrolyte multilayers: temperature and charge dependence. *Langmuir*, **27** (13), 8241–8247.

110 Salloum, D.S. and Schlenoff, J.B. (2004) Rectified ion currents through ultrathin polyelectrolyte complex: Toward chemical transistors. *Electrochem. Solid State Lett.*, **7** (11), E45–E47.

111 Donath, E. *et al.* (1998.) Novel hollow polymer shells by colloid-templated assembly of polyelectrolytes. *Angew. Chem. Int. Ed*, **37** (16), 2202–2205.

112 Caruso, F. *et al.* (1998) Electrostatic self-assembly of silica nanoparticle - Polyelectrolyte multilayers on polystyrene latex particles. *J. Am. Chem. Soc.*, **120** (33), 8523–8524.

113 Caruso, F., Caruso, R.A., and Möhwald, H. (1998) Nanoengineering of inorganic and hybrid hollow spheres by colloidal templating. *Science*, **282** (5391), 1111–1114.

114 Farhat, T.R. and Schlenoff, J.B. (2002) Corrosion control using polyelectrolyte multilayers. *Electrochem. Solid State Lett.*, **5** (4), B13–B15.

# 14
# Conductivity Spectra of Polyelectrolyte Multilayers Revealing Ion Transport Processes

*Monika Schönhoff and Cornelia Cramer*

## 14.1
### Introduction to Conductivity Studies of LbL Films

In electrochemical applications there is a tremendous demand for novel materials with tailored properties, in particular due to the current search for high capacity energy storage devices. Fuel cells and Li ion batteries are competing devices promising solutions to energy management, provided that suitable materials can be developed. In both technologies a key element is the polymer electrolyte membrane which separates the electrodes and has to fulfill the following requirements: (i) exhibit a high ionic ($Li^+$ or protons) and negligible electronic conductivity; (ii) allow thin film processing to reach low overall resistances and (iii) provide a high mechanical stability. In fuel cells Nafion is the benchmark material, showing a microphase separation into hydrophobic, stabilizing domains and hydrated channels with charged interfaces for proton transport [1]. In Li ion batteries state-of-the-art polymeric materials are often salt-in-polymer electrolytes based on poly(ethyleneoxide) with the aim to achieve a compromise between high mechanical stability and high ionic conductivity. A typical approach is for example comb-shaped copolymers with a cross-linkable backbone to provide mechanical stability and short oligoether side chains to solubilize and transport the ions [2–4]. These bulk polymers can be prepared as membranes as thin as, at best, micrometers.

In this context polyelectrolyte multilayers (PEMs) prepared by layer-by-layer-assembly (LbL) [5] have attracted interest, since they already fulfill two of the three main requirements described above by their generic material properties: Concerning requirement (ii), PEM layer thickness can be tuned in the nm to μm range. Concerning requirement (iii), LbL films are mechanically very stable in spite of their low thickness, as demonstrated by Young's moduli of the order of GPa [6, 7].

Requirement (i), however, a sufficiently large ionic conductivity, remains a challenge. In several publications DC conductivities of PEM ranging from $10^{-12}$ S cm$^{-1}$ to $10^{-5}$ S cm$^{-1}$ have been reported [8–11]. Studies of the conductivities of PEM started with the seminal work by Durstock and Rubner, where the classical constituents of LbL films were employed, that is, poly(acrylic acid) (PAA),

---

*Multilayer Thin Films: Sequential Assembly of Nanocomposite Materials*, Second Edition.
Edited by Gero Decher and Joseph B. Schlenoff.
© 2012 Wiley-VCH Verlag GmbH & Co. KGaA. Published 2012 by Wiley-VCH Verlag GmbH & Co. KGaA.

poly(styrene sulfonate sodium salt) (PSS), and poly(allylamine hydrochloride) (PAH). DC conductivities, $\sigma_{DC}$, ranged from $10^{-12}$ S cm$^{-1}$ to $10^{-7}$ S cm$^{-1}$ [8], where the maximum of $10^{-7}$ S cm$^{-1}$ was achieved only at strong hydration. Further optimization of the polymeric constituents was promising: [9, 10] with poly(2-acrylamido-2-methyl-1-propansulfonic acid) (PAMPS) as a polyanion, the DC conductivity at high humidity reached the value $\sigma_{DC} \approx 10^{-5}$ S cm$^{-1}$, already in a realistic range for applications [9].

In further studies, polyion components in LbL films tested for their conductivity involved polymers already known as good ion conductors as bulk material, such as Nafion as an established proton conductor [9], poly(ethylene oxide) (PEO) as a classical polymer electrolyte [10], or polyphosphazenes which provide flexible main chains as polymer electrolytes [11, 12].

In the last few years, in addition to these studies aiming at optimization of conductivities, fundamental aspects governing the ion transport in LbL films were investigated. A major problem in the interpretation of conductivity data is the lack of knowledge about the composition of the films: formed by self-assembly, the stoichiometry of the films is controlled by the compensation of surface charges upon chain adsorption. Full compensation of the excess charge of the previous layer by polyion charges ("intrinsic charge compensation") may occur, or, if this is sterically not favorable, small counterions might incorporate into the film in order to partially compensate the polyion charges ("extrinsic charge compensation"). Thus, PEMs are a material of *a priori* unknown stoichiometry. In a first approximation, the translational entropy of the small counterions favors intrinsic charge compensation, however, in several polyion combinations a deviation from a 1 : 1 stoichiometry of the polyions was found and a substantial degree of extrinsic charge compensation concluded [13–15].

In hydrated multilayers even protons can contribute to the conductivity, in studies employing Nafion in multilayers they are discussed as the main contributors to the DC conductivity [16–19]. Indeed, tremendous differences in conductivity were found comparing dry and completely hydrated films [8, 9, 11, 16, 20]. A review of the activities up to 2007 is given by Lutkenhaus and Hammond [19]. There, it is argued that the films are proton conductors, however, an analysis of the contributions of other small counterions suffers from the lack of systematic knowledge about the composition of the PEMs.

As a reference system with a controlled stoichiometry, polyelectrolyte complexes (PECs) are an interesting material. Their similarity to PEMs was demonstrated by the same short-range interactions of the charged groups [21] and comparable regions of stability in the phase diagrams [22, 23]. Most importantly for the interpretation of conductivities, in PECs the content of small ions is known, as it is defined by the mixing ratio of the polyions. Furthermore, systems with mainly one type of counterion can be prepared, when excess salt is removed by dialysis. Thus, composition-dependent conductivities can be related to the contribution of a single ion type [24–26]. For this purpose, solid PEC samples have to be prepared by drying and pressing.

We review here the progress made by impedance spectroscopy on polyelectrolyte multilayers, not only in terms of extracting DC conductivities, but also, in particular, with respect to knowledge about transport mechanisms which can be gained from evaluation of the spectral shape and variations of it with composition and relative humidity. We also describe a few properties of PECs for comparison, however, a more detailed review comparing PEC to PEM conductive properties was recently published [27].

## 14.2 PEM Spectra: Overview

Room-temperature impedance spectra of PEMs of different polyanion/polycation combinations were determined at different relative humidities, RH [11, 20]. Representative spectra of PAH/LiPSS are shown in Figure 14.1a. In contrast to the spectra of most bulk materials, polarization effects at very low frequencies are strongly pronounced. This polarization arises due to the low thickness of the layers (about 100 nm for 100 layers [20]), and it is caused by ionic charge carriers accumulating at the blocking electrodes at low frequencies. Therefore, the polarization effects become more pronounced the higher the DC conductivity is (see Figure 14.1). At high frequencies, a dispersive regime occurs, which describes ion movements on a more local scale. In the intermediate frequency regime, a DC plateau can be identified just between the regimes of polarization and dispersion. Even when this plateau is less obviously evident, the DC conductivity can be extracted by suitable evaluation methods [20].

For all frequencies, increasing humidity results in an increase in the conductivity. For various combinations of polycations and polyanions the sets of spectra taken at different RH show similar properties. In the spectra of Figure 14.1b, where multilayers are formed from cationic and anionic polyphosphazene derivates [11], conductivities are 1 to 2 orders of magnitude larger than those of the PAH/LiPSS system of Figure 14.1a. This may very well be an effect of the large main chain flexibility of the

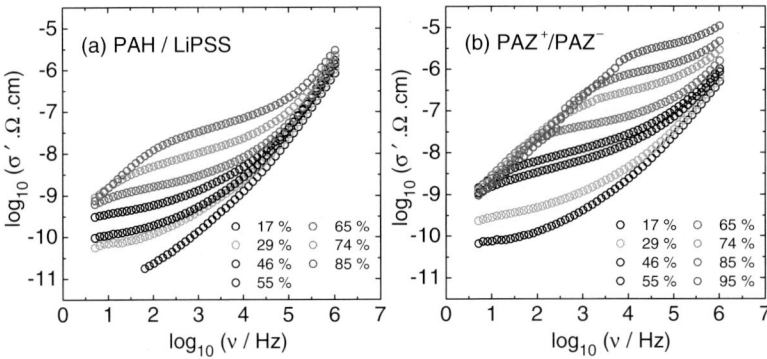

**Figure 14.1** Impedance spectra of (a) PAH/LiPSS multilayers and (b) $PAZ^+/PAZ^-$ multilayers at relative humidities between 17% and 95%. (from Ref. [20]).

polyphosphazenes, though other differences, such as the type and concentration of charge carriers, might additionally contribute.

## 14.3
## DC Conductivities of PEMs

Whereas conductivity *spectra* of PEMs like those presented in Figure 14.1 are very rare in the literature, several authors report on the DC conductivity values of different polyanion/polycation combinations, determined at different relative humidities. Table 14.1 summarizes several examples. In the dry state, DC conductivities of PEMs are generally low, compared to those of classical polymer electrolytes. With increasing relative humidity, however, the values increase significantly and reach values up to $10^{-5}\,\mathrm{S\,cm^{-1}}$. Combination of different types of polycations and polyanions and differences in the preparation (e.g., variation in pH, use of additional salt) also influence the value of $\sigma_{DC}$, however, the relative humidity seems to have the strongest impact.

A difficulty when comparing different values of $\sigma_{DC}$ is the fact that the number density of the charge carriers is not known, and in different films different fractions of counterions of either type might be incorporated. Therefore, the differences in $\sigma_{DC}$ reflected in Table 14.1 might include different contributions: (i) differences in the content of incorporated counterions, acting as charge carriers; (ii) differences in the type and size of relevant charge carriers, causing different charge carrier mobility;

**Table 14.1** Selection of DC conductivities of various PEMs at room temperature.

| PEM System | $\sigma_{DC}/(\mathrm{S\,cm^{-1}})$ | RH (%) | Reference |
|---|---|---|---|
| PAH/PAA (pH 3.5/pH 3.5) | $2 \times 10^{-7}$ | 85–90 | [8] |
| PAH/PSS (pH 3.5/pH 3.5) | $3 \times 10^{-8}$ | 85–90 | [8] |
| PAH/PSS (pH 3.5/pH 3.5) 0.1 M NaCl | $3 \times 10^{-7}$ | 85–90 | [8] |
| (LPEI/NAFION)$_{30}$ 0.1 M NaCl | $(1.2 \pm 0.2) \cdot 10^{-11}$ | 17 | [9] |
| (LPEI/NAFION)$_{30}$ 0.1 M NaCl | $(8.3 \pm 0.4) \cdot 10^{-11}$ | 52 | [9] |
| (LPEI/PAMPS)$_{30}$ no salt | $(9.4 \pm 0.7) \cdot 10^{-6}$ | 52 | [9] |
| (LPEI/PAA)$_{30}$ pH 5 | $(6.4 \pm 0.2) \cdot 10^{-9}$ | 52 | [9] |
| (LPEI/NAFION)$_{30}$ 0.1 M NaCl | $(6.9 \pm 0.4) \cdot 10^{-9}$ | 100 | [9] |
| (LPEI/PAMPS)$_{30}$ no salt | $(1.0 \pm 0.1) \cdot 10^{-5}$ | 100 | [9] |
| (LPEI/PAMPS)$_{30}$ 0.2 M NaCl | $(1.5 \pm 0.2) \cdot 10^{-5}$ | 100 | [9] |
| PEO/PAA, pH 2.5 | $(2.0 \pm 1.0) \cdot 10^{-10}$ | 52 | [10] |
| PEO/PAA, pH 2.5, 1.0 M LiCF$_3$SO$_3$ | $(1.3 \pm 0.1) \cdot 10^{-5}$ | 52 | [10] |
| (PDADMAC/PAMPS)$_{40}$ 0.02 M H$_2$SO$_4$ | $8 \times 10^{-6}$ | 60 | [28] |
| (PDADMAC/PAMPS)$_{40}$ 0.02 M H$_2$SO$_4$ | $4.5 \times 10^{-5}$ | 100 | [28] |
| PDADMAC/NaPSS | $3.5 \times 10^{-6}$ | 74 | [20] |
| PAZ$^+$/PAZ$^-$ | $1.8 \times 10^{-7}$ | 74 | [20] |
| PAH/CsPSS | $7.9 \times 10^{-9}$ | 74 | [20] |
| PAH/LiPSS | $7.1 \times 10^{-9}$ | 74 | [20] |

(iii) differences in the state of hydration, influencing the charge carrier mobility, or (iv) differences in the polyion network mobility. Issue (iii) can be expected to play a role and lead to comparatively large DC conductivities in PEMs with PDADMAC, since PDADMAC with its quaternary ammonium ion group is a very hygroscopic material. Layers formed thereof can be expected to be more strongly hydrated than other polymers. This might enhance the DC conductivity. For the $PAZ^+/PAZ^-$ PEM, we expect a large backbone flexibility (contribution (iv)), however, just from the data of Table 14.1 no separation of the different contributions influencing $\sigma_{DC}$ is possible.

To solve the question about the contribution of different ionic charge carriers, such as residual cations, residual anions, or protons from the hydration water, we look at the results found for analogue polyelectrolyte complexes [24]. In dried PECs the contribution of residual $Na^+$ to the conductivity is dominant over that of $Cl^-$ [24], see also details in Section 14.5. This holds even in PECs with a large fraction of incorporated $Cl^-$ anions (and possibly only traces of $Na^+$). We can thus conclude that even though it might be mainly $Cl^-$ counterions that are incorporated into some PEMs upon layer formation, they will not dominate the ion transport either. This conclusion holds for dry PEMs, but with the assumption that the hydration does not enhance $Cl^-$ mobility much more than $Na^+$ mobility, it holds for hydrated films as well. Thus, the conductivities of Table 14.1 should be dominated by the cations or by protons [20]. In order to determine the influence of the cations, we varied the type of counterion of the polyanion PSS. Our results in Table 14.1, last two rows, show a similar conductivity, irrespective of the type of alkali counterion involved. This implies that either alkali ion incorporation into the film is negligible, or that ion density and mobility are identical for all alkali ions. However, as argued before, the only reasonable conclusion is that the conductivity is not dominated by incorporated counterions, but by protons or hydronium ions [20].

That being the case, the dependence of the DC conductivity on RH, or rather water content, is very interesting. In most studies summarized in Table 14.1, $\sigma_{DC}$ values were only determined for a few RH values, and these vary between different studies. This makes the different data difficult to compare. We have, therefore, systematically determined the influence of RH on the DC conductivity for various PEM materials. The water content of PEMs was already described as increasing linearly or almost linearly with humidity [29, 30].

Figure 14.2 shows the increase in $\log(\sigma_{DC})$ with RH for some examples of PEMs. Remarkably, the increase in $\log(\sigma_{DC})$ is linear over the whole range of RH, and this is valid for all the PEM systems investigated [20]. Thus, there seems to be a general exponential increase in the DC conductivity with RH: $\log(\sigma_{DC}) = a \cdot RH + \text{const}$. This is reminiscent of the exponential dependence of $\sigma_{DC}$ on temperature and we may, therefore, conclude that charge transport in PEMs is a hydration-activated process.

But, although for different PEMs $\sigma_{DC}$ increases exponentially with RH, not only the absolute values of $\sigma_{DC}$, but also the slopes in Figure 14.2 vary significantly for different polyion combinations. The slopes were extracted and are displayed in Figure 14.3 [20]. If we assume that the exponential law of the humidity-dependence of $\sigma_{DC}$ is indeed a general feature, literature data can be included: In studies where DC conductivities were determined for at least two RH values, the parameter $a$ can be

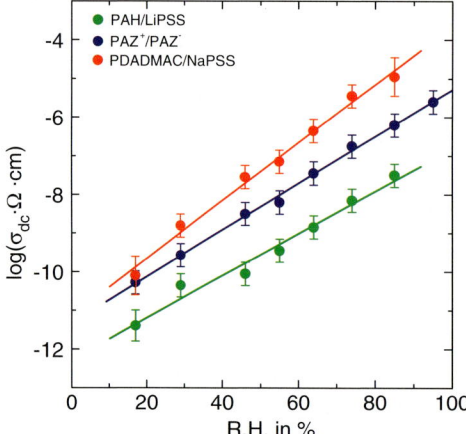

**Figure 14.2** Logarithm of the DC conductivity of multilayers of PAH/LiPSS (green), PAZ$^+$/PAZ$^-$ (blue) and PDADMAC/NaPSS (red) as a function of relative humidity. The solid lines result from linear regression [20].

estimated from the published data points, assuming that the relation $\log(\sigma_{DC}) = a \cdot RH + $ const. is valid over the complete RH range [8, 9, 16]. We note that the open squares in Figure 14.3 do not represent exact values, but lower limits of the slope, since in this case only an upper limit of the dry state conductivity was available [8].

Figure 14.3 clearly shows that the slope values fall into different categories, emphasized by the black frames. Each combination of polyions has its typical slope

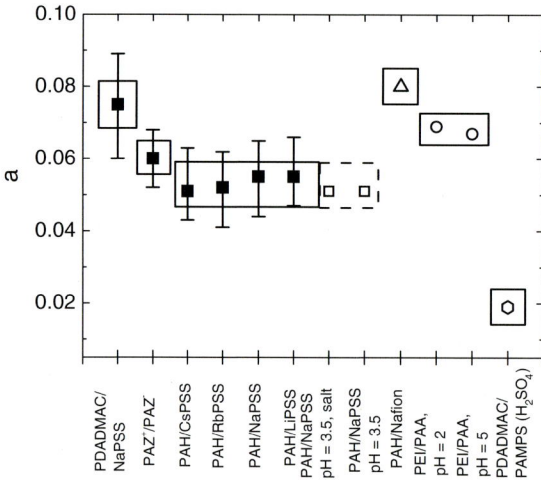

**Figure 14.3** Slopes of $\log_{10}(\sigma_{DC})$ versus RH for different polyelectrolyte multilayer systems of a humidity-dependence study (full symbols) [20] in comparison to slopes derived by interpolation of conductivities; open triangle, Ref. [16]; open circles, Ref. [9]; open hexagon, Ref. [28]; open squares, Ref. [8]; the open squares present lower limits of the slope, see text. Black frames: PEM made from the same polyion combination.

value $a$. The $a$-value can be interpreted as the sensitivity of the respective polymer matrix to hydration. As mentioned before, PEMs with PDADMAC can be expected to be rather hygroscopic. Indeed, here we find the largest value of slope $a$; that is, $\sigma_{DC}$ is most sensitive to hydration. A similar or even larger sensitivity is only found for PEMs containing NAFION, see data point by Daiko [16]. It is also remarkable, that for all PAH/PSS layer systems, irrespective of the type of alkali counterion, and of preparation conditions, the slope $a$ is always the same (see open and filled squares in Figure 14.3). This implies that the sensitivity of $\sigma_{DC}$ to humidity is only controlled by the polymer backbone and its hydration properties. This finding is consistent with the discussion above about the nature of the dominating charge carriers: The mobility of protons (which are the dominant charge carriers in the conduction process) will increase with the amount of water in the films, and the increase in mobility is more pronounced for a material with a stronger hydration behavior.

There might be exceptions where protons are not dominating the charge transport in PEMs, for example specific polyelectrolyte combinations, which yield unusually high extrinsically compensated PEMs. In such PEMs, the counterion content, and thus its contribution to the overall conductivity, might be larger. One might also speculate about very dry PEMs, where incorporated counterions of the polyelectrolyte components might dominate the transport, but in all PEMs presented here, exposed to relative humidities $\geq 17\%$, protons are dominating. Even for much lower humidity there is typically a large water content in PEMs, for example 3.5 water molecules per pair of monomers at 0% humidity, as extracted from neutron reflectivity data [31, 32]. Therefore, even in dry films it might still be protons which dominate the ion transport. In spite of some possible exceptions, the general property of hydrated PEMs is thus that charge transport is governed by protons.

The finding of proton conduction in PEMs is not very surprising in view of the structural similarity of, for example, PSS, employed in many PEM systems, to NAFION, both carrying sulfate groups and hydrophobic moieties. One can speculate about a local separation of hydrophobic and charged groups in PEMs, as is known to exist in NAFION. Due to the disordered structure of PEMs, structural studies at this level of detail are difficult. It has, however, been proven by neutron reflectivity studies that – for suitable preparation conditions – hydrophobic interaction plays a major role in the LbL formation process [31, 33]. It is, therefore, possible that hydrophilic pathways do exist in PEMs, along which sites for protons are available. Such sites should then be closely related to water sites, and the conduction process might follow the Grotthus mechanism. The size of water voids in fully hydrated PEMs was determined to be about 1 nm in diameter [34]. Reducing the water content would also reduce the size of these water voids and, thereby, the proton mobility. The exponential dependence of proton mobility on relative humidity, reported here, is indeed a very strong dependence. Over the whole range of humidity from 0% to 100% the water content in PEMs varies only by about 30% of the total film weight [29, 30]. This implies that the proton mobility increases over orders of magnitude, induced by only small variations in the water content. This is a reasonable finding, since at low degrees of swelling, immobile water, which does not exchange with hydration water was indeed found [32]. The mobility of the protons will be closely connected to that of the hydration water.

Open questions for further, more detailed studies are, whether the water content is the controlling parameter for conductivity, or to what extent structural aspects on the molecular scale might play a role. The size and shape of water voids could have a significant effect on the connectivity of such hydrated pores, and thus on the long-range mobility.

## 14.4
### Modeling of PEM Spectra

This section deals with the shape of the PEM conductivity spectra. The symbols in Figure 14.4 show spectra of $PAZ^+/PAZ^-$ PEM at different RH. The dotted lines represent spectra with a shape typical for conductivity spectra of many ion-conducting materials, including polymers, glasses and crystals [35–37]. This spectral shape is not only found experimentally, but can also be very well reproduced by the MIGRATION model developed by Funke et al. [35, 36]. This model is applicable for ion conductors, such as polymers, inorganic glasses or crystals. It considers ion hopping in a time-dependent energy landscape. The central idea is that, because of their mutual repulsive Coulomb interactions, equally charged mobile ions tend to stay apart from each other. If a mobile ion leaves its site by jumping into a vacant neighboring site, mismatch is created. The system then tends to reduce the mismatch, which can be done either by a correlated backward hop of the ion itself or by a rearrangement of its neighboring ions. In the first case, the previous forward jump of the ion under consideration is unsuccessful, whereas in the second case the ion has successfully moved to a new site. Successful hops are the basis for long-range

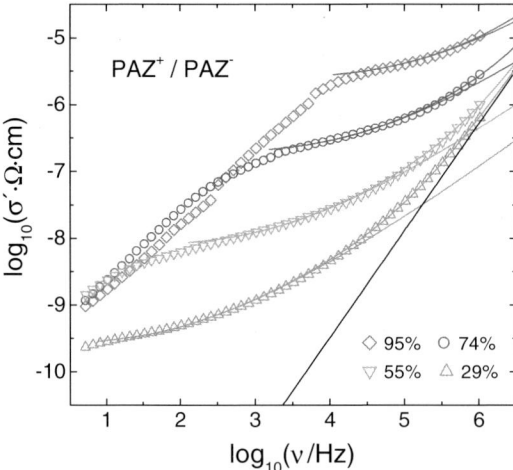

**Figure 14.4** Spectra of $PAZ^+/PAZ^-$ at different RH. Symbols represent experimental spectra. The curved dashed lines were obtained on the basis of the MIGRATION model. Superposition of these model spectra with an additional contribution represented by the straight line yields spectra which are in very good agreement with the experimental results (solid lines).

ion transport and they are probed in the DC regime. In the dispersive regime, all ion hops which have not proven unsuccessful within the time window given by the inverse angular frequency do contribute to the conductivity. The shorter the time window, the more ions have not yet performed a backward jump. This is the reason why the conductivity increases with frequency.

In Figure 14.4 we see that, at sufficiently low frequencies, the shape of PEM spectra agrees very well with that of MIGRATION spectra, but that there are strong deviations at higher frequencies. In the experimentally available frequency window, these deviations become more pronounced the lower the RH. In order to find the origin of the extra contribution to the conductivity, the difference between experimental and MIGRATION spectra was determined [27]. The extra contribution shows a power-law frequency dependence $\sigma'(\nu) = A_q \cdot \nu^q$ with an exponent $q$ larger than 1, but smaller than 2. In PAZ$^+$/PAZ$^-$ PEM, the exponent is 1.6, in other systems like, for example, PAH/PSS it is 1.7 [27].

As shown previously [38], power laws with $q > 1$ imply that the motion under consideration is strictly localized. An exponent of exactly 2 would correspond to a Debye-type process, where a non-interacting dipole moves locally. The other scenario which can be considered is that many dipoles interact strongly with each other, leading to the so-called "nearly constant loss (NCL) behavior" [36, 37, 39–41] with $\sigma' \propto \nu$ over wide ranges in frequency. The exponent $q$ in PEM spectra is between these two scenarios. The dipoles involved in the local motion contribution might involve polarizations of the polyion matrix or even local movements, for example, rotations, of water molecules. In PEMs, these water movements might even be coupled to the proton hopping motions. But in any case, the high-frequency component is of localized type and does not contribute to the long-range transport and, therefore, to the DC conductivity. Within our experimental frequency window, the higher the RH, the less these local motions contribute to the spectra. Due to the absorbed water in the PEMs, the mobility of the small ions – which are not only mobile on a localized scale, but also on a macroscopic scale – increases with RH. This "classical" MIGRATION-type ionic hopping motion then dominates the PEM conductivity spectra for high RH. In the MIGRATION regime, mobile protons in PEMs behave like other mobile ions in various ion-conducting bulk materials. There is a high tendency for the protons which have just moved from one site to another to jump back to the original site. These backward correlations give rise to the dispersive part of the conductivity, well described by the MIGRATION concept. Those protons, which do not jump backwards because they have neighbors that rearrange themselves with respect to the new proton position, determine the long-range ion transport probed by the DC conductivity.

## 14.5
### Ion Conduction in Polyelectrolyte Complexes

Polyelectrolyte complexes have a similar local structure to that of LbL films, however, they provide the advantage that they can be prepared with a defined stoichiometry. After mixing strong polyelectrolytes, that is, polycations and polyanions with

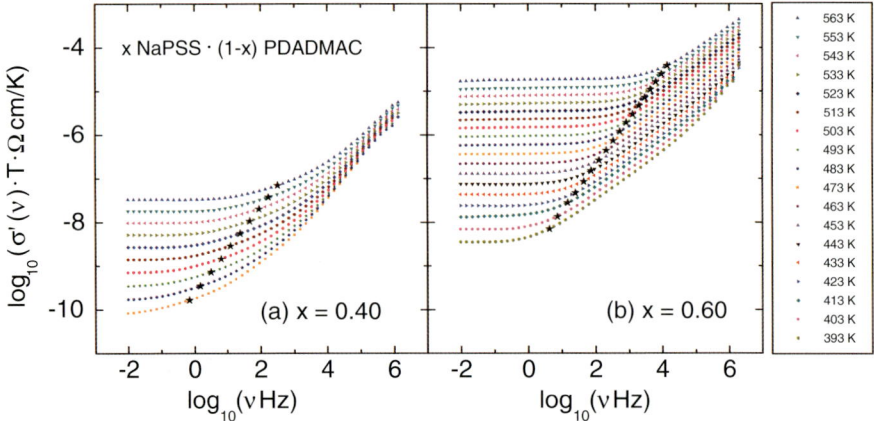

**Figure 14.5** Spectra of the real part of the conductivity of PEC with (a) $x=0.40$ and (b) $x=0.60$, investigated in the dry state at different temperatures. The stars mark the onset frequencies defined by $\sigma'(\nu^*) = 2\sigma_{DC}$. (Spectra from refs [25, 42]).

monomer fractions $x$ and $(1-x)$, respectively, a complex with a defined stoichiometry, and even a defined counterion concentration, results [24]. When excess counterions are removed by dialysis, for $x=0.5$ full intrinsic charge compensation of the polyion charges can be assumed, and nominally no counterions remain in the complex. For a polyanion-rich complex $(x>0.5)$, the extrinsic charge carriers are predominantly the cationic counterions, whereas in polycation-rich complexes $(x<0.5)$, it is the anions. Thus, by varying $x$, complexes with only one type of counterion in defined concentrations can be prepared.

Here, we first focus on the properties of PEC materials in the completely dry state. Typically, these complexes are dialyzed, dried and compressed into solid pellets prior to impedance experiments [24]. Figure 14.5 shows conductivity spectra for $x=0.4$ and $x=0.6$, respectively, in dependence on temperature. In contrast to spectra of PEMs, there is no polarization regime observed, due to the typically larger thickness of the pellets in the mm range. Similar to PEM spectra, however, is the occurrence of a DC plateau and a dispersive regime. With increasing temperature, $T$, the DC conductivity increases, while the onset of conductivity dispersion shifts to higher frequency. This effect is well-known for other ion-conducting materials and is discussed in detail in Section 14.6.2 within the framework of scaling concepts.

The temperature dependence of $\sigma_{DC} \cdot T$ is typically Arrhenius-like over many decades, showing that the ion dynamics in PEC materials is determined by thermally activated hopping processes of the mobile ions. Thus, activation enthalpies can be extracted and compared for different compositions [24]. It is also remarkable that in those PECs for which measurements could be performed at temperatures above *and* below the calorimetric glass transition temperature, there is no change in the temperature dependence of $\sigma_{DC}T$ when passing through the glass transition. This implies that PEC materials are "strong" glasses. Figure 14.6 shows the activation

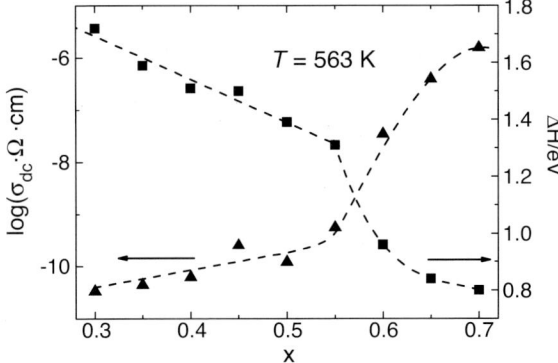

**Figure 14.6** Activation enthalpy (squares, right y-axis) and DC conductivity at 290 °C (triangles, left y-axis) in NaPSS/PDADMAC complexes as a function of composition [24].

enthalpies together with the isothermal DC conductivity as a function of composition. Up to roughly $x = 0.55$ the DC conductivity increases and the activation enthalpy decreases almost linearly. Above $x = 0.55$ the increase in NaPSS content is accompanied by a much stronger increase in $\sigma_{DC}$ and decrease in $\Delta H_{DC}$, respectively. The strong increase in $\sigma_{DC}$ with $x$ was attributed to a strong increase in ion mobility with decreasing polyion network crosslinking density [24].

The asymmetry in Figure 14.6 shows that there are distinct differences between the ion dynamics in PDADMAC- and PSS-rich PECs. Otherwise, $\sigma_{DC}$ as a function of $x$ should show a minimum at $x = 0.5$, which is obviously not observed. The conductivity of PECs with $x \leq 0.5$ could therefore be due to either residual $Na^+$ ions or protons [24]. The fact that the activation enthalpies in dry PECs with different types of small cations, viz. $Li^+$, $Na^+$, $Cs^+$, are very different, points to transport by residual alkali ions rather than by protons [43]. The isothermal DC conductivity increases continuously with NaPSS content, indicating that the chloride ions do not dominate the ion transport, even in PEC materials with an excess of polycations and thus $Cl^-$ as the most abundant mobile charge carrier. Instead, residual $Na^+$ ions seem to govern the ionic conductivity in PDADMAC-rich PECs.

The question why the mobility of $Na^+$ ions in PECs is larger than that of $Cl^-$ ions can partly be explained by the difference in size of the respective ions. However, since the difference in conductivity appears to be orders of magnitude between PSS- and PDADMAC-rich PECs, it is likely that structural differences of the PEC also play a large role: From the asymmetric behavior of the osmotic coefficient of hydrated PECs in dependence on the mixing ratio, $x$, it was concluded that polyanion-rich PECs are homogeneous, while polycation-rich PECs undergo a micro-phase separation of neutral from anion-rich regions [44]. In addition, the shapes of the spectra differ slightly (compare Figure 14.5a and b). In all materials with $x > 0.5$ a "shoulder" in the conductivity spectra is observed, roughly between 0.1 kHz and 10 kHz. A detailed analysis and discussion of the spectral shape and other differences between polycation-rich and polyanion-rich PECs is found in ref. [27]. We conclude that for the Na-rich side of dry PECs the strong increase in the DC conductivity is mainly due to an increase in ion mobility. The latter is due to the fact that the number of available sites increases strongly with ion content [24]. Whereas in systems dilute in Na$^+$ (PDAD-MAC-rich PECs) there are few not well connected ion sites and ions have to travel large distances before they find a new appropriate site, there is a connected system of pathways in PECs with higher Na$^+$ ion content.

In addition to the effect of temperature, the effect of varying humidity on the conductivity spectra of PECs was recently investigated [26]. There are indeed very similar features as compared to the humidity dependence in PEMs, though the charge carriers are alkali ions instead of protons: First, DC conductivities in PEC depend very strongly on relative humidity and also show an exponential increase with RH. Even the slope of log($\sigma_{DC}$) against RH is of the same order as that in LbL films of analogous composition [26]. Thus, the "activation" by humidity given by a smoothing out of the potential energy landscape holds in the same way for protons and alkali ions. There are, however, distinct differences in the humidity-dependent conductivities for different alkali ions: $\sigma_{DC}$ is larger for Cs$^+$ as counterion than for Na$^+$, and it was shown that the ionic mobility is larger for the larger cation with the smaller hydration shell, such that the alkali ions can be considered to be transported with their hydration shells [26].

## 14.6
### Scaling Principles in Conductivity Spectra: From Time–Temperature to Time–Humidity Superposition

### 14.6.1
### General Aspects of Scaling

A general concept in the description of dynamic processes is the "time–temperature superposition principle" (TTSP). This implies that, provided that the mechanisms governing molecular dynamic do not change with temperature, the effect of temperature is a mere enhancement of the rate of all dynamic processes. Indeed, the conductivity spectra of many materials possess a $T$-independent shape and can, therefore, be superimposed to a so-called "mastercurve" [45, 46]. The validity of the TTSP for the real part of the complex conductivity can be presented by a function $\sigma'/\sigma_{DC} = F(\nu/\nu_0)$ where $\nu_0$ is an individual scaling parameter for each conductivity isotherm, often identified with the onset frequency, $\nu^*$, which characterizes the transition from the DC into the dispersive regime.

The special case of Summerfield scaling [47, 48] is fulfilled if $\sigma_{DC} \cdot T$ and $\nu^*$ are proportional to each other. This implies a straight line of slope 1 connecting the onset frequencies in a log–log plot, as in Figure 14.5. The scaling function is then $F(\nu/\nu_0) = F(\nu/(\sigma_{DC} \cdot T))$. Summerfield scaling works for a large number of ion-conducting materials, but fails for some systems presented in the past literature, see references in [27]. When the simple proportionality between $\sigma_{dc} \cdot T$ and $\nu^*$ is no longer valid, other, more general scaling laws might apply. For example, an additional scaling factor for the frequency scale that itself depends on temperature can be included: $\frac{\sigma'(\nu)}{\sigma_{DC}} = F\left(\frac{\nu}{\sigma_{DC} \cdot T} \cdot f(T)\right)$. With $f(T)$ included, all spectra taken at different temperatures can then often be scaled on a master curve. $T^\alpha$ is a common ansatz for $f$ (T) [49, 50]. Note that any of these concepts only apply if TTSP is valid, that is, if the dynamic mechanisms are unaltered by temperature and the spectra can be superimposed to a master curve.

In a plot of log ($\sigma'T$) versus log($\nu$), the slope of a straight line connecting the onset points yields information about the temperature dependence of the number density of mobile ions. While from the validity of Summerfield scaling with a slope of one a constant number density of the charge carriers can be concluded, a slope larger than one implies an increase in number density with temperature. In general, from the evaluation of scaling factors, conclusions about variations of either the number density of mobile ions or the number of available pathways for their transport in dependence on temperature can be drawn [25].

## 14.6.2
### Time–Temperature Superposition Principle in PEC Spectra

We will now review the scaling properties of PEC conductivity spectra with respect to their temperature dependence. On the one hand we will show that the TTSP is valid, indicating that the shapes of the conductivity spectra do not change with temperature for a given composition [25]. In addition, we will discuss these scaling properties as a function of composition.

As marked by the star symbols in Figure 14.5, the frequencies of the dispersion onset, $\nu^*$, can be determined using the relation: $\sigma'(\nu^*) = 2\,\sigma_{DC}$. For both compositions shown there, the onset points are on a straight line, but their slopes differ. For $x = 0.40$ the slope is smaller than one, for $x = 0.60$ it exceeds one. This is one of the marked differences between the spectra of PECs which are rich in PDADMAC and those which are rich in NaPSS.

Nevertheless, as a general feature for all compositions of these PECs, the TTSP holds. Figure 14.7a shows the total agreement of spectra taken at different temperatures after shifting the spectra to make the onset points agree. The conclusion is that there is no change in the dynamic processes with temperature, but temperature enhances all dynamic rates by the same factor. In particular, the ion jumps relevant for long-range transport are accelerated in the same way by temperature as the ionic motions on a more local scale.

A more quantitative evaluation employing the Baranovski–Cordes scaling function $f(T) = T^\alpha$ interestingly yielded a sign and absolute values of $\alpha$ differing strongly with composition [25]. The $\alpha$-exponent values were found to increase from $-1.68$ to $1.58$ when $x$ changes from 0.35 to 0.70. $\alpha = 0$ is equivalent with Summerfield scaling, and the crossover from negative to positive values occurs at $x \approx 0.47$, that is, close to the 1:1 PSS/PDADMAC composition [25]. Following the arguments in Section 14.6.1, a positive $\alpha$-value implies that either the number density of mobile ions increases or the number of available pathways for the ions decreases with increasing temperature. This is the case for PSS-rich PECs with $x \geq 0.50$. The PDADAMAC-rich PECs behave differently, showing a negative $\alpha$ and thus the opposite trend with temperature.

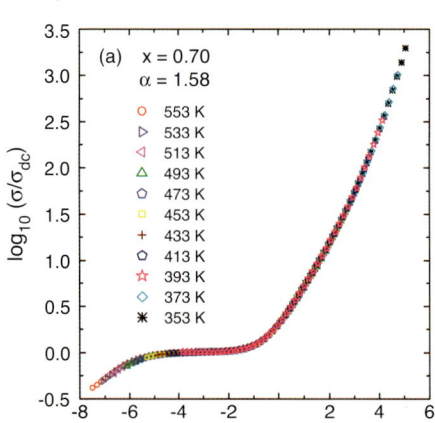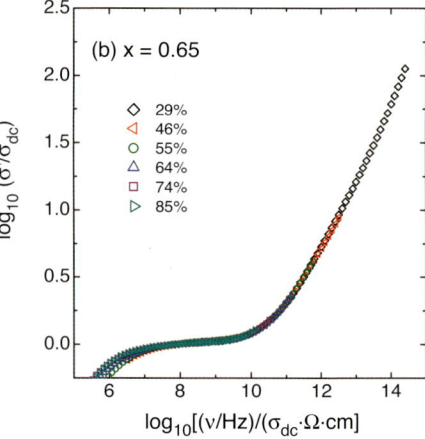

**Figure 14.7** (a) Temperature-dependent spectra of dry NaPSS/PDADMAC PEC superimposed by Baranovski and Cordes scaling with $\alpha = 1.58$ for $x = 0.70$ (according to Ref. [25]) and (b) Humidity-dependent spectra of hydrated PEC for $x = 0.65$ superimposed by Summerfield-scaling.

These differences in scaling behavior support the large differences in the basic conductivity behavior described in Section 14.5 and are due to different structures of PSS- and PDADMAC-rich PECs.

### 14.6.3
### Establishment of a Time–Humidity Superposition Principle

Seeing time–temperature scaling concepts being well verified in polyelectrolyte complexes, it became interesting to investigate whether the shapes of spectra taken at different relative humidity values would also be identical, that is, whether these spectra could also be scaled onto a master curve. Indeed, it turned out that sets of spectra taken at varying RH can be scaled on a master curve, see Figure 14.7b [51]. As the axis legends indicate, the master curve is achieved by simple scaling on the DC conductivity, which represents the case of Summerfield analogue scaling. Here, $\sigma_{DC}$ increases with RH with the same dependence as $\nu^*$. Relative humidity can thus be considered as an activation parameter which enhances long-range transport in the same way as it enhances the local dynamics. The identical shape of all spectra demonstrates that an increase in humidity, and therefore the water content in the sample, does not change the mechanisms of the ion transport. Thus, the role of the water molecules is merely to smooth the potential energy landscape in which the charge carriers are moving, thus enhancing the *rate* of ion jumps, however, the basic *mechanism* of the jumps of a single ion to an adjacent site, which constitutes the long-range transport, remains the same [51]. In particular, the number density of the charge carriers does not increase with increasing water content, implying that the ions contributing to the conductivity – in this case the alkali ions – are already dissociated and mobile at low water content.

The concept of time–humidity scaling was demonstrated for the first time by these data on hydrated non-stoichiometric PECs, and it will be interesting to study in the future whether this concept also applies to PEMs or, more generally, to transport processes in other hydrogels. First indications from unpublished data point to the validity of this concept in PEMs as well.

## References

1. Kreuer, K.D. (2001) *J. Membr. Sci.*, **185**, 29.
2. Cui, M.Z., Li, Z.Y., Zhang, J., and Feng, S.Y. (2008) *Prog. Chem.*, **20**, 1987.
3. Karatas, Y., Kaskhedikar, N., Burjanadze, M., and Wiemhöfer, H.D. (2006) *Macromol. Chem. Phys.*, **207**, 419.
4. Kunze, M., Karatas, Y., Wiemhöfer, H.D., Eckert, H., and Schönhoff, M. (2010) *Phys. Chem. Chem. Phys.*, **12**, 6844.
5. Decher, G., Hong, J.D., and Schmitt, J. (1992) *Thin Solid Films*, **210**, 831.
6. Dubreuil, F., Elsner, N., and Fery, A. (2003) *Eur. Phys. J. E: Soft Mat. Biol. Phys.*, **12**, 215.
7. Picart, C., Senger, B., Sengupta, K., Dubreuil, F., and Fery, A. (2007) *Colloid. Surf. A*, **303**, 30.
8. Durstock, M.F. and Rubner, M.F. (2001) *Langmuir*, **17**, 7865.
9. DeLongchamp, D.M. and Hammond, P.T. (2003) *Chem. Mater.*, **15**, 1165.
10. DeLongchamp, D.M. and Hammond, P.T. (2004) *Langmuir*, **20**, 5403.
11. Akgöl, Y., Hofmann, C., Karatas, Y., Cramer, C., Wiemhöfer, H.D., and Schönhoff, M. (2007) *J. Phys. Chem. B*, **111**, 8532.
12. Argun, A.A., Ashcraft, J.N., Herring, M.K., Lee, D.K.Y., Allcock, H.R., and Hammond, P.T. (2010) *Chem. Mater.*, **22**, 226.
13. Hoogeveen, N., Stuart, M., Fleer, G., and Böhmer, M. (1996) *Langmuir*, **12**, 3675.
14. Jaber, J.A. and Schlenoff, J.B. (2007) *Langmuir*, **23**, 896.
15. Crouzier, T. and Picart, C. (2009) *Biomacromolecules*, **10**, 433.
16. Daiko, Y., Katagiri, K., and Matsuda, A. (2008) *Chem. Mater.*, **20**, 6405.
17. Xi, J.Y., Wu, Z.H., Teng, X.G., Zhao, Y.T., Chen, L.Q., and Qiu, X.P. (2008) *J. Mater. Chem.*, **18**, 1232.
18. Jiang, S.P., Liu, Z.C., and Tian, Z.Q. (2006) *Adv. Mater.*, **18**, 1068.
19. Lutkenhaus, J.L. and Hammond, P.T. (2007) *Soft Mat.*, **3**, 804.
20. Akgöl, Y., Cramer, C., Hofmann, C., Karatas, Y., Wiemhöfer, H.D., and Schönhoff, M. (2010) *Macromolecules*, **43**, 7282.
21. Rodriguez, L.N.J., De Paul, S.M., Barrett, C.J., Reven, L., and Spiess, H.W. (2000) *Adv. Mater.*, **12**, 1934.
22. Kovacevic, D., van der Burgh, S., de Keizer, A., and Stuart, M.A.C. (2002) *Langmuir*, **18**, 5607.
23. Sukhishvili, S.A., Kharlampieva, E., and Izumrudov, V. (2006) *Macromolecules*, **39**, 8873.
24. Imre, Á.W., Schönhoff, M., and Cramer, C. (2008) *J. Chem. Phys.*, **128**, 134905.
25. Imre, Á.W., Schönhoff, M., and Cramer, C. (2009) *Phys. Rev. Lett.*, **102**, 255901.
26. De, S., Cramer, C., and Schönhoff, M. (2011) *Macromolecules*, **44**, 8936.
27. Schönhoff, M., Imre, Á.W., Bhide, A., and Cramer, C. (2010) *Z. Phys. Chem. N.F.*, **224** (10–12), 1555.
28. Farhat, T.R. and Hammond, P.T. (2006) *Adv. Funct. Mater.*, **16**, 433.
29. Kügler, R., Schmitt, J., and Knoll, W. (2002) *Macromol. Chem. Phys.*, **203**, 413.
30. Wong, J.E., Rehfeldt, F., Hanni, P., Tanaka, M., and Klitzing, R.V. (2004) *Macromolecules*, **37**, 7285.
31. Gopinadhan, M., Ivanova, O., Ahrens, H., Gunther, J.U., Steitz, R., and Helm, C.A. (2007) *J. Phys. Chem. B*, **111**, 8426.
32. Ivanova, O., Soltwedel, O., Gopinadhan, M., Koehler, R., Steitz, R., and Helm, C.A. (2008) *Macromolecules*, **41**, 7179.

33 Büscher, K., Graf, K., Ahrens, H., and Helm, C.A. (2002) *Langmuir*, **18**, 3585.
34 Vaca Chávez, F. and Schönhoff, M. (2007) *J. Chem. Phys.*, **126**, 104705.
35 Funke, K., Heimann, B., Vering, M., and Wilmer, D. (2001) *J. Electrochem. Soc.*, **148**, A395.
36 Funke, K. and Banhatti, R.D. (2006) *Solid State Ionics*, **177**, 1551.
37 Funke, K., Banhatti, R.D., Cramer, C., and Wilmer, D. (2009) *Chapter 21* (eds J. Kärger, P. Heitjans, and R. Haberlandt), Springer, Berlin.
38 Cramer, C and Funke, K. (1992) *Ber. Bunsen-Ges. Phys. Chem.*, **96**, 1725.
39 Knödler, D., Dieterich, W., and Petersen, J. (1992) *Solid State Ionics*, **53**, 1135.
40 Rinn, B., Dieterich, W., and Maass, P. (1998) *Philos. Mag. B*, **77**, 1283.
41 Dieterich, W. and Maass, P. (2002) *Chem. Phys.*, **284**, 439.
42 Cramer, C., Akgöl, Y., Imre, A., Bhide, A., and Schönhoff, M. (2009) *Z. Phys. Chem.*, **223**, 1171.
43 Bhide, A., Schönhoff, M., and Cramer, C. (2012) submitted to *Solid State Ionics*.
44 Carrière, D., Dubois, M., Schönhoff, M., Zemb, T., and Möhwald, H. (2006) *Phys. Chem. Chem. Phys.*, **8**, 3141.
45 Dyre, J.C. and Schroder, T.B. (2000) *Rev. Mod. Phys.*, **72**, 873.
46 Dyre, J.C., Maass, P., Roling, B., and Sidebottom, D.L. (2009) *Rep. Prog. Phys.*, **72**.
47 Summerfield, S. (1985) *Philos. Mag. B*, **52**, 9.
48 Summerfield, S. and Butcher, P.N. (1985) *J. Non-Cryst. Solids*, **77–8**, 135.
49 Baranovskii, S.D. and Cordes, H. (1999) *J. Chem. Phys.*, **111**, 7546.
50 Murugavel, S. and Roling, B. (2002) *Phys. Rev. Lett.*, **89**.
51 Cramer, C., De, S., and Schönhoff, M. (2011) *Phys. Rev. Lett.*, **107**, 028301.

# 15
# Responsive Layer-by-Layer Assemblies: Dynamics, Structure and Function

*Svetlana Sukhishvili*

## 15.1
## Introduction

Sequential adsorption of polymers on solid substrates has become an important and widely used tool to produce functional, nanoscopically structured surface coatings, free standing films or capsules. The layer-by-layer (LbL) technique relies on the formation of multiple binding points between self-assembled molecules or particles, often resulting in strong binding of species whose interactions per binding pair of functional groups can be relatively weak. The nature of the interactions between functional groups is not restricted to ionic pairing, but rather may involve non-electrostatic forces, including hydrogen bonding (such as in the case of neutral polymers and protonated polycarboxylic acids [1]) or more complex specific ligand–receptor recognition interactions [2]. For all types of LbL films, molecular binding through multiple binding sites seems to be a unifying feature in LbL film assembly and stability. In this aspect, LbL assembly is generically related to the binding of polymer chains and/or particles in solution within polyelectrolyte complexes (PECs) – an area which developed several decades earlier than the LbL technique [3, 4]. One way of looking at assembled LbL films is as at layered, quasi-2D water-insoluble PECs. Yet such a view is overly simplified. Polymer chains may possess significant mobility, both at segmental and entire-chain levels [5], and idealized layered structure might be significantly compromised. The propensity of polymer chains to intermix within the film is dependent on several factors, including the strength of polymer–polymer segmental pairing (ionic, hydrogen bonding and other interactions), density and total number of binding points per polymer chain, and others. In addition, the degree of polymer intermixing is strongly determined by deposition conditions, including concentration of polymer solutions, pH, ionic strength, and so on, and is generically related to the phase diagram of PECs, that is, solution mixtures of film components. This chapter will discuss the intrinsic similarity and antagonism between PECs and polyelectrolyte multilayers (PEMs), and the consequences of these "love-hate" relationships for LbL film build-up and internal structure. We will also consider intriguing aspects of post-assembly chain dynamics within PEMs, both determined by and affecting unique layered film structures.

*Multilayer Thin Films: Sequential Assembly of Nanocomposite Materials*, Second Edition.
Edited by Gero Decher and Joseph B. Schlenoff.
© 2012 Wiley-VCH Verlag GmbH & Co. KGaA. Published 2012 by Wiley-VCH Verlag GmbH & Co. KGaA.

Specific emphasis will be placed on multilayers of environmentally responsive polymers whose charge and/or hydration state can be varied by applying pH or temperature stimuli. Inclusion of these polymers within LbL films enables the construction of new materials that demonstrate stimuli-triggered changes in swelling, hydration, and small-molecule retention and release. The flexibility of the LbL approach, combined with the possibility of controlling film structure at the nanoscale, afford ample opportunities in engineering response characteristics of such films by varying the chemical and molecular architecture of assembled responsive polymers, as well as the conditions for film assembly. Responsive coatings based on polymer multilayers present significant promise for surface modification of microchannels and surfaces in a confined environment, where alternative techniques for creating such coatings, such as grafting of polymer chains or film stabilization using photo- or e-beam lithography, are hard to apply. The demonstrated high capacity of PEMs for a wide range of functional molecules, including dyes and polypeptides, combined with the engineered response properties of these films, presents an attractive combination for future applications of such films as functional responsive coatings. While the field of responsive LbL fields is broad and continues to grow, this chapter will review a specific type of recently developed assemblies that contain micelles with pH- and temperature-responsive cores. Importantly, assembly conditions (such as pH, temperature and ionic strength), and type of polyelectrolytes involved in LbL film build-up critically affect the response properties of such films. Therefore, correlations between type and strength of molecular interactions, energy and reversibility of polymer–polymer contacts, and the overall chain dynamics and LbL film response will be specifically emphasized.

## 15.2
## Chain Dynamics and Film Layering

### 15.2.1
### Lessons from PECs

#### 15.2.1.1 Phase Diagrams of PECs and LbL Film Deposition

Multilayer growth at surfaces is closely related to the formation of PECs in solution. When a PEM film is brought in contact with a solution of free PE chains during LbL film deposition, PE chains in the added polymer solution are usually in a large excess compared to an oppositely charged polymer within the film. For example, when film is deposited onto a substrate with 5-cm$^2$ surface area in a 50-mL beaker containing 1 mg mL$^{-1}$ polymer solution, the total excess of polymer in solution to that residing at the film surface is five orders of magnitude (assuming the amount of oppositely charged polymer adsorbed within the outermost layer of the film is 1 mg m$^{-2}$). With such a huge excess of PE chains in solution, polymers of an opposite charge deposited within the film can bind and be solubilized by the newly added chains in solution. Polymer chains removed from the LbL film can become bound within water-soluble polyelectrolyte complexes (WPECs) which are solubilized by the excess charge of added polyelectrolyte chains. Thermodynamically, formation of WPECs is always favorable in conditions of large excess of one polyelectrolyte type involved in complex formation, that is, when the molar ratio of positive-to-total charge of mixed polyelectrolyte chains $f^+$ (in molar concentrations of repeat units) is close to zero or to unity. When a polyelectrolyte chain is included within a WPEC, one expects gains in translational and configurational entropy, due to an increase in the number of WPEC species in solution and the greater mobility of dissolved polyelectrolyte chains, respectively. Whether or not such solubilization occurs is strongly determined by the strength of polymer contacts and the equilibration time of interpolyelectrolyte exchange reactions. For polyelectrolyte systems in which chain exchange occurs at the experimental timescale [4], such as for those containing highly hydrated carboxylate or phosphate polyanions and/or tertiary or quaternary ammonium polybases at significant misbalance of positive-to-negative charge in solution (i.e., at $f^+ \ll 0.5$ and $f^+ \gg 0.5$), WPECs form easily and, therefore, chains are removed from surfaces during LbL film construction [5, 6]. Interestingly, contrary to common opinion, polyelectrolytes need not necessarily be of low molecular weight and/or low charge density to exhibit such chain dynamics. For example, polyelectrolytes with matched chain lengths (98% quaternized poly-N-ethyl-4-vinylpyridinium bromide, QPVP, with a polymerization degree $DP_{QPVP}$ of 1600, and poly(methacrylic acid), PMAA, with $DP_{PMAA}$ of 1700) could not be deposited within an LbL film at pH 8.4 (when both components are strongly charged), because of severe chain exchange with solution [5]). Rather than the overall charge density or chain length (both of which do play a role in the shape of phase diagrams of PECs in solution [5]), a low barrier to segmental rearrangements and polymer unit pairing seems to be more important for chain desorption during LbL build-up. High hydration and steric hindrance to ionic pairing both seem to favor faster chain dynamics, exchange and equilibration.

Following work by Cohen Stuart and coworkers [6], who have emphasized the correlation between the phase behavior of PECs in solution and the deposition of PEMs, several groups, including ours, have demonstrated that formation of water-soluble complexes can trigger desorption of polyelectrolytes from multilayers [5–7]. An example of how the phase behavior of PECs in solution controls the deposition of polyelectrolyte chains within LbL films is shown in Figure 15.1 [5]. In experiments with PECs in solution, we have found an interesting phenomenon when the addition of salt to WPEC resulted in the formation of water-insoluble PEC. At lower salt concentrations, WPECs (region I) are formed for fivefold excess of PMAA or QPVP ($f^+ = 0.167$ or $f^+ = 0.833$, respectively). Increased salt concentration triggers redistribution of polymer chains, in region II, and a coexistence of a water-insoluble complex with $f^+ = 0.5$ (resulting in a decrease in the fraction of QPVP remaining in solution, QPVP*$_s$, and a water-soluble complex with $f^+ < 0.167$ or $f^+ > 0.833$ for excess of PMAA or QPVP, respectively). The mechanism for the salt-induced phase separation likely includes contraction of WPEC due to screening of charged units of PMAA (or those of QPVP), followed by cooperative chain rearrangements at the point of phase separation [5]. Importantly, salt-induced redistribution of polymer chains and separation of phases have direct consequences for multilayer film growth.

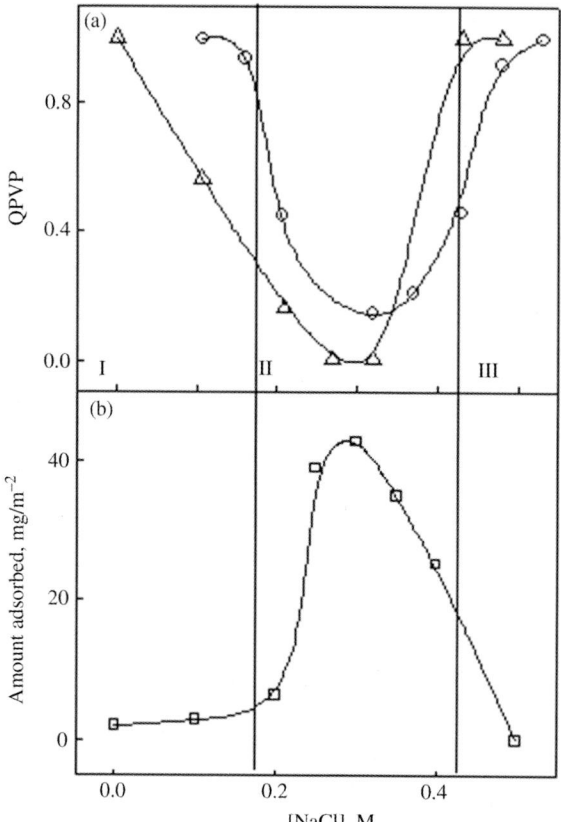

**Figure 15.1** (a) Fraction of QPVP remaining in solution (QPVP*$_s$) of QPVP/PMAA mixtures for PMAA ($M_w$ 350 k, degree of polymerization (DP) of 4000) in excess of PMAA, $f^+ = 0.167$ (circles) and in excess of QPVP ($M_w$ 330 k, DP 1600), $f^+ = 0.833$ (triangles) plotted against concentration of added salt. (b) Total amount adsorbed of nine-layer QPVP/PMAA films deposited from solution with various salt concentrations. Concentrations of polymers were (a) 0.04 M (in monomer units) and (b) 0.1 mg mL$^{-1}$. In all experiments, 0.01 M phosphate buffer at pH 8.4 was used.

Deposition of polymers at surfaces follows trends dictated by PEC in solution (Figure 15.1b). Larger amounts of polymers deposit at surfaces when solution phase separation occurs (region II). At higher salt concentrations (region III), QPVP/PMAA ionic pairs completely dissociate as a result of competition with low molecular weight ions, inhibiting PEC and PEM film formation. Therefore, deposition of LbL films at surfaces is sequentially controlled and "mirrors" the phase diagrams of PECs in solution with excess of added polycation and polyanion [5].

### 15.2.1.2 Polyelectrolyte Type and Equilibrium and Dynamics in PECs

Importantly for the LbL area, polyelectrolyte type critically affects the binding energy and equilibration time of interacting polyelectrolyte pairs in solution, as well as the deposition of polymer chains at surfaces. Many polymers widely used in deposition at surfaces, such as polystyrene sulfonate (PSS) and poly(allylamine hydrochloride) (PAH), show extremely slow equilibration times in solution, and chain desorption from surfaces is kinetically frozen and is usually not observed in the experimental time frame. Figure 15.2 contrasts the phase behavior of polycation/polyanion mixtures in solution for strongly bound and weakly bound polyelectrolyte pairs. With QPVP/PMAA PECs, precipitation occurs only when $f^+$ is close to 0.5, that is, close to an equimolar charge ratio. Excess of PMAA or QPVP (at $f^+ < 0.35$ and $f^+ > 0.6$, respectively) results in WPECs solubilized by either negative or positive charge in unpaired polymer units. In contrast, in mixtures of PMAA with a primary amine polycation, PAH, insoluble complexes are formed over a wide range of polycation-to-polyanion ratios, even at large excess of PMAA or PAH. Similar composition-independent turbidity is observed in mixtures of PSS with polycations. Therefore, the strong affinity of PAH and PSS to oppositely charged chains inhibits equilibration of this system. Strong binding of polycations to $SO_3^-$- or $SO_4^-$-containing polyanions has been demonstrated in studies of interpolyelectrolyte exchange in solutions of WPECs. For example, in mixtures of polyanions, polycations preferentially bind with poly(vinylsulfonate), PSS [8–10] or poly(vinylsulfate) [11]. Even the presence of a small number of sulfate or sulfonate groups in the polycarboxylate polymer chains provides strong selective binding with a polycation in polyanion mixtures [12]. Quantitative studies of such preferential binding revealed that the effective binding constant of the interacting monomeric units, $K_1$, was ~1.2- to 2-fold

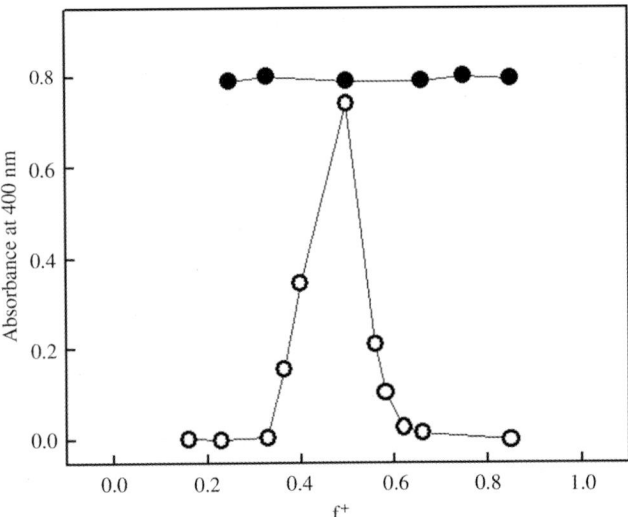

**Figure 15.2** Turbidity of QPVP/PMAA (open circles) and PAH/PMAA (filled circles) mixtures measured as the absorbance of 0.04 M (repeat units) solutions at 400 nm as a function of the mole fraction of positively charged units $f^+$. The pH of 8.4 was supported by 0.01 M Trizma buffer. PMAA, QPVP and PAH with $M_w$ of 72, 200 and 70 k, respectively, were used for the experiment.

larger for the PSS/poly-L-histidine (PLH) system as compared to DNA/PLH, PAA/PLH and PMAA/PLH complexes [13]. These differences in binding constants at the unit-to-unit level translate into enormous differences in the chain-to-chain binding constant $K_n$ through the equation $K_n = K_1^n$, where $n$ is the number of interpolymer contacts per chain. Just as PSS is one of the strongest competitors for binding with polycations, polyamines containing primary amino groups are the strongest binders among polycations [14].

The consequences of the strong binding of PSS-type and PAH-type polyelectrolytes with their partners are very significant for the LbL field. For example, while QPVP/PMAA films could not be deposited at surfaces at pH 8.4, as dictated by solution behavior in Figure 15.2 [5], because of efficient chain exchange and solubilization, robust film growth occurs under the same conditions for strongly interacting systems such as PAH/PMAA or PAH/PSS. Indeed, strong, irreversible ionic pairing in the PAH/PSS system has made this system a favorite choice for multilayer build-up, especially in the early days of the LbL field. Understanding occurrences of PEM erosion is important as it will provide materials scientists with a "map" of PEM growth, enabling them to rationally avoid unstable regimes during film deposition [5]. As the LbL field matures, a better understanding of relationships among strength of interaction, equilibration times and propensity to competitive chain desorption is also developing.

### 15.2.2
**pH-Induced Chain Dynamics and Film Structure**

The use of weak PEs (wPEs) opens ample opportunities for building functional wPEMs [15–17]. When wPEM films are constructed at a specific pH, charge within the films is compensated. However, when pH is varied after film construction it can induce accumulation of excess charge within wPEMs, when the variation occurs in the region close to the apparent $pK_a$ of a wPE within the film.

In extreme cases, when the external pH swings away from the apparent $pK_a$ of an assembled wPE, ionic contacts can completely dissociate, resulting in dissolution of electrostatically assembled films. With hydrogen-bonded films composed of neutral hydrogen-bonding polymers and a protonated form of polyacid, an increase in pH can cause film dissociation at close-to-neutral pH values [1]. Film disruption occurs because of pH-induced accumulation of charge of one sign on wPE chains. The disruptive forces originate from unfavorable interactions of these charges within the film, and the osmotic pressure associated with counterions included within the film to compensate for the emerging charge. On the other hand, hydrophobicity adds to the stability of hydrogen-bonded films. As a result, the critical pH at which film dissociation occurs ($pH_{crit}$) is strongly dependent on the types of polymer involved in film assembly [18]. One example illustrating a significant contribution of film hydrophobicity to stabilization of a hydrogen-bonded film is associated with the behavior of temperature-responsive polymers, such as poly(N-isopropyl acrylamide) (PNIPAM), within assembled films. Figure 15.3 shows the dependence of $pH_{crit}$ on post-assembly temperature for two types of hydrogen-bonded films [19]. First, the

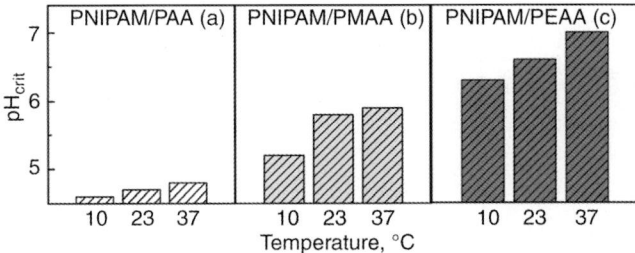

**Figure 15.3** Role of the type of polycarboxylic acid on pH-triggered disintegration of PNIPAM/poly(carboxylic acid) multilayers at various temperatures: PNIPAM/PAA (a), PNIPAM/PMAA (b), and PNIPAM/PEAA (c) films. The five-layer-pair films were deposited at pH 2 and 23 °C.

film stability increases with enhanced hydrophobicity of a polycarboxylic acid, from poly(acrylic acid) (PAA) to PMAA and poly(ethacrylic acid) (PEAA). Secondly, film stability could be modulated through inclusion of two types of neutral polymers – a temperature-insensitive polyvinylpyrrolidone (PVPON), and a temperature-responsive PNIPAM, which loses its solubility in water at temperatures above its lower critical solution temperature (LCST) of ∼32 °C. It is clear that, in contrast to PVPON, the enhanced hydrophobicity of PNIPAM chains at temperatures higher than the LCST improves film stability during pH variations. This allows the use of such pH-responsive films not only as pH-triggered release films [20], but also as release films whose dissolution is triggered by changes in temperature [19]. Figure 15.4 shows an example of such a temperature-triggered preparation for release of a free-standing film from a substrate [19]. The example includes construction of a two-stack LbL film with PNIPAM/PMAA layer pairs assembled close to the substrate surface, and PVPON/PMAA layer pairs deposited on top. As suggested by the data in Figure 15.3, at a constant pH of 5.6, lowering the temperature below PNIPAM's LCST dissolves the PNIPAM/PMAA stack, releasing PVPON/PMAA film (Figure 15.4). A significant appeal of these films is that release occurs in an aqueous environment in mild conditions (neutral pH values and moderate temperature range), safe for mammalian cells. For example, PNIPAM/PMAA release layers have been used to harvest cells with attached LbL patches (or "backpacks") from a substrate [21].

With electrostatically assembled wPEMs, post-assembly pH variations have been used to create excess charge within films, while the films seemed to retain both polymer components. This charge can then be used, for example, to bind and release dyes or drugs [22, 23], to fabricate novel metal-containing inorganic nanocomposite materials [24, 25], or to control electroosmotic flow in microchannels [26]. As a result of pH-induced electrostatic stress within the film, wPEMs swell and change their morphology [27, 28] and pH-induced film porosity can be used advantageously for anti-reflection coatings [29].

In addition to film swelling and morphological changes, pH-induced electrostatic stress can result in complete film dissolution (not necessarily due to dissociation of polymer binding, but rather due to solubilization of PEC in solution), or even result in changes in the film composition. The specific path of film response to pH

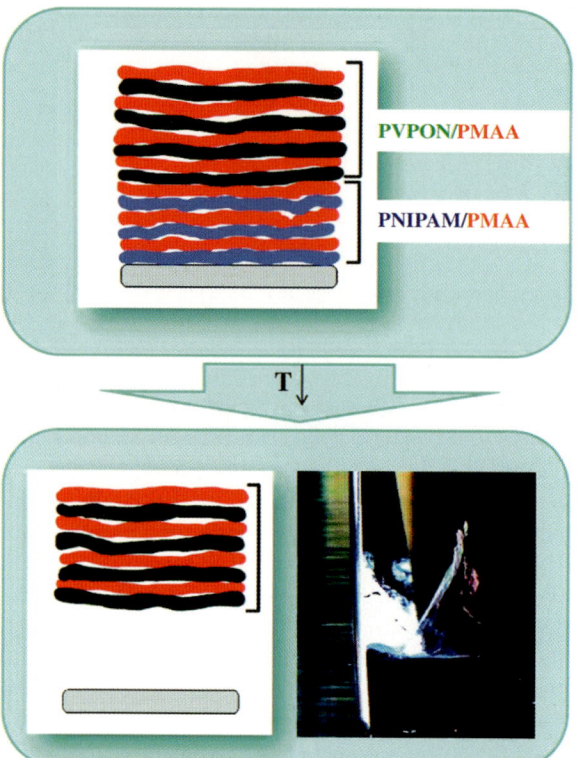

**Figure 15.4** Schematic drawing of temperature-induced release and a snapshot of self-floating hydrogen-bonded (PVPON/PMAA)$_{100}$ film of thickness of ∼340 nm obtained after a 20-min exposure of the (PNIPAM/PMAA)$_{10}$/(PVPON/PMAA)$_{100}$ multilayer to 0.01 M phosphate buffer solutions at pH 5.6 cooled to 10 °C.

variations depends on a delicate balance between strength, number and distribution of individual polymer–polymer contacts, and the magnitude of electrostatic stress within the film. Note that compositional changes within the film are rarely monitored, partly because of instrumental limitations of conventionally used techniques (UV–vis spectroscopy, ellipsometry, quartz crystal microbalance (QCM)) that do not separately monitor individual film components. Our group has explored different scenarios of film response using PMAA/QN multilayers (where QN denotes poly-4-vinylpyridine ($M_w = 200$ kDa) in which N% of units are quaternized with ethyl bromide) deposited at pH 5 and exposed to pH 7.2. Interestingly, Q8- and Q10-containing films showed no mass loss, while PMAA/QN films with quaternization >28 completely dissolved from the substrate after this pH variation. Different modes of film response to pH variations in this case are additionally explained by an increase in solubility of QN polycations with their quaternization degree. An intriguing regime exists for poly-4-vinylpyridine with intermediate degrees of quaternization between 12 and 28. Specifically, chains of one type were expelled from the film, while the film of changed composition remained stable and attached to the substrate.

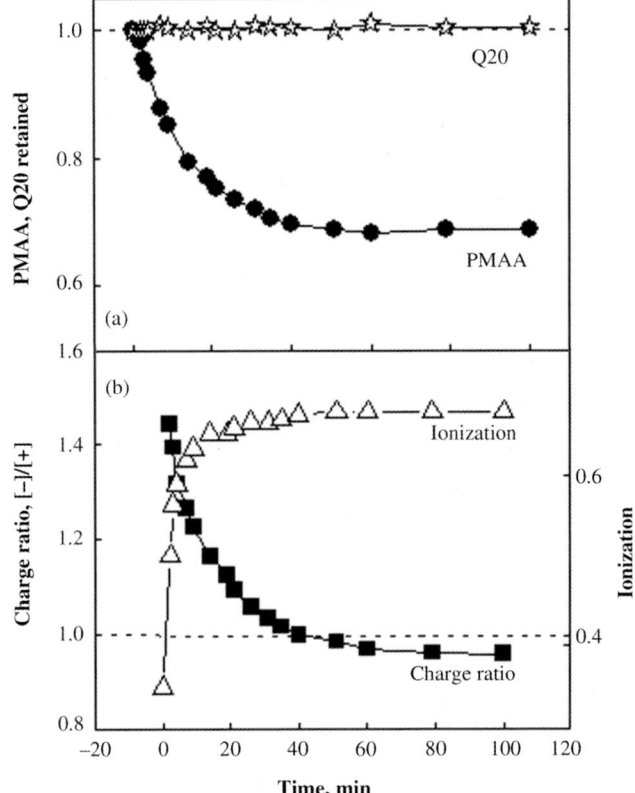

**Figure 15.5** (a) pH-triggered time evolution of Q20 (stars) and PMAA with $M_w = 72$ kDa, (circles) in a 5-layer-pair PMAA/Q20 film deposited at pH 5 and exposed to pH 7.2, as measured by *in situ* ATR-FTIR. (b) Time evolution of PMAA ionization for the PMAA/Q20 film (triangles) and of the charge ratio within the film (squares) at pH 7.2.

Figure 15.5a illustrates that a $(PMAA/Q20)_5$ multilayer releases a fraction of PMAA, while all Q20 remains within the film [30]. A driving force of such chain expulsion, that is, pH-induced imbalance of negative to positive charges, has also been quantified using *in situ* ATR-FTIR. Figure 15.5b shows that when negative charges accumulate within the film after pH variation, PMAA chains with excess charge are released into solution to bring the ratio of " − " to " + " polymer charges back to its original value near unity [30].

Studies of the molecular-weight and film-thickness dependent kinetics of such chain expulsion shed light on the mechanism of PMAA diffusion within wPEMs. With polyacids whose molecular weight was varied from 6 to 350 kDa, the molecular weight of PMAA significantly affected the release kinetics. Figure 15.6 shows that a significantly longer time was required for the release of longer PMAA chains. The characteristic time $\tau_{0.5}$ (the time required to release 50% of PMAA) was almost linearly dependent on the molecular weight of PMAA (inset in Figure 15.6).

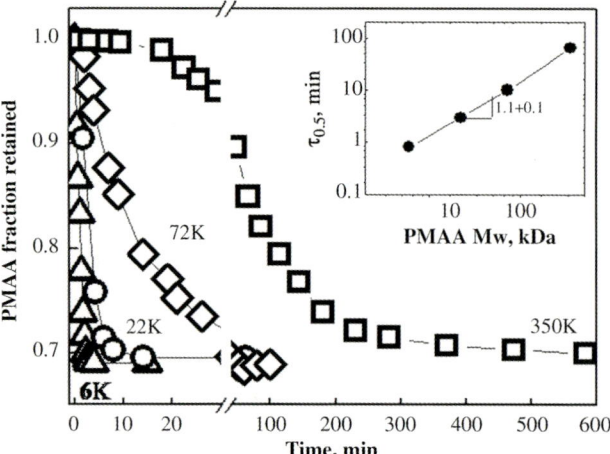

**Figure 15.6** Release rates of PMAA with different molecular weights from a 5-layer-pair PMAA/Q20 film at pH 7.2. Inset shows characteristic time as a function of molecular weight for PMAA release.

Additional experiments with film thickness ($H$) gave a scaling law of $\tau_{0.5} \sim M_w^{1.1 \pm 0.1} H$ [30].

A model consistent with this experimental scaling suggests that the driving force for the release of polyanions from the film is the electric field created by counterions. Specifically, pH-triggered increase in the polyacid charge causes a corresponding release of counterions, which are mobile and can occupy a larger volume than that occupied by the multilayer film. The resulting electric field causes polyanions to move towards the free surface (Figure 15.7a). The polyanions move out of the film by

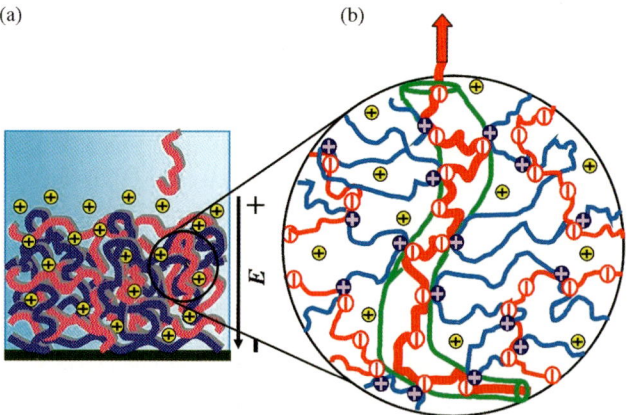

**Figure 15.7** (a) Polycation-polyanion multilayer film with excess charge on polyanions under high pH conditions. The counterions occupy a larger volume than the film, creating a net electric field $E$. (b) "Sticky gel electrophoresis". "Sticky" reptation of a polyanion (bold line) in an electric field occurs through dissociation/association of ionic bonds.

breaking and re-forming ionic bonds with polycations. As in electrophoresis, the polyanions in the multilayer film are entangled within the network and can only move along their confining tubes (Figure 15.7) [30]. In contrast to gel electrophoresis, chain movement within wPEM occurs through dissociation/association of ionic pairs ("stickers"), and the movement of entangled polyanions through the film electric field might be called "sticky gel electrophoresis". As in the conventional gel electrophoresis of entangled polymers, the electrophoretic mobility in the "sticky gel electrophoresis" model is inversely proportional to the molecular weight $M_w$ of moving chains (in this case to the number of "stickers" per chain). Therefore the polyanion release time is proportional to $M_w$, as well as to the distance the polyanions have to move (i.e., the film thickness $H$).

Importantly, the pH-induced expulsion of PMAA from PMAA/Q20 film resulted in drastic changes in the film structure. Figure 15.8 shows neutron reflectivity (NR) data of a 20-layer pair PMAA/Q20 film, where each fifth PMAA layer contained deuterated PMAA (dPMAA), collected with as-deposited film at pH 5 (Figure 15.8a), after exposure of the film to pH 7.5 (Figure 15.8b), and subsequent re-adsorption of

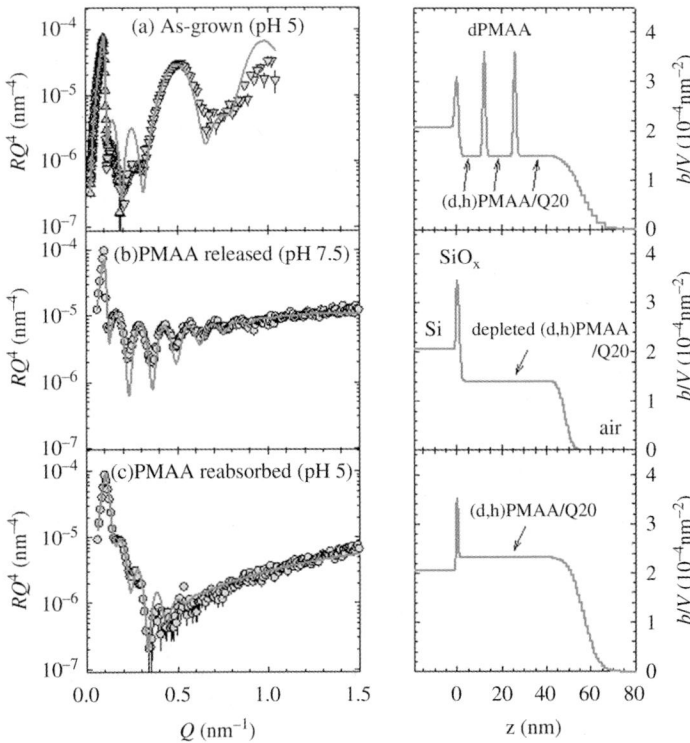

**Figure 15.8** Neutron reflectivity data (left) for air-dried [(PMAA/Q20)$_4$/(dPMAA/Q20)]$_4$ films after deposition at pH 5 (a), after exposure to pH 7.5 (b) and after return to pH 5 (c) and fitted scattering-length-density profiles (right) obtained experimentally. Every fifth PMAA ($M_w = 22$ kDa) layer is deuterated to enhance neutron contrast.

dPMAA at pH 5 (Figure 15.8c). As-assembled films (pH 5) show Bragg peaks characteristic of substrate-mediated layering within the film, with a significant "fussiness" characteristic for PEMs. Strikingly, the multilayer structure completely disappears after pH-induced release of PMAA (Figure 15.8b), as a result of PMAA expulsion. Upon reduction of pH back to its deposition value (pH 5), and exposure to dPMAA solution, dPMAA was completely and homogeneously re-absorbed by the films (Figure 15.8c) [30].

Molecular-level understanding of possible events that might follow the application of environmental stimuli to as-deposited LbL films is important for applications of these films as matrices for release of functional compounds. This knowledge might enable the rational design of responsive coating stabilized exclusively by non-covalent interactions (for example, by choosing polyelectrolytes with stronger ionic pairing), or, alternatively, by a network of covalent crosslinks.

## 15.3
## Responsive Swellable LbL Films

### 15.3.1
### LbL-Derived Hydrogels

#### 15.3.1.1 Preparation and Swelling

The translational dynamics of polymer chains within LbL films can be inhibited through crosslinking. When wPE-containing films are exposed to pH values at which ionic or hydrogen bonding intermolecular contacts dissociate, the films become highly swellable, since macromolecules in these films are held together only by covalent bonds introduced through crosslinking. The LbL hydrogels can be fabricated using sequential chemical crosslinking during self-assembly [31], co-self-assembly of linear polymers with microgel particles [32], or post-assembly treatment of pre-constructed LbL films via thermal-, photo- or chemical cross-linking [33, 34]. Here, we will focus on LbL hydrogels prepared by post-assembly cross-linking of hydrogen-bonded films. Such assembly involves a wPE, such as a polycarboxylic acid, and a neutral polymer [1, 35]. To assure hydrogen bonding, the deposition is performed at acidic pH when the polyacid is totally protonated. As-deposited hydrogen-bonded multilayers, however, readily dissolve at neutral and slightly basic pH values [1]. Crosslinking converts LbL films to surface hydrogels, which retain the polyacid and, therefore, change their swelling in response to pH, and also possess high loading capacity to functional compounds. Hydrogen-bonded multilayers are uniquely suited for constructing responsive hydrogel films since polymer–polymer hydrogen bonds can be easily dissociated by exposure to mild values of pH, often compatible with biological objects. Taken together, these features enable the use of these structures as surface coating/matrices for controlled delivery of bioactive compounds from surfaces.

LbL films can be converted into ultrathin hydrogels via crosslinking between self-assembled polymer chains of the same type, or between two different types of polymers. Using this technique, both single-component and two-component hydrogels materials, as 2D surface-attached coatings or as free-standing 3D hollow capsules, were created [34, 36–38]. Using carbodiimide chemistry, crosslinks were selectively introduced within films of poly(carboxylic acids) and non-functionalized polyvinylpyrrolidone (PVPON), polyvinylcaprolactam (PVCL), or poly(ethylene oxide) (PEO) hydrogen-bonded LbL assemblies. Subsequent exposure to high pH values resulted in complete release of a nonionic polymer from the films, yielding single-component polyacid hydrogel films or capsules. Using this route, single-component poly(acrylic acid) (PAA), PMAA, or poly(ethacrylic acid) (PEAA) hydrogel membranes were prepared. These hydrogels are highly hydrated over a wide range of pH values, and swell up to 300% of their dry weight at pH 7.5 because of the absence of non-covalent binding between units of a poly(carboxylic acid). Hydrogels with strongly enhanced swelling at high pH values were also constructed using PVPON- and PVCL-based copolymers. To that end, PVPON- or PVCL-based copolymers with a fraction of units containing amino groups (PVPON-co-NH$_2$ or PVCL-co-NH$_2$, respectively) [37] were assembled with poly(carboxylic acids) through hydrogen bonding. Following exposure to solutions of water-soluble carbodiimide, amide links formed between PVPON-co-NH$_2$ (or PVCL-co-NH$_2$) and polyacid chains, resulting in two-component hydrogel-like films and capsules [37, 38].

Significant differences were found between single- and two-component hydrogel assemblies [37, 38]. Figure 15.9 shows that when these hydrogels constituted the walls of hollow capsules, both types of capsules reversibly changed their size in response to variations in pH [37]. Unique for the case of two-component hydrogel

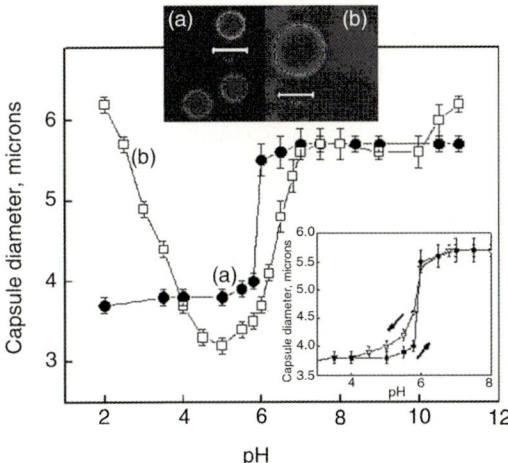

**Figure 15.9** CLSM images of (PVPON-NH$_2$-20/PMAA)$_7$ (a) and (PMAA)$_7$ (b) capsules at pH 2. pH-dependence of the diameter of the (PVPON-NH$_2$-20/PMAA)$_7$ capsules (a, filled circles) and (PMAA)$_7$ capsules (b, open squares) cross-linked for 18 h. The inset shows hysteresis of (PVPON-NH$_2$-20/PMAA)$_7$ capsule size upon increasing (filled circles) and decreasing (open triangles) pH. The pH values were supported by 0.01 M phosphate buffer.

assemblies, hydrogen bonds between a neutral counterpart and poly(carboxylic acid) re-formed at low pH values. PMAA hydrogels additionally swelled at low pH values due to the presence of amino groups, remaining within the membrane as a result of incomplete reaction of a diamine crosslinker [37]. At basic pH values, both types of hydrogels did not contain non-covalent bonds and were highly swollen. An attractive feature of hollow capsules is that their interior volume can be filled with functional cargo and serve as a compartment for controlled delivery, sensing and/or catalysis applications.

#### 15.3.1.2 Mechanical Properties

The highly swollen nature of crosslinked hydrogels derived from hydrogen-bonded LbL assemblies was also reflected in their unique mechanical properties. As revealed by AFM single capsule force spectroscopy measurements, the microcapsule stiffness of one- and two-component hydrogel capsules was strongly pH-dependent (Figure 15.10) [39]. It is remarkable that the deformation properties of the capsules changed in the very narrow range of pH between 5.4 and 6. These changes are in large contrast to the electrostatically stabilized multilayers, where an increase in salt concentration only gradually changes the elastic modulus. Changes in stiffness for both $(PMAA)_7$ and $(PMAA/PVPON\text{-}co\text{-}NH_2)_7$ hydrogels were dramatic (the difference in absolute values of stiffness between the two systems is due to different capsule wall thicknesses) [39]. The values of Young's modulus at pH > 6 were orders

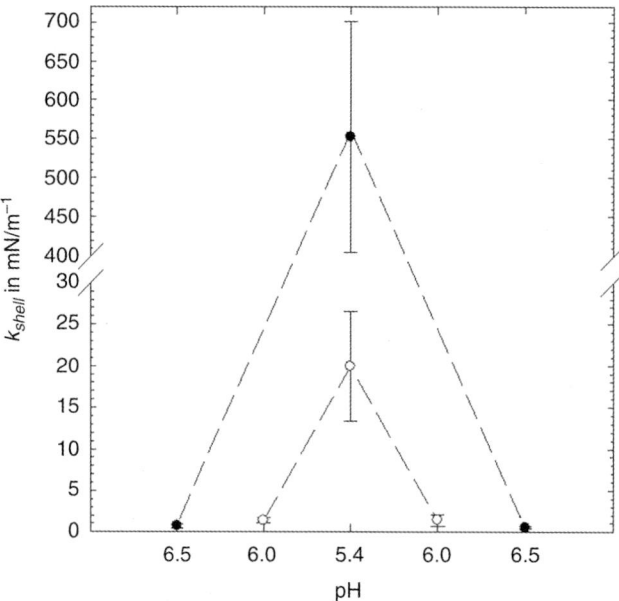

**Figure 15.10** Reversibility of the softening process for crosslinked $(PVPON\text{-}NH_2\text{-}20/PMAA)_7$ (open circles) and cross-linked $(PMAA)_{10}$ (filled circles) microcapsules for pH changes from 5.4 to 6 or 6.5, respectively.

of magnitude lower than those reported in the literature for polymeric electrostatically assembled multilayer systems. The low stiffness of cross-linked hydrogen-bonded capsules at high pH reflects the absence of intermolecular associations in the capsule walls after they are converted to ultrathin hydrogels by a pH variation [39]. Figure 15.9 illustrates that capsule softening is accompanied by drastic swelling of the capsule hydrogel wall as a result of enhanced ionization of carboxylic groups of PMAA. This work showed that the mechanical properties of capsules can be designed to be switchable in response to environmental stimuli and that hydrogen-bonding interactions are highly suitable for this purpose. Importantly, unlike with electrostatically assembled multilayers, the pH-induced dissociation of hydrogen bonds and the resulting drastic microcapsule softening occur at moderate, often compatible with biological materials, pH values.

### 15.3.1.3 LbL Hydrogels as Matrices for Controlled Release of Bioactive Molecules

Hydrogels derived from hydrogen-bonded films are promising materials for pH-controlled loading and release of functional compounds. Loading of proteins and heparin within LbL-derived surface-attached PMAA hydrogels was studied *in situ* using ATR-FTIR and ellipsometry [40]. Large amounts (up to 100% of dry hydrogel weight) of lysozyme (Lys), a protein with an isoelectric point at pH 11.5 could be electrostatically loaded within ultrathin single-component PMAA hydrogels at pH 7.5. In these cases, the hydrogel was relatively sparsely crosslinked (22 monomeric units between covalent crosslinks) [41] to allow transport of protein globules (such as Lys with molecular dimensions $\sim 3 \times 3 \times 4.5 \, nm^3$) through the hydrogel matrix. Importantly, dry thicknesses of Lys-loaded $(PMAA)_n$ films scaled linearly with the number of PMAA layers in the hydrogel matrix, confirming penetration of Lys within the bulk of the film. Adsorption was irreversible towards dilution with buffer solutions at a constant pH, yet Lys was completely released from the hydrogel matrix after the pH was lowered to 4 where the carboxylic groups became protonated. A similar trend of reversible binding of positively charged protein within PMAA hydrogels was obtained with another protein, ribonuclease. At the same time, the most abundant plasma protein, albumin, could not be included into the most sparsely crosslinked PMAA matrix due to its large size and negative charge.

The crosslinker type used for constructing the surface hydrogel appeared to be very important for hydrogel swelling and release of small molecules. This is clearly seen, for example, when comparing hydrogels obtained using two different molecules – ethylenediamine (EDA), or adipic acid dihydrazide (AADH) – for selective crosslinking of PMAA within hydrogen-bonded multilayers [42]. AADH increases the hydrogel hydrophobicity (relative to EDA-crosslinked films) and introduces centers for hydrogen bonding to guest molecules. With hydrogels loaded with antimicrobial agents, AmAs (Figure 15.11), such as gentamicin, and a positively charged peptide L5 [42], retention of AmAs in AADH-crosslinked hydrogels in high-salt solutions was strongly enhanced (of specific interest are the physiologically relevant ionic strengths from 0.15 to 0.2 M NaCl). Moreover, EDA- and AADH-crosslinked hydrogels differed drastically in their ability to release loaded AmAs in response to pH (Figure 15.12) [42]. While EDA-crosslinked hydrogels released AmAs when the pH was lowered

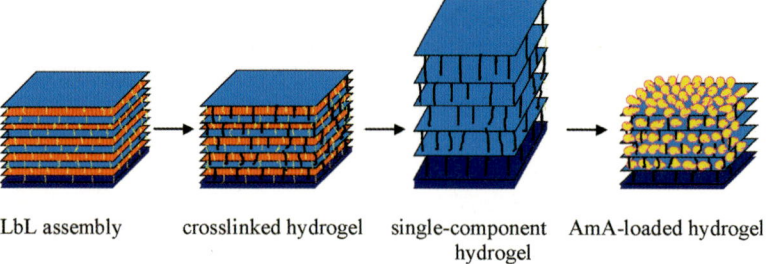

**Figure 15.11** Procedure for preparation of functional antibacterial films.

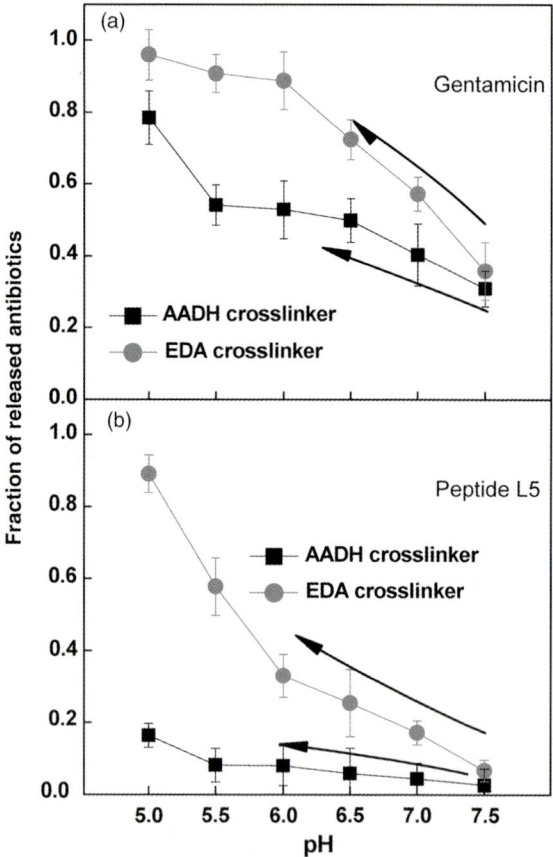

**Figure 15.12** Effect of pH on retention of peptide L5 from EDA- and AADH-stabilized $(PMAA)_{10}$ hydrogels. The fraction released was determined as the ratio of ellipsometric thickness of peptide- loading after pH-triggered release into 0.01 M phosphate buffer containing 0.2 M NaCl thickness to the thickness of antibiotics loaded into $(PMAA)_{10}$ films at pH 7.5. All error bars represent the average standard deviation obtained from three separate experiments.

below 7.5, such release was inhibited with AADH-crosslinked hydrogels. A specifically strong effect was observed with peptide L5, which likely formed hydrogen bonds with amide groups of AADH crosslinker and, therefore, remained bound within the film at low pH values.

The two types of L5-loaded EDA- and AADH-hydrogels (extreme cases of pH-responsive and pH-insensitive systems) also demonstrated distinct antibacterial behaviors. In bacterial culture experiments, pH variations were induced internally rather than externally. Specifically, medium acidification associated with growth of *Staphylococcus epidermidis (S. epidermidis)* was used as an internal trigger to release AmAs from surface coatings. The two types of hydrogels were compared with regard to inhibition of bacterial growth in solution and bacterial attachment at the hydrogel surface (Figure 15.13a and b). A ~40% inhibition of bacterial growth occurred exclusively in the case when bacterial solution was in contact with L5-loaded EDA-crosslinked hydrogel (Figure 15.13a). Consistent with Figure 15.12, AADH-crosslinked hydrogels did not release L5 into solution. At surfaces, a significant amount of bacteria was observed on as-synthesized EDA- and AADH-crosslinked hydrogels, but both types of L5-loaded hydrogel films showed almost complete inhibition of bacterial colonization. However, EDA-crosslinked matrices provide the additional benefit of killing bacteria in solution in response to bacterial growth in the medium surrounding the synthetic surface.

The possibility of tuning the loading capacity, drug retention, and pH-dependent release profiles of therapeutic agents from surface-bound hydrogel matrices has important implications for the development of advanced antibacterial coatings for implanted medical devices. While the LbL hydrogel coating approach is promising for controlled delivery of AmAs from surfaces, it also requires a case-by-case optimization of interactions between a bioactive compound and a polymer matrix – a goal sometimes hard to achieve given the diversity of chemistry, solubility and charge of therapeutic compounds. The use of block copolymer micelles, BCMs, as multilayer building blocks presents a different approach to constructing functional LbL films with hydrophobic "pockets" which can incorporate and release a broad range of hydrophobic bioactive molecules.

### 15.3.2
### LbL Films of Micelles with Responsive Cores

Micelles included within LbL films can be glassy and non-responsive to environmental cues [43–45], and deliver encapsulated bioactive compounds through a passive diffusion mechanism [46, 47]. Strategies also exist to render micelle-containing LbL films pH-responsive by introducing pH-sensitive hydrolyzable linkages between a drug and a micelle [48], or to rely on pH-induced disintegration of hydrogen-bonded LbL films [49]. The following discussion is focused on other types of BCMs which are not permanently hydrophobic, but can undergo collapse/dissolution transitions in aqueous solution in response to pH, temperature [50–52], or light [53]. In solution, such micelles often reversibly transition between micellar and unimer states upon application of environmental stimuli. Assembly of responsive BCMs within LbL films presents an attractive way to stabilize the micellar structure via

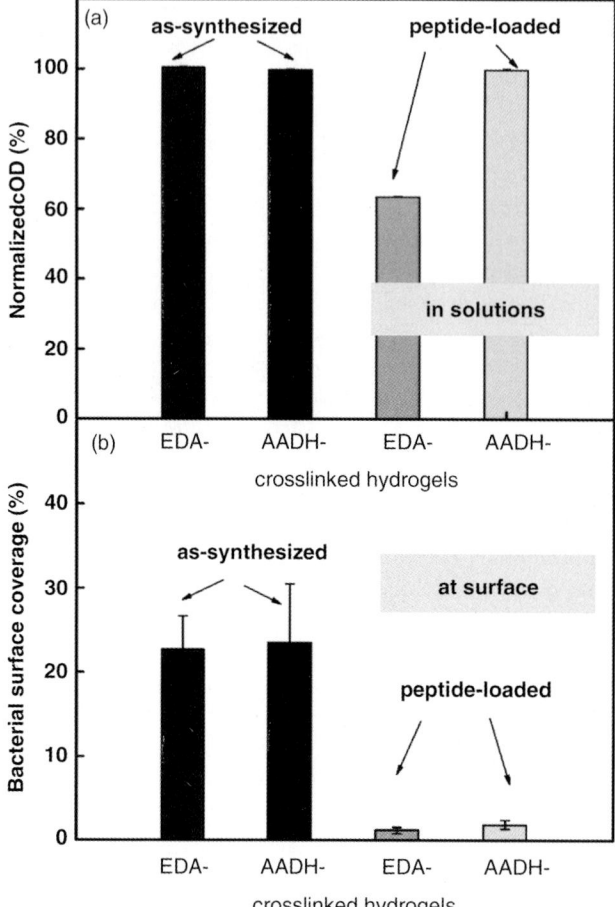

**Figure 15.13** Growth of *S. epidermidis* in the presence of the as-synthesized and L5-loaded (PMAA)$_{10}$ hydrogels. (a) Normalized optical density (OD) at 600 nm of *S. epidermidis* in TSB after incubation for 4 h in the presence of the as-synthesized or peptide-loaded hydrogels. The normalization was done relative to the absorbance of the hydrogel-free bacterial growth in TSB. (b) Surface coverage of *S. epidermidis* after incubation for 4 h in the TSB medium. All error bars represent the average standard deviation obtained from three separate experiments.

polymer binding, as well as to control film swelling, morphology and drug release properties.

Micelles with pH-dissolvable cores were among the first studied as building blocks for constructing responsive LbL films. Electrostatic and hydrogen-bonded LbL assemblies of this type of micelles were reported for several systems [54–56]. The pH-responsive nature of LbL films of micelles with polybasic cores was demonstrated by enhanced release of an encapsulated dye to the bulk solution in acidic environments [55].

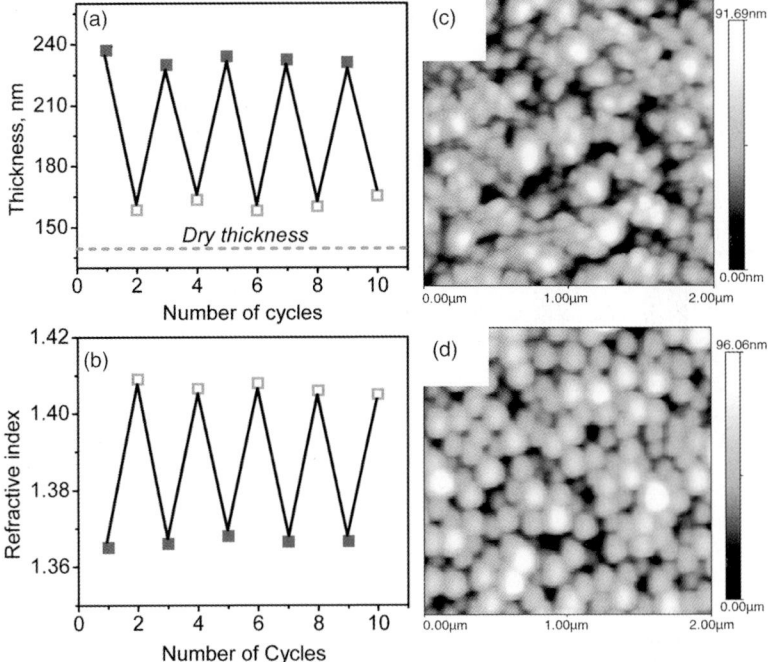

**Figure 15.14** *In situ* ellipsometry measurements of thickness (a) and refractive index (b) of a [BCM/PMAA]$_4$ film in phosphate buffer of pH 5.0 at 20 °C (filled squares) and 45 °C (open squares), and AFM images taken for film dried at 45 °C (c), as well as for wet film at 20 °C (d).

Still other types of micelles incorporated within LbL films contain temperature-responsive cores made of neutral polymers. Our group has reported on hydrogen-bonded films of neutral BCMs of PVPON-*b*-PNIPAM LbL films [57]. After assembly with a polyacid (at 45 °C), PNIPAM cores of these BCMs preserved their ability to collapse and rehydrate in response to variations in temperature, resulting in reversible cycles of film swelling/deswelling (Figure 15.14) [57]. A decrease in temperature below PNIPAM's LCST renders micellar polymer blocks soluble and enhances release of incorporated small hydrophobic molecules [57] (Figure 15.15).

Temperature-responsive micelles can also be included within films via electrostatic assembly. In particular, using pH- and temperature-responsive cationic BCMs of poly (2-(dimethylamino)ethyl methacrylate)-*block*-poly(*N*-isopropylacrylamide) (PDMA-*b*-PNIPAM) and a linear polyanion polystyrene sulfonate (PSS), LbL films were constructed at a temperature above PNIPAM's LCST [58]. Significantly, micelles preserve their morphology when adsorbed at a constant pH. Response of adsorbed BCMs to environmental stimuli was strongly dependent on whether the BCMs were coated with a polyacid layer. Monolayers of uncoated BCMs (also called LBCMs) restructured, irreversibly losing their original dry morphology in response to pH or temperature variations (Figure 15.16). In drastic contrast, depositing a polyanion

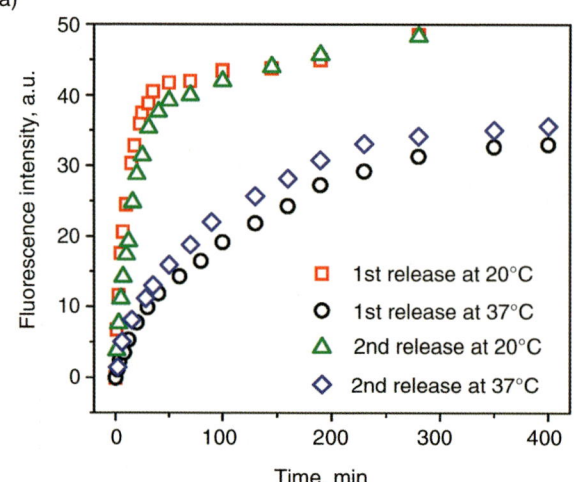

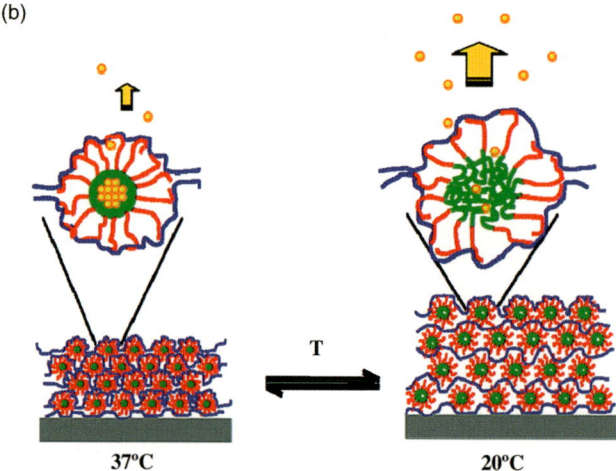

**Figure 15.15** (a) Release kinetics of pyrene from a [BCM/PMAA]$_{10}$ film in 30 mL of pH 5.0 buffer solution at 20 and 37 °C. Pyrene release was monitored by measuring the fluorescence intensity of pyrene accumulated in solution ($\lambda_{ex}=338$ nm, $\lambda_{em}=373$ nm). (b) Schematic representation of reversible temperature-triggered swelling of BCM/PMAA films.

layer atop BCMs inhibited such restructuring (Figure 15.17). Evidently, strong conformal binding of polyacid chains with BCM micelles prevented micellar disintegration after applications of environmental triggers. Such assembly-induced preservation of micellar disintegration within LbL films is central to reversible film swelling in the wet state. Similarly to hydrogen-bonded assemblies, these electrostatic BCM/PSS assemblies showed reversible swelling transitions during temperature cycling below and above PNIPAM's LCST, and exhibited temperature-triggered release of a micelle-loaded hydrophobic dye [58].

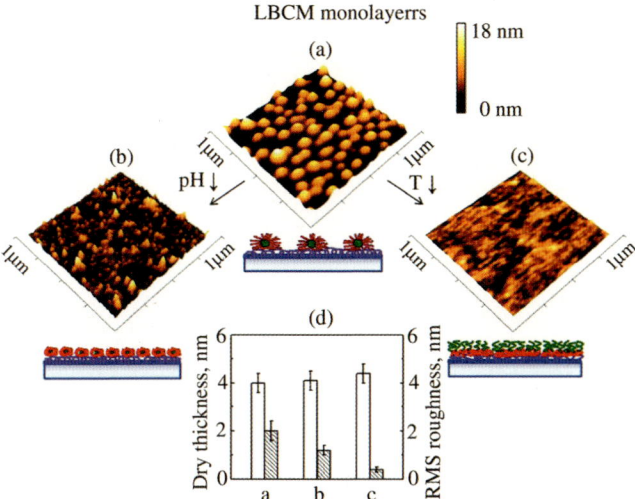

**Figure 15.16** pH and temperature responses of BCM monolayers with dry ellipsometric thickness of 4.0 ± 0.3 nm deposited on a branched poly(ethyleneimine) BPEI/PSS precursor layer measured by AFM of dry films. Films were deposited in 0.01 M phosphate buffer solution at pH 6, 45 °C (a), and then immersed in 0.01 M phosphate buffer solution at pH 4, 45 °C (b) or pH 6, 25 °C (c) for 4 h, respectively. Scale: 1 μm × 1 μm. (d) Comparison of ellipsometric thickness (blank) and RMS roughness (filled) of dry LBCM monolayer films after treatments.

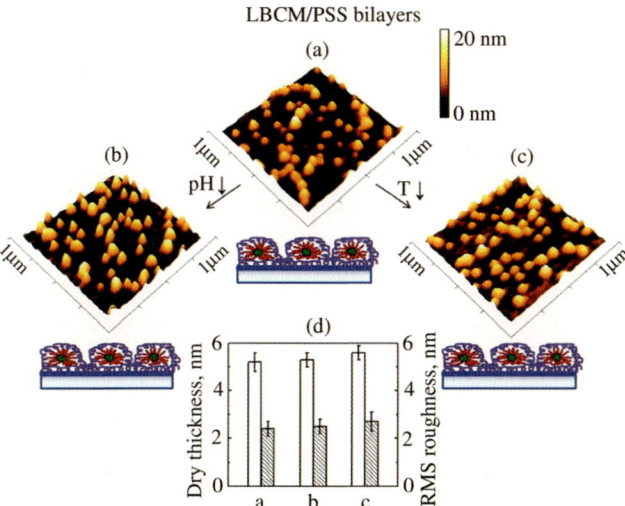

**Figure 15.17** pH and temperature responses of dry LBCM/PSS layer pairs with ellipsometric thickness of 5.2 ± 0.4 nm adsorbed at a BPEI/PSS precursor film as monitored by tapping mode AFM. Films were deposited at pH 6, 45 °C (a), followed by incubation in 0.01 M phosphate buffer solution at pH 4, 45 °C (b) or pH 6, 25 °C (c) for 4 h. Scale: 1 μm × 1 μm. (d) Comparison of ellipsometric thickness (blank) and RMS roughness (filled) of dry LBCM/PSS layer pair films after treatments.

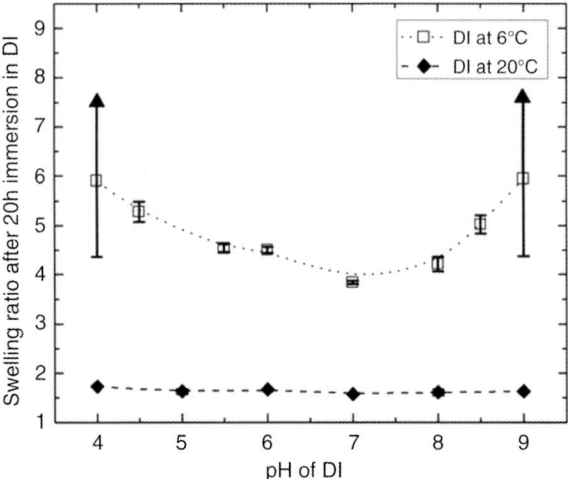

**Figure 15.18** Effect of pH on the swelling of (PD-PPO-PD/PAA)$_{24}$ at 20 and 6 °C. At 20 °C (pH < 3 or >10), as well as at 6 °C (pH < 4.5 or >8.5), the multilayer films become highly swollen (by a factor of 7 times or more).

Finally, the strategy of constructing responsive LbL films is not limited to diblock copolymers. Large-amplitude, reversible temperature-triggered swelling/deswelling transitions were demonstrated over a broad pH range with electrostatically assembled PEMs of triblock copolymer micelles of poly(N,N-dimethylaminoethyl methacrylate) (PD) with poly(propylene oxide) (PPO) cores, assembled with a polyanion (Figure 15.18) [59]. Interestingly, the functionality of these films was dependent on the choice of polyanion. While highly responsive assemblies were obtained with poly(acrylic acid) (PAA), the use of PSS resulted in films that lack temperature responsiveness [59]. Such a dependence of film response on polyanion type makes them interesting candidates for studies of correlations among the strength of intermolecular binding, reversibility of binding contacts, and film response.

## 15.4
### Conclusion and Outlook

The dynamics of polymer chains can first be related to polyelectrolyte exchange between PEMs and PE solutions added during film assembly. While close correlations between PEM and PEC have been overall established, the complexity of the phase diagrams of PEC solutions sometimes impedes prediction of PEM construction using polyelectrolyte solution behavior as a guide. Secondly, chain dynamics is related to post-assembly chain mobility within LbL films. In spite of significant experimental efforts to understand the dynamics of polymer chains in as-assembled PEMs, and correlate their dynamics to film structure, this field is still in its infancy. Simple scaling laws for polymer chain dynamics, such as molecular weight dependence of chain diffusion, established many decades ago for polymers in solution and in melts, have not yet been established for as-assembled PEMs. This is partially because fluorescently labeled, low polydispersed polymers required for such a study are not commercially available. Another open question associated with polymer motions is the degree of chain anisotropy in directions parallel and perpendicular to the substrate. While studies of lateral diffusion of PEs within PEMs have revealed important trends, such as the magnitude of lateral polyelectrolyte diffusion coefficients, and their dependence on polyelectrolyte type, position within the film, as well as solution ionic strength and temperature, it still remains unclear to what degree diffusion of polymer chains is determined by surface-aligned ionic contacts, and to what degree this contributes to layer intermixing. Experiments measuring diffusion in lateral and vertical directions with the same films, performed in identical conditions, might be useful to answer this question.

Finally, there exists a significant potential for endowing LbL films with response to external stimuli. One direction includes constructing LbL-based hydrogel matrices whose interaction with bioactive molecules can be tuned to achieve the desired loading and stimuli-response characteristics. Another promising approach extends the LbL film beyond assemblies of homopolymers, proteins and/or nanoparticles, and includes assembly of containers (such as block copolymer micelles) with responsive cores within LbL films. This latter strategy opens new ways for manipulating film function through environmental stimuli. Importantly, environmentally responsive polymers in such constructs remain highly functional, as their response is not compromised by binding with polymer partners. This strategy of separating response and binding functions presents a versatile principle that can be applied to constructing novel responsive materials. As surface coatings and/or free standing films, these materials might become useful for controlled delivery of functional molecules from surfaces, as well as for microfluidic applications.

## References

1 Sukhishvili, S.A. and Granick, S. (2002) Layered, erasable polymer multilayers formed by hydrogen-bonded sequential self-assembly. *Macromolecules*, **35**, 301–310.

2 Inoue, H., Sato, K., and Anzai, J.I. (2005) Disintegration of layer-by-layer assemblies composed of 2-iminobiotin-labeled poly(ethyleneimine) and avidin. *Biomacromolecules*, **6**, 27–29.

3 Kabanov, V.A. (1994) Basic properties of soluble interpolyelectrolyte complexes applied to bioengineering and cell transformations, in *Macromolecular Complexes in Chemistry and Biology* (eds P.L. Dubin, J. Bock, R.M. Davis, D.N. Schulz, and C. Thies), Springer-Verlag, Berlin, p. 151.

4 Kabanov, V.A. and Zezin, A.B. (1984) Soluble interpolymeric complexes as a new class of synthetic polyelectrolytes. *Pure Appl. Chem.*, **56**, 343–354.

5 Sukhishvili, S.A., Kharlampieva, E., and Izumrudov, V.A. (2006) Where polyelectrolyte multilayers and polyelectrolyte complexes meet. *Macromolecules*, **39**, 8873–8881.

6 Kovačević, D., Van der Burgh, S., de Keizer, A., and Cohen Stuart, M.A. (2002) Kinetics of formation and dissolution of weak polyelectrolyte multilayers: role of

salt and free polyions. *Langmuir*, **18**, 5607–5612.

7  Kovačević, D., Van der Burgh, S., de Keizer, A., and Cohen Stuart, M.A. (2003) Specific ionic effects on weak polyelectrolyte multilayer formation. *J. Phys. Chem. B*, **107**, 7998–8002.

8  Izumrudov, V.A., Bronich, T.K., Zezin, A.B., and Kabanov, V.A. (1985) The kinetics and mechanism of intermacromolecular reactions in polyelectrolyte solutions. *J. Polym. Sci., Polym. Lett. Ed.*, **23**, 439–444.

9  Kabanov, V.A., Zezin, A.B., Izumrudov, V.A., Bronich, T.K., and Bakeev, K.N. (1985) Cooperative interpolyelectrolyte reactions. *Makromol. Chem. Suppl.*, **13**, 137–155.

10  Oupicky, D., Konak, C., Dash, P.R., Seymour, L.W., and Ulbrich, K. (1999) Effect of albumin and polyanion on the structure of DNA complexes with polycation containing hydrophilic nonionic block. *Bioconjugate Chem.*, **10**, 764–772.

11  Kakizawa, Y., Harada, A., and Kataoka, K. (2001) Glutathione-sensitive stabilization of block copolymer micelles composed of antisense DNA and thiolated poly(ethylene glycol)-block-poly(l-lysine): A potential carrier for systemic delivery of antisense DNA. *Biomacromolecules*, **2**, 491–497.

12  Izumrudov, V.A., Chaubet, F., Clairbois, A.-S., and Jozefonvicz, J. (1999) Interpolyelectrolyte reactions in solutions of functionalized dextrans with negatively charged groups along the chains. *Macromol. Chem. Phys.*, **200**, 1753–1763.

13  Zelikin, A.N., Trukhanova, E.S., Putnam, D.A., Izumrudov, V.A., and Litmanovich, A.A. (2003) Competitive reactions in solutions of poly-l-histidine, Calf Thymus DNA, and synthetic polyanions: Determining the binding constants of polyelectrolytes. *J. Am. Chem. Soc.*, **125**, 13693–13699.

14  Izumrudov, V.A., Zhiryakova, M.V., and Kudaibergenov, S. (1999) Controllable stability of DNA-containing polyelectrolyte complexes in water-salt solutions. *Biopolymers*, **52**, 94–108.

15  Shiratori, S.S. and Rubner, M.F. (2000) pH-Dependent thickness behavior of sequentially adsorbed layers of weak polyelectrolytes. *Macromolecules*, **33** (11), 4213–4219.

16  Harris, J.J. and Bruening, M.L. (2000) Electrochemical and in situ ellipsometric investigation of the permeability and stability of layered polyelectrolyte films. *Langmuir*, **16**, 2006–2013.

17  Sukhishvili, S.A. (2005) Responsive polymer films and capsules via layer-by-layer assembly. *Curr. Opin. Colloid Interface.*, **10**, 37–44.

18  Kharlampieva, E. and Sukhishvili, S.A. (2006) Hydrogen-bonded layer-by-layer polymer films. *J. Macromol. Sci., Part C: Polym. Rev.*, **46**, 377–395.

19  Zhuk, A., Pavlukhina, S., and Sukhishvili, S.A. (2009) Temperature-triggered hydrogen-bonded layer-by-layer release films. *Langmuir*, **25**, 14025–14029.

20  Ono, S. and Decher, G. (2006) Preparation of ultrathin self-standing polyelectrolyte multilayer membranes at physiological conditions using pH-responsive film segments as sacrificial layers. *Nano Lett.*, **6**, 592–598.

21  Swiston, A., Cheng, C., Um, S., Irvine, D., Cohen, R., and Rubner, M. (2008) Surface functionalization of living cells with multilayer patches. *Nano Lett.*, **8**, 4446–4453.

22  Chung, A.J. and Rubner, M.F. (2002) Methods of loading and releasing low molecular weight cationic molecules in weak polyelectrolyte multilayer films. *Langmuir*, **18**, 1176–1183.

23  Kharlampieva, E. and Sukhishvili, S.A. (2004) Release of a dye from hydrogen-bonded and electrostatically assembled polymer films triggered by adsorption of a polyelectrolyte. *Langmuir*, **20**, 9677–9685.

24  Wang, T.C., Rubner, M.F., and Cohen, R.E. (2002) Polyelectrolyte multilayer nanoreactors for preparing silver nanoparticle composites: controlling metal concentration and nanoparticle size. *Langmuir*, **18**, 3370–3375.

25  Hammond, P. (2000) Recent explorations in electrostatic multilayer thin film

assembly. *Curr. Opin. Coll. Interface Sci.*, **4**, 430–442.

26 Sui, Zh. and Schlenoff, J.B. (2003) Controlling electroosmotic flow in microchannels with pH-responsive polyelectrolyte multilayers. *Langmuir*, **19** 7829–7831.

27 Mendelsohn, J.D., Barrett, C.J., Chan, V.V., Pal, A.J., Mayes, A.M., and Rubner, M.F. (2000) Fabrication of microporous thin films from polyelectrolyte multilayers. *Langmuir*, **16**, 5017–5023.

28 Sui, Zh. and Schlenoff, J.B. (2004) Phase separations in pH-responsive multilayers: charge extrusion vs. charge expulsion. *Langmuir*, **20**, 6026–6031.

29 Hiller, J., Mendelsohn, J., and Rubner, M.F. (2002) Reversibly erasable nanoporous anti-reflection coatings from polyelectrolyte multilayers. *Nat. Mater.*, **1**, 59–63.

30 Kharlampieva, E., Ankner, J.F., Rubinstein, M., and Sukhishvili, S.A. (2008) pH-Induced release of polyanions from multilayer film. *Phys. Rev. Lett.*, **100**, 128303.

31 Serizawa, T., Nakashima, Y., and Akashi, M. (2003) Stepwise fabrication and characterization of ultrathin hydrogels prepared from poly(vinylamine-co-N-vinylformamide) and poly(acrylic acid) on a solid substrate. *Macromolecules*, **36**, 2072–2078.

32 Serpe, M.J. and Lyon, L.A. (2004) Optical and acoustic studies of pH-dependent swelling in microgel thin films. *Chem. Mater.*, **16**, 4373–4380.

33 Yang, S.Y. and Rubner, M. (2002) Micropatterning of polymer thin films with pH-sensitive and cross-linkable hydrogen-bonded polyelectrolyte multilayers. *J. Am. Chem. Soc.*, **124**, 2100–2101.

34 Kozlovskaya, V., Ok, S., Sousa, A., Libera, M., and Sukhishvili, S.A. (2003) Hydrogen-bonded polymer capsules formed by layer-by-layer self-assembly. *Macromolecules*, **36**, 8590–8592.

35 Sukhishvili, S.A. and Granick, S. (2000) Layered, erasable, ultrathin polymer films. *J. Am. Chem. Soc.*, **122**, 9550–9551.

36 Kozlovskaya, V., Kharlampieva, E., Mansfield, M.L., and Sukhishvili, S.A. (2006) Poly(methacrylic acid) hydrogel films and capsules: response to pH and ionic strength, and encapsulation of macromolecules. *Chem. Mater.*, **18**, 328–336.

37 Kozlovskaya, V. and Sukhishvili, S.A. (2006) pH-Controlled permeability of layered hydrogen-bonded polymer capsules. *Macromolecules*, **39**, 5569–5572.

38 Kozlovskaya, V., Shamaev, A., and Sukhishvili, S.A. (2008) Tuning swelling pH and permeability of hydrogel multilayer capsules. *Soft Matter*, **4**, 1499–1507.

39 Elsner, N., Kozlovskaya, V., Sukhishvili, S.A., and Fery, A. (2006) pH-Triggered softening of crosslinked hydrogen-bonded capsules. *Soft Matter*, **2**, 966–972.

40 Kharlampieva, E., Erel-Unal, I., and Sukhishvili, S.A. (2007) Amphoteric surface hydrogels derived from hydrogen-bonded multilayers: Reversible loading of dyes and macromolecules. *Langmuir*, **23**, 175–181.

41 Kozlovskaya, V.A., Kharlampieva, E.P., Erel-Unal, I., and Sukhishvili, S.A. (2009) Single-component layer-by-layer weak polyelectrolyte films and capsules: Loading and release of functional molecules. *Polym. Sci. – Ser. A*, **51** (6), 719–729.

42 Pavlukhina, S., Lu, Y., Patimetha, A., Libera, M., and Sukhishvili, S.A. (2010) Polymer multilayers with pH-triggered release of antibacterial agents. *Biomacromolecules*, **11**, 3448–3456.

43 Ma, N., Wang, Y., Wang, Z., and Zhang, X. (2006) Polymer micelles as building blocks for the incorporation of azobenzene: enhancing the photochromic properties in layer-by-layer films. *Langmuir*, **22**, 3906–3909.

44 Cho, J., Hong, J., Char, K., and Caruso, F. (2006) Nanoporous block copolymer micelle/micelle multilayer films with dual optical properties. *J. Am. Chem. Soc.*, **128**, 9935–9942.

45 Qi, B., Tong, X., and Zhao, Y. (2006) Layer-by-layer assembly of two different

polymer micelles with polycation and polyanion coronas. *Macromolecules*, **39**, 5714–5719.

46 Nguyen, P.M., Zacharia, N.S., Verploegen, E., and Hammond, P.T. (2007) Extended release antibacterial layer-by-layer films incorporating linear-dendritic block copolymer micelles. *Chem. Mater.*, **19**, 5524–5530.

47 Hu, X. and Ji, J. (2010) Construction of multifunctional coatings via layer-by-layer assembly of sulfonated hyperbranched polyether and chitosan. *Langmuir*, **26**, 2624–2629.

48 Kim, B., Lee, H., Min, Y.H., Poon, Z.Y., and Hammond, P.T. (2009) Hydrogen-bonded multilayer of pH-responsive polymeric micelles with tannic acid for surface drug delivery. *Chem. Commun.*, **28**, 4194–4196.

49 Kim, B.-S., Park, S.W., and Hammond, P.T. (2008) Hydrogen-bonding layer-by-layer-assembled biodegradable polymeric micelles as drug delivery vehicles from surfaces. *ACS Nano*, **2**, 386–392.

50 Gillies, E.R. and Fréchet, J.M. (2005) pH-Responsive copolymer assemblies for controlled release of doxorubicin. *J. Bioconjug. Chem.*, **16**, 361–368.

51 Schilli, C.M., Zhang, M.F., Rizzardo, E., Thang, S.H., Chong, Y.K., Edwards, K., Karlsson, G., and Müller, A.H. (2004) New double-responsive block copolymer synthesized via RAFT polymerization: poly(N-isopropylacrylamide)-block-poly (acrylic acid). *Macromolecules*, **37**, 7861–7866.

52 Rodríguez-Hernández, J. and Lecommandoux, S. (2005) Reversible inside-out micellization of pH-responsive and water-soluble vesicles based on polypeptide diblock copolymers. *J. Am. Chem. Soc.*, **127**, 2026–2027.

53 Lee, H.-I., Wu, W., Oh, J.K., Mueller, L., Sherwood, G., Peteanu, L., Kowalewski, T., and Matyjaszewski, K. (2007) Light-induced reversible formation of polymeric micelles. *Angew. Chem., Int. Edit.*, **46**, 2453–2457.

54 Addison, T., Cayre, O.J., Biggs, S., Armes, S.P., and York, D. (2008) Incorporation of block copolymer micelles into multilayer films for use as nanodelivery systems. *Langmuir*, **24**, 13328–13333.

55 Sakai, K., Webber, G.B., Vo, C.-D., Wanless, E.J., Vamvakaki, M., Bütün, V., Armes, S.P., and Biggs, S. (2008) Characterization of layer-by-layer self-assembled multilayer films of diblock copolymer micelles. *Langmuir*, **24**, 116–123.

56 Erel, I., Zhu, Zh., Zhuk, A., and Sukhishvili, S.A. (2011) Hydrogen-bonded layer-by-layer films of block copolymer micelles with pH-responsive cores. *J. Colloid. Interface Sci.*, **355**, 61–69.

57 Zhu, Z. and Sukhishvili, S.A. (2009) Temperature-induced swelling and small molecule release with hydrogen-bonded multilayers of block copolymer micelles. *ACS Nano.*, **3**, 3595–3605.

58 Xu, L., Zhu, Zh., and Sukhishvili, S.A. (2011) Polyelectrolyte multilayers of diblock copolymer micelles with temperature-responsive cores. *Langmuir*, **27**, 409–415.

59 Tan, W.S., Cohen, R.E., Rubner, M.F., and Sukhishvili, S.A. (2010) Temperature-induced, reversible swelling transitions in multilayers of a cationic triblock copolymer and a polyacid. *Macromolecules*, **43**, 1950–1957.

# 16
# Tailoring the Mechanics of Freestanding Multilayers
*Andreas Fery and Vladimir V. Tsukruk*

## 16.1
## Introduction

Tuning a material's mechanical properties is one of the fundamental aims of material sciences. Mechanical properties, such as elastic modulus, elongation to break, toughness, and ultimate strength, govern the stability of objects under the action of external stresses. Beyond that, mechanical properties relate shape changes to deformation energies. In many cases, mechanical deformation causes symmetry-breakage and/or formation of more complex morphologies. Such mechanical instabilities are universal phenomena observed at all length scales in a wide range of materials in both natural and man-made systems.

As is known, reduction in the elastic energy due to out of plane periodic bending caused by elastic compression of materials manifests itself in a wide range of everyday phenomena, such as wrinkling of the skin, textured cream on milk, and the edges of leaves. Buckling behavior has been extensively investigated in homogenous freely suspended and substrate supported or floating thin films of metals, polymers, and various nanostructures (polymeric and inorganic nanowires, carbon nanotubes) [1–8]. Buckling instabilities have been demonstrated to be valuable in controlling adhesion, enabling flexible electronics, providing means for nanopatterning and optical microgratings [9–14].

Stafford *et al.* have exploited strain-induced elastic buckling instabilities for measuring the mechanical properties (SIBIMM) of uniform thin films [15]. The technique involves a thin polymer layer firmly bound to a compliant substrate, which is subjected to a compressive stress. The periodicity of the uniform buckling pattern is given by:

$$\lambda = 2\pi t \left[ \frac{(1-v_s^2) E_f}{3(1-v_f^2) E_s} \right]^{1/3} \quad (16.1)$$

where $\lambda$ is the wavelength of the periodic buckling pattern, $E_f$ and $v_f$ are the elastic modulus and Poisson's ratio of the film, and $E_s$ and $v_s$ are the elastic modulus and Poisson's ratio of the compliant substrate, and $t$ is the thickness of the film.

---

*Multilayer Thin Films: Sequential Assembly of Nanocomposite Materials,* Second Edition.
Edited by Gero Decher and Joseph B. Schlenoff.
© 2012 Wiley-VCH Verlag GmbH & Co. KGaA. Published 2012 by Wiley-VCH Verlag GmbH & Co. KGaA.

The main focus of this chapter is on summarizing recent results, mostly obtained in the authors' lab, as well as selected examples from the literature on the mechanical behavior of layer by layer (LbL) assemblies in the most common forms of LbL structures, such as planar films and hollow microcapsules. We will analyze the mechanical properties (elastic modulus, elongation to break, toughness, and ultimate strength) of these LbL structures as well as the unique mechanical behavior under external stresses generated by variable environmental conditions. Fine control of mechanical properties is the key issue for many prospective applications of freestanding LbL structures for sensing, drug delivery, and signal transduction, as also will be discussed in this chapter.

## 16.2
### Measurements of Mechanical Properties of Flat LbL Films

To date, numerous different approaches have been exploited to measure the mechanical properties of LbL structures with the main being buckling (compressive properties) and bulging (tensile measurements) tests, as well as surface force spectroscopy, colloidal force spectroscopy, microtensile measurements and resonant frequency measurements, all of which will be briefly discussed in this chapter. The methods discussed here are capable of measuring the elastic modulus, mechanical strength, maximum strain and stress, and mechanical toughness of LbL films with nanometer-scale thickness. The important difference between buckling and bulging techniques is that the buckling method involves compression of the film while bulging involves tensile forces on the film. This fundamental difference in the loading mode might contribute to the systematic differences in the elastic modulus values.

Several examples of collection of different experimental mechanical data and their analysis will be presented below for several traditional LbL films in traditional planar form (this section) as well as in the form of hollow microcapsules (next section).

### 16.2.1
#### Micromechanical Properties from Bulging Experiments

The mechanical properties of the freely suspended LbL films can be measured with the well-known bulging technique introduced by Nix *et al.* [16, 17]. As was demonstrated by Jiang *et al.*, in a bulging test, the deflection of the LbL film suspended across a microscopic opening can be monitored as a function of over (under) pressure applied from one side [18, 19]. The deflection over a wide range of scale from tens of nanometers to tens of microns can be measured with an optical interference set-up. The distance between two rings corresponds to half of the wavelength distance in the $z$-direction. The detailed analysis of the shape of the LbL films allows estimation of stress–strain behavior, elastic modulus, and residual stresses [17].

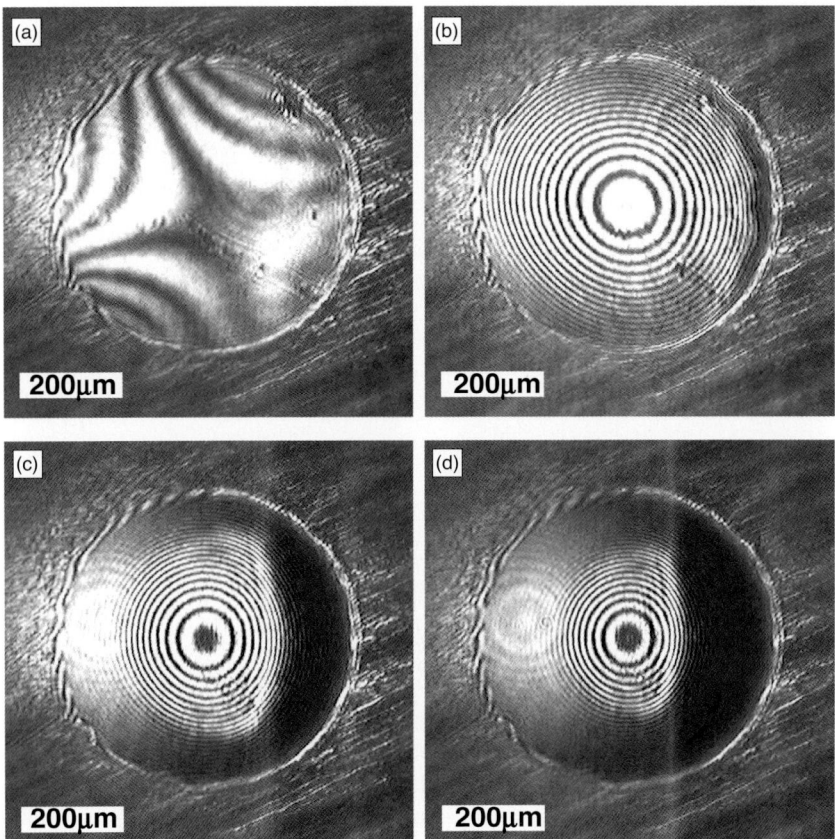

**Figure 16.1** Optical images of the freely suspended 9G9 membrane with 600 μm diameter with variable interference patterns observed at different overpressures, (a) rest state, no pressure applied, (b) 98 Pa, (c) 196 Pa, and (d) 294 Pa. Increasing number and decreasing spacing of the Newton's rings are clearly visible. Reprinted from [19].

The freely suspended LbL film in the rest state displays the fluctuating interference pattern reflecting random deflections due to external acoustic noise (Figure 16.1). Applying an external pressure to the membrane from one side (over- or underpressure) resulted in a series of concentric Newton rings with an increasing number of rings for higher pressures (Figure 16.1). For modest deflections, the spacing of the rings is very uniform and the number of rings is proportional to the film deflection thus simplifying the experimental analysis.

The resulting data from the bulging experiment give the pressure dependence of the membrane's deflection (Figure 16.2). This dependence is highly nonlinear for deflections within a few microns and can be analyzed by using the theoretical mechanical model for a circular elastic plate clamped to the stiff edges of the

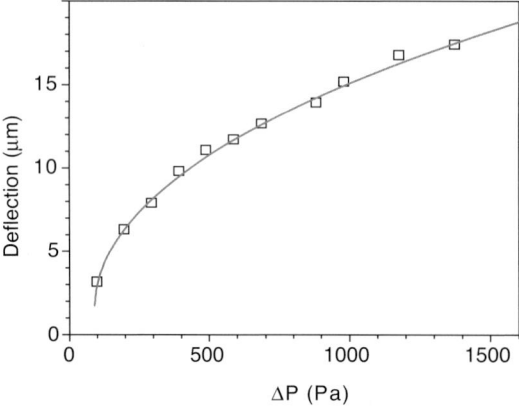

**Figure 16.2** Membrane deflection with varying air pressure and the corresponding theoretical fit (solid line). Reprinted from [19].

opening [20, 21]. According to this model, the membrane deformation can be described by a complex equation [21]:

$$P = P_0 + \left[ C_0 \frac{E}{1-\nu^2} \frac{h^4}{a^4} + C_1 \frac{\sigma_0 h^2}{a^2} \right] \left( \frac{d}{h} \right) + C_2 \frac{E}{1-\nu} \frac{h^4}{a^4} \left( \frac{d}{h} \right)^3 \quad (16.2)$$

where $P$ is over(under)pressure, $P_0$ is initial pressure, $d$ is the deflection of the membrane center, and $\sigma_0$ is the residual stress. The coefficients $C_0$, $C_1$ and $C_2$ depend upon the geometrical shape. The first term corresponds to the initial pressure related to the very small membrane deflection. The second term describes a linear bending regime and includes the residual stresses. The third term describes the regime of deformation that includes tensile stress within the highly deflected film. As is clear from this equation, for large deflections (usually higher than one micron), the third term becomes dominant and the pressure–deflection relationship should follow a cubic law (Figure 16.2). Indeed, the bulging behavior possesses a predominant cubic term contribution, as expected for these regimes where $d$ is much larger then $h$ (below or around a hundred nanometers) (Figure 16.2). The loading and unloading cycles can be repeated multiple times for the same LbL films to test their elastic behavior.

Further analysis can also be conducted by retrieving the stress–strain data from pressure–deflection measurements by converting measured experimental parameters (membrane deflection vs. pressure) to conventional physical parameters: stress $\sigma$, and strain $\varepsilon$, [16] The values of the elastic modulus obtained from these stress–strain data are very similar to those obtained from the independent bulging test. For instance, the elastic modulus measured from stress–strain plots for PSS/PAH LbL films reinforced with gold nanoparticles was around 7 GPa, close to the values derived from direct analysis of the deflection data with Equation 16.2.

As an alternative complementary approach, the frequency measurement of freely suspended LbL films has also been utilized by Markutsya et al. to estimate their mechanical properties under dynamic conditions by measuring the resonant behavior [19]. A mechanical resonance was observed at the high frequencies, about 100 kHz for 400 μm diameter LbL films with the value of the quality factor $Q$ being about 100, indicating highly reversible elastic deformations. The value of the elastic modulus for gold-reinforced PSS/PAH films was found to be in the range 12 to 15 GPa, which is higher than that measured with static experiments and can be related to a well-known time-dependent phenomenon in polymer elastic solids [40].

### 16.2.2
### Buckling Measurements of LbL Films

The buckling technique is widely used for probing the mechanical properties of LbL films of a variety of synthetic polymers, biomacromolecules, and nanoparticles [22].

For LbL films from biomaterials, Jiang et al. have measured the mechanical properties of ultrathin silk films formed by spin coating and spin-assisted LbL assembly [23]. Figure 16.3 shows the buckling patterns in the LbL silk films on PDMS under compressive stresses. From the buckling patterns, the elastic modulus of the methanol-treated LbL silk film with a thickness below 100 nm was found to be much higher than that of the water-treated silk LbL film (6.5 GPa vs. 3.4 GPa), which in turn exhibited a higher elastic modulus than the regular spin-cast silk film (2.8 GPa) (Figure 16.3).

The buckling method has been extensively applied for the determination of the modulus of LbL films filled with inorganic nanostructures (nanoparticles, nanorods, or nanowires) [24–26]. For example, Gunawidjaja et al. have employed this technique to probe the elastic modulus of LbL films with encapsulated silver nanowires [27, 28]. It was found that the elastic modulus increased with the silver nanowire volume fraction from 1.7 GPa (0%) to 5.7 GPa (22.5%). These results were found to be close to the theoretically calculated values for reinforcing nanostructures.

In another related study, Gunawidjaja et al. studied LbL films with uniformly aligned silver nanowires [29]. When the films were compressed in the direction parallel to the silver nanowires, uniform buckling of nanowires was observed over large areas. The modulus was found to be 3.8 GPa along the nanowire orientation. Furthermore, the buckling of the silver nanowires was analyzed using one-dimensional Euler instability and their modulus was calculated to be 118 GPa – close to the known elastic modulus for the (100) silver.

Nolte et al. have extended the metrology technique to vertically stratified layer pair structures [30]. The authors investigated the elastic modulus of polyelectrolyte LbL films assembled on a PS film transferred onto a PDMS substrate. Compression of these structures resulted in buckling of the composite structure with different periodicity. The mechanical contribution of individual layers can be deconvoluted from the experimental data to deduce a Young's modulus of the desired layer. Lee et al. have demonstrated that this phenomenon can be suppressed in LbL films with silica nanoparticles [31]. When the volume fraction of the silica layers increased, the LbL films did not exhibit buckling under external stresses, considering that the presence of the silica particles enhances the higher

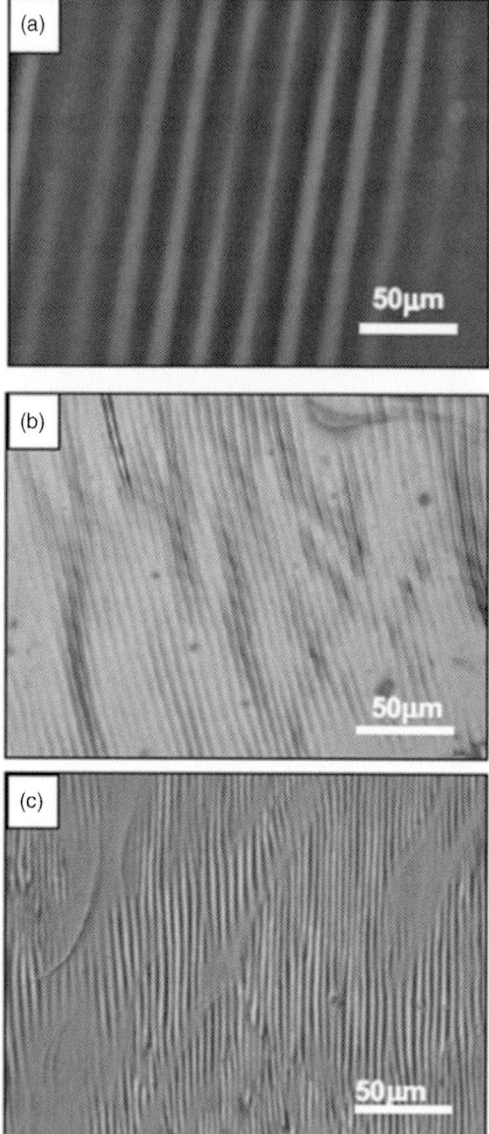

**Figure 16.3** Optical images of the buckling patterns of silk films fabricated by (a) casting, (b) LbL assembly with water treatment, and (c) LbL assembly with methanol treatment. Reprinted from [23].

modulus of the polymer films, thus lowering the critical stress. The authors speculated that the nanoparticles in the film break up, alleviating the stress and thus prevent the film from buckling.

In a more complex example, Jiang *et al.* observed a complicated buckling pattern for LbL films patterned with gold nanoparticles (Figure 16.4) [25]. It has been

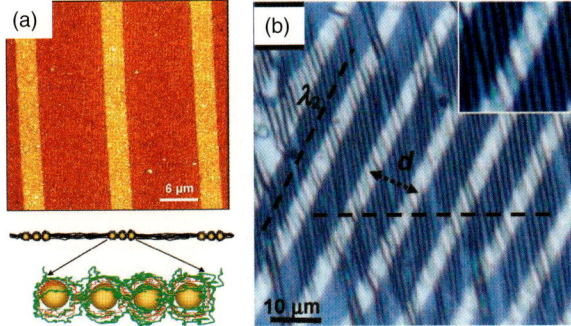

**Figure 16.4** (a) AFM image showing the patterned arrays of gold nanoparticles in the LbL film and the schematic of the nanoparticles incorporated within the LbL films. (b) Optical image showing the buckling of the film in both the regions with and without nanoparticles with different periodicities (inset shows the zigzag buckling at the interface of the two regions). Reprinted from [25].

suggested that the presence of gold nanoparticles in the selected regions results in a different buckling behavior. Indeed, in contrast to one-component LbL films, the composite LbL film shows a complex zigzag-buckling pattern upon compression, with two distinct buckling periodicities (Figure 16.4). The estimation of the elastic properties of two different regions using Equation 16.1 resulted in an elastic modulus of 3.0 GPa for the regions with nanoparticles while the elastic modulus for the non-reinforced regions was found to be lower, 1.6 GPa. The buckling patterns observed for compressed LbL films in this work enable "one-shot" evaluation of the elastic moduli of two compositionally different regions.

In another study, Lu and coworkers reported a novel method to spatially control the buckling of the LbL films on flexible substrates [32]. The authors demonstrated regiospecific wrinkling when a three-step process was applied. In this approach, the PDMS substrate was first modified by oxygen plasma, followed by LbL deposition, which is in turn followed by embossing using a silicone master. The authors noted that, in the cracked regions, the presence of the un-oxidized hydrophobic PDMS surface areas caused the LbL films to de-wet and form an intriguing concentric wrinkle topography confined to specific surface areas.

### 16.2.3
### Local Mechanical Properties Probed with Force Spectroscopies

Alternatively, the nanoscale deformation of freely suspended LbL films under point load was measured directly for the membranes on the TEM grid by Markutsya *et al.* [19]. The authors applied the colloidal probe mode where the AFM tip is replaced with a microscopic glass ball [33]. In this experiment, the force–distance curves (cantilever deflection vs. a normal load) are collected across the grid cell. The bending stiffness of the membrane was calculated at these points from the preliminary calibrated cantilever spring constants and deflections according to the spring-against-spring model [34].

AFM-based force spectroscopies are applied to measure directly the elastic modulus of the LbL films on solid substrates with nanoscale spatial resolution. The loading data are usually analyzed in the framework of the Hertzian or equivalent Snedonn's approach introduced for AFM probing to evaluate the elastic modulus [35–37]. The value of the elastic modulus was found to be in the range of several GPa for PSS/PAH films, which is consistent with the values obtained from a regular bulging test [18].

As an example, the mechanical properties of swollen hydrogen-bonded TA/PVPON LbL film was studied in water with surface force spectroscopy [38]. The surface mapping of the mechanical properties over surface areas of $1\times1\,\mu m^2$ showed a relatively uniform distribution with the standard deviation not exceeding 20%. This loading data for selected locations at the surface shows that elastic deformation of LbL films up to 3 nm is achieved under light normal loads not exceeding 0.1 nN (Figure 16.5). The experimental indentation data follow a linear relationship in Hertzian coordinates, thus indicating the purely elastic nature of the surface deformation under these loading conditions. The value of the elastic modulus of 0.8 MPa estimated from this analysis is close to traditional values known for highly swollen hydrogen-bonded hydrogels.

## 16.2.4
### Summary of Mechanical Properties of Flat LbL Films

A summary of a number of mechanical parameters which have been measured with the different methods listed above for different LbL films has been given in a very recent review on LbL films, as presented in Table 16.1 [39]. Overall, the current set of

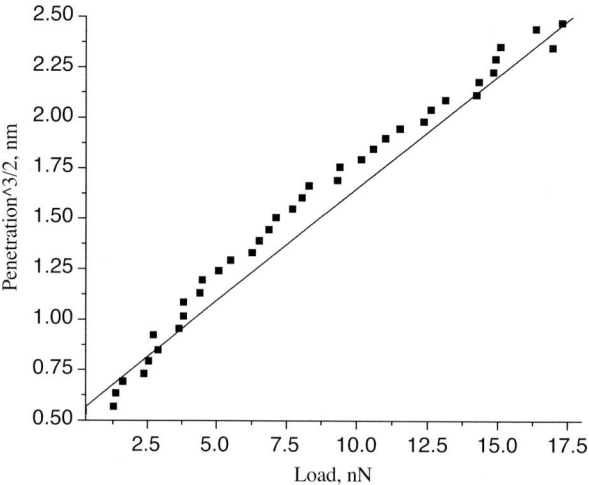

**Figure 16.5** The experimental loading curve in Hertzian coordinates and linear approximation (solid line) for the swollen (TA/PVPON) LbL films.

**Table 16.1** Summary of mechanical parameters of LbL films, as adapted from Reference [39].

| Film constituents | Nature of the interactions | Measurement technique | Elastic modulus $E$ | Ref. |
|---|---|---|---|---|
| PVPON/PMAA | H-bonds and chemical crosslinks | | 610 MPa at pH 2 with strong decrease as the pH increases | [63] |
| (PAH-PSS)$_n$ on PDMS | Electrostatic | SIEBIMM | 4.4 ± 0.7 GPa in the dry state 590 ± 90 MPa in the wet state 300 ± 30 MPa in the presence of 1 M NaCl | [30] |
| (PAH-PAA)$_{25.5}$ | Electrostatic and H-bonds | SIEBIMM | 8.8 ± 0.9 GPa at ambient humidity of 20 ± 4% | [30] |
| PAH-PSS in the form of membranes deposited on PDMS wells | Electrostatic | Deformation of a membrane induced by osmotic pressure difference | 599 ± 52 MPa in the wet state | [111] |
| (PAH-PSS)$_9$-PAH/Au/(PAH-PSS)$_9$-PAH | Electrostatic, spin assited assembly | Bulging test on a freestanding membrane 600 μm in diameter | 6.6 ± 2.3 GPa | [18] |
| (PEI-Prussian Blue)$_n$ | Electrostatic | Nanoindentation | 3.40 GPa in the oxidized state 1.75 GPa in the reduced state. | [112] |
| (PAH-PSS)$_n$ in pure water | Electrostatic | | 590 ± 90 MPa in the presence of water 17 ± 2 MPa in the dry state | [113] |
| (PAH-λ carrageenan)$_n$ | Electrostatic | Colloidal probe AFM | 7 ± 1 MPa | [114] |
| (PAH-ι carrageenan)$_n$ | | | 26 ± 3 MPa | |
| (PEI-single walled carbon nanotubes)$_n$ | Electrostatic | Stretching on freestanding membranes | approx. 220 MPa (ultimate strength) | [115] |
| (PDADMAC-Montmorillonite)$_n$ | | Microtensile tests on freestanding membranes | 11 ± 1 GPa, $n$ = 100 layer pairs, 3% relative humidity 3 ± 0.5 GPa, $n$ = 100 layer pairs, 92% relative humidity approx. 100 MPa (ultimate strength) | [116] |
| (PVA-Montmorillonite)$_n$ | H-bonds and covalent linkages | Microtensile tests on freestanding membranes | 13 ± 2 GPa without crosslinking and 106 ± 11 GPa after crosslinking with glutaraldehyde | [117] |

results for different LbL films briefly summarized below is based upon this review as well as on the analysis of the authors' results.

The elastic moduli measured for different LbL films in the dry state range from several tens to hundreds of MPa for weakly bound (e.g., hydrogen-bonded) LbL films to several GPa for strongly electrostatically bound LbL films, and even higher for LbL films with reinforcing agents (see below). The common range for traditional flat LbL films from conventional polyelectrolyte components (e.g., PAH/PSS) is usually within 1 to 4 GPa, depending upon preparation conditions and humidity, which is common for glassy linear polymers [40]. The level of the residual stress caused by the membrane pre-stretching after transfer to the solid substrate varies significantly from film to film but usually remains within 10–20 MPa. On the other hand, in the highly swollen state, the elastic modulus of these LbL films was shown to decrease dramatically due to the plasticizing effect of the water to the range from a few kPa to several hundred MPa, depending upon the degree of swelling.

The mechanical properties are also strongly dependent upon the nature of the components in the LbL films. In fact, the elastic modulus can be lowered to a few tens of MPa in the dry state, if one or both components are ionic elastomers such as polyurethanes [41]. On the other hand, adding inherently stiff polyelectrolyte components, such as PPV or strong silk fibroin components, can significantly increase the elastic modulus in the dry state to values within 5–9 GPa.

The elastic modulus for LbL films reinforced with inorganic nanostructures (clay particles, metal nanoparticles, silica particles, metal nanowires and nanorods, carbon nanotubes, quantum dots, to name a few) dramatically increases and can routinely reach 6–20 GPa for different specimens and different conditions with a toughness approaching very high values of 1 MJ m$^{-3}$ for many films. For some LbL films with highly crosslinked compositions reinforced with large anisotropic components (e.g., clay nanoplatelets), the value of the elastic modulus might reach record values of 100 GPa, which cannot be achieved for random nanocomposite materials [42].

The elongation to break is usually below 1% for most reinforced LbL films reaching 2% for some purely polymeric films. However, the ultimate elongation can reach 30–200% for some LbL films with the presence of an elastomeric component and in a swollen state. The ultimate mechanical strength of traditional LbL films is usually within 5–10 MPa but it can be as low as a fraction of a MPa for elastomeric LbL films, or as high as 100 MPa for reinforced polyelectrolyte LbL films. Again, some reinforced LbL films with clay nanoplatelets and tailored chemical composition might show outstanding ultimate strength reaching 500 MPa and a toughness of a few MJ m$^{-3}$ [42, 43].

## 16.3
### Mechanical Properties of LbL Microcapsules

Microcapsules are generally defined as thin-walled, hollow structures with typical diameters in the micron and sub-micron range, and typical wall thicknesses in the sub-micron to nanoscale range. Microcapsules have found a large number of applications in areas as diverse as cosmetics, food-design, pharmaceuticals (drug delivery), or even self-healing coatings, to mention several prominent examples [44, 45]. In all of these examples, mechanical properties are relevant for their performance in various ways: They directly govern the stability of the capsules under the action of external forces/stresses, and mechanical instability can provide a fast and efficient release mechanism if it occurs due to external stimuli or triggers. Therefore, controlling the mechanical properties of microcapsules by design is an important step towards intelligent capsule systems. A prerequisite is the availability of techniques that allow direct measurements of a microcapsule's deformation properties.

In the following we will give an overview of techniques which allow measurements of mechanical properties on the single capsule level, with special emphasis on work done on LbL microcapsules. Subsequently, we will discuss some examples, where the control of mechanical properties has been shown to impact on microcapsule performance in biomedical and biotechnological applications.

### 16.3.1
**Theory and Measurement of Mechanical Properties of Microcapsules**

A typical microcapsule deformation experiment yields the deformation force as a function of deformation for a given loading rate. In order to interpret the results in terms of materials constants, and in order to choose meaningful experimental conditions, theoretical considerations are imperative and we will briefly summarize the main findings before turning back to experimental techniques. A more complete treatment with further references can be found in a recent review [46].

### 16.3.2
**Basic Concepts of Shell Theory**

Microcapsules can be described by using shell theory, in which typically the wall material is considered to be a continuum, and the shell is described by an effective two-dimensional membrane located at its middle surface (from which the three-dimensional solution can be derived using the Kirchhoff–Love assumption) [47]. Based on this picture, deformation of shells can be either an in-plane stretching and shear or an out-of-plane bending.

Assuming that the membrane material obeys Hooke's law, and that it is isotropic, that is, characterized by two material constants like Young's modulus $E$ and Poisson ratio $\nu$, the ability of the membrane to resist stretching and bending is characterized by the constants $\eta$ and $\kappa$ [47–49]:

$$\eta = \frac{Eh}{1-\nu^2} \tag{16.3}$$

$$\kappa = \frac{Eh^3}{12(1-\nu^2)} \tag{16.4}$$

where $h$ denotes the shell thickness, $\eta$ is the extensional stiffness, and $\kappa$ the bending stiffness. In the following we will focus on spherical capsules of radius $R$, since those are the most widely investigated (some examples of non-spherical capsules will be covered in Section 16.3). For this case, the deformation energies that occur upon a deformation $d$ associated to stretching and bending, respectively, are

$$E_{\text{stretch}} \propto \eta \left(\frac{d}{R}\right)^2 \tag{16.5}$$

$$E_{\text{bend}} \propto \kappa \left(\frac{d}{R^2}\right)^2 \tag{16.6}$$

Based on these considerations, different regimes of microcapsule deformation can be identified. The ratio $(d/h)$ turns out to be one of the main factors determining the deformation behavior. This is also reflected in the experimental work, where either a small deformation approach ($d$ of the order of $h$) or a large deformation approach ($d \gg h$) has been followed.

### 16.3.3
**Experimental Measurements**

The first qualitative results on LbL-microcapsule deformation properties were obtained by Bäumler and coworkers using micropipette suction, and the first estimates of the wall-material's elastic constants were derived from osmotic pressure induced collapse of microcapsules by Gao and coworkers [50, 51]. However, the flexibility in terms of working conditions (possibility for measurements in air and under solvent, possibility to vary environmental parameters like solvent composition or temperature *in situ*) made AFM the most widely used method. AFM provides a force range between $10^{-12}$ and $10^{-6}$ N combined with a deformation resolution better than 1 nm.

In order to probe the elastic constants of microcapsules with an AFM, the microcapsules have to be immobilized on a solid support. Subsequently, a force–deformation measurement is performed, during which the AFM-probe exerts a known force on the pole of the microcapsule while the deformation is monitored. The force is measured by monitoring the deflection of the AFM-cantilever. The deformation can only be inferred indirectly by reference measurements on non-deformable surfaces (usually the substrate is used as a reference). For a part of the measurements reported in the following, instead of a sharp AFM tip, a colloidal-probe AFM set-up was used. In the colloidal-probe method that was developed earlier by Butt and Ducker, a colloidal particle of several microns diameter is glued to a tip-less AFM cantilever and used as a probe [52, 53]. The main reason for the use of the colloidal-probe technique is that the geometry of the contact between the capsule and the cantilever is well defined and high stresses on the microcapsule as they arise for sharp AFM tips are avoided.

## 16.3.4
### Small Deformation Measurements

Small deformation measurements were first introduced by using a set-up combining AFM with microinterferometry [54]. The latter technique allows monitoring of the capsule pole adjacent to the substrate during deformation, and thus provides information on shape changes of the capsule. Figure 16.6 shows a typical force–deformation curve as obtained for a PE-multilayer microcapsule made from PAH and PSS with 25 nm (dry) thickness and 7.9 μm radius, together with microinterferometry images corresponding to characteristic phases of deformation. One notes a linear

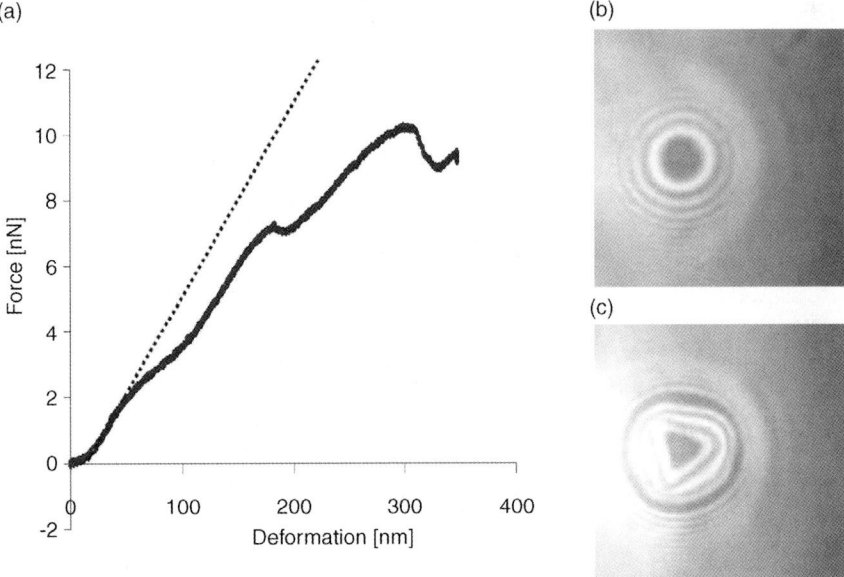

**Figure 16.6** (a) A typical force–deformation curve as obtained for a PE-multilayer microcapsule made from PAH and PSS with 25 nm (dry) thickness and 7.9 μm radius. The analytical solution for small deformations according to the Reissner-model is indicated as a dotted line. Microinterferometry images corresponding to characteristic regions of deformation are displayed: (b) initial spherical contact region, (c) polygonial contact region after buckling. Reprinted from [54] with kind permission from Springer Science + Business Media: Europhys. J. E., Elastic properties of polyelectrolyte capsules studied by atomic force microscopy and RICM, 12(2), 2003, 215–221, Dubreuil F., Elsner N., Fery A.

force–deformation relation for deformations of the order of a few times the wall thickness, followed by a series of stepwise instabilities and an intermittent recovery of the linear force–deformation regimes. The linear region can be explained by the so-called Reissner model of shell deformation [55, 56], which predicts that for a half-dome and deformations of the order of the wall thickness that the force, $F$, should depend on the deformation $d$ via the relationship:

$$F = \frac{4Eh^2}{\sqrt{3(1-\nu^2)}} \frac{d}{R} \tag{16.7}$$

where $h$ denotes the wall thickness, $R$ the radius of curvature.

It is important to discuss the assumptions of this model critically: First, it can only be applied, if the deformation occurs predominately at one pole of the capsule, which can be checked by microinterferometry. The asymmetry is usually caused by the capsule adhesion on the substrate. If deformation occurs symmetrically at both poles, the capsule can be approximately modeled as a system of two serially coupled springs, with spring constants $k = F/d$, according to the equation above. Secondly, this relation is strictly valid only for a point force load at the apex of a shallow spherical cap, but it has turned out to be a good approximation for the deformation by a large colloidal-probe particle, as revealed by finite element modeling [57].

Before discussing the implications of the Reissner-fit to the data in more detail, let us turn our attention to the second feature, the stepwise instabilities in the force–deformation data. These coincide with characteristic shape changes from spherical to polygonal contact regions, as displayed in Figure 16.6. Indeed, theory predicts that, for deformations larger than several times the wall thickness, a so-called buckling transition should occur. Buckling transitions for PE-microcapsules were first observed as a result of osmotic shocks [7]. A buckling transition is characterized by a reversed curvature of the shell in the vicinity of the load-point [58–60]. As a consequence, for a point-load the force–deformation relation should change to [61]

$$F = 1.89 \frac{E}{(1-\nu^2)} \frac{h^{5/2}}{R} \sqrt{d} \tag{16.8}$$

In colloidal-probe AFM experiments, this square-root dependence is not observed, but rather a return a linear slope occurs as a consequence of the contact between the colloidal probe and the rim [11]. The fact that the contact region adapts a polygonal shape is typical for later stages of buckling transitions. While deformations of buckled shells start axisymmetrically, for larger deformations this symmetry is broken.

The experimental results obtained in the small deformation regime have two main implications: first, they allow estimation of the wall's Young's modulus, which was found to be comparable with the values obtained later by other methods for flat freestanding films (see Table 16.1). Indeed, given the fact that the radius of curvature of the microcapsules is large compared to the typical molecular dimensions of the polyelectrolytes used, one does expect consistency between results, which makes flat film measurements a good benchmark for the microcapsule investigations.

Second – and in our opinion more interesting in the long run – they demonstrate impressively the excellent control over deformation properties and allow monitoring of the stimulus responsivity of the microcapsules on the single capsule level. Returning to Equation 16.7, one notices that, for a given material, the stiffness of the microcapsule can be tailored by the wall thickness and the radius of curvature of the capsule. The LbL technique offers control over both parameters: The wall thickness can be tuned by the adsorption conditions and the number of adsorption steps and the radius of curvature is determined by the template-particle used for multilayer assembly (which can be of high monodispersity). Figure 16.7 shows experimental data for the dependence of the average stiffness ($k = F/d$) of microcapsules as a function of the square of the microcapsules wall-thickness [11]. The data are clearly in line with the predictions of Equation 16.7 and similar trends were observed for the scaling with capsule radius. Thus the deformability for small deformations can be predicted and tailored over a large range.

It is well known that the internal structure of polyelectrolyte multilayers can be modified by changes in solvent composition, or by temperature annealing. The impact of these external parameters on microcapsule-deformability has been investigated in detail. Experiments in the small deformation regime offer several advantages as solvent properties, such as salt concentration and pH, or environmental conditions, such as temperature, can be varied during the experiment [62–64]. Due to the non-destructive nature of the measurement, changes in mechanical properties can even be followed *in situ* for individual capsules. The rate dependence of deformation properties merits special attention when studying viscoelastic membranes. For microcapsules made from PDADMAC/PSS, a transition from a glassy to

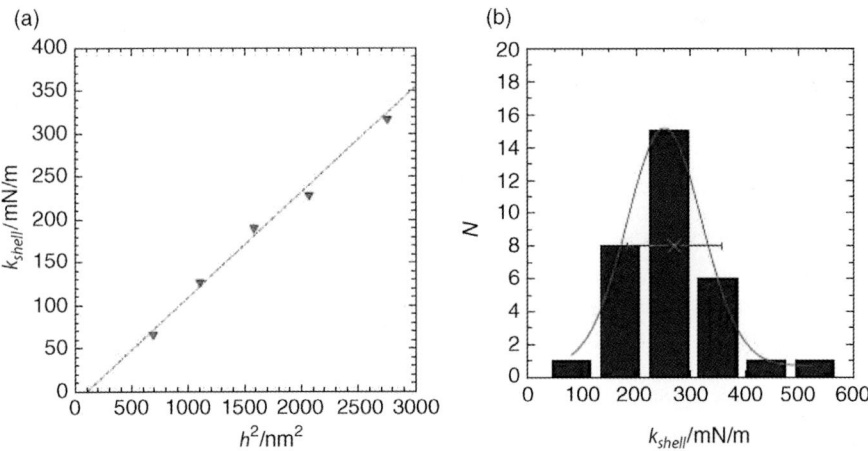

**Figure 16.7** (a) Average stiffness $k$ (defined as ratio of deformation force $F$ versus deformation $d$) of microcapsules with identical radius and variable wall thickness. The (quadratic) scaling meets expectations based on Reissner theory. (b) Distribution of stiffness values for a representative sample, indicating a well-defined mean value, although scatter due to heterogeneities in the wall thickness is considerable. Reprinted from [57].

a viscoelastic regime could be demonstrated. This transition correlates with drastic changes in the (quasi-static) elastic constants of around two orders of magnitude.

Large deformation experiments on polyelectrolyte-multilayer capsules were pioneered by Vinogradova and coworkers [65]. Here, deformations are of the order of up to 50% of the capsule radius and beyond (much larger than the wall thickness $h$). These experiments are in the tradition of parallel plate compression experiments, as first introduced by Cole and coworkers [66–69]. Smith *et al.* first introduced the use of colloidal-probe AFM in this type of experiment.

The physical mechanisms, which determine the deformability of the capsules in this regime differ completely from those in the small deformation regime. While at small deformations only changes in curvature occur, which are localized close to the loading point, the large deformation regime is characterized by global shape changes of the capsules. The energy cost of these shape changes is mainly determined by the membrane stretching due to area increases of the capsule membrane, which accompany these shape changes. Therefore, boundary conditions for the encapsulated volume become important for the interpretation of the results in terms of elastic constants. If the capsule wall material is completely impermeable for the solvent, area stretching can simply be calculated based on volume conservation [70]. However, the situation is more complex for polyelectrolyte-multilayer capsules, since their membrane is well known to be permeable for water (and even larger molecular weight species). Therefore, volume conservation is only partial and should depend on compression speed. This explains why systematically lower elastic constants are derived from experiments using large deformations and assuming volume conservation. Nevertheless, the effects of environmental conditions on microcapsules mechanics agree qualitatively with observations in the small deformation regime (see [71] for a review).

An interesting aspect of measurements in the large deformation regime is the fact that the stability and plasticity of microcapsules can be probed [72]. Fernandes and coworkers investigated polyelectrolyte multilayer-capsules of PDADMAC/PSS filled with fluorescently labeled dextran. Deformation experiments up to 80% of the capsule diameter were carried out while the capsules were monitored *in situ* by fluorescence microscopy. The contrast between fluorescence in the capsule interior and exterior served as a measure for the release of the encapsulated dextran. As shown in Figure 16.8, the data show a well-defined deformation of around 20% relative deformation, upon which release drastically increases, implying membrane rupture. Repeated experiments on the same capsule showed that this threshold corresponds to the onset of plastic deformations.

## 16.4
### Prospective Applications Utilizing Mechanical Properties

#### 16.4.1
### Flat Freestanding LbL Films for Sensing Applications

The peculiar mechanical behavior of ultrathin LbL films can be utilized for different sensing and delivery applications. Further applications such as acoustic microsensors,

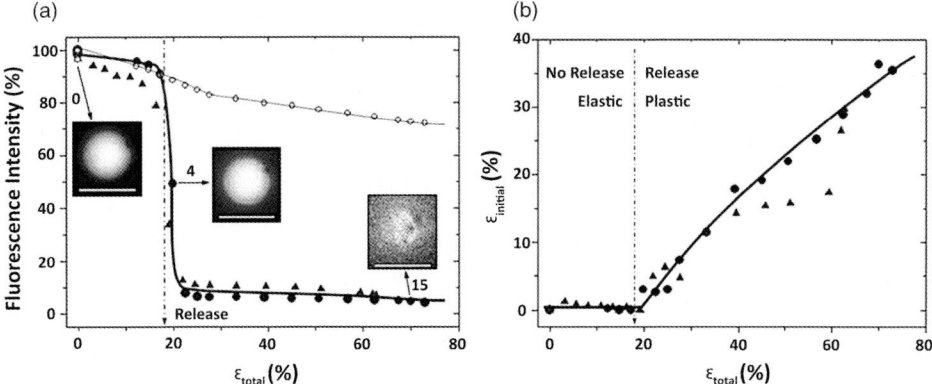

**Figure 16.8** (a) Fluorescence contrast between inside and outside of microcapsules as an indicator for release of fluorescently labeled dextran is recorded as a function of mechanical deformation (full circles, triangles). An abrupt decrease is observed at around 18% deformation, indicating burst release at this deformation. The reference sample (open circles) shows, in the absence of mechanical strains, only gradual bleaching. (b) Plasticity of microcapsules is measured by repeated deformation. Plastic deformations are quantified by the initial deformation encountered for the second deformation cycle. The onset of plasticity corresponds with the deformation at which burst release is found. Reprinted from [72] – Reproduced by permission of The Royal Society of Chemistry.

biological/chemical sensors, micro-transport controllers, microfluidic valves, and air/fluid flow nanosensors, and multiplexed drug storage/release can be envisioned.

The ability of these flexible films to readily deflect upon variable external pressure has been utilized in air pressure/photothermal microfabricated sensing arrays and fluidic osmotic pressure sensors. In an earlier study, Jiang *et al.* demonstrated the extremely high pressure sensitivity of freely suspended LbL membranes which is up to three orders of magnitude better than the threshold hearing of a human ear [18, 73]. They are also two orders of magnitude more sensitive than the existing uncooled thermal sensors based on microcantilever technology, with a theoretical lowest limit of detectable temperature variation well below 1 mK [74].

In a following study, Jiang *et al.* demonstrated that the photothermal behavior can be scaled up to a large (64 × 64) array of microscopic cavities, thus creating an efficient microscopic thermal imaging array [75]. They observed reproducible, multiple transitions from a convex to a concave shape of the flexible nanomembranes caused by air thermal expansion and contraction, with an interesting patterned buckling instability phenomenon which facilitates fast optical response and causes outstanding photothermal sensitivity (Figure 16.9). A strong correlation was found between the large optical responses of the thermal-pneumatic detectors and the trench buckling of LbL films. These patterns are completely reversible and, similar to snap-through membrane behavior, are caused by fast stress relief during the transition between two stable mechanical deformational states [76].

In a related study, McConney *et al.* have exploited the buckling instabilities in freely suspended composite nanomembranes as a highly sensitive platform for sensing minor environmental changes in dynamic mode [77]. The authors demonstrated a

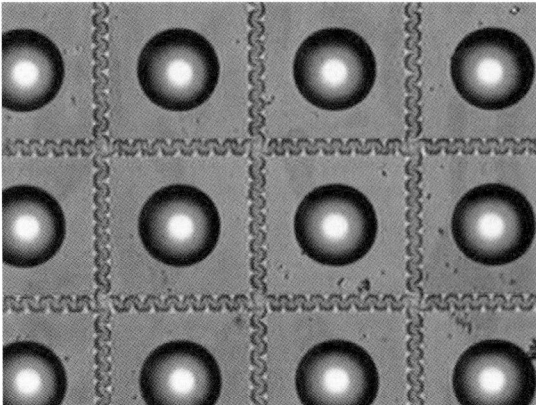

**Figure 16.9** An optical image of a photothermal sensor array with LbL film covering microfabricated substrate with cylindrical cavities and trenches. Notice the worm-like buckling in the trenches and the high contrast of the sealed thermal-pneumatic sensors, which is caused by out-of-plane buckling. Reprinted from [75].

thermal sensitivity of 10 mK by monitoring the relative optical signal from the buckling transformation of freely suspended membranes under fast variation of illumination conditions. Under this dynamic regime, the buckling proved to be very stable with fast response times, better than 25 ms (Figure 16.10). Upon stepwise changing of the temperature by 10 mK from 17.00 to 17.01 °C, a noticeable damping in the optical oscillation signal was observed again in the lower part of the thermal cycle (Figure 16.10). The ability to resolve a 10 mK difference is quite impressive considering that modern photothermal sensors have a temperature sensitivity exceeding 100 mK. We believe that these results pave the way toward a new generation of uncooled, miniature, ultra-sensitive, photothermal imagers for a wide range of sensing applications, from medical *imaging* to fingerprint identification [78, 79].

Alternatively, immersing LbL films into fluids allows the designing of a sensing platform for the monitoring of osmotic pressure. Nolte *et al.* have produced arrays of freestanding LbL membranes by transferring polyelectrolyte multilayers from solid support to PDMS-structures with cylindrical cavities [80]. To detect the influence of osmotic pressure differences across the LbL membrane situated at the fluid/fluid interface, an array was mounted on a confocal microscope, using a sample holder that allows the outer solution to be exchanged (Figure 16.11). This design allows continuous monitoring of the deflection of the membrane for different concentrations of the osmotic active species in outer fluid. When poly(diallyl-dimethyl-ammonium chloride) (PDADMAC) was added to the outer fluid as the osmotic active component to alter the osmotic pressure gradient across the membrane, the membrane deformed.

It has been observed that increasing the polyelectrolyte concentration causes the deflection of the membrane across the hole to increase proportionally [22]. The added

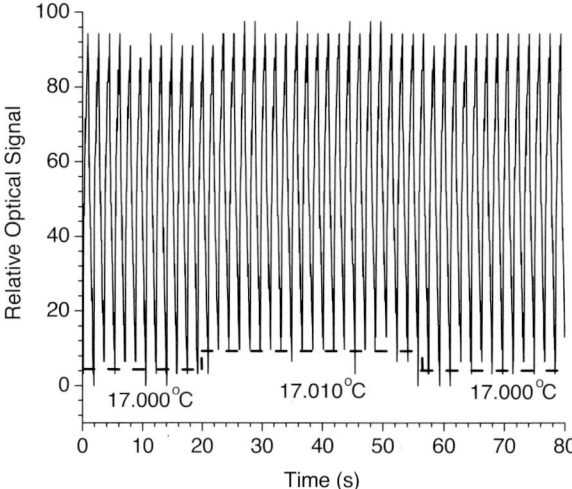

**Figure 16.10** A plot of relative optical signal versus time for a dynamically excited LbL membrane. The surrounding temperature was raised by 10 mK and then returned. A dotted line at the bottom of the graph conveys the noticeable change in the amplitude due to the small temperature change of 10 mK. Reprinted from [77].

polyelectrolyte in the outer phase does not permeate the membrane because of its charge and large molecular weight [23]. A simple linear relation between monomer concentration, $\phi$, and osmotic pressure, $\Pi$, can be assumed [26]:

$$\frac{\Pi}{RT} = \phi \cdot c \tag{16.9}$$

where $c$ is a proportionality constant.

Therefore, the osmotic pressure gradient across the membrane can be tuned and the deflection of the LbL membrane measured as a function of the osmotic pressure.

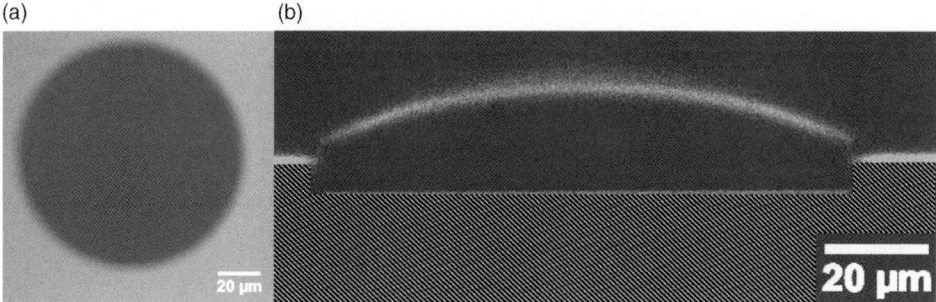

**Figure 16.11** (a) Top-view and (b) side-view confocal micrographs of a freely standing LbL membrane in fluid deformed to the outside due to a negative osmotic pressure difference. Reprinted from [80].

If the concentration difference across the membrane is negative when the concentration inside the cavity is higher than in the outside solution, then the membrane bends away from the PDMS structure (Figure 16.11) [80]. Both positive and "negative" LbL membrane deflection can occur, depending on the concentration difference between the inside and outside of the cavity, thus facilitating continuous monitoring of local osmotic pressure.

### 16.4.2
### Microcapsules for Controlled Delivery Processes

As mentioned in the introduction, mechanical properties of microcapsules are relevant for a broad range of applications, and covering them all would go far beyond the scope of this chapter. Therefore, we will limit the discussion to "mechanical pathways" toward controlling delivery in the biomedical area as a case study, illustrating various fashions in which mechanical properties allow one to gain control over delivery.

#### 16.4.2.1 Exploding Capsules
Since microcapsule membranes are semi-permeable, osmotic pressure was early recognized as one means for causing tension and eventually collapse/bursting of microcapsules [81]. Recently, osmotic pressure has been used for "programming" the mechanical instability of microcapsules: if LbL capsules are templated onto hydrolyzable hydrogel-cores, increasing osmotic pressure can build up in the course of gel hydrolysis and trigger capsule-burst, if the membrane is impermeable for degradation products [82, 83]. Hydrolyzable dextran-based gels are ideally suited for this purpose, since their pressure build-up during swelling and degradation can be controlled [84].

The thickness and composition of the multilayer shell allows fine-tuning of the mechanical stability and resistance toward internal pressure. Thus the timescale on which release occurs can be tuned between seconds and days (see example of fast response in Figure 16.12) [85]. This is especially useful for applications like vaccination, where (multiple) burst deliveries of vaccines are desired, which should be delivered with a single vaccination [86]. The rupture of the membrane occurs usually at one point only, and the ejection of the cargo is driven by the flows induced by the relaxation of the pressure difference. Thus release is directional and not limited by diffusion velocities. As mechanical instability happens at the "weak spot" of the membrane, the release point can be predetermined, by including defects, or even triggered externally by including destabilizing elements [87]. Recently, the mechanical properties and their relevance for release for this class of microcapsules have been investigated using large deformation experiments [88].

#### 16.4.2.2 Asymmetric Capsules
Mechanical deformation properties are naturally dependent not only on elastic constants, but also on microcapsule shape. Thus exploring non-spherical capsule

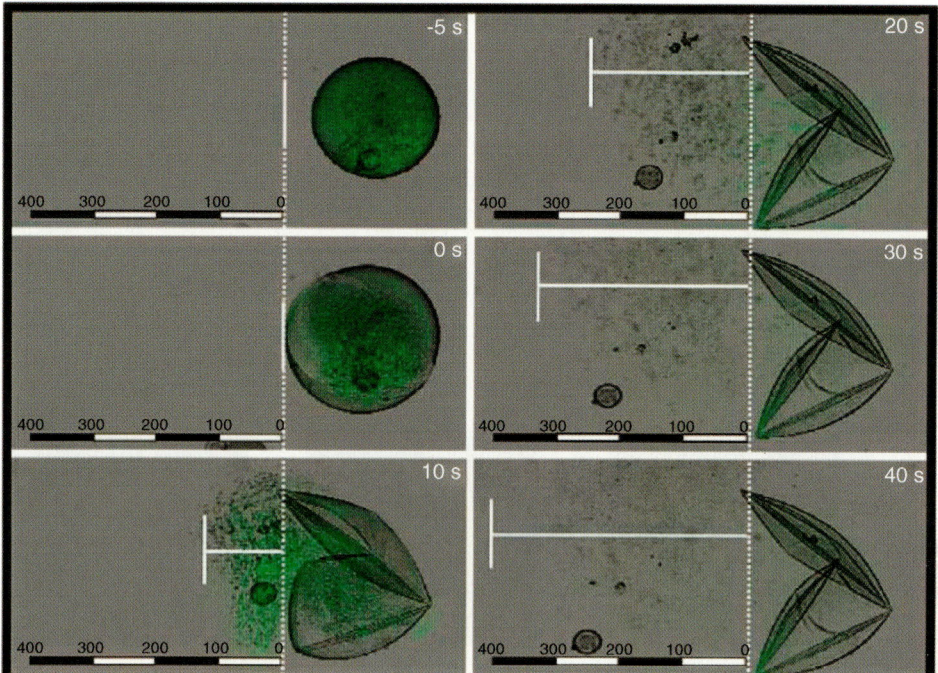

**Figure 16.12** Confocal microscopy snapshots taken at regular time intervals (overlay of green fluorescence channel and transmission channel) of LbL-coated microgels containing fluorescently labeled nanoparticles during the degradation of the microgel core triggered by the addition of sodium hydroxide. The microcapsule explodes 10 s after addition of sodium hydroxide, and subsequently the edge of the propagating front of released nanoparticles is marked by the vertical white line. Reprinted from [101] – Reproduced by permission of The Royal Society of Chemistry.

shapes appears promising. Although the majority of work on LbL microcapsules uses spherical templates, the glassy nature of many polyelectrolyte multilayers allows one to create (and maintain) shapes of non-constant curvature. This is in contrast to alternate approaches, where, especially, sharp-edged shapes are difficult to construct [89–92]. Indeed, a variety of non-spherical microcapsules has been presented in the literature: enzyme and drug crystals have been encapsulated in LbL shells for controlled release applications, cylindrical microcapsules have been obtained by encapsulating fiber fragments [93–96].

Recently, cubic and even tetrahedral microcapsules were prepared by templating inorganic $CdCO_3$ and $SnS$ anisotropic micro- and nanocrystals [97, 98]. The nonspherical shape has interesting implications for the deformation behavior: while spherical capsules react towards compressive pressure with the above-mentioned buckling instability at a random location (typically the mechanically weakest spot of the capsule membrane), cubic or tetrahedral microcapsules maintain their overall shape for a much larger range of pressures, as the facetted structure has a reinforcing effect. Figure 16.13 displays simulation results for cubic and tetrahedral structures under the effect of compressive and expansive pressure gradients. The results are based on computer modeling which combines a lattice Boltzmann model for the fluid dynamics and a lattice spring model of elastic solids [3]. The strain distributions shown in Figure 16.13 point towards an additional feature of non-spherical shapes. The edges allow effective localization of strains, especially under compressive pressures. This localization holds the potential for controlling the position and direction of release. In addition to these advantages, cellular internalization and vascular dynamics govern the benefits of using non-spherical synthetic carriers for targeted drug delivery [99, 100].

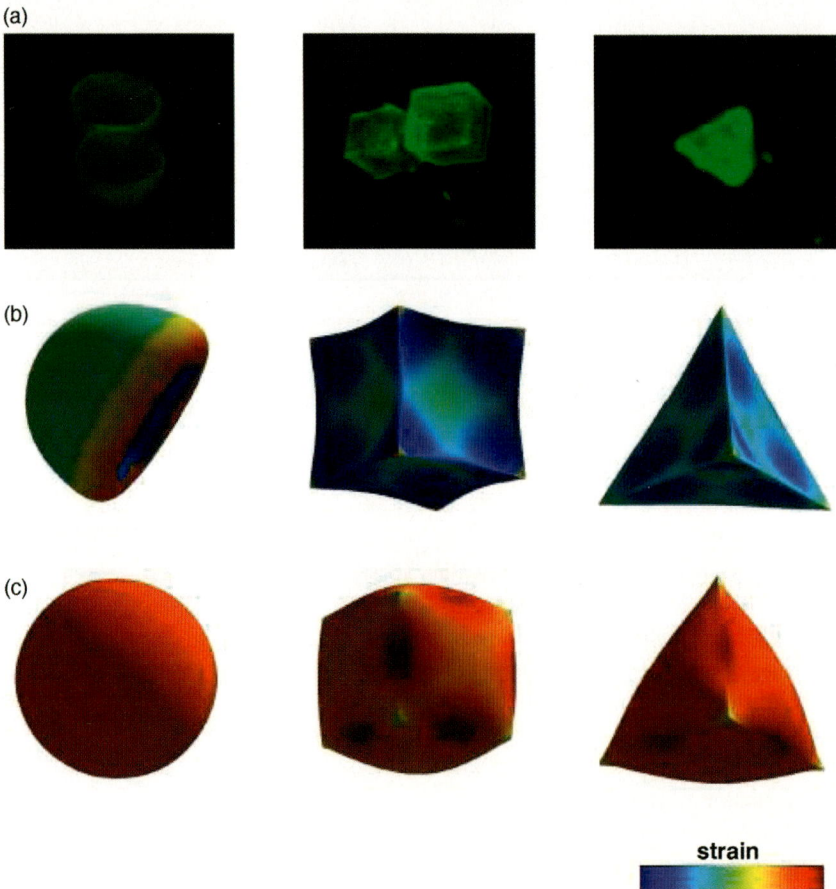

**Figure 16.13** Comparison of spherical with cubic and tetragonal capsules. (a) Confocal laser scanning images. (b) Simulation results for shape changes/strain distributions under the effect of compressive pressure. (c) Simulation results for shape changes/strain distributions under the effect of expansive pressure. Reprinted from [97].

### 16.4.2.3 Mechanical Stability and Intracellular Delivery

Intracellular delivery of proteins, peptides and other biomolecules is of growing importance not only in biomedical research but also in fundamental cell biology [101, 102]. It is essential, however, that the encapsulated cargo is delivered into the cells without losses. Thus, investigation of the mechanical properties of the delivery vehicles, which determine successful delivery, is of interest for both developing carriers capable of withstanding significant pressures, and delivering sufficient intracellular concentration of biomolecules. Although quantitative measurements in living cells are not yet available, it is likely that the intracellular forces exerted by the cytoskeleton are of the order of several hundred nN [103–105]. In this regard, the pressure exerted by a cell upon cellular uptake of a capsule is believed to be one of the key parameters to successful delivery.

Delcea *et al.* have recently investigated the correlations between mechanical stability of LbL-microcapsules and their efficiency in intracellular delivery [106]: Microcapsules filled with fluorescently labeled dextran made from PDADMAC/PSS were exposed to a thermal pretreatment procedure. Depending on the pretreatment temperature, microcapsule stiffness can be varied. The uptake of these microcapsules by cells was subsequently initiated by an electroporation-based method [107].

The mechanical properties of identical microcapsules were investigated using large deformation experiments. It turned out that the mechanical properties and efficiency for intra-cellular delivery correlate strongly. As shown in Figure 16.14, microcapsules pretreated at low temperatures do not survive the uptake process and

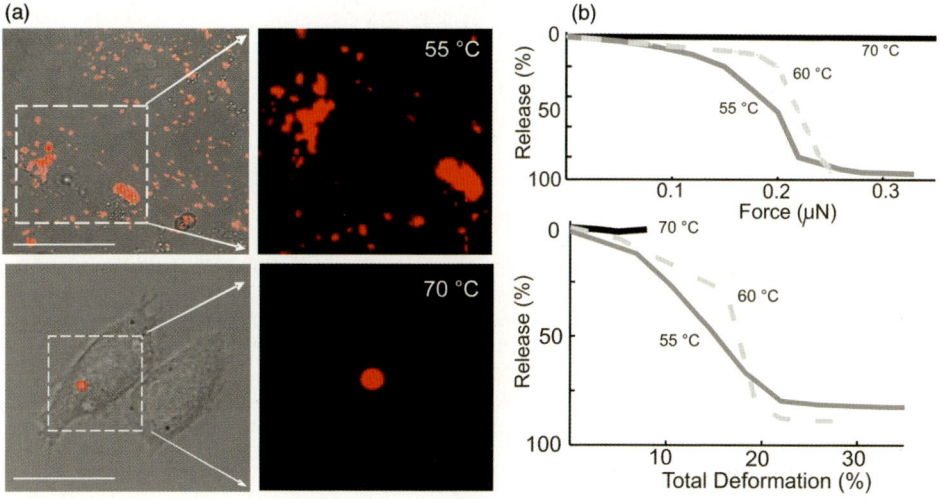

**Figure 16.14** (a) Release characteristics of microcapsules annealed at 55 and 70 °C, respectively. Capsules annealed at low temperature release their cargo prematurely; the fluorescently labeled material is adsorbed on the exterior of the cells. In contrast, the capsules annealed at higher temperature survive the uptake process. (b) Single-capsule force measurements in the large-deformation regime indicate that these differences are due to different mechanical stability [106].

release their cargo prematurely. In contrast, capsules pretreated above a critical temperature are uptaken intact. Thus a threshold of mechanical stability could be identified.

Apart from providing guidelines for the mechanical design of microcapsules for intracellular delivery, these findings can also help in understanding the forces acting on particles during uptake by cells. Microcapsules with well-defined mechanical properties can serve as test objects whose shape changes indicate the forces that they encountered during uptake [108]. The relevance of microcapsule mechanics might go beyond these stability aspects. It is commonly accepted that the mechanical properties of cell substrates have a profound impact on cell–substrate interactions [109]. One could expect similar effects on uptake of, and interactions with particles. Indeed recent studies indicate an impact of particle elasticity on phagocytosis and, consequently, on circulation and lifetime inside living organisms [110].

## Acknowledgements

The Vladimir V. Tsukruk activity in this field is continuously supported with funding provided by the National Science Foundation, the Air Force Office of Scientific Research, the Air Force Research Laboratory, and the Department of Energy. Vladimir V. Tsukruk also acknowledges support from the Alexander von Humboldt Foundation for the preparation of this chapter. Andreas Fery acknowledges support from the Max-Planck Society and from the German Science Foundation. The authors thank Dr. Milana Lizunova, Kyle Anderson, Seth Young, and Dipl. Chem. Christian Kuttner for providing some technical data and proofreading.

## References

1 Bowden, N., Brittain, S., Evans, E.G., Hutchinson, J.W., and Whitesides, G.W. (1998) Spontaneous formation of ordered structures in thin films of metals supported on an elastomeric polymer. *Nature*, **393**, 146–149.

2 Efimenko, K., Rackaitis, M., Manias, E., Vaziri, A., Mahadevan, L., and Genzer, J. (2005) Nested self-similar wrinkling patterns in skins. *Nat. Mater.*, **4**, 293–297.

3 Mahadevan, L. and Rica, S. (2005) Self-organized origami. *Science*, **307**, 1740.

4 Huang, J., Juszkiewicz, M., Jeu, W.H., Cerda, E., Emrick, T., Menon, N., and Russell, T.P. (2007) Capillary wrinkling of floating thin polymer films. *Science*, **317**, 650–653.

5 Chen, J.T., Zhang, M., and Russell, T.P. (2007) Instabilities in nanoporous media. *Nano Lett.*, **7**, 183–187.

6 Sun, Y., Chi, W.M., Jiang, H., Huang, Y.Y., and Rogers, J.A. (2006) Controlled buckling of semiconductor nanoribbons for stretchable electronics. *Nature Nanotechnol.*, **1**, 201–207.

7 Xiao, J., Jiang, H., Khang, D.Y., Wu, J., Huang, Y., and Rogers, J.A. (2008) Mechanics of buckled carbon nanotubes on elastomeric substrates. *J. Appl. Phys.*, **104**, 033543.

8 Khang, D.Y., Xiao, J., Kocabas, C., MacLaren, S., Banks, T., Jiang, H., Huang, Y.Y., and Rogers, J.A. (2008) Molecular scale buckling mechanics in individual aligned single-wall carbon

nanotubes on elastomeric substrates. *Nano. Lett.*, **8**, 124–130.
9. Chan, E.P., Smith, E.J., Hayward, R.C., and Crosby, A.J. (2008) Surface wrinkles for smart adhesion. *Adv. Mater.*, **20**, 711–716.
10. Lin, P., Vajpayee, S., Jagota, A., Hui, C.H., and Yang, S. (2008) Harnessing surface wrinkle patterns in soft matter. *Soft Matter*, **4**, 1830–1835.
11. Sun, Y. and Rogers, J.A. (2007) Structural forms of single crystal semiconductor nanoribbons for high-performance stretchable electronics. *J. Mater. Chem.*, **17**, 832–840.
12. Sun, Y., Choi, W.M., Jiang, H., Huang, Y.Y., and Rogers, J.A. (2006) Controlled buckling of semiconductor nanoribbons for stretchable electronics. *Nat. Nanotech.*, **1**, 201–207.
13. Lu, C.H., Möhwald, H., and Fery, A. (2007) A lithography-free method for directed colloidal crystal assembly based on wrinkling. *Soft Matter*, **3**, 1530–1536.
14. Yoo, P.J., Suh, K.Y., Park, S.Y., and Lee, H.H. (2002) Physical self-assembly of microstructures by anisotropic buckling. *Adv. Mater.*, **14**, 1383–1387.
15. Stafford, C.M., Harrison, C., Beers, K.L., Karim, A., Amis, E.J., Vanlandingham, M.R., Kim, H.C., Volksen, W., Miller, R.D., and Simonyi, E.E. (2004) A buckling-based metrology for measuring the elastic moduli of polymeric thin films. *Nat. Mater.*, **3**, 545–550.
16. Beams, J.W. (1959) *Structure and Properties of Thin Solid Film* (eds C.A. Neugebauer, J.B., Newkirk, and D.A. Vermilyea), John Wiley, New York, p. 183.
17. Vlasak, J.J. and Nix, W.D. (1992) A new bulge test technique for the determination of Young's modulus and Poisson's ratio of thin films. *J. Mater. Res.*, **7**, 3242–3249.
18. Jiang, C., Markutsya, S., Pikus, Y., and Tsukruk, V.V. (2004) Freely suspended nanocomposite membranes as highly-sensitive sensors. *Nature Mater.*, **3**, 721–728.
19. Markutsya, S., Jiang, C., Pikus, Y., and Tsukruk, V.V. (2005) Freestanding multilayered nanomembranes: testing micromechanical properties. *Adv. Funct. Mater.*, **15**, 771–780.
20. Jayaraman, S., Edwards, R.L., and Hemker, K.J. (1999) Relating mechanical testing and microstructural features of polysilicon thin films. *J. Mater. Res.*, **14**, 688–697.
21. Poilane, C., Delobelle, P., Lexcellent, C., Hayashi, S., and Tobushi, H. (2000) Analysis of the mechanical behavior of shape memory polymer membranes by nanoindentation, bulging and point membrane deflection tests. *Thin Solid Films*, **379**, 156–195.
22. Jiang, C. and Tsukruk, V.V. (2006) Free standing nanostructures via Layer-by-Layer Assembly. *Adv. Mater.*, **18**, 829–840.
23. Jiang, C., Wang, X., Gunawidjaja, R., Lin, Y.-H., Gupta, M.K., Kaplan, D.L., Naik, R.R., and Tsukruk, V.V. (2007) Mechanical properties of robust ultrathin silk fibroin films. *Adv. Funct. Mater.*, **17**, 2229–2237.
24. Lu, C., Dönch, I., Nolte, M., and Fery, A. (2006) Au nanoparticle-based multilayer ultrathin films with covalently linked nanostructures: spraying Layer-by-layer assembly and mechanical property characterization. *Chem. Mater.*, **18**, 6204–6210.
25. Jiang, C., Singamaneni, S., Merrick, E., and Tsukruk, V.V. (2006) Thermo-optical arrays of flexible nanomembranes freely suspended over microfabricated cavities as IR microimagers. *Nano Lett.*, **6**, 2254–2259.
26. Lin, Y.H., Jiang, C., Xu, J., Lin, Z., and Tsukruk, V.V. (2007) Sculptured layer-by-layer films. *Adv. Mater.*, **19**, 3827–3832.
27. Gunawidjaja, R., Jiang, C., Ko, H., and Tsukruk, V.V. (2006) Free standing 2D arrays of silver nanorods. *Adv. Mater.*, **18**, 2895–2899.
28. Gunawidjaja, R., Jiang, C., Peleshanko, S., Ornatska, M., Singamaneni, S., and Tsukruk, V.V. (2006) Flexible and robust 2D array of silver nanowires encapsulated within free

standing layer-by-layer films. *Adv. Funct. Mater.*, **16**, 2024–2034.

29 Gunawidjaja, R., Ko, H., Jiang, C., and Tsukruk, V.V. (2007) Buckling behavior of highly oriented silver nanowires encapsulated within LbL film. *Chem. Mater.*, **19**, 2007–2015.

30 Nolte, A.J., Cohen, R.E., and Rubner, M.F. (2006) Applications to polyelectrolyte multilayers. *Macromolecules*, **39**, 4841–4847.

31 Hendricks, T.R. and Lee, I. (2007) Wrinkle-free nanomechanical film: Control and prevention of polym er film buckling. *Nano Lett.*, **7**, 372–379.

32 Lu, C., Möhwald, H., and Fery, A. (2008) Large-scale regioselective formation of well-defined stable wrinkles of multilayered films via embossing. *Chem. Mater.*, **20**, 7052–7059.

33 Zauscher, S. and Klingenberg, D.J. (2000) Normal forces between cellulose surfaces measured with colloidal force microscopy. *J. Colloid Interface Sci.*, **229**, 497–510.

34 Tsukruk, V.V. and Gorbunov, V.V. (2002) Nanomechanical analysis of polymer surfaces. *Probe Microsc.*, **3–4**, 241–247.

35 Chizhik, S.A., Huang, Z., Gorbunov, V.V., Myshkin, N.K., and Tsukruk, V.V. (1998) Micromechanical properties of elastic polymeric materials as probed by scanning force microscopy. *Langmuir*, **14**, 2606–2609.

36 Tsukruk, V.V., Huang, Z., Chizhik, S.A., and Gorbunov, V.V. (1998) Probing of micromechanical properties of compliant polymeric materials. *J. Mater. Sci.*, **33**, 4905.

37 Kovalev, A., Shulha, H., LeMieux, M., Myshkin, N., and Tsukruk, V.V. (2004) Nanomechanical probing of layered nanoscale polymer films with atomic force microscopy. *J. Mater. Res.*, **19**, 716–728.

38 Lisunova, M., Drachuk, I., Shchepelina, O., Anderson, K., and Tsukruk, V.V. (2011) Direct Probing of Micromechanical Properties of Hydrogen-Bonded Layer-by-Layer Microcapsule Shells with Different Chemical Compositions. *Langmuir*, **27**, 11157–11165.

39 Lavalle, P., Voegel, J.C., Vautier, D., Senger, B., Schaaf, P., and Ball, V. (2011) Dynamic aspects of films prepared by a sequential deposition of species: perspectives for smart and responsive materials. *Adv. Mater.*, **23**, 1191–1221.

40 Sperling, L.H. (1997) *Polymeric Multicomponent Materials*, Wiley-VCH, New York.

41 Podsiadlo, P., Arruda, E.M., Kheng, E., Waas, A.M., Lee, J., Critchley, K., Qin, M., Chuang, E., Kaushik, A.K., Kim, H.S., Qi, Y., Noh, S.T., and Kotov, N.A. (2009) LBL assembled laminates with hierarchical organization from nano- to microscale: high-toughness nanomaterials and deformation imaging. *ACS Nano*, **3** (6), 1564–1572.

42 Kotov, N. (2009) The collective behavior of nanoscale building blocks. *ACS Nano*, **3** (6), 1307–1308.

43 Kharlampieva, E., Kozlovskaya, V., Gunawidjaja, R., Shevchenko, V.V., Vaia, R., Naik, R.R., Kaplan, D.L., and Tsukruk, V.V. (2010) Flexible silk-inorganic nanocomposites with transparent to mirror-like optical properties. *Adv. Funct. Mater.*, **20**, 840–846.

44 Sukhorukov, G.B., Fery, A., and Möhwald, H. (2005) Intelligent Micro- and Nanocapsules. *Prog. Polym. Sci.*, **30** (8–9), 885–897.

45 Shchukin, D.G. and Möhwald, H. (2007) Self-repairing coatings containing active nanoreservoirs. *Small*, **3** (6), 926–943.

46 Fery, A. and Weinkamer, R. (2007) Mechanical properties of Micro- and Nanocapsules: Single-capsule measurements. *Polymer*, **48**, 7221–7235.

47 Niordson, I. (1985) *Shell Theory*, Elsevier Science Pub. Co., New York: North-Holland.

48 Timoshenko, S. and Woinowsky-Krieger, S. (1959) *Theory of Plates and Shells*, McGraw-Hill, New York.

49 Ugural, A.C. (1999) *Stresses in Plates and Shells*, WCB/McGraw Hill, Boston.

50 Bäumler, H., Artmann, G., Voigt, A., Mitlöhner, R., Neu, B., and

Kiesewetter, H. (2000) Plastic behaviour of polyelectrolyte microcapsules derived from colloid templates. *J. Microencapsul.*, **17** (5), 651–655.

51 Gao, C., Donath, E., Moya, S., Dudnik, V., and Möhwald, H. (2001) Elasticity of hollow polyelectrolyte capsules prepared by the layer by layer technique. *Eur. Phys. J. E*, **5**, 21–27.

52 Butt, H.J. (1991) Measuring electrostatic, vanderwaals, and hydration forces in electrolyte-solutions with an atomic force microscope. *Biophys J*, **60**, 1438–1444.

53 Ducker, W.A., Senden, T.J., and Pashley, R.M. (1991) Direct measurement of colloidal forces using an atomic force microscope. *Nature*, **353**, 239–241.

54 Dubreuil, F., Elsner, N., and Fery, A. (2003) Elastic properties of polyelectrolyte capsules studied by atomic force microscopy and RICM. *Europhys. J. E*, **12** (2), 215–221.

55 Reissner, E. (1946) Stresses and small displacements of shallow spherical shells. *J. Math. Phys.*, **25**, 80–85.

56 Reissner, E. (1946) Stresses and small displacements of shallow spherical shells. *J. Math. Phys.*, **25** (4), 279–300.

57 Elsner, N., Dubreuil, F., Weinkamer, R., Wasicek, F., Fischer, F.D., and Fery, A. (2006) Mechanical properties of freestanding polyelectrolyte capsules. *Prog. Colloid Polym. Sci.*, **132**, 117–132.

58 Bushnell, D. (1985) *Computerized Buckling Analysis of Shells* (ed. M. Nijhoff), Kluwer Academic, Dordrecht; Boston; Hingham.

59 Kaplan, A. (1974) *Buckling of Spherical Shells* (eds Y.C. Fung and E.E. Sechler), Prentice-Hall, pp. 247–288.

60 Shilkrut, D. and Riks, E. (2002) *Stability of Nonlinear Shells: On the Example of Spherical Shells*, Elsevier Science, New York.

61 Pogorelov, A.V. (1988) *Bending of Surfaces and Stability of Shells*, American Chemical Society, Providence.

62 Heuvingh, J., Zappa, M., and Fery, A. (2005) Salt induced softening transition of polyelectrolyte multilayer capsules. *Langmuir*, **21** (7), 3165–3171.

63 Elsner, N., Kozlovskaya, V., Sukhishvili, S.A., and Fery, A. (2006) pH-triggered softening of crosslinked hydrogen-bonded capsules. *Soft Matter*, **2**, 966–972.

64 Müller, R., Köhler, K., Weinkamer, R., Sukhorukov, G.B., and Fery, A. (2005) Melting of PDADMAC/PSS capsules investigated with AFM force spectroscopy. *Macromolecules*, **38** (23), 9766–9771.

65 Lulevich, V.V., Radtchenko, I.L., Sukhorukov, G.B., and Vinogradova, O.I. (2003) Mechanical properties of polyelectrolyte microcapsules filled with a neutral polymer. *Macromolecules*, **36**, 2832–2837.

66 Bartkowiak, A. and Hunkeler, D. (1999) Alginate–Oligochitosan microcapsules: A mechanistic study relating membrane and capsule properties to reaction conditions. *Chem. Mater.*, **11**, 2486–2492.

67 Carin, M., Barthes-Biesel, D., Edwards-Levy, F., Postel, C., and Andrei, D. (2003) Compression of biocompatible liquid-filled HSA-alginate capsules: Determination of the membrane mechanical properties. *Biotechnol. Bioeng.*, **82**, 207–212.

68 Cole, K.S. (1932) Surface forces of the arbacia egg. *J. Cell. Comp. Physiol.*, **1**, 1–9.

69 Zhang, Z., Saunders, R., and Thomas, C.R. (1999) Mechanical strength of single microcapsules determined by a novel micromanipulation technique. *J. Microencapsul.*, **16**, 117–124.

70 Fery, A., Dubreuil, F., and Möhwald, H. (2004) Mechanics of artificial microcapsules. *New J. Phys.*, **6**, 18.

71 Vinogradova, O.I. (2004) Mechanical properties of polyelectrolyte multilayer microcapsules. *J. Phys.: Condens. Matter*, **16**, 1105–1134.

72 Fernandes, P., Delcea, M., Skirtach, A.G., Möhwald, H., and Fery, A. (2010) Quantification of release from microcapsules upon mechanical

deformation with AFM. *Soft Matter*, **6** (9), 1879–1883.
73 Weng, J. and Guentchev, K.Y. (2001) Three-dimensional sound localization from a compact non-coplanar array of microphones using tree-based learning. *J. Acoust. Soc. Am.*, **110**, 310–323.
74 Rogalski, A. (2003) Infrared detectors: Status and trends. *Prog. Quant. Electron.*, **27**, 59–210.
75 Jiang, C., McConney, M.E., Singamaneni, S., Merrick, E., Chen, Y., Zhao, J., Zhang, L., and Tsukruk, V.V. (2006) Thermo-optical arrays of flexible nanomembranes freely suspended over microfabricated cavities as IR microimagers. *Chem. Mater.*, **18**, 2632–2634.
76 Holmes, D.P. and Crosby, A.J. (2007) Snapping surfaces. *Adv. Mater.*, **19**, 3589–3593.
77 McConney, M.E., Anderson, K.D., Brott, L.L., Naik, R.R., and Tsukruk, V.V. (2009) Bioinspired material approaches to sensing. *Adv. Funct. Mater.*, **19**, 2527–2544.
78 Kayes, R.J. (1977) *Optical and Infrared Detectors;*, Springer-Verlag, Berlin.
79 Rogalski, A. (2003) Infrared detectors: Status and trends. *Prog. Quant. Electron.*, **27**, 59–210.
80 Nolte, C.S., Kurth, D.G., and Fery, A. (2005) Filled microcavity arrays produced by polyelectrolyte multilayer membrane transfer. *Adv. Mater.*, **17**, 1665–1669.
81 Gao, C., Leporatti, S., Moya, S., Donath, E., and Möhwald, H. (2001) Stability and mechanical properties of polyelectrolyte capsules obtained by stepwise assembly of Poly (styrenesulfonate sodium salt) and Poly (diallyldimethyl ammonium)chloride onto melamine resin particles. *Langmuir*, **17**, 3491–3495.
82 De Geest, B.G., Déjugnat, C., Prevot, M., Sukhorukov, G.B., Demeester, J., and De Smedt, S.C. (2007) Self-rupturing and hollow microcapsules prepared from bio-polyelectrolyte-coated microgels. *Adv. Funct. Mater.*, **17** (4), 531–537.
83 De Geest, B.G., Déjugnat, C., Sukhorukov, G.B., Braeckmans, K., De Smedt, S.C., and Demeester, J. (2005) Self-rupturing microcapsules. *Adv. Mater.*, **17** (19), 2357–2361.
84 Stubbe, B.G., Horkay, F., Amsden, B., Hennink, W.E., De Smedt, S.C., and Demeester, J. (2003) Tailoring the swelling pressure of degrading dextran hydroxyethyl methacrylate hydrogels. *Biomacromolecules*, **4** (3), 691–695.
85 De Geest, B.G., McShane, M.J., Demeester, J., De Smedt, S.C., and Hennink, W.E. (2008) Microcapsules ejecting nanosized species into the environment. *J. Am. Chem. Soc.*, **130** (44), 14480–14482.
86 De Cock, L.J., De Koker, S., De Geest, B.G., Grooten, J., Vervaet, C., Remon, J.P., Sukhorukov, G.B., and Antipina, M.N. (2010) Polymeric Multilayer Capsules in Drug Delivery. *Angew. Chem. Int. Edit.*, **49** (39), 6954–6973.
87 Bedard, M.F., De Geest, B.G., Möhwald, H., Sukhorukov, B., and Skirtach, A.G. (2009) Direction specific release from giant microgel-templated polyelectrolyte microcontainers. *Soft Matter*, **5** (20), 3927–3931.
88 Schmidt, S., Fernandes, P.A.L., De Geest, B.G., Delcea, M., Skirtach, A.G., Mohwald, H., and Fery, A. (2011) Release properties of pressurized microgel templated capsules. *Adv. Funct. Mater.*, **21** (8), 1411–1418.
89 Caruso, F. (2000) Hollow capsule processing through colloidal templating and self-assembly. *Chem. Eur. J.*, **6**, 413–419.
90 Decher, G. and Schlenoff, J.B. (2003) *Multilayer Thin Films – Sequential Assembly of Nanocomposite Materials*, Wiley-VCH.
91 Holt, B., Lam, R., Meldrum, F.C., Stoyanov, S., Vesselin, D., and Paunov, N. (2007) Anisotropic nano-papier mache microcapsules. *Soft Matter*, **3**, 188–190.
92 Hua, A., Steven, J.A., and Lvov, Y.M. (2003) Biomedical applications of electrostatic layer-by-layer nanoassembly of polymers, enzymes, and nanoparticles. *Cell Biochem. Biophys.*, **39**, 23–43.
93 Caruso, F., Yang, W., Trau, D., and Renneberg, R. (2000) Microencapsulation of uncharged low

molecular weight organic materials by polyelectrolyte multilayer self-assembly. *Langmuir*, **16** (23), 8932–8936.

94 Ai, H., Jones, S.A., Villiers, M.M., and Lvov, Y.M. (2003) Nano-encapsulation of furosemide microcrystals for controlled drug release. *J. Control Release*, **86**, 59–68.

95 Müller, R., Daehne, L., and Fery, A. (2007) Hollow polyelectrolyte multilayer tubes: Mechanical properties and shape changes. *J. Phys. Chem. B*, **111** (29), 8547–8553.

96 Müller, R., Daehne, L., and Fery, A. (2007) Preparation and mechanical characterization of artificial hollow tubes. *Polymer*, **48**, 2520–2525.

97 Kozlovskaya, V. and Sukhishvili, S.A. (2005) pH-controlled permeability of layered hydrogen-bonded polymer capsules. *Macromolecules*, **38**, 4828–4836.

98 Shchepelina, O., Kozlovskaya, V., Kharlampieva, E., Mao, W., Alexeev, A., and Tsukruk, V.V. (2011) Anisotropic Micro- and Nano-Capsules. *Macromol. Rapid. Commun.*, **31** (23), 2041–2046.

99 Decuzzi, P. and Ferrari, M. (2008) The receptor-mediated endocytosis of nonspherical particles. *Biophys. J*, **94** (10), 3790–3797.

100 Gratton, S.E., Ropp, P.A., Pohlhaus, P.D., Luft, J.C., Madden, V.J., Napier, M.E., and DeSimone, J.M. (2008) The effect of particle design on cellular internalization pathways. *Proc. Natl. Acad. Sci. USA*, **105** (33), 11613–11618.

101 De Geest, B.G., De Koker, S., Sukhorukov, G.B., Kreft, O., Parak, W.J., Skirtach, A.G., Demeester, J., Smedt, S., and Hennink, W.E. (2009) Polyelectrolyte microcapsules for biomedical applications. *Soft Matter*, **5**, 282–291.

102 Skirtach, A.G., Muñoz Javier, A., Kreft, O., Köhler, K., Piera Alberola, A., Möhwald, H., Parak, W.J., and Sukhorukov, G.B. (2006) Laser-induced release of encapsulated materials inside living cells. *Angew. Chem.*, **45**, 4612–4617.

103 Bendix, P.M., Koenderink, G.H., Cuvelier, D., Dogic, Z., Koeleman, B.N., Brichera, W.M., Fielda, C.M., Mahadevan, L., and Weitz, D.A. (2008) A quantitative analysis of contractility in active cytoskeletal protein networks. *Biophys. J.*, **94** (8), 3126–3136.

104 Moreno-Flores, S., Benitez, R., Vivanco, M.dM., and Toca-Herrera, J.L. (2010) Stress relaxation microscopy: Imaging local stress in cells. *J. Biomech.*, **43** (2), 349–354.

105 Schmidt, S., Helfer, E., Carlier, M.F., and Fery, A. (2010) Force generation of curved actin gels characterized by combined AFM–epifluorescence measurements. *Biophys. J.*, **98** (10), 2246–2253.

106 Delcea, M., Schmidt, S., Palankar, R., Fernandes, P.A., Fery, A., Möhwald, H., and Skirtach, A.G. (2010) Mechanobiology: Correlation between mechanical stability of microcapsules studied by AFM and impact of cell-induced stresses. *Small*, **6** (24), 2858–2862.

107 Palankar, R., Skirtach, A.G., Kreft, O., Bédard, M., Garstka, M., Gould, K., Möhwald, H., Sukhorukov, G.B., Winterhalter, M., and Springer, S. (2009) Controlled intracellular release of peptides from microcapsules enhances antigen presentation on MHC Class I molecules. *Small*, **5** (19), 2168–2176.

108 Javier, M., Kreft, O., Alberola, A.P., Kirchner, C., Zebli, B., Susha, A.S., Horn, E., Kempter, S., Skirtach, A.G., Rogach, A.L., Radler, J., Sukhorukov, G.B., Benoit, M., and Parak, W.J. (2006) Combined atomic force microscopy and optical microscopy measurements as a method to investigate particle uptake by cells. *Small*, **2** 394–400.

109 Engler, A.J., Sen, S., Sweeney, H.L., and Discher, D.E. (2006) Matrix elasticity directs stem cell lineage specification. *Cell*, **126** (4), 677–689.

110 Merkel, T.J., Jones, S.W., Herlihy, K.P., Kersey, F.R., Shields, A.R., Napier, M., Luft, J.C., Wu, H., Zamboni, W.C., Wang, A.Z., Bear, J.E., and DeSimone, J.M. (2011) Using mechanobiological mimicry of red blood cells to extend circulation times of hydrogel microparticles. *Proc. Nat. Acad. Sci. USA*, **108** (2), 586–591.

111 Nolte, M., Doench, I., and Fery, A. (2006) Freestanding polyelectrolyte films as sensors for osmotic pressure. *Chemphyschem*, **7** (9), 1985–1989.

112 Schmidt, D.J., Cebeci, F.C., Kalcioglu, Z.I., Wyman, S.G., Ortiz, C., Van Vliet, K.J., and Hammond, P.T. (2009) Electrochemically Controlled Swelling and Mechanical Properties of a Polymer Nanocomposite. *ACS Nano*, **3** (8), 2207–2216.

113 Heuvingh, J., Zappa, M., and Fery, A. (2005) Salt softening of polyelectrolyte multilayer capsules. *Langmuir*, **21** (7), 3165–3171.

114 Schoeler, B., Delorme, N., Doench, I., Sukhorukov, G.B., Fery, A., and Glinel, K. (2006) Polyelectrolyte films based on polysaccharides of different conformations: Effects on multilayer structure and mechanical properties. *Biomacromolecules*, **7** (6), 2065–2071.

115 Mamedov, A., Kotov, N., Prato, M., Guldi, D.M., Wicksted, J.P., and Hirsch, A. (2002) Molecular design of strong single-wall carbon nanotube/polyelectrolyte multilayer composites. *Nat. Mater.*, **1** (3), 190–194.

116 Tang, Z., Kotov, N., Magonov, S., and Ozturk, B. (2003) Nanostructured artificial nacre. *Nat. Mater.*, **2** (6), 413–U8.

117 Podsiadlo, P., Kaushik, A.K., Arruda, E.M., Waas, A.M., Shim, B.S., Xu, J.D., Nandivada, H., Pumplin, B.G., Lahann, J., Ramamoorthy, A., and Kotov, N.A. (2007) Ultrastrong and stiff layered polymer nanocomposites. *Science*, **318** (5847), 80–83.

# 17
# Design and Translation of Nanolayer Assembly Processes: Electrochemical Energy to Programmable Pharmacies

*Md. Nasim Hyder, Nisarg J. Shah, and Paula T. Hammond*

## 17.1
### Introduction

Over the past decade, the layer-by-layer (LbL) assembly approach [1, 2] has advanced rapidly, from the fundamental study of the thermodynamics and kinetics of construction and early incorporation of materials function, to the development of materials with true commercial potential [3], in particular in the areas of functional surface coatings and surface modification, biological and biomedical applications [4–6], passive and reactive membranes [7], and electrochemical energy [8]. The use of electrostatics to build up thin films can provide a number of advantages in these applications. When polymeric species are assembled in these systems, the high degree of interpenetration has yielded nanoscale blends that often exhibit thermal, mechanical and chemical properties that are a combination of the original materials systems. LbL thus provides a way to form homogeneous thin films, often from polymeric backbones that would not form thermodynamic blends in the bulk state without the presence of charge. Furthermore, LbL can be used to incorporate a very broad range of molecular, nano- and micro-scale elements into ultrathin films and coatings, including organic and inorganic nanoparticles, colloidal particles, nanotubes and other nanomaterials, and these materials systems can be introduced at different points during the assembly process, often leading to the ability to create complex graded, stratified or integrated architectures in a manner and with a level of control difficult to create using simple blend approaches. This capability to grow the films "layer-by-layer" essentially allows the design of materials function, and importantly, the tuning of materials interfaces throughout the thickness of the film to impact the transport and interactions of molecules, ions, and electrons. Technologists can use these capabilities to engineer new materials systems with properties inaccessible by other routes.

Perhaps one of the biggest advantages of the LbL method is the ability to process complex materials systems using water at room temperature, and to use processing methods to further extend the application, and manipulate the structure and ordering in films and coatings. By using water-based approaches, it is possible to

*Multilayer Thin Films: Sequential Assembly of Nanocomposite Materials*, Second Edition.
Edited by Gero Decher and Joseph B. Schlenoff.
© 2012 Wiley-VCH Verlag GmbH & Co. KGaA. Published 2012 by Wiley-VCH Verlag GmbH & Co. KGaA.

address the broad range of interesting biological materials, from biomacromolecules, such as proteins and nucleic acids, to actual living entities such as cells and viruses, while maintaining the activity and viability of these sensitive systems. By understanding the nature of electrostatic assembly, as well as hydrogen bonding and other complementary interactions, it is possible to design LbL films that can undergo not only controlled assembly, but disassembly under certain aqueous conditions, thus providing a path to drug delivery and release mechanisms. Furthermore, degradable or responsive polyions can be used as a means of regulating release with a broad range of control. Ultimately, the use of biological species in the multilayer area can also yield a new means of creating materials for a range of applications that include, but also go beyond, biomedicine, such as energy or reactive membrane technology.

The traditional approach for LbL of alternate dipping into water and polyelectrolyte solutions is slow, requiring typically from 5 to 20 min or longer per adsorption step. This slow process, along with some constraints on sample size and shape, had limited strong commercial interest in the method. A new and much more rapid approach – spray-LbL – makes many of these applications more approachable from a commercial standpoint due to speed, ease of application, and the ability to coat a broad range of surfaces [9–12]. By automating this method, we have found that the time to generate an LbL film can be reduced by one or two orders of magnitude, enabling the generation of multilayer films in a matter of seconds per cycle. Furthermore, spray-LbL enables manipulation of the architecture of the final materials system, with the formation of composite membranes, and the manipulation of porosity and surface area that can be gained by coating 3D porous surfaces. Because of the rapid kinetics of the assembly process, spray-LbL can, in many cases, also affect the kinetics of film growth and diffusion of components during assembly, thus leading to new and different film compositions. All of these facets of spray-LbL have led to a new and increasingly more translational range of applications, and have launched an increase in industrial interest. This approach, as well as other rapid assembly methods, will have a large impact on our ability to translate these exciting multilayer systems into everyday use.

In this chapter we address the key areas of growth that have occurred in the multilayer field, including (i) development of membranes and electrodes through the design of selectively permeable systems that can be modulated with respect to ionic and electronic transport for applications that range from water purification to fuel cells to electroactive battery electrodes; (ii) development of biomaterials systems, including thin film LbL systems that can facilitate localized release from surfaces and modulate cell adhesion and growth on or near the surfaces. Potential applications range from drug eluting medical devices to tissue engineering scaffolds for regenerative medicine and therapeutic delivery from biomedical implants. Finally, the recent development of an automated means of spray-LbL, its impact on the design and development of new devices, and the route that this method and similar techniques can have in translation will be described.

## 17.2
## Controlling Transport and Storing Charge in Multilayer Thin Films: Ions, Electrons and Molecules

### 17.2.1
### LbL Systems for Proton Exchange Fuel Cells

Polymer electrolytes are generally thin films that facilitate the transport of a given ion or ions at predetermined operating conditions. Multilayer systems provide the advantages of nanoscale blend capabilities difficult to achieve with blends, and the application of very thin uniform films as readily processed solid state electrolyte or membrane coatings that impart significant properties with relatively small amounts of polymer. With the intent of tuning the transport of ions within an ultrathin film, several multilayer assemblies have been considered, and their structure and properties examined over a range of assembly conditions, such as pH and ionic strength. The studies of proton and ion transport in LbL films include the investigation of hydrogen bond based assembly of poly(ethylene oxide) and poly(acrylic acid) [13, 14] and a range of existing weak and strong polyelectrolyte pairs [15], including organic and inorganic polymer backbones [16]. It was found that certain design rules for the generation of ionically conducting systems could be established. Higher conductivities are generally observed when LbL films are assembled with moderate to extensive amounts of charge shielding via the addition of salt or adjustment of pH solution conditions. Polyion backbones with more hydrophilic characteristics lead to increased proton or ion mobility under hydrated conditions; for this reason, thin films directly assembled with perfluorosulfonated ionomers such as Nafion® and a polycation exhibit much lower ionic conductivities than sulfonated polymers containing hydrogen bonding, and polar groups such as poly(acrylamidosulfonate-2-methyl-1-propane sulfonic acid) can yield much more promising ionic conductivities. Certain LbL candidates that exhibited ionic conductivities in the range of $10^{-4}$–$10^{-5}\,\mathrm{S\,cm^{-1}}$ at high humidity conditions were first introduced into composite membranes for a hydrogen fuel cell in an early demonstration of the viability of the LbL systems in fuel cell applications [17]. These earlier examinations indicated that the ion hopping transport mechanism plays a role in these systems with water-free electrolyte or conditions of low humidity, whereas a Grotthus mechanism of protonic transport is relevant at high humidities or fully wet aqueous conditions [14, 16, 18].

Our group has recently developed a range of multilayer systems that incorporate hydrophilic polymers using electrostatic and hydrogen bonding mechanisms and have shown increases in ionic conductivity of 3 or 4 orders of magnitude [19] from these earlier systems. As can be seen in Figure 17.1, by using a materials design approach, we developed a set of highly selective LbL thin films for proton exchange membranes in direct methanol fuel cells [19, 20]. Sulfonated poly-p-phenyleneoxide (sPPO) was chosen for its inherently low methanol permeability (a common feature of aromatic polyethers) and its highly acidic sulfonic acid functionality.

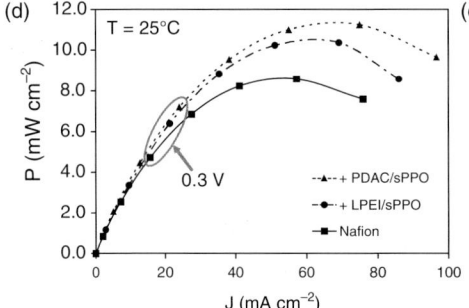

**Figure 17.1** Chemical structures of polyelectrolytes used to assemble multilayer films. (a) PDAC, sPPO, (b) free-standing film of PDAC/sPPO coated on Nafion, (c) SEM cross-sectional image of Nafion 1135 coated with 50 layer pairs of PDAC/sPPO. Nafion 1135 is the amorphous region on the right side, and the LbL film is the lighter band adhered to the surface of the Nafion membrane. Note the sharp transition at the Nafion/LBL film interface, (d) Power curves of single MEA DMFCs comparing unmodified Nafion devices with Nafion membranes coated with LBL films of LPEI/sPPO and PDAC/sPPO, and (e) table comparing the methanol permeability and protonic conductivity of Nafion and two different multilayer systems. Reproduced from Ref. 20.

Because this polymer is fully water soluble in its highly sulfonated form (>85% sulfonation), it had not been fully exploited in terms of ion exchange capacity in traditional cast film fuel cell membranes, but the highly sulfonated version of the polymer is an ideal candidate for the formation of LbL films. The sPPO polymer was synthesized and layered with a series of polycations, including poly(diallyl dimethyl ammonium chloride (PDAC) which is an inexpensive and commercially available polymeric quaternary ammonium salt. It was found that PDAC/sPPO LbL thin films exhibited high ionic conductivity that could be directly tuned by manipulation of the salt concentration in the polyelectrolyte solutions used for assembly. As shown in Table 17.1, the addition of salts in all cases leads to higher ionic conductivity; the mechanism of increased ionic transport appears to be increased shielding of the charged groups on the polymer chains, which ultimately results in lower ionic crosslink density, thicker deposited monolayers in each deposition step, and loopier polymer chain conformations on the adsorbed surface. The effective ionic crosslink density is systematically decreased as fewer ionic groups take part in ion

**Table 17.1** Ionic (proton) conductivity and methanol permeability values of various electrostatic LBL films along with the values obtained from a Nafion 1135 film [20].

| Membrane | Conductivity[a] (mS cm$^{-1}$) | Permeability, $P \times 10^8$ (cm$^2$ s$^{-1}$) |
|---|---|---|
| Nafion 1135 | 97.8 | 282 |
| PDAC/sPPO | 35.3[b] | 2.18 |
| PAH/SPPO | 4.23[c] | 0.57 |
| LPEI/SPPO | 2.12[d] | 1.38 |
| P4VP/SPPO | 1.65[c] | 0.84 |
| LPEI/PAA | 0.01[e] | N/A |

a) At 98% relative humidity.
b) 0.5 M NaCl in the sPPO solution.
c) Assembly pH 2.
d) Assembly pH 1.5.
e) D. M. DeLongchamp, P. T. Hammond, *Chem. Mater.* 2003, **15**, 1165.

complexation with the oppositely charged polymer. Interestingly, the selective addition of salts in the sPPO bath, while leaving the PDAC bath salt-free, leads to even greater increases in protonic conductivity. This result is believed to be due to selective shielding of sPPO sulfonic acid groups, which leads to thicker, denser deposited monolayers of sPPO with respect to PDAC, and could ultimately lead to a higher overall content of shielded sPPO groups. The net results are high ionic conductivities that closely approach the ionic conductivity of Nafion at 100% RH, ranging up to 70 mS cm$^{-1}$ in optimal film systems. In addition, the studied LBL films are highly methanol resistant with permeability values two orders of magnitude lower than Nafion [19]. Coating Nafion membranes with just 2 or 3 layer pairs of PDAC/sPPO – the equivalent of less than 100 nm – resulted in the improvement of DMFC performance by over 50% compared to unmodified Nafion, indicating the power of coupling LbL systems with other high performance existing commercial membranes to achieve a unique combination of properties [20].

One of the advantages of the LbL approach is the ability to create interfaces with a range of different material components. The membrane-electrode assembly (MEA) requires a combination of ionic conductivity and electronic transport, as well as catalytic activity. The assembly of thin-film electrodes for fuel-cell membranes is significantly more demanding than the formation of more traditional electrodes because of the need to maintain these two different kinds of conductivity and access to catalytic surfaces. Layer-by-layer technology that utilizes polyelectrolytes, colloids and nanomaterials is providing new ways to assemble electrochemical devices by enabling the combination of a broad range of organic and inorganic elements within a conformal film architecture [21, 22]. By taking advantage of the ability to develop films with microscopic as well as molecular constituents, we introduced the concept of a complete "soft" membrane electrode assembly (MEA) or "soft fuel cell" based on the LbL technique [23, 24] and its use with traditional fuel-cell devices using forms of amorphous carbon black dispersed in a polyanion or polycation solution to yield charged, stable carbon particles in aqueous suspension.

The multilayer thin film carbon–polymer electrodes exhibit high electronic conductivities of 2–4 S cm$^{-1}$, and their porous structure accommodates ionic conductivities in the range of $10^{-4}$–$10^{-3}$ S cm$^{-1}$ at the high relative humidity operating conditions of a typical fuel cell. These thin electrodes showed remarkable stability towards oxidizing, acidic, or delaminating basic solutions. It was observed that an LbL PDAC/PAMPS/carbon–platinum electrode assembled on a porous stainless steel support yields an open-circuit potential similar to that of a pure platinum electrode. Although the power delivered by the LbL MEA systems must still be optimized, it is an important step forward for LbL technology and the development of thin-film fuel cells for portability.

## 17.2.2
## Charge Storage Using LbL Assemblies

### 17.2.2.1 Carbon Nanotube LbL Electrodes

Multilayer films can also be designed to contain nanomaterials such as polymeric nanofibers, metal or metal oxide nanoparticles and functionalized carbon nanotubes, which retain their reactivity within the film, generating the ability to obtain highly nanoporous electrode morphologies suitable for energy storage applications [25]. Typically, batteries can store a large amount of energy using chemical redox or intercalation reactions; however, batteries cannot be discharged rapidly to gain high power. On the other hand, traditional and electrochemical capacitors rely on pseudocapacitive charge storage and the build-up of the double layer on electrode surfaces; these can deliver energy rapidly for high power, but are limited in the amount of energy that can be stored. In recent years, extensive research has been pursued to develop electrochemical energy storage technologies that can deliver both high energy and power simultaneously. This goal is key to the next significant advances in energy storage, providing promise for automotive and transportation applications, and other key power generation applications to meet future energy needs that cannot be addressed with the current top lithium ion batteries.

Our recent work has demonstrated the design of novel battery cathodes generated from functionalized multiwalled carbon nanotubes (MWNTs) based on LbL assembly [26] that meet the challenge of high energy and power described above. The films are constructed from MWNTs that are oxidized to present a high density of carboxylic acid groups on the outer wall, and a net zeta potential of $-40$ to $-50$ mV. By coupling a large excess of ethylene diamine, MWNTs with similar positive charge can be generated as well. When these two sets of nanotubes are alternated, they exhibit linear growth profiles and form highly uniform thin films. Following a two-step heat treatment of the films at 150 °C followed by 300 °C, they exhibit a high electronic conductivity as well as mechanical stability, enabled by the formation of amide groups between nanotubes. These novel all-carbon multilayer films form an interpenetrating network structure, and the resulting electrodes have a well-developed porous structure that facilitates rapid ion diffusion. As shown in Figure 17.2a, these electrodes allow better utilization of the electrochemically active surface area due to the absence of insulating polymer binder [26]. They exhibit very high packing

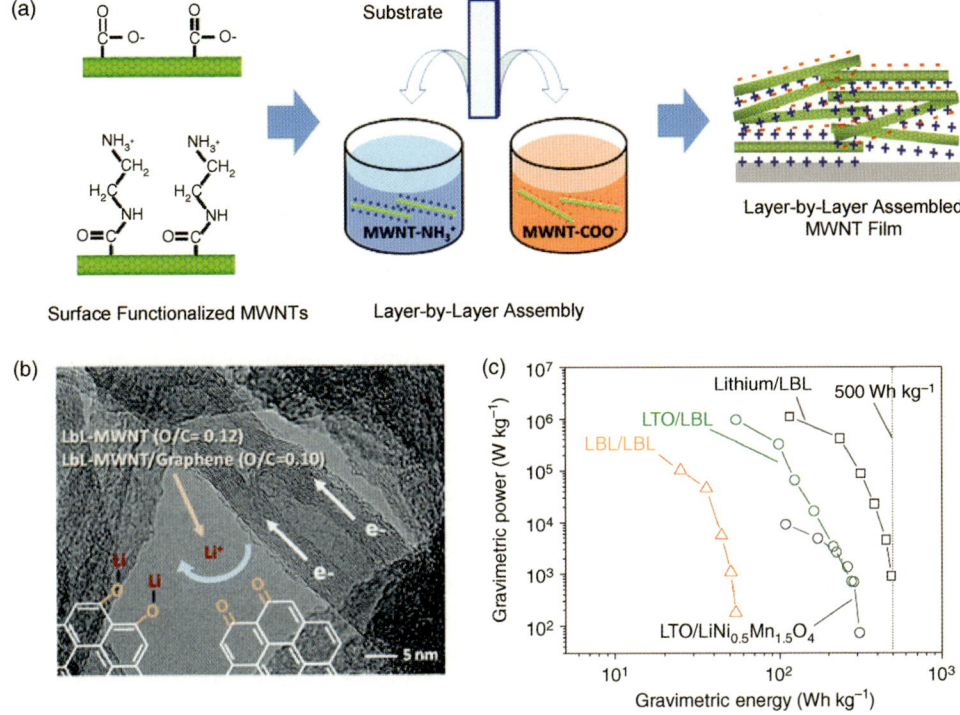

**Figure 17.2** (a) Schematic of LbL assembly of functionalized MWNTs. (reproduced from Ref. 26,31), (b) Schematic of the energy storage mechanism on a TEM image of the LbL-MWNT electrode. White arrows indicate conduction channels of electrons though graphite layers, and (c) Ragone plot for Li/LBL-MWNT (black squares), LTO/LBL-MWNT (green circles), LTO/LiNi0.5Mn1.5O4 (gray circles) and LBL-MWNT/LBL-MWNT (orange triangles) cells with 4.5 V versus Li as the upper-potential limit. The thickness of the LBL-MWNT electrode was 0.3 mm for asymmetric Li/LBL-MWNT and LTO/LBL-MWNT, and 0.4 mm for symmetric LBL-MWNT/LBL-MWNT, (reproduced from Ref. 27).

density (~70%) and provide a high volumetric capacitance (130–150 F cm$^{-3}$) in acidic aqueous electrolyte; however, the true potential for these systems can be observed in their performance as electrodes in lithium battery cells.

When examined for lithium storage capability, we found that the LbL MWNT films can act as cathodes that exhibit the high storage characteristics of a traditional lithium ion battery cathode material, but have the high power of electrochemical capacitors due to a combination of double layer capacitance and Faradaic reactions of surface functional groups. The high energy densities of MWNT electrodes arise from the reactions between Li$^+$ ions and surface functional groups on the MWNT electrodes (rendering high pseudocapacitance), which can be accessed in the voltage window from 1.5 to 4.5 V vs. Li in a nonaqueous electrolyte [27], as illustrated in Figure 17.2b. Batteries consisting of a MWNT positive electrode and either Li metal or a Li$_4$Ti$_5$O$_{12}$

negative electrode exhibit gravimetric energy ($Wh\,kg^{-1}$) densities that are $\sim 5\times$ and $\sim 10\times$ higher than asymmetric and conventional electrochemical capacitors, respectively, but with comparable gravimetric power and cycle life. The MWNT electrodes ($\sim 200\,Wh\,kg^{-1}$ at $100\,kW\,kg^{-1}$) also have higher energy densities in comparison to recently reported state-of-the-art high-power Li storage technologies at high power demands ($\sim 30\,Wh\,kg^{-1}$ at less than $10\,Kw\,kg^{-1}$) [27]. Moreover, the volumetric energy densities of these MWNT electrodes ($\sim 160\,Wh\,l^{-1}$ at high power $\sim 80\,kW\,l^{-1}$) is much higher than that of high-power $LiFePO_4$ electrodes (estimated $\sim 70\,Wh\,l^{-1}$ at $65\,kW\,l^{-1}$) [28]. It should be noted that the voltage window upon which most energy can be obtained from conventional Li storage materials, such as $LiFePO_4$ [27], upon discharge at high power demands is between $\sim 3$ and $\sim 2\,V$ vs. Li, which is similar to the characteristics of MWNT electrodes (Figure 17.2b).

The high energy densities of MWNT electrodes arise from the reactions between $Li^+$ ions and surface functional groups on the MWNT electrodes (rendering high pseudocapacitance), which can be accessed in the voltage window from 1.5 to 4.5 V vs. Li in a nonaqueous electrolyte [27]. The ability to store charge via different mechanisms (both Faradaic and double layer capacitance) in different power regimes without capacity loss or degradation with cycling lends uniquely high versatility to these nanostructured electrodes and opens up a new direction for the development of high-energy, high-power electrodes that bridge the performance gap between batteries and electrochemical capacitor devices. These MWNT electrodes showed no capacity loss up to 2500 cycles at high rates ($\sim C$ rate above) indicating excellent cycling stability of the surface Faradaic reactions and no degradation of the nanostructured electrode.

#### 17.2.2.2 Carbon Nanotube/Conjugated Polymer System

In addition to the all-MWNT based multilayer electrodes, our group has been actively pursuing asymmetric capacitors based on the assembly of MWNTs with other nanomaterials that are psuedocapacitive, or provide other means of Faradaic reaction, including, conjugated polymeric nanofibers, metallic oxide nanoparticles, and graphene oxide. Recently, we fabricated ultrathin conformal films of polyaniline nanofibers (PANi) and oxidized, acid-functionalized MWNTs. For the synthesis of positively charged PANi fibers [29], rapid polymerization in protonic (HCl) media was used, which is a new and powerful approach for a comparatively large amount of aquoeous suspension of nanofibers [30]. Negatively charged MWNTs have been created via chemical functionalization of exterior walls with carboxylic acid groups [26] yielding anionic MWNT. The LbL-PANi/MWNT electrodes show a higher porosity than all carbon MWNT electrodes but with a lower density (Figure 17.3a and b). The as-prepared LbL-PANi/MWNT films were heat treated (at $180\,^\circ C$ for 12 h in vacuum) to improve the mechanical stability of the films via random crosslinking of functional groups. The thermal treatment creates a crosslinked nanostructure (Figure 17.3c) in which the interdiffusion of oligomeric polycationic species becomes greatly hindered, thereby improving the stability of the multilayered films [30]. The LbL PANi/MWNT volumetric energy density in Li non-aqueous cells was higher than the all-MWNT electrodes (as high as $630\,Wh\,l^{-1}$, Figure 17.3d) and these cells

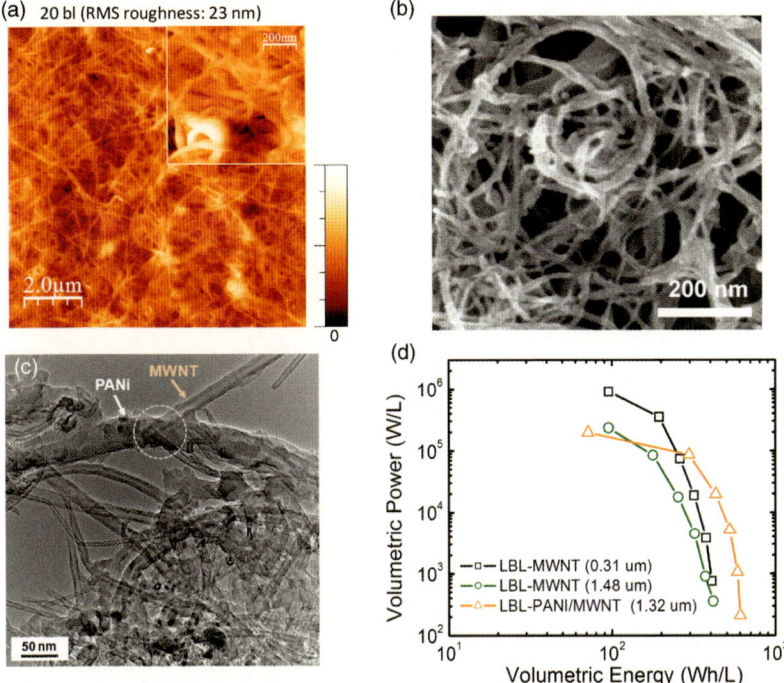

**Figure 17.3** (a) Atomic force microscopy (AFM) height images of LbL-PANi/MWNT thin films. Main image is 10 μm² of LbL-(PANi/MWNT)$_{20}$, inset shows image of 1 μm², (b) scanning electron microscopy (SEM) image of LbL film, (c) high-magnification TEM image of LbL-PANi/MWNT electrode. White and orange arrows in (c) indicate PANi nanofiber and MWNT, respectively. Gray circle in (c) shows one of contact points between PANi nanofibers and MWNTS within the LbL-PANi/MWNT films, and (d) Ragone plot (per electrode volume) comparison for the LbL-PANi/MWNT and the LbL-MWNT electrodes. Reproduced from Ref. 30.

retain comparable power capability and cycle life due to their binder-free, highly homogeneous interpenetrated network morphology. The improved energy metrics are a result of the greater density of the composite material (~1.5 g cm$^{-3}$) and its increased pseudocapacitance (contributions from the reversible redox reactions of PANi around 4 V). There are some opportunities to optimize the number of surface functional groups (increasing pseudocapacitance) and packing density (increasing overall density) of these electrode materials for applications in electrochemical energy storage and conversion devices. It is also possible to use electrochemical deposition methods to deposit metal oxide nanoparticles directly onto the surfaces of nanotubes in the all-MWNT asymmetric electrodes, including RuO$_2$ and MnO$_2$ [25, 31]. These electrodes also show very high capacitance due to the intrinsic highly capacitive nature of the nanoparticles. The nanoparticles have very low conductivities that make them less effective as capacitors alone; however, the presence of the MWNT network makes for rapid transport of electrons and ions within the electrode while achieving much higher energy storage.

## 17.2.3
### Composite Systems

The LbL self-assembly approach enables the juxtapositioning of organic and inorganic nano- and microstructures; it is thus a tool for the integration of materials for hybrid structures, such as light-absorbing and semiconducting nanomaterials, essential for the fabrication of electrodes using a variety of conventional and unconventional materials [1, 12, 32–34]. For example, these hybrid nano-assemblies can act as charge storing or light absorbing semiconducting composites for Li-ion batteries, photovoltaic cells and electrochromic cells [15, 34–36].

#### 17.2.3.1 Virus Battery

One of the key advantages of the LbL method is that the use of water as a medium allows the incorporation of biologically derived materials systems. One case in which this is particularly advantageous is the use of genetically engineered viruses developed in the Belcher lab (MIT) as templates for a variety of metals and metal oxides. For smaller, lightweight flexible batteries, the selection and assembly of materials are key issues to fabricate nanostructured devices through spontaneous non-covalent interactions [32]. Our group has used these genetically altered viruses to synthesize and assemble nanowires (Figure 17.4a) that are subsequently templated with cobalt or cobalt oxide [35]. By exploiting the affinity of the modified M13 virus, nanowires of $Co_3O_4$ were nucleated atop a multilayer polymer electrolyte of linear polyethylenimine (LPEI)/polyacrylic acid (PAA). It is evident from Figure 17.4b that the virus forms a 2D liquid crystalline highly packed structure that results from entropic ordering of the electrodes. This order was achieved through the interdiffusion of the LPEI chains, and its preferential complexation with underlying PAA, thus resulting in an ejection of the more weakly charged virus to the top surface following subsequent adsorption cycles [37]. By harnessing the native assembly behavior of these systems, we can manipulate the interactions between the viruses and LPEI/PAA multilayers, controlling their order and surface density. The virus-templated thin film electrode can store and release Li-ions reversibly [35]; the high surface area provided by the nanowire structures yields cathodes with 60 to 100% higher capacitance than typical $LiCo_3O_4$ systems that are commercially available. Free-standing films achieved by lifting off these films yield electrode thin films that can be rolled and bent readily, demonstrating excellent flexibility and stability. As can be seen in Figure 17.4c, at high current rates, the $Co_3O_4$ electrode exhibits high specific capacity [35] essential for microbattery applications. We have demonstrated that self-assembled nanomaterials based on biological principles can be applied for battery components for high specific capacity and rate capability [35, 38].

#### 17.2.3.2 DSSC LbL Electrodes

The next generation of inexpensive energy conversion and storage devices entails research on novel materials for efficient photovoltaic devices. For solar energy conversion, over the last decade, significant research efforts have been geared toward dye-sensitized solar cells (DSSCs) based on mesoporous $TiO_2$ that have shown the

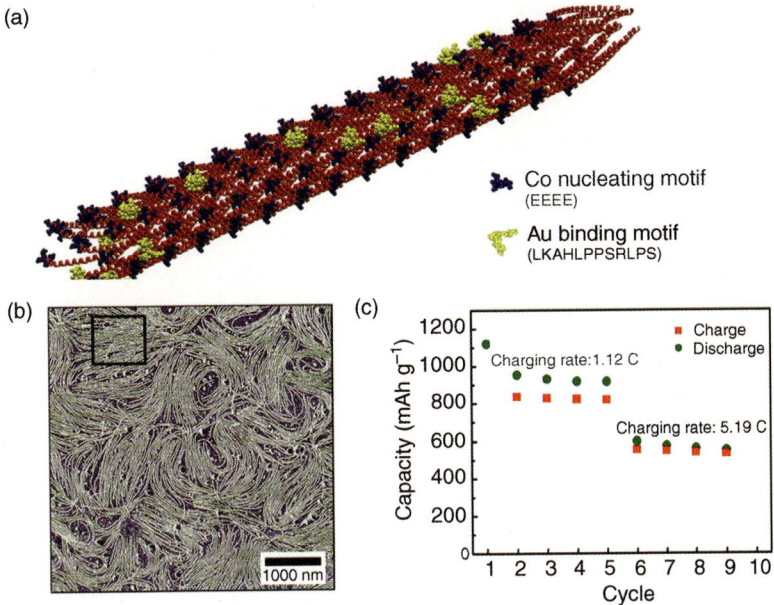

**Figure 17.4** Characterization of the hybrid nanostructure of Au nanoparticles incorporated into $Co_3O_4$ nanowires. (a) visualization of the genetically engineered M13 bacteriophage viruses. P8 proteins containing a gold-binding motif (yellow) were doped by the phagemid method in E4 clones, which can grow $Co_3O_4$, (b) Two-dimensional assembly of $Co_3O_4$ nanowires driven by liquid crystalline ordering of the engineered M13 bacteriophage viruses. Phase-mode atomic force microscope image of macroscopically ordered monolayer of $Co_3O_4$-coated viruses. The Z range is 30°, and (c) capacity for the assembled monolayer of $Co_3O_4$ nanowires/Li cell at two different charging rates. Reproduced from Ref. 34,35,38.

potential to replace silicon as an inexpensive semiconductor material [39]. Multilayer thin films can be designed to contain metal or metal oxide nanoparticles, such as nanostructured titanium dioxide, which retain their reactivity within the multilayer assemblies, and can be used as both photocatalytic and charge transport materials for DSSCs [36]. Furthermore, the development of a more advanced approach to generating conformal, adaptive solid state electrolytes would enable the generation of flexible DSSCs and a solar cell that can be processed using an all-LbL water-based approach, to yield low cost stable photovoltaics.

When LbL thin films constructed of weak polyelectrolytes are assembled under conditions in which a polyion is highly charged, and subsequently exposed to a pH for which the charged state changes dramatically, significant swelling and the formation of highly porous thin film morphologies (Figure 17.5) are observed [40, 41]. These structures can be formed on a broad range of surfaces, and can be used to template a range of organic or inorganic materials. We have previously demonstrated the generation of DSSC cells based on a liquid phase deposition of a $TiO_2$ layer using a porous LbL thin film of LPEI/PAA as a template. The resulting titania film can be annealed at high temperature to achieve the active anatase crystalline phase.

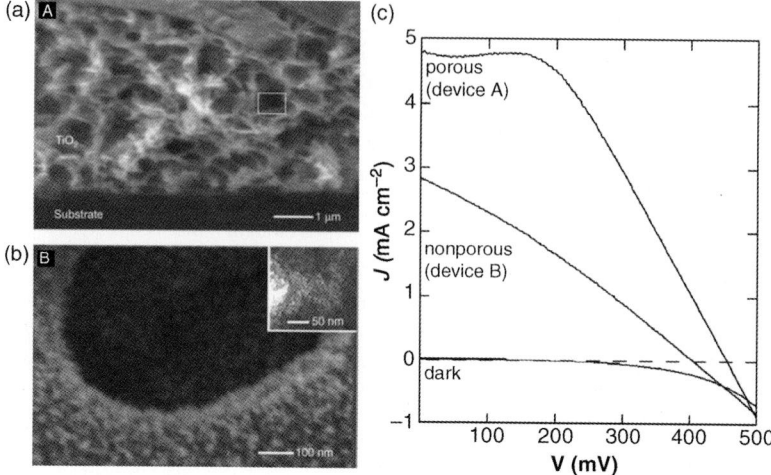

**Figure 17.5** (a) Cross-sectional SEM micrograph of a TiO$_2$ thin film after removal of a (LPEI/PAA)$_{12}$ template, (b) a close-up view of the area indicated in (a), and (c) photovoltaic performance of the device consisting of a dye-sensitized porous TiO$_2$ layer and a (PEO/PAA)$_{45}$ multilayer composite electrolyte. Reproduced from Ref 36.

Following the creation of the titania electrode, a photocharge generator dye was adsorbed onto its surfaces, followed by the deposition of an ionically conducting layer of hydrogen-bonded PEO/PAA. A unique advantage of this approach is the ability of LbL to fill in even very fine gaps; thus the solid state electrolyte effectively presents a very effective pore-filling capacity [42]. The ionic conductivity of such systems can be further enhanced with the addition of short oligoethyleneoxide diacids (OEGDA) as plasticizers. The device showed an impressive performance with an open-circuit voltage(Voc) of 456 mV and short-circuit current of 4.8 mA cm$^{-2}$. Improvements in performance are possible with more advanced multilayer conducting materials.

We have also used other aspects of LbL microfabrication to design titania electrodes with nanoporous structure and micron-scale patterned topography. We used polymer-on-polymer stamping techniques to generate three-dimensional periodic features of the high surface titania, with the goal of increasing light absorption and yield through a constructive scattering effect, and found an optimum efficiency when the pattern size approached the wavelength of visible light. It was observed that the conversion efficiencies were nearly 10 times better for the patterned TiO$_2$ cell compared to a similar device without patterning [39].

### 17.2.3.3 Multicolor Electrochromic Devices

The emerging demand for high contrast, lightweight and flexible displays is an extremely active research field due to their widespread application, from automobiles to building structures to display banners. Electrochromic coatings (ECs) are switchable thin-films applied to glass or plastic that can change appearance reversibly from a clear to a dark tint when a small DC voltage is applied. Compared to other contemporary technologies, an electrochromic device allows a single electrochromic pixel to produce multiple colors in addition to white, depending on the applied potential. LbL assembly allows the incorporation of aqueous-based organic/inorganic materials in these hybrid nanocomposite films to achieve superior contrast and fast switching speeds between various hues with unlimited compositional variations. Our group has demonstrated the successful fabrication of multiple-color electrochromic electrodes using LbL thin films of poly(aniline) (PANi) and Prussian Blue (PB) nanoparticles [43–45]. PB nanocrystals contain potassium inclusions that can dissociate and form stable aqueous dispersions; the negatively charged surfaces of the nanocrystals support their incorporation with cationic PANi using LbL assembly. The multilayer organic/inorganic thin films of PANi/PB films display multiple hues, as shown in Figure 17.6a; in the reduced state (−0.2 V), PANi and PB are both nearly

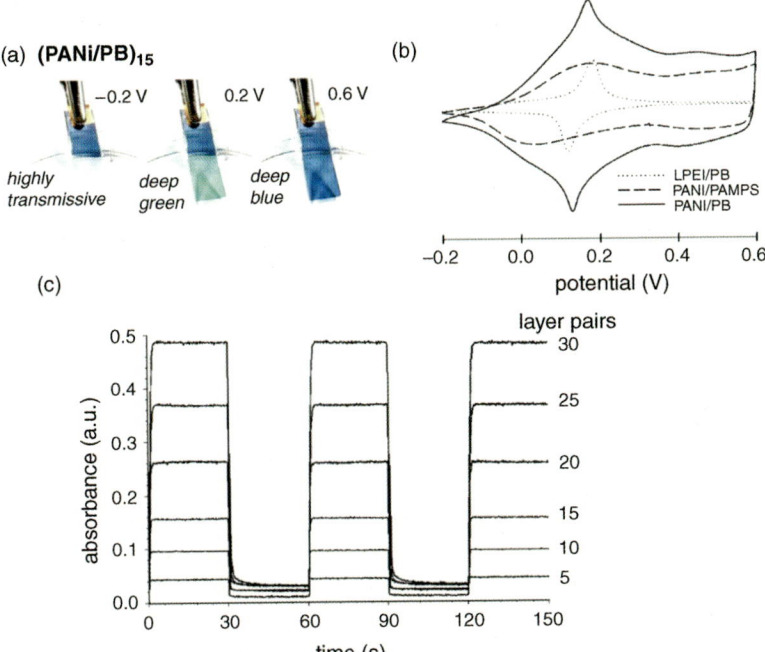

**Figure 17.6** (a) Photographs of the electrochromism of a 15 layer pairs PANi/PB film and the origins of multiple color electrochromism in this dual electrochrome composite. Film was photographed immersed in an electrochemical cell; the electrolyte meniscus is visible on the upper portion of the film. Photographs were collected after a 30-s equilibration at the noted potential. (b) Comparison of cyclic voltammetry of several LBL composites: the LPEI/PB composite exhibits only PB electrochemistry, the PANI/poly (2-acrylamido-2-methyl-1-propanesulfonic acid) (PANI/PAMPS) composite exhibits only PANI electrochemistry, and the PANI/PB composite exhibits both. Scans shown were at a rate of 25 mV s$^{-1}$, and all potentials are vs. K-SCE; the vertical axis is scaled current. (c) The absorbance response at 750 nm of PANi/PB films subjected to a square-wave switching potential. Switching occurred between −0.2 V (low absorbance) and 0.6 V (high absorbance) for 30 s at each potential. Reproduced from Ref. 43,44.

colorless, in the intermediate state (0.2 V), the color is green due to the incomplete oxidation of PB, and in the fully oxidized state(0.6 V) of PB and PANi, the electrode becomes a dark navy blue color. The distinct and noninteracting contributions from the electroactive polymer, PANi and the inorganic nanocrystals, PB are both fully electrochemically accessible, as evident from the oxidation and reduction peaks of the PANi and PB in Figure 17.6b. In addition to the high contrast multiple-hue, the optical switching of different hues appears to be extremely fast (less than 1.5 s) from the sharp appearance of the spectral response to a square-wave potential at a wavelength of 750 nm. The fast optical switching originates from the presence of electroactive conducting PANi in the LbL PANi/PB nanocomposite where the nanostructured high packing density of electrochromic materials contributes to the superior contrast (Figure 17.6c).

By modulating the film composition and morphology LbL-assembly can be applied for the design of multiple-hue flexible nanocomposites in a single electrochemical cell, capable of displaying any visible color on demand. This simple multilayer self-assembly based intermixing of polymer and inorganic nanomaterials provides an excellent platform to make lightweight, flexible high contrast display applications.

## 17.3
### LbL Films for Multi-Agent Drug Delivery – Opportunities for Programmable Release

One of the key needs for drug delivery is the ability to release therapeutics for pre-determined or sustained time periods at specific locations within the body. The development of drug-coated implants such as stents has provided the evidence that controlled release of drugs to targeted areas provides a route to specialized medical treatment relevant to specific tissues or organs on an as-needed basis. By providing localized delivery, it is possible to limit the exposure of the drug to other parts of the body, limiting side-effects, while increasing potency at the desired site. The electrostatic assembly of charged species enables the design of multilayer films that release drugs based on combinations of several different principles of multilayer assembly and disassembly. Drug molecules with charged functional groups or hydrogen-bonding capability are incorporated into LbL thin films via adsorption to groups of complementary functionality. The total drug load may be controlled by simply tuning the number of layers of drug incorporated into the film. These drugs may then exit the film by diffusion out from the LbL matrix, ejection of the charged species due to film swelling mediated by the change in charge density of polyions, or the destabilization of the multilayer film on exposure to changes in pH or ionic strength. Our laboratory has had a particular interest in using passive and active stimuli response to achieve drug release in a more controlled fashion. These approaches include the use of hydrolytically degradable polyelectrolytes, as well as hydrogen bonding and even electrochemical stimulus as a means of controlling release of the drug. Multilayer systems can be made susceptible to each of these stimuli, and sometimes combinations of stimuli, with relative simplicity by manipulating the charge within the layers through the introduction of different conditions. The water-based LbL process allows the incorporation of a range of biologically derived therapeutics, and the combination of multiple kinds of drugs – from small molecules and nucleic acid plasmids to large recombinant proteins – is feasible using this approach. This is a particularly useful aspect of multilayer assembly, and one that offers a great deal of promise for the development of biomedical implant coatings, regenerative tissue scaffolds, and on-demand drug release systems that can be locally embedded.

### 17.3.1
### Passive Controlled Release

By intentionally selecting a hydrolytically degradable polyion, it is possible to design LbL thin films that undergo controlled chemical breakdown at physiological conditions in aqueous media; degradation of these polyelectrolyte multilayer films is abiotic, that is, independent of any external biological action. Release of therapeutic molecules from such systems is not dependent on the cellular mechanisms of breakdown, such as the generation of specific proteases or other enzymes, which may not always be present in sufficient amounts in a given tissue or organ of interest. In collaboration with David Lynn, we demonstrated the approach of controlled release of materials from a hydrolytically biodegradable polyelectrolyte thin film using a poly(β-amino ester), as shown in Figure 17.7 [46].

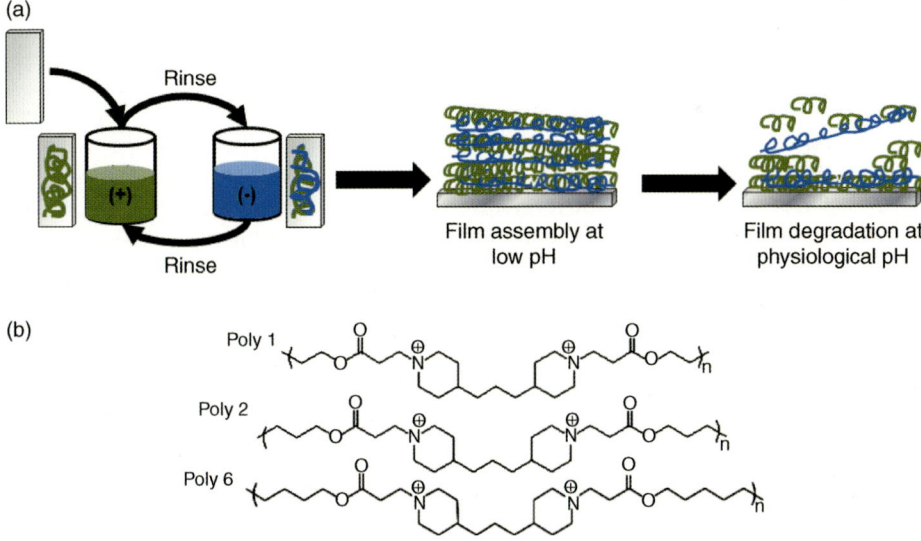

**Figure 17.7** (a) Construction of degradable, layered thin films via LbL deposition of hydrolytically degradable polycations and nondegradable polyanions; (b) variations of the degradable poly(β-amino ester), a polycation. Addition of methyl end groups increases the hydrophobicity of the polymer, allowing tunable degradation of these polymers. Adapted from Ref 46.

The poly(β-amino ester) family described here has a tertiary amine group conferring net positive charge to the polymer that can be designed to exhibit a range of degradation rates depending on chemical composition, with relatively slow degradation rates at acidic pH which enables the construction of films at pH of 6.0 or lower. In early work Poly1 was alternated with common polyanions, including sulfonated polystyrene (SPS) and poly(acrylic acid) (PAA). Vasquez *et al.* observed that the film degraded in a top-down manner and the choice of polyanion affected the rate of film degradation at pH 7.4. The observations made with this early model system suggested that multi-agent drug delivery with specifically tailored drug release profiles might be possible by controlling the architecture of the multilayer films, through specific manipulation of interactions between components of the film as well as controlling the order of introduction of components. By selection of different components, it is possible to incorporate small hydrophilic charged and neutral hydrophobic molecular drugs into the LbL layers using a range of approaches, as well as proteins and other biomacromolecules such as polysaccharides. Thus, nanoscale films with complex architectures containing therapeutic drug molecules can be conformally coated on various geometries and made to adhere to several different surface chemistries. The following sections highlight each of these therapeutic types and the challenges specific to each kind of drug component. Finally, the impact of combining these different drug types together for multifunctional release of drug molecules is discussed, as well as a means of controlling the release profile of each drug independently, for co-release and sequential release of the film components, an approach that is ideally suited to the LbL method, and is of interest for a number of therapeutic applications.

#### 17.3.1.1 Release of Proteins

The ability to entrap proteins and subsequently release them in a controlled manner has led to new treatments for a number of diseases [47]. Biologically derived drug molecules, commonly known as "biologics", are some of the fastest growing therapeutic areas, due to their promise in addressing chronic disease, cancer, wound healing, inflammation, and stem cell and regenerative technologies where proteins or small peptides act as signaling molecules to modulate cell behavior and response. Several new therapies have been developed around the concept of administration of biologics. Therapeutic monoclonal antibodies that specifically target the tumor microenvironment have found application in the clinic, such as bevacizumab, cetuximab and trastuzumab [48]. More recently, application of PDGF-BB has been approved by the FDA to address chronic wound healing [49]. The single biggest issue with most biologic formulations is the instability of the proteins themselves. Proteins are prone to denaturation when exposed to non-physiological pH conditions, solvent, or heat – conditions typical of drug formulation using traditional degradable polymers. Proteins are also susceptible to breakdown in the bloodstream or on exposure to plasma in the presence of a number of proteases, particularly in inflamed or hypoxic regions typical of injured sites on the body. It is particularly challenging to maintain the activity of a protein during its long-term release. For this reason, depot release approaches are somewhat limited, as a large amount of protein is needed to maintain a reasonable efficacy over time. A typical means of protein delivery involves direct bolus injections that yield a greatly limited effect because the protein becomes diluted and cleared or degraded rapidly, thus requiring super-physiologic doses, often at multiple time points. In addition to sharply reducing the frequency of injections, controlled-release formulations of proteins offer protection from *in vivo* degradation, diminished toxicity, improved patient comfort and compliance, localized delivery to a particular site, and more efficient use of the drug, resulting in lower dosages.

Proteins are highly organized, complex structures that must maintain structural and chemical integrity to function. Many formulating methods can jeopardize this by exposing proteins to potentially damaging conditions, such as aqueous/organic interfaces, elevated temperatures, vigorous agitation, hydrophobic surfaces, and detergents. Upon administration, the solid protein becomes hydrated and exposed to physiological temperature within the delivery device for long periods of time. At this stage, the polymer matrix microenvironment and the products of polymer degradation can also reduce protein stability. Thus, proteins are exposed to a variety of stresses over the entire life of a controlled release system, including formulation, long term storage, and release *in vivo*.

Particularly relevant to protein delivery, LbL deposition occurs at mild aqueous processing conditions which preserve fragile protein activity. Because multilayer films are highly conformal and can serve as coatings, it is feasible to coat biomedical devices, implantable scaffolds, and implants with coatings that are nanometers to microns in thickness and deliver significant and biologically relevant quantities of therapeutic protein. Furthermore, LbL can be compatible with existing surface patterning and templating techniques [50–53], opening interesting avenues to explore the cellular response to protein exposure in localized micro-environments.

To demonstrate that proteins can be released in a gradual and controlled fashion through the use of a hydrolytically degradable polymer, we chose the common protein, chicken egg lysozyme, as a model protein. This protein is relatively small and positively charged, with a molecular weight of 14.4 kDa, and pI $\sim$ 9 [54]. The films were constructed in a tetralayer architecture, in which the protein and the positively charged poly($\beta$-amino ester) Poly1 were alternated with a negatively charged biomacromolecular species such as a charged polysaccharide; for example, heparin sulfate, chondroitin sulfate, hyaluronic acid, and dextran sulfate have all been used to construct such tetralayer systems. We demonstrated that these films can be designed to tune and sustain release of active protein quantities from micron scale conformal thin films over periods of several days to several weeks at physiological temperature. The sustained release is of particular interest for proteins, which had often been limited to a few hours in previous systems based on the out-diffusion of proteins, which are often small and globular in size and shape and thus able to diffuse rapidly. For the case of lysozyme, it was shown that release profiles can be achieved over a span of 2 to 34 days. Furthermore, the advantage of the LbL method for drug delivery is that the drug makes up an entire adsorbed layer, yielding systems for which the drug can make up large fractions of the total film composition. In particular, lysozyme quantities on the order of $mg\,cm^{-2}$ were released from these thin coatings, with loadings that scale linearly with number of film layers; control over loading and

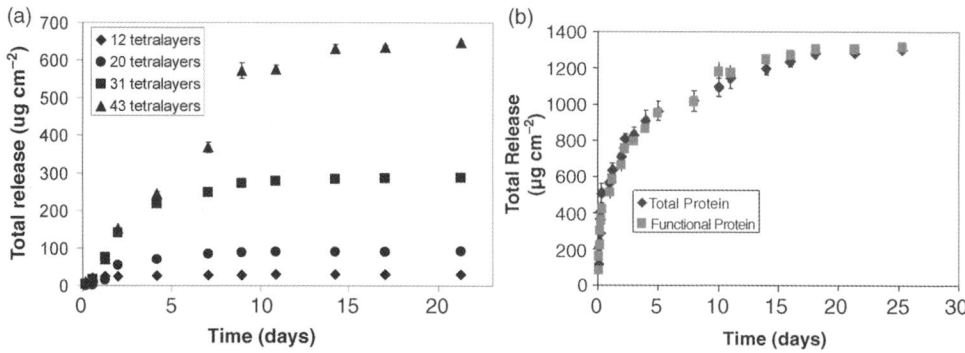

**Figure 17.8** (a) Normalized lysozyme release profiles at 37 °C (Poly1/heparin/lysozyme/heparin)$_n$. (b) The release profile demonstrates a (Poly1/heparin/lysozyme/heparin)$_{43}$ film nearly 100% retains of enzyme activity. Reproduced from Ref 54.

release was demonstrated by tuning the degradability of the polycation by the introduction of an additional methylene group (Poly 2), changing the number of layers used, and choice of the interlayer polyanions, including using heparin or chondroitin sulfate. In each case, functionality, as measured by bacterial lysis activity, was comparable to native protein, as shown in Figure 17.8.

Toward addressing clinically relevant conditions using potent therapeutic molecules, a variety of growth factors were incorporated into similar degradable polyelectrolyte films, including fibroblast growth factor (FGF) [55] and bone morphogenetic protein (BMP) [56]. These soluble factors initiate cell signaling cascades that lead to tissue growth and formation in processes such as wound healing, bone formation, and tissue repair. Fibroblast growth factor 2 (FGF-2) is a potent mitogen for cell proliferation where tuning is desirable. Using the degradable multilayer platform, drug loading was tunable using at least three parameters (number of nanolayers, counter polyanion, and type of degradable polycation) and yielded total drug loading values of 7–45 ng cm$^{-2}$ of FGF-2. Elution time varied between 24 h and 5 days. FGF-2 released from these films retained *in vitro* activity, promoting the proliferation of MC3T3 preosteoblast cells, as illustrated in Figure 17.9. Additionally, the use of counter polyanions, heparin sulfate or chondroitin sulfate, in the tetralayer structures was observed to enhance FGF-2 activity. FGF-2 has binding sites specific to these extracellular matrix polysaccharides; similarly, the CD44 cell surface receptor binds to these polysaccharides and regulates cell adhesion and mobility. By complexing factors such as FGF-2 with these biomacromolecules, it is possible to increase the interactions and bioavailability of the growth factor to cells through specific binding interactions.

Many efforts toward regenerative medicine rely on the co-implantation of autologous derived stem cells with biomaterials that support their proliferation and differentiation. Where possible, however, it would be much more advantageous to directly recruit the native stem cells that exist in the bone marrow or other regions of the body to the desired site of tissue repair using growth factor signaling. An area of

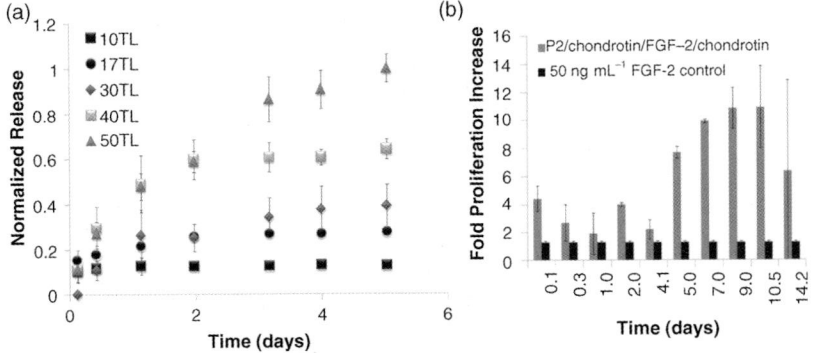

**Figure 17.9** (a) Release is tunable with number of tetralayers, using (P2/heparin/FGF-2/heparin) film as an illustration. (b) MC3T3 preosteoblast cells which were exposed to FGF-2 released from films demonstrated a marked increase in proliferation. Reproduced from Ref 55.

promise for orthopedic applications is the use of rhBMP-2, an osteoinductive growth factor that induces differentiation of preosteoblast cells into osteoblasts [57], which lay down new bone, and is of significant importance in bone regeneration. We were able to introduce 3D osteoconductive scaffolds capable of physiological scale release of BMP-2 using the LbL approach with the incorporation of substantive amounts of protein in the hydrolytically degradable thin films. In this work, tricalcium phosphate/polycaprolactone scaffolds with macroporous channels were coated with the tetralayer architectures containing the BMP-2 protein in alternation with degradable Poly1 or Poly2 and chondroitin sulfate. These scaffolds are capable of directing the host tissue response to create new bone from native progenitor cells without the co-implantation of stem cells. BMP-2 released from LbL films retains its ability to induce bone differentiation in MC3T3 pre-osteoblasts, as demonstrated with several cell culture studies that illustrate quantitative increases in bone production and differentiation for these systems. One of the first *in vivo* studies using LbL films demonstrated bone formation at an ectopic intramuscular site in a rat. The BMP-2 film-coated scaffolds were able to release the protein, which acted to recruit and initiate host progenitor cells to differentiate into bone, which matured and expanded from 4 to 9 weeks. Bone formation was restricted to the scaffold structure, as shown in Figure 17.10, and the bone formation was indicative of native, well-formed lamellar bone. While a specific tissue type consists of the same type of cell, the creation and expansion of tissue typically involves the synergistic cooperation of several different cell types. Formation of new bone tissue is greatly enhanced by angiogenesis, where the proliferation of new blood vessels helps provide a path to enrich the local population of pre-osteoblast cells, which can subsequently differentiate into osteoblast laying down collagen and calcium. rhVEGF has been demonstrated to act synergistically with rhBMP-2 to increase the growth of bone tissue [58]. Localized introduction of such growth factors in specific ratios has been demonstrated to greatly enhance bone tissue regeneration [59]. Using the LbL approach, we were able to independently tune the release of VEGF and

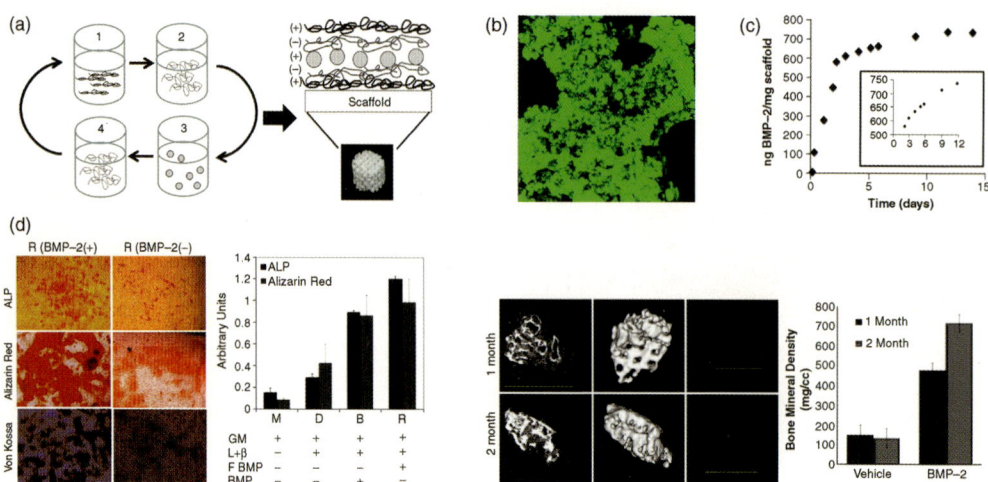

**Figure 17.10** (a) Schematic of LbL deposition on a 3D scaffold which is (b) conformally coated. (c) BMP-2 is released over 2 weeks and retains its activity (d) as quantified by various differentiation assays. *In vivo* scaffolds coated with BMP-2 are able to induce ectopic bone formation, indicating that the system is able to recruit cells and induce differentiation. Reproduced from Ref 56.

BMP-2 [60]. The release of BMP-2 was sustained over a period of 2 weeks, while VEGF eluted from the film over the first 8 days. Released BMP-2 initiated a dose-dependent differentiation cascade in MC3T3-E1S4 pre-osteoblasts, while VEGF upregulated HUVEC proliferation and migration. At an intramuscular ectopic site, bone formed by BMP-2/VEGF PEM films was approximately 33% higher in density than when only rhBMP-2 was introduced, with a higher trabecular thickness. Bone was formed throughout the scaffold when both growth factors were released, which suggests more cellular infiltration due to an increased local vascular network (Figure 17.11). One of the remaining questions about growth factor delivery is whether temporal and spatial cues can be used to more optimally influence wound healing. Ongoing work in this area is facilitated using the LbL approach, as the films can be patterned on 2D and 3D surfaces, and, as discussed further in Section 17.3.2, capabilities to control sequence as well as timing of release continue to advance in this field.

#### 17.3.1.2 Small Molecule Delivery

The delivery of small molecule drugs is relevant to a broad range of biomedical applications, in particular the release of antibiotics to treat infections and infectious disease, anti-inflammatories to regulate pain and inflammation, and to address tissue injury, and statins and antithrombotic molecules that address thrombosis in stents or manage cholesterol. Many of these cases benefit from localized rather than systemic delivery through the blood, airway or oral means, as localized release enables immediate access of the drug to the targeted tissue and limits side-effects that may be undesirable. Biomaterial-mediated delivery schemes offer a unique method to deliver these drugs; however, one challenge with small molecular systems is that

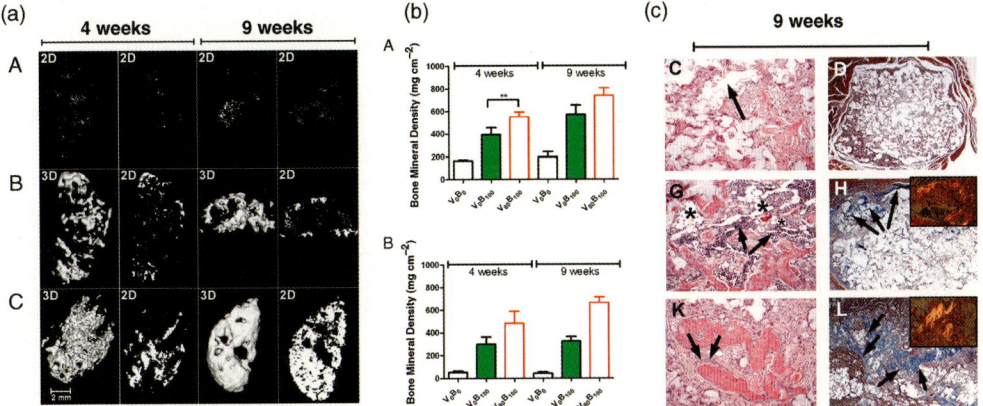

**Figure 17.11** (a) microCT images of PCL-βTCP scaffolds coated with different formulations of combination rhBMP-2 and rhVEGF165 LbL films implanted in an intramuscular site. (A) Control scaffolds lack bone. (B) In BMP-2-only films lacking rhVEGF165, bone formation is restricted to the periphery of the scaffold. (C) Bone grows inward when VEGF is present. Scaffolds releasing rhVEGF-C demonstrate a smooth, continuous profile in the ectopically formed bone that matures from 4 to 9 weeks to fill the entire scaffold. (D) The bone mineral density (BMD) of ectopic bone formed by the rhBMP-2/VEGF combination scaffolds is higher than bone formed by rhBMP-2 scaffolds at 4 and 9 weeks respectively at (A) periphery and (B) interior of the scaffolds. (C) Histology demonstrated recapitulation of the bone environment with fatty marrow when just BMP-2 (G, H) and BMP-2/VEGF 165 (K,L) were introduced. Bone formation restricted to the periphery (H) and throughout the scaffold (L). Control scaffolds demonstrated no bone growth (C,D). Reproduced from Ref 60.

their small size enables rapid diffusion through a polymeric matrix. Additionally, small charged molecules exhibit much lower charge valency for incorporation into polyelectrolyte multilayers, and neutral molecules provide no direct charge-based mechanism for incorporation into multilayer films. Despite these challenges, a diverse range of active small molecules have been successfully incorporated into and released from polyelectrolyte multilayer systems. We have chosen to use two different routes for the incorporation of small drugs. Small charged molecules can be directly adsorbed into multilayer thin films, and the tendency toward diffusion of such molecules can lead to absorption of drug within the polymer matrix, and thus relatively high loadings of drug. On the other hand, neutral and hydrophobic molecular species must be incorporated using polymeric or colloidal carriers that can be incorporated based on electrostatics or hydrogen bonding.

An example of small molecule delivery involving direct adsorption of the drug is the incorporation of the antibiotic gentamicin sulfate, an aminoglycoside used to treat multiple types of gram-negative infections, and favored for orthopedic and other implant infections. Gentamicin is usually administered via intramuscular or intravenous injection; however, because the drug contains five positively charged amines, with a molecular weight of 477 Da, it is possible to incorporate it into heterostructured multilayers to achieve precise tuning of dosage, and control the release rate under aqueous physiological conditions [61]. The released gentamicin from poly (β-amino ester)-based films retains its efficacy against Staphylococcus aureus, as illustrated for a range of polyanion constructs from hyaluronic acid to poly(acrylic acid) (PAA), and the degradation products were nontoxic toward murine osteoblasts MC3T3. Interestingly, the film growth was nonlinear in the early stages of film build-up, but rapidly became linear with large thickness increments per tetralayer, in particular with the use of the high charge density polymer PAA as the polyanionic species. This behavior is highly characteristic of the "in" and "out" diffusion mechanism observed for exponentially growing LbL films [62]; it has been shown that when the film reaches a critical thickness, diffusion limitations take over and the film growth goes from superlinear to linear in such systems. This phenomenon resulted in very thick films ranging from 20 to 80 μm. Drug loadings in the PAA/Poly1/Gentamicin sulfate films were also very high, indicating that the small molecule drug was not only a key substituent in the electrostatic multilayers, but was also absorbed into the film in excess during the assembly process. By ensuring that the final adsorption step is in the gentamicin sulfate dipping solution, it was possible to observe significant increases in drug loading due to direct uptake of the drug within the film. About 70% of the payload eluted over the first three days and the remaining 30% subsequently eluted slowly for more than four additional weeks, reaching a total average release of over 550 mg cm$^{-2}$ from the implant surface. This kind of biphasic release leads to gentamicin concentrations in localized areas that are significantly higher than the minimum inhibitory concentration (MIC) of 0.156 μg ml$^{-1}$ for several hours to days, followed by a lower concentration that is of the same order as the MIC. Rather than simply providing enough antibiotic to prevent infection in an uninfected region, these films are suitable to eradicate an existing infection, a capability that has not yet been achieved for a number of antibiotic release vehicles. For example, it would be highly desirable to eliminate infection in bone and eliminate the need for a two-step surgical process in whole joint replacement surgery, in which the old joint is removed and replaced with antibiotic releasing beads, and then the new joint is inserted in a second surgery weeks later. These polyelectrolyte multilayers were used as a bone implant device coating to address bone infection in a direct implant exchange operation using a rabbit infection model (see Figure 17.12) [63]. Film-coated titanium implants were compared to uncoated implants in an *in vivo* S. aureus rabbit bone infection model, in which the antibiotic-coated devices yielded bone homogenates with a significantly lower degree of infection than uncoated devices for up to one week post infection. These results are very promising, and further tuning of the assembly of the film is expected to lead to even further eradication and the development of new biomedical coatings for multiple surgical applications. It is believed that the release of drug in the case of small molecules such as gentamicin that adsorb directly into the film is based on the combination of two or three mechanisms: (i) drug molecules are loaded in excess in the LbL films due to interdiffusion, and immediately begin to diffuse out when the film is immersed in plasma or buffer solution; (ii) the loss of molecules in the film impacts the charge balance within the multilayer, which in turn leads to polyelectrolyte ejection and erosion of the LbL film; and (iii) hydrolytic degradation of the polymer backbone, which yields surface erosion and release from the surface.

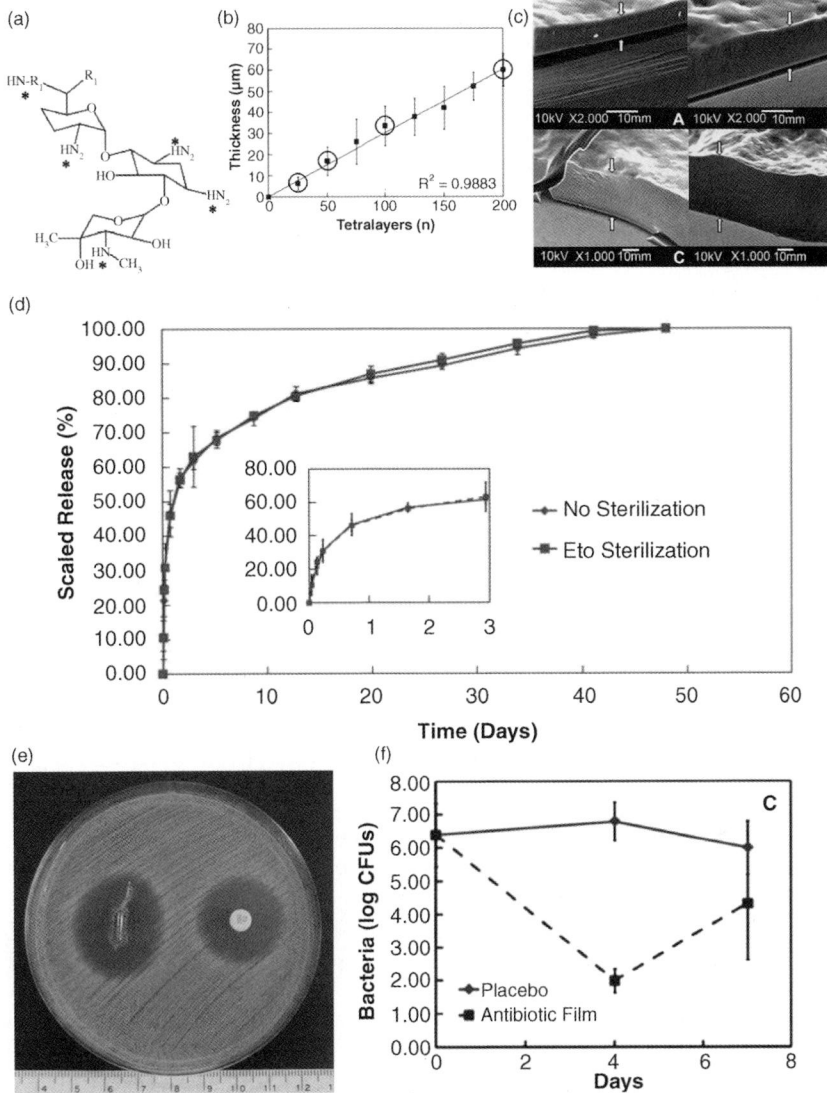

**Figure 17.12** (a) Gentamicin, (b) growth curve, and (c) cross-section SEM images of (polymer 1/PAA/gentamicin/PAA) films. (d) Release curve of gentamicin from the films. (e) Zone of inhibition produced by gentamicin release films (left in figure) comparable to a BD Sensi-Disc control. (f) Log transformed S. aureus CFU data from rabbit femoral condyles treated at day zero with (P1/PAA/gentamicin/PAA) coated Ti implants and after direct exchange. Reproduced from Ref 63.

Although gentamicin is a favored antibiotic for more standard hospital care, vancomycin hydrochloride is a potent multispectrum antibiotic that is capable of killing methicillin resistant (MRSA) and other strains of bacteria. Because this drug has several known side-effects when delivered orally in large doses, it is much more desirable to use delivery technologies to get it to the site of infection while avoiding systemic toxicity [64]. Vancomycin is also a glycopeptide; however, unlike gentamicin, it contains only one net positive charge from a single amine group at pH values of 6 and lower. LbL films directly incorporating vancomycin with Poly1 or Poly2 and a negatively charged polysaccharide can be generated using a combination of electrostatic and other secondary interactions that were discovered to exist between the film components. The importance of film interdiffusion during growth in promoting interactions between film components was critical in the direct incorporation of the weakly charged vancomycin drug in these multilayer films. The resulting coatings were engineered with drug densities ranging from 17–220 $\mu g\, mm^{-3}$ (approximately 20 wt%) for films that are micron to submicron scale in thickness, delivering active vancomycin over timescales of 4 h to 2.5 days.

Although small charged molecules can sometimes be incorporated via alternating adsorption into multilayer thin films, hydrophobic molecules that are typically water-insoluble cannot be directly adsorbed into LbL films. One means of incorporating hydrophobic systems is to use a charged carrier that can complex or encapsulate the hydrophobic drug stably and with high loading efficiency. One recently developed modular approach has been the use of polycyclodextrins, shown in Figure 17.13, which contain hydrophobic pockets within the interiors of the cyclodextrins, toroidally-shaped oligosaccharides that consist of six to eight glucopyranose units, and charged functional groups that reside along the outside of the cyclodextrin rings. These molecules have the capacity to complex with a broad range of clinically relevant drugs, and they have been used as host molecules for inclusion compounds in aqueous drug formulations. While recent work has tried to use a monomeric cyclodextrin ring within multilayer systems, these films still rely on diffusion, and controlled release has not been demonstrated. The use of a polymeric form of cyclodextrin is important, as small charged species can yield less stable layers [65] and exchange of the cyclodextrin molecule out of the film at certain conditions. Smith *et al.* have investigated the ability of these charged polycyclodextrins to act as carriers for various small molecules when incorporated into degradable multilayer thin films [66]. Non-steroidal anti-inflammatory drugs (NSAIDs), ciprofloxacin, and prodan were directly incorporated as molecular complexes with the negatively charged polycyclodextrin (Poly-CD) systems and alternated with the Poly1 and Poly2

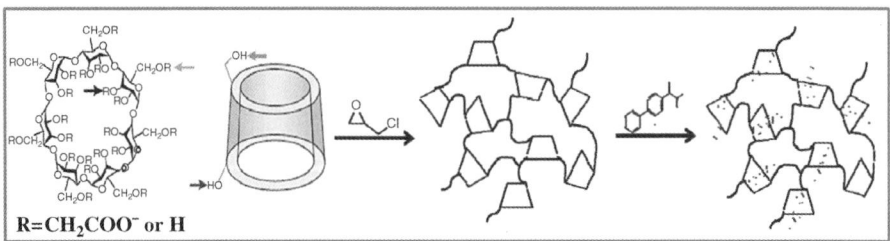

**Figure 17.13** Schematic poly(carboxymethylbeta-cyclodextrin) synthesis and complexation.

series of polycations. The activity of small molecule therapeutics released from the LbL films was then determined by monitoring the effect of released NSAIDs, such as diclofenac, which is a COX-2 inhibitor, on COX-2 activity.

It was found that very thin nanometer-scale multilayer films grew linearly during film assembly of these LbL delivery systems. The drug loadings for these films were sufficient to yield notable COX inhibition with films that were only 20 layer pairs, or in the range of 100 nm in thickness. These systems all exhibited relatively similar growth behavior, and when release profiles were normalized for the PolyCD LbL films, the release curves all collapsed onto a single curve. Examples for the release of two different COX-2 inhibitor drugs are shown in Figure 17.14, along with the inhibition results as a function of time, where the amount eluted each day is isolated and tested for activity. The amount of COX-2 drug in cell culture was sufficient to inhibit over periods of weeks in these systems, a remarkable capability for small molecule drugs in electrostatic thin films. We believe the source of this highly controlled release behavior is the of PolyCD with highly effective in complexing the NSAID drug, and its high charge density creates a highly ionically crosslinked LbL thin film with the degradable PBAE. For these reasons, the macromolecular LbL thin films undergo release at rates more consistent with the slow hydrolytic breakdown of the polycation, without contributions from out-diffusion or polyion ejection that can yield bolus release. The activity of the drug remains unchanged after release, though HPLC studies suggest that some of the drug remains complexed within the PolyCD following release, and is likely eluted from solution even after LbL breakdown. The primary factors for release of these systems are the nature of the polyion pair, Poly1 and PolyCD; thus, the kinetics of film breakdown is not modified as different small molecule drugs are introduced. This approach enables a modular platform to LbL small molecule delivery in which the polymeric carrier independently modulates release rather than the interactions of the drug itself.

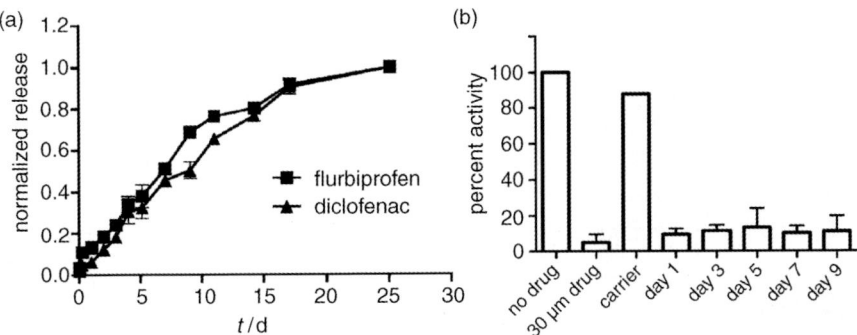

**Figure 17.14** (a) Release of diclofenac and flurbiprofen at 37 °C from (Poly 2/PolyCD-Drug)$_{20}$ (release is independent of drug). (b) Effect of (Poly 3/PolyCD-Diclofenac)$_{20}$ eluent on COX-2 enzyme activity. Reproduced from Ref 66.

## 17.3.2
### Multi-Agent Delivery and Control of Sequence

As mentioned above, one of the key interests in the use of LbL approaches for drug delivery is the ability to load more than one kind of therapeutic into the film based on alternating charge or other interactions. In general, as long as the therapeutic of interest is charged, or can be contained within a charged carrier, it can be introduced into LbL systems; furthermore, hydrogen bonding and hydrophobic interactions can also be used as the basis for drug incorporation. The large range of therapeutics that can be incorporated into LbL systems allows the mixing of components with similar or quite different chemical structure, hydrophobicity, charge and molecular size. This ability to "mix and match" drug formulations with ease, without the need for complex emulsions or solvents, with various bulk polymers is a key advantage to this approach, along with high drug loadings and the ability to create thermodynamically stable films from these mixtures with the aid of electrostatics. We have been able to show that dual combinations of drugs using LbL approaches can be very effective [67]. Multiple tetralayers of vancomycin-containing LbL films can be topped with layer pairs of the PolyCD-diclofenac complex and Poly1 to yield a dual release antibiotic and anti-inflammatory releasing thin film (Figure 17.15). We found that the efficacy

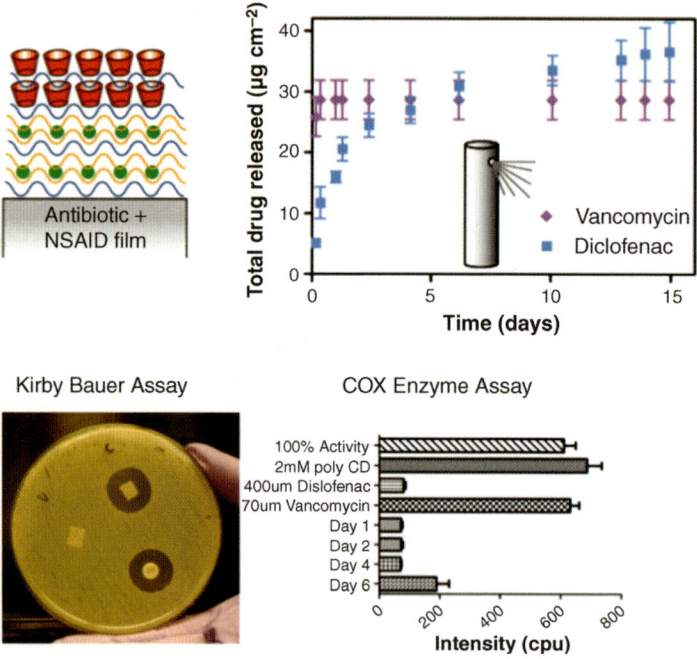

**Figure 17.15** Combination films of vancomycin and diclofenac, a multi-spectrum antibiotic and a Cox-2 inhibitor anti-inflammatory drug, the coincident release profiles of the two drugs, and their relative efficacy as tested by Kirby Bauer Assay and Cox enzyme assay. Reproduced from Ref 67.

of both drugs was maintained, and that, with careful optimization of the placement of the layers as described above, it was possible to avoid undesirable exchange reactions with the small molecular species of the film and the other film components. Ultimately, as will be described in more detail in Section 17.4, the rapid kinetics of assembly using Spray-LbL led to higher drug loadings, and an ability to achieve inhibitory concentrations of both drugs. The rapid release of a high concentration of vancomycin in the first several hours, followed by more extended release of the anti-inflammatory agent over 14 to 15 days also provides an ideal release profile for rapid wound remediation. Other examples of dual release have also been demonstrated in our laboratories via the use of polyelectrolytes to which covalently bound drug molecules can be released from the polymer backbone of micellar species which have been alternated with polyanions with active function, such as heparin sulfate, which acts as an anti-coagulant and has some application in the prevention of thrombosis [68]. Dual protein release and the combination of DNA, proteins and small molecules are active areas of pursuit that are anticipated to lead to new multifunctional drug release depots. In all of the above systems, drug release of each component tends to be coincident; that is, all drug is released starting at approximately the same time, even if the layers are introduced at different points in the film architecture, as was shown in the vancomycin/diclofenac example. These observations suggest that, during assembly, the film components are mixing via diffusive mechanisms. Despite this mixing, we find that the release characteristics of each component in these dual systems tends to be similar to that in the single drug LbL film. It is possible that the film complexes that are formed for a given drug layer pair or tetralayer combination remain intact, but diffuse and complex within the already existing LbL film without disrupting the previously formed complexes. The net result is the retention of release characteristics unique to a given layer pair or tetralayer system, but a lack of time-dependent release with regard to the order of introduction of the components.

A number of applications would also benefit from the ability to control the order of release from an implant, tissue engineering scaffold, or other biomedical device, as well as the drug type and quantity. As demonstrated in Figure 17.16, one ultimate goal of LbL-mediated release is the ability to control the sequence of drug released based primarily on the order in which the drugs are introduced into the film. The example shown is one of an orthopedic implant used for implant replacement surgery, which could be designed to first release a therapeutic quantity of small molecule painkillers, such as NSAIDS, and antibiotics during the early stages following implantation to address inflammation, pain, and infection within the joint; this release might be followed by several days to weeks of growth factor release to modulate the growth of bone tissue around the implant. By providing an impetus for the integration of bone into the implant, the likelihood of future joint loosening and additional surgery is decreased, and the patient is able to heal more rapidly and to generate a more stable mechanical fixture of the joint and the implant [69–71]. Other examples include the elution of drugs to address inflammation and clot formation in stents, and the delivery of different stages of vaccines. One of the great challenges in achieving

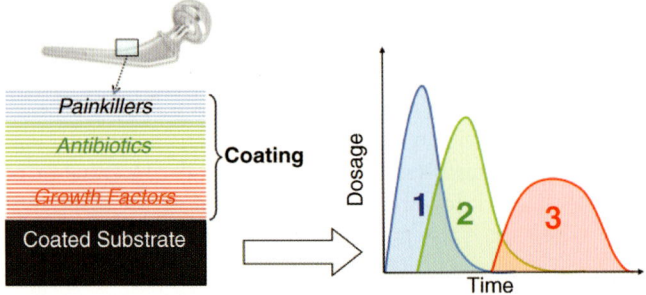

**Figure 17.16** Schematic of concept of sequential drug release based on ordering of drug components within the stratified layers of an LBL thin film.

staged or sequential delivery is the interdiffusion of the adsorbing film components into the LbL matrix, and subsequent mixing of the layers.

We first examined a means of addressing this issue in a study of a series of model polysaccharide therapeutics, including heparin, low molecular weight heparin, and chondroitin sulfate [72]. The film degradation and release of radiolabeled polysaccharide over a range of pH values was observed to follow highly consistent, pseudo-first order degradation kinetics. The pH of the release solution significantly affected the rate and release profile of model polysaccharide heparin, which was also nonlinear and exhibited distinct regimes, which was attributed to the previous observation of top-down degradation of the multilayer film. To determine the mechanistic basis of material release from these hydrolytically degradable polyelectrolyte multilayer films, we considered the phenomena of interlayer diffusion of species within the multilayer film [73].

We systematically evaluated a range of strategies to control the relative positions of multiple species within a single film by constructing physical barriers to separate the two components (Figure 17.17). A base layer of polymer 1/dextran sulfate (DS) coated with a single bilayer of PAH/PAA (covalently crosslinked by heating the films), followed by the deposition of polymer 1/heparin (HEP), resulted in a multistage, serial release of the surface HEP, followed by the underlying DS (Figure 17.10). Here, the use of a single covalently crosslinked PAH/PAA layer was sufficient to separate the two components when deposited onto the surface of the linearly growing polymer 1/DS system, as evidenced by the two-stage release profile. After the 25 h time delay, underlying DS was released with a linear profile. The study demonstrated that while covalently crosslinked barriers can effectively block interlayer diffusion by the creation of compartmentalized structures, even very large numbers of ionically crosslinked barrier layers cannot block interlayer diffusion.

Additional experiments using single and multiple crosslinked PAH/PAA barrier layers demonstrated that the duration of the release delay and the rate of release can be broadly controlled under this approach. On reversing the order of the two labeled components (HEP as the base layer and DS as the surface layer) it was no longer possible to achieve serial release of the two components using crosslinked spacer

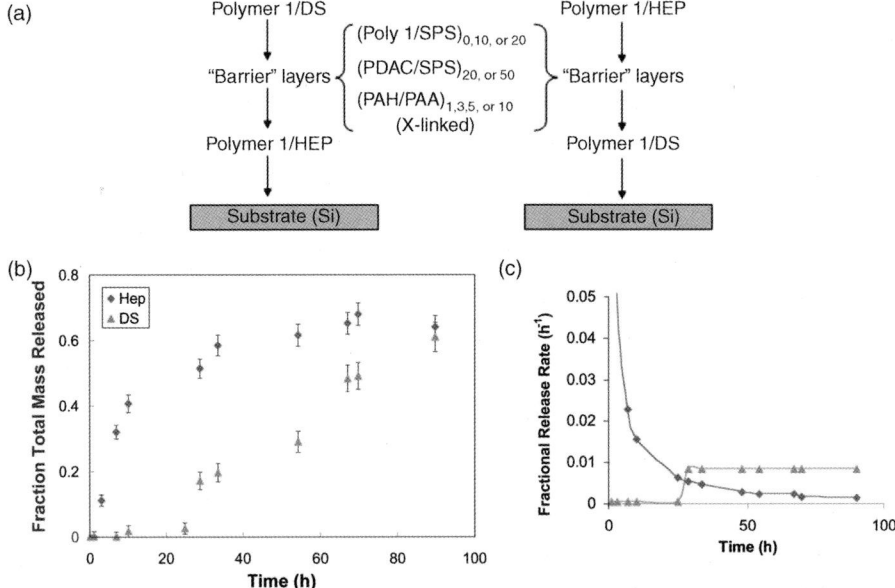

**Figure 17.17** (a) Schematic depicting strategies employed to construct physical barriers to control interlayer diffusion in multicomponent films. (b,c) DS (base layer, ▲) and HEP (surface layer, ◆) loaded layers separated by a single, covalently crosslinked layer of (PAH/PAA) by thermal treatment exhibit sequential release. (b) Fraction of mass released versus degradation time. (c) Fractional release rate versus time. Reproduced from Ref 73.

layers, indicating that the nature of the base film onto which the crosslinked barrier layer is absorbed influences the final properties of the barrier layer.

Unfortunately, the use of amine and acid groups as crosslinking moieties for a barrier layer is not compatible with the incorporation of proteins, peptides and other biologic drugs which contain the same groups, and would undergo denaturation or unintentional crosslinking under chemical or heat conditions that yield amidation or similar reactions. We are currently investigating the use of nanoscale physical barriers to diffusion that can be directly adsorbed into polyelectrolyte multilayer systems. Graphene oxide (GO) is a two-dimensional charged nanomaterial that can be used to create barrier layers in multilayer thin films, trapping molecules of interest for controlled release. Protein-loaded polyelectrolyte multilayer films were fabricated using LbL assembly incorporating cationic Poly with a model protein antigen, ovalbumin (ova) in a bilayer architecture, together with positively and negatively functionalized GO capping layers for the degradable protein films [74]. Ova release without the GO layers takes place in less than 1 h, but can be tuned to release from 30 to 90 days by varying the number of layer pairs of functionalized GO in the multilayer architecture. We demonstrated that proteins can be released in sequence with multi-day gaps between the release of each species by incorporating GO layers between protein-loaded layers. *In vitro* toxicity assays of the individual materials on prolifer-

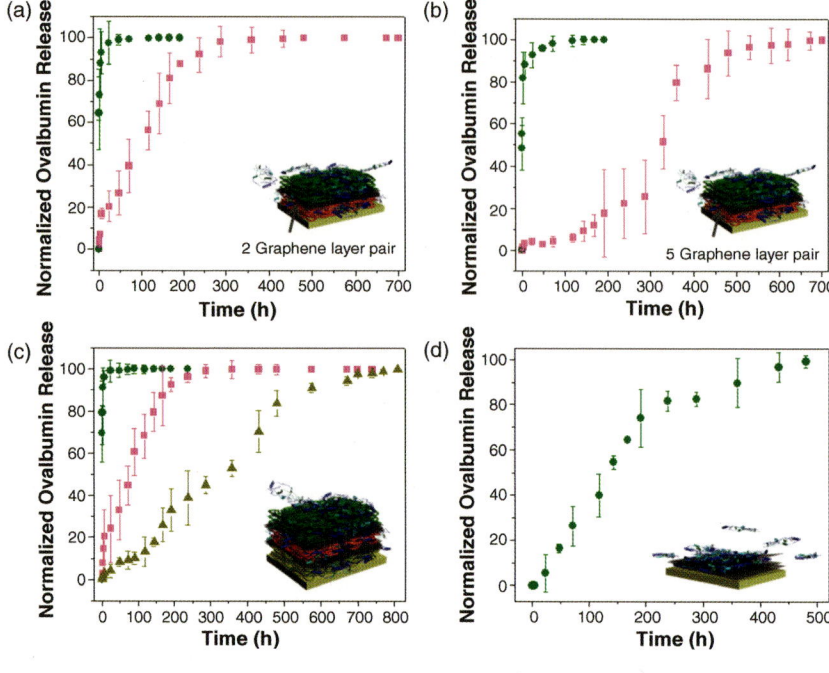

Figure 17.18 Ova release from multilayer films: Normalized cumulative release from: (a) substrate/(Poly1/ova-TR)$_{20}$(GO/GO)$_2$(Poly1/ova-FL)$_{20}$ (b) substrate/(Poly1/ova-TR)$_{20}$ (GO/GO)$_5$ (Poly1/ova-FL)$_{20}$ (c) substrate/(Poly1/ova-AF$^{555}$)$_{20}$(GO/GO)$_5$(Poly1/ova-TR)$_{20}$(GO/GO)$_2$(Poly1/ova-FL)$_{20}$ multilayer films. (d) substrate/(GO/ova-FL)$_{20}$. □-(green line), ■-(pink line) and ▲-(yellow line) indicate the ova released from (Poly1/ova-FL)$_{20}$, (Poly1/ova-TR)$_{20}$ and (Poly1/ova-AF$^{555}$)$_{20}$ respectively. Reproduced from Ref 61.

ating hematopoietic stem cells (HSCs) indicated limited cytotoxic effects (Figure 17.18).

## 17.4
### Automated Spray-LbL – Enabling Function and Translation

One of the most important advances affecting the ability to translate the LbL technique to commercial applications has been the automation of spray-LbL, which involves the alternating application of misted aqueous solutions containing the polyanion, rinse, and polycation directly to the intended substrate under controlled conditions, as shown in Figure 17.19. Previous studies had indicated that LbL films can be generated when applied with hand-misted aerosolized solutions of polyelectrolyte, followed by misted rinse cycles [9, 10]. In our investigations, we found that by automating this method [11, 12], we can precisely tune the time period over which

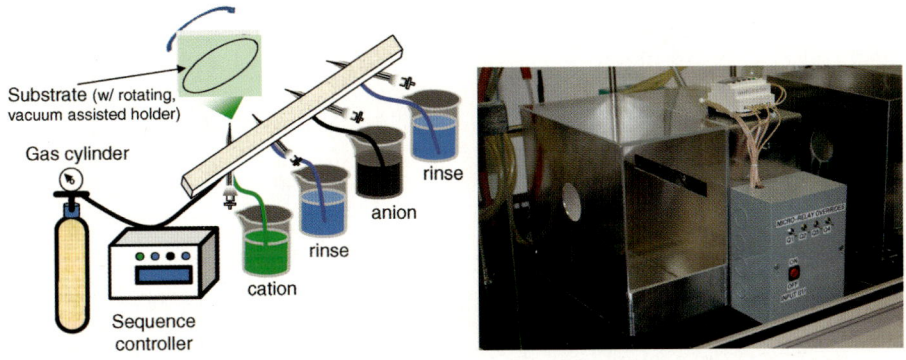

**Figure 17.19** (a) Schematics of spray layer by layer assembly, and (b) automated spray layer by layer assembly developed in Hammond Lab.

mist is applied, the droplet size, and the movement of the intended sample, enabling rotation to achieve uniformity, for example. This approach provides a rapid and lower cost fabrication process. Conventional LbL fabrication via the alternate dipping of a substrate into polyanion, polycation and rinse baths takes of the order of tens of minutes to build a layer pair with thickness of a few nanometers. The automated spray process has been demonstrated with cycle times as short as 10 s for a layer pair, and cuts down the processing time of standard dip-LbL by as much as 50 to 100 times. Ongoing developments of this approach will undoubtedly lead to even shorter cycle times and refined operating conditions, making the ability to generate LbL coatings on a large scale much more accessible. The approach is adaptable to the coating of large and complex surfaces or planar flexible films and devices using methods such as roll-to-roll processing and adjustable spray conditions, and could also be adapted to the creation of free-standing films through the use of a lift-off or release layer. The automated spray-LbL method has served as a foundation to the launch of new commercial ventures involving the application of multilayer thin films.

In general, it has been found that sprayed LbL films are the same thickness, or in some cases, a small percentage thinner, than typical dip-LbL films generated using the same technique. The fact that the spray-LbL process involves rapid contact of polyelectrolytes with the oppositely charged species leads to some advantages in terms of film uniformity. The nature of spray-LbL is that very smooth films can generally be formed, even in the very early stages of film assembly. Typically, the first few layers of polyion adsorbed on the surface exhibit island-like growth in the dip-LbL process until complete surface coverage is achieved, thus leading to the generation of less uniform films in the first few layers and a lower initial rate of film growth. With spray-LbL, we observed that this early limited growth regime appears to go away for many spray-LbL systems (see Figure 17.20); we have hypothesized that this effect is due to the increased spreading of polyelectrolytes on the substrate enabled during the spraying process [11]. Polyelectrolyte multilayer films that undergo traditional linear growth behavior are readily replicated in the spray-LbL approach, and many of the

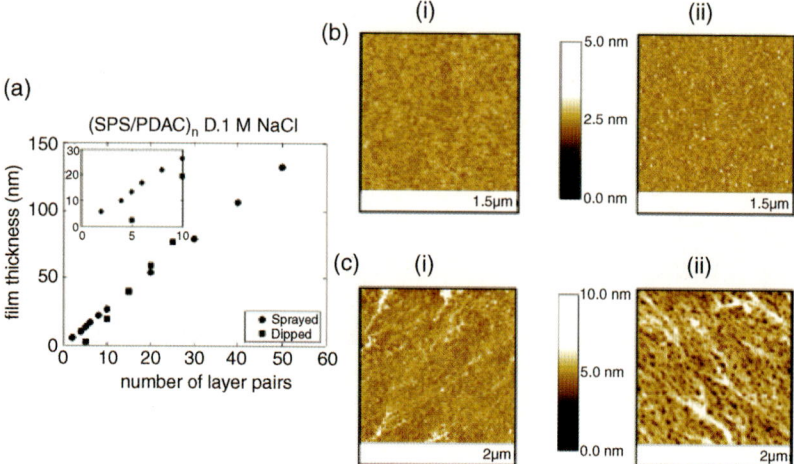

**Figure 17.20** (a) Correlation of total film thickness to layer pair number for the (SPS/PDAC)$_n$ system both by spray deposition and dipping. Thickness was evaluated using ellipsometry and profilometry. Reported values are averages taken from several data points on a silicon wafer and vary by less than 2 nm. Both dipped and sprayed films exhibit linear growth rates above 5 layer pairs, but the sprayed films have no initial nonlinear growth regime. (b) AFM height images of a sprayed (SPS/PDAC)$_n$ (i) 0.5 layer pair, PDAC surface and (ii) 1.0 layer pair, SPS surface. Coverage is thin but uniform. (c) AFM height images of a dipped (SPS/PDAC)$_n$ (i) 0.5 layer pair and (ii) 1.0 layer pair. Initially, "islands" form on the silicon substrate. Reproduced from Ref 11.

properties observed in dipped films can be observed with these thin film coatings as well, confirming that the spray approach follows the same principles of equilibrium-based charge reversal.

Together with the high speed of application and the versatility of processing thin films, there are some unique advantages to the use of spray-LbL. Because we can manipulate the rate of flow of water droplets through porous substrates, we have found this approach to be particularly applicable to the coating of a range of porous materials that can lead to interesting composite film morphologies. We have also found that the rapid timescale of polyelectrolyte deposition in spray-LbL approaches the kinetics for adsorption and equilibrium rearrangements of the polyions in the multilayer, thus impacting the effects of kinetic phenomena such as polymer interdiffusion in exponentially growing LbL films, and polyion and small ion exchange within the film during the adsorption cycle, in a manner that enables access to non-equilibrium structures and provides some unique opportunities. Finally, the use of spray-LbL significantly decreases the exposure of samples to water – this fact is a simple but critical factor in the development of hydrolytically degradable multilayers, and in the coating of water-labile or water-swellable materials systems. In the following sections, examples are provided that highlight each of these unique advantages of spray-LbL.

## 17.4.1
### Means of Achieving New Architectures

The spray-LbL method can be designed to coat porous surfaces by pulling a vacuum across the porous substrate and thus enabling the coating of the fine details of interior surfaces with nanometer scale fidelity. The rate of flow of droplets through the membrane can be controlled with the pressure drop, and when the flow velocity is adjusted appropriately, very conformal coatings are delivered across membranes as thick as hundreds of microns to tens of millimeters. Figure 17.21 illustrates the results of coating a Nylon 6,6 electrospun mat consisting of micron diameter fibers interconnected in a porous network [12]. The individual fibers of the mat are coated throughout the thickness of the substrate with a uniform thin film that increases linearly in thickness with the number of layers. In this particular application, the LbL film is formed from the alternation of the strong polycation, PDAC, with negatively charged titania nanoparticles; these multilayers were first optimized on a planar film substrate to yield highly reactive photocatalytic thin films that act as protective membranes capable of breaking down chemical agents in the presence of UV. When applied to the electrospun mat, the surface area for reaction was significantly increased compared to the planar substrate, thus greatly enhancing the photocatalytic activity. To create an effective photocatalytic film that breaks down significant

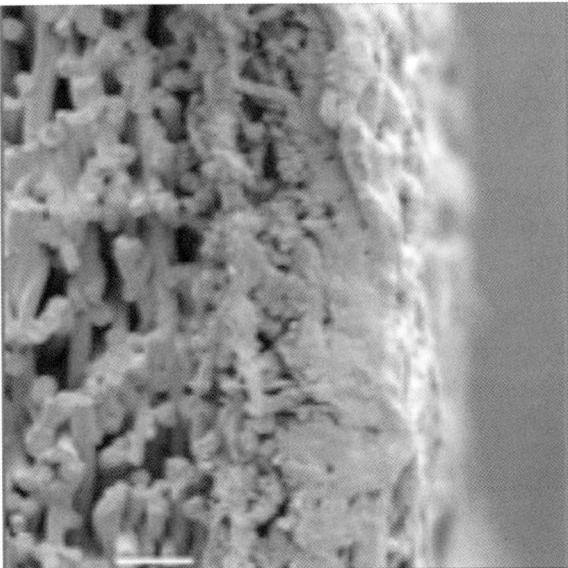

**Figure 17.21** Example of capability to coat porous materials. A microporous electrospun membrane is coated first with a conformal LbL thin film using spray LbL to coat nanofiber surfaces, followed by a separate top surface layer that bridges pores and could act as a bulk release layer. Reproduced from Ref 12.

amounts of agent, it is necessary to retain the agent within the film. By examining the permeability of a series of different polyamine and polyacid LbL systems, the best candidates for retention of the toxic vapor were determined.

Thin condensed multilayer membranes were constructed along the top surface of the electrospun mat. This second set of multilayers was created without a pressure drop applied during the assembly process. The net result is the formation of a thin wetting layer along the surface of the porous membrane which serves to enable the build-up of polyanions and polycations to form a bridged thin film structure. This ultrathin membrane can serve as a barrier layer to the interpenetration of toxic molecules, thus retaining the simulant molecule vapor and providing a much longer retention time within the photocatalytic core of the membrane. The generation of this system and similar hybrid composite membranes illustrates the flexibility of the spray-LbL approach in designing complex architectures through the use of scaffolds and the manipulation of processing conditions. In a similar fashion, we have found that we can coat electrospun mats with sPPO/PDAC to achieve high ionic conductivity, thus leading to the potential use of this approach in methanol fuel cells. Such concepts can be broadened to include composite membrane systems that provide a means of ultrafiltration, water and purification applications, molecular separations, and selective reactive membrane systems.

Other porous scaffolds may also be readily coated using the LbL approach, including membranes with pores ranging in size from tens of nanometers up to microns. By utilizing a vacuum-assisted spray process, the exposure time of the membranes to water is significantly reduced. The constant removal of water provided by the vacuum enables the coating of even highly swellable membrane materials. For example, it is possible to coat gelatin sponges with microporous structures that can absorb up to ten times their weight in blood or plasma. We have applied hydrolytically degradable antibiotic vancomycin [75] or hemostatic thrombin-containing polyelectrolyte multilayers to the gelatin sponges using vacuum-assisted spray-LbL, and have found that the application of multiple layer pairs does not lead to the swelling of the gelatin substrate that would occur during a traditional dip-LbL process involving the immersion of the sponge in aqueous solutions. Instead, the original morphology of the gelatin is effectively preserved, as shown in Figure 17.22 for a 60 [55] tetralayer coating. Interestingly, the application of vancomycin-containing multilayers not only preserves the morphology of the gelatin sponge, but enhances its absorption. The vancomycin LbL films increase the absorbency of the sponges by a factor of over 2.5 for a 120 layer pair coating [75]. The release characteristics from porous media such as sponges, scaffolds, and mats are also often much more extended [55]; sustained release is achieved because it is regulated by the diffusion of water molecules within pores to achieve hydrolytic degradation, and out-diffusion of drug molecules from the pores and channels of the membrane. Finally, drug loadings can also be enhanced by the increased surface area in such substrates, as illustrated in the release of small molecules such as vancomycin [75] and proteins, including growth factors [55].

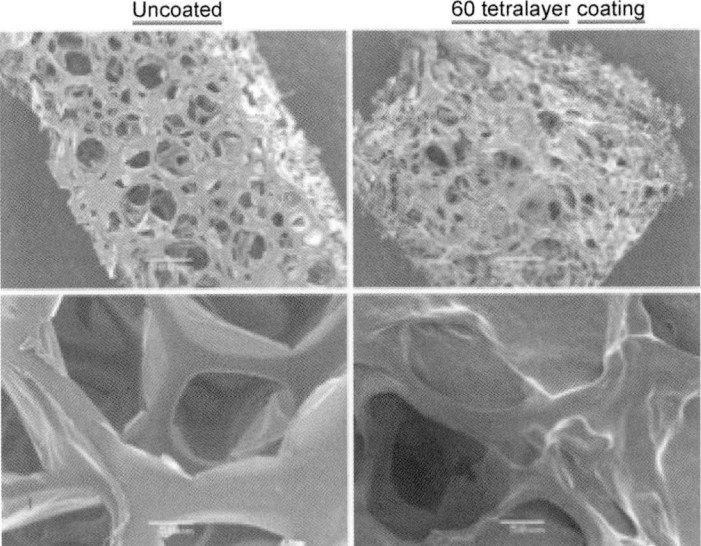

**Figure 17.22** SEM of bovine collagen based Surgifoam® bandages uncoated and spray coated with a (Poly 2/dextran sulfate/vancomycin/dextran sulfate) architecture. Reproduced from Ref 55.

## 17.4.2
### Coating of Complex Surfaces

One of the additional advantages of spray-LbL is the ability to coat not only large areas, but complex surfaces with nanometer to micron scale conformal coatings. We recently reported degradable LbL films that can be loaded with vaccine components, such as antigens or immunomodulators that upregulate the immune response, prepared on flexible substrates for transcutaneous vaccine delivery [76]. However, these planar multilayer transdermal patches required prior disruption of the dense upper layer of skin cells, the stratum corneum, to permit entry of released cargos from LbL films into the epidermis. It is possible to combine these technologies in a unique fashion to facilitate direct entry of the films into the deeper target areas of the skin where immunological dendritic cells reside. Using this pain-free microneedle platform, multiple drugs or vaccine molecules can be effectively released over extended time periods, based on the release rate of the film, without the need to apply multiple doses. This single-step transcutaneous delivery approach is adaptable to a number of different molecules and biomacromolecules, ranging from protein antigens to oligonucleotides, immunostimulatory cytokines, and small molecule drugs. The LbL coating of microneedles is presented in the schematic in Figure 17.23. Degradable poly(lactide-*co*-glycolide) microneedles are molded, and then modified with a spray-LbL coating of degradable polyelectrolyte multilayers loaded with the desired vaccine. PEMs were built through the alternating adsorption of the cationic poly(β-amino ester) Poly1 and anionic plasmid DNA (encoding firefly luciferase,

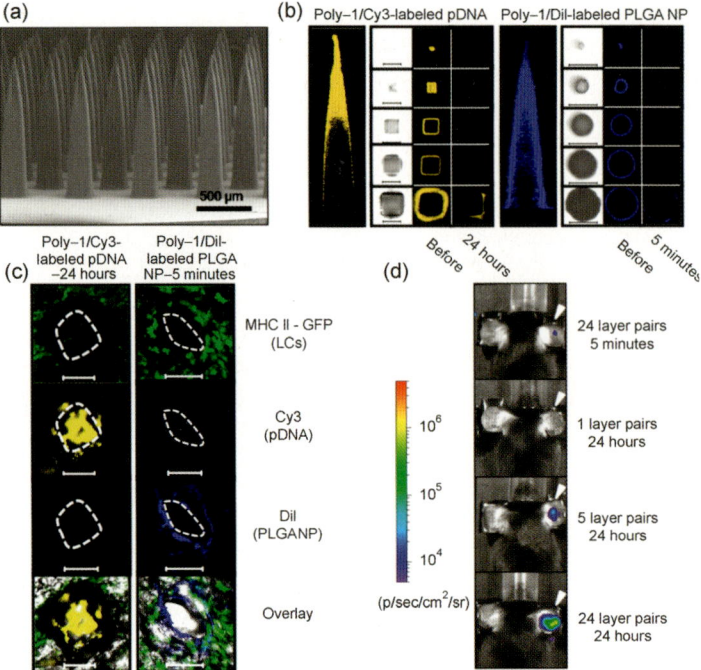

**Figure 17.23** Fabrication and transcutaneous delivery of pDNA and PLGA NP containing PEM films from PLGA microneedles. (a) SEM of PLGA microneedle skin patch array. (b) Confocal microscopy lateral section (left panels) and transverse z-sections (right panels) of microneedles coated with polyelectrolyte multilayer films carrying either Cy3-labeled pLUC or DiI-labeled PLGA nanoparticles. Transverse z-sections are shown before application and after application of the microneedle patches to murine skin for either 24 h or 5 min. Scale bars – 200 μm. (c) Confocal fluorescence images of MHC-II-GFP murine ear skin treated with either Cy3-pLUC- or DiI-PLGA NP-coated microneedles. Scale bar – 200 μm (d) *In vivo* bioluminescent signal observed in mice treated microneedles carrying luciferase-encoding plasmid DNA (1, 5, or 24 layer pairs) for 5 min or 24 h. Arrowheads denote ear where microneedle patches were applied. Reproduced from Ref 77.

pLUC) or PLGA nanoparticles loaded with DNA [77]. As shown in the confocal microscope images, it is clear that the microneedles are uniformly coated, indicating conformal multilayer deposition on the surfaces of coated microneedle arrays.

The microneedles have been shown to effectively penetrate the dorsal skin of mouse ears in an *in vivo* animal model; the DNA contained in the multilayer films is then released, enabling the transfection of luciferase-encoding DNA. The results of this study are summarized in Figure 17.23. We found that the transfection of DNA to the localized regions around the mouse ear was significant, and dose dependent, with higher amounts of DNA delivered and transfected for films with larger numbers of DNA-containing layer pairs. Time-dependent release was also observed over a time period of several days. We found that luciferase gene expression extended to over a week for samples with a range of drug loadings, indicating that the hydrolytic degradation of Poly1, which follows a similar time frame, is likely the primary mechanism of release in these films.

These findings suggest a new paradigm for delivery solutions for the immune system; transcutaneous methods are effective due to the high number of immune cells that reside in the skin, and the ability to use these cells to trigger antibody production and stimulate other immune cells, such as T cells, that may have memory of the antigen. It is also a means of delivering staged release of vaccine molecules, such as an initial dose followed by a booster several weeks later. The ability of PLGA nanoparticles to encapsulate a broad range of small-molecule drugs make this system potentially useful for delivering a variety of therapeutics into the skin without the need for hypodermic needles or trained personnel for injections. Ongoing studies will seek to evaluate this platform for transcutaneous vaccine delivery, specifically the potential to induce protective mucosal immunity.

### 17.4.3
### Using Kinetics to Manipulate Composition

A unique aspect of spray-LbL is the speed of the process of complexation that takes place when a polyion is sprayed onto a surface with oppositely charged polymer. When adsorption takes place in a dip-LbL solution experiment, there is localized depletion of polyelectrolyte during the adsorption process, which lowers the local polyion concentration at the surface. The long time period allowed for adsorption (of the order of several minutes) and relatively low localized surface concentrations experienced under dip equilibrium conditions allow sampling of the polyion with the surface, a high degree of interpenetration, complex formation, and small counterion ejection. In contrast, for the spray-LbL process, the entire surface of the multilayer film is exposed to a relatively high concentration of polyion over the period of a few seconds. Polyions will adsorb immediately to available surface sites, leading to kinetically trapped conformations, and more limited polyion sampling of the surface, rearrangement and exchange of ionic sites for complexation.

We found that we can take advantage of the kinetic nature of spray-LbL to build up films in which specific ionic species are trapped. Typical positively charged counterions used in LbL assembly, such as $Na^+$ or $K^+$, have relatively weak interactions with polyacids, such as polyacrylic acid, due to the differing nature of their electron orbital shells and polarizability; according to Pearson's HSAB classification of acids and bases, $Na^+$ is thought of as a "hard" acid counterion, and the carboxylate anion is considered to be "soft". On the other hand, "soft" counterions such as $Cu^{+2}$, $Fe^{+2}$ and $Ag^+$ are all much more polarizable, and have much stronger interactions with the carboxylate anion. For these reasons, the relative equilibrium constants for the formation of these two different kinds of salt complexes is about five orders of magnitude higher for the "soft" rather than the "hard" metal counterions. When salt complexes of polyacrylic acid with any of these ions are assemble using spray-LbL, it is possible to kinetically trap large amounts of excess metal counterions in the film, due to the kinetic constraints involved in polyelectrolyte adsorption, rearrangement, and complexation, to form intrinsic ion pairs with oppositely charged polymer within the multilayer film. Even following rinsing steps, these ions remain present in the film; the ion content in spray-LbL films is significantly higher than those observed with control films constructed using dip-LbL methods. For example, multilayer films constructed with PAA complexed with $Cu^{2+}$ ($PAA\text{-}Cu^{2+}$) layered with PAH contain approximately 12 wt% copper, compared to 5% in dip-LbL films made under the same conditions. This excess in metal ion content can be used to generate metal nanoparticles, or the free ions can be used to absorb and complex with other molecules; for example, we illustrate the reversible absorption of ammonia vapor with $Cu^{2+}$ loaded films, and the chelation of hydrogen cyanide with $Fe^{2+}$ films. In such systems, the films can be regenerated with appropriate post-treatments, making these systems interesting for a variety of applications. Importantly, this approach might be used to incorporate or embed other ionic species into multilayer films at high loadings.

Another interesting aspect of the kinetic trapping achieved with spray-LbL is the potential impact on the growth behavior of films that undergo interdiffusion. When polyions diffuse into the underlying multilayer film, the rate at which this diffusion process occurs depends on the polyelectrolyte type, charge density, and size. The 1–3 s time period for each polyion deposition step can, in some cases, approach the times required for effective molecular diffusion in the film. One example of this kind of system is the assembly of LbL films containing the antibiotic vancomycin with dextran sulfate and degradable Poly1 in tetralayer architectures. When comparing the dip-LbL and spray-LbL assembled thin films, very different film morphologies and characteristics were found [64]. As shown in Figure 17.24, very thick films are formed

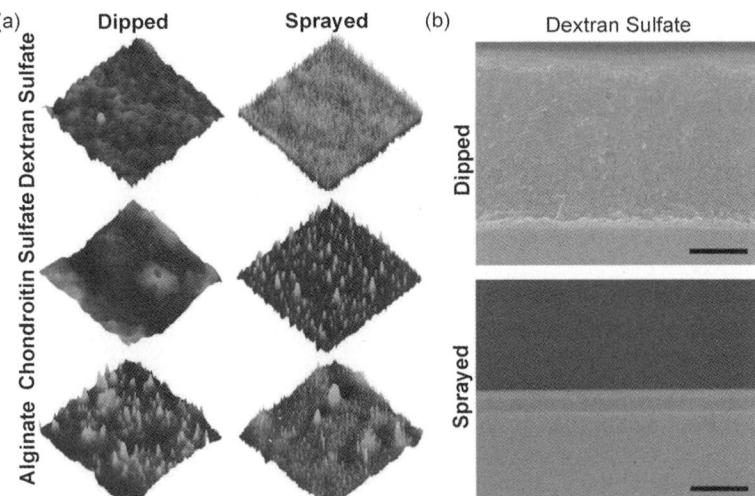

**Figure 17.24** (a) Atomic force microscope images for sprayed and dipped films of architecture (poly 2/polyanion/vancomycin/polyanion) (b) Scanning electron microscope cross-section images for (poly 2/dextran sulfate/vancomycin/dextran sulfate)$_{60}$ dipped and sprayed films (scale bar = 1 μm). Reproduced from Ref 64.

from dip-LbL due to the interdiffusion of film components, including Poly1 and vancomycin; however, the constant exposure to water baths for extended periods also permitted out-diffusion of vancomycin drug molecules, thus leading to thick films with high polyion content but lower drug loading. When these films are assembled using spray-LbL, we find that much thinner films are formed due to the rapid assembly times, which tend to prevent the ability of polyions to interdiffuse during assembly. In this case, vancomycin is also kinetically trapped within the film, yielding much thinner films, but with much higher drug density. By understanding this phenomena further, we can take advantage of the differences in these approaches to further control film form and function.

## 17.5
## Concluding Remarks

As outlined in this chapter, the LbL approach has advanced significantly in the past several years toward the development of new and practical techniques pertinent to surface coatings, functional thin films and active materials for device applications. Remarkably, this water-based assembly process has proven adaptable not only to the biomacromolecular species that are critical to new therapeutics and drug delivery approaches, but also to a broad range of inorganic and organic molecules and nanomaterials that are the core components of electrochemical energy devices, sensors, catalytic supports and reactive membranes. The ability to combine this wide range of materials to achieve mixing on the molecular to nanometer length scale provides a unique route to functional materials fabrication that is not accessible using more conventional means. Finally, the ability to adapt this multi-step process to a commercial platform that is robust and versatile, as well as low cost, is key to the continued advancement of the field as commercial enterprise becomes more interested in using the LbL technique. New automated methods of LbL film application such as the spray-LbL method, that are both rapid and applicable to many different kinds of surfaces and geometries, will lead to invention and further extension of these exciting materials systems into important areas of materials applications.

## References

1 Decher, G. (1997) Fuzzy nanoassemblies: Toward layered polymeric multicomposites. *Science*, **277**, 1232.

2 Decher, G. and Hong, J.-D. (1991) Buildup of ultrathin multilayer films by a self-assembly process, 1) consecutive adsorption of anionic and cationic bipolar amphiphiles on charged surfaces. *Makromol. Chem., Macromol. Symp.*, **46**, 321–327.

3 Hammond, P.T. (2004) Form and function in multilayer assembly: New applications at the nanoscale. *Adv. Mater.*, **16**, 1271–1293.

4 Boudou, T., Crouzier, T., Ren, K.F., Blin, G., and Picart, C. (2010) Multiple functionalities of polyelectrolyte multilayer films: New biomedical applications. *Adv. Mater.*, **22**, 441–467.

5 Detzel, C.J., Larkin, A.L., and Rajagopalan, P. (2011) Polyelectrolyte multilayers in tissue engineering. *Tissue Eng. Part B-Rev.*, **17**, 101–113.

6 Jewell, C.M. and Lynn, D.M. (2008) Multilayered polyelectrolyte assemblies as platforms for the delivery of DNA and other nucleic acid-based therapeutics. *Adv. Drug Deliv. Rev.*, **60**, 979–999.

7 Ji, Y.L., An, Q.F., Qian, J.W., Chen, H.L., and Gao, C.J. (2010) Nanofiltration membranes prepared by layer-by-layer self-assembly of polyelectrolyte. *Prog. Chem.*, **22**, 119–124.

8 Lutkenhaus, J.L. and Hammond, P.T. (2007) Electrochemically enabled polyelectrolyte multilayer devices: from fuel cells to sensors. *Soft Matter*, **3**, 804–816.

9 Schlenoff, J., Dubas, S., and Farhat, T. (2000) Sprayed polyelectrolyte multilayers. *Langmuir*, **16**, 9968–9969.

10 Izquierdo, A., Ono, S.S., Voegel, J.-C., Schaaf, P., and Decher, G. (2005) Dipping versus spraying: exploring the deposition conditions for speeding up layer-by-layer assembly. *Langmuir: ACS J. Surf. Colloids*, **21**, 7558–7567.

11 Krogman, K.C., Zacharia, N.S., Schroeder, S., and Hammond, P.T. (2007) Automated process for improved uniformity and versatility of layer-by-layer deposition. *Langmuir*, **23**, 3137–3141.

12 Krogman, K.C., Lowery, J.L., Zacharia, N.S., Rutledge, G.C., and Hammond, P.T. (2009) Spraying asymmetry into functional membranes layer-by-layer. *Nature Mater.*, **8**, 512–518.

13 Delongchamp, D.M. and Hammond, P.T. (2004) Highly ion conductive Poly (ethylene oxide)-base polymer electrolytes from hydrogen bonding. *Langmuir*, **20**, 5403–5411.

14 Lutkenhaus, J.L., McEnnis, K., and Hammond, P.T. (2007) Tuning the glass transition of and ion transport within hydrogen-bonded layer-by-layer assemblies. *Macromolecules*, **40**, 8367–8373.

15 DeLongchamp, D.M., Kastantin, M., and Hammond, P.T. (2003) High-contrast electrochromism from layer-by-layer polymer films. *Chem. Mater.*, **15**, 1575–1586.

16 Argun, A.A., Ashcraft, J.N., Herring, M.K., Lee, D.K.Y., Allcock, H.R., and Hammond, P.T. (2010) Ion conduction and water transport in polyphosphazene-based multilayers. *Chem. Mater.*, **22**, 226–232.

17 Farhat, T.R. and Hammond, P.T. (2005) Designing a new generation of proton-exchange membranes using layer-by-layer deposition of polyelectrolytes. *Adv. Func. Mater.*, **15**, 945–954.

18 DeLongchamp, D.M. and Hammond, P.T. (2004) Highly ion conductive poly (ethylene oxide)-based solid polymer electrolytes from hydrogen bonding layer-by-layer assembly. *Langmuir*, **20**, 5403–5411.

19 Ashcraft, J.N., Argun, A.A., and Hammond, P.T. (2010) Structure-property studies of highly conductive layer-by-layer assembled membranes for fuel cell PEM applications. *J. Mater. Chem.*, **20**, 6250–6257.

20 Argun, A.A., Ashcraft, J.N., and Hammond, P.T. (2008) Highly conductive, methanol resistant polyelectrolyte multilayers. *Adv. Mater.*, **20**, 1539–1543.

21 Arys, X., Jonas, A.M., Laschewsky, A., and Lagras, R. (2000) *Supramolecular Polymers* (ed. A. Ciferri), Marcel Dekker, pp. 505–564.

22 Bertrand, P., Jonas, A., Laschewsky, A., and Legras, R. (2000) Ultrathin polymer coatings by complexation of polyelectrolytes at interfaces: suitable materials, structure and properties. *Macromol. Rapid. Commun.*, **21**, 319–348.

23 Farhat, T.R. and Hammond, P.T. (2006) Engineering ionic and electronic conductivity in polymer catalytic electrodes using the layer-by-layer technique. *Chem. Mater.*, **18**, 41–49.

24 Farhat, T.R. and Hammond, P.T. (2006) Fabrication of a "soft" membrane electrode assembly using layer-by-layer technology. *Adv. Func. Mater.*, **16**, 433–444.

25 Lee, S.W., Gallant, B.M., Byon, H.R., Hammond, P.T., and Shao-Horn, Y. (2011) Nanostructured carbon-based electrodes: bridging the gap between thin-film lithium-ion batteries and electrochemical capacitors. *Energy Environ. Sci.*, **4**, 1972–1985.

26 Lee, S.W., Kim, B.S., Chen, S., Shao-Horn, Y., and Hammond, P.T. (2009) Layer-by-layer assembly of all carbon nanotube ultrathin films for electrochemical applications. *J. Am. Chem. Soc.*, **131**, 671–679.

27 Lee, S.W., Yabuuchi, N., Gallant, B.M., Chen, S., Kim, B.S., Hammond, P.T., and Shao-Horn, Y. (2010) High-power lithium batteries from functionalized carbon-nanotube electrodes. *Nature Nanotech.*, **5**, 531–537.

28 Kang, B. and Ceder, G. (2009) Battery materials for ultrafast charging and discharging. *Nature*, **458**, 190–193.

29 Huang, J.X. (2006) Syntheses and applications of conducting polymer polyaniline nanofibers. *Pure Appl. Chem.*, **78**, 15–27.

30 Hyder, M.N., Lee, S.W., Cebeci, F.C., Schmidt, D.J., Shao-Horn, Y., and Hammond, P.T. (2011) Layer-by-layer assembled polyaniline nanofiber/multiwall carbon nanotube thin film electrodes for high-power and energy storage applications. *ACS Nano.* vol. 5 (11), 8552–8561.

31 Lee, S.W., Kim, J., Chen, S., Hammond, P.T., and Shao-Horn, Y. (2010) Carbon nanotube/manganese oxide ultrathin film electrodes for electrochemical capacitors. *ACS Nano.*, **4**, 3889–3896.

32 Reches, M. and Gazit, E. (2003) Casting metal nanowires within discrete self-assembled peptide nanotubes. *Science*, **300**, 625–627.

33 Lvov, Y., Ariga, K., Onda, M., Ichinose, I., and Kunitake, T. (1997) Alternate assembly of ordered multilayers of $SiO_2$ and other nanoparticles and polyions. *Langmuir*, **13**, 6195–6203.

34 Lee, J.A., Nam, Y.S., Rutledge, G.C., and Hammond, P.T. (2010) Enhanced photocatalytic activity using layer-by-layer electrospun constructs for water remediation. *Adv. Funct. Mater.*, **20**, 2424–2429.

35 Nam, K.T., Kim, D.W., Yoo, P.J., Chiang, C.Y., Meethong, N., Hammond, P.T., Chiang, Y.M., and Belcher, A.M. (2006) Virus-enabled synthesis and assembly of nanowires for lithium ion battery electrodes. *Science*, **312**, 885–888.

36 Lowman, G.M. and Hammond, P.T. (2005) Solid-state dye-sensitized solar cells combining a porous $TiO_2$ film and a layer-by-layer composite electrolyte. *Small*, **1**, 1070–1073.

37 Yoo, P.J., Nam, K.T., Qi, J.F., Lee, S.K., Park, J., Belcher, A.M., and Hammond, P.T. (2006) Spontaneous assembly of viruses on multilayered polymer surfaces. *Nature Mater.*, **5**, 234–240.

38 Nam, K.T., Wartena, R., Yoo, P.J., Liau, F.W., Lee, Y.J., Chiang, Y.M., Hammond, P.T., and Belcher, A.M. (2008) Stamped microbattery electrodes based on self-assembled M13 viruses. *Proc. Natl. Acad. Sci. USA*, **105**, 17227–17231.

39 Tokuhisa, H. and Hammond, P.T. (2003) Solid-state photovoltaic thin films using $TiO_2$, organic dyes, and layer-by-layer polyelectrolyte nanocomposites. *Adv. Func. Mater.*, **13**, 831–839.

40 Hiller, J. and Rubner, M.F. (2003) Reversible molecular memory and pH-switchable swelling transitions in polyelectrolyte multilayers. *Macromolecules*, **36**, 4078–4083.

41 Zhai, L., Nolte, A.J., Cohen, R.E., and Rubner, M.F. (2004) pH-gated porosity transitions of polyelectrolyte multilayers in confined geometries and their application as tunable Bragg reflectors. *Macromolecules*, **37**, 6113–6123.

42 Lowman, G.M. and Hammond, P.T. (2005) Solid-state dye-sensitized solar cells combining porous $TiO_2$ film and a layer-by-layer composite electrolyte. *Small*, **1**, 1070–1073.

43 DeLongchamp, D.M. and Hammond, P.T. (2004) High-contrast electrochromism and controllable dissolution of assembled Prussian blue/polymer nanocomposites. *Adv. Funct. Mater.*, **14**, 224–232.

44 DeLongchamp, D.M. and Hammond, P.T. (2004) Multiple-color electrochromism from layer-by-layer-assembled polyaniline/Prussian Blue nanocomposite thin films. *Chem. Mater.*, **16**, 4799–4805.

45 DeLongchamp, D.M. and Hammond, P.T. (2005) Electrochromic polyaniline films from layer-by-layer assembly. *Chromogen. Phenom. Polym.*, **888**, 18–33.

46 Vazquez, E., Dewitt, D.M., Hammond, P.T., and Lynn, D.M. (2002) Construction of hydrolytically-degradable thin films via layer-by-layer deposition of degradable polyelectrolytes. *J. Am. Chem. Soc.*, **124**, 13992–13993.

47 Fu, K., Klibanov, A.M., and Langer, R. (2000) Protein stability in controlled-release systems. *Nat. Biotechnol.*, **18**, 24–25.

48 Weiner, L.M., Surana, R., and Wang, S. (2010) Monoclonal antibodies: versatile platforms for cancer immunotherapy. *Nat. Rev. Immunol.*, **10**, 317–327.

49 Goldman, R. (2004) Growth factors and chronic wound healing: past, present, and future. *Adv. Skin. Wound Care*, **17**, 24–35.

50 Hammond, P.T. and Park, J. (2004) Multilayer transfer printing for polyelectrolyte multilayer patterning: Direct transfer of layer-by-layer assembled micropatterned thin films. *Adv. Mater.*, **16**, 520.

51 Hammond, P.T., Jiang, X.P., Zheng, H.P., and Gourdin, S. (2002) Polymer-on-polymer stamping: Universal approaches to chemically patterned surfaces. *Langmuir*, **18**, 2607–2615.

52 Langer, R., Khademhosseini, A., Suh, K.Y., Yang, J.M., Eng, G., Yeh, J., and Levenberg, S. (2004) Layer-by-layer deposition of hyaluronic acid and poly-L-lysine for patterned cell co-cultures. *Biomaterials*, **25**, 3583–3592.

53 Clark, S.L., Montague, M.M., and Hammond, P.T. (1997) Patterning of layer-by-layer polyion films using SAMs as molecular templates. *Abstr. Pap. Am. Chem. S.*, **214**, 88-Pms.

54 Macdonald, M., Rodriguez, N.M., Smith, R., and Hammond, P.T. (2008) Release of a model protein from biodegradable self assembled films for surface delivery applications. *J. Control. Release*, **131**, 228–234.

55 Macdonald, M.L., Rodriguez, N.M., Shah, N.J., and Hammond, P.T. (2010) Characterization of tunable FGF-2 releasing polyelectrolyte multilayers. *Biomacromolecules*, **11**, 2053–2059.

56 Macdonald, M.L., Samuel, R.E., Shah, N.J., Padera, R.F., Beben, Y.M., and Hammond, P.T. (2011) Tissue integration of growth factor-eluting layer-by-layer polyelectrolyte multilayer coated implants. *Biomaterials*, **32**, 1446–1453.

57 Thies, R.S., Bauduy, M., Ashton, B.A., Kurtzberg, L., Wozney, J.M., and Rosen, V. (1992) Recombinant human bone morphogenetic protein-2 induces osteoblastic differentiation in W-20-17 stromal cells. *Endocrinology*, **130**, 1318–1324.

58 Samee, M., Kasugai, S., Kondo, H., Ohya, K., Shimokawa, H., and Kuroda, S. (2008) Bone morphogenetic protein-2 (BMP-2) and vascular endothelial growth factor (VEGF) transfection to human periosteal cells enhances osteoblast differentiation and bone formation. *J. Pharmacol. Sci.*, **108**, 18–31.

59 Peng, H., Wright, V., Usas, A., Gearhart, B., Shen, H.C., Cummins, J., and Huard, J. (2002) Synergistic enhancement of bone formation and healing by stem cell-expressed VEGF and bone morphogenetic protein-4. *J. Clin. Invest.*, **110**, 751–759.

60 Shah, N.J., Macdonald, M.L., Beben, Y.M., Padera, R.F., Samuel, R.E., and Hammond, P.T. (2011) Tunable dual growth factor delivery from polyelectrolyte multilayer films. *Biomaterials*, **32**, 6183–6193.

61 Chuang, H.F., Smith, R.C., and Hammond, P.T. (2008) Polyelectrolyte multilayers for tunable release of antibiotics. *Biomacromolecules*, **9**, 1660–1668.

62 Lavalle, P., Gergely, C., Cuisinier, F.J.G., Decher, G., Schaaf, P., Voegel, J.C., and Picart, C. (2002) Comparison of the structure of polyelectrolyte multilayer films exhibiting a linear and an exponential growth regime: An in situ atomic force microscopy study. *Macromolecules*, **35**, 4458–4465.

63 Moskowitz, J.S., Blaisse, M.R., Samuel, R.E., Hsu, H.-P., Harris, M.B., Martin, S.D., Lee, J.C., Spector, M., and Hammond, P.T. (2010) The effectiveness of the controlled release of gentamicin from polyelectrolyte multilayers in the treatment of Staphylococcus aureus

infection in a rabbit bone model. *Biomaterials*, **31**, 6019–6030.

64 Shukla, A., Avadhany, S.N., Fang, J.C., and Hammond, P.T. (2010) Tunable vancomycin releasing surfaces for biomedical applications. *Small*, **6**, 2392–2404.

65 Jessel, N.B., Schwinte, P., Donohue, R., Lavalle, P., Boulmedais, F., Darcy, R., Szalontai, B., Voegel, J.C., and Ogier, J. (2004) Pyridylamino-beta-cyclodextrin as a molecular chaperone for lipopolysaccharide embedded in a multilayered polyelectrolyte architecture. *Adv. Func. Mater.*, **14**, 963–969.

66 Smith, R.C., Riollano, M., Leung, A., and Hammond, P.T. (2009) Layer-by-layer platform technology for small-molecule delivery. *Angew. Chem. Int. Ed.*, **48**, 8974–8977.

67 Shukla, A., Fuller, R.C., and Hammond, P.T. (2011) Design of multi-drug release coatings targeting infection and inflammation. *J. Control. Release*, **155**, 159–166.

68 Kim, B.S., Lee, H.I., Min, Y., and Poon, Z.Y. (2009) Hydrogen-bonded multilayer of pH-responsive polymeric micelles with tannic acid for surface drug delivery. *Chem. Comm.*, **28**, 4194–4196.

69 Soballe, K. and Overgaard, S. (1996) The current status of hydroxyapatite coating of prostheses. *J. Bone Joint Surg. Br.*, **78**, 689–691.

70 Engh, C.A., Bobyn, J.D., and Glassman, A.H. (1987) Porous-coated hip replacement. The factors governing bone ingrowth, stress shielding, and clinical results. *J. Bone Joint Surg. Br.*, **69**, 45–55.

71 Geesink, R.G. and Hoefnagels, N.H. (1995) Six-year results of hydroxyapatite-coated total hip replacement. *J. Bone Joint Surg. Br.*, **77**, 534–547.

72 Wood, K.C., Boedicker, J.Q., Lynn, D.M., and Hammond, P.T. (2005) Tunable drug release from hydrolytically degradable layer-by-layer thin films. *Langmuir*, **21**, 1603–1609.

73 Wood, K.C., Chuang, H.F., Batten, R.D., Lynn, D.M., and Hammond, P.T. (2006) Controlling interlayer diffusion to achieve sustained, multiagent delivery from layer-by-layer thin films. *Proc. Natl. Acad. Sci. USA*, **103**, 10207–10212.

74 Hong, J., Shah, N.J., Drake, A., Lee, J.B., Chen, J., and Hammond, P.T. (2011) Graphene multilayer dressings for sequential release of proteins. *ACS Nano* 2012 **6** (1), 81–88.

75 Shukla, A., Fang, J.C., Puranam, S., and Hammond, P.T. (2011) Release of vancomycin from multilayer coated absorbent gelatin sponges. *J. Control. Release*, 2012, **157** (1), 64–71.

76 Su, X.F., Kim, B.S., Kim, S.R., Hammond, P.T., and Irvine, D.J. (2009) Layer-by-layer-assembled multilayer films for transcutaneous drug and vaccine delivery. *ACS Nano.*, **3**, 3719–3729.

77 DeMuth, P.C., Su, X.F., Samuel, R.E., Hammond, P.T., and Irvine, D.J. (2010) Nano-layered microneedles for transcutaneous delivery of polymer nanoparticles and plasmid DNA. *Adv. Mater.*, **22**, 4851.

# 18
# Surface-Initiated Polymerization and Layer-by-Layer Films
*Nicel Estillore and Rigoberto C. Advincula*

## 18.1
## Introduction

Since the dissemination of the layer-by-layer (LbL) technique in the early 1990s by Decher and Hong, this technique has evolved into one of the notably regarded methods for the preparation of highly ordered but interpenetrated multilayer ultrathin films [1–3]. The versatility and simplicity of the LbL method are among the attractive traits that drive researchers to further explore the possibilities of this technique. The broad range of commercially available polyelectrolyte pairs provides a useful tool in fabricating multilayered thin films tailored to a specific application. In general, by having an initial charge on the surface, the consecutive deposition of oppositely charged polyelectrolytes is observed, based primarily on strong electrostatic interactions.

For a long time, polymers have been primarily used to protect or improve the function and property of a material with its environment, but with the modern era of advancing technology, polymer surface coatings have extended beyond the decorative and protective aspects. Today, surface modifications by polymer brushes are becoming increasingly important for fundamental research as well as for industrial applications [4–6]. Although there have been countless reviews on the LbL as a stand-alone technique, the combined LbL and surface-initiated polymerization (SIP) approach is a relatively new area of research.

The remainder of this chapter is divided into three major sections. The first provides an overview of polymer brushes and their synthesis. The second focuses on the combined LbL and SIP techniques for the fabrication of stimuli-responsive brush surfaces. Within this section three different strategies for the adsorption of oppositely charged, water-soluble macroinitiators onto various substrates are outlined. Finally, several applications of these LbL-SIP brush modified surfaces are presented.

---

*Multilayer Thin Films: Sequential Assembly of Nanocomposite Materials*, Second Edition.
Edited by Gero Decher and Joseph B. Schlenoff.
© 2012 Wiley-VCH Verlag GmbH & Co. KGaA. Published 2012 by Wiley-VCH Verlag GmbH & Co. KGaA.

## 18.2
### Overview of Surface-Grafted Polymer Brushes

Polymer chains that are tethered by one end to a surface with sufficiently high grafting densities are termed polymer brushes [7, 8]. The steric hindrance, imposed by the close proximity of the grafting points, forces the polymer chains to adopt a more stretched out or extended conformation away from the tethering site [7, 8]. Polymer brushes have been used in various "smart" surface coating applications due to their ability to undergo reversible conformational changes as a result of an external stimulus, that is, solvent, temperature, light, pH, and electric field. In the last decade or so, polymer brushes have become increasingly popular for a variety of potential applications ranging from materials science to biomedical [4–6].

At low grafting densities, the pancake and mushroom morphologies are also possible, depending on the strength of interaction between the polymer and substrate/solvent, as illustrated in Figure 18.1 [7]. Therefore, careful control of the initiator density is crucial in assuring that a high grafting density is achieved for the polymer chains to grow in the brush regime. Although experimental parameters such as polymerization time, monomer concentration, solvent medium, and so on, can all be varied to optimize the brush growth, the most convenient way is to control the initiator density. Perhaps the easiest way of controlling the initiator density is to vary the initiator concentration [9, 10] but other more advanced methods, including photodecomposition [11, 12], gradient approach [13, 14], and Langmuir-Schaefer techniques [15], have been reported. Recently, the combined LbL and surface-initiated polymerization (SIP) techniques have been shown to effectively regulate the initiator density through the repetitive deposition of oppositely charged macroinitiators [16].

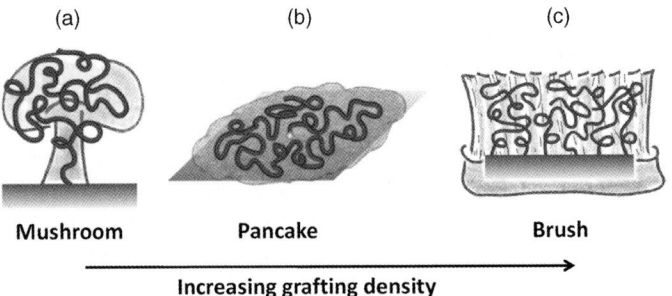

**Figure 18.1** Morphologies of surface-grafted polymer chains: (a) mushroom, (b) pancake, and (c) brush. With increasing grafting density, the brush regime is attained while at the low grafting densities, the mushroom and pancake are possible [7].

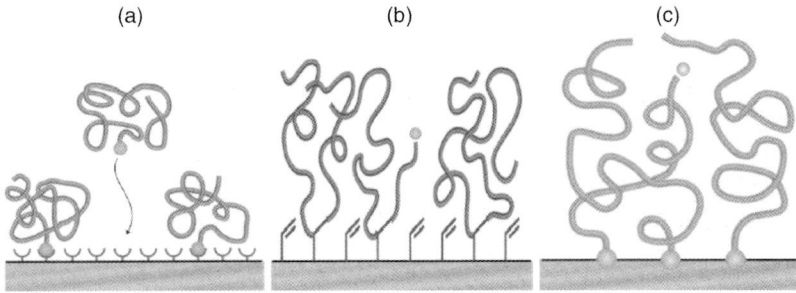

**Figure 18.2** Different strategies for the growth of polymer brushes: (a) "grafting to", (b) "grafting through", and (c) "grafting from" or surface-initiated polymerization (SIP) [7].

### 18.2.1
### Synthesis of Polymer Brushes

Polymer brushes are synthesized either through physisorption or chemisorption methods [7, 17]. Because the chemisorption method, namely "grafting to" and "grafting from", results in the covalent attachment of the polymers to the surfaces, it is more widely used for the synthesis of robust polymer brush films (Figure 18.2). Although the "grafting to" is a suitable method for the covalent attachment of polymer brushes to reactive groups inherent to a surface, this method is self-limiting in that as soon as the preformed polymer attaches to the surface, steric hindrance starts to accumulate, thereby preventing the diffusion of the incoming polymer molecules to the available anchor groups on the surface [7].

In order to circumvent the diffusion problem associated with the "grafting to" method, the "grafting from", more commonly known as the surface-initiated polymerization (SIP) method has been widely explored for the growth of thick polymer brushes [7, 18]. In this method, monomers grow directly from surfaces functionalized with initiators, thereby producing thick and dense polymer brushes.

### 18.2.2
### Surface-Initiated Atom Transfer Radical Polymerization (SI-ATRP)

Atom transfer radical polymerization (ATRP) is arguably the most widely used controlled or living radical polymerization (CLRP) technique for the synthesis of well-defined polymer brushes with controlled molecular weights and narrow polydispersity index (PDI) [18, 19]. Furthermore, ATRP allows the formation of block copolymers by simply reinitiating the dormant chain ends and subsequently extending the polymer chains. One of the components essential in conducting ATRP is an initiator, which is typically a low molecular weight alkyl halide (i.e., ethyl-2-bromoisobutyrate). However, when the initiating moiety is already attached to a polymer chain, it is referred to as a macroinitiator. As will be discussed in the later sections, several groups have synthesized water-soluble ATRP macroinitiators to allow their surface attachment to be conducted under aqueous conditions.

## 18.3
## Layer-by-Layer (LbL) Self-Assembly

The LbL method has evolved into a well-established technique for the fabrication of multilayered ultrathin films [1–3]. Coating of this type can be applied to any surfaces that can tolerate a water-based LbL adsorption process. Because both planar and colloidal substrates can be utilized a wider spectrum of applications, ranging from electronics to biotechnology, is now realized [20]. The LbL technique is based on the consecutive deposition of complementary polyelectrolyte pairs which are held together through strong ionic interactions. This repetitive process gives one control over the film composition and thickness from the angstrom to the micrometer scale. Other types of non-covalent forces of interaction, like H-bonding, ion–dipole interaction, complexation, and so on, should enable the alternate adsorption of complementary polymers, small molecules, or nanoparticles [1–3]. The complimentary polyelectrolyte pairing can be depicted ideally as a stratified structure. However, significant interpenetration of the polyelectrolyte chains has been observed, and the resulting multilayer film has been described as an interdigitated yet stratified structure, exhibiting some order of the internal composition [1–3].

#### 18.3.1.1 Water-Soluble ATRP Macroinitiators

Typically, low molecular weight initiators are employed for SIP studies and are mainly based on silane and thiol anchoring groups [7, 18]. Although these types of surface-bound initiators are widely accepted, they often suffer from several problems. The coupling method is usually specific for an initiator/substrate combination. For example, oxide substrates, including silicon (Si) wafer, glass, quartz, and indium tin oxide (ITO), have to be coupled with silane-based initiators, while thiol-based initiators are utilized for gold (Au) substrates. In addition, the synthesis of these initiators involves expensive and/or toxic reagents, such as $H_2PtCl_6$ and $HSiCl_3$ [21]. Also, alkoxy silanes have a limited shelf life due to possible hydrolysis. Furthermore, the Au–S bonds are unstable over time due to possible redox cleavage and the preparation of Au substrates is an expensive process [22]. In general, the formation of initiators from a self-assembled monolayer (SAM) necessitates their surface attachment to be conducted under anhydrous conditions [21, 23]. In other instances, the surface attachment of initiators is performed in multiple reaction steps occurring *in situ* on the substrate which can be time consuming and not amenable to polymeric substrates [24].

Therefore, water-soluble macroinitiators provide an attractive alternative to the conventional surface-bound initiators [21]. In particular, the synthesis of polyelectrolyte ATRP macroinitiators is increasingly popular due to its simple reaction and post modification steps [21]. Armes and other groups have reported the synthesis of

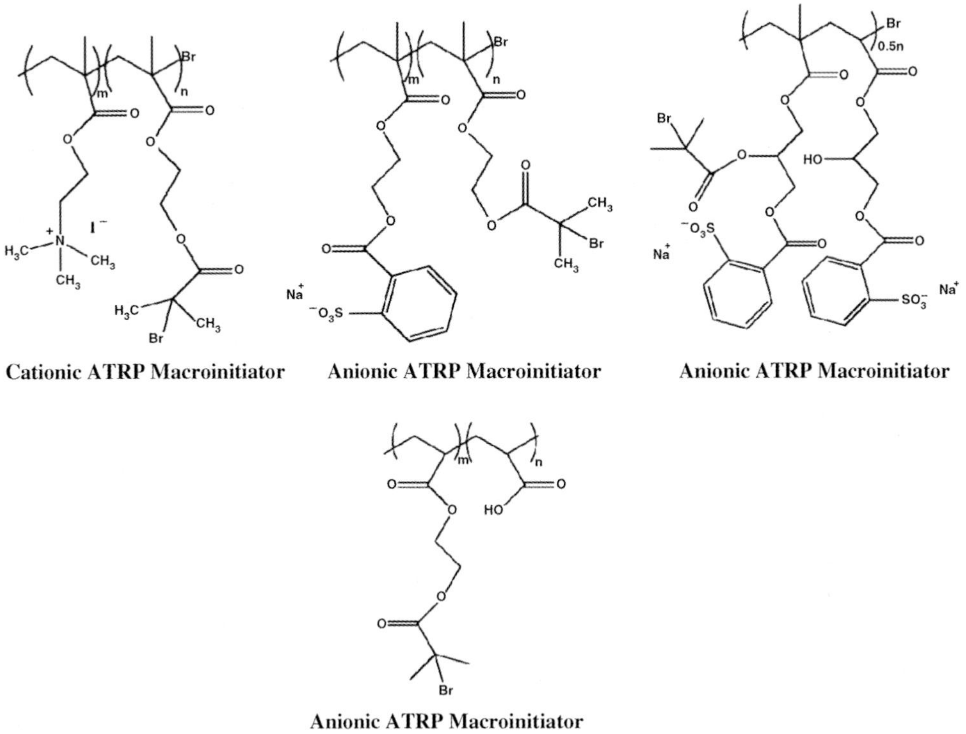

**Cationic ATRP Macroinitiator**   **Anionic ATRP Macroinitiator**   **Anionic ATRP Macroinitiator**

**Anionic ATRP Macroinitiator**

**Figure 18.3** Structures of the cationic and anionic ATRP macroinitiators utilized for the LbL deposition process [25–29].

both cationic and anionic ATRP macroinitiators, as shown in Figure 18.3 [21, 25–32]. The use of water-soluble macroinitiators can be applied to coat various substrates, including aqueous dispersions of particles [25–27]. By having an initial surface charge, the deposition of a single or multiple layer(s) of polyelectrolyte macroinitiators is possible. Thus, the surface attachment of polyelectrolyte macroinitiators can be done under ambient conditions without the need for expensive reagents or experimental set-up.

## 18.4
## Combined LbL-SIP Approach

Recently, the combined LbL and SIP techniques have become increasingly attractive for the fabrication of stimuli-responsive brush surfaces [21]. The ability to fine tune the multilayer assembly and decorate the outer surface with polymer brushes enables interesting "smart" coatings to be prepared. The novelty in the LbL-SIP approach lies in the preparation of dual responsive surfaces where the polyelectrolyte assembly and the polymer brushes behave differently to an external stimulus. Moreover, the LbL technique allows the brush thickness to be easily and precisely controlled for specific applications. Scheme 18.1 illustrates a general scheme for the preparation of brush-modified surfaces through the combined LbL-SIP approach [33].

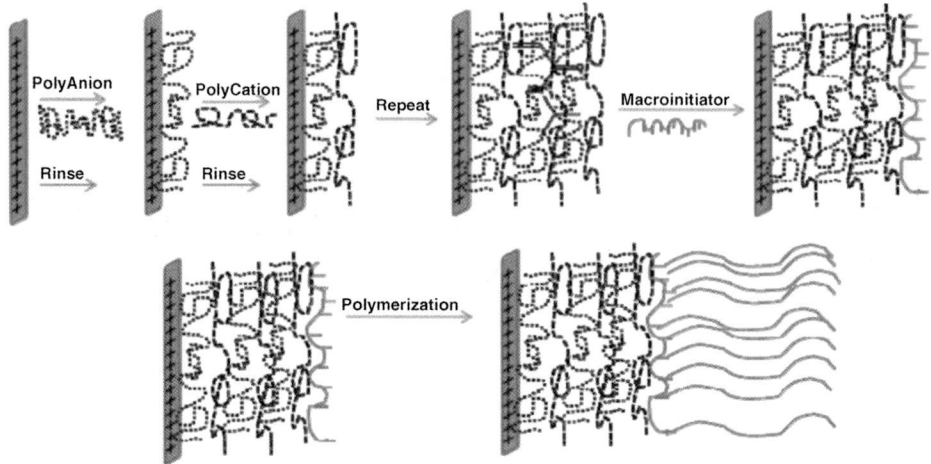

**Scheme 18.1** General scheme for the combined LbL-SIP approach on a flat substrate [33].

There are three different strategies for the deposition of polyelectrolyte ATRP macroinitiators onto planar and colloidal surfaces: (i) adsorption of a single layer of a cationic or anionic macroinitiator onto a charged substrate; (ii) adsorption of a single or multiple layer(s) of oppositely charged macroinitiators onto a pre-existing multilayer assembly; and (iii) sole adsorption of polyelectrolyte macroinitiators onto a charged substrate.

### 18.4.1
**Single Adsorption of a Polyelectrolyte ATRP Macroinitiator onto a Charged Surface**

In order to probe the versatility and efficiency of these newly synthesized ATRP macroinitiators, a single layer of either the cationic or anionic macroinitiator was directly deposited onto charged substrates (both flat and spherical) [25–28]. For example, Armes and coworkers reported the deposition of a cationic ATRP macroinitiator onto ultrafine inorganic silica sols followed by the aqueous ATRP of various methacrylate and acrylamide monomers (Scheme 18.2) [25, 26]. The same group has also reported the use of ultrafine cationic Ludox (CL) silica sols for the direct attachment of an anionic ATRP macroinitiator followed by the polymerization of various hydrophilic monomers in aqueous media at room temperature [27]. These sterically stabilized particles were characterized using techniques such as dynamic light scattering, transmission electron microscopy (TEM), aqueous electrophoresis, fourier transform infrared (FTIR) spectroscopy, and thermal analysis. The surface

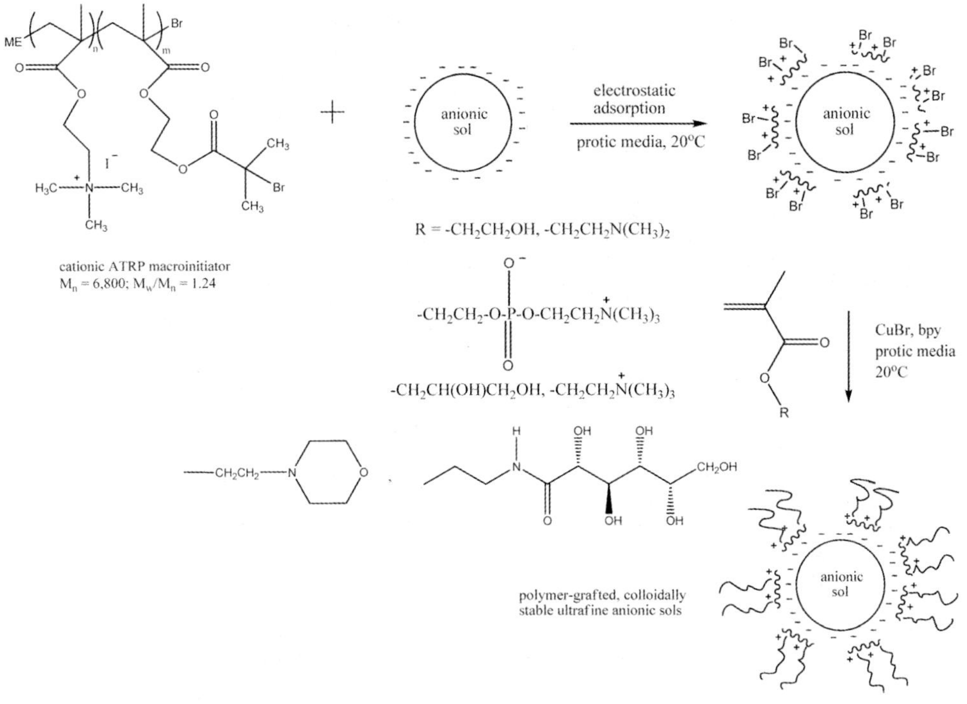

**Scheme 18.2** Surface attachment of a cationic ATRP macroinitiator onto a spherical particle [25, 26].

attachment of an anionic macroinitiator onto an aminated planar substrate and the SI-ATRP growth of poly(2-hydroxyethyl methacrylate) (PHEMA) brushes has also been reported [28]. This macroinitiator-brush modified substrate was further applied toward the fabrication of patterned surfaces which displayed micrometer-sized features, as shown in Figure 18.4.

In all these cases, the polyelectrolyte ATRP macroinitiators proved to be efficient in producing well-defined and controlled brush composition and thickness. However, the control of initiator density is crucial in assuring that the polymer chains have a sufficient grafting density such that they will be forced to grow or fully extend away from the tethering site [7]. Therefore, several groups have exploited the combined LbL and SIP techniques for growing well-controlled polymer brushes.

### 18.4.2
### Adsorption of Oppositely Charged ATRP Macroinitiator(s) onto an LbL Deposited Multilayer Assembly

In 2006, our group reported one of the earliest studies demonstrating the combined LbL-SIP approach which involved the synthesis of fuzzy particles based on the SI-ATRP onto the LbL polyelectrolyte-modified macroinitiator PS particles

**Figure 18.4** Optical microscope image of the patterned PHEMA brushes grown from a microcontact-printed cationic ATRP macroinitiator. The inset shows the AFM image of the PHEMA brush surface (z-scale 150 nm, feature height is ca. 54 nm) [28].

(Scheme 18.3) [29]. The multilayer build-up of the poly(diallyl-dimethylammonium chloride)/poly(acrylic acid) (PDADMAC/PAA) core shell was monitored using the ζ-potential. Figure 18.5 shows the alternating surface charges after the deposition of PDADMAC and PAA with variations in the potential ranging from ±40 to 45 mV. In order to incorporate the ATRP initiator, a simple esterification reaction of the PAA with 2-bromoisobutyryl bromide (ATRP initiator) yielded the esterified PAA macro-

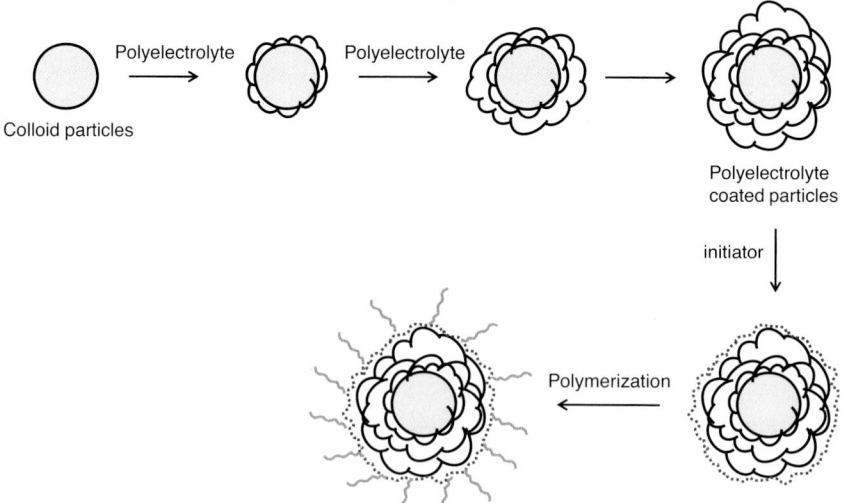

**Scheme 18.3** Schematic illustration for the LbL multilayer assembly of the polyelectrolyte core shell decorated with an anionic ATRP macroinitiator [29].

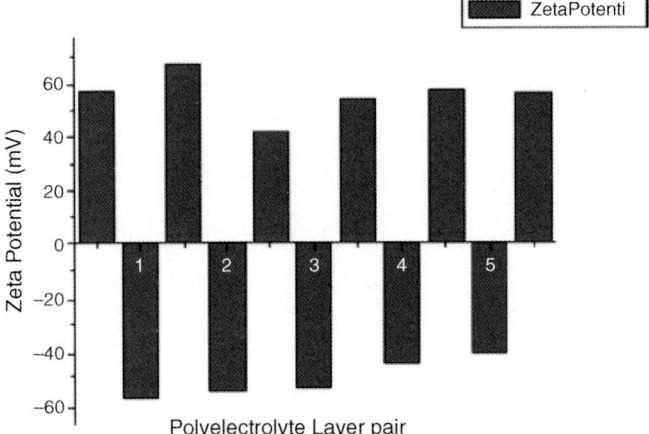

**Figure 18.5** Electrophoretic measurements displaying the alternating charges after each deposited polyelectrolyte pair [29].

initiator. The PAA was only 30% functionalized with the bromoisobutyryl initiator so as to allow it to be adsorbed onto the prior PDADMAC layer. By having the ATRP macroinitiator as the outermost layer, the polymerization of methyl methacrylate (MMA) was conducted successfully. TGA measurements were performed on the PS-modified polyelectrolyte/PMMA brush particles. The thermograms clearly showed the decomposition of the PS core at 413 °C while at 192 °C they revealed a 7% mass increase in the particles after the deposition of 12 layers. Moreover, the peak maximum at 387 °C was attributed to the decomposition of the PMMA brush. TEM was also utilized to observe changes in the particle shape and size before and after the depositions and brush growth. Although there was no change in the shape of the particles after the LbL depositions, the TEM images confirmed the deformation of the particles after the SIP growth of the PMMA brushes, as displayed in Figure 18.6. Furthermore, there was a 50 nm increase in diameter size of the PS particles after the deposition of six layer pairs of PDADMAC/PAA. More interestingly, the size of the particle changed dramatically changed between 1100 and 1500 nm after the grafting of the PMMA brushes.

Several other groups have also demonstrated similar LbL-SIP strategies. For instance, the LbL deposition of poly(styrene sulfonate)/PDADMAC (PSS/PDADMAC) and a polycationic ATRP macroinitiator layer onto polyethersulfone (PES) porous membranes was reported by Bruening and coworkers [30]. They prepared two films with different compositions: (i) alternating deposition of (polycationic macroinitiator/PSS)$_n$ and (2) (PDADMAC/PSS)$_n$ and a single layer of the polycationic macroinitiator. With the alternate deposition of the PSS and polycationic macroinitiator they concluded that each layer with the macroinitiator throughout the film could act as an initiation site for the polymer brush growth. A brush thickness of 160 nm was obtained after the ATRP growth of PHEMA brushes in a relatively short

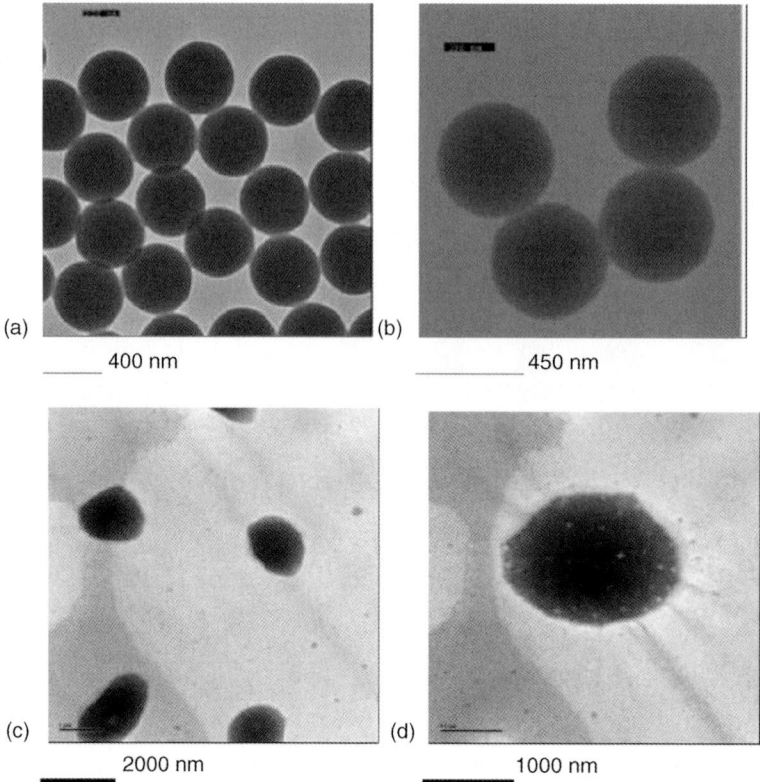

**Figure 18.6** TEM images of the (a) pristine PS particles 400 nm, (b) PS particles with a six layer pair shell of ~450 nm showing a roughened surface, and (c, d) PS particles after the pMMA brush polymerization showing an increase in particle size of ~1100–1500 nm [29].

time of 2 h. Kinetic studies for the brush growth were also performed on the single deposition of the polycationic macroinitiator on top of the (PDADMAC/PSS)$_n$ layers.

By employing complementary polyelectrolytes with weakened electrostatic interactions, the adsorption of a single layer of an anionic ATRP macroinitiator and the subsequent brush growth can be effectively controlled [31, 32]. Laschewsky and coworkers investigated how a nonlinear growth of the interlayer assembly composed of various polyelectrolyte pairs affected the adsorption of a single macroinitiator and the thickness of the resulting polymer brush [31, 32]. When a fluorescently labeled PDADMAC was incorporated into the interlayer assembly with a sequence of {PEI/PSS/F-PDADMAC/PSS)$_n$}, a nonlinear growth was observed, as shown in Figure 18.7a. The diffusion of a polyelectrolyte in and out of the previously deposited layer seems to be involved during the adsorption process, leading to the nonlinear growth [32, 34]. It was concluded that the thickness of the adsorbed macroinitiator increases with the number of cycles $n$ of the interlayer assembly (Figure 18.7b). Consequently, the thickness of the grafted thermoresponsive polymer brushes also increased proportionally (Figure 18.7c and d).

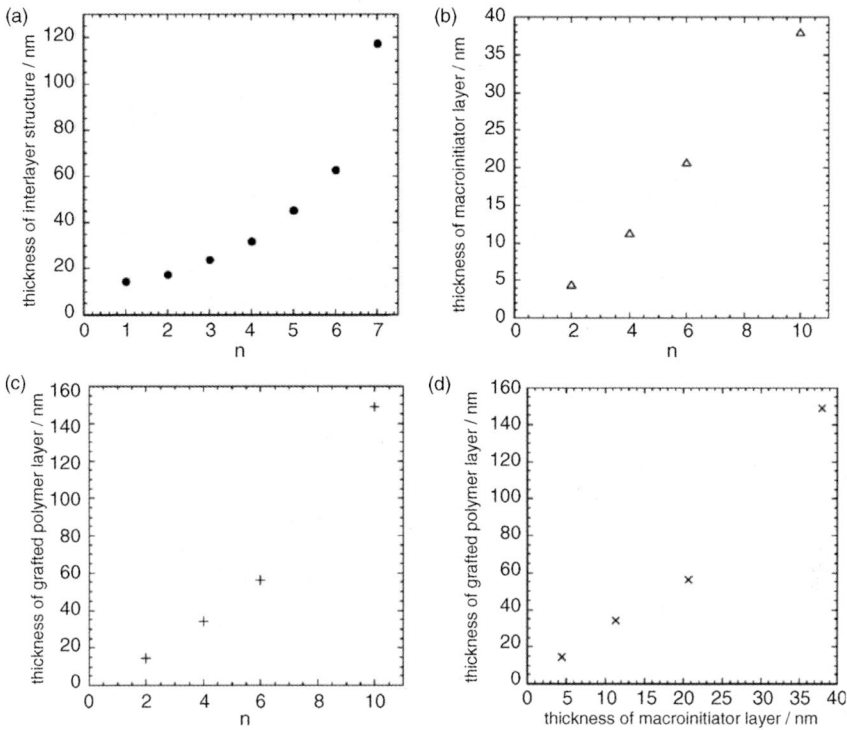

**Figure 18.7** Ellipsometry plots of the layer system {PEI/PSS/(F-PDADMAC/PSS)$_n$/F-PDADMAC/MA01/graft-MEO$_2$MA-co-OEGMA 475} (n is the number of interlayer pairs deposited): (a) nonlinear (or exponential) growth of the interlayer assembly, {PEI/PSS/(F-PDADMAC/PSS)$_n$. (b) Dependence of the thickness of the macroinitiator (MA01) on n. (c) Dependence of the thickness of the grafted polymer brush on n. (d) Correlation of the grafted polymer brush thickness with the thickness of the macroinitiator [32].

### 18.4.3
### Multiple Adsorption of Oppositely Charged ATRP Macroinitiators for Enhanced Initiator Density

Critical in achieving dense and thick polymer brushes is a high initiator density [7]. Armes and coworkers demonstrated the sole deposition of oppositely charged ATRP macroinitiators on functionalized planar substrates (Scheme 18.4) [16]. By having multiple layers of the macroinitiators deposited on the surface, the graft density increases, thus achieving highly thick and dense polymer brushes. Based on the thickness data, a linear relationship was observed between the LbL film thickness and the number of layers deposited on the surface (Figure 18.8a). In addition, the use of dual polarization interferometry (DPI) confirmed that the adsorbed amount of macroinitiators was directly proportional to the surface density of the 2-bromoester groups. Specifically, the bromoester initiator density was estimated to be

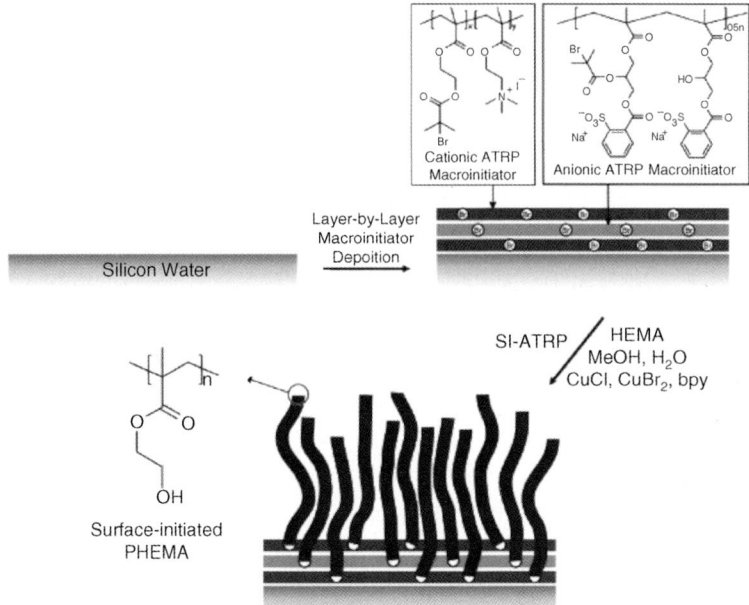

**Scheme 18.4** General scheme illustrating the deposition of oppositely charged ATRP macroinitiators onto a planar substrate [16].

$4.0 \pm 0.2\,\text{nm}^{-2}$ for 17 layers of macroinitiators deposited. The X-ray photoelectron spectroscopy (XPS) data further confirmed the increase in the bromine signal with increasing layers of the macroinitiators deposited (Figure 18.8b). For a fixed polymerization time and monomer concentration, the PHEMA brush grew more dense and thick with increasing macroinitiator layers deposited, indicating that each deposited layer was capable of initiating the poly-

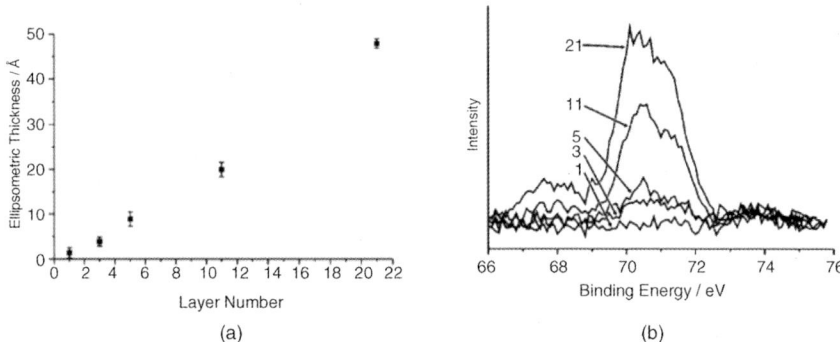

**Figure 18.8** (a) Multilayer film assembly monitored with ellipsometry. (b) High resolution XPS scan of the bromine (Br3d) signal from the different LbL films [16].

merization. For 21 layers of macroinitiators deposited, the PHEMA brush thickness was 111 nm, as compared to only 9 nm for a single macroinitiator layer deposited.

## 18.5
## Applications of the Combined LbL-SIP Approach

The LbL self-assembly method has provided a way of controlling the initiator density by virtue of depositing alternating charges of water-soluble macroinitiators. Moreover, the binary film architecture resulting from the combined LbL and SIP techniques allows one to gain control of the stimuli response of the interlayer assembly as well as that of the outer brushes.

### 18.5.1
### Dual Stimuli-Response of the LbL-SIP Brush Modified Surfaces

A dual intelligent surface exhibiting both pH and temperature responses was demonstrated by our group [33]. The multilayered film assembly was composed of weak polyelectrolytes, PAA and poly(allylamine hydrochloride) (PAH), and several layers of oppositely charged ATRP macroinitiators on top (Scheme 18.1). The deposition of PAA/PAH layers was conducted using pH 3/6 in order to allow for the pH response to occur [33]. Several layers of the ATRP macroinitiators were deposited to ensure that a relatively thick and dense poly(n-isopropylacrylamide) (PNIPAM) brushes was grown. The multilayered build-up exhibited a stepwise increase in the film thickness as a function of the amount of layers deposited. However, the build-up of the PAA/PAH revealed a higher average thickness per layer pair deposited than the ATRP macroinitiators. The ATRP macroinitiators employed in this study possessed higher charge densities than the weak polyelectrolytes [16, 33].

The pH response of the PAA:PAH/macroinitiator films displayed a tunable swelling of the LbL assembly at pH 3/9. This swelling behavior was attributed to the degree of ionization for both PAA and PAH. Once the pH response of the PAA:PAH/macroinitiator film was demonstrated, the ATRP of NIPAM was conducted. The successful grafting of highly dense and thick PNIPAM brushes was confirmed by FTIR and XPS. Moreover, the wetting of the film as a function of the lower critical solution temperature (LCST) showed close to a 45° increase in the water contact angle going from 21 to 40 °C, indicating a more hydrophobic surface.

The dual stimuli-response of the PAA:PAH/PNIPAM brush film was demonstrated using cyclic voltammetry (CV). In particular, the permeability of $Fe(CN_6)^{3-}$ in the film was measured as a function of both pH and temperature. Figure 18.9 plots the CV scans for the pH-dependent response of the PAA:PAH:PNIPAM LbL brush film taken at 21 and 60 °C. At either temperature, the highest current density was observed at pH 3 where the PAA:PAH layered assembly was in the most swelled state. Thus, the anionic probe molecule permeated freely through the film. However, a

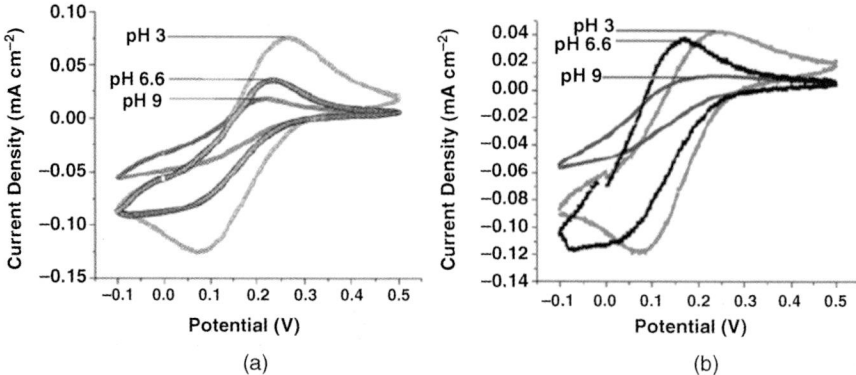

**Figure 18.9** Cyclic voltammetry scans of Fe(CN$_6$)$^{3-}$ for the PAA:PAH/PNIPAM brush films as a function of pH taken at (a) room temperature and (b) 60 °C [33].

closer examination of the film taken at pH 3 at 21 and 60 °C revealed that the maximum current density decreased by 45% at 60 °C, indicating the collapsed conformation of the PNIPAM brushes.

## 18.5.2
### Functional Free-Standing Brush Films

The groups of Advincula and Takeoka further explored the combined LbL and SIP techniques for the fabrication of a free-standing brush nanosheet [35]. The spin-assisted (SA) LbL film was prepared by the alternating deposition of chitosan and alginate. In order to introduce the initiating layers into the LbL assembly, the hydroxy and amine groups of the polysaccharides were reacted with a 2-bromoisobutyryl bromide ATRP initiator, yielding multilayers of the initiator covalently attached to the polysaccharide LbL assembly. The appearance of the bromine signal, which was not present prior to the reaction, was confirmed with XPS. As a result, an 88 nm thick PNIPAM brush was obtained after just 30 min of polymerization. The temperature response of the free-standing PNIPAM nanosheets was investigated by first detaching the polymer film from the solid substrate using water-soluble poly(vinyl alcohol) (PVA) as the sacrificial layer. After the dissolution of the PVA sacrificial layer in water, the free-standing PNIPAM nanosheet was released. The photographic images of the free-standing PNIPAM nanosheet in response to changes in temperature are provided in Figure 18.10. At room temperature, the nanosheet was observed to have a pale blue color but as the temperature was elevated to 40 °C, the nanosheet showed a yellow color. The reversibility of the temperature response was tested by returning the temperature to room temperature, and indeed the appearance of the pale blue color reappeared. The changing color was attributed to the structural thickness change of the PNIPAM as a result of its LCST. Moreover, the interfacial hydrodynamics of the PNIPAM nanosheet was concluded to depend on the concentration of an anionic surfactant, sodium $n$-dodecylsulfate (SDS).

## 18.5 Applications of the Combined LbL-SIP Approach

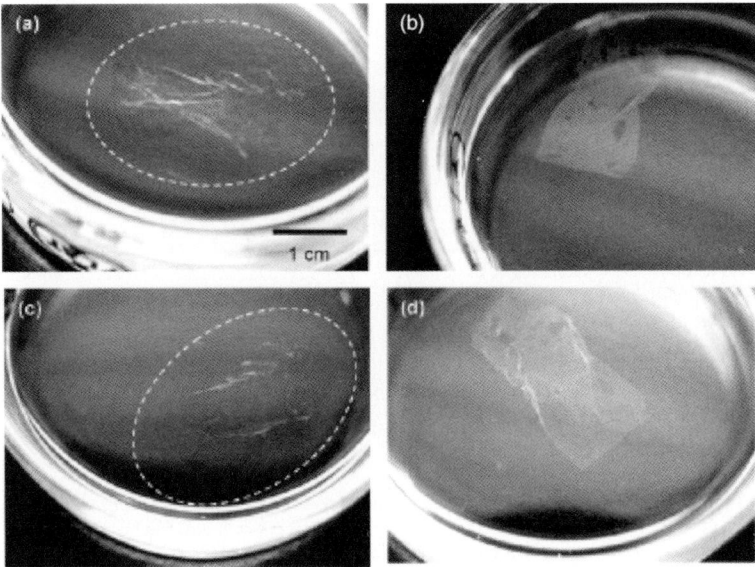

**Figure 18.10** Photographic images of the PNIPAM nanosheet suspended in water at temperatures: (a) 25 °C, (b) 40 °C, (c) 25 °C, and (d) 40 °C [35].

### 18.5.3
### Cell Adhesion

The combined LbL and SIP techniques have also been applied to biosurfaces. Laschewsky and coworkers demonstrated the regulation of cell adhesion on thermo-responsive grafted polymer brushes [31, 32]. Although a single layer of the macroinitiator was deposited, its thickness as well as the resulting brush thickness was easily controlled by the nonlinear growth of the interlayer assembly. For a layer sequence {PEI/PSS/(F-PDADMAC/PSS)$_n$/F-PDADMAC/MA01/graft-MEO$_2$MA-co-OEGMA 475} with $n = 2, 4$, and 10, only $n = 4$ displayed the optimum conditions for the cell adhesion studies. Figure 18.11 shows the images for the adhesion of L 929 mouse fibroblast on the surfaces of $n = 2, 4$, and 10. At 37 °C, the fibroblasts cells spread and grew on the surface but as the temperature was lowered to 22 °C cell rounding occured in order to minimize contact with the surface.

### 18.5.4
### Stimuli Responsive Free-Standing Films

Recently, the formation of free-standing films utilizing the LbL macroinitiator layers was demonstrated in block copolymer brushes [36] and mixed polymer brushes [37]. First, the fabrication of free-standing semifluorinated polymer brushes from LbL macroinitiators was demonstrated. The LbL deposition of macroinitiators on solid-support substrates was followed by surface-initiated atom transfer radical polymer-

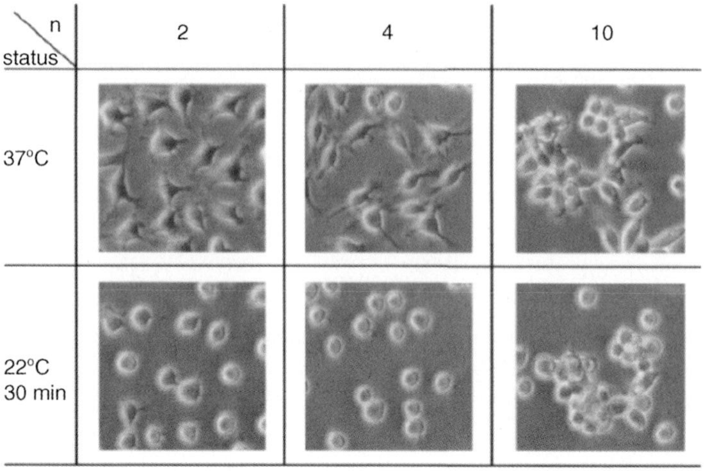

| n status | 2 | 4 | 10 |
|---|---|---|---|
| 37°C | | | |
| 22°C 30 min | | | |

**Figure 18.11** Adhesion of L 929 mouse fibroblast on {PEI/PSS/(F-PDADMAC/PSS)$_n$/F-PDADMAC/MA01/graft-MEO$_2$MA-co-OEGMA 475} brush surface. For $n = 4$, the cell adhesion at 37 °C and rounding at 22 °C were optimum as compared to the $n = 2$ and 10 surfaces [32].

ization (SI-ATRP). Increasing the number of LbL macroinitiator layers increased the grafting density and thickness, as expected [36]. The SI-ATRP was conducted by first polymerizing a block of polystyrene followed by the formation of the semifluorinated poly-2,2,2-trifluoroethyl methacrylate block. Solvent selective response was investigated by AFM imaging and water contact angle measurements. Specifically, the solvent response of the semifluorinated block copolymer brushes exhibited alternate selectivity to cyclohexane and trifluorotoluene. Solvent responsive free-standing films were then fabricated and also showed interesting solvent response behavior. The Advincula group also recently reported a facile approach to preparing binary mixed polymer brushes and free-standing films by the LbL-SIP technique [37]. In this instance, the grafting of mixed polymer brushes of poly(n-isopropylacrylamide) and polystyrene (pNIPAM-pSt) onto LbL-macroinitiator-modified planar substrates was demonstrated with ATRP and free radical polymerization for pNIPAM and pSt brushes, respectively. The composition of the two polymers was controlled by varying the number of macroinitiator layers deposited on the substrate (i.e., LbL layers = 4, 8, 12, 16, and 20); consequently, mixed brushes with different thicknesses and composition ratios were obtained. Moreover, the switching behavior of the LbL mixed brush films as a function of solvent and temperature was also demonstrated and evaluated by water contact angle and AFM experiments. It was found that both the solvent and temperature stimuli responses were a function of the mixed brush composition and thickness ratio, where the dominant component played a larger role in the response behavior. Furthermore, the ability to obtain free-standing films was exploited, showing interesting swelling and contraction behavior based on the solvent or temperature. Thus the LbL technique provided the macroinitiator density variation necessary for the preparation of stable free-standing mixed brush films. The free-standing films exhibited the rigidity to withstand changes in the solvent and temperature environment and, at the same time, were flexible enough to respond accordingly to the external stimuli.

## 18.6
## Concluding Remarks

The versatility of the combined LbL and SIP techniques has so far been beneficial in the tuning of a wide range of surface coatings. Different polyelectrolyte combination pairs can be used to fine tune the overall film thickness and internal composition and, at the same time, the single or multiple deposition of polyelectrolyte macroinitiators can be used to graft "smart" or responsive polymer brushes. It should be obvious that the technique allows the possibility of getting the best of the combined properties of the two films or methods of preparation into one coating. This means that small molecules, nanoparticles, and other functional polyelectrolytes can be prepared on the LbL film, and another side of the film will be composed of high density polymer brushes of a different functionality or role. This means that the stimuli response can be a function of both the LbL or the SIP films and their ratio. The nuances of the technique at least require the presence of a macroinitiator that can be deposited as a layer/s and, from this, polymer brushes can be grown selectively by a number of polymerization mechanisms (CLRP is highlighted here). In the future, it should be possible to extend the combinations to previously reported techniques for both LbL and SIP. Studies are underway.

## Acknowledgment

We acknowledge the contributions of past and present members of the Advincula Research Group and collaborators on work related to LbL and SIP (www.nanostructure.uh.edu). Funding from the National Science Foundation (NSF) DMR-10-06776, CBET-0854979, CHE-10-41300, and Texas NHARP 01846, and Robert A. Welch Foundation, E-1551 is also acknowledged. Technical support from KSV Instruments, MetroOhm, Agilent Technologies and Optrel is also acknowledged.

## References

1 Decher, G. (1997) *Science*, **277**, 1232.
2 Bertrand, P., Jonas, A., Laschewsky, A., and Legras, R. (2000) *Macromol. Rapid Commun.*, **21**, 319.
3 Decher, G. and Schlenoff, J. (2003) *Multilayer Thin Films: Sequential Assembly of Nanocomposite Materials*, Wiley-VCH, Weinheim, Germany.
4 Ayres, N. (2010) *Polym. Chem.*, **1**, 769.
5 Ionov, L. (2010) *J. Mater. Chem.*, **20**, 3382.
6 Chen, T., Ferris, R., Zhang, J., Ducker, R., and Zauscher, S. (2010) *Prog. Polym. Sci.*, **35**, 94.
7 Advincula, R.C., Brittain, W.J., Caster, K.C. and Rühe, J. (eds) (2004) *Polymer Brushes: Synthesis,*

8 Brittain, W.J. and Minko, S. (2007) *J. Polym. Sci. Part A: Polym. Chem.*, **45**, 3505.

9 Jones, D.M., Brown, A.A., and Huck, W.T.S. (2002) *Langmuir*, **18**, 1265.

10 Bao, Z., Bruening, M.L., and Baker, G.L. (2006) *Macromolecules*, **39**, 5251.

11 Yamamoto, S., Ejaz, M., Tsujii, Y., and Fukuda, T. (2000) *Macromolecules*, **33**, 5608.

12 Tsujii, Y., Ejaz, M., Yamamoto, S., Fukuda, T., Shigeto, K., Mibu, K., and Shinjo, T. (2002) *Polymer*, **43**, 3837.

13 Wu, T., Efimenko, K., and Genzer, J. (2002) *J. Am. Chem. Soc.*, **124**, 9394.

14 Liu, Y., Klep, V., Zdyrko, B., and Luzinov, I. (2005) *Langmuir*, **21**, 11806.

15 Estillore, N.C., Park, J.Y., and Advincula, R.C. (2010) *Macromolecules*, **43**, 6588.

16 Edmonson, S., Vo, C.D., Armes, S.P., Unali, G.F., and Weir, M.P. (2008) *Langmuir*, **24**, 7208.

17 Belder, G.F., Brinke, G.T., and Hadziioannou, G. (1997) *Langmuir*, **13**, 4102.

18 Barbey, R., Lavanant, L., Paripovic, D., Schüwer, N., Sugnaux, C., Tugulu, S., and Klok, H.-A. (2009) *Chem. Rev.*, **109**, 5437.

19 Matyjaszewski, K. and Xia, J. (2001) *Chem. Rev.*, **101**, 2921.

20 Quinn, J.F., Johnston, A.P.R., Such, G.K., Zelikin, A.N., and Caruso, F. (2007) *Chem. Soc. Rev.*, **36**, 707.

21 Edmondson, S. and Armes, S.P. (2009) *Polym. Int.*, **58**, 307.

22 Lee, M.T., Hsueh, C.C., Freund, M.S., and Ferguson, G.S. (1998) *Langmuir*, **14**, 6419.

23 Ista, L.K., Mendez, S., Pérez-Luna, V., and López, G.P. (2001) *Langmuir*, **17**, 2552.

24 Wang, Y., Hu, S., and Brittain, W.J. (2006) *Macromolecules*, **39**, 5675.

25 Chen, X. and Armes, S.P. (2003) *Adv. Mater.*, **15**, 1558.

26 Chen, X.Y., Armes, S.P., Greaves, S.J., and Watts, J.F. (2004) *Langmuir*, **20**, 587.

27 Vo, C.-D., Schmid, A., Armes, S.P., Sakai, K., and Biggs, S. (2007) *Langmuir*, **23**, 408.

28 Edmondson, S., Vo, C.-D., Armes, S.P., and Unali, G.-F. (2007) *Macromolecules*, **40**, 5271.

29 Fulghum, T.M., Patton, D.L., and Advincula, R.C. (2006) *Langmuir*, **22**, 8397.

30 Jain, P., Dai, J., Grajales, S., Saha, S., Baker, G.L., and Bruening, M.L. (2007) *Langmuir*, **23**, 11360.

31 Wischerhoff, E., Uhlig, K., Lankenau, A., Börner, H.G., Laschewsky, A., Duschl, C., and Lutz, J.-F. (2008) *Angew. Chem. Int. Ed.*, **47**, 5666.

32 Wischerhoff, E., Glatzel, S., Uhlig, K., Lankenau, A., Lutz, J-F., and Laschewsky, A. (2009) *Langmuir*, **25**, 5949.

33 Fulghum, T.M., Estillore, N.C., Vo, C.-D., Armes, S.P., and Advincula, R.C. (2008) *Macromolecules*, **41**, 429.

34 Porcel, C., Lavalle, P., Decher, G., Senger, B., Voegel, J.-C., and Schaaf, P. (2007) *Langmuir*, **23**, 1898.

35 Fujie, T., Park, J.Y., Murata, A., Estillore, N.C., Tria, M.C.R., Takeoka, S., and Advincula, R.C. (2009) *ACS Appl. Mater. Interfaces*, **1**, 1404.

36 Estillore, N. and Advincula, R. (2011) *Langmuir*, **27** (10), 5997.

37 Estillore, N. and Advincula, R. (2011) *Macromolecular Chemistry and Physics* **212** (15), 1552–1566, macp.201100066.

# 19
# Quartz Crystal Resonator as a Tool for Following the Build-up of Polyelectrolyte Multilayers

*Mikko Salomäki and Jouko Kankare*

## 19.1
## Introduction

Utilization of thin polymer films in a large number of applications, such as sensors, coatings, actuators, energy storage devices, displays, and so on, demands the development of a considerable arsenal of characterization methods. Acoustic wave devices offer unique means of studying both the thickness and elastic properties of thin films *in situ* during the deposition process. They provide methods of studying thin films in a non-destructive and non-interrupted manner and are capable of providing measurement data even in optically opaque solutions. Thickness-shear mode (TSM) quartz crystal resonator (QCR) has become a routine tool when studying the film formation on various surfaces. Usually the integrated measurement instrument is called a quartz crystal microbalance (QCM) to emphasize its traditional use as a balance for weighing.

The QCM is widely used in polyelectrolyte multilayer study. In most cases it is used simply for estimating the amount of material collected on the surface of the resonator. In the simplest case the quartz resonator is utilized in an oscillator-mode simply to determine the resonant frequency. Most of the reports concentrated on polyelectrolyte multilayers and QCM utilize the Sauerbrey equation [1] to analyze the mass growth. There are not very many methods which allow a continuous and controlled growth of films and simultaneous monitoring of their build-up. Vacuum evaporation of metals is one of those methods. In fact, it was the original application of a quartz crystal microbalance. Much more recent are the methods where the layer is grown under a liquid, for example in the electropolymerization of conductive polymers. QCM equipment and analysis offer nanogram precision per added layer, which gives valuable opportunities for developing the layer-by-layer (LbL) assembly. Presently, the build-up of polyelectrolyte multilayers forms an especially interesting problem because of the dual growth mechanism. Some polyelectrolyte systems seem to grow linearly with the number of deposited layers in the LbL process, whereas others grow exponentially, that is, each layer pair formed by the anionic and cationic polyelectrolyte is always thicker than the previous one by a constant factor.

*Multilayer Thin Films: Sequential Assembly of Nanocomposite Materials*, Second Edition.
Edited by Gero Decher and Joseph B. Schlenoff.
© 2012 Wiley-VCH Verlag GmbH & Co. KGaA. Published 2012 by Wiley-VCH Verlag GmbH & Co. KGaA.

## 19.2
## Basic Concepts

The already classical application of the quartz crystal resonator (QCR) is to use it as a quartz crystal microbalance (QCM). An AT cut quartz plate has mechanical thickness–shear mode resonant frequencies, that is, the fundamental frequency and its odd harmonics. In the first approximation, when the resonator is coated with a thin laterally homogeneous layer of any material, the relative decrease in the resonant frequency is equal to the relative increase of the areal mass density of the resonator:

$$-\frac{\Delta f_{res}}{f_{res}} \simeq \frac{\Delta m}{m_q} \tag{19.1}$$

Here $m_q = d_q \varrho_q$ where $d_q$ is the thickness of the quartz plate and $\varrho_q$ is the density of quartz. This equation – a slightly different form of the Sauerbrey equation – is a very good approximation in vacuum where the QCM was first used. It is also a good approximation in the liquid phase provided that (i) the added layer is thin enough and homogeneously distributed on the resonator surface, (ii) the viscoelastic properties of the layer material and liquid are not too close, and (iii) there is no slip between the film and QCR surface or film and liquid. We mainly treat case (i), that is, we see what happens to the equations as the layers on the QCR become thick. The correction due to the violation of condition (ii) is briefly treated later in Equation 19.15. The case (iii) is very rare [2] and we do not treat it here.

If the layer is in contact with liquid, the effect of viscoelasticity is stronger and depends on the density–viscosity product of the liquid and the complex shear modulus of the layer material [3]. In principle, the elastic properties of the layer material have first order influence on the estimated mass, but usually the ratio of the shear moduli of the liquid and the layer is so small that the effect is negligible within the conventional range of the areal mass density. However, if the thickness of the layer is continuously increased, the point where the acoustic thickness of the layer approaches the quarter wavelength of the shear wave may be finally reached. If one is measuring the change in the resonant frequency as a function of the increasing layer thickness, this point appears as a minimum, and further increase of thickness now causes increase in frequency and, in certain cases, the resonant frequency may even exceed that of an unloaded resonator. This quite rare occurrence may be falsely interpreted as negative mass. The stationary point is called the *acoustic film resonance* (AFR, Figure 19.1.). With further growth of the layer AFR is repeated at every odd multiple of the quarter-wave. Near to and after the AFR the classical Sauerbrey equation is no longer valid. It is often mentioned in the literature that a film with excessive damping generates a "viscoelastic effect" and, therefore, the QCM-analysis is not reliable. However, valuable measurable information is unquestionably also available in the neighborhood of the AFR [4].

The coating material on the surface of a QCR is subjected to sinusoidally oscillating shear stress. If the material is purely elastic, the induced strain is strictly in phase with the stress. On the other hand, if the material is a purely viscous liquid, "Newtonian" liquid, the strain is 90° out of phase. The properties of macromolecular compounds

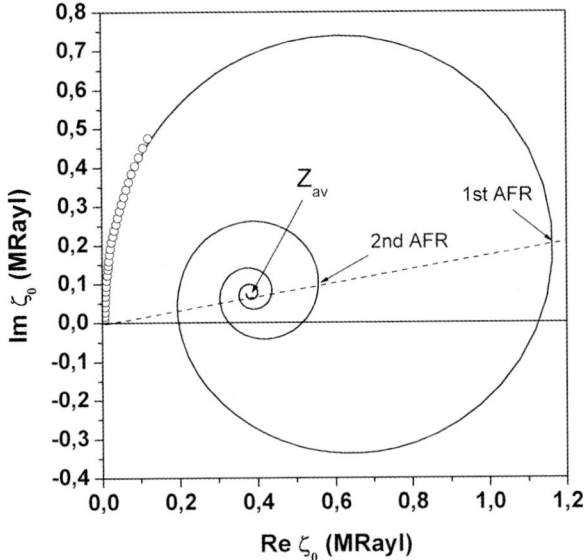

**Figure 19.1** Impedance spiral as an Argand diagram of surface acoustic impedance [5]. The polar line is drawn through the origin and spiral center. Its intercepts with the right-hand branches of the spiral give the locations of the acoustic film resonance (AFR). The small circles are experimental values of surface acoustic impedance of a PSS-PDADMA system. Every tenth data point is shown.

are between these two extremes; they are viscoelastic. The phase shift between stress and strain is between 0° and 90°. In analogy with electric AC voltage and current, it is advantageous to express the phase shift by using complex quantities. The ratio of shear stress $\tau$ to strain $\gamma$ is the complex shear modulus G:

$$\frac{\tau}{\gamma} = G = G' + jG''; \quad j^2 = -1 \tag{19.2}$$

The real part $G'$ is the shear storage modulus and the imaginary part $G''$ the shear loss modulus. The inverse of the shear modulus is the complex shear compliance J.

$$J = J' - jJ'' = 1/G \tag{19.3}$$

The loss angle $\delta$ is defined as

$$\tan \delta = \frac{G''}{G'} = \frac{J''}{J'} \tag{19.4}$$

For Newtonian liquids $G = j\omega\eta$ (where $\eta$ is viscosity and $\omega$ angular frequency) and consequently $\delta = {}^{1/2}\pi$, and for elastic solids $\delta = 0$. An important quantity is also the characteristic or bulk acoustic impedance Z of the material:

$$Z^2 = \varrho G = \varrho/J \tag{19.5}$$

where $\varrho$ is the density of the material. At every point of the material undergoing acoustic vibrations one can define the local acoustic impedance (LAI) $\zeta$ as the ratio of stress and the transverse displacement velocity $v$ of particles (not to be mistaken with the propagation velocity):

$$\zeta = -\frac{\tau}{v} \tag{19.6}$$

The velocity is also a complex quantity because of the phase shift versus stress. Local acoustic impedance is measurable only at the surface of the resonator and it is denoted as $\zeta_0$ and called the surface acoustic impedance or surface load impedance.

It should be noted that the AT cut quartz crystal undergoes transverse vibrations and consequently induces transverse acoustic waves to the surrounding medium. Transverse acoustic waves are very strongly attenuated in liquids and propagate into liquids only a very short distance. For instance, in water at 10 MHz the damping distance is only about 170 nm. This means that a QCR in contact with a layer of liquid senses the characteristic acoustic impedance as the surface acoustic impedance, even if the liquid layer is only a few micrometers thick:

$$\zeta_0 = Z_{liq} = \sqrt{\varrho_{liq} G_{liq}} = \sqrt{j\omega \varrho_{liq} \eta_{liq}} = \sqrt{\frac{\omega \eta_{liq} \varrho_{liq}}{2}}(1+j) \tag{19.7}$$

An important assumption is that LAI is continuous over the interface, meaning that there is no slippage between two materials. Both $Z$ and $\zeta$ have the same dimension: $kg\,m^{-2}\,s^{-1}$, corresponding to the acoustic unit *Rayl*.

The main benefit in using the LAI in the problems of a loaded QCR is that it can be obtained as a solution to a first-order nonlinear differential equation, the Riccati equation [3]. In this equation differentiation is done in terms of areal mass density $m$. The solution allows one to calculate LAI at the point $m = m_2$ when LAI at point $m = m_1$ is known, by using linear fractional transformation or Möbius transformation [5]:

$$\zeta(m_2) = \frac{a\zeta(m_1) + b}{c\zeta(m_1) + d} = \mathbf{A} \circ \zeta(m_1) \tag{19.8}$$

Coefficients $a$, $b$, $c$ and $d$ can be represented as a Möbius matrix

$$\mathbf{A} = \begin{pmatrix} a & b \\ c & d \end{pmatrix} = \begin{pmatrix} \cos\frac{\omega \Delta m}{Z} & jZ\sin\frac{\omega \Delta m}{Z} \\ jZ^{-1}\sin\frac{\omega \Delta m}{Z} & \cos\frac{\omega \Delta m}{Z} \end{pmatrix} \tag{19.9}$$

Here $\Delta m = m_1 - m_2$. This matrix is not unique because the elements can be multiplied by the same factor without changing the value of the Möbius transformation (19.8). The use of this equation requires that the impedance $Z$ is constant between $m_1$ and $m_2$. For instance, if we have a thin homogeneous layer of some material with areal mass density $m$ and characteristic acoustic impedance $Z$ on the surface of a QCR in contact with air ($Z_{air} \approx 0$), we obtain

$$\zeta(m_2 = 0) = \zeta_0 = jZ \tan\frac{\omega m}{Z} \tag{19.10}$$

On the other hand, if the material is in contact with a liquid, we have

$$\zeta_0 = \frac{Z_{\text{liq}} \cos(\omega m/Z) + jZ \sin(\omega m/Z)}{jZ_{\text{liq}} Z^{-1} \sin(\omega m/Z) + \cos(\omega m/Z)} \tag{19.11}$$

In most cases the first term in the denominator is small and we can write

$$\zeta_0 \approx Z_{\text{liq}} + jZ \tan(\omega m/Z) \tag{19.12}$$

In many practical cases very thin layers are studied and then $\omega m \ll |Z|$. Then we can take only the first term in the Taylor expansion of the tangent function and

$$jZ \tan(\omega m/Z) \approx j\omega m \tag{19.13}$$

We see that in this approximation the acoustic impedance of the material has no influence and we can form a modified "Sauerbrey equation":

$$\frac{\Delta f_{\text{res}}}{f_{\text{res}}} \approx -\frac{\Delta \zeta''_0}{\omega \varrho_q d_q} \tag{19.14}$$

where $\Delta \zeta''_0$ is the change in the imaginary part of the surface acoustic impedance when the QCR is loaded with a thin layer of some material. With the approximation (19.13) we obtain Equation 19.1. As a matter of fact, a better approximation is obtained if the viscoelastic properties of the material are taken into account. If the imaginary part of the shear compliance of this coating material is denoted by $J''$ and its density by $\varrho$, we have [3]

$$\frac{\Delta f_{\text{res}}}{f_{\text{res}}} = -\frac{\Delta m}{\varrho_q d_q}\left(1 - \frac{\omega \varrho_{\text{liq}} \eta_{\text{liq}}}{\varrho} J''\right) \tag{19.15}$$

In most practical cases this correction factor can be neglected.

A big advantage in using LAI is the way of handling polyelectrolyte multilayers on a QCR. Suppose that we have two successive layers, A and B, on a QCR, with A in contact with the QCR surface and B in contact with liquid. The areal mass densities are $m_a$ and $m_b$ and characteristic impedances $Z_a$ and $Z_b$. The matrix (19.9) corresponding to this bilayer can be shown to be the matrix product **AB**. and the surface acoustic impedance is obtained from

$$\zeta_0 = \mathbf{AB} \circ Z_{\text{liq}} \tag{19.16}$$

Here, we have used a short-hand notation for the linear fractional transformation (19.8). Matrices **A** and **B** do not commute unless the impedances are equal. However, if the layers are thin enough the impedance does not contribute and the only observed change to the unloaded QCR in liquid is

$$\Delta \zeta_0 = j\omega(m_a + m_b) \tag{19.17}$$

If the layer system consists of a large number $n$ of these layer pairs, such as polyelectrolyte multilayers, we have

$$\zeta_0 = (\mathbf{AB})^n \circ Z_{\text{liq}} \tag{19.18}$$

If we now measure $\zeta_0$ after each addition of a layer pair during the LbL procedure and plot the points in the complex plane, we obtain an Argand diagram of surface acoustic impedance, an impedance spiral (Figure 19.1). The convergence point of the spiral, the spiral center, corresponds to the bulk impedance of the multilayer material, the surface acoustic impedance of an "infinite" number of layer pairs.

By applying the Lie product formula [6], the bulk impedance or apparent characteristic acoustic impedance $Z_{\text{app}}$ of the infinite multilayer is given as[1]:

$$\frac{1}{Z_{\text{app}}^2} \cong \frac{m_a}{m_a + m_b} \frac{1}{Z_a^2} + \frac{m_b}{m_a + m_b} \frac{1}{Z_b^2} \tag{19.19}$$

This can also be written in terms of the compliance:

$$J_{\text{app}} = \frac{h_a}{h_a + h_b} J_a + \frac{h_b}{h_a + h_b} J_b \tag{19.20}$$

Here $h_a$ and $h_b$ are the thicknesses of the layers and $J_{\text{app}}$ is the apparent shear compliance of the multilayer.

The main assumption is the presence of distinct layers of equal thickness. However, it is known that most PE layers are strongly interpenetrated and there are no clear borderlines between them [7]. Still, these equations indicate that densely stratified thin layers in a polyelectrolyte multilayer (PEM) system can be treated as a single thick layer, presuming that the characteristic impedance of the layers is constant within the region. An important question is: When can we expect the impedance to be constant? An obvious answer is that in the close proximity of interfaces we may expect that the characteristic impedance differs from the value in the bulk. A common assumption is that the PEM can be divided into three zones [8]. Zone I is in contact with the QCR surface, zone III in contact with the solution and zone II is the main growth region in the middle. The real boundary lines between the zones are not sharp, but yet in terms of approximations a useful assumption is the presence of dividing lines which confine the regions of constant impedance. The surface acoustic impedance measured on a QCR loaded with a PEM becomes

$$\zeta_0 = \mathbf{L}_1 \mathbf{L}_2 \mathbf{L}_3 \circ Z_{\text{med}} \tag{19.21}$$

where $\mathbf{L}_i$ s are the Möbius matrices of zones I, II and III. It is also customary to assume that zone I in contact with the resonator surface is very thin, only a few layer pairs, and consequently the contribution of zone I is limited only to a small mass increase. Zone III in contact with liquid is assumed to be a fuzzy, ill-organized and highly hydrated region, allowing relatively fast diffusion of even macromolecular species. Its thickness depends on the nature of the polyelectrolytes, being sometimes a dominant region.

In addition to following the mass growth in the LbL techniques, it is in most cases useful to characterize the viscoelastic properties of the PEM material. For this

---

1) Matrices in equation 19.9 form a special Lie group SL(2; $\mathbb{C}$).

purpose the complex shear modulus G or its inverse, complex compliance J (Equation 19.5) are the parameters of choice. One measurement results in two parameters, the real and imaginary parts of surface acoustic impedance or equivalent. However, the full characterization requires three parameters, the areal mass density and the real and imaginary parts of the shear modulus or compliance. Hence, we have three unknowns[2] but only two equations. A common remedy is to make additional measurements at higher overtones. Obviously this increases the number of equations, but it implies also that the frequency-dependence of the viscoelastic parameters is known. This is not the case as the functional form of this dependence varies between different polymers. A common assumption is that the parameters follow the Voigt model. According to this model the complex shear modulus can be written

$$G = G' + jG'' = \mu + j\omega\eta \qquad (19.22)$$

where elastic shear modulus $\mu$ and viscosity $\eta$ do not depend on the frequency. However, this is a considerable simplification. Real polymers seldom follow this simple model. Many materials of PEMs are very soft and hydrous, resembling more or less polymer solutions which do not obey the simple frequency dependence of the Voigt model. An alternative method for determining the viscoelastic parameters is based on the growth process of PEMs.

## 19.3
## Growth Processes

There are two main build-up regimes that are usually found in PEMs: linear and exponential. Linearity means that the areal mass density $m$ is proportional to the number $n$ of layer pairs:

$$m = An \qquad (19.23)$$

whereas the exponential build-up means:

$$m = Ae^{\beta n} \qquad (19.24)$$

The linear build-up can be explained by the shallow interpenetration range of polyelectrolytes. Each new addition of polyelectrolyte occupies only the surface region, partially interpenetrating into a few underlying layers. The origin of the exponential build-up is still somewhat debatable. One explanation is the free diffusion of at least one of the polyelectrolytes within the layers. Polyelectrolyte does not remain in the surface region but is diluted into the entire volume of the multilayer. Consequently, more material is transferred from the solution until the chemical potentials match in the interface. A further consequence is the increase in the multilayer volume with the concomitant increase in the next portion of

---

[2] Actually four unknowns, because we have to know also the density in order to estimate the shear modulus. However, in the case of PEMs we can estimate the density to be 1200–1400 kg m$^{-3}$ without making a major error.

incorporated PE. The linear build-up in a polyelectrolyte multilayer is usually observed after the nonlinear initial build-up [8, 9]. This change in the growth mechanism may be explained by the reduced diffusion range. Diffusion may occur only in zone III whose thickness stays constant after the initial stages. The internal diffusion has been proposed and demonstrated for some polyelectrolytes of biological nature [10–13].

According to another suggestion the nonlinear build-up is caused by the continuous increase in surface roughness, which leads to the increase of physical surface area available for adsorption. That is observed in the case of lowly charged polyelectrolytes [14, 15] and in high ionic strength [16].

In most cases the layer-by-layer build-up is self-limited to constant increments and can be sustained in the linear regime for hundreds or even thousands of layers.

## 19.3.1
### Effect of Temperature

The resonant frequency of a 10 MHz QCR in water changes by about 22 Hz K$^{-1}$ at room temperature, mainly due to the temperature dependence of viscosity. This corresponds approximately to the effect of a 1 nm polymer layer. Consequently, if we are supposed to do accurate measurements on a QCR and the measurement takes a long time, thermostatting is necessary. Temperature has an important role in the layer-by-layer build-up [17, 18]. Actually, the general mechanism of the build-up process can be changed simply by increasing the temperature. While the build-up observed for the deposition at room temperature is mainly linear, the increasing temperature brings along a clear progression in the multilayer build-up. The build-up changes to exponential at higher temperatures. The trend has been observed and studied in detail using PSS/PDADMA films (see Figure 19.2 [18]).

The thickness of zone III can be estimated from the total mass of the layers at the end of the exponential regime by assuming a plausible density for the material. The zone III thickness values for PSS/PDADMA film, deposited in NaBr, show really strong temperature dependence. The estimated values are: 69, 88, 120, 360 and 6000 nm for deposition temperatures 15, 25, 35, 45, 55 °C, respectively [18]. It is noteworthy that the values increase exponentially as the temperature increases, reaching an extremely high value at 55 °C. Also the layer-pair mass density in the linear regime exhibits a strong increase as the temperature is raised. In addition, even though the PSS/PAH -pair has been generally considered as a model example of linear build-up [19, 20], this system also follows the same trend [18]. It can be stated that the effect of the deposition temperature in this polyelectrolyte system is enormous.

A common feature in all of the studied systems is the extension of the exponential build-up regime with increasing temperature.

The build-up is in most cases divided into distinct steps. First, there is an initial build-up which can be related to the formation of zone I. The measurements carried out at five different temperatures show almost identical build-up during the first four layer pairs. The build-up is exponential but it is independent of temperature. After the initial stage there exists an exponential build-up regime with a temperature-depen-

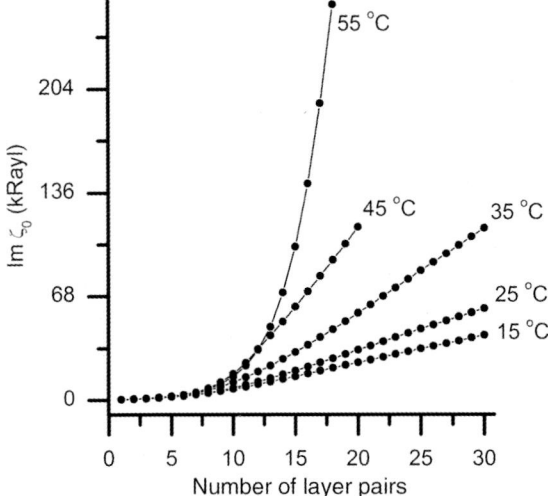

**Figure 19.2** PSS/PDADMA multilayers deposited in 0.1 M NaBr at different temperatures [18]. (Reprinted with permission from *Langmuir* **2005**, *21*, 11232–11240. Copyright 2005 American Chemical Society.)

dent rate. The basic assumption is that now there are zones I and III present and at least one of the polymers is rather freely diffusing within the polymer blend at the polymer/solution interface. It is also assumed that at the end of each deposition step there exists equilibrium between the polymer concentration in solution and the composition of the film. Within these conditions the build-up can be expressed using the growth exponent $\beta$ (cf. Equation 19.24). The growth exponent is the essential parameter for determining the temperature dependence of the multilayer build-up. The relationship between the growth exponent and the deposition temperature leads to an Arrhenius-like function [18]. Under these conditions the layer-by-layer deposition is exponential if the diffusion is able to carry polymer within the entire film during the applied contact time. Eventually, when the film becomes adequately thick, the complete mixing of the polymer blend within the contact time becomes impossible and the depositing polymer will have a finite blending depth. The build-up is no longer exponential and is independent of the thickness of the underlying film. The zones I and III are now separated by zone II and the multilayer adopts its final configuration, providing a base for the linear build-up.

## 19.4
## Experimental Techniques

The QCM has been used in the liquid phase for about 30 years. Since those days a number of different methods have been used to characterize QCR and the material on its surface. The classical and most simple method is to connect the quartz crystal as the frequency-controlling component in an oscillator circuit. From the change in the resonant frequency the areal mass density of the film on the QCR surface can be estimated at certain conditions. Commercial instruments based on this principle are still available. However, polyelectrolyte multilayers are lossy viscoelastic materials and, if relatively thick layers are being measured, a more elaborate instrument is needed. Two types of instrumentation are in use:

a) Electrical impedance measurement of the QCR. There is a functional relationship between the electrical impedance and surface acoustical impedance [21–24]. There are several instruments, network analyzers, for the general impedance measurement available from the major electronic instrument companies. A few instruments devoted solely to the QCR measurements are also commercially available (e.g., Resonant Probes GmbH). These instruments are mostly suitable for the measurements of PEMs although some difficulties may occur with strongly lossy thick coatings measured in the liquid phase. Previously, the impedance measurement of a QCR was analyzed by using the Butterworth–Van Dyke equivalent circuit, consisting of an inductance, capacitance and resistance in series, forming together the so-called motional impedance, and a static capacitance in parallel with the motional impedance. The resonant frequency is then mainly determined by the inductance and capacitance and the losses by the resistance. Presently, the necessary parameters are obtained directly from the impedance/admittance spectrum. Areal mass densities are obtained from the frequency shift and losses are estimated from the half-band-half-width of the conductance spectrum. Let the half-band-half-width be denoted by $\Delta\omega_w$. The connection to the real part of surface acoustic impedance is

$$\frac{\Delta\omega_w}{\omega} \simeq \frac{\zeta'_0}{\omega \rho_q d_q} \tag{19.25}$$

We have developed a new method whereby the surface acoustic impedance of a loaded QCR is measured by using a heterodyne principle with modulated carrier frequency. The main advantage compared with the commercial instruments is higher sensitivity allowing thick and highly lossy layers [24].

b) Impulse excitation and the free decay of oscillation. In this method the quartz crystal oscillation is started by an RF pulse, the crystal is disconnected from the oscillator circuit and the decay time $\tau$ of oscillation is measured. Instruments based on this principle are commercially available ("QCM-D," Biolin Scientific AB). The decay rate is inversely proportional to the losses of the resonator. The energy dissipation $D$ is defined as

$$D = \frac{2}{\omega \tau} = \frac{E_{\text{dissipated}}}{2\pi E_{\text{stored}}} \tag{19.26}$$

Another dimensionless quantity is the relative change in the resonant frequency. These quantities are related to the surface acoustic impedance by

$$D \simeq \frac{2\zeta'_0}{\omega \rho_q d_q} \; ; \; -\frac{\Delta f}{f} \simeq \frac{\zeta''_0}{\omega \rho_q d_q} \tag{19.27}$$

The numerical value of the acoustical impedance of quartz plate in Equations 19.14, 19.25 and 19.27 at the nominal resonant frequency becomes

$$\omega_{nom}\varrho_q d_q = \sqrt{\bar{c}_{66}\varrho_q(\pi^2 - 8K_0^2)} = 2.768 \times 10^7 \text{ kg m}^{-2}\text{ s}^{-1} \quad (19.28)$$

Here $\bar{c}_{66}$ is the piezoelectrically stiffened elastic constant and $K_0^2$ the electromechanical coupling factor of lossless quartz.

The use of a QCR in connection with PEMs incurs certain requirements on the instrumentation. A flow cell is necessary in order to introduce polyelectrolyte solutions in the build-up procedure. The geometry of the cell interior should be optimized: a small volume for minimizing the dead volume, but still a suitable shape for minimizing the spurious reflections of longitudinal acoustic waves. Practical experience shows that the dome-shaped hemispherical cell interior with the quartz crystal as a "floor" is good in this sense. The construction material of the cell should be chosen according to the intended use. For instance, if the measurements are to be done at several temperatures, good thermal conductivity of the material is advantageous in order to speed thermal equilibration. Metals are good in this sense but unfortunately subject to corrosion, even gold-coated copper. Presently, the cell used in the authors' laboratory is made of PFA-coated Shapal M, machinable aluminum nitride ceramics with high thermal conductivity

For the room temperature work Kel-F or PTFE are suitable materials. One should keep in mind that the alternate application of cationic and anionic polyelectrolytes gathers multilayers also on any unintentional surfaces, to some extent even on fluoropolymers. Sometimes this even leads to blocking of the capillary PTFE tubing and, hence, common conduits are to be avoided.

An automated system for delivering solutions into the QCR cell is advisable if the number of layers in the build-up procedure exceeds 10 to 20. The system in the authors' laboratory consists of several helium-pressurized storage bottles containing the polyelectrolyte solutions and washing liquids. The bottles are connected to the solenoid valves and the QCR cell with PTFE capillary tubing. Each solution is brought into the cell in separate tubings. The outlet from the cell conducts liquids into a vessel on a digital balance which controls the amount of each solution passing through the cell. The valves, balance and QCR are controlled by using a program written in LabView. The schematic diagram of the system is shown in Figure 19.3.

## 19.5
### Analysis of QCR Data

The collection of experimental data and the analysis of these data by some mathematical technique depends on the method of measurement. However, the data used for conclusions or further analysis are in most cases either surface acoustic impedances or proportional to them, such as presented in Equation 19.27. If the data have been collected during the LbL process, they can be used to estimate viscoelastic parameters and mass growth, with certain assumptions. The methods of

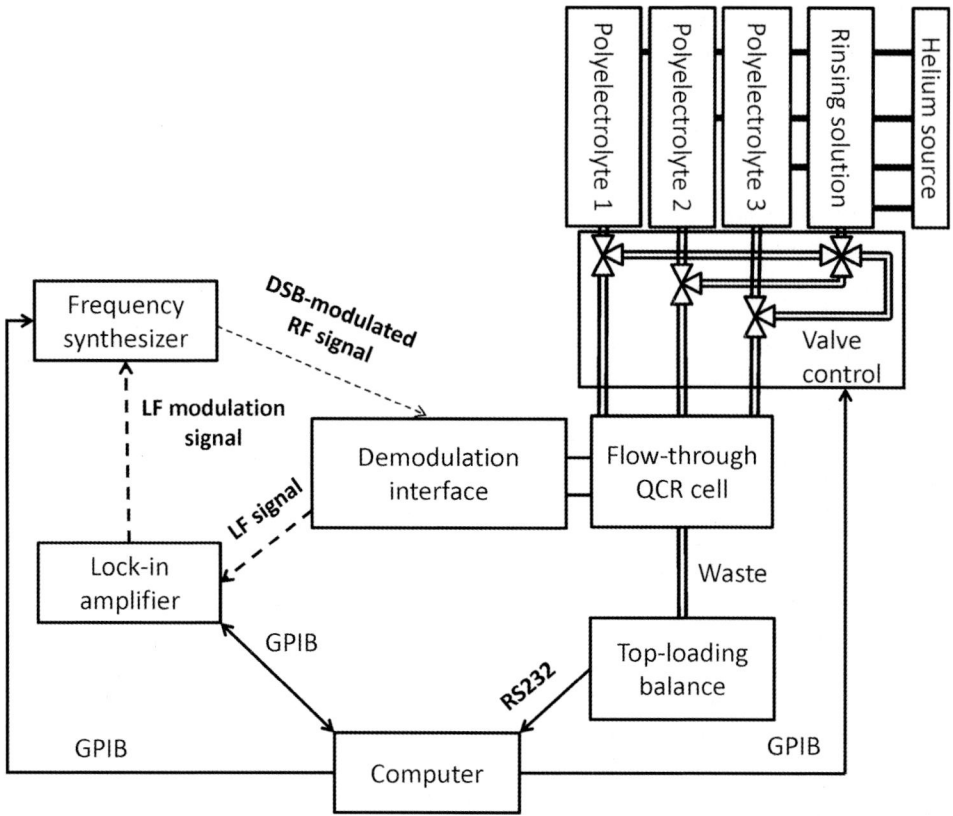

**Figure 19.3** Schematic diagram of the fully automatic computer-controlled instrument for making polyelectrolyte multilayers by using the layer-by-layer technique. The QCR measurement system is described in ref. 24.

analysis are essentially different, depending on whether the build-up process is either linear or nonlinear. If the build-up is linear, we may assume that the layer pairs, that is, two layers formed from cationic and anionic polyelectrolytes, are identical. This leads to the use of the special properties of the Möbius transformation. According to Equation 19.18 the surface acoustic impedance of an $n$-layer system is obtained by repeating the Möbius transformation **AB** $n$ times. However, in this case we are making an assumption that the impedance of the layer material is constant throughout the multilayer. As stated before, a multilayer can be roughly divided into three zones (cf. Equation 19.21). Zone I, closest to the QCR, surface is thin and remains constant during the build-up. Its contribution to the surface acoustic impedance is a small constant addition to the areal mass density. Also, zone III, in contact with solution, is assumed to stay constant in the case of linear growth. Hence, the linear growth occurs only in zone II and the multilayers can be presented in the form:

$$\zeta_0^{(n)} = \mathbf{L}_1(\mathbf{AB})^n\mathbf{L}_3 \circ \zeta_{aq} \tag{19.29}$$

A fundamental property of Möbius transformations is the invariance of the cross ratio $r$. The cross ratio is defined in terms of four points in the complex plane:

$$r = \frac{(z_1-z_4)(z_3-z_2)}{(z_1-z_3)(z_2-z_4)} = \frac{\left(\zeta_0^{(k)}-\zeta_0^{(k+3m)}\right)\left(\zeta_0^{(k+2m)}-\zeta_0^{(k+m)}\right)}{\left(\zeta_0^{(k)}-\zeta_0^{(k+2m)}\right)\left(\zeta_0^{(k+m)}-\zeta_0^{(k+3m)}\right)} \quad (19.30)$$

Each point is obtained from the previous one by applying the same Möbius transformation:

$$z_n = \mathbf{N} \circ z_{n-1}; n = 2, 3, 4 \quad (19.31)$$

Let us now take four experimental values of surface acoustic impedance corresponding to the multilayers consisting of $k$, $k + m$, $k + 2m$ and $k + 3m$ layer pairs. In this case

$$\mathbf{N} = (\mathbf{AB})^m \quad (19.32)$$

and matrices $\mathbf{N}$ and $\mathbf{AB}$ can be solved relatively easily. The weakness in this procedure is that only four points in the data set are utilized. It is also rather sensitive to the choice of the points and, especially, the four points chosen should not be on the same straight line in the Argand diagram. This is actually a general rule whatever method is used: the viscoelastic parameters are obtained with a reasonable accuracy only if there is enough curvature in the measured part of the spiral. In practice, this means that, in the case of linear growth, a very large number, even hundreds of layers should be laid before a sufficient amount of data has been collected for reasonably accurate viscoelastic parameters.

The cross ratio invariance is a unique technique in the sense that the Möbius matrix is obtained without any prior assumption on the layer pair. However, we have reasons to expect that separate layers in the "layer pair" do not exist in most cases. The layers are generally strongly interpenetrated and the "layer pair" can be treated as a layer of single material. The Möbius matrix of this layer is not then given as a product of two matrices but simply as in Equation 19.9, that is, a matrix with three adjustable parameters $m$ and the real and imaginary parts of the bulk acoustic impedance $Z$. An alternative method of data treatment, matrix iteration, is to try to find these parameters and the corresponding matrix which gives the best least-squares fit $F$:

$$F = \sum_{n=1}^{N} \left|\zeta_0^{(k+n)} - \mathbf{A}^n \circ \zeta_0^{(k)}\right|^2 \quad (19.33)$$

If the covariance matrix of the real and imaginary parts of surface acoustic impedance is known, a slight variation is to minimize the chi-square function instead. The initial multilayer with $k$ layer pairs is chosen to be far enough from zone I, that is, $k$ should be greater than 10. The total number of layers, $N$, should be high enough and the track of data points in the complex plane should be clearly curved in order to obtain reliable values for the parameters. As an example, in Figure 19.1 150 layer pairs of poly(diallyldimethylammonium chloride) and poly(sodium 4-styrenesulfonate) have been laid and every 10th shown. Still, at least 700 layer pairs would be needed in order to achieve the close neighborhood of the spiral center [5].

Typical for the exponential growth is the extremely fast growth rate of the layer thickness. Only in this case is it possible to reach the complete spiral form of the Argand diagram within a moderate number of layers. Another distinctive feature of the exponential growth is splitting of the impedance spiral into two branches with even and odd numbered layers. In the even-numbered layers the number of cationic and anionic polyelectrolyte layers is equal, whereas in the odd branch the numbers of these layers are unequal. Sometimes the even and odd branches are widely separated, showing even separate convergence centers. This shows that the polyelectrolytes are completely mixed after each addition. In Figure 19.4a a good example of exponential growth is shown [4]. The build-up of the multilayer formed by hyaluronic acid and the polysaccharide amine chitosan is exponential at room temperature and in 0.5 M NaCl.

If the build-up process is nonlinear, the basic assumption on the constant size of the "layer pair" is not valid. The matrix iteration and the least-squares procedure should be modified by taking the $N$ values of mass increase $\Delta m$ as the new adjustable parameters. This sounds like too many parameters for simple modeling but, because the data points are complex numbers with real and imaginary parts, the actual number of data is $2N$. Hence the number of degrees of freedom is still $N-2$. The data of Figure 19.4a have been analyzed using this method. The main assumption is that the bulk acoustic impedance is constant within the fitting range but no assumption on the growth process is made. The excellent fit extending up to the spiral center supports the assumption on the constant impedance. In Figure 19.4b the logarithm of areal mass density is plotted against the layer pair number. The linear fit is excellent up to the 18th layer pair with $\beta = 0.275$.

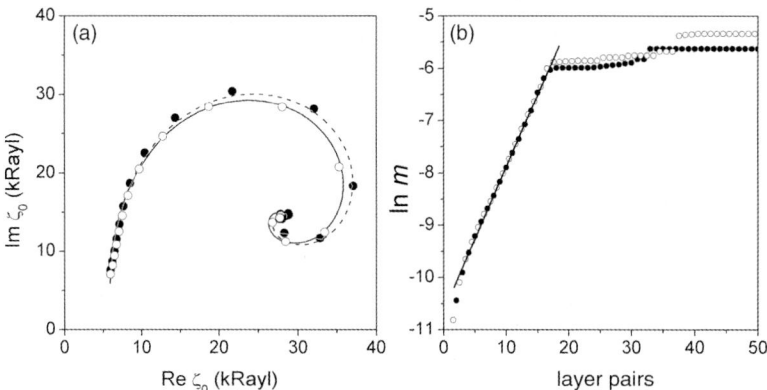

**Figure 19.4** (a) Argand diagram of the hyaluronan/chitosan multilayer build-up in 0.5 M NaCl at 25 °C: dashed curve and filled circles, even branch; solid curve and open circles, odd branch. Curves are calculated values. (b) Logarithm of areal mass density (in kg m$^{-2}$) as a function of the number of layer pairs. Areal mass density was calculated from the data of (a) as described in ref. [4]. (Reprinted with permission from *J. Phys. Chem.* **2007**, *111*, 8509–8519. Copyright 2007 American Chemical Society.)

## List of Abbreviations

| | |
|---|---|
| DSB | double sideband |
| GPIB | general purpose interface bus |
| LbL | layer-by-layer |
| LF | low frequency |
| PAH | poly(allyl amine hydrochloride) |
| PDADMA | poly(diallyldimethylammonium) |
| PE | polyelectrolyte |
| PEM | polyelectrolyte multilayer |
| PSS | poly(sodium 4-styrenesulfonate) |
| QCM | quartz crystal microbalance |
| QCR | quartz crystal resonator |
| RF | radiofrequency |
| RS-232 | standard serial connection |

## References

1 Sauerbrey, G. (1959) *Z. Phys.*, **155**, 206–222.
2 Salomäki, M. and Kankare, J. (2004) *Langmuir*, **20**, 7794–7801.
3 Kankare, J. (2002) *Langmuir*, **18**, 7092–7094.
4 Salomäki, M. and Kankare, J. (2007) *J. Phys. Chem. B*, **111**, 8509–8519.
5 Salomäki, M., Loikas, K., and Kankare, J. (2003) *Anal. Chem.*, **75**, 5895–5904.
6 Hall, B.C. (2004) *Lie Groups, Lie Algebras, and Representations*, Springer, p. 35.
7 Decher, G. (1997) *Science*, **277**, 1232–1237.
8 Ladam, G., Schaad, P., Voegel, J.C., Schaaf, P., Decher, G., and Cuisinier, F. (2000) *Langmuir*, **16**, 1249–1255.
9 Schlenoff, J.B. and Dubas, S.T. (2001) *Macromolecules*, **34**, 592–598.
10 Picart, C., Lavalle, P., Hubert, P., Cuisiner, F.J.G., Decher, G., Schaaf, P., and Voegel, J.C. (2001) *Langmuir*, **17**, 7414–7424.
11 Picart, C., Mutterer, J., Richert, L., Luo, Y., Prestwich, G.D., Schaaf, P., Voegel, J.C., and Lavalle, P. (2002) *Proc. Natl. Acad. Sci. USA*, **99**, 12531–12535.
12 Richert, L., Lavalle, P., Payan, E., Zheng, X.S., Prestwich, G.D., Stolz, J.F., Schaaf, P., Voegel, J.C., and Picart, C. (2004) *Langmuir*, **20**, 448–458.
13 Boulmedais, F., Ball, V., Schwinte, P., Frisch, B., Schaaf, P., and Voegel, J.C. (2003) *Langmuir*, **19**, 440–445.
14 Schoeler, B., Poptoshev, E., and Caruso, F. (2003) *Macromolecules*, **36**, 5258–5264.
15 DeLongchamp, D.M., Kastantin, M., and Hammond, P.T. (2003) *Chem. Mater.*, **15**, 1575–1586.
16 McAloney, R.A., Sinyor, M., Dudnik, V., and Goh, M.C. (2001) *Langmuir*, **17**, 6655–6663.
17 Tan, H.L., McMurdo, M.J., Pan, G., and Van Patten, P.G. (2003) *Langmuir*, **19**, 9311–9314.
18 Salomäki, M., Vinokurov, I.A., and Kankare, J. (2005) *Langmuir*, **21**, 11232–11240.
19 Caruso, F., Niikura, K., Furlong, D.N., and Okahata, Y. (1997) *Langmuir*, **13**, 3427–3433.
20 Ramsden, J.J., Lvov, Y.M., and Decher, G. (1995) *Thin Solid Films*, **254**, 246–251.
21 Martin, S.J., Granstaff, V.E., and Frye, G.C. (1991) *Anal. Chem.*, **63**, 2272–2281.
22 Granstaff, V.E. and Martin, S.J. (1994) *J. Appl. Phys.*, **75**, 1319–1329.
23 Lucklum, R., Behling, C., Cernosek, R.W., and Martin, S.J. (1997) *J. Phys. D: Appl. Phys.*, **30**, 346–356.
24 Kankare, J., Loikas, K., and Salomäki, M. (2006) *Anal. Chem.*, **78**, 1875–1882.

*Edited by*
*Gero Decher and*
*Joseph B. Schlenoff*

**Multilayer Thin Films**

## Related Titles

Pacchioni, G., Valeri, S. (eds.)

**Oxide Ultrathin Films**

Science and Technology

2012

ISBN: 978-3-527-33016-4

Friedbacher, G., Bubert, H. (eds.)

**Surface and Thin Film Analysis**

A Compendium of Principles, Instrumentation, and Applications

2011

ISBN: 978-3-527-32047-9

Knoll, W., Advincula, R. C. (eds.)

**Functional Polymer Films**

2 Volume Set

2011

ISBN: 978-3-527-32190-2

Kumar, C. S. S. R. (ed.)

**Polymeric Nanomaterials**

2011

ISBN: 978-3-527-32170-4

Urban, M. W. (ed.)

**Handbook of Stimuli-Responsive Materials**

2011

ISBN: 978-3-527-32700-3

Fernandez-Nieves, A., Wyss, H., Mattsson, J., Weitz, D. A. (eds.)

**Microgel Suspensions**

Fundamentals and Applications

2011

ISBN: 978-3-527-32158-2

Samori, P., Cacialli, F. (eds.)

**Functional Supramolecular Architectures**

for Organic Electronics and Nanotechnology

2011

ISBN: 978-3-527-32611-2

Chujo, Y. (ed.)

**Conjugated Polymer Synthesis**

Methods and Reactions

2011

ISBN: 978-3-527-32267-1

Leclerc, M., Morin, J.-F. (eds.)

**Design and Synthesis of Conjugated Polymers**

2010

ISBN: 978-3-527-32474-3

*Edited by*
*Gero Decher and Joseph B. Schlenoff*

# Multilayer Thin Films

Sequential Assembly of Nanocomposite Materials

*Second, Completely Revised and Enlarged Edition*

WILEY-VCH Verlag GmbH & Co. KGaA

**The Editors**

*Prof. Gero Decher*
Université de Strasbourg
Insitut Charles Sadron, UPR 22
23, rue du Loess
67034 Strasbourg Cedex 2
France

*Prof. Joseph B. Schlenoff*
Florida State University
Chemistry Department
Tallahassee, FL 32306-4390
USA

All books published by **Wiley-VCH** are carefully produced. Nevertheless, authors, editors, and publisher do not warrant the information contained in these books, including this book, to be free of errors. Readers are advised to keep in mind that statements, data, illustrations, procedural details or other items may inadvertently be inaccurate.

**Library of Congress Card No.:** applied for

**British Library Cataloguing-in-Publication Data**
A catalogue record for this book is available from the British Library.

**Bibliographic information published by the Deutsche Nationalbibliothek**
The Deutsche Nationalbibliothek lists this publication in the Deutsche Nationalbibliografie; detailed bibliographic data are available on the Internet at http://dnb.d-nb.de.

© 2012 Wiley-VCH Verlag & Co. KGaA, Boschstr. 12, 69469 Weinheim, Germany

All rights reserved (including those of translation into other languages). No part of this book may be reproduced in any form – by photoprinting, microfilm, or any other means – nor transmitted or translated into a machine language without written permission from the publishers. Registered names, trademarks, etc. used in this book, even when not specifically marked as such, are not to be considered unprotected by law.

**Composition**   Thomson Digital, Noida, India
**Printing and Binding**   betz-druck GmbH, Darmstadt
**Cover Design**   Schulz Grafik-Design, Fußgönheim

Printed in the Federal Republic of Germany
Printed on acid-free paper

**Print ISBN:**   978-3-527-31648-9
**ePDF ISBN:**   978-3-527-64677-7
**oBook ISBN:**   978-3-527-64674-6
**ePub ISBN:**   978-3-527-64676-0

## Contents

**List of Contributors** XXV

**Volume 1**

| | | |
|---|---|---|
| **1** | **Layer-by-Layer Assembly (Putting Molecules to Work)** 1 | |
| | *Gero Decher* | |
| 1.1 | The Whole is More than the Sum of its Parts 1 | |
| 1.2 | From Self-Assembly to Directed Assembly 1 | |
| 1.3 | History and Development of the Layer-by-Layer Assembly Method 4 | |
| 1.4 | LbL-Assembly is the Synthesis of Fuzzy Supramolecular Objects 6 | |
| 1.5 | Reproducibility and Choice of Deposition Conditions 7 | |
| 1.6 | Monitoring Multilayer Build-up 10 | |
| 1.7 | Spray- and Spin-Assisted Multilayer Assembly 13 | |
| 1.8 | Recent Developments 14 | |
| 1.8.1 | Self-patterning LbL-Films 14 | |
| 1.8.2 | Deposition of LbL-Films on Very Small Particles 15 | |
| 1.8.3 | Purely Inorganic LbL-"Films" 17 | |
| 1.9 | Final Remarks 18 | |
| | References 19 | |
| **Part I** | **Preparation and Characterization** 23 | |
| **2** | **Layer-by-Layer Processed Multilayers: Challenges and Opportunities** 25 | |
| | *Michael F. Rubner and Robert E. Cohen* | |
| 2.1 | Introduction 25 | |
| 2.2 | Fundamental Challenges and Opportunities 25 | |
| 2.2.1 | LbL Assembly on Nanoscale Elements and in Confined Geometries 25 | |
| 2.2.2 | Living Cells as Functional Elements of Polyelectrolyte Multilayers 28 | |
| 2.2.3 | Multilayer Cellular Backpacks 28 | |
| 2.2.4 | Direct LbL Processing of Living Cells 29 | |
| 2.3 | Technological Challenges and Opportunities 31 | |
| 2.3.1 | Improving Processing Time and Versatility 31 | |
| 2.3.2 | Towards Mechanically Robust Multilayer Coatings 32 | |
| 2.4 | The Path Forward 36 | |
| | References 36 | |

| 3 | **Layer-by-Layer Assembly: from Conventional to Unconventional Methods** 43 |
|---|---|
| | *Guanglu Wu and Xi Zhang* |
| 3.1 | Introduction 43 |
| 3.2 | Conventional LbL Methods 44 |
| 3.2.1 | Electrostatic LbL Assembly 44 |
| 3.2.2 | Hydrogen-Bonded LbL Assembly 47 |
| 3.2.3 | LbL Assembly Driven by Coordination Interaction 50 |
| 3.2.4 | To Combine LbL Assembly and Post-Chemical Reaction for the Fabrication of Robust Thin Films 50 |
| 3.3 | Unconventional LbL Methods 52 |
| 3.3.1 | Electrostatic Complex for Unconventional LbL Assembly 52 |
| 3.3.1.1 | Nanoreactors with Enhanced Quantum Yield 53 |
| 3.3.1.2 | "Ion Traps" for Enhancing the Permselectivity and Permeability 53 |
| 3.3.1.3 | Surface Imprinted LbL Films 54 |
| 3.3.1.4 | Cation-Selective μCP Based on SMILbL Film 58 |
| 3.3.2 | Hydrogen-Bonded Complex for Unconventional LbL Assembly 58 |
| 3.3.3 | Block Copolymer Micelles for Unconventional LbL Assembly 61 |
| 3.3.4 | π–π Interaction Complex for Electrostatic LbL Assembly 62 |
| 3.4 | Summary and Outlook 64 |
| | References 64 |

| 4 | **Novel Multilayer Thin Films: Hierarchic Layer-by-Layer (Hi-LbL) Assemblies** 69 |
|---|---|
| | *Katsuhiko Ariga, Qingmin Ji, and Jonathan P. Hill* |
| 4.1 | Introduction 69 |
| 4.2 | Hi-LbL for Multi-Cellular Models 70 |
| 4.3 | Hi-LbL for Unusual Drug Delivery Modes 72 |
| 4.4 | Hi-LbL for Sensors 75 |
| 4.4.1 | Mesoporous Carbon Hi-LbL 75 |
| 4.4.2 | Mesoporous Carbon Capsule Hi-LbL 76 |
| 4.4.3 | Graphene/Ionic-Liquid Hi-LbL 78 |
| 4.5 | Future Perspectives 79 |
| | References 80 |

| 5 | **Layer-by-Layer Assembly Using Host-Guest Interactions** 83 |
|---|---|
| | *Janneke Veerbeek, David N. Reinhoudt, and Jurriaan Huskens* |
| 5.1 | Introduction 83 |
| 5.2 | Supramolecular Layer-by-Layer Assembly 84 |
| 5.3 | 3D Patterned Multilayer Assemblies on Surfaces 85 |
| 5.4 | 3D Supramolecular Nanoparticle Crystal Structures 88 |
| 5.5 | Porous 3D Supramolecular Assemblies in Solution 90 |
| 5.6 | Conclusions 95 |
| | References 95 |

| 6 | **LbL Assemblies Using van der Waals or Affinity Interactions and Their Applications** *99* |
|---|---|
| | *Takeshi Serizawa, Mitsuru Akashi, Michiya Matsusaki, Hioharu Ajiro, and Toshiyuki Kida* |
| 6.1 | Introduction *99* |
| 6.2 | Stereospecific Template Polymerization of Methacrylates by Stereocomplex Formation in Nanoporous LbL Films *100* |
| 6.2.1 | Introduction *100* |
| 6.2.2 | LbLit-PMMA/st-PMMA Stereocomplex Uultrathin Film *102* |
| 6.2.3 | Fabrication of Template Nanospaces in Films *104* |
| 6.2.4 | Polymerization within Template Nanospaces Using a Double-Stranded Assembly *106* |
| 6.2.5 | Studies on the Porous Structure Obtained by LbL Assembly *111* |
| 6.3 | Preparation and Properties of Hollow Capsules Composed of Layer-by-Layer Polymer Films Constructed through van der Waals Interactions *113* |
| 6.3.1 | Introduction *113* |
| 6.3.2 | Preparation of Hollow Nanocapsules Composed of Poly(methyl methacrylate) Stereocomplex Films *114* |
| 6.3.3 | Preparation and Fusion Properties of Novel Hollow Nanocapsules Composed of Poly(lactic acid)s Stereocomplex Films *116* |
| 6.4 | Fabrication of Three-Dimensional Cellular Multilayers Using Layer-by-Layer Protein Nanofilms Constructed through Affinity Interaction *120* |
| 6.4.1 | Introduction *120* |
| 6.4.2 | Hierarchical Cell Manipulation *121* |
| 6.4.3 | High Cellular Activities Induced by 3D-Layered Constructs *124* |
| 6.4.4 | Quantitative 3D-Analyses of Nitric Oxide Diffusion in a 3D-Artery Model *124* |
| 6.5 | Conclusion *129* |
| | References *129* |
| | |
| 7 | **Layer-by-Layer Assembly of Polymeric Complexes** *135* |
| | *Junqi Sun, Xiaokong Liu, and Jiacong Shen* |
| 7.1 | Introduction *135* |
| 7.2 | Concept of LbL Assembly of Polymeric Complexes *136* |
| 7.2.1 | LbL Assembly of Polyelectrolyte Complexes for the Rapid and Direct Fabrication of Foam Coatings *136* |
| 7.2.2 | LbL Assembly of Hydrogen-Bonded Polymeric Complexes *138* |
| 7.2.3 | LbL Assembly of Polyelectrolyte-Surfactant Complexes *139* |
| 7.3 | Structural Tailoring of LbL-Assembled Films of Polymeric Complexes *140* |
| 7.3.1 | Mixing Ratio of PECs *140* |
| 7.3.2 | LbL Codeposition of PECs and Free Polyelectrolytes *142* |
| 7.3.3 | Salt Effect *143* |
| 7.4 | LbL-Assembled Functional Films of Polymeric Complexes *144* |

| 7.4.1 | Self-Healing Superhydrophobic Coatings  144 |
| --- | --- |
| 7.4.2 | Mechanically Stable Antireflection- and Antifogging-Integrated Coatings  146 |
| 7.4.3 | Transparent and Scratch-Resistant Coatings with Well-Dispersed Nanofillers  147 |
| 7.5 | Summary  149 |
| | References  149 |
| | |
| **8** | **Making Aqueous Nanocolloids from Low Solubility Materials: LbL Shells on Nanocores**  151 |
| | *Yuri Lvov, Pravin Pattekari, and Tatsiana Shutava* |
| 8.1 | Introduction  151 |
| 8.2 | Formation of Nanocores  153 |
| 8.2.1 | Stabilizers versus Layer-by-Layer Polyelectrolyte Shell  153 |
| 8.3 | Ultrasonication-Assisted LbL Assembly  154 |
| 8.3.1 | Top-Down Approach  154 |
| 8.3.2 | Bottom-Up Approach  158 |
| 8.4 | Solvent-Assisted Precipitation Into Preformed LbL-Coated Soft Organic Nanoparticles  159 |
| 8.5 | Washless (Titration) LbL Technique  161 |
| 8.6 | Formation of LbL Shells on Nanocores  163 |
| 8.7 | Drug Release Study  165 |
| 8.8 | Conclusions  168 |
| | References  168 |
| | |
| **9** | **Cellulose Fibers and Fibrils as Templates for the Layer-by-Layer (LbL) Technology**  171 |
| | *Lars Wågberg* |
| 9.1 | Background  171 |
| 9.2 | Formation of LbLs on Cellulose Fibers  172 |
| 9.3 | The use of LbL to Improve Adhesion between Wood Fibers  176 |
| 9.4 | The Use of LbL to Prepare Antibacterial Fibers  179 |
| 9.5 | The use of NFC/CNC to Prepare Interactive Layers Using the LbL Approach  182 |
| 9.6 | Conclusions  185 |
| | References  186 |
| | |
| **10** | **Freely Standing LbL Films**  189 |
| | *Chaoyang Jiang and Vladimir V. Tsukruk* |
| 10.1 | Introduction  189 |
| 10.2 | Fabrication of Freely Standing Ultrathin LbL Films  189 |
| 10.2.1 | LbL Films with Gold Nanoparticles and Silver Nanowires  191 |
| 10.2.2 | LbL Films with Quantum Dots  193 |
| 10.2.3 | LbL Films with Nanocrystals  194 |
| 10.2.4 | LbL Films with Graphene Oxides and Carbon Nanotubes  195 |

| | | |
|---|---|---|
| 10.2.5 | LbL with Biomaterials  *198* | |
| 10.2.6 | Mechanical and Optical Properties  *199* | |
| 10.2.7 | Surface Morphology and Internal Microstructure  *201* | |
| 10.3 | Porous and Patterned Freely Standing LbL Films  *204* | |
| 10.3.1 | Nanoporous LbL Films  *204* | |
| 10.3.2 | Patterned LbL Films  *205* | |
| 10.3.3 | Sculptured LbL Films  *207* | |
| 10.4 | Freely Standing LbL Films with Weak Interactions  *208* | |
| 10.4.1 | Responsive LbL Films  *208* | |
| 10.4.2 | Anisotropic LbL Microcapsules  *209* | |
| 10.4.3 | LbL-Shells for Cell Encapsulation  *210* | |
| | References  *214* | |
| **11** | **Neutron Reflectometry at Polyelectrolyte Multilayers**  *219* | |
| | *Regine von Klitzing, Ralf Köhler, and Chloe Chenigny* | |
| 11.1 | Introduction  *219* | |
| 11.2 | Neutron Reflectometry  *219* | |
| 11.2.1 | Specular Reflectometry  *220* | |
| 11.2.2 | Contrast Variation: Interesting Features for PEMs  *222* | |
| 11.2.2.1 | Superlattice Structure  *222* | |
| 11.2.2.2 | Block Structure  *223* | |
| 11.2.3 | Fitting  *223* | |
| 11.3 | Preparation Techniques for Polyelectrolyte Multilayers  *224* | |
| 11.3.1 | Dipping  *224* | |
| 11.3.2 | Spraying  *228* | |
| 11.3.3 | Spin-Coating  *231* | |
| 11.3.4 | Comparison of the Techniques and Model  *231* | |
| 11.4 | Types of Polyelectrolytes  *233* | |
| 11.4.1 | A Strong Polyanion and a Weak Polycation  *233* | |
| 11.4.2 | Two Strong Polyelectrolytes  *235* | |
| 11.4.3 | Two Weak Polyelectrolytes  *237* | |
| 11.4.4 | Multilayers with Non-Electrostatic Interactions  *238* | |
| 11.5 | Preparation Parameters  *238* | |
| 11.5.1 | Ion Effects  *238* | |
| 11.5.1.1 | Ionic Strength  *238* | |
| 11.5.1.2 | Counterions  *239* | |
| 11.5.2 | pH  *241* | |
| 11.5.3 | Temperature  *241* | |
| 11.6 | Influence of External Fields After PEM Assembly  *242* | |
| 11.6.1 | Scalar Fields: Ionic Strength  *242* | |
| 11.6.2 | Scalar Fields: Temperature  *244* | |
| 11.6.3 | Scalar Fields: Water  *246* | |
| 11.6.3.1 | Swelling in Water  *247* | |
| 11.6.3.2 | Models for Determination of the Water Content  *252* | |

| 11.6.3.3 | Effect of Polymer Charge Density and Ions During Preparation on Swelling in Water  *254* |
|---|---|
| 11.6.3.4 | Odd-Even Effect  *256* |
| 11.6.4 | Directed External Fields: Mechanical and Electrical Fields  *257* |
| 11.6.4.1 | Mechanical Stress  *257* |
| 11.6.4.2 | Electric Field  *258* |
| 11.7 | PEM as a Structural Unit  *259* |
| 11.7.1 | Cushion  *259* |
| 11.7.1.1 | Cell-Membrane Mimic Systems  *260* |
| 11.7.1.2 | Multilayers Beyond Polyelectrolytes  *261* |
| 11.7.2 | Matrix  *261* |
| 11.8 | Conclusion and Outlook  *261* |
|  | References  *262* |

**12 Polyelectrolyte Conformation in and Structure of Polyelectrolyte Multilayers**  *269*
*Stephan Block, Olaf Soltwedel, Peter Nestler, and Christiane A. Helm*

| 12.1 | Introduction  *269* |
|---|---|
| 12.2 | Results  *270* |
| 12.2.1 | The First Polyelectrolyte Layer: Brush and/or Flatly Adsorbed Chains  *270* |
| 12.2.2 | Adsorption of Additional Layers and Interdiffusion  *275* |
| 12.2.3 | The Outermost Layer: From Odd–Even to Even–Odd Effect  *277* |
| 12.3 | Conclusion and Outlook  *279* |
|  | References  *280* |

**13 Charge Balance and Transport in Ion-Paired Polyelectrolyte Multilayers**  *281*
*Joseph B. Schlenoff*

| 13.1 | Introduction  *281* |
|---|---|
| 13.2 | Association Mechanism: Competitive Ion Pairing  *283* |
| 13.2.1 | Intrinsic vs. Extrinsic Charge Compensation  *285* |
| 13.2.1.1 | Key Equilibria  *285* |
| 13.2.1.2 | Interaction Energies  *287* |
| 13.2.1.3 | Multilayer Decomposition  *290* |
| 13.2.1.4 | Doping-moderated Mechanical Properties  *292* |
| 13.3 | Surface versus Bulk Polymer Charge  *292* |
| 13.3.1 | Distribution of Surface Charge in Layer-by-Layer Build-up: Mechanism  *297* |
| 13.4 | Polyelectrolyte Interdiffusion  *301* |
| 13.4.1 | Equilibrium versus Non-Equilibrium Conditions for Salt and Polymer Sorption  *304* |
| 13.5 | Ion Transport Through Multilayers: the "Reluctant" Exchange Mechanism  *305* |

| | | |
|---|---|---|
| 13.5.1 | Practical Consequences: Trapping and Self-Trapping | *313* |
| 13.6 | Concluding Remarks | *315* |
| | References | *315* |

**14 Conductivity Spectra of Polyelectrolyte Multilayers Revealing Ion Transport Processes** *321*
*Monika Schönhoff and Cornelia Cramer*
14.1 Introduction to Conductivity Studies of LbL Films *321*
14.2 PEM Spectra: Overview *323*
14.3 DC Conductivities of PEMs *324*
14.4 Modeling of PEM Spectra *328*
14.5 Ion Conduction in Polyelectrolyte Complexes *329*
14.6 Scaling Principles in Conductivity Spectra: From Time–Temperature to Time–Humidity Superposition *332*
14.6.1 General Aspects of Scaling *332*
14.6.2 Time–Temperature Superposition Principle in PEC Spectra *333*
14.6.3 Establishment of a Time–Humidity Superposition Principle *334*
References *335*

**15 Responsive Layer-by-Layer Assemblies: Dynamics, Structure and Function** *337*
*Svetlana Sukhishvili*
15.1 Introduction *337*
15.2 Chain Dynamics and Film Layering *338*
15.2.1 Lessons from PECs *338*
15.2.1.1 Phase Diagrams of PECs and LbL Film Deposition *338*
15.2.1.2 Polyelectrolyte Type and Equilibrium and Dynamics in PECs *340*
15.2.2 pH-Induced Chain Dynamics and Film Structure *342*
15.3 Responsive Swellable LbL Films *348*
15.3.1 LbL-Derived Hydrogels *348*
15.3.1.1 Preparation and Swelling *348*
15.3.1.2 Mechanical Properties *350*
15.3.1.3 LbL Hydrogels as Matrices for Controlled Release of Bioactive Molecules *351*
15.3.2 LbL Films of Micelles with Responsive Cores *353*
15.4 Conclusion and Outlook *358*
References *359*

**16 Tailoring the Mechanics of Freestanding Multilayers** *363*
*Andreas Fery and Vladimir V. Tsukruk*
16.1 Introduction *363*
16.2 Measurements of Mechanical Properties of Flat LbL Films *364*
16.2.1 Micromechanical Properties from Bulging Experiments *364*

| 16.2.2 | Buckling Measurements of LbL Films  367 |
| 16.2.3 | Local Mechanical Properties Probed with Force Spectroscopies  369 |
| 16.2.4 | Summary of Mechanical Properties of Flat LbL Films  370 |
| 16.3 | Mechanical Properties of LbL Microcapsules  372 |
| 16.3.1 | Theory and Measurement of Mechanical Properties of Microcapsules  373 |
| 16.3.2 | Basic Concepts of Shell Theory  373 |
| 16.3.3 | Experimental Measurements  374 |
| 16.3.4 | Small Deformation Measurements  375 |
| 16.4 | Prospective Applications Utilizing Mechanical Properties  378 |
| 16.4.1 | Flat Freestanding LbL Films for Sensing Applications  378 |
| 16.4.2 | Microcapsules for Controlled Delivery Processes  382 |
| 16.4.2.1 | Exploding Capsules  382 |
| 16.4.2.2 | Asymmetric Capsules  382 |
| 16.4.2.3 | Mechanical Stability and Intracellular Delivery  385 |
| | References  386 |

**17 Design and Translation of Nanolayer Assembly Processes: Electrochemical Energy to Programmable Pharmacies**  393
*Md. Nasim Hyder, Nisarg J. Shah, and Paula T. Hammond*

| 17.1 | Introduction  393 |
| 17.2 | Controlling Transport and Storing Charge in Multilayer Thin Films: Ions, Electrons and Molecules  395 |
| 17.2.1 | LbL Systems for Proton Exchange Fuel Cells  395 |
| 17.2.2 | Charge Storage Using LbL Assemblies  398 |
| 17.2.2.1 | Carbon Nanotube LbL Electrodes  398 |
| 17.2.2.2 | Carbon Nanotube/Conjugated Polymer System  400 |
| 17.2.3 | Composite Systems  402 |
| 17.2.3.1 | Virus Battery  402 |
| 17.2.3.2 | DSSC LbL Electrodes  402 |
| 17.2.3.3 | Multicolor Electrochromic Devices  404 |
| 17.3 | LbL Films for Multi-Agent Drug Delivery – Opportunities for Programmable Release  406 |
| 17.3.1 | Passive Controlled Release  407 |
| 17.3.1.1 | Release of Proteins  408 |
| 17.3.1.2 | Small Molecule Delivery  412 |
| 17.3.2 | Multi-Agent Delivery and Control of Sequence  418 |
| 17.4 | Automated Spray-LbL – Enabling Function and Translation  422 |
| 17.4.1 | Means of Achieving New Architectures  425 |
| 17.4.2 | Coating of Complex Surfaces  427 |
| 17.4.3 | Using Kinetics to Manipulate Composition  429 |
| 17.5 | Concluding Remarks  431 |
| | References  431 |

| | | |
|---|---|---|
| **18** | **Surface-Initiated Polymerization and Layer-by-Layer Films** *437* | |
| | *Nicel Estillore and Rigoberto C. Advincula* | |
| 18.1 | Introduction *437* | |
| 18.2 | Overview of Surface-Grafted Polymer Brushes *438* | |
| 18.2.1 | Synthesis of Polymer Brushes *439* | |
| 18.2.2 | Surface-Initiated Atom Transfer Radical Polymerization (SI-ATRP) *439* | |
| 18.3 | Layer-by-Layer (LbL) Self-Assembly *440* | |
| 18.3.1.1 | Water-Soluble ATRP Macroinitiators *440* | |
| 18.4 | Combined LbL-SIP Approach *441* | |
| 18.4.1 | Single Adsorption of a Polyelectrolyte ATRP Macroinitiator onto a Charged Surface *442* | |
| 18.4.2 | Adsorption of Oppositely Charged ATRP Macroinitiator(s) onto an LbL Deposited Multilayer Assembly *443* | |
| 18.4.3 | Multiple Adsorption of Oppositely Charged ATRP Macroinitiators for Enhanced Initiator Density *447* | |
| 18.5 | Applications of the Combined LbL-SIP Approach *449* | |
| 18.5.1 | Dual Stimuli-Response of the LbL-SIP Brush Modified Surfaces *449* | |
| 18.5.2 | Functional Free-Standing Brush Films *450* | |
| 18.5.3 | Cell Adhesion *451* | |
| 18.5.4 | Stimuli Responsive Free-Standing Films *451* | |
| 18.6 | Concluding Remarks *453* | |
| | References *453* | |
| | | |
| **19** | **Quartz Crystal Resonator as a Tool for Following the Build-up of Polyelectrolyte Multilayers** *455* | |
| | *Mikko Salomäki and Jouko Kankare* | |
| 19.1 | Introduction *455* | |
| 19.2 | Basic Concepts *456* | |
| 19.3 | Growth Processes *461* | |
| 19.3.1 | Effect of Temperature *462* | |
| 19.4 | Experimental Techniques *463* | |
| 19.5 | Analysis of QCR Data *465* | |
| | References *469* | |

**Volume 2**

| | | |
|---|---|---|
| **Part II** | **Applications** *471* | |
| | | |
| **20** | **Electrostatic and Coordinative Supramolecular Assembly of Functional Films for Electronic Application and Materials Separation** *473* | |
| | *Bernd Tieke, Ashraf El-Hashani, Kristina Hoffmann, and Anna Maier* | |
| 20.1 | Introduction *473* | |
| 20.2 | Polyelectrolyte Multilayer Membranes *474* | |
| 20.2.1 | Water Desalination *474* | |
| 20.2.2 | Size- and Charge-Selective Transport of Aromatic Compounds *476* | |

| | | |
|---|---|---|
| 20.2.3 | Ion Separation from Polyelectrolyte Blend Multilayer Membranes *479* | |
| 20.2.4 | Membranes Containing Macrocyclic Compounds *480* | |
| 20.2.4.1 | p-Sulfonato-Calix[n]arene-Containing Membranes *480* | |
| 20.2.4.2 | Azacrown Ether-Containing Membranes *483* | |
| 20.2.4.3 | Membranes Containing Hexacyclen-Hexaacetic Acid *486* | |
| 20.2.5 | LbL-Assembled Films of Prussian Blue and Analogues *487* | |
| 20.2.6 | Coordinative Assembly of Functional Thin Films *493* | |
| 20.2.6.1 | Films of Coordination Polymers Based on Schiff-Base-Metal Ion Complexes *493* | |
| 20.2.6.2 | Films of Coordination Polymers Based on Terpyridine-Metal Ion Complexes *495* | |
| 20.2.6.3 | Films of Coordination Polymers Based on Bisimidazolylpyridine-Metal Ion Complexes *502* | |
| 20.3 | Summary and Conclusions *504* | |
| | References *506* | |

**21 Optoelectronic Materials and Devices Incorporating Polyelectrolyte Multilayers** *511*

*H.D. Robinson, Reza Montazami, Chalongrat Daengngam, Ziwei Zuo, Wang Dong, Jonathan Metzman, and Randy Heflin*

| | |
|---|---|
| 21.1 | Introduction *511* |
| 21.2 | Second Order Nonlinear Optics *512* |
| 21.3 | Plasmonic Enhancement of Second Order Nonlinear Optical Response *515* |
| 21.4 | Nonlinear Optical Fibers *519* |
| 21.5 | Optical Fiber Biosensors *521* |
| 21.6 | Antireflection Coatings *525* |
| 21.7 | Electrochromic Devices *527* |
| 21.8 | Electromechanical Actuators *530* |
| | References *533* |

**22 Nanostructured Electrodes Assembled from Metal Nanoparticles and Quantum Dots in Polyelectrolytes** *539*

*Lara Halaoui*

| | |
|---|---|
| 22.1 | Introduction *539* |
| 22.2 | Nanostructured Pt Electrodes from Assemblies of Pt Nanoparticles in Polyelectrolytes *540* |
| 22.2.1 | Assembly of Polyacrylate-capped Pt NPs in Polyelectrolytes *540* |
| 22.2.2 | Surface Characterization by Hydrogen Underpotential Deposition *541* |
| 22.2.3 | $H_2O_2$ Sensing at Arrays of Pt NPs in Polyelectrolyte at Low Surface Coverage *545* |
| 22.2.4 | Biosensing at Assemblies of Glucose Oxidase Modified Pt NPs in Polyelectrolytes *548* |

| | | |
|---|---|---|
| 22.2.5 | Surface Oxidation and Stability of Pt NP Assemblies | 549 |
| 22.2.6 | Oxygen Reduction at Pt NP Assemblies | 550 |
| 22.3 | Nanostructured Photoelectrodes from Assemblies of Q-CdS in Polyelectrolytes | 552 |
| 22.3.1 | Dip versus Dip–Spin Assembly of Q-CdS in Polyelectrolytes | 553 |
| 22.3.2 | Photoelectrochemistry at PDDA/Q-CdS Assembly in the Presence of Hole Scavengers | 555 |
| 22.3.3 | Photocurrent Polarity-Switching at Q-CdS Photoelectrodes | 556 |
| 22.4 | Conclusions | 558 |
| | References | 559 |
| | | |
| **23** | **Record Properties of Layer-by-Layer Assembled Composites** | **573** |
| | *Ming Yang, Paul Podsiadlo, Bong Sup Shim, and Nicholas A. Kotov* | |
| 23.1 | Introduction | 573 |
| 23.2 | LbL Assemblies of Clays | 574 |
| 23.2.1 | Structure and Properties of Clay Particles | 574 |
| 23.2.2 | Structural Organization in Clay Multilayers | 575 |
| 23.2.3 | Clay Multilayers as High-Performance Nanocomposites | 576 |
| 23.2.4 | Applications of Clay Multilayers in Biotechnology | 578 |
| 23.2.5 | Anisotropic Transport in Clay Multilayers | 580 |
| 23.2.6 | Clay Multilayers for Optical and Electronic Applications | 581 |
| 23.3 | LBL Assemblies of Carbon Nanotubes | 582 |
| 23.3.1 | Structure and Properties of CNTs | 582 |
| 23.3.2 | Structural Organization and Mechanical Properties in Multilayers of Carbon Nanotubes | 583 |
| 23.3.3 | Electrical Conductor Applications | 584 |
| 23.3.4 | Sensor Applications | 585 |
| 23.3.5 | Fuel Cell Applications | 587 |
| 23.3.6 | Nano-/Micro-Shell LbL Coatings | 587 |
| 23.3.7 | Biomedical Applications | 588 |
| 23.4 | Conclusions and Perspectives | 589 |
| | References | 590 |
| | | |
| **24** | **Carbon Nanotube-Based Multilayers** | **595** |
| | *Yong Tae Park and Jaime C. Grunlan* | |
| 24.1 | Introduction | 595 |
| 24.2 | Characteristics of Carbon Nanotube Layer-by-Layer Assemblies | 596 |
| 24.2.1 | Growth of Carbon Nanotube-Based Multilayers | 596 |
| 24.2.2 | Electrical Properties | 600 |
| 24.2.3 | Mechanical Behavior | 601 |
| 24.3 | Applications of Carbon Nanotube Layer-by-Layer Assemblies | 602 |
| 24.3.1 | Transparent Electrodes | 602 |
| 24.3.2 | Sensor Applications | 604 |
| 24.3.3 | Energy-Related Applications | 607 |

| | | |
|---|---|---|
| 24.3.4 | Biomedical Applications | 608 |
| 24.4 | Conclusions | 609 |
| | References | 609 |

**25 Nanoconfined Polyelectrolyte Multilayers: From Nanostripes to Multisegmented Functional Nanotubes** 613

*Cécile J. Roy, Cédric C. Buron, Sophie Demoustier-Champagne, and Alain M. Jonas*

| | | |
|---|---|---|
| 25.1 | Introduction | 613 |
| 25.2 | Estimation of the Size of Polyelectrolyte Chains in Dilute Solutions | 614 |
| 25.3 | Confining LbL Assembly on Flat Surfaces | 618 |
| 25.3.1 | LbL Assembly Templated by Chemically Nanopatterned Surfaces | 619 |
| 25.3.2 | LbL Lift-Off | 621 |
| 25.4 | Confining LbL Assembly in Nanopores | 624 |
| 25.4.1 | Peculiarities of LbL Assembly in Nanopores | 627 |
| 25.4.2 | Multisegmented LbL Nanotubes | 630 |
| 25.5 | Conclusions | 633 |
| | References | 634 |

**26 The Design of Polysaccharide Multilayers for Medical Applications** 637

*Benjamin Thierry, Dewang Ma, and Françoise M. Winnik*

| | | |
|---|---|---|
| 26.1 | Introduction | 637 |
| 26.2 | Polysaccharides as Multilayered film Components: An Overview of Their Structure and Properties | 638 |
| 26.2.1 | Polycations | 639 |
| 26.2.2 | Polyanions | 641 |
| 26.3 | Multilayers Formed by Assembly of Weak Polyanions and Chitosan or Chitosan Derivatives | 642 |
| 26.3.1 | Hyaluronan | 642 |
| 26.3.2 | Polygalacturonic Acid | 646 |
| 26.3.3 | Alginate | 647 |
| 26.4 | Multilayers Formed by Assembly of Strong Polyanions and Chitosan or Chitosan Derivatives | 647 |
| 26.4.1 | Heparin | 647 |
| 26.4.2 | Chondroitin Sulfate | 649 |
| 26.5 | Cardiovascular Applications of Polysaccharide Multilayers | 650 |
| 26.5.1 | Anti-Thrombogenic Properties of Polysaccharide Multilayers | 650 |
| 26.5.2 | Preparation of Polysaccharide Multilayers on Blood Vessels | 651 |
| 26.5.3 | Drug Delivery to the Vascular Wall from Polysaccharide Multilayers | 652 |
| 26.6 | Conclusions | 654 |
| | References | 655 |

| 27 | **Polyelectrolyte Multilayer Films Based on Polysaccharides: From Physical Chemistry to the Control of Cell Differentiation** *659* |
|---|---|
| | *Thomas Boudou, Kefeng Ren, Thomas Crouzier, and Catherine Picart* |
| 27.1 | Introduction *659* |
| 27.2 | Film Internal Composition and Hydration *660* |
| 27.2.1 | Film Growth *660* |
| 27.2.2 | Hydration and Swellability *662* |
| 27.2.3 | Internal Composition *665* |
| 27.3 | Film Cross-Linking: Relation Between Composition and Mechanical Properties *666* |
| 27.3.1 | Ionic Pairing *666* |
| 27.3.2 | Covalent Amide Bonds *668* |
| 27.3.3 | Mechanical Properties of Films and Correlation with Cross-Link Density *669* |
| 27.4 | Cell Adhesion onto Cross-Linked Films: Cell Adhesion, Cytoskeletal Organization and Comparison with Other Model Materials *671* |
| 27.4.1 | Film Cross-Linking Modulates Myoblast Attachment, Proliferation and Cytoskeletal Organization *671* |
| 27.4.2 | Correlation Between the Cell Spreading Area and Young's Modulus $E_0$ *675* |
| 27.4.3 | Comparison with Other Material Substrates: Polyacrylamide (PA) and Polydimethysiloxane (PDMS) *677* |
| 27.5 | Cell Differentiation: ESC and Myoblasts *679* |
| 27.5.1 | Film Cross-Linking Influences the Myogenic Differentiation Process Morphologically but not the Expression of Muscle-Specific Proteins *679* |
| 27.5.2 | Film Cross-Linking Drives the Fate of Mouse Embryonic Stem *681* |
| 27.6 | Conclusions *684* |
| | References *685* |

| 28 | **Diffusion of Nanoparticles and Biomolecules into Polyelectrolyte Multilayer Films: Towards New Functional Materials** *691* |
|---|---|
| | *Marc Michel and Vincent Ball* |
| 28.1 | Introduction *691* |
| 28.2 | LBL Films in Which Nanoparticles are Incorporated Step-By-Step *693* |
| 28.3 | LBL Films Made Uniquely From Nanoparticles *693* |
| 28.4 | Nanoparticles Produced by Post-treatment of Deposited Films *694* |
| 28.5 | Diffusion of Colloids in Already Deposited Films *698* |
| 28.5.1 | Permeability of LBL Films Towards Ions, Small Drugs and Dyes *698* |
| 28.5.2 | Diffusion of Nanoparticles in and out of LBL Films *700* |
| 28.6 | Emerging Properties of Films Filled with Nanoparticles by the Post-incubation Method *705* |
| 28.6.1 | Production of Anisotropically-Coated Colloids *705* |

| | | |
|---|---|---|
| 28.6.2 | Sensing and Controlled Release Applications | 705 |
| 28.6.3 | Electrical Conductivity | 705 |
| 28.7 | Conclusions and Perspectives | 706 |
| | References | 707 |

**29 Coupling Chemistry and Hybridization of DNA Molecules on Layer-by-Layer Modified Colloids** 711

*Jing Kang and Lars Dähne*

| | | |
|---|---|---|
| 29.1 | Introduction | 711 |
| 29.2 | Materials and Methods | 712 |
| 29.2.1 | Materials | 712 |
| 29.2.2 | Methods | 713 |
| 29.2.2.1 | Coupling of Oligonucleotides onto LbL-Coated Colloidal Particles | 713 |
| 29.2.2.2 | Hybridization of Complementary Oligos Onto the LbL-Oligo Particles | 716 |
| 29.3 | Results | 716 |
| 29.3.1 | One-Step versus Two-Step Coupling Process by EDC | 716 |
| 29.3.2 | Selection of Polyelectrolytes | 716 |
| 29.3.2.1 | Selection of Polycations | 717 |
| 29.3.2.2 | Selection of Polyanions | 718 |
| 29.3.3 | Non-Specific Binding and Its Minimization | 718 |
| 29.3.4 | Coupling by 1.1′-Carbonyldiimidazol in Organic Solvent | 721 |
| 29.3.5 | Number of Coating Layers | 722 |
| 29.3.6 | Coupling and Hybridization Efficiency | 722 |
| 29.3.6.1 | Coupling Efficiency | 722 |
| 29.3.6.2 | Hybridization Efficiency | 723 |
| 29.3.7 | Comparison of LbL Particles with Conventional Carboxylated Particles | 724 |
| 29.3.8 | Arrangement of the Coupled Oligo Molecules | 724 |
| 29.3.8.1 | Analysis by FRET Investigations | 724 |
| 29.3.8.2 | FRET of dsOligos Free in Solution | 725 |
| 29.3.8.3 | FRET of dsOligos on LbL Particles | 726 |
| 29.3.9 | Stability of the LbL-Oligo Particles | 727 |
| 29.4 | Summary | 727 |
| | References | 729 |

**30 A "Multilayered" Approach to the Delivery of DNA: Exploiting the Structure of Polyelectrolyte Multilayers to Promote Surface-Mediated Cell Transfection and Multi-Agent Delivery** 731

*David M. Lynn*

| | | |
|---|---|---|
| 30.1 | Introduction | 731 |
| 30.2 | Surface-Mediated Delivery of DNA: Motivation and Context, Opportunities and Challenges | 732 |
| 30.3 | Films Fabricated Using Hydrolytically Degradable Cationic Polymers | 734 |

| | | |
|---|---|---|
| 30.4 | Toward Spatial Control: Release of DNA from the Surfaces of Implants and Devices *736* | |
| 30.5 | Toward Temporal Control: Tunable Release and Sequential Release *739* | |
| 30.5.1 | Approaches Based on Incorporation of Different Hydrolytically Degradable Polyamines *740* | |
| 30.5.2 | Approaches Based on Incorporation of Cationic "Charge-Shifting" Polymers *742* | |
| 30.5.3 | New Approaches to Rapid Release *744* | |
| 30.6 | Concluding Remarks *745* | |
| | References *746* | |

**31 Designing LbL Capsules for Drug Loading and Release** *749*
*Bruno G. De Geest and Stefaan C. De Smedt*

| | |
|---|---|
| 31.1 | Introduction *749* |
| 31.2 | Engineering Microparticulate Templates to Design LbL Capsules for Controlled Drug Release *750* |
| 31.3 | Engineering the Shell to Design LbL Capsules for Controlled Drug Release *753* |
| 31.4 | Interaction of LbL Capsules with Living Cells *In Vitro* and *In Vivo* *759* |
| 31.5 | Conclusions *761* |
| | References *761* |

**32 Stimuli-Sensitive LbL Films for Controlled Delivery of Proteins and Drugs** *765*
*Katsuhiko Sato, Shigehiro Takahashi, and Jun-ichi Anzai*

| | |
|---|---|
| 32.1 | Introduction *765* |
| 32.2 | Avidin-Containing LbL Films *765* |
| 32.3 | Concanavalin A-containing LbL Films *768* |
| 32.4 | Dendrimer-Containing LbL Films *771* |
| 32.5 | Insulin-Containing LbL Films *772* |
| 32.6 | Conclusions *774* |
| | References *776* |

**33 Assembly of Multilayer Capsules for Drug Encapsulation and Controlled Release** *777*
*Jinbo Fei, Yue Cui, Qiang He, and Junbai Li*

| | |
|---|---|
| 33.1 | Introduction *777* |
| 33.2 | Magnetically Sensitive Release *779* |
| 33.3 | Ultrasound-Stimulated Release *780* |
| 33.4 | Photo-Stimulated Release *781* |
| 33.5 | Thermo-Stimulated Release *783* |
| 33.6 | pH-Sensitive Release *785* |
| 33.7 | Redox-Controlled Release *787* |

| | | |
|---|---|---|
| 33.8 | Bio-Responsive Release | 788 |
| 33.9 | Extension | 792 |
| 33.10 | Concluding Remarks | 794 |
| | References | 794 |

**34 Engineered Layer-by-Layer Assembled Capsules for Biomedical Applications** 801

*Angus P.R. Johnston, Georgina K. Such, Sarah J. Dodds, and Frank Caruso*

| | | |
|---|---|---|
| 34.1 | Introduction | 801 |
| 34.2 | Template Selection | 801 |
| 34.3 | Material Assembly | 804 |
| 34.4 | Loading | 809 |
| 34.4.1 | Preloading on/in Template | 810 |
| 34.4.2 | Loading Within Layers | 811 |
| 34.4.3 | Post-Loading | 813 |
| 34.5 | Degradation and Release | 813 |
| 34.6 | Applications | 816 |
| 34.6.1 | Microreactors | 816 |
| 34.6.2 | Targeting | 818 |
| 34.6.3 | Therapeutic Delivery | 819 |
| 34.6.3.1 | Small Molecules | 819 |
| 34.6.3.2 | Vaccines | 821 |
| 34.6.3.3 | DNA | 822 |
| 34.7 | Conclusions | 823 |
| | References | 824 |

**35 Assembly of Polymer Multilayers from Organic Solvents for Biomolecule Encapsulation** 831

*Sebastian Beyer, Jianhao Bai, and Dieter Trau*

| | | |
|---|---|---|
| 35.1 | Introduction | 831 |
| 35.1.1 | Bio-Template-Based LbL Encapsulation in the Aqueous Phase | 831 |
| 35.1.2 | Loading-Based LbL Biomolecule Encapsulation in the Aqueous Phase | 832 |
| 35.1.3 | Diffusion-Based LbL Biomolecule Encapsulation in the Aqueous Phase | 833 |
| 35.2 | Limitations of LbL-Based Biomolecule Encapsulation in Aqueous Phase | 834 |
| 35.3 | LbL Biomolecule Encapsulation in the Organic Phase | 835 |
| 35.3.1 | Reverse-Phase LbL | 836 |
| 35.3.1.1 | Mechanism | 836 |
| 35.3.1.2 | Technique | 839 |
| 35.3.1.3 | Encapsulation of Biomolecules | 840 |
| 35.3.2 | "Inwards Build-Up Self-Assembly" of Polymers for Biomolecule Encapsulation in the Organic Phase | 845 |
| 35.3.2.1 | Mechanism and Technique | 845 |

| | | |
|---|---|---|
| 35.3.2.2 | Encapsulation of Biomolecules | 846 |
| 35.4 | Conclusion and Outlook | 847 |
| | References | 849 |

**36 Stimuli-Responsive Polymer Composite Multilayer Microcapsules and Microchamber Arrays** *851*

*Maria N. Antipina, Maxim V. Kiryukhin, and Gleb B. Sukhorukov*

| | | |
|---|---|---|
| 36.1 | Introduction | 852 |
| 36.2 | Fabrication of Stimuli-Responsive LbL Microcapsules | 853 |
| 36.2.1 | pH-Responsive Capsules | 855 |
| 36.2.2 | Salt-Responsive Capsules and Capsule Fusion | 858 |
| 36.2.3 | Redox-Responsive Capsules | 862 |
| 36.2.4 | Chemical-Responsive Capsules | 862 |
| 36.2.4.1 | Solvent | 862 |
| 36.2.4.2 | Glucose | 863 |
| 36.2.4.3 | $CO_2$ | 863 |
| 36.2.4.4 | Enzymes | 864 |
| 36.2.5 | Temperature-Responsive Capsules | 865 |
| 36.2.6 | Remote Responsive Capsules | 866 |
| 36.2.6.1 | Magnetic LbL Capsules | 866 |
| 36.2.6.2 | Ultrasound-Triggered Release | 867 |
| 36.2.6.3 | Optically Addressable Capsules | 868 |
| 36.2.7 | Mechanical Addressing of Individual Capsules | 869 |
| 36.2.8 | Patterning Polyelectrolyte Capsules | 872 |
| 36.3 | Microchamber Arrays | 873 |
| 36.3.1 | Fabrication of Microchambers by LbL Assembly on Imprinted Surfaces | 873 |
| 36.3.1.1 | LbL assembly of the PEMs in confined geometries | 874 |
| 36.3.1.2 | Dissolving the template and making a free-standing PEM film with an array of standing hollow microchambers | 874 |
| 36.3.2 | Microchamber Loading with Substances of Interest | 877 |
| 36.3.3 | Responsiveness of Chambers to Light and Mechanical Load | 879 |
| 36.4 | Conclusion | 881 |
| | References | 882 |

**37 Domain-Containing Functional Polyelectrolyte Films: Applications to Antimicrobial Coatings and Energy Transfer** *891*

*Aurélie Guyomard, Bernard Nysten, Alain M. Jonas, and Karine Glinel*

| | | |
|---|---|---|
| 37.1 | Introduction | 891 |
| 37.2 | Polyelectrolyte Films Incorporating Randomly Distributed Hydrophobic Nanodomains for Antimicrobial Applications | 893 |
| 37.2.1 | Hydrophobic Nanodomains in Hydrophilic PEMs with Amphiphilic Macromolecules | 893 |
| 37.2.2 | Entrapment of an Antibacterial Peptide in Hydrophobic Nanodomains | 895 |

| | | |
|---|---|---|
| 37.3 | Multicompartmentalized Stratified Polyelectrolyte Films for Control of Energy Transfer *898* | |
| 37.3.1 | How Precisely Can the Stratification of a LbL Film Be Controlled? *898* | |
| 37.3.2 | Fabrication of Dye-Impermeable Polyelectrolyte Barriers *901* | |
| 37.3.3 | Control of a Cascade of Events in a Multicompartmentalized Stratified Polyelectrolyte Film *902* | |
| 37.4 | Conclusions and Perspectives *903* | |
| | References *904* | |
| | | |
| **38** | **Creating Functional Membranes Through Polyelectrolyte Adsorption** *907* | |
| | *Merlin L. Bruening* | |
| 38.1 | Introduction *907* | |
| 38.2 | Functionalization of the Interior of Membranes *908* | |
| 38.2.1 | Deposition of PEMs in Porous Media *908* | |
| 38.2.2 | Functionalization of Membranes with Proteins *910* | |
| 38.2.2.1 | Protein Adsorption in Membranes *910* | |
| 38.2.2.2 | Trypsin-Containing Polyelectrolyte Films for Protein Digestion *911* | |
| 38.2.3 | Catalytic Films and Membranes *914* | |
| 38.2.3.1 | Catalytic, Nanoparticle-Containing Films *914* | |
| 38.2.3.2 | Catalytic Membranes *915* | |
| 38.3 | LBL Films as Membrane Skins *918* | |
| 38.3.1 | Early Studies *918* | |
| 38.3.2 | Removal of Dyes and Small Organic Molecules from Water *918* | |
| 38.3.3 | Selective Rejection of $F^-$ and Phosphate *920* | |
| 38.3.4 | Variation of PSS/PDADMAC film Properties with the Number of Adsorbed Layers *921* | |
| 38.4 | Challenges *922* | |
| | References *922* | |
| | | |
| **39** | **Remote and Self-Induced Release from Polyelectrolyte Multilayer Capsules and Films** *925* | |
| | *Andre G. Skirtach, Dmitry V. Volodkin, and Helmuth Möhwald* | |
| | References *940* | |
| | | |
| **40** | **Controlled Architectures in LbL Films for Sensing and Biosensing** *951* | |
| | *Osvaldo N. Oliveira Jr., Pedro H.B. Aoki, Felippe J. Pavinatto, and Carlos J.L. Constantino* | |
| 40.1 | Introduction *951* | |
| 40.2 | LbL-Based Sensors and Biosensors *952* | |
| 40.2.1 | Optical Detection Methods *952* | |
| 40.2.2 | Mass Change Methods *954* | |
| 40.2.3 | Electrochemical Methods *955* | |

| | | |
|---|---|---|
| 40.2.4 | Methods Involving Electrical Measurements | 956 |
| 40.2.5 | E-Tongues and E-Noses | 958 |
| 40.2.6 | Extending the Concept of E-Tongue to Biosensing | 963 |
| 40.3 | Special Architectures for Sensing and Biosensing | 964 |
| 40.4 | Statistical and Computational Methods to Treat the Data | 969 |
| 40.4.1 | Artificial Neural Networks and Regression Methods | 970 |
| 40.4.2 | Optimization of Biosensing Performance Using Multidimensional Projections | 971 |
| 40.5 | Conclusions and Perspectives | 977 |
| | References | 978 |
| | | |
| **41** | **Patterned Multilayer Systems and Directed Self-Assembly of Functional Nano-Bio Materials** 985 | |
| | *Ilsoon Lee* | |
| 41.1 | New Approaches and Materials for Multilayer Film Patterning Techniques | 985 |
| 41.2 | Cell Adhesion and Patterning Using PEMs | 988 |
| 41.3 | PEMs Incorporating Proteins and Their Patterning | 990 |
| 41.4 | Metal/Graphene Conductive Patterning via PEM Films | 992 |
| 41.5 | Ordered and Disordered Particles on PEMs | 995 |
| 41.6 | Mechanical Aspects of PEM Films and Degradable Films | 997 |
| | References | 999 |
| | | |
| **42** | **Electrochemically Active LbL Multilayer Films: From Biosensors to Nanocatalysts** 1003 | |
| | *Ernesto. J. Calvo* | |
| 42.1 | Introduction | 1003 |
| 42.2 | Electrochemical Response | 1004 |
| 42.3 | Dynamics of Charge Exchange | 1012 |
| 42.3.1 | Propagation of Redox Charge (Electron Hopping) | 1012 |
| 42.3.2 | Ion Exchange | 1018 |
| 42.3.3 | Applications | 1024 |
| 42.3.4 | Biosensors | 1026 |
| 42.3.5 | Core–Shell Nanoparticles | 1030 |
| 42.3.6 | Nanoreactors | 1030 |
| 42.3.7 | Biofuel Cell Cathodes | 1031 |
| 42.4 | Conclusions | 1033 |
| | References | 1034 |
| | | |
| **43** | **Multilayer Polyelectrolyte Assembly in Feedback Active Coatings and Films** 1039 | |
| | *Dmitry G. Shchukin and Helmuth Möhwald* | |
| 43.1 | Introduction. The Concept of Feedback Active Coatings | 1039 |
| 43.2 | Polyelectrolyte-Based Self-Healing Anticorrosion Coatings | 1040 |
| 43.2.1 | Passive Protection Activity of Polyelectrolyte Multilayers | 1042 |

| | | |
|---|---|---|
| 43.2.2 | Controlled Release of Inhibiting Agents | *1044* |
| 43.3 | Coatings with Antibacterial Activity | *1045* |
| 43.4 | Conclusions and Outlook | *1050* |
| | References | *1050* |

**Index** *1053*

# List of Contributors

**Rigoberto C. Advincula**
University of Houston
Departments of Chemistry and
Chemical and Biomolecular
Engineering
136 Fleming Building
Houston, TX 77204-5003
USA

**Hioharu Ajiro**
Osaka University
Department of Applied Chemistry
2-1 Yamada-oka
Suita, Osaka 565-0871
Japan

**Mitsuro Akashi**
Osaka University
Department of Applied Chemistry
2-1 Yamada-oka
Suita, Osaka 565-0871
Japan

**Maria N. Antipina**
Agency for Science, Technology and
Research (A*STAR)
Institute of Materials Research and
Engineering
3 Research Link
Singapore 117602
Singapore

*Jun-ichi Anzai*
Tohoku University
Graduate School of
Pharmaceutical Sciences
Aramaki, Aoba-ku
Sendai 980-8578
Japan

*Pedro H.B. Aoki*
Universidade Estadual Paulista
(UNESP)
Faculdade de Ciências e Tecnologia
19060-900 Presidente Prudente, SP
Brazil

*Katsuhiko Ariga*
National Institute for Materials Science
(NIMS)
World Premier International (WPI)
Research Center for Materials
Nanoarchitectonics (MANA)
1-1 Namiki
Tsukuba 305-0044
Japan

and

JST
CREST
1-1 Namiki
Tsukuba 305-0044
Japan

**Jianhao Bai**
National University of Singapore
Division of Bioengineering
Singapore 117576
Singapore

**Vincent Ball**
Centre de Recherche Public
Henri Tudor
Department of Advanced Materials and Structures
66 rue de Luxembourg
4002 Esch-sur-Alzette
Luxembourg

**Sebastian Beyer**
National University of Singapore
Division of Bioengineering
Singapore 117576
Singapore

and

National University of Singapore
Graduate School for Integrative Sciences and Engineering
Singapore 11756
Singapore

**Stephan Block**
Ernst-Moritz-Arndt Universität
Institut für Physik
Felix-Hausdorff-Str. 6
17487 Greifswald
Germany

**Thomas Boudou**
Grenoble Institute of Technology and Centre National de la Recherche Scientifique
CNRS UMR 5628, LMGP, MINATEC
3 Parvis Louis Néel
38016 Grenoble
France

**Merlin L. Bruening**
Michigan State University
Department of Chemistry
East Lansing, MI 48824
USA

**Cédric C. Buron**
Université Catholique de Louvain
Institute of Condensed Matter and Nanosciences – Bio & Soft Matter
Croix du Sud 1
1348 Louvain-la-Neuve
Belgium

**Ernesto J. Calvo**
Universidad de Buenos Aires
Departamento de Química Inorgánica
Electrochemistry Group, INQUIMAE
Pabellón 2, Ciudad Universitária
Buenos Aires 1428
Argentina

**Frank Caruso**
The University of Melbourne
Department of Chemical and Biomolecular Engineering
Building 173
Melbourne, Victoria 3010
Australia

**Chloe Chevigny**
TU Berlin
Department of Chemistry
Straße des 17. Juni 124
10623 Berlin
Germany

**Robert E. Cohen**
Massachusetts Institute of Technology
Departments of Materials Science and Engineering and Chemical Engineering
77 Massachusetts Avenue
Cambridge, MA 02139
USA

*Carlos J.L. Constantino*
Universidade Estadual Paulista
(UNESP)
Faculdade de Ciências e Tecnologia
19060-900 Presidente Prudente, SP
Brazil

*Cornelia Cramer*
University of Münster
Institute of Physical Chemistry
Corrensstr. 28/30
48149 Münster
Germany

*Thomas Crouzier*
Grenoble Institute of Technology and
Centre National de la Recherche
Scientifique
CNRS UMR 5628, LMGP, MINATEC
3 Parvis Louis Néel
38016 Grenoble
France

*Yue Cui*
Chinese Academy of Sciences
Institute of Chemistry
Beijing National Laboratory for
Molecular Sciences (BNLMS)
Zhongguancun North First Street 2
Beijing 100190
China

*Chalongrat Daengngam*
Virginia Polytechnic Institute and
State University
Department of Physics
Robeson Hall (0435)
Blacksburg, VA 24061-0435
USA

*Bruno G. De Geest*
Ghent University
Department of Pharmaceutics
Harelbekestraat 72
9000 Ghent
Belgium

*Sophie Demoustier-Champagne*
Université Catholique de Louvain
Institute of Condensed Matter and
Nanosciences – Bio & Soft Matter
Croix du Sud 1
1348 Louvain-la-Neuve
Belgium

*Stefaan C. De Smedt*
Ghent University
Department of Pharmaceutics
Harelbekestraat 72
9000 Ghent
Belgium

*Wang Dong*
Virginia Polytechnic Institute and
State University
Department of Physics
Robeson Hall (0435)
Blacksburg, VA 24061-0435
USA

*Ashraf El-Hashani*
University of Cologne
Department of Chemistry
Luxemburger Str. 116
50939 Köln
Germany

*Nicel Estillore*
University of Houston
Departments of Chemistry and
Chemical and Biomolecular
Engineering
136 Fleming Building
Houston, TX 77204-5003
USA

*Jinbo Fei*
Chinese Academy of Sciences
Institute of Chemistry
Beijing National Laboratory for
Molecular Sciences (BNLMS)
Zhongguancun North First Street 2
Beijing 100190
China

*Andreas Fery*
University of Bayreuth
Department of Physical Chemistry II
95440 Bayreuth
Germany

*Karine Glinel*
Université de Rouen, CNRS
Laboratoire Polymères, Biopolymères, Surfaces
Bd M. de Broglie
7681 Mont Saint Aignan
France

and

Université Catholique de Louvain
Institute of Condensed Matter and Nanosciences – Bio & Soft Matter
Croix du Sud 1
1348 Louvain-la-Neuve
Belgium

*Jaime C. Grunlan*
Texas A&M University
Department of Mechanical Engineering
College Station, TX 77843-3123
USA

*Aurélie Guyomard*
Université de Rouen, CNRS
Laboratoire Polymères, Biopolymères, Surfaces
Bd M. de Broglie
7681 Mont Saint Aignan
France

*Lara Halaoui*
American University of Beirut
Department of Chemistry
Beirut
Lebanon

*Paula T. Hammond*
Massachusetts Institute of Technology
Department of Chemical Engineering
77 Massachusetts Avenue
Cambridge, MA 02139
USA

*Qiang He*
Harbin Institute of Technology
Micro/Nano Technology Research Centre
Harbin 150080
China

*Randy Heflin*
Virginia Polytechnic Institute and State University
Department of Physics
Robeson Hall (0435)
Blacksburg, VA 24061-0435
USA

*Christiane A. Helm*
Ernst-Moritz-Arndt Universität
Institut für Physik
Felix-Hausdorff-Str. 6
17487 Greifswald
Germany

*Jonathan P. Hill*
National Institute for Materials Science (NIMS)
World Premier International (WPI)
Research Center for Materials Nanoarchitectonics (MANA)
1-1 Namiki
Tsukuba 305-0044
Japan

and

JST
CREST
1-1 Namiki
Tsukuba 305-0044
Japan

*Kristina Hoffmann*
University of Cologne
Department of Chemistry
Luxemburger Str. 116
50939 Köln
Germany

*Jurriaan Huskens*
University of Twente
MESA+ Institute for Nanotechnology
Molecular Nanofabrication Group
7500 AE Enschede
The Netherlands

*Md Nasim Hyder*
Massachusetts Institute of Technology
Department of Chemical Engineering
77 Massachusetts Avenue
Cambridge, MA 02139
USA

*Qingmin Ji*
National Institute for Materials Science (NIMS)
World Premier International (WPI)
Research Center for Materials Nanoarchitectonics (MANA)
1-1 Namiki
Tsukuba 305-0044
Japan

*Chaoyang Jiang*
University of South Dakota
Chemistry Department
414 East Clark Street
Vermillion, SD 57069
USA

*Alain M. Jonas*
Université Catholique de Louvain
Institute of Condensed Matter and Nanosciences – Bio & Soft Matter
Croix du Sud 1
1348 Louvain-la-Neuve
Belgium

*Jouko Kankare*
University of Turku
Department of Chemistry
Laboratory of Materials Chemistry and Chemical Analysis
20014 Turku
Finland

*Toshiyuki Kida*
Osaka University
Department of Applied Chemistry
2-1 Yamada-oka
Suita, Osaka 565-0871
Japan

*Maxim V. Kiryukhin*
Agency for Science, Technology and Research (A*STAR)
Institute of Materials Research and Engineering
3 Research Link
Singapore 117602
Singapore

*Ralf Köhler*
TU Berlin
Department of Chemistry
Straße des 17. Juni 124
10623 Berlin
Germany

*Nicholas A. Kotov*
University of Michigan
Departments of Chemical Engineering, Materials Science and Engineering, and Biomedical Engineering
2300 Hayward Street
Ann Arbor, MI 48109
USA

*Ilsoon Lee*
Michigan State University
Department of Chemical Engineering
and Materials Science
2527 Engineering Building
East Lansing, MI 48824
USA

*Junbai Li*
Chinese Academy of Sciences
Institute of Chemistry
Beijing National Laboratory for
Molecular Sciences (BNLMS)
Zhongguancun North First Street 2
Beijing 100190
China

*Xiaokong Liu*
Jilin University
College of Chemistry
State Key Laboratory of Supramolecular
Structure and Materials
Changchun 130012
China

*Yuri Lvov*
Louisiana Tech University
Institute for Micromanufacturing
911 Hergot Avenue
Ruston, LA 71272
USA

*David M. Lynn*
University of Wisconsin – Madison
Department of Chemical and Biological
Engineering
1415 Engineering Drive
Madison, WI 53706
USA

*Dewang Ma*
Université de Montréal
Faculté de Pharmacie
Succursale Centre-Ville
Montréal, Quebec H3C 3J7
Canada

*Anna Maier*
University of Cologne
Department of Chemistry
Luxemburger Str. 116
50939 Köln
Germany

*Michiya Matsusaki*
Osaka University
Department of Applied Chemistry
2-1 Yamada-oka
Suita, Osaka 565-0871
Japan

*Jonathan Metzman*
Virginia Polytechnic Institute and
State University
Department of Materials Science and
Engineering
Robeson Hall (0435)
Blacksburg, VA 24061-0435
USA

*Marc Michel*
Centre de Recherche Public
Henri Tudor
Department of Advanced Materials and
Structures
66 rue de Luxembourg
4002 Esch-sur-Alzette
Luxembourg

*Helmuth Möhwald*
Max-Planck Institute of
Colloids and Interfaces
Fraunhofer Institute of
Biomedical Technology
Research Campus Golm
Am Mühlenberg 1
14424 Potsdam-Golm
Germany

*Reza Montazami*
Virginia Polytechnic Institute and
State University
Department of Materials Science and
Engineering
Robeson Hall (0435)
Blacksburg, VA 24061-0435
USA

*Peter Nestler*
Ernst-Moritz-Arndt Universität
Institut für Physik
Felix-Hausdorff-Str. 6
17487 Greifswald
Germany

*Bernard Nysten*
Université Catholique de Louvain
Institute of Condensed Matter and
Nanosciences – Bio & Soft Matter
Croix du Sud 1
1348 Louvain-la-Neuve
Belgium

*Osvaldo N. Oliveira Jr.*
Universidade de São Paulo
Instituto de Física de São Carlos
13560-970 São Carlos, SP
Brazil

*Yong Tae Park*
Texas A&M University
Department of Mechanical Engineering
College Station, TX 77843-3123
USA

*Pravin Pattekari*
Louisiana Tech University
Institute for Micromanufacturing
911 Hergot Avenue
Ruston, LA 71272
USA

*Felippe J. Pavinatto*
Universidade de São Paulo
Instituto de Física de São Carlos
13560-970 São Carlos, SP
Brazil

*Catherine Picart*
Grenoble Institute of Technology and
Centre National de la Recherche
Scientifique
CNRS UMR 5628, LMGP, MINATEC
3 Parvis Louis Néel
38016 Grenoble
France

*Paul Podsiadlo*
University of Michigan
Department of Chemical Engineering
2300 Hayward Street
Ann Arbor, MI 48109
USA

*David N. Reinhoudt*
University of Twente
MESA+ Institute for Nanotechnology
Molecular Nanofabrication Group
7500 AE Enschede
The Netherlands

*Kefeng Ren*
Grenoble Institute of Technology and
Centre National de la Recherche
Scientifique
CNRS UMR 5628, LMGP, MINATEC
3 Parvis Louis Néel
38016 Grenoble
France

*H.D. Robinson*
Virginia Polytechnic Institute and
State University
Department of Physics
Robeson Hall (0435)
Blacksburg, VA 24061-0435
USA

**Cécile J. Roy**
Université Catholique de Louvain
Institute of Condensed Matter and
Nanosciences – Bio & Soft Matter
Croix du Sud 1
1348 Louvain-la-Neuve
Belgium

**Michael F. Rubner**
Massachusetts Institute of Technology
Departments of Materials Science and
Engineering and Chemical Engineering
77 Massachusetts Avenue
Cambridge, MA 02139
USA

**Mikko Salomäki**
University of Turku
Department of Chemistry
Laboratory of Materials Chemistry and
Chemical Analysis
20014 Turku
Finland

**Katsuhiko Sato**
Tohoku University
Graduate School of
Pharmaceutical Sciences
Aramaki, Aoba-ku
Sendai 980-8578
Japan

**Joseph B. Schlenoff**
Florida State University
Department of Chemistry and
Biochemistry
95 Chieftan Way
Tallahassee, FL 32306-4390
USA

**Monika Schönhoff**
University of Münster
Institute of Physical Chemistry
Corrensstr. 28/30
48149 Münster
Germany

**Takeshi Serizawa**
University of Tokyo
Research Center for Advanced Science
and Technology
4-6-1 Komaba, Meguro-ku
Tokyo 153-8904
Japan

**Nisarg Shah**
Massachusetts Institute of Technology
Department of Chemical Engineering
77 Massachusetts Avenue
Cambridge, MA 02139
USA

**Dmitry G. Shchukin**
Max-Planck Institute of
Colloids and Interfaces
Research Campus Golm
Am Mühlenberg 1
14424 Potsdam-Golm
Germany

**Jiacong Shen**
Jilin University
College of Chemistry
State Key Laboratory of Supramolecular
Structure and Materials
Changchun 130012
China

**Bong Sup Shim**
University of Michigan
Department of Chemical Engineering
2300 Hayward Street
Ann Arbor, MI 48109
USA

**Tatsiana Shutava**
Louisiana Tech University
Institute for Micromanufacturing
911 Hergot Avenue
Ruston, LA 71272
USA

**Andre G. Skirtach**
Max-Planck-Institute of
Colloids and Interfaces
Fraunhofer Institute of
Biomedical Technology
Research Campus Golm
Am Mühlenberg 1
14476 Potsdam-Golm
Germany

**Olaf Soltwedel**
Ernst-Moritz-Arndt Universität
Institut für Physik
Felix-Hausdorff-Str. 6
17487 Greifswald
Germany

**Svetlana Sukhishvili**
Stevens Institute of Technology
Department of Chemistry, Chemical
Biology and Biomedical Engineering
1 Castle Point on Hudson
Hoboken, NJ 07030
USA

**Gleb B. Sukhorukov**
Queen Mary University of London
School of Engineering and
Materials Science
Mile End Road
London E1 4NS
UK

**Junqi Sun**
Jilin University
College of Chemistry
State Key Laboratory of Supramolecular
Structure and Materials
Changchun 130012
China

**Shigehiro Takahashi**
Tohoku University
Graduate School of
Pharmaceutical Sciences
Aramaki, Aoba-ku
Sendai 980-8578
Japan

**Benjamin Thierry**
University of South Australia
Ian Wark Research Institute
Mawson Lakes Campus
Mawson Lakes, South Australia 5095
Australia

**Bernd Tieke**
University of Cologne
Department of Chemistry
Luxemburger Str. 116
50939 Köln
Germany

**Dieter Trau**
National University of Singapore
Division of Bioengineering
Singapore 117576
Singapore

and

National University of Singapore
Department of Chemical &
Biomolecular Engineering
Singapore 11756
Singapore

**Vladimir V. Tsukruk**
Georgia Institute of Technology
Materials Science and Engineering
771 Ferst Drive N.W.
Atlanta, GA 30332-0245
USA

**Janneke Veerbeek**
University of Twente
MESA+ Institute for Nanotechnology
Molecular Nanofabrication Group
7500 AE Enschede
The Netherlands

**Dmitry V. Volodkin**
Max-Planck-Institute of
Colloids and Interfaces
Fraunhofer Institute of
Biomedical Technology
Research Campus Golm
Am Mühlenberg 1
14476 Potsdam-Golm
Germany

**Regine von Klitzing**
TU Berlin
Department of Chemistry
Straße des 17. Juni 124
10623 Berlin
Germany

**Lars Wågberg**
KTH – Royal Institute of Technology
School of Chemical Science and
Engineering
Fibre and Polymer Technology and
The Wallenberg Wood Science Centre
Teknikringen 56–58
10044 Stockholm
Sweden

**Françoise M. Winnik**
Université de Montréal
Faculté de Pharmacie and
Département de Chimie
Succursale Centre-Ville
Montréal, Quebec H3C 3J7
Canada

and

National Institute for Materials Science
(NIMS)
World Premier International (WPI)
Research Center for Materials
Nanoarchitectonics (MANA)
1-1 Namiki
Tsukuba 305-0044
Japan

**Guanglu Wu**
Tsinghua University
Department of Chemistry
116 Hetian Building
Beijing 100084
China

**Ming Yang**
University of Michigan
Department of Chemical Engineering
2300 Hayward Street
Ann Arbor, MI 48109
USA

**Xi Zhang**
Tsinghua University
Department of Chemistry
308 Hetian Building
Beijing 100084
China

**Ziwei Zuo**
Virginia Polytechnic Institute and
State University
Department of Physics
Robeson Hall (0435)
Blacksburg, VA 24061-0435
USA

# Part II
# Applications

# 20
# Electrostatic and Coordinative Supramolecular Assembly of Functional Films for Electronic Application and Materials Separation

*Bernd Tieke, Ashraf El-Hashani, Kristina Hoffmann, and Anna Maier*

## 20.1
### Introduction

Since the pioneering work of Kuhn *et al.* [1–3] in the 1960s, the preparation and study of functional organized films with thickness control in the nanometer range has become an interesting and attractive research topic. Up to the 1990s, organized films were mainly prepared by the Langmuir–Blodgett technique [4–7] while adsorption techniques studied by Sagiv *et al.* [8], Mallouk *et al.* [9, 10], and Ulman [11] were of minor importance. In 1991, the situation changed when Decher *et al.* [12–14] first reported on alternating electrostatic layer-by-layer (LbL) assembly of bolaamphiphiles and/or polyelectrolytes of opposite charge on pretreated substrates. Since then the LbL assembly has become the most important and frequently used method for preparation of organized films. A variety of potential applications in surface modification, as anti-reflective coatings, as electroactive films, sensors, separation membranes, or for preparation of hollow microcapsules useful as drug carriers or microreactors have been reported since, and a number of review articles have appeared [15–22].

The purpose of this chapter is to review our recent activities in the field of LbL assembled films. Our work is mainly concerned with the preparation of highly selective and efficient separation membranes, and with the fabrication of functional films exhibiting electroactive and optoelectronic properties. Since the first compilation in 2003 [23], we have studied the following topics:

- Selective transport of ions and molecules across polyelectrolyte multilayer membranes: water softening and desalination were studied under nanofiltration (NF) and reverse osmosis (RO) conditions; size-selective transport of aromatic compounds was investigated, and ion separation was studied using polyelectrolyte blend multilayer membranes.
- Highly selective separation membranes of polyelectrolytes and macrocyclic compounds: membranes containing calixarenes or azacrown ethers were studied with respect to their ion separation and rejection under diffusion dialysis, NF- and RO-conditions.

*Multilayer Thin Films: Sequential Assembly of Nanocomposite Materials*, Second Edition.
Edited by Gero Decher and Joseph B. Schlenoff.
© 2012 Wiley-VCH Verlag GmbH & Co. KGaA. Published 2012 by Wiley-VCH Verlag GmbH & Co. KGaA.

- LbL-assembled films of Prussian Blue and analogues on substrates: films were prepared upon sequential assembly of transition metal ions and complex anions, and their optical and electrochemical properties, structure and composition, and their potential use for ion sieving were studied.
- Coordinative supramolecular assembly of films of linear and cross-linked metallo-polymers: films of linear and cross-linked metal Schiff-base coordination polymers, cross-linked coordination polymers with metal ion-terpyridine and -bis (imidazolyl-pyridine) complexes were prepared, and their electrochemical, electrochromic or fluorescent properties studied.

Some of our studies have already been compiled in short review articles [24–29], while in the present chapter our studies are comprehensively reviewed.

## 20.2
## Polyelectrolyte Multilayer Membranes

### 20.2.1
### Water Desalination

From previous studies [30–37] it was known that the LbL assembly of polyelectrolyte multilayer membranes (PEMs) on porous supports leads to a new type of composite membrane with high separation capability. Gas separation [30, 31], separation of alcohol–water mixtures under pervaporation conditions [32, 33], and ion separation under diffusion dialysis conditions [34–38] were investigated. Since pressure-driven separation processes, like nanofiltration (NF) and reverse osmosis (RO) are technically important, it was interesting to study the use of PEMs as NF- and RO-membranes for water softening and desalination in greater detail. While NF is restricted to separation processes at low pressure (<30 bar), RO is carried out at higher pressure. In a first study [39] we investigated the pressure-driven transport of aqueous NaCl, $Na_2SO_4$, $MgCl_2$, and $MgSO_4$ solutions in different concentration across polyvinylamine/polyvinylsulfate (PVA/PVS) membranes built up on porous polyacrylonitrile/polyethylene terephthalate (PAN/PET) supporting membranes. Sixty layer pairs were always adsorbed, to ensure that the pores were covered and the separation layer was free of defects. The home-made apparatus worked under dead-end conditions. In order to avoid concentration polarization it was important to rapidly stir the feed solution.

The hydraulic permeability of the composite membrane was 113.7 mL m$^{-2}$ h$^{-1}$ bar$^{-1}$, while the bare support had a permeability of 237 L m$^{-2}$ h$^{-1}$ bar$^{-1}$. This means that the multilayered polyelectrolyte coating reduced the permeability by a factor of 2000. For $MgCl_2$ and $MgSO_4$, a complete rejection was found, which was independent from the applied pressure and the concentration of the feed solution. The rejections $R$ of NaCl and $Na_2SO_4$ were pressure-dependent. They increased from 84 and 96% at 5 bar to 93.5 and 98.5% at 40 bar, respectively (Figure 20.1) [39, 40]. The salt rejection increased in the series 1,1- < 1,2- < 2,1- ≤ 2,2-electrolyte, because

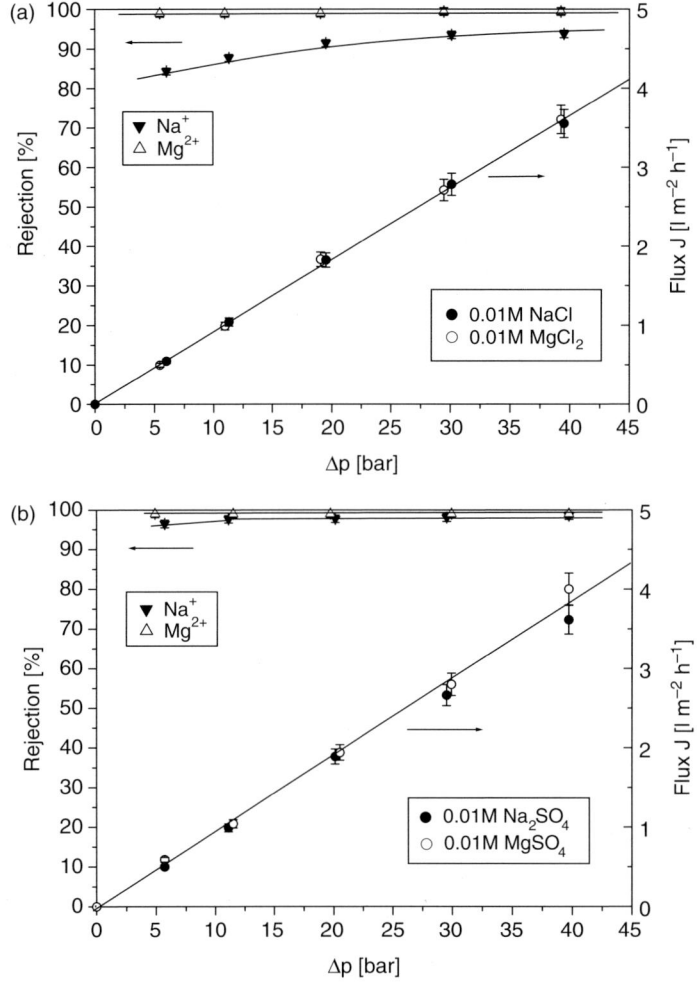

**Figure 20.1** Plot of cation rejection and permeation flux of 10 mM aqueous solutions of sodium and magnesium chloride (a), and sodium and magnesium sulfate (b) as a function of the operative pressure $\Delta p$. Room temperature, feed solution stirred at 700 rpm. Membrane: 60 layer pairs PVA/PVS on porous PAN/PET support (from [25] with permission of Elsevier, Amsterdam).

the permeating divalent ions interact more strongly with the membrane-bound charges than the monovalent ones [34]. Sodium sulfate is more strongly rejected than magnesium chloride because the permeating magnesium ions exhibit a stronger affinity to the sulfate groups of PVS than do the permeating sulfate ions to the ammonium groups of PVA. The good $R$ values indicate that the PEMs are suitable for water softening at low pressure, and water desalination at high pressure. Only the flux was rather low and still has to be improved by using supporting membranes with higher flux.

Motivated by the good results from the study on ion rejection, we also investigated the use of PEMs for sea water desalination under RO conditions [40]. The permeate solution was analyzed using a HPLC apparatus equipped with a conductivity detector. Again composite membranes consisting of 60 layer pairs PVA/PVS adsorbed on a PAN/PET support were studied. The sea water was either used as non-diluted feed solution, or after dilution with pure water in the ratio (sea water: water) 1 : 10, 1 : 100 and 1 : 1000. Using diluted sea water (1 : 10, 1 : 100, 1 : 1000) it was possible to reject 99% $Mg^{2+}$, 97% $Ca^{2+}$, and 92.5% $Na^+$ at 40 bar, while from non-diluted sea water 98% $Mg^{2+}$, 96.4% $Ca^{2+}$, and 74.5% $Na^+$ were rejected. The flux increased linearly with the pressure and reached $4\,L\,m^{-2}\,h^{-1}$ at the highest investigated pressure of 40 bar. A study of the influence of the number of deposited polyelectrolyte layer pairs on the rejection showed that a useful rejection of 90% $Mg^{2+}$ and 80% $Na^+$ could already be reached, if 30 layer pairs were adsorbed.

### 20.2.2
### Size- and Charge-Selective Transport of Aromatic Compounds

Since the LbL-assembly of oppositely charged polyelectrolytes proceeds under ion pairing, the PEMs exhibit a highly cross-linked structure, which might act as a molecular sieve able to separate small and large molecules from each other. This could be of technical interest for applications such as controlled release, water purification, and purification and separation of organic compounds. The porosity of the network structure can be tailored by the use of polyelectrolytes of either high or low charge density for membrane formation. With polyelectrolytes of high charge density, such as PVA/PVS, a dense network should be formed, whereas a more porous structure should be obtained with poly(diallyldimethyl-ammonium chloride) and poly(styrenesulfonate) (PDADMA/PSS) (Figure 20.2). Size-selective transport across PEMs was first reported by Liu and Bruening [41] studying the separation of mono- and polyfunctional alcohol derivatives. A clear size selective discrimination between glucose and sucrose was found and a transport model was derived.

In our group, the transport of aromatic compounds across PEMs was investigated [42]. The pore size and the size of the permeating compounds were varied. First, the transport of aromatic compounds of different size in ethanolic solutions was studied across the PDADMA/PSS membrane (Figure 20.3a). It was found that the smaller compounds phenol (ph), hydroquinone (hq) and naphthalene (np) with the longest molecular axes below 0.8 nm exhibit high permeation rates $P_R$ being above $10^{-5}$ $cm\,s^{-1}$. The larger compounds pyrene (py) and perylene (pe) with the longest molecular axes above 0.9 nm showed lower permeation rates, the $P_R$ values were below $2 \times 10^{-6}\,cm\,s^{-1}$. From this result, a mean pore diameter of about $0.82 \pm 0.09$ nm could be estimated for the PDADMA/PSS membrane. The transport across the PAH/PSS membrane (PAH: polyallylamine hydrochloride) was also studied. Since PAH has a higher charge density than PDADMA, a smaller pore size could be expected. In fact, high $P_R$ values of about $5 \times 10^{-6}\,cm\,s^{-1}$ were only observed for the two

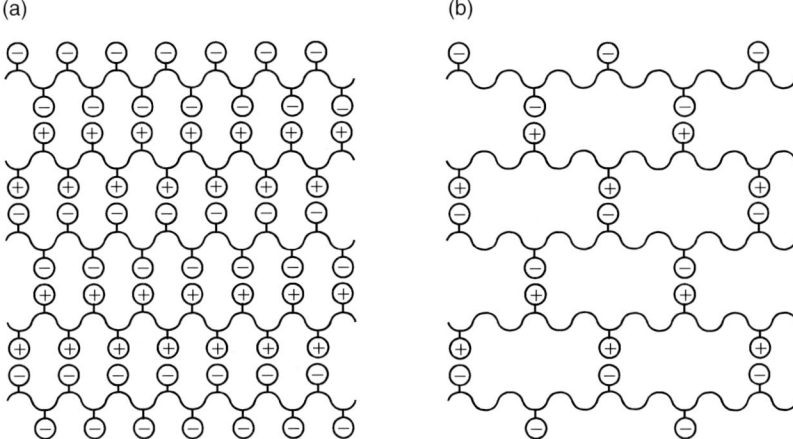

**Figure 20.2** Simplified structure model of polyelectrolyte complex membrane of high (a) and low charge density (b). Possible chain interdigitation and incomplete neutralization are not considered (from [25] with permission of Elsevier, Amsterdam).

smallest compounds ph and hq, whereas the permeation rates of the larger ones np, py and pe ($P_R$ being about $2 \times 10^{-7}$ cm s$^{-1}$) were much lower (Figure 20.3b). From the difference, a mean pore diameter of about $0.67 \pm 0.04$ nm could be derived. For the same kind of membranes, Liu and Bruening determined a pore size of 0.4–0.5 nm [41]. The difference may arise from different preparation conditions of the membrane and the fact that they used water as solvent, while we used ethanol. Finally the PVA/PVS membrane with even higher charge density of the polyelectrolytes was studied. It was impermeable for the five aromatic compounds, even phenol was rejected. Thus it could be assumed that the pore diameter was below 0.54 nm, the maximal length of the phenol molecule.

After the study of size-selective transport we investigated a possible charge-selective transport of organic compounds. The transport of differently charged aromatic compounds of similar size was investigated [42]. The compounds were ph, benzene sulfonate (bs), benzene-1,2-disulfonate (1,2-bs), and 1,3-disulfonate (1,3-bs). In Figure 20.3c, the $P_R$ values of the four compounds across the PDADMA/PSS membrane are plotted. Although size-selective transport can be ruled out because of the large pores, largely different $P_R$ values were obtained, the ratio of the values of ph:bs:1,2-bs:1,3-bs being 47.5 : 20.9 : 2.8 : 1. With PAH/PSS an even higher ratio of ph:bs:1,3-bds of 263 : 64 : 6:1 was found. This indicates that the organic ions with the highest charge density receive the strongest electrostatic (Donnan) rejection from the membrane, quite similar to the inorganic ions [34]. The difference in the $P_R$ values of the two isomeric bs-derivatives originates from their different dipole moments. Because of its larger moment (1,2-bs: 6.5 Da; 1,3-bs: 3.8 Da) the 1,2-isomer receives a stronger rejection. A size-dependent transport of organic ions was also studied [42]. The permeation of three organic salts with monovalent ions of different size of the organic part (bs, naphthalene sulfonate (ns), and methyl orange

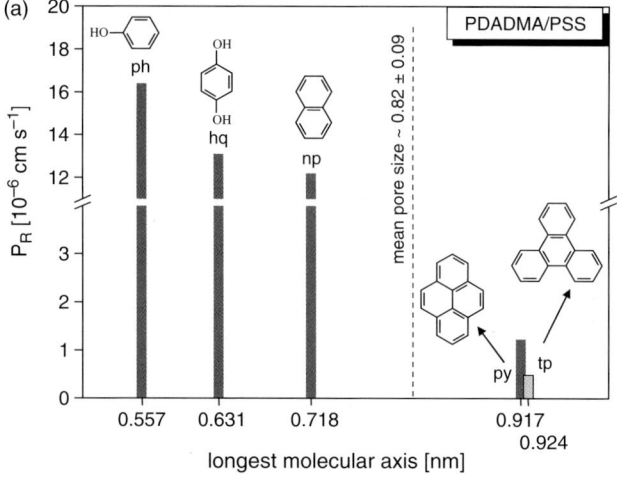
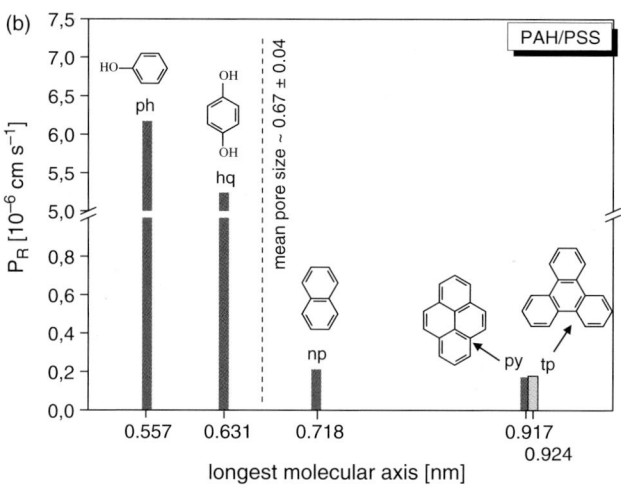
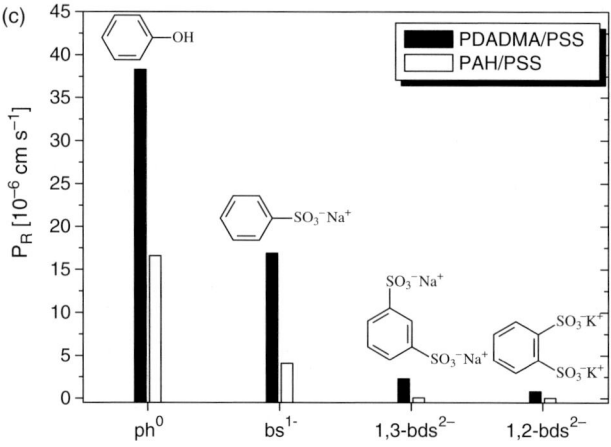

(MO)) was studied. It was found that the $P_R$ values decrease in the series bs > ns >> MO, that is, in the direction of increasing size of the anion. However, the size-based effect was small compared with the charge-based separation. For the PAH/PSS membrane, the separation factor $\alpha$ (bs/ns) for size-based separation was only 4, whereas $\alpha$ (bs/1,3-bs) for charge-based separation was about 65.

### 20.2.3
### Ion Separation from Polyelectrolyte Blend Multilayer Membranes

Recent studies indicated that the morphology of multilayered polyelectrolyte films can be decisively influenced if blends of weak and strong polyelectrolytes are adsorbed instead of single component polymers [43–46]. Films of PAH as cationic polyelectrolyte and a mixture of PSS and PAA (polyacrylic acid) as strong and weak anionic polyelectrolytes were investigated [47]. Since the permeability of a membrane is dependent on the morphology, it was interesting to study the influence of a blend formation on the ion separation. As an example, the ion transport across the PAH/PSS-PAA membrane was studied under conditions of diffusion dialysis, NF and RO [48].

Infrared spectroscopic studies [48] of the blend-PEMs indicated that PSS and PAA were not adsorbed in the same molar ratio as they were present in solution, in agreement with an earlier study [49]. The adsorption of the week polyelectrolyte PAA was lowered in favor of the strong polyelectrolyte PSS. Furthermore, film thickness and composition were also found to depend on the pH and salt content of the dipping solution.

Ion permeation was studied across the blend films of PAH/PSS-PAA prepared from dipping solutions of different PSS:PAH weight ratio. For comparison, the single-component films of PAH/PSS and PAH/PAA were also studied [48]. It was found that the PSS/PAA blend membranes containing 84% PSS prepared from a PAH solution of pH 7.5 and a mixed PSS–PAA solution of pH 3.5 showed a slightly improved selectivity $\alpha(NaCl/MgCl_2)$ of $39 \pm 3$ (PAH/PSS: $22 \pm 2$; PAH/PAA $2 \pm 1$). Since PAH was adsorbed at high pH, it is likely that the adsorbed polymer chains contained many non-dissociated amino groups, which were only protonated when the ion permeation took place at pH 4 or 5. The protonation created positive excess charges in the membrane, from which the permeating magnesium ions received a strong rejection. The reverse effect was found for PSS–PAA blend membranes containing 83% PSS prepared from dipping solutions of low pH of 1.7. For these membranes, an increased anion selectivity was found, the $\alpha(NaCl/Na_2SO_4)$ value was $197 \pm 10$ (PAH/PSS: 45; PAH/PAA: $10 \pm 6$). In this case it is likely that the adsorbed

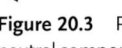

**Figure 20.3** Plot of the permeation rates $P_R$ of neutral compounds in ethanolic solution through PDADMA/PSS (a), and PAH/PSS membranes (b) as a function of the longest molecular axes of the compounds. The plot demonstrates the size-selective transport of the aromatic hydrocarbons. In (c) the permeation rates $P_R$ of neutral and differently charged compounds across the PDADMA/PSS and PAH/PSS membranes (60 layer pairs) are plotted showing the charge-selective transport of aromatic anions (from [25] with permission of Elsevier, Amsterdam).

PAA contained many non-dissociated carboxylic acid groups, which only became ionized when the ion permeation took place at pH 4 or 5. The deprotonation created negative excess charges in the membrane, from which the permeating sulfate ions received a strong rejection.

Under NF and RO conditions, the rejection of NaCl and $Na_2SO_4$ of the blend membrane was clearly superior to the PAH/PSS membrane [48]. For blend membranes prepared from a PAH solution and a mixed PSS–PAA solution of 1 : 1 ratio at pH 1.7, the $R$ values were 84 and 97% at 40 bar, respectively. For the PAH/PSS membrane, the corresponding salt rejections were only 15 and 27% at 40 bar.

### 20.2.4
### Membranes Containing Macrocyclic Compounds

#### 20.2.4.1 p-Sulfonato-Calix[n]arene-Containing Membranes

In order to improve the separation behavior of PEMs, we also tried to modify the membranes by replacing one of the two oppositely charged polyelectrolytes with charged macrocyclic compounds. Macrocycles such as calixarenes [50, 51] and crown ethers [52, 53] are known as building blocks in supramolecular chemistry, able to form host–guest complexes with a variety of molecules and ions. The chemical structures of the macrocycles used in our studies are shown in Scheme 20.1.

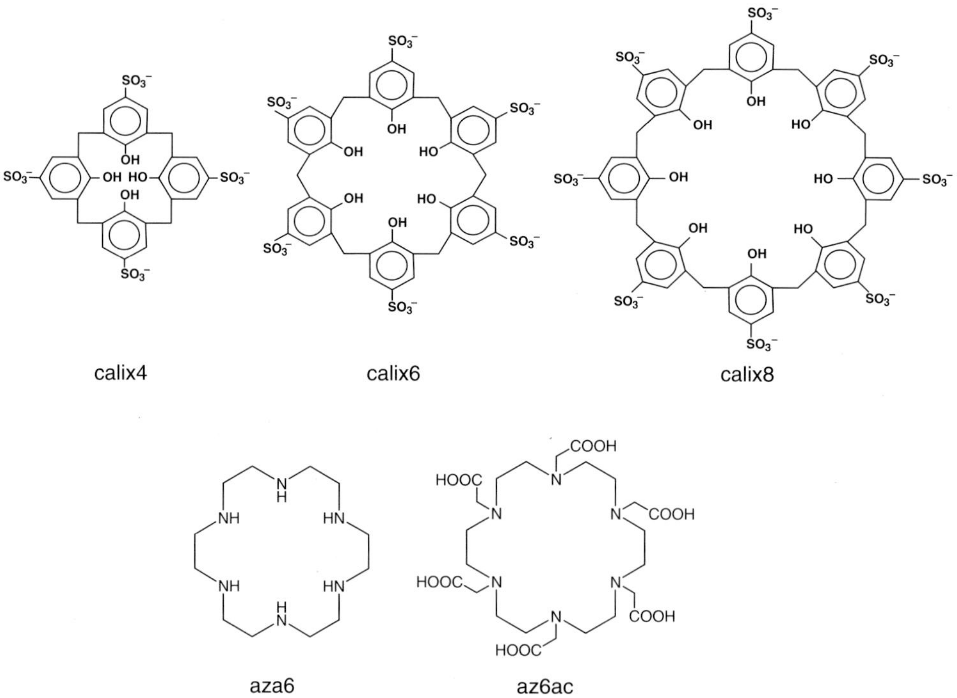

**Scheme 20.1** Chemical structures of macrocyclic compounds used for membrane preparation.

Calix[*n*]arenes are known for their complex formation with certain metal ions and small molecules [54]. However, only a few attempts to incorporate the compounds in LbL assemblies have been reported [55, 56], and the preparation of ion-selective membranes was not studied at all. For film preparation (Scheme 20.2), *p*-sulfonatocalix[*n*]arenes [57–59] with different ring sizes, *n* being 4, 6, and 8 (denoted as calix4, calix6 and calix8) were used [60]. The counter-polyelectrolyte was PVA. UV–Vis absorption studies indicated that dipping solutions of pH 6.5 to 7 without any supporting electrolyte were most suited for film formation. Calix8 was adsorbed in higher concentration than calix6 or calix4, probably because desorption was less pronounced. However, different conformations of the three macrocyclic compounds may also have influenced the absorption behavior.

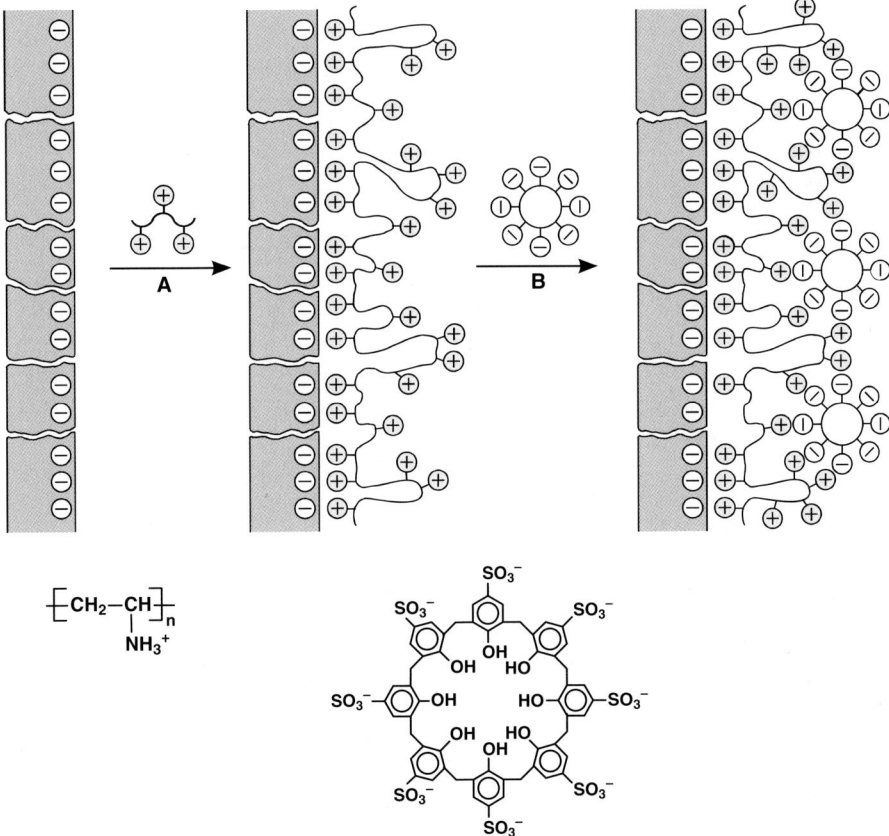

**Scheme 20.2** LbL-assembly of PVA and calix8 on porous supporting membrane. The actual orientation of the macrocycles is unknown.

In Figure 20.4, the permeation rates of various metal salts across the PVA/calixarene membranes are listed. For comparison, a PVA/PSS control membrane

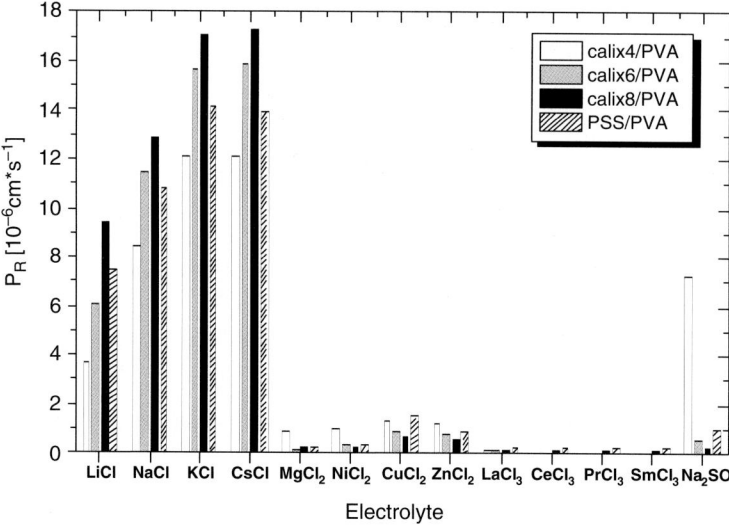

**Figure 20.4** Plot of permeation rates $P_R$ of various electrolyte salts across PVA/calixarene membranes and the PVA/PSS control membrane (always 60 dipping cycles) (data taken from [60]).

prepared under the same conditions was also investigated. Several important results were obtained. First, it turned out that the membranes were more permeable for monovalent than for divalent ions, quite similar to the all-polyelectrolyte membranes. The reason is that the di- and trivalent ions receive stronger repulsive forces from the membrane-bound charges of equal sign than the monovalent ions. For the same reason the permeation rates of alkali metal ions increase in the direction of lower charge density of the ions, that is, in the series $Li^+ < Na^+ < K^+ \leq Cs^+$. Secondly, there is a clear effect of the ring size $n$ of the calixarenes on the permeation rates of the ions. The $P_R$-values of the alkali metal ions increase with increasing ring size $n$, while for magnesium ions a minimum $P_R$ value is found for $n = 6$. For transition and rare earth metal ions, the $P_R$ values decrease with increasing ring size $n$. Thirdly, for nearly all metal salts investigated a calixarene-based membrane could be found exhibiting higher selectivities than the PVA/PSS control membrane. It can be concluded that the ion transport across the PVA/calixarene membranes is first determined by electrostatic interactions between the permeating ions and the membrane-bound charged groups, but is also influenced by a complex formation with the calixarene units. Arguments for host–guest interactions are based on the ring-size dependence of the $P_R$ values, and the observation that the $P_R$ values of some of the metal salts are up to three times lower than for the polyelectrolyte control membrane. Especially low transport rates were found for lithium ions through calix4-containing membranes, magnesium ions through calix6-based membranes, and rare earth metal ions through calix8-based membranes. A detailed study on the use of PVA/calix8 membranes for selective enrichment of rare earth metal ions revealed large separation factors $\alpha(NaCl/LaCl_3)$ of 138, $\alpha(NaCl/CeCl_3)$ of 128, $\alpha(NaCl/PrCl_3)$ of 130, $\alpha(NaCl/SmCl_3)$ of 130, and $\alpha(NaCl/YCl_3)$ of 160 [61].

#### 20.2.4.2 Azacrown Ether-Containing Membranes

Besides calixarenes, crown ether compounds are also among the most important host compounds in supramolecular chemistry. Because of their positive charge in the protonated state, and their ability to form complexes with a variety of organic and inorganic ions [62–69], azacrown ethers represent interesting building blocks for the preparation of LbL-assembled ion separation membranes. Among several azacrown ether compounds, which we tested for LbL-assembly, 1,4,7,10,13,16-hexaazacyclooctadecane (18-azacrown-6, hexacyclen, aza6) [53] turned out to be most useful.

In a first study the formation of LbL assemblies of aza6 and PSS, calix6, calix8, and hexacyanoferrate(II) ($HCF^{II}$) was demonstrated [70]. Since the protonation of aza6 takes place over a wide pH range [53, 62], the film formation was strongly pH-dependent. Linear growth of aza6/PSS films was found at pH 1.7 (where aza6 is fully protonated) and at pH 6. Superlinear growth was found in between, especially at pH 3.7. At this pH, aza6 is mainly present in a fourfold protonated state, that is, it can be desorbed and re-adsorbed, which would lead to the observed exponential layer growth. It was also possible to prepare LbL assemblies consisting entirely of aza6 and calix6 or calix8 [70]. The film growth was linear at pH 1.7 and 6 although at pH 6 the surface coverage was six times lower than at pH 1.7 due to the weak protonation of aza6. In Scheme 20.3, the LbL assembly of aza6/PSS and aza6/calix6 macrocycles is

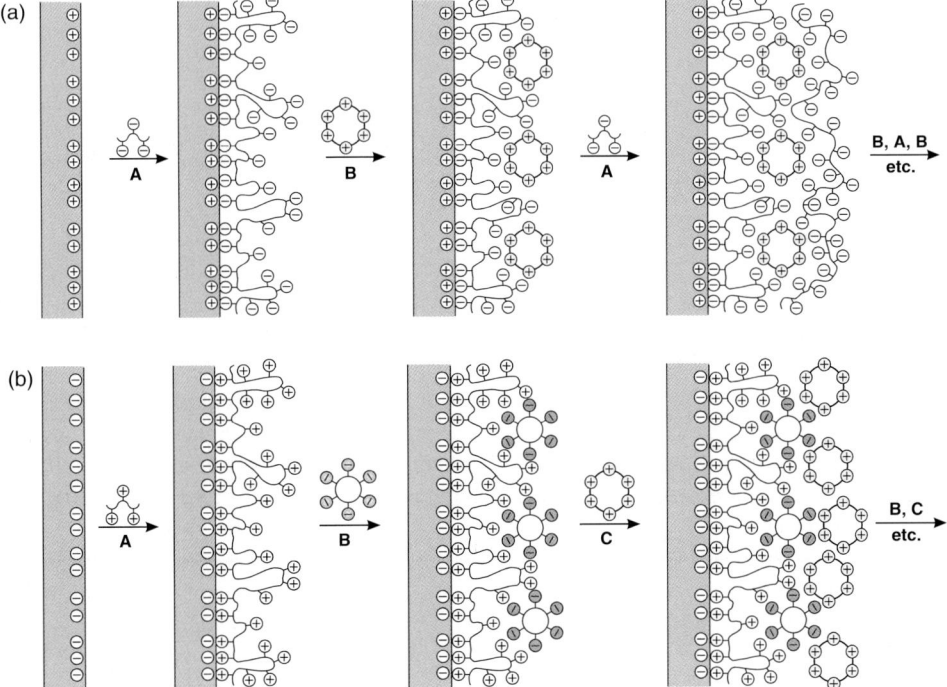

**Scheme 20.3** LbL-assembly of aza6 and PSS (a) and aza6/calix6 (b) on a solid support. The mutual orientation of the macrocycles is unknown.

represented. Due to the high affinity of aza6 towards anions, it was also possible to build up LbL assemblies of protonated aza6 and HCF$^{II}$ [70]. The resulting films were highly sensitive to Fe(III) ions. In the presence of Fe(III) they immediately turned blue due to formation of Prussian Blue particles. The effect is so strong that the films may serve as indicators for Fe(III) ions.

Having optimized the LbL growth of aza6-based films, composite membranes containing 60 layers of aza6 and PVS were built up on porous PAN/PET supports [71]. A control membrane of branched PEI (polyethyleneimine) and PVS was also prepared. The aza6/PVS membranes were highly permeable for electrolyte solutions, but the selectivity in ion transport was only poor ($\alpha$ (Cl$^-$/SO$_4^{2-}$) = 2). Since aza6 is known to form numerous host–guest complexes with metal ions and anions, we immersed the membrane in a 0.1 M aqueous copper(II) acetate solution and studied the ion permeation after the copper salt treatment. The selectivity of the membrane was strongly increased (Figure 20.5). Separation factors $\alpha$(Cl$^-$/SO$_4^{2-}$) of 110 and $\alpha$(Cl$^-$/SO$_3^{2-}$) of 1420 were obtained, while the cation separation remained very poor. UV- and IR-spectroscopic studies [71] indicated the formation of a binuclear complex [72] of aza6 and copper acetate according to Equation 20.1. The complex formation was also indicated by the deep blue coloration of the membrane (see the inset of Figure 20.5).

$$\text{aza6} + 2\, Cu(CH_3COO^-)_2 \longrightarrow [\text{aza6-Cu}_2(CH_3COO)_2]^{2+} + 2\, CH_3COO^- \quad (1)$$

As indicated in Scheme 20.4 the complex formation likely proceeds under release of protons from the triply protonated aza6. After formation of the binuclear complex, only two positive charges are left per macrocycle. As a consequence, some of the sulfate groups of PVS are no longer needed for charge neutralization and represent excess negative charges, which are responsible for the Donnan rejection of the permeating anions. The concept of enhancing the ion transport selectivity in PEMs by creating excess charges in the membrane was first introduced by Bruening and Sullivan [73].

Stimulated by the good performance of the copper-complexed aza6-PVS membrane in ion separation, we also studied the use of the membrane for water desalination under NF conditions [74]. The pressure-driven transport of a mixed aqueous solution of NaCl and Na$_2$SO$_4$ in 1 : 1 molar ratio (concentration: 10 mM each) was investigated. The total flux of a membrane consisting of 60 layer pairs was 0.85 L m$^{-2}$ h$^{-1}$; the rejections $R$ of chloride and sulfate ions were 56.5 and 98.6% at 14 bar, and 45.7 and 91.9% at 25 bar. On increasing the operating pressure the $R$

## 20.2 Polyelectrolyte Multilayer Membranes

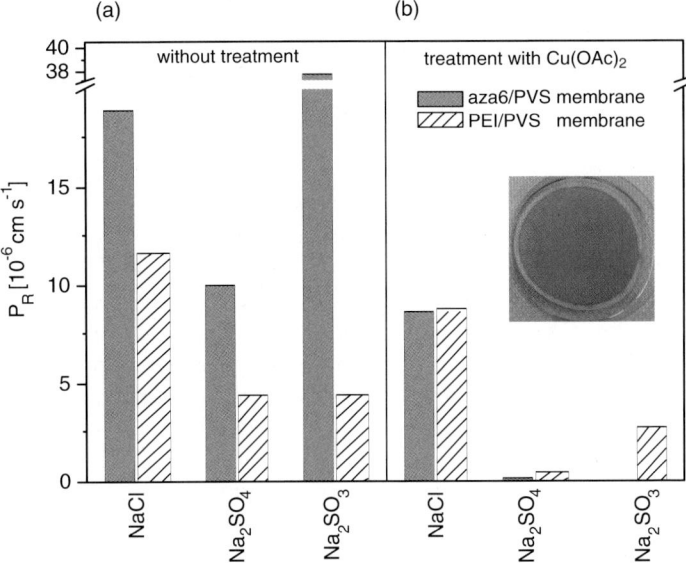

**Figure 20.5** Permeation rates $P_R$ of 0.1 M aqueous solutions of sodium chloride, sulfate and sulfite across aza6/PVS membranes and PEI/PVS control membrane (always 60 layer pairs) (a) before and (b) after treatment with aqueous copper(II) acetate solution (from [26] with permission of Elsevier, Amsterdam). The inset shows the bluish colored aza6/PVS membrane (60 layer pairs) after treatment with 0.1 M aqueous copper(II)acetate and careful washing with water (diameter: 5 cm) (from ref. [74] with permission by Elsevier, Amsterdam).

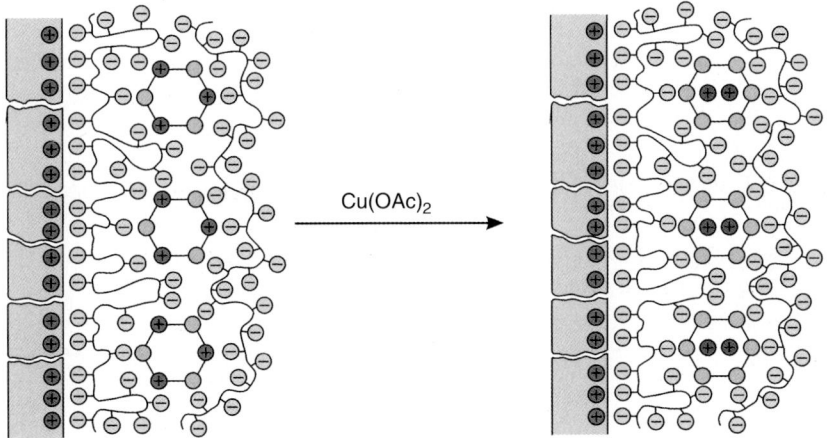

[**aza6** ·3 H]$^{3+}$ ·3 PVS$^-$ + 2 Cu(OAc)$_2$ → [**aza6** ·2 CuOAc]$^{2+}$ ·2 PVS$^-$ + 2 HOAc + H$^+$ + PVS$^-$

**Scheme 20.4** Chemical equation and model of complex formation between triply protonated aza6 and copper(II)acetate in aza6/PVS membrane.

values decreased, but the decrease was reversible. The decrease either originated from concentration polarization, or from a structural change in the separation layer induced by the higher pressure applied. The sulfate rejection of the copper-treated aza6/PVS membrane was comparable with the best rejection values reported previously for all-polyelectrolyte PVA/PVS membranes [39], whereas the chloride rejection was lower.

### 20.2.4.3 Membranes Containing Hexacyclen-Hexaacetic Acid

After the study of membranes containing anionic or cationic macrocycles, we also became interested in membranes containing zwitterionic macrocycles. N-Carboxymethyl-substituted azacrown ethers such as 1,4,7,10,13,16-hexacarboxymethyl 1,4,7,10,13,16-hexaazacyclo-octadecane (hexacyclen-hexaacetic acid, az6ac) [75, 76] are known as receptors for alkaline earth metal ions [77], lanthanide and actinide ions [78, 79]. Thus it was interesting to prepare membranes containing az6ac and study their ion separation properties [80].

Since it was not clear if the zwitterionic az6ac is preferentially adsorbed as cation or anion, we tried to build up LbL assemblies with PSS, PVS and PVA. The best films were obtained if PVA and az6ac were alternately adsorbed. Since 12 protonation states exist for az6ac [75, 76], the film formation was highly pH dependent. As indicated from ATR-IR-studies, a linear film growth only occurred at pH 6. At that pH, az6ac is most likely present in the fourfold protonated state [80]. In Scheme 20.5, the molecular structure of az6ac and a structure model of the PVA/az6ac membrane prepared at pH 6 are shown. According to the scheme, LbL assembly most likely proceeds via a salt formation between the carboxylate groups of az6ac and the ammonium groups of PVA. Protonation of the N atoms in the rings creates positive excess charges, which can lead to an improved cation rejection, as discussed in the previous section.

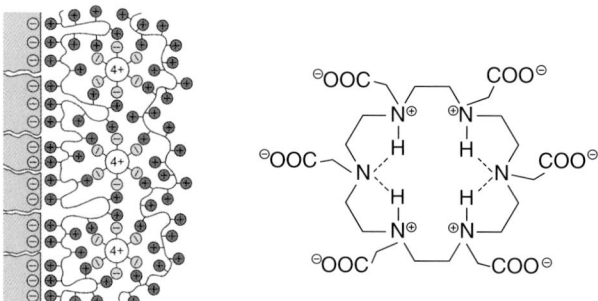

**Scheme 20.5** Molecular structure of az6ac at pH 6 and structure model of PVA/az6ac membrane prepared at pH 6. The actual orientation of the macrocycles is unknown.

For the study of ion transport, a PVA/az6ac membrane (60 layer pairs) was used [80]. The separation factors for cations were relatively high, for example, $\alpha(NaCl/MgCl_2)$ was 28, whereas the anion separation was poor, for example,

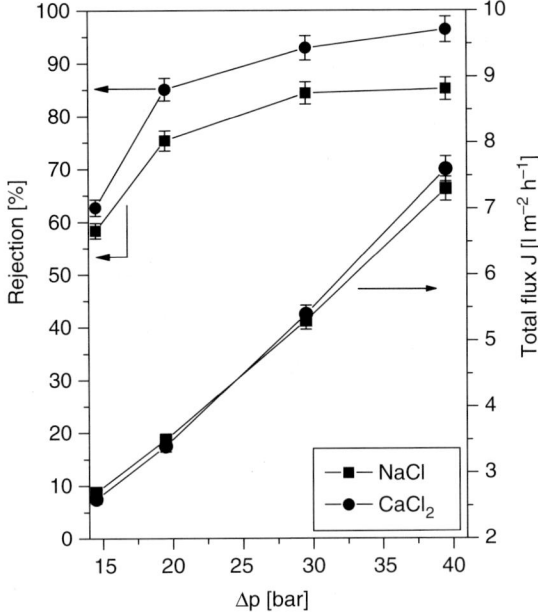

**Figure 20.6** Plot of rejection $R$ and total flux $J$ of 0.1 M aqueous NaCl and CaCl$_2$ solution as a function of operating pressure $\Delta p$. Sample: PVA/aza6ac membrane (60 dipping cycles). Dipping solutions contained 0.5 M NaCl, the pH was 6 (from [80] with permission of Elsevier, Amsterdam).

$\alpha(\text{NaCl}/\text{Na}_2\text{SO}_4)$ was only 2.3. The high cation selectivity can be related to the presence of positive excess charges in the membrane, as discussed above.

Under RO conditions at 40 bar, the PAH/az6ac membranes were very efficient in salt rejection [80]. High rejection values were found for sodium chloride (85%) and calcium chloride (95%), and the total flux was three times higher than for the corresponding all-polyelectrolyte membrane PVA/PVS (Figure 20.6).

In a further experiment we compared the transport behavior of membranes containing the macrocyclic az6ac and its linear chain analogue. For this purpose poly(ethyleneimine N-acetic acid) (LPEI-ac) was synthesized, and the ion transport across PVA/LPEI-ac membranes was studied. The PVA/LPEI-ac membrane was clearly less selective ($\alpha(\text{NaCl}/\text{MgCl}_2)$ was only 10) and salt rejection was lower and less efficient [80]. This might be due to the fact that the positive excess charges responsible for the improved ion separation are much better stabilized by the macrocyclic structure of az6ac than by the linear chain structure of LPEI-ac.

### 20.2.5
### LbL-Assembled Films of Prussian Blue and Analogues

LbL assembly of organized thin films is not only restricted to small organic and polymeric compounds, but may also occur with organic and inorganic compounds, or may take place exclusively with inorganic compounds. One of the first reports on electrostatic and coordinative LbL assembly of inorganic ions was concerned with films of Prussian Blue (PB) and analogous complex salts [81–83]. In a number of studies we reported on growth, morphology, elemental composition, optical and electrochemical properties, and on the ion transport across the films of PB and analogues [81, 84, 85].

Films of PB can be prepared upon alternate dipping of pretreated charged substrates into aqueous solutions of potassium hexacyanoferrate(III) (HCF$^{III}$) and ammonium ferrous sulfate, or by dipping into solutions of potassium HCF$^{II}$ and ferric chloride [81]. In the former case films of Fe$^{II}$HCF$^{III}$, in the latter one of Fe$^{III}$HCF$^{II}$ are formed, which are chemically identical, but for Fe$^{III}$HCF$^{II}$ the charge numbers of the adsorbed ions are higher ($z^+ = 3$, $z^- = 4$) than for Fe$^{II}$HCF$^{III}$ ($z^+ = 2$, $z^+ = 3$). Therefore, the surface charge of the films is higher and more oppositely charged ions are adsorbed. Consequently, after the same number of dipping cycles, the optical density of the Fe$^{III}$HCF$^{II}$ film is 1.4 times higher (Figure 20.7a, b) [81].

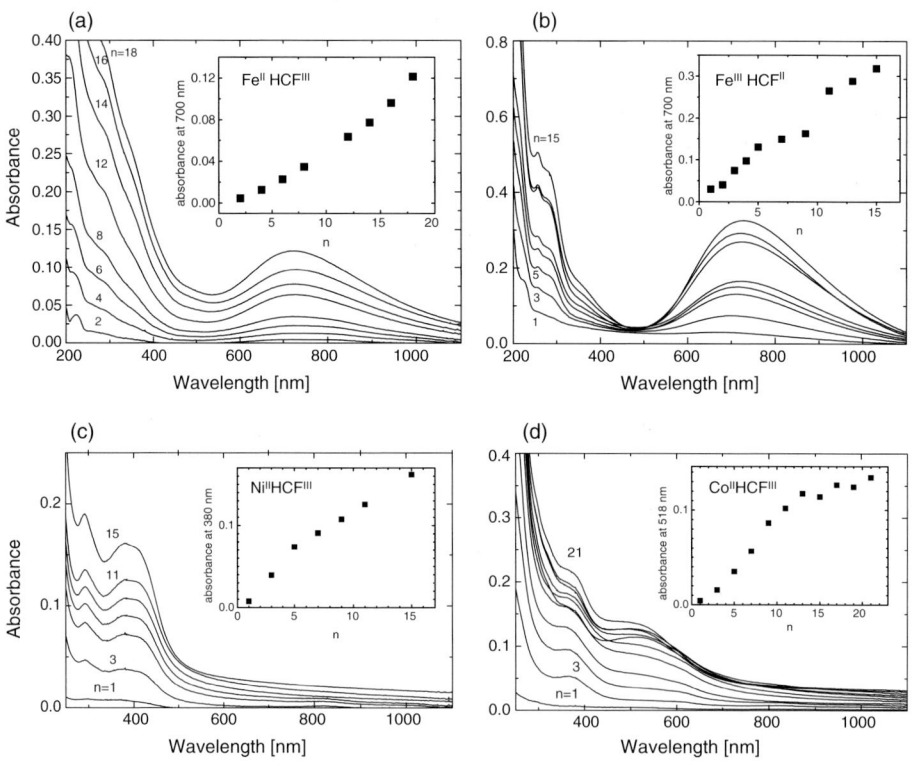

**Figure 20.7** UV–Vis absorption spectra of self-assembled films of Fe$^{II}$HCF$^{III}$ (a), Fe$^{III}$HCF$^{II}$ (b), Ni$^{II}$HCF$^{III}$ (c), and Co$^{II}$HCF$^{III}$ (d) after different number of dipping cycles. The insets show plots of the maximum absorbance versus $n$ (from ref. [81,84]. Copyright 2001 and 2003 American Chemical Society. Reproduced with permission).

If Ni(II) or Co(II) ions are adsorbed instead of Fe(II), the analogous complex salts $Ni^{II}HCF^{III}$ and $Co^{II}HCF^{III}$ are formed [84]. $HCF^{II}$ can also be replaced by hexacyanocobaltate, $HCC^{II}$, and films of $Fe^{III}HCC^{II}$ are obtained. Films of $Ni^{II}HCF^{III}$ are yellow because the $Ni^{II}$–NC–$Fe^{III}$ sequence only exhibits a weak mixed valence character. Films of $Co^{II}HCF^{III}$ are purple–brown. The 518 nm absorption can be ascribed to the charge-transfer transition of the $Fe^{III}$–CN–$Co^{II}$ sequence. Films of $Fe^{III}HCC^{II}$ are brownish-yellow, a charge transfer band does not occur (Figure 20.7c, d). The plots of the maximum absorbance versus the number of dipping cycles $n$ in the insets of Figure 20.7a–d indicate that linear film growth dominates, but nonlinear growth also occurs.

For the study of the electrochemical behavior, the films were adsorbed on ITO-coated glass substrates. Voltammetric studies (Figure 20.8a–c) indicate reversible oxidation behavior [84]. In the potential range between $-0.2$ and $+1.2$ V vs. SCE the PB films exhibit two anodic waves at 0.20 and 0.88 V, which can be ascribed to the oxidation of the two iron centers. Films of $Ni^{II}HCF^{III}$ and $Co^{II}HCF^{III}$, however, only show single anodic waves with peak potentials at 0.67 and 0.71 V, respectively. For the Co-based film an insulating effect of the substrate precoating is also shown. The peak current is much lower, if three PSS/PAH layer pairs are used instead of a single

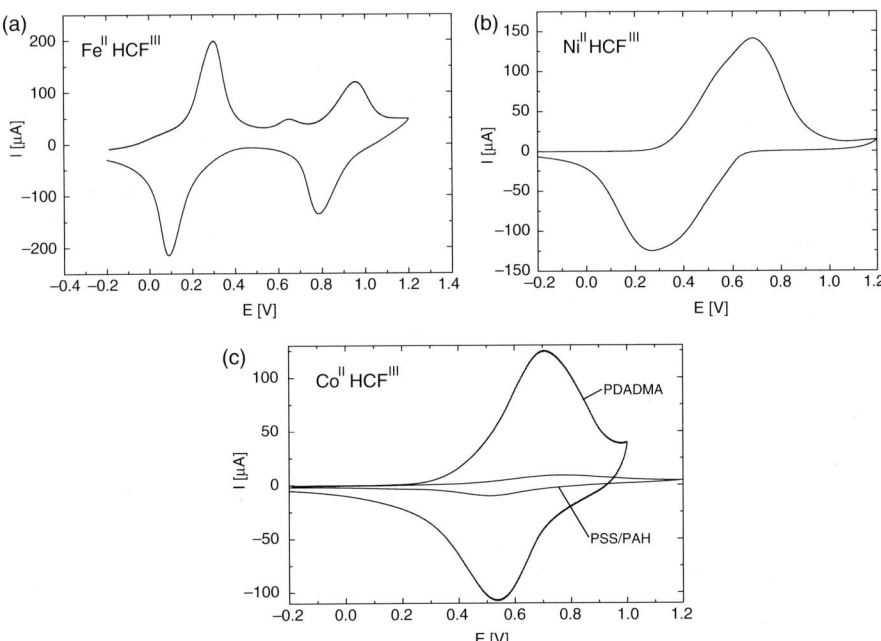

**Figure 20.8** Cyclic voltammograms of self-assembled films of $Fe^{II}HCF^{III}$ (a), $Ni^{II}HCF^{III}$ (b) and $Co^{II}HCF^{III}$ (c). Scan rate 2.3 mV s$^{-1}$, $T = 20\,^\circ$C. Substrates ITO-coated glass with a monolayer of PDADMA (a–c) or three layer pairs PSS/PAH (c) (from ref. [81,84]. Copyright 2001 and 2003 American Chemical Society. Reproduced with permission).

PDADMA layer. Using LbL assembly it was also possible to prepare films directly in the fully oxidized or reduced state, such as $Fe^{II}HCF^{II}$ (Everitt's salt, Prussian White) and $Fe^{III}HCF^{III}$ (Prussian Yellow) [84].

PB exhibits an open, zeolite-like structure, which in the past already led to studies of size-selective ion transport through PB-type films to an underlying electrode [86–88]. The studies indicated that the films are useful as ion-selective electrodes, electrocatalytic surfaces, and ion exchangers [89, 90]. In these studies the ion transport was driven by a potential gradient, while we tried to apply a concentration gradient. Before the ion transport was investigated, the growth conditions of the films were optimized [85]. The dipping time was varied, and absorbance, thickness, and morphology of the films were determined after 5, 10 and 20 dipping cycles. To our surprise, it was found that a dipping time of only 1 min into each of the solutions led to films with the highest absorbance. The increase in absorbance was fairly linear, although at a higher number of dipping cycles some scattering occurred. We believe that long dipping times favored desorption of PB particles and led to the observed effects.

AFM and EDX studies revealed that the PB films consist of a multitude of small, densely packed particles of $KFe[Fe(CN)_6]$, 10–100 nm in diameter. AFM studies also indicated that long dipping times of 30 min favor deposition of isolated, large PB particles on the substrate, while after short dipping times of only 1 min, and only a few dipping cycles applied, a fairly homogeneous surface coverage was reached (Figure 20.9) [85]. If more dipping cycles were applied, larger particles were formed in addition. Based on the AFM study we believe that two adsorption processes are involved in the film formation (Figure 20.10). The first comprises a rapid adsorption

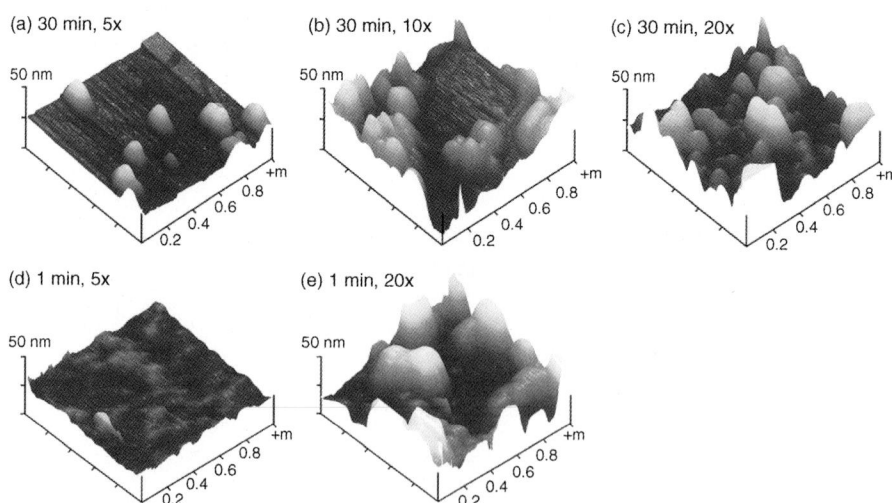

**Figure 20.9** AFM images of $Fe^{II}HCF^{III}$ films as a function of dipping cycles and dipping time into each solution. Dipping times are 1 min (d, e) and 30 min (a–c). Numbers of dipping cycles are 5 (a, d), 10 (b) and 20 (c, e) (from ref. [85]. Copyright 2003 American Chemical Society. Reproduced with permission).

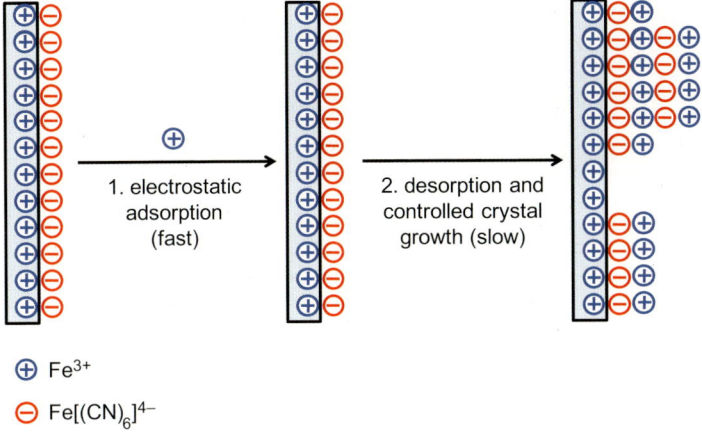

⊕ $Fe^{3+}$

⊖ $Fe[(CN)_6]^{4-}$

**Figure 20.10** Schematic representation of film growth of PB involving two processes: 1. electrostatic adsorption of ions within short dipping time, and 2. desorption/readsorption of ions under formation of particles (controlled crystal growth) at long dipping time.

of the ions at the substrate surface leading to homogeneous film growth. The second, much slower process comprises desorption, re-adsorption and formation of 3D crystallites, and leads to particle formation and heterogeneous structures. Therefore, homogeneous films can only be obtained if the dipping times are short, and the number of dipping cycles is small.

Having optimized the film growth conditions it was possible to build up defect-free films of PB and analogues on porous PAN/PET membranes. Up to 100 dipping cycles were applied. The macroscopic appearance of various metal hexacyanoferrate membranes on the porous supports is shown in Figure 20.11a. Since the pores of the PB structure have a radius of 0.16 nm, they are large enough to allow small hydrated ions such as $Cs^+$ and $K^+$ (hydrodynamic radius $r_h$: 0.119 and 0.125 nm, respectively) [91] to pass the membrane, whereas $Na^+$, $Li^+$ and $Mg^{2+}$ ($r_h$: 0.184, 0.239, and 0.347 nm, respectively) [91] should be rejected. In Figure 20.11b, the flux rates of various metal chloride salts are plotted and the Stokes radii of the hydrated ions are indicated. It can be seen that only the salts with small ions, such as $Cs^+$, $K^+$ and $NH_4^+$, are able to pass the membrane, while the others are rejected. High separation factors $\alpha(Cs^+/Na^+)$ of 7.7 and $\alpha(K^+/Na^+)$ of 5.9 were found [85]. For membranes of PB analogues, the separation factors were lower. The experiments demonstrate that the PB lattice acts as an ion sieve. It can be used, for example, to separate radioactive cesium ions from nuclear waste water. The ion transport mechanism is quite different from the PEMs, where non-hydrated ions are separated according to their charge density, and not by the sieving of hydrated ions.

In a subsequent study by Jin et al. [92] an alternative, facile preparation method for PB films on a Pt-electrode by direct aerosol deposition was reported. Aqueous solutions of potassium $HCF^{II}$ and ferric chloride were added to ultrasonic nebulizers. First, the $HCF^{II}$ was deposited from the aerosol on the electrode in an air-tight

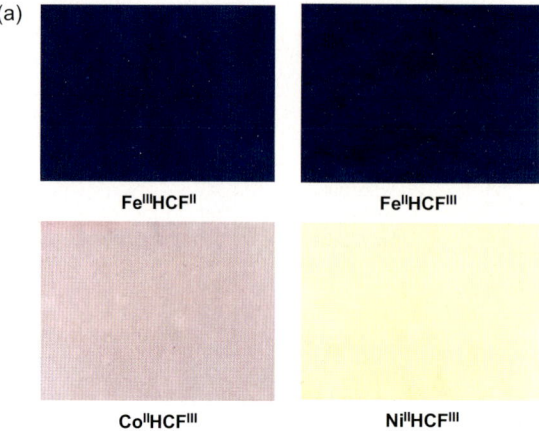

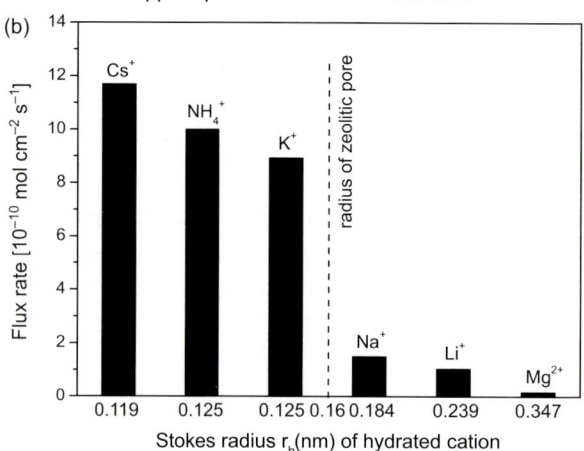

**Figure 20.11** (a) Macroscopic view of metal hexacyanoferrate membranes on PAN/PET support prepared with 60 dipping cycles. Size of membrane sections: $7 \times 5$ cm$^2$, and (b) plot of flux rates of various electrolyte salts across Fe$^{II}$HCF$^{III}$ membranes (100 dipping cycles) versus the Stokes radius $r_n$ of the hydrated cations (b) (from ref. [85]. Copyright 2003 American Chemical Society. Reproduced with permission).

organic glass container for 2 h. Subsequently, the coated Pt-electrode was subjected to the FeCl$_3$ aerosol and a PB film formed on the electrode. After washing and heating to 100 °C a very homogeneous film was deposited. The films can be applied in biosensors with high selectivity. For example, the detection of hydrogen peroxide with high sensitivity (1163 mA M$^{-1}$ cm$^{-2}$), and with an excellent anti-interference to ascorbic acid was demonstrated [93].

## 20.2.6
### Coordinative Assembly of Functional Thin Films

#### 20.2.6.1 Films of Coordination Polymers Based on Schiff-Base-Metal Ion Complexes

LbL assemblies are usually prepared upon alternating electrostatic adsorption of oppositely charged compounds, but deposition may also occur via hydrogen bonding [94–96] or coordinative interactions between the compounds [97]. Coordinative assembly is advantageous because films of organic–inorganic coordination polymers and metallopolymers can be directly built up in a step-by-step process. Among the first examples reported were films of zinc–bisquinoline [98], zirconium–tetrasalicylidene–diaminobenzidine [99], metal tetrathiooxalates [100], and others [101–104]. Furthermore, it was demonstrated that functional films with interesting photo- and electroactive properties can be fabricated [98, 100, 103].

In a continuation of this work, we investigated the supramolecular assembly of Schiff-base ligands and metal ions on solid supports [105]. As the ligand compounds the ditopic 5,5′-methylene-bis(N-methylsalicylidene amine) (MBSA), tetratopic N,N′,N″,N‴-tetra-salicylidene polyamidoamine (TSPA), and the polytopic poly(N-salicylidene vinylamine) (PSVA) were used (Scheme 20.6). The metal ions were Cu(II), Zr(II), Fe(II), Fe(III) and Ce(IV). For film formation, a charged substrate was alternately dipped into an ethanolic solution of the metal salt, and an ethanolic (or chloroform) solution of the ligand compound. The LbL assembly of MBSA-metal complex films is represented in Scheme 20.7a.

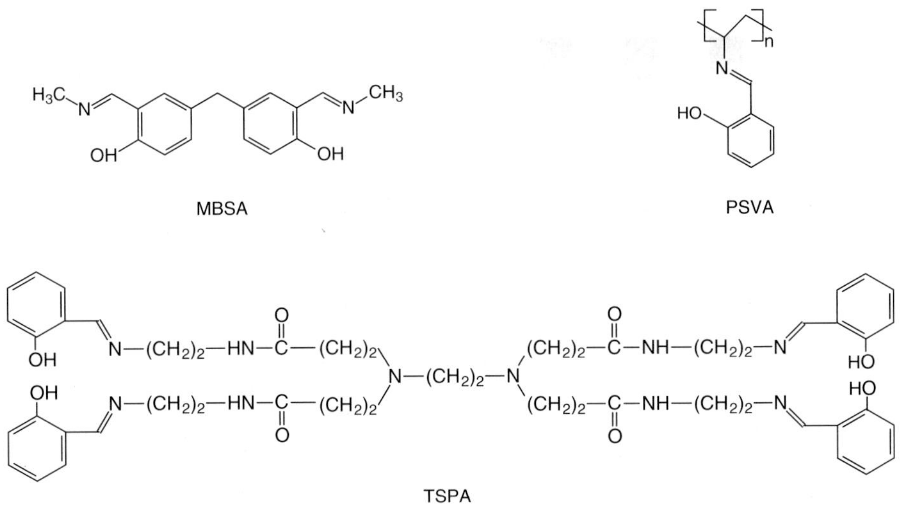

**Scheme 20.6** Chemical structures of Schiff-base ligands for multiple sequential assembly with metal ions.

In Figure 20.12, optical spectra of films of copper complexes of the three Schiff-base ligands are compared. All films exhibit a brownish color with three absorption bands near 370, 270 and 225 nm, which are typical for the salen–copper complex

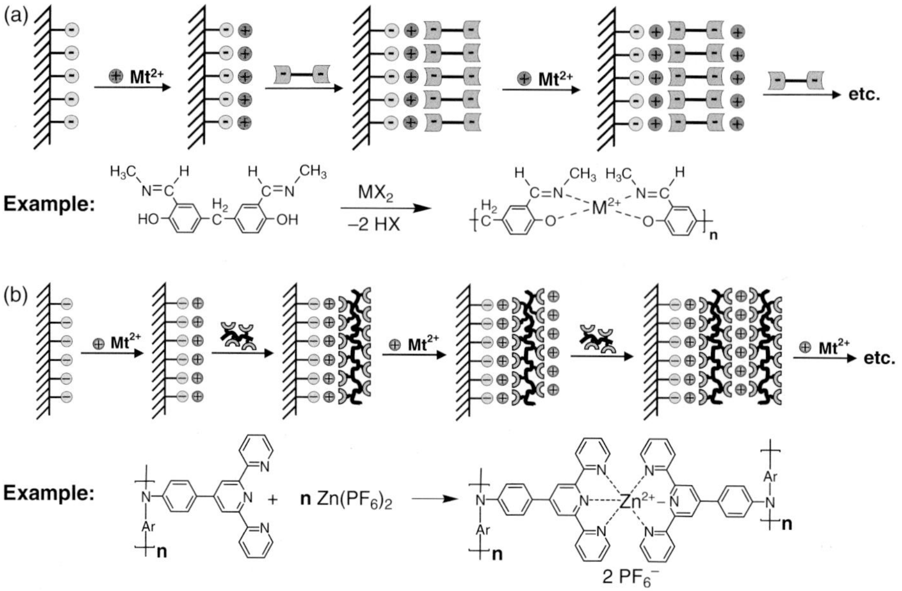

**Scheme 20.7** Coordinative LbL-assembly of ditopic (a) and polytopic (b) ligands with divalent metal ions on solid substrates. Examples for the formation of MBSA-M(II) (a) and polyiminoarylene-tpy-zinc ion complexes (b) are also shown.

chromophore. The insets of Figure 20.16 indicate linear or nearly linear film growth. After 12 dipping cycles, the absorbance of the films containing copper complexes of MBSA, TSPA and PSVA is quite different. For the three complexes it increases in the ratio 1 : 1.2 : 7.1. This indicates that the adsorption of PSVA is favored, very likely due to the large number of Schiff-base binding sites attached to the polymer chain, and due to the fact that a much higher density of salen groups can be reached at the substrate surface than with the ditopic MBSA or tetratopic TSPA. For the latter ones desorption is more likely. While the MBSA-based films contain linear coordination polymers with 2 : 1 salen:metal(II) complexes in the main chain (Scheme 20.7a), the PSVA-based films are cross-linked and contain 2 : 1 salen:metal(II) complexes, or 4 : 1 salen:Ce(IV) complexes in the side chains (Scheme 20.8). The metal–salen binding is based on electrostatic interactions with the phenolic OH-groups, accompanied by the release of protons, and coordinative interactions with the nitrogen atoms of the aldimine groups.

Using the same deposition conditions, it was also possible to build up films with other metal salts. Film preparation was most successful with Cu(II) and Ni(II) ions but not very successful with Zn(II) ions. In the former case, square planar complexes are obtained, while in the latter case tetragonal, voluminous complexes would be formed, which, however, is sterically unfavorable. The electrochemical properties of films of Cu(II)-PSVA on ITO-coated glass supports were studied. Voltammetric studies indicated a partially reversible anodic oxidation with two peak potentials at 0.6 and 0.9 V vs. Pt [105].

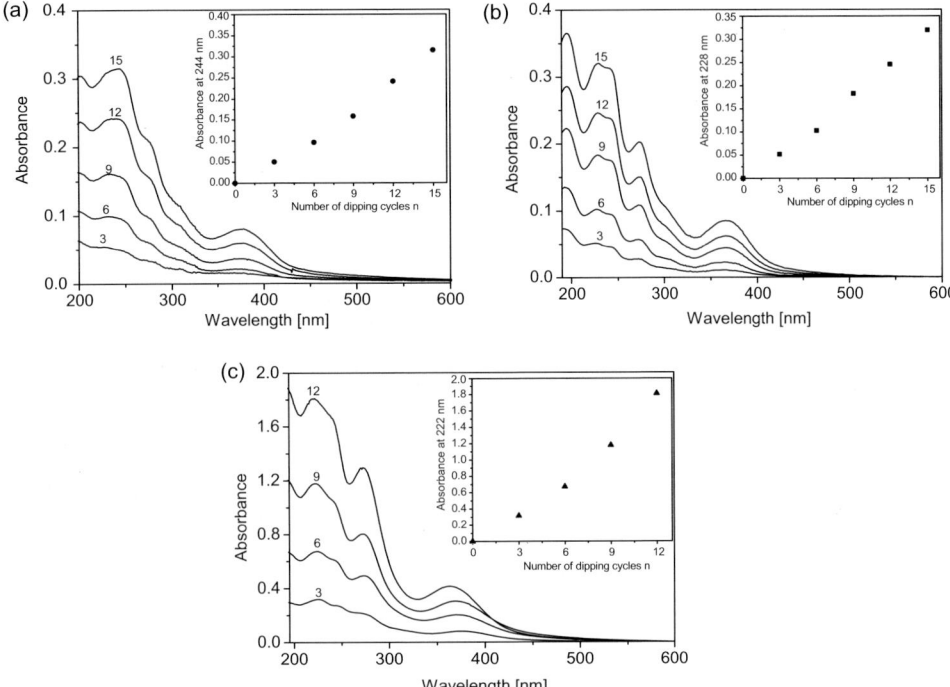

**Figure 20.12** UV–Vis spectra of Cu-MBSA (a), Cu-TSPA (b), and Cu-PSVA films (c) prepared upon multiple sequential adsorption. In the insets the increase in maximum absorbance with the number of dipping cycles is shown (from ref. [105]. Copyright 2007 American Chemical Society. Reproduced with permission).

### 20.2.6.2 Films of Coordination Polymers Based on Terpyridine-Metal Ion Complexes

The formation of metal-terpyridine complexes is exclusively based on coordinative interactions. The coordinative assembly of di- or trivalent metal ions with 2,2′:6,2″-terpyridine (tpy) ligands leads to positively charged octahedral complexes of 2:1 stoichiometry and $D_{2d}$-symmetry [106]. Stimulated by the good results on coordinative LbL assembly of polymers with salen ligands as substituent groups and metal ions (see Section 20.2.6.1), we synthesized the corresponding polymers with tpy substituent groups, and studied the coordinative assembly of these compounds and metal ions. The chemical structures of the polymers are compiled in Scheme 20.9. Most of the polymers consist of a polyiminoarylene main chain with easily oxidizable nitrogen atoms, which may induce hole-conducting and electrochromic behavior, and tpy substituent groups for metal ion complexation and sequential assembly of LbL films. Details of the polymer synthesis have been previously described [107].

With all polymers shown in Scheme 20.9 it was possible to build up LbL assemblies with homogenous surface structure. In a first communication [108], films of P1 and zinc ions were investigated. Subsequent studies were concerned with zinc, cobalt and nickel complexes of P1 [109], P2 and P3 [110]. A study of LbL assemblies of metal ion

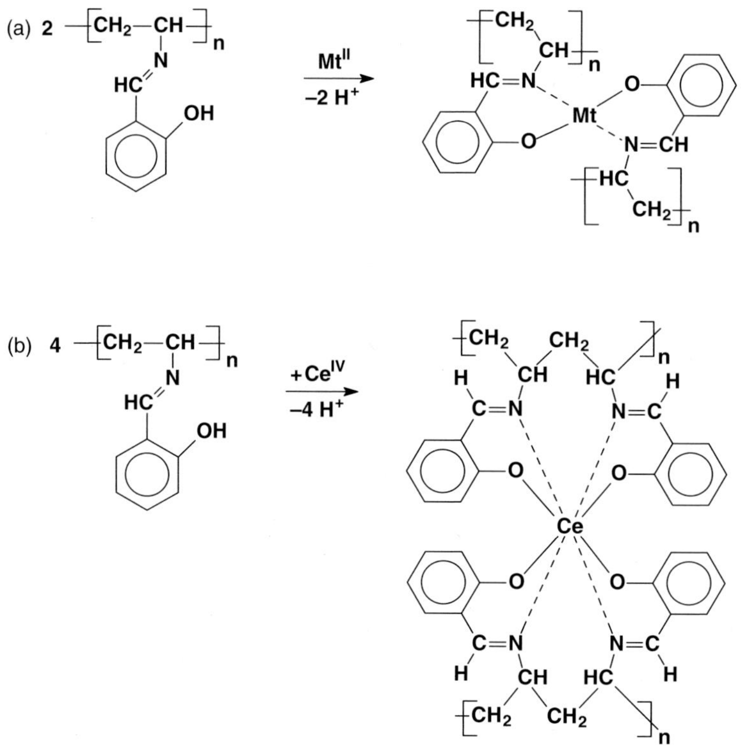

**Scheme 20.8** Formation of (a) PSVA-Mt(II) and (b) PSVA-Ce(IV) coordination polymer networks.

complexes of P4 and P5 was published very recently [111] and will be the subject of a future publication. Films of the P1-zinc ion complex were prepared upon alternate dipping of pretreated substrates into solutions of P1 and zinc hexafluorophosphate, as outlined in Scheme 20.7b. For this system, the growth conditions were optimized by adjusting the parameters of film preparation [109]. Concentration of the dipping solutions, dipping time and composition of the organic solvent mixture of the dipping solution were varied. In order to lower the solubility of the metal ligand complex and to favor its deposition on the substrate, metal hexafluorophosphate salts were used, and a non-solvent of the complex, such as n-hexane, was added to the dipping solution. As a result of the optimization, dipping times could be shortened from 10 min to 5 s, while the film thickness and surface homogeneity were essentially maintained. In Figure 20.13 optical spectra taken after different numbers of dipping cycles, and by application of different deposition conditions are compared. SEM pictures of films subjected to 12 dipping cycles are also shown. Working near to the precipitation limit of the complex, it was possible to prepare films of 100 nm thickness in less than 6 min. If too much hexane was added, uncontrolled precipitation set in, as shown in Figure 20.13. The films became inhomogeneous and very thick [109].

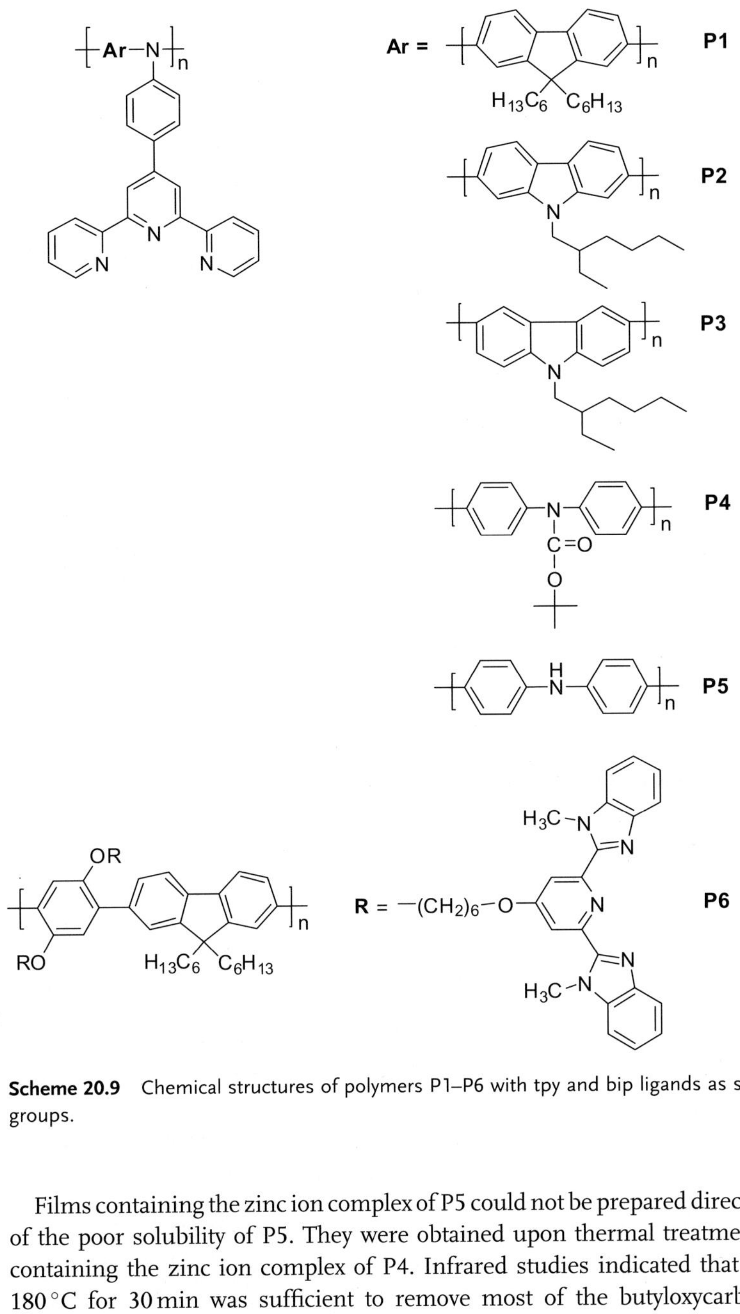

**Scheme 20.9** Chemical structures of polymers P1–P6 with tpy and bip ligands as substituent groups.

Films containing the zinc ion complex of P5 could not be prepared directly because of the poor solubility of P5. They were obtained upon thermal treatment of films containing the zinc ion complex of P4. Infrared studies indicated that heating at 180 °C for 30 min was sufficient to remove most of the butyloxycarbonyl (boc) protecting groups from P4.

The optical and electrochemical properties of films of the various metal ion complexes of P1 to P5 were studied. The UV–Vis absorption spectra in Figure 20.14 show that the films of the zinc and nickel ion complexes exhibit yellowish colors with

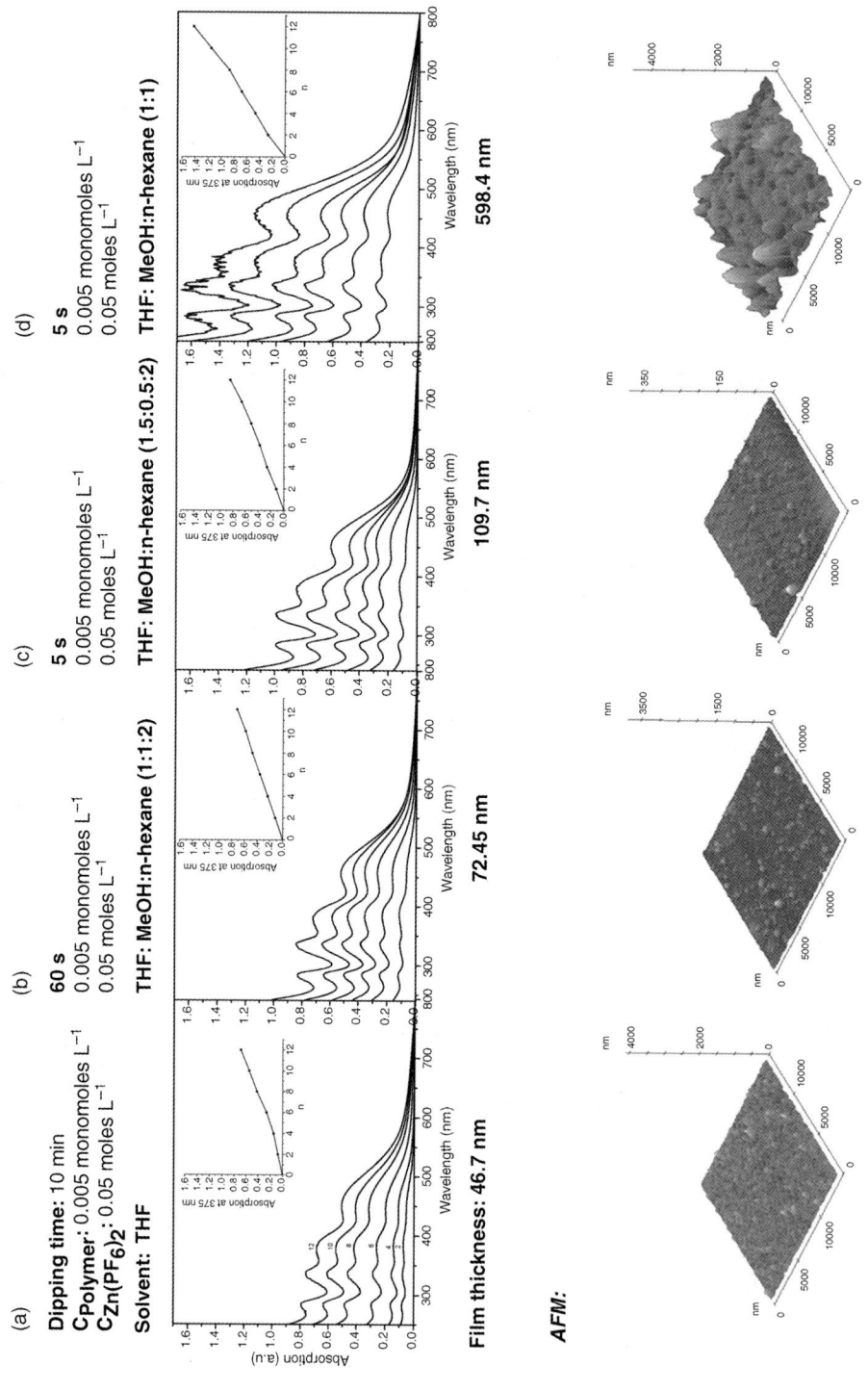

**Figure 20.13** Optical spectra and AFM pictures of films of the Zn-P1 complex prepared under different preparation conditions (solvent and concentration of dipping solutions). Optical spectra were monitored after different numbers of dipping cycles (from ref. [109]. Copyright 2009 American Chemical Society. Reproduced with permission).

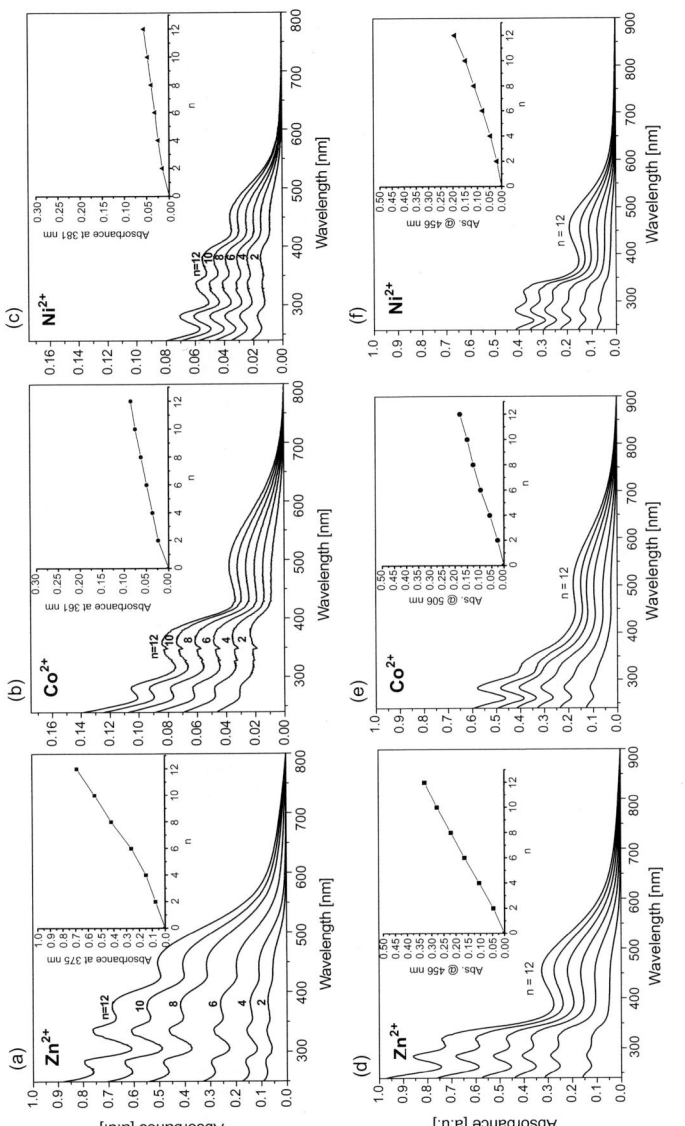

**Figure 20.14** UV–Vis absorption spectra of coordination polymer films of zinc (a, d), cobalt (b, e) and nickel ion complexes (c, f) of P1 (a–c) and P2 (d–f) monitored after $n$ dipping cycles. The insets show plots of the maximum absorbance versus $n$ ((a–c) from ref. [109]. Copyright 2009 American Chemical Society. Reproduced with permission; (d–f) from ref. [110] with permission from Royal Society of Chemistry).

the longest wavelength maxima around 460 nm, whereas those with cobalt are pink, purple, or brown ($\lambda_{max}$ around 490 nm). Films of copper(II) and iron(II) complexes of P1 are brown–red ($\lambda_{max}$: 496 nm) or olive green ($\lambda_{max}$: 589 nm) [111]. The ionochromism originates from differences in the metal-to-ligand charge transfer (MLCT) for the individual complexes, leading to different wavelength positions of the corresponding MLCT bands in the optical spectra.

Cyclovoltammetric studies indicate a highly reversible anodic oxidation of metal ion complexes of P1, P4 and P5 and a quasi-reversible oxidation behavior of films containing P2 and P3 (Figure 20.15a–d). Upon cycling between −0.2 and +1.0 V (in some examples 1.2 V) vs. FOC, either two anodic waves or a single broad wave appear, which can be related to the oxidation of the polymer with formation of the cation radical and dication state as outlined for P1 in Scheme 20.10 [109]. Depending on whether one or two anodic waves occur the oxidation is accompanied by one or two color transitions. A spectroelectrochemical study of a number of films is shown in Figure 20.15 (e–h) [109, 110]. In this figure, UV–Vis spectra of films (12 dipping cycles) on ITO-coated glass substrates are shown, which were monitored while different potentials were applied to the films. In all cases, oxidation is accompanied by color transitions to blue, green, or gray colors involving isosbestic points. The shift of the

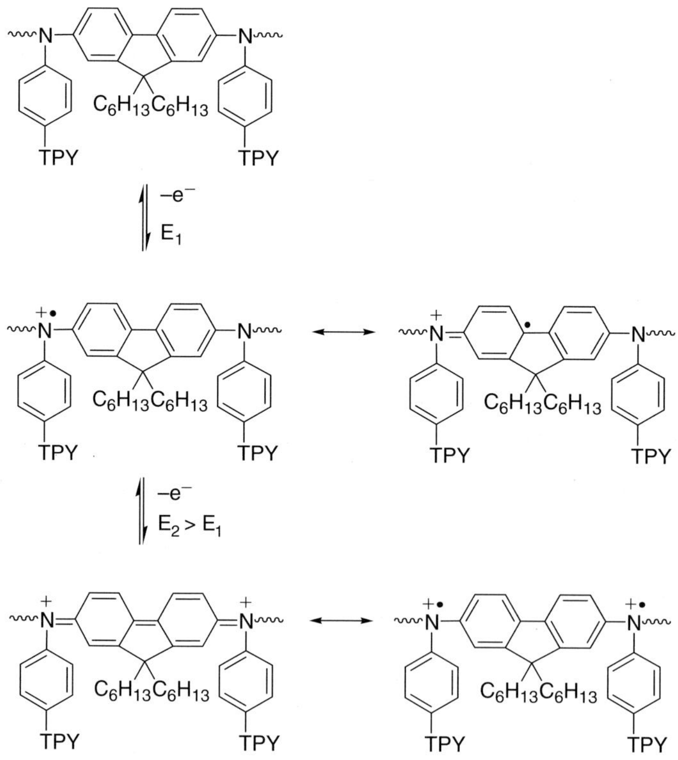

**Scheme 20.10** Chemical oxidation states of P1.

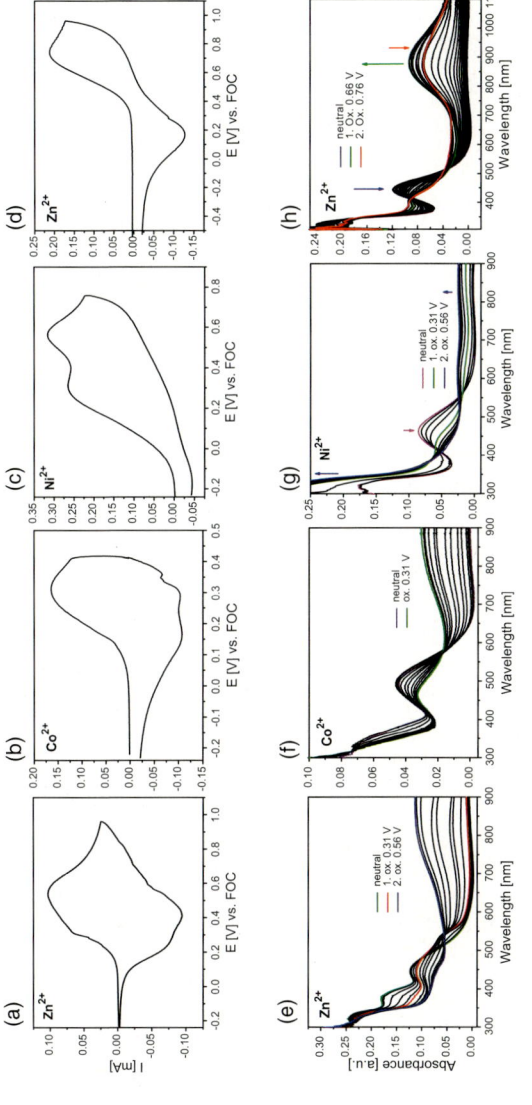

**Figure 20.15** Oxidative electrochemical cycles (a–d) and spectroelectrochemistry (e–h) of coordination polymer films on ITO-coated glass supports (always 12 dipping cycles). Scan rate: 50 mV s$^{-1}$. Samples: Zn-P1 (a, e), Co-P1 (b, f), Ni-P2 (c, g), and Zn-P4 complex (d, h) ((a,b,d–f,h) from ref. [109, 111b]. Copyright 2009 and 2012 American Chemical Society. Reproduced with permission; (c,g) from ref. [110] with permission by Royal Society of Chemistry).

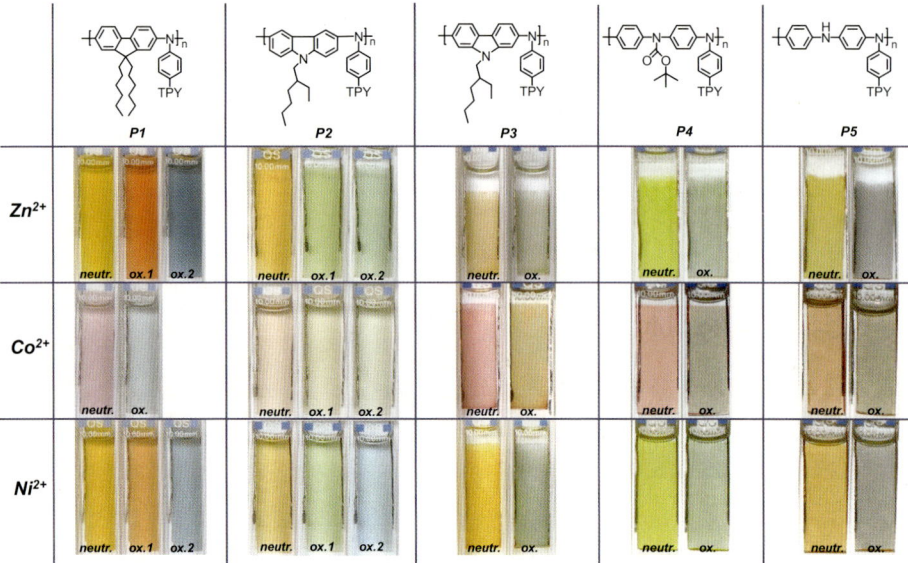

**Figure 20.16** Electrochromism of LbL-assembled coordination polymer films of zinc, cobalt, and nickel ion complexes of P1–P5. Colors of the neutral, partially and fully oxidized film are shown (from ref. [111a]).

absorption to longer wavelengths (and in some cases even to the infrared) originates from the formation of quinoid structures in the polymer backbone, as pointed out for P1 in Scheme 20.10. Oxidation of the metal ions does not occur at the applied potentials. Typical colors of a variety of coordination polymer films with different metal ions and polymers in the neutral and oxidized state are shown in Figure 20.16 [111].

The color changes are highly reversible [108]. At a scan rate of 200 mV s$^{-1}$, the loss in anodic current after 100 cycles is about 8%, if the films are cycled under ambient conditions. Films of 50 nm in thickness usually exhibit response times between 300 and 500 ms for switching between the neutral and fully oxidized state, and a contrast of 12 to 20% (in Δ% transmittance). This indicates that the films compare well with previously published electrochromic LbL assemblies of polyaniline and PB [112], polyethyleneimine and PB [113], and PEDOT and PAH [114], although it must be admitted that a direct comparison is difficult because of different device geometries and experimental conditions.

### 20.2.6.3 Films of Coordination Polymers Based on Bisimidazolylpyridine-Metal Ion Complexes

In a recent study [115], the coordinative LbL assembly of metal ions and the new conjugated polymer P6 with benzimidazolylpyridine ligands [116] in the side chain (Scheme 20.11) resulting in fluorescent films was described. The polymer (Scheme 20.9) contains a poly(phenylene-alt-fluorene) backbone, the bip units are attached to the phenylene units via alkyl spacer groups. Due to the aliphatic spacer

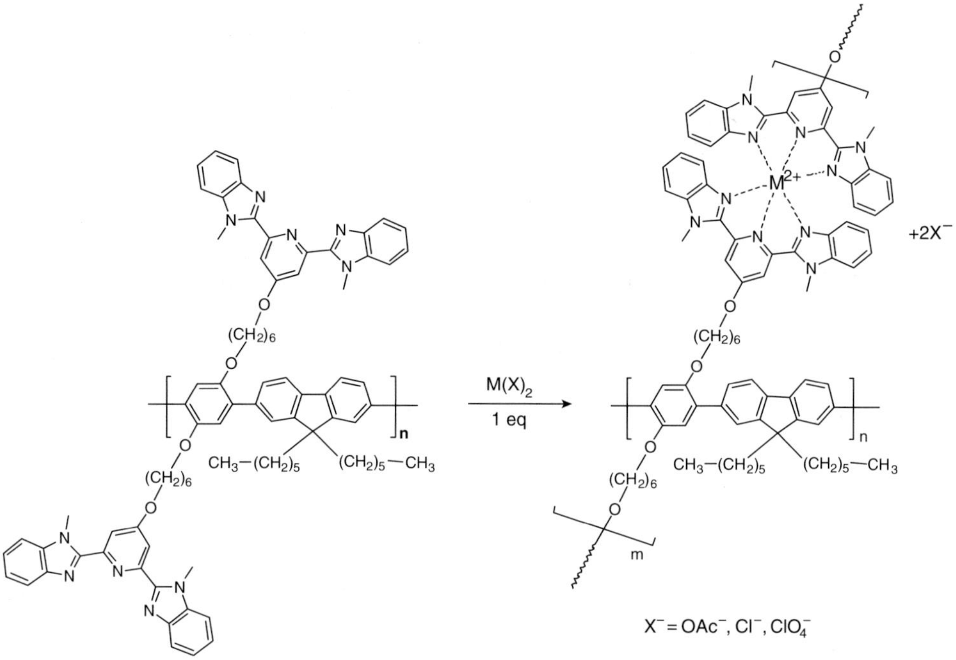

Scheme 20.11 Metal ion coordination of P6 involving cross-linking of side chains and supramolecular sequential assembly of coordination polymer network films based on P6 and divalent metal ions.

groups, the backbone and ligand chromophores are electronically decoupled, and metal ion coordination during film formation (see Scheme 20.11) only quenches the ligand fluorescence, whereas the blue backbone fluorescence is partially retained. LbL assembly of thin films on quartz supports was carried out by alternate dipping of

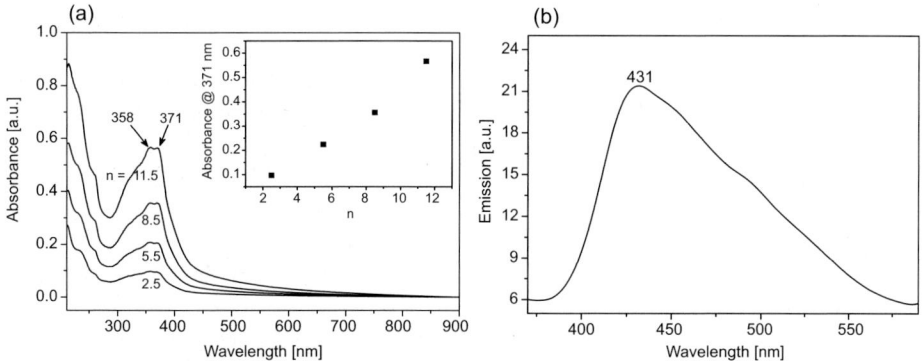

Figure 20.17 Spectroscopic study of LbL-assembled films of the copper ion complex of P6. (a) UV–Vis absorption spectra monitored after $n$ dipping cycles, (b) fluorescence spectrum of the film after 12 dipping cycles (from ref. [115]. Copyright 2011 American Chemical Society. Reproduced with permission).

the substrate into solutions of zinc(II) or copper(II) salts and the polymer. Homogeneous films were obtained in a nearly linear growth reaction (Figure 20.17a). The resulting films of P6 and zinc or copper ions exhibited a blue fluorescence with maximum around 425 nm (Figure 20.17b) [115]. The new materials could be useful for metal ion sensing and the preparation of fluorescent coatings.

## 20.3
## Summary and Conclusions

Our studies indicate the broad application potential of LbL-assembled films, ranging from highly selective membranes to inorganic films and membranes of Prussian Blue (PB) and analogues, and to functional films of coordination polymer network structures with reversible electrochromic behavior and fluorescence. A variety of compounds was used for LbL-assembly, such as polyelectrolytes, macrocycles, transition metal ions, metal complex anions, organic and polymeric ligands.

Polyelectrolyte multilayered membranes have been proven useful for water softening under nanofiltration conditions, and water and sea-water desalination under reverse osmosis conditions. Size-selective separation of aromatic compounds was demonstrated and the mean pore diameters of membranes were determined to be about 0.5 to 0.8 nm depending on the charge density of the polyelectrolytes. In order to improve the ion separation, polyelectrolyte blend membranes, and membranes consisting of polyelectrolytes and macrocycles were prepared. Unfortunately, the composition of the blend membranes always differed from the mixture composition of the dipping solution, and was also influenced by the pH and ion content of the dipping solution. Nevertheless, our study indicated that after a proper adjustment of the preparation conditions blend membranes became accessible, which showed a higher selectivity in ion transport than the corresponding single component polyelectrolyte membranes. With membranes of cationic polyelectrolytes and *p*-sufonatocalix[n]arenes a highly selective ion transport was found, which primarily originates from the strong electrostatic rejection that the multivalent ions receive from equally charged parts of the membrane, but is also caused by complex formation with the macrocyclic compounds. Due to host–guest interactions, the transport rates of lithium ions through calix4-containing membranes, magnesium ions through calix6-based membranes, and rare earth metal ions through calix8-based membranes were especially low. The strong interaction between *p*-sufonatocalix[n]arenes and rare earth metal ions allows the use of the calixarene/polyelectrolyte membranes for enrichment of rare earth metal ions. If azacrown ethers such as hexacyclen are incorporated in the membranes and converted into copper(II) complexes *in situ*, membranes with negative excess charges become accessible, which exhibit a highly selective ion transport with separation factors for mono- and divalent anions up to $10^3$. LbL-assembled membranes containing a zwitterionic azacrown ether derivative, such as hexacyclen hexaacetic acid az6ac, exhibit a highly selective cation transport. The high selectivity likely originates from positive excess charges created in the membrane during ion permeation. Under nanofiltration and reverse osmosis conditions, PVA/az6ac membranes exhibit a high flux and ion rejection, and, therefore, are very efficient in water softening and desalination.

The multiple sequential adsorption of di- or trivalent transition metal ions and metal complex anions such as hexacyanoferrate or cobaltate allowed one to build up films of Prussian Blue and analogous complex salts on charged substrates. Ultrathin films of a variety of inorganic complex salts became accessible and their optical and electrochemical properties could be studied. The high stability and excellent reversibility of the oxidation behavior led to application in biosensors with high selectivity. Structure studies of the Prussian Blue films indicated that they consist of a multitude of small, densely packed particles of $KFe[Fe(CN)_6]$ with diameters in the 10–100 nm range. Two processes contribute to the film formation: a rapid process of regular ion adsorption at the surface, and a slower process of ion desorption and re-adsorption under particle and crystallite formation. Therefore, homogeneous films can only be obtained if the second process is suppressed, that is, by applying short dipping times and only a few dipping cycles. Because of the porous, zeolitic structure of PB and its analogues it was possible to use the films as membranes for ion separation. If deposited on a porous support, the PB membranes were permeable for ions with small Stokes radius, such as cesium, potassium, and chloride, whereas large hydrated ions such as sodium, magnesium and sulfate were blocked. High separation factors of 7.7 for cesium/sodium separation, and 5.9 for potassium/sodium separation were found.

Further studies were concerned with functional films of metallopolymers and metallopolymer networks containing metal-Schiff-base complexes, metal ion-terpyridine (tpy), and metal ion-bisimidazolylpyridine (bip) complexes. The deposition of the films containing metal-Schiff-base complexes is driven by electrostatic interactions of the metal ions with the OH-groups of the ligands and coordinative interactions with the nitrogen atoms of the aldimine groups, whereas the deposition of the films with metal ion-tpy and -bip complexes is exclusively driven by coordinative interactions with the pyridine-N-atoms of the ligands. Using di-, tetra- and polytopic Schiff-base ligands for film formation it could be shown that polytopic ligands, that is, polymers with the ligand groups attached to the polymer main chain, are much better suited for film formation than the di- or tetratopic ligands. The reasons are that desorption is less likely, and a higher ligand concentration can be reached at the substrate surface. Sequential adsorption of metal ions and polyiminoarylenes with terpyridine substituent groups results in colored films, the color depending on the nature of the metal ions and the nature of the aromatic groups in the main chain. The ionochromism might be useful for ion sensing, for example. Since the oxidation potential of the polyiminoarylenes is very low the films can be cycled under ambient conditions with good reversibility. Electrochemical cycling is accompanied by electrochromic effects. In the neutral state yellow shades dominate, while in the oxidized state blue or green colors occur. Switching times below 500 ms and contrasts of about 20% for films of 50 nm in thickness compare well with previously reported LbL assemblies of conjugated polymers and Prussian Blue. Finally, it is shown that polymers with bip ligands can be assembled on pretreated substrates via complex formation with divalent transition metal ions. Since the conjugated polymer main chain is fluorescent, and is electronically decoupled from the bip ligands via alkyl spacer units, the fluorescence is not completely quenched upon metal ion complexation, and a fluorescent film is obtained.

In future work, we will continue studies on the coordinative assembly of films using polymers with other ligand groups, and metal salts with organic and functional (electrochromic) anions. Furthermore, the coordination polymer network structure might be interesting for membrane applications. The pore sizes are adjustable via the length of the spacer units between the ligand groups and the main chain, and the positive charge of the network structure originating from the complexed metal ions may induce ion exchanger properties, which might enable a highly selective ion transport.

## Acknowledgment

Financial support by the Deutsche Forschungsgemeinschaft (projects TI 219/6-3, 7-1, 7-2, 7-3, 11-1 and 11-2) is gratefully acknowledged.

## References

1 Kuhn, H., Möbius, D., and Bücher, H. (1972) Spectroscopy of Monolayer Assemblies, in *Physical Methods of Chemistry*, vol. **i**, Part IIIB (eds A. Weissberger and B.W. Rossiter), John Wiley & Sons, Inc, New York, pp. 577–702.
2 Kuhn, H. and Möbius, D. (1971) *Angew. Chem. Int. Ed.*, **10**, 620–637.
3 Bücher, H., Drexhage, K.H., Fleck, M., Kuhn, H., Möbius, D., Schäfer, F.P., Sondermann, J., Sperling, W., Tillmann, P., and Wiegand, J. (1967) *Mol. Cryst.*, **2**, 199–230.
4 Blodgett, K.B. (1935) *J. Am. Chem. Soc.*, **57**, 1007–1022.
5 Tredgold, R.H. (1994) *Order in Thin Organic Films*, Cambridge University Press, Cambridge.
6 See for example: Proceedings of International Conferences in Langmuir–Blodgett Films, in *Thin Solid Films*, 1983, **99**; 1985, **132**; 1987, **159**; 1989, **178**; 1992, **210/211**; 1994, **242/244**; 1996, **284/285**; 1998, **327/329**.
7 Roberts, G.G. (1990) *Langmuir–Blodgett Films*, Plenum Press, New York.
8 (a) Netzer, L. and Sagiv, J. (1983) *J. Am. Chem. Soc.*, **105**, 674–676; (b) Netzer, L., Iscovici, R., and Sagiv, J. (1983) *Thin Solid Films*, **99**, 235–241.
9 Cao, G., Hong, H.G., and Mallouk, T.E. (1992) *Acc. Chem. Res.*, **25**, 420–427.
10 Lee, H., Keplay, L.J., Hong, H.G., and Mallouk, T.E. (1988) *J. Am. Chem. Soc.*, **110**, 618–6200.
11 Ulman, A. (1996) *Chem. Rev.*, **9**, 1533–1554.
12 Decher, G. and Hong, J.D. (1991) *Makromol. Chem. Macromol. Symp.*, **46**, 321–327.
13 Decher, G. (1997) *Science*, **277**, 1232–1237.
14 Decher, G., Eckle, M., Schmitt, J., and Struth, B. (1998) *Curr. Opin. Colloid. Interface Sci.*, **3**, 32–39.
15 Boudou, T., Crouzier, T., Ren, K., Blin, G., and Picart, C. (2010) *Adv. Mater.*, **22**, 441–467.
16 Lichter, J.A., VanVliet, K.J., and Rubner, M.F. (2009) *Macromolecules*, **42**, 8573–8586.
17 Tung, Z., Wang, Y., Podsiadlo, P., and Kotov, M.A. (2006) *Adv. Mater.*, **18**, 3203–3224.
18 Ariga, K., Hill, J.P., and Ji, Q. (2007) *Phys. Chem. Chem. Phys.*, **9**, 2319–2340.
19 Jaber, J.A. and Schlenoff, J.B. (2006) *Curr. Opin. Colloid. Interface Sci.*, **11**, 324–329.
20 Hammond, P.T. (2004) *Adv. Mater.*, **16**, 1271–1293.

21 Schönhoff, M. (2003) *Curr. Opin. Colloid. Interface Sci.*, **8**, 86–95.
22 Lutkenhaus, J.L. and Hammond, P.T. (2007) *Soft Matter*, **3**, 804–816.
23 Decher, G. and Schlenoff, J.B. (eds) (2003) *Multilayer Thin Films. Sequential Assembly of Nanocomposite Materials*, Wiley VCH Verlag GmbH, Heidelberg.
24 Klitzing, R.v. and Tieke, B. (2004) *Adv. Polym. Sci.*, **165**, 177–210.
25 Tieke, B., Toutianoush, A., and Jin, W. (2005) *Adv. Colloid. Interface Sci.*, **116**, 121–131.
26 Tieke, B., El-Hashani, A., Toutianoush, A., and Fendt, A. (2008) *Thin Solid Films*, **516**, 8814–8820.
27 Tieke, B. and Toutianoush, A. (2009) Chapter 7, Electrostatic layer-by-layer fabrication of ultrathin separation membranes, in *Bottom-up Nanofabrication: Supramolecules, Self-Assemblies, and Organized Films*, vol. **5** (eds K. Ariga and H.S. Nalwa), American Scientific Publishers, Stevenson Ranch, California, USA, pp. 201–217.
28 Hoffmann, K., El-Hashani, A., and Tieke, B. (2010) *Macromol. Symp.*, **287**, 22–31.
29 Tieke, B. (2011) *Curr. Opin. Colloid Interface Sci.*, **16**, 499–507.
30 Stroeve, P., Vasquez, V., Coelho, M.A.N., and Rabolt, J.F. (1996) *Thin Solid Films*, **284–285**, 708–712.
31 Ackern, F.v., Krasemann, L., and Tieke, B. (1998) *Thin Solid Films*, **327–329**, 762–766.
32 Krasemann, L., Toutianoush, A., and Tieke, B. (2001) *J. Membr. Sci.*, **181**, 221–228.
33 Meier-Haack, J., Lenk, W., Lehmann, D., and Lunkwitz, K. (2001) *J. Membr. Sci.*, **184**, 233–243.
34 Krasemann, L. and Tieke, B. (2000) *Langmuir*, **16**, 287–290.
35 Harris, J.J., Stair, J.L., and Bruening, M.L. (2000) *Chem. Mater.*, **12**, 1941–1946.
36 Tieke, B., Ackern, F.v., Krasemann, L., and Toutianoush, A. (2001) *Eur. Phys. J. E*, **5**, 29–39.
37 Stair, J.L., Harris, J.J., and Bruening, M.L. (2001) *Chem. Mater.*, **13**, 2641–2648.
38 Toutianoush, A. and Tieke, B. (2001) Ultrathin self-assembled polyvinylamine/polyvinylsulfate membranes for separation of ions, in *Studies in Interface Science, Vol. 11 Novel Methods to Study Interfacial Layers* (eds D. Möbius, and R. Miller), Elsevier, Amsterdam, pp. 415–425.
39 Jin, W., Toutianoush, A., and Tieke, B. (2003) *Langmuir*, **19**, 2550–2553.
40 Toutianoush, A., Jin, W., Deligöz, A., and Tieke, B. (2005) *Appl. Surf. Sci.*, **246**, 437–443.
41 Liu, X. and Bruening, M.L. (2004) *Chem. Mater.*, **16**, 351–357.
42 Jin, W., Toutianoush, A., and Tieke, B. (2005) *Appl. Surf. Sci.*, **246**, 444–450.
43 Debreczeny, M., Ball, V., Boulmedais, F., Szalontai, B., Voegel, J.L., and Schaaf, P. (2003) *J. Phys. Chem. B*, **107**, 12734–12739.
44 Quinn, J.F., Yeo, J.C.C., and Caruso, F. (2004) *Macromolecules*, **37**, 6537–6543.
45 Yap, H.P., Quinn, J.F., Ng, S.M., Cho, J., and Caruso, F. (2005) *Langmuir*, **21**, 4328–4333.
46 Quinn, A., Such, G.K., Quinn, J.F., and Caruso, F. (2008) *Adv. Funct. Mater.*, **18**, 17–26.
47 Quinn, A., Tjipto, E., Yu, A., Gengenbach, T.R., and Caruso, F. (2007) *Langmuir*, **23**, 4944–4949.
48 Hoffmann, K., Friedrich, T., and Tieke, B. (2011) *Polym. Eng. Sci.*, **51**, 1497–1506.
49 Cho, J., Quinn, J.F., and Caruso, F. (2004) *J. Am. Chem. Soc.*, **126**, 2270–2271.
50 Böhmer, V. (1995) *Angew. Chem. Int. Ed.*, **34**, 713–745.
51 Asfari, Z., Böhmer, V., Harrowfield, J., and Vicens, J. (eds) (2001) *Calixarenes 2001*, Kluwer, Dordrecht, The Netherlands.
52 Gokel, G.W. (1991) *Crown Ethers and Cryptands: Monographs in Supramolecular Chemistry*, Royal Society of Chemistry, Cambridge, UK.
53 Kimura, E., Sakonaka, A., Yatsunami, T., and Kodama, M. (1981) *J. Am. Chem. Soc.*, **103**, 3041–3045.
54 Li, H.B., Chen, Y.Y., and Liu, S.L. (2003) *J. Appl. Polym. Sci.*, **89**, 1139–1144.

55 Yang, X., Johnson, S., Shi, J., Holesinger, T., and Swanson, B. (1997) *Sens. Actuators B*, **45**, 87–92.

56 Yang, Z.H. and Cao, W.Y. (2003) *Chin. J. Polym. Sci.*, **21**, 473.

57 Shinkai, S., Mori, S., Koneishi, H., Tsubaki, T., and Manabe, O. (1986) *J. Am. Chem. Soc.*, **108**, 2409–2416.

58 Atwood, J.L., Barhour, L.J., Hardie, M.J., and Raston, C.L. (2001) *Coord. Chem. Rev.*, **222**, 3–32.

59 Sonoda, A., Hayashi, K., Nishida, M., Ishii, D., and Yoshida, I. (1998) *Anal. Sci.*, **14**, 493–499.

60 Toutianoush, A., Schnepf, J., El-Hashani, A., and Tieke, B. (2005) *Adv. Funct. Mater.*, **15**, 700–708.

61 Toutianoush, A., El-Hashani, A., Schnepf, J., and Tieke, B. (2005) *Appl. Surf. Sci.*, **246**, 430–436.

62 Kimura, E. and Koike, T. (1998) *Chem. Commun.*, 1495–1500.

63 Cullinane, J., Gelb, R.I., Margulis, T.N., and Zompa, L.J. (1982) *J. Am. Chem. Soc.*, **104**, 3048–3053.

64 Gelb, R.I., Schwartz, L.M., and Zompa, L.J. (1986) *Inorg. Chem.*, **25**, 1527–1535.

65 Dietrich, B., Hosseini, M.W., Lehn, J.M., and Sessions, R.B. (1981) *J. Am. Chem. Soc.*, **103**, 1282–1283.

66 Warden, A.C., Warren, M., Hearn, M.T.W., and Spiccia, L. (2004) *Inorg. Chem.*, **43**, 6936–6943.

67 Kodama, M., Kimura, E., and Yamaguchi, S. (1980) *J. Chem. Soc., Dalton Trans.*, 2536–2538.

68 Royer, D.J., Grant, G.J., van der Veer, D.G., and Castillo, M.J. (1982) *Inorg. Chem.*, **21**, 1902–1908.

69 Chandrasekhar, S., Fortier, D.G., and McAuley, A. (1993) *Inorg. Chem.*, **32**, 1424–1429.

70 El-Hashani, A. and Tieke, B. (2006) *J. Nanosci. Nanotechnol.*, **6**, 1710–1717.

71 El-Hashani, A., Toutianoush, A., and Tieke, B. (2007) *J. Phys. Chem. B*, **111**, 8582–8588.

72 Barker, J.E., Liu, Y., Martin, N.D., and Ren, T. (2003) *J. Am. Chem. Soc.*, **125**, 13332–13333.

73 Bruening, M.L. and Sullivan, D.M. (2002) *Chem. Eur. J.*, **8**, 3832–3837.

74 El-Hashani, A., Toutianoush, A., and Tieke, B. (2008) *J. Membr. Sci.*, **318**, 65–70.

75 Kodama, M., Koike, T., Mahatma, A.B., and Kimura, E. (1991) *Inorg. Chem.*, **30**, 1270–1273.

76 Kimura, E., Fujioka, H., Yatsunami, A., Nihira, H., and Kodama, M. (1985) *Chem. Pharm. Bull.*, **33**, 655–661.

77 Clarke, E. and Martell, A.E. (1991) *Inorg. Chim. Acta*, **190**, 27–36.

78 Choi, K.Y. and Hong, C.P. (1994) *Bull. Kor. Chem. Soc.*, **15**, 293–297.

79 Deal, K.A., Davis, I.A., Mirzadeh, S., Kennel, S.J., and Brechbiel, M.W. (1999) *J. Med. Chem.*, **42**, 2988–2992.

80 Hoffmann, K. and Tieke, B. (2009) *J. Membr. Sci.*, **341**, 261–267.

81 Pyrasch, M. and Tieke, B. (2001) *Langmuir*, **1**, 7706–7709.

82 Millward, R.C., Madden, C.E., Sutherland, I., Mortimer, R., Fletcher, S., and Marken, F. (2001) *Chem. Commun.*, 1994–1995.

83 Barathi, S., Nogami, M., and Ikeda, S. (2001) *Langmuir*, **17**, 7468–7471.

84 Pyrasch, M., Toutianoush, A., Jin, W., Schnepf, J., and Tieke, B. (2003) *Chem. Mater.*, **15**, 245–254.

85 Jin, W., Toutianoush, A., Pyrasch, M., Schnepf, J., Gottschalk, H., Rammensee, W., and Tieke, B. (2003) *J. Phys. Chem. B*, **107**, 12062–12070.

86 Koncki, R. (2002) *Crit. Rev. Anal. Chem.*, **32**, 79–96.

87 Karyakin, A.A. (2001) *Electroanalysis*, **13**, 813–819.

88 de Tacconi, N.R., Rajeshwar, K., and Lezna, R.O. (2003) *Chem. Mater.*, **15**, 3046–3062.

89 Shankaran, D.R. and Narayanan, S.S. (1999) *Fresenius J. Anal. Chem.*, **365**, 663–665.

90 Jeerage, K.M. and Schwartz, D.T. (2000) *Sep. Sci. Technol.*, **35**, 2375–2392.

91 Pau, P.C.F., Berg, J.O., and McMillan, W.G. (1990) *J. Phys. Chem.*, **94**, 2671–2679.

92 Chu, Z., Liu, Y., Jin, W., Xu, N., and Tieke, B. (2009) *Chem. Commun.*, 3566–3567.

93 Chu, Z., Zhang, Y., Dong, X., Jin, W., Xu, N., and Tieke, B. (2010) *J. Mater. Chem.*, **20**, 7815–7820.
94 Stockton, W.B. and Rubner, M.F. (1997) *Macromolecules*, **30**, 2717–2725.
95 Wang, L., Wang, Z., Zhang, X., Shen, J., Chi, L., and Fuchs, H. (1997) *Macromol. Rapid Commun.*, **18**, 509–514.
96 Kharlampieva, E. and Sukhishvili, S.A. (2006) *J. Macromol. Sci.: Polym. Rev.*, **46**, 377–395.
97 Sarno, D.M., Jiang, B., Grosfeld, D., Afriyie, J.O., Mationzo, W., and Jones, W.E.Jr. (2000) *Langmuir*, **16**, 6191–6199.
98 Thomsen, D.L. III, Phely-Bobin, T., and Papadimitrakopoulos, F. (1998) *J. Am. Chem. Soc.*, **120**, 6177–6178.
99 Byrd, H., Holloway, C.E., Pogue, J., Kircus, S., Advincula, R.C., and Knoll, W. (2000) *Langmuir*, **16**, 10322–10328.
100 Pyrasch, M., Amirbeyki, D., and Tieke, B. (2001) *Adv. Mater.*, **13**, 1188–1191.
101 Wanunu, M., Vaskevich, A., Cohen, S.R., Cohen, H., Arad-Yellin, R., Shanzer, A., and Rubinstein, I. (2005) *J. Am. Chem. Soc.*, **127**, 17877–17887.
102 Krass, H., Papastravou, G., and Kurth, D.G. (2003) *Chem. Mater.*, **15**, 196–203.
103 Zhao, W., Tong, B., Pan, Y., Shen, J., Zhi, J., Shi, J. et al. (2009) *Langmuir*, **25**, 11796–11801.
104 Greenstein, M., Ishay, R.B., Maoz, B.M., Leader, H., Vaskevich, A., and Rubinstein, I. (2010) *Langmuir*, **26**, 7277–7284.
105 Belghoul, B., Welterlich, I., Maier, A., Toutianoush, A., Rabindranath, A.R., and Tieke, B. (2007) *Langmuir*, **23**, 5062–5069.
106 Constable, E.C. and Cargill Thompson, A.M.W. (1992) *J. Chem. Soc., Dalton Trans.*, 3467–3475.
107 Rabindranath, A.R., Maier, A., Schäfer, M., and Tieke, B. (2009) *Macromol. Chem. Phys.*, **210**, 659–668.
108 Maier, A., Rabindranath, A.R., and Tieke, B. (2009) *Adv. Mater.*, **21**, 959–963.
109 Maier, A., Rabindranath, A.R., and Tieke, B. (2009) *Chem. Mater.*, **21**, 3668–3676.
110 Maier, A., Fakhrnabavi, H., Rabindranath, A.R., and Tieke, B. (2011) *J. Mater. Chem.*, **21**, 5795–5804.
111 (a) Maier, A. (2010) doctoral thesis, University of Cologne, ISBN 978-3-942109-48-2. (b) Maier, A. and Tieke, B. (2012) *J. Phys. Chem. B*, **116**, 925–934.
112 Delongchamp, D.M. and Hammond, P.T. (2004) *Chem. Mater.*, **16**, 4799–4805.
113 Delongchamp, D.M. and Hammond, P.T. (2004) *Adv. Funct. Mater.*, **14**, 224–232.
114 Cutler, C.A., Bouguettaya, M., and Reynolds, J.R. (2002) *Adv. Mater.*, **14**, 684–688.
115 Welterlich, I. and Tieke, B. (2011) *Macromolecules*, **44**, 4194–4203.
116 (a) Petoud, S., Bünzli, J.-C., Schenk, K.J., and Piguet, C. (1997) *Inorg. Chem.*, **36**, 1345–1353; (b) Piguet, C., and Bünzli, J.-C. (1996) *Eur. J. Solid State Inorg. Chem.*, **33**, 165.

# 21
# Optoelectronic Materials and Devices Incorporating Polyelectrolyte Multilayers

*H.D. Robinson, Reza Montazami, Chalongrat Daengngam, Ziwei Zuo, Wang Dong, Jonathan Metzman, and Randy Heflin*

## 21.1
### Introduction

The nanoscale control of thickness and composition afforded by polyelectrolyte multilayers (PEMs), which are fabricated by layer-by-layer (LbL) deposition, and are also referred to as ionic self-assembled multilayer (ISAM) films, enables novel or enhanced properties of thin films. Since the initial demonstration [1] by Decher and colleagues of polymeric PEMs, following the early work by Iler on charged colloids [2], there has been a burgeoning number of extensive studies on the fundamental properties and potential applications of PEM films. In this chapter, we overview some of our work that has been centered on the utilization of PEMs in a wide range of optoelectronic applications. The work is focused on the manner in which the exquisite control of thickness, composition, and orientational order that can be obtained in PEMs containing polymers and nanoparticles can enhance the properties and performance.

For example, the ionic interactions between layers can induce highly stable, polar ordering of the films. This noncentrosymmetric order is essential for second order nonlinear optical responses and is, in general, not straightforward to achieve in a simple, inexpensive processing approach. In order to further increase the degree of polar order, we developed a new hybrid covalent/ionic self-assembly method that enables the fabrication of films with nonlinear optical responses comparable to those of commercial inorganic crystals. Furthermore, the combination of these self-assembled films with metallic nanoparticles leads to orders of magnitude plasmonic enhancement of the nonlinear optical susceptibilities. Taking advantage of the ability of PEMs to grow conformally on surfaces of arbitrary shape and size, polar self-assembled films have been deposited on the surface of optical fibers in order to imbue the fibers with second order nonlinear optical properties that they do not innately possess. The ability to deposit PEMs on optical fibers has been further utilized to develop a novel biosensor platform in which the binding of analytes to a

*Multilayer Thin Films: Sequential Assembly of Nanocomposite Materials*, Second Edition.
Edited by Gero Decher and Joseph B. Schlenoff.
© 2012 Wiley-VCH Verlag GmbH & Co. KGaA. Published 2012 by Wiley-VCH Verlag GmbH & Co. KGaA.

functionalized PEM film on the fiber surface results in a dramatic change in the attenuation spectrum of an optical fiber long-period grating. As an example, methicillin-resistant Staphylococcus aureus has been detected at a level of 100 cells/ml.

The combination of silica nanoparticles with a polyelectrolyte leads to a highly porous film that has an exceptionally low refractive index. The low refractive index combined with nanoscale control of thickness enables a highly effective antireflection coating that can have a transmittance of >99.7% on a glass substrate. Numerous polymeric and nanoparticle PEM films have been demonstrated to have strong electrochromic responses in which the transmittance of the film is controlled by an applied voltage. We have demonstrated PEM electrochromic devices with transmittance changes >80% and response times of the order of 30 ms. Lastly, gold nanoparticles have been employed in electromechanical actuators to provide porous, conductive surfaces that allow large ion accumulation that leads to large and fast bending motion under applied voltage.

## 21.2
### Second Order Nonlinear Optics

In order to possess nonzero second order nonlinear optical susceptibilities and exhibit effects such as frequency doubling and the electro-optic effect, *a material must lack a center of inversion at the macroscopic level* [3]. The macroscopic second order susceptibility, $\chi^{(2)}$ governs the nonlinear polarization $P$ of the medium at frequency $\omega_3$ in response to (optical) electric fields $E$ at frequencies $\omega_1$ and $\omega_2$ through

$$P_i^{\omega_3} = \chi_{ijk}^{(2)}(-\omega_3;\omega_1,\omega_2) E_j^{\omega_1} E_k^{\omega_2} \tag{21.1}$$

where the subscripts refer to the directions of polarization of the fields. Second order nonlinear effects governed by Equation 21.1 include second harmonic generation (where $\omega_1 = \omega_2 = \omega$, so that radiation $\omega_3 = 2\omega$ is generated at twice the frequency of the incident radiation) and the electro-optic (EO) effect (where $\omega_1 = \sim 0$, so that $\omega_2 = \omega_3 = \omega$). The most common second order NLO materials are ferroelectric, inorganic crystals, such as potassium dihydrogen phosphate (KDP), beta-barium borate (BBO), and lithium niobate, this last having a second harmonic generation $\chi^{(2)} = 200 \times 10^{-9}$ esu and electro-optic coefficient $r_{33} = 30$ pm V$^{-1}$ [4]. Growth of such high-quality inorganic crystals, however, is difficult, time-consuming, and expensive. We and others have demonstrated that PEM films can be fabricated into noncentrosymmetric structures for $\chi^{(2)}$ applications as an alternative to inorganic crystals [5–9].

In our nonlinear optical work, PEM films are generally deposited on glass microscope slide substrates. Noncentrosymmetric, PEM $\chi^{(2)}$ films have been primarily produced using two different polyanions: the polymeric dyes Poly S-119 (Sigma-Aldrich), which consists of a poly(vinylamine) backbone with an ionic azo dye chromophore, and poly{1-[4-(3-carboxy-4-hydroxyphenylazo)benzensulfonamido]-1,2-ethanediyl, sodium salt} (PCBS, from Sigma-Aldrich). While either Poly S-119 or PCBS served as the polyanion for the ISAM fabrication, poly(allylamine hydrochloride) (PAH), which has no $\chi^{(2)}$ response, was used for the polycation. The formation of each monolayer is exceptionally rapid with these polymers. Through measurements of absorbance and of film thickness (by ellipsometry) as a function of immersion time, each monolayer was found to be fully deposited in less than 20 s of immersion in the polyelectrolyte. This allows the rapid build-up of self-assembled, multilayer films.

Second harmonic generation (SHG) experiments were carried out using the 1200 nm output from a broadband, BBO optical parametric oscillator (OPO) on Poly S-119/PAH [5]. By comparison to a quartz wedge, the $\chi^{(2)}_{zzz}$ value for Poly S-119/PAH was found to be 0.70 times the value of quartz, or $1.34 \times 10^{-9}$ esu [5]. To beyond 100 layer pairs, the films were found to exhibit a quadratic growth in the SHG intensity with film thickness as expected theoretically. This demonstrates that the degree of polar orientation of the chromophores is maintained for each successive layer. More recently, it has been shown that even films with 600 layer pairs and a total thickness of 740 nm continue to exhibit the quadratic dependence of SHG on thickness, demonstrating that ISAM films are also suitable as nonlinear optical materials for waveguide structures where thicknesses of the order of 1 μm are required [10].

While PEM films are remarkably stable, their NLO performance is too low for practical devices. The degree of net polar ordering that can be achieved with a polymeric dye PEM film is limited by: (i) orientation of dipoles in opposite directions above and below the polymeric dye layer, resulting in partial cancellation; (ii) random orientation of dipoles within thicker monolayers due to polymer loops; and (iii) steric constraints on orientation since the relatively bulky chromophores are separated along the polymer backbone by only two carbon–carbon bonds.

We have developed a novel methodology for achieving a high degree of polar ordering of organic chromophores in thin films – a hybrid covalent-electrostatic self-assembled multilayer process [11, 12]. Our first example used Procion Red MX-5B, a common textile reactive dye having a dichlorotriazine moiety and two sulfonate moieties. Polar order was achieved by *alternating the mechanism of layer adsorption between covalent reaction and electrostatic attraction for a single pair of materials.* LbL films were constructed by alternately immersing the substrate into aqueous solutions of an NLO-inactive polymeric "glue" (PAH) and the reactive dye (illustrated in Figure 21.1 for the dye Procion Brown MX-GRN). The dichlorotriazine moiety is susceptible to nucleophilic attack and reacts readily with unprotonated amine groups at pH > 9 at 25 °C. [13] Covalent attachment is rapid and essentially complete within 2 min in unstirred solutions at 25 °C. A PAH layer is then deposited *electrostatically* from a solution at a pH less than the $pK_a$ (~9), such that the amine groups are protonated. The key to obtaining a high degree of net polar order is that the covalent interaction occurs preferentially at the high pH of the dye solution, while the electrostatic interaction occurs preferentially at the lower pH of the PAH solution, resulting in an orientation of the chromophore with the dichlorotriazine end directed towards the substrate and the sulfonates directed away.

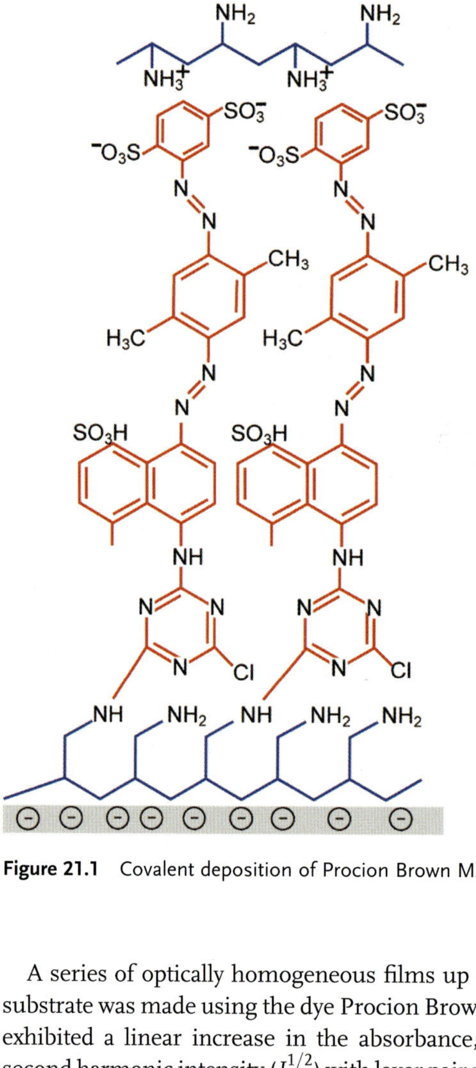

**Figure 21.1** Covalent deposition of Procion Brown MX-GRN onto an adsorbed layer of PAH.

A series of optically homogeneous films up to 30 layer pairs on each side of the substrate was made using the dye Procion Brown MX-GRN and PAH [14]. The films exhibited a linear increase in the absorbance, thickness, and square root of the second harmonic intensity ($I_{2\omega}^{1/2}$) with layer pair number. The effects of ionic strength on the chromophore density and polar ordering were also examined. The $\chi_{zzz}^{(2)}$ for Procion Brown films made with no added NaCl was $20 \times 10^{-9}$ esu, and the average tilt angle of the Procion Brown chromophores with respect to the surface normal was 43°. As the ionic strength of the Procion Brown deposition solution was increased to 0.5 M NaCl, the $\chi_{zzz}^{(2)}$ increased by approximately 150% to $50 \times 10^{-9}$ esu. This increase in $\chi_{zzz}^{(2)}$ was concomitant with a decrease in the average chromophore tilt angle to 38° with respect to the surface normal. The data suggest that, as the ionic strength of the Procion Brown solution is increased, charge shielding of the sulfonate moieties enables the chromophores to pack closer together on the PAH monolayers. The electro-optic coefficient component ($r_{33}$) for Procion Brown/PAH films was measured to have an average value of $r_{33} = 14 \pm 2$ pm V$^{-1}$ for films of 50 layer pairs

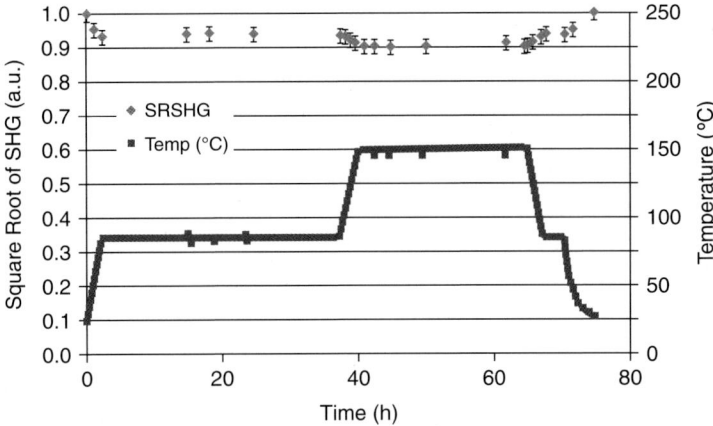

**Figure 21.2** The square root of the SHG intensity (left axis, diamonds, normalized to 1.0 at the beginning of the experiment) of a Procion Brown/PAH film as a function of time and temperature (right axis, squares) during a heating cycle. Reprinted with permission from Ref. [14]. © American Chemical Society.

of Procion Brown with 0.5 M NaCl, which compares favorably to 30 pm V$^{-1}$ for lithium niobate.

Temporal and thermal stability of electro-optic materials are critical for NLO device applications. Procion Brown/PAH films stored at ambient conditions have exhibited no measurable decay of $\chi^{(2)}$ over a period of more than four years. Moreover, the films made with 0.5 M NaCl also have exceptional stability at elevated temperatures (Figure 21.2). Using an apparatus that allowed *in situ* measurement of the SHG intensity, $(I^{2\omega})^{1/2}$ was found to be reduced by 7% after 36 h at 85 °C. The sample was then kept at 150 °C for 24 h, and an additional 3% decrease in $(I^{2\omega})^{1/2}$ was measured, for an overall 10% loss in $(I^{2\omega})^{1/2}$ for the entire process. Upon cooling to room temperature, $(I^{2\omega})^{1/2}$ returned to its initial value over a period of several hours, indicating complete recovery of polar ordering of the chromophores in the film.

## 21.3
### Plasmonic Enhancement of Second Order Nonlinear Optical Response

In 1999, Pendry *et al.* [15] proposed that structures with large effective NLO coefficients could be made by combining small quantities of NLO-active material with metallic structures that are resonant with the incident light. This is based on the observation that metal nanostructures and particles possess surface plasmon resonances that can concentrate electromagnetic energy into small volumes, known as hot spots, located at sharp points or narrow gaps of the structures. These resonances are collective oscillations of the conduction electrons in the metal structures, coupled to an electromagnetic field that is maximal at the metal surface, and decays exponentially both into the metal and the surrounding dielectric. Surface plasmons in noble metals, such and silver and gold, have the smallest losses, and the choice of materials also sets an upper limit on the frequency of the plasmon resonance. Beyond this, the resonance is strongly affected by the shape of the metal nanostructure, and it is possible to engineer gold and silver particles and structures with plasmon resonances located at essentially any visible or infrared wavelength [16]. On resonance, the light intensity at the hot spots can be several orders of magnitude larger in the hot spots than in the incident wave [17, 18], which is key to the enhanced NLO properties. The efficiency of NLO processes increases with light intensity, so if the volume of the hot spots is filled with NLO-active material, this can lead to higher effective nonlinear coefficients than if the light had been evenly distributed throughout space, while simultaneously requiring much smaller quantities of the active material.

As described in the previous section, polyelectrolyte multilayers of PAH/PCBS or PAH/Procion Brown can have large and stable second order non-linear optical susceptibilities. However, fabricating films that are thick enough for conventional NLO devices, such as waveguides, can be time consuming. By contrast, the films are ideally suited as the active material in plasmon-enhanced metal-based NLO materials of the type proposed by Pendry. PEMs deposit well on metal surfaces, and conform easily to the shape of complicated metal structures. In addition, the plasmon hot spots typically only extend a few nanometers from the metal surface, so even quite thin PEMs will take full advantage of the plasmonic NLO enhancement. The freedom to design the metal-PEM structure is quite large; the only restriction is that the assembly must lack global inversion symmetry, otherwise $\chi^{(2)} = 0$.

We have demonstrated plasmon-enhanced second harmonic generation (SHG) in a PAH/PCBS NLO PEM film combined with triangular silver nanoparticles arranged as shown schematically in Figure 21.3a, where the particles are arranged in an array on top of a planar PEM film deposited on a glass slide [19]. The planar configuration of the film takes maximum advantage of the asymmetry of the PEM film, and since the dominant plasmon hot spots tend to be displaced toward the higher index material [20], the configuration also quite effectively takes advantage of the field concentration inside the hot spots, which are located at the apexes of the triangular particles.

The glass slides were cleaned using the standard RCA cleaning process, after which a PAH/PCBS PEM film was deposited at pH 7. Arrays of silver nanoparticles were fabricated on top of the films using nanosphere lithography (NSL) [21], a technique that allows rapid deposition of large arrays of triangular nanoparticles. In this technique, a suspension of monodisperse polystyrene nanospheres is allowed to dry on the substrate, which causes areas of close-packed monolayers of spheres to form. In our case, 720 nm diameter spheres conjugated with carboxyl groups and suspended in DI water were used. Silver was then deposited onto the substrate by electron beam evaporation, after which the spheres were dissolved away using ultrasonication in dichloromethane. Because the deposited metal only reaches the surface through the gaps in the close-packed colloidal crystal, this results in a hexagonal array of triangular nanoparticles, as shown in Figure 21.3b. The thickness of the silver was 72 nm, which results in a principal plasmon resonance near 1064 nm, resonant with the 10 Hz Q-switched Nd:YAG laser used in the SHG measurements.

(a)

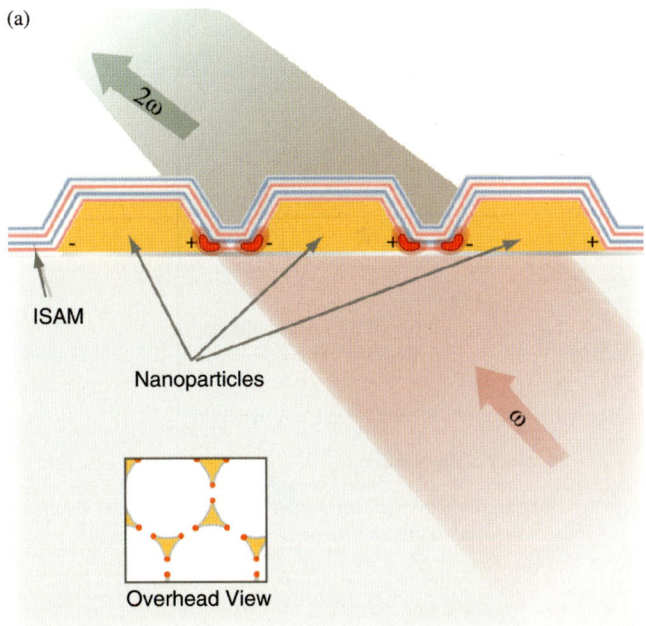

(b)

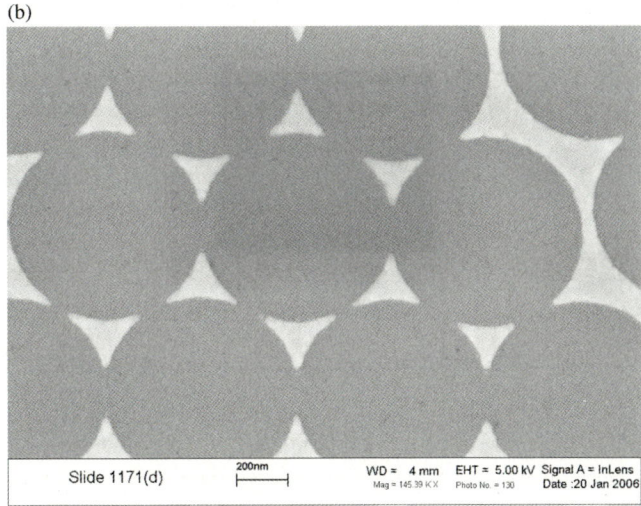

**Figure 21.3** (a) Schematic sideview of a plasmonically enhanced NLO ISAM film. A beam of light (red) traverses a layer of triangular nanoparticle (yellow). The localized intensity enhancement due to the LSPRs (red spots) leads to a commensurate increase in SHG light (green) from the ISAM film (red/white/blue). (b) SEM image of triangular silver nanoparticles (bright phase) fabricated using nanosphere lithography. Reprinted with permission from Ref. [19]. © American Chemical Society.

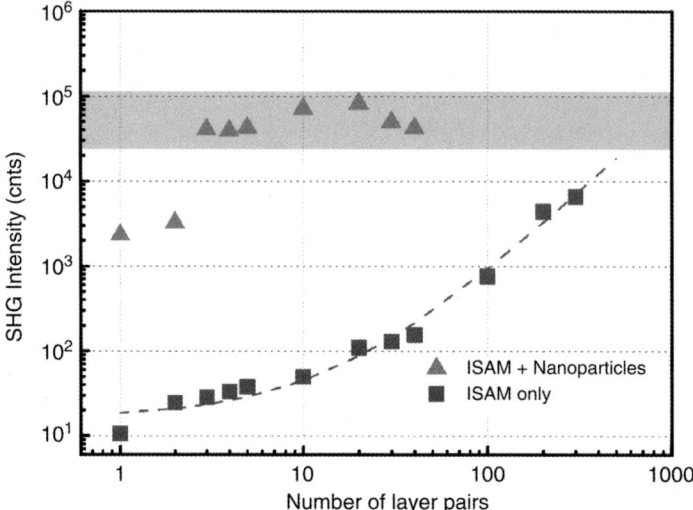

**Figure 21.4** Comparison of SHG efficiencies in unmodified (squares) and plasmonically enhanced (triangles) PAH/PCBS ISAM films as a function of film thickness. Adapted from Ref. [19]. © American Chemical Society.

The pulsed laser light was weakly focused on the sample in a 1 mm spot and at an incident angle of about 45°. The SHG light was collected in a forward-transmission geometry, filtered to remove the fundamental laser wavelength, and detected with a photomultiplier tube. The measured SHG intensities for a number of different film thicknesses, either with or without sliver nanotriangles deposited on top, are plotted in Figure 21.4. The SHG signal from the film without metal particles scales as the square of the film thickness for sufficiently thick films, which is the expected behavior. For films with less than about 20 layer pairs, each layer pair being approximately 1 nm thick, the signal scales more slowly, which reflects that the SHG contains a significant contribution from the glass/film and film/air interfaces, which are independent of thickness. The dashed line is a fit to a model taking this effect into account.

When silver nanotriangles are deposited on the film, the SHG signal increases rapidly for the first three layer pairs and then plateaus, with some variation in signal strength occurring due to variations in nanoparticle quality and orientation. This is precisely what one would expect, given that the decay length of the plasmon mode hot spots is typically of the order of 3–4 nm, which is the thickness of a 3-layer-pair PEM film. Thicker films, therefore, see no additional plasmonic enhancement, and do not add appreciably to the signal. Notably, the SHG from the 3-layer-pair plasmon-enhanced film is 1600 times higher than from an unadorned film, corresponding to an increase in effective $\chi^{(2)}$ by a factor of 40. Stated differently, because the plasmonic enhancement takes advantage of interface SHG, a plain NLO PEM would need 700–1000 layer pairs in order to be as efficient an SHG emitter as the plasmon-enhanced 3-layer-pair film.

## 21.4 Nonlinear Optical Fibers

Large as these numbers are, the structure has not been optimized in any way. The metal nanostructures could be chosen better and deposited more densely, and rather than depositing the PEM film on a planar surface, it could, for instance, be deposited on an optical fiber to create a high efficiency fiber-optical upconversion device.

### 21.4
### Nonlinear Optical Fibers

The attempt to generate strong and thermodynamically stable second-order nonlinear optical properties in a silica fiber has been of increasing interest since it enables many potential applications such as SHG and optical parametric oscillation (OPO). Significant second-order susceptibility $\chi^{(2)}$ in a silica fiber could be accomplished by breaking the material symmetry using various poling techniques; however, the poor stability of the produced $\chi^{(2)}$ is still problematic [22, 23]. In contrast to current methods, we have theoretically [24] and experimentally [25] demonstrated that it is possible to generate significant second-order nonlinearity in silica fiber by coating it with layers of radially-aligned nonlinear molecules. The fiber is coated with nonlinear optical PEMs, where the alignment of the nonlinear chromophores is maintained by electrostatic interaction and is, therefore, thermodynamically stable.

The schematic structure of a nonlinear optical fiber is illustrated in Figure 21.5a. The uniform silica core is covered by PEMs of radially-aligned nonlinear chromophores. The origin of the second-order nonlinearity comes from the interaction of the propagating beam with the surface nonlinear layers. The great advantage of the fiber waveguide configuration for nonlinear processes is that it can provide a very long

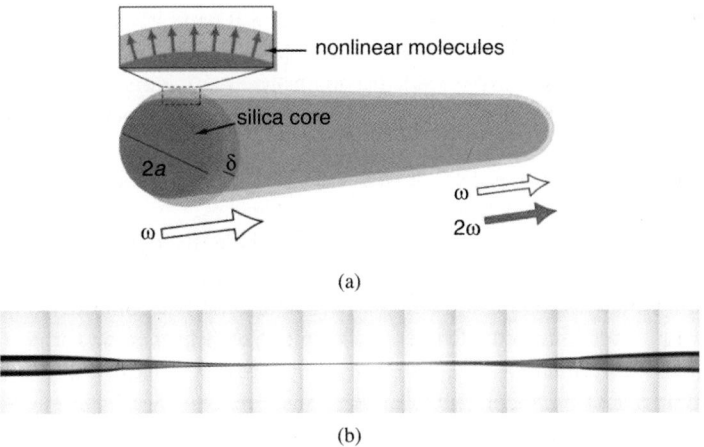

**Figure 21.5** Nonlinear optical PEM coating on a silica fiber. (a) A schematic structure of nonlinear fiber with full cylindrical symmetry. The radially aligned nonlinear chromphores provide a second-order susceptibility tensor dominated by the $\chi^{(2)}_{rrr}$ component. (b) A composite optical microscope image of a nonlinear fiber taper. Reprinted with permission from Ref. [25]. © Optical Society of America.

interaction length between the light and nonlinear layers, allowing the amplification of the nonlinear output power if the phase matching condition is satisfied, and hence high conversion efficiency. In practice, most commercial optical fibers are designed to be insensitive to the outside environment, that is, the mode field completely propagates inside the fiber. Therefore, the fabrication of the nonlinear optical fiber requires a fiber taper with a size of several microns in which the mode field extends into the outside region, enabling strong interaction with the nonlinear coating. The taper can be produced by slowly stretching a common optical fiber from both sides while heating at temperature $\sim$1600 °C [26]. As shown in Figure 21.5b, a typical fiber taper consists of an approximately uniform waist (with a taper radius that ranges from 2.8 to 7.5 μm) and shallow transition regions to assure adiabatic coupling with minimum insertion loss. The taper length and shape can be controlled by the heating profile [27]. The optical transmission of a freshly produced fiber taper normally exceeds 90% in the near-IR region.

The fiber tapers are coated with nonlinear molecules in a LbL fashion, utilizing the novel hybrid covalent/ionic self-assembled multilayer technique described in Section 21.2 by which high polar ordering can be achieved through the alternation of adsorption mechanism between covalent interaction and electrostatic attraction between nonlinear molecules and a polyelectrolyte layer. Briefly, the fiber tapers are immersed into a polycationic solution containing PAH at pH 7.0 and 10 mM concentration, whereupon a monolayer of positively charged PAH film grows uniformly on the negatively charged taper surface. Following PAH deposition, the fiber sample is dipped into an aqueous solution containing nonlinear molecules Procion Brown MX-GRN (PB) at pH 10.5, 5 mg mL$^{-1}$ concentration, and 0.5 M NaCl added. At this pH, the dichlorotriazine moieties of PB form covalent bonds with the unprotonated amines of PAH. Upon deposition of the sequential PAH layer at pH 7.0, the negatively-charged sulfonates of PB bind electrostatically with the protonated amine groups of PAH. The deposition of each monolayer requires less than 2 min and can be repeated as many times as desired to produce multilayer nonlinear optical films.

The self-assembled layer of the nonlinear molecules is very uniform and smooth with surface roughness less than $\pm$3 nm as revealed by AFM [25]. This smooth film surface minimizes the scattering loss of the nonlinear taper.

For the nonlinear characterization, SHG measurements were performed using a pump beam with $\lambda = 1294$ nm, pulsewidth $= 10$ ns, and repetition rate of 20 Hz, obtained from an OPO system. The pump light was coupled into the nonlinear fiber taper through an untapered multimode fiber. As the pump pulse propagates through the fiber taper region, it interacts with the nonlinear coating, giving rise to SHG light which then couples back into the multimode fiber. The output signals are collimated into free space, where the remaining pump beam is filtered, leaving only the SHG to be detected by a photomultiplier tube (PMT).

Figure 21.6 shows the measurement of SHG power versus pump pulse energy for the same nonlinear taper sample with different thickness of PAH/PB coating, that is, 0, 10, 20, 30 and 40 layer pairs. The second harmonic power exhibits quadratic dependence on pump pulse energy, a characteristic feature of SHG. Furthermore, an

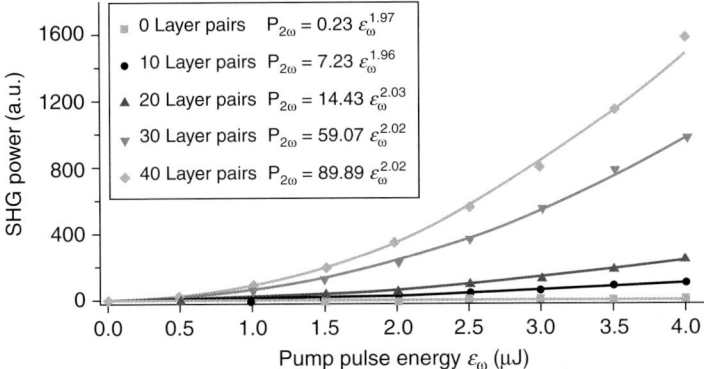

**Figure 21.6** SHG power as a function of pump pulse energy. The measurement was taken for a 3.8-μm-radius taper with 0, 10, 20, 30 and 40 layer pairs of nonlinear film. The lines are quadratic fits to the data. Reprinted with permission from Ref. [25]. © Optical Society of America.

increase in nonlinear film thickness directly increases the SHG power. For a 40-layer-pair coating, a 400-fold enhancement in SHG power is observed compared to the bare taper case. This result confirms that the measured second harmonic component is indeed generated by the nonlinear PAH/PB coating. The enhancement in SHG is even more drastic for a smaller taper size. Typically the SHG power is proportional to $1/a^4$ (where $a$ is the taper waist radius) [25].

Quasi-phase-matching in the nonlinear taper can possibly be achieved by periodic coating of nonlinear film with period $\Lambda = 2l_{coh}$ (where $l_{coh}$ is the coherence build-up length for the SHG power). Several fabrication techniques are promising to produce such a periodic structure of nonlinear film on a taper surface, for example, laser ablation on the film surface using a phase mask, direct electron-beam patterning, and photochemical patterning. Furthermore, intermodal phase-matching can be accomplished by adjusting the taper radius down to the submicron scale [28]. Under the phase-matched condition, just a 100-μm-long nonlinear taper with 40 layer pairs of PB/PAH coating could achieve 10% SHG efficiency with a pump peak power of 400 W.

## 21.5
## Optical Fiber Biosensors

Long-period gratings (LPGs) are a type of photonic technology that consists of a periodic variation in the refractive index of an optical fiber. While optical fibers generally have very high optical transmittance, the coupling of light from the fundamental mode in the fiber core to higher order cladding modes will lead to the attenuation of the transmission spectrum at a specific wavelength, depending on the period of the LPG $\Lambda_{LPG}$ and the values of the effective refractive indices of the core and cladding. The transmittance remains high at other wavelengths. In contrast to fiber Bragg gratings (FBGs) with a period length at the sub-micron level, LPGs typically involve grating periods from 100 to 1000 μm.

Several papers [29, 30] have discussed the computational methods for quantitative determination of the characteristics of LPGs with chosen parameters. One direct approach for understanding the grating properties is to consider the phase-matching condition of the LPG. For an LPG fiber with period $\Lambda_{\text{LPG}}$, the coupling between the fundamental mode in the higher-index core surrounded by lower-index cladding and a forward-propagating mode within the cladding surrounded by an infinite medium, for example, air, will cause resonance at the wavelength $\lambda^{(m)}$, defined by the relationship:

$$\lambda^{(m)} = \left(n_{\text{eff}}^{\text{core}} - n_{\text{eff}}^{\text{clad}(m)}\right) \cdot \Lambda_{\text{LPG}} \tag{21.2}$$

where $n_{\text{eff}}^{\text{core}}$ is the effective refractive index of the core mode, and $n_{\text{eff}}^{\text{clad}(m)}$ is the effective refractive index of the $m$-th cladding mode, which is also a function of the incident light's wavelength $\lambda^{(m)}$. The strength of the coupling is described using the detuning parameter δ, which can be expressed as the phase-matching condition:

$$\delta(\lambda) = \frac{1}{2}\left(\beta_1 - \beta_2 - \frac{2\pi}{\Lambda_{\text{LPG}}}\right) \tag{21.3}$$

where the propagation constant $\beta = \frac{2\pi}{\lambda} n_{\text{eff}}$. The smaller δ becomes, the stronger the coupling that will be induced.

For a given LPG, $n_{\text{eff}}^{\text{core}}$ and $\Lambda_{\text{LPG}}$ are predetermined and remain constant. However, by coating the fiber with materials with different thickness and refractive index, it is possible to change the effective index of the cladding mode $n_{\text{eff}}^{\text{clad}(m)}$ and, therefore, produce a shift of the location of the phase-matched wavelength.

The cladding has a much larger radius compared to the radius of the core, which guarantees the existence of a large number of cladding modes. Analysis [31] has proven that a large number difference between the core mode and cladding mode is necessary to observe efficient coupling, for example, $LP_{0,5}$ or $LP_{0,12}$. This fact has a significant importance. If only the coupling between the fundamental mode in the core and the cladding mode is considered, calculations have found that for some higher order coupling modes, such as $LP_{0,9} \sim LP_{0,12}$, equation (21.2) becomes a concave function. In other words, the resonant wavelength $\lambda^{(m)}$ no longer increases monotonically with the increment of the grating period $\Lambda_{\text{LPG}}$ in these modes. Instead, The extremum where $d\Lambda_{\text{LPG}}/d\lambda^{(m)} = 0$ is referred to as the turnaround point (TAP).

Therefore, for a TAP-LPG, it is possible to see two attenuation peaks occurring at different wavelengths for coupling to the same mode. Moving towards the region of the TAP, the two resonant wavelengths will get closer. From the perspective of the transmission spectrum, the two attenuation peaks will move towards each other and eventually merge at the TAP. However, when the value of $\Lambda_{\text{LPG}}$ falls slightly beyond the TAP, there will be no solution to equation (21.2), which means the phase mismatch will grow dramatically and the merged attenuation peak will disappear very quickly as $\Lambda_{\text{LPG}}$ moves further away from TAP. This phenomenon indicates a promising application of TAP-LPGs as an ultra-sensitive sensor and a huge advantage of this type of LPG over the conventional LPGs. Not only do TAP-LPGs have higher sensitivity than conventional LPGs, they also show their response as a change in the strength of the attenuation at a fixed wavelength rather than as a shift of the attenuation wavelength, which makes detection simpler and more cost-effective.

Practically, a TAP-LPG has a constant value of $\varLambda_{LPG}$, and by varying the value of $n_{eff}^{clad(m)}$, the phase matching curve (which is the solution curve to equation (21.2)) will be pushed up or down to vary the strength of the attenuation. If we are able to tune the fiber just beyond the TAP, then a very small change of the thickness or the refractive index of the coating, which will alter the value of $n_{eff}^{clad(m)}$, will be reflected as a large change in the strength of the attenuation peak in the transmission spectrum.

In order to perform biological target detection and sensing tasks with TAP-LPGs, a thin medium with highly controlled thickness and refractive index that can be used to tune the phase-matching condition, and which should also have the capacity to couple with biomolecules, is required. PEMs fulfill this role exquisitely.

The conformal deposition of PEMs onto the thin silica fiber is easy to achieve. Since polyanion/polycation layers have a distinct refractive index that can be controlled by varying the relative thickness of the polycation and polyanion through adjustment of pH or ionic strength, the accumulating thickness will alter the effective refractive index of the cladding mode, and hence shift the phase-matching curve of the LPGs. We have examined experimentally [32] and theoretically [33] the effect of PEMs of varying thickness and refractive index on the location of the resonant wavelength in conventional LPGs. The incremental wavelength shift is found to increase with increased thickness and refractive index of the PEM coating on the LPG. Figure 21.7 shows the shift of the attenuation wavelength as a function of the number of layer

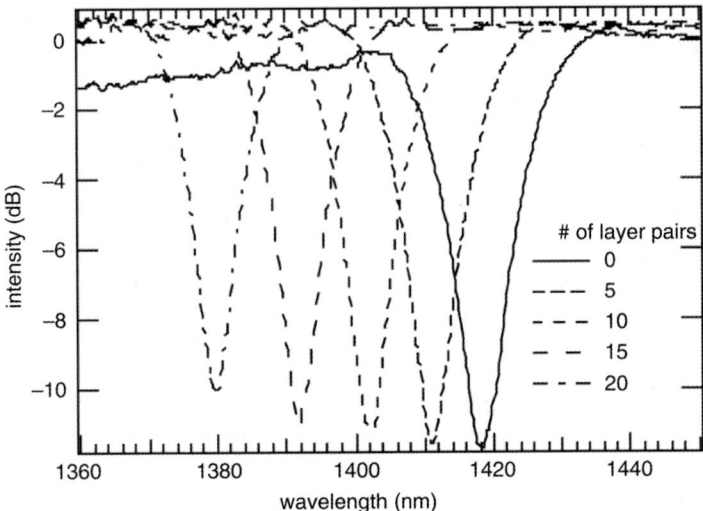

**Figure 21.7** Experimental results of ISAM film coatings on an LPG. LPG transmission spectra with 0, 5, 10, 15 and 20 layer pairs of PAH/PCBS ISAM film for PAH solution pH of 9 and PCBS solution pH of 8. Reprinted with permission from Ref. [32]. © American Institute of Physics.

pairs of PAH (10 mM, pH 9.0) and PCBS (10 mM, pH 8.0), which results in films with a relatively high refractive index of 1.67.

A biosensor is a type of analytic device that can transform a response to the existence, type, and concentration of a specific biological analyte into a physical signal. An LPG biosensor is able to take full advantage of the LPGs' ultra-sensitivity to the variation of its coating's thickness, and link the shifting of the attenuation strength with the concentration of the analyte in the sample solution.

Our TAP-LPG PEM biosensor consists of an LPG fiber, which is usually 30 mm in length; a PEM film surrounding the LPG that will tune the fiber near the TAP of a high order coupling mode; and a receptor layer (e.g., antibody) that is cross-linked with the PEM film. The receptor layer can be anything that has specific binding only with the desired analyte. When the receptors successfully recognize and bind the analyte, the thickness of the coating of the TAP-LPG will be increased, which will result in an increase in the attenuation of the light intensity at the resonant wavelength. The higher the concentration of the analyte, the stronger will be the attenuation observed in the transmission spectrum. This provides a method to not only differentiate cells of close genus due to the exclusive cell-selecting characteristic of antibodies, but also has the ability to quantitatively determine the concentration of the analyte in the solution from the percentage decrease of the resonant wavelength attenuation.

This type of biosensor has the potential to perform quick diagnosis of bacterial infections. One of our studies has investigated the application of the TAP-LPG PEM biosensor to detect and differentiate Methicillin-resistant *Staphylococcus aureus* (MRSA) from Methicillin-sensitive *Staphylococcus aureus* (MSSA). MRSA is a bacterium that can cause several difficult-to-treat human infections and results in thousands of deaths annually in the U.S. It is, by definition, any sort of strain of *Staphylococcus aureus* that has developed a resistance to traditional antibiotics, such as penicillin and cephalosporins. A patient suffering from an MRSA infection may show very similar symptoms to an MSSA infection. Early diagnoses of MRSA can help a physician prescribe appropriate treatment before the infection becomes serious. The conventional method to differentiate these two types of infections requires culturing of the bacteria and can take several days to obtain a result.

In contrast, our TAP-LPG PEM biosensor is able to give answers on whether the infection is induced by MRSA and how bad the infection is within 1 h, which is a typical time period required for the MRSA antibodies to capture the cells. Figure 21.8 shows our sequential test with different concentrations of aqueous solution of MRSA cells [34]. For each test, 500 µl of sample solution is applied to the fiber sensor with a MRSA antibody as the outermost layer of the coatings. The fiber sensor is immersed in each of the sample solutions for 1 h in a dark environment and then rinsed with phosphate-buffered saline (PBS) with 0.5% of Tween-20 for 10 s. All measurements are done in 1 M PBS at pH 7. A MRSA sample with concentration of 100 cells/ml caused a 26% decrease in the transmitted power at the resonant wavelength of 1546.2 nm. Exposure of the fiber to MSSA samples of equivalent concentrations yields negligible changes in transmittance.

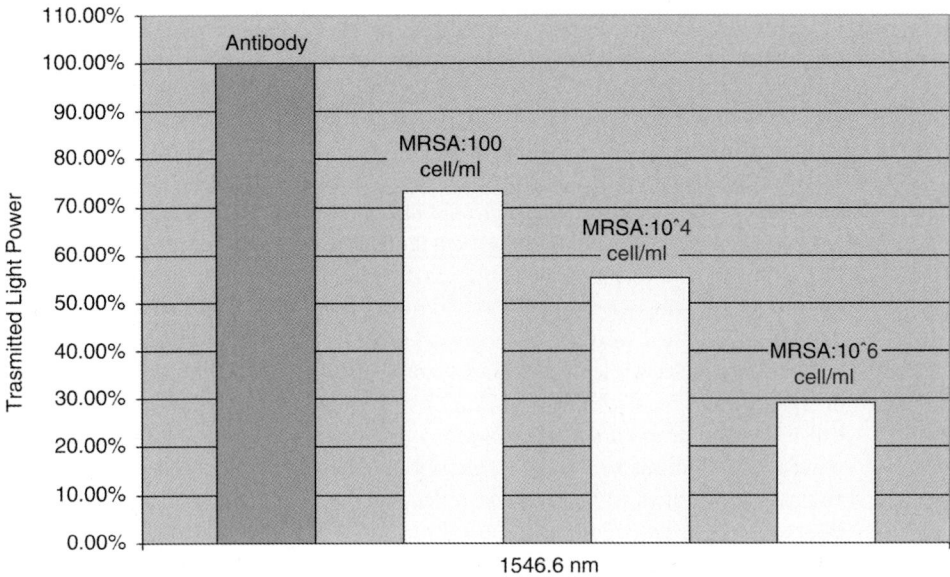

**Figure 21.8** Relative transmittance at 1546.6 nm for exposure of the LPG biosensor to 100, $10^4$, and $10^6$ MRSA cells/ml.

## 21.6
## Antireflection Coatings

Antireflection coatings (ARCs) are commonly used to reduce substrate reflection, while maximizing transmission, in the visible wavelength range of incident light. Along with a decrease in loss of transmitted light, ARCs diminish glare created from high levels of reflected light. These coatings are commonly utilized for applications such as eyeglasses, flat-screen monitors, and hand-held electronics.

ARCs must fulfill two conditions to exhibit zero reflectance due to destructive interference at the desired wavelength $\lambda$. The optical thickness of the coating (product of the physical thickness and refractive index) must be equal to one-quarter of the wavelength of incident light. This creates a $\pi$-phase shift between the light reflected off the substrate and the light reflected off the coating, resulting in destructive interference. Secondly, the refractive index of the coating must be equal to $\sqrt{n_1 n_2}$, where $n_1$ and $n_2$ refer to the refractive indices of the medium on either side of the coating. In typical circumstances, the substrate is composed of glass ($n = 1.50$) and the surrounding medium is air ($n = 1.0$) so that the refractive index of the coating should be $n_c = 1.22$. When this condition is met, the amplitude of the two reflected waves will match to yield complete destructive interference. The former condition is far less stringent to accomplish, as modern methods of film growth can precisely control thickness. The latter condition requires control of the composition of the coating, which is not typically variable. ARCs can consist either of one transparent dielectric material for low reflectance in a relatively narrow range of wavelengths

(V-type ARC) or a stack of two or more dielectrics for low reflectance in a wide range of wavelengths (broadband ARC). $MgF_2$ is typically used for V-type ARCs due to its low index of refraction at $n = 1.38$. However, as this is still significantly larger than the ideal refractive index value, $MgF_2$ ARCs reduce the reflectance on a glass substrate from approximately 4 to 1.2% per surface.

ARCs have been fabricated by a variety of techniques including e-beam evaporation [35], sol–gel synthesis [36], high vacuum deposition [37], spin coating [38], and chemical etching [39]. Although these methods feature a range of pros and cons, the most common drawbacks relate to high cost and complexity in growth (especially for concave or flexible substrates). PEMs address these issues due to the versatility of deposition and their relatively inexpensive nature. Films can be grown conformally and homogenously while the thickness can be easily controlled through the number of layer pairs. Several groups have demonstrated ARCs from PEMs using polyelectrolytes and/or charged colloidal solutions. Rubner and coworkers [40] created antireflective ISAM films by inducing nanoporosity into the weak polyelectrolytes PAH and poly(acrylic acid) (PAA). The controlled and reversible nanoporosity obtained by pH-induced swelling allows continuous tuning of the refractive index. We demonstrated ARCs formed by PEMs of silica nanoparticles and PAH in which the void space between the nanoparticles provides the low refractive index [41]. Rubner and colleagues fabricated ARCs from PEMs of silica and titania nanoparticles and employed calcination to achieve robustness of the films [42].

Figure 21.9 shows an SEM image of our 40 nm diameter silica nanoparticle/PAH ARCs on a glass substrate. Due to the random close packed (RCP) structure, the silica nanoparticles (and thin PAH surface coating) occupy 64% of the volume, leaving 36% void space. This lowers the overall average refractive index of the film from $n = 1.50$

**Figure 21.9** SEM image of a PEM film comprising PAH and 40 nm silica nanoparticles.

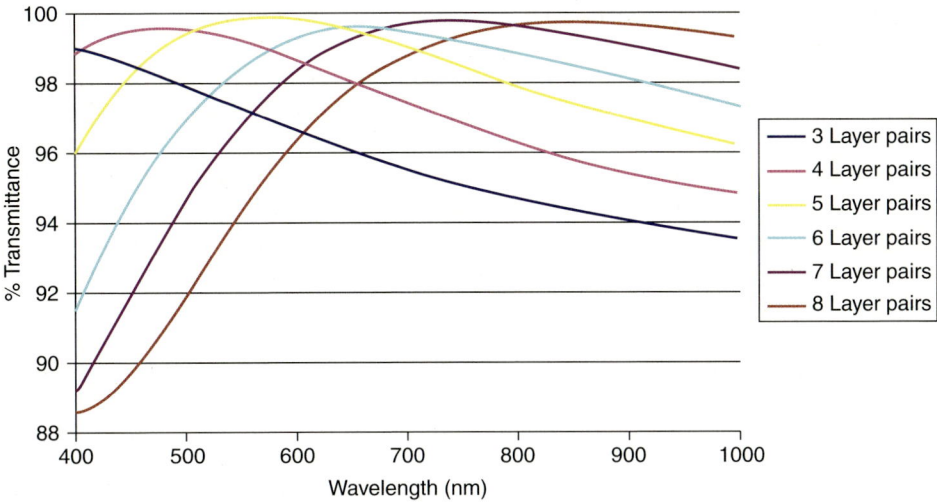

**Figure 21.10** Transmission spectra of PAH/SiO$_2$ antireflection coatings on glass as a function of the number of layer pairs.

(refractive index of silica) to approximately $n = 1.30$, as determined by ellipsometry. The transmission spectra for these ARCs, with transmittance levels reaching greater than 99.7%, are shown in Figure 21.10 as a function of the number of layer pairs.

Nanoparticle PEM films typically do not have sufficient mechanical robustness as fabricated. As mentioned above, calcination is a successful approach for imparting mechanical robustness for ARCs on glass substrates. However, transparent plastics, such as polycarbonate, cannot withstand typical calcination temperatures. As an alternative, we are currently examining several different crosslinking methods to improve robustness in ARC PEM films. A UV-curable diazo-resin oligomer that can crosslink with PSS is being examined with silica nanoparticles as a means of strengthening the mechanical stability in the films. We are also exploring thermal crosslinking of the weak polyelectrolytes, PAH and PAA, in an ISAM matrix with silica nanoparticles as a method to create both V-type and broadband films.

## 21.7
## Electrochromic Devices

Electrochromic devices (ECDs) exhibit reversible change in their optical transmittance when subjected to an external electric field. The change is due to the occurrence of electrochemical reactions in the EC materials within the device. ECDs based on a wide variety of materials, including inorganic compounds [43], phthalocyanines [44, 45] and polymers [46, 47], have been fabricated. EC polymers have been among the most heavily studied EC materials in recent years. Changes in the optical transmittance of EC polymers are primarily due to variations in their redox states. When activated, EC polymers undergo an ion exchange with their environment, resulting in reduction or oxidation of the polymer, which has a different optical transmittance than the polymer in the neutral state. The process can be reversed to neutral or the opposite redox state by termination or reversal of the applied electric field. Electrochemical reactions can occur between the EC materials and a dilute electrolyte solution or viscous conductive gels. EC polymers are usually deposited in the form of a thin film on a transparent or reflective substrate with precise control over the thickness and homogeneity of the thin film.

ECDs have a wide range of potential applications, including smart glass and heads-up displays. Smart glass can be used in green buildings, for example, to control transmittance of visible and/or infrared light from the sun to optimize heating/cooling efficiency of the building, as well as a means to provide privacy.

ECDs consisting of one or two EC polymers (single or dual [48, 49] electrochrome ECDs, respectively) have been fabricated from PEMs with the general goals of increasing the optical contrast and the switching speed. High optical contrast between redox states and fast switching speed are among the desired characteristics of ECDs that have been reported for polymer-based ECDs. Several groups have studied ECDs fabricated from PEM films. The redox states of PEM films were studied by Schlenoff and colleagues using cyclic voltammetry and spectroelectrochemistry [50, 51]. DeLongchamp and Hammond performed extensive studies of the electrochromic properties of PEM films. They fabricated dual electrochrome films with PEDOT:PSS and polyaniline as complementary switching materials and observed transmittance changes of up to 30% near 650 nm with a coloration time of 1.22 s and a decoloration time of 0.37 s [52]. They fabricated similar devices from a polyviologen and PEDOT:PSS and observed a transmittance change of 82% at 525 nm and switching times of 1–4 s [53]. Reynolds and coworkers synthesized a water-soluble sulfonated derivative of PEDOT (PEDOT-S) and demonstrated electrochromism of PEDOT-S/PAH films. They observed changes in transmittance of up to 40% for a 40-layer-pair film in $Na_2SO_4$ with a switching time of several seconds [54].

We have examined a number of aspects of PEM EC films and devices. A cationic polyviologen (PV) was combined with poly(2-acrylamido 2-methylpropanesulfonic acid) (PAMPS) and a transmittance change of 61% at 515 nm between 0.0 and 0.6 V for a 40-layer-pair film in 0.1 M $NaClO_4$ [55]. The PV/PAMPS film was combined with a polyaniline (PANI)/PAMPS film with PAMPS gel between the two films to make a quasi-solid state device that exhibited 25% contrast at 515 nm with coloration and decoloration times of 100 and 250 ms, respectively. Solid-state ECDs fabricated from pairs of PEM films on ITO containing PAH and a sulfonated 3,4-propylene-dioxythiophene derivative demonstrated 40% contrast with a coloration time of 100 ms and a decoloration time of 50 ms [56]. PEM films containing novel viologen-bridged polysilsesquioxane nanoparticles and PAMPS showed EC contrast of 50% at 550 nm [57]. We have also fabricated ECDs with two electrochromic polymers incorporated into a single PEM film. Devices fabricated from PANI and polyaniline sulfonic acid (PASA) were compared to those incorporating each of the two polymers individually. The PANI/PASA devices showed more uniform transmittance changes with a contrast of >49% across the visible spectrum as well as faster

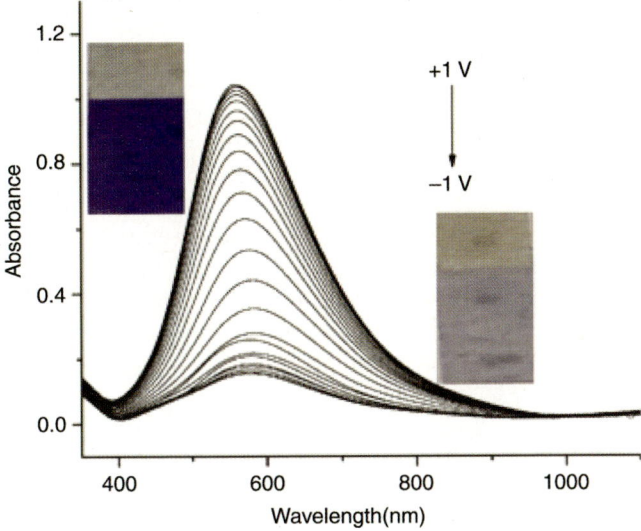

**Figure 21.11** Spectroelectrochemistry data of the absorbance of a 40-layer-pair RuP film in 0.1 M LiClO$_4$–ACN solution on increasing the voltage from +1 V to −1 V in 0.1 V steps. Reprinted with permission from Ref. [60]. © Royal Society of Chemistry.

switching than devices from either of the polymers with an EC-inactive polyelectrolyte [58]. Furthermore, we have examined ECDs employing single-wall carbon nanotube networks as alternative transparent, conducting substrates to ITO [59].

With respect to high contrast, we have achieved a transmittance change >84% at 560 nm with 20–30 nm Ruthenium purple (RuP) nanoparticle PEM EC devices [60]. The 40-layer-pair RuP/poly(ethylene imine) (PEI) film was switched in 0.1 M LiClO$_4$–acetonitrile and showed a very high coloration efficiency of 205 cm$^2$ C$^{-1}$. The absorption spectrum of the RuP/PEI film from −1 to +1 V is shown in Figure 21.11. CV data confirmed that the high optical contrast results because the thin film is fully electrochemically accessible, such that the entire thickness of the film participates in the redox reaction.

Exceptionally fast switching has been obtained in quasi-solid-state PEM EC devices containing PEDOT: PSS [61]. Two PEDOT: PSS/PAH films on ITO were sandwiched together with PAMPS gel between them. In this symmetric EC device, the PEDOT: PSS film at the cathode changes color while the one at the anode remains transparent. The temporal response of the devices was monitored with a He–Ne laser and photodiode as a square wave voltage (0–1.4 V) was applied to the electrochromic device. Figure 21.12 shows the electrochromic film response for an 80-layer-pair (two 40-layer-pair films) 0.6 cm$^2$ area PAH/PEDOT device. The coloration and decoloration times (to 90% of equilibrium value) are 31 ms and 6 ms, respectively. Furthermore, the film response time decreases linearly with the active area of the device, similar to the response of an RC circuit. Pixel areas associated with active displays are often < 0.05 cm$^2$. The data suggest that switching times for 0.05 cm$^2$ area devices should be of the order of 3 ms for coloration and 0.6 ms for decoloration.

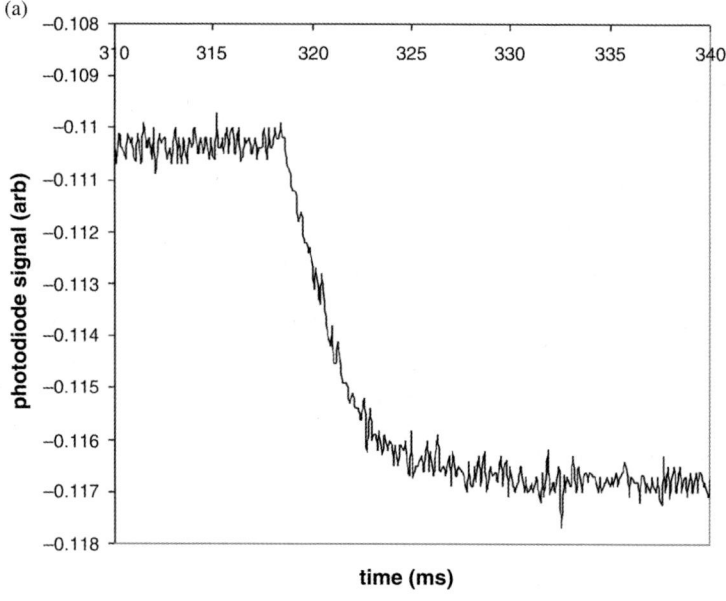

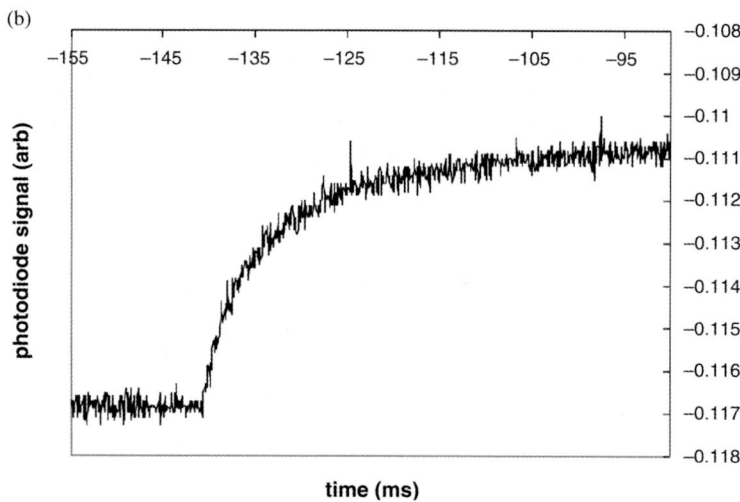

**Figure 21.12** (a) Decoloration and (b) coloration of 0.6 cm$^2$ device consisting of two 40-layer-pair PAH/PEDOT films with applied voltage of 0–1.4 V. Reprinted with permission from Ref. [61]. © American Institute of Physics.

## 21.8
### Electromechanical Actuators

Electromechanical actuators can expand, contract, or bend under applied voltage. By providing both force and motion, they convert electrical energy to mechanical energy.

Compared to other commonly used actuator materials, such as piezoelectric ceramics, electrostatic elastomers, and electrostrictive polymers, ionic polymer conductive network composite (IPCNC) actuators have light weight, high durability, and can be operated under low voltage, all of which give them promising prospects in robotic, medical, and other electromechanical applications.

IPCNC bending actuators consist of an ion exchange membrane (usually Nafion) for ion conduction, a conductive network composite (CNC) to provide a porous and conductive framework through which ions can move and accumulate, and outer metal electrodes for electrical contact. Under applied voltage, ions within the actuator will move toward and accumulate near the electrode of opposite charge, causing that side of the actuator to swell and resulting in bending. While actuators can function without the CNC, the presence of the CNC generally increases the amplitude and speed of the motion through the increased volume in which the ions can accumulate.

Conventionally, the ions in the actuator have been provided by using water to dissolve the ions associated with the pendant groups in Nafion. However, water has a low electrolysis voltage ($\sim$1.23 V) and will easily evaporate, which causes the performance of the actuator to degrade quickly over time when operated in air. Bennett and Leo showed that ionic liquids (ILs), such as 1-ethyl-3-methylimidazolium trifluoromethanesulfonate (EMI-Tf), could be used as a novel electrolyte in actuator design [62]. ILs, which are salts that are liquid at or near room temperature, are often composed of organic cations and inorganic anions. ILs not only provide both cations and anions for the actuation, but also have higher electrochemical and thermal stability window (4.1 V and 400 °C for EMI-Tf, for example), as well as nearly zero vapor pressure. So they are excellent candidates to replace water as the electrolyte in actuator applications.

Typically, the CNC layer has been fabricated by electro/electroless plating or direct assembly [63] methods. We have recently demonstrated that CNCs fabricated from PEMs can significantly enhance actuator performance through the nanoscale control over thickness and composition. PAH has been used as the polycation to build the inactive polymer matrix network while anionic colloidal gold nanoparticle (Au NP) suspension fills into and attaches onto the PAH network, which results in a relatively smooth, porous, and conductive CNC structure. This structure provides the actuator with large (the porous structure provides volume for ion accumulation) and fast (Au nanoparticles provide high electrical conductivity) bending actuation.

As an example of an IPCNC bending actuator, two CNC layers of PAH and Au NPs are coated by the LbL technique on both sides of a 25 μm thick Nafion extruded membrane [64]. The coated film is then soaked in EMI-Tf IL to 40 wt%. Finally, gold foils are hot-pressed onto both sides of the film, acting as the electrodes for connection to the electrical power supply.

As shown in Figure 21.13, the bending performance (in terms of the time response of the normalized bending strain) can be fitted with a typical capacitor charging curve. The characteristic response time $\tau$ represents the time needed by the actuator to reach approximately 63% ($e^{-1}$) of its maximum bending. A very fast ($\tau = 0.18$ s) bending response can be achieved by the IPCNC bending actuator with 0.4 μm PAH/Au NPs

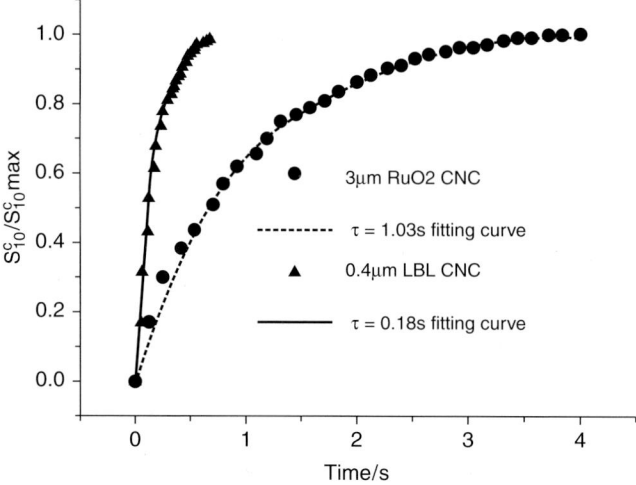

**Figure 21.13** Bending response of a Nafion IPCNC actuator (1 × 8 mm²) with 0.4 μm PEM CNC to a 4 V step input, compared with one with 3 μm RuO₂ CNC fabricated by direct assembly. Reprinted with permission from Ref. [64]. © American Institute of Physics.

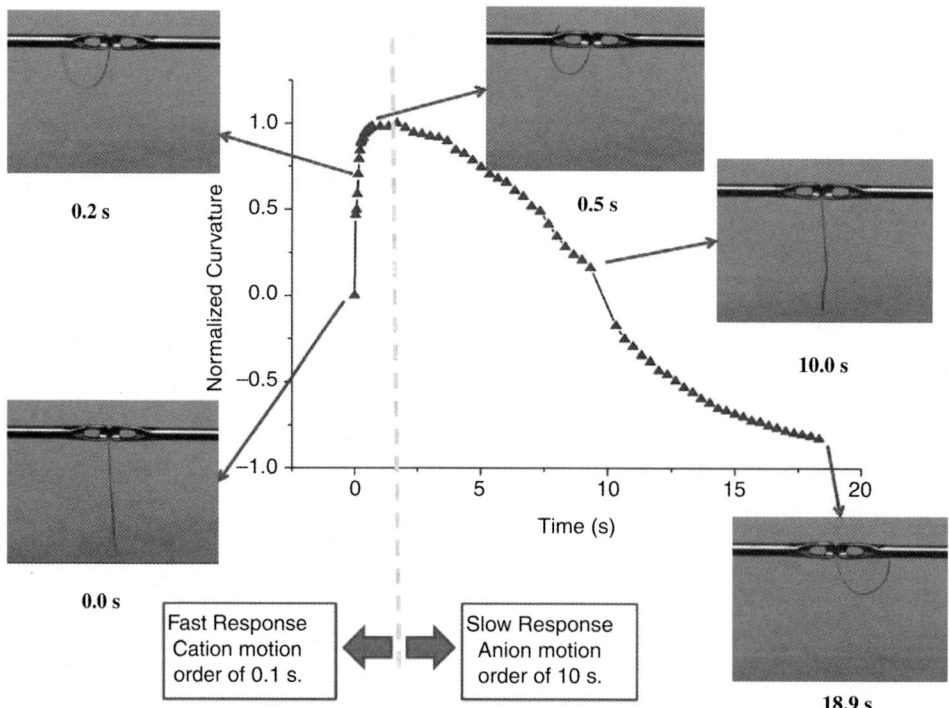

**Figure 21.14** Bidirectional bending response of a Nafion IPCNC actuator (1 × 8 mm²) with 0.1 μm PAH/Au NPs PEM CNC to a 4 V step input. Anode is on the left and cathode is on the right.

(~2 nm diameter, Purest Colloids Inc.) LbL CNC. For comparison, an actuator with a 3 μm direct-assembled $RuO_2$ CNC has significantly slower bending response.

Since both cations and anions from the IL in the actuator contribute to bending, the actuator shows a bidirectional motion on a longer timescale, as shown in Figure 21.14 [65]. The bidirectional bending is caused by the difference in size and speed of the cations and anions of EMI-Tf ionic liquid. When the cations have smaller size but faster speed than the anions, a large number of cations will quickly accumulate on the cathode side of the CNC after switching on the applied step voltage, while the anions move more slowly towards the anode. Initially, the swelling mostly happens on the cathode side of the actuator, which makes the actuator bend towards the anode. This corresponds to the fast bending response. After a certain time, the swelling caused by the anions starts to take over. Because of their larger size, their accumulation in the CNC at the anode side has a larger swelling effect, which will overcome the swelling on the cathode side caused by cations, resulting in slower motion in the opposite direction. The actuation has also been studied as a function of the number of bilayers, and it has been shown that curvature is linear in the thickness of the PAH/Au NP CNC, such that the large net intrinsic strain of 6.1% is independent of the thickness [66].

## Acknowledgements

This work was supported in part by National Science Foundation grant ECS-9907747, Air Force Office of Scientific Research grant FA9550-07-0357, National Institutes of Health grant NIAID 1R21A1081130-01, the Virginia Tech Carilion Research Institute, and the Institute for Critical Technology and Applied Science at Virginia Tech. This work was also supported in part by U.S. Army Research Office under Grant No. W911NF-07-1-0452 Ionic Liquids in Electro-Active Devices (ILEAD) MURI and by the MultiTASC Materials Center of Excellence sponsored by the Army Research Laboratory accomplished under Agreement Number W911NF-06-2-0014.

## References

1 Decher, G. and Hong, J.-D. (1991) Buildup of ultrathin multilayer films by a self-assembly process: 1. consecutive adsorption of anionic and cationic bipolar amphiphiles on charged surfaces. *Makromol. Chem, Macromol. Symp.*, **46**, 321–327.

2 Iler, R.K. (1966) Multilayers of colloidal particles. *J. Colloid Interface Sci.*, **21**, 569–594.

3 See, for example Boyd, R.W. (1992) *Nonlinear Optics*, Academic Press, Boston.

4 Sutherland, R.L. (2003) *Handbook of Nonlinear Optics*, Marcel Dekker, New York.

5 Heflin, J.R., Liu, Y., Figura, C., Marciu, D., and Claus, R.O. (1999) Thickness dependence of second harmonic generation in thin films fabricated from ionically self-assembled monolayers. *Appl. Phys. Lett.*, **74**, 495–497.

6 Lenahan, K.M., Wang, Y.X., Liu, Y., Claus, R.O., Heflin, J.R., Marciu, D., and Figura, C. (1998) Novel polymer dyes for nonlinear optical applications using

ionic self-assembled monolayers. *Adv. Mater.*, **10**, 853–855.
7 Lvov, Y., Yamada, S., and Kinitake, T. (1997) Nonlinear optical effects in layer-by-layer alternate films of polycations and an azobenzene-containing polyanion. *Thin Solid Films*, **300**, 107–110.
8 Wang, X., Balasubramanian, S., Li, L., Jiang, X., Sandman, D.J., Rubner, M.F., Kumar, J., and Tripathy, S.K. (1997) Self-assembled second order nonlinear optical multilayer azo polymer. *Macromol. Rapid Commun.*, **18**, 451–459.
9 Roberts, M.J., Lindsay, G.A., Herman, W.N., and Wynne, K.J. (1998) Thermally stable nonlinear optical films by alternating polyelectrolyte deposition on hydrophobic substrates. *J. Am. Chem. Soc.*, **120**, 11202–11203.
10 Garg, A., Davis, R.M., Durak, C., Heflin, J.R., and Gibson, H.W. (2008) Polar orientation of a pendant anionic chromophore in thick layer-by-layer self-assembled polymeric films. *J. Appl. Phys.*, **104**, 053116:1-8.
11 Van Cott, K.E., Heflin, J.R., Gibson, H.W., and Davis, R.M. (2005) Polar Ordering of Reactive Chromophores in Layer-by-Layer Nonlinear Optical Materials, U.S. Patent #6,953,607.
12 Van Cott, K.E., Guzy, M., Neyman, P., Brands, C., Heflin, J.R., Gibson, H.W., and Davis, R.M. (2002) Layer-by-layer deposition and ordering of low molecular weight dye molecules for second order nonlinear optics. *Angew. Chem. Int. Ed.*, **41**, 3236–3238.
13 Hermanson, G.T. (1996) *Bioconjugate Techniques*, Academic Press, San Diego.
14 Heflin, J.R., Guzy, M.T., Neyman, P.J., Gaskins, K.J., Brands, C., Wang, Z., Gibson, H.W., Davis, R.M., and Van Cott, K.E. (2006) Efficient, thermally-stable, second order nonlinear response in organic hybrid covalent/ionic self-assembled films. *Langmuir*, **22**, 5723–5727.
15 Pendry, J.B., Holden, A.J., Robbins, D.J., and Stewart, W.J. (1999) Magnetism from conductors and enhanced nonlinear phenomena. *IEEE Trans. Microwave Theory Tech.*, **47** (11), 2075–2084.
16 Liz-Marzan, L.M. (2006) Tailoring surface plasmons through the morphology and assembly of metal nanoparticles. *Langmuir*, **22** (1), 32–41.
17 Nie, S. and Emory, S.R. (1997) Probing single molecules and single nanoparticles by surface-enhanced raman scattering. *Science*, **275**, 1102–1106.
18 Hao, E. and Schatz, G.C. (2004) Electromagnetic fields around silver nanoparticles and dimers. *J. Chem. Phys.*, **120** (1), 357–366.
19 Chen, K., Durak, C., Heflin, J.R., and Robinson, H.D. (2007) Plasmon enhanced second-harmonic generation from ionically self-assembled multilayers film. *Nano Lett.*, **7**, 254–258.
20 Sherry, L.J., Chang, S.H., Schatz, G.C., Van Duyne, R.P., Wiley, B.J., and Xia, Y.N. (2005) Localized surface plasmon resonance spectroscopy of single silver nanocubes. *Nano Lett.*, **5**, 2034–2038.
21 Haynes, C.L. and Van Duyne, R.P. (2001) Nanosphere lithography: A versatile nanofabrication tool for studies of size-dependent nanoparticle optics. *J. Phys. Chem. B*, **105**, 5599–5611.
22 Moura, A.L., de Araujo, V.D., Vermelho, M.V.D., and Aitchison, J.S. (2006) Improved stability of the induced second-order nonlinearity in soft glass by thermal poling. *J. Appl. Phys.*, **100**, 033509:1-5.
23 Fujiwara, T., Takahashi, M., and Ikushima, A.J. (1997) Decay behaviour of second-order nonlinearity in $GeO_2$-$SiO_2$ glass poled with UV-irradiation. *Electron. Lett.*, **33**, 980–982.
24 Xu, Y., Wang, A., Heflin, J.R., and Liu, Z. (2007) Proposal and analysis of a silica fiber with large thermodynamically stable second order nonlinearity. *Appl. Phys. Lett.*, **90**, 21110:1-3.
25 Daengngam, C., Hofmann, M., Liu, Z., Wang, A., Heflin, J.R., and Xu, Y. (2011) Demonstration of a cylindrically symmetric second-order nonlinear fiber with self-assembled organic surface layers. *Opt. Express*, **19**, 10326–10335.
26 Xue, S., van Eijkelenborg, M., Barton, G.W., and Hambley, P. (2007) Theoretical, numerical, and experimental

analysis of optical fiber tapering. *J. Lightwave. Technol.*, **25**, 1169–1176.

27 Brisk, T.A. and Li, Y.W. (1992) The shape of fiber tapers. *J. Lightwave. Technol.*, **10**, 432–438.

28 Wiedemann, U., Karapetyan, K., Dan, C., Pritzkau, D., Alt, W., Irsen, S., and Meschede, D. (2010) Measurement of submicrometre diameters of tapered optical fibers using harmonic generation. *Opt. Express*, **18**, 7693–7704.

29 Vasiliev, S.A., Dianov, E.M., Kurkov, A.S., Medvedkov, O.I., and Protopopov, V.N. (1997) Photoinduced in-fibre refractive-index gratings for core-cladding mode coupling. *Quantum Electron.*, **27**, 146–149.

30 Erdogan, T. (1997) Cladding-mode resonances in short- and long-period fiber grating filters. *J. Opt. Soc. Am. A*, **14**, 1760–1773.

31 Erdogan, T. (1997) Cladding mode resonances in short and long period fibre grating filters. *J. Opt. Soc. Am.*, **14**, 1760–1773;Erdogan, T. (2000) Cladding mode resonances in short and long period fibre grating filters: errata. *J. Opt. Soc. Am.*, **17**, 2113.

32 Wang, Z.Y., Heflin, J.R., Stolen, R.H., and Ramachandran, S. (2005) Highly sensitive optical response of optical fiber long period gratings to nanometer-thick ionic self-assembled multilayers. *Appl. Phys. Lett.*, **86**, **223104**, 1–3.

33 Wang, Z., Heflin, J.R., Stolen, R.H., and Ramachandran, S. (2005) Analysis of optical response of long-period fiber gratings to nm-thick thin-film coatings. *Opt. Express*, **13**, 2808–2813.

34 Zuo, Z., Bandara, A., Michalenka, A., Inzana, T., and Heflin, J.R.(to be published).

35 Tesar, A., Balooch, M., Shotts, K., and Siekhaus, W. (1991) Morphology and laser damage studies by atomic force microscopy of e-beam evaporation deposited AR and HR coatings. *SPIE Proc.*, **1441**, Laser-Induced Damage in Optical Materials, 228–236.

36 Uhlmann, D., Suratwala, T., Davidson, K., Boulton, J., and Teowee, G. (1997) Sol-gel derived coatings on glass. *J. Non-Cryst. Solids*, **218**, 113–122.

37 Peterson, R.E. (1975) Thin Film coatings in solar thermal-power systems. *J. Vac. Sci. Technol.*, **12**, 174–181.

38 Alexieva, Z.I., Nenova1, Z.S., Bakardjieva1, V.S., Milanova, M.M., and Dikov1, H.M. (2010) Antireflection coatings for GaAs solar cell applications. *J. Phys. Conf. Ser.*, **223**, 012045:1-4.

39 Schirone, L. and Sotgiu, G. (1997) Chemically etched porous silicon as an anti-reflection coating for high efficiency solar cells. *Thin Solid Films*, **297**, 296–298.

40 Hiller, J.A., Mendelsohn, J.D., and Rubner, M.F. (2002) Reversibly erasable nanoporous anti-reflection coatings from polyelectrolyte multilayers. *Nat. Mater.*, **1**, 59–63.

41 Yancey, S.E., Zhong, W., Heflin, J.R., and Ritter, A.L. (2006) The influence of void space on antireflection coatings of silica nanoparticle self-assembled films. *J. Appl. Phys.*, **99**, 034313:1-10.

42 Lee, D., Rubner, M.F., and Cohen, R.E. (2006) All-nanoparticle Thin-film Coatings. *Nano Lett.*, **6**, 2305–2312.

43 Mortimer, R.J. (1997) Electrochromic materials. *Chem. Soc. Rev.*, **26**, 147–156.

44 Riou, M.T. and Clarisse, C. (1988) The rare-earth substitution effect on the electrochemistry of diphthalocyanine films in contact with an acidic aqueous-medium. *J. Electroanal. Chem.*, **249**, 181–190.

45 Kobayashi, T., Fujita, K., and Muto, J. (1996) Characteristics of electrochromic magnesium phthalocyanine films for repeated oxidation-reduction cycling. *J. Mater. Sci. Lett.*, **15**, 1276–1278.

46 Sonmez, G., Meng, H., and Wudl, F. (2004) Organic polymeric electrochromic devices: polychromism with very high coloration efficiency. *Chem. Mater.*, **16**, 574–580.

47 DeLongchamp, D.M. and Hammond, P.T. (2004) Multiple-color electrochromism from layer-by-layer-assembled polyaniline/Prussian blue nanocomposite thin films. *Chem. Mater.*, **16**, 4799–4805.

48 DeLongchamp, D.M., Kastantin, M., and Hammond, P.T. (2003) High-contrast electrochromism from layer-by-layer

49 Montazami, R., Jain, V., and Heflin, J.R. (2010) High-contrast asymmetric solid state electrochromic devices based on layer-by-layer deposition of polyaniline and poly (aniline sulfonic acid). *Electrochim. Acta*, **56**, 990–994.

50 Laurent, D. and Schlenoff, J.B. (1997) Multilayer assemblies of redox polyelectrolytes. *Langmuir*, **13**, 1552–1557.

51 Schlenoff, J.B., Ly, H., and Li, M. (1998) Charge and mass balance in polyelectrolyte multilayers. *J. Am. Chem. Soc.*, **120**, 7626–7634.

52 DeLongchamp, D. and Hammond, P.T. (2001) Layer-by-layer assembly of PEDOT/polyaniline electrochromic devices. *Adv. Mater.*, **19**, 1455–1459.

53 DeLongchamp, D., Katantin, M., and Hammond, P.T. (2003) High-contrast electrochromism from layer-by-layer polymer films. *Chem. Mater.*, **15**, 1575–1586.

54 Cutler, C.A., Bouguettaya, M., and Reynolds, J.R. (2002) PEDOT polyelectrolyte based electrochromic films via electrostatic adsorption. *Adv. Mater.*, **14**, 684–688.

55 Jain, V., Yochum, H., Wang, H., Montazami, R., Hurtado, M.A.V., Mendoza-Galvan, A., Gibson, H.W., and Heflin, J.R. (2008) Solid-state electrochromic devices via ionic self-assembled multilayers of a polyviologen. *Macromol. Chem. Phys.*, **209**, 150–157.

56 Jain, V., Sahoo, R., Mishra, S.P., Sinha, J., Montazami, R., Yochum, H.M., Heflin, J.R., and Kumar, A. (2009) Synthesis and characterization of regioregular water-soluble propylenedioxythiophene derivative and its application in the fabrication of high-contrast solid-state eelectrochromic devices. *Macromolecules*, **42**, 135–140.

57 Jain, V., Kiterer, M., Montazami, R., Yochum, H.M., Shea, K.J., and Heflin, J.R. (2009) High-contrast solid-state electrochromic devices of viologen-bridged polysilsesquioxane nanoparticles fabricated by layer-by-layer assembly. *Appl. Mater. Interf.*, **1**, 83–89.

58 Montazami, R., Jain, V., and Heflin, J.R. (2010) High contrast asymmetric solid state electrochromic devices based on layer-by-layer deposition of polyaniline and poly(aniline sulfonic acid). *Electrochim. Acta*, **56**, 990–994.

59 Jain, V., Yochum, H.M., Montazami, R., Heflin, J.R., Hu, L., and Gruner, G. (2008) Modification of single-walled carbon nanotube electrodes by layer-by-layer assembly for electrochromic devices. *J. Appl. Phys.*, **103**, 074504:1-5.

60 Jain, V., Sahoo, R., Jinschek, J., Montazami, R., Yochum, H.M., Beyer, F.L., Kumar, A., and Heflin, J.R. (2008) High contrast solid state electrochromic devices based on ruthenium purple nanocomposites fabricated by layer-by-layer assembly. *Chem. Comm.*, **2008**, 3663–3665.

61 Jain, V., Yochum, H.M., Montazami, R., and Heflin, J.R. (2008) millisecond switching in sold state electrochromic polymer devices fabricated from ionic self-assembled multilayers. *Appl. Phys. Lett.*, **92**, 033304:1-3.

62 Bennett, M.D. and Leo, D.J. (2004) Ionic liquids as stable solvents for ionic polymer transducers. *Sens. Actuat. A*, **115**, 79–90.

63 Akle, B.J., Bennett, M.D., Leo, D.J., Wiles, K.B., and McGrath, J.E. (2007) Direct assembly process: a novel fabrication technique for large strain ionic polymer transducers. *J. Mater. Sci.*, **42**, 7031–7041.

64 Liu, S., Montazami, R., Liu, Y., Jain, V., Lin, M., Heflin, J.R., and Zhang, Q.M. (2009) Layer-by-layer self-assembled conductive network composites in ionic polymer metal composite actuators with high strain response. *Appl. Phys. Lett.*, **95**, 023505:1-3.

65 Liu, Y., Liu, S., Lin, J., Wang, D., Jain, V., Montazami, R., Heflin, J.R., Li, J., Madsen, L., and Zhang, Q.M. (2010) Ion transport and storage of ionic

liquids in ionic polymer conductor network composites. *Appl. Phys. Lett.*, **96**, 223503:1-3.

66 Montazami, R., Liu, S., Liu, Y., Wang, D., Zhang, Q.M., and Heflin, J.R. (2011) Thickness dependence of curvature, strain, and response time in ionic electroactive polymer actuators fabricated via layer-by-layer assembly. *J. Appl. Phys.*, **109**, 104301:1-5.

# 22
# Nanostructured Electrodes Assembled from Metal Nanoparticles and Quantum Dots in Polyelectrolytes

*Lara Halaoui*

## 22.1
## Introduction

Some very interesting questions in chemistry and physics today revolve around effects of quantum confinement in semiconductors on enhancing solar energy conversion [1–3], and effects of the size and shape of small metal particles on catalysis [4]. In addition to fundamental curiosity, research is driven by global energy needs and aims to reduce cost and rare material requirements in alternative clean energy production and utilization. Reducing the semiconductor size to dimensions comparable to the exciton diameter results in unique optoelectronic properties that placed quantum dots (QD) [5, 6] at the center of research to develop (low-cost high-efficiency) third generation solar cells [1–3]. At metal nanoparticles (NP), the surface-to-mass ratio increases with reduced dimension, and the distribution of surface sites becomes dependent on size and shape, which may lead to enhanced kinetics for some structure-sensitive reactions. Catalysis at metal NP can thus potentially offer control over reactivity (and selectivity) for the rational design of catalysts in several technologies, including fuel cells. Addressing several of the outstanding questions in this field necessitates varying NP size, shape and surface modification, assembly allowing architecture manipulation in two and three dimensions, and investigations of electrochemical and photoelectrochemical behavior.

NPs can be deposited on surfaces either by transfer of colloids or by direct deposition, however, separating synthesis and assembly steps may allow a finer control of NP size, shape and surface distribution. Examples of metal NP (viz., Pt) assembly on electrodes include electrooxidative coupling of NPs encapsulated in dendrimers [7], deposition using $C_{60}$ linker [8], open circuit potential deposition [9] and electrophoretic deposition [10], while direct growth methods include metal evaporation [11, 12], chemical deposition [13], electroless deposition [14], underpotential deposition redox replacement [15] and potentiostatic deposition [16–19]. Multilayers of (Pt) NPs have been fabricated by NP adsorption on polyelectrolytes [20–25], reduction of Pt salt impregnated in polyelectrolytes [26, 27], or as Langmuir Blodgett films [28]. Semiconductor NPs were incorporated in solar

*Multilayer Thin Films: Sequential Assembly of Nanocomposite Materials*, Second Edition.
Edited by Gero Decher and Joseph B. Schlenoff.
© 2012 Wiley-VCH Verlag GmbH & Co. KGaA. Published 2012 by Wiley-VCH Verlag GmbH & Co. KGaA.

cells by sensitizing wide band-gap semiconductors via surface linkers [29–34] or by direct deposition (e.g., by SILAR (successive ionic layer adsorption and reaction) [35]), dispersion in organic hole-transport polymer [36–38], assembly of metal/QD Schottky junction [39, 40], or as photoelectrodes with a liquid interface as Langmuir–Blodgett films [41, 42], self-assembled monolayers [43–50], chemically deposited films [51, 52], and multilayers in polyelectrolyte [53–57].

Ideally, the assembly should afford architecture control of NP coverage and interlayer spacing, and should allow incorporation of various materials for build-up of heteronanostructures with multiple functional units. To serve as electrodes, assembled films should naturally be stable in electrolyte and must not present high barriers to mass and charge transport or block catalytic sites. The layer-by-layer (LbL) assembly of NPs in polyelectrolyte, based on initial work by Iler on the assembly of colloids [58], and later work on polyelectrolyte multilayers [59–63], provides an attractive method successfully employed to transfer a wide range of NPs on surfaces, including CdS, CdSe, CdTe, $TiO_2$, Au, Ag, Pt, $SiO_2$, $Fe_2O_3$ [20–25, 53–57, 64–98]. It is simple, does not require major instrumentation, yields stable photo- or electrochemically active films, and can be readily extended to assemble a multitude of NPs and materials with tunability of coverage and thickness by varying deposition parameters, such as adsorption time and ionic strength [20, 24, 25, 62, 96].

The subject of this chapter is the fabrication of metal and semiconductor nanostructured electrodes from polyacrylate-capped Pt NPs [20–24, 98] and CdS quantum dots (Q-CdS) [54, 55, 67] assembled in polyelectrolyte, either as multilayers using a LbL dip or dip–spin self-assembly, or at submonolayer coverage, and a review of the electrochemical or photoelectrochemical characteristics of these electrodes with applications in sensors [24, 98], catalysis [22, 23] and energy conversion [54, 55].

## 22.2
### Nanostructured Pt Electrodes from Assemblies of Pt Nanoparticles in Polyelectrolytes

Pt is a catalyst for several significant structure-sensitive reactions, such as oxygen reduction, methanol oxidation, and hydrogen oxidation [99–117], and questions remain on structural and electronic factors and those of surface modification and coverage on catalysis at Pt nanoparticles [23, 105–107, 116–127]. We present herein the fabrication of nanostructured electrodes from polyacrylate-capped 2.5 nm Pt NPs assembled in polyelectrolyte at varied coverage, and electrochemical characterization of surface, catalytic, and sensing properties of these electrodes.

### 22.2.1
### Assembly of Polyacrylate-capped Pt NPs in Polyelectrolytes

Polyacrylate (PAC)-capped Pt NPs with average size $2.5 \pm 0.6$ nm were prepared by reducing Pt(IV) with citrate in the presence of PAC [20] and were assembled in poly(diallyldimethylammonium) chloride, PDDA, by consecutively dipping electrodes (ITO, glassy carbon) or other surfaces (silicon, quartz) in solutions of 10 mM

PDDA/0.1 M NaCl, followed by rinsing, drying, then immersing in the NP solution (1 h for 1 layer pair), and repeating this cycle (n-times) to form a multilayer [20–24]. The NP surface was modified with polyacrylate, an anionic polyelectrolyte which, in addition to terminating growth, provided the basis of interactions with the cationic polyelectrolyte. Films are referred to as $(PDDA/PAC-Pt\ NP)_n$.

For TEM imaging, films were assembled on $SiO_x$-Cu grids by floating on the solution/air interface [21, 23] and exhibited a fractal NP distribution at submonolayer coverage in 1 layer (Figure 22.1). The NP density increased per layer and reached almost full coverage at 4 layer pairs [23]. The observed fractal framework with NP proximity in some areas could have been caused by the morphology of PDDA assembled from a high ionic strength medium [23, 24]. The UV–visible absorbance increased linearly with $n$ [20], indicating reproducible surface charge reversal despite less than full NP coverage per layer, which could be due to "defect" coverage by the polyelectrolyte.

The assembly dynamics of Pt NP on PDDA was followed by UV–visible absorbance and was found to exhibit two kinetics regimes, an initial faster adsorption rate was recorded in the first 40 min, followed by a considerably slower adsorption with pseudo-saturation at $t$ longer than about 60 min, whether adsorption was interrupted and films dried every $\Delta t = 10$ or 2 min (Figure 22.2) [20]. The presence of two kinetics regimes conforms to the observed adsorption dynamics of polymers and polyelectrolytes [20, 128–131]. At low coverage, the adsorption rate depends on the rate of NP transport or their availability at the surface, while, at long times, adsorption sites become scarce and adsorption is thought to be governed by the rate of NP/polyelectrolyte rearrangement to open sites [20].

## 22.2.2
### Surface Characterization by Hydrogen Underpotential Deposition

Recent advances in preparing Pt NPs with different sizes and preferential surface orientation paved the way to systematic studies of nanoscale catalysis [132–144].

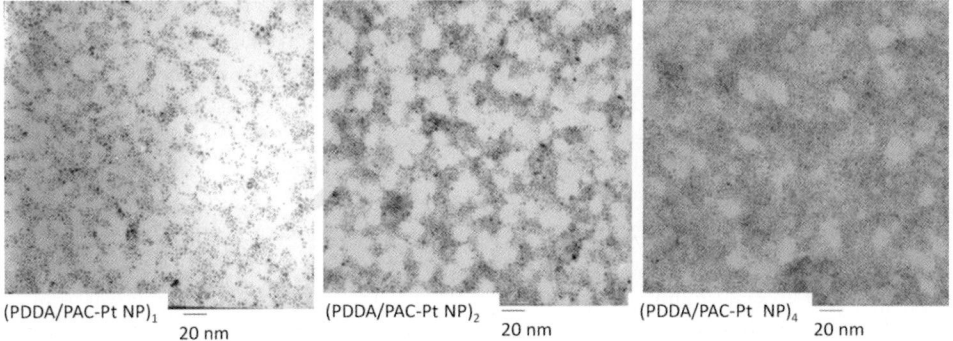

**Figure 22.1** TEM images of 1, 2, and 4 layer pairs of PDDA/PAC-Pt NP assembled on $SiO_x$/Cu grid. The scale bars are 20 nm. Reprinted with permission from Estephan, Z., Alawiye, L., Halaoui, L. I. *J. Phys. Chem. C*, 2007, **111**, 8060. Copyright 2007 American Chemical Society.

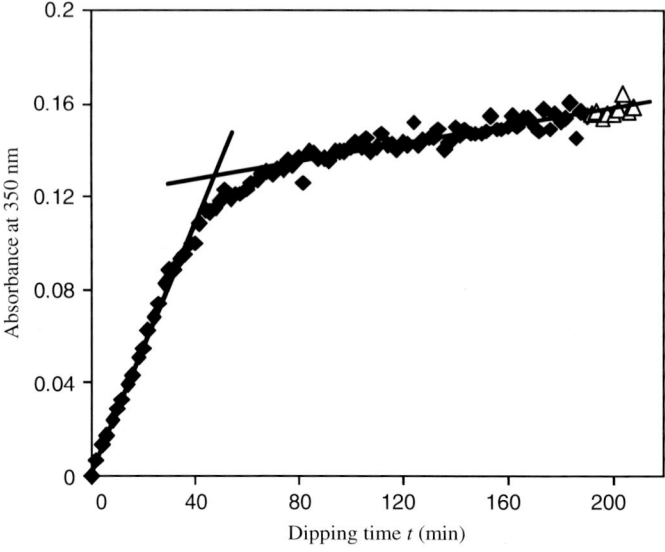

**Figure 22.2** Absorbance at 350 nm versus total dipping time $t$ in NP solution of PDDA/(PAC-Pt NP)$_t$ assembled by dipping a PDDA-primed quartz substrate at 2 min intervals in the PAC-Pt NP solution, followed by 1 min rinse and 15 min drying. The open symbols (△) correspond to data acquired after changing the NP solution. Reprinted with permission from Ghannoum, S., Xin, Y., Jaber, J., Halaoui, L. I. *Langmuir*, 2003, *19*, 4804–4811. Copyright 2003 American Chemical Society.

The NP surface is normally encapsulated with a capping molecule, such as polyacrylate [132–136, 142], polyvinylpyrrolidone [135–139, 144] or dendrimer [7, 145]. Understanding nanoscale catalysis thus requires determining active sites of surface-modified NPs. Surface de-capping can be done, for instance, by base [144] or heat treatment [143], but this could alter the NP size or shape and the surface ligand may be desired, beyond controlling NP growth, to provide interaction forces for assembly, and could also play a part in reactivity or selectivity by tuning surface access. TEM provides size and shape distribution but does not indicate if surface ligands are blocking adsorption sites. Electrochemical atom adsorption is a technique that yields a statistical average of active surface sites over a large electrode area, and can show structural changes *in situ* during electrocatalysis.

We studied hydrogen underpotential deposition ($H_{upd}$) at PDDA/PAC-Pt NPs films to characterize active sites and active Pt surface area [21]. $H_{upd}$ is a structure-sensitive adsorption at Pt whose potential depends on the adsorption site [146–154]. Active surface sites are assigned in accordance with studies on Pt single crystals [146], and the distribution of sites and real surface area are determined by integration of areas under the peaks. Figure 22.3 shows cyclic voltammograms (CVs) at $n = 2, 4,$ and 14 and at bulk Pt in 1 M $H_2SO_4$ at 20 mV s$^{-1}$ [21]. I–V curves exhibited two well-resolved $H_{upd}$ peaks at Pt NP: a strongly-adsorbed state H(S) attributed to (100) sites and a weakly-adsorbed state H(W) attributed to (110) sites, followed by hydrogen

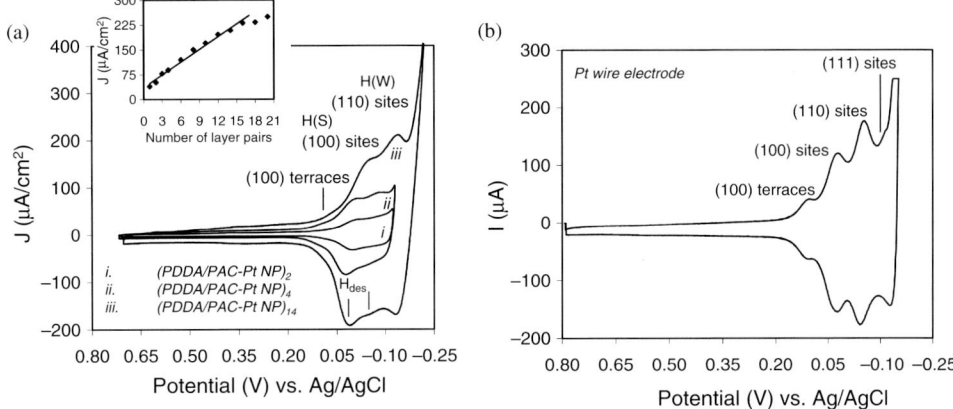

**Figure 22.3** (a) Cyclic voltammograms at 2, 4, and 14 layer pairs of PDDA/PAC-Pt NP on ITO and (b) at a polycrystalline Pt wire electrode in deoxygenated 1 M $H_2SO_4$. Scan rate is 20 mV s$^{-1}$. Inset of (a) shows H(W) peak current density at 20 mV s$^{-1}$ as a function of the number of layer pairs. Reprinted with permission from Markarian, M. Z., El Harakeh, M., Halaoui, L. I. *J. Phys. Chem. B* 2005, *109*, 11616. Copyright 2005 American Chemical Society.

evolution [21, 146]. The shoulder prior to H(S) is ascribed to long-range adsorption at (100) terraces [21, 146]. A total charge of 677 µC cm$^{-2}$ is computed at (PDDA/PAC-Pt NP)$_4$, corresponding to 3.22 cm$^2_{real}$/cm$^2_{geom}$ assuming the charge of 210 µC cm$^{-2}$ at polycryst-Pt [21]. We calculated that this charge is equivalent to a NP coverage exceeding 1 ($\theta = 1.37$) relative to a hypothetical closely-packed assembly of 2.5 nm Pt NPs capped with a 1.5 Å-thick organic layer, which was qualitatively consistent with the almost full coverage observed by TEM, and indicated that a significant fraction of PAC-Pt NP sites remained open. This could be rationalized since (110) sites lie below the Pt surface and some (100) sites could be fourfold hollow sites [154] that, therefore, can remain active in the presence of a surface ligand. The active sites of surface-modified Pt NPs can thus be identified upon assembly in polyelectrolyte, providing needed structural information in studying structure–reactivity relations at the nanoscale.

Reported I–V curves at Pt NPs on electrodes varied in resolving $H_{upd}$ features which can be due to differences in surface modification, adsorbates, or oxidation state [8, 13, 26, 27, 126, 127, 141, 142, 144, 155]. In any case, in some studies the surface ligand was removed [144], or it may not be indicated if films were electrochemically activated [23, 26, 27, 126, 127]. It is noteworthy that to identify active sites of *as*-prepared Pt NPs, films should not be subjected to electrode activation, in order not to cause surface oxidation that may lead to restructuring (or decapping).

Nanostructured Pt electrodes can be assembled at varying NP coverage on a PDDA layer, or in multilayers [21–24]. Figure 22.4 presents the variation in $H_{upd}$ total charge ($Q$) and real surface area at dipping time $t$ from 2 to 120 min, or in 2 to 12 layer pairs [23]. The increase in $Q$ with $t$ on one PDDA layer was consistent with the

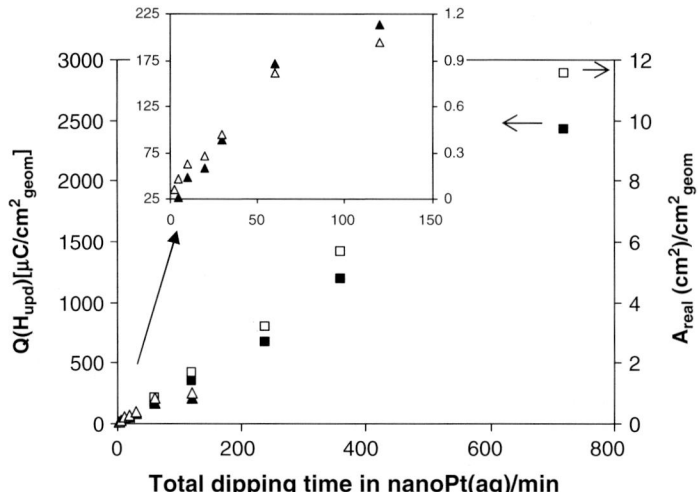

**Figure 22.4** Total charge Q for H$_{upd}$ and active surface area versus the total dipping time in the Pt NP solution for assemblies of PDDA/(PAC-Pt NP)$_t$ at $t = 2$ to 120 min (triangles), and at multilayered assembly of (PDDA/PAC-Pt NP)$_n$ for $n = 1$ to 12 (squares). Data was acquired from CVs in deoxygenated 1 M H$_2$SO$_4$ at scan rate of 20 mV s$^{-1}$. Reprinted with permission from Estephan, Z., Alawiye, L., Halaoui, L. I. J. Phys. Chem. C, 2007, **111**, 8060. Copyright 2007 American Chemical Society.

presence of two kinetics adsorption regimes. The charge density increased linearly with $n$ up to 16 layers in measurements at 20 mV s$^{-1}$ [21]. It follows that NP charging must be taking place, thus Pt NPs are electrochemically addressed and their active sites are accessible in the thickness of polyelectrolyte multilayers.

The electrochemical behavior of Pt NP assemblies was compared to polycryst-Pt to elucidate the effects of the polyelectrolyte capping and matrix on the electrode properties [21]. Atomic hydrogen adsorption peaks were negatively shifted at the NP assemblies compared to the bulk, and the shift increased with the number of layer pairs, but an anodic shift in desorption was not observed at 20 mV s$^{-1}$ (Figure 22.3). H$_{upd}$ at these films involves several processes (Scheme 22.1) [21]. Charging of Pt NPs must be taking place at an applied potential by charge hopping between NPs, driven by an electrochemical potential gradient possibly facilitated by networks of interconnected particles at short proximity, and charge must be tunneling across a capping layer [21]. At a feasible electrochemical potential, an electron is transferred to H$^+$ in the film with adsorption and this may be accompanied by capping agent/polyelectrolyte and solvent reorganization [21]. Mass transport of H$^+$ in the films, and possibly of counterions (HSO$_4^-$ and SO$_4^{2-}$), to solution maintains the films electroneutrality [21]. Mass transport and electrochemistry at polyelectrolyte multilayers have been reported to depend on the type of polyelectrolyte and assembly conditions, the nature and charge of the redox species and the surface charge of the outermost layer [156–165], and mass transport in these systems has been modeled by a capillary membrane model [156] or via an

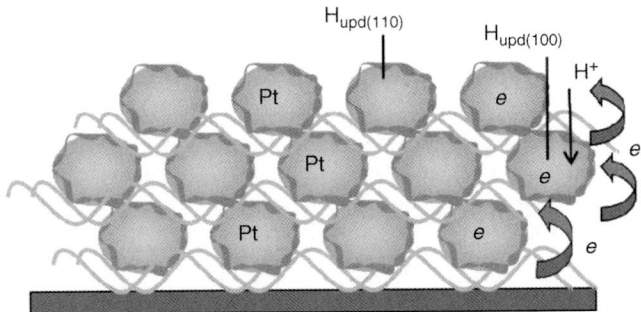

**Scheme 22.1** Sketch of main processes leading to $H_{upd}$ at PAC-capped Pt NP assembled in PDDA: film charging via charge hopping between Pt NP across the polyelectrolyte and reduction of H$^+$ in the films with adsorption on a (100) or (110) site. Reprinted with permission from Markarian, M. Z., El Harakeh, M., Halaoui, L. I. *J. Phys. Chem. B* 2005, *109*, 11616. Copyright 2005 American Chemical Society.

ion-hopping mechanism [157–159]. Some of the same factors affecting mass transport in polyelectrolyte multilayers may be at work in the NP/polyelectrolyte assembly. The resistances that can limit the rate of $H_{upd}$ are, therefore, of charge hopping ($R_{ch}$), mass transport ($R_{mt}$), and electron transfer, which includes the adsorption/desorption terms ($R_{et}$) [21]. The cathodic shift in $H_{upd}$ at the assemblies relative to bulk Pt was attributed to an activation barrier in electron transfer/H-adsorption affected by the capping ligand/polyelectrolyte environment, rather than to charge hopping or mass transfer resistances at this timescale (20 mV s$^{-1}$) because then the absence of an anodic shift could not be rationalized [21]. Electronic factors due to size variations could also affect adsorption, but would also affect desorption potentials [127]. Upon increasing the scan rate, shifts in anodic and cathodic peaks occurred, reflecting the effects of mass transport and/or charge hopping limiting kinetics [21].

### 22.2.3
### H$_2$O$_2$ Sensing at Arrays of Pt NPs in Polyelectrolyte at Low Surface Coverage

Miniaturizing the size of an electrode can offer several advantages [24, 166–169]. At an ultramicroelectrode (UME), the ratio of the Faradaic current to area ($i/A$) is inversely proportional to the electrode apparent radius of curvature ($a_o$), leading to improved signal-to-background current ratio with decreased dimension [24, 166–168], and the current is almost unaffected by convection variations [166, 169]. The small surface area limits the current magnitude and thus UME arrays have been used with an ideal response (of multiple single-UME) achieved at large inter-electrode separation distances relative to the electrode dimension $r_o$ ($d \gg 2\, r_o$) [166–170].

An interesting question is the behavior of such electrochemical arrays upon shrinking the size of the active electrode element to the nanoscale [24]. H$_2$O$_2$ sensing is significant in several industries [171–183], and electrochemical detection based on $H_2O_2$ oxidation uses Pt because of its catalytic activity [184–188]. A problem with a bulk Pt electrode is its detection limit, and deactivation by thick oxide formation at anodic potential, and several alternatives have been explored (including Pt black [189], mesoporous Pt UME [190], Pt NPs with carbon nanotubes [191], or Pt NPs in a carbon film [180]). We explored electrochemical detection in random arrays of NPs by studying amperometric sensing of $H_2O_2$ at PDDA/PAC-Pt NP films prepared by varying the adsorption time ($t$) of PDDA-primed ITO electrode in Pt NP solution [24]. Films are referred to as PDDA/(PAC-Pt NP)$_t$ ($t=60$ min equivalent to $n=1$). An average sensitivity to $H_2O_2$ of $0.500 \pm 0.019$ AM$^{-1}$ cm$^{-2}$ was measured at PDDA/(PAC-Pt NP)$_{t=60\,min}$ ($N=8$ films) with a limit of detection of 42 nM and a linear range up to 0.16 mM (Figure 22.5). Decreasing coverage at submonolayer at $t<60$ min reduced both the current and the upper linear range limit (to 8 µM) but films still detected 42 nM [24]. We measured the intrinsic (or real) sensitivity, which is the sensitivity (AM$^{-1}$) per real surface area determined from $H_{upd}$, and found it increased with decreasing loading, reaching 0.9 AM$^{-1}$ cm$^{-2}$$_{real}$ at the lowest coverage (0.2–0.4 cm$^2_{real}$/cm$^2_{geom}$) (Figure 22.6). It was notable that at a PDDA/(PAC-Pt NP)$_{t=20\,min}$ array, the current output was only 15% lower than at PDDA/(PAC-Pt NP)$_{t=60\,min}$, while the intrinsic sensitivity was 200% higher, showing that a highest effective turnover rate per active site under kinetic and diffusion control is reached at a more loosely packed assembly [24]. The PDDA/(PAC-Pt NP)$_{t=20\,min}$ corresponded to 0.4 cm$^2_{real}$/cm$^2_{geom}$ or an estimated coverage $\theta=0.17$ relative to a hypothetical monolayer of fully active and closely packed 2.5 nm Pt NPs capped with an organic layer (or Pt mass of ~380 ng cm$^{-2}$) compared to $\theta=0.39$ at $t=60$ min.

The development of the diffusion field of an array of small size electrodes from the individual diffusion fields with increasing loading is sketched in Scheme 22.2 [24, 167]. Whereas at loosely-packed arrays the diffusion fields of NPs (or islands of

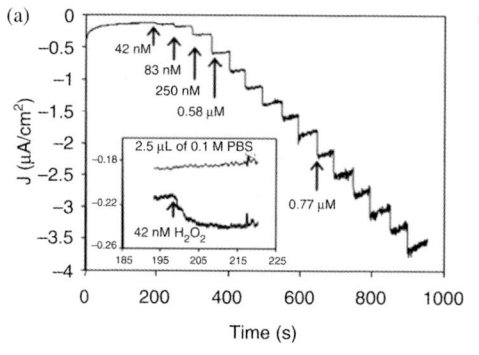
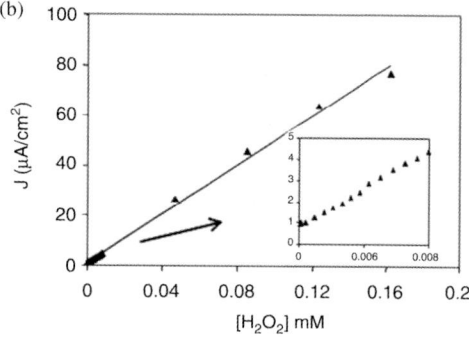

**Figure 22.5** (a) Continuous amperometry plot in response to $H_2O_2$ additions at a PDDA/(Pt NP)$_{t=60\,min}$ biased at 0.6 V vs Ag/AgCl in air-saturated 0.1 M phosphate buffer solution (PBS) at pH 7, the inset shows the current response to 42 nM $H_2O_2$ and the same volume of 0.1 M PBS without the analyte. (b) Steady-state current density versus $H_2O_2$ concentration. Reprinted with permission from Karam, P., Halaoui, L. I. Anal. Chem. 2008, 80, 5441. Copyright 2008 American Chemical Society.

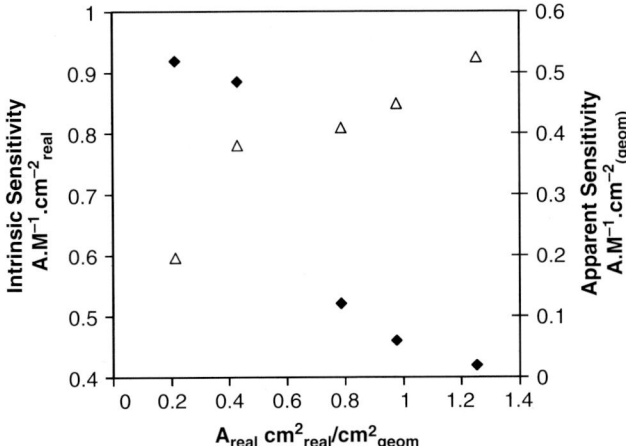

**Figure 22.6** Plots of intrinsic (real) sensitivity per real surface area (◆) and apparent sensitivity per geometric surface area (△, secondary axis) to $H_2O_2$ at a PDDA/PAC-Pt NP array as a function of the real Pt surface area. Reprinted with permission from Karam, P., Halaoui, L. I. *Anal. Chem.* 2008, *80*, 5441. Copyright 2008 American Chemical Society.

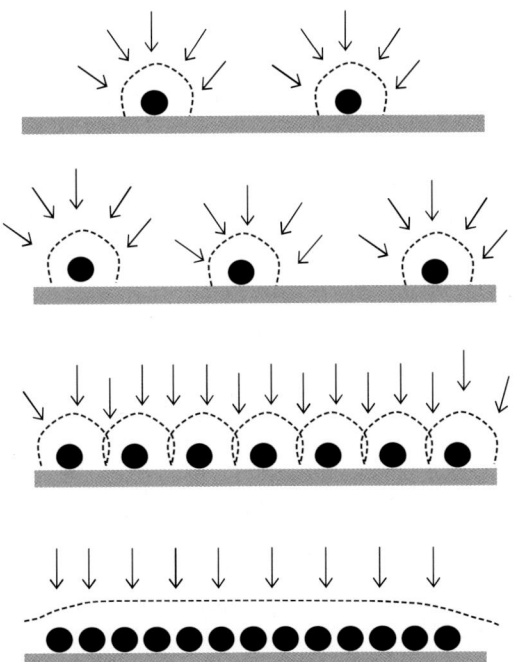

**Scheme 22.2** Sketch of the effective diffusion area at an array of small size electrodes assumed as spherical particles adsorbed on an insulating matrix with decreasing interparticle separation distance. Reprinted with permission from Karam, P., Halaoui, L. I. *Anal. Chem.* 2008, *80*, 5441. Copyright 2008 American Chemical Society.

NPs) are independent, at high coverage or short interparticle distances the individual (hemi-) spherical diffusion fields overlap and access to some sites becomes blocked to the analyte, resulting in a current not increasing proportionally with the microscopic area. TEM images showed a fractal surface coverage with some particles adsorbing at close distances. NPs adsorbing near others with increasing $t$ would not contribute to the *effective* catalytic and diffusion area in proportion to their active sites. Under kinetic and diffusion control it can, therefore, be understood that the highest rate per surface site would be reached at the lowest loading [24]. This behavior agreed with the reported dependence of UME arrays response on packing density [169]. The limiting intrinsic sensitivity of NP arrays of $\sim 1\,\text{AM}^{-1}\,\text{cm}^{-2}_{\text{real}}$ is one order of magnitude greater than the intrinsic sensitivity of a Pt disk electrode (2 mm diameter, $0.07\,\text{AM}^{-1}\,\text{cm}^{-1}_{\text{real}}$ over a range from LOD 0.08–3.8 mM [24]), and 30 times larger than the real sensitivity we calculated for a reported mesoporous Pt UME (with a surface area 100 times greater than the geometric area [190]). It is, therefore, a significant minimization of the use of Pt when an electrode is structured as an array of Pt NPs in polyelectrolyte at low coverage.

It is noteworthy, considering the deactivation of bulk Pt in $H_2O_2$ sensors with anodic polarization, that another advantage of reducing the size of the active element to the nanoscale is increased stability for continuous $H_2O_2$ detection [24]. Applying an anodic potential (0.6 V vs. Ag/AgCl) for 3 h caused no measurable loss of activity at the NP arrays [24]. This stability is believed to be rooted in a more difficult oxidation of Pt NPs, as detailed in Section 22.2.5.

### 22.2.4
**Biosensing at Assemblies of Glucose Oxidase Modified Pt NPs in Polyelectrolytes**

Electrochemical detection of $H_2O_2$ provides signal transduction for biomolecular recognition of molecules such as glucose [176–178, 182], cholesterol [180], and acetylcholine [180–183]. At glucose oxidase (GOx)-modified Pt electrodes the enzyme catalyzes the oxidation of β-D-glucose by oxygen, producing gluconic acid and $H_2O_2$ which is oxidized at Pt. We prepared a nanoscale biosensing element consisting of Pt NP stabilized with GOx, a homodimer glycoprotein of dimensions $7.0 \times 5.5 \times 8.0\,\text{nm}^3$ [192, 193], and assembled the NPs on PDDA [194]. In addition to gains from miniaturizing the Pt size, including enhanced S/N ratio and stability against oxidation of the catalyst surface (*vide infra*) [24], linking an enzyme to a NP may protect the enzyme from denaturation on flat surfaces [195, 196].

GOx-modified Pt NPs of size $4.05 \pm 0.62$ nm were prepared by reducing Pt(IV) with hydrogen in the presence of GOx at a GOx:Pt ratio of 1:880, corresponding to 2.6:1 GOx per Pt NP [194]. The enzyme surface groups, such as COOH of aspartic and glutamic acid, $NH_2$ of lysine and arginine [192], or other O–H or C=O groups, must have bound to the NP surface or precursor, terminating growth and stabilizing the NP [194]. The response of PDDA/(GOx-Pt NP)$_{t\,=\,60\,\text{min}}$ to $H_2O_2$ shows its catalytic oxidation at Pt NPs stabilized with GOx (Figure 22.7a) [194]. The film exhibited a sensitivity to glucose equal to $0.42\,\mu\text{A}\,\text{mM}^{-1}\,\text{cm}^{-2}$ over a linear range from 0.56 to 16 mM and a limit of detection about 30 μM (Figure 22.7b) [194], revealing retention

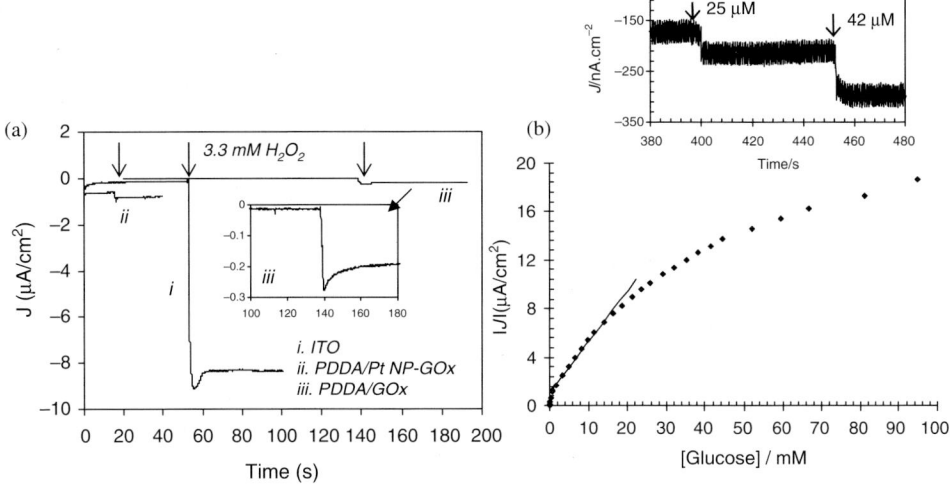

**Figure 22.7** (a) Current response to (3.3 mM) H$_2$O$_2$ in air-saturated 0.1 M PBS at PDDA/Pt NP-GOx on ITO (i), ITO (ii), and a PDDA/GOx film on ITO (iii) at 0.6 V vs. Ag/AgCl. (b) Steady-state current density as a function of glucose concentration at PDDA/GOx-Pt NP at 0.6 V vs. Ag/AgCl. Inset of (b) shows the current response to 25 and 42 µM glucose. Reprinted with permission from Karam, P., Xin, Y., Jaber, S., Halaoui, L. I. *J. Phys. Chem. C* 2008, *112*, 13846. Copyright 2008 American Chemical Society.

of catalytic activity at both NP and enzyme, and charge hopping to the NP core capped with GOx.

## 22.2.5
### Surface Oxidation and Stability of Pt NP Assemblies

A significant difference between the Pt NP assembled electrode and bulk Pt was a more difficult Pt oxidation at the former. OH adsorption on Pt is of interest because of the effect of OH$_{adlayer}$ on oxygen reduction and CO oxidation at Pt [104, 106, 116, 126, 127]. OH adsorption at polycryst-Pt took place with an onset at 480 mV and another at 700 mV, while OH adsorption at Pt NP assembly started at 800 mV and the reduction peak was negatively shifted by 100 mV, indicating a more irreversible process (Figure 22.8) [23]. Increased oxygenated species adsorption strength with decreasing Pt particle size, reflected by the negative potential shift in OH-desorption, agreed with reported studies [17, 127, 130, 196]. Mukerjee *et al.* attributed a stronger OH adsorption at Pt particles smaller than 5 nm to a greater fraction of low coordination sites [196], and Mayrhofer *et al.* attributed a negative shift in desorption potential from 30 to 1 nm of Pt-on-C to a negative shift in the pztc (but shifts in anodic potentials were not clearly apparent) [127]. On the other hand, a positive potential shift in Pt-oxide formation was also observed at unmodified Pt nanocrystals on HOPG [17] and at Pt NPs deposited in graphite [180]. You *et al.* ascribed the positive potential shift at 2.5 nm Pt NPs in graphite to a (0.9 eV) greater binding energy of Pt 4f electrons relative to the bulk [180], and a higher 4f electrons binding energy was also

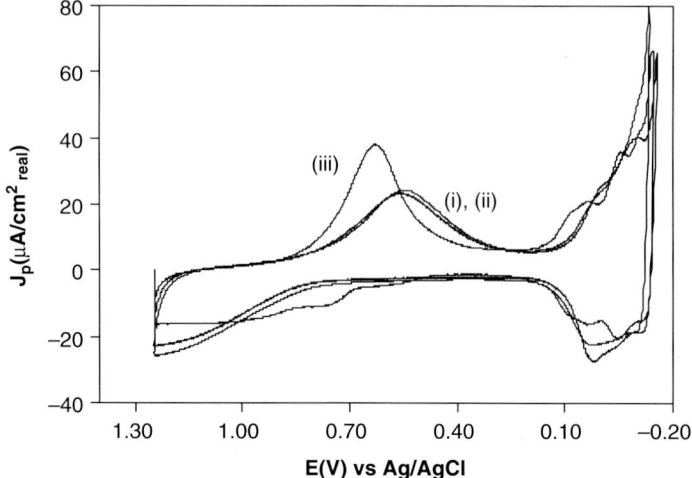

**Figure 22.8** CVs at (i) (PDDA/PAC-Pt NP)$_1$, (ii) (PDDA/PAC-Pt NP)$_2$ assembled on ITO and polycryst-Pt disk electrode (iii) in 1 M H$_2$SO$_4$ at 20 mV s$^{-1}$. Currents were normalized to the real Pt surface area of each electrode ($A_{real}$) computed from the charge of H$_{upd}$. Reprinted with permission from Estephan, Z., Alawiye, L., Halaoui, L. I. *J. Phys. Chem. C*, 2007, *111*, 8060. Copyright 2007 American Chemical Society.

measured at 1.7 nm tris(4-phosphonatophenyl phosphine)-stabilized Pt NPs [25]. In addition to electronic and structural factors at small Pt crystals, it may be possible that a surface ligand can also affect OH adsorption, as we observed in recent work a significantly more hindered oxide formation at larger polyvinylpyrrolidone-capped Pt NPs [197]. A more difficult oxide formation at Pt NPs could have implications in sensing and catalysis where OH is a site blocking species [105, 106, 116], or where a thick oxide layer deactivates the surface [24, 180].

### 22.2.6
### Oxygen Reduction at Pt NP Assemblies

The Pt crystallite-size effect on the oxygen reduction reaction has been a long studied problem [104, 105, 107, 116, 118–127]. Structural factors caused by size variations [107, 118–120], surface defects [119, 126], electronic factors [127], and surface distribution [117, 123, 124] have been proposed to contribute to the observed activity. A structural model was proposed by Kinoshita to explain the maximum mass activity at ~3.5–5.5 nm Pt and the lowering of activity per surface area with decreasing particle size [107], but still cannot reconcile all observations at Pt NPs with studies at single crystals [104–106, 116, 121, 122]. Watanabe *et al.* hypothesized a "territory model" that implicates interparticle distances in the observed maximum catalytic activity [117, 123, 124]. It seems, therefore, imperative that tuning particle size and shape should be decoupled from control of surface distribution in order to understand the determining factors.

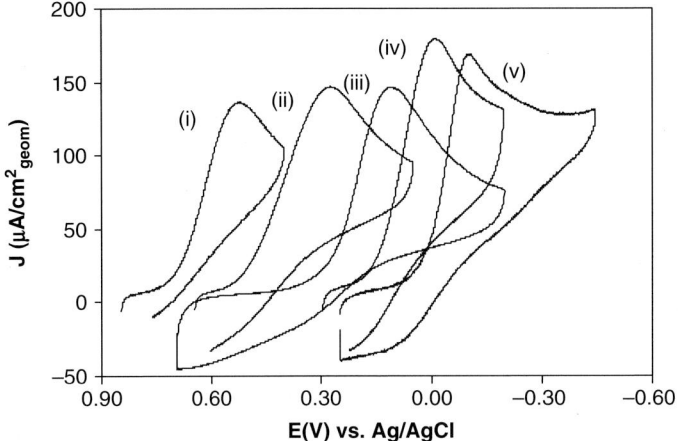

**Figure 22.9** CVs at (PDDA/PAC-Pt NP)$_4$ in air-saturated 0.1 M HClO$_4$ (i), 0.25 M phosphate buffer at pH 4 (ii), 7 (iii), 11 (iv), and 0.1 M KOH (v). Scan rate at 20 mV s$^{-1}$. Reprinted with permission from Estephan, Z., Alawiye, L., Halaoui, L. I. *J. Phys. Chem. C*, 2007, *111*, 8060. Copyright 2007 American Chemical Society.

We investigated oxygen reduction by CV at Pt NP/polyelectrolyte assemblies at varied coverage [23]. CVs at (PDDA/PAC-Pt NP)$_4$ in air-saturated electrolytes of different pH exhibited an irreversible peak corresponding to oxygen reduction catalyzed at Pt NP (Figure 22.9). The films were stable over a wide pH range, and the peak potential ($E_{peak}$) shifted by −50 mV/pH unit, in agreement with the expected potential shift with pH for this reaction [23]. Reduction of oxygen to water is a four-electron process proposed to take place at Pt either via a direct mechanism, or a series mechanism with a first two-electron transfer to form H$_2$O$_{2ads}$ [104, 116, 198]. Pt NP assemblies also catalyzed H$_2$O$_2$ reduction in a similar potential region as oxygen reduction [23]. PAC-Pt NPs assembled in polyelectrolyte thus retained the catalytic activity of the metal despite the presence of the surface ligand, in agreement with H$_{upd}$ and H$_2$O$_2$ sensing results.

We studied the effect of coverage and multilayer architecture by varying Pt NP coverage on ITO or GC at submonolayer or in multilayers [23]. Figure 22.10 shows CVs at PDDA/(PAC-Pt NP)$_{t=2\ or\ 10\ min}$, and at (PDDA/PAC-Pt NP)$_{n=2\ or\ 12}$ on ITO in deaerated and air-saturated 1 M H$_2$SO$_4$ (at 20 mV s$^{-1}$) [23]. The oxygen reduction peak potential ($E_p$) shifted positively with increasing NP coverage, reflecting the greater catalytic surface (from 104 mV vs. Ag/AgCl at PDDA/(PAC-Pt NP)$_{t=2\ min}$ to 475–492 mV at 4–12 layer pairs on ITO, or 590 mV at 12 layer pairs on GC) [23]. A significant kinetic limitation was not apparent at this timescale, as $E_p$ at 12 layer pairs on GC was similar to polycryst-Pt. However, we observed that the diffusion-limited peak current did not increase proportionally with real surface area, and Figure 22.10 presents the decrease in peak current per $A_{real}$ with increasing coverage. Even though the diffusion area increased with NP loading, the increase was not in proportion with the increase in microscopic Pt surface [23]. This behavior was again explained by picturing that as the diffusion limiting rate is reached the overall diffusion field is an

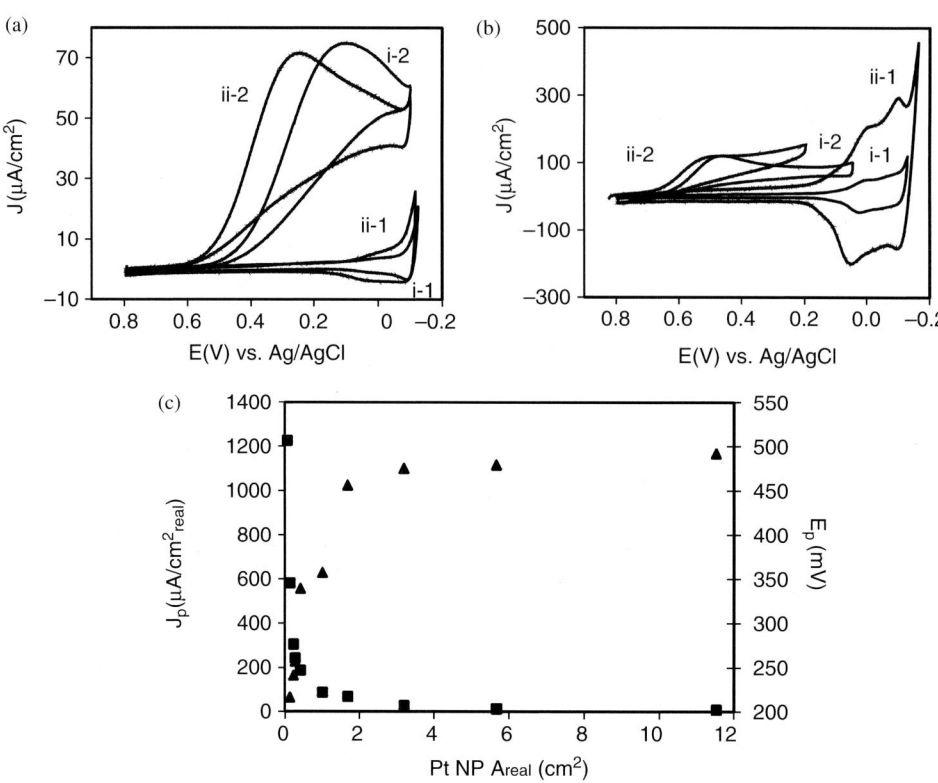

**Figure 22.10** CVs in deoxygenated (1) and in air-saturated (2) 1 M $H_2SO_4$ at the following assemblies on ITO electrodes: PDDA/(PAC-Pt NP)$_{t=2\,min}$ [i] and PDDA/(PAC-Pt NP)$_{t=10\,min}$ [ii] (a), and at layer pairs (PDDA/PAC-Pt NP)$_2$ [i] and (PDDA/PAC-Pt NP)$_{12}$ [ii] in (b). (c) Peak current density per real Pt surface area (■) and $E_{peak}$ (▲, vs. Ag/AgCl) versus real Pt surface area for oxygen reduction in air-saturated 1 M $H_2SO_4$ at assemblies on ITO. Data from CVs collected at 20 mV s$^{-1}$ scan rate. Reprinted with permission from Estephan, Z., Alawiye, L., Halaoui, L. I. *J. Phys. Chem. C*, 2007, *111*, 8060. Copyright 2007 American Chemical Society.

overlap or a sum of the individual diffusion fields of NPs or islands of NPs, thus NPs adsorbing very close to others with increasing $t$ may not contribute to the diffusion area, while those adsorbing further apart will. This leads to an increase in the *effective* diffusion area less than the increase in real area, with the largest rate per site at the lowest NP loading [23]. The trend is consistent with the effect of coverage on the amperometry response [24], and shows the effect of NP coverage on the electrochemical behavior of nanostructured electrodes [23, 24, 117, 123, 124, 155, 199].

## 22.3
### Nanostructured Photoelectrodes from Assemblies of Q-CdS in Polyelectrolytes

In QDs, spatial confinement results in a quantized size-dependent eigenspectrum [5, 6] and the extinction coefficient increases with decreasing dimension [200],

## 22.3 Nanostructured Photoelectrodes from Assemblies of Q-CdS in Polyelectrolytes

which allows fine tuning of band edge positions and enhances light harvesting. Furthermore, quantum size effects offer the potential of exceeding in QD-solar cells the thermodynamic limit [201] for light conversion at a bulk semiconductor (by improving photovoltage via hot-carrier collection or enhancing photocurrent via multiple exciton generation) [3, 202–204]. Several problems would need to be overcome before this potential is realized [55, 203, 205, 206], and in this the QD surface and assembly will play a vital role in determining the fate of photogenerated carriers [55, 205]. Understanding the photoelectrochemistry at assemblies of QDs is thus fundamental to developments in this area. In this section we review the assembly of photoelectrodes from Q-CdS in polyelectrolyte and their photoelectrochemistry in the presence of hole scavengers [55].

### 22.3.1
### Dip versus Dip–Spin Assembly of Q-CdS in Polyelectrolytes

PAC-capped Q-CdS particles of $3.6 \pm 0.5$ nm size [67] were assembled in PDDA via dip self-assembly (SA) or dip–spin SA [55]. To build a layer pair by dip SA, ITO electrodes were immersed in 10 mM PDDA/0.4 M NaCl (aq) for 30 min (followed by rinsing in water), dipped in PAC-Q-CdS solution for 1 h, rinsed in water and air-dried [55, 67]. In the dip–spin SA each dipping and rinsing step was followed by spinning the electrode for 1 min at 5000 rpm [67].

UV–visible absorbance revealed a 1.5-fold greater Q-CdS adsorbed amount per PDDA layer with spinning (Figure 22.11) [55]. TEM images of dip SA films showed

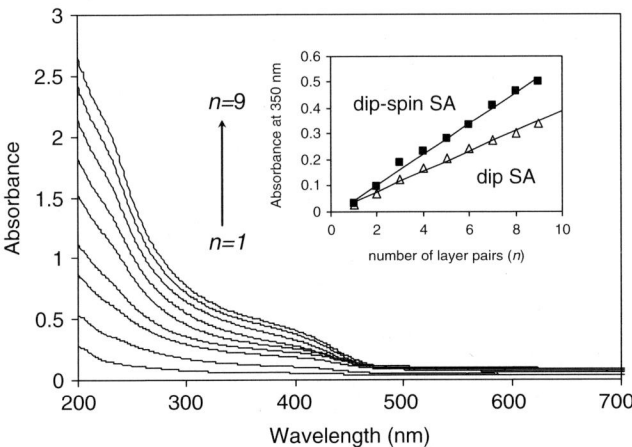

**Figure 22.11** UV–visible absorbance spectra of (PDDA/PAC-Q-CdS)$_n$ layer pairs by dip–spin SA. The inset shows plots of absorbance at 350 nm versus the number of layer pairs ($n$) for assembled films by dip SA and dip–spin SA. [El Harakeh, M., Alawieh, L., Saouma, S., Halaoui, L. I. *Phys. Chem. Chem. Phys.* 2009, 11, 5962] – Reproduced by permission of the PCCP Owner Societies.

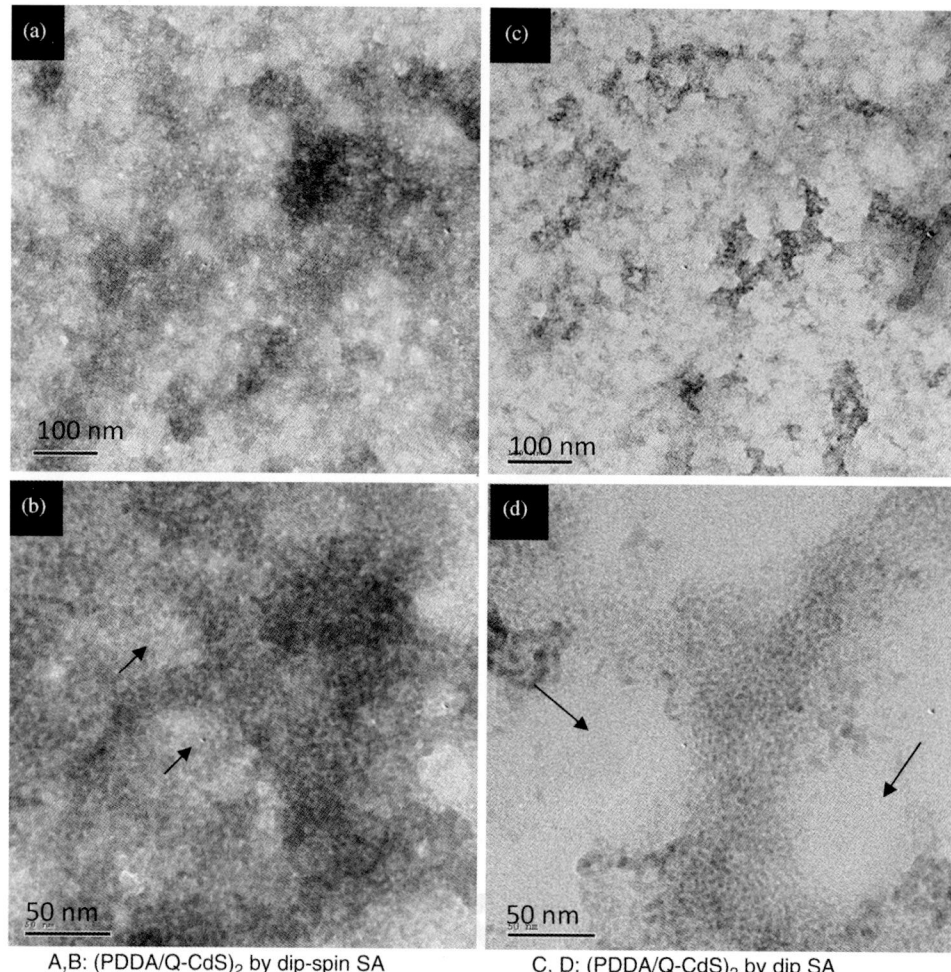

A,B: (PDDA/Q-CdS)$_2$ by dip-spin SA    C, D: (PDDA/Q-CdS)$_2$ by dip SA

**Figure 22.12** TEM images of (PDDA/PAC-Q-CdS)$_2$ films assembled by dip–spin SA (a,b), or dip SA (c,d). [El Harakeh, M., Alawieh, L., Saouma, S., Halaoui, L. I. *Phys. Chem. Chem. Phys.* 2009, *11*, 5962] – Reproduced by permission of the PCCP Owner Societies.

a fractal surface distribution similar to Pt NP layers, while a greater and more uniform coverage was observed in dip–spin SA films (Figure 22.12) and was confirmed in AFM images [55]. Islands that appeared naked of NPs at the former films were evidently covered by Q-CdS upon spinning (arrows in TEM images) [55]. Cho et al. reported deposition of Q-CdS/poly(diallylamine hydrochloride) films by casting followed by spinning (spin SA) and observed greater QD adsorption and an ordered internal structure upon spinning, ascribed to faster water removal and greater material concentration, and to the air sheer force flattening films [207]. We proposed that the kinetics of NP adsorption on PDDA becomes limited at longer times by slow surface rearrangement to open sites [20]. Substrate spinning and

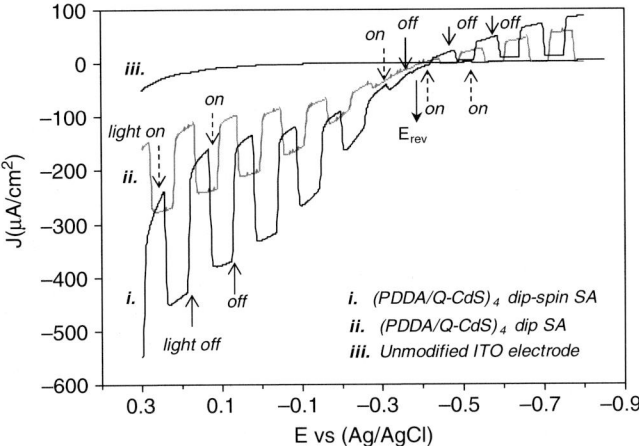

**Figure 22.13** Current–potential plots acquired under chopped illumination at (PDDA/PAC-Q-CdS)$_4$ films assembled on ITO by dip–spin SA (i) or dip SA (ii) and at an ITO electrode (iii) in deoxygenated 0.1 M Na$_2$S/0.2 M NaOH at 20 mV s$^{-1}$. [El Harakeh, M., Alawieh, L., Saouma, S., Halaoui, L. I. *Phys. Chem. Chem. Phys.* 2009, *11*, 5962] – Reproduced by permission of the PCCP Owner Societies.

fast drying could be facilitating NP/polyelectrolyte reorganization opening adsorption sites and smoothing films, leading to a greater adsorbed amount and more uniform coverage [55].

## 22.3.2
### Photoelectrochemistry at PDDA/Q-CdS Assembly in the Presence of Hole Scavengers

Figure 22.13 shows *I–V* curves under chopped white-light illumination at (PDDA/PAC-Q-CdS)$_4$ assembled via dip or dip–spin SA in alkaline sulfide electrolyte [55]. Sulfide is a hole scavenger that enhances charge separation and protects cadmium chalcogenides against photooxidation by scavenging photogenerated holes that otherwise attack the lattice [208, 209]. The curves exhibited a switch in photocurrent polarity from anodic to cathodic at a switching potential of −350 to −400 mV vs. Ag/AgCl, in a photoelectrochemical behavior different from n-type bulk CdS where anodic photocurrents flow [55]. In the presence of ascorbic acid, the photoelectrochemistry at QD films was again notably different from the bulk. While anodic photocurrents were recorded at the CdS thin film in the presence of this hole scavenger, predominantly photocathodic currents were measured at Q-CdS films (Figure 22.14) [55].

The photocurrent magnitude in sulfide increased with increasing *n* up to 6 layer pairs [55]. The monochromatic incident photon-to-current conversion efficiency (%IPCE at 100 mV vs. Ag/AgCl) was 2.8 to 6.5-fold (at 360–440 nm) larger at 4- compared to 2-layer pairs, greater than the difference in absorbance, possibly because of enhanced charge hopping due to denser NP coverage. The efficiency of

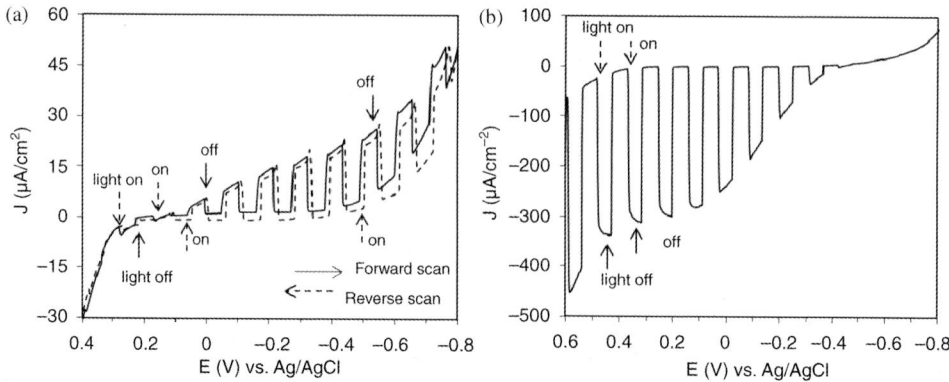

**Figure 22.14** Current–potential plots acquired under chopped illumination at (PDDA/PAC-Q-CdS)$_4$ film assembled on ITO by dip–spin SA (a) and at an electrodeposited thin CdS film on ITO (b) in deoxygenated 0.1 M ascorbic acid at 20 mV s$^{-1}$. The arrows indicating light on-off are for the forward I-V scan. [El Harakeh, M., Alawieh, L., Saouma, S., Halaoui, L. I. *Phys. Chem. Chem. Phys.* 2009, *11*, 5962] – Reproduced by permission of the PCCP Owner Societies.

photocurrent generation at Q-CdS multilayers depends on the extent of photon absorbance and the yields of hole scavenging by the redox species, and of charge transport and injection to the electrode before electron–hole recombination. In films thicker than 6 layer pairs recombination losses limited the photocurrents. The same trend was observed with white light illumination. Assembling NPs with thinner intercalating polyelectrolyte layer by decreasing ionic strength could provide a way to enhance charge hopping in multilayered films, but this needs study.

Photocurrents were at some potentials more than twofold greater at dip–spin SA films than at the same layer pairs of dip SA films, reflecting an effect of structure on photocurrent generation beyond the increase in light absorbance [55]. Charge transport could have been facilitated by a more uniform NP coverage and different NP/polyelectrolyte structure upon spinning that may have provided closer hopping distances between NPs in underlying layers, thus enhancing charge separation [55].

### 22.3.3
### Photocurrent Polarity-Switching at Q-CdS Photoelectrodes

The phenomenon of photoelectrochemical polarity switching has gained attention as a basis for nanoscale switches and logic devices, and the interested reader is directed to a review of the literature on the subject in ref. 210. In bulk semiconductors, the field created in the space-charge region causes separation of photogenerated carriers. At potentials positive of the flat band potential of an n-type semiconductor (upward band bending) the majority of carriers are depleted (depletion layer) and photogenerated holes flow to the surface and can transfer to a reduced species, resulting in photoanodic currents [211]. Photocathodic currents at p-type semiconductors flow by a similar mechanism. Anomalous photocurrent polarity behavior at semiconductors (e.g., $TiO_2$, Se, CdS, CdTe, and CdSe) [212–220] including, for example, switching from n-type to p-type behavior at CdSe nanocrystalline films in polysulfide following an acid etch [220], has been attributed to energy levels in the gap, such as surface states. On the other hand, both photogenerated carriers are available at a QD surface to undergo charge transfer, and the photocurrent polarity at QD assemblies would be determined by the redox species and applied potential, and, in general, photoanodic currents were reported in the presence of a hole scavenger, or photocathodic in the presence of an electron acceptor, and a potential-dependent polarity was observed with both redox species [41–52, 221, 222]. Examples are photocurrent polarity switching at Q-PbS monolayer in $S^{2-}/S_x^{2-}$, photoanodic current with a hole scavenger ($MV^{+\cdot}$), or photocathodic with an electron acceptor (benzoquinone) [44], and photoanodic or photocathodic at Q-CdS/DNA assembly with an electron donor (triethanolamine) or electron acceptor (oxygen), respectively [48]. At PDDA/Q-CdS assembly, we observed earlier photocurrent polarity switching in oxygenated and deoxygenated phosphate buffer but with significantly smaller currents than in sulfide [54], and a similar observation was reported at L-lysine/CdSe or L-lysine/CdSe@CdS multilayers in 1,2-dichloroethane [58].

The various processes leading to photoanodic and photocathodic currents in sulfide are presented in Scheme 22.3 from ref. 55. A sulfur species was believed to be directly involved in the photocurrent polarity switching, since only photoanodic currents were measured in its absence (in 0.1 M NaOH). Photoanodic currents result

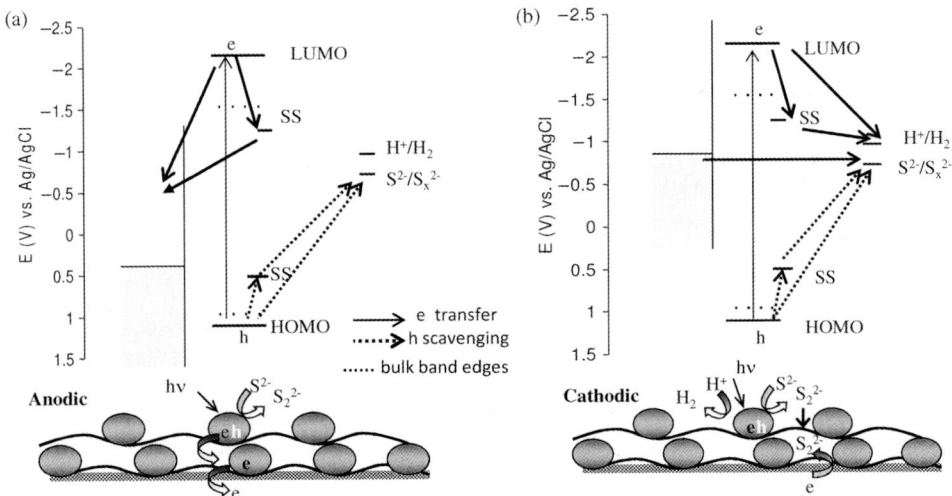

**Scheme 22.3** Sketches of the predominant charge transfer processes at (PDDA/PAC-Q-CdS)$_n$ film assembled on ITO in 0.1 M Na$_2$S/0.2 M NaOH electrolyte leading to photoanodic (a) or photocathodic (b) current. The HOMO and LUMO of a 3.6 nm Q-CdS particle are approximated using the effective mass model [6], the CdS bulk band edges are also shown. [El Harakeh, M., Alawieh, L., Saouma, S., Halaoui, L. I. Phys. Chem. Chem. Phys. 2009, 11, 5962] – Reproduced by permission of the PCCP Owner Societies.

from scavenging of photogenerated holes by $S^{2-}$ and collection of photogenerated electrons at the substrate electrode at potentials positive of the switching potential [55]. Photocathodic currents at more negative potentials were attributed to reduction at the underlying electrode of a photo-oxidized sulfur species ($S_2^{2-}$) trapped in the film upon scavenging of the photogenerated hole by $S^{2-}$. The photogenerated electron was proposed to either become trapped at surface states or to photocatalyze hydrogen evolution. Ascorbic acid was also involved in the photocathodic current generation since only photoanodic currents were recorded in its absence at the same pH [55]. The photocathodic current was ascribed by the same mechanism to the electrode acting as a collector of a photo-oxidized species trapped in the film after ascorbic acid scavenges the photogenerated hole [55, 223–225].

## 22.4
## Conclusions

Chemistry at the size scale between the molecular and the bulk is an attractive area of science fundamentally, and for its potential technological impact. The transition between colloidal preparation and device fabrication, for catalytic, photocatalytic, sensing, plasmonic, and photovoltaic applications, necessitates assembly that permits manipulation of architecture and an understanding of ensemble nanoparticle behavior. Assembly of metal and semiconductor NPs in polyelectrolyte resulted in photoelectrochemically or electrochemically active films, surface-modification and polyelectrolyte matrix notwithstanding. Electrocatalytic surfaces, sensing arrays, and photoelectrodes have thus been fabricated. The electrochemical behavior of Pt NP/polyelectrolyte assemblies showed retention of electrocatalytic activity for reactions catalyzed at Pt, such as oxygen reduction, hydrogen evolution and oxidation, and hydrogen peroxide oxidation, and a dependence on surface coverage in what can lead to optimizing the use of rare metals in sensing and catalysis. This would allow, in future work, fabrication of electrodes from NPs of varied sizes and shapes and surface coverage – albeit in this work randomly – for fundamental studies of reactivity at the nanoscale. Photoelectrochemistry at QD/polyelectrolyte films revealed appreciable charge separation, leading to potential-dependent photocurrent-polarity switching, an interesting phenomenon at the nanosize semiconductor. The photoelectrochemical behavior depended on the assembly method, layer pair number, redox species, and electrode potential. Future studies in this area will be directed toward assembling heteronanostructures with different NPs or other functionalities in polyelectrolytes, such as QD with metal NP or metal NP with protein, for cascades of catalytic reactions, biomolecular recognition, or possible light field enhancement, and control of film architecture such as interparticle separation, NP order, and interlayer spacing, and studying effects on light harvesting, photogenerated charge separation and charge transfer and transport in these nanostructured films.

## Acknowledgments

I thank the students (in alphabetical order) who have contributed to this work over the years: Leen Alawiyeh, Maysaa El Harakeh, Zaki Estephan, Mohammad Houry, Pierre Karam, Marie Zabel Markarian, Sarah Jaber, Samer Saouma, and I thank Dr. Yan Xin at the NHMFL at Florida State University Tallahasee for her work in TEM imaging, and thank NSF (DMR-9625692) and NHMFL under cooperative agreement DMR-0084173. Acknowledgment is made to the donors of the American Chemical Society Petroleum Research Fund (PRF #40548-B10) and to the American University of Beirut University Research Board (URB) for financial support of this work, and to the Lebanese National Council for Scientific Research (LNCSR, grant 2004–05) for partial financial support of this research.

## References

1 Lewis, N.S., Crabtree, G.W., Nozik, A.J., Wasielewski, M.R., and Alivisatos, A.P. (2005) Basic Energy Science Report on Basic Research Needs for Solar Energy Utilization, Office of Science, US Department of Energy, Washington DC, April 18–21.

2 Kamat, P.V. (2007) Meeting the clean energy demand: Nanostructure architectures for solar energy conversion. *J. Phys. Chem. C*, **111**, 2834–2860.

3 Nozik, A.J. (2002) Quantum dot solar cells. *Phys. E.*, **14**, 115–120.

4 Burda, C., Chen, X., Narayanan, R., and El-Sayed, M.A. (2005) The chemistry and properties of nanocrystals of different shapes. *Chem. Rev.*, **105** (4), 1025–1102.

5 Brus, L.E. (1984) Electron–electron and electron–hole interactions in small semiconductor crystallites: The size dependence of the lowest excited electronic state. *J. Chem. Phys.*, **80**, 4403–4409.

6 Brus, L. (1986) Electronic wave functions in semiconductor clusters: Experiment and theory. *J. Phys. Chem.*, **90**, 2555–2560.

7 Ye, H. and Crooks, R.M. (2005) Electrocatalytic $O_2$ reduction at glassy carbon electrodes modified with dendrimer-encapsulated Pt nanoparticles. *J. Am. Chem. Soc.*, **127**, 4930–4934.

8 Roth, C., Hussain, I., Bayati, M., Nichols, R.J., and Schiffrin, D.J. (2004) Fullerene-linked Pt nanoparticle assemblies. *Chem. Comm.*, 1532–1533.

9 Friedrich, K.A., Henglein, F., Stimming, U., and Unkauf, W. (2000) Size dependence of the CO monolayer oxidation on nanosized Pt particles supported on gold. *Electrochim. Acta*, **45**, 3283–3293.

10 Teranishi, T., Hosoe, M., Tanaka, T., and Miyake, M. (1999) Size control of monodispersed Pt nanoparticles and Their 2D organization by electrophoretic deposition. *J. Phys. Chem. B*, **103**, 3818–3827.

11 Takasu, Y., Ohashi, N., Zhang, X.-G., Murakami, Y., Minigawa, H., Sato, S., and Yahikozawa, K. (1996) Size effects of platinum particles on the electroreduction of oxygen. *Electrochim. Acta*, **41**, 2595–2600.

12 Yahikozawa, K., Fujii, Y., Matsuda, Y., Nishimura, K., and Takasu, Y. (1991) Electrocatalytic properties of ultrafine platinum particles for oxidation of methanol and formic acid in aqueous solutions. *Electrochim. Acta*, **36**, 973–978.

13 Cherstiouk, O.V., Simonov, P.A., and Savinova., E.R. (2003) Model approach to evaluate particle size effects in electrocatalysis: preparation and properties of Pt nanoparticles supported

on GC and HOPG. *Electrochim. Acta*, **48**, 3851–3860.

14 Kokkinidis, G., Papoutsis, A., Stoychev, D., and Milchev, A.J. (2000) Electroless deposition of Pt on Ti — catalytic activity for the hydrogen evolution reaction. *Electroanal. Chem.*, **486**, 48–55.

15 Park, S., Yang, P., Corredor, P., and Weaver, M.J. (2002) Transition metal-coated nanoparticle Films: Vibrational characterization with surface-enhanced raman scattering. *J. Am. Chem. Soc.*, **124**, 2428–2429.

16 Cui, H.-F., Ye, J.-S., Zhang, W.-D., Wang, J., and Sheu, F.-S. (2005) Electrocatalytic reduction of oxygen by a platinum nanoparticle/carbon nanotube composite electrode. *J. Electroanal. Chem.*, **577**, 295–302.

17 (29) Zoval, J.V., Lee, J., Gorer, S., and Penner, R.M. (1998) Electrochemical preparation of platinum nanocrystallites with size selectivity on basal plane oriented graphite surfaces. *J. Phys. Chem. B*, **102**, 1166–1175.

18 Antoine, O. and Durand, R. (2001) In situ electrochemical deposition of Pt nanoparticles on carbon and inside nafion. *Electrochem. Solid State Lett.*, **4**, A55–A58.

19 Chang, G., Oyama, M., and Hirao, K. (2006) In situ chemical reductive growth of platinum nanoparticles on indium tin oxide surfaces and their electrochemical applications. *J. Phys. Chem. B*, **110**, 1860–1865.

20 Ghannoum, S., Xin, Y., Jaber, J., and Halaoui, L.I. (2003) Self-assembly of polyacrylate-capped platinum nanoparticles on a polyelectrolyte surface: Kinetics of adsorption and effect of ionic strength and deposition protocol. *Langmuir*, **19**, 4804–4811.

21 Markarian, M.Z., El Harakeh, M., and Halaoui, L.I. (2005) Adsorption of atomic hydrogen at a nanostructured electrode of polyacrylate-capped Pt nanoparticles in polyelectrolyte. *J. Phys. Chem. B*, **109**, 11616–11621.

22 Karam, P., Estephan, Z.G., El Harakeh, M., Houry, M., and Halaoui, L.I. (2006) Electrocatalytic activity at surface-modified Pt nanoparticles assembled in polyelectrolyte. *Electrochem. Solid-State Lett.*, **9** (3), A144–A146.

23 Estephan, Z., Alawiye, L., and Halaoui, L.I. (2007) Oxygen reduction at nanostructured electrodes assembled from polyacrylate-capped Pt nanoparticles in polyelectrolytes. *J. Phys. Chem. C*, **111**, 8060–8068.

24 Karam, P. and Halaoui, L.I. (2008) Sensing of $H_2O_2$ at low surface density assemblies of Pt nanoparticles in polyelectrolyte. *Anal. Chem.*, **80**, 5441–5448.

25 Kostelansky, C.N., Pietron, J.J., Chen, M.-S., Dressick, W.J., Swider-Lyons, K.E., Ramaker, D.E., Stroud, R.M., Klug, C.A., Zelakiewicz, B.S., and Schull, T.L. (2006) Triarylphosphine-stabilized platinum nanoparticles in three hyphen;dimensional nanostructured films as active electrocatalysts. *J. Phys. Chem. B*, **110**, 21487–21496.

26 Huang, M., Shao, Y., Sun, X., Chen, H., Liu, B., and Dong, S. (2005) Alternate assemblies of platinum nanoparticles and metalloporphyrins as tunable electrocatalysts for dioxygen reduction. *Langmuir*, **21**, 323–329.

27 Shen, Y., Liu, J., Wu, A., Jiang, J., Bi, L., Liu, B., Li, Z., and Dong, S. (2003) Preparation of multilayer films containing Pt nanoparticles on a glassy carbon electrode and application as an electrocatalyst for dioxygen reduction. *Langmuir*, **19**, 5397–5401.

28 Cavaliere-Jaricot, S., Etcheberry, A., Herlem, M., Nöel, V., and Perez, H. (2007) Electrochemistry at capped platinum nanoparticle Langmuir Blodgett films: A study of the influence of platinum amount and of number of LB layers. *Electrochim. Acta*, **52**, 2285–2293.

29 Vogel, R., Pohl, K., and Weller, H. (1990) Sensitization of highly porous, polycrystalline TiO2 electrodes by quantum sized CdS. *Chem. Phys. Lett.*, **174**, 241–246.

30 Vogel, R., Hoyer, P., and Weller, H. (1994) Quantum-sized PbS, CdS, $Ag_2S$, $Sb_2S_3$, and $Bi_2S_3$ particles as sensitizers for various nanoporous wide-bandgap semiconductors. *J. Phys. Chem.*, **98**, 3183–3188.

31 Nasr, C., Kamat, P.V., and Hotchandani, S. (1997) Photoelectrochemical behavior of coupled SnO$_2$CdSe nanocrystalline semiconductor films. *J. Electroanal. Chem.*, **420**, 201–207.

32 Zaban, A., Mićić, O.I., Gregg, B.A., and Nozik, A.J. (1998) Photosensitization of nanoporous TiO$_2$ electrodes with InP quantum dots. *Langmuir*, **14**, 3153–3156.

33 Peter, L.M., Wijayantha, K.G.U., Riley, D.J., and Waggett, J.P. (2003) Band-edge tuning in self-assembled layers of Bi$_2$S$_3$ nanoparticles used to photosensitize nanocrystalline TiO$_2$. *J. Phys. Chem. B*, **107**, 8378–8381.

34 Robel, I., Subramanian, V., Kuno, M., and Kamat, P.V. (2006) Quantum dot solar cells. harvesting light energy with CdSe nanocrystals molecularly linked to mesoscopic TiO$_2$ films. *J. Am. Chem. Soc.*, **128**, 2385–2393.

35 Lee, H., Wang, M., Chen, P., Gamelin, D.R., Zakeeruddin, S.M., Grätzel, M., and Nazeeruddin, Md.K. (2009) Efficient CdSe Quantum Dot-Sensitized Solar Cells Prepared by an Improved Successive Ionic Layer Adsorption and Reaction Process. *Nano Lett.*, **9**, 4221–4227.

36 Huynh, W.U., Peng, X.G., and Alivisatos, A.P. (1999) CdSe nanocrystal rods/poly(3-hexylthiophene) composite photovoltaic devices. *Adv. Mater.*, **11**, 923–927.

37 Huynh, W.U., Dittmer, J.J., and Alivisatos, A.P. (2002) Hybrid nanorod-polymer solar cells. *Science*, **295**, 2425–2427.

38 Greenham, N.C., Peng, X., and Alivisatos, A.P. (1996) Charge separation and transport in conjugated-polymer/semiconductor-nanocrystal composites studied by photoluminescence quenching and photoconductivity. *Phys. Rev. B: Condens. Matter Mater. Phys.*, **54**, 17628–17637.

39 Luther, J.M., Law, M., Beard, M.C., Song, Q., Reese, M.O., Ellignson, R.J., and Nozik, A.J. (2008) Schottky solar cells based on colloidal nanocrystal films. *Nano Lett*, **8**, 3488–3492.

40 Johnston, K.W., Parrantyus-Abraham, A.G., Clifford, J.P., Myrskog, S.H., MacNeil, D.D., Levina, L., and Sargent, E.H. (2008) Schottky-quantum dot photovoltaics for efficient infrared power conversion. *Appl. Phys. Lett.*, **92**, 151115-1–3.

41 Mansur, H.S., Grieser, F., Marychurch, M.S., Biggs, S., Urquhart, R.S., and Furlong, D.N. (1995) Photoelectrochemical properties of "Q-state" CdS particles in arachidic acid Langmuir– Blodgett films. *J. Chem. Soc., Faraday Trans.*, **91**, 665–672.

42 Mansur, H.S., Grieser, F., Urquhart, R.S., and Furlong, D.N. (1995) Photoelectrochemical behavior of Q-state CdS$_x$Se$_{(1-x)}$ particles in arachidic acid Langmuir–Blodgett films. *J. Chem. Soc., Faraday Trans.*, **91**, 3399–3404.

43 Ogawa, S., Fan, F.F., and Bard, A.J. (1995) Scanning tunneling microscopy, tunneling spectroscopy, and photoelectrochemistry of a film of Q-CdS particles incorporated in a self-assembled monolayer on a gold surface. *J. Phys. Chem.*, **99**, 11182–11189.

44 Ogawa, S., Fan, F.F., and Bard, A.J. (1997) Photoelectrochemistry of films of quantum size lead sulfide particles incorporated in selfassembled monolayers on gold. *J. Phys. Chem. B*, **101**, 5707–5711.

45 Nakanishi, T., Ohtani, B., and Uosaki, K. (1998) Fabrication and characterization of CdS-nanoparticle mono- and multilayers on a selfassembled monolayer of alkanedithiols on gold. *J. Phys. Chem. B*, **102**, 1571–1577.

46 Bakkers, E.P.A.M., Reitsma, E., and Kelly, J.J. (1999) Vanmaeckelbergh, D. Excited-state dynamics in CdS quantum dots adsorbed on a metal electrode. *J. Phys. Chem. B*, **103**, 2781–2788.

47 Bakkers, E.P.A.M., Kelly, J.J., and Vanmaeckelbergh, D. (2000) Time resolved photoelectrochemistry with size-quantized PbS adsorbed on gold. *J. Electroanal. Chem.*, **482**, 48–55.

48 Gill, R., Patolsky, F., Katz, E., and Willner, I. (2005) Electrochemical control of the photocurrent direction in intercalated DNA/CdS nanoparticle systems. *Angew. Chem., Int. Ed.*, **44**, 4554–4557.

49 Sheeney-Haj-Ichia, L. and Willner, I. (2002) Enhanced photoelectrochemistry in supramolecular CdS-nanoparticle-

stoppered pseudorotaxane monolayers assembled on electrodes. *J. Phys. Chem. B*, **106**, 13094–13097.

50. Sheeney-Haj-Ichia, L., Basnar, B., and Willner, I. (2005) Efficient generation of photocurrents by using CdS/carbon nanotube assemblies on electrodes. *Angew. Chem., Int. Ed.*, **44**, 78–83.

51. Hickey, S.G. and Riley, D.J. (1999) Photoelectrochemical studies of CdS nanoparticle-modified electrodes. *J. Phys. Chem. B*, **103**, 4599–4602.

52. Hickey, S.G., Riley, D.J., and Tull, E.J. (2000) Photoelectrochemical studies of CdS nanoparticle modified electrodes: Absorption and photocurrent investigations. *J. Phys. Chem. B*, **104**, 7623–7626.

53. Kotov, N.A., Dekany, I., and Fendler, J.H. (1995) Layer-by-layer self assembly of polyelectrolyte-semiconductor nanoparticle composite films. *J. Phys. Chem.*, **99**, 13065–13069.

54. Halaoui, L.I. (2003) Photoelectrochemistry in aqueous media at polyacrylate-capped Q-CdS assembled in polyelectrolyte matrix on electrode surfaces. *J. Electrochem. Soc.*, **150**, E455–E460.

55. El Harakeh, M., Alawieh, L., Saouma, S., and Halaoui, L.I. (2009) Charge separation and photocurrent polarity-switching at CdS quantum dots assembly in polyelectrolyte interfaced with hole scavengers. *Phys. Chem. Chem. Phys.*, **11**, 5962–5973.

56. Hojeiji, M., Eugstar, N., Su, B., and Girault, H.H. (2006) CdSe sensitized thin aqueous films: Probing the potential distribution inside multilayer assemblies. *Langmuir*, **22**, 10652–10658.

57. Hojeiji, M., Su, B., Tan, S., Meriguet, G., and Girault, H.H. (2008) Nanoporous photocathode and photoanode made by multilayer assembly of quantum dots. *ACS Nano*, **2**, 984–992.

58. Iler, R.K. (1966) Multilayers of colloidal particles. *J. Colloid Interface Sci.*, **21**, 569–594S.

59. Decher, G., Hong, J.D., and Schmitt, J. (1992) Buildup of ultrathin multilayer films by a self-assembly process: III. Consecutively alternating adsorption of anionic and cationic polyelectrolytes on charged surfaces. *Thin Solid Films*, **210/211**, 831–835.

60. Lvov, Y., Decher, G., and Moehwald, H. (1993) Assembly, structural characterization, and thermal behavior of layer-by-layer deposited ultrathin films of poly(vinyl sulfate) and poly (allylamine). *Langmuir*, **9**, 481–486.

61. Decher, G. (1996) Layered nanoarchitectures via directed assembly of anionic and cationic molecules, in *Comprehensive Supramolecular Chemistry, Vol. 9, Templating, Self-Assembly and Self-Organization* (eds J.P. Sauvage and M.W. Hosseini), Pergamon Press, New York, pp. 507–528.

62. Decher., G. (2002) Polyelectrolyte multilayers, an overview, in *Multilayer Thin Films* (eds G. Decher and J.B. Schlenoff), Wiley-VCH Verlag GmbH & Co. KGaA, pp. 1–46.

63. Decher, G. (1997) Fuzzy nanoassemblies: Toward layered polymeric multicomposites. *Science*, **277**, 1232–1237.

64. Kotov, N.A., Dekany, I., and Fendler, J.H. (1995) Layer-by-layer self-assembly of polyelectrolyte-semiconductor nanoparticle composite films. *J. Phys. Chem.*, **99**, 13065–13069.

65. Lvov, Y., Ariga, K., Onda, M., Ichinose, I., and Kunitake, T. (1997) Alternate assembly of ordered multilayers of $SiO_2$ and other nanoparticles and polyions. *Langmuir*, **13**, 6195–6203.

66. Liu, Y., Wang, A., and Claus, R. (1997) Molecular self-assembly of $TiO_2$/polymer nanocomposite films. *J. Phys. Chem. B*, **101**, 1385–1388.

67. Halaoui, L.I. (2001) Layer-by-layer assembly of polyacrylate-capped CdS nanoparticles in poly (diallyldimethylammonium chloride) on solid surface. *Langmuir*, **17**, 7130–7136.

68. Ostrander, J.W., Mamedov, A.A., and Kotov, N.A. (2001) Two modes of linear layer-by-layer growth of nanoparticle-polyelectrolyte multilayers and different interactions in the layer-by-layer deposition. *J. Am. Chem. Soc.*, **123**, 1101–1110.

69 Feldheim, D.L., Grabar, K.C., Natan, M.J., and Mallouk, T.C. (1996) Electron transfer in self-assembled inorganic polyelectrolyte/metal nanoparticle heterostructures. *J. Am. Chem. Soc.*, **118**, 7640–7641.

70 Cassagneau, T. and Fendler, J.H. (1999) Preparation and layer-by-layer self-assembly of silver nanoparticles capped by graphite oxide nanosheets. *J. Phys. Chem.*, **103**, 1789–1793.

71 Hicks, J.F., Seok-Shon, Y., and Murray, R.W. (2002) Layer-by-layer growth of polymer/nanoparticle films containing monolayer-protected gold clusters. *Langmuir*, **18**, 2288–2294.

72 Malikova, N., Pastoriza-Santos, I., Schierhorn, M., Kotov, N.A., and Liz-Marzán, L.M. (2002) Layer-by-layer assembled mixed spherical and planar gold nanoparticles: Control of interparticle interactions. *Langmuir*, **18**, 3694–3697.

73 Caruso, F. and Möhwald, H. (1999) Preparation and characterization of ordered nanoparticle and polymer composite multilayers on colloids. *Langmuir*, **15**, 8276–8281.

74 Sennerfors, T., Bogdanovic, G., and Tiberg, F. (2002) Formation, chemical composition, and structure of polyelectrolyte-nanoparticle multilayer films. *Langmuir*, **18**, 6410–6415.

75 Kotov, N.A. (2002) Layer-by-layer assembly of nanoparticles and nanocolloids: Intermolecular interactions, structure and materials perspectives, in *Multilayer Thin Films*, (ed G. Decher and J.B. Schlenoff), Wiley-VCH Verlag GmbH & Co. KGaA, pp. 207–243.

76 Srivastava, S. and Kotov, N. (2008) Composite layer-by-layer (LBL) assembly with inorganic nanoparticles and nanowires. *Acc. Chem. Res.*, **41** (12), 1831–1841.

77 Lesser, C., Gao, M., and Kirstein, S. (1999) Highly luminescent thin films from alternating deposition of CdTe nanoparticles and polycations. *Mater. Sci. Eng. C*, **8–9**, 159–162.

78 Hao, E., Yang, B., Ren, H., Qian, X., Xie, T., Shen, J., and Li, D. (1999) Fabrication of composite film comprising $TiO_2$/CdS and polyelectrolytes based on ionic attraction. *Mater. Sci. Eng., C*, **10**, 119–122.

79 Hao, E., Qian, X., Yang, B., Wang, D., and Shen, J. (1999) Assembly and photoelectrochemical studies of $TiO_2$/CdS nanocomposite film. *Mol. Cryst. Liq. Cryst. Sci. Technol., Sect. A*, **337**, 181–184.

80 Gao, M.Y., Richter, B., Kirstein, S., and Möhwald, H. (1998) Electroluminescence studies on self-assembled films of PPV and CdSe nanoparticles. *J. Phys. Chem.B*, **102**, 4096–4103.

81 Cassagneau, T., Mallouk, T.E., and Fendler, J.H. (1998) Layer-by-layer assembly of thin film zener diodes from conducting polymers and CdSe nanoparticles. *J. Am. Chem. Soc.*, **120**, 7848–7859.

82 Hao, E. and Lian, T. (2000) Layer-by-layer assembly of CdSe nanoparticles based on hydrogen bonding. *Langmuir*, **16**, 7879–7881.

83 Gao, M., Lesser, C., Kirstein, S., Möhwald, H., Rogach, A.L., and Weller, H. (2000) Electroluminescence of different colors from polycation/CdTe nanocrystal self-assembled films. *J. Appl. Phys.*, **87**, 2297–2302.

84 Sun, J., Hao, E., Sun, Y., Zhang, H., Yang, B., Zou, S., Shen, J., and Wang, S. (1998) Multilayer assemblies of colloidal ZnS doped with silver and polyelectrolytes based on electrostatic interaction. *Thin Solid Films*, **327–329**, 528–531.

85 Sun, Y., Hao, E., Zhang, X., Yang, B., Shen, J., Chi, L., and Fuchs, H. (1997) Buildup of composite films containing $TiO_2$/PbS nanoparticles and polyelectrolytes based on electrostatic interaction. *Langmuir*, **13**, 5168–5174.

86 Hao, E., Yang, B., Zhang, J., Zhang, X., Sun, J., and Shen, J. (1998) Assembly of alternating $TiO_2$/v\CdS nanoparticle composite films. *J. Mater. Chem.*, **8**, 1327–1328.

87 Rosidian, A., Liu, Y.J., and Claus, R.O. (1998) Ionic self-assembly of ultrahard $ZrO_2$/polymer nanocomposite thin films. *Adv. Mater.*, **10**, 1087–1091.

88 Pastoriza-Santos, I., Koktysh, D.S., Mamedov, A.A., Giersig, M., Kotov, N.A., and Liz-Marzan, L.M. (2000) One-pot synthesis of Ag@TiO$_2$ core–shell nanoparticles and their layer-by-layer assembly. *Langmuir*, **16** 2731–2735.

89 Schrof, W., Rozouvan, S., Vankeuren, E., Horn, D., Schmitt, J., and Decher, G. (1998) Nonlinear optical properties of polyelectrolyte thin films containing gold nanoparticles investigated by wavelength dispersive femtosecond degenerate four wave mixing (DFWM). *Adv. Mater.*, **10**, 338–341.

90 He, J.A., Valluzzi, R., Yang, K., Dolukhanyan, T., Sung, C., Kumar, J., Tripathy, S.K., Samuelson, L., Balogh, L., and Tomalia, D.A. (1999) Electrostatic multilayer deposition of a gold–dendrimer nanocomposite. *Chem. Mater.*, **11**, 3268–3274.

91 Hao, E. and Lian, T. (2000) Buildup of polymer/Au nanoparticle multilayer thin films based on hydrogen bonding. *Chem. Mater.*, **12**, 3392–3396.

92 Lvov, Y., Ariga, K., Onda, M., Ichinose, I., and Kunitake, T. (1997) Alternate assembly of ordered multilayers of SiO$_2$ and other nanoparticles and polyions. *Langmuir*, **13**, 6195–6203.

93 Lvov, Y.M., Rusling, J.F., Thomsen, D.L., Papadimitrakopoulos, F., Kawakami, T., and Kunitake, T. (1998) High-speed multilayer film assembly by alternate adsorption of silica nanoparticles and linear polycation. *Chem. Commun.*, 1229–1230.

94 Ichinose, I., Tagawa, H., Mizuki, S., Lvov, Y., and Kunitake, T. (1998) Formation process of ultrathin multilayer films of molybdenum oxide by alternate adsorption of octamolybdate and linear polycations. *Langmuir*, **14**, 187–192.

95 Rogach, A.L., Koktysh, D.S., Harrison, M., and Kotov, N.A. (2000) Layer-by-layer assembled films of HgTe nanocrystals with strong infrared emission. *Chem. Mater.*, **12**, 1526–1528.

96 Mamedov, A., Ostrander, J., Aliev, F., and Kotov, N.A. (2000) Stratified assemblies of magnetite nanoparticles and montmorillonite prepared by the layer-by-layer assembly. *Langmuir*, **16**, 3941–3949.

97 Liu, Y.J., Wang, A.B., and Claus, R.O. (1997) Layer-by-layer electrostatic self-assembly of nanoscale Fe$_3$O$_4$ particles and polyimide precursor on silicon and silica surfaces. *Appl. Phys. Lett.*, **71**, 2265–2267.

98 Karam, P., Xin, Y., Jaber, S., and Halaoui, L.I. (2008) Active Pt nanoparticles stabilized with glucose oxidase. *J. Phys. Chem. C*, **112**, 13846–13850.

99 Conway, B.E. and Bockris, J.O'M. (1957) Electrolytic hydrogen evolution kinetics and its relation to the electronic and adsorptive properties of the metal. *J. Chem. Phys.*, **26**, 532–542.

100 Parsons, R. (1958) The rate of electrolytic hydrogen evolution and the heat of adsorption of hydrogen. *Trans. Faraday Soc.*, **54**, 1053–1063.

101 Marković, N.M., Grgur, B.N., and Ross, P.N. (1997) Temperature-dependent hydrogen electrochemistry on platinum low-index single-crystal surfaces in acid solutions. *J. Phys. Chem. B*, **101**, 5405–5413.

102 Marković, N.M., Sarraf, T.S., Gasteiger, H.A., and Ross, P.N. (1996) Hydrogen electrochemistry on platinum low-index single-crystal surfaces in alkaline solution. *J. Chem. Soc., Faraday Trans.*, **92**, 3719–3725.

103 Schmidt, T.J., Marković, N.M., and Ross, P.N. Jr. (2002) Temperature dependent surface electrochemistry on Pt single crystals in alkaline electrolytes: Part 2. The hydrogen evolution/oxidation reaction. *J. Electroanal. Chem.*, **524–525**, 252–260.

104 Marković, N.M., Schmidt, T.J., Stamenkovic, V., and Ross, P.N. (2001) Oxygen reduction reaction on Pt and Pt bimetallic surfaces: A selective review. *Fuel Cells*, **1**, 105–116.

105 Marković, N.M., Gasteiger, H.A., and Ross, P.N. (1995) Oxygen reduction on platinum low-index single-crystal surfaces in sulfuric acid solution: Rotating ring-Pt(hkl) disk studies. *J. Phys. Chem.*, **99**, 3411–3415.

106 Marković, N.M., Gasteiger, H.A., and Ross, P.N. (1996) Oxygen reduction on platinum low-index single-crystal surfaces in alkaline solution: Rotating ring disk Pt(hkl) studies. *J. Phys. Chem.*, **100**, 6715–6721.

107 Konishita, K. (1990) Particle size effects for oxygen reduction on highly dispersed platinum in acid electrolytes. *J. Electrochem. Soc.*, **137**, 845–848.

108 Climent, V., Marković, N.M., and Ross, P.N. (2000) Kinetics of oxygen reduction on an epitaxial film of palladium on Pt(111). *J. Phys. Chem. B*, **104**, 3116–3120.

109 Paulus, U.A., Wokaun, A., Scherer, G.G., Schmidt, T.J., Stamenkovic, V., Radmilovic, V., Marković, N.M., and Ross, P.N. Jr. (2002) Oxygen Reduction on Carbon-Supported Pt–Ni and Pt–Co Alloy Catalysts. *J. Phys. Chem. B*, **106**, 4181–4191.

110 Schmidt, T.J., Stamenkovic, V., Arenz, M., Marković, N.M., and Ross, P.N. Jr. (2002) Oxygen electrocatalysis in alkaline electrolyte: Pt(hkl), Au(hkl) and the effect of Pd-modification. *Electrochim. Acta*, **47**, 3765–3776.

111 Toda, T., Igarashi, H., and Watanabe, M. (1998) Role of electronic property of Pt and Pt alloys on electrocatalytic reduction of oxygen. *J. Electrochem. Soc.*, **145**, 4185–4188.

112 Toda, T., Igarashi, H., and Watanabe, M. (1999) Enhancement of the electrocatalytic $O_2$ reduction on Pt–Fe alloys. *J. Electroanal. Chem.*, **460**, 258–262.

113 Toda, T., Igarashi, H., Uchida, H., and Watanabe, M. (1999) Enhancement of the electroreduction of oxygen on Pt alloys with Fe, Ni, and Co. *J. Electrochem. Soc.*, **146**, 3750–3756.

114 Takasu, Y., Ohashi, N., Zhang, X.-G., Murakami, Y., Minagawa, H., Sato, S., and Yahikozawa, K. (1996) Size effects of platinum particles on the electroreduction of oxygen. *Electrochim. Acta*, **41**, 2595–2600.

115 Kabbabi, A., Gloaguen, F., Andolfatto, F., and Durand, R. (1994) Particle size effect for oxygen reduction and methanol oxidation on Pt/C inside a proton exchange membrane. *J. Electroanal. Chem.*, **373**, 251–254.

116 Marković, N.M., Radmilovic, V., and Ross, P.N. Jr. (2003) chapter 9. Physical and electrochemical characterization of bimetallic nanoparticle electrocatalysts, in *Catalysis and Electrocatalysis at Nanoparticle Surfaces* (eds A. Wieckowski, E.R. Savinova, and C.G. Vayenas), Marcel Dekker, Inc., New York, pp. 311–342.

117 Watanabe, M., Sei, H., and Stonehart, P. (1989) The influence of platinum crystallite size on the electroreduction of oxygen. *J. Electroanal. Chem. Interfac. Electrochem.*, **261**, 375–387.

118 Bregoli, L.J. (1978) The influence of platinum crystallite size on the electrochemical reduction of oxygen in phosphoric acid. *Electrochim. Acta*, **23**, 489–492.

119 Sattler, M.L. and Ross, P.N. (1986) The surface structure of Pt crystallites supported on carbon black. *Ultramicroscopy*, **20**, 21–28.

120 Peuckert, M., Yoneda, T., Dalla Betta, R.A., and Boudart, M. (1986) Oxygen reduction on small supported platinum oarticles. *J. Electrochem. Soc.*, **133**, 944–947.

121 Schmidt, T.J., Stamenkovic, V., Ross, P.N. Jr., and Markovic, N.M. (2003) Temperature dependent surface electrochemistry on Pt single crystals in alkaline electrolyte Part 3. The oxygen reduction reaction. *Phys. Chem. Chem. Phys.*, **5**, 400–406.

122 Geniès, L., Faure, R., and Durand, R. (1998) Electrochemical reduction of oxygen on platinum nanoparticles in alkaline media. *Electrochim. Acta*, **44**, 1317–1327.

123 Watanabe, M., Saegusa, S., and Stonehart, P. (1988) Electro-catalytic activity on supported platinum crystallites for oxygen reduction in sulphuric acid. *Chem. Lett.*, **9**, 1487–1490.

124 Watanabe, M. (2003) Chapter 22, Design of electrocatalysts for fuel cells, in *Catalysis and Electrocatalysis at Nanoparticle Surfaces* (eds A. Wieckowski, E.R. Savinova, and C.G. Vayenas), Marcel Dekker Inc, New York, pp. 827–846.

125 Giordano, N., Passalacqua, E., Pino, L., Arico, A.S., Antonucci, V., Vivaldi, M., and Kinoshita, K. (1991) Analysis of platinum particle size and oxygen reduction in phosphoric acid. *Electrochim. Acta*, **36**, 1979–1984.

126 Arenz, M., Mayrhofer, K.J.J., Stamenkovic, V., Blizanac, B.B., Tomoyuki, T., Ross, P.N., and Markovic, N.M. (2005) The Effect of the particle size on the kinetics of CO electrooxidation on high surface area Pt catalysts. *J. Am. Chem. Soc.*, **127**, 6819–6829.

127 Mayrhofer, K.J.J., Blizanac, B.B., Arenz, M., Stamenkovic, V.R., Ross, P.N., and Markovic, N.M. (2005) The impact of geometric and surface electronic properties of Pt-catalysts on the particle size effect in electrocatalysis. *J. Phys. Chem. B*, **109**, 14433–14440.

128 Miyano, K., Asano, K., and Shimomura, M. (1991) Adsorption kinetics of water-soluble polymers onto a spread monolayer. *Langmuir*, **7**, 444–445.

129 Motschmann, H., Stamm, M., and Toprakcioglu, Ch. (1991) Adsorption kinetics of block copolymers from a good solvent: a two-stage process. *Macromolecules*, **24**, 3681–3688.

130 Hoogeveen, N.G., Cohen Stuart, M.A., and Fleer, G.J. (1996) Polyelectrolyte adsorption on oxides: I. Kinetics and adsorbed amounts. *J. Colloid Interface Sci.*, **182**, 133–145.

131 Fleer, G.J., Cohen Stuart, M.A., Scheutjens, J.M.H.M., Cosgrove, T., and Vincent, B. (1993) *Polymers at Interfaces*, Chapman & Hall, London, pp. 343–373.

132 Ahmadi, T.S., Wang, Z.L., Green, T.C., Henglein, A., and El-Sayed, M.A. (1996) Shape-controlled synthesis of colloidal platinum nanoparticles. *Science*, **272**, 1924–1925.

133 Ahmadi, T.S., Wang, Z.L., Henglein, A., and El-Sayed, M.A. (1996) "Cubic" Colloidal Platinum Nanoparticles. *Chem. Mater.*, **8**, 1161–1163.

134 Petrovski, J.M., Wang, Z.L., Green, T.C., and El-Sayed, M.A. (1998) Kinetically controlled growth and shape formation mechanism of platinum nanoparticles. *J. Phys. Chem. B*, **102**, 3316–3320.

135 Narayanan, R. and El-Sayed, M. (2004) Effect of nanocatalysis in colloidal solution on the tetrahedral and cubic nanoparticle SHAPE: Electron-transfer reaction catalyzed by platinum nanoparticles. *J. Phys. Chem. B*, **108** (18), 5726–5733.

136 Narayanan, R. and El-Sayed, M.A. (2004) Changing catalytic activity during colloidal platinum nanocatalysis due to shape changes: Electron-transfer reaction. *J. Am. Chem. Soc.*, **126** (23), 7194–7195.

137 Narayanan, R. and El-Sayed, M.A. (2003) Effect of catalytic activity on the metallic nanoparticle size distribution: Electron-transfer reaction between $Fe(CN)_6$ and thiosulfate ions catalyzed by PVP-platinum nanoparticles. *J. Phys. Chem. B.*, **107**, 12416–12424.

138 Narayanan, R. and El-Sayed, M.A. (2004) Shape-dependent catalytic activity of platinum nanoparticles in colloidal solution. *Nano Lett.*, **4** (7), 1343–1348.

139 Narayanan, R. and El-Sayed, M.A. (2005) Effect of colloidal nanocatalysis on the metallic nanoparticle shape: The suzuki reaction. *Langmuir*, **21** (5), 2027–2033.

140 Narayanan, R. and El Sayed, M. (2005) Catalysis with transition metal nanoparticles in colloidal solution: Nanoparticle shape dependence and stability. *J. Phys. Chem. B*, **109**, 12663–12676.

141 Solla-Gullón, J., Vidal-Iglesias, F.J., Rodríguez, P., Herrero, E., Feliu, J.M., Clavilier, J., and Aldaz, A. (2004) In situ surface characterization of preferentially oriented platinum nanoparticles by using electrochemical structure sensitive adsorption reactions. *J. Phys. Chem. B*, **108**, 13573–13575.

142 Susut, C., Chapman, G.B., Samjeske', G., Osawa, M., and Tong, Y. (2008) An unexpected enhancement in methanol electro-oxidation on an ensemble of Pt (111) nanofacets: a case of nanoscale single crystal ensemble electrocatalysis. *Phys. Chem. Chem. Phys.*, **10**, 3712–3721.

143 Lee, I., Delbecq, F., Morales, R., Albiter, R., and Zaera, F. (2009) Tuning selectivity in catalysis by controlling particle shape. *Nat. Mater.*, **8**, 132–138.

144 Solla-Gullón, J., Vidal-Iglesias, F.J., López-Cudero, A., Garnier, E., Feliu, J.M., and Aldaza, A. (2008) Shape-dependent electrocatalysis: methanol and formic acid electrooxidation on preferentially oriented Pt nanoparticles. *Phys. Chem. Chem. Phys.*, **10**, 3689–3698.

145 Zhao, M. and Crooks, R.M. (1999) Dendrimer-encapsulated Pt nanoparticles: Synthesis, characterization, and applications to catalysis. *Adv. Mater*, **11** 217–220.

146 Clavilier, J. (1988) Electrochemical surface characterization of platinum electrodes using elementary electrosorption processes at basal and stepped surfaces, in *Electrochemical Surface Science: Molecular Phenomena at Electrode Surfaces* (ed. M.P. Soriaga), ACS Symposium Series 378, American Chemical Society, Washington, DC, pp. 202–215.

147 Will, F.G. (1965) Hydrogen adsorption on platinum single crystal electrodes. *J. Electrochem. Soc.*, **112**, 451–455.

148 Duarte, M.Y., Martins, M.E., and Arvia, A. (1980) The electrosorption and the electrodesorption of hydrogen and oxygen at the polycrystalline platinum-alkaline carbonate interface. *J. Electrochim. Acta*, **25**, 1613–1618.

149 Kita, H., Ye, S., Aramata, A., and Furuya, N. (1990) Adsorption of hydrogen on platinum single crystal electrodes in acid and alkali solutions. *J. Electroanal. Chem. Interfac. Electrochem.*, **295**, 317–331.

150 Ye, S., Kita, H., and Aramata, A. (1992) Hydrogen and anion adsorption at platinum single crystal electrodes in phosphate solutions over a wide range of pH. *J. Electroanal. Chem.*, **333**, 299–312.

151 Kinoshita, K., Ferrier, D.R., and Stonehart, P. (1978) Effect of electrolyte environment and Pt crystallite size on hydrogen adsorption—V. *Electrochim. Acta*, **23**, 45–54.

152 Gómez, R., Orts, J.M., Álvarez-Ruiz, B., and Feliu, J.M. (2004) Effect of temperature on hydrogen adsorption on Pt(111), Pt(110), and Pt(100) electrodes in 0.1M $HClO_4$. *J. Phys. Chem. B*, **108**, 228–238.

153 Bard, A.J. and Faulkner, L.R. (2001) Chapters 13 and 14, in *Electrochemical Methods: Fundamentals and Applications*, 2nd edn, John Wiley & Sons, New York.

154 Markovic, N.M. and Ross, P.N. Jr. (1999) Electrocatalysis at well-defined surfaces: Kinetics of oxygen reduction and hydrogen oxidation/evolution on Pt(hkl) electrodes, in *Interfacial Electrochemistry: Theory, Experiment, and Applications* (ed. A. Wieckowski), Marcel Dekker, Inc, New York, Basel, pp. 821–841.

155 Baret, B., Aubert, P.-H., Mayne-L'Hermitea, M., Pinault, M., Reynauda, C., Etcheberry, A., and Pereza., H. (2009) Nanocomposite electrodes based on pre-synthesized organically capped platinum nanoparticles and carbon nanotubes. Part I: Tuneable low platinum loadings, specific Hupd feature and evidence for oxygen reduction. *Electrochim. Acta*, **54**, 5421–5430.

156 Barreira, S.V.P., Garcia-Morales, V., Pereira, C.M., Manzanares, J.A., and Silva, F. (2004) Electrochemical impedance spectroscopy of polyelectrolyte multilayer modified electrodes. *J. Phys. Chem. B*, **108**, 17973–17982.

157 Farhat, T.R. and Schlenoff, J.B. (2001) Ion transport and equilibria in polyelectrolyte multilayers. *Langmuir*, **17**, 1184–1192.

158 Farhat, T.R. and Schlenoff, J.B. (2003) Doping-controlled ion diffusion in polyelectrolyte multilayers: Mass transport in reluctant exchangers. *J. Am. Chem. Soc.*, **125**, 4627–4636.

159 Rmaile, H.H. and Farhat, T.R. (2003) pH-Gated permeability of variably charged species through polyelectrolyte multilayer membranes. *J. Phys. Chem. B*, **107**, 14401–14406.

160 Krasemann, L. and Tieke, B. (2000) Selective ion transport across self-assembled alternating multilayers of cationic and anionic polyelectrolytes. *Langmuir*, **16**, 287–290.

161 Lebedev, K., Ramírez, P., Mafé, S., and Pellicer, J. (2000) Modeling of the salt permeability in fixed charge multilayer membranes. *Langmuir*, **16**, 9941–9943.

162 Antipov, A.A., Sukhorukov, G.B., and Möhwald, H. (2003) Influence of the ionic strength on the polyelectrolyte multilayers' permeability. *Langmuir*, **19**, 2444–2448.

163 Han, S. and Lindholm-Sethson, B. (1999) Electrochemistry at ultrathin polyelectrolyte films self-assembled at planar gold electrodes. *Electrochim. Acta*, **45**, 845–853.

164 Harris, J.J. and Bruening, M.L. (2000) Electrochemical and in situ ellipsometric investigation of the permeability and stability of layered polyelectrolyte films. *Langmuir*, **16**, 2006–2013.

165 Pardo-Yissar, V., Katz, E., Lioubashevski, O., and Willner, I. (2001) Layered polyelectrolyte films on Au electrodes: Characterization of electron-transfer features at the charged polymer interface and application for selective redox reactions. *Langmuir*, **17**, 1110–1118.

166 Morf, W.E. (1996) Theoretical treatment of the amperometric current response of multiple microelectrode arrays. *Anal. Chim. Acta*, **330**, 139–149.

167 Bard, A.J. and Faulkner, L.R. (2001) Chapter 5, in *Electrochemical Methods. Fundamentals and Applications*, 2nd edn, John Wiley & Sons, New York.

168 Feeney, R. and Kounaves, S.P. (2000) Microfabricated ultramicroelectrode arrays: Developments, advances, and applications in environmental analysis. *Electroanalysis*, **12**, 677–684.

169 Morf, W.E. and de Rooij, N.F. (1997) Performance of amperometric sensors based on multiple microelectrode arrays. *Sens. Actuators, B*, **44**, 538–541.

170 Weber, S.G. (1989) Signal-to-noise ratio in microelectrode-array-based electrochemical detectors. *Anal. Chem.*, **61**, 295–302.

171 Kriz, K., Anderlund, M., and Kriz, D. (2001) Real-time detection of ascorbic acid and hydrogen peroxide in crude food samples employing a reversed sequential differential measuring technique of the SIRE-technology based biosensor. *Biosens. Bioelectrochem.*, **16**, 363–369.

172 Kriz, K., Kraft, L., Krook, M., and Kriz, D. (2002) Amperometric determination of L-Lactate based on entrapment of lactate oxidase on a transducer surface with a semi-permeable membrane using a SIRE technology based biosensor. Application: Tomato paste and baby food. *J. Agric. Food Chem.*, **50**, 3419–3424.

173 Chai, P.C., Long, L.H., and Halliwel, B. (2003) Contribution of hydrogen peroxide to the cytotoxicity of green tea and red wines. *Biochem. Biophys. Res. Commun.*, **304**, 650–654.

174 Wang, Y., Huang, J., Zhang, C., Wei, J., and Zhou, X. (1998) Determination of hydrogen peroxide in rainwater by using a polyaniline film and platinum particles co-modified carbon fiber microelectrode. *Electroanalysis*, **10**, 776–778.

175 Vione, D., Maurino, V., Minero, C., Borghesi, D., Lucchiari, M., and Pelizzetti, E. (2003) New processes in the environmental chemistry of nitrite. 2. The role of hydrogen peroxide. *Environ. Sci. Technol.*, **37**, 4635–4641.

176 Clark, L.C. Jr. (1965) Membrane polarographic electrode system and method with electrochemical compensation. U. S. Patent 3539455.

177 Armstrong, F.A. and Wilson, G.S. (2000) Recent developments in faradaic bioelectrochemistry. *Electrochim. Acta*, **45**, 2623–2645.

178 Thome-Duret, V., Reach, G., Gangnerau, M.N., Lemonnier, F., Klein, J.C., Zhang, Y., Hu, Y., and Wilson, G.S. (1996) Use of a subcutaneous glucose sensor to detect decreases in glucose concentration prior to observation in blood. *Anal. Chem.*, **68**, 3822–3826.

179 Yang, M., Yang, Y., Yang, H., Shen, G., and Yu, R. (2006) Layer-by-layer self-assembled multilayer films of carbon nanotubes and platinum nanoparticles with polyelectrolyte for the fabrication of biosensors. *Biomaterials*, **27**, 246–255.

180 You, T., Niwa, O., Tomita, M., and Hirono, S. (2003) Characterization of platinum nanoparticle-embedded carbon film electrode and its detection of hydrogen peroxide. *Anal. Chem.*, **75**, 2080–2085.

181 Niwa, O., Horiuchi, T., Kurita, R., and Torimitsu, K. (1998) On-line

electrochemical sensor for selective continuous measurement of acetylcholine in cultured brain tissue. *Anal. Chem.*, **70**, 1126–1132.

182 Yang, L., Janle, E., Huang, T., Gitsen, J., Kissinger, P.T., Vreeke, M., and Heller, A. (1995) Applications of "wired" peroxidase electrodes for peroxide determination in liquid chromatography coupled to oxidase immobilized enzyme reactors. *Anal. Chem.*, **67**, 1326–1331.

183 Ruiz, B.L., Dempsey, E., Hua, C., Smyth, M.R., and Wang, J. (1993) Development of amperometric sensors for choline, acetylcholine and arsenocholine. *Anal. Chim. Acta*, **273**, 425–430.

184 Hall, S.B., Khudaish, E.A., and Hart, A.L. (1998) Electrochemical oxidation of hydrogen peroxide at platinum electrodes. Part II: effect of potential. *Electrochim. Acta*, **43**, 2015–2024.

185 Hall, S.B., Khudaish, E.A., and Hart, A.L. (1999) Electrochemical oxidation of hydrogen peroxide at platinum electrodes. Part III: Effect of temperature. *Electrochim. Acta*, **44**, 2455–2462.

186 Hall, S.B., Khudaish, E.A., and Hart, A.L. (1999) Electrochemical oxidation of hydrogen peroxide at platinum electrodes. Part IV: phosphate buffer dependence. *Electrochim. Acta*, **44**, 4573–4582.

187 Hall, S.B., Khudaish, E.A., and Hart, A.L. (2000) Electrochemical oxidation of hydrogen peroxide at platinum electrodes. Part V: inhibition by chloride. *Electrochim. Acta*, **45**, 3573–3579.

188 Pravda, M., Jungar, C.M., Iwuoha, E.I., Smyth, M.R., Vytras, K., and Ivaska, A. (1995) Evaluation of amperometric glucose biosensors based on co-immobilisation of glucose oxidase with an osmium redox polymer in electrochemically generated polyphenol films. *Anal. Chim. Acta*, **304**, 127–138.

189 Niwa, O., Horiuchi, T., Marita, M., Huang, T., and Kissinger, P.T. (1996) Determination of acetylcholine and choline with platinum-black ultramicroarray electrodes using liquid chromatography with a post-column enzyme reactor. *Anal. Chim. Acta*, **318**, 167–173.

190 Evans, S.A.G., Elliott, J.M., Andrews, L.M., Bartlett, P.N., Doyle, P.J., and Denuault, G. (2002) Detection of hydrogen peroxide at mesoporous platinum microelectrodes. *Anal. Chem.*, **74**, 1322–1326.

191 Hrapovic, S., Liu, Y., Male, K.B., and Luong, J.H.T. (2004) Electrochemical biosensing platforms using platinum nanoparticles and carbon nanotubes. *Anal. Chem.*, **76**, 1083–1088.

192 Hecht, H.J., Schomburg, D., Kalisz, H., and Schmid, R.D. (1993) The 3D structure of glucose oxidase from *Aspergillus niger*. Implications for the use of GOD as a biosensor enzyme. *Biosens. Bioelectron.*, **8**, 197–203.

193 Hecht, H.J., Kalisz, H.M., Schmid, R.D., and Schomburg, D. (1993) Crystal structure of glucose oxidase from *Aspergillus niger* refined at 2·3 Å resolution. *J. Mol. Biol.*, **229**, 153–172.

194 Crumbliss, A.L., Perine, S.C., Stonehuerner, J., Tubergen, K.R., Zhao, J.G., and Henkens, R.W. (1992) Colloidal gold as a biocompatible immobilization matrix suitable for the fabrication of enzyme electrodes by electrodeposition. *Biotechnol. Bioeng.*, **40**, 483–490.

195 Stonehuerner, J.G., Zhao, J., O'Daly, J.P., Crumbliss, A.L., and Henkens, R.W. (1992) Comparison of colloidal gold electrode fabrication methods: the preparation of a horseradish peroxidase enzyme electrode. *Biosens. Bioelectron.*, **7**, 421–428.

196 Mukerjee, S. and McBreen, J. (1998) Effect of particle size on the electrocatalysis by carbon-supported Pt electrocatalysts: an in situ XAS investigation. *J. Electroanal. Chem.*, **448**, 163–171.

197 Jaber, S., Nasr, P., Xin, Y., Sleem, F., and Halaoui, L.I. (2012) Electrochemical Surface Characterization and Stability of Polyvinylpyrrolidone-capped Tetrahedral Pt Nanoparticles Assembled in a Polyelectrolyte. unpublished work. Submitted.

198 Wang, Y. and Balbuena, P. (2005) Ab initio molecular dynamics simulations of the oxygen reduction reaction on a Pt (111) surface in the presence of hydrated hydronium $(H_3O)^+(H_2O)_2$: Direct or series pathway? *J. Phys. Chem. B*, **109**, 14896–14907.

199 Perez part 2: Marcha, G., Volatron, F., Lachauda, F., Chenga, X., Bareta, B., Pinaulta, M., Etcheberry, A., and Perez, H. (2011) Nanocomposite electrodesbasedonpre-synthesizedorganicallycapped platinum nanoparticlesandcarbonnanotubes. Part II: Determination of diffusion area for oxygen reduction reflects platinum accessibility. *Electrochim. Acta*, **56**, 5151–5157.

200 Trindale, T., O'Brien, P., and Pickett, N.L. (2001) Nanocrystalline semiconductors: Synthesis, properties, and perspectives. *Chem. Mater*, **13** 3843–3858.

201 Shockley, W. and Queisser, H.J. (1961) Detailed balance limit of efficiency of p–n junction solar cells. *J. Appl. Phys.*, **32**, 510–519.

202 Nozik, A.J. (2001) Spectroscopy and hot electron relaxation dynamics in semiconductor quantum wells and quantum dots. *Annu. Rev. Phys. Chem.*, **52**, 193–231.

203 Nozik, A.J. (2008) Multiple exciton generation in semiconductor quantum dots. *Chem. Phys. Lett.*, **457**, 3–11, and references therein.

204 Blackburn, J.L., Ellingson, R.J., Mićić, O.I., and Nozik, A.J. (2003) Electron relaxation in colloidal InP quantum dots with photogenerated excitons or chemically injected electrons. *J. Phys. Chem. B*, **107**, 102–109.

205 Beard, M.C., Midgett, A.G., Law, M., Semonin, O.E., Ellingson, R.J., and Nozik, A.J. (2009) Variations in the quantum efficiency of multiple exciton generation for a series of chemically treated PbSe nanocrystal films. *Nano Lett.*, **9**, 836–845.

206 Law, M., Luther, J.M., Song, Q., Hughes, B.K., Perkins, C.L., and Nozik, A.J. (2008) Structural, optical, and electrical properties of PbSe nanocrystal solids treated thermally or with simple amines. *J. Am. Chem. Soc.*, **130**, 5974–5985.

207 Cho, J., Char, K., Hong, J., and Lee, K. (2001) Fabrication of highly ordered multilayer films using a spin self-assembly method. *Adv. Mater.*, **13**, 1076–1078.

208 Ellis, A.B., Kaiser, S.W., Bolts, J.M., and Wrighton, M.S. (1977) Study of n-type semiconducting cadmium chalcogenide-based photoelectrochemical cells employing polychalcogenide electrolytes. *J. Am. Chem. Soc.*, **99**, 2839–2848.

209 Ellis, A.B., Kaiser, S.W., and Wrighton, M.S. (1976) Optical to electrical energy conversion. characterization of cadmium sulfide and cadmium selenide based photoelectrochemical cells. *J. Am. Chem. Soc.*, **98**, 6855–6866.

210 Gaweda, S., Podborska, A., Macyk, W., and Szaciłowski, K. (2009) Nanoscale optoelectronic switches and logic devices. *Nanoscale*, **1**, 299–316.

211 Bard, A.J. and Faulkner, L.R. (2001) Chapter 18, in *Electrochemical Methods: Fundamentals and Applications*, 2nd edn, John Wiley & Sons, New York, pp. 745–760.

212 Morisaki, H., Hiraya, M., and Yazawa, K. (1977) Anomalous photoresponse of n-$TiO_2$ electrode in a photoelectrochemical cell. *Appl. Phys. Lett.*, **30**, 7–9.

213 Gissler, W. (1980) Photoelectrochemical investigation on trigonal selenium film electrodes. *J. Electrochem. Soc.*, **127**, 1713–1716.

214 Agostinelli, G. and Dunlop, E.D. (2003) Local inversion of photocurrent in cadmium telluride solar cells. *Thin Solid Films*, **431–432**, 448–452.

215 Zhang, P., Cheng, C., Jiao, P., Li, Y., He, Z., and Zhang, H. (2008) Well improved photoswitching characteristic of CdSe nanorods via CdS nanoparticle-decoration. *Mater. Lett.*, **62**, 1151.

216 Minoura, H. and Tsuiki, M. (1978) Cathodic photo-effect at CdS electrode in aqueous polysulfide solution. *Chem. Lett.*, 205–208.

217 Müller, N., Hodes, G., and Vainas, B. (1984) Cathodic current photoenhancement at mechanically

damaged CdS electrodes. *J. Electroanal. Chem. Interfac. Electrochem.*, **172**, 155–165.
218 Podborska, A., Gaweł, B., Pietrzak, Ł., Szymańska, I.B., Jeszka, J.K., Łasocha, W., and Szaciłowski, K. (2009) Anomalous photocathodic behavior of CdS within the Urbach tail region. *J. Phys. Chem. C*, **113**, 6774–6784.
219 VaInAs, B., Hodes, G., and Dubow, V. (1981) A photocathodic effect at the CdS-electrolyte interface. *J. Electroanal. Chem.*, **130**, 391–394.
220 Kronik, L., Ashkenasy, N., Leibovitch, M., Fefer, E., Shapira, Y., Gorer, S., and Hodes, G. (1998) Surface states and photovoltaic effects in CdSe quantum dot films. *J. Electrochem. Soc.*, **145**, 1748–1755.
221 Baron, R., Huang, C., Bassani, D.M., Onopriyenko, A., Zayats, M., and Willner, I. (2005) Hydrogen-bonded CdS nanoparticle assemblies on electrodes for photoelectrochemical applications. *Angew. Chem., Int. Ed.*, **44**, 4010–4015.
222 Miyake, M., Torimoto, T., Nishizawa, M., Sakata, T., Mori, H., and Yoneyama, H. (1999) Effects of surface charges and surface states of chemically modified cadmium sulfide nanoparticles immobilized to gold electrode substrate on photoinduced charge transfers. *Langmuir*, **15**, 2714–2718.
223 Wehmeyer, K.R. and Wightman, R.M. (1985) Cyclic voltammetry and anodic stripping voltammetry with mercury ultramicroelectrodes. *Anal. Chem.*, **57**, 1989–1993.
224 Perone, S.P. and Kretlow, W.J. (1966) Application of controlled potential techniques to study of rapid succeeding chemical reaction coupled to electro-oxidation of ascorbic acid. *Anal. Chem.*, **38**, 1760–1763.
225 Steenken, S. and Neta, P. (1982) One-electron redox potentials of phenols. hydroxy- and aminophenols and related compounds of biological interest. *J. Phys. Chem.*, **86**, 3661–3667.

# 23
# Record Properties of Layer-by-Layer Assembled Composites
*Ming Yang, Paul Podsiadlo, Bong Sup Shim, and Nicholas A. Kotov*

## 23.1
## Introduction

The development of composite materials provides more freedom to engineer structural and functional properties (mechanical, electrical, thermal, optical, electrochemical, catalytic, etc.) markedly different from those of the component materials. Scientists continue to seek the path to better combinations of different materials resulting in composites with record properties [1]. The flourishing of nanotechnology can be seen in basically all technical disciplines, among which a prominent area is polymer matrix based nanocomposites [2]. New physical properties, novel behavior and enhanced performances that are absent in the matrices could be introduced by the nanoscale dispersion of controlled nanostructures in the composite. The area of the interface between the matrix and nanoscale building blocks is typically an order of magnitude larger than that for conventional composite materials which means an observable effect on the macroscale properties of the composite can be expected from a relatively small amount of nanoscale reinforcement. However, so far only layer-by-layer assembly (LbL), a new method for preparation of nanocomposites based on the sequential layering of inorganic and organic components, has been able to demonstrate that this property resource can be tapped into effectively, including composites with high inorganic contents. This distinct feature of using a nanoscale component in the LBL composites enables the blue-print design and tuning of high performance materials with a variety of finely controlled functional characteristics, which was very difficult to achieve by the traditional methods of composite preparation.

Nanoscale components or building blocks can nowadays be synthesized with great control with respect to their composition, for example, inorganic, organic, polymeric, biological, as well as structure and function, making a much broader spectrum of nanoscale components with properties substantially different or better than those of the polymer matrix. While the high aspect ratio and surface area of nanoscale components can be critical for effective stress transfer based on a large interface, this also brings about the problem of achieving nearly perfect dispersion of nanoscale components in a polymer matrix (i.e., when the particles are uniformly dispersed over the entire volume following either a perfect statistical distribution or a specific

*Multilayer Thin Films: Sequential Assembly of Nanocomposite Materials*, Second Edition.
Edited by Gero Decher and Joseph B. Schlenoff.
© 2012 Wiley-VCH Verlag GmbH & Co. KGaA. Published 2012 by Wiley-VCH Verlag GmbH & Co. KGaA.

order) due to the strong tendency of nanomaterials to agglomerate, negating any benefit associated with the nanoscale dimension, and resulting in a typical low component percolation threshold of nanocomposites. The challenge of this field is how to synergistically combine the materials from almost limitless combinations, or how to effectively transfer the nanoscale properties of these materials into macroscale structures, which has so far been nearly impossible to achieve by the existing composite preparation techniques, such as extrusion, solution blending, melt mixing, *in situ* polymerization and others.

This challenge requires a new method of composite synthesis with capabilities to manipulate molecular scale and nanoscale components in the large range of concentrations and ratios without disturbing the perfection of dispersion or the ordering in the composite. The LbL assembly technique is one of the most versatile methods for the preparation of organic–inorganic materials. The underlying mechanism of LbL is sequential adsorption of compounds with oppositely charged components, allowing optimization of the interactions between the components, nanometer-level structural control, and combinations of different functionalities. From the perspective of mechanical properties, LbL is advantageous due to the intrinsically strong interactions between the components and the great uniformity of the resulting materials. Wide range tuning of different properties can be based on the multiscale control of the structure and components during LbL assembly, namely the number, sequence, and the structure of individual layers, which can be modified virtually at will. The LbL building of the interface could guarantee the good dispersion of components and their effective communication with the matrix. LbL assembly holds promise for the fabrication of hierarchically ordered materials with tailored structures and functionalities spanning multiple length scales and dimensions.

Since the first demonstration of LbL assembly by Iler [3] and later by Decher *et al.* in the 1990s [4], the LbL field has experienced rapid growth covering a vast number of molecular species and architectures. In this chapter, we will focus on the preparation and properties of LbL nanocomposites, especially with two of the most investigated nanoparticles – clay and carbon nanotubes (CNTs) as nanoscale components. The predominant use of these two nanomaterials in LbL nanocomposites is due to their commercial availability, excellent properties, ready-to-use dispersion technique, and high aspect ratios. Some new emerging areas in this field will also be discussed. We will detail the technology involved in LbL-assembled multilayered composites, and also include the likely applications of these high performance materials.

## 23.2
## LbL Assemblies of Clays

### 23.2.1
### Structure and Properties of Clay Particles

Clays are well-known for their layered structure. Exfoliated clays are characterized by two-dimensional sheets of corner-sharing $SiO_4$ and $AlO_4$ tetrahedra, always bonded to octahedral sheets with Al or Mg at their center and coordinated by six oxygen atoms. Depending on the way that the tetrahedral and octahedral sheets are packaged into layers, clays can be classified into 1 : 1 and 2 : 1 clay. Montmorillonite (MTM), primarily used in LbL assembly, is a 2 : 1 clay, meaning that it has 2 tetrahedral sheets sandwiching a central octahedral sheet.

MTM can swell in aqueous conditions due to water penetrating the interlayer molecular spaces and concomitant adsorption. The amount of expansion depends mainly on the type of interlayer exchangeable cations. The presence of monovalent cations, such as $Na^+$ or $Li^+$, as the predominant exchangeable cation can result in the clay swelling to several times its original volume. The exfoliated clay may be individual dispersed lamellae with a thickness $\sim$0.96 nm and lateral (in-plane) dimensions of 100–1000 nm. The high aspect ratios of exfoliated clay are generally advantageous for a number of composite properties. The anisotropic, sheet-like structure is of great importance for control of transport properties through the films.

The negatively charged exfoliated clay sheets can be very stable in aqueous solution and thus a good partner for many polymers during LbL assembly. The wide utilization of clay in LbL assembly is also due to its low cost imparted by the natural abundance, and exceptional mechanical properties associated with its atomic framework and defect-free structure, with in-plane modulus of elasticity ($E$) calculated to be of the order of 270 GPa [5].

## 23.2.2
### Structural Organization in Clay Multilayers

The first demonstration of LbL assembly of clay nanosheets with polyelectrolytes (PEs) was based on sequential adsorption of a cationic polyelectrolyte, poly(diallyldimethylammonium chloride) (PDDA) and individual sheets of the silicate mineral, hectorite (Laponite RD) [6]. Stepwise formation of multilayered films via an ion-exchange mechanism has allowed structural order in few-hundred-nanometers-thick films with an average linear increment per layer pair of $\sim$3.6 nm. The large lateral extent of the silicate sheets allows each layer to cover any packing defects in the underlying layer, thus preserving structural order in the growing film.

More detailed understanding of the adsorption kinetics, organization, and control of the internal structure of the clay/PEs films (adsorption time, concentration of cationic polymer, amount of clay in the dispersion and pH) has been the subject of several research groups. Lvov *et al.* extended the preparation of clay multilayers to PDDA/MTM and poly(ethyleneimine) (PEI)/MTM systems [7]. They showed that the assembly process is based on saturated adsorption of the two components and drying is not obligatory for the film growth. A saturation time (5–6 min) was observed in the condition of low concentration of MTM, which was shown to be essential to produce a single, uniform layer in each adsorption step. van Duffel *et al.* investigated the influence of pH, polymer concentration, and clay particle size on the building and roughness of LbL-assembled natural and synthetic clay–polymer films [8]. The thickness increase per cycle depends on the clay particle size and the concentration of polymer. The film roughness is affected by the same parameters and increases with the particle size of the clay and PDDA concentration. They suggested that at low polymer concentration the polymer chains can bind strongly with clay platelets, resulting in stretched chains and a small number of unbound polymer units, resulting in low surface roughness. For high PE concentrations, multiple polymer chains are only partially bonded to the substrate and stick out of the surface. Bundles of clay platelets can adsorb under these conditions, being more or less stacked on top of each, other resulting in a rough film surface. Optimal conditions for the deposition of smooth multilayer films consist of the combined use of a 0.05–0.5% (w/w) aqueous solution of PDDA and a 0.05% (w/w) clay suspension at pH 9–10.

The synthetic clays offer new opportunities for high-quality hybrid thin films with multifunctionality since different combinations of ions can be artificially introduced into the nanosheet gallery. For example, Li et al. for the first time showed LbL assembly of magnesium–aluminum synthetic clays called "layered double hydroxides" (LDH) with anionic polymer poly(styrenesulfonate) (PSS) [9]. Recently, a series of free-standing, strong, transparent, and functional layered organic–inorganic hybrid films reinforced with LDH micro- and nanoplatelets have been fabricated through the LbL assembly procedure using a series of LDH platelets as building blocks [10].

### 23.2.3
### Clay Multilayers as High-Performance Nanocomposites

One of the unique perspectives for clay multilayers is their potential as high-performance nanocomposites, considering the exceptional mechanical properties of clay sheets, comparable with steel and its alloys. Kotov et al. have shown for the first time that PDDA/MTM multilayers have unusually high strength, flexibility, and resistance to crack proliferation [11]. LbL films of PDDA/MTM were shown to self-assemble on flexible PET substrates and AFM tests demonstrated that the individual alumosilicate sheets are unexpectedly flexible, and oriented in parallel to the substrate. Fan et al. used the nanoindentation test on PDDA/MTM composite films to show the hardness ($H$) and modulus of the films to be $H = 0.46$ GPa and $E = 9.5$ GPa, respectively [12]. The utilization of the flexibility and high strength of clay layers in LbL assembly of PDDA and magnetite nanoparticles resulted in the substantial improvement of the strength, and the effect of strengthening can be traced to the ability to lift-off films of different architecture [13]. The free-standing membrane of this multifunctional composite could be further patterned into a magnetic free-standing microstrip with a root anchored on the substrate, with potential sensing applications [14].

An interesting "nanostructured artificial nacre" with alternating PDDA and MTM was demonstrated by Tang et al. [15]. The tensile strength of the prepared multilayers approached that of nacre ($\sigma_{UTS} = 100 \pm 10$ MPa), whereas their ultimate Young's modulus was similar to that of lamellar bones ($E = 11 \pm 2$ GPa). The macromolecular folding effect reveals itself in the unique saw-tooth pattern of differential stretching curves, attributed to the gradual breakage of ionic crosslinks in the PE chains.

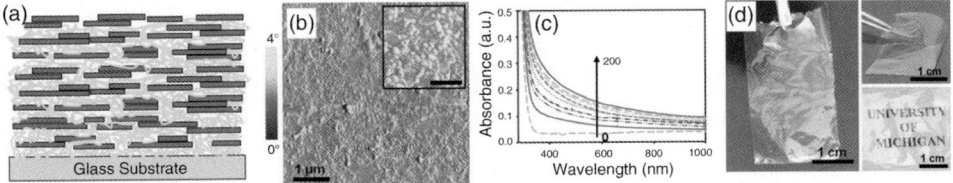

**Figure 23.1** Preparation of PVA/MTM nanocomposites. (a) Schematic representation of the internal architecture of the PVA/MTM nanocomposite (picture shows 8 layer pairss). (b) ATM phase image of a single PVA/MTM layer pair adsorbed on top of a silicon wafer. (Inset) Close-up of the main image showing individual MTM platelets more clearly. Scale bar in inset, 400 nm. (c) Compilation of UV–Vis absorbance spectra collected after multiples of 25 layer pairs of PVA/MTM composite deposited on both sides of a microscope glass slide up to 200 layer pairs. a.u., arbitrary units. (d) Free-standing, 300-layer pair PVA/MTM composite film showing high flexibility and transparency. The lower image was taken at an angle to show diffraction colors.

Structural and functional resemblance makes this kind of clay–PE multilayers a close replica of natural biocomposites. To better control the interfacial properties of clay/polymer nanocomposites, we turned our attention to another polymer, poly(vinyl alcohol) (PVA) which allows further covalent cross-linking chemistry with glutaraldehyde (GA) [16] (Figure 23.1). The resulting tensile strength was, $\sigma_{UTS} = 400 \pm 40$ MPa and $E = 106 \pm 11$ GPa, approaching the theoretical maxima. The explanation of the effective stiffening of the PVA matrix lies in its close proximity to, and many interactions with, the MTM platelets.

From the above examples, we can see several advantages of the LbL technique for the preparation of high-strength clay nanocomposites: (i) the nanocomposites are constructed by alternately depositing nanometer-thick layers of polymer and clay, thus allowing for nanometer-level control of preparation; (ii) alternating the layers of clay nanosheets with a few-nanometer-thick layers of polymers translates into volume fractions upwards of 50 vol%; (iii) the colloidal self-assembly process restricts the adsorption of clay to well exfoliated sheets; and (iv) sandwiching the nanosheets between polymer layers and strong interfacial bonding prevent phase segregation of the nanocomponent.

Despite the unique ability of molecular scale component manipulation, slow deposition speeds are currently limiting the LbL technique to applications in coatings and thin membranes. As alternative approaches to the PE/clay films preparation we should mention two most recent cases which deviate from traditional dipping and/or monolayer deposition and have important implications for the structural organization of the clay and PE layers. The first case is spin-assisted LbL assembly in which, instead of dipping of the substrate into solutions of the constituents, the solutions are alternately spin-coated onto the substrate with intermediate rinsing steps with pure water. Recently, Lee *et al.* showed the first successful preparation of clay multilayers by spin-assisted self-assembly [17] of poly (*p*-phenylene vinylene) (PPV) and layered silicate. The combination of LbL with spin-coating makes it more time-efficient and practical for various needs of the coatings. In the second case, we recently showed that MTM nanosheets can be successfully incorporated into the so-called "exponential"

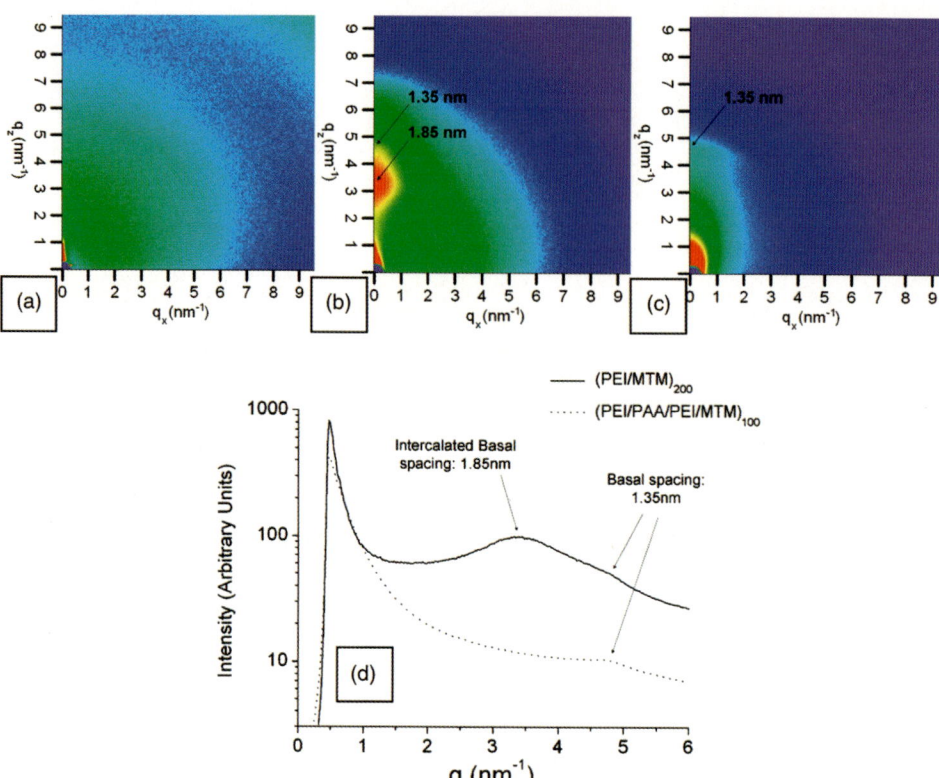

**Figure 23.2** 2D SAXS patterns of free-standing films of (a) (PEI/PAA)$_{200}$, (b) (PEI/MTM)$_{200}$, and (c) (PEI/PAA/PEI/MTM)$_{100}$. The scattering features of interest are indicated by arrows and the corresponding spacings are noted. (d) 1D SAXS patterns of free-standing films of (PEI/MTM)$_{200}$ and (PEI/PAA/PEI/MTM)$_{100}$. These plots are radial integrations of the 2D images shown in (b) and (c). The intensities were shifted for clarity.

LbL assembly (e-LBL) of PEI and poly(acrylic acid) (PAA) which leads to novel architectures [18]. A proof of an "in-and-out" diffusion of the PEs during e-LBL, responsible for the exponential growth of the film thickness, was given by Picart et al. [19] and Lavalle et al. [20]. Although the e-LBL mode offers new opportunities for the preparation of thick multilayers with unique internal organization in a short period of time, the internal ordering of the film was substantially decreased, as revealed by small-angle X-ray scattering (SAXS) (Figure 23.2).

### 23.2.4
### Applications of Clay Multilayers in Biotechnology

The appreciable surface area, ordered structure, intercalation properties, natural origins, high stability, and high exchange capacity of clay films impart them with additional advantages for biomedical constructs. The first application of clay multilayers for this purpose was demonstrated by Lvov *et al.* [7, 21]. They used water-soluble proteins in combination with positively charged PEI or with negatively charged PSS, as well as MTM. This biomolecular architecture opened a way to construct artificially orchestrated protein systems that could carry out complex enzymatic reactions.

Later, stable films of clay and hemoglobin (Hb) were assembled LbL on various solid substrates by alternate adsorption of negatively charged clay platelets from their aqueous dispersions and positively charged Hb from pH 4.5 buffers [22]. The electroactivity of Hb was found to be extended to six clay/Hb layer pairs. Electrochemical reductions of trichloroacetic acid, oxygen, and hydrogen peroxide were catalyzed by Hb in the (Clay/Hb)$_6$ films.. In further studies, myoglobin (Mb) or horseradish peroxidase (HRP) was introduced into clay/protein films [23]. Effective electron transfer rates involving the protein heme Fe(III)/Fe(II) redox couple were greatly enhanced in the microenvironment of clay/protein films compared with that of bare electrodes in the protein solutions. These LbL clay/protein films had reasonable stability and the protein could essentially retain its native state at medium pH, suggesting a promising approach to the fabrication of mediator-free biosensors or bioreactors.

The clay multilayers have also been used as interfaces/substrates for cell culture. Antibacterial activity can be introduced by alternating clay layers with starch-stabilized silver nanoparticles in combination with PDDA [24]. The resulting composite showed excellent structural stability with no detectable levels of silver lost over a 1 month period. Evaluation of the antibacterial properties showed almost complete growth inhibition of *E. coli* over an 18 h period with excellent biocompatibility with the human osteoblast cell line. Later, our group demonstrated that LbL-assembled clay/PDDA multilayers can be used to modify the surface of inverted colloidal crystal (ICC) scaffolds and to enhance cell adhesion [25]. Similarly, Mehta *et al.* modified the surface of PDMS, based on LBL deposition of PDDA, clay, type IV collagen and fibronectin, to promote bone marrow cell attachment and spreading [26]. Adherent primary bone marrow cells attached and spread best on a surface with composition of (PDDA/clay)$_5$ (Collagen/Fibronectin)$_2$ with negatively charged fibronectin exposed on the top, remaining well spread and proliferating for at least two weeks.

Besides the biocompatibility of clay multilayers, their high dielectric constant and excellent ionic conductivity can help generate sufficient potential difference at the clay membrane interface and ionic currents through the membrane. Pappas *et al.* engineered a neuronal interface for photoelectric stimulation of the cells [27]. The authors used multicomponent LbL assemblies incorporating HgTe nanoparticles capped with PDDA/MTM film (Figure 23.3). The PDDA/MTM layer was used to improve biocompatibility and neuronal cell attachment, as well as being a reservoir of Na$^+$ cations which were necessary for cell depolarization and firing of neurons. Increased cathodic capacitive spike at the light "turn-on" point was detected with a (PDDA/HgTe)$_{12}$(PDDA/Clay)$_2$ film.

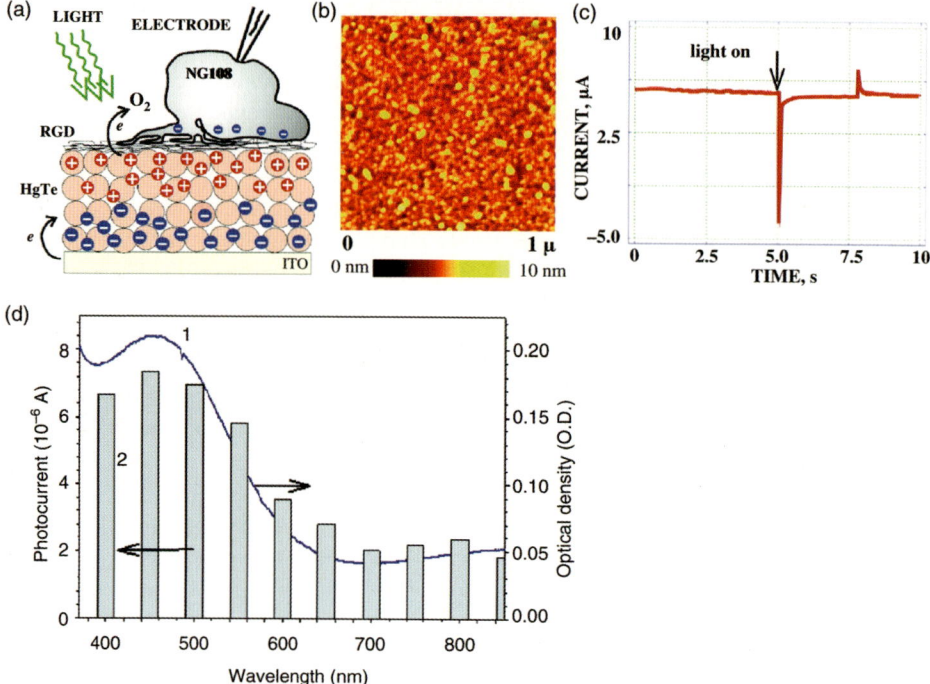

**Figure 23.3** Photovoltaic properties of LbL films from HgTe. (a) Schematic of coupling between NP and the neuron. (b) Atomic force microscopy image of the (HgTe/PDDA)$_1$ on Si wafer. (c) Photocurrent response measured at 0.0 V bias for (PDDA/HgTe)$_{12}$ under monochromatic 550 nm illumination. (d) Light absorption characteristics of HgTe NPs and LbL films. (1, solid line) UV – Vis absorption spectrum on HgTe NP dispersion stabilized by thioglycerol used for fabrication of LBL films. (2, bars) Dependence of photogenerated voltage in (PDDA/HgTe)$_{12}$PLP multilayer on the wavelength of incident light.

### 23.2.5
### Anisotropic Transport in Clay Multilayers

The stratified organizations of clay multilayers with respect to the substrate, the inorganic composition, and the flat morphology of the platelets, have also imparted the clay multilayers with unique anisotropic transport properties. Kotov et al. realized the potential of the clay multilayers to preclude gas diffusion through defects and to design highly selective ultrathin membranes [11]. This fact made the PE/MTM multilayers stand out among other thin films for which the diffusion through defects was the primary mechanism of gas permeation.

LbL assembly of MTM and cationic polyacrylamide on a PET film resulted in a transparent film (transparency greater than 90%) having an oxygen transmission rate (OTR) below the detection limit of commercial instrumentation ($<0.005$ cc m$^{-2}$ d$^{-1}$ atm$^{-1}$). This low OTR is believed to be due to a brick wall nanostructure comprised of completely exfoliated clay in polymeric mortar [28]. A more detailed study by Priolo

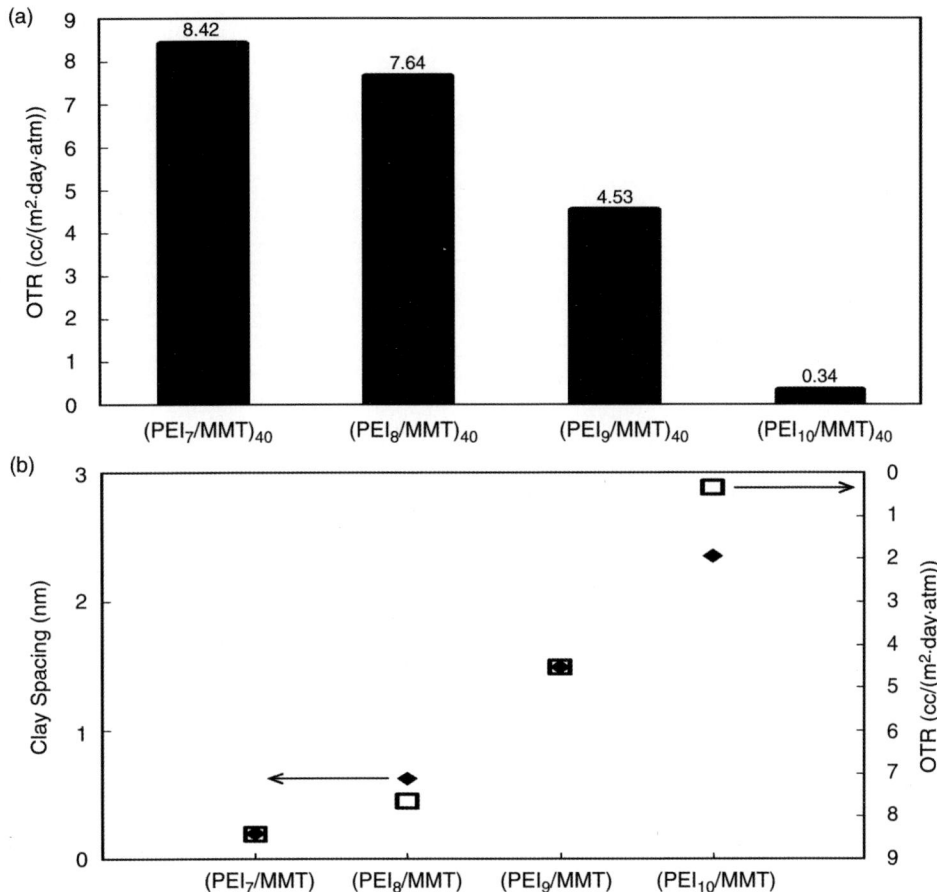

**Figure 23.4** (a) Oxygen transmission rates of 40-layer pair films as a function of their PEI pH values, and (b) clay spacing as a function of PEI pH with OTR of 40-layer pair films.

et al. presented the possibility of tailoring OTR by changing the pH of branched PEI in the assembly with clay [29] (Figure 23.4). Most recently, the treatment of cotton fabric by LbL-assembled PEI/clay has been shown to impart flame-retardant behavior to cotton fabric ([30].). This work provides an effective way for using polymer/clay thin film assemblies to make a variety of complex substrates (foam, fabrics, etc.) flame resistant. A new application for clay multilayers was demonstrated by Kim et al., using saponite multilayers as coatings for fuel cells membranes to enhance resistance to methanol diffusion through a Nafion membrane [31].

## 23.2.6
### Clay Multilayers for Optical and Electronic Applications

The high surface area and high exchange capacity of clay nanosheets were also beneficial for incorporation of optically and electronically active molecules and polymers. Ru(bpy)$_3^{2+}$ immobilized in the multilayer films showed good electrochemical and ECL activity [32]. The sensitivity of the {clay/Ru(bpy)$_3^{2+}$}$_n$ multilayer films modified electrode for determination of TPA and oxalate was improved by about one order of magnitude. With respect to incorporation of dye molecules, an ionene-type polycation incorporating a nonlinear optical chromophore was employed for the assembly of alternating PE multilayers, using different organic polyanions as well as MTM clay as anionic counterparts [33]. In contrast to the organic polyanions, the use of MTM as inorganic polyanion in the assembly gave rise to a red shift compared to the aqueous solution spectrum of nonlinear optical dyes. Different kinds of aggregation were observed for different species of polyanions employed. The degree of orientation of the chromophores was investigated by second-harmonic generation and the nonlinear response was found to depend strongly on the anionic species. Optically active polymers PPV have also been incorporated into clay multilayers by spin-assisted LbL assembly using the electrostatic forces between the negatively charged surface of layered silicate and the cationic PPV precursor which finally thermally converted to (PPV/Laponite)$_n$ film [17].

## 23.3
## LBL Assemblies of Carbon Nanotubes

### 23.3.1
### Structure and Properties of CNTs

Nanotubes are categorized as single-walled carbon nanotubes (SWNTs) and multi-walled carbon nanotubes (MWNTs). Individual nanotubes naturally align themselves into "ropes", held together by van der Waals forces. The structure of a SWNT can be conceptualized by wrapping a one-atom-thick layer of graphite, called grapheme, into a seamless cylinder. MWNTs consist of multiple rolled layers (concentric tubes) of graphite. CNTs represent one of the most important areas of materials science today for their ultrahigh aspect ratio, extraordinary strength and unique electrical properties, making them ideal candidates as components in lightweight polymer composites. The stiffness and toughness of CNTs are arguably known as about up to 1 TPa and 300 GPa, respectively [34]. CNTs tend not to bond strongly with their host matrix, resulting in limitations of potential increases and improvements in the mechanical properties of a nanotube composite. Chemical functionalization and surface modification of CNTs are used to remedy this problem, aiming at achieving effective load transfer. Sonication of CNTs within a solvent in the presence of polymers is one of the most common methods to overcome the collective van der Waals forces, which are exerted in a CNT bundle and prevent exfoliation, dramatically increasing as the length increases. Currently, only limited length, under 100 μm, CNTs are employed in solution processing of CNT.

## 23.3.2
### Structural Organization and Mechanical Properties in Multilayers of Carbon Nanotubes

The effective utilization of CNTs in composite applications depends strongly on the ability to homogeneously disperse them throughout the matrix without destroying their integrity. The individual nanoscale building blocks underperform when they are incorporated into composites due to inefficient stress transfer with the polymer matrix. LbL assembly provides possibilities of careful structural design of the materials and control of the polymer nanotube interface, addressing the challenging problems of good dispersion and effective stress transfer. First, the alternate generation of a polymer–nanotube interface promises a uniform dispersion of CNTs into a composite, preventing phase segregation; second, the intermediate rinsing stages of the LbL process stimulate efficient interfacial bonding between the components of the composites by removal of polymeric chains loosely attached to SWNTs; third, accurately controlled multi-component nanolayers enable tunable multi-functional properties of a composite.

The early pioneering work of CNT LbL assembly is the introduction of successful conquest over the dispersion challenges of CNTs in polymer composites. Mamedov et al. [35] reported exceptionally uniform dispersion of SWNTs in nano-thin layered structures $((PEI/PAA)(PEI/SWNT)_5)_n$. The ultimate tensile strength was found to be $220 \pm 40$ MPa with some readings as high as 325 MPa. Notable mechanical functionalities of the SWNT LbL composite are attributed to uniform dispersion of CNTs in a polymer matrix, high loading of CNTs without compromising the homogeneity of the composite at the nanometer level, and functional activated interfacial bonding between CNTs and the matrix (PEI). The high structural homogeneity and interconnectivity of the structural components of the LbL films combined with high SWNT loading from CNT multilayers has drawn broad attention from various disciplines.

Successive experimental reports have confirmed that LbL assembly of SWNTs indeed allowed exceptional exfoliation and homogeneous dispersion in a polymeric composite [36] and nano-scale stepwise deposition of CNTs. The loading of CNTs in LbL-assembled composites can be controlled in the range 10% [37] to 75% [38], which depends on many variables of LbL assembly such as the polymer matrix, the stabilizer of the CNTs, the process conditions, and so on. Most recently, SWNT nanocomposites were made by LbL assembly with PVA with optimized connectivity at the grapheme/polymer interface [39]. The resulting SWNT-PVA composites demonstrated a tensile strength of $504.5 \pm 67.3$ MPa, stiffness $15.6 \pm 3.8$ GPa, and toughness $121.2 \pm 19.2$ J g$^{-1}$, representing the strongest and stiffest nonfibrous SWNT composites made to date (Figure 23.5). The toughness of SWNT-PVA composites exceeds that of Kevlar by threefold. The successful combination of high strength, stiffness, and toughness in LbL composites is due to a synergistic multiscale action of intermolecular interactions, improved load transfer by crosslinking, high crack surface roughening, and energy dissipation in microcracks.

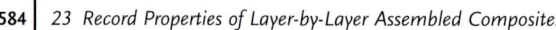

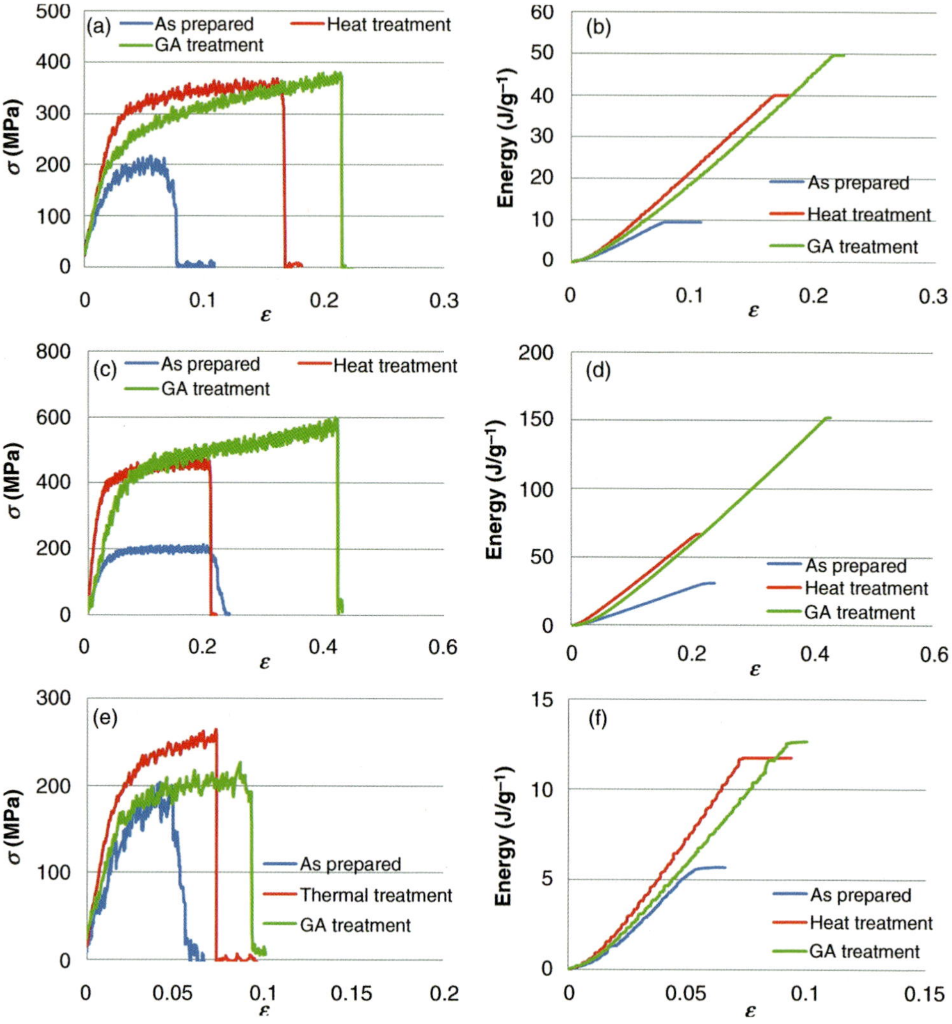

**Figure 23.5** Tension test results of SWNT LbL nanocomposites. (a,c,e) Stress – strain and (b,d,f) toughness – strain curves of (a,b) [PVA/SWNT$_{-COOH}$(15.8) + PSS]$_{200}$, (c,d) [PVA/SWNT$_{-COOH}$(7.9) + PSS]$_{200}$, and (e,f) [PVA/SWNT$_{-COOH}$(0) + PSS]$_{200}$. Additional processing steps of the composites are indicated on the graphs.

### 23.3.3
### Electrical Conductor Applications

Electrically, a SWNT can be either semiconducting or metallic, with the general presumption that one-third of bulk SWNTs are metallic and the rest are semiconducting. The electrical conductivities of metallic SWNT ropes are 1 to $\sim 3 \times 10^6$ S m$^{-1}$. The combination of the exceptional nano-organization of LbL assembly and the unique electrical properties of CNTs has opened wide research opportunities.

One advantage of LbL assemblies over conventional solution casting is conservation of the dispersion quality of nanobuilding blocks from a solution to a solid state. Once we prepare exfoliated SWNT dispersions, multifunctional properties can be transferred and tailored to a macroscale film by LbL assembly with high precision.

Highly anisotropic electrical properties of CNTs LbL composites were expected because of the layered structures. Kovtyukhova *et al.* reported that the conductivity differs between the in-plane and out-of-plane directions by more than a factor of three, so that great potential for property modulation was suggested, with ultrathin film thickness controls [40]. As a way to improve the electrical properties of LbL-assembled CNT nanocomposites, Shim *et al.* reported that the electrical conductivities of [PVA/(SWNT + PSS)]$_n$ film are correlated with a micro-/nano-scale continuous conductive path by analysis specific to molecular structures of LbL composites [37]. Thermal treatment of the film at 300 °C boosted the conductivity to $4.15 \times 10^4$ S m$^{-1}$ due to a gradual closing of intrinsic nano- and angstrom-scale voids/pores, and relaxation of the polymer conformation to one that adheres to nanotubes better. The controllability of the thickness with an increment of 2–3 nm implies precise tuning of the light transmission and suggests the potential of these films as high performance transparent conductors (TCs). Further achievement in this system was demonstrated by Shim *et al.* recently [41]. A superacid (120% $H_2SO_4$) treatment at room temperature alters the micro- and nanostructure of a CNT LbL film by strong dehydration and modification of nanotube–polymer interactions, leading to higher effective charge carrier concentrations in the SWNT networks. They display $80.2 \pm 0.1\%$ light transmittance ($T$ at $\lambda = 550$ nm) and $86 \pm 1$ Ω/sq surface resistivity ($R_s$), which matches or even exceeds those obtained for analogous SWNT TCs reported so far. Compared to ITO and bucky-paper-type TCs, a 100 times improvement in the maximum bending strain before electrical failure was achieved, due to the excellent flexibility. A new type of SWNT-doping based on electron transfer from the valence bands of the nanotubes to unoccupied levels of sulfonated polyetheretherketone (SPEEK) through π–π interactions was identified in (hydroethyl cellulose/SPEEK-SWNT)$_n$ composite coatings [42]. The low energy of the LUMO in SPEEK, the presence of another unoccupied orbital above it, and tight wrapping of the polymer around the nanotubes provide all the necessary conditions for effective electron transfer from SWNT to SPEEK (Figure 23.6). The conductivity of $1.1 \times 10^5$ S m$^{-1}$ at 66 wt% loadings of SWNT outperforms other polymer/SWNT composites and translates into a surface conductivity of 920 Ω/sq and transmittance of 86.7% at 550 nm. These discoveries highlight the advantages of LbL in satisfying seemingly impossible to combine multiple requirements, including transparency, charge transport, environmental resilience, and bending strain. Interestingly, SWNT LbL films can also be used as thin film transistors (TFTs) [43] using $SiO_2$ nanoparticles as the gate dielectric material.

### 23.3.4
**Sensor Applications**

The compliant combination of mechanical and electrical properties of CNT LbL film makes it very suitable for monitoring physical and chemical state changes.

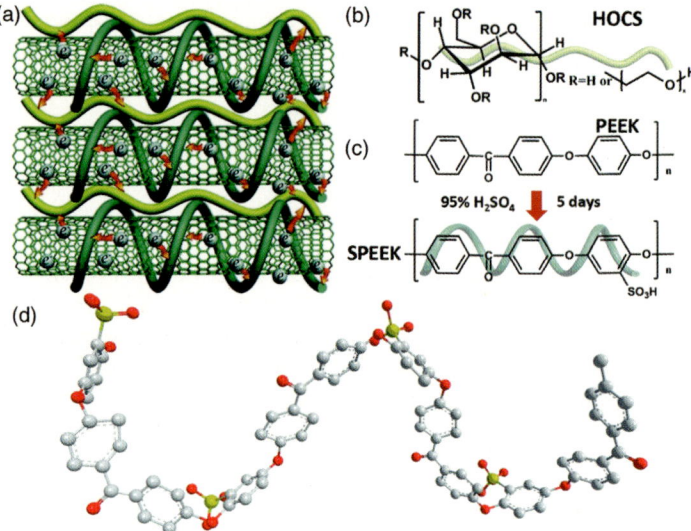

**Figure 23.6** (a) Proposed ideal architecture for LBL films. The red arrows indicate hole-doping from the surrounding polymers. SPEEK, green macromolecules; HOCS, yellow macromolecules. (b) Chemical structure of hydroxyethyl cellulose (HOCS) and (c) poly (etheretherketone) (PEEK). (c) Chemical structure of sulfonated PEEK (SPEEK). (d) Ball-and-stick model of SPEEK in the minimal energy state calculated by the molecular mechanics (MM2) algorithm. Gray spheres are assigned to carbon atoms, red to oxygen atoms, and yellow to sulfur atoms. Hydrogen atoms are omitted for greater clarity.

Loh et al. investigated the ability to tailor strain sensor sensitivities by varying the LbL fabrication parameters of a PSS-SWNT/PVA thin film [44]. They found that increasing CNT and/or PSS concentration during LbL fabrication increases the overall film strain sensitivity, due to creation of more nanotube-to-nanotube junctions. Simply by adjusting the weight fraction of SWNT solutions and film thickness, strain sensitivities between 0.1 and 1.8 have been achieved.

A robust and reliable fabrication method for a freely suspended nanomembrane that is subsequently released by using conventional micromachining processes was developed based on spin-assisted LbL assembly [45]. The integration of LbL assembled composites into micro-/nano-electrical mechanical system (MEMS/NEMS) devices is new fusion technology of top-down and bottom-up nano-processes. Nanomembrane arrays with a thickness of 7–26 nm, freely suspended over large square openings with side dimensions of 50–500 mm can be fabricated with controlled electrical conductivity and outstanding mechanical stability.

The detection of biological materials such as DNA, glucose, dopamine, uric acid, and toxic materials such as arsenic, and phenols has often been achieved by amperometric measurements of CNT LbL nanocomposites, utilizing the combinations of electrocatalytic activities, electrochemical sensitivities of CNTs, and immobilization of various materials, given by versatile selections of LbL assembly. Environmental changes like humidity are easily monitored by CNT LbL composites [46].

## 23.3.5
### Fuel Cell Applications

The LbL technique affords the preparation of composites from CNTs with loadings of above 50% without phase segregation of dissimilar materials, with controlled porosity and free selection of the ionic matrix. Fuel cell proton exchange membranes (PEMs) were constructed by LbL-assembled CNT nanocomposites to give highly conductive and chemically durable electrodes. Michel *et al.* reported that CNTs wrapped with Nafion and Pt catalysts [PEI/(Pt/SWNT + Nafion)]$_n$ showed unusual fuel cell performances when they were LbL assembled on a fuel cell PEM, featuring high peak power densities (195 mW cm$^{-2}$) and Pt utilization (408 mW mg(Pt)$^{-1}$) for [PEI/(Pt/SWNT + Nafion)]$_{400}$ catalysts [47]. Interestingly, biofuel cell applications were also developed by CNT/poly-L-(lysine)/fungal laccase LbL assembly with electrical potential generation by stable enzyme immobilization [48]. The multilayer structured cathode exhibited high power output (329 μW cm$^{-2}$) in the glucose/O$_2$ biofuel cell.

## 23.3.6
### Nano-/Micro-Shell LbL Coatings

The diversity of LbL assembly allows coating not only on planar substrates but also on micro-/nano- complex objects. For these complex coatings, two design schemes associated with CNTs are introduced. The first is assembling CNT LbL films over microspheres. Hollow spherical cages made of nested SWNTs were first reported by Sano *et al.* using amine-terminated silica gels in solution as a template which was etched away to give hollow cages [49]. The following efforts have been directed at improving the properties of micro-spherical shells, such as strength using infiltrated silica and titania [50]. Other types of microsphere templates include sulfonated PS particles [51] and poly(methacrylic acid) (PMAA) microspheres [52]. Various types of micro-objects can be used as templates for LbL assemblies. Coating over soft-microcapsules was developed to protect liposome by adding to the liposomal suspension aliquots of the aqueous suspension of negatively charged SWCNT and positively charged polyallylammine hydrochloride (PAC) [53]. Hybrid hollow MWNTs/PE nanofibers were prepared by a combination of an electrospinning method and the LbL technique, and selectively removal of PS [54]. Our group recently used a patterned PDMS substrate derived from a lithographically patterned silicon mold as a substrate for the growth of 100 layer pairss of PVA and MTM (Andres *et al.*, manuscript in preparation). Due to the conformal and limiting nature of LbL assembly, the film takes on the shape of the PDMS template and does not fill in the cavities. With the aid of polymer to polymer transfer, microcapsules of PVA/MTM film can be successfully fabricated.

### 23.3.7
### Biomedical Applications

LbL assembly has several advantages significant for applications in the biomedical field, including simple integration of biological species into the resulting material, remarkable mechanical properties necessary for device engineering, high loading of CNTs and controllable nanotube–nanotube contacts with high electrical conductivity, and biocompatibility. Gheith *et al.* demonstrated that SWNT LbL films can support the growth, viability, and differentiation of neuronal NG108-15 neuroblastoma/glioma hybrid cells. The free-standing SWNT/PE multilayers can serve as a prototype for a variety of nanotube-based implants [55]. Later, they studied the possibility of utilizing the electrical conductivity of LbL-assembled, modified SWNT films (PAA/SWNT)$_{30}$ to stimulate the neurophysiological activity of NG108-15 cells, as an important evidence of electrical coupling between SWNT LbL films and NG108-15 cells in the lateral electrical configuration, which can be quite convenient for medical applications [56] (Figure 23.7). Furthermore, these neural cell tests were extended to the differentiation of embryonic neural stem cells on CNT LbL nanocomposites [57]. The SWNT LbL composites have recently been used for *in situ* gene delivery with an additional layer of PEI–pDNA deposited on top of the PVA/SWNT multilayers-(PVA/SWNT)$_3$(PEI–pDNA) [58]. The multilayer method was found to be more efficient than the conventional solution-mediated delivery. The successful generation of neurons on conductive SWNT films was triggered by multilayer-mediated delivery of neural bHLH protein expression vectors. These live cell interface experiments corroborate that CNT LbL composites are indeed one of the most viable functional materials options for biomedical implants, prosthetic devices, and stem cell growth platforms.

**Figure 23.7** Scanning electron microscopy image of a differentiated NG108-15 cell on a SWNT LBL film showing the outgrowth of neurites and branches that attach to the surface.

Another type of biocompatible, blood-compatible surface test was performed on LbL-assembled poly(lactic-co-glycolic-acid) (PLGA) and CNT nanocomposites for the aim of thromboresistance suppression to foreign artificial blood prostheses [59]. A significant reduction in platelet adhesion with the absence of platelet activation on PLGA-CNT composite was observed. Nepal *et al.* designed large-scale robust biomimetic SWNT coatings with significant antimicrobial activity [60]. LbL assembly of DNA-SWNT and lysozyme-SWNT dispersions resulted in nonleaching coatings with long-term stability and protection against bacterial colonization.

## 23.4
## Conclusions and Perspectives

The LbL technique is currently one of the most widely used methods for the preparation of multifunctional, nanostructured thin films, with applications ranging from nanocomposites, optical and electrical devices, biotechnology, battery, and environmental detections. Its popularity stems from its simplicity and universality, as well as robustness and versatility in combining a plethora of available colloids and macromolecules into finely tuned architectures with nanometer-scale control. The possibility to control the structure of composites at the nanometer scale with a high degree of accuracy is the pathway to attaining the record properties of materials.

Further development of the LbL technique relies addressed several issues: (i) Integration with other material processing techniques. We recently applied direct-write lithography on LbL films which is applicable to a wide range of functional materials (Bai *et al.*, manuscript in preparation). The integrated approach may provide multifunctional molecular assemblies of tailored architectures and material properties for various versatile device materials. (ii) Scale up of the LbL materials. One of the key factors is to accelerate the composite accumulation process. Some modified LbL methods are very promising, including spraying [61] and spin-coating LbL. An ink-jet LbL method was developed recently in our laboratory using a standard inkjet printer as a practical method to alternately deliver precise amounts of each component for the formation of LbL films without intermediate rinsing [62] (Figure 23.8). (iii) Non-aqueous LbL deposition. Elimination of water could prevent many chemical processes that lead to corrosion of substrates or electro/photo active films. The absence of water is also important to avoid the hydration of the previous coatings, which can alter the electrical properties of the films and their key functional characteristics. Nakashima *et al.* used room-temperature ionic liquids (RTILs) as media for LbL assembly which proved fairly advantageous to the further expansion of the LbL technique and the applications of LbL films [63]. (iv) Realization of greater mechanical properties in LbL composites. Currently, the ideal stress transfer has only been realized for clay platelets. CNTs will require much greater research effort to achieve such conditions. Alternatively, new nanoscale building blocks having structural and functional similarities with CNTs could be useful. Our group recently discovered a new kind of strong flexible polymeric nanofiber – aramid or Kevlar nanofibers [64]. These nanofibers can be successfully processed into films using

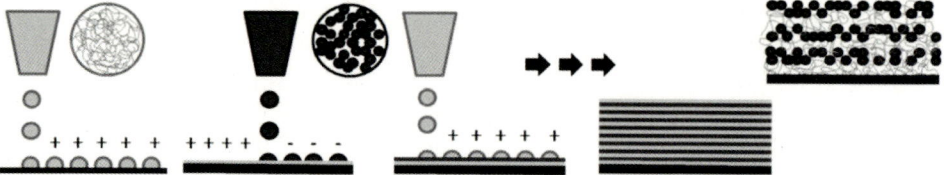

**Figure 23.8** Schematic of multilayer film production by inkjet LbL assembly.

LbL assembly and building blocks of other high performance materials, in place of or in combination with CNTs.

Some of the applications of the LbL technique are still in their infancy. Currently, we are just beginning to develop a new generation of composite biomaterials made by the LbL technique that are likely to provide enabling technologies for many applications in medicine, from implants to imaging. Further exploration of LbL assembly in various new technical disciplines can be expected with the development of the technique itself.

## References

1 Hofmann, D.C., Suh, J.-Y., Wiest, A., Duan, G., Lind, M.-L., Demetriou, M.D., and Johnson, W.L. (2008) Designing metallic glass matrix composites with high toughness and tensile ductility. *Nature*, **451**, 1085-U3.
2 Fischer, H. (2003) Polymer nanocomposites: from fundamental research to specific applications. *Mater. Sci. Eng. C-Biomim.*, **23**, 763–772.
3 Iler, R.K. (1966) Multilayers of colloidal particles. *J. Colloid Interface Sci.*, **21**, 569.
4 Decher, G. (1997) Fuzzy nanoassemblies: Toward layered polymeric multicomposites. *Science*, **277**, 1232–1237.
5 Manevitch, O.L. and Rutledge, G.C. (2004) Elastic properties of a single lamella of montmorillonite by molecular dynamics simulation. *J. Phys. Chem. B*, **108**, 1428–1435.
6 Kleinfeld, E.R. and Ferguson, G.S. (1994) Stepwise formation of multilayered nanostructural films from macromolecular precursors. *Science*, **265**, 370–373.
7 Lvov, Y., Ariga, K., Ichinose, I., and Kunitake, T. (1996) Formation of ultrathin multilayer and hydrated gel from montmorillonite and linear polycations. *Langmuir*, **12**, 3038–3044.
8 van Duffel, B., Schoonheydt, R.A., Grim, C.P.M., and de Schryver, F.C. (1999) Multilayered clay films: Atomic force microscopy study and modeling. *Langmuir*, **15**, 7520–7529.
9 Li, L., Ma, R.Z., Ebina, Y., Iyi, N., and Sasaki, T. (2005) Positively charged nanosheets derived via total delamination of layered double hydroxides. *Chem. Mater.*, **17**, 4386–4391.
10 Yao, H.-B., Fang, H.-Y., Tan, Z.-H., Wu, L.-H., and Yu, S.-H. (2010) Biologically inspired, strong, transparent, and functional layered organic-inorganic hybrid films. *Angew. Chem. Int. Ed.*, **49**, 2140–2145.
11 Kotov, N.A., Magonov, S., and Tropsha, E. (1998) Layer-by-layer self-assembly of alumosilicate-polyelectrolyte composites: Mechanism of deposition, crack resistance, and perspectives for novel membrane materials. *Chem. Mater.*, **10**, 886–895.
12 Fan, X.W., Park, M.K., Xia, C.J., and Advincula, R. (2002) Surface structural characterization and mechanical testing by nanoindentation measurements of

hybrid polymer/clay nanostructured multilayer films. *J. Mater. Res.*, **17**, 1622–1633.
13 Mamedov, A.A. and Kotov, N.A. (2000) Free-standing layer-by-layer assembled films of magnetite nanoparticles. *Langmuir*, **16**, 5530–5533.
14 Hua, F., Cui, T.H., and Lvov, Y.M. (2004) Ultrathin cantilevers based on polymer-ceramic nanocomposite assembled through layer-by-layer adsorption. *Nano Lett.*, **4**, 823–825.
15 Tang, Z.Y., Kotov, N.A., Magonov, S., and Ozturk, B. (2003) Nanostructured artificial nacre. *Nature Mater.*, **2**, 413-U8.
16 Podsiadlo, P., Kaushik, A.K., Arruda, E.M., Waas, A.M., Shim, B.S., Xu, J.D., Nandivada, H., Pumplin, B.G., Lahann, J., Ramamoorthy, A., and Kotov, N.A. (2007) Ultrastrong and stiff layered polymer nanocomposites. *Science*, **318**, 80–83.
17 Lee, H.C., Lee, T.W., Kim, T.H., and Park, O.O. (2004) Fabrication and characterization of polymer/nanoclay hybrid ultrathin multilayer film by spin self-assembly method. *Thin Solid Films*, **458**, 9–14.
18 Podsiadlo, P., Michel, M., Lee, J., Verploegen, E., Kam, N.W.S., Ball, V., Qi, Y., Hart, A.J., Hammond, P.T., and Kotov, N.A. (2008) Exponential growth of LBL films with incorporated inorganic sheets. *Nano Lett.*, **8**, 1762–1770.
19 Picart, C., Mutterer, J., Richert, L., Luo, Y., Prestwich, G.D., Schaaf, P., Voegel, J.C., and Lavalle, P. (2002) Molecular basis for the explanation of the exponential growth of polyelectrolyte multilayers. *Proc. Natl. Acad. Sci. USA*, **99**, 12531–12535.
20 Lavalle, P., Gergely, C., Cuisinier, F.J.G., Decher, G., Schaaf, P., Voegel, J.C., and Picart, C. (2002) Comparison of the structure of polyelectrolyte multilayer films exhibiting a linear and an exponential growth regime: An in situ atomic force microscopy study. *Macromolecules*, **35**, 4458–4465.
21 Lvov, Y.M. and Sukhorukov, G.B. (1997) Protein architecture: Assembly of ordered films by means alternated adsorption of opposite charged macromolecules. *Biol. Membr.*, **14**, 229–250.
22 Zhou, Y.L., Li, Z., Hu, N.F., Zeng, Y.H., and Rusling, J.F. (2002) Layer-by-layer assembly of ultrathin films of hemoglobin and clay nanoparticles with electrochemical and catalytic activity. *Langmuir*, **18**, 8573–8579.
23 Li, Z. and Hu, N.F. (2003) Direct electrochemistry of heme proteins in their layer-by-layer films with clay nanoparticles. *J. Electroanal. Chem.*, **558**, 155–165.
24 Podsiadlo, P., Paternel, S., Rouillard, J.M., Zhang, Z.F., Lee, J., Lee, J.W., Gulari, L., and Kotov, N.A. (2005) Layer-by-layer assembly of nacre-like nanostructured composites with antimicrobial properties. *Langmuir*, **21**, 11915–11921.
25 Lee, J., Shanbhag, S., and Kotov, N.A. (2006) Inverted colloidal crystals as three-dimensional microenvironments for cellular co-cultures. *J. Mater. Chem.*, **16**, 3558–3564.
26 Mehta, G., Kiel, M.J., Lee, J.W., Kotov, N., Linderman, J.J., and Takayama, S. (2007) Polyelectrolyte-clay-protein layer films on microfluidic PDMS bioreactor surfaces for primary murine bone marrow culture. *Adv. Func. Mater.*, **17**, 2701–2709.
27 Pappas, T.C., Wickramanyake, W.M.S., Jan, E., Motamedi, M., Brodwick, M., and Kotov, N.A. (2007) Nanoscale engineering of a cellular interface with semiconductor nanoparticle films for photoelectric stimulation of neurons. *Nano Lett.*, **7**, 513–519.
28 Jang, W.S., Rawson, I., and Grunlan, J.C. (2008) Layer-by-layer assembly of thin film oxygen barrier. *Thin Solid Films*, **516**, 4819–4825.
29 Priolo, M.A., Gamboa, D., and Grunlan, J.C. (2010) Transparent clay-polymer nano brick wall assemblies with tailorable oxygen barrier. *ACS Appl. Mater. Interface.*, **2**, 312–320.
30 Li, Y.-C., Schulz, J., Mannen, S., Delhom, C., Condon, B., Chang, S., Zammarano, M., and Grunlan, J.C. (2010) Flame retardant behavior of polyelectrolyte-clay thin film assemblies on cotton fabric. *ACS Nano.*, **4**, 3325–3337.
31 Kim, D.W., Choi, H.S., Lee, C., Blumstein, A., and Kang, Y. (2003)

Investigation on methanol permeability of Nafion modified by self-assembled clay-nanocomposite multilayers, in *1st International Conference on Polymer Batteries and Fuel Cells (PBFC-1)*, Jeju Isl, Pergamon-Elsevier Science Ltd, South Korea.

32 Guo, Z.H., Shen, Y., Zhao, F., Wang, M.K., and Dong, S.J. (2004) Electrochemical and electrogenerated chemiluminescence of clay nanoparticles/Ru(bpy)(3)(2+) multilayer films on ITO electrodes. *Analyst.*, **129**, 657–663.

33 Laschewsky, A., Wischerhoff, E., Kauranen, M., and Persoons, A. (1997) Polyelectrolyte multilayer assemblies containing nonlinear optical dyes. *Macromolecules*, **30**, 8304–8309.

34 Yu, M.F., Lourie, O., Dyer, M.J., Moloni, K., Kelly, T.F., and Ruoff, R.S. (2000) Strength and breaking mechanism of multiwalled carbon nanotubes under tensile load. *Science*, **287**, 637–640.

35 Mamedov, A.A., Kotov, N.A., Prato, M., Guldi, D.M., Wicksted, J.P., and Hirsch, A. (2002) Molecular design of strong single-wall carbon nanotube/polyelectrolyte multilayer composites. *Nature Mater.*, **1**, 190–194.

36 Rouse, J.H. and Lillehei, P.T. (2003) Electrostatic assembly of polymer/single walled carbon nanotube multilayer films. *Nano Lett.*, **3**, 59–62.

37 Shim, B.S., Tang, Z.Y., Morabito, M.P., Agarwal, A., Hong, H.P., and Kotov, N.A. (2007) Integration of conductivity transparency, and mechanical strength into highly homogeneous layer-by-layer composites of single-walled carbon nanotubes for optoelectronics. *Chem. Mater.*, **19**, 5467–5474.

38 Xue, W. and Cui, T.H. (2007) Characterization of layer-by-layer self-assembled carbon nanotube multilayer thin films. *Nanotechnology*, **18**, 7.

39 Shim, B.S., Zhu, J., Jan, E., Critchley, K., Ho, S., Podsiadlo, P., Sun, K., and Kotov, N.A. (2009) Multiparameter structural optimization of single-walled carbon nanotube composites: toward record strength, stiffness, and toughness. *ACS Nano.*, **3**, 1711–1722.

40 Kovtyukhova, N.I. and Mallouk, T.E. (2005) Ultrathin anisotropic films assembled from individual single-walled carbon nanotubes and amine polymers. *J. Phys. Chem. B*, **109**, 2540–2545.

41 Shim, B.S., Zhu, J., Jan, E., Critchley, K., and Kotov, N.A. (2010) Transparent conductors from layer-by-layer assembled swnt films: importance of mechanical properties and a new figure of merit. *ACS Nano.*, **4**, 3725–3734.

42 Zhu, J., Shim, B.S., Di Prima, M., and Kotov, N.A. (2011) Transparent conductors from carbon nanotubes LBL-assembled with polymer dopant with pi-pi electron transfer. *J. Am. Chem. Soc.*, **133**, 7450–7460.

43 Xue, W., Liu, Y., and Cui, T.H. (2006) High-mobility transistors based on nanoassembled carbon nanotube semiconducting layer and $SiO_2$ nanoparticle dielectric layer. *Appl. Phys. Lett.*, **89**, 3.

44 Loh, K.J., Lynch, J.P., Shim, B.S., and Kotov, N.A. (2008) Tailoring piezoresistive sensitivity of multilayer carbon nanotube composite strain sensors. *J. Intel. Mater. Syst. Struct.*, **19**, 747–764.

45 Kang, T.J., Cha, M., Jang, E.Y., Shin, J., Im, H.U., Kim, Y., Lee, J., and Kim, Y.H. (2008) Ultra-thin and conductive nanomembrane arrays for nanomechanical transducers. *Adv. Mater.*, **20**, 3131–3137.

46 Yu, H.H., Cao, T., Zhou, L.D., Gu, E.D., Yu, D.S., and Jiang, D.S. (2006) Layer-by-layer assembly and humidity sensitive behavior of poly(ethyleneimine)/multiwall carbon nanotube composite films. *Sensor Actuat. B-Chem.*, **119**, 512–515.

47 Michel, M., Taylor, A., Sekol, R., Podsiadlo, P., Ho, P., Kotov, N., and Thompson, L. (2007) High-performance nanostructured membrane electrode, assemblies for fuel cells made by layer-by-layer assembly of carbon nanocolloids. *Adv. Mater.*, **19**, 3859.

48 Deng, L., Shang, L., Wang, Y.Z., Wang, T., Chen, H.J., and Dong, S.J. (2008) Multilayer structured carbon nanotubes/poly-L-lysine/laccase composite cathode

for glucose/O-2 biofuel cell. *Electrochem. Commun.*, **10**, 1012–1015.
49. Sano, M., Kamino, A., Okamura, J., and Shinkai, S. (2002) Noncovalent self-assembly of carbon nanotubes for construction of "cages". *Nano Lett.*, **2**, 531–533.
50. Ji, L.J., Ma, J., Zhao, C.G., Wei, W., Wang, X.C., Yang, M.S., Lu, Y.F., and Yang, Z.Z. (2006) Porous hollow carbon nanotube composite cages. *Chem. Commun.*, 1206–1208.
51. Kim, B.S., Kim, B., and Suh, K.D. (2008) Electrorheological properties of carbon nanotube/polyelectrolyte self-assembled polystyrene particles by layer-by-layer assembly. *J. Polym. Sci. Polym. Chem.*, **46**, 1058–1065.
52. Shi, J.H., Chen, Z.Y., Qin, Y.J., and Guo, Z.X. (2008) Multiwalled carbon nanotube microspheres from layer-by-layer assembly and calcination. *J. Phys. Chem. C*, **112**, 11617–11622.
53. Angelini, G., Boncompagni, S., de Maria, P., de Nardi, M., Fontana, A., Gasbarri, C., and Menna, E. (2007) Layer-by-layer deposition of shortened nanotubes or polyethylene glycol-derivatized nanotubes on liposomes: A tool for increasing liposome stability. *Carbon*, **45**, 2479–2485.
54. Pan, C., Ge, L.-Q., and Gu, Z.-Z. (2007) Fabrication of multi-walled carbon nanotube reinforced polyelectrolyte hollow nanofibers by electrospinning. *Compos. Sci. Technol.*, **67**, 3271–3277.
55. Gheith, M.K., Sinani, V.A., Wicksted, J.P., Matts, R.L., and Kotov, N.A. (2005) Single-walled carbon nanotube polyelectrolyte multilayers and freestanding films as a biocompatible platform for neuroprosthetic implants. *Adv. Mater.*, **17**, 2663.
56. Gheith, M.K., Pappas, T.C., Liopo, A.V., Sinani, V.A., Shim, B.S., Motamedi, M., Wicksted, J.R., and Kotov, N.A. (2006) Stimulation of neural cells by lateral layer-by-layer films of single-walled currents in conductive carbon nanotubes. *Adv. Mater.*, **18**, 2975–.
57. Jan, E. and Kotov, N.A. (2007) Successful differentiation of mouse neural stem cells on layer-by-layer assembled single-walled carbon nanotube composite. *Nano Lett.*, **7**, 1123–1128.
58. Jan, E., Pereira, F.N., Turner, D.L., and Kotov, N.A. (2011) In situ gene transfection and neuronal programming on electroconductive nanocomposite to reduce inflammatory response. *J. Mater. Chem.*, **21**, 1109–1114.
59. Koh, L.B., Rodriguez, I., and Zhou, J.J. (2008) Platelet adhesion studies on nanostructured poly(lactic-co-glycolic-acid)-carbon nanotube composite. *J. Biomed. Mater. Res. A*, **86**, 394–401.
60. Nepal, D., Balasubramanian, S., Simonian, A.L., and Davis, V.A. (2008) Strong antimicrobial coatings: Single-walled carbon nanotubes armored with biopolymers. *Nano Lett.*, **8**, 1896–1901.
61. Krogman, K.C., Lowery, J.L., Zacharia, N.S., Rutledge, G.C., and Hammond, P.T. (2009) Spraying asymmetry into functional membranes layer-by-layer. *Nature Mater.*, **8**, 512–518.
62. Andres, C.M. and Kotov, N.A. (2010) Inkjet Deposition of Layer-by-Layer Assembled Films. *J. Am. Chem. Soc.*, **132**, 14496–14502.
63. Nakashima, T., Zhu, J.A., Qin, M., Ho, S.S., and Kotov, N.A. (2010) Polyelectrolyte and carbon nanotube multilayers made from ionic liquid solutions. *Nanoscale*, **2**, 2084–2090.
64. Yang, M., Cao, K., Sui, L., Qi, Y., Zhu, J., Waas, A., Arruda, E.M., Kieffer, J., Thouless, M.D., and Kotov, N.A. (2011) Dispersions of aramid nanofibers: A new nanoscale building block. *ACS Nano.*, **5**, 6945–6954.

# 24
# Carbon Nanotube-Based Multilayers
Yong Tae Park and Jaime C. Grunlan

## 24.1
## Introduction

Intense study has followed the discovery of carbon nanotubes (CNTs) [1, 2] because of their impressive properties. CNTs have been one of the most promising materials in the growing field of nanotechnology over the last two decades. Single-walled carbon nanotubes (SWNTs), consisting of one layer of the hexagonal graphite lattice rolled to form a seamless cylinder with a radius up to a few nanometers, are especially promising. CNT research has made tremendous strides in synthesis, purity, and analysis of this new nanoscale material and numerous studies are now focused on the electrical, optical, thermal, mechanical, and chemical properties of large-area thin films. Recently, CNT-based thin films have been investigated as components in electronic devices, including high mobility transistors [3, 4], sensors [5, 6], solar and fuel cells [7, 8], integrated circuits [9, 10], and transparent electrodes [11]. CNT-based thin films are a more recent alternative to existing transparent conductive layers. However, one of the major obstacles to introducing CNTs to macroscale composite applications is the reduction in their unique properties. When random networks of CNTs are formed on a large substrate, the electrical performance of the resulting film decreases dramatically due to disconnection of each individual nanotube. This undesirable effect necessitates the development of more optimal methods of large-scale production.

Several methods to prepare CNT thin films, such as vacuum filtration [12], air-spraying [13], transfer printing [14], rod coating [15], spin coating [16], dip-coating [17], direct CVD growth [18], and electrophoretic deposition [19] have been explored. Although transfer printing is one of the best methods for producing CNT thin films, there has been difficulty with scale-up, breakage of the film during transfer, and comparatively brittle final films [20]. These challenges have opened the door for new methods of production, including the use of layer-by-layer (LbL) assembly to produce these important CNT films. LbL assembly is one of the best nanoscale coating techniques due to its simplicity and versatility, which makes ultrathin films (1~100 nm) in a homogeneous and controlled manner. One of the

*Multilayer Thin Films: Sequential Assembly of Nanocomposite Materials*, Second Edition.
Edited by Gero Decher and Joseph B. Schlenoff.
© 2012 Wiley-VCH Verlag GmbH & Co. KGaA. Published 2012 by Wiley-VCH Verlag GmbH & Co. KGaA.

major advantages of LbL assemblies is the ability to incorporate several different materials into precise levels within sub-micron films. In addition to conventional polyelectrolytes, various functional materials have been assembled by electrostatic LbL assembly. These unconventional ingredients include biomaterials, especially DNA [21] and proteins [22], and charged objects like colloidal nanoparticles [23], metal oxides [24], clay [25], nanosheets [26], and nanotubes [27]. Photonic crystals have been built by altering the size of quantum dots (QDs) in successive layers [28], which highlights the power of this technique. Carbon nanotubes require the use of stabilizers to impart the surface charge needed for effective LbL assembly. This chapter focuses on state-of-the-art LbL assembly with CNTs, and properties of these multilayer films, along with a survey of various applications. The survey begins with the preparation and procedure of LbL assembly with carbon nanotubes.

## 24.2
## Characteristics of Carbon Nanotube Layer-by-Layer Assemblies

### 24.2.1
### Growth of Carbon Nanotube-Based Multilayers

LbL assembly typically produces thin films through electrostatic interactions, by alternately exposing a substrate to positively and negatively-charged materials in aqueous solutions, as shown in Figure 24.1. Film thickness is determined by the

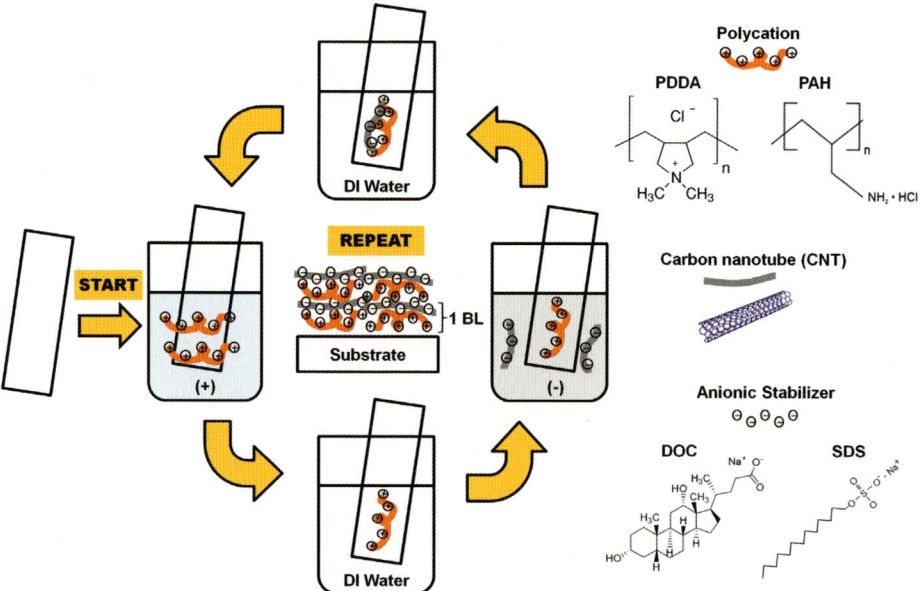

**Figure 24.1** Schematic of the deposition process for CNT-based multilayers. Cationic and anionic ingredients shown are examples taken from the literature.

number of deposition cycles, in which each positive and negative pair deposited is referred to as a layer pair. For assemblies made with CNTs, it is most common to stabilize with a negatively charged polymer or surfactant [29–31]. These negatively-stabilized tubes are then alternately deposited with a cationic polymer.

Ultrathin conductive assemblies made with CNTs can exhibit high transparency, low sheet resistance, and improved mechanical properties, making them an important topic of study. The principal advantages of CNT-based LbL assemblies include the simple process, done at room-temperature under ambient conditions, and the tunable properties achieved by altering the number of deposition cycles and/or sequence of layers. Another major advantage of LbL deposition of CNTs is versatile combination of complementary materials as a counterpart of the assembly process. These ingredients include: (i) conventional polymers, such as poly(diallyldimethyl ammonium chloride) (PDDA) [29], poly(allylamine hydrochloride) (PAH) [30], poly(aniline) (PANI) [31], polyethyleneimine (PEI) [32], poly(vinyl alcohol) (PVA) [33], poly(acrylic acid) (PAA) [34], poly(styrene sulfonate) (PSS) [35]; (ii) functional polymers such as poly(viologen) (POV) [36], polypyrrole (PPy) [37], poly-L-lysine (PLL) [38], and diazoresin (DR) [39]; (iii) bio-related macromolecules, such as biopolymer chitosan [40], polyamidoamine (PAMAM) dendrimers [41], hemoglobin (Hb) and myoglobin (Mb) [42], and enzyme horseradish peroxidase (HRP) [43]; (iv) inorganic nanoparticles, such as cetylpyridinium bromide (CPB) [44], platinum [45], gold [46], iron(II)phthalocyanine [47], titanium oxide nanoparticles [48], and graphene oxide nanosheets [49]. The most common materials used for LbL assembly are polymers. Polymeric LbL is much less dependent on the substrate, or the substrate charge density, than films made with smaller molecules. Completely nanoparticle-based LbL assemblies have also been successfully produced [50], but nanoparticle-polymer assemblies are much more common.

As previously mentioned, CNTs typically make use of stabilizers to impart the surface charge needed for effective LbL assembly. Stabilized nanotubes behave as rigid polyelectrolytes as a result of imparted charge. One of the most challenging issues in CNT LbL deposition is to fully disperse individual nanotubes in the solution, because nanotubes tend to form bundles. Poor dispersion also leads to aggregates of nanotubes in the resulting thin films, which results in reduced electrical conductivity and initiation of cracks in these composite thin films. Nevertheless, several methods have been successfully used to obtain excellent dissolution of CNTs in organic or aqueous solutions due to chemical modification of the side walls, including covalent attachment of chemical groups and the use of dispersing agents such as surfactants, dispersants, or other solubilizing agents [51–53]. Most LbL studies have used amphiphilic surfactants as the dispersing agents [29–31, 54–56], because of their ability to stabilize the hydrophobic CNTs in water. Unfortunately, covalent modification degrades the electronic and mechanical properties due to loss of their high aspect ratio from lattice damage. Noncovalent interactions are milder than covalent modification, but excess stabilizer can also cause reduction in electrical conductivity by providing individual nanotubes with an insulating layer.

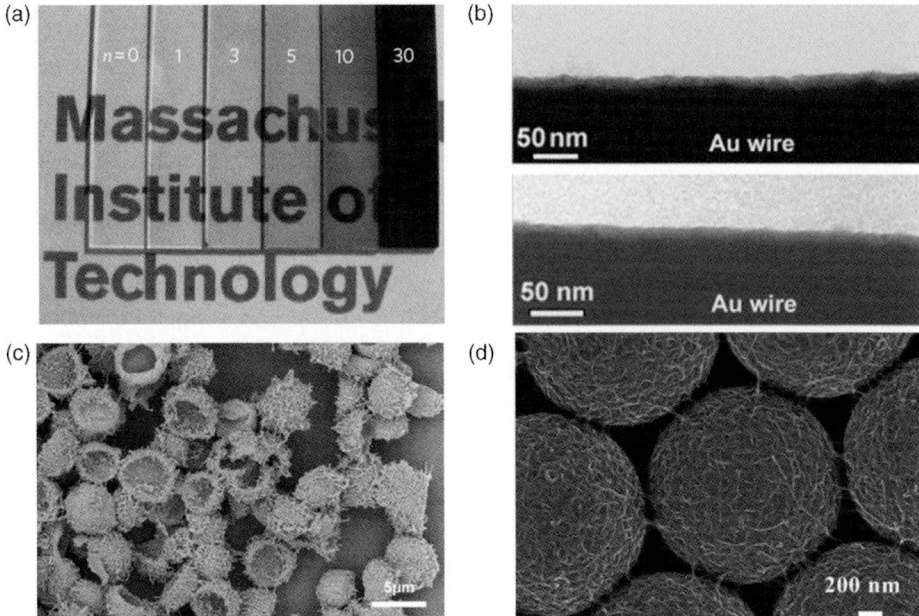

**Figure 24.2** CNT-based LbL coatings on the surfaces of (a) ITO-coated glass slides (the number on each image indicates the number of layer pairs ($n$) of (MWNT–NH$_2$/MWNT–COOH)$_n$) [57], (b) gold wires [60], (c) PS beads (ambient-dried PDDA/CNT hollow spheres after template removal, mostly collapsed) [61], and (d) PS templates (image of PS particles coated with one layer of CNTs) [63].

LbL deposition of CNTs is compatible with most substrates and there are no restrictions with respect to size, shape, and topology of the substrate. CNT-based assemblies have been grown on two-dimensional substrates, such as glass slides, indium tin oxide (ITO)-coated substrates [57], glassy carbon electrodes [58], gold disc electrodes [59], semiconductor wafers [41], and PET and PS films [29]. The conformal nature of this coating process typically results in a uniform coating of every three-dimensional surface of polyurethane foams, fabrics, gold wire [60], spherical beads [61, 62], PS hemispherical templates [63], and PDMS stamps [64]. Figure 24.2 highlights several examples of these assemblies on various substrates. Several studies used surface treatment of the substrates and/or a primer layer for charge enhancement. Corona treatment is used to oxidize the surface of polymer films, which leads to the formation of hydrophilic groups that help the polycationic species to adhere better. The glass slides were treated with an oxygen plasma etcher prior to use [27]. A precursor, or primer layer also facilitates adhesion, with strong electrostatic attraction to the underlying substrate as well as to the layers above. Several types of positively and negatively charged primer layers have been deposited on the surfaces of glass or silicon [30, 46], gold electrodes [43, 47, 56, 65], and patterned substrates [66].

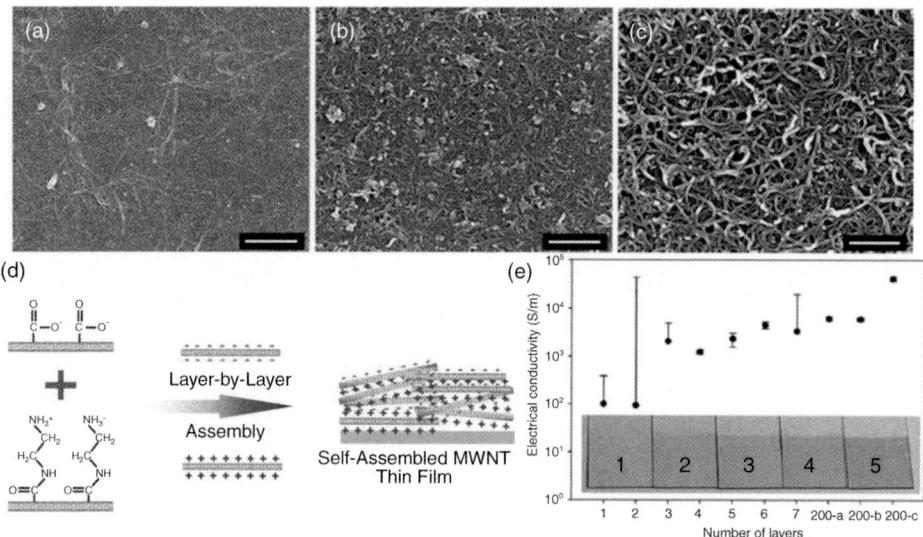

**Figure 24.3** SEM morphology of (a) [PVA/PSS]$_3$[PVA/(SWNT + PSS)]$_{15}$, (b) [PVA/PSS]$_3$[PVA/(TWNT + PSS)]$_{25}$, and (c) [PVA/PSS]$_3$[PVA/(MWNT-PSS)]$_{17}$ (reproduced from [68]). (d) MWNT thin film assembled with positively- and negatively-charged MWNTs (reproduced from [70]). (e) Electrical conductivity of [PVA/(SWNT + PSS)]$_n$ as a function of the number of bilayers deposited (reproduced from [69]).

Following the creation of the first SWNT-based LbL multilayer in 2002 [32], several research groups prepared CNT assemblies using a variety of recipes: oxidized individual SWNTs and PEI (denoted as PEI/SWNT$_{ox}$) [32], PANI/SWNT$_{ox}$ [67], PVA/(CNT + PSS) [68], PDDA/(CNT + DOC) [27], PDDA/(SWNT + SDS) [54], negatively and positively functionalized MWNTs (MWNT-NH$_2$/MWNT-COOH) [57], PDDA/(SWNT + ND) and (SWNT + PD)/PSS [30], where DOC, SDS, MWNT, ND and PD are sodium deoxycholate, sodium dodecyl sulfate, multiwalled carbon nanotube, naphthalene derivatives, and pyrene derivatives, respectively (see examples in Figure 24.3). These nanotube-based assemblies exhibit linear growth of thickness, mass, and absorbance, as a function of the number of layer pairs deposited (Figure 24.4) [27, 32, 54, 55, 67, 69–71], which suggests successful combination of CNTs and their counterions within the resulting composite, and constant composition of each component in each cycle. The growth of these thin films is also affected by the number of nanotube walls [27, 68], ionic strength of the CNT solution (by pH control [70] or choice of polyions or dispersant [30, 43]), CNT concentration [71], addition of electrolyte [40], and processing temperature [54]. Most commonly, SWNT-based films are much thinner than MWNT, suggesting that the film thickness depends on the radial size of the CNT. On the macroscopic scale, the surface roughness of the CNT-based multilayers depends on the components, and increases with the film thickness. It is this control of growth and composition that allows precise tuning of multilayer properties.

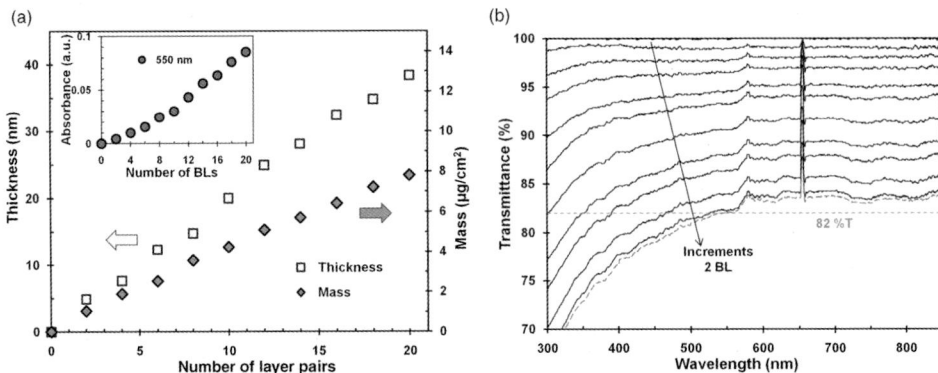

**Figure 24.4** Growth of PDDA/(SWNT + DOC) thin films as a function of the number of layer pairs deposited. Film thickness was obtained by ellipsometry, mass by QCM, and absorbance (at 550 nm) by UV-vis spectroscopy (inset), respectively [29].

## 24.2.2
### Electrical Properties

Four-point probe and current–voltage ($I$–$V$) measurements have been used to study the electrical properties of the CNT thin films. The in-plane and out-of-plane resistances of several LbL films were measured and the influence of CNT type on electrical conductivity was compared [27, 33, 49, 72, 73]. In Figure 24.5a, the conductivity of SWNT-based thin films is shown to gradually increase from 40 to 150 S cm$^{-1}$ as the number of layer pairs increases, but not as dramatically as a percolation transition. This suggests that the PDDA/(SWNT + DOC) system has a CNT concentration over the percolation threshold, and the density of intersecting pathways increases as the network transitions from two- to three-dimensional. SWNTs are largely deposited parallel to the substrate and generate a more two-dimensional structure, especially at low layer pair numbers. As more layers are deposited, the density of intersecting pathways can increase due to interconnection between SWNTs in the upper and lower layers. In contrast, the MWNT films have relatively constant conductivity, which suggests that homogeneous three-dimensional networks are formed from the beginning due to greater rigidity (from larger radius and lower aspect ratio) that would not allow the tubes to fully lay down on the substrate [27]. $I$–$V$ studies indicate the anisotropic electrical properties of LbL films are due to their layered structure that is dominated by the nanotubes lying in the horizontal plane (Figure 24.5b, c) [67]. It has been suggested that the charge transfer behavior through the CNT LbL film is controlled by quantum mechanical tunneling of carriers between the nanotubes [74]. Additionally, the in-plane $I$–$V$ behaviors were found to be dependent on the concentration of CNTs, exhibiting a nonlinear $I$–$V$ regime at low concentration and ohmic conduction behavior with higher CNT loading in the thin films [75].

Several studies have reported that the electrical properties of CNT-based LbL nanocomposites were improved by increasing the density of electrical pathways.

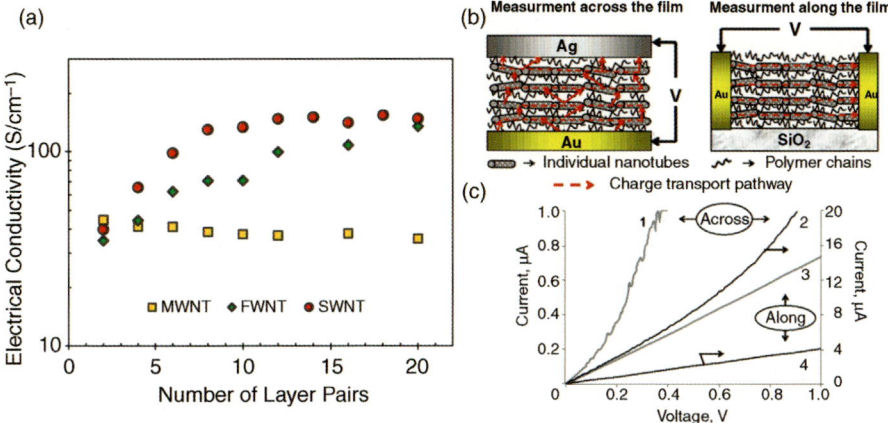

**Figure 24.5** (a) Electrical conductivity of [PDDA/(CNT + DOC)]$_n$ thin films as a function of the number of layer pairs deposited on PS substrates [27]. (b) Schematic representation of the SWNTox/polymer multilayer film structure and electrode positioning for electrical measurements across and along the films. Red arrows indicate charge pathways through the films. (c) I–V characteristics of [PANI/SWNTox]$_{19}$PANI (1, 3) and [PEI/SWNTox]$_{15}$PEI (2, 4) thin film devices measured across (1, 2) and along (3, 4) the films [67].

Heat treatment at 200 and 300 °C was used to form tighter connections between individual nanotubes, due to removal of water molecules [74] and polymer in the resulting films [69, 73, 76]. Moreover, a series of post-treatments, involving heating and nitric acid doping, were used to increase conductivity to 1430 from 160 S cm$^{-1}$, with no change in transparency, owing to the removal of insulating molecules and charge transfer doping [27].

Electrochemical properties of CNT-based LbL assemblies were also investigated, showing good capacitive performance and sensitive electrochemical behavior. An additive-free and functionalized MWNT-based LbL assembly exhibited reversible capacitance for lithium ions when cycled thousands of times [57]. The well-developed nanoporous structure between the MWNTs may create fast ion diffusion channels to facilitate ion transport. Addition of MnO$_2$ to MWNT multilayers yielded a higher capacitance, with good retention, due to faster electron and ion transport [77]. Other CNT-based hybrid multilayer films have also exhibited electrochemical activity towards specific molecules or pH [78–80]. These electrochemical properties were highly dependent on the number of layers. The thickness-dependent behavior demonstrates the tailorability of electrochemical properties in these assemblies.

### 24.2.3
### Mechanical Behavior

As mentioned above, the properties of these multilayer assemblies are strongly affected by the strength of interactions between nanotubes in the adjacent layers. CNT-based multilayers can be thought of as fiber- reinforced composites consisting of a soft polymer matrix encapsulating stiffer and electrically conductive fibers. The high aspect ratio of CNTs leads to high strength composites due to good load transfer from the matrix to the nanotubes. Conventional CNT-polymer composites produced by mixing, or *in situ* polymerization, often fail to show high strength due to difficulty forming strong interfaces between nanotubes and the surrounding polymer matrix [51]. In contrast, LbL multilayer composites exhibit remarkable mechanical properties due to high nanotube content, uniform distribution, and strong interfacial bonding between CNTs and the polyelectrolyte matrix. Additional mechanical modification by CNT-polymer crosslinking was performed with heat treatment and addition of a crosslinking agent [32]. 40-layer-pair PEI/SWNT films were shown to exhibit a large ultimate strength of 220 MPa and Young's modulus of 35 GPa [32]. Some other studies also reported high ultimate tensile strength and Young's modulus, with a relatively high strain (1~5%), for MWNT-based free-standing films [55] and SWNT films [33, 54, 69]. In addition, nanoindentation and nanoscratch techniques were used to evaluate the hardness (1 GPa) and coefficient of friction (0.66) of CNT-based multilayers [63, 71, 76, 81–83]. These values are much better than those of the other CNT–polymer composites reported, which leads to the potential for stiff nanoscale electronic devices (e.g., cantilevers) [84]. The high mechanical strength and electrical conductivity of CNT-based multilayers create the opportunity for a new generation of electrical, mechanical, and optical applications.

## 24.3
### Applications of Carbon Nanotube Layer-by-Layer Assemblies

CNT-based LbL films have recently emerged as a useful new electronic material, for use in applications with macroscopic dimensions. These thin films, with random CNT networks, can be used in applications ranging from transparent electrodes to sensors. In this section, potential applications are illustrated by the demonstration of (i) good optoelectronic performance for transparent electrodes; (ii) sensitive detection for physical/chemical sensors; (iii) superior charge storage for energy-related devices; and (iv) biocompatible or antimicrobial behavior for bioelectronics. Each of the highlighted applications use LbL assembled multilayers made with CNTs and counterparts, such as polymers and inorganic nanoparticles.

### 24.3.1
**Transparent Electrodes**

As the cost of carbon nanotubes comes down, and high speed/volume LbL deposition is demonstrated, the superior mechanical and electro-optical performance of these thin films is spurring continued research toward a fully organic ITO replacement. Optoelectronic devices, such as touch screens and flexible displays, require electrodes that exhibit high transparency and low sheet resistance. CNT thin films are an interesting alternative to existing transparent conductive layers, such as brittle and chemically unstable ITO and low transparency poly(3,4-ethylenedioxythiophene)-

poly(styrenesulfonate) [PEDOT-PSS]. Layer-by-layer assembly has been shown to generate highly dense nanotube networks with good transparency and electrical conductivity [29, 33]. Although other processes are also used for producing CNT thin films [85, 86], there has been difficulty with scale-up, breakage of the film during transfer, and comparatively brittle final films. These challenges have opened the door for the use of LbL assembly to produce these important electrodes. The thickness tailorability, down to an increment of 2–4 nm, allows precise tuning of light transmission, which is vital for optoelectronic devices.

Following the first study of an LbL-produced transparent electrode, with 2.5 k$\Omega$/sq and 87% average light transmittance [87], a series of studies has been performed to obtain thin films with low sheet resistance and high transparency [27, 29, 33, 49, 68, 69]. Figure 24.6 compares sheet resistance as a function of visible light transmission for a variety of CNT LbL assemblies. Many of these systems approach the performance characteristics of ITO. Very recently, a double walled carbon nanotube (DWNT)-based LbL electrode, together with nitric acid treatment, achieved a sheet resistance of 100 $\Omega$/sq and visible light transmittance greater than 85%, which is in the middle of the performance range for commercial ITO-coated PET (Figure 24.7). Highly transparent thin film electrodes were assembled through the alternate exposure of flexible substrates to aqueous mixtures of positively-charged PDDA and DWNTs, stabilized with negatively-charged DOC. DWNT thin films exhibit a significant increase in electrical conductivity after exposure to nitric acid vapor. Strong acid dopes the individual nanotubes and removes insulating material (polymer and surfactant). Additionally, these DWNT LbL coatings on PET substrates have excellent flexibility without any loss of conductivity after 100 bending cycles,

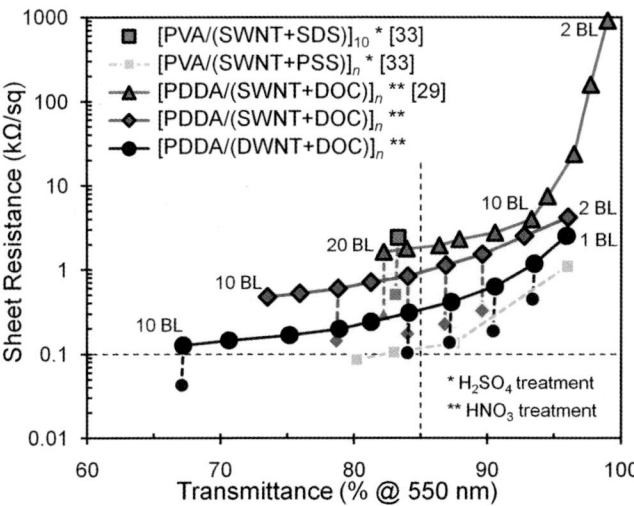

**Figure 24.6** Optoelectronic performance of several CNT LbL systems before and after acid treatment. Points with the black outline are values of as-assembled LbL films and those with no outline are acid-treated.

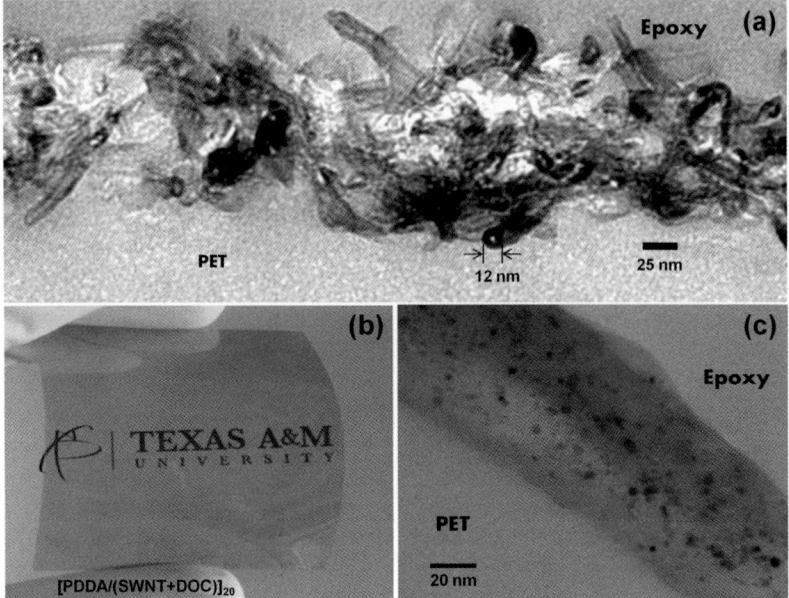

**Figure 24.7** (a) TEM cross-section of a [PDDA/(MWNT + DOC)]$_{20}$ thin film [27]. (b) Optical image of a [PDDA/(SWNT + DOC)]$_{20}$ coating on both sides of a 7.5 × 7.5 cm$^2$ PS film. (c) TEM cross-section of a [PDDA/(SWNT + DOC)]$_{20}$ thin film [29].

unlike ITO. These highly dense CNT networks with low sheet resistance, good transparency, superior mechanical flexibility, and electrochemical stability meet the criteria for ITO replacement in most electronics applications.

### 24.3.2
### Sensor Applications

In addition to producing simple electrodes, the LbL technique can be used to prepare nanostructured physical, chemical, or biological sensors by the sequential exposure of cationic and anionic ingredients. The source of this sensing behavior can be: electrical property change of CNTs by external stimuli, such as pH, deformation, or humidity; electrochemical sensitivity of counterpart materials of CNTs; or hybrid multilayered structures with sensing layers on the CNT-based conductive electrode.

CNT-based LbL thin films have been fabricated that exhibit changes in their electrical properties due to humidity [88], strain [5, 89], and pH [5, 66, 80]. An example of this is a highly sensitive, quick response, and reversible humidity sensor, based on PDDA/CNT composite films, whose resistance was shown to increase exponentially with an increase in humidity (from 20 to 98%) [88]. The CNT junction resistance dominates the humidity sensing in these nanocomposite films (Figure 24.8a). In other studies, a strain sensor based upon a piezoresistive response was produced by LbL deposition of PVA/SWNT [5] and highly sensitive pH sensors were used to detect

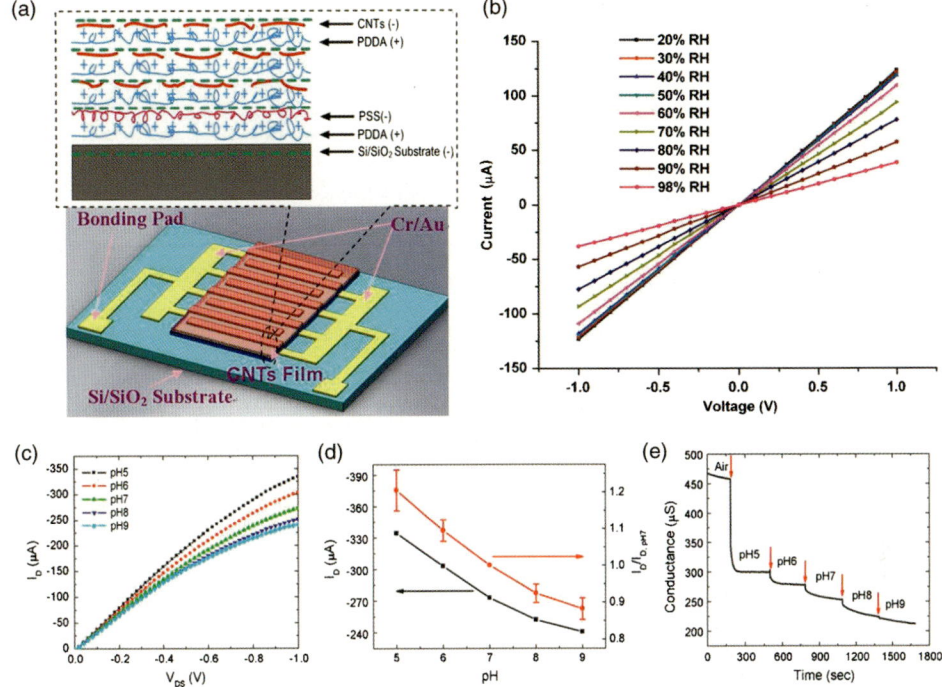

**Figure 24.8** (a) Structure of a PDDA/CNT multilayer composite humidity sensor. The LbL-assembled film consists of a (PDDA/PSS)$_2$ precursor and (PDDA/CNT)$_2$ layers. (b) I–V curves of a typical eight-layer PDDA/SWNT sensor, with varying RH from 20 to 98%, measured at 25 °C [88]. (c–e) The pH response of the flexible SWNT sensor: (c) representative $I_D$–$V_{DS}$ curves under different pH buffers at fixed $V_G = -1.5$ V, (d) pH-responsive drain current and normalized $I_D$ by the current at pH 5 buffer at fixed $V_{DS} = -1.0$ V and $V_G = -1.5$ V, and (e) time response of the flexible SWNT multilayer conductance to pH without gate voltage applied [80].

the corrosion of metallic structural materials by using PANI in conjunction with nanotubes (Figure 24.8b–e) [80].

This pH-dependent response of CNT multilayers has also been used to make biosensors, based on changes in electrical properties caused by a pH change due to electrocatalytic activity of enzymes [66, 80, 90]. Many biosensors have utilized the response of CNTs to a change in hydrogen peroxide (H$_2$O$_2$) concentration [35, 37, 42, 47, 59, 65, 91–93]. CNT-based assemblies have been applied to detect the reduction or oxidation of various biomolecules electrochemically, by using various polymers [91–95], proteins [42], and nanoparticles [43, 44, 47, 59] in the recipes. Biosensors for glucose [37, 59, 64, 65, 92–94], proteins [42], DNA [95, 96], cysteine [43, 97], cholesterol [35], dopamine [44], acetylcholine [90], and penicillin [41] have been developed with CNT-based multilayered electrodes. Most biosensors have utilized a specific enzyme as a bio-receptor, such as glucose oxidase (GOx) [37, 59, 64, 65, 92, 93], horseradish peroxidase (HRP) [43], penicillinase [41], acetylcholinesterase [90],

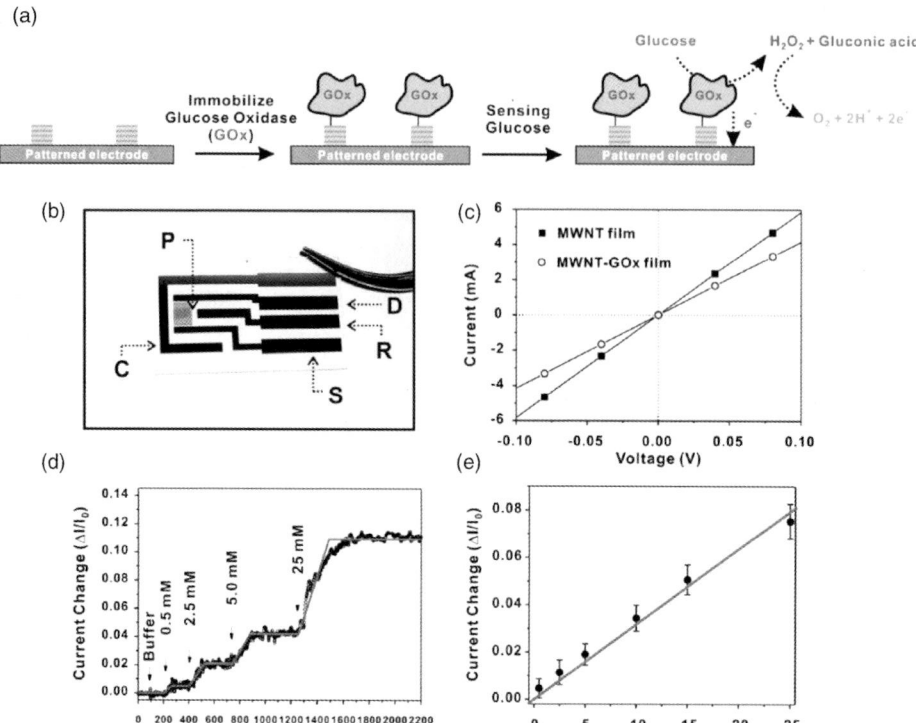

**Figure 24.9** (a) Schematic of patterned enzyme-MWNT film for glucose sensing. (b) Photograph of patterned enzyme-MWNT film on microelectrode array (the patterned MWNT film (P), source (S), drain (D), reference (R), and counter (C) electrodes are individually labeled). (c) Characteristic I–V curve of MWNT film and MWNT–GOx film. (d) Real-time response of the sensor upon a serial addition of 0.5–25.0 mM$_{glucose}$ (the arrow indicates the addition of analyte solutions), and (e) calibration curve of the sensor (sensitivity was determined as the current change measured when the saturated value is reached after the addition of glucose) [64].

cholesterol oxidase [35], or choline oxidase [91]. For glucose sensors, the CNT-based multilayer film has a catalytic effect towards $H_2O_2$, with remarkable current and rapid response. GOx was immobilized on the CNT LbL electrode and exhibited excellent catalytic behavior toward glucose oxidation to form $H_2O_2$, which shows good ability to detect glucose (Figure 24.9). Contrary to amperometric glucose sensors, ion-sensitive field-effect transistors were fabricated as conductometric sensors, composed of three different layers: (PDDA/PSS)$_2$ as a precursor layer for charge enhancement, (PDDA/SWNT)$_5$ as an electrochemical transducer, and (PDDA/GOx)$_3$ as a bioreceptor [66, 80]. This tailorable LbL assembly of CNTs and enzymes on transparent and flexible substrates lends itself to the creation of various chemical and biological sensors suitable for *in vivo* application.

### 24.3.3
### Energy-Related Applications

The photovoltaic/electrochemical applications described below are limited to recent advancements in the application of CNT-based LbL assemblies. CNTs facilitate charge transport by assembling active layers for an efficient pathway. Assembling chromophores and CNTs is a recent approach for photovoltaic and electron storage devices, because the chromophore and CNT function as an electron donor and acceptor, respectively [56, 98]. For the controlled organization of SWNTs and electron donor molecules onto the electrodes, a PDDA/PSS/(SWNT + pyrene)/ZnP$^{8-}$ LbL film was assembled onto ITO substrates (ZnP$^{8-}$: negatively charged zinc porphyrin) [99]. Electrons transfer from the reduced SWNTs to the ITO electrodes through the PDDA/PSS layer, and the sodium ascorbate reduces the oxidized porphyrins, as shown in Figure 24.10a. Furthermore, repetitive deposition steps allow modification of device efficiency with (SWNT + pyrene)/ZnP$^{8-}$ multilayers. In addition, a 5,10,15,20-tetrakis(1-methyl-4-pyridinio)porphyrin tetra(p-toluenesulfonate) (TMPyP)–SWNT/sodium copper chlorophyllin (SCC)–SWNT LbL film was introduced by dissolving SWNTs in water-soluble cationic TMPyP and anionic SCC [56]. The incorporation of MWNTs was found to enhance the photocatalytic activity of TiO$_2$ [48]. These enhanced photocatalytic activities demonstrate the benefit of introducing CNTs to highly efficient photoelectrochemical devices, such as dye-sensitized solar cells. In this manner, LbL assemblies can be extended to the combinations of CNTs with various photoactive molecules together with optimization of the device structure.

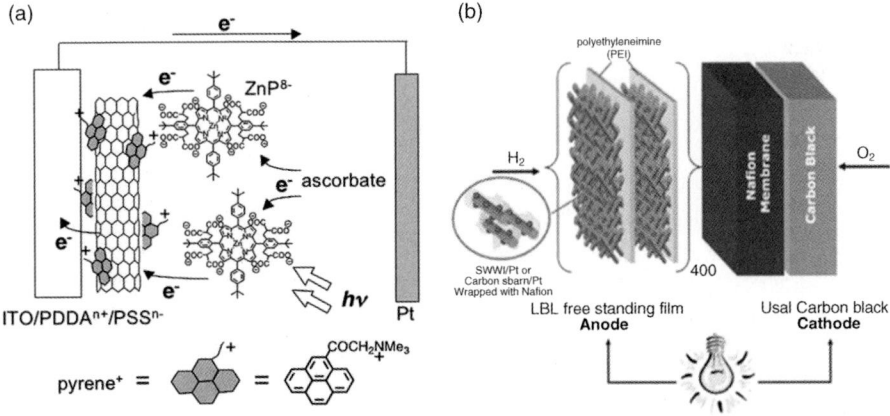

**Figure 24.10** (a) Schematic of photocurrent generation with a photoelectrochemical device using a PDDA/PSS/(SWNT + pyrene)/ZnP$^{8-}$ electrode [99]. (b) High-performance nanostructured membrane electrode assemblies for fuel cells made by LbL assembly of carbon nanocolloids (i.e., SWNT and carbon fiber) functionalized by platinum nanoparticles [101].

CNT-based multilayer films have also been used in fuel cells. Due to their ultrathin and nanoporous structure, uniform integrity, ease of preparation, and readily controllable platinum loading, PDDA/(Pt–CNT) multilayers exhibit the potential for application to electrodes in direct methanol fuel cells [100]. These carbon-supported platinum nanocomposite materials were also incorporated into catalyst and gas diffusion layers with polyelectrolytes, using LbL assembly techniques (i.e., (PANI or PEI)/(Pt–SWNTs + Nafion)) [31, 101]. These LbL-deposited catalyst layers yielded much higher platinum utilization than membrane electrode assemblies produced using conventional methods, when applied to the anode and cathode of a polymer electrolyte membrane fuel cell, as shown in Figure 24.10b. Additionally, CNT-based assemblies have been used as electrodes for biofuel cell (BFC) and microbial fuel cell (MFC) applications. The (MWNT/poly-L-lysine/laccase/poly-L-lysine) quadlayer shows potential for use as a cathode for improving BFC performance by promoting the electron transfer of lactase and poly-L-lysine [38]. The CNT-based MFC anode enabled good anodic bacteria anchoring and electron transfer due to more specific surface area and less interfacial resistance from the LbL-assembled CNTs [102].

### 24.3.4
### Biomedical Applications

CNT-based assemblies have been constructed as favorable substrates for neural differentiation and attachment, and electrical stimulation of neural cells [103]. It was reported that the properties of freestanding SWNT films, such as flexibility, non-biodegradability, and durability, make them potential candidates for extracellular implants and neuroprosthetic elements, due to favorable cell interactions. One of the key challenges to engineering neural interfaces is to minimize their immune response toward implanted electrodes. Also, electrical stimulation of excitable neural cells was evaluated for neurite outgrowth using SWNT multilayers with sufficiently high electrical conductivity [34]. Following the studies of neural cells, the differentiation of environment-sensitive neural stem cells was successfully performed on a SWNT–polyelectrolyte multilayer [104]. A multilayer implanted electrode, made with SWNTs and protein (i.e., laminin) for greater structural resemblance to living tissues, was also studied, to show viability for humans [73]. The SWNT/laminin thin films were suitable for excitation of NSCs and the products of their differentiation. In addition to the neural cells, the effects of a CNT-based multilayer on bone tissue *in vivo* in rats were studied and found to be biocompatible and to promote bone formation [82]. The unique electrical and mechanical properties of CNTs make them a promising substrate for bone implants.

In contrast to the biocompatible studies just described, significant antimicrobial activity was measured from large scale biomimetic SWNT-based coatings. Non-leaching lysozyme–SWNT/DNA–SWNT multilayers exhibit robust mechanical properties and long-term antimicrobial efficacy due to preservation of antimicrobials over time [71]. Additionally, a novel delivery system was developed with polyoxometalate (POM)-modified SWNTs and chitosan [40]. Addition of SWNTs suppressed the initial release of POMs, a promising antitumor drug, and guaranteed higher loading and prolonged release of POMs due to strong interaction with SWNTs. The conformal nature of LbL assemblies makes this a promising technique for imparting antimicrobial and drug delivery behavior to a variety of complex substrates, such as wound dressings and patches.

## 24.4
## Conclusions

This chapter surveyed the preparation, properties, and applications of carbon nanotube-based thin films made using layer-by-layer assembly. Most CNT-based multilayers grow linearly as a function of the layer pairs deposited, which allows precise tailoring of thickness and various properties. The combination of exceptional nanotube properties and the unique characteristics of LbL assembly provides a number of opportunities for important applications. CNTs impart high electrical conductivity (or low sheet resistance) and significant mechanical strength to multilayer thin films. These useful properties originate with the nanoscale control and homogeneity afforded by the LbL process. The applications described in this chapter demonstrate the benefits of these films when used as transparent electrodes, physical/chemical/biological sensors, and energy-related devices. Further work is likely to improve both the electrical and mechanical properties of these assemblies and open the door to more applications (and eventual commercial use).

## References

1 Oberlin, A., Endo, M., and Koyama, T. (1976) *J. Cryst. Growth*, **32**, 335.
2 Iijima, S. (1991) *Nature*, **354**, 56.
3 Kang, S.J., Kocabas, C., Ozel, T., Shim, M., Pimparkar, N., Alam, M.A., Rotkin, S.V., and Rogers, J.A. (2007) *Nat. Nanotechnol.*, **2**, 230.
4 Xue, W., Liu, Y., and Cui, T.H. (2006) *Appl. Phys. Lett.*, **89**, 163512.
5 Loh, K.J., Kim, J., Lynch, J.P., Kam, N.W.S., and Kotov, N.A. (2007) *Smart Mater. Struct.*, **16**, 429.
6 Modi, A., Koratkar, N., Lass, E., Wei, B.Q., and Ajayan, P.M. (2003) *Nature*, **424**, 171.
7 Lowman, G.M. and Hammond, P.T. (2005) *Small*, **1**, 1070.
8 Endo, M., Hayashi, T., Kim, Y.A., Terrones, M., and Dresselhaus, M.S. (2004) *Philos. Trans. R. Soc. London, Ser. A*, **362**, 2223.
9 Cao, Q., Kim, H.S., Pimparkar, N., Kulkarni, J.P., Wang, C.J., Shim, M., Roy, K., Alam, M.A., and Rogers, J.A. (2008) *Nature*, **454**, 495.
10 Friedman, R.S., McAlpine, M.C., Ricketts, D.S., Ham, D., and Lieber, C.M. (2005) *Nature*, **434**, 1085.
11 Wu, Z.C., Chen, Z.H., Du, X., Logan, J.M., Sippel, J., Nikolou, M., Kamaras, K., Reynolds, J.R., Tanner, D.B., Hebard, A.F., and Rinzler, A.G. (2004) *Science*, **305**, 1273.
12 Chen, P.C., Shen, G., Sukcharoenchoke, S., and Zhou, C. (2009) *Appl. Phys. Lett.*, **94**, 043113.
13 Mo, C.B., Hwang, J.W., Cha, S.I., and Hong, S.H. (2009) *Carbon*, **47**, 1276.
14 Zhang, D.H., Ryu, K., Liu, X.L., Polikarpov, E., Ly, J., Tompson, M.E., and Zhou, C.W. (2006) *Nano Lett.*, **6**, 1880.

15 Dan, B., Irvin, G.C., and Pasquali, M. (2009) *ACS Nano*, **3**, 835.
16 Williams, Q.L., Liu, X., Walters, W., Zhou, J.G., Edwards, T.Y., and Smith, F.L. (2007) *Appl. Phys. Lett.*, **91**, 143116.
17 Spotnitz, M.E., Ryan, D., and Stone, H.A. (2004) *J. Mater. Chem.*, **14**, 1299.
18 Ma, W.J., Song, L., Yang, R., Zhang, T.H., Zhao, Y.C., Sun, L.F., Ren, Y., Liu, D.F., Liu, L.F., Shen, J., Zhang, Z.X., Xiang, Y.J., Zhou, W.Y., and Xie, S.S. (2007) *Nano Lett.*, **7** 2307.
19 Lima, M.D., de Andrade, M.J., Bergmann, C.P., and Roth, S. (2008) *J. Mater. Chem.*, **18**, 776.
20 Yu, X., Rajamani, R., Stelson, K.A., and Cui, T. (2008) *Surf. Coat. Technol.*, **202**, 2002.
21 Ma, H., Zhang, L., Pan, Y., Zhang, K., and Zhang, Y. (2008) *Electroanalysis*, **20**, 1220.
22 Seo, H.S., Kim, S.E., Park, J.S., Lee, J.H., Yang, K.Y., Lee, H., Lee, K.E., Han, S.S., and Lee, J. (2010) *Adv. Funct. Mater.*, **20**, 4055.
23 Park, J., Fouché, L., and Hammond, P. (2005) *Adv. Mater.*, **17**, 2575.
24 Park, Y.T. and Grunlan, J.C. (2010) *Electrochim. Acta*, **55**, 3257.
25 Priolo, M.A., Gamboa, D., and Grunlan, J.C. (2010) *ACS Appl. Mater. Interfaces*, **2**, 312.
26 Hendricks, T.R., Lu, J., Drzal, L.T., and Lee, I. (2008) *Adv. Mater.*, **20**, 2008.
27 Park, Y.T., Ham, A.Y., and Grunlan, J.C. (2010) *J. Phys. Chem. C*, **114**, 6325.
28 Mamedov, A.A., Belov, A., Giersig, M., Mamedova, N.N., and Kotov, N.A. (2001) *J. Am. Chem. Soc.*, **123**, 7738.
29 Park, Y.T., Ham, A.Y., and Grunlan, J.C. (2011) *J. Mater. Chem.*, **21**, 363.
30 Paloniemi, H., Lukkarinen, M., Ääritalo, T., Areva, S., Leiro, J., Heinonen, M., Haapakka, K., and Lukkari, J. (2006) *Langmuir*, **22**, 74.
31 Taylor, A.D., Michel, M., Sekol, R.C., Kizuka, J.M., Kotov, N.A., and Thompson, L.T. (2008) *Adv. Func. Mater.*, **18**, 3003.
32 Mamedov, A.A., Kotov, N.A., Prato, M., Guldi, D.M., Wicksted, J.P., and Hirsch, A. (2002) *Nat. Mater.*, **1**, 257.
33 Shim, B.S., Zhu, J., Jan, E., Critchley, K., and Kotov, N.A. (2010) *ACS Nano*, **4**, 3725.
34 Gheith, M.K., Pappas, T.C., Liopo, A.V., Sinani, V.A., Shim, B.S., Motamedi, M., Wicksted, J.R., and Kotov, N.A. (2006) *Adv. Mater.*, **18**, 2975.
35 Yang, M.H., Yang, Y., Yang, H.F., Shen, G.L., and Yu, R.Q. (2006) *Biomaterials*, **27**, 246.
36 Wang, X., Huang, H.X., Liu, A.R., Liu, B., Wakayama, T., Nakamura, C., Miyake, J., and Qian, D.J. (2006) *Carbon*, **44**, 2115.
37 Shirsat, M.D., Too, C.O., and Wallace, G.G. (2008) *Electroanalysis*, **20**, 150.
38 Deng, L., Shang, L., Wang, Y.Z., Wang, T., Chen, H.J., and Dong, S.J. (2008) *Electrochem. Commun.*, **10**, 1012.
39 Tang, M.X., Qin, Y.J., Wang, Y.Y., and Guo, Z.X. (2009) *J. Phys. Chem. C*, **113**, 1666.
40 Zhao, Q.C., Feng, X.D., Mei, S.L., and Jin, Z.X. (2009) *Nanotechnology*, **20**, 105101.
41 Siqueira, J.R., Abouzar, M.H., Poghossian, A., Zucolotto, V., Oliveira, O.N., and Schoning, M.J. (2009) *Biosens. Bioelectron.*, **25**, 497.
42 Zhao, L.Y., Liu, H.Y., and Hu, N.F. (2006) *Anal. Bioanal. Chem.*, **384**, 414.
43 Liu, L.J., Xi, F.N., Zhang, Y.M., Chen, Z.C., and Lin, X.F. (2009) *Sensor Actuat. B-Chem.*, **135**, 642.
44 Zhang, Y.Z., Pan, Y., Sit, S., Zhang, L.P., Li, S.P., and Shao, M.W. (2007) *Electroanalysis*, **19**, 1695.
45 Wang, L., Guo, S.J., Huang, L.J., and Dong, S.J. (2007) *Electrochem. Commun.*, **9**, 827.
46 Kim, J., Lee, S.W., Hammond, P.T., and Shao-Horn, Y. (2009) *Chem. Mater.*, **21**, 2993.
47 Pillay, J. and Ozoemena, K.I. (2009) *Electrochim Acta*, **54**, 5053.
48 Tettey, K.E., Yee, M.Q., and Lee, D. (2010) *ACS Appl. Mater. Interfaces*, **2**, 2646.
49 Hong, T.K., Lee, D.W., Choi, H.J., Shin, H.S., and Kim, B.S. (2010) *ACS Nano*, **4**, 3861.
50 Lee, D., Rubner, M.F., and Cohen, R.E. (2006) *Nano Lett.*, **6**, 2305.

51 Ajayan, P.M. and Tour, J.M. (2007) *Nature*, **447**, 1066.
52 Huang, W., Lin, Y., Taylor, S., Gaillard, J., Rao, A.M., and Sun, Y.-P. (2002) *Nano Lett.*, **2**, 231.
53 Haggenmueller, R., Rahatekar, S.S., Fagan, J.A., Chun, J.H., Becker, M.L., Naik, R.R., Krauss, T., Carlson, L., Kadla, J.F., Trulove, P.C., Fox, D.F., DeLong, H.C., Fang, Z.C., Kelley, S.O., and Gilman, J.W. (2008) *Langmuir*, **24** 5070.
54 Shim, B.S., Zhu, J., Jan, E., Critchley, K., Ho, S.S., Podsiadlo, P., Sun, K., and Kotov, N.A. (2009) *Acs Nano*, **3**, 1711.
55 Olek, M., Ostrander, J., Jurga, S., Mohwald, H., Kotov, N., Kempa, K., and Giersig, M. (2004) *Nano Lett.*, **4**, 1889.
56 Baba, A., Matsuzawa, T., Sriwichai, S., Ohdaira, Y., Shinbo, K., Kato, K., Phanichphant, S., and Kaneko, F. (2010) *J. Phys. Chem. C*, **114**, 14716.
57 Lee, S.W., Yabuuchi, N., Gallant, B.M., Chen, S., Kim, B.S., Hammond, P.T., and Shao-Horn, Y. (2010) *Nat. Nanotechnol.*, **5**, 531.
58 Zhang, M.N., Yan, Y.M., Gong, K.P., Mao, L.Q., Guo, Z.X., and Chen, Y. (2004) *Langmuir*, **20**, 8781.
59 Liu, Y., Wu, S., Ju, H.X., and Xu, L. (2007) *Electroanalysis*, **19**, 986.
60 Kovtyukhova, N.L. and Mallouk, T.E. (2005) *Adv. Mater.*, **17**, 187.
61 Ji, L.J., Ma, J., Zhao, C.G., Wei, W., Ji, L.J., Wang, X.C., Yang, M.S., Lu, Y.F., and Yang, Z.Z. (2006) *Chem. Commun.*, 1206.
62 Sugikawa, K., Numata, M., Kaneko, K., Sada, K., and Shinkai, S. (2008) *Langmuir*, **24**, 13270.
63 Firkowska, I., Olek, M., Pazos-Perez, N., Rojas-Chapana, J., and Giersig, M. (2006) *Langmuir*, **22**, 5427.
64 Kim, B.S., Lee, S.W., Yoon, H., Strano, M.S., Shao-Horn, Y., and Hammond, P.T. (2010) *Chem. Mater.*, **22**, 4791.
65 Yan, X.B., Chen, X.J., Tay, B.K., and Khor, K.A. (2007) *Electrochem. Commun.*, **9**, 1269.
66 Lee, D. and Cui, T.H. (2009) *IEEE Sens. J.*, **9**, 449.
67 Kovtyukhova, N.I. and Mallouk, T.E. (2005) *J. Phys. Chem. B*, **109**, 2540.
68 Park, H.J., Oh, K.A., Park, M., and Lee, H. (2009) *J. Phys. Chem. C*, **113**, 13070.
69 Shim, B.S., Tang, Z.Y., Morabito, M.P., Agarwal, A., Hong, H.P., and Kotov, N.A. (2007) *Chem. Mater.*, **19**, 5467.
70 Lee, S.W., Kim, B.S., Chen, S., Shao-Horn, Y., and Hammond, P.T. (2009) *J. Am. Chem. Soc.*, **131**, 671.
71 Nepal, D., Balasubramanian, S., Simonian, A.L., and Davis, V.A. (2008) *Nano Lett.*, **8**, 1896.
72 Palumbo, M., Lee, K.U., Ahn, B.T., Suri, A., Coleman, K.S., Zeze, D., Wood, D., Pearson, C., and Petty, M.C. (2006) *J. Phys. D Appl. Phys.*, **39**, 3077.
73 Kam, N.W.S., Jan, E., and Kotov, N.A. (2009) *Nano Lett.*, **9**, 273.
74 Lee, K.U., Cho, Y.H., Petty, M.C., and Ahn, B.T. (2009) *Carbon*, **47**, 475.
75 Jombert, A.S., Coleman, K.S., Wood, D., Petty, M.C., and Zeze, D.A. (2008) *J. Appl. Phys.*, **104**, 094503.
76 Xue, W. and Cui, T.H. (2007) *Nanotechnology*, **18**, 145709.
77 Lee, S.W., Kim, J., Chen, S., Hammond, P.T., and Shao-Horn, Y. (2010) *ACS Nano*, **4**, 3889.
78 Kong, B., Zeng, J.X., Luo, G.M., Luo, S.L., Wei, W.Z., and Li, J. (2009) *Bioelectrochemistry*, **74**, 289.
79 Zhai, J.F., Zhai, Y.M., Wen, D., and Dong, S.J. (2009) *Electroanalysis*, **21**, 2207.
80 Lee, D. and Cui, T.H. (2010) *Biosens. Bioelectron.*, **25**, 2259.
81 Lu, H.B., Huang, G., Wang, B., Mamedov, A., and Gupta, S. (2006) *Thin Solid Films*, **500**, 197.
82 Bhattacharya, M., Wutticharoenmongkol-Thitiwongsawet, P., Hamamoto, D.T., Lee, D., Cui, T.H., Prasad, H.S., and Ahmad, M. (2011) *J. Biomed. Mater. Res. A*, **96**, 75.
83 Lee, D. and Cui, T.H. (2011) *Nanotechnology*, **22**, 165601.
84 Xue, W. and Cui, T.H. (2007) *Sens. Actuat. A-Phys.*, **136**, 510.

85 Geng, H.Z., Kim, K.K., So, K.P., Lee, Y.S., Chang, Y., and Lee, Y.H. (2007) *J. Am. Chem. Soc.*, **129**, 7758.
86 Hu, L., Hecht, D.S., and Gruner, G. (2004) *Nano Lett.*, **4**, 2513.
87 Yu, X., Rajamani, R., Stelson, K.A., and Cui, T. (2006) *Sens. Actuat. A-Phys.*, **132**, 626.
88 Liu, L.T., Ye, X.Y., Wu, K., Zhou, Z.Y., Lee, D.J., and Cui, T.H. (2009) *IEEE Sens. J.*, **9**, 1308.
89 Loh, K.J., Hou, T.C., Lynch, J.P., and Kotov, N.A. (2009) *J. Nondestruct. Eval.*, **28**, 9.
90 Xue, W. and Cui, T.H. (2008) *Sens. Actuat. B-Chem.*, **134**, 981.
91 Qu, F.L., Yang, M.H., Jiang, J.H., Shen, G.L., and Yu, R.Q. (2005) *Anal Biochem.*, **344**, 108.
92 Liu, G.D. and Lin, Y.H. (2006) *Electrochem. Commun.*, **8**, 251.
93 Wang, Y., Wei, W.Z., Liu, X.Y., and Zeng, X.D. (2009) *Mater. Sci. Eng. C-Biomim. Supramol. Syst.*, **29**, 50.
94 Wang, Y.D., Joshi, P.P., Hobbs, K.L., Johnson, M.B., and Schmidtke, D.W. (2006) *Langmuir*, **22**, 9776.
95 Du, M., Yang, T., Zhang, Y.C., and Jiao, K. (2009) *Electroanalysis*, **21**, 2521.
96 Kang, T.J., Lim, D.K., Nam, J.M., and Kim, Y.H. (2010) *Sens. Actuat. B-Chem.*, **147**, 691.
97 Chen, X., Yang, Y., and Ding, M.Y. (2006) *Anal. Chim. Acta*, **557**, 52.
98 Sgobba, V., Rahman, G.M.A., Guldi, D.M., Jux, N., Campidelli, S., and Prato, M. (2006) *Adv. Mater.*, **18**, 2264.
99 Umeyama, T. and Imahori, H. (2008) *Energy Environ. Sci.*, **1**, 120.
100 Yuan, J.H., Wang, Z.J., Zhang, Y.J., Shen, Y.F., Han, D.X., Zhang, Q., Xu, X.Y., and Niu, L. (2008) *Thin Solid Films*, **516**, 6531.
101 Michel, M., Taylor, A., Sekol, R., Podsiadlo, P., Ho, P., Kotov, N., and Thompson, L. (2007) *Adv. Mater.*, **19**, 3859.
102 Sun, J.J., Zhao, H.Z., Yang, Q.Z., Song, J., and Xue, A. (2010) *Electrochim. Acta*, **55**, 3041.
103 Gheith, M.K., Sinani, V.A., Wicksted, J.P., Matts, R.L., and Kotov, N.A. (2005) *Adv. Mater.*, **17**, 2663.
104 Jan, E. and Kotov, N.A. (2007) *Nano Lett.*, **7**, 1123.

# 25
# Nanoconfined Polyelectrolyte Multilayers: From Nanostripes to Multisegmented Functional Nanotubes[1]

*Cécile J. Roy, Cédric C. Buron, Sophie Demoustier-Champagne, and Alain M. Jonas*

## 25.1
## Introduction

Layer-by-layer (LbL) assembly is by now a mature technique of surface functionalization. The simplicity of the methodology, the wide variety of components that can be included in the layers, and the frequent responsiveness of LbL assemblies to stimuli, such as pH, ionic strength or biological signals, have made LbL a method of choice for the rational design of complex functional films. Such films can be easily stratified in the vertical direction [1], and the degree of control over their vertical composition is currently so good that it was even possible to create directional cascades of energy transfer between a succession of stacked compartments only a few nm in thickness [2].

While the control over the structure of LbL films is excellent in the vertical direction, it is much more difficult to achieve the same level of precision in the two other spatial dimensions. It is possible to micropattern polyelectrolyte multilayers by a range of techniques, from optical lithography followed by lift-off or adsorption [3–6], to soft lithography [7, 8] and micro-embossing [9–11]; but reports on nanopatterning remain scarcer [12, 13]. Likewise, growing LbL assemblies over microparticles is a well-established technique [14, 15]; but deposition on nanoparticles of diameter below 100 nm is much more demanding [16]. This is because most polyelectrolytes have radii of gyration in the 10 to 50 nm range, depending on molar mass and ionic strength; therefore, when attempting to control the tridimensional structure at this level, entropic elasticity and intra-chain electrostatic repulsions become significant forces to overcome. A similar problem arises when attempting to assemble multilayers inside the porosity of a template (such as a nanoporous membrane); it is by now clear that the laws governing LbL assembly in nanopores are different from those valid for flat surfaces [17–19].

The relaxation of polyelectrolyte chains after adsorption is one key to understanding nanoconfinement effects on LbL assembly. When a polyelectrolyte chain adsorbs

---

1) This chapter is dedicated to the memory of Professor Jean-Louis Habib-Jiwan, who was working with us on dye-loaded LbL assemblies.

*Multilayer Thin Films: Sequential Assembly of Nanocomposite Materials*, Second Edition.
Edited by Gero Decher and Joseph B. Schlenoff.
© 2012 Wiley-VCH Verlag GmbH & Co. KGaA. Published 2012 by Wiley-VCH Verlag GmbH & Co. KGaA.

on a surface, its time-average conformation evolves from the one in solution to the final one adopted on the surface [20, 21]. The chain first establishes a few contact points with the surface, then increases their number by progressively relaxing and spreading towards a more stable state [22–24]. During this relaxation process, the chain may become trapped in a kinetically-frozen state, where the total strength of already-established surface/monomer contacts and steric hindrances resulting from neighboring chains prevent complete relaxation towards the most thermodynamically stable state. When adsorbing on nanopatterned surfaces, the spatial confinement of the adsorption process to tiny regions will limit even more strongly the possible relaxation of the chains, for instance their flattening. Likewise, in small cavities, the initial random attachment of a few monomer units on the concave surface of the cavity may strongly constrain the subsequent possible relaxation pathways of the adsorbing chain. Strong perturbations of LbL assembly are thus to be expected under conditions of confinement, the extent of confinement being defined with respect to the natural size of the chains in solution.

The field of nanoconfined LbL assembly is still largely in its infancy, and work to understand the phenomena occurring when adsorbing polyelectrolytes in nanopores or over nanopatterned surfaces is ongoing. Yet, going to the nanoscale is expected to open new perspectives for LbL assembly. For instance, nanopatterned LbLs would allow us to fabricate surfaces capable of interacting with biological cells at a spatial scale similar to the natural dimensions of the fibers of the extracellular matrix, which is relevant for the regulation of collective cell behavior (adhesion, proliferation, differentiation, gene expression) [25]; or to direct the assembly of hybrid components, such as DNA, nanoparticles, and synthetic polymers, in order to obtain multifunctional smart surfaces. Likewise, the fabrication of LbL nanowires and nanotubes is of interest for the design of new delivery vehicles resembling elongated viruses; or to prepare soft functional nanostructures without having to use the harsh and expensive standard nanolithography techniques.

This chapter thus focuses on polyelectrolyte multilayers grown under conditions where spatial confinement of the chains cannot be neglected. After a brief discussion of the size of polyelectrolyte chains in solution – which is required to define the length scales at which confinement effects may arise – we will consider LbL assembly over chemically or topographically nanopatterned surfaces. Then, assembly in nanoporous templates will be presented, with a bias towards the fabrication of biomacromolecule-containing nanotubes and nanowires. This chapter is not intended to be an exhaustive review of the field, but instead aims at providing general concepts and illustrating them by selected examples, mostly taken from our own work in the field.

## 25.2
### Estimation of the Size of Polyelectrolyte Chains in Dilute Solutions

Three different regimes characterize polyelectrolyte chains in *dilute* aqueous solutions, depending on the linear charge density of the chain, $\tau$, and on the relative values of the bare persistence length, $l_0$, and the Debye screening length, $\kappa^{-1}$ [26, 27].

The bare persistence length is the persistence length of the chain in the absence of charges; it defines the average distance over which the local orientation of the chain persists. Typical values for standard neutral synthetic polymers are in the range 0.5–1 nm [28], but $l_0$ increases to a few nm for semi-rigid polymers such as poly(n-hexyl isocyanate) [29], and can reach tens of nm for some biomacromolecules such as double-stranded DNA [28]. The Debye screening length $\kappa^{-1}$ is the distance over which electrostatic interactions are felt between two unit charges in the solution; it depends essentially on the ionic strength of the solution, because small mobile ions accumulate around the charged chain and, thereby, screen its charge. Mathematically, the inverse Debye length is [30]:

$$\kappa = (e^2 \Sigma_j c_j Z_j^2 / \varepsilon_r \varepsilon_0 k T)^{1/2} \tag{25.1}$$

where $c_j$ is the concentration of ions of type $j$ and charge $Z_j$, $\varepsilon_r$ is the relative permittivity of water and $\varepsilon_0$ the permittivity of vacuum, $e$ is the charge of the electron, $k$ is the Boltzmann constant and $T$ the temperature. The Debye length of a 0.1 M aqueous NaCl solution is ∼1 nm at room temperature, close to that of human blood plasma at 36 °C, which is ∼0.8 nm [30]. It is generally considered that distilled water has a Debye length of ∼100 nm, due to the presence of residual ionic impurities or a pH slightly different from 7 [30]. The linear charge density $\tau$ is the number of unit electrostatic charges per unit length along the contour of the chain. In general, one has to take into account the Manning condensation of counterions [31, 32] to evaluate an effective linear charge density $\tau^*$ [27], since the average distance between two charges along the chain should not exceed $l_B$. The Bjerrum length $l_B$ is the distance at which the electrostatic energy between two unit charges is equal to the thermal energy $kT$ (∼0.7 nm in water at 20 °C).

The boundaries of the three regimes of polyelectrolyte chains in dilute solution are defined in Table 25.1 [27]. When the Debye screening length is larger than the average distance between two charges along the chain, the electrostatic repulsion between neighboring monomers is strong, and the chain enters the "persistent regime" in which it is locally stiffened by electrostatic repulsion. The persistence length $l$ then becomes the sum of $l_0$ and of an electrostatic contribution, $l_B \tau^{*2}/(4\kappa^2)$ [27], which can be much larger than $l_0$ for solutions of low ionic strength. For instance, the experimentally-determined total persistence length of poly(styrene sulfonate) in aqueous solutions $10^{-3}$ M in NaCl is 7–12 nm, much larger than the intrinsic persistence length $l_0$ which is ∼1 nm [33]. In contrast, in the Gaussian

Table 25.1 Domains of existence of the three regimes of polyelectrolyte chains in dilute solution.

| Scaling Regime | $\tau^* (l_B l_0)^{1/2}$ | $\tau^* \kappa^{-1} (l_B/l_0)^{1/2}$ | $\tau^* \kappa^{-3/2} (l_B/l_0^2)^{1/2}$ |
|---|---|---|---|
| Persistent regime | >1 | >1 | |
| Gaussian-persistent regime | <1 | | >1 |
| Gaussian regime | | <1 | <1 |

regime, the electrostatic repulsion is strongly screened, and the chain is not stiffened electrostatically at any length scale. It behaves as a standard Gaussian chain with $l_0$ as the persistence length. Finally, in the Gaussian-persistent regime, the chain is not stiffened by electrostatic repulsions at a local scale but, nevertheless, electrostatic stiffening exists at a larger scale. In this regime, the chain behaves as a string of electrostatically-repulsive blobs, with a persistence length intermediate between the two other regimes. Note that the previous considerations apply for dilute solutions, for which the concentration in polyelectrolyte is lower than the overlap concentration $c^*$ [28].

From knowledge of the total persistence length $l$ and of the contour length $L$ of the chain (which is the length of the chain measured along its trajectory), the root-mean-square end-to-end distance $R_E$ of the chain can be computed using:

$$R_E = (2l(L-l[1-e^{-L/l}]))^{1/2} \tag{25.2}$$

which reduces to $L$ when $L \ll l$, and to $2lL$ when $L \gg l$ [34]. The end-to-end distance provides a good estimation for the average size of the macromolecules, since it is proportional to $6^{1/2} R_g$ where $R_g$ is the radius of gyration. (This is strictly valid for a Gaussian chain only.) Figure 25.1 presents the estimated values of $R_E$ as a

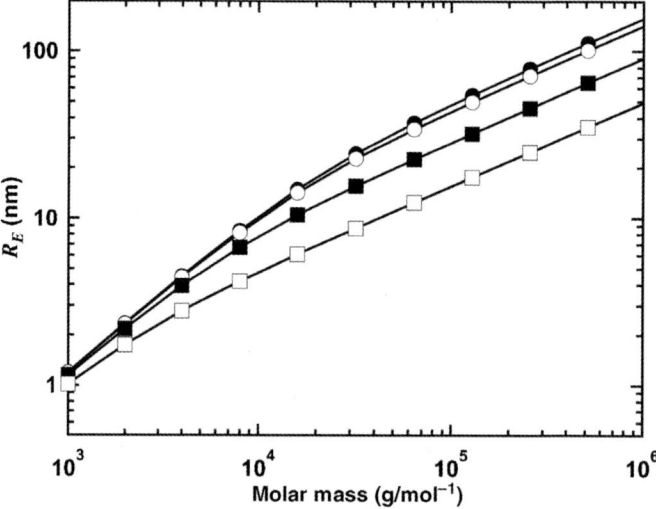

**Figure 25.1** Root-mean-square end-to-end distance $R_E$ of NaPSS chains in dilute aqueous solutions of increased molarity in NaCl, as a function of the NaPSS molar mass. The concentration in NaPSS repeat units is 0.01 M. The values of $R_E$ are roughly estimated based on the standard theory of polyelectrolyte solutions, neglecting excluded volume interactions and taking as bare persistence length $l_0 \sim 1$ nm. Closed circles: $10^{-4}$ M NaCl; open circles: $10^{-3}$ M NaCl; closed squares: $10^{-2}$ M NaCl; open squares: $10^{-1}$ and 1 M NaCl, fused together.

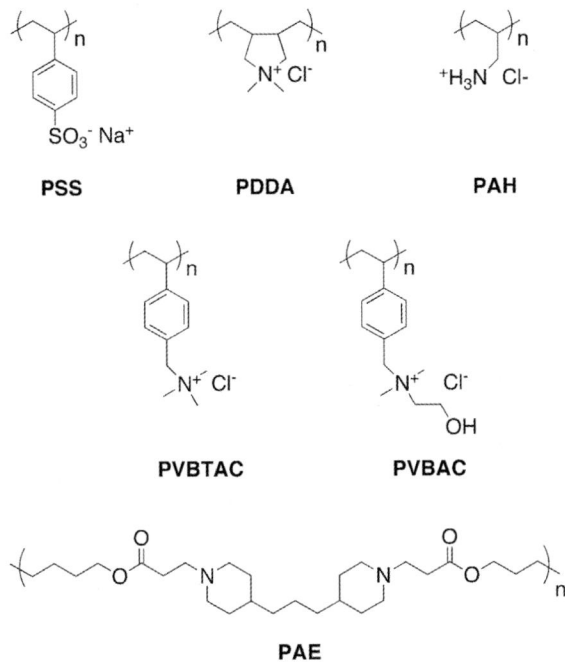

**Figure 25.2** Chemical structures of the synthetic polycations and polyanions of this chapter.

function of molar mass for poly(styrene sodium sulfonate) (NaPSS, Figure 25.2) in aqueous solutions of increasing NaCl concentration. The NaPSS concentration is 0.01 M in monomer units (a typical concentration for LbL experiments), and the NaCl concentration varies between $10^{-4}$ and 1 M. The computed values of $R_E$ are only rough estimations, because excluded volume interactions are not taken into account, and because the theory of polyelectrolyte solutions is not fully quantitative. Nevertheless, they provide a reasonable idea of the size of NaPSS chains in solution.

At low salt concentration (below $10^{-2}$ M in NaCl), the chains are in the persistent regime and have larger dimensions than at higher salt concentration where the chains are in the Gaussian regime. There is very little difference between the size of the chains for $10^{-4}$ and $10^{-3}$ M NaCl, because the ionic strength is dominated by the free (not condensed) ions of the NaPSS chain. Likewise, at high salt concentrations, the chain size converges to the value of the corresponding uncharged chain. For a typical NaPSS molar mass of 70 000 g mol$^{-1}$, $R_E$ ranges from ~10 to ~40 nm, depending on the ionic strength; for a molar mass of 250 000 g mol$^{-1}$, it ranges from ~25 to ~80 nm. This shows that polyelectrolyte chains are relatively large entities in solution, that will start to experience confinement when adsorbed on patches of size below ~100 nm, or in pores of size of this order of magnitude, or for which the concavity cannot be neglected at this length scale.

## 25.3
## Confining LbL Assembly on Flat Surfaces

There are basically two methodologies that allow one to confine LbL assembly laterally (Scheme 25.1). The first is the binary chemical patterning of a substrate in order to create regions that favor or prevent polyelectrolyte adsorption. Immersion of the substrate in polyelectrolyte solutions results in guided – *aka* templated – LbL assembly. The second method is a direct transposition of the classical lift-off procedures used in semiconductor technology. Here, a protective mask on a substrate is etched away at specific locations by a suitable lithography, and LbL assembly is performed on this etched mask. When the mask is dissolved, the LbL film grown over the mask is "lifted-off", leaving a film on the substrate in the open cavities of the mask. Both methodologies have been tested at the micrometer scale, where no special confinement effects are expected [3–5, 7, 8]. It is possible to transpose them at the sub-micrometer scale; however, because the required techniques are much more demanding, reports are scarce and basically limited to our own work.

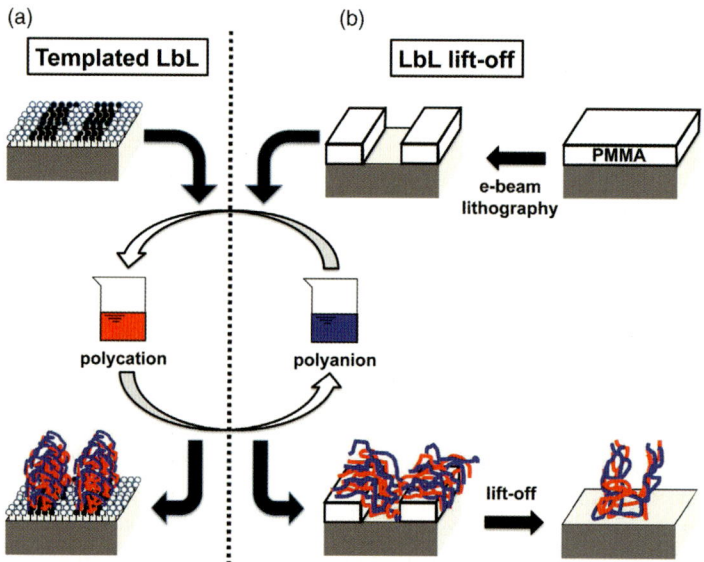

**Scheme 25.1** Two methods of fabrication of nanopatterned LbL films. (a) Templated LbL assembly over chemically-nanopatterned surfaces. (b) LbL assembly over patterned resists, followed by lift-off.

Interestingly, it is also possible to micro- and nano-pattern LbL films by embossing them with hard molds, thereby creating a surface relief [9–11, 13]. This is a simple application of nanoimprint lithography to LbL films, which will not be discussed here because it is more related to the mechanical properties of LbL films than to confined assembly.

### 25.3.1
### LbL Assembly Templated by Chemically Nanopatterned Surfaces

Nanopatterned surfaces able to direct LbL growth require (i) a patterning methodology suitable for small scale, and (ii) the selection of two surface functionalities able to, respectively, attract and repel polyelectrolytes. As regards nanopatterning methods capable of offering a wide range of surface functionalities, microcontact printing [35], which was used before to pattern LbL at the micrometer scale [7, 8], cannot be easily applied to smaller scales. In contrast, dip pen lithography and its variants allow the localized direct deposition of functional thiols on gold surfaces [36–38]. However, gold is not a very attractive substrate for high resolution work because its relatively high roughness, resulting from its intrinsically grainy nature, prevents smooth feature edges [39]. We therefore usually prefer to use electron-beam or nanoimprint lithography to draw openings into polymer masks on silicon wafers, through which local deposition of silanes is then performed from the gas phase, with resolutions down to 10 nm [40]. Removal of the polymer mask by dissolution, followed by a subsequent second silanation, provides access to binary chemical nanopatterned silicon wafers.

It is rather easy to select a surface chemical functionality from which LbL assemblies can be grown. Indeed, it is well-known that LbL assembly can be performed on almost any surface [41], because a range of interactions are involved in the adsorption of the first layer in water, including electrostatic, hydrophobic, van der Waals, or H-bonding interactions. In addition, even for a polyelectrolyte-repelling surface, once a few chains are adsorbed on surface defects they will create anchoring points for the deposition of subsequent layers and initiate LbL assembly (however, this does not mean that adhesion in air will be large). For synthetic polyelectrolytes, which usually have a hydrophobic backbone, it is thus convenient to use hydrophobic patches to guide LbL assembly.

In contrast, finding surfaces that resist polyelectrolyte adsorption is much more difficult, a fact that is considered advantageous by most researchers but which is a nagging problem when attempting to control LbL assembly spatially. Hydrophilic oligo(ethylene glycol) (OEG) monolayers are well-known to resist protein adsorption [42–44], and self-assembled monolayers of OEG-thiols on gold were shown to resist polyelectrolyte adsorption under specific conditions [7, 45, 46]. The resistance of OEG-silane monolayers on silicon towards polyelectrolyte adsorption was also investigated using different pairs of polyelectrolytes of varying molar mass, and for aqueous solutions of different pH [47, 48]. Although the OEG-silane monolayers were previously proved to resist protein adsorption, they fail in many cases to prevent the growth of polyelectrolyte multilayers. At basic pH, OEG-silane monolayers undergo partial hydrolysis, leading to the formation of negatively charged $SiO^-$ defects on which polycation adsorption occurs [49]. In the absence of hydrolytic degradation, unreacted free silanols also contribute to negative charging of the surface, since the isoelectric point of silicon oxide is $\sim 2$ [50, 51]. Once the polycation is adsorbed, further adsorption results in the formation of a discontinuous multilayer in the form of blobs, or of an open network of adsorbed strands. However, at pH 3, OEG-silane monolayers exhibit good polyelectrolyte repellency provided that the polyelectrolytes do not bear moieties capable of hydrogen bonding to the OEG ether groups [47].

On this basis, nanopatterned substrates were fabricated and used as guiding templates for LbL deposition at pH 3, with hydrophobic adsorbing stripes or dots of size below 100 nm in a repelling background of hydrophilic OEG-silanes [12]. A lateral force AFM image of 50 nm-wide hydrophobic stripes is shown in Figure 25.3a; the reversal of the contrast depending on scan direction indicates that the contrast originates from variations of friction on the surface, as expected for a hydrophobic/hydrophilic pattern. The topography of this sample after five cycles of deposition of poly(diallyldimethylammonium chloride) (PDDA) and poly(styrene sulfonate) (PSS) from aqueous solutions (at pH 3) is shown in Figure 25.3b. LbL assembly is clearly restricted to the hydrophobic stripes, despite the fact that their width is in the same range as the root-mean-square (rms) end-to-end distances of the polyelectrolyte chains (~40 and ~85 nm for PSS and PDDA, respectively).

As stated in Section 25.1, the adsorption of polyelectrolyte chains is a two-step process, consisting first of the rapid attachment of the chain onto the surface, followed by a progressive structural relaxation of the chain toward a more spread and stable state, within the specific constraints of the adsorption experiment. A perturbation of this relaxation process appears when adsorbing polyelectrolytes on nanopatches because chain spreading after adsorption is severely hampered. As a result, the chains keep a more coiled conformation, thereby occupying smaller footprints on the surface. Therefore, the layer thickness increases. This effect is only observable for nanopatches of size comparable to or smaller than the typical size of the macromolecules. This is shown in Figure 25.4, which reports the thickness of PDDA/PSS multilayers (five cycles of adsorption) versus the size of the adsorbing nanopatches

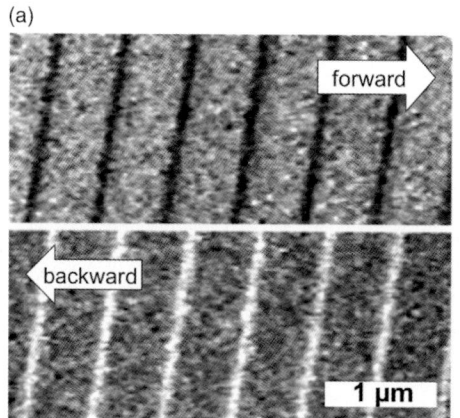 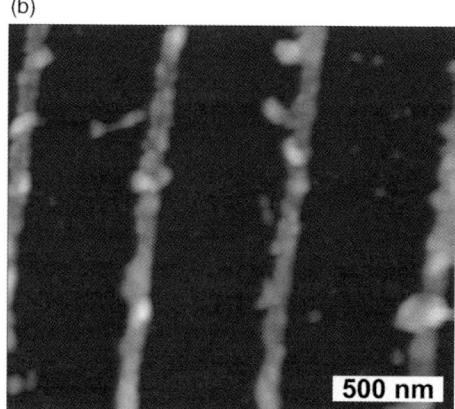

**Figure 25.3** (a) AFM images (lateral force mode) of 50 nm-wide stripes of a hydrophobic alkanesilane in a background of hydrophilic OEG-silane. The contrast, which originates from differences in friction, depends on the scan direction (arrows). (b) AFM topographical image of the same sample, after five cycles of adsorption of PDDA/PSS from aqueous solutions (pH 3). The greyscale covers 45 nm of height variation.

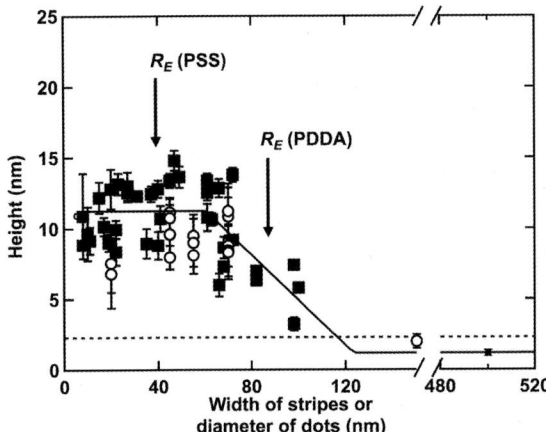

**Figure 25.4** Experimental height of (PDDA/PSS)$_5$ LbL multilayers, grown at pH 3 on nanopatterned flat substrates with hydrophobic stripes (squares) or dots (circles) drawn in an OEG-silane hydrophilic background. The dashed line is the experimental thickness of the laterally-infinite multilayer. Arrows indicate the average end-to-end distance of the PSS and PDDA chains in solution. Adapted from [12].

(stripes and dots) [12]. The more confined multilayers have a thickness four to five times larger than that measured for the laterally-infinite multilayer (or on patches of large size), indicating that the perturbation of LbL by the nanoconfinement is very strong. Significantly, the perturbations occur once the size of the nanopatches enters the range below the average end-to-end distance of the PDDA chains, as expected.

These experiments provide direct evidence for the possibility to manipulate the characteristics of LbL assembly by designing substrates at the nanometer scale. Furthermore, they provide strong support to the notion of relaxation of polyelectrolytes after adsorption. Finally, they indicate the feasibility of building soft nanostructures by LbL assembly. In particular, because polyelectrolyte multilayers can serve as containers for a large range of functional molecules [52], it becomes possible to use LbL assembly on nanopatterned surfaces to accumulate functional molecules at specific locations on a surface, which opens tremendous perspectives for bottom-up assembly of complex architectures. However, the main remaining difficulty is to find a more universal background that better resists the adsorption of polyelectrolytes, which is far from trivial.

### 25.3.2
### LbL Lift-Off

The problem of finding a universal polyelectrolyte-repelling background can be relaxed by using alternative well-established methods of semiconductor fabrication technology. One could, for instance, spin-cast an electron-sensitive resist atop a LbL film, pattern the resist by electron beam lithography or another convenient process, and finally plasma-etch the LbL film through the open windows of the mask resist before dissolving the mask. However, this process is intrinsically destructive and will result in significant perturbations of the chemical composition and molar mass of the chains at the edges of the nanostructure. For nanostructures, this perturbation could be overwhelmingly large since it is related to the surface-to-volume ratio of the nanostructure. Another possibility would be to use LbL films containing an electron-sensitive polyelectrolyte, such as poly(methacrylic acid). But here again, the destructive nature of the process could be a problem.

An interesting alternative, non-destructive patterning methodology was proposed by Lvov and others [3, 4]. It consists of first patterning a mask of resist spin-coated on a substrate, then depositing a LbL multilayer on the patterned mask, before dissolving the mask by immersion in a solvent (Scheme 25.1b). During the dissolution of the resist, the osmotic pressure generated by the difference in polymer resist concentration above and below the LbL film may be sufficient to fracture the LbL film; as a result, the LbL film is lifted-off at the places where it lies over the resist. In contrast, in the open cavities of the resist, the film is in contact with the substrate and is not displaced. The main difficulty in the process is that the LbL film must fracture neatly on the vertical walls of the open windows, whereas the substrate/LbL film interface must be strong enough to avoid unwanted lift-off. The first parameter is related to the mechanical properties of the multilayer, whereas the second is associated with the strength of the surface/polyelectrolyte interactions.

We have transposed this methodology to a smaller scale, by growing LbL films of poly(vinylbenzyltrimethylammonium chloride) (PVBTAC, $M_w$: 100 000 g mol$^{-1}$) and sodium poly(styrene sulfonate) (PSS, $M_w$: 70 000 g mol$^{-1}$) onto 95 nm-thick masks of poly(methyl methacrylate) (PMMA) patterned by electron beam lithography, from aqueous solutions 10$^{-2}$ M in monomer units and 0 or 0.1 M in NaCl at pH 7. A dye (fluorescein disodium salt dihydrate) was added in all baths at a concentration of $3 \times 10^{-4}$ M to dope the multilayer film during LbL growth [52]. The LbL-coated nanolithographed Si wafers were finally immersed in acetone under sonication for 2 min to remove the PMMA resist and lift-off the adsorbed polyelectrolyte layer.

Figure 25.5 shows AFM topographic images (b) and laser-scanning confocal microscopy (LSCM) images (a) of a fluorescein-doped (PVBTAC/PSS)$_5$ sample obtained in this way. As for micropatterned multilayer films, the quality of the nanopatterned multilayer depends on the experimental conditions during polyelectrolyte adsorption and during the dissolution of the mask. We found that the drying step at the end of each adsorption cycle is a key parameter for production of well-defined edges after lift-off, with 20 s being the optimum. When the drying step was omitted or increased, no pattern was obtained after lift-off. This is presumably related to the amount of residual water in the film, which controls its mechanical properties. As previously noted for micropatterns [3], ultrasonic treatment during lift-off was also required to obtain a proper reproduction of the patterns.

Importantly, the AFM topography and fluorescence maps indicate the presence of excess material at the borders of the stripes and dots (Figure 25.5b top figure), which are remains of the LbL film initially resting on the vertical walls of the mask (Scheme 25.1b). These parts of the LbL film are torn during lift-off and finally flip

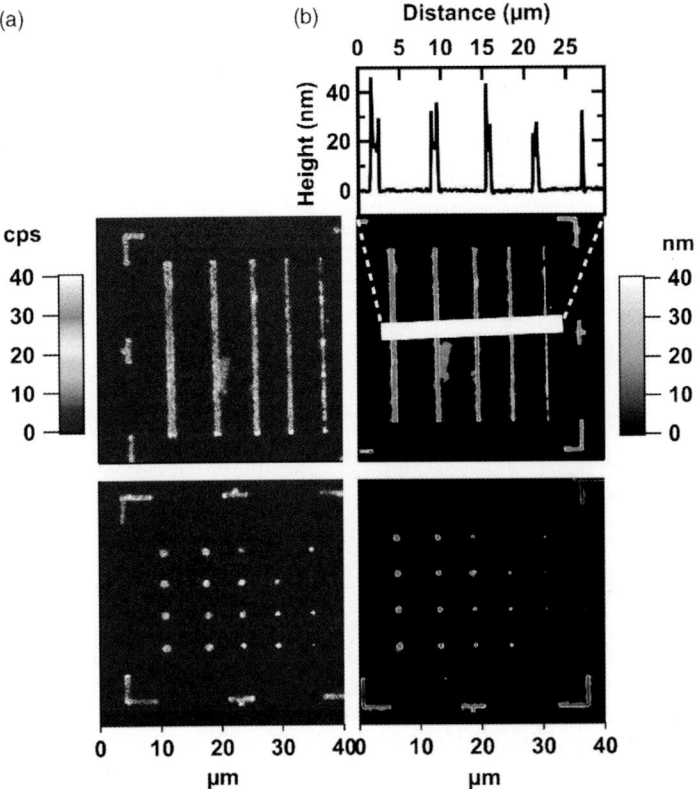

**Figure 25.5** (a) Laser-scanning confocal images of a lifted-off (PVBTAC/PSS)$_5$ LbL film, doped with fluorescein ($\lambda_{ex}$ = 495 nm). The film was grown from aqueous solutions 0.1 M in NaCl and $3 \times 10^{-4}$ M in fluorescein. (b) AFM topographic images of the same regions. The top figure is a height profile drawn along the thick white line.

over the edges of the patterned features. There is, however, no significant difference of the average height in the center of the stripes or dots compared to the laterally-homogeneous multilayer (whose thickness when grown under identical conditions as in Figure 25.5 is ~22 nm). This is logical, since the polyelectrolyte chains hardly experience confinement when adsorbing over the PMMA mask. However, when the stripes become very narrow, the two mounds of torn material merge, which results in an apparent higher thickness for the LbL assembly (last stripe of Figure 25.5). This increase of thickness should not be confused with the one observed for LbL assembly on chemically-patterned flat substrates; both phenomena arise from entirely different causes.

The amount of dye included in the nanopatterned features was estimated by integrating the fluorescence intensity over the stripes (Figure 25.6). The intensity increases linearly with the width of the stripes. However, the lines do not converge to the origin; the non-zero intercept is the amount of dye in the torn material

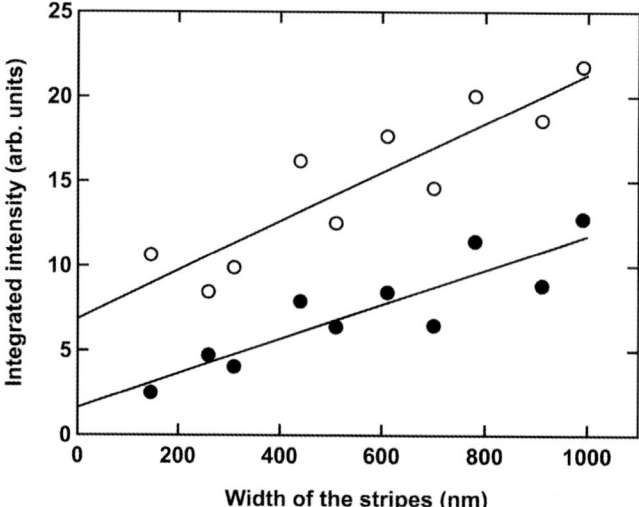

**Figure 25.6** Integrated fluorescence intensity of fluorescein entrapped in (PVBTAC/PSS)$_5$ multilayers confined by a lift-off process in stripes of varying width, for two different NaCl concentrations (0 M, closed circles; 0.1 M, open circles).

accumulating at the edge of the stripes, which does not depend on stripe width. It was found that this edge material is less voluminous when the LbL assembly is performed in the absence of salt (Figure 25.6). Similar observations were made for pyranine-doped multilayers.

The fabrication of nanoscale features by LbL lift-off is thus certainly feasible, but the accumulation of torn material at the edges of the nanostructures is inevitable. It is also obvious that reducing the size of the stripes to the range of the dimensions of the chains will create serious problems during the lift-off process. The lowest stripe width that we succeeded to fabricate by LbL lift-off was ~150 nm, well above the average rms distance between the chain ends of the polyelectrolytes. In this respect, LbL lift-off appears as useful for not-too-confined multilayers, but will certainly not allow the sort of confinement that was achieved by the direct adsorption of polyelectrolytes over chemically-nanopatterned surfaces, or when building the multilayers in cylindrical nanopores, as will be discussed in the next section.

## 25.4
### Confining LbL Assembly in Nanopores

Another example of multilayers built under polyelectrolyte confinement is provided by LbL assembly in porous templates, with pores in the same size range as the size of the macromolecular chains. By replicating the cylindrical open porosity of alumina or track-etched polymer membranes, multilayer nanotubes or nanowires of well-controlled dimensions can readily be obtained after dissolution of the membrane

[15, 53–62]. The macromolecular chains can either be filtered through the porous template, or allowed to diffuse freely within the pores of the membrane which is then immersed in the polyelectrolyte baths, as for flat surfaces (Scheme 25.2).

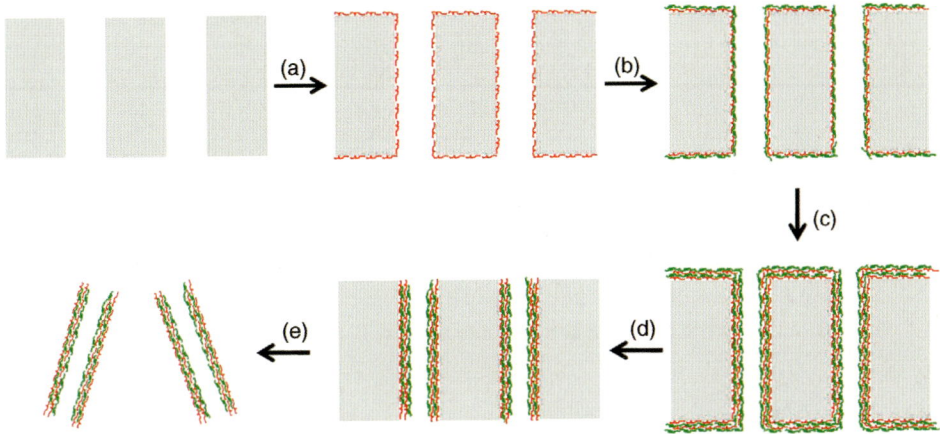

**Scheme 25.2** Synthesis procedure of multilayer nanotubes via the membrane-templated LbL assembly. The scheme shows a cross-sectional view of the porous membrane. (a) Adsorption of the first polyelectrolyte layer. (b) Adsorption of the second polyelectrolyte layer of opposite charge. (c) Adsorption of subsequent polyelectrolyte layers. (d) Removal of the multilayer film deposited on the top and bottom surfaces of the template. (e) Dissolution of the template and liberation of the nanotubes.

To remove the multilayer film deposited on the top and bottom surfaces of the membrane, various techniques, such as plasma etching, polishing [63] or wiping off the surfaces with alumina powder [55], sand paper [64] or a cotton swab [58] have been reported. We developed another softer strategy based simultaneously on the disruption and removal of the multilayer film with a cotton swab saturated with a strong basic aqueous solution of high ionic strength [18]. An interesting alternative would be to prevent the adsorption of polyelectrolytes on the surfaces of the template with an appropriate coating though, as previously noted, this is far from trivial. Martin's group reports on one specific example of this strategy where the adsorption of phosphonate-based compounds was prevented by sputtering a thin Au film on both faces of an alumina membrane [65–67].

Using this membrane-templated LbL assembly process, several research groups succeeded in obtaining multilayer nanotubes composed of a wide range of materials, such as synthetic polyelectrolytes [54, 55, 64, 68], dendrimers [56], nanoparticles [53], proteins [58–60, 66, 69–75], peptides [63], polysaccharides [57], DNA [67], or even viruses [61]. Applications are envisioned in tissue engineering [59, 75], analyte trapping [60, 61, 71], selective filtration [76, 77], enzymatic nanoreactors [66, 73, 78], micro- and nanofluidic devices [79], sensor devices [70], controlled drug release [57, 74] or gene delivery [63, 67]. For some of these applications, the LbL nanotubes can be kept within the templating membrane; for others, they have to be freed and dispersed in a suitable solvent (most often water). This extraction process is still a limiting factor. When using alumina membranes as templates, the harsh conditions required for the dissolution of alumina (strongly basic or acidic solutions) are, for instance, incompatible with the use of biological compounds in the nanotubes. When using polymer membrane templates, the dissolution is much easier but the separation of the LbL nanotubes from the dissolved polymer is much more difficult by centrifugation. With both types of template, a filtration process allows us to collect nanotubes on a porous substrate, but the interactions between the liberated nanomaterials and the filter are often so strong that redispersion of the nanotubes in a solvent is difficult. However, a promising method was recently reported by Komatsu et al. [60], who collected the nanotubes by lyophilization after the dissolution of the membrane, and could subsequently readily redisperse them in aqueous solutions. Promising results were also obtained recently in our group by collecting the nanotubes at the interface between two immiscible liquids, one of which is a solvent of the PC membrane.

As an example of the type of LbL nanotubes that can be obtained, Figure 25.7 displays DNA-containing LbL nanotubes grown within the pores of track-etched polycarbonate (PC) membranes of 100 nm pore diameter. A poly(aminoester), PAE (Figure 25.2), was used as a biodegradable polycation to condense the DNA through electrostatic interactions. Polymers of the PAE family can be synthesized by Michael addition polymerization of diacrylate and amine monomers [80]; under physiological conditions, they undergo hydrolytic degradation at a rate that is tuned by their chemical composition [81–86]. In addition, because the degradation products of PAE

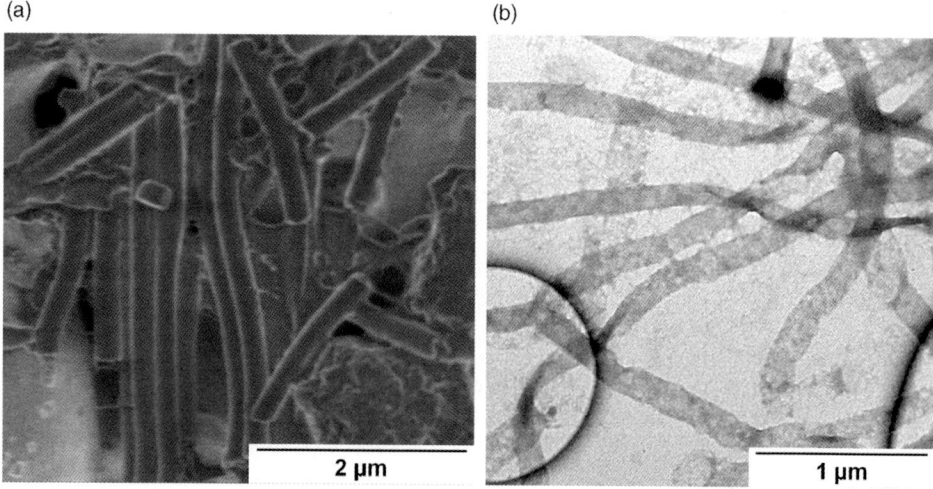

**Figure 25.7** (a) SEM image of (PAE/DNA)$_{10}$ multilayer nanotubes. (b) TEM image of (PAE/DNA)$_5$ multilayer nanotubes.

are biocompatible and noncytotoxic [80–82, 87], the PAE/DNA nanotubes of Figure 25.7 are thus of direct interest for gene delivery applications.

The LbL nanostructures of Figure 25.7 have a maximum length corresponding to the thickness of the membrane, demonstrating that the LbL process is efficient over the entire thickness of the membrane. Some shorter tubes are also observed, arising from rupture during the sample preparation process. The outer diameter of the nanotubes differs slightly from the pore diameter of the membrane because nanotubes tend to flatten when adsorbed on the TEM grids, or to shrink when freed from the membrane. These effects make difficult the observation of the hollow structure and the measurement of reliable dimensions of the nanostructures by microscopy; therefore, other techniques, such as gas flow porometry or conductimetry have usually to be used [18].

### 25.4.1
**Peculiarities of LbL Assembly in Nanopores**

Although multilayer growth was demonstrated to occur within cylindrical nanopores, it is frequently reported to differ from that taking place on planar surfaces, independently of the method used for the LbL assembly (filtration or immersion) and of the types of polyelectrolytes (synthetic or biological compounds). The wall thickness of LbL nanotubes composed of (semi-)flexible polyelectrolytes is, for instance, found to be larger than the thickness of corresponding films fabricated on a flat substrate under identical deposition conditions. For very small pores, this behavior can be attributed to the perturbation of chain conformation resulting from confinement, and to the associated decreased configurational entropy of the chains. In addition, for larger pores, the concave geometry of the nanopores favors the initial formation of multiple interaction points between the macromolecular chains and the walls of the cavity, with a topological distribution quite different from that on flat surfaces. The perturbation of the ensuing relaxation process by the initial topological constraints then leads to a final equilibrium conformation that differs markedly from that attained on planar surfaces, and results in an increase in the thickness. Importantly, this behavior, which has some similarity with that observed for multilayers adsorbed on chemically nanopatterned surfaces, is not noticed for rigid polyelectrolytes such as spherical dendrimers [19]. For such rigid objects, the conformational freedom is intrinsically much more restricted than that of (semi-)flexible chains; therefore, the spatial extent of their interaction with pore walls is limited, as well as their relaxation by lateral spreading, leading to a layer thickness comparable to that obtained on flat surfaces.

The conformational change is, however, not the only parameter affected by the confinement. LbL studies performed with rigid polymers, or within the nanoscale interstice of two planar charged surfaces, have highlighted that confinement also hinders macromolecular transport due to steric effects and electrostatic interactions, and consequently leads to a slowing down of the multilayer growth kinetics or even to an inhibition of the multilayer growth [19, 79]. According to the scaling theory of Rubinstein [88], the polyelectrolyte diffusion coefficient in dilute solution, $D_0$, varies

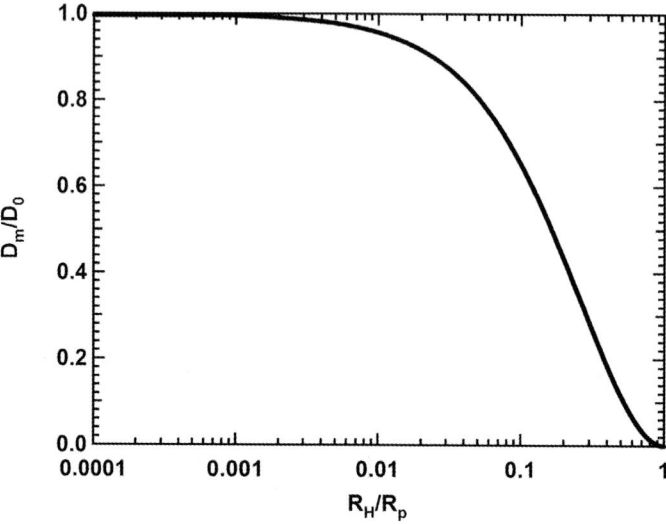

**Figure 25.8** Ratio of the pore-to-bulk diffusivities as a function of the ratio between the hydrodynamic radius of a rigid sphere ($R_H$) and the radius of the pore ($R_p$) derived from the Renking model. Larger values of $R_H/R_p$ correspond to more confined spheres.

as $D_0 \sim N^{-3/5}$, where $N$ is the degree of polymerization. In salted solution, $D_0$ is of the order of $\sim 10^{-7}$ cm$^2$ s$^{-1}$ for a typical molar mass; hence, the characteristic diffusion time for polyelectrolyte chains to travel unhindered through cylindrical pores of about 10–20 μm length is about 10 s [89–91]. This value is also valid for the transport of polymer chains through pores of radius $R_p$, much larger than the hydrodynamic radius of the chains, $R_H$, in which case the polymer chains can be considered as rigid, spherical solutes. The Renking model [92] can be used to describe the diffusion of small spherical, non-deformable solutes when $R_p > 3R_H$ (Figure 25.8): for small enough pore sizes ($R_p < 10R_H$), the diffusion coefficient, $D_m$, becomes significantly reduced due to the hydrodynamic interaction between the sphere and the pore walls [90, 91, 93, 94]. For polyelectrolyte chains, the situation is more complex due to the deformability of the random coil and electrostatic effects. In the so-called semidilute-unentangled regime ($5R_H/3 < R_p < 10R_H$), the polymer chains are deformed and adopt an elongated conformation; their diffusion follows the Rouse behavior which is characterized by a diffusion coefficient that varies as $D_m \sim N^{-1}$. For even stronger confinement ($R_p < 5R_H/3$), the chains further elongate and enter the semidilute-entangled regime for which reptation behavior becomes the main diffusion mechanism, with an even slower diffusion coefficient $D_m \sim N^{-2}$. Various more complex models, such as the Davidson and Deen theory [95], were derived from the Renking model to express the diffusion of polyelectrolytes under strong confinement. Yet these models are often limited to a specific polymer and are, therefore, not generally applicable. On the basis of these considerations, it is thus expected that multilayer growth under confinement should also be progressively limited by the diffusion of the chains in the nanopores.

(a) **Regime 1**   (b) **Regime 2**

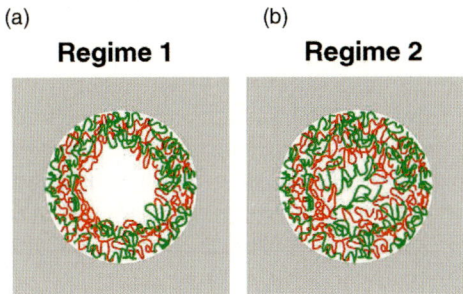

**Scheme 25.3** Schematic representation of polyelectrolytes multilayer structures in nanopores corresponding to the dilute diffusion regime (a) and the semi-dilute diffusion regime (b). Adapted from [18].

The mechanism of multilayer growth under polyelectrolyte confinement was studied by following the stepwise construction of PAH/PSS multilayers within pores of diameter ranging from 100 to 500 nm [18]. In addition to the geometrical constraint (pore diameter), the size of the macromolecular chains was tuned by varying the molar mass of the polyelectrolytes and the ionic strength of the solutions. Two regimes were observed in the kinetics of multilayer growth in nanopores (Scheme 25.3). The first regime occurs at the beginning of the multilayer build-up when the size of the polyelectrolyte chains is significantly smaller than the pore diameter. In this dilute diffusion regime, the polymer chains diffuse within the nanopores like rigid spheres. LbL assembly then occurs as on flat surfaces, and with a similar influence of the ionic strength, but with a larger multilayer thickness resulting from the curvature of the pore wall. The second regime takes place when the macromolecular chains are able to make interconnections with each other across the nanopores, that is, when the polyelectrolyte size is in the same range as the pore size. In this semi-dilute diffusion regime, the macromolecular transport within the nanopores is hindered by chain entanglement, leading to the formation of an increasingly denser gel. As a result, the multilayer growth kinetics is slowed down compared to that observed in the first regime. Formation of clogged nanotubes, or even nanowires, is finally observed for a large number of deposition cycles thanks to the reptation of the polyelectrolytes which diffuse in this gel. This gel model is in full agreement with the few other studies on multilayer deposition in confined media carried out by filtration or immersion of rigid or flexible polyelectrolytes within planar or tubular nanochannels [17, 19, 79].

When using small pore diameters or relatively bulky chains, LbL deposition can occur in the semi-dilute diffusion regime from the very start of the assembly process. This is, for example, the case for the PAE/DNA nanotubes of Figure 25.7: in nanopores of 100 nm diameter, the DNA chains, which have a radius of gyration of about 12 nm, (120 base pairs, $l \sim 50$ nm and $L \sim 40$ nm [96]) are in the semi-dilute regime ($R_p \sim 4.2 R_H$) from the very beginning of the multilayer construction. The pore diameter of the templating membranes was measured by gas flow porometry after each step of the construction process (Figure 25.9a): the pore diameter decreases after deposition of the first layer pair, but then hardly decreases for the next deposition cycles. This is a direct signature for the growth occurring in the semi-dilute diffusion regime. Figure 25.9 also shows that the saturation of the growth apparently happens before the pore is fully filled, since a roughly constant value of ~60 nm is measured for the pore diameter after more than one deposition cycle. This is because the pore diameters are measured in the dry state, after collapse of the swollen gel which fully fills the pores in the wet state. A similar behavior was observed for collagen/PSS multilayers built in PC membranes with pore diameters of 200 and 500 nm [75].

The actual amount of DNA included in the nanotubes can be estimated by immersing the filled PC membrane in a phosphate buffer solution, PBS, at $37\,°C$; under these conditions, the PAE component hydrolyzes and DNA is released in solution. The quantification of the released DNA was carried out with a DNA probe, $[Ru(phen)_2ddpz]^{2+}$, on the basis of its strong binding to double helical DNA and a practically negligible background emission of any unbound complex [97]. As shown in Figure 25.9b, the amount of DNA released from PAE-DNA nanotubes increases proportionally to the number of adsorption cycles, similarly to what is observed with PAE/DNA multilayer films deposited on flat surfaces. This is surprising, since the gas flow porometry results (Figure 25.9a) show that the thickness of the multilayer adsorbed in the pores is almost independent of the number of adsorption cycles. The only way to reconcile these seemingly contradictory results is to accept that the relative amount of DNA adsorbed in the nanopores increases with the number of deposition cycles, at the expense of PAE. Therefore, diffusion and redissolution certainly occur during multilayer formation in the nanopores – in the present case in favor of DNA – which indicates that LbL deposition in nanopores may involve processes similar to those encountered for exponentially-growing flat multilayers [98, 99]. More research is, however, needed before the peculiarities of LbL assembly in nanopores can be fully understood.

### 25.4.2
### Multisegmented LbL Nanotubes

The LbL templating method is certainly a powerful and versatile tool to study the confinement of polyelectrolytes and to fabricate a large variety of potentially useful anisotropic nanostructures. One of its unique features is the ability to control the dimensions of the resulting nanotubes over a wide range. The outer diameter of the nanotubes, determined by the diameter of the template, can indeed be varied from a few microns down to only 30 nm and the length of the nanotubes, determined by the thickness of the template, can range anywhere from 1 to 50 μm. Moreover, nanotubes presenting various degrees of rigidity can easily be obtained by varying the number of layers incorporated into the nanostructures. Furthermore, the distinct inner and outer surfaces can be differentially functionalized allowing, for instance, the loading of therapeutic agents within the nanotubes and, simultaneously, the modification of the outer surface with biochemical compounds that enhance the biocompatibility, recognition and/or targeting properties of the nanotubes. Finally, the large aspect

## 25.4 Confining LbL Assembly in Nanopores

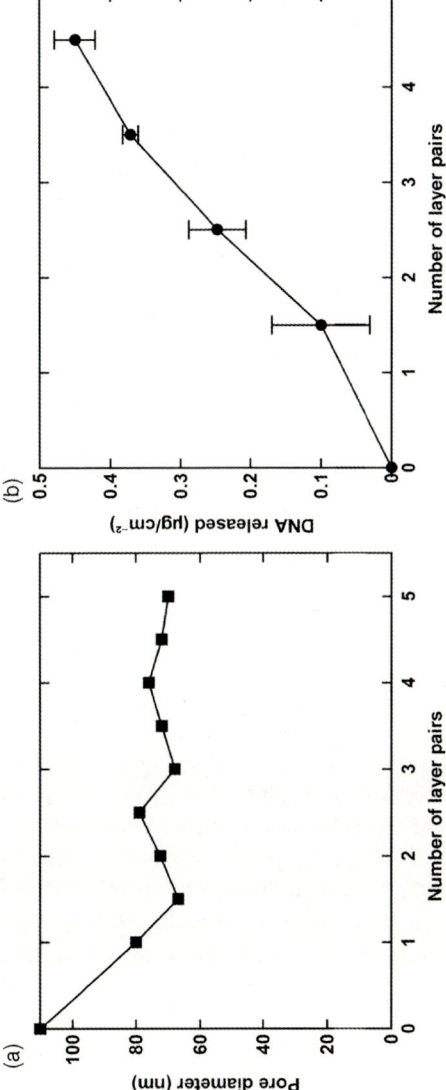

**Figure 25.9** (a) Evolution of the dry pore diameter estimated by gas flow porometry as a function of the number of PAE/DNA layer pairs deposited within the pores of a PC membrane (~110 nm initial diameter). Half-integer values of the number of layer pairs coincide with the adsorption of a PAE layer, integer ones with the adsorption of a DNA layer. (b) Amount of DNA released from (PAE/DNA) nanotubes embedded in a polycarbonate membrane and immersed in PBS at 37 °C (average values obtained from three experiments, expressed per area of nanopores).

ratio of LbL nanotubes provides multiple attachment sites for various functions along the nanotube axis. This provides the possibility to fabricate complex nanostructures which could even eventually mimic sophisticated natural assemblies, such as filamentous viruses.

By combining LbL assembly with other templating methodologies, it even becomes possible to fabricate multisegmented nanowires. For instance, we have synthesized nanowires containing four successive segments, namely gold, polypyrrole (PPy), PAE/DNA, and a glass colloid (Figure 25.10a). A short gold nanorod of 6 µm length was first electrodeposited within a PC track-etched membrane of ~10 µm thickness. Next, a PPy segment of 0.5–1 µm length was electropolymerized from this base; by suitable selection of the conditions of electropolymerization, a nanotube open at one side could be obtained [100]. Five layer pairs of PAE/DNA were subsequently adsorbed in the remaining length of the nanopores. After surface cleaning, a glass colloid was adsorbed on the tip of the LbL nanotube.

To follow the location of the LbL multilayer, the PAE was replaced by PAH labeled with fluorescein. Figure 25.10b is a fluorescent image of such a multisegmented nanowire, without a glass colloid at its tip. The fluorescence profile traced along a nanostructure allows the identification of the three segments of the nanowire: the multilayer nanotube exhibits the brightest green emission due to the fluorescein dyes (FITC) coupled to the PAH chains, the gold nanorod emits a light green color due to the reflection of the emission light, and the polypyrrole nanocontainer appears in black in-between. Importantly, the fluorescence profile indicates the presence of the multilayer inside the PPy nanocontainers: while an abrupt decrease in the fluorescence intensity is observed at the left edge of the LbL nanotube, the intensity is only decreasing progressively on the side of the polypyrrole segment.

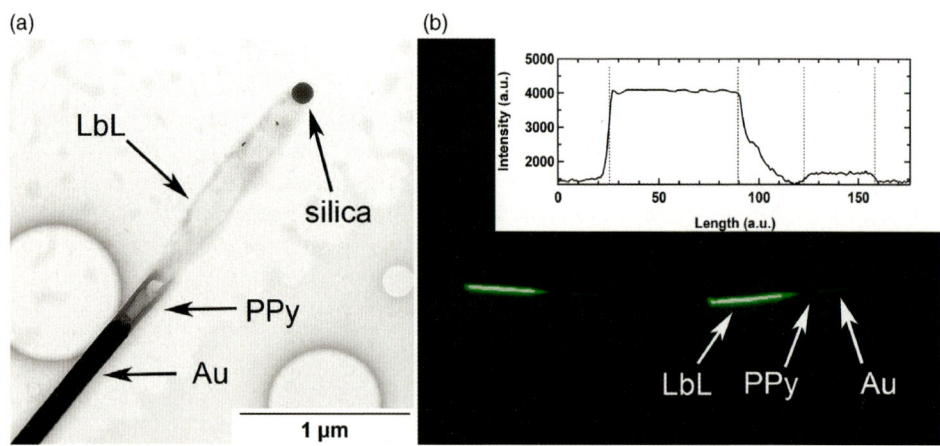

**Figure 25.10** (a) TEM image of a polypyrrole nanotube electrosynthesized atop a gold nanorod and further functionalized with a (PAE/DNA)$_5$ multilayer and a terminal silica nanobead. (b) Fluorescence microscopy image of tri-segmented Au/PPy/(PAH$^{FITC}$/DNA)$_5$ nanorods. The inset is a fluorescence profile along a multisegmented nanorod.

This example shows the level of complexity one can currently reach using templated assembly. It is to be noted that further surface modifications can be performed on specific segments of such nanostructures, for example by grafting amino-poly(ethylene glycol) over the PPy segment (if carboxylic acid-functionalized) [100], or by silanation of the terminal glass beads. Such multifunctional nanostructures could thus certainly be used for advanced targeted gene delivery.

## 25.5
## Conclusions

In this chapter, we have provided general concepts and illustrations of polyelectrolyte multilayer assembly in conditions of steric or hydrodynamic confinement. Compared to macro- and microscopic flat surfaces, the behavior of polyelectrolytes differs markedly when they adsorb on substrates of dimensions comparable to the size of the macromolecular chains. This is mainly attributed to a perturbation of the relaxation process of the polyelectrolytes that is spatially restricted when the chains adsorb either on laterally-confined substrates (such as chemically-nanopatterned surfaces), or on substrates presenting a concave geometry (such as nanopores). Consequently, the polymer chains are prevented from adopting an equilibrium flat-on conformation (as observed for laterally-infinite surfaces) which results in a higher thickness for confined multilayers. Moreover, the transport of polyelectrolytes through nanopores is hindered by hydrodynamic interactions, which progressively limit chain diffusion and the growth kinetics of the multilayers.

Although these studies already provide a consistent picture of the effects of confinement on LbL assembly, further work should be performed to understand in finer detail the peculiar behavior of polyelectrolytes under confinement conditions. This should ultimately lead to a precise control of LbL assembly in the three spatial dimensions required for advanced nanotechnology applications. Nevertheless, it is already possible to fabricate functional nanostructures by LbL assembly, and we have provided a small number of examples, such as dye-loaded nanostripes, or multi-segmented nanotubes containing biomacromolecules. This shows that LbL assembly complements nicely the already vast arsenal of nanofabrication technologies, and that it can be combined with patterning, templating and deposition methodologies to create complex functional hybrid nanostructures. It is, therefore, to be expected that LbL assembly will progressively find niche applications in fields beyond its original background of surface functionalization.

## Acknowledgments

It is appropriate to acknowledge the dedication of people who contributed to the work reported here: H. Alem, G. Baralia, N. Chorine, A. De Boulard, F. Denis, N. Frederich, J. Landoulsi, B. Muls, A. Pallandre, E. Nicol, S. Peralta, S. Saghazadeh.

The authors also benefited from the expertise of, and scientific discussions with, B. De Geest, S. De Smedt, C. Dupont-Gillain, B. Elias, K. Glinel, A. Laschewsky, B. Nysten, as well as many researchers from the LbL community. The authors are thankful to E. Ferain and the it4ip company for supplying polycarbonate membranes. CJR is a research fellow of the F.R.S.-FNRS and SD-C is senior research associate of the F.R.S.-FNRS. Financial support was provided by the Communauté Française de Belgique (ARC 06-11/339), the Belgian Federal Science Policy (IAP-PAI P6/27), the F.R.S.-FNRS, and the Wallonia Region (Nanotic-Feeling).

## References

1 Glinel, K., Jonas, A.M., Laschewsky, A., and Vuillaume, P.Y. (2003) in *Multilayer Thin Films* (eds G. Decher and J. Schlenoff,), Wiley-VCH, Weinheim, pp. 177–205.
2 Peralta, S., Habib-Jiwan, J.-L., and Jonas, A.M. (2009) *ChemPhysChem*, **10**, 137–143.
3 Hua, F., Cui, T., and Lvov, Y. (2002) *Langmuir*, **18**, 6712–6715.
4 Cho, J., Jang, H., Yeom, B., Kim, H., Kim, R., Kim, S., Char, K., and Caruso, F. (2006) *Langmuir*, **22**, 1356–1364.
5 Chen, X., Sun, J., and Shen, J. (2009) *Langmuir*, **25**, 3316–3320.
6 Hua, F., Shi, J., Lvov, Y., and Cui, T. (2002) *Nano Lett.*, **2**, 1219–1222.
7 Hammond, P.T. and Whitesides, G.M. (1995) *Macromolecules*, **28**, 7569–7571.
8 Hammond, P.T. (2003) in *Multilayer Thin Films* (eds G. Decher and J. Schlenoff), Wiley-VCH, Weinheim, pp. 271–299.
9 Gao F C., Wang, B., Feng, J., and Shen, J. (2004) *Macromolecules*, **37**, 8836–8839.
10 Lu, Y., Chen, X., Hu, W., Lu, N., Sun, J., and Shen, J. (2007) *Langmuir*, **23**, 3254–3259.
11 Ladhari, N., Hemmerlé, J., Haikel, Y., Voegel, J.-C., Schaaf, P., and Ball, V. (2008) *Appl. Surf. Sci.*, **255**, 1988–1995.
12 Pallandre, A., Moussa, A., Nysten, B., and Jonas, A.M. (2006) *Adv. Mater.*, **18**, 481–486.
13 Lu, Y., Hu, W., Ma, Y., Zhang, L., Sun, J., Lu, N., and Shen, J. (2006) *Macromol. Rapid Commun.*, **27**, 505–510.
14 Peyratout, C. and Dähne, L. (2004) *Angew. Chem. Int. Ed.*, **43**, 3762–3783.
15 Wang, Y., Angelatos, A.S., and Caruso, F. (2008) *Chem. Mater.*, **20**, 848–858.
16 Gittins, D.I. and Caruso, F. (2001) *J. Phys. Chem. B*, **105**, 6846–6852.
17 Alem, H., Blondeau, F., Glinel, K., Demoustier-Champagne, S., and Jonas, A.M. (2007) *Macromolecules*, **40**, 3366–3372.
18 Roy, C.J., Dupont-Gillain, C., Demoustier-Champagne, S., Jonas, A.M., and Landoulsi, J. (2010) *Langmuir*, **26**, 3350–3355.
19 Lazzara, T.D., Lau, K.H.A., Abou-Kandil, A.I., Caminade, A.-M., Majoral, J.-P., and Knoll, W. (2010) *ACS Nano*, **4**, 3909–3920.
20 Pefferkron, E. and Elaissari, A. (1990) *J. Colloid. Interface Sci.*, **138**, 187–194.
21 Fleer, G.J., Cohen Stuart, M.A., Scheutjens, J.M.H.M., Cosgrove, T., and Vincent, B. (1993) *Polymer at Interfaces*, Chapman & Hall, London.
22 Cohen Stuart, M.A. and Tamai, H. (1988) *Macromolecules*, **21**, 1863–1866.
23 Elaissari, A. and Pefferkron, E. (1991) *J. Colloid. Interface Sci.*, **143**, 85–91.
24 McAloney, R.A. and Goh, M.C. (1999) *J. Phys. Chem. B*, **103**, 10729–10732.
25 Alves, N.M., Pashkuleva, I., Reis, R.L., and Mano, J.F. (2010) *Small*, **6**, 2208–2220.
26 Mandel, M. (1988) in *Encyclopedia of Polymer Science and Engineering*, 2nd edn (eds H.F. Mark, N.M. Bikales, C.G. Overberger, and G. Menges), John Wiley & Sons, New York, pp. 739–829.

27 Netz, R.R. and Andelman, D. (2003) *Phys. Rep.*, **380**, 1–95.
28 Rubinstein, M. and Colby, R.H. (2003) *Polymer Physics*, Oxford University Press, Oxford.
29 Teraoka, I. (2002) *Polymer Solutions. An Introduction to Physical Properties*, John Wiley & Sons, New York.
30 Butt, H.-J., Graf, K., and Kappl, M. (2006) *Physics and Chemistry of Interfaces*, 2nd edn, Wiley-VCH, Weinheim.
31 Manning, G.S. (1969) *J. Chem. Phys.*, **51**, 924–931.
32 Manning, G.S. and Ray, J. (1998) *J. Biomol. Struct. Dyn.*, **16**, 461–476.
33 Borochov, N. and Eisenberg, H. (1994) *Macromolecules*, **27**, 1440–1445.
34 Grosberg, A.Y. and Khokhlov, A.R. (1994) *Statistical Physics of Macromolecules*, American Institute of Physics, New York.
35 Kumar, A. and Whitesides, G.M. (1993) *Appl. Phys. Lett.*, **63**, 2002–2004.
36 Piner, R.D., Zhu, J., Xu, F., Hong, S.H., and Mirkin, C.A. (1999) *Science*, **283**, 661–663.
37 Huo, F.W., Zheng, Z.J., Zheng, G.F., Giam, L.R., Zhang, H., and Mirkin, C.A. (2008) *Science*, **321**, 1658–1660.
38 Shim, W., Braunschweig, A.B., Liao, X., Chai, J., Lim, J.K., Zheng, G.F., and Mirkin, C.A. (2011) *Nature*, **469**, 516–520.
39 Baralia, G.G., Pallandre, A., Nysten, B., and Jonas, A.M. (2006) *Nanotechnology*, **17**, 1160–1165.
40 Pallandre, A., Glinel, K., Jonas, A.M., and Nysten, B. (2004) *Nano Lett.*, **4**, 365–371.
41 Arys, X., Jonas, A.M., Laschewsky, A., Legras, R., and Mallwitz, F. (2005) in *Supramolecular Polymers*, 2nd edn (ed. A. Ciferri), Taylor & Francis, Boca Raton, pp. 651–710.
42 Golander F C-.G., Herron, J.N., Lim, K., Claesson, P., Stenius, P., and Andrade, J.D. (1992) in *Poly(Ethylene Glycol) Chemistry: Biotechnical and Biomedical Applications* (ed. J.M. Harris), Plenum Press, New York, pp. 221–245.
43 Prime, K.L. and Whitesides, G.M. (1993) *J. Am. Chem. Soc.*, **115**, 10714–10721.
44 Zhang, M., Desai, T., and Ferrari, M. (1998) *Biomaterials*, **19**, 953–960.
45 Clark, S.L., Montague, M.F., and Hammond, P.T. (1997) *Macromolecules*, **30**, 7237–7244.
46 Clark, S.L. and Hammond, P.T. (2000) *Langmuir*, **16**, 10206–10214.
47 Buron, C.C., Callegari, V., Nysten, B., and Jonas, A.M. (2007) *Langmuir*, **23**, 9667–9673.
48 Alvarez, M., Bocchio, N.L., and Kreiter, M. (2009) *Langmuir*, **25**, 1097–1102.
49 Dekeyser, C.M., Buron, C.C., Mc Evoy, K., Dupont-Gillain, C.C., Marchand-Brynaert, J., Jonas, A.M., and Rouxhet, P.G. (2008) *J. Colloid. Interface Sci.*, **324**, 118–126.
50 Baudrant, A., Tardif, F., and Wyon, C. (2003) *Caractérisation et nettoyage du silicium*, éditions Lavoisier, Paris.
51 Hoogeveen, N.G., Cohen Stuart, M.A., and Fleer, G.J. (1996) *J. Colloid. Interface Sci.*, **182**, 133–145.
52 Nicol, E., Habib-Jiwan, J.-L., and Jonas, A.M. (2003) *Langmuir*, **19**, 6178–6186.
53 Liang, Z., Susha, A.S., Yu, A., and Caruso, F. (2003) *Adv. Mater.*, **15**, 1849–1853.
54 Ai, S., Lu, G., He, Q., and Li, J. (2003) *J. Am. Chem. Soc.*, **125**, 11140–11141.
55 Ai, S., He, Q., Tao, C., Zheng, S., and Li, J. (2005) *Macromol. Rapid Commun.*, **26**, 1965–1969.
56 Kim, D.H., Karan, P., Göring, P., Julien Leclaire, J., Caminade, A.-M., Majoral, J.-P., Gösele, U., Steinhart, M., and Knoll, W. (2005) *Small*, **1**, 99–102.
57 Yang, Y., He, Q., Duan, L., Cui, Y., and Li, J. (2007) *Biomaterials*, **28**, 3083–3090.
58 Qu, X., Lu, G., Tsuchida, E., and Komatsu, T. (2008) *Chem. Eur. J.*, **14**, 10303–10308.
59 Landoulsi, J., Roy, C.J., Dupont-Gillain, C., and Demoustier-Champagne, S. (2009) *Biomacromolecules*, **10**, 1021–1024.
60 Qu, X. and Komatsu, T. (2009) *ACS Nano*, **4**, 563–573.
61 Komatsu, T., Qu, X., Ihara, H., Fujihara, M., Azuma, H., and Ikeda, H. (2011) *J. Am. Chem. Soc.*, **133**, 3246–3248.
62 He, Q., Cui, Y., Ai, S., Tian, Y., and Li, J. (2009) *Curr. Opin. Colloid Interface Sci.*, **14**, 115–125.

63 He, Q., Tian, Y., Cui, Y., Möhwald, H., and Li, J. (2008) *J. Mater. Chem.*, **18**, 748–754.
64 Tian, Y., He, Q., Tao, C., and Li, J. (2006) *Langmuir*, **22**, 360–362.
65 Hou, S., Harrell, C.C., Trofin, L., Kohli, P., and Martin, C.R. (2004) *J. Am. Chem. Soc.*, **126**, 5674–5675.
66 Hou, S., Wang, J., and Martin, C.R. (2005) *Nano Lett.*, **5**, 231–234.
67 Hou, S., Wang, J., and Martin, C.R. (2005) *J. Am. Chem. Soc.*, **127**, 8586–8587.
68 He, Q., Song, W., Möhwald, H., and Li, J. (2008) *Langmuir*, **24**, 5508–5513.
69 Tian, Y., He, Q., Cui, Y., and Li, J. (2006) *Biomacromolecules*, **7**, 2539–2542.
70 Lu, G., Ai, S., and Li, J. (2005) *Langmuir*, **21**, 1679–1682.
71 Lu, G., Komatsu, T., and Tshuchida, E. (2007) *Chem. Commun.*, 2980–2982.
72 Dougherty, S.A., Zhang, D., and Liang, J. (2009) *Langmuir*, **25**, 13232–13237.
73 Qu, X., Kobayashi, N., and Komatsu, T. (2010) *ACS Nano*, **4**, 1732–1738.
74 Tao, C., Yang, S., and Zhang, J. (2010) *Chinese J. Chem.*, **28**, 325–328.
75 Landoulsi, J., Demoustier-Champagne, S., and Dupont-Gillain, C. (2011) *Soft Matter*, **7**, 3337–3347.
76 Lee, D., Nolte, A.J., Kunz, A.L., Rubner, M.F., and Cohen, R.E. (2006) *J. Am. Chem. Soc.*, **128**, 8521–8529.
77 Hollman, A.M. and Bhattacharyya, D. (2004) *Langmuir*, **20**, 5418–5424.
78 Komatsu, T., Terada, H., and Kobayashi, N. (2011) *Chem. Eur. J.*, **17**, 1849–1854.
79 DeRocher, J.P., Mao, P., Han, J., Rubner, M.F., and Cohen, R.E. (2010) *Macromolecules*, **43**, 2430–2437.
80 Lynn, D.M. and Langer, R. (2000) *J. Am. Chem. Soc.*, **122**, 10761–10768.
81 Zhang, J., Chua, L.S., and Lynn, D.M. (2004) *Langmuir*, **20**, 8015–8021.
82 Jewell, C.M. and Lynn, D.M. (2008) *Adv. Drug Deliver. Rev.*, **60**, 979–999.
83 Lynn, D.M., Anderson, D.G., Putnam, D., and Langer, R. (2001) *J. Am. Chem. Soc.*, **123**, 8155–8156.
84 Anderson, D.G., Lynn, D.M., and Langer, R. (2003) *Angew. Chem. Int. Ed.*, **42**, 3153–3158.
85 Zhang, J., Fredin, N.J., Janz, J.F., Sun, B., and Lynn, D.M. (2006) *Langmuir*, **22**, 239–245.
86 Smith, R.C., Leung, A., Kim, B.-S., and Hammond, P.T. (2009) *Chem. Mater.*, **21**, 1108–1115.
87 Jewell, C.M., Zhang, J., Fredin, N.J., and Lynn, D.M. (2005) *J. Control. Release*, **106**, 214–223.
88 Dobrynin, A.V., Colby, R.H., and Rubinstein, M. (1995) *Macromolecules*, **28**, 1859–1871.
89 Cong, R., Temyanko, E., and Russo, P.S. (2005) *Macromolecules*, **38**, 10627–10630.
90 Cannell, D.S. and Rondelez, F. (1980) *Macromolecules*, **13**, 1599–1602.
91 Bohrer, M.P., Patterson, G.D., and Carroll, P.J. (1984) *Macromolecules*, **17**, 1170–1173.
92 Renkin, E.M. (1954) *J. Gen. Physiol.*, **38**, 225–243.
93 Gu, H., Faucher, S., and Zhu, S. (2010) *AIChE J.*, **56**, 1684–1692.
94 Shao, J. and Baltus, R.E. (2000) *AIChE J.l*, **46**, 1149–1156.
95 Davidson, M.G. and Deen, W.M. (1988) *J. Membr. Sci.*, **35**, 167–192.
96 Huertas, M.L., Navarro, S., Lopez Martinez, M.C., and García de la Torre, J. (1997) *Biophys. J.*, **73**, 3142–3153.
97 Hiort, C., Lincoln, P., and Norden, B. (1993) *J. Am. Chem. Soc.*, **115**, 3448–3454.
98 Picart, C., Mutterer, J., Richert, L., Luo, Y., Prestwich, G.D., Schaaf, P., Voegel, J.-C., and Lavalle, P. (2002) *PNAS*, **99**, 12531–12535.
99 Haynie, D.T., Cho, E., and Waduge, P. (2011) *Langmuir*, **27**, 5700–5704.
100 Roy, C.J., Leprince, L., De Boulard, A., Landoulsi, J., Callegari, V., Jonas, A.M., and Demoustier-Champagne, S. (2011) *Electrochim. Acta*, **56**, 3641–3648.

# 26
# The Design of Polysaccharide Multilayers for Medical Applications
*Benjamin Thierry, Dewang Ma, and Françoise M. Winnik*

## 26.1
## Introduction

Polysaccharides form an important class of natural polymers with critical structural functions and biological activities in plants and all living organisms. It comes as no surprise that polysaccharides are key players in the area of biomaterials, where they have found applications in fields as diverse as drug delivery, tissue engineering and regenerative medicine, biosensing and biodiagnostics [1]. Nonetheless, in the early phase of polyelectrolyte multilayers development, up to ~2002, there were only a few studies of layer-by-layer (LbL) assemblies containing polysaccharides. These natural polymers, intrinsically, are more difficult to handle than synthetic polymers or proteins that can be defined precisely in terms of structure, architecture, and molecular weight. Polysaccharides consist of carbohydrate units linked by glycosidic bonds and occur naturally with a range of molecular weights. Specific substituents, such as hydroxyls, sulfates, carboxylates, and amines linked to the glycoside rings, directly control the physicochemical and biological properties of polysaccharides. They also affect the multilayer assembly process, often in unpredictable ways. In addition, polysaccharides undergo facile degradation in strong acidic or alkaline environments. Hence, the multilayer construction conditions need to be selected carefully in order to avoid undesired side-reactions.

In spite of these difficulties, the field of polysaccharide multilayers has been flourishing over the last decade. A 2010 review by Crouzier *et al.* on this topic lists nearly 100 references to articles reporting multilayers containing one or two polysaccharides [2]. The surge in activity in this area has been triggered by two key results, which pointed to the unique properties of polysaccharide multilayers. First, from the mechanistic viewpoint, researchers noticed the unique propensity of polysaccharides to promote *exponential multilayer growth* [3, 4], unlike most synthetic polyelectrolytes for which multilayer growth occurs linearly as a function of the number of layers deposited [5, 6]. This observation led to a better understanding of the LbL process in general and generated further fundamental studies of polysaccharide multilayers. The practical implications of polysaccharide multilayers became apparent at the same

---

*Multilayer Thin Films: Sequential Assembly of Nanocomposite Materials*, Second Edition.
Edited by Gero Decher and Joseph B. Schlenoff.
© 2012 Wiley-VCH Verlag GmbH & Co. KGaA. Published 2012 by Wiley-VCH Verlag GmbH & Co. KGaA.

time. Serizawa *et al.* reported, in 2000, the use of polysaccharide multilayers as anti- or pro-coagulants of human blood [7]. In 2003, Thierry *et al.* described the construction of polysaccharide multilayers onto damaged arteries as a means to protect them against blood coagulation during the revascularization procedure [8]. The same group indicated also that the artery healing process can be accelerated by incorporating a releasable drug within the multilayer [9]. These results, which demonstrated the exceptional and varied functional properties of natural polysaccharides, led the way to the design of numerous polysaccharides multilayers for biomedical applications.

In this chapter, we give an overview of the work on polysaccharide multilayers carried out over the last decade, up to mid-2011. We chose, somewhat arbitrarily, to focus our attention on films composed exclusively of polysaccharides. In specific cases we refer to systems in which one of the polyelectrolytes is of a different class, either a synthetic polymer or a protein, to contrast their properties to those of all-carbohydrate multilayers. Throughout the chapter, we stress the importance of the chemical composition and structure of the polysaccharides, a point often neglected by researchers who focus their interest on the physical or biomedical properties of multilayer films. As a help to the reader not familiar with the chemistry of polysaccharides, we describe briefly in the first section of the chapter the structure and properties of commonly used polysaccharides. The core of the chapter is devoted to the physico-chemical properties of the most important current systems, from the fundamental view point. A final section is devoted to specific applications of polysaccharide multilayers in the treatment of cardiovascular diseases.

## 26.2
### Polysaccharides as Multilayered film Components: An Overview of Their Structure and Properties

The physico-chemical properties of natural polysaccharides used in multilayer assemblies are listed in Table 26.1. Figures 26.1 and 26.2 present, respectively, the chemical structure of cationic and anionic polysaccharides.

**Table 26.1** Physico-chemical properties of polysaccharides used in multilayer assemblies.

|  | Abbreviation | Polyelectrolyte nature | $pK_a$ | Persistence length (nm) |
|---|---|---|---|---|
| Chitosan | CH | weak | ~6.5 [10] | 6–12 [10] |
| Hyaluronan | HA | weak | ~3.0 [10] | 6 [10] |
| Alginate | AL | weak | ~3.5 [11] | — |
| Dextran sulfate | DEXS | strong | — | — |
| Heparin | HEP | strong | — | — |
| Chondroitin sulfate | CS | strong | — | — |
| Polygalacturonic acid | PGaLA | weak | 3.0–4.5 [12] | 10–12 [13] |
| κ-Carrageenan | — | strong | — | ~ 5 [14] |
| ι-Carrageenan | — | strong | — | 74 [15] |
| λ-Carrageenan | — | strong | — | 13 [14] |

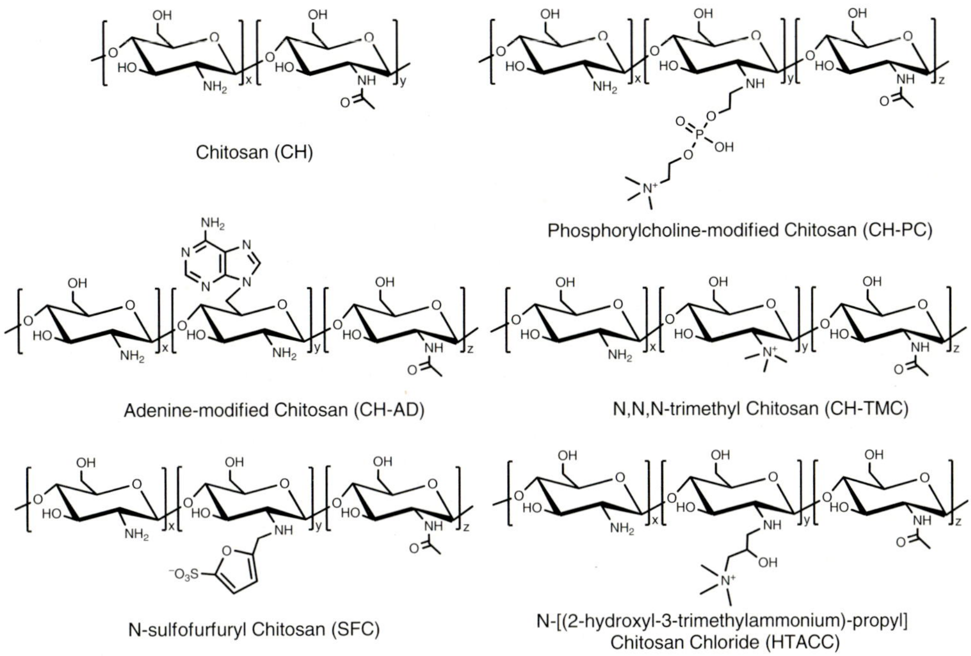

## 26.2 Polysaccharides as Multilayered film Components: An Overview of Their Structure and Properties | 639

**Figure 26.1** Chemical structures of cationic polysaccharides used in multilayer assemblies.

### 26.2.1
### Polycations

Strictly speaking, there are no positively charged polysaccharides in nature. Chitosan, a linear random copolymer of N-acetyl-β-D-glucosamine and β-D-glucosamine linked (1 → 4) is often considered to be "natural" since it is obtained by partial deacetylation of chitin, poly(N-acetyl-β-D-glucosamine), a natural constituent of shell fish [16–19]. Chitosans vary in molecular weight, degree of deacetylation (DDA, defined as the average mol% of glucosamine units/chain), and sequence, namely whether the N-acetyl-glucosamine residues are distributed randomly along the chain or in a "blocky" manner. As a result of the structural differences, the physicochemical properties of chitosan, such as the $pK_a$, vary from sample to sample. Chitosans of DDA ≥80% are used in most studies of multilayer assemblies. They are weak polycations. At low pH, the primary amines of the glucosamine residues are protonated, imparting to chitosan its solubility and cationic properties. At neutral or alkaline pH, the amines are deprotonated; the polymer becomes neutral and loses its solubility in water. The soluble to insoluble transition occurs for pH values between 6 and 6.5 for chitosan of DDA ∼ 80% [20]. In order for chitosan to have a sufficiently high charge density, the LbL construction should be carried out with solutions of pH 6 or lower. This places limitations on the polyanion partner, which has to be charged in solutions of pH ≤ 6 and should not be degraded under these conditions.

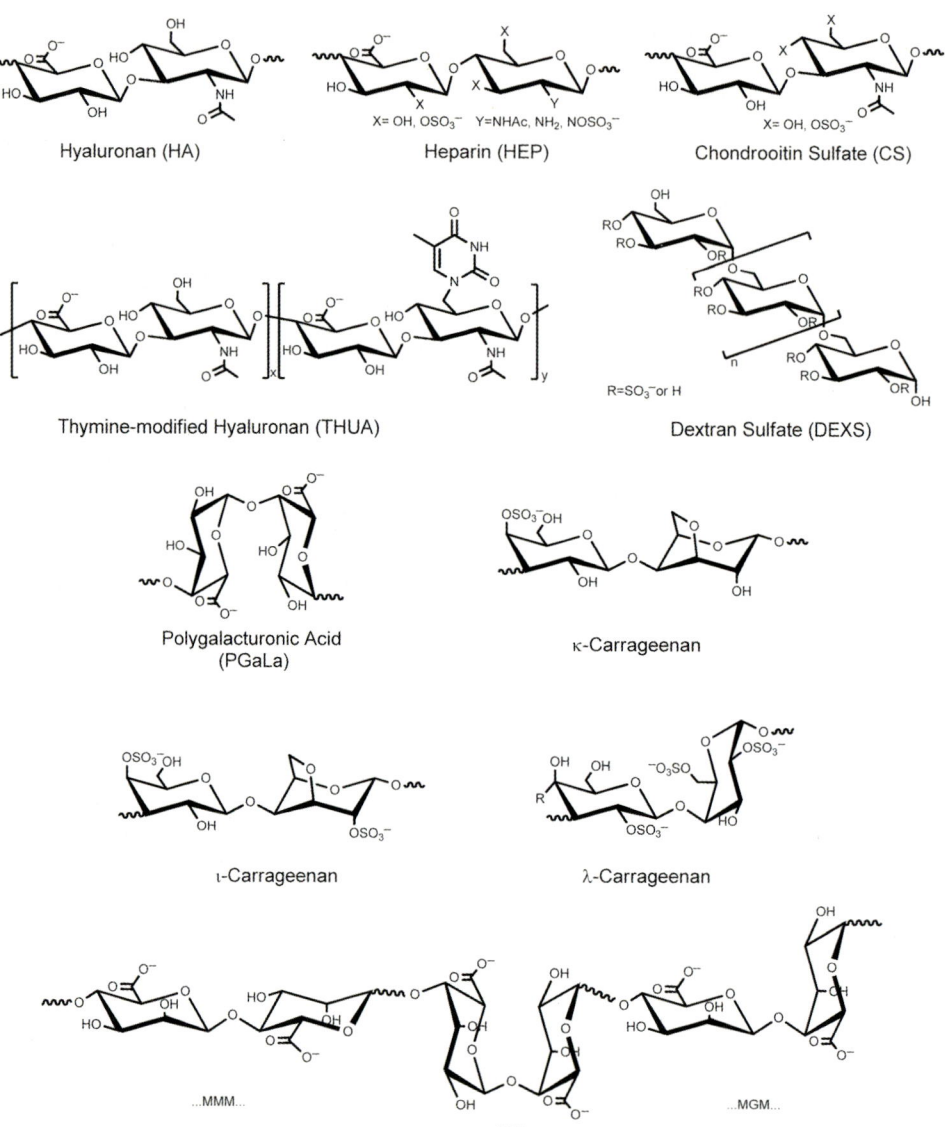

Figure 26.2 Chemical structures of anionic polysaccharides used in multilayer assemblies.

The insolubility of chitosan under physiological conditions (pH 7.4) is particularly problematic when one envisages *in vitro* and *in vivo* applications. There exist currently a number of chitosan derivatives that are soluble in neutral pH, or higher. Tiera *et al.* reported the preparation of phosphorylcholine (PC) modified chitosans, which proved to be non-toxic and soluble under physiological conditions, even with a modest level of

PC incorporation (~ 20 mol% PC per glucosamine unit) [21]. Like chitosan, chitosan-PC (CH-PC, Figure 26.1) is a polycation under acidic conditions, but the solution properties of the two polymers are different as a consequence of the conversion of a fraction of the primary amines into secondary amines ($pK_a \sim 7.20$) upon covalent linkage of the PC-groups. It is also possible to convert chitosan into a strong polyelectrolyte by methylation of the amine moieties. Thus, N,N,N-trimethyl chitosan (CH-TMC, Figure 26.1) with a degree of quaternization of 18% was employed successfully in LbL assembly [22]. Manna et al. described the preparation of adenine-modified chitosan (CH-AD, Figure 26.1), a water-soluble polymer, which together with a thymine-modified polysaccharide was used in the construction of multilayered films held together by hydrogen bonds between adenine and thymine [23]. Other researchers prepared a quaternary ammonium modified chitosan, N-[(2-hydroxy-3-trimethylammonium)-propyl] chitosan chloride (HTACC, Figure 26.1), which is soluble in water at all pH values and formed polyelectrolyte multilayers with a modified chitosan bearing negative charges, N-sulfofurfuryl chitosan (SFC, Figure 26.1) [24]. For the sake of completeness, we signal to the readers the commercial availability of (i) a chitosan hydrochloride that is soluble in neutral conditions and forms multilayered films with polyanions under pH conditions ranging from 5.5 to 7.0 [25] and (ii) a quaternized hydroxyethyl cellulose ethoxylate reported to form LbL films with anionic poysaccharides [26].

### 26.2.2
### Polyanions

Hyaluronan (HA, Figure 26.2), a glycosaminoglycan, has been used extensively in fundamental and applied studies of polysaccharide multilayers. It is a linear polymer consisting of alternating N-acetyl-β-D-glucosamine and β-D-glucuronic acid residues linked (1 → 3) and (1 → 4), respectively [27–29]. It is a weak polyelectrolyte with a $pK_a \sim 3.0$. Originally, commercial HA was isolated from natural sources, such as the rooster comb. This situation has changed considerably over the last decade in view of the increasing use of HA in commercial formulations. Currently, several companies offer well characterized HA samples of controlled molecular weight (from ~ 5000 to $10^6$ g mol$^{-1}$) [30, 31]. Alginate, which is extracted from seaweed, is also a weak polyelectrolyte ($pK_a \sim 3.5$). It is a copolymer of two monosaccharides linked (1 → 4), β-D manuronic acid and its C-5 epimer α-L-guluronic acid (AL, Figure 26.2) [18, 32, 33]. Polygalacturonic acid (PGaLA, Figure 26.2) is obtained by de-esterification of pectin extracted from the peel of citrus fruit. It is a linear weak polyelectrolyte consisting of (1 → 4) linked α-galacturonysyl units. Both pectin and PGaLA were assessed as components of polysaccharide multilayers [12, 13].

Sulfated glycosaminoglycans (sGAG) are heteropolysaccharides, that contain both aminosugar units and sulfonate residues. Hence, they are strong polyanions, remaining negatively charged in water over the entire pH range. They are mostly of animal origin. They constitute most of the non-collagenous biomacromolecules of the extracellular matrix. Two sGAG, heparin and chondroitin sulfate, have been used extensively as constituents of polyelectrolyte multilayer films. Heparin (HEP) is a complex macromolecule containing D-glucuronic acid, L-iduronic acid, and D-glucosamine units. The glucosamine units may be *N*-acetylated or *N*-sulfonated, such that HEP contains, on average, 2.5 sulfate groups per disaccharide unit (Figure 26.2) [34–38]. The glucuronic acid and iduronic acid are not distributed randomly, but occur in blocks. Heparin acts as an anticoagulant and is widely used in medicine for this purpose. Chondroitin sulfates are found in bone, skin, and cartilage in the form of proteoglycans that surround collagen fibers, lending them rigidity and incompressibility. The chondroitin sulfate chains are linear alternating copolymers of D-glucuronic acid and *N*-acetyl-D-galactosamine. They contain, on average, 1 sulfate per disaccharide unit linked either to carbon-4 or carbon-6 of the *N*-acetyl-D-galactosamine unit (Figure 26.2) (see Ref. [39] and references therein). Chondroitin sulfate is a non-toxic polymer sold as a health supplement.

Carrageenans are sulfated linear galactans, consisting of digalactopyranoside units joined by alternating (1 → 4) and (1 → 3) bonds [40, 41]. They are extracted from red seaweeds and are widely used in food, pharmaceutical and cosmetic formulations. Several carrageenans are commercially available. They have the same carbohydrate backbone, but differ by the position and number of sulfate groups along the chain. The approximate repeat units of three carrageenans are presented in Figure 26.2. The sulfate contents of λ- and *l* carrageenans are similar (42 and 37%, respectively), while κ-carrageenan has a lower sulfate content (~ 20%). λ-Carrageenan exists as a random coil in water [40]. It does not form gels. Both κ- and *l*-carrageenans, which have a 3,6-anhydrobridge in each disaccharide repeat unit, adopt a helical conformation in water [14, 15]. The sulfate groups are situated at the outside of the helix with the helical structure stabilized by internal hydrogen bonds. Helical carrageenans form highly viscous solutions or gels in water. They are widely used as food additives and in cosmetic formulations. Dextran sulfates (DEXS) are sulfated derivatives of dextran, a glucose homopolymer synthesized by bacteria [7]. The glucose units are linked (1 → 6), with branches linked to the backbone at the carbons 2 or 5 of the glucose units.

## 26.3
### Multilayers Formed by Assembly of Weak Polyanions and Chitosan or Chitosan Derivatives

### 26.3.1
**Hyaluronan**

Richert *et al.* reported in 2003 that the growth of polyelectrolyte multilayer films consisting of two oppositely charged polysaccharides, the polyanion HA and the polycation CH, follows the exponential mode previously observed for films containing poly(L-lysine) [42]. On the basis of a confocal laser scanning microscopy investigation with fluorescently-labeled polysaccharides, they attributed the exponential growth of HA/CH films to the diffusion through the multilayer of CH, not

HA. An *in situ* atomic force microscopy (AFM) investigation revealed that the film build-up proceeds in two stages: the surface is initially covered with isolated islets, which grow in size and coalesce as new polyelectrolyte is added. In the second stage, the islets interconnect, leading to a vermiculate morphology. Kujawa *et al.* [10] carried out AFM imaging of hydrated HA/CH multilayers throughout the build-up process (from layer pair 1 to 12) using polyelectrolyte pairs of different molecular weights (HA: $M_n$ 31 and 360 kDa; CH: $M_n$ 30 and 160 kDa). For assemblies of 12 layer pairs, films of (HA-360K/CH-160 K) were twice as thick as films of HA-31K/CH-30K (820 ± 150 nm vs. 448 ± 202 nm) (Figure 26.3). Full coverage of the surface with polyelectrolyte islets and the progression to interconnected

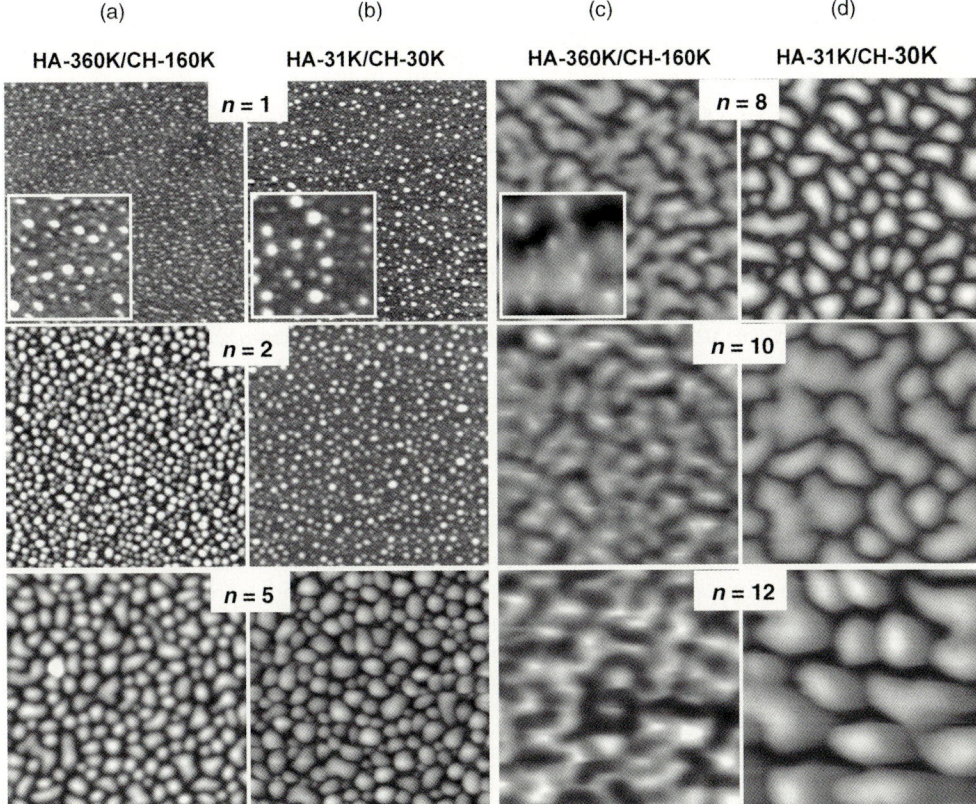

**Figure 26.3** AFM topography images (7.5 μm × 7.5 μm) acquired in a 0.15 M NaCl solution (pH 4.5) for different steps of the alternate deposition of high-MW (a, c) and low-MW (b, d) onto a PEI-coated MUA-Au substrate. *n* indicates the number of HA/CH layer pairs deposited. The insets shown for *n* = 1 are 1.5 μm × 1.5 μm magnifications, while that for *n* = 8 is 2.5 μm × 2.5 μm magnification. Maximum height or z-ranges of high-MW are 35 nm (*n* = 1), 90 nm (*n* = 2), 225 nm (*n* = 5), 500 nm (*n* = 8), 500 nm (*n* = 10), and 600 nm (*n* = 12). Maximum height or z-ranges of low-MW are 35 nm (*n* = 1), 150 nm (*n* = 2), 400 nm (*n* = 5), 400 nm (*n* = 8), 700 nm (*n* = 10), and 850 nm (*n* = 12) [10].

domains occurred after the deposition of fewer layers in the case of high-MW pairs, compared to low-MW pairs. Interestingly, much larger domains were present in the vermiculate morphology exhibited by the low-MW pair, compared to the high-MW pair. Film assembly of HA and chitosan-phosphoryl choline (CH-PC, PC: 20 mol% relative to the number of saccharide units) led to rougher and thicker films than those formed by HA/CH of similar MW [43]. The HA/CH-PC multilayers adopted a vermiculate morphology in the early construction steps (after deposition of the second layer pair) contrary to HA/CH films for which rounded densely packed globules formed preferentially.

Surface plasmon resonance (SPR) measurements with HA/CH pairs of high- and low-MW indicated that in the initial stage of multilayer construction, that is, layer pairs 1 to 3, the film thickness increment per deposition is small and of the same magnitude (~3 nm per layer) regardless of the MW of the polyelectrolytes [10]. This situation changed, however, as the build-up continued and a steep growth regime (exponential) prevailed. The crossover occurred after the deposition of ~2 layer pairs for the HA-360K/CH-160K pair, but only after the deposition of ~3 layer pairs in the case of the HA-31K/CH-30K pair. The number of layer pairs for which the crossover occured for the high (~2) and low (~3) MW pairs coincides with the stage in the multilayer build-up at which the vermiculate morphology appears, resulting in a more effective CH diffusion. In the exponential regime, the growth rate did not depend on the MW of the polysaccharides. Hence, the existence of thicker films for the high MW polysaccharides at a given layer pair number in the regime of steep growth, was attributed to an earlier crossover from the first to second growth regime, rather than a larger incremental increase in the layer thickness *per se*, as reported in a previous QCM study [42]. SPR measurements of the multilayer build-up with HA/CH-PC revealed that the surface coverage at each step is the same as in the case of HA/CH, although the thickness of the films measured by AFM was different (*vide supra*) [43]. Since the AFM scratch test yields the total thickness of the films, whereas SPR provides a measure of the polymer mass in the film, the discrepancy between the results of the two techniques was taken as an indication of large differences in the degree of hydration of HA/CH and HA/CH-PC. Throughout the growth of eight layer pairs, both the loss and storage moduli of the films, $G'$ and $G''$, derived from an impedance analysis of QCM data, were one order of magnitude lower in the case of HA/CH-PC films, compared to HA/CH films (Figure 26.4).

Richert *et al.* showed that the thickness of CH/HA films increases with the ionic strength ($10^{-4}$ to 0.15 M NaCl) [42]. Films constructed with HA and CH in aqueous 0.15 M NaCl solutions (pH 4.5) exposed to further increase in ionic strength unraveled rapidly as the ionic strength exceeded 0.8 M [44]. This salt concentration is higher than that observed for multilayers of HA and the weak polyelectrolyte poly (allylamine hydrochloride) assembled at pH 4.0 which decomposed as the ionic strength exceeded 0.20 M (NaCl). The enhanced stability of HA/CH films may reflect differences between the energies of interpolymer association: the HA/CH pair may undergo hydrophobic interactions in addition to electrostatic association. Multilayers of HA/CH (4 layer pairs) formed at pH 4.5 (0.15 M NaCl) were subjected to changes in pH by addition of buffer of increasing or decreasing pH, starting from

## 26.3 Multilayers Formed by Assembly of Weak Polyanions and Chitosan or Chitosan Derivatives

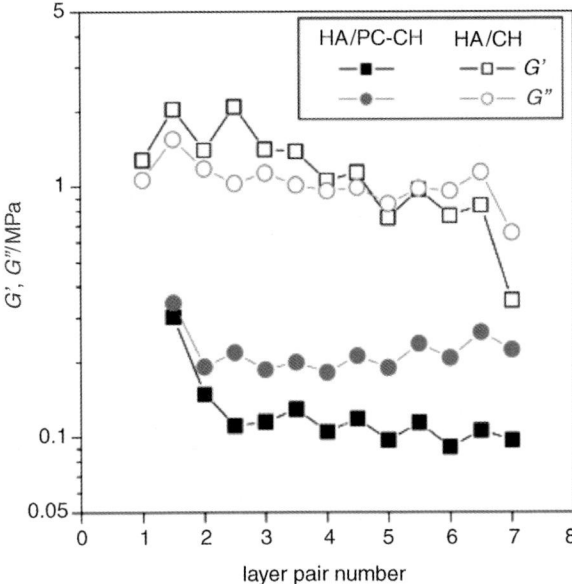

**Figure 26.4** Storage modulus ($G'$, square) and loss modulus ($G''$, circle) as a function of the number of layer pairs during the alternate adsorption of HA and CH (open symbols) or HA and PC-CH (full symbols). Data were calculated from dissipative QCM data [43].

pH 4.5 [44]. QCM measurements indicated that the multilayers were stable over the 2 to 9.0 pH range, independently of the molecular weight of the polymers. AFM observations of multilayers prepared on mica under the same conditions indicated that complete unraveling of the multilayers took place at pH 9.0 and 2.5 (Figure 26.5). Multilayers of HA and CH-PC remained intact until the pH of the contacting solution exceeded 11.0 [43]. The disintegration of the multilayers in the high pH range is governed by the decrease in the ionization degree of CH. Incorporation of PC modifies the $pK_a$ of the polymer (CH: $pK_a \sim 6.7$, CH-PC: $pK_a = 7.20$) [21]; thus the coatings become more stable in neutral and alkaline pH. The stability of the multilayers in the low pH range is governed by the ionization of HA. Both HA/CH and HA/CH-PC decompose in media of pH $\leq 2.5$.

An altogether different approach towards the formation of HA/CH multilayers consists in linking moieties able to form strong hydrogen bonds to HA and CH. The feasibility of this concept, which mimics the formation of the DNA double helix, was demonstrated in a study of the assembly of CH bearing adenine groups (CH-AD, Figure 26.1) with HA bearing thymine groups (THUA, Figure 26.2) [23]. Multilayers were obtained by sequential dipping of a substrate in aqueous solutions of CH-AD and THUA containing 0.15 M NaCl. The substrate used to initiate the assembly was prepared by dipping a quartz substrate into acidic CH followed by electrostatically-driven adsorption of THUA. The following layers were held together via H-bonds only. The CH-AD/THUA were used as delivery systems for drugs, such as doxorubicin.

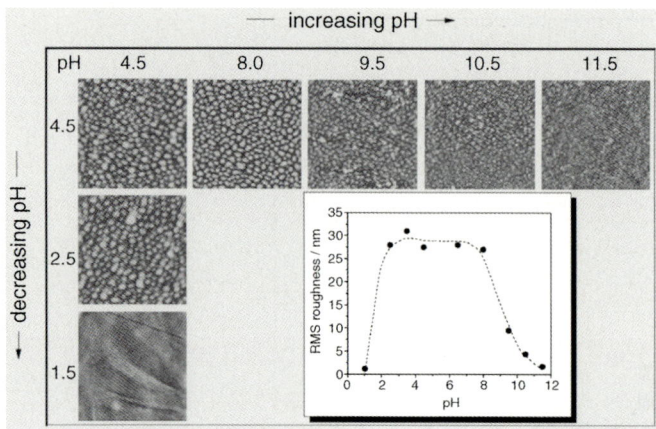

**Figure 26.5** AFM topography images (7.5 μm × 7.5 μm) acquired in a 0.15 M NaCl solution (pH 4.5) for different steps of the treatment of a PEI-(HA-360 K/CH-160 K)$_4$ multilayer with solutions of different pH. The film surface was rinsed with a 0.15 M NaCl solution (pH 4.5) between each step. Maximum heights or z-ranges are: 300 nm (pH 2.5, 4.5 and 8.0), 100 nm (pH 9.5), 50 nm (pH 10.5), and 25 nm (pH 1.5 and 11.5). Inset: RMS roughness versus pH dependence [44].

## 26.3.2
## Polygalacturonic Acid

Multilayered constructs between chitosan and poly(galacturonic acid) (PGaLA, Figure 26.2) were reported by Westwood *et al.*, who monitored the build-up process using three complementary techniques: QCM-D, dual polarization interferometry (DPI), and Fourier transform infrared spectrometry in attenuated total reflectance mode (FTIR-ATR) [45]. The multilayers were formed in a 10-step deposition of polyelectrolyte solutions (0.6 mg mL$^{-1}$) in a pH 7.0, 10 mM phosphate buffer containing 30 mM NaCl. Under these conditions, PGaLA is fully charged (p$K_a \leq 5.3$), while CH is only partially ionized. The chitosan employed had a DDA of ∼ 50% [46], hence only about half the monosaccharide units had a protonable amine group. Moreover, due to the deposition pH, only about half of these are predicted to be charged. Hence, the fraction of charged monosaccharide units in CH was ∼ 0.25%. PGaLA/CH multilayer growth followed the exponential mode, reaching an optical thickness of 32 nm (5 layer pairs). FTIR-ATR indicated that the films are richer in PGaLA, the fully charged polyelectrolyte. Hence, the films were far from achieving intrinsic charge balance, suggesting that electrostatic forces involving the buffer and salts and non-electrostatic interactions play an important part in the formation and characteristics of the films. QCM-D measurements showed that the multilayers were more viscous when CH was the top layer and more elastic when PGaLA was the uppermost layer. A similar pattern was observed also during the construction of HA/CH multilayers [43].

The pH of the fluid in contact with a 10-layered PGaLA/CH multilayer was decreased gradually from 7.0 to 2.0, then increased up to 7.0 to assess the stability of the films [47]. The multilayer shrank from 32 to 24 nm as the pH was lowered from 7.0 to 3.6, and then underwent a slight re-swelling with further decrease in pH to 2.0. On increasing the pH back to 7.0, the multilayer shrank further, reaching a final thickness of 17 nm. The refractive index of the multilayer was not affected by these changes in pH, implying that the reduction of the film thickness was due to the dissolution of the polyelectrolytes. By assessing the composition of the resulting films via FTIR-ATR, it was determined that chitosan redissolved preferentially. The losses in CH and PGaLA during the treatment were 50 and 10%, respectively. The authors suggest that the self-association of PGaLA in the acid form is responsible for the stability of the film at low pH.

### 26.3.3
### Alginate

The alginate/chitosan pair has been used extensively to encapsulate particles and cells and to form physically-crosslinked hydrogels [48]. Multilayered AL/CH constructs on flat surfaces have also been investigated, but to a much lesser extent. Of note is a study of the effect of pH on the construction of multilayers with a solution of CH (medium molecular weight, pH 3) and solutions of AL (sodium salt, 20–40 cP) of pH 3.0, 4.0, or 5.0 [49]. From SPR measurements, it was estimated that each layer pair contained excess CH relative to AL, and the excess increased with the pH of the AL solution used in the build-up. The multilayer morphology observed by AFM was also affected by the pH of the AL solution, changing from well-defined contiguous small microglobules (pH 3.0) to large polydispersed globules (pH 5.0). Dried AL/CH films (AL: medium viscosity, in 0.5 M NaCl; CH: DDA 90%; in a pH 3.02 citric acid buffer, 0.5 M NaCl) were deposited on silicon wafers [11]. Their composition was compared to that of AL/CH mixed films and of precipitated, preformed AL/CH complexes. This study emphasizes the importance of using two complementary tools (here XPS and FTIR) to analyze the composition of films.

## 26.4
## Multilayers Formed by Assembly of Strong Polyanions and Chitosan or Chitosan Derivatives

### 26.4.1
### Heparin

The elaboration of HEP/CH multilayers has been motivated by the possible use of the films on medical implants, to which they may impart the antibacterial properties of CH and the antiadhesive characteristics of HEP. Recent studies of this pair of polysaccharides have focused on the effects of the pH and the ionic strength of the deposition solution on the growth, structure, and bacterial adhesion properties of the multilayers. Fu et al. reported that HEP/CH multilayers deposited onto poly (ethylene terephthalate) surfaces using polyelectrolyte solutions of pH 4.0 were more effective bactericides than films obtained using solutions of pH 6.0 [50]. They also observed that an increase in solution pH leads to an increase in the amount of CH and HEP in each layer, which was ascribed to the weak polyelectrolyte nature of CH. Increase in the solution ionic strength (from 0.1 to 0.5 M) also triggers an increase in the multilayer thickness, as demonstrated by Boddohi *et al.* in a study also aimed at mapping the relationship between the HEP/CH multilayer thickness and both the pH and ionic strength of the deposition solutions [51]. In this study, the HEP/CH multilayers were deposited on a mercaptoundecanoic acid monolayer adsorbed on a gold surface. The multilayer thickness was derived from FT-SPR *in situ* measurements and *ex situ* ellipsometry. The largest change with pH (from 4.6 to 5.8) in multilayer thickness was recorded for the intermediate ionic strength (0.2 M, from 2 nm/layer to 4 nm/layer).

Lundin *et al.* studied HEP/CH multilayers deposited on a 50-nm silica layer coated onto quartz crystals for *in situ* QCM-D analysis, complemented by AFM imaging and DPI [52]. The authors mapped the effect of pH (4.0 to 5.8) and ionic strength (0.1 to 150 mM) on multilayer growth and thickness. In all cases the multilayer growth followed the exponential mode. This mode is generally believed to be a consequence of the diffusion of at least one of the two polyelectrolytes within the film. This was visualized by the distance-dependent fluorescence resonance energy transfer (FRET) technique using fluorescently labeled polyelectrolytes, which showed that CH diffused vertically (out of the plane) in all multilayers, independently of the pH and ionic strength of the solutions (Figure 26.6) [53]. The authors also established that, of

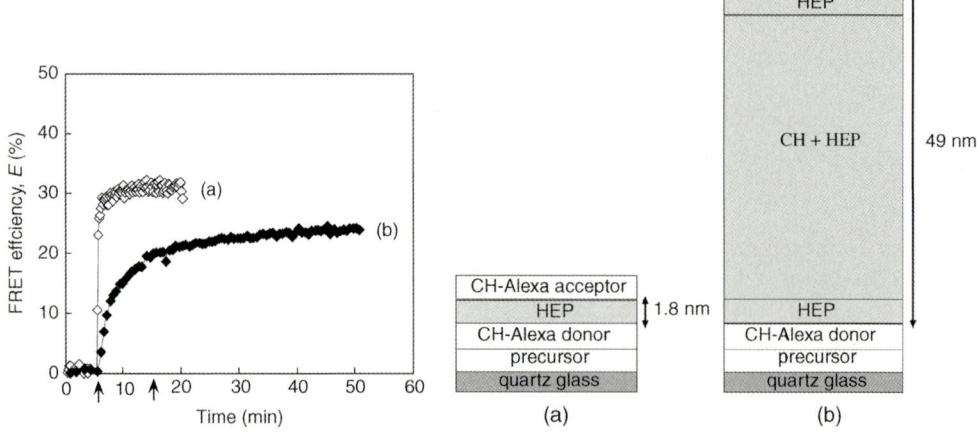

**Figure 26.6** Energy transfer efficiency between CH-Alexa 488 and CH-Alexa 555 dye molecules adsorbed in/on a CH-HEP multilayer film prepared in a solution of 150 mM NaCl at pH 5.8. Donor and acceptor were initially separated either by (a) 1 HEP layer (open diamonds) or (b) 13 layers of CH and HEP (filled diamonds). These two situations are illustrated schematically. CH-Alexa 488 was present in layer 5. Arrows mark the addition of CH-Alexa 555 (acceptor dye) and the beginning of a rinse with polymer-free electrolyte solution. Lines are drawn to guide the eye [53].

the two polysaccharides, CH is the dominant species in all multilayers. The fact that the HEP/CH ratio is not stoichiometric was taken as an indication that the association of the two polysaccharides involves short-range interactions that are not of electrostatic origin. The QCM-D data were used also to gain information on the viscoelastic properties of the films [52]. On the basis of the Voigt viscoelasticity model for an adsorbed layer, the authors showed that it is the film viscosity and shear that define the layer viscoelasticity of a film and not the absolute value of the dissipation energy. This point is indeed applicable to QCM-D studies of all viscoelastic films, beyond HEP/CH or polysaccharides.

### 26.4.2
### Chondroitin Sulfate

Chondroitin sulfate (CS) is a component of the non-collagenous extracellular matrix of lower charge density than HEP (see Figure 26.2). Multilayers formed by CS and several polycations have been studied extensively in view of potential applications in tissue engineering. Two recent studies have assessed the construction of multilayers between either CH or poly-(L-lysine) (PLL) as polycation and HA, CS, and HEP as polyanions [22, 54]. Both studies stress the importance of ion pair interactions, and for PLL-based multilayers the ion pairing between the polycation increases with increasing charge density of the polyanion (HA < CS < HEP). Moreover, the increase in ion pairing within polyanion/PLL films was associated with a decrease in the film hydration. For CH-based films, assemblies with the strong polyelectrolytes CS and HEP were thinner in the wet form (from *in situ* FT-SPR, 10 layers) and upon drying (from *ex situ* ellipsometry, 20 layers) and less hydrophilic (from contact angle determinations) than HA/CH films. Moreover, when the weak polycation, CH, was replaced by a stronger polycation (TMC, Figure 26.1), all multilayers were thinner, the largest change being recorded in the case of HA/polycation multilayers (HA/CH: wet thickness = 28 nm; HA/TMC: wet thickness = 13 nm). For multilayers constructed with CS, the change in thickness fell within experimental uncertainty for measurements by FT-SPR (wet), but it was significant in the case of dry films: CS/CH: 11 nm; CS/TMC: 4 nm. Contact angle measurements revealed that the HA/CH, CS/TMC, and HEP/TMC films were the most hydrophilic, while HA/TMC, CS/CH, and HEP/CH were the most hydrophobic. These observations were taken as an indication that multilayers formed between weak polyelectrolytes tend to have reduced ion-pairing interactions, hence can be highly hydrated. Polyelectrolytes of higher charge density, in contrast, readily undergo inter-chain ion pairing and, consequently, have a reduced ability to swell in water and are macroscopically more hydrophobic than films formed by weak polyelectrolytes. Conversion of CH to CH-TMC introduces strong cationic charges along the chain, without affecting significantly the $pK_a$ of the polymer. In the presence of strong polyelectrolytes, such as CS or HEP, the charge density of CH-TMC is not affected strongly. In this case, a significant amount of positive counterions must be present in the multilayers to achieve charge neutrality. The large amount of salts in the film imparts significant osmotic pressure to the films, allowing them to swell in water and to exhibit a higher hydrophilicity, compared to the corresponding CH films.

## 26.5
### Cardiovascular Applications of Polysaccharide Multilayers

#### 26.5.1
##### Anti-Thrombogenic Properties of Polysaccharide Multilayers

As stated in the introduction, the publication of Serizawa *et al.* on the alternating anti- and pro-coagulant activities of DEXS/CH multilayers opened the way toward numerous applications of polysaccharide multilayers in therapeutics and tissue engineering [7]. Laboratories worldwide have reported remarkably thromboresistant coatings using anionic polysaccharides other than DEXS, such as HEP and HA, assembled with cationic polyelectrolytes, such as polyethyleneimine (PEI) and PLL [55, 56]. The impact of the LbL approach is particularly impressive in cardiovascular therapy, as described in a recent review by Kerdjoudj *et al.* [57]. Thierry *et al.* demonstrated for the first time the validity of the approach, coating nickel-titanium (NiTi) stents with HA/CH [8]. A PEI primer layer was used to initiate the LbL build-up on the metallic surface of a multilayer, which was studied in the dry form using time-of-flight secondary ions mass spectrometry (ToF-SIMS), contact angle measurements, and atomic force microscopy (AFM). The anti-thrombogenic characteristics of the coatings were investigated *in vitro*: a significant decrease of platelet adhesion was observed in comparison to the unmodified metallic alloy (38% reduction; $p = 0.036$). An extracorporeal procedure in *ex vivo* perfusion chambers was further used to determine the adhesion of polymorphonuclear neutrophils radiolabeled with $^{111}$In. Despite their anti-thrombogenic properties, the polysaccharide coatings did not prevent fouling by neutrophils which could be attributed to specific adhesion through ligand–receptor mediated interactions such as CD44 that are expressed on polymorphonuclear neutrophils. Recently, the high binding affinity of nonadherent B lymphocytes for HA-terminated HA/CH multilayers has been taken advantage of to prepare surfaces with selective binding abilities of lymphocytes [58]. HA deposition conditions that favor loops and tails, such as low pH and with added salt, result in more available CD44 binding ligands and higher cell binding efficiency.

Promising results have also been obtained *in vivo*. Chitosan and heparin multilayer coatings displayed a good hemocompatibility and no significant toxicity. In a pig model of vascular injury, HEP/CH films were effective in promoting re-endothelialization and intimal healing after stent implantation (Figure 26.7) [59]. While the blood compatibility of polysaccharide multilayer coating is important towards their use in vascular applications, their long term ability to promote healthy endothelial cell function is also critical, especially when combined with a drug eluting approach which is typically associated with increased risks of late angiographic stent thrombosis. Heng *et al.* investigated the adhesion, proliferation, and gene expression of human umbilical vein endothelial cell (HUVEC) on polyelectrolyte coatings composed of an outermost layer of hyaluronan, heparin, or chondroitin sulfate and an underlying layer of either poly-L-lysine or chitosan [60]. Seven genes related to the adhesion, migration, and endothelial function of HUVEC (VWF, VEGFR, VEGFA, endoglin, integrin-5, ICAM1, and ICAM2) were investigated and found to be

## 26.5 Cardiovascular Applications of Polysaccharide Multilayers

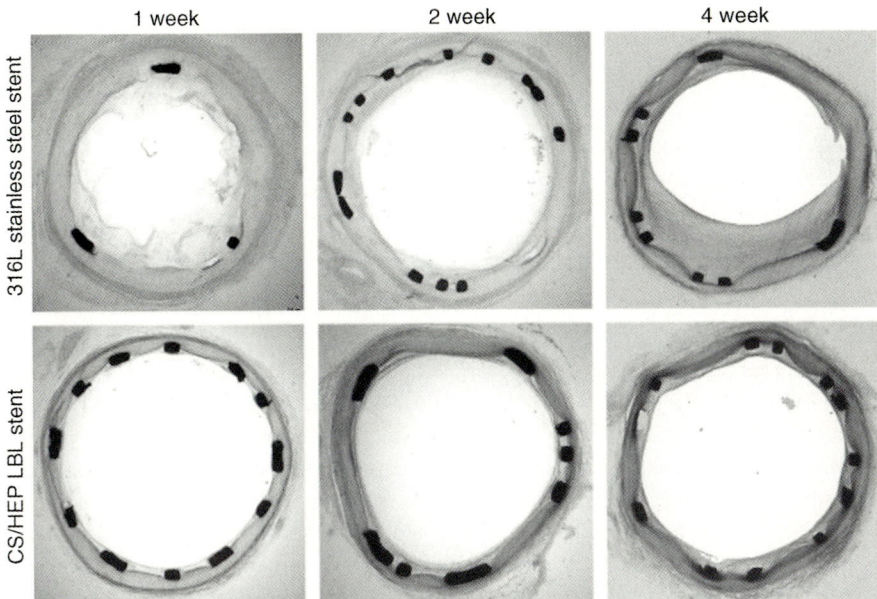

**Figure 26.7** Typical optical photographs (25 ×) of the cross-section slices of porcine arteries with bare metal stent and chitosan/heparin coated stents after implantation for 1, 2 and 4 weeks, respectively [59].

modulated by the underlying polysaccharide layer. HA/CH multilayers displayed the highest level of expression of VWF, reflecting the important role of hyaluronan in vascular biology.

### 26.5.2
### Preparation of Polysaccharide Multilayers on Blood Vessels

Since the pioneering work of Hubbell's group [5], the feasibility of self-assembling polyelectrolyte multilayers on biological substrates and, in turn, to drastically modulate their properties, has been explored. Owing to the broad range of polysaccharides compatible with LbL processes and the resulting biological functions of these films, innovative applications have advanced this field considerably. In particular, the ability to prepare polysaccharide multilayers on blood vessels has attracted much attention. Minimally invasive revascularization procedures, such as coronary angioplasty, are often compromised by complications related to the re-occlusion of the treated blood vessel, that is, restenosis. These complications result, at least in part, from the injury to the vascular wall during the angioplasty procedure which denudes the protective endothelium and initiates excessive vascular cell proliferation within the artery. The angioplasty-damaged blood vessels may lose their thromboresistance due to the exposure of underlying media to the blood. The ability of polysaccharide multilayers to alleviate the thrombogenicity of damaged blood vessels has been demonstrated *in vitro* using chitosan and hyaluronan multilayers [61]. In this work,

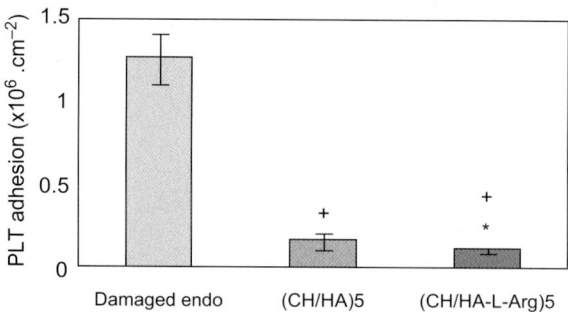

**Figure 26.8** Platelet adhesion ($n \geq 4$) on damaged endothelium (damaged endo), (CH/HA)5 and (CH/HA-L-Arg)5 coated damaged endothelium ((CH/HA)5 and (CH/HA-L-Arg)5); *$p< 0.05$ Vs. Damaged endo; + $p< 0.05$ Vs. (CH/HA)5 by paired Student-t-test [61].

HA/CH multilayers were constructed on damaged and healthy aortic porcine arteries using a perfusion chamber matching. Linear build-up profiles were obtained on the damaged blood vessels with lower amounts of polymers being deposited on healthy endothelium compared to the damaged vessels. The thrombogenicity of the damaged arterial surfaces was significantly reduced by 4 layer pair HA/CH films, as shown by a drastic reduction in platelet adhesion (87%, $p < 0.05$) (Figure 26.8).

A potential drawback of polysaccharide-based multilayers for the protective coating of biological materials is their susceptibility to enzymatic degradation *in vivo*, which may limit their long term efficiency. Using a rat mouth model, the enzymatic degradation of HA/CH films, native or cross-linked using carbodiimide chemistry, was studied. While rapid degradation of the non-cross-linked films was observed, much slower degradation kinetics was observed for the cross-linked ones, demonstrating the possibility to tune the degradation of such polysaccharide multilayers. However, chemical cross-linking may not be compatible with *in vivo* deposition of polysaccharide multilayers on damaged blood vessels. Voegel's group reported recently the use of poly(styrene sulfonate)/poly(allylamine hydrochloride) (PSS/PAH) multilayers instead of HA/CH. Human umbilical arteries treated with PSS/PAH multilayers demonstrated a high graft patency after 3 months of implantation, suggesting that such modified arteries could constitute a useful approach for small vascular replacement [57].

### 26.5.3
### Drug Delivery to the Vascular Wall from Polysaccharide Multilayers

Important efforts have been devoted toward the incorporation of active agents within polyelectrolyte multilayers. Several reviews have been written on this topic, particularly in the case of drug delivery systems [62–64]. The use of polysaccharide multilayers as a drug-carrier was demonstrated first in the case of HA/CH multilayers assembled onto damaged vascular vessels that were loaded with the cationic peptide L-arginine (amount incorporated: 213 ng cm$^{-2}$ within 5 layer pairs) [61]. The release of L-arginine occurred rapidly (burst release), nonetheless the incorpo-

## 26.5 Cardiovascular Applications of Polysaccharide Multilayers | 653

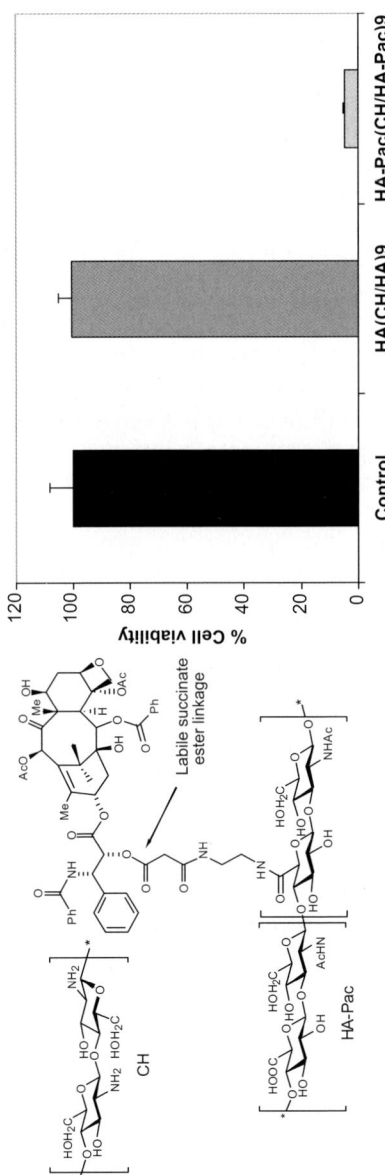

**Figure 26.9** Viability of J774 macrophages cultured onto HA(CH/HA)$_9$ and HA-Pac(CH/HA-Pac)$_9$ multilayers for 4 days.

ration of L-arginine in the CH/HA multilayers resulted in a reduction of 30% in the amount of platelet adhesion after 90 min ($p < 0.05$), compared to arteries protected by CH/HA multilayers devoid of the cationic peptide.

In order to prevent the burst release typically associated with the physical loading of low molecular weight drugs, Thierry *et al.* devised an approach that combined standard LbL assembly and prodrug synthesis [9]. Paclitaxel (Pac), a microtubule-stabilizing agent, was covalently linked to HA via a labile succinate ester moiety with incorporation levels of 3 and 6 mol% versus HA disaccharide units. The HA-Pac prodrug was successfully employed in the construction of multilayers with CH, as shown by QCM-D and AFM imaging. The *in vitro* Pac release was monitored upon immersion of the multilayers in water. The drug half-life in the multilayers was 3 h. Murine J774 macrophages plated at a density of 50 000 cells/well onto paclitaxel loaded multilayers HA-Pac (HA-Pac/CH)$_9$ displayed a 95% reduction in viability ($p < 0.05$ by ANOVA-Dunnett), while cells cultured onto (HA/CH)$_9$ were healthy (100.2% cell survival), demonstrating the highly potent cytotoxicity of the modified polysaccharide films (Figure 26.9). It is important to note that the release of the drug from the multilayer, as a consequence of the succinate ester hydrolysis, does not trigger the unraveling of the multilayer, which continues to act as thromboresistant agent over much longer time periods. Hence the mode of action of Pro-drug multilayers is quite different from the approach proposed by Lynn *et al.* [65], who pioneered the design of multilayered films that erode in physiological environments and, in turn, enable the release in a controlled fashion of macromolecular therapeutics.

## 26.6
## Conclusions

The field of functionalized films and coatings for medical applications is evolving rapidly. Multilayered polyelectrolyte films play a pivotal role among the various approaches assessed currently. Although polysaccharides are excellent components of multilayers, in view of their inherent biocompatibility, nontoxicity, and inherent biological activity, there are limitations, however, to the use of this class of polyelectrolytes. As stated in Section 26.5, polysaccharides are biodegradable, so *in vivo* they will eventually decompose by enzymatic degradation. Materials design has to take into consideration the fact that polysaccharide multilayers disintegrate with time unless they are crosslinked chemically. A second problem associated with the design of all-polysaccharide multilayer films is the limited choice of natural polycations. The construction of films between synthetic and natural polyelectrolytes poses no problem *per se*, but for *in vivo* applications the well-documented toxicity elicited by most polycations poses a significant challenge. This issue may be solved by fundamental mechanistic studies on the *in vitro* and *in vivo* interactions between carefully designed polycations and cells or organs coupled with conceptually robust multilayer architectures, as indicated by recent publications [66].

## References

1. Dove, A. (2001) The bittersweet promise of glycobiology. *Nat. Biotechnol.*, **19** (10), 913–917.
2. Crouzier, T., Boudou, T., and Picart, C. (2010) Polysaccharide-based polyelectrolyte multilayers. *Curr. Opin. Colloid Interface Sci.*, **15** (6), 417–426.
3. Picart, C., Lavalle, P., Hubert, P., Cuisinier, F.J.G., Decher, G., Schaaf, P., and Voegel, J.C. (2001) Buildup mechanism for poly(L-lysine)/hyaluronic acid films onto a solid surface. *Langmuir*, **17** (23), 7414–7424.
4. Picart, C. (2008) Polyelectrolyte multilayer films: From physico-chemical properties to the control of cellular processes. *Curr. Med. Chem.*, **15** (7), 685–697.
5. Elbert, D.L., Herbert, C.B., and Hubbell, J.A. (1999) Thin polymer layers formed by polyelectrolyte multilayer techniques on biological surfaces. *Langmuir*, **15** (16), 5355–5362.
6. von Klitzing, R. (2006) Internal structure of polyelectrolyte multilayer assemblies. *Phys. Chem. Chem. Phys.*, **8** (43), 5012–5033.
7. Serizawa, T., Yamaguchi, M., Matsuyama, T., and Akashi, M. (2000) Alternating bioactivity of polymeric layer-by-layer assemblies: Anti- vs procoagulation of human blood on chitosan and dextran sulfate layers. *Biomacromolecules*, **1** (3), 306–309.
8. Thierry, B., Winnik, F.M., Merhi, Y., Silver, J., and Tabrizian, M. (2003) Bioactive coatings of endovascular stents based on polyelectrolyte multilayers. *Biomacromolecules*, **4** (6), 1564–1571.
9. Thierry, B., Kujawa, P., Tkaczyk, C., Winnik, F.M., Bilodeau, L., and Tabrizian, M. (2005) Delivery platform for hydrophobic drugs: Prodrug approach combined with self-assembled multilayers. *J. Am. Chem. Soc.*, **127** (6), 1626–1627.
10. Kujawa, P., Moraille, P., Sanchez, J., Badia, A., and Winnik, F.M. (2005) Effect of molecular weight on the exponential growth and morphology of hyaluronan/chitosan multilayers: A surface plasmon resonance spectroscopy and atomic force microscopy investigation. *J. Am. Chem. Soc.*, **127** (25), 9224–9234.
11. Lawrie, G., Keen, I., Drew, B., Chandler-Temple, A., Rintoul, L., Fredericks, P., and Grondahl, L. (2007) Interactions between alginate and chitosan biopolymers characterized using FTIR and XPS. *Biomacromolecules*, **8** (8), 2533–2541.
12. Ralet, M.C., Dronnet, V., Buchholt, H.C., and Thibault, J.F. (2001) Enzymatically and chemically de-esterified lime pectins: Characterisation, polyelectrolyte behaviour and calcium binding properties. *Carbohydr. Res.*, **336** (2), 117–125.
13. Cros, S., Garnier, C., Axelos, M.A.V., Imberty, A., and Perez, S. (1996) Solution conformations of pectin polysaccharides: Determination of chain characteristics by small angle neutron scattering, viscometry, and molecular modeling. *Biopolymers*, **39** (3), 339–352.
14. Vanneste, K., Slootmaekers, D., and Reynaers, H. (1996) Light scattering studies of the dilute solution behaviour of kappa-, iota- and lambda-carrageenan. *Food Hydrocolloid*, **10** (1), 99–107.
15. Reynaers, H. (2003) Light scattering study of polyelectrolyle polysaccharides - the carrageenans. *Fibres Text. East. Eur.*, **11** (5), 88–96.
16. Yi, H.M., Wu, L.Q., Bentley, W.E., Ghodssi, R., Rubloff, G.W., Culver, J.N., and Payne, G.F. (2005) Biofabrication with chitosan. *Biomacromolecules*, **6** (6), 2881–2894.
17. Kumar, M., Muzzarelli, R.A.A., Muzzarelli, C., Sashiwa, H., and Domb, A.J. (2004) Chitosan chemistry and pharmaceutical perspectives. *Chem. Rev.*, **104** (12), 6017–6084.
18. Rinaudo, M. (2006) Chitin and chitosan: Properties and applications. *Prog. Polym. Sci.*, **31** (7), 603–632.
19. Payne, G.F. and Raghavan, S.R. (2007) Chitosan: a soft interconnect for hierarchical assembly of nano-scale components. *Soft Matter*, **3** (5), 521–527.

20 Schoeler, B., Delorme, N., Doench, I., Sukhorukov, G.B., Fery, A., and Glinel, K. (2006) Polyelectrolyte films based on polysaccharides of different conformations: Effects on multilayer structure and mechanical properties. *Biomacromolecules*, **7** (6), 2065–2071.

21 Tiera, M.J., Qiu, X.P., Bechaouch, S., Shi, Q., Fernandes, J.C., and Winnik, F.M. (2006) Synthesis and characterization of phosphorylcholine-substituted chitosans soluble in physiological pH conditions. *Biomacromolecules*, **7** (11), 3151–3156.

22 Almodovar, J., Place, L.W., Gogolski, J., Erickson, K., and Kipper, M.J. (2011) Layer-by-layer assembly of polysaccharide-based polyelectrolyte multilayers: A spectroscopic study of hydrophilicity, composition, and ion pairing. *Biomacromolecules*, **12** (7), 2755–2765.

23 Manna, U., Bharani, S., and Patil, S. (2009) Layer-by-layer self-assembly of modified hyaluronic acid/chitosan based on hydrogen bonding. *Biomacromolecules*, **10** (9), 2632–2639.

24 Channasanon, S., Graisuwan, W., Kiatkamjornwong, S., and Hoven, V.P. (2007) Alternating bioactivity of multilayer thin films assembled from charged derivatives of chitosan. *J. Colloid. Interface Sci.*, **316** (2), 331–343.

25 Martins, G.V., Mano, J.F., and Alves, N.M. (2010) Nanostructured self-assembled films containing chitosan fabricated at neutral pH. *Carbohyd. Polym.*, **80** (2), 570–573.

26 Liu, Y., Liu, H.Y., Guo, X.H., and Hu, N.F. (2010) Influence of ionic strength on electrochemical responses of myoglobin loaded in polysaccharide layer-by-layer assembly films. *Electroanalysis*, **22** (19), 2261–2268.

27 Morra, M. (2005) Engineering of biomaterials surfaces by hyaluronan. *Biomacromolecules*, **6** (3), 1205–1223.

28 Toole, B.P. (2004) Hyaluronan: From extracellular glue to pericellular cue. *Nat. Rev. Cancer*, **4** (7), 528–539.

29 Hargittai, I. and Hargittai, M. (2008) Molecular structure of hyaluronan: an introduction. *Struct. Chem.*, **19** (5), 697–717.

30 Brown, S.H. and Pummill, P.E. (2008) Recombinant production of hyaluronic acid. *Curr. Pharm. Biotechnol.*, **9** (4), 239–241.

31 DeAngelis, P.L. (2008) Monodisperse hyaluronan polymers: Synthesis and potential applications. *Curr. Pharm. Biotechnol.*, **9** (4), 246–248.

32 Augst, A.D., Kong, H.J., and Mooney, D.J. (2006) Alginate hydrogels as biomaterials. *Macromol. Biosci.*, **6** (8), 623–633.

33 Yang, J.S., Xie, Y.J., and He, W. (2011) Research progress on chemical modification of alginate: A review. *Carbohyd. Polym.*, **84** (1), 33–39.

34 Salmivirta, M., Lidholt, K., and Lindahl, U. (1996) Heparan sulfate: A piece of information. *Faseb J.*, **10** (11), 1270–1279.

35 Ruppert, R., Hoffmann, E., and Sebald, W. (1996) Human bone morphogenetic protein 2 contains a heparin-binding site which modifies its biological activity. *Eur. J. Biochem.*, **237** (1), 295–302.

36 Lindahl, U., Kusche, M., Lidholt, K., and Oscarsson, L.G. (1989) Biosynthesis of heparin and heparan-sulfate. *Ann. NY Acad. Sci.*, **556**, 36–50.

37 Comper, W.D. (1981) *Heparin (and Related Polysaccharides)*, Gordon and Breach, New York.

38 Lane, D. (1992) *Heparin and Related Polysaccharides*, Plenum, New York.

39 Seyrek, E. and Dubin, P. (2010) Glycosaminoglycans as polyelectrolytes. *Adv. Colloid Interface*, **158** (1–2), 119–129.

40 Therkelsen, G.H. (1993) Chapter 7, in *Industrial Gums - Polysaccharides and their Derivatives*, 3rd edn, vol. Academic Press, New York.

41 van der Velde, F. and De Ruiter, G.A. (2002) Chapter 9, in *Biopolymers - Polysaccharides*, vol. **II**, Wiley-VCH Verlag GmbH, Weinheim.

42 Richert, L., Lavalle, P., Payan, E., Shu, X.Z., Prestwich, G.D., Stoltz, J.F., Schaaf, P., Voegel, J.C., and Picart, C. (2004) Layer by layer buildup of polysaccharide films: Physical chemistry and cellular adhesion aspects. *Langmuir*, **20** (2), 448–458.

43 Kujawa, P., Schmauch, G., Viitala, T., Badia, A., and Winnik, F.M. (2007) Construction of viscoelastic biocompatible films via the layer-by-layer assembly of hyaluronan and

phosphorylcholine-modified chitosan. *Biomacromolecules*, **8** (10), 3169–3176.

44 Kujawa, P., Sanchez, J., Badia, A., and Winnik, F.M. (2006) Probing the stability of biocompatible sodium hyaluronate/chitosan nanocoatings against changes in salinity and pH. *J. Nanosci. Nanotechnol.*, **6** (6), 1565–1574.

45 Westwood, M., Noel, T.R., and Parker, R. (2010) The characterisation of polygalacturonic acid-based layer-by-layer deposited films using a quartz crystal microbalance with dissipation monitoring, a dual polarization interferometer and a Fourier-transform infrared spectrometer in attenuated total reflectance mode. *Soft Matter*, **6** (21), 5502–5513.

46 Sannan, T., Kurita, K., and Iwakura, Y. (1976) Studies on chitin. 2. Effect of deacetylation on sulibility. *Makromol. Chem.*, **177**, 3589–3600.

47 Westwood, M., Noel, T.R., and Parker, R. (2011) Environmental Responsiveness of Polygalacturonic Acid-Based Multilayers to Variation of pH. *Biomacromolecules*, **12** (2), 359–369.

48 Shenoy, D.B., Antipov, A.A., Sukhorukov, G.B., and Mohwald, H. (2003) Layer-by-layer engineering of biocompatible, decomposable core-shell structures. *Biomacromolecules*, **4** (2), 265–272.

49 Yuan, W.Y., Dong, H., Li, C.M., Cui, X.Q., Yu, L., Lu, Z.S., and Zhou, Q. (2007) pH-Controlled construction of chitosan/alginate multilayer film: Characterization and application for antibody immobilization. *Langmuir*, **23** (26), 13046–13052.

50 Fu, J.H., Ji, J., Yuan, W.Y., and Shen, J.C. (2005) Construction of anti-adhesive and antibacterial multilayer films via layer-by-layer assembly of heparin and chitosan. *Biomaterials*, **26** (33), 6684–6692.

51 Boddohi, S., Killingsworth, C.E., and Kipper, M.J. (2008) Polyelectrolyte multilayer assembly as a function of pH and ionic strength using the polysaccharides chitosan and heparin. *Biomacromolecules*, **9** (7), 2021–2028.

52 Lundin, M., Solaqa, F., Thormann, E., Macakova, L., and Blomberg, E. (2011) Layer-by-Layer Assemblies of Chitosan and Heparin: Effect of Solution Ionic Strength and pH. *Langmuir*, **27** (12), 7537–7548.

53 Lundin, M., Blomberg, E., and Tilton, R.D. (2010) Polymer Dynamics in Layer-by-Layer Assemblies of Chitosan and Heparin. *Langmuir*, **26** (5), 3242–3251.

54 Grohmann, S., Rothe, H., Frant, M., and Liefeith, K. (2011) Colloidal force spectroscopy and cell biological investigations on biomimetic polyelectrolyte multilayer coatings composed of chondroitin sulfate and heparin. *Biomacromolecules*, **12** (6), 1987–1997.

55 Acharya, G. and Park, K. (2006) Mechanisms of controlled drug release from drug-eluting stents. *Adv. Drug Deliv. Rev.*, **58** (3), 387–401.

56 Takahashi, H., Letourneur, D., and Grainger, D.W. (2007) Delivery of large biopharmaceuticals from cardiovascular stents: A review. *Biomacromolecules*, **8** (11), 3281–3293.

57 Kerdjoudj, H., Berthelemy, N., Rinckenbach, S., Kearney-Schwartz, A., Montagne, K., Schaaf, P., Lacolley, P., Stoltz, J.F., Voegel, J.C., and Menu, P. (2008) Small vessel replacement by human umbilical arteries with polyelectrolyte film-treated arteries in vivo behavior. *J. Am. Coll. Cardiol.*, **52** (19), 1589–1597.

58 Vasconcellos, F.C., Swiston, A.J., Beppu, M.M., Cohen, R.E., and Rubner, M.F. (2010) Bioactive polyelectrolyte multilayers: Hyaluronic acid mediated B Lymphocyte adhesion. *Biomacromolecules*, **11** (9), 2407–2414.

59 Meng, S., Liu, Z.J., Shen, L., Guo, Z., Chou, L.S.L., Zhong, W., Du, Q.G., and Ge, J. (2009) The effect of a layer-by-layer chitosan-heparin coating on the endothelialization and coagulation properties of a coronary stent system. *Biomaterials*, **30** (12), 2276–2283.

60 Heng, B.C., Bezerra, P.P., Meng, Q.R., Chin, D.W.L., Koh, L.B., Li, H., Zhang, H., Preiser, P.R., Boey, F.Y.C., and Venkatraman, S.S. (2010) Adhesion, proliferation, and gene expression profile of human umbilical vein endothelial cells cultured on bilayered polyelectrolyte coatings composed of

glycosaminoglycans. *Biointerphases*, **5** (3), FA53–FA62.

61 Thierry, B., Winnik, F.M., Merhi, Y., and Tabrizian, M. (2003) Nanocoatings onto arteries via layer-by-layer deposition: Toward the in vivo repair of damaged blood vessels. *J. Am. Chem. Soc.*, **125** (25), 7494–7495.

62 Ariga, K., McShane, M., Lvov, Y.M., Ji, Q.M., and Hill, J.P. (2011) Layer-by-layer assembly for drug delivery and related applications. *Expert Opin. Drug Deliv.*, **8** (5), 633–644.

63 Ariga, K., McShane, M., Lvov, Y.M., Ji, Q.M., and Hill, J.P. (2011) Layer-by-layer self-assembled shells for drug delivery. *Adv. Drug. Deliv. Rev.*, **63** (9), 762–771.

64 Zelikin, A.N. (2010) Drug releasing polymer thin films: New Era of surface-mediated drug delivery. *ACS Nano*, **4** (5), 2494–2509.

65 Lynn, D.M. (2006) Layers of opportunity: nanostructured polymer assemblies for the delivery of macromolecular therapeutics. *Soft Matter*, **2** (4), 269–273.

66 Wilson, J.T., Cui, W.X., Kozovskaya, V., Kharlampieva, E., Pan, D., Qu, Z., Krishnamurthy, V.R., Mets, J., Kumar, V., Wen, J., Song, Y.H., Tsukruk, V.V., and Chaikof, E.L. (2011) Cell Surface Engineering with Polyelectrolyte Multilayer Thin Films. *J. Am. Chem. Soc.*, **133** (18), 7054–7064.

# 27
# Polyelectrolyte Multilayer Films Based on Polysaccharides: From Physical Chemistry to the Control of Cell Differentiation

*Thomas Boudou, Kefeng Ren, Thomas Crouzier, and Catherine Picart*

## 27.1
## Introduction

The internal structure and growth properties of polyelectrolyte multilayers (PEM) have been an intensive field of research for the past decade [1–4]. Differences in growth modes, that is, linear versus exponential, have been evidenced, depending on the build-up conditions and on the polyelectrolyte intrinsic properties [5–7]. An important question that remains under debate is how the charge balance in the films is achieved. In a pioneering work by Schlenoff [8, 9], the concept of intrinsic (i.e., polyelectrolyte of opposite charge) versus extrinsic (i.e., counterions) charge matching was introduced. Since then, studies have mostly focused on the widely investigated synthetic polyelectrolyte system poly(allylamine)hydrochloride/poly(styrenesulfonate) (PAH/PSS) [10–12] that is mainly found to be "intrinsically" compensated, but little is known about the internal charge balance in films made of natural polyelectrolytes [13]. During the past years, an increasing interest has emerged for the investigation of natural-based films [14, 15] because these natural polyelectrolytes possess specific properties (enzymatic biodegradability, low immune reactivity and bioactivity...). Furthermore, these natural polyelectrolytes are already widely employed in the biomedical field and in the food sector under various conditions (gels, capsules, membranes...) [16]. Also, and importantly, polyanions like polysaccharides (hyaluronan, chondroitin sulfate, heparan sulfate and heparin), also called glycosaminoglycans, are natural components of the extra-cellular matrices (ECM) of soft tissues where they associate with proteins via covalent and electrostatic interactions [17]. They play a role in the release of growth factors [18] and in the control of tissue hydration [19]. This motivates their use in the design of tissue engineered constructs [20, 21]. Self-assembly of natural biopolymers thus gives the possibility to mimic the organization of tissues and to controllably prepare thin reservoirs for bioactive molecules, such as growth factors to guide cell proliferation and differentiation [22], or to deliver locally drugs such as anti-tumoral agents [23].

In this chapter, we will first show that we can obtain insights into the internal composition, hydration and ion pairing in polysaccharide and polypeptide PEM films

*Multilayer Thin Films: Sequential Assembly of Nanocomposite Materials*, Second Edition.
Edited by Gero Decher and Joseph B. Schlenoff.
© 2012 Wiley-VCH Verlag GmbH & Co. KGaA. Published 2012 by Wiley-VCH Verlag GmbH & Co. KGaA.

by means of the quartz crystal microbalance with dissipation monitoring (QCM-D) and infrared spectroscopy in attenuated total reflection mode (ATR-FTIR). We will then focus on the controlled cross-linking of the films by means of a water-soluble carbodiimide, which leads to a change in the films' mechanical properties. We will then investigate the effect of cross-linking on cell adhesion, spreading and proliferation. More particularly, we will compare our data to those obtained on other types of well-acknowledged model substrates, such as polyacrylamide gels and polydimethysiloxane. Finally, we will focus on a much more complex cellular process – cell differentiation. We will examine how the differentiation of myoblasts into myotubes depends on the extent of cross-linking of the film, with a more efficient differentiation in fused and elongated myotubes being observed on the stiffest films. Very interestingly, embryonic stem cell (ESC) differentiation also depended on the film's cross-linking, as only cells on native poly(L-lysine)/hyaluronan (PLL/HA) films secured their pluripotency. Conversely, those grown on cross-linked films drove the cells from the inner cellular mass state to a primary specification toward a fate of epiblast, an early stage of differentiation.

## 27.2
## Film Internal Composition and Hydration

### 27.2.1
### Film Growth

Polysaccharide-based PEM films were at the origin of the discovery of a new growth mode, namely the exponential growth of film thickness based on the diffusion of at least one of the polyelectrolyte species in the films. Polysaccharide multilayer films containing alginate (ALG) and HA in combination with PLL (PLL/ALG) [24] and (PLL/HA) [4] were the first exponentially growing films to be reported. These greatly contrasted with the widely studied and linearly growing (PSS/PAH) synthetic films. This type of growth was initially mostly observed in films based on polysaccharides and polyaminoacids [4, 6, 24] but nowadays it is widely recognized that many different types of systems, including synthetic PEM films, can grow exponentially [7]. The QCM-D is frequently used to follow film growth. In our group, we have shown that films made of (PLL/HA), (PLL/chondroitin sulfate) (PLL/CSA) and (chitosan/HA) (CHI/HA) grow exponentially (Figure 27.1 and data not shown). Other films that do not contain polysaccharides, like (PAH/poly(L-glutamic acid) (PAH/PGA) films also grow exponentially. In contrast, PLL/heparin (PLL/HEP) films grow linearly. It is noteworthy that some films, including (PAH/PGA) and (CHI/HA) exhibit purely elastic behavior, which is characterized by a good overlap of the normalized frequency shifts delta $(f)/\nu$ curves over the whole range. Conversely, the normalized frequency shifts did not overlap for (PLL/HA) and (PLL/CSA), which is characteristic of viscoelastic behavior that can be described by the Voinova model [25].

Polyelectrolyte diffusion was found to be a key feature of this type of growth. Evidenced for the first time on (PLL/HA) films [4, 7], some PLL chains labeled with

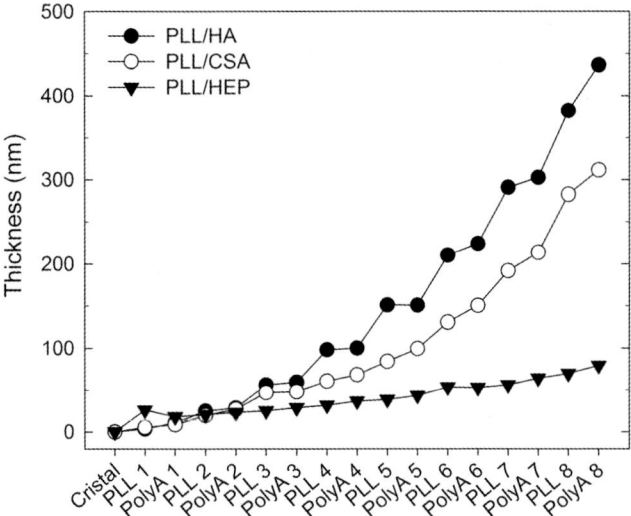

**Figure 27.1** Film thickness as a function of the number of layer pairs deduced from the QCM-D data for (PLL/HA) (●) (PLL/CSA) (○) and (PLL/HEP) (▼) films. Films were built in a 20 mM Hepes/150 mM NaCl buffer at pH 7.4 with PLL concentration at 0.5 mg mL$^{-1}$ and polycations at 1 mg mL$^{-1}$. PolyA stands for each polyanion deposition for all three types of films. (reproduced with permission from Crouzier et al., Biomacromolecules 2009, copyright ACS 2009).

fluorescein isothiocyanate (FITC) were found to remain free in the films. These were able to diffuse toward the upper part of the films upon deposition of the HA layer. A significant contribution to film growth thus came from these free chains that were able to interact with the incoming HA chains. Watery and very thick (PLL/HA) films have become a model for deeper understanding of these phenomena. Later studies brought a better view of the diffusion process, in particular of its limitation to a certain zone [26]. Better understanding of the different growth mechanisms is also emerging [27]. For instance, Porcel et al. showed that a transition from exponential to linear growth occurs at a certain level in film build-up [26]. It also appears that, even for synthetic polyelectrolyte films, exponential growth becomes dominant when NaCl concentrations increase [5] or when the temperature is increased [28]. Interestingly, isothermal titration microcalorimetry investigations indicate that the linear growth regime is associated with exothermic complexation, whereas the exponential growth regime relates to endothermic complexation [29].

Simulations are also only in the early stages. According to Holm et al. in their recent review [30], no simulation study has been able to reproduce the exponential growth of PEMs. The interesting predictions made by Hoda and Larson [31] about exponential growth require a test via numerical simulations. One of the reasons for the lack of results may be that, so far, numerical studies have never explored the influence of a mismatch on the degree of charges between two different polyelectrolytes. As indicated earlier, and as will be shown below, this is, however, quite probably often the case for exponentially growing films.

## 27.2.2
### Hydration and Swellability

Usually, the known properties of polysaccharides in solution will directly influence the properties of the PEM films formed. Hydration is a critical parameter for PEM films, impacting film thickness, swellability and diffusion of the film's components. Neutron scattering is a major technique to study the distribution of salt ions and water in synthetic polyelectrolyte multilayer films [27]. However, due to the inherent difficulty in drying the polysaccharide-based films without changing their structure, and due to the fact that deuterated polysaccharides are difficult to prepare, there are, to our knowledge, no data available on polysaccharide-based PEM films probed by neutrons. For such films, hydration is usually probed by comparing hydrated adsorbed masses measured by QCM-D and dried masses measured by optical methods such as surface plasmon resonance (SPR), ellipsometry or ATR-FTIR. In our group, we have compared QCM-D data with FTIR spectra for films built *in situ* on the ATR crystal. As FTIR provides a specific signature of the different peaks, internal composition can thus be determined [32]. An increase of specific bands was noticeable after each polyelectrolyte deposition (Figure 27.2). In (PLL/HA) films, this was particularly visible for amine I and saccharide bands as well as for the $COO^-$ group at $1610\,cm^{-1}$ (Figure 27.2c). Very similar features showed up for (CHI/HA) films (Figure 27.2b). For (PAH/PGA) films, the amide I band in the $1640\,cm^{-1}$ region and the $COO^-$ peaks at 1563 and $1405\,cm^{-1}$ were particularly visible. For (PLL/CSA) films a similar pattern emerges, but with an additional $SO_3^-$ peak showing up at $1063\,cm^{-1}$ (data not shown). For (PLL/HEP) films, deposition of PLL increased the amide I band ($1635\,cm^{-1}$) whereas deposition of HEP led to an increase in the $COO^-$ peak ($1610\,cm^{-1}$) and $SO_3^-$ peak ($1000\,cm^{-1}$) (data not shown).

From each individual spectrum and from the known calibration constants measured for these polyelectrolytes individually [3], the adsorbed amounts of each polyanion and polycation were calculated for each deposited layer. In consistency with QCM-D data, we found that film growth was exponential for HA and CSA-based films and deviated only slightly from linearity for the HEP films. However, the adsorbed amounts calculated from FTIR data, corresponding to the dry mass of the films, were systematically smaller than those estimated from the QCM-D measurements, corresponding to their hydrated mass (Table 27.1). Consequently, comparison between FTIR and QCM-D data allowed film hydration to be estimated. We found the highest water contents in HA-based films: ~86% in (PLL/HA) films [33], 81% for (CHI/HA) [34], 63% in (PLL/CSA), 29% in (PAH/PGA) and 20% for (PLL/HEP) films. It has to be noted that the reproducibility of the experiments was in the range 5–15%, depending on the type of film (three independent experiments for each type of film). $(PAH/PGA)_8$ and $(PLL/HEP)_8$ films were thus clearly different with a much lower hydration degree. Several chemical side groups appear to interact with water molecules via hydrogen bonds. Acetamido groups present on CSA and HA may be partially responsible for the high hydration of (PLL/CSA) and (PLL/HA) films. Interestingly, although HA and CSA have differently

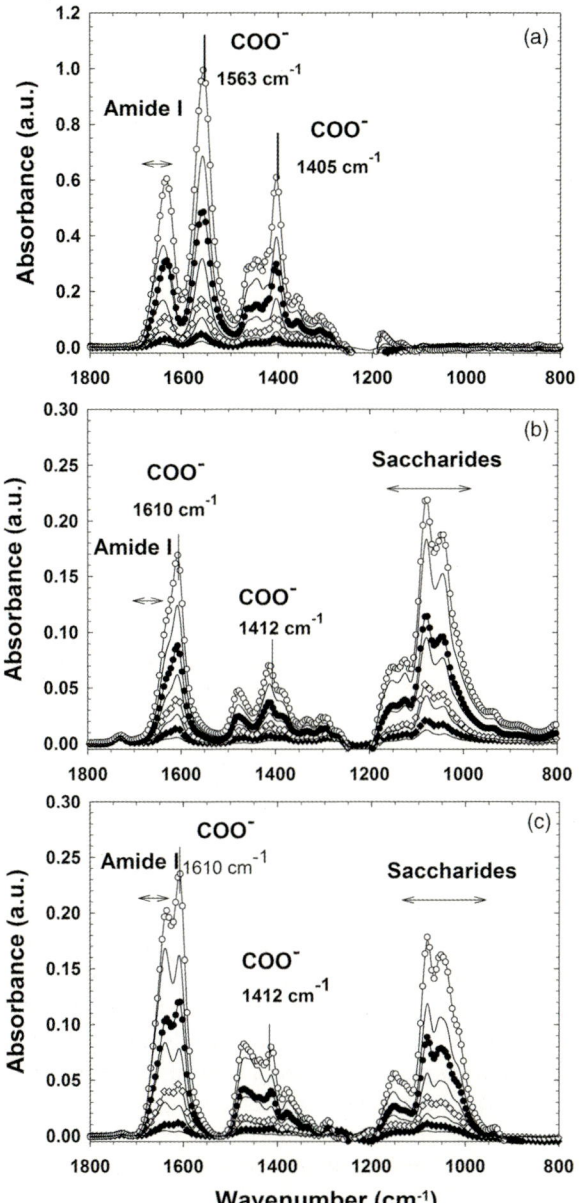

**Figure 27.2** ATR-FTIR spectra acquired during the build-up of: (a) $(PAH/PGA)_8$, (b) $(CHI/HA)_8$ and (c) $(PLL/HA)_8$ films. For the (CHI/HA) films, the polyelectrolyte solutions were used at 1 mg mL$^{-1}$ in a 10 mM sodium acetate/0.15 M NaCl buffer at pH 4.6. For the two other films, the polyelectrolytes were dissolved at 1 mg mL$^{-1}$ (except for PLL at 0.5 mg mL$^{-1}$) in a 20 mM Hepes/150 mM NaCl buffer at pH 7.4. For clarity, only the absorbance spectra of the polyanion-ended films obtained after deposition of an even number of layers are shown: 2 (◆), 4 (◇), 6 (●) and 8 (○). The spectra of the polycation-ended films obtained after deposition of an odd number of layers are represented within continuous lines. (reproduced with permission from Boudou et al., Langmuir 2009, copyright ACS 2009).

Table 27.1 Adsorbed amounts of polyelectrolyte determined from ATR-FTIR spectroscopy (or by microfluorimetry (µfluo) for PAH) and QCM-D data. Film hydration was calculated from the ratio of total adsorbed amounts. (adapted from Crouzier and Picart, *Biomacromolecules* 2009 and Boudou et al., *Langmuir* 2009).

|            | Technique   | Polycation µg cm$^{-2}$ | Polyanion µg cm$^{-2}$ | Total mass µg cm$^{-2}$ | Hydration (% w/w) |
|------------|-------------|-------------------------|------------------------|--------------------------|-------------------|
| (PLL/HA)$_8$ | FTIR        | 3.3                     | 4.3                    | $7.6 \pm 0.8$            | 86%               |
|            | QCM-D       | 47.6                    | 5.8                    | $53.4 \pm 9.5$           |                   |
| (CHI/HA)$_8$ | FTIR        | 3.1                     | 4.0                    | $7.0 \pm 1.0$            | 81%               |
|            | QCM-D       | 19.1                    | 18.6                   | $37.7 \pm 4.9$           |                   |
| (PLL/CSA)$_8$ | FTIR       | 3.5                     | 5.8                    | $9.3 \pm 1.7$            | 63%               |
|            | QCM-D       | 15.5                    | 9.5                    | $25.0 \pm 0.6$           |                   |
| (PLL/HEP)$_8$ | FTIR       | 2                       | 3.1                    | $5.1 \pm 0.6$            | 20%               |
|            | QCM-D       | 2.2                     | 4.2                    | $6.4 \pm 2.2$            |                   |
| (PAH/PGA)$_8$ | µfluo-FTIR | 13.8                    | 12.7                   | $26.5 \pm 1.2$           | 29%               |
|            | QCM-D       | 16.5                    | 20.9                   | $37.4 \pm 7.8$           |                   |

charged chemical functions and charge densities, they both have the acetamido group and both yield thick, hydrated films in combination with PLL [33]. The low hydration of PGA-containing films as compared to HA ones may be due to the different properties of these polyelectrolytes. HA is known to be highly hydrated [35] and in a random coil, whereas PGA can be structured in a more compact α-helix. Boulmedais et al. effectively showed for (PAH/PGA) films that 30% of PGA is structured in an α-helix [36]. Moreover, the charge density of PGA is higher than that of HA. Heparin, with a similar structure to HA and CSA but without this acetamido group, also yielded thin and dense films.

In relation to this high hydration, film refractive indices are usually low (~1.38–1.40) for polysaccharide-based films [37]. The film swelling properties (i.e., its ability to change volume and thickness with changing environmental conditions, such as pH, ionic strength, hydration) have been found to depend on build-up conditions. In the initial studies on HA-based PEM films, the build-up was performed in physiological conditions with no intermediate drying steps. The films are thus already highly hydrated and do not swell further in physiological solution (pH 6.5–7.5 and ionic strength of 0.1–0.15 M NaCl). The film was found to shrink to ~50% upon dehydration in ethanol baths of increased concentration [38]. Film cross-linking also induced a slight film swelling (~10%) as cross-linking is performed at a slightly lower pH (5.5) [39]. On the contrary, when the films are built by alternate dipping in the polyelectrolyte solution followed by intermediate drying steps, the swelling of the film between the dried and wet states can be very high (hundreds of %). This is indeed what we observed for (PLL/HA) films built in pH-amplified conditions, that is, with PLL deposited at pH 9.5 and HA deposited at pH 2.9 [40].

## 27.2.3
### Internal Composition

Film hydration is related to the affinity between the charged groups of the two polyelectrolytes. A polyelectrolyte couple with high affinity will chase water during complexation and form very dense and highly cross-linked networks, resulting in films close to a "glassy state" (i.e., "frozen" chains without mobility) [41]. The interactions between the polyelectrolytes in terms of stoichiometry and affinity are thus critical parameters. A question arises from these facts: how does polyelectrolyte chemistry, and in particular charge density, affect the film composition and ionic pairing of the polymers?

It has to be noted that ionic pairing in polysaccharide films has been investigated very little. The question was assessed by two different approaches. Ring *et al.* investigated systematic changes in the degree of de-acetylation of pectin in (PLL/pectin) films [13, 42]. The ratio of PLL/pectin monomers was slightly below 1, between 0.83 and 0.97 depending on the degree of pectin de-esterification. In our group, we chose a series of structurally similar polysaccharides that bear an increasing charge density. To this end, HA, CSA and HEP, with their increasing sulfate contents (from 0 to 2.5) and charge density, were selected. The internal compositions for (PLL/HA), (PLL/CSA), (PLL/HEP) were probed by FTIR [33]. The PLL/polyanion monomer ratio oscillated and tended to stabilize around 0.5 after the deposition of five layer pairs. It was systematically slightly higher when the polyanion layer was deposited. Very interestingly, the oscillation pattern and ratio values are very similar for the three types of films, whatever the nature of the polyanion. This indicates that, although the absolute charges of the disaccharide units are very different, the overall ratio of disaccharide unit to lysine monomer is constant and does not depend on the polysaccharide monomer charge.

First, it seems that sterical hindrance is an important driver for polysaccharide assembly, which is, in a certain way, independent of the charge position or density. Secondly, because of this approximately constant ratio, the charge is directly influenced by the charge density of the polyelectrolytes used. Thus (PLL/pectin) [42], (PLL/HA) and (CHI/HA) films were found to be positively charged, (PLL/CSA) is almost neutral whereas (PLL/HEP) films are negatively charged. This excess of charges is compensated by mobile counterions to ensure the overall electro-neutrality of the film. Thirdly, the growth mode was affected by charge density, more linear growth being observed for highly charged polysaccharides (highly de-acetylated pectin or heparin), whereas growth was exponential for the less charged polysaccharides (HA, CSA, highly acetylated pectin).

In our recent work, we showed that sulfate groups are much more likely to interact with the ammonium groups of the polycation than the carboxylic groups [33]. This must be related to the preferential incorporation observed in blended films of HA and HEP in combination with PLL as polycation.

## 27.3
### Film Cross-Linking: Relation Between Composition and Mechanical Properties

The internal structure of films can be manipulated by introducing covalent cross-links into the films. To this end, we have examined, for different types of multilayer films, how they responded to the presence of a cross-linking agent (ethylenediaminecarbodiimide, EDC) brought into contact with the film after build-up. In our initial studies, we have shown that this reaction, which involves only carboxylic groups $COO^-$ with amine groups $NH_3^+$, leads to the creation of covalent amide bonds [43, 44]. Such bonds can be quantified by ATR-FTIR spectroscopy by following the disappearance of the $COO^-$ group.

### 27.3.1
### Ionic Pairing

The cross-linking reaction can be employed to selectively quantify the $COO^-$ groups interacting with $NH_3^+$ groups thanks to the spectral separation of $COO^-$ versus $SO_3^-$ in the FTIR spectra. No change in the infrared absorbance of sulfate groups or the saccharide ring was noticed, thus confirming that the reaction is specific to amine and carboxylic groups. In all cases, a clear increase in the amide band (around $1635\,cm^{-1}$) and a decrease in the carboxylic peak ($1610\,cm^{-1}$) were observed, although this was less pronounced for HEP-containing films. From the FTIR spectra of the cross-linking reactions, the percentage of $COO^-$ groups involved in the cross-linking reaction was quantified (Figure 27.3a). It was assumed to be equal to the percentage of $COO^-$ groups paired in electrostatic interactions with $NH_3^+$ groups before cross-linking. This percentage was 89% for (PLL/HA) films and 74% for (PLL/CSA) films. In (PLL/HEP) films, only 49% were paired with $NH_3^+$ groups. The remaining $COO^-$ groups are presumably involved in weaker interactions with the mobile $Na^+$ counterions, due to the presence of 0.15 M NaCl in the rinsing solutions and since the charge neutrality in the film has to be respected. However, as $Cl^-$ and $Na^+$ are not IR-active, it was not possible to get direct proof of their presence. The presence of counterions in polysaccharide and polypeptide PEM films is indeed strongly suggested by the effect of ionic strength on film growth and stability [45, 46]. Few carboxylate groups may become protonated, but we verified that the percentage of protonated carboxylic groups was always below 7%. From these FITR data, ionic pairing could also be deduced for the ammonium groups. $NH_3^+$ groups are paired either with counterions ("free" groups), carboxylic groups or sulfate groups for HEP and CSA-based films. One can first notice that the percentage of "free" $NH_3^+$ groups decreases when the sulfate content (and the overall negative charge) is increased. Thus, while 46% of the $NH_3^+$ groups in (PLL/HA) films are paired with counterions (or "free"), this percentage falls to 21% for (PLL/CSA) films and to almost zero for (PLL/HEP) films (Figure 27.3b), that is, all the $NH_3^+$ are engaged in electrostatic interactions and are probably not highly mobile. These differences between the three polysaccharides

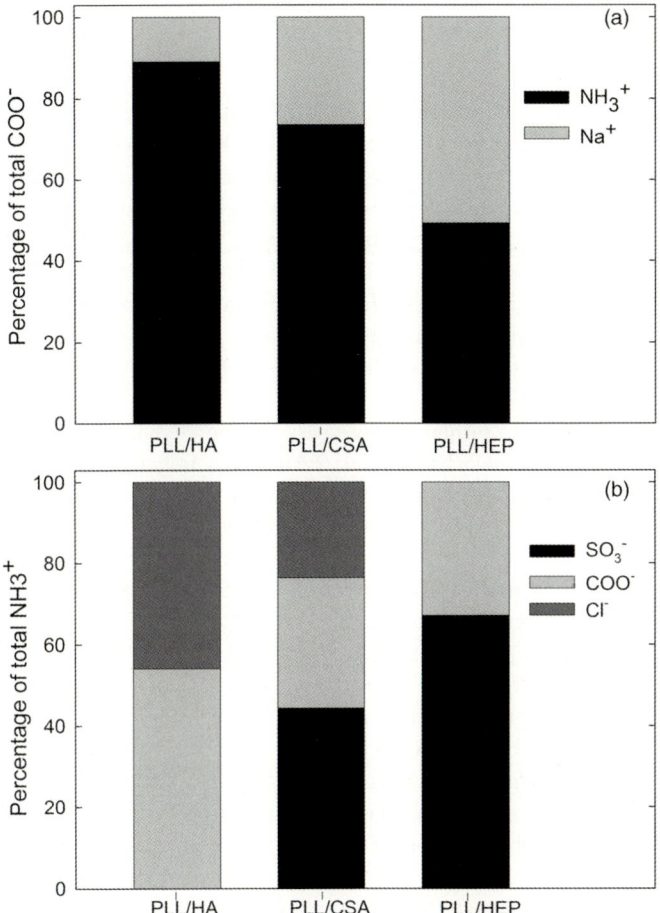

**Figure 27.3** Ion pairing within the polysaccharide-based films. From the quantitative analysis of the FTIR data, pairing of the carboxylic groups (a) and of the amine groups (b) was determined. For the COO$^-$ groups, they can either be paired with an amine group or with a positive counterion (Na$^+$ in the case of NaCl). For the amine groups, they can be paired with a SO$_3^-$ group, a COO$^-$ group or a negative counterion (Cl$^-$ in the case of NaCl). (Same buffer condition and polyelectrolyte concentrations as for Figure 27.1) (reproduced with permission from Crouzier et al., Biomacromolecules 2009, copyright ACS 2009).

may thus be related to the differences in the PLL diffusion coefficient measured by FRAP, that is, very little if no diffusion of PLL in HEP-based film, intermediate diffusion in CSA-based films and more diffusion in HA-based films. NH$_3^+$ groups interacting with COO$^-$ groups represent 54% of the total amines in (PLL/HA) films, and only 33% (respectively 30%) for (PLL/CSA) (respectively (PLL/HEP)) films. Finally, the interaction between the NH$_3^+$ and SO$_3^-$ groups accounts for 46% of total NH$_3^+$ in (PLL/CSA) films and 70% in (PLL/HEP) films.

## 27.3.2
### Covalent Amide Bonds

Film cross-linking can be achieved at different EDC concentrations, allowing the formation of a tunable amount of covalent amide bonds [47]. This was first evidenced for (PLL/HA) films, for which the percentage of carboxylic groups engaged in cross-linking steadily increased when the EDC concentration was increased [48]. In addition, it correlated well with the consumption of EDC molecules. As one carboxylic group forms a covalent amide bond with one amine group, the overall charge density in the film bulk should decrease, whereas the film surface should remain negative (due to the remaining free carboxylic groups). This was indeed verified by zeta potential measurements [38].

Recently, we quantified and compared the cross-linking densities of three PEM films, namely (PLL/HA), (CHI/HA) and (PAH/PGA) [49] (Figure 27.4). We systematically measured the decrease in the carboxylic peak of PGA or HA, which was directly related to the formation of amide bonds as the reaction had one-to-one stoichiometry between carboxylic and amine groups [43]. All curves of amide bond formation showed a similar trend characterized by an initial increase, followed by stabilization toward a plateau value at high EDC concentrations (data not shown). The $(PAH/PGA)_8$ films had the highest absolute decrease in molar density of the $COO^-$ groups, whereas the consumption of these groups expressed as a percentage is lower than for the other films. Thus, whereas 60% of the $COO^-$ groups were engaged in covalent amide bonds for an EDC concentration of $100\,\mathrm{mg\,mL^{-1}}$, this fraction rose to ~80% for $(CHI/HA)_8$ films and ~90% for $(PLL/HA)_8$ films. In all cases, a fraction of the $COO^-$ groups did not react, presumably because of the steric hindrance and entanglements of the polymer chains. Considering that the charge neutrality must be respected in the film, we assumed that the remaining "free" $COO^-$ groups were involved in weaker interactions with $Na^+$ counterions due to the presence of 0.15 M NaCl in the rinsing solutions. Similarly, the total number of ammonium groups was much higher for the $(PAH/PGA)_8$ films than for the other films. Only 30–50% of the ammonium groups, depending on film type, reacted during the cross-linking reaction, which suggested that these groups may interact to a large extent with $Cl^-$ counterions. However, for technical reasons (the counterions present here did not absorb infrared light), it was not possible to obtain direct proof of their presence, even though the effect of ionic strength on film growth and stability strongly suggested it [50, 51].

The molar density of the amide bonds created is shown in Figure 27.4a. It was much higher for $(PAH/PGA)_8$ films ($16.2\,\mathrm{\mu mol\,cm^{-3}}$) than for $(PLL/HA)_8$ ($2.2\,\mathrm{\mu mol\,cm^{-3}}$) and $(CHI/HA)_8$ films ($2.4\,\mathrm{\mu mol\,cm^{-3}}$) at an EDC concentration of $100\,\mathrm{mg\,mL^{-1}}$. This is a ~sevenfold increase for the former as compared to the latter. Thus, due to the large initial amount of charged $COO^-$ groups incorporated into the films, the (PAH/PGA) films allowed a variation of the covalent amide bonds over a wider range.

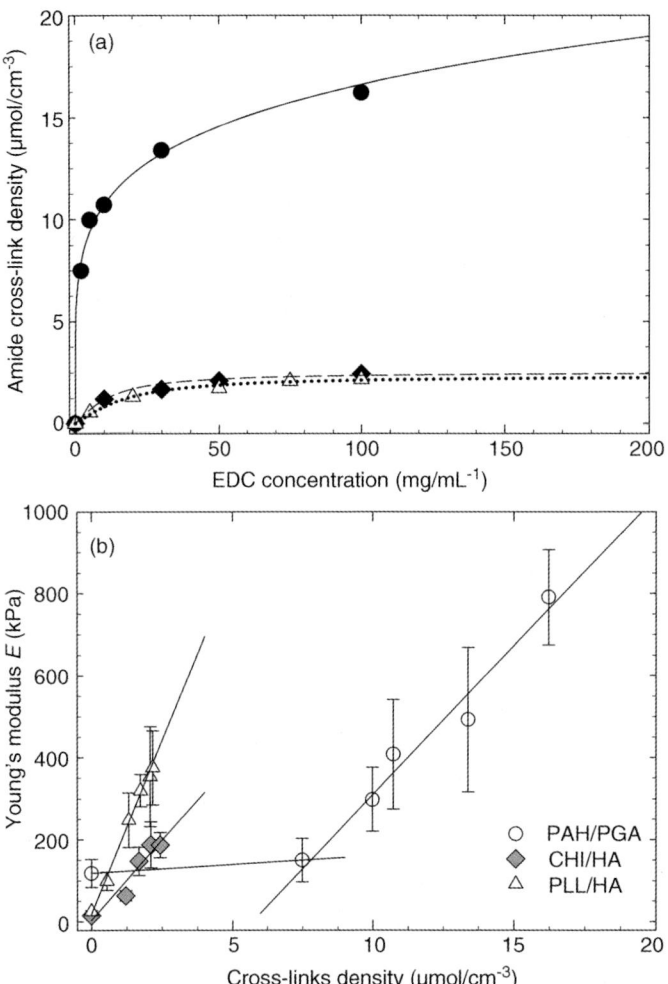

**Figure 27.4** (a) Molar density of covalent amide bonds created upon film cross-linking for (PAH/PGA) (○), (CHI/HA) (◇) and (PLL/HA) (△) films. The lines are drawn to guide the eyes (same conditions as for Figure 27.2) (b) Young's modulus $E_0$ plotted as a function of cross-link density (lower x-axis) and mean distance between cross-links $L_{CL}$ (upper x-axis) for the three types of films. The lines correspond to the best linear regression. (reproduced with permission from Boudou et al., Langmuir 2009, copyright ACS 2009).

### 27.3.3
### Mechanical Properties of Films and Correlation with Cross-Link Density

We subsequently investigated whether there is a correlation between the formation of the amide cross-links and mechanical properties. Such a relationship is indeed already known to exist for macroscopic gels in solution, for which the elasticity (or shear modulus) derived from rheological measurements is found to depend on the cross-linking density [52]. However, to our knowledge, such a relation had never been established for polyelectrolyte multilayer films that are only few micrometers in thickness. To obtain a quantitative estimate of the films' mechanical properties, we performed AFM nano-indentation experiments in a buffered medium (Hepes-NaCl buffer). A colloidal probe of radius $R = 2.5\,\mu m$ was used as the indenter to avoid damaging the films. This technique consists in indenting the films by a probe sphere in the small deformation regime and recording the force versus indentation curve. By using the finite thickness corrected Hertz sphere model [53] to analyze the force measurement data, the average Young's moduli ($E_0$) of the PEM films were estimated. This method has already been recognized as being an accurate means of probing the nanomechanical properties of thin films and gels in an aqueous environment, including polyvinylalcohol gels [53], polyacrylamide gels [54, 55] and multilayer films [56–58].

As anticipated from their polymer density estimated by ATR-FTIR spectroscopy analysis, we found that native (PAH/PGA) films (i.e., uncross-linked) were considerably more rigid, with $E_{Nat} = 118 \pm 28$ kPa, than native (PLL/HA) and (CHI/HA) films with $E_{Nat} = 25 \pm 2$ kPa and $E_{Nat} = 15 \pm 3$ kPa, respectively. All the films became stiffer when the concentration of EDC ([EDC]) increased and reached a plateau value, but with significant differences between the three systems. In order to quantitatively compare the three systems, we fitted the $E_0$ as a function of [EDC] curves using a simple hyperbola equation:

$$E_0 = E_{Nat} + E_\infty \frac{[EDC]}{[EDC]_{1/2} + [EDC]} \qquad (27.1)$$

where $E_\infty$ is the asymptotic modulus (i.e., the maximal Young's modulus for a given type of film) and $[EDC]_{1/2}$ represents the [EDC] at half maximum rigidity.

The best fits were obtained for $[EDC]_{1/2} = 20$ mg mL$^{-1}$ for the three types of films and $E_\infty = 460$ kPa for (PLL/HA) films, $E_\infty = 240$ kPa for (CHI/HA) films and $E_\infty = 710$ kPa for (PAH/PGA) films.

For a given EDC concentration, (CHI/HA) films thus appeared to be $\sim$2 times softer than (PLL/HA) films and $\sim$3 times softer than (PAH/PGA) films, as anticipated from their cross-linking density. Native (PAH/PGA) had relatively high stiffness that may be due to the presence of α-helices in 30% of the PGA chains [59] as the secondary structure of polyelectrolytes is known to influence the mechanical properties of films [60]. The tightly ionically cross-linked network resulting from this very dense film, plus the low hydration, may also play an important role. The PLL in (PLL/HA) films was much less structured and had only a random or turn structure [33].

The $E_0$ values were also compared to the molar densities of covalent cross-links (Figure 27.4b). It should be noted that this graph only takes into account the covalent cross-links that were created upon EDC cross-linking. Interestingly, two different kinds of behavior emerged. For the (PLL/HA) and (CHI/HA) films, $E_0$ varied linearly with the cross-link density. Conversely, for (PAH/PGA) films, a two-step behavior was observed. For a cross-link density ($d_{CL}$) lower than $d^*_{CL} \sim 8\,\mu mol\,cm^{-3}$, $E_0$ increased only slightly with the cross-link density (slope of 4.3 kPa/(μmol cm$^{-3}$)). However, for higher densities, a sharp increase was visible with a slope of 72.6 kPa/(μmol cm$^{-3}$). This kind of behavior is somewhat similar to the mechanical behavior of semiflexible polymer networks. Semiflexible polymers are so called because their characteristic bending length is comparable to other length scales in the problem, such as the contour length or the network mesh size, and thus cannot be neglected [52]. These semiflexible polymer networks are characterized by the existence of qualitatively distinct regimes in the elastic response, separated by a rigidity percolation threshold. Below a critical cross-link density, the elastic deformation of the gel is spatially heterogeneous over long length scales and is dominated by filament bending, leading to a low bulk stiffness. When the cross-link density exceeds the critical value, the strain is guided by the extension/contraction of the polymer chains and becomes uniform throughout the sample [52]. The mechanical behavior of (PAH/PGA) films thus presents a sort of "percolation threshold", except that, in the present case, the $E$ value of the native films is not zero.

Noticeably, our $E_0$ values are of the same order of magnitude as those measured for other PEM films, in particular by Van Vliet et al. for (PAH/PAA) films and by Vautier et al. for (PLL/HA) films capped with PSS/PAH layers [61]. The $E_0$ values for PEM also appear to fall within a physiological range, as soft tissues have stiffness ranging from hundreds of Pa (such as brain tissue) to MPa (cartilage tissue) [62, 63].

## 27.4
### Cell Adhesion onto Cross-Linked Films: Cell Adhesion, Cytoskeletal Organization and Comparison with Other Model Materials

### 27.4.1
### Film Cross-Linking Modulates Myoblast Attachment, Proliferation and Cytoskeletal Organization

Adhesion is usually the very first step when a cell encounters a material surface, followed by cell proliferation if the conditions are satisfactory. We have investigated cell adhesion in contact with different types of PEM films, especially polysaccharide-based PEM films and exponentially growing films. One of our working models is that of $(PLL/HA)_{12}$ films, whose stiffness can be modulated as described above. In the past years, we have been working on C2C12 skeletal myoblast as cell models. These cells are a common model for studying muscle cell differentiation [64]. Usually, these cells have first to be grown in a growth medium (GM), where they attach and proliferate. In a second step, the medium has to be changed to a differentiation medium (DM) that contains a lower concentration of serum (2% as compared to 10% for the growth medium) to induce cell differentiation.

C2C12 cells were thus cultured in GM on $(PLL/HA)_{12}$ films cross-linked at various EDC concentrations. Few cells attached on the native (e.g., uncross-linked) films, whereas the cells cultured on cross-linked films adhered and spread. The cell number increased with cross-linking and significant differences exist between low and high cross-linked films (Figure 27.5). However, for EDC concentrations higher than 50 mg mL$^{-1}$, all the seeded cells attached as well as on plastic. We measured cell

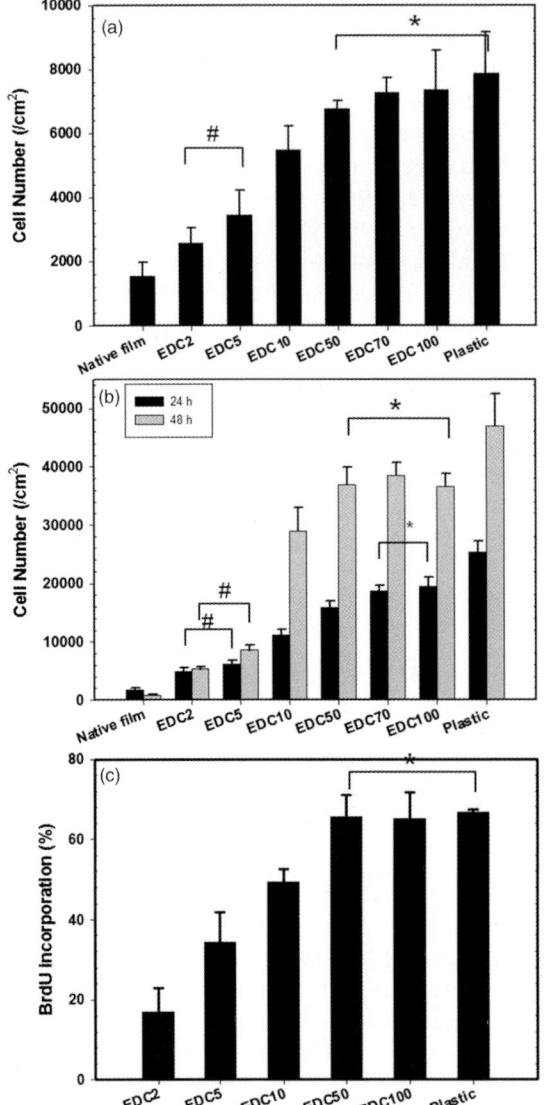

**Figure 27.5** Cell adhesion and proliferation depend on film cross-linking. Quantification of the number of adherent cells on cross-linked (PLL/HA)$_{12}$ films for increasing EDC concentrations, after 4 h (a), 24 and 48 h (b) in growth medium (GM). BrdU incorporation was also quantified (c). Cells were seeded at 7500 cells cm$^{-2}$ in GM. (data are means ± standard deviation, SD of three independent wells or slides). ANOVA revealed film-dependent adhesion (conditions belonging to the same groups are represented by * and #, the other conditions being different from one another and from the groups). (reproduced with permission from Ren et al., Adv; Funct. Mat 2008, copyright Wiley 2008).

proliferation by direct cell counting and by a Bromodeoxyuridine (BrdU) proliferation assay. For the native films, the cell number decreased as a function of time. Cells cultured on films crosslinked with EDC at 5 mg/mL (noted EDC5) films show little proliferation. Cells grown on EDC10 cross-linked films exhibit a slightly higher proliferation rate, but still significantly lower than for stiffer films, for which the proliferation rate levels off for EDC concentrations equal to or greater than 50 mg mL$^{-1}$ (Figure 27.5b). These results were confirmed by the BrdU quantification test (Figure 27.5c).

We also observed by immuno-fluorescence the cell cytoskeleton by staining actin and the formation of focal adhesions by labeling vinculin. We observed that cells cultured on soft films did not exhibit clear vinculin plaques or F-actin stress fibers. In contrast, cells cultured on the stiffest EDC50 and EDC100 films showed numerous, large, elongated focal adhesions and well organized F-actin stress fibers. Vinculin was clearly localized at the tip of elongated structures. The density of focal adhesions was also significantly increased on the stiffest films compared to the low cross-linked films. Obviously the cells developed more numerous and better organized adhesion structures on the more cross-linked films. Cell/film contacts were also observed by using the scanning electron microscopy (SEM) technique at high resolution. Interestingly, short and small microspikes or filopodia were visible on the softest films, some of them being as long as 5 μm. Conversely, cells cultured on the stiffest films (EDC50 and EDC100) showed numerous, very thin and long filopodia and/or cell extensions, whose length could reach about 10 μm (data not shown). Lamellipodia were visible [65] and dorsal microvilli were also observed on the stiffest films.

All these observations suggested that it was difficult for the cells to anchor and spread on the softest films whereas the highly cross-linked films are much more favorable for cell attachment and anchoring. In a subsequent step, we investigated the possible effect of integrin activation for cells cultured on soft and stiff films [66]. Integrins are trans-membrane adhesion receptors whose role in mechano-transduction is now well established [67]. It is known that the change in integrin conformation from a "low affinity" (i.e., inactive) to a "high-affinity" (i.e., activated) state [68] can be obtained by activation by manganese ($Mn^{2+}$) ions [69, 70]. We thus investigated the effects of $Mn^{2+}$ on integrin organization for cells cultured on soft and stiff (PLL/HA)$_{12}$ films. In particular, we visualized the expression and organization of vinculin and of F-actin (Figure 27.6). On soft films, the cells exhibited nascent focal contacts (FCs) and were more spread after just 1 h of adhesion in the presence of $Mn^{2+}$ compared to untreated cells. After 4 h, mature FAs and organized F-actin fibers were observed. The cells were also well spread (Figure 27.6d). Conversely, the untreated cells exhibited only nascent FCs. On stiff films, the effect of $Mn^{2+}$ was of much less than on soft films, as the cells were already spread without $Mn^{2+}$ (Figure 27.6e–h). Nevertheless, an effect on FA formation at the cell periphery was visible after 1 h of treatment (Figure 27.6g). Quantitative measurements of cell spreading areas (Figure 27.6i) and vinculin spot size (Figure 27.6j) confirmed that there are significant differences between cells on soft films without or with $Mn^{2+}$ added to the culture medium for 1 or 4 h. Both parameters were statistically different on treatment with $Mn^{2+}$. This effect was particularly visible on soft films, precisely because the

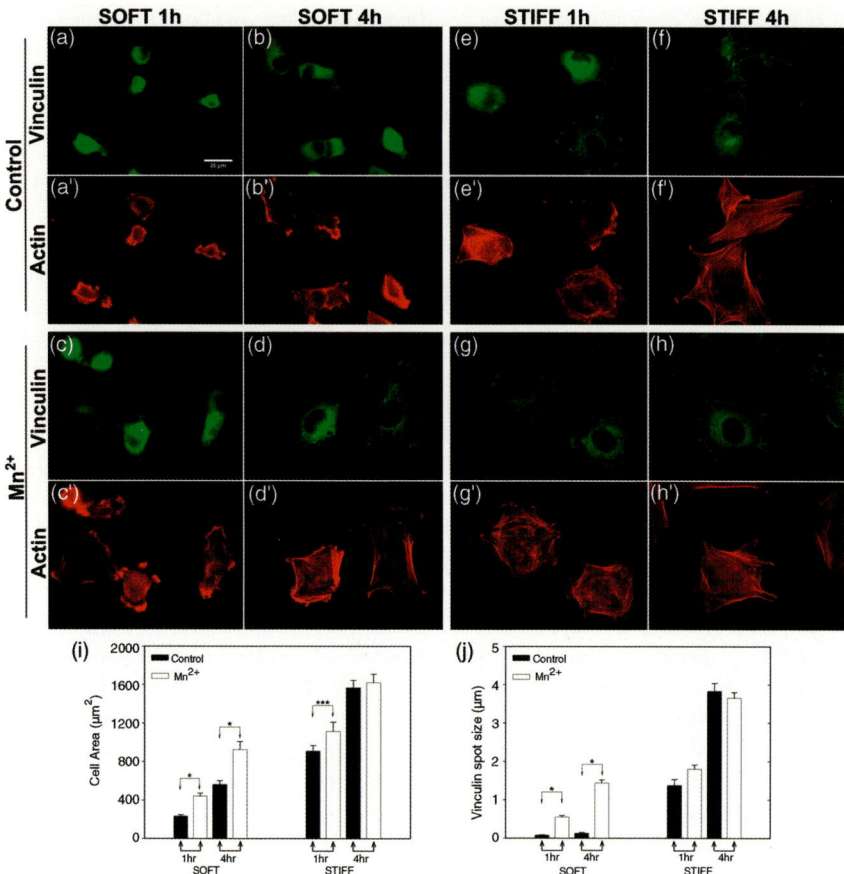

**Figure 27.6** C2C12 myoblasts cultured on soft (a–d) or stiff (e–h) films were stimulated for 1 or 4 h without (a–f and a′–f′) or with $Mn^{2+}$ (c–h and c′–h′) added to the culture medium. After fixation, cells were stained for vinculin (green, a–f and c–h), a focal adhesion marker and actin (red, a′–f′ and C′–H′). Scale bar: 20 μm. The cell spreading area (i) and vinculin spot size (j) were quantified. Data are means from at least 35 cells (on 4 different areas) per condition (means ± SE). *$P < 0.05$. (reproduced with permission from Ren et al., Acta Biomater. 2010, copyright Elsevier 2010).

cells were initially round. On stiff films, the difference in spreading was significant only at a short time point (1 h), whereas the vinculin spot sizes were similar. The effects were much smaller as the cells already had the potential for high spreading on these films, a process that was accompanied by focal adhesion formation.

The promotion of focal adhesions on soft films via integrin activation by $Mn^{2+}$ confirmed that integrins play a key role in the cell response on the different (PLL/HA) cross-linked films. Very interestingly, a morphological switch in the state of the adhesion sites can be achieved by manipulating integrin conformation and converting round cells to highly spread ones.

## 27.4.2
**Correlation Between the Cell Spreading Area and Young's Modulus $E_0$**

As shown above, besides (PLL/HA) films, (CHI/HA) and (PAH/PGA) films can be cross-linked to different extents. Thus, we investigated cell behavior on these three types of films by following the attachment and spreading of C2C12 myoblasts. After C2C12 myoblast seeding in serum-containing media on the (PLL/HA)$_8$, (CHI/HA)$_8$ and (PAH/PGA)$_8$ films cross-linked at various [EDC], cell adhesion and proliferation were quantified. Here again, we found that cell proliferation depends on the extent of film cross-linking, the more cross-linked films leading to the highest proliferation [71]. Similar trends were observed for the three types of films whatever their internal composition and chemistry.

In order to investigate whether cell spreading could be correlated to $E_0$, we plotted the mean cell spreading area for C2C12 myoblasts as a function of the measured Young's moduli for all films (Figure 27.7a). Very interestingly, we observed that all experimental points were distributed over a unique curve. The round cells, characterized by a small cell area of about 350 μm$^2$, were found on soft films ($E_0 < 90$ kPa). Cell spreading was observed to increase steadily with an adhesion area of ~1000 μm$^2$ for $E_0 \sim 200$ kPa. A high plateau value at ~2000 μm$^2$ was reached for $E_0 > 400$ kPa. This plateau also corresponded to an almost infinitely hard substrate (glass or tissue culture polystyrene). We also tried to quantify myoblast spreading at 4 h in a serum-free medium, to further discard any possible role of serum proteins in the initial adhesion process. However, as serum contains several vital nutrients, these sensitive cells did not adhere or proliferate and cell spreading could not be quantified (data not shown). To confirm our findings, we investigated whether a more robust cell type, 3T3 fibroblasts, which are also used for mechano-sensitivity studies [72, 73], exhibited similar sensitivity to the film's Young's moduli $E_0$. For this purpose, 3T3 fibroblasts were seeded on the two most tunable films, that is, (PLL/HA) and (PAH/PGA) films with different degrees of cross-linking, in the presence or absence of serum. The spreading areas of the fibroblasts are shown as a function of $E_0$ in Figure 27.7b.

Interestingly, the curve for 3T3 exhibited a very similar trend to that for C2C12, although with lower maximum spreading. This indicates that the "unique mechano-sensitivity" curve remained valid for 3T3 fibroblasts. The fibroblasts were round on the softest films and widely spread cells on stiffer films. The spreading areas of 3T3 cells, cultured with or without serum, were very similar, indicating that serum proteins have a negligible influence in this case. Indeed, we quantified by microfluorimetry the adsorption of fluorescent fibronectin on the three types of films after 2 h at 37 °C. We observed a low adsorption that did not depend on the film type or on the extent of cross-linking. Furthermore, both C2C12 and 3T3 cells seeded on the rough (PAH/PGA) films exhibited the same behavior when seeded on rather smooth (PLL/HA) or (CHI/HA) films with similar rigidity. This was further proof that random nanotopography in the range of tens of nanometers on PEM films does not seem to contribute to initial cell adhesion and proliferation. Thus, the spreading of C2C12 myoblasts and 3T3 fibroblasts on all the PEM films appeared to depend mainly on the film's stiffness.

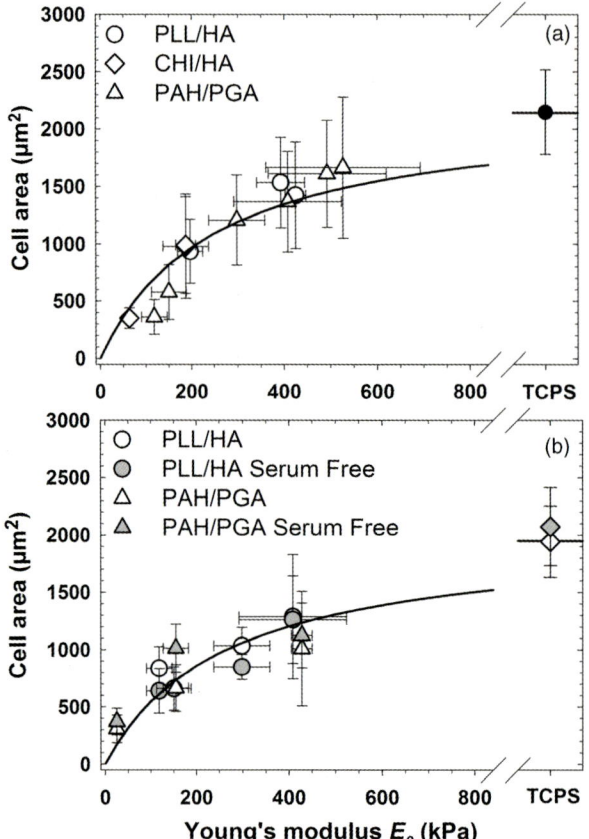

**Figure 27.7** (a) C2C12 cell spreading area after 1 day in GM measured for cell cultured on the three different types of films having different Young's moduli, in comparison with TCPS (tissue culture polystyrene). Data are plotted on a unique curve for (PLL/HA) (○), (CHI/HA) (◇) and (PAH/PGA) (△) films. (b) Spreading area of NIH-3T3 fibroblasts after 4 h either in growth medium containing serum at 10% (open symbols) or in a serum-free medium (gray filled symbols). The experimental data were fitted according to Equation 27.2 and the best fits were obtained for parameters given in Table 3. Data are means ± SE of 50 to 100 cells. (reproduced with permission from Boudou et al., Macromol. Bioscience 2011, copyright Wiley 2011).

We also observed on these three types of films that focal adhesions and actin stress fiber organization were affected by film stiffness. In all cases, cells cultured on the softest films were round. Actin was distributed at their periphery and there was no stress fiber formation [71]. Localization of vinculin was diffuse. On the contrary, cells cultured on the stiffest films were highly spread. They showed well-organized F-actin stress fibers as well as numerous, large and elongated focal adhesions. Vinculin was clearly localized at the tip of elongated structures. A common trend was thus observed for all films, regardless of their internal chemistry. First, C2C12 myoblasts appeared to proliferate faster on the stiffest films and secondly, spreading and cytoskeleton organization were enhanced on the stiffest films while cells remained round on soft films.

In order to highlight the relative importance of mechano-sensitivity over material-specific interactions, the relationship between adhesion area and film stiffness for different materials will be now discussed.

### 27.4.3
### Comparison with Other Material Substrates: Polyacrylamide (PA) and Polydimethysiloxane (PDMS)

To date, polyacrylamide has been widely employed for mechano-sensitivity studies of cells [74]. It is now acknowledged that cells are able to sense their micro-environment and respond by adjusting the organization of their cytoskeleton [75]. In particular, actin assembly was altered on soft polyacrylamide gels [54, 76]. In the literature, we identified another study by Engler *et al.* reporting data on C2C12 cell spreading [55], as well as two other studies investigating NIH-3T3 cell spreading on collagen-coated PA gels [72, 73]. A study of 3T3 fibroblasts on fibronectin-coated PDMS substrates gave quantitative measurement of cell spreading [77]. Figure 27.8 shows the comparison between our data and those from the literature for cell spreading areas as a function of $E_0$ for C2C12 cells (a) and 3T3 fibroblasts (b). On PDMS gels of increased stiffness, vascular smooth muscle cells and 3T3 cells also showed differences in cell spreading between 500 and 1000 $\mu m^2$ [77, 78].

Interestingly, the trends observed here for C2C12 and 3T3 cells were similar for PEM, PA and PDMS gels, that is, increased cell spreading with increased Young's moduli. Similar trends have been observed previously for smooth muscle cells on PA [54] and PDMS gels [78]. On PEM films, a recent study on another cell type, that is, epithelial cells, reported increased spreading areas that depended on the stiffness of a composite (PLL/HA) film capped with (PSS/PAH) layers [61].

It can be noted that the $E_0$ values for PEM films are about one order of magnitude higher than those reported for PA gels (5–30 kPa) and closer to values reported for PDMS substrates (50 kPa to 1.7 MPa by varying the base to cross-linker ratio). Thus, the effect of substrate stiffness seems to be material-dependent.

In order to compare the different studies quantitatively, we fitted the cell areas as a function of $E_0$ curves using a simple hyperbolic equation, as already proposed by Engler *et al.* for smooth muscle cells [54]:

$$\text{Area} = aE_0/(E_{1/2} + E_0) \tag{27.2}$$

where $a$ is the asymptotic area (i.e., the maximal adhesion area for a given cell line) and $E_{1/2}$ represents an intermediate Young's modulus at half maximum spreading. On substrates with $E_0 \sim E_{1/2}$, a cell spread more or less depending on external stimulations. The most suitable parameters for $a$ and $E_{1/2}$ have been added to the data in Figure 27.8. We found $E_{1/2} \sim 2.5$–$7$ kPa for both cell lines on PA gels and $E_{1/2}$ to be around 300 kPa for 3T3 and C2C12 on PEM films and PDMS gels. Thus, whereas a similar offset in cell behavior is observed for cell spreading on substrates with

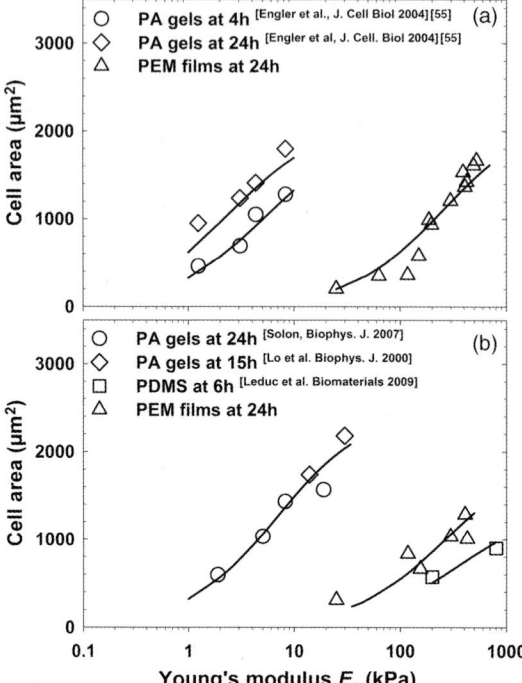

**Figure 27.8** C2C12 (a) and 3T3 (b) cells spreading area measured for cells cultured on different types of substrates having different Young's moduli. The experimental data were fitted according to Equation 27.2. (reproduced with permission from Boudou et al., Macromol. Bioscience 2011, copyright Wiley 2011).

increasing stiffness, the characteristic $E_{1/2}$ values differ and are about 40 times higher for PEM and PDMS gels than for PA gels. There is very limited overlap between the elastic moduli for PA gels (maximum 20 kPa) and those measured for PEM and PDMS gels (minimum 20 kPa). It thus appears that evolution in the spreading area is material-dependent. One hypothesis is that the differences are the result of extracellular matrix protein grafting on to PA gels, whereas PEM films are uncoated. Engler et al. effectively provided evidence for collagen-coated PA gels which, for a given value of $E_0$, had a smooth muscle cell spreading area that increased twofold when the collagen density was increased from 5 to 1000 ng cm$^{-2}$ [54]. In other words, a PA gel with an intermediate stiffness and a low amount of grafted collagen would correspond to a soft PA gel with a high amount of grafted collagen. This would also correspond to a stiff PEM film not pre-coated with serum proteins. Another hypothesis is that the mechanical properties of the films, which are necessarily measured by means of nano-indentations due to low film thickness (a few hundred nanometers to a few μm), could be anisotropic and differ in the tangential direction. $E_0$ is measured in the vertical direction, normal to the film's surface, whereas cells pull normally as well as tangentially to the substrates [79]. The possible anisotropy of the film's properties is, however, very difficult to check experimentally due to their thickness and the lack of appropriate tools.

An important point is that our values lie in the range of values measured by other groups for other types of PEM film [61, 80]. Also, these groups reported mechanosensitivity for different types of cells grown on PEM films. Thus, although trends are similar between different material substrates, whether they are PA, PDMS or PEM films, it is important to take into account that similar cell processes occur on these different materials in different $E_0$ ranges.

## 27.5
## Cell Differentiation: ESC and Myoblasts

### 27.5.1
### Film Cross-Linking Influences the Myogenic Differentiation Process Morphologically but not the Expression of Muscle-Specific Proteins

The development of skeletal muscle is a multistep process, including initial cell proliferation, subsequent withdrawal from the cell cycle and differentiation into multinucleated myotubes [81]. Cell differentiation is a much more complex process than cell adhesion or proliferation. During normal myogenesis, myoblasts must exit the cell cycle and follow an ordered set of cellular events, including cell adhesion, migration and membrane fusion [82]. Mammalian myoblast fusion occurs in two phases [83]. Initially, myoblasts fuse with one another to form small, nascent myotubes. Additional myoblasts subsequently fuse with nascent myotubes, leading to the formation of large, mature myotubes. Myoblast differentiation requires much more than just adhesion and proliferation and the material surface appears extremely important. For instance, it has previously been shown that C2C12 cells can adhere but neither proliferate nor differentiate on cross-linked chitosan gels [84]. When cultured on N-isopropylacrylamide (NiPAM) polymers, they even differentiate into osteoblasts (instead of differentiating into myotubes) [85].

To determine the effect of (PLL/HA) film stiffness on C2C12 cell differentiation, cells were first allowed to proliferate close to confluency in GM and then switched to DM to induce differentiation. We took care that the cell density was the same for all films when switching to DM. Cell differentiation was followed by phase-contrast microscopy, by immuno-staining and by immuno-blots. Tissue culture polystyrene (TCPS) was always used as a control surface as C2C12 cells are known to not differentiate optimally on bare glass [55].

During differentiation, C2C12 cells align and fuse together to form myotubes, when cultured on control TCPS, as usually observed [86]. We observed cell differentiation qualitatively, that is, cell alignment, fusion and formation of myotubes, on all types of films, and noticed important differences in the fate of the cells and in the kinetics of the events. We followed the analysis of the expression of muscle-specific proteins like myogenin and troponin T. For control TCPS surfaces, these proteins are known to be expressed after about 1–2 days in DM [86, 87] and until at least day 6.

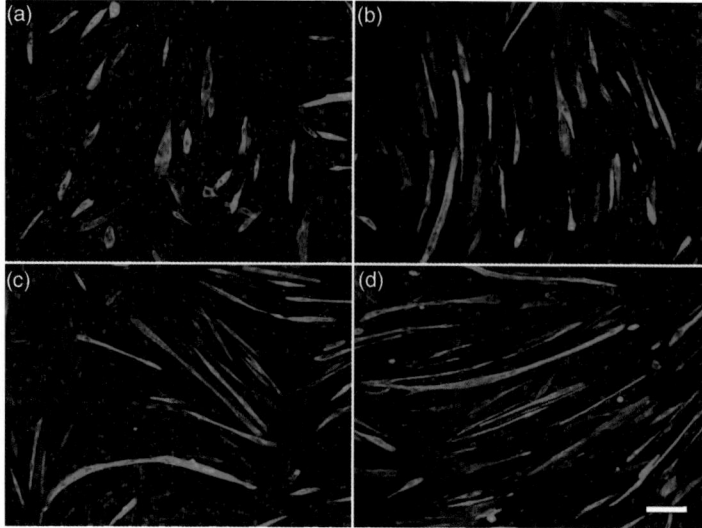

**Figure 27.9** Film cross-linking affects myotube morphology. Immuno-fluorescence images of troponin T (green) and DNA (blue) of C2C12 cells cultured on (PLL/HA)$_{12}$ films cross-linked at various EDC concentration after 3 days in DM: EDC30 (a), EDC50 (b), EDC100 (c), as compared to tissue-culture plastic (d). (Scale bar is 100 μm) (reproduced with permission from Ren et al., Adv. Funct. Mater. 2008, copyright Wiley 2008).

Figure 27.9 shows immuno-staining of troponin T for cells grown for 3 days on the different films. We noticed that cells on soft films (EDC5, 10) fused but began to detach after 2–3 days in DM. In this case, the cells formed large aggregates or "clumps" and started to lose connectivity with the films. This phenomenon could be observed for stiffer films, although it was greatly delayed. For EDC30, cells differentiated and formed myotubes after 1–2 days in DM but they started to detach afterwards. The more cross-linked the films, the longer the time period over which differentiation progresses and myotubes could remain adherent on the films. Thus, cells grown on EDC50 films form myotubes (visible at day 3 in Figure 27.9b) that begin to clump between day 5 and day 7. For films grown on EDC100 films, myotubes that formed after 2 days orientated, elongated and thickened over the course of the experiment (Figure 27.9). They remained adherent until they also began to clump and detach between day 8 and day 10 (data not shown). Thus, the kinetics of detachment occurred in time periods from days to weeks, depending on the film stiffness. We also noticed that the largest variability from one experiment to another was observed for the soft films (EDC $\leq$ 30 mg mL$^{-1}$).

Importantly, it has to be noticed that this detachment was not attributed to a film rupture, as was checked by confocal laser scanning microscopy using fluorescently labeled PLL for visualizing the film [7]. Such aggregate formation and spontaneous detachment after 2–3 days has indeed already been observed for C2C12 cells cultured on alginate matrices of various compositions [88]. In this case, the cells that tended to aggregate were apparently the non-fused ones. C2C12 detachment was also evidenced for cells plated on a silicone or PDMS substrate [89], or when they were plated at a high density on wavy micropatterned silicone substrates [90]. In this latter study, the detachment was attributed to cell overgrowth from the initial high plating density and it was mentioned that the balance between cell–cell interactions and cell–substrate interactions is particularly important.

Furthermore, we observed by immuno-blots that cells on films cross-linked at the highest degree (EDC50 and EDC100) expressed both myogenin and troponin T at day 3 and day 6.

Notably, our results showed that early stages of the differentiation process are affected by (PLL/HA)$_{12}$ film stiffness. For soft films, the absence of well organized adhesion structures may lead to a deregulated myofibril assembly, which may hinder further myotube growth. Myosin II, a molecular motor known to play an important role in the generation of traction forces and in contractility [62, 91], might also be involved in the cell clumping observed on soft films. Another hypothesis is that film stiffness might affect myoblast motility, a parameter that was recently shown to be very important in the growth of myotubes after their initial formation [92] and which is currently under study.

Notably, the morphological differences between the films were even increased at a later stage in cell differentiation, when myotubes are striated. At this stage, only reached by the stiffest films (EDC $\geq$ 50, $E_0 \sim$ 350 kPa), a relatively small difference in film stiffness induces an appreciable effect on the percentage of striated myotubes, the tissue culture polystyrene surface remaining, however, the most favorable for myotube striation. We thus observed a correlation between film stiffness and myotubes striation, increased stiffness leading to a higher striation percentage.

Our results were somewhat different from the results obtained on patterned PA gels for which striation was optimal at an intermediate Young's modulus of 12 kPa [55]. As myotubes are exerting increasing forces upon formation, while they are contracting, the cell-film adhesion has to be strong enough to support the whole differentiation process [93]. The contractile force caused by myotubes could thus be the main reason for the late detachment observed on the stiffest films. The more mature the myotubes, the higher the force they will produce [94]. In fact, Engler *et al.* showed that the adhesion force of isolated mature myotubes cultured on micropatterned polyacrylamide gels increases exponentially with substrate stiffness [55]. We are currently further exploring the role of film stiffness in the early stages of the formation of myotubes, especially on the formation of adhesion complexes.

## 27.5.2
### Film Cross-Linking Drives the Fate of Mouse Embryonic Stem

In our recent studies, we have investigated whether (PLL/HA)$_{12}$ films at different cross-linking degrees could guide stem cell fate. To this end, we used embryonic stem cells (ESCs), which are derived from the inner cell mass (ICM) of the blastocyst at an early stage of embryonic development following the segregation of the embryo into the ICM and the trophectoderm [95]. In contrast to the cells located in the inner cell mass, mouse ESC lines remain pluripotent *in vitro* and self-renew in the presence of the cytokine leukemia inducible factor (LIF). ESCs commute between metastable states from the ICM to the epiblast stage. These reversible states are associated with a distinct potential of differentiation [96–98]. This is thus a highly dynamic self-renewing cell population which responds to environmental cues to maintain their pluripotency or to differentiate. The latter cues are several and include growth factors in the culture medium surrounding ESCs colonies or secreted within the colonies, signals arising from adhesion to the substrate (i.e., extracellular matrices, ECM), and/or stiffness of the substrate [99]. Little attention has been paid so far to the impact of the substrate on cell pluripotency as most ESCs lines are grown on feeder cells or in feeder-free conditions on gelatin-coated dishes.

The cellular environment can be changed by changing the adhesive state of the cells, the substrate's chemistry or the substrate stiffness. A pioneering work by Discher and coworkers suggested that multipotent stem cells might be driven in their differentiation by their tight or loose attachment, depending upon the stiffness of the substrate [62], in this case a polyacrylamide gel. Usually, adhesive ESCs generate two-dimensional flat colonies while cells in suspension form more three-dimensional and compact structures featuring a spherical shape. Such a change in morphology might affect the ESCs microenvironment and, in turn, their phenotype.

We investigated the effect of $(PLL/HA)_{12}$ film cross-linking on ESCs gene expression. We first analyzed the expression of mRNAs encoding pluripotency genes in ESCs grown on native or cross-linked nanofilms. The expression level of Oct4, Sox2, Nanog, three master genes gatekeepers of stem cell pluripotency [95], were unchanged in ESCs whatever the nanofilms they grew on (Figure 27.10). We next surmised that cross-linking the nanofilms and, in turn, modifying their adhesive properties might drive the cells from the ICM state to a primary specification toward a fate of epiblast, an early stage of differentiation. Accordingly, real time PCR was used to monitor the expression of genes specific for both stages of development, the epiblast (i.e., *Brachyury, Goosecoïd*) or the ICM (i.e., *Rex1, stella, Tbx3*) in ESCs cultured on native or cross-linked nanofilms. While ESCs in suspension cultured on native films expressed high levels of *Rex1, Stella* and *Tbx3*, expression of the ICM genes was dramatically decreased as soon as the film was even slightly cross-linked (EDC of $10\,mg\,mL^{-1}$). This also corresponded to conditions where more cells attached to the film. ICM genes expression remained stable in cells that have adhered on films cross-linking with higher amounts of EDC (40, 70 and $100\,mg\,mL^{-1}$) (data not shown). In addition, this transcriptional event was accompanied by a concomitant increase in *Brachyury* and *Goosecoïd*, which reached a maximum level for cells cultured on films crosslinked with $40\,mg\,mL^{-1}$ EDC (Figure 27.10). It is noteworthy at or below this level of cross-linking, that monitoring gene expression in adherent cells only, or in the whole adherent and non-adherent cell population, gave the same results. To further investigate the switch of ESCs from ICM to epiblast stage of pluripotency, we used an ESC clone engineered to express GFP under the transcriptional control of the Rex promoter. Expression of Rex1 is restricted in the ICM and its monitoring in ESCs reflects the metastable state of the cells [98]. RexGFP cells

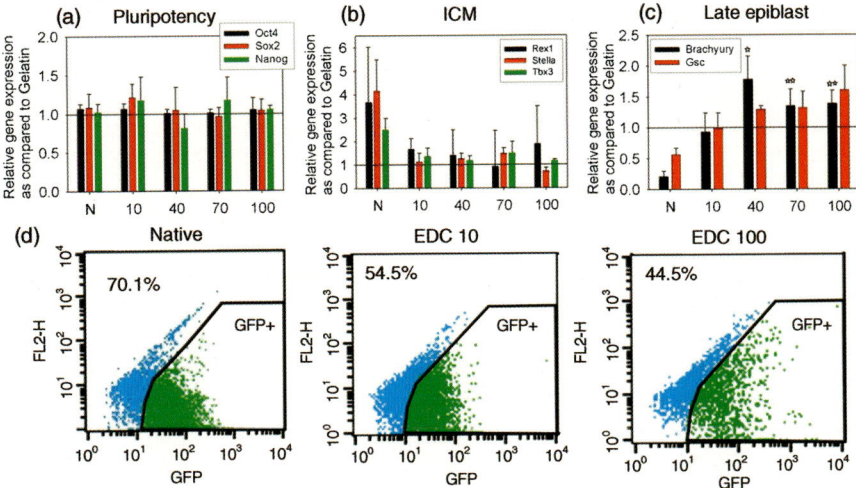

**Figure 27.10** Developmental stage of ESCs grown on native and cross-linked films. (a–c) Quantitative polymerase chain reactions (PCR) experiments showing the relative gene expression profile of cells grown on films for 4 days as compared to cells grown on gelatin. Error bars represent S.E.M ($n = 4$). (*$p < 0.5$, **$p < 0.05$). (d) Flow cytometry quantification of Rex1 expression within the cell populations obtained after culture on native, EDC10 and EDC100 films, from left to right. The range axis represents the fluorescence intensity in the orange channel and the domain axis represents the fluorescence intensity in the green channel. (reproduced with permission from Blin et al., Biomaterials 2010, copyright Elsevier 2010).

cultured in suspension on native films remained GFP positive for 70% of them, as measured by FACS analysis. On the other hand, the proportion of GFP+ cells on EDC10 and 100 cross-linked films decreased to 54 and 44%, respectively, suggesting that the cells underwent differentiation.

Thus, ESCs secured their pluripotency when grown on native (PLL/HA) films, which prevented their adhesion. Their phenotype was more reminiscent of the ICM stage of embryogenesis. We predicted that such a stage of development should be associated with a better potential of differentiation than cells which reached the epiblast stage, with a more limited repertoire of differentiation.

We subsequently investigated whether the phenotype of undifferentiated ESCs and, in turn, their potential of differentiation depended upon the film stiffness, their adhesion prior to differentiation and/or differences in the chemical microenvironment. To address this point, we compared the behavior of ESCs cultured on native nanofilms composed of PLL and HA or on agarose, a chemically inert substrate as soft as native films. In both experimental conditions, ESCs did not attach to the surface and formed colonies in suspension. However, ESCs on native films expressed much more strongly gene markers of ICM (Rex1, Stella, Tbx3) than ESCs on agarose. As a counterpart, the pattern of genes markers of the epiblast (Brachyury, Goosecoïd) was opposite to that of ICM. While both Brachyury and Goosecoïd were readily expressed in ESCs grown on agarose, the genes were poorly detected in ESCs grown on native nanofilms. These findings pointed to a chemical effect of the native nanofilms. We indeed found that a very small release of PLL ($< 3\,\mu g\,mL^{-1}$) was noted for the native films.

## 27.6
## Conclusions

In summary, physico-chemical properties of polysaccharide and polypeptide multilayer PEM films can be unraveled using QCM-D and ATR-FTIR to probe the internal composition of the film and ion pairing. Film cross-linking by means of the carbodiimide chemistry appears to be simple strategy to create covalent amide in the film and to modulate film stiffness over a large range. Advantageously, this cross-linking procedure is selective to carboxylic and ammonium groups, which allowed us to identify clearly the cross-linking of $COO^-$ groups in polyelectrolytes containing both carboxylic and sulfate groups, such as chondroitin sulfate and heparin. Film stiffness expressed as the Young's modulus $E_0$ was found to be correlated to the molar charge density of the ionic groups within the films. $E_0$ was varied over at least two orders of magnitude, depending on the film type and cross-linking extent. We have shown that the film mechanical properties have a strong influence on cell behavior, cell adhesion and spreading, being higher when the cross-linking extent is increased. Adhesion receptors from the integrin family are involved in the early adhesive process. Thus, induction of integrin activation led to the formation of focal adhesions on the otherwise poorly adhesive films (i.e., soft $(PLL/HA)_{12}$ films). Moreover, we have shown that C2C12 myoblasts can differentiate on (PLL/HA) films cross-linked at various degrees, but the efficiency of myotube formation depended on the cross-linking degree: stiffer films led to better myotube formation without any detachment of cell sheets. For a different application in the field of cardiovascular therapy, ESCs were found to adhere and proliferate on the films in a very similar manner to myoblasts: the stiffer the film, the higher the adhesion and the subsequent proliferation. Notably, native films allowed the cells to secure their pluripotency, whereas cross-linked films directed their fate toward the epiblast.

All together, our results open the perspective for the use of PEM films to direct stem cell fate and to manipulate the cell adhesive behavior of various cell types. Such films could thus be employed as a tool to unravel early cell adhesive events.

## Acknowledgements

C.P. is indebted to the Institut Universitaire de France, to the Association Française contre les Myopathies (AFM, grant no. 12671 and fellowship to T.C.), to the Agence Nationale pour la Recherche (grant ANR-06-NANO-006, ANR-07-NANO-002) and to the Association Recherche contre le Cancer (ARC, fellowship to K.R.) for financial support.

# References

1 Decher, G. (1997) Fuzzy nanoassemblies: Toward layered polymeric multicomposites. *Science*, **277** (5330), 1232–1237.

2 Ladam, G., Schaad, P., Voegel, J.-C., Schaaf, P., Decher, G., and Cuisinier, F.J.G. (2000) In situ determination of the structural properties of initially deposited polyelectrolyte multilayers. *Langmuir*, **16** (3), 1249–1255.

3 Sukhishvili, S. and Granick, S. (2002) Layered, erasable polymer multilayers formed by hydrogen-bonded sequential self-assembly. *Macromolecules*, **35** (1), 301–310.

4 Picart, C., Lavalle, P., Hubert, P., Cuisinier, F.J.G., Decher, G., Schaaf, P., and Voegel, J.-C. (2001) Buildup mechanism for poly(L-lysine)/hyaluronic acid films onto a solid surface. *Langmuir*, **17** (23), 7414–7424.

5 McAloney, R.A., Sinyor, M., Dudnik, V., and Goh, M.C. (2001) Atomic force microscopy studies of salt effects on polyelectrolyte multilayer film morphology. *Langmuir*, **17** (21), 6655–6663.

6 Lavalle, P., Gergely, C., Cuisinier, F., Decher, G., Schaaf, P., Voegel, J.-C., and Picart, C. (2002) Comparison of the structure of polyelectrolyte multilayer films exhibiting a linear and an exponential growth regime: An in situ atomic force microscopy study. *Macromolecules*, **35** (11), 4458–4465.

7 Picart, C., Mutterer, J., Richert, L., Luo, Y., Prestwich, G.D., Schaaf, P., Voegel, J.-C., and Lavalle, P. (2002) Molecular basis for the explanation of the exponential growth of polyelectrolyte multilayers. *Proc. Natl. Acad. Sci. U. S. A.*, **99** (20), 12531–12535.

8 Schlenoff, J., Ly, H., and Li, M. (1998) Charge and mass balance in polyelectrolyte multilayers. *J. Am. Chem. Soc.*, **120** (30), 7626–7634.

9 Schlenoff, J.B. and Dubas, S.T. (2001) Mechanism of polyelectrolyte multilayer growth: Charge overcompensation and distribution. *Macromolecules*, **34** (3), 592–598.

10 Riegler, H. and Essler, F. (2002) Polyelectrolytes. 2. Intrinsic or extrinsic charge compensation? Quantitative charge analysis of PAH/PSS multilayers. *Langmuir*, **18** (17), 6694–6698.

11 Lourenco, J.M., Ribeiro, P.A., Botelho do Rego, A.M., Braz Fernandes, F.M., Moutinho, A.M., and Raposo, M. (2004) Counterions in poly(allylamine hydrochloride) and poly(styrene sulfonate) layer-by-layer films. *Langmuir*, **20** (19), 8103–8109.

12 Jaber, J.A. and Schlenoff, J.B. (2007) Counterions and water in polyelectrolyte multilayers: a tale of two polycations. *Langmuir*, **23** (2), 896–901.

13 Krzeminski, A., Marudova, M., Moffat, J., Noel, T.R., Parker, R., Wellner, N., and Ring, S.G. (2006) Deposition of pectin/poly-L-lysine multilayers with pectins of varying degrees of esterification. *Biomacromolecules*, **7** (2), 498–506.

14 Tang, Z., Wang, Y.l., Podsiadlo, P., and Kotov, N.A. (2006) Biomedical applications of layer-by-layer assembly: from biomimetics to tissue engineering. *Adv. Mater.*, **18**, 3203–3224.

15 Picart, C. (2008) Polyelectrolyte multilayer films: from physico-chemical properties to the control of cellular processes. *Curr. Med. Chem.*, **15**, 685–697.

16 Nishinari, K. and Takahashi, R. (2003) Interaction in polysaccharide solutions and gels. *Curr. Opin. Colloid Interface.*, **8** (4–5), 396–400.

17 Schlessinger, J., Lax, I., and Lemmon, M. (1995) Regulation of growth factor activation by proteoglycans: what is the role of the low affinity receptors? *Cell*, **83** (3), 357–360.

18 Taipale, J. and Keski-Oja, J. (1997) Growth factors in the extracellular matrix. *FASEB J.*, **11** (1), 51–59.

19 Wiig, H., Gyenge, C., Iversen, P.O., Gullberg, D., and Tenstad, O. (2008) The role of the extracellular matrix in tissue distribution of macromolecules in normal and pathological tissues: potential therapeutic consequences. *Microcirculation*, **15** (4), 283–296.

20 Drury, J.L. and Mooney, D.J. (2003) Hydrogels for tissue engineering: scaffold design variables and applications. *Biomaterials*, **24** (24), 4337–4351.

21 Almond, A. (2007) Hyaluronan. *Cell Mol. Life Sci.*, **64** (13), 1591–1596.

22 Ma, L., Zhou, J., Gao, C., and Shen, J. (2007) Incorporation of basic fibroblast growth factor by a layer-by-layer assembly technique to produce bioactive substrates. *J. Biomed. Mater. Res.: Appl. Biomater.*, **83** (1), 285–292.

23 Schneider, A., Vodouhê, A., Richert, L., Francius, G., Le Guen, E., Schaaf, P., Voegel, J.-C., Frisch, F., and Picart, C. (2007) Multi-functional polyelectrolyte multilayer films: combining mechanical resistance, biodegradability and bioactivity. *Biomacromolecules*, **8** (1), 139–145.

24 Elbert, D.L., Herbert, C.B., and Hubbell, J.A. (1999) Thin polymer layers formed by polyelectrolyte multilayer techniques on biological surfaces. *Langmuir*, **15** (16), 5355–5362.

25 Voinova, M.V., Rodahl, R., Jonson, R., and Kasemo, B. (1999) Viscoelastic acoustic response of layered polymer films at fluid-solid interfaces: Continuum mechanics approach. *Phys. Scr.*, **89**, 391–396.

26 Porcel, C., Lavalle, P., Ball, V., Decher, G., Senger, B., Voegel, J.C., and Schaaf, P. (2006) From exponential to linear growth in polyelectrolyte multilayers. *Langmuir*, **22** (9), 4376–4383.

27 von Klitzing, R. (2006) Internal structure of polyelectrolyte multilayer assemblies. *Phys. Chem. Chem. Phys.*, **8** (43), 5012–5033.

28 Salomaki, M., Vinokurov, I.A., and Kankare, J. (2005) Effect of temperature on the buildup of polyelectrolyte multilayers. *Langmuir*, **21** (24), 11232–11240.

29 Laugel, N., Betscha, C., Winterhalter, M., Voegel, J.C., Schaaf, P., and Ball, V. (2006) Relationship between the growth regime of polyelectrolyte multilayers and the polyanion/polycation complexation enthalpy. *J. Phys. Chem. B*, **110** (39), 19443–19449.

30 Cerda, J.J., Qiao, B.F., and Holm, C. (2009) Understanding polyelectrolyte multilayers: an open challenge for simulations. *Soft Matter*, **5** (22), 4412–4425.

31 Hoda, N. and Larson, R.G. (2009) Modeling the buildup of exponentially growing polyelectrolyte multilayer films. *J. Phys. Chem. B*, **113** (13), 4232–4241.

32 Xie, A.F. and Granick, S. (2002) Local electrostatics within a polyelectrolyte multilayer with embedded weak polyelectrolyte. *Macromolecules*, **35** (5), 1805–1813.

33 Crouzier, T. and Picart, C. (2009) Ion pairing and hydration in polyelectrolyte multilayer films containing polysaccharides. *Biomacromolecules*, **10** (2), 433–442.

34 Boudou, T., Crouzier, T., Auzely-Velty, R., Glinel, K., and Picart, C. (2009) Internal composition versus the mechanical properties of polyelectrolyte multilayer films: the influence of chemical cross-linking. *Langmuir*, **25**, 13809–13819.

35 Haxaire, K., Maréchal, Y., Milas, M., and Rinaudo, M. (2003) Hydration of hyaluronan polysaccharide observed by IR spectrometry. II. Definition and quantitative analysis of elementary hydration spectra and water uptake. *Biopolymers*, **72** (3), 149–161.

36 Boulmedais, F., Ball, V., Schwinte, P., Frisch, B., Schaaf, P., and Voegel, J.-C. (2003) Buildup of exponentially growing multilayer polypeptide films with internal secondary structure. *Langmuir*, **19** (2), 440–445.

37 Picart, C. (2008) Polyelectrolyte multilayer films: from physico-chemical properties to the control of cellular processes. *Curr. Med. Chem.*, **15** (7), 685–697.

38 Richert, L., Boulmedais, F., Lavalle, P., Mutterer, J., Ferreux, E., Decher, G., Schaaf, P., Voegel, J.-C., and Picart, C. (2004) Improvement of stability and cell adhesion properties of polyelectrolyte multilayer films by chemical cross-linking. *Biomacromolecules*, **5** (2), 284–294.

39 Richert, L., Schneider, A., Vautier, D., Jessel, N., Payan, E., Schaaf, P., Voegel, J.-C., and Picart, C. (2006) Imaging cell interactions with native and cross-linked polyelectrolyte multilayers. *Cell Biochem. Biophys.*, **44** (2), 273–276.

40 Shen, L., Chaudouet, P., Ji, J., and Picart, C. (2011) pH-Amplified multilayer films based on hyaluronan: influence of HA molecular weight and concentration on film growth and stability. *Biomacromolecules*, **12** (4), 1322–1331.

41 Kovacevic, D., von der Burgh, S., De Keizer, A., and Cohen Stuart, M. (2003) Specific ionic effects on weak polyelectrolyte multilayer formation. *J. Phys. Chem. B*, **107** (32), 7998–8002.

42 Moffat, J., Noel, T.R., Parker, R., Wellner, N., and Ring, S.G. (2007) The environmental response and stability of pectin and poly-L-lysine multilayers. *Carbohydr. Polym.*, **70**, 422–429.

43 Richert, L., Boulmedais, F., Lavalle, P., Mutterer, J., Ferreux, E., Decher, G., Schaaf, P., Voegel, J.-C., and Picart, C. (2004) Improvement of stability and cell adhesion properties of polyelectrolyte multilayer films by chemical cross-linking. *Biomacromolecules*, **5** (2), 284–294.

44 Schneider, A., Francius, G., Obeid, R., Schwinte, P., Hemmerle, J., Frisch, B., Schaaf, P., Voegel, J.-C., Senger, B., and Picart, C. (2006) Polyelectrolyte multilayers with a tunable young's modulus:  Influence of film stiffness on cell adhesion. *Langmuir*, **22** (3), 1193–1200.

45 Richert, L., Arntz, Y., Schaaf, P., Voegel, J.-C., and Picart, C. (2004) pH dependent growth of poly(L-lysine/poly(L-glutamic) acid multilayer films and their cell adhesion properties. *Surf. Sci.*, **570**, 13–29.

46 Richert, L., Lavalle, P., Payan, E., Stoltz, J.-F., Shu, X.Z., Prestwich, G.D., Schaaf, P., Voegel, J.-C., and Picart, C. (2004) Layer-by-layer buildup of polysaccharide films: Physical chemistry and cellular adhesion aspects. *Langmuir*, **1** (2), 284–294.

47 Schneider, A., Francius, G., Obeid, R., Schwinté, P., Frisch, B., Schaaf, P., Voegel, J.-C., Senger, B., and Picart, C. (2006) Polyelectrolyte multilayer with tunable Young's modulus: influence on cell adhesion. *Langmuir*, **22** (3), 1193–1200.

48 Ren, K., Crouzier, T., Roy, C., and Picart, C. (2008) Polyelectrolyte multilayer films of controlled stiffness modulate myoblast differentiation. *Adv. Func. Mater.*, **18** (9), 1378–1389.

49 Boudou, T., Crouzier, T., Auzely-Velty, R., Glinel, K., and Picart, C. (2009) Internal composition versus the mechanical properties of polyelectrolyte multilayer films: The influence of chemical cross-linking. *Langmuir*, **25** (24), 13809–13819.

50 Richert, L., Arntz, Y., Schaaf, P., Voegel, J.-C., and Picart, C. (2004) pH dependent growth of poly(L-lysine)/poly(L-glutamic) acid multilayer films and their cell adhesion properties. *Surf. Sci.*, **570** (1–2), 13–29.

51 Richert, L., Lavalle, P., Payan, E., Shu, X.Z., Prestwich, G.D., Stoltz, J.-F., Schaaf, P., Voegel, J.-C., and Picart, C. (2004) Layer-by-layer buildup of polysaccharide films: physical chemistry and cellular adhesion aspects. *Langmuir*, **20** (2), 448–458.

52 Head, D.A., Levine, A.J., and MacKintosh, F.C. (2003) Distinct regimes of elastic response and deformation modes of cross-linked cytoskeletal and semiflexible polymer networks. *Phys. Rev. E*, **68** (6), 061907.

53 Dimitriadis, E.K., Horkay, F., Maresca, J., Kachar, B., and Chadwick, R.S. (2002) Determination of elastic moduli of thin layers of soft material using the atomic force microscope. *Biophys. J.*, **82** (5), 2798–2810.

54 Engler, A., Bacakova, L., Newman, C., Hategan, A., Griffin, M., and Discher, D.E. (2004) Substrate compliance versus ligand density in cell on gel responses. *Biophys. J.*, **86** (1), 617–628.

55 Engler, A.J., Griffin, M.A., Sen, S., Bonnemann, C.G., Sweeney, H.L., and Discher, D.E. (2004) Myotubes differentiate optimally on substrates with tissue-like stiffness: pathological implications for soft or stiff microenvironments. *J. Cell Biol.*, **166** (6), 877–887.

56 Mermut, O., Lefebvre, J., Gray, D.G., and Barrett, C.J. (2003) Structural and mechanical properties of polyelectrolyte multilayer films studied by AFM. *Macromolecules*, **36** (23), 8819–8824.

57 Richert, L., Engler, A.J., Discher, D.E., and Picart, C. (2004) Elasticity of native and cross-linked polyelectrolyte multilayers. *Biomacromolecules*, **5** (5), 1908–1916.

58 Francius, G., Hemmerle, J., Ohayon, J., Schaaf, P., Voegel, J.-C., Picart, C., and Senger, B. (2006) Effect of cross-linking on the elasticity of polyelectrolyte multilayer films measured by colloidal probe AFM. *Microsc. Res. Tech.*, **69** (2), 84–92.

59 Boulmedais, F., Ball, V., Schwinté, P., Frisch, B., Schaaf, P., and Voegel, J.-C. (2003) Buildup of exponentially growing multilayer polypeptide films with internal secondary structure. *Langmuir*, **19** (2), 440–445.

60 Schoeler, B., Delorme, N., Doench, I., Sukhorukov, G.B., Fery, A., and Glinel, K. (2006) Polyelectrolyte films based on polysaccharides of different conformations: effects on multilayer structure and mechanical properties. *Biomacromolecules*, **7** (6), 2065–2071.

61 Kocgozlu, L., Lavalle, P., Koenig, G., Senger, B., Haikel, Y., Schaaf, P., Voegel, J.C., Tenenbaum, H., and Vautier, D. (2010) Selective and uncoupled role of substrate elasticity in the regulation of replication and transcription in epithelial cells. *J. Cell Sci.*, **123** Pt (1), 29–39.

62 Engler, A.J., Sen, S., Sweeney, H.L., and Discher, D.E. (2006) Matrix elasticity directs stem cell lineage specification. *Cell*, **126** (4), 677–689.

63 Bernal, R., Pullarkat, P.A., and Melo, F. (2007) Mechanical properties of axons. *Phys. Rev. Lett.*, **99** (1), 018301.

64 Silberstein, L., Webster, S.G., Travis, M., and Blau, H.M. (1986) Developmental progression of myosin gene expression in cultured muscle cells. *Cell*, **46** (7), 1075–1081.

65 Adams, J.C. (2002) Regulation of protrusive and contractile cell-matrix contacts. *J. Cell Sci.*, **115** (2), 257–265.

66 Ren, K., Fourel, L., Gauthier-Rouviere, C., Albiges-Rizo, C., and Picart, C. (2010) Manipulation of the adhesive behaviour of skeletal muscle cells on soft and stiff polyelectrolyte multilayers. *Acta Biomater.*, **6** (11), 4238–4248.

67 Roca-Cusachs, P., Gauthier, N.C., Del Rio, A., and Sheetz, M.P. (2009) Clustering of alpha(5)beta(1) integrins determines adhesion strength whereas alpha(v)beta(3) and talin enable mechanotransduction. *Proc. Natl. Acad. Sci. U. S. A.*, **106** (38), 16245–16250.

68 Hynes, R.O. (2002) Integrins: bidirectional, allosteric signaling machines. *Cell*, **110** (6), 673–687.

69 Mould, A.P., Askari, J.A., Barton, S., Kline, A.D., McEwan, P.A., Craig, S.E., and Humphries, M.J. (2002) Integrin activation involves a conformational change in the alpha 1 helix of the beta subunit A-domain. *J. Biol. Chem.*, **277** (22), 19800–19805.

70 Wipff, P.J., Rifkin, D.B., Meister, J.J., and Hinz, B. (2007) Myofibroblast contraction activates latent TGF-beta1 from the extracellular matrix. *J. Cell Biol.*, **179** (6), 1311–1323.

71 Boudou, T., Crouzier, T., Nicolas, C., Ren, K., and Picart, C. (2011) Polyelectrolyte multilayer nanofilms used as thin materials for cell mechanosensitivity studies. *Macromol. Biosci.*, **11** (1), 77–89.

72 Lo, C.M., Wang, H.B., Dembo, M., and Wang, Y.L. (2000) Cell movement is guided by the rigidity of the substrate. *Biophys. J.*, **79** (1), 144–152.

73 Solon, J., Levental, I., Sengupta, K., Georges, P.C., and Janmey, P.A. (2007) Fibroblast adaptation and stiffness matching to soft elastic substrates. *Biophys. J.*, **93** (12), 4453–4461.

74 Beningo, K.A., Lo, C.M., and Wang, Y.L. (2002) Flexible polyacrylamide substrata for the analysis of mechanical interactions at cell-substratum adhesions. *Methods Cell Biol.*, **69**, 325–339.

75 Geiger, B., Spatz, J.P., and Bershadsky, A.D. (2009) Environmental sensing through focal adhesions. *Nat. Rev. Mol. Cell Biol.*, **10** (1), 21–33.

76 Deroanne, C.F., Lapiere, C.M., and Nusgens, B.V. (2001) In vitro tubulogenesis of endothelial cells by relaxation of the coupling extracellular matrix-cytoskeleton. *Cardiovasc. Res.*, **49** (3), 647–658.

77 Chou, S.Y., Cheng, C.M., and LeDuc, P.R. (2009) Composite polymer systems with control of local substrate elasticity and their effect on cytoskeletal and morphological characteristics of adherent cells. *Biomaterials*, **30** (18), 3136–3142.

78 Brown, X.Q., Ookawa, K., and Wong, J.Y. (2005) Evaluation of polydimethylsiloxane scaffolds with physiologically-relevant elastic moduli: interplay of substrate mechanics and surface chemistry effects on vascular smooth muscle cell response. *Biomaterials*, **26** (16), 3123–3129.

79 Chien, H.W., Chang, T.Y., and Tsai, W.B. (2009) Spatial control of cellular adhesion using photo-crosslinked micropatterned polyelectrolyte multilayer films. *Biomaterials*, **30** (12), 2209–2218.

80 Chen, A.A., Khetani, S.R., Lee, S., Bhatia, S.N., and Van Vliet, K.J. (2009) Modulation of hepatocyte phenotype in vitro via chemomechanical tuning of polyelectrolyte multilayers. *Biomaterials*, **30** (6), 1113–1120.

81 Andres, V. and Walsh, K. (1996) Myogenin expression, cell cycle withdrawal, and phenotypic differentiation are temporally separable events that precede cell fusion upon myogenesis. *J. Cell Biol.*, **132** (4), 657–666.

82 Knudsen, K.A. and Horwitz, A.F. (1977) Tandem events in myoblast fusion. *Dev. Biol.*, **58** (2), 328–338.

83 Horsley, V. and Pavlath, G.K. (2004) Forming a multinucleated cell: molecules that regulate myoblast fusion. *Cells Tissues Organs*, **176** (1–3), 67–78.

84 Yeo, Y., Geng, W., Ito, T., Kohane, D.S., Burdick, J.A., and Radisic, M. (2007) Photocrosslinkable hydrogel for myocyte cell culture and injection. *J. Biomed. Mater. Res. B*, **81** (12), 312– 312.

85 Smith, E., Yang, J., McGann, L., Sebald, W., and Uludag, H. (2005) RGD-grafted thermoreversible polymers to facilitate attachment of BMP-2 responsive C2C12 cells. *Biomaterials*, **26** (35), 7329–7338.

86 Charrasse, S., Comunale, F., Fortier, M., Portales-Casamar, E., Debant, A., and Gauthier-Rouviere, C. (2007) M-cadherin activates Rac1 GTPase through the Rho-GEF trio during myoblast fusion. *Mol. Biol. Cell*, **18** (5), 1734–1743.

87 Dedieu, S., Mazeres, G., Cottin, P., and Brustis, J.J. (2002) Involvement of myogenic regulator factors during fusion in the cell line C2C12. *Int. J. Dev. Biol.*, **46** (2), 235–241.

88 Rowley, J.A. and Mooney, D.J. (2002) Alginate type and RGD density control myoblast phenotype. *J. Biomed. Mater. Res.*, **60** (2), 217–223.

89 Dennis, R.G., Kosnik, P.E., Gilbert, M.E., and Faulkner, J.A. (2001) Excitability and contractility of skeletal muscle engineered from primary cultures and cell lines. *Am. J. Physiol. Cell Physiol.*, **280** (2), C288–C295.

90 Lam, M.T., Sim, S., Zhu, X., and Takayama, S. (2006) The effect of continuous wavy micropatterns on silicone substrates on the alignment of skeletal muscle myoblasts and myotubes. *Biomaterials*, **27** (24), 4340–4347.

91 Guo, W.H., Frey, M.T., Burnham, N.A., and Wang, Y.L. (2006) Substrate rigidity regulates the formation and maintenance of tissues. *Biophys. J.*, **90** (6), 2213–2220.

92 Jansen, K.M. and Pavlath, G.K. (2006) Mannose receptor regulates myoblast motility and muscle growth. *J. Cell Biol.*, **174** (3), 403–413.

93 Neumann, T., Hauschka, S.D., and Sanders, J.E. (2003) Tissue engineering of skeletal muscle using polymer fiber arrays. *Tissue Eng.*, **9** (5), 995–1003.

94 Clemmens, E.W. and Regnier, M. (2004) Skeletal regulatory proteins enhance thin filament sliding speed and force by skeletal HMM. *J. Muscle Res. Cell. Motil.*, **25** (7), 515–525.

95 Niwa, H. (2007) How is pluripotency determined and maintained? *Development*, **134** (4), 635–646.

96 Hayashi, K., Lopes, S.M., Tang, F., and Surani, M.A. (2008) Dynamic equilibrium and heterogeneity of mouse pluripotent stem cells with distinct functional and epigenetic states. *Cell Stem Cell*, **3** (4), 391–401.

97 Pelton, T.A., Sharma, S., Schulz, T.C., Rathjen, J., and Rathjen, P.D. (2002) Transient pluripotent cell populations during primitive ectoderm formation:

correlation of in vivo and in vitro pluripotent cell development. *J. Cell Sci.*, **115** Pt (2), 329–339.

98 Toyooka, Y., Shimosato, D., Murakami, K., Takahashi, K., and Niwa, H. (2008) Identification and characterization of subpopulations in undifferentiated ES cell culture. *Development*, **135** (5), 909–918.

99 Discher, D.E., Mooney, D.J., and Zandstra, P.W. (2009) Growth factors, matrices, and forces combine and control stem cells. *Science*, **324** (5935), 1673–1677.

# 28
# Diffusion of Nanoparticles and Biomolecules into Polyelectrolyte Multilayer Films: Towards New Functional Materials

*Marc Michel and Vincent Ball*

## 28.1
## Introduction

The first films processed according to a step-by-step (SBS) scheme, a name we prefer to the usual "layer-by-layer" (LBL) one, were indeed made from modified silica particles 45 years ago [1]. The field of step-by-step deposition was then neglected for 35 years, with some exceptions in the deposition of films made from amphiphilic molecules using the Langmuir–Blodgett deposition method [2]. The step-by-step deposition method was the subject of a "renaissance" in the early 1990s. The long ignorance of Iler's contribution [1] was most probably due to the lack of analytical tools able to characterize films a few nanometers in thickness. The first investigated systems to build-up LBL films were oppositely charged polyelectrolyes and polyelectrolytes and bolaamphiphiles [3, 4]. These films were called polyelectrolyte multilayer films (PEMs). Very rapidly in the history of step-by-step deposited films, colloids other than synthetic polymers were inserted during the alternated deposition of interacting species: DNA [5, 6] proteins [7], viruses [8, 9] lipid vesicles [10–13] or analogues of vesicles [14], and clays [15], to cite only a few. The layer-by-layer deposition incorporating polymers and inorganic nanoparticles started very soon in the development of SBS films, after 1994–1995. The aim was to produce hybrid films with controlled optical, electronic and mechanical properties. The use of clay nanoparticles [15] allowed the production of films behaving as diodes [16] and biomimetic structures having outstanding mechanical properties [17, 18]. A similar concept was applied with carbon nanotubes (CNTs) [19], other kinds of nanowires [20] and graphene oxide [21]. The properties of composite films made from polymers and CNTs, or from polymers and clays have been recently reviewed [22]. The outstanding mechanical properties of such films are reviewed in this book by Fery and Tsukruk and also by Kotov.

Very interestingly, LBL films incorporating graphene oxide and poly(diallyldimethyl ammonium chloride) (PDADMAC) appeared already in 1996 [23]. The electrical conductivity of such films could be switched on or off by reversible reduction of graphene oxide or oxidation of graphene, respectively. Many fascinating studies showed the opportunity to use films incorporating nanoparticles as components of electronic devices [24] and for solar energy conversion [25, 26]. Other studies showed the possibility to use films in which enzymes were deposited in a step-by-step manner with other species as potential biosensors [27, 28].

In all these films the position of the incorporated colloids reflects more or less, with some small interlayer interpenetration, the initial distance from the substrate at which the colloids were deposited. The situation is, hence, identical to that of films made from the alternated deposition of strongly interacting and, hence, weakly interpenetrating polymers. This is the reason why the term "LBL" deposition is so frequently used: by using such a term one implies that the polymers or deposited nanoparticles remain localized in the film at a location close (at the nm scale) to where they were initially adsorbed. In many cases this in not true because of high mobility in the film, as we will see later. This holds not only for polyelectrolytes [29] but also for polymers interacting via hydrogen bonding [30–32].

The fact that some combinations of oppositely charged polyelectrolytes lead to films displaying a supralinear increase in their thickness with the number of layer pairs [33–37] led to the question of their deposition mechanism. The finding that fluorescently labeled poly(-L-lysine) (PLL) diffuses through the whole film thickness of $(PLL-HA)_n$ films (where HA designates poly(sodium hyaluronate), as observed by laser confocal scanning microscopy (LCSM), highlighted the high mobility of such chains along the direction perpendicular to the substrate. In these films, the presence of a charged surface acts a barrier which impedes the diffusion of the free polyelectrolytes from the multilayer film into the bulk of the solution [38]. In addition, these films are characterized by a strong chain mobility in the direction parallel to the plane of the film, as has been measured by means of fluorescence recovery after photobleaching [39].

A few years ago we asked the question, in collaboration with Prof. Nicholas A. Kotov, as to whether nanoparticles could replace polyelectrolytes and diffuse through the whole thickness of a multilayer film, growing according to an exponential growth regime. This was indeed possible with thioglycollic acid-capped CdTe nanoparticles [40] and opened new avenues to the production of composite films.

The aim of this chapter is not to review the properties of films made by the step-by-step deposition of colloids and synthetic polymers, even if we will describe briefly the main trends, but to describe the major advances that have been made recently in films made from inorganic nanoparticles only, in films that have been incubated in a solution containing precursor ions of the nanoparticles to be synthesized in the films or by "post-incubation" of the film in solutions containing already synthesized inorganic colloids, proteins or CNTs. The three main approaches used up to now to introduce nanoparticles or colloids into LBL films are represented in Scheme 28.1. This chapter is not an exhaustive review but aims to show that PEM films in which nanoparticles, nanowires or proteins are allowed to diffuse after the deposition of the film are a fantastic platform for new applications as well as for exciting fundamental studies.

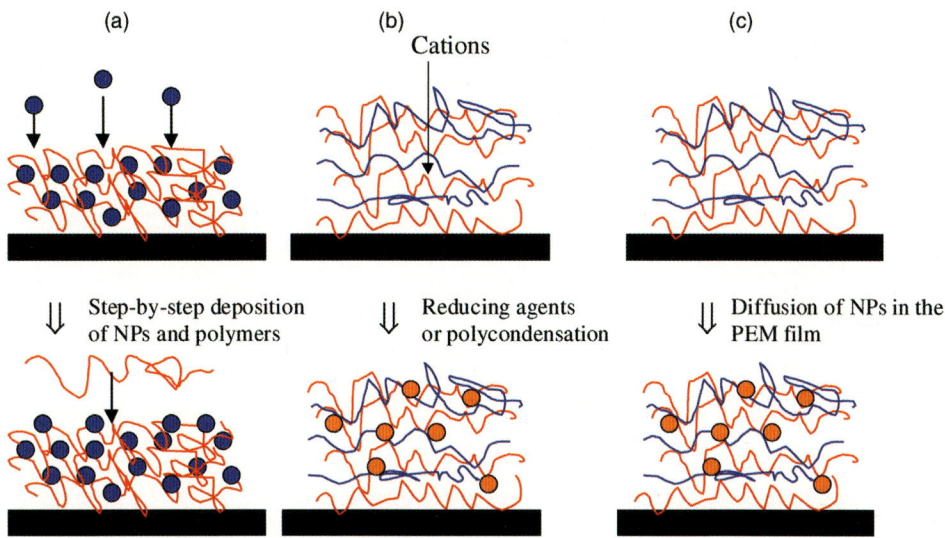

**Scheme 28.1** Schematic representation of the three ways to incorporate nanoparticles in films deposited in a step-by-step manner. (a) Step-by-step deposition of polymers and nanoparticles. (b) Diffusion of cations or of metal complexes in films produced by a step-by-step deposition method and their subsequent transformation into nanoparticles by means of a reduction or a polycondensation reaction. (c) Diffusion of nanoparticles in already deposited films, produced in a step-by-step manner.

## 28.2
### LBL Films in Which Nanoparticles are Incorporated Step-By-Step

The most described way to produce nanoparticles containing films using the step-by-step deposition method is represented in Scheme 28.1a. A plethora of studies have shown broad applications for such composite films. As an example, films made from the alternated deposition of clays and polyelectrolytes may act as efficient barriers for the diffusion of gases [41], in particular oxygen [42]. The anisotopic orientation of the laponite clays in the (PEI-Clay-PEO)$_n$ films translates into an anisoptropic proton conductivity of the hybrid films ($7.2 \times 10^{-8}$ S cm$^{-1}$ in the direction parallel to the substrate and $6.8 \times 10^{-10}$ S cm$^{-1}$ in the direction perpendicular to the substrate in the dry state and at 401 K) [43] Many other examples of films produced by a step-by-step deposition of nanoparticles and polymers are available. The interested reader should refer to the excellent review by Srivastava and Kotov [44].

## 28.3
### LBL Films Made Uniquely From Nanoparticles

The field of multilayered films made uniquely from nanoparticles regained interest a few years ago under the impulsion of Rubner's group [45]. A step-by-step approach was also used to produce Prussian Blue (PB) based coatings following the sequential adsorption steps of hexacyanoferrate anions and $Fe^{3+}$ cations [46]. An even more elegant approach consists in the alternated spray deposition [47] of interacting inorganic species to yield insoluble salts, such as $CaF_2$ (fluorine), $CaC_2O_4$ (calcium oxalate) and $CaHPO_4$ (brushite) in the form of thin films [48]. The deposition of films made from inorganic clusters can also be achieved through the simultaneous spray deposition method [49, 50], in which the film thickness grows linearly with the spraying time, the notion of an elementary deposition step being replaced by an infinitesimal spray deposition time.

## 28.4
### Nanoparticles Produced by Post-treatment of Deposited Films

The global strategy to produce nanoparticles in PEM films by post-treatment with a solution of precursor cations or metal complexes is represented in Scheme 28.1b. Owing to the relative permeability of PEM films with respect to metallic cations, as found by Miller and Bruening [51] and Krasemann and Tieke [52], Joly et al. introduced the concept of reduction of metal cations inside the films [53, 54]. Cobalt hydroxide [55] and iron oxy-hydroxide magnetic nanoparticles have also been synthesized in PEM films [56]. The controlled incorporation of hexacyanoferrate anions in exponentially growing films made from poly(sodium-L-glutamic acid) (PGA) and poly(allylamine hydrochloride) (PAH) made possible the further reaction of these anions with $Fe^{3+}$ cations to produce films containing Prussian Blue (PB). This control was achieved through the post-treatment of the films with their constituent polyelectrolytes (Figure 28.1). This post-treatment had the effect of changing the value of the Donnan potential of the PEM films.

The PB-containing films exhibited the magnetic properties expected for PB nanoparticles [57].

In their initial report, Joly et al. reported that "The problem for the most part has been the use of strong polyelectrolytes, such as poly(styrene sulfonic acid) (PSS), which tend to form layers in which all of the acid groups are partnered with the cationic groups of the surrounding layers, thereby leaving little or no active sites available for subsequent chemistry" [53]. This is directly related to the mode of charge compensation in the PEM films: in the case of "intrinsic charge" compensation all the charges carried by the polyanion are compensated by those of the polycation in the bulk of the film. There are "noncompensated charges" only at the substrate/film and film/solution interfaces. In the case of "intrinsic charge" compensation the films do not incorporate ions from the electrolyte solution [58]. However, in this pioneering work, it was already demonstrated that some ion influx from the solution in the film can occur through reduction of viologen groups in [poly (viologen)-(sodium poly-4-styrene sulfonate]$_n$ films. Hence, such films can undergo an externally triggered transition from "intrinsic charge compensation" to "extrinsic charge compensation". On the other hand some films made from other polyelectrolytes are "extrinsically charge compensated": they have to incorporate counter-ions from the solution to

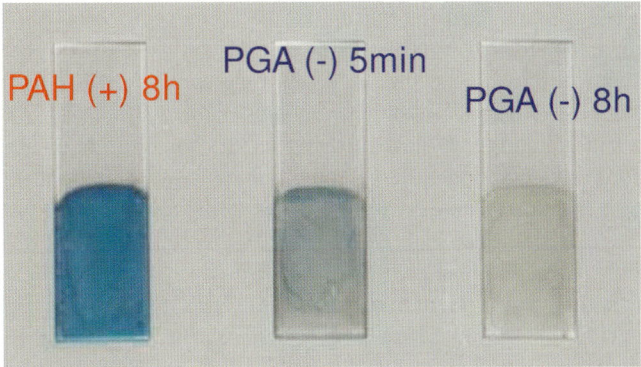

**Figure 28.1** Controlled incorporation of PB in (PAH-PGA)$_{10}$ films deposited on quartz slides and subsequently put in contact with PAH (at 1 mg mL$^{-1}$ in the presence of Tris-NaCl buffer) for 8 h, with PGA (at 1 mg mL$^{-1}$ in the presence of Tris-NaCl buffer) for 5 min and 8 h. The films were subsequently put in the presence of potassium hexacyanoferrate at 1 mM in the presence of the same buffer for 5 h, rinsed with buffer and exposed to a FeCl$_3$ aqueous solution at 1 mM for 1 h. Adapted from reference [57], with copyright from the American Chemical Society.

ensure their electroneutrality during their deposition. Most of the exponentially growing films belong in such a category and can, hence, be used as ion exchange membranes [59, 60] allowing incorporation of ionic species and subsequent chemical processes allowing production of the nanomaterials of interest. Joly *et al.* achieved the production of Ag nanoparticles inside PEM films through infiltration of Ag$^+$ cations into (PAH-PAA)$_n$ films (PAA stands for poly(acrylic acid)) and their further reduction to Ag using H$_2$ as the reducing agent. The obtained nanoparticles had an average diameter of 2 nm. The pH of the solution used to deposit the PEM film was of prime importance: a high density of nanoparticles was obtained when PAA was deposited at pH values between 2.5 and 3.5 where it is only weakly ionized, whereas a low density of nanoparticles was obtained when PAA was dissolved at pH values higher than 3.5 [53]. Using the concept that nonbound negatively charged groups constitute binding sites for metallic cations, whereas films such as (PAH-PSS)$_m$ with pure intrinsic charge compensation do not posses such sites, the same group used [(PAH-PAA)$_n$-(PAH-PSS)$_m$] stacks to produce Bragg reflectors after selective reduction of silver in the (PAH-PAA)$_n$ stacks [61]. The size of the nickel nanoparticles produced through electroless deposition on palladium seeds present in (PAA-PAH)$_n$ films could be controlled by means of manipulation of the rate of electroless deposition [62]. The palladium seeds were produced through diffusion of tetraaminepalladium cations in the (PAH-PAA)$_n$ films and the subsequent reduction of these cations with hydrogen. Interestingly, when the PEM films were made from 15 layer pairs and an additional deposition of PAH (PAH was deposited at pH 7.5 whereas PAA was dissolved at pH 3.5), a gradient in nanoparticle size was observed with larger nanoparticles (14 nm in diameter) present at the film solution interface and a progressive decrease in size down to 3 nm near the substrate.

Dai and Bruening used a somewhat different approach to deposit composite PEM films containing silver nanoparticles: silver cations were precomplexed with poly (ethyleneimine) (PEI) via tertiary amino groups of the polycation. The PEI-$Ag^+$ complexes were then deposited in a step-by-step manner with polyacrylic acid (PAA). The $Ag^+$ cations were subsequently reduced to Ag using sodium borohydride [63]. In this approach the cations do not have to diffuse through the multilayer film because they are homogeneously distributed across the whole film, being initially bound to the PEI chains.

As another approach, it has been shown that a layer-by-layer process with a polyamine and a water soluble precursor of titanium dioxide, Ti(IV) (bis ammonium lactato) dihydroxyde (TiBisLac) allows production of a film containing titanium dioxide [64]. This work was inspired by the fact that polyamines are able to catalyze biosilica formation [65] as well as the formation of $TiO_2$ in solution [66]. Surprisingly, even if the whole process was performed at ambient temperature it was found that the obtained $TiO_2$ was partially crystalline, with a crystalline domain size of about 5 nm, as calculated from the width at half maximum of the diffraction peak using Scherrer's equation. The obtained anatase particles are monodisperse, which was attributed to a polycondensation process restricted in the loops of the adsorbed polyamine [67]. Similarly, the polycondensation of TiBislac was performed on a layer of sillafin proteins adsorbed at the surface of a $(PSS-PAH)_n$ film [68, 69].

The same concept of "reactive layer-by-layer" deposition also works to produce composite films containing silica. In this case, sodium silicate was deposited in an SBS manner with a whole set of polyamines [70]. Very recently, it was shown that composite films containing both silica and $TiO_2$ could be deposited using the reactive step-by-step deposition from blends containing sodium silicate and TiBisLac at different molar fractions but at a constant total concentration of inorganic precursor. Using UV–vis spectroscopy and X-ray photoelectron spectroscopy, it was found that silica was preferentially deposited in the LBL films with respect to $TiO_2$ [71]. The ionic strength at which the "reactive step-by-step" deposition is performed is of major importance, not only for the thickness increment per deposited "layer pair", which is common for films containing polyelectrolytes, but also for the properties of the films: they become progressively impermeable to hexacyanoferrate anions when the ionic strength of the dipping solutions increases. This was attributed to an increase in the mass fraction of $TiO_2$ in the film leading to a decrease in the amount of free amino groups of the PEI chains and, hence, a reduced amount of binding sites available for the negatively charged redox probe. Additionally, it was shown by means of electron microscopic tomography, that the distribution of $TiO_2$ particles is not uniform along the direction perpendicular to the plane of the film. This finding is highly surprising: the film growth being linear with the number of processing steps, one would have expected to have a regular and homogeneous distribution of $TiO_2$ in the film. On the basis of a time decrease in the zeta potential of the films ending with the PEI, the inhomogeneous distribution of $TiO_2$ in the film was attributed to a progressive migration, with a time scale of the order of a few hours, of the nanoparticles to the film/solution interface [72].

Finally, we also showed that $TiO_2$ can be produced in $(PLL-HA)_n$ films (HA being sodium hyaluronate and PLL being poly(L-lysine hydrobromide)) by simply allowing its precursor complex anion, TiBisLac, to diffuse in the architecture of the PEM film. Polycondensation then occurs through the presence of unpaired free amino groups in the film [73]. Again, the fact that the charge compensation of the PEM film is of extrinsic nature plays a major role in its ability to induce the polycondensation of TiBisLac into $TiO_2$.

PEM films have also been used as templates for biomineralization processes [74, 75]. Of particular interest, was the finding that the average conformation of the polypeptides in PEM films made from PLL as the polycation and a blend of poly(L-glutamic acid) (PGA) and poly(L-aspartic acid) (PAsp) can be modulated, and that this modulation has a marked influence on both the induction time preceding crystal growth and the morphology of the obtained crystals. Namely, $PEI-(PGA-PLL)_n$ films are rich in $\beta$ sheets whereas $PEI-(PAsp-PLL)_n$ are rich in $\alpha$ helixes. The films made from blends of PGA and PAsp contain a percentage of $\beta$ sheets that is higher than expected on the basis of the mole fraction of each polyanion in the film [76]. This mole fraction could be estimated by means of infrared spectroscopy in the totally attenuated reflection mode, ATR-FTIR spectroscopy (Figure 28.2). When the fraction of PGA in the film is above 50%, the amount of $\beta$ sheets remains constant but higher than in films made from PGA as the unique polyanion. At this critical concentration in PGA, the induction time preceding crystal growth is maximal and the shape of the obtained crystals is that of sand roses (Figure 28.2).

The scanning electron micrographs shown above the graph display the morphology of the calcium phosphate crystals obtained on the different films. The FTIR spectra of the calcium phosphates (not shown here) are consistent with the presence of hydroxyapatite, at all values of the PGA percentage in the film. Another fascinating investigation showing the possibility to use biomineralization concepts in PEM films is the work of Nam et al. [77]. In this investigation, aimed to produce lithium ion battery electrodes, M13 viruses were genetically modified so that all their main coat proteins carried an N-terminal tetraglutamate motif. This allowed easy sequestration of cobalt cations whose subsequent reduction and spontaneous re-oxidation in water allowed conversion in monodisperse $Co_3O_4$ oxide nanoparticles. The virus decorated nanowires were then deposited on multilayer films made from linear poly(ethyleneimine) and polyacrylic acid to produce films which have excellent ionic conductivity [78].

The addition of $(PLL-HA)_n$ films to dopamine solutions in the presence of basic oxygenated solutions allowed the introduction of melanine particles into the films. These melanine particles allowed a significant decrease of the chain mobility in the film, as quantified by means of fluorescence recovery after photobleaching, and a subsequent detachment of the whole film as a free standing membrane (Figure 28.3) [79]. The apparent change in the mechanical properties of such melanine-doped films, not yet quantified, is believed to be due to chemical crosslinking between free amino groups from PLL chains in the film and catechol groups at the surface of the melanine grains [80].

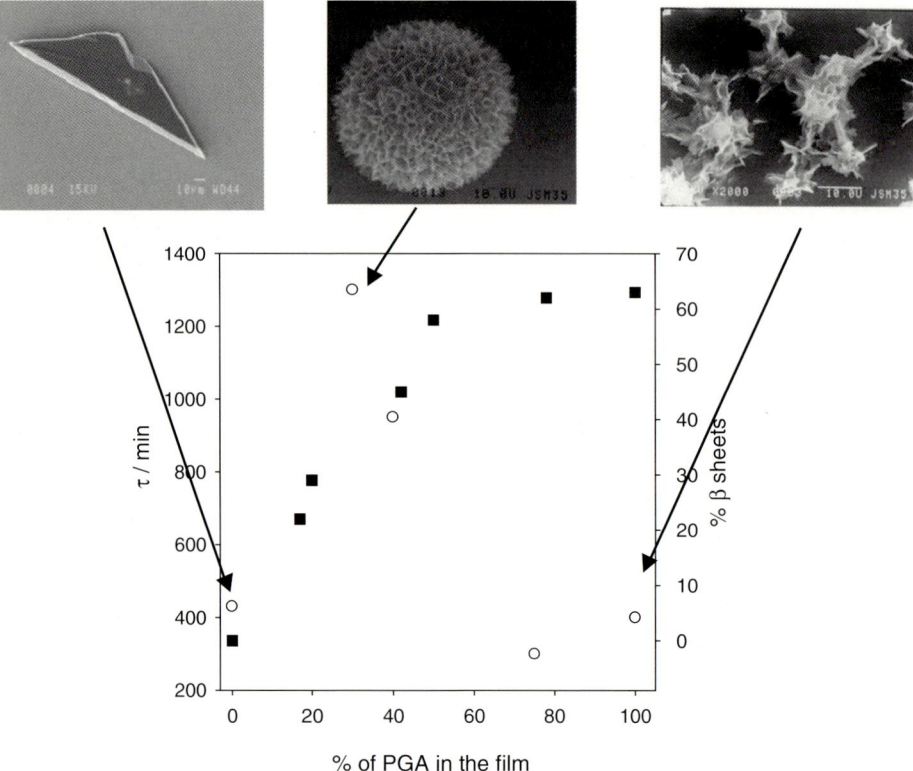

**Figure 28.2** (○) evolution of the induction time preceding crystal growth on PEI-(PGA$_x$-PAsp$_{1-x}$-PLL)$_6$ PEM films in contact with a flowing Tris (10 mM) + NaCl (0.15 M) solution at pH 7.5 containing 6 mM CaCl$_2$ and 6 mM sodium phosphate salts as a function of the percentage of PGA in the blended films. Such films were deposited on substrates from the alternated adsorption of poly(ethylene imine) (PEI), a mixture of poly-L-glutamic acid (PGA) (molar fraction $x$ in monomer units) and poly(-L-aspartic acid) (molar fraction 1-$x$ in monomer units). The film deposition and the nucleation kinetics were followed by means of a quartz crystal microbalance with dissipation monitoring. (■) evolution of the % of the amide band intensity (1600–1700 cm$^{-1}$ in ATR-FTIR spectroscopy) attributed to β sheets. The percentage of PGA in the film and the % of β sheets in the amide I band were determined from ATR-FTIR spectroscopy experiments. Adapted from reference [75], with copyright from the American Chemical Society.

## 28.5
## Diffusion of Colloids in Already Deposited Films

### 28.5.1
### Permeability of LBL Films Towards Ions, Small Drugs and Dyes

As already explained, PEM films can be characterized by the presence of small ions from the solution in contact with the film. Such films are "extrinsically charge

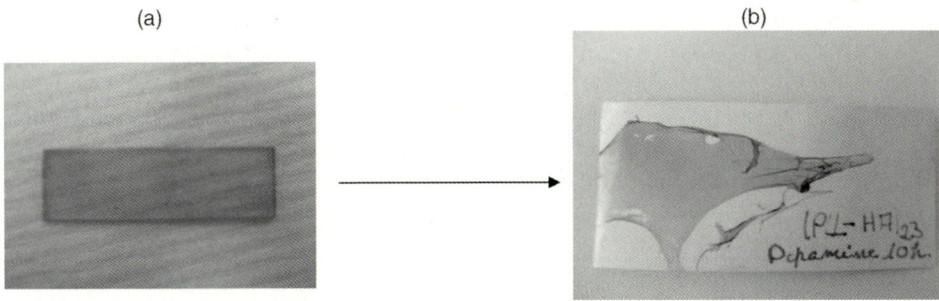

**Figure 28.3** (a) A (PLL-HA)$_{23}$ film deposited on a quartz slide and put in contact with an aerated dopamine (2 mg mL$^{-1}$) solution in the presence of 50 mM Tris buffer at pH 8.5 for 10 h and its subsequent detachment as a free-standing film after immersion in a 0.1 M hydrochloric acid containing solution (b). Adapted from reference [79] with copyright from the Royal Society of Chemistry.

compensated" as opposed to "intrinsically compensated" films where no detectable amounts of ions are present [58]. At first glance one could thus expect that only "extrinsically compensated" films could act as ion exchange membranes to accommodate new ionic species and to perform some chemistry with them. Unfortunately, the situation is not so clear-cut: "intrinsically compensated films" can be ion doped through an increase in the ionic strength of the solution in contact with the film [81]. This ion doping allows a significant increase in the chain mobility [82] in the film and a decrease in chain elasticity [83]. These findings mean that almost all linearly growing and internally compensated PEM films can be converted to externally compensated ones. It has also to be noted that, for a given set of polyelectrolytes, transition from linear to supralinear growth regimes can be achieved by either increasing the ionic strength [84] or the temperature [85]. In the case of films already incorporating small electrolytes from the external solution, a further increase [86] or decrease [87] in the ionic strength in the solution bathing the film induces some film swelling or contraction with the fascinating appearance of transient water cavities.

Another way to create some nonintrinsically compensated charges in PEM films is to remove some sacrificial moieties by post-treatment. Hence, PAA can be partially complexed with $Cu^{2+}$ cations to produce $Cu^{2+}$-templated PAA-PAH films [88]. Upon treatment with a ligand complexing $Cu^{2+}$, the films contain an excess of uncompensated carboxylic groups, increasing the selectivity of the membranes for $Cl^-$ versus $SO_4^{2-}$ ions in separation membranes made from such films [88].

Hence, it is possible to create "extrinsically compensated" sites in PEM films not containing small ions at the stage of their deposition. This offers many possibilities for the incorporation of charged dyes and drugs.

Burke and Barrett [89] showed that the loading as well as the release kinetics of the positively charged indoine blue and of the negatively charged chromotrope 2R in (PAH-HA)$_n$ films is strongly pH dependent.

In a similar way, we investigated the loading of two oppositely charged probes of similar size, $Cu^{2+}$ phtalocyanines, in (PLL-PGA)$_n$ or in (PLL-PGA)$_n$-PLL films at constant pH, namely 7.5. We found a strong preferential incorporation of the negatively charged tetrasulfonated Cu(II) phtalocyanine (CuPc-SO$_4$) with respect to the positively charged tetrapyridine analogue (by a factor of at least 60–70) This finding was independent of the nature of the last deposited polyelectrolyte and thus reflects the fact that the investigated PEM films possess a positive Donnan potential and behave as anion exchange membranes [90]. We concluded that such kinds of "extrinsically charge compensated films" behave similarly to hydrogels made from charged polyaminoacids [91]. In the same investigation it was shown that, as soon as the "binding sites" of (PLL-HA)$_n$ films were saturated with either CuPC-SO$_4$ or TiO$_2$ (produced from the polycondensation of TiBisLac as previously explained [73]), the films were no longer able to host the other competing species. Both species, CuPC-SO$_4$ and TiO$_2$ could be incorporated in the PEM films when present simultaneously in the buffer solution in contact with the (PLL-HA)$_{10}$ architecture [90].

It has to be noted that even if the influence of the charge of the molecule to be incorporated in the PEM film is of major importance it is not the only parameter playing a role. Indeed the water insoluble antiproliferative drug, paclitaxel, could be incorporated in (PLL-HA)$_n$ films [92]. In addition, LCSM showed that the PEM film was homogeneously filled with fluorescently labeled drug. The drug release rate from the film was controlled by means of a capping layer made from dye-impermeable layers made from PSS and PAH [92].

## 28.5.2
### Diffusion of Nanoparticles in and out of LBL Films

A study performed by Salloum and Schlenoff showed that proteins such as Bovine serum albumin (BSA) are able not only to adsorb on the surface of PEM films but also to diffuse in their bulk [93]. This observation was made by means of infrared spectroscopy in the attenuated total internal reflection mode, a technique that is able to sense molecules in the region of the substrat/-film interface in which an evanescent wave penetrates from the substrate. The (PDADMAC-PSS)$_n$ films were significantly thicker than the penetration depth of the IR wave from the crystal in the film. Hence, the observation of an absorption at 1600–1700 cm$^{-1}$, typical of the amide I band of proteins but absent from the IR signature of the pristine film, was proof that the protein, dissolved in a solution in contact with the film, was able to diffuse down to the substrate/film interface. The diffusion of proteins in PEM films, further confirmed in the case of bovine insulin diffusing in PEM films made from PDADMAC and PAA [94], illustrates the possibility of embedding proteins by incubation of the films in a protein-containing solution as a post-deposition treatment. This could be an elegant way to produce biosensors since many proteins have been shown to acquire increased thermal stability during their step-by-step embedding in PEM films [95] and to keep their enzymatic activity for a longer time than the same enzyme in solution [96]. It has been shown that many redox active proteins can be addressed electrochemically when deposited step-by-step in alternation with polyelectrolytes or redox active polymers [97, 98]. It has to be checked, in forthcoming studies, whether such proteins will also keep their conformation and can be electrochemically addressed after diffusion in already deposited PEM films. However, to reach such a goal one has to understand the basic parameters allowing the diffusion of a colloid in an already constituted PEM film. One can assume, at this level of knowledge, taking all the acquired knowledge in step-by-step assembled films into account [99, 100], that the most important parameters may be:

1) The internal structure of the PEM film, namely its ability to swell upon incorporation of charged or uncharged species.
2) The size and the shape of the colloidal particles to be incorporated in the PEM film
3) Their charge density and eventually the charge distribution.
4) The influence of external parameters like the ionic strength, pH and temperature of the solution containing the nanoparticles, particularly if the values of these parameters are different from those of the solutions used to deposit the PEM template.

Many proteins are known to have a very heterogeneous distribution of their surface charge and may hence not be ideal candidates for fundamental investigations aimed at investigating the diffusion of nanoparticles in PEM films. Nanoparticles, oxides, and quantum dots are more appropriate for such a goal because they can be produced in a way in which their size distribution is monomodal with a good monodispersity and a controlled surface charge density.

The first study reporting some partial engulfment of colloids in PEM films was that of Picart et al. [101]. This investigation showed that even very compact and linearly growing (PAH-PSS)$_n$ films can progressively integrate negatively charged latex particles. In this case, however, the influence of gravity cannot be ruled out as an explanation of the incorporation of the micrometer-sized colloids in the PEM films. In our opinion this surprising finding leads to the question as to whether it is possible to have diffusion of colloids even in strongly interdigitated films characterized by an "intrinsic charge compensation".

The first systematic study investigating the incorporation of either negatively or positively charged quantum dots, namely CdTe nanoparticles, into exponentially growing PEMs made from poly(diallyldimethyl ammonium chloride) (PDADMAC) and poly(acrylic acid) (PAA, deposited from 1% w/w solutions at pH 3) was performed by Srivistava et al. [40]. LCSM experiments showed that (PDAPMAC-PAA)$_{100}$ films, about 5 μm thick before contact with the nanoparticles, can be homogeneously filled with quantum dots after a few hours of incubation with a solution of the negatively charged CdTe particles (Figure 28.4a). The loading of the PEM film with such nanoparticles is always accompanied by a pronounced swelling of the film (Figure 28.4b). The film loading–unloading cycle in the presence of water at pH 9 is a fully reversible process (at least for three loading–release cycles). When the films are loaded in the presence of the NPs at pH 9 and the release kinetics is performed at pH 7, almost no nanoparticle release in the surrounding solution is observed. This has been explained by a significant reduction in the fraction of dissociated groups of PAA upon contact with a solution at pH 7, which has the effect of increasing the interaction strength between the nanoparticles and the film carrying a constant density of quaternary ammonium groups (the charge density

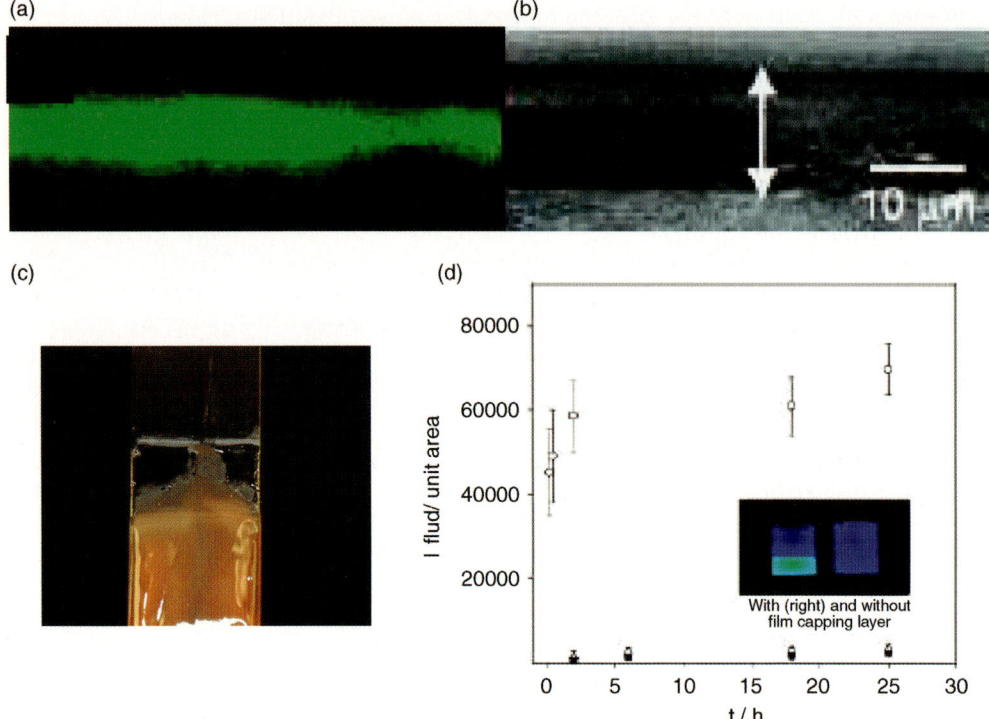

**Figure 28.4** (a) Laser confocal scanning microscopy image of a (PDADMAC-PAA)$_{100}$ film put in contact for 6 h with the solution containing the negatively charged CdTe nanoparticles at pH 9. (b) Bright film image of the same film after 24 h contact with water at pH 9 allowing a quantitative release of the nanoparticles. (c) Picture of a (PDADMAC-PAA)$_{100}$ film in which the lower part was immersed for 6 h in a solution containing the negatively charged CdTes whereas the upper part was left in air, showing that the incorporation of nanoparticles induces an important swelling of the PEM template. (d) Release kinetics of the negatively charged CdTe nanoparticles from (PDADMAC-PAA)$_{100}$ films as measured by fluorescence spectroscopy in contact with water at pH 9 (○), pH 7 (■), pH 9 but the film being capped with an impermeable (PDADMAC-PSS)$_{10}$ film before the beginning of the release kinetics (△). The inset shows the fluorescence of the solutions at pH 9 in contact with the films without (left) and with a capping layer (right) at the end of the release kinetics. Adapted from reference [40] with copyright from the American Chemical Society.

of PDADMAC is pH independent). The CdTe release observed at pH 9 could be almost totally suppressed when the nanoparticle-loaded films were capped with a linearly growing and impermeable (PDADMAC-PSS)$_{10}$ film (Figure 28.4c) [40].

Similar results were obtained by Wang *et al.* who showed that the negatively charged dye Methyl orange, as well as mercaptoacetic capped CdTe nanoparticles can be loaded up to a high concentration in PEM films made from dextran-modified PAH and PSS. The loaded dye and nanoparticles could be released slowly in 0.154 M sodium chloride solutions [102].

Similarly, silver colloids can diffuse in (PDADMAC-PAA)$_{30}$ films. The obtained composite films have been shown to act as highly sensitive substrates for surface enhanced Raman scattering, not only because of the known surface enhancement effect of silver but also because the molecules to be analyzed are able to concentrate in the film [103]. Note that, in order to have correct organization of the used silver nanoparticles (5 nm in diameter) in the film, it was mandatory to immerse the film in an aqueous solution of trisodium citrate before incubation in the nanoparticles-containing solution. The diffusion of the silver nanoparticles in the (PDADMAC-PAA)$_n$ films was slow, as followed by UV–vis spectroscopy: the steady state absorbance value of the localized surface plasmon resonance band of silver at 434 nm was only found after 24 h. Transmission electron microcopy images of the cross-sectional view of the film showed that the silver nanoparticles invaded the film only in its first 150 nm at the film/solution interface. This finding of inhomogeneous film filling with silver nanoparticles (the pristine film thickness was about 1 μm) is in contrast to the homogeneous filling of the same kind of PEM films with thioglycolic acid-capped CdTe nanoparticles [40].

Single-walled carbon nanotubes capped with either sodium dodecyl sulfate (NT1) with poly(acrylic acid) (NT2) or with positively charged cetyltrimethylammonium bromide (NT3) can also diffuse in (PDADMAC-PAA)$_n$ films. The incorporation of the NT1 dissolved at pH 4.8 in the films is extremely slow and even after 7 days of incubation in the NT1 suspension, a gradient in nanotubes was found through the film's section. No release from the film was observed at any pH, contrary to the CdTe nanoparticles that are released from the (PDADMAC-PAA)$_n$ films at pH 9 but not at pH 7 [40]. To obtain an almost homogeneous filling of the films, they had to be detached from the silica substrates as free-standing membranes to ensure incorporation of the modified SWNT from both sides of the film. The incorporation of the NT1 in the PEM films induced tremendous swelling as well as a marked change in the electrical resistance versus temperature behavior (Figure 28.5) [104]. In strong analogy with the positively charged CdTe nanoparticles, no detectable penetration of the positively charged NT3 in the PEM film was found, confirming again that the positive Donnan potential of these films plays a major role for the incorporation of nanoparticles. It was also found that rigid and negatively charged CdTe nanowires do not penetrate the films, meaning that the shape as well as the rigidity and the charge density contribute to explaining the penetration of anisotropic colloids in such exponentially growing PEM films [104].

Many investigations have described the possibility of using step-by-step assembly to immobilize functional polyoxometallates in thin films [105, 106] to obtain films with many possible applications, for instance electrochromic devices with fast response times [107]. In order to investigate the possibility of loading PEM films by post-incubation with a polyoxometallate-containing solution, Ball et al. [108] used the negatively charged heteropolytungstate $[H_7P_8W_{48}O_{184}]^{33-}$ (which will be denoted as $P_8W_{48}$ for simplification) and let it interact with an exponentially growing PEM known to have a positive Donnan potential, namely films made from HA and PLL. They showed that the film loading of PEI-(HA-PLL)$_9$ -HA films with $P_8W_{48}$ ($10.8 \times 10^{-9}$ A nm$^{-1}$ as determined by cyclic voltammetry) is similar to those of

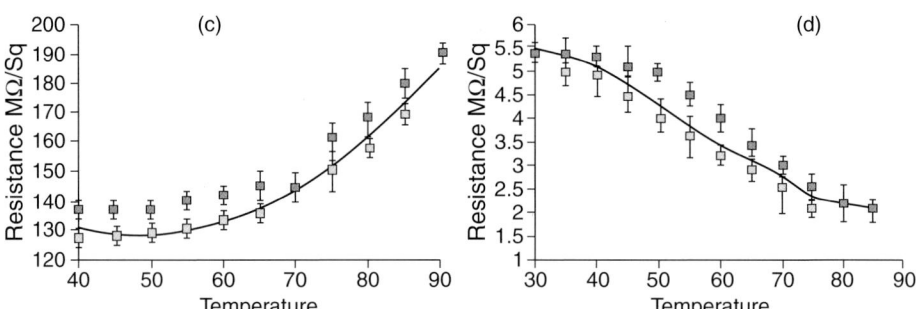

**Figure 28.5** (a) Top view of a (PDADMAC-PAA)$_{200}$ film put in contact with the NT1-containing solution for 1 d. (b) The same film in side view showing the important swelling of the film after incorporation of NT1. (c, d) Temperature dependence of the area resistance of films incubated with NT1 and NT2, respectively. The grey symbols correspond to the experiments performed with increasing temperature whereas the dark grey symbols correspond to those performed with decreasing temperature. Adapted from reference [104], with copyright from the American Chemical Society.

(PLL-P$_8$W$_{48}$)$_n$ films (37.4 × 10$^{-9}$ A nm$^{-1}$). Of the highest interest was the finding that the incorporation of P$_8$W$_{48}$ reached a maximal value when the film was put in contact with a solution at 2.4 × 10$^{-5}$ M whereas the incorporation was lower in contact with P$_8$W$_{48}$ solutions at 1.2 × 10$^{-6}$ and 1.2 × 10$^{-4}$ M. This finding suggests that the filling of the PEM matrix is not homogeneous across the whole thickness of the film. This needs to be investigated further. Anyway, films presenting a gradient in their sensing molecules, P$_8$W$_{48}$ is a useful catalyst for the reduction of nitrite anions, may present fascinating perspectives for new generations of sensors. In addition the immobilization of polyoxometallates in thin films offers a plethora of other possible applications [109].

## 28.6 Emerging Properties of Films Filled with Nanoparticles by the Post-incubation Method

### 28.6.1 Production of Anisotropically-Coated Colloids

It has been shown that the partial embedding of silica spheres (4.8 µm in diameter) coated with a thin and rigid (PDADMAC-PSS)$_4$ PEM in a (PLL-HA)$_{12}$-PLL film allows the production of anisotropic multicompartment capsules. With that aim, the partially embedded particles were put in contact with smaller (PDADMAC-PSS)$_4$-PDAMAC-coated silica particles (0.5 µm in diameter). The (PLL-HA)$_{12}$-PLL film acted as a mask, allowing the adsorption of the smaller colloids only on the solution-exposed part of the larger, partially embedded colloids. The obtained Janus particles could then be released by desorption from the (PLL-HA)$_{12}$-PLL template in diluted (0.01 M) sodium hydroxide solution. Fine tuning of the pH allows detachment of the capsules without decomposition of the underlying matrix, eventually allowing multistep production of Janus particles [110].

### 28.6.2 Sensing and Controlled Release Applications

(PDADMAC-PAA)$_{30}$ films pretreated with sodium citrate before the incorporation of silver nanoparticles enabled the acquisition of the first Raman spectra of 2-benzoyl-dibenzo-$p$-dioxin with a detection level down to $10^{-8}$ M in solution. This exceptional sensitivity was due to the ability of the dioxin to concentrate in the PEM film in close proximity to the silver nanoparticles [103].

### 28.6.3 Electrical Conductivity

The electrical resistance per square area of (PDADMAC-PAA)$_{200}$ films progressively decreased when the incubation time of the PEM films with the NT1 suspension increased from 6 h (260 kΩ sq$^{-1}$) to 7 d (3 kΩ sq$^{-1}$) [104]. The replacement of NT1 (nanotubes coated with SDS) by NT2 (nanotubes coated with PAA) had a less marked effect on the increase in the film's conductivity. This was due to a lower penetration of NT2 (about 2–3 µm) than NT1 in the film depth (20–25 µm). This difference in penetration depth was attributed to the higher flexibility of the SDS-coated nanotubes (NT1).

Most interestingly, the temperature dependence of the area resistance of the (PDADMAC-PAA)$_n$ films loaded with NT1 and NT2 was markedly different (see

Figure 28.5c and d). The films loaded with NT1 were characterized by a positive $dR/dT$ value and hence behave as a resistor, whereas the films loaded with NT2 seem to behave as a semiconductor with a negative $dR/dT$ coefficient. However $I$ versus $V$ curves for both films displayed an ohmic behavior. The authors ascribed the difference in the temperature dependence of the area resistance to a difference in the thermal activation modes of the films depending on their loading with NT1 or NT2 [104].

## 28.7
## Conclusions and Perspectives

This chapter has shown that it is possible to load polyelectrolyte multilayer films with nanoparticles through a post-incubation method either by letting precursor ions diffuse in the film [53–56] and then be further transformed (either by reduction, controlled precipitation or sol–gel chemistry) into nanoparticles or by diffusion of the already synthesized colloids in the bulk of the film [40, 102–104, 108]. In the first case, it is now well established that the binding of ions from the solution to available binding sites of the film is possible if the film carries groups whose charge is compensated by counter ions from the solution ("extrinsic charge compensation") and not from charged groups of the oppositely charged polyelectrolyte ("intrinsic charge compensation"). It appears, however, that the reasons why colloids, even very anisoptropic ones such as single-walled carbon nanotubes, diffuse in PEM films are not well known. The charge of the nanoparticles to be incorporated is important [40, 104] but is not the only reason for successful diffusion and retention of the species of interest. Hence, in order to use this alternative way to produce composite films in a step-by-step manner and subsequent functionalization with nanoparticles, nanowires or proteins (and eventually DNA or RNA), some systematic studies will have to be performed to better understand which are the most important properties of a PEM film for accommodation of nanoparticles by diffusion. What are the properties required for the molecules of interest to be accommodated and retained in such films? How can we trigger the controlled release from such films without disintegration of the template film?

It is not sure that the answer to these questions will be trivial: initially one could have expected that nanoparticles or proteins would only diffuse in exponentially growing films characterized by a high level of hydration, a high mobility of the constituent chains, and the presence of "extrinsically compensated sites". The finding that bovine serum albumin is able to diffuse in linearly growing (PDADMAC-PSS)$_n$ films [93] contradicts the general validity of the assumptions made previously.

Of the highest interest will be the possibility to control the diffusion rate of colloids in PEM films in order to produce composite films displaying *a gradient in their composition along the direction perpendicular to the substrate*. Such films will be of high interest, not only for bio-applications but also for the development of new generations of sensors. The control of such gradients in thin films will need collaboration with experts in analytical depth profiling techniques (for instance dynamic time of flight secondary ion mass spectrometry).

**Acknowledgements**

The authors' research is supported by the "Suprabat" and "Captochem" grants, Luxembourg.

**References**

1. Iler, R.K. (1966) *J. Colloid Interface Sci.*, **21**, 569–582.
2. Nuzzo, R.G. and Allara, D.L. (1983) *J. Am. Chem. Soc.*, **105**, 4481.
3. Decher, G., Hong, J.D., and Schmitt, J. (1992) *Thin Solid Films*, **210**, 831–835.
4. Decher, G. and Hong, J.D. (1991) *Ber. Bunsenges. Phys. Chem.*, **95**, 1430–1434.
5. Sukhorukov, G.B., Mohwald, H., Decher, G., and Lvov, Y.M. (1996) *Thin Solid Films*, **284**, 220–226.
6. Jessel, N., Oulad-Abdelhani, N., Meyer, F., Lavalle, P., Haïkel, Y., Schaaf, P., and Voegel, J.-C. (2006) *Proc. Natl. Acad. Sci. U.S.A.*, **103**, 8618–8621.
7. Lvov, Y., Ariga, K., Ichinose, I., and Kunitake, T. (1995) *J. Chem.Soc., Chem. Commun.*, 2313–2314.
8. Lvov, Y., Haas, H., Decher, G., Mohwald, H., Mikhailov, A., Mtchedlishvily, B., Morgunova, E., and Vainshtein, B. (1994) *Langmuir*, **10**, 4232.
9. Yoo, P.J., Nam, K.T., Qi, J.F., Lee, S.K., Park, J., Belcher, A.M., and Hammond, P.T. (2006) *Nature Mater.*, **5**, 234–240.
10. Michel, M., Vautier, D., Voegel, J.-C., Schaaf, P., and Ball, V. (2004) *Langmuir*, **20**, 4835–4839.
11. Michel, M., Izquierdo, A., Decher, G., Voegel, J.-C., Schaaf, P., and Ball, V. (2005) *Langmuir*, **21**, 7854–7859.
12. Volodkin, D.V., Schaaf, P., Mohwald, H., Voegel, J.C., and Ball, V. (2009) *Soft Matter*, **5**, 1394–1405.
13. Katagiri, K., Hamasaki, R., Ariga, K., and Kicuchi, J. (2002) *J. Am. Chem. Soc.*, **124**, 7892–7893.
14. Katagiri, K., Hamasaki, R., Ariga, K., and Kikuchi, J. (2002) *Langmuir*, **18**, 6709–6711.
15. Kleinfeld, E.R. and Ferguson, G.S. (1994) *Science*, **265**, 370–372.
16. Eckle, M. and Decher, G. (2001) *Nano Lett.*, **1**, 45–49.
17. Tang, Z.Y., Kotov, N.A., Magonov, S., and Ozturk, B. (2003) *Nature Mater.*, **2**, 413–418.
18. Podsiadlo, P., Kaushik, A.K., Arruda, E.M., Waas, A.M., Shim, B.S., Xu, J., Nandivada, H., Pumplin, B.G., Lahann, J., Ramamoorthy, A., and Kotov, N.A. (2007) *Science*, **318**, 80–83.
19. Mamedov, A.A., Kotov, N.A., Prato, M., Guldi, D.M., Wicksted, J.P., and Hirsch, A. (2002) *Nature Mater.*, **1**, 190–194.
20. Gunawidjaja, R., Ko, H., Jiang, C.Y., and Tsukruk, V.V. (2007) *Chem. Mater.*, **19**, 2007–2015.
21. Kulkarni, D.D., Choi, I., Singameneni, S.S., and Tsukruk, V.V. (2010) *ACS Nano*, **4**, 4667–4676.
22. Podsiadlo, P., Shim, B.S., and Kotov, N.A. (2009) *Coord. Chem. Rev.*, **253**, 2835–2851.
23. Kotov, N., Dékány, I., and Fendler, J.H. (1996) *Adv. Mater.*, **8**, 637–641.
24. Kotov, N.A., Dekany, I., and Fendler, J.H. (1995) *J. Phys. Chem.*, **99**, 13065.
25. Guldi, D.M., Rahman, G.M.A., Prato, M., Jux, N., Qin, S., and Ford, W. (2005) *Angew. Chem. Int. Ed.*, **44**, 2015–2018.

26 Agrios, A.G., Cesar, I., Comte, P., Nazeerudin, M.K., and Grätzel, M. (2006) *Chem. Mater.*, **18**, 5395–5397.

27 Forzani, E.S., Pérez, M.A., Teijelo, M.L., and Calvo, E.J. (2002) *Langmuir*, **18**, 9867–9873.

28 Ferryra, N., Coche-Guérante, L., and Labbé, P. (2004) *Electrochim. Acta*, **49**, 477–484.

29 Decher, G. (1997) *Science*, **277**, 1232–1237.

30 Stockton, W.B. and Rubner, M.F. (1997) *Macromolecules*, **30**, 2717–2725.

31 Laschewsky, A., Wischerhoff, E., Denzinger, S., Ringsdorf, H., Delcorte, A., and Bertrand, P. (1997) *Chem-Eur. J.*, **3**, 34–38.

32 Sukhishvili, S.A. and Granick, S. (2000) *J. Am. Chem. Soc.*, **122**, 9550–9551.

33 Elbert, D.L., Herbert, C.B., and Hubbell, J.A. (1999) *Langmuir*, **15**, 5355–5362.

34 Picart, C., Lavalle, P., Hubert, P., Cuisinier, F.J.G., Decher, G., Schaaf, P., and Voegel, J.C. (2001) *Langmuir*, **17**, 7414–7424.

35 Picart, C., Mutterer, J., Richert, L., Luo, Y., Prestwich, G.D., Schaaf, P., Voegel, J.C., and Lavalle, P. (2002) *Proc. Natl. Acad. Sci. U.S.A.*, **99**, 12531–12535.

36 Lavalle, P., Gergely, C., Cuisinier, F.J.G., Decher, G., Schaaf, P., Voegel, J.C., and Picart, C. (2002) *Macromolecules*, **35**, 4458–4465.

37 Boulmedais, F., Ball, V., Schwinté, P., Frisch, B., Schaaf, P., and Voegel, J.-C. (2003) *Langmuir*, **19**, 440–445.

38 Lavalle, P., Picart, C., Mutterer, J., Gergely, C., Reiss, H., Voegel, J.C., Senger, B., and Schaaf, P. (2004) *J. Phys. Chem. B*, **108**, 635–648.

39 Picart, C., Mutterer, J., Arntz, Y., Voegel, J.-C., Schaaf, P., and Senger, B. (2005) *Microsc. Res. Tech.*, **66**, 43–57.

40 Srivastava, S., Ball, V., Podsiadlo, P., Lee, J., Ho, P., and Kotov, N.A. (2008) *J. Am.Chem. Soc.*, **130**, 3748–3749.

41 Kotov, N.A., Magonov, S., and Tropsha, E. (1998) *Chem. Mater.*, **10**, 886–895.

42 Priolo, M.A., Gamboa, D., Holder, K.M., and Grunlan, J.C. (2011) *Nano Lett.*, **10**, 4970–4974.

43 Lutkenhaus, J.L., Olivetti, E.A., Verploegen, E.A., Cord, B.M., Sadoway, D.R., and Hammond, P.T. (2007) *Langmuir*, **23**, 8515–8521.

44 Srivastava, S. and Kotov, N.A. (2008) *Acc. Chem. Res.*, **41**, 1831–1841.

45 Lee, D., Rubner, M.F., and Cohen, R.E. (2006) *Nano Lett.*, **6**, 2305–2312.

46 Pyrash, M. and Tieke, B. (2001) *Langmuir*, **17**, 7706–7709.

47 Schlenoff, J.B., Dubas, S.T., and Farhat, T. (2000) *Langmuir*, **16**, 9968–9969.

48 Popa, G., Boulmedais, F., Zhao, P., Hemmerlé, J., Vidal, L., Mathieu, E., Félix, O., Schaaf, P., Decher, G., and Voegel, J.-C. (2010) *ACS Nano*, **4**, 4792–4798.

49 Porcel, C.H., Izquierdo, A., Ball, V., Decher, G., Voegel, J.-C., and Schaaf, P. (2005) *Langmuir*, **21**, 800–802.

50 Lefort, M., Popa, G., Seyrek, E., Szamocki, R., Félix, O., Hemmerlé, J., Vidal, L., Voegel, J.-C., Boulmedais, F., Decher, G., and Schaaf, P. (2010) *Angew. Chem. Int. Ed.*, **49**, 10110–10113.

51 Miller, M.D. and Bruening, M.L. (2004) *Langmuir*, **20**, 11545–11551.

52 Krasemann, L. and Tieke, B. (2000) *Langmuir*, **16**, 287–290.

53 Joly, S., Kane, R., Radzilowski, L., Wang, T., Wu, A., Cohen, R.E., Thomas, E.L., and Rubner, M.F. (2000) *Langmuir*, **16**, 1354–1359.

54 Wang, T.C., Rubner, M.F., and Cohen, R.E. (2003) *Chem. Mater.*, **15**, 299–304.

55 Zhang, L., Dutta, A.K., Jarero, G., and Stroeve, P. (2000) *Langmuir*, **16**, 7095–7100.

56 Dante, S., Hou, Z., Risbud, S., and Stroeve, P. (1999) *Langmuir*, **15**, 2176–2182.

57 Laugel, N., Boulmedais, F., El Haitami, A.E., Rabu, P., Rogez, G., Voegel, J.C., Schaaf, P., and Ball, V. (2009) *Langmuir*, **25**, 14030–14036.

58 Schlenoff, J.B., Ly, H., and Li, M. (1998) *J. Am. Chem. Soc.*, **120**, 7626–7634.

59 Hübsch, E., Fleith, G., Fatisson, J., Labbé, P., Voegel, J.-C., Schaaf, P., and Ball, V. (2005) *Langmuir*, **21**, 3664–3669.

60 Noguchi, T. and Anzai, J.-i. (2006) *Langmuir*, **22**, 2870–2875.

61 Nolte, A.J., Rubner, M.F., and Cohen, R.E. (2004) *Langmuir*, **20**, 3304–3310.

62 Wang, T.C., Rubner, M.F., and Cohen, R.E. (2003) *Chem. Mater.*, **15**, 299–304.

63 Dai, J. and Bruening, M.L. (2002) *Nano Lett.*, **2**, 497–501.

64 Shi, X., Cassagneau, T., and Caruso, F. (2002) *Langmuir*, **18**, 904–910.

65 Coradin, T., Durupthy, O., and Livage, J. (2002) *Langmuir*, **18**, 2331–2336.

66 Sumerel, J.L., Yang, W., Kisailus, D., Weaver, J.C., Choi, J.H., and Morse, D.E. (2003) *Chem. Mater.*, **15**, 4804–4809.

67 Laugel, N., Hemmerlé, J., Ladhari, N., Arntz, Y., Gonthier, E., Haikel, Y., Voegel, J.-C., Schaaf, P., and Ball, V. (2008) *J. Colloid Interface Sci.*, **324**, 127–133.

68 Kharlampieva, E., Slocik, J.M., Singamaneni, S., Poulsen, N., Kroger, N., Naik, R.R., and Tsukruk, V.V. (2009) *Adv. Funct. Mater.*, **19**, 2303–2311.

69 Kharlampieva, E., Jung, C.M., Kozlovskaya, V., and Tsukruk, V.V. (2010) *J. Mater. Chem.*, **20**, 5242–5250.

70 Laugel, N., Hemmerle, J., Porcel, C., Voegel, J.C., Schaaf, P., and Ball, V. (2007) *Langmuir*, **23**, 3706–3711.

71 Ball, V., Dahéron, L., Arnoult, C., Toniazzo, V., and Ruch, D. (2011) *Langmuir*, **27**, 1859–1866.

72 Ladhari, N., Ringwald, C., Ersen, O., Ileana, F., Hemmerlé, J., and Ball, V. (2011) *Langmuir*, **27**, 7934–7943.

73 Zouari, R., Michel, M., Di Martino, J., and Ball, V. (2010) *Mater. Sci. Eng. C*, **30**, 1291–1297.

74 Ngankam, P.A., Lavalle, P., Voegel, J.C., Szyk, L., Decher, G., Schaaf, P., and Cuisinier, F.J.G. (2000) *J. Am. Chem. Soc.*, **122**, 8998.

75 Ball, V., Michel, M., Boulmedais, F., Hemmerlé, J., Haikel, Y., Schaaf, P., and Voegel, J.-C. (2006) *Cryst. Growth Des.*, **6**, 327–334.

76 Debreczeny, M., Ball, V., Boulmedais, F., Szalontai, B., Voegel, J.C., and Schaaf, P. (2003) *J.Phys. Chem. B*, **107**, 12734–12739.

77 Nam, K.T., Kim, D.W., Yoo, P.J., Chiang, C.-Y., Meethong, N., Hammond, P.T., Chiang, Y.-M., and Belcher, A.M. (2006) *Science*, **312**, 885–888.

78 Delongchamp, D.M. and Hammond, P.T. (2003) *Chem. Mater.*, **15**, 1165.

79 Bernsmann, F., Richert, L., Senger, B., Lavalle, P., Voegel, J.C., Schaaf, P., and Ball, V. (2008) *Soft Matter*, **4**, 1621–1625.

80 Lee, H., Rho, J., and Messersmith, P.B. (2009) *Adv. Mater.*, **21**, 431–434.

81 Fahrat, T.R. and Schlenoff, J.B. (2003) *J. Am. Chem. Soc.*, **125**, 4627–4636.

82 Jomma, H.W. and Schlenoff, J.B. (2005) *Macromolecules*, **38**, 8473–8480.

83 Jaber, J.A. and Schlenoff, J.B. (2006) *J. Am. Chem. Soc.*, **128**, 2940–2947.

84 Laugel, N., Betscha, C., Winterhalter, M., Voegel, J.C., Schaaf, P., and Ball, V. (2006) *J. Phys. Chem. B*, **110**, 19443–19449.

85 Salomäki, M., Vinokourov, I.A., and Kankare, Y. (2005) *Langmuir*, **21**, 11232–11240.

86 Mjahed, H., Voegel, J.-C., Senger, B., Chassepot, A., Rameau, A., Ball, V., Schaaf, P., and Boulmedais, F. (2009) *Soft Matter*, **5**, 2269–2276.

87 Mjahed, H., Cado, G., Boulmedais, F., Senger, B., Ball, V., and Voegel, J.-C. (2011) *J. Mater. Chem.*, **21**, 8416–8421.

88 Balachandra, A.M., Dai, J., and Bruening, M.L. (2002) *Macromolecules*, **35**, 3171–3178.

89 Burke, S.E. and Barrett, C.J. (2004) *Macromolecules*, **37**, 5375–5384.

90 Sorrenti, E., Ball, V., Del Frari, D., Arnoult, C., Toniazzo, V., and Ruch, D. (2011) *J. Phys Chem. C*, **115**, 8248–8259.

91 Malmsten, M., Bysell, H., and Hansson, P. (2010) *Curr. Opin. Colloid Interface Sci.*, **15**, 435–444.

92 Vodouhê, C., LeGuen, E., Mendez Garza, J., Francius, G., Déjugnat, C., Ogier, J., Schaaf, P., Voegel, J.-C., and Lavalle F Ph. (2006) *Biomaterials*, **27**, 4149–4156.

93 Salloum, D.S. and Schlenoff, J.B. (2004) *Biomacromolecules*, **5**, 1089–1096.

94 Ladhari, N., Hemmerlé, J., Haikel, Y., Voegel, J.-C., and Ball, V. (2010) *Bio-Med. Mater. Eng.*, **20**, 217–223.

95 Schwinté, P., Ball, V., Szalontai, B., Haikel, Y., Voegel, J.-C., and Schaaf, P. (2002) *Biomacromolecules*, **3**, 1135–1143.

96 Derbal, L., Lesot, H., Voegel, J.-C., and Ball, V. (2003) *Biomacromolecules*, **4**, 1255–1263.

97 Dronov, R., Kurth, D.G., Moehwald, H., Scheller, F.W., and Lisdat, F. (2008) *Angew. Chem. Int. Ed.*, **47**, 3000–3003.

98 Sarauli, D., Tanne, J., Schafer, D., Schubart, I.W., and Lisdat, F. (2009) *Electrochem. Commun.*, **11**, 2288–2291.

99 Ariga, K., Hill, J.P., and Ji, Q. (2007) *Phys. Chem. Chem. Phys.*, **9**, 2319–2340.

100 Lavalle, P., Voegel, J.-C., Vautier, D., Senger, B., Schaaf, P., and Ball, V. (2011) *Adv. Mater.*, **23**, 1191–1221.

101 Picart, C., Sengupta, K., Schilling, J., Maurstad, G., Ladam, G., Bausch, A.R., and Sackmann, E. (2004) *J.Phys. Chem. B*, **108**, 7196–7205.

102 Wang, L., Wang, X., Xu, M., Chen, D., and Sun, J. (2008) *Langmuir*, **24**, 1902–1909.

103 Abalde Cela, S., Ho, S., Rodriguez-Gonzalez, B., Correa-Duarte, M.A., Alvarez-Puebla, R.A., Liz-Marzan, L.M., and Kotov, N.A. (2009) *Angew. Chem. Int. Ed.*, **48**, 5326–5329.

104 Srivastava, S., Podsiadlo, P., Critchley, K., Zhu, J., Qin, M., Shim, B.S., and Kotov, N.A. (2009) *Chem. Mater.*, **21**, 4397–4400.

105 Caruso, F., Kurth, D.G., Volkmer, D., Koop, M.J., and Müller, A. (1998) *Langmuir*, **14**, 3462–3465.

106 Liu, S., Kurth, D.G., Bredenkötter, B., and Volkmer, D. (2002) *J. Am. Chem. Soc.*, **124**, 12279–12287.

107 Liu, S., Möhwald, H., Volkmer, D., and Kurth, D.G. (2006) *Langmuir*, **22**, 1949–1951.

108 Ball, V., Bernsmann, F., Werner, S., Voegel, J.-C., Piedra-Garza, L.F., and Kortz, U. (2009) *Eur. J. Inorg. Chem.*, 5115–5124.

109 Liu, S. and Tang, Z. (2010) *Nano Today*, **5**, 267–281.

110 Delcea, M., Madaboosi, N., Yashchenok, A.M., Subedi, P., Volodkin, D.V., De Geest, B.G., Möhwald, H., and Skirttach, A.G. (2011) *Chem Comm.*, **47**, 2098–2100.

# 29
# Coupling Chemistry and Hybridization of DNA Molecules on Layer-by-Layer Modified Colloids
*Jing Kang and Lars Dähne*

## 29.1
## Introduction

Colloidal particles in the micrometer range are often applied in liquid suspensions for diagnostic and sensing applications [1, 2]. Compared to planar systems they offer a larger surface area per volume and a better mobility and distribution in the analyte solution, which ensures faster and geometrically unlimited specific binding of the target molecules. Simple separation procedures, like centrifugation or magnetic collection of the particles, enable their concentration at the point of signal detection. Therefore, particle-based methods have reached a high degree of sensitivity [3]. A desirable further increase requires a higher concentration of binding sites per particle volume (Figure 29.1a). This can be achieved either by using smaller particles, in the nanometer size range, or porous particles with a large inner surface area. However, while the first case yields increasing difficulties in the separation procedures, the second case limits the application, with decreasing pore size, to smaller molecules, thus excluding important target molecules such as proteins (antibodies) or DNA due to difficulties in diffusion [4].

Hence, core–shell particles are often applied, exhibiting a porous and rough surface area in the nanometer scale and a solid core (Figure 29.1b). Such particles have been prepared by coating solid particles with spacer molecules, like brushes, or with hydrogels [5, 6]. Due to the porous and rough coating in the nanometer range a large number of binding sites can be assembled and the diffusion of the target molecules is not hindered. It can be assumed that layer-by-layer (LbL) films [7, 8], that were developed 20 years ago on planar surfaces and transferred to colloidal templates just 13 years ago, may show similar properties [9–12].

The LbL coating, as described in other chapters of this book, results in three zones of the polyelectrolyte film [13–15]. While zone 1 on the substrate will not be taken into consideration, zone 2 in the middle of the LbL film still shows porosity due to a water content above 30%, dependent on the polyelectrolyte combination [16]. Therefore, small and medium-sized molecules will penetrate the zone[17]. In contrast, the

---

*Multilayer Thin Films: Sequential Assembly of Nanocomposite Materials*, Second Edition.
Edited by Gero Decher and Joseph B. Schlenoff.
© 2012 Wiley-VCH Verlag GmbH & Co. KGaA. Published 2012 by Wiley-VCH Verlag GmbH & Co. KGaA.

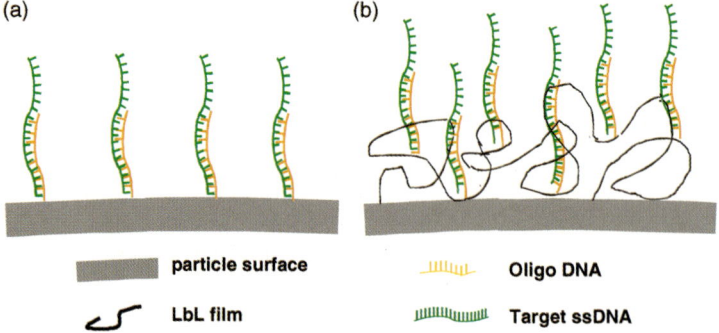

**Figure 29.1** Scheme of (a) hybridized DNA molecules on a solid particle surface and (b) hybridized DNA molecules on a core–shell structure with a nanoporous shell structure. (ssDNA = single stranded DNA).

outermost zone on the surface is highly charged and strongly swollen in water and can be accessed even by large protein or DNA molecules [18]. Therefore the LbL coating should offer in zone 3 a nanoporous and rough surface, similar to a coating with spacer molecules or with hydrogels. However, compared to these methods the LbL technology allows easy control of the surface parameters, like layer thickness, charge, binding sites, roughness and porosity, and it can be applied on almost every colloidal template surface [19–21]. Despite these advantages LbL particles have not been reported yet as diagnostic particles, probably due to the instability of an electrostatic assembly of the probe molecules and to increased nonspecific binding of proteins to the LbL surface.

We have tried to exploit LbL-coated particles for the covalent linkage of small oligonucleotide molecules (Oligos) as DNA diagnostic particles. We will report important parameters for their application, such as coupling method, density of coupled probe molecules, efficiency of DNA hybridization, non-specific binding, and stability under application conditions. From the viewpoint of chemistry the findings should also be valid for RNA molecules, although they have less stability due to enzymatic destruction.

## 29.2
## Materials and Methods

### 29.2.1
### Materials

Different polyelectrolyte combinations have been investigated for the coating and coupling procedure. Their structures are shown in Figure 29.2.

The oligonucleotides for the coupling procedure and for the hybridization process are listed in Table 29.1. The aminolinker is marked with –NH$_2$ and the dye labeling with rhodamine (Rho) or fluorescein (Flu) in either the 5′ or 3′ position.

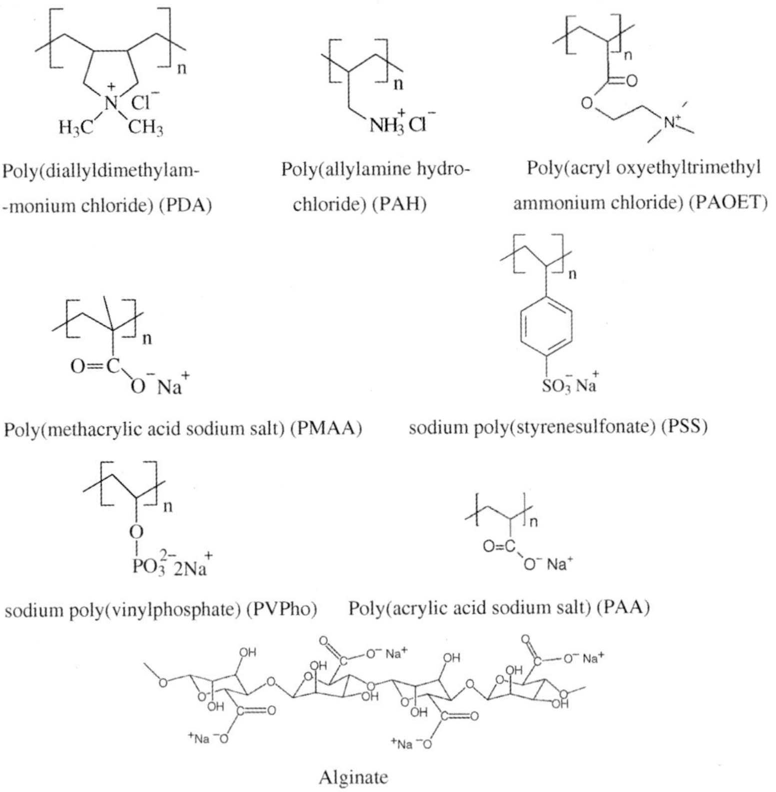

Poly(diallyldimethylam-
-monium chloride) (PDA)

Poly(allylamine hydro-
chloride) (PAH)

Poly(acryl oxyethyltrimethyl
ammonium chloride) (PAOET)

Poly(methacrylic acid sodium salt) (PMAA)

sodium poly(styrenesulfonate) (PSS)

sodium poly(vinylphosphate) (PVPho)  Poly(acrylic acid sodium salt) (PAA)

Alginate

**Figure 29.2** Structures of used polyelectrolytes.

## 29.2.2
## Methods

### 29.2.2.1 Coupling of Oligonucleotides onto LbL-Coated Colloidal Particles

DNA is a polyanion and can be assembled easily on cationic surfaces via electrostatic forces [22]. In order to suppress nonspecific binding of DNA molecules the LbL

**Table 29.1** Overview of Oligos used.

| Probe Oligos for covalent coupling | Abbreviation |
|---|---|
| 5'-NH$_2$-AAA AAA AAA AAA AAA AAA AAA-3' | A$_{21}$ |
| 5'-NH$_2$-AAT(Flu)AAA AAA AAA AAA AAA AAA-3' | FluA$_{21}$ |
| 5'-NH$_2$-CACTACGGTGCTGAAGCGACAAA-3' | — |
| Oligos for DNA hybridization | Abbreviation |
| 5'-TTT TTT TTT TTT TTT TTT TT-3' | T$_{20}$ |
| 5'-T(Rho)TT TTT TTT TTT TTT TTT TT-3' | RhoT$_{20}$ |
| 5'-TTT TTT TTT TTT TTT TTT TT(Rho)-3' | T$_{20}$Rho |
| 5'-TTT TTG TCG CTT CAG CAC CGT AGT G-3' | — |

surface has to be completely negatively charged; this excludes several coupling techniques based on primary amino or aldehyde groups on the particle surface [23]. Negatively charged LbL surfaces can be realized by polyelectrolytes based on sulfate, sulfonate, phosphate or carboxylate groups. While the first three functionalities show low reactivity and low p$K$ values, the carboxylate group loses its negative charge due to protonation at decreasing pH value. For covalent linkage of amino- to carboxyl groups the latter one has to be activated to form an amide bond. The activation can be performed in aqueous solution by 1-(3-dimethylaminopropyl)-3-ethylcarbodiimide hydrochloride (EDC) (Figure 29.3) or in water-free organic solvents by 1.1'-carbonyldiimidazol or diisopropyl carbodiimide (Figure 29.4). The subsequent reaction with the Oligos can be in the presence of water because of the lower reactivity toward

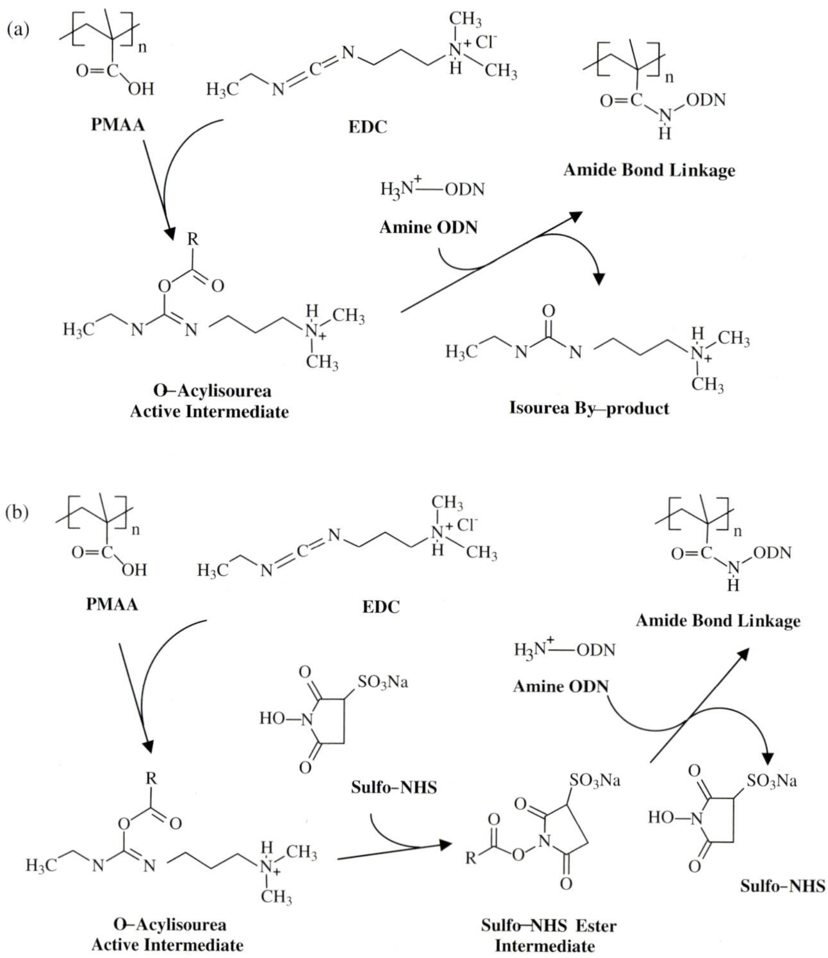

**Figure 29.3** Scheme of coupling mechanism of amino-Oligos(ODN) to PMAA in aqueous solution: (a) by crosslinker EDC only; (b) by EDC plus sulfo-NHS.

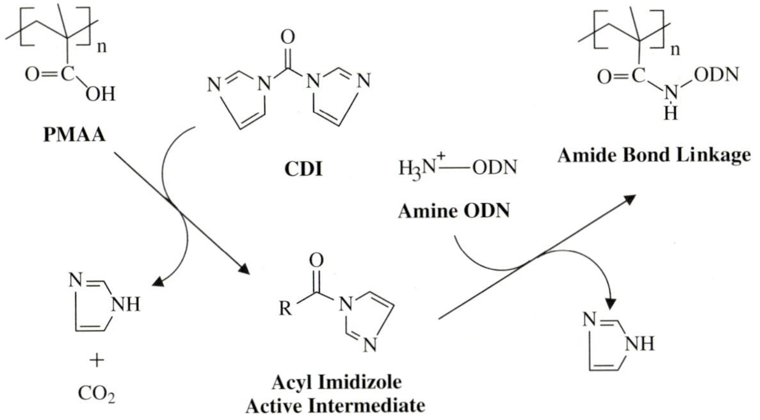

**Figure 29.4** Scheme of coupling mechanism of amino-Oligos (ODN) to PMAA in organic solvent by crosslinker CDI.

hydroxy groups of the activated acid groups than that of the activation agents. For the linkage of synthetic Oligos they were equipped on the end with an amino-function (Table 29.1).

**29.2.2.1.1. Coupling of Probe Oligos Onto LbL Particles by EDC Activation** Amino-modified Oligos were covalently bound to carboxylic groups on 4.3 µm LbL-coated silica particles by EDC alone, or together with N-hydroxysulfosuccinimide sodium salt (sulfo-NHS) (Figure 29.3), which should give higher coupling yield due to the higher stability of the formed succinimide [23]. In a one-step coupling process, amino-Oligos, EDC, sulfo-NHS and the LbL particles were mixed together at the same time. In a two-step coupling procedure, as the first step 10 µl EDC (50 mg ml$^{-1}$) and 10 µl sulfo-NHS (50 mg ml$^{-1}$) was used to activate 50 µl silica particles (2 wt%, 4.3 µm) coated with two layer pairs of (PDA/PMAA). After 30 min activation the particles were washed and amino-Oligos were added to the LbL particle suspension as the second step. The reaction concentration of Oligos was kept at 1.5 µM and the mixture was incubated at room temperature for 1 h before thorough washing with water. 50 mM borate buffer (pH 8.0) was used in the coupling step.

**29.2.2.1.2. Coupling of Probe Oligos Onto LbL Particles by 1.1′-carbonyldiimidazol (CDI) Activation** Amino-Oligos were also covalently bound to carboxylate groups on the LbL particles by mediation of 1.1′-carbonyldiimidazol in organic solvent DMSO (Figure 29.4). The LbL particles were washed with water-free DMSO five times for solvent exchange. 10 µl of 1.1′-carbonyldiimidazol solution (100 mg ml$^{-1}$ dissolved in water-free DMSO) was used to activate 50 µl silica particles (2 wt%, 4.3 µm) coated with two layer pairs of (PDA/PMAA). The activation of the LbL particles was allowed to proceed for 60 min at 35 °C before washing and addition of amino-Oligos (dissolved in DMSO and water mixture due to the poor solubility of the Oligos in water-free DMSO). The reaction concentration of Oligos was kept at 1.5 µM and the mixture was incubated at room temperature for 2 h before thorough washing with water.

#### 29.2.2.2 Hybridization of Complementary Oligos Onto the LbL-Oligo Particles

For hybridization between simple polyadenosine/polythymidine (poly dA/T) Oligos, the LbL-Oligo particles were mixed with the complementary Oligo in 10 mM HEPES buffer (pH 7.4, 500 mM NaCl) at room temperature ($c_{particle} = 2.5\%$) for 30 min. For hybridization of the Oligo sequence from Escherichia coli (*E. coli.*) the same buffer was used but first a temperature of 60 °C was applied for 15 min before cooling to room temperature.

For storage of the LbL-Oligo particles, 10 mM TRIS buffer containing 0.1% EDTA (pH 7.0) was used to prevent cleavage of the Oligo sequence by DNases that might be present in the suspension.

## 29.3 Results

### 29.3.1 One-Step versus Two- Step Coupling Process by EDC

In one-step coupling, the LbL particles, crosslinking agent EDC and the probe Oligo FluA$_{21}$ were mixed together. However, using this simple method the coupling efficiency was rather low. In a control experiment, it was observed that the aminolinked FluA$_{21}$ in solution reacted already with EDC in the absence of the particles causing strong reduction of the fluorescence intensity of the FluA$_{21}$ solution. This could be caused by activation of the carboxylate group in fluorescein or the phosphate groups from the Oligos by EDC, leading to a self-crosslinking of the Oligo via the amine group, thus reducing the coupling efficiency. Therefore, a two-step coupling process was developed, in which the probe Oligos were added in a second step after washing away the excess EDC when the activation on the particle surface was completed.

It is well known that EDC and the EDC-activated intermediate groups are hydrolyzed quickly in aqueous solution. Thus freshly prepared EDC was used and the time of washing between activation and Oligo addition in the two-step process was kept as short as possible. It is reported that in the presence of sulfo-NHS with EDC, a more stable NHS-ester is formed, which increases the coupling yield (Figure 29.3b). Therefore, the NHS stabilization was tried in the beginning. But as described later, this leads to increased unspecific binding.

### 29.3.2 Selection of Polyelectrolytes

Selection of the polyelectrolyte for LbL coating on the particles was considered under the following aspects:

1) Maximal coupling efficiency of probe Oligos
2) Negligible non-specific binding of non-target DNA on the LbL surface
3) Stability of polyelectrolyte layers under the hybridization conditions.

#### 29.3.2.1 Selection of Polycations

PDA ($M_w = 200-350$ kDa), PAOET ($M_w = 80-100$ kDa) and PAH ($M_w = 70$ kDa), which are commonly used in LbL film assembly, were tested. The former two consist of quaternary amine groups that do not react with activated carboxylate groups. Also the charge density of them is independent of the pH. However, LbL films containing PDA showed lower thermal stability, as reported by Köhler et al. [24]. PAH consists of primary amine groups that generally yield more stable LbL films due to additional hydrogen bonding. However, there could be a competition between the amine groups on PAH and the amino-modified probe Oligos to react with the activated carboxylate groups on the particle surface.

The different polycations have been combined with PMAA and the coupling efficiency has been investigated. 100 pmol of FluA$_{21}$ was used to couple onto 1 mg of LbL particles coated with four polyelectrolyte layers of (PAOET/PMAA)$_2$, (PDA/PMAA)$_2$ (PAH/PMAA)$_2$ and (PAH/PMAA/PDA/PMAA), respectively. The amount of coupled Oligo was calculated from non-coupled Oligo in the supernatant, determined by UV absorption spectroscopy. After an identical coupling process, the (PAOET/PMAA)$_2$ and (PDA/PMAA)$_2$ surfaces showed similar coupling capacity to probe Oligos. On the contrary, only one third of FluA$_{21}$ was coupled onto the particles coated with (PAH/PMAA)$_2$. This is consistent with the measurement of fluorescence intensity by CLSM (confocal laser scanning microscopy), which shows that the fluorescence intensity of (PAH/PMAA)$_2$-based LbL-FluA$_{21}$ particles was remarkably lower than that of the (PAOET/PMAA)$_2$ or (PDA/PMAA)$_2$-coated ones. (Figure 29.5)

The remarkably reduced coupling efficiency of FluA$_{21}$ onto the PAH-containing LbL film can be ascribed to the consumption of activated carboxylate groups due to crosslinking between PAH and PMAA within the LbL layers. This effect can be especially pronounced for the two-step coupling process used, because the NH$_2$-Oligos were added after the EDC activation process, whereas the crosslinking

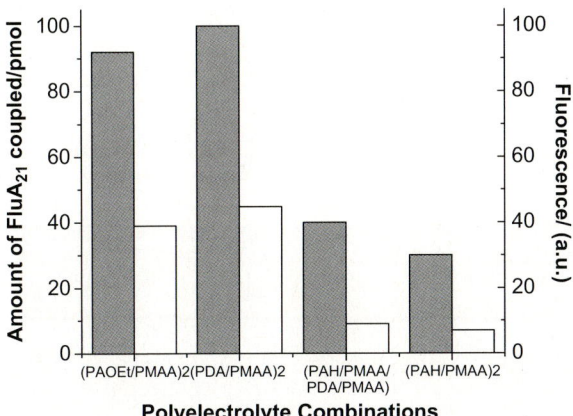

**Figure 29.5** Amount of FluA$_{21}$ that can be coupled onto particles coated with different polyelectrolyte combinations (gray) and fluorescence intensity of the respective LbL-FluA$_{21}$ particles measured by CLSM (white).

of LbL film already took place during the activation. It is worth mentioning that, even when PAH was assembled 3 layers below (i.e., only used as the first coating layer) the effect was still remarkable, pointing either to a rather strong interpenetration of the layers or to a coupling of the Oligos also to the second PMAA layer inside the film. This could be possible in respect to the Oligo size and permeability of the LbL film. Due to the strong decrease in activated carboxylate groups available for covalent linkage to probe Oligos, PAH in the LbL film is not suitable for Oligo coupling. Hence, quaternary polyamines have to be used for the LbL coating. (PDA/PMAA) coating has been selected because the alternative (PAOET/PMAA)$_2$ particles showed more agglomeration.

#### 29.3.2.2 Selection of Polyanions

For selection of polyanions, different polymers with carboxylate groups like alginate, PAA and PMAA, and also with other anionic functionalities such as PVPho and PSS (see Figure 19.2), were LbL coated alternately with the selected polycation PDA. The following polyelectrolyte combinations resulted in covalent coupling of probe Oligos: (PDA/PMAA)$_2$, (PDA/PSS/PDA/PMAA), (PDA/PSS)$_2$, (PDA/PVPho)$_2$. The last two were used in the first instance as control samples because no reaction with EDC was expected. 500 pmol of FluA$_{21}$ was used for coupling on 1 mg of each type of LbL particles under identical process conditions. Surprisingly, FluA$_{21}$ was coupled on all four types of LbL particles, including (PDA/PSS)$_2$ and (PDA/PVPho)$_2$-coated ones. It seems that the phosphate and sulfonate groups are also activated by EDC, though we have, so far, not found reports on activation of these groups by EDC. All four samples yielded similar coupling efficiency (250 pmol per mg particles). The fluorescence intensity of each sample was quantified by CLSM. Unfortunately the brightness of the particles in Figure 29.6 looks different, but that is due to the different voltages taken for the images. Taking that into account, all particles exhibit similar fluorescence intensities.

### 29.3.3
### Non-Specific Binding and Its Minimization

One important requirement of a DNA diagnostic kit is negligible non-specific binding, that is, where no non-target molecules bind onto the sensor. The

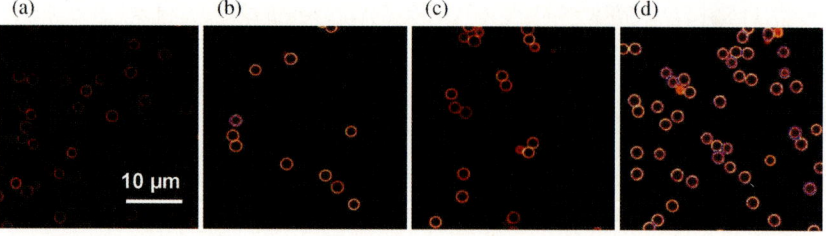

**Figure 29.6** CLSM images of FluA$_{21}$ coupled onto LbL particles coated with four polyelectrolyte layers of (a) (PDA/PVPho)$_2$, (photo multiplier voltage, PMV = 500); (b) (PDA/PSS)$_2$, (PMV = 600); (c) (PDA/PSS/PDA/PMAA), (PMV = 550); (d) (PDA/PMAA)$_2$, (PMV = 600).

non-specific binding test of LbL-Oligo particles was carried out in three ways: (i) Affinity of Oligos to the LbL particles without EDC activation; (ii) affinity of Oligos (without amine-linker modification) to the LbL particles after EDC (plus sulfo-NHS) activation; (iii) affinity of non-complementary Oligos to the LbL particles coupled with probe Oligos by EDC (plus sulfo-NHS).

First, the affinity of $RhoT_{20}$ (without amine-linker) to the LbL particles without any crosslinking agents was tested. No adsorption of Oligos on LbL film was expected due to the electrostatic repulsion between the negatively charged Oligo and the LbL particles with a polyanion as the outermost coating layer. However, non-specific binding of $RhoT_{20}$ on $(PDA/PVPho)_2$ coated particles was observed (Figure 29.7a), while no binding was found on the other polyelectrolyte combinations tested. Therefore, polyanions containing phosphate groups could not be used in the LbL film for specific coupling of Oligos.

Secondly, the affinity test of $RhoT_{20}$ (without amine-linker) on particles coated with $(PDA/PSS)_2$ or $(PDA/PMAA)_2$ after EDC activation was tested by CLSM. Strong affinity of $RhoT_{20}$ was found on both surfaces (Figure 29.7b and c). This can be ascribed to the positive charges on the surface produced by the crosslinking intermediate groups.

Thirdly, the affinity of $RhoA_{20}$ and $RhoT_{20}$ (both without amine-linker) to the LbL-$FluA_{21}$ particles, that is, particles coated with $(PDA/PSS)_2$ or $(PDA/PMAA)_2$ and covalently coupled with $FluA_{21}$ was tested (Figure 29.8).

Figure 29.8 shows that $RhoA_{20}$ also bound non-specifically onto the LbL-$A_{21}$ particles, which indicates that the complete binding of $RhoT_{20}$ onto the LbL-$A_{21}$ particles is not only via specific DNA hybridization but also partially by non-specific interactions. This strong non-specific affinity of Oligos on LbL particles after EDC plus sulfo-NHS activation was mostly caused by remaining activated groups that were not used by coupling the probe Oligos. It is known that the EDC active intermediates undergo slow hydrolysis in aqueous solution and return to carboxylate groups. As reported, this process should be rather fast, within minutes to one hour. But even after several days, non-specific binding was still observed for the particles activated by EDC plus sulfo-NHS. In order to prevent this non-specific binding, efforts were made to accelerate this process and to reverse the activated groups back to carboxylate groups after sufficient probe Oligos were coupled.

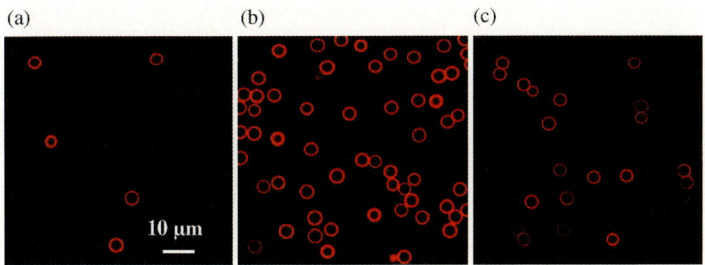

**Figure 29.7** Non-specific binding of $RhoT_{20}$ on particles coated with (a) $(PDA/PVPho)_2$ without EDC activation (PMV = 700); (b) $(PDA/PSS)_2$ after EDC activation (PMV = 700); (c) $(PDA/PMAA)_2$ after EDC (plus sulfo-NHS) activation (PMV = 650).

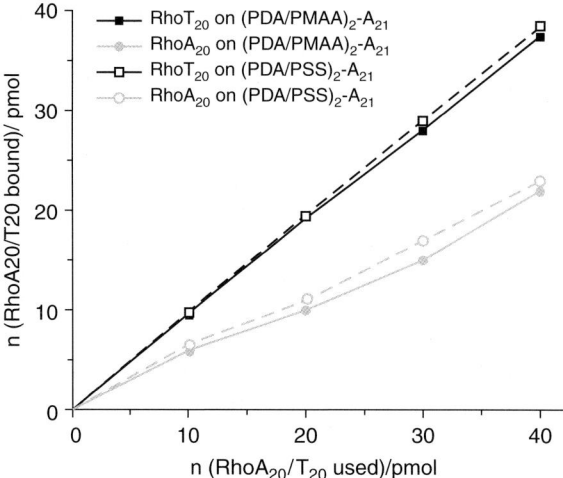

**Figure 29.8** Amount of RhoA$_{20}$ (mainly specific via DNA hybridization, squares) or RhoT$_{20}$ (non-specific, circles) bound onto LbL-A$_{21}$ particles coated with polyelectrolyte layers of (PDA/PMAA)$_2$ (solid line) and (PDA/PSS)$_2$ (dashed line) and activated by EDC plus sulfo-NHS.

Several substances to deactivate the EDC/sulfo-NHS ester groups was tried. The efficacy of deactivation was investigated by affinity test of NH$_2$-FluA$_{21}$ onto the LbL particle surface after the deactivation. 100 pmol NH$_2$-FluA$_{21}$ was used for coupling onto 1 mg of particles coated with certain polyelectrolyte combinations, activated by EDC (with or without sulfo-NHS) and then treated for 14 h with different solutions (shown in Table 29.2) for deactivation. The amount of NH$_2$-FluA$_{21}$ coupled onto each type of LbL particle after treatment was analyzed.

As shown in Table 29.2, complete quenching of the activated carboxyl groups was only found on NHS-ester free (PDA/PMAA)$_2$ particles that were treated with acetic acid (0.1 M, pH 3.0). In addition, the deactivation process was monitored by the ζ-potential of the particles at pH 7 (in 10 mM Tris buffer). The original (PDA/PMAA)$_2$ particles showed a potential of −58 mV. After activation with only EDC the ζ-potential

**Table 29.2** Amount of NH$_2$-FluA$_{21}$ bound (per pmol) on treated LbL particles[a].

| Deactivation Solution (pH) | LbL film activated by EDC with sulfo-NHS | | | LbL film activated by EDC only |
|---|---|---|---|---|
| | 1 M glutamic acid | 0.5 M NH$_2$OH·HCl | 0.5 M CH$_3$COOH | 0.1 M CH$_3$COOH |
| Polyelectrolytes in the LbL film | | | | |
| (PDA/PSS)$_2$ | 34 | 35 | 76 | 9.5 |
| (PDA/PSS/PDA/PMAA) | 39 | 29.5 | 10 | 2.5 |
| (PDA/PMAA)$_2$ | 40 | 27 | 5 | 0 |

a) The LbL particles were first activated by EDC (alone or with sulfo-NHS) and then treated with the listed solution overnight. ($c_{particles} = 0.5$ mg ml$^{-1}$).

reversed to +50 mV, explaining the high non-specific binding with the negatively charged Oligos. After incubation of the activated particles in 0.1 M acetic acid for 1 h it decreased to +19 mV, after 4 h to −23 mV and after 2 days to −30 mV. Longer incubation did not recover the original potential of the pure PMAA surface.

A further test revealed that a 4 h treatment with acetic acid solution was sufficient to achieve complete deactivation. In order to avoid non-specific binding, the prepared Oligo-particles were treated with 0.1 M acetate buffer at pH 3.0 for 4 h and then washed with water and hybridization buffer. Later, their affinity to $RhoT_{20}$ and $RhoA_{20}$ was tested. After addition of the $LbL-A_{21}$ particles to the solution of a non-complementary Oligo ($RhoA_{20}$), the supernatant fluorescence did not change, meaning that no coupling to the $LbL-A_{21}$ particles took place. In contrast, addition of complementary Oligo $RhoT_{20}$ to $LbL-A_{21}$ particles resulted in negligible fluorescence of the supernatant up to 100 nM $RhoT_{20}$, indicating efficient specific binding via DNA hybridization (Figure 29.9).

The activation of the carboxylate groups, especially without addition of SulfoNHS, by EDC yielded for smaller particles severe problems of irreversible agglomeration, because attachment of the cationic EDC neutralizes the anionic charges of the PMAA surface (see Figure 29.3) and destroys the electrostatic stabilization of the particles. Hence Oligoparticles below 1 µm in diameter could not be prepared by this technique.

### 29.3.4
### Coupling by 1,1′-Carbonyldiimidazol in Organic Solvent

Besides EDC, there are alternative activation agents for the coupling of amine-containing compounds to carboxylate groups, such as $N,N'$-carbonyldiimidazole,

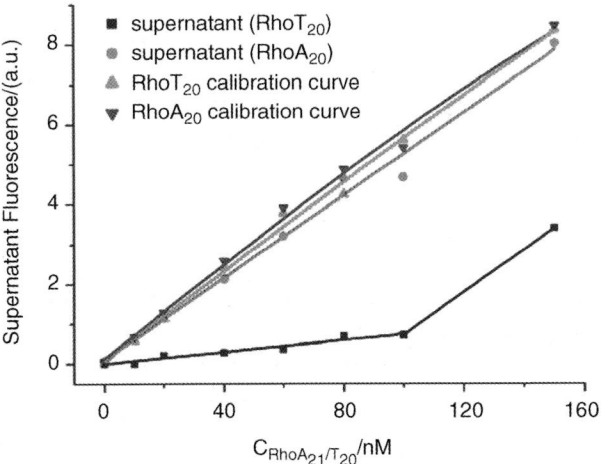

**Figure 29.9** Fluorescence intensity of supernatant after addition of different concentrations of $RhoA_{20}$ (circle) or $RhoT_{20}$ (square) to $LbL-A_{21}$ particles ($c_{particle} = 1$ mg ml$^{-1}$). The corresponding calibration curves of $RhoA_{20}$ (inverted triangle) or $RhoT_{20}$ (triangle) solutions without particles are shown.

diisopropyl carbodiimide and dicyclohexylcarbodiimide. All three coupling agents require water-free solvent for solubilization to avoid almost instant hydrolysis. The LbL particles were washed with water-free DMSO five times and activated with 1.1′-carbonyldiimidazol. After washing away the 1.1′-carbonyldiimidazol solution, the $NH_2$-Oligos were added in a solution of 80% DMSO and 20% water, due to the low Oligo solubility in pure DMSO. The $NH_2$-Oligos were successfully coupled to the particle surface. This method has the advantage that deactivation of excess esters is not necessary because the activated esters undergo faster hydrolysis than in the case of the EDC activation when the solvent is shifted back to aqueous solution. In addition, the problem of agglomeration of 1.1′-carbonyldiimidazol-activated particles is less pronounced. LbL-Oligo particles with diameter smaller than 1.0 μm could be prepared only by the 1.1′-carbonyldiimidazol method.

The coupling efficiency was slightly lower than that achieved by EDC activation in aqueous solution, most likely due to the presence of water molecules in the LbL film that hydrate the remaining charges. Therefore, the solvent exchange inside the layers might not be complete. In addition, the higher reactivity of the 1.1′-carbonyldiimidazol activated LbL particles could result in a less selective reaction with the amine groups, meaning that more groups are deactivated by the hydroxy groups of the 20% water in the second step.

### 29.3.5
**Number of Coating Layers**

The amount of probe Oligos that can be coupled onto the LbL particles increases with the number of coating layers. One double layer (PDA/PMAA) couples maximal 150 pmol Oligos per mg particles, two double layers couple 250 pmol and three double layers couple 325 pmol. The coupled amount increases. It is not quite clear if the reason for that is an exponential growth of the film, leading to more carboxylate groups on the particle outside, or if the layers underneath also couple Oligos due to the high permeability of the layers for such rather small molecules.

Although use of the polyelectrolyte combination (PDA/PMAA) yielded the least non-specific binding, which is essential for the LbL-Oligo system, aggregation of the particles during the LbL film assembly was observed. The aggregation level increased with increasing number of (PDA/PMAA) layers coated on the particle surface. Since a sufficient amount of probe Oligos can be coupled already onto particles coated with two layer pairs of these polyelectrolytes, the $(PDA/PMAA)_2$ system was selected to modify the particle surface for further experiments.

### 29.3.6
**Coupling and Hybridization Efficiency**

#### 29.3.6.1 Coupling Efficiency
The maximal amount of coupled Oligos was determined by measuring the consumption of Oligos from the Supernatant during the coupling process (Figure 29.10). Up to 250 pmol Amine-FluA$_{21}$ were covalently coupled onto 1 mg

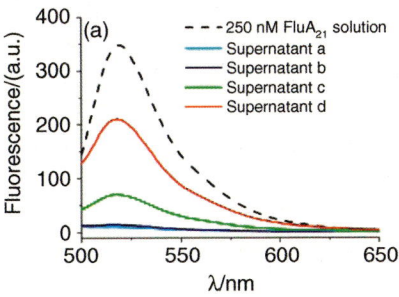
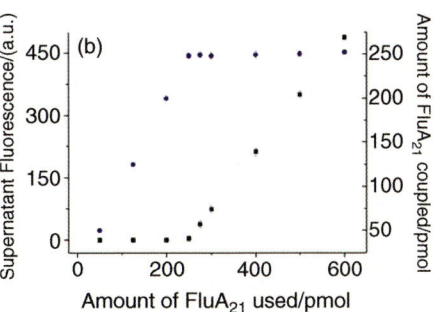

**Figure 29.10** (a) Fluorescence spectra of supernatant solution after coupling procedure. Amount of FluA$_{21}$ used for coupling: 125 pmol (cyan line), 250 pmol (blue line), 300 pmol (green line) and 400 pmol (red line). Fluorescence of 250 nM FluA$_{21}$ solution is shown as dashed line as reference. (b) Supernatant fluorescence (squares) and the coupled amount of FluA$_{21}$ (circles) in dependence on the amount of FluA$_{21}$ used. (Note: a constant amount of 1 mg LbL particles was used for each sample.)

of the 4.3 μm LbL particles. This corresponds to $2 \times 10^{-17}$ mole Oligos (or $1.2 \times 10^7$ Oligo molecules) per particle. Possibly, by increasing the amount of EDC for the activation, the coupling capacity of probe Oligos could be further increased. However, for such high density of probe Oligos on the particle surface (approximately one Oligo molecule per 5 nm$^2$) the hybridization efficiency decreased, which was most likely caused by steric hindrance for DNA hybridization (a double-stranded DNA helix has a diameter of 2 nm) [25].

During further work unlabeled probe Oligos were also coupled onto the particles. When relatively large batches were prepared, the amount of Oligos coupled could also be determined by UV/Vis spectroscopy due to the distinctive absorption of Oligos at 260 nm. The coupling efficiency for the unlabeled Oligos was consistent with the value obtained by the fluorescence labeled ones. Hence, the label on the Oligos had no influence on the coupling efficiency. However, the fluorescence measurement offers much higher sensitivity than the absorption method and requires less sample. Therefore, it was preferentially used in this work for quantification of the immobilized Oligos.

### 29.3.6.2 Hybridization Efficiency

Single-stranded (ss)DNA containing the complementary sequence to the probe Oligo can form double helices on the LbL-Oligo particles via DNA hybridization. The hybridization efficiency is defined as the ratio of the amounts of complementary ssDNA hybridized to the coupled Oligos per particle. When A$_{21}$ was used as the probe Oligo, RhoT$_{20}$ or T$_{20}$Rho was used as target ssDNA to test the hybridization. The amount of hybridized Oligos was determined in the same manner as that for the coupling efficiency, by measuring the fluorescence or absorption intensity in the supernatant. As mentioned, a single 4.3 μm LbL particle can carry up to $2 \times 10^{-17}$ mole probe Oligos. However, at such a high Oligo density, the hybridization efficiency of the complementary Oligos was below 50%. In contrast, a coupling amount of only 125 pmol per mg particles results in a hybridization yield of approximately 80% for both RhoT$_{20}$ and T$_{20}$Rho (100 pmol T$_{20}$ per mg LbL-A$_{21}$ particles).

### 29.3.7
### Comparison of LbL Particles with Conventional Carboxylated Particles

The LbL particles were compared with commercially available carboxylated polystyrene particles from Polyscience Inc. (4 μm) and MagSense Inc. (1.6 μm), respectively. By an identical coupling procedure, 25 pmol mg$^{-1}$ FluA$_{21}$ could be coupled on the Polyscience particles, which is only 20% of the LbL particle amount. In the case of the MagSense particles it was below 10% when the higher surface area was taken into account. This was in agreement with the analysis by CLSM: the fluorescence of the Polyscience-Oligo particles and the MagSense-Oligo particles was much weaker than the LbL-Oligo ones (Figure 29.11). (Note that the photomultiplier voltage (PMV) of 700 (used for Polyscience-FluA$_{21}$ particles) and 800 (used for MagSense-FluA$_{21}$ particles) yields a 3.6- and 10.0-fold signal amplification compared to 600 PMV (used for LbL-FluA$_{21}$ particles). The uneven distribution of fluorescence on the Polyscience-FluA$_{21}$ particles revealed that the Oligos were immobilized onto the particle surface inhomogeneously or as clusters.

The results clearly show that LbL films exhibit superior properties for probe molecule immobilization. Due to the nanoroughness of the LbL-surface and even distribution of the functional groups on the polyanions, higher amounts of probe Oligos were coupled onto the particle surface. Furthermore, the excellent hybridization efficiency despite the high Oligo concentration can be explained by the high flexibility of the LbL film.

### 29.3.8
### Arrangement of the Coupled Oligo Molecules

#### 29.3.8.1 Analysis by FRET Investigations
For the analysis of the Oligo arrangement in the LbL-film fluorescence resonance energy transfer measurements (FRET) were applied. If, for two different fluorescent

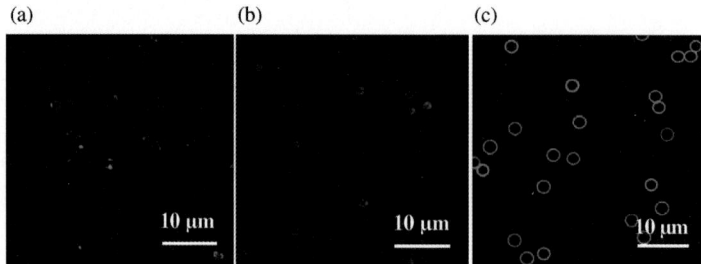

**Figure 29.11** CLSM images of FluA$_{21}$ coupled under same coupling protocols (taking the different particle diameters into account) onto (a) carboxylated polystyrene particles from Polyscience Inc., PMV: 700; (b) MagSense Inc., PMV: 800; (c) LbL particles, PMV: 600.

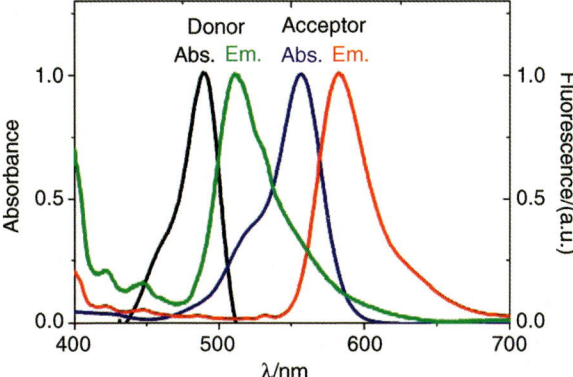

**Figure 29.12** Spectral overlap between Rho absorption and Flu emission as FRET pair: the absorption and emission spectra of Flu are shown as black and green lines, and the absorption and emission spectra of Rho are shown as blue and red lines, respectively.

dyes, the emission energy of the donor overlaps the absorption energy of the acceptor (Figure 29.12) and the molecules are in close proximity (typically 10–100 Å) [26] then energy is transferred from the excited donor dye to the acceptor through transition–dipole interaction, without emission of a photon. The FRET efficiency is highly dependent on the separation distance between the donor and acceptor in the nanometer range. Therefore, FRET has been used as a molecular ruler for distance analysis, and to reveal the interpenetration between polyelectrolyte layers in the LbL films [27]. We applied the FRET pair rhodamine and fluorescein for the investigation of the hybridization process and the arrangement of the coupled Oligo molecules on the LbL films. [28].

### 29.3.8.2 FRET of dsOligos Free in Solution

As the first step to study the FRET between Flu–Rho, the behavior of the dye pair-labeled Oligos free in solution was investigated. By using the combination of Oligo FluA$_{21}$ and T$_{20}$Rho or RhoT$_{20}$, the distance between the FRET pair was well defined for a convenient FRET analysis. To take into account the possible quantum yield changes of the fluorophores by DNA hybridization, FluA$_{21}$ was also hybridized with unlabeled Oligo T$_{20}$ and T$_{20}$Rho was hybridized with unlabeled Oligo A$_{21}$ as control samples, so that the quantification was always based on double-stranded helices. The fluorescence spectra of the hybridized dsOligos solution with varying dye position on the Oligo sequence are shown in Figure 29.13.

The evaluation of the spectra showed a decrease in donor emission by 26.0% when Flu and Rho dye molecules were 20 bps apart (i.e., 6.8 nm separation distance) on FluA$_{21}$–RhoT$_{20}$ hybrids (approximately 130% of the Förster radius of Flu–Rho) [29]. As a comparison, when the FRET pair were directly next to each other (i.e., on FluA$_{21}$–T$_{20}$Rho hybrids), the quenching of the donor increased to 87.5%. Simultaneously with decreasing donor emission, an increase in the acceptor emission was observed at 580 nm, which clearly indicated the occurrence of FRET.

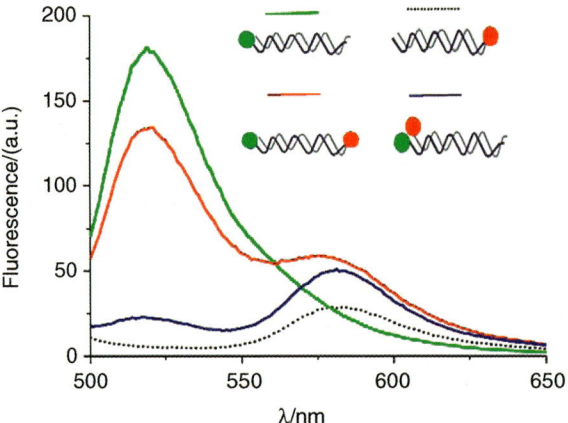

**Figure 29.13** Fluorescence spectra of Oligo hybrids free in solution ($C_{dsOligos} = 50$ nM). FluA$_{21}$-T$_{20}$ (—), FluA$_{21}$-T$_{20}$Rho (—), FluA$_{21}$-RhoT$_{20}$ (—), A$_{21}$-T$_{20}$Rho (—). Spectra were taken at $\lambda_{ex} = 480$ nm, slit width: 5/10 nm.

#### 29.3.8.3 FRET of dsOligos on LbL Particles

For FRET analysis on LbL films, Oligos RhoT$_{20}$ or T$_{20}$Rho were hybridized on the LbL-FluA$_{21}$ particles and the fluorescence spectra of LbL-Oligo particles were measured (Figure 29.14). The spectra show that, although the arrangement of the donor–acceptor dyes on each dsOligo helix is the same as the measurement in solution, the difference in FRET induced by the different distance between the donor and acceptor (i.e., Oligo hybrids FluA$_{21}$–RhoT$_{20}$ and FluA$_{21}$–T$_{20}$Rho) vanished almost completely. Even with a large distance of 20 base pairs in the case of FluA$_{21}$ hybridized to RhoT$_{20}$, the FRET signal was almost the same as for the FluA$_{21}$–T$_{20}$Rho scenario.

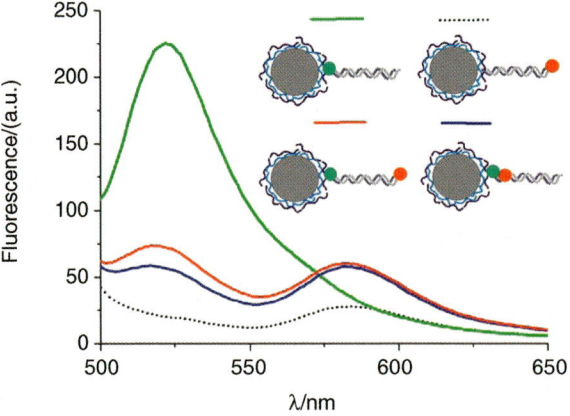

**Figure 29.14** Fluorescence spectra of Oligo hybrids coupled to the LbL particle surface ($c_{particle} = 1$ mg ml$^{-1}$ with 50 nM T$_{20}$Rho/RhoT$_{20}$). FluA$_{21}$-T$_{20}$ (—), FluA$_{21}$-T$_{20}$Rho (—), FluA$_{21}$-RhoT$_{20}$ (—), A$_{21}$-T$_{20}$Rho (—). Spectra were taken at $\lambda_{ex} = 480$ nm, slits width: 5/10 nm.

While in solution both dye molecules on the hybridized double-stranded chain interact only with each other, the high donor dye concentration on the LbL surface leads also to multiple donor–acceptor interactions between dyes on different double strands in the neighborhood. This screens out the distance difference between the dyes on the two different hybrids (FluA$_{21}$–RhoT$_{20}$ and FluA$_{21}$–T$_{20}$Rho, see also Figure 29.1b), having Oligos at different heights on a rough surface.

### 29.3.9
### Stability of the LbL-Oligo Particles

The DNA hybridization assay usually includes an initial incubation step above 60 °C in order to ensure complete melting of any existing dimer or duplex DNAs and to ensure the energetically lowest hybridization state in the right order during the slow cooling process. High salt concentration (500 mM) is used for efficient shielding of the repulsive Oligo charges. In addition, other substances, such as surfactants like SDS and proteins like BSA, are often added in hybridization buffers in order to prevent non-specific binding. The LbL-Oligo particles need to be stable under these conditions. LbL films are known to be responsive to environmental changes such as temperature, ionic strength and surfactants, with possible conformational changes or partial dissolution of the polyelectrolyte layers in the film [30].

Therefore the stability of the LbL particles was investigated under different hybridization conditions by varying the following parameters: room temperature to 70 °C, HEPES buffer (pH 7.4), 0.5 M NaCl, 0.1% SDS and 0.1 mg ml$^{-1}$ BSA. The LbL-Oligo particles remained stable during hybridization at room temperature with the aforementioned parameters; whereas for particles incubated at 70 °C and in the presence of SDS, a partial release of the PMAA from LbL-Oligo particles coated with (PDA/PMAA)$_2$ was observed.

Different treatments were performed on the LbL particles prior to the Oligo-coupling process in order to achieve the most stable LbL films. An most effective method was their incubation in 50 mM borate buffer (pH 8.0) at 70 °C for 2 h, by which all the excess PMAA was released during the treatment and a leakage of polyelectrolytes in the hybridization step was no longer observed. This treatment of the LbL particles did not influence the coupling efficiency for probe Oligos. However, heating for a longer time or at higher temperature caused further loss of polyelectrolyte molecules from the film and led to increasing aggregation of the particles.

### 29.4
### Summary

We have evaluated the LbL technology for the preparation of DNA-sensing particles and compared the results to the state of the art. DNA-sensing particles are covalently linked with oligonucleotides of defined sequence that bind selectively target DNA of complementary sequence. Compared to conventional carboxylated particles with a plane surface normally used for this purpose, the LbL particles showed several peculiarities. They required a specific adaptation of known coupling and preparation methods for stable and efficient attachment of Oligos, for high hybridization rates and for low non-specific binding.

Only a few polyelectrolyte combinations were found to be suitable for these requirements. Good coupling results could only be achieved on LbL films composed of polycations having quaternary amines and poly(methacrylic acid), which has to be the outermost layer. The use of primary or secondary amines in the LbL film reduced the coupling efficiency remarkably, even when the amines were assembled several layers below the outermost layer. The use of the common LbL component PSS yielded strong non-specific binding, even if it was not used for the last layer.

Carboxylate groups were used for the linkage of amine-modified Oligos. The necessary activation was preferably performed either by EDC in aqueous solution or by 1.1'-carbonyldiimidazol in organic solution. The often applied intermediate step of a succinimide formation using sulfoNHS leads to increased non-specific binding and cannot be recommended.

In general a two-step process is required in which, after activation of the carboxylic groups, the excess of activation agent is washed away. Although the EDC coupling in water can be easily performed, it leads to problems of particle aggregation and non-specific binding. The latter could be suppressed by post-treatment at pH 3. However the irreversible aggregation problem for particles smaller than 1–2 $\mu$m could not be solved for the coupling method using EDC.

The coupling via 1.1'-carbonyldiimidazol activation requires more effort due to complete solvent exchange to a water-free environment, but the problems of aggregation and non-specific binding were prevented. Therefore, it is especially recommended for small silica particles or DNA/RNA molecules that are sensitive to longer treatment at a pH value around 3. However, it cannot be applied to solvent-sensitive materials such as PS particles.

Overall, taking the optimization points into account, specific DNA sensing could be achieved by LbL particles, which exceeded in most parameters the performance of the best commercially available particles. These parameters are high coupling density of Oligos, efficient hybridization of target DNA, very low non-specific binding and a spatially extended FRET sensing of the coupling process. These advantages are based on the intrinsic properties of LbL films, namely a rough, porous and swollen surface coating in the nanometer-range. The surface structure is comparable to surfaces modified with long spacer molecules, exhibiting high charge density and number of coupling sites and low hydrophobic interactions, as in the case of PMAA.

Although the polymers in the LbL-film are not covalently linked with each other they are quite robust under the quite harsh conditions of coupling chemistry and the hybridization process. The stability is most likely remarkably increased by using PDA that contains few primary amine groups due to crosslinking the polyelectrolytes. Last but not least, the possibility of modifying the LbL film with fluorescent dyes with nanometer precision offers the efficient detection of binding events via FRET [31].

Therefore, having the merits of the LbL films, the technology is well suited for efficient coupling of synthetic DNA/RNA sequences for selective binding of defined target DNA by hybridization. A big advantage of the method is that it can be applied on almost every planar, rough or colloidal surface, independent of the material for covalent linkage of Oligos.

This work was supported by a grant of the Federal Ministry of Education and Research (BMBF project INUNA FKZ 0312027A)

## References

1. Wang, Y.J., Price, A.D., and Caruso, F. (2009) *J. Mater. Chem.*, **19**, 6451–6464.
2. Velev, O.D. and Gupta, S. (2009) *Adv. Mater.*, **21**, 1897–1905.
3. Tkachov, R., Senkovskyy, V., and Oertel, U. (2010) *Macromol. Rapid Commun.*, **31**, 2146–2150.
4. Ho, K.M., Li, W.Y., Wong, C.H., and Li, P. (2010) *Colloid Polym. Sci.*, **288**, 1503–1523.
5. Edmondson, S., Osborne, V.L., and Huck, W.T. (2004) *Chem Soc. Rev.*, **33**, 14–22.
6. Acosta, M.A., Ymele-Leki, P., and Kostov, Y.V. (2009) *Biomaterials*, **30**, 3068–3074.
7. Decher, G. (1997) *Science*, **277**, 1232–1237.
8. Decher, G. and Hong, J.D. (1991) *Macromol. Chem. Macromol. Symp.*, **46**, 321–327.
9. Sukhorukov, G.B., Donath, E., and Lichtenfeld, H. (1998) *Colloids Surf. A.*, **137**, 253–266.
10. Donath, E., Sukhorukov, G.B., Caruso, F., Davis, S.A., and Möhwald, H. (1998) *Angew. Chem., Int. Ed.*, **37**, 2202–2205.
11. Sukhorukov, G.B., Donath, E., and Davis, S. (1998) *Polym. Adv. Tech.*, **10/11**, 759–767.
12. Jiang, X.P. and Hammond, P.T. (2000) *Langmuir*, **16**, 8501–8509.
13. Schlenoff, J.B. and Dubas, S.T. (2001) *Macromol.*, **34**, 592–598.
14. Ladam, G., Schaad, P., Vögel, J.C., Schaff, P., Decher, G., and Cuisinier, F. (2000) *Langmuir*, **16**, 1249–1255.
15. Lavalle, Ph., Gergely, C., Cuisinier, F.J.G., Decher, G., Schaaf, P., and Voegel, J.C., and Picart, C. (2002) *Macromolecules*, **35**, 4458–4465.
16. Loesche, M., Schmitt, J., Decher, G., Bouwman, W.G., and Kjaer, K. (1998) *Macromolecules*, **31**, 8893–8906.
17. Qiu, X.P., Donath, E., and Mohwald, H. (2001) *Macromol. Mater. Eng.*, **286**, 591–597.
18. Caruso, F., Furlong, D.N., and Ariga, K. (1998) *Langmuir*, **14**, 4559–4565.
19. Ariga, K., Hill, J.P., and Ji, Q. (2007) *Phy. Chem. Chem. Phys.*, **9**, 2319–2340.
20. Picart, C. (2008) *Curr. Med. Chem.*, **15**, 685–697.
21. Quinn, J.F., Johnston, A.P.R., Such, G.K., Zelikin, A.N., and Caruso, F. (2007) *Chem. Soc. Rev.*, **36**, 707–718.
22. Pei, R.J., Cui, X.Q., and Yang, X.R. (2001) *Biomacromolecules*, **2**, 463–468.
23. Hermanson, G.T. (1996) Ch. 2, in *Bioconjugate Techniques*, Elsevier Science, San Diego.
24. (a) Köhler, K., Shchukin, D.G., Möhwald, H., and Sukhorukov, G.B. (2005) *J. Phys. Chem. B.*, **109**, 18250–18259; (b) Köhler, K., Möhwald, H., and Sukhorukov, G.B. (2006) *J. Phys. Chem. B.*, **47**, 24002–24010.
25. Voet, D. and Voet, J.G. (2004) Ch. 5, in *Biochemistry*, 3rd edn, Wiley-VCH, Weinheim.
26. Lakowicz, J.R. (2006) *Principles of Fluorescence Spectroscopy*, 3rd edn, Springer, New York.
27. Baur, J.W., Rubner, M.F., and Reynolds, J.R. (1999) *Langmuir*, **19**, 6460–6469.
28. Kang, J., Loew, M., Arbuzova, A., Andreou, I., and Dähne, L. (2010) *Adv. Mater.*, **22**, 3548–3552.
29. Parkhurst, K.M. and Parkhurst, L. (1995) *Biochemistry*, **34**, 285–292.
30. Kang, J. and Dähne, L. (2011) *Langmuir*, **27**, 4627–4634.
31. Kang, J. (2011) PhD thesis, FU Berlin, www.diss.fu-berlin.de

# 30
# A "Multilayered" Approach to the Delivery of DNA: Exploiting the Structure of Polyelectrolyte Multilayers to Promote Surface-Mediated Cell Transfection and Multi-Agent Delivery

*David M. Lynn*

## 30.1
## Introduction

Although the earliest reports on layer-by-layer assembly were focused on fundamental aspects related to the growth and structure of films fabricated using synthetic building blocks [1–3], it took only a few short years to recognize that this entirely aqueous and generally "forgiving" approach could also be used to assemble films using biologically relevant nucleic acid-based materials, such as DNA. A seminal report on the incorporation of DNA constructs into polyelectrolyte multilayers in 1993 provided robust proof of concept [4] and the basis of a powerful and general approach to the design of thin films and interfaces of interest in myriad biotechnological and biomedical contexts.

One particularly intriguing possibility arising from that early work was that methods for the layer-by-layer (LbL) incorporation of DNA *into* thin films might also lead to well-defined methods for the subsequent *release* of DNA from surfaces coated with these materials (and, thus, to new approaches to the localized or surface-mediated delivery of DNA in therapeutic contexts). This conceptually straightforward idea has been a focus of many different research groups, including our own, over the last 10 years [5]. The collective results of all of these efforts have provided, in a relatively short period of time, many different and complimentary approaches to the design of DNA-containing thin films that release and/or deliver functional DNA to cells and tissues *in vitro* and *in vivo*.

This chapter provides an overview and account of specific approaches pursued in our laboratory for the design of films that promote the surface-mediated release of DNA. It is also intended to provide some broader perspective on the development of these materials as well as insight into currently unresolved issues and opportunities for further development in view of some potential applications. Although work reported by other groups is discussed briefly where additional context is helpful, we note at the outset that many other important approaches to polyelectrolyte multilayers (PEMs) designed to release or deliver DNA have been reviewed comprehensively elsewhere [5] and are not discussed in detail in the limited space provided here.

*Multilayer Thin Films: Sequential Assembly of Nanocomposite Materials*, Second Edition.
Edited by Gero Decher and Joseph B. Schlenoff.
© 2012 Wiley-VCH Verlag GmbH & Co. KGaA. Published 2012 by Wiley-VCH Verlag GmbH & Co. KGaA.

Readers interested in the incorporation and release of other agents (e.g., proteins, peptides, viruses, small molecules, etc.) from PEMs in the context of other biomedical and biotechnological applications will also find additional detail and broader context in numerous other reviews [5–14] and in several other chapters of this book. Finally, we note that in addition to work focused on drug delivery and controlled release, the incorporation of DNA into stable PEMs has been investigated extensively as a platform for the development of biosensors [15].

## 30.2
### Surface-Mediated Delivery of DNA: Motivation and Context, Opportunities and Challenges

The majority of the work in our laboratory on the design of DNA-containing PEMs has been motivated broadly by the idea of gene therapy, or the notion that functional nucleic acid-based constructs could be used to treat or prevent certain types of diseases. While this is an attractive approach and, in principle, a conceptually simple idea, it has proven to be very difficult to achieve. Although many materials-based platforms (e.g., cationic polymers, cationic lipids, etc. [16, 17]) have been developed to deliver DNA to cells, it has become clear that many important and complicated issues will need to be addressed before it is possible to advance from the relatively simple idea of "DNA delivery" to the application of these materials in therapeutic contexts. Our work has been motivated by the idea that methods for the fabrication of DNA-containing PEMs on surfaces might provide opportunities to address particular challenges associated with achieving spatial and/or temporal control over the delivery of DNA (that is, delivering DNA where it needs to be, when it needs to be there, and, ultimately, rendering it available in a form that cells or tissues are able to process efficiently).

In the context of DNA delivery, PEMs and LbL methods of assembly provide several potential practical advantages relative to conventional methods for the encapsulation and surface-mediated release of DNA. These practical advantages range from the aqueous nature of the assembly process and the ultrathin and conformal nature of the resulting films to the fact that LbL assembly can provide precise control over the loading of DNA by managing the number of layers of DNA deposited. LbL assembly also provides opportunities to deposit different layers of several different DNA constructs, making it possible to fabricate multilayers with hierarchical structures capable of promoting multi-agent release. Finally, it is important to remember that the assembly of PEMs using double-stranded DNA (a negatively charged polymer) as a building block also requires the alternate adsorption of layers of cationic polymers, a class of materials used widely in conventional approaches to the delivery of DNA (e.g., by formation of self-assembled polymer/DNA "polyplexes" that can promote entry into cells [18–20]). In the context of DNA delivery, then, the juxtaposition and comingling of DNA with layers of cationic polymers provides compelling and potentially powerful opportunities to design films that not only release DNA, but also promote more efficient internalization and intracellular trafficking of DNA. In view of these and other potential advantages, this "multilayered" approach to

## 30.2 Surface-Mediated Delivery of DNA: Motivation and Context, Opportunities and Challenges

assembly could contribute broadly to the design of thin films and coatings capable of delivering precise and well-defined quantities of DNA (or combinations of DNA and other agents) of interest in a broad range of fundamental and applied biomedical contexts.

LbL assembly provides clear opportunities to incorporate DNA constructs into the structure of a PEM [4]. We note here, however, that PEMs are also often quite stable in aqueous media, and that, at the beginning of our own studies 10 years ago, there were relatively few methods that could be used to promote the controlled disassembly of these ionically-crosslinked assemblies in physiologically relevant media (i.e., to promote physical film erosion and the subsequent release of incorporated DNA that is too large to diffuse passively through an otherwise intact film). While several groups had demonstrated, at that point, that changes in pH and ionic strength could be used to disrupt PEMs in aqueous media [21–25] (including an example of the salt-induced deconstruction of DNA-containing films [24]), we sought to develop approaches to film disruption that could be used in biological environments for which pH and ionic strength are either constant or difficult to manipulate.

Our work has centered on three primary approaches. The first is based on fabrication of PEMs using hydrolytically degradable cationic polymers (Scheme 30.1a) to promote controlled film disruption. PEMs fabricated using these polymers erode gradually and promote surface-mediated DNA delivery and cell transfection *in vitro* and *in vivo*, and changes in polymer structure can provide tunable control over film disassembly. The second approach is based upon the design of a new class of "charge-shifting" cationic polymers (Scheme 30.1b) that undergo gradual reductions in net charge and introduce destabilizing interactions in ionically crosslinked PEMs. Charge-shifting polyelectrolytes can be used to design films that promote long-term release of DNA or to fabricate films that permit control over the release of multiple different DNA constructs with separate, distinct, and predictable release profiles. A third, more recent, approach leverages the pH-dependent

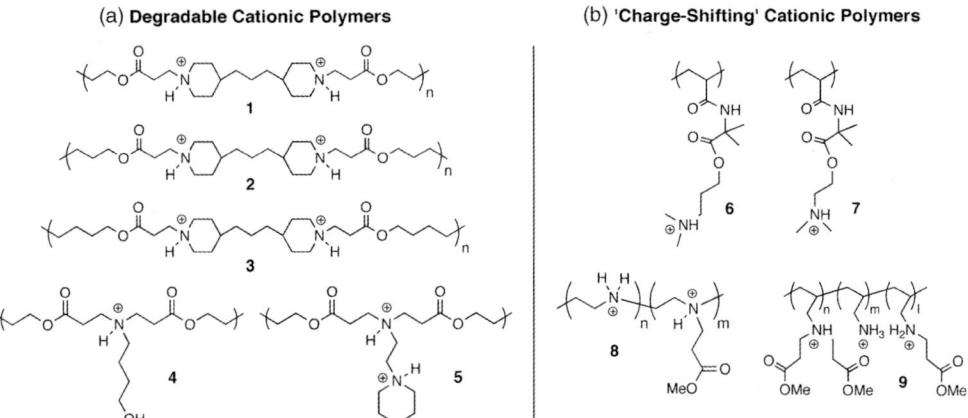

**Scheme 30.1** Chemical structures of (a) hydrolytically degradable cationic polymers and (b) "charge-shifting" cationic polymers used to fabricate PEMs that degrade in aqueous media.

properties of additional layers of weak polyelectrolytes to design assemblies that promote more rapid release of DNA.

In the sections below, we describe fundamental aspects related to the design, characterization, and potential applications of PEMs fabricated using these approaches. Before concluding this section, however, we note again that many other methods for the disruption of PEMs (and the release/delivery of DNA [5] or other agents) have been developed and exploited by other groups in parallel with our own efforts over the last 10 years [5–14]. These methods include, but are not limited to (i) the assembly of PEMs using enzymatically and reductively degradable polyelectrolytes (as well as other approaches using hydrolytically degradable polymers), (ii) the creative use of reversible crosslinking reactions and supramolecular interactions, and (iii) the use of other physical or chemical stimuli (e.g., light, ultrasound, electrochemical potentials, etc.) to promote the disruption of PEMs. More extensive discussions of these and other approaches are included in other chapters of this book, and leading references can again also be found in a variety of helpful reviews [5–14].

## 30.3
### Films Fabricated Using Hydrolytically Degradable Cationic Polymers

Our first efforts to design films that could promote the release of DNA and other anionic polymers were focused on the fabrication of PEMs using hydrolytically degradable cationic polymers as positively charged building blocks [26]. These early investigations were conducted in collaboration with Professor Paula Hammond, and were designed to test the hypothesis that the slow and gradual chemical hydrolysis of the cationic layers of a PEM could be used to promote gradual physical erosion and film disassembly (that is, that a cationic polymer could be used as both a structural element (Figure 30.1a) and a transient element (Figure 30.1b) of these assemblies).

Initial experiments demonstrated that films fabricated using polymer 1, a model member of a larger class of hydrolytically degradable synthetic cationic polymers known as poly($\beta$-amino ester)s [27–30], could be used to fabricate PEMs using several different anionic polymers [e.g., sodium poly(styrene sulfonate) (SPS) or poly(acrylic acid) (PAA)] [26]. These studies also demonstrated that films fabricated using this degradable polymer eroded gradually and released SPS and PAA into solution when they were incubated under physiologically relevant conditions (e.g., phosphate-buffered saline (PBS), 37 °C). Subsequent studies by our group revealed that films fabricated using SPS and a structural analog of polymer 1 having amide bonds, rather than ester bonds, in the backbone of the polymer do not erode under otherwise identical conditions [31]. These results, and those of other experiments, established the important role of the ester functionality of polymer 1 in promoting film erosion and are, in general, consistent with a mechanism of film disassembly that involves the gradual hydrolytic degradation of polymer 1.

We also demonstrated that polymer 1 could be used to fabricate PEMs that promote the surface-mediated release of DNA [32]. Although numerous past studies had demonstrated the ability to fabricate PEMs using model (non-functional) DNA

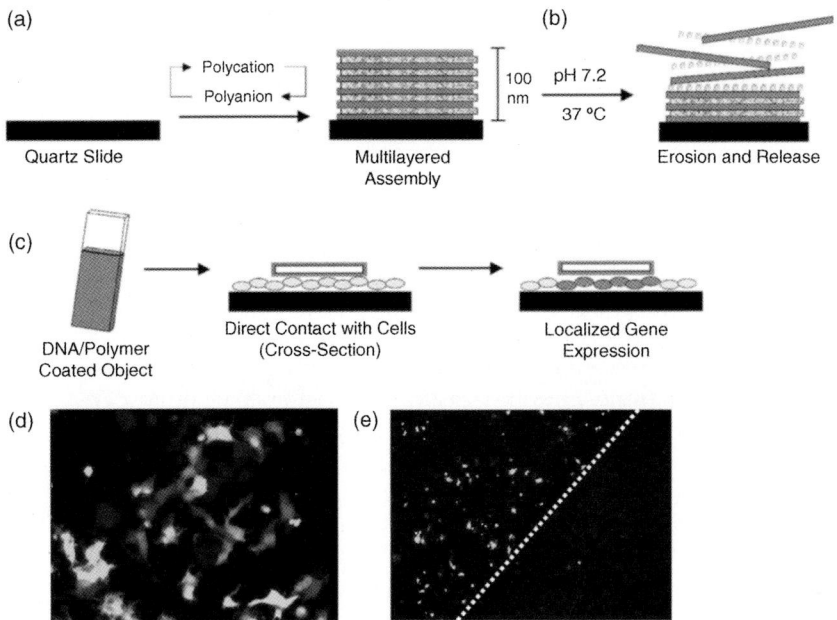

**Figure 30.1** (a) LbL assembly of a PEM fabricated from alternating layers of cationic polymer (light gray) and anionic polymer (dark gray). (b) Idealized scheme showing erosion of a PEM fabricated using a hydrolytically degradable polyamine; hydrolysis of the polyamine results in film erosion and release of anionic layers. (c) Scheme depicting surface-mediated transfection of cells using an object coated with a degradable DNA-containing PEM. (d and e) Fluorescence microscopy images showing expression of EGFP in COS-7 cells promoted by a glass slide coated with a PEM fabricated using polymer 1 and plasmid DNA encoding EGFP. Images show views near the center (d) and near the edge (e) of the film-coated slide (dotted line indicates the slide edge). Adapted from figures previously published in refs [75] (a–c; reproduced by permission of the Royal Society of Chemistry (RSC)) and [38] (d and e; reprinted with permission of Elsevier).

constructs (e.g., salmon sperm DNA, etc. [4]), this study was, to our knowledge, the first to incorporate functional (i.e., biologically/transcriptionally active) plasmid DNA constructs. The results of this study demonstrated that LbL film growth proceeded in a stepwise and linear manner, thus providing straightforward control over the loading of DNA in a film by control over the number of layers of DNA deposited. Films ~100 nm thick fabricated using polymer 1 and a plasmid DNA construct encoding enhanced green fluorescent protein (EGFP) eroded and released DNA gradually into solution over a period of ~30 h when incubated in PBS [32]. A series of subsequent studies (described in further detail below) demonstrated approaches that could be used to accelerate, slow down, or otherwise tune release profiles of PEMs using poly (β-amino ester)s [31, 33–35], but this initial study established clearly the feasibility of this degradable PEM-based approach as a platform for the gradual and surface-mediated release of DNA from film-coated objects. More importantly, at least from the standpoint of potential biomedical applications of this approach, characterization of DNA released from these PEMs revealed it to remain capable of promoting the expression of EGFP when administered to mammalian cells *in vitro* (e.g., when co-delivered to cells using a cationic lipid transfection agent) [32]. Additional detailed physical characterization of the surfaces of these films using atomic force microscopy (AFM) and scanning electron microscopy (SEM) demonstrated that these DNA-containing assemblies undergo large-scale changes in morphology (e.g., from smooth and relatively defect-free to a more complex nanoparticulate morphology) in the first several hours after incubation in PBS [36, 37]. We return to a discussion of these observations again in the sections below.

We demonstrated, in a subsequent study, that films fabricated using polymer 1 and plasmid DNA encoding EGFP could be used to promote the localized and surface-mediated delivery of DNA directly to cells *in vitro* (e.g., by placing film-coated objects in contact with cells; Figure 30.1c) [38]. The images in Figure 30.1d,e show cells growing under (d) or near the edge (e) of a film-coated glass slide for 48 h. The observation of green fluorescence in the cells in these images demonstrates that these degradable PEMs are able to promote the localized internalization and expression of EGFP by cells directly (that is, without the addition of any exogenous agents used to promote transfection). In the context of this latter observation, we note again the attractive possibility that the presence of cationic polymer 1 in these films could play a role in promoting more efficient internalization and processing of DNA by cells (e.g., as demonstrated in past studies on the use of polymer 1 [27–30] and other polymers [18–20] to promote polyplex-mediated gene delivery). Subsequent studies using dynamic light scattering revealed the presence of nanoscale aggregates (ranging from 100 to 600 nm in size) in samples of PBS used to incubate polymer 1/DNA films [39]. These observations are consistent with the idea that DNA could be released in a form that is bound to fragments of polymer 1, but the compositions of these aggregates and the extent to which this might occur in the presence of cells has not yet been established. Ongoing studies in our laboratory are focused on developing approaches (including the synthesis of fluorescently end-labeled derivatives of polymer 1 [40, 41]) that can be used to identify the presence of such aggregates and provide additional important insight into the roles that polymer 1 might play in promoting surface-mediated transfection. In addition to exploiting the role of degradable polymer 1, it could, of course, also prove possible to incorporate additional layers of other cationic polymers (either degradable or non-degradable) used more conventionally for the delivery of DNA [18–20] into these degradable films to promote more effective formation of polymer/DNA aggregates, or to address other important intracellular barriers to transfection.

## 30.4
### Toward Spatial Control: Release of DNA from the Surfaces of Implants and Devices

The ability to use LbL assembly to fabricate conformal multilayers on the surfaces of topographically and topologically complex objects creates opportunities to coat the surfaces of interventional devices or other implantable objects and characterize the

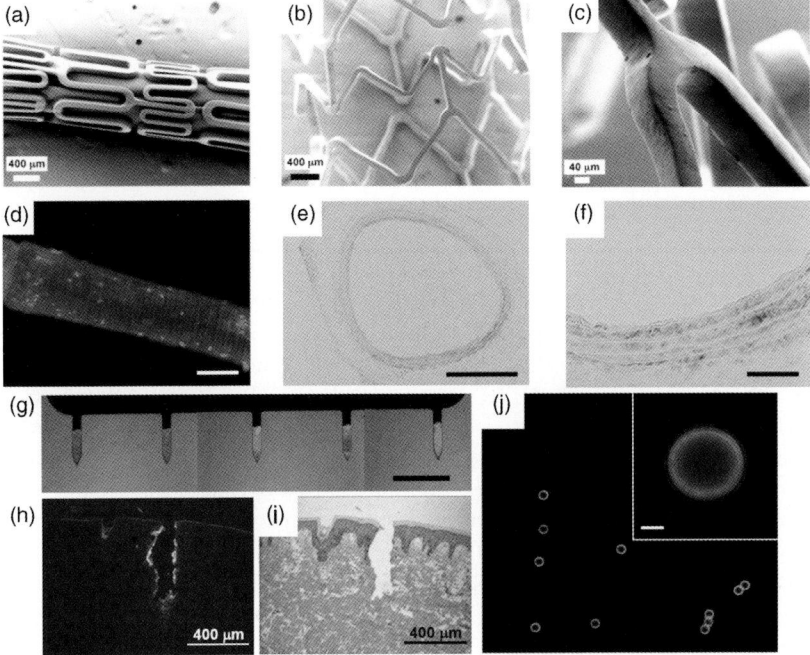

**Figure 30.2** SEM images of intravascular stents coated with PEMs fabricated using polymer 1 and plasmid DNA. Images are of stents imaged as-coated on a balloon assembly (a) or after balloon expansion (b and c). (d) Fluorescence microscopy image of an inflatable catheter balloon coated with a PEM fabricated using polymer 1 and fluorescently labeled plasmid DNA (scale bar = 500 μm). (e and f) Optical micrographs of cross-sectional samples of the carotid artery of a rat (with X-gal staining; blue) treated with a catheter balloon coated with a film fabricated using polymer 1 and plasmid DNA encoding beta-galactosidase. Scale bars for E and F are 500 μm and 200 μm, respectively. (g) Fluorescence microscopy images of a microneedle array coated with PEMs fabricated using polymer 1 and fluorescently labeled DNA. Scale bar = 1 mm. (h and i) Fluorescence microscopy image (h) and histology image (i) of porcine cadaver skin after insertion and removal of a film-coated microneedle. (j) Confocal microscopy image showing cross-sections of polymer microparticles coated with PEMs fabricated using polymer 1 and fluorescently labeled plasmid DNA. Scale bar in inset = 2 μm. Individual images adapted with permission from figures previously published in refs [39] (a–c; copyright American Chemical Society), [45] (d–f; reprinted with permission of Elsevier), [46] (g–i; copyright American Chemical Society), and [52] (j; reprinted with permission of Elsevier).

ability of DNA-containing PEMs to deliver DNA and transfect cells in more complex *in vivo* environments. Our group reported on the fabrication of polymer 1/DNA films on the surfaces of stainless steel intravascular stents [39] as part of an approach to the localized delivery of DNA to the vascular wall (and, by extension, as a step toward the development of potential gene-based treatments for complications that can arise when these interventional devices are implanted [42–44]). The images in Figure 30.2a–c are SEM images of two film-coated stents: one shown mounted on an inflatable balloon assembly (a), and one that was coated and then expanded prior to imaging (b, c) [39]. These images highlight two features of this film system, and this general approach, that are likely to be important in the context of this potential application. First, these images reveal the presence of a smooth, uniform, and defect-free film that conforms to the surface of the curved stent struts. Second, these images reveal that these ultrathin films (~120 nm thick) do not crack, peel, or delaminate when the stent is manipulated and subjected to mechanical forces representative of those encountered during the deployment of these devices (e.g., during balloon expansion).

We have also demonstrated that this LbL approach can be used to coat the surfaces of inflatable catheter balloons, and that film-coated balloons can be used to promote the localized transfection of vascular tissue *in vivo* [45]. Figure 30.2d shows a fluorescence microscopy image of a film-coated balloon prepared using this approach (this film was fabricated using fluorescently labeled DNA to permit imaging and characterization of film uniformity). To characterize the ability of these films to promote transfection *in vivo*, we used catheter balloons coated with films fabricated using polymer 1 and plasmid DNA encoding either EGFP or β-galactosidase and an *in vivo* rat model of balloon-mediated arterial injury used previously to study intimal hyperplasia. The images in Figure 30.2e, f show representative optical microscopy images of cross-sections of the carotid artery of a rat treated with a film-coated balloon for 20 min (tissue was harvested 72 h after initial treatment) and shows the presence of significant levels of cell transfection (here, indicated by the presence of blue cells) in the medial layers of the artery. These and other results of this study constitute a proof-of-concept demonstration of the ability of these degradable PEMs to promote localized tissue transfection *in vivo* [45]. Additional studies will be required to investigate the potential for toxicity, inflammation, or a range of other important issues. However, these results suggest the basis of a more general platform that could be used for the balloon-mediated delivery of therapeutically relevant DNA constructs (or other agents) targeted specifically to prevent, suppress, or study fundamental mechanisms of intimal hyperplasia or other vascular diseases and conditions.

Finally, we have also demonstrated an approach to the delivery of DNA across the skin using arrays of micrometer-scale needles coated with thin polymer 1/DNA films [46]. Figure 30.2g shows an image of a stainless steel microneedle array consisting of five needles (each ~750 μm long) coated with films fabricated using fluorescently labeled DNA. The images in Figure 30.2h,i show a fluorescence microscopy image (h) and a corresponding histology image (i) of a needle track resulting from the insertion of a film-coated microneedle into porcine cadaver skin for 2 h. The image in Figure 30.2h reveals that DNA was deposited in the skin along the edges of the needle track to depths of up to ~500 to 600 μm. Additional results demonstrated that the films were removed completely from the surfaces of the microneedles over this 2-h insertion period [46]. This period of time is significantly shorter than the erosion and release of DNA from polymer 1/DNA films incubated in PBS (which, in general, takes place on the order of days [32, 39]) and suggests that the behavior of these films in defined media may differ substantially from that observed in more complex tissue environments, where many different physical, chemical, and mechanical factors could also influence film behavior significantly. Although more rapid release might prove useful in the context of microneedle insertion and removal, the potential for differences in film behavior *in vivo* will, of course, need to be borne in mind with respect to the design of these and other PEMs for specific biomedical applications.

Microneedle-based approaches to transdermal delivery have been investigated broadly as a means for the pain-free delivery of small and large molecules across the skin [47–50]. The results of the study described above demonstrate that LbL approaches to the design of PEM-coated microneedles can provide a useful platform for the delivery of DNA (or protein [46]) across the skin, and thus suggest opportunities for the administration of therapeutically relevant doses of DNA-based (or protein-based) vaccines to immune cells that reside in the dermis [47–50]. The potential of this LbL approach to delivery is supported further by the results of recent studies by DeMuth *et al.* [51] also demonstrating methods for the transdermal delivery of DNA using a 3D array of microneedles coated with polymer 1/DNA films. In the broader context of new approaches to the delivery of DNA vaccines, our group has also reported on polymer microparticles coated with these erodible films (Figure 30.2j; using methods similar to those reported by other groups for the fabrication of PEMs on colloidal substrates [7, 14]) toward a size-based means of targeting delivery of DNA to macrophages or other cells of the immune system [52].

## 30.5
**Toward Temporal Control: Tunable Release and Sequential Release**

The section above highlights the range of different size scales and the complexity of surfaces that can be coated using LbL assembly. The examples above also illustrate one important way in which this approach can be used to exert spatial control over the delivery of DNA—that is, that coating an object, and then placing that object in contact with cells or with tissues, provides a conceptually straightforward and practical means of assuring that release is localized in the vicinity of that object. In this section, we highlight ways that this "multilayered" approach to materials design can also provide useful levels of temporal control over the rates at which DNA is released from a film-coated surface.

The importance of control over the timescale of delivery in the context of potential therapeutic applications cannot, of course, be overstated, and a relatively simple framework for discussion emerges from just the small number of examples provided in the previous section. For example, whereas a stent-based approach to delivery could potentially benefit from gradual and sustained release of DNA over a period of weeks or months, catheter balloon-based interventions are inherently limited to short surgical procedures (e.g., <20 min), and rapid or instantaneous release might be most desirable for certain applications of film-coated microneedle arrays. In addition, whereas some applications may require short-, medium-, or long-term release of a single agent, many biological processes and therapeutic approaches could potentially benefit from materials that provide control over the relative rates at which *multiple* different agents are released (e.g., either simultaneously or in sequence; see

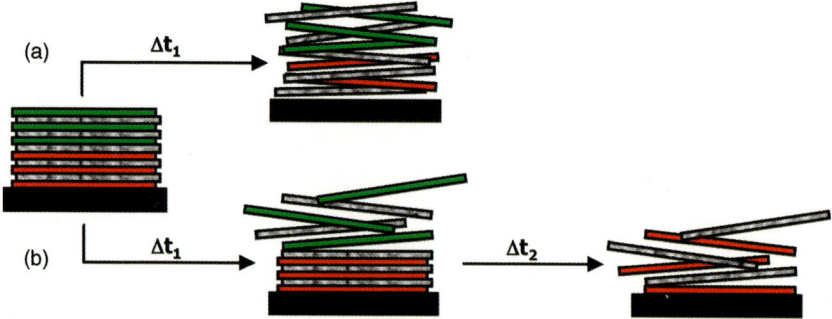

**Scheme 30.2** Idealized schematic illustration depicting the simultaneous (path a) or sequential (path b) release of two different components in a PEM (depicted here in green and red) upon erosion of a film with a hierarchical structure fabricated from different layers of two different macromolecular components.

Scheme 30.2). The ability to define the locations of different layers and design multicomponent PEMs with "hierarchical" structures renders approaches based on LbL assembly well suited to address these and many other important aspects of temporal control (and, ultimately, does so in ways that dovetail nicely with the inherent approaches to spatial control described above).

### 30.5.1
### Approaches Based on Incorporation of Different Hydrolytically Degradable Polyamines

One approach to exerting temporal control over film erosion is to alter the structures of the degradable cationic polymers used to assemble the films. Scheme 30.1a shows the structures of several examples of different hydrolytically degradable poly(β-amino ester)s that we have investigated for this purpose: for polymers **1–3**, the hydrophobicity of the polymer backbone (and, thus, the rates at which the ester bonds hydrolyze in aqueous solution [33]) has been varied systematically; polymers **4** and **5** are examples for which the side chain structure has been varied [35].

Several past studies from our group [31, 33–35] and others [53–56] demonstrate that such changes to the structure of a poly(β-amino ester) can result in large differences in the rates at which PEMs erode and release a range of different small molecules and macromolecules (e.g., over periods ranging from days to several weeks or months). In the specific context of DNA delivery, films fabricated using polymer **4** (which possesses a hydroxy-functionalized, non-ionic side chain) erode and release plasmid DNA over a period of ~2 days, but otherwise identical films fabricated using polymer **5** (which possesses a protonatable, amine-functionalized side chain) erode and release DNA over ~2 weeks (Figure 30.3a) [35]. Whereas differences in the erosion of films fabricated using polymers **1–3** is, in general, thought to arise from differences in rates of backbone hydrolysis [31, 33], large differences in the release of DNA using polymers **4** and **5** are likely to arise, at least in

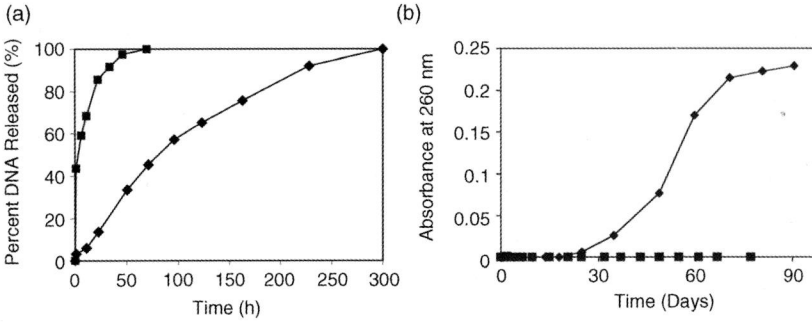

**Figure 30.3** (a) Plot showing percentage of DNA released versus time for two PEMs fabricated using plasmid DNA and either polymer **4** (■) or polymer **5** (■). Films were fabricated on planar silicon substrates and incubated in PBS at 37 °C. Figure replotted from data reported in ref [35]. (b) Plot of absorbance at 260 nm versus time showing the release of DNA from films fabricated from polymer **6** (■) and an analog of this polymer containing side chain amide bonds (■; see text). Adapted from ref [60] (copyright Wiley-VCH Verlag GmbH & Co. KGaA; reproduced with permission).

part, from differences in the amount of ionic crosslinking present in these films (as dictated by differences in side chain structure). Results of other experiments using polymers **1–3** have also demonstrated that it is possible to design PEMs with broadly and systematically tunable erosion/release rates simply by using different blends of these individual polymers during film fabrication [34]. Although many other poly(β-amino ester) structures exist [28, 29], this "blending" approach makes it possible to exploit the availability of a limited number of polymers and can, at least in some cases, obviate the need to design new structures to achieve other desired release profiles.

While approaches based on the incorporation of hydrolytically degradable polyamines can be used to exert useful levels of temporal control over the erosion of PEMs, this basic approach has proven to be limited, at least thus far, in at least two important ways. First, while this approach is useful for extending the release of DNA over periods of days to weeks [32, 35, 38], the design of new degradable polymers that promote the release of DNA more rapidly (e.g., over seconds, minutes, or hours) or more slowly (e.g., over months) has proven difficult. Second, while this degradable poly(β-amino ester) approach permits incorporation of different layers of two different plasmid DNA constructs into a film [35, 38], the resulting assemblies generally exhibit simultaneous or substantially overlapping release profiles (that is, it has been difficult to design films that permit the release of two DNA constructs with distinct and non-overlapping release profiles using this approach). The reasons for this are not entirely understood, but are likely a result of (i) the interpenetration that is inherent between different layers of a PEM [57], (ii) potential interdiffusion of film components during assembly (leading to degradable films with more blended, non-hierarchical structures [58]), and/or (iii) the demonstrated propensity of DNA-containing films fabricated using polymer **1** to undergo large-scale physical transformations upon incubation in physiologically relevant media (which could result in the mixing of layers in films that may otherwise have hierarchical structures) [36, 37].

## 30.5.2
### Approaches Based on Incorporation of Cationic "Charge-Shifting" Polymers

To address these and other issues, we have also developed alternative approaches to the design of erodible PEMs based on the concept of cationic "charge-shifting" polymers originally developed by our group [59–63] and others [64–66] for the disruption of solution-based aggregates of polymer and DNA. Scheme 30.1b shows the structures of several different cationic charge-shifting polymers developed and investigated by our group. These cationic polymers were designed to undergo "shifts" in their net charge (e.g., to become either less cationic or net anionic) as a result of side chain hydrolysis. In the case of polymers **6** and **7**, side chain ester hydrolysis results in the loss of cationic (amine) functionality and the unmasking of anionic carboxylate groups (e.g., Eq. (30.1)) [60, 63]; in the limit of time, these cationic polymers ultimately become completely anionic. Polymers **8** and **9** are examples resulting from a second approach, in which ester-functionalized side chains are conjugated to the backbones of a cationic polymer; gradual hydrolysis of the ester groups again results in unmasking of carboxylates that reduce the net charge of these polymers (e.g., Eq. (30.2) [59, 61, 62]. In both cases, time-dependent ester hydrolysis provides time-dependent control over the disruption of ionic interactions in PEMs in ways that can be accomplished at otherwise constant (and physiologically relevant) pH and ionic strength.

The rate of hydrolysis of the side chain esters in polymer **6** is slow (with a half-life of ~200 days in aqueous solution) [60]. This polymer can thus be used to fabricate films that promote prolonged release (e.g., over months). Figure 30.3b shows a plot of the release of DNA from a film ~100 nm thick fabricated using polymer **6** and plasmid DNA encoding EGFP; release is observed to occur over a period of ~3 months [60]. Further inspection of this release profile reveals the presence of an initial lag period of several weeks prior to the onset of the release of DNA. Although the reasons for this lag period are not completely understood, it is likely related to the time required to reach a threshold number of side chain hydrolysis events sufficient to allow film disruption to occur (the results of control experiments using an analog of polymer **6** having amide groups, rather than ester groups, support the view that side chain hydrolysis plays an important role in promoting film erosion [60]). This lag period is not observed for the release of DNA from films fabricated using hydrolytically degradable cationic polymers, and is a feature of this charge-shifting approach that could potentially be exploited, if it can be controlled, in applications for which the immediate release of DNA is neither required nor desired. Plasmid DNA released over the complete 3-month course of these experiments remained transcriptionally active, suggesting the feasibility of approaches to long-term delivery using these assemblies.

One particularly attractive outcome of approaches to the design of charge-shifting polymers based on conjugation of ester functionality to cationic polymer backbones (e.g., polymers **8** and **9**) is that it is possible to control both the extent of the resulting charge-shift and the rate at which it occurs by control over the number of ester-functionalized side chains that are added [59, 61, 62]. This approach can thus be used to design polymers that promote the erosion of DNA-containing films rapidly (e.g., over ~30 min) or slowly (e.g., over ~2 days), and, as a result, to design multicomponent, hierarchical assemblies that promote the release of two different DNA constructs with separate and distinct (non-overlapping) release profiles (e.g., by depositing several layers of one construct to form the bottommost layers of a film, followed by several layers of a second construct to form the topmost layers; e.g., see Scheme 30.2). As one example, we demonstrated that it is possible to use different derivatives of polymer **9** to design assemblies that promote the rapid and complete release of a first DNA construct over ~30 min, followed by the slower and sustained release of a different DNA construct over a period of several days [62]. The results of additional experiments demonstrated that this multilayered approach to assembly could also be used to control the order with which these constructs were released by changing the order in which they were deposited (i.e., a "first on/last off" approach).

Additional experiments will be required to characterize the internal, molecular-level structures of these multicomponent assemblies. The results of the study above, however, are consistent with the presence of hierarchical structures, and the degree of segregation of the two DNA constructs (e.g., along the z-axis of a film) is at least sufficient to provide distinct control over the release of each individual component [62]. Approaches to the deposition of defined numbers of intermediate "barrier" layers to further separate individual layers and/or provide barriers to interdiffusion of different components in PEMs have been reported [58, 67] (including a report from Jessel *et al.* on the design of DNA-containing films that promote sequential cell transfection [67]). This general approach could provide additional means to control or tune the release of multiple different film components. Other work in our group has demonstrated that it is also possible to expand on this approach and broaden the temporal "window" over which two different DNA constructs can be released by making judicious changes to polymer structure. For example, we reported that hierarchical films fabricated using polymer **6** (which promotes release over a period of months; see above) and polymer **7** (a structural variant of polymer **6** that promotes release over a period of days) can be designed to sustain the initial release of a first DNA construct over 3 days, followed by the extended release of a second construct over a period of months [63]. Additional experiments demonstrate that it is possible to use blends of these two charge-shifting polymers to design intermediate release profiles (similar to the approaches described above) [63], suggesting that it should be possible to design hierarchical films that can be tuned to release multiple DNA constructs (or DNA constructs in combination with other agents) over periods of time ranging from several hours to several days or months to meet the requirements of a range of potential applications.

### 30.5.3
### New Approaches to Rapid Release

One aspect of temporal control that has, thus far, been difficult to address through the design and incorporation of new hydrolytically degradable polyamines or charge-shifting polymers alone is the fabrication of PEMs that promote the rapid release of DNA (e.g., over timescales of the order of seconds, minutes, or a few hours, as opposed to several days, weeks, or months). As outlined above, films that promote rapid release could be particularly useful in DNA delivery applications that are time-limited (such as balloon-mediated interventions) or for which extended delivery times are either not needed or impractical (e.g., microneedle-based approaches to vaccine delivery).

We recently reported an approach that can be used to accelerate the erosion of DNA-containing PEMs considerably. This approach is based on the LbL incorporation of additional layers of weak poly(acid)s, such as PAA, into DNA-containing PEMs [68] and builds upon a range of past studies demonstrating that the pH-dependent properties of these polymers can be used to promote the pH-dependent disruption of ionic interactions in these assemblies (in various ways, and to varying extents [8, 21–23, 69–72]). Whereas films fabricated using alternating layers of polymer **1** and DNA typically erode over a period of several days upon incubation in PBS (see above [32, 39]), substitution of every other layer of DNA with PAA resulted in films that eroded and released DNA over a period of 3–6 h when incubated under the same conditions [68]. Additional experiments demonstrated that the incorporation of PAA into PEMs fabricated using plasmid DNA and linear poly(ethyleneimine) (LPEI) – a non-degradable polyamine that leads to polymer/DNA films that generally erode very slowly – can also be used to promote the complete release of DNA in as little as 3 h. The results of this study are consistent with the view that pH-dependent changes in the ionization of PAA upon introduction to PBS lead to destabilizing interactions that promote rapid film disassembly.

These results, when combined, are of interest for several reasons. First, in addition to simply promoting more rapid release, this approach does so in a way that removes the need for incorporation of a degradable or charge-shifting cationic polymer to promote film erosion. This approach thus opens the door to incorporation of many other types of non-degradable cationic polymers that are designed specifically for

DNA delivery [18–20] and/or may be more effective at directing the intracellular trafficking of DNA (see above for additional discussion). Second, this approach provides a platform for the co-incorporation of other weak poly(acid)s, such as poly(propylacrylic acid) (PPAA), that have been demonstrated in other contexts [73, 74] to have properties that help promote more efficient intracellular processing of therapeutic agents. Third, films that erode rapidly upon exposure to conditions of physiological pH and ionic strength could be well suited for applications (such as balloon-mediated delivery) in which other types of stimuli developed to promote the "triggered" disassembly of PEMs may be difficult to apply. Finally, we note that, in addition to potential therapeutic applications, new approaches to PEMs that promote rapid release, very rapid release, or rapid "on-demand" release of DNA (e.g., on timescales of seconds and minutes) could contribute significantly to the development of new research tools and may be useful in other fundamental contexts. Exciting examples from several groups have demonstrated redox-triggered release mechanisms, electrochemically-stimulated release mechanisms, and other approaches that have great potential in this regard (for additional discussion of recent examples of these latter approaches, see ref. [12]).

## 30.6
## Concluding Remarks

The sections above highlight several different ways that LbL assembly can be used to fabricate PEMs that erode in physiologically relevant environments and promote the release of functional DNA from film-coated surfaces. This chapter has also highlighted several ways that this "multilayered" approach can provide opportunities to exert spatial and/or temporal control over the release of DNA and promote cell transfection *in vitro* and *in vivo*. In closing, however, it is important to note again that this chapter presents one laboratory's perspective on this interesting and rapidly growing area of research. In drawing on the results of our own research, we have endeavored to detail some of the specific opportunities that exist, some of the materials properties that can potentially be exploited, and some of the unresolved issues, gaps in knowledge, and opportunities for further development that remain in both fundamental and applied contexts. Additional insights, potential solutions, and creative new directions, however, are evident in the collective body of work arising from the efforts of many other groups working on the fabrication, characterization, and applications of DNA-containing PEMs [5]. Finally, it is also important to note that while these materials have evolved very rapidly since the first report on the incorporation of DNA into a PEM [4], the application of this general approach in the areas of gene and drug delivery remains in its infancy – it remains too early, in many important ways, to determine the range of different applications for which these materials may or may not be well suited. What does seem clear, however, is that the development of these materials as platforms for potential therapeutic applications will continue to benefit considerably from the influx of fresh ideas and the interdisciplinary experience of researchers in biology, medicine, and the pharmaceutical sciences.

## Acknowledgments

The author is grateful to the many students, collaborators, and colleagues whose contributions to this work are acknowledged by reference in many of the citations included above. Work from the author's laboratory was supported generously by the National Institutes of Health (R21 EB02746 and R01 EB006820) and the Arnold and Mabel Beckman Foundation.

## References

1 Decher, G. and Hong, J.D. (1991) *Makromol. Chem.-Macromol. Symp.*, **46**, 321.
2 Decher, G. and Hong, J.D. (1991) *Ber. Bunsen-Ges. Phys. Chem.*, **95**, 1520.
3 Decher, G., Hong, J.D., and Schmitt, J. (1992) *Thin Solid Films*, **210**, 831.
4 Lvov, Y., Decher, G., and Sukhorukov, G. (1993) *Macromolecules*, **26**, 5396.
5 Jewell, C.M. and Lynn, D.M. (2008) *Adv. Drug Deliv. Rev.*, **60**, 979.
6 Ai, H., Jones, S.A., and Lvov, Y.M. (2003) *Cell Biochem. Biophys.*, **39**, 23.
7 Peyratout, C.S. and Dahne, L. (2004) *Angew Chem. Int. Ed. Engl.*, **43**, 3762.
8 Sukhishvili, S.A. (2005) *Curr. Opin. Colloid Interface.*, **10**, 37.
9 Angelatos, A.S., Katagiri, K., and Caruso, F. (2006) *Soft Matter*, **2**, 18.
10 Tang, Z.Y., Wang, Y., Podsiadlo, P., and Kotov, N.A. (2006) *Adv. Mater.*, **18**, 3203.
11 De Geest, B.G., Sanders, N.N., Sukhorukov, G.B., Demeester, J., and De Smedt, S.C. (2007) *Chem. Soc. Rev.*, **36**, 636.
12 Lynn, D.M. (2007) *Adv. Mater.*, **19**, 4118.
13 Boudou, T., Crouzier, T., Ren, K.F., Blin, G., and Picart, C. (2010) *Adv. Mater.*, **22**, 441.
14 De Cock, L.J., De Koker, S., De Geest, B.G., Grooten, J., Vervaet, C., Remon, J.P., Sukhorukov, G.B., and Antipina, M.N. (2010) *Angew. Chem. Int. Ed.*, **49**, 6954.
15 Rusling, J.F., Hvastkovs, E.G., Hull, D.O., and Schenkman, J.B. (2008) *Chem. Commun.*, 141.
16 Kabanov, A.V., Felgner, P.L., and Seymour, L.W. (1998) *Self-Assembling Complexes for Gene Delivery: From Laboratory to Clinical Trial*, John Wiley and Sons, New York.
17 Amiji, M.M. (2004) *Polymeric Gene Delivery: Principles and Applications*, CRC Press, New York.
18 Pack, D.W., Hoffman, A.S., Pun, S., and Stayton, P.S. (2005) *Nat. Rev. Drug Discov.*, **4**, 581.
19 Park, T.G., Jeong, J.H., and Kim, S.W. (2006) *Adv. Drug Delivery Rev.*, **58**, 467.
20 Putnam, D., (2006) *Nat. Mater.*, **5**, 439.
21 Sukhishvili, S.A. and Granick, S. (2000) *J. Am. Chem. Soc.*, **122**, 9550.
22 Dubas, S.T., Farhat, T.R., and Schlenoff, J.B. (2001) *J. Am. Chem. Soc.*, **123**, 5368.
23 Dubas, S.T. and Schlenoff, J.B. (2001) *Macromolecules*, **34**, 3736.
24 Schuler, C. and Caruso, F. (2001) *Biomacromolecules*, **2**, 921.
25 Sukhishvili, S.A. and Granick, S. (2002) *Macromolecules*, **35**, 301.
26 Vazquez, E., Dewitt, D.M., Hammond, P.T., and Lynn, D.M. (2002) *J. Am. Chem. Soc.*, **124**, 13992.
27 Lynn, D.M. and Langer, R. (2000) *J. Am. Chem. Soc.*, **122**, 10761.
28 Lynn, D.M., Anderson, D.G., Putnam, D., and Langer, R. (2001) *J. Am. Chem. Soc.*, **123**, 8155.
29 Anderson, D.G., Lynn, D.M., and Langer, R. (2003) *Angew. Chem. Int. Ed.*, **42**, 3153.
30 Lynn, D.M., Anderson, D.G., Akinc, A.B., and Langer, R. (2004) *Polymeric Gene Delivery: Principles and Applications* (ed. M. Amiji), CRC Press, New York.
31 Zhang, J., Fredin, N.J., and Lynn, D.M. (2006) *J. Polym. Sci. Polym. Chem.*, **44**, 5161.
32 Zhang, J., Chua, L.S., and Lynn, D.M. (2004) *Langmuir*, **20**, 8015.

33 Zhang, J., Fredin, N.J., Janz, J.F., Sun, B., and Lynn, D.M., (2006) *Langmuir*, **22**, 239.
34 Zhang, J. and Lynn, D.M. (2006) *Macromolecules*, **39**, 8928.
35 Zhang, J.T., Montanez, S.I., Jewell, C.M., and Lynn, D.M. (2007) *Langmuir*, **23**, 11139.
36 Fredin, N.J., Zhang, J., and Lynn, D.M. (2005) *Langmuir*, **21**, 5803.
37 Fredin, N.J., Zhang, J., and Lynn, D.M. (2007) *Langmuir*, **23**, 2273.
38 Jewell, C.M., Zhang, J., Fredin, N.J., and Lynn, D.M. (2005) *J. Control. Release*, **106**, 214.
39 Jewell, C.M., Zhang, J., Fredin, N.J., Wolff, M.R., Hacker, T.A., and Lynn, D.M. (2006) *Biomacromolecules*, **7**, 2483.
40 Fredin, N.J., Flessner, R.M., Jewell, C.M., Bechler, S.L., Buck, M.E., and Lynn, D.M. (2010) *Microsc. Res. Tech.*, **73**, 834.
41 Bechler, S.L. and Lynn, D.M. (2011) *J. Polym. Sci. Polym. Chem.*, **49**, 1572.
42 Klugherz, B.D., Song, C., DeFelice, S., Cui, X., Lu, Z., Connolly, J., Hinson, J.T., Wilensky, R.L., and Levy, R.J. (2002) *Hum. Gene Ther.*, **13**, 443.
43 Fishbein, I., Stachelek, S.J., Connolly, J.M., Wilensky, R.L., Alferiev, I., and Levy, R.J. (2005) *J. Control. Release*, **109**, 37.
44 Fishbein, I., Alferiev, I.S., Nyanguile, O., Gaster, R., Vohs, J.M., Wong, G.S., Felderman, H., Chen, I.W., Choi, H., Wilensky, R.L., and Levy, R.J. (2006) *Proc. Natl. Acad. Sci. USA*, **103**, 159.
45 Saurer, E.M., Yamanouchi, D., Liu, B., and Lynn, D.M. (2011) *Biomaterials*, **32**, 610.
46 Saurer, E.M., Flessner, R.M., Sullivan, S.P., Prausnitz, M.R., and Lynn, D.M. (2010) *Biomacromolecules*, **11**, 3136.
47 Coulman, S., Allender, C., and Birchall, J. (2006) *Crit. Rev. Ther. Drug.*, **23**, 205.
48 Gill, H.S. and Prausnitz, M.R. (2007) *J. Control. Release*, **117**, 227.
49 Prausnitz, M.R., Mikszta, J.A., Cormier, M., and Andrianov, A.K. (2009) *Curr. Top. Microbiol. Immunol.*, **333**, 369.
50 Donnelly, R.F., Singh, T.R.R., and Woolfson, A.D. (2010) *Drug Deliv.*, **17**, 187.
51 Demuth, P.C., Su, X., Samuel, R.E., Hammond, P.T., and Irvine, D.J. (2010) *Adv. Mater.*, **22**, 4851.
52 Saurer, E.M., Jewell, C.M., Kuchenreuther, J.M., and Lynn, D.M. (2009) *Acta Biomater.*, **5**, 913.
53 Chuang, H.F., Smith, R.C., and Hammond, P.T. (2008) *Biomacromolecules*, **9**, 1660.
54 Macdonald, M., Rodriguez, N.M., Smith, R., and Hammond, P.T. (2008) *J Control. Release*, **131**, 228.
55 Macdonald, M.L., Rodriguez, N.M., Shah, N.J., and Hammond, P.T. (2010) *Biomacromolecules*, **11**, 2053.
56 Shukla, A., Fleming, K.E., Chuang, H.F., Chau, T.M., Loose, C.R., Stephanopoulos, G.N., and Hammond, P.T. (2010) *Biomaterials*, **31**, 2348.
57 Decher, G. (1997) *Science*, **277**, 1232.
58 Wood, K.C., Chuang, H.F., Batten, R.D., Lynn, D.M., and Hammond, P.T. (2006) *Proc. Natl. Acad. Sci. USA*, **103**, 10207.
59 Liu, X., Yang, J.W., Miller, A.D., Nack, E.A., and Lynn, D.M. (2005) *Macromolecules*, **38**, 7907.
60 Zhang, J. and Lynn, D.M. (2007) *Adv. Mater.*, **19**, 4218.
61 Liu, X.H., Yang, J.W., and Lynn, D.M. (2008) *Biomacromolecules*, **9**, 2063.
62 Liu, X.H., Zhang, J.T., and Lynn, D.M. (2008) *Adv. Mater.*, **20**, 4148.
63 Sun, B. and Lynn, D.M. (2010) *J. Control. Release*, **148**, 91.
64 Funhoff, A.M., van Nostrum, C.F., Janssen, A.P.C.A., Fens, M.H.A.M., Crommelin, D.J.A., and Hennink, W.E. (2004) *Pharm. Res.*, **21**, 170.
65 Veron, L., Ganee, A., Charreyre, M.T., Pichot, C., and Delair, T. (2004) *Macromol. Biosci.*, **4**, 431.
66 Luten, J., Akeroyd, N., Funhoff, A., Lok, M.C., Talsma, H., and Hennink, W.E. (2006) *Bioconjugate Chem.*, **17**, 1077.
67 Jessel, N., Oulad-Abdelghani, M., Meyer, F., Lavalle, P., Haikel, Y., Schaaf, P., and Voegel, J.C. (2006) *Proc. Natl. Acad. Sci. USA*, **103**, 8618.
68 Flessner, R.M., Yu, Y., and Lynn, D.M. (2011) *Chem. Comm.*, **47**, 550.
69 Mendelsohn, J.D., Barrett, C.J., Chan, V.V., Pal, A.J., Mayes, A.M., and Rubner, M.F. (2000) *Langmuir*, **16**, 5017.

70 Hiller, J.A., Mendelsohn, J.D., and Rubner, M.F. (2002) *Nat. Mater.*, **1**, 59.
71 Cho, J. and Caruso, F. (2003) *Macromolecules*, **36**, 2845.
72 Sui, Z.J. and Schlenoff, J.B. (2004) *Langmuir*, **20**, 6026.
73 Seki, K. and Tirrell, D.A. (1984) *Macromolecules*, **17**, 1692.
74 Murthy, N., Robichaud, J.R., Tirrell, D.A., Stayton, P.S., and Hoffman, A.S. (1999) *J. Control. Release*, **61**, 137.
75 Lynn, D.M. (2006) *Soft Matter*, **2**, 269.

# 31
# Designing LbL Capsules for Drug Loading and Release
*Bruno G. De Geest and Stefaan C. De Smedt*

## 31.1
## Introduction

Layer-by-layer (LbL) [1] assembly is a powerful tool for engineering microparticulate structures. On the one hand it allows one to tailor the surface chemistry of microparticles, rendering them responsive to physicochemical stimuli such as pH, ionic strength, light, and so on, to allow bio-specific recognition or just to prevent adsorption of unwanted species [2]. On the other hand, using microparticles as sacrificial templates, one can fabricate spherically-shaped free-standing polymeric multilayer films, forming hollow capsules [3–5]. These capsules are formed in multiple steps. First, alternating polymeric layers are deposited onto core microparticles. Typically electrostatics or H-bonding [6–8] between the successive layers is used as the driving force for this multilayer build-up. Secondly, the core microparticles are decomposed into low molecular weight degradation products which can freely diffuse through the LbL membrane. The resulting hollow capsules have walls with a thickness of typically a few tens of nanometers and surround an aqueous void.

From this conceptual point of view, LbL capsules form a binary system in which both the LbL membrane and the hollow void can be exploited to perform a specific function. As the primary focus of our research laboratories lies in the field of drug delivery, we will present in this chapter several approaches that we have developed in applying LbL technology to the design of drug delivery systems. For drug delivery purposes, two aspects are crucial: (i) something should be encapsulated inside the capsules and (ii) the encapsulated content should be released. The most evident route for encapsulation is without doubt the incorporation of drug molecules within the hollow void of LbL capsules, while engineering the LbL membrane to be responsive to specific physicochemical stimuli looks a straightforward way to release encapsulated material. In collaboration with the Sukhorukov group we developed several strategies that allowed encapsulation of macromolecular drugs into capsules using pre-loaded microparticulate templates. Furthermore, we explored several possibilities to design LbL capsules that could release their payload in a controlled fashion, not only by engineering the LbL membrane, but also by engineering the core templates.

*Multilayer Thin Films: Sequential Assembly of Nanocomposite Materials*, Second Edition.
Edited by Gero Decher and Joseph B. Schlenoff.
© 2012 Wiley-VCH Verlag GmbH & Co. KGaA. Published 2012 by Wiley-VCH Verlag GmbH & Co. KGaA.

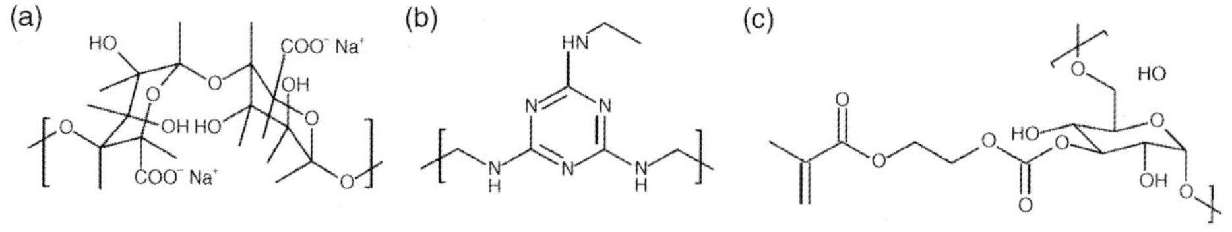

Figure 31.1 Molecular structure of (a) alginate, (b) melamine formaldehyde and (c) dex-HEMA.

## 31.2
### Engineering Microparticulate Templates to Design LbL Capsules for Controlled Drug Release

Engineering the capsule core templates allows one to incorporate drug molecules and provides opportunities to equip the capsules with a drug releasing mechanism. Hydrogel microspheres are a well established class of materials used in drug delivery. For example, calcium alginate microspheres, produced by pouring an aqueous alginate solution into a solution of divalent calcium ions, causing ionic gelation of the glucuronic and mannuronic acid (Figure 31.1a) residues by the $Ca^{2+}$ ions, has been used widely for drug and cell encapsulation. Furthermore, in order to stabilize these calcium alginate beads, poly-L-lysine coatings have been applied onto their surface to prevent the beads from dissolving when the $Ca^{2+}$ ions become exchanged by monovalent $Na^+$ ions in physiological media [9]. These coatings were also observed to form a diffusional barrier against diffusion of macromolecules into or out of the gel beads. These reports appeared in the early 1980s and calcium alginate beads were also among the first non-spherical substrates to be LbL coated. In 1994 Pommersheim et al. reported on the deposition of a multilayer coating onto calcium alginate beads with a diameter of several hundred micrometers [10]. By varying the number of deposited layers, the authors observed that, upon dissolution of the calcium alginate beads in the presence of $Na^+$ ions, either stable capsules were obtained or the capsule membrane ruptured due to the osmotic pressure of the dissolving hydrogel cores. The McShane group elaborated further on the use of calcium alginate as core templates from drug loaded LbL capsules [11]. As calcium alginate bears an anionic charge at physiological pH, it can absorb relatively high amounts of oppositely charged molecules and serve as a drug reservoir. Subsequent LbL coating of these drug-loaded alginate beads allowed stabilization of the beads and could also play a role in controlling subsequent drug release.

This concept of electrostatic loading was first described by Mohwald and coworkers using melamine formaldehyde microparticles as the sacrificial template [12]. Upon decomposition of melamine formaldehyde (Figure 31.2b) in acidic media, not all melamine oligomers diffused through the capsules' LbL membrane and a cationic gel-like structure that could be loaded with oppositely charged molecules remained within the capsule membrane. Just as observed by Pommersheim et al., these authors also observed that decomposition of the core templates could lead to capsule rupturing due to an osmotic shock of dissolved core template components [13]. De Geest et al. elaborated on this phenomenon to equip LbL capsules with a

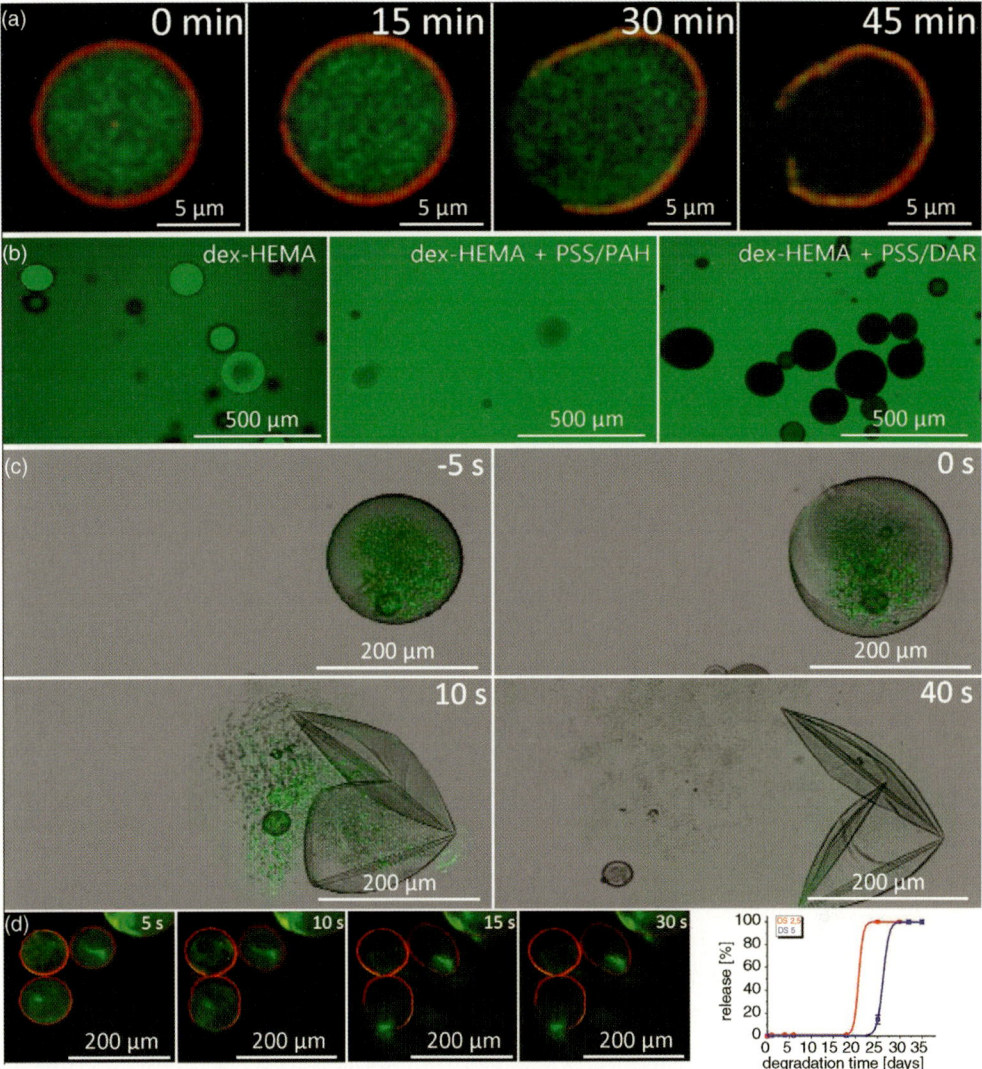

**Figure 31.2** (a) Confocal microscopy snapshots of self-exploding microcapsules. Dex-HEMA microgels were coated with a (PSS/PAH)$_3$ multilayer membrane and the degradation of the microgel core was accelerated by elevating the pH of the medium to 9 and heating to 37 °C. (b) Confocal microscopy images and fluorescence intensity profiles of bare dex-HEMA microgels, dex-HEMA microgels coated with (PSS/PAH)$_4$, and dex-HEMA microgels coated with (PSS/DAR)2 incubated in a 1 mg mL$^{-1}$ solution of FITC-dex. (c) Confocal microscopy snapshots taken at regular time intervals (overlay of green fluorescence channel and transmission channel) of (PSS/DAR)$_2$-coated microgels containing FITC-NP during the degradation of the microgel core triggered by the addition of sodium hydroxide. (d) Confocal microscopy snapshots of exploding capsules triggered by the addition of sodium hydroxide. The corresponding cumulative release curves of encapsulated 50 nm green fluorescent latex beads show that by varying the DS of the dex-HEMA (i.e., a DS of respectively 2.5 (red curve) and 5 (blue curve)) the onset of release upon incubation at physiological conditions can be tailored.

time-controlled release mechanism [14]. Figure 31.1c shows the molecular structure of these "dex-HEMA" hydrogels. Dextran was substituted with methacrylate groups that are connected to the dextran backbone by a carbonate ester. Polymerization of the methacrylates allows the formation of a crosslinked 3D network while hydrolysis of the carbonate esters allows this network to degrade into dextran and methacrylate oligomers. The use of dex-HEMA hydrogels offers several advantages: (i) the net charge can be tailored by incorporation of acidic (e.g., methacrylic acid) or basic (e.g., dimethylaminoethyl methacrylate (DMAEMA)), (ii) the degradation rate of the hydrogels can be tailored by varying the amount of crosslinks, and (iii) encapsulation of proteins into spherical dex-HEMA microgels with diameters ranging from 1 to 1000 μm is easily achieved using a water-in-water emulsion method based on the immiscibility of an aqueous poly(ethylene glycol) (PEG) and an aqueous dextran phase. This approach is of particular interest for the encapsulation of proteins as these often show enhanced affinity for a dextran phase compared to a PEG phase, in addition to the fact that organic solvents which might cause protein denaturation are avoided. The first paper in a series on this topic reported on the use of cationic microgels obtained by copolymerization of dex-HEMA with DMAEMA followed by coating of these microgels with three PSS/PAH polyelectrolyte bilayers. To prove the concept that the swelling pressure exerted by the degrading hydrogel core could be used as a trigger to rupture the $(PSS/PAH)_3$ membrane, the capsules were incubated at elevated pH (i.e., pH 9) and physiological temperature in order to accelerate the degradation of the dex-HEMA hydrogels. As shown in Figure 31.2a, at a certain moment, when the swelling pressure exceeds the tensile strength of the LbL membrane, the capsule suddenly ruptures and releases its payload.

In a number of subsequent papers, De Geest et al. further elaborated on this concept and elucidated that both the LbL membrane and the size and composition of the microgel templates played a critical role in controlling the release properties of these so-called "self-exploding capsules" [15]. An important observation was that, depending on the type of polyelectrolyte, either exploding or intact hollow capsules were obtained upon degradation of the microgel core. Furthermore, using FITC-labeled albumin or FITC-labeled dextran as a model drug, it was found that a significant fraction of the encapsulated payload was able to diffuse through the capsule membrane prior to capsule explosion [16]. These observations indicated the need to reinforce the capsule membrane to avoid premature leakage. This was tackled by applying a covalently stabilized LbL membrane, comprising PSS and diazoresin (DAR), that formed a covalent bond with the PSS's sulfonate groups [17]. The resulting capsules exhibited a dramatic decrease in permeability (Figure 31.2b) but, in order to allow rupturing of the capsule membrane upon degradation of the microgel core, it was necessary to synthesize microgels with a diameter larger than 100 μm to reduce the pressure required to overcome the capsules' tensile strength. As shown in Figure 31.2c, $(PSS/DAR)_2$-coated microgels are literally slashed prior to releasing their payload. By encapsulating fluorescent latex beads, it was possible to follow the trajectory of the released payload. Interestingly, it was observed that this type of "ejecting capsule" gives the latex beads such a momentum that they are propelled with an 800-fold increase in speed compared to mere Brownian motion.

Besides reinforcing the LbL membrane, another strategy to reduce drug diffusion is to use nano- or microparticles as the model drug rather than soluble macromolecules, as it could be hypothesized that the mesh size of an LbL membrane would be too small to allow premature release of particle-like materials. Furthermore, this also allowed the use of bio-polyelectrolytes, such as polysaccharides and polypeptides, instead of synthetic polymers such as PSS, PAH or DAR. Figure 31.2d shows confocal microscopy images of exploding capsules containing 50-nm sized latex beads [18]. An LbL membrane consisting of four layer pairs of dextran sulfate and poly-L-arginine was used to coat the dex-HEMA microgels. Upon dissolution of the microgel core, these capsules exploded and released their payload. Importantly, it was also demonstrated that by varying the crosslink density of the microgels (Figure 31.2d; DS: degree of substitution, that is, the amount of methacrylate groups per dextran backbone), it is possible to tailor the onset of burst release from the capsules. To further assess the versatility of this approach, LbL-coated 3-µm sized calcium carbonate microparticles and hollow LbL capsules were also demonstrated to be encapsulated and released from self-exploding capsules [19]. This allowed the construction of multi-compartment particles with the potential to load and release a wide variety of substances.

## 31.3
## Engineering the Shell to Design LbL Capsules for Controlled Drug Release

Due to their polyionic nature, polyelectrolytes are inherently stimuli-responsive. In an aqueous medium at low ionic strength, in the absence of salt, polyelectrolytes adopt an elongated "rod-like" conformation, while addition of salt leads to a more compact "coiled" conformation. It had already been observed in the early days of LbL research that salt had a tremendous effect on both polyelectrolyte multilayer assembly and on pre-formed polyelectrolyte multilayers. Salt ions are capable of screening electrostatic charges and inducing swelling of the multilayers, increasing their permeability [20]. This also holds true for LbL capsules, and the Mohwald group demonstrated that LbL empty capsules could be loaded with macromolecules at elevated salt concentration, be "closed" at low ionic strength, and subsequently release their payload by again raising the salt concentration [21]. Besides responsiveness to salt, pH is also often an inherent trigger to change the behavior of a polyelectrolyte [22]. Especially, weak polyelectrolytes, such as PAA and PAH, exhibit a pronounced charge-shifting behavior which allows them to be uncharged, partly charged or fully charged, depending on their apparent $pK_a$ and the pH of the surrounding medium. Capsules consisting of one or more weak polyelectrolytes have also been demonstrated to exhibit reversible swelling/shrinking and loading/unloading by cycling the pH around the apparent $pK_a$ of the polyelectrolyte complexes [23].

However, although both of these mechanisms, that is, varying ionic strength and pH, clearly allow controlled release from LbL capsules, therapeutic applications of these concepts are scarce as large variations in pH or ionic strength are predominantly encountered in the gastro-intestinal tract where competition with other established drug delivery systems intended for oral intake is difficult. In order to find their way in the field of drug delivery it is most likely that LbL capsules will have to be administered through parenteral injection, where they could act as a depot that releases its content after a specific stimulus, or after cellular uptake. Keeping this in mind our research laboratories, in collaboration with the Sukhorukov group, have attempted to work towards the design of LbL capsules that could release their payload under physiologically relevant conditions.

Glucose is a common metabolite and patients suffering from diabetes mellitus fail to secrete sufficient levels of insulin to lower glucose levels in the blood stream. Rather than injecting fixed doses of insulin at fixed time intervals, it could be advantageous to design drug delivery systems that release insulin on demand, that is, when glucose levels pass a certain threshold. This inspired us to synthesize polymers that could shift their overall charge and their charge density depending on the concentration of glucose in the medium. Phenylboronic acids are known to form anionic complexes with glucose and copolymerization of a phenylboronic acid containing monomer with a basic monomer yielded a polyelectolyte which had a net positive charge in the absence of glucose, while addition of glucose induced anionic charges on the polymer backbone, which lowered the net charge density and changed the conformation of the polyelectrolyte. Assembling this polymer with PSS onto sacrificial polystyrene microtemplates, followed by decomposition of the polystyrene in THF, resulted in hollow capsules that contained glucose responsive moieties. Addition of glucose effectively induced disassembly of the capsules [24]. This concept could potentially be used for glucose-induced insulin release. However, the type of phenyl boronic acid used to construct these capsules has a $pK_a$ between 8 and 9, which means that glucose induced charge-shifting only takes place in this pH range. Recently, several groups have reported on the synthesis of phenylboronic acids that are responsive under physiological conditions, which could be of interest for application in our concept of glucose responsive capsules.

Another common stimulus that can be provided by the human body, and which is not restricted to diseased patients, is enzymatic hydrolysis. Enzymes are omnipresent in body fluids and actively phagocyting cells contain an abundance of proteases in their lysosomes. Pioneering work by Picart *et al.* on planar films composed of hyaluronic acid and poly-L-lysine has demonstrated that living cells could attach to these LbL films, invade and gradually digest them [25]. Therefore, multilayer capsules built from polypeptides in their shell should also be prone to enzymatic hydrolysis and could thus serve as carriers for intracellular release of encapsulated therapeutics. This concept was explored by De Geest and coworkers using calcium carbonate ($CaCO_3$) coated with a polyelectrolyte multilayer film of dextran sulfate and poly-L-arginine as a sacrificial template [26]. The use of porous microparticles such as $CaCO_3$, which was introduced by the Sukhorukov group [27], as well as the use of mesoporous silica particles, which was introduced by the Caruso group [28], offers tremendous potential for the encapsulation of macromolecular drugs such as proteins and polynucleic acids. $CaCO_3$ is cheap, non-toxic and biocompatible. It is synthesized under ambient conditions by mixing aqueous solutions of sodium carbonate and calcium chloride, and the resulting precipitate forms fairly monodisperse particles with a diameter of, typically, 3 μm and a high surface to volume ratio. By adding proteins during the precipitation reaction, these become incorporated within the pores with a nearly 100% encapsulation efficiency. Moreover, due to their porous nature, exhibiting a much higher surface roughness than typically "smooth" particles, such as PS, MF or silica, significantly higher amounts of polyelectrolytes are adsorbed during each deposition cycle, creating thick walled capsules which are more robust and less prone to buckling instabilities. This is of particular interest for intracellular drug delivery following parenteral administration, as the surrounding tissue will cause a certain mechanical pressure which should be withstood by the capsules prior to cellular internalization. Finally, $CaCO_3$ is dissolved under mild conditions in water by complexation with EDTA, leading to non-toxic low molecular weight degradation products, such as $CO_2$ and $Ca^{2+}$. As an alternative to enzymatic hydrolysis, De Geest and coworkers also explored the use of degradable charge-shifting polycations (i.e., poly(hydroxypropyl methacrylamide-dimethylaminoethyl) (poly(HPMA-DMAE))) developed by the Hennink group [29]. These polycations were based on a polymethacrylamide backbone which was substituted with tertiary amine groups that were linked to the polymer backbone through a hydrolyzable carbonate ester. The use of degradable polycations – based on polyamines synthesized by Michael addition of dimethacrylates to diamines – to construct "erodable" LbL films was pioneered by Lynn and coworkers, and has been shown to be an effective approach for surface-mediated drug delivery [30]. As shown in Figure 31.3, two types of degradable capsules, as well as non-degradable synthetic PSS/PAH capsules were incubated with an *in vitro* cultured cell line (i.e., VERO cells). All three types of capsules were efficiently internalized by this cell line and through co-localization with "LysoTracker" (a fluorescent marker which stains intracellular acidic vesicles) and several endocytotic inhibitors it was found that these LbL capsules enter the cell through caveolae-mediated endocytosis and end up in endo/lyso/phago-somal vesicles. Whereas PSS/PAH capsules remaine intact over several days within lysosomal compartments of the cells, capsules based on dextran sulfate poly-L-arginine, or capsules containing the degradable polycation poly(HPMA-DMAE) exhibited intracellular degradation and, after several days of incubation, only debris of degraded capsules could be observed.

Besides enzymatic hydrolysis, another intracellular stimulus is the reductive environment which is encountered upon cellular internalization. This offers the possibility for redox-responsive capsules to deliver their payload selectively inside living cells. The Caruso group developed an elegant approach to the construction of capsules that could disassemble through reduction of disulfides in the presence of physiologically relevant glutathione concentrations [31]. First, poly(methacrylic acid) was substituted with thiol groups ($PMA^{SH}$) and assembled onto sacrificial silica microtemplates with PVP through hydrogel bonding. Subsequently, the $PMA^{SH}$ layers were crosslinked by oxidative disulfide formation and the silica microparticles were dissolved in diluted HF solution. The resulting capsules were stable under normal physiological conditions but decomposed in an oxidative medium. The introduction of both redox-sensitive and enzymatically degradable capsules has

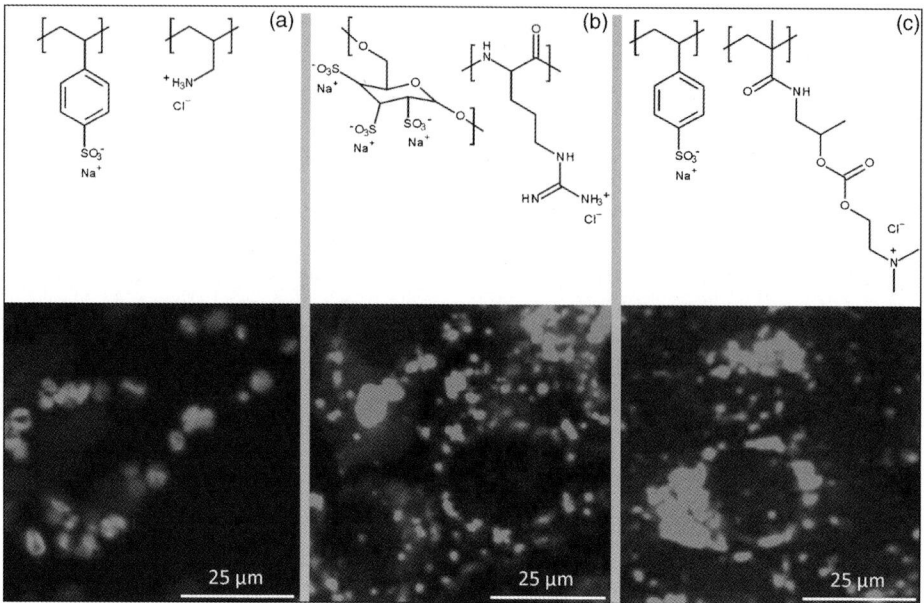

**Figure 31.3** Molecular structure of different polyelectrolytes used to construct LbL capsules. Confocal microscopy images of PSS/PAH, dextran sulfate/poly-L-arginine and PSS/poly (HPMA-DMAE) capsules that were subsequently incubated with VERO cells for 60 h. The green fluorescence originates from FITC-dextran that was encapsulated.

paved the road for the use of LbL capsules in a therapeutic setting and they are currently being evaluated in a number of drug delivery applications.

An alternative strategy to the use of physiological stimuli to induce drug release from LbL capsules, is the use of external triggers, such as light, ultrasound, magnetism or radiofrequency fields. Multilayer capsules susceptible to one of these triggers might be used as a drug depot – either extracellular or intracellular – and only release their payload after application of the specific physico-chemical stimulus. Light-triggered release from LbL capsules was elaborated on by both the Caruso and Sukhorukov groups by incorporating gold nanoparticles within the LbL shell [32, 33]. Upon irradiation with IR light, the gold nanoparticles absorb the energy and transform it into thermal energy. As a consequence, the capsules are heated far above their glass transition temperature ($T_g$) and break, releasing their encapsualetd payload. This process has been shown to be well tolerated by living cells, and it was even shown that triggered release from capsules that were first fagocyted by living cells ruptured the lysosomal compartment, and released the capsules' payload within the cellular cytoplasm [34]. Preliminary experiments showed no effect on cell viability and still allowed intracellular processes such as MHC-I presentation of peptides, released from the capsules, to take place.

Our research laboratories have, in collaboration with the Sukhorukov group, also contributed to the field of triggered release, focusing on the release of encapsulated macromolecules from calcium carbonate template capsules. As mentioned earlier, due to its high porosity, CaCO$_3$ tends to adsorb higher amounts of polyelectrolyte than other template particles. This leads to thicker shells that exhibit enhanced mechanical stability and which will render the capsules less prone to small distortions caused by external triggers, such as light and ultrasound. To cope with these issues, hybrid nanoparticle/polyelectrolyte capsules were constructed. The rationale behind this was to reduce the capsule elasticity to render the capsules more susceptible to fracture. Mercaptosuccinic acid stabilized 1 nm gold nanoparticles with an anionic surface charge were assembled with the polycation PAH onto CaCO$_3$ microparticles, without any additional polyanion [35]. As shown in Figure 31.4a, this leads to capsules with a high content of gold nanoparticles which are literally glued together by PAH. These capsules were evaluated for their responsiveness to IR laser

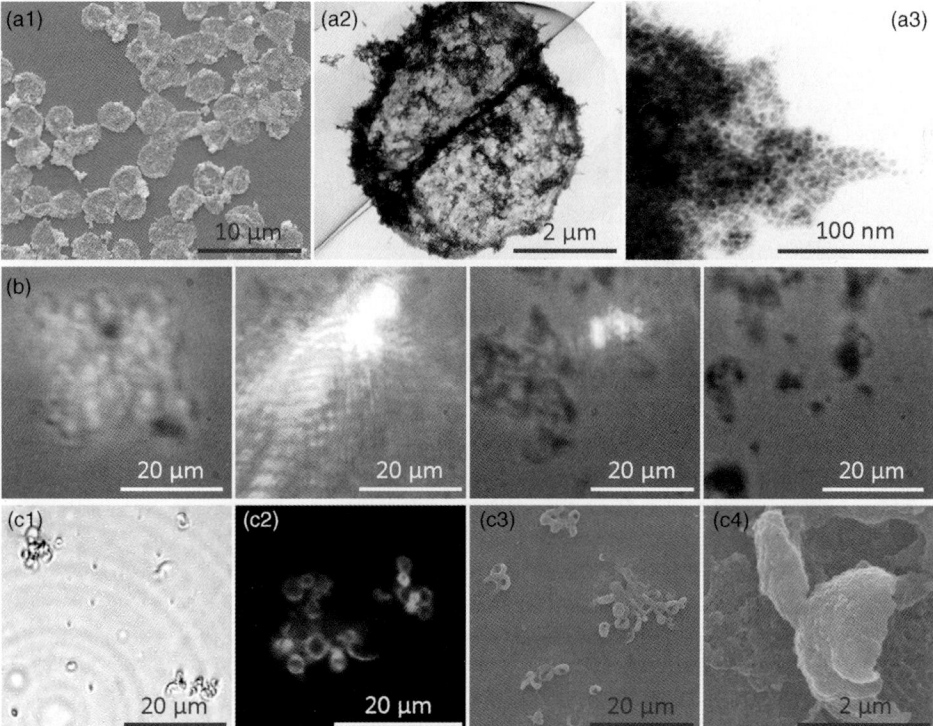

**Figure 31.4** (a) Scanning and (a2, a3) transmission electron microscopy images of hybrid nanoparticle/polyelectrolyte capsules composed of gold nanoparticles and PAH. (b) Optical transmission, confocal microscopy and scanning electron microscopy images of hybrid capsules after ultrasonic treatment. (c) Fluorescence microscopy images of a cluster of FITC-dextran (bright color)-filled nanoparticle/polyelectrolyte capsules upon irradiation with IR laser light. The time interval between the successive images is 2 s. (c1) Optical transmission, (c2) confocal fluorescence images and (c3,c4) scanning electron microscopy of hybrid nanoparticle/polyelectrolyte capsules after ultrasonic treatment.

# 758 | 31 Designing LbL Capsules for Drug Loading and Release

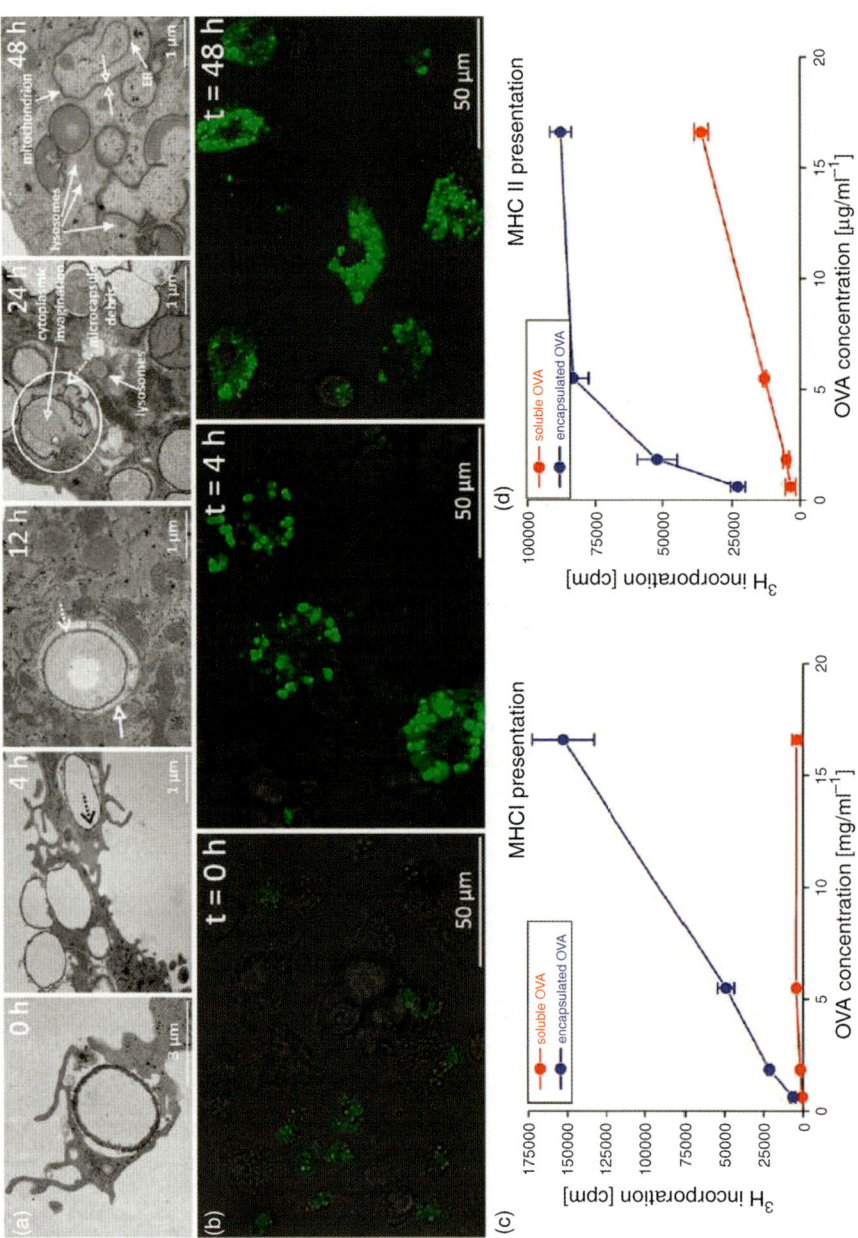

irradiation and ultrasound. Exposure to a focused 30 mW laser beam caused a whole agglomerate of capsules to explode and release their fluorescent payload (Figure 31.4b). A similar result was obtained when a capsule suspension was subjected to 10 s of a 20 W ultrasonic treatment with a frequency of 20 kHz (Figure 31.4c) [36]. These data clearly demonstrated the potential of laser and ultrasound irradiation for on-demand drug release from LbL capsules. However, there is still a long way to go for these systems towards clinical applications. Several issues regarding the penetration depth of light and ultrasound required to address the capsules have to be resolved, as well as the tissue reaction to injected capsules.

## 31.4
### Interaction of LbL Capsules with Living Cells *In Vitro* and *In Vivo*

For applications in drug delivery, it is of the utmost importance that LbL capsules can be designed in such a way that they are non-toxic to living cells. This issue has been addressed by several groups, so far *in vitro*, and a general consensus is that at moderate capsule to cell ratios no acute cytotoxicity is observed [37, 38]. LbL capsules could be of particular interest for intracellular drug delivery. Several cancer cell lines, as well as immune cells, such as macrophagues and dendritic cells, have been shown to be capable of internalizing LbL capsules. The Caruso group is currently performing pioneering work on engineering LbL capsules to load both hydrophilic and hydrophobic low molecular weight anticancer drugs. Recently, this group was able to demonstrate highly specific targeting and uptake of antibody functionalized LbL capsules by receptor recognition with extreme precision [39].

Our research laboratories have been active in evaluating LbL capsules for vaccine delivery to dendritic cells, which are the most potent antigen presenting cells. As mentioned earlier, upon cellular uptake, LbL capsules end up in intracellular acidic vesicles. The mechanism through which dendritic cells internalize LbL capsules composed of dextran sulfate and poly-L-arginine was investigated using various inhibitors of different endocytotic pathways. Blocking of actin polymerization appeared to completely abolish capsule uptake, suggesting an important role for

**Figure 31.5** (a) TEM images of BM-DCs that have internalized dextran sulfate/poly-L-arginine microcapsules at the indicated time intervals. Microcapsule shell: dotted arrows; membranes surrounding the microcapsules: open arrows. In the encircled area, microcapsule rupture and cytoplasmic invagination are clearly distinguishable. Lysosomes, endoplasmic reticulum (ER), and a mitochondrion are indicated by the solid arrows. (b) Processing of dextran sulfate/poly-L-arginine microcapsule encapsulated OVA was analyzed using DQ-OVA. Confocal microscopy images of BM-DCs incubated with OVA-DQ microcapsules for 0, 4 and 48 h (overlay of green fluorescence and DIC). (DQ-OVA is ovalbumin oversaturated with BODIPY dyes. Upon proteolytic cleavage, quenching is relieved and green fluorescence appears. (c) Antigen presentation by BM-DCs after uptake of soluble and encapsulated OVA. Proliferation of OT-I cells was used as a measure for MHC-I-mediated cross-presentation of OVA (d), proliferation of OT-II cells as a measure for MHC-II mediated presentation.

cytoplasmic engulfment [40]. This was confirmed by transmission electron microscopy, and actin staining with fluorescence microscopy proved the role of cytoplasmic protrusions in the process of capsule internalization. Transmission electron microscopy was further used to assess the intracellular fate of the internalized capsules, and it was observed that the capsules remained surrounded by a lipid membrane. However, over time, the capsule shell ruptured and cytoplasmic content protruded into the capsule core (Figure 31.5a), which can most likely be attributed to a combination of enzymatic degradation by endo/lysosomal proteases and mechanical force exerted by the surrounding cytoplasm. Furthermore, several cellular organelles, such as lysosomes, mitochondria and endoplasmatic reticulum, were recruited towards the ruptured capsules, which will likely play a role in the processing and presentation of peptide fragments from encapsulated vaccine antigens.

To assess the intracellular fate of encapsulated antigens, ovalbumin (OVA) was encapsulated as model protein antigen. Antigen processing was investigated using a fluorogenic substrate of OVA (i.e., DQ-OVA); comprising OVA that is oversaturated with BODIPY dyes, thereby forcing the fluorescence in a quenched state. When DQ-OVA is degraded into small peptide fragments, the quenching is relieved and a bright green fluorescence emerges. This technique demonstrated that LbL capsules consisting of degradable polyelectrolytes (i.e., dextran sulfate and poly-L-arginine) allowed ready processing of the encapsulated proteins, as confocal microscopy and flow cytometry showed that antigen processing started in less than 4 h after cellular uptake (Figure 31.5b). As a control, DQ-OVA was encapsulated in non-degradable PSS/PAH capsules and found not to be processed upon cellular internalization. These observations indicated the crucial influence of capsule design in order to grant access of proteases to encapsulated protein antigens. To assess whether this fast antigen processing was accompanied with enhanced presentation of the OVA CD4 and CD8 peptide fragments to CD4, respectively CD8 T-cells, dendritic cells that were pulsed with OVA-loaded capsules were co-cultured with OT-I, respectively OT-II cells. OT-I cells are CD8 T cells with a transgenic T-cell receptor that specifically recognizes the OVA peptide SIINFEKL presented by MHC-I, whereas OTII cells are transgenic CD4 T cells that specifically recognize the OVA peptide LSQAVHAAHAEINEAGR presented by MHC-II. As shown in Figure 31.5c, LbL capsule-mediated OVA delivery dramatically induces T-cell proliferation – as a measure of antigen presentation of dendritic cells to T-cells – compared to soluble OVA. This was especially found to be the case for presentation to CD8 T-cells, which is referred to as cross-presentation and believed to be a crucial step in the induction of cellular immunity against insidious intracellular pathogens, as well as cancer.

So far, few studies on the *in vivo* performance of LbL capsules have been reported. De Koker and coworkers have assessed the *in vivo* fate of LbL capsules composed of dextran sulfate and poly-L-arginine after subcutaneous injection and pulmonary delivery [41, 42]. Both studies were performed in mice. The pro-inflammatory response to subcutaneously injected capsules was characterized by the recruitment of polymorphonuclear cells and monocytes, and found to be within the same range as FDA approved vaccine adjuvants such as aluminum hydroxide. The injected capsules behaved as a porous implant with cellular infiltration emerging from the periphery and proceeding over time over the whole injected volume [41]. No ulceration was observed and the inflammatory response remained confined to the injection site, which became surrounded by several layers of fibroblasts. Tissue sections obtained from mice that were injected with fluorescently labeled capsules revealed that the capsules remained intact before becoming phagocyted by infiltrating cells. Two weeks post injection, all capsules were found to be inside cells and to have lost their spherical shape. One month post injection, only capsule debris could be observed within the cells. Taken together, these experiments demonstrate that LbL capsules fabricated from degradable polyelectrolytes are well tolerated *in vivo* and could serve as a drug carrier towards phagocyting cells.

Instillation of OVA-loaded capsules into the lungs of mice revealed a transient inflammation and promoted strong humoral and cellular immune responses [42]. This was attributed to the ability of LbL capsules to restrict the antigen to actively phagocyting cells, such as dendritic cells, whereas non-encapsulated soluble antigen would readily diffuse into the surrounding tissue. Moreover, capsule-mediated antigen delivery also resulted in an increased activation state of antigen presenting cells, through complement activation.

## 31.5
### Conclusions

In the early years of LbL technology, since it was introduced by Gero Decher in 1991, and extended to hollow LbL capsules by the Mohwald group in 1998, a first objective has been to explore the potential of LbL technology for a wide range of drug delivery applications and to assess whether there was potential to compete with existing technologies, or even if there was an opportunity for LbL technology to offer an advantage. In this context, several concepts have been developed and evaluated, mainly in chemistry labs without direct applications being readily at hand. Nowadays, the field has moved more and more towards the development of LbL capsules specifically engineered for a well defined drug delivery purpose, for example, the delivery of cancer therapeutics, vaccine delivery, and so on. Furthermore, another emerging trend is simplification of the fabrication procedure. Whereas LbL technology inherently suffers from a multistep assembly, involving many time- and product consuming batch operations, more and more groups, both in the fields of planar LbL films as well as LbL capsules, are making efforts to drastically reduce the number of steps needed to generate capsules while aiming to keep the versatility of the LbL approach.

## References

1 Decher, G. (1997) *Science*, **277**, 1232.
2 De Geest, B.G., Sanders, N.N., Sukhorukov, G.B., Demeester, J., and De Smedt, S.C. (2007) *Chem. Soc. Rev.*, **36**, 636.
3 Caruso, F., Caruso, R.A., and Mohwald, H. (1998) *Science*, **282**, 1111.

4 Donath, E., Sukhorukov, G.B., Caruso, F., Davis, S.A., and Mohwald, H. (1998) *Angew Chem. Int. Edit.*, **37**, 2202.
5 Sukhorukov, G.B., Donath, E., Davis, S., Lichtenfeld, H., Caruso, F., Popov, V.I., and Mohwald, H. (1998) *Polym. Adv. Technol.*, **9**, 759.
6 Such, G.K., Johnston, A.P., and Caruso, F. (2011) *Chem. Soc. Rev.*, **40**, 19.
7 Quinn, J.F., Johnston, A.P., Such, G.K., Zelikin, A.N., and Caruso, F. (2007) *Chem. Soc. Rev.*, **36**, 707.
8 Sukhishvili, S. (2005) *Curr. Opin. Colloid Interface Sci.*, **10**, 37.
9 Jarvis, A., Grinda, T., Chipura, W., Sullivan, M., and Koch, G. (1983) *Abstr. Pap. Am. Chem. Soc.*, **185**, 8.
10 Pommersheim, R., Schrezenmeir, J., and Vogt, W. (1994) *Macromol. Chem. Phys.*, **195**, 1557.
11 Zhu, H., Srivastava, R., and McShane, M. (2005) *Biomacromolecules*, **6**, 2221.
12 Gao, C., Donath, E., Mohwald, H., and Shen, J. (2002) *Angew Chem. Int. Edit.*, **41**, 3789.
13 Gao, C., Moya, S., Donath, E., and Mohwald, H. (2002) *Macromol. Chem. Phys.*, **203**, 953.
14 De Geest, B., Dejugnat, C., Sukhorukov, G., Braeckmans, K., Demeester, J., and De Smedt, S. (2005) *Adv. Mater.*, **17**, 2357.
15 De Geest, B., Dejugnat, C., Prevot, M., Sukhorukov, G., Demeester, J., and De Smedt, S. (2007) *Adv. Func. Mater.*, **17**, 531.
16 De Geest, B., De Koker, S., Demeester, J., De Smedt, S., and Hennink, W. (2010) *Polym. Chem.*, **1**, 137.
17 De Geest, B., McShane, M., Demeester, J., De Smedt, S., and Hennink, W. (2008) *J. Am. Chem. Soc.*, **130**, 14480.
18 De Geest, B., De Koker, S., Demeester, J., De Smedt, S., and Hennink, W. (2009) *J. Control Release*, **135**, 268.
19 De Geest, B., De Koker, S., Immesoete, K., Demeester, J., De Smedt, S., and Hennink, W. (2008) *Adv. Mater.*, **20**, 3687.
20 Schlenoff, J., Ly, H., and Li, M. (1998) *J. Am. Chem. Soc.*, **120**, 7626.
21 Ibarz, G., Dahne, L., Donath, E., and Mohwald, H. (2002) *Macromol. Rapid. Commun.*, **23**, 474.
22 Shiratori, S. and Rubner, M. (2000) *Macromolecules*, **33**, 4213.
23 Mauser, T., Dejugnat, C., and Sukhorukov, G. (2004) *Macromol. Rapid. Commun.*, **25**, 1781.
24 De Geest, B., Jonas, A., Demeester, J., and De Smedt, S. (2006) *Langmuir*, **22**, 5070.
25 Picart, C., Schneider, A., Etienne, O., Mutterer, J., Schaaf, P., Egles, C., Jessel, N., and Voegel, J. (2005) *Adv. Func. Mater.*, **15**, 1771.
26 De Geest, B., Vandenbroucke, R., Guenther, A., Sukhorukov, G., Hennink, W., Sanders, N., Demeester, J., and De Smedt, S. (2006) *Adv. Mater.*, **18**, 1005.
27 Volodkin, D., Larionova, N., and Sukhorukov, G. (2004) *Biomacromolecules*, **5**, 1962.
28 Wang, Y., Yu, A., and Caruso, F. (2005) *Angew Chem. Int. Edit.*, **44**, 2888.
29 Funhoff, A., van Nostrum, C., Janssen, A., Fens, M., Crommelin, D., and Hennink, W., (2004) *Pharm. Res.*, **21**, 170.
30 Jewell, C., and Lynn, D. (2008) *Adv. Drug Deliv. Rev.*, **60**, 979.
31 Zelikin, A., Quinn, J., and Caruso, F. (2006) *Biomacromolecules*, **7**, 27.
32 Radt, B., Smith, T., and Caruso, F. (2004) *Adv. Mater.*, **16**, 2184.
33 Skirtach, A., Dejugnat, C., Braun, D., Susha, A., Rogach, A., Parak, W., Mohwald, H., and Sukhorukov, G. (2005) *Nano Lett.*, **5**, 1371.
34 Skirtach, A., Javier, A., Kreft, O., Kohler, K., Alberola, A., Mohwald, H., Parak, W., and Sukhorukov, G. (2006) *Angew Chem. Int. Edit.*, **45**, 4612.
35 De Geest, B., Skirtach, A., De Beer, T., Sukhorukov, G., Bracke, L., Baeyens, W., Demeester, J., and De Smedt, S. (2007) *Macromol. Rapid. Commun.*, **28**, 88.
36 De Geest, B., Skirtach, A., Mamedov, A., Antipov, A., Kotov, N., De Smedt, S., and Sukhorukov, G. (2007) *Small*, **3**, 804.
37 Javier, A., Kreft, O., Alberola, A., Kirchner, C., Zebli, B., Susha, A., Horn, E., Kempter, S., Skirtach, A., Rogach, A., Radler, J., Sukhorukov, G., Benoit, M., and Parak, W. (2006) *Small*, **2** 394.
38 Kirchner, C., Javier, A., Susha, A., Rogach, A., Kreft, O., Sukhorukov, G.,

and Parak, W. (2005) *Talanta*, **67**, 486.
39 Kamphuis, M., Johnston, A., Such, G., Dam, H., Evans, R., Scott, A., Nice, E., Heath, J., and Caruso, F. (2010) *J. Am. Chem. Soc.*, **132**, 15881.
40 De Geest, B., De Koker, S., Singh, S., De Rycke, R., Naessens, T., Van Kooyk, Y., Demeester, J., De Smedt, S., and Grooten, J. (2009) *Angew Chem. Int. Edit.*, **48**, 8485.
41 De Koker, S., De Geest, B., Cuvelier, C., Ferdinande, L., Deckers, W., Hennink, W., Mertens, N., and De Smedt, S. (2007) *Adv. Func. Mater.*, **17**, 3754.
42 De Koker, S., Naessens, T., De Geest, B., Bogaert, P., Demeester, J., De Smedt, S., and Grooten, J. (2010) *J. Immunol.*, **184**, 203.

# 32
## Stimuli-Sensitive LbL Films for Controlled Delivery of Proteins and Drugs

*Katsuhiko Sato, Shigehiro Takahashi, and Jun-ichi Anzai*

### 32.1
### Introduction

Layer-by-layer (LbL) deposited thin films have attracted much attention in the development of controlled delivery systems for drugs and proteins. In most cases, LbL films are constructed by the alternate and repeated adsorption of polycations and polyanions on the surface of a solid support from solution. The driving force of LbL deposition is not limited to electrostatic forces of attraction; binding interactions such as hydrogen bonding and biological affinity can also be utilized. As such, a variety of materials have been employed as components for LbL films, including dyes, proteins, DNA, and nanoparticles. Thus, desired components and functionalities can be incorporated into LbL films, and this forms the basis for the development of stimuli-sensitive LbL films. This chapter focuses on LbL films that can be disintegrated in response to external stimuli, such as pH changes and the addition of biological compounds. As examples of such stimuli-sensitive systems, LbL films and microcapsules composed of binding proteins (avidin and concanavalin A), poly (amidoamine) dendrimers, and insulin are discussed in relation to their future application to the controlled delivery of proteins and drugs.

### 32.2
### Avidin-Containing LbL Films

Avidin is a glycoprotein found in egg white and is isolated as a tetramer of identical polypeptide subunits. The subunit contains a strong binding site for biotin and its analogues (binding constant for biotin, $\sim 10^{15}\,\mathrm{M}^{-1}$) (Figure 32.1) [1]. The binding sites for biotin are arranged in two pairs on the outer faces of the tetrad of avidin subunits, whose molecular dimensions are approximately $4.0 \times 5.0 \times 5.5\,\mathrm{nm}^3$ [1]. This arrangement of binding sites suggests that avidin-based LbL films could be constructed using biotin-tagged polymers and/or proteins. Indeed, avidin has previously been employed as a building block for the construction of LbL films in combination with biotin-modified polymers and proteins [2–5]. The avidin-containing

*Multilayer Thin Films: Sequential Assembly of Nanocomposite Materials*, Second Edition.
Edited by Gero Decher and Joseph B. Schlenoff.
© 2012 Wiley-VCH Verlag GmbH & Co. KGaA. Published 2012 by Wiley-VCH Verlag GmbH & Co. KGaA.

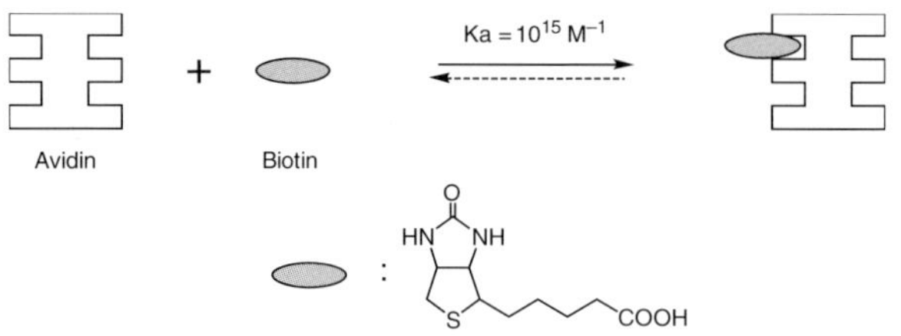

**Figure 32.1** Binding equilibrium between avidin and biotin.

LbL films were found to be highly stable because of the strong affinity between avidin and biotin.

In this context, we have used 2-iminobiotin-labeled poly(ethyleneimine) (ib-PEI) in place of biotin-labeled polymer to construct LbL films consisting of avidin and ib-PEI, and studied stimuli-induced disintegration of the films. Avidin binds 2-iminobiotin less strongly than biotin and the binding constant is pH-dependent. The binding constant of avidin for 2-iminobiotin is $2.9 \times 10^{10}\,M^{-1}$ in basic media while the binding constant of avidin for the protonated form of 2-iminobiotin in acidic media is $\sim 10^3\,M^{-1}$ (Figure 32.2). Thus, it is reasonable to assume that LbL films composed of ib-PEI would be sensitive to the pH environment. In fact, avidin/ib-PEI films were decomposed by changing the environmental pH from basic to weakly acidic or by adding biotin to the solution in which the LbL film was immersed [6, 7]. Figure 32.3a shows the disintegration of a 10 layer pair (avidin/ib-PEI)$_{10}$ film induced by pH changes. The LbL film was highly stable at pH 8.0 or higher, while at pH 7.0 the film was partly degraded. In contrast, the film completely decomposed within a few minutes at pH 5.0 or 6.0 due to the reduced affinity of ib-PEI for avidin as a result of protonation of the 2-iminobiotin moiety in the film. This decomposition is not due to the denaturation of avidin in the acidic media as avidin is known to retain its binding affinity for biotin and its analogues in the pH range 3–7. These results clearly show that the avidin/ib-PEI LbL film can be disintegrated by changing the solution pH from basic to weakly acidic. The pH response of the avidin/ib-PEI film is in clear contrast to that of hydrogen bond-mediated LbL films which are decomposed in neutral and basic solutions [8].

Another interesting feature of avidin/ib-PEI LbL films is that they are sensitive to biotin and its analogues (Figure 32.3b). The (avidin/ib-PEI)$_{10}$ LbL film completely decomposed within a few minutes of addition of biotin to a concentration of $1 \times 10^{-5}\,M$ in solution at pH 8.0 due to the preferential binding of added biotin

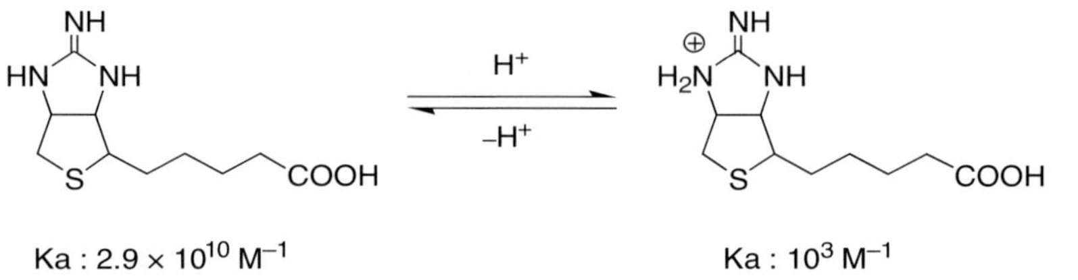

**Figure 32.2** pH-dependent binding constant of 2-iminobiotin.

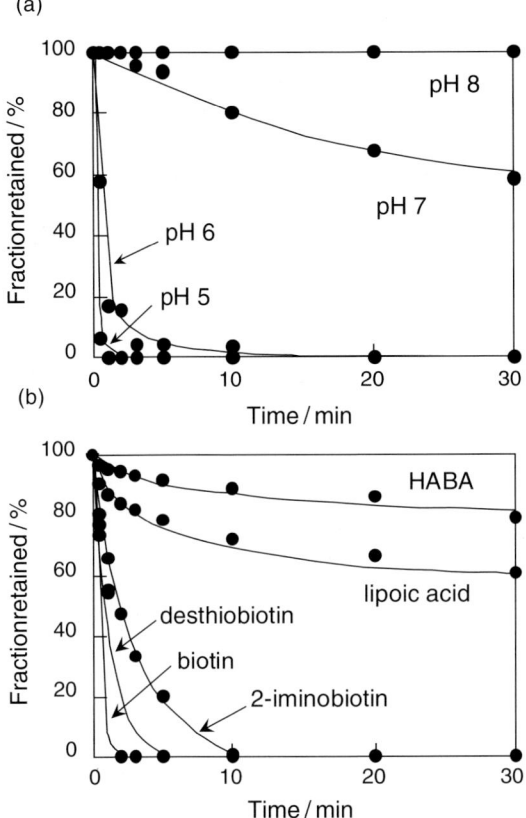

**Figure 32.3** pH-induced (a) and biotin analog-induced (b) disintegration of (avidin/ib-PEI)$_{10}$ film. Concentration of biotin and analogues: $10^{-5}$ M (pH 8.0). (Reprinted with permission from Ref. [6]. Copyright 2005, The American Chemical Society).

to the binding sites of the avidin. Biotin analogues also induced the disintegration of the LbL film to varying extents depending on the binding affinity of the analog for avidin. The rate of film decomposition correlated positively with the concentration of biotin analogues added. For example, the addition of $1 \times 10^{-3}$ M lipoic acid or HABA induced the complete decomposition of the LbL film in a few minutes. In short, the decomposition rate of the avidin/ib-PEI film can be rationally controlled by changing the type of stimulant and its concentration.

The avidin/ib-PEI films can also be disintegrated in response to electrochemical stimuli (i.e., electrode potential). Figure 32.4 illustrates schematically the electrochemical disintegration of an LbL film deposited on the surface of a platinum (Pt) film-coated quartz crystal microbalance (QCM) probe [9]. The decomposition of the LbL film was dependent on the magnitude of the electrode potential applied and the concentration and pH of the buffer. The avidin/ib-PEI film decomposed instantly when +0.9 or +1.0 V of electrode potential was applied to the Pt layer in 1 mM buffer

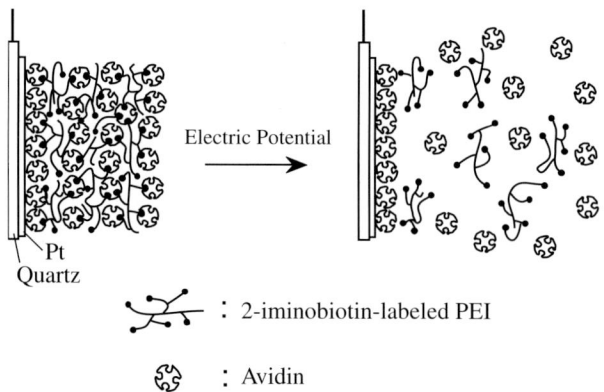

**Figure 32.4** Disintegration of (avidin/ib-PEI) LbL film induced by the application of electric potential. (Reprinted with permission from Ref. [9]. Copyright 2006, The American Chemical Society).

at pH 9.0. This is probably due to a local pH change induced by the electrolysis of water in the vicinity of the surface of the Pt electrode. It is likely that the applied potential caused water to oxidize, generating $H^+$ ions on the Pt electrode and resulting in the acidification of the medium around the film/electrode interface. Thus, the binding affinity of the 2-iminobiotin residues for avidin in the LbL film was lowered. The effects of the buffering capacity of the medium were also significant. Film decomposition was more rapid in a lower concentration buffer (1 mM) than in higher concentration (10 or 100 mM) buffers with identical pH, which further supports the supposition that the film decomposed as a result of the local pH change at the electrode. In this system, the application of electrode potential was sufficient to decompose the LbL film, with no reagent addition required. This is an advantage over other systems which require chemical stimuli such as acids, bases, ions, or other molecules to induce film decomposition. The electrochemical decomposition of the LbL film was highly accelerated in the presence of hydrogen peroxide ($H_2O_2$) as a result of electrolysis of $H_2O_2$ on the electrode surface according to Equation 32.1 [10]. $H_2O_2$ can be oxidized at a lower electrode potential ($+0.5$–$0.6$ V) compared to the rather high potential needed for the electrolysis of water. This phenomenon may be useful for developing electrically controlled delivery systems by coupling the avidin/ib-PEI films with oxidase enzymes that produce $H_2O_2$ via a catalytic reaction, such as glucose oxidase.

$$H_2O_2 \xrightarrow{-2e^-} O_2 + 2H^+ \tag{32.1}$$

## 32.3
### Concanavalin A-containing LbL Films

Concanavalin A (Con A) is a member of the lectin protein family which is found in the Jack bean and is known to contain four binding sites to sugars such as D-mannose

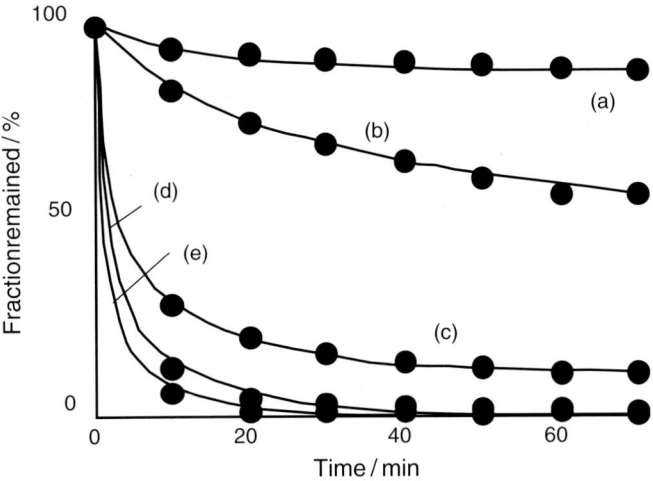

**Figure 32.5** Disintegration of the (Con A/glycogen)$_{10}$ film in the presence of 10 mM of D-galactose (a), D-glucose (b), D-mannose (c), Me-Man (d), and Me-Glu (e) at pH 7.4. (Reprinted with permission from Ref. [14]. Copyright 2005, The American Chemical Society).

and D-glucose [11]. Kunitake and coworkers first used Con A for developing LbL films in combination with glycogen, that is, a branched polysaccharide composed of D-glucose units [12]. The Con A/glycogen LbL films may be sensitive to sugars because these films rely on the reversible binding between Con A and D-glucose units in glycogen. From that viewpoint, Con A-containing LbL films were recently studied for the development of optical glucose sensors [13].

We have studied sugar-induced disintegration of LbL films composed of Con A and glycogen and found that these films are indeed decomposed in response to the addition of various sugars [14]. Figure 32.5 shows the disintegration of a 10 layer pair (Con A/glycogen)$_{10}$ film in the presence of different sugars at 10 mM, pH 7.4. The LbL film was almost completely decomposed in 10–20 min in the presence of methyl-α-mannose (Me-Man) or methyl-α-glucose (Me-Glu) because of the high affinity of the methylated sugars for Con A. The binding constants of Me-Man and Me-Glu are reported to be $2.1 \times 10^4$ and $4.9 \times 10^3$ M$^{-1}$, respectively [15]. It is likely that the added sugars diffuse into the film interior to displace the glycogen from the binding sites of Con A, resulting in disintegration of the film. D-Mannose was more effective than D-glucose in inducing decomposition of the film due to the higher affinity of D-mannose than that of D-glucose for Con A; the binding constants of D-mannose and D-glucose are $2.2 \times 10^3$ and $0.8 \times 10^3$ M$^{-1}$, respectively [16]. In contrast, the LbL film was scarcely decomposed in the presence of D-galactose because D-galactose is known not to bind to Con A. The film decomposition depends on the concentration of the sugar as well as the sugar type. For example, 70–80% of the LbL film remained intact in the presence of 5 mM D-glucose while only 10–20% remained intact upon exposure to 50 mM D-glucose. Con A-containing LbL films may be useful for the development of sugar-sensitive drug delivery systems. Sugar-bearing synthetic

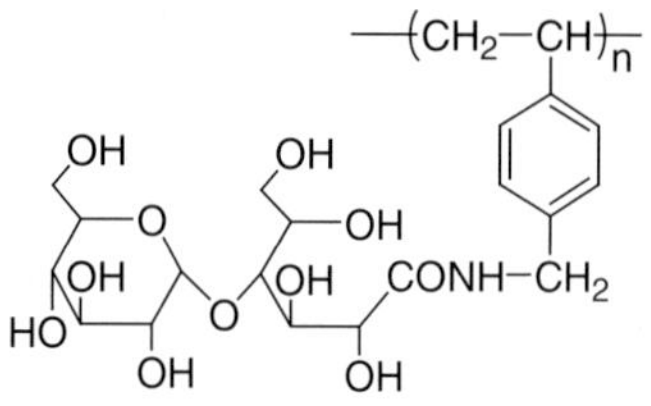

**Figure 32.6** The chemical structure of maltose-bearing polymer.

polymer (Figure 32.6) has also been used successfully for the construction of Con A LbL films, and films constructed using these polymers were sensitive to sugars [17].

The Con A/glycogen LbL films have also been used for the electrochemical sensing of sugars [18]. To this end, an LbL film was constructed on the surface of a glassy carbon (GC) electrode using ferrocene-labeled glycogen (Fc-glycogen, Figure 32.7) and Con A. A cyclic voltammogram (CV) for the Con A/Fc-glycogen film-coated GC electrode exhibited redox peaks at ~0.3 V and the peak height increased with the increasing number of the Con A/Fc-glycogen layers, suggesting the redox-active ferrocene residues were accumulating on the electrode surface. The oxidation and reduction peak potentials are nearly identical to one another for the 1–3 layer pair films, as is typical of CVs for surface-confined redox species. It is to be expected that the intensity of the redox current would be reduced upon addition of sugars because the Con A/Fc-glycogen film would be decomposed in part by the added sugars and thus dissociate from the electrode surface. In fact, the redox current was decreased in the presence of D-glucose (10–100 mM), D-mannose (5–20 mM), and methylated sugars (1–10 mM). A limitation of the present system for sugar determination arises from the irreversible decomposition of the LbL film; the LbL film-coated electrodes are useful only as single-use devices.

The controlled release of insulin for the treatment of diabetes mellitus is one of the ultimate goals for the development of sugar-sensitive devices. Thus, a variety of systems have been studied which would afford glucose-triggered insulin delivery [19]. We developed insulin-containing microcapsules by the LbL deposition of Con A and glycogen onto a calcium carbonate ($CaCO_3$) microparticle doped with insulin, followed by dissolution of the $CaCO_3$ core [20]. The surface of the microcapsules was reinforced with a polymer layer composed of poly(ethyleneimine) and poly(styrene sulfonate) because the microcapsules were too fragile without this surface layer.

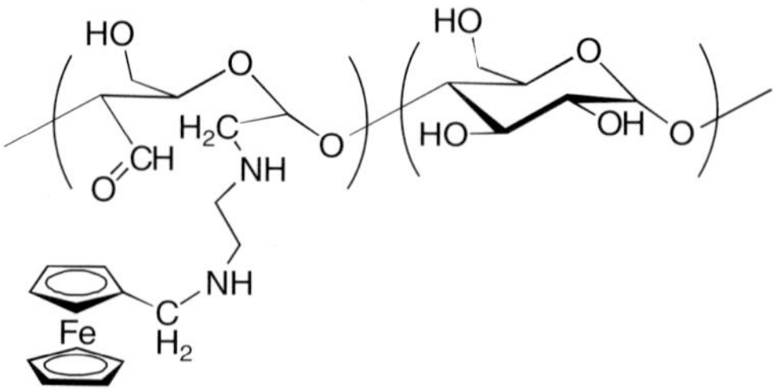

**Figure 32.7** The chemical structure of ferrocene-labeled glycogen.

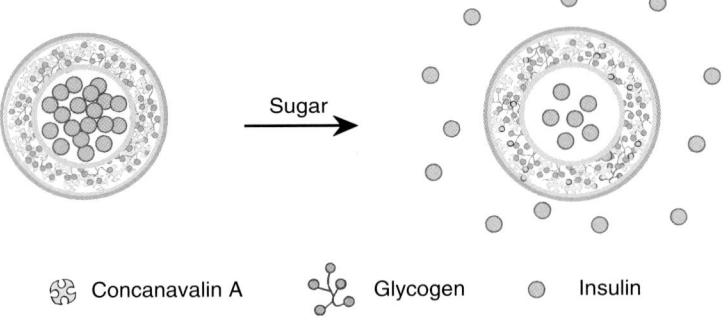

**Figure 32.8** Sugar-induced release of insulin from a (Con A/glycogen) microcapsule.

Figure 32.8 illustrates the concept of glucose-triggered release of insulin from the Con A/glycogen LbL microcapsule. We found that the release rate of insulin was enhanced to some extent in the presence of 100 mM D-glucose. The result was, however, still unsatisfactory in view of the fact that the blood level of D-glucose of diabetic patients is 5–10 mM.

## 32.4
## Dendrimer-Containing LbL Films

Dendrimers have been widely studied for the development of molecular containers or host compounds that bind and release small molecules [21]. In addition, dendrimers have been used as a building block of LbL films. Poly(amidoamine) (PAMAM) dendrimers bearing $-NH_2$ terminal groups were assembled into LbL films through their electrostatic force of attraction to polyanions [22]. Carboxy-terminated PAMAM dendrimers (PAMAM-COOH) have also been used as anionic components for constructing LbL films [23].

We have previously prepared LbL films using PAMAM-COOH (generation 3.5) and poly(methacrylic acid) (PMA) or poly(acrylic acid) (PAA) and studied their pH-sensitive disassembly [24]. Figure 32.9 shows the QCM response to the deposition of five layer pair (PMA/PAMAM-COOH)$_5$ film on the quartz resonator at pH 4.0 and its disintegration at pH 7.0. The QCM results clearly demonstrate successful deposition of the LbL film in the acidic medium, and its decomposition at pH 7.0. The threshold pH for the decomposition of the (PMA/PAMAM-COOH)$_5$ film was pH 5.0; the film was decomposed at pH 5.5 or higher but was stable at pH 5.0. These results suggest that dissociation of –COOH residues into the anionic forms is responsible for the decomposition of the film. In other words, the (PMA/PAMAM-COOH)$_5$ film was stabilized mainly through hydrogen bonding between the −COOH residues in PMA and PAMAM-COOH. A separate experiment revealed that, unexpectedly, the (PMA/PAMAM-COOH)$_5$ film was decomposed at pH 2.0 or lower, where hydrogen bonding should occur. The results suggest an essential role for electrostatic forces of attraction in addition to hydrogen bonding in stabilizing the (PMA/PAMAM-COOH)$_5$

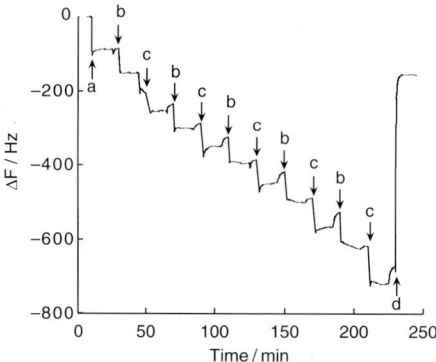

**Figure 32.9** A QCM response for the construction of (PMA/PAMAM-COOH)$_5$ film at pH 4.0 and its decomposition at pH 7.0. The quartz resonator was exposed to polyethyleneimine (a), PMA (b), PAMAM-COOH (c), and pH 7.0 medium (d). (Reprinted with permission from Ref. [24]. Copyright 2008, Elsevier).

film. It is likely that PMA contains a small fraction of carboxylate anion, which assists PMA in binding to PAMAM-COOH because the tertiary amino groups in PAMAM-COOH are positively charged at pH 3.0–5.0. At pH 2.0 or lower, PMA chains are fully protonated and, therefore, lose electrostatic affinity for the dendrimer in the film. Thus, the (PMA/PAMAM-COOH)$_5$ film is stabilized only when both hydrogen bonding and electrostatic affinity are available simultaneously. This is in clear contrast to the pH-stability of hydrogen bond-mediated LbL films consisting of poly(carboxylic acid) and acceptor polymers such as poly(vinylpyrrolidone) and poly(ethylene oxide), which are stable at pH 2.0. The pH-stability of (PAA/PAMAM-COOH)$_5$ film was slightly different from that of the PMA-based film [25]. The (PAA/PAMAM-COOH)$_5$ film was stable at pH 2.0–4.5, while the pH-stable range for the PMA-based film was 3.0–5.0. Thus, the pH-stability of these two LbL films differs due to the difference in acidity between PAA and PMA.

The (PMA/PAMAM-COOH)$_5$ film may be promising for future application to pH-controlled drug delivery because dendrimers are known to bind small molecules in their cavities or on their surfaces. Therefore, uptake and pH-sensitive release by the (PMA/PAMAM-COOH)$_5$ film was studied using Rose Bengal and sulfonated tetraphenylporphirin [24]. Both dyes were released from the LbL film slowly at pH 4.0; the complete release of the dyes took more than 20 h at this pH. In contrast, the dyes were released in a few minutes at pH 7.0 as a result of film decomposition.

## 32.5
### Insulin-Containing LbL Films

Much effort has been devoted to the development of glucose-sensitive formulations that release insulin in response to elevated blood glucose levels, for the treatment of diabetes mellitus [26]. Glucose-sensitive formulations of insulin would eliminate the

**Table 32.1** Loading of insulin in LbL films.

| LbL film | Loading of insulin in LbL films/$10^{-6}$ g cm$^{-2}$ | | |
| --- | --- | --- | --- |
| | $n=5$ | $n=10$ | $n=15$ |
| (PVS-insulin)$_n$ | 5.4 | 14 | 19 |
| (DS-insulin)$_n$ | 1.9 | 4.9 | 15 |
| (PAA-insulin)$_n$ | 2.9 | 36 | 61 |

The average values of three or four preparations are listed. These values contain 10–30% errors. (Reprinted with permission from Ref. [29]. Copyright 2010, The Royal Society of Chemistry).

need for repeated insulin injections. Insulin formulations that can be orally administered as an alternative to subcutaneous injection are also under extensive study [27]. Recently, LbL films and microcapsules have been employed for the development of insulin delivery systems [28].

We have prepared insulin-containing LbL films by the alternate deposition of insulin and polyanions such as PAA, poly(vinyl sulfate) (PVS), and dextran sulfate (DS) at pH 3.0 through the electrostatic forces of attraction between positively-charged insulin and the polyanions [29]. Table 32.1 summarizes the amounts of insulin loaded in the LbL films thus prepared. The insulin loading increased with the increasing number of layers for all films. The type of polyanion used had a significant influence on the loading of insulin, suggesting different conformations of the polymer chains in each of these LbL films. Presumably, PAA chains form a flexible coiled conformation on the film surface because the number of negative charges in the PAA chains is rather limited at pH 3.0, resulting in a thicker layer that can accommodate larger amounts of insulin than PVS and DS. It seems, in view of the fact that the monolayer coverage of insulin on a flat surface with close packing corresponds to $1.1 \times 10^{-7}$ g cm$^{-2}$, that insulin may form aggregates in the LbL films [30]. Consequently, the loading of insulin listed in Table 32.1 is much higher than that expected for monomolecular deposition in each layer.

The pH-dependent decomposition of (PVS/insulin)$_{15}$ film was evaluated at 20 and 37 °C, based on the changes in absorbance of the solutions in which the LbL film was immersed at different pH (Figure 32.10). The LbL film decomposed at pH 6.0 and 7.4, while the film was stable at pH 4.0 and 5.0. The pH threshold observed for film decomposition corresponds qualitatively with the isoelectric point of insulin (pI, 5.4). Therefore, it is clear that the decomposition of the film was induced by electrostatic repulsion between the insulin and polyanion in the film. The effect of temperature on film decomposition was negligible. The rapid film decomposition response at neutral pH is a key benefit of the PVS-containing LbL film.

The structural integrity of insulin released from the LbL film was studied by means of circular dichroism (CD) spectroscopy. The CD spectrum of released insulin exhibited typical α-helical characteristics with a double minimum that was nearly identical to that of native insulin, suggesting that no conformational changes were induced in the released insulin. In other words, the released insulin is likely to retain its biological activity. In contrast, the CD spectrum of insulin within the LbL film

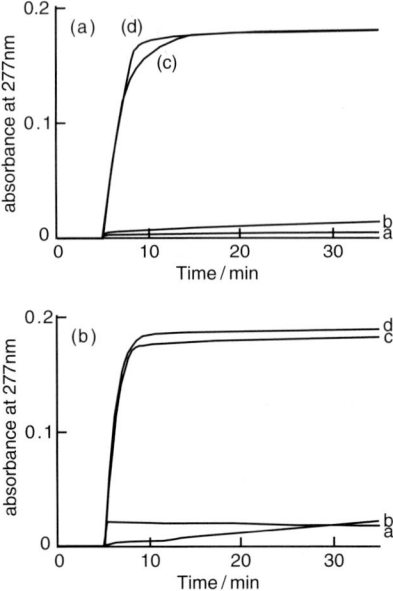

**Figure 32.10** pH-dependent decomposition of the (PVS/insulin)$_{15}$ film at 20 °C (A) and 37 °C (B). The absorbance of the solutions in which the LbL film was immersed was recorded for pH changes from pH 3.0 to pH 4.0 (a), pH 5.0 (b), pH 6.0 (c), and pH 7.4 (d). (Reprinted with permission from Ref. [29]. Copyright 2010, The Royal Society of Chemistry).

deviated slightly from that of native insulin, probably due to the electrostatic effects of PVS chains in the film. Thus, insulin released from the LbL film was intact, although its three-dimensional structure within the LbL film may have differed slightly from that of native insulin.

The stability of the LbL fims in the gastric juice environment in the stomach (pH 1.4) was evaluated at 37 °C in the presence and absence of digestive enzyme (pepsin). Figure 32.11 shows the stability of PVS- and DS-based films. Both LbL films were reasonably stable at pH 1.4 in the absence of pepsin, while in the presence of the enzyme insulin was digested to some extent. The PVS-based films were more durable than the DS-based films. It is likely that DS and insulin are loosely packed in the LbL films, due to the limited flexibility and lower charge density of DS, compared to the dense packing in the PVS-insulin film. Both of these insulin-containing LbL films may be promising for future development of oral formulations of insulin.

## 32.6
## Conclusions

Stimuli-sensitive LbL films and microcapsules have been successfully constructed using dendrimers and proteins for the development of controlled delivery systems. LbL films constructed with avidin and ib-PEI can be decomposed by pH changes and

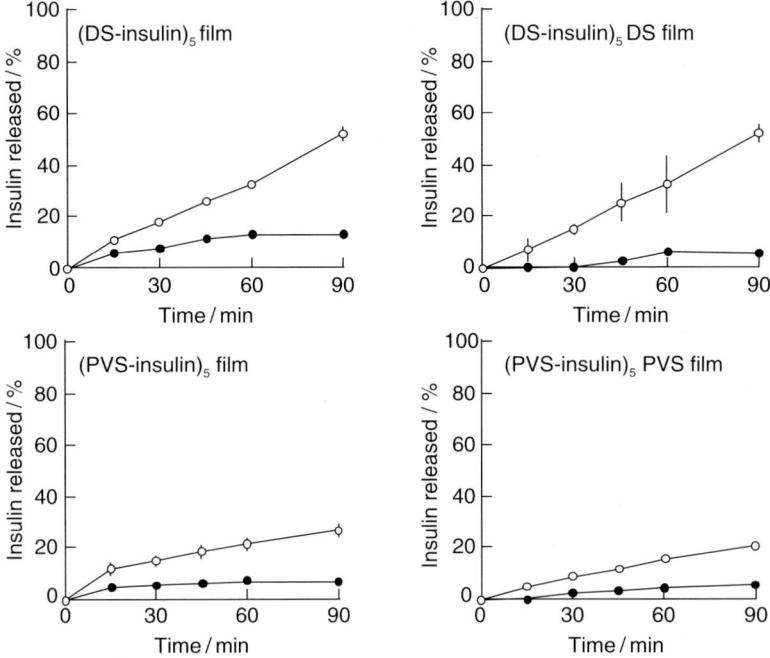

**Figure 32.11** Stability of insulin-containing LbL films in the presence (○) and absence of 0.5 mg mL$^{-1}$ pepsin (•) in acidic solution (pH 1.4) at 37 °C. (Reprinted with permission from Ref. [29]. Copyright 2010, The Royal Society of Chemistry).

by adding biotin and its analogues. In addition, an electrochemical technique can be employed to decompose these LbL films without the addition of reagents. Con A can be used to construct LbL films and microcapsules that are sensitive to sugars. These Con A/glycogen LbL films can be disintegrated by adding sugars, depending on the binding affinity of the sugars for Con A. Con A-containing films and capsules may be useful for the future development of glucose-sensitive delivery systems. PAMAM-COOH can be built into LbL films by alternate adsorption with PMA or PAA, mainly through hydrogen bonding. The PAMAM-COOH/(PMA or PAA) LbL films can bind small dyes and release these in response to pH changes. Insulin-containing LbL films have also been constructed by the alternate deposition of insulin and polyanions such as PVS and DS through electrostatic forces of attraction. These insulin-containing LbL films are stable at pH 1.4 (i.e., stomach pH) even in the presence of digestive enzymes, while insulin is released at pH 7.4, suggesting a potential use of such LbL films for the development of oral formulations of insulin.

### Acknowledgments

Support of this work by the Japan Society for the Promotion of Science (JSPS) is gratefully acknowledged.

## References

1 Wilcheck, M. and Bayer, E.A. (1990) *Methods Enzymol.*, **184**, 14–45.
2 Rao, S.V., Anderson, K.W., and Bachas, L.G. (1999) *Biotechnol. Bioeng.*, **65**, 389–396.
3 Anzai, J., Kobayashi, Y., Nakamura, N., Nishimura, M., and Hoshi, T. (1999) *Langmuir*, **15**, 221–226.
4 Pieczonka, N.P.W., Goulet, P.J.G., and Aroca, R.F. (2006) *J. Am. Chem. Soc.*, **128**, 12626–12627.
5 Dai, Z., Wilson, J.T., and Chaikof, E.L. (2007) *Mater. Sci. Eng. C*, **27**, 402–408.
6 Inoue, H., Sato, K., and Anzai, J. (2005) *Biomacromolecules*, **6**, 27–29.
7 Inoue, H. and Anzai, J. (2005) *Langmuir*, **21**, 8354–8359.
8 Zhuk, A., Pavlukhina, S., and Sukhishvili, S.A. (2009) *Langmuir*, **25**, 14025–14029.
9 Sato, K., Kodama, D., Naka, Y., and Anzai, J. (2006) *Biomacromolecules*, **7**, 3302–3305.
10 Sato, K., Naka, Y., and Anzai, J. (2007) *J. Colloid Interface Sci.*, **315**, 396–399.
11 Becker, J.W., ReekeJr., G.N., Cunnigham, B.A., and Edelman, G.M. (1976) *Nature*, **259**, 406–409.
12 Lvov, Y., Ariga, K., Ichinose, I., and Kunitake, T. (1995) *J. Chem. Soc., Chem. Commun.*, 2313–2314.
13 Chinnayelka, S. and Macshane, M.J. (2004) *J. Fluoresc.*, **14**, 585–595.
14 Sato, K., Imoto, Y., Sugama, J., Seki, S., Inoue, H., Odagiri, T., Hoshi, T., and Anzai, J. (2005) *Langmuir*, **21**, 797–799.
15 Schwarz, F.P., Puri, K.D., Bhat, R.G., and Surolia, A. (1993) *J. Biol. Chem.*, **268**, 7668–7677.
16 Mandel, D.K., Kishore, N., and Brewer, C.F. (1994) *Biochemistry*, **33**, 1149–1156.
17 Sato, K., Kodama, D., and Anzai, J. (2006) *Anal. Sci.*, **21**, 1375–1378.
18 Sato, K., Kodama, D., and Anzai, J. (2006) *Anal. Bioanal. Chem.*, **386**, 1899–1904.
19 Qi, W., Yan, X., Rei, J., Wang, A., Cui, Y., and Li, J. (2009) *Biomaterials*, **30**, 2799–2806.
20 Sato, K., Kodama, D., Endo, Y., and Anzai, J. (2009) *J. Nanosci. Nanotechnol.*, **9**, 386–390.
21 Tomalia, D.A. (2005) *Prog. Polym. Sci.*, **30**, 294–324.
22 Khopade, A.J. and Caruso, F. (2002) *Biomacromolecules*, **3**, 1154–1162.
23 Tsukruk, V.V., Rinderspacher, F., and Bliznyuk, V.N. (1997) *Langmuir*, **13**, 2171–2176.
24 Tomita, S., Sato, K., and Anzai, J. (2008) *J. Colloid Interface Sci.*, **326**, 35–40.
25 Tomita, S., Sato, K., and Anzai, J. (2009) *Kobunshi Ronbunshu*, **66**, 75–78.
26 Zhao, Y., Trewyn, B.G., Slowing, I.I., and Lin, V.S.-Y. (2009) *J. Am. Chem. Soc.*, **131**, 8398–8400.
27 Kim, S.K., Lee, S., Jin, S., Moon, H.T., Jeon, O.C., Lee, D.Y., and Byun, Y. (2010) *Mol. Pharm.*, **7**, 708–717.
28 Zheng, J., Yue, X., Dai, Z., Wang, Y., Liu, S., and Yan, X. (2009) *Acta Biomater.*, **5**, 1499–1507.
29 Yoshida, K., Sato, K., and Anzai, J. (2010) *J. Mater. Chem.*, **20**, 1546–1552.
30 Mollmann, S.H., Jorgensen, L., Bukrinsky, J.T., Elofsson, U., Norde, W., and Frokjaer, S. (2006) *Eur. J. Pharm. Sic.*, **27**, 194–204.

# 33
# Assembly of Multilayer Capsules for Drug Encapsulation and Controlled Release
*Jinbo Fei, Yue Cui, Qiang He, and Junbai Li*

## 33.1
## Introduction

In general, the main objectives in developing drug encapsulation and controlled release systems are to avoid biological barriers, enhance the *in vivo* bioavailability of drugs and realize targeted drug delivery [1, 2]. Nanotechnology can provide strong and powerful platforms for these purposes. During the past 30 years, much attention has been paid to nanomedicine in relation to the development of drug carriers. By using different methods, many kinds of inorganic and organic nanostructured materials, including 0D nanoparticles [3–6], 1D nanotubes [7, 8], and 2D nanofilms [9], have been prepared or assembled for drug delivery systems. These nanomaterials have exhibited interesting and attractive properties, such as controllable shape and size, and adjustable chemical, physical and biological properties. However, great challenges in the development of advanced drug formulations are how to increase programmed longevity and stability of the carrier in the circulation, how to target to the pathological site via specific biorecognition, how to realize stimuli sensitivity to external trigger (such as ultrasound, magnetic field, and laser irradiation) or to the local microenvironment of the disease site (such as pH, temperature and electrochemical potential), how to enhance intracellular delivery of the drug cargo, and how to integrate multifunctional contrast agents with high efficiency and sensitivity for intracellular imaging of the delivery systems and real-time measurement of analytes *in vivo* [10]. To fulfill the demand mentioned above, multilayer microcapsules assembled via the layer-by-layer (LbL) technique have emerged as one of most promising multifunctional carrier platforms.

Introduced by Decher in 1991, the LbL technique was originally exploited for ultrathin films by sequential adsorption of oppositely charged polymers on a charged planar substrate [11, 12]. Over two decades, the LbL technique has been extended from polyelectrolyte systems to almost any kind of charged building block, such as atom and molecular clusters, low-dimensional nanostructures, organic dyes, dendrimers, polysaccharides, peptides, DNA, proteins and viruses. The simplicity and universality of the LbL process has led to the wide and rapid development of the related biomedical applications. Assemblies via electrostatic interaction [13],

---

*Multilayer Thin Films: Sequential Assembly of Nanocomposite Materials*, Second Edition.
Edited by Gero Decher and Joseph B. Schlenoff.
© 2012 Wiley-VCH Verlag GmbH & Co. KGaA. Published 2012 by Wiley-VCH Verlag GmbH & Co. KGaA.

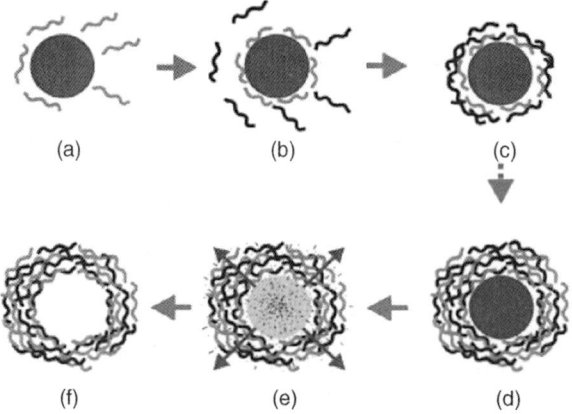

**Figure 33.1** Schematic illustration of the polyelectrolyte deposition process and subsequent core decomposition. Reprinted with permission from [20]. Copyright 1998, Wiley-VCH.

hydrogen bonding [14], charge transfer [15], covalent bonding and biological recognition [16–19], have been investigated. The availability of a wide spectrum of fabrication components, the variety of substrates, and the versatility of assembly methods dramatically enriches the biological applications of LbL films. Furthermore, the LbL technique has been developed from 2D films to nonplanar multilayer capsules and nanotubes by many outstanding research groups.

Multilayer microcapsules were first constructed by Möhwald's group by depositing polyelectrolytes onto charged colloidal microparticles, which were used as sacrificial templates (Figure 33.1) [20]. Since then, these microcapsules have attracted a lot of research interest based on the fact that their properties, such as size, shape, composition, shell thickness, permeability, and stiffness, can be tailored easily for controlled encapsulation and release in many fields, including food, cosmetics and drug delivery.

Usually, most LbL-based multilayer microcapsules are composed of two different compartments: the cavity and the shell. The cavity can encapsulate many kinds of materials, and the shells can be devoted to different chemical, physical and biological properties. In terms of drug encapsulation and controlled release, compared with conventional strategies, LbL assembly offers great potential advantages of controlling the order and location of building blocks with nano-sized precision, and defining the concentrations of incorporated materials simply by changing the number of layers. Recently, several excellent articles have reviewed the progress of LbL multilayer microcapsules with relation to physicochemical and mechanical properties [21–23], permeability [24], and smart loading and release [25–28]. In addition, Gao's group reviewed the recent progress in LbL microcapsules with respect to manipulation of their properties by chemical cross-linking, fabrication of microcapsules based on novel driving forces, such as covalent interaction, base pair interaction, host–guest interaction and van der Waals interaction, and their bio-related functions and applications [29]. In particular, Mercato et al. reviewed LbL multilayer capsules with regard to recent progress and the future outlook for their use in life sciences [30]. In this chapter, we will give a review of some of the very recent progress achieved by several active groups in the design of novel microcapsules and their use as drug delivery vehicles stimulated by many different pathways. In addition, while considering the benefits and limits of current systems, we firmly believe there is a bright future for the emerging strategies for engineering multifunctional drug encapsulation and controlled release vehicles.

## 33.2
## Magnetically Sensitive Release

Recently, magnetically sensitive alginate-templated polyelectrolyte multilayer microcapsules were successfully assembled by combining emulsification and LbL techniques [31]. The as-synthesized microcapsules were superparamagnetic with a saturation magnetization of 14.2 emu g$^{-1}$, and contained about 30 wt% Fe$_3$O$_4$ nanoparticles. The drug (doxorubicin) encapsulation efficiency was 56.4% and loading content was 3.5%. The authors investigated the *in vitro* release behavior under a high-frequency magnetic field (HFMF) and the results indicated that the applied HFMF could accelerate drastically the drug release from the microcapsules. Lu *et al.* also used a magnetic field to modulate the permeability of polyelectrolyte multilayer microcapsules [32]. It should be noted that ferromagnetic gold-coated cobalt (Co@Au) core–shell nanoparticles were incorporated inside the capsule shells. External alternating magnetic fields were applied to rotate the embedded ferromagnetic nanoparticles, which subsequently disturbed and distorted the capsule wall and significantly increased its permeability to macromolecules like FITC-labeled dextran. Katagiri *et al.* studied magnetoresponsive multilayer capsules composed of polyelectrolytes, lipid bilayers and magnetic nanoparticles [33]. They demonstrated that the magnetically induced release was attributed to the phase transition of the lipid membrane, caused by the heat of Fe$_3$O$_4$ nanoparticles under magnetic stimuli (see Figure 33.2).

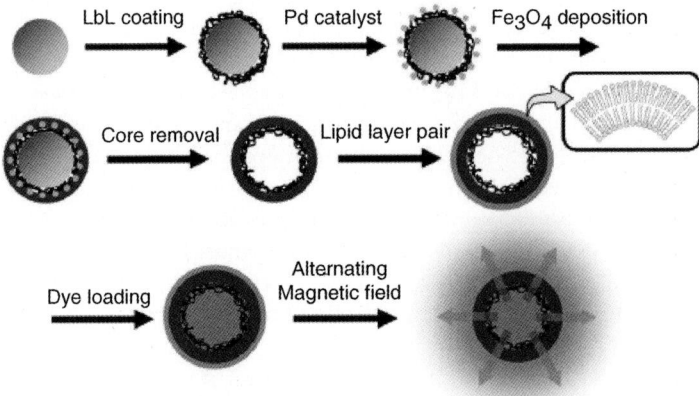

**Figure 33.2** Schematic representation of the preparation process and release of substances from magnetoresponsive hybrid capsules. Reprinted with permission from [33]. Copyright 2009, American Chemical Society.

Furthermore, Gomes et al. embedded superparamagnetic nanoparticles encapsulated by liposomes into multilayer capsules [34]. They found that liposomes coated with magnetic nanoparticles could be separated rapidly from unbound polyelectrolytes in an external magnetic field.

Interestingly, using the LbL technique, Han et al. developed a new targeted delivery system by depositing magnetic nanoparticles on multilayer protein containers [35]. The results in this study indicated that the vesicles containing dye were stable and could sustain the deposition treatment without loss of dye due to the protection of protein nanoshells.

## 33.3
### Ultrasound-Stimulated Release

Generally, ultrasound (US) is cyclic sound pressure with a frequency greater than 20 kHz (see Figure 33.3). In nature, bats use US for hunting by penetrating a medium and measuring the reflection signature. It is well known that with sonography, pictures of fetuses in the human womb can be produced by employing ultrasound technology [36, 37]. In fact, US has been used widely in the biomedical field, including for the destruction and fragmentation of contrast agents [38], gas release, and destruction of polymer and albumin shells of microbubbles [39, 40]. Moreover, US with low frequency can be used to promote drug delivery and fracture healing by its cavitational effect [41]. However, up to now, there have been few reports on drug encapsulation and controlled release by multilayer microcapsules based on the LbL technique.

Very recently, our group reported that procainamide hydrochloride, a water-soluble drug, could be loaded in large amount and concentrated by more than two orders of magnitude in the assembled multilayer capsules of PDADMAC/PSS through heat treatment (shown in Figure 33.4) [42]. Additionally, the encapsulated amount could be quantitatively controlled via the drug concentration in the bulk. The unloading rate of drugs could be controlled by using ultrasonic treatment.

Gou et al. introduced silica@Zn–Sr@silica multilayer nanospheres assembled by the LbL technique. Controlled release of active trace ions (Zn and Sr) could be realized by US with low frequency [43]. The authors demonstrated that this strategy mentioned above was potentially advantageous for administering multicomponents and exhibiting a synergistic effect for promoting tissue regeneration.

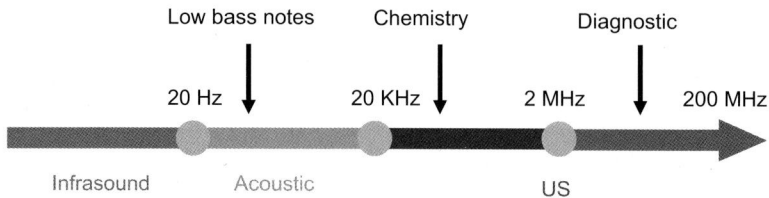

**Figure 33.3** Approximate frequency ranges with respect to US and some applications.

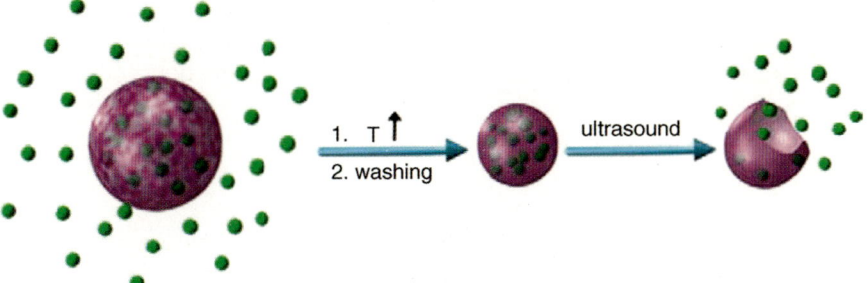

**Figure 33.4** Scheme showing encapsulation of materials into assembled capsules and unloading of the encapsulated materials using US. Reprinted with permission from [42]. Copyright 2009, Elsevier.

## 33.4
## Photo-Stimulated Release

As is well known, light is a portion of electromagnetic radiation. Visible light has a wavelength ranging from 380 or 400 nm to about 760 or 780 nm, with a frequency range of about 405 to 790 THz. In physics, the term light often includes the adjacent radiation regions of infrared (at lower frequencies) and ultraviolet (at higher), not visible to the human eye [44]. It should be noted that in biological systems, most tissues show negligible absorption in the wavelength range 800–1200 nm [45].

Controlled encapsulation and triggered release of bioactive substances by light is a very hot and interesting topic in designing drug delivery systems. It is worth mentioning that, very recently, Sukhorukov et al. have given an excellent review on light-sensitive polymer-based multilayer microcapsules in general terms [46]. Their comprehensive review referred to the use of microcapsules sensitive to light over a wide wavelength range as delivery systems for biological applications by incorporating many kinds of building blocks, including light-responsive polymers, functional dyes, metal and metal oxide nanoparticles. In this section, we will focus mainly on controlled release in the near-in-infrared (NIR) by assembling hybrid multilayer microcapsules.

At present, to our best knowledge, it is still a great challenge to develop chromophores that absorb in the biologically friendly NIR range. Fortunately, the surface plasmon resonance signal can be modulated to shift into the IR region above 800 nm by controlling the morphology and aggregation state of gold nanostructures on the surface of multilayer capsules. Halas and coworkers first reported that gold nanoshells had tunable optical resonances and strong absorption in the NIR [47]. In their subsequent studies, the authors demonstrated that, by embedding gold nanoshells into thermo-sensitive polymer films, the entrapped substance could be released upon illumination at the nanoshell resonance wavelength [48, 49]. Since then, this approach has been explored extensively for rupture of the assembled multilayer microcapsules and release of the encapsulated materials. For instance, Caruso's group reported the preparation of NIR-responsive capsules based on gold nanoparticles incorporated within a LbL-assembled polyelectrolyte multilayer shell [50].

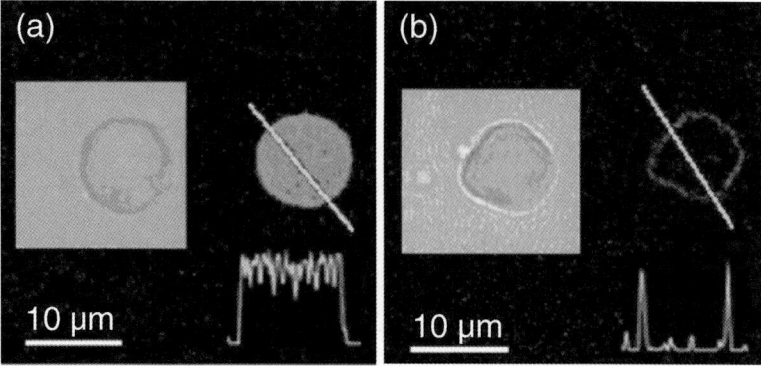

**Figure 33.5** Confocal microscope images demonstrating remote release of encapsulated rhodamine-labeled PSS polymers from a polyelectrolyte multilayer capsule containing gold sulfide core/gold shell nanoparticles in its walls. Fluorescence intensity profiles along the line through the capsule show that it is filled with fluorescent polymers before (a) and empty after (b) laser illumination. After the release of encapsulated polymers, the leftover fluorescent intensity is observed only in the walls of the capsule, (b). Insets show black and white transmission microscope images of the same capsule. Incident intensity of laser diode operating at 830 nm was set at 50 mW. Reprinted with permission from [53]. Copyright 2005, American Chemical Society.

Furthermore, they introduced light-responsive microcapsules composed of multiple polyelectrolyte layers and gold nanoparticles [51]. The encapsulated material, fluorescein isothiocyanate-labeled dextran, could be released upon irradiation of the nanoengineered hybrid systems with short (10 ns) laser pulses in the NIR (1064 nm).

In addition, Javier et al. studied polyelectrolyte capsules with metal nanoparticles in their walls and fluorescently labeled polymers as cargo inside their cavity [52]. The authors found that photo-induced heating of the metal nanoparticles in the capsule walls led to rupture of the capsule walls, and the polymeric cargo was released to the whole cytosol. Sukhorukov and coworkers reported laser-mediated remote release of encapsulated fluorescently labeled polymers from polyelectrolyte multilayer capsules containing gold sulfide core/gold shell nanoparticles in their walls [53]. Figure 33.5 showed that the work was studied in real time on a single capsule level. They developed a simple method for measuring the temperature increase and quantitatively investigated the influence of absorption, size, and surface density of complex nanoparticles. They demonstrated that the treatment presented in the study could be extended and was applicable to any system where nanoparticles were used as absorbent.

Recently, Skirtach et al. reported the spontaneous embedding of gold nanoparticle (NP) aggregates or polyelectrolyte microcapsules modified with NPs in biocompatible hyaluronic acid/poly (L-lysine) (HA/PLL) films [54]. The films functionalized with gold NPs became active in response to a NIR laser at a power of about 20 mW. The activation was characterized by a localized temperature increase in the film, allowing conversion of light energy to heat in confined volumes. Microcapsules

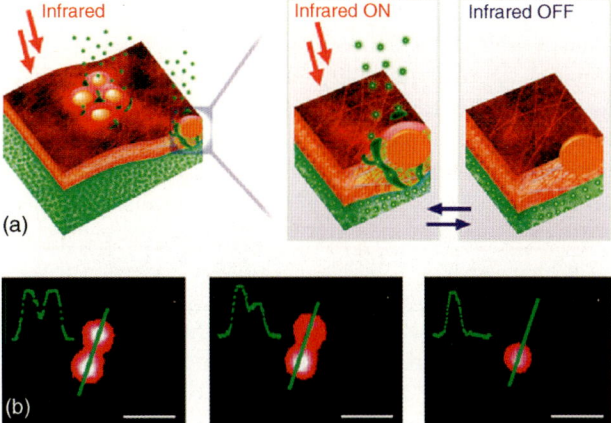

**Figure 33.6** Remote release from microcapsules: (a) schematics of nanoparticle functionalized polymeric nanomembranes opening channels upon laser illumination; (b) a polymeric microcapsule shell acts as a reversible nanomembrane. Upon laser light illumination the microcapsule (left image) partially releases encapsulated polymers and reseals (middle). After the second illumination the microcapsule completely releases its content (right). Profiles in the left upper corner are drawn along the green line. Scale bars correspond to 5 µm. Reprinted with permission from [55]. Copyright 2008, American Chemical Society.

adsorbed onto the film could release their cargo under NIR light because of localized permeability changes in their walls.

Moreover, further fine-tuning the composition of the polymer multilayer capsules allowed the capsule shell to be made reversibly permeable by IR irradiation. To our interest, this system below could release only discrete portions of encapsulated material while keeping the whole capsule (as shown in Figure 33.6) [55]. This approach was applied to drug delivery by the same group and results indicated that laser-triggered opening could be performed within living cells without impairing their viability [56]. Furthermore, they also reported a much milder approach to light addressable systems using gold nanorods as building blocks [57].

## 33.5
## Thermo-Stimulated Release

A temperature-responsive polymer is one which undergoes a physical change when external thermal stimuli are presented [58]. The ability to undergo such changes under easily controlled conditions makes this kind of polymer fall into the category of intelligent materials. These physical changes can be exploited for many analytical techniques, especially for separation chemistry.

Poly (N-isopropylacrylamide) (PNIPAM) is one of the most widely investigated thermoresponsive polymers. In most cases, the polymer undergoes a reversible volume phase transition at a lower critical solution temperature (LCST) of about

32 °C, where the hydrogel hydrophobically collapses upon itself, expelling water in an entropically favored fashion [59]. Using the LbL technique, novel thermoresponsive multilayer systems, exhibiting thermal tunability in uptake and release, have been developed for controlling the encapsulation and release of various model drugs [60–62]. For example, Prevot et al. presented the uptake of PNIPAM inside polyelectrolyte microcapsules, and described the dependence of the LCST on the nature and concentration of different salts [62]. With the related results supported, they demonstrated that it was possible to tune and finely control the collapse of encapsulated PNIPAM.

Recently, through direct covalent LbL assembly using popular click chemistry, Huang et al. reported the preparation of azido- and acetylene-functionalized PNIPAM copolymers and their use in the fabrication of ultrathin thermoresponsive microcapsules (see Figure 33.7) [63]. In detail, the clickable copolymers poly[N-isopropylacrylamide- co-(trimethylsilyl)propargylacrylamide] and poly(N-isopropylacrylamide-co-3-azideopropylacrylamide) were prepared through atom transfer radical polymerization (ATRP). After removing the protective trimethylsilyl groups, these clickable PNIPAM copolymers were assembled alternately onto azido-modified silica particles in aqueous media through highly efficient click reactions catalyzed by a mixture of copper sulfate and sodium ascorbate. After removal of the template, the microcapsules remained whole in the presence of the stable covalently bonded triazole units and underwent thermoreversible swelling/deswelling when the temperature was changed.

Kharlampieva et al. deposited poly(N-vinylcaprolactam) (PVCL) and poly(vinyl methyl ether) (PVME) onto particulate substrates by using LbL alternative adsorption with poly(methacrylic acid) (PMAA) [64]. The authors demonstrated that PMAA/PVME and PMAA/PVCL assembled multilayers held promise as temperature-responsive containers for controlled delivery applications.

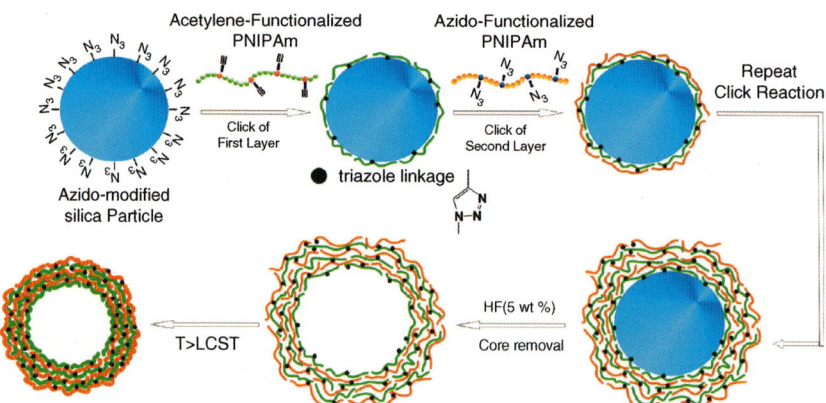

**Figure 33.7** Schematic representation of the preparation of covalently stabilized thermoresponsive microcapsules through LbL assembly using popular click chemistry. Reprinted with permission from [63]. Copyright 2005, American Chemical Society.

## 33.6
## pH-Sensitive Release

The values of pH in different parts of the human body are not the same. For example, pH in the stomach is about 2.0 and pH in intestines is from 5.0 to 8.0. Generally, pH in a tumor is lower than that of normal organisms. Especially, the pH in lysosome and endosome are about 4.5 and 5.8, respectively. Based on the differences mentioned above, many pH-sensitive sheddable drug delivery nano-sized systems have been developed by using pH-sensitive covalent bonds. Usually, pH-sensitive linkers are orthoester, hydrazone, thiopropionate, phosphoramidate and benzoic-imine. However, to our best knowledge, most of them are related to polymer micelles and few reports describe the use of the LbL technique.

Up to now, most pH-sensitive multilayer films have been assembled via electrostatic interaction or hydrogen bonds. At the same time, pH-stimulated multilayer microcapsules driven by the same forces for drug release have been reported in excellent articles [65–72]. For instance, Teng et al. developed a novel "oil-in-water" multifunctional lipophilic drug carrier by combining an ultrasonic technique and LbL assembly [67]. In detail, polyglutamate/polyethyleneimine/poly(acrylic acid) multilayer nanocontainers loaded with the lipophilic drug, rifampicin, dissolved in soybean oil were fabricated. The experimental results indicated that the drug could be released by changing the pH of the media because of the pH-responsive properties of the shell.

In addition, Manna introduced a multilayer microcapsule of poly(vinyl alcohol) (PVA)-borate complex and chitosan (CHI) loaded with the anticancer drug doxorubicin by the LbL technique [68]. Release rates of this delivery system were measured at different pH values. Shu et al. reported biodegradable hollow capsules encapsulating protein drugs prepared via LbL assembly of water-soluble chitosan and dextran sulfate on protein-entrapping amino-functionalized silica particles, and the subsequent removal of the silica [69]. The CHI-integrated microcapsules were loaded with a model drug, bovine serum albumin labeled with fluorescein isothiocyanate (FITC-BSA), and it was found that pH was an effective trigger for controlling the loading and release of FITC-BSA. Very recently, our group fabricated biocompatible, biodegradable, and nontoxic microcapsules by LbL assembly of CHI and alginate dialdehyde (ADA) [70]. The stability of the microcapsules was effectively improved in the presence of covalent Schiff's bases between CHI and ADA. Meanwhile, the Schiff's bases enabled the microcapsules autofluorescence property and pH-response (see Figures 33.8 and 33.9). Experiments proved that the strategy could be extended to other polysaccharides, such as heparin and starch. These series of polysaccharide microcapsules would have great potential for the applications of biological tracing and drug delivery.

Chaturbedy et al. prepared pH stimuli-responsive organic–inorganic hybrid spheres by coating the colloidal polystyrene spheres with polyelectrolyte-protected aminoclay, Mg phyllo (organo) silicate layers through a LbL method [71]. The hybrid spheres underwent a size change up to 60% as the pH was modified from 9 to 4. The stimuli-responsive property of the hybrid spheres was used for the controlled release

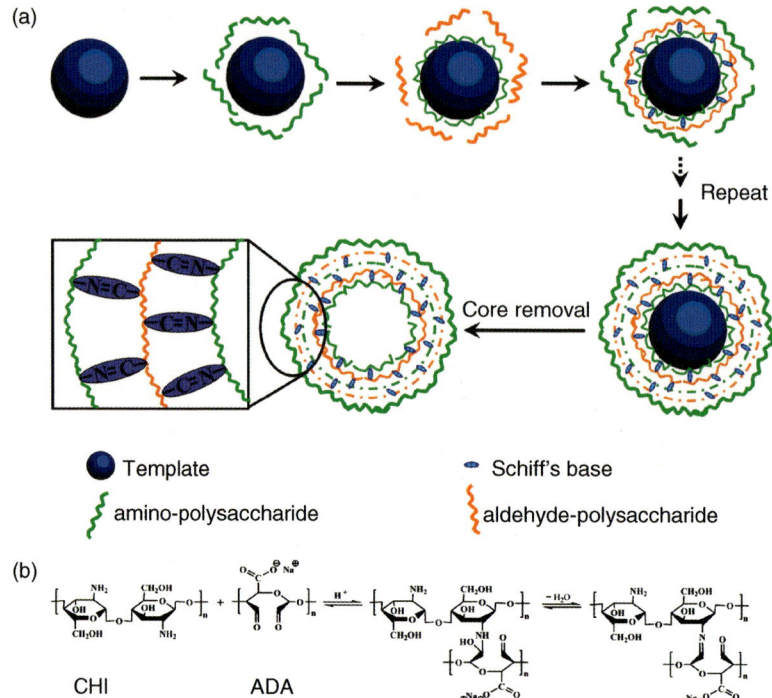

**Figure 33.8** Schematic illustrations to show (a) the fabrication process of covalent assembly of polysaccharide-based microcapsules and (b) the reaction between (i) CHI and (ii) ADA. Reprinted with permission from [70]. Copyright 2011, the Royal Society of Chemistry.

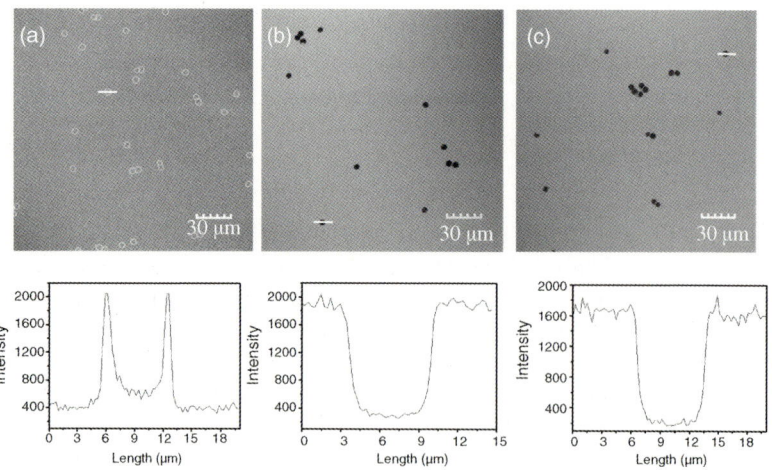

**Figure 33.9** CLSM images of (CHI/ADA)$_5$/CHI microcapsules in different pH media with FITC-dextran (20 kDa) as a reporting molecule: (a) pH 5, (b) pH 7 and (c) pH 9. The corresponding fluorescence intensity profiles are shown below. Reprinted with permission from [70]. Copyright 2011, the Royal Society of Chemistry.

of eosin and ibuprofen at different pH. In addition, the superparamagnetic multilayer hybrid hollow microspheres have been fabricated using the LbL assembly technique by the electrostatic interaction between CHI and the hybrid anion citrate-modified $Fe_3O_4$ nanoparticles ($Fe_3O_4$-CA) onto the sacrificial polystyrene sulfonate microspheres [72]. Results of this work indicated that the hybrid hollow microspheres show pH-sensitive characteristics.

## 33.7
## Redox-Controlled Release

Disulfide or diselenide bonds can act as the cross-links that can be cleaved to single thiols or selenols upon reduction [73–76]. The transition between oxidative and reductive states has been exploited to trigger the destruction or decomposition of multilayer capsules following cellular take. Haynie et al. were the first to stabilize multilayer capsules by disulfide bonds through the design of oppositely charged 32-mer peptides modified with cysteine moieties [73, 74]. Furthermore, Caruso's group assembled poly(methacrylic acid) (PMA) modified with cysteamine capsules via hydrogen bonding. In detail, the polyelectrolyte microcapsules were obtained through sequential deposition of poly(vinylpyrrolidone) (PVPON) and thiolated poly(methacrylic acid) ($PMA^{SH}$) onto sacrificial silica core templates, followed by oxidative cross-linking of the thiol moieties and removal of the silica cores with HF [77]. The microcapsules as-prepared were broken down in vitro under reductive conditions which were very similar to those in the intracellular space. The same group has extended this technology to encapsulate oligonucleotides, peptides, and low-molecular-weight anticancer drugs [78]. They demonstrated that these capsules, using the strategy above, held great promise for in vivo applications, because the reducing intercellular environment could facilitate cleavage of the disulfide bonds to degrade the capsules. They also adopted this strategy to encapsulate single-stranded (ss), linear double-stranded (ds), and plasmid DNA in disulfide-cross-linked PMA microcapsules [79]. As shown in Figure 33.10, the encapsulation procedure involved four steps: adsorption of DNA onto silica particles modified with amine; sequential deposition of thiolated $PMA^{SH}$ and PVPON to form multilayers; cross-linking of the thiol groups of the $PMA^{SH}$ in the multilayers into disulfide linkages; and removal of the sacrificial silica particles. The encapsulation strategy could be applied to nucleic acids with different size and conformation, and concentrated DNA over 100-fold from solutions into capsules. The release of encapsulated DNA was successfully employed in a polymerase chain reactions (PCR) as both templates and primer sequences.

In addition, Höhwald's group introduced redox-active organometallic polyelectrolyte multilayer capsules, prepared by LbL self-assembly onto colloidal templates, followed by core removal [80]. In particular, incorporation of polyanions and polycations of poly(ferrocenylsilane) (PFS) into the system was helpful to investigate the effects of changing the redox state on capsule wall permeability. The permeability of

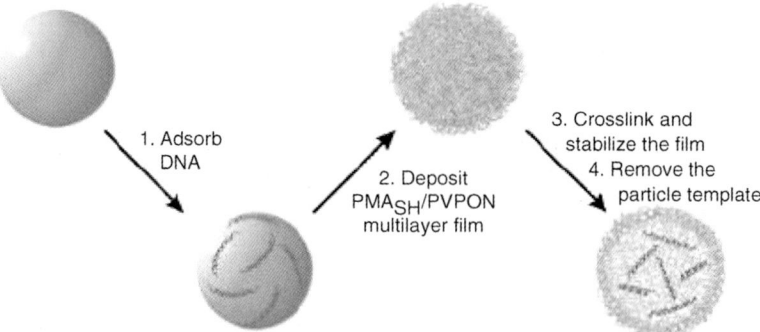

**Figure 33.10** Encapsulation of DNA into degradable polymer capsules. Adsorption of DNA onto amine-functionalized silica particles (1) is followed by the assembly of a thin polymer film prepared via the alternating deposition of PMA$^{SH}$ and PVPON (2). Oxidation of the PMA$^{SH}$ thiol groups into bridging disulfide linkages (3) and removal of the core particles (4) result in stable polymer capsules. DNA chains are confined within the capsule interior and are released in a reducing environment. Reprinted with permission from [79]. Copyright 2007, American Chemical Society.

these capsules could be sensitively adjusted by adding a little weak oxidant (FeCl$_3$) into the system, which led to a drastic permeability increase following a fast capsule expansion.

## 33.8
### Bio-Responsive Release

Enzymatically degradable polymer multilayer capsules based on oppositely charged polypeptides and/or polysaccharides have now been studied by various research groups. For instance, De Geest et al. demonstrated that polymer multilayer capsules consisting of dextran sulfate and poly-L-arigine could be degraded by proteases [81]. They were the first to report intracellularly degradable polyelectrolyte capsules. The *in vivo* conditions used in this work appeared to be able to induce this decomposition. The capsules had the outstanding advantage of requiring no external trigger for their decomposition (see Figure 33.11). These capsules might have great potential applications for the intracellular delivery of therapeutic nucleic acids (DNA, siRNA) and proteins.

Similar findings were reported later by different research groups, who used hyaluronidase and chitinase to decompose capsules containing hyaluronic acid and chitisan, respectively. For example, Itoh et al. studied encapsulation and controlled release of a model protein using hollow capsules composed of chitosan and dextran sulfate [82]. They found that chitosanase could degrade the chitosan component, resulting in the gradual deformation and final decomposition of the capsules and sustained release of the encapsulated proteins. In their further study, multilayer microcapsules with the same biodegradable polyelectrolytes mentioned above

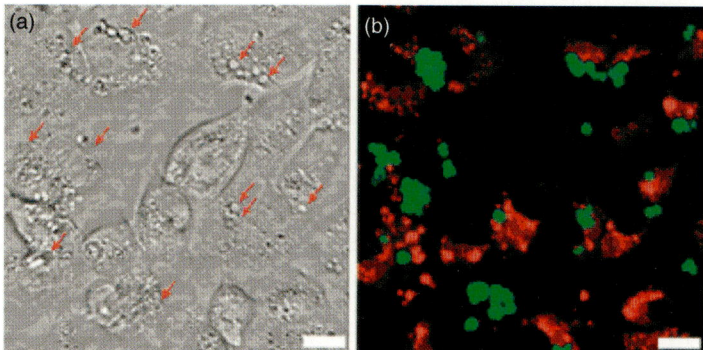

**Figure 33.11** Transmission (a) and confocal (b) images of DEXS/pARG capsules filled with FITC-dextran taken up by VERO-1 cells after 20 h incubation. The transmission image clearly shows the presence of intact capsules a few micrometers in size in the cells. The red arrows indicate the presence of intact capsules. In the confocal image Lyso-Tracker Red, which stains the lysosomes in the cytosol red, was used. The green structures are the DEXS/pARG capsules filled with FITC-dextran present in the cells. Scale bars represent 10 μm. Reprinted with permission from [81]. Copyright 2006, Wiley-VCH.

encapsulating basic fibroblast growth factor (bFGF) severed as acytokine release carrier [83]. The bFGF was encapsulated into the multilayered capsules by changing the pH in the solution systems. Surprisingly, about 30% of the encapsulated bFGF was released in a serum-free medium within a few hours and the release could be sustained over 70 h. In addition, HA/PLL polyelectrolyte microcapsules with a cross-linked shell by carbodiimide chemistry were prepared by LbL adsorption and subsequent core removal [84]. By changing the permeability by adjusting the pH, a model protein drug (bovine serum albumin (BSA)) could be loaded. Adding an HA digesting enzyme (hyaluronidase) into the incubation medium could readily modulate BSA release profiles from the microcapsules.

Becker and coworkers used PNAs to impart new properties to multilayered thin films, while retaining the advantages of sequence-directed assembly [85]. Changing the amount of DNA incorporated into the films could tune the nuclease degradation rate of thin films. In their subsequent study, high concentrations of uncomplexed, short oligonucleotide chains confined within monodisperse, degradable microcapsules could be obtained by a polycation-free encapsulation method [86]. In detail, the encapsulation method exploited amine-functionalized silica particles to adsorb oligonucleotides, followed by the multilayer assembly of PMA$^{SH}$ and PVPON. Removal of the template particles with HF produced degradable capsules filled with oligonucleotides. Furthermore, researchers in the same group reported covalently stabilized, biodegradable, and drug-loaded microcapsules through the assembly of a polymer-drug conjugate (shown in Figure 33.12) [87]. First, DOX was grafted to alkyne-functionalized poly(L-glutamic acid) (PGAAlk) via amide bond formation. Then, PGAAlk and PGAAlk-DOX were assembled with PVPON via hydrogen bonding onto colloidal silica templates. Furthermore, the shells were subsequently covalently stabilized using diazide cross-linkers via click chemistry, and PVPON was

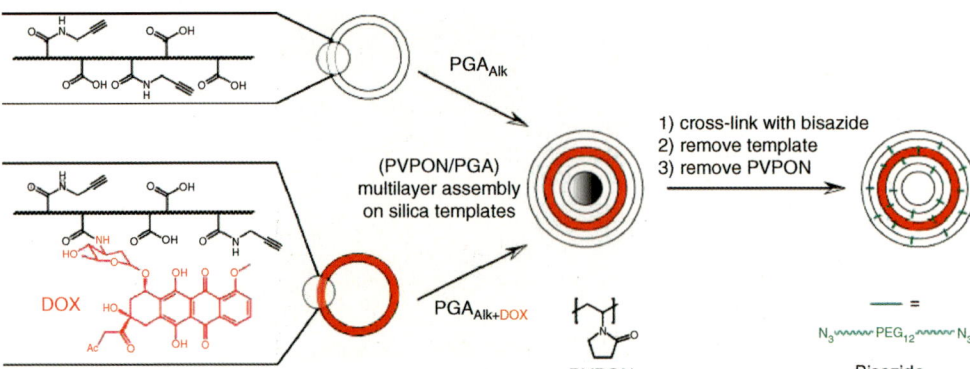

**Figure 33.12** Modular assembly of PGAAlk (white) click capsules with drug-loaded multilayers. The polymer_drug conjugate (PGAAlk-DOX, red) can be incorporated in defined positions and with controlled dose, after which the multilayer films are crosslinked via click chemistry using a bisazide crosslinker (green bars). Reprinted with permission from [87]. Copyright 1998, Wiley-VCH.

removed from the multilayers by changing the pH in solution to disrupt hydrogen bonding. Finally, single-component PGAAlk capsules were obtained after removal of the sacrificial template. Results in this work indicated that the drug-loaded capsules could be degraded by enzyme and the sustained release time of DOX was over 2 h.

As shown in Figure 33.13, novel self-disintegrating microcapsules were prepared by incorporating a highly active mixture of proteases (Pronase) into biodegradable polyelectrolyte shells [88]. In detail, Pronase was localized by calcium carbonate microparticles that were subsequently embedded into multilayer shells of poly(L-arginine) and poly(L-aspartic acid). After removal of the calcium carbonate constituents from the resulting core–shell particles by EDTA treatment, the release of Pronase led to degradation of the polyelectrolyte shell. The author demonstrated that

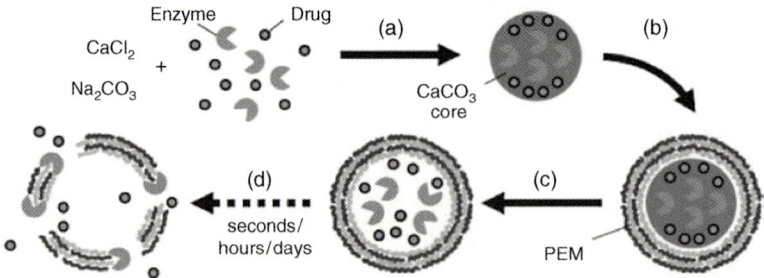

**Figure 33.13** Principle of time-delayed drug release from enzyme-degradable polyelectrolyte capsules. (a) Co-immobilization of a polyelectrolyte- degrading enzyme and drug molecules in CaCO₃ microparticles ("coprecipitation"). (b) LbL coating with specific, enzyme-degradable polyelectrolytes. (c) Core-dissolution (EDTA) leads to release of enzyme and drug into the inner void of the capsule. (d) Enzyme digests capsule-shell and releases drug. Reprinted with permission from [88]. Copyright 2007, Wiley-VCH.

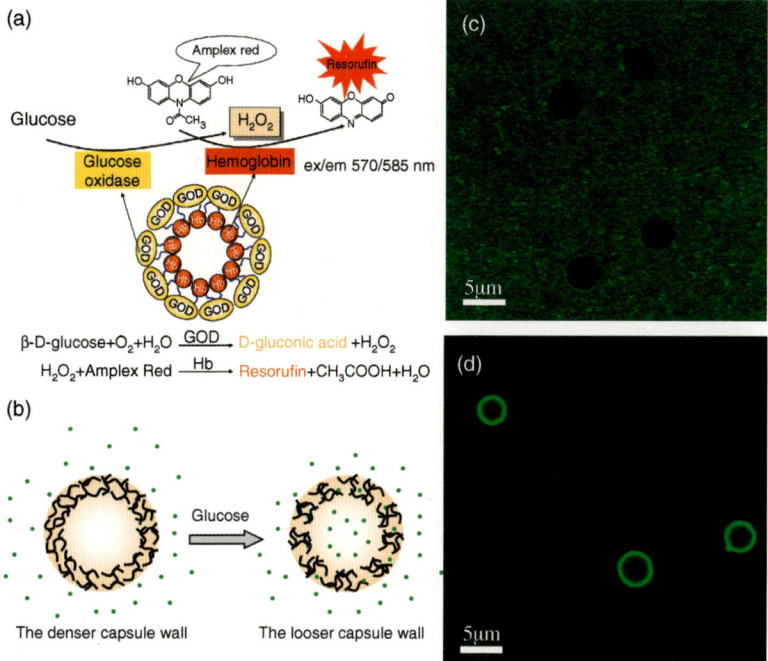

**Figure 33.14** Schematic illustration of (a) coupled enzymatic process based on glucose oxidase (GOD) and hemoglobin (Hb) co-immobilized as LbL microcapsules components; (b) glucose-stimulated enhancement of the Hb/GOD capsules wall permeability. CLSM images of (Hb/GOD)$_5$ microcapsules mixed with FITC-dextran (2000 kDa); (c) without glucose, (d) after adding glucose for 3 h. Reprinted with permission from [89]. Copyright 2009, American Chemical Society.

varying the amount of Pronase encapsulated could adjust the lifetimes of such self-disintegrating capsules from seconds, to hours to days.

In addition, enzymatic proteins have also been employed as layer materials to form multilayer shells for specific reactions. Recently, our group developed glucose-sensitive microcapsules by an enzymatic cooperation of Hb and GOD through LbL assembly [89]. Glucose-stimulated enhancement of the wall permeability could be observed because of the decrease in the local pH and the loosening of the multilayer structure (Figure 33.14). In a further work, this stimulation behavior was used to realize the controlled release of insulin [90]. Briefly, GOD and catalase (CAT) were assembled onto insulin microparticles alternately via glutaraldehyde (GA) cross-linking. The release ratio of insulin from the enzymatic multilayers increased remarkably after addition of external glucose (Figure 33.15).

Phenylboronic acids are well known to form covalent complexes with polyol compounds. A novel polyelectrolyte with phenylboronic acid has been synthesized for the fabrication of glucose-sensitive multilayer capsules [91]. The results showed that after being brought into contact with a glucose-containing medium, a rather fast

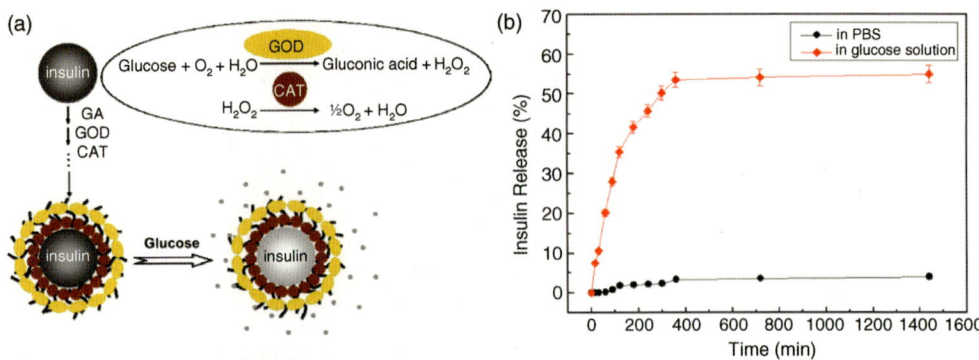

**Figure 33.15** Controlled release of insulin from glucose-sensitive enzyme multilayer shells. (a) Schematic representation of coupled reactions of glucose oxidase (GOD) and catalase (CAT) assembled onto insulin particles, followed by the enhanced permeability of the capsule for release of insulin. (b) Release profiles of coated insulin particles before (black line) and after (red line) external application of glucose solution, respectively. Reprinted with permission from [90]. Copyright 2009, Elsevier.

dissolution of the capsules could be observed readily. The authors demonstrated that these polyelectrolyte capsules assembled were the first polyelectrolyte capsules capable of responding to a stimulus provided by the human body, and the strategy used had promising applications for the controlled delivery of insulin. In addition, Levy et al. reported carbohydrate-sensitive polymer multilayers assembled onto colloidal $CaCO_3$ particles via reversible covalent ester formation between the polysaccharide mannan and phenylboronic acid moieties grafted onto poly(acrylic acid) (PAA) [92]. The multilayer films obtained were sensitive to many kinds of carbohydrates, and exhibited the highest sensitivity to fructose. Very recently, Manna and Patil reported a novel approach for glucose-triggered anticancer drug delivery from the LbL self-assembly of borate-modified PVA and CHI [93]. They investigated the disassembly of microcapsules obtained in the presence of glucose. They also examined the effect of glucose on the release behavior of DOX embedded in multilayer microcapsules. The related results indicated that this multilayer thin film was very efficient for controlled release of DOX molecules above a certain concentration of glucose.

## 33.9
### Extension

It is worth mentioning that, very recently, nanometer-sized colloidal drug delivery systems prepared by the LbL technique have also been developed for cancer and other therapies [94]. As shown in Figure 33.16, in these systems, various multifunctional building blocks can be engineered rationally to obtain highly stable dispersions and nano-objects can be assembled stealthily with various smart polymers. The authors demonstrated that this strategy makes such delivery systems promising to enhance

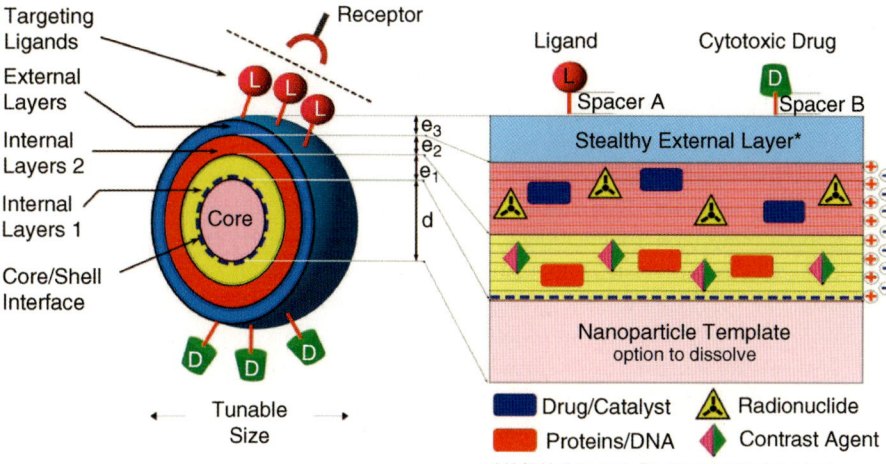

**Figure 33.16** Schematic depiction of nanoparticles coated with multilayer shells as a new drug delivery system. The multifunctionality arises from the stepwise construction of the shell that is assembled by the LBL method. The internal layers are arbitrarily split in two compartments (yellow (1) and red (2)) in order to indicate that different functionalities can be integrated in a modular way in different layers. Theoretically there is no limit with respect to the number of different internal compartments. The yellow compartment (1) serves primarily to compatibilize between the core and the external layers enabling the use of different core materials while maintaining the same functionalization process. However, the yellow compartment (1) as well as the red one (2) may also serve to incorporate additional functional entities, such as drugs, catalysts for biochemical reactions, radionuclids for radiotherapy, proteins/nucleotides for bioactivity, or contrast agents for detection. In the present study this was not realized, but such options are well established from LBL films on flat substrates. The external layers carry functionalities such as enzymatically cleavable drugs or ligands for receptor-mediated targeting, both of which must be accessible on the outside. Only the functionalization with an attached drug was chosen in the present study. The attachment of additional ligands, however, follows the same principle which has already been described for a variety of different functional groups in the past. Reprinted with permission from [94]. Copyright 2009, American Chemical Society.

the accumulation of active drug in the interesting disease sites. Elbakry et al. prepared LbL-engineered gold nanoparticles coated with siRNA and poly(ethylenimine) (PEI) [95]. The assembled nanoparticles with caveolae-like structures demonstrated effective knockdown of enhanced green fluorescent protein (EGFP) expression in stably transfected Chinese hamster ovary cells (CHO-K1). In another study, Poon et al. introduced LbL-coated quantum dot (QD) nanoparticles based on the electrostatic interactions between dextran sulfate (DS) and PLL [96]. These LbL nanoparticles in vivo were evaluated and the result indicated that the number of layers and the surface chemistry are key factors for layer stability and nanoparticle biodistribution. Interestingly, the circulation time of DS/PLL LbL nanoparticles could be extended by incorporating HA as the outer layer. Poon et al. also reported the formation and use of pH-sensitive LbL engineered nanoparticles for systemic targeting of hypoxic tumors in vivo [97].

## 33.10
## Concluding Remarks

In this review, we have highlighted some recent progress in designing novel types of LbL-based multilayer microcapsules for drug encapsulation and controlled release. Their intelligently adjustable permeability allows the loading of many different kinds of drug molecules. They can be easily modified with various biologically friendly natural molecules (such as lipids, nucleotides and proteins) or inorganic nanoparticles (such as Au and $Fe_3O_4$ nanoparticles) to create multifunctional shells with desirable combined or synergetic properties. It should be noted that capsules with various physicochemical and biological properties have been assembled for wide applications, ranging from stimulative imaging and ready delivery of biologically active molecules to bio-microreactors. Nevertheless, there are still some limits left to be overcome for further development of LbL capsule delivery systems for drug encapsulation and controlled release.

In most cases, the fabrication of multilayered capsules is relatively time and cost consuming and still stays at the laboratory scale because of their essential multistep coating and separation. New efficient methods for the industrial production and application have to be developed if they are to be a powerful alternative to encapsulate and release traditional drug molecules. Moreover, when multilayered capsules are used for intravenous administration, we have to focus on the design of multilayer capsules with a size of less than 500 nm, considering the well-known EPR (enhanced permeability and retention) effect of the blood vessels. Developing new molecules and studying new active targeting and controlled release mechanisms are necessary in order to integrate new complex multifunctional systems transferable to the clinic. Personally speaking, they could be designed and constructed according to a traditional Chinese medicine (TCM) formula [98]. TCM follows the tenet that a formula should have four major ingredients (emperor, minister, assistant and delivering servant), each playing its unique role while working together synergistically, to achieve the optimum therapy. Certainly, to overcome the limitations of the currently available LbL systems needs sincere and effective cooperation between different scientific communities.

## Acknowledgments

We acknowledge the financial support of this research by the National Nature Science Foundation of China (No.20833010), and National Basic Research Program of China (973 program) 2009CB930101.

## References

1 Langer, R. and Tirrell, D.A. (2004) Designing materials for biology and medicine. *Nature*, **428**, 487–492.
2 Peer, D., Karp, J.M., Hong, S., Farokhzad, O.C., Margalit, R., and Langer, R. (2007) Nanocarriers as an emerging platform for cancer therapy. *Nat. Nanotechnol.*, **2**, 751–760.
3 Boisselier, E. and Astruc, D. (2009) Gold nanoparticles in nanomedicine: preparations, imaging, diagnostics,

therapies and toxicity. *Chem. Soc. Rev.*, **38**, 1759–1782.

4 Arruebo, M., Fernandez-Pacheco, R., Ibarra, M.R., and Santamaria, J. (2007) Magnetic nanoparticles for drug delivery. *Nano Today*, **2**, 22–32.

5 Zhang, L., Gu, F.X., Chan, J.M., Wang, A.Z., Langer, R.S., and Farokhzad, O.C. (2008) Nanoparticles in medicine: therapeutic applications and developments. *Clin. Pharmacol. Ther.*, **83**, 761–769.

6 Sperling, R.A., Rivera-Gil, P., Zhang, F., Zanella, M., and Parak, W.J. (2008) Biological applications of gold nanoparticles. *Chem. Soc. Rev.*, **37**, 1896–1908.

7 Son, S.J., Bai, X., and Lee, S. (2007) Inorganic hollow nanoparticles and nanotubes in nanomedicine: Part 2: Imaging, diagnostic, and therapeutic applications. *Drug Discov. Today*, **12**, 657–663.

8 Son, S.J., Bai, X., and Lee, S. (2007) Inorganic hollow nanoparticles and nanotubes in nanomedicine: Part 1. Drug/gene delivery applications. *Drug Discov. Today*, **12**, 650–656.

9 Yang, K., Zhang, S., and Zhang, G.X. (2010) Graphene in mice: ultrahigh in vivo tumor uptake and efficient photothermal therapy. *Nano Lett.*, **10**, 3318–3323.

10 Torchilin, V. (2009) Multifunctional and stimuli-sensitive pharmaceutical nanocarrier. *Eur. J. Pharm. Biopharm.*, **71**, 431–444.

11 Decher, G. and Hong, J.D. (1991) Buildup of ultrathin multilayer films by a self-assembly process. 1 Consecutive adsorption of anionic and cationic bipolar amphiphiles on charged surfaces. *Makromol Chem. Macromol. Symp.*, **46**, 321–327.

12 Decher, G. (1997) Fuzzy nanoassemblies: toward layered polymeric multicomposites. *Science*, **277**, 1232–1237.

13 Cheung, J.H., Stockton, W.B., and Rubner, M.F. (1997) Molecular-level processing of conjugated polymers. 3. layer-by-layer manipulation of polyaniline via electrostatic interactions. *Macromolecules*, **30**, 2712–2716.

14 Wang, L.Y., Wang, Z.Q., Zhang, X., and Shen, J.C. (1997) A new approach for the fabrication of alternating multilayer film of poly (4-vinylpyridine) and polyacrylic acid based on hydrogen bonding. *Macromol. Rapid Commun.*, **18**, 509–514.

15 Shimazaki, Y., Mitsuishi, M., Ito, S., and Yamamoto, M. (1997) Preparation of the layer-by-layer deposited ultrathin film based on the charge-transferinteraction. *Langmuir*, **13**, 1385–1387.

16 Onitsuka, O., Fou, A.C., Ferreira, M., Hsieh;, B.R., and Rubner, M.F. (1996) Enhancement of light emitting diodes based on self-assembled heterostructures of poly(phenylenevinylene). *J. Appl. Phys.*, **80**, 4067–4071.

17 Zhang, X.M., He, Q., Yan, X.H., Boullanger, P., and Li, J.B. (2007) Glycolipid patterns supported by human serum albumin for E. coli recognition. *Biochem. Biophys. Res. Commun.*, **358**, 424–428.

18 Zhang, X.M., He, Q., Cui, Y., Duan, L., and Li, J.B. (2006) Human serum albumin supported lipid patterns for the targeted recognition of microspheres coated by membrane based on ss-DNA hybridization. *Biochem. Biophys. Res. Commun.*, **349**, 920–924.

19 Hua, D.Y., Yang, Y., Zhang, X.M., Zhu, P.L., Fei, J.B., and Li, J.B. (2010) Biotinylated lipid membrane patterns supported by proteins for the recognition of streptavidined polystyrene microspheres. *J. Nanosci. Nanotechnol.*, **10**, 6318–6323.

20 Donath, E., Sukhorukov, G.B., Caruso, F., Davis, S.A., and Möhwald, H. (1998) Novel hollow polymer shells by colloid-templated assembly of polyelectrolytes. *Angew. Chem. Int. Ed.*, **37**, 2201–2205.

21 Johnston, A.P., Cortez, R.C., Angelatos, A.S., and Caruso, F. (2006) Layer-by-layer engineered capsules and their applications. *Curr. Opin. Colloid Interface Sci.*, **11**, 203–209.

22 Peyratout, C.S. and Daehne, L. (2004) Tailor-made polyelectrolyte microcapsules: From multilayers to smart containers. *Angew. Chem. Int. Ed.*, **43**, 3762–3783.

23 Fery, A. and Weinkamer, R. (2007) Mechanical properties of micro- and nanocapsules: single-capsule measurements. *Polymer*, **48**, 7221–7235.

24 Antipov, A. and Sukhorukov, G.B. (2004) Polyelectrolyte multilayer capsules as vehicles with tunable permeability. *Adv. Colloid Interface Sci.*, **111**, 49–61.

25 Sukhorukov, G.B., Fery, A., Brumen, M., and Möhwald, H. (2004) Physical chemistry of encapsulation and release. *Phys. Chem. Chem. Phys.*, **6**, 4078–4089.

26 Sukhorukov, G.B., Fery, A., and Möhwald, H. (2005) Intelligent micro and nanocapsules. *Prog. Polym. Sci.*, **30**, 885–897.

27 De Geest, B.G., Sanders, N.N., Sukhorukov, G.B., Demeestera, J., and Smedt, S.C.D. (2007) Release mechanisms for polyelectrolyte capsules. *Chem. Soc. Rev.*, **36**, 636–649.

28 De Cock, L.J., De Koker, S., De Geest, B.G., Grooten, J., Vervaet, C., Remon, J.P., Sukhorukov, G.B., and Antipina, M.N. (2010) Polymeric multilayer capsules in drug delivery. *Angew. Chem. Int. Ed.*, **49**, 6954–6973.

29 Tong, W.J. and Gao, C.Y. (2008) Multilayer microcapsules with tailored structures for bio-related applications. *J. Mater. Chem.*, **18**, 3799–3812.

30 del Mercato, L.L., Rivera-Gil, P., Abbasi, A.Z., Ochs, M., Ganas, C., Zins, I., Sonnichsen, C., and Parak, W.J. (2010) LbL multilayer capsules: recent progress and future outlook for their use in life sciences. *Nanoscale*, **2**, 458–467.

31 Liu, J.W., Zhang, Y., Wang, C.Y., Xu, R.Z., Chen, Z.P., and Gu, N. (2010) Magnetically sensitive alginate-templated polyelectrolyte multilayer microcapsules for controlled release of doxorubicin. *J. Phys. Chem. C*, **114**, 7673–7679.

32 Lu, Z.H., Prouty, M.D., Lu, Z.H., Prouty, M.D., Guo, Z.H., Golub, V.O., Kumar, C.S.S.R., and Lvov, Y.M. (2005) Magnetic switch of permeability for polyelectrolyte microcapsules embedded with Co@Au nanoparticles. *Langmuir*, **21**, 2042–2050.

33 Katagiri, K., Nakamura, M., and Koumoto, K. (2010) Magnetoresponsive smart capsules formed with polyelectrolytes, lipid bilayers and magnetic nanoparticles. *ACS Appl. Mater. Interfaces*, **2**, 768–773.

34 Silva Gomes, J.F.P., Rank, A., Kronenberger, A., Jurgen, F., Winterhalter, M., and Ramaye, Y. (2009) Polyelectrolyte-coated unilamellar nanometer-sized magnetic liposomes. *Langmuir*, **25**, 6793–6799.

35 Han, Y.S., Radziuk, D., Shchukin, D., and Möhwald, H. (2008) Sonochemical synthesis of magnetic protein container for targeted delivery. *Macromol. Rapid Commun.*, **29**, 1203–1207.

36 http://en.wikipedia.org/wiki/Ultrasound (last accessed 26/02/2011).

37 Novelline, R. (1997) *Squire's Fundamentals of Radiology*, 5th edn, Harvard University Press, pp. 34–35.

38 Chen, W.S., Matula, T.J., Brayman, A.A., and Crum, L.A. (2003) Comparison of the fragmentation thresholds and inertial cavitation dose of different ultrasound contrast agents. *J. Acoust. Soc. Am.*, **113**, 643–651.

39 Postema, M., Bouakaz, A., and de Jong, N. (2005) Ultrasound-induced gas release from contrast agent microbubbles. *IEEE Trans. Ultrason. Ferroelectr. Freq. Control*, **52**, 1035–1041.

40 Bloch, S.H., Wan, M., Dayton, P.A., and Ferrara, K.W. (2004) Optical observation of lipid- and polymer-shelled ultrasound microbubble contrast agents. *Appl. Phys. Lett.*, **84**, 631–633.

41 Mitragotri, S. (2005) Healing sound: the use of ultrasound in drug delivery and other therapeutic applications. *Nat. Rev. Drug Discovery*, **4**, 255–260.

42 Song, W.X., He, Q., Möhwald, H., Yang, Y., and Li, J.B. (2009) Smart polyelectrolyte microcapsules as carriers for water-soluble molecular drug. *J. Controlled Release*, **139**, 160–166.

43 Gou, Z.R., Weng, W.J., Du, P.Y., and Han, G.R. (2006) An Efficient US-enhanced controllable release of biologically active trace elements on bioactive silica-gel-based material. *Chem. Lett.*, **35**, 1214–1215.

44 http://en.wikipedia.org/wiki/Light (last accessed 27/02/2011).

45 Vogel, A. and Venugopalan, V. (2003) Mechanisms of pulsed laser ablation of

biological tissues. *Chem. Rev.*, **103**, 577–644.

46 Bédard, M.F., De Geest, B.G., Skirtach, A.G., Möhwald, H., and Sukhorukov, G.B. (2010) Polymeric microcapsules with light responsive properties for encapsulation and release. *Adv. Colloid Interface Sci.*, **158**, 2–14.

47 Oldenburg, S.J., Jackson, J.B., Westcott, S.L., and Halas, N.J. (1999) Infrared extinction properties of gold nanoshells. *App. Phys. Lett.*, **75**, 2897–2899.

48 Sershen, S.R., Westcott, S.L., Halas, N.J., and West, J.L. (2000) Temperature-sensitive polymer-nanoshell composites for photothermally modulated drug delivery. *J. Biomed. Mater. Res.*, **51**, 293–298.

49 Sershen, S.R., Westcott, S.L., Halas, N.J., and West, J.L. (2002) Independent optically addressable nanoparticle-polymer optomechanical composites. *Appl. Phys. Lett.*, **80**, 4609–4611.

50 Radt, B., Smith, T.A., and Caruso, F. (2004) Optically addressable nanostructured capsules. *Adv. Mater.*, **16**, 2184–2189.

51 Angelatos, A.S., Radt, B., and Caruso, F. (2005) Light-responsive polyelectrolyte/gold nanoparticle microcapsules. *J. Phys. Chem. B*, **109**, 3071–3076.

52 Javier, A.M., del Pino, P., Bedard, M.F., Ho, D., Skirtach, A.G., Sukhorukov, G.B., Plank, C., and Parak, W.J. (2008) Photoactivated release of cargo from the cavity of polyelectrolyte capsules to the cytosol of cells. *Langmuir*, **24**, 12517–12520.

53 Skirtach, A.G., Dejugnat, C., Braun, D., Susha, A.S., Rogach, A.L., Parak, W.J., Möhwald, H., and Sukhorukov, G.B. (2005) The role of metal nanoparticles in remote release of encapsulated materials. *Nano Lett.*, **5**, 1371–1377.

54 Volodkin, D.V., Delcea, M., Möhwald, H., and Skirtach, A.G. (2009) Remote near-IR light activation of a hyaluronic acid/poly (L-lysine) multilayered film and film-entrapped microcapsules. *ACS Appl. Mater. Interfaces*, **11**, 1705–1710.

55 Skirtach, A.G., Karageorgiev, P., Bédard, M.F., Sukhorukov, G.B., and Möhwald, H. (2008) Reversibly permeable nanomembranes of polymeric microcapsules. *J. Am. Chem. Soc.*, **130**, 11572–11573.

56 Skirtach, A.G., Javier, A.M., Kreft, O., Kohler, K., Alberola, A.P., Möhwald, H., Parak, W.J., and Sukhorukov, G.B. (2006) Laser-induced release of encapsulated materials inside living cells. *Angew. Chem. Int. Ed.*, **45**, 4612–4617.

57 Skirtach, A.G., Karageorgiev, P., De Geest, B.G., Pazos-Perez, N., Braun, D., and Sukhorukov, G.B. (2008) Nanorods as wavelength-selective absorption centers in the visible and near-infrared regions of the electromagnetic spectrum. *Adv. Mater.*, **20**, 506–510.

58 http://en.wikipedia.org/wiki/Temperature-responsive_polymer (last accessed 26/02/2011).

59 Kaneko, Y., Nakamura, S., Sakai, K., Aoyagi, T., Kikuchi, A., Sakurai, Y., and Okano, T. (1998) Rapid deswelling response of poly (N-isopropylacrylamide) hydrogels by the formation of water release channels using poly (ethylene oxide) graft chains. *Macromolecules*, **31**, 6099–6105.

60 Peppas, N.A. and Langer, R. (1994) New challenges in biomaterials. *Science*, **263**, 1715–1720.

61 Peppas, N.A., Huang, Y., Torres-Lugo, M., Ward, J.H., and Zhang, J. (2000) Physicochemical, foundations and structural design of hydrogels in medicine and biology. *Annu. Rev. Biomed. Eng.*, **2**, 9–29.

62 Prevot, M., Dějugnat, C., Möhwald, H., and Sukhorukov, G.B. (2006) Behavior of temperature-sensitive PNIPAM confined in polyelectrolyte capsules. *Chem. Phys. Chem.*, **7**, 2497–2502.

63 Huang, C.J. and Chang, F.C. (2009) Using click chemistry to fabricate ultrathin thermoresponsive microcapsules through direct covalent layer-by-layer assembly. *Macromolecules*, **42**, 5155–5166.

64 Kharlampieva, E., Kozlovskaya, V., Tyutina, J., and Sukhishvili, S.A. (2005) Hydrogen-bonded multilayers of thermoresponsive polymers. *Macromolecules*, **38**, 10523–10531.

65 Zhao, Q.H. and Li, B.Y. (2008) pH-controlled drug loading and release

from biodegradable microcapsules. *Nanomed: Nanotechnol., Biol. Med.*, **4**, 302–310.

66 Usov, D. and Sukhorukov, G.B. (2010) Dextran coatings for aggregation control of layer-by-layer assembled polyelectrolyte microcapusles. *Langmuir*, **26**, 12575–12584.

67 Teng, X.R., Shchukin, D.G., and Möhwald, H. (2008) A novel drug carrier: lipophilic drug-loaded polyglutamate/polyelectrolyte nanocontainers. *Langmuir*, **24**, 383–389.

68 Manna, U. and Patil, S. (2009) Borax mediated layer-by-layer self-assembly of neutral poly (vinyl alcohol) and chitosan. *J. Phys. Chem. B*, **113**, 9137–9142.

69 Shu, S.J., Sun, C.Y., Zhang, X., Wu, Z.M., Wang, Z., and Li, C.X. (2010) Hollow and degradable polyelectrolyte nanocapsules for protein drug delivery. *Acta Biomater.*, **6**, 210–217.

70 Jia, Y., Fei, J.B., Cui, Y., Yang, Y., Gao, L., and Li, J.B. (2011) pH-responsive polysaccharides microcapsules through covalent bonding assembly. *Chem. Commun.*, **47**, 1175–1177.

71 Chaturbedy, P., Jagadeesan, D., and Eswaramoorthy, M. (2010) pH-sensitive breathing of clay within the polyelectrolyte matrix. *ACS Nano*, **4**, 5921–5929.

72 Mu, B., Liu, P., Dong, Y., Lu, C.Y., and Wu, X.L. (2010) Superparamagnetic pH-sensitive multilayer hybrid hollow microspheres for targeted controlled release. *J. Polym. Sci.: Part A: Polym. Chem.*, **48**, 3135–3134.

73 Li, B.Y. and Haynie, D.T. (2004) Multilayer biomimetics: reversible covalent stabilization of a nanostructured biofilm. *Biomacromolecules*, **5**, 1667–1670.

74 Haynie, D.T., Palath, N., Liu, Y., Li, B.Y., and Pargaonkar, N. (2005) Biomimetic nanostructured materials:inherent reversible stabilization of polypeptide microcapsules. *Langmuir*, **21**, 1136–1138.

75 Ma, N., Li, Y., Xu, H.P., Wang, Z.Q., and Zhang, X. (2010) Dual redox responsive assemblies formed from diselenide block copolymers. *J. Am. Chem. Soc.*, **132**, 442–443.

76 Ma, N., Li, Y., Ren, H.F., Xu, H.P., Li, Z.B., and Zhang, X. (2010) Selenium-containing block copolymers and their oxidation-responsive aggregates. *Polym. Chem.*, **1**, 1609–1614.

77 Zelikin, A.N., Li, Q., and Caruso, F. (2008) Disulfide-stabilized poly (methacrylic acid) capsules: formation, cross-linking, and degradation behavior. *Chem. Mater.*, **20**, 2655–2661.

78 Wang, Y.J., Bansal, V., Zelikin, A.N., and Caruso, F. (2008) Templated synthesis of single-component polymer capsules and their application in drug delivery. *Nano Lett.*, **8**, 1741–1745.

79 Zelikin, A.N., Becker, A.L., Johnston, A.P.R., Wark, K.L., Turatti, F., and Caruso, F. (2007) A general approach for DNA encapsulation in degradable polymer microcapsules. *ACS Nano*, **1**, 63–69.

80 Ma, Y.J., Dong, W.F., Hempenius, M.A., Höwald, H., and Vancso, G.J. (2005) Redox-controlled molecular permeability of composite-wall microcapsules. *Nat. Mater.*, **5**, 724–729.

81 De Geest, B.G., Vandenbroucke, R.E., Guenther, A.M., Sukhorukov, G.B., Hennink, W.E., Sanders, N.N., Demeester, J., and De Smedt, S.C. (2006) Intracellularly degradable polyelectrolyte microcapsules. *Adv. Mater.*, **18**, 1005–1009.

82 Itoh, Y., Matsusaki, M., Kida, T., and Akashi, M. (2006) Enzyme-responsive release of encapsulated proteins from biodegradable hollow capsules. *Biomacromolecules*, **7**, 2715–2718.

83 Itoh, Y., Matsusaki, M., Kida, T., and Akashi, M. (2008) Locally controlled release of basic fibroblast growth factor from multilayered capsules. *Biomacromolecules*, **9**, 2202–2206.

84 Lee, H., Jeong, Y., and Park, T.G. (2007) Binding and release of consensus peptides by poly(acrylic acid) microgels. *Biomacromolecules*, **8**, 3705–3711.

85 Zelikin, A.N., Li, Q., and Caruso, F. (2006) Degradable polyelectrolyte capsules filled with oligonucleotide sequences. *Angew. Chem. Int. Ed.*, **45**, 7743–7745.

86 Becker, A.L., Johnston, A.P.R., and Caruso, F. (2010) Peptide nucleic acid films and capsules: assembly and enzymatic degradation. *Macromol. Biosci.*, **10**, 488–495.

87 Ochs, C.J., Such, G.K., Yan, Y., van Koeverden, M.P., and Caruso, F. (2010) Biodegradable click capsules with engineered drug-loaded multilayers. *ACS Nano*, **4**, 1653–1663.

88 Borodina, T., Markvicheva, E., Kunizhev, S., Möhwald, H., Sukhorukov, G.B., and Kreft, O. (2007) Controlled release of DNA from self-degrading microcapsules. *Macromol. Rapid Commun.*, **28**, 1894–1899.

89 Qi, W., Yan, X.H., Duan, L., Cui, Y., Yang, Y., and Li, J.B. (2009) Glucose-sensitive microcapsules from glutaraldehyde cross-linked hemoglobin and glucose oxidase. *Biomacromolecules*, **10**, 1212–1216.

90 Qi, W., Yan, X.H., Fei, J.B., Wang, A.H., Cui, Y., and Li, J.B. (2009) Triggered release of insulin from glucose-sensitive enzyme multilayer shells. *Biomaterials*, **30**, 2799–2806.

91 De Geest, B.G., Jonas, A.M., Demeester, J., and De Smedt, S.C. (2006) Glucose-responsive polyelectrolyte capsules. *Langmuir*, **22**, 5070–5074.

92 Levy, T., Déjugnat, C., and Sukhorukov, G.B. (2008) Polymer microcapsules with carbohydrate-sensitive properties. *Adv. Funct. Mater.*, **18**, 1586–1594.

93 Manna, U. and Patil, S. (2010) Glucose-triggered drug delivery from borate mediated layer-by-layer self-assembly. *ACS Appl. Mat. Interfaces*, **2**, 1521–1527.

94 Schneider, G.F., Subr, V., Ulbrich, K., and Decher, G. (2009) Multifunctional cytotoxic stealth nanoparticles. a model approach with potential for cancer therapy. *Nano Lett.*, **9**, 636–642.

95 Elbakry, A., Zaky, A., Liebl, R., Rachel, R., Goepferich, A., and Breunig, M. (2009) Layer-by-layer assembled gold nanoparticles for siRNA delivery. *Nano Lett.*, **9**, 2059–2064.

96 Poon, Z., Lee, J.B., Morton, S.W., and Hammond, P.T. (2011) Controlling in vivo stability and biodistribution in electrostatically assembled nanoparticles for systemic delivery. *Nano Lett.*, **11**, 2096–2103.

97 Poon, Z., Chang, D., Zhao, X., and Hammond, P.T. (2011) Layer-by-layer nanoparticles with a pH sheddable lLayer for in vivo targeting of tumor hypoxia. *ACS Nano*, **5**, 4284–4292.

98 Wang, L., Zhou, G.B., Liu, P., Song, J.H., Liang, Y., Xu, F., Wang, B.S., Mao, J.H., Shen, Z.X., Chen, S.J., and Chen, Z. (2008) Dissection of mechanisms of Chinese medicinal formula Realgar-Indigo naturalis as an effective treatment for promyelocytic leukemia. *Pro. Natl. Acad. Sci. U. S. A.*, **105**, 4826–4831.

# 34
# Engineered Layer-by-Layer Assembled Capsules for Biomedical Applications

*Angus P.R. Johnston, Georgina K. Such, Sarah J. Dodds, and Frank Caruso*

## 34.1
### Introduction

Over the past decade, there has been a convergence of materials science with biology to develop advanced materials for drug delivery, biomolecule detection and regenerative medicine [1, 2]. In particular, particle-based systems (also known as carriers or capsules) have the potential to improve the treatment of a number of diseases, including cancer, HIV and type I diabetes. However, there are still significant challenges with integrating these engineered carriers with biological systems [3]. Over the last 20 years, the layer-by-layer (LbL) assembly process has been extensively used for the preparation of advanced materials, largely because it is an inexpensive, highly adaptable and facile solution-based assembly method, thus allowing materials to be designed and assembled with specific properties and nanoscale precision. Significant progress has been made in the field since the publication of "Multilayer Thin Films: Sequential Assembly of Nanocomposite Materials" [4]. In this chapter we explore some of the recent advances made with delivering therapeutic compounds using layer-by-layer (LbL)-engineered particles and capsules (Figure 34.1). The key considerations for engineering these systems are: (i) template selection (Section 34.2), (ii) material assembly (Section 34.3), (iii) drug loading (Section 34.4), and (iv) capsule degradation and cargo release (Section 34.5). In addition, we highlight recent scientific progress aimed at the application of these nanoengineered particles to current biomedical challenges, such as cancer treatment and vaccine delivery (Section 34.6).

## 34.2
### Template Selection

One significant advantage of the LbL assembly method is that it can be performed on a range of substrates. LbL assembly was initially developed for planar surfaces and research in this area continues to generate developments for applications in biomedicine [5, 6]. However, for drug delivery, LbL assembly on particle templates

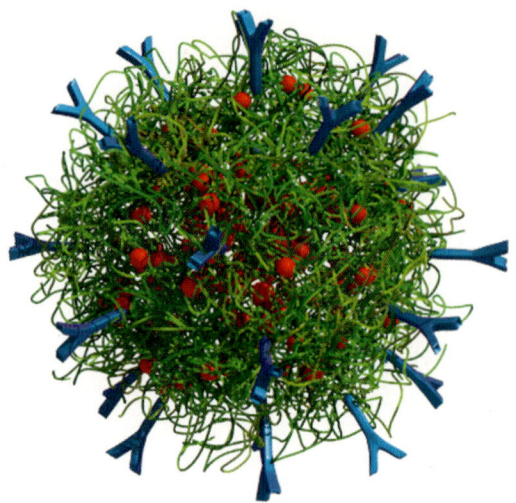

**Figure 34.1** Schematic diagram of an LbL-assembled capsule (green) loaded with therapeutic cargo (red) and functionalized with targeting antibodies (blue).

offers a unique approach to controlling the size, surface functionality, morphology and degradability of the delivery vehicles [7].

Another key advantage of LbL assembly is that a suite of particle templates can be chosen for the preparation of capsules, with sizes ranging from nanoparticles (~13 nm [8]) through to particles larger than tens of micrometers. It is generally accepted that particle size and monodispersity play an important role in the properties of drug delivery vehicles, therefore, these characteristics are fundamental in the design of polymeric carriers [3]. Developing an understanding of how particle size affects interactions with biological systems is key to advancing LbL carriers for biomedical applications. Recent literature on particle carriers has demonstrated that flexibility and shape can also influence the interaction of carriers with cells [3]. Both the size and shape of LbL-engineered particles can be readily tailored through use of an appropriate template [7]. There are two types of LbL-engineered particle systems: (i) core–shell particles, where the core is retained after multilayer assembly; and (ii) capsules, where the core is removed.

LbL capsules have been prepared using particle templates of diverse composition, including polystyrene, melamine formaldehyde (MF), silica and calcium carbonate [9]. Organic particle templates, such as polystyrene or MF, require the use of organic solvents or highly acidic conditions for their removal, and therefore there are limitations with the encapsulation of sensitive biomolecules. Over recent years silica particles (both nonporous and porous) have become widely used templates. One of the significant advantages of porous particle templates is that they can be used to load high amounts of cargo (e.g., DNA or proteins). The cargo can remain entrapped in subsequently generated capsules following LbL assembly and removal of the porous particles [10]. Silica particle templates are dissolved using buffered dilute hydrofluoric acid (HF). A variety of cargos, including proteins, DNA and peptides, can be loaded into porous silica spheres, and after LbL coating and HF-induced particle template removal, these loaded materials remain active and functional [10, 11]. Calcium carbonate ($CaCO_3$) particles are also commonly used templates, as they can be dissolved using aqueous ethylenediaminetetraacetic acid. However, unlike silica particles, it is challenging to make uniformly sized and submicrometer-sized $CaCO_3$ particles.

While silica and calcium carbonate particles have emerged as the templates of choice, the versatility of LbL assembly has led to its application to a host of templates, including gold nanoparticles, emulsions, protein crystals/aggregates and cells [7]. Gold nanoparticle templates are particularly attractive, as small-sized LbL capsules (<100 nm) may lead to improved *in vivo* circulation times for these particle carriers. Recent work by Decher and coworkers demonstrated the preparation of a multifunctional cytotoxic stealth core–shell system based on gold nanoparticles [8].

Liquid crystal (LC) emulsions have been used as templates for LbL assembly; for example, poly(styrene sulfonate) (PSS) and poly(allylamine hydrochloride) (PAH) multilayers were assembled onto thermotropic LC emulsions [12]. It was found that the surfactant-induced transition that occurs in the uncoated emulsions (bipolar to radial) was retarded by two orders of magnitude when the emulsions were coated with the multilayers. These LC emulsions are potentially useful in three-dimensional molecular sensing, offering higher sensitivity and faster response. In related work, it was also demonstrated that polymer multilayers could be assembled onto LC emulsions by using microfluidic devices [13]. This was achieved by stabilizing the emulsion droplets with poly(*N*-vinyl pyrrolidone) (PVPON) and surfactant. As the emulsion droplets were transported through the microfluidic channels, the first polymer was selectively removed and replaced with a rinse solution and then thiolated poly(methacrylic acid) ($PMA_{SH}$) to facilitate LbL assembly on the emulsions. The advantage of this approach is that it significantly reduces the LbL assembly times for deposition of each layer, as centrifugation steps are not required. The automated system can deposit three polymer layers in ~2 min, compared with >30 min for the conventional assembly approach that requires centrifugation–supernatant-resuspension cycles to remove excess polymer from solution. The use of emulsion templates also allows high cargo loadings, a feature that has been applied to the synthesis of LbL microreactors and to load low molecular weight drugs [14].

Earlier work reported that protein crystals, crystals of low molecular weight compounds, and protein aggregates can also be used as templates for LbL assembly, and offer the advantage of allowing efficient and high cargo loadings [15, 16]. However, the polydispersity and limited number of proteins that can be crystallized (or aggregated into stable structures) limits the wide applicability of these templates. A recent study demonstrated the use of recrystallized taxol as a template for coating with poly(*N*-isopropylacrylamide) (PNIPAM) and alginate acid [17].

Liposomes, which are spherical vesicles formed from phospholipids, have also been applied as templates for LbL assembly [18]. In a recent study, liposomes were loaded with acetylcholinesterase and its degradation was investigated using proteinase K. It was found that the stability of the protein loaded in the liposomes was significantly increased for the liposomes coated with a cross-linked polyelectrolyte shell [18]. While LbL assembly on liposomes can be used to alter the physicochemical and surface properties of the systems, challenges associated with purification of the coated liposomes can make their isolation difficult.

A range of cells have also been used as templates, which is of interest as the mutilayers can potentially modify cell function and protect them from harmful conditions. However, modification techniques must still allow the diffusion of nutrients to the cells and use non-toxic materials. A number of different studies have been conducted using different cell types [19]. Recently, a study conducted using yeast cells demonstrated long-term cell survivability (over seven days) at 79% using hydrogen-bonded tannic acid (TA)/PVPON multilayers. Cells coated with PSS/PAH had a survivability of 20%.

## 34.3
## Material Assembly

When materials are designed for biomedical applications, consideration needs to be given to both the biocompatibility of the material and its ability to respond to biological stimuli. The flexibility of the LbL process (Figure 34.2) makes it possible to incorporate a wide variety of materials within the layers, as well as to control the thickness, morphology and structure of the films. This is achieved by altering (i) the specific materials being assembled, (ii) the number of layers deposited, and/or (iii) the specific adsorption conditions used [20]. For example, by varying the temperature of adsorption [21], the ionic strength of the adsorption and rinse solutions [22], and the solvent polarity [23], the thickness and morphology of each adsorbed layer can be tuned with nanometer precision. Film assembly can be driven via electrostatic interactions, hydrogen bonding, covalent bonding, or hydrophobic interactions, depending on the desired properties of the films [20].

LbL assembly of various materials on planar surfaces can be followed by using established surface characterization techniques such as quartz crystal microgravimetry (QCM), surface plasmon resonance (SPR), ellipsometry and UV–vis/IR spectroscopy. However, these techniques cannot be directly applied to quantify film assembly on particles [25]. Dynamic light scattering (DLS) cannot readily distinguish the deposition of nanometer-thick layers onto most particle templates. Although microelectrophoresis has been used extensively to qualitatively follow the reversal of surface charge for the assembly of electrostatically associated LbL materials, no such surface charge reversal occurs for many hydrogen-bonded and covalently assembled multilayers. A fluorescence-based method for quantifying the assembly of fluorescently labeled polymers onto particles was recently developed [25]. A flow cytometer can be used to measure the fluorescence intensity of individual particles (as small as 500 nm), and by correlating the fluorescence signal with confocal microscopy and QCM, it was possible to detect the adsorption of <1 fg of material onto a particle [25]. This method allows the analysis of thousands of particles per second with high throughput, and the fluorescence measured is independent of the particle concen-

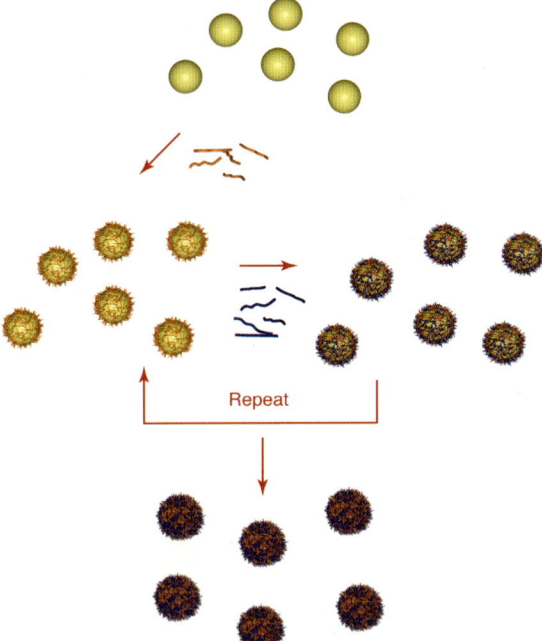

**Figure 34.2** LbL assembly of polymers on spherical particle templates. Particles are incubated in a solution of polymer to deposit the first polymer layer. Following several centrifugation/rinsing steps to remove excess polymer, the coated particles are incubated in another polymer solution containing an interacting polymer. The process is repeated until the desired number of layers (and hence thickness) is achieved. The particles can be subsequently removed to yield polymer capsules [24].

tration. In contrast, if fluorescence spectroscopy is used to determine the amount of fluorescent polymers adsorbed onto particles, it is necessary to determine the concentration of particles accurately. This presents a significant challenge when working with submicron-sized particles and does not readily lend itself to precise, quantitative measurements.

For capsule assembly, oppositely charged strong polyelectrolytes have been extensively employed in the LbL process, as they often lead to robust films due to the strong ionic interactions between the polymers [26]. Electrostatic interactions can also be used to load macromolecular cargo (such as proteins or DNA) by exploiting the inherent charge of the biomolecules (see Section 34.4.2). While these systems are well suited for materials science applications, the highly charged nature of the films can lead to issues with fouling in biological systems and it can be challenging to design materials that are responsive within biologically-relevant conditions (e.g., pH 4–8 and 37 °C). To impart specific responsive characteristics, a number of weak polyelectrolytes, such as poly(acrylic acid) (PAA) [27], poly(L-glutamic acid) (PGA), poly(L-lysine) (PLL) [28], and poly(L-arginine) (pARG) [29], and more advanced polymer structures, such as polyrotaxanes [30], have been incorporated into the films. Films assembled from polypeptides such as PLL, pARG and PGA are susceptible to enzymatic degradation, which provides a useful trigger for drug release [29, 31]. Furthermore, the structure of films containing weak acids, such as PGA and PAA, can be controlled by varying the degree of ionization in a physiologically relevant pH range (pH 4–8) [32].

On planar surfaces, films assembled from weak polyelectrolytes show excellent stability and responsive properties. In contrast, weak polyelectrolyte multilayers deposited on particles can result in decreased colloidal stability of the coated particles, presenting issues with subsequent processing and application of these systems. The decrease in colloidal stability can be, in part, attributed to the migration of the weakly charged polymer species through the multilayer film. Weak polyelectrolytes, such as PAA, have been blended with strong polyelectrolytes (e.g., PSS) to improve the colloidal stability of PAA-coated particles; however, these particles remain less stable than PSS/PAH-coated particles and their colloidal stability decreases as the PAA content in the films is increased [33].

Covalent interactions can also be used to form multilayer assemblies. This was initially demonstrated with the LbL assembly of small diamine and diisocyanate compounds by Kohli and Blanchard [34], and extended to LbL polymer assembly by Akashi and coworkers [35]. There is a potentially limitless array of covalent reactions that can be used to form LbL-assembled materials; however, some of the most versatile and widely applicable are those based on "click chemistry" reactions [36]. The advent of click chemistry is one of the most significant advances in chemical science over the past decade. Click chemistry refers to a class of chemical reactions that rapidly form covalent bonds under mild reaction conditions while having no nonspecific side reactions. The most commonly employed reaction involves the copper-catalyzed cycloaddtion between an alkyne and an azide functionality to form a 1,2,3-triazole linkage (i.e., the 1,3- Huisgen dipolar cycloaddition) [36]. We demonstrated the first example of LbL assembly using click chemistry by forming single-component, pH-responsive PAA films [37] and capsules [38]. PAA was modified with ~10% azide or alkyne functionality (denoted $PAA_{Az}$ and $PAA_{Alk}$, respectively) and sequentially assembled in the presence of a Cu(I) catalyst. LbL assembly has been subsequently demonstrated with a number of different click chemistries/reactions, such as thiol-ene [39] and with a number of different polymer systems [40, 41].

An attractive alternative to electrostatic and covalent assembly is hydrogen-bond facilitated film build-up, which is based on the alternate deposition of polymers containing a hydrogen bond acceptor and a hydrogen bond donor [20]. Hydrogen bonding opens up new possibilities for the incorporation of a variety of uncharged polymers that cannot be used in electrostatic-based assembly; for example, PVPON and poly(ethylene glycol) (PEG) [42, 43]. These polymers are of significant interest for biomedical applications, as they have been demonstrated to be low fouling in biological systems and show good biocompatibility [42, 43]. Hydrogen-bond acceptors, like PVPON and PEG, are typically co-layered with polyacids such as PMA, PAA and tannic acid the protonated carboxylic acid groups act as the hydrogen-bond donors to the lone pair electrons on the oxygen of PVPON and PEG [24].

A limitation with some early hydrogen-bonded systems was the relatively narrow pH range in which the films remained stable. For biological applications where stability in the pH range of 6.5–7.5 is important, this can be overcome by cross-linking the films, either thermally [44] or through the use of a cross-linking agent [45], or by using a polyacid with a p$K_a$ above physiological pH (~7.5) [46]. The addition of a cross-linking agent opens up a wide range of possibilities for stabilizing films with nondegradable or degradable cross-linkers that respond to biological stimuli such as redox potential. Sukishivili and coworkers demonstrated that PMA/PVPON films assembled at pH 4 on sacrificial $CaCO_3$ templates could be used to form stable capsules at physiological conditions by cross-linking the carboxylic acid groups on PMA with ethylenediamine [45]. Similarly, using thiolated PMA ($PMA_{SH}$), we have shown that $PMA_{SH}$ assembled with PVPON can be cross-linked to form LbL capsules that are stable in physiological conditions, but are susceptible to degradation in simulated reducing conditions of the cytoplasm of the cell [47]. At pH 7 the PMA is protonated and the hydrogen-bonding interaction between PMA and PVPON is disrupted, causing the PVPON to be expelled from the film, leaving single-component PMA capsules. Capsules formed from low fouling materials can also be prepared by click chemistry approaches using polymers such as PVPON [45] or PEG [46] that have been co-polymerized with a small percentage of a cross-linkable group (Figure 34.3). If a small percentage (2–10%) of a click reactive group, such as an alkyne, is incorporated into the polymer, the films can be stabilized by a bifunctional azide cross-linker that can either be nondegradable, or may be engineered to contain a degradable group, such as a disulfide. Capsules assembled from $PVPON_{Alk}$ and $PEG_{Alk}$ have been shown to have significantly lower fouling properties than those assembled from charged polymers [48, 49]. A related approach was also used to assemble PGA multilayers by assembly of $PGA_{Alk}$ with PVPON and then cross-linking the films using a biazide linker [50].

More recently, using the same approach of layering a polymers containing a small percentage of a cross-linkable component, we demonstrated the synthesis of capsules assembled from the charge shifting polymer poly(2-(diisopropylamino)ethyl methacrylate) ($PDPA_{Alk}$) [51]. $PDPA_{Alk}$ was cross-linked with a redox-cleavable disulfide-containing linker. At physiological pH (~7), the PDPA is hydrophobic, which causes the capsules to shrink significantly to approximately two thirds of their assembled size. This shrinkage, which is also reversible with pH, is advantageous when trying to assemble small drug carriers that are designed to evade the immune system (see Section 34.5).

A different class of hydrogen-bonded assembly is driven by the specific interaction of DNA base pairs [52]. DNA can be readily incorporated into LbL systems by exploiting the inherent negative charge on the phosphate backbone [53]; however, this does not capitalize on one of the most interesting characteristics of DNA – the programmable nature of DNA interactions. Multilayer films can be assembled from short DNA sequences (oligonucleotides) if the oligonucleotide contains a sequence of bases that recognizes a complementary DNA sequence in the film. In the simplest case, oliognucleotides with two homopolymeric blocks (diblock) of nucleotides can be used, such as the alternating deposition of $polyA_{15}G_{15}$ and $polyT_{15}C_{15}$ [52]. If an oligonucleotide with a single homopolymeric block is used, then only limited assembly occurs as, following hybridization between the first two layers, there is insufficient free single-stranded (ss) DNA to allow hybridization of subsequent

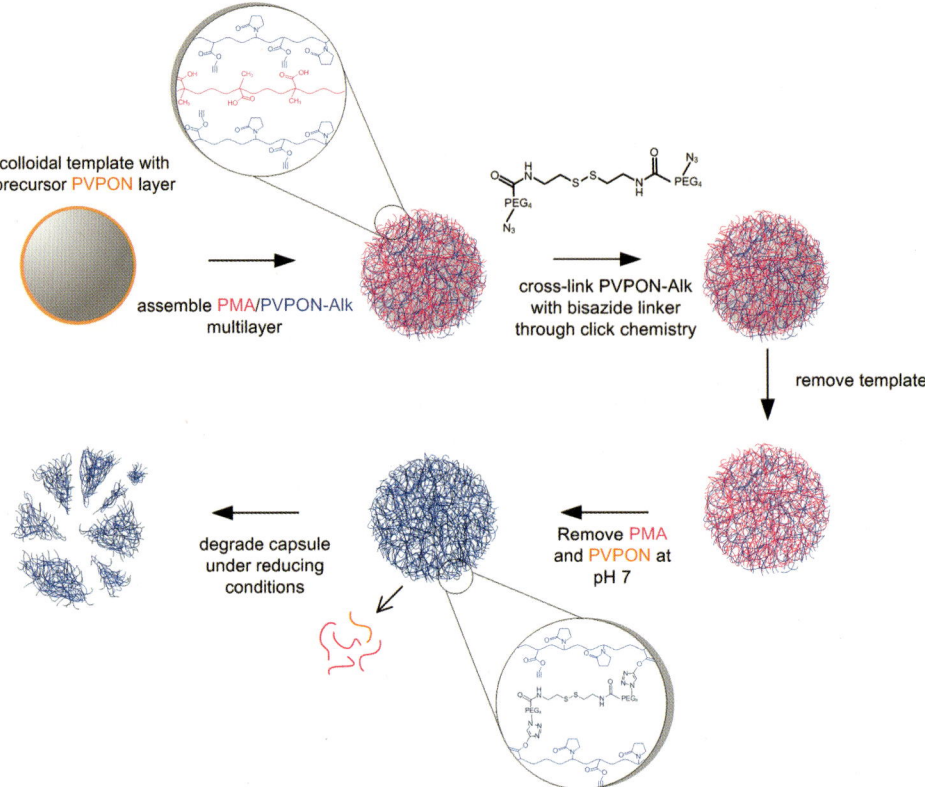

**Figure 34.3** Hydrogen-bonded multilayers of PMA and PVPON$_{Alk}$ are assembled onto silica particle templates coated with a precursor layer of PVPON. Alkyne groups in the PVPON$_{Alk}$ are used to covalently cross-link the multilayers by click chemistry with a biazide linker containing a disulfide bond. After removal of the template particles and subsequent release of PMA at pH 7, low-biofouling, degradable, PVPON capsules are obtained. Taken from [48].

layers. Control over the assembly can be achieved through varying the length of the hybridizing blocks [54], and controlling the repulsion between the phosphate backbones of the oligonucleotides by varying the salt concentration of the hybridizing solution [55]. Both of these variables can be modeled using molecular and phenomenological approaches to predict the structure of the films that are formed [56, 57].

Greater complexity can be engineered into DNA systems by increasing the number of interacting "blocks" within the nucleotide, and by exploiting the specific nature of DNA interactions (Figure 34.4) [58]. The hybridization between two DNA sequences is directional, which means that the two strands of the double helix are oriented in opposite directions [59]. This property can be exploited in the assembly of films to control the morphology of the multilayer structures. We have shown that capsules assembled from sequences that cause the direction of the helix to change after each layer shrink significantly more than capsules assembled from sequences that promote a continually growing helix. The degree of capsule shrinkage can also be

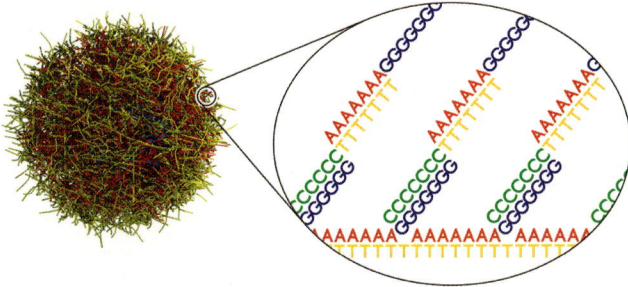

**Figure 34.4** Schematic representation of an idealized orientation of oligonucleotides assembled from homopolymeric blocks of repeating nucleotides (polyTC and polyAG) in DNA capsules. A primer layer of polyT is also shown [60].

controlled by the sequence used, with capsules assembled from "random" DNA blocks shrinking more than those assembled from homopolymeric blocks [59].

In addition to conventional LbL assembly, there has recently been interest in spontaneous film formation, including the use of dopamine to facilitate layer growth [61]. Dopamine is a small molecule which undergoes oxidative self-polymerization and has been shown to form thin films on planar substrates [62]. Postma *et al.* reported that the dopamine polymerization approach could be applied to coat particles [61]. It was shown that the thickness of the dopamine layer could be controlled by the dopamine polymerization time and the number of dopamine exposure steps. Further, subsequent removal of the sacrificial particle templates resulted in polydopamine capsules. In related work, monodisperse dimethyldiethoxysilane emulsion droplets with a range of sizes (400 nm to 2.4 μm) were coated with polydopamine [63]. These droplets were loaded with a range of cargo, including magnetic $Fe_3O_4$ nanoparticles, fluorescent QDs and the anticancer drug thiocoroline. By removing the emulsion cores using ethanol, polydopamine capsules with aqueous centers could be achieved.

## 34.4
## Loading

Loading therapeutic cargos within LbL capsules is fundamental for developing these systems for biomedical applications. A wide range of molecules, including small anticancer drugs, oligonucleotides, plasmids, peptides and whole proteins, have been encapsulated within LbL capsules [7]. The different chemical and physical properties of such therapeutics suggest that a number of different approaches are required for the successful encapsulation of each therapeutic. In particular, low molecular weight species (such as anticancer drugs) can freely diffuse through LbL multilayer films; hence careful design of the encapsulation strategy is required. Encapsulation techniques that have been most commonly used fall broadly into three categories: preloading the cargo either in or onto the template (Section 34.4.1),

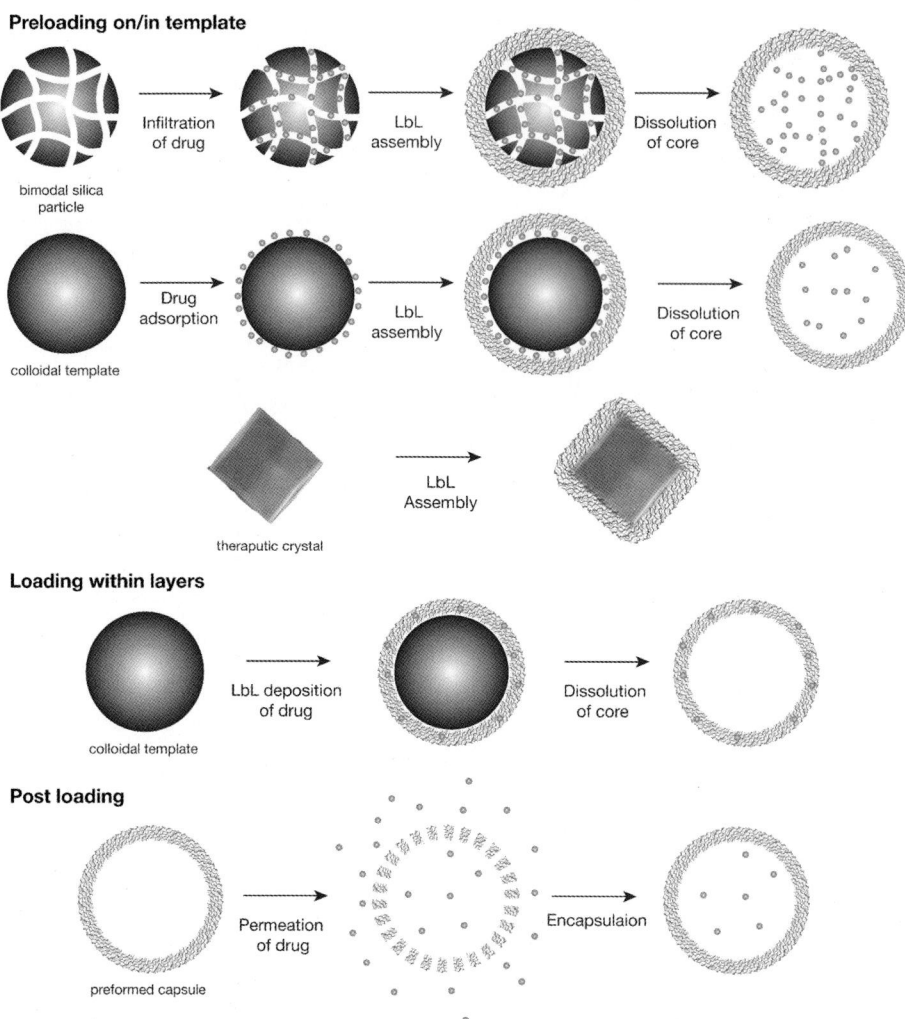

**Figure 34.5** Therapeutic loading strategies: Preloading the cargo by either adsorption onto the template surface, infiltration into a porous template, or assembling layers directly onto a therapeutic crystal; Loading within layers; Post-loading a preformed capsule. Adapted from [9].

incorporating the cargo with the film during multilayer assembly (Section 34.4.2), or post-loading the cargo into preformed capsules (Section 34.4.3) (Figure 34.5).

### 34.4.1
### Preloading on/in Template

Mesoporous templates have been commonly used as a means of infiltrating various cargos. In 2004, Wang *et al.* demonstrated the ability of multilayered mesoporous silica particles to load biomolecules. A model enzyme, catalase, was used to demonstrate that cargo retention was increased by approximately 25% when the protein-loaded silica particles were coated with a LbL film [64]. This encapsulation strategy also enhanced the stability of the catalase to both pH and protease. Later studies demonstrated this approach could be used to load DNA molecular beacons (MB) [65], and that these beacons could be used as sensors to probe the permeability of LbL capsules using single-stranded DNA. It is typically difficult to encapsulate low molecular weight cargo within mesoporous materials without significant leakage. However, when this small cargo is hydrophobic it can be retained in the carrier by interaction with a hydrophobic host material. A recent example involved the use of mesoporous silica particles to load a range of hydrophobic small cargo, including thiocoraline and paclitaxel, which were then encapsulated in $PMA_{SH}$ multilayers. It was demonstrated that the cargo was retained within the capsules and retained activity to kill model cancer cells [66]. In an analogous approach, biomolecules can be loaded into porous $CaCO_3$ particles through a similar infiltration process, or by coprecipitation during the template synthesis [67].

Mesoporous templates can also be used to synthesize polymeric replicas by infiltration of a polymer and subsequent cross-linking or stabilization [68]. These replicas can then be used to load materials based on interactions with the polymeric component. One such study infiltrated a hydrophobic poly-3-hydroxybutyrate polymer into mesoporous silica particles [69]. The polymer formed replicas on template removal due to the strong associations between the polymer chains. These replicas were then used to load a rare-earth complex with potential application in diagnostics or fluorescence immunoassays. The materials exhibited luminescence over a broad pH range and were stable over several months.

A related technique to preload cargo is by deposition on a solid silica template before multilayer assembly. This approach, which uses a positively charged silica template to efficiently adsorb the negatively charged cargo, has been successfully used to load single-stranded oligonucleotides, double-stranded DNA and plasmids [11, 70]. One advantage of this technique is the high loading efficiency that can be achieved: the high surface area-to-volume ratio of particles (e.g., below 1 μm in diameter) allowed a significant amount of DNA (~10 000 30b single-stranded DNA, ~1000 800 bp double-stranded DNA and ~300 3 kb plasmid) to be encapsulated by a single DNA adsorption step. After DNA loading, PMA/PVPON multilayers can be assembled on the DNA-coated particles and then the core removed, yielding DNA-loaded capsules [11, 70, 71]. A similar technique can be used to load proteins (ovalbumin (OVA) [72]) and siRNA [73].

### 34.4.2
**Loading Within Layers**

Loading within the layers of LbL capsules can be achieved by covalently conjugating the intended cargo to the assembly material before film build-up. This approach is particularly useful in the case of low molecular weight cargo that would otherwise pass through the polymer shell. Ochs *et al.* reported the conjugation of doxorubicin

(DOX) to alkyne-modified PGA (PGA$_{Alk+DOX}$) and subsequent LbL assembly based on the hydrogen-bonding interaction of PGA$_{Alk+DOX}$ with PVPON [50]. The use of a DOX-conjugated polymer has also been reported by Schneider et al. for N-(2-hydroxypropyl)methacrylamide (HPMA) copolymers with oligopeptide-spaced DOX groups [8].

Sub-compartments can also be loaded into LbL capsules during film assembly, which leads to hybrid materials that offer the properties of both the parent carrier and the sub-units. Capsosomes, for example, are LbL capsules that contain liposomes, and combine the loading capacity of the liposomes (which are able to host both hydrophilic and hydrophobic cargo) with the stability and adaptability of polymer capsules [74]. These systems have been used to load both enzymes and hydrophobic drug cargo [75]. Cholesterol-modified polymers such as poly(L-lysine)-cholesterol (PLL$_C$) and poly(methacrylic) acid-cholesterol (PMA$_C$) have been used as primer and capping layers, respectively, for depositing liposomes (Figure 34.6) [76]. The pendant cholesterol moieties provide stability by being incorporated into the lipid bilayer of the liposomes. Model enzymatic cargo, β-lactamase and luciferase, were used to demonstrate cargo retention throughout the assembly and subsequent core removal steps. Liposome retention was shown to be independent of the number of PVPON/PMA$_{SH}$ bilayers used to encapsulate the liposomal cargo. Subsequent studies have demonstrated that multiple liposome layers can be incorporated into capsosomes, allowing up to 160 000 sub-compartments to be layered alternately with polymer separation layers on 3 μm-diameter silica particle templates [77]. Loading of hydrophobic cargo into capsosomes has also been demonstrated by incorporation of the anticancer drug thiocoraline into the liposomal sub-compartments [75].

Polymersomes, which are vesicles assembled from block copolymers, have the advantage of having thicker, more robust, less permeable membranes than

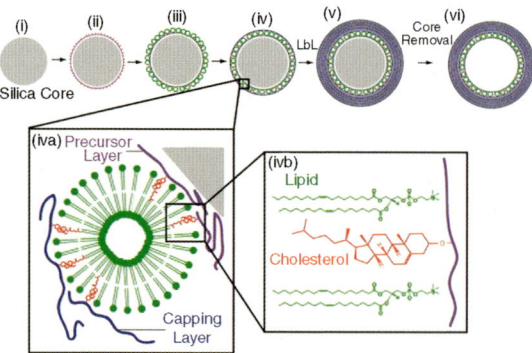

Figure 34.6 Schematic illustration of capsosome assembly. A silica template particle (i) is coated with a cholesterol-functionalized polymer precursor layer (ii), liposomes (iii), and a cholesterol-functionalized polymer capping layer (iv), followed by subsequent LbL assembly (v), and removal of the silica core (vi). Cholesterol-modified polymers are used to noncovalently adsorb liposomes onto the polymer surface (iva). Cholesterol is spontaneously incorporated into the lipid bilayer (ivb). Taken from [76].

liposomes. Due to these properties the incorporation of polymersomes into hydrogen-bonded LbL films for the loading of plasmid DNA was investigated [78]. A pH-tunable, amphiphilic diblock copolymer, poly(oligoethylene glycol methacrylate)-*block*-poly(2-(diisopropylamino)ethyl methacrylate), was used to form polymersomes that were stable at physiological pH but disassemble to release cargo when acidified within cellular compartments.

LbL capsules have also been used to load multiple polymer capsules within one larger capsule [79]. In this case, 300 nm-diameter thiol-modified PMA capsules were loaded into 3 μm-diameter thiol-modified PMA capsules. Three different cross-linking strategies were used to demonstrate control over the release mechanisms of a model peptide cargo from sub-compartments, as well as the release of the sub-compartments from the parent carriers. Loading of the smaller capsules with a peptide vaccine demonstrated a potential application of this hybrid capsule system in therapeutic delivery.

### 34.4.3
### Post-Loading

Post-loading of LbL capsules exploits the permeability of the walls of preformed capsules and how this can be varied by changing external stimuli. By altering pH [80] or ionic strength [81], capsules assembled from suitably chosen materials can be made to swell, resulting in an increase in their membrane porosity. If the swollen capsules are exposed to a solution containing a molecule previously unable to permeate through the membrane, the increased porosity can allow it to pass through into the capsule void. Reversal of the conditions back to the original pH or ionic strength causes the capsules to shrink, restoring the reduced permeability and entrapping the molecules within the capsules. Temperature-induced shrinkage has also been used to trap molecules within LbL capsules. Köhler *et al.* demonstrated the encapsulation of 10 and 70 kDa fluorescently labeled dextran molecules within poly(diallyldimethylammonium chloride) (PDADMAC)/PSS capsules by subjecting them to temperatures above the glass transition temperature ($T_g$) in the presence of the dextran polymers [82]. However, using high temperatures to achieve capsule shrinkage is not suitable for the encapsulation of many biomolecules.

### 34.5
### Degradation and Release

Effective drug carriers typically include a mechanism to release the encapsulated drug. This can be achieved by using an external stimulus, such as temperature, light or magnetic field, or by using an inherent biological stimulus such as a change in pH, redox conditions or enzymatic environment [83, 84].

Release induced by an external stimulus, such as light or magnetic field, is an attractive mechanism, as there is minimal chance of premature release of the cargo due to nonspecific interactions within biological systems. One unique method for inducing release is to incorporate gold nanoparticles into the capsule walls. These nanoparticles can adsorb light in the near-infrared spectrum and induce heating of the capsule wall [85, 86]. The high localized temperature in the capsule wall causes the capsules to rupture, but causes minimal damage to the encapsulated cargo, such as lysozyme. Similar heat-induced release can be triggered by incorporating superparamagnetic iron oxide nanoparticles into the films and exposing them to an oscillating magnetic field [87].

Release from capsosomes can also be induced by using heat or by exploiting the phospholipid membrane phase transition temperature ($T_m$). Chandrawati et al. reported the enzymatic hydrolysis of nitrocefin by dipalmitoylphosphatidylcholine (DPPC) liposome-encapsulated β-lactamase as a colorimetric assay to demonstrate release from the liposomal compartments [77]. Below $T_m$, in this case 41 °C, nitrocephin in solution was unable to cross the liposome membrane. However, when the temperature was raised above the $T_m$, the membrane permeability increased due to a phase transition, and a color change was observed due to nitrocefin hydrolysis. The β-lactamase remained encapsulated and could be used in three further hydrolysis cycles without a change in reaction rate (Figure 34.7).

An alternative approach is to exploit cellular processes to promote cargo release. One phenomenon that is widely used in drug delivery is the difference in the redox potential between the cytoplasm of the cell and the extra-cellular environment. The environment outside of the cell is oxidizing and as such stabilizes disulfide bonds. In contrast, the cellular cytoplasm is a reducing environment and thus promotes the degradation of disulfide bonds. With the aim to exploit such differences in redox potential, a range of disulfide-stabilized polymer capsules have been prepared (e.g.,

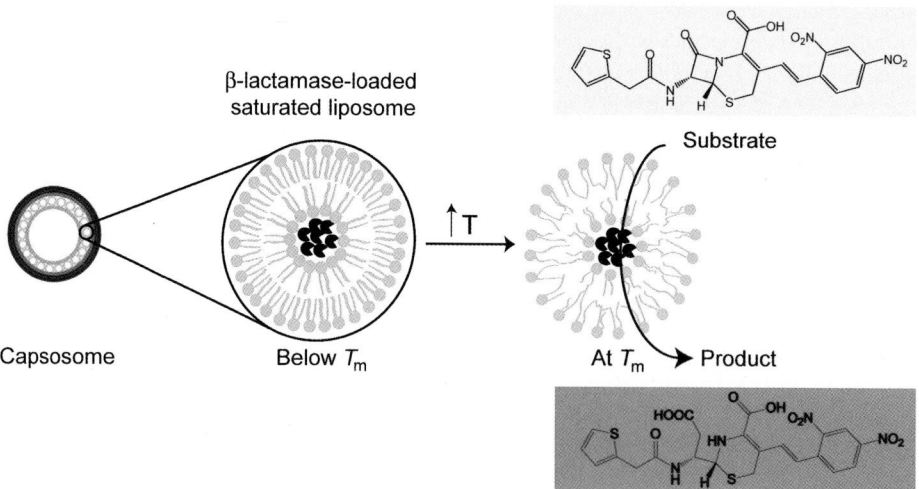

**Figure 34.7** Temperature-triggered enzymatic conversion within a capsosome. An increase in temperature to the phase transition temperature ($T_m$) of the liposomes results in the lipid membrane becoming disordered, allowing nitrocefin to cross the membrane and be hydrolyzed while the β-lactamase remains inside the compartments. Taken from [77].

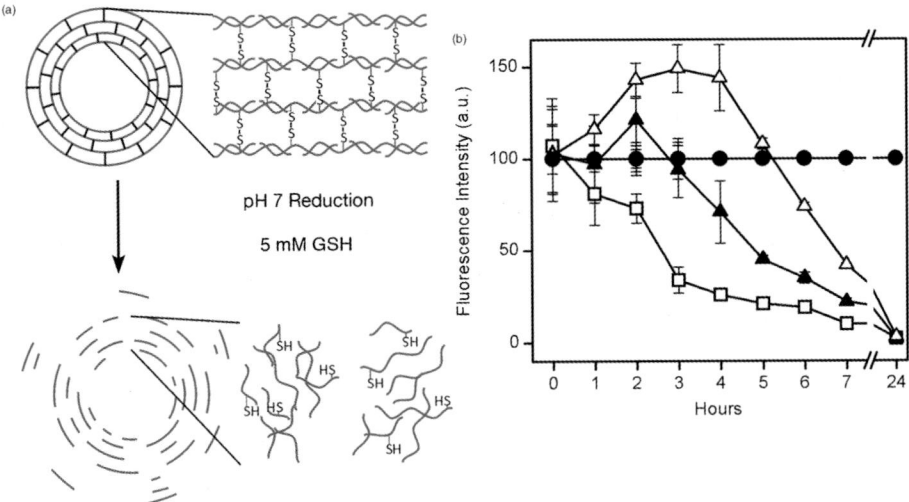

**Figure 34.8** (a) Schematic illustration of the degradation of PMA$_{SH}$ capsules by GSH. (b) Degradation of PMA$_{SH}$ capsules with varying degrees of thiolation upon exposure to 5 mM GSH. Open squares correspond to 10% thiolation, closed triangles to 15% thiolation, and open triangles to 20% thiolation. Closed circles correspond to capsules in PBS buffer. Taken from [88].

PMA$_{SH}$ [47], PVPON [48], PEG [49], and PDPA [51]). These capsules are resistant to degradation in conditions simulating the oxidizing environment of the bloodstream but readily degrade in simulated cytoplasmic conditions (5 mM glutathione (GSH)) (Figure 34.8). The rate of degradation can be controlled by the degree of cross-linking [88], the length of the cross-linker [71], and the polymers used in the assembly.

Another stimulus that has been extensively exploited for cargo release is pH. When, for example, particles are internalized into a cell they typically enter via an endo/lysosomal pathway [3]. In this process the foreign material is compartmentalized into an endocytic compartment that is gradually acidified from physiological pH (~7.4) to between pH 4.5 and 5. This large change in pH has been exploited for drug release. The charge shifting polymer PDPA has a p$K_a$ of ~6.5, which makes it ideal for responding to endocytic pH changes [51]. At pH 7.4 PDPA is uncharged and hydrophobic. However, below pH 6, the PDPA is protonated and hydrophilic, causing the PDPA capsules to swell to more than double their original size. Furthermore, at physiological pH PDPA capsules were found to be resistant to degradation in reducing environments, despite the presence of a disulfide linker. This is likely to be due to the hydrophobic nature of the PDPA limiting the accessibility of the reducing species such as GSH. However, at pH 6 in the presence of 0.5 mM GSH (one tenth the simulated physiological concentration) the capsules are rapidly degraded. This phenomenon is likely to be highly beneficial for therapeutic delivery with these systems, as both a decrease in pH and a reducing environment are required for the drug to be released.

Perhaps the most widely exploited cellular release mechanism employed is enzymatic degradation. Materials assembled from polypeptides, DNA and carbohydrates are susceptible to nonspecific degradation by proteases, nucleases and carbohydrases, respectively. De Geest and coworkers demonstrated that pARG/dextran sulfate (DEXS) capsules and (PSS)/poly(hydroxypropylmethacrylamide dimethylaminoethyl ester) (pHPMA-DMAE) capsules are degraded intracellularly, whereas PSS/PAH capsules remain intact [29]. Similarly, we have shown that PGA capsules assembled using click chemistry are degraded intracellularly and can release a drug cargo (see Section 34.6.3.1) [50]. Degradation induced by specific enzymes can also be engineered into LbL capsules. This was demonstrated using LbL DNA capsules incorporating the DNA sequence GAATTC. The films rapidly degraded and released a model cargo (ovalbumin) in the presence of the restriction enzyme EcoRI, which recognizes the GAATTC sequence [89]. If the GAATTC sequence was not present in the film, the assembly remained stable in the presence of EcoRI, but was still susceptible to degradation by the nonspecific nuclease DNAse. Decher and coworkers demonstrated the enzymatic release of DOX from LbL gold nanoparticles using a cathepsin degradable linker [8]. The cathepsin enzyme is present in lysosomal compartments and provides a specific trigger for the release of internalized cargo.

## 34.6
## Applications

By combining the aforementioned four aspects of material design it has been demonstrated that LbL assembly can be used for engineering microreactors and carriers that can deliver therapeutics, such as anticancer drugs, peptide-based vaccines, and nucleic acids. In the following we discuss recent progress in the application of the various LbL capsule systems in these areas.

### 34.6.1
### Microreactors

The application of polymeric carriers as microreactors is a growing area of interest within materials science, as they can effectively combine multiple simultaneous reactions within one multicompartment reaction vessel [14]. LbL microreactors based on emulsion templates have been used as PCR reaction containers [90] and the inherent permeability of LbL capsules has been used to facilitate triggered enzymatic degradation [91]. The latter study [91] used porous silica particles to co-encapsulate double-stranded DNA (dsDNA) and the enzyme DNase I. PVPON/PMA$_{SH}$ multilayers were then assembled onto the template, cross-linked and then the sacrificial templates were removed. DNase I is an endonuclease with a strong affinity for dsDNA and requires divalent cations to be active. It was demonstrated that the dsDNA is retained within the multilayers; however, when the capsules were incubated with divalent ions at 35 °C, degradation of the dsDNA was facilitated. Due to their small size the degradation products then diffused out of the capsules.

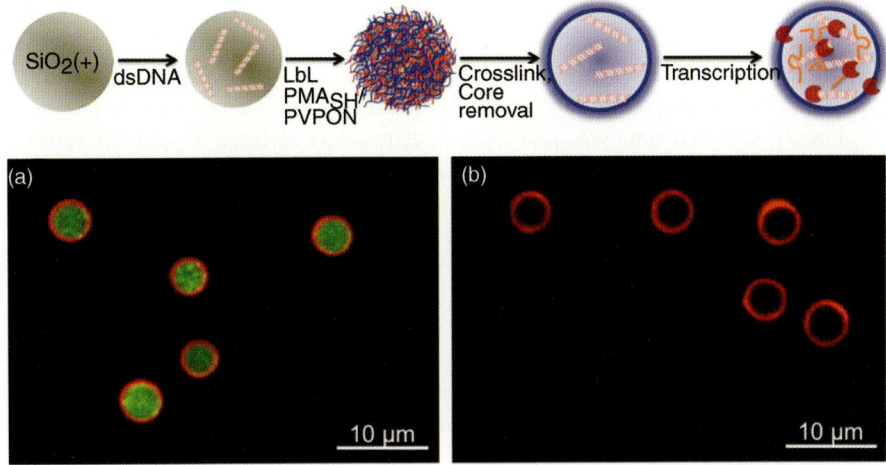

**Figure 34.9** Schematic illustration of an RNA microreactor. dsDNA is encapsulated within a PMA$_{SH}$ polymer capsule by preloading onto a silica template. (a) In the presence of the enzyme (T7 RNA polymerase) and nucleotides, RNA is synthesized inside the capsules (green). (b) Without the dsDNA template, no RNA is synthesized. Taken from [92].

A related approach was used to synthesize RNA oligomers; RNA is known to be highly susceptible to degradation and thus strategies for *in situ* synthesis are desirable (Figure 34.9) [92]. A dsDNA sequence was loaded onto a silica particle template using the loading technique described in Section 34.4.1. The DNA contained a specific promoter sequence, which is required for enzyme binding and initiation of RNA transcription. After PMA$_{SH}$ capsule formation, the loaded capsules were incubated in a solution containing T7 RNA polymerase. Due to the selective permeability of LbL capsules, this enzyme can diffuse into the capsules and can bind to the dsDNA. When the dsDNA loaded capsules were incubated with individual ribonucleotide triphosphates, the T7 RNA polymerized the nucleotides into single-stranded RNA polymers. After synthesis, the RNA was retained in the capsules due to its size and charge. The RNA synthesis was monitored using flow cytometry and it was shown that the synthesized RNA was active. The potential to deliver such RNA-loaded capsules could have interesting applications for the modification of cellular function.

The coupled reaction of horseradish peroxidase (HRP) and glucose oxidase (GOD) is one of the classical bienzymatic systems used for the detection of glucose [93]. GOD catalyses the oxidation of glucose producing hydrogen peroxide ($H_2O_2$) which, in the presence of an electron donor, acts as a substrate for HRP. The reaction by HRP is ideally proportional to the amount of $H_2O_2$ produced in the first reaction. When encapsulated within a single capsule, a coupled catalytic reaction takes place within a limited space. This potentially increases the sensitivity of the reaction to glucose detection while creating a diffusion barrier for the glucose substrate. In one example, HRP and GOD were co-encapsulated in PSS/PAH capsules using calcium carbonate particles. The enzymes successfully performed the coupling reaction with glucose, however, at a significantly reduced level (85% reduction) [93]. A related system using shell-in-shell capsules based on GOD and peroxidase (POD) was reported [94]. In this system, POD was encapsulated in $CaCO_3$ particle templates and then PSS/PAH multilayers were assembled on the surface. Then a $CaCO_3$ layer was deposited containing GOD and additional PSS/PAH multilayers were deposited. The sacrificial $CaCO_3$ was then removed. Due to the permeability of the LbL wall, the glucose diffused into the LbL capsule allowing glucose to be converted to hydrogen peroxide. Peroxide then diffused into the inner capsule, which reacted with POD in the presence of the dye Amplex Red to generate a fluorescent signal.

### 34.6.2
### Targeting

The ability to functionalize particle-based drug delivery vehicles with targeting ligands is one of the significant advantages of engineered delivery systems. Targeting high doses of therapeutics to a particular site would have the combined advantage of reducing side effects to healthy cells and increasing therapeutic efficiency [3]. LbL capsules are highly amenable to surface modification with targeting ligands due to the high degree of control achievable over surface chemistry.

In 2006 Cortez et al. demonstrated successful targeting of A33 antigen expressing LIM1215 human colorectal cells with humanized A33 monoclonal antibody (huA33 mAb)-functionalised PSS/PAH core–shell particles [95]. The layers were assembled based on electrostatic interactions and then post-functionalized with antibody through electrostatic adsorption [95, 96]. Flow cytometry analysis showed that the number of cells with bound particles was approximately four times greater when the particles were functionalized with the huA33 mAb.

Successful targeting of cells relies on the carrier exhibiting low non-specific binding to cells in order that the carriers have the opportunity to reach the antigen expressing cells. As charged materials are more prone to biofouling than certain uncharged polymers, Kamphuis et al. extended the huA33 mAb approach to targeting human colorectal cancer cells with PVPON capsules [97]. The uncharged PVPON core–shell particles were shown to exhibit low protein fouling behavior compared to core–shell particles with charged terminating layers [48]. However, due to the low fouling behavior of the PVPON, covalent functionalization was required to couple the Ab to the capsules. Click chemistry-stabilized PVPON capsules were post-functionalized with a huA33 mAb modified with a linear PEG chain terminating in an azide group, in the presence of a $Cu^I$ catalyst. A chelating ligand was used to prevent aggregation of the $Cu^I$ with the antibody. It was shown that huA33 mAb-functionalized capsules were associated with 90% of the total human A33 antigen expressing LIM2405 + cell population compared with 5% association with LIM2405- cells which do not express the antigen (Figure 34.10). Low nonspecific binding of IgG-functionalized capsules was observed with less than 15% association with both LIM2405 + and LIM2405- cells. The specificity of the targeting was demonstrated by incubating capsules with LIM2405 + and LIM 2405- cells at a ratio of 0.1 : 99.9 respectively. Even at this very low concentration, 50% of LIM2405 + cells were associated with huA33 mAb-functionalized capsules.

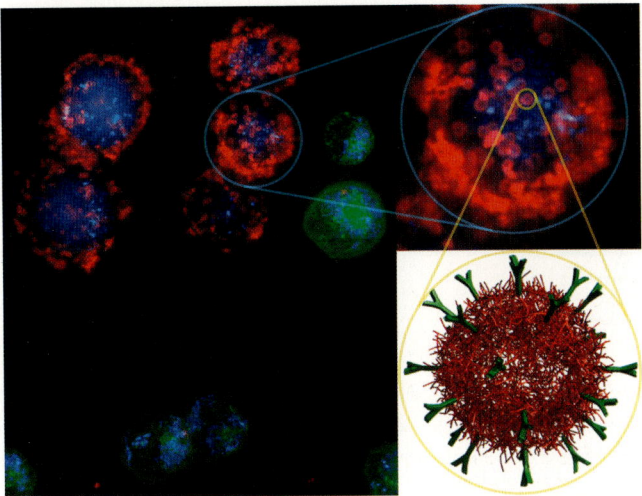

**Figure 34.10** Targeting antibody-functionalized LbL-assembled capsules to cancer cells. Fluorescence microscopy image of cells expressing the huA33 antigen (blue) and cells that do not express the huA33 antigen (green) were incubated with capsules functionalized with the huA33 antibody (red). Taken from [97].

### 34.6.3
### Therapeutic Delivery

The encapsulation and delivery of therapeutic cargo requires careful engineering of the LbL carrier based on the characteristics of the cargo. Recently, a number of intelligent strategies to achieve encapsulation and release have been demonstrated.

#### 34.6.3.1 Small Molecules

The encapsulation of small molecules within LbL capsules presents a significant challenge due to the permeability of the capsule wall. Efforts to circumvent this limitation have been driven by the low molecular weight of most chemotherapeutic drugs. One approach to bypassing diffusion through the carrier film is to conjugate the drug candidate to the assembly material. Ochs *et al.* reported the LbL assembly and click stabilization of $PGA_{Alk}$ capsules to which the anticancer drug DOX was covalently conjugated through an amide bond ($PGA_{Alk+DOX}$) [50]. The release of DOX occurred through enzymatic degradation of the polypeptide backbone. MTT assays showed that, when delivered in capsule form, $PGA_{Alk+DOX}$ reduced the cell viability of LIM1899 cells to 32% (compared with 100% for untreated cells). This decrease was significantly greater than that observed for the free polymer–drug conjugate in the presence of protease, which decreased cell viability and proliferation to 78%. Confocal laser scanning microscopy showed the presence of DOX associated fluorescence in the cytoplasm, vesicles, and, significantly, the nucleus of LIM1899 cells when DOX was delivered via $PGA_{Alk+DOX}$ capsules over 24 h. In contrast, LIM1899 cells incubated with the equivalent amount of DOX as the free

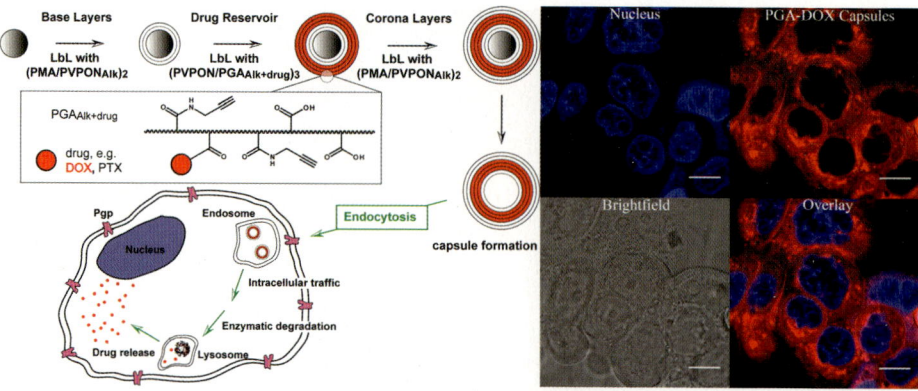

**Figure 34.11** (a) Schematic diagram showing the assembly of drug-loaded capsules and the mechanism by which MDR is bypassed. The structure of the polymer–drug conjugate is shown (expanded). The drug-loaded capsules are internalized by MDR cancer cells via endocytosis. Subsequently incorporated drugs are released by enzymatic degradation in lysosomes. (b) CLSM images showing the intracellular distribution of PGA-DOX capsules in LIM1899MDR cells. Nuclei were counterstained with Hoechst 33342 (blue). Scale bars, 10 μm. Taken from [98].

polymer–drug conjugate showed much lower DOX-associated fluorescence after 24 h, and no DOX was found to be localized with the cell nuclei.

$PGA_{Alk + DOX}$ and $PGA_{Alk + Paclitaxel}$ capsules have also been used to overcome multidrug resistance (MDR) in cancer cells. MDR contributes to the failure of metastatic cancer treatment in more than 90% of patients. LIM1899 cancer cells were specifically generated to be resistant to the anticancer drugs DOX (LIM1899MDR) (Figure 34.11) [98]. Using $PGA_{Alk + DOX}$ capsules led to internalization via endocytosis and thus the efflux mechanism, which is stimulated in MDR cells, was bypassed. Drug resistance in cells is measured by the resistance index. LIM1899MDR cells have a resistance index of 14.1 towards free DOX (i.e., 14.1 times more free DOX is required to kill LIM1899MDR cells compared to LIM1899 cells). In comparison to free DOX, when $PGA_{Alk + DOX}$ capsules were used, the resistance index dropped to 2.7.

Encapsulation of low molecular weight drugs within a stabilized emulsion has also been demonstrated [99]. Sivakumar et al. reported the infiltration of DOX-loaded oleic acid into the cavity of LbL assembled, disulfide-stabilized PMA capsules. Encapsulating the emulsion within the disulfide-stabilized carrier provides a controlled release mechanism that can be exploited in the reducing environment of the cytosol. The templated LbL approach to assembling the capsules gives rise to monodispersity, and hence a high degree of control over dose. The DOX-loaded PMA capsules were shown to be efficiently internalized by LIM1215 cancer cells, likely via a macropinocytotic process [100]. While the $PMA_{SH}$ carriers were found to localize in the acidic lysosomal compartments of the cell, the DOX was able to escape from these compartments into the cytosol and nucleus. This resulted in enhanced killing of LIM1215 cells compared with free DOX and DOX-loaded carriers with no engineered

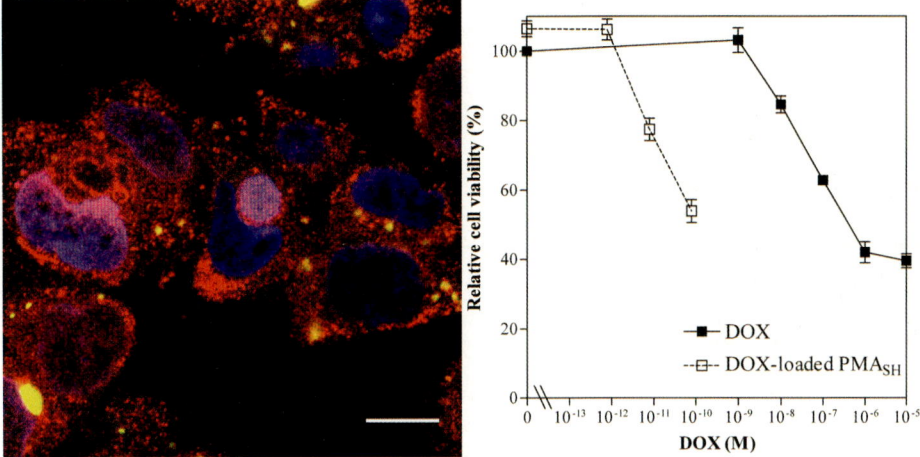

**Figure 34.12** Anticancer drug delivery. (a) DOX (red) release from PMA$_{SH}$ capsules (green) in LIM1215 capsules. Nucleus stained blue with DAPI. (b) Dose response curve of free DOX (solid squares) and DOX-loaded capsules (open squares). Taken from [100].

release mechanism, namely nondegradable PSS/PAH capsules (Figure 34.12). Given that the amount of DOX delivered via the loaded PMA carriers was more than three orders of magnitude less than the concentration of free DOX used, these nanoengineered carriers have potential as efficient drug delivery systems [100].

Another post-loading approach to the encapsulation of DOX and daunorubicin (DNR) was reported by Zhao *et al.* [101]. The electrostatic interaction between the positively charged drug molecules and negatively charged PSS, encapsulated within a PAH/PSS shell, was used to sequester drug from solution. Controlled release by diffusion was demonstrated. This principle was subsequently extended to a more biologically relevant system in which layers were used to encapsulate the polyanion carboxylmethyl cellulose [102]. Compared to free DOX, DOX-loaded chitosan/alginate capsules showed increased cell apoptosis *in vitro* and decreased tumor growth *in vivo*.

### 34.6.3.2 Vaccines

Peptides have been shown to be successful candidates for inducing T cell immunity in the fight against infectious diseases such as HIV/AIDS, which cannot be treated using classical vaccine technologies. However, the susceptibility of peptides to enzymatic degradation in the body, and the volume of peptide material needed to administer peptide treatments *ex vivo* currently make using peptides in the clinic unfeasible. In contrast, the protection afforded to cargo encapsulated within a carrier system makes this method of delivery an attractive prospect. In particular, the multiple strategies for controlled release offered by polymer capsule systems make them attractive candidates for the delivery of peptides to antigen presenting cells. De Rose *et al.* reported the successful loading and delivery of an oligopeptide vaccine and demonstrated that this led to an immune response [103]. Loading of the KP9 9-amino

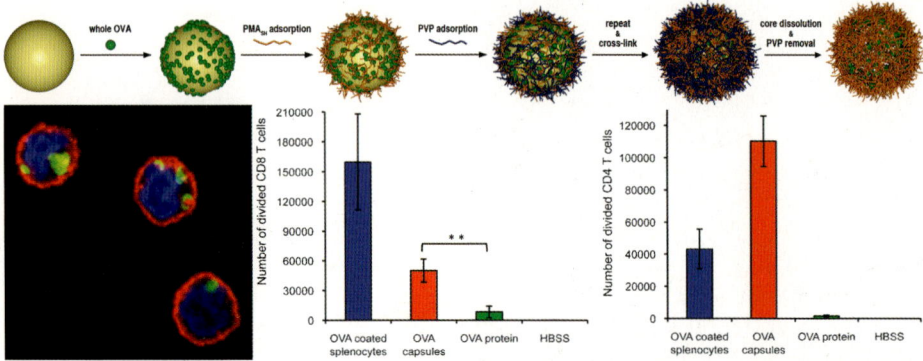

**Figure 34.13** (a) Schematic illustration of the loading and assembly of proteins into PMA$_{SH}$ capsules. (b) CLSM image of dendritic cells with internalized capsules (nucleus – blue, cell membrane – red, capsules – green). Number of divided (c) CD8 T-cells (MHC class I response) and (d) CD4 T-cells (MHC class II response) treated with OVA-coated splenocytes (positive control), HBSS (saline – negative control), OVA-loaded capsules and free OVA protein. Taken from [72, 103].

acid peptide into disulfide-stabilized PMA$_{SH}$ capsules was achieved by conjugation of KP9, modified at the N-terminus with a cysteine residue to the PMA$_{SH}$ polymer prior to assembly. In this approach, intracellular reduction of disulfide bonds provides the mechanism for both capsule disassembly and KP9 release inside the target antigen presenting cells. The KP9 loaded capsules were found to stimulate KP9-specific T cells to produce cytokines, which supports replication of the *ex vivo* behavior. The capsule supernatant and non-loaded capsules showed no stimulation.

Subsequent work by Sexton *et al.* extended this system to an *in vivo* study in mice [72]. Encapsulated ovalbumin (OVA) was found to be more effectively delivered to mouse antigen presenting cells than OVA delivered alone (Figure 34.13). It was also found that substantially less protein or peptide antigen was needed to stimulate an equivalent immune response when the cargo was encapsulated. While these initial results are promising, the MHC class I response (essential for effective vaccination) was significantly lower than the MHC class II response. This is most likely due to the intracellular processing of the capsule and the cargo through a lysosomal pathway [3]. Model antigen delivery to dendritic cells using LbL capsules was also reported by De Koker *et al.*, who successfully delivered OVA encapsulated in DEXS/pARG capsules to dendritic cells from mouse bone marrow [104]. Uptake by macropinocytosis into lysomomal compartments was suggested. The enzymatic degradation of encapsulated OVA was demonstrated through loading of OVA-DQ, which is a boron-dipyrromethene (BODIPY)-labeled protein that is self-quenched as an intact protein but exhibits fluorescence when it is degraded into peptides.

### 34.6.3.3 DNA

Gene therapies for the treatment of a range of serious diseases are currently undergoing clinical trials [105]. However, there are still a number of barriers that stand in the way of the widespread use of this technology across the range of applicable diseases. One of the issues facing the field is vector transport and uptake. The instability of nucleic acids in the presence of enzymes within the body makes them obvious candidates for protection within carriers. DNA has been used as a building block in electrostatic LbL assembly since the early days of the technique [53]. Numerous examples of loading DNA into the walls and cavities of capsules have been reported, with release mechanisms based on changing ionic strength and co-encapsulation of capsule-degrading enzymes reported. Recently, disulfide-stabilized PMA capsules have been used to encapsulate single-stranded, linear double-stranded and plasmid DNA, which were preloaded onto positively charged amine-modified silica templates [11, 70]. The PMA$_{SH}$ layers that are stable in simulated oxidizing conditions of the bloodstream but degrade under simulated reducing conditions such as those in the intracellular environment, offer both cargo protection and stimuli-induced release.

A small interfering RNA (siRNA), siSurv, which inhibits the expression of survivin, a protein key to the proliferation and viability of almost all human cancer cells, has also been loaded using this approach [73]. The ability of the PMA$_{SH}$ capsules to inhibit the expression of survivin was compared to that of Lipofectamine 2000, a standard lipid-mediated transfer system. It was shown that siSurv delivered in LbL capsules resulted in twice the inhibition as the same amount delivered using Lipofectamine 2000. Interestingly, the capsules without siRNA also showed inhibition of survivin and other autophagy markers, highlighting the need for further studies into LbL capsule–cell interactions. In another example, the electrostatic interaction between siRNA and PEI was used to assemble PEI/siRNA/PEI onto gold nanoparticle templates [106]. Delivery of the core–shell particles to CHO-K1 cells led to a dose-dependent silencing of EGFP (enhanced green fluorescent protein). However, the number of particles needed to achieve the effect was regarded as high and was attributed to the nondegradable nature of the electrostatic layers leading to insufficient siRNA release.

## 34.7
## Conclusions

LbL technology has significant potential for application in therapeutic delivery due to the ability to finely control the size, composition, degradation and surface functionality of therapeutic-loaded capsules. In the last five years, the loading of a range of therapeutic compounds, such as anticancer drugs, peptides, proteins, DNA and siRNA, has been demonstrated for application in cancer therapy, vaccination and gene therapy. Importantly, the loading of these therapeutics has been coupled with the development of intelligent release mechanisms, which enables the release of drugs based on changes in pH, redox potential, temperature and enzymatic environment. The first steps towards controlling the targeting and uptake of capsules in specific cells has been made by functionalizing capsules with targeting molecules such as antibodies. LbL technology has also been applied to prepare compartmentalized bioreactors that are capable of synthesizing RNA and detecting biomolecules such as glucose. However, for *in vivo* applications, challenges include overcoming the natural clearance mechanisms of the body to increase the bioavailability of encapsulated drugs as well as controlling how the drugs/capsules are processed once they are internalized by cells. We anticipate that over the next decade significant progress will be made to address such challenges and that the LbL technology will be at the forefront of scientific innovations in biomedical research.

## References

1 Davis, M.E., Chen, Z., and Shin, D.M. (2008) Nanoparticle therapeutics: an emerging treatment modality for cancer. *Nat. Rev. Drug Disc.*, **7** (9), 771–782.
2 Peer, D., Karp, J.M., Hong, S., Farokhzad, O.C., Margalit, R., and Langer, R. (2007) Nanocarriers as an emerging platform for cancer therapy. *Nat. Nanotechnol.*, **2** (12), 751–760.
3 Johnston, A.P.R., Such, G.K., Ng, S.L., and Caruso, F. (2011) Challenges facing colloidal delivery systems: From synthesis to the clinic. *Curr. Opin. Colloid Interface Sci.*, **16** (3), 171–181.
4 Schlenoff, J.B. and Decher, G. (eds) (2003) *Multilayer Thin Films: Sequential Assembly of Nanocomposite Materials*, Wiley-VCH, Weinheim.
5 Boudou, T., Crouzier, T., Ren, K.F., Blin, G., and Picart, C. (2010) Multiple functionalities of polyelectrolyte multilayer films: New biomedical applications. *Adv. Mater.*, **22** (4), 441–467.
6 Zelikin, A.N. (2010) Drug releasing polymer thin films: New era of surface-mediated drug delivery. *ACS Nano*, **4** (5), 2494–2509.
7 Becker, A.L., Johnston, A.P.R., and Caruso, F. (2010) Layer-by-layer-assembled capsules and films for therapeutic delivery. *Small*, **6** (17), 1836–1852.
8 Schneider, G.F., Subr, V., Ulbrich, K., and Decher, G. (2009) Multifunctional cytotoxic stealth nanoparticles. A model approach with potential for cancer therapy. *Nano Lett.*, **9** (2), 636–642.
9 Johnston, A.P.R., Cortez, C., Angelatos, A.S., and Caruso, F. (2006) Layer-by-layer engineered capsules and their applications. *Curr. Opin. Colloid Interface Sci.*, **11** (4), 203–209.
10 Wang, Y., Angelatos, A.S., and Caruso, F. (2008) Template synthesis of nanostructured materials via layer-by-layer assembly. *Chem. Mater.*, **20** (3), 848–858.
11 Zelikin, A., Becker, A.L., Johnston, A.P.R., Wark, K.L., Turatti, F., and Caruso, F. (2007) A general approach for DNA encapsulation in degradable polymer microcapsules. *ACS Nano*, **1** (1), 63–69.
12 Tjipto, E., Cadwell, K.D., Quinn, J.F., Johnston, A.P.R., Abbott, N.L., and Caruso, F. (2006) Tailoring the interfaces between nematic liquid crystal emulsions and aqueous phases via layer-by-layer assembly. *Nano Lett.*, **6** (10), 2243–2248.
13 Priest, C., Quinn, A., Postma, A., Zelikin, A.N., Ralston, J., and Caruso, F. (2008) Microfluidic polymer multilayer absorption on liquid crystal droplets for microcapsule synthesis. *Lab Chip*, **8**, 2182–2187.
14 Price, A.D., Johnston, A.P.R., Such, G.K., and Caruso, F. (2010) Reaction vessels assembled by the sequential adsorption of polymers, in *Modern Techniques for Nano- and Microreactors/-Reactions* (ed. F. Caruso), Springer-Verlag, Heidelberg, pp. 155–179.
15 Caruso, F., Yang, W., Trau, D., and Renneberg, R. (2000) Microencapsulation of uncharged low molecular weight organic materials by polyelectrolyte multilayer self-assembly. *Langmuir*, **16** (23), 8932–8936.
16 Caruso, F., Trau, D., Möhwald, H., and Renneberg, R. (2000) Enzyme

encapsulation in layer-by-layer engineered polymer multilayer capsules. *Langmuir*, **16** (4), 1485–1488.

17 Wang, A., Tao, C., Cui, Y., Duan, L., Yang, Y., and Li, J. (2009) Assembly of environmental sensitive microcapsules of PNIPAAm and alginate acid and their application in drug release. *J. Colloid Interface Sci.*, **332** (2), 271–279.

18 Germain, M., Grube, S., Carriere, V., Richard-Foy, H., Winterhalter, M., and Fournier, D. (2006) Composite nanocapsules: Lipid vesicles covered with several layers of crosslinked polyelectrolyte. *Adv. Mater.*, **18** (21), 2868–2871.

19 Kozlovskaya, V., Harbaugh, S., Drachuk, I., Shchepelina, O., Kelley-Loughnane, N., Stone, M., and Tsukruk, V.V. (2011) Hydrogen-bonded LbL shells for living cell surface engineering. *Soft Matter*, **7** (6), 2364–2372.

20 Quinn, J.F., Johnston, A.P.R., Such, G.K., Zelikin, A.N., and Caruso, F. (2007) Next generation, sequentially assembled ultrathin films: Beyond electrostatics. *Chem. Soc. Rev.*, **36** (5), 707–718.

21 Tan, H.L., McMurdo, M.J., Pan, G.Q., and van Patten, P.G. (2003) Temperature dependence of polyelectrolyte multilayer assembly. *Langmuir*, **19** (22), 9311–9314.

22 Dubas, S.T. and Schlenoff, J.B. (1999) Factors controlling the growth of polyelectrolyte multilayers. *Macromolecules*, **32** (24), 8153–8160.

23 Poptoshev, E., Schoeler, B., and Caruso, F. (2004) Influence of solvent quality on the growth of polyelectrolyte multilayers. *Langmuir*, **20** (3), 829–834.

24 Such, G.K., Johnston, A.P.R., and Caruso, F. (2011) Engineered hydrogen-bonded polymer multilayers: From assembly to biomedical applications. *Chem. Soc. Rev.*, **40** (1), 19–29.

25 Johnston, A.P.R., Zelikin, A.N., Lee, L., and Caruso, F. (2006) Approaches to quantifying and visualizing polyelectrolyte multilayer film formation on particles. *Anal. Chem.*, **78** (16), 5913–5919.

26 Donath, E., Sukhorukov, G.B., Caruso, F., Davis, S.A., and Möhwald, H. (1998) Novel hollow polymer shells by colloid-templated assembly of polyelectrolytes. *Angew. Chem. Int. Ed.*, **37** (16), 2202–2205.

27 Choi, J. and Rubner, M.F. (2005) Influence of the degree of ionization on weak polyelectrolyte multilayer assembly. *Macromolecules*, **38** (1), 116–124.

28 Lavalle, P., Gergely, C., Cuisinear, F.J.G., Decher, G., Schaaf, P., Voegel, J.C., and Picart, C. (2002) Comparison of the structure of polyelectrolyte multilayer films exhibiting a linear and an exponential growth regime: An in situ atomic force microscopy study. *Macromolecules*, **35** (11), 4458–4465.

29 De Geest, B.G., Vandenbroucke, R.E., Guenther, A.M., Sukhorukov, G.B., Hennink, W.E., Sanders, N.N., Demeester, J., and De Smedt, S.C. (2006) Intracellularly degradable polyelectrolyte microcapsules. *Adv. Mater.*, **18** (8), 1005–1009.

30 Dam, H.H. and Caruso, F. (2011) Construction and degradation of polyrotaxane multilayers. *Adv. Mater.*, **23** (27), 3026–3029 doi: 10.1002/adma201101210

31 Ochs, C.J., Such, G.K., Staedler, B., and Caruso, F. (2008) Low-fouling, biofunctionalized and biodegradable click capsules. *Biomacromolecules*, **9** (12), 3389–3396.

32 Park, S.Y., Barrett, C.J., Rubner, M.F., and Mayes, A.M. (2001) Anomalous adsorption of polyelectrolyte layers. *Macromolecules*, **34** (10), 3384–3388.

33 Yap, H.P., Quinn, J.F., Johnston, A.P.R., and Caruso, F. (2007) Compositional engineering of polyelectrolyte blend capsules. *Macromolecules*, **40** (21), 7581–7789.

34 Kohli, P. and Blanchard, G.J. (2000) Design and demonstration of hybrid multilayer structures: Layer-by-layer mixed covalent and ionic interlayer linking chemistry. *Langmuir*, **16** (22), 4655–4661.

35 Serizawa, T., Nanameki, K., Yamamoto, K., and Akashi, M. (2002) Thermoresponsive ultrathin hydrogels prepared by sequential chemical reactions. *Macromolecules*, **35** (6), 2184–2189.

36 Kolb, H.C., Finn, M.G., and Sharpless, K.B. (2001) Click chemistry: diverse chemical function from a few good reactions. *Angew. Chem. Int. Ed.*, **40** (11), 2004–2021.

37 Such, G.K., Quinn, J.F., Quinn, A., Tjipto, E., and Caruso, F. (2006) Assembly of ultrathin polymer multilayer films by click chemistry. *J. Am. Chem. Soc.*, **128** (29), 9318–9319.

38 Such, G.K., Tjipto, E., Postma, M., Johnston, A.P.R., and Caruso, F. (2007) Ultrathin, responsive polymer click capsules. *Nano Lett.*, **7** (6), 1706–1710.

39 Connal, L.A., Kinnane, C.R., Zelikin, A.N., and Caruso, F. (2009) Stabilization and functionalization of polymer multilayers and capsules via thiol-ene click chemistry. *Chem. Mater.*, **21** (4), 576–578.

40 Huang, C.-J. and Chang., F.-C. (2009) Using click chemistry to fabricate ultrathin thermoresponsive microcapsules through direct covalent layer-by-layer assembly. *Macromolecules*, **42** (14), 5155–5166.

41 De Geest, B.G., van Camp, W., Du Prez, F.E., De Smedt, S.C., Demeester, J., and Hennink, W.E. (2008) Degradable multilayer films and hollow capsules via a "click" strategy. *Macromol. Rapid Commun.*, **29** (12–13), 1111–1118.

42 Stockton, W.B. and Rubner, M.F. (1997) Molecular-level processing of conjugated polymers. 3. Layer-by-layer manipulation of polyaniline via electrostatic interactions. *Macromolecules*, **30** (9), 2717–2725.

43 Wang, L., Wang, Z.Q., Zhang, X., Shen, J.C., Chi, F., and Fuchs, H. (1997) Hydrogen bonded layer-by-layer assembly of poly(2-vinylpyridine) and poly(acrylic acid): Influence of molecular weight on the formation of microporous film by post-base treatment. *Macromol. Rapid Commun.*, **18** (7), 509–514.

44 Yang, S.Y. and Rubner, M.F. (2002) Micropatterning of polymer thin films with ph-sensitive and cross-linkable hydrogen-bonded polyelectrolyte multilayers. *J. Am. Chem. Soc.*, **124** (10), 2100–2101.

45 Kozlovskaya, V., Ok, S., Sousa, A., Libera, M., and Sukhishvili, S.A. (2003) Hydrogen-bonded polymer capsules formed by layer-by-layer self-assembly. *Macromolecules*, **36** (23), 8590–8592.

46 Erel-Unal, I. and Sukhishvili, S.A. (2008) Hydrogen-bonded hybrid multilayers: Film architecture controls release of macromolecules. *Macromolecules*, **41** (22), 3962–3970.

47 Zelikin, A.N., Quinn, J.F., and Caruso, F. (2006) Disulfide cross-linked polymer capsules: en route to biodeconstructible systems. *Biomacromolecules*, **7** (1), 27–30.

48 Kinnane, C.R., Such, G.K., Antequera-García, G., Yan, Y., Dodds, S.J., Liz-Marzan, L.M., and Caruso, F. (2009) Low-fouling poly(N-vinyl pyrrolidone) capsules with engineered degradable properties. *Biomacromolecules*, **10** (10), 2839–2846.

49 Leung, M.K.M., Such, G.K., Johnston, A.P.R., Biswas, D.P., Zhu, Z., Yan, Y., Lutz, J.-F., and Caruso, F. (2011) Assembly and degradation of low-fouling click-functionalized poly(ethylene glycol)-based multilayer films and capsules. *Small*, **7** (8), 1075–1085.

50 Ochs, C.J., Such, G.K., Yan, Y., van Koeverden, M.P., and Caruso, F. (2010) Biodegradable click capsules with engineered drug-loaded multilayers. *ACS Nano*, **4** (3), 1653–1663.

51 Liang, K., Such, G.K., Zhu, Z., Yan, Y., Lomas, H., and Caruso, F., Charge shifting click capsules with dual-responsive cargo release mechanisms. *Adv. Mater.*, **23** (36), H273–H277.

52 Johnston, A.P.R., Read, E.S., and Caruso, F. (2005) DNA multilayer films on planar and colloidal supports: Sequential assembly of like-charged polyelectrolytes. *Nano Lett.*, **5** (5), 953–956.

53 Lvov, Y., Decher, G., and Sukhorukov, G.B. (1993) Assembly of thin films by means of successive deposition of alternate layers of DNA and poly(allylamine). *Macromolecules*, **26** (20), 5396–5399.

54 Johnston, A.P.R., Mitomo, H., Read, E.S., and Caruso, F. (2006) Compositional and structural engineering of DNA multilayer films. *Langmuir*, **22** (7), 3251–3258.

55 Lee, L., Cavalieri, F., Johnston, A.P.R., and Caruso, F. (2010) Influence of salt

concentration on the assembly of DNA multilayer films. *Langmuir*, **26** (5), 3415–3422.

56 Kato, N., Lee, L., Chandrawati, R., Johnston, A.P.R., and Caruso, F. (2011) Optically characterized DNA multilayered assemblies and phenomenological modeling of layer-by-layer hybridization. *J. Phys. Chem. C*, **113** (50), 21185–21195.

57 Singh, A., Snyder, S., Lee, L., Johnston, A.P.R., Caruso, F., and Yingling, Y.G. (2010) Effect of oligonucleotide length on the assembly of DNA materials: Molecular dynamics simulations of layer-by-layer DNA films. *Langmuir*, **26** (22), 17339–17347.

58 Johnston, A.P.R. and Caruso, F. (2008) Stabilization of DNA multilayer films through oligonucleotide crosslinking. *Small*, **4** (5), 612–618.

59 Johnston, A.P.R. and Caruso, F. (2007) Exploiting the directionality of DNA: Controlled shrinkage of engineered oligonucleotide capsules. *Angew. Chem. Int. Ed.*, **46** (15), 2677–2680.

60 Johnston, A.P.R., Zelikin, A.N., and Caruso, F. (2007) Assembling DNA into advanced materials: From nanostructured films to biosensing and drug delivery. *Adv. Mater.*, **19** (21), 3727–3730.

61 Postma, A., Yan, Y., Wang, Y., Zelikin, A.N., Tjipto, E., and Caruso, F. (2009) Self-polymerization of dopamine as a versatile and robust technique to prepare polymer capsules. *Chem. Mater.*, **21** (14), 3042–3044.

62 Lee, H., Dellatore, S.H., Miller, W.M., and Messersmith, P.H. (2007) Mussel-inspired surface chemistry for multifunctional coatings. *Science*, **318** (5849), 426–430.

63 Cui, J., Wang, Y., Postma, A., Hao, J.C., Hosta-Rigau, L., and Caruso, F. (2010) Monodisperse polymer capsules: Tailoring size, shell thickness and hydrophobic cargo loading via emulsion templating. *Adv. Funct. Mater.*, **20** (10), 1625–1631.

64 Wang, Y. and Caruso, F. (2005) Mesoporous silica spheres as supports for enzyme immobilization and encapsulation. *Chem. Mater.* **17** (5), 953–961.

65 Angelatos, A.S., Johnston, A.P.R., Wang, Y., and Caruso, F. (2007) Probing the permeability of polyelectrolyte multilayer capsules via a molecular beacon approach. *Langmuir*, **23** (8), 4554–4562.

66 Wang, Y., Yan, Y., Cui, J., Hosta-Rigau, L., Heath, J.K., Nice, E.C., and Caruso, F. (2010) Encapsulation of water-insoluble drugs in polymer capsules prepared using mesoporous silica templates for intracellular drug delivery. *Adv. Mater.*, **22** (38), 4293–4297.

67 Petrov, A.I., Volodkin, D.V., and Sukhorukov, G.B. (2005) Protein-calcium carbonate coprecipitation: A tool for protein encapsulation. *Biotechnol. Prog.*, **21** (3), 918–925.

68 Wang, Y., Yu, A.M., and Caruso, F. (2005) Nanoporous polyelectrolyte spheres prepared by sequentially coating sacrificial mesoporous silica spheres. *Angew. Chem. Int. Ed.*, **44** (19), 2888–2892.

69 Cui, J., Wang, Y., Hao, J., and Caruso, F. (2009) Mesoporous silica-templated assembly of luminescent polyester particles. *Chem. Mater.*, **21** (18), 4310–4315.

70 Zelikin, A.N., Li, Q., and Caruso, F. (2006) Degradable polyelectrolyte capsules filled with oligonucleotide sequences. *Angew. Chem. Int. Ed.*, **45** (46), 7743–7745.

71 Ng, S.L., Such, G.K., Johnston, A.P.R., Antequera-García, G., and Caruso, F. (2011) Controlled release of DNA from polymer capsules using cleavable linkers. *Biomaterials*, **32** (26), 6277–6284.

72 Sexton, A., Whitney, P.G., Chong, S.-F., Zelikin, A.N., Johnston, A.P.R., De Rose, R., Brooks, A.G., Caruso, F., and Kent, S.J. (2009) A protective vaccine delivery system for in vivo T cell stimulation using nanoengineered polymer hydrogel capsules. *ACS Nano*, **3** (11), 3391–3400.

73 Becker, A.L., Orlotti, N.I., Folini, M., Cavalieri, F., Zelikin, A.N., Johnston, A.P.R., Zaffaroni, N., and Caruso, F. (2011) Redox-active polymer microcapsules for the delivery of a

survivin-specific siRNA in prostate cancer cells. *ACS Nano*, **5** (2), 1335–1344.

74 Städler, B., Chandrawati, R., Price, A.D., Chong, S.-F., Breheney, K., Postma, A., Connal, L.A., Zelikin, A.N., and Caruso, F. (2009) A microreactor with thousands of subcompartments: Enzyme-loaded liposomes within polymer capsules. *Angew. Chem. Int. Ed.*, **48** (24), 4359–4362.

75 Hosta-Rigau, L., Städler, B., Yan, Y., Nice, E.C., Heath, J.K., Albericio, F., and Caruso, F. (2010) Capsosomes with multilayered subcompartments: Assembly and loading with hydrophobic cargo. *Adv. Funct. Mater.*, **20** (1), 59–66.

76 Chandrawati, R., Städler, B., Postma, A., Connal, L.A., Chong, S.-F., Zelikin, A.N., and Caruso, F. (2009) Cholesterol-mediated anchoring of enzyme-loaded liposomes within disulfide-stabilized polymer carrier capsules. *Biomaterials*, **30** (30), 5988–5998.

77 Chandrawati, R., Hosta-Rigau, L., Vanderstraaten, D., Lokuliyana, S.A., Städler, B., Albericio, F., and Caruso, F. (2010) Engineering advanced capsosomes: Maximizing the number of subcompartments, cargo retention, and temperature-triggered reaction. *ACS Nano*, **4** (3), 1351–1361.

78 Lomas, H., Johnston, A.P.R., Such, G.K., Zhu, Z., Liang, K., van Koeverden, M.P., Alongkornchotikul, S., and Caruso, F. (2011) Polymersome-loaded capsules for controlled release of DNA. *Small*, **7** (14), 2109–2119.

79 Kulygin, O., Price, A.D., Chong, S.-F., Städler, B., Zelikin, A.N., and Caruso, F. (2010) Subcompartmentalized polymer hydrogel capsules with selectively degradable carriers and subunits. *Small*, **6** (14), 1558–1564.

80 Shutava, T., Prouty, M., Kommireddy, D., and Lvov, Y. (2005) pH responsive decomposable layer-by-layer nanofilms and capsules on the basis of tannic acid. *Macromolecules*, **38** (7), 2850–2858.

81 Ibarz, G., Dähne, L., Donath, E., and Möhwald, H. (2001) Smart micro- and nanocontainers for storage, transport, and release. *Adv. Mater.*, **13** (17), 1324–1327.

82 Köhler, K. and Sukhorukov, G.B. (2007) Heat treatment of polyelectrolyte multilayer capsules: A versatile method for encapsulation. *Adv. Funct. Mater.*, **17** (13), 2053–2061.

83 Johnston, A.P.R., Such, G.K., and Caruso, F. (2010) Triggering release of encapsulated cargo. *Angew. Chem. Int. Ed.*, **49** (15), 2664–2666.

84 De Geest, B.G., Sanders, N.N., Sukhorukov, G.B., Demeester, J., and De Smedt, S.C. (2007) Release mechanisms for polyelectrolyte capsules. *Chem. Soc. Rev.*, **36** (4), 636–649.

85 Radt, B., Smith, T.A., and Caruso, F. (2004) Optically addressable nanostructured capsules. *Adv. Mater.*, **16** (23–24), 2184–2189.

86 Angelatos, A.S., Radt, B., and Caruso, F. (2005) Light-responsive polyelectrolyte/gold nanoparticle microcapsules. *J. Phys. Chem. B*, **109** (7), 3071–3076.

87 Thorek, D.L.J., Chen, A.K., Czupryna, J., and Tsourkas, A. (2006) Superparamagnetic iron oxide nanoparticle probes for molecular imaging. *Ann. Biomed. Eng.*, **34** (1), 23–38.

88 Becker, A.L., Zelikin, A.N., Johnston, A.P.R., and Caruso, F. (2009) Tuning the formation and degradation of layer-by-layer assembled polymer hydrogel microcapsules. *Langmuir*, **25** (24), 14079–14085.

89 Johnston, A.P.R., Lee, L., Wang, Y., and Caruso, F. (2009) Controlled degradation of DNA capsules with engineered restriction-enzyme cut sites. *Small*, **5** (12), 1418–1421.

90 Mak, W.C., Cheung, K.Y., and Trau, D. (2008) Diffusion controlled and temperature stable microcapsule reaction compartments for high-throughput microcapsule-PCR. *Adv. Funct. Mater.*, **18** (19), 2930–2937.

91 Price, A.D., Zelikin, A.N., Wang, Y., and Caruso, F. (2009) Triggered enzymatic degradation of DNA within selectively permeable polymer capsule microreactors. *Angew. Chem. Int. Ed.*, **48** (2), 329–332.

92 Price, A.D., Zelikin, A.N., Wark, K.L., and Caruso, F. (2010) A biomolecular "ship-in-a-bottle": Continuous RNA

synthesis within hollow polymer hydrogel assemblies. *Adv. Mater.*, **22** (6), 720–723.

93 Stein, E.W., Volodkin, D.V., McShane, M.J., and Sukhorukov, G.B. (2006) Real-time assessment of spatial and temporal coupled catalysis within polyelectrolyte microcapsules containing coimmobilized glucose oxidase and peroxidase. *Biomacromolecules*, **7** (3), 710–719.

94 Kreft, O., Prevot, M., Möhwald, H., and Sukhorukov, G.B. (2007) Shell-in-shell microcapsules: A novel tool for integrated, spatially confined enzymatic reactions. *Angew. Chem. Int. Ed.*, **46** (29), 5605–5608.

95 Cortez, C., Tomaskovic-Crook, E., Johnston, A.P.R., Radt, B., Cody, S.H., Scott, A.M., Nice, E.C., Heath, J.K., and Caruso, F. (2006) Targeting and uptake of multilayered particles to colorectal cancer cells. *Adv. Mater.*, **18** (15), 1998–2003.

96 Cortez, C., Tomaskovic-Cook, E., Johnston, A.P.R., Scott, A.M., Nice, E.C., Heath, J.K., and Caruso, F. (2007) Influence of size, surface, cell line and kinetic properties on the specific binding of A33 antigen-targeted multilayered particles to colorectal cancer cells. *ACS Nano*, **1** (2), 93–102.

97 Kamphuis, M.M.J., Johnston, A.P.R., Such, G.K., Dam, H.H., Evans, R.A., Scott, A.M., Nice, E.C., Heath, J.K., and Caruso, F. (2010) Targeting of cancer cells using click-functionalized polymer capsules. *J. Am. Chem. Soc.*, **132** (45), 15881–15883.

98 Yan, Y., Ochs, C.J., Such, G.K., Heath, J.K., Nice, E.C., and Caruso, F. (2010) Bypassing multidrug resistance in cancer cells with biodegradable polymer capsules. *Adv. Mater.*, **22** (47), 5398–5403.

99 Sivakumar, S., Bansal, V., Cortez, C., Chong, S.-F., Zelikin, A.N., and Caruso, F. (2009) Degradable, surfactant-free, monodisperse polymer-encapsulated emulsions as anticancer drug carriers. *Adv. Mater.*, **21** (18), 1820–1824.

100 Yan, Y., Johnston, A.P.R., Dodds, S.J., Kamphuis, M.M.J., Ferguson, C., Parton, R.G., Nice, E.C., Heath, J.K., and Caruso, F. (2010) Uptake and intracellular fate of disulfide-bonded polymer hydrogel capsules for doxorubicin delivery to colorectal cancer cells. *ACS Nano*, **4** (5), 2928–2936.

101 Zhao, Q., Zhang, S., Tong, W., Gao, C., and Shen, J. (2006) Polyelectrolyte microcapsules templated on poly(styrene sulfonate)-doped $CaCO_3$ particles for loading and sustained release of daunorubicin and doxorubicin. *Eur. Polym. J.*, **42** (12), 3341–3351.

102 Zhao, Q., Han, B., Wang, Z., Gao, C., Peng, C., and Shen, J. (2007) Hollow chitosan-alginate multilayer microcapsules as drug delivery vehicle: doxorubicin loading and in vitro and in vivo studies. *Nanomed. Nanotechnol. Biol. Med.*, **3** (1), 63–74.

103 De Rose, R., Zelikin, A.N., Johnston, A.P.R., Sexton, A., Chong, S.-F., Cortez, C., Mulholland, W., Caruso, F., and Kent, S.J. (2008) Binding, internalization, and antigen presentation of vaccine-loaded nanoengineered capsules in blood. *Adv. Mater.*, **20** (24), 4698–4703.

104 De Koker, S., De Geest, B.G., Singh, S.K., De Rycke, R., Naessens, T., Van Kooyk, Y., Demeester, J., De Smedt, S.C., and Grooten, J. (2009) Polyelectrolyte microcapsules as antigen delivery vehicles to dendritic cells: Uptake, processing, and cross-presentation of encapsulated antigens. *Angew. Chem. Int. Ed.*, **48** (45), 8485–8489.

105 Kay, M.A. (2011) State-of-the-art gene-based therapies: The road ahead. *Nat. Rev. Genet.*, **12** (5), 316–328.

106 Elbakry, A., Zaky, A., Liebl, R., Rachel, R., Goepferich, A., and Breunig, M. (2009) Layer-by-layer assembled gold nanoparticles for siRNA delivery. *Nano Lett.*, **9** (5), 2059–2064.

# 35
# Assembly of Polymer Multilayers from Organic Solvents for Biomolecule Encapsulation

*Sebastian Beyer, Jianhao Bai, and Dieter Trau*

## 35.1
### Introduction

Through the introduction of microencapsulation many life-sciences and biomedical engineering fields have made tremendous progress, such as cellular [1] or biomolecular [2] therapy, drug delivery [3], enzymatic bioreactors [4] and biosensors [5]. The popularity of microencapsulation in these fields stems from many reasons including: (i) The minute nature of microcapsules allows for the efficient exchange of materials between the microcapsules and their environment; (ii) the prevention of infinite dilution, and (iii) the protection of encapsulated materials (e.g., proteins and DNA) from undesired external agents (e.g., proteases and nucleases). Many microcapsule fabrication techniques have been developed, including the self-assembly of microparticles [6] or polymers [7], solvent evaporation [8], and spray drying [9].

The layer-by-layer (LbL) polymer self-assembly technique has proven to be a very reproducible and versatile approach for microcapsule fabrication and biomolecule encapsulation. Initial approaches of biomolecule encapsulation, via the LbL technique, were conventionally conducted in the aqueous phase and we have generalized these approaches into three main categories: (i) bio-template-based [10], (ii) loading-based [11] and (iii) diffusion-based encapsulation [12].

### 35.1.1
#### Bio-Template-Based LbL Encapsulation in the Aqueous Phase

Biomolecule encapsulation within LbL microcapsules can be achieved via the LbL self-assembly of polymers onto biomolecule microcrystals (bio-template) as depicted in Figure 35.1. The solubility of biomolecule microcrystals was reduced through the use of chilled high salt aqueous solution [10]. Significant advantages of this approach are the very high biomolecule loading and encapsulation efficiency. By using, for example, enzyme crystals with the highest possible packing density of molecules per volume element, the highest probable encapsulation densities can be achieved. However, encapsulating mixtures of biomolecules with defined

---

*Multilayer Thin Films: Sequential Assembly of Nanocomposite Materials*, Second Edition.
Edited by Gero Decher and Joseph B. Schlenoff.
© 2012 Wiley-VCH Verlag GmbH & Co. KGaA. Published 2012 by Wiley-VCH Verlag GmbH & Co. KGaA.

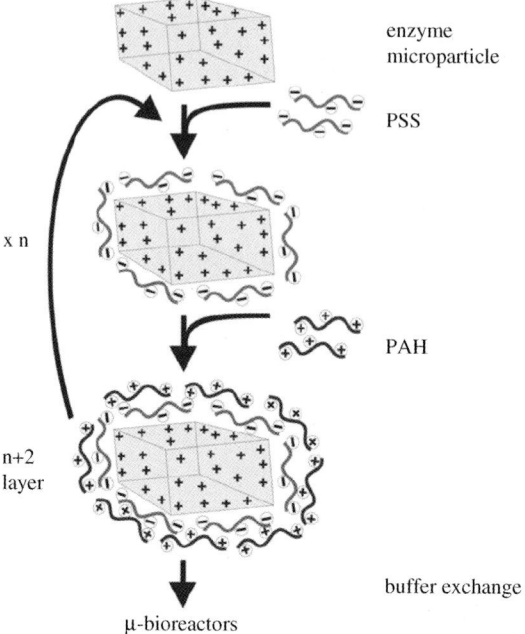

**Figure 35.1** Schematic illustration of the bio-template-based LbL encapsulation within aqueous phase. A microparticle composed of amorphous or crystalline biomolecules is used as a template dispersed in the aqueous phase under conditions which render the biomolecule water insoluble. The polyelectrolytes poly (styrene sulfonate, sodium salt) (PSS) and poly (allylamine hydrochloride) (PAH) are deposited sequentially to form a polymeric capsule around the solid bio-template (figure reprinted with permission from [10b]).

concentration using this approach is limited by difficult fabrication of suitable template microparticles. In addition, most biomolecules will still dissolve in chilled high salt aqueous solution, thus preventing the formation of a solid bio-template and limiting this approach.

### 35.1.2
### Loading-Based LbL Biomolecule Encapsulation in the Aqueous Phase

Another approach involves the loading of biomolecules within a template matrix material prior to fabrication of an LbL multilayer polymer shell around the template (Figure 35.2). Examples of such loading techniques include the adsorption of biomolecules within silica [11a, 11b] or carbonate [11c] microparticles, or pre-mixing of biomolecules during template formation of agarose microbeads [11d, 11e] or carbonate [11f, 11g] microparticles. Although loading of biomolecules via adsorption is a simple procedure, the quantity of biomolecules to be loaded cannot be controlled easily. Also, not all biomolecules will be loaded, resulting in low initial loading efficiency. Biomolecule loading by mixing them with carbonate precursors during

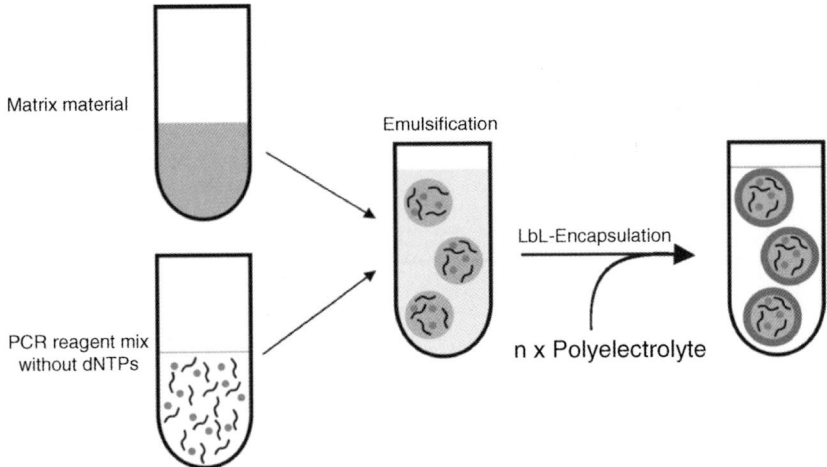

**Figure 35.2** Schematic illustration of the "matrix-assisted LbL encapsulation" technique; an example of an aqueous LbL preloading biomolecule encapsulation approach. From left to right: Mixing of matrix material (e.g., agarose) and biomolecules (e.g., polymerases and primers) followed by emulsification in an oil phase to form a water-in-oil emulsion at elevated temperature. Solidification by cooling of the matrix material forms microbead templates. Fabrication of LbL multilayer polymer shells for the encapsulation and retention of the biomolecules within the microbead templates (figure reprinted with permission from [11d]).

microparticle synthesis also faces similar problems. By using a water in oil emulsification approach to load biomolecules into emulsified agarose microbead templates, quantitative biomolecule loading can be achieved. In this approach the agarose acts as a matrix material in which any desired biomolecules, or mixture thereof, can be loaded. However, after transfer of the biomolecule-loaded agarose templates into an aqueous phase for LbL encapsulation a considerable amount of biomolecules leaches out and results in relatively low (~50%) encapsulation efficiency for biomolecules [11d].

### 35.1.3
### Diffusion-Based LbL Biomolecule Encapsulation in the Aqueous Phase

Another approach to the encapsulation of biomolecules within LbL microcapsules is by diffusion of biomolecules from an external solution into hollow LbL microcapsules. This diffusion process is driven by the biomolecule concentration gradient between the external solution and the interior of the hollow LbL microcapsules. Retention of biomolecules within the hollow LbL microcapsules can be achieved through a number of methods. For example, by changing (reducing) the permeability of the LbL "semi-permeable membrane" shells through pH [12a], solvent [12b], drying [12c], ionic strength [12d], or ultraviolet irradiation [12e] means after the biomolecules had diffused into the core. Unfortunately for this diffusion driven

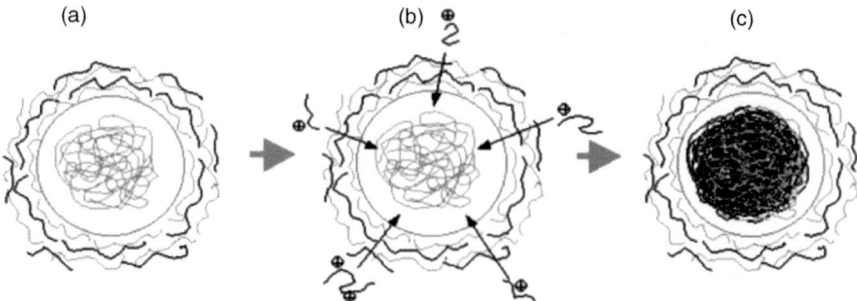

**Figure 35.3** Schematic illustration of an example of protein encapsulation into hollow LbL microcapsules. (a) LbL microcapsule made from a melamine-formaldehyde (MF) template and containing polystyrenesulfonate (PSS)/MF complex. (b) Positively charged proteins diffusing into the microcapsules. (c) Due to the negative charge of the PSS complex within the microcapsule, an insoluble protein/PSS complex is further formed and is retained within the microcapsule (figure reprinted with permission from [12g]).

loading approach, the concentration of loaded biomolecules is limited to the concentration of biomolecules in the external solution and is also difficult to control. Therefore, only a fraction of biomolecules will be encapsulated and this method results in low encapsulation efficiency. In a variation of this approach, as illustrated in Figure 35.3 a complex is formed within the microcapsules and a higher loading and better encapsulation efficiency can be achieved [12f, 12g].

## 35.2
### Limitations of LbL-Based Biomolecule Encapsulation in Aqueous Phase

However, performing LbL in the aqueous phase has some limitations: the encapsulation efficiency for biomolecules is relatively low and water-soluble biomolecules cannot be used in a direct templating approach. In addition, any water-sensitive materials cannot be encapsulated (e.g., biomolecules that are hydrolyzed rapidly in water, or materials such as sodium borohydride). Furthermore, it has been pointed out recently that common entrapment mechanisms, such as reversibly changing the membrane permeability, heat-induced membrane densification, and spontaneous accumulation due to electrostatic interaction between the molecule of interest and the matrix material inside the microcapsules have one great common limitation. It was stated by De Cock and coworkers that the most prevailing limitation of those methods is that *"it suffers from very low encapsulation efficiency (that is the amount of protein that becomes encapsulated within the capsules, relative to the amount of protein that was initially added), possible loss of bioactivity and low integrity of therapeutic macromolecules because of the harsh conditions required to make the PMLC (polymeric multilayer capsules) membrane permeable [3a]."* A recent study by De Smedt [13] and coworkers stressed that understanding parameters which influence encapsulation yield for aqueous LbL using the calcium carbonate sacrificial template method is the essential requirement to pave the way for pharmaceutical application of protein-filled polyelectrolyte microcapsules. The same study revealed that the entrapment efficiencies of proteins within calcium carbonate particles are highly dependent on their isoelectric point, and that only highly negatively charged proteins can be entrapped with a suitable yield of around 90%, and that positively charged proteins are excluded from this method by their very low encapsulation yield. Neutral biomacromolecules, such as polysaccharides, were not included in this study; it might be assumed that entrapment of those molecules within calcium carbonate particles is relatively low due to the absence of charged groups. Also, calcium carbonate as the template material for the encapsulation of biomolecules is limited to cargo that is not pH sensitive and which might not lose function upon interaction with EDTA [3a]. The latter applies especially to di-or trivalent cations that might be necessary as a cofactor for enzymes. Reverse phase LbL (RP-LbL), or the "inwards build-up technique" overcome most of these limitations. Both methods use polymers that are also frequently used in aqueous-based LbL. For direct encapsulation of protein crystals with polyelectrolyte multilayer using the RP-LbL technique, an initial entrapment of 100% can be achieved due to the insolubility of proteins in organic solvents. In cases where hydrogel microbead template materials are used for RP-LbL or the inwards build-up method, the initial biomolecule concentration within the aqueous phase can be precisely controlled. Microbead template material preparation by emulsion approaches and subsequent RP-LbL coating or inwards build-up of a polymer layer should lead to very high encapsulation yields. This should be valid for all water-soluble biomolecules due to their insolubility in the organic phase that surrounds the template and later the core shell material during the entire manufacturing process. The retention (upon transfer to the aqueous phase) of biomolecules within the core shell structures prepared by RP-LbL or the inwards build-up technique was demonstrated to be similar to that of core shell materials prepared by conventional methods. Retention of biofunctionality for various enzymes was also demonstrated within the same studies.

## 35.3
### LbL Biomolecule Encapsulation in the Organic Phase

Performing LbL in the organic phase opens some interesting new avenues; a variety of new templates and polymers can be used in organic solvents to create an even larger pool of LbL-derived core–shell materials. LbL templates can be extended into biomolecule-based templates such as amorphous or microcrystalline proteins (e.g., enzyme crystals), or organic materials (e.g., sodium ascorbate). Two techniques were developed by the Trau Group at the National University of Singapor (NUS) over recent years. The first technique is RP-LbL and the second technique is the "inwards build-up self-assembly". Both techniques are performed in organic solvents to minimize loss of biomolecules during template formation and microencapsulation by polymer multilayer build-up. In addition, the use of organic solvents allows the encapsulation of water-sensitive materials. This section will be divided into two subsections. The first covers the physico-chemical similarities and differences between polymers in aqueous and organic solvents; also different mechanisms for aqueous LbL, hydrogen-bonded LbL (HB-LbL) and the newly established RP-LbL are discussed, and some applications demonstrated. The second reports on the "inwards build-up self-assembly" method and its applications.

## 35.3.1
### Reverse-Phase LbL

#### 35.3.1.1 Mechanism

This section focuses on the distinct differences between conventional aqueous LbL, HB-LbL and organic phase RP-LbL for polymer multilayer assembly. The RP-LbL technique is capable of creating multilayers of polymers onto template materials like the other LbL techniques that have been developed. Similar to water based LbL, RP-LbL layers can serve as a "semi-permeable" membrane. In contrast to aqueous LbL that commonly uses polyelectrolyte salts in aqueous solvents, the RP-LbL technique employs polyamines and polyacids dissolved in organic solvents [14]. To the best of our knowledge the use of organic solvents in LbL can be divided into two fields: (i) LbL multilayer post-processing with organic solvents to alter their internal structure [15], and (ii) the use of organic solvents or solvent/water mixtures as the deposition medium for polymers. The latter employs organic solvents for four major purposes, to alter polyelectrolyte multilayer build-up by changing the solvent polarity [16], for dissolution of water-insoluble polyelectrolyte salts [17], to perform HB-LbL [18], and to protect water-sensitive template materials [19]. Polymer interaction and multilayer build-up in the above-mentioned methods can be grouped into two fields, the hydrogen bonding interaction of polymers and the electrostatic interaction of polymers. The first uses polymers that do not dissociate in charged species, multilayer build-up and interaction is based primarily on hydrogen bonding. The latter is based on electrostatic interaction during the build-up process using polyelectrolyte salts dissolved in polar organic solvents that allow ionic dissociation [20].

In contrast, the RP-LbL method employs polyamines that do not carry a charge while in solution and polyacids that only carry a very low charge due to auto-protolysis in some solvents while in solution [21].

**Properties of Polymers in Solution:** Polyelectrolytes used in aqueous-based LbL are most commonly alkali salts of poly carboxylic acids, poly sulfonic acids and poly ammonium or quaternary poly(alkyl ammonium) salts with a halogenide counter ion. The nature of these polymers implies a high charge density per polymer molecule in aqueous solution. Since the charge is the same between the monomer units [22], charge–charge interaction within a polyelectrolyte is repulsive. It is easy to visualize that a polyelectrolyte thus appears to be in a stretched form when dissolved in pure water. Shielding of the charges around each charged group is possible by addition of salts, gradually changing the polyelectrolyte conformation from a stretched to a coiled form [22]. The addition of water-miscible organic solvents leads to a decrease in solvent quality, leading to a more coiled polymer conformation. So far, only polyelectrolyte salts able to dissociate to a large extent in aqueous-based solutions, with a significantly decreased ionic bond dissociation causing polyelectrolyte precipitation at high solvent concentration, have been reviewed. In the case of the RP-LbL and the inwards build-up techniques, poly amines in free base form and poly acids as free acid are dissolved in organic solvents. These polymers exhibit excellent solubility in solvent mixtures and in pure aliphatic alcohols with various polarities. The solvent quality might thus not decrease that dramatically with decreasing permittivity, as is the case for polymers used in aqueous LbL. Furthermore, amines do not have a significant charge in organic solvents and can be considered neutral, whereas acids have a certain autoprotolysis and thus have a significant negative charge in aqueous solution and organic solvents. Decreasing the polarity of the solvent or changing the whole solvent from water-based mixtures to lower alcohols such as methanol or ethanol, results in a significant increase of the apparent pKa for poly carboxylic acids. An increased apparent pKa value leads to less charged units per molecule. The consequence for organic phase systems might be that the polymers have a less stretched conformation in solution compared to aqueous solutions. Furthermore, especially in the case of poly amines, inter- and intra- molecular hydrogen bonding might play an important role in the solution behavior of the polymers as hydrogen bonding is not possible between poly ammonium cations in aqueous phase. This, however, has just begun to be understood and is the subject of current investigations.

**RP-LbL Multilayer Build-Up:** It was postulated that, once the non-charged polymers come in close proximity to the surface during the assembly process, a *Brønsted acid–base* reaction (e.g., the non-dissociated proton of the poly acid protonates the amine to form the respective ammonium salt) takes place at the surface and the multilayer holds together due to electrostatic interactions. This mechanism is significantly distinct from previously introduced mechanisms such as aqueous LbL based on electrostatic interaction and HB-LbL as depicted in Figure 35.4. Polyelectrolyte multilayers consisting of polystyrene sulfonic acid (PSS) and poly allylamine (PA) assembled from aliphatic alcohols have a significantly increased layer thickness, which is consistent with the observation that polyelectrolyte multilayer thickness increases

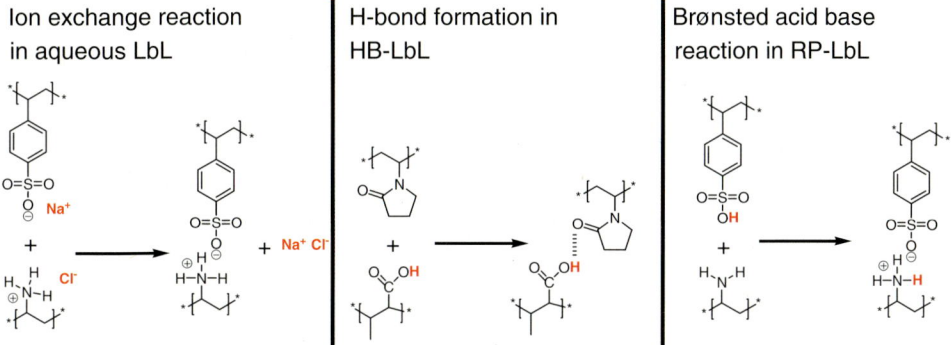

**Figure 35.4** Molecular interactions between polymers in the aqueous LbL, HB-LbL and RP-LbL techniques.

significantly but to a less great extent upon addition of organic solvents (data not shown). Zeta potential studies of the sequential absorption of polymers by the RP-LbL method show a clear reversal of zeta potential upon absorption of each layer. Given that the poly amine itself carries no charge in solution, it must acquire positive charge while absorbed on the surface. This might be possible since protons of poly acids have a certain mobility in organic solvents, especially aliphatic alcohols. To what extent polymer absorption is based on electrostatic interaction, diffusional motion and other types of interaction remains to be elucidated.

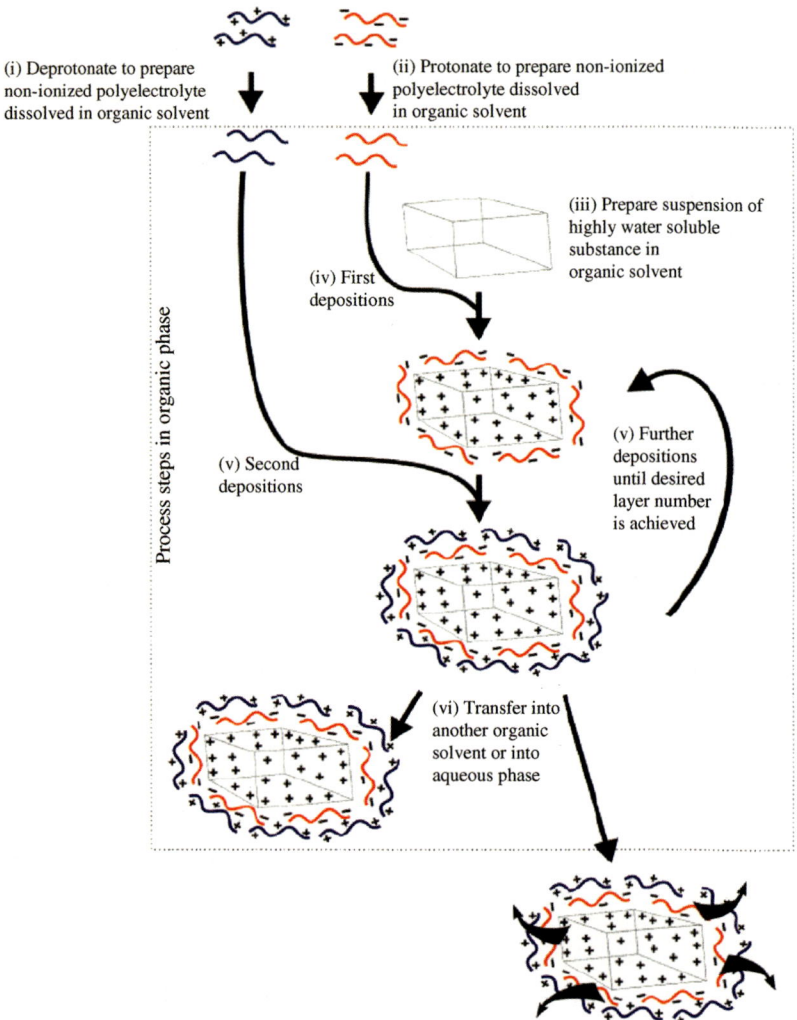

**Figure 35.5** Schematic diagram of the RP-LbL technique. Significant differences, as highlighted in this figure, are the use of an organic phase and Brønsted acid–base reactions taking place between polymers or polymers with the template (Reprinted with permission from [14]).

Earlier investigations have revealed that, in aqueous LbL, the driving force to form water-insoluble polyelectrolyte complexes is entropic due to the release of small molecular weight counter ions [23]. This might also be the case for methods in which common polyelectrolyte salts are dissolved in very polar organic solvents, such as formamide [20a] or ionic liquids [20b], in which ionic bond dissociation is possible. In this case the interaction between a polymer pair is electrostatic and far-reaching and layer formation is via an ionic bond. In contrast to electrostatic interaction, HB-LbL relies on hydrogen bonding interaction. Occasionally, the use of organic solvents has been reported to facilitate HB-LbL. The main reasons for using organic solvents in HB-LbL was the insolubility of certain polymers in water, and the fact that hydrogen bonds have a stronger interaction in nonpolar solvents. RP-LbL was postulated to be based on an acid–base reaction as the driving force, which differs in nature from those of aqueous LbL and HB-LbL. In RP-LbL no small molecular weight counter ions exist that could be released upon polyelectrolyte complex formation. The acidic strength of the poly acids used in combination with the strong proton-accepting ability of primary amines resulting in the formation of ionic bonds by an acid–base reaction. In contrast to HB-LbL, the nature of the polymer complex interaction is electrostatic. This hypothesis is supported by previous studies in the field of ion exchange resins, in which thermodynamic studies have been performed to prove that monomolecular amine bases absorb onto ion exchange resins in acid form to produce a salt [24]. Another study of polymer interaction using spectroscopic methods has been demonstrated by Borodina *et al.* for RP-LbL [25]. In this study it was clearly shown that a poly carboxylic acid forms ionic bonds to some extent upon interaction with primary poly amines (good proton acceptors) when both polymers are dissolved and mixed in dichloromethane (DCM). For relatively weak poly carboxylic acids IR-spectroscopy has revealed that hydrogen bonding interaction occurs between poly acids and relatively weak proton acceptors [26]. Hydration of sulfonic acids makes it difficult to investigate proton dissociation from the acid group as they resemble sulfonates in the hydrated form. However, it might be reasonable to assume a higher degree of salt formation with increasing acidic strength. Again, the most important implication for the encapsulation of biomolecules is probably that microcapsules based on electrostatic interaction can be assembled while guaranteeing a very high encapsulation efficiency and control over biomolecule concentration within microcapsules. Electrostatically bound microcapsules have certain advantages over HB-LbL-based microcapsules in terms of stability against environmental influences.

### 35.3.1.2 Technique

The implication for biomolecule encapsulation is the combination of a relatively strong electrostatic interaction, and thus a stable resulting microcapsule, with high encapsulation efficiency as no biomolecules can dissolve in the organic phase. The RP-LbL technique works similarly to other techniques to prepare microcapsules, employing sequential absorption of a polymer layer followed by washing off excess non-absorbed polymer after each deposition step.

Table 35.1 Examples of polymers used for the RP-LbL technique.

| Polymer | Solvent | $M_W$ (kDa) | Concentration used (mg mL$^{-1}$) | Reference |
|---|---|---|---|---|
| **Amine functional polymers** | | | | |
| Polydiallyldimethyl ammonium chloride | Ethanol | 100–200 | 10 | [14] |
| Polyethyleneimine, linear | Chloroform | 250 | Saturation | [14] |
| Poly(allylamine) | 1-Butanol | 65 | 1 | [27] |
| Polyethyleneimine, branched | Dichloromethane | 10–25 | 4 | [25] |
| **Acid functional polymers** | | | | |
| Poly(methacrylic acid) | Ethanol | 100 | 10 | [14] |
| Poly(acrylic acid) | 1-Butanol | 450 | 1 | [27] |
| Poly(acrylonitrile-co-butadieneco-acrylic acid) | Dichloromethane | 3.6 | 4 | [25] |
| Poly(styrene sulfonic acid) | 1-Butanol | 70 | 5 | [28] |

**Polymers:** The polymers used for the RP-LbL technique are usually poly amines or poly acids. Many suitable polymers are commercially available but may also be prepared from their polyelectrolyte form by acid–base chemistry in which small molecular weight counterions are removed, in analogy to ion exchange resin chemistry. Polymers prepared in this way are termed non-ionized or non-ionic to indicate their polyelectrolyte origin. Non-ionic polymers are preferred for the RP-LbL technique. This allows polymer dissolution in organic solvents. A summary of polymers that can be used for RP-LbL and their solubility in organic solvents can be found in Table 35.1.

**Solvents:** Solvents for RP-LbL range over various classes of solvents, the main criteria being sufficient solubility for poly amines and poly acids. Although the solubility of polymers has been shown for various solvents, such as dimethyl formamide, dimethyl sulfoxide, formamides and dichloromethane, our group favors the use of aliphatic alcohols as they are less toxic, biofriendly, and some are even considered Generally Recognized as Safe (GRAS) by the US Food And Drug Administration (FDA).[1]

**Template materials:** Template materials can be chosen from a broad range of materials. Special attention has been paid to highly water-soluble materials that cannot be encapsulated by "direct bio-templating" using common aqueous LbL methods, such as saccharides, poly saccharides, proteins, nucleic acids, organic or inorganic salts, drugs and hydrophilic vitamins.

### 35.3.1.3 Encapsulation of Biomolecules

Demonstrated methods of encapsulating biomolecules via the RP-LbL technique include the direct encapsulation of biomolecule crystals [14] (Figure 35.6a), self-

---

1) Database of National Archives and Records Administration of the United States of America, Sec. 172.864 Synthetic fatty alcohols, Sec. 178.3480 fatty alcohols, synthetic.

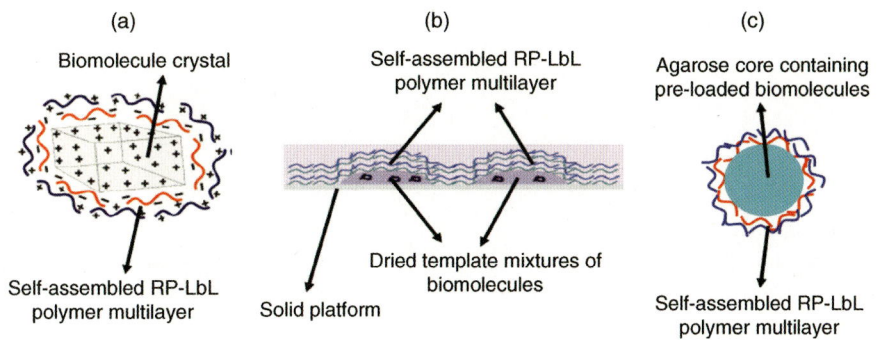

**Figure 35.6** Schematic diagram illustrating various methods for encapsulation of biomolecules via the RP-LbL technique. (a) Direct encapsulation of biomolecule crystals, (b) self-assembly of polymers onto dried template mixtures of biomolecules on a flat solid substrate to form surface-bound microenclosures, and (c) encapsulation of biomolecules loaded into agarose microbeads (Images taken and modified from [14, 29]).

assembly of polymers onto biomolecules dried onto a flat solid substrate to form surface-bound microenclosures [29] (Figure 35.6b), and encapsulation of biomolecules loaded into agarose microbeads [27, 30] (Figure 35.6c). Encapsulation is achieved by dispersing the colloidal template material or surface-bound template in an organic solvent, such as aliphatic alcohols, containing the polymer. Next, polymers with good absorption properties at the template interface will be absorbed until the surface is covered with the polymer. Excess polymer is washed off, followed by incubation of the polymer-coated template into a complementary polymer solution. This process can be repeated to achieve the desired number of LbL polymer multilayers.

Direct encapsulation of water-soluble biomolecules bears the advantage that no intermediate steps, such as biomolecule loading or removal of a sacrificial template, are necessary. However, upon transfer into the aqueous phase encapsulated macromolecular biomolecules might build up sufficient osmotic pressure to rupture the capsule [14]. Figure 35.7 depicts the release of encapsulated BSA-FITC ($M_W$ 65 kDa), obtained via the RP-LbL technique, when transferred into an aqueous phase. Fortunately, this problem can easily be circumvented through the simultaneous encapsulation of saccharides [29]. Certain organic solvents (e.g., ethanol) are inherently much more volatile than water. By using such solvents to encapsulate biomolecules via the RP-LbL technique, powders are easily prepared by evaporation of the solvent. Preferentially, the solvent is evaporated under vacuum at room temperature under very mild conditions to prepare powders of biomolecule-filled microcapsules. It was demonstrated that the powder can be reconstituted in water or buffer to form a microcapsule suspension. Powders have many applications and can usually be more easily stored than liquid suspensions.

Another application to encapsulate biomolecules by RP-LbL was demonstrated by forming "surface bound microenclosures" (SBMEs) [29] on flat substrate surfaces. Instead of directly encapsulating biomolecule crystals, a solution of biomolecules and

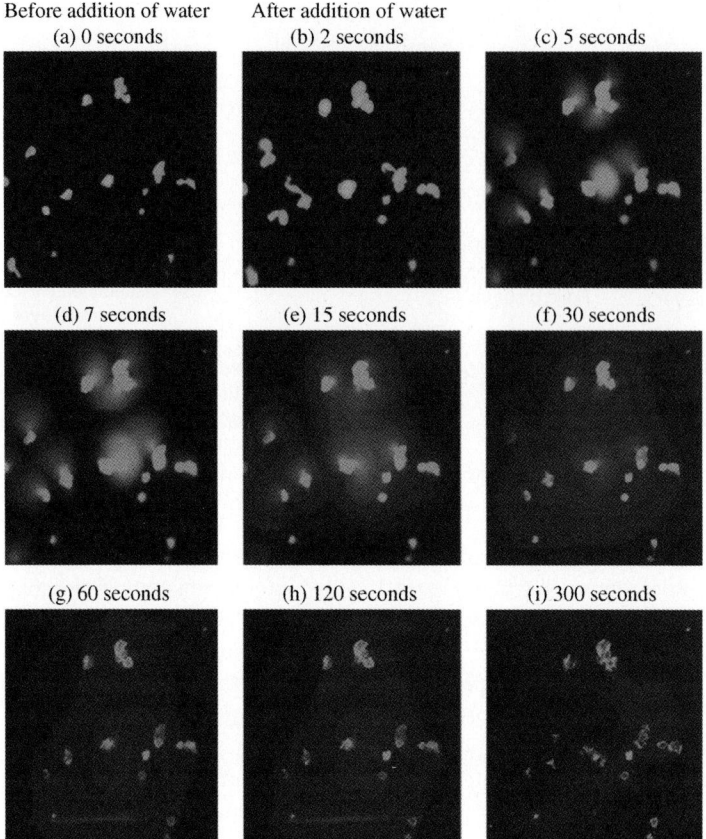

**Figure 35.7** Fluorescent micrographs demonstrating the RP-LbL encapsulation of the highly water soluble protein bovine serum albumin (BSA): (a) before addition of water. (b) Capsule volume increases due to build up of osmotic pressure in capsules. Some capsules show "small jets" releasing BSA. (c to h) Further release of BSA forming "clouds" and diffusion of BSA away from capsules. (i) Remaining capsule material. BSA-FITC was used to demonstrate release. The particle size was 5–20 μm. Non-encapsulated BSA dissolves immediately in the first 2 s (Image taken from [14]).

glucose was dried on a solid substrate, forming a solid bound template. Then, sequential immersion of the substrate with the dried template into ethanol with dissolved polymers caused the deposition of RP-LbL membranes and formation of SBMEs. By encapsulating the desired biomolecules within the SBMEs, various physical properties and biological reactions could be studied in a closed confinement. This is especially useful for parallel and multiplexed observations of biochemical reactions and biophysical processes by an "easy to handle" protocol. By using this method with encapsulated NeutrAvidin, the membrane permeability of fluorescence-labeled biotin and its interaction with NeutrAvidin was studied. Also, the control of enzymatic reaction within the SBMEs was demonstrated by DNase activity

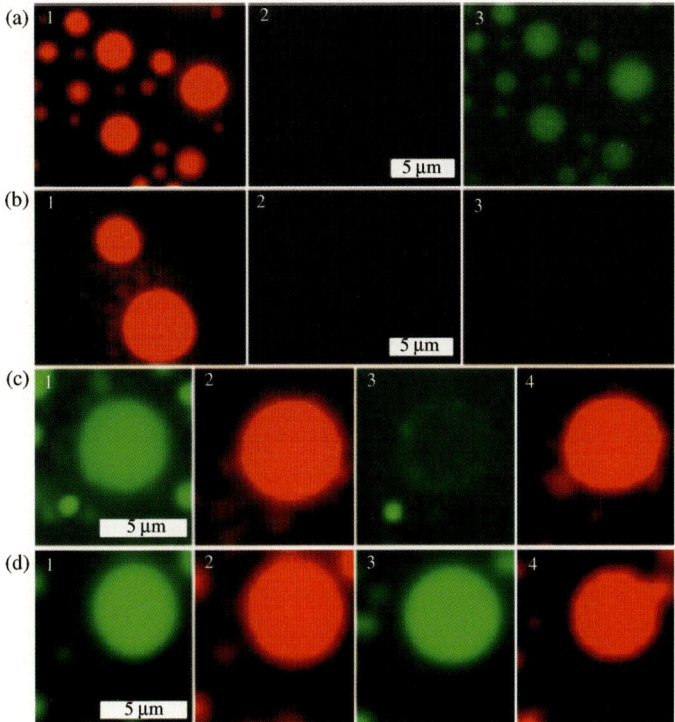

**Figure 35.8** Applications of SBMEs observed with fluorescence microscopy. (a) Interaction of enclosed NeutrAvidin molecules with FITC-biotin diffusing in from outside the SBMEs. (1) SBMEs filled with Texas Red labeled BSA and NeutrAvidin (red channel); (2) Image of the same area before adding FITC-biotin (green channel); (3) same area 3 min after adding FITC-biotin (green channel). (b) Control experiment without NeutrAvidin. Same sequence of events as in (a). (c) Enzymatic DNA digestion. (1) SBME filled with Alexa Fluor 488 labeled oligonucleotide, DNase I, and Texas Red/BSA (green channel); (2) same area as in (1) (red channel); (3) same area 10 min after adding buffer containing $Mg^{2+}$ and $Ca^{2+}$ ions to start the enzymatic reaction (green channel) (4) same area as in (3) (red channel). (d) Control experiment of enzymatic DNA digestion with SBMEs containing no DNase. Same sequence as in (c) (Image taken from [29]).

that could be switched on and off by addition or removal of $Mg^{2+}$ ions to the external solution, indicating selective permeability of the RP-LbL membrane for small molecular weight materials (Figure 35.8).

Core–shell materials employing an agarose core are particularly useful for capsule-based bio-reactor and biosensor applications. The high water content of the agarose core provides a good environment for biochemical reactions. Mixtures of biomolecules can be loaded within the agarose with high control over the absolute concentration and the ratio of biomolecules for encapsulation. However, before polymers can be self-assembled from the organic phase onto agarose templates to form core–shell materials, it is necessary to prevent aggregation and shrinking of the agarose microbead core. Shrinking could occur due to loss of water into the organic phase, and aggregation caused by the relatively hydrophobic organic phase. Two approaches to prevent shrinking and aggregation have been demonstrated. The first entails the use of polystyrene microparticles to form colloidosomes [30] and the second is based on ADOGEN® 464 detergent (cationic) [27, 31]. After stabilization of the agarose bead, polymers can be self-assembled to form biomolecule-filled agarose core based core–shell materials.

By using the RP-LbL self-assembly of polymers in organic solvents onto stabilized agarose core templates containing biomolecules, (Figure 35.6c) the encapsulation yields for biomolecules were at least doubled compared to fabrication via the conventional aqueous LbL approach. It was also shown that these core–shell materials have high retention stability (~100%) of encapsulated BSA when dispersed in an aqueous phase for a period of 7 days [30]. This ability to achieve a high encapsulation efficiency and retention stability for biomolecules with the RP-LbL technique allows the fabrication of better performing core–shell materials for various applications.

Interestingly, additional loading of small molecules such as tris(hydroxymethyl) aminomethane (TRIS) and sucrose within the agarose-polymer core–shell materials resulted in the fabrication of "inflated" microcapsules with a unique "bead-in-a-capsule" morphology (Figure 35.9) [27]. The inflation of the RP-LbL shell is driven by a difference in osmotic pressure between the core–shell materials' interior and exterior environment when the agarose-polymer core–shell materials were transferred from an organic phase into an aqueous phase. The different distribution of

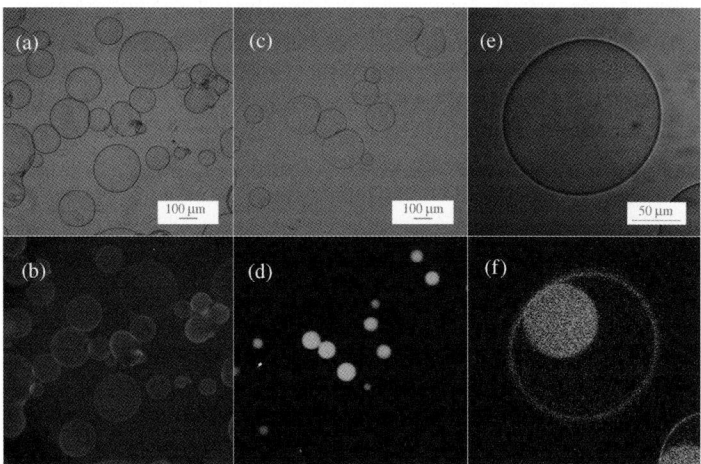

**Figure 35.9** (a) Optical and corresponding (b) fluorescence micrograph of inflated microcapsules fabricated with poly(acrylic acid)–Rhodamine 123 conjugate and by the RP-LbL method. Fluorescence is observed from the RP-LbL polymeric shell. (c) Optical and corresponding (d) fluorescence micrograph of inflated microcapsules fabricated with agarose– Rhodamine 123. The agarose microbeads are clearly fluorescent. (e) Confocal optical and corresponding (f) fluorescence micrograph of inflated microcapsules fabricated with both Polyacrylic acid–Rhodamine 123 and agarose–Rhodamine 123. The agarose microbead is observed to be partially attached to the LbL capsular wall (Image taken from [27]).

materials within the multiphase interior of "inflated" microcapsules was demonstrated. Small molecules could diffuse out of the bead, occupying the entire interior, whereas entrapped particles remained in the agarose bead, creating a distinct localization. This demonstrates that "inflated" microcapsules could permit control over localized chemical or enzymatic reactions for future studies.

## 35.3.2
## "Inwards Build-Up Self-Assembly" of Polymers for Biomolecule Encapsulation in the Organic Phase

### 35.3.2.1 Mechanism and Technique

An alternative and recently developed novel method for polymer self-assembly from organic solvents to form polymer multilayers is termed the "inwards buildup self-assembly" technique [28, 31]. Unlike conventional LbL techniques where polymers are built up outwards from the template, the inwards build-up self-assembly technique generates well-defined polymer multilayers through the deposition of polyamines into and within the porous matrices of agarose microbeads that "grow inwards". This novel process is driven by a poly amine (e.g., poly(allylamine) and branched polyethyleneimine) concentration gradient between the exterior and interior environments of agarose microbeads, causing the polymers to diffuse into the microbeads. The details of the interaction between the polyamines and the agarose polymers are not fully known and are still under investigation. However, an initial assessment is that certain distribution coefficients for the polymer exist between the organic solvent and the aqueous agarose matrix, causing the deposition of the well-defined shells on the agarose matrix. After a first shell is formed, further addition of polymer will again result in in-diffusion of poly amines and the formation of another inner shell within the agarose microbeads. It should be noted that the shells are formed by the same polymer and not, as in LbL, from complementary polymers in charge or HB-bonding capabilities. Finally, these self-assembled poly amines can be "immobilized" within agarose microbeads through the use of bi-functional amino cross-linkers or polymers of opposite charge (e.g., poly(styrene sulfonic acid)) (Figure 35.10) [28]. Although it appears that the inwards build-up self-assembly technique is limited to hydrogel type templates such as agarose, this alternative technique can, however, fabricate unique core–shell materials and perform biomolecular encapsulation tasks that are not feasible with the conventional LbL technology. Typically a self-assembled LbL polymer layer is $\sim$3–5 nm thick, due to its self-limiting characteristic; in contrast, polymer layers assembled by the inwards build-up self-assembly technique are limited by the amount of poly amines available or the incubation time. As a consequence, the thickness of self-assembled polymer layers of the inwards build-up self-assembly type is easily tunable by varying the incubation time and poly amine concentration, and micrometer thickness of polymer layers was demonstrated to be easily achievable within minutes, and in a single step. In addition, the density of the self-assembled polymer layers obtained via the inwards build-up self-assembly technique is tunable by the density of the hydrogel material forming the microbead template [28].

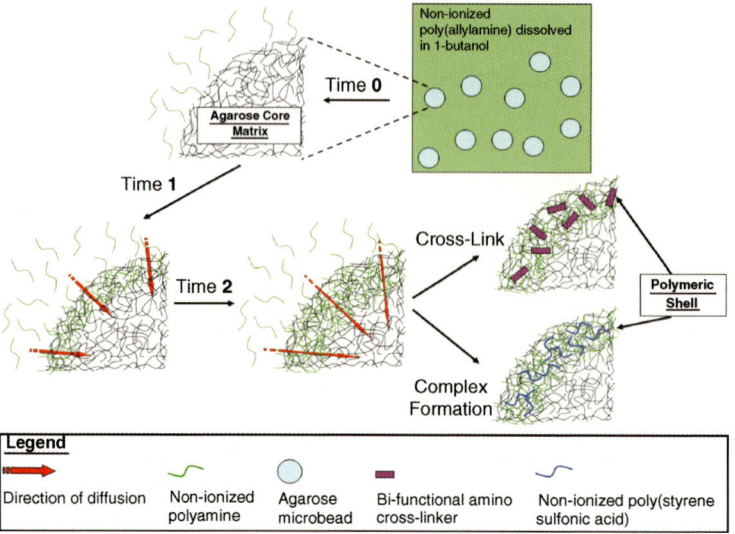

**Figure 35.10** Schematic diagram illustrating the self-assembly of polymers from an organic solvent, via the inwards build-up self-assembly technique, into porous agarose core templates for the formation of agarose–polymer core–shell materials.

### 35.3.2.2 Encapsulation of Biomolecules

LbL encapsulation and retention of macromolecules of relatively low $M_W$ is a challenge due to their relatively good permeability through self-assembled polymeric shells. Reduction of permeability for successful biomolecule encapsulation was demonstrated by a post-treatment of the LbL layers [32]. Advantageously, by employing the inwards build-up self-assembly technique, poly(allylamine) shells of micrometer thickness could be easily achieved within 2 h to encapsulate and retain relatively low $M_W$ dextran ($M_W = 4$ kDa) [28]. Interestingly, by using branched polyethyleneimine of similar concentrations, an increase in shell thickness was observed, and by using a lower percentage of agarose a decrease in shell density was obtained, in both cases the permeability was increased, thereby providing an easy means to control the permeability. Core–shell materials obtained by our method are good candidates for passive release applications such as drug delivery. A "redox responsive" core–shell material was obtained by using cleavable disulfide cross-linkers (-SS-) for poly amine fixation during fabrication. An immediate release of biomolecules was demonstrated after addition of a reductive agent (dithiothreitol) to emulate certain physiological conditions.

Encoding of polymer self-assembled capsules has been demonstrated using encapsulated CdTe nanocrystals of different fluorescence emission wavelengths in previous studies [33]. Advantageously, the inwards build-up self-assembly technique provides two modes for encoding: by permutation of the layer color and/or the layer thickness (Figure 35.11). Permutation of layer color was achieved through incubation of poly amines, conjugated with different dyes in a desired sequence, while the permutation of layer thickness was achieved by tuning the incubation time and poly amine amount (by volume or concentration variations).

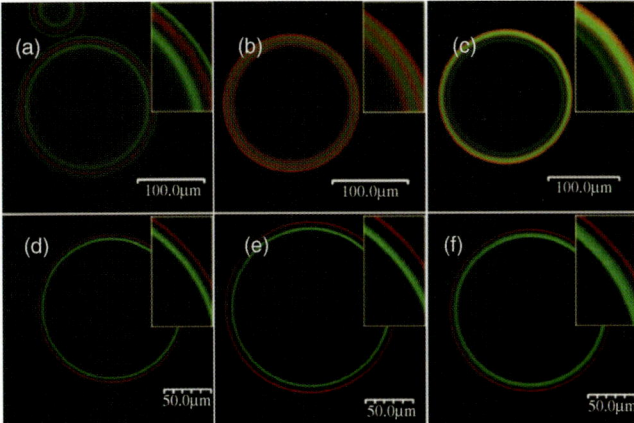

**Figure 35.11** (a–c) Confocal images of agarose microbeads in 0.01x PBS with 5 concentric layers of different color coding permutations (R – RED, G – GREEN, B – BLANK). Fabrication was done in the following order: Layers 1/2/3/4/5 (a) G/B/R/B/G (b) R/G/R/G/R (C) R/G/R/B/G. (d–f) Confocal images of agarose microbeads in 0.01x PBS with 3 concentric layers (Layer 1/2/3) of the same color encoding permutation (R/B/G) but with different thickness permutations due to the variation of the polymer amount by using different volumes. (d) 500 μL/500 μL/500 μL (e) 500 μL/1 mL/500 μL (f) 500 μL/500 μL/1 mL. The insets in the confocal images are magnified images of the fluorescence layers (Image taken from [31]).

Core–shell materials with multiplexing analytical capabilities based on encapsulated enzymes were fabricated using the inwards build-up technique. Encapsulation of three different biomolecules (GOx, HRP and BSA) was performed within three different color-coded (green, red and green/red fluorescence, respectively) core–shell materials. Then, the three capsule batches were mixed and split into three samples (Figure 35.12a–c) in which the capsules look similar in white light microscopy. Through de-coding of the stained fluorescence dyes (Figure 35.12d–f) by fluorescence microscopy it was possible to distinguish the different types of core–shell materials in the samples. To confirm the de-coding HRP and GOx specific substrates were added, causing typical staining of capsules (Figure 35.12g–i). The results obtained in this work indicate the potential and capabilities of the inwards build-up self-assembly technique to create multifunctional layer assemblies for encapsulation and decoding.

## 35.4 Conclusion and Outlook

The encapsulation of biomolecules within core–shell materials prepared by RP-LbL or inwards build-up was demonstrated to be superior to aqueous LbL in terms of biomolecule loading and encapsulation efficiency during the encapsulation process due to the insolubility of biomolecules in the organic phase. Organic solvents have

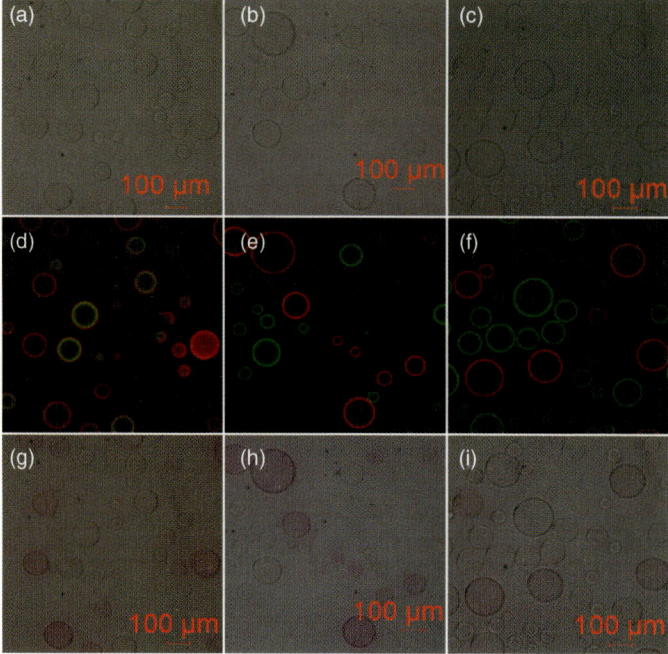

**Figure 35.12** Demonstration of enzymatic viability in core–shell materials encapsulating HRP (labeled red only), encapsulating GOx (labeled green only) and encapsulating BSA (labeled green and red) used as a control. Optical transmission images of (a) HRP and BSA filled capsules, (b, c) HRP and GOx filled capsules, and corresponding overlapping FITC and TRITC fluorescence images of (d) HRP and BSA-filled capsules and (e, f) HRP and GOx-filled capsules before addition of substrates. (g, h) Addition of $H_2O_2$ and Ampliflu Red (AR) to the HRP and BSA-filled capsules (g) and HRP and GOx-filled capsules (h). After 10 s, only the HRP-filled capsules were observed to turn purple. (i) Addition of glucose and AR to the HRP and GOx-filled capsules. After 2 min, only the HRP-filled capsules turned purple. A longer time was required for the $H_2O_2$ produced by the GOx-filled capsules to diffuse to the HRP-filled capsules to cause a color change (Image taken from [31]).

often been claimed to be unfavorable for biomedical applications or industrial processes due to their toxicity, or for economic reasons. In the case of aliphatic alcohols, that are mainly used for RP-LbL and inwards build-up, toxicity is not an issue as some might be consumed and are "Generally Recognized as Safe" (GRAS) by the FDA. The use of organic solvents might even be economically favorable when one takes into consideration that biomolecule loss using water-based LbL encapsulation can be up to 90%. Encapsulation efficiency might be close to 100% when employing the organic phase based RP-LbL. Furthermore, in the case of lower aliphatic alcohols, evaporation has inherently less energy requirement, fostering easier recycling processes compared to water as solvent. Denaturation of biomacromolecules by organic solvents might be another concern in the biomedical or diagnostic fields. This, however, only applies to proteins, for which the tertiary structure is crucial;

macromolecules, such as DNA, RNA, sacchharides, peptides, hormones (even insulin), and macromolecular drugs are generally not affected by organic solvents. The encapsulation of model enzymes (glucose oxidase (GOx) and horseradish peroxidase (HRP)) with retained biological activity [30, 31] further highlights the potential of using the RP-LbL technique to fabricate core–shell materials for biomedical or diagnostic applications. Much groundwork has been done to establish the RP-LbL and inwards build-up methods as alternative methods to aqueous-based LbL and other LbL encapsulation methods. The focus of this work was to extend the spectrum of template materials to generate LbL microcapsules to those that are not feasible for other methods (e.g., small water-soluble molecules and water-soluble biomacromolecules). Challenges for the future will be to create uniform, ideally spherical, microcapsules by the RP-LbL method from water-soluble template materials. Another interesting development is to perform the LbL process in microfluidic devices; up to date only water-based microfluidics LbL has been demonstrated [42]. Microfluidics-based LbL in water or organic solvents is potentially very useful due to the high degree of control over the LbL process, microcapsule morphology and its speed. The relatively young RP-LbL technique has yet to prove its applicability in biomedicine and other fields, as the 20 years old aqueous LbL technique did in recent years. Since RP-LbL and conventional aqueous LbL can often be used interchangeably for many applications, the RP-LbL might prove more suitable for applications in which minimum loss of often expensive biomacromolecules is crucial.

## References

1 (a) Murua, A., Portero, A., Orive, G., Hernandez, R.M., de Castro, M., and Pedraz, J.L. (2008) *J. Control Release*, **132**, 76–83; (b) Hernandez, R.M., Orive, G., Murua, A., and Pedraz, J.L. (2010) *Adv. Drug Deliver. Rev.*, **62**, 711–730.

2 Becker, A.L., Johnston, A.P.R., and Caruso, F. (2010) *Small*, **6**, 1836–1852.

3 (a) De Cock, L.J., De Koker, S., De Geest, B.G., Grooten, J., Vervaet, C., Remon, J.P., Sukhorukov, G.B., and Antipina, M.N. (2010) *Angew. Chem. Int. Ed.*, **49**, 6954–6973; (b) Hoffman, A.S. (2008) *J. Control Release*, **132**, 153–163.

4 van Dongen, S.F.M., de Hoog, H.P.M., Peters, R.J.R.W., Nallani, M., Nolte, R.J.M., and van Hest, J.C.M. (2009) *Chem. Rev.*, **109**, 6212–6274.

5 Tong, W.J. and Gao, C.Y. (2008) *J. Mater. Chem.*, **18**, 3799–3812.

6 (a) Cayre, O.J., Noble, P.F. and Paunov, V.N. (2004) *J. Mater. Chem.*, **14**, 3351–3355; (b) Laib, S. and Routh, A.F. (2008) *J. Colloid Interface Sci.*, **317**, 121–129.

7 Peyratout, C.S. and Dahne, L. (2004) *Angew. Chem. Int. Ed.*, **43**, 3762–3783.

8 Kim, J.W., Jung, M.O., Kim, Y.J., Ryu, J.H., Kim, J., Chang, I.S., Lee, O.S., and Suh, K.D. (2005) *Macromol. Rapid. Commun.*, **26**, 1258–1261.

9 (a) Sinha, V.R., Singla, A.K., Wadhawan, S., Kaushik, R., Kumria, R., Bansal, K., and Dhawan, S. (2004) *Int. J. Pharm.*, **274**, 1–33; (b) Vehring, R. (2008) *Pharm. Res.*, **25**, 999–1022.

10 (a) Caruso, F., Trau, D., Mohwald, H., and Renneberg, R. (2000) *Langmuir*, **16**, 1485–1488; (b) Trau, D. and Renneberg, R. (2003) *Biosens. Bioelectron.*, **18**, 1491–1499.

11 (a) Yu, A.M., Wang, Y.J., Barlow, E., and Caruso, F. (2005) *Adv. Mater.*, **17**, 1737; (b) Wang, Y.J. and Caruso, F. (2004) *Chem. Commun.*, 1528–1529; (c) Volodkin, D.V., Larionova, N.I., and Sukhorukov, G.B. (2004) *Biomacromolecules*, **5**, 1962–1972;

(d) Mak, W.C., Cheung, K.Y., and Trau, D. (2008) *Adv. Func. Mater.*, **18**, 2930–2937; (e) Mak, W.C., Cheung, K.Y., and Trau, D. (2008) *Chem. Mater.*, **20**, 5475–5484; (f) Petrov, A.I., Volodkin, D.V., and Sukhorukov, G.B. (2005) *Biotechnol. Progr.*, **21**, 918–925; (g) Kreft, O., Prevot, M., Mohwald, H., and Sukhorukov, G.B. (2007) *Angew. Chem. Int. Ed.*, **46**, 5605–5608.

12 (a) Ghan, R., Shutava, T., Patel, A., John, V.T., and Lvov, Y. (2004) *Macromolecules*, **37**, 4519–4524; (b) Lvov, Y., Antipov, A.A., Mamedov, A., Mohwald, H., and Sukhorukov, G.B. (2001) *Nano Lett.*, **1**, 125–128; (c) Kreft, O., Georgieva, R., Baumler, H., Steup, M., Muller-Rober, B., Sukhorukov, G.B., and Mohwald, H. (2006) *Macromol. Rapid Commun.*, **27**, 435–440; (d) Georgieva, R., Moya, S., Hin, M., Mitlohner, R., Donath, E., Kiesewetter, H., Mohwald, H., and Baumler, H. (2002) *Biomacromolecules*, **3**, 517–524; (e) Zhu, H.G., and McShane, M.J. (2005) *Langmuir*, **21**, 424–430; (f) Gao, C.Y., Donath, E., Mohwald, H., and Shen, J.C. (2002) *Angew. Chem. Int. Ed.*, **41**, 3789–3793; (g) Gao, C.Y., Liu, X.Y., Shen, J.C., and Mohwald, H. (2002) *Chem. Commun.*, 1928–1929.

13 De Temmerman, M.L., Demeester, J., De Vos, F., and De Smedt, S.C. (2011) *Biomacromolecules*, **12**, 1283–1289.

14 Beyer, S., Bai, J., Blocki, A.M., Kantak, C., Xue, Q., Raghunath, M., and Trau, D. (2012) Assembly of Biomacromolecule Loaded Polyelectrolyte Multilayer Capsules by Using Water Soluble Sacrificial Templates Soft Matter, 8(9), 2760–2768.

15 Kim, B.S., Lebedeva, O.V., Koynov, K., Gong, H.F., Glasser, G., Lieberwith, I., and Vinogradova, O.I. (2005) *Macromolecules*, **38**, 5214–5222.

16 Poptoshev, E., Schoeler, B., and Caruso, F. (2004) *Langmuir*, **20**, 829–834.

17 Tuo, X.L., Chen, D., Cheng, H., and Wang, X.G. (2005) *Polym. Bull.*, **54**, 427–433.

18 (a) Wang, L.Y., Wang, Z.Q., Zhang, X., Shen, J.C., Chi, L.F., and Fuchs, H. (1997) *Macromol. Rapid Commun.*, **18**, 509–514; (b) Kharlampieva, E., Kozlovskaya, V., and Sukhishvili, S.A. (2009) *Adv. Mater.*, **21**, 3053–3065; (c) Such, G.K.,

Johnston, A.P.R., and Caruso, F. (2011) *Chem. Soc. Rev.*, **40**, 19–29.

19 (a) Kamineni, V.K., Lvov, Y.M., and Dobbins, T.A. (2007) *Langmuir*, **23**, 7423–7427; (b) Nakashima, T., Zhu, J.A., Qin, M., Ho, S.S., and Kotov, N.A. (2010) *Nanoscale*, **2**, 2084–2090.

20 (a) Dobbins, T.A., Kamineni, V.K., and Lvov, Y.M. (2007) *Langmuir*, **23**, 7423–7427; (b) Kotov, N.A., Nakashima, T., Zhu, J.A., Qin, M., and Ho, S.S. (2010) *Nanoscale*, **2**, 2084–2090.

21 Klooster, N.T.M., Van der Touw, F., and Mandel, M. (1984) *Macromolecules*, **17**, 2070–2078.

22 Kitano, T., Taguchi, A., Noda, I., and Nagasawa, M. (1980) *Macromolecules*, **13**, 57–63.

23 Bharadwaj, S., Montazeri, R., and Haynie, D.T. (2006) *Langmuir*, **22**, 6093–6101.

24 (a) Arnett, E.M., Haaksma, R.A., Chawla, B., and Healy, M.H. (1986) *J. Am. Chem. Soc.*, **108**, 4888–4896; (b) Hart, M., Fuller, G., Brown, D.R., Dale, J.A., and Plant, S. (2002) *J. Mol. Catal. A-Chem.*, **182**, 439–445.

25 Borodina, T.N., Grigoriev, D.O., Andreeva, D.V., Mohwald, H., and Shchukin, D.G. (2009) *ACS Appl. Mater. Interface.*, **1**, 996–1001.

26 (a) Stockton, W.B., and Rubner, M.F. (1997) *Macromolecules*, **30**, 2717–2725; (b) Sukhishvili, S.A., and Granick, S. (2001) *Macromolecules*, **35**, 301–310.

27 Bai, J., Beyer, S., Mak, W.C., and Trau, D. (2009) *Soft Matter*, **5**, 4152–4160.

28 Bai, J., Beyer, S., Toh, S.Y., and Trau, D.W. (2011) *ACS Appl. Mater. Interface.*, **3**, 1665–1674.

29 Lin, L.Y., Beyer, S., Wohland, T., Trau, D., and Lubrich, D. (2010) *Angew. Chem. Int. Ed.*, **49**, 9773–9776.

30 Mak, W.C., Bai, J., Chang, X.Y., and Trau, D. (2009) *Langmuir*, **25**, 769–775.

31 Bai, J.H., Beyer, S., Mak, W.C., Rajagopalan, R., and Trau, D. (2010) *Angew. Chem. Int. Ed.*, **49**, 5189–5193.

32 Kohler, K. and Sukhorukov, G.B. (2007) *Adv. Funct. Mater.*, **17**, 2053–2061.

33 Gaponik, N., Radtchenko, I.L., Sukhorukov, G.B., Weller, H., and Rogach, A.L. (2002) *Adv. Mater.*, **14**, 879–882.

# 36
## Stimuli-Responsive Polymer Composite Multilayer Microcapsules and Microchamber Arrays

*Maria N. Antipina, Maxim V. Kiryukhin, and Gleb B. Sukhorukov*

**Abbreviations**

| | |
|---|---|
| LbL | layer-by-layer |
| PEM | polyelectrolyte multilayer |
| PMMA | poly(methyl methacrylate) |
| EDTA | ethylenediaminetetraacetic acid |
| PAH | poly(allylamine hydrochloride) |
| PSS | poly(styrene sulfonate) |
| PDDA | poly(dimethyldiallylamide) |
| PVPON | poly(N-vinylpyrrolidone) |
| PMA | poly-(methacrylic acid) |
| TA | tannic acid |
| FITC | fluorescein isothiocyanate |
| PVCL | poly(N-vinylcaprolactam) |
| PNIPAM | poly(N-isopropylacrylamide) |
| DNA | deoxyribonucleic acid |
| PDADMAC | poly(diallyldimethylammonium chloride) |
| TRITC | tetramethylrhodamine isothiocyanate |
| PFS | poly(ferrocenylsilane) |
| pARG | poly-L-arginine |
| DEXS | dextran sulfate |
| HFMF | high frequency magnetic field |
| MF | melamine formaldehyde |
| PS | polystyrene |

*Multilayer Thin Films: Sequential Assembly of Nanocomposite Materials*, Second Edition.
Edited by Gero Decher and Joseph B. Schlenoff.
© 2012 Wiley-VCH Verlag GmbH & Co. KGaA. Published 2012 by Wiley-VCH Verlag GmbH & Co. KGaA.

## 36.1
## Introduction

Stimuli-responsive materials are under extensive investigation due to emerging demand for different sensing devices, controlling systems and actives delivery.

As a rule, stimulus response of a polymer material is generated either via interactions at the molecular level causing macroscopic structural change in a single component system, or via different response of constituents to an external trigger in two- or multicomponent materials, resulting in mechanical stress and macroscopic structural change. For instance, a light-responsive functional polymer incorporated into a "base" polymer may be able to cause shape response of the entire material.

A majority of the responsive materials developed so far are susceptible to such triggers as pH and redox potential, chemicals (salt, glucose), temperature and mechanical load. Double component pH-, redox-, and chemical-responsive materials are mainly designed in a way that a triggering stimulus or compound interferes and weakens the interaction between the constituents. More recent development is polymer materials that can be remotely triggered by physical impact, for example, magnetic field, or laser light irradiation. On the submicron level there is quite a limited number of options to introduce a stimuli-responsiveness, apart from linking of functional groups or incorporation of nanoparticles. In submicron-sized systems, like liposomes or block-co-polymer vesicles, the responsiveness can be achieved by incorporation of light- or pH-sensitive compounds in their structures. Changes in external pH or light irradiation result in structure unbalancing and affect the integrity of the whole construct. These systems have already been applied in drug delivery. Incorporation of magnetic nanoparticles into vesicles or liposomes adds another important functionality to the system; it becomes susceptible to magnetic field and can be remotely navigated. Despite substantial advances in the field, liposomes and vesicles have limited versatility and are not of multiple use. There is no universal component to build in the constructs enabling a larger variety of applications. Thus, the systems possessing multiple functionalities are under development.

At present, the most intensive ongoing elaboration of the multifunctional carriers employs layer-by-layer (LbL) technology which allows tailoring of different functionalities through assembly of functional constituents. LbL technology appears as a generic and versatile approach to achieve many functions in one entity. Moreover, the uniqueness of the LbL technique is the possibility to assemble multilayers on various substrates of different shapes and curvature, such as organic and inorganic colloids, flat wafers and patterned surfaces. Multilayers repeat surface topology in almost nanometer precision. This opens up plenty of opportunities to design micron and submicron confined carriers of any desired geometry. In this chapter we focus on recent advances in the fabrication of multilayer capsules and chamber arrays with stimuli-responsive restricting walls, including response to mechanical load, to chemicals, and to externally induced impact by ultrasound, light and magnetic field. We particularly aim at describing the design strategies, loading routines and control release properties of the multilayer capsules and chambers and discuss their potential application areas.

Besides stimuli-responsive particles and capsules, there is a demand in the art for materials enabling permeation of certain compounds in response to changeable conditions of the surrounding media. Indeed, a number of responsive coatings, such as those with open/closed pores in the underlying membrane, or those changing the surface properties once subjected to chemical modification have been reported. These materials are out of the scope of this chapter, which rather focuses on microcarriers for packaging and control release delivery of actives.

## 36.2
## Fabrication of Stimuli-Responsive LbL Microcapsules

The basic principle of multilayer polymeric composite film assembly is the alternate adsorption and complexation of complementary macromolecules and/or nanosized species on a substrate surface [1]. So far, a variety of interactions and mechanisms have been utilized for fabrication of multilayer films. Among those, electrostatic interactions of oppositely charged polyelectrolytes [2], hydrogen bonding [3–5], covalent bonding [6–11], and specific recognition [12, 13] can be named. Formation of confined capsules is usually performed on a sacrificial template of water-dispersed colloids, such as organic and inorganic solid microparticles, and hydrogel microbeads [2, 14–16]. LbL coating of water-dispersed oil microdroplets [17–19] attracts considerable attention as a multilayer film improves the stability of the oil microdroplets towards coalescence, protects them from oxidative degradation [20] and slows down the speed of the gravitational separation in a double phase system [21]. Recently, patterned poly(methyl methacrylate) (PMMA) substrates have been utilized as a two-dimensional template to obtain an array of polymeric multilayer microcapsules on a flat support [22].

The fabrication of an encapsulating system of any kind would not make sense without elaborating a certain mechanism for capsule loading and subsequent release of the payload. In other words, control over capsule integrity and/or permeability for the payload is required. Ideal capsules are highly permeable in the stage of actives infiltration and loading, tightly sealed on the way to a target, and then again permeable for the loaded species at the target site or they simply disassemble releasing the payload.

Synthetic and bio-originated polymers are the most popular building blocks used for assembly of multilayers arranged as a coating film on a planar or colloidal substrate. Properties of polymers, such as chemical structure, melting and glass transition temperature, $pK_a$, and solubility in different solvents predetermine the structural stability of the multilayer complex in response to changes in temperature, pH, ionic strength, redox potential, solvent polarity, and concentration of specific chemicals, for instance, glucose or $CO_2$. These changes in external conditions, the so-called external stimuli, may increase the capsule permeability for payload by weakening interlayer interactions, or cause total decomposition of the multilayer structure. For instance, pH is an obvious stimulus to alter the integrity of capsules assembled from weak polyelectrolytes and/or hydrogen-bonded multilayers. Ionic strength shields charges that weaken electrostatic interactions. Redox potential influences the integrity of the multilayers with redox sensitive bonds or groups.

The responsiveness to an external stimulus plays a key role upon loading of the polymeric multilayer capsules with actives. Generally, molecules of interest can be loaded into polymer multilayer capsules by means of two different strategies: "preloading" [23–26] and "postloading" [27, 28] (Figure 36.1). Preloading is usually done by absorbing payload molecules on a porous inorganic template, such as $CaCO_3$ or mesoporous silica microparticles. The saturated template is subsequently coated with a polymeric multilayer shell and then dissolved by an appropriate solvent. For instance, ethylenediaminetetraacetic acid (EDTA) and HF treatment can be used to dissolve $CaCO_3$ and mesoporous silica, respectively. The pH in a capsule suspension changes substantially in both cases. Thus, the pH becomes acidic upon HF treatment of silica, and alkaline upon $CaCO_3$ decomposition by EDTA. Apparently, it is essential that the capsule multilayer complex does not decompose in the corresponding pH region to avoid a considerable loss of the preloaded active. In the "postloading" approach, alternated permeability of prefabricated capsules is utilized to infiltrate and then entrap the molecules of interest. In such a way, postloading of polymeric multilayer capsules has been performed in response to external changes in pH, ionic strength, temperature and solvent polarity [29–32]. The composition of a multilayer film has to be decided in accordance with the stimulus the capsules will undergo in each particular application. Thus, capsules to be used in consumer care and foodstuff

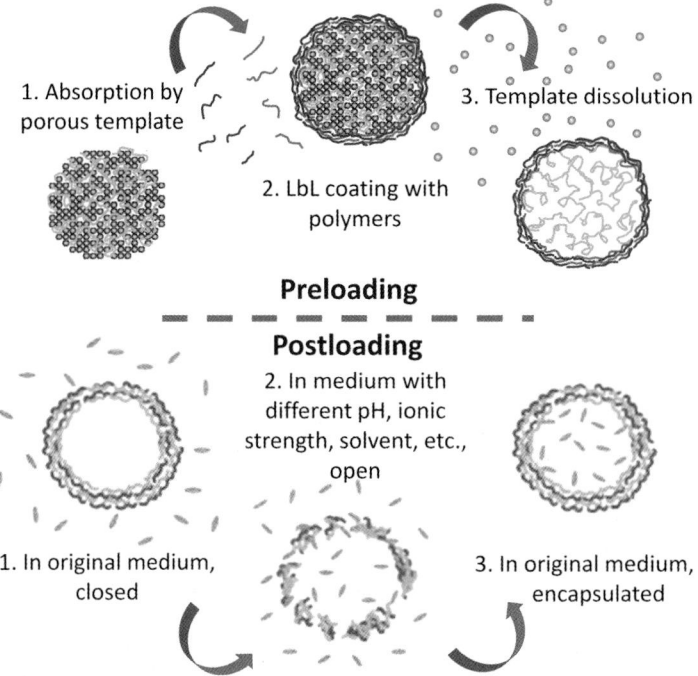

**Figure 36.1** General scheme of capsule loading.

products which are subsequently diluted with water (e.g., detergent powder, fabric softener, shower gels and shampoos, soluble beverages, etc.) could be designed salt-responsive, pH-responsive and water-soluble. Glucose-responsive capsules may also be considered for food related applications. Another extensively used release stimulus in food is a temperature increase, which happens, for instance, upon heating of tertiary processed foods. The changes in physical and chemical conditions in leaving systems have insufficient ranges to induce release from the majority of polymeric capsules developed so far. However, redox- and pH-responsive capsules may find some applications in intranuclear gene delivery and oral drug delivery to the gut. Apparently, biocompatibility is an essential requirement for the capsules to be used in foodstuffs, skin care products, and *in vitro* and *in vivo* drug delivery. DNA [33], polypeptides and polysaccharides [34] and proteins [35, 36] have been used as building blocks for LbL film assembly to produce biodegradable capsules.

Another type of sensitivity available for polymeric multilayer capsules involves remote physical stimuli, such as light and magnetic field. To produce a light- or magnetic field-sensitive capsule, susceptible species (usually metal nanoparticles) are incorporated within the capsule polymeric network. In such a way, one may not only influence the permeability to guide the capsules but also control their location [37]. Patterning of capsules on a flat solid substrate allows direct rupturing of them individually by a microneedle [38].

The scope of this chapter is to give an overview of LbL-formed capsules with permeability controlled by exposure to external conditions, bioactive molecules and physical impact, as well as to describe a mechanism of film response to each particular stimulus.

### 36.2.1
### pH-Responsive Capsules

In the case of polyelectrolyte multilayer (PEM) capsules, pH-responsiveness is attributed to weak polyelectrolytes used as wall constituents. As the ionization properties of the weak polyelectrolytes are changeable, a shift in the environmental pH can cause uncompensated positive or negative charges within the multilayer at pH close to the $pK_a$ of the polymer due to protonation or deprotonation of the charged groups. Accumulation of like-charges leads to stronger repulsion in the polymer network. Thus, capsules undergo reversible swelling [27, 39, 40], increase in permeability for macromolecules [41], and changes in the thickness of the walls [42]. Further increase of pH in the case of polybase, or decrease in the case of polyacid, leads to capsule disassembly.

Being $pK_a$-dependent, the regions of permeable/impermeable states of PEM capsules are, therefore, different for each polyelectrolyte couple, which opens up a possibility to match the pH of the highest release rate required in a specific application. However, as for the majority of currently available PEM capsules, the pH regions of increased permeability are shifted to a very high or very low value. This happens because the actual $pK_a$ values of the polymer in multilayer films may be substantially different from those in solution. For instance, poly(allylamine hydrochloride) (PAH), one of the most frequently used weak polyelectrolytes for fabrication of polyelectrolyte capsules has a $pK_a$ of 8.7 in salt-free solution. When PAH is used in alternation with poly(styrene sulfonate) (PSS), its apparent $pK_a$ changes to 10.7 [43]. Presumably, the lower dielectric environment within the multilayer film results in suppressed deprotonation of polyelectrolyte [4]. PSS/PAH capsules start to swell when the pH is approaching 11 and disassemble when the medium pH exceeds 12 [27]. The pH-dependent integrity of the PEM shell enables infiltration and loading of high molecular weight substances through multiple pores visualized by AFM. [29]. In such a way, the PSS/PAH polyelectrolyte capsules were filled with various macromolecules in the acidic region, where PAH molecules are highly ionized [29, 44]. PSS/PAH capsules did not swell at low pH values, presumably because of the high ionization degree of PSS. In another example, PMA/PAH capsules [39] composed of a couple of weak polyelectrolytes were characterized by pH-responsive swelling at both low and high pH. Capsule swelling and dissolution were observed at pH values below 2.5 and above 11.5, due to the lack of electrostatic interactions between the polymers destabilizing the capsules. A comparative permeability study of capsules based on tannic acid (TA) showed a clear difference in pH-dependent behavior, if a strong polyelectrolyte (poly(dimethyldiallylamide) (PDDA)) or a weak polyelectrolyte (PAH) was coupled with the polyphenolic compound [45].

Hydrogen-bonded multilayer films of neutral polymers have gained considerable interest as offering the possibility to avoid the use of cytotoxic polycations. The most studied nonionic polymer with strong hydrogen-accepting carbonyl groups is poly(N-vinylpyrrolidone) (PVPON). Protonated hydroxy groups of weak polyacids, for example, poly(methacrylic acid) (PMA) [46] and TA [4] become excellent hydrogen donors for binding with carbonyl groups of pyrrolidone rings (Figure 36.2). LbL assembly of hydrogen-bonded multilayers is performed at low pH (pH 2). These films disintegrate on pH increase due to ionization of coupled weak polyacid and electrostatic repulsion caused by ionization.

Similar to polyelectrolyte films, hydrogen-bonded multilayers are stable at pH values higher than the $pK_a$ of the coupled polyacid. For instance, the critical dissolution pH of PVPON/TA multilayer film is almost 1 pH unit higher than the $pK_a$ of TA in solution [4].

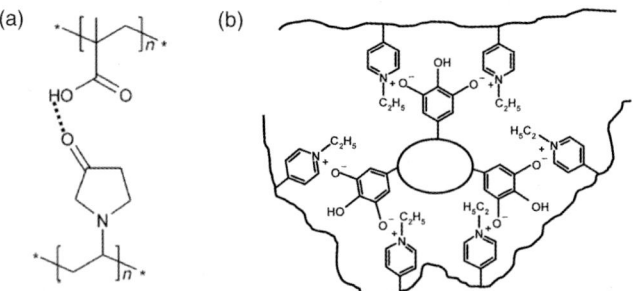

**Figure 36.2** Structures of the polymers PMA and PVPON, which form a complex through hydrogen bonding (a); schematic presentation of hydrogen-bonding interactions between carbonyl groups of PVPON and phenolic hydroxyl groups of TA (b) [4].

The pH stability intervals of hydrogen-bonded multilayers can be controlled through selection of hydrogen donor polyacids with different $pK_a$ values, and by varying the chemical nature and hydrophobicity of a neutral polymer component [4]. Covalent stabilization of the hydrogen-bonded film prevents its irreversible decomposition. In such a way, carbodiimide was coupled with the carboxy groups of the PMA by using ethylenediamine as a cross-linking agent [47]. The resulting cross-linked capsules were stable over the whole pH range, and released their PVPON fraction at alkaline pH values, since it was no longer retained by the now deprotonated PMA. The resulting single-component capsules exhibited interesting pH-dependent shrinking/swelling, with a steep reversible transition from a shrunken to a swollen state when the pH value increased above 6. Furthermore, carbodiimide-assisted cross-linking of hydrogen-bonded multilayers was utilized to produce pH-responsive two-component capsules of amino-containing PVPON and PMA (PVPON-NH2-20/PMA) [48]. These capsules reversibly responded to changes in environmental pH by variations in the capsule diameter (Figure 36.3). As for the capsule permeability, (PVPON-NH2-20/PMAA)$_7$ shells excluded fluorescein isothiocyanate (FITC) labeled dextrans of $M_w = 4$ kDa and above at pH 4.6, and were able to selectively control permeability of macromolecules, exclusively allowing permeation of FITC-dextran with $M_w = 4$ kDa and rejecting higher molecular weight permeants. Similar to the PVPON/PMA system, the permeability of capsules of TA assembled

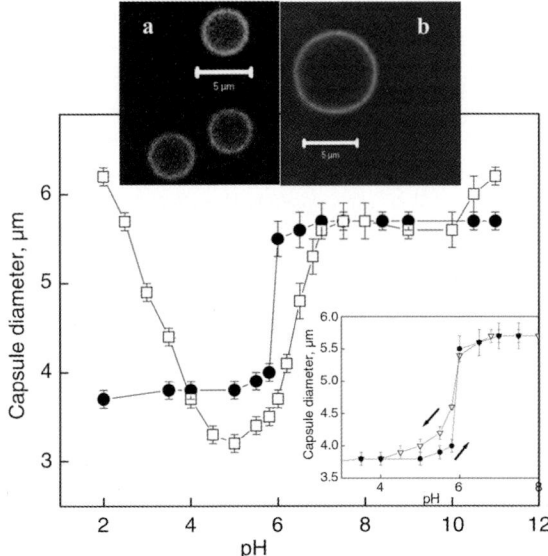

**Figure 36.3** CLSM images of (PVPON-NH2-20/PMAA)$_7$ (a) and (PMAA)$_7$ (b) capsules at pH 2. pH dependence of the diameter of the (PVPONNH2-20/PMAA)$_7$ capsules (filled circles) and (PMAA)$_7$ capsules (open squares) cross-linked for 18 h. The inset shows hysteresis of (PVPON-NH2-20/PMAA)$_7$ capsule size upon increasing (filled circles) and decreasing (open triangles) pH. The pH values were supported by 0.01 M phosphate buffer [48].

with a range of neutral polymers, PVPON, poly(N-vinylcaprolactam) (PVCL) or poly (N-isopropylacrylamide) (PNIPAM) can be controlled by changing the environmental pH [49]. These shells show reversible pH-triggered changes in surface charge and permeability towards FITC-labeled polysaccharide molecules.

The pH-responsive erasable capsules are envisioned to be applied for rapid release of actives as is required for instance in some foodstuff and laundry products, whereas the stable capsules with pH-dependent release profile can find an application in drug delivery [50].

### 36.2.2
### Salt-Responsive Capsules and Capsule Fusion

The effect of salt on the permeability and morphology of PEM films and capsules is explained by a shielding of the charges on the film-forming polyelectrolytes [30, 51–54]. Polyelectrolyte capsules made of deoxyribonucleic acid (DNA) and spermidine were the first reported system decomposable by salt [55]. It is known that DNA–spermidine interactions are reduced at higher ionic strength. When the DNA/spermidine capsules were immersed in a solution containing 5 M of salt, the multilayers completely dissolved, leading to destruction of the capsules.

Addition of salt remarkably increases the permeability of the PEM membrane for macromolecules due to breakage of bonds between oppositely charged polyelectrolyte segments which allows one to postload the capsules [29, 30] (Figure 36.4). The influence of ionic strength on PEM capsules was traced by measuring dye molecule diffusion through PEMs as a function of varying salt concentration [29, 31]. The permeability coefficient was found to have a very strong nonlinear dependence on salt. In the presence of salt, capsules can thus be filled with molecules of interest; the encapsulated molecules can be later released by similar treatment with salt. However, the filling efficiency is rather low and that might be a limitation for this type of encapsulation.

Although addition of salt increases the permeability of the polyelectrolyte layer, an opposite effect can be achieved if one polymer in the pair possesses hydrophobic groups, for example PSS. Upon increase in ionic strength, the electrostatic interactions between the charged groups decrease, and that allows hydrophobic groups of PSS to shrink the capsules. Based on this principle, encapsulation of FITC-dextran $M_w = 2000$ kD by opening and reclosing capsules made of PSS and PDADMAC was performed [51].

Salt-responsive polyelectrolyte capsules will probably not find any application in food and consumer care products, which mainly require capsule stability at high salt concentration and disintegration at low ionic strength (release upon dilution). However, some salt responsive drug delivery systems have been proposed [56–59]. The main obstacle to salt-responsive release is apparently there are no ionic strength variations in the human body. The exception is the intracellular concentration of a number of ions, which differs significantly from the extracellular one. For example, the concentration of sodium, calcium and potassium inside the living cells (respectively, ~11 mM, ~230 nM, ~115 mM) is significantly different from the

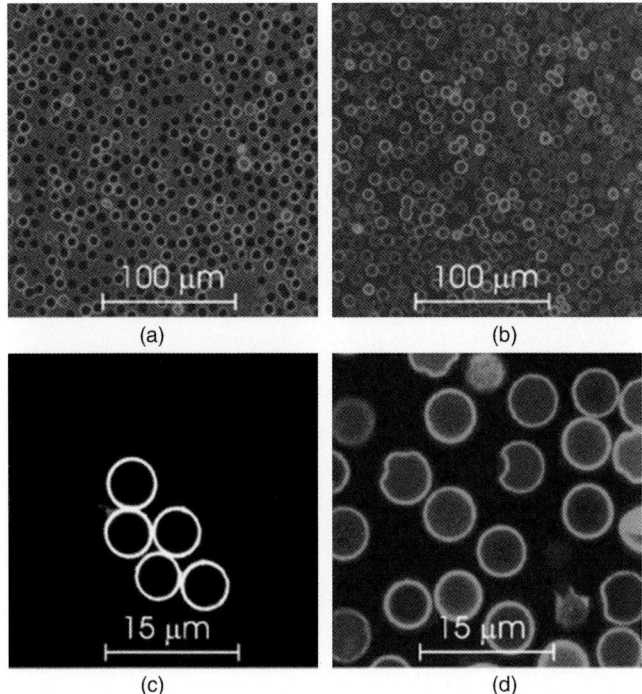

**Figure 36.4** Confocal laser scanning microscopy images showing: (a) capsules in a solution of $5 \times 10 \pm 3$ M PAH ± rhodamine and $10 \pm 2$ M NaCl after 24 h incubation time; (b) capsules in a solution of $5 \times 10 \pm 3$ M PAH ± Rhodamine ($M_w$ 70 000 g mol$^{-1}$) after 24 h incubation time; (c) capsules, made of 8 layers of PSS/PAH, colored by Rhodamine 6G; (d) capsules, loaded with $5 \times 10 \pm 3$ M PAH ± Rhodamine [30].

concentration in serum (respectively, ~140 mM, ~2 mM, ~4.5 mM) [60–62]. This phenomenon gives room to explore an application of salt-responsive capsules.

Additionally, salt-responsiveness can be utilized for significant improvement of encapsulation efficiency performed upon capsule heating [31] (Figure 36.5).

As has already been mentioned, salt concentration changes the morphology of the planar PEM films drastically [54]. In a 3D system, such as PEM capsules, these morphology changes open up the possibility of capsule fusion at high concentration of salt in an aqueous medium.

Salt-induced capsule fusion has been observed for poly(diallyldimethylammonium chloride) (PDADMAC) and PSS ((PDADMAC/PSS)$_4$) microcapsules during the evaporation of NaCl solution [63]. Figure 36.6 presents a series of snapshots after addition of an excess of 3 M NaCl to a suspension of hollow (PDADMAC/PSS)$_4$ capsules (a) and (PDADMAC/PSS)$_4$ capsules filled with FITC and tetramethylrhodamine isothiocyanate (TRITC) labeled dextran, respectively (b). The elevated concentration of salt creates defects in the membranes, decreases the electrostatic repulsion of polyelectrolytes in the membranes, causes a conformational change

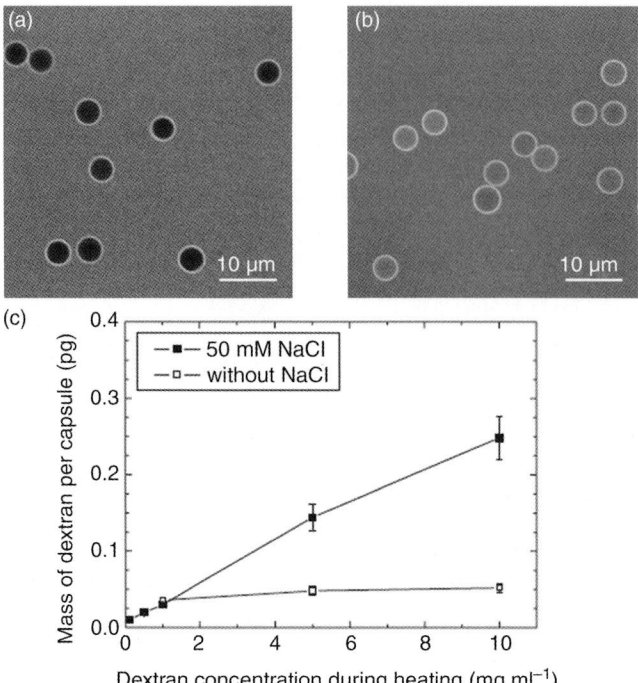

**Figure 36.5** Confocal micrographs of (PDADMAC/PSS)$_4$ capsules incubated for 1 h in 1 mg ml$^{-1}$ FITC-dextran 70 kDa (a) without NaCl and (b) with 50 mM NaCl. (c) Comparison of the encapsulated amount of FITC-dextran 70 kDa per (PDADMAC/PSS)$_4$ capsule as a function of dextran concentration during heating in 0 and 50 mM NaCl [31].

of the polyelectrolytes, and increases the osmotic pressure gradient across the microcapsule membranes. Due to the surface tension of the microcapsules and the hydrophobic interaction of the polyelectrolytes, microcapsules decrease the surface area and finally fuse. During capsule fusion, the inner contents of two differently labeled neutral molecules can diffuse freely and mix (Figure 36.6b). At the same time, the PEM membranes remain impermeable for both probes, enabling fusion without leakage. Fusion of fluorescently labeled and unlabeled capsules displayed "Janus" membranes (Figure 36.7), showing that polyelectrolyte molecules in membranes do not mix considerably during fusion. This observation can be explained by conformation changes of polyelectrolyte molecules in concentrated salt solutions. The polyelectrolyte molecules change conformation from relatively extended to coiled in a more and more concentrated salt solution. Due to the loss of charges and the hydrophobic interaction, the polyelectrolyte molecules entangle, which prevents their free dispersion in a capsule membrane.

The phenomenon of capsule fusion allows one to foresee application of PEM capsules in intracellular delivery, gene transfection and fabrication of artificial cells [64].

## 36.2 Fabrication of Stimuli-Responsive LbL Microcapsules

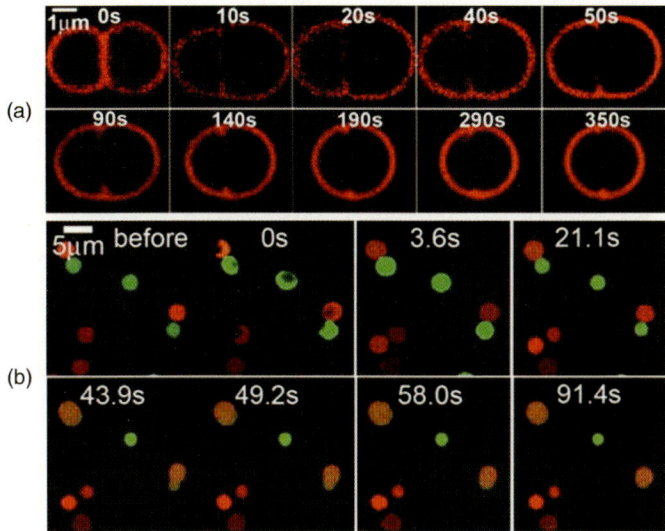

**Figure 36.6** A series of snapshots of the salt-induced microcapsules fusion in an open chamber after 3 M NaCl was added (at $t=0$). Fusion of hollow (PDADMAC/PSS)$_4$ capsules (a) and (PDADMAC/PSS)$_4$ capsules filled with FITC- and TRITC-dextran, respectively (b). The overlaid fluorescence images (b) clearly prove that the orange inner content ($t=22.9$–$91.4$ s) of the fused microcapsules is due to the free diffusion and mixing of the original FITC-dextran (green) and TRITC-dextran (red) [63].

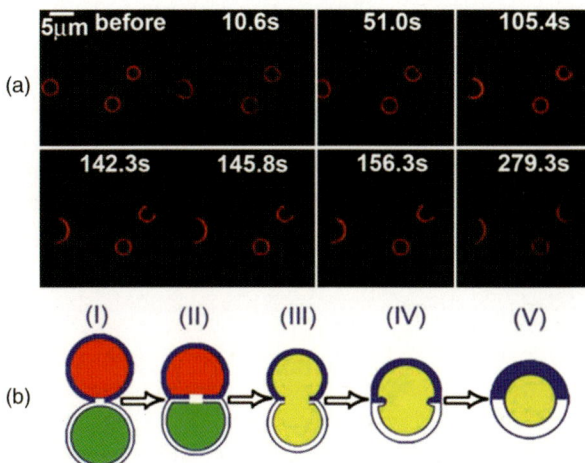

**Figure 36.7** A series of snapshots of the salt-induced microcapsule fusion (a). At $t=0$ a solution of 3 M NaCl was added to a mixture of unlabeled and labeled (PDADMAC/PSS)$_4$ microcapsules (4th layer MRho-PSS). The microcapsules are filled with dextran 70 kDa. Fusion mechanism of microcapsules (b). (I) Salt creates defects and pores in the contacting membranes; (II) pores enlarge by lateral tension; (III) the inner contents of neutral polymers mix, and (IV–V) the polyelectrolyte molecules in the membranes do not mix [63].

### 36.2.3
### Redox-Responsive Capsules

The first redox-responsive PEM capsules were fabricated of anionic and cationic polypeptides [65] containing cysteine groups [66]. Cross-linking of the cysteine thiol groups leads to disulfide bond formation, and the capsules were found to be stable at both neutral and acidic pH. However, after reducing the disulfide bonds, the capsules disassembled at a pH lower than the $pK_a$ of the anionic polypeptides as the multilayers were then no longer stabilized by electrostatic or covalent bonds. In the pioneering work [66], disulfide bond formation in cysteine-containing polypeptide capsules was reversible, suggesting that the controlled formation and cleavage of disulfides could be an element of the polypeptide capsule design and engineering process for various projected applications, including *in vivo* drug delivery.

Another example of the redox-responsive polyelectrolyte capsules is the organometallic capsules based on the water-soluble polyanion and polycation of poly (ferrocenylsilane) (PFS) [67]. The capsules were fabricated by the electrostatic LbL assembly of these polyelectrolytes onto colloidal templates. After template removal, the capsules were incubated in 1 mM ferric chloride ($FeCl_3$) solution as $FeCl_3$ is known for its effectiveness in oxidizing PFS. On exposure to $Fe^{3+}$ ions originating from $FeCl_3$, the PFS capsules exhibited a continuous expansion accompanied by an increasing permeability of the capsules. Expansion of the capsules proceeds in a continuous manner until their final disappearance.

Disulfide bonds were employed for fabrication of the redox-responsive hydrogen-bonded polymer multilayer capsules. Capsules were assembled by hydrogen bond interactions between PMA and PVPON on protein-filled mesoporous silica [14]. Cross-linking through disulfide bonding was accomplished using PMA functionalized with cysteinamine moieties. Capsules cross-linked through disulfide bonds were found to be stable under physiological conditions. However, when the disulfide bonds were reduced, the capsules disassembled readily and released their content. This approach has potential for intracellular drug delivery as thiol–disulfide exchange leading to deconstruction of the capsules can occur with the assistance of intracellular proteins.

### 36.2.4
### Chemical-Responsive Capsules

#### 36.2.4.1 Solvent
Although water is the most common solvent for the assembly and application of capsules, organic solvents can also be used for both purposes. 1 : 1 ethanol/water mixture was found to cause pores in PSS/PAH capsules, accompanied by an increasing permeability of the capsules to macromolecules as, for instance, molecules of urease (545 kDa) (Figure 36.8) [32]. The solvent-induced pore formation was reversible, so that enzyme was retained by the membrane after washing out the ethanol.

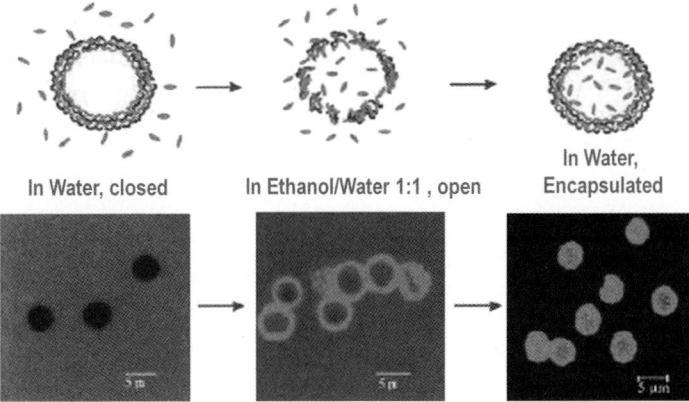

**Figure 36.8** Schematic and CLSM images illustrating permeation and encapsulation of urease–FITC into polyion multilayer capsules in water, in water/ethanol mixture 1 : 1, and the capsule with encapsulated urease again in the water [32].

The role of ethanol in pore creation is not fully understood. It may be related to a partial removal of the hydration water between the polyelectrolytes, resulting in segregation of the polyion network in the presence of ethanol. It is important to mention that solvent can affect the activity of the payload. In the case of urease, the presence of ethanol influences its activity, which was found to be lower than the activity of free urease in bulk solution.

#### 36.2.4.2 Glucose

Glucose-responsive polymer multilayers capsules are very promising candidates for insulin delivery and self-regulated release. In the ideal case, polymer multilayer shells encapsulating insulin should be triggered to release the drug when the glucose level in the blood of a diabetic patient reaches $2\,\mathrm{mg\,ml^{-1}}$ [68]. Release in response to a relevant glucose concentration was demonstrated from the capsules fabricated from polyelectrolytes containing phenylboronic acid as a glucose-sensitive moiety [69, 70]. So far, the pH conditions required for either glucose-dependent permeability [69] or stability [70] of the microcapsules were substantially above physiological values.

#### 36.2.4.3 $CO_2$

Synthesis of a new class of well-defined, homodisperse oligoamine structures based on natural basic amino acids allowed the formation of $CO_2$-responsive capsules [71]. These novel oligoamine "patches" contain exactly 11 amine groups and their overall charge can be reversibly switched in the ordinary pH range (5.5–7.5) between cationic and anionic through interaction with $CO_2$ and carbamate formation. PEM capsules built with these patches as wall constituents showed overall stability of the architecture, but a reversible permeation of large water-soluble polymer species (70 kDa dextran) when bubbled with $CO_2$ (Figure 36.9). Importantly, the oligoamine structures were found to be non-cytotoxic which opens an avenue

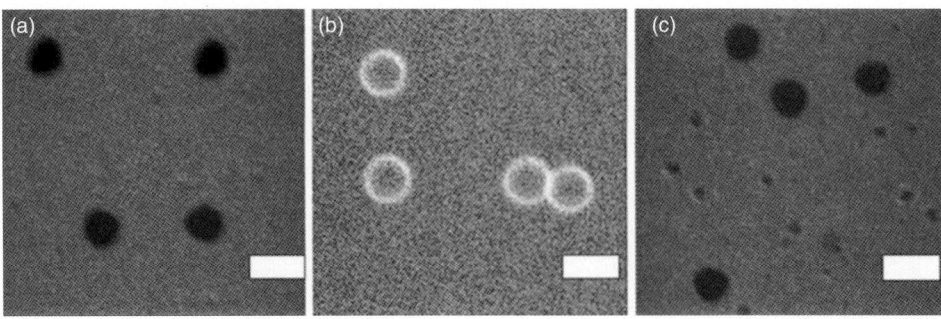

**Figure 36.9** Confocal scanning microscopy images of (PDDA/(PSS/II)5PDDA/PSS) capsules in a solution of FITC-labeled dextran (70 kDa) before (a) and after bubbling $CO_2$ and $N_2$ gas for 20 min (b and c, respectively). In order to visualize the capsules in (b), the orange fluorescence from the Rhodamine-labeled PSS contained in the shell is kept. The scale bars are 2 mm [71].

for a variety of interesting applications in the fields of biomedicine, gene- and drug delivery.

### 36.2.4.4 Enzymes

The development of biodegradable polymer multilayer capsules is of high significance for applications in intracellular drug delivery and gene transfection, drug delivery *in vivo*, bioengineering, and biotechnology [72, 73]. In the pioneering study enzymatic degradation of two different types of capsules was observed [74]. These are the capsules composed of (i) poly-L-arginine (pARG) as the polycation and dextran sulfate (DEXS) as the polyanion, and (ii) the capsules composed of poly(hydroxypropylmethacrylamide dimethylaminoethyl) (p(HPMA-DMAE) and PSS used as the polycation and polyanion, respectively. It was shown, that polyelectrolyte capsules that contain an enzymatically or hydrolytically degradable polycation (DEXS/pARG and PSS/pHPMA-DMAE) are subject to intracellular degradation in VERO-1 cells, while those made from synthetic PSS/PAH are intact upon intracellular uptake. More recently, release of DNA from DEXS/pARG capsules decomposable by a mixture of proteases (Pronase), was also shown *in vitro* [75].

Degradation of capsules made of biopolymers after internalization by cells was also demonstrated for the hyaluronic acid/poly-L-lysine composition [76], suggesting their potential as a carrier for intracellular drug delivery. In another example, enzyme-responsive capsules constructed of DEXS and chitosan were degradable by chitosanase [77]. This type of capsule is not suitable for application in mammals but can be used as a degradable carrier in a large variety of other living organisms.

So far, besides light-responsive capsules (see below), the enzymatically degradable capsules are the only polymer multilayer delivery system that has been tested for intracellular drug delivery and release. The biggest advantage of enzyme degradable capsules is that no external trigger is required for their opening. However, establishing a precise initial time of release and release rate is difficult as the release caused by the capsule disassembly may be also overlaid by the spontaneous release (leakage) of encapsulated molecules [78, 79].

## 36.2.5
### Temperature-Responsive Capsules

The effect of temperature on PEMs is closely related to polymer mobility and interdiffusion [80]. Polymer multilayers are kinetically stable, so an increase in temperature is expected to provide enough thermal energy for the film rearrangements. Surprisingly, PEM films deposited on flat substrates exhibit only negligible changes in thickness upon heating [81]. However, they shrink noticeably if heated at 100% humidity, indicating that water desorption takes place [82]. In the case of LbL-assembled capsules dispersed in water, shrinking in response to thermal treatment above the glass transition temperature, $T_g$, of the polyelectrolyte complex has been observed [28]. Molecular rearrangements enabled by heating significantly densify the capsule membrane and reduce its permeability [31]. In such a way, leakage-free encapsulation of low molecular weight dextran (10 kDa) has been achieved. Dextran molecules freely diffused through the capsule membrane at room temperature and were entrapped within the capsule interior after exposure to heating above the $T_g$ of the polyelectrolyte complex (Figure 36.10).

Similar to the effect of salt on the integrity and structure of the PEMs, temperature-induced shrinkage was explained by the interplay of electrostatic and hydrophobic forces [51]. Thermal shrinking of the PEM capsules is accompanied by stiffening of the walls and, therefore, increasing their mechanical strength. Heating (PDADMAC/PSS)$_n$ capsules (where $n$ is the number of layer pairs) has been shown to induce the reorganization of loosely arranged polyelectrolyte layers into a denser structure [83]. The shrinking behavior can be adjusted by the total number of layer pairs, or by the charge balance: microcapsules with an odd number of layers swell, while those with an even number of layers shrink [28]. Embedding gold nanoparticles within the multilayer enables the thermal capsule shrinking to be shifted to higher

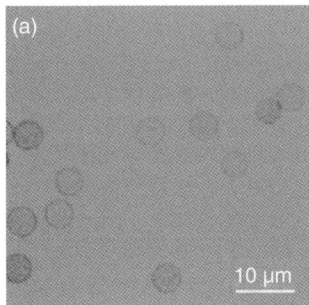
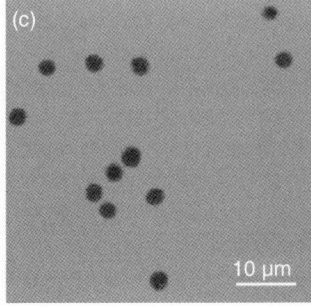
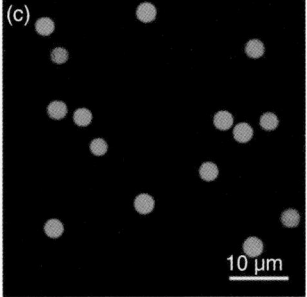

**Figure 36.10** Confocal micrographs of (PDADMAC/PSS)$_4$ capsules (a) before and (b) after heating at 52 °C for 20 min and subsequent incubation in 2 mg ml$^{-1}$ FITC-dextran 10 kDa for 1 h. (c) Confocal micrograph of (PDADMAC/PSS)$_4$ capsules filled with FITC-dextran 10 kDa. Encapsulation was performed in 2 mg ml$^{-1}$ FITC-dextran at 52 °C for 20 min [31].

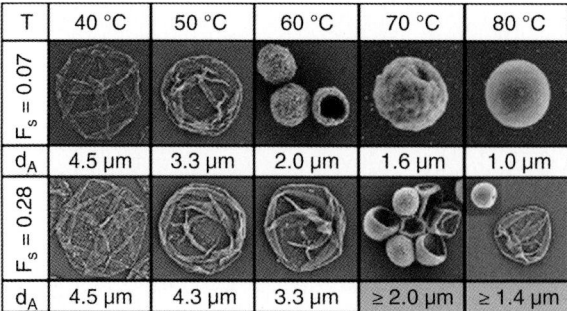

**Figure 36.11** Representative SEM images of thermally shrunk capsules with a gold nanoparticle surface coverage of 7 and 28% ($F_S = 0.07$ and $0.28$), and their average diameter ($d_A$) [85].

temperatures, indicating the importance of the thermal behavior of the multilayer constituents [84]. It was found that capsules containing gold nanoparticles needed significantly higher temperatures to shrink to the same diameter compared with those that were particle-free. A remarkable difference in thermal shrinking profile was observed for capsules containing low (7%) and high (28%) fractions of gold nanoparticles (Figure 36.11) [85]. Capsules with a higher amount of embedded nanoparticles treated at low temperatures (40 and 50 °C) were visibly bigger and displayed denser walls than those containing smaller fraction of nanoparticles. Exposed to 60 °C, the walls of the capsules containing 7% of nanoparticles were dense enough to support themselves and not collapse in the dried state, whereas shells carrying a higher gold load needed further heating before this could be partially achieved.

Responsiveness to the thermal treatment, which leads to densification of the capsule and permeability decrease, is an important finding. This feature of the PEM capsules can be employed for postloading of low molecular weight compounds. As will be described below, a fraction of embedded nanoparticles makes the thermally shrunk capsules susceptible to remote stimuli, such as light, magnetic field and ultrasound, enabling *in situ* controlled release of payload.

## 36.2.6
### Remote Responsive Capsules

#### 36.2.6.1 Magnetic LbL Capsules

Microcapsules with magnetic properties can be employed as promising delivery vehicles to tackle the problem of remotely navigated delivery and controlled release of payload. A common way to introduce the magnetic properties to the PEM capsules is the tailoring of magnetic iron oxides. So far, a few strategies have been developed to incorporate $Fe_3O_4$ nanoparticles into the polymer multilayer capsules. These are the use of stabilized and charged magnetic nanoparticles as a shell constituent [86–89], impregnation from water-based magnetic fluid [90], *in situ* synthesis inside a capsule [91–94], and pre-loading by means of co-precipitation with $CaCO_3$ core followed by LbL coating [95]. Another type of magnetic PEM capsule is represented by the surfactant-stabilized emulsion of ferrofluid [96] additionally covered by the polyelectrolyte shell [97].

In the magnetic field of a permanent magnet, polymer multilayer capsules with incorporated magnetic substances will align [86] or move, driven by the magnetic field gradient [98, 99]. Such remotely driven and controlled motions of capsules seem to be of high technological importance, for example, for separation of the capsules from dispersion [95] or addressed intracellular delivery *in vitro* [98].

Besides the capability to be remotely navigated, magnetic polyelectrolyte multilayer microcapsules can be remotely triggered to release their payload by the high frequency magnetic field (HFMF) [100, 101]. Among the suggested release mechanisms is the alignment of the magnetic particles embedded in the polyelectrolyte shell structure along the direction of the magnetic field creating stress inside the polyelectrolyte network, which becomes loose and facilitates penetration and desorption of macromolecules. Another reason for the increased permeability can be the local heating caused by HFMF, leading to relaxation within the PEM film, followed by pore formation and rupturing of the capsules after prolonged exposure.

The magnetic PEM capsule is a universal delivery system regarding the options for remote manipulation, in that it can be both directed and triggered to release payload by means of remotely applied external physical stimulus.

### 36.2.6.2 Ultrasound-Triggered Release

The effect of ultrasound as a release trigger is attributed to acoustic cavitation in liquids under ultrasonic vibrations with a frequency of more than 20 kHz. As the ultrasonic waves propagate from the sonicator probe tip, microbubbles are generated from the air which was initially dissolved in the fluid. The microbubbles start to oscillate due to the ultrasonic waves and finally collapse, the so-called cavitation. Even at low input power, the collapse of microbubbles in liquid results in an enormous concentration of energy. When the capsules are subjected to ultrasound, a morphological change in the capsule wall should occur due to the creation of shear forces between the successive fluid layers, which results in the disruption of the capsule membrane and release of its payload [102].

The feasibility to use the ultrasonic treatments for opening of polyelectrolyte microcapsules was investigated for the example of PSS/PAH shells [103–106]. Among those, the capsules of $(PAH/PSS)_6$ and $(PAH/PSS)_8$ formed on a polystyrene (PS) template, the melamine formaldehyde (MF) templated $(PSS/PAH)_4$ capsules, and the $(PSS/PAH)_4$ capsules assembled on porous $CaCO_3$ particles were studied. The microcapsules were subjected to a treatment with an ultrasonic probe operating at a frequency of 20 kHz and different values of power output (between 40 and 500 W). The treatment duration varied in different studies from very short (1 s) [105] to quite prolonged (over 1 min) [103]. Regardless of the applied power, treatment duration and amount of polyelectrolyte layers in the shell, ultrasound had a mild effect on the integrity of the capsules formed on organic microparticles [103, 104, 106]. Thus, both the MF and PS templated PSS/PAH shells were just deformed upon ultrasonic treatment. In contrast, almost 100% of the capsules formed on the porous $CaCO_3$ core were ruptured after 10 s exposure to 20–100 W ultrasound [105].

The observed resistance to the ultrasonic-induced rupturing of the microcapsules formed on organic templates compared to those formed on an inorganic core can be possibly explained by incomplete dissolution of the MF [107, 108] and PS [109] core materials. The remnants of the organic templates can provide an additional firmness and mechanical stability to the capsule polyelectrolyte network.

The effect of ultrasonic treatment can be enhanced by passing the acoustic waves through media of different densities. Indeed, the sensitivity of the PSS/PAH capsules assembled on the organic templates was remarkably increased if one or several monolayers of metallic (magnetite or silver) nanoparticles was/were embedded in the polyelectrolyte network [103, 104, 106]. It was shown that a higher amount of embedded nanoparticles leads to decrease in the Young's modulus and shell elasticity. As a result, the capsules with lower Young's modulus were highly sensitive to ultrasound [106].

The capsule sensitivity to ultrasonic power depends on the stiffness and elasticity of the shell network, and thus can be tuned to operate under bio- and medically friendly conditions. Therefore, the application of ultrasound as an external remote trigger to induce the release of payload from the polyelectrolyte microcapsules has an optimistic outlook after further improvement of their sensitivity to the stimulus.

### 36.2.6.3 Optically Addressable Capsules

The first light-responsive capsules were fabricated from PSS/PAH with an embedded monolayer of silver nanoparticles (average diameter ca. 8 nm) and addressed remotely by an 830 nm laser beam [110]. Generally, the effect of light on polyelectrolyte capsules is attributed to absorption and accumulation of light energy by a chromophore embedded in a non-absorbing polyelectrolyte network, heating the chromophore locally and inducing an increase in permeability or rupture of the shell. The mechanism of light-controlled tunable permeability of the polyelectrolyte multilayer shells is shown schematically in Figure 36.12a [111]. Upon illumination, the temperature of the nanoparticles rises as light energy is converted into heat. The heat produced by the absorbing nanoparticles transiently melts the surrounding polymeric network, locally increasing the membrane permeability. If the local temperature rises just slightly above the glass transition temperature ($T_g$) of the polyelectrolyte complex (40 °C in the case of the PSS/PAH film) extensive release of the payload will occur leaving the capsules undamaged. Switching off the laser was proved to cause self-sealing of the polyelectrolyte network, bringing the capsules back to the impermeable stage (Figure 36.12b). By choosing the appropriate nanoparticles and adjusting their size, one can control their absorption properties, regulate heat production, and the capsules response to the irradiating power. Embedded chromophores (metal nanoparticles, nanorods, organic fluorophores etc.) can have either a narrow adsorption spectrum (like gold nanoparticles) or a broad one (silver nanoparticles) enabling them to be addressed in the near-IR region which is the most "friendly" region of the electromagnetic spectrum regarding the influence on biological tissue [111–116]. Tuning of the absorption spectra of nanoparticles can be done through control of their size and assemblies [111, 117–119].

Laser-induced release of molecular cargo from the microcapsules internalized by the living cells, such as MDA-MB-435s breast cancer cells [120, 121], and African

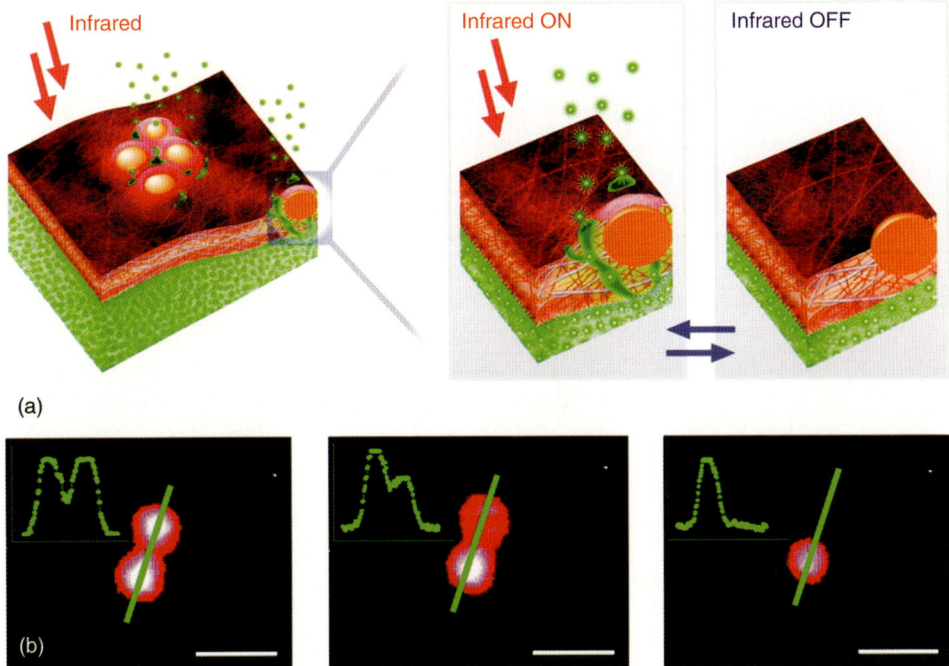

**Figure 36.12** Remote release from microcapsules: (a) schematics of nanoparticle functionalized polymeric nanomembranes opening channels upon laser illumination; (b) a polymeric microcapsule shell acts as a reversible nanomembrane. Upon laser light illumination the microcapsule (left image) partially releases encapsulated polymers and reseals (middle). After the second illumination the microcapsule completely releases its content (right). Profiles in the left upper corner are drawn along the green line. Scale bars correspond to 5 μm [111].

green monkey (Vero) and Chinese hamster ovary (CHO) fibroblasts [122] have been reported. Irradiation of the internalized capsules with near-IR laser light was shown to rupture their shells, releasing the model drug without any major destructive effect on the cell at the particular laser power (2.3 mW, 830 nm wavelength).

Composite microcapsules with iron oxide and gold nanoparticles in the shell can be both remotely delivered to target and triggered to release the payload in the predetermined way [99].

### 36.2.7
### Mechanical Addressing of Individual Capsules

For a number of applications addressable opening of individual selected capsules is crucial. As we wrote above the individual capsules can be located by optical microscopy. They can also be probed mechanically by a tip of the AFM or nanoindentation system. Comprehensive reviews on the force-induced deformation of PEM capsules in aqueous media have been published recently [123, 124].

The response of a capsule toward applied force depends significantly on the permeability of the capsule shell. If the drainage of the inner solution (water and loaded molecules) through the shell is rapid compared with the selected compression speed, decrease of the capsule volume occurs, that is, the capsule deflates under the applied force and gradually releases its content into the surrounding media. Note that the rate of drainage is not related to the diffusion of molecules and ions through the capsule shells discussed in the previous sections. This drainage is caused by excess hydrostatic pressure upon capsule compression, and occurs through the nanopores existing in the multilayer shell or formed by local rupturing of the defects or weak points in the shell (Figure 36.13a). In this case the energy of elastic deformation of the capsule shell is mainly the energy of the shell bending, which is concentrated in a narrow line separating the capsule and the probe, called the bending line. The theory of elasticity predicts the following dependence of the reaction force due to shell bending, $F$, on the relative deformation of spherical capsules, $\varepsilon$:

$$F = \frac{1.8 E h^{\frac{5}{2}}}{R^{\frac{1}{2}}(1-v^2)^{\frac{3}{4}}} \varepsilon^{\frac{1}{2}} = a\varepsilon^{1/2} \tag{36.1}$$

$$= a\, \varepsilon^{1/2} \text{(for } R \gg R_s\text{) [125]}$$

$$F = \frac{\pi E h^2}{2\sqrt{2}} \varepsilon^{\frac{1}{2}} = b\varepsilon^{1/2} \tag{36.2}$$

(for $R \ll R_s$) [126]

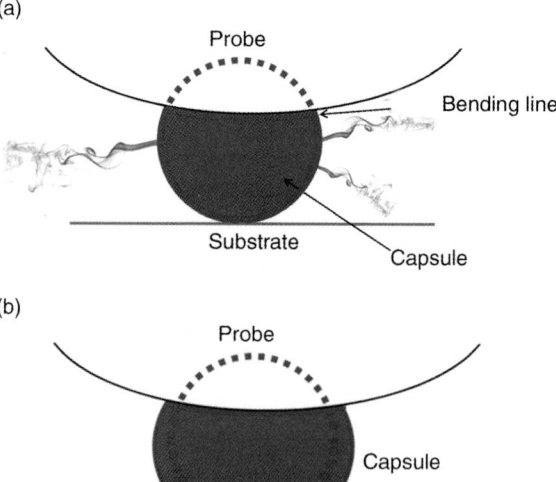

**Figure 36.13** Sketch of a hollow microcapsule in the undeformed and compressed states for highly permeable (a) and impermeable (b) shells.

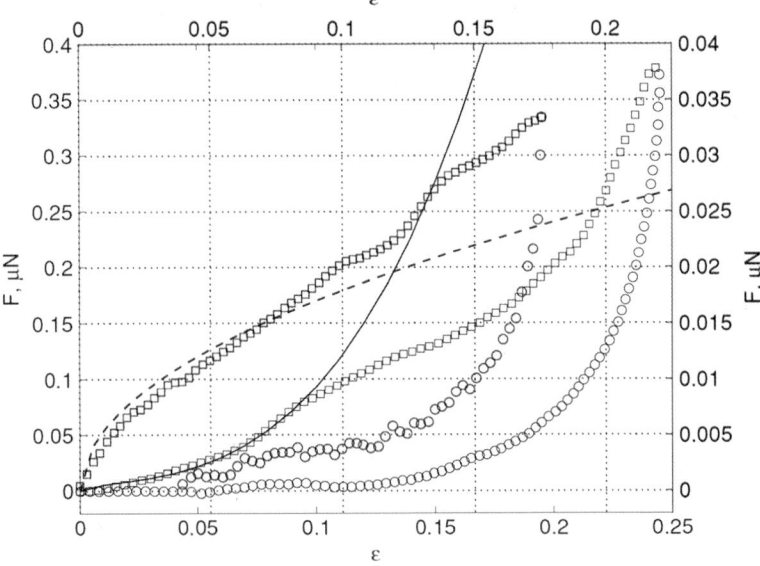

**Figure 36.14** Typical loading (squares)/unloading (circles) curves obtained for impermeable PSS/PAH (blue) and permeable PSS/phosphorus dendrimer (black) capsules. Only every fourth point is shown. Fit to Equation 36.2 and to Equation 36.3 gives an estimate for the Young's modulus $E = 250$ MPa and $E = 70$ MPa, correspondingly [124].

where $E$ is the Young's modulus of the shell material, $v$ its Poisson's ratio, $h$ the shell thickness, $R$ the radius of the capsule, and $R_s$ the radius of the probe. Such behavior under deformation is characteristic for highly porous or fragile capsules, like those made of DNA/PAH or phosphorus dendrimer/PAH multilayer [127, 128]. Figure 36.14 (black) shows a typical force–deformation curve for highly permeable capsules probed with a large glass sphere, the dotted line is the fit to Equation 36.2.

The other scenario happens if the drainage of the inner solution through the capsule shell is slow enough compared with the speed of capsule compression. In this case the shell is considered as almost impermeable, the capsule volume remains constant and the capsule compression leads to stretching of the capsule shell (Figure 36.13b). As a result the capsule reacts towards deformation with a restoring force of an elastic spring, $F_{spring}$, and membrane stretching, $F_{stretch}$.

$$F = F_{spring} + F_{stretch} = \frac{4Eh^2}{3}\varepsilon + \frac{16\pi EhR}{3}\varepsilon^2 \ [123] \tag{36.3}$$

A good example of such capsules with impermeable shells is those made of PSS-PAH multilayers [129–131]. Their deformation is completely reversible and follows the above law until $\varepsilon \sim 0.1$–$0.2$ (Figure 36.14 (blue)). The volume of the capsule does not change in this regime and the concentration of molecules

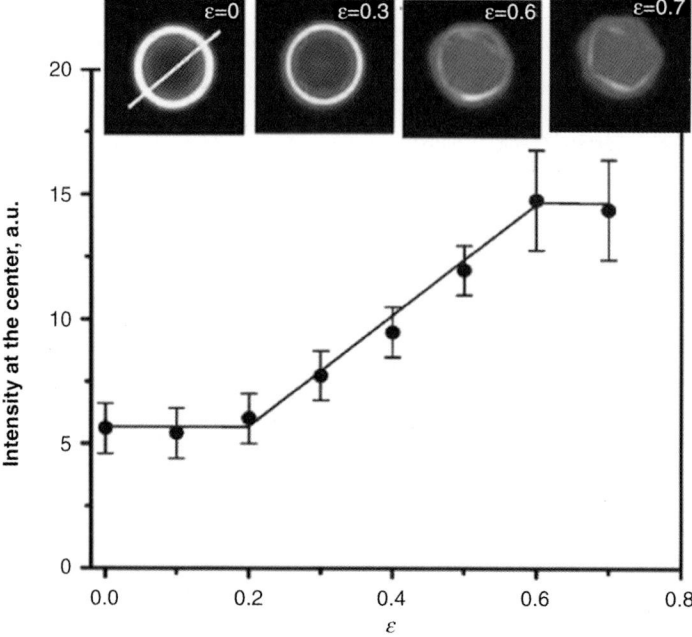

**Figure 36.15** Relative fluoresecence intensity at the center of a PAH-PSS capsule filled with RBITC-labeled PSS molecules ($C_{PSS} \sim 0.10$ mol L$^{-1}$) at different stages of deformation and corresponding images from confocal scanning of the capsule [131].

inside the capsule remains constant, as it is shown for PSS-loaded PAH-PSS capsules (Figure 36.15) [131]. Further increase in the relative deformation from 0.2 to 0.6 induces buckling of the capsules and their volume starts to shrink. However, the shell is still impermeable as the concentration of loaded PSS molecules in the capsules increases. The gradual release of loaded molecules starts only at relative deformation $\varepsilon > 0.6$, which causes major damage of the capsule.

### 36.2.8
### Patterning Polyelectrolyte Capsules

All methods mentioned above rely on micromanipulation of individual capsules that is quite time consuming and technically challenging. Fabrication of a highly-ordered pattern of PEM capsules on a substrate surface could simplify the task significantly. Such patterning of the capsules has been achieved by their selective adsorption on chemically patterned surfaces [132–134] or physical entrapment in the arrays of imprinted microwells [44], as shown in Figure 36.16. It allows addressing of the individual capsules and release of loaded substances on-demand in a programmed way.

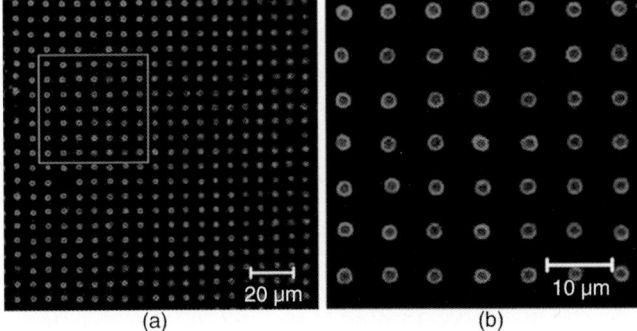

**Figure 36.16** Confocal laser scanning microscopy (CLSM) images of PSS-PAH microcapsules entrapped inside the microwells imprinted on an ETFE film. (a) and (b) are the corresponding zoomed in and zoomed out images [44].

## 36.3
## Microchamber Arrays

### 36.3.1
### Fabrication of Microchambers by LbL Assembly on Imprinted Surfaces

Responsive multilayer capsules could be addressed individually with a number of stimuli releasing a payload on-demand, as was discussed in the previous section. Patterning of the capsules on a substrate is beneficial, easing the positioning of capsules and enabling automatic release in a programmed way. However capsules have certain drawbacks. In most cases they are made of very thin shells in the range of a few nm, so they are rather fragile and can remain intact only while kept under solvent. Drying usually causes irreversible collapse of capsules. Moreover, capsule loading, by pre-adsorption of a payload in the pores of template particles (like porous silica, $CaCO_3$) [24, 135, 136] or post-loading of empty capsules with a payload using responsive permeability of the shells [32, 137] may result in degradation of an active in the harsh conditions required to increase the capsule permeability, or during the core dissolution process. Thus, some other methods of making microcompartments from responsive multilayer films have been developed recently.

In one method a flat PEM film assembled on a sacrificial substrate was brought in contact with a PDMS stamp having an array of microcavities on its surface, coated with an adhesive layer. Dissolving the sacrificial substrate gives an array of microcavities sealed with responsive PEM film [138, 139].

In a second method the PEM film is assembled on a sacrificial template having an array of imprinted microwells on its surface. Then the PEM-coated template is transferred onto a support coated with adhesive layer and the template is dissolved to reveal the structure of a microbubble wrap PEM film, as shown in Figure 36.17 [22]. In this case the whole microcompartment is made of responsive multilayer shell that could be sealed on any substrate: transparent, conductive, flexible, and so on, thus significantly broadening possible applications.

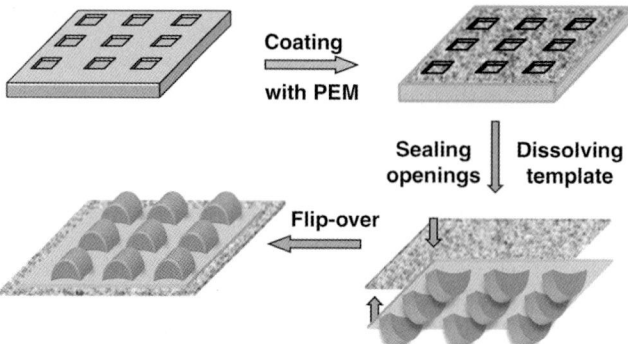

**Figure 36.17** Schematic illustration of the route for patterned PEM films [22].

Let us discuss each of the steps leading to patterned PEM microcompartments in more detail.

#### 36.3.1.1 LbL assembly of the PEMs in confined geometries

Even deposition of polyelectrolytes both on the outer surface of the sacrificial template and the inner surface of the microwells is crucial for successful fabrication of the microchambers. The following factors could cause detrimental nonuniformity of the film thickness:

A thinner PEM film is formed inside the wells if:

- there is poor wetting of these wells with polyelectrolyte solutions;
- the size of the wells is smaller than the dimensions of the polyelectrolyte coils and the physical exclusion of polyelectrolytes occurs;
- there is a depletion of polyelectrolyte concentration across the wells from the surface to the bottom due to the electrostatic interaction between polyelectrolyte coils and charged surfaces confining the wells, or due to the slow diffusion of coils into the wells.

Conversely, a thicker PEM film is formed inside the wells if:

- non-adsorbed polyelectrolytes are incompletely drained out of the wells during the washing steps.

In practice, special precautions should be taken to avoid air bubbles from being trapped inside the wells, for example, by applying the ultrasound prior to LbL assembly. Using polyelectrolyte solutions at high ionic strength helps to overcome the surface charge-induced depletion of polyelectrolyte concentration due to electrostatic screening. Also, the time of the polyelectrolyte adsorption step as well as the time and number of the washing steps should be chosen properly.

#### 36.3.1.2 Dissolving the template and making a free-standing PEM film with an array of standing hollow microchambers

The condition to retain the structural shape of the standing chambers puts additional requirements on the mechanical properties of the PEM shell. The theory of elastic stability predicts the existence of certain critical stress, $\sigma_{cr}$, and load, $P_{cr}$, above which elastic shells collapse by buckling. For cylindrical shells they are [140]:

$$\sigma_{cr} = \frac{Eh}{a\sqrt{3(1-v^2)}} \qquad (36.4)$$

$$P_{cr} = \sigma_{cr} A = 2\pi a h \sigma_{cr} = \frac{2\pi E h^2}{\sqrt{3(1-v^2)}} \qquad (36.5)$$

where $E$ is Young's modulus, $v$ Poisson's ratio, $A$ the shell cross-section, $h$ the shell thickness, and $a$ the radius of a cylinder.

Now let us consider, what kind of stresses PEM microchambers are subjected to (Figure 36.18):

- the load on a chamber due to its roof weight only, $P_{gr} = \pi a^2 h \varrho g$, where $\varrho$ is the PEM film mass density (~1200 kg m$^{-3}$), $g$ is gravitational acceleration (9.81 m s$^{-2}$). One can easily estimate that the load on a chamber made of tens to hundreds nm thick PEM film and having several microns in diameter, $P_{gr} \sim 10^{-15} - 10^{-14}$ N;
- the bending force acting on the chamber walls upon drying. It originates from the difference between the pressures of air outside the microchamber, $P_{air}$, and solvent inside the microchamber, $P_s$, in accordance with the Laplace equation: $P_{surf} = P_{air} - P_s = \gamma \cos\theta \, H/2 \, a^{-1}$, where $\gamma$ is the surface tension of the solvent, $\theta$ the contact angle of the solvent and PEM film, and $H$ is the height of the microchamber. In the case of a common solvent like toluene, with surface tension $\gamma \sim 30$ mN m$^{-1}$, a rough estimation gives bending stress, $P_{surf} \sim 10^{-8}$ N;
- finally, capillary forces could bring the chamber roof into contact with the underlying substrate and cause it to adhere. The adhesion of the chamber's roof to the underlying support, $P_{ad} = \sigma_{ad} \pi a^2$, where $\sigma_{ad}$ is the adhesion strength of PEMs to the support, for example, SiO$_2$ (~ 6 MPa) [141]. This gives $P_{ad} \sim 10^{-5}$ N.

Comparing the stresses one can see that gravity plays a negligible part in the collapse of microchambers. Capillary forces could induce chamber deformation, but

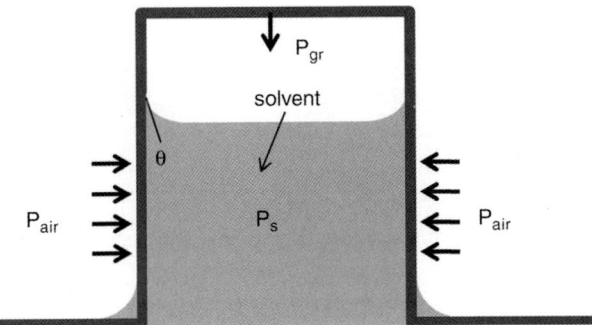

**Figure 36.18** The cross-section of a microchamber depicting the stresses it is subjected to upon dissolving of the template.

the main mechanism behind the collapse is the adhesive contact between the chamber roof and the support. The structure is stable if it is able to spring back to its original standing position. Irreversible collapse occurs if the adhesion strength is larger than the critical stress that the chambers could withstand. The criterion for microchamber collapse could be written as:

$$\sigma_{ad}\pi \cdot a^2 = \frac{2\pi E h^2}{\sqrt{3(1-v^2)}} \tag{36.6}$$

So the PEM shell needs to have certain critical thickness, $h_{cr}$, to retain the structural shape of the standing chambers:

$$h_{cr}^2 = \frac{a^2 \sigma_{ad} \sqrt{3(1-v^2)}}{2E} \tag{36.7}$$

For example, the reported Young's modulus, $E$, for wet PAH-PSS multilayer is $\sim 500$–$750$ MPa [142, 143]. Then the critical thickness for the cylindrical shell 7 μm in diameter standing on a silicon oxide is $\sim 300$–$400$ nm. Stable 25 μm chambers should be made of a $\sim 1$–$1.5$ μm thick shell. Softer PDADMAC-PSS film has Young's modulus of $\sim 100$ MPa [144], giving the critical thickness of $\sim 740$ nm for 7 μm chambers. These rough estimations work well for predicting the mechanical stability of PEM chambers. Figure 36.19 shows SEM images of a number of chambers of different geometries and made of different PEMs, clearly demonstrating collapsed chambers with their roofs contacting the support if made of PEM films thinner than critical thickness, and standing chambers if made of thicker films.

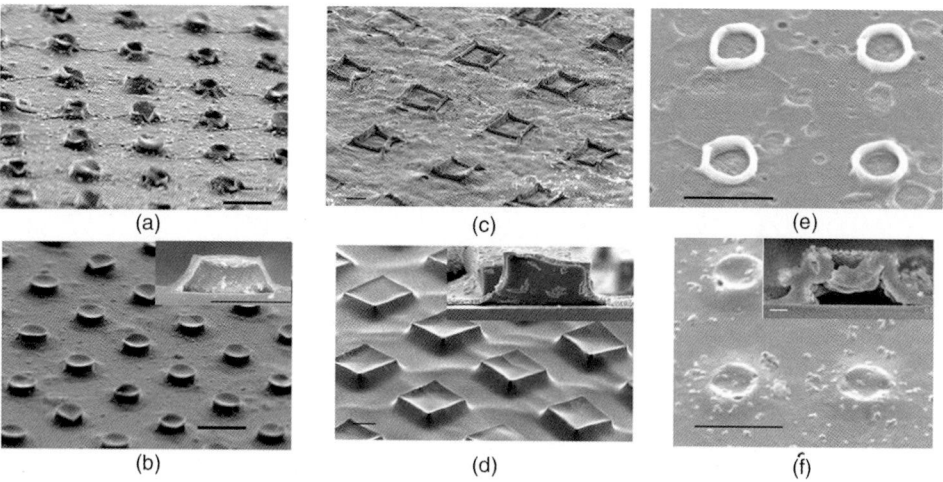

**Figure 36.19** SEM images of microchambers of different geometries made of PAH-PSS (a–d) and PDADMAC-PSS (e, f) multilayers. Thickness of the PEM film is 220 nm (a), 370 nm (b, c, e), 740 nm (d), 540 nm (f). Scale bars: 10 μm [22, 38].

## 36.3.2
### Microchamber Loading with Substances of Interest

The hollow PEM microcompartments described in the previous section could be used for housing a cargo. One way to fill these compartments with a cargo could be by using the tunable permeability of PEM shells, as was shown above for PEM capsules. However, much thicker PEM shells that ensure their mechanical stability (typically hundreds of nm) could hinder diffusion of molecules and the feasibility of each particular technique for loading the microchambers should be verified. Successful filling of PEM microchambers with oil by a solvent-exchange method has been demonstrated [38]. Composite CRM images containing information from both oil (red) and PEM (green) bands demonstrate that oil droplets are specifically located inside the PEM chambers and completely fill them (Figure 36.20). Another possible strategy lies in putting a cargo into the wells before the transfer process and removal of the sacrificial template. A number of approaches could be exploited, depending on the nature of cargo. For example, the wells could be filled with a solution containing high-molecular-weight molecules, these molecules remain entrapped inside after the PEM film transfer, as shown in Figure 36.20b [138]. Selective microwells could be filled with chosen solution using the discontinuous dewetting process developed by Whitesides et al. [145]. When the liquid front meets a microwell, the abrupt change in angle at the edge of the well causes the drop to be pinned. As the bulk liquid continues to recede, the pinned drop hinges on the edge of the well, and the drop tends toward its equilibrium contact angle at the top edge of the well. Drainage of the liquid at the lower edge of the well causes the film to thin until, finally, it ruptures and pins at the bottom lip of the well. This process leaves each of the microwells filled with an equal volume of liquid.

Colloid particles could be selectively deposited into the wells using the template-assisted self-assembly of colloids developed by Xia et al. [146]. The aqueous

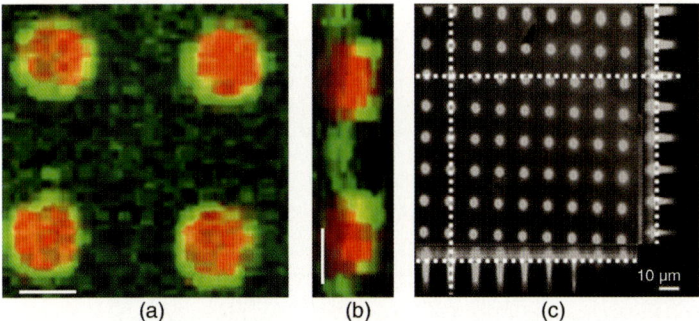

**Figure 36.20** Confocal Raman microscope images of PAH-PSS 7 μm chambers filled with oil: top view (a) and cross-section (b). Color code of the image: red = oil (1660 cm$^{-1}$ band is assigned to double bond, C=C stretching); green = PEM (1604 cm$^{-1}$ band is assigned to C—C stretching in the aromatic ring of PSS). All scale bars are 5 mm [38]. (c) CLSM image of microcavity array filled with 500 kDa dextran [138].

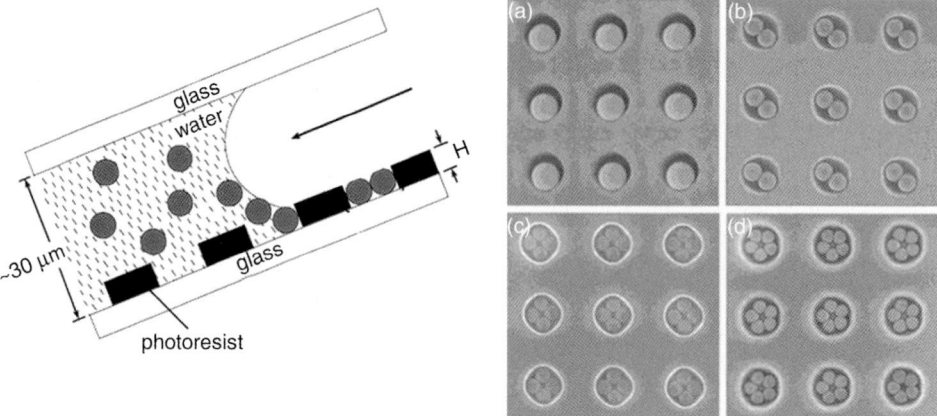

**Figure 36.21** A schematic illustration of the procedure to assemble spherical colloids by the 2D array of cylindrical wells patterned in a thin film of photoresist (a). The size, shape, and structure of the resultant clusters could be easily controlled by changing the ratios between the dimensions of the wells and the diameter of the beads (b) [146].

dispersion of colloid particles is allowed to move slowly in a thin layer confined between the substrate with an array of wells and the glass above, as depicted in Figure 36.21a. As the rear front of the liquid moves, the capillary forces exerted on this interface drag colloid particles across the surface until they are physically trapped by the wells. Depending on the ratio of the well dimensions to the diameter of the particles a number of self-assembled clusters could be obtained (Figure 36.21b).

This approach was successfully used to fill the PEM chambers with colloid particles, as shown in the Figure 36.22a. Porous colloids could serve as carriers for absorbed proteins or polymers to bring them into the microchambers (Figure 36.22b).

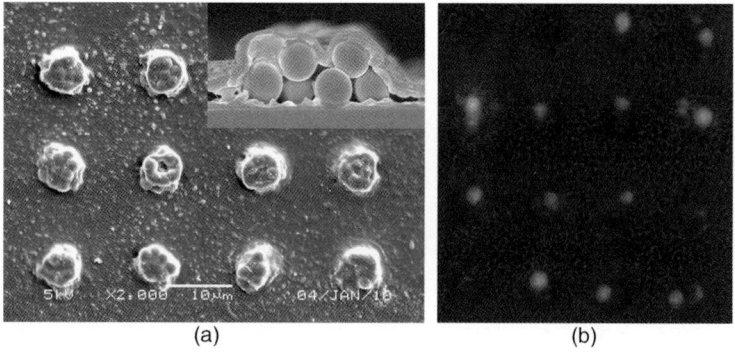

**Figure 36.22** SEM images of PAH-PSS microchambers loaded with 1.87 μm MF particles (a). Optical micrograph of PAH-PSS microchambers filled with $CaCO_3$ porous particles saturated with Dextran-RITC (b).

## 36.3.3
### Responsiveness of Chambers to Light and Mechanical Load

Highly-ordered patterns of PEM microchambers allow the addressing of specific chambers in a programmed way, releasing the loaded active cargo on-demand. The methods to trigger release from individual chambers are mechanical probing and remote addressing with a focused laser beam.

Mechanical probing of chambers was done by the nanoindentation system using a diamond sharp Berkovich indenter (tip diameter $\sim$50 nm) [38]. The tip was positioned over the chamber roof using a 40× objective. The microchambers were probed at different loading speeds. The indenter tip does not leave any footprint on a chamber's roof if loaded at a speed $V = 0.07$ mN s$^{-1}$. At $V = 0.33$ mN s$^{-1}$ a tiny scratch was detected. At higher loading speeds, the tetrahedral indenter starts to make a triangular hole in the roof. The higher the $V$, the larger the hole in the chamber, indicating deeper penetration of the indenter (Table 36.1).

The force–deformation curves of the PEM chambers for low loading speeds (0.03 and 0.07 mN s$^{-1}$) are similar to those of PEM capsules with highly porous shells, when fast drainage of the solution from the capsules is possible (see above). In the low deformation regime ($\varepsilon < \sim$0.25) the data falls well into a straight line in coordinates $F(\varepsilon^{1/2})$ indicating deflation of air from the chambers and shell bending with no stretching. At $\varepsilon > \sim$0.25 severe plastic buckling of the chambers' walls develops.

The force–deformation curves of PEM microchambers at higher loading speeds ($> 0.33$ mN s$^{-1}$) are shown in Figure 36.23. First, significant deformation develops with negligible applied force; that corresponds to the chamber roof being pierced by the tetrahedral indenter. As it penetrates deeper, the contact area with the chamber roof increases and, at a certain point, plastic deformation of the chamber starts to develop, requiring a drastic increase in applied force in the second regime. The higher the loading speed, the deeper the indenter could penetrate. The critical loading speed $V = 0.33$ mN s$^{-1}$ should correspond to a relaxation rate of loaded PEM shell.

Thus, mechanical piercing could be applied to open the selected PEM microchambers and the release kinetics of pre-loaded active cargo could be precisely controlled by changing the size of the hole in the chamber's roof.

Remote rupture of individual chambers could be achieved using focused laser radiation. Incorporation of metal nanoparticles in a PEM shell makes it sensitive towards irradiation within the plasmon absorption band of those nanoparticles.

**Table 36.1** The altitudes of triangle holes in the roofs of microchambers loaded at different speed.

|   | Loading speed, mN s$^{-1}$ | Altitude of triangle hole, μm |
| --- | --- | --- |
| 1 | 0.66 | 1.0 |
| 2 | 1.33 | 3.0 |
| 3 | 3.33 | 1.0 |
| 4 | 6.67 | 4.5 |
| 5 | 13.33 | 6.0 |

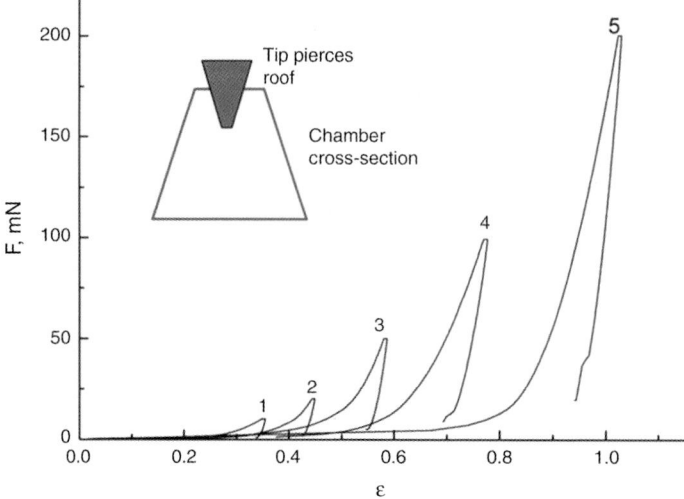

**Figure 36.23** Force–deformation curves for (PAH-PSS) microchambers addressed at loading speed 0.66 (1), 1.33 (2), 3.33 (3), 6.67 (4), and 13.33 (5) mN s$^{-1}$, respectively. The sketch demonstrates piercing of the chamber's roof with the sharp indenter [38].

The energy of absorbed light dissipates as heat, resulting in a highly localized temperature increase that destroys the surrounding few hundred nm thick PEM film in the same way as was shown above for few nm thick PEM capsules.

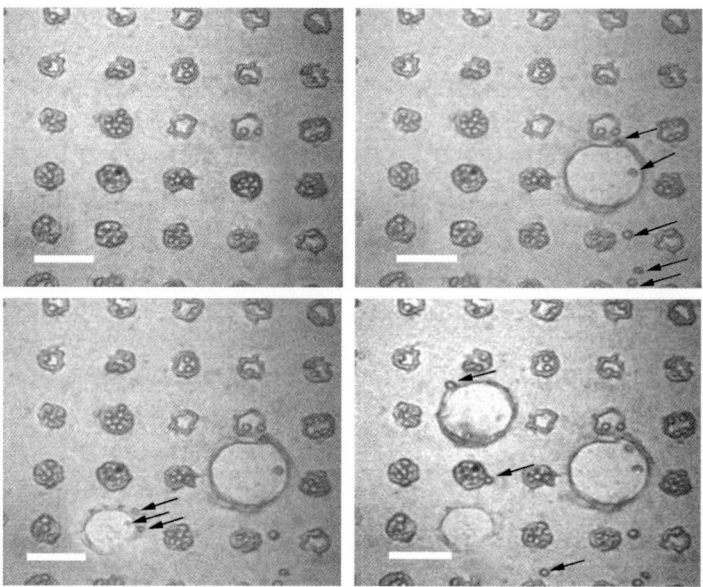

**Figure 36.24** Optical microscope images of PEM chambers loaded with MF particles after sequential addressing of three different chambers with a single pulse of focused Nd-YAG laser beam. Arrows show released MF particles. Scale bar – 10 μm.

Figure 36.24 demonstrates sequential addressing of three different microchambers with entrapped MF particles, releasing the particles into surrounding water.

## 36.4
## Conclusion

The development of biomedicine, bioengineering, consumer care and foodstuff products creates a strong demand for microcarriers enabling packaging and control release delivery of various actives. In this chapter we have illustrated how the LbL technique for thin film assembly can be applied to achieve carriers possessing changeable permeability and integrity in response to a variety of external stimuli, and discussed the perspectives of stimuli-responsive capsules in some practical applications. Stimuli-responsive permeability of multilayers has been first utilized for capsule loading. More recent advances in the field of research allow controlled delivery of capsules and *in situ* release of their contents by remote physical stimuli, such as magnetic field or light. Reported data on the application of capsules in drug delivery has been extensively reviewed elsewhere [147]. Time- and site-controlled biochemical reactions inside the light-responsive capsule have been demonstrated, with the aim to model artificial organelles [95].

LbL assembly appears as a generic and versatile approach to introduce capsule responsiveness to pH, redox potential, ionic strength, chemicals (glucose and $CO_2$) and active enzymes, as well as to a number of physical stimuli, such as temperature, light, magnetic field, ultrasound, and mechanical load. The unique feature of the LbL technique for the microcapsule fabrication – the possibility to use templates of different materials, shape and dimension – has been utilized to achieve a novel type of capsule represented by an array of sealed chambers on a solid substrate. The approaches to introduce stimuli-responsiveness are universal for multilayer capsules of different geometry. However, the relatively thick walls of the chambers if compared to the microcapsules might possess quite limited permeability in response to such stimuli as pH or salt, but rather could be ruptured and opened by mechanical load or laser light. Nevertheless, the arrays of microchambers have their own advantage over capsules assembled on colloid particles. A colloid core template has to be removed by an appropriate solvent, and some remnants of a core material often remain in the capsule interior and get entangled within the PEM shell [148]. Another challenge for the colloid capsules is encapsulation of small water soluble molecules. To some extent, the microchambers may prove to be more versatile for housing different compounds, and also to be more robust in processing. Although more research has to be carried out, we can propose that the mechanical properties of the microchambers will be different for different couples of forming polymers [22, 38]. Another prediction is the feasibility of multiple loading and release of actives from the chambers. Controlled release of different bioactives in a sequence or with a certain time delay would be a powerful tool for biomedical researchers. Another anticipated application of the microchamber array is a drug eluting surface. Indeed, stent or implants with deposited microchambers can be activated inside the body by any available remote stimuli to release the encapsulated antibiotics or other drugs in a controlled manner.

Obviously, each strategy for fabrication of stimuli-responsive capsules using multilayer assembly has its own niche of applications. The aqueous dispersed capsules have clear advantages over the available alternatives, such as liposomes and polymeric vesicles, due to multimodal functionality for delivery of actives in biomedical applications, food and consumer care products. The microchamber array has a good perspective for application as a drug eluting surface or a multiple sensor. Further investigations are envisaged in this area.

Concluding this chapter we would like to highlight that the LbL technique for film assembly represents a powerful tool for the design of versatile stimuli-responsive carriers with defined micron- and submicron-sized geometries on 2D and 3D levels, from the aqueous suspension of functional capsules to arrays of microchambers on planar surfaces. These capsules can perform preserved storage and be capable of time- and site-specific controlled release delivery of encapsulated actives.

## References

1 Decher, G. (1997) Fuzzy nanoassemblies: Toward layered polymeric multicomposites. *Science*, **277** (5330), 1232–1237.

2 Donath, E., Sukhorukov, G.B., Caruso, F., Davis, S.A., and Moehwald, H. (1998) Novel hollow polymer shells by colloid-templated assembly of polyelectrolytes. *Angew. Chem., Int. Ed.*, **37** (16), 2202–2205.

3 Stockton, W.B. and Rubner, M.F. (1997) Molecular-level processing of conjugated polymers. 4. Layer-by-layer manipulation of polyaniline via hydrogen-bonding interactions. *Macromolecules*, **30** (9), 2717–2725.

4 Erel-Unal, I. and Sukhishvili, S. (2008) Hydrogen-bonded multilayers of a neutral polymer and a polyphenol. *Macromolecules*, **41** (11), 3962–3970.

5 Kharlampieva, E., Kozlovskaya, V., and Sukhishvili, S.A. (2009) Layer-by-layer hydrogen-bonded polymer films: From fundamentals to applications. *Adv. Mater.*, **21** (30), 3053–3065.

6 Tong, W.J., Gao, C.Y., and Moehwald, H. (2006) Single polyelectrolyte microcapsules fabricated by glutaraldehyde-mediated covalent layer-by-layer assembly. *Macromol. Rapid Commun.*, **27** (24), 2078–2083.

7 Tong, W.J., Gao, C.Y., and Moehwald, H. (2006) Stable weak polyelectrolyte microcapsules with pH-responsive permeability. *Macromolecules*, **39** (1), 335–340.

8 Tong, W.J., Gao, C.Y., and Moehwald, H. (2005) Manipulating the properties of polyelectrolyte microcapsules by glutaraldehyde cross-linking. *Chem. Mater.*, **17** (18), 4610–4616.

9 Feng, Z.Q., Fan, G.Q., Wang, H.X., Gao, C.Y., and Shen, J.C. (2009) Polyphosphazene microcapsules fabricated through covalent assembly. *Macromol. Rapid Commun.*, **30** (6), 448–452.

10 Such, G.K., Quinn, J.F., Quinn, A., Tjipto, E., and Caruso, F. (2006) Assembly of ultrathin polymer multilayer films by click chemistry. *J. Am. Chem. Soc.*, **128** (29), 9318–9319.

11 Such, G.K., Tjipto, E., Postma, A., Johnston, A.P.R., and Caruso, F. (2007) Ultrathin, responsive polymer click capsules. *Nano Lett.*, **7** (6), 1706–1710.

12 Wang, Z.P., Feng, Z.Q., and Gao, C.Y. (2008) Stepwise assembly of the same polyelectrolytes using host-guest interaction to obtain microcapsules with multiresponsive properties. *Chem. Mater.*, **20** (13), 4194–4199.

13 Kida, T., Mouri, M., and Akashi, M. (2006) Fabrication of hollow capsules composed of poly(methyl methacrylate) stereocomplex films. *Angew. Chem. Int. Ed.*, **45** (45), 7534–7536.

14 Zelikin, A.N., Quinn, J.F., and Caruso, F. (2006) Disulfide cross-linked polymer capsules: En route to biodeconstructible systems. *Biomacromolecules*, **7** (1), 27–30.

15 De Geest, B.G., Déjugnat, C., Prevot, M., Sukhorukov, G., Demeester, J., and De Smedt, S.C. (2007) Self-rupturing and hollow microcapsules prepared from bio-polyelectrolyte-coated microgels. *Adv. Funct. Mater.*, **17** (4), 531–537.

16 De Geest, B.G., Déjugnat, C., Sukhorukov, G.B., Braeckmans, K., De Smedt, S.C., and Demeester, J. (2005) Self-rupturing microcapsules. *Adv. Mater.*, **17** (19), 2357–2361.

17 Grigoriev, D.O., Bukreeva, T., Moehwald, H., and Shchukin, D.G. (2008) New method for fabrication of loaded micro- and nanocontainers: Emulsion encapsulation by polyelectrolyte layer-by-layer deposition on the liquid core. *Langmuir*, **24** (3), 999–1004.

18 Wackerbarth, H., Schoen, P., and Bindrich, U. (2009) Preparation and characterization of multilayer coated microdroplets: droplet deformation simultaneously probed by atomic force spectroscopy and optical detection. *Langmuir*, **25** (5), 2636–2640.

19 Szczepanowicz, K., Dronka-Góra, D., Para, G., and Warszyński, P. (2010) Encapsulation of liquid cores by layer-by-layer adsorption of polyelectrolytes. *J. Microencapsul.*, **27** (3), 198–204.

20 Lomova, M.V., Sukhorukov, G.B., and Antipina, M.N. (2010) Antioxidant coating of micronsize droplets for prevention of lipid peroxidation in oil-in-water emulsion. *ACS Appl. Mater. Interface*, **2** (12), 3669–3676.

21 Sadovoy, A.V., Kiryukhin, M.V., Sukhorukov, G.B., and Antipina, M.N. (2011) Kinetic stability of water-dispersed oil droplets encapsulated in a polyelectrolyte multilayer shell. *Phys. Chem. Chem. Phys.*, **13** (9), 4005–4012.

22 Kiryukhin, M.V., Man, S.M., Sadovoy, A.V., Low, H.Y., and Sukhorukov, G.B. (2011) Peculiarities of polyelectrolyte multi layer assembly on patterned surfaces. *Langmuir*, **27** (13), 8430–8436.

23 Sukhorukov, G.B., Volodkin, D.V., Günther, A., Petrov, A.I., Shenoy, D.B., and Moehwald, H. (2004) Porous calcium carbonate microparticles as templates for encapsulation of bioactive compounds. *J. Mater. Chem.*, **14** (14), 2073–2081.

24 Petrov, A.I., Volodkin, D.V., and Sukhorukov, G.B. (2005) Protein-calcium carbonate coprecipitation: A tool for protein encapsulation. *Biotechnol. Prog.*, **21** (3), 918–925.

25 Volodkin, D.V., Larionova, N.I., and Sukhorukov, G.B. (2004) Protein encapsulation via porous $CaCO_3$ microparticles templating. *Biomacromolecules*, **5** (5), 1962–1972.

26 Volodkin, D.V., Petrov, A.I., Prevot, M., and Sukhorukov, G.B. (2004) Matrix polyelectrolyte microcapsules: New system for macromolecule encapsulation. *Langmuir*, **20** (8), 3398–3406.

27 Déjugnat, C. and Sukhorukov, G.B. (2004) pH-responsive properties of hollow polyelectrolyte microcapsules templated on various cores. *Langmuir*, **20** (17), 7265–7269.

28 Koehler, K., Shchukin, D.G., Moehwald, H., and Sukhorukov, G.B. (2005) Thermal behavior of polyelectrolyte multilayer microcapsules. 1. The effect of odd and even layer number. *J. Phys. Chem. B*, **109** (39), 18250–18259.

29 Antipov, A.A., Sukhorukov, G.B., Leporatti, S., Radtchenko, I.L., Donath, E., and Moehwald, H. (2002) Polyelectrolyte multilayer capsule permeability control. *Colloids Surf., A*, **198**, 535–541.

30 Ibarz, G., Daehne, L., Donath, E., and Moehwald, H. (2001) Smart micro- and nanocontainers for storage, transport, and release. *Adv. Mater.*, **13** (17), 1324–1327.

31. Koehler, K. and Sukhorukov, G.B. (2007) Heat treatment of polyelectrolyte multilayer capsules: A versatile method for encapsulation. *Adv. Funct. Mater.*, **17** (13), 2053–2061.

32. Lvov, Y., Antipov, A., Mamedov, A., Moehwald, H., and Sukhorukov, G.B. (2001) Urease encapsulation in nanoorganized microshells. *Nano Lett.*, **1** (3), 125–128.

33. Lvov, Y., Decher, G., and Sukhorukov, G. (1993) Assembly of thin-films by means of successive deposition of alternate layers of DNA and poly(allylamine). *Macromolecules*, **26** (20), 5396–5399.

34. De Koker, S., De Geest, B.G., Cuvelier, C., Ferdinande, L., Deckers, W., Hennink, W.E., De Smedt, S., and Mertens, N. (2007) In vivo cellular uptake, degradation, and biocompatibility of polyelectrolyte microcapsules. *Adv. Funct. Mater.*, **17** (18), 3754–3763.

35. Keller, S.W., Kim, H.N., and Mallouk, T.E. (1994) Layer-by-layer assembly of intercalation compounds and heterostructures on surfaces – toward molecular beaker epitaxy. *J. Am. Chem. Soc.*, **116** (19), 8817–8818.

36. Lvov, Y., Ariga, K., Ichinose, I., and Kunitake, T. (1995) Assembly of multicomponent protein films by means of electrostatic layer-by-layer adsorption. *J. Am. Chem. Soc.*, **117** (22), 6117–6123.

37. Antipina, M.N. and Sukhorukov, G.B. (2011) Remote control over guidance and release properties of composite polyelectrolyte based capsules. *Adv. Drug Deliv. Rev.*, **63** (9), 716–729.

38. Kiryukhin, M.V., Man, S.M., Gorelik, S.R., Subramanian, G.S., Low, H.Y., and Sukhorukov, G.B. (2011) Fabrication and mechanical properties of microchambers made of polyelectrolyte multilayers. *Soft Matter.*, **7** (14), 6550–6556.

39. Mauser, T., Déjugnat, C., and Sukhorukov, G. (2004) Reversible pH-dependent properties of multilayer microcapsules made of weak polyelectrolytes. *Macromol. Rapid Commun.*, **25** (20), 1781–1785.

40. Mauser, T., Déjugnat, C., and Sukhorukov, G.B. (2006) Balance of hydrophobic and electrostatic forces in the pH response of weak polyelectrolyte capsules. *Phys. Chem. B*, **110** (41), 20246–20253.

41. Déjugnat, C., Halozan, D., and Sukhorukov, G. (2005) Defined picogram dose inclusion and release of macromolecules using polyelectrolyte microcapsules. *Macromol. Rapid Commun.*, **26** (12), 961–967.

42. Shiratori, S.S. and Rubner, M.F. (2000) pH-dependent thickness behavior of sequentially adsorbed layers of weak polyelectrolytes. *Macromolecules*, **33** (11), 4213–4219.

43. Petrov, A.I., Antipov, A.A., and Sukhorukov, G.B. (2003) Base-acid equilibria in polyelectrolyte systems: From weak polvelectrolytes to interpolyelectrolyte complexes and multilayered polyelectrolyte shells. *Macromolecules*, **36** (26), 10079–10086.

44. Antipina, M.N., Kiryukhin, M.V., Chong, K., Low, H.Y., and Sukhorukov, G.B. (2009) Patterned microcontainers as novel functional elements for µTAS and LOC. *Lab Chip*, **9** (10), 1472–1475.

45. Shutava, T., Prouty, M., Kommireddy, D., and Lvov, Y. (2005) pH responsive decomposable layer-by-layer nanofilms and capsules on the basis of tannic acid. *Macromolecules*, **38** (7), 2850–2858.

46. Sukhishvili, S.A. and Granick, S. (2000) Layered, erasable, ultrathin polymer films. *J. Am. Chem. Soc.*, **122** (39), 9550–9551.

47. Kozlovskaya, V., Kharlampieva, E., Mansfield, M.L., and Sukhishvili, S.A. (2006) Poly(methacrylic acid) hydrogel films and capsules: Response to pH and ionic strength, and encapsulation of macromolecules. *Chem. Mater.*, **18** (2), 328–326.

48. Kozlovskaya, V. and Sukhishvili, S.A. (2006) pH-Controlled permeability of layered hydrogen-bonded polymer capsules. *Macromolecules*, **39** (16), 5569–5572.

49. Kozlovskaya, V., Kharlampieva, E., Drachuk, I., Cheng, D., and Tsukruk, V.V. (2010) Responsive microcapsule reactors based on hydrogen-bonded tannic acid

layer-by-layer assemblies. *Soft Matter.*, **6** (15), 3596–3608.

50 Yun, J. and Kim, H.I. (2010) Control of release characteristics in pH-sensitive poly(vinylalcohol)/poly(acrylic acid) microcapsules containing chemically treated alumina core. *J. Appl. Polym. Sci.*, **115** (3), 1853–1858.

51 Gao, C., Moehwald, H., and Shen, J. (2004) Enhanced biomacromolecule encapsulation by swelling and shrinking procedures. *ChemPhysChem*, **5**, 116–120.

52 Buscher, K., Graf, K., Ahrens, H., and Helm, C.A. (2002) Influence of adsorption conditions on the structure of polyelectrolyte multilayers. *Langmuir*, **18**, 3585–3591.

53 Fery, A., Scholer, B., Cassagneau, T., and Caruso, F. (2001) Nanoporous thin films formed by salt-induced structural changes in multilayers of poly(acrylic acid) and poly (allylamine). *Langmuir*, **17**, 3779–3783.

54 McAloney, R.A., Sinyor, M., Dudnik, V., and Goh, M.C. (2001) Atomic force microscopy studies of salt effects on polyelectrolyte multilayer film morphology. *Langmuir*, **17**, 6655–6663.

55 Schuler, C. and Caruso, F. (2001) Decomposable hollow biopolymer-based capsules. *Biomacromolecules*, **2** (3), 921–926.

56 Park, T.G. and Hoffman, A.S. (1993) Sodium chloride-induced phase-transition in nonionic poly(N-isopropylacryamide) gel. *Macromolecules*, **26** (19), 5045–5048.

57 Chang, Y.K., Powell, E.S., and Allcock, H.R. (2005) Environmentally responsive micelles from polystyrene-poly[bis(potassium carboxylatophenoxy) phosphazenel block copolymers. *J. Polym. Sci., Part A: Polym. Chem.*, **43** (13), 2912–2920.

58 Solomatin, S.V., Bronich, T.K., Bargar, T.W., Eisenberg, A., Kabanov, V.A., and Kabanov, A.V. (2003) Environmentally responsive nanoparticles from block ionomer complexes: Effects of pH and ionic strength. *Langmuir*, **19** (19), 8069–8076.

59 Zhang, R.S., Tang, M.G., Bowyer, A., Eisenthal, R., and Hubble, J. (2005) A novel pH- and ionic-strength-sensitive carboxy methyl dextran hydrogel. *Biomaterials*, **26** (22), 4677–4683.

60 Balkay, L., Marian, T., Emri, M., Krasznai, Z., and Tron, L. (1997) Flow cytometric determination of intracellular free potassium concentration. *Cytometry*, **28** (1), 42–49.

61 Erstad, B.L. (2002) Laboratory monitoring: Back to basics. *Am. J. Pharm. Educ.*, **66** (2), 199–203.

62 Fry, C.H., Hall, S.K., Blatter, L.A., and McGuigan, J.A.S. (1990) Analysis and presentation of intracellular measurements obtained with ion-selective microelectrodes. *Exp. Physiol.*, **75** (2), 187–198.

63 Zhang, R.J., Koehler, K., Kreft, O., Skirtach, A., Moehwald, H., and Sukhorukov, G. (2010) Salt-induced fusion of microcapsules of polyelectrolytes. *Soft Matter.*, **6**, 4742–4747.

64 Riske, K. and Dimova, R. (2005) Timescales involved in electro-deformation, poration and fusion of giant vesicles resolved with fast digital imaging. *Biophys. J.*, **88** (2), 1143–1155.

65 Li, B. and Haynie, D.T. (2004) Multilayer biomimetics: Reversible covalent stabilization of a nanostructured biofilm. *Biomacromolecules*, **5** (5), 1667–1670.

66 Haynie, D.T., Palath, N., Liu, Y., Li, B.Y., and Pargaonkar, N. (2005) Biomimetic nanostructured materials: Inherent reversible stabilization of polypeptide microcapsules. *Langmuir*, **21** (3), 1136–1138.

67 Ma, Y.J., Dong, W.F., Hempenius, M.A., Moehwald, H., and Vancso, G.J. (2006) Redox-controlled molecular permeability of composite-wall microcapsules. *Nat. Matter.*, **5** (9), 724–729.

68 De Geest, B.G., Sanders, N.N., Sukhorukov, G.B., Demeester, J., and De Smedt, S.C. (2007) Release mechanisms for polyelectrolyte capsules. *Chem. Soc. Rev.*, **36** (4), 636–649.

69 De Geest, B.G., Jonas, A.M., Demeester, J., and De Smedt, S.C. (2006)

Glucose-responsive polyelectrolyte capsules. *Langmuir*, **22** (11), 5070–5074.

70 Levy, T., Déjugnat, C., and Sukhorukov, G.B. (2008) Polymer microcapsules with carbohydrate-sensitive properties. *Adv. Funct. Mater.*, **18** (10), 1586–1594.

71 Hartmann, L., Bedard, M., Börner, H.G., Moehwald, H., Sukhorukov, G.B., and Antonietti, M. (2008) $CO_2$-switchable oligoamine patches based on amino acids and their use to build polyelectrolyte containers with intelligent gating. *Soft Matter*, **3**, 534–539.

72 She, S.J., Zhang, X.G., Wu, Z.M., Wang, Z., and Li, C.X. (2010) Gradient cross-linked biodegradable polyelectrolyte nanocapsules for intracellular protein drug delivery. *Biomaterials*, **31**, 6039–6049.

73 He, Q., Cui, Y., and Li, J.B. (2009) Molecular assembly and application of biomimetic microcapsules. *Chem. Soc. Rev.*, **38**, 2292–2303.

74 De Geest, B.G., Vandenbroucke, R.E., Guenther, A.M., Sukhorukov, G.B., Hennink, W.E., Sanders, N.N., Demeester, J., and De Smedt, S.C. (2006) Intracellularly degradable polyelectrolyte microcapsules. *Adv. Mater.*, **18** (8), 1005–1009.

75 Borodina, T., Markvicheva, E., Kunizhev, S., Moehwald, H., Sukhorukov, G.B., and Kreft, O. (2007) Controlled release of DNA from self-degrading microcapsules. *Macromol. Rapid Commun.*, **28** (18–19), 1894–1899.

76 Szarpak, A., Cui, D., Dubreuil, F., De Geest, B.G., De Cock, L.J., Picart, C., and Auzely-Velty, R. (2010) Designing hyaluronic acid-based layer-by-layer capsules as a carrier for intracellular drug delivery. *Biomacromolecules*, **11** (3), 713–720.

77 Itoh, Y., Matsusaki, M., Kida, T., and Akashi, M. (2006) Enzyme-responsive release of encapsulated proteins from biodegradable hollow capsules. *Biomacromolecules*, **7**, 2715–2718.

78 She, Z., Antipina, M.N., Li, J., and Sukhorukov, G.B. (2010) Mechanism of protein release from polyelectrolyte multilayer microcapsules. *Biomacromolecules*, **11** (5), 1241–1247.

79 Pavlov, A.M., Sapelkin, A.V., Huang, X., Ping, K.M.Y., Bushby, A.J., Sukhorukov, G.B., and Skirtach, A.G. (2011) Neuron cells uptake of polymeric microcapsules and subsequent intracellular release. *Macromol. Biosci.*, **11** (6), 846–852.

80 Soltwedel, O., Ivanova, O., Nestler, P., Muller, M., Kohler, R., and Helm, C.A. (2010) Interdiffusion in polyelectrolyte multilayers. *Macromolecules*, **43**, 7288–7293.

81 Steitz, R., Leiner, V., Tauer, K., Khrenov, V., and von Klitzing, R. (2002) Temperature-induced changes in polyelectrolyte films at the solid–liquid interface. *Appl. Phys. A Mater. Sci. Proc.*, **74**, S519–S521.

82 Ahrens, H., Büscher, K., Eck, D., Förster, S., Luap, C., Papastavrou, G., Schmitt, J., Steitz, R., and Helm, C.A. (2004) Poly(styrene sulfonate) adsorption: electrostatic and secondary interactions. *Macromol. Symp.*, **211**, 93–106.

83 Dejugnat, C., Koehler, K., Dubois, M., Sukhorukov, G.B., Moehwald, H., Zemb, T., and Guttmann, P. (2007) Membrane densification of heated polyelectrolyte multilayer capsules characterized by soft X-ray microscopy. *Adv. Mater.*, **19**, 1331–1336.

84 Bedard, M.F., Braun, D., Sukhorukov, G.B., and Skirtach, A.G. (2008) Towards self-assembly of nanoparticles on polymeric capsules: release threshold and permeability. *ACS Nano*, **2**, 1807–1816.

85 Bedard, M.F., Munoz-Javier, A., Mueller, R., del Pino, P., Fery, A., Parak, W.J., Skirtach, A.G., and Sukhorukov, G.B. (2009) On the mechanical stability of polymeric microcontainers functionalized with nanoparticles. *Soft Matter.*, **5**, 148–155.

86 Caruso, F., Susha, A.S., Giersig, M., and Moehwald, H. (1999) Magnetic core-shell particles: Preparation of magnetite multilayers on polymer latex microspheres. *Adv. Mater.*, **11** (11), 950–953.

87 Caruso, F., Spasova, M., Susha, A., Giersig, M., and Caruso, R. (2001) Magnetic nanocomposite particles and hollow spheres constructed by a sequential layering approach. *Chem. Mater.*, **13** (1), 109–116.

88 Andreeva, D.V., Gorin, D.A., Shchukin, D.G., and Sukhorukov, G.B. (2006) Magnetic microcapsules with low permeable polypyrrole skin layer. *Macromol. Rapid Commun.*, **27** (12), 931–936.

89 Sadasivan, S. and Sukhorukov, G.B. (2006) Fabrication of hollow multifunctional spheres containing MCM-41 nanoparticles and magnetite nanoparticles using layer-by-layer method. *J. Colloid Interface Sci.*, **304** (2), 437–441.

90 Shchukin, D.G., Radtchenko, I.L., and Sukhorukov, G.B. (2003) Micron-scale hollow polyelectrolyte capsules with nanosized magnetic $Fe_3O_4$ inside. *Mater. Lett.*, **57** (11), 1743–1747.

91 Radtchenko, I.L., Giersig, M., and Sukhorukov, G.B. (2002) Inorganic particle synthesis in confined micron-sized polyelectrolyte capsules. *Langmuir*, **18** (21), 8204–8208.

92 Shchukin, D.G. and Sukhorukov, G.B. (2004) Nanoparticle synthesis in engineered organic nanoscale reactors. *Adv. Mater.*, **16** (8), 671–682.

93 Shchukin, D.G., Radtchenko, I.L., and Sukhorukov, G.B. (2003) Micron-scale hollow polyelectrolyte capsules with nanosized magnetic $Fe_3O_4$ inside. *Mater. Lett.*, **57** (11), 1743–1747.

94 Shchukin, D.G., Radtchenko, I.L., and Sukhorukov, G.B. (2003) Synthesis of nanosized magnetic ferrite particles inside hollow polyelectrolyte capsules. *J. Phys. Chem. B*, **107** (1), 86–90.

95 Kreft, O., Prevot, M., Moehwald, H., and Sukhorukov, G.B. (2007) Shell-in-shell microcapsules: a novel tool for integrated, spatially confined enzymatic reactions. *Angew. Chem. Int. Ed.*, **46** (29), 5605–5608.

96 Bibette, J. (1993) Monodisperse ferrofluid emulsions. *J. Magn. Magn. Mater.*, **122** (1–3), 37–41.

97 Veyret, R., Delair, T., and Elaissari, A. (2005) Preparation and biomedical application of layer-by-layer encapsulated oil in water magnetic emulsion. *J. Magn. Magn. Mater.*, **293** (1), 171–176.

98 Zebli, B., Susha, A.S., Sukhorukov, G.B., Rogach, A.L., and Parak, W.J. (2005) Magnetic targeting and cellular uptake of polymer microcapsules simultaneously functionalized with magnetic and luminescent nanocrystals. *Langmuir*, **21** (10), 4262–4265.

99 Gorin, D.A., Portnov, S.A., Inozemtseva, O.A., Luklinska, Z., Yashchenok, A.M., Pavlov, A.M., Skirtach, A.G., Moehwald, H., and Sukhorukov, G.B. (2008) Magnetic/gold nanoparticle functionalized biocompatible microcapsules with sensitivity to laser irradiation. *Phys. Chem. Chem. Phys.*, **10** (45), 6899–6905.

100 Lu, Z.H., Prouty, M.D., Guo, Z.H., Golub, V.O., Kumar, C.S.S.R., and Lvov, Y.M. (2005) Magnetic switch of permeability for polyelectrolyte microcapsules embedded with Co@Au nanoparticles. *Langmuir*, **21** (5), 2042–2050.

101 Hu, S.H., Tsai, C.H., Liao, C.F., Liu, D.M., and Chen, S.Y. (2008) Controlled rupture of magnetic polyelectrolyte microcapsules for drug delivery. *Langmuir*, **24** (20), 11811–11818.

102 Rae, J., Ashokkumar, M., Eulaerts, O., von Sonntag, C., Reisse, J., and Grieser, F. (2005) Estimation of ultrasound induced cavitation bubble temperatures in aqueous solutions. *Ultrason. Sonochem.*, **12** (2), 325–329.

103 Shchukin, D.G., Gorin, D.A., and Moehwald, H. (2006) Ultrasonically induced opening of polyelectrolyte microcontainers. *Langmuir*, **22** (17), 7400–7404.

104 Skirtach, A.G., De Geest, B.G., Mamedov, A., Antipov, A.A., Kotov, N.A., and Sukhorukov, G.B. (2007) Ultrasound stimulated release and catalysis using polyelectrolyte multilayer capsules. *J. Mater. Chem.*, **17** (11), 1050–1054.

105 De Geest, B.G., Skirtach, A.G., Mamedov, A., Antipov, A.A., Kotov, N.A.,

De Smedt, S.C., and Sukhorukov, G.B. (2007) Ultrasound-triggered release from multilayered capsules. *Small*, **3** (5), 804–808.

106 Kolesnikova, T.A., Gorin, D.A., Fernandes, P., Kessel, S., Khomutov, G.B., Fery, A., Shchukin, D.G., and Moehwald, H. (2010) Nanocomposite microcontainers with high ultrasound sensitivity. *Adv. Funct. Mater.*, **20** (7), 1189–1195.

107 Tiourina, O.P., Antipov, A.A., Sukhorukov, G.B., Larionova, N.I., Lvov, Y., and Moehwald, H. (2001) Entrapment of alpha-chymotrypsin into hollow polyelectrolyte microcapsules. *Macromol. Biosci.*, **1** (5), 209–214.

108 Gao, C.Y., Donath, E., Moehwald, H., and Shen, J.C. (2002) Spontaneous deposition of water-soluble substances into microcapsules: Phenomenon, mechanism, and application. *Angew. Chem., Int. Ed.*, **41** (20), 3789–3793.

109 Ye, S.Q., Wang, C.Y., Liu, X.X., and Tong, Z. (2005) Multilayer nanocapsules of polysaccharide chitosan and alginate through layer-by-layer assembly directly on PS nanoparticles for release. *J. Biomater. Sci., Polym. Ed.*, **16** (7), 909–923.

110 Skirtach, A.G., Antipov, A.A., Shchukin, D.G., and Sukhorukov, G.B. (2004) Remote activation of capsules containing Ag nanoparticles and IR dye by laser light. *Langmuir*, **20** (17), 6988–6992.

111 Skirtach, A.G., Karageorgiev, P., Bedard, M.F., Sukhorukov, G.B., and Moehwald, H. (2008) Reversibly permeable nanomembranes of polymeric microcapsules. *J. Am. Chem. Soc.*, **130** (35), 11572–11573.

112 Vogel, A. and Venugopalan, V. (2003) Mechanisms of pulsed laser ablation of biological tissues. *Chem. Rev.*, **103** (2), 577–644.

113 Bukreeva, T.V., Parakhonsky, B.V., Skirtach, A.G., Susha, A.S., and Sukhorukov, G.B. (2006) Preparation of polyelectrolyte microcapsules with silver and gold nanoparticles in a shell and the remote destruction of microcapsules under laser irradiation. *Crystallogr. Rep.*, **51** (5), 863–869.

114 Skirtach, A.G., Karageorgiev, P., De Geest, B.G., Pazos-Perez, N., Braun, D., and Sukhorukov, G.B. (2008) Nanorods as wavelength-selective absorption centers in the visible and near-infrared regions of the electromagnetic spectrum. *Adv. Mater.*, **20** (3), 506–510.

115 Bedard, M.F., Sadasivan, S., Sukhorukov, G.B., and Skirtach, A. (2009) Assembling polyelectrolytes and porphyrins into hollow capsules with laser-responsive oxidative properties. *J. Mater. Chem.*, **15**, 2226–2233.

116 Wang, K.W., He, Q., Yan, X.H., Cui, Y., Qi, W., Duan, L., and Li, J.B. (2007) Encapsulated photosensitive drugs by biodegradable microcapsules to incapacitate cancer cells. *J. Mater. Chem.*, **17** (38), 4018–4021.

117 Skirtach, A.G., Déjugnat, C., Braun, D., Susha, A.S., Rogach, A.L., and Sukhorukov, G.B. (2007) Nanoparticles distribution control by polymers: aggregates versus nonaggregates. *J. Phys. Chem. C*, **111** (2), 555–564.

118 Bedard, M.F., Braun, D., Sukhorukov, G.B., and Skirtach, A.G. (2008) Toward self-assembly of nanoparticles on polymeric microshells: near-IR release and permeability. *ACS Nano*, **2** (9), 1807–1816.

119 Radziuk, D., Shchukin, D.G., Skirtach, A., Moehwald, H., and Sukhorukov, G. (2007) Synthesis of silver nanoparticles for remote opening of polyelectrolyte microcapsules. *Langmuir*, **23** (8), 4612–4617.

120 Skirtach, A.G., Muñoz Javier, A., Kreft, O., Koehler, K., Alberola, A.P., Moehwald, H., Parak, W.J., and Sukhorukov, G.B. (2006) Laser-induced release of encapsulated materials inside living cells. *Angew. Chem., Int. Ed.*, **45** (28), 4612–4617.

121 Muñoz Javier, A., del Pino, P., Bedard, M.F., Ho, D., Skirtach, A.G., Sukhorukov, G.B., Plank, C., and Parak, W.J. (2008) Photoactivated release of cargo from the cavity of polyelectrolyte capsules to the cytosol of cells. *Langmuir*, **24** (21), 12517–12520.

122 Palankar, R., Skirtach, A.G., Kreft, O., Bedard, M., Garstka, M., Gould, K., Moehwald, H., Sukhorukov, G.B., Winterhalter, M., and Springer, S. (2009) Controlled intracellular release of peptides from microcapsules enhances antigen presentation on MHC Class I molecules. *Small*, **5** (19), 2168–2176.

123 Fery, A., Dubreuil, F., and Mohwald, H. (2004) Mechanics of artificial microcapsules. *New J. Phys.*, **6**, 18.

124 Vinogradova, O.I., Lebedeva, O.V., and Kim, B.S. (2006) Mechanical behavior and characterization of microcapsules. *Annu. Rev. Mater. Res.*, **36**, 143–178.

125 Landau, L.D. and Lifshitz, E.M. (1987) Deformation of shells, in *Theory of Elasticity*, vol. 7 of Course of Theoretical Physics, 4th edn, Nauka, Moscow, pp. 80–84. (In Russian).

126 Lulevich, V.V., Andrienko, D., and Vinogradova, O.I. (2004) Elasticity of polyelectrolyte multilayer microcapsules. *J. Chem. Phys.*, **120** (8), 3822–3826.

127 Vinogradova, O.I., Lebedeva, O.V., Vasilev, K., Gong, H., Garcia-Turiel, J., and Kim, B.S. (2005) Multilayer DNA/Poly(allylamine hydrochloride) microcapsules: assembly and mechanical properties. *Biomacromolecules*, **6** (3), 1495–1502.

128 Kim, B.S., Lebedeva, O.V., Kim, D.H., Caminade, A.M., Majoral, J.P., Knoll, W., and Vinogradova, O.I. (2005) Assembly and mechanical properties of phosphorous dendrimer/polyelectrolyte multilayer microcapsules. *Langmuir*, **21** (16), 7200–7206.

129 Dubreuil, F., Elsner, N., and Fery, A. (2003) Elastic properties of polyelectrolyte capsules studied by atomic-force microscopy and RICM. *Eur. Phys. J. E*, **12**, 215–221.

130 Lulevich, V.V., Radtchenko, I.L., Sukhorukov, G.B., and Vinogradova, O.I. (2003) Deformation properties of nonadhesive polyelectrolyte microcapsules studied with the atomic force microscope. *J. Phys. Chem. B*, **107** (12), 2735–2740.

131 Lebedeva, O.V., Kim, B.S., and Vinogradova, O.I. (2004) Mechanical properties of polyelectrolyte-filled multilayer microcapsules studied by atomic force and confocal microscopy. *Langmuir*, **20** (24), 10685–10690.

132 Feng, J., Wang, B., Gao, C., and Shen, J. (2004) Selective adsorption of microcapsules on patterned polyelectrolyte multilayers. *Adv. Mater.*, **16** (21), 1940–1944.

133 Nolte, M. and Fery, A. (2004) Coupling of individual polyelectrolyte capsules onto patterned substrates. *Langmuir*, **20** (8), 2995–2998.

134 Wang, B., Zhao, Q., Wang, F., and Gao, C. (2006) Biologically driven assembly of polyelectrolyte microcapsule patterns to fabricate microreactor arrays. *Angew. Chem. Int. Ed.*, **45**, 1560–1563.

135 Yu, A., Wang, Y., Barlow, E., and Caruso, F. (2005) Mesoporous silica particles as templates for preparing enzyme-loaded biocompatible microcapsules. *Adv. Mater.*, **17**, 1737–1741.

136 Wang, Y., Yu, A., and Caruso, F. (2005) Nanoporous polyelectrolyte spheres prepared by sequentially coating sacrificial mesoporous silica spheres. *Angew. Chem. Int. Ed.*, **44**, 2888–2892.

137 Sukhorukov, G.B., Antipov, A.A., Voigt, A., Donath, E., and Mohwald, H. (2001) pH-controlled macromolecule encapsulation in and release from polyelectrolyte multilayer nanocapsules. *Macromol. Rapid Commun.*, **22** (1), 44–46.

138 Nolte, M., Schoeler, B., Peyratout, C.S., Kurth, D.G., and Fery, A. (2005) Filled microcavity arrays produced by polyelectrolyte multilayer membrane transfer. *Adv. Mater.*, **17**, 1665–1669.

139 Nolte, M. and Fery, A. (2006) Freestanding polyelectrolyte multilayers as functional and construction elements. *IEE Proc.-Nanobiotechnol.*, **153** (4), 112–120.

140 Timoshenko, S.P. and Gere, J.M. (1963) Buckling of shells, in *Theory of Elastic Stability*, 2nd edn, McGraw-Hill Book Co., Singapore, pp. 457–521.

141 Matsukuma, D., Aoyagi, T., and Serizawa, T. (2009) Adhesion of two physically contacting planar substrates coated with layer-by-layer assembled films. *Langmuir*, **25** (17), 9824–9830.

142 Nolte, A.J., Rubner, M.F., and Cohen, R.E. (2005) Determining the Young's modulus of polyelectrolyte multilayer films via stress-induced mechanical buckling instabilities. *Macromolecules*, **38** (13), 5367–5370.

143 Gao, C., Donath, E., Moya, S., Dudnik, V., and Mohwald, H. (2001) Elasticity of hollow polyelectrolyte capsules prepared by the layer-by-layer technique. *Eur. Phys. J. E*, **5**, 21–27.

144 Mueller, R., Kohler, K., Weinkamer, R., Sukhorukov, G., and Fery, A. (2005) Melting of PDADMAC/PSS capsules investigated with AFM force spectroscopy. *Macromolecules*, **38** (23), 9766–9771.

145 Jackman, R.J., Duffy, D.C., Ostuni, E., Willmore, N.D., and Whitesides, G.M. (1998) Fabricating large arrays of microwells with arbitrary dimensions and filling them using discontinuous dewetting. *Anal. Chem.*, **70** (11), 2280–2287.

146 Yin, Y. and Xia, Y. (2001) Self-assembly of monodispersed spherical colloids into complex aggregates with well-defined sizes, shapes, and structures. *Adv. Mater.*, **13** (4), 267–271.

147 De Cock, L.J., De Koker, S., De Geest, B.G., Grooten, J., Vervaet, C., Remon, J.P., Sukhorukov, G.B., and Antipina, M.N. (2010) Polymeric multilayer capsules in drug delivery. *Angew. Chem. Int. Ed.*, **49** (39), 6954–6973.

148 Sukhorukov, G.B. and Moehwald, H. (2007) Multifunctional cargo systems for biotechnology. *Trends Biotechnol.*, **25** (3), 93–98.

# 37
# Domain-Containing Functional Polyelectrolyte Films: Applications to Antimicrobial Coatings and Energy Transfer

*Aurélie Guyomard, Bernard Nysten, Alain M. Jonas, and Karine Glinel*

## 37.1
## Introduction

Polyelectrolyte multilayers (PEMs) are ideally suited to the fabrication of functional coatings with specific optical, electrical or bioactive properties. Indeed, they can be used as reservoirs to trap various functional or bioactive molecules, such as dyes, enzymes, DNA, proteins, growth factors, or nanoparticles, to produce thin films for biosensors and implanted medical devices, to protect material surfaces against bacterial adhesion or corrosion, or, more generally, to provide a new function to the surface of a material [1–4]. It is, therefore, important to develop methodologies able to load functional molecules within PEMs, and to be able to control their release or to make sure they are trapped irreversibly in the multilayers, depending on the envisioned application.

A series of methods has been developed to dope PEMs with functional molecules. Charged components can be directly embedded in the multilayers by adsorption or co-adsorption during LbL (layer-by-layer) construction. For instance, enzymes [5, 6], cationic antibacterial peptides [7, 8], hydrophilic dyes [9–12] and DNA molecules [5, 13, 14] were successfully inserted into PEMs to fabricate (bio)active coatings. However, this approach is restricted to hydrophilic charged substances that are able to interact through electrostatic interactions with the polyelectrolyte chains constituting the PEM. Moreover, strong electrostatic interactions between the active substance and the polyelectrolyte matrix can reduce considerably the activity of the guest molecules, which can be particularly detrimental in the case of bioactive compounds. Another critical issue remains the accessibility of the immobilized active substance, notably when the activity requires interactions or exchanges with the surrounding environment. This is especially the case for embedded biological compounds, such as proteins or enzymes, which interact with high molar mass substances or large entities like cells. In this particular case, the embedding of the active substance into the PEM has to be performed during the last deposition steps in order to locate it closer to the external surface.

*Multilayer Thin Films: Sequential Assembly of Nanocomposite Materials*, Second Edition.
Edited by Gero Decher and Joseph B. Schlenoff.
© 2012 Wiley-VCH Verlag GmbH & Co. KGaA. Published 2012 by Wiley-VCH Verlag GmbH & Co. KGaA.

Another approach to incorporate a functional agent into a PEM involves first building the LbL assembly, followed by the post-diffusion of the guest molecule into the polyelectrolyte matrix [15–18]. Again, this method is suitable only for water-soluble substances having enough affinity with the polyelectrolyte film via charged or hydrogen-bonding groups. Moreover, this method is intrinsically slow and can be used only for low molar mass compounds which can diffuse through the PEM free volume. Finally, it is not easily applied to the construction of multicompartmentalized films, wherein different molecules are stacked in a spatial sequence of isolated functional compartments.

Therefore, it remains challenging to incorporate poorly-charged or hydrophobic (bio)active substances into traditional multilayers in order to prepare functional coatings, notably due to their lack of affinity with the polyelectrolyte matrix. A possible approach to overcome this issue is to incorporate into the structure of the PEM, nanodomains showing a specific affinity for the guest molecules. For instance, micellar aggregates of amphiphilic block copolymers with charged coronas were successfully used as building blocks for the fabrication of LbL films [19–21]. Incorporation of hydrophobic dyes in such films was achieved by preloading the micelles before LbL assembly [20, 22]. Liposomes [23], dendrimers [24–26], cyclodextrins [27] and microgels [28] were also explored to serve as nanocontainers for guest agents to be incorporated into PEM films. Moreover, we have developed an alternative strategy based on the pre-assembly of a hydrophobic guest substance with amphiphilic charged biopolymers to produce complex building blocks that can be incorporated into PEMs [29–31]. This method was used to incorporate apolar (macro) molecules into polyelectrolyte films [30, 31].

Once the functional agent is incorporated into polyelectrolyte films, a second challenge is to prevent or, in contrast, to control its slow diffusion through the assembly. The slow release of the active component from the films is often required for biological applications in which the bioactivity is based on a close interaction between the active molecules and biological entities like cells or proteins. In contrast, some applications require a permanent immobilization of the active molecules into the films. This is notably the case for coatings developed for optical, electronic or biosensing devices.

The control of the diffusion of guest molecules through the PEM films can be achieved by changing the porosity of the layer. Different post-treatments based on the variation of physical and chemical surrounding conditions were explored [3, 32]. Crosslinking of the pre-assembled functional films was also developed [3, 33]. However, all these methodologies are not completely satisfying since they can affect the intrinsic activity of the guest substances and, depending on the size of the guest molecules, do not permit a fine control of their diffusion into the multilayers.

In this context, other strategies based on the control of the microstructure of PEMs are being explored. The idea is to fabricate more or less permeable physical barriers to control the entrapment or the diffusion of guest molecules from the films. A possible approach is, for instance, to cap the entire film with a semipermeable or impermeable layer, such as a nonpolar polymer layer [34], a lipid bilayer [35] or a layer of wax particles [36], which results in the construction of single-compartment films. More sophisticated multilayers composed of several stratified compartments can also be prepared by incorporating internal barriers into the film structure. These barriers can be based on polyelectrolyte layers whose chemical composition differs from that of the polyelectrolyte compartments [37, 38], or on inorganic particles such as clay platelets [39–41]. The incorporation of an internal barrier within the films is achieved by varying the LbL depositing sequence during multilayer build-up, which is after all the main interest of LbL assembly compared to other coating methodologies.

Here we report the preparation of polyelectrolyte multilayers whose structure contains separated compartments which can be advantageously used to trap guest molecules in order to fabricate functional coatings. Two different systems differing by their internal structure are described (Scheme 37.1): (i) polyelectrolyte films containing hydrophobic nanodomains randomly distributed in a hydrophilic polyelectrolyte matrix and (ii) stratified multilayers incorporating a succession of polyelectrolyte compartments separated by organic or hybrid polyelectrolyte barriers. The elaboration as well as the internal characterization of these different systems will be described. Moreover, practical examples showing the fabrication of functional films will be reported.

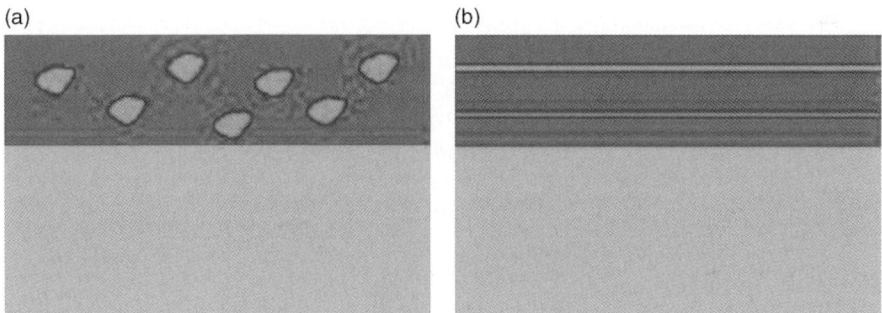

**Scheme 37.1** Description of the polyelectrolyte multilayer morphologies described in this chapter: (a) films incorporating hydrophobic nanodomains; (b) multi-compartment stratified films incorporating organic or hybrid barriers.

## 37.2
## Polyelectrolyte Films Incorporating Randomly Distributed Hydrophobic Nanodomains for Antimicrobial Applications

### 37.2.1
### Hydrophobic Nanodomains in Hydrophilic PEMs with Amphiphilic Macromolecules

In order to incorporate hydrophobic nanodomains in PEMs, we developed amphiphilic derivatives by grafting alkyl chains on carboxymethylpullulan (CMP), an anionic flexible polysaccharide [29]. These hydrophobically modified polysaccharides were shown to develop spontaneous intra- and inter-molecular hydrophobic

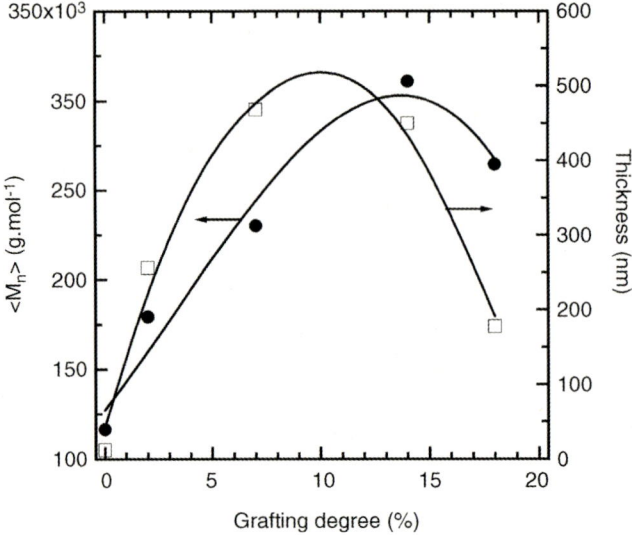

**Figure 37.1** Variation of the apparent number-average molar mass ($<M_n>$) of decyl-grafted CMP derivatives in solution, and of the thickness of (PEI/decyl grafted-CMP) 10-layer pair films, versus the degree in grafted decyl chains of the decyl-grafted CMP derivatives.

associations in aqueous solution, depending on the content and on the length of the grafted alkyl chains. This self-associating behavior results in the formation of hydrophobic aggregates which can advantageously be exploited to entrap various nonpolar compounds such as dyes, peptides or even proteins.

The growth of polyelectrolyte films based on CMP derivatives grafted by decyl groups was shown to be in good agreement with the associating behavior measured for these amphiphilic polymers in aqueous solution. Indeed, the apparent molar mass of the amphiphilic aggregates, measured as a function of the grafting degree in decyl chains, displayed a maximum which corresponds to the formation of large intermolecular aggregates in solution (Figure 37.1). By comparison, the polyelectrolyte multilayers based on the same amphiphilic derivatives and various polycations showed also a maximum of thickness as a function of the grafting degree in alkyl chains, for a given number of deposited layer pairs (Figure 37.1). The grafting degree inducing the formation of larger macromolecular aggregates in solution, and the one leading to thicker films in LbL assembly, are close to each other. This suggests that the large amphiphilic aggregates present in solution are adsorbed and mainly preserved during the LbL deposition process.

The incorporation of these aggregates into the films results in the presence of hydrophobic clusters which can trap hydrophobic molecules such as poorly water-soluble Nile Red dye molecules. Two different methodologies based on either the diffusion of the dye in pre-assembled amphiphilic polyelectrolyte films, or on the pre-assembly of amphiphilic derivatives with the dye before film build-up, led to the successful incorporation of the hydrophobic dye molecules in the films, as confirmed by UV spectroscopy [30]. Moreover, the higher the grafting degree in alkyl chains, the

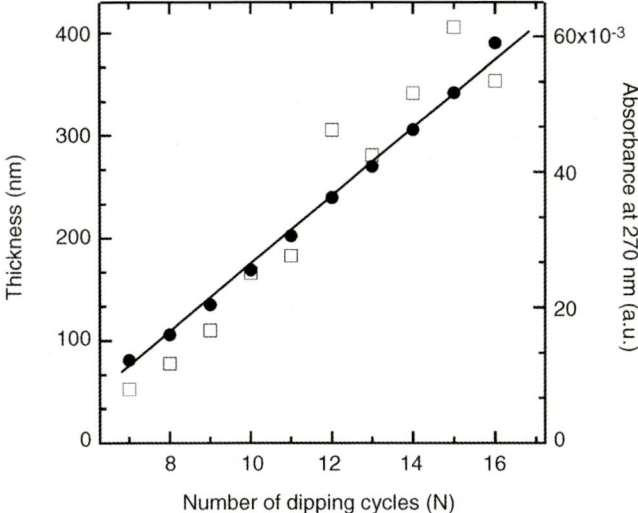

**Figure 37.2** Variation of the thickness (circle) and UV absorbance (square) of (PEI/decyl-grafted CMP(18%) + Nile Red) films as a function of the number of dipping cycles (N).

higher the content of dye entrapped in the films. The amount of dye inserted into the films can also be finely tuned since it increases linearly with the number of polycation/anionic amphiphilic CMP layer pairs deposited (Figure 37.2).

### 37.2.2
### Entrapment of an Antibacterial Peptide in Hydrophobic Nanodomains

Polyelectrolyte films containing hydrophobic nanodomains were tested to trap a poorly water-soluble antibacterial peptide, namely gramicidin A, to prepare bactericidal coatings [31]. Gramicidin molecules, which consist of 15 amino acid residues, were first pre-assembled in aqueous solution with a decyl-grafted CMP derivative to form an anionic complex (Scheme 37.2). The efficiency of CMP derivatives for solubilizing the hydrophobic peptide was systematically investigated as a function of the grafting degree in the decyl chains and of the molar mass of the CMP derivatives. UV measurements performed on the resulting solutions showed that the higher the alkyl group content, the higher the efficiency to trap gramicidin A [31]. This result confirmed the previous results obtained with Nile Red dye [30]. In contrast to the entrapment of low molar mass molecules, the efficiency of the pre-assembly of amphiphilic CMP with gramicidin was shown to depend also on the molar mass of the amphiphilic chains. Indeed, a higher content of gramicidin A could be entrapped by alkyl-grafted CMP of lower molar mass [31]. As a consequence, the macromolecular characteristics of the amphiphilic derivatives have to be adjusted for each considered guest compound in order to optimize its entrapment in polyelectrolyte films. Interestingly, circular dichroism measurements performed on a solution

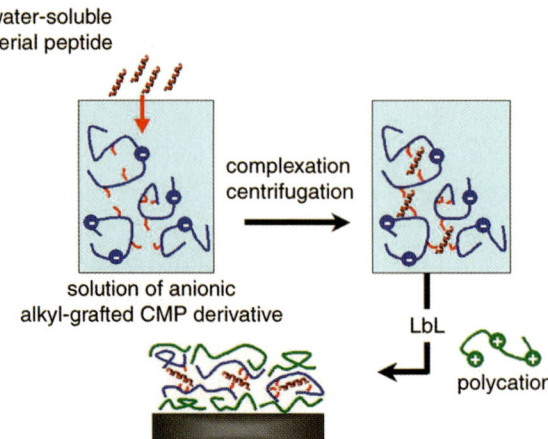

**Scheme 37.2** Preparation of amphiphilic polyelectrolyte films containing a poorly water-soluble antibacterial peptide.

containing gramicidin A pre-assembled with a low molar mass decyl-grafted CMP, confirmed that the helical conformation of the peptide molecules was not affected by the presence of the amphiphilic macromolecules, which is an essential requirement to keep the peptide active against bacteria.

The anionic pre-assembly complex of decyl-grafted CMP and gramicidine A was subsequently layer-by-layer assembled with cationic poly(L-lysine) (PLL) to form peptide-functionalized films (Scheme 37.2). UV and circular dichroism measurements performed on these multilayers confirmed that the poorly soluble peptide molecules were successfully entrapped without denaturation of their secondary helical structure [31]. Bacterial assays performed on these peptide-functionalized films revealed a high surface biocidal activity against gram positive bacteria like *Enterococcus faecalis*. Indeed the presence of lyzed bacterial cells was clearly observed on the surface of gramicidin-loaded films while large microcolonies were detected on non-loaded films (Figure 37.3).

Moreover, a slow release of the peptide molecules in the surrounding solution was evidenced by monitoring the growth of planktonic cells in solutions put in contact with gramicidin-loaded films. Not surprisingly, the efficiency of the films in inhibiting bacterial growth in their surrounding environment increases with the number of gramicidin-loaded layers deposited during the LbL process, as shown by the strong decrease in the optical density measured for bacterial suspensions in contact with peptide-loaded films (Figure 37.4). Interestingly, it was also shown that peptide molecules were only partially released from the films, even after extensive rinsing of the polyelectrolyte multilayers. The gramicidin molecules remained partially entrapped in the coatings, which contributed to prevention of the formation of biofilms on the film surface with time. Only a few individual bacteria were observed on the film surface, even after a 7-day test [31].

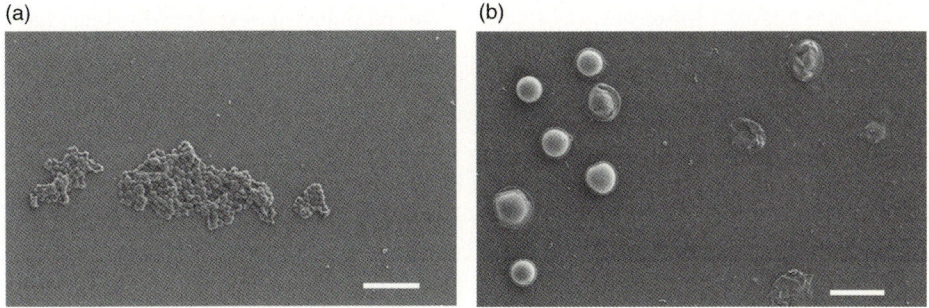

**Figure 37.3** SEM images of E. faecaelis bacteria adhered on polylectrolyte films without (a) and with (b) gramicidin A. Scale bar represents 10 μm.

These results demonstrate that the development of LbL films containing amphiphilic charged polyectrolytes is a powerful strategy to incorporate poorly water-soluble guest molecules into multilayers in order to fabricate active surfaces. However, it is important to note that this general approach is not universal. Although antibiotic molecules, antibacterial peptides and poorly soluble dyes have been

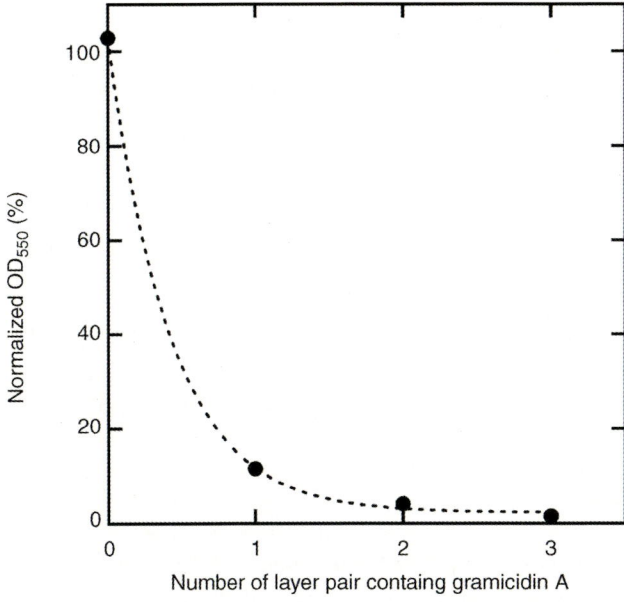

**Figure 37.4** Variation of the normalized optical density measured at 550 nm ($OD_{550}$) for bacterial suspensions after 6 h incubation in the presence of polyelectrolyte films containing a given number of (PLL/decyl-grafted CMP (18%) + gramicidin A) layer pairs. The normalized $OD_{550}$ was computed by comparison with the $OD_{550}$ measured for a bacterial suspension in the absence of polyelectrolyte multilayer.

successfully entrapped in PEMs of amphiphilic polyelectrolytes that form hydrophobic clusters, a very different behavior was observed for amphiphilic polyelectrolytes which bear hydrophobic chromophore groups [42]. Indeed, the complete disruption of the hydrophobic aggregates formed in solution was observed upon LbL deposition of these systems, which precludes the formation of hydrophobic nanodomains into the films. This illustrates that the molecular characteristics of amphiphilic polyelectrolytes have to be finely tuned to attain an efficient entrapment of guest substances into the coatings.

## 37.3
### Multicompartmentalized Stratified Polyelectrolyte Films for Control of Energy Transfer

The preparation of coatings composed of a succession of stratified compartments offers the possibility to confine active guest molecules at a specific vertical location in the films. Such an internal organization is of interest for the preparation of multifunctional coatings in which an oriented cascade of events such as enzymatic reactions, or transfers of energy or charge can occur. To achieve this goal, the active compounds have to be trapped in successive compartments without possible diffusion into neighboring compartments, but nevertheless with the possible diffusion of substrates and products between compartments, or with the possible transfer of energy or charge between them. In addition, a high degree of control over the distance between the successive compartments may be needed in the nanometer range, in order to avoid a disruption of the cascade. This is especially true for charge or energy transfers, which most often occur over relatively small distances [43]. Therefore, methods to generate, by LbL, ultra-thin barriers of well controlled permeability at precise locations need to be developed.

### 37.3.1
### How Precisely Can the Stratification of a LbL Film Be Controlled?

LbL assembly is *per se* a methodology that permits one to structure films in the vertical direction, simply based on the proper selection of the adsorption sequence [37, 41]. However, due to interpenetration of successive layers, together with, in some cases, polyelectrolyte exchange or diffusion during growth [44, 45], there might be a difference between the actual stratification of a multilayer and its adsorption sequence. We have therefore investigated systems having as structure $((A/B)_m/(A/C)_n)_p$, where the $(A/B)_m$ sequence is a hybrid barrier and the $(A/C)_n$ sequence is an organic compartment, and checked conditions to obtain multi-compartmentalized PEMs having a structure replicating the sequence of adsorption steps used to create them. The polycation A of both the compartments and the barriers was poly(diallylpyrrolidinium bromide) (PDAPB). The polyanions B of the barriers were clay platelets (exfoliated Laponite), and the polyanion C of the organic compartments

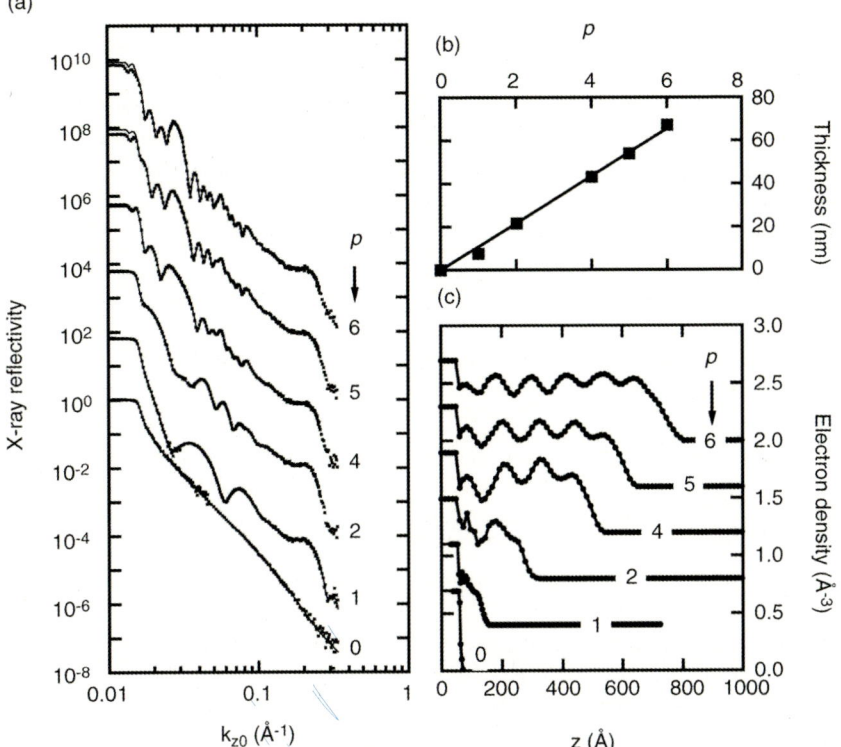

**Figure 37.5** X-ray reflectometry results on multicompartmentalized multilayers grown on Si wafers. The adsorption sequence is $((PDAPB/clay)_2/(PDAPB/PSS)_{10})_p$. (a) X-ray reflectivity (dots) and fits (lines); the data are multiplied by integral powers of 100 for clarity. (b) Total film thickness versus $p$, the number of deposition cycles of the $(PDAPB/clay)_2/(PDAPB/PSS)_{10}$ sequence. (c) Electron density profiles obtained from the fit of the XRR data; the profiles are shifted vertically by multiples of 0.4 Å$^{-3}$ for clarity.

was poly(styrene sulfonate) (PSS). Because the silicate clay platelets are of larger electron density than organic materials, it becomes possible to have a direct view of the stratification by X-ray reflectometry (XRR), without having to deuterate specific layers and use neutron reflectometry, as is usually done [46–48].

As an example, Figure 37.5a displays the reflectivity of a series of $((PDAPB/clay)_2/(PDAPB/PSS)_{10})_p$ samples grown from salt-free aqueous solutions on Si wafers, with $p$ ranging from 0 to 6. The electron density profiles of these samples were obtained by fitting the XRR data by methods described elsewhere [49], and are displayed in Figure 37.5c. The regular modulation of electron density with $\sim$12 nm repeat period testifies for the proper stratification of the films, which consist of a succession of organic compartments of lower density separated by denser clay-rich barriers. The transition between the compartments and the barriers is not sharp, due to the

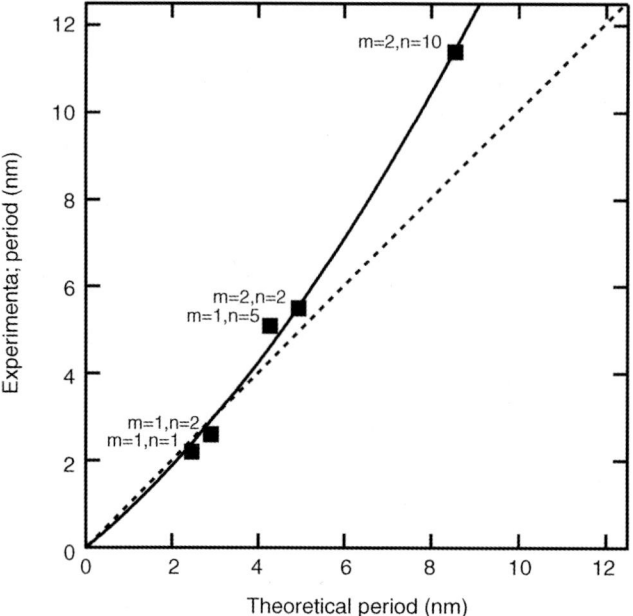

**Figure 37.6** Experimental repeat period of multilayers of adsorption sequence ((PDAPB/clay)$_m$/(PDAPB/PSS)$_n$)$_p$, grown on Si wafers from salt-free solutions, versus the predicted repeat period based on the increment of thickness of (PDAPB/clay) and (PDAPB/PSS) simple multilayers. The dashed line corresponds to the experimental period being equal to the predicted one.

interpenetration of successive layers [45] coupled to the intrinsic roughness of clay-based multilayers [39]; this results in a very slight progressive blurring of the stratification as the distance from the substrate increases. In addition, the total thickness of the multilayer scales linearly with $p$ (Figure 37.5b).

Both the (PDAPB/clay) and (PDAPB/PSS) systems grow linearly from salt-free water solutions, with an increment of thickness per cycle of deposition of, respectively, 2.02 and 0.45 nm. If this linearity is preserved in multicompartmentalized films having ((PDAPB/clay)$_m$/(PDAPB/PSS)$_n$)$_p$ as adsorption sequence, the expected repeat period of the stratified multilayer should simply be $2.02m + 0.45n$ nm. Figure 37.6 displays the experimentally determined repeat period versus the predicted one. The equivalence is preserved for repeat periods below ~6 nm; above this value, corresponding to relatively large values of $m$ and/or $n$, the experimental period is larger than predicted by ~25%. The simple additive rule is thus only valid when the number of deposition cycles of the repeating unit is limited, which illustrates further the intrinsic complexity of LbL assembly. Obviously, deviations from additivity are more probable for pairs of polycations and polyanions that do not exhibit linear growth when assembled in a PEM. This was actually checked by performing similar experiments, where the (PDAPB/PSS) component was grown from aqueous solutions 0.1 M in NaCl. In this case, the (PDAPB/PSS) multilayers no longer grow linearly; as a result, larger deviations, up to 50%, were found between predicted and experimental values of the thickness of the repeating unit of $((PDAPB/clay)_m/(PDAPB/PSS)_n)_p$ assemblies.

Nevertheless, these results indicate that it is possible to build internally-stratified multilayers with a high degree of control over the internal structure, for repeat periods that are between 3 and 6 nm, provided the conditions of assembly are properly selected to favor linearity. Since these distances are in the range into which nonradiative fluorescence resonance energy transfer (FRET) occurs [43], it should be possible to fabricate by LbL a multicompartmentalized system into which light energy would be absorbed in a first compartment, then driven in a series of cascades towards a final compartment, that is, to fabricate a "light diode" where the flow of energy occurs only in one direction. However, this requires being able to confine a succession of complementary dyes in each compartment in turn, which is only possible if dye-impermeable barriers can be developed.

### 37.3.2
### Fabrication of Dye-Impermeable Polyelectrolyte Barriers

The deposition of a polyelectrolyte stratum whose nature differs from the main polyelectrolyte sequence is an easy way to construct an internal barrier. For instance, the poly(allylamine) (PAH)/polystyrene (PSS) pair was successfully used to construct a barrier between hyaluronic acid (HA)/poly(L-lysine) (PLL) compartments [37]. This barrier was shown to prevent the diffusion of "free" polyelectrolyte chains between two consecutive compartments while allowing the diffusion of protons. However, purely organic LbL barriers are generally inefficient at preventing the diffusion of small molecules such as dyes, as was shown by FRET experiments on a poly (vinylbenzylammonium chloride) PVBAC/PSS barrier separating two compartments loaded with complementary dyes [50]. More efficient barriers could be obtained by replacing PSS by negatively-charged clay platelets, in which case the increased tortuosity of possible dye diffusion pathways was shown to slow down dye mixing during multilayer build-up [50]. However, this effect was moderate and did not prevent partial dye mixing, even when increasing the thickness of the barrier in the 10 nm range, above which FRET does not occur.

Dye-impermeable barriers could be created in PEMs by using PAH/poly(acrylic acid) (PAA) layer pairs crosslinked thermally [50]. Experimentally, the multilayers were heated in an inert atmosphere after deposition of each $(PAH/PAA)_m$ layer pair sequence onto $(PVBAC/PSS)_n$ compartments doped with a succession of complementary dyes. FRET measurements performed on these systems revealed that nonradiative energy transfer occurred from compartment to compartment, but with a decreasing efficiency as the thickness of the PAH/PAA barrier increased: the FRET totally vanished for a $(PAH/PAA)_m$ barrier thickness larger than 6 nm. These results confirm that a strongly-crosslinked PAH/PAA barrier is able to prevent the diffusion of dyes between two consecutive compartments since FRET would have been measured if diffusion of the dyes through the barriers had occurred.

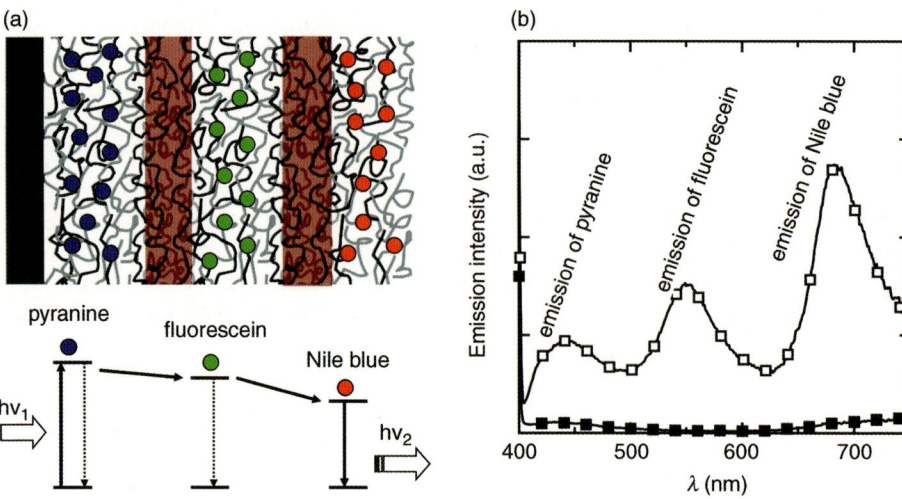

**Figure 37.7** Cascade of FRET in a three-compartment dye-loaded multilayer (barriers are in brown). (a) Scheme of the structure of the multilayer. (b) Emission intensity upon excitation at 380 nm, for a simple one-compartment Nile Blue-loaded multilayer (black squares), and for the three-compartment system (open squares).

### 37.3.3
### Control of a Cascade of Events in a Multicompartmentalized Stratified Polyelectrolyte Film

The strategy based on the incorporation of crosslinked polyelectrolyte barriers in a multicompartment multilayer was finally explored to fabricate an optical device mimicking a light-harvesting antenna system. For this, multilayers composed of three stacked (PVBAC/PSS)$_5$ compartments doped by co-adsorption [10] with pyranine, fluorescein and Nile Blue dyes, respectively, and separated by crosslinked (PAH/PAA)$_5$ barriers, were tested to produce a FRET cascade [50]. The first donor compartment is loaded with pyranine, which transfers energy to the next, fluorescein-loaded compartment, which in turn transfers energy to a last, Nile Blue-loaded acceptor compartment (Figure 37.7a). This triplet of dyes respects the conditions required for FRET, with the emission and absorption peaks of each neighboring dye (in the multilayers) having a substantial overlap. However, the emission of pyranine and the absorption of Nile Blue hardly overlap, and the distance between the compartments containing these two dyes is too large to allow direct FRET between the first and third compartments to occur. Therefore, energy transfer has to be relayed by the intermediate, fluorescein-loaded compartment. This can only occur if the distance between successive compartments is small enough, which is the case since the thickness of the crosslinked (PAH/PAA)$_5$ barriers was selected to be 2.5 nm.

Figure 37.7b is the fluorescence emission spectrum of the three-compartment multilayer when excited at 380 nm. At this excitation wavelength, Nile Blue is not excited, as demonstrated by the absence of fluorescence of a Nile Blue-loaded single compartment multilayer (black squares). However, it emits strongly in the three-compartment system (open squares), which proves the existence of a cascade of energy transfers. The fluorescence of the pyranine and fluorescein compartments is strongly decreased, although not zero, which indicates incomplete transfer of energy. Nevertheless, these results confirm that a unidirectional FRET cascade exists in the multilayer under proper illumination conditions, resulting in a concentration of light energy at the outer surface of the multilayer. This demonstrates that LbL assembly, combined with doping, can fabricate functional systems to control structure in the 5 nm range. This proof-of-concept demonstration in the field of energy transfer could be easily transposed to other systems which require cascades of events to occur at separate locations in the film.

## 37.4
## Conclusions and Perspectives

Polyelectrolyte multilayers can be loaded with functional molecules, providing them with new properties. When the functional molecules are hydrosoluble, they may simply be co-adsorbed; by selecting suitable conditions of deposition, and specific assembly sequences, it is possible to construct multicompartmentalized multilayers where each compartment contains a different functional molecule separated by impermeable barriers. When the functional molecules are hydrophobic, they may be complexed with amphipilic polyelectrolytes before being deposited in the films, which then consist in part of hydrophobic domains. In both cases, the multilayers are thus heterogeneous, although the spatial distribution and the shape of the nanodomains is entirely different.

Examples of functional multilayers were also given in this chapter. By loading the hydrophobic nanodomains of a PEM with an antibacterial peptide, efficient biocidal coatings have been produced, that were stable for prolonged immersion in aqueous solutions. By loading three compartments with complementary dyes, it was possible to control the flow of light energy in a multilayer, and to concentrate light energy at its outer surface. These are just two examples, among the many possibilities offered by LbL assembly, which illustrate how the internal structure of PEM films can be tuned at a very small scale to attain complex properties. Other examples are emerging, and it is to be expected that much more complex functional coatings will appear in the coming years, enabled by the powerful control of LbL assembly over structure and thus function.

## Acknowledgments

The authors are indebted to a series of researchers who contributed to the results reported here, mainly E. Dé, T. Jouenne, A. Moussa, G. Muller, E. Nicol, and S. Peralta. The authors also benefited from discussions with A. Laschewsky and J.-L. Habib-Jiwan. Financial support was provided by the Communauté Française de

Belgique (ARC 06-11/339), the Belgian Federal Science Policy (IAP-PAI P6/27), the F.R.S.-FNRS (MIS-Ulysse), the French "Matériaux Polymères, Plasturgie" network and the French National Research Agency (Grant ANR-06-BLAN-0196-01). BN is Senior Research Associate of the F.R.S.-FNRS.

## References

1 Bertrand, P., Jonas, A.M., Laschewsky, A., and Legras, R. (2000) *Macromol. Rapid Commun.*, **21**, 319–348.

2 Hammond, P.T. (1999) *Curr. Opin. Colloid Interface Sci.*, **4**, 430–442.

3 Tang, Z., Wang, Y., Podsiadlo, P., and Kotov, N.A. (2006) *Adv. Mater.*, **18**, 3203–3224.

4 Boudou, T., Crouzier, T., Ren, K., Blin, G., and Picart, C. (2010) *Adv. Mater.*, **22**, 441–467.

5 Lvov, Y.M., Lu, Z., Schenkman, J.B., Zu, X., and Rusling, J.F. (1998) *J. Am. Chem. Soc.*, **120**, 4073–4080.

6 Onda, M., Lvov, Y., Ariga, K., and Kunitake, T. (1996) *Biotechnol. Bioeng.*, **51**, 163–167.

7 Etienne, O., Picart, C., Taddei, C., Haikel, Y., Dimarcq, J.L., Schaaf, P., Voegel, J.C., Ogier, J.A., and Egles, C. (2004) *Antimicrob. Agents Chemother.*, **48**, 3662–3669.

8 Shuklaa, A., Fleminga, K.E., Chuanga, H.F., Chaua, T.M., Loosea, C.R., Stephanopoulosa, G.N., and Hammond, P.T. (2010) *Biomaterials*, **31**, 2348–2357.

9 Nicol, E., Moussa, A., Habib-Jiwan, J.-L., and Jonas, A.M. (2004) *J. Photochem. Photobiol. A: Chemistry*, **167**, 31–35.

10 Nicol, E., Habib-Jiwan, J.-L., and Jonas, A.M. (2003) *Langmuir*, **19**, 6178–6186.

11 Ariga, K., Lvov, Y., and Kunitake, T. (1997) *J. Am. Chem. Soc.*, **119**, 2224–2231.

12 Tedeschi, C., Caruso, F., Mohwald, H., and Kirstein, S. (2000) *J. Am. Chem. Soc.*, **122**, 5841–5848.

13 Lvov, Y.M., Decher, G., and Sukhorukov, G.B. (1993) *Macromolecules*, **26**, 5396–5399.

14 Shi, X., Sanedrin, R.J., and Zhou, F. (2002) *J. Phys. Chem. B*, **106**, 1173–1180.

15 Chung, A.J., and Rubner, M.F. (2002) *Langmuir*, **18**, 1176–1183.

16 Wang, B., Gao, C., and Liu, L. (2005) *J. Phys. Chem. B*, **109**, 4887–4892.

17 Schneider, A., Vodouhê, C., Richert, L., Francius, G., Le Guen, E., Schaaf, P., Voegel, J.-C., Frisch, B., and Picart, C. (2007) *Biomacromolecules*, **8**, 139–145.

18 Berg, M.C., Zhai, L., Cohen, R.E., and Rubner, M.F. (2006) *Biomacromolecules*, **7**, 357–364.

19 Xu, L., Zhu, Z., and Sukhishvili, S.A. (2011) *Langmuir*, **27**, 409–415.

20 Ma, N., Zhang, H., Song, B., Wang, Z., Zhang, X. (2005) *Chem. Mater.*, **17**, 5065–5069.

21 Zhang, X., Chen, H., and Zhang, H. (2007) *Chem. Commun.*, **14**, 1395–1405.

22 Kim, B.-S., Park, S.W., and Hammond, P.T. (2008) *ACS Nano*, **2**, 386–392.

23 Michel, M., Arntz, Y., Fleith, G., Toquant, J., Haikel, Y., Voegel, J.-C., Schaaf, P., and Ball, V. (2006) *Langmuir*, **22**, 2358–2364.

24 Watanabe, S. and Regen, S.L. (1994) *J. Am. Chem. Soc.*, **116**, 8855–8856.

25 Khopade, A.J. and Caruso, F. (2002) *Nano Lett.*, **2**, 415–418.

26 Jiao, Q., Yi, Z., Chen, Y., and Xi, F. (2008) *Polymer*, **49**, 1520–1526.

27 Jessel, N., Oulad-Abdelghani, M., Meyer, F., Lavalle, P., Haïkel, Y., Schaaf, P., and Voegel, J.-C. (2006) *Proc. Natl. Acad. Sci. USA*, **103**, 8618–8621.

28 Wang, L., Wang, X., Xu, M., Chen, D., and Sun, J. (2008) *Langmuir*, **24**, 1902–1909.

29 Guyomard, A., Muller, G., and Glinel, K. (2005) *Macromolecules*, **38**, 5737–5742.

30 Guyomard, A., Nysten, B., Muller, G., and Glinel, K. (2006) *Langmuir*, **22**, 2281–2287.

31 Guyomard, A., Dé, E., Jouenne, T., Malandain, J.-J., Muller, G., and Glinel, K. (2008) *Adv. Funct. Mater*, **18**, 758–765.

32 Zelikin, A.N. (2010) *ACS Nano*, **4**, 2494–2509.
33 Crouzier, T., Ren, K., Nicolas, C., Roy, C., and Picart, C. (2009) *Small*, **5**, 598–608.
34 Rouse, J.H. and Ferguson, G.S. (2002) *Langmuir*, **18**, 7635–7640.
35 Moya, S., Donath, E., Sukhorukov, G.B., Auch, M., Bäumler, H., Lichtenfeld, H., and Möhwald, H. (2000) *Macromolecules*, **33**, 4538–4544.
36 Glinel, K., Prevot, M., Krustev, R., Sukhorukov, G.B., Jonas, A.M., and Möhwald, H. (2004) *Langmuir*, **20**, 4898–4902.
37 Méndez Garza, J., Schaaf, P., Muller, S., Ball, V., Stoltz, J.-F., Voegel, J.-C., and Lavalle, P. (2004) *Langmuir*, **20**, 7298–7302.
38 Dubas, S.T., Farhat, T.R., and Schlenoff, J.B. (2001) *J. Am. Chem. Soc.*, **123**, 5368–5369.
39 Glinel, K., Jonas, A.M., and Laschewsky, A. (2002) *J. Phys. Chem. B*, **106**, 11246–11252.
40 Glinel, K., Laschewsky, A., and Jonas, A.M. (2001) *Macromolecules*, **34**, 5267–5274.
41 Glinel, K., Jonas, A.M., Laschewsky, A., and Vuillaume, P.Y. (2003) *Multilayer Thin Films* (eds G. Decher and J.B. Schlenoff), Wiley-VCH, Weinheim, pp. 177–205.
42 Laschewsky, A., Mallwitz, F., Baussard, J.-F, Cochin, D., Fisher, P., Habib-Jiwan, J.-L., and Wischeroff, E. (2004) *Macromol. Symp.*, **211**, 135–156.
43 Lakowicz, J.R. (2006) *Principles of Fluorescence Spectroscopy*, 3rd ed., Springer, New York.
44 Picart, C., Mutterer, J., Richert, L., Luo, Y., Prestwich, G.D., Schaaf, P., Voegel, J.-C., and Lavalle, P. (2002) *Proc. Natl. Acad. Sci. USA*, **99**, 12531–12535.
45 Decher, G. (1997) *Science*, **277**, 1232–1237.
46 Schmitt, J., Grunewald, T., Decher, G., Pershan, P.S., Kjaer, K., and Lösche, M. (1993) *Macromolecules*, **26**, 7058–7053.
47 Kellogg, G.J., Mayes, A.M., Stockton, W.B., Ferreira, M., Rubner, M.F., and Satija, S.K. (1996) *Langmuir*, **12**, 5109–5113.
48 Lösche, M., Schmitt, J., Decher, G., Bouwman, W.G., and Kjaer, K. (1998) *Macromolecules*, **31**, 8893–8906.
49 Arys, X., Laschewsky, A., and Jonas, A.M. (2001) *Macromolecules*, **34**, 3318–3330.
50 Peralta, S., Habib-Jiwan, J.-L., and Jonas, A.M. (2009) *ChemPhysChem*, **10**, 137–143.

# 38
# Creating Functional Membranes Through Polyelectrolyte Adsorption
*Merlin L. Bruening*

## 38.1
## Introduction

Since the initial wave of fundamental research on layer-by-layer (LbL) polyelectrolyte deposition [1], many studies have focused on potential applications of polyelectrolyte multilayers (PEMs) in areas ranging from photovoltaics to drug delivery [2–5]. PEMs are attractive because the LbL technique affords control over film thickness at the nanometer scale, and the wide variety of species that can form these coatings provides an extensive range of film properties and functions. Moreover, LBL adsorption can occur on substrates with a wide range of geometries.

This chapter examines modification of porous substrates with PEMs to create membranes that capture and digest proteins, catalyze reactions, and separate ions. These different applications have very different coating requirements. In catalytic reactors the films should coat the membrane pores without blocking them and facilitate attachment of catalysts such as enzymes or nanoparticles (Figure 38.1a), whereas when creating ion-separation membranes, the PEM should form a continuous, ultrathin skin on the surface of the support (Figure 38.1b). Thus, this chapter first reviews studies of polyelectrolyte deposition in membrane pores and then examines applications that may benefit from such PEMs. A subsequent section explores separation membranes that contain PEMs as selective skins on porous substrates. The functions of PEM-modified membranes, which range from rapid protein digestion to complete removal of textile dyes from simulated waste streams, are quite remarkable. A final section discusses important challenges for creating practical PEM-modified membranes.

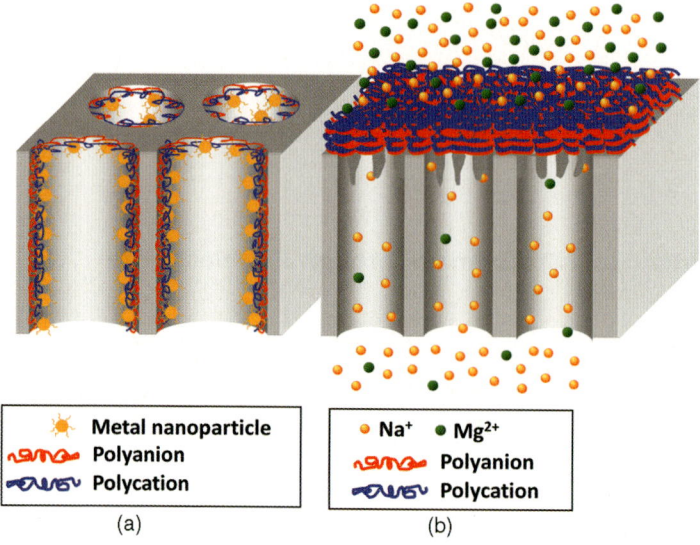

**Figure 38.1** Schematic drawings of (a) a catalytic membrane formed by depositing a polyelectrolyte/nanoparticle film in the pores of a membrane and (b) an ion-separation membrane prepared by deposition of a polyelectrolyte multilayer on the surface of a porous substrate. Used by permission of the American Chemical Society [8].

## 38.2
## Functionalization of the Interior of Membranes

### 38.2.1
### Deposition of PEMs in Porous Media

Several research groups have examined adsorption of PEMs in porous materials. A few early studies employed LbL deposition in membrane pores followed by dissolution of the porous membrane to create polyelectrolyte nanotubes [6–8]. Figure 38.2 shows nanotubes prepared by adsorption of poly(acrylic acid) (PAA)/protonated poly(allylamine) (PAH) films in porous alumina and subsequent dissolution of the alumina in base. (Heat-induced cross-linking of the PAA/PAH film makes it stable in the basic solution that dissolves the alumina.) The interior of the tubes is clearly open, so the films do not entirely fill the membrane pores.

Ai and coworkers found that the thicknesses of nanotubes deposited in alumina membranes were an order of magnitude greater than that those of films deposited on flat surfaces [6]. In contrast, nanotubes prepared by Liang et al. had much smaller thicknesses per layer pair [7]. The disparity in these studies may stem from differences in the film preparation methods. Thicker nanotubes often form when forcing the polyelectrolyte solutions through a membrane, whereas allowing the polyelectrolyte to diffuse into the pores leads to films with thicknesses closer to those on flat substrates. Alem and coworkers reported the formation of extremely thick

**Figure 38.2** SEM image of polyelectrolyte nanotubes prepared by deposition of a [PAA/PAH]$_3$ film in porous alumina and subsequent alumina dissolution after cross-linking of the film. Used by permission of the American Chemical Society [8].

films (up to 100-fold greater than on flat surfaces) when passing polyelectrolyte solutions through track-etched membranes [9]. In that case, swollen films with only a few polyelectrolyte layer pairs essentially occluded the pores, even for pore diameters as high as 800 nm. The authors suggested that entanglement of chains during filtration led to much thicker films in these membranes. Such entanglement is likely less significant when preparing polyelectrolyte/nanoparticle films by filtration. We coated porous alumina, track-etched polycarbonate, and nylon membranes (pore sizes between 0.2 and 0.45 μm) with polyanion/PAH/Au nanoparticle films. SEM images (e.g., Figure 38.3) clearly show the presence of the nanoparticles, but the pores are open, and the films do not greatly restrict flow [8, 10, 11]. Similar deposition can also occur in hollow fiber membranes [12].

**Figure 38.3** Cross-sectional SEM image of a porous alumina membrane coated with a PAA/PAH/Au nanoparticle film. The nanoparticles are well-separated, and the film does not significantly occlude the membrane pores. Used by permission of the American Chemical Society [10].

Roy and coworkers recently deposited PAH/poly(styrene sulfonate) (PSS) films in track-etched membranes using diffusion to carry polyelectrolytes into the pores [13]. They found an initial rapid growth of the films followed by slow growth as the pore diameter decreased, excluding polyelectrolytes. In the rapid growth, films were again much thicker than on flat surfaces.

Other studies report that polyelectrolyte deposition by diffusion into porous substrates yields either modest increases or modest decreases in film thickness relative to flat supports. Lee and coworkers prepared PAA/PAH films in track-etched membranes and found that the film thickness inside the pores is about 40% greater than the thickness on the surface [14]. They suggested that the increased thickness stems from incomplete draining of polyelectrolyte solutions in pores. These PAA/PAH films undergo an order of magnitude increase in percent swelling on going from low to high pH, which allows gating of flow through the membrane.

In 2010, Derocher and coworkers examined the deposition of PAH/PSS films in channels with widths of around 400 nm. In this case the coating on the flat surface above the channels is 1.4- to 1.8-fold thicker than the films inside the channels. Electrostatic exclusion of polyelectrolytes from the nanochannels may decrease the rate of their deposition. This result is consistent with studies of polyelectrolyte deposition in photonic crystals, where thicknesses within the crystal are sometimes five-fold lower than thicknesses on the crystal surface [15]. At some point the film on the surface can cover underlying pores to restrict deposition [16].

Ali and coworkers deposited polyelectrolyte multilayers in conical nanopores to control ion transport through these systems [17]. The surface charge created by the polyelectrolyte film creates a rectified ion current through the conical pores. However, the net surface charge decreased with the number of layers in the film, suggesting that the film structure may change in these confined geometries. In some cases, the zeta potentials of polyelectrolyte films deposited on flat surfaces can also decrease with the number of deposited layers, particularly in the case of exponentially growing films [18].

In summary, coating of pores during convective flow of polyelectrolyte solutions through membranes often results in films that are much thicker than those on flat surfaces, perhaps because polymers become entangled during filtration. In the case of diffusion of polyelectrolytes into pores, the thicknesses of PEMs may be somewhat less than on flat surfaces due to exclusion of polyelectrolytes from the pores, but the literature is not consistent on this issue. Deposition of polyelectrolyte/nanoparticle films with a few layer pairs does not occlude membrane pores.

## 38.2.2
**Functionalization of Membranes with Proteins**

### 38.2.2.1 Protein Adsorption in Membranes
In the early development of the LbL method, several studies demonstrated that proteins can serve as components of these films [19–21]. More recently, Smuleac and coworkers immobilized avidin, glucose oxidase, and alkaline phosphatase in membranes modified with PEMs [22]. The accessibility of the biotin-binding sites in the

immobilized avidin is as high as 90%, although the amount of avidin immobilized is only 7 mg cm$^{-3}$. For comparison, random covalent immobilization gives accessibilities around 50%. Similarly, the activities of glucose oxidase and alkaline phosphatase are around 70% of those for the enzymes in solution, demonstrating that the electrostatic adsorption procedure does not greatly alter the enzymes or their accessibility. Earlier work by Wang and Caruso revealed significantly increased enzyme stability using electrostatic adsorption in macroporous zeolite membranes [23]. The Caruso group also showed increasing catalytic activity with the number of enzyme/polyelectrolyte layer pairs in track-etched polycarbonate membranes [24]. However, after deposition of 5–7 layers, the activity of the membrane drops or plateaus, depending on the pore size, perhaps because the films plug the membrane pores.

Liu and coworkers examined lysozyme binding to nylon membranes (nominal 5 μm pore size) modified with polyelectrolyte films [25]. In these large pores, deposition of polyelectrolyte films with 3.5 layer pairs decreases water flux by only 20%. The amount of adsorbed lysozyme is similar for membranes containing PEMs terminated with poly(acrylic acid) (PAA) and PSS. Additionally, membranes modified with (PSS/polyethyleneimine, PEI)$_3$/PSS, (PSS/PAH)$_3$/PSS, and (PSS/poly(diallyldimethylammonium chloride, PDADMAC)$_3$/PSS films all bind comparable amounts of protein (around 10 mg cm$^{-3}$). Increasing the ionic strength from 0.5 to 2.0 M during deposition of the outer PSS layer of (PSS/PEI)$_3$/PSS films increases the binding capacity to 16 mg cm$^{-3}$. Presumably more loops and tails form in the outer layer of films adsorbed at high ionic strength, giving more binding sites. Although the binding capacities of membranes modified with polyelectrolyte multilayers are only ~50% of those of commercial ion-exchange membranes, stacking of membranes is a viable method for increasing capacity. Simultaneous modification of three stacked membranes gives a binding capacity that is essentially 3-fold greater than the capacity of a single membrane.

One major advantage of membranes as substrates for protein immobilization is their high surface area compared to flat supports. Using the procedure in Figure 38.4, Dai and coworkers exploited PAA-containing PEMs to covalently immobilize antibodies, specifically antig-IgG, in porous alumina [26]. Because immobilization happens throughout the membrane pores, the surface area available for antibody attachment is approximately 500-fold larger than that on a flat surface. When such membranes serve as substrates for arrays of spotted antibodies, this enhanced area and the ability to flow antigen solutions through the membrane leads to detection limits that are two orders of magnitude lower than for antibody arrays prepared on a flat surface. Moreover, the PAA-terminated polyelectrolyte film resists non-specific protein adsorption at physiological ionic strength, so antibody arrays on these substrates have very low background signals.

### 38.2.2.2 Trypsin-Containing Polyelectrolyte Films for Protein Digestion

We are particularly interested in employing trypsin-modified membranes to digest proteins for mass spectrometry analysis (Figure 38.5). Trypsin catalyzes protein cleavage after arginine and lysine residues to create small peptides that are readily

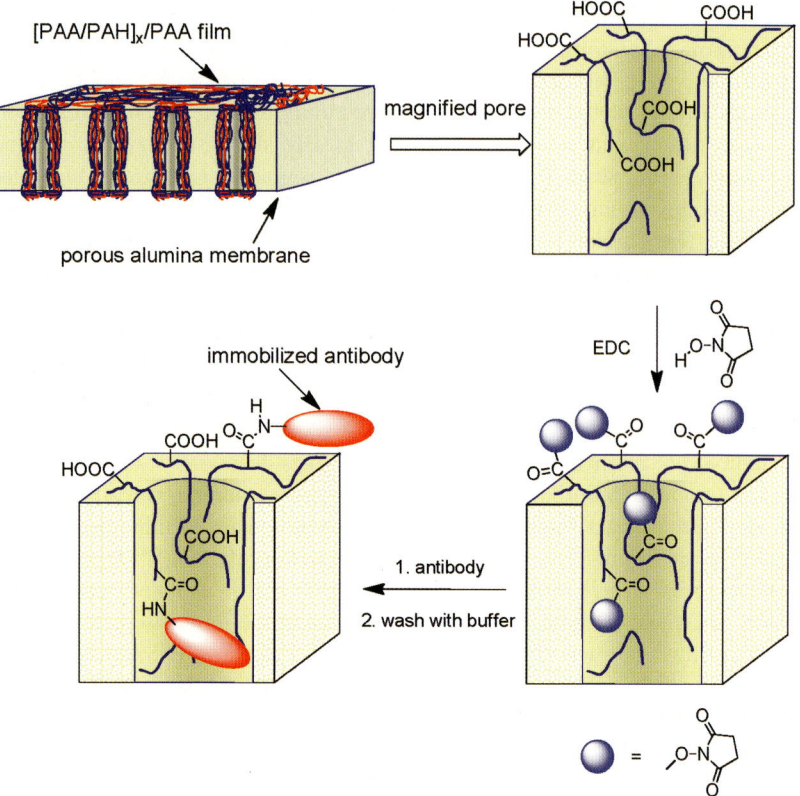

**Figure 38.4** Procedure for immobilization of antibodies by covalent binding to PAA-terminated PEMs in membrane pores. EDC = 1-[3-(dimethylamino)propyl]-3-ethylcarbodiimide hydrochloride. Used by permission of the American Chemical Society [26].

amenable to interrogation by common mass spectrometers. Nevertheless, for in-solution tryptic digestion, the ratio of trypsin to the protein of interest must be low to avoid digestion of trypsin by itself, and this low amount of trypsin can lead to digestion times as long as 16 h. In contrast, anchoring of trypsin to beads or

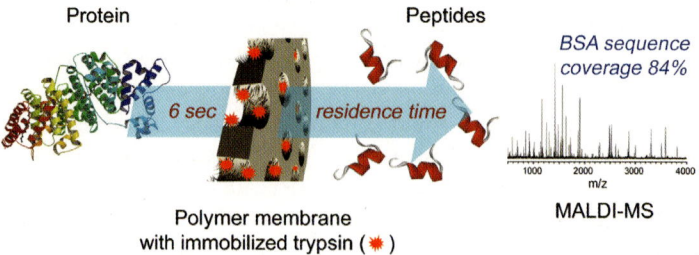

**Figure 38.5** Conceptual drawing of protein digestion in membranes containing adsorbed trypsin. Used by permission of the American Chemical Society [36].

membranes results in high enzyme densities, and immobilization limits self-digestion. A number of studies demonstrate that a high density of trypsin bound to solid supports can catalyze digestion in a few minutes or less [27–29]. Unfortunately, the covalent immobilization procedures that anchor enzymes in substrates such as monoliths typically require a high concentration of enzyme or relatively long times (up to 19 h) [29–31]. Random covalent immobilization also frequently decreases enzyme activity [32].

LbL adsorption provides a simple way to immobilize trypsin in membranes, and, as mentioned above, enzymes in PEMs frequently retain most of their activity. In microfluidic channels, Ji and coworkers deposited (PDADMAC/zeolite)$_3$ films and then adsorbed trypsin in the zeolites [33]. Such devices can digest bovine serum albumin (BSA) in just 6 s. Similarly, Liu and coworkers adsorbed trypsin in microchips containing PDADMAC/Au nanoparticle films [34]. Digestion in the chips for times as short as 5 s facilitates protein identification by matrix-assisted laser desorption/ionization mass spectrometry (MALDI-MS). However, adsorption of trypsin in microchannels with dimensions $>100\,\mu m$ likely leads to significant diffusion limitations. Assuming a protein diffusion coefficient, $D$, of $6 \times 10^{-7}\,cm^2\,s^{-1}$ [35], Equation 38.1 reveals that the average distance a protein can diffuse, $l$, in a time, $t$, of 5 s is only 25 μm. Thus in 5 s, much of the protein in channels with dimensions $>100\,\mu m$ will not reach the walls.

$$l = \sqrt{2Dt} \qquad (38.1)$$

To overcome diffusion limitations, we anchor trypsin to nylon membranes with nominal pore sizes of 0.45 μm [36]. In this case, diffusion of protein to the pore walls is not a significant limitation until residence times reach values around 1 ms. Immobilization of the trypsin takes place during formation of a PSS/trypsin layer pair by passing PSS, water, and trypsin solutions through the membrane. Addition of 2.7 mM HCl to the trypsin deposition solution reversibly inactivates the enzyme, to avoid self-digestion, and increases the charge on the protein to promote adsorption to negatively charged PSS. Analysis of the trypsin deposition solution before and after circulating through the membrane suggests an immobilization density of 11 mg of trypsin per cm$^3$ of membrane pores. This density is about 450-fold greater than typical trypsin concentrations used during in-solution digestion.

The PSS/trypsin-modified membranes are very effective in digesting proteins prior to MALDI-MS analysis. In the case of α-casein, sodium dodecyl sulfate polyacrylamide gel electrophoresis reveals no undigested protein for membrane residence times as short as 0.79 s. Additionally, membrane-based protein digestion yields high amino acid sequence coverage in MALDI-MS analysis of the resulting proteolytic peptides. For BSA, after membrane digestion the peptides detected by MALDI-MS account for 84% of the amino acids in the protein. High sequence coverages are vital for identifying protein modifications such as phosphorylation. BSA digestion in the membrane results in 52 detectable proteolytic peptides, whereas in-solution digestion gives detectable MALDI-MS signals for only 37 peptides. Moreover, the membrane retains much of its efficacy in the presence of 0.05 wt% sodium dodecyl sulfate, whereas in-solution digestion is ineffective under these conditions. Digestion in the presence of surfactants is important for studying membrane proteins that require surfactants to maintain their solubility.

One particular advantage of membranes for protein digestion is that their small thicknesses allow fine control of residence time through variation of the flow rates. At fast enough flow rates, incomplete digestion will lead to large peptides that are important for "middle-down" proteomics [37], where investigation of correlations between protein modifications proceeds through fragmentation of large peptides in the mass spectrometer. Larger peptides may also more readily identify proteins than several smaller peptides. Our current studies suggest that ms residence times give rise to large peptides for such studies. Overall, LbL deposition in membranes is a convenient method for creating highly active enzyme reactors, and variation of flow rates to control residence times can make this technique especially versatile.

### 38.2.3
### Catalytic Films and Membranes

#### 38.2.3.1 Catalytic, Nanoparticle-Containing Films

Metal nanoparticles are attractive catalysts because of their high surface area and unique electronic properties, and polyelectrolyte films present a versatile platform for forming or immobilizing such nanoparticles. Synthesis of nanoparticles in LbL films occurs by depositing metal ions in the film, either during film growth or after film deposition, and subsequently reducing the ions to form nanoparticles. For example, Joly and coworkers grew PAA/PAH films at pH values where a significant fraction of the –COOH groups are protonated. After deprotonation, complexation of metal ions by these groups and reduction of the bound metal ions yield metal nanoparticles encapsulated in polyelectrolyte films [38]. In a related method, we employed PEI-metal ion complexes as polycations for LbL adsorption, and subsequently reduced the metal ions to form nanoparticles [39]. The alternative strategy is to first form metal nanoparticles and then use the preformed particles as one of the alternating species in multilayer films [40, 41].

PEMs can both stabilize nanoparticles and potentially impart selectivity by restricting access to catalytic sites. Nanoparticles formed by reduction of Pd(II) in PEMs exhibit alkene hydrogenation selectivities based on the substitutents in the α-position with respect to the double bond. For example, hydrogenation of allyl alcohol is 12-fold faster than hydrogenation of 3-methyl-1-penten-3-ol (Figure 38.6) [42]. Related films show hydrogenation selectivity based on the degree of substitution on a double bond. Remarkably, the selectivity for hydrogenation of monosubstituted over trisubstituted double bonds can reach a value of 100, which is higher than the selectivities of related homogeneous catalysts [43].

Polyelectrolyte films are also a useful medium for controlling nanoparticle size and investigating how catalytic selectivity varies with particle diameter. To control particle size, we simply vary the ratio of metal ion to polyelectrolyte in the solutions used to adsorb metal ion-containing films. Studies of catalysis as a function of the average particle size in TEM images reveal dramatic increases in selectivity as the particle diameter decreases [44]. For example, as Figure 38.7 presents graphically, for films

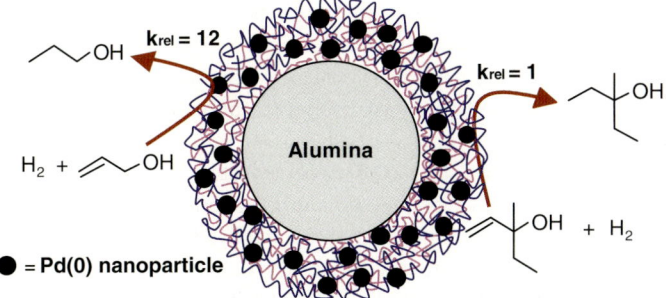

**Figure 38.6** Graphical representation of the selective hydrogenation of allyl alcohol by Pd nanoparticles embedded in PEMs. Used by permission of the American Chemical Society [42].

formed with a high concentration of Pd(II), larger nanoparticles (3.4 nm average diameter) have a selectivity of only 3 for hydrogenation of allyl alcohol over 2-methyl-2-propen-1-ol. In contrast, films containing nanoparticles with an average diameter of 2.2 nm display a selectivity of 240, although this is somewhat complicated by isomerization. The higher selectivities on smaller nanoparticles indicate that highly active sites on these particles are much less accessible to multisubstituted double bonds.

### 38.2.3.2 Catalytic Membranes

Deposition of catalysts in membranes results in efficient flow-through reactors with high surface areas, and if the membrane pores are sufficiently small, diffusion will not limit the reaction rate. Additionally, after reaction the catalyst does not have to be separated from the product stream. Membrane reactors also provide a means for controlling the extent of reaction and rapidly examining reaction kinetics. Because residence time in the membrane is directly proportional to flow rate, variation of the flow rate controls the reaction time.

As demonstrated above with enzymes, LBL adsorption is a very effective method for immobilizing catalysts in membrane pores. Figure 38.3 clearly shows that LBL adsorption of a PAA/PAH/Au nanoparticle films yields separated nanoparticles,

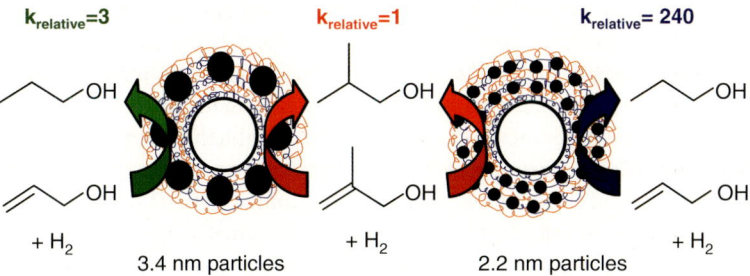

**Figure 38.7** Relative rates for hydrogenation catalyzed by large (3.4 nm) and small (2.2 nm) Pd nanoparticles embedded in PEMs on alumina. The smaller nanoparticles show significantly higher selectivity. Used by permission of the American Chemical Society [44].

which is vital for preserving their high surface area and unique catalytic activity. Moreover, the LBL process simply involves passing polyelectrolyte, nanoparticle, and rinsing solutions through the membrane and is very convenient. (In modifying these membranes, we usually first form the nanoparticles and subsequently immobilize them in the film).

Initially, reduction of p-nitrophenol by NaBH$_4$ served as a model reaction for examining the catalytic activity of membranes containing gold nanoparticles [11]. By measuring the percent conversion of p-nitrophenol to p-aminophenol as a function of the flux through the membrane, we were able to determine the rate constants of the catalytic nanoparticles. Remarkably, the nanoparticles in the PEMs exhibit the same rate constant as nanoparticles in solution. Thus the PEM does not significantly limit access to catalytic sites on the nanoparticles. Alumina, track-etched polycarbonate, and nylon membranes all give similar results [10]. Deposition of two layers of nanoparticles increases the catalytic activity, as expected.

Variation of the flow rate through catalytic membranes can potentially control the reaction products. For example, in the reduction of o-nitrotoluene by NaBH$_4$ (Scheme 38.1), on increasing the flux through the membrane by an order of magnitude, the conversion to o-nitrosotoluene increases from 5 to 61%, while the conversion to o-aminotoluene drops from 94 to 34%. We could not further increase the flux due to the strength of the membrane, but at larger fluxes the yield of the nitroso compound may be even higher. However, at some point significant amounts of starting material will remain. In addition to providing control over the reaction,

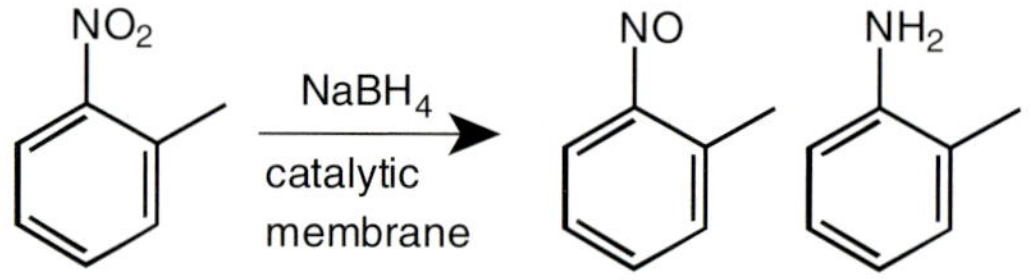

**Scheme 38.1** Reduction of o-nitrotoluene.

studies of the reaction as a function of flow rate indicate that the nitroso compound is an intermediate in the reduction to the amine.

Nanoparticle-modified membranes can also function as contactors for reactions such as wet air oxidation of pollutants (Figure 38.8). Oxidation of pollutants by air is attractive for treating waste streams that are toxic to remediating bacteria and too dilute for effective incineration [45, 46]. Deposition of the catalyst near the air/water interface, where most of the oxidation takes place, is vital for efficient use of expensive precious metals. LbL adsorption is a convenient method for depositing catalyst only near the interior wall of the membrane. In the case of tubular ceramic membranes, with the narrowest pores near their lumen, water wets only the interior (lumen) wall of the membrane, restricting the reactive volume to this location [47]. To confine polyelectrolyte deposition near the lumen interface, we simply limit the amount of nanoparticle and polyelectrolyte solutions we pass from the lumen to the shell of the tubular ceramic membrane. Nanoparticle adsorption is rapid, so all of the precious metal adsorbs near the membrane surface.

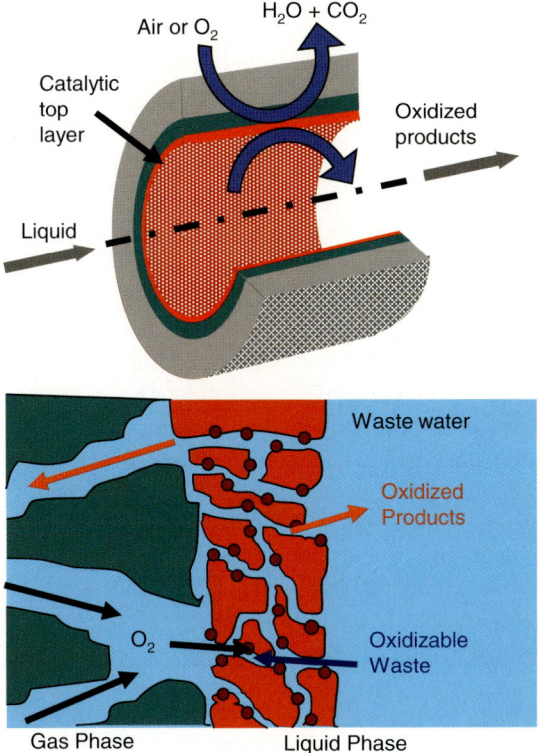

**Figure 38.8** Diagram of catalytic wet-air oxidation in a membrane contactor. The catalyst resides near the air/liquid interface to enhance its utilization. Used by permission of Elsevier [48].

LbL deposition yields remarkably active catalysts for wet air oxidation [48]. For rates of catalysis normalized to the amount of deposited Pt, the LbL method gives formic acid oxidation catalytic activities that are two- to three-fold greater than the corresponding activities of catalysts deposited by anionic impregnation/reduction and evaporation/recrystallization/reduction. In acetic acid and phenol oxidation, the LbL films show activities an order of magnitude greater than those of traditional catalysts. The enhanced activity likely stems from confinement of the catalyst at the interface in the LbL method.

Most recently, we have begun modifying polymeric hollow fibers with metal nanoparticles [12]. These membranes are much cheaper than ceramic membranes, and their smaller inner diameter should minimize diffusion distances. Unfortunately, these fibers are not designed for wetting to occur solely near the lumen interface. In preliminary studies, we deposited Pt throughout the hollow fiber and achieved much higher conversions of formic acid to $CO_2$ than with ceramic membranes, but the activity with respect to the amount of Pt is lower than in ceramic membranes because the Pt is distributed throughout the fiber. Still, the minimal cost of polymeric hollow fibers and high conversions may make these membranes attractive.

## 38.3
## LBL Films as Membrane Skins

### 38.3.1
### Early Studies

Practical separation membranes consist of an ultrathin, selective skin on a porous support (Figure 38.1b). The minimal thickness of the skin affords high flux together with selectivity, and the highly permeable support provides mechanical stability and minimal resistance to transport. Typical membrane fabrication methods include phase inversion casting of membranes and interfacial polymerization. The phase inversion process produces a porous polymer with a dense skin on its surface. In contrast, interfacial polymerization results in a composite membrane, where the skin layer may have a completely different composition than the support.

Complementary to interfacial polymerization, the LbL method can also deposit an ultrathin skin on a highly permeable support. Early studies found that a film with 10 layers or less could cover the 20 nm-diameter pores of porous alumina supports to yield membranes that reject divalent ions or molecules as small as glucose [49–51]. Tieke's group showed that membranes with 60 layer pairs can desalinate water, but the permeability of these systems is low, and depositing such a large number of layers is cumbersome [52, 53]. Polyelectrolyte films are also highly selective as pervaporation membranes [54, 55].

### 38.3.2
### Removal of Dyes and Small Organic Molecules from Water

Our more recent studies focused on the unique selectivities that are possible with PEM membranes, together with gaining an understanding of some of the unusual properties of PSS/PDADMAC films. PEMs are especially effective in rejecting dyes that must be removed from textile waste streams. Equation 38.2 defines rejection, $R$, of a given species, where $C_{perm}$ is the concentration of the species in the permeate stream and $C_{feed}$ is its concentration in the feed. Equation 38.3 defines the selectivity, $S$, for passage of species A relative to species B.

$$R = \left(1 - \frac{C_{perm}}{C_{feed}}\right) \times 100\% \tag{38.2}$$

$$S = \frac{C_{A,perm}}{C_{A,feed}} \frac{C_{B,feed}}{C_{B,perm}} = \frac{100\% - R_A}{100\% - R_B} \tag{38.3}$$

As Table 38.1 shows, [PSS/PAH]$_4$PSS films on porous alumina exhibit >99.9% rejection of a number of dyes, while allowing 60–80% passage (passage is $1 - R$) of NaCl [56]. This translates into a NaCl/dye selectivity of more than 2000. Moreover, the flux through these membranes is around 2 m$^3$/(m$^2$ d) at 4.8 bar, which is at the high end of fluxes for nanofiltration (NF) membranes. Studies with commercial

**Table 38.1** Dye and NaCl rejections and NaCl/dye selectivities during nanofiltration through a [PSS/PAH]₄PSS film on a porous alumina support [56]. The feed solution contained 1 g L⁻¹ dye and 0.01 M NaCl.

| Dye | NaCl Rejection | Dye Rejection | NaCl/dye Selectivity |
|---|---|---|---|
| Reactive Blue | $40 \pm 4$ | $99.98 \pm 0.01$ | $2900 \pm 1100$ |
| Reactive Orange | $19 \pm 3$ | $99.96 \pm 0.02$ | $2700 \pm 1200$ |
| Reactive Black | $15 \pm 3$ | $99.92 \pm 0.07$ | $2200 \pm 1500$ |

NF membranes report typical dye rejections of 98–99% with fluxes around 1 m³/(m² d) [57, 58]. Thus, the dye passage is an order of magnitude lower with the PEM than with the commercial membranes. The PEMs also display >99% rejection of sucrose along with NaCl/sucrose selectivities of >100, so salt removal or dewatering of sugar solutions is feasible [56].

Recent studies by Vankelecom and coworkers demonstrate that PEMs are effective at concentrating dyes from organic solvents [59, 60]. Removal of small molecules from organic solvents is challenging because many membrane materials are soluble and unstable in nonaqueous systems. Nevertheless, recovery of catalysts or rejection of organic molecules, such as dyes in organic solvents, is important for recycling.

Because many PEMs swell less in water than in organic solvents [61], they may have low permeabilities to these solvents. However, deposition of sulfonated poly(ether ether ketone)/PDADMAC membranes from 1.0 M NaCl solutions gives membranes with reasonable permeabilities to isopropanol and THF, together with rose Bengal rejections from 85–99% [60].

Tieke's group prepared a number of PEM membranes using complex molecules such as calixarenes, azamacrocycles, and cyclodextrins [62–67]. In ethanolic solutions, the cyclodextrin-containing membranes afford size-based selective passage of naphthalene over pyrene and perylene, although the mechanism of selectivity is not yet clear [65, 68]. PEMs with calixarenes likely form complexes with lanthanide ions, and have $Na^+$/lanthanide ion selectivities >100 [67, 69].

### 38.3.3
### Selective Rejection of $F^-$ and Phosphate

Typically, PEMs display low selectivities among monovalent ions, but [PSS/PDADMAC]$_4$PSS films exhibit $Cl^-/F^-$ selectivities of around 3 [70]. For monovalent ions, this selectivity is quite high, as commercial nanofiltration membranes have corresponding selectivities around 1 [70]. The selectivity of the [PSS/PDADMAC]$_4$PSS coatings holds at $Cl^-/F^-$ ratios in the feed solution from 1 to 40, and at pressures from at least 2.5 to 6 bar. Removal of $F^-$ from drinking water is important in some regions because high levels of $F^-$ may cause dental and bone fluorosis, and relatively high rejections of $F^-$ and low rejections of $Cl^-$ are desirable to minimize the osmotic pressure across the membrane. The [PSS/PDADMAC]$_4$PSS films had $F^-$ rejections of 70% and $Cl^-$ rejections around 10%. The rejection of $F^-$ is not especially high, but it should be sufficient in many cases where $F^-$ levels need to be reduced from 3 to <1 ppm.

One unusual feature of the [PSS/PDADMAC]$_4$PSS membranes is that selectivity decreases with the addition of more layers. The $Cl^-/F^-$ selectivity of [PSS/PDADMAC]$_6$PSS films is only 1.1. Additionally, [PSS/PAH]$_4$PSS membranes, which are less permeable than [PSS/PDADMAC]$_4$PSS, have selectivities of only 1.1. The next section discusses how the properties of PSS/PDADMAC films vary with the number of layers.

Phosphate removal from water is an important issue in some parts of the world because of the emerging scarcity of phosphorus and legal limits on the levels of phosphorus in discharge streams [71]. Because of the high charge of phosphate at basic pH values, PEMs are quite effective at rejecting this species [72]. For example, at pH 8.4 [PSS/PDADMAC]$_4$PSS films on porous alumina membranes show 98% phosphate rejection and only 21% $Cl^-$ rejection. (We use the term phosphate to include phosphate and all of its protonated forms.) The low $Cl^-$ rejection minimizes osmotic pressure, which is attractive when desalination is unnecessary. For comparison, commercial NF90 nanofiltration membranes exhibit 99.7% phosphate rejection at pH 8.4, but $Cl^-$ rejection is 98%.

Phosphate rejection by [PSS/PDADMAC]$_x$PSS membranes decreases with pH because more of the phosphate becomes the singly charged anion $H_2PO_4^-$. Nevertheless, even at pH 5.6, [PSS/PDADMAC]$_4$PSS films have phosphate rejections of

86%. Notably, phosphate rejections decrease on going from [PSS/PDADMAC]₄PSS to [PSS/PDADMAC]₅PSS films, even though the latter better covers the underlying substrate.

### 38.3.4
### Variation of PSS/PDADMAC film Properties with the Number of Adsorbed Layers

To understand why transport properties vary with the number of layers in PSS/PDADMAC films, we examined both anion-exchange capacities and zeta potentials as a function of the number of adsorbed layers in these coatings [18]. Infrared spectroscopy of films on a germanium crystal revealed no detectable $SCN^-$ binding in PSS-terminated $[PDADMAC/PSS]_x$ films with less than 6 layer pairs. In contrast, coatings with more than 7 layer pairs have an ion-exchange capacity of about 0.5 M. The high concentration of anion-exchange sites in these films suggests a significant excess of PDADMAC.

The zeta potentials of these films also suggest an excess of PDADMAC for coatings with more than 5 or 6 layers. Figure 38.9 shows how zeta potentials of PDADMAC/PSS films on 100 kDa polyethersulfone membranes vary with the number of deposited layers. Initially, the zeta potential is negative after deposition of PSS and positive after deposition of PDADMAC. However, after adsorption of 5 or 6 layer pairs, the zeta potential is positive even after deposition of PSS. Both the ion-exchange results and zeta potential data are consistent with an exponential growth mechanism where PDADMAC diffuses throughout the film, and upon exposure to PSS, the PDADMAC may diffuse to the surface to form a thick layer. Evidently, the PSS does not completely compensate the charge on PDADMAC. Ellipsometric data

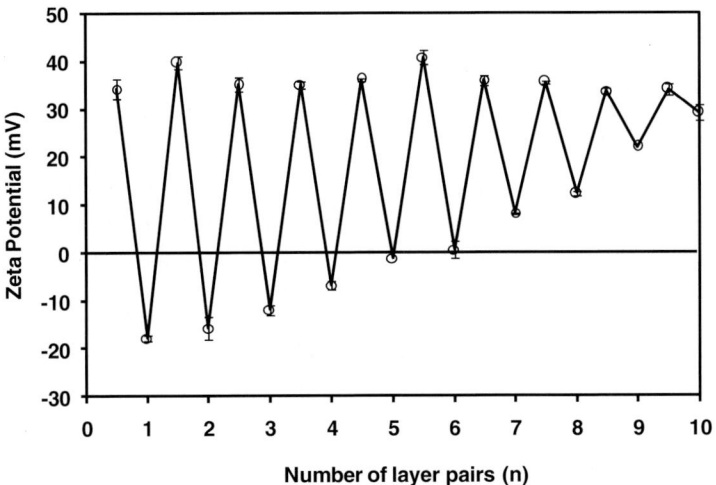

**Figure 38.9** Zeta potentials of 100 kDa polyethersulfone membranes after deposition of $[PDADMAC/PSS]_n$ coatings. Films with noninteger numbers of layer pairs terminate in PDADMAC. Used by permission of the American Chemical Society [18].

confirm that the thickness of PSS/PDADMAC films increases nonlinearly when they are deposited at high ionic strength (1.0 M in this case) [18].

The excess of PDADMAC in the film after deposition of 5–6 layer pairs likely explains the drops in selectivity that occur when adding more layers to [PSS/PDADMAC]$_4$PSS films. In addition to the examples of F$^-$ and phosphate rejections described above, [PSS/PDADMAC]$_4$PSS films on porous alumina membranes display Cl$^-$/SO$_4^{2-}$ selectivities >30, whereas [PSS/PDADMAC]$_6$PSS films exhibit selectivities of only 3 [18]. The excess PDADMAC reduces the negative charge in the film to decrease rejections of divalent ions and may also increase swelling to lower size-based selectivity.

## 38.4
## Challenges

For large scale applications such as water softening [73] or fuel cell membranes [74], the many steps in the LbL process may prevent adoption of the technique. Although one can envision rolling a substrate through a number of alternating solutions, the performance of PEMs must be far superior to current membranes to justify the LbL technique. Several studies describe spray-coating of multilayer films [75–77], but in many cases the solution still must drip from the substrate, and waste solutions will result. Deposition of polyelectrolyte complexes provides an alternative method to form polyelectrolyte films [78]. However, this method does not afford the control of thickness and composition available with LbL adsorption.

In contrast, for smaller scale, high value applications the fine control of the LbL method should justify its multiple steps. In creating functional membranes, sensors, or biocompatible materials with spatially organized functionalities, LbL assembly may well play an important role.

## Acknowledgments

I thank the many students and postdocs who participated in this work. The Department of Energy Office of Basic Energy Sciences, the US National Science Foundation (OIS-0530174), and the US National Institutes of Health (GM080511) generously funded this work.

## References

1 Decher, G. (1997) *Science*, **277**, 1232.
2 Shim, B.S., Zhu, J.A., Jan, E., Critchley, K., and Kotov, N.A. (2010) *ACS Nano*, **4**, 3725.
3 Ariga, K., McShane, M., Lvov, Y.M., Ji, Q.M., and Hill, J.P. (2011) *Expert Opin. Drug. Deliv.*, **8**, 633.
4 Becker, A.L., Johnston, A.P.R., and Caruso, F. (2010) *Small*, **6** 1836.
5 Briscoe, J., Gallardo, D.E., Hatch, S., Lesnyak, V., Gaponik, N., and Dunn, S. (2011) *J. Mater. Chem.*, **21**, 2517.

6 Ai, S.F., Lu, G., He, Q., and Li, J.B. (2003) *J. Am. Chem. Soc.*, **125**, 11140.
7 Liang, Z.J., Susha, A.S., Yu, A.M., and Caruso, F. (2003) *Adv. Mater.*, **15**, 1849.
8 Bruening, M.L., Dotzauer, D.M., Jain, P., Ouyang, L., and Baker, G.L. (2008) *Langmuir*, **24**, 7663.
9 Alem, H., Blondeau, F., Glinel, K., Demoustier-Champagne, S., and Jonas, A.M. (2007) *Macromolecules*, **40**, 3366.
10 Dotzauer, D.A., Bhattacharjee, S., Wen, Y., and Bruening, M.L. (2009) *Langmuir*, **25**, 1865.
11 Dotzauer, D.M., Dai, J.H., Sun, L., and Bruening, M.L. (2006) *Nano Lett.*, **6**, 2268.
12 Lu, O.Y., Dotzauer, D.M., Hogg, S.R., Macanas, J., Lahitte, J.F., and Bruening, M.L. (2010) *Catal. Today*, **156**, 100.
13 Roy, C.J., Dupont-Gillain, C., Demoustier-Champagne, S., Jonas, A.M., and Landoulsi, J. (2010) *Langmuir*, **26**, 3350.
14 Lee, D., Nolte, A.J., Kunz, A.L., Rubner, M.F., and Cohen, R.E. (2006) *J. Am. Chem. Soc.*, **128**, 8521.
15 Arsenault, A.C., Halfyard, J., Wang, Z., Kitaev, V., Ozin, G.A., Manners, I., Mihi, A., and Miguez, H. (2005) *Langmuir*, **21**, 499.
16 Harris, J.J., Stair, J.L., and Bruening, M.L. (2000) *Chem. Mater.*, **12**, 1941.
17 Ali, M., Yameen, B., Cervera, J., Ramirez, P., Neumann, R., Ensinger, W., Knoll, W., and Azzaroni, O. (2010) *J. Am. Chem. Soc.*, **132**, 8338.
18 Adusumilli, M. and Bruening, M.L. (2009) *Langmuir*, **25**, 7478.
19 Lvov, Y., Ariga, K., Ichinose, I., and Kunitake, T. (1995) *J. Am. Chem. Soc.*, **117**, 6117.
20 Lvov, Y., Ariga, K., and Kunitake, T. (1994) *Chem. Lett.*, 2323.
21 Kong, W., Wang, L.P., Gao, M.L., Zhou, H., Zhang, X., Li, W., and Shen, J.C. (1994) *J. Chem. Soc., Chem. Commun.*, 1297.
22 Smuleac, V., Butterfield, D.A., and Bhattacharyya, D. (2006) *Langmuir*, **22**, 10118.
23 Wang, Y.J. and Caruso, F. (2004) *Adv. Funct. Mater.*, **14**, 1012.
24 Yu, A.M., Liang, Z.J., and Caruso, F. (2005) *Chem. Mater.*, **17**, 171.
25 Liu, G.Q., Dotzauer, D.M., and Bruening, M.L. (2010) *J. Membr. Sci.*, **354**, 198.
26 Dai, J.H., Baker, G.L., and Bruening, M.L. (2006) *Anal. Chem.*, **78**, 135.
27 Kim, J., Kim, B.C., Lopez-Ferrer, D., Petritis, K., and Smith, R.D. (2010) *Proteomics*, **10**, 687.
28 Gao, J., Xu, J.D., Locascio, L.E., and Lee, C.S. (2001) *Anal. Chem.*, **73**, 2648.
29 Krenkova, J., Lacher, N.A., and Svec, F. (2009) *Anal. Chem.*, **81**, 2004.
30 Dulay, M.T., Baca, Q.J., and Zare, R.N. (2005) *Anal. Chem.*, **77**, 4604.
31 Spross, J. and Sinz, A. (2010) *Anal. Chem.*, **82**, 1434.
32 Brady, D. and Jordaan, J. (2009) *Biotechnol. Lett.*, **31**, 1639.
33 Ji, J., Zhang, Y., Zhou, X., Kong, J., Tang, Y., and Liu, B. (2008) *Anal. Chem.*, **80**, 2457.
34 Liu, Y., Xue, Y., Ji, J., Chen, X., Kong, J., Yang, P.Y., Girault, H.H., and Liu, B.H. (2007) *Mol. Cell. Proteomics*, **6**, 1428.
35 Raj, T. and Flygare, W.H. (1974) *Biochem.*, **13**, 3336.
36 Xu, F., Wang, W.H., Tan, Y.J., and Bruening, M.L. (2010) *Anal. Chem.*, **82**, 10045.
37 Meyer, B., Papasotiriou, D.G., and Karas, M. (2011) *Amino Acids*, **41**, 291.
38 Joly, S., Kane, R., Radzilowski, L., Wang, T., Wu, A., Cohen, R.E., Thomas, E.L., and Rubner, M.F. (2000) *Langmuir*, **16**, 1354.
39 Dai, J.H. and Bruening, M.L. (2002) *Nano Lett.*, **2**, 497.
40 Schrof, W., Rozouvan, S., Van Keuren, E., Horn, D., Schmitt, J., and Decher, G. (1998) *Adv. Mater.*, **10**, 338.
41 Mamedov, A.A., Belov, A., Giersig, M., Mamedova, N.N., and Kotov, N.A. (2001) *J. Am. Chem. Soc.*, **123**, 7738.
42 Kidambi, S., Dai, J.H., Li, J., and Bruening, M.L. (2004) *J. Am. Chem. Soc.*, **126**, 2658.
43 Bhattacharjee, S. and Bruening, M.L. (2008) *Langmuir*, **24**, 2916.
44 Bhattacharjee, S., Dotzauer, D.M., and Bruening, M.L. (2009) *J. Am. Chem. Soc.*, **131**, 3601.

45 Guibelin, E. (2004) *Water Sci. Technol.*, **49**, 209.
46 Matatov-Meytal, Y.I. and Sheintuch, M. (1998) *Ind. Eng. Chem. Res.*, **37**, 309.
47 Uzio, D., Miachon, S., and Dalmon, J.A. (2003) *Catal. Today*, **82**, 67.
48 Dotzauer, D.M., Abusaloua, A., Miachon, S., Dalmon, J.A., and Bruening, M.L. (2009) *Appl. Catal. B Environ.*, **91**, 180.
49 Liu, X.Y. and Bruening, M.L. (2004) *Chem. Mater.*, **16**, 351.
50 Stanton, B.W., Harris, J.J., Miller, M.D., and Bruening, M.L. (2003) *Langmuir*, **19**, 7038.
51 Harris, J.J. and Bruening, M.L. (2000) *Langmuir*, **16**, 2006.
52 Jin, W.Q., Toutianoush, A., and Tieke, B. (2003) *Langmuir*, **19**, 2550.
53 Toutianoush, A., Jin, W.Q., Deligoz, H., and Tieke, B. (2005) *Appl. Surf. Sci.*, **246**, 437.
54 Krasemann, L., Toutianoush, A., and Tieke, B. (2001) *J. Membr. Sci.*, **181**, 221.
55 Toutianoush, A., Krasemann, L., and Tieke, B. (2002) *Colloid. Surf. A. Phys. Eng. Asp.*, **198**, 881.
56 Hong, S.U., Miller, M.D., and Bruening, M.L. (2006) *Ind. Eng. Chem. Res.*, **45**, 6284.
57 Koyuncu, I., Topacik, D., and Yuksel, E. (2004) *Sep. Purif. Technol.*, **36**, 77.
58 Tang, C. and Chen, V. (2002) *Desalination*, **143**, 11.
59 Li, X.F., De Feyter, S., Chen, D.J., Aldea, S., Vandezande, P., Du Prez, F., and Vankelecom, I.F.J. (2008) *Chem. Mater.*, **20**, 3876.
60 Li, X.F., Goyens, W., Ahmadiannamini, P., Vanderlinden, W., De Feyter, S., and Vankelecom, I. (2010) *J. Membr. Sci.*, **358**, 150.
61 Miller, M.D. and Bruening, M.L. (2005) *Chem. Mater.*, **17**, 5375.
62 El-Hashani, A., Toutianoush, A., and Tieke, B. (2007) *J. Phys. Chem. B*, **111**, 8582.
63 El-Hashani, A., Toutianoush, A., and Tieke, B. (2008) *J. Membr. Sci.*, **318**, 65.
64 Hoffmann, K. and Tieke, B. (2009) *J. Membr. Sci.*, **341**, 261.
65 Tieke, B., El-Hashani, A., Toutianoush, A., and Fendt, A. (2008) *Thin Solid Films*, **516**, 8814.
66 Tieke, B., Toutianoush, A., and Jin, W.Q. (2005) *Adv. Colloid Interface Sci.*, **116**, 121.
67 Toutianoush, A., El-Hashani, A., Schnepf, J., and Tieke, B. (2005) *Appl. Surf. Sci.*, **246**, 430.
68 Jin, W.Q., Toutianoush, A., and Tieke, B. (2005) *Appl. Surf. Sci.*, **246**, 444.
69 Toutianoush, A., Schnepf, J., El Hashani, A., and Tieke, B. (2005) *Adv. Funct. Mater.*, **15**, 700.
70 Hong, S.U., Malaisamy, R., and Bruening, M.L. (2007) *Langmuir*, **23**, 1716.
71 van Voorthuizen, E.M., Zwijnenburg, A., and Wessling, M. (2005) *Water Res.*, **39**, 3657.
72 Hong, S.U., Ouyang, L., and Bruening, M.L. (2009) *J. Membr. Sci.*, **327**, 2.
73 Ouyang, L., Malaisamy, R., and Bruening, M.L. (2008) *J. Membr. Sci.*, **310**, 76.
74 Ashcraft, J.N., Argun, A.A., and Hammond, P.T. (2010) *J. Mater. Chem.*, **20**, 6250.
75 Izquierdo, A., Ono, S.S., Voegel, J.C., Schaaf, P., and Decher, G. (2005) *Langmuir*, **21**, 7558.
76 Krogman, K.C., Lowery, J.L., Zacharia, N.S., Rutledge, G.C., and Hammond, P.T. (2009) *Nat. Mat.*, **8**, 512.
77 Lefort, M., Boulmedais, F., Jierry, L., Gonthier, E., Voegel, J.C., Hemmerle, J., Lavalle, P., Ponche, A., and Schaaf, P. (2011) *Langmuir*, **27**, 4653.
78 Ji, Y.L., An, Q.F., Zhao, Q., Chen, H.L., Qian, J.W., and Gao, C.J. (2010) *J. Membr. Sci.*, **357**, 80.

# 39
# Remote and Self-Induced Release from Polyelectrolyte Multilayer Capsules and Films

*Andre G. Skirtach, Dmitry V. Volodkin, and Helmuth Möhwald*

Polyelectrolyte multilayer capsules and films stimulate various application areas, including drug delivery, surface coatings, optoelectronics, and corrosion protection. In all these areas one needs to design smart micro- and nano-sized containers which protect their content from external influences and release it only in response to certain environmental conditions at the desired destination. Mimicking both encapsulation and release processes would be especially interesting for the targeted delivery of drugs to cells. There are different approaches to encapsulate substances in the capsule interior and to release them at the specific target by external stimuli. Selective permeability is an essential attribute of a capsule wall regulating transport of molecules. Polymeric multilayer capsules assembled using the layer-by-layer technique (LbL) [1] are promising candidates for more complex tasks of storage, encapsulation and release which can be readily engineered and functionalized with the desired properties. Their mechanical stability, elasticity, morphology, biocompatibility, permeability, and surface characteristics can also be adjusted accordingly. The most significant advantages of polyelectrolyte multilayer (PEM) capsules are multifunctionality and availability of various stimuli to affect and control their properties. Thus, recent efforts in the field have been devoted to designing "smart" micro- and nanocontainers which respond to these external stimuli. This chapter is devoted to remote and self-induced release from microcapsules and films, Figure 39.1. A brief introduction is presented on the building blocks of capsules and nanoshells, such as polymers, and on the basic principles of stimuli-responsive encapsulation, release and manipulation. Some applications are further presented, focusing on intracellular delivery and release, enzymatic reactions, sensors and multicompartment capsules. In the case of films, their build-up processes and applications in delivery are described.

Polyelectrolytes have been used as building blocks of microcapsules and films because of their high versatility and that of the assembly procedures. The main difference between capsules and films is that with films polymers are immobilized on the substrate instead of a sacrificial nano-or microparticle. PEM capsules fabricated by the LbL technique have been an active area of research [1–22]. Release of the contents of microcapsules [23–25] is an area that has experienced rapid growth. It is possible to assemble LbL films using uncharged colloids [26], however, assembling

*Multilayer Thin Films: Sequential Assembly of Nanocomposite Materials*, Second Edition.
Edited by Gero Decher and Joseph B. Schlenoff.
© 2012 Wiley-VCH Verlag GmbH & Co. KGaA. Published 2012 by Wiley-VCH Verlag GmbH & Co. KGaA.

MICROCAPSULES and FILMS

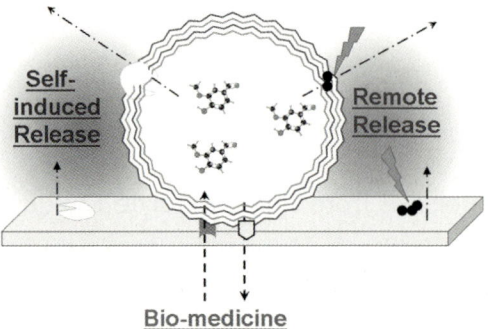

**Figure 39.1** Schematics of remote and self-induced release on polyelectrolyte multilayer capsules and films.

the films based on charges remains a method of choice in the LbL film area [27]. Centrifugation and removal of the non-adsorbed excess of polyelectrolytes are typically used as steps in the fabrication. Although assembling on nanometer-sized templates is possible [28], micrometer-sized capsules are attractive due to simplicity in observation, characterization and imaging, prevention of aggregation, and high loading. The main challenge of assembling nanometer-sized capsules is aggregation, and although some progress has been achieved [29], it remains challenging [30]. As an example of the application of larger capsules, several hundreds of micrometers in size, [31] one can consider direction-specific release [32].

Various templates (also called cores) can be used for the fabrication of capsules [33]: melamine formaldehyde (MF) [34], polystyrene (PS) [35], gold NP (NP) [28], carbonate cores [36], silica particles [37, 38], and cells [39, 40]. MF and PS raise concerns because of some left-over oligomers in the interior of the capsules; they are dissolved in organic solvents but residuals in the shell may alter the structure and stability of the polyelectrolyte multilayers [35]. For dissolution, the coated cells are incubated in a basic hypochlorite solution; here oxidation of the shell-constituting polyelectrolytes remains of concern [39]. In contrast to organic cores, inorganic templates dissolve well. In this context, carbonate cores exhibit good biocompatibility but inherit high polydispersity. They are attractive, however, due to their high porosity ($CaCO_3$) [41, 42]. $SiO_2$ particles are smooth and monodisperse; one significant disadvantage is the necessity of dissolution in HF (hydrogen fluoride) [37, 38, 43]. Various materials have been demonstrated for incorporation into PEM: synthetic polyelectrolytes, biocompatible polymers [38, 44], proteins [45], deoxyribonucleic acid (DNA) [46, 47], lipids [48, 49], molecules [50, 51], charged nanoparticles [52–56], dendrimers [57], viruses [58], nanotubes, nanowires and nanoplates [59, 60], as well as differently functionalized polyelectrolytes [61, 62] have been incorporated into the capsule wall. This diversity of possible building-blocks with respect to functionalization, together with the precisely tunable wall thickness makes the LbL capsules attractive for such applications as drug-resistant bio-films and fungal cells [63], textile industries [64], healing of corrosion [65], and a broad range of biomedical applications [66].

PEM capsules comprised of weak polyelectrolytes are pH-responsive due to protonation/deprotonation of charged groups; the biggest advantage of this approach is its reversibility. pH allows control of the charges of the polymer repeat units and, as a consequence, the interaction between the charges of polymers, by adjusting the acidity of the surrounding solution [67]. Even the thickness of the PEM can be controlled by pH [68]. When the pH is altered relative to the $pK_a$ of the polyelectrolytes constituting the capsule walls, protonation/deprotonation of charged groups takes place. Accumulation of additional charges on the polyelectrolyte then leads to stronger mutual repulsion, which causes capsule swelling (and increased permeability). Thus the shells show reversible pH-triggered changes in surface charge and permeability, for example, towards FITC-labeled polysaccharide molecules. It can be noted that sensitivity to pH can also be used for patterning of polymer thin films [69]. pH-responsive spheres made of organo-clays were used to release ibuprofen and eosin [70], which were demonstrated to expand and shrink (in size) over a wide range of pH (4 to 9). Poly(allylamine hydrochloride) (PAH), and the polyacids poly (acrylic acid) (PAA) and poly(methacrylic acid) (PMA) are some representative and widely studied weak polyelectrolytes [71]. Besides electrostatic stabilization, hydrogen bonding between uncharged carboxylic acids, amino functions, and hydroxy groups could also contribute to the stability of the microcapsules [72].

Thermal treatment allows the production of mechanically strengthened capsules [73]. Shrinking of LbL-assembled microcapsules, which can be used for encapsulation and enhancement of mechanical properties, in response to thermal treatment has been reported by Köhler *et al.* [74]. PEMs are kinetically stable, so it is logical to expect that a temperature increase can provide enough thermal energy to surpass the barrier necessary for polymeric film rearrangements. PEM films deposited on flat substrates exhibit only negligible changes in thickness upon heating [75]. However, they shrink noticeably if heated at 100% humidity, indicating that water desorption takes place [76]. In the case of capsules, a significant reduction in permeability for low molecular-weight compounds accompanies heat treatment [77]. This is due to the interplay of electrostatic and hydrophobic forces; in this way thermal encapsulation is similar to salt-induced encapsulation [78]. It was reported that stimuli affect block-co-polymers, micelles, nanogels, and core–shell nanoparticles [79–82]. Another distinct feature of PEM response to thermal treatment is that the temperature has to be raised above the glass transition temperature, $T_g$, of the polyelectrolyte complex [74]. Subsequent response to thermal shrinking of microcapsules is accompanied by stiffening of the walls and, therefore, increase in the mechanical strength of microcapsules. Heating Polydiallyldimethylammoniumchloride/poly-(4-styrenesulfonate)/polystyrene sulfonate (PDADMAC/PSS)$n$ microcapsules (where $n$ is the number of layer pairs) has been shown to induce the reorganization of loosely arranged polyelectrolyte layers into a denser structure [83]. The shrinking behavior can be adjusted by the total number of layer pairs, or by the charge balance: microcapsules with an odd number of layers swell, while those with an even number of layers shrink [74]. It can be noted that understanding polymer mobility and inter-diffusion [84] is essential not only for construction of capsules. Bedard and coworkers found that gold nanoparticles strengthen PEM capsules [85].

External fields [86] provide remote means for stimulating the permeability of capsules and affecting the films. In the case of laser light it is essential to operate in the so-called biologically "friendly" near-IR window [87]. Light-sensitive capsules have been reported by several groups [88–92]. The mechanism of affecting the permeability is based on localized heating of metal nanoparticles (NPs) upon laser exposure. Therefore, the concentration and aggregation of NPs and high and low concentrations [93, 94] play an important role in this process. The interaction of NPs accomplished through their aggregation increases the absorption in the near-IR part of the spectrum. The release of macromolecules upon IR irradiation of polyelectrolyte capsules functionalized with gold NP has been the subject of extensive research [52, 53, 90, 91, 95–98]. On the other hand, laser light can also be used for encapsulation. Bedard and coworkers showed light-initiated shrinking and encapsulation of a fluorescently labeled polymer in a (PAH/poly[1-[4-(carboxy-4-hydroxyphenylazo) benzenesulfoamido]-1,2-ethanediyl, sodium salt] (PAzo))$_3$/PAH/poly(vinylsulfonate) (PVS) microcapsule [99], while the effectiveness of such encapsulation increases with time [100]. Nanomembranes with controllable shell permeability [101, 102] have been implemented using polymeric microcapsules as a model system [96]. Wavelength selectivity is possible by tuning the absorption of nanorods [103]. Light-responsive polymeric microcapsules have been intensively studied [104–108] paving the way to such new applications as improved photodynamic therapy (PDT) [106], intracellular delivery and time-programmed release [53, 109], or remotely controllable bioreactors [110]. In addition to light, ultrasound and magnetic fields represent means of remote release from capsules and films. Ultrasound is another method which is used in drug delivery [111]. High power ultrasound (in the range 100–500 W, 20 kHz) was shown to be capable of destroying PEM capsules [112], and thus initiating remote release. In this area, NPs were shown to enhance this effect [112]. Polyelectrolyte microcapsules with ZnO NPs embedded in their shell proved to have potential as drug-delivery systems, with the possibility of opening under the action of ultrasound [113]. Encapsulation is also possible using ultrasound [114]. With regard to reduction of the power of ultrasound necessary for remote release, a step forward has been achieved by attaching liposomes to microcapsules [115]. An oscillating magnetic field is another candidate for remote permeability control [116, 117]. For instance, Lu and coworkers showed that this may change the permeability of microcapsules [118] (100–300 Hz; strength of 1200 Oe). Katagiri and coworkers showed release from magneto-responsive smart capsules formed with polyelectrolytes, lipid bilayers and magnetic NPs [119]. Functionalization by magnetic and metallic NPs produces capsules which respond to multiple stimuli [120].

Mechanical stability of microcapsules is required for successful delivery. Structural [122] and mechanical [85, 123, 124] characterization techniques have been applied to investigate the properties of microcontainers. Microcapsules are deformed upon intracellular uptake [125], which leads to loss of encapsulated cargo. Although enhancement of mechanical strength can be achieved by adsorbing gold NPs [85] or carbon nanotubes [126], further enhancement of the polymeric shell is required. Mechanical stability studies were made by applying mechanical deformation on PEM

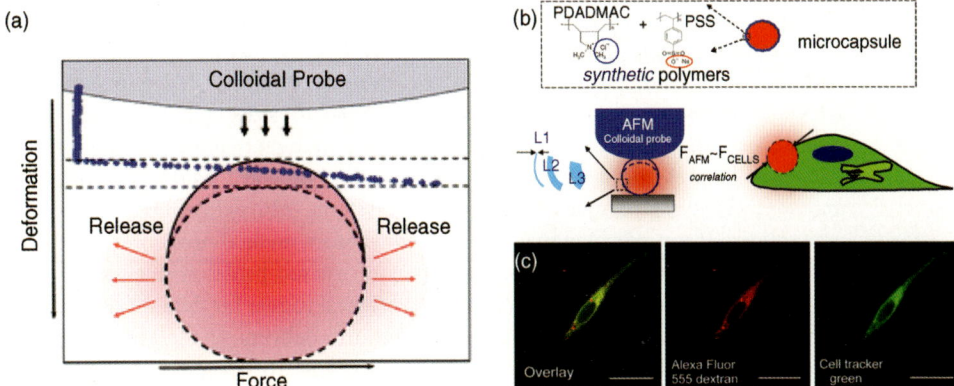

**Figure 39.2** (a) Push force-curve on a typical capsule. Schematics of the capsule before contact with the colloidal probe and at maximum deformation are shown. (b) Average fluorescence intensity from a typical microcapsule subjected to mechanical deformation (filled circles) and a control microcapsule not subjected to mechanical deformation (open circles), calculated from images taken after each push–pull cycle as a function of total capsule deformation. The dashed-dotted line indicates the threshold of total deformation (18%) beyond which release is triggered [121]. Reproduced by permission of The Royal Society of Chemistry. (b) Schematics representing capsules made of synthetic polymers acting as sensors for estimation of forces exerted by cells upon intracellular incorporation. (c) Fluorescence images of a cell after uptake of microcapsules shrunk at 55 °C and containing encapsulated Alexa Fluor 555 dextran. The left panel shows the overlay depicting the release of Alexa Fluor 555 dextran molecules (the red channel, middle panel) introduced into the intracellular environment (the green channel, right panel). The scale bars correspond to 30 μm [25]. Copyright Wiley-VCH Verlag GmbH & Co. KGaA. Reproduced with permission.

capsules through colloidal probe AFM, and combining this experiment with fluorescence microscopy [121], Figure 39.2a. By heat treatment we increase the mechanical stability, although even after heating to 55 °C capsules are still soft enough for rupture. Other studies of the mechanical stability of the capsules included probing hardness [127, 128] and deformation [129]. The studies described above can be also applied for probing cell-induced stresses; there the forces inducing the release of encapsulated material from polymeric microcapsules can be estimated in an *ex situ* experiment by the colloidal probe AFM. Comparing AFM-induced release with the release observed in living cells provides the possibility of estimating the pressure of cells upon uptake, which was found to be above 0.2 μN [25].

Self-induced release is closely associated with biodegradable PEM capsules, for example, enzymatic degradation of (poly-L-arginine (pARG)/dextran sulfate (DEXS) capsules [130], which can be used for delivery of drugs, nucleic acids and proteins [131]. In this area the release of oligonucleotides has been demonstrated from microcapsules assembled by the polycation-free method [132].

Enzymatic degradation has also been demonstrated for pARG and DEXS systems [130]. There, the degradation of these microcapsules was demonstrated by using the non-specific degradation by Pronase, a mixture of proteases that cleaves proteins and peptides. Degradation of microcapsules made of hyaluronic acid (HA)/PLL demonstrated the potential of such systems [133]. Orozco et al. reported the design and enzymatic degradation of multilayer assemblies by α-chymotrypsin [134]. In that study, the biodegradation process was monitored with UV–Vis spectroscopy. DEXS/Chitosan microcapsules are another example of biodegradable systems [135]. Release of transferrin from disulfide cross-linked polymer capsules made of poly (vinylpyrrolidone) and poly(methacrylic acid) functionalized with cysteamine are another interesting example of biodegradable systems [136] with potential for *in vivo* applications. It can be noted that this is the only other method used for intracellular delivery and release besides laser activation of microcapsules inside living cells. Enzyme-degradable capsules do not need an external trigger for their decomposition and may have high potential for the intracellular delivery of therapeutics, other biomimetic applications [137] and *in vivo* applications. Their disadvantages include difficulties in establishing a precise initial time of release [138]. Another representation of self-induced release is achieved by fusion of microcapsules (upon mixing of their contents). Microcapsule fusion provides opportunities for potential applications in gene transfection and drug transport across multilayers [139]. The salt-induced fusion of (PDADMAC/PSS)$_4$ microcapsules has been demonstrated [140]. It was established that, due to the surface tension and the hydrophobic interaction of polyelectrolytes, microcapsules decrease in surface area and fuse. Encapsulated molecules diffuse and mix but do not escape the bigger, fused microcapsule. Also, the polyelectrolytes in the membranes do not mix. The fluorescence images clearly prove that the orange inner content of the fused microcapsules is due to the free diffusion and mixing of the original FITC-dextran (green) and tetramethyl rhodamine isothiocyanate (TRITC)-dextran (red). Fusion of microcapsules is important to approach promising artificial microcontainers. Fusion of microcapsules is a promising approach for mimicking cellular functions.

Self-induced release is closely related to the area of enzyme-catalyzed reactions. These represent important biological functionality. Multicompartmental concentric microcapsules [18] or capsule-based capsosomes [141] can be used for monitoring reactions under a microscope; they can also be used to study coupled enzyme reactions [142, 143] or target cells [144]. Recent developments in enzyme-encapsulated LbL devices and their related functions, such as medical applications [145] and reactor sensors [146, 147] have been described recently. Encapsulation and functionality of glucose oxidase (GOX) and horseradish peroxidase (HRP) within concentric colored polymeric layers for the fabrication of striated multicolored spherical shells within agarose microbeads was also demonstrated [148]. These shells can simultaneously encapsulate biomolecules and can be used for encoding. Georgieva and Bäumler studied a coupled chain reaction in concentrically-built CaCO$_3$ capsules with three enzymes separated from each other by PEM walls: HRP, GOX and β-glucosidase (β-Glu) [142]. CaCO$_3$ microparticles have also been suggested for enzyme-catalyzed reactions [115], where the reaction was shown to take place in porous CaCO$_3$ particles upon disruption of attached substrate-filled liposomes (it should be noted that the enzyme is still protected by a PEM shell). Upon ultrasound triggered release from liposomes containing the peroxidase substrate, and Amplex Red, attached on the periphery in liposomes, the substrate was observed to diffuse

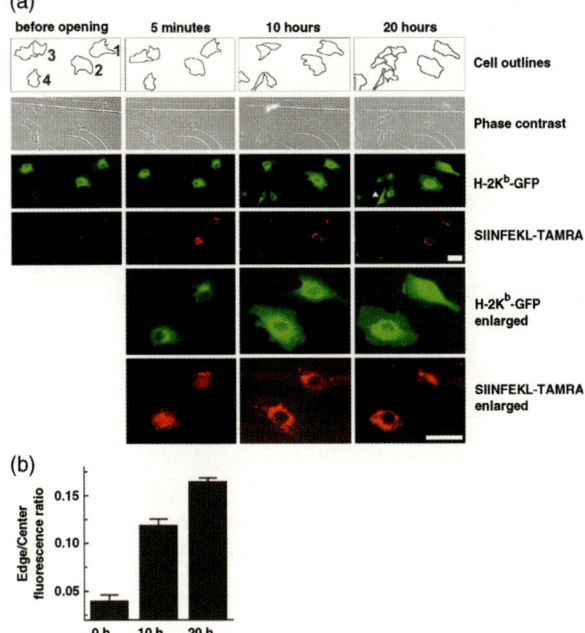

**Figure 39.3** Release of SIINFEKL peptides from microcapsules leads to surface transport of H-2K$^b$ –GFP. (a) Time course of capsule opening and class I transport. The scale bar is 50 μm. (b) Ratio of cell edge (i.e., surface) over cell center (total) fluorescence of the complex at three time points (an average of five cells; error bars show standard deviation) [109]. Copyright Wiley-VCH Verlag GmbH & Co. KGaA. Reproduced with permission.

into the inner part of the multicompartment containers. The same enzymatic reaction occurs faster in microcapsules [149] than inside particles.

Intracellular delivery and cell studies are a promising application of remote or self-induced release. Indeed, once sufficient mechanical stability of microcapsules is achieved, they can be placed inside living cells for studying intracellular processes [109, 150]. It was shown that four layer pairs PSS/PDADMAC capsules thermally shrunk above 60–65 °C can be used for this purpose. Palankar and coworkers investigated major histocompatibility complex (MHC) class I-mediated antigen presentation to the cell surface [109]. Capsules were incorporated into Vero and CHO cells and opened by a near-IR laser pulse from inside the cells. The kinetics of peptide-protein complexing was observed, Figure 39.3. The peptide-MHC Class I complex intracellular transport and the surface presentation were verified by adding antibodies which bind (from outside cells) to the peptide–protein complex, Figure 39.4. It can be noted that the laser-induced release from polymer microcapsules does not need chemical modification of the microcapsules and thus opens new approaches in biology and drug delivery. Some of these new approaches have already been demonstrated inside MDA-MB-435S cancer cells [53], African green monkey (Vero) cells [109] and Chinese hamster ovary (CHO) [109] and CHO-M1 [150]

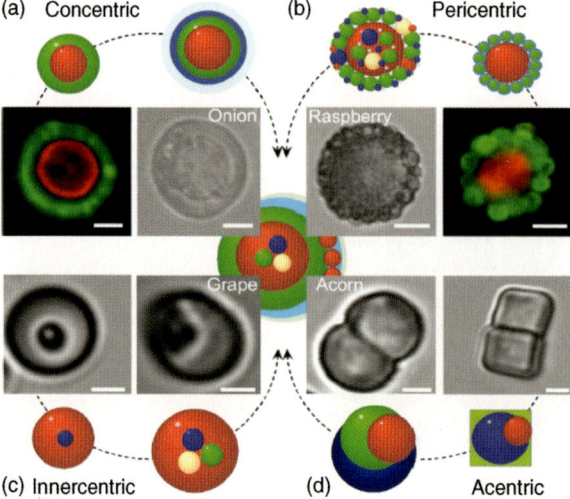

**Figure 39.4** Overview of multicompartment microcapsules. Four different approaches are identified: (a) concentric, (b) pericentric, (c) innercentric, and (d) acentric [149]. Copyright Wiley-VCH Verlag GmbH & Co. KGaA. Reproduced with permission.

fibroblasts, and neurons [138]. The temperature rise in such remote release experiments has been measured [52], and cell viability has been studied [151]. Extending biomedical research, microcapsules have been used in siRNA delivery [152], stem cells research [153], mucosal irritation [154], and bone formation [155].

Following the general trend of increasing the degree of complexity and designing more sophisticated capsules and particles, PEM microcapsules can be fabricated with more than one compartment. This can be achieved by multicompartmentalization, which can be used for simultaneous multiple drug delivery, both *in vitro* and *in vivo*, for conducting biochemical reactions in protected volumes, and gives the possibility to control various subcompartments of the capsules separately. In addition to increasing functionalities and augmenting a number of possibilities for studying reactions in protected volumes [110], multicompartmentalization can become particularly desirable in simultaneous delivery of several molecules, simultaneous diagnostics, sensing and delivery, and in response to multiple stimuli [143]. Four different approaches to construct multicompartment capsules: concentric ("onion"-like), pericentric ("raspberry"-like), innercentric ("grape"-like), and acentric ("acorn"-like), Figure 39.4, were recently presented [149]. A significant advantage of the structures shown in (a) and (b) is simplicity of fabrication. It can be noted that the acentric structure is different from the three previously described approaches because it represents an anisotropic structure. As such, anisotropic nano-, micro- and mesoscale natural and synthetic structures have been widely investigated [156–158], they are particularly important for pulmonary drug delivery [159]. Combining multicompartmentalization and anisotropic geometry can be used for improved control of biochemical reactions [110] or, if combined with direction-specific release [32], targeting specific organelles inside cells.

A novel strategy of fabricating anisotropic capsules consists in embedding larger containers into a soft film, for example PLL/HA, followed by adsorption of smaller containers on top of the unmasked surface [160]. This approach is viewed as a combination of capsules and films, and it opens extensive opportunities for research in biomedicine, specifically for controlled transport, delivery and separation of biomolecules [161–165]. Exponentially grown thick multilayer films [166, 167] exhibit interesting dynamic [168] properties as reservoirs of biomolecules [169]. They represent a class of relatively soft films, therefore direct deposition of particles or capsules leads to their partial entrapment and embedding in the films. Direct embedding of particles and capsules into the hydrogel films upon their sedimentation (due to gravity) leads to their protrusion into the films. Upon this process, the lower part of the particles or capsules is masked by the film, while their upper part is available for subsequent functionalization. Since only the upper part is available for functionalization, addition and adsorption of smaller containers, particles or molecules leads to the formation of anisotropic capsules. Figure 39.5 shows schematics where larger microcapsules loaded with Alexa Fluor 488 dextran (in green) are functionalized with smaller $SiO_2$ nano-containers labeled with TRITC-PAH (in red). It can be noted that although diffusion of molecules into the films poses certain limitations for retaining the masking, the soft "hydrogel"-like nature of the PLL/HA multilayers, combined with controllable thickness, the high loading capacity of the films, and high polymer diffusion, present advantages of their application. The

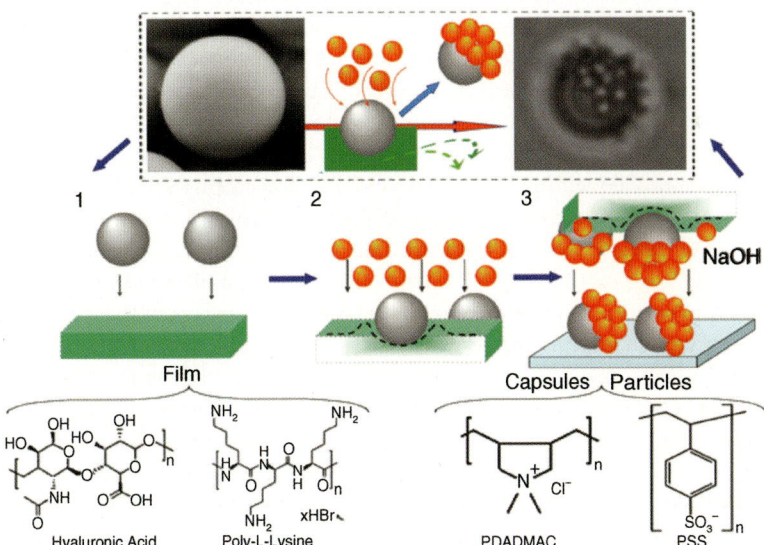

**Figure 39.5** Formation of anisotropic capsules and particles: (1) entrapping large silica capsules/particles (gray spheres) in biocompatible polymeric PLL/HA films (12 layers, PLL-terminated and labeled in green), followed by (2) adsorption of small silica containers (in red), and (3) film destruction by NaOH followed by extraction on a glass substrate [160]. Reproduced by permission of The Royal Society of Chemistry.

kinetics of particle/capsule sedimentation is essential for efficient functionalization; our studies showed that 10 min sedimentation was sufficient for particles. Controllable spatial self-embedding of microcapsules can be achieved by varying the stiffness of the films. Once particles and capsules are embedded into the film and functionalized by materials of choice, anisotropic particles/capsules need to be extracted from the film. This is done by slightly loosening the interaction of the films with the particles/capsules, that is, by adding a solution of NaOH. For prevention of aggregation and increased efficiency of recovery, the films are flipped over upside-down so that particles/capsules are located on the lower part of the films, Figure 39.5. Addition of NaOH, which basically acts as a self-inducing release agent for films, at this point leads to extraction of particles/capsules. Anisotropic structures open vast opportunities for diverse biomedical applications as they can be loaded with different biomolecules and used as reactors for enzyme-catalyzed reactions. On the other hand, functionalization of films with capsules and control over their interaction is of interest for coatings.

The results described above represent an area where microcapsules are used together with films. Compared to planar films, microcapsules are an ideal system since the PEMs are surrounded by liquid on both sides. On the other hand, modification of the surface of films by microcapsules allows one to introduce additional functionality to the film itself. Recently, Volodkin et al. proposed and studied modification of the surface of films by nanoparticles and microcapsules [170]. It was shown that nanoparticles are located at the interface of the film with the buffer solution in which the film is immersed. In the case of microcapsules, embedding was observed to be similar to the case of particles/capsules described above. Once the films are functionalized with particles or capsules, remote activation or remote release can be conducted. Laser activation of the films functionalized with gold NPs has been shown to affect the upper coating of the films. The surface of the film can thus be locally cross-linked by laser–NP interaction [170]. Stable adsorption of PEM microcapsules modified with gold NPs onto biocompatible HA/PLL films has also been demonstrated. In these experiments, PEM microcapsules loaded with dextran adsorbed onto the film were shown to release their cargo under stimulation with near-IR light, Figure 39.6. It can be seen from Figure 39.6 that, first, a microcapsule located on the left-hand side of the panel releases its contents upon exposure to laser light, then, sequentially, the second capsule is exposed and releases its content. The area on the film where these experiments took place is shown in the white rectangle, Figure 39.6d, while profiles of the capsules before all release experiments, and after the first and second release are shown in Figure 39.6e. Transmission images of both microcapsules at the end of the experiments are shown in Figure 39.6f. The above-described combination of films and capsules, both of which could be functionalized with light absorbing NPs, represents a combination of self-induced and remote release methods. As such, a wide range of future applications in this area for studying films is envisaged, including those with controlled release functions, bio-coatings, and cell detachment, as well as surfaces with antibacterial properties.

The area of modification of films by microcapsules and NPs is closely related to modification of surfaces by liposomes (Lip), which are widely used in medical

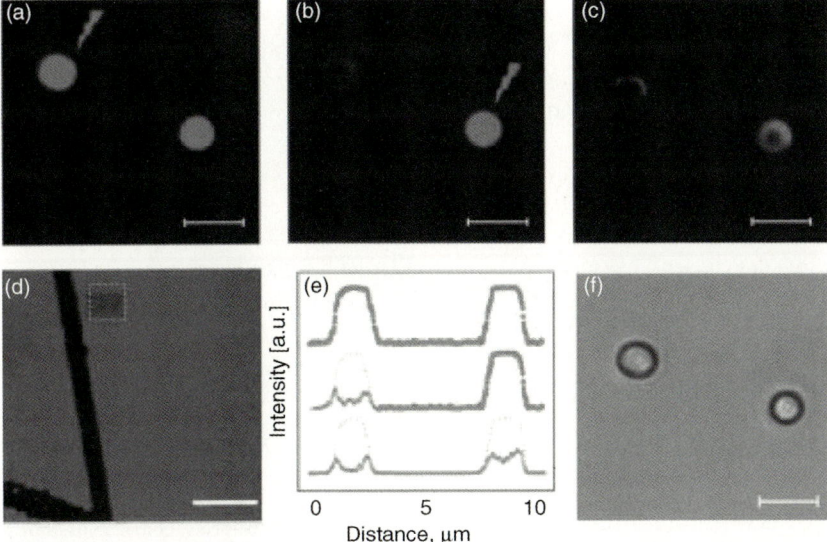

**Figure 39.6** Adsorption of microcapsules onto (PLL/HA)24/PLL films. (a)–(c) CLSM images of the capsules exposed to near-IR light irradiation. (d) CLSM image of the film surface (the film is prepared with PLL-FITC; black lines are scratches made by a needle for easier film imaging). (e) Cross-sectional profile of the capsules after step-by-step laser exposure (the sections from top to bottom correspond to the images (a–c), respectively). (f) TEM images of the capsules after light irradiation. The scale bar in (a–c) and (f) is 4 μm, that in (d) is 25 μm [170]. Copyright the American Chemical Society. Reproduced with permission.

applications [171–174]. They allow encapsulation and release of small amphiphilic and water-soluble molecules. Therefore, surfaces with immobilized vesicles are perfect candidates for drug delivery applications [175]. In addition, biocompatibility of the liposomes makes them attractive candidates for preserving trans-membrane proteins or water-soluble biomolecules. It can be noted that the main difficulty in immobilization of intact liposomes is vesicle destabilization or fusion upon contact with solid/fluid interfaces [176, 177]. Immobilization of liposomes on solid surfaces has been realized by covalent or other specific interactions [178, 179]. It has been shown that the PEM films can be functionalized with liposomes containing encapsulated molecules inside the internal aqueous compartment (Figure 39.7a–c) [180–184]. The vesicles were stabilized with polylysine (Lip-PLL) by surface adsorption, and were further embedded into the film. Control over PLL stabilization is important for avoiding vesicle aggregation [185–187]. The vesicles of 130 nm diameter were made by extrusion and composed of 1,2-dipalmitoyl-sn-glycero-3-phosphocholine, 1,2-dipalmitoyl-sn-glycero-3-[phospho-rac-(1-glycerol)], and cholesterol.

Interpolyelectrolyte and lipid–polyelectrolyte interactions are critically important for the embedding process. Optimal polyelectrolyte combination and lipid charge allow successful vesicle adsorption. Embedding without rupture can be identified based on the interaction between PLL-Lip and the outermost polyanion layer of the film. An experimental framework for selection of combinations of polyelectrolytes

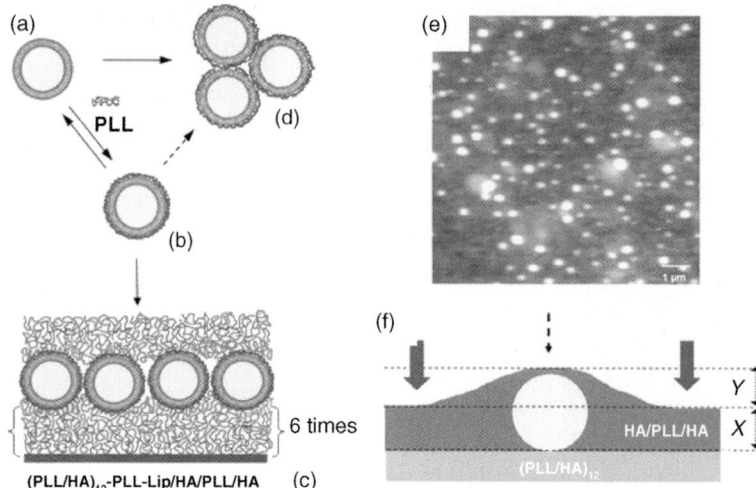

**Figure 39.7** Scheme of vesicle complexation with PLL resulting in the formation of stabilized single liposomes, noted PLL-Lip (a, b) or liposome aggregates (a–d). Liposome-containing films (PLL/HA)$_{12}$/PLL-Lip/HA/PLL/HA are formed upon adsorption of PLL-Lip on the (PLL/HA)$_{12}$ films followed by additional deposition of HA/PLL/HA (b, c). PLL is in blue and HA is in red. (e) AFM image of liposome-containing films. (f) Representation of PLL/HA films with incorporated liposomes or polystyrene latex particles, both stabilized by PLL coating. X represents the thickness increase over the (PLL/HA)$_{12}$ film upon deposition of HA/PLL/HA in a location without deposited vesicles and (X + Y) represents the thickness increase upon adsorption of HA/PLL/HA in a location atop the deposited particles, assuming that they deposit on the (PLL/HA)$_{12}$ stratum without penetrating into its architecture. Reproduced by permission of The Royal Society of Chemistry [180].

has been constructed by investigating the interactions between PLL-Lip and the polyanion of the film [180]. The state of liposomes in the film with regard to embedding can be predicted by taking into account the interaction enthalpies between polymers and lipids. Highly hydrated LbL films, such as HA/PLL films, are suitable reservoirs for embedding intact vesicles, while synthetic films such as PSS/PAH are not suited because the PSS–PLL interaction is stronger than that of PLL–DPPG. A fine detail of functionalization of films by liposomes stems from the fact that gravity does not play a role in sedimentation of these nanometer-sized carriers, the process is totally driven by diffusion of liposomes in solution above the films. As they come into contact with the surface of the film, particles are immersed similarly to micrometer-sized capsules [188] which are embedded to approximately one half of their diameter. This can be assigned to the increase in the number of ionic contacts during immersion into the film. Figures 39.7e,f show that additional coating after vesicle adsorption can result in almost complete liposome embedding into the film (as indicated by a reduction in the vesicle height in the film). It can be noted that, in solution, vesicles have a diameter of 130 nm, while after even additional coating the height is 21 nm. The embedding is attributed to the difference in the growth regime on top of and in between the entrapped vesicles (polymer diffusion into the film is restricted by the presence of vesicles). The behavior of soft vesicles in contact with the film was found to be similar to that of latex particles [181]. Both types of particles preserve their spherical shapes and settle down into the film similarly to adenoviral vectors [189].

Surface-immobilized liposome-based systems could be considered for controlled drug delivery. Localized delivery can be pursued if loaded liposomes are immobilized and are kept intact on films (for example, for contact lenses). Cargo of the film-entrapped vesicles (carboxyfluorescein, CF) is not released if the lipids are in the solid state below the transition temperature (41 °C), but pronounced dye release can be induced by temperature increase over the transition temperature, when the lipid bilayer is in the liquid state possessing higher permeability (Figure 39.8a). The ATR-FTIR experiment revealed that CF release is not due to thermally induced structural changes within the film (that can also affect liposome integrity), but due to permeabilization of the lipidic membrane upon heating (Figure 39.8b–d). Slightly faster release kinetics of film-embedded vesicles in comparison with that for vesicles in solution is due to destabilization of the lipid bilayer interacting with the

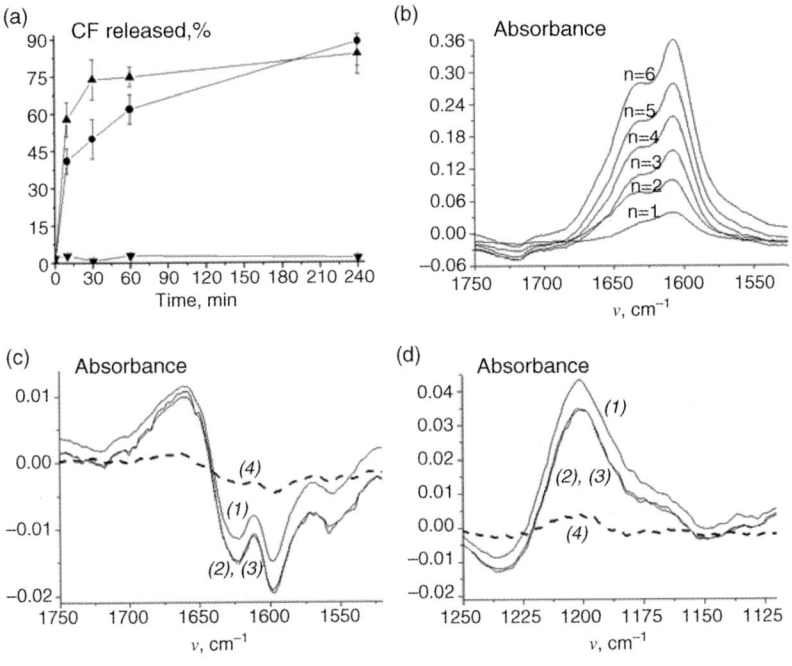

**Figure 39.8** (a) Time evolution of the cumulative CF release from vesicles embedded inside a (PLL/HA)$_{12}$/Lip-PLL/HA/PLL/HA film architecture, when the film is maintained at ambient temperature (▼) or heated and maintained at 45 °C (▲). These release kinetics are compared to the release kinetics obtained for the same PLL-covered vesicles in aqueous solution at 45 °C (●). (b) IR absorption spectra of (PLL/HA)$_6$ film build-up. Spectra (c) and (d) show the wavelength regions corresponding to the HA and D$_2$O absorption: spectra of the (PLL/HA)$_7$ film upon heating at 55 °C for (1) 10, (2) 30, and (3) 60 min, and after cooling to 25 °C (4). Reproduced by permission of The Royal Society of Chemistry [180, 181].

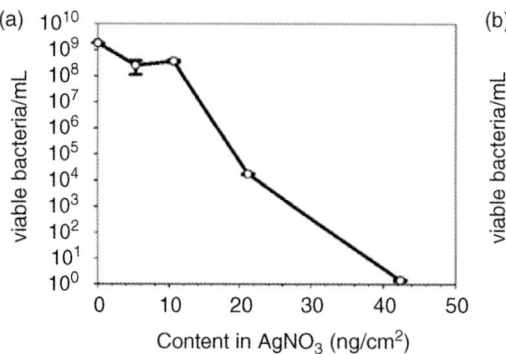
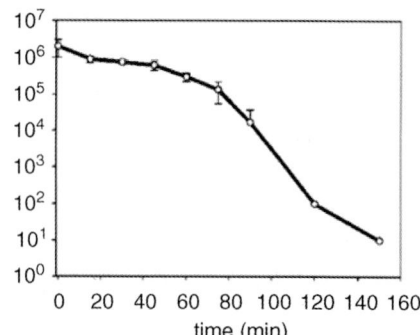

**Figure 39.9** (a) Number of viable E. coli after incubation for 20 h at 37 °C versus AgNO$_3$ concentration in the AgNO$_3$ coating. The initial concentration of bacteria put into contact with AgNO$_3$ coatings was around 10$^6$ cfu mL$^{-1}$. The AgNO$_3$ content is expressed as the amount of AgNO$_3$ encapsulated per cm$^2$. (b) Evolution of the number of viable E. coli cells as a function of the incubation time in contact with AgNO$_3$ coating (120 ng cm$^{-2}$ in AgNO$_3$). Copyright the American Chemical Society. Reproduced from [194] with permission.

polyelectrolyte film network. This study demonstrates the feasibility of a general strategy aimed at encapsulating solutes in the internal aqueous compartment of vesicles embedded inside highly hydrated PEM films, without spontaneous release but with the possibility to trigger the release of hydrophilic compounds with external stimuli tolerated by the environment.

PEM polymer films are promising candidates for antibacterial coatings of biomedical devices. Bacterial infections at a site of an implanted medical device present a serious source of problems, leading, if untreated, to chronic microbial infection, inflammation, tissue necrosis, and so on. Multilayers containing antimicrobial agents, such as peptides, silver nanoparticles, and hydrophobic bactericides, have been reported [190–193]. Temperature-induced release of small molecules from liposome-containing films could be used for the preparation of antibacterial coatings responsive to temperature. HA/PLL films with embedded AgNO$_3$-loaded liposomes showed a strong and rapid antibacterial effect against E. coli at 37 °C (Figure 39.9). Two-hour contact with an AgNO$_3$ coating (120 ng cm$^{-2}$ AgNO$_3$) induces a 4-log reduction of the bacterial population. Although PLL/HA films could not be directly loaded with a significant amount of silver ions to induce bacterial death, embedded vesicles provide a large amount of a bactericidal agent. The developed coatings present a solvent-based antibacterial system through release of bactericidal agent from the film under external stimulation. Shi reported PEMs having antimicrobial properties obtained by in situ wet phase reduction of Ag$^+$ ions in silver nanoparticles [192]. These films can release silver ions into the surroundings and show about 2-log reduction of the number of viable E. coli cells after 2 h incubation. The liposome-containing films have fast and strong performance against bacteria growth compared to other multilayer coatings.

HA/PLL films demonstrate reservoir properties, that is, nano- and micro-carriers, such as liposomes or polymeric capsules as well as biomolecules (DNA, proteins, peptides, etc.), can be incorporated in the film [180, 181, 195, 196]. Interaction of

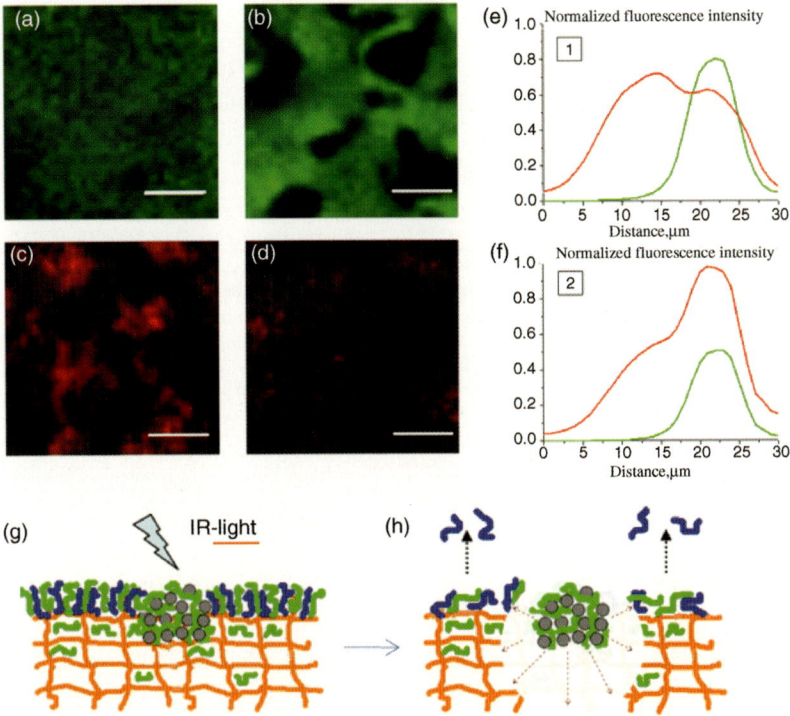

**Figure 39.10** (PLL/HA)$_{24}$/PLL film with embedded DNA before (a,c) and after (b,d) irradiation with IR-laser light. PLL in the film is labeled with FITC (in green), and DNA (in red) is labeled with EtBr. The scale bar is 5 μm. Stack profiles obtained by CLSM for (PLL/HA)$_{24}$/PLL with embedded DNA labeled with EtBr. The film was prepared with PLL-FITC before (e) and after (f) heating to 70 °C for 30 min. The green curve corresponds to the PLL-FITC stack profile, and the red one corresponds to the DNA-EtBr stack profile. The distance on the y-axis is from the top of the film to the bottom. Schematic presentation of the suggested mechanism of DNA release induced by distortion of DNA-PLL interaction due to partial thermal film decomposition around nanoparticle aggregates (g, h). Copyright the American Chemical Society. Reproduced from [195] with permission.

nano- and micrometer-sized species with LbL-assembled films depends significantly on the polymer mobility in the films. The HA/PLL films with high polymer mobility [197] (PLL diffusion coefficient up to $1\,\mu m^2\,s^{-1}$) have very high loading capacity for macromolecules due to complexation of adsorbing material with the polymer molecules doped onto the film surface [195]. This results in accumulation of a large amount of adsorbed materials (significantly larger than that for low polymer mobility PSS/PAH films). Diffusion of embedded DNA into the HA/PLL films can be triggered by heating, as indicated by the shift of the DNA profile (red line) relative to the film (green line) (Figure 39.10e–f) [195]. Adsorption of macromolecules such as DNA on the film is driven by complexation with PLL doping from the film interior to the surface. Interruption of this interaction by, for instance, local heating can result in increase in DNA mobility and release from the film surface (Figure 39.10a–d and g–h).

## References

1. Decher, G. (1997) Fuzzy nanoassemblies: Toward layered polymeric multicomposites. *Science*, **277**, 1232–1237.
2. Donath, E., Sukhorukov, G.B., Caruso, F., Davis, S.A., and Möhwald, H. (1998) Novel hollow polymer shells by colloid-templated assembly of polyelectrolytes. *Angew. Chem. Int. Ed.*, **37**, 2202–2205.
3. Caruso, F., Caruso, R.A., and Möhwald, H. (1998) Nanoengineering of inorganic and hybrid hollow spheres by colloidal templating. *Science*, **282**, 1111–1114.
4. Klitzing, R., Wong, J.E., Jaeger, W., and Steitz, R. (2004) Short range interaction in polyelectrolyte multilayers. *Curr. Opin. Colloid Interface Sci.*, **9**, 158–162.
5. Georgieva, R., Dimova, R., Sukhorukov, G.B., Ibarz, G., and Möhwald, H. (2005) Influence of different salts on micro-sized polyelectrolyte hollow capsules. *J. Mater. Chem.*, **15**, 4301–4310.
6. Sukhishvili, S.A. and Granick, S. (1998) Adsorbed monomer analog of a common polyelectrolyte. *Phys. Rev. Lett.*, **80**, 3646–3649.
7. Tong, W.J. and Gao, C.Y. (2008) Multilayer microcapsules with tailored structures for bio-related applications. *J. Mater. Chem.*, **18**, 3799–3812.
8. De Cock, L.J., De Koker, S., De Geest, B.G., Grooten, J., Vervaet, C., Remon, J.P., Sukhorukov, G.B., and Antipina, M.N. (2010) Polymeric multilayer capsules in drug delivery. *Angew. Chem. Int. Ed.*, **49**, 6954–6973.
9. Zelikin, A.N., Price, A.D., and Stadler, B. (2010) Poly(Methacrylic Acid) polymer hydrogel capsules: Drug carriers, sub-compartmentalized microreactors, artificial organelles. *Small*, **6**, 2201–2207.
10. Qi, W., Wang, A.H., Yang, Y., Du, M.C., Bouchu, M.N., Boullanger, P., and Li, J.B. (2010) The lectin binding and targetable cellular uptake of lipid-coated polysaccharide microcapsules. *J. Mater. Chem.*, **20**, 2121–2127.
11. Decher, G., Hong, J.D., and Schmitt, J. (1992) Buildup of ultrathin multilayer films by a self-assembly process: III. Consecutively alternating adsorption of anionic and cationic polyelectrolytes on charged surfaces. *Thin Solid Films*, **210**, 831–835.
12. Lvov, Y., Decher, G., and Möhwald, H. (1993) Assembly, structural characterization and thermal behavior of layer-by-layer deposited ultrathin films of Poly(vinylsulfate) and Poly(allylamine). *Langmuir*, **9**, 481–486.
13. Sukhishvili, S.A. (2005) Responsive polymer films and capsules via layer-by-layer assembly. *Curr. Opin. Colloid Interface Sci.*, **10**, 37–44.
14. Delongchamp, D.M. and Hammond, P.T. (2003) Fast ion conduction in layer-by-layer polymer films. *Chem. Mater.*, **15**, 1165–1173.
15. Tanchak, O.M., Yager, K.G., Fritzsche, H., Harroun, T., Katsaras, J., and Barrett, C.J. (2006) Water distribution in multilayers of weak polyelectrolytes. *Langmuir*, **22**, 5137–5143.
16. Schönhoff, M. (2003) Layered polyelectrolyte complexes: physics of formation and molecular properties. *J. Phys. Condens. Matter*, **15**, R1781–R1808.
17. Lo, M.Y., Lay, C.L., Lu, X.H., and Liu, Y. (2008) Finer structures of polyelectrolyte multilayers reflected by solution H-1 NMR. *J. Phys. Chem. B*, **112**, 13218–13224.
18. Kreft, O., Prevot, M., Mohwald, H., and Sukhorukov, G.B. (2007) Shell-in-shell microcapsules: A novel tool for integrated, spatially confined enzymatic reactions. *Angew. Chem. Int. Ed.*, **46**, 5605–5608.
19. Kato, N. and Caruso, F. (2005) Homogeneous, competitive fluorescence quenching immunoassay based on gold nanoparticle/polyelectrolyte coated latex particles. *J. Phys. Chem. B*, **109**, 19604–19612.
20. Stein, E.W., Volodkin, D.V., McShane, M.J., and Sukhorukov, G.B.

(2006) Real-time assessment of spatial and temporal coupled catalysis within polyelectrolyte microcapsules containing coimmobilized glucose oxidase and peroxidase. *Biomacromolecules*, **7**, 710–719.

21 Jewell, C.M. and Lynn, D.M. (2008) Multilayered polyelectrolyte assemblies as platforms for the delivery of DNA and other nucleic acid-based therapeutics. *Adv. Drug Deliv. Rev.*, **60**, 979–999.

22 Peteiro-Cattelle, J., Rodriguez-Pedreira, M., Zhang, F., Rivera Gil, P., del Mercato, L.L., and Parak, W.J. (2009) One example on how colloidal nano- and microparticles could contribute to medicine. *Nanomedicine*, **4**, 967–979.

23 Verberg, R., Alexeev, A., and Balazs, A.C. (2006) Modeling the release of nanoparticles from mobile microcapsules. *J. Chem. Phys.*, **125**, 2247121–22471210.

24 Tavera, E.M., Kadali, S.B., Bagaria, H.G., Liu, A.W., and Wong, M.S. (2009) Experimental and modeling analysis of diffusive release from single-shell microcapsules. *AIChE J.*, **55**, 2950–2965.

25 Delcea, M., Schmidt, S., Palankar, R., Fernandes, P.A.L., Fery, A., Moehwald, H., and Skirtach, A.G. (2010) Mechanobiology: Corrrelation of mechanical stability of microcapsules studied by AFM. *Small*, **6**, 2858–2862.

26 Ariga, K., Hill, J.P., and Ji, Q. (2007) Layer-by-layer assembly as a versatile bottom-up nanofabrication technique for exploratory research and realistic application. *Phys. Chem. Chem. Phys.*, **9**, 2319–2340.

27 Caruso, F. and Sukhorukov, G.B. (2003) *Multilayer Thin Films* (eds G. Decher and J.B. Schlenoff), Wiley-VCH Verlag GmbH, Weinheim.

28 Schneider, G. and Decher, G. (2004) From functional core/shell nanoparticles prepared via layer-by-layer deposition to empty nanospheres. *Nano Lett.*, **4**, 1833–1839.

29 Bantchev, G., Lu, Z.H., and Lvov, Y. (2009) Layer-by-layer nanoshell assembly on colloids through simplified washless process. *J. Nanosci. Nanotech.*, **9**, 396–403.

30 Patra, D., Sanyal, A., and Rotello, V.M. (2010) Colloidal microcapsules: Self-assembly of nanoparticles at the liquid-liquid interface. *Chem. Asian J.*, **5**, 2442–2453.

31 De Geest, B.G., McShane, M.J., Demeester, J., De Smedt, S.C., and Hennink, W.E. (2008) Microcapsules ejecting nanosized species into the environment. *J. Am. Chem. Soc.*, **130**, 14480–14482.

32 Bedard, M.F., De Geest, B.G., Möhwald, H., Sukhorukov, G.B., and Skirtach, A.G. (2009) Direction specific release from giant microgel-templated polyelectrolyte microcontainers. *Soft Matter*, **5**, 3927–3931.

33 Peyratout, C.S. and Dähne, L. (2004) Tailor-made polyelectrolyte microcapsules: from multilayers to smart containers. *Angew. Chem. Int. Ed.*, **43**, 3762–3783.

34 Sukhorukov, G.B., Donath, E., Davis, S., Lichtenfeld, H., Caruso, F., Popov, V.I., and Möhwald, H. (1998) Stepwise polyelectrolyte assembly on particle surfaces: a novel approach to colloid design. *Polym. Adv. Technol.*, **9**, 759–767.

35 Dejugnat, C. and Sukhorukov, G.B. (2004) pH-responsive properties of hollow polyelectrolyte microcapsules templated on various cores. *Langmuir*, **20**, 7265–7269.

36 Antipov, A.A., Shchukin, D., Fedutik, Y.A., Petrov, A.I., Sukhorukov, G.B., and Möhwald, H. (2003) Carbonate microparticles for hollow polyelectrolyte capsules fabrication. *Colloids Surf. A*, **224**, 175–183.

37 Adalsteinsson, T., Dong, W.F., and Schönhoff, M. (2004) Dynamics of 77kDa dextran in sub-micron polyelectrolyte capsule dispersion measured using PFG-NMR. *J. Phys. Chem. B*, **108**, 20056–20063.

38 Itoh, Y., Matsusaki, M., Kida, T., and Akashi, M. (2004) Preparation of biodegradable hollow nanocapsules by silica template method. *Chem. Lett.*, **33**, 1552–1553.

39 Moya, S., Dähne, L., Voigt, A., and Möhwald, H. (2001) Polyelectrolyte multilayer capsules templated on biological cells: core oxidation influences

layer chemistry. *Colloids Surf. A*, **183**, 27–40.

**40** Neu, B., Voigt, A., Mitlöhner, R., Leporatti, S., Gao, C.Y., Donath, E., Kiesewetter, H., Möhwald, H., Meiselman, H.J., and Bäumler, H. (2001) Biological cells as templates for hollow microcapsules. *J. Microencapsulation*, **18**, 385–395.

**41** Volodkin, D.V., Petrov, A.I., Prevot, M., and Sukhorukov, G.B. (2004) Matrix polyelectrolyte microcapsules: new system for macromolecule encapsulation. *Langmuir*, **20**, 3398–3406.

**42** Petrov, A.I., Volodkin, D.V., and Sukhorukov, G.B. (2005) Protein-calcium carbonate coprecipitation: a tool for protein encapsulation. *Biotechnol. Prog.*, **21**, 918–925.

**43** Schuetz, P. and Caruso, F. (2003) Copper-assisted weak polyelectrolyte multilayer formation on microspheres and subsequent film crosslinking. *Adv. Funct. Mater.*, **13**, 929–937.

**44** Shenoy, D.B., Antipov, A.A., Sukhorukov, G.B., and Möhwald, H. (2003) Layer-by-layer engineering of biocompatible, decomposable core-shell structures. *Biomacromolecules*, **4**, 265–272.

**45** Caruso, F. and Möhwald, H. (1999) Protein multilayer formation on colloids through a stepwise self-assembly technique. *J. Am. Chem. Soc.*, **121**, 6039–6046.

**46** Vinogradova, O.I., Lebedeva, O.V., Vasilev, K., Gong, H.F., Garcia-Turiel, J., and Kim, B.S. (2005) Multilayer DNA/Poly (Allylamine Hydrochloride) microcapsules: assembly and mechanical properties. *Biomacromolecules*, **6**, 1495–1502.

**47** Schüler, C. and Caruso, F. (2001) Decomposable hollow biopolymer-based capsules. *Biomacromolecules*, **2**, 921–926.

**48** Moya, S., Donath, E., Sukhorukov, G.B., Auch, M., Bäumler, H., Lichtenfeld, H., and Möhwald, H. (2000) Lipid coating on polyelectrolyte surface modified colloidal particles and polyelectrolyte capsules. *Macromolecules*, **33**, 4538–4544.

**49** An, Z.H., Tao, C., Lu, G., Möhwald, H., Zheng, S., Cui, Y., and Li, J.B. (2005) Fabrication and characterization of human serum albumin and L-α-dimyristoylphosphatidic acid microcapsules based on template technique. *Chem. Mater.*, **17**, 2514–2519.

**50** Radtchenko, I.L., Sukhorukov, G.B., Leporatti, S., Khomutov, G.B., Donath, E., and Möhwald, H. (2000) Assembly of alternated multivalent ion/polyelectrolyte layers on colloidal particles. Stability of the multilayers and encapsulation of macromolecules into polyelectrolyte capsules. *J. Colloid Interface Sci.*, **230**, 272–280.

**51** Dai, Z.F., Voigt, A., Leporatti, S., Donath, E., Dähne, L., and Möhwald, H. (2001) Layer-by-layer self-assembly of polyelectrolyte and low molecular weight species into capsules. *Adv. Mater.*, **13**, 1339–1342.

**52** Skirtach, A.G., Dejugnat, C., Braun, D., Susha, A.S., Rogach, A.L., Parak, W.J., Mohwald, H., and Sukhorukov, G.B. (2005) The role of metal nanoparticles in remote release of encapsulated materials. *Nano. Lett.*, **5**, 1371–1377.

**53** Skirtach, A.G., Munoz Javier, A., Kreft, O., Kohler, K., Piera Alberola, A., Mohwald, H., Parak, W.J., and Sukhorukov, G.B. (2006) Laser-induced release of encapsulated materials inside living cells. *Angew. Chem. Int. Ed.*, **45**, 4612–4617.

**54** Bédard, M.F., Braun, D., Sukhorukov, G.B., and Skirtach, A.G. (2008) Towards self-assembly of nanoparticles on polymeric capsules: release threshold and permeability. *ACS Nano.*, **2**, 1807–1816.

**55** Caruso, F., Lichtenfeld, H., Giersig, M., and Möhwald, H. (1998) Electrostatic self-assembly of silica nanoparticle–polyelectrolyte multilayers on polystyrene latex particles. *J. Am. Chem. Soc.*, **120**, 8523–8524.

**56** Kotov, N.A., Dekany, I., and Fendler, J.H. (1995) Layer-by-layer self-assembly of polyelectrolyte-semiconductor nanoparticle composite. *J. Phys. Chem.*, **99**, 13065–13069.

**57** Khopade, A.J. and Caruso, F. (2002) Stepwise self-assembled Poly (amidoamine) dendrimer and Poly (styrenesulfonate) microcapsules as

58 Yoo, P.J., Nam, K.T., and Qi, J. (2006) Spontaneous assembly of viruses on multilayered polymer surfaces. *Nat. Mater.*, **5**, 234–240.

59 Jiang, C., Ko, H., and Tsukruk, V.V. (2005) Strain-sensitive raman modes of carbon nanotubes in deflecting freely suspended nanomembranes. *Adv. Mater.*, **17**, 2127–2131.

60 Kleinfeld, E.R. and Ferguson, G.S. (1994) Stepwise formation of multilayered nanostructural films from macromolecular precursors. *Science*, **265**, 370–373.

61 Antipov, A.A., Sukhorukov, G.B., Fedutik, Y.A., Hartmann, J., Giersig, M., and Möhwald, H. (2002) Fabrication of a novel type of metallized colloids and hollow capsules. *Langmuir*, **18**, 6687–6693.

62 Li, L.D., Tedeschi, C., Kurth, D.G., and Möhwald, H. (2004) Synthesis of a pyrene-labeled polyanion and its adsorption onto polyelectrolyte hollow capsules functionalized for electron transfer. *Chem. Mater.*, **16**, 570–573.

63 Karlsson, A.J., Flessner, R.M., Gellman, S.H., Lynn, D.M., and Palecek, S.P. (2010) Polyelectrolyte multilayers fabricated from antifungal beta-peptides: design of surfaces that exhibit antifungal activity against candida albicans. *Biomacromolecules*, **11**, 2321–2328.

64 Gowri, S., Almeida, L., Amorim, T., Carneiro, N., Souto, A.P., and Esteves, M.F. (2010) Polymer nanocomposites for multifunctional finishing of textiles – a review. *Text. Res. J.*, **80**, 1290–1306.

65 Verberg, R., Dale, A.T., Kumar, P., Alexeev, A., and Balazs, A.C. (2007) Healing substrates with mobile, particle-filled microcapsules: designing a 'repair and go' system. *J. R. Soc. Interface*, **4**, 349–357.

66 Skirtach, A.G. and Kreft, O. (2009) Nanotechnology in drug delivery, in *Nanotechnology in Drug Delivery* (eds M. de Villiers, P. Aramwit, and G.S. Kwon), Springer, New York, p. 545.

67 Sui, Z.J. and Schlenoff, J.B. (2004) Phase separations in pH-responsive polyelectrolyte multilayers: Charge extrusion versus charge expulsion. *Langmuir*, **20**, 6026–6031.

68 Shiratori, S.S. and Rubner, M.F. (2000) pH-dependent thickness behavior of sequentially adsorbed layers of weak polyelectrolytes. *Macromolecules*, **33**, 4213–4219.

69 Yang, S.Y. and Rubner, M.F. (2002) Micropatterning of polymer thin films with pH-sensitive and cross-linkable hydrogen-bonded polyelectrolyte multilayers. *J. Am. Chem. Soc.*, **124**, 2100–2101.

70 Chaturbedy, P., Jagadeesan, D., and Eswaramoorthy, M. (2010) pH-sensitive breathing of clay within the polyelectrolyte matrix. *ACS Nano.*, **4**, 5921–5929.

71 Burke, S.E. and Barrett, C.J. (2003) pH-responsive properties of multilayered poly(L-lysine)/hyaluronic acid surfaces. *Biomacromolecules*, **4**, 1773–1783.

72 Stockton, W.B. and Rubner, M.F. (1997) Molecular-level processing of conjugated polymers. 4. Layer-by-layer manipulation of polyaniline via hydrogen-bonding interactions. *Macromolecules*, **30**, 2717–2725.

73 Buscher, K., Graf, K., Ahrens, H., and Helm, C.A. (2002) Influence of adsorption conditions on the structure of polyelectrolyte multilayers. *Langmuir*, **18**, 3585–3591.

74 Köhler, K., Shchukin, D.G., Möhwald, H., and Sukhorukov, G.B. (2005) Thermal behavior of polyelectrolyte multilayer microcapsules. 1.The effect of odd and even layer number. *J. Phys. Chem. B*, **109**, 18250–18259.

75 Steitz, R., Leiner, V., Tauer, K., Khrenov, V., and von Klitzing, R. (2002) Temperature-induced changes in polyelectrolyte films at the solid-liquid interface. *Appl. Phys. A: Mater. Sci. Process.*, **74**, S519–S521.

76 Ahrens, H., Büscher, K., Eck, D., Förster, S., Luap, C., Papastavrou, G., Schmitt, J., Steitz, R., and Helm, C.A. (2004) Poly(styrene sulfonate) adsorption: Electrostatic and Secondary

Interactions Macromol. *Symp.*, **211**, 93–106.

77 Ibarz, G., Dähne, L., Donath, E., and Möhwald, H. (2002) Controlled permeability of polyelectrolyte capsules via defined annealing. *Chem. Mater.*, **14**, 4059–4062.

78 Gao, C., Möhwald, H., and Shen, J. (2004) Enhanced biomacromolecule encapsulation by swelling and shrinking procedures. *Chem. Phys. Chem.*, **5**, 116–120.

79 Mendes, P.M. (2008) Stimuli-responsive surfaces for bio-applications. *Chem. Soc. Rev.*, **37**, 2512–2529.

80 Stuart, M.A.C., Huck, W.T.S., Genzer, J., Muller, M., Ober, C., Stamm, M., Sukhorukov, G.B., Szleifer, I., Tsukruk, V.V., Urban, M., Winnik, F., Zauscher, S., Luzinov, I., and Minko, S. (2010) Emerging applications of stimuli-responsive polymer materials. *Nature Mater.*, **9**, 101–113.

81 Lee, H.I., Pietrasik, J., Sheiko, S.S., and Matyjaszewski, K. (2010) Stimuli-responsive molecular brushes. *Prog. Polym. Sci.*, **35**, 24–44.

82 Aznar, E., Coll, C., Marcos, M.D., Martinez-Manez, R., Sancenon, M., Soto, J., Amoros, P., Cano, J., and Ruiz, E. (2009) Borate-driven gatelike scaffolding using mesoporous materials functionalised with saccharides. *Chem.-Eur. J.*, **15**, 6877–6888.

83 Dejugnat, C., Köhler, K., Dubois, M., Sukhorukov, G.B., Mohwald, H., Zemb, T., and Guttmann, P. (2007) Membrane densification of heated polyelectrolyte multilayer capsules characterized by soft X-ray microscopy. *Adv. Mater.*, **19**, 1331–1336.

84 Soltwedel, O., Ivanova, O., Nestler, P., Muller, M., Kohler, R., and Helm, C.A. (2010) Interdiffusion in polyelectrolyte multi layers. *Macromolecules*, **43**, 7288–7293.

85 Bedard, M.F., Munoz-Javier, A., Mueller, R., del Pino, P., Fery, A., Parak, W.J., Skirtach, A.G., and Sukhorukov, G.B. (2009) On the mechanical stability of polymeric microcontainers functionalized with nanoparticles. *Soft Matter*, **5**, 148–155.

86 Timko, B.P., Dvir, T., and Kohane, D.S. (2010) Remotely triggerable drug delivery systems. *Adv. Mater.*, **22**, 4925–4943.

87 Roggan, A., Friebel, M., Dorschel, K., Hahn, A., and Muller, G. (1999) Optical properties of circulating human blood in the wavelength range 400–2500nm. *J. Biomed. Opt.*, **4**, 36–46.

88 Tao, X., Li, J., and Möhwald, H. (2004) Self-assembly, optical behavior, and permeability of a novel capsule based on an azo dye and polyelectrolytes. *Chem. Eur. J.*, **10**, 3397–3403.

89 Ning, L., Kommireddy, D.S., Lvov, Y., Liebenberg, W., Tiedt, L.R., and De Villiers, M.M. (2006) Nanoparticle multilayers: Surface modification of photosensitive drug microparticles for increased stability and in vitro bioavailability. *J. Nanosci. Nanotechnol.*, **6**, 3252–3260.

90 Skirtach, A.G., Antipov, A.A., Shchukin, D.G., and Sukhorukov, G.B. (2004) Remote activation of capsules containing Ag nanoparticles and IR dye by laser light. *Langmuir*, **20**, 6988–6992.

91 Radt, B., Smith, T.A., and Caruso, F. (2004) Optically addressable nanostructured capsules. *Adv. Mater.*, **16**, 2184–2189.

92 Koo, H.Y., Lee, H.J., Kim, J.K., and Choi, W.S. (2010) UV-triggered encapsulation and release from polyelectrolyte microcapsules decorated with photoacid generators. *J. Mater. Chem.*, **20**, 3932–3937.

93 Skirtach, A.G., Dejugnat, C., Braun, D., Susha, A.S., Rogach, A.L., and Sukhorukov, G.B. (2007) Nanoparticles distribution control by polymers: Aggregates versus nonaggregates. *J. Phys. Chem. C*, **111**, 555–564.

94 Parakhonskiy, B.V., Bedard, M.F., Bukreeva, T.V., Sukhorukov, G.B., Mohwald, H., and Skirtach, A.G. (2010) Nanoparticles on polyelectrolytes at low concentration: controlling concentration and size. *J. Phys. Chem. C*, **114**, 1996–2002.

95 Angelatos, A.S., Radt, B., and Caruso, F. (2005) Light-responsive polyelectrolyte/gold nanoparticle microcapsules. *J. Phys. Chem. B*, **109**, 3071–3076.

96 Skirtach, A.G., Karageorgiev, P., Bedard, M.F., Sukhorukov, G.B., and Möhwald, H. (2008) Reversibly permeable nanomembranes of polymeric microcapsules. *J. Am. Chem. Soc.*, **130**, 11572–11573.

97 Angelatos, A.S., Smith, S., and Caruso, F. (2006) Bioinspired colloidal systems via layer-by-layer assembly. *Soft Matter*, **2**, 18–23.

98 De Geest, B.G., Skirtach, A.G., De Beer, T.R.M., Sukhorukov, G.B., Bracke, L., Bacyens, W.R.G., Demeester, J., and De Smedt, S.C. (2007) Stimuli-responsive multilayered hybrid nanoparticle/polyelectrolyte capsules. *Macromol. Rapid Commun.*, **28**, 88–95.

99 Bedard, M., Skirtach, A.G., and Sukhorukov, G.B. (2007) Optically driven encapsulation using novel polymeric hollow shells containing an azobenzene polymer. *Macromol. Rapid Commun.*, **28**, 1517–1521.

100 Koo, H.Y., Lee, H.J., Kim, J.K., and Choi, W.S. (2010) UV-triggered encapsulation and release from polyelectrolyte microcapsules decorated with photoacid generators. *J. Mater. Chem.*, **20**, 3932–3937.

101 von Klitzing, R. and Tieke, B. (2004) Polyelectrolyte membranes, in *Polyelectrolytes with Defined Molecular Architecture I*, Springer-Verlag Berlin, Berlin, pp. 177–210.

102 Jin, W.Q., Toutianoush, A., and Tieke, B. (2005) Size- and charge-selective transport of aromatic compounds across polyelectrolyte multilayer membranes. *Appl. Surf. Sci.*, **246**, 444–450.

103 Skirtach, A.G., Karageorgiev, P., De Geest, B.G., Pazos-Perez, N., Braun, D., and Sukhorukov, G.B. (2008) Nanorods as wavelength-selective absorption centers in the visible and near-infrared regions of the electromagnetic spectrum. *Adv. Mater.*, **20**, 506–510.

104 Rosenbauer, E.M., Wagner, M., Musyanovych, A., and Landfester, K. (2010) Controlled release from polyurethane nanocapsules via pH-, UV-Light- or temperature-induced stimuli. *Macromolecules*, **43**, 5083–5093.

105 Katagiri, K., Koumoto, K., Iseya, S., Sakai, M., Matsuda, A., and Caruso, F. (2009) Tunable UV-responsive organic-inorganic hybrid capsules. *Chem. Mater.*, **21**, 195–197.

106 Bedard, M.F., Sadasivan, S., Sukhorukov, G.B., and Skirtach, A. (2009) Assembling polyelectrolytes and porphyrins into hollow capsules with laser-responsive oxidative properties. *J. Mater. Chem.*, **19**, 2226–2233.

107 Erokhina, S., Benassi, L., Bianchini, P., Diaspro, A., Erokhin, V., and Fontana, M.P. (2009) Light-driven release from polymeric microcapsules functionalized with bacteriorhodopsin. *J. Am. Chem. Soc.*, **131**, 9800–9804.

108 Tao, X. and Su, J. (2008) Confined photoreaction in nano-engineered multilayer microshells. *Curr. Nanosci.*, **4**, 308–313.

109 Palankar, R., Skirtach, A.G., Kreft, O., Bedard, M., Garstka, M., Gould, K., Mohwald, H., Sukhorukov, G.B., Winterhalter, M., and Springer, S. (2009) Controlled intracellular release of peptides from microcapsules enhances antigen presentation on MHC class I molecules. *Small*, **5**, 2168–2176.

110 Kreft, O., Skirtach, A.G., Sukhorukov, G.B., and Möhwald, H. (2007) Remote control of bioreactions in multicompartment capsules. *Adv. Mater.*, **19**, 3142–3145.

111 Unger, E. (1997) Drug and gene delivery with ultrasound contrast agents, in The Leading Edge in Diagnostic Ultrasound, Conference, First Annual International Symposium of Contrast Agents in Diagnostic Ultrasound; sponsored by the Thomas Jefferson University and Jefferson Medical College, Atlantic City, NJ, 1995.

112 Skirtach, A.G., De Geest, B.G., Mamedov, A., Antipov, A.A., Kotov, N.A., and Sukhorukov, G.B. (2007) Ultrasound stimulated release and catalysis using polyelectrolyte multilayer capsules. *J. Mater. Chem.*, **17**, 1050–1054.

113 Kolesnikova, T.A., Gorin, D.A., Fernandes, P., Kessel, S., Khomutov, G.B., Fery, A., Shchukin, D.C., and Mohwald, H. (2010) Nanocomposite microcontainers with high ultrasound sensitivity. *Adv. Funct. Mater.*, **20**, 1189–1195.

114 Han, Y.S., Shchukin, D., Yang, J., Simon, C.R., Fuchs, H., and Möhwald, H. (2010) Biocompatible protein nanocontainers for controlled drugs release. *ACS Nano*, **4**, 2838–2844.

115 Yashchenok, A.M., Delcea, M., Videnova, K., Jares-Erijman, E.A., Jovin, T.M., Konrad, M., Mohwald, H., and Skirtach, A.G. (2010) Enzyme reaction in the pores of CaCO3 particles upon ultrasound disruption of attached substrate-filled liposomes. *Angew. Chem. Int. Ed.*, **49**, 8116–8120.

116 Wang, W., Liu, L., Ju, X.J., Zerrouki, D., Xie, R., Yang, L.H., and Chu, L.Y. (2009) A novel thermo-induced self-bursting microcapsule with magnetic-targeting property. *ChemPhysChem*, **10**, 2405–2409.

117 Hu, S.H., Tsai, C.H., Liao, C.F., Liu, D.M., and Chen, S.Y. (2008) Controlled rupture of magnetic polyelectrolyte microcapsules for drug delivery. *Langmuir*, **24**, 11811–11818.

118 Lu, Z., Prouty, M.D., Guo, Z., Golub, V.O., Kumar, C.S.S.R., and Lvov, Y.M. (2005) Magnetic switch of permeability for polyelectrolyte microcapsules embedded with Co@Au nanoparticles. *Langmuir*, **21**, 2042–2050.

119 Katagiri, K., Nakamura, M., and Koumoto, K. (2010) Magnetoresponsive smart capsules formed with polyelectrolytes, lipid bilayers and magnetic nanoparticles. *ACS Appl. Mater Interface.*, **2**, 768–773.

120 Gorin, D.A., Portnov, S.A., Inozemtseva, O.A., Luklinska, Z., Yashchenok, A.M., Pavlov, A.M., Skirtach, A.G., Mohwald, H., and Sukhorukov, G.B. (2008) Magnetic/gold nanoparticle functionalized biocompatible microcapsules with sensitivity to laser irradiation. *Phys. Chem. Chem. Phys.*, **10**, 6899–6905.

121 Fernandes, P.A.L., Delcea, M., Skirtach, A.G., Möhwald, H., and Fery, A. (2010) Quantification of release from microcapsules upon mechanical deformation with AFM. *Soft Matter*, **6**, 1879–1883.

122 Tzvetkov, G., Graf, B., Fernandes, P., Fery, A., Cavalieri, F., Paradossi, G., and Fink, R. (2008) In situ imaging of gas-filled microballoons by soft X-ray microscopy. *Soft Matter*, **4**, 510–514.

123 Fery, A. and Weinkamer, R. (2007) Mechanical properties of micro- and nanocapsules: Single-capsule measurements. *Polymer*, **48**, 7221–7235.

124 Dubreuil, F., Elsner, N., and Fery, A. (2003) Elastic properties of polyelectrolyte capsules studied by atomic-force microscopy and RICM. *Eur. Phys. J. E*, **12**, 215–221.

125 Munoz-Javier, A., Kreft, O., Semmling, M., Kempter, S., Skirtach, A.G., Bruns, O.T., del Pino, P., Bedard, M.F., Raedler, J., Kaes, J., Planck, C., Sukhorukov, G.B., and Parak, W.J. (2008) Uptake of colloidal polyelectrolyte-coated particles and polyelectrolyte multilayer capsules by living cells. *Adv. Mater.*, **20**, 4281–4287.

126 Yashchenok, A.M., Bratashov, D.N., Gorin, D.A., Lomova, M.V., Pavlov, A.M., Sapelkin, A.V., Shim, B.S., Khomutov, G.B., Kotov, N.A., Sukhorukov, G.B., Möhwald, H., and Skirtach, A.G. (2010) Carbon nanotubes on polymeric microcapsules: free-standing structures and point-wise laser openings. *Adv. Funct. Mater.*, **20**, 3136–3142.

127 Müller, R., Köhler, K., Weinkamer, R., Sukhorukov, G., and Fery, A. (2005) Melting of PDADMAC/PSS capsules investigated with AFM force spectroscopy. *Macromolecules*, **38**, 9766–9771.

128 Lulevich, V.V., Nordschild, S., and Vinogradova, O.I. (2004) Investigation of molecular weight and aging effect on the stiffness of polyelectrolyte multilayer microcapsules. *Macromolecules*, **37**, 7736–7741.

129 Sukhorukov, G.B., Rogach, A.L., Zebli, B., Liedl, T., Skirtach, A.G., Kohler, K., Antipov, A.A., Gaponik, N., Susha, A.S., Winterhalter, M., and Parak, W.J. (2005) Nanoengineered polymer capsules: Tools for detection, controlled delivery and site specific manipulation. *Small*, **1**, 194–200.

130 De Geest, B.G., Vandenbroucke, R.E., Guenther, A.M., Sukhorukov, G.B., Hennink, W.E., Sanders, N.N., Demeester, J., and De Smedt, S.C. (2006) Intracellularly degradable polyelectrolyte

microcapsules. *Adv. Mater.*, **18**, 1005–1009.

131 She, S.J., Zhang, X.G., Wu, Z.M., Wang, Z., and Li, C.X. (2010) Gradient cross-linked biodegradable polyelectrolyte nanocapsules for intracellular protein drug delivery. *Biomaterials*, **31**, 6039–6049.

132 Zelikin, A.N., Li, Q., and Caruso, F. (2006) Degradable polyelectrolyte capsules filled with oligonucleotide sequences. *Angew. Chem. Int. Ed.*, **45**, 7743–7745.

133 Szarpak, A., Cui, D., Dubreuil, F., De Geest, B.G., De Cock, L.J., Picart, C., and Auzely-Velty, R. (2010) Designing hyaluronic acid-based layer-by-layer capsules as a carrier for intracellular drug delivery. *Biomacromolecules*, **11**, 713–720.

134 Orozco, V.H., Kozlovskaya, V., Kharlampieva, E., Lopez, B.L., and Tsukruk, V.V. (2010) Biodegradable self-reporting nanocomposite films of poly(lactic acid) nanoparticles engineered by layer-by-layer assembly. *Polymer*, **51**, 4127–4139.

135 Itoh, Y., Matsusaki, M., Kida, T., and Akashi, M. (2006) Enzyme-responsive release of encapsulated proteins from biodegradable hollow capsules. *Biomacromolecules*, **7**, 2715–2718.

136 Zelikin, A.N., Quinn, J.F., and Caruso, F. (2006) Disulfide cross-linked polymer capsules: En route to biodeconstructible systems. *Biomacromolecules*, **7**, 27–30.

137 He, Q., Cui, Y., and Li, J.B. (2009) Molecular assembly and application of biomimetic microcapsules. *Chem. Soc. Rev.*, **38**, 2292–2303.

138 Pavlov, A.M., Sapelkin, A.V., Huang, X., P'ing, K.M.Y., Bushby, A.J., Sukhorukov, G.B., and Skirtach, A.G. (2011) Neuron cells uptake of polymeric microcapsules and subsequent intracellular release. *Macromol. Biosci.* doi: 10. 1002/mabi.201000494

139 Riske, K. and Dimova, R. (2005) Timescales involved in electro-deformation, poration and fusion of giant vesicles resolved with fast digital imaging. *Biophys. J.*, **88**, 1143–1155.

140 Zhang, R.J., Kohler, K., Kreft, O., Skirtach, A., Mohwald, H., and Sukhorukov, G. (2010) Salt-induced fusion of microcapsules of polyelectrolytes. *Soft Matter*, **6**, 4742–4747.

141 Stadler, B., Chandrawati, R., Price, A.D., Chong, S.F., Breheney, K., Postma, A., Connal, L.A., Zelikin, A.N., and Caruso, F. (2009) A microreactor with thousands of subcompartments: enzyme-loaded liposomes within polymer capsules. *Angew. Chem. Int. Ed.*, **48**, 4359–4362.

142 Bäumler, H. and Georgieva, R. (2010) Coupled Enzyme Reactions in Multicompartment Microparticles. *Biomacromolecules*, **11**, 1480–1487.

143 van Dongen, S.F.M., Nallani, M., Cornelissen, J.L.L.M., Nolte, R.J.M., and van Hest, J.C.M. (2009) A three-enzyme cascade reaction through positional assembly of enzymes in a polymersome nanoreactor. *Chem. Eur. J.*, **15**, 1107–1114.

144 Yu, J., Javier, D., Yaseen, M.A., Nitin, N., Richards-Kortum, R., Anvari, B., and Wong, M.S. (2010) Self-assembly synthesis, tumor cell targeting, and photothermal capabilities of antibody-coated indocyanine green nanocapsules. *J. Am. Chem. Soc.*, **132**, 1929–1938.

145 Ariga, K., Ji, Q.M., and Hill, J.P. (2010) Enzyme-encapsulated layer-by-layer assemblies: current status and challenges toward ultimate nanodevices. *Adv. Polym. Sci.*, **229**, 51–87.

146 Zhang, H.Y., Wang, Z.Q., Zhang, Y.Q., and Zhang, X. (2004) Hydrogen-bonding-directed layer-by-layer assembly of poly(4-vinylpyridine) and poly(4-vinylphenol): Effect of solvent composition on multilayer buildup. *Langmuir*, **20**, 9366–9370.

147 Zhang, F., Wu, Q., Chen, Z.C., Zhang, M., and Lin, X.F. (2008) Hepatic-targeting microcapsules construction by self-assembly of bioactive galactose-branched polyelectrolyte for controlled drug release system. *J. Colloid Interface Sci.*, **317**, 477–484.

148 Bai, J.H., Beyer, S., Mak, W.C., Rajagopalan, R., and Trau, D. (2010) Inwards buildup of concentric polymer layers: A method for biomolecule encapsulation and microcapsule

encoding. *Angew. Chem. Int. Ed.*, **49**, 5189–5193.

149 Delcea, M., Yashchenok, A., Videnova, K., Kreft, O., Möhwald, H., and Skirtach, A.G. (2010) Multicompartmental micro- and nanocapsules: hierarchy and applications in biosciences. *Macromol. Biosci.*, **10**, 465–474.

150 Gregersen, K.A.D., Hill, Z.B., Gadd, J.C., Fujimoto, B.S., Maly, D.J., and Chiu, D.T. (2010) Intracellular delivery of bioactive molecules using light-addressable nanocapsules. *ACS Nano*, **4**, 7603–7611.

151 Javier, A.M., del Pino, P., Bedard, M.F., Ho, D., Skirtach, A.G., Sukhorukov, G.B., Plank, C., and Parak, W.J. (2008) Photoactivated release of cargo from the cavity of polyelectrolyte capsules to the cytosol of cells. *Langmuir*, **24**, 12517–12520.

152 Raemdonck, K., Van Thienen, T.G., Vandenbroucke, R.E., Sanders, N.N., Demeester, J., and De Smedt, S.C. (2008) Dextran microgels for time-controlled delivery of siRNA. *Adv. Func. Mater.*, **18**, 993–1001.

153 Veerabadran, N.G., Goli, P.L., Stewart-Clark, S.S., Lvov, Y.M., and Mills, D.K. (2007) Nanoencapsulation of stem cells within polyelectrolyte multilayer shells. *Macromol. Biosci.*, **7**, 877–882.

154 De Cock, L.J., Lenoir, J., De Koker, S., Vermeersch, V., Skirtach, A.G., Dubruel, P., Adriaens, E., Vervaet, C., Remon, J.P., and De Geest, B.G. (2011) Mucosal irritation potential of polyelectrolyte multilayer capsules. *Biomaterials*. doi: 10.1016/j.biomaterials.2010.1011.1012

155 Facca, S., Cortez, C., Mendoza-Palomares, C., Messadeq, N., Dierich, A., Johnston, A.P.R., Mainard, D., Voegel, J.C., Caruso, F., and Benkirane-Jessel, N. (2010) Active multilayered capsules for in vivo bone formation. *Proc. Natl. Acad. Sci. USA*, **107**, 3406–3411.

156 Shchepelina, O., Kozlovskaya, V., Singamaneni, S., Kharlampieva, E., and Tsukruk, V.V. (2010) Replication of anisotropic dispersed particulates and complex continuous templates. *J. Mater. Chem.*, **20**, 6587–6603.

157 Chen, R.T., Muir, B.W., Such, G.K., Postma, A., McLean, K.M., and Caruso, F. (2010) Fabrication of asymmetric "Janus" particles via plasma polymerization. *Chem. Commun.*, **46**, 5121–5123.

158 Shchepelina, O., Kozlovskaya, V., Kharlampieva, E., Mao, W.B., Alexeev, A., and Tsukruk, V.V. (2010) Anisotropic Micro- and Nano-Capsules. *Macromol. Rapid Commun.*, **31**, 2041–2046.

159 Simone, E.A., Dziubla, T.D., and Muzykantov, V.R. (2008) Polymeric carriers: role of geometry in drug delivery. *Expert Opin. Drug Deliv.*, **5**, 1283–1300.

160 Delcea, M., Madaboosi, N., Yashchenok, A.M., Subedi, P., Volodkin, D.V., De Geest, B.G., Möhwald, H., and Skirtach, A.G. (2011) Anisotropic multicompartment micro- and nano-capsules produced via embedding into biocompatible PLL/HA films. *Chem. Commun.*, **47**, 2098–2100.

161 Tokarev, I. and Minko, S. (2010) Stimuli-responsive porous hydrogels at interfaces for molecular filtration, separation, controlled release, and gating in capsules and membranes. *Adv. Mater.*, **31**, 3446–3462.

162 Thierry, B., Kujawa, P., Tkaczyk, C., Winnik, F.M., Bilodeau, L., and Tabrizian, M. (2005) Delivery platform for hydrophobic drugs: Prodrug approach combined with self-assembled multilayers. *J. Am. Chem. Soc.*, **127**, 1626–1627.

163 Volodkin, D.V., Madaboosi, N., Blacklock, J., Skirtach, A.G., and Möhwald, H. (2009) Self-assembled polyelectron multilayers: Structure and function perspective. *Langmuir*, **25**, 14037–14043.

164 Mjahed, H., Voegel, J.C., Senger, B., Chassepot, A., Rameau, A., Ball, V., Schaaf, P., and Boulmedais, F. (2009) Hole formation induced by ionic strength increase in exponentially growing multilayer films. *Soft Matter*, **5**, 2269–2276.

165 Blacklock, J., Mao, G.Z., Oupicky, D., and Mohwald, H. (2010) DNA release dynamics from bioreducible layer-by-layer films. *Langmuir*, **26**, 8597–8605.

166 Picart, C., Mutterer, J., Richert, L., Luo, Y., Prestwich, G.D., Schaaf, P., Voegel, J.C., and Lavalle, P. (2002) Molecular basis for the explanation of the exponential growth of polyelectrolyte multilayers. *Proc. Natl. Acad. Sci. USA*, **99**, 12531–12535.

167 Picart, C., Lavalle, P., Hubert, P., Cuisinier, F.J.G., Decher, G., Schaaf, P., and Voegel, J.C. (2001) Buildup mechanism for poly(L-lysine)/hyaluronic acid films onto a solid surface. *Langmuir*, **17**, 7414–7424.

168 Lavalle, P., Voegel, J.C., Vautier, D., Senger, B., Shchaaf, P., and Ball, V. (2011) Dynamic aspects of films prepared by a sequential deposition of species: Perspective for smart and responsive materials. *Adv. Mater.* doi: 10.1002/adma.201003309

169 Laugel, N., Boulmedais, F., El Haitami, A.E., Rabu, P., Rogez, G., Voegel, J.C., Schaaf, P., and Ball, V. (2009) Tunable synthesis of prussian blue in exponentially growing polyelectrolyte multilayer films. *Langmuir*, **25**, 14030–14036.

170 Volodkin, D.V., Delcea, M., Möhwald, H., and Skirtach, A.G. (2009) Remote near-IR light activation of a hyaluronic Acid/Poly(l-lysine) multilayered film and film-entrapped microcapsules. *ACS Appl. Mater. Interface.*, **1**, 1705–1710.

171 Lasic, D.D. (1993) *Liposomes: From Physics to Applications*, Elsevier, Amsterdam.

172 Lasic, D.D. and Papahadjopoulos, D. (1998) *Medical Applications of Liposomes*, Elsevier, Amsterdam.

173 Graff, A., Winterhalter, M., and Meier, W. (2001) Nanoreactors from polymer-stabilized liposomes. *Langmuir*, **17**, 919–923.

174 Barenholz, Y. (2001) Liposome application: Problems and prospects. *Curr. Opin. Colloid Interface Sci.*, **6**, 66–77.

175 Christensen, S.M. and Stamou, D. (2007) Surface-based lipid vesicle reactor systems: Fabrication and applications. *Soft Matter*, **3**, 828–836.

176 Richter, R.P., Berat, R., and Brisson, A.R. (2006) Formation of solid-supported lipid bilayers: An integrated view. *Langmuir*, **22**, 3497–3505.

177 Reviakine, I. and Brisson, A. (2000) Formation of supported phospholipid bilayers from unilamellar vesicles investigated by atomic force microscopy. *Langmuir*, **16**, 1806–1815.

178 Yoshina-Ishii, C., Miller, G.P., Kraft, M.L., Kool, E.T., and Boxer, S.G. (2005) General method for modification of liposomes for encoded assembly on supported bilayers. *J. Am. Chem. Soc.*, **127**, 1356–1357.

179 Chifen, A.N., Forch, R., Knoll, W., Cameron, P.J., Khor, H.L., Williams, T.L., and Jenkins, A.T.A. (2007) Attachment and phospholipase A2-induced lysis of phospholipid bilayer vesicles to plasmapolymerized maleic anhydride/SiO2 multilayers. *Langmuir*, **23**, 6294–6298.

180 Volodkin, D., Schaaf, P., Mohwald, H., Voegel, J.-C., and Ball, V. (2009) Effective embedding of liposomes into polyelectrolyte multilayered films. The relative importance of lipid-polyelectrolyte and interpolyelectrolyte interactions. *Soft Matter*, **5**, 1394–1405.

181 Volodkin, D.V., Arntz, Y., Schaaf, P., Mohwald, H., Voegel, J.-C., and Ball, V. (2008) Composite multilayered biocompatible polyelectrolyte films with intact liposomes: stability and triggered dye release. *Soft Matter*, **4**, 122–130.

182 Michel, M., Izquierdo, A., Decher, G., Voegel, J.-C., Schaaf, P., and Ball, V. (2005) Layer-by-Layer self-assembled polyelectrolyte multilayers with embedded phospholipid vesicles obtained by spraying: integrity of the vesicles. *Langmuir*, **21**, 7854–7859.

183 Michel, M., Vautier, D., Voegel, J.-C., Schaaf, P., and Ball, V. (2004) Layer-by-Layer self-assembled polyelectrolyte multilayers with embedded phospholipid vesicles. *Langmuir*, **20**, 4835–4839.

184 Volodkin, D.V., Michel, M., Schaaf, P., Voegel, J.-C., Mohwald, H., and Ball, V. (2008) Liposome embedding into polyelectrolyte multilayers: a new way to create drug reservoirs at solid-liquid interfaces, in *Advances in Planar Lipid Bilayers and Liposomes* (ed. A.L. Liu) Elsevier, Amsterdam.

185 Volodkin, D., Ball, V., Schaaf, P., Voegel, J.-C., and Mohwald, H. (2007) Complexation of phosphocholine

liposomes with polylysine. Stabilization by surface coverage versus aggregation. *Biochim. Biophys. Acta.*, **1768**, 280–290.

186 Volodkin, D., Mohwald, H., Voegel, J.-C., and Ball, V. (2007) Stabilization of negatively charged liposomes by polylysine surface coating. Drug release study. *J. Control. Release*, **117**, 111–120.

187 Volodkin, D.V., Ball, V., Voegel, J.-C., Möhwald, H., Dimova, R., and Marchi-Artzner, V. (2007) Control of the interaction between membranes and vesicles: Adhesion, fusion and release of dyes. *Colloids Surf. A*, **303**, 89–96.

188 Delcea, M., Madaboosi, N., Yashchenok, A.M., Subedi, P., Volodkin, D.V., De Geest, B.G., Mohwald, H., and Skirtach, A.G. (2011) Anisotropic multicompartment micro- and nano-capsules produced via embedding into biocompatible PLL/HA films. *Chem. Commun.*, **47**, 2098–2100.

189 Dimitrova, M., Arntz, Y., Lavalle, P., Meyer, F., Wolf, M., Schuster, C., Haïkel, Y., Voegel, J.-C., and Ogier, J. (2007) Adenoviral gene delivery from multilayered polyelectrolyte architectures. *Adv. Funct. Mater.*, **17**, 233–245.

190 Etienne, O., Picart, C., Taddei, C., Haikel, Y., Dimarcq, J.L., Schaaf, P., Voegel, J.C., Ogier, J.A., and Egles, C. (2004) Multilayer polyelectrolyte films functionalized by insertion of defens in: a new approach to protection of implants from bacterial colonization. *Antimicrob Agents Chemother*, **48**, 3662–3669.

191 Nguyen, P.M., Zacharia, N.S., Verploegen, E., and Hammond, P.T. (2007) Extended release antibacterial layer-by-layer films incorporating linear- dendritic block copolymer micelles. *Chem. Mater.*, **19**, 5524–5530.

192 Shi, Z., Neoh, K.G., Zhong, S.P., Yung, L.Y.L., Kang, E.T., and Wang, W. (2006) In vitro antibacterial and cytotoxicity assay of multilayered polyelectrolyte-functionalized stainless steel. *J. Biomed. Mater. Res. Part A*, **76**, 826–834.

193 Lee, D., Rubner, M.F., and Cohen, R.E. (2005) Formation of nanoparticle-loaded microcapsules based on hydrogen-bonded multilayers. *Chem. Mater.*, **17**, 1099–1105.

194 Malcher, M., Volodkin, D., Heurtault, B., Andre, P., Schaaf, P., Mohwald, H., Voegel, J.-C., Sokolowski, A., Ball, V., Boulmedais, F., and Frisch, B. (2008) Embedded silver ions-containing liposomes in polyelectrolyte multilayers: Cargos films for antibacterial agents. *Langmuir*, **24**, 10209–10215.

195 Volodkin, D.V., Madaboosi, N., Blacklock, J., Skirtach, A.G., and Mohwald, H. (2009) Surface-supported multilayers decorated with bio-active material aimed at light-triggered drug delivery. *Langmuir*, **25**, 14037–14043.

196 Volodkin, D.V. and Mohwald, H. (2009) Polyelectrolyte multilayers for drug delivery, in *Encyclopedia of Surface and Colloid Science* (ed. P. Somasundaran) Taylor & Francis Group, LLC, p. 14.

197 Jourdainne, L., Lecuyer, S., Arntz, Y., Picart, C., Schaaf, P., Senger, B., Voegel, J.C., Lavalle, P., and Charitat, T. (2008) Dynamics of poly(L-lysine) in hyaluronic acid/poly(L-lysine)multilayer films studied by fluorescence recovery after pattern photobleaching. *Langmuir*, **24**, 7842–7847.

# 40
# Controlled Architectures in LbL Films for Sensing and Biosensing

*Osvaldo N. Oliveira Jr., Pedro H.B. Aoki, Felippe J. Pavinatto, and Carlos J.L. Constantino*

## 40.1
## Introduction

The layer-by-layer (LbL) method [1, 2] has as one key feature the possible control of molecular architectures, including the adsorption of various materials in a single film. Unique properties of LbL films arise from their nanostructured nature, with molecules arranged in a multilayered fashion in ultrathin films. This has made it a popular method for sensing and biosensing, for the small thickness is advantageous owing to the increased surface area/volume ratios, with very large surface areas being attained for rough films. Furthermore, these tasks require the films to respond to external stimuli, which is normally achieved with functional materials. For biosensing, in particular, there is the additional, stringent requirement that biomolecules should be immobilized with their activity preserved. Preservation of bioactivity in LbL films containing biomolecules is associated with the mild conditions under which the films are fabricated, in addition to the presence of hydration water that helps prevent denaturing [3, 4]. In most cases, however, this can only be reached upon optimizing the film fabrication conditions, such as immersion time, temperature, and concentration of the solutions, and the film architectures with adequate choice of components [5]. For instance, antigens immobilized in LbL films have their biological activity preserved for a long period of time when dendrimers are used as the supporting layer. If dendrimers are replaced by polyelectrolytes, the antigen biological activity can no longer be preserved [3]. There is no rule for predicting which template materials will be appropriate for a given biomolecule [5]. It is known, however, that chitosan, dendrimers [5], carbon nanotubes (CNTs) [6] and phospholipids [7, 8] are considered biocompatible when used as supporting layers for adsorption of molecules with biological activity.

In this chapter we shall review the many types of sensors and biosensors made with LbL films. Though we shall try to cover the main contributions in this field, our survey will by no means be exhaustive as the number of papers published is very high. By way of illustration, a search in the Web of Science with the keywords LbL (or layer-by-layer) and biosens* or sensing or sensor*) retrieved over 1180 papers in

---

*Multilayer Thin Films: Sequential Assembly of Nanocomposite Materials*, Second Edition.
Edited by Gero Decher and Joseph B. Schlenoff.
© 2012 Wiley-VCH Verlag GmbH & Co. KGaA. Published 2012 by Wiley-VCH Verlag GmbH & Co. KGaA.

September, 2011. The chapter is organized as follows: The bulk of the literature will be described according to the methods of detection, in Section 40.2. Some examples, however, will be discussed in Section 40.3 dedicated to special film architectures where synergy was sought to improve device performance. Before presenting the conclusions and outlook in Section 40.5, we discuss the use of information visualization methods to enhance sensitivity in Section 40.4.

## 40.2
## LbL-Based Sensors and Biosensors

With the large variety of materials amenable to fabrication in LbL films, any detection principle mentioned in the literature for sensing and biosensing can also be used in conjunction with LbL-based devices. The most employed ones are optical methods, electrical and electrochemical methods, and methods involving measurements of mass changes. Sensors and biosensors are used to detect a variety of analytes in liquids and vapors, with applications in many areas, such as clinical diagnosis, environmental control, and quality monitoring, either in the food or pharmaceutical industries.

### 40.2.1
### Optical Detection Methods

The versatility of the LbL technique allows one to combine distinct materials for tailored applications, for example by choosing materials whose optical properties can be varied upon interaction with the liquid or vapor to be sensed. Among the optical methods one may include those based on absorption and reflection of light in the UV–Vis range, fluorescence spectroscopy, Raman scattering and surface plasmon resonance. In some cases, the measurements are straightforward, as in low-cost sensors based on colorimetry, but in others the optical sensors involve special film architectures, such as those containing metallic nanoparticles to enhance plasmonic effects [9].

Surface plasmon resonance (SPR) has been applied to detect copper, nickel, and iron ions in aqueous solutions down to $10^{-5}$ M using LbL films of poly(ethyleneimine)/poly(ethylene-alt-maleic acid) (PEI/PMAE), and poly(ethyleneimine)/poly(styrene sulfonate) (PEI/PSS) [10]. The SPR spectra were found to depend on both the LbL architecture, especially the number of layers and outer layers, and on the humidity [11]. This was performed with LbL nanocomposite films containing 8 nm anionic Au nanoparticles (AuNPs) and M13 viruses genetically engineered to display a positively charged peptide (tetraArg) on their side walls. M13 bacteriophage is a filamentous virus that specifically infects bacterial cells, being nontoxic to human beings, and can carry different functionalities. The results have opened the possibility of organizing M13 viruses and nanocomponents into solid devices by simply integrating virus-templated nanoassembly and LbL films.

Biosensors made with LbL films containing three-dimensional Au nanoparticles (AuNPs) and multiwalled carbon nanotubes (MWCNTs) were used with SPR. The detection limit was $0.5 \times 10^{-9}$ M for streptavidin and $3.33 \times 10^{-9}$ M for anti-human serum albumin (HSA) [12]. DNA and Ag nanoparticle multilayer LbL films were applied to detect gentamicin, with a good linear relationship in the concentration range from $5 \times 10^{-8}$ to $1 \times 10^{-4}$ M, and a detection limit of $1 \times 10^{-9}$ M [13]. The plasmonic effects on colloidal particles were exploited in surface-enhanced Raman scattering (SERS) to detect small analyte concentrations, as reviewed by Abalde-Cela et al. [14]. Details regarding special metal nanoparticle structures in LbL films are discussed in Section 40.3.

With regard to fluorescence used as a detection tool in sensing, various reports have been made. The concentration of potassium was determined in a device where a potassium ion indicator was adsorbed within poly(styrene sulfonate)/poly(allylamine hydrochloride) (PSS/PAH) LbL films, with a coating of fluorescent europium nanoparticles, leading to a passive intracellular nanosensor [15]. The separation of the fluorescent indicator from the cellular environment is attractive because it may prevent cytotoxicity and probe compartmentalization. Stubbe et al. [16] reported a sophisticated preparation process of LbL films containing PAH/PSS layer pairs coating fluorescent microparticles carrying fluorescently labeled protease substrates (peptidic trypsin) for detecting protease.

LbL films of 1,3,6,8-pyrenetetrasulfonat acid tetrasodium salt (PTS) and Zn–Al LDH nanosheets on quartz substrates were applied by Shi et al. [17] as fluorescence chemosensors for copper ions ($Cu^{2+}$), with a linear response in fluorescence quenching from 0.6 to $50 \times 10^{-3}$ M $Cu^{2+}$ concentration. The quenching was attributed to either electron transfer or energy transfer between the transition metal and the excited fluorophore. The detection limit was $0.2 \times 10^{-3}$ M, which is sufficient to determine the $Cu^{2+}$ concentration in the human blood system (from 15.7 to $23.6 \times 10^{-3}$ M) and in drinking water within the US regulations limit ($20 \times 10^{-3}$ M). Fluorescence quenching was also applied by Ma et al. [18] to detect formaldehyde vapor using LbL films of CdTe quantum dots/poly(dimethyldiallylemmonium chloride) (PDDA). The fluorescence intensity decreased linearly with increasing formaldehyde concentration from 5 to 500 ppb, with a detection limit of 1 ppb. Yang et al. [19] applied fluorescence quenching to detect $Hg^{2+}$ using LbL films of oppositely charged CdTe quantum dots and PDDA, with the outermost layer cross-linked to bovine serum albumin (BSA). The quenching of the photoluminescence intensity varied almost linearly with the $Hg^{2+}$ concentration from $1.0 \times 10^{-8}$ to $1.0 \times 10^{-6}$ M, with a detection limit of $4.5 \times 10^{-9}$ M.

The catalysis processes involving enzymes immobilized in LbL films may, in some cases, be monitored via changes in optical absorption and reflection, particularly in cases where the products from the reaction exhibit optical properties that neither of the film components do. This was the case for biosensors to detect cathecol using LbL films containing Cl-catechol 1,2-dioxygenase (CCD) layers alternated with poly(amidoamine) generation 4 (PAMAM G4) dendrimer [4]. The film architecture is shown in Figure 40.1a. Detection was carried out following the absorbance at 260 nm, assigned to the formation of cis–cis muconic acid, resulting from the reaction between CCD and catechol, after immersion of the LbL film in catechol solutions down to $10^{-7}$ M, as shown in Figure 40.1b.

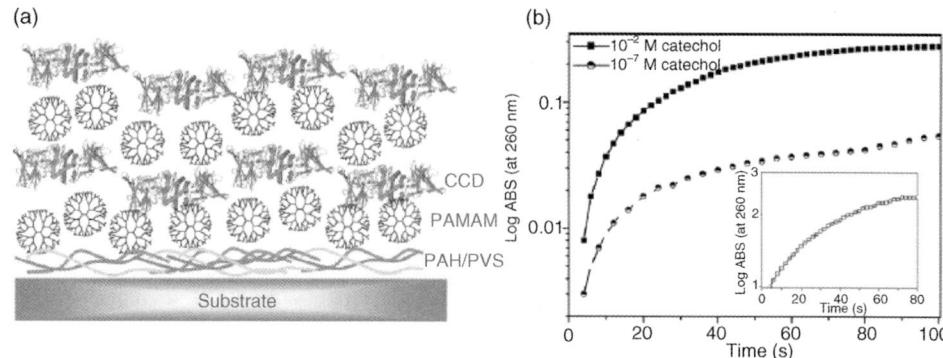

**Figure 40.1** (a) Illustration of the LbL film molecular architecture containing alternating layers of PAMAM/CCD deposited onto a first LbL layer pair of poly(allylamine hydrochloride)/poly(vinylsulfonic acid) (PAH/PVS). (b) Absorbance at 260 nm versus time for a 10-layer-pair PAMAM/CCD LbL film immersed into $10^{-7}$ and $10^{-2}$ M catechol solutions. Also shown in the inset is the absorbance at 260 nm versus time for CCD solution in the presence of catechol, obtained to evaluate the CCD activity. Reproduced with permission from: V. Zucolotto, A. P. A. Pinto, T. Tumolo, M. L. Moraes, M. S. Baptista, A. Riul Jr., A. P. U. Araujo and O. N. Oliveira Jr., Biosens. Bioelectron., 2006, 21, 1320–1326.

Optical changes arising from changes in the photochemical properties of a film component were exploited by Egawa et al. [20], who applied LbL films of poly (allylamine) and Brilliant Yellow as pH sensors. They varied the solution under study from pH 10.0 to 5.0, which caused shifts in the UV–Vis spectra of Brilliant Yellow. The pH response in the neutral range is useful for biological sensors and environmental applications. Another phenomenon used for sensing was the change in refractive index of LbL films containing zinc peroxide/poly(acrylic acid), $ZnO_2$/PAA [21]. The effective refractive index was modified by the water molecules adsorbed on the LbL films, leading to changes in the reflection properties.

### 40.2.2
### Mass Change Methods

The monitoring of mass change with a quartz crystal microbalance (QCM) has been used as a detection method with LbL films. Nakane and Kubo [22] deposited liposome/bacteriorhodopsin (bR, membrane protein) LbL films onto a QCM crystal to detect nonylphenol in solution, which is one of the endocrine disrupting chemicals (EDCs). Its presence was detected for concentrations between 0.1 and 10 ppm from a decrease in mass of the deposited LbL films, which was attributed to the disruption of liposomes by the nonylphenol. Xie et al. [23] developed a methane gas sensor by depositing LbL films of polyaniline/palladium oxide nanoparticles (PANI/PdO) onto a QCM support. The concern with methane comes from the need to ensure safety in homes, industries and mines. QCM results allowed the real-time response of the PANI/PdO sensor to different concentrations of methane gas.

## 40.2.3
### Electrochemical Methods

Electrochemistry has been largely used for sensing with distinct mechanisms being exploited. In most cases the LbL films contain an electroactive material whose properties are affected upon interacting with the analyte. In other instances, the analyte itself may be electroactive, and there are still other situations where electroactive species are generated from the reaction between a film component and the analyte. With these various possibilities, there are a huge number of sensors and biosensors made with LbL films employing electrochemistry as the principle of detection, where the specific technique may be cyclic voltammetry, amperometry, chronoamperometry, and electrochemical impedance spectroscopy. Reviews on this subject can be found in Wei et al. [24] for LbL films applied as biosensors, in Siqueira et al. [25] for LbL and Langmuir–Blodgett (LB) films in sensors and biosensors, and in Riul et al. [26] for electronic tongues.

Most of the electrochemical biosensors made with LbL films contain immobilized enzymes, and a few examples will be highlighted here. Caseli et al. [27] reported on biosensors for detecting glucose where glucose oxidase (GOD) layers were alternated with chitosan. The latter was proven to be a good scaffolding material, preserving the GOD catalytic activity toward glucose oxidation. They found a detection limit of $0.2\,\text{mmol}\,L^{-1}$ and an activity of $40.5\,\mu A\,\text{mmol}^{-1}\,L\,\mu g^{-1}$. Concerning the role played by the enzyme in the LbL film, the highest sensitivity was achieved when only the top layer contained GOD, indicating that the enzyme in inner layers did not contribute to glucose oxidation, probably because they hampered the analyte diffusion and electron transport through the LbL film.

An unconventional method for assembling graphene multilayer LbL films was developed by Zeng et al. [28]. The graphene sheets were modified by pyrene-grafted PAA in an aqueous solution, and then used for assembling the LbL films with poly (ethyleneimine) (PEI). These graphene LbL films were modified with GOD and glucoamylase (GA), being applied as transducers for detecting glucose and maltose in amperometric experiments ("bienzyme system"). Basically, in the LbL film containing graphene, GOD, and GA, the GA catalyzes the hydrolysis of maltose to produce glucose, which is then oxidized by GOD to produce $H_2O_2$ that is detected by the graphene.

Amperometric biosensors were developed by Guo et al. [29] to detect cholesterol using LbL films containing the cationic PDDA and cholesterol oxidase (ChOx) on MWCNTs, which were coated with the nonconducting poly(o-phenylenediamine) (PPD), as the protective layer. A linear response up to $6.0 \times 10^{-3}$ M was observed, with a detection limit of $0.2 \times 10^{-3}$ M. Good distinction between dopamine and ascorbic acid, which acts as a natural interferent in biological fluids, was achieved by Zucolotto et al. [30] via cyclic voltammetry using LbL films containing PANI and tetrasulfonated iron phthalocyanine (FeTsPc). Detection of ascorbic acid was also reported by Ragupathy et al. [31] for LbL films comprising poly(diphenylamine) (PDPA) and phosphotungstic acid (PTA) using cyclic voltammetry and amperometry. The enhanced electrocatalytic activity of $(PDPA/PTA)_n$ LbL film was attributed to the tungsten atoms within the interlayers.

Amperometric biosensors have been obtained with LbL films containing different enzymes capable of reducing hydrogen peroxide. Li et al. [32] used LbL films of negatively charged horseradish peroxidase (HRP) and a quaternized poly(4-vinylpyridine) complex of $(Os(bpy)_2Cl)^{+/2+}$ (PVP–Os), yielding a biosensor stable for 3 weeks, with a detection limit of $3 \times 10^{-6}$ M and a linear relationship from $3 \times 10^{-6}$ to $3.7 \times 10^{-3}$ M of $H_2O_2$. Chen et al. [33] used alternating layers of chitosan and manganese oxide nanoflakes to produce a sensor whose linear response covered the range from $2.5 \times 10^{-6}$ to $1.05 \times 10^{-3}$ M, with a sensitivity of $0.038$ A $M^{-1}$ $cm^{-2}$. Kafi et al. [34] combined hemoglobin (Hb) and poly-allylamine hydrochloride (PAH) in LbL films, which also exhibited high electrocatalytic response for reducing hydrogen peroxide, with a linear range from $2.5 \times 10^{-6}$ to $5 \times 10^{-4}$ M and detection limit of $0.2 \times 10^{-6}$ M.

LbL films containing alternating layers of mesoporous $SiO_2$ and PDDA were applied by Shi et al. [35] to detect nitroaromatic compounds such as TNT, TNB, DNT and DNB, with a linear current response ranging from $10^{-9}$ to $10^{-7}$ M. This high sensitivity was achieved using differential pulse voltammetry, which was dependent on the number of layers, pH and ionic strength of the electrolyte. High sensitivity was also reached by Liu et al. [36] for phenolic compounds using LbL films containing PAH-wrapped MWCNTs (PAH-MWCNTs) and horseradish peroxidase (HRP). The biosensor was applied to detect a series of 17 phenolic samples via electrochemical impedance spectroscopy, displaying a linear response for catechol, for instance, from 0.1 to $20.4 \times 10^{-6}$ M, with a detection limit of $0.06 \times 10^{-6}$ M.

The strategy of Du et al. [37] to produce LbL films with ferrocene-appended poly (ethyleneimine), Fc-PEI, CNTs and aptamers led to a high performance in detecting proteins via differential pulse voltammetry (DPV). Thrombin and lysozyme were applied as target molecules. The strategy with LbL films to detect thrombin was as follows. The aptamer on the outermost layer of the LbL film caught the target on the electrode interface, making a barrier for electrons and inhibiting electron transfer, resulting in the decreased DPV signals of Fc-PEI. A similar sensing strategy was applied for detecting lysozyme. For thrombin, the detection limit was 0.14 ng $mL^{-1}$ for the range between 0.3 and 165 ng $mL^{-1}$, and for lysozyme the detection limit was 0.17 ng $mL^{-1}$ between 0.2 ng $mL^{-1}$ and 1.66 µg $mL^{-1}$.

Dong et al. [38] used cyclic voltammetry (CV), chronoamperometry (i–t), and electrochemical impedance spectroscopy (EIS) to detect target DNA down to $0.7 \times 10^{-15}$ M. Basically, a high content of HRP was adsorbed on $Fe_3O_4$ nanoparticles (NPs) using the LbL technique. Then, the signal probe and diluting probe were immobilized on the HRP-functionalized $Fe_3O_4$ NPs through the bridge of AuNPs. The resulting DNA–Au–HRP–$Fe_3O_4$ (DAHF) bioconjugates were anchored to the Au nanofilm (GNF) modified electrode surface, forming the sandwich-type electrochemical DNA biosensor.

## 40.2.4
### Methods Involving Electrical Measurements

Electrical parameters such as resistance, current and capacitance are widely applied as detection methods in sensing and biosensing. The incorporation of conducting materials to the LbL films often causes the output signal and sensitivity to increase. Two classes of material extensively used for this purpose are metallic nanoparticles and nanowires/nanotubes [39–41]. Gold and silver NPs are now the most used [13, 42–47]. Papers dealing with biosensors containing gold NPs were reviewed by Pingarrón et al. [45]. The use of inorganic layered materials (e.g., clay materials) and CNTs was the focus of a survey by Podsiadlo and coworkers [48]. Here we shall discuss some of the work in sensing and biosensing with LbL films using DC conductivity and capacitance measurements, and then elaborate upon electronic tongues (e-tongues) and electronic noses (e-noses), which have also received considerable attention from the community of LbL films.

Nohria et al. [49] compared LbL and spin-coated films of poly(anilinesulfonic acid) (SPANI) as humidity sensors via electrical resistance changes. They found that the LbL films presented better sensing performance in terms of response time, sensitivity and repeatability. The electrical resistance was also applied by Loh et al. [50] for strain sensors with LbL films made with SWCNTs, PSS and poly(vinyl alcohol) (PVA), leading to the molecular architecture SWCNT-PSS/PVA. The idea was to correlate changes in LbL fabrication parameters to strain sensing properties of the bulk material. For instance, they found that by adjusting the weight fraction of SWNT solutions and film thickness, strain sensitivities between 0.1 and 1.8 were achieved.

The combination of electrical resistance and spray-LbL films made with polycarbonate and MWCNTs was explored by Lu et al. [51] to detect volatile organic compounds. The vapor sensing performance was investigated as a function of MWCNT content, film thickness, vapor flow and vapor solubility, leading to the ranking Ar(toluene) > Ar(methanol) > Ar(water). A novel flexible $H_2$ gas sensor was fabricated by Su and Chuang [52] using LbL films containing surface-oxidized MWCNTs on a polyester (PET) substrate. The Pd-based complex was self-assembled in situ on the as-prepared MWCNTs layer, which was reduced to form an MWCNT-Pd thin film. The flexible $H_2$ gas sensor showed strong response and high linearity between 200 and 10 000 ppm, fast response (300 s), high reproducibility and long-term stability (84 days at least).

Current–time curves at a given voltage were applied by Hornok and Dékány [53] to detect hydrogen peroxide and acetic acid or hydrochloric acid vapors, using as transducer LbL films of Prussian Blue (PB) nanoparticles and PAH. Capacitance–voltage curves were used by Siqueira et al. [25, 54] to detect penicillin G using field-effect (bio-)chemical sensors. The latter contained LbL films with SWNTs and polyamidoamine (PAMAM) dendrimers, onto which the penicillinase enzyme was immobilized. The structure of the device is of the capacitive electrolyte–insulator–semiconductor (EIS) type, as shown in Figure 40.2. Penicillin concentrations from $5.0 \times 10^{-6}$ to $25 \times 10^{-3}$ M were evaluated, achieving a sensitivity of about 116 mV/decade. Many features of this biosensor (as well as the synergy between them) deserve attention owing to the advantages of using LbL films in biosensing. The incorporation of each film component imparted an important feature to the sensor, as follows: (i) PAMAM dendrimers confer porosity to LbL films and facilitate diffusion of $H^+$ ions to the chip surface. (ii) SWCNTs increased the film conductivity, thus facilitating

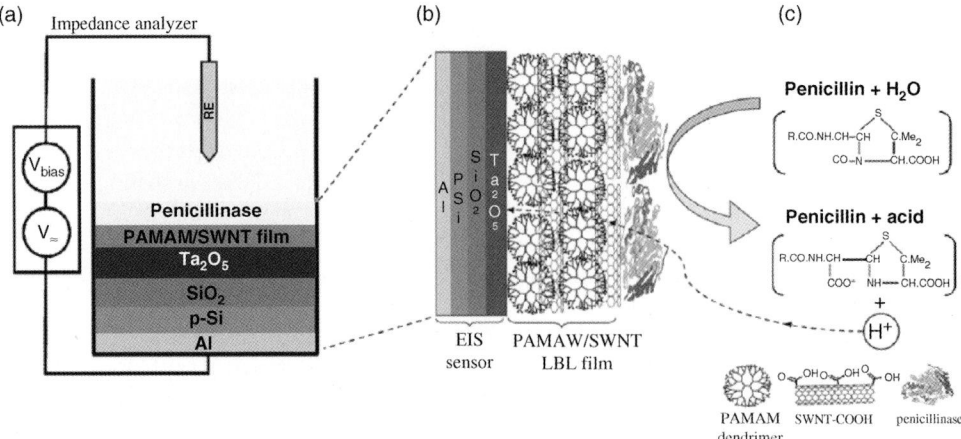

**Figure 40.2** Illustration of the biosensor for detecting penicillin based on electrolyte–insulator–semiconductor (EIS) structure functionalized with PAMAM/SWNT LbL film and its operation principle. In (a) the architecture is depicted, which included the semiconductor chip made with aluminum coated with layers of p-type silicon, $SiO_2$ and $Ta_2O_5$, onto which the LbL film was deposited. A side view of the device structure is shown in (b) and (c) depicts the reactions involving penicillin which are catalyzed by penicillinase. Reproduced with permission from: J. R. Siqueira Jr., L. Caseli, F. N. Crespilho, V. Zucolotto and O. N. Oliveira Jr., Biosens. Bioelectron., 2010, 25, 1254–1263.

the detection through potentiometry. (iii). The enzyme penicillinase conferred specificity to the target analyte. Moreover, using technologies of microelectronics, as this is a kind of field-effect device, the authors opened the way for miniaturization.

Lee and Cui [55] developed flexible pH and glucose sensors based on LbL films containing SWCNTs and PDDA deposited on a polyethylene terephthalate substrate. The glucose sensor showed sensitivity from 18 to 45 $\mu A\,mM^{-1}$ and a linear range from 2 to $10 \times 10^{-3}$ M.

## 40.2.5
### E-Tongues and E-Noses

Many of the sensors developed nowadays represent an attempt to mimic the smell and taste functions of human beings, leading to the so-called electronic noses (e-noses) and tongues (e-tongues), respectively. The taste we feel is produced by a chemical or physical interaction of absorbed material with the papillae on the tongue surface. Transduction is carried out by the papillae and an associated complex network of neurons. Thus, papillae, or more specifically some receptors inserted in the plasmatic membranes of the epithelial cells, are the receptors for sensing taste [56]. Though controversies exist about the mechanism of taste, the signal detection theory (SDT) is one of the most accepted. It postulates that a specific substance is tasted by the human tongue only if its concentration is above a certain minimum, known as the threshold concentration [57]. Ideally, to mimic the human tongue efficiently one should develop tables or databases for threshold concentrations of specific substances, but this cannot be done accurately since disturbances in the threshold level of the substances are provoked by factors such as people's mood or illnesses [58].

The lack of reproducibility and the limits of detection imposed by the threshold concentration represent an opportunity for sensors and biosensors as "artificial tongues" (e-tongues) [26]. Their applicability is not based on the qualitative and quantitative detection of analytes, in contrast to typical chemical or biological sensors. Instead, it exploits the classification of substances within the global selectivity principle [59], analogously to what occurs in the human brain to classify the different tastes. The development of e-tongues has been pursued for a long time, with several architectures, transducers and principles of measurements being described in the literature [26, 60–62].

The first electronic tongue system for liquid analyses based on an array of sensing units was introduced in 1985 by Otto and Thomas [63]. E-tongues met nanostructured films in 2002 [64], with the array of sensing units being made of Langmuir–Blodgett (LB) and LbL films, and using impedance spectroscopy as the method of detection. The films are normally deposited onto interdigitated electrodes. In the seminal work on e-tongue employing impedance spectroscopy, the data were analyzed using equivalent circuits containing capacitors and resistors [65]. One of such equivalent circuits is shown in Figure 40.3 together with a typical capacitance versus frequency curve for a metal/film/solution system.

A comparison among many samples with similar capacitance versus frequency curves is obviously very difficult. One normally takes the value at a single frequency and the comparison with many samples is done via a statistical method. The most used is principal component analysis (PCA) [66], a linear technique in which a new orthogonal basis of lower dimension is obtained via linear combinations of the data variance. The PCA plot in Figure 40.4 was obtained from the capacitance values for samples representing the basic tastes, viz. salty (with NaCl), sour (HCl), sweet (sucrose) and bitter (quinine), and pure water (Milli-Q), measured by several sensing units. The values were grouped by similarities of some specific feature, and therefore the distance between points represents dissimilarity among the samples.

Treating the data with PCA [67] it was possible to detect trace amounts of small molecules and heavy metals in aqueous solutions, in addition to distinguishing similar samples of coffee and coconut water. Since this pioneering work, other papers and patents have been reported on e-tongues based on LbL films, whose main advantages are the wide choice of materials and the low thickness. For instance, Dos Santos et al. [68] reported the distinction of red wines, with no need to pretreat the samples or carry out complex laboratory analysis, using only LbL films of chitosan alternated with sulfonated polystyrene (PSS) and a bare electrode.

In order to reach a high distinguishing ability, several points need to be considered. Perhaps the most important is an adequate choice of materials for the sensing units, and their combination. Normally, a sensor array contains from four to six sensing units. Numerous materials have been used in LbL films for sensing units in e-tongues. Borato et al. were the first to use nanoparticles [69], with 10-layer LbL films with successive deposition of POEA alternated with chitosan or nanoparticles of

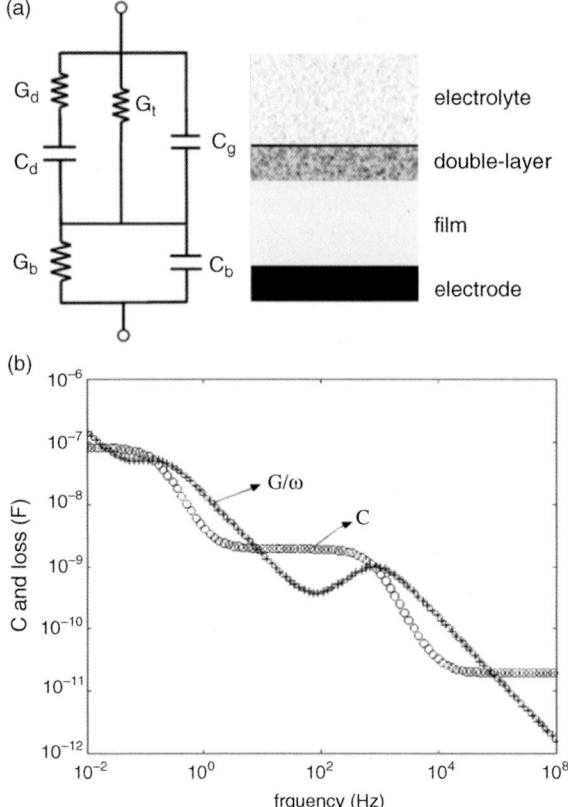

**Figure 40.3** (a) Equivalent circuit used to model the electrical properties of the metal-film/solution interface of a typical sensing unit of an e-tongue. $C_g$ is the geometric capacitance of the electrodes; $G_t$ is the conductance associated with charge transfer through the film electrolyte interface; $C_d$ is the double-layer capacitance; $G_d$ is the electrolytic conductance; $G_b$ and $C_b$ are film conductance and capacitance, respectively. (b) Sketch of typical capacitance and loss versus frequency plots for metallic electrodes covered with low conductivity thin film immersed in an electrolyte. Reproduced with permission from: D. M. Taylor and A. G. Macdonald, *J. Phys. D-Appl. Phys.*, 1987, **20**, 1277–1283.

chitosan in a poly(methacrylic acid) matrix (CS-PMA), as illustrated in Figure 40.5. These LbL films were adsorbed onto interdigitated chrome electrodes. Advantage was taken of the very distinct properties of these films, for the LbL films from chitosan-PMA nanoparticles behaved differently from chitosan. Films of POEA co-deposited with the nanoparticles are much thicker, with a higher amount of conducting polymer adsorbed. With such sensing units it was possible to detect copper ions in aqueous solutions from 1 to 50 mM. Even bare electrodes could classify samples by forming clusters in PCA graphs, but for some specific films, like the POEA-chitosan LbL film, sensitivity was increased when the nanostructured film was present.

POEA-PSS LbL films with 1 to 20 layer pairs were employed in a systematic study of the influence of film thickness and electrode geometry over sensor and e-tongue

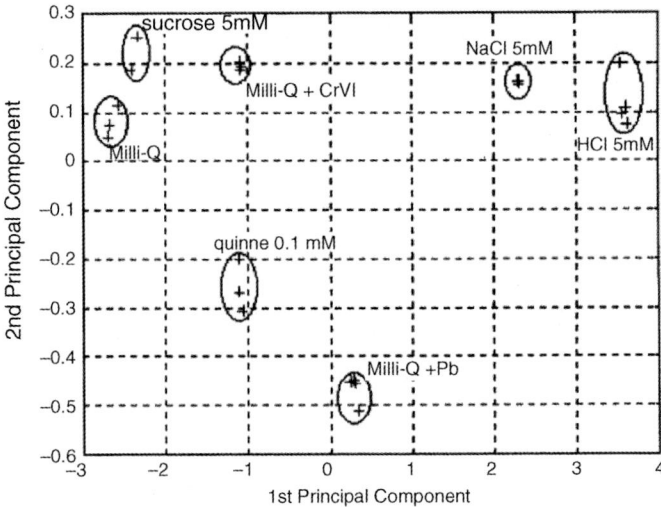

**Figure 40.4** PCA plot generated by the analysis of capacitance values at 1 kHz obtained from an array of four sensors containing LB and LbL films to distinguish salty, sour, sweet, and bitter tastes, in addition to an inorganic contaminant (Cr) in ultrapure water. Reproduced with permission from: A. Riul Jr., D. S. dos Santos Jr., K. Wohnrath, R. Di Tommazo, A. Carvalho, F. J. Fonseca, O. N. Oliveira Jr., D. M. Taylor and L. H. C. Mattoso, *Langmuir*, 2002, **18**, 239–245.

properties [70]. The sensitivity to NaCl in solution was increased for thicker films, as evaluated for two frequencies (100 Hz and 1 kHz). Moreover, depending on the range of NaCl concentration, a specific sensor was more sensitive. This was unexpected since the sensing units were made of the same material. The apparent discrepancy was attributed to the increased roughness, which caused the area in contact with the ions in solution to be higher. POEA adsorbs as globules and many agglomerates and voids are formed. The POEA-PSS LbL films with different numbers of layers were grouped in an array and used as an e-tongue. Such an e-tongue was innovative since it

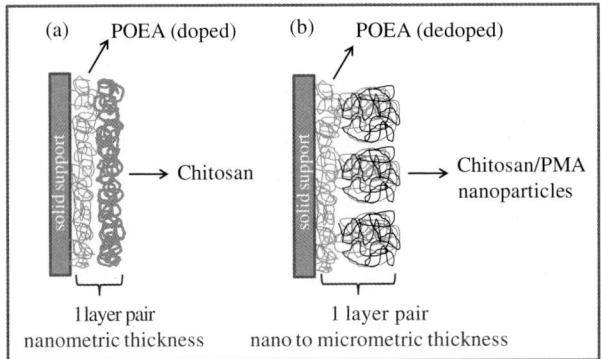

**Figure 40.5** Scheme of the structure of a layer pair formed with alternating deposition of POEA and chitosan (a) or POEA and chitosan-PMA nanoparticles (b).

was composed of four identical sensing units differing only in film thickness. Its detection limit for NaCl in solution was as low as 10 µM.

The use of a conducting polymer with different doping states as active material [71] is another example of how the tuning of LbL film properties can dramatically affect the sensing features. POEA was employed in alternating films with PSS and several acids were used to dope it. The heterogeneity of the counter ions imparted different structural and morphological properties to the films. Heavier, larger counter ions, such as $NO_3^-$ and $HSO_4^-$, were less solvated by water and bound closer to the amine groups of POEA, increasing charge screening effects along the chain [72]. As a result, the polymer chains adopted a more compact structure and were adsorbed in a larger amount than POEA chains undoped or doped with smaller acids, for example, HCl. These sensors with a larger mass of POEA adsorbed had their capacitance signal increased. An e-tongue was obtained from an array of sensing units containing the same materials (POEA/PSS). The electrical response of this e-tongue was very fast, of the order of seconds, and the detection results were highly reproducible.

The versatility of the sensing units to distinguish both basic tastes and wines was investigated by Borato et al. [73] combining LbL films of POEA and sulfonated lignin (LS) and Langmuir–Blodgett (LB) films. Olivati et al. [74] detected trace levels of phenolic compounds (phenol; 2-chloro-4-methoxyphenol; 2-chlorophenol; 3-chlorophenol) in water down to $10^{-9}$ M also combining LbL films of tetrasulfonated metallic phthalocyanines (Fe and Cu)/PAH layer pairs and LB films. Volpati et al. [75] applied sensing units based on LbL, LB, and physical vapor deposited (PVD) films of iron phthalocyanine to detect copper ions ($Cu^{2+}$) in aqueous solutions down to 0.2 mg L$^{-1}$, not only in ultrapure water but also in distilled and tap water, which is sufficiently sensitive for the quality control of water for human consumption. Significantly, a sensor array could be made from sensing units with identical or similar materials by simply varying the molecular architectures of the thin films, revealing the importance of fabrication processes.

Brominated trihalomethanes were detected in water by using an e-tongue with 10 sensing units (LbL films of POEA, PANI, aquatic humic substances (AHS) and sulfonated lignins (SL), in several combinations) deposited on Au interdigitated electrodes, using AC impedance at 1 kHz, and PCA to treat the data [76]. Sensing units in which POEA was combined in LbL films with AHS or SL showed the lowest detection limits. Three brominated compounds could be distinguished in different regions of the PCA plot, and their increased concentration in solution could also be characterized by the e-tongue system. These compounds are carcinogenic byproducts formed when drinking water is disinfected by the chlorination method, and their detection is important for public health.

LbL films have also been used in sensing units for gases and vapors, in which the film permeability to gases was exploited to reach high sensitivities. Materials such as dendrimers are often advantageous in this regard [77, 78], for their three-dimensional structure leads to porous LbL films that are more permeable to gases. Therefore, improved gas sensors [79–81], in addition to sensors for molecules in solution [4, 54, 82–84], can be obtained with dendrimers. In some cases, a sensor array is used for combining the response of different materials, yielding the so-called electronic noses (e-noses), whose functioning is very similar to e-tongues. Lu et al. [85] produced stable sensors made with hybrid networks of PANI and carbon nanotubes deposited onto interdigitated electrodes, which were used to detect volatile organic compounds (VOCs) and may be suitable for e-noses. The same applied to LbL films obtained from chitosan and carbon nanotubes [86], where the quantitative responses fitted with the Langmuir–Henry clustering (LHC) model allowed the electrical signal to be correlated with the vapor content.

### 40.2.6
### Extending the Concept of E-Tongue to Biosensing

The sensing units reported in the last subsection are based on the concept of global selectivity, with no need for specific interactions between the analytes and the transducers. However, a step forward was taken in e-tongue systems by including sensing units with molecular recognition capability and specific interaction, characteristic of biological systems. In these sensing units, LbL films of enzymes or antigens were produced. One such example has already been mentioned in Section 2.1, where Zucolotto et al. [4] detected catechol by immobilizing Cl-catechol 1,2-dioxygenase (CCD) in LbL films with PAMAM generation 4 dendrimer. In addition to the optical detection mentioned in Section 2.1, catechol in solutions at concentrations as low as $10^{-10}$ M could be determined using impedance spectroscopy in the frequency range from 1 Hz to 1 kHz. The same strategy was applied to detect ethanol, immobilizing the enzyme alcohol dehydrogenase (ADH) with layers of PAMAM G4 onto interdigitated electrodes using the LbL technique [84]. The latter allowed detection of ethanol concentrations at 1 part per million by volume (ppmv). Cholesterol concentrations in aqueous solutions down to $10^{-6}$ M were detected by Moraes et al. [87] via LbL films containing PAH/ChOx layers.

Zucolotto et al. [3] immobilized antigen-containing liposomes (proteoliposomes) in alternating layers of PAMAM G4, as shown in Figure 40.6. In order to get increased

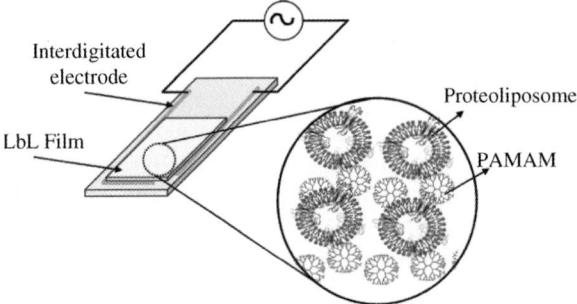

**Figure 40.6** Illustration of the biosensor based on an interdigitated electrode coated by an LbL film of PAMAM/proteoliposome, which is immersed into solutions containing buffer and the IgG to be detected. Reproduced with permission from: V. Zucolotto, K. R. P. Daghastanli, C. O. Hayasaka, A. Riul Jr., P. Ciancaglini and O. N. Oliveira Jr., Anal. Chem., 2007, **79**, 2163–2167.

performance, a sensor array was used which contained one bare electrode, an electrode coated with 5-layer-pairs PAMAM/PVS LbL film, and an electrode coated with 5-layer-pairs PAMAM/proteoliposome LbL film. Because of the preserved activity of the antigen, it was possible to detect immunoglobulin G (IgG) against Pasteurellosis at concentrations as low as nanograms per milliliter. In addition, due to the molecular recognition capability, a distinction could be made between specific and nonspecific IgG.

In another work by Perinoto et al. [88], a sensor array made with LbL films containing different types of antigen was used to distinguish between two tropical diseases, viz. Chagas' disease and Leishmaniasis, again with the concept of an extended e-tongue. These results will be described in detail when the use of information visualization methods is discussed in Section 40.4.

## 40.3
### Special Architectures for Sensing and Biosensing

The possibility of immobilizing nanoparticles in a controlled, well-defined manner using the LbL technique has been extensively exploited for producing sensors and biosensors. This is the case for nanostructures such as spherical-like nanoparticles [89], nanowires [90], nanorods [91–93], and carbon nanotubes [94, 95]. Moreover, the unique optical properties that result from surface plasmon resonance in the visible range of the electromagnetic spectrum make them particularly attractive for optical sensing applications, which are directly related to NP size, shape, composition, aggregation and local surroundings [89]. Indeed, studies involving NPs have been carried out to optimize the synthesis and reaction conditions [96], control of shape and size [97, 98] and select reaction reagents [99]. Figure 40.7a and b display AFM images of AuNPs with different shapes, including large polygons (mainly triangles and hexagons) and smaller structures immobilized in a supporting layer of PDDA with the LbL technique [89], which can be used as surface plasmon resonance substrates for sensing. Figure 40.7c shows the high-magnification TEM micrographs of Au nanorods in poly(sodium-4-styrenesulfonate) (PSS) [91], while an AFM image of an LbL film containing Ag nanowires, shaped particles and dendrimers [100] is shown in Figure 40.7d.

Ag and Au nanostructures can sustain localized surface plasmon resonances (LSPR) in the visible range of the electromagnetic spectrum, providing the basis for plasmon-enhanced spectroscopy [101]. The enhancements of visible and IR radiation absorption [102–104], fluorescence radiation [105], and Raman scattering from analytes are well known [101] and represent a different way of obtaining information about both enhancing nanoparticles and analyte. Surface-enhanced Raman scattering (SERS) and surface-enhanced resonance Raman scattering (SERRS) are optical techniques that have become widely used in the characterization of nanomaterials. The enhancement of the Raman signal on metal nanoparticles allows one to obtain scattering cross sections to achieve single molecule detection [106, 107], which makes SERS one of the ultimate analytical tools. Single-molecule SERS has been reported

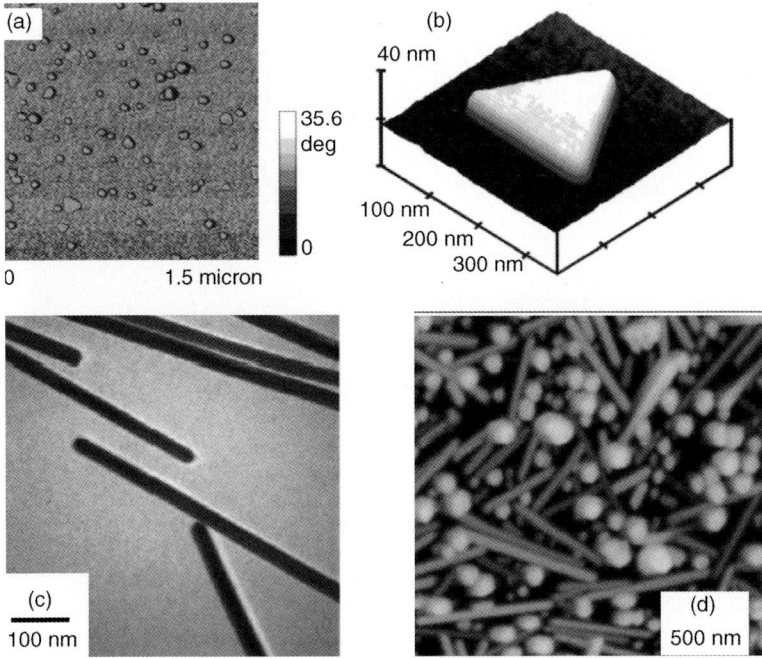

**Figure 40.7** (a) AFM topographic image of PDDA/Au layer pair. (b) Zoom of a triangular particle observed in (a). (c) TEM micrographs of gold nanorods deposited on a layer of PSS. (d) AFM topographic images of Ag nanowire/G5 DAB-Am dendrimer LbL film. Reproduced with permission from respectively: N. Malikova, I. Pastoriza-Santos, M. Schierhorn, N. A. Kotov and L. M. Liz-Marzan, *Langmuir*, 2002, **18**, 3694–3697.; A. Gole and C. J. Murphy, *Chem. Mater.*, 2005, **17**, 1325–1330.; R. F. Aroca, P. J. G. Goulet, D. S. dos Santos Jr., R. A. Alvarez-Puebla and O. N. Oliveira Jr., *Anal. Chem.*, 2005, **77**, 378–382.

from different environments [108], including under electrochemical conditions [109], and in single living cells [110].

SERS is useful for investigation and structural characterization of interfacial and thin film systems [101]. For instance, Aoki et al. [111] reported the surface-enhanced phenomenon to investigate LbL films, taking advantage of the strong SERRS signal of phenothiazine methylene blue (MB). The detection of this pharmaceutical drug was made with LbL films containing the anionic phospholipid dipalmitoyl phosphatidyl glycerol (DPPG). The Ag NPs played a key role in achieving the MB SERRS signal, allowing the detection of low amounts of material deposited even in a single LbL bilayer. Figure 40.8a shows the MB Raman spectra collected for two LbL films containing 1-tetralayer each, following the sequence shown in the insets. The Raman intensity was much weaker when the DPPG was placed between MB and Ag NPs, showing the strong dependence of the SERRS effect on the distance between the analyte and the NPs. The enhancement factor was estimated as $10^4$, in good agreement with the factors for the electromagnetic mechanism [101]. The Ag NPs distribution along the LbL films was determined from SEM images for 10 layer pairs of PAH/(DPPG + AgNP + MB) film, as shown in Figure 40.8b.

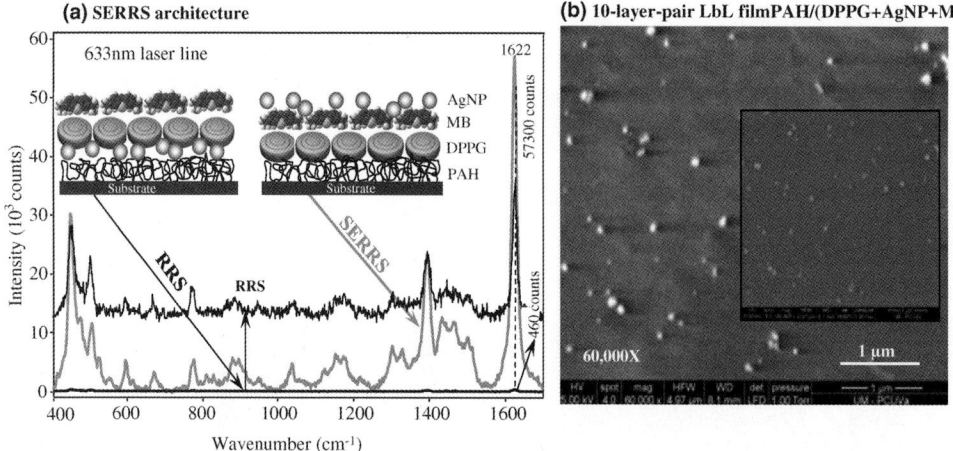

**Figure 40.8** (a) MB Raman spectra collected for LbL films containing the two molecular architectures appearing in the inset. (b) SEM image collected for a 10-layer-pair LbL film of PAH/(DPPG + AgNP + MB). The inset show the details of AgNP distribution using the back scattering electron detector. Reproduced with permission from: P. H. B. Aoki, P. Alessio, J. A. De Saja and C. J. L. Constantino, *J. Raman Spectrosc.*, 2010, **41**, 40–48.

Analytical techniques have been combined with impedance spectroscopy to improve the understanding of molecular-level interactions involved in sensing. SERS spectra were obtained by Aoki *et al.* [111] with LbL films incorporating Ag NPs and containing PAH layers alternated with vesicles made with negative or zwitterionic phospholipids, such as DPPG or dipalmitoyl phosphatidyl choline (DPPC). The combined use of an e-tongue and SERS was made [7] for the negative phospholipid cardiolipin (CLP) forming PAH/CLP LbL films containing Ag NP. The film structure is represented in Figure 40.9, together with a SEM image of the film and a scheme of the approach. The sensing unit was able to detect MB down to $10^{-11}$ M via impedance spectroscopy and the SERS spectra could be obtained directly from the sensing unit after the impedance spectroscopy experiment, adding chemical information to the transducer–analyte interaction.

Li *et al.* [112] reported a new way of obtaining SERS using Au NPs encapsulated in an ultrathin silica shell (2–4 nm), which separates them from direct contact with the analyte. For a fluorophore located on an enhancing nanostructure, increasing the shell thickness causes a continuous transition from fluorescence quenching to fluorescence enhancement [113]. Since the fluorescence-enhanced cross section is larger than SERS, the improvement of experimental methods allowing the positioning of fluorophores at well defined distances near metal nanoparticles is of great interest for optical sensing applications. For instance, this system may find applications in biomedical diagnostics [114, 115], where information is frequently conveyed by fluorescence. In this context, Schneider *et al.* [116] reported the preparation, characterization, and photophysical study of new fluorescent core/shell nanoparticles fabricated with the LbL technique. Oppositely charged polyelectrolytes were

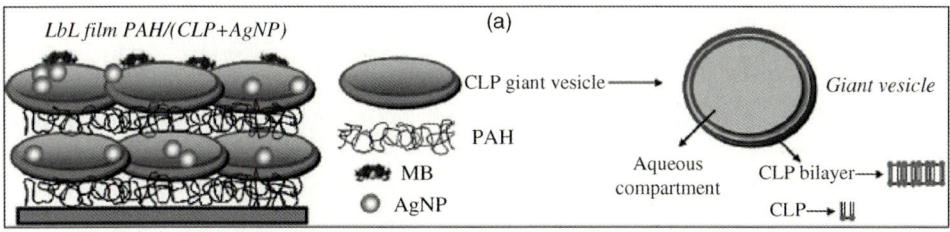
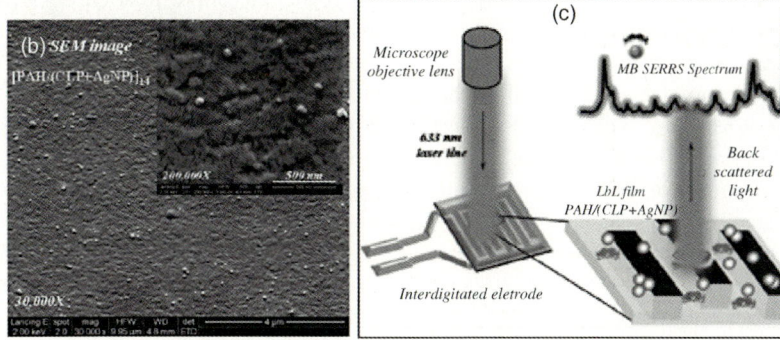

**Figure 40.9** (a) Schematic illustration of LbL films containing the CLP giant vesicles, AgNP nanoparticles immobilized with alternating layers of PAH. (b) SEM image collected for a 14-bilayer PAH/(CLP + AgNP) LbL film. The inset shows a zoom with details of the AgNP. (c) Experimental set-up applied for coupling SERRS and e-tongue. Reproduced with permission from: P. H. B. Aoki, P. Alessio, A. Riul Jr., J. A. D. Saez and C. J. L. Constantino, *Anal. Chem.*, 2010, **82**, 3537–3546.

deposited onto 13-nm diameter Au colloids, with the fluorescent organic dyes fluorescein isothiocyanate and lissamine rhodamine B (referred to as FITC and LISS, respectively) placed at various distances from the Au core. A clear increase in the fluorescence intensity with the increasing number of spacer layer pairs was found, which confirms the distance dependence of the fluorescence quenching. Guerrero et al. [117] reported a similar work using Au NP recovered with $SiO_2$ forming the so-called shell-isolated nanoparticle (SHIN), which was applied for surface-enhanced fluorescence (SEF), leading to the so-called shell-isolated nanoparticle enhanced fluorescence (SHINEF) by using a single LB monolayer that contains fluorescent probes.

Nanostructured materials combined with biomolecules often exhibit interesting physical and chemical properties for applications such as mimetic biomaterials [118] and sensors [6, 119]. For instance, nanostructures and biomolecules deposited with the LbL technique have been applied as electric sensors for detecting proteins [55, 120], enzymes [94] and pesticides [121]. Carbon nanotubes [122] and biopolymers may be assembled in films with suitable electrical, mechanical and optical properties [123–125]. CNT is particularly interesting for electrode materials for sensing [126], by enhancing the electron transfer of enzyme redox centers [6, 94]. CNTs can also serve as a scaffold for biomolecule immobilization, due to their high accessible surface area [94, 127]. Mantha et al. [6] reported a novel nanocomposite based on modified MWCNT LbL films for sensing. For the growth of LbL films, a supporting layer of MWCNT modified with cationic polyethyleneimine (MWCNT-PEI) was deposited, followed by a MWCNT modified with anionic DNA (MWCNT-DNA) layer. The deposition of this first layer pair allows further deposition of MWCNT modified with the cationic enzyme organophosphorus hydrolase (MWCNT-OPH). Such films were applied as biosensors in detecting paraoxon (parasympathomimetic) by flow-injection data for the concentration range from 0.5 to $10 \times 10^{-6}$ M. The biosensor presented both higher sensitivity and lower detection limit due to the large specific surface area of the oxidized MWNT [6]. Moreover, the structure of LbL films containing the MWCNT nanocomposites preserved the biological activity of proteins for a long time.

Crespilho et al. [128] demonstrated a strategy to obtain an efficient enzymatic bioelectrochemical device, in which urease was immobilized by drop-coating on electroactive nanostructured membranes (ENMs) made with polyaniline and Ag NPs stabilized in polyvinyl alcohol (PANI/PVA–AgNP). The modified electrode architecture, shown in Figure 40.10, promoted an efficient catalytic conversion of urea into ammonium and bicarbonate ions. After enzymatic conversion, the highly permeable PVA-AgNP layer allowed fast diffusion of ammonia ions, leading to an efficient response toward urea detection. Moreover, the modified electrode architecture was suitable for new enzymatic devices with high bioactivity preservation.

A similar strategy was used by Deng et al. [129] to fabricate a sensitive and stable glucose biosensor by LbL self-assembling GOD on a multiwall CNT-modified glassy carbon (GC) electrode. Crespilho et al. [83] also demonstrated the possibility of fabricating a sensitive and stable glucose biosensor but now using enzymatic devices based on the deposition of GOD on aligned and highly oriented CoNiMo metallic

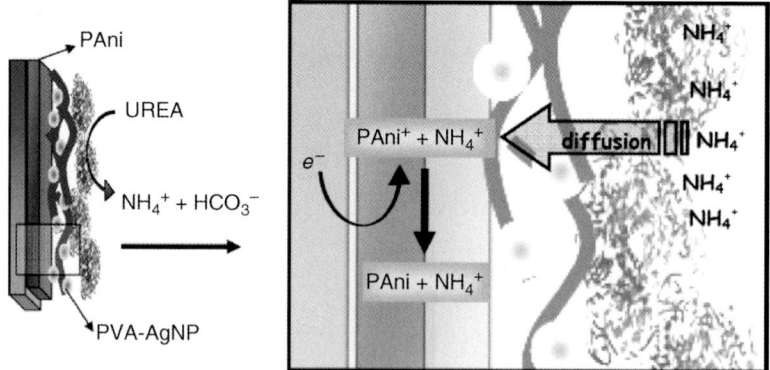

**Figure 40.10** Illustration of ITO/PANI/PVA–AgNP/urease electrode molecular architecture. The hydrolysis reaction of urea is given in the zoom. Reproduced with permission from: F. N. Crespilho, R. M. Iost, S. A. Travain, O. N. Oliveira Jr. and V. Zucolotto, *Biosens. Bioelectron.*, 2009, **24**, 3073–3077.

nanowires. Gu et al. [92] deposited alternated layers of PSS and HRP on the ZnO nanorods to detect $H_2O_2$.

Biomolecules are frequently incorporated into LbL films to produce sophisticated biosensors [25, 130]. Residual water molecules usually stay entrapped in the LbL architecture, which helps to preserve the biological activity of proteins, enzymes and other biological molecules. Since the pioneering work of Lvov and coworkers [131], several examples of films with enzymes and other molecules with specific interactions are available in the literature [3, 4, 54, 82–84]. Such films are highly versatile for dedicated, efficient sensors, as is the case for the detection of femtomoles of a protein in solution [132].

A paper from Kotov's group illustrates the importance of the nanostructured nature of the LbL multilayer, not only for sensor conductivity but also for the final property of the sensor [133]. Silver NPs with a shell of titania had their core dissolved with ammonium hydroxide, resulting in a nanoshell of titania. Such nanoshells were alternately co-deposited with PAA onto glassy carbon electrodes. As the nanoshells had an average inner diameter of 10–30 nm, they formed a network of voids and channels with dimensions comparable to the extent of a typical electrical double-layer for porous materials. This network allowed the "filtration" of ions and molecules by simply controlling pH and ionic strength, which consequently regulated the accessibility of analytes to the electrodes. This switchable system could be used to detect dopamine, an important neurotransmitter. Under physiological conditions, the electrical double layer of the porous structure was tuned to allow the penetration of the positively charged dopamine and to hinder the accessibility of the negatively charged ascorbic acid (interfering) to the sensor. Thus, the film provided a high selectivity, confirmed by the similar electrochemical response produced by the sensor when exposed to dopamine and to a mixture of dopamine and ascorbic acid. Moreover, the compatibility of the sensors covered with the PAA/nanoshells films with neuronal cells was demonstrated, with which one may envisage real biological applications.

## 40.4
### Statistical and Computational Methods to Treat the Data

The large amounts of data generated in experiments with sensors and biosensors have brought the need to use statistical and computational methods to treat the data. Indeed, many issues of sensor design and classification of similar samples are now dealt with using the concepts of chemometrics [134], which is a science dedicated to extracting information (and knowledge) in data-driven discovery approaches. Within the realms of chemometrics, perhaps the method most relevant for sensing is multivariate analysis, which includes PCA [66] and classical scaling [135]. A review of multivariate data analysis for biosensors can be found in Lindholm-Sethson et al. [136]. With the high throughput achieved in miniaturized systems, such as those in lab-on-a-chip technologies [137] one now needs to go beyond and explore further statistical and computational methods.

New paradigms for treating the data have been suggested, such as those belonging to data-intensive discovery or e-science [138]. By e-science one means a computationally intensive science making use of a highly distributed computer network, or a science dealing with large amounts of data for which grid computing is used. Working within the e-science paradigm normally involves cloud computing and parallel processing, which are helpful to handle the massive amounts of data. Also important are the computational methods of data mining, information retrieval and knowledge generation. We have not been able to find e-science initiatives specific for sensing and biosensing, but researchers are becoming aware of the need (and convenience) of using more sophisticated data treatment methods.

Here, we summarize findings where the performance of sensors or biosensors was considerably enhanced by using statistical or computational methods, going beyond PCA and other multivariate data analyses. The latter have been widely used for sensing for many years [136]. They belong to the category of information visualization methods, where multidimensional data instances are represented in 1D, 2D or 3D spaces. In most cases each data instance is mapped to a graphical marker, which can be a pixel, a line or an icon. Examples of these methods are *scatterplots* [139], *table lens* [140], *parallel coordinates* [141], *pixel oriented techniques* [142], *self-organizing maps* [143], *point placement or multidimensional projection techniques* [144]. A detailed description of these methods can be found in [145, 146]. The PCA plots for the e-tongue results in Figure 40.8 (Section 40.2) in this chapter are good examples of these multidimensional projections.

### 40.4.1
#### Artificial Neural Networks and Regression Methods

In searching the literature for sensing and biosensing based on LbL films, we could only find application of more sophisticated analysis methods in work by Brazilian groups, and, therefore, this section is limited to reviewing contributions from these groups. The distinction of six types of red wine was obtained using artificial neural networks (ANNs) [147] to treat data provided by impedance spectroscopy on an array of sensing units made with LbL films of chitosan and LB films. While the distinction with regard to vintage, type of grape, and brand of the red wine could be made with PCA, the identification of wine samples stored under different conditions was possible via ANNs. In the study, 900 wine samples were considered, being obtained from 30 measurements for each of five bottles of the six wines. Since ANNs can learn from data by using training algorithms [148], upon choosing the appropriate algorithm for training, 100% accuracy could be achieved.

In the e-tongue results discussed in this chapter, we mentioned the high sensitivity to detect trace amounts of analytes, and the ability to distinguish between very similar samples. However, one may go one step further to correlate the data from an e-tongue to the taste as perceived by human experts. This was performed for coffee samples from the Brazilian Association for the Coffee Industry, which have been evaluated by human experts who assigned scores for the different samples. Using a hybrid regression approach based on the random subspace method and PCA, Ferreira *et al.* [149] obtained a high correlation, with a Pearson coefficient of 0.964 (1 would be total correlation) between the predicted scores and those of the experts.

## 40.4.2
## Optimization of Biosensing Performance Using Multidimensional Projections

The challenge to distinguish between tropical diseases, for which many false positives occur even in sophisticated immunoassays [150, 151], has led to the use of further multidimensional projections for impedance spectroscopy data. Multidimensional projection techniques are useful to convey global similarity relationships among data instances, by placing similar instances close to each other on a plane or 3D plot. PCA, for instance, is a projection technique in which a new orthogonal basis of lower dimension is obtained via linear combinations of the data variance. In a PCA plot the axes are labeled as first, second or third principal components, which represent most of the variance in the data. Other multidimensional projection techniques are based on defining the similarity/dissimilarity by some distance function in the high-dimensional data space. In a mathematical formulation of this type of projection, $X = \{x_1, x_2, \ldots, x_n\}$ is the set of data instances, and $\delta(x_i, x_j)$ is a dissimilarity (distance) function defined between two instances. By defining $Y = \{y_1, y_2, \ldots, y_n\}$ as the set of visual markers corresponding to X, and $d(y_i, y_j)$ a distance function among them, the projection technique amounts to using an injective function $f: X \to Y$ aimed at making $|\delta(x_i, x_j) - d(f(x_i), f(y_j))| \approx 0$, $\forall x_i, x_j \in X$ [152]. In other words, one tries to preserve in the projected 2D or 3D space the similarity of the data instances in the original, multidimensional space.

These projection techniques may be classified into *linear* and *non-linear* methods, depending on the mapping function $f$ [144]. In attempts to improve the performance of biosensors, Paulovich et al. [153, 154] found that linear techniques may fail to capture data features that depend on non-linear relationships with the analyte concentrations. On the other hand, better results were obtained with non-linear techniques, though the precise reasons for this finding have not been determined. Particularly useful non-linear methods have been the Sammon's Mapping [155] and interactive document map (IDMAP) [156]. The cost function in Sammon's mapping is

$$S = \frac{1}{\sum_{i<j} \delta(x_i, x_j)} \sum \frac{(d(y_i, y_j) - \delta(x_i, x_j))^2}{\delta(x_i, x_j)}$$

where $\delta$ is a dissimilarity measure between samples $x_i$ and $x_j$, while $d$ is the distance among their projections $y_i$ and $y_j$ onto a 2D plot. S represents the information lost in the projection procedure.

For IDMAP, the cost function is

$$S_{IDMAP} = \frac{\delta(x_i, x_j) - \delta_{min}}{\delta_{max} - \delta_{min}} - d(y_i, y_j)$$

where $\delta_{min}$ and $\delta_{max}$ are the minimum and maximum distances between the samples.

For the work on the tropical diseases Leishmaniasis and Chagas' disease mentioned before in Section 40.2, optimization of device performance was essential for the following reasons. The concept of an extended e-tongue was used, with a sensor array containing four sensing units: the bare interdigitated gold electrode, and electrodes coated with 5-layer-pair LbL films made with PAMAM/PVS and with proteoliposomes containing immobilized antigens with molecular recognition capability toward anti-Leishmania and anti-T. Cruzi (for Chagas' disease) antibodies. In the LbL films with proteoliposomes, the alternating layers contained PAMAM dendrimers. The schematic diagram for a biosensing unit is shown in Figure 40.11, together with capacitance versus frequency data for antibody solutions at $10^{-5}$ mg mL$^{-1}$ for three of the sensing units. While it is virtually impossible to distinguish the samples with the two units that did not have specific interactions, good distinction could be made with the (biosensor) unit containing PAMAM/proteoliposome LbL films. This demonstrates the importance of specific interactions for biosensing.

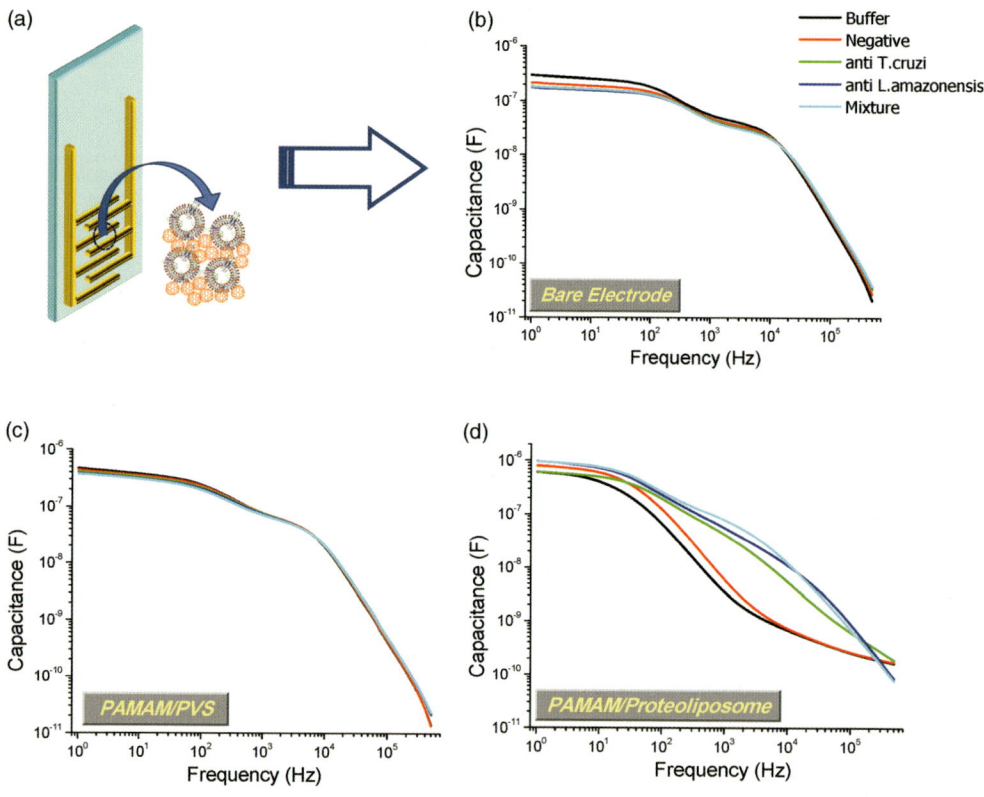

**Figure 40.11** (a) A schematic diagram for the sensing device, where an LbL film containing antigens in proteoliposomes is deposited onto an interdigitated electrode. (b)–(d) Capacitance versus frequency curves for three electrodes immersed in $10^{-5}$ mg mL$^{-1}$ antibody solutions, as indicated. Note that distinction between the samples is superior with the electrode containing a 5-layer-pair LbL film of PAMAM/proteoliposome (d). Reproduced with permission from: A. C. Perinoto, R. M. Maki, M. C. Colhone, F. R. Santos, V. Migliaccio, K. R. Daghastanli, R. G. Stabeli, P. Ciancaglini, F. V. Paulovich, M. C. F. de Oliveira, O. N. Oliveira and V. Zucolotto, *Anal. Chem.*, 2010, **82**, 9763–9768.

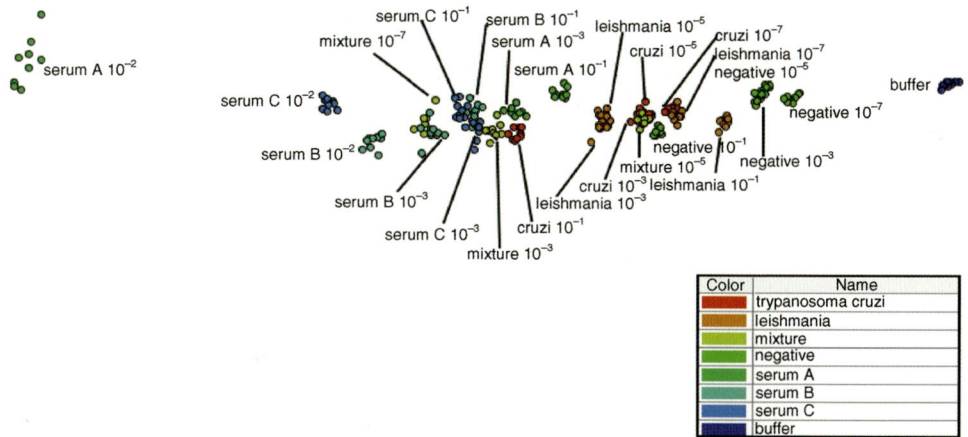

**Figure 40.12** Projection using PCA of the electrical impedance data obtained with the bare electrode for *L. amazonensis* and *T. Cruzi* samples with different concentrations, as follows. Serum A contained negative antibodies, serum B contained anti-*Leishmania* antibodies, serum C contained anti-*T. Cruzi* antibodies. The other samples were the buffer, and the so-called synthetic samples made with the buffer to which anti-*Leishmania*, anti-*T. Cruzi* and negative antibodies were added. The mixtures were synthetic samples with anti-*Leishmania*, anti-*T. Cruzi* antibodies together. Reproduced with permission from: F. V. Paulovich, R. M. Maki, M. C. F. de Oliveira, M. C. Colhone, F. R. Santos, V. Migliaccio, P. Ciancaglini, K. R. Perez, R. G. Stabeli, A. C. Perinoto, O. N. Oliveira, Jr. and V. Zucolotto, *Anal. Bioanal. Chem.*, 2011, **400**, 1153–1159.

Good distinction of the capacitance data in Figure 40.11 could be obtained with PCA, as only samples with antibodies added to the buffer were included. However, when all the samples were considered, which were obtained also from the blood serum of infected animals for the diseases under study, it was no longer possible to achieve a good distinction with PCA. In order to demonstrate the superior performance of a non-linear technique, we initially take the impedance spectroscopy data obtained with the bare electrode for all the samples. With the lack of specificity of the sensing unit toward the analytes (antibodies), the distinction was rather poor for any projection method. However, a direct comparison between Figures 40.12 and 40.13 indicates that Sammon's mapping is more suitable than PCA.

When the data from all four sensing units mentioned above were used, full distinction could be achieved with Sammon's mapping, as shown in Figure 40.14 [153].

The superiority of non-linear methods for biosensing was studied by plotting the data from the four sensors with PCA in subsidiary studies. Though the distinction was good, it was not perfect as it was with Sammon's mapping. Our hypothesis for the better performance of the non-linear methods is based on the highly non-linear fashion with which the electrical properties of the sensing units should depend on the specific interactions involved in molecular recognition processes toward analytes.

In addition to improving the distinguishing ability in a dataset with many samples, through the use of different projection techniques, one may further optimize the

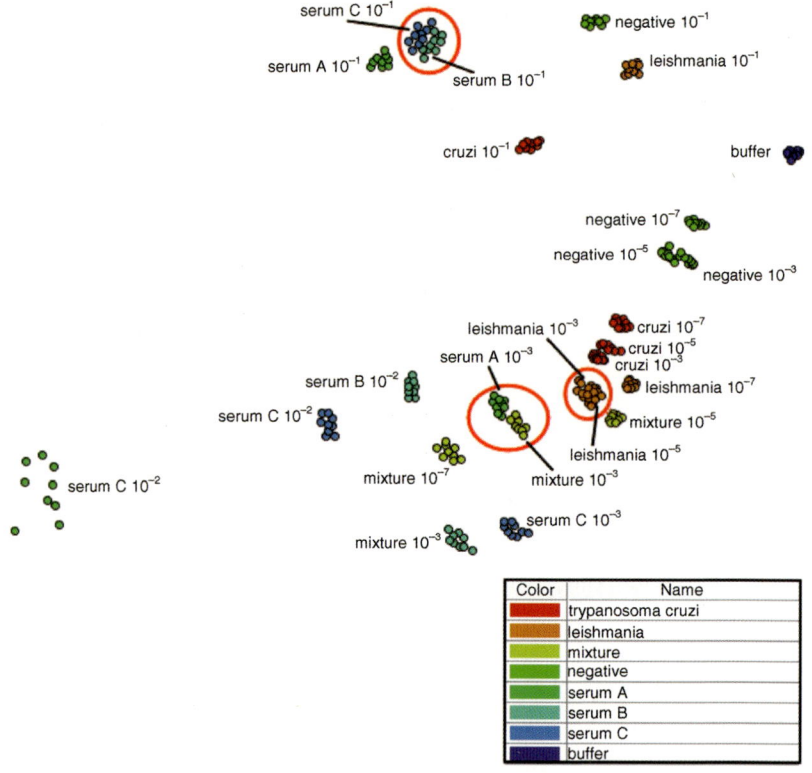

**Figure 40.13** Projection using Sammon's mapping of the same samples in Figure 40.12. Though data points from different samples are still mixed (circled in red), the distinction is better than in Figure 40.12 where PCA was used. Reproduced with permission from: F. V. Paulovich, R. M. Maki, M. C. F. de Oliveira, M. C. Colhone, F. R. Santos, V. Migliaccio, P. Ciancaglini, K. R. Perez, R. G. Stabeli, A. C. Perinoto, O. N. Oliveira, Jr. and V. Zucolotto, *Anal. Bioanal. Chem.*, 2011, **400**, 1153–1159.

sensing performance by selecting specific features in the dataset. A piece of work in this direction was reported by Moraes *et al.* [157] and Paulovich *et al.* [153], where the method known as parallel coordinates (PC) [141] was used. In PC the attributes are not mapped onto orthogonal axes of a Cartesian plane, but are rather mapped to axes arranged in parallel on the plane. The data instances are then represented as lines crossing these axes at a point determined by the corresponding attribute value. With PC it is possible to visualize a relatively large number of attributes on a single plane, which is useful to highlight patterns on the data and to infer functional dependences among the data attributes.

The PC plots in Figure 40.15 were obtained from the capacitance data for a sensing unit made with LbL films of PAH/PVS layers used to detect phytic acid in aqueous solutions. Owing to the lack of specificity in the interaction between the polyelectrolytes in the film and phytic acid, distinction of the different concentrations

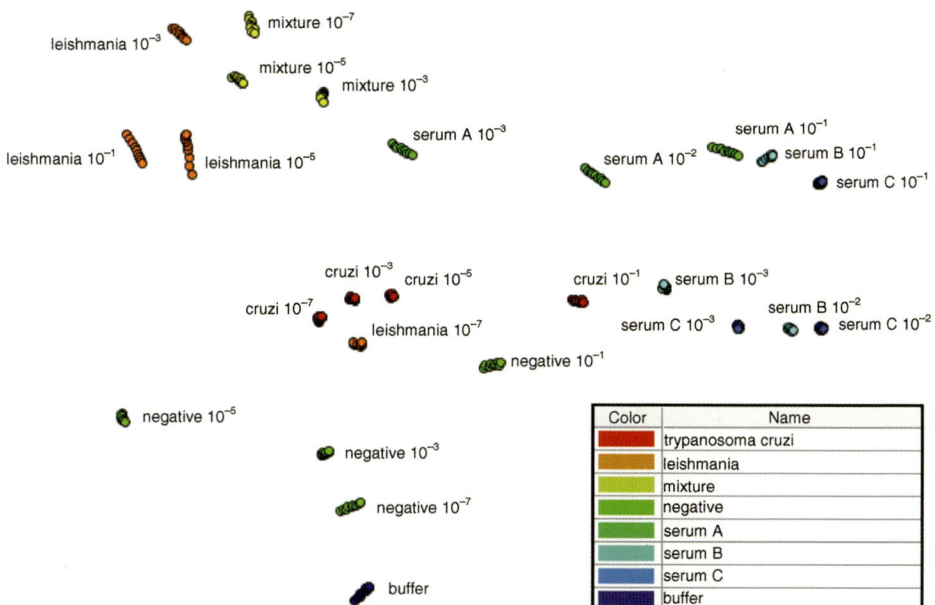

**Figure 40.14** Projection using Sammon's mapping of the capacitance data obtained with four sensors for all the samples shown in Figure 40.12. All samples can now be clearly separated. Reproduced with permission from: F. V. Paulovich, R. M. Maki, M. C. F. de Oliveira, M. C. Colhone, F. R. Santos, V. Migliaccio, P. Ciancaglini, K. R. Perez, R. G. Stabeli, A. C. Perinoto, O. N. Oliveira, Jr. and V. Zucolotto, *Anal. Bioanal. Chem.*, 2011, **400**, 1153–1159.

is poor, as is clear from the plots. In fact, this sensing unit was chosen because it was known to lead to a poor performance, and the objective of the study was to evaluate how the performance could be improved with optimization.

A better performance was achieved by selecting frequencies automatically from the dataset. This was carried out by computing the silhouette coefficient [158], which is a metric varying from $-1$ to 1 for evaluating the quality of a data cluster. Higher coefficient values, better cluster quality. The silhouette coefficient is:

$$S = \frac{1}{n}\sum_{i=1}^{n}\frac{(b_i - a_i)}{\max(a_i, b_i)}$$

where $a_i$ is the average value for the distances between the $i$th data point and the remaining points of the same cluster, while $b_i$ is the minimum distance between the $i$th data point and the points from the other clusters. In order to scan the whole data space of silhouette coefficient values, a genetic algorithm was used to select the frequencies leading to the best distinction [153]. The results from this procedure are illustrated in Figure 40.16 where the data are visualized in a parallel coordinates plot. A better distinction capability is readily observed in comparison with Figure 40.15. Significantly, the silhouette coefficient values, depicted by the color in the little boxes at the top of the plot – with blue representing high coefficient values and red

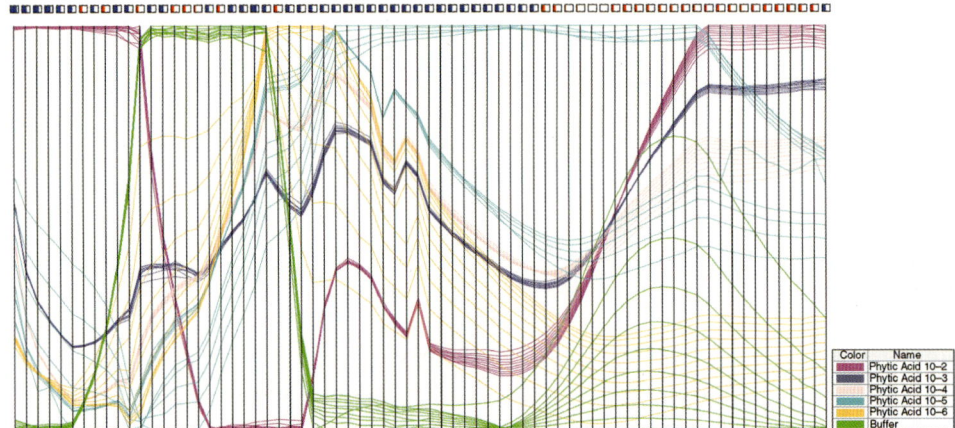

**Figure 40.15** Visualization of capacitance data with a sensing unit made of a PAH/PVS LbL film deposited onto an interdigitated gold electrode, using the parallel coordinates technique. The x-axis is the frequency and the y-axis gives normalized values for the capacitance. Note that for some small concentrations of phytic acid (denoted by different colors), there is overlap of the graphs. The little boxes on the top of the figure represent the silhouette coefficient for each data attribute. Blue boxes indicate frequencies that are useful for distinguishing the samples whereas the opposite applies for the red boxes. Reproduced with permission from: F. V. Paulovich, M. L. Moraes, R. M. Maki, M. Ferreira, O. N. Oliveira and M. C. F. de Oliveira, *Analyst*, 2011, **136**, 1344–1350.

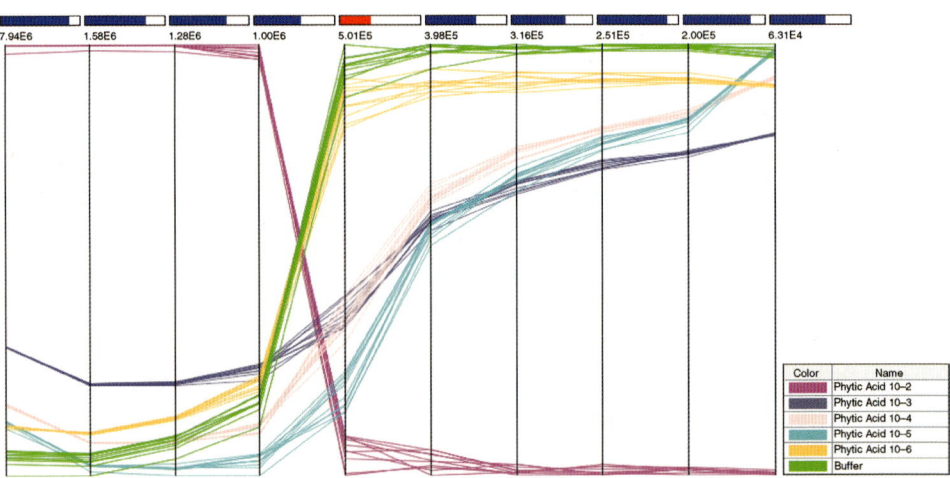

**Figure 40.16** Visualization with parallel coordinates of the same data in Figure 40.15, but now only with 10 selected frequencies to improve the distinguishing ability. The boxes representing the silhouette coefficients are almost all blue, for an optimization procedure was performed. Reproduced with permission from: F. V. Paulovich, M. L. Moraes, R. M. Maki, M. Ferreira, O. N. Oliveira and M. C. F. de Oliveira, *Analyst*, 2011, **136**, 1344–1350.

representing low values, are much higher in Figure 40.16. Indeed, most of the frequencies selected had high coefficients. The only exception was a particular frequency denoted with a red box, which was not good for distinguishing between clusters when used in isolation, but together with other frequencies led to an improvement in the overall distinguishing ability.

## 40.5
## Conclusions and Perspectives

The versatility of the LbL technique has been extensively exploited in producing novel architectures for sensing and biosensing, as we hope to have emphasized in this chapter. Several points are worth recalling. First, sensing units may be produced with LbL films for virtually any type of sensor, that is, with any type of principle of detection and kind of application. Indeed, sensing and biosensing may be performed with LbL films for different liquids and vapors. The second important point is that LbL films are normally advantageous in comparison with films produced with other methods because they are ultrathin and may have their architecture controlled precisely. Moreover, the LbL technique is especially suitable for biosensors because it allows the activity of biomolecules to be preserved. In this context, the third feature to be mentioned is the synergy among distinct materials that are deposited in a single LbL film. Throughout the chapter we have presented a long list of examples where polymers, nanoparticles, biomolecules, nanotubes, shells, and so on, were combined in LbL films to enhance the sensitivity and selectivity of the sensors and biosensors.

With so many positive features in the use of the LbL technique one may wonder where to go from here, and what are the challenges to be faced. As far as future developments are concerned, one may suggest several possibilities:

1) Combination of the LbL technique with microfluidics and other methods to yield high-throughput measurements, which are essential for some applications with biological samples.
2) Exploiting surface-enhanced methods that may lead to widespread detection of single molecules and/or single cells, where LbL films may be useful for functionalizing the substrates employed in the surface-enhancing measurements.
3) New developments in device engineering in order to take the sensors and biosensors produced in the lab to the real world. There are several platforms that could be employed in this context, including the use of field-effect devices based on conventional microelectronics. In fact, in this chapter we did mention successful examples of LbL films deposited on field-effect devices, and this can be pursued further.
4) Use of more sophisticated data analysis methods, which will be crucial for demanding applications where large amounts of data are collected and need to be processed. There has been, in this context, a timid development over the last few years, for example, with information visualization methods, but much more

needs to be done. In order to produce a fully-fledged diagnosis system, for instance, it will be necessary to merge information not only from sensors and biosensors but also from other data sources, such as clinical data of patients and information about diseases. This demanding task can only be performed with a suite of computational methods from areas such as artificial intelligence, data mining and information retrieval.

Last but not least, considerable progress in sensing and biosensing using LbL films could be achieved if one could understand the molecular-level interactions prevailing in the sensing task. In this chapter we have touched upon some surface specific experimental methods that serve this purpose, but most still needs to be done in theoretical simulations. Because the molecular-level interactions occur in very complex systems, the simulation and theoretical methods at present cannot cope with the variety of interaction forces, especially in biosensing where molecular recognition is expected. This is perhaps the main challenge in sensing and biosensing for the next decades.

## Acknowledgments

This work was supported by FAPESP, CNPq, CAPES and nBioNet (Brazil).

## References

1 Decher, G. (1997) *Science*, **277**, 1232–1237.
2 Decher, G., Hong, J.D., and Schmitt, J. (1992) *Thin Solid Films*, **210**, 831–835.
3 Zucolotto, V., Daghastanli, K.R.P., Hayasaka, C.O., Riul, A. Jr., Ciancaglini, P., and Oliveira, O.N. Jr. (2007) *Anal. Chem.*, **79**, 2163–2167.
4 Zucolotto, V., Pinto, A.P.A., Tumolo, T., Moraes, M.L., Baptista, M.S., Riul, A. Jr., Araujo, A.P.U., and Oliveira, O.N. Jr. (2006) *Biosens. Bioelectron.*, **21**, 1320–1326.
5 Schmidt, T.F., Caseli, L., dos Santos, D.S. Jr., and Oliveira, O.N. (2009) *Mat. Sci. Eng. C*, **29**, 1889–1892.
6 Mantha, S., Pedrosa, V.A., Olsen, E.V., Davis, V.A., and Simonian, A.L. (2010) *Langmuir*, **26**, 19114–19119.
7 Aoki, P.H.B., Alessio, P., Riul, A. Jr., Saez, J.A.D., and Constantino, C.J.L. (2010) *Anal. Chem.*, **82**, 3537–3546.
8 Aoki, P.H.B., Volpati, D., Riul, A. Jr., Caetano, W., and Constantino, C.J.L. (2009) *Langmuir*, **25**, 2331–2338.
9 Schmid, G. (2005) *Nanoparticles. From Theory to Applications*, Wiley-VCH Verlag GmbH, Weinheim.
10 Palumbo, M., Nagel, J., and Petty, M.C. (2005) *IEEE Sens. J.*, **5**, 1159–1164.
11 Liu, A.H., Abbineni, G., and Moo, C.B. (2009) *Adv. Mater.*, **21**, 1001–1005.
12 Guo, L.H., Chen, G.N., and Kim, D.H. (2010) *Anal. Chem.*, **82**, 5147–5153.
13 Chen, Y.L., Dong, B.Y., and Zhou, W.Y. (2010) *Appl. Surf. Sci.*, **257**, 1021–1026.
14 Abalde-Cela, S., Aldeanueva-Potel, P., Mateo-Mateo, C., Rodriguez-Lorenzo, L., Alvarez-Puebla, R.A., and Liz-Marzan, L.M. (2010) *J. R. Soc. Interface*, **7**, S435–S450.
15 Brown, J.Q. and McShane, M.J. (2005) *IEEE Sens. J.*, **5**, 1197–1205.
16 Stubbe, B.G., Gevaert, K., Derveaux, S., Braeckmans, K., De Geest, B.G.,

Goethals, M., Vandekerckhove, J., Demeester, J., and De Smedt, S.C. (2008) *Adv. Func. Mater.*, **18**, 1624–1631.

17 Shi, W.Y., Lin, Y.J., Kong, X.G., Zhang, S.T., Jia, Y.K., Wei, M., Evans, D.G., and Duan, X. (2011) *J. Mater. Chem.*, **21**, 6088–6094.

18 Ma, Q., Cui, H.L., and Su, X.G. (2009) *Biosens. Bioelectron.*, **25**, 839–844.

19 Yang, F.P., Ma, Q.A., Yu, W., and Su, X.G. (2011) *Talanta*, **84**, 411–415.

20 Egawa, Y., Hayashida, R., and Anzai, J.I. (2006) *Anal. Sci.*, **22**, 1117–1119.

21 Pal, E., Sebok, D., Hornok, V., and Dekany, I. (2009) *J. Colloid. Interface Sci.*, **332**, 173–182.

22 Nakane, Y. and Kubo, I. (2009) *Thin Solid Films*, **518**, 678–681.

23 Xie, G.Z., Sun, P., Yan, X.L., Du, X.S., and Jiang, Y.D. (2010) *Sens. Actuat. B-Chem.*, **145**, 373–377.

24 Zhao, W., Xu, J.J., and Chen, H.Y. (2006) *Electroanalysis*, **18**, 1737–1748.

25 Siqueira, J.R. Jr., Caseli, L., Crespilho, F.N., Zucolotto, V., and Oliveira, O.N. Jr. (2010) *Biosens. Bioelectron.*, **25**, 1254–1263.

26 Riul, A. Jr., Dantas, C.A.R., Miyazaki, C.M., and Oliveira, O.N. Jr. (2010) *Analyst.*, **135**, 2481–2495.

27 Caseli, L., dos Santos, D.S. Jr., Foschini, M., Goncalves, D., and Oliveira, O.N. Jr. (2006) *J. Colloid. Interface Sci.*, **303**, 326–331.

28 Zeng, G.H., Xing, Y.B., Gao, J.A., Wang, Z.Q., and Zhang, X. (2010) *Langmuir*, **26**, 15022–15026.

29 Guo, M.L., Chen, J.H., Li, J., Nie, L.H., and Yao, S.Z. (2004) *Electroanalysis*, **16**, 1992–1998.

30 Zucolotto, V., Ferreira, M., Cordeiro, M.R., Constantino, C.J.L., Moreira, W.C., and Oliveira, O.N. Jr. (2006) *Sens. Actuat. B-Chem.*, **113**, 809–815.

31 Ragupathy, D., Gopalan, A.I., and Lee, K.P. (2009) *Microchim. Acta*, **166**, 303–310.

32 Li, W.J., Wang, Z., Sun, C.Q., Xian, M., and Zhao, M.Y. (2000) *Anal. Chim. Acta*, **418**, 225–232.

33 Chen, X., Zhang, X., Yang, W.S., and Evans, D.G. (2009) *Mater. Sci. Eng. C-Biomim. Supramol. Syst.*, **29**, 284–287.

34 Kafi, A.K.M., Lee, D.Y., Park, S.H., and Kwon, Y.S. (2007) *Thin Solid Films*, **515**, 5179–5183.

35 Shi, G., Qu, Y., Zhai, Y., Liu, Y., Sun, Z., Yang, J., and Jin, L. (2007) *Electrochem. Commun.*, **9**, 1719–1724.

36 Liu, L.J., Zhang, F., Xi, F.N., and Lin, X.F. (2008) *Biosens. Bioelectron.*, **24**, 306–312.

37 Du, Y., Chen, C.G., Li, B.L., Zhou, M., Wang, E.K., and Dong, S.J. (2010) *Biosens. Bioelectron.*, **25**, 1902–1907.

38 Dong, X.Y., Mi, X.N., Wang, B., Xu, J.J., and Chen, H.Y. (2011) *Talanta*, **84**, 531–537.

39 Balasubramanian, K. and Burghard, M. (2006) *Anal. Bioanal. Chem.*, **385**, 452–468.

40 Srivastava, S. and Kotov, N.A. (2008) *Acc. Chem. Res.*, **41**, 1831–1841.

41 Kim, S.N., Rusling, J.F., and Papadimitrakopoulos, F. (2007) *Adv. Mater.*, **19**, 3214–3228.

42 Shu, K.W., Qin, Y.J., Luo, H.X., Zhang, P., and Guo, Z.X. (2011) *Mater. Lett.*, **65**, 1510–1513.

43 Chen, S.H., Chuang, Y.C., Lu, Y.C., Lin, H.C., Yang, Y.L., and Lin, C.S. (2009) *Nanotechnology*, **20**, 215501.

44 Lin, J.H., He, C.Y., Zhao, Y., and Zhang, S.S. (2009) *Sens. Actuat. B-Chem.*, **137**, 768–773.

45 Pingarron, J.M., Yanez-Sedeno, P., and Gonzalez-Cortes, A. (2008) *Electrochim. Acta*, **53**, 5848–5866.

46 Crespilho, F.N., Ghica, M.E., Florescu, M., Nart, F.C., Oliveira, O.N. Jr., and Brett, C.M.A. (2006) *Electrochem. Commun.*, **8**, 1665–1670.

47 Lu, L.P., Wang, S.Q., and Lin, X.Q. (2004) *Anal. Chim. Acta*, **519**, 161–166.

48 Podsiadlo, P., Shim, B.S., and Kotov, N.A. (2009) *Coordin. Chem. Rev.*, **253**, 2835–2851.

49 Nohria, R., Khillan, R.K., Su, Y., Dikshit, R., Lvov, Y., and Varahramyan, K. (2006) *Sens. Actuat. B-Chem.*, **114**, 218–222.

50 Loh, K.J., Lynch, J.P., Shim, B.S., and Kotov, N.A. (2008) *J. Intel. Mat. Syst. Str.*, **19**, 747–764.

51 Lu, J.B., Kumar, B., Castro, M., and Feller, J.F. (2009) *Sens. Actuat. B-Chem.*, **140**, 451–460.

52 Su, P.G. and Chuang, Y.S. (2010) *Sens. Actuat. B-Chem.*, **145**, 521–526.

53 Hornok, V. and Dekany, I. (2007) *J. Colloid. Interface Sci.*, **309**, 176–182.

54 Siqueira, J.R. Jr., Abouzar, M.H., Poghossian, A., Zucolotto, V., Oliveira, O.N. Jr., and Schoning, M.J. (2009) *Biosens. Bioelectron.*, **25**, 497–501.

55 Lee, D. and Cui, T.H. (2010) *Biosens. Bioelectron.*, **25**, 2259–2264.

56 Dulac, C. (2000) *Cell*, **100**, 607–610.

57 Paredes-Olay, C., Moreno-Fernandez, M.M., Rosas, J.M., and Ramos-Alvarez, M.M. (2010) *Food Qual. Prefer.*, **21**, 562–568.

58 Heath, T.P., Melichar, J.K., Nutt, D.J., and Donaldson, L.F. (2006) *J. Neurosci.*, **26**, 12664–12671.

59 Toko, K. (1996) *Mat. Sci. Eng. C-Biomim.*, **4**, 69–82.

60 Vlasov, Y.G., Legin, A.V., and Rudnitskaya, A.M. (2008) *Russian J. Gen. Chem.*, **78**, 2532–2544.

61 Escuder-Gilabert, L. and Peris, M. (2010) *Anal. Chim. Acta*, **665**, 15–25.

62 Zeravik, J., Hlavacek, A., Lacina, K., and Skladal, P. (2009) *Electroanalysis*, **21**, 2509–2520.

63 Otto, M. and Thomas, J.D.R. (1985) *Anal. Chem.*, **57**, 2647–2651.

64 Riul, A. Jr., dos Santos, D.S. Jr., Wohnrath, K., Di Tommazo, R., Carvalho, A., Fonseca, F.J., Oliveira, O.N. Jr., Taylor, D.M., and Mattoso, L.H.C. (2002) *Langmuir*, **18**, 239–245.

65 Taylor, D.M. and Macdonald, A.G. (1987) *J. Phys. D Appl. Phys.*, **20**, 1277–1283.

66 Gorban, A.K., Kegl, B., Wunsch, D., and Zinovyev, A. (2007) *Principal Manifolds for Data Visualisation and Dimension Reduction*, Springer, Berlin.

67 Joliffe, I. (2002) *Principal Component Analysis*, Springer-Verlag Press, New York.

68 Dos Santos, D.S. Jr., Riul, A. Jr., Malmegrim, R.R., Fonseca, F.J., Oliveira, O.N. Jr., and Mattoso, L.H.C. (2003) *Macromol. Biosci.*, **3**, 591–595.

69 Borato, C.E., Leite, F.L., Mattoso, L.H.C., Goy, R.C., Campana, S.P., de Vasconcelos, C.L., Neto, C., Pereira, M.R., Fonseca, J.L.C., and Oliveira, O.N. Jr. (2006) *IEEE T. Dielect. El. In.*, **13**, 1101–1109.

70 Wiziack, N.K.L., Paterno, L.G., Fonseca, F.J., and Mattoso, L.H.C. (2007) *Sens. Actuat. B-Chem.*, **122**, 484–492.

71 Brugnollo, E.D., Paterno, L.G., Leite, F.L., Fonseca, F.J., Constantino, C.J.L., Antunes, P.A., and Mattoso, L.H.C. (2008) *Thin Solid Films*, **516**, 3274–3281.

72 Paterno, L.G. and Mattoso, L.H.C. (2002) *J. Appl. Polym. Sci.*, **83**, 1309–1316.

73 Borato, C.E., Riul, A. Jr., Ferreira, M., Oliveira, O.N. Jr., and Mattoso, L.H.C. (2004) *Instrum. Sci. Technol.*, **32**, 21–30.

74 Olivati, C.A., Riul, A. Jr., Balogh, D.T., Oliveira, O.N. Jr., and Ferreira, M. (2009) *Bioproc. Biosystems. Eng.*, **32**, 41–46.

75 Volpati, D., Alessio, P., Zanfolim, A.A., Storti, F.C., Job, A.E., Ferreira, M., Riul, A. Jr., Oliveira, O.N. Jr., and Constantino, C.J.L. (2008) *J. Phys. Chem. B*, **112**, 15275–15282.

76 Carvalho, E.R., Filho, N.C., Venancio, E.C., Osvaldo, N.O., Mattoso, L.H.C., and Martin-Neto, L. (2007) *Sensors*, **7**, 3258–3271.

77 Krasteva, N., Besnard, I., Guse, B., Bauer, R.E., Mullen, K., Yasuda, A., and Vossmeyer, T. (2002) *Nano Lett.*, **2**, 551–555.

78 Vossmeyer, T., Guse, B., Besnard, I., Bauer, R.E., Mullen, K., and Yasuda, A. (2002) *Adv. Mater.*, **14**, 238–.

79 Joseph, Y., Besnard, I., Rosenberger, M., Guse, B., Nothofer, H.G., Wessels, J.M., Wild, U., Knop-Gericke, A., Su, D.S., Schlogl, R., Yasuda, A., and Vossmeyer, T. (2003) *J. Phys. Chem. B*, **107**, 7406–7413.

80 Krasteva, N., Guse, B., Besnard, I., Yasuda, A., and Vossmeyer, T. (2003) *Sens. Actuat. B-Chem.*, **92**, 137–143.

81 Krasteva, N., Krustev, R., Yasuda, A., and Vossmeyer, T. (2003) *Langmuir*, **19**, 7754–7760.

82 Siqueira, J.R. Jr., Maki, R.M., Paulovich, F.V., Werner, C.F., Poghossian, A., de Oliveira, M.C.F., Zucolotto, V., Oliveira, O.N. Jr., and Schoning, M.J. (2010) *Anal. Chem.*, **82**, 61–65.

83 Crespilho, F.N., Esteves, M.C., Sumodjo, P.T.A., Podlaha, E.J., and Zucolotto, V. (2009) *J. Phys. Chem. C*, **113**, 6037–6041.

84 Perinotto, A.C., Caseli, L., Hayasaka, C.O., Riul, A. Jr., Oliveira, O.N. Jr., and

Zucolotto, V. (2008) *Thin Solid Films*, **516**, 9002–9005.

85 Lu, J., Park, B.J., Kumar, B., Castro, M., Choi, H.J., and Feller, J.-F. (2010) *Nanotechnology*, **21**, 255501.

86 Kumar, B., Feller, J.-F., Castro, M., and Lu, J. (2010) *Talanta*, **81**, 908–915.

87 Moraes, M.L., de Souza, N.C., Hayasaka, C.O., Ferreira, M., Rodrigues, U.P., Riul, A. Jr., Zucolotto, V., and Oliveira, O.N. Jr. (2009) *Mater. Sci. Eng. C-Biomim. Supramol. Syst*, **29**, 442–447.

88 Perinoto, A.C., Maki, R.M., Colhone, M.C., Santos, F.R., Migliaccio, V., Daghastanli, K.R., Stabeli, R.G., Ciancaglini, P., Paulovich, F.V., de Oliveira, M.C.F., Oliveira, O.N. Jr., and Zucolotto, V. (2010) *Anal. Chem.*, **82**, 9763–9768.

89 Malikova, N., Pastoriza-Santos, I., Schierhorn, M., Kotov, N.A., and Liz-Marzan, L.M. (2002) *Langmuir*, **18**, 3694–3697.

90 MacKenzie, R., Fraschina, C., Sannomiya, T., Auzelyte, V., and Voros, J. (2010) *Sensors*, **10**, 9808–9830.

91 Gole, A. and Murphy, C.J. (2005) *Chem. Mater.*, **17**, 1325–1330.

92 Gu, B.X., Xu, C.X., Zhu, G.P., Liu, S.Q., Chen, L.Y., Wang, M.L., and Zhu, J.J. (2009) *J. Phys. Chem. B*, **113**, 6553–6557.

93 Wang, Z.Y., Zong, S.F., Yang, J., Song, C.Y., Li, J., and Cui, Y.P. (2010) *Biosens. Bioelectron.*, **26**, 241–247.

94 Chen, H.A., Xi, F.N., Gao, X., Chen, Z.C., and Lin, X.F. (2010) *Anal. Biochem.*, **403**, 36–42.

95 Hu, Z.C., Xu, J.J., Tian, Y.A., Peng, R., Xian, Y.Z., Ran, Q., and Jin, L.T. (2010) *Carbon*, **48**, 3729–3736.

96 Gu, J., Fan, W., Shimojima, A., and Okubo, T. (2008) *J. Solid. State Chem.*, **181**, 957–963.

97 Zhao, Y.Y., Qi, W., Li, W., and Wu, L.X. (2010) *Langmuir*, **26**, 4437–4442.

98 Kan, C.X., Wang, C.S., Zhu, J.J., and Li, H.C. (2010) *J. Solid. State Chem.*, **183**, 858–865.

99 Park, J.-I., and Cheon, J. (2001) *J. Am. Chem. Soc.*, **123**, 5743–5746.

100 Aroca, R.F., Goulet, P.J.G., dos Santos, D.S. Jr., Alvarez-Puebla, R.A., and Oliveira, O.N. Jr. (2005) *Anal. Chem.*, **77**, 378–382.

101 Aroca, R. (2006) *Surface-Enhanced Vibrational Spectroscopy*, John Wiley & Sons Ltd, Chichester.

102 Glass, A.M., Liao, P.F., Bergman, J.G., and Olson, D.H. (1980) *Opt. Lett.*, **5**, 368–370.

103 Garoff, S., Weitz, D.A., Gramila, T.J., and Hanson, C.D. (1981) *Opt. Lett.*, **6**, 245–247.

104 Osawa, M. (1997) *Bull. Chem. Soc. Jpn.* **70**, 2861–2880.

105 Aroca, R., Kovacs, G.J., Jennings, C.A., Loutfy, R.O., and Vincett, P.S. (1988) *Langmuir*, **4**, 518–521.

106 Constantino, C.J.L., Lemma, T., Antunes, P.A., and Aroca, R. (2001) *Anal. Chem.*, **73**, 3674–3678.

107 Kneipp, K., Wang, Y., Kneipp, H., Perelman, L.T., Itzkan, I., Dasari, R., and Feld, M.S. (1997) *Phys. Rev. Lett.*, **78**, 1667–1670.

108 Pieczonka, N.P.W., and Aroca, R.F. (2008) *Chem. Soc. Rev.*, **37**, 946–954.

109 Dos Santos, D.P. Jr., Andrade, G.F.S., Temperini, M.L.A., and Brolo, A.G. (2009) *J. Phys. Chem. C*, **113**, 17737–17744.

110 Scaffidi, J.P., Gregas, M.K., Seewaldt, V., and Vo-Dinh, T. (2009) *Anal. Bioanal. Chem.*, **393**, 1135–1141.

111 Aoki, P.H.B., Alessio, P., De Saja, J.A., and Constantino, C.J.L. (2010) *J. Raman Spectrosc.*, **41**, 40–48.

112 Li, J.F., Huang, Y.F., Ding, Y., Yang, Z.L., Li, S.B., Zhou, X.S., Fan, F.R., Zhang, W., Zhou, Z.Y., Wu, D.Y., Ren, B., Wang, Z.L., and Tian, Z.Q. (2010) *Nature*, **464**, 392–395.

113 Gersten, J. and Nitzan, A. (1981) *J. Chem. Phys.*, **75**, 1139–1152.

114 Demers, L.M., Mirkin, C.A., Mucic, R.C., Reynolds, R.A., Letsinger, R.L., Elghanian, R., and Viswanadham, G. (2000) *Anal. Chem.*, **72**, 5535–5541.

115 Hamad-Schifferli, K., Schwartz, J.J., Santos, A.T., Zhang, S.G., and Jacobson, J.M. (2002) *Nature*, **415**, 152–155.

116 Schneider, G., Decher, G., Nerambourg, N., Praho, R., Werts, M.H.V., and Blanchard-Desce, M. (2006) *Nano. Lett.*, **6**, 530–536.

117 Guerrero, A.R. and Aroca, R.F. (2011) *Angew Chem. Int. Ed.*, **50**, 665–668.
118 Palin, E., Liu, H.N., and Webster, T.J. (2005) *Nanotechnology*, **16**, 1828–1835.
119 Yan, J., Pedrosa, V.A., Simonian, A.L., and Revzin, A. (2010) *ACS Appl. Mater. Interface.*, **2**, 748–755.
120 Wang, J. (2007) *Electroanalysis*, **19**, 769–776.
121 Qu, Y.H., Sun, Q., Xiao, F., Shi, G.Y., and Jin, L.T. (2010) *Bioelectrochemistry*, **77**, 139–144.
122 Iijima, S. (1991) *Nature*, **354**, 56–58.
123 Liu, T.X., Phang, I.Y., Shen, L., Chow, S.Y., and Zhang, W.D. (2004) *Macromolecules*, **37**, 7214–7222.
124 Raravikar, N.R., Schadler, L.S., Vijayaraghavan, A., Zhao, Y.P., Wei, B.Q., and Ajayan, P.M. (2005) *Chem. Mater.*, **17**, 974–983.
125 Du, F.M., Scogna, R.C., Zhou, W., Brand, S., Fischer, J.E., and Winey, K.I. (2004) *Macromolecules*, **37**, 9048–9055.
126 Cui, H.F., Cui, Y.H., Sun, Y.L., Zhang, K., and Zhang, W.D. (2010) *Nanotechnology*, **21**, 215601.
127 Xi, F.N., Liu, L.J., Chen, Z.C., and Lin, X.F. (2009) *Talanta*, **78**, 1077–1082.
128 Crespilho, F.N., Iost, R.M., Travain, S.A., Oliveira, O.N. Jr., and Zucolotto, V. (2009) *Biosens. Bioelectron.*, **24**, 3073–3077.
129 Deng, C.Y., Chen, J.H., Nie, Z., and Si, S.H. (2010) *Biosens. Bioelectron.*, **26**, 213–219.
130 Zhao, W., Xu, J.-J., and Chen, H.-Y. (2006) *Electroanalysis*, **18**, 1737–1748.
131 Lvov, Y., Decher, G., and Sukhorukov, G. (1993) *Macromolecules*, **26**, 5396–5399.
132 Kumpumbu-Kalemba, L. and Leclerc, M. (2000) *Chem. Commun.*, **36**, 1847–1848.
133 Koktysh, D.S., Liang, X., Yun, B.G., Pastoriza-Santos, I., Matts, R.L., Giersig, M., Serra-Rodríguez, C., Liz-Marzán, L.M., and Kotov, N.A. (2002) *Adv. Funct. Mater.*, **12**, 255–265.
134 Geladi, P., Nelson, A., and Lindholm-Sethson, B. (2007) *Anal. Chim. Acta*, **595**, 152–159.
135 Torgeson, W.S. (1965) *Psychometrika*, **30**, 379–393.
136 Lindholm-Sethson, B., Nystrom, J., Malmsten, M., Ringstad, L., Nelson, A., and Geladi, P. (2010) *Anal. Bioanal. Chem.*, **398**, 2341–2349.
137 Trietsch, S.J., Hankemeier, T., and van der Linden, H.J. (2011) *Chemometr. Intell. Lab.*, **108**, 64–75.
138 Hey, T., Tansley, S., and Tolle, K. (2009) *The Fourth Paradigm – Data Intensive Scientific Discovery Microsoft Research*, Redmond.
139 Cleveland, W.S. (1993) *Visualizing Data*, Hobart Press, New Jersey.
140 Rao, R. and Card, S.K. (1994) Proceedings of Human Factors in Computing Systems (CHI'94), pp. 318–322.
141 Inselberg, A. and Dimsdale, B. (1990) Proceedings of the IEEE Visualization (Vis'90), pp. 361–337.
142 Keim, D.A. (2000) *IEEE T. Vis. Comput. Gr.*, **6**, 59–78.
143 Kohonen, T. (1990) *P. IEEE*, **78**, 1464–1480.
144 Paulovich, F.V. and Minghim, R. (2008) *IEEE T. Vis. Comput. Gr.*, **14**, 1229–1236.
145 de Oliveira, M.C.F. and Levkowitz, H. (2003) *IEEE T. Vis. Comput. Gr.*, **9**, 378–394.
146 Grinstein, G., Trutschl, M., and Cvek, U. (2001) Proceedings 7th Data Mining Conference KDD Workshop, pp. 7–19.
147 Riul, A. Jr., de Sousa, H.C., Malmegrim, R.R., dos Santos, D.S. Jr., Carvalho, A., Fonseca, F.J., Oliveira, O.N. Jr., and Mattoso, L.H.C. (2004) *Sens. Actuat. B-Chem.*, **98**, 77–82.
148 Haykin, S. (1999) *Neural Networks: A Comprehensive Foundation*, Prentice-Hall, Englewood-Cliffs.
149 Ferreira, E.J., Pereira, R.C.T., Delbem, A.C.B., Oliveira, O.N. Jr., and Mattoso, L.H.C. (2007) *Electron. Lett.*, **43**, 1138–1140.
150 Ben Nouir, N., Gianinazzi, C., Gorcii, M., Mueller, N., Nouri, A., Babba, H., and Gottstein, B. (2009) *Trans. R. Soc. Trop. Med. Hyg.*, **103**, 355–364.
151 Singh, S. and Sivakumar, R. (2003) *J. Postgrad. Med.*, **49**, 55–60.
152 Tejada, E., Minghim, R., and Nonato, L.G. (2003) *Inf. Vis.*, **2**, 218–231.
153 Paulovich, F.V., Moraes, M.L., Maki, R.M., Ferreira, M., Oliveira, O.N.

Jr., and de Oliveira, M.C.F. (2011) *Analyst.*, **136**, 1344–1350.

154 Paulovich, F.V., Maki, R.M., de Oliveira, M.C.F., Colhone, M.C., Santos, F.R., Migliaccio, V., Ciancaglini, P., Perez, K.R., Stabeli, R.G., Perinoto, A.C., Oliveira, O.N. Jr., and Zucolotto, V. (2011) *Anal. Bioanal. Chem.*, **400**, 1153–1159.

155 Sammon, J.W. (1969) *IEEE T. Comput.*, **C18**, 401.

156 Minghim, R., Paulovich, F.V., and Lopes, A.A. (2006) IS&T/SPIE Symposium on Electronic Imaging – Visualization and Data Analysis, pp. S1–S12.

157 Moraes, M.L., Maki, R.M., Paulovich, F.V., Rodrigues, U.P., de Oliveira, M.C.F., Riul, A. Jr., de Souza, N.C., Ferreira, M., Gomes, H.L., and Oliveira, O.N. Jr. (2010) *Anal. Chem.*, **82**, 3239–3246.

158 Tan, P.N., Steinbach, M., and Kumar, V. (2005) *Introduction to Data Mining*, Addison-Wesley Longman Publishing Co., Boston.

# 41
# Patterned Multilayer Systems and Directed Self-Assembly of Functional Nano-Bio Materials

*Ilsoon Lee*

## 41.1
### New Approaches and Materials for Multilayer Film Patterning Techniques

Patterned and controlled layer-by-layer (LbL) films on surfaces can serve as excellent molecular templates for applications in optoelectronic display materials, biosensor arrays, and drug screening devices. The ionic LbL assembly technique, introduced by Decher in 1991 [1, 2], is among the most exciting developments in this area. Films formed by electrostatic interactions between oppositely charged poly-ion species to create alternating layers of sequentially adsorbed poly-ions are called "polyelectrolyte multilayers" (PEMs). One of the soft lithographic methods, microcontact printing (μCP), has been used in physics, chemistry, materials science, and biology to transfer patterned thin organic films to surfaces with sub-micron resolution [3, 4]. Unlike other fabrication methods that merely provide topographic contrast between the feature and the background, μCP also allows chemical contrast to be achieved by selection of an appropriate ink. Microcontact printing offers advantages over conventional photolithographic techniques because it is simple to perform and is not diffraction limited.

Lee and coworkers at the Michigan State University have challenged new patterning approaches for the last several years. For the first time, they have demonstrated self-assembled monolayer (SAM) patterning on PEMs, as opposed to on gold or silicon substrates [5]. In the work, the process of creating chemically patterned and physically structured surfaces was described by stamping polyethylene acid molecules on PEM coated surfaces, as illustrated in Figure 41.1. The activated carboxylate functional group binds ionically to the topmost positive surface of the PEM surfaces, and the other end (PEG units) resists the deposition of subsequent polymer (polyelectrolyte) layers. To deposit thin uniform PEG SAMs on PEMs, ionic interactions were capitalized. The deposited PEG patterns acted like an efficient resisting area which resists non-specific adsorption of polyelectrolytes, charged particles, and biomolecules and cells. The exposed PEM regions served as active surfaces attracting a variety of functional species (Figure 41.2) [5].

*Multilayer Thin Films: Sequential Assembly of Nanocomposite Materials*, Second Edition.
Edited by Gero Decher and Joseph B. Schlenoff.
© 2012 Wiley-VCH Verlag GmbH & Co. KGaA. Published 2012 by Wiley-VCH Verlag GmbH & Co. KGaA.

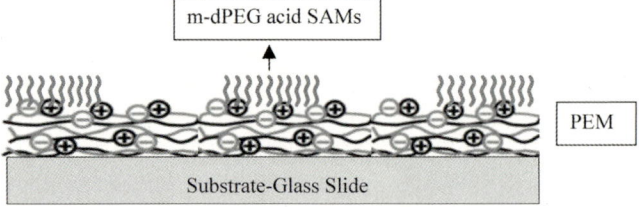

**Figure 41.1** Illustration of patterned SAMs on PEM. Reproduced with permission from [5].

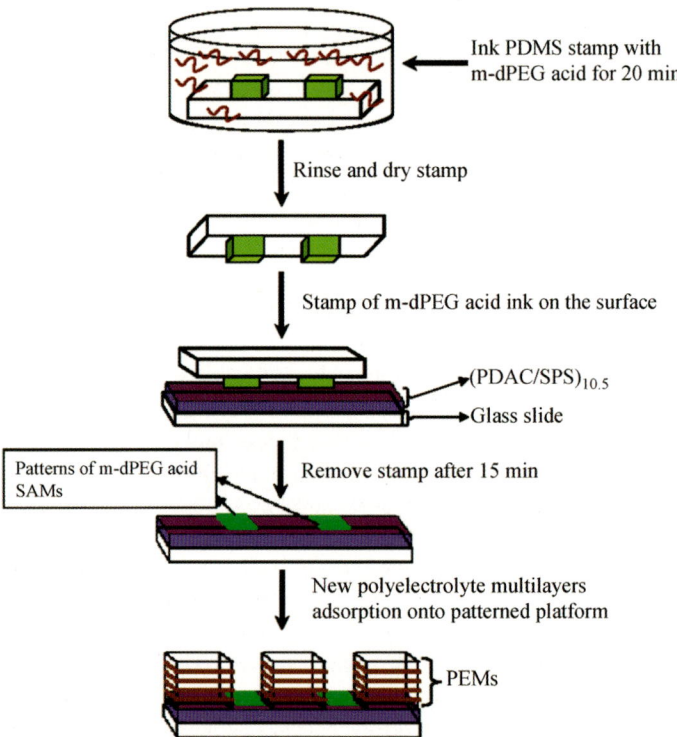

**Figure 41.2** Diagram illustrating the stamping process of m-dPEG acid on a PDAC/SPS multilayer platform. Reproduced with permission from [5].

In addition, they developed novel salt tunable resistive PEG SAMs on PEM surfaces that can provide a tunable template to design numerous sorted surfaces [6]. The previous PEG SAMs on PEMs system where the carboxylate functional group in the PEG molecule binds ionically to the topmost positive surface of the PEM surfaces, and the other end of the PEG molecule resists the deposition of subsequent macromolecules, including polyelectrolyte layers, was utilized for this purpose. The PEG patterns were tuned under mild conditions (salt concentration = 0.25 M) to reveal active regions that can be used to create multicomponent systems. This study extended the tunable PEG surfaces to engineer multicomponent systems of macromolecules with similar physical and chemical properties. Such resistive and removable PEG patterns on PEMs facilitated the directed deposition of various macromolecules, such as polymers, dyes, colloidal particles, proteins, liposomes, nucleic acids, and cells.

LbL directed assembly of 3D structures onto pre-patterns can be hindered by a lack of chemical contrast between features and background, making it difficult to deposit additional layers cleanly onto only the features or the background. This problem can be particularly challenging when a layer of amphiphilic molecules can adsorb to both hydrophilic and hydrophobic surfaces. To overcome this, Lee and coworkers developed a new multilayer patterning approach, intact pattern transfer (IPT), especially to establish well-defined, 3D, layered bionanocomposite patterns containing alternating layers of polyelectrolytes, dendrimers, and amphiphilic proteins [7, 8]. Unlike the directed selective depositions, this new approach allows the preparation of high quality, 3D patterned LbL films on substrates whose surface properties are incompatible with existing self-assembly methods. Using this new approach, Lee and coworkers presented a simple method of creating patterned conductive multilayered polymer/graphene films on a non-conductive substrate [9]. In the work, multilayered graphite was exfoliated, followed by milling to create size-controlled graphenes. The detailed procedure is described in Section 41.4.

Recently, Lee and coworkers have demonstrated that PEM films can be transferred from a stamp to the base substrate under aqueous conditions, whereby the two surfaces are in a noncontact mode [10]. Using this new approach, they showed an alternative method for creating a sandwiched 3D cell co-culture by transferring polyelectrolyte multilayers onto a charged "base" substrate under aqueous conditions, in a noncontact mode. In this noncontact transfer mode, the base substrate and the stamp do not contact each other during the multilayer transfer. Noncontact multilayer transfer can be useful for creating a 3D cellular co-culture with a permeable polymer layer sandwiched between two monolayers of cells. An advantage is that deposition of the polymer over the cells is achieved without the need for removing the culture media.

Lee and coworkers also reported the first application of microcontact printing of the amphiphilic and crosslinkable poly(amidoamine organosilicon) dendrimers having dimethoxymethylsilyl endgroups (PAMAMOS-DMOMS) dendrimers [11] and poly(amidoamine) (PAMAM) dendrimers [12] on glass slides, silicon wafers, and PEMs in which the pattern average thickness was controlled by spin self-assembly (i.e., spin-inking). The results provided a framework for controlling the geometry of the deposited patterns. The lateral footprint of the pattern was controlled by the shape of the elastomeric stamp, and the thickness of the patterns was controlled by adjusting the spin coating method, the surface properties of the stamp, and the substrate used. The results also confirmed the well-known influences of spin speed, concentration, and solvent on the thickness of spin-coated films. They presented methods to fabricate arrays of bilayer lipid membranes (BLMs) and liposomes on PEMs [13]. Arrays of BLMs were created by exposing PDAC patterns, polyethylene glycol (m-dPEG acid) patterns, and PAH patterns on PEMs to liposomes of various compositions. The formation of a novel biomimetic interface consisting of an electrolessly deposited gold film overlaid with a tethered BLM was demonstrated by Lee and coworkers [14]. BLMs were tethered to the gold film by first depositing an inner molecular leaflet using a mixture of 1,2-dipalmitoyl-sn-glycero-3-phosphoethanolamine-N-[3-(2-pyridyldithio)propionate],1,2-di-O-phytanyl-sn-glycero-3-phosphoethanolamine (DPGP), and cystamine in ethanol onto a freshly prepared electrolessly deposited gold surface. The outer leaflet was then formed by the fusion of liposomes made from DPGP or 1,2-dioleoyl-sn-glycero-3-phosphocholine on the inner leaflet. To provide functionality, two membrane biomolecules were also incorporated into the tethered BLMs. Microcontact printing was used to form arrays of electrolessly deposited gold patterns on glass slides. Subsequent deposition of lipids yielded arrays of tethered BLMs. They also demonstrated for the first time how to control the 2D polyelectrolyte aggregates created by microcontact printing [15]. A key feature of the work is the thin film morphology study of microcontact printed polyelectrolyte aggregates. The polyelectrolyte ink and stamping processes were designed for the formation of polyelectrolyte aggregates (e.g., tree-like ramified structures). They reported the coarsening of the ramified structures of polyelectrolyte by confining the stamp's contact area to a size in which the pattern is smaller than that of the ramified structures. In addition, the ramified structures can be directed by directional stamping without conformal contact at the interface.

## 41.2
### Cell Adhesion and Patterning Using PEMs

Cell–substrate interactions are important to many biological phenomena. One of the major challenges in studying the mechanism of cell–substrate interactions on synthetic surfaces is discerning the relative role of the chemical functional groups in this interaction. Elucidating these interactions and how they may be controlled is crucial to understanding how to manipulate and design better biological systems and medical devices. The physical and chemical properties of a substrate affect the attachment and growth of cells. PEMs have become excellent candidates for biomaterial applications due to (i) their biocompatibility and bioinertness, (ii) their ability to incorporate biological molecules, such as proteins, and (iii) the high degree of molecular control of the film structure and thickness, providing a much simpler approach to the construction of complex 3D surfaces as compared with photolithography. Lee and coworkers described the successful attachment and spreading of primary hepatocytes on PEM films without the use of adhesive proteins such as collagen or fibronectin [16]. They demonstrated, for the first time, that primary hepatocytes attached, spread, and maintained differentiated function, such as albumin and urea production, on a synthetic PEM surface with poly(4-styrenesulfonic acid) (SPS) as the topmost surface. The aim of this study was to characterize the attachment, spreading, and function of primary rat hepatocytes cultured on PEM surfaces in which the PEMs were used to produce defined cell-resistant and cell-adhesive properties, depending on the topmost surface and the type of cells used. This was extended to describe the formation of patterned cell co-cultures using the

LbL deposition of synthetic ionic polymers and without the aid of adhesive proteins/ ligands such as collagen or fibronectin [17]. To create patterned co-cultures on PEMs, they capitalized on the preferential attachment and spreading of primary hepatocytes on SPS as opposed to PDAC surfaces. In contrast, fibroblasts readily attached to both PDAC and SPS surfaces, and as a result, they were able to obtain patterned co-cultures of fibroblast and primary hepatocytes on synthetic PEM surfaces.

Lee and coworkers demonstrated that PEM-coated PDMS surfaces with different topographies affect the attachment, spreading, and proliferation of three types of mammalian cells: transformed 3T3 fibroblasts (3T3s), HeLa (transformed epithelial) cells, and primary hepatocytes. The PEMs were built using LbL assembly of the polyelectrolytes poly(diallyldimethylammonium chloride) (PDAC, the poly-cation) and sulfonated poly(styrene) sodium salt (SPS, the poly-anion) [18]. After cell seeding, they observed differences in cell attachment and spreading depending on the grooves and patterns on the PDMS surfaces. Cell morphology and attachment varied depending on the pattern geometries. Using imaging techniques, they showed that changes in the surface topographical features alter the attachment and spreading of cells, suggesting a physical means of controlling the interaction between the cell and its environment.

In addition, they demonstrated patterned co-cultures of primary neurons and astrocytes on PEM films without the aid of adhesive proteins/ligands to study the oxidative stress mediated by astrocytes on neuronal cells (Figure 41.3) [19]. In their study, they used synthetic polymers, namely PDAC and SPS as the polycation and polyanion, respectively, to build the multilayers. Primary neurons attached and

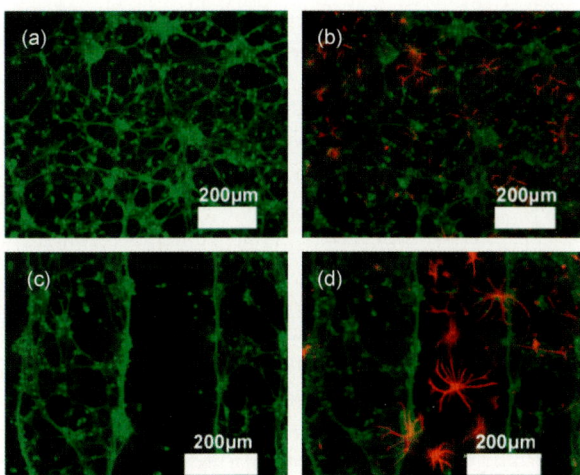

**Figure 41.3** Fluorescent images of primary neurons and astrocytes co-culture on SPS surfaces (a) Random neuron monocultures (green) after 7 days in culture, (b) Random co-culture of neurons (green) and astrocytes (red) after seeding astrocytes. (c) Patterned primary neurons on SPS patterns after 7 days in culture (d) Patterned co-culture of neurons (green) and astrocytes (red) after seeding astrocytes (Scale bars: 200 m). Reproduced with permission from [19].

spread preferentially on SPS surfaces, while primary astrocytes attached to both SPS and PDAC surfaces. SPS patterns were formed on PEM surfaces, either by microcontact printing SPS onto PDAC surfaces or vice versa, to obtain patterns of primary neurons and patterned co-cultures of primary neurons and astrocytes. They also demonstrated an alternative method of cell–cell patterned co-culture system by utilizing the salt tunable PEG patterns on PEMs [6]. The removable PEG SAMs on PEMs provide surfaces that can be readily switched from cell repulsive to cell adhesive using cell-friendly conditions. The approach is advantageous since they can be used to form patterned co-cultures irrespective of the types of cell or the order of seeding of the different types of cells.

Recently, Lee and coworkers have shown that increasing the number of layer pairs (deposition cycles) of PDAC/SPS films from 10 to 20, corresponding to a film thickness of 37.6 nm (40 nm) to 95.9 nm (100 nm), respectively, switches the films from a cytophilic to a cytophobic surface [20]. They demonstrated this effect with bone marrow mesenchymal stem cells (MSCs) and NIH3T3 fibroblasts. The thickness increases linearly as the number of layer pairs increases, causing a shift to cytophobic behavior with a concomitant decrease in cell spreading and adhesion. They implemented a finite element analysis to help elucidate the observed trends in cell spreading. The simulation results suggest that cells maintain a constant level of energy consumption (energy homeostasis) during active probing, and thus respond to changes in the film stiffness as the film thickness increases by adjusting their morphology and the number of focal adhesions recruited, and thereby their attachment to a substrate. They also demonstrated that a 3D celluar scaffold using a noncontact, aqueous-phase multilayer (NAM) transfer method [10]. A multilayer transfer process was performed under physiological conditions that involved transferring an assembly of PDAC and SPS onto a charged base-substrate without the stamp being in full contact with the base substrate. The successful transfer of the NAM depended upon the number of layer pairs in the PEM to be transferred, and on the height of the stamping surface from the base substrate.

## 41.3
### PEMs Incorporating Proteins and Their Patterning

Biological activity can be imparted to such arrays by incorporating proteins or other biomolecules. The biological activity can be augmented by co-immobilization of macromolecular adjuvants, such as dendrimers or PEMs. PEMs are thin films formed when two oppositely charged polyelectrolytes are alternately adsorbed onto a surface one layer at a time. PEMs are robust, easy to fabricate, and have tunable architectures (i.e., film composition and physical and chemical microstructure). Polyelectrolytes can be used to immobilize hydrophobic membrane proteins onto hydrophilic substrates, entrap ionic or polar small molecules needed by the proteins, and act as an ion-selective barrier to screen out interfering molecules.

Lee and coworkers have compared a direct self-assembly of amphiphilic biomacromolecules and a novel direct multilayer transfer method that overcomes the above-mentioned difficulties in establishing well-defined, 3D, layered bionanocomposite patterns containing alternating layers of polyelectrolytes, dendrimers, and amphiphilic proteins [7, 8]. The approach entails combining spin self-assembly and layer-by-layer self-assembly to pre-establish a multilayered structure on an elastomeric stamp, and then using μCP to transfer the 3D structure intact to the target surface. While μCP was recently used to transfer preformed PEMs to a substrate by the Hammond group at MIT [21], the papers presented, for the first time, conclusive evidence of the formation of bionanocomposite layered structures on a micropatterned stamp, and subsequent transfer of the structures intact to a target substrate.

The preserved activity of proteins transferred onto a substrate such as bioelectronic interfaces is crucial to fabricating highly sensitive and fast responsive biosensors. Lee and coworkers presented the first continuous, electrochemical biosensor for real-time, rapid measurement of neuropathy target esterase (NTE or NEST) activity [7]. The biosensor was fabricated by co-immobilizing NEST and tyrosinase on an electrode using the LbL assembly approach. Potential applications of the biosensor include detecting the presence of neuropathic agents that target NTE, screening industrial and agricultural Organophosphorus compounds for NTE inhibition, studying the fundamental reaction kinetics of NTE, and investigating the effects of NTE mutations on its enzymatic properties. They also developed a theoretical model to analyze the bi-enzyme electrode containing NEST and tyrosinase [22]. The LbL molecular architecture of the bi-enzyme electrode is shown in Figure 41.4. The NEST

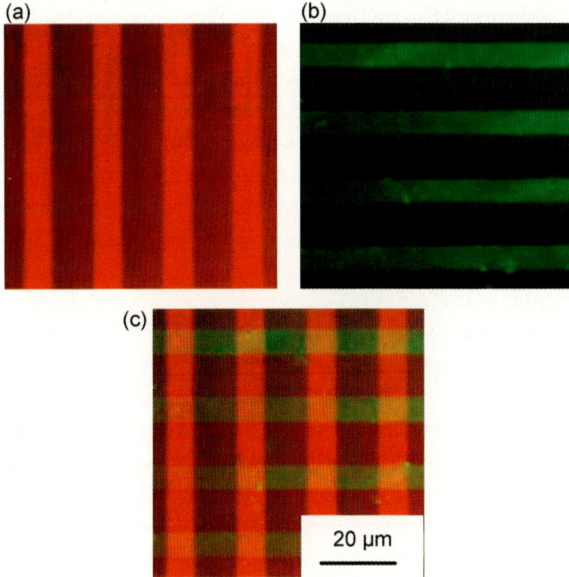

**Figure 41.4** (a) Red fluorescence from vertically printed (Alexa Fluor labeled-sDH)$_1$(PDAC/SPS)$_{20}$ lines on a PEM (10.5 PDAC/SPS layer pairs) coated glass substrate. (b) Green fluorescence from horizontally printed (FITC labeled-sADH)$_1$(PDAC/SPS)$_{20}$ lines, on a PEM (10.5 PDAC/SPS layer pairs) coated glass substrate. (c) Digitally combined fluorescence image showing both red and green fluorescence. Reproduced with permission from [23].

protein converts phenyl valerate to phenol, which is converted to o-quinone by tyrosinase. The o-quinone is electrochemically reduced to catechol at the electrode's surface, resulting in a current. A portion of the catechol produced is then converted to o-quinone by tyrosinase. Catechol thus serves as a shuttle analyte that can undergo successive cycles of enzymatic oxidation–electrochemical reduction (substrate recycling), resulting in an amplification of the biosensor's response. The theoretical model includes the influence of the mass transport, permeation through the enzyme layers, and enzyme kinetics. This model is expressed in dimensionless form to minimize the number of constants that must be evaluated. The biosensor was assembled on a rotating disk electrode, and the biosensor's performance was measured at a variety of rotational velocities and substrate concentrations to evaluate the constants and validate the model.

Lee and coworkers also presented a novel method based on LbL self-assembly to fabricate a renewable bioelectronic interface in which the enzyme and cofactor can be removed and replaced [24]. The polycation poly(ethylenimine) (PEI) was used to couple the electron mediator, cofactor, and enzyme to a carboxylic acid-modified gold electrode in such a way that mediated electron transfer was achieved. Decreasing the pH of the solution protonated the surface-bound carboxylic acid groups, disrupting the ionic bonds and releasing the enzyme and cofactor. After neutralization, fresh enzyme and cofactor could be bound, allowing the interface to be reconstituted. In addition, they presented a novel and convenient bench-top fabrication method to deposit gold, and then overlay high-performance bioelectronic interfaces containing dehydrogenase enzymes, on a variety of nonconductive surfaces [25]. The method integrates LbL deposition of polyelectrolytes, colloidal gold deposition, and molecular self-assembly.

Degradable multilayer assemblies, based on sequential embedding of drugs during the fabrication, can incorporate any drug, independent of its molecular weight. Fabrication of hydrogen bond (H-bond)-based LbL multilayer films was initially reported by Rubner and coworkers [26]. Lee and coworkers present a simple approach for controlled delivery of proteins from agarose gels, where the proteins are incorporated within the degradable LbL multilayer coatings formed over the agarose [27]. Carboxylic acid (−COOH)-based weak polyelectrolytes form H-bond interactions at low pH (e.g., pH < 3.5 in the case of poly(acrylic acid (PAA)) and deprotonate to carboxylate ions (−COO$^-$) at high pH, which degrades the H-bonded multilayer assembly. H-bonded PAA/poly(ethylene oxide) (PEO) multilayer films when built on a planar substrate are known to degrade in about 30 min upon exposure to a pH of 3.5 or higher. However, they showed that the H-bonded films when prepared over agarose as the substrate provided sustained release of the incorporated protein under physiological conditions for a period of more than 4 weeks.

## 41.4
### Metal/Graphene Conductive Patterning via PEM Films

Inexpensive metal or other conductive patterning techniques with high selectivity have been the focus of current research in displays, radio frequency identification

(RFID) transponders, sensors and other nano- and microelectronic device fabrication. Recently, many techniques have been developed to pattern metals on surfaces. Photolithography based top-down methods are the standard industrial patterning technique in microelectronics. However, this process is an expensive step in device fabrication, limits the functionality of substrates and other materials, and has an inability to work with curved substrates or the complex 3D structures needed for new electronic devices. Microcontact printing, a soft lithographic patterning technique, combined with PEM coatings offers a multitude of cost-effective routes for creating functional 3D structures on plastic and other flexible substrates.

Lee and Hendricks presented a new process for creating versatile and selective copper patterns by combining PEM coatings, μCP, and electroless deposition (ELD) (see Figure 41.5) [28]. For the first time μCP was used to pattern a charged palladium catalyst onto oppositely charged PEM-coated substrates. PEMs, unlike silanes and thiols, can be stably coated onto virtually any substrate, including hydrophobic polymer surfaces. This resulted in a highly selective, electrostatically bound charged palladium ion complex on the PEM-coated substrates. The substrate was then placed into an ELD bath where copper selectively plated only at the catalyzed regions. Their system which involves PEMs as the stable adhesion layer is more versatile, economical and works over a larger range of substrates than previous approaches. The combination of PEMs and μCP allows the control of 3D features on the micron and submicron scale. Using the process it was possible to create stable and selective copper patterns with nanometer dimensions on flexible substrates, which can result in lower fabrication costs to produce flexible display electronic circuits, sensors, RFID transponders, and other nano- or microelectronic devices.

**Figure 41.5** Reflected light optical micrographs of selective copper lines on PEM coated substrates. Parts (a) and (b) have glass substrates while (c) is on a polystyrene substrate. (d) Transmitted light optical micrograph of polystyrene particles deposited on the active unpatterned regions of the PEM surface next to the black copper lines. (e) A PEM-coated flexible polyester transparency film substrate with electroless copper patterns which are 30 nm thick. Reproduced with permission from [28].

For a similar purpose, Lee and coworkers demonstrated a simple method of creating patterned conductive multilayered polymer/graphene nanocomposite films by using the LBL assembly of graphene and the intact pattern transfer (IPT) [7, 8] of these films to a substrate [9]. The graphene was coated with a negatively charged polymer to form a stable aqueous solution. The solution was used for electrostatic LBL assembly, with a positively charged polyelectrolyte as the counter ion, onto the surface of an uncharged hydrophobic elastomeric stamp. Once the film was formed, it was placed in direct contact with a substrate of the opposite charge to directly transfer the multilayer film. If enough layers of graphene were adsorbed onto the stamp, the LBL film became conductive. Before LBL assembly, the elastomeric stamp is coated with a layer of polyelectrolyte using relatively weak hydrophobic interactions between the stamp and film. When the stamp is removed from the substrate, the strong electrostatic interactions between the oppositely charged films on the stamp and substrate hold the multilayer film on the substrate surface. The process is illustrated in Figure 41.6.

Lee and coworkers have built a versatile new bench-top method to form bioelectronic interfaces by combining LbL deposition of polyelectrolytes, electroless metal deposition, and directed molecular self-assembly on non-conductive substrates such as polystyrene and glass [25]. This method involves colloidal gold deposition on the LbL deposited polyelectrolytes on a flexible substrate to form a high performance bioelectronic interface containing dehydrogenase enzymes. They also presented

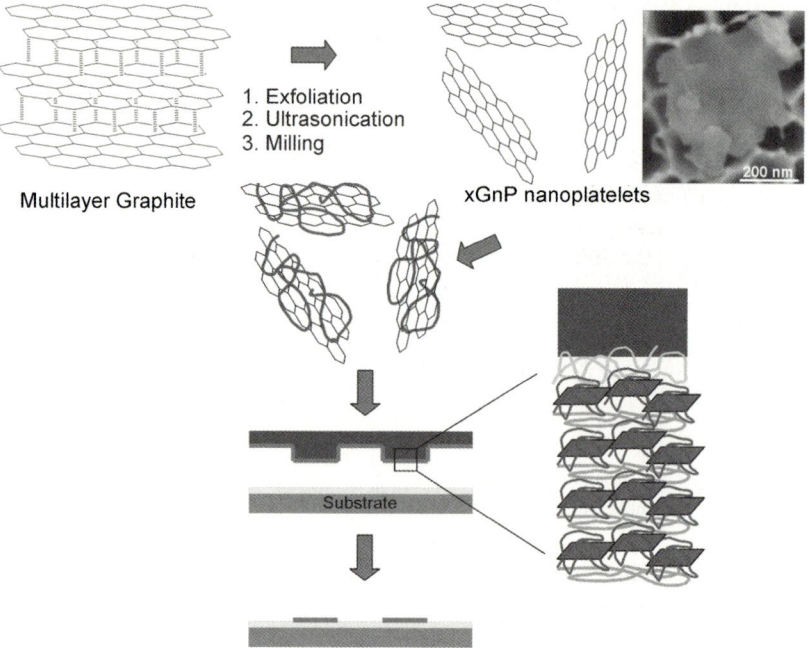

**Figure 41.6** Illustration of the process to form graphenes, Subsequent film formation on poly(dimethylsiloxane) and transfer to a PEM-coated substrate. Reproduced with permission from [9].

the fabrication of a novel biomimetic interface consisting of a conductive, electrolessly deposited gold film overlaid with a tethered BLM. Self-assembly of colloidal gold particles was used to create an electrolessly deposited gold film on a glass slide [14].

Lee and coworkers have utilized selective electroless metal plating techniques on a variety of arrayed or assembled particles on PEMs. Selective electroless nickel plating was demonstrated at the micron scale on 3D surface microstructures using a two-step method. The 3D patterned and functional surface microstructures consisted of microparticles, polyelectrolyte multilayers, and self-assembled monolayers, which were built by microcontact printing and directed self-assemblies of polyelectrolytes and particles [29]. They extended the work to control specular and diffuse reflection of light [30]. In the work, charged polystyrene colloidal particles, ranging in size from 100 nm to 5 μm, were adsorbed from solution onto oppositely charged PEMs. The monodisperse particle monolayers were coated with nickel in a two-step electroless plating process using palladium catalysts. The surfaces can be used as diffusive metal reflectors with a uniformly controlled surface roughness due to the uniform size of the deposited particles.

They also reported a step-edge-like methodology for the fabrication of polyelectrolyte supported nanowires [31]. Polyelectrolytes provided flexible support to metallic nanowires and prevented them from falling apart. An alumina membrane was functionalized with hydrophobic molecules and then broken to expose freshly cleaved hydrophilic edges along the broken pore walls. Then, PEMs were built on the hydrophilic edge of the pore membrane and an electroless nickel bath was used for the deposition of nickel onto the multilayers to form nickel nanowires. After dissolving the membrane, free standing nickel nanowires were obtained. They studied the effects of catalyst introduction methods using poly(amidoamine) (PAMAM) dendrimers on the nickel patterning of PEM-coated substrates [12]. PEM films were fabricated and used as platforms for electroless nickel patterning. μCP was used to pattern the palladium catalyst on the PEM platforms. After applying the catalyst, the samples were placed into an electroless bath. In the electroless bath, the initial nickel plating rate, nickel morphology, and nickel pattern selectivity (i.e., relative amount of metal deposition in desired places versus undesired places) were all affected by the method of catalyst introduction. Also, the number of PEM layer pairs required to remove the substrate effect on nickel patterning was investigated.

## 41.5
### Ordered and Disordered Particles on PEMs

The precise positioning of particles in 2D/3D structures can be crucial for optoelectronic devices, photonic bandgap materials, and biochip devices and sensor applications. For the fabrication of 2D photonic band gap materials, Lee and coworkers demonstrated how to obtain 2D arrays of single particles, and groups of particles on polyelectrolyte surfaces using patterned LbL thin films as functional templates, in which the size of the template is as small as that of the colloid [32]. The precise control of the number of colloids on each isolated patterned polyelectrolyte region was discussed. PEM film was used to increase adhesion between the particle and the surface template; surface wettability and adhesion characteristics of the multilayer were tuned by varying the adsorption conditions. Such selectively adsorbed particle arrays on PEMs were also selectively coated with metal such as nickel using a two-step electroless plating technique (Figure 41.7) [29]. Particle arrays on surfaces offer great potential in optical, electrical, and magnetic sensors and devices, and in photonic and bio-chip applications. The preparation of these metal particle arrays and a means of selectively plating different surface components of the arrays were discussed. On the other hand, they presented a unique random particle deposition on PEM films [33]. Monolayers of charged polystyrene latex particles ranging in size from 100 nm to 10 μm were deposited on oppositely charged polyelectrolyte multilayers (PEMs) by electrostatic interactions and capillary forces. Ultrathin PEMs formed on a glass slide provided an excellent underlying adhesive layer. As the sample surface was being dried, strong capillary forces between particles resulted in a unique pattern of particle

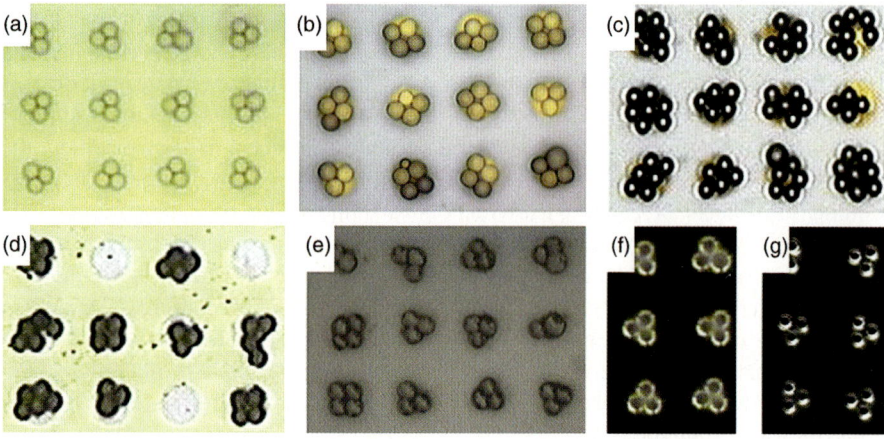

**Figure 41.7** Demonstration of various two-step selective nickel-platings of carboxylated polystyrene bead ($D = 4.3$ μm) arrays deposited on patterned polyelectrolyte multilayer templates, (PDAC/SPS)$_{10.5}$. Catalyst 1 ([Pd(NH$_3$)$_4$]Cl$_2$) was used in (b), (c), and (g), and catalyst 2 (Na$_2$[PdCl$_4$]) was used in (d) and (e). A patterned PS bead array on patterned polyelectrolyte templates before nickel plating. (b) Nickel plating on both the carboxylated PS beads and the EG-SAM area (outside circle) was partially made. (c) Nickel plating was completed, and incomplete particle array shows the uncoated multilayer template (yellow). (d) After several dehydration steps of the EG-SAM area before step 1 (dipping samples into 0.5 M NaOH solution followed by air-drying), the EG-SAM area became inert toward the nickel deposition due to the dehydration of EG functional groups. The incomplete particle array was chosen to show that nickel was also plated on the multilayer platform (blue). (e) Nonselective nickel-plating was made using catalyst 2. (f) Dark-field image of uncoated particle array of sample (a). (g) Shows the dark-field image of a nickel-coated array which reflects the incident light. Reproduced with permission from [29].

monolayers (i.e., 2D particle aggregates). The resulting topographical structure of the coatings strongly influenced the transmission of visible light through the slides. The total and specular transmittances showed three different characteristics as a function of particle size: (i) anti-reflection, when the particle diameter is around a quarter of the wavelength of the incident visible light, (ii) diffraction, when the particle diameter is equivalent to the wavelength of the incident beam, and (iii) diffusive scattering when the particle diameter is bigger than the wavelength of the incident beam. Additionally, for the first time, monolayer coverage and fractal-dimension analyses have been reported over a wide range of particle sizes. This work was extended to control the specular and diffuse reflection after metal coating on the deposited particle monolayers, instead of transmission of light [30]. Lee and coworkers also reported a novel method to produce polymer supported nickel nanowires using a step-edge like fabrication technique along with anodized alumina membranes [31]. The prepared nanowires are asymmetric in shape and function. First the membrane was treated with fluorosilanes to make the surface completely hydrophobic. It was then cleaved to expose the freshly prepared alumina edges, which are hydrophilic. These freshly cleaved hydrophilic edges were used as templates for the deposition of polyelectrolyte multilayers and nickel. They selectively deposited polyelectrolyte multilayers on those hydrophilic edges. Then the electroless deposition of nickel was used to create nickel nanowires. After dissolving the membranes, nanostructured particles were obtained.

Lee and coworkers have prepared a variety of nanocomplex particles with PEI and small interfering RNA (siRNA) in the size range 50–250 nm for gene delivery [34]. They described the application of a LbL-assembled degradable multilayer film for patterned delivery of siRNA using a forward transfection approach. The transfection process involved the following steps. (i) pH controlled, biocompatible, and degradable multilayers were fabricated using LbL assembly under acidic conditions. (ii) Nanoparticles of vector–siRNA complexes prepared at physiological pH conditions were used as "ink" in μCP to form patterns on multilayer substrates. (iii) Multilayer substrate containing patterned nanoparticles laid on top of cells at physiological pH conditions degraded the multilayer and formed patterns of transfected cells. The method is a variation of the forward transfection technique; denoted as the multilayer mediated forward transfection (MFT) method. Quantification of MFT efficiencies with linear polyethylenimine (LPEI)-siRNA nanoparticles found nitrogen/phosphate (N/P) ratios of 30 gave significant transfection (60%). MFT of patterned siRNA provides an efficient and simple approach to spatially controlled siRNA delivery for tissue engineering applications.

## 41.6
### Mechanical Aspects of PEM Films and Degradable Films

The morphology and mechanical properties of PEMs have recently become of interest. For example, the natural folding and wrinkling of materials is a ubiquitous

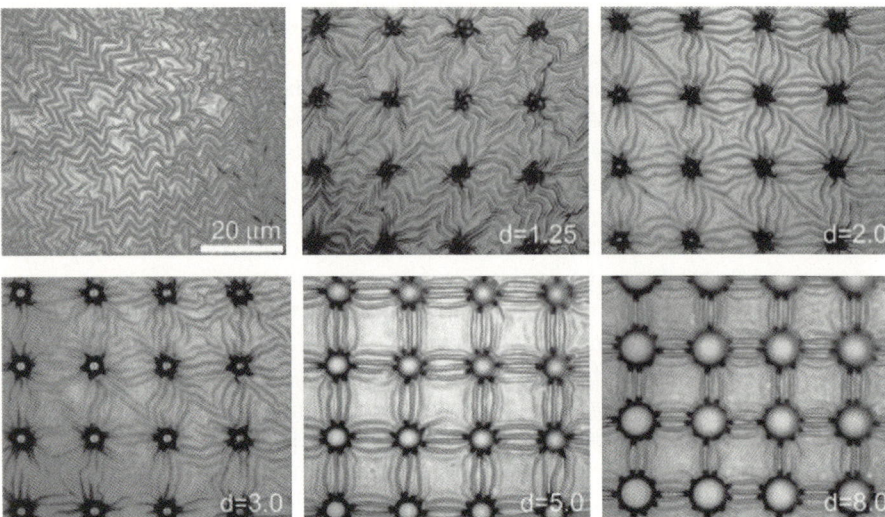

**Figure 41.8** Optical microscope images of a (PAH/PAA)$_{5.5}$ film on a topographically patterned PDMS substrate after thermal processing. The column diameter determines whether the buckled film is ordered ($d > \lambda$) or randomly oriented ($d < \lambda$). Reproduced with permission from [35].

occurrence observed in many facets of daily life. Wrinkling is observed in nature during the formation of mountain ranges, in smiles or frowns on a human face, the drying of fruits or vegetables, and the heating or cooling of thin films. Lee and Hendricks reported methods to create, spatially control, and prevent permanently buckled PEM films on flat PDMS substrates (Figure 41.8) [35, 36]. The wrinkling is caused by the release of compressive stress created by heating the substrate and allowing it to cool. This thermal cycling can be an irreversible process creating internal stress which induces a permanently buckled morphology. Control over the film buckling morphology was demonstrated by varying the film thickness and the surface topography of the substrate. Additionally, the effects of plasma treatment and the critical temperature for permanent buckling have been studied. With the knowledge they obtained, the prevention of permanent buckling by adding nanoparticles to the films was demonstrated for the first time. Lee and coworkers showed that the self-assembled particles at the polymer and metal interface deflected the internal stresses that build up at the interface while the metal is being deposited [30]. This allowed a thicker metal film to be deposited before delamination occurred.

In contrast, LbL films can be weakly built to be degradable under specific conditions. These degradable PEM films co-deposited with drug molecules like proteins can be used for gene or drug delivery. Lee and coworkers presented a simple approach for controlled delivery of proteins from agarose gels, where the proteins are incorporated within the degradable LbL multilayer coatings formed over the agarose [27]. They also presented a method to deliver patterns of siRNA that capitalize on a forward transfection method (transfection by introducing siRNA transfection reagent complexes onto plated cells); herein denoted as multilayer mediated forward transfection (MFT) [34]. This method separates the substrate-mediated delivery from the cell adhesive properties of the surface. pH responsive layer-by-layer (LbL) assembled multilayers were used as the delivery platform and the microcontact printing technique was used to pattern nanoparticles of transfection reagent–siRNA complexes onto degradable multilayers.

## Acknowledgment

Financial support from the National Science Foundation (0609164, 0832730, and 0928835), the University Research Corridor, the Michigan Initiative for Innovation & Entrepreneurship, and the MSU Foundation are greatly appreciated.

## References

1 Decher, G. and Hong, J.D. (1991) Buildup of ultrathin multilayer films by a self-assembly process.1. Consecutive adsorption of anionic and cationic bipolar amphiphiles on charged surfaces. *Makromol. Symposia*, **46**, 321–327.

2 Decher, G. and Hong, J.D. (1991) Buildup of ultrathin multilayer films by a self-assembly process.2. Consecutive adsorption of anionic and cationic bipolar amphiphiles and polyelectrolytes on charged surfaces. *Ber. Bunsen-Ges. Phys. Chem.*, **95**, 1430–1434.

3 Kumar, A., Biebuyck, H.A., and Whitesides, G.M. (1994) Patterning self-assembled monolayers - applications in materials science. *Langmuir*, **10**, 1498–1511.

4 Wilbur, J.L., Kumar, A., Biebuyck, H.A., Kim, E., and Whitesides, G.M. (1996) Microcontact printing of self-assembled monolayers: Applications in microfabrication. *Nanotechnology*, **7**, 452–457.

5 Kidambi, S., Chan, C., and Lee, I.S. (2004) Selective depositions on polyelectrolyte multilayers: Self-assembled monolayers of m-dPEG acid as molecular template. *J. Am. Chem. Soc.*, **126**, 4697–4703.

6 Kidambi, S., Chan, C., and Lee, I. (2008) Tunable resistive m-dPEG acid patterns on polyelectrolyte, multilayers at physiological conditions: Template for directed deposition of biomacromolecules. *Langmuir*, **24**, 224–230.

7 Kohli, N. et al. (2007) Nanostructured biosensor for measuring neuropathy target esterase activity. *Anal. Chem.*, **79**, 5196–5203.

8 Kohli, N., Worden, R.M., and Lee, I. (2005) Intact transfer of layered, bionanocomposite arrays by microcontact printing. *Chem. Commun.*, 316–318.

9 Hendricks, T.R., Lu, J., Drzal, L.T., and Lee, I. (2008) Intact pattern transfer of conductive exfoliated graphite nanoplatelet composite films to polyelectrolyte multilayer platforms. *Adv. Mater.*, **20**, 2008–2012.

10 Mehrotra, S., Lee, I., Liu, C., and Chan, C. (2011) Polyelectrolyte multilayer stamping in aqueous phase and non-contact mode. *Ind. Eng. Chem. Res.*, 8851–8858.

11 Kohli, N., Dvornic, P.R., Kaganove, S.N., Worden, R.M., and Lee, I. (2004) Nanostructured crosslinkable micropatterns by amphiphilic dendrimer stamping. *Macromol. Rapid. Commun.*, **25**, 935–941.

12 Hendricks, T.R., Dams, E.E., Wensing, S.T., and Lee, I. (2007) Effects of

catalyst introduction methods using PAMAM dendrimers on selective electroless nickel deposition on polyelectrolyte multilayers. *Langmuir*, **23**, 7404–7410.

13 Kohli, N., Vaidya, S., Ofoli, R.Y., Worden, R.M., and Lee, I. (2006) Arrays of lipid bilayers and liposomes on patterned polyelectrolyte templates. *J. Colloid Interface Sci.*, **301**, 461–469.

14 Kohli, N. et al. (2006) Tethered lipid bilayers on electrolessly deposited gold for bioelectronic applications. *Biomacromolecules*, **7**, 3327–3335.

15 Lee, I., Ahn, J.S., Hendricks, T.R., Rubner, M.F., and Hammond, P.T. (2004) Patterned and controlled polyelectrolyte fractal growth and aggregations. *Langmuir*, **20**, 2478–2483.

16 Kidambi, S., Lee, I., and Chan, C. (2004) Controlling primary hepatocyte adhesion and spreading on protein-free polyelectrolyte multilayer films. *J. Am. Chem. Soc.*, **126**, 16286–16287.

17 Kidambi, S. et al. (2007) Patterned co-culture of primary hepatocytes and fibroblasts using polyelectrolyte multilayer templates. *Macromol. Biosci.*, **7**, 344–353.

18 Kidambi, S. et al. (2007) Cell adhesion on polyelectrolyte multilayer coated polydimethylsiloxane surfaces with varying topographies. *Tissue Eng.*, **13**, 2105–2117.

19 Kidambi, S., Lee, I., and Chan, C. (2008) Primary neuron/astrocyte co-culture on polyelectrolyte multilayer films: A template for studying astrocyte-mediated oxidative stress in neurons. *Adv. Func. Mater.*, **18**, 294–301.

20 Mehrotra, S. et al. (2010) Cell adhesive behavior on thin polyelectrolyte multilayers: Cells attempt to achieve homeostasis of its adhesion energy. *Langmuir*, **26**, 12794–12802.

21 Park, J. and Hammond, P.T. (2004) Multilayer transfer printing for polyelectrolyte multilayer patterning: Direct transfer of layer-by-layer assembled micropatterned thin films. *Adv. Mater.*, **16**, 520.

22 Kohli, N., Lee, I., Richardson, R.J., and Worden, R.M. (2010) Theoretical and experimental study of bi-enzyme electrodes with substrate recycling. *J. Electroanal. Chem.*, **641**, 104–110.

23 Kohli, N., Worden, R.M., and Lee, I. (2007) Direct transfer of preformed patterned bio-nanocomposite films on polyelectrolyte multilayer templates. *Macromol. Biosci.*, **7**, 789–797.

24 Hassler, B.L., Kohli, N., Zeikus, J.G., Lee, I., and Worden, R.M. (2007) Renewable dehydrogenase-based interfaces for bioelectronic applications. *Langmuir*, **23**, 7127–7133.

25 Hassler, B.L., Amundsen, T.J., Zeikus, J.G., Lee, I., and Worden, R.M. (2008) Versatile bioelectronic interfaces on flexible non-conductive substrates. *Biosens. Bioelectron.*, **23**, 1481–1487.

26 Stockton, W.B. and Rubner, M.F. (1997) Molecular-level processing of conjugated polymers.4. Layer-by-layer manipulation of polyaniline via hydrogen-bonding interactions. *Macromolecules*, **30**, 2717–2725.

27 Mehrotra, S. et al. (2010) Time controlled protein release from layer-by-layer assembled multilayer functionalized agarose hydrogels. *Adv. Func. Mater.*, **20**, 247–258.

28 Hendricks, T.R. and Lee, I. (2006) A versatile approach to selective and inexpensive copper patterns using polyelectrolyte multilayer coatings. *Thin Solid Films*, **515**, 2347–2352.

29 Lee, I.S., Hammond, P.T., and Rubner, M.F. (2003) Selective electroless nickel plating of particle arrays on polyelectrolyte multilayers. *Chem. Mater.*, **15**, 4583–4589.

30 Ahn, J.S., Hendricks, T.R., and Lee, I. (2007) Control of specular and diffuse reflection of light using particle self-assembly at the polymer and metal interface. *Adv. Func. Mater.*, **17**, 3619–3625.

31 Srivastava, D., Hendricks, T.R., and Lee, I. (2007) Step-edge like template fabrication of polyelectrolyte supported nickel nanowires. *Nanotechnology*, **18**, 245305–245310.

32 Lee, I., Zheng, H.P., Rubner, M.F., and Hammond, P.T. (2002) Controlled cluster size in patterned particle arrays via

directed adsorption on confined surfaces. *Adv. Mater.*, **14**, 572–577.

33 Ahn, J.S., Hammond, P.T., Rubner, M.F., and Lee, I. (2005) Self-assembled particle monolayers on polyelectrolyte multilayers: particle size effects on formation, structure, and optical properties. *Colloid. Surf. A*, **259**, 45–53.

34 Mehrotra, S., Lee, I., and Chan, C. (2009) Multilayer mediated forward and patterned siRNA transfection using linear-PEI at extended N/P ratios. *Acta Biomater.*, **5**, 1474–1488.

35 Hendricks, T.R. and Lee, I. (2007) Wrinkle-free nanomechanical film: Control and prevention of polymer film buckling. *Nano Lett.*, **7**, 372–379.

36 Hendricks, T.R., Wang, W., and Lee, I. (2010) Buckling in nanomechanical films. *Soft Matter*, **6**, 3701–3706.

# 42
# Electrochemically Active LbL Multilayer Films: From Biosensors to Nanocatalysts
*Ernesto. J. Calvo*

## 42.1
## Introduction

In this chapter we describe a particular case of layer-by-layer (LbL) multilayer films containing electrochemically or redox active components which can undergo exchange of electrons with an underlying conducting substrate, and ion and solvent exchange with the bathing electrolyte adjacent to the film.

In 1997 Laurent and Schlenoff [1] described the electrochemical response of viologen groups in poly(styrenesulfonate)/poly(butanylviologen) LbL thin films and Hodak *et al.* [2] showed the biosensor response to glucose with the enzyme glucose oxidase (GOx) wired by ferrocene-derivatized poly-(allylamine) LbL multilayers. Since then a large number of publications have described electrochemically active polyelectrolyte multilayers (r-PEMs) including osmium pyridyl-bipyridyl modified poly(allylamines) [3–15], poly(vinylpyridine) osmium complexes [16, 17], ferrocene poly(allylamines [2, 18–20], polyalkylviologens [21–24], sulfonated polyaniline [25], polythiophenes [26], polyoxo-metalates [27–29], enzymes [2, 9–11, 16, 24, 30–33] and redox proteins [34–38], ferrocene-containing poly(amidoamine) dendrimers [39–44], Prussian Blue [45–48], carbon nanotubes [49], metal nanoparticles [50], and so on. The topic has been reviewed elsewhere [51–53].

Electrochemically or redox active polyelectrolyte multilayers (r-PEMs) deposited on conductive substrates (electrodes) and immersed in electrolyte solutions are of particular interest since they have electron acceptors/donors that can exchange electrons with the underlying conductive substrate and propagate redox changes along the direction normal to the substrate. The charge imbalance due to the exchange of electrons between the electrode and the r-PEM results in the transfer of charge across the LbL layers and exchange of solvent and ions with the external electrolyte. In layered thin films the stratified structure results in characteristic behavior and the fuzzy nature of interpenetrated layers facilitates the transfer of charges along the film.

From a practical point of view, LbL redox polyelectrolyte multilayers find application as model systems in electrochemical sensors, biosensors, electrochromic devices, actuators, corrosion protection layers, antibacterial films, nanoreactors for

*Multilayer Thin Films: Sequential Assembly of Nanocomposite Materials*, Second Edition.
Edited by Gero Decher and Joseph B. Schlenoff.
© 2012 Wiley-VCH Verlag GmbH & Co. KGaA. Published 2012 by Wiley-VCH Verlag GmbH & Co. KGaA.

the formation of nanocatalysts by confining ions and thus preventing ionic diffusion that results in large crystals, and so on.

Some fundamental interest in redox LbL multilayers on conductive substrates is given by the stratified structure yet fuzzy interpenetrating layers, and the unique ability to change the nature and charge of the topmost layer in multilayer films. For instance, the polyion–polyion intrinsic charge compensation can be broken by ion exchange induced by the electrochenical oxidation and reduction forming extrinsic polyion–counter ion pairing, which will be discussed in the treatment of the dynamics of these systems.

The possibility to vary the fixed charges in the PEMs by application of an external electrical potential across the multilayer results in unique and interesting features, both for the equilibrium electrochemical response and the dynamics of these systems.

The apparent redox potential of redox couples entrapped (either by charge or covalent binding to the polyelectrolyte backbone) in a charged matrix depends strongly on the concentration and pH of the external electrolyte, in terms of Donnan or membrane potential with respect to the reference electrode in the external solution; the effect of solution pH on the apparent redox potential for pH-independent redox couples is also a consequence of charges in ionizable groups at the polyelectrolyte backbone [21].

During electrochemically induced changes of redox charge in r-PEMs, the exchange of ion and solvent between the polyelectrolyte film and the external electrolyte is observed by different experiments. The exchange of solvent and ions results in swelling/deswelling and viscoelastic changes of the polymer film by uptake and release of the solvent, with mechanical work terms contributing to the free energy of the system.

Charge transfer at the electrode/r-PEM interface results in diffusion-like propagation of charge by electron hopping of electrons between neighboring redox sites and segmental motion, with operational diffusion coefficients very dependent on the concentration of redox sites, film structure, pH and the nature of the charge in the topmost layer. Thus, the existence of a mixed mechanism of electron hopping due to polymer segmental motion and physical diffusion of counter ions to compensate charge should be considered [34].

## 42.2
### Electrochemical Response

The most commonly used tool to investigate the electrochemical behavior of redox active films on conducting surfaces is cyclic voltammetry (CV) [54]. In this technique the electrode potential with respect to a suitable reference electrode in the electrolyte solution, is linearly scanned between initial and final values while recording the current that circulates in the external circuit.

Figure 42.1 shows typical cyclic voltammograms of 0.5 mM $Fe(CN)_6^{-4}$ in 0.1 M $Na_2SO_4$ at a bare Au electrode, and at a thin $PEI-(PGA-PAH)_5$ self-assembled multilayer [55].

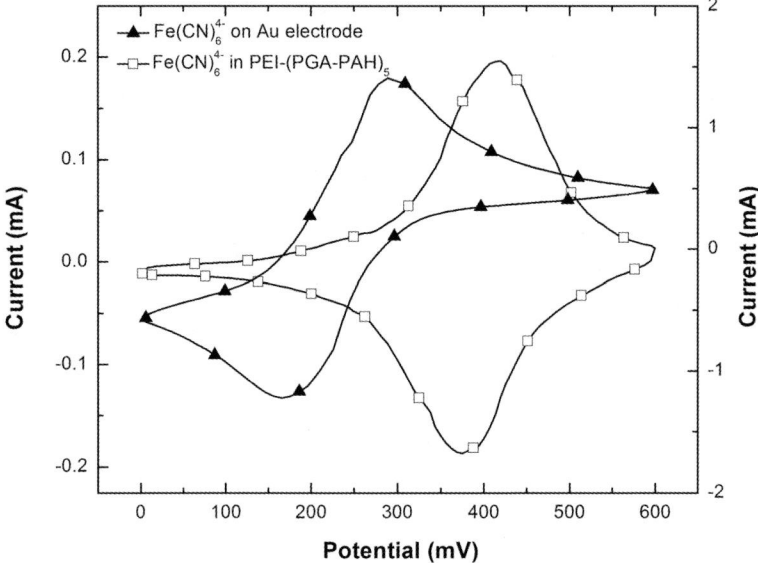

**Figure 42.1** CV of Fe(CN)$_6^{4-}$/Fe(CN)$_6^{3-}$ recorded at a scan rate of 50 mV s$^{-1}$ on a gold electrode via EQCM-D. A clear shift in apparent redox potential is visible between the voltammogram of free ferrocyanide (triangles) and ferrocyanide containing PEI-(PGA-PAH)5 film (squares). Taken from ref. [55].

The main features of the electrochemical response are (i) the electrode potential peak position and its dependence on electrolyte composition, (ii) anodic and cathodic peak separation, (iii) the peak full width at half height (FWHH), and (iv) the peak current dependence on potential sweep rate. After double layer charging correction, the integrated current under the curve yields the redox charge exchanged with the electrode, and from this the moles of redox groups per specific area can be calculated by using Faraday's law.

Figures 42.2 and 42.3 depict the CV for the enzyme GOx and osmium poly-allylamine, PAH-Os/(GOx) self-assembled multilayer [11], and PAH-Os/PVS multilayer [5], respectively, with almost symmetrical oxidation reduction peaks and 90 mV peak width at half height. For fast electron transfer and low sweep rate, when the timescale of the experiment is shorter than the characteristic timescale for charge transport in the film ($l^2/D_{app}$, where $l$ is the film thickness and $D_{app}$ the diffusion coefficient), all redox sites within the film are accessible for oxidation or reduction and thus can be regarded as in electrochemical equilibrium.

We describe this thin film voltammetry as ideal or Nernstian behavior and the following analytical expression for the current–potential curve describes the experimental data:

$$i = i_{dl} + \frac{F^2 v \Gamma_{Os}}{RT} \frac{\exp\left[\left(\frac{F}{RT}\right)(E-E^{o\prime})\right]}{\left[1+\exp\left(\frac{F}{RT}\right)(E-E^{o\prime})\right]^2} \qquad (42.1)$$

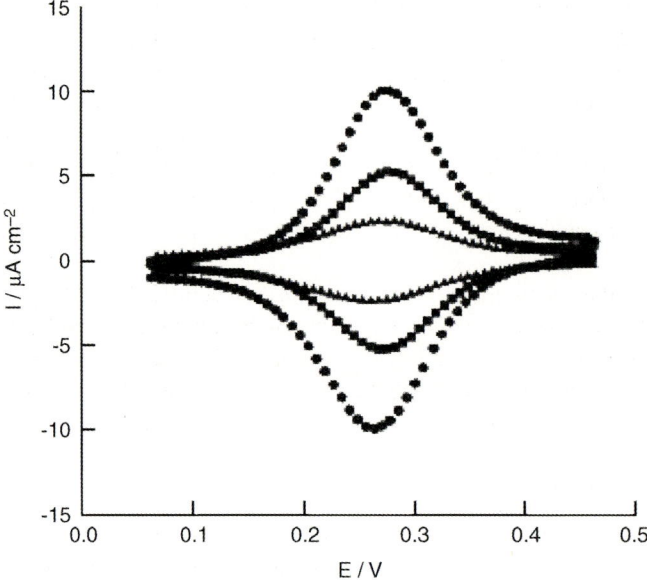

**Figure 42.2** CV curves of (PAH-Os)$_n$(GOx)$_n$ for different numbers of layer pairs with $n = 2, 4, 6$ in 0.1 M KNO$_3$, TRIS buffer of pH 7 at 50 mV s$^{-1}$. Taken from Ref. [11].

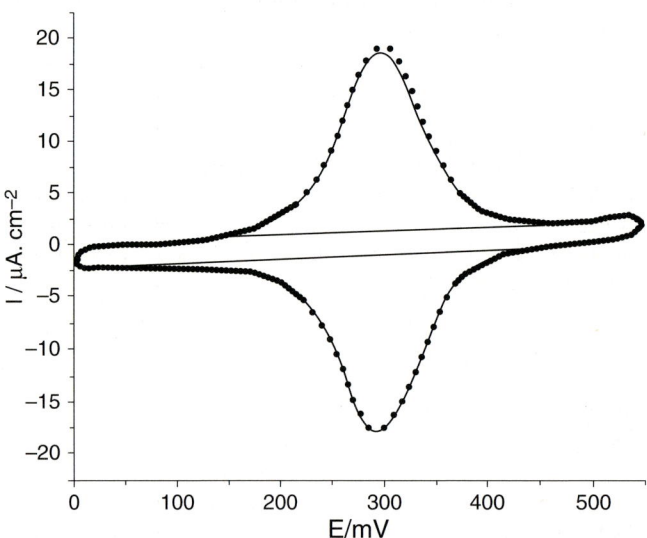

**Figure 42.3** CV of (PAH-Os)$_5$(PVS)$_4$ in 0.2 M K$_{NO_3}$, 20 mM TRIS buffer of pH 7.4 (dots) and best fit CV curve for ideal reversible thin layer voltammetry (equation 42.1) (solid line). Taken from Ref. [5].

where $\Gamma_{Os}$ is the total osmium surface coverage, $v\,(=dE/dt)$ is the scan rate and $i_{dl}$ is the double layer charging current under these conditions, anodic and cathodic current–potential waves are mirror images (zero peak splitting) and the current is proportional to the scan rate [54].

As shown in Figure 42.1, due to the presence of interactions in the film, the apparent redox potential of a redox couple inside a polyelectrolyte film may differ from the standard formal redox potential of the redox couple in solution [54], since the free energy required to oxidize a mole of redox sites within the film differs from that needed in solution.

In particular, when the interactions are due to the presence of fixed charged groups in the polymer backbone, a potential difference between this phase and the solution (the reference electrode is in the electrolyte), known as the Donnan or membrane potential, contributes to the apparent potential of the redox couple and contributes to the peak position [4, 55–58]:

$$E_{app}^{1/2} = E^0 + \Delta\phi_{DONNAN} \tag{42.2}$$

where $E_{app}^{1/2}$ is the apparent redox potential measured by cyclic voltammetry, $E^0$ is the standard redox potential of the redox couple and $\Delta\phi_D$ is the Donnan potential given by:

$$\Delta\phi_D = \frac{RT}{F}\ln\left[\frac{C_F + (C_F^2 + 4C_S^2)^{1/2}}{2C_S}\right] \tag{42.3}$$

and strongly dependent on salt concentration, $C_S$, in the electrolyte, and protonation-dependent charge concentration in the polyelectrolyte backbone, $C_F$ [58]. This potential difference results when the concentration of immobile charges inside the film is larger than the electrolyte concentration, salt ions of the same charge as the immobile groups are excluded from the film, while those of equal charge are incorporated to keep electroneutrality. The film is said to be permselective to these ions. The Donnan potential can be regarded as the electrostatic potential difference required to maintain this imbalance in the concentration of mobile ions. In Equation 42.3 $C_F > 0$ implies positive Donnan potentials and anion permselectivity, whereas $C_F < 0$ yields negative Donnan potentials and cation permselectivity. At the other limit, when the electrolyte concentration is larger than the concentration of immobile charges in the film, the concentration of mobile ions is similar in the film and the solution and thus the Donnan potential vanishes (Donnan breakdown).

From Equations 42.2 and 42.3 we obtain:

$$E_{app}^{1/2} = E^0 + \frac{RT}{F}\ln\left[\frac{C_F + (C_F^2 + 4C_S^2)^{1/2}}{2C_S}\right] \tag{42.4}$$

from which the concentration and sign of the immobile charges inside the film can be determined. Figure 42.4 shows CV curves of a PAH-Os modified electrode at 25 mV s$^{-1}$ with 4 mM and 1.2 M salt concentrations, respectively. A peak shift due to the presence of an interfacial potential is seen.

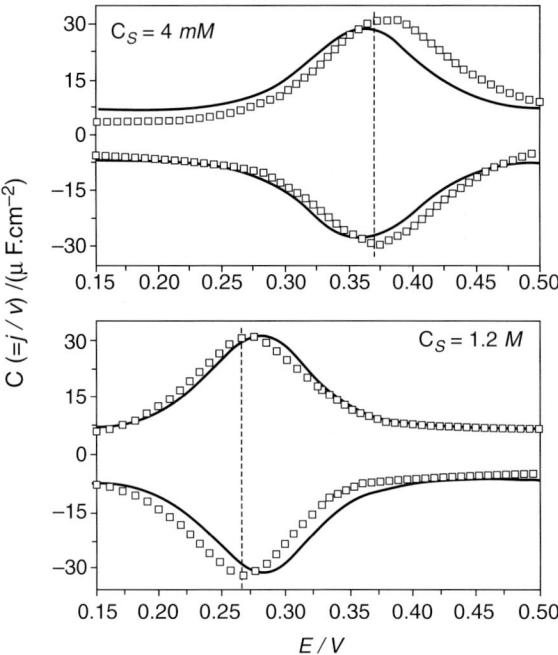

**Figure 42.4** Comparison of experimental (squares, $v = 0.025\,\text{V}\,\text{s}^{-1}$) and theoretical (solid lines) current–potential plots measured for an Au/MPS/PAH-Os electrode in solutions of different ionic strength and 1 mM $HNO_3$ (pH 3). Taken from Ref. [63].

Tagliazucchi et al. have recently shown that, due to the possibility to protonate/deprotonate both positive and negative weak polyions in redox PEMs, the polarity of net immobile charge can be controlled at will by the choice of the outermost layer and the relationship between assembly and testing solution pH [58]. Figure 42.5 shows four Donnan graphs for films self-assembled in solutions of pH 8.3, 7.3, 5.5 and 3.5 and measured in a solution of pH 7 so that protonation and deprotonation of the films occurred in the measuring electrolyte. The concentration of fixed charge in the film, $C_F$, calculated from the best fit of the experimental results to Equation 42.4, increases monotonically with the increase in assembly pH, since protonation in solutions of pH 7 increases, and thus films assembled at pH 8.3 are always anion exchangers, while films self-assembled at pH 3.5 are always cation exchangers. For films self-assembled and measured at the same electrolyte pH no change in polymer backbone protonation is expected, and the nature of the topmost layer charge (positive or negative) determines the ion permselectivity (anion or cation, respectively). Therefore, there are two contributions to the excess of fixed charges in the film which arise, respectively, from the topmost layer charge in contact with the liquid electrolyte and the protonation equilibrium of ionizable groups, that is, $NH_3^+$ and $COO^-$ in weak polyelectrolytes.

It should be noticed that there are several approximations in Equation 42.4: (i) the distribution of non-balanced immobile charges in a PEM is inhomogeneous and

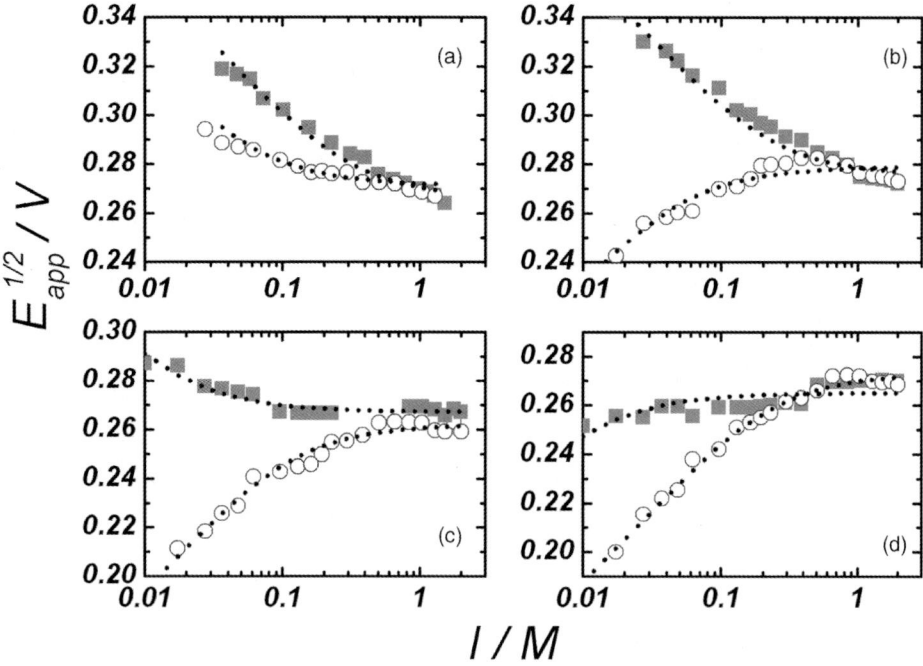

**Figure 42.5** Apparent formal redox potential for (■) (PAH-Os)$_4$(PVS)$_4$PAHOs (positively charged capping layer) and (○) (PAH-Os)$_5$(PVS)$_5$ (negatively charged capping layer) modified electrodes self-assembled from PAH-Os/PVS solutions of different pH, (a) 8.3, (b) 7.3, (c) 5.5, and (d) 3.5, and measured in pH 7.3 solutions of different KNO$_3$ concentration, C$_S$. Dotted lines are best fits with Equation 42.4. Taken from Ref. [58].

mainly located in the film/electrolyte interface, thus $C_F$ should be regarded as an effective averaged charge concentration; (ii) it is assumed that for pH-dependent groups the degree of ionization does not depend on the electrolyte concentration; and (iii) during film oxidation and reduction redox charges are created and destroyed in the film, respectively, but these are neglected in $C_F$.

The Donnan partition model assumes two well-defined phases (polyelectrolyte film and electrolyte solution) with a sharp electrostatic potential drop between them. This approximation is valid when the characteristic Debye length is much shorter than the film thickness. Otherwise, when most charges are located close to the outermost plane and the Debye length is larger than the film thickness, the potential distribution is given by the surface potential model:

$$\Delta\phi_S = \frac{2RT}{F} \text{arcsin} \, h \left( \frac{\sigma_M}{[8c_s \varepsilon_r \varepsilon_0 kT]^{1/2}} \right) \quad (42.5)$$

where $\Delta\phi_S$ is the surface–solution potential difference and $\sigma_M$ is the charge density of the surface. For film thicknesses comparable to the Debye length numerical solutions should be used [59].

A molecular theory of chemically modified electrodes has recently been developed and applied to the study of redox polyelectroyte modified electrodes [60–63]. The molecular approach explicitly includes the size, shape, charge distribution, and conformations of all of the molecular species in the system, as well as the chemical equilibria (redox and acid–base) and intermolecular interactions. For the osmium pyridine–bipyridine complex covalently bound to poly(allylamine) backbone (PAH-Os) adsorbed onto mercapto-propane sulfonate (MPS) thiolated gold electrode the potential and electrolyte composition dependent redox and non-redox capacitance have been calculated in very good agreement with voltammetric experiments under reversible conditions without the use of a freely adjustable parameter, as shown in Figure 42.4 for two electrolyte concentrations [61, 63].

The new molecular theory is based on the minimization of the free energy functional of the system, and thus includes the different existing models employed to describe the electrochemical response of chemically modified electrodes: Laviron redox pseudo-capacitance [64, 65], interaction broadening or narrowing of peaks [66–68], peak position due to Donnan equilibrium [56], and so on. Unlike existing phenomenological models, the theory links the electrochemical behavior with the structure of the polymer layer and predicts a highly inhomogeneous distribution of acid–base and redox states that strongly couple with the spatial arrangement of the molecular species in the nanostructured redox film.

Figure 42.5 depicts a comparison of experimental results and the prediction of the molecular theory for the dependence of peak potential on electrolyte concentration. [61] The experimental data can also be fitted to Equations 42.4 or 42.5 to extract the adjustable parameters, and to compare to the predictions of the molecular theory to see whether these limiting cases are applicable. We can compare, for example, the best fitted value of $\Delta\phi_S$ ($3.02\,\mu\mathrm{C\,cm^{-2}}$) in Equation 42.5 with that obtained by summing the charge density of the negatively charged MPS and the positively charged PAH-Os. This calculation for a fully ionized thiol layer results in a surface charge ($-7.14\,\mu\mathrm{C\,cm^{-2}}$) that does not correspond to the fitted value. However, if we consider the protonation of sulfonate groups predicted by the theory, which is a consequence of strong charge regulation [61], then the fitted value agrees with the expected one ($3.18\,\mu\mathrm{C\,cm^{-2}}$).

Charge regulation also results in the strong coupling of redox and acid–base equilibria in the redox film [62], as shown in Figure 42.6 for the experimental redox peak positions versus electrolyte pH, which are compared to the prediction of the molecular theory. There are two regions to be considered: for pH < 10 the positive charge in the film (protonated amines and osmium complexes) remains constant as the peak potential. Between 10 < pH < 12 amine deprotonation on increasing the pH leads to a decrease in peak potential, and at pH > 12 all the amines become neutral and the peak position should be pH-independent again.

Besides the dependence of the redox couple apparent redox potential on solution pH for a pH-independent redox couple, such as the osmium pyridine-bypyridine complex, the theory also predicts amine deprotonation during film oxidation as a response to the creation of positive Os(III) sites which increment the charge in the film [62]. The maximum effect is observed at high pH and low salt concentration.

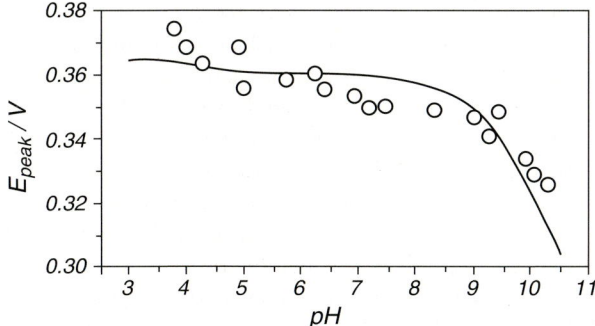

**Figure 42.6** Comparison between theoretical (solid line) and experimental ($v = 0.025$ V s$^{-1}$, circles) voltammetric peak potential for electrolyte solutions of different pH (prepared from 1 mM NaOH and HNO$_3$ solutions containing 4 mM NaNO$_3$). Taken from ref. [58].

For a one-electron Nernstian behavior Equation 42.1 predicts a peak full width at half height (FWHH) of 90.6 mV. However, most experimental data show a FWHH which differs from that value. Brown and Anson [66] and Albery et al. [67] have explained the departure from ideality by the presence of lateral interactions between oxidized and reduced sites, so that the free energy for oxidation depends on the fraction of oxidized sites on the surface:

$$E = E^{0\prime} + \frac{RT}{F}\ln\left(\frac{f_{\mathrm{Ox}}}{1-f_{\mathrm{Ox}}}\right) - \frac{RT}{F}\Gamma f_{\mathrm{Ox}} r \qquad (42.6)$$

where $E^{0\prime}$ is the standard redox potential, $f_{\mathrm{Ox}}$ is the fraction of oxidized sites, $\Gamma_{\mathrm{Os}}$ is the total redox site coverage and $r$ is an interaction parameter (equal to $r_{\mathrm{O}} + r_{\mathrm{R}}$ in Anson's original work) [66]. The apparent redox potential ($E^{0\prime} - RTF^{-1}\Gamma f_{\mathrm{Ox}} r$) depends on the electrode potential. Albery's model [67], on the other hand, considers that not all the redox sites in the film have the same redox potential, but a Gaussian distribution of apparent redox potential. Both models rely on a fitting parameter: $r$ in the case of Brown and Anson and the width of the Gaussian distribution in Albery's model. Other models, such as that of Smith and White, predict peak broadening due to interfacial potential distribution in monolayer-coated electrodes [69]. Redepenning's model takes into account the effect of charges created during a potential scan that contribute to the Donnan potential [70].

The molecular theory described predicts FWHH for a single PAH-Os layer on an electrode without the need for any adjustable fitting parameters [61]. It has been shown that the models of Brown and Anson, and of Albery to describe interactions are limiting cases of the molecular theory prediction, and that the fitting parameters in these models can be obtained from the molecular theory. For instance, at low ionic strength the average Os(III)/Os(II) apparent formal potential in the film, defined as [63]:

$$E^0_{\mathrm{app}} = E - \frac{RT}{F}\ln\left(\frac{\langle f_{\mathrm{Ox}}\rangle}{1-\langle f_{\mathrm{Ox}}\rangle}\right) \qquad (42.7)$$

where $<f_{Ox}>$ is the z-averaged oxidation fraction at an electrode potential $E$, as predicted by the molecular theory, depends on the osmium sites oxidation fraction in the film, in line with the peak broadening based on lateral interactions, while at high ionic strength it is almost constant for all redox concentrations, as expected for an ideal system [71].

## 42.3
## Dynamics of Charge Exchange

### 42.3.1
### Propagation of Redox Charge (Electron Hopping)

Both the initial work of Laurent and Schlenoff [1] with poly(butanylviologen) and poly(styrensulfonate) and the early work of Hodak et al. with poly(allylamine) ferrocene and glucose oxidase [2] have shown by CV nearly ideal surface waves of viologen and ferrocene, respectively, with charge independent of sweep rate, and the redox surface concentration was obtained from integration of the voltammetric peaks. One of the interesting features of electrochemically active LbL films is that the redox charge increases in step with the number of layers deposited. The amperometric enzyme electrode response also increases with the number of layers, since glucose oxidase could be wired by ferrocinium propagating charge in the multilayer [2].

In the seminal work of Laurent and Schlenoff [1], multilayers containing electrochemically active viologen units were constructed using the LbL deposition technique with poly(styrenesulfonate) and poly(butanylviologen) as polyanion and polycation, respectively. Varying the distance between the redox active layer and the electrode by intercalating PAH/PSS redox inert layer pairs, at least four inert layer pairs were needed to offset the electrochemical response due to interpenetration of the layers. Therefore, redox active material in electrochemically active LbL multilayers can be addressed through the film via electron hopping between neighboring redox sites.

Direct proof of diffusion type behavior for the propagation of charge has been provided by Wolosiuk [9] who intercalated GOx layers with inactive FAD-free apo-GOx layers in four enzyme/osmium poly(allylamine) layer pairs structures, where the active enzyme layer position was changed from the first layer adjacent to the metal electrode to the fourth layer in contact with the electrolyte. The generation of an electrical biosensor signal from the molecular recognition of glucose by GOx involves the following sequence of events: $FADH_2$ re-oxidation, $k_M$, and propagation of electrons from the active layer to the underlying electrode through the multilayer, $k_D$. The experimental data could be described by expressing the observed rate of $FADH_2$ re-oxidation under glucose saturation, $k_M(obs)$ as the sum of the reciprocals of the re-oxidation rate $k_M$ in the absence of electron propagation limitations and the electron hopping diffusion constant $k_D$. This propagation rate constant was further expressed as the reciprocal of the characteristic diffusion time given by the Einstein equation: $\tau = \Delta x^2 / 2 D_e$ with $\Delta x = n \cdot d_{layer}$, where $n$ is the layer number, $d_{layer}$ the average interlayer distance and $D_e$ the electron-hopping diffusion coefficient. From the linear dependence of $1/k_M(obs)$ versus $n^2$, both $k_M = 4500\,s^{-1}$ and $D_e = 10^{-9}\,cm^2\,s^{-1}$ for an estimated $d_{layer} = 5\,nm$ were obtained, in excellent agreement with the electron hopping diffusion constant measured by ultra microelectrode chronoamperometry in random hydrogels of the same composition.

It has been shown by chronoamperometry that charge propagation in ultrathin polymer films fabricated by LbL deposition of poly(acrylic acid), (PAA) and a polycation bearing ferrocene (Fc) moieties (P(CM-Fcx)), shows diffusion behavior. When the characteristic charge diffusion time is lower than the experiment timescale, not all redox sites in the film can be oxidized or reduced, and an apparent diffusion coefficient for charge propagation, $D_{app}$, can be obtained, with typical values around $10^{-10}\,cm^2\,s^{-1}$ [5, 72].

Figure 42.7 depicts a scheme of the electrochemical process at a redox polyelectrolyte multilayer. The electrode potential perturbation causes electron transfer from a reduced Os(II) site adjacent to the underlying metal electrode at the metal/r-PEM interface. Subsequently, redox charge propagates by bimolecular electron transfer from an adjacent Os(III) site, and gives rise to electron hopping between Os(II)/Os (III) sites in the multilayer. At the same time, at the r-PEM/electrolyte interface anions balance charge by ion and solvent exchange from the electrolyte.

We define two characteristic parameters: the heterogeneous rate constant for electron transfer between the electrode and the adjacent redox sites, $k_o$, with units of $cm^{-1}$, since $C^*$ is the volume redox concentration in the film, and $D_e$ a phenomenological apparent diffusion coefficient for the electron-hopping mechanism between neighboring redox sites in the multilayer.

Laviron introduced a CV method to determine the electron transfer rate constant in surface-confined redox systems by recording the peak potential and peak current as a function of the potential scan rate [64, 65]. Electrochemical impedance spectroscopy

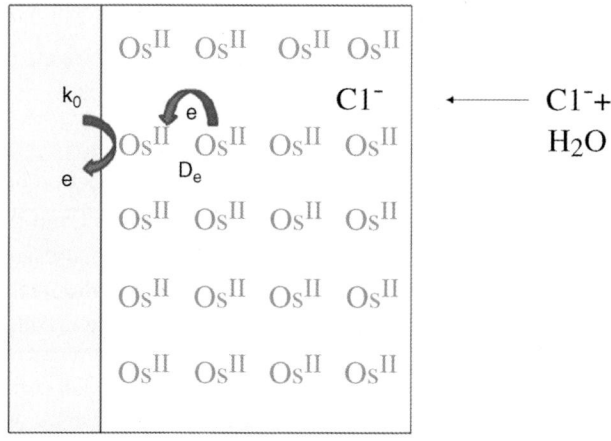

**Figure 42.7** Scheme of electron transfer and charge propagation in a redox PEM under electrochemical perturbation.

(EIS) [73] and chronoamperometry [5] have also been employed to measure charge transport in redox polyelectrolyte multilayers.

Charge propagation in redox active films can be described either by the physical displacement of the electron acceptor/donor sites or electron hopping by bimolecular collision of neighboring reduced/oxidized centers, or by a combination of both processes. In the case of freely diffusing redox couples entrapped in oppositely charged polyelectrolytes multilayers, both processes occur and an apparent diffusion coefficient can be defined and measured [74, 75]:

$$D_{app} = D_{phys} + D_e \qquad (42.8)$$

with $D_{phys}$ the coefficient of physical diffusion and $D_e$ the diffusion coefficient for the electron hopping process.

For an immobile assembly of entrapped redox centers, it follows that:

$$D_e = \frac{k_{ex} C^* \delta^2}{6} \qquad (42.9)$$

where $k_{ex}$ is the bimolecular rate constant for electron self-exchange, $\delta$ the center-to-center separation during self-exchange, and $C^*$ the volume concentration of redox sites.

For redox sites covalently bound to a polymer backbone, only $D_e$ contributes significantly to charge transport, and Equation 42.9 has systematically failed to explain the redox concentration dependence of $D_{app}$ [76]. Blauch and Savéant have shown that charge transport can be described by a percolation process for randomly distributed immobile isolated clusters of electrochemically connected sites [77, 78].

When the redox sites in the polymeric structure can move around their equilibrium position, they can become in contact by dynamic rearrangement and charge transport occurs under bound diffusion. If the motion is approximated by harmonic displacements, it follows that:

$$D_{app} = \frac{k_{bim}(\delta^2 + 3\lambda^2)C^*}{6} \qquad (42.10)$$

where $k_{bim}$ is the bimolecular activation-limited rate constant for electron hopping $k_{bim} = \frac{k_{ex} k_D}{(k_{ex} + k_D)}$, $k_D$ is the diffusion-limited bimolecular rate constant and $\delta$ is the distance between adjacent redox centers in the film, while $\lambda$ represents the mean displacement of a redox center out of its equilibrium position. This distance can be related to an imaginary spring $\lambda = (2k_B T/f_s)^{1/2}$, where $k_B$ is the Boltzmann constant, $T$ the temperature, and $f_s$ the spring force constant.

Savéant and Blauch further investigated the bound diffusion model using analytical methods and Monte Carlo simulations [79]. Long linkers between the polymer and the electroactive groups can dramatically help by allowing the redox sites to explore larger spatial regions around their equilibrium positions. Experimental validation of this model was achieved by Mano and Heller [80], who reported an unprecedented diffusion coefficient of $5.8 \times 10^{-6}\,cm^2\,s^{-1}$ for a redox polymer bearing a long 13 atom spacer, which is more than five orders of magnitude higher that those typically reported for redox polymers and very close to diffusion coefficients of freely diffusing molecules.

While the simple models discussed above have shed light on the mechanism of charge transport, experimental results are affected by the structure and composition of the multilayer polymeric matrix and the electrolyte composition. For example, polymer cross-linking hinders charge transport due to the slowdown of segmental motions, with the consequent failure of redox sites to come close enough to allow efficient electron transfer [81]. Charge propagation can be enhanced by film swelling, giving rise to more flexible structures [82].

The effect of solution composition, due to the exchange of mobile ions and solvent, on electron transport in an electrode chemically modified by redox films has shown that ionic cross-linking of polymer films also results in a decrease in the charge diffusion coefficient [83, 84]. For osmium-containing poly(allylamines)/poly(vinylsulfonate) LbL multilayers, dramatic effects have been observed for specific ions, with a $D_{app}$ reduction from $1.8 \times 10^{-10}\,cm^2\,s^{-1}$ in chloride solutions to $6.6 \times 10^{-15}\,cm^2\,s^{-1}$ in perchlorate electrolyte supporting the ionic crosslinking hypothesis [6].

The effect of the structure of multilayer films on the electrochemical response has been shown by Flexer with GOx osmium poly(allylamines) multilayers assembled at different electrolyte pHs and, therefore, with different film structure and thickness [3]. More irreversible voltammperograms are observed for PAH-Os/PVS films assembled at low pH (pH 3.5–5.5) when compared to those assembled at high pH (pH 8.3). Since PAH-Os/PVS films assembled at lower pH are thinner and more compact than those obtained at high pH, diffusion limitations would explain the different apparent electron hopping diffusion coefficient that decreases with the assembly pH. This can arise from slower segmental motions and the mobility of the redox sites in the more compact structures assembled at low pH as compared to the rather floppy films obtained at higher pH.

In a seminal work Hodak et al. [2] reported that the redox property of ferrocene modified poly(allylamine) multilayers depends significantly on the number of layers, with a decrease in the redox charge for negatively charged GOx topmost layers. Furthermore, it has been shown that with the addition of a single layer of osmium complex polycation, PAH-Os(+), adsorbed on top of the terminal negatively charged GOx(−) layer with the same enzyme concentration, the catalytic response was significantly larger and the amount of "wired enzyme" increased from 2.1 to 24% of the total enzyme present on the surface [11]. Anzai demonstrated that the redox properties of ferrocene-containing PEMs depend on the deposition of outer layers, even if these outer layers contain no redox sites [18]. The redox charge changes, depending on the nature of the topmost layer, either PAH(+) or PVS(−), with lower charge densities for PVS-terminated layers, and the effect was apparent for both thin and thick multilayer films. Xie and Granick reported that the dissociation equilibrium of poly(metacrylic acid) in the inner PEMs layers depends significantly on the polarity of the strong polyelectrolyte deposited on the outermost surface of the film [85].

Unlike random polymers, LbL films exhibit the unique ability to control the nature and charge of the topmost layer at the polymer/solution interface. With polyanion and polycation constructs, it has been shown by EIS that polyanion-terminated films exhibit a hindrance for redox switching as compared to polycation-capped ones [73], but the nature of the slow process could not be unveiled with the EIS technique. While the adsorption of the polyanion induces a decrease in the redox charge, increasing the ionic strength completely offsets the effect, which suggests an electrostatic nature of the phenomenon.

A modified Laviron diffusion model has been presented by Tagliazucchi to account for the experimentally observed dependence of the average peak potential on the scan rate in the CV of $(PAH\text{-}Os)_n(PVS)_m$ LbL multilayer films [86]. The effect of the nature of the charge in the topmost layer in the LbL multilayers and the electrolyte ionic strength have been studied with this method. For instance, Figure 42.8 shows CVs recorded for PAH-Os/PVS LbL multilayers deposited at pH 8.3. While the CVs for PAH-Os-capped films present an almost ideal thin-layer behavior, the voltammograms of PVS-capped

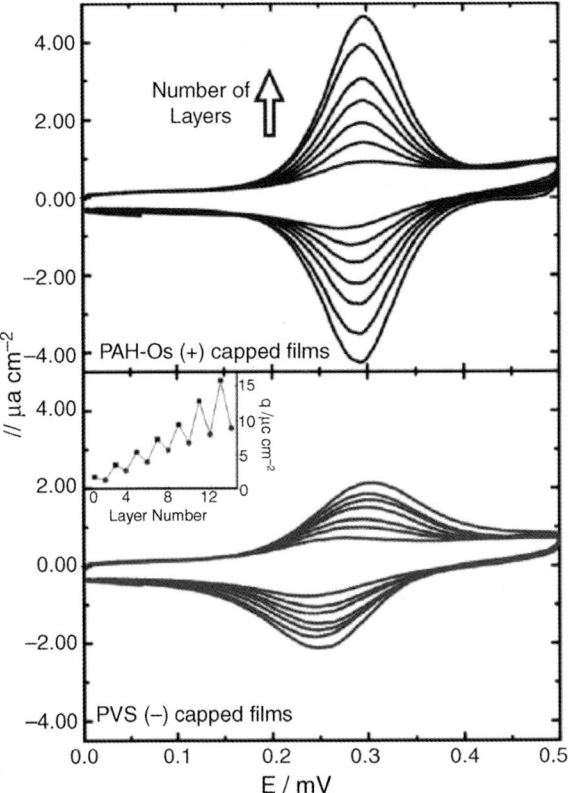

**Figure 42.8** Current–potential curves recorded in buffer TRIS (20 mM, pH 7.3) + KNO3 (0.2 M) after the successive assembly steps of a $(PAH\text{-}Os/PVS)_7$ film. The pH value of the assembly solutions was 8.3 and the scan rate was 25 mV s$^{-1}$. Inset in lower panel: redox charge versus number of layers for PAH-Os-capped (squares), PVS-capped (circles) films. Taken from Ref. [86].

**Table 42.1** Best-fit values of De and ko obtained from cyclic voltammetry experiments. Taken from ref. [86].

| | m/M | $D/\text{cm}^2\,\text{s}^{-1}$ | $k_o/\text{cm}\,\text{s}^{-1}$ |
|---|---|---|---|
| (PAH-Os)$_5$(PVS)$_5$(−) | 0.2 | $2 \times 10^{-13}$ | $1.2 \times 10^4$ |
| (PAH-Os)$_5$(PVS)$_5$(−) | 0.5 | $5 \times 10^{-13}$ | $0.21 \times 10^4$ |
| (PAH-Os)$_5$(PVS)$_5$(−) | 1.0 | $2.8 \times 10^{-12}$ | $0.52 \times 10^4$ |
| (PAH-Os)$_5$(PVS)$_5$(−) | 2.0 | $1.5 \times 10^{-10}$ | $2.2 \times 10^4$ |
| (PAH-Os)$_5$(PVS)$_5$ (PAH-Os) (+) | 0.2 | $2.0 \times 10^{-10}$ | $2.2 \times 10^4$ |

multilayer films exhibit a rather large peak separation, indicative of irreversibility in the redox process that hinders complete oxidation/reduction of all osmium sites. EIS studies have clarified this behavior, as explained below.

While the charge transfer rate constant, $k_o$, does not vary appreciably with the ionic strength or the nature of the outermost layer, the electron hopping diffusion coefficient, $D_e$, at low ionic strength (ca. 0.2 M) drops by three orders of magnitude for polyanion-capped films ($k_o = 0.12 \times 10^4\,\text{cm}^2\,\text{s}^{-1}$ and $D_e = 2.2 \times 10^{-13}\,\text{cm}^2\,\text{s}^{-1}$) with respect to positively charged PAH-Os-capped films ($k_o = 2.2 \times 10^4\,\text{cm}^2\,\text{s}^{-1}$ and $D_e = 2.0 \times 10^{-10}\,\text{cm}^2\,\text{s}^{-1}$).

The effect of electrolyte ionic strength on negatively charged polyanion-capped films is shown in Table 42.1. A dramatic decrease, by three orders of magnitude, in $D_e$ is produced at low ionic strength for negatively charged topmost layers, while films capped with PAH-Os or PAH-Fc exhibit fast redox kinetics. The results in Table 42.1 indicate that the polyanion topmost layer affects the charge transport through the multilayer, but not the charge transfer at the metal/polymer interface. This has been further confirmed by either:

1) Alternating electro-inactive polyelectrolytes PAH and PVS on top of PAH-Os/PVS multi-layers, or
2) Alternate adsorption of PAH and PVS on a thiol partly derivatized self-assembled osmium pyridine-bipyridine complex monolayer. For this diffusionless redox system with osmium sites located on a well-defined plane at the surface, within tunneling distance of the electrode, direct electron transfer is the only redox mechanism and no oscillations in the redox charge with the nature of the topmost layer are observed, supporting the idea that charge transfer is not affected by the outermost layer. Therefore, the phenomenon observed with redox LbL multilayers terminated in negatively charged layers can be solely ascribed to changes in the diffusion of redox charge, $D_{app}$, due to the polyion character of these polyelectrolytes rather than to their redox activity.

A complete explanation of the effect of the outermost layer in the electron hopping diffusion has not been given. Limitations to ion transport across the polymer/electrolyte interface due to the negative surface charge introduced by the outermost layer should be ruled out, since ion fluxes are needed to keep electroneutrality during the redox process, and hence redox reversibility would also be affected, which is not the case.

However, it has been suggested that the dynamics of polyelectrolyte segments carrying the redox groups would be altered by the outermost layer charge. Quartz crystal shear resonator dissipation studies of the viscoelastic properties of LbL films have contributed to the understanding of the topmost layer effect [87]. While the mass increases stepwise during LbL self-assembly of alternate layers, viscoelastic effects are reported to be strongly dependent on the nature of the topmost layer in poly (allylamine)-poly(acrylic acid) built at pH 7.5, as shown by shear wave energy dissipation at 10 MHz through the multilayer as a function of adsorption steps. An increase in dissipation is observed as the PAH is adsorbed while the dissipation decreases with the adsorption of PAA. The evidence points to changes in the diffusion rate of mobile species within the film.

### 42.3.2
### Ion Exchange

Lindholm-Sethson [88] has measured, by CV and EIS, the permeability of the electrochemically active anion $Fe(CN)_6^{4-}$ and the cation, $Os(bipy)_3^{3+}$ in self-assembled multilayer films formed on positively-charged gold electrode surfaces modified by consecutive adsorption of poly(styrene sulfonate),PSS, and protonated poly(allyl amine), PAH. Electrochemical and *in situ* ellipsometric studies on the permeability and stability of poly-(allylamine hydrochloride)/poly(styrenesulfonate), (PAH/PSS) and PAH/poly(acrylic acid), (PAA) films were also reported by Bruening et al. [89, 90]. The permeability of these layered polyelectrolyte films to $Fe(CN)_6^{3-}$ and $Ru(NH_3)_6^{3+}$ depends on the solution pH, the number of layer pairs in the film, whether supporting electrolyte is present during film deposition, and the nature of the polycations and polyanions. In these experiments the electron transfer reaction takes place at the underlying metal electrode, and the polyelectrolyte multilayer simply behaves as a selective membrane to the transport of electrochemically active ions. Using CV at a rotating disc electrode, Schlenoff studied the transport of redox active probe ions through multilayers of highly charged polyelectrolytes [91].

Unlike exchange membranes the "as-prepared" PEM films are *"reluctant"* amphoteric exchangers due to the intrinsic charge balance between polyion–polyion segment ion pairs [51]. It has been found that, in excess salt in the external bathing electrolyte solution, exchange of small ions is forced and the exchanger site concentration is proportional to the solution salt concentration, while the population of redox ions within the multilayer remains constant. The initial intrinsic PEM then becomes extrinsically compensated by small ions after breaking the polyion–polyion interactions and making polyion–small ion ones as shown in Scheme 42.1.

A similar phenomenon occurs when the PEM contains redox sites covalently bound to the polymer backbone and is subject to an imbalance of charge produced at the metal/PEM interface by an electrochemical reaction. The excess charge created in the film needs to be compensated by exchanging ions with the bathing electrolyte. The ionic exchange during the initial conditioning of the film (called "break-in" [8]) undergoing oxidation–reduction cycles, and recovery after equilibration in the reduced state, has been studied by electrochemical quartz crystal microbalance (EQCM) and probe beam deflection [7, 8] (Scheme 42.1).

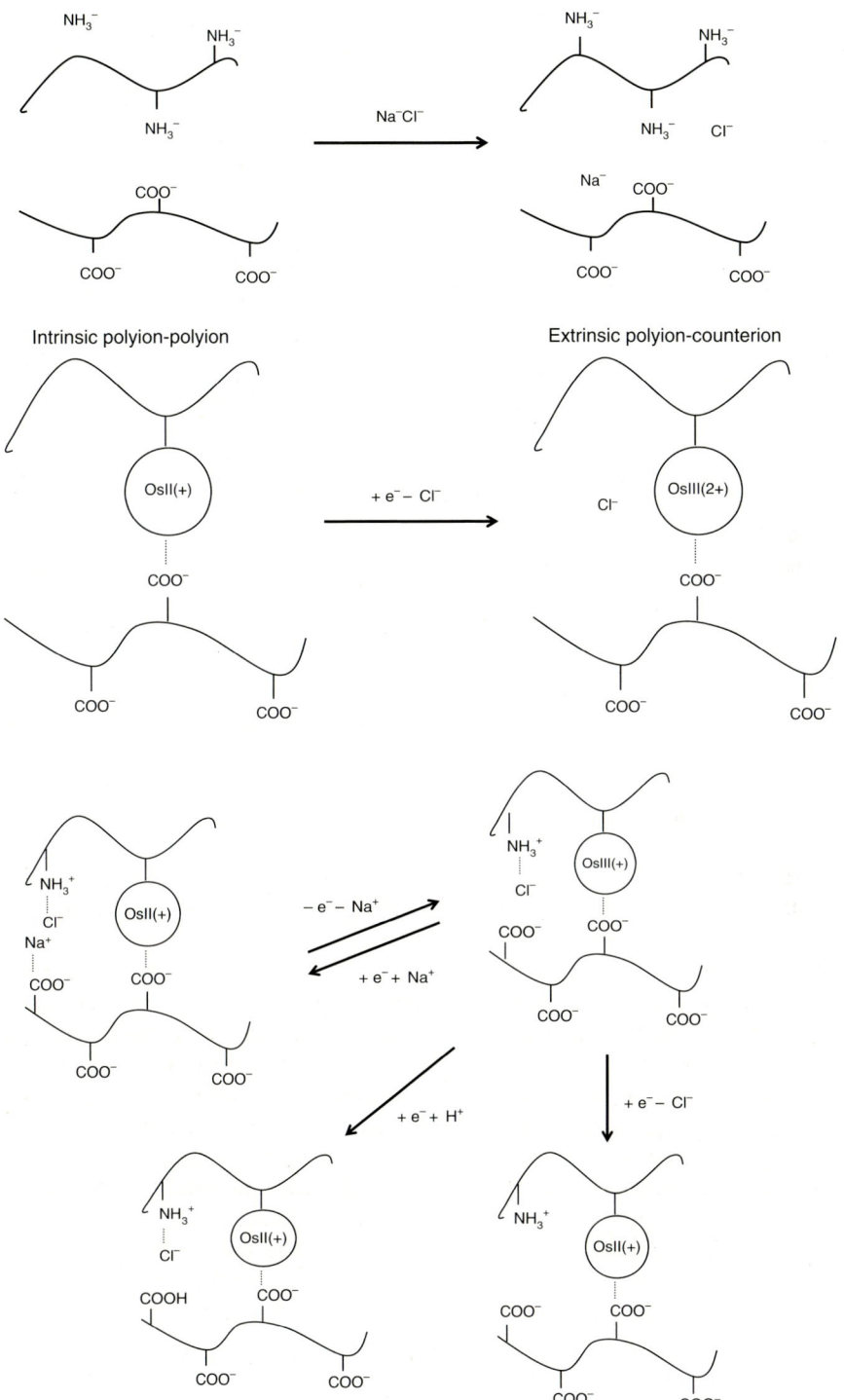

**Scheme 42.1** Different mechanism for charge compensation in extrinsic redox polyelectrolyte multilayers.

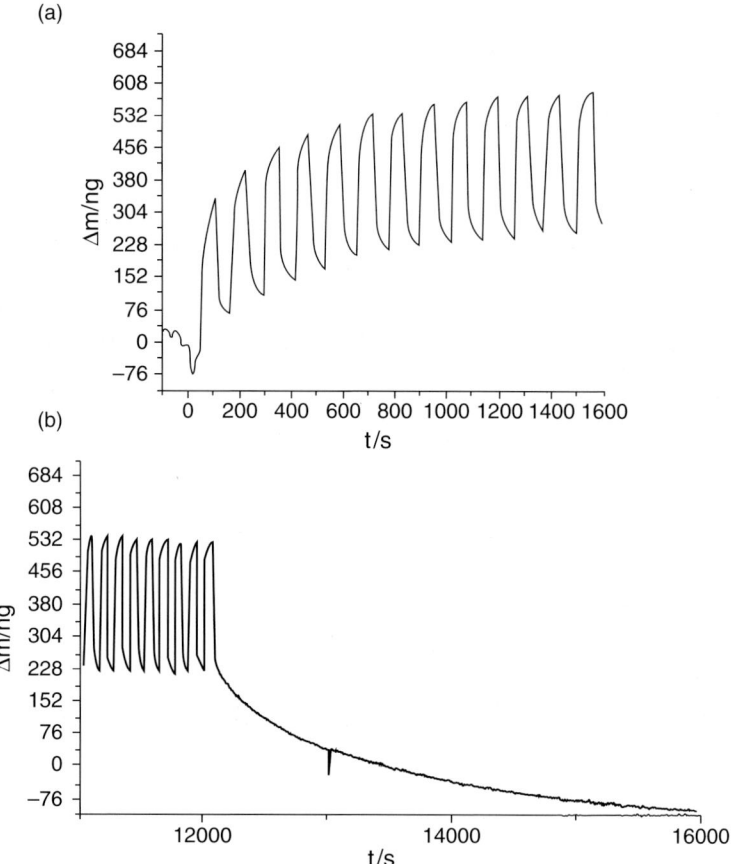

**Figure 42.9** EQCM gravimetric response of (PAH-Os)$_{15}$(PSS)$_{14}$ film in 10 mM NaCl for repetitive potential steps from 0.1 to 6 V. (a) First switching cycles. (b) Electrode mass transient after switching off to 0.10 V. Taken from Ref. [5].

A similar "break-in" effect has been observed in poly(vinylferrocene), PVF, films due to the incorporation of solvent and ions into the film in the first oxidation–reduction cycles, thus decreasing its resistivity [92]. This effect has been observed for several polyelectrolyte and polymer-modified electrodes, for example, polyaniline [93]. The first voltammetry curve of a series always shows a larger anodic to cathodic peak separation than the subsequent ones [5]; while *in situ* EQCM simultaneous to an electrode potential square wave perturbation results in mass transients, shown in Figure 42.9a for a freshly prepared (PAHOs)$_{15}$(PSS)$_{14}$ during the first few redox switching cycles. An increase in the mass is observed in every oxidation step, and a decrease in mass in the subsequent reduction step to yield Os(II) film. However, the original mass of the fully reduced film is never recovered and a continuous mass build up in the film is apparent during several short oxidation–reduction cycles. Thus, a fraction of the species that gets into the film from the electrolyte solution during oxidation, remains in the film in the following reduction cycle. Full recovery of the original mass is only observed if the electrode potential is kept constant at full reduction values (i.e., 0.1 V), remaining in the fully reduced state for more than 1 h, as shown in the transient of Figure 42.9b.

The EQCM measures the total mass of ions and neutrals (solvent and salt) exchanged at the film/electrolyte interface simultaneously with the injection of charge at the electrode/film interface in the potential step experiments. Plots of the change in mass per unit area versus the charge density during an end-to-end oxidation–reduction cycle are linear, and from their slope the calculated molar mass results always in large excess with respect to the molar mass of the mobile ions that can be exchanged (i.e., $Na^+$ and $Cl^-$ respectively). Furthermore, the mass-to-charge ratio is very dependent on the anion in the electrolyte but not on the cation, as expected for anion exchanger films [6].

Since the increase in film mass observed by EQCM during the "break-in" process and steady state could have an origin in the accumulation of mobile ions and solvent, experiments with external reflection infrared spectroscopy (SNIFTIRS – subtractively normalized interfacial Fourier transform infrared spectroscopy) while modulating the electrode potential were carried out, and showed direct evidence of water content in the film increasing during oxidation–reduction cycles and accumulation during "break-in" with a similar pattern to the mass accumulation (see Figure 42.10) [94].

The role of mobile ions in redox polyelectrolyte multilayers was first described by Schlenoff et al. using radio-labeled $^{45}Ca^{2+}$ ions with poly(butylviologen)/PSS multilayers with exchange during the reduction scan, but with accumulation of a small

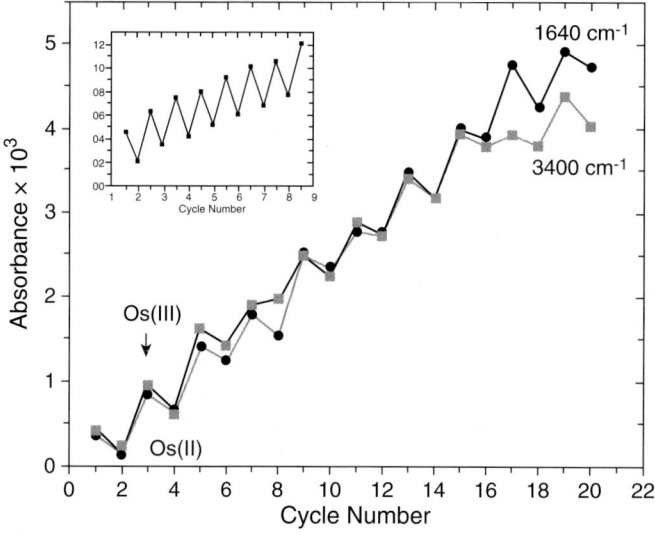

**Figure 42.10** Infrared absorbance at 1640 and 3400 $cm^{-1}$ as a function of the number of oxidation–reduction cycles of PAH-Os and PSS multilayer film. Inset: Variation of mass (EQCM) during repetitive oxidation–reduction cycles. Taken from Ref. [94].

fraction in the film during oxidation, and increase in $Ca^{2+}$ content in subsequent cycles [95]. The authors suggested that simultaneous anion uptake occurs during reduction, in the same proportion to the fraction of $Ca^{2+}$ that was retained in the film, in order to balance charge.

A laser probe beam deflection (PBD) technique is used to measure how the refractive index gradient in the solution adjacent to the electrode/electrolyte interface changes due to ionic fluxes [25, 26]. In this technique positive deflections indicate the flux of ions towards the electrode and, therefore, for anodic potential steps it implies anion uptake. Cation release results, under these conditions, in a negative deflection. This technique provided direct evidence of ion exchange in PAH-Os/PSS multilayers [25, 26]. PBD has also shown differences in both the deflection signal and the electrical current transients between the first and successive oxidation–reduction cycles for freshly prepared PAH-Os/PSS films in 10 mM HCl [28]. Both the PBD signal and the electrode current decrease with increasing number of cycles and this is not a consequence of electroactive material loss, as proven by EQCM gravimetric experiments. The electrochemical response is recovered after holding the system at a reducing potential for a rest period.

PBD oxidation transients (Figure 42.11) show the presence of a positive peak due to anion uptake, but also a negative pre-peak originating from the release of cations. Due to the difference in diffusion of proton and chloride, fitting the transients in Figure 42.11 to the PBD analytical expressions yields the ratio of exchanged anions to exchanged cations ($Cs_{anion}/Cs_{cation}$), which is 4.2 for positively capped multilayers and 1.4 for negatively capped ones. As expected, anion exchange contributes more to charge compensation for positively capped films than for those with a negative topmost layer. Direct molecular evidence of water uptake/release during the anodic/cathodic scans has been obtained by observing the associated water IR-bands at 1540 and 3450 $cm^{-1}$, respectively, by SNIFTIRS [5, 94].

A combination of EQCM, PBD and integrated charge has revealed the nature of exchanged ions, and the population of ions and solvent during the redox switching of r-PEMs.

The amount of exchanged species resulting in mass change can only be explained by water fluxes; which is well above ionic hydration [5, 6] (around 30 water molecules per electron, see Table 42.2).

Indirect evidence for solvent exchange has also been gained by "*in situ*" ellipsometry, which measures film thickness and optical properties as a function of the electrode potential (linear sweep voltammetry) or time (chronoamperometry). Figure 42.12 shows redox driven swelling of a LbL enzyme-redox polyelectrolyte multilayer during oxidation and shrinking during reduction, consistent with EQCM and infrared spectroscopy measurements [96]. A simultaneous decrease in the refractive index from 1.42 to 1.38 during oxidation also supports the uptake of water, which has a refractive index of 1.33.

Zambelli has recently reviewed the swelling of electrochemically active PEMs [97]. Upon electrochemical stimulus, the influx of counter ions and solvent increases the osmotic pressure in the film, resulting in swelling. The electrochemically driven volume expansion can be as large as 15%.

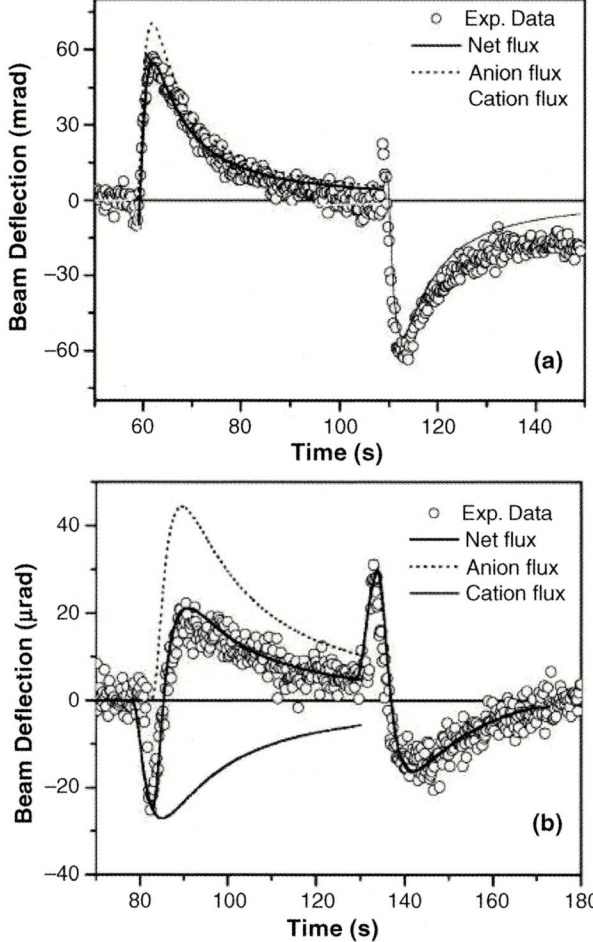

**Figure 42.11** Experimental steady-state PBD transients (circles) of the oxidation and reduction of (PAH-Os(II))n(PSS)m in HCl 0.01 M (full lines), best fits to Equation 42.2 and calculated individual contributions of Cl⁻ and H⁺ (dashed lines): (a) (PAH-Os)$_{15}$(PSS)$_{14}$ and (b) (PAH-Os)$_{15}$-(PSS)$_{15}$. The same d$n$/d$C$ value was assumed for Cl⁻ and H⁺ species, $D_{Cl^-} = 1.5 \times 10^{-5}$ and $D_{H^+} = 9.3 \times 10^{-5}$. Taken from Ref. [7].

Film swelling and contraction in ferrocyanide-containing PEMs upon application of an electric potential has been studied by Boulmedais and co-workers [55, 98–100] using CV, electrochemical quartz crystal microbalance with dissipation (EQCM-D), and electrochemical atomic force microscopy (EC-AFM) in relation to biomedical applications for drug release [99]. They subsequently studied the effect of supporting electrolyte anion on the thickness of PSS/PAH multilayer films and ionic permeability [100]. Both the Buenos Aires [6] and the Strasbourg groups [100] found that the film hydration and thickness, for a given number of

Table 42.2 End-to-end mass/charge ratios taken from ref. [6]..

| Sodium salt | $(\Delta m/\Delta q^1)/g\ C^{-1}$ (x $10^3$) | $\Delta m^1/g\ mol^{-1}$ (x $10^3$) | $N_{H_2O}$ |
|---|---|---|---|
| NaCl | 64 ± 9 | 615 ± 90 | 32 |
| NaF | 72 ± 9 | 692 ± 90 | 37 |
| NaNO$_3$ | 19 ± 9 | 185 ± 90 | 7 |
| NaBF$_4$ | 20 ± 9 | 195 ± 90 | 6 |
| NaClO$_4$ | 6 ± 9 | 58 ± 90 | −2.3 |
| Chloride salt | | | |
| NaCl | 62 ± 9 | 598 ± 90 | 31 |
| LiCl | 62 ± 9 | 596 ± 90 | 31 |
| CsCl | 58 ± 9 | 564 ± 90 | 29 |
| HCl | 69 ± 9 | 662 ± 90 | 35 |

deposition steps, follows the Hofmeister series from cosmotropic to chaotropic anions (F$^-$, Cl$^-$, NO$_3^-$, ClO$_4^-$) where the monovalent anions are ranked according to their hydration entropy.

Hammond and coworkers described the electrochemically controlled swelling and mechanical properties of a PEM nanocomposite thin film containing cationic linear poly-(ethyleneimine), (LPEI), and anionic Prussian Blue nanoparticles, with electrochemical control over film thickness and mechanical properties [47]. They found reversible changes in the film thickness (2–10%) and Young's elastic modulus (50%) of the hydrated composite film. Viscoelastic studies with the EQCM have also shown important changes in the shear moduli ($G'$ storage modulus and $G''$ the loss modulus) at 10 MHz under oxidation–reduction cycles (this frequency is characteristic of ionic atmosphere relaxation in the polyions). The electrostatically self-assembled (PAH-Os)$_n$(GOx)$_n$ enzyme multilayers behave as viscoelastic films at 10 MHz with $G'$ and $G''$ of the order of $10^6$ Pa in the reduced Os(II) state. Upon oxidation a film with $G' = 5.9$ MPa and $G'' = 1.1$ MPa swells (10% thickness increase) and exhibits a change to $G' = 4.8$ MPa and $G'' = 5.2$ MPa [13, 101]. Figure 42.13 shows both components of the Butterworth–Van Dyke electrical equivalent circuit elements of the surface redox enzyme-PEM film for a different number of self-assembled layers as a function of the applied potential [101]. It should be noted that for more than 7 layer pairs (thickness > 100 nm), there are considerable viscoelastic losses in the LbL film, as shown by the increase in the shear wave dissipation parameter $R_S$ with potential.

### 42.3.3
### Applications

The LbL electrostatic assembly technique is a rich, versatile, and significantly inexpensive approach to building up electroactive thin films by alternate adsorption of positively and negatively charged polyelectrolytes from aqueous solutions,

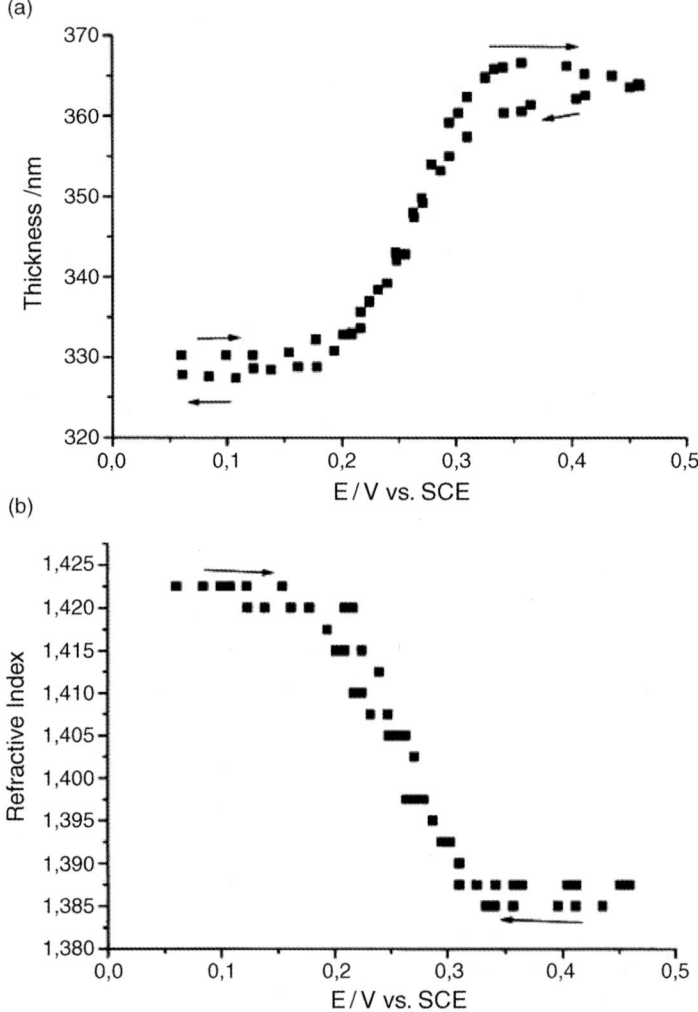

**Figure 42.12** Plots of (a) $d_f$–$E$ and (b) $n_f$–$E$ dependences obtained for a (PAH-Os)$_7$(GOx)$_7$ multilayer film under potentiodynamic conditions ($v = 5$ mV s$^{-1}$), at $\lambda = 632.8$ nm. Optical properties obtained by simulation of $\psi$–$\Delta$ values with an isotropic single layer model on thiolated gold $n = 0.767$, $k = 2.540$, assuming $k$ (632.8 nm) $= 0.000$. Taken from Ref. [96].

with applications to biosensors, electrochromics, actuators, nanoreactors, and so on [53].

Functional polyelectrolytes are synthetic polymers or copolymers that contain ionizable electrolyte groups, as well as additional functionality imparted by the molecular (polymer) structure: redox active, optical active, catalytic, molecular recognition, and so on [102].

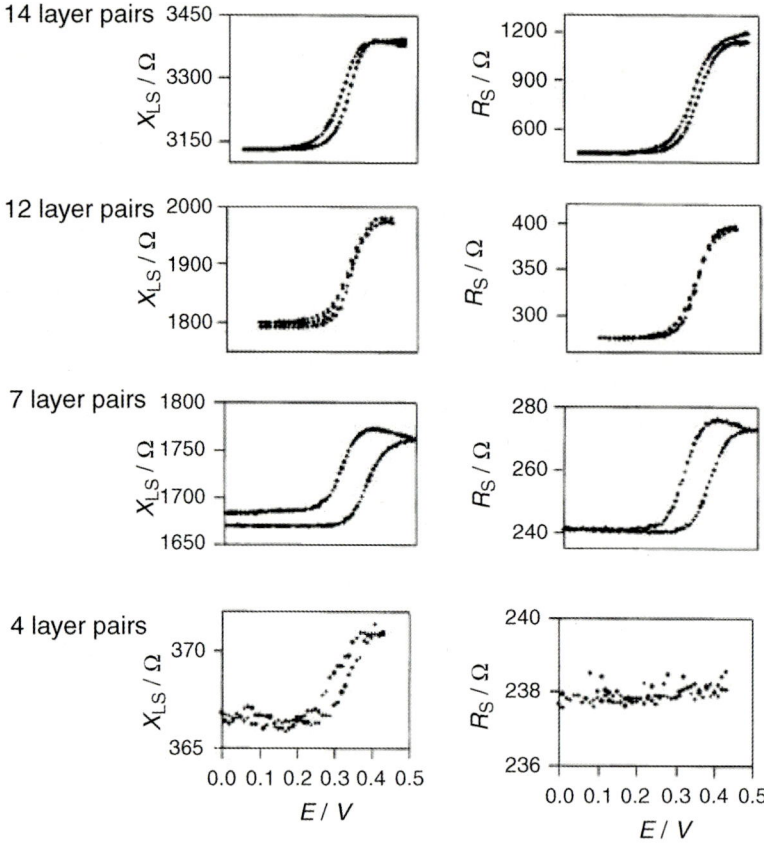

**Figure 42.13** Real ($R_S$) and imaginary ($XL_S$) components of the surface acoustic impedance at 10 MHz for (PAH-Os)$_n$(GOx)$_n$ in contact with 0.12 M KNO$_3$ and 10 mM Tris buffer electrolyte of pH 7.5 for different film thicknesses with $n = 4, 7, 12$ and 14 during voltammetric oxidation of the film under CV conditions at 5 mV s$^{-1}$. Taken from Ref. [101].

## 42.3.4
### Biosensors

Biosensors can be defined as analytical devices incorporating a biological material (e.g., enzymes, antibodies, nucleic acids, tissues, microorganisms, organelles, cell receptors, etc.), a biologically derived material, or a biomimetic one intimately associated with or integrated within a physicochemical transducer microsystem [103]. In particular, biosensors and molecular devices can take advantage of electrically "wired" enzymes. Molecular recognition with enzymes and with enzyme-labeled immuno and genomic electrodes based, respectively, on the antigen–antibody interaction and single-strand DNA (ss-DNA) hybridization with self-contained redox relays to generate an electrical signal, can be integrated in circuits. In an amperometric enzyme biosensor electrode there are three consecutive steps: molecular recognition of the target substrate molecule by the enzyme that changes its redox state, generation of an electrical signal by propagation of charge in the sensing surface layer film, and amplification of the electrical signal in the external circuit.

Direct electron transfer from the electrode surface to the prosthetic group, $FADH_2$, buried inside large enzymes, such as GOx- (186 kDa), is hindered. Heller demonstrated that electrical communication between the $FADH_2$ in GOx and electrodes can be achieved by electrostatically complexing the negatively charged enzyme in a solution of pH above the isoelectric point (4.2) with a cationic quaternized poly(vinylpyridine) and poly(vinylpyridine) $Os(bpy)_2Cl$ redox mediator polyelectrolyte copolymer. He further introduced a two-component epoxy technique combining GOx and other oxidase enzymes with the polycationic redox mediator cross-linked with a bifunctional reagent [104].

Unlike redox hydrogels, with random distribution of components without control of the molecular orientation, spatially ordered enzyme assemblies offer several advantages over random polymers for molecular recognition by the enzyme, redox mediation by the redox polymer, and signal generation at the electrode surface. Enzymes deposited in ordered monolayers and multilayer systems have been described using different assembling techniques for protein immobilization, such as Langmuir–Blodgett, [105] self-assembled monolayers [106–108], step-by-step electrostatic adsorption of alternate multilayers [2, 109–111], antigen–antibody [112–114] and avidin–biotin interactions [115–117], surfactant films [118], electrostatic adsorption of hyper-branched polyelectrolytes [119], redox functionalized dendrimers [42], and so on.

In 1997 Hodak [2] introduced an integrated biosensor based on the redox mediation of GOx in self-assembled structures of cationic poly(allylamine) modified by ferrocene and anionic GOx deposited step-wise in alternate polymer enzyme multilayers. The redox charge, the amount of enzyme and the catalytic biosensor signal increases with each adsorption step. An extensive literature has followed this pioneering work [9–11, 31, 32, 42, 115, 120–126].

In the specific case of the glucose bio-electrode using GOx and poly(allyl-amine) with an osmium complex covalently tethered (PAA-Os) in LbL multilayers, formed by sequential immersion in GOx solutions of suitable pH so that the enzyme carries a net negative charge, and the positively charged "wiring" osmium-containing polyelectrolyte, the reactions are [11]:

$$GOx_{RED}(FADH_2) + 2PAH-Os(III) \xrightarrow{k} GOx_{OX}(FAD) + 2PAH-Os(II) + 2H^+ \tag{42.11}$$

$$\beta-(D)-glucose + GOx_{OX}(FAD) \underset{}{\overset{K_{MS}}{\rightleftharpoons}} S\text{-}GOx_{OX}(FAD) \xrightarrow{k_{cat}} \text{Gluconic acid} + GOx_{RED}(FADH) \tag{42.12}$$

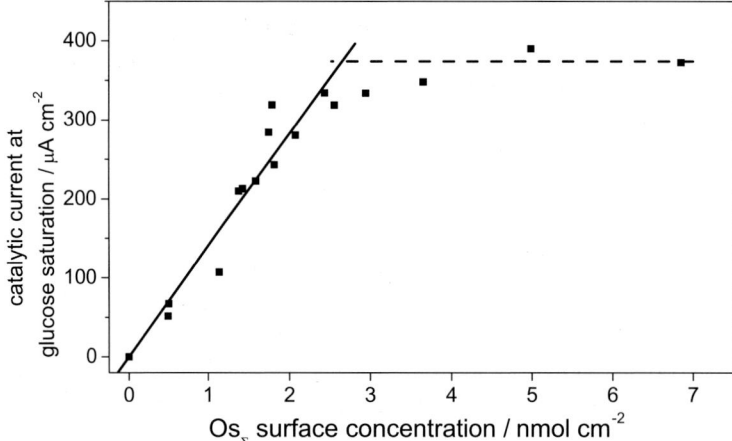

**Figure 42.14** Catalytic current at glucose saturation as a function of total Os surface charge concentration. Full line is the best linear fit of the experimental data for the thinner electrodes (case I). The broken line is the average plateau current for the thicker electrodes (case II). Taken from Ref. [130].

In the LbL film the oxidized mediator, PAH-Os(III) regenerates the oxidized form of the enzyme with a second order rate constant $k$, according to the conventional "ping-pong" mechanism, while the oxidized enzyme reacts with its substrate β-(D)-glucose to yield the reaction product gluconic acid. The double enzyme-redox cycle is completed at the film/electrode interface where electrons are exchanged with the Os (II) polymer:

$$PAH-Os(II) \rightarrow PAH-Os(III) + e \qquad (42.13)$$

In the steady state, when the diffusion of substrate and redox mediator (electron hopping) balances the enzyme kinetics, the concentration of enzyme, substrate and redox mediator can be described by nonlinear second order differential equations with boundary conditions at the film/electrode and film/electrolyte interfaces, which cannot be solved analytically [127]. Pratt and Bartlett [128, 129] made approximate assumptions and obtained analytical expressions for seven different limiting cases. A test of this kinetic case diagram has been recently reported by Flexer [130] using LbL self-assembled enzyme multilayer electrodes with fine control over the enzyme film thickness, enzyme loading, osmium and glucose concentrations. The kinetic analysis and kinetic data extraction were based on the approximate analytical solutions, which were used to obtain initial estimates for the kinetic parameters, followed by application of the relaxation method with automated grid point allocation [131], combined with the simplex algorithm to fit a large number of experimental data points to the full model, and to extract best fit estimates for the kinetic parameters. Flexer has visited experimentally four cases in the diffusion-kinetics case diagram by systematically varying the film thickness, glucose concentration and redox potential and, hence, redox mediator concentration in the enzyme multilayer films.

The method proved successful and concentration profiles of the oxidized enzyme and substrate inside the enzyme multilayer thin film have been calculated.

Figure 42.14 shows the catalytic response in excess glucose as a function of the number of self-assembled polymer-enzyme layer pairs. For thin enzyme films the biocatalytic current increases proportionally to the number of enzyme layers, as expected for kinetic case I in the Pratt–Bartlett case diagram:

$$I = nFA[Os(III)]ke_{\Sigma}L \tag{42.14}$$

where $n$ is the number of electrons exchanged, $F$ the Faraday constant, $A$ the electrode area, [Os(III)] the concentration of oxidized osmium, $e_{\Sigma}$ the total surface enzyme concentration, and $L$ the film thickness.

However, for thicker enzyme films, the biocatalytic current is no longer proportional to the film thickness and reaches a plateau when the conditions for kinetic case II are fulfilled:

$$I = nFA[Os(III)](D_e k e_{\Sigma})^{1/2} \tag{42.15}$$

with $D_e$ the electron hopping diffusion coefficient. The boundary conditions between cases I and II correspond to $L = (D_e k e_{\Sigma})^{1/2}$ and this critical thickness is reached after some ten dipping cycles in the LbL method or 770 nm in Figure 42.13 for $D_e = 1.2 \times 10^{-9}$ cm$^2$ s$^{-1}$, $k = 2 \times 10^4$ M$^{-1}$ s$^{-1}$ and $e_{\Sigma} = 5 \times 10^{-4}$ M. Notice the different enzyme concentration dependence in these two kinetic cases.

An important finding of the early studies with LbL "wired enzyme multilayers" [2, 11, 30, 132] was that, for thin films, only a small fraction of the active assembled GOx molecules are "electrically wired" by the redox polymer, even though all of the enzyme molecules can be oxidized by a soluble mediator [2, 11, 30]. However, it has recently been shown that the wiring efficiency increases with the number of layers and for films thicker than 300 nm the LbL enzyme multilayer approaches the behavior of redox hydrogels [133].

In a seminal work Hodak [2] reported that the redox property of ferrocene-modified poly-(allyamine) multilayers depends significantly on the number of layers, with a decrease in redox charge for negatively charged GOx topmost layers. Furthermore, in ref. [30] it has been shown that, upon addition of a single layer of osmium complex tethered to polyallylamine, PAH-Os, adsorbed on top of the terminal GOx layer with the same enzyme concentration, the catalytic response was significantly larger and the amount of "wired enzyme" increased from 2.1 to 24% of the total enzyme present on the surface [30]. Unlike random polymers, LbL films exhibit the unique ability to control the nature and charge of the topmost layer at the polymer/solution interface.

As described above, another unique possibility with LbL enzyme multilayer films is the ability to control the position of each component in stratified structures: Wolosiuk [9] intercalated GOx layers with inactive FAD-free apo-GOx layers. In amperometric bienzyme electrodes, LbL strategy allowed the separation in space of two enzymes working in cascade, that is, soy beam peroxidase (SBP) and GOx or glucose dehydrogenase (GDH) and GOx [30].

### 42.3.5
### Core–Shell Nanoparticles

An interesting application of redox active LbL films containing enzymes is the biosensor described by Scodeller and coworkers. The authors reported a core–shell system comprised of gold nanoparticles with a LbL film shell containing both GOx and the osmium polymer PAH-Os molecular wire. The system exploited both the Raman scattering and electrochemical properties of core–shell gold–GOx nanoparticles to develop a surface-enhanced Raman-based sensor for glucose in the low millimolar concentration range [32]. The proof of concept with a wired-enzyme core–shell Au nanoparticle biosensor has extended the work done with wired enzymes in LbL planar surfaces to core–shell nanoparticles.

The core–shell functional nanoparticles were characterized by EQCM, spectroscopy, and electrochemical techniques which showed that the catalytically active shell has a structure as designed and all components are active in the self-assembled multilayer shell. Furthermore, amperometric reagentless detection of glucose and contactless photonic biosensing by the Os(II) resonant Raman signal were demonstrated. The enzymatic reduction of FAD by glucose, and further reduction of the Raman silent Os(III) by $FADH_2$, yields a characteristic enzyme–substrate calibration curve in the milimolar range. Furthermore, coupling of electronic resonant Raman of the osmium complex with the SERS amplification by Au NPs plasmon resonance has been demonstrated, which leads to an extra enhancement of the biosensor signal.

### 42.3.6
### Nanoreactors

An interesting application of LbL polyelectrolyte films acting as nanoreactors is to entrap ions, which subsequently react to yield interesting nanoparticles inside the polyelectrolyte multilayer by confining the ions in the nanoscale, thus preventing aggregation of large crystals. For instance, multiple-color electrochromism has been described in layer-by-layer-assembled poly-aniline/Prussian Blue nanocomposite thin films [32, 45], which leads to electrochemical control of swelling and mechanical properties [47]. Tunable synthesis of Prussian Blue in exponentially growing polyelectrolyte multilayer films has been reported by Boulmedais and coworkers [48].

Vago and coworkers described a highly efficient and selective material for electrocatalytic hydrogenation prepared by depositing monodisperse $(6 \pm 1)$ nm palladium nanoparticles by electrochemical reduction of $PdCl_4^{2-}$ confined in a PEM structure [134]. This work demonstrated that homogeneous metal nanoparticles can be electrodeposited within PEMs, resulting in electrodes with enhanced performance as compared to those containing microparticles. The crystal size and structure seem to determine the product selectivity while improving the efficiency due to a large area to volume ratio.

While this was the first electrochemical reduction of confined metal ions in the LbL nanostructure, previous work by Rubner, Cohen and coworkers had described a method that used LbL PEMs as nanoreactors to grow inorganic nanocrystals from coordinated metal ions [135–137]. In that strategy, electrostatically self-assembled multilayers were immersed in a solution of cations $Ag^+$, $Pb^{2+}$ or $Pd(NH_3)_4^{2+}$, which bind to carboxylic groups in poly(acrylic acid) (PAA); the ion-exchanged cations incorporated in the film were subsequently converted into nanoparticles through a chemical reaction (i.e., reduction or sulfidation). Bruening et al. developed an alternative method where palladium ions were co-deposited during multilayer formation and the system was applied to selective hydrogenation catalysis [138–140].

### 42.3.7
### Biofuel Cell Cathodes

Szamocki and coworkers studied high potential purified Trametes trogii laccase as a biocatalyst for the four-electron oxygen reduction in cathodes composed of layer-by-layer fungal enzyme laccase in anionic form and a polycation osmium redox mediator on mercaptopropane sulfonate modified gold, to produce an all integrated "wired" enzymatic oxygen cathode [33]. The adsorption of laccase was followed by monitoring the mass uptake with a quartz crystal microbalance and the oxygen reduction electrocatalysis was studied by linear scan voltammetry using a rotating disc electrode. For the fully integrated enzyme–mediator system a catalytic reduction of oxygen could be recorded at different oxygen partial pressures. The authors further reported an important role of hydrogen peroxide on the kinetics of the oxygen cathode using the LbL self-assembly strategy for biofuel cells [121]: detection of traces of $H_2O_2$, intermediate in the $O_2$ reduction, with scanning electrochemical microscopy (SECM) and the inhibition effect of peroxide on the biocatalytic current.

Dong and coworkers [141] prepared LbL self-assembled laccase, poly-L-lysine (PLL) and multi-walled carbon nanotube (MWNT) films and obtained a uniform growth of the multilayer, as shown by UV–vis spectroscopy and SECM. The catalytic behavior of the $(MWNTs/PLL/laccase)_n$ modified electrode was investigated for the four-electron reduction of $O_2$ to water without a mediator. The authors then applied the laccase-catalyzed $O_2$ reduction to a glucose/$O_2$ biofuel that exhibited an open-circuit voltage of 0.7 V, and achieved a maximum power density of $329\,\mu W\,cm^{-2}$ at cell voltage 470 mV.

The use of carbon nanotubes (CNTs) in this strategy allows one to extend a conducting CNT network in the direction normal to the substrate electrode and, therefore, to increase the electrochemical response. The unique properties of CNTs, particularly their huge surface area-to-weight ratio, make them extremely attractive amplification platforms, as shown by Wang and coworkers in 2005 [142]. Likewise, the strategy has been employed in LbL enzyme biosensing multilayers of multiwall CNTs modified with ferrocene-derivatized poly(allyl-amine) redox polymer and glucose oxidase by electrostatic self-assembly [143]. The incorporation of redox-polymer-functionalized CNTs into enzyme films resulted in a 6-10-fold increase in the glucose electrocatalytic current while the bimolecular rate constant of $FADH_2$ oxidation (wiring efficiency) was increased up to 12-fold.

Hammond [53] reported Au nanoparticle/multiwall carbon nanotubes (AuNP/MWNT) thin films fabricated by using the LbL assembly technique. The microstructure of the film characterized by TEM and AFM, showed that the MWNTs in the AuNP/MWNT LBL films are randomly entangled to form a dense network, in which the AuNPs are intercalated. Surface plasmon resonance of the AuNP/MWNT films increased linearly with the number of deposition cycles and exhibited a red shift with increasing number of layer pairs as a result of increasing sizes of the AuNP agglomerates in the films. The AuNP/MWNT films were electrochemically active toward methanol oxidation reaction in alkaline solution, and linear sweep voltammetry revealed a systematic increase in the methanol oxidation current with increasing number of layer pairs.

A very innovative concept recently introduced by the Hammond group has been all-MWCNT LbL-deposited thin films using surface functionalized MWNTs carrying, respectively, negative and positive charges [144]. The pH-dependent surface charge on the MWNTs gives these systems unique characteristics over weak polyelectrolytes LbL-assembled CNTs, controlling thickness and morphology with assembly pH conditions. The electrode arrays prepared with all-MWCNT LbL films, which do not use polyelectrolytes, exhibit higher conductivity than those assembled with polyelectrolytes, full electrochemical functionality, and have been employed in soft membranes [49, 145, 146] and stamped battery electrodes [147]. The versatile approach for fabricating and positioning electrodes may provide greater flexibility for implementing advanced battery designs, such as those with interdigitated microelectrodes or 3D architectures. These structures will find application in new generation fuel cells with membrane-electrode assemblies (MEAs) that are several times thinner, assume multiple geometries and, hence, will be more compact. Applications are also foreseen in high power lithium batteries from functionalized CNT electrodes [148].

An electrochromic material changes color at an applied potential, since the visible spectrum depends on the oxidation state and, therefore, can be modulated electrochemically. The first redox active electrochromic PEM was disclosed by Schlenoff using poly(butylviologen)/PSS films [149]. DeLongchamp and Hammond [150] reported the first solid state electrochromic device using LbL assembly of PEDOT/polyaniline with a fully functional switchable chromic device in which complementary coloration is achieved from films containing poly-(aniline) (PANI) and poly(3,4-ethylenedioxythiophene) (PEDOT). This first device achieved a maximum optical contrast of 30% and a switching time of 1 s. Replacing PANI with poly(hexylviologen) within the LbL assembly resulted in a high transmittance change of 82.1%, the highest reported in any LbL film structure, and among the highest reported in the literature for any polymeric electrochromic device [151]. A number of electrochromic and photovoltaic oxide systems has been proposed [152].

A multiple-color electrochromism from LbL-assembled PANI/Prussian Blue nanocomposite thin films has been reported by Longchamp and Hammond

[45, 153]. The electrochromic LbL electrode films were formed by electrostatic attraction between polycationic PANI and polyanions Prussian Blue nanoparticles dispersion. The non-interacting PANI and PB were fully electrochemically accessible in thick films with high contrast and fast switching from colorless to green to blue over a moderate potential range.

Finally, we shall comment on a LbL multilayer film particularly responsive to the presence of trace amounts of heavy metal ions in solution. Tagliazucchi reported a chemical sensor responsive to $Zn^{2+}$, $Cd^{2+}$, $Ni^{2+}$ and $Cu^{2+}$ soluble ions but insensitive to $Ca^{2+}$, $Pb^{2+}$, $La^{3+}$ [154]. The LbL films, composed of alternate layers of a cyano-osmium complex of pyridine tethered to poly(allyl-amine) and poly(acrylic acid) respectively and poly(vinylsulfonate), PVS, showed an amperometric response that was completely offset by a micromolar quantity of bivalent ions in solution, but could be restored reversibly by treating the electrode with EDTA solution. X-ray photoelectron spectroscopy and ATR IR studies demonstrated the interaction of the bivalent inhibiting cations with CN groups in the polymers. It is suggested that the blocking effect of the cations arises from a hindrance of polymer chain mobility due to coordination cross-linking.

## 42.4
## Conclusions

The molecular structure and gradient properties in the direction normal to the electrode/PEM plane interface results from the interplay of chemical and electrochemical equilibria with strong charge regulation due to long-range electrostatic and osmotic forces. Redox and acid–base equilibria are strongly coupled in LbL films containing weak polyelectrolytes, which can modify the degree of protonation with the electrolyte solution pH or the state of redox charge.

Permselectivity to small ions in solution arises from Donnan or surface potential due to fixed charges in the reluctant ion exchanger films where the surface charge plays a key role.

Under electrical or chemical gradients, the electron donor and acceptor chemical groups in polyelectrolytes self-assembled LbL multilayers give rise to exchange of electrons, ions and solvent through these thin films. The resulting dynamics of redox-PEM electrodes are characterized by ion and solvent exchange with the bathing electrolyte solution, which also results in swelling and mechanical responses to the perturbing gradients.

The ability to control the LbL organized structure and the nature of the charge in the topmost layer provides unique possibilities to design systems and devices for a variety of applications.

Redox LbL self assembled multilayers on conducting substrates are a type of chemically modified electrode with a large range of applications, such as biosensors, nanocatalysts, proton exchange membranes, electrochromic devices, battery electrodes, fuel cells and biofuel cells, photovoltaics and dye-sensitized solar cells, and so on.

## References

1. Laurent, D. and Schlenoff, J.B. (1997) *Langmuir*, **13**, 1552.
2. Hodak, J., Etchenique, R., Calvo, E.J., Singhal, K., and Bartlett, P.N. (1997) *Langmuir*, **13**, 2708.
3. Flexer, V., Forzani, E.S., Calvo, E.J., Luduenä, S.J., and Pietrasanta, L.I. (2000) *Anal. Chem.*, **78**, 399.
4. Calvo, E.J. and Wolosiuk, A. (2002) *J. Am. Chem. Soc.*, **124**, 8490.
5. Tagliazucchi, M., Grumelli, D., Bonazzola, C., and Calvo, E.J. (2006) *J. Nanosci. Nanotech.*, **6**, 1731.
6. Tagliazucchi, M., Grumelli, D., and Calvo, E.J. (2006) *Phys. Chem. Chem. Phys.*, **8**, 5086.
7. Grumelli, D.E., Garay, F., Barbero, C.A., and Calvo, E.J. (2006) *J. Phys. Chem. B*, **110**, 15345.
8. Grumelli, D.E., Wolosiuk, A., Forzani, E., Planes, G.A., Barbero, C., and Calvo, E.J. (2003) *Chem. Commun.*, 3014.
9. Calvo, E.J., Danilowicz, C., and Wolosiuk, A. (2002) *J. Am. Chem. Soc.*, **124**, 2452.
10. Calvo, E.J., Danilowicz, C.B., and Wolosiuk, A. (2005) *Phys. Chem. Chem. Phys.*, **7**, 1800.
11. Calvo, E.J., Etchenique, R., Pietrasanta, L., Wolosiuk, A., and Danilowicz, C. (2001) *Anal. Chem.*, **73**, 1161.
12. Calvo, E.J., Forzani, E., and Otero, M. (2002) *J. Electroanal. Chem.*, **538–539**, 231.
13. Calvo, E.J., Forzani, E.S., and Otero, M. (2002) *Anal. Chem.*, **74**, 3281.
14. Calvo, E.J. and Wolosiuk, A. (2004) *Chemphyschem*, **5**, 235.
15. Calvo, E.J. and Wolosiuk, A. (2005) *Chemphyschem*, **6**, 43.
16. Narváez, A., Suárez, G., Popescu, I.C., Katakis, I., and Domínguez, E. (2000) *Biosens. Bioelectron.*, **15**, 43.
17. Sun, J., Sun, Y., Zou, S., Zhang, X., Sun, C., Wang, Y., and Shen, S. (1999) *Macromol. Chem. Phys.*, **200**, 840.
18. Liu, A. and Anzai, J. (2003) *Langmuir*, **19**, 4043.
19. Liu, A., Kashiwagi, Y., and Anzai, J. (2003) *Electroanalysis*, **15** (13), 1139–1142.
20. Liu, A.H. and Anzai, J. (2004) *Anal. Bioanal. Chem.*, **380**, 98.
21. Schlenoff, J.B., Laurent, D., Ly, H., and Stepp, J. (1998) *Adv. Mater.*, **10**, 347.
22. Schlenoff, J.B., Ly, H., and Li, M. (1998) *J. Am. Chem. Soc.*, **120**, 7626.
23. Jason, S. and Schlenoff, B.J. (1997) *J. Electrochem. Soc.*, **144**, 155.
24. Ferreyra, N.F., Coche-Guerente, L., Labbe, P., Calvo, E.J., and Solis, V.M. (2003) *Langmuir*, **19**, 3864.
25. Sarkar, N., KuRam, M., Sarkar, A., Narizzano, R., Paddeu, S., and Nicolini, C. (2000) *Nanotechnology*, **11**, 30.
26. Lukkari, J., Salomäki, M., Viinikanoja, A., Ääritalo, T., Paukkunen, J., Kocharova, N., and Kankare, J. (2001) *J. Am. Chem. Soc.*, **123**, 6083.
27. Liu, S., Kurth, D.G., Bredenkötterand, B., and Volkmer, D. (2002) *J. Am. Chem. Soc.*, **124**, 12279.
28. Cheng, L., Niu, L., Gong, J., and Dong, S. (1999) *Chem. Mater.*, **11**, 1465.
29. Girina, G.P., Ovsyannikova, E.V., and Alpatova, N.M. (2007) *Russ. J. Electrochem.*, **43**, 1026.
30. Calvo, E.J., Battaglini, F., Danilowicz, C., Wolosiuk, A., and Otero, M. (2000) *Faraday Discuss.*, **116**, 47.
31. Calvo, E.J., Flexer, V., Tagliazucchi, M., and Scodeller, P., (2010) *Phys. Chem. Chem. Phys.*, **12**, 10033–10039.
32. Scodeller, P., Flexer, V., Szamocki, R., Calvo, E.J., Tognalli, N., Troiani, H., and Fainstein, A. (2008) *J. Am. Chem. Soc.*, **130**, 12690.
33. Szamocki, R., Flexer, V., Levin, L., Forchiasin, F., and Calvo, E.J. (2009) *Electrochim. Acta*, **54**, 1970.
34. Bonk, S.M. and Lisdat, F. (2009) *Biosens. Bioelectron.*, **25**, 739.
35. Balkenhohl, T., Adelt, S., Dronov, R., and Lisdat, F. (2008) *Electrochem. Commun.*, **10**, 914.
36. Spricigo, R., Dronov, R., Rajagopalan, K.V., Lisdat, F., Leimkuhler, S., Scheller, F.W., and Wollenberger, U. (2008) *Soft Matter*, **4**, 972.

37 Beissenhirtz, M.K., Kafka, B., Schafer, D., Wolny, M., and Lisdat, F. (2005) *Electroanalysis*, **17**, 1931.

38 Beissenhirtz, M.K., Scheller, F.W., Stocklein, W.F.M., Kurth, D.G., Mohwald, H., and Lisdat, F. (2004) *Angew. Chem. Int. Edit.*, **43**, 4357.

39 Crespilho, F.N., Zucolotto, V., Brett, C.M.A., Oliveira, O.N., and Nart, F.C. (2006) *J. Phys. Chem. B*, **110**, 17478.

40 Crespilho, F.N., Huguenin, F., Zucolotto, V., Olivi, P., Nart, F.C., and Oliveira, O.N. (2006) *Electrochem. Commun.*, **8**, 348.

41 Crespilho, F.N., zucolotto, V., Oliveira, O.N., and Nart, F.C. (2006) *Int. J. Electrochem. Sci.*, **1**, 194.

42 Yoon, H.C., Hong, M.Y., and Kim, H.S. (2000) *Anal. Chem.*, **72**, 4420.

43 Yoon, H.C., Hong, M.Y., and Kim, H.S. (2000) *Anal. Biochem.*, **282**, 121.

44 Yoon, H.C. and Kim, H.S. (2000) *Anal. Chem.*, **72**, 922.

45 DeLongchamp, D.M. and Hammond, P.T. (2004) *Chem. Mater.*, **16**, 4799.

46 Zhao, W., Xu, J.J., Shi, C.G., and Chen, H.Y. (2005) *Langmuir*, **21**, 9630.

47 Schmidt, D.J., Cebeci, F.C., Kalcioglu, Z.I., Wyman, S.G., Ortiz, C., Van Vliet, K.J., and Hammond, P.T. (2009) *ACS Nano*, **3**, 2207.

48 Laugel, N., Boulmedais, F., El Haitami, A.E., Rabu, P., Rogez, G., Voegel, J.C., Schaaf, P., and Ball, V. (2009) *Langmuir*, **25**, 14030.

49 Farhat, T.R. and Hammond, P.T. (2006) *Adv. Func. Mater.*, **16**, 433.

50 Ferreyra, N., Coche-Guerente, L., Fatisson, J., Teijelo, M.L., and Labbe, P. (2003) *Chem. Commun.*, 2056.

51 Schlenoff, J.B. (2003) in *Multilayer Thin Films Charge Balance and Transport in Polyelectrolyte Multilayers* (eds G. Decher and J.B. Schlenoff), Wiley-VCH Verlag GmbH, Weinheim, p. 99.

52 Tagliazucchi, M. and Calvo, E.J. (2009) in *Chemically Modified Electrodes Vol. 11: Electrochemically Active Polyelectrolyte-Modified Electrodes* (eds R.C. Alkire, D.M. Kolb, J. Lipkowski, and P.N. Ross), Wiley-VCH Verlag GmbH, Weinheim, p. 57.

53 Hammond, P.T. (2004) *Adv. Mater.*, **16**, 1271.

54 Bard, A.J. and Faulkner, L.R. (2001) *Electrochemical Methods*, John Wiley and Sons, New York.

55 Zahn, R., Boulmedais, F., Voros, J., Schaaf, P., and Zambelli, T. (2010) *J. Phys. Chem. B*, **114**, 3759.

56 Doblhofer, K. and Vorotyntsev, M. (1994) in *Electroactive Polymer Electrochemistry. Fundamentals* (ed. M.E.G. Lyons), Plenum, New York, p. 375.

57 Redepenning, J. and Anson, F.C. (1987) *J. Phys. Chem.*, **91**, 4549.

58 Tagliazucchi, M., Williams, F.J., and Calvo, E.J. (2007) *J. Phys. Chem. B*, **111**, 8105.

59 Ohshima, H. and Ohki, S. (1985) *Biophys. J.*, **47**, 673.

60 Tagliazucchi, M., Calvo, E.J., and Szleifer, I. (2010) *AICHE J.*, **56**, 1952.

61 Tagliazucchi, M., Calvo, E.J., and Szleifer, I. (2008) *J. Phys. Chem. C*, **112**, 458.

62 Tagliazucchi, M., Calvo, E.J., and Szleifer, I. (2008) *Langmuir*, **24**, 2869.

63 Tagliazucchi, M., Calvo, E.J., and Szleifer, I. (2008) *Electrochim. Acta*, **53**, 6740.

64 Laviron, E. (1979) *J. Electroanal. Chem.*, **101**, 19.

65 Laviron, E., Roullier, L., and Degrand, C. (1980) *J. Electroanal. Chem*, **112**, 11.

66 Brown, A.P. and Anson, F.C. (1977) *Anal. Chem.*, **49**, 1589.

67 Albery, W.J., Boutelle, M.G., Colby, P.J., and Hillman, A.R. (1982) *J. Electroanal. Chem.*, **133**, 135.

68 Chidsey, C.E.D. and Murray, R.W. (1986) *J. Phys. Chem.*, **90**, 1479.

69 Smith, C.P. and White, H.S. (1992) *Anal. Chem.*, **64**, 2398.

70 Redepenning, J., Miller, B.R., and Burnham, S. (1994) *Anal. Chem.*, **66**, 1560.

71 Tagliazucchi, M. and Calvo, E.J. (2009) in *Chemically Modified Electrodes Vol. 11: Electrochemically Active Polyelectrolyte-Modified Electrodes* (eds R.C. Alkire, D.M. Kolb, J. Lipkowski, and P.N. Ross),

Wiley-VCH Verlag GmbH, Weinheim, p. 57.

72 Fushimi, T., Oda, A., Ohkita, H., and Ito, S. (2005) *Thin Solid Films*, **484**, 318.

73 Tagliazucchi, M.E. and Calvo, E.J. (2007) *J. Electroanal. Chem.*, **599**, 249.

74 Ruff, I. and Botar, L. (1985) *J. Chem. Phys.*, **83**, 1292.

75 Dahms, H. (1968) *J. Phys. Chem.*, **72**, 362.

76 Lyons, M.E.G. (1994) *Electroactive Polymer Electrochemistry*, Plenum Press, New York.

77 Blauch, D.N. and Savéant, J.M. (1993) *J. Phys. Chem.*, **97**, 6444.

78 Blaunch, D.N. and Saveant, J.M. (1992) *J. Am. Chem. Soc.*, 3323.

79 Anne, A., Demaille, C., and Moiroux, J. (2001) *J. Am. Chem. Soc.*, **123**, 4817.

80 Mao, F., Mano, N., and Heller, A. (2003) *J. Am. Chem. Soc.*, **125**, 4951.

81 Aoki, A. and Heller, A. (1993) *J. Phys. Chem.*, **97**, 11014.

82 Aoki, A., Rajagopalan, R., and Heller, A. (1995) *J. Phys. Chem.*, **99**, 5102.

83 Oh, S.M. and Faulkner, L.R. (1989) *J. Electroanal. Chem.*, **269**, 77.

84 Oh, S.M. and Faulkner, L.R. (1989) *J. Am. Chem. Soc.*, **111**, 5613.

85 Xie, A.F. and Granick, S. (2002) *Macromolecules*, **35**, 1805.

86 Tagliazucchi, M. and Calvo, E.J. (2010) *Chemphyschem*, **11**, 2957.

87 Notley, S.M., Eriksson, M., and Wagberg, L. (2005) *J. Colloid Interface Sci.*, **292**, 29.

88 Lindholm-Sethson, B. (1996) *Langmuir*, **12**, 3305.

89 Harris, J.J., Stair, J.L., and Bruening, M.L. (2000) *Chem. Mater.*, **12**, 1941.

90 Harris, J.J. and Bruening, M.L. (2000) *Langmuir*, **16**, 2006.

91 Farhat, T.R. and Schlenoff, J.B. (2001) *Langmuir*, **17**, 1184.

92 Daum, P. and Murray, R.W. (1981) *J. Phys. Chem.*, **85**, 389.

93 Barbero, C., Kotz, R., Kalaji, M., Nyholm, L., and Peter, L.M. (1993) *Synth. Met.*, **55**, 1545.

94 Grumelli, D., Bonazzola, C., and Calvo, E.J. (2006) *Electrochem. Commun.*, **8**, 1353.

95 Schlenoff, J.B., Ly, H., and Li, M. (1998) *J. Am. Chem. Soc.*, **120**, 7626.

96 Forzani, E.S., Perez, M.A., Teijelo, M.L., and Calvo, E.J. (2002) *Langmuir*, **18**, 9867.

97 Zahn, R., Voros, J., and Zambelli, T., (2010) *Curr. Opin. Colloid Interface.*, **15** (6), 427–434.

98 Guillaume-Gentil, O., Graf, N., Boulmedais, F., Schaaf, P., Voros, J., and Zambelli, T. (2010) *Soft Matter*, **6**, 4246.

99 Grieshaber, D., Voros, J., Zambelli, T., Ball, V., Schaaf, P., Voegel, J.C., and Boulmedais, F. (2008) *Langmuir*, **24**, 13668.

100 El Haitami, A.E., Martel, D., Ball, V., Nguyen, H.C., Gonthier, E., Labbe, P., Voegel, J.C., Schaaf, P., Senger, B., and Boulmedais, F. (2009) *Langmuir*, **25**, 2282.

101 Calvo, E.J., Forzani, E., and Otero, M. (2002) *J. Electroanal. Chem.*, **538**, 231.

102 Schanze, K.S. and Shelton, A.H. (2009) *Langmuir*, **25**, 13698.

103 Stipek, S. and Calvo, E.J. (2008) in *Piezoelectric Transducers and Applications Biosensors: Natural Systems and Machines* (ed. A. Arnau), Springer Verlag, Berlin, p. 259.

104 Heller F A. (1990) *Acc. Chem. Res.*, **23**, 128.

105 Sun, S.C., Hosi, P.H., and Harrison, D.J. (1991) *Langmuir*, **7**, 727.

106 Mizutani, F., Sato, Y., Yabuki, S., and Hirata, Y. (1996) *Chem. Lett.*, 251.

107 Kinnear, K.T. and Monbouquette, H.G. (1993) *Langmuir*, **9**, 2255.

108 Guiomar, A.J., Guthrie, J.T., and Evans, S.D. (1999) *Langmuir*, **15**, 1198.

109 Onda, M., Lvov, Y., Ariga, K., and Kunitake, T. (1996) *J. Ferment. Bioeng.*, **82**, 502.

110 Lvov, Y. (2000) in *Protein Architecture*, vol. ch. 6 (eds Y. Lvov and H. Mohwald), Marcel Dekker, New York, p. 125.

111 Lvov, Y., Ariga, K., Ichinose, I., and Kunitake, T. (1995) *J. Am. Chem. Soc.*, **117**, 6117.

112 Bourdillon, C., Demaille, C., Moiroux, J., and Saveant, J.M. (1996) *Acc. Chem. Res.*, **29**, 529.

113 Bourdillon, C., Demaille, C., Moiroux, J., and Saveant, J.M. (1995) *J. Am. Chem. Soc.*, **117**, 11499.

114 Bourdillon, C., Demaille, C., Moiroux, J., and Saveant, J.M. (1994) *J. Am. Chem. Soc.*, **116**, 10328.

115 Anicet, N., Anne, A., Moiroux, J., and Saveant, J.M. (1998) *J. Am. Chem. Soc.*, **120**, 7115.
116 Anicet, N., Bourdillon, C., Moiroux, J., and Saveant, J.M. (1999) *Langmuir*, **15**, 6527.
117 Anicet, N., Bourdillon, C., Moiroux, J., and Saveant, J.M. (1998) *J. Phys. Chem. B*, **102**, 9844.
118 Rusling, J.F. (2000) in *Protein Architecture*, vol. **ch. 13** (eds Y. Lvov and H. Mohwald), Marcel Dekker, New York, p. 337.
119 Franchina J.G., Lackowski, W.M., Dermody, D.L., Crooks, R.M., Bergbreiter, D.E., Sirkar, K., Russell, R.J., and Pishko, M.V. (1999) *Anal. Chem.*, **71**, 3133.
120 Flexer, V., Forzani, E.S., Calvo, E.J., Luduenä, S.J., and Pietrasanta, L.I. (2006) *Anal. Chem.*, **78**, 399.
121 Scodeller, P., Carballo, R., Szamocki, R., Levin, L., Forchiassin, F., and Calvo, E.J. (2010) *J. Am. Chem. Soc.*, **132**, 11132.
122 Zhang, S.X., Yang, W.W., Niu, Y.M., Li, Y.C., Zhang, M., and Sun, C.Q. (2006) *Anal. Bioanal. Chem.*, **384**, 736.
123 Dominguez, E., Suarez, G., and Narvaez, A. (2006) *Electroanalysis*, **18**, 1871.
124 Narvaez, A., Suarez, G., Popescu, I.C., Katakis, I., and Dominguez, E. (2000) *Biosens. Bioelectron.*, **15**, 43.
125 Sirkar, K., Revzin, A., and Pishko, M.V. (2000) *Anal. Chem.*, **72**, 2930.
126 Cui, X.Q., Pei, R.J., Wang, X.Z., Yang, F., Ma, Y., Dong, S.J., and Yang, X.R. (2003) *Biosens. Bioelectron.*, **18**, 59.
127 Bartlett, P.N., Toh, C.S., Calvo, E.J., and Flexer, V. (2008) in *Bioelectrochemistry: Fundamentals, Experimental Techniques and Applications Modelling Biosensor Responses* (ed. P.N. Bartlett), John Wiley & Sons, New York, p. 266.
128 Bartlett, P.N. and Pratt, K.F.E. (1995) *J. Electroanal. Chem.*, **397**, 61.
129 Bartlett, P.N. and Pratt, K.F.E. (1993) *Biosens. Bioelectron.*, **8**, 451.
130 Flexer, V., Calvo, E.J., and Bartlett, P.N. (2010) *J. Electroanal. Chem.*, **646**, 24.
131 Flexer, V., Pratt, K.F.E., Garay, F., Bartlett, P.N., and Calvo, E.J. (2008) *J. Electroanal. Chem.*, **616**, 87.
132 Calvo, E.J., Battaglini, F., Danilowicz, C., Wolosiuk, A., and Otero, M. (2000) *Faraday Discuss.*, **116**, 47–65.
133 Flexer, V., Calvo, E.J., and Bartlett, P.N. (2009) *J. Electroanal. Chem.*, **646**, 24.
134 Vago, M., Tagliazucchi, M., Williams, F.J., and Calvo, E.J. (2008) *Chem. Commun.*, 5746.
135 Joly, S., Kane, R., Radzilowski, L., Wang, T., Wu, A., Cohen, R.E., Thomas, E.L., and Rubner, M.F. (2000) *Langmuir*, **16**, 1354.
136 Wang, T.C., Rubner, M.F., and Cohen, R.E. (2002) *Langmuir*, **18**, 3370.
137 Wang, T.C., Rubner, M.F., and Cohen, R.E. (2003) *Chem. Mater.*, **15**, 299.
138 Dai, J.H. and Bruening, M.L. (2002) *Nano Lett.*, **2**, 497.
139 Kidambi, S. and Bruening, M.L. (2005) *Chem. Mater.*, **17**, 301.
140 Kidambi, S., Dai, J.H., Li, J., and Bruening, M.L. (2004) *J. Am. Chem. Soc.*, **126**, 2658.
141 Deng, L., Shang, L., Wang, Y.Z., Wang, T., Chen, H.J., and Dong, S.J. (2008) *Electrochem. Commun.*, **10**, 1012.
142 Munge, B., Liu, G.D., Collins, G., and Wang, J. (2005) *Anal. Chem.*, **77**, 4662.
143 Deng, L., Liu, Y., Yang, G.C., Shang, L., Wen, D., Wang, F., Xu, Z., and Dong, S.J. (2007) *Biomacromolecules*, **8**, 2063.
144 Lee, S.W., Kim, B.S., Chen, S., Shao-Horn, Y., and Hammond, P.T. (2009) *J. Am. Chem. Soc.*, **131**, 671.
145 Farhat, T.R. and Hammond, P.T. (2006) *Chem. Mater.*, **18**, 41.
146 Farhat, T.R. and Hammond, P.T. (2005) *Adv. Func. Mater.*, **15**, 945.
147 Nam, K.T., Wartena, R., Yoo, P.J., Liau, F.W., Lee, Y.J., Chiang, Y.M., Hammond, P.T., and Belcher, A.M. (2008) *Proc. Natl. Acad. Sci. U. S. A.*, **105**, 17227.
148 Lee, S.W., Yabuuchi, N., Gallant, B.M., Chen, S., Kim, B.S., Hammond, P.T., and Shao-Horn, Y., (2010) *Nat. Nanotechnol.*, **5** (7), 531–537.

149 Stepp, J. and Schlenoff, J.B. (1997) *J. Electrochem. Soc.*, **144**, L155.
150 DeLongchamp, D. and Hammond, P.T. (2001) *Adv. Mater.*, **13**, 1455.
151 DeLongchamp, D.M., Kastantin, M., and Hammond, P.T. (2003) *Chem. Mater.*, **15**, 1575.
152 Lutkenhaus, J.L. and Hammond, P.T. (2007) *Soft Matter*, **3**, 804.
153 DeLongchamp, D.M. and Hammond, P.T. (2004) *Adv. Func. Mater.*, **14**, 224.
154 Tagliazucchi, M., Williams, F.J., and Calvo, E.J. (2010) *Chem. Commun.*, **46**, 9004.

# 43
# Multilayer Polyelectrolyte Assembly in Feedback Active Coatings and Films

*Dmitry G. Shchukin and Helmuth Möhwald*

## 43.1
### Introduction. The Concept of Feedback Active Coatings

Development of a new generation of multifunctional coatings and films, which will possess not only passive functionality but also active and rapid feedback activity in response to changes in local environment, is a key technology for the fabrication of future high-tech products and functional surfaces. These new multifunctional coatings should combine passive components inherited from "classical" ones (barrier, color, adhesion), and active components which provide fast response of the coating properties to changes occurring either in the matrix of multifunctional coatings and films (e.g., cracks, local pH change, bacteria activity), or in the local environment surrounding the coating (temperature, humidity). From this perspective, the coatings should also have several functionalities (e.g., antireflection, antifungal and anticorrosion) exhibiting synergistic effects.

The coatings should provide sustained or immediate release of the active material on demand, within a short time after changes in the environment or in film integrity. The active part of the coating has to be incorporated into a passive matrix or form a layered structure together with the passive matrix. Recent developments in surface science and technology show new opportunities for modern engineering concepts for the fabrication of active feedback coatings through the integration of nanoscale layers (carriers) loaded with the active compounds (e.g., inhibitor, lubricant) into existing "classical" films, thus designing completely new coating systems of the "passive" host–"active" guest structure.

The most important part in the design of new active films is the development of active layers or carriers incorporated into the coating matrix. There are several approaches demonstrated so far for the design of the active components of the feedback active coatings: (i) polymer containers [1], (ii) polymer or glass fibers [2], (iii) nanocontainers with a polyelectrolyte shell [3], (iv) layered double hydroxides and mesoporous inorganic materials [4] and, finally, design of the coatings employing polyelectrolyte layer-by-layer assembly (LbL) multilayer approach [5]. All of the mentioned methods have their own advantages and drawbacks concerning the

---

*Multilayer Thin Films: Sequential Assembly of Nanocomposite Materials*, Second Edition.
Edited by Gero Decher and Joseph B. Schlenoff.
© 2012 Wiley-VCH Verlag GmbH & Co. KGaA. Published 2012 by Wiley-VCH Verlag GmbH & Co. KGaA.

upscaling possibility, performance, feasibility to employ different active materials, and so on, whose detailed description is out of the scope of the present chapter focusing on the most versatile, in our opinion, approach for the formation of feedback active coatings – polyelectrolyte LbL assembly.

The LbL templating technique attracted significant interest as a simple, highly variable approach that was widely spread to prepare nanostructured materials with tailored properties. The process involves the sequential deposition of species, such as polymers, nanoparticles, lipids, proteins and dye molecules, onto various templates, which are subsequently removed to yield free-standing structures. Since its introduction [6] the LbL assembly technique has expanded rapidly to become a popular method for the preparation of nanoscale films. Classically, the LbL deposition procedure involves the step-wise electrostatic assembly (usually by dipping into the polyelectrolyte-containing solution, which is time-consuming) of oppositely charged species (e.g., polyelectrolytes, proteins, nanoparticles) on the substrate surface with nanometer scale precision, forming a coating with multiple functionality. The coating properties, such as composition, thickness, and function, can be controlled by the number of deposition cycles, the conditions employed during the assembly process and the types of materials used. The versatility of the LbL approach allowed the formation of a broad range of films due to electrostatic interactions, hydrogen bonding, hydrophobic interactions, covalent bonding and complementary base pairing [7, 8].

Polyelectrolyte multilayer deposition can be simplified and speeded up enormously by applying the polyion-containing liquid by spray [9] rather than by bringing the receiving surface into contact with the solution by dipping. The quality of films with a deposition time of 60 s per layer is equal or superior to films obtained by dipping with a deposition time of 1520 s per layer (classical dipping conditions). This corresponds to an acceleration of the LbL deposition by a factor of about 25. LbL films prepared by dipping are always thicker than films prepared by spraying. The practice of spraying polyelectrolyte solutions onto a substrate in order to construct thin films via the LbL technique was extended to the automated approach [10]. Film growth is shown to be similar to that in conventional "dipped" LbL assembly, whereas the reported technology allows one to realize additional 25-fold decreases in process times.

Summarizing this short introduction, the LbL approach for deposition of various charged organic and inorganic species on a substrate offers a very versatile tool for surface modification, bringing the various functional properties (both active and passive) to the surface coatings and films in a controlled (on the nanoscale) manner.

## 43.2
### Polyelectrolyte-Based Self-Healing Anticorrosion Coatings

High-value materials, for example, in the automotive and aerospace industry require increasingly sophisticated coating for improved performance, self-healing and durability, and, in this respect, recent developments of nanotechnology are most promising. The need for a new generation of protective coating systems is important, especially taking into consideration the banning of carcinogenic Cr(VI) [11]. The anticorrosion coatings developed so far passively prevent the interaction of corrosive species with the metal. Corrosion processes develop rapidly after disruption of the protective barrier and are accompanied by a number of reactions that change the composition and properties of both the metal surface and the local environment (e.g., the formation of oxides, diffusion of metal cations into the coating matrix, and local changes in pH and electrochemical potential) [12].

An approach to active prevention of corrosion propagation on metal surfaces is the suppression of these physico-chemical reactions. The self-healing activity of the polyelectrolyte LbL-assembled multilayer system (Figure 43.1) could be based on three mechanisms arising from their inherent properties: (i) the polyelectrolytes

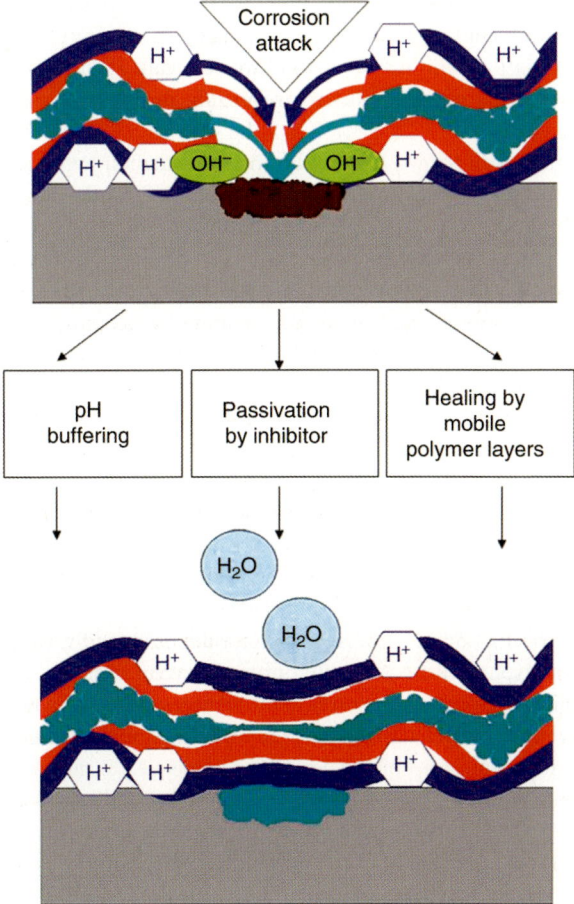

**Figure 43.1** Schematic mechanism of corrosion protection by LbL-assembled polyelectrolyte structures. Reprinted from Shchukin D.G. et al. (2010) *Appl. Mater. & Interfaces*, 2, 1954–1962. Copyright the WILEY VCH.

have pH-buffering activity and can stabilize the pH between 5 and 7.5 (the usual pH passivation range) at the metal surface in corrosive media; (ii) the inhibitors are released from polyelectrolyte multilayers only after the start of the corrosion process preventing the corrosion propagation directly in the rusted area; (iii) polyelectrolytes forming the coating are relatively mobile and have the tendency to seal and eliminate the mechanical cracks in the coating.

### 43.2.1
**Passive Protection Activity of Polyelectrolyte Multilayers**

One of the important characteristics of corrosion is the local changes in pH in the corroding area. Depending on the corrosion mechanism, intermetallic inclusions and surface properties of the metal substrate, the local pH shift could occur both to acidic and alkaline regions [13]. Therefore, local pH neutralization will prevent the corrosion process. Neutralization could be achieved by formation of a coating with pronounced pH-buffering activity on the metal surface, which could stabilize the pH between 5 and 7.5 at the metal surface in corrosive media, thus terminating the corrosive reactions.

Polyelectrolyte or composite polyelectrolyte/clay multilayers, due to their polyacid–polybase nature, can act as a buffering coating when LbL assembled on the metal surface [14, 15]. The mechanism of such a coating is to buffer the pH value at a metal surface leaving the metal in a passivated state and thus suppressing the appearance of corrosion. The films can be enriched in one polymer compared to the other by fabrication in a pH-regime where one of the polymers is weakly charged while the other is strongly charged.

A method of corrosion protection based on the buffering activity of the combination of a strong negative polyelectrolyte, poly(styrene sulfonate) (PSS) and a weak positive polyelectrolyte, poly(ethyleneimine) (PEI), deposited on aluminum alloy AA2024 in a LbL manner was presented in [16]. The PEI and PSS layers (of thickness about 8 nm, as estimated by ellipsometry) were deposited on degreased and etched aluminum alloy AA2024 by the spin coating method from a $2\,\text{mg}\,\text{ml}^{-1}$ solution of polyelectrolytes in water/ethanol (1 : 1, v/v) mixture. The corrosion behavior was studied by dipping the samples in aqueous NaCl solution at $20\,^\circ\text{C}$.

The introduced coatings possess active and passive corrosion protection, combining effective barrier properties with the possibility of termination of corrosion propagation in the coating. Figure 43.2 shows pictures of the samples coated with two PEI/PSS layer pairs (a) and without polyelectrolyte coating (b). Corrosion defects can be observed after 12 h of immersion in 0.1 M NaCl on the unmodified aluminum, whereas the sample with only two layer pairs does not exhibit any visible signs of corrosion attack, even after 7 days of immersion, without any detachment of the polyelectrolyte multilayers. So, the deposition of the polyelectrolyte coating consisting of two strong–weak polyelectrolyte layer pairs is enough to demonstrate the corrosion protection efficiency of the polyelectrolyte coating. This pronounced difference shows the very high buffering activity of the multilayer protection system against corrosion.

**Figure 43.2** (a) Aluminum alloy coated with (PEI/PSS)$_2$ polyelectrolyte complex in 0.1 M NaCl after 21 days of immersion. (b) Unmodified aluminum alloy in 0.1 M NaCl after 12 h of immersion. Reprinted from Shchukin, D.G. et al. (2008) *J. Mater. Chem.* **18**, 1738–1740. Copyright the Royal Society of Chemistry.

The self-healing efficiency of the PEI/PSS polyelectrolyte coatings was investigated by the scanning vibrating electrode technique (SVET). The SVET method obtains current density maps over the selected surface of the sample, thus allowing the monitoring of local cathodic and anodic activity in the corrosion zones. A small anodic activity was observed on the aluminum coated with PEI/PSS multilayers after defect formation. However, the defect zone on the polyelectrolyte-coated sample was passivated with time and there was no sign of the corrosion process detected after 3 h of immersion in the NaCl solution. Thus, the polyelectrolyte coating deposited on the surface of aluminum alloy heals the defect in the corrosion zone of the sample and terminates corrosion propagation after 3 h of immersion without addition of any functional inhibitors. Samples with more, 3–5, layer pairs exhibited faster self-healing of the coating in response to the damage.

The most probable mechanism for the PEI/PSS multilayer coating system is based on local neutralization of the pH in the area of the corrosion process by neutral ($(R)_3N$-, R-NH-R, R-NH$_2$) and charged ($(R)_3$-NH$^+$, $(R)_2$-NH$_2^+$-, R-NH$_3^+$) amine groups of PEI, which are able to bind the free hydroxide ions produced during the corrosion process in the corrosive area.

Combinations of strong–weak, weak–weak and strong–strong polyelectrolytes (positively charged poly(ethyleneimine), poly(diallyldimethylammonium) chloride (PDADMAC) and negatively charged poly(styrene sulfonate) and poly(acrylic acid) (PAA) were chosen to reveal their contribution to each corrosion–protection mechanism of polyelectrolytes: pH buffering, regeneration, and barrier to aggressive species.[5] Both PEI–PAA and PEI–PSS form a continuous nanonetwork on the aluminum surface differing slightly in the surface roughness. In contrast, the PDADMAC–PSS coating consists of approximately 20–40 nm aggregates homogeneously distributed on the surface. Formation of the aggregates could be caused by poor adhesion of the PDADMAC on the aluminum surface due to weak affinity of this polyelectrolyte to the aluminum. These results are in good agreement with the spectroscopic analysis and confirm the formation of a stable and continuous coating by PEI–PSS and PEI–PAA pairs and agglomerates of PDADMAC–PSS. As well as the PEI–PSS nanonetwork, the film of two weak polyelectrolytes PEI and PAA is able to protect the aluminum surface in aggressive media. The PEI–PAA buffering system suppresses propagation of corrosion degradation within 30 min. On the contrary, both anodic and catholic peaks appear on the aluminum surface covered by a strong–strong PDADMAC–PSS nanonetwork, resulting in defect propagation throughout the whole surface of the sample. The negative effect of strong–strong polyelectrolyte multilayers can be explained by their strong acidic or alkaline activity, carrying high charge densities over wide pH ranges. Therefore, interaction between them is too strong to be altered by pH change. Furthermore, due to the formation of the stable complex between PDADMAC and PSS, the interaction of PDADMAC with the substrate is very weak.

### 43.2.2
### Controlled Release of Inhibiting Agents

A multilayer anticorrosion system consisting of polyelectrolyte and inhibitor layers deposited on aluminum alloy surfaces pre-treated by sonication was demonstrated for the protection of aluminum and iron surfaces [17]. The design of our novel anticorrosion system is shown schematically in Figure 43.3. The poly(ethyleneimine) (MW $\sim$ 600–1000 kDa), poly(styrene sulfonate) (MW $\sim$ 70 kDa) and 8-hydroxyquinoline (8HQ) nanolayers are deposited on the pre-treated aluminum alloy AA2024 by spray drying from a 2 mg ml$^{-1}$ solution of polyelectrolytes in water/ethanol (1 : 1, v/v) mixture. The thickness of each layer was about 5–10 nm, as measured

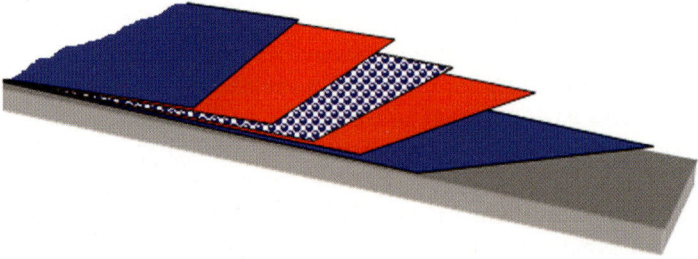

**Figure 43.3** Schematic diagram of the sandwich-like polyelectrolyte/inhibitor corrosion protection system on aluminum alloy. Reprinted from Shchukin D.G. et al. (2010) Appl. Mater. & Interfaces, 2, 1954–1962. Copyright the WILEY- VCH.

by ellipsometry. 8-Hydroxyquinoline was deposited between two PSS layers from 10 wt% solution of 8HQ in ethanol.

The formed LbL film was characterized by infrared absorption bands, which can be assigned to all film components.

Even the nanometer-thick polyelectrolyte/inhibitor coating provided effective corrosion protection for the aluminum alloy. The samples with polymer/inhibitor coating exhibited neither anodic activity nor corrosion propagation after their artificial scratching with the knife and placement in 0.1 M NaCl solution. The corrosion inhibitor incorporated, as a component of the LbL film, into the protective coating is responsible for the most effective mechanism of corrosion suppression. 8-HQ was found to prevent the adsorption of chloride ions and to improve the corrosion resistance due to the formation of an insoluble chelate of aluminum, which protects the oxide film [18]. Either mechanical (mechanical scratch) or chemical (polymer swelling due to changes in local pH) rupture of the polymer film causes the release of encapsulated inhibitor, which occurs in the damaged part of the metal surface. Therefore, the inhibitor release occurs in response to corrosion attack. This results in termination of the corrosion process and prolongation of corrosion protection. The long-term coating stability in a corrosive environment was also studied by dipping the samples in aqueous NaCl solution at 20 °C. Corrosion defects can be observed after 12 h of immersion in 0.1 M NaCl on the unmodified aluminum whereas the sample with the polymer/inhibitor complex does not exhibit any visible signs of corrosion attack even after 21 days of immersion.

## 43.3
### Coatings with Antibacterial Activity

Biofouling or biocontamination is relevant in a wide range of applications [19]: surgical equipment and protective apparel in hospitals [20], medical implants [21], biosensors [22], textiles [23], food packaging [24] and food storage [25], water purification systems [26], and marine and industrial equipment [27]. Surfaces that resist the nonspecific adsorption of protein and microbes are also vital in catheters, prosthetic devices, contact lenses, and in immunological assays [28].

The attachment of bacteria to a surface leads to subsequent colonization, resulting in the formation of a biofilm [29]. Fouling is also a concern in the case of surfaces subjected to aquatic environments where marine microorganisms can bind to a surface and form a conditioning layer, which then provides an easily accessible platform for other aquatic species, such as diatoms and algae, to attach and proliferate [30]. The critical issues associated with biofouling include increased operational and maintenance cost due to fouling of water conduits and ship hulls and the degradation of abiotic materials, among others [27]. The key strategies for preparing antifouling coatings are to resist the adhesion of biocontaminants, or to degrade or kill them [31]. There are several approaches for this, like the immobilization of polymers, such as poly(ethylene glycol), on surfaces, photoactivated self-cleaning coatings, incorporation of biocidal agents (silver, antibiotics, nanoparticles, polycations, enzymes, and antimicrobial peptides) and the use of structured surfaces. With regard to marine antifouling coatings, restrictions on the use of biocide-releasing coatings make the generation of nontoxic antifouling surfaces more important. While considerable progress has been made in the design of antifouling coatings, ongoing research in this area should result in the development of even better antifouling materials in the future.

The first report of silver nanoparticle-containing polyelectrolyte multilayers that exhibit antibacterial properties was published by Dai and Bruening [32]. Grunlan *et al.* also studied the antimicrobial activity of silver ion and/or cetrimide loaded polyelectrolyte multilayers [33]. Etienne *et al.* reported the antimicrobial properties of multilayers that contain an antibacterial peptide, defensin [34]. Antibacterial coatings based on hydrogen-bonded multilayers containing *in situ* synthesized Ag nanoparticles were created on planar surfaces [35]. The zone of inhibition determined by the disk-diffusion test increases as the thickness of the multilayer film is increased. This observation suggests that, in order to incrementally increase the zone of inhibition, an exponentially increasing amount of Ag is required within the multilayers. The duration of sustained release of antibacterial Ag ions from these coatings, however, could be prolonged by increasing the total supply of zerovalent silver in the films via multiple loading and reduction cycles. Both Gram-positive strain (S. epidermidis) and Gram-negative strain (E. coli) bacteria were susceptible to the biocidal activity of Ag nanoparticle-loaded multilayer thin films.

The layer-by-layer assembly of a biomimetic nanostructured composite from $Na^+$-montmorillonite clay nanosheets, starch-stabilized silver nanoparticles and poly(diallylmethylammonium chloride) was demonstrated [36]. The resulting composite showed excellent structural stability with no detectable levels of silver lost over a 1 month period. Evaluation of the antibacterial properties showed almost complete growth inhibition of E. coli over an 18 h period. The amount of silver eluted from the LbL composite over a 1 month period was determined to be only $0.5–3.0\,\mu g\,L^{-1}$.

Recently, Hammond and coworkers used the LbL deposition technique to fabricate polyelectrolyte multilayer films that incorporate gentamicin (GS), an antibiotic effective against biofilms of several staphylococci [37]. One of the salient features of this approach is the incorporation of charged small molecules without physical or chemical modification, rendering the construction of the film simple and efficient. The LbL heterostructure consisted of a hydrolytically degradable poly(β-aminoester), a biocompatible polyanionic hyaluronic acid (HA) and GS. For poly(β-aminoester), with a higher charge density and hence stronger electrostatic interactions, 95% of the GS release occurred over prolonged time periods (10 to 15 h) compared to the burst release of the drug observed in the case of a lower charge density (95% drug release within 4 h). Additionally, the release rate was shown to be tunable by varying the film architecture (the number and composition of multilayers).

Lichter *et al.* suggested that highly swelled multilayer films can efficiently reduce bacterial attachment [38]. By tuning the pH of assembly of the multilayer and postassembly conditions, polycationic sites can be generated in the multilayer (PSS/PAH, PAA/PAH), giving rise to biocidal activity, even without the incorporation of external biocidal agents. The multilayers showed biocidal activity at neutral pH against E. coli.

Li *et al.* fabricated dual-function antimicrobial materials based on silver ions or nanoparticles and immobilized quaternary ammonium salts (QAS) [39]. A reservoir region made of 20 layer pairs of poly(allylamine hydrochloride) (assembly pH 8.5) and poly(acrylic acid) (assembly pH 3.5) and a cap region made of 10 layer pairs of PAH (assembly pH 7.5) and silica nanoparticles (20 nm in diameter, assembly pH 9.5) can be constructed successfully through polyelectrolyte LbL deposition.

The silica nanoparticles had modified surface chemistry, which was exploited to tether QAS. QAS is a contact-based antimicrobial agent, as opposed to silver which produces its antimicrobial effect through a release-based mechanism and thus is depleted slowly. This fabrication strategy is illustrated in Figure 43.4. In the test conditions studied, killing efficiencies were greater than 99.0%. This coating scheme is not limited only to silver and quaternary ammonium salts but should also allow the incorporation of a variety of other antibacterial agents, such as antibiotics and titanium oxide.

A PEO-b-PHEMA block copolymer conjugate of the chemotherapeutic drug doxorubicin through acid-labile carbamate was synthesized and self-assembled to form a polymeric micelle [40]. Through hydrogen-bonding interactions between the corona of the polymeric micelles and biologically active tannic acid, bioactive multilayers were prepared. Release of the drug from the multilayers upon hydrolysis

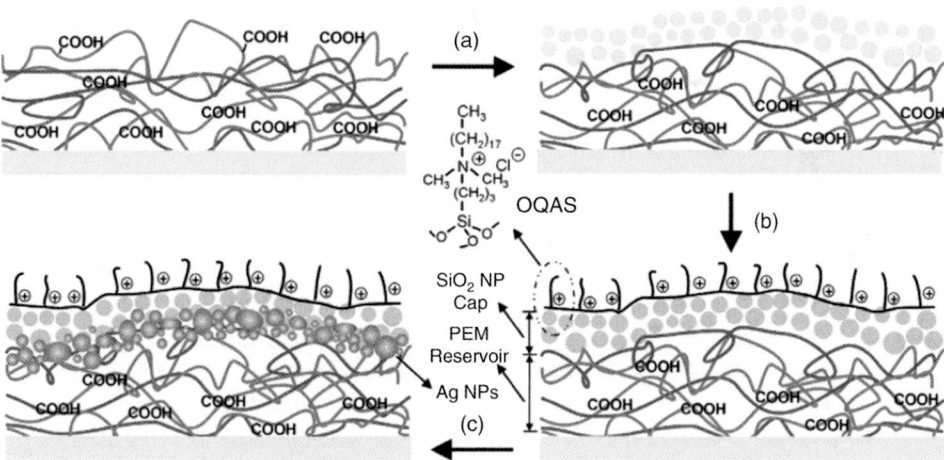

**Figure 43.4** Scheme showing the design of a two-level dual-functional antibacterial coating with both quaternary ammonium salts and silver. The coating process begins with LbL deposition of a reservoir made of layer pairs of PAH and PAA. (a) A cap region made of layer pairs of PAH and SiO$_2$ nanoparticles is added to the top. (b) The SiO$_2$ nanoparticles cap is modified with a quaternary ammonium silane, OQAS. (c) Ag$^+$ can be loaded inside the coating using the available unreacted carboxylic acid groups in the LbL multilayers. Reprinted from Rubner M.F. *et al.* (2006) *Langmuir* **22**, 9820–9823. Copyright the ACS.

of the pH-sensitive carbamate linkage resulted in a drastic cancer cell death, indicating that the cytotoxic effect of doxorubicin is preserved in the film construct.

Recently, it was shown that a layer of catalase or $Fe_3O_4$ nanoparticles deposited via LbL assembly on the outermost surface of PAS/PAH multilayered microcapsules can significantly inhibit oxidation of an encapsulated protein in hydrogen peroxide solutions [41]. The influence of a catalase (Cat) layer located at different depths in the LbL hemoglobin/polystyrene sulfonate films with an $(Hb/PSS)_{20-x}/(Cat/PSS)/(Hb/PSS)_x$ ($x$: 0–20) architecture on the kinetics of hemoglobin degradation was studied under treatment with hydrogen peroxide solutions of different concentrations and features of $H_2O_2$ decay in the surrounding solutions [42]. While assembled on the top of the multilayers, the catalase layer shows the highest activity in hydrogen peroxide decomposition. Hemoglobin in such films retains its native state for a longer period of time. The effect of catalase layers is compared with that of protamine, horseradish peroxidase, and inactivated catalase. Positioning an active layer with catalytic properties as an outer layer is the best protection strategy for LbL assembled films in aggressive media. A protection based on targeted decomposition of pollutants by the catalysts deposited on the outermost layer is preferable because the protection (decomposition) action can be repeated many times.

Modification of wood surfaces through noncovalent attachment of amine-containing water soluble polyelectrolytes provides a path to create functional surfaces in a controlled manner [43]. Polycation adsorption was maximized under basic pH without the addition of electrolyte. Polyelectrolyte adsorption could be modeled by both Langmuir and Freundlich equations, although the wood surface is known to be heterogeneous. Ionic interactions play a dominant role in the adsorption of polycations to wood surfaces. Polyelectrolyte adsorption onto wood surfaces is also facilitated by a higher concentration of polymers than what is typically reported for ionically assembled monolayers. Layered films masked ultrastructural features of the cell wall while leaving the microscale features of wood (cut lumen walls and openings) evident. So, the nanoscale films on wood can be deposited without changing the microscopic and macroscopic texture. Functionalized wood surfaces created by nanoscale films have a role in adhesives systems for wood composites as well as wood protection.

For timely control of the biological activity of cells in contact with a substrate, multicompartment films made of different polyelectrolyte multilayers deposited sequentially on the wood substrate constitute a promising new approach. Such multicompartment films can be designed by alternating exponentially growing polyelectrolyte multilayers acting as reservoirs, and linearly growing ones acting as barriers. However, these barriers composed of synthetic polyelectrolytes are not degraded, despite the presence of phagocytic cells. An alternative approach was presented where exponentially growing poly-(L-lysine)/hyaluronic acid (PLL/HA) multilayers, used as reservoirs, are alternated with biodegradable polymer layers consisting of poly(lactic-co-glycolic acid) (PLGA) and acting as barriers for PLL chains that diffuse within the PLL/HA reservoirs [44]. They are able to contain, for example, specific peptides or proteins. By changing the thickness of the PLGA layer, the time delay of degradation can be tuned. Such mixed architectures made of polyelectrolyte multilayers and hydrolyzable polymeric layers could act as coatings allowing us to induce a time-scheduled cascade of their biological activities.

As was mentioned before, the antibacterial activity of the protective coating could appear not only to be due to the bactericide effect of the coating components, but also to the surface structure (making it more hydrophobic), which can resist adhesion of the living organisms. A two-level surface mimicking the texture of the well-known lotus leaf was suitably designed from porous polyelectrolyte multilayers, silica nanoparticles and a simple hydrophobic surface treatment [45]. The net result is a stable superhydrophobic surface with high contact angles and low contact angle hysteresis. Hydrophilic patterns on superhydrophobic surfaces were created with water/2-propanol solutions of a polyelectrolyte to produce surfaces with extreme hydrophobic contrast [46]. Selective deposition of multilayer films onto the hydrophilic patterns introduces different properties to the area including superhydrophilicity. A structure was built by depositing an array of hydrophilic spots, size 750 μm, onto a superhydrophobic surface from polyacrylic acid water/2-propanol solution. The advancing water contact angle in the patterned region was 144°, whereas the receding contact angle was 12°.

Figure 43.5 demonstrates the behavior of small water droplets on a PAA-patterned superhydrophobic surface. Spraying a mist of water onto the surface leads to water droplets of 250 μm that do not wet the superhydrophobic surface and form nearly

**Figure 43.5** (a) Small water droplets sprayed on a (PAA/PAH/silica nanoparticle/semi-fluorosilane) superhydrophobic surface with an array of hydrophilic domains patterned with a 1% PAA water/2-propanol solution (scale bar 5 mm). (b) Sprayed small water droplets accumulate on the patterned hydrophilic area shown in (a) (scale bar 750 μm). Reprinted from R.E. Cohen et al. (2006) Nano Lett. **6**, 1213–1217. Copyright the ACS.

perfect spheres. Most of the droplets bounced and rolled on the superhydrophobic regions and eventually stuck to the patterned hydrophilic regions where large water droplets were formed. Potential applications of such surfaces include water harvesting surfaces, controlled drug release coatings, open-air microchannel devices, and lab-on-chip devices. In addition, specific functional groups or LbL-assembled polyelectrolyte multilayers can be selectively and easily introduced onto the patterned areas.

## 43.4
## Conclusions and Outlook

In this brief chapter we have demonstrated with examples of self-healing anticorrosion coatings and coatings with antibacterial activity that multilayer polyelectrolyte assembly can be used as a host for various active materials to fabricate a new family of feedback active coatings, which possess quick response to changes in the coating environment or coating integrity. The release of corrosion inhibitors encapsulated into LbL-assembled coatings as well as overall feedback activity of the polyelectrolyte multilayer coatings is triggered by the corrosion process. This prevents leakage of the corrosion inhibitor out of the coating. Moreover, the coating can have several active functionalities (e.g., antibacterial, hydrophobic and anticorrosion) when several types of layers in a multilayer system are replaced by (or entrap) different corresponding active agents. The demonstrated universal approach for the fabrication of active coatings on the other hand is also a great challenge for the development of multifunctional LbL-assembled polyelectrolyte film able to entrap active material, retain it for a long period, and immediately release it on demand. This requires the investigation of kinetic and structural properties of the sandwich-like multilayer structures and diffusion of the released active material inside the film matrix. This will surely be a matter of future intense research, which, as a result, may lead to a new generation of highly sophisticated multifunctional coatings.

## References

1 White, S.R., Sottos, N.R., Geubelle, P.H., Moore, J.S., Kessler, M.R., Sriram, S.R., Brown, E.N., and Viswanathan, S. (2001) Autonomic healing of polymer composites. *Science*, **409**, 794–797.
2 Hansen, C.J., Wu, W., Toohey, K.S., Sottos, N.R., White, S.R., and Lewis, J.A. (2009) Self-healing materials with interpenetrating microvascular networks. *Adv. Mater.*, **21**, 4143–4146.
3 Shchukin, D.G. and Möhwald, H. (2007) Self-repairing coatings containing active nanoreservoirs. *Small*, **3**, 926–943.
4 Zheludkevich, M.L., Poznyak, S.K., Rodrigues, L.M., Raps, D., Hack, T., Dick, L.F., Nunes, T., and Ferreira, M.G.S. (2010) Active protection coatings with layered double hydroxide nanocontainers of corrosion inhibitor. *Corros. Sci.*, **52**, 602–611.
5 Andreeva, D.V., Skorb, E.V., Möhwald, H., and Shchukin, D.G. (2010) Layer-by-Layer polyelectrolyte/inhibitor nanostructures for metal corrosion protection. *Appl. Mater. Interfaces*, **2**, 1954–1962.

6 Decher, G. (1997) Fuzzy nanoassemblies: Towards layered polymeric multicomposites. *Science*, **277**, 1232–1237.

7 Peyratout, C.S. and Dähne, L. (2004) Tailor-made polyelectrolyte microcapsules: from multilayers to smart containers. *Angew. Chem., Int. Ed.*, **43**, 3762–3783.

8 Ariga, K., Hill, J.P., and Ji, Q. (2007) Layer-by-layer assembly as a versatile bottom-up nanofabrication technique for exploratory research and realistic application. *Phys. Chem. Chem. Phys.*, **9**, 2319–2340.

9 Izquierdo, A., Ono, S.S., Voegel, J.-C., Schaaf, P., and Decher, G. (2005) Dipping versus spraying: exploring the deposition conditions for speeding up layer-by-layer assembly. *Langmuir*, **21**, 7558–7567.

10 Krogman, K.C., Zacharia, N.S., Schroeder, S., and Hammond, P.T. (2007) Automated process for improved uniformity and versatility of layer-by-layer deposition. *Langmuir*, **23**, 3137–3141.

11 (2000) Directive 2000/53/EC of the European Parliament and the council of 18 September 2000 on end-life vehicles. *Official J. Eur. Communities*, **L269**, 34.

12 Osborne, J.H. (2001) Observation on chromate conversion coatings from a sol-gel perspective. *Prog. Org. Coat.*, **41**, 280–286.

13 Schmutz, P. and Frankel, G.S. (1998) Characterization of AA2024-T3 by scanning Kelvin probe microscopy. *J. Electrochem. Soc.*, **145**, 2285–2295.

14 Farhat, T.R. and Schlenoff, J.B. (2002) Corrosion control using polyelectrolyte multilayers. *Electrochem. Solid State Lett.*, **5**, B13–B15.

15 Kachurina, O., Knobbe, E., Metroke, T.L., Ostrander, J.W., and Kotov, N.A. (2004) Corrosion protection with synergistic LbL/Ormosil nanostructured thin films. *Int. J. Nanotech.*, **1**, 347–265.

16 Andreeva, D.V., Fix, D., Möhwald, H., and Shchukin, D.G. (2008) Buffering polyelectrolyte multilayers for active corrosion protection. *J. Mater. Chem.*, **18**, 1738–1740.

17 Andreeva, D.V., Fix, D., Möhwald, H., and Shchukin, D.G. (2008) Self-healing anticorrosion coatings based on pH-sensitive polyelectrolyte/inhibitor sandwich-like nanostructures. *Adv. Mater.*, **20**, 2789–2794.

18 Cicileo, G.P., Rosales, B.M., Farela, F.E., and Vilche, J.R. (1998) Inhibitory action of 8-hydroxyquinoline on the copper corrosion process. *Corros. Sci.*, **40**, 1915–1926.

19 Banerjee, I., Pangule, R.C., and Kane, R.S. (2011) Antifouling coatings: recent developments in the design of surfaces that prevent fouling by proteins, bacteria and marine organisms. *Adv. Mater.*, **23**, 690–718.

20 Donlan, R.M. (2001) Biofilm formation: A clinically relevant microbiological process. *Clin. Infect. Dis.*, **33**, 1387–1392.

21 Vasilev, K., Cook, J., and Griesser, H.J. (2009) Antibacterial surfaces for biomedical devices. *Exp. Rev. Med. Dev,*, **6**, 553–567.

22 Wisniewski, N. and Reichert, M. (2000) Methods for reducing biosensor membrane biofouling. *Colloids Surf., B*, **18**, 197–219.

23 Bozja, J., Sherrill, J., Michielsen, J., and Stojiljkovic, I. (2003) Porphyrin-based, light-activated antimicrobial materials. *J. Polym. Sci., Part A: Polym. Chem.*, **41**, 2297–2303.

24 Conte, A., Buonocore, G.G., Bevilacqua, A., Sinigaglia, M., and Del Nobile, M.A. (2006) Immobilization of lysozyme on polyvinylalcohol films for active packaging applications. *J. Food Prot.*, **69**, 866–870.

25 Kenawy, E.R., Worley, S.D., and Broughton, R. (2007) The chemistry and applications of antimicrobial polymers: A state-of-the-art review. *Biomacromolecules*, **8**, 1359–1384.

26 Asuri, P., Karajanagi, S.S., Kane, R.S., and Dordick, J.S. (2007) Polymer-nanotube-enzyme composites as active antifouling films. *Small*, **3**, 50–53.

27 Chambers, L.D., Stokes, K.R., Walsh, F.C., and Wood, R.J.K. (2006) Modern approaches to marine antifouling coatings. *Surf. Coat. Technol.*, **201**, 3642–3652.

28 Kane, R.S., Deschatelets, P., and Whitesides, G.M. (2003) Kosmotropes

29. Costerton, J.W., Stewart, P.S., and Greenberg, E.P. (1999) Bacterial biofilms: A common cause of persistent infections. *Science*, **284**, 1318–1322.
30. Olsen, S.M., Pedersen, L.T., Laursen, M.H., Kiil, S., and Dam-Johansen, K. (2007) Enzyme-based antifouling coatings: A review. *Biofouling*, **23**, 369–383.
31. Glinel, K., Jonas, A.M., Jouenne, T., Leprince, J., Galas, L., and Huck, W.T.S. (2009) Antibacterial and antifouling polymer brushes incorporating antimicrobial peptide. *Bioconj. Chem.*, **20**, 71–77.
32. Dai, J.H. and Bruening, M.L. (2002) Catalytic nanoparticles formed by reduction of metal ions in multilayered polyelectrolyte films. *Nano Lett.*, **2**, 497–501.
33. Grunlan, J.C., Choi, J.K., and Lin, A. (2005) Antimicrobial behavior of polyelectrolyte multilayer films containing cetrimide and active silver. *Biomacromolecules*, **6**, 1149–1153.
34. Etienne, O., Picart, C., Taddei, C., Haikel, Y., Dimarcq, J.L., Schaaf, P., Voegel, J.C., Ogier, J.A., and Egles, C. (2004) Multilayer polyelectrolyte films functionalized by insertion of defensin: A new approach to protection of implants from bacterial colonization. *Antimicrob. Agents Chemother.*, **48**, 3662–3669.
35. Lee, D., Cohen, R.E., and Rubner, M.F. (2005) Antibacterial properties of Ag nanoparticle loaded multilayers and formation of magnetically directed antibacterial microparticles. *Langmuir*, **21**, 9651–9659.
36. Podsiadlo, P., Paternel, S., Rouillard, J.-M., Zhang, Zh., Lee, J., Lee, J.-W., Gulari, E., and Kotov, N.A. (2005) Layer-by-layer assembly of nacre-like nanostructured composites with antimicrobial properties. *Langmuir*, **21**, 11915–11921.
37. Chuang, H.F., Smith, R.C., and Hammond, P.T. (2008) Polyelectrolyte multilayers for tunable release of antibiotics. *Biomacromolecules*, **9**, 1660–1668.
38. Lichter, J.A., Thompson, M.T., Delgadillo, M., Nishikawa, T., Rubner, M.F., and Van Vliet, J.K. (2008) Substrata mechanical stiffness can regulate adhesion of viable bacteria. *Biomacromolecules*, **9**, 1571–1578.
39. Li, Z., Lee, D., Sheng, X., Cohen, R.E., and Rubner, M.F. (2006) Two-level antibacterial coating with both release-killing and contact-killing capabilities. *Langmuir*, **22**, 9820–9823.
40. Kim, B.-S., Lee, H., Min, Y., Poon, Z., and Hammond, P.T. (2009) Hydrogen-bonded multilayer of pH-responsive polymeric micelles with tannic acid for surface drug delivery. *Chem. Commun.*, 4194–4196.
41. Shchukin, D.G., Shutava, T., Shchukina, E., Sukhorukov, G.B., and Lvov, Y.M. (2004) Modified polyelectrolyte microcapsules as smart defense systems. *Chem. Mater.*, **16**, 3446–3451.
42. Shutava, T.G., Kommireddy, D.S., and Lvov, Y.M. (2006) Layer-by-layer enzyme/polyelectrolyte films as a functional protective barrier in oxidizing media. *J. Am. Chem. Soc.*, **128**, 9926–9934.
43. Renneckar, S. and Zhou, Y. (2009) Nanoscale coatings on wood: Polyelectrolyte adsorption and layer-by-layer assembled film formation. *ACS Appl. Mater. Interfaces*, **1**, 559–566.
44. Garza, J.M., Jessel, N., Ladam, G., Dupray, V., Muller, S., Stoltz, J.-F., Schaaf, P., Voegel, J.-C., and Lavalle, P. (2005) Polyelectrolyte multilayers and degradable polymer layers as multicompartment films. *Langmuir*, **21**, 12372–12377.
45. Zhai, L., Cebeci, F.C., Cohen, R.E., and Rubner, M.F. (2004) Stable superhydrophobic coatings from polyelectrolyte multilayers. *Nano Lett.*, **4**, 1349–1353.
46. Zhai, L., Berg, M.C., Cebeci, F.C., Kim, Y., Milwid, J.M., Rubner, M.F., and Cohen, R.E. (2006) Patterned superhydrophobic surfaces: Toward a synthetic mimic of the Namib desert beetle. *Nano Lett.*, **6**, 1213–1217.

# Index

## a
AADH (adipic acid dihydrazide)  351
abrasion test  34
acentric multicompartment
   microcapsules  932
acids
– acrylic  47–55
– adipic  351
– Brønsted acid–base reactions  838
– hexacyclen-hexaacetic  486–487
– hyaluronic  30
– Pearson's HSAB classification  429
– phenyl boronic  754
– polyacid layers  355
– polycarboxylic  343
– polygalacturonic  646–647
– tannic  210–213
– weak polyacids  856
acoustic impedance  457, 460, 1026
activation
– CDI  715
– EDC  715
– enthalpy  331
activity
– antibacterial  1045–1050
– biological  951, 1048
– catalytic  546
– passive protection  1042–1044
actuators  530–533
adamantyl-functionalized dendrimers  85, 92–94
addressable capsules  868–872
adhesion
– cell  451, 671–679, 988–990
– platelet  652
– wood fibers  176–179
adipic acid dihydrazide (AADH)  351
ADOGEN® 464 detergent  843
adsorption  3
– additional layers  275
– alternate  1017
– electrostatic  491
– flatly adsorbed chains  270–275
– kinetics  232
– microcapsules  935
– multiple  447–449
– number of adsorbed layers  921–922
– polyelectrolytes  664, 907–924
– polymers  232
– proteins  910–911
– sequence  899
– single  442–443
– time  8
affinity interactions  99–133
Ag, see silver
agarose  841, 846
AIDS  821
air shear force  225
Alexander–de Gennes theory  271
alginate  647
– molecular structure  750
– structures  713
alkali counterions  325
alkali metal ions  482
alkane silane  620
allyl alcohol  915
alternate adsorption  1017
alternate deposition  141, 643
alternating charges  445
alumina  909
aluminum alloys  1043
amide bonds  668–669
amine functionality  742
amine groups  667
amino-oligonucleotides  714
amorphous biomolecules  832
amphiphiles  43
amphiphilic dendrimers  987

amphiphilic macromolecules   893–895
amphiphilic molecules   691
amphiphilic polyelectrolyte films   896
amphoteric exchangers   308, 1018
anionic ATRP macroinitiators   441
anionic carboxylate groups   742
anionic cerasomes/polyelectrolytes   72
anionic phospholipid   965–966
anionic polysaccharides   640
anionic surfactants   450
anions, permselectivity   1007
anisotropic capsules/particles   933
anisotropic electrical properties   583
anisotropic LbL microcapsules   209–210
anisotropic transport   580–581
anisotropically-coated colloids   705
annealing   302
– temperature   377
antibacterial coatings   938, 1045–1050
antibacterial fibers   179–182
antibacterial films   352
antibacterial peptides   895–898
antibiotics   352
antibodies   525, 711, 973
– immobilization   912
– monoclonal   818
– targeting   802
anticoagulants   638
anticorrosion coatings   1040–1045
antigens   760, 951
antiinflammatory drugs   418–419
antimicrobial coatings   891–905
antireflection coatings (ARCs)   35, 525–527
antireflection-integrated coatings   146–147
antithrombogenic properties   650–651
apparent diffusion coefficient   1014
apparent redox potential   1004–1005
applications
– biomedical   588–589, 816–823
– biotechnology   578–580
– cardiovascular   650–654
– CNT-based LbL assembly   602
– controlled release   705
– electrical conductor   584–585
– electrochemically active multilayers   1024–1026
– electronic   473–509, 581–582
– energy-related   607–608
– flat LbL films   378–382
– functional domain-containing films   891–905
– LbL-SIP approach   449

– optical   581–582
– orthopedic   411
– sensing   585–587, 604–606, 705
aqueous nanocolloids   151–170
aqueous phase   831
aqueous solutions   614
architectures
– controlled   951–983
– SERRS   965–966
ARCs (antireflection coatings)   525–527
areal charge density   296
Argand diagram   457, 460, 468
aromatic chromophore   11
aromatic compounds   476–479
arrays
– carboxylated PS bead   994
– microchamber   873–890
– Pt nanoparticles   545–548
arteries
– 3D model   124–128
– porcine   651
artificial cell membranes   70
artificial nacre   576
artificial neural networks   970
assembly
– capsosome   812
– chitosan   642–650
– coordinative   493–504
– dip–spin   553–555
– directed   1–4
– double-stranded   106–111
– drug-loaded capsules   820
– dynamics   541
– feedback active coatings/films   1039–1050
– GOx-modified Pt nanoparticles   548–549
– helical   114
– hierarchical   743
– hydrogen-bonded   807
– immobile   1014
– LbL, see LbL assembly
– material   804–809
– membrane-electrode   397–398
– modular   790
– multilayer capsules   777–799
– multiple sequential   493
– nanolayer   393–435
– PAC-capped Pt nanoparticles   540–541
– PDDA/Q-CdS   555–556
– PEM   242–259
– polymer multilayers   831–850
– proteins   822
– Q-CdS   552–558
– self-, see self-assembly
– spherical colloids   878

- spray-/spin-assisted multilayer  13–14
- stepwise  103
- strong polyanions  647–650
- supramolecular  473–509
- weak polyanions  642–647
association constant  291
association mechanism  283–292
astrocytes  989
asymmetric capsules  382–384
atom transfer radical polymerization (ATRP)  439–441
Au, *see* gold
automated spray-LbL  422–431
average stiffness  377
avidin  911
avidin-containing LbL films  765–768
azacrown ether  483–486
- *N*-carboxymethyl-substituted  486

## b

bis-azide linker  808
backbone
- poly (phenylene-alt-fluorene)  502
- segments  202
backpacks  28–29
bacteria
- antibacterial fibers  181
- biofilm  1045
balance
- Cahn technique  174
- charge  281–320
balloons, inflatable catheter  738
band bending  556
bandages  427
barriers
- biological  777
- dye-impermeable polyelectrolyte  901–902
- hybrid  893
- organic  893
- physical  421
bases
- Brønsted acid–base reactions  838
- DNA base pairs  807
- Pearson's HSAB classification  429
basic fibroblast growth factor (bFGF)  789
BBO (beta-barium borate)  512–513
BCMs (block copolymer micelles)  61–62, 353
bead arrays  994
bending  374
bending line  870
bending response  532
benzoquinone  557
beta-barium borate (BBO)  512–513
bFGF (basic fibroblast growth factor)  789

bidirectional bending response  532
binary systems  749
binding
- covalent  912
- equilibrium  766
- non-specific  718–721
- pH-dependent constant  766
- *see also* bonds
binuclear complex  484
bio-template-based LbL encapsulation  831–832
bioactive molecules  351–353
biocatalytic current  1029
biocompatibility, PMMA  114
biodegradable porous scaffolds  120
biofilm  1045
biofuel cell cathodes  1031–1033
biological activity  951, 1048
biological barriers  777
biomaterials  1
- functional nano-  985–1001
- LbL films  198–199
biomedical applications  588–589, 608–609, 816–823
biomimetic SWNT coatings  589
biomolecules
- amorphous  832
- biomacromolecules  394
- crystalline  832
- diffusion  691–710
- encapsulation  831–850
bioresponsive release  788–792
biosensors
- controlled architectures  951–983
- e-noses  963–964
- electrochemically active multilayers  1026–1029
- GOx-modified Pt nanoparticles  548–549
- optical fiber  521–525
- optimization  971–977
biotechnology, clay multilayers  578–580
biotin  766
bipolar amphiphiles  43
bis-azide linker  808
bis-imidazolylpyridine-metal ion complexes  502–504
bis-triazine (DTA)  59–60
bleached softwood fibers  171
blend-PEMs  479–480
block copolymer brushes  452
block copolymer micelles (BCMs)  61–62, 353
block structure  223, 236
blocks, homopolymeric  809
blood, human  638

blood vessels 651–652
bonds
– covalent amide 668–669
– disulfide/diselenide 787
– *see also* binding
bone mineral density (BMD) 413
bottom-up approach 158–159
bovine collagen 427
bovine serum albumin (BSA) 842
1-box model 249
Bragg peaks 234
Bragg stacks 31–32
branched precursor layer 357
"break-in" effect 1020
bridging flocculation 16
bromodeoxyuridine (BrdU) 672–673
2-bromoester groups 447
Brønsted acid–base reactions 838
brushes 270–275
– block copolymer 452
– free-standing films 450–451
– surface-grafted 438–440
buckling measurements 367–369
buckling pattern 363
build-up 461–462
building blocks, self-assembly 44
bulging experiments 364–367
bulk polymer charge 292–301
burst release 654

c
cadmium sulfide (CdS) quantum dots (Q-CdS) 552–558
cadmium telluride (CdTe) nanoparticles 702
Cahn balance technique 174
calcination 146
– thermal 33
calcium fluoride ($CaF_2$) films 17
cancer cells 819
cancer therapies 792–793
cap region 1047
capacitance 960
capillary electrophoresis 295
capping 179, 1009
capsosome 814
– assembly 812
capsules
– anisotropic 933
– asymmetric 382–384
– drug-loaded 820
– engineered LbL 801–829
– exploding 382
– fusion 858–861
– hollow 113–120
– hybrid nanoparticle/polyelectrolyte 757
– LbL 749–763
– loading 854
– magnetoresponsive 779
– mechanical addressing 869–872
– mechanically strengthened 927
– micro-, *see* microcapsules
– multilayer 777–799
– optically addressable 868–869
– patterning 872–873
– polyion multilayer 863
– push force-curve 929
– salt-induced fusion 930
– self-exploding 752
– self-induced/remote release 925–950
– shell design 753–759
– silica 73
– size hysteresis 349, 857
– structures 384
captivities 194
carbon dioxide 863–864
carbon Hi-LbL 75–78
carbon nanocolloids 607
carbon nanotubes (CNTs)
– biofuel cell cathodes 1031–1032
– biomedical applications 588–589
– (bio)sensing 967–968, 1031–1032
– coatings 598
– conjugated polymer system 400–401
– LbL electrodes 398–400
– LbL films 195–198
– multilayers 582–589, 595–612
– MWNTs 398–400
– SWNTs 582–589, 608
carbonyl vibration bands 106
1.1′-carbonyldiimidazol (CDI) 715, 721–722
carboxyfluorescein (CF) 937
carboxyl groups 173
carboxyl-terminated polyether dendrimer (DEN-COOH) 47
carboxylate groups 75, 742
carboxylated particles 724
carboxylated polystyrene bead arrays 994
carboxylic groups 667
*N*-carboxymethyl-substituted azacrown ethers 486
carboxymethylpullulan (CMP) 893–894
cardiolipin (CLP) 966–967
cardiovascular applications 650–654
cascade of events 902–903
catalysts, nano- 1003–1038
catalytic activity 546
catalytic current 1028

catalytic films   914–915
catalytic membranes   908, 915–917
catechol   953–954
catheter balloons   738
cation-selective μCP   58
cationic ATRP macroinitiators   441
cationic cerasomes/polyelectrolytes   72
cationic peptides   654
cationic polymers
– charge-shifting   742–744
– chemical structures   733
– degradable   734–736
cationic polysaccharides   639
cations, permselectivity   1007
cavities   778
– cylindrical   380
C2C12 myoblasts   674–675
CCD (Cl-catechol 1,2-dioxygenase)   953–954
CD (cyclodextrin), SAMs   84
CD-functionalized polystyrene (PS)   88
CdS (cadmium sulfide) quantum dots
   (Q-CdS)   552–558
cell-induced stresses   929
cells
– flow-through QCR   466
– fuel, see fuel cells
– solar, see solar cells
cells (biological)
– adhesion   451, 671–679, 988–990
– artificial membranes   70
– biological activity   1048
– cancer   819
– cell-membrane mimic systems   260–261
– dendritic   28
– differentiation control   659–690
– encapsulation   210–213
– endothelial   124–127
– fibroblasts   123, 452
– HAECs   127–128
– hierarchical manipulation   121–124
– HUVEC   124
– living   28–31, 394, 759–761
– NG108-15   588
– programmable pharmacies   394
– smooth muscle   124–127, 677
– stem   410, 660, 679–684
– surface-mediated transfection   731–748
– UASMC   124
– VERO   755, 864
– yeast   210–212
cellular activities   124–128
cellular backpacks   28–29
cellular multilayers, 3D   120
cellulose fibers/fibrils   171–187

cellulose nanocrystals (CNC)   172, 182–184
cellulose PEMs   238
center of inversion   512
centrifugation   152, 805
cerasomes   71–72
cetyltrimethylammonium bromide (CTAB)
   micelles   139–140
CF (carboxyfluorescein)   937
chains
– complexed   292
– cross-linking   503
– dynamics   338–348
– flatly adsorbed   270–275
– frozen   665
– macromolecular   624
– polyelectrolyte   614–617
– post-assembly dynamics   337
charge
– alternating   445
– carriers   324
– compensation   239, 285–292, 322, 1019
– density   254, 296
– end-to-end mass/charge ratios   1024
– equilibrium-based reversal   424
– exchange dynamics   1012
– hopping   545
– balance   281–320
– metal-to-ligand transfer   500
– redox   1012–1018
– regulation   1010
– storage   395–406
– surface/bulk   292–301
– transfer   557
charge-selective transport   476–479
charge-shifting polymers   742–744
charged capping layer   1009
charged colloids   511
charged groups   327
– (de)protonation   927
charged surfaces   442–443
chemical reactions, see reactions
chemical-responsive capsules   862–864
chemical structures
– anionic polysaccharides   640
– cationic polymers   733
– cationic polysaccharides   639
– chitosan   639
– ferrocene-labeled glycogen   770
– macrocyclic compounds   480
– maltose-bearing polymers   770
– PMMA   101
– polyelectrolytes   396–397
– poorly-soluble organic compounds   155
– Schiff-base ligands   493

– synthetic polycations/-anions  617
– *see also* structures
chemically modified electrodes  1010, 1015
chemically nanopatterned surfaces  619–621
chemistry
– click  784, 790
– DNA  711–729
chitosan (CT)  30, 126
– chemical structure  639
– derivatives  642–650
– hyaluronan/chitosan multilayer  468
– polysaccharide-based microcapsules  785–786
chondroitin sulfate (CS)  649
– chemical structure  640
chopped illumination  555–556
chromophores  11
α-chymotrypsin  930
Cl-catechol 1,2-dioxygenase (CCD)  953–954
clay multilayers  575–582
– anisotropic transport  580–581
– structural organization  575–576
clays  898–901
– exfoliated  574
– LbL assembly  574–582
click chemistry  784, 790
CLP (cardiolipin)  966–967
CLSM (confocal laser scanning microscopy)  717–719
CMK-3 (mesoporous carbon)  75–78
CMP (carboxymethylpullulan)  893–894
CNC (cellulose nanocrystals)  172, 182–184
CNTs (carbon nanotubes)
– biofuel cell cathodes  1031–1032
– biomedical applications  588–589
– (bio)sensing  967–968, 1031–1032
– coatings  598
– conjugated polymer system  400–401
– LbL electrodes  398–400
– LbL films  195–198
– multilayers  582–589, 595–612
– MWNTs  398–400
– SWNTs  582–589, 608
co-culture, neurons/astrocytes  989
coacervates  281
coagulants  638
coated silica particles  114–115, 117
coatings
– AgNO$_3$  938
– anisotropically-coated colloids  705
– antibacterial  938, 1045–1050
– antimicrobial  891–905
– antireflection  146–147, 525–527
– biomimetic SWNT  589

– CH/heparin stents  651
– CNT-based LbL assembly  598
– complex surfaces  427–429
– conformal nanochannel  27
– degradable polymers  735
– feedback active  1039–1050
– fluorescent  504
– foam  136–138
– glass substrates  991
– graded-index antireflection  35
– nano-/micro-shell LbL  587
– nickel-titanium (NiTi) stents  650
– number of layers  722
– oligonucleotide coupling  713–716
– periodic  521
– porous materials  425
– robust  32–35
– scratch-resistant  147–148
– self-healing  146–147, 1040–1045
– spin-  231
– stimuli-responsive polymer  208
– thermally cross-linked  148
– transparent  147–148
cobalt ion complexes  502
codeposition, LbL  142–143
collagen  988
– bovine  427
colloid probe tapping mode (CPTM)  271
colloids
– anisotropically-coated  705
– charged  511
– core  113
– diffusion  698–704
– LbL-modified  711–729
– probe  929
– spherical  878
coloration  530
compensation, charge  285–292, 322, 1019
competitive ion pairing  283–292
complementary oligonucleotides  716
complementary polyelectrolytes  446
complex shear modulus  457
complex surfaces  427–429
complexation
– polycation–polyanion  239
– polyelectrolyte segments  284
– theories and models  485
– vesicles  936
complexed chains  292
complexes
– binuclear  484
– hydrogen-bonded  138–139
– π–π interaction  62–64
– interpolyelectrolyte  166

- PECs   136–144
- polyelectrolyte–surfactant   136, 139–140
- polymeric, *see* polymeric complexes
- polymeric–inorganic   136
- Schiff-base-metal ion   493–495
- step-by-step deposition   693
- stereo-   100–120
- supramolecular   44
- terpyridine-metal ion   495–502

complexing efficiency   105
composites   402–406
- free-standing   94
- LbL assembly   573–593
- membranes   484
- multilayer microcapsules   851–873
- synthesis   574

composition
- and mechanical properties   666–671
- composition-independent turbidity   341
- internal   660–666
- manipulation   429–431

compressive properties   364
computational methods   969
concanavalin A (Con A)-containing LbL films   768–769

concentration
- minimum inhibitory   414
- overlap   616

concentric multicompartment microcapsules   932
conductive network composite (CNC)   531–533
conductive patterning   992–995

conductivity
- electrical   192, 601, 705–706
- isothermal   331
- spectra   321–336

confined geometries   25–28
- LbL assembly   874

confinement
- LbL assembly   618–633
- spatial   552

confocal microscopy   381–383, 717–719
conformal films   400
conformal nanochannel coatings   27

conformation
- polyelectrolytes   269–280
- time-average   614

conjugated polymer system   400–401
contact angle   175
contact zone   177
contrast variation   222–223

control
- architectures   951–983

- cell differentiation   659–690
- energy transfer   898
- sequence   418–422
- spatial   736–739
- temporal   739
- transport   395–406

controlled delivery   382–386, 765–776
controlled release
- applications   705
- bioactive molecules   351–353
- drugs   750–759
- inhibiting agents   1044–1045
- insulin   792
- multilayer capsules   777–799
- passive   407–417

conventional LbL methods   44–52
coordination cross-linking   1033
coordination interactions   50
coordination polymers   493–504
- networks   496

coordinative assembly   493–504
coordinative interactions   495
coordinative supramolecular assembly   473–509

copolymers
- block, *see* block copolymer ...
- synthetic   1025

copper-complexed aza6-PVS   484

core
- agarose   841
- decomposition   778
- responsive   353–358
- self-induced/remote release   926

core–shell materials   835
- agarose–polymer   846

core–shell nanoparticles   1030
core–shell particles   711–712
corrosion propagation   1042
cost function   971
counter-polyelectrolyte   481

counterions
- alkali   325
- counterionic polyelectrolytes   77
- Manning condensation   615
- mobility   325
- PEM preparation   239–241
- release   284

coupling
- efficiency   722–723
- oligonucleotides   713–716
- organic solvents   721–722

covalent amide bonds   668–669
covalent binding   912
covalent cross-links   670

## Index

covalent deposition 514
covalent interactions 806
covalently stabilized microcapsules 784
COX enzyme assay 418
µCP (microcontact printing) 49
– cation-selective 58
cross-linking 348
– coordination 1033
– covalent 670
– cross-link density 669–671
– dendrimers 987
– ESCs 679–684
– films 666–671
– ion pairing 292
– myogenic differentiation 679–681
– side chains 503
– thermal 33, 148
cross ratio invariance 467
crystal structures 88–90
crystalline biomolecules 832
crystalline PMMA 112
crystallinity 183
crystals, tetrahedral SnS 211
CS (chondroitin sulfate) 649
– chemical structure 640
CT (chitosan) 30, 126
– chemical structure 639
– derivatives 642–650
– hyaluronan/chitosan multilayer 468
– polysaccharide-based microcapsules 785–786
CTAB (cetyltrimethylammonium bromide) micelles 139–140
Cu, see copper
cubic microcapsules 210, 384
curcumin 154
current–voltage measurements 600, 1005–1006
cushion structures 259–261
cyclodextrin (CD) 84
cyclohexane 77
cylindrical cavities/trenches 380
cylindrical microscopic captivities 194
cylindrical wells 878
cytoplasmic protrusions 760
cytoskeletal organization 671–679
cytotoxic drug 793

### d

3D artery model 124–128
3D cellular multilayers 120
3D objects 70
3D patterned multilayer assemblies 85–88
3D supramolecular structures 91
DADMAC (diallyldimethyl ammonium chloride) 230
damaged endothelium 652
DAR (diazoresin) 51, 55, 137
data-intensive discovery 970
DC conductivity 324–328
– isothermal 331
DCs (dendritic cells) 28
Debye length 270, 283
– inverse 615
– surface potential model 1009
Debye–Waller factor 221
decoloration 530
decomposition 1048
– core 778
– multilayers 290–292
– pH-dependent 773–774
decyl-grafted CMP 894–895
defensin 1046
deflection, membrane 366
deformation 879
degradation
– cationic polymers 734–736
– engineered LbL capsules 813–816
– enzymatic 930
– films 407, 997–999
– hydrolytic 740–741
– polyamines 740–741
delivery
– controlled 382–386, 765–776
– DNA 731–748
– drug, see drug delivery
– intracellular 384–386
– multi-agent 406–422, 731–748
– proteins 765–776
– small molecule 412–417
– targeted gene 633
– therapeutic 819–823
– transcutaneous 428
– see also release
DEN-COOH (carboxyl-terminated polyether dendrimer) 47
dendrimers
– adamantyl-functionalized 85, 92–94
– amphiphilic 987
– cross-linkable 987
– DEN-COOH 47
– LbL films 769–770
– PAMAM 771–772, 957–958
– spherical 627
dendritic cells (DCs) 28
density

– bone mineral   413
– charge   254, 296
– cross-link   669–671
– density–viscosity product   456
– SLD   203–204, 220, 250, 303
depletion layer   556
deposition
– alternate   141, 643
– CNT multilayers   596
– co-   142–143
– conditions   7–10
– covalent   514
– film   3
– hydrogen underpotential   541–545
– Langmuir–Blodgett   691
– LbL films   15–17, 338–340
– PEMs in porous media   908–910
– polyelectrolytes   778
– post-treatment   694–698
– reactive LbL   696
– step-by-step   693
deprotonation   1008
– charged groups   927
desalination   474–476
design
– LbL capsules   749–763
– nanolayer assembly processes   393–435
– polysaccharide multilayers   637–658
– shell   753–759
desorption   491
deuterated layers   223
deuterated PMAA   347
dex-HEMA   750
dextran sulfate (DS)   126, 165, 430
– chemical structure   640
diallyldimethyl ammoniumchloride (DADMAC)   230
diameter
– dry pore   631
– hydrodynamic   143
diazoresin (DAR)   51, 55, 137
diclofenac   416, 418
dielectric mirrors   31–32
differentiation   412
– myogenic   679–681
diffusion
– apparent coefficient   1014
– colloids   698–704
– diffusion-limited flux   307
– dilute diffusion regime   629–630
– effective area   547
– "in-and-out"   147

– inter-   275, 301–305
– interlayer   421
– Laviron model   1010, 1016
– LbL encapsulation   833–834
– LbL films   700–705
– nanoparticles and biomolecules   691–710
– nitric oxide   124–128
– nonlinear control   311
– pore-to-bulk diffusivity   628
– post-   892
– (semi-)dilute regime   629–630
digestion, proteins   911–914
digestive enzymes   774
dilute solutions   614–617
1-(3-dimethylaminopropyl)-3-ethylcarbodiimide hydrochloride (EDC)   714–715
dipalmitoyl phosphatidyl glycerol (DPPG)   965–966
dipping   224–228
– cycles   488
– dip-assisted LbL assembly   204
– dip-LbL films   423
– dip–spin assembly   553–555
– solution temperature   242
– time   542
direct encapsulation   841
direct LbL processing   29–31
directed assembly   1–4
directed external fields   257
directed self-assembly   985–1001
disassembly, triggered   744
diselenide bonds   787
disintegration
– LbL films   767–768
– pH-triggered   343
disk electrodes   550
– rotating   305–306
disordered particles   995–997
dispersion   118
displacement model   250
dissolution, templates   874–876
disulfide bonds   787
ditopic ligands   494
divalent metal ions   494
DLVO theory   270
DNA   4
– base pairs   807
– chemistry and hybridization   711–729
– delivery   731–748
– direct LbL processing   30
– DNA-containing LbL nanotubes   626
– double-stranded   816–817
– embedded   939

– mobility   939
– plasmid   741
– self-assembled polymer/DNA polyplexes   732
– single-stranded   723
– therapeutic delivery   822–823
$n$-dodecanethiol   46
domain-containing functional polyelectrolyte films   891–905
Donnan potential   1007, 1011
doping
– doping-moderated mechanical properties   292
– fluorescein   623
– hole-   586
– level   289
double hydroxides   576
double layers   247
double-stranded assembly   106–111
double-stranded DNA (dsDNA)   816–817
double walled carbon nanotubes (DWNTs)   603
doxorubicin   811–812, 819–821
DPPG (dipalmitoyl phosphatidyl glycerol)   965–966
DQ-OVA   760
drug delivery   765–776
– cancer therapies   792–793
– unusual modes   72–75
– vascular wall   652–654
drugs
– antiinflammatory   418–419
– controlled release   750–759
– cytotoxic   793
– drug-loaded capsules   820
– drug-loaded multilayers   790
– encapsulation   777–799
– loading   749–763
– nanocrystal formulations   153
– release   165–167, 749–763
– sequential release   420
– small   698–700
– time-delayed release   790
dry pore diameter   631
drying   228
DS (dextran sulfate)   126, 165
– chemical structure   640
DSSCs (dye-sensitized solar cells)   402–404
DTA (bis-triazine)   59–60
dual-functional antibacterial coating   1047
dual protein release   419
dual stimuli-response   449–450
DWNTs (double walled carbon nanotubes)   603

dyes
– DSSCs   402–404
– dye-impermeable polyelectrolyte barriers   901–902
– encapsulated   354
– Nile Blue   902
– pair-labeled oligonucleotides   725–726
– permeability   698–700
– pigment Orange 13   155–156, 158
– Procion Brown MX-GRN   513–515
– Prussian Blue   487–492, 694–695, 957
– removal   918–920
– rhodamine   116, 725, 782, 843
– stained fluorescence   847
dynamically excited LbL membrane   381
dynamics
– assembly   541
– chain   338–348
– charge exchange   1012
– PECs   340–342

e
e-noses   958–964
e-science   970
e-tongues   958–963
ECDs (electrochromic devices)   527–530
– multicolor   404–406
ECM (extracellular matrix)   120–121
– nano-   100
ECs (endothelial cells)   124–127
EDC (1-(3-dimethylaminopropyl)-3-ethylcarbodiimide hydrochloride)   714–715
EDCs (endocrine disrupting chemicals)   954
effective diffusion area   547
EGFP (enhanced green fluorescent protein)   735–736
eigenspectrum, quantized size-dependent   552
EIS (electrolyte–insulator–semiconductor) structure   958
electrical conductivity   192, 601
– post-incubation method   705–706
electrical conductor applications   582–589
electrical fields   258–259
electrical measurements   464, 600, 956–958
electrical properties
– anisotropic   583
– CNT multilayers   600–601
electro-optic effect   512
electrocatalytic activity   62
electrochemical cycles   501
electrochemical energy   393–435
electrochemical perturbation   1013

electrochemical potential 258
electrochemical QCM (EQCM) 1018–1021
electrochemical response 1004–1012
electrochemical sensors 955–956
electrochemical swelling 259
electrochemically active multilayers 1003–1038
electrochromic devices (ECDs) 527–530
– multicolor 404–406
electrochromism 502
electrodes
– biofuel cell cathodes 1031–1033
– chemically modified 1010, 1015
– CNTs 398–400
– disk 305–306, 550
– DWNT-based 603
– geometric capacitance 960
– graphene-modified 62
– interdigitated 963, 972
– LbL 398–402
– nanostructured 539–369
– nanostructured photoelectrodes 552–558
– platinum 540–552
– Q-CdS photoelectrodes 556–558
– transparent 602–604
– wire 543
electrolytes
– EIS structure 958
– poly-, see polyelectrolytes
– salt-in-polymer 321
electromagnetic energy 515
electromechanical actuators 530–533
electron hopping 1012–1018
electron transfer 1005
– electrochemical perturbation 1013
electronic applications 473–509
– clay multilayers 581–582
electroosmotic flow 295
electrophoresis
– capillary 295
– sticky gel 346–347
electrospinning 120
electrostatic adsorption 491
electrostatic interactions 596
electrostatic LbL assembly 44–47, 52–58
– π–π interaction complex 62–64
electrostatic supramolecular assembly 473–509
electrosynthesis 632
ellipsometric thickness 290
ellipsometry 278
embedded DNA 939
embryonic stem cells (ESCs) 660, 679–684
emulsification 833

emulsions, LC 803
encapsulation 74
– antigens 760
– bio-template-based 831–832
– biomolecules 831–850
– cells 210–213
– diffusion-based 833–834
– direct 841
– dyes 354
– efficiency 834
– loading-based 832–833
– organic phase 835
– OVA 822
– proteins 834
– rhodamine-labeled PSS polymers 782
– SLbL 151
end-to-end mass/charge ratios 1024
endocrine disrupting chemicals (EDCs) 954
endothelial cells (ECs) 124–127
endothelium 652
energy
– applications 607–608
– electrochemical 393–435
– electromagnetic 515
– FRET 724–727
– interaction 287–290
– pump pulse 521
– transfer efficiency 648
energy transfer
– control 898
– efficiency 648
– functional polyelectrolyte films 891–905
engineered LbL capsules 801–829
enhanced green fluorescent protein (EGFP) 735–736
enhanced quantum yield 53
entrapment, antibacterial peptides 895–898
entrapped redox centers 1014
enzymes
– chemical-responsive capsules 864
– COX enzyme assay 418
– digestive 774
– enzymatic conversion 814
– enzymatic degradation 930
– films 1029
– GA 64
– GOx 64
– HRP 127, 847–849
– kinetics 992
– luciferase 428
– patterned enzyme-MWNT film 606
epiblast 682
EQCM (electrochemical QCM) 1018–1021
equations, see laws and equations

equilibrium
– binding   766
– charge compensation   285–287
– equilibrium constant   286
– equilibrium-based charge reversal   424
– PECs   340–342
– salt/polymer sorption   304–305
erodable LbL films   755
ESCs (embryonic stem cells)   660, 679–684
ester functionality   743
etched gold patterns   59
even–odd effect   278–279
Everitt's salt   490
exchange
– charge   1012
– ion   1018–1024
– polyelectrolyte   358
– proton exchange fuel cells   395–398
– "reluctant" mechanism   305
exchangers, amphoteric   308, 1018
excitation, impulse   464–465
exfoliated clays   574
– Laponite   898
exploding capsules   382
exponential build-up   461–462
exponential growth   300–301, 637
expression
– muscle-specific proteins   679–681
– vectors   588
external fields   242–259
– directed   257
extracellular matrix (ECM)   120–121
– nano-   100
extrinsic charge compensation   285–292, 322
– films   698–699
extrinsic redox PEMs   1019

## f

Faraday's law   1005
Fc-SiO$_2$ (ferrocenyl-functionalized silica)   87
feedback active coatings/films   1039–1050
ferrocyanide (FC)   259
– ion-paired PEMs   309–312
ferrocene-labeled glycogen   770
ferrocenyl-functionalized silica (Fc-SiO$_2$)   87
fibers
– antibacterial   179–182
– cellulose   171–187
– nonlinear optical   519–525
– silica   519
– wood   176–179
fibroblasts   123, 452
fibroin   198–199

fibronectin (FN)   120–122, 988
– fibronectingelatin films   100
fillers   190
films
– amphiphilic polyelectrolyte   896
– bacterial biofilms   1045
– CaF$_2$   17
– catalytic   914–915
– charging   545
– conductivity spectra   321–323
– conformal   400
– cross-linking   666–671
– degradable   734–736, 997–999
– deposition   3
– emerging properties   705
– enzyme-MWNT   606
– ESCs   679–684
– exponentially growing   301
– feedback active   1039–1050
– free-standing   189–218, 450–453, 578, 874–876
– functional   144, 352, 473–509, 891–905
– growth   660–661
– GS–IL   78–79
– hydration   660–666
– hydrogen-bonded   342
– internal composition   660–666
– ISAM   511, 517
– Langmuir-Blodgett   5–6, 955
– layering   338–348
– LbL, see LbL films
– mesoporous nanocompartment   73
– metal film/solution interface   960
– microporous   48
– multicomponent   7, 421
– myogenic differentiation   679–681
– nanoparticle-containing   914–915
– nanoporous   204–205
– patterning techniques   985–988
– polyelectrolyte multilayer, see polyelectrolyte multilayers
– polymer   113–120,
– post-treatment   694–698
– PSS/PDADMAC   921–922
– remote release   925–950
– responsive free-standing   451–453
– self-assembled   488
– self-induced release   925–950
– silk   367–368
– SMILbL   54–58
– stability   118
– stratified polyelectrolyte   898
– structural tailoring   140–144
– structure   342–348

- template nanospaces 104–106
- thickness 10
- thin, *see* thin films
- trypsin-containing polyelectrolyte 911–914
- ultrathin 102–104, 189–204, 400
- *see also* nanofilms
filtration 152
fitting 223–224
flat LbL films 364–372
- sensing applications 378–382
flat substrates 442
flat surfaces 618–624
flatly adsorbed chains 270–275
flexible nanocomposites 406
flexible substrates 369
flocculation 16
Flory–Huggins fit 249
flow-through QCR cell 466
fluorescein 725
- doping 623
- FITC 857–860, 863
fluorescence dyes 847
fluorescence microscopy 116
fluorescence resonance energy transfer (FRET) 724–727
fluorescent coatings 504
fluorescent multilayer backpacks 29
fluoride 920–921
flurbiprofen 416
FN (fibronectin) 120–122, 988
foam coatings 136–138
force
- force–deformation curves 375–376, 879
- force extension curves 49
- pull-off 178
- spectroscopy 369–370
- virous 225
four-point probe measurements 600
fracturing, LbL films 207
free decay of oscillation 464–465
free radical initiator 106
free-standing films 189–218
- brush 450–451
- clay multilayers 578
- PEM 874–876
- responsive 451–453
free-standing hydride particle composites 94
free-standing multilayers 363–392
- sensing applications 378–382
frequency
- resonant 456
- synthesizer 466
- ultrasound ranges 780

frozen chains 665
frozen segments 301
fuel cells
- biofuel cell cathodes 1031–1033
- CNT multilayers 587
- proton exchange 395–398
functional films 473–509
- antibacterial 352
- free-standing brush 450–451
- polymeric complexes 144
- thin 493–504
functional groups 9
- *see also* groups
functional materials 691–710
- nano-bio- 985–1001
functional membranes 907–924
functional nanoscale fillers 190
functional nanotubes 613–636
functional polyelectrolytes 1025
- films 891–905
functional polyoxometallates 703
functionalities
- amine 742
- ester 743
- multiple 852
functionalization
- membrane interior 908–918
- MWNTs 398–400, 601
- polymeric nanomembranes 783, 869
fusion
- capsules 858–861
- salt-induced 930
fuzzy supramolecular objects 6–7

## g

Gaussian regime 615
gelatin 122
gelatin-based nanoparticles 160
gels
- hydro-, *see* hydrogels
- micro- 383
gene delivery 633
gene expression, luciferase 428
gentamicin 414–415
geometric capacitance 960
geometries, confined 25–28, 874
giant vesicles 967
glass substrates, coated 489, 991
glass transition 330
glassy state 665
glucoamylase (GA) 64
glucose 63, 754
- chemical-responsive capsules 863
- sensing 606

glucose oxidase (GOx) 64
– modified Pt nanoparticles 548–549
glucose saturation 1028
glutathione (GSH) 815
glycogen, ferrocene-labeled 770
glycopeptides 415
glycoproteins 765
gold
– etched patterns 59
– nanostructures 964
– (super)hydrophobic 45–47
gold nanoparticles 191–193
– patterned arrays 369
– sandwiched 201
graded-index antireflection coatings 35
graphene
– graphene/ionic-liquid Hi-LbL 78
– graphene-modified electrodes 62
– immobilization 62
– patterning 992–995
– sheets 78, 197
– synthesis 992
graphene oxide (GO), LbL films 195–198
graphite 992
gratings 521–524
groups
– amine 667
– 2-bromoester 447
– carboxylate 742
– carboxylic 173, 667
– charged 927
– functional, see functional groups
– hydrophobic 327
– pH-sensitive 26
– prosthetic 1027
– protonated hydroxy 856
– side 202
– substituent 497
– substrate-binding 26
– unmasking 742
growth
– CNT multilayers 596–600
– exponential 637
– films 660–661
– QCR monitoring 461–462
GS (graphene sheets) 78–79, 197
GSH (glutathione) 815
guest molecules 892
– active 1039

**h**

HA (hyaluronan) 642–646
– chemical structure 640
– hyaluronan/chitosan multilayer 468

– PLL/HA films 660–664
HAECs (human aortic endothelial cells) 127–128
harmonic displacements 1014
HB-LbL (hydrogen-bonded LbL) 836–839
heating, microcapsules 865–866
helical assembly 114
hemicellulose 173
hemoglobin 1048
heparin 647–649
– chemical structure 640
– layers 162
Hertzian coordinates 370
hexacyanoferrate (HCF) 483–484, 488–492
hexacyclen-hexaacetic acid 486–487
hierarchic LbL assembly 69–81
– sensors 75
hierarchical assemblies 743
hierarchical cell manipulation 121–124
hierarchical structure 740
high-performance nanocomposites 576–578
history of LbL assembly 4–6
HIV 821
holes
– hole-doping 586
– hole scavengers 540–557
– photogenerated 558
– triangle 879
hollow capsules 113–120
hollow microchambers 874–876
hollow structures 372
homogeneous swelling 249
homopolymeric blocks 809
Hooke's law 373–374
hopping
– charge 545
– electron 1012–1018
hopping model 311
host–guest interactions 83–97
HRP (horseradish peroxidase) 127
human aortic endothelial cells (HAECs) 127–128
human blood 638
human umbilical vascular endothelial cells (HUVEC) 124
humidity
– relative 248, 324
– sensor 605
– time–humidity superposition 332–335
HUVEC (human umbilical vascular endothelial cells) 124
hyaluronan (HA) 642–646

– chemical structure 640
– hyaluronan/chitosan multilayer 468
– hyaluronic acid 30
– PLL/HA films 660–664
hybrid barriers 893
hybrid capsules 779
hybrid materials 91
hybrid nanostructures 403, 757
hybridization
– complementary oligonucleotides 716
– DNA 711–729
– efficiency 722–723
hydrated ions 491
hydrated region 460
hydration 285
– films 660–666
– polymer matrix 327
– swellability 662–664
hydration number 290
hydraulic permeability 474
hydride particle composites 94
hydrodynamic diameter 143
hydrogels
– dex-HEMA 750–752
– LbL-derived 348–353
– pH-sensitive 209
hydrogen-bonded assembly 807
hydrogen-bonded films 342
– self-floating 344
hydrogen-bonded LbL (HB-LbL) 47–50, 58–60, 836–839
hydrogen-bonded multilayers 808
hydrogen-bonded polymeric complexes 138–139
hydrogen peroxide ($H_2O_2$)-sensing 545–548
hydrogen underpotential deposition 541–545
hydrogenation, selective 915
hydrolysis, urea 968
hydrolytic degradation 407, 734–736, 740–741
hydrophilic OEG-silane 620
hydrophilic patterns 1049
hydrophilic PEMs 893–895
hydrophilic polymers 395
hydrophobic alkane silane 620
hydrophobic gold 47
hydrophobic groups 327
hydrophobic nanodomains 893–898
hydrophobic pockets 353
hydrophobically modified polyvinylamine (PVAm) 180
hydroquinone 312
hydrothermal treatment 33
hydroxides, layered double 576

hydroxy groups, protonated 856
hyperbolic equation 670, 677
hyperswelling 288
hysteresis, capsule size 349, 857

*i*

IDMAP (interactive document map) 971
illumination, chopped 555–556
bis-imidazolylpyridine-metal ion complexes 502–504
2-iminobiotin 766
immobile assembly 1014
immobilization
– antibodies 912
– avidin 911
– functional polyoxometallates 703
– graphene 62
– liposomes 934–935
immunoblots 679–681
immunofluorescence images 680
immunostaining 125
impedance 464
– spectra 323
– spiral 457, 460
– surface acoustic 457, 460, 1026
impermeable shells 870
imprinted surfaces 873
impulse excitation 464–465
"in-and-out" diffusion 147
*in vitro* release 167
*In vivo* scaffolds 412
incorporation constant 105
indium tin oxide (ITO) 45–46, 440
– ITO-coated glass substrates 489
inflatable catheter balloons 738
inflated microcapsules 843–844
inhibiting agents 1044–1045
inhibitory concentration 414
inkjet LbL assembly 590
innercentric multicompartment microcapsules 932
inorganic LbL films 17
insulin 164, 754, 770–774
interactions
– affinity 99–133
– coordinative 50, 495
– covalent 806
– electrostatic 596
– energies 287–290
– host–guest 83–97
– molecular 837
– monomer/substrate 274
– non-electrostatic 238–242
– π–π 62–64

– polycation–protein  167
– polyion–polyion  1018
– van der Waals  99–133
– weak  208–213
interactive document map (IDMAP)  971
interactive layers  182–184
intercalating layers  226
interdiffusion  275
– polyelectrolyte  301–305
interdigitated electrodes  963, 972
interfaces
– metal film/solution  960
– metal/polymer  1017
– multilayer/solution  298
interference colors  184
interlayer diffusion  421
interlayer interpenetration  296, 692
internal composition, films  660–666
internal microstructure  201–204
interpolyelectrolyte complexes  166
intracellular delivery  384–386
intramuscular site  413
intravascular stents  737
intrinsic charge compensation  239, 285–292, 322
– films  698–699
inverse Debye length  615
inversion, center of  512
inwards build-up self-assembly  845–847
ion conduction  329–332
ion pairing  666–667
– association constant  291
– competitive  283–292
– cross-links  292
– PEMs  281–320
ion transport  305
– conductivity spectra  321–336
ionic liquids (ILs)  78
– electromechanical actuators  531–533
ionic polymer CNC (IPCNC) actuators  531–533
ionic self-assembled multilayer (ISAM) films  511, 517
ionic strength  143, 238–239, 1008
– scalar fields  242–244
– water swelling  255
ions
– dynamics  331
– electrochemical energy  395–406
– exchange  1018–1024
– hydrated  491
– ion effects  238–242
– ion traps  53–54
– organic  477

– permeability  698–700
– separation  479–480, 908
isotactic (it) PMMA  100–104
isothermal DC conductivity  331
isothermal titration microcalorimetry  661
isotopic effect  254
ITO (indium tin oxide)  45–46, 440
– ITO-coated glass substrates  489

## j
J776 macrophages  653

## k
key equilibria, charge compensation  285–287
Kiessig fringe  244
kinetic trapping  430
kinetics
– adsorption  232
– composition manipulation  429–431
– enzyme  992
– release  356
– SMILbL films  56
Kirby Bauer assay  418

## l
$\beta$-lactamase  814
Langmuir–Blodgett (LB) films  5–6, 955
– deposition  691
Langmuir plot  105
lanthanide-doped nanocrystals  194
Laponite  898
laser-induced release  868
Laviron diffusion model  1010, 1016
laws and equations
– complex shear modulus  457
– cost function  971
– Donnan potential  1007, 1011
– equilibrium constant  286
– Faraday's law  1005
– Hooke's law  373–374
– hyperbolic equation  670, 677
– interactive document map  971
– inverse Debye length  615
– Levich equation  306
– Lie product formula  460
– Möbius transformation  458, 467
– nonlinear polarization  512
– phase-matching condition  522
– refractive index  220
– Sammon's mapping  971, 974–975
– scattering length density  220
– silhouette coefficient  975
– Young's modulus  373–374
layer-by-layer . . ., see LbL . . .

layered double hydroxides (LDH)   576
layers
– adsorbed   921–922
– branched   357
– capping   1009
– depletion   556
– deuterated   223
– double/quad-   247
– fibroblast cell   123
– film layering   338–348
– heparin   162
– interactive   182–184
– intercalating   226
– interpenetration   296, 692
– lipid layer pair   779
– loading within   811–813
– mobile polymer   1041
– mono-   619
– multi-, see multilayers
– organic–metallic   92
– phospholipid   122
– poly-L-lysine   162
– polyacid   355
– protonated   223
– S-   260
– selectively labeled   203–204
– stratified   420
– submonolayers   551
– substrate-mediated layering   348
– see also multilayers
LbL assembly   1–21
– automated spray-   422–431
– cellulose fibers/fibrils   171–187
– charge storage   398–402
– clays   574–582
– CNT-based   596–602
– composites   573–593
– confined geometries   25–28, 618–633, 874
– dip-assisted   204
– electrostatic   44–47, 52–58, 62–64
– functionalized MWNTs   399–400
– hierarchic   69–81
– host–guest interactions   83–97
– hydrogen-bonded   47–50, 58–60
– imprinted surfaces   873
– inkjet   590
– membrane-templated   625
– methods   43–67
– multi-agent delivery   735
– nanoscale elements   25–28
– polymeric complexes   135–150
– porous structure   111–113
– proton exchange fuel cells   395–398
– responsive   337–362

– sensors   75
– spin-assisted   203–204
– structural tailoring   140–144
– supramolecular   84–85
– surface charge distribution   297–301
– templates   171–187, 619–621
– ultrasonication-assisted   154
– van der Waals/affinity interactions   99–133
– washless   161–163
LbL capsules
– design   749–763
– engineered   801–829
– living cells   759–761
– shell design   753–759
LbL coatings
– microgels   383
– nano-/micro-shell   587
– nanoparticles   159–161
LbL codeposition   142–143
LbL deposition   696
LbL electrodes
– CNTs   398–400
– DSSCs   402–404
– DWNT-based   603
LbL encapsulation
– bio-template-based   831–832
– diffusion-based   833–834
– hydrogen-bonded   836–839
– limitations   834–835
– loading-based   832–833
– organic phase   835
– reverse-phase   836–840
LbL films
– avidin-containing   765–768
– (bio)sensing   951–983
– buckling measurements   367–369
– CNTs   195–198
– concanavalin A (Con A)-containing   768–769
– conductivity spectra   321–323
– controlled architectures   951–983
– dendrimer-containing   769–770
– deposition   338–340
– diffusion   700–705
– dip-   423
– disintegration   767–768
– electrochemically active   1003–1038
– erodable   755
– flat   364–372
– fracturing   207
– freely standing   189–218
– hydrophobic pockets   353
– inorganic   17
– insulin-containing   770–774

– membrane skins   918
– methanol-treated   367
– multi-agent drug delivery   406–422
– nanoparticles   693–694
– nanoporous   100–113
– patterning   14–15, 204–208
– PB   487–492
– permeability   698–700
– photovoltaic properties   580
– porous   204–208
– sculptured   207–208
– step-by-step deposition   693
– stimuli-sensitive   765–776
– stratification   898–901
– surface-initiated polymerization   437–454
– swellable   348–358
LbL hydrogels   348–353
– pH-sensitive   209
LbL lift-off   621–624
LbL membrane   381
LbL microcapsules
– anisotropic   209–210
– magnetic   866
– mechanical properties   372–378
– stimuli-responsive   853–873
LbL-modified colloids   711–729
LbL nanotubes   626, 630–633
LbL polyelectrolyte shells   153–154
LbL polymer films   113–120
LbL processed multilayers   25–41
LbL processing, direct   29–31
LbL protein nanofilms   120
LbL self-assembly   440–441
LbL shells   151–170
– cell encapsulation   210–213
LC (liquid crystal) emulsions   803
lectin   768
length scale   2
2-level antibacterial coating   1047
Levich equation   306
Lie product formula   460
lift-off LbL   621–624
ligands
– di-/polytopic   494
– metal-to-ligand charge transfer   500
– Schiff-base   493
– targeting   793
light harvesting   553
light-responsive microchambers   879–881
light-sensitive microcapsules   781
linear build-up   461–462
linear growth   300
linker, bis-azide   808
lipid layer pair   779

liposomes   803–804, 814
– immobilization   934–935
– proteo-   963, 972
– stabilized   936
liquid crystal (LC) emulsions   803
liquids, ionic, see ionic liquids
lithography
– nanoimprint   83, 85–88, 92–93, 618
– nanosphere   516
living cells   28
– direct LbL processing   29–31
– LbL capsules   759–761
– programmable pharmacies   394
loading
– capacity   60
– capsules   854
– drugs   749–763
– loading curve   370
– loading-based LbL encapsulation   832–833
– load-responsive microchambers   879–881
– microchambers   877–878
– proteins   822
– therapeutic cargos   809–813
– within layers   811–813
local mechanical properties   369–370
localized micro-environments   409
long-period gratings (LPGs)   521–524
long-range transport   334
loss modulus   645
low solubility materials   151–170
LPEI-ac (poly (ethyleneimine $N$-acetic acid))   487
luciferase   428
luminescence, suppression   193–194
lymphocytes   28
lysozyme release   410

**m**

macrocyclic compounds   480–487
macroinitiators   440–441
macromolecular chains   624
macromolecules, amphiphilic   893–895
macrophages   29
– J776   653
magnetic LbL capsules   866
magnetically sensitive release   779–780
major histocompatibility complex (MHC)   931
maltose   63, 770
manipulation
– composition   429–431
– hierarchical   121–124
Manning condensation   615

mass change methods 954
mass/charge ratio 1024
material assembly 804–809
materials science 2
materials separation 473–509
matrices
– extracellular, see ECM
– hydration 327
– LbL-derived hydrogels 351–353
– PEM 261
MBSA (5,5'-methylene-bis(N-methylsalicylidene amine)) 493–494
MDR (multidrug resistance) 820
MEA (membrane-electrode assembly) 397–398
mechanical fields 257
mechanical properties
– and composition 666–671
– CNT multilayers 583–584, 601–602
– cross-link density 669–671
– doping-moderated 292
– free-standing multilayers 363–392
– LbL-derived hydrogels 350–351
– LbL films 199–201
– LbL microcapsules 372–378
– local 369–370
– measurement 364–372
– PEMs 997–999
– tailoring 363–392
mechanical stability 384–386
mechanical stress 257–258
mechanically robust coatings 32–35, 146–147
mechanically strengthened capsules 927
mediated forward transfection (MFT) 999
medical applications 637–658
– CNT-based LbL assembly 608–609
– CNT multilayers 588–589
– engineered LbL capsules 816–823
melamine formaldehyde (MF) 802, 867–868
– molecular structure 750
membrane-electrode assembly (MEA) 397–398
membranes
– catalytic 908, 915–917
– cell 260–261
– composite 484
– contactor 917
– deflection 366
– dynamically excited 381
– functional 907–924
– functionalized polymeric nanomembranes 783
– ion-separation 908
– macrocyclic compounds 480–487
– membrane flux 310
– membrane-templated LbL assembly 625
– PEM 474–504
– polyethersulfone 921
– preparation 480
– semi-permeable 836
– skins 918
– see also nanomembranes
MEMS (microelectromechanical systems) 90, 586
mesoporous carbon (CMK-3) Hi-LbL 75–78
– capsules 76–78
mesoporous nanocompartment films 73
mesoporous templates 810
metabolites 754
metal complexes, step-by-step deposition 693
metal film/solution interface 960
metal/graphene conductive patterning 992–995
metal ions
– alkali 482
– divalent 494
– metal ion complexes 493–504
– rare earth 482
metal nanoparticles 539–369
metal/polymer interfaces 1017
metal-to-ligand charge transfer (MLCT) 500
metastable pitting region 314
methacrylates 100–113
methanol-treated LbL films 367
methicillin-resistant Staphylococcus aureus (MRSA) 524–525
methods and techniques
– buckling measurements 367–369
– Cahn balance technique 174
– colloid probe tapping mode 271
– computational 969
– cross ratio invariance 467
– electrochemical sensing 955–956
– impulse excitation 464–465
– Langmuir–Blodgett method 691
– LbL assembly 43–67
– MEA 397–398
– microcantilever technology 379
– nanosphere lithography 516
– non-destructive patterning 622
– optical detection methods 952–954
– parallel coordinates technique 976
– patterning techniques 985–988
– PEM preparation 224–233
– post-incubation method 705
– regression 970

– reverse-phase LbL   836–840
– stamping process   986
– statistical   969
– step-edge-like methodology   993
– thickness-shear mode   455
5,5′-methylene-bis(N-methylsalicylidene amine) (MBSA)   493–494
MF (melamine formaldehyde)   802, 867–868
– molecular structure   750
MHC (major histocompatibility complex)   931
micelles
– BCMs   353
– block copolymer   61–62
– CTAB   139–140
– responsive cores   353–358
micro-environments, localized   409
micro-shell LbL coatings   587
microcalorimetry   661
microcantilever technology   379
microcapsules
– adsorption   935
– anisotropic LbL   209–210
– chemical-responsive   862–864
– controlled delivery processes   382–386
– covalently stabilized   784
– cubic   210
– heating   865–866
– inflated   843–844
– light-sensitive   781
– magnetic LbL   866
– mechanical properties   372–378
– multicompartment   932
– permeability   928
– pH-responsive   855–858
– polymer composite multilayer   851–873
– polysaccharide-based   786
– redox-responsive   862
– salt-responsive   858–861
– self-induced/remote release   925–950
– stimuli-responsive   853–873
– thermo-responsive   865–866
– see also capsules; encapsulation
microchambers
– arrays   873–890
– loading   877–878
– responsive   879–881
– roofs of   879
– standing hollow   874–876
microcontact printing (μCP)   58
– cation-selective   49
microcontainers   925
microcores   164

microcrystals
– paclitaxel   157
– pigment Orange 13   158
microelectromechanical systems (MEMS)   90, 586
microencapsulation   831
– see also capsules; encapsulation
microenclosures   841–843
microfluidic channels   27
microgels, LbL-coated   383
microinterferometry   375–376
micromechanical properties   364–367
microparticulate templates   750–753
microporous films   48
microreactors   816–818
microscopic captivities   194
microscopy
– confocal   381–383, 717–719
– fluorescence   116
microshells   151
microstructure   201–204
"middle-down" proteomics   914
MIGRATION model   328–329
mimicry, cell-   71
minimum inhibitory concentration (MIC)   414
mixing ratio, PECs   140–141
MLCT (metal-to-ligand charge transfer)   500
mobility
– charge carriers   324
– DNA   939
– polymer layers   1041
Möbius transformation   458, 467
model materials   671–679
models, see theories and models
modified polyvinylamine (PVAm)   180
modular assembly   790
modulus
– complex shear   457
– storage/loss   645
– Young's, see Young's modulus
molar mass, number-average   894
molecular contact zone   177
molecular interactions   837
molecular motion   113
molecular structures
– alginate   750
– dex-HEMA   750
– melamine formaldehyde   750
– polyelectrolytes   756
– see also structures
molecules
– bio-, see biomolecules
– bioactive   351–353

Index | 1073

- electrochemical energy 395–406
- guest 892, 1039
- small organic 918–920
- therapeutic delivery 819–821
monoclonal antibodies 818
monolayers, OEG 619
monomer/substrate interactions 274
montmorillonite (MTM) 575
- PVA/MTM nanocomposites 577
morphologies
- myotube 680
- PEMs 893, 997–999
- surface 201–204
- surface-grafted polymer chains 438
- transition 119
motifs, host–guest 83
motion, molecular 113
mouse fibroblasts 452
MRSA (methicillin-resistant *Staphylococcus aureus*) 524–525
multiagent delivery 731–748
- LbL films 406–422
multicellular models 70–72
multicolor electrochromic devices 404–406
multicompartment microcapsules 932
multicompartmentalized polyelectrolyte films 898
multicomponent films 7, 421
multidimensional projections 971–977
multidrug resistance (MDR) 820
multifunctional nanostructures 633
multilayer capsules 777–799
multilayered three-dimensional objects 70
multilayers 642–650
- buildup 10–13
- capping 179
- cellular backpacks 28–29
- clay 575–582
- CNT 583–584, 595–612
- 3D cellular 120
- 3D patterned assemblies 85–88
- decomposition 290–292
- drug-loaded 790
- electrochemically active 1003–1038
- exponential growth 637
- free-standing 363–392, 578
- graphite 992
- hyaluronan/chitosan 468
- hydrogen-bonded 808
- ion transport 305
- LbL processed 25–41
- multilayer/solution interface 298
- neutron reflectometry 219–268, 275
- Nile Blue-loaded 902

- non-electrostatic interactions 238–242
- patterned systems 985–1001
- polyelectrolyte, *see* polyelectrolyte multilayers
- polyion capsules 863
- polymer 831–850
- polymer composite microcapsules 851–873
- polysaccharide 637–658
- reverse-phase LbL build-up 837–838
- robust coatings 32–35
- shear compliance 460
- SI-ATRP 443–447
- spray-/spin-assisted assembly 13–14
- stress–strain curves 293
- structural organization 575–576
- thin films 69–81, 395–406
- *see also* layers
multiple adsorption 447–449
multiple functionalities 852
multiple-hue flexible nanocomposites 406
multiple sequential assembly 493
multisegmented functional nanotubes 613–636
multistack heterostructures 32
multiwalled carbon nanotubes (MWNTs) 582
- biofuel cell cathodes 1032
- (bio)sensing 968
- functionalized 398–400, 601
- layer growth 599
- patterned enzyme-MWNT film 606
muscle cells, smooth 124–127, 677
muscle-specific proteins 679–681
mushroom morphology 438
myoblasts 679–684
- attachment 671–674
- C2C12 674–675
myogenic differentiation 679–681
myotube morphology 680

## n

Na, *see* sodium
nacre, nanostructured artificial 576
Nafion 322, 327
- bending response 532
nano-electrical mechanical system (NEMS) 586
nano-extracellular matrix (ECM) 100
nano-/micro-shell LbL coatings 587
nanobiomaterials 985–1001
nanocapsules 114–120
nanocatalysts 1003–1038
nanochannel coatings 27
nanocolloids

- aqueous 151–170
- carbon 607
- paclitaxel 157, 160
nanocompartment films 73
nanocomposites
- high-performance 576–578
- multiple-hue flexible 406
- PVA/MTM 577
nanocontainers 925
nanocores 151–170
- surface modification 163
nanocrystals
- cellulose 172, 182–184
- drug formulations 153
- lanthanide-doped 194
- LbL films 194–195
nanodomains 893–898
nanofibers
- nanofibrillated cellulose 172, 182–184
- polyaniline 400–401, 405–406
- scaffolds 120
nanofillers 147–148
nanofilms
- protein 120
- see also films
nanofiltration 919
nanoimprint lithography (NIL) 83, 85–88, 92–93, 618
nanolayer assembly processes 393–435
nanomembranes 196
- functionalized polymeric 783, 869
- silk fibroin 198–199
- see also membranes
nanoparticles
- bridging flocculation 16
- catalytic nanoparticle-containing films 914–915
- CdTe 702
- core–shell 1030
- diffusion 691–710
- gelatin-based 160
- gold 191–193, 201, 369
- hybrid nanoparticle/polyelectrolyte capsules 757
- LbL films 693–694
- metal 539–369
- nonlinear optics 517
- PAC-capped 540–541
- PB 957
- PLGA 428–429
- semiconductor 539
- silica 526
- soft organic 159–161
- supramolecular crystal structures 88–90

nanopatterned surfaces 619–621
nanoporous LbL films 100–113, 204–205, 624–633
nanoporous shell structure 711–712
nanoreactors 1030–1031
- enhanced quantum yield 53
nanoscale elements 25–28
nanoscale fillers 190
nanosheets
- graphene 78, 197
- PNIPAM 451
nanospaces 104–106
nanosphere lithography (NSL) 516
nanostripes 613–636
nanostructured artificial nacre 576
nanostructured electrodes 539–571
nanostructured photoelectrodes 552–558
nanostructures
- gold 964
- hybrid 403
- multifunctional 633
- silver 964
nanotransfer printing (nTP) 85
nanotubes
- carbon, see carbon nanotubes
- DNA-containing LbL 626
- multisegmented functional 613–636
- polypyrrole 632
nanowires 403
- silver 191–193
- synthesis 993
near-infrared (NIR) irradiation 781
nearly constant loss (NCL) behavior 329
NEMS (nano-electrical mechanical system) 586
networks
- coordination polymer 496
- neural 970
neurites 588
neutralization 1042
neutron reflectometry 203–204, 219–268, 275
Newton's rings 365
NFC (nanofibrillated cellulose) 172, 182–184
NG108-15 cell 588
nickel ion complexes 497–499, 502
nickel-platings, selective 994
nickel-titanium (NiTi) stents 650
NIL (nanoimprint lithography) 83, 85–88, 92–93, 618
Nile Blue-loaded multilayer 902
NIR (near-infrared) irradiation 781
nitric oxide diffusion 124–128

o-nitrotoluene 916
NMR (nuclear magnetic resonance), PMMA 107
noncharged polymer–polymer complexes 136
nondegradable polyanions 407
nondestructive patterning 622
nonelectrostatic interactions 238–242
nonlinear build-up 461–462
nonlinear control, diffusion 311
nonlinear optical fibers 519–525
nonlinear polarization 512
nonlinear swelling 249
nonspecific binding 718–721
NSL (nanosphere lithography) 516
nTP (nanotransfer printing) 85
nuclear magnetic resonance (NMR), PMMA 107
null-ellipsometry 278
number-average molar mass 894

## o

odd–even effect 256–257, 278–279
oligo(ethylene glycol) (OEG) monolayers 619
oligonucleotides
– arrangement 724–727
– coupling 713–716
– idealized orientation 809
one-step coupling 716
optical applications, clay multilayers 581–582
optical detection methods 952–954
optical fibers 519–525
optical properties, LbL films 199–201
optically addressable capsules 868–869
optoelectronic devices 602
optoelectronic materials 511–537
ordered particles 995–997
organic barriers 893
organic compounds, poorly-soluble 155
organic ions 477
organic–metallic layers 92
organic nanoparticles, soft 159–161
organic phase
– inwards build-up self-assembly 845–847
– LbL encapsulation 835
organic solvents 721–722, 831–850
orthopedic applications 411
osmium-containing polyelectrolyte 1027
osmotic pressure 246
– electrochemical swelling 259
ovalbumin (OVA) 758–761
– encapsulated 822
overlap concentration 616
oxidation

– ferro cyanide (FC) 259
– surface 549–550
– wet-air 917
oxidative electrochemical cycles 501
oxygen reduction 550–552
oxygen transmission rate 581

## p

π–π interaction complex 62–64
PA (polyacrylamide) 677–679
PAA (poly (acrylic acid)) 47–55, 695–698
– cellulose fibers/fibrils 174–179
– PECs 137–139
PAC (polyacrylate-capped) nanoparticles 540–541
paclitaxel 154–155
– drug release 166
– microcrystals/nanocolloids 157
– nanocolloids 160
PAH (poly (allylamine hydrochloride)) 57, 174–179
– aqueous nanocolloids 156–158
– biomolecule encapsulation 832
– diffusion of nanoparticles 694–698
– LbL films 190–191
– optoelectronic materials 513–514
– PAH-PSS capsules 871–873
– PEMs 226–227, 234–235
pairing, ion, see ion pairing
PAMAM (poly (amidoamine)) 963
– dendrimers 771–772, 957–958
PAMPS (poly (2-acrylamido 2-methylpropanesulfonic acid)) 528–529
pancake morphology 438
PANi (polyaniline nanofibers) 400–401, 405–406
parallel coordinates technique 976
particles
– anisotropic 933
– carboxylated 724
– clay 574–575
– core–shell 711–712
– LbL film deposition 15–17
– nano-, see nanoparticles
– spherical 443
– triangular 965
passive controlled release 407–417
passive host–active guest structure 1039
passive protection activity 1042–1044
patterned arrays, gold nanoparticles 369
patterned enzyme-MWNT film, for glucose sensing. 606
patterned LbL films, freely standing 205–207
patterned multilayer systems 985–1001

patterned resists 618
patterned target 89
patterning
- buckling 363
- conductive 992–995
- graphene 992–995
- hydrophilic 1049
- LbL films 14–15
- metal 992–995
- non-destructive 622
- PEMs 988–990
- polyelectrolyte capsules 872–873
- proteins 990–995
- self-assembled monolayers 985
PB (Prussian Blue) 487–492, 694–695
- nanoparticles 957
PBD (probe beam deflection) 1022
PCA (principal component analysis) 959
PCR (polymerase chain reactions) 682–683
PDAC (poly (diallyldimethyl ammonium chloride)) 396–397
PDADMA (poly (diallyldimethyl ammonium)) 287, 291–292, 295–299
- QCR monitoring 457, 462–463
- selective transport 476–478
PDADMAC (poly (diallyldimethyl ammonium chloride)) 227–230, 235–237, 443–447
- capsule fusion 859
- counterions 240
- diffusion of nanoparticles 701–705
- feedback active coatings/films 1043–1044
- PSS/PDADMAC films 921–922
PDAPB (poly (diallyl pyrrolidinium bromide)) 898–901
PDDA (poly (diallyldimethyl ammonium chloride)) 45, 71, 139, 146
- clay multilayers 575–576
- nanoconfined multilayers 620
- nanostructured electrodes 540–557
- PDDA/Q-CdS assembly 555–556
PDMS (polydimethylsiloxane) 677–679
- stamp 89–90
Pearson's HSAB classification 429
PECs (polyelectrolyte complexes) 136–144
- ion conduction 329–332
- mixing ratio 140–141
- phase diagrams 338–340
- quasisoluble 281
- water-soluble 338
PEI (polyethyleneimine) 63
- cellulose fibers/fibrils 184
PEMs (polyelectrolyte multilayers) 14
- blend- 479–480

- cell differentiation control 659–690
- conductivity spectra 321–336
- confined geometries 874
- diffusion into 691–710
- (dis)ordered particles 995–997
- feedback active coatings/films 1039–1050
- hydrophilic 893–895
- ion-paired 281–320
- living cells 28
- mechanical properties 997–999
- membranes 474–504
- morphologies 997–999
- multi-agent delivery 731–748
- nanoconfined 613–636
- neutron reflectometry 219–268, 275
- optoelectronic materials 511–537
- passive protection activity 1042–1044
- patterning 988–995
- polysaccharides 659–690
- porous media 908–910
- preparation techniques 224–233, 238–242
- QCR 455–469
- redox active 1003
- self-induced/remote release 925–950
- structural units 259
- structure 269–280
- triggered disassembly 744
penicillin 958
PEO (polyethyleneoxide) 175
pepsin 774
peptides
- antibacterial 895–898, 1046
- cationic 654
- SIINFEKL 931
percolation 252–254, 671
pericentric multicompartment microcapsules 932
periodic buckling pattern 363
periodic coating 521
permeability 53–54
- hydraulic 474
- LbL films 698–700
- microcapsules 928
- shell 213, 870
permeation rate 482
permselectivity 53–54, 1007
persistent regime 615
perturbation, electrochemical 1013
PGA (poly (sodium-L-glutamic acid)) 694–698
pH-buffering 1041
pH-dependent binding constant 766
pH-dependent decomposition 773–774
pH-dependent response 605

pH-effects 241
pH-induced chain dynamics 342–348
pH-responsive capsules 855–858
pH-sensitive LbL hydrogels 209
pH-sensitive release 785–787
pH-sensitive substrate-binding groups 26
pH-triggered disintegration 343
phagocytes 28
pharmacies, programmable 393–435
phase diagrams, PECs 338–340
phase-matching condition 522
π-phase shift 525
PHEMA (poly (2-hydroxyethyl methacrylate)) 443–444
phenyl boronic acid 754
phosphate, selective rejection 920–921
phospholipids 805
– anionic 965–966
– layer 122
photo-responsive microchambers 879–881
photo-stimulated release 781–783
photocurrent generation 607
photocurrent polarity-switching 556–558
photocycloaddition 54
photocyclomers 54
photoelectrochemistry, PDDA/Q-CdS assembly 555–556
photoelectrodes 552–558
photogenerated holes 558
photothermal sensor array 380
photovoltaic performance 404
photovoltaic properties, LbL films 580
phthalocyanate derivatives 58
physical barriers 421
physical chemistry, polysaccharides 659–690
piezoresistive response 604
pigment Orange 13 155–156
– microcrystals 158
pitting region 314
PLA (poly (lactic acid)) 116–120
– morphology transition 119
plasmid DNA 741
plasmon resonance, SPR 201, 644, 952
plasmonic enhancement 515–519
plasmonic properties 190
platelet adhesion 652
platinum electrodes 540–552
platinum nanoparticles 545–548
PLGA nanoparticles 428–429
PLL (poly-L-lysine) 156–157, 163–164
– layers 162
– PLL/HA films 660–664
pluripotency 682

PMA (poly (methacrylic acid)), structure 856
PMAA (poly (methacrylic acid)) 339–345
– deuterated 347
– LbL-oligonucleotide particles 714–718, 727
PMMA (poly (methyl methacrylate))
– chemical structure 101
– crystalline 112
– host–guest interactions 86, 92–94
– it/st 100–104, 114–116
– stereocomplex films 114–116
PNIPAM (poly (N-isopropyl acrylamide)) 245–246, 342–343, 355–356
– SIP 449–451
POEA (polyethoxylated tallow amine) 959–962
point-load 376
Poisson ratio 373–374, 871
polarity-switching 556–558
poly (2-acrylamido 2-methylpropanesulfonic acid) (PAMPS) 528–529
poly (acrylic acid) (PAA) 47–55, 695–698
– cellulose fibers/fibrils 174–179
– PECs 137–139
poly (allylamine hydrochloride) (PAH) 57
– aqueous nanocolloids 156–158
– biomolecule encapsulation 832
– cellulose fibers/fibrils 174–179
– diffusion of nanoparticles 694–698
– LbL films 190–191
– optoelectronic materials 513–514
– PAH-PSS capsules 871–873
– PEMs 226–227, 234–235
poly (amidoamine) (PAMAM) 963
– dendrimers 771–772, 957–958
poly (carboxymethylbeta-cyclodextrin) 415
poly (diallyldimethyl ammonium chloride) (PDAC) 396–397
poly (diallyldimethyl ammonium) (PDADMA) 287, 291–292, 295–299
– QCR monitoring 457, 462–463
– selective transport 476–478
poly (diallyldimethyl ammonium chloride) (PDADMAC) 227–230, 235–237, 443–447
– capsule fusion 859
– counterions 240
– diffusion of nanoparticles 701–705
– feedback active coatings/films 1043–1044
– PSS/PDADMAC films 921–922
poly (diallyldimethyl ammonium chloride) (PDDA) 45, 71, 139, 146
– clay multilayers 575–576
– nanoconfined multilayers 620
– nanostructured electrodes 540–557
– PDDA/Q-CdS assembly 555–556

poly (diallyl pyrrolidinium bromide)
  (PDAPB)   898–901
poly (ethyleneimine *N*-acetic acid)
  (LPEI-ac)   487
poly (ethyleneimine) (PEI)   63
poly (2-hydroxyethyl methacrylate)
  (PHEMA)   443–444
poly (*N*-isopropyl acrylamide)
  (PNIPAM)   245–246, 342–343, 355–356
– SIP   449–451
poly L-lysine (PLL)   156–157, 163–164
– layers   162
– PLL/HA films   660–664
poly (lactic acid) (PLA)   116–120
– morphology transition   119
poly (methacrylic acid) (PMA), structure   856
poly (methacrylic acid) (PMAA)   339–345
– deuterated   347
– LbL-oligonucleotide particles   714–718, 727
poly (methyl methacrylate) (PMMA)
– chemical structure   101
– crystalline   112
– host–guest interactions   86, 92–94
– it/st   100–104, 114–116
– stereocomplex films   114–116
poly (phenylene-alt-fluorene) backbone   502
poly (*N*-salicylidene vinylamine)
  (PSVA)   493–496
poly (sodium-L-glutamic acid) (PGA)   694–698
poly (sodium phosphate) (PSP)   13–15
poly (sodium salt) (PSS), biomolecule encapsulation   832
poly (sodium styrene sulfonate) (PSS)   76
– aqueous nanocolloids   156–158
– conformation   269–280
– LbL films   190–191
– PEMs   226–227, 234–235
poly (styrene-b-acrylic acid)   61
poly (styrene sulfonate) (PSS)   10–13
– biomolecule encapsulation   832
– ion-paired PEMs   287
– nanoconfined multilayers   620
– PAH-PSS capsules   871–873
– PSS/PDADMAC films   921–922
– rhodamine-labeled PSS polymers   782
– selective transport   476–478
poly (vinyl alcohol) (PVA), PVA/MTM nanocomposites   577
poly (4-vinylpyridine) (PVP)   47–49
poly (vinylpyrrolidone) (PVPON)   138–139, 210–213, 343–344, 349–350
– engineered LbL capsules   803–808

– redox-controlled release   787–788
– structure   856
polyacid layers   355
polyacids   856
polyacrylamide (PA)   677–679
polyacrylate-capped (PAC) nanoparticles   540–541
polyamines   740–741
polyaniline nanofibers (PANi)   400–401, 405–406
polyanions
– nondegradable   407
– polysaccharides   641–642
– selection   718
– strong   647–650
– synthetic   617
– weak   642–647
polycarboxylic acid   343
polycations
– hydrolytically degradable   407
– polycation–protein interaction   167
– polysaccharides   639–641
– selection   717–718
– synthetic   617
– toxic   213
polycondensation   693
polycrystalline Pt electrode   543, 550
polydimethylsiloxane (PDMS)   677–679
– stamp   89–90
polyelectrolyte complexes (PECs)   136–144
– ion conduction   329–332
– membranes   477
– mixing ratio   140–141
– phase diagrams   338–340
– quasisoluble   281
– water-soluble   338
polyelectrolyte multilayers (PEMs)   14, 28
– blend-   479–480
– cell differentiation control   659–690
– conductivity spectra   321–336
– confined geometries   874
– diffusion into   691–710
– (dis)ordered particles   995–997
– feedback active coatings/films   1039–1050
– hydrophilic   893–895
– ion-paired   281–320
– living cells   28
– mechanical properties   997–999
– membranes   474–504
– morphologies   997–999
– multi-agent delivery   731–748
– nanoconfined   613–636
– neutron reflectometry   219–268, 275
– optoelectronic materials   511–537

– passive protection activity 1042–1044
– patterning 988–995
– polysaccharides 659–690
– porous media 908–910
– preparation techniques 224–233, 238–242
– QCR 455–469
– redox active 1003
– self-induced/remote release 925–950
– structural units 259
– structure 269–280
– triggered disassembly 744
polyelectrolyte-repelling background 621
polyelectrolyte–sufactant complexes 136, 139–140
polyelectrolytes 43
– adsorption 664, 907–924
– amphiphilic films 896
– anionic/cationic 72
– ATRP macroinitiators 442–443
– capsule patterning 872–873
– chains 614–617
– chemical structures 396–397
– codeposition 142–143
– complementary 446
– complexation segments 284
– conformation 269–280
– counter-polyelectrolyte 481
– counterionic 77
– deposition process 778
– dip–spin assembly 553–555
– dye-impermeable barriers 901–902
– exchange 358
– functional domain-containing films 891–905
– hybrid nanoparticle/polyelectrolyte capsules 757
– interdiffusion 301–305
– LbL films 190
– LbL shells 153–154
– molecular structure 756
– multicompartmentalized films 898
– osmium-containing 1027
– quantum dots 539–369
– segment movement 304
– selection 716–718
– self-healing anticorrosion coatings 1040–1045
– stratified films 898
– structures 713
– trypsin-containing films 911–914
– types 233–238, 340–342
– viscosity 179
polyether dendrimers 47
polyetheretherketone 585

polyethersulfone membranes 921
polyethoxylated tallow amine (POEA) 959–962
polyethyleneimine (PEI), cellulose fibers/fibrils 184
polyethyleneoxide (PEO) 175
polygalacturonic acid 646
polyion multilayer capsules 863
polyion–polyion interactions 1018
polymerase chain reactions (PCR) 682–683
polymerases 833
polymeric carriers 816
polymeric complexes 135–150
– functional films 144
– hydrogen-bonded 138–139
polymeric–inorganic complexes 136
polymeric nanomembranes 783, 869
polymerization
– double-stranded assembly 106–111
– SI-ATRP 439–440
– stereospecific template 100–113
– surface-initiated 437–454
polymers
– adsorption 232
– agarose–polymer core–shell materials 846
– charge density 254
– charge-shifting 742–744
– CNT/conjugated system 400–401
– degradable cationic 734–736
– electrochromic 527
– films 113–120
– hydrophilic 395
– in solution 836–837
– inwards build-up self-assembly 845–847
– maltose-bearing 770
– matrix hydration 327
– metal/polymer interfaces 1017
– mobile layers 1041
– molecular interactions 837
– multilayers 831–850
– reverse-phase LbL 840
– rhodamine-labeled PSS 782
– self-assembled polymer/DNA polyplexes 732
– solutions 99
– sorption 304–305
– stimuli-responsive 208, 851–873
– surface/bulk charge 292–301
– surface-grafted brushes 438–440
– synthetic 1025
polyoxometalate (POM) 608–609
– functional 703
polyphenol 161
polyplexes 732

polypyrrole nanotube   632
polysaccharide-based microcapsules   786
polysaccharide multilayers   637–658
– antithrombogenic properties   650–651
– cardiovascular applications   650–654
– drug delivery   652–654
– preparation   651–652
polystyrene (PS)
– carboxylated   994
– CD-functionalized   88
– TCPS   675–676
polytopic ligands   494
polyvinylamine (PVA)   180, 474–476
polyvinylcaprolactam (PVCL)   349
polyviologen (PV)   528
poorly-soluble inorganic compounds   159
poorly-soluble organic compounds   155
porcine arteries   651
pore-to-bulk diffusivity   628
porous alumina   909
porous LbL films   204–208
porous materials, coating   425
porous media, PEM deposition   908–910
porous scaffolds   120
porous structure   111–113
porous supramolecular assemblies   90–94
porphyrin derivatives   55
post-assembly chain dynamics   337
post-assembly processing   297
post-chemical reaction   50–52
post-diffusion   892
post-incubation method   705
post-loading   813, 854
Pratt–Bartlett case diagram   1029
pre-loading   854
precipitation, solvent-assisted   159–161
preformed LbL-coated nanoparticles   159–161
preloading, on/in template   810–811
preparation
– amphiphilic polyelectrolyte films   896
– hollow capsules   113–120
– LbL-derived hydrogels   348–350
– membranes   480
– polysaccharide multilayers   651–652
pressure
– osmotic   246, 259
– pressure–deflection relationship   366
primary neurons   989
primers   833
principal component analysis (PCA)   959
printing
– μCP   49, 58
– imprinted surfaces   873
– nanotransfer   85

– NIL   83, 85–88, 92–93, 618
– SMILbL films   54–58
probe beam deflection (PBD)   1022
processing time   31–32
Procion Brown MX-GRN   513–515
procoagulants   638
programmable pharmacies   393–435
programmable release   406–422
projections, multidimensional   971–977
proliferation   671–674
PROMAXX® insulin microcores   164
propagation
– corrosion   1042
– redox charge   1012
prosthetic group   1027
proteases   864
protection activity   1042–1044
proteins
– adsorption   910–911
– avidin   765–768
– collagen   988
– Con A   768–769
– delivery   765–776
– digestion   911–914
– dual release   419
– EGFP   735–736
– encapsulation   834
– expression vectors   588
– fibronectin   988
– glycoproteins   765
– lectin   768
– loading and assembly   822
– membrane functionalization   910–914
– muscle-specific   679–681
– patterning   990–995
– polycation–protein interaction   167
– release   408–412
proteoliposomes   963, 972
proteomics, "middle-down"   914
proton exchange fuel cells   395–398
protonated hydroxy groups   856
protonated layers   223
protonation   1008
– charged groups   927
protrusions, cytoplasmic   760
Prussian Blue (PB)   487–492, 694–695
– nanoparticles   957
PS (polystyrene)
– carboxylated   994
– CD-functionalized   88
– TCPS   675–676
PSP (poly (sodium phosphate))   13–15
PSS (poly (sodium salt)), biomolecule encapsulation   832

PSS (poly (sodium styrene sulfonate)) 76
– aqueous nanocolloids 156–158
– conformation 269–280
– LbL films 190–191
– PEMs 226–227, 234–235
PSS (poly (styrene sulfonate)) 10–13
– biomolecule encapsulation 832
– ion-paired PEMs 287
– nanoconfined multilayers 620
– PAH-PSS capsules 871–873
– PSS/PDADMAC films 921–922
– rhodamine-labeled PSS polymers 782
– selective transport 476–478
PSVA (poly (N-salicylidene vinylamine)) 493–496
pull-off force 178
pump pulse energy 521
push force-curve 929
PV (polyviologen) 528
PVA (poly (vinyl alcohol)), PVA/MTM nanocomposites 577
PVA (polyvinylamine), membranes 474–476
PVAm (modified polyvinylamine) 180
PVCL (polyvinylcaprolactam) 349
PVP (poly (4-vinylpyridine)) 47–49
PVPON (poly (vinylpyrrolidone)) 138–139, 210–213, 343–344, 349–350
– engineered LbL capsules 803–808
– redox-controlled release 787–788
– structure 856
pyrene 139–140, 356

## q

Q-CdS (CdS quantum dots) 552–558
quadlayers 247
quantitative abrasion test 34
quantized size-dependent eigenspectrum 552
quantum dots 193–194
– CdS 552–558
– in polyelectrolytes 539–369
quantum yield, enhanced 53
quartz crystal microbalance (QCM) 11–12, 455
– electrochemical 1018–1021
– Hi-LbL 71, 74–77
– PECs 141
quartz crystal resonator (QCR) 455–469
– flow-through cell 466
quasi-phase-matching 521
quasisoluble PECs (qPECs) 281
quaternized poly-N-ethyl-4-vinylpyridinium bromide (QPVP) 339–342

## r

randomly distributed hydrophobic nanodomains 893–898
rapid release 744–745
rare earth metal ions 482
RDE (rotating disk electrode) 305–306
reactions
– Brønsted acid–base 838
– oxidation, see oxidation
– oxygen reduction 550–552
– PCR 682–683
– photocycloaddition 54
– polycondensation 693
– polymerization, see polymerization
– post-chemical 50–52
– selective hydrogenation 915
– urea hydrolysis 968
reactive LbL deposition 696
reactors
– nano- 53, 1030–1031
– RNA micro- 816–817
readsorption 491
redox active PEMs (r-PEMs) 1003
– extrinsic 1019
redox centers, entrapped 1014
redox charge propagation 1012–1018
redox-controlled release 787–788
redox potential 1004–1005
redox-responsive capsules 862
reducing agents 693
reduction
– o-nitrotoluene 916
– oxygen 550–552
reflectivity, neutron 203–204
refractive index 220
– profiles 35
regenerative medicine 410
regression methods 970
Reissner-model 375–376
relative humidity 248, 324
relaxation, segment 276
release
– bio-responsive 788–792
– burst 654
– CF 937
– controlled 351–353, 750–759, 777–799, 1044–1045
– DNA 736–739
– drugs 165–167, 749–763
– dual protein 419
– engineered LbL capsules 813–816
– in vitro 167
– insulin 754, 792
– kinetics 356

– laser-induced  868
– lysozyme  410
– magnetically sensitive  779–780
– passive controlled  407–417
– pH-sensitive  785–787
– photo-stimulated  781–783
– programmable  406–422
– proteins  408–412
– pyrene  356
– rapid  744–745
– redox-controlled  787–788
– remote  782–783, 925–950
– self-induced  925–950
– sequential  420, 739
– simultaneous  740
– spatial control  736–739
– stimuli-free controlled  73
– sugar-induced  771
– temperature-induced  344
– thermo-stimulated  783–784
– time-delayed  790
– tunable  739
– ultrasound-stimulated  780–781, 867–868
– see also delivery
"reluctant" exchange mechanism  305
remote responsive capsules  866–869
removal, dyes  918–920
reproducibility, LbL assembly  7–10
resists, patterned  618
resonant frequencies  456
response
– bending  532
– dual stimuli-  449–450
– electrochemical  1004–1012
– nonlinear optical  515–519
– pH-dependent  605
– piezoresistive  604
responsive cores, micelles  353–358
responsive free-standing films  451–453
responsive LbL assemblies  337–362
– swellable  348–358
responsive LbL films  208–209
responsive multilayer microcapsules  851–873
reverse-phase LbL  836–840
reversibility, softening process  350
rhodamine  725
rhodamine 123  843
rhodamine 6G  116
rhodamine-labeled PSS polymers  782
rinsing  8, 805
RNA  823, 997–999
RNA microreactor  816–817
robust multilayer coatings  32–35
robust thin films  50–52

roofs of microchambers  879
rotating disk electrode (RDE)  305–306
Ruthenium purple (RuP)  529

s
S-layers  260
salts
– Everitt's  490
– saloplastics  281
– salt effect  143–144
– "salt-gated" transistor  312–313
– salt-in-polymer electrolytes  321
– salt-induced fusion  930
– salt-responsive capsules  858–861
– sorption  304–305
Sammon's mapping  971, 974–975
SAMs (self-assembled monolayers)
– CD  84
– patterning  985
SAMs (self-assembled multilayers)  511, 517
sandwich-like structure  1044
sandwiched gold nanoparticles  201
SANP (sodium 9-anthracenepropionate)  52–54
SBMEs (surface-bound microenclosures)  841–843
scaffolds
– In vivo  412
– nanofiber  120
– porous  426
scalar fields  242
scaling
– conductivity spectra  332–335
– regimes  615
– Summerfield  332–334
scattering length density (SLD)  203–204, 220, 250
– ion-paired PEMs  303
Schiff-base-metal ion complexes  493–495
scratch-resistant coatings  147–148
sculptured LbL films  207–208
SDS (sodium n-dodecylsulfate)  450
SDT (signal detection theory)  958
SEC (size exclusion chromatography)  108
second harmonic generation (SHG)  513–518
second order nonlinear optics  512–515
segments
– backbone  202
– complexation  284
– frozen  301
– movement  304
– relaxation  276
selection

– polyelectrolytes 716–718
– template 801–804
selective extraction 109
selective hydrogenation 915
selective nickel-platings 994
selective rejection 920–921
selective separation 473
selective transport 473
selectively labeled layers 203–204
self-assembled films 488
self-assembled monolayers (SAMs)
– CD 84
– patterning 985
self-assembled multilayers (SAMs) 511, 517
self-assembly 1–4
– building blocks 44
– directed 985–1001
– inwards build-up 845–847
– LbL 440–441
– polymer/DNA polyplexes 732
– spontaneous 69
self-embedding, spatial 934
self-exploding capsules 752
self-floating hydrogen-bonded films 344
self healing coatings 1040–1045
– anticorrosion 1040–1045
– superhydrophobic 144–146
self-limited build-up 461–462
self-trapping 313–315
semi-dilute diffusion regime 629–630
semi-permeable membrane 836
semiconductor nanoparticles 539
semifluorinated block copolymer brushes 452
sensors
– bio- 548–549
– CNT-based LbL assembly 604–606
– CNT multilayers 585–587
– controlled architectures 951–983
– electrochemical 955–956
– flat LbL films 378–382
– glucose 606
– hierarchic LbL assembly 75
– humidity 605
– optical 521–525, 952–954
– post-incubation method 705
separation
– ion 479–480
– materials 473–509
– selective 473
sequential addressing 879
sequential assembly, multiple 493
sequential release 420, 739
SERRS, architecture 965–966

shear compliance 460
shear force, air 225
shear modulus, complex 457
β-sheet formation 198–199
shells
– core–shell materials 835, 846
– core–shell particles 711–712, 1030
– design 753–759
– (im)permeable 870
– LbL 151–170
– permeability 213
– polyelectrolyte 153–154
– theory 373–374
SHG (second harmonic generation) 513–518
side groups 202
sieving, hydrated ions 491
signal detection theory (SDT) 958
SIINFEKL peptides 931
silanes 619–620
silhouette coefficient 975
silica capsules 73
silica fiber 519
silica nanoparticles, PEM films 526
silica particles 114–115, 117
silicon wafer 15
silk fibroin 198–199
silk films 367–368
silver nanostructures 964
silver nanowires 191–193
silver nitrate 938
simultaneous release 740
single adsorption 442–443
single-stranded (ss) DNA 723
single-walled carbon nanotubes (SWNTs) 582–589
– SWNT/laminin thin films 608
SIP (surface-initiated polymerization) 437–454
size-dependent eigenspectrum 552
size-dependent transport 477
size estimation, polyelectrolyte chains 614–617
size exclusion chromatography (SEC) 108
size-selective transport 476–479
skins, membrane 918
SLbL (sonication-assisted LbL) encapsulation 151
SLD (scattering length density) 203–204, 220, 250
– ion-paired PEMs 303
small deformation measurements 375–378
small drugs, permeability 698–700
small interfering RNA (siRNA) 823, 997–999
small molecules

– delivery   412–417
– organic   918–920
– therapeutic delivery   819–821
small particles, LbL film deposition   15–17
smart micro-/nanocontainers   925
SMILbL (surface molecular imprinted LbL) films   54–58
smooth muscle cells (SMCs)   124–127, 677
SnS crystals, tetrahedral   211
sodium
– PSS, see PSS
– sodium 9-anthracenepropionate (SANP)   52–54
– sodium *n*-dodecylsulfate (SDS)   450
soft organic nanoparticles   159–161
softening, reversibility   350
softwood fibers, bleached   171
solar cells   402–404, 553
solubility   151
– low   151–170
– organic compounds   155
solutions
– dilute   614–617
– dipping temperature   242
– metal film/solution interface   960
– multilayer/solution interface   298
– polymer   99
– porous supramolecular assemblies   90–94
solvents
– chemical-responsive capsules   862–863
– organic   721–722, 831–850
– reverse-phase LbL   840
– solvent-assisted precipitation   159–161
sonication-assisted LbL (SLbL) encapsulation   151
sorption, salt/polymer   304–305
spatial confinement   552
spatial control   736–739
spatial self-embedding   934
specular reflectometry   220–222
SPEEK (sulfonated polyetheretherketone)   585
spherical capsules   384
spherical colloids, assembly   878
spherical dendrimers   627
spherical particles   443
– templates   805
spherical vesicles   803–804
spin-assisted LbL assembly   203–204
spin-coating   231
spontaneous self-assembly   69
sPPO (sulfonated poly-*p*-phenyleneoxide)   395–396
SPR (surface plasmon resonance)   201

– (bio)sensing   952
– polysaccharide multilayers   644
spray-/spin-assisted multilayer assembly   13–14
spraying   228–231
spreading area   675–677
ss (single-stranded) DNA   723
stability
– antireflection-integrated coatings   146–147
– LbL-oligonucleotide particles   727
– mechanical   384–386
– Pt nanoparticles   549–550
– stereocomplex polymer films   118
stabilized liposomes   936
stabilizers   153–154, 597
stacks, Bragg   31–32
stained fluorescence dyes   847
stamp, PDMS   89–90
stamping process   986
standing hollow microchambers   874–876
*Staphylococcus aureus*   512
– methicillin-resistant   524–525
*Staphylococcus epidermidis*   353–354
statistical methods   969
– principal component analysis   959
steady-state diffusion-limited flux   307
steady-state growth   300
steady-state PBD transients   1023
stem cells   410
– embryonic   660, 679–684
stents
– CH/heparin   651
– intravascular   737
– nickel-titanium (NiTi)   650
step-by-step deposition   693
step-by-step laser exposure   935
1-step coupling   716
2-step coupling   716
step-edge-like methodology   993
stepwise assembly   103
stereocomplex films
– PLAs   116–120
– PMMA   114–116
– stability   118
– ultrathin   102–104
stereocomplex formation   100–113
stereospecific template polymerization   100–113
sterilization   415
sticky gel electrophoresis   346–347
stiffness, average   377
stimuli-free controlled release   73
stimuli-response

- dual 449–450
- free-standing films 451–453
stimuli-responsive microcapsules 851–873
stimuli-responsive polymer coatings 208
stimuli-sensitive LbL films 765–776
Stokes radius 492
storage charge 395–406
storage modulus 645
stratification, LbL films 898–901
stratified layers 420
stratified polyelectrolyte films 898
stress 176
- cell-induced 929
- mechanical 257–258
- stress–strain curves 293
stretching 374
- vibrations 115
strong polyanions 233–237, 647–650
strong polycations 235–237
structural organization 575–576
- CNT multilayers 583–584
structural tailoring 140–144
structures
- alginate 713, 750
- ATRP macroinitiators 441
- block 223, 236
- capsules 384
- clay particles 574–575
- CNTs 582
- cushion 259–261
- DAR 137
- dex-HEMA 750
- EIS 958
- fibrillar 173
- film 342–348
- hetero- 32
- hierarchical 740
- hollow 372
- hybrid material 91
- melamine formaldehyde 750
- nano-, see nanostructures
- passive host–active guest 1039
- PEMs 259, 269–280
- PMA 856
- polyelectrolyte complex membrane 477
- polyelectrolytes 713
- polymeric complexes 136
- polysaccharides 638–642
- porous 111–113
- PVPON 856
- sandwich-like 1044
- superlattice 222–223
- supramolecular crystal 88–90

- thin-walled 372
- zeolite-like 490
styrene 11
- see also PS, PSS
submonolayers 551
substituent groups 497
substrates
- flat 442
- flexible 369
- glass 991
- ITO-coated glass 489
- pH-sensitive substrate-binding groups 26
- substrate-mediated layering 348
sugar-induced release 771
sulfonate 289
sulfonated poly-$p$-phenyleneoxide (sPPO) 395–396
sulfonated polyetheretherketone (SPEEK) 585
$p$-sulfonato-calix[$n$]arene 480–482
Summerfield scaling 332–334
superhydrophobic coatings, self-healing 146–147
superhydrophobic surfaces 45–46, 1049
superlattice structure 222–223
supernatant 721–723
supramolecular assembly 473–509
supramolecular complex 44
supramolecular crystal structures 88–90
supramolecular hybrid material structures 91
supramolecular LbL assembly 84–85
- porous 90–94
supramolecular objects, fuzzy 6–7
surface acoustic impedance 457, 460, 1026
surface-bound microenclosures (SBMEs) 841–843
surface charge 442–443
- distribution 297–301
- polymers 292–301
- reversal 294–295
surface-grafted polymer brushes 438–440
surface-initiated atom transfer radical polymerization (SI-ATRP) 439–440
surface-initiated polymerization (SIP) 437–454
surface-mediated cell transfection 731–748
- multi-agent delivery 731–748
surface molecular imprinted LbL (SMILbL) films 54–58
surface plasmon resonance (SPR) 201
- (bio)sensing 952
- polysaccharide multilayers 644

surfaces
- characterization 541–545
- chemically nanopatterned 619–621
- coating 427–429
- 3D patterned multilayer assemblies 85–88
- flat 618–624
- implants 736–739
- imprinted 873
- low coverage 545–548
- modification 163
- morphology 201–204
- oxidation 549–550
- potential model 1009
- stereocomplex assembly 104
- superhydrophobic 45–46, 1049
surfactants 139–140, 450
Surgifoam® bandages 427
susceptibility tensor 519
sweep rate 1005
swellability 662–664
swelling
- antibacterial activity 1046
- electrochemical 259
- hyperswelling 288
- in water 247–256
- LbL-derived hydrogels 348–350
- LbL films 348–358
- PSS layers 273
- r-PEMs 1022, 1025
- volume swelling coefficient 287
SWNTs (single-walled carbon nanotubes) 582–589
- SWNT/laminin thin films 608
syndiotactic (st) PMMA 100–104
synthesis
- composites 574
- graphene 992
- multilayer nanotubes 625
- nanowires 993
- polymer brushes 439
- supramolecular objects 6–7
synthetic polycations/-anions, chemical structures 617
synthetic polymers 1025
systemic toxicity 415

*t*
tailoring
- mechanics 363–392
- structural 140–144
tamoxifen 155–156
tannic acid (TA) 210–213
taper 519–520
targeting 818–819

- antibodies 802
- gene delivery 633
- ligands 793
- materials 69
- topographically patterned 89
TCPS (tissue culture polystyrene) 675–676
techniques, *see* methods and techniques
tellurium, CdTe nanoparticles 702
temperature
- annealing 377
- dipping 242
- scalar fields 244–246
- temperature-dependent spectra 334
- temperature-induced release 344
- temperature-triggered enzymatic conversion 814
- time–temperature superposition 332–335
temperature effect
- PEM preparation 241–242
- QCR monitoring 462–463
templates
- bio- 831–832
- cellulose fibers/fibrils 171–187
- dissolvation 874–876
- mesoporous 810
- microparticulate 750–753
- nanospaces 104–106
- preloading 810–811
- selection 801–804
- self-induced/remote release 926
- spherical particle 805
- stereospecific polymerization 100–113
- templated LbL assembly 619–621
temporal control 739
tensile index 176
tensile measurements 364
terpyridine-metal ion complexes 495–502
tests
- abrasion 34
- tension 582
$N,N',N'',N'''$-tetra-salicylidene polyamidoamine (TSPA) 493–494
tetragonal capsules 384
tetrahedral SnS crystals 211
theories and models
- Alexander–de Gennes theory 271
- complexation 485
- 3D artery model 124–128
- displacement model 250
- DLVO 270
- hopping model 311
- Laviron diffusion model 1010, 1016

- microcapsule mechanical properties 373–378
- MIGRATION model 328–329
- model materials 671–679
- multi-cellular 70–72
- one-box model 249
- PEM conductivity spectra 328–329
- PEM preparation 231–233
- percolation model 252–254
- Reissner-model 375–376
- shell theory 373–374
- signal detection theory 958
- SLD model 250
- surface potential model 1009
- volume model 252
- water content 252–254
therapeutic delivery 819–823
thermal calcination 33
thermal cross-linking 33
thermal processing 996
thermal treatment 927
thermally cross-linked coatings 148
thermo-responsive capsules 865–866
thermo-stimulated release 783–784
thermodynamics, SMILbL films 56
thickness
- ellipsometric 290
- films 10
- thickness-shear mode (TSM) 455
thin films
- degradable 407
- electrostatic interactions 596
- functional 493–504
- functional polyoxometallates 703
- multilayer 69–81, 395–406
- robust 50–52
- SWNT/laminin 608
thin-walled structures 372
thiolation 815
three-dimensional..., see 3D...
time-average conformation 614
time-delayed drug release 790
time–humidity/temperature superposition 332–335
tissue culture polystyrene (TCPS) 675–676
titanium 650, 696–697
titration 161–163
- isothermal microcalorimetry 661
top-down approach, ultrasonication 154–158
topographically patterned target 89
toxic polycations 213
toxicity, systemic 415
transcutaneous delivery 428
transfection

- MFT 999
- surface-mediated 731–748
transfer
- electron 1005, 1013
- energy 648, 891–905
- FRET 724–727
- MLCT 500
- nanotransfer printing (nTP) 85
transformation, Möbius 458
transistor, "salt-gated" 312–313
transition-states 304
transparent coatings 147–148
transparent electrodes 602–604
transport
- anisotropic 580–581
- controlling 395–406
- ion-paired PEMs 281–320
- long-range 334
- multilayers 305
- selective 473
- size-/charge-selective 476–479
- size-dependent 477
trapping 313–315
- antibacterial peptides 895–898
- kinetic 430
- redox centers 1014
- self- 313–315
trenches, cylindrical 380
triangle holes 879
triangular particles 965
bis-triazine (DTA) 59–60
triggered disassembly 744
trypsin-containing polyelectrolyte films 911–914
TSM (thickness-shear mode) 455
TSPA ($N,N',N'',N'''$-tetra-salicylidene polyamidoamine) 493–494
tunable release 739
turbidity 341
two-level antibacterial coating 1047
two-step coupling 716

*u*

UASMC (umbilical artery smooth muscle cells) 124
UC (upconversion) 194
ultrasonication-assisted LbL assembly 154
ultrasound-stimulated release 780–781, 867–868
ultrathin films 102–104
- conformal 400
- freely standing 189–204
umbilical artery smooth muscle cells (UASMC) 124

unmasking, anionic carboxylate groups   742
upconversion (UC)   194
upward band bending   556
urea, hydrolysis   968
urease–FITC   863

## v
VA-044   106
vaccines, therapeutic delivery   821–822
van der Waals interactions   99–133
vancomycin   415, 418
vascular wall, drug delivery   652–654
vectors, protein expression   588
VERO cells   755, 864
vesicles
– CF release   937
– complexation   936
– giant   967
– spherical   803–804
vessels, blood   651–652
vibrations
– carbonyl bands   106
– stretching   115
virous force   225
virus battery   402
viscoelasticity   456
viscosity   179
volume model   252
volume rinsing   8
volume swelling coefficient   287

## w
washless LbL assembly   161–163
water
– content   252–254
– desalination   474–476
– dispersion   118
– dye removal   918–920
– scalar fields   246–257
– swelling in   247–256
– water-soluble ATRP macroinitiators   440–441
– water-soluble PECs (WPECs)   338
weak interactions, LbL films   208–213
weak polyacids   856
weak polyanions   237, 642–647
weak polycations   233–235, 237
well-dispersed nanofillers   147–148
wells, cylindrical   878
wet-air oxidation   917
wire electrode, polycryst-Pt   543
wood fibers   176–179
WPECs (water-soluble PECs)   338

## x
X-ray diffraction (XRD), PMMA   111–112
$x,y$ control   88
xyloglucan   238

## y
yeast cells   210–212
yEGFP-reporter   212–213
yield, quantum   53
Young's modulus   200, 321
– shell theory   373–374
– spreading area   675–677
– stimuli-responsive LbL microcapsules   871

## z
ς-potential   162, 445
– polyethersulfone membranes   921
zeolite crystals   76
zeolite-like structure   490
zinc ion complexes   497–499, 502